CW00369058

Civil Aircraft and Glider Registers of United Kingdom and Ireland 2005

Air-Britain supports the fight against terrorism and the efforts of the Police and other Authorities in protecting airports and airfields from criminal activity.

If you see anything suspicious do not hesitate to call the
Anti-Terrorist Hotline 0800 789321
or alert a Police Officer.

CIVIL AIRCRAFT & GLIDER REGISTERS OF UNITED KINGDOM & REPUBLIC OF IRELAND 2005

41st Year of Publication – Collated and Edited by Barrie Womersley

CONTENTS **PAGE**

Published by: Air-Britain (Historians) Limited
Sales Department: 41 Penshurst Road, Leigh, Tonbridge, Kent TN11 8HL
Membership Enquiries: 1 Rose Cottages, 179 Penn Road, Hazlemere, Bucks HP15 7NE
Web-site: http://www.air-britain.com
Sales e-mail: mike@absales.demon.co.uk

Front Cover: Desoutter I G-AAPZ performing a fly-past at Old Warden on 5.9.2004. (David W Yates)

Rear Cover: Twin Pioneer 3 G-APRS displays STOL characteristics at Biggin Hill 19.9.2004 (Roger Birchall)

Provost T.1 G-BKFW/XF597 gets airborne for a fly-past at Brimpton 20.6.04 (Dave Partington)

Sikorsky S-61N of Stornoway Coastguard in action at Loch Broom 25.8.04 (Richard F Hewitt)

ISBN 0 85130 359 5 ISSN 0264-5270

Printed by The Cromwell Press Ltd, Trowbridge

INTRODUCTION AND EDITORIAL

For forty one consecutive years Air-Britain has produced a highly detailed account of the happenings relative to the United Kingdom and Ireland Civil Aircraft Registers. Glider details were introduced in 1983 and so this information has now appeared for twenty three of those years.

With every year there brings a challenge to improve the presentation and content. I believe that mostly we succeed but occasionally we cause a minor upset. I have always attempted to respond to the latter and revisit the topic and this year is no exception. Three issues have been illuminated in recent times - the absence of an Index covering the numerous overseas aircraft based here, the difficulty of relating the museum entries in Section 1 to the data published elsewhere whilst some members felt strongly that the main Indices should be at the very end of the book. I hope you find that this year's edition addresses all these issues.

I must comment on the second of those issues mentioned above.The introduction of Museum entries (SECTION 3) into the main Sections has alarmed some of our older readers (just as the retention of some cancelled aircraft within the main text did for some of us over twenty years ago). It is an issue which divides the readership and strong support is expressed for both views. Whilst I appeciate that such "cancelled" information seems to get in the way we must remember the Air-Britain register has never sought to be a clone of the CAA's registration data. I acknowledge now that the annual Register should seek to be a comprehensive revelation of current activities and this should include our dynamic museum scene where it impinges on UK and Irish registered aircraft.

Once again we highlight a number of people behind this year's production. So my thanks to Bernard Martin for his work on the monthly UK Changes of Ownership and "Round and About" section, Phil Butler for updating the register information from the British Gliding Association's records. Peter Hornfeck for working on the Irish Register and, finally, Paul Hewins who now compiles the Non-UK Registered Section and has also revived the dedicated Index. Special thanks to Graham Slack, Co-ordinating Editor of Air Britain News and Alan Johnson (UK Register) for passing on the official Civil Aviation Authority data and to all those members who supply regular updates either direct or via ABIX and/or the A-B Sightings web-site.

I am grateful to all the following contributors, including those who assist in the background. This year thanks go to Phil Ansell, Rob Barlex, Pete Bish Jim Brazier, Peter Budden, Phil Butler, Ian Burnett, Mike Cain, Ian Callier, Richard Cawsey, Chris Chatfield, Noel Collier, Terry Dann, John M Davis, Graham Duke, Phil Dunnington, Ken Ellis, Bill Fisher, Bryan Foster, Wal Gandy, Stuart Greer, Dave Haines, Kenneth Hearn, Paul Hewins, Les Hitchings, Nigel Hitchman, Peter Hornfeck, Pete Hughes, Paul Jackson, Kenneth James, Alan Johnson, Mark Jones, Nigel Kemp, Phil Kemp, Bob Kent, Anne Lindsay, Stuart MacConnacher, Peter J Marson, Bernard Martin, Tony Morris, Alistair Ness, Ken Parfitt, Dave Partington,Tony Pither, Nigel Ponsford, Geoff Pott, Dave Reid, Robin Sauvery, Trevor Sexton, Graham Slack, Colin M Smith, Terry Smith, Tony Smith, Dave Sparrow, Trevor Spedding, Martyn Steggalls, Mike Stroud, Barry V Taylor, Gerard Terry, Dave Thompson, John Tietjen and Pete Webber. Finally, very special thanks to my wife, Angela, for another year of patience!

The CAA's annual UK statistical information at 31st December 2004 is shown at Page xl. The size of the UK Register continues to increase, last year by over 3%. However, whilst there has been growth in almost all areas the number of aircraft not certificated has increased by over 14%. All official UK aircraft registration and ownership information shown is published by the CAA and includes the issue and status of each Cerftificate of Airworthiness. All CAA and BGA registration data is correct to the middle of February 2005. Irish Aviation Authority data is correct to 28th January 2005.

Comments and further information for the next edition should reach me no later than 31st January 2006, please.

BARRIE WOMERSLEY
19 The Pastures, Westwood
Bradford-on-Avon
Wiltshire BA15 2BH
email: barrie.womersley@air-britain.co.uk

February 2005

REGISTRATION OF UK AIRCRAFT

There were no international regulations controlling the registration of civil aircraft within the United Kingdom at the end of the First World War. Consequently the Air Ministry's Civil Air Department specified a system of temporary registration marks in May 1919 but this practice prevailed only until July of that year. Two temporary registers were established for (i) military aircraft sold for civil purposes and already bearing Service serials - existing aircraft would be allocated their serials as registration marks with the Service ring markings obliterated and (ii) new aircraft and those built from spares - they were allocated marks in a special Service sequence commencing at K100. Subsequently, a number of these aircraft were re-allocated registrations in the first permanent register which replaced the two temporary registers. This was inaugurated on 31st July 1919 and ran until 29th July 1928 when the registration G-EBZZ had been issued - see SECTION 1 Part 1. Meantime, the civil use of Airships and Balloons came under the supplementary air traffic regulations of the Air Navigation Act 1911-1919. and a separate lighter-than-air register series, G-FAAA- to FAAZ, was established until the end of 1928.

With the growth of international civil aviation the Director-General of Civil Aviation decided to terminate the first series and he authorised a new sequence of registration marks commencing at G-AAAA including airships and balloons. This second permanent register was introduced retrospectively from 30th July 1928 and, in effect, continues today - see SECTION 1 Part 2. Registrations were usually allocated in alphabetical sequence until the late 1970s although there have been numerous sporadic exceptions to this rule throughout this time. The G-AAAA-AZZZ series was allocated by July 1972 and a new series G-BAAA onwards was used in the same month: this series became exhausted in June 2001. Notwithstanding this, and commencing in 1974, many registrations were issued ahead of the natural alphabetical sequence and came from all of the forthcoming G-Bxxx to G-Zxxx series, examples being Concorde G-BSST (May 1968), Accountant G-BTEL (August 1957) and Harrier G-VSTO (June 1971). The advance G-Bxxx registrations were eventually subsumed within the proper sequence and we have moved on to the G-Cxxx series which has been in regular use since June 2001.

Small parts of an original G-Cxxx series were allocated to Canadian civilian aircraft from 20th April 1920 until 31st December 1928 commencing with G-CAAA. Registration G-CAWI was the last to be issued although G-CAXP had been formally allocated earlier in May 1928 being ex G-EBXP. On 1st January 1929 Canada adopted the nationality marks CF, followed by a hyphen, and three registration marks running from AAA to ZZZ. Those aircraft registered previously with G-CAxxx markings continued to display them until they were retired from service. In addition, two separate series of quasi-military markings were created for aircraft operated on Canadian Government Air Operations by the Royal Canadian Air Force (RCAF). Series G-CYAA to G-CYHD was used from 18th June 1920 to 11th June 1926 and G-CYUR to G-CYZZ, but not sequentially, from 15th March 1927 to 18th July 1931. The batch from G-CYHE to G-CYUQ were not allotted as the RCAF were using a numerical series for registration marks on aircraft in their use by this time and marks were being displayed in an abbreviated form, that is the last two letters only. Aircraft still wearing these marks were absorbed into the existent RCAF numerical series by 1939. Consequently, the Civil Aviation Authority has not allocated any further registrations from the G-CAxx range although nine advance registrations were issued between December 1977 and March 1999 and remain extant. These were not allocated previously although two were reserved in 1928.

In addition, a number of special registrations series have been used as shown below:

a) During 1979 and 1980 specific alpha-numeric registrations G-N81AC and G-N94AA to G-N94AE which were used for British Airways' Concordes,

b) From January 1982 until 1997 registrations G-FYAA to G-FYZZ which were dedicated for minimum lift balloons, and

c) From 1981 until 1998 registration batches G-MBAA to G-MBZZ, G-MGAA to G-MGZZ, G-MJAA to G-MJZZ, G-MMAA to G-MNZZ, G-MTAA to G-MTZZ and G-MYAA to G-MZZZ which were allocated originally for microlight aircraft use. Subsequently, other aircraft can now be found within these batches.

USER GUIDE

There are a number of purposes to this annual volume. The primary one is to list all aircraft on the current civil aircraft registers, giving full details of types and previous identities, registered ownership and/or operator, probable home base and certification of airworthiness status. This information reproduces, expands upon and amplifies the official country registers. However, we go well beyond that; included are all other known, but no longer currently registered, UK and Irish aircraft noted in a reasonably identifiable condition and which are displayed or on/for rebuild. Some of these are held for instructional, fire or spares use. The majority of these are in located in the UK and Ireland but a few are resident abroad. In addition we detail the extensive numbers of foreign registered aircraft now located in both the UK and Ireland, some of which may aspire to G- and EI- registrations in due course. Finally, we include a comprehensive listing of all gliders in use throughout the area in order to present an all-inclusive guide of the civil aviation scene.

Secondly, on a more technical level, we cater for the aviation specialist and historian who wishes to know more about a particular aircraft by providing detailed information such as the reasons for that aircraft's non-airworthy state where known and if applicable. Also, it has been our policy for many years to record engine details in the text, for example the modification of an airframe to receive a non standard engine has to be approved by the CAA and is then recorded in their Register. When the first microlights appeared in 1981 there was little standardisation and, consequently, we recorded all engine information. Now that the majority are produced commercially, either in whole or kit form, engine fitment is standard to type in many cases. Accordingly, a default fit is shown in the Section 6 Index whilst variations remain identified in Section 1. A similar situation exists for PFA and BMAA approved, and other miscellaneous imported, home-build types.

A guide to the main text is as follows:

Registration Registrations are set out in alphabetic order. Aircraft no longer currently registered are marked with an asterisk (*). A few aircraft, either real or static reproductions are identified in fictitious UK civil marks for display purposes and are identified in Section 6. Those marks which have been re-issued or re-allotted, particularly either if the first holder or allottee did not use the marks or they were not allocated at the time, are shown with a suffix, for example (2) after the registration.

Type We adopt the official type description as set down by the manufacturer or designer. Where there is doubt, reference is made to the relevant issue of "Jane's All the Worlds Aircraft", "Airlife's World Aircraft" and the annual "World Directory of Leisure Aviation". Indication is given if the manufacturer is a successor company or a licence builder - although not always if it is merely a sub-contractor. Under this column, we show engine types in parenthesis if the engine is non-standard for all PFA and BMAA approved and other home-build types and for those vintage and classic aircraft where engines do vary. Standard fit engines for most PFA and BMAA approved types are shown in the respective Indices in Section 6 . In addition, and especially for all PFA and BMAA types, we show the manufacturer. Final explanatory notes comment on the airframe's identity and where the true position is at variance with official records.

Construction Number Sometimes referred to as the Manufacturers' Serial Number and generally quoted in official registers, the construction number should be a constant traceable reference throughout an airframe's life by means of a unique identification plate. Homebuilt and kit build examples should have a home-builder's personal reference number. Some aircraft have more than one construction number, for example PFA and BMAA approved types. Both bodies use special series for their approved designs. These are, in effect, project numbers but are often accepted as construction numbers in the absence of a more specific reference.

Many PFA and BMAA examples have a manufacturers' Plan or Kit number; the latter sometimes identified by a "K..." prefix. It is known that main components carry adhesive labels with individual kit numbers when packed but often after assembly these are discarded and there is no longer any visible identity on a particular airframe therefore, all traceability is lost.unfortunately.

Although manufacturers often identify weightshift microlights with separate construction numbers for the Trike unit and for the Wing respectively the CAA usually only records the latter number. For example, manufacturers Hornet, Mainair and Medway issue a composite construction number comprising both units. Both numbers are identified, if known, with the construction number for the Trike unit preceding that for the Wing.

Previous Identities These are set out in reverse order with the most recent identity first. Some of these may have been issued several times and are noted with a suffix as stated above. Registrations shown in parenthesis have been allotted but were never officially used to the best of our knowledge. The nationality of foreign military serials is indicated only where it may not be apparent. Manufacturers' test marks, also known as "B Conditions" identities, are given where known.

Registration Date This is the date of the original registration for those particular marks even where subsequently removed and restored.

Owner(*Operator*) This is the registered owner for current aircraft as recorded in CAA records. Where the operator is known to be different this is shown in parenthesis. Included under this column are details of the latest reported status, for example if the C of A is not current or the aircraft is known to be under repair plus details of any names and, in particular, any military colour schemes and marks worn. Some aircraft are shown as temporary un-registered ("Temp unregd"); this is where the CAA has not received an application from a new owner following a sale. Usually the CAA gives a period of discretion and if no response is received the Certificate is cancelled and the aircraft is not permitted to fly. Such action usually stimulates the new owner to produce the relevant documentation.

(Unconfirmed) Base The information in this column is not guaranteed: there is no official information. Reports by members and other readers who visit airfields and strips are perused to compile this column. Aircraft change base frequently. Balloons are generally shown as being based at the owner's registered address and details are in parenthesis. Bases for active flexwing aircraft can present a dilemma as Trikes are often stored at owners' homes after flying whilst corresponding Sailwings are pegged out at base airfields. When the location is uncertain the owner's hometown is shown in parenthesis.

Readers are reminded that the identification of a base, particularly if it is a private strip, is not an invitation to visit and in a number of cases visiting is actively discouraged because of previous abuses. We recognise the need for privacy in this area and consequently not all information held is published. Readers are also directed to the AB-IX Bases database which has current details of airfields and listings of occupants. This is updated frequently.

CofA Expiry Date Information is taken from information published by the CAA. The date indicates the currency of the aircraft's certificate and details of the coding suffix letters applied are set out below. Where a CofA has expired or lapsed and an aircraft has been reported since that date further details are shown.

CERTIFICATES OF AIRWORTHINESS (CofA) STATUS

The formation of the European Aviation Safety Agency (EASA) and the implementation of the associated European legislation has changed the responsibilities and procedures for the regulation of continuing airworthiness. Consistent with these developments, substantial changes have been made to CAA mandatory airworthiness information.

As a direct result of the European legislation, and with effect from 28th September 2004, UK-registered aircraft are now divided into two groups:

- *"EASA aircraft"*; that is, aircraft subject to regulation by EASA; and
- *"Non-EASA aircraft"*; that is, aircraft that remain subject to regulation by the CAA.

The specific aircraft are detailed in CAP747. Therefore, a new "EASA Standard Certificate of Airworthiness" has been introduced for those UK-registered aircraft now defined as *"EASA aircraft"*. CAA definitions for Private, Aerial Work and Transport categories remain for *"Non-EASA aircraft"*

These changes are in transition but it is already affecting the way we show CofA information. Two specific points to note are (i) all type certificated balloons are now deemed EASA aircraft and (ii) all gliders with a structutral mass of 80kg or more (single seater) or 100 kg or more (two seater) will now need to be issued with a CofA.

In Section 1 codings shown after the date of expiry indicate the CofA Certification category. The absence of a code letter means an aircraft holds a Private category CofA of either one or three years duration. Code are as follows:-

E: EASA
Denotes new EASA Standard CofA

T: Transport
Issued to any Passenger aircraft operated for hire or reward, usually for either one or three years duration, and

T(C): Transport (Cargo)
Issued to any aircraft operated for carrying cargo for hire or reward only, usually for either one or three years duration. Very few aircraft remain in this category.

P: Permit to Fly
Introduced in 1950 this covers homebuilder, vintage and microlight aircraft and is issued by the CAA, on the recommendation of the PFA. It is the "prime document" that legalises flight of a Permit aircraft. The Permit to Fly is issued for the life of the aircraft but is only valid when certain conditions are fulfilled. The most fundamental of these is the Certificate of Validity. The C of V is issued by the PFA following a satisfactory report concerning the annual inspection and test flying of the aircraft.

AW: Aerial Work
This category normally indicates aerial advertising, mainly by hot air balloons, and a few banner towing aircraft.

Exemption:
These non-expiring exemptions were issued originally to early microlights and hot air balloons. They are being progressively upgraded

S: Special
Mainly lapsed now and replaced by Permit to Fly.

GLOSSARY of TERMS and ABBREVIATIONS

AA	Automobile Association
AB	Aktiebolaget (1)
AAC	Army Air Corps
AAIB	Air Accidents Investigation Branch
AC	Awaiting Certification
AERONCA	Aeronautical Corpn of America
AESL	Aero Engine Services Ltd
AIA	Atelier Industriel de l'Aéronautique d'Alger
AIRCO	Aircraft Manufacturing Co
ALAT	Aviation Légère de l'Armée de Terre
AMD-BA	Avions Marcel Dassault-Breguet and Aviation
ANEC	Air Navigation and Engineering Co
ANG	Air National Guard
APSS	Aviation Preservation Society of Scotland
ASS	Air Signals School
AVIA	Azionara Vercellese Ind.Areo
A/c	Aircraft
Aka	also known as
Assn	Association
BA	British Aircraft Manufacturing Co Ltd
BAC	British Aircraft Company
BAC	British Aircraft Corporation
BAT	British Aerial Transport Co Ltd
BoBMF	Battle of Britain Memorial Flight
BMAA	British Microlight Aircraft Association
BV	Besloten Vennootschap (2)
CAARP	Coopérative des Ateliers Aéronautiques de la Région Parisienne
CAB	Constructions Aéronautiques de Béarn
CAB	Constructions Aéronautiques de Bourgogne
CAF	Candian Air Force
CASA	Construcciones Aeronáuticas SA
CC	County Council
CCF	Canadian Car and Foundry
CEA	Centre Est Aviation
Cobelavia	Compagnie Belge d'Aviation
CSS	Centralne Studium Samolotów
CZAL	Ceskoslovenske Zavody Automobilove a Letecke
Corp	Corporation
c	circa
C of A	Certificate of Airworthiness
C of R	Certificate of Registration
c/s	Colour scheme
DBF	Destroyed by fire
DEFRA	Department for Environment, Food and Rural Affairs
DEFTS	Defence Elementary Flying Training School
DOSAAF	Dobrovol'noe Obshchestvo Sodeistviya Armii, Aviasii i Flotu
EASA	European Aviation Safety Agency
EKW	Eidgenössiche Konstruktions Werkstätte
EMBRAER	Empresa Brasileira de Aeronautica SA
EoN	Elliotts of Newbury Ltd
ERCO	Engineering and Research Corporation
ETPS	Empire Test Pilots' School
FAA	Fleet Air Arm
FLPH	Foot Launched Propelled Hang-glider
FTS	Flying Training School
f/f	First flight
fsm	Full Scale Model
GAF	Government Aircraft Factory
GC	Gliding Club
GmbH	Gesellschaft mit beschrankter Haftung (3)
HAB	Hot-Air Balloon
IAC	Irish Air Corps
IAA	Irish Aviation Authority
IAR	Industria Aeronautica Romania
IAV	Intreprinderea deAvioane
ICA	Intreprinderea de Constructii Aeronautice
ICAO	International Civil Aviation Organisation
III	Iniziative Industriali Italiane
IMCO	Intermountain Manufacturing Co
IMES	Irish Marine Emergency Services
IPTN	Industri Pesawat Terbang Nusantara
IWM	Imperial War Museum
Intl	International
JAA	Joint Aviation Authorities
JAR	Joint Aviation Regulations
KG	Kommanditgesellschaft (4)
KK	Kabushiki Kaisha (5)
LAK	Litovskaya Aviatsyonnaya Konstruktsiya
LET	Letecky Narodny Podnik
LLC	Limited Liability Corporation (6)
LLP	Limited Liability Partnership (7)
LVG	Luft-Verkehrs Gesellschaft
Lsg	Leasing
Ltd	Limited (8)
MBA	Micro Biplane Aviation
MBB	Messerschmitt-Bölkow-Blohm
MLB	Minimum Lift Balloon
MPA	Man Powered Aircraft
NV	Naamloze Vennootschap (9)
NK	Not known
Ntu	Not taken up
n/w	nose-wheel

OGMA	Oficinas Gerais de Material Aeronautico
Op	Operated by
PFA	Popular Flying Association
PIK	Polytecknikkojen Ilmailukerho
PLC	Public Limited Company (10)
PRC	Peoples' Republic of China
PT	Pesawat Terbang (11)
PWFU	Permanently WFU
PZL	Panstwowe Zaklady Lotnicze (State Aviation Works)
Qv	Which See
R	Reservation
RAF	Royal Aircraft Factory
RAF	Royal Air Force
RAFC	RAF College
RCAF	Royal Canadian Air Force
Rep	Reproduction
RFC	Royal Flying Corps
RJAF	Royal Jordanian Air Force
RN	Royal Navy
RNAS	Royal Naval Air Service
RSAF	Royal Saudi Air Force
RTS	Reduced to spares
SA	Société Anonyme (12)
SA	Sociedad Anónima (13)
SA	Spoika Akeyjna (14)
SAAC	Society of Amateur Aircraft Constructors
SAI	Skandinavsk Aero Industri
SAN	Société Aéronautique Normande
SAR	Search and Rescue
SAS	Société par Actions Simplifiées (15)
SEAE	School of Electrical and Aeronautical Engineering
SIPA	Société Industrielle pour l'Aéronautique
SNCAC	Société Nationale de Constructions Aéronautiques du Centre
SNCAN	Société Nationale de Constructions Aéronautiques du Nord
SOCATA	Société de Construction d'Avions de Tourisme et d'Affaires
SoS	Secretary of State
SRCM	Société de Recherches et de Constructions Mécaniques
SRL	Société Anonyme à Responsabilité Limitée (16)
SpA	Societa per Azioni (17)
SPP	Strojirny Prvni Petiletky
SEAE	School of Electrical and Aeronautical Engineering
Srs.	Series
SS	Special Shape
SZD	Szybowcowy Zaklad Dowswiadczalny
TAD	Technical Aid and Demonstrator
TBA	To be advised
TEAM	Tennessee Engineering and Manufacturing
TWU	Tactical Weapons Unit
t/a	Trading as
tr	Trustee of
t/w	tail-wheel
UAS	University Air Squadron
USAAC	United States Army Air Corps
Vw	Volkswagen
WACO	Weaver Aircraft Corp
WFU	Withdrawn from Use
WSK	Wytwornia Sprzetu Komunikacy Jnego Okecie

Company Constitution Notes

(1)	Sweden	Joint Stock
(2)	Netherlands	Private
(3)	Germany	Private Limited
(4)	Germany	Limited Partnership
(5)	Japan	
(6)	USA	Limited Partnership
(7)	UK	Limited Partnership
(8)	UK	Private Limited
(9)	Belgian/Netherlands	
(10)	UK	Public Limited
(11)	Indonesia	
(12)	France/Romania	Public Limited
(13)	Spain	Public Limited
(14)	Poland	
(15)	France	Joint Stock Company
(16)	Italy	Public Limited Company
(17)	Italy	Public Limited Company

CAA DATA

UNITED KINGDOM REGISTER OF CIVIL AIRCRAFT

	CERTIFICATION STATUS @ 1st JANUARY 2005							
CLASS	**TCP**	**TCC**	**AW**	**PRIVATE**	**SPECIAL**	**PERMIT**	**NOT CERTIFICATED**	**TOTAL**
FIXED WING mtow > 5700 KG	**883** *(935)*	**7** *(10)*	**18** *(16)*	**8** *(13)*	- -	**26** *(27)*	**240** *(153)*	**1180** *(1154)*
FIXED WING mtow > 2730 KG <= 5700 KG	**159** *(154)*	- -	**10** *(9)*	**31** *(39)*	- -	**63** *(66)*	**114** *(109)*	**377** *(377)*
FIXED WING mtow <= 2730 KG excl Microlights	**2264** *(2329)*	- -	**20** *(19)*	**2270** *(2421)*	**2** *(2)*	**1549** *(1489)*	**2478** *(2148)*	**8583** *(8408)*
MICROLIGHTS incl FLPH	- -	- -	- -	- -	- -	**2303** *(2328)*	**1762** *(1496)*	**4065** *(3824)*
GLIDERS	-	-	-	-	-	-	**2** (1)	**2** *(1)*
HELICOPTERS	**810** *(816)*	- -	-	**111** *(139)*	- -	**48** *(53)*	**269** *(169)*	**1238** *(1159)*
GYROPLANES	-	-	-	**1** *(-)*-	-	**68** *(62)*	**182** *(185)*	**251** *(247)*
GAS FILLED AIRSHIPS	**2** *(2)*	- -	- -	- -	- -	- -	**3** *(3)*	**5** *(5)*
HOT AIR AIRSHIPS	- -	- -	**5** *(9)*	- -	- -	- -	**19** *(16)*	**24** *(25)*
GAS FILLED BALLOONS	**1** *(-)*	- -	- *(2)*	- -	- -	- -	**12** *(11)*	**13** *(13)*
HOT AIR BALLOONS	**203** *(185)*	- -	**515** *(525)*	**32** *(40)*	- -	- -	**994** *(931)*	**1744** (1681)
GAS/HOT AIR BALLOONS	- -	- -	- -	- -	- -	- -	**6** *(6)*	**6** (6)
MINIMUM LIFT BALLOONS	- -	- -	- -	- -	- -	- -	**99** *(99)*	**99** *(99)*
TOTALS	**4322** *(4429)*	**7** *(10)*	**566** *(562)*	**2453** *(2624)*	**2** *(2)*	**4057** *(4025)*	**6160** *(5360)*	**17587** *(17012)*

Key: TCP - Transport Category Passenger, TCC - Transport Category Cargo, AW - Aerial Work

Data produced by CAA Aircraft Registration Section, Kingsway, London WC2 @ 1st January 2005

SECTION 1 - UNITED KINGDOM

PART 1 – FIRST PERMANENT REGISTER

G-EAAA - G-EAZZ

Registration	Type	Construction No	Previous Identity	Date	Registered Owner(*Operator*)	(Unconfirmed) Base	CofA Expy
G-EACN*	BAT FK.23 Bantam 1				See SECTION 3 Part 1	Schiphol, The Netherlands	
G-EACQ*	Avro 534 Baby				See SECTION 3 Part 1	Brisbane, Australia	
G-EAFN*	BAT FK.23 Bantam				See SECTION 3 Part 1	Lelystad, The Netherlands	
G-EAML*	Airco DH.6				See SECTION 3 Part 1	Pretoria, South Africa	
G-EAOU*	Vickers FB.27A Vimy IV				See SECTION 3 Part 1	Adelaide, Australia	
G-EAQM*	Airco DH.9				See SECTION 3 Part 1	Canberra, Australia	
G-EASD	Avro 504L	E.5	S-AHAA	26. 3.20	G.M.New	Woodford	
	(Le Clerget @ 130hp)		S-AAP, G-EASD, (RAF)		*(Under restoration 3.04)*		

G-EBAA - G-EBZZ

Registration	Type	Construction No	Previous Identity	Date	Registered Owner(*Operator*)	(Unconfirmed) Base	CofA Expy
G-EBHX	de Havilland DH.53 Humming Bird	98	No.8*	22. 9.23	Richard Shuttleworth Trustees	Old Warden	21. 8.04P
	(ABC Scorpion II)		*(*Lympne 1923 trials)*		*"L'Oiseau-Mouche"*		
G-EBIA	Royal Aircraft Factory SE.5A	654/2404	F904	26. 9.23	Richard Shuttleworth Trustees	Old Warden	5. 5.05P
	(Built Wolseley Motors Ltd) (Wolseley Viper @ 200hp)		"D7000", G-EBIA, F904		*(As "F904/H" in 56 Sqdn RFC c/s)*		
G-EBIB*	Royal Aircraft Factory SE.5A				See SECTION 3 Part 1	South Kensington, London	
G-EBIC*	Royal Aircraft Factory SE.5A				See SECTION 3 Part 1	Hendon	
G-EBIR	de Havilland DH.51	102	VP-KAA	22. 1.24	Richard Shuttleworth Trustees	Old Warden	26. 5.05P
	(ADC Renault @ 120hp)		G-KAA, G-EBIR		*"Miss Kenya"*		
G-EBJE*	Avro 504K				See SECTION 3 Part I	Hendon	
G-EBJG*	Parnall Pixie III				See SECTION 3 Part 1	Coventry	
G-EBJO	Air Navigation and Engineering Co ANEC II	1	No.7*	17. 7.24	Richard Shuttleworth Trustees	Old Warden	30.11.35
			*(*Lympne 1924 trials)*		*(Rebuild nearing completion 9.03)*		
G-EBKY	Sopwith Pup	w/o 3004/14	"N5180"	27. 3.25	Richard Shuttleworth Trustees	Old Warden	7. 5.04P
	(Le Rhone 80hp) *(ex Sopwith Dove)*		"N5184", G-EBKY		*(As "N6181" in 3 Sqdn RNAS c/s) "Happy"*		
G-EBLV	de Havilland DH.60 Moth	188		22. 6.25	BAE SYSTEMS (Operations) Ltd	Old Warden	21. 6.05P
	(ADC Cirrus III)				*(On loan to Richard Shuttleworth Trustees)*		
G-EBMB*	Hawker Cygnet I				See SECTION 3 Part 1	RAF Cosford	
G-EBNO*	de Havilland DH.60 Moth				See SECTION 3 Part 1	Halmstad, Sweden	
G-EBNU*	Avro 504K				See SECTION 3 Part 1	Jyväskylä, Finland	
G-EBNV	English Electric S.1 Wren	4	(BAPC.11)	9. 4.26	Richard Shuttleworth Trustees	Old Warden	23. 6.87P
	(ABC @ 398cc)		G-EBNV	*(Composite - principally c/n 3 rebuilt 1955/56: as "No.4": no marks carried: noted 5.01)*			
G-EBOV*	Avro 581E Avian				See SECTION Part I	South Brisbane, Australia	
G-EBQP	de Havilland DH.53 Humming Bird	114	J7326	?. 4.27	M.C.Russell	Salisbury Hall, London Colney	AC
	(Bristol Cherub III)				tr G-EBQP Syndicate		
			(Fuselage on rebuild 5.04: will use wings ex Martin Monoplane G-AEYY and carry "J7326")				
G-EBUB*	Westland Widgeon III				See SECTION 3 Part 1	Wangaratta, Victoria, Australia	
G-EBWD	de Havilland DH.60X Moth	552		2. 3.28	Richard Shuttleworth Trustees	Old Warden	24. 4.05P
	(ADC Hermes 2)						
G-EBXU	de Havilland DH.60X Moth	627		2. 5.28	G.G.Pugh	Stapleford	21. 6.05P
	(DH Gipsy II)						
G-EBYY*	Cierva C.8L Mk.2				See SECTION 3 Part 1	Le Bourget, Paris, France	
G-EBZM*	Avro 594A Avian IIIA				See SECTION 3 Part 1	Manchester	

PART 2 – SECOND PERMANENT REGISTER

G-AAAA - G-AAZZ

Registration	Type	Construction No	Previous Identity	Date	Registered Owner(*Operator*)	(Unconfirmed) Base	CofA Expy
G-AAAH*	de Havilland DH.60G Moth				See SECTION 3 Part 1	South Kensington, London	
G-AACN*	Handley Page HP.39 Gugnunc				See SECTION 3 Part 1	Wroughton	
G-AADR (2)	American Moth Corporation DH.60GM Moth	138	NC939M	2. 6.86	Edwina V. Moffatt	Woodlow Farm, Bosbury	16. 8.02P
	(DH Gipsy I)				*(New owner 7.04)*		
G-AAEG	de Havilland DH.60G Moth	1027	D-EUPI	4. 2.29	I.B.Grace	(Ada, MI, USA)	
			D-1599, G-AAEG		*(New owner 1.02)*		
G-AAHD*	Avro 594 Avian IV				See SECTION 3 Part 1	Stockholm	
G-AAHI	de Havilland DH.60G Moth	1082		25. 5.29	N.J.W.Reid	Lee-on-Solent	5. 4.05P
	(DH Gipsy I) *(Original fuselage used in 1953 rebuild of G-AAWO)*						
G-AAHY	de Havilland DH.60M Moth	1362	HB-AFI	10. 5.29	D.J.Elliott	Thruxton	13. 8.05P
	(DH Gipsy I)		G-AAHY		*(Brooklands Flying Club titles)*		
G-AAIN	Parnall Elf II	2 & J.6		11. 6.29	Richard Shuttleworth Trustees	Old Warden	16. 7.05P
	(ADC Hermes 2)						
G-AALY	de Havilland DH.60G Moth	1175	F-AJKM	9. 9.29	K M Fresson	Lee-on-Solent	15. 5.05T
	(DH Gipsy I) *(Composite: rebuilt from components*		G-AALY				
G-AAMX (2)*	American Moth Corporation DH.60GM Moth				See SECTION 3 Part 1	Hendon	
G-AAMY (2)	American Moth Corporation DH.60GMW Moth	86	N585M	2. 5.80	Totalsure Ltd	Seppe-Hoeven, The Netherlands	11. 8.04P
	(Wright Gipsy L320)		NC585M				
G-AANF (2)*	American Moth Corporation DH.60GMW Moth				See SECTION 3 Part 1	Mandeville, Gore, New Zealand	
G-AANG (2)	Bleriot Type XI	14	BAPC.3	29.11.81	Richard Shuttleworth Trustees	Old Warden	26. 5.05P
	(1910 original) (Anzani @ 25hp)				*(No external marks)*		
G-AANH (2)	Deperdussin Monoplane	43	BAPC.4	29.10.81	Richard Shuttleworth Trustees	Old Warden	26. 5.05P
	(Anzani Y @ 35hp) *(Possibly c/n 143)*				*(No external marks)*		

Registration	Type	C/n	Previous identity	Reg date	Owner/operator	Base	CofA expiry
G-AANI (2)	Blackburn Single Seat Monoplane	"9"	BAPC.5	29.10.81	Richard Shuttleworth Trustees	Old Warden	15. 6.05P
	(Gnome @ 50hp no.683)		No.9		*(No external marks)*		
G-AANJ (2)*	Luft-Verkehrs Gesellschaft C.VI				See SECTION 3 Part 1	Hendon	
G-AANL (2)	de Havilland DH.60M Moth	1446	OY-DEH	26. 6.87	A L.Berry	Perth	13. 7.04P
	(DH Gipsy II) *(Composite rebuild)*		RDAF S-357, S-107		*(National Flying Services titles)*		
G-AANM (2)	Bristol F.2b Fighter composite	"67626"	BAPC.166	16. 7.87	Aero Vintage Ltd	Duxford	AC
	(Built British and Coloniial Aero Co 1917) (RR Falcon)				*(As "D7889") (Noted 1.05)*		
G-AANO (2)	American Moth Corporation DH.60GMW Moth	165	N590N	3. 3.88	A W.and M.E.Jenkins	(Comberton, Cambridge)	
	(Composite rebuild 11.91)		NC590N		*(New CofR 3.03)*		
G-AANV (2)	de Havilland DH.60M Moth	13	HB-OBU	8. 3.84	R.A Seeley	Longwood Farm, Morestead	9. 9.05P
	(Built Morane Saulnier) (DH Gipsy I)		CH-349, F-AJNY				
G-AAOK (2)	Curtiss-Wright Travel Air CW-12Q		N370N	18.11.81	Shipping and Airlines Ltd	Rushett Farm, Chessington	18. 1.84P
	(Warner Scarab @ 125hp)	12Q-2026	NC370N, NC352M		*(Damaged in gales Rijeka, Yugoslavia 21.10.83: noted 2004)*		
G-AAOR (2)	de Havilland DH.60G Moth	1075	EC-AAO	15. 4.85	B.R.Cox and N.J.Stagg	(Bristol)	17.11.05P
	(DH Gipsy I) *(C/n uncertain: probably a composite)*						
G-AARO (2)*	Arrow Sport A2-60				See SECTION 3 Part 1	Washington, DC, USA	
G-AAPZ	Desoutter I	D.25		?. ?.31	Richard Shuttleworth Trustees	Old Warden	12. 4.05P
	(ADC Hermes)				*(National Flying Services titles)*		
G-AATC*	de Havilland DH.80A Puss Moth				See SECTION 3 Part 1	Mandeville, Gore, New Zealand	
G-AAUP	Klemm L 25-1a	145		19. 2.30	J I.Cooper	Denford Manor, Hungerford	21.11.84P
	(Salmson AD9)				t/a Newbury Aeroplane Co *"Clementine" (On rebuild 10.01)*		
G-AAWO	de Havilland DH.60G Moth	1235		2. 5.30	N.J.W.Reid	Lee-on-Solent	10. 2.05P
	(DH Gipsy I) *(1953 rebuild substituted original fuselage of G-AAHI)*						
G-AAXG*	de Havilland DH.60M Moth				See SECTION 3 Part 1	Mandeville, Gore, New Zealand	
G-AAYT	de Havilland DH.60G Moth	1233	DR606	?. 5.30	P.Groves	Lee-on-Solent	
			G-AAYT		*(Noted 9.04)*		
G-AAYX	Southern Martlet	202		14. 5.30	Richard Shuttleworth Trustees	Old Warden	8. 5.05P
	(AS Genet Major 1A)						
G-AAZG	de Havilland DH.60G Moth	1253	EC-AAE	23. 5.30	C.C, Jennifer M.Lovell and J.A Pothecary	Chilbolton	
			EC-MMA, M-CMMA, MW-133, G-AAZG			*(On rebuild 9.04)*	
G-AAZP	de Havilland DH.80A Puss Moth	2047	HL537	4. 6.30	R.P.Williams	Denford Manor, Hungerford	17. 6.06
	(DH Gipsy Major)		G-AAZP, SU-AAC, G-AAZP		*"British Heritage"*		

G-ABAA - G-ABZZ

Registration	Type	C/n	Previous identity	Reg date	Owner/operator	Base	CofA expiry
G-ABAG	de Havilland DH.60G Moth	1259		23. 6.30	A and P.A Wood		
	(DH Gipsy I)				Woodleys Farm, Gainsford End, Toppesfield, Halstead		2. 5.05P
G-ABBB*	Bristol 105A Bulldog IIA				See SECTION 3 Part 1	Hendon	
G-ABDW*	de Havilland DH.80A Puss Moth				See SECTION 3 Part 1	East Fortune	
G-ABDX	de Havilland DH.60G Moth	1294	HB-UAS	22. 8.30	M.D.Souch	Hill Farm, Durley	13.10.05P
	(DH Gipsy I)		G-ABDX				
G-ABEV (2)	de Havilland DH.60G Moth	1823	N4203E	10. 3.77	S.L.G.Darch	East Chinnock, Yeovil	9. 9.05P
	(DH Gipsy I)		G-ABEV (2), HB-OKI, CH-217				
G-ABHC*	de Havilland DH.80A Puss Moth				See SECTION 3 Part 1	Mandeville, Gore, New Zealand	
G-ABIH*	de Havilland DH.80A Puss Moth				See SECTION 3 Part 1	Rhinebeck, New York, USA	
G-ABLF*	Avro 616 Avian Sport				See SECTION 3 Part 1	Salisbury, Australia	
G-ABLK*	Avro 616 Avian V				See SECTION 3 Part 1	South Brisbane, Australia	
G-ABLM*	Cierva C.24				See SECTION 3 Part 1	Salisbury Hall, London Colney	
G-ABLS	de Havilland DH.80A Puss Moth	2164		7. 5.31	R.I.and D.E.Souch	Hill Farm, Durley	2.10.04P
	(DH Gipsy Major)						
G-ABMR*	Hawker Hart				See SECTION 3 Part 1	Hendon	
G-ABNT	Civilian CAC.1 Coupe	O.2.3		10. 9.31	Shipping and Airlines Ltd	Biggin Hil	9. 9.05P
	(AS Genet Major 1A) *(C/n also quoted as O.3)*						
G-ABNX	Robinson Redwing 2	9		2. 7.31	R.J.Burgess tr Redwing Syndicate	Redhil	12. 5.03P
G-ABOI*	Wheeler Slymph				See SECTION 3 Part 1	Coventry	
G-ABOX (2)	Sopwith Pup	-	N5195	2. 9.84	C.M.D.and A P.St.Cyrien	AAC Middle Wallop	22. 4.93P
	(Le Rhone 80hp)				*(On loan to Museum of Army Flying as "N5195")*		
G-ABSD	de Havilland DH.60G Moth	1883	A7-96	21.11.31	M.E.Vaisey	(Hemel Hempstead)	
	(Identity unconfirmed)		VH-UTN, G-ABSD		*(On rebuild following import from USA in 1985)*		
G-ABTC*	Comper CLA.7 Swift	S.32/1		1. 1.32	Not known	Trevissick Farm, Porthtowan	18. 7.84P
					(Cancelled 22.2.99 by CAA) (Noted 6.04)		
G-ABUS	Comper CLA.7 Swift	S.32/4		27. 2.32	R.C.F.Bailey	(Bosbury, Ledbury)	19. 6.79P
	(Pobjoy Niagara 3)				*(Current status unknown)*		
G-ABUU*	Comper CLA.7 Swift				See SECTION 3 Part 1	Madrid, Spain	
G-ABVE	Arrow Active 2	2		19. 3.32	R.A Fleming	Breighton	18.11.05P
	(DH Gipsy III)						
G-ABWP	Spartan Arrow 1	78		?. 4.32	R.E.Blain	Redhil	18. 6.04P
	(Cirrus Hermes 2)						
G-ABXH	Cierva C.19 Mk.IVP				See SECTION 3 Part 1	Madrid, Spain	
G-ABXL	Granger Archaeopteryx	3A		3. 6.32	J.R.Granger	Radcliffe-on-Trent	22. 9.82P
	(Cherub III)				*(New owner 4.02)*		
G-ABYA	de Havilland DH.60G Moth	1906		?. 7.32	D A Hay and J.F.Moore	Biggin Hill	17. 2.05
	(DH Gipsy I)						
G-ABYN*	Spartan 3-Seater II				See SECTION 3 Part 1	Mandeville, Gore, New Zealand	
G-ABZB (2)	de Havilland Moth Major	5011	SE-AEL	11. 9.80	R.Earl and B.Morris	Manor Farm, Garford	12.11.04P
	(DH Gipsy Major 1C)		OY-DAK				

G-ACAA - G-ACZZ

Registration	Type	C/n	Previous identity	Reg date	Owner/operator	Base	CofA expiry
G-ACAA (2)	Bristol F.2b	7434	F4516	25.10.91	Patina Ltd	Duxford	25. 4.01P
	(RR Falcon) *(Restored as original but rebuilt from various components)*				*(Op The Fighter Collection as "D8084/S")*		
G-ACCB	de Havilland DH.83 Fox Moth	4042		24. 1.33	E.A Gautrey	(Nuneaton)	20. 7.57
					(Crashed off Southport 25.9.56 and on rebuild 10.95: current status unknown)		

G-ACDA	de Havilland DH.82A Tiger Moth	3175	BB724	6. 2.33	B.D.Hughes	Denford Manor, Hungerford	26. 6.82
			G-ACDA		*(Crashed and burned out near Cirencester 27.6.79 - fuselage reported stored 10.01)*		
G-ACDC	de Havilland DH.82A Tiger Moth	3177	BB726	6. 2.33	The Tiger Club (1990) Ltd	Headcorn	15. 5.05T
	(Composite airframe)		G-ACDC				
G-ACDI	de Havilland DH.82A Tiger Moth	3182	BB742	6. 2.33	J.A Pothecary	Newton Toney, Salisbury	AC
	(Composite rebuild)		G-ACDI		*(On rebuild 2002)*		
G-ACDJ	de Havilland DH.82A Tiger Moth	3183	BB729	6. 2.33	de Havilland School of Flying Ltd	White Waltham	18. 8.07T
			G-ACDJ				
G-ACEJ	de Havilland DH.83 Fox Moth	4069		21. 4.33	Janice I.Cooper	Brickhouse Farm, Frogland Cross	14. 8.07
					t/a Newbury Aeroplane Co *(Scottish Motor Traction titles)*		
G-ACET	de Havilland DH.84 Dragon	6021	2779M	21. 4.33	M.D.Souch	(Hedge End, Southampton)	
			AW171, G-ACET		*(On rebuild 1.00: composite based on original wings)*		
G-ACGO*	Saunders-Roe A.19 Cloud				See SECTION 3 Part 1	Prague, Czech Republic	
G-ACGR*	Percival P.1C Gull Four IIA				See SECTION 3 Part 1	Brussels, Belgium	
G-ACGZ	de Havilland Moth Major	5038	VT-AFW	30. 5.33	N.H.Lemon	(Maidenhead)	
			G-ACGZ		*(Restored 28.9.99: on rebuild 2000)*		
G-ACIT*	de Havilland DH84 Dragon 1	6039		24. 7.33	See SECTION 3 Part 1	Wroughton	
G-ACLL	de Havilland DH.85 Leopard Moth	7028	AW165	16. 1.34	D.C.M. and V.M.Stiles	Jurby, Isle of Man	6.12.95P
			G-ACLL		*(New owners 1.03)*		
G-ACMA	de Havilland DH.85 Leopard Moth	7042	BD148	14. 3.34	S.J.Filhol	Headcorn	3. 2.94P
			G-ACMA		*(Stored 2.95: current status unknown)*		
G-ACMD (2)	de Havilland DH.82A Tiger Moth	3195	N182DH	20. 1.88	M.J.Bonnick	Rectory Farm, Abbotsley	18. 7.05
			EC-AGB, Sp AF 33-5		*(Provenance doubtful: EC-AGB had interim Spanish AF serial 30-104)*		
G-ACMN	de Havilland DH.85 Leopard Moth	7050	X9381	?. 4.34	M.R.and K.E.Slack	Duxford	25. 7.06
			G-ACMN				
G-ACNS	de Havilland Moth Major	5068	ZS-???	?. 3.34	R.I.and D.Souch	Hill Farm, Durley	12. 8.05
			G-ACNS				
G-ACOJ (2)	de Havilland DH.85 Leopard Moth	7035	F-AMXP	5. 6.87	A J.Norman	Rendcomb	20. 9.07
	(Composite with wings from HB-OXO)				tr Norman Aeroplane Trust		
G-ACPP*	de Havilland DH.89 Dragon Rapide				See SECTION 3 Part 1	Wetaskiwin, Alberta, Canada	
G-ACSP	de Havilland DH.88 Comet	1994	CS-AAJ	21. 8.34	T.M., Margaret L., D.A.and P.M.Jones	Egginton	
			G-ACSP, E-1		*(New owners 8.04)*		
G-ACSS	de Havilland DH.88 Comet	1996	K5084	4. 9.34	Richard Shuttleworth Trustees	Old Warden	2. 6.94P
	(DH Gipsy Queen 2)		G-ACSS		*"Grosvenor House/34" (Active 2002)*		
	(i) Three static replicas exist, all as "G-ACSS" - see SECTION 3, Part 1 and two with identities BAPC.216 and BAPC.257 - see SECTION 4.						
	(ii) Flying replica built Repeat Aircraft, Riverside CA, USA 1993 for T J Wathen regd N88XD [T7] is also as "G-ACSS".						
G-ACTF	Comper CLA.7 Swift	S.32/9	VT-ADO	24. 5.34	Richard Shuttleworth Trustees	Old Warden	7. 9.05P
	(Pobjoy Niagara 2)				*"The Scarlet Angel"*		
G-ACTH*	de Havilland DH.85 Leopard Moth				See SECTION 3 Part 1	Trento, Italy	
G-ACUP*	Percival P.3 Gull Six				See SECTION 3 Part 1	Wangaratta, Victoria, Australia	
G-ACUS (2)	de Havilland DH.85 Leopard Moth	7082	HB-OXA	17.11.77	R.A and V.A Gammons	RAF Henlow	18.12.05
	(Composite including parts ex HB-OXO c/n 7045)		(G-ACUS)				
G-ACUU*	Cierva C.30A				See SECTION 3 Part 1	Duxford	
G-ACUX*	Short S.16 Scion 1				See SECTION 3 Part 1	Belfast	
G-ACVA*	Kay Gyroplane 33/1				See SECTION 3 Part 1	East Fortune	
G-ACWM*	Cierva C.30A				See SECTION 3 Part 1	Weston-super-Mare	
G-ACWP*	Cierva C.30A				See SECTION 3 Part 1	South Kensington, London	
G-ACXA*	Cierva C.30A				See SECTION 3 Part 1	Milan, Italy	
G-ACXB (2)	de Havilland Moth Major	5098	EC-ABY	24. 1.89	D.F.Hodgkinson	(Gravesend)	
			EC-BAX, Sp AF 30-53, EC-YAY			*(On rebuild 2000)*	
G-ACXE	British Klemm L 25c1 Swallow	21		29.10.34	J.G.Wakeford	Bexhill-on-Sea	7. 4.40
	(On rebuild since 1989 using components to re-draw plans and produce a substantially "new-build" airframe: progress continuing 2002 with some mainspar build sub-contracted to Denford Manor)						
G-ACYK*	Spartan Cruiser III				See SECTION 3 Part 1	East Fortune	
G-ACYR*	de Havilland DH.89 Dragon Rapide				See SECTION 3 Part 1	Madrid, Spain	
G-ACYZ*	Miles M.2H Hawk Major				See SECTION 3 Part 1	Markham, Ontario, Canada	
G-ACZE	de Havilland DH.89A Dragon Rapide	6264	G-AJGS	20.11.34	Wessex Aviation and Transport Ltd	Haverfordwest	18. 8.95
			G-ACZE, Z7266, G-ACZE		*(Stored 10.02)*		

G-ADAA - G-ADZZ

G-ADAH*	de Havilland DH.89 Dragon Rapide				See SECTION 3 Part 1	Manchester	
G-ADCG*	de Havilland DH.82 Tiger Moth				See SECTION 3 Part 1	Wevelghem, Belgium	
G-ADEV (2)	Avro 504K	R3/LE/61400	G-ACNB	18. 4.84	Richard Shuttleworth Trustees	Old Warden	24. 4.05P
	(Le Rhone 110hp)		"E3404"		*(As "H5199")*		
	(P/i not confirmed but, if correct, full p/i is 3118M, BK892, G-ADEV, H5199)						
G-ADGP	Miles M.2L Hawk Speed Six	160	G-ADGP	20. 5.35	R.A Mills *"8"*	White Waltham	7. 5.05P
G-ADGT	de Havilland DH.82A Tiger Moth	3338	BB697	23. 5.35	M.F.Dalton	Andrewsfield	10.11.05T
			G-ADGT				
G-ADGV	de Havilland DH.82A Tiger Moth	3340	(D-E...)	23. 5.35	K.J. and P.J.Whitehead	Whitchurch Hill, Reading	23. 7.05
			G-ADGV, (G-BACW), BB694, G-ADGV				
G-ADGZ*	de Havilland DH.82A Tiger Moth	3344		23. 5.35	Not known	Hill Farm, Durley	
					(Cancelled 17.9.40: to RAF as BB700 in 9.40) (Reported 5.02)		
G-ADHA (2)*	de Havilland DH.83 Fox Moth				See SECTION 3 Part 1	Mandeville, Gore, New Zealand	
G-ADHD (2)	de Havilland Moth Major	5105	EC-...	17. 2.88	M.E.Vaisey	RAF Henlow	
	(Rebuild of ex Spanish components ex USA)		Spanish AF 34-5, EC-W32		*(Noted 7.02)*		
G-ADIA	de Havilland DH.82A Tiger Moth	3368	BB747	13. 8.35	S.J. Beaty	Wold Lodge, Finedon	5. 5.05
			G-ADIA				
G-ADJJ	de Havilland DH.82A Tiger Moth	3386	BB819	29. 8.35	J.M.Preston	Chilbolton	20. 3.75
			G-ADJJ		*(Noted on rebuild 11.04)*		
G-ADKC	de Havilland DH.87B Hornet Moth	8064	X9445	27. 3.36	A J.Davy	White Waltham	20.12.04
			G-ADKC				

G-ADKK	de Havilland DH.87B Hornet Moth	8033	W5749	9.11.35	R M and D R Lee	Kemble	17. 5.07
			G-ADKK				
G-ADKL	de Havilland DH.87B Hornet Moth	8035	F-BCJO	?.11.35	P.R.and M.J.F.Gould		
			G-ADKL, W5750, G-ADKL			Coulommiers, Seine-et-Marne, France	8. 7.05
G-ADKM	de Havilland DH.87B Hornet Moth	8037	W5751	12.11.35	L.V.Mayhead	Hill Farm, Durley	6. 7.01
			G-ADKM				
G-ADLS*	Miles M.3C Falcon Six				See SECTION 3 Part 1	Madrid, Spain	
G-ADLV*	de Havilland DH.82A Tiger Moth				See SECTION 3 Part 1	Helsingborg, Sweden	
G-ADLY	de Havilland DH.87B Hornet Moth	8020	W9388	5.10.35	Totalsure Ltd	Seppe-Hoeven, The Netherlands	9. 8.07
			G-ADLY				
G-ADMT	de Havilland DH.87B Hornet Moth	8093		8. 5.36	S and Helen Roberts *"Curlew"*	(Cirencester)	13. 6.07
G-ADMW*	Miles M.2H Hawk Major				See SECTION 3 Part 1	RAF Stafford	
G-ADND	de Havilland DH.87B Hornet Moth	8097	W9385	4. 8.36	Richard Shuttleworth Trustees	Old Warden	12. 4.05P
			G-ADND		*(As "W9385/YG-L/3" in 502 Sqdn c/s)*		
G-ADNE	de Havilland DH.87B Hornet Moth	8089	X9325	10. 3.36	R Felix	Oaksey Park	25. 4.06
			G-ADNE		tr G-ADNE Group *"Ariadne"*		
G-ADNL	Miles M.5 Sparrowhawk	239		12. 8.35	Angela Patrice Pearson	(Ramsbottom, Bury)	13. 5.58S
	(Reconstructed c1953 as M.77 Sparrowjet)				*(On rebuild with Tim Cox in Bristol area 4.04 as replica/reproduction M.5)*		
G-ADNZ (2)	de Havilland DH.82A Tiger Moth	85614	6948M	10.10.74	D.C.Wall	Old Buckenham	12. 8.06
			DE673		*(As "DE673")*		
G-ADOT*	de Havilland DH.87B Hornet Moth				See SECTION 3 Part 1	Salisbury Hall, London Colney	
G-ADPC	de Havilland DH.82A Tiger Moth	3393	BB852	24. 9.35	D J Marshall	Charity Farm, Baxterley	2. 8.06
			G-ADPC				
G-ADPJ	BAC Drone 2	7		21. 8.35	N.H.Ponsford	(Selby)	17. 5.55
	(Douglas Sprite)				*(Crashed Leicester 3.4.55: on rebuild 12.99 using parts from G-AEJR c/n 22)*		
G-ADPR*	Percival P.3 Gull Six				See SECTION 3 Part 1	Auckland, New Zealand	
G-ADPS	BA Swallow 2	410		4. 9.35	J.F.Hopkins	Watchford Farm, Yarcombe	5. 9.05P
	(Pobjoy Cataract 2)						
G-ADRA (2)	Pietenpol AirCamper	PFA 1514		10. 4.78	A J.Mason	Bicester	12. 4.05P
	(Built A J.Mason) (Continental A65)				*"Edna May"*		
G-ADRH (2)	de Havilland DH.87B Hornet Moth	8038	G-ADRH	6. 8.82	R G Grocott	Mandeville, Gore, New Zealand	
	(Originally regd with c/n IMC/8164, amended 20.5.85)		F-AQBY, HB-OBE		*(On rebuild 2000 - reserved as ZK-ANR)*		
G-ADRR (2)	Aeronca C.3	A.734	N17423	6. 9.88	S.J.Rudkin	Roughay Farm, Bishops Waltham	
			NC17423		*(Stored 1992: wings @ Skycraft/Spalding 11.01)*		
G-ADSK*	de Havilland DH.87B Hornet Moth				See SECTION 3 Part 1	Mandeville, Gore, New Zealand	
G-ADUR	de Havilland DH.87B Hornet Moth	8085	N9026Y	10. 3.36	R.A.Seeley	Phoenix Farm, Lower Upham	1. 8.04
			G-ADUR				
G-ADWJ	de Havilland DH.82A Tiger Moth	3450	BB803	9.12.35	C.Adams	(Madley)	
			G-ADWJ		*(Under restoration 3.97)*		
G-ADWO*	de Havilland DH.82A Tiger Moth				See SECTION 3 Part 1	Southampton	
G-ADWT	Miles M.2W Hawk Trainer	215	CF-NXT	18.11.35	R.Earl and B.Morris	Manor Farm, Garford	27. 5.07P
			G-ADWT, NF750, G-ADWT				
G-ADXT	de Havilland DH.82A Tiger Moth	3436		9.12.35	J.R.Hanauer	Goodwood	26. 5.06
	(Mainly rebuild of components)						
G-ADYS	Aeronca C.3	A.600		?. 1.36	Janice I Cooper	Brickhouse Farm, Frogland Cross	6. 5.04P
	(Aeronca E113C)				*(London Air Park Flying Club titles)*		

G-AEAA - G-AEZZ

G-AEBB	Mignet HM.14 Pou-Du-Ciel	KWO.1		24. 1.36	Richard Shuttleworth Trustees	Old Warden	31. 5.39
	(Built K W Owen)						
G-AEBJ	Blackburn B.2	6300/8		4. 2.36	BAE SYSTEMS (Operations) Ltd	Warton	30. 6.05
	(DH Gipsy Major)						
G-AEDB	BAC Drone 2	13		18. 3.36	P.L.Kirk and R E Nerou	Hucknall	26. 5.87P
	(Cherub III)				*(Registered as BGA2731 31.3.81: composite with wings of G-AEJH and tail of G-AEEN: noted 7.01)*		
G-AEDT*	de Havilland DH.90 Dragonfly				See SECTION 3 Part 1	Mandeville, Gore, New Zealand	
G-AEDU (2)	de Havilland DH.90A Dragonfly	7526	N190DH	4. 6.79	A J.Norman	Langham	22. 7.05
			G-AEDU, ZS-CTR, CR-AAB		tr Norman Aeroplane Trust		
G-AEEG	Miles M.3A Falcon Major	216	SE-AFN	14. 3.36	P.R.Holloway	Old Warden	18. 6.07
			Fv913, SE-AFN, G-AEEG, U-20				
G-AEEH*	Mignet HM.14 Pou-Du-Ciel				See SECTION 3 Part 1	RAF Cosford	
G-AEFG*	Mignet HM.14 Pou-du-Ciel	JN.1	BAPC.75	27. 3.36	Anne Lindsay and N.H.Ponsford	Selby	
	(Built J Nolan)				*(Cancelled in 31.3.38 census) (Under restoration 2.04)*		
G-AEFT	Aeronca C.3	A.610		17. 4.36	N.C.Chittenden	(Herodsfoot, Liskeard)	8. 5.04P
	(JAP J.99)				*(Rebuilt in 1976 with major parts of G-AETG (qv) and carries c/n AB.110 thereof)*		
G-AEGV*	Mignet HM.14 Pou-Du-Ciel				See SECTION 3 Part 1	Coventry	
G-AEHM*	Mignet HM.14 Pou-Du-Ciel				See SECTION 3 Part 1	Wroughton	
G-AEKR*	Mignet HM.14 Pou-Du-Ciel				See SECTION 3 Part 1	Doncaster	
G-AEKV*	Kronfeld (BAC) Drone de luxe				See SECTION 3 Part 1	Brooklands	
G-AEKW*	Miles M.12 Mohawk				See SECTION 3 Part 1	RAF Cosford	
G-AELO	de Havilland DH.87B Hornet Moth	8105	AW118	30. 7.36	M.J.Miller	Audley End	27. 5.06
			G-AELO				
G-AEML	de Havilland DH.89 Dragon Rapide	6337	X9450	1. 9.36	Amanda Investments Ltd	Rendcomb	22. 5.05
			G-AEML		*"Proteus"*		
G-AENP (2)	Hawker Afghan Hind	41H/81902	(BAPC.78)	29.10.81	Richard Shuttleworth Trustees	Old Warden	3. 6.05P
	(Kestrel V)		R.Afghan AF		*(As "K5414 "in 15 Sqdn c/s)*		
G-AEOA	de Havilland DH.80A Puss Moth	2184	ES921	1.10.36	A and P A Wood	Audley End	27. 6.95P
	(DH Gipsy Major)		G-AEOA, YU-PAX, UN-PAX		t/a P and A Wood *(Current status unknown)*		
G-AEOF (2)	Rearwin 8500 Sportster	462	N15863	1.12.81	Shipping and Airlines Ltd	Biggin Hill	5.10.01P
	(Le Blond 5DF @ 85hp)		NC15863				
G-AEPH	Bristol F.2b	7575	D8096	13.11.36	Richard Shuttleworth Trustees	Old Warden	12. 4.05P
	(RR Falcon 3) *(Original c/n 3746 and rebuilt c.1931)*		G-AEPH, D8096		*(As "D8096")*		

G-AERV	Miles M.11A Whitney Straight	307	EM999	30.12.36	R.A Seeley	Hill Farm, Durley	9. 4.66
			G-AERV		*(New owner 5.02)*		
G-AERD*	Percival P.3 Gull Six				See SECTION 3 Part 1	Canberra, Australia	
G-AERN*	de Havilland DH.89A Dragon Rapide				See SECTION 3 Part 1	Caernarfon	
G-AESB (2)	Aeronca C.3	A.638	N15742	5. 8.88	R.J.M.Turnbull	Rydinghurst Farm, Cranleigh	
			NC15742		*(For rebuild 2002)*		
G-AESE	de Havilland DH.87B Hornet Moth	8108	W5775	13. 1.37	J.G.Green	(White Ox Mead, Bath)	4.10.01
			G-AESE		*"Sheena" (Current status unknown)*		
G-AESZ	Chilton DW.1	DW.1/1		?. 1.37	R.E.Nerou	Rendcomb	13. 8.05P
	(Carden Ford @ 32hp)						
G-AETA*	Caudron G.III				See SECTION 3 Part 1	Hendon	
G-AETL*	Miles M.14A Hawk Trainer III				See SECTION 3 Part 1	Albany, Auckland, New Zealand	
G-AEUJ	Miles M.11A Whitney Straight	313		19. 2.37	R.E.Mitchell	RAF Cosford	4. 6.70
					(Stored 4.02)		
G-AEVS	Aeronca 100	AB.114		3.37	R.A.Fleming	Breighton	5. 9.05P
	(JAP J.99) *(Composite including parts of original G-AEXD)*				*"Jeeves"*		
G-AEWU*	Aeronca 100	AB.116		3.37	(B Cox)	Brickhouse Farm, Frogland Cross	
					(DBR when portable hangar collapsed Farnborough 2.55) (Stored 2003)		
G-AEXD	Aeronca 100	AB.124		1. 4.37	M A and R W Mills	(London W7)	20. 4.70P
	(JAP J.99) *(Mainly comprises parts of G-AESP after rebuild in 1958)*				*(Stored 2003)*		
G-AEXF	Percival P.6 Mew Gull	E.22	ZS-AHM	18. 5.37	R.A Fleming	Breighton	9. 8.05P
	(Rebuilt as c/n PFA 13-10020)				*(Noted 1.05)*		
G-AEXT	Dart Kitten II	123		?. 4.37	A J.Hartfield	Marsh Hill Farm, Aylesbury	2. 7.05P
	(JAP J.99)						
G-AEXZ	Piper J-2 Cub	997		5. 2.38	J.R. and Mrs M.Dowson	(Leicester)	2.11.78S
	(Built Taylor Aircraft Co Inc) (Continental A75)				*(On rebuild: by 3.00 wings only and remainder with owner at home)*		
G-AEYF*	General Aircraft Monospar ST.25 Ambulance				See SECTION 3 Part 1	Billund, Denmark	
G-AEZJ	Percival P.10 Vega Gull	K.65	SE-ALA	2. 7.37	D.P.H.Hulme	Biggin Hill	10. 6.07
			D-IXWD, PH-ATH, G-AEZJ				
G-AEZX (2)	Bücker Bü.133C Jungmeister	1018	N5A	10. 5.88	T J Reeve	North Lopham, Diss	27. 7.00P
			PP-TDP		*(As "LG+03" in Luftwaffe c/s) (New owner 3.04)*		

G-AFAA - G-AFZZ

G-AFAX	BA Eagle 2	138	VH-ACN	26.10.37	J.G.Green and M.J.Miller	Audley End	9. 5.05
	(DH Gipsy Major)		G-AFAX				
G-AFBS*	Miles M.14A Hawk Trainer 3				See SECTION 3 Part 1	Duxford	
G-AFCL	BA L.25c Swallow II	462		3.11.37	M.Mordue and C.P.Bloxham	Shotteswell	16. 8.04P
	(Pobjoy Niagara 3)						
G-AFDO (2)	Piper J-3C-65 Cub	2593	N21697	7. 6.88	R.Wald	Hill Farm, Durley	27. 7.99P
	(Frame No.2633)		NC21697		*"Butter Cub" (On rebuild 12.02)*		
G-AFDX*	Hanriot HD.1				See SECTION 3 Part 1	Hendon	
G-AFEL (2)	Monocoupe 90A	A.782	N19432	7. 6.82	M.Rieser	(Germany)	22.12.05P
	(Lambert R266)		NC194323				
G-AFFD	Percival P.16A Q-Six	Q.21	(G-AIEY)	12. 2.38	B.D.Greenwood	(Sywell)	31. 8.56
			X9407, G-AFFD		*(On rebuild 2003)*		
G-AFFH	Piper J-2 Cub	1166	EC-ALA	26. 3.38	M.J.Honeychurch	(Pewsey)	29. 8.53
	(Built Taylor Aircraft Co Inc) (Continental A40)		G-AFFH		*(On restoration at owner's home 10.01)*		
G-AFGC	BA L.25c Swallow II	467	BK893	4. 4.38	G E.Arden	Thorns Cross Farm, Chudleigh	20. 3.51
	(Pobjoy Niagara 3)		G-AFGC		*(Stored 1.98: current status unknown)*		
G-AFGD*	BA L.25c Swallow II	469	BK897	4. 4.38	A T.Williams, B.Arden, C.A Cook, J.Hughes and M.Barnby		9. 4.01P
	(Pobjoy Cataract 3)		G-AFGD		tr South Wales Swallow Group	Shobdon	
					(Cancelled 6.4.01 by CAA) (Noted 12.04)		
G-AFGE	BA L.25c Swallow II	470	BK894	4. 4.38	G.R.French	Benson's Farm, Laindon	27. 7.98P
	(Pobjoy Niagara 2)		G-AFGE		*(Noted 2.04)*		
G-AFGH	Chilton DW.1	DW.1/2		20. 3.38	M.L. and G.L.Joseph	Denford Manor, Hungerford	7. 7.83P
	(Lycoming O-145-A2) *(To be re-engined with Carden-Ford)*				*(On rebuild 10.01)*		
G-AFGI	Chilton DW.1	DW.1/3		30. 3.38	J.E and K.A A McDonald	White Waltham	9. 8.05P
	(Walter Mikron 2)						
G-AFGK*	Miles M.11A Whitney Straight				See SECTION 3 Part 1	Wetaskiwin, Alberta, Canada	
G-AFGM (2)	Piper J-4A Cub Coupé	4-943	N26895	30.12.81	P.H.Wilkinson	Carlisle	10. 4.05P
			NC26895				
G-AFGZ	de Havilland DH.82A Tiger Moth	3700	G-AMHI	9. 5.38	M.R.Paul	Lee-on-Solent	25. 3.06
			BB759, G-AFGZ				
G-AFHA (2)	Moss MA.1	MA.1/2		27. 2.67	C.V.Butler	(Allesley, Coventry)	
					(Small components only stored)		
G-AFIN	Chrislea LC.1 Airguard	LC.1	BAPC.203	7. 7.38	N H Wright	Queach Farm, Bury St.Edmunds	
					(On rebuild 5.03 using original wings, tailplane and metal fittings)		
G-AFIR	Luton LA-4 Minor	JSS.2		7. 7.38	A J.Mason	(Aylesbury)	30. 7.71
	(Built J S Squires) (JAP J-99)				*(Damaged near Cobham 14.3.71 and on rebuild 2000)*		
G-AFJA	Taylor-Watkinson Dingbat	DB.100		2. 8.38	K.Woolley	(Berkswell, Coventry)	23. 6.75
	(Built Taylor Watkinson Aircraft Co) (Carden-Ford 32hp)				*(Damaged Headcorn 19.5.75 and partially rebuilt: stored 12.01)*		
G-AFJB	Foster-Wikner GM.1 Wicko	5	DR613	15. 8.38	J.Dibble	Hill Farm, Durley	12. 7.63
	(DH Gipsy Major 1)		G-AFJB		*(On rebuild 5.02)*		
G-AFJR*	Tipsy Trainer 1				See SECTION 3 Part 1	Brussels, Belgium	
G-AFJU*	Miles M.17 Monarch				See SECTION 3 Part 1	East Fortune	
G-AFJV (2)	Moss MA.2	MA.2/2		27. 2.67	C.V.Butler	(Allesley, Coventry)	
					(Small components only stored)		
G-AFLW*	Miles M.17 Monarch	792		2.11.38	N.I.Dalziel	White Waltham	30. 7.98
					(Cancelled 3.5.01 by CAA)		
G-AFNG	de Havilland DH.94 Moth Minor	94014	AW112	2. 5.39	D.Saunders	Carnmore, Galway	21.10.98P
			G-AFNG		tr The Gullwing Trust *(Noted 8.04)*		

Registration	Type	C/n	Previous identity	Regd	Owner/Operator	Base	CofA
G-AFNI	de Havilland DH.94 Moth Minor	94035	W7972	11. 5.39	J.Jennings	Fenland	26. 5.67
			G-AFNI		*(On rebuild 2002)*		
G-AFNJ*	de Havilland DH.94 Moth Minor				See SECTION 3 Part 1	La Ferté-Alais, Essonne, France	
G-AFOB	de Havilland DH.94 Moth Minor	94018	X5117	16. 5.39	K.Cantwell	(Royston)	11. 5.93P
			G-AFOB		*(New owner 2.02)*		
G-AFOJ	de Havilland DH.94 Moth Minor	9407	E-1	21. 7.39	R.M.Long *"Bugs 2"*	Salisbury Hall, London Colney	27. 8.69P
	(Cabin)		E-0236, G-AFOJ		*(On loan to de Havilland Heritage Museum) (Noted 5.04)*		
G-AFON*	de Havilland DH.94 Moth Minor				See SECTION 3 Part 1	Albany, Auckland, New Zealand	
G-AFOR*	de Havilland DH.94 Moth Minor				See SECTION 3 Part 1	Caboolture, Queensland, Australia	
G-AFOW*	de Havilland DH.94 Moth Minor				See SECTION 3 Part 1	Albury, New South Wales, Australia	
G-AFPN	de Havilland DH.94 Moth Minor	94044	X9297	23. 5.39	J.W and A R.Davy	Redhill	30. 5.05
	(Now regd with c/n 94016)		G-AFPN				
G-AFPR*	de Havilland DH.94 Moth Minor				See SECTION 3 Part 1	Mandeville, Gore, New Zealand	
G-AFRV*	Tipsy Trainer I				See SECTION 3 Part 1	Brussels, Belgium	
G-AFRZ	Miles M.17 Monarch	793	G-AIDE	24. 3.39	R.E.Mitchell	RAF Cosford	29. 6.70
			W6463, G-AFRZ		*(Believed stored 4.02)*		
G-AFSC	Tipsy Trainer 1 (Walter Mikron 2)	11		15. 7.39	D.M.Forshaw	Panshanger	7.10.04P
G-AFSH*	de Havilland DH.82A Tiger Moth				See SECTION 3 Part 1	Masterton, New Zealand	
G-AFSV	Chilton DW.1A	DW.1A/1		5. 4.39	R.E.Nerou	(Coventry)	12. 7.72
	(Train 45hp)				*(For rebuild 8.02)*		
G-AFSW*	Chilton DW.2	DW.2/1		6. 4.39	R.I.Souch	(Hedge End, Southampton)	
			(Not completed originally: fuselage box in poor condition: stored 1.00: current status unknown)				
G-AFTA	Hawker Tomtit	30380	K1786	26. 4.39	Richard Shuttleworth Trustees	Old Warden	31. 7.05P
	(Mongoose 3C)		G-AFTA, K1786		*(As "K1786")*		
G-AFTN*	Taylorcraft Plus C2				See SECTION 3 Part 1	Coalville	
G-AFUP (2)	Luscombe 8A Master	1246	N25370	7. 6.88	R.Dispain	(Fordingbridge)	12. 3.97P
	(Continental A65)		NC25370				
G-AFUU*	de Havilland DH.94 Moth Minor				See SECTION 3 Part 1	Mandeville, Gore, New Zealand	
G-AFVE (2)	de Havilland DH.82A Tiger Moth	83720	T7230	1. 2.78	W.N.Gibson	Booker	30. 4.07T
	(Built Morris Motors Ltd)				t/a Tigerfly *(As "T-7230")*		
G-AFVN	Tipsy Trainer 1	12		15. 7.39	D.F.Lingard	Fenland	2. 1.03P
	(Walter Mikron 2)						
G-AFWH (2)	Piper J-4A Cub Coupé	4-1341	N33093	14. 1.82	C.W.Stearn and R.D.W.Norton	(Ely)	2. 7.01P
	(Continental A65)		NC33093		*(Stored 4.04)*		
G-AFWI	de Havilland DH.82A Tiger Moth	82187	BB814	19. 7.39	E.Newbigin		
			G-AFWI			Brown Shutters Farm, Norton St.Philips, Somerset	12. 8.06
G-AFWN*	Auster J/1 Autocrat				See SECTION 3 Part 1	Stauning, Denmark	
G-AFWT	Tipsy Trainer 1	13		1. 8.39	N.Parkhouse	Chelwood Gate, West Sussex/Redhill	25. 4.05P
	(Walter Mikron 2)						
G-AFYD (2)	Luscombe 8AF Silvaire	1044	N25120	29. 7.75	J.D.Iliffe	Chilbolton	12. 9.07
	(Continental C90)		NC25120				
G-AFYO (2)	Stinson HW-75 Voyager	7039	F-BGQP	25. 4.77	M.Lodge	Westfield Farm, Hailsham	24. 5.05P
			NC22586 *(Probably ex Fr.Mil with identity "22586")*				
G-AFZA (2)	Piper J-4A Cub Coupé	4-873	N26198	27. 6.84	R.A.Benson	Trenchard Farm, Eggesford	2.12.02P
	(Continental A65)		NC26198		*(New owner 8.04)*		
G-AFZK (2)	Luscombe 8A Master	1042	N25118	24.10.88	M.G.Byrnes	Haverfordwest	19. 5.05P
	(Continental A65)		NC25118				
G-AFZL (2)	Porterfield CP-50	581	N25401	18. 3.82	P.G.Lucus and S.H.Sharpe	White Waltham	5. 7.05P
	(Continental A50)		NC25401				
G-AFZN (2)	Luscombe 8A Master	1186	N25279	5.10.81	J.L.Truscott	Henstridge	25. 7.05P
	(Continental A65)		NC25279				

G-AGAA - G-AGZZ

Registration	Type	C/n	Previous identity	Regd	Owner/Operator	Base	CofA
G-AGAT (2)	Piper J-3F-50 Cub	4062	N26126	17. 7.87	O.T.Taylor and C.J.Marshall	Haverfordwest	15.10.05P
	(Franklin 4AC-150)		NC26126				
G-AGBN*	General Aircraft GAL.42 Cygnet 2				See SECTION 3 Part 1	East Fortune	
G-AGCN*	Lockheed 18-56 Lodestar II (C-56D-LO)				See SECTION 3 Part 1	Auckland, New Zealand	
G-AGEG (2)	de Havilland DH.82A Tiger Moth	82710	N9146	16. 8.82	A J.Norman tr Norman Aeroplane Trust	Rendcomb	15. 5.07
			D-EDIL, R.Neth AF A-32, PH-UFK, A-32, R4769				
G-AGFT (2)	Avia FL.3	176	I-TOLB	21. 8.84	K.Joynson and K.Cracknell	Breighton	7. 1.05P
	(CNA D4S)		MM....		*(As "W7" in Italian Co-Belligerent AF c/s)*		
G-AGHY (2)	de Havilland DH.82A Tiger Moth	82292	N9181	17. 2.88	P.Groves	Lee-on-Solent	22.12.06
G-AGIJ*	Lockheed 18-56 Lodestar II				See SECTION 3 Part 1	Stockholm, Sweden	
G-AGIP*	Douglas C-47A-1-DK Dakota				See SECTION 3 Part 1	Florida, USA	
G-AGIV (2)	Piper J-3C-65 Cub (L-4J-PI)	12676	OO-AFI	13. 8.82	P.C.and F.M.Gill	Waits Farm, Belchamp Walter	7. 1.04P
	(Frame No.12506)		OO-GBA, 44-80380				
G-AGJG	de Havilland DH.89A Dragon Rapide	6517	X7344	25.10.43	M.J. and D.J.T.Miller	Duxford	29. 6.07T
G-AGLK	Taylorcraft J Auster 5D	1137	RT475	25. 8.44	C R Harris	Rochester	28. 4.07
G-AGMI (2)	Luscombe 8A Master	1569	N28827	15.11.88	P.R.Bush	RAF Kinloss	9. 4.05P
	(Continental A65)		NC28827				
G-AGNJ (2)	de Havilland DH.82A Tiger Moth	660	VP-YOJ	21. 2.89	B.P, A J. and P.J.Borsberry	Kidmore End, Reading	
	(Built de Havilland Aircraft Pty Ltd, Australia)		ZS-BGF, SAAF 2366		*(On rebuild 6.95) (Current status unknown)*		
G-AGNV*	Avro 685 York C.1				See SECTION 3 Part 1	Cosford	
G-AGOH	Auster V J/1 Autocrat	1442		19. 4.45	Leicestershire County Council Museums	Winthorpe	24. 8.95
					(On loan to Newark Air Museum)		
G-AGOS*	Reid and Sigrist RS.4 Desford Trainer				See SECTION 3 Part 1	Coalville	
G-AGOY	Miles M.48 Messenger 3	4690	EI-AGE	5. 6.45	P.A Brook	West Chiltington, Pulborough	25.11.53
			G-AGOY, HB-EIP, G-AGOY, U-0247 *(On rebuild 4.92: as "U-0247")*				
G-AGPG*	Avro 652A Anson 19 Series 2				See SECTION 3 Part 1	Hooton Park	
G-AGPK (2)	de Havilland DH.82A Tiger Moth	86566	N657DH	27.10.88	Delta Aviation Ltd	Sywell	23. 4.05T
			F-BGDN, Fr AF, PG657		*(Fitted with wings, tailplane, fin and rudder ex G-ANLH 2002)*		
G-AGRU*	Vickers 657 Viking 1				See SECTION 3 Part 1	Brooklands	
G-AGRW*	Vickers 639 Viking 1				See SECTION 3 Part 1	Vienna, Austria	

Regn	Type	C/n	Previous identity	Regd	Owner / operator	Base	CA expiry
G-AGSH	de Havilland DH.89A Dragon Rapide 6 *(Built Brush Coachworks Ltd)*	6884	EI-AJO G-AGSH, NR808	25. 7.45	Techair London Ltd *(BEA titles) "Jemma Meeson"*	Bournemouth	20. 9.07
G-AGTB*	Percival P.44 Proctor V				See SECTION 3 Part 1	Kogarah, New South Wales, Australia	
G-AGTM	de Havilland DH.89A Dragon Rapide 6 *(Built Brush Coachworks Ltd)*	6746	JY-ACL D-ABP, G-AGTM, NF875	19. 9.45	Aviation Heritage Ltd	Coventry	13. 5.06T
G-AGTO	Auster V J/1 Autocrat	1822		2.10.45	M.J.Barnett and D.J.T.Miller	Duxford	9. 3.06
G-AGTT	Auster V J/1 Autocrat	1826		2.10.45	R.Farrer *(Stored 12.97: current status unknown)*	(Bromham, Bedford)	11. 2.93
G-AGVG	Auster V J/1 Autocrat (Lycoming O-360-A2A) *(Modified tail surfaces)*	1858		7.12.45	S.J.Riddington	Leicester	15. 6.06
G-AGVN	Auster V J/1 Autocrat	1873	EI-CKC G-AGVN	18. 1.46	R J Bentley	Pallas West, Toomyvara, County Tipperary	28. 9.05
G-AGVV (2)	Piper J-3C-65 Cub (L-4H-PI)	11163	F-BCZK Fr.AF, 43-29872	19. 2.81	M.Molina-Ruano	(Malaga, Spain)	2. 9.04P
G-AGWE*	Avro 19 Series 2				See SECTION 3 Part 1	Tico, Florida, USA	
G-AGXN	Auster J/1N Alpha	1963		22. 1.46	I.M.Godfrey-Davies tr Gentleman's Aerial Touring Carriage Group	Barton Ashes	16. 5.05
G-AGXS*	Auster V J/1 Autocrat				See SECTION 3 Part 1	Sils-Girona, Spain	
G-AGXU	Auster J/1N Alpha	1969		24. 1.46	B.H.Austen	Oaksey Park	13. 8.06
G-AGXV	Auster V J/1 Autocrat	1970		1. 2.46	B.S.Dowsett and I.M.Oliver *"Pamela IV"*	Little Gransden	18. 9.06
G-AGYD	Auster J/1N Alpha	1985		4. 2.46	P.R.Hodson *(Damaged near Felthorpe 25.11.90: on rebuild 4.94: current status unknown)*	Little Gransden	24.11.90
G-AGYH	Auster J/1N Alpha *(Built Taylor Aeroplanes)*	1989		4. 2.46	I.M.Staves *(New owner 1.05)*	(Northallerton)	10.10.72S
G-AGYK	Auster V J/1 Autocrat	2002		4. 2.46	M.C.Hayes tr Autocraft Syndicate	Shobdon	17. 8.07
G-AGYT	Auster J/1N Alpha	1862		18. 1.46	P.J.Barrett *(On overhaul 6.94: current status unknown)*	(Lightwater)	27. 2.91
G-AGYU	de Havilland DH.82A Tiger Moth *(Built Morris Motors Ltd)*	85265	DE208	10. 1.46	P.L.Jones *(As "DE208")*	Tuam, Galway, County Galway	14. 9.07
G-AGYX*	Douglas C-47A-10DK Dakota 3				See SECTION 3 Part 1	Hendon	
G-AGYY (2)	Ryan ST3KR (PT-21-RY) (Kinner R56)	1167	N56792 41-1947	15. 6.83	G.G.Smit tr Dutch Nostalgic Wings *(As "27" in USAAC c/s)*	Hoogeveen, The Netherlands	3. 7.05P
G-AGZF*	Douglas C-47A-1-DL Dakota				See SECTION 3 Part 1	Val-de-Marne, France	
G-AGZI*	Consolidated-Vultee CV.32-3 (LB-30) Liberator II				See SECTION 3 Part 1	Anchorage, Alasaka, USA	
G-AGZZ (2)	de Havilland DH.82A Tiger Moth *(Built de Havilland Aircraft Pty Ltd, Australia)*	T256 & 926	N3862 VH-BTU, VH-RNM, VH-BMY, A17-503	14. 5.82	R.C.Mercer	(Pewsey)	13. 6.07

G-AHAA - G-AHZZ

Regn	Type	C/n	Previous identity	Regd	Owner / operator	Base	CA expiry
G-AHAA	Miles M.28 Mercury 6				See SECTION 3 Part 1	Stauning, Denmark	
G-AHAG	de Havilland DH.89A Dragon Rapide *(Built Brush Coachworks Ltd)*	6926	RL944	31. 1.46	Pelham Ltd *(On rebuild 10.01)*	Membury	15. 7.73
G-AHAL	Auster J/1N Alpha *(Built Taylor Aeroplanes)*	1870		31. 1.46	J.W.Frecklington and R.Merwood t/a Wickenby Aviation	Wickenby	4. 5.07T
G-AHAM	Auster V J/1 Autocrat	1885		21. 1.46	C.P.L.Jenkins	Church Farm, Shotteswell	26. 3.06
G-AHAN (2)	de Havilland DH.82A Tiger Moth *(Built Morris Motors Ltd)*	86553	N90406 F-BGDG, Fr.AF, PG644	31. 5.85	Tiger Associates Ltd	White Waltham	9. 6.07T
G-AHAP	Auster V J/1 Autocrat (Rover V-8)	1887		8. 2.46	W.D.Hill *(New owner 6.04)*	(March)	20. 2.91P
G-AHAT*	Auster J/1N Alpha				See SECTION 3 Part 1	Dumfries	
G-AHAU	Auster V J/1-160 Autocrat (Lycoming O-320-A) *(Built-up fin/fuselage fillet)*	1850	(HB-EOL)	11. 2.46	A C.Webber	Andreas, Isle of Man	21. 4.06
G-AHAV	Auster V J/1N Alpha *(Built Taylor Aeroplanes)*	1863	(HB-EOM)	13. 2.46	J.S.Antrum *(New CofR 10.04)*	Headcorn	21. 6.75
G-AHAY*	Auster V J/1 Autocrat				See SECTION 3 Part 1	Beersheba, Israel	
G-AHBL	de Havilland DH.87B Hornet Moth	8135	P6786 CF-BFN	6. 2.46	H.D.Labouchere	Blue Tile Farm, Langham	17. 6.06
G-AHBM	de Havilland DH.87B Hornet Moth	8126	P6785 CF-BFJ, (CF-BFO), CF-BFJ	6. 2.46	P.A and E.P.Gliddon	Redhill	21. 6.02
G-AHCF*	Auster V J/1 Autocrat				See SECTION 3 Part 1	Burgos, Spain	
G-AHCL	Auster J/1N Alpha *(Originally regd as J/1 Autocrat)* (Lycoming 0-320-A2B)	1977	G-OJVC G-AHCL	13. 5.46	Electronic Precision Ltd	RAF Mona	16. 9.07
G-AHCN	Auster J/1N Alpha *(Originally regd as J/1 Autocrat)*	1980	OY-AVM G-AHCN, OY-AVM, G-AHCN	25. 3.46	G.L.Brown and E.Martinsen *(Current status unknown)*	(Peterborough/Hornbaek, Denmark	27. 3.79
G-AHCR	Gould-Taylorcraft Plus D Special (Continental C90)	211	LB352	15. 4.46	D.E.H.Balmford and D.R.Shepherd tr Wagtail Flying Group	Dunkeswell	14.10.04P
G-AHDI*	Percival P.28 Proctor I				See SECTION 3 Part 1	Melbourne, Victoria, Australia	
G-AHEC (2)	Luscombe 8A Silvaire (Continental A65)	3428	N72001 NC72001	28.10.88	P.G.Baxter	Hill Farm, Nayland	14. 7.05P
G-AHED*	de Havilland DH.89A Dragon Rapide 6				See SECTION 3 Part 1	RAF Stafford	
G-AHGD	de Havilland DH.89A Dragon Rapide *(Built Brush Coachworks Ltd)*	6862	NR786	1. 4.46	R.Jones t/a Southern Sailplanes *(Destroyed near.Audley End 30.6.91: components for possible rebuild 8.97)*	Membury	20. 9.92
G-AHGW	Taylorcraft Plus D	222	LB375	2. 9.46	C.V.Butler *(Op Military Auster Flight as "LB375")*	Shenington	3. 5.96P
G-AHGZ	Taylorcraft Plus D	214	LB367	24. 4.46	M.Pocock *(As "LB367") (Op Stephen White)*	Duxford	13.10.05
G-AHHE*	Auster V J/1 Autocrat				See SECTION 3 Part 1	Hamilton East, New Zealand	
G-AHHH	Auster J/1N Alpha	2011	F-BAVR G-AHHH	11. 5.46	H.A Jones	RAF Coltishall	21.12.05P
G-AHHK*	Auster J/1 Autocrat	2014		11. 5.46	C.J.Baker *(Cancelled 3.4.89 as WFU) (Derelict frame noted 1.05)*	Carr Farm, Thorney, Newark	22. 3.70
G-AHHT	Auster J/1N Alpha	2022		11. 5.46	A C.Barber and N.J.Hudson tr Southdowns Auster Group	Durleighmarsh Farm, Rogate	31. 5.04

G-AHIP (2)	Piper J-3C-65 Cub (L-4H-PI)	12122	OO-GEJ (2)	3. 7.85	A R.M.Cot-Croft	Rayne Hall Farm, Braintree	11.12 03P
	(Frame No.11950)		OO-ALY, 44-79826		*(Noted 1.05)*		
	(Officially regd with c/n 12008: see G-AJAD)						
G-AHIZ	de Havilland DH.82A Tiger Moth	86533	PG624	23. 4.46	CFG Flying Ltd	Cambridge	1. 6.06T
	(Built Morris Motors Ltd) (Regd with fuselage no.4610)						
G-AHJR (2)*	Short S.25 Sunderland MR.5				See SECTION 3 Part 1	Auckland, New Zealand	
G-AHKO*	Taylorcraft Plus D				See SECTION 3 Part 1	Stauning, Denmark	
G-AHKX	Avro 19 Series 2	1333		18. 5.46	BAE SYSTEMS (Operations) Ltd	Old Warden	21. 4.05P
G-AHKY*	Miles M.18 Series 2				See SECTION 3 Part 1	Eastt Fortune	
G-AHLK	Taylorcraft E Auster III	700	NJ889	1. 5.46	E.T.Brackenbury *(On rebuild 1.05)*	Leicester	21. 9.97
G-AHLO*	de Havilland DH.80A Puss Moth				See SECTION 3 Part 1	Winnipeg, Manitoba, Canada	
G-AHLT	de Havilland DH.82A Tiger Moth	82247	N9128	2. 5.46	M P Waring	(London W10)	13 8.06
G-AHLX*	Douglas C-47A-30-DK Dakota				See SECTION 3 Part 1	Belgrade, Serbia	
G-AHMJ*	Cierva C.30A				See SECTION 3 Part 1	Polk City, Florida ,USA	
G-AHNR (2)	Taylorcraft BC-12D	7204	N43545	15.11.88	T.P.Hancock	Leicester	2. 6.05P
	(Continental A65)		NC43545				
G-AHOO (2)	de Havilland DH.82A Tiger Moth	86150	6940M	6. 6.85	J.T.and A D.Milsom	Little Farm, Hampstead Marshall	10. 8.06
	(Built Morris Motors Ltd) (Regd with c/n 86149)		EM967				
G-AHOT*	Vickers 498 Viking 1A				See SECTION 3 Part 1	Johannesburg, South Africa	
G-AHPB*	Vickers 639 Viking 1				See SECTION 3 Part 1	Winterthur, Switzerland	
G-AHPZ	de Havilland DH.82A Tiger Moth	83794	EI-AFJ	22. 5.46	N.J.Wareing	Lee-on-Solent	19. 5.07
			G-AHPZ, T7280				
G-AHRI	de Havilland DH.104 Dove 1B				See SECTION 3 Part 1	Newark	
G-AHRO (2)	Cessna 140	8069	N89065	25. 1.82	R.H.Screen	Ashcroft Farm, Winsford	20. 7.06
			NC89065				
G-AHSA	Avro 621 Tutor	-	K3215	21. 6.46	Richard Shuttleworth Trustees	Old Warden	30. 8.05P
	(Lynx IVM)		G-AHSA, K3215		*(As "K3241" in RAF Central Flying School 1930s c/s)*		
G-AHSD	Taylorcraft Plus D	182	LB323	1. 7.46	A L.Hall-Carpenter	(Shipdham)	10. 9.62
					(On rebuild 2.99: current status unknown)		
G-AHSO	Auster J/1N Alpha	2123		8. 8.46	W.P.Miller	Mavis Enderby	6. 4.95T
					(On rebuild 8.98: current status unknown)		
G-AHSP	Auster V J/1 Autocrat	2134	F-BGRO	8. 8.46	R.M.Weeks	Earls Colne	1. 6.07
			G-AHSP				
G-AHSS	Auster J/1N Alpha	2136		8. 8.46	A M.Roche *"Sunday Sierra"*	Great Massingham	21. 7.06
G-AHTE	Percival P.44 Proctor V	Ae58		26. 6.46	D.K.Tregilgas	New Farm House, Great Oakley	10. 8.61
					(Stored for rebuild 7.02 - new CofR 10.04)		
G-AHTN*	Percival P.28 Proctor I				See SECTION 3 Part 1	Canberra, Australia	
G-AHTW*	Airspeed AS.40 Oxford 1				See SECTION 3 Part 1	Duxford	
G-AHUF (2)	de Havilland DH.82A Tiger Moth	86221	A2123	26. 2.85	Dream Ventures Ltd	Full Sutton	22. 8.05T
	(Built Morris Motors Ltd)		NL750		*(As "T7997") (Also see G-AOBH)*		
G-AHUG	Taylorcraft Plus D	153	LB282	5. 6.46	D.Nieman *(Current status unknown)*	(Thame)	12. 7.70
G-AHUJ*	Miles M.14A Hawk Trainer 3	1900	R1914	6. 6.46	Not known	Strathallan	9. 7.98P
					(Cancelled 19.11.99 as WFU) (Stored as "R1914" 2004)		
G-AHUN (2)	Globe GC-1B Swift	3536/766	EC-AJK	24. 7.86	R J Hamlet	Kings Farm, Thurrock	4. 8.95P
			OO-KAY, NC77764		*(Noted 2.00)*		
G-AHUV	de Havilland DH.82A Tiger Moth	3894	N6593	24. 6.46	A D.Gordon	Blair Athol	10. 7.06
G-AHVG*	Percival P.30 Proctor II				See SECTION 3 Part 1	Alice Springs, Australia	
G-AHVU	de Havilland DH.82A Tiger Moth	84728	T6313	14. 8.46	R.A L.Hubbard *(As "T6313")*	West Meon	19. 3.06
G-AHVV	de Havilland DH.82A Tiger Moth	86123	EM929	24. 6.46	B.M.Pullen	Barton Ashes	14. 10.06
	(Built Morris Motors Ltd)						
G-AHWJ	Taylorcraft Plus D	165	LB294	20. 6.46	M.Pocock *(New owner 1.03)*	Kemble	30. 6.71
G-AHXE	Taylorcraft Plus D	171	LB312	9. 7.46	Jenny M.C.Pothecary *(As "LB312")*	AAC Netheravon	17. 7.05P
G-AHXW*	de Havilland DH.89A Dragon Rapide				See SECTION 3 Part 1	Oshkosh, Wisconsin, USA	
G-AHZY*	Percival P.44 Proctor V				See SECTION 3 Part 1	Brussels, Belgium	

G-AIAA - G-AIZZ

G-AIBA*	Douglas C-47A-40-DL Dakota				See SECTION 3 Part 1	Reykjavik, Iceland	
G-AIBE*	Fairey Fulmar 2				See SECTION 3 Part 1	Yeovilton	
G-AIBH	Auster J/1N Alpha	2113		19. 8.46	M.J.Bonnick	Standalone Farm, Meppershall	18. 4 98
G-AIBM	Auster V J/1 Autocrat	2148		2. 9.46	D.G.Greatrex	Colthrop Manor, Thatcham	9.11.04
G-AIBR	Auster V J/1 Autocrat	2151		2. 9.46	P.R.Hodson	Felthorpe	22. 7.06T
G-AIBV*	Auster J/1N Alpha				See SECTION 3 Part 1	Ardmore, Auckland, New Zealand	
G-AIBW	Auster J/1N Alpha	2158		2. 9.46	W.B.Bateson *(Stored 12.01)*	Blackpool	4. 5.97T
G-AIBX	Auster V J/1 Autocrat	2159		2. 9.46	B.H.Beeston tr The Wasp Flying Group	Little Gransden	5.12.05
G-AIBY	Auster V J/1 Autocrat	2160		2. 9.46	D.Morris	Sherburn-in-Elmet	13. 4.81
					(Stored 6.97: current status unknown)		
G-AICH*	Bristol 170 Freighter Mk.1A				See SECTION 3 Part 1	Buenos Aires, Argentina	
G-AICX (2)	Luscombe 8A Silvaire	2568	N71141	27. 1.88	R.V.Smith	Dunkeswell	8. 7.05P
	(Continental A65)		NC71141		*"Easy Grace"*		
G-AIDL	de Havilland DH.89A Dragon Rapide 6	6968	TX310	23. 8.46	Atlantic Air Transport Ltd	Coventry	13. 6.05T
	(Built Brush Coachworks Ltd)						
G-AIDN*	Vickers Supermarine 502 Spitfire T.8				See SECTION 3 Part 1	Tillamook, Oregon, USA	
G-AIDS	de Havilland DH.82A Tiger Moth	84546	T6055	22. 8.46	K.D.Pogmore and T.Dann	Benson's Farm, Laindon	11. 8.06
					"The Sorcerer"		
G-AIEK	Miles M.38 Messenger 2A	6339	U-9	27. 8.46	J.Buckingham	New Farm, Felton	7. 9.06
					(As "RG333" in 2 TAF Comm Sqdn c/s)		
G-AIFZ	Auster J/1N Alpha	2182		2.11.46	M.D.Anstey	Rushett Farm, Chessington	20. 8.01
					(Noted 2004)		
G-AIGD	Auster V J/1 Autocrat	2186		2.11.46	R.B.Webber	Trenchard Farm, Eggesford	5. 6.04P
	(Officially regd incorrectly as J/1N Alpha)						
G-AIGF	Auster J/1N Alpha	2188		5.11.46	A R.C.Mathie	(Southampton)	17.11.05
G-AIGH*	Auster V J/1 Autocrat				See SECTION 3 Part 1	Burgos, Spain	

Regn	Type	C/n	Previous identity	Regd	Owner	Base	CofA
G-AIGT	Auster J/1N Alpha	2176		12.10.46	R.R.Harris *(New owner 4.02)*	(Crowfield)	22.10.76
G-AIIH	Piper J-3C-65 Cub (L-4H-PI)	11945	44-79649	14. 9.46	J.A de Salis	Oxford	4. 7.05P
G-AIJI*	Auster J/1N Alpha	2307		15. 4.47	C.J.Baker	Carr Farm, Thorney, Newark	
	(Originally regd as J/1 Autocrat)	*(Damaged in gales Humberside 12.1.75: cancelled 12.3.75 as WFU) (Dismantled frame and tail section noted 1.05)*					
G-AIJK*	Auster V J/4				See SECTION 3 Part 1	Coalville	
G-AIJM	Auster V J/4	2069	EI-BEU	13.11.46	N.Huxtable *"Priscilla"*	Cheddington	28. 3.97
			G-AIJM		*(Damaged near Tring 5.1.97: stored pending overhaul/repairs)*		
G-AIJT	Auster V J/4 Series 100	2075		13.11.46	J.L.Thorogood	Pittrichie Farm, Whiterashes	9. 5.05
	(Continental O-200-A)				tr The Aberdeen Auster Flying Group		
G-AIKE	Taylorcraft J Auster 5	1097	NJ728	15.11.46	C.J.Baker	Carr Farm, Thorney, Newark	3. 2.66
	(Built Taylorcraft Aeroplanes Ltd) (Frame No.TAY 2450)				*(Crashed Luton 1.9.65 and dismantled 1.05)*		
G-AIKR*	Airspeed AS.65 Consul				See SECTION 3 Part 1	Ottawa, Ontario, Canada	
G-AIMI*	Bristol 170 Freighter 21E				See SECTION 3 Part 1	Point Cook, Victoria,Australia	
G-AINT*	Bristol 170 Freighter 31MNZ				See SECTION 3 Part 1	Christchurch, New Zealand	
G-AIPE*	Taylorcraft J Auster 5				See SECTION 3 Part 1	Rosendaal, The Netherlands	
G-AIPR	Auster V J/4	2084		9. 1.47	R.W. and Mrs M.A Mills	Popham	27. 5.00P
					tr The MPM Flying Group		
G-AIPV	Auster V J/1 Autocrat	2203		9. 1.47	W.P.Miller *"Buttercup"*	Mavis Enderby	7. 2.05
G-AIPW*	Taylorcraft Auster 5A Series 160				See SECTION 3 Part 1	Amman, Jordan	
G-AIRC	Auster V J/1 Autocrat	2215		13. 1.47	A Noble	Perth	2. 8.07
G-AIRI*	de Havilland DH.82A Tiger Moth	3761	N5488	22.10.46	E.R.Goodwin	Little Gransden	9.11.81
					(Cancelled 3.4.89 as WFU) (Stored 10.01)		
G-AIRK	de Havilland DH.82A Tiger Moth	82336	N9241	22.10.46	R.C.Teverson, R.W.Marshall and C.E.McKinney		
					tr The Belchamp Tiger Moth Group		
						Waits Farm, Belchamp Walter	1. 8.07
G-AISA	Tipsy B Series 1	17		24. 4.47	A A M.and C.W.N.Huke	(Salisbury)	13. 4.05P
G-AISB*	Tipsy B Series 1				See SECTION 3 Part 1	Keiheuvel, Belgium	
G-AISC	Tipsy B Series 1	19		24. 4.47	D.R.Shepherd	(Pretswick)	23. 5.79P
					tr Wagtail Flying Group *(Stored 5.02)*		
G-AISD*	Miles M.65 Gemini 1A	6285	OO-RLD	26.11.46	J E Homewood	(Spanhoe)	
			VP-KDH, G-AISD		*(Cancelled 7.1.88 - to OO-RLD: noted as OO-RLD 5.04)*		
G-AISS (2)	Piper J-3C-65 Cub (L-4H-PI)	12077	D-ECAV	3. 9.85	K.W.Wood and F.Watson	Insch	25. 6.97P
	(Frame No.11904)		SL-AAA, 44-79781		*(Under restoration 5.04 as "479781" in USAAF Army Air Corps c/s)*		
G-AIST	Vickers Supermarine 300 Spitfire IA	WASP/20/2	AR213	25.10.46	Sheringham Aviation UK Ltd	Booker	26. 3.02P
	(Built Westland Aircraft Ltd) (Also Heston Aircraft Company c/n HA1 6S/5 139)				*(As "AR213/PR-D" in 609 Sqdn c/s)*		
G-AISU*	Vickers Supermarine 349 Spitfire LF.VB				See SECTION 3 Part 1	RAF Coningsby	
G-AISX	Piper J-3C-85 Cub (L-4H-PI)	11663	43-303722	8.10.46	A M Turney tr Cubfly	Booker	21. 5.05P
	(Frame No.11489) (Rebuilt with ex Spanish airframe)						
G-AITB*	Airspeed AS.40 Oxford 1				See SECTION 3 Part 1	Hendon	
G-AITF*	Airspeed AS.40 Oxford 1				See SECTION 3 Part 1	Pretoria, South Africa	
G-AIUA	Miles M.14A Hawk Trainer 3	2035	T9768	11.11.46	K.P.Hunt and D.L.Hunt *(Noted 1.04)*	Redhill	13. 7.67P
	(Wings fitted 1960s fromG-ANWO: crashed Roborough 26.9.65 and original centre section used to rebuild G-AKPF: fuselage stored)						
G-AIVG*	Vickers 610 Viking 1B				See SECTION 3 Part 1	Mulhouse, France	
G-AIWY*	de Havilland DH.89A Dragon Rapide				See SECTION 3 Part 1	Billund, Denmark	
G-AIXA*	Taylorcraft Plus D				See SECTION 3 Part 1	Hendon	
G-AIXJ	de Havilland DH.82A Tiger Moth	85434	DE426	28.11.46	D.Green	(Pulborough)	13. 8.06
	(Built Morris Motors Ltd) (Probably composite airframe rebuilt by Newbury Aeroplane Co 1991)						
G-AIXN	Automobilove Zavody Mraz M.1C Sokol	112	OK-BHA	22. 4.47	A J.Wood	Breighton	10. 2.05P
G-AIYG (2)	SNCAN Stampe SV-4B	21	OO-CKZ	31. 8.89	R.Lageirse and H.Ewaut	(Drongen, Belgium)	7. 6.03
	(DH Gipsy Major 10)		F-BCKZ, Fr Mil				
G-AIYR	de Havilland DH.89A Dragon Rapide	6676	HG691	11.12.46	Spectrum Leisure Ltd	Duxford	1. 5..05T
	(Built Brush Coachworks Ltd)				*(Op Classic Wings as "HG691")*		
G-AIYS	de Havilland DH.85 Leopard Moth	7089	YI-ABI	16.12.46	R.A and V.A Gammons	RAF Henlow	18. 5.07
			SU-ABM				
G-AIYT*	Douglas C-47A-10-DK Dakota				See SECTION 3 Part 1	Beersheba,.Israel	
G-AIZE*	Fairchild F.24W-41A Argus II				See SECTION 3 Part 1	RAF Cosford	
G-AIZG*	Vickers Supermarine VS.236 Walrus 1				See SECTION 3 Part 1	Yeovilton	
G-AIZU	Auster V J/1 Autocrat	2228		31. 1.47	C.J.and J.G.B.Morley	Popham	26. 6.06
G-AIZW*	Auster V J/1 Autocrat				See SECTION 3 Part 1	Halmstad, Sweden	
G-AIZY	Auster V J/1 Autocrat	2233		31. 1.47	B.J.Richards	(Brunel Technical College, Bristol)	20. 9.78S
					(Damaged Portskewett, Caldicot 8.89: on rebuild 6.91: current status unknown)		

G-AJAA - G-AJZZ

Regn	Type	C/n	Previous identity	Regd	Owner	Base	CofA
G-AJAD (2)	Piper J-3C-65 Cub (L-4H-PI)	12008	OO-GEJ (1)	26. 6.84	C R Shipley	(Bristol)	12. 7.05P
	(Frame No.11835) (Regd with c/n 11700)		44-79712 *(Airframe has original fuselage of OO-GEJ discarded in a rebuild in 1970s: OO-GEJ rebuilt with Frame No.11950 (c/n 12122) ex OO-ALY/44-79826 and now G-AHIP: OO-ALY rebuilt from c/n 11700 ex OO-TON/43-30409)*				
G-AJAE	Auster J/1N Alpha	2237		4. 2.47	J Cooke tr Lichfield Auster Group	Streethay	31. 7.06
G-AJAJ	Auster J/1N Alpha	2243		4. 2.47	R.B.Lawrence	Watchford Farm, Yarcombe	1. 4.07
G-AJAL*	Auster V J/1 Autocrat				See SECTION 3 Part 1	Burgos, Spain	
G-AJAM	Auster V J/2 Arrow	2371		8. 2.47	D.A Porter	Griffins Farm, Temple Bruer	17. 6.05P
G-AJAP (2)	Luscombe 8A Silvaire	2305	N45778	26. 1.89	R J Thomas	Hamilton Farm, Bilsington	31. 8.04P
	(Continental A65)		NC45778				
G-AJAS	Auster J/1N Alpha	2319		14. 3.47	C.J.Baker	Carr Farm, Thorney, Newark	11. 4.90
					(Noted complete 1.05)		
G-AJAV*	Douglas C-47A-5-DK Dakota				See SECTION 3 Part 1	Aragua, Maracay, Venezuela	
G-AJAZ*	Douglas C-47A-50-DL Dakota				See SECTION 3 Part 1	Schwäbisch-Gmund, Germany	
G-AJBJ*	de Havilland DH.89A Dragon Rapide	6765	NF894	20. 1.47	John Pierce Aviation Ltd	Ley Farm, Chirk	14. 9.61T
					(Cancelled 16.12.91 by CAA) (Noted 8.04 w/o marks)		
G-AJCL (2)*	de Havilland DH.89A Dragon Rapide	6722	NF851	7. 9.48	John Pierce Aviation Ltd	Ley Farm, Chirk	
					(WFU Shobdon 1.71: cancelled 24.5.71) (Noted 8.04 w/o marks)		
G-AJCP (2)	Druine D.31 Turbulent	PFA 512		9. 2.59	B.R.Pearson	Eaglescott	4. 9.78S
	(Built Rollason Aircraft and Engines) (Ardem 4C02)				tr Turbulent Group *(Stored 10.95: current status unknown)*		

G-AJEB*	Auster J/1N Alpha				See SECTION 3 Part 1	Hooton Park	
G-AJEC*	Auster V J/1 Autocrat				See SECTION 3 Part 1	Wanaka, New Zealand	
G-AJEH	Auster J/1N Alpha	2312		14. 3.47	J.T.Powell-Tuck *(Current status unknown)*	(Pontypool)	28. 5.90
G-AJEI	Auster J/1N Alpha	2313		14. 3.47	J.Siddall	(Messingham, Scunthorpe)	13. 8.94T
	(Originally regd as Auster J/1 Autocrat) (Composite rebuilt 1976 with fuselage of F-BFUT c/n 3357: original fuselage stored 1.97) (New owner 8.04)						
G-AJEM	Auster V J/1 Autocrat	2317	F-BFPB	14. 3.47	C.D.Wilkinson	(Taunton)	13. 5.07
			G-AJEM				
G-AJES (2)	Piper J-3C-65 Cub (L-4H-PI)	11776	OO-ACB	21. 9.84	G.W.Jarvis	Shifnal	24. 7.05P
	(Frame No.11602)		43-30485		*(As "330485/44-C" in USAAC c/s)*		
G-AJGJ	Taylorcraft J Auster 5	1147	RT486	31. 1.47	D.Gotts and E.J.Downing	Lee-on-Solent	25. 5.07
					tr Auster RT486 Flying Group *(As "RT486/PF-A")*		
G-AJHM*	SAI KZ.VII UA Laerke				See SECTION 3 Part 1	Stauning, Denmark	
G-AJHR*	de Havilland DH.82A Tiger Moth				See SECTION 3 Part 1	Ardmore, Auckland, New Zealand	
G-AJHS	de Havilland DH.82A Tiger Moth	82121	N6866	12. 2.47	J.M.Voeten and H.Van Der Paauw	Seppe, The Netherlands	22. 7.06
					(Op Vliegend Museum)		
G-AJHU	de Havilland DH.82A Tiger Moth	83900	T7471	12. 2.47	G.Valenti *(New owner 8.04*	(Parma, Italy)	23. 8.98T
G-AJHW*	Sikorsky S-51				See SECTION 3 Part 1	Point Cook, Victoria,Australia	
G-AJIH	Auster V J/1 Autocrat	2318		2. 4.47	D.G.Curran	(Weyarn, Germany)	2. 6.07
G-AJIS	Auster J/1N Alpha	2336		30. 4.47	J.D.Smith and J.M.Hodgson	Baxby Manor, Husthwaite	17.6.06
					tr Husthwaite Auster Group		
G-AJIT	Auster V J/1 Kingsland	2337		30. 4.47	A J.Kay	Netherthorpe	17.6.05P
	(Continental O-200-A)				tr G-AJIT Group		
G-AJIU	Auster V J/1 Autocrat	2338		30. 4.47	M.D.Greenhalgh	Netherthorpe	20. 6.03
G-AJIW	Auster J/1N Alpha	2340		30. 4.47	Truman Aviation Ltd	Tollerton	22. 8.04
G-AJJP*	Fairey FB.2 Jet Gyrodyne				See SECTION 3 Part 1	Woodley	
G-AJJS (2)	Cessna 120	13047	8R-GBO	7. 1.87	R.W.Marchant, I.D.Ranger and S.C.Parsons		
	(Continental O-200-A)		VP-GBO		tr Robhurst Flying Group		
	(Rebuilt 1994 with new airframe?)		VP-TBO, N1106M, YV-T-CTA, NC2786N			Little Robhurst Farm, Woodchurch	6. 5.05P
G-AJJT (2)	Cessna 120	12881	N2621N	27. 1.88	J.S.Robson	Franklyn's Field, Chewton Mendip	7. 7.04P
	(Continental C85)		NC2621N				
G-AJJU (2)	Luscombe 8E Silvaire	2295	N45768	10. 1.89	S.C.Weston and R.J.Hopcraft	Enstone	20. 8.02P
	(Continental C85)		NC45768				
G-AJKB (2)	Luscombe 8E Silvaire	3058	N71631	4. 1.89	A F.Hall and S.P.Collins	Tibenham	1. 7.05P
	(Continental C85)		NC71631				
G-AJKF (2)*	de Havilland DH.84 Dragon III				See SECTION 3 Part 1	Point Cook, Victoria,Australia	
G-AJKG (2)*	Miles M.38 Messenger 2A				See SECTION 3 Part 1	Melbourne, Victoria, Australia	
G-AJLR*	Airspeed AS.65 Consul				See SECTION 3 Part 1	Singapore	
G-AJMC*	Bristol 156 Beaufighter TF.X				See SECTION 3 Part 1	Beersheba, Israel	
G-AJOA*	de Havilland DH.82A Tiger Moth	83167	T5424	29. 4.47	Aero Antiques	Durley	22. 5.03
	(Badly damaged landing Lotmead Farm, Wanborough, Swindon 13.5.01: sold 10.01 for rebuild) (Cancelled 19.3.03 as wfu)						
G-AJOC*	Miles M.38 Messenger 2A				See SECTION 3 Part 1	Belfast	
G-AJOE	Miles M.38 Messenger 2A	6367		28. 4.47	P.W.Bishop	Kemble	18.12.07
G-AJON (2)	Aeronca 7AC Champion	7AC-2633	OO-TWH	3. 1.86	A Biggs and D.S.Moores	Shenington	15. 5.05P
G-AJOZ*	Fairchild 24W-41A Argus 1				See SECTION 3 Part 1	Elvington	
G-AJPI	Fairchild 24R-46A Argus III	851	HB614	26. 4.47	K.A Doornbos	(Eelde, The Netherlands)	9. 5.07
	(UC-61A-FA)		43-14887		*(As "314887" in USAAF c/s)*		
G-AJRB	Auster V J/1 Autocrat	2350		12. 5.47	B.Maurer tr G-AJRB Flying Group	Sywell	8. 3.04T
G-AJRC	Auster V J/1 Autocrat	2601		12. 5.47	M Barker	Octon Lodge Farm, Thwing	14. 7.02
G-AJRE	Auster V J/1 Autocrat	2603		12. 5.47	Pauline Slater	Wellesbourne Mountford	5. 5.07
G-AJRH*	Auster J/1N Alpha				See SECTION 3 Part 1	Loughborough	
G-AJRS	Miles M.14A Hawk Trainer 3	1750	P6382	30. 4.47	Richard Shuttleworth Trustees	Old Warden	8. 9.05P
	(Composite a/c which flew as "G-AJDR" 1.54/3.71)		G-AJDR, G-AJRS, P6382		*(As "P6382/C" in 16 EFTS c/s)*		
G-AJTI*	Miles M.65 Gemini 1A				See SECTION 3 Part 1	Pretoria, South Africa	
G-AJTW	de Havilland DH.82A Tiger Moth	82203	N6965	21. 5.47	J.A Barker *(As "N6965/FL-J")*	Tibenham	9. 9.00
			(Crashed landing Raydon near Ipswich 7.6.99 and extensively damaged: wreck noted again 7.02)				
G-AJUD*	Auster V J/1 Autocrat	2614		5. 6.47	C.L.Sawyer	(Bromham, Bedford)	18. 5.74
					(On rebuild 12.97: cancelled 31.3.99 by CAA) (Noted for sale 2000)		
G-AJUE	Auster V J/1 Autocrat	2616		5. 6.47	P.H.B.Cole	Craysmarsh Farm, Melksham	22. 6.06
G-AJUL	Auster J/1N Alpha	2624		18. 6.47	M.J.Crees	Halstead, Essex	11. 9.81
					(On rebuild 12.90)		
G-AJVE	de Havilland DH.82A Tiger Moth	85814	DE943	28. 5.47	R.A Gammons	RAF Henlow	10. 6.06
				(Composite 1981 rebuild including substantial parts of G-APGL c/n 86460/NM140)			
G-AJVH*	Fairey Swordfish II	Not known	LS326	28. 5.47	Royal Navy Historic Flight	RNAS Yeovilton	
					(Cancelled 30.4.59 as restored to RN)		
					"City of Liverpool" (Noted as "LS326/L2" in 836 Sqdn c/s 11.03)		
G-AJVL*	Miles M.38 Messenger 2A				See SECTION 3 Part 1		
						Caboolture, Queensland, Australia	
G-AJWB	Miles M.38 Messenger 2A	6699		17. 6.47	G.E.J.Spooner	Fanners Farm, Great Waltham	23. 1.05
G-AJXC*	Taylorcraft J Auster 5	1409	TJ343	11. 6.47	J.Graves	Scotland Farm, Hook	2. 8.82
			(Damaged Hook in gales 16.10.87: cancelled 3.4.89 by CAA) (Stored 9.94: fuselage noted 10.02)				
G-AJXV	Taylorcraft G Auster 4	1065	F-BEEJ	8. 9.47	Barbara A Farries	Carr Farm, Thorney, Newark	13. 7.06
			G-AJXV, NJ695		*(As "NJ695") "Little Lulu"*		
G-AJXY	Taylorcraft G Auster 4	792	MT243	4. 5.48	D.A Hall	(Melton Mowbray)	10.11.70
					(On rebuild 1993: new owner 7.00)		
G-AJYB	Auster J/1N Alpha	847	MS974	3. 2.49	P.J.Shotbolt	Ingthorpe Farm, Great Casterton, Lincoln	25. 7.05

G-AKAA - G-AKZZ

G-AKAA*	Piper J-3C-65 Cub (L-4H-PI)				See SECTION 3 Part 1	Madrid, Spain	
G-AKAO*	Miles M.38 Messenger 2A				See SECTION 3 Part 1	Halmstad, Sweden	
G-AKAT	Miles M.14A Hawk Trainer 3	2005	F-AZOR	2. 7.47	R.A.Fleming	Breighton	31. 7.05P
			G-AKAT, T9738		*(As "T9738")*		

G-AKAZ (2)	Piper J-3C-65 Cub (L-4A-PI)	AN.1 & 8499	F-BFYL	19. 4.82	Frazerblades Ltd	Duxford	11. 8.05P
	(Frame No.8616)		Fr Mil, 42-36375		*(As "54884/57-H" in 83rd FS/78th FG USAAC c/s)*		
G-AKBO*	Miles M.38 Messenger 2A	6378		15. 7.47	P.R.Holloway *(Cancelled 24.8.04 by CAA)*	Enstone	3. 8.03
G-AKBT*	Piper J-3C-85 Cub				See SECTION 3 Part 1	Alverca, Portugal	
G-AKCO*	Short S.25 Sandringham 7				See SECTION 3 Part 1	Le Bourget, Paris, France	
G-AKDK*	Miles M.65 Gemini 1A				See SECTION 3 Part 1	Billund, Denmark	
G-AKDN	de Havilland DHC-1A Chipmunk 10	11		14. 8.47	P.S.Derry	Bagby	31. 5.07
G-AKDW	de Havilland DH.89A Dragon Rapide	6897	F-BCDB	25. 8.47	de Havilland Aircraft Museum Trust Ltd		
	(Built Brush Coachworks Ltd)		G-AKDW, YI-ABD, NR833		*"City of Winchester"*	Salisbury Hall, London Colney	8. 5.59
					(On long term restoration 5.04)		
G-AKEE*	de Havilland DH.82A Tiger Moth				See SECTION 3 Part 1	Ratamalana, Sri Lanka	
G-AKEL*	Miles M.65 Gemini 1A				See SECTION 3 Part 1	Belfast	
G-AKEZ	Miles M.38 Messenger 2A	6707		27. 8.47	P.G.Lee	Fanners Farm, Great Waltham, Essex	15.11.68
					(Noted on rebuild 10.01)		
G-AKGE*	Miles M.65 Gemini 3C				See SECTION 3 Part 1	Belfast	
G-AKGV*	de Havilland DH.89A Dragon Rapide				See SECTION 3 Part 1		
						Sault Sainte Marie, Ontario, Canada	
G-AKHP	Miles M.65 Gemini 1A	6519		3.10.47	P.A Brook	Shoreham	11. 6.06
G-AKHW*	Miles M.65 Gemini 1A				See SECTION 3 Part 1	Albany, Auckland, New Zealand	
G-AKIB (2)	Piper J-3C-90 Cub (L-4H-PI)	12311	OO-RAY	18. 4.84	M.C.Bennett	Bodmin	20. 8.05P
	(Frame No.12139)		44-80015		*(As "480015/44-M" in USAAC c/s)*		
G-AKIF	de Havilland DH.89A Dragon Rapide	6838	LN-BEZ	24. 9.47	Airborne Taxi Services Ltd	Duxford/Booker	19. 8.06T
			G-AKIF, NR750				
G-AKIN	Miles M.38 Messenger 2A	6728		19. 9.47	D.L.Sentance tr Sywell Messenger Group	Sywell	10.10.05
G-AKIS*	Miles M.38 Messenger 2A				See SECTION 3 Part 1	Brussels, Belgium	
G-AKIU	Percival P.44 Proctor V	Ae129		20. 2.48	Air Atlantique Ltd	Seaton Ross	24. 1.65
					(Under restoration by Hornet Aviation 2.04)		
G-AKKA*	Miles M.65 Gemini 1A				See SECTION 3 Part 1	Stavangar, Norway	
G-AKKB	Miles M.65 Gemini 1A	6537		28.10.47	J.Buckingham *(Air Total c/s)*	New Farm, Felton	29. 4.05
G-AKKH	Miles M.65 Gemini 1A	6479	OO-CDO	23. 7.48	J.S.Allison	Bicester	8. 9.06T
G-AKKR*	Miles M.14A Hawk Trainer 3				See SECTION 3 Part 1	Middle Wallop	
G-AKKY*	Miles M.14A Hawk Trainer 3				See SECTION 3 Part 1	Woodley	
G-AKLL*	Douglas C-47A-30-DK Dakota				See SECTION 3 Part 1	Schleissheim, Germany	
G-AKLS*	Short SA.6 Sealand 1				See SECTION 3 Part 1	Belgrade, Serbia	
G-AKLW*	Short SA.6 Sealand 1				See SECTION 3 Part 1	Belfast	
G-AKNP*	Short S.45 Solent 3				See SECTION 3 Part 1	Oakland, California, USA	
G-AKNV*	de Havilland DH.89A Dragon Rapide				See SECTION 3 Part 1	Brussels, Belgium	
G-AKOE*	de Havilland DH.89A Dragon Rapide 4	6601	X7484	3.12.47	J.E.Pierce	Ley Farm, Chirk	25. 7.82
					(Cancelled 18.6.02 by CAA) (Noted 8.04 in BEA c/s)		
G-AKOW*	Taylorcraft J Auster 5				See SECTION 3 Part 1		
G-AKPF	Miles M.14A Hawk Trainer 3	2228	V1075	27. 1.48	P.R.Holloway *(As "N3788" in RAF c/s)*	Old Warden	9. 4.05P
	(i) Rebuilt 1955 as composite with centre-section from G-AIUA, fuselage from G-ANLT, wings from G-AHYL)						
	(ii) Rebuilt 1970/80 with about 10% of fuselage from G-AKPF: tail unit also from G-ANLT)						
G-AKRA (2)	Piper J-3C-65 Cub (L-4H-PI)	11255	I-FIVI	15. 6.84	W.R.Savin	(Cambridge)	
	(Frame No.11080)		43-29964		*(On rebuild 9.00)*		
G-AKRP	de Havilland DH.89A Dragon Rapide 4	6940	CN-TTO	26. 1.48	R.H.Ford	Gamlingay	26. 6.03T
			(F-DAFS), G-AKRP, RL958		*"Northamptonshire Rose" (Noted 2004)*		
G-AKRS*	de Havilland DH.89A Dragon Rapide				See SECTION 3 Part 1	Beersheba, Israell	
G-AKSY	Taylorcraft J Auster 5	1567	F-BGOO	10. 2.48	A Brier	Breighton	28. 4.07
			G-AKSY, TJ534		*(As "TJ534")*		
G-AKSZ	Taylorcraft J Auster 5C	1503	F-BGPQ	10. 2.48	P.W.Yates and R.G.Darbyshire	(Chorley)	9. 5.02
	(DH Gipsy Major 1) *(Large fin and rudder)*		G-AKSZ, TJ457				
G-AKTH (2)	Piper J-3C-65 Cub (L-4J-PI)	13211	OO-AGL	14. 7.86	G.J.Harry, The Viscount Goschen		
	(Frame No.13041) (Regd with incorrect c/n 13047)		PH-UCR, 45-4471			Bradleys Lawn, Heathfield	27. 4.05P
G-AKTI (2)	Luscombe 8A Silvaire	4101	N1374K	27. 5.87	M.W.Olliver	Farley Farm, Romsey	16. 5.05P
	(Continental A65)		NC1374K				
G-AKTK (2)	Aeronca 11AC Chief	11AC-1017	N9379E	13. 3.89	R.W.Marshall	Waits Farm, Belchamp Walter	11.9.03P
	(Continental A65)		NC9379E		tr Aeronca Tango Kilo Group		
G-AKTN (2)	Luscombe 8A Silvaire	3540	N77813	22. 7.88	D.Taylor	Clacton	1. 6.04P
	(Continental A65)		NC77813				
G-AKTO (2)	Aeronca 7BCM Champion	7AC-940	N8515X	19. 5.88	D.C.Murray	Lee-on-Solent	26. 8.04P
	(Continental A75) *(Modified ex 7AC standard 1950)*		N82311, NC82311				
G-AKTP (2)	Piper PA-17 Vagabond	17-82	N4683H	24. 6.88	P.J.B.Lewis	Swansea	5. 8.03P
	(Continental C85)		NC4683H		tr Golf Tango Papa Group		
G-AKTR (2)	Aeronca 7AC Champion	7AC-3017	N58312	19. 6.89	C.Fielder	Oaksey Park	17. 6.04P
			NC58312		*"Eddie" (Noted 1.05)*		
G-AKTS (2)	Cessna 120	11875	N77434	26. 5.88	M Isterling	Insch	1. 9.05P
			NC77434				
G-AKTT (2)	Luscombe 8A Silvaire	3279	N71852	21. 7.88	S.J.Charters	Octon Lodge Farm, Thwing, Eddsfield	23. 6.92P
	(Continental A65)		NC71852		*(Crashed 6.7.91: stored 1.96) (Current status unknown)*		
G-AKUE (2)	de Havilland DH.82A Tiger Moth	P.68	ZS-FZL	12. 2.86	D.F.Hodgkinson	Maypole Farm, Chislet	21.10.07T
	(Built OGMA)		CR-AGM, FAP				
G-AKUF (2)	Luscombe 8F Silvaire	4794	N2067K	1. 8.88	E.J.Lloyd	(Caterham)	7. 8.03P
	(Continental C90)		NC2067K		*(Undershot landing Guestling, near Hastings 25.5.03 and substantially damaged)*		
G-AKUG (2)	Luscombe 8A Silvaire	3689	N77962	21. 7.88	(R S Bird)	(Chilton Foliat, Hungerford)	16. 3.05P
	(Continental A65)		NC77962		*(Damaged on take off Stockton, Wiltshire 17.7.04: under repair with completion 2005/2006)*		
G-AKUH (2)	Luscombe 8E Silvaire	4644	N1917K	24.10.88	E.J.Lloyd	Cardington	27. 7.05P
	(Continental O-200-A)		NC1917K				
G-AKUI (2)	Luscombe 8E Silvaire	2464	N45937	24.10.88	D.A Sims	Yeatsall Farm, Abbots Bromley	22. 3.05P
	(Continental O-200-A)		NC45937				
G-AKUJ (2)	Luscombe 8E Silvaire	5282	N2555K	4. 8.88	R.C.Green	Coventry	12.12.04P
	(Continental C85)		NC2555K				
G-AKUK (2)	Luscombe 8A Silvaire	5793	N1166B	28.10.88	O.R.Watts	(Eastleigh)	15. 9.05P
	(Continental A65)		NC1166B				

Registration	Type	C/n	Previous identity	Date	Owner / Operator	Location	CofA
G-AKUL (2)	Luscombe 8A Silvaire	4189	N1462K	9. 2.89	E.A Taylor	Southend	21. 5.90P
	(Continental A65)		NC1462K		*(Noted stored 2.05)*		
G-AKUM (2)	Luscombe 8F Silvaire	6452	N2025B	17. 2.88	D A Young	North Weald	23.5.05P
	(Continental C90)						
G-AKUN (2)	Piper J-3C-85 Cub	6914	N38304	13. 1.89	W.R.Savin	Coldharbour Farm, Willingham	27. 5.05P
			NC38304				
G-AKUO (2)	Aeronca 11AC Chief	11AC-1376	N9730E	16. 1.89	L.W.Richardson	White Waltham	17. 9.05P
			NC9730E				
G-AKUP (2)	Luscombe 8E Silvaire	5501	N2774K	9. 5.89	D.A Young	North Weald	
	(Lycoming O-320)		NC2774K		*(Dismantled awaiting rebuild 1.05)*		
G-AKUR (2)	Cessna 140	13819	N1647V	26. 1.89	J.Greenaway and C.A Davis	Popham	21. 9.95
			NC1647V		*(Current status unknown)*		
G-AKUW	Chrislea CH.3 Series 2 Super Ace	105		8. 3.48	J and S Rickett	North Coates	15. 2.05P
G-AKVD*	Chrislea CH.3 Series 2 Super Ace				See SECTION 3 Part 1	Tokyo, Japan	
G-AKVF	Chrislea CH.3 Series 2 Super Ace	114	AP-ADT	8. 3.48	B.Metters	Bourne Park, Hurstbourne Tarrant	15. 4.05P
			G-AKVF				
G-AKVM (2)	Cessna 120	13431	N3173N	10. 1.89	N.and Susan Wise	Croft Farm, Croft-on-Tees	17. 6.05P
			NC3173N				
G-AKVN (2)	Aeronca 11AC Chief	11AC-469	N3742B	13. 1.89	C.E.Ellis	Priory Farm, Tibenham	27 5.05P
			N86047, NC86047		tr Breckland Aeronca Group *(Carries "3742B" on fin)*		
G-AKVO (2)	Taylorcraft BC-12D	9845	N44045	10. 1.89	M.Gibson	Breighton	29 6.05P
	(Continental A65)		NC44045				
G-AKVP (2)	Luscombe 8A Silvaire	5549	N2822K	21. 7.48	J M Edis	Charity Farm, Baxterley	5. 4.05P
	(Continental A65)		NC2822K				
G-AKVR	Chrislea CH.3 Series 4 Skyjeep	125	VH-OLD	8. 3.48	C.L.Needham	Old Manor Farm, Anwick, Seaford	AC
			VH-RCD, VH-BRP, G-AKVR		*(Flown 30.5.04 after rebuild)*		
G-AKVZ	Miles M.38 Messenger 4B	6352	RH427	25. 6.48	Shipping and Airlines Ltd	Biggin Hill	4.10.06
G-AKWS	Auster 5A-160	1237	RT610	1. 4.48	A J.Collins	Bidford	4. 6.06
	(Lycoming O-320)				t/a Interesting Aircraft Company *(As "RT610")*		
G-AKWT*	Taylorcraft J Auster 5	998	MT360	1. 4.48	C.J.Baker	Carr Farm, Thorney, Newark	22. 7.49
					(Crashed Tollerton 7.8.48) (Derelict frame stored 1.05)		
G-AKXP	Taylorcraft J Auster 5	1017	NJ633	13. 4.48	M.Pocock	RAF Keevil	19.12.70
					(Crashed St.Mary's, Isles of Scilly 9.4.70: noted as "NJ633" 1.04)		
G-AKXS	de Havilland DH.82A Tiger Moth	83512	T7105	13. 4.48	J. and G.J.Eagles	Oaksey Park	21. 3.03T
					(Spun into ground White Waltham 21.7.02 and badly damaged)		
G-AKZN*	Percival P.34A Proctor III	K.386	8380M	24. 5.48	Royal Air Force Museum Reserve Collection	RAF Stafford	29.11.63
			Z7197		*(Cancelled 27.9.63 as Marks WFU and reverted to Military Marks)*		
					(Noted 2.04 as "Z7197")		
G-AKZY*	Messerschmitt Bf.108D-1 Taifun				See SECTION 3 Part 1	Albuquerque, New Mexico, USA	

G-ALAA - G-ALZZ

Registration	Type	C/n	Previous identity	Date	Owner / Operator	Location	CofA
G-ALAD (2)*	de Havilland DH.82A Tiger Moth				See SECTION 3 Part 1	Ardmore, Auckland, New Zealand	
G-ALBD	de Havilland DH.82A Tiger Moth	84130	T7748	27. 5.48	C.H.Schoonbeek	Midden Zeeland, The Netherlands	31.10.81
		(Damaged Leopoldsburg, Belgium 24.5.81: noted 9.02 with Gyrocopter Aviation being rebuilt for static display)					
G-ALBJ	Taylorcraft J Auster 5	1831	TW501	3. 6.48	P.N.Elkington	Bloxholm, Sleaford	23. 8.07
G-ALBK	Taylorcraft J Auster 5	1273	RT644	3. 6.48	S.J.Wright	Swayfield, Grantham	29. 6.06
G-ALBN*	Bristol 173 Mk.1				See SECTION 3 Part 1	Kemble	
G-ALBP*	Miles M.38 Messenger 4A				See SECTION 3 Part 1	Wangaratta, Victoria, Australia	
G-ALCK*	Percival P.34A Proctor III				See SECTION 3 Part 1	Duxford	
G-ALCS (2)*	Miles M.65 Gemini 3C				See SECTION 3 Part 1	Leitrim, County Leitrim	
G-ALCT*	Taylorcraft J Auster 5				See SECTION 3 Part 1	Helsingborg, Sweden	
G-ALCU*	de Havilland DH.104 Dove 2B				See SECTION 3 Part 1	Coventry	
G-ALDG*	Handley Page HP.81 Hermes IV				See SECTION 3 Part 1	Duxford	
G-ALEH (2)	Piper PA-17 Vagabond	17-87	N4689H	17. 8.81	A D.Pearce	White Waltham	28. 3.05P
	(Continental A65)		NC4689H				
G-ALFA	Taylorcraft J Auster 5	1236	RT607	20.10.48	S.P.Barrett	(Irby-in-the-Marsh, Skegness)	7.10.07
	(P/i uncertain as c/n 1236 considered sold as HB-EOC 4.48: reported as c/n 826 (MS958) but doubtful)						
G-ALFT*	de Havilland DH.104 Dove 6				See SECTION 3 Part 1	Caernarfon	
G-ALFU*	de Havilland DH.104 Dove 6				See SECTION 3 Part 1	Duxford	
G-ALGA (2)	Piper PA-15 Vagabond	15-348	N4575H	3.12.86	G.A Brady	Enstone	19. 1.05P
	(Lycoming O-145)		NC4575H				
G-ALGB*	de Havilland DH.89A Dragon Rapide				See SECTION 3 Part 1	Le Bourget, Paris, France	
G-ALGT	Vickers Supermarine 379 Spitfire F.XIVc	6S/432263	RM689	9. 2.49	Rolls-Royce plc	Filton	31 .7.92P
			"RM619", G-ALGT, RM689		*(Destroyed in crash Woodford 27.6.92: under restoration Sandown 8.04)*		
G-ALIJ (2)	Piper PA-17 Vagabond	17-166	N4866H	13. 2.87	A S.Cowan	Popham	21. 4.05P
	(Continental A65)				tr Popham Flying Group G-ALIJ		
G-ALIS*	Percival P.34A Proctor III				See SECTION 3 Part 1	Bull Creek, Western Australia	
G-ALIW (2)	de Havilland DH.82A Tiger Moth	82901	N27WB	17. 8.81	D.I.M.Geddes and F.R.Curry	White Waltham	14. 8.06
			ZK-ATI, NZ899, R5006				
G-ALJF	Percival P.34A Proctor III	K.427	Z7252	3. 3.49	J.F.Moore	Biggin Hill	26. 7.07
G-ALJL	de Havilland DH.82A Tiger Moth	84726	T6311	7. 3.49	D and R I Souch	Hill Farm, Durley	28. 9.50
	(Built Morris Motors Ltd)				*(On long term rebuild from components 8.00)*		
G-ALMB*	Westland-Sikorsky S-51 Dragonfly Mk.1A				See SECTION 3 Part 1	Vigna di Valle. Italy	
G-ALNA	de Havilland DH.82A Tiger Moth	85061	T6774	11. 4.49	R.J.Doughton	Vendee Air Park, France	12. 5.07T
					(Brooklands Aviation titles)		
G-ALND	de Havilland DH.82A Tiger Moth	82308	N9191	12. 4.49	J.T.Powell-Tuck	Abergavenny	11. 4.82
					(As "N9191" in RN c/s) (Crashed Panshanger 8.3.81 and on rebuild 3.96)		
G-ALNV*	Taylorcraft J Auster 5	1216	RT578	21. 4.49	C.J.Baker	Carr Farm, Thorney, Newark	4. 7.50
					(WFU and silver frame stored 1.05)		
G-ALOD (2)	Cessna 140	14691	N2440V	14.10.83	J.R.Stainer	Whitehall Farm, Benington	11. 1.08E
G-ALRI	de Havilland DH.82A Tiger Moth	83350	ZK-BAB	2. 5.51	Mark Squared Ltd	(Chepstow)	22. 6.07
	(Built Morris Motors Ltd)		G-ALRI, T5672		*(As "T5672" in RAF c/s)*		

Registration	Type	c/n	Previous identity	Regd	Owner	Base	CofA
G-ALRX*	Bristol 175 Britannia Series 101				See SECTION 3 Part 1	Kemble	
(G-ALSP)*	Bristol 171 Sycamore 3				See SECTION 3 Part 1	Rochester	
G-ALSX*	Bristol 171 Sycamore 3				See SECTION 3 Part 1	Weston-super-Mare	
G-ALTO (2)	Cessna 140	14253	N2040V	19. 1.82	T.M., P.M. and Margaret L.Jones	Egginton	3. 7.04
	(Continental C85)						
G-ALUA*	Winter LF-1 Zaunkönig				See SECTION 3 Part 1	Munich, Germany	
G-ALUC	de Havilland DH.82A Tiger Moth	83094	R5219	28. 6.49	D.R.and Mrs M.Wood	Fowle Hall Farm, Paddock Wood	7.11.04
G-ALVP*	de Havilland DH.82A Tiger Moth	82711	R4770	26. 9.49	D and R Leatherland	Tollerton	15. 2.61
					(Cancelled as WFU 2.61 and stored) (Noted dismantled 11.04)		
G-ALWB	de Havilland DHC-1 Chipmunk 22A	C1/0100	OE-ABC	28.12.49	D J Neville and P A Dear-Neville	(Royston)	18. 5.03
			G-ALWB				
G-ALWC*	Douglas C-47A-25-DK Dakota				See SECTION 3 Part 1	Toulouse, France	
G-ALWF*	Vickers 701 Viscount				See SECTION 3 Part 1	Duxford	
G-ALWS	de Havilland DH.82A Tiger Moth	82415	N9328	24. 1.50	A P.Beynon	Trehelig, Welshpool	
	(Officially regd with c/n 82413)				*(On rebuild 8.00)*		
G-ALWW	de Havilland DH.82A Tiger Moth	86366	NL923	24. 1.50	D.E.Findon	Bidford	11. 3.06
	(Built Morris Motors Ltd)				tr Stratford-upon-Avon Tiger Moth Group		
G-ALXT*	de Havilland DH.89A Dragon Rapide				See SECTION 3 Part 1	Wroughton	
G-ALXZ	Taylorcraft J Auster 5-150	1082	D-EGOF	1. 2.50	M.F.Cuming	Sackville Lodge Farm, Riseley	4. 8.06
	(Lycoming O-320) *(Frame No.TAY24070)*		PH-NER, G-ALXZ, NJ689				
G-ALYB*	Taylorcraft J Auster 5				See SECTION 3 Part 1	Doncaster	
G-ALYG	Taylorcraft J Auster 5D	835	MS968	14. 3.50	A L.Young	Henstridge	19. 1.70
	(Officially regd with incorrect identity MT968:)				*(Frame stored 8.02: for rebuild as Auster 5)*		
G-ALZA*	de Havilland DH.82A Tiger Moth				See SECTION 3 Part 1	Mandeville, Gore, New Zealand	
G-ALZE*	Britten-Norman BN-1F				See SECTION 3 Part 1	Southampton	
G-ALZF*	de Havilland DH.89A Dragon Rapide				See SECTION 3 Part 1	La Ferté-Alais, Essonne, France	
G-ALZL*	de Havilland DH.114 Heron Series 1				See SECTION 3 Part 1	Bull Creek, Western Australia	
G-ALZO (2)*	Airspeed AS.57 Ambassador 2				See SECTION 3 Part 1	Duxford	

G-AMAA - G-AMZZ

Registration	Type	c/n	Previous identity	Regd	Owner	Base	CofA
G-AMAU*	Hawker Hurricane IIc	Not known			See SECTION 3 Part 1	RAF Coningsby	
G-AMAW	Luton LA-4 Minor	SA.I		29. 4.50	R.H.Coates	Breighton	6. 8.88P
	(Built J R Coates - c/n JRC.1) (Bristol Cherub 3) *(Aka as Swalesong SA.I)*				*(Stored 1.05)*		
G-AMBB	de Havilland DH.82A Tiger Moth	85070	T6801	1. 5.50	J.Eagles	Oaksey Park	
	(Composite rebuild - parts to "G-MAZY" ? - see SECTION 7, Part 5)				*(On rebuild 6.95: current status unknown)*		
G-AMBY*	Beech D-17S (YC-43) Traveler				See SECTION 3 Part 1	Eindhoven, The Netherlands	
G-AMCA*	Douglas C-47B-30DK Dakota 3				See SECTION 3 Part 1	Lelystad, The Netherlands	
G-AMCK	de Havilland DH.82A Tiger Moth	84641	N65N	15. 6.50	Avia Special Ltd	Leicester	5. 5.07T
			C-GBBF, SLN-05, D-EGXY, HB-UAC, G-AMCK, T6193				
G-AMCM	de Havilland DH.82A Tiger Moth	85295	DE249	14.12.50	A K and J.I Cooper	Denford Manor, Hungerford	28. 5.56
	(Regd with c/n "89259")		*(Crashed near Somerton 25.9.55: rear fuselage frame on restoration 10.01 but not original G-AMCM!)*				
G-AMDA*	Avro 652A Anson 1				See SECTION 3 Part 1	Duxford	
G-AMDD*	de Havilland DH.104 Dove 6				See SECTION 3 Part 1	Leitrim, County Leitrim	
G-AMEN (2)	Piper PA-18 Super Cub 95 *(L-18C-PI)*	18-1998	(G-BJTR)	29.12.81	A Lovejoy and W.Cook tr Sierra Golf Flying Group		
	(Frame No.18-1963) (Italian rebuild c/n OMA.71-08)		MM52-2398 "EI.71", I-EIAM, MM52-2398, 52-2398			Popham	26. 7.03P
G-AMEX*	de Havilland DH.82A Tiger Moth				See SECTION 3 Part 1	Levin, New Zealand	
G-AMHB*	Westland-Sikorsky S-51 Dragonfly Mk.1B				See SECTION 3 Part 1	Bebeduoro, Brasil	
G-AMHF	de Havilland DH.82A Tiger Moth	83026	R5144	6. 2.51	Wavendon Social Housing Ltd	Sywell	14. 9.03
	(Rebuilt with components from G-BABA c/n 86584 ex F-BGDT/PG687)						
G-AMHJ*	Douglas C-47A-25DK Dakota 6				See SECTION 3 Part 1	RAF Shawbury	
G-AMIU	de Havilland DH.82A Tiger Moth	83228	T5495	9. 4.51	M D Souch	Hill Farm, Durley	9. 9.71
			(Crashed Booker 15.10.69: frame reported on restoration Denford Manor 10.01: now departed for completion)				
G-AMJD*	de Havilland DH.82A Tiger Moth				See SECTION 3 Part 1	Brussels, Belgium	
G-AMJW*	Westland-Sikorsky S-51 Mk.1A				See SECTION 3 Part 1	Bangkok, Thailand	
G-AMJX*	Douglas C-47B-20-DK Dakota				See SECTION 3 Part 1	Scotia, New York, USA	
G-AMJY*	Douglas C-47B-40-DK Dakota				See SECTION 3 Part 1	Ratmalana, Sri Lanka	
G-AMKL*	Auster B.4	2983	XA177	3. 7.51	C.J.Baker	Carr Farm, Thorney, Newark	
			G-AMKL, G-25-2		*(Dismantled Rearsby 1956 and cancelled 24.9.58) (New fuselage 1.05)*		
G-AMKU	Auster 5 J/1B Aiglet	2721	ST-ABD	10. 7.51	P.G.Lipman	Romney Street Farm, Sevenoaks	7. 7.06
			SN-ABD, G-AMKU				
G-AMLZ*	Percival P.50 Prince 6E				See SECTION 3 Part 1	Caernarfon	
G-AMMA*	de Havilland DHC-1 Chipmunk 21				See SECTION 3 Part 1	Stauning, Denmark	
G-AMMC*	Miles M.14A Hawk Trainer III				See SECTION 3 Part 1	Auckland, New Zealand	
G-AMMK*	de Havilland DH.82A Tiger Moth				See SECTION 3 Part 1	Mandeville, Gore, New Zealand	
G-AMMS	Auster J/5K Aiglet Trainer	2745		11.10.51	R.B.Webber	Trenchard Farm, Eggesford	19.10.98
					(On rebuild 11.04)		
G-AMNL*	Douglas C-47B-35-DK Dakota				See SECTION 3 Part 1	Montevideo, Uruquay	
G-AMNN	de Havilland DH.82A Tiger Moth	86457	NM137	24.12.51	M.Thrower *"Spirit of Pashley"*	Shoreham	20. 7.03T
	(Composite rebuild with unidentified airframe: original G-AMNN posssibly absorbed into G-BPAJ qv)				t/a Northbrook College of Aeronautical Engineering		
G-AMOG (2)*	Vickers 701 Viscount				See SECTION 3 Part 1	RAF Cosford	
G-AMOI*	Vickers 701 Viscount				See SECTION 3 Part 1	Bebeduoro, Brasil	
G-AMOU*	de Havilland DH.82A Tiger Moth				See SECTION 3 Part 1	Bangkok, Thailand	
G-AMPG (2)	Piper PA-12 Super Cruiser	12-985	N2647M	25. 3.85	N.Sutton	Preston Court, Ledbury	28. 9.05P
	(Hoerner wing-tips)		NC2647				
G-AMPI (2)	SNCAN Stampe SV-4C	213	N6RA	13. 2.84	T.W.Harris	Booker	7. 8.06
	(Renault 4P)		F-BCFX				
G-AMPM*	de Havilland DH.82A Tiger Moth				See SECTION 3 Part 1	Mackay, Queensland, Australia	
G-AMPN*	de Havilland DH.82A Tiger Moth				See SECTION 3 Part 1	Tyabb, Victoria, Australia	
G-AMPO*	Douglas C-47B-30-DK Dakota 3				See SECTION 3 Part 1	RAFLlyneham	
G-AMPP*	Douglas C-47B-15-DK Dakota 3				See SECTION 3 Part 1	Noord-Brabant, The Netherlands	
G-AMPY	Douglas C-47B-15-DK Dakota 3	15124/26569	(EI-BKJ)	8. 3.52	Atlantic Air Transport Ltd	Coventry	5. 1.02A
			G-AMPY, N15751, G-AMPY, TF-FIO, G-AMPY, JY-ABE, G-AMPY, KK116, 43-49308 *(On restoration 1.05)*				

G-AMRA	Douglas C-47B-15DK Dakota 6	15290/26735	XE280	8. 3.52	Atlantic Air Transport Ltd	Coventry	13. 7.05T
			G-AMRA, KK151, 43-49474		*(Atlantic c/s with Air Atlantique titles)*		
G-AMRF	Auster J/5F Aiglet Trainer	2716	VT-DHA	20. 3.52	D.A.Hill	Fenland	31. 3.07
			G-AMRF				
G-AMRK	Gloster Gladiator I	-	L8032	16. 5.52	Richard Shuttleworth Trustees	Old Warden	18. 8.05P
	(Bristol Mercury XXX)		"K8032", G-AMRK, L8032		*(As "423-Port/427-Starboard" in R.Norwegian AF c/s)*		
G-AMRM*	de Havilland DH.82A Tiger Moth				See SECTION 3 Part 1	Auckland, New Zealand	
G-AMSG	SIPA 903	77	OO-VBL	25.11.81	S.W.Markham	Valentine Farm, Odiham	1. 5.05P
			F-BGHB				
G-AMSM*	Douglas C-47B-20-DK Dakota				See SECTION 3 Part 1	Brenzett	
G-AMSN*	Douglas C-47B-35DK Dakota IV				See SECTION 3 Part 1	North Weald	
G-AMSV	Douglas C-47B-25DK Dakota 3	16072/32820	(F-BSGV)	15. 5.52	Atlantic Air Transport Ltd	Coventry	13. 8.03A
			G-AMSV, KN397, 44-76488				
			(Pollution Control - MCA logo on rudder: minus engines, WFU and in open store 1.05)				
G-AMTA	Auster J/5F Aiglet Trainer	2780		24. 5.52	J.D.Manson	Tollerton	3. 8.06
G-AMTD*	Auster J/5F Aiglet Trainer	2783	EI-AVL	24. 5.52	Not known	Leicester	
			G-AMTD *(Damaged landing Hayrish Farm, Okehampton 7.8.93: cancelled 15.1.99 as WFU) (Noted 5.04)*				
G-AMTK	de Havilland DH.82A Tiger Moth	3982	N6709	18. 6.52	S.W.McKay and M.E.Vaisey	(Berkhamsted)	27. 5.66
					(Stored 12.99: CofR @ 4.03)		
G-AMTL*	de Havilland DH.82A Tiger Moth				See SECTION 3 Part 1	Zaventem, Belgium	
G-AMTM	Auster V J/1 Autocrat	3101	G-AJUJ	3. 7.52	R.J.Stobo	Oaklands Farm, Stonesfield, Oxon	27. 6.04P
	(Auster rebuild - originally c/n 2622)						
G-AMTP*	de Havilland DH.82A Tiger Moth				See SECTION 3 Part 1	Zaventem, Belgium	
G-AMTV	de Havilland DH.82A Tiger Moth	3858	OO-SOE	5. 8.52	M.A Wray	Oaksey Park	4. 3.07
			G-AMTV, N6545		tr Tango Victor Flying Group		
G-AMUF	de Havilland DHC-1 Chipmunk 21	C1/0832		2. 9.52	Redhill Tailwheel Flying Club Ltd	Redhill	5. 2.05
G-AMUH*	de Havilland DHC-1 Chipmunk 21				See SECTION 3 Part 1	Wanaka, New Zealand	
G-AMUI	Auster J/5F Aiglet Trainer	2790		29. 8.52	R B Webber *(Stored 11.04)*	Trenchard Farm, Eggesford	15. 2.66T
G-AMVD	Taylorcraft J Auster 5	1565	F-BGTF	6.10.52	M.Hammond	Airfield Farm, Hardwick	25. 5.07
			G-AMVD, TJ565		*(As "TJ565")*		
G-AMVP	Tipsy Junior	J.111	OO-ULA	23.10.52	A R.Wershat	Sandown	22. 6.94P
	(Walter Mikron 2)				*(Damaged Wroughton 4.7.93: stored 6.03)*		
G-AMVS	de Havilland DH.82A Tiger Moth	82784	OO-SOJ	12.11.52	J.T.Powell-Tuck	(Pontypool)	21.12.53
			G-AMVS, R4852		*(On rebuild 8.92: current status unknown)*		
G-AMWI*	Bristol 171 Sycamore 4				See SECTION 3 Part 1		
						Kogarah, New South Wales, Australia	
G-AMXR*	de Havilland DH.104 Dove 6				See SECTION 3 Part 1	Salisbury Hall, London Colney	
G-AMXS	de Havilland DH.104 Dove 2A				See SECTION 3 Part 1	Maracay, Venezuela	
G-AMXX*	de Havilland DH.104 Dove 2A				See SECTION 3 Part 1	Hermeskeil, Trier, Germany	
G-AMYD	Auster J/5L Aiglet Trainer	2773		13. 2.53	G.H.Maskell	Duckend Farm, Wilstead, Bedford	3. 9.04
G-AMYJ*	Douglas C-47B-25DK Dakota 6				See SECTION 3 Part 1	Elvington	
G-AMYL (2)	Piper PA-17 Vagabond	17-30	N4613H	24. 4.87	P.J.Penn-Sayer	Scaynes Hill, Haywards Heath	20. 6.89P
	(Continental C75)		NC4613H		t/a The Fun Airplane Co *"Yankee Lady" (Stored 2004)*		
G-AMZH*	Douglas C-47B-20-DK Dakota				See SECTION 3 Part 1		
						Port Moresby, Papua-New Guinea	
G-AMZI	Auster J/5F Aiglet Trainer	3104		4. 5.53	J.F.Moore	Rexden, Rye	16. 2.07
G-AMZO*	de Havilland DH.87B Hornet Moth				See SECTION 3 Part 1	Stauning, Denmark	
G-AMZT	Auster J/5F Aiglet Trainer	3107		28. 5.53	D.Hyde, J.W.Saull and J.C.Hutchinson		
						Standalone Farm, Meppershall	25. 5.07
G-AMZU	Auster J/5F Aiglet Trainer	3108		28. 5.53	J.A Longworth, A R.M. and C.B.A Eagle		
					tr Flying Flicks	White Waltham	25. 9.05
G-AMZW*	Douglas C-47B-20-DK Dakota				See SECTION 3 Part 1	Johannesburg, South Africa	

G-ANAA - G-ANZZ

G-ANAF	Douglas C-47B-35DK Dakota 3	16688/33436	N170GP	17. 6.53	Atlantic Air Transport Ltd	Coventry	27. 2.03A
			G-ANAF, KP220, 44-77104		*(Noted with "THALES" titles - special radar fit under nose 1.05)*		
G-ANAV*	de Havilland DH.106 Comet 1A				See SECTION 3 Part 1	Wroughton	
G-ANCF*	Bristol 175 Britannia Series 308F				See SECTION 3 Part 1	Kemble	
G-ANCS	de Havilland DH.82A Tiger Moth	82824	R4907	12. 9.53	C.E.Edwards and E.A Higgins	Rush Green	7.10.05
G-ANCX	de Havilland DH.82A Tiger Moth	83719	T7229	15. 9.53	D.R.Wood	Fowle Hall Farm, Paddock Wood	28. 7.02
	(Built Morris Motors Ltd)						
G-ANCY*	de Havilland DH.82A Tiger Moth				See SECTION 3 Part 1	Stauning, Denmark	
G-ANDM	de Havilland DH.82A Tiger Moth	3946	EI-AGP	23. 9.53	N.J.Stagg	Bristol	14. 8.06
			G-ANDM, EI-AGP, G-ANDM, (G-ANDI), N6642				
G-ANDP	de Havilland DH.82A Tiger Moth	82868	D-EBEC	22. 9.53	A H.Diver	Newtownards	27. 3.05
			N9920F, G-ANDP, R4960				
G-ANEF*	de Havilland DH.82A Tiger Moth				See SECTION 3 Part 1	Västerås, Sweden	
G-ANEH	de Havilland DH.82A Tiger Moth	82067	N6797	29. 9.53	G.J.Wells *(As "N-6797")*	(Booker)	14. 7.07
G-ANEJ*	de Havilland DH.82A Tiger Moth				See SECTION 3 Part 1	Kuala Lumpur, Malaysia	
G-ANEL	de Havilland DH.82A Tiger Moth	82333	N9238	1.10.53	Totalsure Ltd	Seppe-Hoeven, The Netherlands	29.10.05
G-ANEM	de Havilland DH.82A Tiger Moth	82943	EI-AGN	1.10.53	P.J.Benest	Hamstead Marshall	16. 7.05
			G-ANEM, R5042				
G-ANEN	de Havilland DH.82A Tiger Moth	85418	OO-ACG	2.10.53	A J.D.Douglas-Hamilton	Goodwood	13. 4.05
			G-ANEN, DE410				
G-ANEW	de Havilland DH.82A Tiger Moth	86458	NM138	6.10.53	A L.Young	Henstridge	18. 6.62T
	(Built Morris Motors Ltd)				*(Frame stored 8.02)*		
G-ANEZ	de Havilland DH.82A Tiger Moth	84218	T7849	20.10.53	C.D.J.Bland	Sandown	18. 8.05
G-ANFC	de Havilland DH.82A Tiger Moth	85385	DE363	13.10.53	J.E.Pierce	Ley Farm, Chirk	9.10.03T
	(Built Morris Motors Ltd)				*(Noted 8.04)*		
G-ANFH*	Westland WS.55 Whirlwind 1				See SECTION 3 Part 1	Weston-super-Mare	
G-ANFI	de Havilland DH.82A Tiger Moth	85577	DE623	16.10.53	G.P.Graham *(As "DE623")*	Shobdon	28. 4.06
	(Built Morris Motors Ltd) (Tiger Moth D-EDON as "DE623" is displayed @ Auto und Technik Museum, Sinsheim, Germany)						

G-ANFL	de Havilland DH.82A Tiger Moth	84617	T6169	22.10.53	Felthorpe Tiger Group Ltd	Felthorpe	6.10.07
	(Built Morris Motors Ltd)						
G-ANFM	de Havilland DH.82A Tiger Moth	83604	T5888	22.10.53	L.S.Mitton, A J.Coker and N.H.Lemon		
	(Built Morris Motors Ltd)				tr Reading Flying Group	White Waltham	15. 8.04
G-ANFP	de Havilland DH.82A Tiger Moth	82530	N9503	28.10.53	G D Horn *(Frame only 1.00)*	(Fordingbridge)	1. 7.63
G-ANFU*	Taylorcraft J Auster 5				See SECTION 3 Part 1	Newcastle upon Tyne	
G-ANFV	de Havilland DH.82A Tiger Moth	85904	DF155	1.12.53	R.A L.Falconer	Shempston Farm, Lossiemouth	31. 7.04
	(Built Morris Motors Ltd)				*(As "DF155")*		
G-ANFW*	de Havilland DH.82A Tiger Moth				See SECTION 3 Part 1	Ta'Qali, Malta	
G-ANGK (2)	Cessna 140A	15396	N9675A	10. 3.89	G.A Copeland	Popham	12. 8.07
G-ANGV*	AusterV J/1B Aiglet				See SECTION 3 Part 1	Queenstown, New Zealand	
G-ANHD*	Vickers 701C Viscount				See SECTION 3 Part 1	Bebeduoro, Brasil	
G-ANHK	de Havilland DH.82A Tiger Moth	82442	F-BHIM	4.12.53	J.D.Iliffe	Hampstead Norreys	23. 2.07
			G-ANHK, N9372				
G-ANHM*	Taylorcraft G Auster 4				See SECTION 3 Part 1	Murwillumbah, New South Wales, Australia	
G-ANHR	Taylorcraft J Auster 5	759	MT192	5.12.53	H.L.Swallow *(Stored 6.04)*	Hibaldstow	20. 7.86
G-ANHS	Taylorcraft G Auster 4	737	MT197	5.12.53	R.G.Tomlinson tr Tango Uniform Group	Spanhoe	22. 8.04
G-ANHU	Taylorcraft G Auster 4	799	EC-AXR	5.12.53	D.J.Baker	Carr Farm, Thorney, Newark	22.10.66
			G-ANHU, MT255		*(Noted dismantled, camouflaged 1.05)*		
G-ANHW*	Taylorcraft J Auster 5D	1396	TJ320	5.12.53	D.J.Baker	Carr Farm, Thorney, Newark	9. 3.70
	(Originally regd as Auster 5)		*(Forced landed Carlton Manor, Norfolk 1970: WFU 15.12.71) (Derelict fuselage noted stored 1.05)*				
G-ANHX	Taylorcraft J Auster 5D	2064	TW519	5.12.53	D.J.Baker	Carr Farm, Thorney, Newark	2.11.73
					(Crashed 28.3.70: noted dismantled 1.05)		
G-ANIE	Taylorcraft J Auster 5	1809	TW467	5.12.53	M J Whitwell	Kemble	27.11.05
					(As "TW467/ROD-F" in 664 Sqdn c/s)		
G-ANIJ	Taylorcraft J Auster 5D	1680	TJ672	5.12.53	M.Pocock	Kemble	5. 5.71
					(As "TJ672" in 657 Sqdn c/s: noted 12.00)		
G-ANIU*	Taylorcraft J Auster 5				See SECTION 3 Part 1	Stockhol, Sweden	
G-ANJA	de Havilland DH.82A Tiger Moth	82459	N9389	7.12.53	P.Aukland *(As "N9389")*	Seething	6. 6.05
G-ANJD	de Havilland DH.82A Tiger Moth	84652	T6226	8.12.53	I.Laws	Audley End	8. 7.06
	(Built Morris Motors Ltd)						
G-ANJG*	de Havilland DH.82A Tiger Moth				See SECTION 3 Part 1	La Ferté-Alais, Essonne, France	
G-ANJV*	Westland WS-55 Whirlwind 3				See SECTION 3 Part 1	Weston-super-Mare	
G-ANKK	de Havilland DH.82A Tiger Moth	83590	T5854	24.12.53	P A Cambridge	Charity Farm, Baxterley	20. 6.04
					tr Halfpenny Green Tiger Group *(As "T5854")*		
G-ANKN*	de Havilland DH.82A Tiger Moth				See SECTION 3 Part 1	Columbus, Indiana, USA	
G-ANKT	de Havilland DH.82A Tiger Moth	85087	T6818	24.12.53	Richard Shuttleworth Trustees*(As "T6818")*	Old Warden	1. 9.05
G-ANKV*	de Havilland DH.82A Tiger Moth				See SECTION 3 Part 1	Croydon	
G-ANKZ	de Havilland DH.82A Tiger Moth	3803	(N)	30.12.53	D.W.Graham	Bossington	10. 6.06
			F-BHIO, G-ANKZ, N6466		*(As "N-6466")*		
G-ANLD	de Havilland DH.82A Tiger Moth	85990	OO-DPA	30.12.53	K.Peters	White Waltham	17.12.05
			G-ANLD, EM773				
G-ANLS	de Havilland DH.82A Tiger Moth	85862	DF113	7. 1.54	P.A Gliddon	Great Fryup, Egton, Whitby	29. 6.03
	(Built Morris Motors Ltd)						
G-ANLW*	Westland WS-51 Series 2 Widgeon				See SECTION 3 Part 1	Flixton, Bungay	
G-ANMO	de Havilland DH.82A Tiger Moth	3255	F-BHIU	22. 1.54	E. and K.M.Lay	White Waltham	26. 8.06
			G-ANMO, K4259		*(As "K-4259/71")*		
G-ANMV	de Havilland DH.82A Tiger Moth	83745	F-BHAZ	22. 1.54	B.P.Sanders	Booker	26. 6.01T
	(Built Morris Motors Ltd)		G-ANMV, T7404		tr Tigerfly *(Dismantled 7.01 and sold to Germany 2002)*		
G-ANMY	de Havilland DH.82A Tiger Moth	85466	OO-SOL	22. 1.54	F.P.Le Coyte	Manor Farm, Garford	28. 9.07
	(Built Morris Motors Ltd)		"OO-SOC", G-ANMY, DE470		tr Lotmead Flying Group *(As "DE470/16" in RAF c/s)*		
G-ANNB	de Havilland DH.82A Tiger Moth	84233	N6037	22. 1.54	G.M.Bradley	(Colchester)	12. 6.58
			D-EGYN, G-ANNB, T6037		*(On rebuild 2002)*		
G-ANNC*	de Havilland DH.82A Tiger Moth				See SECTION 3 Part 1	Santa Teresa, New Mexico, USA	
G-ANNE (2)	de Havilland DH.82A Tiger Moth	"83814"		15. 4.94	C.R.Hardiman	Shobdon	30. 5.58
	(Composite airframe)				*(On rebuild 9.02)*		
G-ANNG	de Havilland DH.82A Tiger Moth	85504	DE524	22. 1.54	P.F.Walter	Farnborough	18. 5.01
	(Built Morris Motors Ltd)						
G-ANNI	de Havilland DH.82A Tiger Moth	85162	T6953	22. 1.54	C E, O C and M E Ponsford *(As "T-6953)*	Goodwood	6. 9.06
G-ANNK	de Havilland DH.82A Tiger Moth	83804	F-BFDO	22. 1.54	D.R.Wilcox	Sywell	25. 9.87
	(Built Morris Motors Ltd)		G-ANNK, T7290		*(New owner 5.02: wings only noted Spanhoe 4.03)*		
G-ANNW*	Auster J/5F Aiglet Trainer				See SECTION 3 Part 1	Kuwait	
G-ANOH	de Havilland DH.82A Tiger Moth	86040	EM838	22. 2.54	N.Parkhouse	Redhill	30.10.05
	(Built Morris Motors Ltd)						
G-ANOK*	SAAB 91C Safir	91311	SE-CAH	22. 4.54	A F.Galt and Co Ltd	(Yarrow Ford)	5. 2.73
					(Cancelled 15.10.81 by CAA) (Stored 10.01)		
G-ANOM	de Havilland DH.82A Tiger Moth	82086	N6837	2. 3.54	A L.Creer	(Bristol)	3. 5.62T
					(Crashed Fairoaks 17.12.61: on rebuild 6.00)		
G-ANON	de Havilland DH.82A Tiger Moth	84270	T7909	4. 3.54	R.C.Hields	Sherburn-in-Elmet	22. 4.06T
	(Built Morris Motors Ltd)				t/a Hields Aviation *(As "T7909")*		
G-ANOO	de Havilland DH.82A Tiger Moth	85409	DE401	11. 3.54	R.K.Packman	Compton Abbas	16. 1.06
	(Built Morris Motors Ltd)						
G-ANOS*	de Havilland DH.82A Tiger Moth				See SECTION 3 Part 1	Brandon, Manitoba, Canada	
G-ANOV*	de Havilland DH.104 Dove 6				See SECTION 3 Part 1	East Fortune	
G-ANPE	de Havilland DH.82A Tiger Moth	83738	G-IESH	27. 3.54	I.E.S.Hudleston	Duxford	6. 8.06T
	(Built Morris Motors Ltd)		G-ANPE, F-BHAT, G-ANPE, T7397				
G-ANPK*	de Havilland DH.82A Tiger Moth	3571	L6936	5. 4.54	A D.Hodgkinson	Thruxton	10. 7.97T
			(Damaged Jaywick Sands, Clacton-on-Sea 18.8.96: on rebuild 2002: cancelled 7.1.03 by CAA)				
G-ANPV (2)*	de Havilland DH.114 Heron 2D				See SECTION 3 Part 1	Melbourne, Victoria, Australia	
G-ANRF	de Havilland DH.82A Tiger Moth	83748	T5850	24. 5.54	C.D.Cyster	Glenrothes	24. 8.07
	(Built Morris Motors Ltd)						
G-ANRM	de Havilland DH.82A Tiger Moth	85861	DF112	8. 6.54	Spectrum Leisure Ltd	Duxford	28. 7.07T
	(Built Morris Motors Ltd)				*(Op Classic Wings as "DF112")*		
G-ANRN	de Havilland DH.82A Tiger Moth	83133	T5368	24. 5.54	J.J.V.Elwes	Rush Green	19. 5.07

Regn	Type	C/n	Previous identity	Regd	Owner/operator	Base	CofA
G-ANRP	Taylorcraft J Auster 5	1789	TW439	21. 5.54	I.C.Naylor and P.G.Wood *(As "TW439")*	Bagby	3. 4.06
G-ANRX*	de Havilland DH.82A Tiger Moth				See SECTION 3 Part 1	Salisbury Hall, London Colney	
G-ANSE*	de Havilland DH.82A Tiger Moth				See SECTION 3 Part 1	Kjeller, Oslo, Norway	
G-ANSJ*	de Havilland DH.82A Tiger Moth				See SECTION 3 Part 1	Masterton, New Zealand	
G-ANSM	de Havilland DH.82A Tiger Moth	82909	R5014	3. 6.54	Northamptonshire School of Flying Ltd	Sywell	28. 8.06T
	(Built Morris Motors Ltd)						
G-ANSO*	Gloster Meteor T.7				See SECTION 3 Part 1	Halmstad, Sweden	
G-ANSU*	de Havilland DH.82A Tiger Moth				See SECTION 3 Part 1	Masterton, New Zealand	
G-ANTE	de Havilland DH.82A Tiger Moth	84891	T6562	20. 9.54	P Reading	Sywell	16. 7.05T
	(Built Morris Motors Ltd)				*(Fuselage noted 1.05)*		
G-ANTK*	Avro 685 York C.1				See SECTION 3 Part 1	Duxford	
G-ANTV*	de Havilland DH.82A Tiger Moth				See SECTION 3 Part 1	Kungsangen, Sweden	
G-ANUO*	de Havilland DH.114 Heron 2D				See SECTION 3 Part 1	Croydon	
G-ANUW*	de Havilland DH.104 Dove 6	04458		16. 5.55	Jet Aviation Preservation Group	Long Marston	22. 7.81
					(Cancelled 5.6.96 as WFU) (Noted 2.04)		
G-ANVU*	de Havilland DH.104 Dove 1B				See SECTION 3 Part 1	Linköping, Sweden	
G-ANWB	de Havilland DHC-1 Chipmunk 21	C1/0987	G-5-17	15. 2.55	G.Briggs	Blackpool	17.12.04T
G-ANWO	Miles M.14A Hawk Trainer 3	718	L8262	31.12.58	A G.Dunkerley	(Bristol)	18. 4.63
	(DBR Kirton-in-Lindsey 21.4.62 and cancelled as damaged: restored 24.6.87 but unlikely little residue: wings to G-AIUA in 1960s and fuselage remnants slight of substance: for possible incorporation into rebuild of G-ADNL (qv))						
G-ANXB*	de Havilland DH.114 Heron 1B				See SECTION 3 Part 1	Newark	
G-ANXC	Auster J/5R Alpine	3135	5Y-UBD	4.12.54	R.B.Webber	Trenchard Farm, Eggesford	14. 9.06
			VP-UBD, G-ANXC, (AP-AHG), G-ANXC		tr Alpine Group		
G-ANXP*	Piper J-3C-65 Cub (L-4H-PI)				See SECTION 3 Part 1	Liorac-sur-Louyre, Dordogne, France	
G-ANXR	Percival P.31C Proctor IV	H.803	RM221	14.12.54	L.H.Oakins *(As "RM221")*	Rochester	4. 3.07
G-ANZS*	de Havilland DH.82A Tiger Moth				See SECTION 3 Part 1	Yering, Victoria, USA	
G-ANZT	Thruxton Jackaroo	84176	T7798	4. 3.55	D.J.Neville and P.J.Dear	Booker	14. 8.05
G-ANZU	de Havilland DH.82A Tiger Moth	3583	L6938	9. 3.55	P.A Jackson	Brookfield Farm, Great Stukeley	17. 3.91
					(Stored 1994: current status unknown)		
G-ANZZ	de Havilland DH.82A Tiger Moth	85834	DE974	14. 3.55	J.I.B.Bennett and P.P.Amershi	(Hatfield)	28. 2.69T
	(Built Morris Motors Ltd)						

G-AOAA - G-AOZZ

Regn	Type	C/n	Previous identity	Regd	Owner/operator	Base	CofA
G-AOAA	de Havilland DH.82A Tiger Moth	85908	DF159	14. 3.55	R.C.P.Brookhouse	Thruxton	8.12.91T
	(Built Morris Motors Ltd)				*(Damaged Redhill 4.6.89: under restoration 2002)*		
G-AOAF*	de Havilland DH.82A Tiger Moth				See SECTION 3 Part 1	Masterton, New Zealand	
G-AOAI*	Blackburn Beverley C.1				See SECTION 3 Part 1	Beverley	
G-AOAO*	de Havilland DH.89A Dragon Rapide				See SECTION 3 Part 1	La Ferté-Alais, Essonne, France	
G-AOBG*	Somers-Kendall SK-1	1		30. 3.55	A J.E.Smith	Breighton	26. 6.58
					(WFU after engine turbine failure 11.7.57: stored 1.05)		
G-AOBH	de Havilland DH.82A Tiger Moth	84350	T7997	31. 3.55	P.Nutley	Trenchard Farm, Eggesford	2.12.06
	(Built Morris Motors Ltd) (Regd with c/n 83818 ex T7439)				*(As "NL750" which belongs to G-AHUF once regd to this owner)*		
G-AOBO	de Havilland DH.82A Tiger Moth	3810	N6473	23. 4.55	J.S.and J.V.Shaw	Cubert, Newquay	28. 8.69T
					(On rebuild 10.97: current status unknown)		
G-AOBU	Hunting Percival P.84 Jet Provost T.1	P84/6	XM129	2. 5.55	T.J.Manna	North Weald	2. 3.05P
			G-AOBU, G-42-1		t/a Kennet Aviation *(As "XD693/Z-Q" in 2 FTS c/s)*		
G-AOBX	de Havilland DH.82A Tiger Moth	83653	T7187	26. 4.55	S.Bohill-Smith	White Waltham	4.12.05
	(Built Morris Motors Ltd)				tr David Ross Flying Group		
G-AOCP (2)*	Taylorcraft J Auster 5	1800	W462	25. 5.56	C.J.Baker	Carr Farm, Thorney, Newark	
	(Composite using G-AKOT)				*(WFU 22.6.68) (Fuselage frame noted 1.05)*		
G-AOCR (2)	Taylorcraft J Auster 5D	1060	EI-AJS	25. 5.56	G.J.McDill	Bagby	2. 5.05
			G-AOCR, NJ673		*(As "NJ673")*		
G-AOCU (2)	Taylorcraft J Auster 5	986	MT349	8. 6.56	S.J.Bal	Leicester	7. 3.07
G-AODA*	Westland WS-55 Whirlwind 3				See SECTION 3 Part 1	Weston-super-Mare	
G-AODR	de Havilland DH.82A Tiger Moth	86251	G-ISIS	4. 8.55	L.A Groves	Lee-on-Solent	29. 3.62
	(Built Morris Motors Ltd)		G-AODR, NL779		*(New CofR 9.03)*		
G-AODT	de Havilland DH.82A Tiger Moth	83109	R5250	4. 8.55	R.A Harrowven	Tibenham	30. 4.01
G-AOEH	Aeronca 7AC Champion	7AC-2144	N79854	8. 9.55	R.A and S.P.Smith	Great Yeldham	7. 4.05P
	(Continental A65)		OO-TWF				
G-AOEI	de Havilland DH.82A Tiger Moth	82196	N6946	14. 9.55	CFG Flying Ltd	Cambridge	11. 7.05T
	(Regd with fuselage no.MCO/DH3409 which should correspond to ex DE298 [85332]: a/c is probably composite airframe)						
G-AOEL*	de Havilland DH.82A Tiger Moth				See SECTION 3 Part 1	East Fortune	
G-AOES	de Havilland DH.82A Tiger Moth	84547	T6056	6.10.55	K.A and A J.Broomfield	Charity Farm, Baxterley	15. 6.02
	(Built Morris Motors Ltd)						
G-AOET	de Havilland DH.82A Tiger Moth	85650	DE720	7.10.55	Techair London Ltd	Bossington	29. 9.06
	(Built Morris Motors Ltd)						
G-AOEX	Thruxton Jackaroo	86483	NM175	10.10.55	A T.Christian	Walkeridge Farm, Overton	3. 2.68T
					(On rebuild 10.01)		
G-AOFE	de Havilland DHC-1 Chipmunk 22A	C1/0150	WB702	13. 9.56	W.J.Quinn *(As "WB702")*	Goodwood	8.10.04
G-AOFJ (2)	Auster Alpha 5	3401		3.10.56	N.Fraser	Carnmore, Galway	6. 7.07
G-AOFM	Auster J/5P Autocar	3178		16. 6.55	S.J.Cooper	Sturgate	13. 5.05
G-AOFR*	de Havilland DH.82A Tiger Moth				See SECTION 3 Part 1	Egeskov, Denmark	
G-AOFS	Auster J/5L Aiglet Trainer	3143	EI-ALN	28.10.55	P.N.A Whitehead	Leicester	26. 4.04
			G-AOFS		*(Noted 1.05)*		
G-AOFX (2)*	Vickers 701C Viscount				See SECTION 3 Part 1	Sao Paulo, Brasil	
G-AOGA*	Miles M.75 Aries 1				See SECTION 3 Part 1	Leitrim, County Leitrim	
G-AOGE*	Percival P.34A Proctor III	H.210	BV651	24.11.55	Not known	Biggin Hill	21. 5.84
					(Cancelled 19.1.99 by CAA) (Stored 4.04)		
G-AOGI	de Havilland DH.82A Tiger Moth	85922	(N)	14.12.55	W.J.Taylor	(Skegness)	23. 8.91
	(Built Morris Motors Ltd)		OO-SOA, G-AOGI, DF186		t/a Lincs Aerial Spraying Co *(Stored 10.92)*		
G-AOGR	de Havilland DH.82A Tiger Moth	84566	XL714	20. 1.56	M.I.Edwards	Grange Farm, Boughton	21.12.00
	(Built Morris Motors Ltd)		G-AOGR, T6099		*(As "XL714": noted 5.02)*		

Registration	Type	C/n	Previous identity	Date	Owner/operator	Location	Expiry
G-AOGV	Auster J/5R Alpine	3302		2. 2.56	R.E.Heading	Walnut Tree Farm, Thorney, Whittlesey	17. 7.72
					(Stored 12.97: current status unknown)		
G-AOGW*	de Havilland DH.114 Heron 2E				See SECTION 3 Part 1	Sault Sainte Marie, Ontario, Canada	
G-AOHD (2)*	Hunting Percival P.84 Jet Provost T.2				See SECTION 3 Part 1	Point Cook, Victoria,Australia	
G-AOHY	de Havilland DH.82A Tiger Moth	3850	N6537	23. 2.56	R.H.and Samantha J.Cooper	(Stow, Lincoln)	20. 8.60
					(On rebuild 9.00: new owners 7.04)		
G-AOHZ	Auster J/5P Autocar	3252		28. 2.56	A D.Hodgkinson *(Noted 1.05)*	Farley Farm, Romsey	25. 9.03
G-AOIE*	Douglas DC-7C				See SECTION 3 Part 1	Dromod, Leitrim, County Leitrim	
G-AOIL	de Havilland DH.82A Tiger Moth	83673	XL716	20. 8.56	C.D.Davidson	Compton Abbas	17. 9.05T
	(Built Morris Motors Ltd)		G-AOIL, T7363		*(As "XL-716")*		
G-AOIM	de Havilland DH.82A Tiger Moth	83536	T7109	27. 8.56	C.R.Hardiman	Shobdon	8. 4.04
	(Built Morris Motors Ltd)				*(New owner 8.04)*		
G-AOIR	Thruxton Jackaroo	82882	R4972	13. 1.56	K.A and A J.Broomfield	Charity Farm, Baxterley	1. 4.05
G-AOIS	de Havilland DH.82A Tiger Moth	83034	R5172	13. 1.56	J.K.Ellwood *(As "R5172/FIJ-E")*	Sherburn-in-Elmet	10. 7.04
G-AOIY	Auster J/5V-160 Autocar	3199		1. 3.56	R.A Benson	Trenchard Farm, Eggesford)	15. 1.07
	(Lycoming O-320)						
G-AOJG*	Hunting-Percival P.66 President I				See SECTION 3 Part 1	Billund, Denmark	
G-AOJH	de Havilland DH.83C Fox Moth	FM.42	AP-ABO	29. 3.56	Connect Properties Ltd	Kemble	12. 1.06
G-AOJJ	de Havilland DH.82A Tiger Moth	85877	DF128	5. 4.56	E.and K.M.Lay *(As "DF128/RCO-U")*	White Waltham	26. 7.03
	(Built Morris Motors Ltd)	*(Swung on take-off Goodwood 8.7.01, stuck parked aircraft, wings and tail broken off: new owners 10.01)*					
G-AOJK	de Havilland DH.82A Tiger Moth	82813	R4896	5. 4.56	R.J.Willies	Top Farm, Croydon, Royston	18. 8.05
G-AOJR	de Havilland DHC-1 Chipmunk 22	C1/0205	SE-BBS	9. 4.56	G.J.G-H.Caubergs and N.Marien	Grimbergen, Belgium	28. 7.05
			OY-DFB, D-EGIM, G-AOJR, D-EGIM, G-AOJR, WB756				
G-AOJT*	de Havilland DH.106 Comet IXB				See SECTION 3 Part 1	Salisbury Hall, London Colney	
G-AOJX*	de Havilland DH.82A Tiger Moth				See SECTION 3 Part 1	Brussels, Belgium	
G-AOKL	Percival P.40 Prentice T.1	PAC/208	VS610	13. 4.56	Richard Shuttleworth Trustees	Old Warden	20. 9.96
					(As "VS610/K-L") (Under restoration 3.04)		
G-AOKO*	Percival P.40 Prentice T.1				See SECTION 3 Part 1	Doncaster	
G-AOKZ*	Percival P.40 Prentice T.1				See SECTION 3 Part 1	Coventry	
G-AOLK	Percival P.40 Prentice T.1	PAC/225	VS618	25. 4.56	A Hilton	Southend	14. 6.07
G-AOLP*	Percival P.40 Prentice T.1				See SECTION 3 Part 1		
G-AOLU	Percival P.40 Prentice T.1	B3/1A/PAC/283	EI-ASP	25. 4.56	N.J.Butler	(Montrose)	8. 5.76
	(Regd with c/n 5830/3)		G-AOLU, VS356		*(As "VS356" Montrose Air Station Museum)*		
G-AOMF*	Percival P.40 Prentice T.1				See SECTION 3 Part 1	Camarillo, California, USA	
G-AOPL*	Percival P.40 Prentice T.1				See SECTION 3 Part 1	Pretoria, South Africa	
G-AOPO*	Percival P.40 Prentice T.1				See SECTION 3 Part 1	Brussels, Belgium	
G-AORB (2)	Cessna 170B	20767	OO-SIZ	13. 2.84	A R.Thompson	Hawley Farm, Tadley	9. 4.06
			N2615D		tr Hawley Farm Group		
G-AORG	de Havilland DH.114 Heron	214101	XR441	1. 5.56	Duchess of Brittany (Jersey) Ltd	Jersey	23. 4.05
	(Built as Sea Heron C.1)		G-AORG, G-5-16		*(Jersey Airlines titles) "Duchess of Brittany"*		
G-AORW	de Havilland DHC-1 Chipmunk 22A	C1/0130	WB682	28. 5.56	Skylark Aviation Ltd	(Sandford, Strathaven)	19. 1.06T
G-AOSF	de Havilland DHC-1 Chipmunk 22	C1/0023	D-EIIZ	25. 6.56	T.S.Olsen	(Luxembourg)	6. 4.06
			G-AOSF, HB-TUA, G-AOSF, WB571 *(As "WB571/34")*				
G-AOSK	de Havilland DHC-1 Chipmunk 22A	C1/0178	WB726	26. 6.56	L.J.Irvine	RAF Halton	5. 3.06
					(As "WB726/E" in Cambridge UAS c/s)		
G-AOSP*	de Havilland DHC-1 Chipmunk 22				See SECTION 3 Part 1	Scone, New South Wales, Australia	
G-AOSU	de Havilland DHC-1 Chipmunk 22	C1/0217	WB766	28. 6.56	T.Holloway tr RAFGSA	Easterton	28. 6.06
	(Lycoming O-360)		*(Op Fulmar Gliding Club) (Force landed in nearby field 18.1.204 and suffered substantial damage)*				
G-AOSW*	de Havilland DHC-1 Chipmunk 22				See SECTION 3 Part 1	Ardmore, Auckland, New Zealand	
G-AOSY	de Havilland DHC-1 Chipmunk 22	C1/0037	WB585	29. 6.56	B.A Webster	Seething	26. 6.05
					tr WFG Chipmunk Group *(As "WB585/M)*		
G-AOTD	de Havilland DHC-1 Chipmunk 22	C1/0040	WB588	30. 6.56	S.Piech	Old Sarum	4.10.06
					(As "WB588/D" in Oxford UAS c/s)		
G-AOTF	de Havilland DHC-1 Chipmunk 23	C1/0015	WB563	2. 7.56	T.Holloway	AAC Dishforth	28.10.07
	(Lycoming O-360)				tr RAFGSA *(Op Clevelands Gliding Club)*		
G-AOTI*	de Havilland DH.114 Heron 2D				See SECTION 3 Part 1	Salisbury Hall, London Colney	
G-AOTK	Druine D.53 Turbi	PFA 230		1.11.56	T.J.Adams	RAF Henlow	14. 9.05P
	(Built TK Flying Group - c/n 1) (Walter Mikron 3)						
G-AOTR	de Havilland DHC-1 Chipmunk 22	C1/0045	HB-TUH	12. 7.56	M.R.Woodgate	Newtownards	12. 6.06
			D-EGOG, G-AOTR, WB604				
G-AOTY	de Havilland DHC-1 Chipmunk 22	C1/0522	WG472	12. 7.56	A A Hodgson	Bryn Gwyn Bach	5. 4.07
					tr Bryn Gwyn Bach Chipmunk Group *(As "WG472" in RAF c/s)*		
G-AOUF*	de Havilland DH.104 Dove 6				See SECTION 3 Part 1	Billund, Denmark	
G-AOUO	de Havilland DHC-1 Chipmunk 22	C1/0179	WB730	10. 8.56	T.Holloway	RAF Cosford	16. 3.06
	(Lycoming O-360)				tr RAFGSA *(Op Wrekin Gliding Club)*		
G-AOUP	de Havilland DHC-1 Chipmunk 22	C1/0180	WB731	10. 8.56	A R.Harding	Leicester	3. 8.06
G-AOUR*	de Havilland DH.82A Tiger Moth				See SECTION 3 Part 1	Belfast	
G-AOVF*	Bristol 175 Britannia Series 312F				See SECTION 3 Part 1	RAF Cosford	
G-AOVT*	Bristol 175 Britannia Series 312				See SECTION 3 Part 1	Duxford	
G-AOVU*	de Havilland DH.106 Comet 4C				See SECTION 3 Part 1	Seattle, Washington, USA	
G-AOVW	Taylorcraft J Auster 5	894	MT119	16.11.59	B.Marriott	Wilsford, Sleaford	8.10.06
G-AOXG*	de Havilland DH.82A Tiger Moth				See SECTION 3 Part 1	RNAS Yeovilton	
G-AOXL*	de Havilland DH.114 Heron 1B				See SECTION 3 Part 1	Stavangar, Norway	
G-AOXN	de Havilland DH.82A Tiger Moth	85958	EM727	31.10.56	S.L.G.Darch	East Chinnock, Yeovil	21.12.01
	(Built Morris Motors Ltd)						
G-AOYM*	Vickers 806 Viscount				See SECTION 3 Part 1	Tenerife,.Spain	
G-AOYU*	de Havilland DH.82A Tiger Moth				See SECTION 3 Part 1	Wetaskiwin, Alberta, Canada	
G-AOZE*	Westland WS-51/2 Widgeon				See SECTION 3 Part 1	Weston-super-Mare	
G-AOZH	de Havilland DH.82A Tiger Moth	86449	NM129	18. 1.57	M.H.Blois-Brooke	White Waltham	10. 6.05T
	(Built Morris Motors Ltd)				*(As "K2572")*		
G-AOZL	Auster J/5Q Alpine	3202		5. 2.57	R.M.Weeks *(On rebuild 7.02)*	Earls Colne	28. 5.88
G-AOZP	de Havilland DHC-1 Chipmunk 22A	C1/0183	WB734	14. 2.57	S.J.Davies	Sandtoft	27. 3.05
G-AOZZ*	Armstrong-Whitworth 650 Argosy Series 100				See SECTION 3 Part 1	Belleville, Michigan, USA	

G-APAA - G-APZZ

Registration	Type	C/n	Previous identity	Reg date	Owner/operator	Probable base	CofA expiry
G-APAD*	Edgar Percival EP.9				See SECTION 3 Part 1	Greenock, South Australia	
G-APAF	Auster Alpha 5	3404	G-CMAL	25. 3.57	J J J Mostyn	Henstridge	1. 8.05
			G-APAF		*(As "TW511")*		
G-APAH	Auster Alpha 5	3402		29. 3.57	T.J.Goodwin	New Farm House, Great Oakley	5. 4.04
G-APAJ	Thruxton Jackaroo	83314	VH-KRK	3. 4.57	J.T.H. and Suna J. Page *(New CofR 12.04)*		
			G-APAJ, T5616			Hurstbourne Tarrant, Bourne Park	
G-APAL	de Havilland DH.82A Tiger Moth	82102	N6847	3. 4.57	Avia Special Ltd *(As "N6847")*	Barton	9. 7.06T
G-APAM	de Havilland DH.82A Tiger Moth	3874	N6580	3. 4.57	R.P.Williams	Denford Manor, Hungerford	17. 8.07
					tr Myth Group *"Myth"*		
	(Caught by crosswind landing Radley Farm, Hungerford 5.4.04 and left wing tip struck ground: damage to upper wing, rudder, engine cowling and propeller)						
G-APAO	de Havilland DH.82A Tiger Moth	82845	R4922	3. 4.57	H.J.Maguire *(Op Classic Wings)*	Duxford	24. 9.05T
G-APAP	de Havilland DH.82A Tiger Moth	83018	R5136	3. 4.57	J C Wright *(As "R5136")*	RAF Henlow	17. 7.06
G-APAS*	de Havilland DH.106 Comet 1A				See SECTION 3 Part 1	RAF Cosford	
G-APBE	Auster Alpha 5	3403		7. 5.57	R.J.Napp	Brock Farm, Billericay	6.12.04
G-APBI	de Havilland DH.82A Tiger Moth	86097	EM903	16. 5.57	A Wood	Halstead, Essex	19. 4.82
	(Built Morris Motors Ltd)				*(Damaged Audley End 7.7.80: on rebuild 12.90: current status unknown)*		
G-APBO	Druine D.53 Turbi	PFA 229		3. 6.57	R.C.Hibberd	(Devizes)	1. 7.05P
	(Built F Roche) (Continental C75)						
G-APBW	Auster Alpha 5A	3405		23. 5.57	N.Huxtable	Cheddington	1. 6.06
G-APCB	Auster J/5Q Alpine	3204		5. 6.57	A A Beswick *(Noted 8.04)*	Thruxton	11. 7 07
G-APCC	de Havilland DH.82A Tiger Moth	86549	PG640	11. 6.57	L.J.Rice	Quebec Farm, Knook, Warminster	8. 7.06
G-APCN*	Boulton Paul P.108 Balliol T.2				See SECTION 3 Part 1	Negombo, Sri Lanka	
G-APCU*	de Havilland DH.82A Tiger Moth				See SECTION 3 Part 1	Glize-Rijen, The Netherlands	
G-APDB*	de Havilland DH.106 Comet 4				See SECTION 3 Part 1	Duxford	
G-APDF*	de Havilland DH.106 Comet 4	6407		2. 2.57	(Not known)	(Chipping Campden)	
					(To RAE as XV814 @ 3.67 and cancelled) (Nose preserved 2002)		
G-APEJ*	Vickers 953C Vanguard Merchantman				See SECTION 3 Part 1	Brooklands	
G-APEP*	Vickers 953C Vanguard Merchantman				See SECTION 3 Part 1	Brooklands	
G-APES*	Vickers 953C Vanguard Merchantman				See SECTION 3 Part 1	Nottingham East Midlands	
G-APFA	Druine D.52 Turbi	PFA 232		5. 2.57	F.J.Keitch	Smiths Farm, Brixham	22. 9.92P
	(Built Britten-Norman Ltd) (Continental A65)				*(Current status unknown)*		
G-APFG*	Boeing 707-436	17708	N5094K	7. 8.59	G.Spoors	Cove	24. 5.81T
					(Cancelled 11.68 as WFU) (Nose only 5.03)		
G-APFJ	Boeing 707-436				See SECTION 3 Part 1	Duxford	
G-APFU	de Havilland DH.82A Tiger Moth	86081	EM879	28. 8.57	Leisure Assets Ltd	Goodwood	11. 4.06T
G-APGL	de Havilland DH.82A Tiger Moth	86460	NM140	6. 9.57	K.A Broomfield	Charity Farm, Baxterley	
	(Built Morris Motors Ltd)				*(Not previously converted: on rebuild 3.97: see G-AJVE)*		
G-APGM*	de Havilland DH.82A Tiger Moth				See SECTION 3 Part 1	Nepi, Rome, Italy	
G-APHV*	Avro 652A Anson C.19 Series 2				See SECTION 3 Part 1	East Fortune	
G-APHY*	Scottish Aviation Twin Pioneer Series 1				See SECTION 3 Part 1	Quesnel, British Columbia, Canada	
G-APHZ*	Thruxton Jackaroo				See SECTION 3 Part 1	Kitchener, Ontario, Canada	
G-APIE	Tipsy Belfair	535	(OO-TIE)	22.10.57	D.Beale	Witchford	15. 4.05P
	(Walter Mikron 2)						
G-APIH	de Havilland DH.82A Tiger Moth	82981	N111DH	25.10.57	K.Stewering	Borken-Gemen, Germany	8. 4.06
			OY-DGJ, D-EMEX, G-APIH, R5086				
G-APIK	Auster Alpha	3375		11.11.57	J.H.Powell-Tuck	Gloucestershire	6. 2.06
G-APIM*	Vickers 806 Viscount				See SECTION 3 Part 1	Brooklands	
G-APIT*	Percival P.40 Prentice T.1				See SECTION 3 Part 1	Lasham	
G-APIY*	Percival P.40 Prentice T.1				See SECTION 3 Part 1	Newark	
G-APIZ	Druine D.31 Turbulent	PFA 478		22.11.57	E J I Musty	White Waltham	29.10.04P
	(Built Rollason Aircraft and Engines) (Volkswagen 1600)				*"Witch Lady"*		
G-APJB	Percival P.40 Prentice T.1	PAC/086	VR259	28.11.57	Atlantic Air Transport Ltd	Coventry	10. 7.05T
					(As "VR259/M" in 2 ASS c/s)		
G-APJJ (2)*	Fairey Ultralight Helicopter				See SECTION 3 Part 1	Coventry	
G-APJO	de Havilland DH.82A Tiger Moth	86446	NM126	23.12.57	D.R.and Mrs M.Wood	Tunbridge Wells	27. 3.59T
	(Built Morris Motors Ltd) (C/n quoted as "17712":)				*(Crashed Ross-on-Wye 5.8.58: on rebuild and may include components from G-APJR)*		
G-APJP*	de Havilland DH.82A Tiger Moth				See SECTION 3 Part 1	Billund, Denmark	
G-APJT*	Scottish Aviation Twin Pioneer Series 1				See SECTION 3 Part 1	Kuala Lumpur, Malaysia	
G-APJZ	Auster Alpha	3382	5N-ACY	3. 1.58	P.G.Lipman	Romney Street Farm, Sevenoaks	15. 7.77
			(VR-NDR), G-APJZ		*(Damaged Thornicombe 10.11.75: on rebuild 12.97: current status unknown)*		
G-APKM*	Auster Alpha	3385		27. 1.58	C.J.Baker	Carr Farm, Thorney, Newark	9. 1.89
					(Cancelled 9.10.91 as TWFU) (Noted dismantled 1.05)		
G-APKN	Auster Alpha	3387		27. 1.58	P.R.Hodson tr The Felthorpe Auster Group	Felthorpe	8. 9.05
					(Badly damaged in arson attack Felthorpe 8.2.03)		
G-APLG*	Auster J/5L Aiglet Trainer				See SECTION 3 Part 1	Carlisle	
G-APLI*	de Havilland DH.82A Tiger Moth				See SECTION 3 Part 1	Helsingborg, Sweden	
G-APLK*	Miles M.100 Student 2				See SECTION 3 Part 1	Woodley	
G-APLO	de Havilland DHC-1 Chipmunk 22	C1/0144	EI-AHU	1. 5.58	Lindholme Aircraft Ltd	Jersey	2.11.06T
			WB696		*(As "WD379/K" in Cambridge UAS c/s)*		
G-APLR*	de Havilland DH.82A Tiger Moth				See SECTION 3 Part 1	Nepi, Rome	
G-APLU	de Havilland DH.82A Tiger Moth	85094	VR-AAY	2. 4.58	R.A Bishop and M.E.Vaisey	Rush Green	14. 8.04
	(Built Morris Motors Ltd)		F-OBKK, G-APLU, T6825				
G-APMH	Auster J/1U Workmaster	3502	F-OBOA	15. 4.58	J.L.Thorogood	Insch	2. 6.07
			G-APMH				
G-APML*	Douglas C-47B-1-DK Dakota	14175/25620	KJ836	17. 4.58	(Aces High)	North Weald	27. 7.84T
			43-48359		*(Cancelled 20.3.01 as WFU) (Rear fuselage only 1.05)*		
G-APMX	de Havilland DH.82A Tiger Moth	85645	DE715	9. 5.58	M.A Broughton	Popham	13. 7.07
	(Built Morris Motors Ltd)				*(Noted 10.04)*		
G-APMY*	Piper PA-23-160 Apache				See SECTION 3 Part 1	Doncaster	
G-APNJ*	Cessna 310				See SECTION 3 Part 1	Newark	
G-APNN*	Auster 5 Alpha				See SECTION 3 Part 1	Helsingborg, Sweden	

G-APNS	Garland-Bianchi Linnet	001		17. 6.58	P.M.Busaidy	Scaynes Hill, Haywards Heath	6.10.78S
	(Continental C90)				*(Stored 2004)*		
G-APNT	Currie Wot	HAC-3		18. 6.58	B.J.Dunford	Longwood Farm, Morestead	29. 6.05P
	(Built Hampshire Aero Club) (Continental PC60) *(Regd with c/n P.6399)*				*"Airymouse"*		
G-APNV*	Saunders-Roe P.531-1				See SECTION 3 Part 1	RAF Yeovilton	
G-APNZ	Druine D.31 Turbulent	PFA 482		17. 4.58	J.Knight	(Hailsham)	13.12.95P
	(Built Rollason Aircraft and Engines) (Ardem 4C02)				*(Damaged River Rother near Iden 3.9.95: on rebuild)*		
G-APOD*	Tipsy Belfair	536	(OO-TIF)	16. 7.58	L.F.Potts	(Culloden)	23. 8.88P
	(Walter Mikron 2)				*(Cancelled 6.9.00 by CAA) (Under restoration 2.01)*		
G-APOI	Saunders-Roe Skeeter Series 8	S2/5081		29. 7.58	B.Chamberlain	Otley, Ipswich	2. 8.00P
G-APOL*	Druine D.31 Turbulent	PFA 439		31. 7.58	A Gregori and S.Tinker	(Morpeth)	18. 6.94P
	(Built P J Houston) (Ardem 4C02)				*(Damaged Charterhall 24.7.93: cancelled 13.9.00 as WFU) (Stored 2.03)*		
G-APOU*	de Havilland DH.82A Tiger Moth				See SECTION 3 Part 1	Västerås, Sweden	
G-APPA	de Havilland DHC-1 Chipmunk 22	C1/0792	N5073E	11. 9.58	D.M.Squires	(Wellesbourne Mountford)	14. 7.85
			G-APPA, WP917		*(On slow rebuild 1.03)*		
G-APPL	Percival P.40 Prentice 1	PAC/013	VR189	7.10.58	Susan J.Saggers	Biggin Hill	2. 6.07
G-APPM	de Havilland DHC-1 Chipmunk 22	C1/0159	WB711	14.10.58	S.D.Wilch *(As "WB711")*	Sywell	15. 8.05
G-APPN	de Havilland DH.82A Tiger Moth	83839	T7328	26. 9.03	John Colours SRL	Phoenix Farm, Lower Upham	4.11.07
					(As "T7328")		
G-APPT*	de Havilland DH.82A Tiger Moth				See SECTION 3 Part 1	Brussesl, Belgium	
G-APRF	Auster Alpha 5	3412	VR-LAF	8.12.58	W.B.Bateson	Blackpool	14.11.00
			G-APRF		*(Stored 8.02)*		
G-APRJ	Avro 694 Lincoln B.2	-	RF342	29.12.58	D.Copley	(Belton, Doncaster)	
			G-36-3, G-29-1, G-APRJ, RF342		*(Fuselage as "G-29-1" and wings as "RF342" - current status unknown)*		
G-APRK*	de Havilland DH.114 Heron 2D				See SECTION 3 Part 1	Langley, British Columbia, Canada	
G-APRL*	Armstrong-Whitworth 650 Argosy Series 101				See SECTION 3 Part 1	Coventry	
G-APRR	CZL Super Aero 45 Series 04	04-014	OK-KFQ	5. 1.59	R.H.Jowett	Ronaldsway	21. 7.07
G-APRS	Scottish Aviation Twin Pioneer 3	561	G-BCWF	9. 1.59	Bravo Aviation Ltd	Coventry	15. 7.05T
			XT610, G-APRS, (PI-C430)		*(ETPS titles)*		
G-APRT	Taylor JT.1 Monoplane	PFA 537		15. 1.59	D.A Slater	Redhill	26. 5.05P
	(Built J Taylor) (Ardem 4C02)						
G-APSA	Douglas DC-6A	45497	4W-ABQ	12. 2.59	Atlantic Air Transport Ltd	Coventry	11. 4.05T
			HZ-ADA, G-APSA, CF-MC		*(Air Atlantique - old style white c/s)*		
G-APSO*	de Havilland DH.104 Dove 5	04505	(N1046T)	16. 2.59	Cormack (Aircraft Services) Ltd	Kemble	8. 7.78T
			G-APSO		*(To Devonair) (Cancelled 2.5.01 as WFU)*		
			(Wings fitted to G-BWWC: forward part of fuselage fitted with stub wings and used as engine test-bed 4.03)				
G-APSR	Auster J/1U Workmaster	3499	OO-HXA	22. 4.59	D and K Aero Services Ltd	Namur-Temploux, Belgium	20.10.05A
			G-APSR, VP-JCD, G-APSR, (F-OBHR)		*(Op P.De Liens)*		
G-APSY*	Bensen B.7Mc	2		22. 2.59	J.Howell	(Copthorne, Sussex)	
	(Built J.Howell - c/n JH/001)				*(Cancelled 13.4.73 as WFU) (Stored 10.03)*		
G-APTR	Auster Alpha	3392		15. 4.59	C.J.and D.J.Baker	Carr Farm, Thorney, Newark	11. 4.87
					(WFU but noted complete 1.05)		
G-APTS*	de Havilland DHC-1 Chipmunk 22A				See SECTION 3 Part 1	Adelaide, South Australia	
G-APTU	Auster Alpha 5	3413		20. 4.59	A J. and J.M.Davis	Leicester	8. 6.98
					tr G-APTU Flying Group *(On rebuild 3.00)*		
G-APTW*	Westland WS-51/2 Widgeon				See SECTION 3 Part 1	Usworth, Sunderland	
G-APTY	Beech G35 Bonanza	D-4789	EI-AJG	4. 6.59	G.E.Brennand	Blackpool	11. 6.06
G-APTZ	Druine D.31 Turbulent	PFA 508		18. 3.59	Tiger Club (1990) Ltd	Headcorn	19. 5.05P
	(Built Rollason Aircraft and Engines) (Volkswagen 1600)						
G-APUD*	Bensen B-7Mc				See SECTION 3 Part 1	Manchester	
G-APUE	Orlican L-40 Meta-Sokol	150708	OK-NMB	2. 6.59	S.E.and M.J.Aherne	Top Farm, Croydon, Royston	20. 5.06
G-APUP*	Sopwith Pup rep				See SECTION 3 Part 1	Hendon	
G-APUR	Piper PA-22-160 Tri-Pacer	22-6711		3. 7.59	S.T.A.Hutchinson	Rathfriland	20. 8.07
G-APUW	Auster J/5V Series 160 Autocar	3273		23. 6.59	E.A J.Hibbard	Hill Farm, Nayland	18.12.03
G-APUY	Druine D.31 Turbulent	PFA 509		24. 6.59	C.Jones	Barton	10. 6.86P
	(Built K F W Turner) (Volkswagen 1300)				*(Stored 1.04)*		
G-APUZ	Piper PA-24-250 Comanche	24-1094	N6000P	3. 7.59	Tatenhill Aviation	Tatenhill	30. 9.07E
G-APVF	Putzer Elster B	006	D-EEQX	29.12.83	A and E.A Wiseman	Breighton	19. 5.05P
	(Continental O-200-A)		97+04, D-EJUH		*(As "97+04" in Luftwaffe c/s)*		
G-APVG	Auster J/5L Aiglet Trainer	3306	(ZK-BQW)	10. 7.59	R.Farrer	Cranfield	20. 3.00
G-APVL	Saunders-Roe P531-2	S2/5311	XP166	23. 7.59	R.E.Dagless	Yaxham, Dereham	
			G-APVL		*(Noted 7.02)*		
G-APVN	Druine D.31 Turbulent	PFA 511		24. 7.59	R.Sherwin	Swanborough Farm, Lewes	24. 6.94P
	(Built J P Knight) (Volkswagen 1600)				*(Stored 6.04)*		
G-APVS	Cessna 170B	26156	N2512C	7. 8.59	N.Simpson *"Stormin' Norman"*	East Kirkby	23. 6.06
G-APVU	Orlican L-40 Meta-Sokol	150706	OK-NMI	21. 8.59	S.A and M.J.Aherne	(St.Albans)	27. 6.79
					(Damaged Manchester 12.9.78: on rebuild 1993)		
G-APVV*	Mooney M.20A				See SECTION 3 Part 1	Newark	
G-APVZ	Druine D.31 Turbulent	PFA 545		23. 7.59	The Tiger Club 1990 Ltd	Headcorn	28. 4.04P
	(Built Rollason Aircraft and Engines) (Ardem 4C02)						
G-APWA*	Handley Page HPR.7 Dart Herald 100				See SECTION 3 Part 1	Woodley	
G-APWJ*	Handley Page HPR.7 Dart Herald 201				See SECTION 3 Part 1	Duxford	
G-APWL	EoN AP.10 460 Standard Series 1A	EoN/S/001	BGA.1172(BRK)	2. 9.59	D.G.Andrew	Eaglescott	
			G-APWL, RAFGSA.268, G-APWL		*(Noted 2000)*		
G-APWN*	Westland WS-55 Whirlwind 3				See SECTION 3 Part 1	Coventry	
G-APWP	Druine D.31 Turbulent	PFA 497		14. 9.59	C.F.Rogers	(Wheathamstead)	27. 6.67
	(Built C F Rogers)				*(Current status unknown)*		
G-APWY*	Piaggio P.166				See SECTION 3 Part 1	Wroughton	
G-APXA*	Westland-Sikorsky S-55 Whirlwind Series 2				See SECTION 3 Part 1	Kuwait	
G-APXB*	Westland-Sikorsky S-55 Whirlwind Series 2				See SECTION 3 Part 1	Kuwait	
G-APXG*	de Havilland DH.114 Heron 2D				See SECTION 3 Part 1	Winnipeg, Manitoba, Canada	
G-APXJ	Piper PA-24-250 Comanche	24-291	VR-NDA	11.12.59	T.Wildsmith	Sheffield City	29. 1.06
			N10F				
G-APXR	Piper PA-22-160 Tri-Pacer	22-7172	N10F	29. 1.60	A Troughton	Belfast Aldergrove	2.12.04

Registration	Type	C/n	Previous identity	Date	Owner	Base	Date
G-APXT	Piper PA-22-150 Tri-Pacer	22-3854	N4545A	16. 2.60	A E Cuttler	(Wokingham)	5. 7.87T
			(Damaged Southend 26.12.85 and on rebuild to PA-20 Pacer configuration: new owner 12.01)				
G-APXU	Piper PA-22-150 Tri-Pacer	22-474	N1723A	10. 2.60	The Scottish Aero Club Ltd	Perth	20. 2.85
					"The Cloth Bomber" (Noted 10.03)		
G-APXW*	Lancashire Aircraft EP-9 Prospector				See SECTION 3 Part 1	Middle Wallop	
G-APXX*	de Havilland DHA.3 Drover 2				See SECTION 3 Part 1	Lasham	
G-APXY	Cessna 150	17711	N7911E	15. 1.60	The Merlin Flying Club Ltd	Hucknall	1. 4.05T
G-APYB	Tipsy Nipper T.66 Series 3	T66/S/39		28. 1.60	B.O.Smith	Yearby	12. 6.96P
	(Built Avions Fairey SA) (Volkswagen 1834)				*(On rebuild 1.02)*		
G-APYD*	de Havilland DH.106 Comet 4B				See SECTION 3 Part 1	Wroughton	
G-APYG	de Havilland DHC-1 Chipmunk 22	C1/0060	OH-HCB WB619	11.11.60	E.J.I.Musty tr PYGS Flying Group	White Waltham	22. 7.07
G-APYI	Piper PA-22-135 Tri-Pacer	22-2218	N8031C	8. 2.60	B.T. and J.Cullen	Ballyboy, County Meath	9. 5.07
	(Modified to PA-20 Pacer configuration)						
G-APYN	Piper PA-22-160 Tri-Pacer	22-6797	N2804Z	24. 2.60	S.J.Raw	Morgansfield, Fishburn	8. 7.07
G-APYT	Champion 7FC Tri-Traveler	387		9. 5.60	B.J.Anning	Watchford Farm, Yarcombe	26. 1.04P
G-APZJ	Piper PA-18-150 Super Cub	18-7233	N10F	29. 1.60	R.Jones t/a Southern Sailplanes	Lasham	27. 5.06
					(Rebuilt 1986 after accident 12.6.83 using un-identified new fuselage frame)		
G-APZL	Piper PA-22-160 Tri-Pacer	22-7054	EI-ALF N10F	27. 1.60	B.Robins	Dunkeswell	14. 5.99
G-APZX	Piper PA-22-150 Tri-Pacer	22-5181	N7420D	28. 4.60	Applied Signs Ltd	Streethay	3. 9.06
	(Modified to PA-20 Pacer configuration)						

G-ARAA - G-ARZZ

Registration	Type	C/n	Previous identity	Date	Owner	Base	Date
G-ARAD*	Phoenix Luton LA-5A Major				See SECTION 3 Part 1	Usworth, Sunderland	
G-ARAI	Piper PA-22-160 Tri-Pacer	22-7421	N10F	17. 5.60	S.T.A Hutchinson	Rathfriland	14.11.04
G-ARAM	Piper PA-18-150 Super Cub	18-7312	N10F	17. 5.60	Spectrum Leisure Ltd	Clacton	22. 6.02T
G-ARAN	Piper PA-18-150 Super Cub	18-7307	N10F	28. 4.60	A P.Docherty	Clacton	8. 6.07
G-ARAO	Piper PA-18 Super Cub 95	18-7327	N10F	17. 5.60	R.G.Manton *(As "607327/09-L" in USAAC c/s)*	Denham	30. 1.05
G-ARAS	Champion 7FC Tri-Traveler	7FC-396		12. 9.60	G.J.Taylor tr Alpha Sierra Flying Group *(Noted 9.03)*	Yeatsall Farm, Abbots Bromley	22. 6.01P
G-ARAT	Cessna 180C	50827	N9327T	18. 5.60	S.D.Pryke and J.Graham	Old Buckenham	27.10.07E
G-ARAW	Cessna 182C Skylane	52843	N8943T	18. 5.60	R.P.Beck, G.and R.L.McLean t/a Ximango UK	Sherburn-in-Elmet	8.12.05T
G-ARAX	Piper PA-22-150 Tri-Pacer	22-3830	N4523A	22. 4.60	J.W.Iliffe	(Melbourne, Derby)	14. 4.06
G-ARAZ	de Havilland DH.82A Tiger Moth	82867	R4959	25. 3.60	D.A Porter *(As "R4959/59" in RAF c/s)*	Griffins Farm, Temple Bruer	1. 6.07
G-ARBE	de Havilland DH.104 Dove 8	04517		6. 5.60	M.Whale and M.W.A Lunn	Kemble	3.10.02
G-ARBF*	Bensen B.7M Gyrocopter	004		6. 5.60	(D R Shepherd)	(Prestwick)	
					(Crashed Tangmere 29.4.61 and cancelled) (Airframe stored at home 2003)		
G-ARBG	Tipsy Nipper T.66 Series 2	ABAC.1 & 57		11. 5.60	D.Shrimpton	Compton Abbas	17. 8.84P
	(Built Avions Fairey SA) (Volkswagen 1834 Acro)				*(New owner 7.04)*		
G-ARBM*	Auster 5 J/1B Aiglet	2792	EI-AMO G-ARBM, VP-SZZ, VP-KKR	8. 6.60	A D.Hodgkinson *(Cancelled 30.7.04 by CAA)*	Dunkirk Farm, Canterbury	6. 6.03
G-ARBO	Piper PA-24-250 Comanche	24-2117	N10F	15. 6.60	Arrow Aviation Services Ltd	Farley Farm, Romsey	27. 5.84
			(Force landed Morecambe Bay 27.4.83: new owner 6.01: unmarked fuselage noted 11.02)				
G-ARBP	Tipsy Nipper T.66 Series 2	54		7. 6.60	F.W.Kirk	Seighford	13. 6.05P
	(Built Avions Fairey SA) (Volkswagen 1834)						
G-ARBS	Piper PA-22-160 Tri-Pacer	22-6858	N2868Z	24. 8.60	S.D.Rowell	Valley Farm, Winwick	22. 1.05
	(Modified to PA-20 Pacer configuration)				*"Greta"*		
G-ARBV	Piper PA-22-160 Tri-Pacer	22-5836	N8633D	29. 6.60	Lorraine M.Williams	(Needham Market, Ipswich)	3. 7.06
	(Rebuilt 1983/84 using fuselage of G-ARDP ex N7004B [22-4254]						
G-ARBZ	Druine D.31 Turbulent	PFA 553		6. 5.60	G.Richards	Headcorn	15.10.99P
	(Built Rollason Aircraft and Engines) (Ardem 4C02)				*(New owner 2.03)*		
G-ARCC	Piper PA-22-150 Tri-Pacer	22-4006	N4853A	23. 6.60	A S.Cowan tr Popham Flying Group G-ARCC	Popham	31. 8.06
G-ARCF	Piper PA-22-150 Tri-Pacer	22-4563	N5902D	28. 6.60	M.J.Speakman	North Coates	20. 5.05
G-ARCS	Auster D.6 Series 180	3703		4. 7.60	E.A Matty	Shobdon	3. 9.03
G-ARCT	Piper PA-18 Super Cub 95	18-7375	EI-AVE G-ARCT, N10F	6. 7.60	C.F.O'Neil *(Damaged Mullaghmore 29.3.87: on rebuild 11.04)*	Newtownards	21. 4.86
G-ARCV	Cessna 175A Skylark (Continental O-300D)	17556757	N8057T	7.11.60	R.Francis and C.Campbell	Sandtoft	3. 9.05
G-ARCW	Piper PA-23 Apache *(Modified to PA-23-160 standard)*	23-796	N2187P	7. 7.60	F W Ellis	Clough Farm, Croft, Skegness	24. 8.07T
G-ARCX*	Gloster Meteor NF.14				See SECTION 3 Part 1	East Fortune	
G-ARDB	Piper PA-24-250 Comanche	24-2166	PH-RON G-ARDB, N7019P	15. 8.60	P.Crook	Andrewsfield	21. 6.07
G-ARDD	Scintex CP.301C1 Emeraude *(Rebuilt EMK Aeroplanes - c/n EMK.004)*	549		4. 7.60	M.W.Bodger and M.H.Hoffmann	Yeatsall Farm, Abbots Bromley	5. 8.05P
G-ARDE*	de Havilland DH.104 Dove 6				See SECTION 3 Part 1	Sharjah, United Arab Emirates	
G-ARDG*	Lancashire Aircraft EP-9 Prospector 2	47		14. 7.60	G.Pearce *(Cancelled 28.5.82 as WFU) (Cockpit hulk noted 4.04)*	Washington, West Sussex	
G-ARDJ	Auster D.6 Series 180	3704		15. 7.60	R.E.Neal t/a RN Aviation (Leicester Airport) *(Damaged near Leicester 30.5.86: noted 5.04)*	Leicester	7. 7.88T
G-ARDO	Jodel D.112J *(Built Etablissement Couesnon)*	146	F-PBTE F-BBTE, F-WBTE	22. 8.60	W.R.Prescott *(Composite with fuselage of G-AYEO ex F-BIGG [684] c 1974)*	Kilkeel, County Down	27. 4.05P
G-ARDS	Piper PA-22-150 Caribbean	22-7154	N3214Z	4. 9.60	N.P.McGowan	Newtownards	28. 3.05
G-ARDT	Piper PA-22-160 Tri-Pacer	22-6210	N9158D	15. 9.60	B.W.Haston	Cheyene Farm, Stonehaven	29. 6.05
G-ARDV	Piper PA-22-160 Tri-Pacer	22-7487	EI-APA G-ARDV, N10F	28. 7.60	R.W.Christie *(Damaged Ballymena 10.7.98: noted inder repair 10.04)*	Newtownards	2. 1.99
G-ARDY	Tipsy Nipper T.66 Series 2 *(Built Avions Fairey SA)* (Martlet Volkswagen)	55		10. 8.60	Deborah House	Enstone	12.12.00P

Registration	Type	C/n	Previous identity	Reg date	Owner / operator	Base	CofA
G-AREA*	de Havilland DH.104 Dove 8				See SECTION 3 Part 1	Salisbury Hall, London Colney	
G-AREH	de Havilland DH.82A Tiger Moth	85287	(G-APYV)	4. 7.60	C.D.Cyster and A J.Hastings	Glenrothes	19. 4.66T
	(Built Morris Motors Ltd)		6746M, DE241		*(On rebuild and new owners 2.02)*		
G-AREI	Taylorcraft E Auster III	518	9M-ALB	14.12.60	R.B.Webber	Trenchard Farm, Eggesford	12. 5.06
			VR-RBM, VR-SCJ, MT438		*"Akyab" (As "MT438" in SEAC c/s)*		
G-AREL	Piper PA-22-150 Caribbean	22-7284	N3344Z	14. 9.60	H.H.Cousins t/a Fenland Aeroservices	Fenland	8. 1.05
G-AREO	Piper PA-18-150 Super Cub	18-7407	N10F	24. 8.60	The Vale of the White Horse Gliding Club		
						Sandhill Farm, Shrivenham	17. 9.07
G-ARER*	Vickers 708 Viscount				See SECTION 3 Part 1	Clermont-Ferrand, Puy-de-Dôme, France	
G-ARET	Piper PA-22-160 Tri-Pacer	22-7590	N10F	2. 9.60	I.S.Runnalls	(Norfolk)	20. 5.83T
					(On rebuild 2004)		
G-AREV	Piper PA-22-160 Tri-Pacer	22-6540	N9628D	25.10.60	D.J.Ash *"Smart Cat"*	Barton	29. 1.07
G-AREX	Aeronca 15AC Sedan	15AC-61	CF-FNM	12. 9.60	R.J.M.Turnbull	Rydinghurst Farm, Cranleigh	12.10.07E
G-ARFB	Piper PA-22-150 Caribbean	22-7518	N3625Z	8. 9.60	D.Shaw	Egginton	3. 4.06
G-ARFD	Piper PA-22-160 Tri-Pacer	22-7565	N3667Z	8. 9.60	J.R.Dunnett	Priory Farm, Tibenham	7. 6.07E
G-ARFG	Cessna 175AX Skylark	56505	N7005E	15.11.60	P.K.Blair	(Little Hadham, Ware)	4. 11.07E
	(Rebuilt to Cessna 172 standard 1988)				tr Foxtrot Golf Group		
G-ARFH	Piper PA-24-250 Comanche	24-2240	N7087P	13.10.60	A B.W.Taylor	Great Massingham	13. 6.07
G-ARFI	Cessna 150A	15059100	N41836	1. 2.61	Ann R.Abrey	(Wotton-under-Edge)	19. 7.06
			G-ARFI, N7000X				
G-ARFO	Cessna 150A	15059174	N7074X	23. 3.61	Breakthrough Aviation Ltd	Leicester	3. 8.06T
G-ARFT	SAN Jodel DR.1050 Ambassadeur	170		27.10.60	R.Shaw	(Sowerby Bridge)	13.10.84
					(Damaged Prestwick 15.6.84: current status unknown)		
G-ARFV	Tipsy Nipper T.66 Series 2	44		5.10.60	C.J.Pidler	(Wellington)	9. 7.05P
	(Built Avions Fairey SA) (Volkswagen 1834)						
G-ARGB*	Auster 6A Tugmaster	2593	VF635	12.10.60	C.J.Baker	Carr Farm, Thorney, Newark	21. 6.74
					(Dismantled fuselage noted 1.05)		
G-ARGG	de Havilland DHC-1 Chipmunk 22	C1/0247	WD305	19.10.60	D.Curtis	Prestwick	11. 2.07T
G-ARGI (2)*	Auster 6A Tugmaster				See SECTION 3 Part 1	Doncaster	
G-ARGO	Piper PA-22-108 Colt	22-8034		18. 1.61	D.R.Smith	Sleap	18. 7.05
G-ARGV	Piper PA-18-150 Super Cub	18-7559	N10F	20.12.60	Wolds Gliding Club Ltd	Pocklington	17. 3.05
	(Lycoming O-360-A4)						
G-ARGY	Piper PA-22-160 Tri-Pacer	22-7620	G-JEST	20.12.60	D.H.and R.T.Tanner	Wellesbourne Mountford	16. 3.06T
	(Modified to PA-20 configuration)		G-ARGY, N10F				
G-ARGZ	Druine D.31 Turbulent	PFA 562		7.11.60	The Tiger Club (1990) Ltd	Headcorn	13.10.04P
	(Built Rollason Aircraft and Engines) (Volkswagen 1600)						
G-ARHB	Forney F-1A Aircoupe	5733		17. 4.61	K.J.Peacock and S.F.Turner	Earls Colne	6. 1.06
G-ARHC	Forney F-1A Aircoupe	5734		26. 5.61	A P.Gardner	Little Gransden	8. 9.05
G-ARHI	Piper PA-24-180 Comanche	24-2260	N7299P	20.12.60	D.D.Smith	Norwich	27. 7.06
			N10F				
G-ARHM	Auster 6A	2515	VF557	5. 1.61	R.C.P.Brookhouse	(London SW10)	9.12.01
G-ARHN	Piper PA-22-150 Caribbean	22-7514	N3622Z	10. 1.61	I.S.Hodge and S.Haughton	Field Farm, Hillington	5. 3.07
	(Rebuilt with parts of G-ATXB which was damaged beyond repair 26.8.74)						
G-ARHP	Piper PA-22-160 Tri-Pacer	22-7549	N3652Z	10. 1.61	R.N.Morgan	Andrewsfield	7. 4.07
G-ARHR	Piper PA-22-150 Caribbean	22-7576	N3707Z	10. 1.61	A R.Wyatt	Cottered, Hertford	22.10.04
G-ARHW	de Havilland DH.104 Dove 8	04512		10. 1.61	Pacelink Ltd	Fairoaks	19. 5.06T
G-ARHX*	de Havilland DH.104 Dove 8				See SECTION 3 Part 1	Doncaster	
G-ARHZ	Druine D.62A Condor	PFA 247		13.12.60	E.Shouler	Beeches Farm, South Scarle	26. 7.95P
	(Built Rollason Aircraft and Engines - c/n RAE/602)				*(Damaged Damyn's Hall, Upminster 4.9.94: new owner 3.04)*		
G-ARID	Cessna 172B Skyhawk	48209	N7709X	2. 2.61	L.M.Edwards	Sleap	10. 7.06
G-ARIF	Ord-Hume O-H 7 Minor Coupe	PAL/1401		22. 8.60	N.H.Ponsford	Wigan	
	(Built A W J G Ord-Hume - c/n O-H 7) (Modified Luton LA-4C Minor)				*(Stored incomplete 3.96)*		
G-ARIH	Auster 6A	2463	TW591	23. 1.61	R.J.Griffin	Compton Abbas	14. 6.07
					(As "TW591" in 664 (AOP) Sqdn c/s)		
G-ARIK	Piper PA-22-150 Caribbean	22-7570	N3701Z	26. 1.61	A.Taylor	Manor Farm, Binham	5. 6.06
G-ARIL	Piper PA-22-150 Caribbean	22-7574	N3705Z	26. 1.61	A.Fergusson and A.W.Mcblain	Kilkerran	14.10.06
G-ARIM	Druine D.31 Turbulent	PFA 510		27. 2.61	R.M.White	High Kype Farm, Strathaven	
	(Built A Schima)				*(Noted 3.04)*		
G-ARJB	de Havilland DH.104 Dove 8	04518		29. 9.60	M.Whale and M.W.A Lunn	Kemble	10.12.73T
					(JCB titles) "Exporter" (Stored 5.02)		
G-ARJE	Piper PA-22-108 Colt	22-8184		29. 3.61	C.I.Fray	(Disley)	29. 4.73
					(On rebuild 1993: new owner 10.00)		
G-ARJF	Piper PA-22-108 Colt	22-8199		23. 3.61	Tandycel Co Ltd	Dunkeswell	16. 3.07
G-ARJH	Piper PA-22-108 Colt	22-8249		29. 3.61	F.Vogels	Goodwood	8.10.06
G-ARJS	Piper PA-23-160 Apache G	23-1977	N10F	3. 3.61	Bencray Ltd	Blackpool	3. 2.07T
					(Op Blackpool and Fylde Aero Club)		
G-ARJT	Piper PA-23-160 Apache G	23-1981	N10F	3. 3.61	J.A Cole	Leicester	18. 8.07T
G-ARJU	Piper PA-23-160 Apache G	23-1984	N10F	3. 3.61	G.R.Manley	Andrewsfield	8. 5.06T
G-ARJV	Piper PA-23-160 Apache G	23-1985	N10F	3. 3.61	Metham Aviation Ltd	Thruxton	17. 6.05
G-ARKG	Auster J/5G Cirrus Autocar	3061	AP-AHJ	22. 2.61	B.C.C.Harrison	Church Farm, Shotteswell	27. 4.07
			VP-KKN				
G-ARKJ	Beech N35 Bonanza	D-6736		5. 5.61	P.D.and Jane L.Jenkins	Goodwood	20. 1.05
G-ARKK	Piper PA-22-108 Colt	22-8290		12. 4.61	R.D.Welfare	Rochester	20. 1.07
G-ARKM	Piper PA-22-108 Colt	22-8313		12. 4.61	D.Dytch and J.Moffat	Perth	1.12.04T
G-ARKN	Piper PA-22-108 Colt	22-8327		9. 5.61	R.Redfern	Egginton	8. 8.05
G-ARKP	Piper PA-22-108 Colt	22-8364		19. 5.61	C.J.and J.Freeman	Headcorn	20.12.04T
G-ARKS	Piper PA-22-108 Colt	22-8422		7. 6.61	R.A Nesbitt-Dufort	Bradleys Lawn, Heathfield	28.11.07E
	(Lycoming O-320)						
G-ARLB	Piper PA-24-250 Comanche	24-2352	G-BUTL	21. 3.61	D.Heater	Blackbushe	28. 7.05
			G-ARLB, N10F				
G-ARLG	Auster D.4/108	3606		4. 4.61	R.D.Helliar-Symons	Bourne Park, Hurstbourne Tarrant	14. 5.05P
					tr Auster D4 Group		
G-ARLK	Piper PA-24-250 Comanche	24-2433	EI-ALW	25. 5.61	I.Kazi	Headcorn	2. 4.05
			G-ARLK, N10F				

Registration	Type	C/n	Previous identity	Date	Owner	Base	CofA
G-ARLP (2)	Beagle A 61 Terrier 1	3724 (1)	VX123	11. 4.61	D.R.Whitby tr Gemini Flying Group	(Fakenham)	31.10.91
	(C/n officially quoted as 2573/VF631 which became G-ARLM(2)/G-ASDK)				*(Damaged Truleigh Farm, Edburton 4.8.91: on rebuild 2000)*		
G-ARLR	Beagle A 61 Terrier 2	3721 & B.601	VW996	11. 4.61	M. Palfreman	Bagby	9. 9.01
G-ARLX	SAN Jodel D.140B Mousquetaire II	66		12. 4.61	J.S and S.V.Shaw	Dunkeswell	7. 9.07
G-ARLY*	Auster J/5P Autocar	3271	(VH-...)	14. 4.61	Not known	(Switzerland)	6. 6.71
					(Sold in Switzerland 11.87 and cancelled 25.2.99 by CAA)		
	(On rebuild to D.6/180 standard using parts ex Airedale G-ARNR and wings ex J/5R G-APAA 1.03)						
G-ARLZ	Druine D.31A Turbulent	RAE/578		7. 4.61	Little Bear Ltd	Exeter	16. 9.05P
	(Built Rollason Aircraft and Engines) (Ardem 4C02)						
G-ARMA	Piper PA-23-160 Apache G	23-1967	N4448P	8. 5.61	C J Hopewell *(Stored 10.01)*	Sibson	22. 7.77
G-ARMC	de Havilland DHC-1 Chipmunk 22A	C1/0151	WB703	26. 4.61	J.T.H.Henderson	White Waltham	7. 6.02
					(As "WB703" in RAF c/s)		
G-ARMD	de Havilland DHC-1 Chipmunk 22A	C1/0237	WD297	26. 4.61	D.M.Squires *(Stored 1.03)*	(Redditch)	5. 6.76
G-ARMF	de Havilland DHC-1 Chipmunk 22A	C1/0394	WG322	26. 4.61	D.M.Squires	(Redditch)	12.10.98
					(As "WZ868") (Damaged 1996: stored 1.03)		
G-ARMG	de Havilland DHC-1 Chipmunk 22A	C1/0575	WK558	26. 4.61	J.Archer tr MG Group	Bidford	1. 3.07
G-ARML	Cessna 175B Skylark	17556995	N8295T	12. 7.61	R.W.Boote	RAF Lyneham	1. 9.07T
G-ARMN	Cessna 175B Skylark	17556994	N8294T	18. 8.61	B.R.Nash	Lower Wasing Farm, Brimpton	13. 6.05
G-ARMO	Cessna 172B Skyhawk	48560	N8060X	12. 6.61	I.M.Latiff	Little Staughton	23.12.05
G-ARMR	Cessna 172B Skyhawk	48566	N8066X	12. 6.61	Sunsaver Ltd	Barton	3. 9.06
G-ARMZ	Druine D.31 Turbulent	PFA 565		2. 5.61	J.Mickleburgh and D.Clark	Headcorn	18. 3.05P
	(Built Rollason Aircraft and Engines) (Volkswagen 1500)						
G-ARNB	Auster J/5G Cirrus Autocar	3169	AP-AHL VP-KNL	18. 5.61	R.F.Tolhurst	Lenham, Maidstone	19. 2.77
					(Possibly on rebuild 1995: current status unknown)		
G-ARND	Piper PA-22-108 Colt	22-8484		6. 6.61	E.J.Clarke	Seighford	4. 8.99
G-ARNE	Piper PA-22-108 Colt (Lycoming O-320)	22-8502		15. 6.61	T.D.L.Bowden	Knettishall	11. 7.07
G-ARNG	Piper PA-22-108 Colt (Lycoming O-320)	22-8547		26. 6.61	F.B.Rothera	Stoneacre Farm, Farthing Corner	18. 3.07
G-ARNJ	Piper PA-22-108 Colt	22-8587		3. 8.61	R.A Keech	Woodvale	18. 1.06
G-ARNK	Piper PA-22-108 Colt	22-8622		5. 9.61	A.T.J.and D.Hyatt		
	(Reported as converted to PA-20 configuration with Lycoming O-320 as "Super Colt")					(Freulleville, France/Tunbridge Wells)	5.11.04
G-ARNL	Piper PA-22-108 Colt	22-8625		3. 8.61	Miss J.A Dodsworth	White Waltham	22. 3.07
G-ARNN	Globe GC-1B Swift	1272	VP-YMJ VP-RDA, ZS-BMX, NC3279K	11. 5.61	K.E.Sword	(Leicester)	11. 7.74
					(Crashed Hucknall 1.9.73: current status unknown)		
G-ARNO	Beagle A 61 Terrier 1	3722	VX113	8. 5.61	R.Webber	Trenchard Farm, Eggesford	19. 6.81
	(Official p/i now shown as VX115)				*(On rebuil d 11.04)*		
G-ARNP	Beagle A 109 Airedale	B.503		10. 5.61	S.T. and M.Isbister	North Weald	27. 4.06
	(Original regd with c/n A109-P1)						
G-ARNY	SAN Jodel D.117	595	F-BHXQ	13. 6.61	P.Jenkins	Inverness	9. 4.05P
G-ARNZ	Druine D.31 Turbulent	PFA 579		28. 6.61	The Tiger Club (1990) Ltd	Headcorn	25.11.05P
	(Built Rollason Aircraft and Engines) (Volkswagen 1600)						
G-AROA	Cessna 172B Skyhawk	48628	N8128X	19. 9.61	D.E.Partridge tr The D and P Group	Rayne Hall Farm, Braintree	4. 9.06T
G-AROC	Cessna 175BX Skylark	17556997	G-OTOW G-AROC, N8297T	2.10.61	A J.Symms	High Ham, Yeovil	13. 3.06
	(Modified to 172 configuration)						
G-AROI*	de Havilland DH.104 Dove 5				See SECTION 3 Part 1	Billund, Denmark	
G-AROJ*	Beagle A 109 Airedale	B.508	HB-EUC G-AROJ	17. 5.61	C.J.Baker	Carr Farm, Thorney, Newark	8. 1.76
	(Originally regd with c/n A.109-1)				*(Cancelled 21.1.80) (Noted dismantled 1.05)*		
G-ARON	Piper PA-22-108 Colt	22-8822		23.11.61	R.Gibson tr The G-ARON Flying Group *(New owners 11.04)*	White Waltham	5. 7.01
G-AROO	Forney F-1A Aircoupe	5750	N25B	3.11.61	W.J.McMeekan	Newtownards	21.11.04
G-AROW	SAN Jodel D.140B Mousquetaire II	71		13. 9.61	A R.Crome	New Farm House, Great Oakley	5. 6.06
G-AROY	Boeing Stearman A75N1 (PT-17) Kaydet (Pratt and Whitney R985)	75-4775	N56418 42-16612	6. 6.61	Abbey Security Services Ltd	Old Buckenham	19. 7.07
G-ARPH*	de Havilland DH.121 Trident 1C				See SECTION 3 Part 1		
G-ARPK*	de Havilland DH.121 Trident 1C	2111		13. 4.61	Fire School	Manchester	17. 5.82T
					(WFU 24.5.82: cancelled.1.98) (Noted @ fire dump 2.04)		
G-ARPO (2)*	de Havilland DH.121 Trident 1C	2116	(G-ARPP)	23. 3.64	International Fire Training Centre	Durham Tees Valley	12. 1.86T
					(WFU 12.12.83) (Noted 2.04)		
G-ARRD	SAN Jodel DR.1051 Ambassadeur	274		20. 7.61	D.J.Taylor and J.P.Brady	RAF Halton	5. 8.05P
G-ARRE	SAN Jodel DR.1050 Ambassadeur	275		20. 7.61	G.and Rosemary Ward	(Holbeach, Spalding)	16. 7.05
G-ARRI	Cessna 175B Skylark	17557001	N8301T	5.10.61	R.D.Fowden	Haverfordwest	1. 6.07
G-ARRL	Auster 5 J/1N Alpha	2115	VP-KFK VP-KPF, VP-KFK, VP-UAK	13. 6.61	A C.Ladd	Romney Street Farm, Sevenoaks	22. 7.05
G-ARRM	Beagle B.206X				See SECTION 3 Part 1	Kemble	
G-ARRO	Beagle A.109 Airedale	B.507	EI-AYL (2) G-ARRO, (EI-AVP), G-ARRO	16. 6.61	M.and S.W.Isbister	Spanhoe	17. 1.74
	(Originally regd with c/n A.109-P5)				*(Stored 4.03 as "EI-AYL")*		
G-ARRS	Menavia Piel CP.301A Emeraude	226	F-BIMA	29. 6.61	Julia P.Drake	Sherburn-in-Elmet	8. 7.05P
G-ARRT	Wallis WA-116/Mc	2		28. 6.61	K.H.Wallis	Reymerston Hall, Norfolk	26. 5.83P
	(McCulloch 4318A) *(Originally regd as a Wallis Gyroplane then became WA-116/Mc) (Noted 8.01)*						
G-ARRU	Druine D.31 Turbulent	PFA 502		28. 6.61	D.G.Huck	Rugby	27. 2.97P
	(Built J O'Connor) (Volkswagen 1600)				*(New CofR 8.03)*		
G-ARRX	Auster 6A	2281	VF512	4. 7.61	J.E.D.Mackie	Popham	17. 6.06
					(As "VF512/PF-M" in 43 OTU c/s) "Peggy Too"		
G-ARRY	SAN Jodel D.140B Mousquetaire II	72		13. 9.61	Fictionview Ltd	Bodmin	11.11.04
G-ARRZ	Druine D.31 Turbulent	PFA 580		21. 8.61	C.I.Jefferson *"Tarzan"*	(Liverpool)	21.12.90P
	(Built Rollason Aircraft and Engines) (Ardem 4C02)				*(Damaged Horley, Surrey 21.7.90: rear fuselage Priory Farm, Tibenham 6.04)*		
G-ARSG	Roe Triplane Type IV rep	TRI-1	(BAPC.1)	29.10.81	Richard Shuttleworth Trustees	Old Warden	21. 8.05P
	(Built Hampshire Aero Club) (ADC Cirrus III)				*(Built for "Those Magnificent Men in Their Flying Machines" film) (No external marks)*		
G-ARSL	Beagle A.61 Terrier 2	2539	VF581	13. 7.61	D.J.Colclough *(As "VF581")*	Trenchard Farm, Eggesford	10. 9.06
G-ARSU	Piper PA-22-108 Colt	22-8835	EI-AMI G-ARSU	23.11.61	D.P.Owen	Thruxton	7.12.06

G-ARTH	Piper PA-12 Super Cruiser	12-3278	EI-ADO	22. 9.61	R.I.Souch and B.J.Dunford	Hill Farm, Durley	21. 4.95P
					(Stored 12.01)		
G-ARTJ*	Bensen B.8M	7		22. 9.61	M A Stewart	(Wilkieston Farm, Peat Inn)	
	(Volkswagen 1600) *(Originally regd with c/n 8-104-100)*				*(Cancelled 6.6.75 as WFU) (Stored 2003?)*		
G-ARTL	de Havilland DH.82A Tiger Moth		"T7281"	22. 9.61	F G Clacherty	Longwood Farm, Morestead	25. 5.06
	(P/i is doubtful - if correct the c/n is 83795)				*(As "T7281" in RAF c/s)*		
G-ARTM*	Beagle A.61 Terrier 1	3723	WE536	9.10.61	C.J.Baker	Carr Farm, Thorney, Newark	13.11.71
	(Crashed Priory Farm, Turvey, Bedford 28.5.70 and cancelled 12.9.73 as WFU) (Fuselage, part rebuilt. noted 1.05)						
G-ARTR*	de Havilland DHC-2 Beaver 1				See SECTION 3 Part 1	Papakura, Auckland, New Zealand	
G-ARTZ (1)*	McCandless M.2				See SECTION 3 Part 1	Belfast	
G-ARTZ (2)	McCandless M.4	M4/1		24.10.61	W.R.Partridge	St.Merryn	13.10.69P
	(Volkswagen 1500)				*(Noted 5.03)*		
G-ARUE (2)*	de Havilland DH.104 Dove 7				See SECTION 3 Part 1	Schleissheim, Gemany	
G-ARUG	Auster J/5G Cirrus Autocar	3272		2. 1.62	D.P.H.Hulme	Biggin Hill	18. 4.03
G-ARUI	Beagle A.61 Terrier 1	2529	VF571	9. 3.62	T.W.J.Dann	Southend	18. 2.05
G-ARUL	LeVier Cosmic Wind	103	N22C	28.11.61	P.G.Kynsey	Duxford	4. 7.05P
	(Originally built Tony LeVier Associates Inc) (Continental O-200-A)				*"Ballerina"*		
	(Rebuilt 1973 as c/n PFA 1511: contains little of original as fuselage, wings and data plate for original held elsewhere)						
G-ARUO*	Piper PA-24-180 Comanche	24-2427	N7251P	16. 1.62	Not known		
						Walton New Road Business Park, Bruntingthorpe	22. 8.00
					(Cancelled 18.7.00 by CAA) (Fuselage noted in scrapyard compound 2.04)		
G-ARUV	Piel CP.301 Emeraude Series 1	PFA 700		2. 2.62	P.O'Fee	RAF Keevil	25. 4.05P
	(Built M N Harrison) (Continental C90)				*"Emma"*		
G-ARUY	Auster J/1N Alpha	3394		2. 2.62	D.Burnham	Andrewsfield	27.11.05
G-ARUZ	Cessna 175C Skylark	17557080	N8380T	23. 2.62	M.Lowe and L.E.Brown tr Cardiff Skylark Group	Cardiff	18. 5.06
G-ARVF*	Vickers VC-10 Series 1101				See SECTION 3 Part 1	Hermeskeil, Trier, Germany	
G-ARVM (2)*	Vickers VC-10 Series 1101				See SECTION 3 Part 1	Duxford	
G-ARVN (2)*	Servotec Rotorcraft Grasshopper 1				See SECTION 3 Part 1	Weston-super-Mare	
G-ARVO	Piper PA-18 Super Cub 95	18-7252	D-ENFI	18. 1.83	Northamptonshire School of Flying Ltd	Sywell	20.11.04T
			N3376Z				
G-ARVT	Piper PA-28-160 Cherokee	28-379		21. 3.62	Red Rose Aviation Ltd	Liverpool	15. 3.07
G-ARVU	Piper PA-28-160 Cherokee	28-410	PH-ONY	30. 3.62	Barton Mudwing Ltd	Barton	3. 6.07
			G-ARVU				
G-ARVV	Piper PA-28-160 Cherokee	28-451		11. 7.62	G.E.Hopkins	Shobdon	10. 4.04
G-ARVZ	Druine D.62B Condor	RAE/606		6.12.61	Anna A M.Huke	Dinton, Wiltshire	6. 5.05P
	(Built Rollason Aircraft and Engines)						
G-ARWB	de Havilland DHC-1 Chipmunk 22A	C1/0621	WK611	2. 1.62	P.G.Alston	Thruxton	4.11.05
					tr Thruxton Chipmunk Flying Group *(As "WK611")*		
G-ARWO	Cessna 172C Skyhawk	49187	N1487Y	10. 4.62	R.Taggart	(Enfield, County Meath	15. 6.06
G-ARWR	Cessna 172C Skyhawk	49172	N1472Y	13. 4.62	M.McCann tr Devanha Flying Group	Insch	26. 7.04
G-ARWS	Cessna 175C Skylark	17557102	N8502X	12. 4.62	M.D.Fage	Egginton	28.10.07
G-ARXB	Beagle A.109 Airedale	B.509	EI-BBK	5. 2.62	M.Isbister	Spanhoe	9. 9.76
	(Originally regd with c/n A.109-2)		G-ARXB, EI-ATE, G-ARXB		*(Stored as "EI-BBK" 4.03)*		
G-ARXC*	Beagle A.109 Airedale	B.510	EI-ATD	9. 4.62	C.J.Baker	Carr Farm, Thorney, Newark	27. 6.76
	(Originally regd with c/n A.109-3)		G-ARXC		*(Cancelled 12.4.89 as WFU) (Fuselage on rebuild 1.05)*		
G-ARXD	Beagle A.109 Airedale	B.511		19. 4.62	D.Howden	Lumphanan	13. 6.86
	(Originally regd with c/n A.109-4)				*(Under restoration 6.00)*		
G-ARXG	Piper PA-24-250 Comanche	24-3154	N10F	21. 2.62	R.F.Corstin t/a Fairoaks Comanche	Fairoaks	24. 7.05
G-ARXH	Bell 47	G 40	N120B	13. 2.62	A.B.Searle	Cranfield	6. 7.90
			NC120B		*(Noted 7.00)*		
G-ARXP	Phoenix Luton LA-4A Minor	PFA 816		23. 2.62	E.Evans	Benson's Farm, Laindon	17.10.95P
	(Built W C Hymas - c/n PAL/1119) (Walter Mikron 3)				*(Fuselage stored 2.04)*		
G-ARXT	SAN Jodel DR.1050 Ambassadeur	355		14. 3.62	M.F.Coy	Wellesbourne Mountford	13.11.07
					tr CJM Flying Group		
G-ARXU	Auster 6A	2295	VF526	5. 3.62	E.C.Tait and M.Pocock	AAC Netheravon	16. 9.05
					(As "VF526/T" in Army c/s)		
G-ARXW	Morane Saulnier MS.885 Super Rallye	100		30. 3.62	M.J.Kirk	Haverfordwest	4. 5.04
G-ARYB*	de Havilland DH.125 Series 1				See SECTION 3 Part 1	Coventry	
G-ARYC*	de Havilland DH.125 Series 1				See SECTION 3 Part 1	Salisbury Hall, London Colney	
G-ARYD*	Auster AOP.6				See SECTION 3 Part 1	Middle Wallop	
G-ARYF	Piper PA-23-250 Aztec B	27-2065	N10F	11. 4.62	D.A Hitchcock *(New owner 6.02)*	Biggin Hill	18. 6.99
G-ARYH	Piper PA-22-160 Tri-Pacer	22-7039	N3102Z	9. 3.62	C.Watt	Crosland Moor	11. 6.05
G-ARYI	Cessna 172C	49260	N1560Y	13. 7.62	Joyce Rhodes	Blackbushe	10. 8.03T
G-ARYK	Cessna 172C	49288	N1588Y	13. 7.62	G.W.Goodban	Lydd	26. 4.07T
G-ARYR	Piper PA-28-160 Cherokee B	28-770		12. 7.62	R.P.Synge and C.S.Wilkinson	Turweston	28. 4.06
					tr GARYR Flying Group		
G-ARYS	Cessna 172C Skyhawk	49291	N1591Y	13. 7.62	Squires Gear and Engineering Ltd	Coventry	8.10.06
G-ARYV	Piper PA-24-250 Comanche	24-2516	N7337P	17. 4.62	D.C.Hanss	Elstree	30. 7.05T
G-ARYZ	Beagle A.109 Airedale	B.512		9. 4.62	C.W.Tomkins	Spanhoe	26. 2.01
G-ARZB	Beagle-Wallis WA-116 Series 1 Agile	B.203	XR943	18. 4.62	K.H.Wallis	Reymerston Hall, Norfolk	29. 6.93P
	(McCulloch 4318A)		G-ARZB		*"Little Nellie"*		
	(Flown as "XR943" for evaluation 1962 and remained on UK Register: used in 1966 for James Bond film "You Only Live Twice") (Noted 8.01)						
G-ARZN	Beech N35 Bonanza	D-6795	N215DM	23. 5.62	D.W.Mickleburgh	Leicester	20.12.07
G-ARZS	Beagle A 109 Airedale	B.515	EI-BAL	11. 5.62	M.and S.W.Isbister	Spanhoe	23. 5.75
			G-ARZS		*(Under rebuild 4.03)*		
G-ARZW	Phoenix Currie Wot	1		25. 5.62	B.R.Pearson	Eaglescott	7. 1.89P
	(Built J H B Urmston) (Walter Mikron 3)				*(Damaged near Headcorn 12.2.88: on rebuild 10.99 as Pfalz D.VII scale rep)*		

G-ASAA - G-ASZZ

G-ASAA	Phoenix Luton LA-4A Minor	O-H/4		19. 4.62	M.J.Aubrey	(Kington, Hereford)	7. 6.01P
	(Built P D Lea and Partners) (JAP J.99)				*(Noted 2002)*		
G-ASAI	Beagle A 109 Airedale	B.516		26. 6.62	K.R.Howden	(Lumphanan)	20. 5.77S
					(On rebuild 6.00: current status unknown)		

G-ASAJ	Beagle A 61 Terrier 2	B.605	WE569	26. 6.62	R.Skingley tr G-ASAJ Flying Group	Bassingbourn	19. 8.01
	(Initially allocated c/n 3732)				*(Op Military Auster Flight as Auster T.7 "WE569") (Current status unknown)*		
G-ASAK	Beagle A 61 Terrier 2	B.604	WE591	26. 6.62	Igleszas Frank *(As "WE591/Y")*	(Swanley)	8. 6.05
G-ASAL (2)	Scottish Aviation Bulldog Series 100/101	BH120/239	(G-BBHF) G-31-17	5. 9.73	Pioneer Flying Co Ltd	Prestwick	25. 4.05P
G-ASAT	Morane Saulnier MS.880B Rallye Club	178		21. 6.62	M.Cutovic	Croft Farm, Defford	2.11.03
G-ASAU	Morane Saulnier MS.880B Rallye Club	179		21. 6.62	L.O.Queen	Blackpool	10. 9.06
G-ASAX	Beagle A 61 Terrier 2	B.609	TW533	12. 6.62	P.G.Morris	Cheyene Farm, Stonehaven	1. 9.96
	(Converted from Auster 6 c/n 1911)				*"The Jacobite Air Force" (Under restoration 6.00: current status unknown)*		
G-ASAZ	Hiller UH-12E-4	2070	N5372V	18. 6.62	R.C.Hields t/a Hields Aviation *(As "XS165/37")*	Sherburn-in-Elmet	24. 7.05T
G-ASBA	Phoenix Currie Wot	PFA 3005		16. 8.62	J.C.Lister	Valley Farm, Winwick	9. 5.05P
	(Built A Etherbridge - c/n AE.1) (Continental C90)						
G-ASBH	Beagle A 109 Airedale	B.519		26. 6.62	D.T.Smollett	Bratton Clovelly, Okehampton	19. 2.99
G-ASBU*	Beagle A 61 Terrier 2	3733 (1) & B.613	WE570	12. 7.62	P.G.Morris	Cheyene Farm, Stonehaven	5. 7.82
			(Damaged Netherley 12.8.80: cancelled 16.10.85 as WFU) (Stored for spares 6.00: current status unknown)				
G-ASBY	Beagle A 109 Airedale	B.523		23. 7.62	F.A Forster *(New owner 4.02)*	Lydd	22. 3.80
G-ASCB*	Beagle A 109 Airedale				See SECTION 3 Part 1	Alverca, Portugal	
G-ASCC	Beagle E.3 Mk.11	B.701	(G-25-12) XP254	23. 7.62	P.T.Bolton *(As "XP254" in 'ARMY' camouflage c/s)*	South Lodge Farm, Widmerpool	31. 7.05P
G-ASCD*	Beagle A 61 Terrier 2				See SECTION 3 Part 1	Elvington	
G-ASCF*	Beagle A 61 Terrier 2				See SECTION 3 Part 1	Gardermoen, Norway	
G-ASCJ*	Piper PA-24-250 Comanche	24-2368	5N-AEB N7197P	2. 8.62	Not known	Walton New Road Business Park, Bruntingthorpe	2.10.88
					(DBR landing Bournemouth 10.9.86: cancelled 8.1.87 as WFU) (Fuselage in scrapyard 2.04)		
G-ASCM	Isaacs Fury II	PFA 2002/1B		1. 8.62	E.C.and P.King	Eastbach Farm, Coleford	24.10.02P
	(Built J O Isaacs - c/n 1) (Lycoming O-290) *(PFA c/n = Builder's membership no.?) (As "K2050" in pre-war RAF c/s)*						
G-ASCT*	Bensen B.7Mc				See SECTION 3 Part 1	Weston-super-Mare	
G-ASCU	Piper PA-18A-150 Super Cub	18-6797	VP-JBL	31. 8.62	D.J.O'Mahony	(Dublin)	16. 7.06
G-ASCX*	de Havilland DH.114 Heron 2D				See SECTION 3 Part 1	Launceston, Tasmania, Australia	
G-ASCY*	Phoenix Luton LA-4A Minor				See SECTION 3 Part 1	Miami, Florida, USA	
G-ASCZ	Menavia Piel CP.301A Emeraude	233	F-BIMG	1.10.62	P.Johnson	Goodwood	18.12.05P
G-ASDF*	Edwards Gyrocopter	NAFE.1		17.10.62	M.J.Aubrey	(Kington, Hereford)	
	(Built N A F Edwards) (Triumph T110) *(Mod. Adams-Wilson XH-1 Hobbycopter)*				*(Cancelled in 1963 as not completed) (Noted 2002)*		
G-ASDK	Beagle A 61 Terrier 2	B.702	G-ARLM (2) G-ARLP(1), VF631	26.10.62	J.Swallow	Hibaldstow	3. 6.06
	(Converted from Auster AOP.6 c/n 2573)						
G-ASDS*	Vickers 843 Viscount				See SECTION 3 Part 1	Beijing, China	
G-ASDY	Beagle-Wallis WA-116/F	B.205	XR944 (G-ARZC (1))	9.11.62	K.H.Wallis *(Noted 8.01)*	Reymerston Hall, Norfolk	28.10.97P
	(Franklin 2A-120-B)						
	(Regd with c/n B.204 as Beagle-Wallis WA.116 Srs 1 and powered by McCulloch 4318A: fitted with 990cc Hillman Imp engine 1965 and known as WA.119: re-engined 1971 with 60hp 2-cylinder Franklin 2A-120-A and re-designated)						
G-ASEA	Phoenix Luton LA-4A Minor	PAL/1154		14.11.62	C.Willmot *(New owner 12.03)*	(Street)	16. 8.89P
	(Built G P Smith and M Fawkes) (JAP J.99) *(Regd with c/n PFA 1154*						
G-ASEB	Phoenix Luton LA-4A Minor	PAL/1149		26.11.62	S.R.P.Harper *(Under restoration 10.01)*	Walkeridge Farm, Overton	29.10.82P
	(Built J A Anning) (Lycoming O-145)						
G-ASEF*	Auster 6A Tugmaster	---	VW985	17.12.62	(M.Thomason)	Not known	19.12.66
					(Damaged Bicester 1966 and cancelled 13.1.67) (Stored 10.03)		
G-ASEO	Piper PA-24-250 Comanche	24-3367	(G-ASDX) N10F	23. 1.63	M.Scott t/a Pixies Day Nursery	Southampton	5. 3.07
G-ASEP	Piper PA-23-235 Apache	27-541		28. 1.63	Air Warren Ltd *(Maxim titles)*	Denham	10. 7.05
G-ASEU	Druine D.62A Condor	RAE/607		12. 2.63	W.M.Grant	Inverness	13. 5.05P
	(Built Rollason Aircraft and Engines) (Continental C90-8F)						
G-ASFA	Cessna 172D Skyhawk	17250182	N2582U	21. 2.63	D.Halfpenny	Maypole Farm,Chislet	18. 6.04
G-ASFD	LET L-200A Morava	170808	OK-PHH	26. 2.63	M.Emery *(New CofR 12.02)*	(Redhill)	12. 7.84T
G-ASFI*	de Havilland DH.114 Heron 2D				See SECTION 3 Part 1	Bankstown, New South Wales, Australia	
G-ASFK	Auster J/5G Cirrus Autocar	3276		7. 3.63	T.D.G.Lancaster	Manor Farm, Garford	25. 6.06
G-ASFL	Piper PA-28-180 Cherokee B	28-1170		7. 3.63	J.Simpson and D.Kennedy	Lee-on-Solent	28. 1.07
G-ASFR	Bölkow Bö.208C Junior	522	D-EGMO	12. 3.63	S.T.Dauncey *(Stored 1.02)*	Yearby	29. 3.90P
G-ASFX	Druine D.31 Turbulent	PFA 513		18. 3.63	E.F.Clapham and W.B.S.Dobie	Oldbury-on-Severn	25.11.05P
	(Built E.F.Clapham) (Volkswagen 1600)						
G-ASGC*	Vickers Super VC-10 Series 1151				See SECTION 3 Part 1	Duxford	
G-ASHD*	Brantly B.2A				See SECTION 3 Part 1	Weston-super-Mare	
G-ASHH	Piper PA-23-250 Aztec	27-63	N455SL N4557P	25. 3.63	C.Fordham and L.Barr	Leicester	29. 8.03
G-ASHS	SNCAN Stampe SV-4C	265	F-BCFN	23. 4.63	D G Girling	Liverpool	6. 2.05T
	(DH Gipsy Major 10) *(Original fuselage for rebuild of G-AWEF 1980: rebuilt 1984 with fuselage of G-AZIR c/n 452 ex F-BCXR)*						
G-ASHT	Druine D.31 Turbulent	PFA 1610		23. 4.63	C.W.N.Huke	Dinton, Wiltshire	13. 4.05P
	(Built Rollason Aircraft and Engines) (Volkswagen 1600)						
G-ASHU	Piper PA-15 Vagabond	15-46	N4164H NC4164H	1. 5.63	T.J.Ventham tr The Calybe Flying Group *"Calybe"*	Little Bredy	19. 5.05P
	(Rotax 912-UL)						
G-ASHX	Piper PA-28-180 Cherokee B	28-1266	N7382W	3. 5.63	Powertheme Ltd	Barton	29. 5.05
G-ASIB	Cessna F172D Skyhawk	F172-0006	F-WLIR	9. 5.63	D.A.Smart and A.Jones tr G-ASIB Flying Group	Hawarden	26. 4.07
	(Built Reims Aviation SA) (Wichita c/n 17250091)						
G-ASII	Piper PA-28-180 Cherokee B	28-1264		21. 5.63	T.R.Hart and R.W.S.Matthews	Exeter	25. 5.07
G-ASIJ	Piper PA-28-180 Cherokee B	28-1333	N7445W	21. 5.63	G.R.Moore tr G-ASIJ Group	Andrewsfield	29. 4.07T
G-ASIL	Piper PA-28-180 Cherokee B	28-1350	N7461W	21. 5.63	C.D.Powell	Leicester	8.11.04
G-ASIP*	Auster 6A Tugmaster	2549	VF608	22. 5.63	(C.Applegarth)	Not known	19. 7.73
					(Damaged by fire Nympsfield 7.5.73) (On rebuild 9.03)		
G-ASIS	Wassmer Jodel D.112	1166	EI-CKX G-ASIS, F-BKNR	24. 2.81	W.R.Prescott *(New CofR 11.03)*	Riverstown	21. 3.95P
G-ASIT	Cessna 180	32567	N7670A	24. 5.63	R.A.Seeley	(London SW13)	20. 5.07
G-ASIX*	Vickers VC-10 Series 1103				See SECTION 3 Part 1	Duxford	
G-ASIY	Piper PA-25-235 Pawnee	25-2446		30. 5.63	T.Holloway tr RAFGSA	RAF Halton	18. 1.05

G-ASJL	Beech H35 Bonanza	D-5132	N5582D	14. 6.63	A J.Orchard	Biggin Hill	24. 7.05
G-ASJO	Beech 23 Musketeer	M-518		18. 6.63	K A Boon tr G-ASJO Syndicate	Bembridge	12. 2.01
G-ASJV	Vickers Supermarine 361 Spitfire LF.IXb	CBAF.IX.552	OO-ARA	3. 7.63	Merlin Aviation Ltd *(Op The Old Flying Machine Co)*	Duxford	19. 5.05P
			Belgian AF SM-41, Fokker B-13, R Neth H-68, H-105, MH434 *(As "MH434/ZD-B" in 316 Sqdn c/s)*				
G-ASJY	Sud-Aviation Gardan GY-80-160 Horizon	13		9. 7.63	No 6 Group Aviation Ltd	Bagby	2. 7.05
G-ASJZ	SAN Jodel D.117A	826	F-BITD	5. 7.63	W.J.Siertsema	Bicester	12. 7.05P
G-ASKB*	de Havilland DH.98 Mosquito TT.35				See SECTION 3 Part 1	Miami, Florida, US	
G-ASKC*	de Havilland DH.98 Mosquito TT.35				See SECTION 3 Part 1	Duxford	
G-ASKK*	Handley Page HPR.7 Dart Herald 211				See SECTION 3 Part 1	Norwich	
G-ASKL	SAN Jodel D.150 Mascaret	27		18. 7.63	J.M.Graty	Nuthampstead	17. 9.04P
G-ASKP	de Havilland DH.82A Tiger Moth	3889	N6588	22. 7.63	The Tiger Club (1990) Ltd	Headcorn	6. 3.06T
G-ASKT	Piper PA-28-180 Cherokee B	28-1410	N7497W	24. 7.63	T.J.Herbert	Biggin Hill	10. 9.05
G-ASLH	Cessna 182F Skylane	18254905	N3505U	19. 8.63	A.L.Brown and A.L.Butcher	Bourn	9.10.06T
G-ASLP*	Bensen B.7	11		3. 9.63	R.Light and T.Smith	Stockport	
					(Cancelled 4.9.73 as WFU) (Stored 12.00)		
G-ASLR*	Agusta-Bell 47J-2 Ranger	2057		3. 9.63	Flamingo Club and Harleys American Diner		
					Avenida Espana, Adeje, Playa de Las Americas, Tenerife		7. 3.96T
	(Damaged on take-off Bristol 31.1.96: cancelled 28.2.97 as WFU) (Noted @ 2.04 minus tail rotor: regn not carried but visible on c/n plate in canopy)						
G-ASLV	Piper PA-28-235 Cherokee	28-10048		11. 9.63	I.L.Harding	Sackville Lodge Farm, Riseley	10. 2.05
					tr Sackville Flying Group		
G-ASLX	Menavia Piel CP.301A Emeraude	292	F-BISV	12. 9.63	J.J.Reilly	(Thurles, County Tipperary)	8. 2.05P
G-ASMA	Piper PA-30 Twin Comanche	30-143	N10F	17. 9.63	K.Cooper	Wolverhampton	5. 1.03
	(Modified to PA-39 C/R status)						
G-ASME	Bensen B.8M	12		24. 9.63	R.M. Harris	Tollerton	12. 8.04P
	(Rotax 582)						
G-ASMF	Beech D95A Travel Air	TD-565		26. 9.63	M.J.A Hornblower	Southend	11. 8.06T
G-ASMJ	Cessna F172E	F172-0029		25.10.63	Aeroscene Ltd	Sherburn-in-Elmet	26. 5.07T
	(Built Reims Aviation SA) (Wichita c/n 17250584)						
G-ASML	Phoenix Luton LA-4A Minor	PFA 802		28.10.63	S.Slater	Bicester	31. 3.05P
	(Built R M Kirby - c/n PAL/1148) (Volkswagen 1600)						
G-ASMM	Druine D.31 Turbulent	PFA 1611		31.10.63	W.J. Browning	Redhill	9. 5.05P
	(Built Rollason Aircraft and Engines) (Ardem 4C02)				*"Mouche Miel"*		
G-ASMS	Cessna 150A	15059204	N7104X	18.11.63	R.N.Ainsworth and M.A.Brown	Barton	4. 4.07
G-ASMT	Fairtravel Linnet 2	004		20.11.63	A F.Cashin	Maypole Farm, Chislet	20.12.04P
G-ASMW	Cessna 150D	15060247	N4247U	26.11.63	Aviation Business Centres Ltd	Netherthorpe	3. 8.07T
G-ASMY	Piper PA-23-160 Apache H	23-2032	N4309Y	3.12.63	R.D.Forster *(New CofR 9.03)*	Beccles	25.11.95T
G-ASMZ	Beagle A 61 Terrier 2	B.629	G-35-11	4.12.63	B.Andrews	Trenchard Farm, Eggesford	7. 4.06
	(Conversion of Auster AOP.10 c/n 2285)		VF516		*(As "VF516")*		
G-ASNB	Auster 6A Tugmaster	3725 (2)	VX118	6.12.63	B.C.C.Harrison *(As "VX118")*	Shotteswell	29. 4.05T
G-ASNC	Beagle D.5/180 Husky	3678		9.12.63	Peterborough and Spalding Gliding Club Ltd	Crowland	11. 1.07
G-ASNI	Scintex CP.1310-C3 Super Emeraude	925		20.12.63	D.Chapman	Wickenby	13. 2.06
G-ASNK	Cessna 205	205-0400	N8400Z	27.12.63	Justgold Ltd *(Op Blackpool Air Centre)*	Blackpool	29. 6.06T
G-ASNW	Cessna F172E	F172-0031		13. 1.64	B.M.Tremain	Draycott Farm, Chiseldon	14.10.07E
	(Built Reims Aviation SA) (Wichita c/n 17250613)				tr G-ASNW Group		
G-ASOC	Auster 6A Tugmaster	2544	VF603	21. 1.64	M.J.Kirk *(New owner 4.03)*	Haverfordwest	18. 5.02
G-ASOH	Beech 95-B55A Baron	TC-656		31. 1.64	G.S.Goodsir tr GMD Group	Biggin Hill	21. 9.07
G-ASOI	Beagle A 61 Terrier 2	B.627	G-35-11	31. 1.64	G.D.B.Delmege	Kemble	19. 6.98
			WJ404		*(Noted 7.04)*		
G-ASOK	Cessna F172E	F172-0057		31. 1.64	D.W.Disney	Egginton	18 9.05
	(Built Reims Aviation SA)						
G-ASOL*	Bell 47D-1				See SECTION 3 Part 1	Usworth, Sunderland	
G-ASOM	Beagle A 61 Terrier 2	B.622	G-JETS	3. 2.64	D.Humphries	Spanhoe	11. 3.06
			G-ASOM, G-35-11, VF505				
G-ASOX*	Cessna 205A	205-0556	N4856U	3. 2.64	A Turnbull	Bournemouth	1. 8.92
					(Cancelled 15.8.00 by CAA) (Noted 2.04)		
G-ASPF	Wassmer Jodel D.120 Paris-Nice	02	F-BFNP	26. 2.64	T.J.Bates	Dairy House Farm, Worleston	19. 6.05P
G-ASPK	Piper PA-28-140 Cherokee	28-20051		28. 2.64	Westward Airways (Lands End) Ltd	St.Just	10. 4.05T
G-ASPP	Bristol Boxkite rep	BOX.1 & BM.7279	(BAPC.2)	29.10.81	Richard Shuttleworth Trustees	Old Warden	27. 7.05P
	(Built Miles Aviation) (Continental O-200-B)				*(Built for "Those Magnificent Men in Their Flying Machines" film) (As No."12A")*		
G-ASPS	Piper J-3C-65 Cub Special	22809	N3571N	2. 3.64	A J.Chalkley	Rhoshirwaun, Pwllheli	27. 5.05P
	(Frame No.21971)		NC3571N				
G-ASPU	Druine D.31 Turbulent	PFA 1623		4. 3.64	M.K.Field	Sleap	14.10 05P
	(Built A S Usherwood) (Volkswagen 1500)						
G-ASPV (2)	de Havilland DH.82A Tiger Moth	84167	T7794	5. 3.64	Z J Rockey	(Exeter)	31. 8.97
	(Built Morris Motors Ltd) (P/i obscure - original G-ASPV sold Norway 7.75 and rebuilt as LN-MAX) (New owner 3.04)						
G-ASRB	Druine D.62B Condor	RAE/608		11. 3.64	R.J.Bentley Pallas West, Toomywvara, County Tipperary		1.11.98
	(Built Rollason Aircraft and Engines)				*(New owners 7.04)*		
G-ASRC	Druine D.62C Condor	RAE/609		11. 3.64	C.R.Isbell	Andrewsfield	29. 4.05P
	(Built Rollason Aircraft and Engines) (Continental O-240-A)						
G-ASRF*	Gowland GWG.2 Jenny Wren				See SECTION 3 Part 1	Flixton, Bungay	
G-ASRK	Beagle A 109 Airedale	B.538		26. 3.64	Bio Pathica Ltd	Lydd	3. 8.07
G-ASRL*	Beagle A 61 Terrier 2				See SECTION 3 Part 1	Krugersdorp, South Africa	
G-ASRO	Piper PA-30 Twin Comanche	30-395	N10F	31. 3.64	D.W.Blake tr Five Star Flying Group	Gloucestershire	14. 7.05
G-ASRT	SAN Jodel D.150 Mascaret	45		6. 4.64	P.Turton *(Stored 6.94 - current status unknown)*	(Crewe)	3. 6.94P
G-ASRW	Piper PA-28-180 Cherokee B	28-1606	N11C	21. 4.64	G.N.Smith	(Dickleburgh, Diss)	12. 5.06T
G-ASSB*	Piper PA-30 Twin Comanche	30-432	N10F	22. 4.64	Brooklands Technical College	Weybridge	11. 3.93T
					(Cancelled 25.8.92 as WFU) (Instructional airframe 10.03)		
G-ASSF	Cessna 182G Skylane	18255593	N2493R	5. 5.64	J.D.Bingham	(Doncaster)	31. 3.07
G-ASSM*	Hawker Siddeley HS.125 Series 1/522				See SECTION 3 Part 1	South Kensington, London	
G-ASSP	Piper PA-30 Twin Comanche	30-458	N10F	7. 5.64	P.H.Tavener	Redhill	22.10.06
G-ASSS	Cessna 172E	17251467	N5567T	7. 5.64	D.H.N.Squires and P.R.March	Filton	27. 5.06
G-ASST	Cessna 150D	15060630	N5930T	7. 5.64	F.R.H.Parker	Pear Tree Farm, Marsh Gibbon, Bicester	1. 8.07
G-ASSV	Kensinger KF	PFA 168-13923	N23S	11. 5.64	C.I.Jefferson	(Liverpool)	30. 7.69P
	(Built N Kensinger - c/n 2) (Continental C85)				*(Crashed Wolverhampton 2.7.69: stored Priory Farm, Tibenham 8.04)*		

Reg	Type	C/n	Previous identity	Regd	Owner/operator	Probable base	CofA expiry
G-ASSW	Piper PA-28-140 Cherokee	28-20055	N11C	11. 5.64	G.S.Stone	Biggin Hill	22. 7.07
G-ASSY	Druine D.31 Turbulent	PFA 586		12. 5.64	M.N.King and J A Thomas	Branscombe	20. 4.84P
	(Built F J Parker) (Volkswagen 1500)				*(On long term rebuild 1.04)*		
G-ASTA	Druine D.31 Turbulent	152	F-PJGH	12. 5.64	P.A Cooke	(Kimbolton, Huntingdon)	13.11.97P
	(Built M Barboni) (Ardem 4C02)				*(Current status unknown)*		
G-ASTG	Nord 1002 Pingouin II	183	F-BGKI	21. 5.64	L.M.Walton	Duxford	26.10.73S
			Fr.AF 183		*(On rebuild and unmarked 4.03)*		
G-ASTI	Auster 6A Tugmaster	3745	WJ359	27. 5.64	C.C.Burton	Headcorn	31. 7.06
G-ASTL*	Fairey Firefly 1				See SECTION 3 Part 1	RNAS Yeovilton	
G-ASTP*	Hiller UH-12C				See SECTION 3 Part 1	Weston-super-Mare	
G-ASUB	Mooney M.20E Super 21	397	N7158U	24. 6.64	S.C.Coulbeck	North Coates	24. 8.07
G-ASUD	Piper PA-28-180 Cherokee B	28-1654	N7673W	29. 6.64	S.J and A.Rogers	Andrewsfield	29. 1.07
					tr G-ASUD Flying Group		
G-ASUE	Cessna 150D	15060718	N6018T	30. 6.64	D.Huckle	West Thurrock	1. 8.90
					(Stored 6.94: current status unknown)		
G-ASUG*	Beech E18S-9700				See SECTION 3 Part 1	East Fortune	
G-ASUI	Beagle A 61 Terrier 2	B.641	VF628	6. 7.64	R.J.Bentley		
	(Conversion of Auster AOP.10 c/n 2570)					Pallas West, Toomyvara, County Tipperary	11. 1.06
G-ASUP	Cessna F172E	F172-0071		22. 7.64	P.T. and L.E.Trivett	Cardiff	3. 8.06
	(Built Reims Aviation SA)				t/a Gasup Air		
G-ASUR	Dornier Do.28A-1	3051	D-IBOM	28. 7.64	P.R.Dyson	Thruxton	12.11.06
G-ASUS	Jurca MJ.2E Tempete	PFA 2001		28. 7.64	D.G.Jones	Coventry	20. 8.04P
	(Built D.G.Jones) (Continental O-200-A)						
G-ASVC*	de Havilland DH.114 Heron 2D				See SECTION 3 Part 1	Caloundra, Queensland, Australia	
G-ASVG	Rousseau Piel CP.301B Emeraude	109	F-BILV	7. 8.64	K.S.Woodard *"Emma II"*	Priory Farm, Tibenham	31. 5.05P
G-ASVM	Cessna F172E	F172-0077		11. 8.64	R.Seckington	Lee-on-Solent	9. 2.06T
	(Built Reims Aviation SA)						
G-ASVN	Cessna 206 Super Skywagon	206-0275	N5275U	12. 8.64	N.D.Johnston t/a British Skysports Paracentre	Langar	27. 2.06
G-ASVO*	Handley Page HPR.7 Dart Herald 214	185	PP-SDG	13. 8.64	Not known	Glenrothes	14. 1.00T
			G-ASVO, G-8-3		*(WFU after collision Hurn 8.4.97: cancelled 25.9.01 by CAA) (Cockpit only 1.05)*		
G-ASVP	Piper PA-25-235 Pawnee	25-2978	N10F	17. 8.64	Aquila Gliding Club Ltd	Hinton-in-the-Hedges	27. 3.05
G-ASVZ	Piper PA-28-140 Cherokee	28-20357	N11C	24. 8.64	J.S.Garvey	Sleap	8. 10.06
G-ASWF*	Beagle A 109 Airedale	B.537		26. 8.64	C.J.Baker	Carr Farm, Thorney, Newark	24. 7.83
					(Cancelled 3.2.89 by CAA) (Dismantled but sold in "London", awaiting collection 1.05)		
G-ASWL	Cessna F172F	F172-0087		10. 9.64	Ensiform Aviation Ltd	Haverfordwest	30. 6.07T
	(Built Reims Aviation SA)						
G-ASWN	Bensen B.8M	14		15. 9.64	D.R.Shepherd	(Prestwick)	AC
	(Built D.R.Shepherd)				*(Components stored 2003)*		
G-ASWW	Piper PA-30 Twin Comanche	30-556	N7531Y	1.10.64	R.Jenkins	Bournemouth	19. 8.06
			N10F		t/a RJ Motors		
G-ASWX	Piper PA-28-180 Cherokee C	28-1932	N11C	1.10.64	A F.Dadds	Biggin Hill	16. 4.06
G-ASXD	Brantly B.2B	435		7.10.64	Lousada plc	Crawley Park, Husborne Crawley, Bedford	2. 7.05
G-ASXF*	Brantly 305	1014		7.10.64	Not known	Amen Corner, Binfield, Bracknell	16 .2.79
					(Cancelled 24.5.82 as WFU) (Noted 12.04)		
G-ASXI	Tipsy Nipper T.66 Series 3	56	VH-CGH	13.10.64	B.Dixon	Bagby	12. 7.05P
	(Built Avions Fairey SA) (Jabiru 2200A)		OO-KOC, (VH-CGC)				
G-ASXJ	Phoenix Luton LA-4A Minor	PFA 801		14.10.64	P.N.Stacey	Sandown	30. 4.05P
	(Built P D Lea and E A Linguard) (Lycoming O-145)				*"Pride and Joy"*		
G-ASXM*	Armstrong-Whitworth 650 Argosy Series 222				See SECTION 3 Part 1	Marlborough, New Zealand	
G-ASXN*	Armstrong-Whitworth 650 Argosy Series 222				See SECTION 3 Part 1	Marlborough, New Zealand	
G-ASXR	Cessna 210	57532	5Y-KPW	16.10.64	A Schofield	Barton	3. 1.93
			VP-KPW, N6532X		*(Noted dismantled 11.04)*		
G-ASXS	SAN Jodel DR.1050 Ambassadeur	133	F-BJNG	19.10.64	R.A Hunter	Finmere	17.12.06
G-ASXU	Wassmer Jodel D.120A Paris-Nice	196	F-BKAG	19.10.64	M.Ferid	Stoneacre Farm, Farthing Corner Corner	31. 3.05P
					tr G-ASXU Group		
G-ASXX*	Avro 683 Lancaster B.VII				See SECTION 3 Part 1	East Kirkby	
G-ASXY	SAN Jodel D.117A	914	F-BIVA	27.10.64	P.A., R.A.Davies and D.G.Claxton	Cardiff	9. 3.05P
G-ASXZ	Cessna 182G Skylane	18255738	N3238S	28.10.64	Last Refuge Ltd	Gedney Marsh Farm, Gedney, Wells	2.10.06
G-ASYD*	British Aircraft Corporation One-Eleven 475AM				See SECTION 3 Part 1	Duxford	
G-ASYG	Beagle A 61 Terrier 2	B.637	VX927	3.11.64	R.H. and S.J.Cooper	Gamston	19. 2.70T
					(Noted silver overall 1.05: being rebuilt to T.7 standard as "VX927")		
G-ASYJ	Beech D95A Travel Air	TD-595	N8675Q	6.11.64	Crosby Aviation (Jersey) Ltd	Jersey	31.10.07E
G-ASYK*	Piper PA-30-160 Twin Comanche	30-573	N7543Y	6.11.64	Not known		
						Walton New Road Business Park, Bruntingthorpe	28. 7.97T
					(Crashed on take-off from Sandown, IoW 11.5.96) (Cancelled 30.10.96 as WFU) (Noted 2.04)		
G-ASYP	Cessna 150E	15060794	N6094T	23.11.64	A C.Melmore tr Henlow Flying Group	RAF Henlow	10.10.06
G-ASZB	Cessna 150E	15061113	N3013J	16.12.64	R.J.Scott	Popham	19. 4.07
G-ASZD	Bölkow Bö.208A-2 Junior	563	D-ENKI	16.12.64	M.J.Ayres	Full Sutton	23. 6.03P
G-ASZE	Beagle A 61 Terrier 2	B.636	VF552	17.12.64	P.J.Moore	Lee-on-Solent	1. 9.05
	(Conversion of Auster 6 c/n 2510)						
G-ASZR	Fairtravel Linnet 2	005		5. 1.65	R.Palmer and D.Scott	Swanborough Farm, Lewes	9. 6.05P
G-ASZS	Sud-Aviation Gardan GY-80-160 Horizon	70		6. 1.65	L.R.Burton tr ZS Group	Wellesbourne Mountford	25.11.04
G-ASZU	Cessna 150E	15061152	N3052J	13. 1.65	T.H.Milburn	Sywell	15. 5.05
G-ASZV	Tipsy Nipper T.66 Series 2	45	5N-ADE	14. 1.65	J.M.Gough	(Sale)	23. 5.90P
	(Built Avions Fairey SA) (Volkswagen 1835)		5N-ADY, VR-NDD		*(Stored 9.97: new owner 5.01)*		
G-ASZX	Beagle A 61 Terrier 1	3742	(SE-ELO)	18. 1.65	R.B.Webber	Trenchard Farm, Eggesford	8. 7.07
			WJ368		*(On rebuild 11.03)*		

G-ATAA - G-ATZZ

Reg	Type	C/n	Previous identity	Regd	Owner/operator	Probable base	CofA expiry
G-ATAF	Cessna F172F	F172-0135		25. 1.65	Summit Media Ltd	(Norwich)	7. 7.07T
	(Built Reims Aviation SA)						
G-ATAG	CEA Jodel DR.1050 Ambassadeur	226	F-BKGG	25. 1.65	T.M.Dawes-Gamble	Oxford	4.10.02

G-ATAS	Piper PA-28-180 Cherokee C	28-2137	N11C	4. 2.65	R Osborn tr Atlas Group	Andrewsfield	17. 8.06
G-ATAU	Druine D.62B Condor	RAE/610		10. 2.65	M.A Peare	Siege Cross Farm, Thatcham	16.11.02
	(Built Rollason Aircraft and Engines)				tr Golf Alpha Uniform Group		
G-ATAV	Druine D.62C Condor	RAE/611		10. 2.65	Agri Air Services Ltd *(New owner 1.05)*		
	(Built Rollason Aircraft and Engines) (Continental O-240-A)					(Newbold on Stour,Stratford-upon-Avon)	6. 8.94
G-ATBG	Nord 1002 Pingouin II	121	F-BGVX	24. 2.65	T.W.Harris	Booker	17. 5.05P
			F-OTAN-5, Fr.Mil		*(As "NJ+C11" in Luftwaffe c/s)*		
G-ATBH	CZL Aero 145	172015		24. 2.65	P.D.Aviram *(On rebuild 11.04)*	Redhill	26.10.81
	(Built Strrojirny Prvni Petilesky)						
G-ATBI	Beech A23 Musketeer II	M-696		26. 2.65	A C.Dent tr Three Musketeers Flying Group	Oxford	20. 5.06
G-ATBJ	Sikorsky S-61N	61-269	N10043?	12. 3.65	Veritair Ltd t/a British International	(Sherborne)	2. 6.06T
G-ATBL	de Havilland DH.60G Moth	1917	HB-OBA	2. 3.65	J.M.Greenland	Blackacre Farm, Holt, Trowbridge	23. 7.05P
	(DH Gipsy I)		CH-353				
G-ATBP	Fournier RF3	59		11. 3.65	D.McNicholl	Inverness	12. 3.06
G-ATBS	Druine D.31 Turbulent	PFA 1620		16. 3.65	J.A Lear	Wigtown	17. 8.05P
	(Built C R Shilling) (Volkswagen 1500)				*"Fly Baby Fly"*		
G-ATBU	Beagle A 61 Terrier 2	B.635	VF611	17. 3.65	D.M.Snape	Hucknall	2. 6.05
	(Conversion of Auster 6 c/n 2552)				tr K9 Flying Group		
G-ATBW	Tipsy Nipper T.66 Series 2	52	OO-MAG	19. 3.65	S.Bloomfield and C.Firth	Stapleford	23 2.05P
	(Built Cobelavia) (Volkswagen 1834 Acro)				tr Stapleford Nipper Group		
G-ATBX	Piper PA-20-135 Pacer	20-904	VP-KRX	19. 3.65	G.D.and P.M.Thomson		
			VR-TCH, VP-KKE			Standalone Farm, Meppershall	17. 6.05
G-ATBZ	Westland WS-58 Wessex 60 Series 1				See SECTION 3 Part 1	Weston-super-Mare	
G-ATCC	Beagle A 109 Airedale	B.542		25. 3.65	J.R.Bowden	Headcorn	24. 3.05
G-ATCD	Beagle D.5/180 Husky	3683		25. 3.65	D.J.O'Gorman	Enstone	2. 4.06
G-ATCE	Cessna U206 Super Skywagon	U2060380	N2180F	25. 3.65	C M J Fitzmaurice	Tilstock	15. 7.05
G-ATCJ	Phoenix Luton LA-4A Minor	PFA 812		5. 4.65	P.R.Diffey	Top Farm, Croydon, Royston	5. 8.03P
	(Built R M Sharphouse - c/n PAL/1163) (Volkswagen 1600)						
G-ATCL	Victa Airtourer 100	93		5. 4.65	A D.Goodall	Cardiff	25. 7.05
G-ATCU	Cessna 337 Super Skymaster	3370133	N2233X	22. 4.65	The Committee for Aerial Photography, University of Cambridge	Cambridge	25. 4.05A
G-ATCX	Cessna 182H Skylane	18255848	N3448S	26. 4.65	Craft Associates Worldwide Ltd	(London W1G)	21.12.06
					(Note fuselage of cancelled G-OLSC is also marked as "G-ATCX")		
G-ATDA	Piper PA-28-160 Cherokee	28-206	EI-AME	27. 4.65	Portway Aviation Ltd	Shobdon	9. 1.06
			(G-ARUV)				
G-ATDB	SNCAN 1101 Noralpha	186	F-OTAN-6	27. 4.65	J.W.Hardie	Prestwick	22.11.78S
			Fr.Mil		*(Note "F-OTAN-6" used on G-BAYV) (Fuselage noted 2003)*		
G-ATDD*	Beagle B.206 Series 1				See SECTION 3 Part 1	Kemble	
G-ATDN	Beagle A 61 Terrier 2	B.638	TW641	7. 5.65	Susan J.Saggers	Biggin Hill	11. 7.07T
	(Conversion of Auster 6 c/n 2499)				*(As "TW641")*		
G-ATDO	Bölkow Bö.208C Junior	576	D-EGZU	10. 5.65	P.Thompson	Netherthorpe	26. 9.05P
G-ATEF	Cessna 150E	15061378	N3978U	25. 5.65	A J.White and B.M.Scott t/a Swans Aviation	Blackbushe	12.11.05
G-ATEM	Piper PA-28-180 Cherokee C	28-2329	N11C	26. 5.65	G D Wyles	Bovingdon	15. 4.07
G-ATEP*	EAA Biplane	PFA 1301		28. 5.65	E.L.Martin	Sausmarez Park, Guernsey	18. 6.73
	(Built E.L.Martin) (Continental C75)				*(Cancelled 14.7.86 by CAA) (Frame stored 5.03)*		
G-ATES*	Piper PA-32-260 Cherokee Six	2-20		31. 5.65	Stirling Parachute Centre	(Easter Poldar Farm, Thornhill)	11. 6.83
	(Crashed near Kinglassie 8.2.81 and cancelled 22.10.84 as WFU) (Used as para-trainer 6.00: ceased operations 2000: current status unknown)						
G-ATEV	CEA Jodel DR.1050 Ambassadeur	18	F-BJHL	31. 5.65	J.C.Carter and J.L.Altrip	(Cambridge)	13. 8.71
					(On rebuild 9.00: current status unknown)		
G-ATEW	Piper PA-30 Twin Comanche	30-719	N7640Y	3. 6.65	Air Northumbria (Woolsington) Ltd	Newcastle	10.11.07E
G-ATEX	Victa Airtourer 100	110	(VH-MTU)	3. 6.65	D.R.Henson	RAF Halton	29. 9.06
					tr Halton Victa Group *"Matilda"*		
G-ATEZ	Piper PA-28-140 Cherokee	28-21044	N11C	8. 6.65	EFI Aviation Ltd	Norwich	29. 5.06T
G-ATFD	CEA Jodel DR.1050 Ambassadeur	311	F-BKIM	14. 6.65	V.Usher	Wickenby	21. 2.04
G-ATFF	Piper PA-23-250 Aztec C	27-2898	N5769Y	16. 6.65	T.J.Wassell	Tatenhill	15. 5.05
G-ATFG*	Brantly B.2B				See SECTION 3 Part 1	East Fortune	
G-ATFK*	Piper PA-30-160 Twin Comanche	30-721	N7642Y	17. 6.65	Not known		
						Walton New Road Business Park, Bruntingthorpe	22.11.90
			(Damaged landing White Waltham 12.6.89) (Cancelled 21.10.92 by CAA) (Noted 2.04)				
G-ATFM	Sikorsky S-61N Mk.II	61-270	CF-OKY	21. 6.65	British International Ltd	Plymouth	1.10.06T
			N10052 *(US p/i not confirmed)*				
G-ATFR	Piper PA-25 Pawnee	25-135	OY-ADJ	28. 6.65	Borders (Milfield) Gliding Club Ltd	Milfield	29. 5.06
			N10F				
G-ATFU*	de Havilland DH.85 Leopard Moth				See SECTION 3 Part 1	Mandeville, Gore, New Zealand	
G-ATFV*	Agusta-Bell 47J-2A Ranger	2093	9J-ACX	1. 7.65	Not known	Ley Farm, Chirk	8. 8.92T
			G-ATFV, MM80-417		*(Cancelled 22.12.92 by CAA) (Stored 9.00)*		
G-ATFY	Cessna F172G	F172-0199		8. 7.65	Jaguar Aviation Ltd	Errol	11.12.06
	(Built Reims Aviation SA)				*(Op Fife Parachute Centre)*		
G-ATGE	SAN Jodel DR.1050 Ambassadeur	114	F-BJJF	9. 7.65	J.Turner	(Thatcham)	14. 2.07
G-ATGJ*	Riley Dove 400				See SECTION 3 Part 1	Ballarat, Victoria, Australia	
G-ATGK*	Riley Dove 400				See SECTION 3 Part 1	Ammam, Jordan	
G-ATGN*	Thorn K-800 Coal Gas Balloon				See SECTION 3 Part 1	Newbury	
G-ATGO	Cessna F172G	F172-0181		12. 7.65	Poetpilot Ltd	Leicester	22.11.07E
	(Built Reims Aviation SA)						
G-ATGY	Sud-Aviation Gardan GY-80-160 Horizon	121		20. 7.65	D Cowen	Newcastle	23.11.06
G-ATGZ	Griffiths GH-4 Gyroplane	G.1		20. 7.65	R.W.J.Cripps	(Shardlow, Derby)	
	(Built G Griffiths)				*(Stored 7.91: current status unknown)*		
G-ATHD	de Havilland DHC-1 Chipmunk 22	WP971	G-ATHD	26. 7.65	O L Cubitt and N Keveren	Denham	30. 6.06
		C1/0837	WP971		tr Spartan Flying Group *(As "WP971")*		
G-ATHK	Aeronca 7AC Champion	7AC-971	N82339	2. 8.65	T.P.Mcdonald and T.Crawley	Crosland Moor	9. 4.05P
	(Continental A75)		NC82339				
G-ATHM	Wallis WA-116/F	402 & 211	4R-ACK	3. 8.65	Wallis Autogyros Ltd	Reymerston Hall, Norfolk	23. 5.93P
	(Originally McCulloch - 60hp Franklin fitted 1974)		G-ATHM		*(Noted 8.01)*		

G-ATHN*	SNCAN 1101 Noralpha	84	F-BFUZ	5. 8.65	E.L.Martin	St.Peter Port, Guernsey	27. 6.75S
			Fr.Mil		*(Cancelled 16.12.91 by CAA) (Stored 5.03)*		
G-ATHR	Piper PA-28-180 Cherokee C	28-2343	EI-AOT	11. 8.65	Britannia Airways Ltd	Luton	10. 8.07T
			N11C				
G-ATHT	Victa Airtourer 115	120		16. 8.65	D.A Beese	Badminton	23. 9.06
G-ATHU	Beagle A 61 Terrier 1	AUS/127/FM	7435M	16. 8.65	J.A L.Irwin	Park Farm, Eaton Bray	24. 9.04
			WE539				
G-ATHV	Cessna 150F	15062019	N8719S	16. 8.65	S.Greenwood	Sherburn-in-Elmet	24. 2.06
					tr Cessna Hotel Victor Group		
G-ATHZ	Cessna 150F	15061586	(EI-AOP)	20. 8.65	R.D.Forster	Beccles	27.3.98T
			N6286R		*(New CofR 9.03)*		
G-ATIA	Piper PA-24-260 Comanche	24-4049	N8650P	20. 8.65	L.A Brown	Enstone	22.10.04
			N10F		tr The India Alpha Partnership		
G-ATIC	CEA Jodel DR.1050 Ambassadeur	6	F-BJCJ	23. 8.65	R.E.Major	Trevissick Farm, Porthtowan	13.10.06
G-ATIG*	Handley Page HPR.7 Dart Herald	214	PP-SDI	25. 8.65	Nordic Oil Services Ltd	Norwich	14.10.97T
		177	G-ATIG		*(Cancelled 29.10.96 as WFU) (Stored 4.04)*		
G-ATIN	SAN Jodel D.117	437	F-BHNV	8. 9.65	A.Ayre	(St. Andrews)	18. 4.96P
					(New owner 7.04)		
G-ATIR	AIA Stampe SV-4C	1047	F-BNMC	9. 9.65	Austin Trueman Ltd	(St.Albans)	29. 6.03
	(Renault 4P)		G-ATIR, F-BMKQ, Aéronavale, F-BCDM, Aéronavale				
G-ATIS	Piper PA-28-160 Cherokee C	28-2713	N11C	9. 9.65	M.J.Barton	Lee-on-Solent	10. 2.06
G-ATIZ	SAN Jodel D.117	636	F-BIBR	15. 9.65	D.K.Shipton	Deenethorpe	6.12.05P
G-ATJA	SAN Jodel DR.1050 Ambassadeur	378	F-BKHL	15. 9.65	D.A Head and G.W.Cunningham	Bicester	27. 3.06
					tr Bicester Flying Group		
G-ATJC	Victa Airtourer 100	125		16. 9.65	Aviation West Ltd	Cumbernauld	15. 1.07T
G-ATJG	Piper PA-28-140 Cherokee	28-21299		20. 9.65	Christine A McGee and L.K.G.Manning	Biggin Hill	8.12.05T
G-ATJL	Piper PA-24-260 Comanche	24-4203	N8752P	23. 9.65	M.J.Berry and T.R.Quinn	Blackbushe	17. 7.06
			N10F				
G-ATJM	Fokker Dr.1 Triplane rep	002	N78001	23. 9.65	R.J.Lamplough	Manor Farm, East Garston	10. 9.93P
	(Built Bitz Flugzeugbau GmbH) (Siemens SH-14A-165)		EI-APY, G-ATJM		*(Noted as "152/17" 6.02)*		
G-ATJN	Dormois Jodel D.119	863	F-PINZ	23. 9.65	R.L.Wademan tr Oxenhope Flying Group	Oxenhope	27. 6.05P
G-ATJT	Sud-Aviation Gardan GY-80-160 Horizon	108		4.10.65	N.Huxtable	Cheddington	28. 5.05
G-ATJV	Piper PA-32-260 Cherokee Six	32-103	TF-GOS	7.10.65	Wingglider Ltd	Hibaldstow	14. 7.07A
			G-ATJV, N11C				
G-ATKF	Cessna 150F	15062386	N3586L	20.10.65	P.Asbridge	Sleap	17.11.06T
G-ATKH	Phoenix Luton LA-4A Minor	PFA 809		25.10.65	H.E.Jenner	Brenchley, Kent	1. 7.05P
	(Built E B W Woodhall) (Lycoming O-145)						
G-ATKI	Piper J-3C-65 Cub	17545	N70536	25.10.65	C.O'Donnell	(Kinsale, County Cork)	23. 4.05P
	(Continental A75)		NC70536		tr KI Group		
G-ATKT	Cessna F172G	F172-0206		9.11.65	P.J.Megson	Goodwood	15. 5.05
	(Built Reims Aviation SA)						
G-ATKV*	Westland WS-55 Whirlwind 3				See SECTION 3 Part 1	Weston-super-Mare	
G-ATKX	SAN Jodel D.140C Mousquetaire III	163		19.11.65	I.V.Sharman tr Kilo Xray Syndicate	Redhill	24. 6.07
G-ATLA	Cessna 182J Skylane	18256923	N2823F	24.11.65	J.W.and J.T.Whicher	Full Sutton	10.12 05
G-ATLB	SAN Jodel DR.1050M Excellence	78	F-BIVG	29.11.65	D.J.Gibson	Hill Farm, Nayland	24. 8.05
					tr Le Syndicate du Petit Oiseau		
G-ATLH*	Fewsdale Tigercraft Gyroplane	F.T5		6.12.65	R.Light	(Stockport)	
					(Cancelled 10.2.82 as WFU) (Stored 7.04)		
G-ATLM	Cessna F172G	F172-0252		6.12.65	Airfotos Ltd	Newcastle	30 3.06T
	(Built Reims Aviation SA)						
G-ATLP	Bensen B.8M	17		9.12.65	R.F.G.Moyle	(Penryn)	19. 5.97P
	(Built C D Julian) (McCulloch Motors 4318F)				*(Current status unknown)*		
G-ATLT	Cessna U206A Super Skywagon	U2060523	N4823F	13.12.65	A.I.M. and A.J.Guest	Dunkeswell	10. 6.05T
G-ATLV	Wassmer Jodel D.120 Paris-Nice	224	F-BKNQ	15.12.65	L.S.Thorne	(Lichfield)	8. 7.05P
G-ATMC	Cessna F150F	F150-0020		28.12.65	G.H.Farrah and D.Cunnane		
	(Built Reims Aviation SA) (Wichita c/n 15062849)					Abbeyshrule, County Longford	22. 8.07
G-ATMH	Beagle D.5/180 Husky	3684		3. 1.66	Dorset Gliding Club Ltd	Gallows Hill, Bovington	16. 7.06
G-ATMI*	Hawker Siddeley HS.748 Series 2A/225	1592	VP-LIU	4. 1.66	Emerald Airways Ltd	Blackpool	18. 5.00T
			G-ATMI, VP-LIU, G-ATMI, VP-LIU, G-ATMI, VP-LIU, G-ATMI				
					(Cancelled 30.7.01 as WFU) (Fuselage on fire dump 1.04)		
G-ATMJ	Hawker Siddeley HS.748 Series 2A/225	1593	VP-LAJ	4. 1.66	Emerald Airways Ltd	Liverpool	7. 9.06T
			G-ATMJ, 6Y-JFJ, G-ATMJ				
G-ATML	Cessna F150F	F150-0014		6. 1.66	G.I.Smith	Octon Lodge Farm, Thwing	4.10.04
	(Built Reims Aviation SA) (Wichita c/n 15062722)						
G-ATMM	Cessna F150F	F150-0016		6. 1.66	Skytrax Aviation Ltd	Crowfield	19. 5.07T
	(Built Reims Aviation SA) (Wichita c/n 15062775)						
G-ATMN (2)*	Cessna F150F	F150-0060	(G-ATNE)	6. 1.66	C.R.Hardiman	Egginton	2. 7.84T
	(Built Reims Aviation SA) (Wichita c/n 15063526)				*(Noted 8.04: cancelled 21.10.04 as WFU)*		
G-ATMT	Piper PA-30 Twin Comanche	30-439	XW938	10. 1.66	Montagu-Smith and Co Ltd	Hinton-in-the-Hedges	11. 7.05
			G-ATMT, N7385Y				
G-ATMW	Piper PA-28-140 Cherokee	28-21486		11. 1.66	Bencray Ltd	Blackpool	27. 5.07T
					(Op Blackpool and Fylde Aero Club)		
G-ATMY	Cessna 150F	15062642	SE-ETD	13. 1.66	A.Dobson	(Newark)	20.10.06
			N8542G				
G-ATNB	Piper PA-28-180 Cherokee C	28-3057	N11C	20. 1.66	K.N.Macdonald t/a Ken Macdonald and Co	Stornoway	31. 7.06
G-ATNE	Cessna F150F	F150-0042		20. 1.66	A.D.Revill	Leicester	10. 2.07
	(Built Reims Aviation SA) (Wichita c/n 15063252)						
G-ATNL	Cessna F150F	F150-0066		25. 1.66	G.A.Lauf	Lower Upham Farm, Chiseldon	24. 7.05
	(Built Reims Aviation SA) (Wichita c/n 15063652)				tr G-ATNL Flying Group		
G-ATNV	Piper PA-24-260 Comanche	24-4350	N8896P	28. 1.66	A.Heydn and K.Powell	King's Farm, Thurrock	14.11.04
G-ATOA	Piper PA-23-160 Apache G	23-1954	N4437P	31. 1.66	Oscar Alpha Ltd	Stapleford	13. 6.03
G-ATOD	Cessna F150F	F150-0003		1. 2.66	D.Lugg	(Helston)	10.10.05
	(Built Reims Aviation SA) (Wichita c/n 15062342)						

G-ATOH	Druine D.62B Condor	RAE/612		3. 2.66	J.Cooke	Streethay	17. 6.05P
	(Built Rollason Aircraft and Engines)				tr Three Spires Flying Group		
G-ATOI	Piper PA-28-140 Cherokee	28-21556	N11C	3. 2.66	R.W.Nash	RAF Brize Norton	27. 5.05
G-ATOJ	Piper PA-28-140 Cherokee	28-21584	N11C	3. 2.66	A Flight Aviation Ltd	Prestwick	23.10.06T
					(Op Prestwick Flying Club)		
G-ATOK	Piper PA-28-140 Cherokee	28-21612	N11C	3. 2.66	G.T.S.Done and P.R.Harrison	White Waltham	8. 3.07
					tr ILC Flying Group		
G-ATOL	Piper PA-28-140 Cherokee	28-21626	N11C	3. 2.66	L.J.Nation and G.Alford tr G-ATOL Flying Group	Cardiff	23. 1.98
G-ATOM	Piper PA-28-140 Cherokee	28-21640	N11C	3. 2.66	A Flight Aviation Ltd	Prestwick	27. 8.07T
					(Op Prestwick Flying Club)		
G-ATON	Piper PA-28-140 Cherokee	28-21654	N11C	3. 2.66	R.G.Walters	Shobdon	14.11.07E
G-ATOO	Piper PA-28-140 Cherokee	28-21668	N11C	3. 2.66	A K.Komosa	Shoreham	20.11.05
G-ATOP	Piper PA-28-140 Cherokee	28-21682	N11C	3. 2.66	P.R.Coombs tr The Aero 80 Flying Group	Popham	22. 5.05
G-ATOR	Piper PA-28-140 Cherokee	28-21696	N11C	3. 2.66	D.Palmer tr Aligator Group	Shobdon	22. 6.06
G-ATOT	Piper PA-28-180 Cherokee C	28-3061	N11C	3. 2.66	Totair Ltd *"Totty"*	Shipdham	31. 7.06T
G-ATOU	Mooney M.20E Super 21	961	N5946Q	3. 2.66	A D.Morgan	Sherburn-in-Elmet	23. 7.06
					tr Mooney M20 Flying Group		
G-ATOY*	Piper PA-24-260 Comanche B				See SECTION 3 Part 1	East Fortune	
G-ATOZ	Bensen B.8M	18		7. 2.66	N.C.White	Sorbie Farm, Kingsmuir	9.12.05P
	(Built J D M Wilson) (Rotax 503 *(Substantially rebuilt in 1986, original airframe stored Wimborne)*						
G-ATPD*	Hawker Siddeley HS.125 Series 1B/522	25085	5N-AGU	11. 2.66	Wessex Air (Holdings) Ltd	Bournemouth	14.10.98T
			G-ATPD		*(WFU 1997 and cancelled 2.12.03 as wfu) (Noted on dump 2.04)*		
G-ATPN	Piper PA-28-140 Cherokee	28-21899	N11C	18. 2.66	R.W.Harris, M.F.Hatt, P.E.Preston and A Jahanfar	Southend	25. 3.05T
					(Op Southend Flying Club)		
G-ATPT	Cessna 182J Skylane	18257056	N2956F	22. 2.66	G.B.Scholes tr Papa Tango Group	Elstree	12. 8.07
G-ATPV	Gardan GY-20 Minicab	JB-01	F-PJKA	22. 2.66	C.F.O'Neill	Newtownards	28. 4.99P
	(Continental C90) *(Rebuild of GY-20 F-PHUC c/n A 155 by J Barritault-Bauge as JB.01 Minicab) (Current status unknown)*						
G-ATRG	Piper PA-18-150 Super Cub	18-7764	5B-CAB	1. 3.66	Lasham Gliding Society Ltd	Lasham	31. 5.07
	(Lycoming O-360-A4)		N4985Z				
G-ATRI	Bölkow Bö.208C Junior	602	D-ECGY	3. 3.66	H.P.Brooks	Deanland	23. 9.06
G-ATRK	Cessna F150F	F150-0049	(G-ATNC)	4. 3.66	G.G.and J.G.Armstrong	Carlisle	7. 8.06T
	(Built Reims Aviation SA) (Wichita c/n 15063381)				t/a Armstrong Aviation		
G-ATRL*	Cessna F150F	F150-0050		4. 3.66	S.S.Delwarte tr G-ATRL Flying Group	Shoreham	21. 2.98
	(Built Reims Aviation SA) (Wichita c/n 15063382)		*(Wings used in rebuild of G-AVHM qv: cancelled 23. 5.01 by CAA) (Fuselage unidentifiable on fire dump 2004)*				
G-ATRM	Cessna F150F	F150-0053	(G-ATNJ)	4. 3.66	J.Redfearn	Morgansfield, Fishburn	14. 7.07T
	(Built Reims Aviation SA) (Wichita c/n 15063454)						
G-ATRO	Piper PA-28-140 Cherokee	28-21871	N11C	4. 3.66	M.A.Smith	Seething	4. 8.06
G-ATRP*	Piper PA-28-140 Cherokee	28-21885	N11C	4. 3.66	JRB Aviation Ltd	Southend	20. 9.84.
		(Damaged Boughton Monchelsea 16.10.81: cancelled 10.11.86 as WFU) (Wreck stored dismantled 1.05)					
G-ATRR	Piper PA-28-140 Cherokee	28-21892	N11C	4. 3.66	Marham Investments Ltd	Ronaldsway	1. 7.07T
					(Op Manx Flyers Aero Club)		
G-ATRW	Piper PA-32-260 Cherokee Six	32-360	N11C	8. 3.66	Moxley and Frankl Ltd and J.Pringle	Biggin Hill	20.10.07E
G-ATRX	Piper PA-32-260 Cherokee Six	32-390	N11C	8. 3.66	A M., A C.M. and M.R.Harrhy	Bembridge	4.10.04T
G-ATSI	Bölkow Bö.208C Junior	605	D-EFNU	14. 3.66	R.S.Jordan tr G-ATSI Group	Shipdham	16. 4.06
G-ATSL	Cessna F172G	F172-0260		16. 3.66	G.F.Robinson	Movenis	14. 9.07
	(Built Reims Aviation SA)				tr Alpha Aviation		
G-ATSM	Cessna 337A Super Skymaster	337-0434	N5334S	23. 3.66	I.J. and H.R.Jones	Thruxton	10. 7.97T
					t/a Landscape and Ground Maintenance *(Current status unknown)*		
G-ATSR	Beech M35 Bonanza	D-6236	EI-ALL	29. 3.66	C.B.Linton tr G-ATSR Group	Gloucestershire	7. 2.05
G-ATSX	Bölkow Bö.208C Junior	608	D-EJUC	7. 4.66	Little Bear Ltd *(New owner 3.04)*	Exeter	1. 7.02
G-ATSY	Wassmer WA.41 Super Baladou IV	117		12. 4.66	R.L.and K.P.McLean	Rufforth	23.11.91
					t/a McLean Aviation *(Spares use for G-ATZS 5.01)*		
G-ATSZ	Piper PA-30 Twin Comanche B	30-1002	EI-BPS	13. 4.66	Sierra Zulu Aviation Ltd	Cambridge	16. 7.05T
			G-ATSZ, (AN-...), G-ATSZ, (EI-BBS), G-ATSZ, N7912Y				
G-ATTB	Wallis WA-116/F	214		19. 4.66	D.A Wallis	Reymerston Hall, Norfolk	27. 5.05P
	(Franklin 2A) *(Rebuild of WA-116 G-ARZC(2)/XR944 c/n 205)*				*(As "XR944")*		
	(Originally regd as Wallis WA.116 Srs 1 with McCulloch engine then Franklin 2A 1981: redesignated WA-116/F: to XR944 for Service participation)						
G-ATTD	Cessna 182J Skylane	18257229	N3129F	19. 4.66	Batesons Hotels (1958) Ltd	(Warton)	1. 4.05
G-ATTI	Piper PA-28-140 Cherokee	28-21951	N11C	24. 4.66	T.Marsh tr G-ATTI Flying Group	Bristol	28.10 07T
G-ATTK	Piper PA-28-140 Cherokee	28-21959	N11C	25. 4.66	D.J.E.Fairburn tr The G-ATTK Flying Group	Southend	3. 6.07
G-ATTM	CEA Jodel DR.250/160 Capitaine	65		26. 4.66	R.W.Tomkinson	Seletar, Singapore	19.12.05
G-ATTN*	Piccard HAB (62,000 cu ft)				See SECTION 3 Part 1	Wroughton	
G-ATTR	Bölkow Bö.208C Junior	612	D-EHEH	28. 4.66	S.Luck	Audley End	9.10.06
G-ATTV	Piper PA-28-140 Cherokee	28-21991	N11C	2. 5.66	N.E.Leech tr G-ATTV Group	Andrewsfield	14. 2.05
G-ATTX	Piper PA-28-180 Cherokee C	28-3390	PH-VDP	2. 5.66	IPAC Aviation Ltd	Earls Colne	29. 1.06
			(G-ATTX), N11C				
G-ATUB	Piper PA-28-140 Cherokee	28-21971	N11C	2. 5.66	R.H.Partington and M.J.Porter	Wombleton	18. 3.05
G-ATUD	Piper PA-28-140 Cherokee	28-21979	N11C	2. 5.66	J.J.Ferguson	Belle Vue Farm, Yarnscombe	25. 9.06T
G-ATUF	Cessna F150F	F150-0040		4. 5.66	D.P.Williams	Weeley Heath	16. 5.05
	(Built Reims Aviation SA) (Wichita c/n 15063229)				*"Honeysuckle"*		
G-ATUG	Druine D.62B Condor	RAE/614		4. 5.66	R.Crosby	Watchford Farm, Yarcombe	22. 6.05P
	(Built Rollason Aircraft and Engines) (Continental C90-14F)						
G-ATUH	Tipsy Nipper T.66 Series 1	6	OO-NIF	4. 5.66	M.D.Barnard and C.Voelger	RAF Henlow	28. 4.04P
	(Built Avions Fairey SA) (Volkswagen 1600)						
G-ATUI	Bölkow Bö.208C Junior	611	D-EHEF	4. 5.66	A W.Wakefield	Stapleford	26.11.05
G-ATUL	Piper PA-28-180 Cherokee C	28-3033	N9007J	6. 5.66	Barry Fielding Aviation Ltd	Ronaldsway	21. 6.05
G-ATVF	de Havilland DHC-1 Chipmunk 22	C1/0265	WD327	25. 5.66	T.M.Holloway	RAF Syerston	26. 5.07
	(Lycoming AEIO-360)				tr RAFGSA *(Op Four Counties Gliding Club)*		
G-ATVK	Piper PA-28-140 Cherokee	28-22006	N11C	27. 5.66	Broadland Flyers Ltd	Norwich	11.12.04T
G-ATVO	Piper PA-28-140 Cherokee	28-22020	N11C	27. 5.66	G.R.Bright	Little Gransden	20. 3.06T
G-ATVP*	Vickers FB.5 Gunbus rep				See SECTION 3 Part 1	Hendon	
G-ATVS	Piper PA-28-180 Cherokee C	28-3041	N9014J	1. 6.66	T.A.Buckley	Sandown	16.11.06T
G-ATVW	Druine D.62B Condor	RAE/615		7. 6.66	G.G.Roberts	Rayne Hall Farm, Braintree	24. 5.07
	(Built Rollason Aircraft and Engines)						

G-ATVX	Bölkow Bö.208C Junior	615	D-EHER	9. 6.66	A.V.Hurley and J.Farquhar	RAF Henlow	19.12.05P
					tr Henlow Juniors		
G-ATWA	SAN Jodel DR.1050 Ambassadeur	296	F-BKHA	10. 6.66	C.R.Elliott tr One Twenty Group	Tollerton	13.12.05
G-ATWB	SAN Jodel D.117	423	F-BHNH	10. 6.66	D.P.Ash	Andrewsfield	5. 8.05P
					tr Andrewsfield Whisky Bravo Group		
G-ATWJ	Cessna F172F	F172-0095	EI-ANS	21. 6.66	C.J.and J.Freeman	Headcorn	10. 6.07T
	(Built Reims Aviation SA)				t/a Weald Air Services		
G-ATWR*	Piper PA-30-160 Twin Comanche B	30-1134	N8025Y	30. 6.66	Not known		
					Walton New Road Business Park, Bruntingthorpe		22.12.94T
	(Damaged in crash Crosland Moor 14.9.93: cancelled 18.4.95 as TWFU) (Fuselage in scrapyard compound 3.04)						
G-ATWS*	Phoenix Luton LA-4A Minor	PAL/1195		30. 6.66	W M Grant	(Tain)	26. 3.69P
	(Built D H Handley) (Incorporates c/n PFA/818)				*(Cancelled 8.2.82) (Noted 2003)*		
G-ATXA	Piper PA-22-150 Tri-Pacer	22-3730	N4403A	8. 7.66	S.Hildrop	Top Farm, Croydon, Royston	17. 5.07
	(Modified to PA-20 Super Pacer configuration)						
G-ATXD	Piper PA-30 Twin Comanche B	30-1166	N8053Y	12. 7.66	P.A.Brook	Shoreham	21. 8.06
G-ATXL*	Avro 504K rep				See SECTION 3 Part 1	Rhinebeck, New York, US	
G-ATXM	Piper PA-28-180 Cherokee C	28-2759	N8809J	19. 7.66	M.J.Stack tr G-ATXM Flying Group	Stapleford	12.10.02
G-ATXN	Mitchell-Procter Kittiwake 1	PFA 1306		19. 7.66	R.G.Day	Biggin Hill	23. 5.05P
	(Built R Procter - c/n 1) (Lycoming O-290)						
G-ATXO	SIPA 903	41	F-BGAP	19. 7.66	D.F.Hurn	Sandown	29. 7.05P
G-ATXR*	Abingdon Gas/HAB				See SECTION 3 Part 1	Newbury	
G-ATXX*	McCandless M.4				See SECTION 3 Part 1	Belfast	
G-ATXZ	Bölkow Bö.208C Junior	624	D-ELNE	28. 7.66	M.R.Kaye tr G-ATXZ Group	Tatenhill	12. 7.05P
G-ATYM	Cessna F150G	F150-0074		15. 8.66	P.D'Costa	Rochester	15. 1.07
	(Built Reims Aviation SA)						
G-ATYS	Piper PA-28-180 Cherokee C	28-3296	N9226J	19. 8.66	E.Baker tr G-ATYS Flying Group	Lydd	24. 5.06
G-ATZK	Piper PA-28-180 Cherokee C	28-3128	N9090J	21. 9.66	I.A Eddy	Oaksey Park	21. 4.06T
			(D-EFUN), N9090J		tr G-ZK Group		
G-ATZM	Piper J-3C-90 Cub Special	20868	N2092M	26. 9.66	N.D.Marshall	RAF Haltonl	3.12.05P
	(Frame No.21310)		NC2092M				
G-ATZS	Wassmer WA.41 Super Baladou IV	128		30. 9.66	Little Bear Ltd	Exeter	16. 4.06
G-ATZY	Cessna F150G	F150-0135		14.10.66	Prestwick Flight Centre Ltd	Prestwick	6. 2.06T
	(Built Reims Aviation SA)						

G-AVAA - G-AVZZ

G-AVAA*	Cessna F150G				See SECTION 3 Part 1	Doncaster	
G-AVAR	Cessna F150G	F150-0122		27.10.66	J.A Rees	Haverfordwest	3.10.07E
	(Built Reims Aviation SA)						
G-AVAU	Piper PA-30 Twin Comanche B	30-1328	N8230Y	8.11.66	Enrico Ermano Ltd	Fairoaks	19. 6.05
			N10F				
G-AVAV*	Vickers Supermarine 509 Spitfire Trainer IX				See SECTION 3 Part 1	Mesa, Arizona, US	
G-AVAW	Druine D.62C Condor	RAE/617		10.11.66	S.Banyard	Tibenham	25. 5.06
	(Built Rollason Aircraft and Engines) (Continental O-240-A				tr Condor Aircraft Group		
G-AVAX	Piper PA-28-180 Cherokee C	28-3798	N11C	11.11.66	J.J.Parkes	Wolverhampton	30. 5.05
G-AVBG	Piper PA-28-180 Cherokee C	28-3801	N11C	11.11.66	R.A Cayless and R.D.B.Severn	White Waltham	24. 3.06
					tr G-AVBG Flying Group		
G-AVBH	Piper PA-28-180 Cherokee C	28-3802	N11C	11.11.66	T.R.Smith (Agricultural Machinery) Ltd		
						New Lane Farm, North Elmham	18. 5.06
G-AVBS	Piper PA-28-180 Cherokee C	28-3938	N11C	14.11.66	A G.Arthur	Perranporth	7. 7.07T
G-AVBT	Piper PA-28-180 Cherokee C	28-3945	N11C	14.11.66	J.F.Mitchell	Shoreham	1. 7.07T
G-AVCM	Piper PA-24-260 Comanche B	24-4520	N9054P	5.12.66	R.F.Smith	Stapleford	26. 6.05
G-AVCN	Britten-Norman BN-2A-8 Islander	3	N290VL	6.12.66	Airstream International Group Ltd	Bembridge	5.11.76T
	(Originally regd as BN-2)		F-OGHG, G-AVCN		*(For restoration by Britten-Norman Aircraft Preservation Society)*		
G-AVCV	Cessna 182J Skylane	18257492	N3492F	15.12.66	University of Manchester, School of Earth, Atmospheric and		
					Environmental Sciences	Woodford	29. 3.07
G-AVCY*	Piper PA-30 Twin Comanche B	30-1367	N8241Y	16.12.66	Not known		
					Walton New Road Business Park, Bruntingthorpe		26. 7.93
	(Crashed on take-off Cardiff 9.3.91 and cancelled 17.7.91 as WFU) (Noted 2.04)						
G-AVDA	Cessna 182K Skylane	18257959	N2759Q	16.12.66	F.W.Ellis	Water Leisure Park, Skegness	17. 6.01
G-AVDF*	Beagle B.121 Pup Series 100	B.121-001		28.12.66	D Collings	Lower Wasing Farm, Brimpton	22. 5.68
	(Originally registered as B.121C c/n B.151, became B.121 Series 100 2.69)				*(Cancelled 22.5.68 as WFU) (Stored 1.03)*		
G-AVDG	Wallis WA-116 Series 1 Agile	215		28.12.66	K.H.Wallis	Reymerston Hall, Norfolk	23. 5.92P
	(Variously powered by McCulloch: Fuji 440, Norton twin-rotor Wankel and now Rotax 532) (Stored 8.01)						
G-AVDS*	Beech 65-B80 Queen Air	LD-337	A40-CS	5. 1.67	Brunel Technical College	Filton	26. 8.77
			G-AVDS		*(Cancelled 1.3.89 as WFU) (Dumped 2.04)*		
G-AVDT	Aeronca 7AC Champion	7AC-6932	N3594E	5. 1.67	D.Cheney and G.Moore	(Newry, County Armagh)	27. 6.05P
			NC3594E				
G-AVDV	Piper PA-22-150 Tri-Pacer	22-3752	N4423A	5. 1.67	S.C.Brooks	Wellcross Grange, Slinfold	23.10.03
	(Modifed to PA-20 Super Pacer configuration)						
G-AVDY	Phoenix Luton LA-4A Minor	PFA 808		10. 1.67	R.Targonski	(Birmingham B18)	9. 8.00P
	(Built M E Pendlebury - c/n PAL/1183) (Lycoming O-145)				*(Damaged landing Stapleford 18.12.99: new owner 4.04)*		
G-AVEC	Cessna F172H	F172-0405		13. 1.67	Quick Flight Images LLP	Fairoaks	11. 5.05
	(Built Reims Aviation SA)						
G-AVEF	SAN Jodel D.150 Mascaret	16	F-BLDK	19. 1.67	Prop-Air Corporation Ltd	Headcorn	20. 6.05T
G-AVEH	SIAI-Marchetti S.205-20R	346		20. 1.67	R.E.Gretton, K.Fear, A D.F.Flintoft and R.L.F.Darby		
						Shipdham	5. 6.06
G-AVEM	Cessna F150G	F150-0198		23. 1.67	T.D.and J.A Warren	Redhill	30. 5.05
	(Built Reims Aviation SA)						
G-AVEN	Cessna F150G	F150-0202		23. 1.67	R.A.Lambert	Bourn	8.12.05
	(Built Reims Aviation SA)						
G-AVER	Cessna F150G	F150-0206		23. 1.67	LAC (Enterprises) Ltd	Barton	19.12.04T
	(Built Reims Aviation SA)				*(Op Lancashire Aero Club)*		

G-AVEU	Wassmer WA.41 Super Baladou IV	136		27. 1.67	H. and S.Roberts	Enstone	10.10.05
G-AVEX	Druine D.62B Condor	RAE/616		31. 1.67	C.A.Macleod	Hinton-in-the-Hedges	17. 6.05P
	(Built Rollason Aircraft and Engines)						
G-AVEY	Phoenix Currie Super Wot	PFA 3006		31. 1.67	B.J.Anning	Watchford Farm, Yarcombe	4. 9.04P
	(Built K Sedgwick - c/n SE.100) (Pobjoy "R")						
G-AVEZ*	Handley Page HPR.7 Dart Herald 210	169	PP-ASW	31. 1.67	Not known	Norwich	
			G-AVEZ, HB-AAH		*(WFU on 5.1.81 and cancelled 4.1.83) (On fire dump 4.04)*		
G-AVFB*	Hawker Siddeley HS.121 Trident 2E				See SECTION 3 Part 1	Duxford	
G-AVFE*	Hawker Siddeley HS.121 Trident 2E	2144		1. 2.67	Belfast Airport Fire Service	Belfast Aldergrove	6. 5.85T
					(WFU 20.3.85: noted 2.04)		
G-AVFG*	Hawker Siddeley HS.121 Trident 2E	2146		1. 2.67	Fire Station	Manchester	2. 7.85T
					(WFU 24.5.85) (Noted as "G-SMOKE" 2.04)		
G-AVFH*	Hawker Siddeley HS.121 Trident 2E				See SECTION 3 Part 1	Salisbury Hall, London Colney	
G-AVFJ*	Hawker Siddeley HS.121 Trident 2E	2149		1. 2.67	International Fire Training Centre	Durham Tees Valley	18. 9.83T
					(WFU 6.82: cancelled 9.7.82) (Front fuselage extant 11.03)		
G-AVFP	Piper PA-28-140 Cherokee	28-22652	N11C	1. 2.67	Rebecca L.Howells	Barton	6. 8.05
G-AVFR	Piper PA-28-140 Cherokee	28-22747	N11C	1. 2.67	R.R.Orr	Newtownards	5. 6.05
G-AVFU	Piper PA-32-300 Cherokee Six	32-40182	N11C	1. 2.67	M.J.Hoodless	Kirkbride	30. 4.03T
					(Noted less engine 5.03)		
G-AVFX	Piper PA-28-140 Cherokee	28-22757	N11C	1. 2.67	J.Watson	Perth	18.11.07E
G-AVFZ	Piper PA-28-140 Cherokee	28-22767	N11C	1. 2.67	C.M.Toyne tr G-AVFZ Flying Group	Yeovil	14.10.07E
G-AVGA	Piper PA-24-260 Comanche B	24-4489	N9027P	31. 1.67	J.T.M.Ball	Biggin Hill	26. 2.06
					tr Conram Aviation *"C'est Si Bon"*		
G-AVGC	Piper PA-28-140 Cherokee	28-22777	N11C	31. 1.67	R.Dagg-Heston	Popham	16. 5.07
G-AVGD	Piper PA-28-140 Cherokee	28-22782	N11C	31. 1.67	T.Akeroyd	Bourn	10. 2.08E
G-AVGE	Piper PA-28-140 Cherokee	28-22787	N11C	31. 1.67	A.J.Cutler	Bournemouth	22. 4.07
G-AVGI	Piper PA-28-140 Cherokee	28-22822	N11C	31. 1.67	R.D.A.Gilchrist tr GI Group	Liverpool	16.12.07E
G-AVGK	Piper PA-28-180 Cherokee C	28-3639	N9516J	2. 2.67	I.R.Chaplin	Andrewsfield	13. 8.06T
G-AVGU	Cessna F150G	F150-0199		8. 2.67	Coulson Flying Services Ltd	Cranfield	13. 2.07T
	(Built Reims Aviation SA)						
G-AVGY	Cessna 182K Skylane	18258112	N3112Q	17. 2.67	R.M.C.Sears	Stoke Ferry	3. 8.06
G-AVGZ	CEA Jodel DR.1050 Sicile	341	F-BKPR	14. 2.67	D.C.Webb *(Stored 10.00)*	Bagby	13. 7.97
G-AVHE*	Vickers 812 Viscount				See SECTION 3 Part 1	Stuttgart, Germany	
G-AVHH	Cessna F172H	F172-0337		20. 2.67	Avon Aviation Ltd	Netherthorpe	2. 2.05T
	(Built Reims Aviation SA)						
G-AVHL	SAN Jodel DR.105A Ambassadeur	90	F-BIVY	23. 2.67	P.J.McMahon	Ludham	17. 8.06
G-AVHM	Cessna F150G	F150-0181		24. 2.67	M.Murphy	Goodwood	31. 1.05T
	(Built Reims Aviation SA) (Rebuilt 1997 with wings from G-ATRL qv)				tr M and N Flying Group		
G-AVHT	Beagle E.3 (Auster AOP.9M)	Not known	WZ711	1. 3.67	J.Pyett	Spanhoe	29. 4.01
	(Lycoming O-360)				*(Noted for restoration 6.04)*		
G-AVHY	Fournier RF4D	4009		10. 3.67	I.K.G.Mitchell	Halesland	5. 4.05P
G-AVIA	Cessna F150G	F150-0184		10. 3.67	Cheshire Air Training Services Ltd	Liverpool	5. 5.07T
	(Built Reims Aviation SA)						
G-AVIB	Cessna F150G	F150-0180		10. 3.67	Far North Aviation	Wick	6.11.06T
	(Built Reims Aviation SA)						
G-AVIC	Cessna F172H	F172-0320	N17011	10. 3.67	Leeside Flying Ltd	Cork, County Cork	23. 5.07
	(Built Reims Aviation SA)						
G-AVID	Cessna 182K	18257734	N2534Q	10. 3.67	Jaguar Aviation Ltd *(Op Fife Parachute Centre)*	Eroll	18. 4.06
G-AVII	Agusta-Bell 206B JetRanger II	8011		10. 3.67	Bristow Helicopters Ltd *"Brighton Belle"*	Redhill	4. 1.07T
G-AVIL	Alon A-2 Aircoupe	A 5	N5471E	14. 3.67	D.J.Hulks *(As "VX147" in RAF c/s)*	Headcorn	4. 11.07E
G-AVIN	SOCATA MS.880B Rallye Club	884		14. 3.67	R.Bunce	Compton Abbas	6. 8.05
G-AVIP	Brantly B.2B	471		14. 3.67	W.G.B.Yard	(Aberdesach, Caernarfon)	18.10.01
					(New owner 6.03)		
G-AVIS	Cessna F172H	F172-0413		14. 3.67	J.P.A.Freeman	Headcorn	20. 1.05T
	(Built Reims Aviation SA)						
G-AVIT	Cessna F150G	F150-0217		14. 3.67	P.J.Mitchell	Hinton-in-the-Hedges	20.12.07E
	(Built Reims Aviation SA)						
G-AVIZ	Scheibe SF25A Motorfalke	4552	(D-KOFY)	21. 3.67	T.J.Wiltshire	Spilsby	19. 9.91
	(Hirth F10A)				tr Splisby Gliding Trust		
G-AVJB*	Vickers 815 Viscount				See SECTION 3 Part 1	Hillerstorp, Sweden	
G-AVJE	Cessna F150G	F150-0219		29. 3.67	T.F.Fisher	Hinton-in-the-Hedges	5. 8.07
	(Built Reims Aviation SA)				tr G-AVJE Syndicate		
G-AVJF	Cessna F172H	F172-0393		31. 3.67	J.A and G.M.Rees	Haverfordwest	10. 2.07T
	(Built Reims Aviation SA)						
G-AVJH*	Druine D.62 Condor	PFA 603		31. 3.67	R Chapman	East Grinstead	4.11.83P
	(Built J Norton)		*(Crashed Nefyn, Gwynedd 31.7.83: cancelled 5.1.89) (As spares 4.00 for rebuild of G-AXGU qv)*				
G-AVJJ	Piper PA-30 Twin Comanche B	30-1420	N8285Y	7. 4.67	A H.Manser	Gloucestershire	8. 9.07T
G-AVJK	SAN Jodel DR.1050M Excellence	453	F-BLJH	7. 4.67	A.A.Robertson and D.S.Spillane		
	(Originally built as DR.1051)					Sackville Lodge Farm, Riseley	20. 7.05
G-AVJO	Fokker E.III rep	PPS/FOK/1 & PPS/REP/6		12. 4.67	Bianchi Aviation Film Services Ltd	Compton Abbas	5. 4.04P
	(Built Personal Plane Services Ltd - c/n PPS/FOK/6) (Continental C85) *(As "E.III 422/15" in German c/s with Flying Aces Movie Aeroplane Collection)*						
G-AVJV	Wallis WA-117 Series 1	K/402/X		12. 4.67	K.H.Wallis	Reymerston Hall, Norfolk	21. 4.89P
	(RR Continental O-200-B) *(Used major parts of G-ATCV c/n 301)*				*(Stored 8.01)*		
G-AVJW	Wallis WA-118/M Meteorite	K/502/X		12. 4.67	K.H.Wallis	Reymerston Hall, Norfolk	21. 4.83P
	(Meteor Alfa 1) *(Originally regd as Wallis WA.118 Srs 2: used major components of G-ATPW c/n 401) (Stored 8.01)*						
G-AVKB	Brochet MB.50 Pipistrelle	02	F-PFAL	17. 4.67	W.B.Cooper	Goodwood	30.10.96P
	(Walter Mikron 3)				*(Noted 1.05)*		
G-AVKD	Fournier RF4D	4024		19. 4.67	R.E.Cross tr Lasham RF4 Group	Lasham	30. 5.04P
G-AVKE*	Gadfly HDW-1				See SECTION 3 Part 1	Weston-super-Mare	
G-AVKG	Cessna F172H	F172-0345		21. 4.67	Springbank Aviation Ltd	(Castletown, Isle Of Man)	10.10.03
	(Built Reims Aviation SA) (Rebuilt with fuselage of G-AVDC c/n F172-0382 @ 1986)						
G-AVKI	Nipper T.66 RA.45 Series 3	S.102/1586		24. 4.67	J.M.Greenway	(Wolverhampton)	7. 8.91P
	(Built Slingsby Sailplanes Ltd from Tipsy c/n 31)(Ardem 4C02)				*(Current status unknown)*		

Registration	Type	C/n	Previous identity	Date	Owner/operator	Base	CofA expiry
G-AVKK	Nipper T.66 RA.45 Series 3	S.104/1588	EI-BJH	24. 4.67	C.Watson	Newtownards	6. 4.05P
	(Built Slingsby Sailplanes Ltd from Tipsy c/n 74) (Ardem 4C02) G-AVKK						
G-AVKL	Piper PA-30 Twin Comanche B	30-1418	OY-DHL	25. 4.67	Bravo Aviation Ltd	Jersey	11. 6.05
			G-AVKL, N8284Y				
G-AVKN	Cessna 401	401-0082	(N3282Q)	26. 4.67	Law Leasing Ltd	Rochester	1. 7.06
G-AVKP	Beagle A 109 Airedale	B.540	SE-EGA	26. 4.67	D.R.Williams	Peplow	26. 9.03
G-AVKR	Bölkow Bö.208C Junior	648	D-EGRA	28. 4.67	C.H.Morris	Deanland	30. 4.07
G-AVKT*	Tipsy Nipper T.66 Series 3	70	OO-HEL	1. 5.67	Not known *(Frame noted 1.02)*	Yearby	
	(Built Cobelavia)		OO-DEL		*(Crashed Constable Burton, Paull, Yorkshire 19.9.72: cancelled 14.2.73 as destroyed)*		
G-AVLB	Piper PA-28-140 Cherokee	28-23158	N11C	8. 5.67	M.Wilson	Sywell	2.12.06T
G-AVLC	Piper PA-28-140 Cherokee	28-23178	N11C	8. 5.67	C.M.Tyers *(New owner 4.04)*	Spanhoe	25. 9.98
G-AVLE	Piper PA-28-140 Cherokee	28-23223	N11C	8. 5.67	G.E.Wright	South Lodge Farm, Widmerpool	23.12.07E
					t/a Video Security Services		
G-AVLF	Piper PA-28-140 Cherokee	28-23268	N11C	8. 5.67	G.H.Hughesdon	White Waltham	29. 3.07T
G-AVLG	Piper PA-28-140 Cherokee	28-23358	N11C	8. 5.67	C.H.R.Hewitt	Poplar Hall Farm, Elmsett	23. 8.06
G-AVLH*	Piper PA-28-140 Cherokee	28-23368		8. 5.67	Not known	North Weald	18. 8.00
					(Cancelled 23.4.02 by CAA) (Noted derelict 1.05		
G-AVLI	Piper PA-28-140 Cherokee	28-23388	N11C	8. 5.67	Lima India Aviation Ltd	Southend	20. 5.07
G-AVLJ	Piper PA-28-140 Cherokee	28-23393	9H-AAZ	8. 5.67	Cherokee Aviation Holdings Jersey Ltd	Jersey	5. 8.05T
			G-AVLJ, N11C				
G-AVLM	Beagle B.121 Pup Series 2	B121-003		8. 5.67	T.M.and D.A Jones	Egginton	29. 4.69S
					(On slow restoration 1.03)		
G-AVLN	Beagle B.121 Pup Series 2	B121-004		8. 5.67	A P.Marks Tr Dogs Flying Group	Sywell	9. 5.07S
G-AVLO	Bölkow Bö.208C Junior	650	D-EGUC	8. 5.67	P.J.Swain	Sandford Hall, Knockin	8. 6.04P
G-AVLT	Piper PA-28-140 Cherokee	28-23328	G-KELC	9. 5.67	B.Maurer	Wellesbourne Mountford	20. 9.04T
			G-AVLT, N11C		*(Noted 1.05)*		
G-AVLY	Wassmer Jodel D.120A Paris-Nice	331		11. 5.67	N.V.de Candole	Loders Hill Farm, Bridport	3. 5.05P
G-AVMA	Sud-Aviation Gardan GY-80-180 Horizon	196		12. 5.67	B.R.Hildick	Shenstone Hall Farm, Shenstone	14. 7.07
G-AVMB	Druine D.62B Condor	RAE/621		12. 5.67	L.J.Dray	Watchford Farm, Yarcombe	9. 9.05P
	(Built Rollason Aircraft and Engines) (Continental C90-14F)				*"Spirit of Silver City"*		
G-AVMD	Cessna 150G	15065504	N2404J	16. 5.67	T.A White t/a Bagby Aviation	Bagby	16. 8.04
G-AVMF	Cessna F150G	F150-0203		17. 5.67	J.F.Marsh	Newton Farm, Sudbury	19. 8.06
	(Built Reims Aviation SA)						
G-AVMJ*	British Aircraft Corporation One-Eleven 510ED	BAC.138		11. 5.67	European Aviation Ltd	Bournemouth	17.11.94T
					(WFU 6.94 and cancelled 11.5.01 by CAA) (Forward fuselage only 1.04)		
G-AVMN	British Aircraft Corporation One-Eleven 510ED	BAC.142		11. 5.67	European Aviation Ltd	Bournemouth	21. 6.00T
					(In open store awaiting scrapping 1.05)		
G-AVMO*	British Aircraft Corporation One-Eleven 510ED				See SECTION 3 Part 1	RAF Cosford	
G-AVMT*	British Aircraft Corporation One-Eleven 510ED	BAC.147		11. 5.67	European Aviation Air Charter Ltd	Cardiff	5.12.03T
					(Cancelled 17.12.04 as WFU) (To Fire Section 11..04)		
G-AVMU*	British Aircraft Corporation One-Eleven 510ED				See SECTION 3 Part 1	Duxford	
G-AVMZ*	British Aircraft Corporation One-Eleven 510ED	BAC.153	(5N-OSA)	11. 5.67	European Aviation Ltd	Not known	17.10.02T
			G-AVMZ		*(Cancelled 10.6.03 by CAA)*		
					(Wings and tail removed: remaining fuselage sold for use as restaurant and left by road 2003)		
G-AVNC	Cessna F150G	F150-0200		18. 5.67	J.R.Alderson	Popham	24. 5.04
	(Built Reims Aviation SA)						
G-AVNE*	Westland WS-58 Wessex 60 Series 1				See SECTION 3 Part 1	Weston-super-Mare	
G-AVNN	Piper PA-28-180 Cherokee C	28-4049	N11C	26. 5.67	J.Acres tr G-AVNN Flying Group	Eaglescott	7. 4.06
G-AVNO	Piper PA-28-180 Cherokee C	28-4105	N11C	26. 5.67	Allister Flight Ltd	Southend	11.10.07E
G-AVNP*	Piper PA-28-180 Cherokee C	28-4113	N11C	26. 5.67	(Southend Airport Fire Services)	Southend	21.10.01T
					(Force landed near Nayland 28.4.01: cancelled 27.11.01 as destroyed) (Wreck used for rescue training 1.05		
G-AVNS	Piper PA-28-180 Cherokee C	28-4129	N11C	26. 5.67	J.G.O'Brien	North Weald	27. 6.06T
G-AVNU	Piper PA-28-180 Cherokee C	28-4153	N11C	26. 5.67	O.Durrani	Lydd	31. 3.07T
G-AVNW	Piper PA-28-180 Cherokee C	28-4210	N11C	26. 5.67	Len Smith's School and Sports Ltd	Fairoaks	3. 8.06T
G-AVNZ	Fournier RF4D	4030		26. 5.67	C.D.Pidler	Franklyn's Field, Chewton Mendip	18. 6.05
G-AVOA	SAN Jodel DR.1050 Ambassadeur	195	F-BJYY	31. 5.67	D.A Willies	Anwick	2.10.06
G-AVOC	CEA Jodel DR.221 Dauphin	67		2. 6.67	J.P.Coulter and J.Chidley	Nuthampstead	21. 3.05
					tr Alpha One Flying Group		
G-AVOH	Druine D.62B Condor	RAE/622		6. 6.67	Halegreen Associates Ltd	Hinton-in-the-Hedges	1. 5.05T
	(Built Rollason Aircraft and Engines)						
G-AVOM	CEA Jodel DR.221 Dauphin	65		6. 6.67	C.J.S.Drewett tr Avon Flying Group	Bidford	29. 7.06
G-AVOO	Piper PA-18-150 Super Cub	18-8511	N10F	7. 6.67	Dublin Gliding Club Ltd	Gowran Grange, County Kildare	1. 5.06
	(Lycoming O-360-A4)						
G-AVOZ	Piper PA-28-180 Cherokee C	28-3711	N9574J	13. 6.67	P.Hoskins and R.Flavell	Booker	28. 5.07
					tr Oscar Zulu Flying Group		
G-AVPD	Jodel D.9 Bébé	521 & PFA 927		15. 6.67	S.W.McKay	(Berkhamsted)	6. 6.75S
	(Built S.W.McKay - c/n MAC.1) (Volkswagen 1500)				*(Stored 12.99)*		
G-AVPI*	Cessna F172H	F172-0409		20. 6.67	R.W.Cope	Water Leisure Park, Skegness	30. 5.03
	(Built Reims Aviation SA)				*(Cancelled 12.2.02 as destroyed).(On rebuild using fuselage and parts ex EI-AOK 4.04)*		
G-AVPJ	de Havilland DH.82A Tiger Moth	86326	NL879	20. 6.67	C C.Silk	Bericote Farm, Blackdown, Leamington Spa	19. 8.07
	(Built Morris Motors Ltd)						
G-AVPM	SAN Jodel D.117	593	F-BHXO	20. 6.67	J.C.Haynes	Breighton	14. 7.05P
G-AVPN*	Handley Page HPR.7 Dart Herald 213				See SECTION 3 Part 1	Elvington	
G-AVPO	Hindustan HAL-26 Pushpak	PK-127	9M-AOZ	31. 3.83	M.B.Johns	(Leamington Spa)	14. 9.05P
	(Continental C90)		VT-DWL				
G-AVPS	Piper PA-30 Twin Comanche B	30-1548	N8393Y	27. 6.67	J.M.Bisco	Farley Farm, Romsey	11.11.05
					(Apparently written off in landing accident Farley Farm 11.04)		
G-AVPV	Piper PA-28-180 Cherokee C	28-2705	9J-RBP	27. 6.67	K.A Passmore	Rayne Hall Farm, Braintree	8. 3.03
			N11C		*(Derelict 1.05)*		
G-AVPY	Piper PA-25-235 Pawnee C	25-4330	N4636Y	7. 7.67	Southdown Gliding Club Ltd	Parham Park	28. 6.06
			N10F				
G-AVRK	Piper PA-28-180 Cherokee C	28-4041	N11C	11. 7.67	J.Gama	Tatenhill	2. 4.06
G-AVRP	Piper PA-28-140 Cherokee	28-23153	N11C	14. 7.67	K.J.Bowers tr Trent 199 Flying Group	Tatenhill	10. 7.06
G-AVRS	Sud-Aviation Gardan GY-80-180 Horizon	224		14. 7.67	C.Clark-Monks and B.A Heath	Alderney	2. 7.06

G-AVRU	Piper PA-28-180 Cherokee C	28-4025	N11C	17. 7.67	D.J.Rowell tr G-AVRU Partnership	Clacton	11.12.05
G-AVRW	Gardan GY-20 Minicab	PFA 1800		18. 7.67	D.J.Smith	Hucknall	2. 9.05P
	(Built R Hart to JB.01 Minicab standard - c/n OH-1549) (Continental C90)				tr Kestrel Flying Group		
G-AVRY	Piper PA-28-180 Cherokee C	28-4089	N11C	24. 7.67	Brigfast Ltd	Popham	9. 4.06
G-AVRZ	Piper PA-28-180 Cherokee C	28-4137	N11C	24. 7.67	Mantavia Group Ltd	Guernsey	26.11.05
G-AVSA	Piper PA-28-180 Cherokee C	28-4184	N11C	24. 7.67	J.Walker tr G-AVSA Flying Group	Barton	5. 5.05
G-AVSB	Piper PA-28-180 Cherokee C	28-4191	N11C	24. 7.67	D.L.Macdonald	Denham	2. 5.05
G-AVSC	Piper PA-28-180 Cherokee C	28-4193	N11C	24. 7.67	MSC019 Ltd	White Waltham	22. 6.06T
G-AVSD	Piper PA-28-180 Cherokee C	28-4195	N11C	24. 7.67	Landmate Ltd	Haverfordwest	22.11.07E
G-AVSE	Piper PA-28-180 Cherokee C	28-4196	N11C	24. 7.67	F.Glendon *(Noted 7.04)*	Kilrush, County Kildare	30. 4.00T
G-AVSF	Piper PA-28-180 Cherokee C	28-4197	N11C	24. 7.67	S.E.Pick and D.A Rham tr Monday Club	Blackbushe	28. 4.07
G-AVSI	Piper PA-28-140 Cherokee	28-23148	N11C	24. 7.67	C.M.Royle tr G-AVSI Flying Group	White Waltham	1. 4.07
G-AVSP	Piper PA-28-180 Cherokee C	28-3952	N11C (PJ-ACT)	8. 8.67	Airways Flight Training (Exeter) Ltd	Exeter	28. 1.07T
G-AVSR	Beagle D.5/180 Husky	3689		8. 8.67	A.L.Young	Henstridge	6. 4.06A
G-AVSZ	Agusta-Bell 206B JetRanger II	8032	VH-BEQ	8. 8.67	Patriot Aviation Ltd	Cranfield	16. 6.99T
			PK-HBZ, VR-BCR, PK-HBD, VR-BCR, G-AVSZ *(Amended CofR 3.02)*				
G-AVTC	Nipper T.66 RA.45 Series 3	S.106/1583		9. 8.67	M.W.Bodger	Yeatsall Farm, Abbots Bromley	1. 6.79P
	(Built Slingsby Aircraft Co Ltd) (Ardem Mk.10)						
G-AVTL*	Brighton Ax7-65 HAB				See SECTION 3 Part 1	Newbury	
G-AVTP	Cessna F172H	F172-0458		17. 8.67	M.J.Green and Virginia Ann Newbold	White Waltham	16. 8.07
	(Built Reims Aviation SA)				tr Tango Papa Group		
G-AVTT*	Ercoupe 415D	4399	SE-BFZ	21. 8.67	Wright Farm Eggs Ltd	Cherry Tree Farm, Monewden	20. 1.86
	(Continental C85)		NC3774H		*(Stored 6.00) (Cancelled 12.4.02 as temporarily WFU)*		
G-AVTV	SOCATA MS.893A Rallye Commodore 180	10725		24. 8.67	Staffordshire Gliding Club Ltd	Seighford	6. 8.06
G-AVUD	Piper PA-30 Twin Comanche B	30-1515	N8422Y N9???N	5. 9.67	P.M.Fox	Biggin Hill	4.11.07E
G-AVUG	Cessna F150H	F150-0234		11. 9.67	V.J.Larkin and P.Woodburn	Gamston	9. 6.05
	(Built Reims Aviation SA)				tr Skyways Flying Club		
G-AVUH	Cessna F150H	F150-0244		11. 9.67	C.M.Chinn	Strubby	26. 8.07
	(Built Reims Aviation SA)						
G-AVUO	Phoenix Luton LA.4A Minor	PAL/1313		21. 9.67	M.E.Vaisey	(Hemel Hempstead)	
			(Initially not completed: parts used in construction of G-AXKH - possible long-term build project)				
G-AVUS	Piper PA-28-140 Cherokee	28-24065	(G-AVUT) N11C	25. 9.67	D.J.Hunter	Shipdham	6.12.04T
G-AVUT	Piper PA-28-140 Cherokee	28-24085	(G-AVUU) N11C	25. 9.67	Bencray Ltd	Blackpool	20. 7.07T
					(Op Blackpool and Fylde Aero Club)		
G-AVUU	Piper PA-28-140 Cherokee	28-24100	(G-AVUS) N11C	25. 9.67	R.W.Harris, A Jahanfar, P.E.Preston and M.F.Hatt		
					(Op Southend Flying Club)	Southend	11. 5.06T
G-AVUZ	Piper PA-32-300 Cherokee Six	32-40302	N11C	29. 9.67	Ceesix Ltd	Jersey	23. 4.06
G-AVVC	Cessna F172H	F172-0443		29. 9.67	M.Turnbull	Eshott	17. 6.06T
	(Built Reims Aviation SA)						
G-AVVI*	Piper PA-30-160 Twin Comanche B	30-1613	EI-AVD G-AVVI, N8454Y	5.10.67	Not known	Walton New Road Business Park, Bruntingthorpe	7. 3.94
					(Cancelled as WFU 29.9.97: fuselage in scrapyard 2.04)		
G-AVVJ	SOCATA MS.893A Rallye Commodore 180	10752		6.10.67	M.Powell	Tibenham	29. 7.05
G-AVVL	Cessna F150H	F150-0257		6.10.67	Team Arena Ltd	Gamston	29.10.06T
	(Built Reims Aviation SA) (Wilksch WAM-120)				*"Samurai"*		
G-AVVO*	Avro 652A Anson 19 Series 2				See SECTION 3 Part 1	Newark	
G-AVVR*	Avro 652A Anson 19 Series 2				See SECTION 3 Part 1	Bilbrook, Wolverhampton	
G-AVWA	Piper PA-28-140 Cherokee	28-23660	N11C	19.10.67	SFG Ltd	Shipdham	27. 2.06T
G-AVWD	Piper PA-28-140 Cherokee	28-23700	N11C	19.10.67	C.Bentley and B.Marlowe t/a Evelyn Air	Leeds-Bradford	30. 9.04T
G-AVWE*	Piper PA-28-140 Cherokee	28-23720	N11C	19.10.67	Not known	Blackpool	22. 4.82T
			(WFU and cancelled 8.6.89 by CAA) (Fuselage noted 3.00: current status unknown)				
G-AVWI	Piper PA-28-140 Cherokee	28-23800	N11C	19.10.67	Mrs.L.M.Middleton	Cranfield	17. 2.06
G-AVWJ	Piper PA-28-140 Cherokee	28-23940	N11C	19.10.67	A.M.Harrhy	Sandown	29. 7.04
G-AVWL	Piper PA-28-140 Cherokee	28-24000	N11C	19.10.67	S.H.and Claire Louise Maynard	Durham Tees Valley	4.10 07
G-AVWM	Piper PA-28-140 Cherokee	28-24005		19.10.67	A.Jahanfar, P.E.Preston, M.F.Hatt and R.W.Harris		
					(Op Southend Flying Club)	Southend	14. 7.07T
G-AVWN	Piper PA-28R-180 Cherokee Arrow	28R-30170	N11C	19.10.67	Vawn Air Ltd	Jersey	10. 4.05
G-AVWO	Piper PA-28R-180 Cherokee Arrow	28R-30205	N11C	19.10.67	I.P.Scobell	(South Godstone)	9. 2.07
G-AVWR	Piper PA-28R-180 Cherokee Arrow	28R-30242	N11C	19.10.67	S.J.French tr SJ French and Partners	Dunkeswell	14. 6.07
G-AVWT	Piper PA-28R-180 Cherokee Arrow	28R-30362	N11C	19.10.67	Cloudbase Aviation Ltd	Barton	21. 5.06
G-AVWU	Piper PA-28R-180 Cherokee Arrow	28R-30380	N11C	19.10.67	A M.Alam	Elstree	23. 4.06T
G-AVWV	Piper PA-28R-180 Cherokee Arrow	28R-30404	N11C	19.10.67	R.V.Thornton and R.Barron	Perth	18. 6.05
					tr Strathtay Flying Group		
G-AVWY	Fournier RF4D	4031		26.10.67	P.Turner	(Gloucestershire)	4. 7.04P
			(U/c raised too early on take-off Halesland 24.8.03 and badly damaged: damaged remains noted 9.04)				
G-AVXA	Piper PA-25-235 Pawnee C	25-4244	N4576Y	26.10.67	South Wales Gliding Club Ltd	Usk	21. 5.06
	(Re-built using new frame - c/n unknown)						
G-AVXC	Nipper T.66 RA.45 Series 3	S.108/1605		26.10.67	P.A Gibbs	(Inverness)	2. 5.04P
	(Built Slingsby Aircraft Co Ltd) (Ardem 4C02)						
G-AVXD	Nipper T.66 RA.45 Series 3	S.109/1606		26.10.67	R.L.Fraser	Dundee	19.12.03P
	(Built Slingsby Aircraft Co Ltd) (Volkswagen 1834 Acro)				tr Tayside Nipper Group		
G-AVXF	Piper PA-28R-180 Cherokee Arrow	28R-30044	N11C	26.10.67	A D.C.McNeile tr GAVXF Group	North Weald	24. 6.07
G-AVXI*	Hawker Siddeley HS.748 Series 2A/238	1623		2.11.67	Hanningfield Metals	Templewood, Stock	30. 8.98T
					(Cancelled 24.10.01 by CAA) (Fuselage noted 2.04)		
G-AVXV*	Bleriot XI				See SECTION 3 Part 1	La Baule, Loire-Atlantique, France	
G-AVXW	Druine D.62B Condor	RAE/625		3.11.67	A J.Cooper	Rochester	30. 9.01
	(Built Rollason Aircraft and Engines)						
G-AVXY	Auster AOP.9	AUS.10/92	XK417	7.11.67	G.J. Siddall	South Lodge Farm, Widmerpool	9. 7.00P
	(Officially regd as c/n AUS/120)				*(New owner 8.04) (As "XK417" in Army c/s)*		
G-AVYL	Piper PA-28-180 Cherokee D	28-4622	N11C	24.11.67	N.E.Binner	Full Sutton	23. 5.05
G-AVYM	Piper PA-28-180 Cherokee D	8-4638	N11C	24.11.67	Carlisle Aviation (1985) Ltd	Carlisle	14. 5.07T

Registration	Type	C/n	Previous identity	Date	Owner/Operator	Base	CofA
G-AVYP*	Piper PA-28-140 Cherokee	28-24211	N11C	4.11.67	K.Hobbs	Belfast Aldergrove	14. 2.04T
					t/a Aldergrove Flight Training Centre *(Cancelled 3.7.03 by CAA)*		
G-AVYR	Piper PA-28-140 Cherokee	28-24226	N11C	24.11.67	P.J. Huxley tr SAS Flying Group	Compton Abbas	18. 6.06
G-AVYS	Piper PA-28R-180 Cherokee Arrow	28R-30456	N11C	24.11.67	Musicbank Ltd	Ludham	5. 3.06T
G-AVYT	Piper PA-28R-180 Cherokee Arrow	28R-30472	N11C	24.11.67	J.R.Tindale	Blackpool	15. 7.06
G-AVYV	Wassmer Jodel D.120A Paris-Nice	252	F-BMAM	27.11.67	A J.Sephton	Brickhouse Farm, Frogland Cross	20. 7.04P
G-AVZB*	LET Z-37 Cmelak				See SECTION 3 Part 1	Wroughton	
G-AVZI	Bölkow Bö.208C Junior	673	D-EGZF	19.12.67	C.F.Rogers *(Stored 10.00)*	(Wheathampstead)	24. 7.76
G-AVZN	Beagle B.121 Pup Series 1	B121-006		19.12.67	B.L.Elvy	Shipdham	16. 8.04
					tr Shipdham Aviators Flying Group		
G-AVZP	Beagle B.121 Pup Series 1	B121-008		19.12.67	T.A White	Bagby	21. 9.07
G-AVZR	Piper PA-28-180 Cherokee C	28-4114	N4779L	19.12.67	Lincoln Aero Club Ltd	Sturgate	12. 6.06T
G-AVZU	Cessna F150H	F150-0283		29.12.67	R.D.Forster	Beccles	23.12.05T
	(Built Reims Aviation SA)				*(Op Norfolk and Norwich Aero Club)*		
G-AVZV	Cessna F172H	F172-0511		29.12.67	E.L.and D.S.Lightbown	Crosland Moor	12. 1.07
	(Built Reims Aviation SA)						
G-AVZW	EAA Biplane Model P	PFA 1314		29.12.67	R.G.Maidment and G.R.Edmondson		
	(Built R.G.Maidment) (Lycoming O-290)					Lower Wasing Farm, Brimpton	17. 6.05P
G-AVZX	SOCATA MS.880B Rallye Club	1165		29.12.67	J.Nugent	Kilrush, County Kildare	19.11.06

G-AWAA - G-AWZZ

Registration	Type	C/n	Previous identity	Date	Owner/Operator	Base	CofA
G-AWAC	Sud-Aviation Gardan GY-80-180 Horizon	234		29.12.67	P.B.Hodgson *"Le Fantome"*	(Tewkesbury)	11. 6.04
					(Force landed wheels-up Semley, Wilts 22.7.03 and badly damaged: new owner 3.04)		
G-AWAH	Beech D55 Baron	TE-540		1. 1.68	B.J.S.Grey	Duxford	1. 4.07
G-AWAJ	Beech D55 Baron	TE-536		1. 1.68	Aflex Hose Ltd	Blackpool	29. 7.07E
G-AWAT	Druine D.62B Condor	RAE/627		8. 1.68	B.Woolford	Sandown	16. 7.04
	(Built Rollason Aircraft and Engines)						
G-AWAU*	Vickers FB.27A Vimy rep				See SECTION 3 Part 1	Hendon	
G-AWAW*	Reims/Cessna F150F				See SECTION 3 Part 1	South Kensington, London	
G-AWAX	Cessna 150D	15060153	OY-TRJ	5. 1.68	Joyce Haunch	Fenland	1.12.07E
	(Tail-wheel conversion)		N4153U				
G-AWAZ	Piper PA-28R-180 Cherokee Arrow	28R-30512	N11C	8. 1.68	Aero Anglia Ltd	Poplar Hall Farm, Elmsett	28. 8.05
G-AWBA	Piper PA-28R-180 Cherokee Arrow	28R-30528	N11C	8. 1.68	A.Taplin and G.A Dunster	Stapleford	20. 2.06
					tr March Flying Group		
G-AWBB	Piper PA-28R-180 Cherokee Arrow	28R-30552	N11C	8. 1.68	D.L.Claydon	(Sudbury)	24. 6.05
G-AWBC	Piper PA-28R-180 Cherokee Arrow	28R-30572	N11C	8. 1.68	Anglo Aviation (UK) Ltd	Bournemouth	28.12.06
G-AWBE	Piper PA-28-140 Cherokee	28-24266	N11C	8. 1.68	B.E.Boyle	Shenington	24.11.05
G-AWBG	Piper PA-28-140 Cherokee	28-24286	N11C	8. 1.68	G.D.Cooper	Rochester	5. 5.07T
G-AWBH	Piper PA-28-140 Cherokee	28-24306	N11C	8. 1.68	Proofgolden Ltd t/a Mainstreet Aviation	Newcastle	30.12.04T
G-AWBJ	Fournier RF4D	4055		12. 1.68	J.M.Adams	RAF Syerston	25. 4.05P
G-AWBM	Druine D.31A Turbulent	PFA 1647		17. 1.68	A D.Pratt	North Coates	20. 7.95P
	(Built J R D Bygrave) (Volkswagen 1700)				*(New owner 8.04)*		
G-AWBN	Piper PA-30 Twin Comanche B	30-1472	N8517Y	18. 1.68	Stourfield Investments Ltd	Jersey	2.12.05
G-AWBS	Piper PA-28-140 Cherokee	28-24331	N11C	22. 1.68	M.A English and T.M.Brown	Little Snoring	18.12.05
G-AWBU	Morane Saulnier Type N Rep	PPS/REP/7		22. 1.68	Bianchi Aviation Film Services Ltd	Compton Abbas	28. 4.04P
	(Built D E Bianchi) (Continental C90-8F)				*(As "MS824" in French AF c/s with Flying Aces Movie Aeroplane Collection)*		
G-AWBX	Cessna F150H	F150-0286		22. 1.68	J.Meddings	Tatenhill	17. 9.06
	(Built Reims Aviation SA)						
G-AWCN	Reims FR172E Rocket	FR17200020		25. 1.68	R.C.Lunnon and A J.Speight	Stapleford	4. 7.07
G-AWCP	Cessna F150H	F150-0354		29. 1.68	C.E.Mason	Sleap	12. 2.06
	(Built Reims Aviation SA) (Tail-wheel conversion)						
G-AWCR*	Piccard Ax6 HAB				See SECTION 3 Part 1	Newbury	
G-AWDA	Nipper T.66 RA.45 Series 3	S.117/1624		7. 2.68	J.A Cheesbrough	Ottringham	4. 6.05P
	(Built Slingsby Aircraft Co Ltd) (Volkswagen Acro 1834)						
G-AWDO	Druine D.31 Turbulent	PFA 1649		21. 2.68	R.N.Crosland	Deanland	9. 5.05P
	(Built R Watling-Greenwood) (Volkswagen 1600)						
G-AWDP	Piper PA-28-180 Cherokee D	28-4870	N11C	21. 2.68	B.H. and P.M.Illston	Norwich	12. 2.05T
					(Op Norwich School of Flying)		
G-AWDR	Reims FR172E Rocket	FR17200004		21. 2.68	B.A Wallace	Nuthampstead	9. 4.07
G-AWDU	Brantly B.2B	481		23. 2.68	B.M.Freeman	(Stourport-on-Severn)	17. 3.06
G-AWDW	Campbell-Bensen CB.8MS	DS.1330		26. 2.68	M.R.Langton	(Taplow)	7.10.71P
	(Built D J C Summerfield) (McCulloch.Motors 4318C)				*(Stored 12.00)*		
G-AWEF	SNCAN Stampe SV-4C	549	F-BDCT	29. 3.68	R.A.F.Buchanan	Headcorn	19.12.07T
	(DH Gipsy Major 10)						
G-AWEI	Druine D.62B Condor	RAE/628		6. 3.68	Alison M.Noble	Thruxton	10.11.98T
	(Built Rollason Aircraft and Engines)				*(New owner 12.04)*		
G-AWEL	Fournier RF4D	4077		7. 3.68	A B.Clymo	Wolverhampton	1. 8.05P
G-AWEM	Fournier RF4D	4078		7. 3.68	B.J.Griffin	Wickenby	11. 6.05P
G-AWEP	Gardan GY-20 Minicab	PFA 1801		12. 3.68	A.Louth	(Boston)	17. 8.04P
	(Built S Jackson to JB.01 Minicab standard) (Continental C90)						
G-AWES	Cessna 150H	15068626	N22933	20. 3.68	P.Montgomery-Stuart	Netherthorpe	5. 8.84
					(Noted re-built 9.04)		
G-AWET	Piper PA-28-180 Cherokee D	28-4871	N11C	21. 3.68	Broadland Flying Group Ltd	Old Buckenham	25. 5.06
G-AWEV	Piper PA-28-140 Cherokee	28-24460	N11C	21. 3.68	Norflight Ltd *(Noted 4.04)*	Shipdham	6. 1.01
G-AWEX	Piper PA-28-140 Cherokee	28-24472	N11C	21. 3.68	N.D. Wyndow	Coventry	27. 5.07
					tr Sir W.G.Armstrong-Whitworth Flying Group		
G-AWEZ	Piper PA-28R-180 Cherokee Arrow	28R-30592	N11C	21. 3.68	T.R.Leighton, R.G.E.Simpson and D.A C.Clissett	Stapleford	14. 1.06
G-AWFB	Piper PA-28R-180 Cherokee Arrow	28R-30689	N11C	21. 3.68	J.C.Luke	Filton	17. 4.05
G-AWFC	Piper PA-28R-180 Cherokee Arrow	28R-30670	N11C	21. 3.68	B.J.Hines	White Waltham	30. 9.07E
G-AWFD	Piper PA-28R-180 Cherokee Arrow	28R-30669	N11C	21. 3.68	D.J.Hill	Moorlands Farm, Farway Common	26. 5.05
G-AWFF	Cessna F150H	F150-0280		25. 3.68	Westflight Aviation Ltd	Gloucestershire	17.10.05T
	(Built Reims Aviation SA)						

G-AWFJ	Piper PA-28R-180 Cherokee Arrow	28R-30688	N11C	26. 3.68	Parplon Ltd	Barton	24 3.05
G-AWFM*	de Havilland DH.104 Dove 6				See SECTION 3 Part 1	Johannesburg, South Africa	
G-AWFN	Druine D.62B Condor *(Built Rollason Aircraft and Engines)*	RAE/629		27. 3.68	P.B.Lowry	Deanland	3. 7.05P
G-AWFO	Druine D.62B Condor *(Built Rollason Aircraft and Engines)*	RAE/630		27. 3.68	R.E.Major	Trevissick Farm, Porthtowan	5.11.05P
G-AWFP	Druine D.62B Condor *(Built Rollason Aircraft and Engines)*	RAE/631		27. 3.68	D.J.Taylor tr Blackbushe Flying Club	White Waltham	6. 7.07
G-AWFT	Jodel D.9 Bébé *(Built W.H.Cole)* (Volkswagen 1200)	PFA 932		29. 3.68	W.H.Cole *(Noted 5.01)*	Spilsteads Farm, Sedlescombe	22. 7.69P
G-AWFW	SAN Jodel D.117	599	PH-VRE F-BHXU	2. 4.68	C.J.Rodwell	Hawksbridge Farm, Oxenhope	30. 8.05P
G-AWFZ	Beech 19A Musketeer Sport	MB-323	N2811B	3. 4.68	Bob Crowe Aircraft Sales Ltd	Cranfield	18.12.05T
G-AWGA*	Beagle A 109 Airedale	B.535	EI-ATA G-AWGA, D-ENRU	3. 4.68	(J.R.Bowden) *(Cancelled 30.9.86 as WFU) (Stored for spares 2004?)*	(Headcorn)	
G-AWGD	Cessna F172H *(Built Reims Aviation SA)*	F172-0503		5. 4.68	R.P.Vincent	Shoreham	18. 7.06T
G-AWGK	Cessna F150H *(Built Reims Aviation SA)*	F150-0347		8. 4.68	G.E.Allen	(Saxilby, Lincoln)	1. 5.07
G-AWGN	Fournier RF4D	4084		9. 4.68	M.P.Barley	(Huntingdon)	9. 6.05P
G-AWGZ	Taylor JT.1 Monoplane *(Built J Morris - c/n M.1)* (Ardem 4C02)	PFA 1406		17. 4.68	R.L.Sambell	Stoke Golding	21. 7.05P
G-AWHA*	CASA 2111D				See SECTION 3 Part 1	Munich, Germany	
G-AWHB*	CASA 2111D (Heinkel 111H-16) *(Officially quoted as c/n 167 ex Spanish AF B2I-37)* *(Fuselage noted 9.02 - to create flyable Heinkel 111 for Flying Heritage Collection, Seattle, US with wings ex CASA 2111 B21-39 and Junkers Jumo engines)*	049	Spanish AF B2I-57	14. 5.68	Not known *(Cancelled 27.4.01 as sold US)*	Great Massingham	
G-AWHE*	Hispano HA.1112-MIL Buchon				See SECTION 3 Part 1	Mesa, Arizona, US	
G-AWHG*	Hispano HA.1112-MIL Buchon				See SECTION 3 Part 1	La Ferté-Alais, Essonne, France	
G-AWHJ*	Hispano HA.1112-MIL Buchon				See SECTION 3 Part 1	Kalamazoo, Michigan, US	
G-AWHL*	Hispano HA.1112-MIL Buchon				See SECTION 3 Part 1	Mesa, Arizona, US	
G-AWHN*	Hispano HA.1112-MIL Buchon				See SECTION 3 Part 1	Tillamook, Oregon, US	
G-AWHO*	Hispano HA.1112-MIL Buchon				See SECTION 3 Part 1	Oshkosh, Wisconsin, US	
G-AWHS*	Hispano HA.1112-MIL Buchon				See SECTION 3 Part 1	Schleissheim, Germany	
G-AWHX	Rollason Beta B.2	RAE/04	(G-ATEE)	17. 4.68	S.G.Jones *"Vertigo" (On rebuild 10.01)*	Membury	14. 6.87P
G-AWHY	Falconar F-11-3 *(Built A E Pritchard and A E Riley-Gale)* (Continental C90)	PFA 1322	G-BDPB (G-AWHY)	17. 4.68	J.Porthouse tr Why Fly Group	Goodwood	29.10.04P
G-AWID*	Britten-Norman BN-2A Islander				See SECTION 3 Part 1	Khlong Luang, Thailand	
G-AWIF	Brookland Mosquito *(Built Brooklands Aero Ltd)*	LC-1 & 3		17. 4.68	C.A Reeves *(Noted 10.03)*	Henstridge	7. 1.82P
G-AWII	Vickers Supermarine 349 Spitfire LF.Vc *(Built Westland Aircraft Ltd)*	WASP/20/223	AR501	25. 4.68	Richard Shuttleworth Trustees *(As "AR501/NN-A" in 310 Sqdn c/s)*	Old Warden	31. 3.05P
G-AWIJ*	Vickers Supermarine 329 Spitfire IIA				See SECTION 3 Part 1	RAF Coningsby	
G-AWIP	Phoenix Luton LA-4A Minor *(Built T Reagan c/n PAL/1308)* (Continental A65)	PFA 830		30. 4.68	J.Houghton *(Damaged near Holme-on-Spalding Moor 20.7.88: stored 2000)*	(North Ferriby)	8. 5.89P
G-AWIR	Bushby-Long Midget Mustang *(Built A F Jarman and Co Ltd)* (Continental O-200-A)	PFA 1315		30. 4.68	K.E.Sword *(On overhaul 1991: current status unknown)*	Leicester	6. 3.90P
G-AWIT	Piper PA-28-180 Cherokee D	28-4987	N11C	30. 4.68	Cherry Orchard Aparthotel Ltd *(Op Manx Flyers Aero Club)*	Ronaldsway	11. 3.07T
G-AWIV	Storey TSR.3 *(Built J M Storey)* (Continental PC60) *(Off icially regd as c/n "1325")*	PFA 1325		30. 4.68	F.R.Hutchings *"Stor"*	Polgear Farm, Four Lanes, Polgear	17. 6.05P
G-AWIW	SNCAN Stampe SV-4B (DH Gipsy Major 10)	532	F-BDCC	2. 5.68	R.E.Mitchell *(Believed stored 4.02)*	RAF Cosford	6. 5.73
G-AWJA*	Cessna 182L Skylane	18258883	N1658C	3. 5.68	Not known *(Cancelled as destroyed 29.11.88) (Wreck noted 5.04)*	Movenis	21.4.85
G-AWJB*	Brighton MAB-65 HAB				See SECTION 3 Part 1	Newbury	
G-AWJE	Nipper T.66 RA.45 Series 3 *(Built Slingsby Aircraft Co Ltd)* (Volkswagen 1834)	S.121/1628		8. 5.68	Katrine G.G.Howe	Barton	14.10.05P
G-AWJF*	Nipper T.66 RA.45 Series 3 *(Built Slingsby Aircraft Co Ltd)*	S.122/1629		8. 5.68	S.Maric *(Cancelled 17.9.91 by CAA) (Stored 2004)*	(Stewarton)	7. 6.88P
G-AWJV*	de Havilland DH.98 Mosquito TT.35				See SECTION 3 Part 1	Salisbury Hall, London Colney	
G-AWJX	Moravan Zlin Z-526 Trener Master	1049		22. 5.68	P.A Colman *(New owner 11.02)*	Luxters Farm, Hambleden, Henley-on-Thames	29. 5.85A
G-AWJY	Moravan Zlin Z-526 Trener Master	1050		22. 5.68	M.Gainza	White Waltham	26. 4.03
G-AWKD	Piper PA-17 Vagabond (Continental A65)	17-192	F-BFMZ N4892H	27. 5.68	A T.and Mrs.M.R.Dowie	Scotland Farm, Hook	12. 9.05P
G-AWKM*	Beagle B.121 Pup Series 1	B121-017		11. 6.68	(D.M.G.Jenkins) *(Damaged Swansea 7.91: cancelled 28.5.02 by CAA) (Fuselage only 2.05)*	Bourne Park, Hurstbourne Tarrant	29. 6.84
G-AWKO	Beagle B.121 Pup Series 1	B121-019		11. 6.68	Sarah E.Ford	(Ely)	7. 6.04T
G-AWKT	SOCATA MS.880B Rallye Club	1235		17. 6.68	A.Ringland and P.Keating	Movenis	26. 3.06
G-AWLA	Cessna F150H *(Built Reims Aviation SA)*	F150-0269	N13175	27. 6.68	T.A White t/a Bagby Aviation	Bagby	19. 7.07T
G-AWLF	Cessna F172H *(Built Reims Aviation SA)*	F172-0536		27. 6.68	H.Sharp and A.Mackey tr Gannet Aviation	City of Derry	5. 1.07
G-AWLG	SIPA 903	82	F-BGHG	27. 6.68	S.W.Markham *(Stored 1997: current status unknown)*	Valentine Farm, Odiham	22. 8.79P
G-AWLI	Piper PA-22-150 Tri-Pacer	22-5083	N7256D	1. 7.68	J.S.Lewery *"Little Peach"*	Shoreham	15. 8.02
G-AWLO	Boeing Stearman E75 (PT-13D) Kaydet (Pratt and Whitney R985)	75-5563	5Y-KRR VP-KRR, 42-17400	9. 7.68	N.D.Pickard *(Op Sky High Advertising Ltd)*	Gloucestershire	16 5.05
G-AWLP	Mooney M.20F Executive 21	680200		9. 7.68	I.C.Lomax *(Noted 1.05)*	Gamston	7. 7.00
G-AWLR	Nipper T.66 RA.45 Series 3 *(Built Slingsby Aircraft Co Ltd)* (Ardem 4C02)	S.125/1662		9. 7.68	T.D.Reid	Newtownards	16. 5.05P
G-AWLS	Nipper T.66 RA.45 Series 3 *(Built Slingsby Aircraft Co Ltd)* (Ardem Mk.10)	S.126/1663		9. 7.68	G.A Dunster and B.Gallagher *(Damaged Stapleford 14.1.88: on rebuild 1995: current status unknown)*	(Loughton, Essex)	25. 3.88P
G-AWLZ	Fournier RF4D	4099		12. 7.68	J.H.Taylor tr Nympsfield RF4 Group	Nympsfield	19. 2.05P

Registration	Type	C/n	Previous identity	Reg date	Owner	Location	CofA
G-AWMD	Jodel D.11	PFA 904		19. 7.68	Rachel M.Worth	Leicester	25. 4.05P
	(Built F H French) (Continental C90)				*(Noted 9.04)*		
G-AWMF	Piper PA-18-150 Super Cub	18-8674	N4356Z	23. 7.68	Booker Gliding Club Ltd	Booker	26. 8.06
	(Lycoming O-360-A4)						
G-AWMI	AESL Airtourer T2 (115)	505		24. 7.68	M.Furse	Cardiff	10. 5.04
G-AWMN	Phoenix Luton LA-4A Minor	PFA 827		30. 7.68	B.J.Douglas	Kilrush, County Kildare	8. 8.05P
	(Built R Wilks) (Volkswagen 1800)						
G-AWMO*	Omega O-84 HAB				See SECTION 3 Part 1	Newbury	
G-AWMP	Cessna F172H	F172-0488		31. 7.68	R.J.D.Blois	Yoxford, Saxmundham	23.12.05
	(Built Reims Aviation SA)						
G-AWMR	Druine D.31 Turbulent	PFA 1661		1. 8.68	M.J.Freeman	Shobdon	24. 2.05P
	(Built S J Hargreaves- c/n 43) (Volkswagen 1390)				*"Demelza"*		
G-AWMZ*	Cessna F172H	F172-0554		2. 8.68	Not known	Cark-in-Cartmel	
	(Built Reims Aviation SA)	*(Hit ground Bucknarrowbridge, Bootle 18.1.76: cancelled 1.9.81 as WFU) (Used as parachute club training aid 7.03)*					
G-AWNT	Britten-Norman BN-2A Islander	32		2. 8.68	Sterling Helicopters Ltd	Norwich	2. 9.07A
G-AWOA	SOCATA MS.880B Rallye Club	1258		2. 8.68	J.A Rimmer	RAF Mona	19.11.05
G-AWOE	Aero Commander 680E	680E-753-41	N3844C	5. 8.68	J.M.Houlder t/a Elstree Flying Club	Elstree	19. 3.06
G-AWOF	Piper PA-15 Vagabond	15-227	F-BETF	6. 8.68	C.M.Hicks	Barton	8. 7.05P
	(Continental C90)						
G-AWOH	Piper PA-17 Vagabond	17-191	F-BFMY	6. 8.68	I.M.Callier	(Hampshire)	24. 7.03P
	(Continental C90)		N4891H	*(On rebuild 2004 after suffering fire damage High Flatts Farm, Chester-le-Street 2003?)*			
G-AWOK*	Sussex Gas (Free) Balloon				See SECTION 3 Part 1	Newbury	
G-AWON*	English Electric Lightning F.53	95291	G-27-56	9. 8.68	See SECTION 3 Part 1	Norwich	
G-AWOT	Cessna F150H	F150-0389		14. 8.68	M.J.Willoughby	Turweston	28. 1.07T
	(Built Reims Aviation SA)						
G-AWOU	Cessna 170B	25829	VQ-ZJA	16. 8.68	S.Billington	Ashcroft Farm, Winsford	10. 4.06
			ZS-CKY, CR-ADU, N3185A				
G-AWOX*	Westland Wessex 60 Series 1	WA/686	G-17-2	28. 8.68	Paintball Adventure West	Bristol	13. 1.83
			G-AWOX, 5N-AJO, G-AWOX, 9Y-TFB, G-AWOX, VH-BHE(3), G-AWOX, VR-BCV, G-AWOX, G-17-1				
					(Cancelled 23.11.82 as TWFU) (Extant 2.04)		
G-AWPH	Percival P.56 Provost T.1	PAC/F/003	WV420	6. 9.68	J.A.D.Bradshaw	Three Mile Cross, Reading	28. 6.05P
G-AWPJ	Cessna F150H	F150-0376		9. 9.68	W.J.Greenfield	Humberside	24. 4.05T
	(Built Reims Aviation SA)				*(Op Humberside Flying Club)*		
G-AWPN	Shield Xyla	PFA 1320		13. 9.68	K.R.Snell	Deanland	23. 6.05P
	(Built G W Shield - c/n 2) (Continental A65)						
G-AWPS	Piper PA-28-140 Cherokee	28-20196	5N-AEK	16. 9.68	A R.Matthews	Sittles Farm, Alrewas	17. 3.07
G-AWPU	Cessna F150J	F150-0411		18. 9.68	LAC (Enterprises) Ltd	Barton	12. 2.07T
	(Built Reims Aviation SA)				*(Op Lancashire Aero Club)*		
G-AWPW	Piper PA-12 Super Cruiser	12-3947	N78572	23. 9.68	AK Leasing (Jersey) Ltd	Jersey	5. 4.07
			NC78572				
G-AWPY	Campbell-Bensen B.8M	CA/314		20. 9.68	J.Jordan	Melrose Farm, Melbourne	
					(Current status unknown)		
G-AWPZ	Andreasson BA-4B	1	SE-XBS	24. 9.68	J.M.Vening	(Lyminster, Littlehampton)	5.11.03P
	(Built B Andreasson) (Continental O-200A)				*(Current status unknown)*		
G-AWRK	Cessna F150J	F150-0410		8.10.68	Systemroute Ltd	Shoreham	23. 7.06T
	(Built Reims Aviation SA)				*(Op Southern Strut Flying Group)*		
G-AWRP*	Cierva Rotorcraft CR.LTH.1 Grasshopper III				See SECTION 3 Part 1	Weston-super-Mare	
G-AWRS*	Avro 652A Anson C.19 Series 2				See SECTION 3 Part 1	Usworth, Sunderland	
G-AWRY	Hunting Percival P.56 Provost T.1	PAC/F/339	XF836	29.10.81	Sylmar Aviation and Services Ltd		
			8043M		*(As "XF836")*	Lower Wasing Farm, Brimpton	22. 8.88P
					(Damaged near Newbury 28.7.87: on rebuild 6.94: current status unknown)		
G-AWSA*	Avro 652A Anson C.19/2				See SECTION 3 Part 1	Flixton, Bungay	
G-AWSH	Moravan Zlin Z-526 Trener Master	1052	OK-XRH	23.11.68	Avia Special Ltd	White Waltham	23.12.04T
			G-AWSH				
G-AWSL	Piper PA-28-180 Cherokee D	28-4907	N11C	30.10.68	Fascia Services Ltd	King's Farm, Thurrock	14.12.06
G-AWSM	Piper PA-28-235 Cherokee C	28-11125	N11C	30.10.68	N.A.Wright t/a Aviation Projects	Shoreham	13. 5.07T
G-AWSN	Druine D.62B Condor	RAE/632		31.10.68	M.K.A Blyth	Little Gransden	19. 1.05P
	(Built Rollason Aircraft and Engines)						
G-AWSP	Druine D.62B Condor	RAE/634		31.10.68	R.Q. and A S.Bond	Enstone	23. 1.95
	(Built Rollason Aircraft and Engines)				*(Stored 6.03)*		
G-AWSS	Druine D.62B Condor	RAE/636		31.10.68	N.J. and D.Butler	(Fordoun)	19.10.94P
	(Built Rollason Aircraft and Engines)				*(Stored 3.98:current status unknown - at owner's home?)*		
G-AWST	Druine D.62B Condor	RAE/637		31.10.68	T.P.Lowe	Fenland	31. 5.05P
	(Built Rollason Aircraft and Engines)						
G-AWSW	Beagle D.5/180 Husky	3690	XW635	4.11.68	C.Tyers	Spanhoe	25. 5.05T
			G-AWSW		t/a Windmill Aviation *(As "XW635")*		
G-AWTD*	Percival P.56 Provost T.1				See SECTION 3 Part 1	Gweru, Zimbabwe	
G-AWTJ	Cessna F150J	F150-0419		8.11.68	P.L.Jameson	Elstree	8.12.04T
	(Built Reims Aviation SA)						
G-AWTL	Piper PA-28-180 Cherokee D	28-5068	N11C	12.11.68	I.R.Chaplin	Andrewsfield	28. 7.07T
G-AWTS	Beech 19A Musketeer Sport	MB-412	OO-BGN	14.11.68	Cinque Ports Flying Group Ltd	Lydd	15. 8.05T
			G-AWTS, N2763B				
G-AWTV	Beech 19A Musketeer Sport	MB-424	N2770B	14.11.68	J.Whittaker	Trehelig, Welshpool	20.10.06
G-AWTX	Cessna F150J	F150-0404		18.11.68	R.D.Forster	Clacton	25. 6.95T
	(Built Reims Aviation SA)				*(New CofR 9.03)*		
G-AWUB	Gardan GY-201 Minicab	A 205	F-PERX	22.11.68	R.A Hand	RAF Barkston Heath	5. 4.05P
	(Built Aéronautique Havraise) (Continental A65)						
G-AWUE	SAN Jodel DR.1050 Ambassadeur	299	F-BKHE	22.11.68	K.W. and F.M.Wood *(On rebuild 4.97: noted 3.02)*	Insch	17.10.87
G-AWUG	Cessna F150H	F150-0299		25.11.68	Prestwick Flight Centre Ltd	Prestwick	6. 2.06T
	(Built Reims Aviation SA)						
G-AWUH*	Cessna F150H	F150-0307		25.11.68	Not known	Phoenix Farm, Lower Upham	16. 7.94
	(Built Reims Aviation SA)				*(Cancelled 8.7.97 as WFU) (Fuselage in poor condition 8.04)*		
G-AWUJ	Cessna F150H	F150-0332		25.11.68	S.R.Hughes	Netherthorpe	18. 2.06
	(Built Reims Aviation SA)						

Registration	Type	C/n	Previous identity	Date	Owner / Operator	Base	CofA
G-AWUK*	Cessna F150H				See SECTION 3 Part 1	Caernarfon	
G-AWUL	Cessna F150H	F150-0346		25.11.68	C.A and L.P.Green		
	(Built Reims Aviation SA)					Drayton Manor, Drayton St.Leonard	15. 9.05
G-AWUN	Cessna F150H	F150-0377		25.11.68	D.Valentine	Sturgate	22. 1.05
	(Built Reims Aviation SA)				tr C150 Group		
G-AWUO	Cessna F150H	F150-0380		25.11.68	S.Stevens	Popham	22. 7.07
	(Built Reims Aviation SA)				tr SAS Flying Group		
G-AWUT	Cessna F150J	F150-0405		25.11.68	S.J.Black	Sherburn-in-Elmet	6.11.06
	(Built Reims Aviation SA)						
G-AWUU	Cessna F150J	F150-0408	EI-BRA	25.11.68	A.L.Grey	Armshold Farm, Kingston, Cambridge	15. 6.97
	(Built Reims Aviation SA)		G-AWUU				
G-AWUX	Cessna F172H	F172-0577		25.11.68	B.J.Portch	St. Just	23. 5.07
	(Built Reims Aviation SA)				tr G-AWUX Group		
G-AWUZ	Cessna F172H	F172-0587		25.11.68	I.R.Judge	Shoreham	27.10.04
	(Built Reims Aviation SA)				tr G-BUJU Flying Group		
G-AWVA	Cessna F172H	F172-0597		25.11.68	Barton Air Ltd	Barton	12.10.06
	(Built Reims Aviation SA)						
G-AWVB	SAN Jodel D.117	604	F-BIBA	26.11.68	H.Davies	Swansea	31. 5.05P
G-AWVC	Beagle B.121 Pup Series 1	B121-026	(OE-CUP)	27.11.68	J.J.West	Sturgate	17. 7.07
G-AWVE	CEA Jodel DR.1050/M1 Sicile Record	612	F-BMPQ	27.11.68	E.A Taylor *(Noted 2.05)*	Southend	18. 5.00
G-AWVF	Hunting Percival P.56 Provost T.1	PAC/F/375	XF877	28.11.68	A.J.House	Lower Wasing Farm, Brinpton	10.10.07P
					(As "XF877/JX")		
G-AWVG	AESL Airtourer T2 (115)	513	OO-WIC	29.11.68	C.J.Scholfield	Top Farm, Croydon, Royston	20. 9.07
			G-AWVG				
G-AWVN	Aeronca 7AC Champion	7AC-6005	N2426E	4.12.68	P.K.Brown	Rush Green	6. 4.05P
			NC2426E		tr Champ Flying Group		
G-AWVZ	Jodel D.112	898	F-PKVL	12.12.68	D.C.Stokes	Dunkeswell	1. 8.05P
G-AWWE	Beagle B.121 Pup Series 2	B121-032	G-35-032	12.12.68	J.N.Randle	Coventry	10.10.07
G-AWWI	SAN Jodel D.117	728	F-BIDU	13.12.68	W.J.Evans	Rhigos	13. 6.04P
G-AWWM	Gardan GY-201 Minicab	A 195	F-BFOQ	1. 1.69	P J Brayshaw	Haddock Stone Farm, Markington	10.12.92P
	(Built M Heron) (Continental A65)				*(Current status unknown)*		
G-AWWN	SAN Jodel DR.1050 Sicile	398	F-BLJA	8. 1.69	R.A J.Hurst	Nuthampstead	20. 7.07
G-AWWO	CEA Jodel DR.1050 Sicile	552	F-BLOI	8. 1.69	A R.Grimshaw and A A Macleod	Barton	15. 5.06
					tr The Whiskey Oscar Group		
G-AWWP	Aerosport Woody Pusher Mk.3	PFA 1323		7. 1.69	M.S.and Mrs R.D.Bird	Pepperbox, Salisbury	
	(Built Woods Aeroplanes - c/n WA/163)				*(Stored 6.93)*		
G-AWWT	Druine D.31 Turbulent	PFA 1653		15. 1.69	E.L.Phillips	Andrewsfield	23. 4.97P
	(Built D A Levermore) (Volkswagen 1600)				*(Damaged Andrewsfield 7.10.96: current status unknown)*		
G-AWWU	Reims FR172F Rocket	FR17200111		15. 1.69	Westward Airways (Lands End) Ltd	St.Just	16. 4.06T
G-AWWW	Cessna 401	401-0294	N8446F	19.12.68	Treble Whiskey Aviation Ltd	Blackpool	5. 6.04T
					(Op Westair Flying Services)		
G-AWXR	Piper PA-28-180 Cherokee D	28-5171	N11C	24. 1.69	Aero Club de Portugal	(Lisbon, Portugal)	11. 5.07
G-AWXS	Piper PA-28-180 Cherokee D	28-5283	N11C	24. 1.69	J.A.Hardiman	Shobdon	6. 1.06T
G-AWXZ	SNCAN Stampe SV-4C	360	F-BHMZ	30. 1.69	Bianchi Aviation Film Services Ltd	Booker	1.10.05A
	(Renault 4P)		Fr.Mil, F-BCOI				
G-AWYB	Reims FR172F Rocket	FR17200075		30. 1.69	J.R.Sharpe	Southend	22. 9.07
G-AWYI*	Royal Aircraft Factory BE.2c rep				See SECTION 3 Part 1	Rhinebeck, New York, US	
G-AWYJ	Beagle B.121 Pup Series 2	B121-038	G-35-038	10. 2.69	H.C.Taylor	Popham	28. 7.07
G-AWYL	CEA Jodel DR.253B Régent	143		11. 2.69	K.Gillam	Radley, Hungerford	18. 5.06
G-AWYO	Beagle B.121 Pup Series 1	B121-041	G-35-041	11. 2.69	B.R.C.Wild	Popham	12.12.05
G-AWYV*	British Aircraft Corporation One-Eleven 501EX	BAC.178		11. 2.69	European Aviation Air Charter Ltd	Bournemouth	24. 6.04T
					(Cancelled 17.12.04 as wfu) (In open storage 12.04)		
G-AWYX	SOCATA MS.880B Rallye Club	1311		11. 2.69	Marjorie J.Edwards *(Noted 8.03)*	St.Just	27. 6.86
G-AWYY*	Slingsby T.57 Sopwith Camel F.1 rep				See SECTION 3 Part 1	RNAS Yeovilton	
G-AWZI*	Hawker Siddeley HS.121 Trident 3B Series 101				See SECTION 3 Part 1	Farnborough	
G-AWZJ*	Hawker Siddeley HS.121 Trident 3B Series 101				See SECTION 3 Part 1	Dumfries	
G-AWZK*	Hawker Siddeley HS.121 Trident 3B Series 101				See SECTION 3 Part 1	Manchester	
G-AWZM*	Hawker Siddeley HS.121 Trident 3B Series 101				See SECTION 3 Part 1	Wroughton	
G-AWZP*	Hawker Siddeley HS.121 Trident 3B Series 101				See SECTION 3 Part 1	Manchester	
G-AWZR*	Hawker Siddeley HS.121 Trident 3B Series 101	2318		14. 1.69	International Fire Training Centre	Durham Tees Valley	9. 4.86T
					(WFU 27.9.85 and cancelled 26.3.86 as WFU) (Noted 11.03)		
G-AWZS*	Hawker Siddeley HS.121 Trident 3B Series 101	2319		14. 1.69	International Fire Training Centre	Durham Tees Valley	9. 9.86T
					(WFU 5.12.85 and cancelled 18.3.86 as WFU) (Noted 11.03)		
G-AWZU*	Hawker Siddeley HS.121 Trident 3B Series 101	2321		14. 1.69	Not known	(Basingstoke)	3. 7.86T
					(Cancelled 18.3.86 as WFU) (Nose only noted 9.03) "Tina"		

G-AXAA - G-AXZZ

Registration	Type	C/n	Previous identity	Date	Owner / Operator	Base	CofA
G-AXAB	Piper PA-28-140 Cherokee	28-20238	EI-AOA	17. 2.69	Bencray Ltd	Blackpool	25. 6.07T
			N6206W		*(Op Blackpool and Fylde Aero Club)*		
G-AXAN	de Havilland DH.82A Tiger Moth	85951	F-BDMM	21. 2.69	Leading Edge Marketing Ltd	Little Gransden	27. 8.06T
			Fr.AF, EM720		*(As "EM720")*		
G-AXAS	Wallis WA-116-T/Mc	217		25. 2.69	K.H.Wallis *(Noted 8.01)*	Reymerston Hall, Norfolk	13. 4.05P
	(McCulloch 4318A @ 72hp) *(Originally registered as Wallis WA-116-T two-seater tandem version: used major components from G-AVDH c/n 216)*						
G-AXAT	SAN Jodel D.117A	836	F-BITJ	26. 2.69	P.S.Wilkinson	Garton, Insch	29. 4.05P
G-AXBF	Beagle D.5/180 Husky	3691	OE-DEW	17.10.84	J.H.Powell-Tuck	Popham	17. 3.07
G-AXBG	Bensen B.8M	RC.1		12. 3.69	R.Curtis	(Bury St.Edmunds)	
	(Built R Curtis)				*(Current status unknown)*		
G-AXBH	Cessna F172H	F172-0571		12. 3.69	D.F.Ranger	Popham	20. 3.06T
	(Built Reims Aviation SA)						
G-AXBJ	Cessna F172H	F172-0573		12. 3.69	S.E.Goodman	Leicester	19. 2.07
	(Built Reims Aviation SA)				tr BJ Flying Group		

G-AXBW	de Havilland DH.82A Tiger Moth	83595	6854M	12. 3.69	Hunter Wing Ltd	Frensham, Wilshanger	5. 4.07
			T5879		*(As "T-5879/RUC-W")*		
G-AXBZ	de Havilland DH.82A Tiger Moth	86552	F-BGDF	14. 3.69	W.J.de Jong Cleyndert	(Dereham)	3. 4.05
	(Built Morris Motors Ltd)		Fr.AF, PG643				
G-AXCA	Piper PA-28R-200 Cherokee Arrow	28R-35053	N11C	18. 3.69	W.H.Nelson	Southend	7. 5.06
G-AXCG	SAN Jodel D.117	510	PH-VRA	19. 3.69	C.A White	Andrewsfield	19. 4.06P
			F-BHXI		tr The Charlie Golf Group		
G-AXCM	SOCATA MS.880B Rallye Club	1322		25. 3.69	D.C.Maniford	Bidford	18.12.04
G-AXCX	Beagle B.121 Pup Series 2	B121-046	G-35-046	31. 3.69	L.A.Pink	Sandown	10. 7.94
					(On active restoration in Wiltshire 1.03)		
G-AXCY	SAN Jodel D.117A	499	F-BHXB	31. 3.69	S.Marom	Whitehall Farm, Benington	10. 2.05P
G-AXCZ	SNCAN Stampe SV-4C	186	ZS-VFW	31. 3.69	J.Price	Trenchard Farm, Eggesford	10. 7.83
	(Renault 4P)		G-AXCZ, F-BCFG		*(Fuselage noted 11.04)*		
G-AXDC	Piper PA-23-250 Aztec D	27-4169	N6829Y	8. 4.69	N.J.Lilley *(New CofR 3.03)*	Bodmin	24. 8.98
G-AXDI	Cessna F172H	F172-0574		14. 4.69	M.F. and J.R.Leusby	Maypole Farm, Chislet	17. 2.06
	(Built Reims Aviation SA)				t/a Jeanair		
G-AXDK	CEA Jodel DR.315 Petit Prince	378		16. 4.69	M.R.Weatherhead and T.J.Thomas	Sywell	3.12.05
					tr Delta Kilo Flying Group		
G-AXDN*	British Aircraft Corporation-Aérospatiale Concorde				See SECTION 3 Part 1	Duxford	
G-AXDU*	Beagle B.121 Pup Series 2	B121-048	G-35-048	18. 4.69	Not known	Near Parish Hall, Bennington, Stevenage	14. 5.98T
			(Damaged North Weald 20.9.97: cancelled 12.1.98 as destroyed) (Battered fuselage noted 1.04)				
G-AXDV	Beagle B.121 Pup Series 1	B121-049		18. 4.69	T.A White	Bagby	28. 6.04
G-AXDW	Beagle B.121 Pup Series 1	B121-053		18. 4.69	I.Beaty, P.J.Abbott and J.R.A Stevens	Cranfield	27. 3.05
					tr Cranfield Delta Whiskey Group		
G-AXED	Piper PA-25-235 Pawnee B	25-3586	OH-PIM	24. 4.69	Wolds Gliding Club Ltd	Pocklington	18. 5.06
			OH-CPY, N7540Z				
G-AXEE*	English Electric Lightning F.53				See SECTION 3 Part 1	Kuwait City, Kuwait	
G-AXEH*	Beagle B.125 Bulldog 1				See SECTION 3 Part 1	East Fortune	
					(Also see G-CCOA below)		
G-AXEI*	Ward P.45 Gnome	P.45		25. 4.69	A J.E.Smith and N.H.Ponsford	Breighton	
	(Built M Ward)				*(Cancelled 30.5.84 as WFU) (Noted 1.05)*		
G-AXEO	Scheibe SF2525B Falke	4645	D-KEBC	1. 5.69	The Borders (Milfield) Gliding Club Ltd	Millfield	28. 6.07
	(Stark-Stamo MS1500)						
G-AXEV	Beagle B.121 Pup Series 2	B121-070		6. 5.69	G.Benson and D.S.Russell	Gloucestershire	6. 8.06
G-AXFM*	Cierva Rotorcraft CR.LTH.1 Grasshopper III				See SECTION 3 Part 1	Weston-super-Mare	
G-AXFN	Jodel D.119	980	F-PHBU	19. 5.69	D.M.Jackson	Netherthorpe	25.11.05P
	(Built M Ganu)						
G-AXGE	SOCATA MS.880B Rallye Club	1353		23. 5.69	R.P.Loxton	(Bridport)	25. 9.04
G-AXGG	Cessna F150J	F150-0440		28. 5.69	S.G.Moores and A.J.Simpson	Maypole Farm, Chislet	17. 9.06
	(Built Reims Aviation SA)						
G-AXGP	Piper J-3C-90 Cub (L-4J-PI)	12544	F-BGPS	2. 6.69	J.R.Howard	Liverpool	12. 2.04P
	(Frame No.12374) (Reported as c/n 9542 ex 43-28251)		F-BDTM, 44-80248		tr G-AXGP Group		
G-AXGR	Phoenix Luton LA-4A Minor	PAL/1125		2. 6.69	B.A.Schlussler	Cawthorpe	23. 9.05P
	(Built R Spall) (JAP J.99)						
G-AXGS	Druine D.62B Condor	RAE/638		3. 6.69	P.A Kirkham	Wellcross Grange, Slinfold	19. 6.05P
	(Built Rollason Aircraft and Engines)				tr G-AXGS Condor Group		
G-AXGV	Druine D.62B Condor	RAE/641		3. 6.69	S.B.Robson	Watchford Farm, Yarcombe	23. 9.05P
	(Built Rollason Aircraftibn and Engines)						
G-AXGZ	Druine D.62B Condor	RAE/643		3. 6.69	A.J.Cooper	Rochester	16. 9.04P
	(Built Rollason Aircraft and Engines)						
G-AXHA	Cessna 337A Super Skymaster	3370484	(EI-ATH)	5. 6.69	I.M.Latiff	Little Staughton	30. 8.02
			N5384S		*(New owner 3.03)*		
G-AXHC	SNCAN Stampe SV-4C	293	F-BCFU	6. 6.69	D.L.Webley	Wickenby	8. 5.06
	(Renault 4P)						
G-AXHO	Beagle B.121 Pup Series 2	B121-077		9. 6.69	L.H.Grundy	King's Farm, Thurrock	28. 6.07
G-AXHP	Piper J-3C-65 Cub (L-4J-PI)	12932	F-BETT	9. 6.69	Witham (Specialist) Vehicles Ltd	(Grantham)	1. 1.04P
	(Frame No.12762)		NC74121, 44-80636		*(As "480636/58-A" in USAAC c/s)*		
	(Regd with c/n "AF36506" which is USAAC contract no)						
G-AXHR	Piper J-3C-65 Cub (L-4H-PI)	10892	F-BETI	9. 6.69	K.B.Raven and E.Cundy	Hill Farm, Nayland	15. 8.05P
			43-29601		tr G-AXHR Cub Group *(As "329601/44-D" in US Army c/s)*		
G-AXHS	SOCATA MS.880B Rallye Club	1357		9. 6.69	W.B.and A Swales	Bagby	14. 7.06
G-AXHT	SOCATA MS.880B Rallye Club	1358		9. 6.69	K.J.Duthie tr Hotel Tango Group	Oxford	19. 4.07
G-AXHV	SAN Jodel D.117A	695	F-BIDF	9. 6.69	J.S.Ponsford tr Derwent Flying Group	Hucknall	1. 6.05P
G-AXIA	Beagle B.121 Pup Series 1	B121-078		17. 6.69	P.S.Shuttleworth	Leicester	28.10.07T
G-AXIE	Beagle B.121 Pup Series 2	B121-087		17. 6.69	G.McD.Moir	Tatenhill	7. 6.07
G-AXIF	Beagle B.121 Pup Series 2	B121-088	(SE-FGV)	17. 6.69	J.R.Faulkner	Egginton	16. 9.05T
G-AXIG	Scottish Aviation Bulldog Series 100/104	BH120/002		24. 6.69	A.A.Douglas-Hamilton	Archerfield Estate, Dirleton	6. 5.05
G-AXIO	Piper PA-28-140 Cherokee B	28-25764	N11C	26. 6.69	Just Plane Trading Ltd	Top Farm, Royston	20. 3.05T
G-AXIR	Piper PA-28-140 Cherokee B	28-25795	N11C	26. 6.69	A.G.Birch	Weston Zoyland	15. 6.07
G-AXIW	Scheibe SF25B Falke	4657	(D-KABJ)	3. 7.69	M.B.Hill	Nympsfield	30. 1.06
	(Stark-Stamo MS1500)						
G-AXIX	AESL Airtourer T4 (150)	A 527		3. 7.69	J.C.Wood	Shobdon	22 .4.07
G-AXJB	Omega 84 HAB	04		9. 7.69	Semajan Ltd	(Romsey)	20. 8.73S
	(Initially flown as G-AXDT)				tr Southern Balloon Group *"Jester"*		
G-AXJH	Beagle B.121 Pup Series 2	B121-089		11. 7.69	D.Collings tr The Henry Flying Group	Popham	2. 5.07
G-AXJI	Beagle B.121 Pup Series 2	B121-090		11. 7.69	J.J.Sanders	Egginton	3. 6.05
G-AXJJ	Beagle B.121 Pup Series 2	B121-091		11. 7.69	M.L., T.M., D.A and P.M.Jones	Egginton	23.10.06T
G-AXJO	Beagle B.121 Pup Series 2	B121-094		11. 7.69	J.A D.Bradshaw *"Joey"*	Three Mile Cross, Reading	9. 8.06
G-AXJR	Scheibe SF25B Falke	4652	D-KICD	4. 7.69	R.I.Hey	Nympsfield	6. 8.06
	(Stark-Stamo MS1500)				tr The Falke Syndicate		
G-AXJV	Piper PA-28-140 Cherokee B	28-25572	N11C	14. 7.69	ATC (Lasham) Ltd	Lasham	7. 6.07T
G-AXJX	Piper PA-28-140 Cherokee B	28-25990	N11C	14. 7.69	Patrolwatch Ltd	Barton	4.11.07E
G-AXKH	Phoenix Luton LA-4A Minor	PFA 823		21. 7.69	M.E.Vaisey	(Hemel Hempstead)	18. 4.84P
	(Built M.E.Vaisey - c/n PAL/1316) (Volkswagen 1600)				*(Current status unknown)*		

Regn	Type	C/n	Previous identity	Regd	Owner	Base	CofA
G-AXKJ	Jodel D.9 Bébé	PFA 928B & PFA 941		22. 7.69	P.D.Smalley	Spanhoe	12.10.04P
	(Built Southdown Aero Services Ltd - c/n SAS.002) (Volkswagen 1600)						
G-AXKO	Westland-Bell 47G-4A	WA/720	G-17-5	22. 7.69	G.P.Hinkley	Channons Hall, Tibenham	21. 1.07
G-AXKS*	Westland-Bell 47G-4A				See SECTION 3 Part 1	Middle Wallop	
G-AXKX	Westland-Bell 47G-4A	WA/728	G-17-13	22. 7.69	South Yorkshire Aviation Ltd	Gamston	1. 8.07
G-AXKY	Westland-Bell 47G-4A	WA/729	G-17-14	22. 7.69	C.P.Golborne	(Whittlesey, Peterborough)	15. 5.06
G-AXLG	Cessna 310K	310K0204	N3804X	25. 7.69	Smiths (Harlow) Aerospace Ltd	Andrewsfield	26.10 07E
G-AXLI	Nipper T.66 RA.45 Series 3	S.131/1707		25. 7.69	D.and Margaret Shrimpton	RAF Keevil	2. 6.05P
	(Built Slingsby Aircraft Co Ltd) (Ardem Mk.10)						
G-AXLS	SAN Jodel DR.105A Ambassadeur	86	F-BIVR	31. 7.69	J.J.Boon tr Axle Flying Club	Popham	6. 4.07
G-AXLZ	Piper PA-18 Super Cub 95	18-2052	PH-NLB	31. 7.69	R.J.Quantrell	Low Farm, South Walsham	23. 4.00
	(L-18C-PI) *(Frame No.18-2065)*		R.Neth.AF R-45, 8A-45, 52-2452 *(Damaged Low Farm 14.8.97: current status unknown)*				
G-AXMA	Piper PA-24-180 Comanche	24-3467	N8214P	5. 8.69	J.A Fletcher	(Louth)	18. 6.04
G-AXMD*	Omega O-20 HAB				See SECTION 3 Part 1	Newbury	
G-AXMN	Auster J/5B Autocar	2962	F-BGPN	14. 8.69	J.A and S.M.Fletcher	(Louth)	15. 4.04
G-AXMT	Bücker Bü.133C Jungmeister	46	N133SJ	19. 8.69	R.A Fleming	Breighton	23. 5.05P
	(Built AG Fur Dornier-Flugzeuge)		G-AXMT, HB-MIY, Swiss AF U-99 *(As "U-99")*				
G-AXMW	Beagle B.121 Pup Series 1	B121-101		19. 8.69	DJP Engineering (Knebworth) Ltd	Cambridge	22. 7.07
G-AXMX	Beagle B.121 Pup Series 2	B121-103	VH-UPT	19. 8.69	Susan A Jones	(Cannes, France)	5.10.06T
			G-AXMX, G-35-103				
G-AXNJ	Wassmer Jodel D.120 Paris-Nice	52	F-BHYO	29. 8.69	D.I.Vernon tr Clive Flying Group	Sleap	15. 9.05P
G-AXNL	Beagle B.121 Pup Series 1	B121-113		3. 9.69	CAVOK Ltd	White Waltham	9. 5.07T
G-AXNM	Beagle B.121 Pup Series 1	B121-114		3. 9.69	T.W.Anderson	Bembridge	23.1.06
G-AXNN	Beagle B.121 Pup Series 2	B121-104		3. 9.69	Gabrielle Aviation Ltd *"Gabrielle"*	Shoreham	20. 8.06
G-AXNP	Beagle B.121 Pup Series 2	B121-106		3. 9.69	J.W.Ellis and R.J.Hemmings	Ashcroft Farm, Winsford	24. 6.05
G-AXNR	Beagle B.121 Pup Series 2	B121-108		3. 9.69	J.R.Clegg	Raby's Farm, Great Stukeley	25. 3.05
					tr November Romeo Group		
G-AXNS	Beagle B.121 Pup Series 2	B121-110		3. 9.69	D.Beckwith and D.Long tr Derwent Aero Group	Gamston	8. 7.07
G-AXNW	SNCAN Stampe SV-4C	381	F-BFZX	11. 9.69	Carolyn S.Grace	Blooms Farm, Sible Hedingham	20. 5.06
	(Renault 4P)		Fr.Mil				
G-AXNX	Cessna 182M	18259322	N70606	16. 9.69	D.B.Harper	Biggin Hill	12. 5.05T
G-AXNZ	Pitts S-1C	PFA 1383		16. 9.69	W.A Jordan	(Sandy)	30. 8.91P
	(Built W Berry and B Etheridge - c/n EB.1) (Official c/n shown as EBX2)				*(Stored 12.97: current status unknown)*		
G-AXOH	SOCATA MS.894A Rallye Minerva 220	11062	D-EAGU	17. 9.69	Bristol Cars Ltd	White Waltham	25. 9.06
G-AXOJ	Beagle B.121 Pup Series 2	B121-109	G-35-109	24. 9.69	T.J.Martin tr Pup Flying Group	Rochester	5. 5.07
G-AXOM*	Penn-Smith Gyroplane	DJPS.1		26. 9.69	See SECTION 3 Part 1	Lower Stondon	
G-AXOS	SOCATA MS.894A Rallye Minerva 220	11079		3.10.69	M.N.Stevens	Popham	26. 5.06
					tr Rallye Racers Flying Group		
G-AXOT	SOCATA MS.893A Rallye Commodore 180	11433		3.10.69	P.Evans and J.C.Graves	Doncaster	26. 3.06
G-AXOZ	Beagle B.121 Pup Series 1	B121-115	N70290	7.10.69	R.J.Ogborn	Hawarden	14. 1.07
			G-AXOZ, G-35-115				
G-AXPA	Beagle B.121 Pup Series 1	B121-116	D-EATL	7.10.69	D.G.Lewendon	Spanhoe	4.11.87
			G-AXPA, G-35-116		*(On rebuild 6.04)*		
G-AXPB	Beagle B.121 Pup Series 1	B121-117	G-35-117	7.10.69	M.J.K.Seary and R.T.Austin	Leicester	17. 1.05
					tr Beagle Flying Group		
G-AXPC	Beagle B.121 Pup Series 1	B121-119	PH-VRS	7.10.69	T.A White	Bagby	22. 8.07
			G-AXPC				
G-AXPF	Reims/Cessna F150K	F15000543		14.10.69	D.R.Marks *(Noted 5.04)*	Enstone	22. 4.02
G-AXPG	Mignet HM.293	PFA 1333		14.10.69	W.H.Cole	Spilsteads Farm, Sedlescombe	20. 1.77P
	(Built W H Cole) (Volkswagen 1300)				*(Noted 5.01)*		
G-AXPM	Beagle B.121 Pup Series 1	B121-122	G-35-122	20.10.69	S.C.Stanton	North Weald	29. 5.05
G-AXPN	Beagle B.121 Pup Series 2	B121-123	G-35-123	20.10.69	A.Richardson	Egginton	28. 2.05
G-AXPZ	Campbell Cricket	CA/320		3.11.69	W.R.Partridge	St.Merryn	27. 4.99P
	(Rotax 582)				*(Noted 10.00)*		
G-AXRC	Campbell Cricket	CA/323		3.11.69	L.R.Morris	(Newry, County Armagh)	18. 5.78S
	(Volkswagen 1600)		*(Damaged Wittering 22.10.77: stored Tattershall Thorpe 7.91: new owner 12.02)*				
G-AXRL*	Piper PA-28-160 Cherokee	28-324	PH-CHE	5.11.69	T.W.Clarke	(Rye)	20. 2.86
			D-EFRI, N11C		*(WFU 8.84 due to corrosion and cancelled by CAA 12.8.94) (Noted on M25 18.7.04)*		
G-AXRP	SNCAN Stampe SV-4C	554	F-BDCZ	7.11.69	C.C.Manning	Rotary Farm, Hatch	20.11.06
	(Originally regd as SV-4C (Renault 4P): damaged Gransden 19.10.74: restored 2.85 as SV-4A G-BLOL with c/n SS-SV-R1: NTU and restored 9.94 as G-AXRP but stored as "G-BLOL" 10.99: on rebuild with DH Gipsy Major 2 @ 2.01: type re-regd as SV-4C 10.03)						
G-AXRR	Auster AOP.9	AUS.178 & B5/10/178	XR241	7.11.69	R.B.Webber	Trenchard Farm, Eggesford	11.12.04P
			G-AXRR, XR241		*(As "XR241" in Army yellow c/s)*		
G-AXRT	Reims/Cessna FA150K Aerobat	FA1500018		12.11.69	C.C.Walley	Elstree	25. 1.06T
	(Tail-wheel conversion)						
G-AXSC	Beagle B.121 Pup Series 1	B121-138	G-35-138	13.11.69	R.J.MacCarthy	Swansea	28. 4.07
G-AXSD	Beagle B.121 Pup Series 1	B121-139	G-35-139	13.11.69	AURS Aviation Ltd	Prestwick	1. 4.05
G-AXSF	Nash Petrel	PFA 1516		17.11.69	Nash Aircraft Ltd	Lasham	?. 4.94P
	(Built Nash Aircraft Ltd - c/n P.003) (Lycoming O-360) *(Second allocation of PFA 1516 has no connection with G-BACA) (Stored 10.95: current status unknown)*						
G-AXSG	Piper PA-28-180 Cherokee E	28-5605	N11C	17.11.69	Admiral Property Ltd	Old Buckenham	26. 6.05
G-AXSI	Reims/Cessna F172H	F17200687	G-SNIP	19.11.69	R.Collins	(Cleethorpes)	8..9.07
			G-AXSI				
G-AXSM	CEA Jodel DR.1051 Sicile	512	F-BLRH	20.11.69	T.R.G.Barnby and M.S.Regendanz	Headcorn	1. 6.06
G-AXSW	Reims/Cessna FA150K Aerobat	FA1500003		25.11.69	R.Mitchell	(Chalfont St.Giles)	2. 3.07
G-AXSZ	Piper PA-28-140 Cherokee B	28-26188	N11C	26.11.69	R.Gibson and B.Collins	White Waltham	24. 4.06
					tr The White Wings Flying Group		
G-AXTA	Piper PA-28-140 Cherokee B	28-26301	N11C	26.11.69	P.J.Farrell tr G-AXTA Aircraft Group	Shoreham	25. 5.07
G-AXTC	Piper PA-28-140 Cherokee B	28-26265	N11C	26.11.69	W.J.Knott tr G-AXTC Group	Strubby	29.10.05
G-AXTJ	Piper PA-28-140 Cherokee B	28-26241	N11C	26.11.69	K.Patel	Elstree	26. 2.07T
G-AXTL	Piper PA-28-140 Cherokee B	28-26247	N11C	26.11.69	Pegasus Aviation (Midlands) Ltd	Tatenhill	1.12.07E
G-AXTO	Piper PA-24-260 Comanche C	24-4900	N9449P	28.11.69	Jean L.Richardson	Turweston	4. 8.06
			N9705N		*"Betsy Baby"*		
G-AXTP	Piper PA-28-180 Cherokee C	28-3791	OH-PID	1.12.69	M.Whyte	(Williamstown, County Galway)	10. 1.07
G-AXTY*	de Havilland DH.82A Tiger Moth				See SECTION 3 Part 1	Oshkosh, Wisconsin, US	

Registration	Type	C/n	Previous identity	Regd	Owner/Operator	Base	CofA
G-AXUA	Beagle B.121 Pup Series 1	B121-150	G-35-150	4.12.69	P.Wood	Bourn	12. 6.03
G-AXUB	Britten-Norman BN-2A Islander	121	5N-AIJ	4.12.69	Headcorn Parachute Club Ltd	Headcorn	18. 5.05
			G-AXUB, N859JA, G-51-47				
G-AXUC	Piper PA-12 Super Cruiser	12-621	5Y-KFR	5.12.69	J.J.Bunton	Maypole Farm, Chislet	21. 4.05
			VP-KFR, ZS-BIN				
G-AXUF	Reims/Cessna FA150K Aerobat	FA1500043		9.12.69	W.B.Bateson	Blackpool	6. 5.07T
G-AXUJ	Auster V J/1 Autocrat	1957	G-OSTA	11.12.69	C.J.Harrison and S.J.McKenna	(Warwick)	6. 5.07
			G-AXUJ, PH-OTO				
G-AXUK	SAN Jodel DR.1050 Ambassadeur	292	F-BJYU	11.12.69	G.J. Keegan	Deanland	23. 8.06
					tr Downland Flying Group (2KI)		
G-AXVB	Reims/Cessna F172H	F17200703		22.12.69	R.and J.Turner	Charlton Park, Malmesbury	22. 6.07
G-AXVK	Campbell Cricket	CA/327		1. 1.70	B.Jones	Melrose Farm, Melbourne	24. 4.05P
	(Volkswagen 1600)						
G-AXVM	Campbell Cricket	CA/329		1. 1.70	D.M.Organ	Gloucestershire	29.10.05P
	(Volkswagen 1834)						
G-AXVN	McCandless M.4	M4/6		5. 1.70	W.R.Partridge	St.Merryn	
	(Volkswagen 1700)				*(Stored 5.03)*		
G-AXVU*	Omega 84 HAB				See SECTION 3 Part 1	Newbury	
G-AXWA	Auster AOP.9	B5/10/133	XN437	13. 1.70	C M Edwards	North Weald	
					(Unmarked on slow rebuild 1.05)		
G-AXWT	Jodel D.11	PFA 911		26. 1.70	R.C.Owen	Danehill	2. 6.00P
	(Built C King and R C Owen) (Continental C90)						
G-AXWV	CEA Jodel DR.253 Régent	104	F-OCKL	2. 2.70	R Friedlander and D C Ray	Grateley, Andover	26.11.04
G-AXWZ	Piper PA-28R-200 Cherokee Arrow	28R-35605	N11C	3. 2.70	P.Walkley	Bournemouth	7.10.05P
G-AXXC	Rousseau Piel CP.301B Emeraude	117	F-BJAT	4. 2.70	L.F.Clayton	Wellesbourne Mountford	19. 1.05P
G-AXXP*	Bradshaw 76 (Ax7) HAB				See SECTION 3 Part 1	Newbury	
G-AXXV	de Havilland DH.82A Tiger Moth	85852	F-BGJI	24. 2.70	C.N.Wookey	Membury	1. 7.07
	(Built Morris Motors Ltd)		Fr.AF, DE992		*(As "DE992")*		
G-AXXW	SAN Jodel D.117	632	F-BIBN	26. 2.70	D.F.Chamberlain and M.A Hughes	Haverfordwest	31. 8.05P
G-AXYK	Taylor JT.1 Monoplane	PFA 1409		2. 3.70	G.D.Bailey	(Otterbourne, Winchester)	7.12.05P
	(Built C Oakins) (Volkswagen 1600)						
G-AXYU	Jodel D.9 Bébé	547	EI-BVE	5. 3.70	P.Turton and H.C.Peake-Jones	Ashcroft Farm, Winsford	13. 9.01P
	(Volkswagen 1600)		G-AXYU		*(Current status unknown)*		
G-AXYZ	WHE Airbuggy	1005		10. 3.70	B.Gunn	Not known	22.12.92P
	(Volkswagen 1600) *(Originally regd as McCandless M.4)*				*(Noted on trailer M62 westbound 4 .04)*		
G-AXZB	WHE Airbuggy	1007		10. 3.70	B.Gunn	Melrose Farm, Melbourne	18.11.86P
	(Originally regd as McCandless M.4)				*(New CofR 6.02)*		
G-AXZD	Piper PA-28-180 Cherokee E	28-5609	N11C	12. 3.70	G.M.Whitmore	High Cross, Ware	15.10.04
G-AXZF	Piper PA-28-180 Cherokee E	28-5688	N11C	12. 3.70	E.P.C.and W.R.Rabson	Compton Abbas	23. 7.07T
					(Op Carill Aviation)		
G-AXZK	Britten-Norman BN-2A-26 Islander	153	V2-LAD	12. 3.70	P.Johnson	Hinton-in-the-Hedges	16. 4.06
			VP-LAD, G-AXZK, G-51-153		t/a Global Parachute Club		
G-AXZM	Nipper T.66 RA.45 Series 3A	PFA 1378		16. 3.70	G.R.Harlow	Newcastle	24. 8.89P
	(Built S J Booth and A Young fromSlingsby kit c/n S.133/1709) (Volkswagen 1600)				*(Damaged near Eshott 21.8.89: possible rebuild 5.90: current status unknown)*		
G-AXZO	Cessna 180	31137	N3639C	17. 3.70	J.C.King	Bourne Park, Hurstbourne Tarrant	17.10.05
					tr Bourne Park Flyers		
G-AXZP	Piper PA-E23-250 Aztec D	27-4464	N13819	17. 3.70	D.M.Harbottle	(Ottershaw)	22. 8.04T
G-AXZT	SAN Jodel D.117A	607	F-BIBD	17. 3.70	N.Batty	Bagby	25. 7.05P
G-AXZU	Cessna 182N Skylane	18260104	N92233	19. 3.70	C.D.Williams	Goodwood	31. 3.05

G-AYAA - G-AYZZ

Registration	Type	C/n	Previous identity	Regd	Owner/Operator	Base	CofA
G-AYAB	Piper PA-28-180 Cherokee E	28-5804	N11C	24. 3.70	J.R.Green	Turweston	21. 8.06
G-AYAC	Piper PA-28R-200 Cherokee Arrow	28R-35606	N11C	24. 3.70	G.A J.Smith-Bosanquet	Knettishall	21. 4.07
					tr The Fersfield Flying Group		
G-AYAG*	Boeing 707-321				See SECTION 3 Part 1	Leitrim, County Leitrim	
G-AYAJ*	Cameron O-84 HAB				See SECTION 3 Part 1	Newbury	
G-AYAK*	Yakovlev Yak C.11				See SECTION 3 Part 1	Polk City, Florida, US	
G-AYAL*	Omega 56 HAB				See SECTION 3 Part 1	Newbury	
G-AYAN	Slingsby Cadet III	PFA 1385	BGA1224	6. 4.70	D.C.Pattison	Brunton	13. 4.04P
	(Built P J Martin and D R Wilkinson - c/n 003) (Volkswagen 1600) RAFGSA.223				*"Thermal Hopper"*		
	(Converted from T.31B [Frame No.SSK/FF776])						
G-AYAR	Piper PA-28-180 Cherokee E	28-5797	N11C	8. 4.70	A Jahanfar *(Op Seawing Flying Club)*	Southend	13. 2 05T
G-AYAT	Piper PA-28-180 Cherokee E	28-5801	N11C	8. 4.70	A Goodchild tr G-AYAT Flying Group	Seething	18. 3.07E
G-AYAW	Piper PA-28-180 Cherokee E	28-5805	N11C	14. 4.70	R.C.Pendle tr GAYAW Group	Denham	19. 6.05T
G-AYAZ*	Britten-Norman BN-2A-7 Islander				See SECTION 3 Part 1	Hong Kong, China	
G-AYBD	Reims/Cessna F150K	F15000583		7. 4.70	Apollo Aviation Advisory Ltd	Shoreham	13.10.07E
					(Op Ace Aviation)		
G-AYBG	Scheibe SF25B Falke	4696	(D-KECJ)	13. 4.70	H.H.T.Wolf	Gallows Hill, Bovington	4. 4.97
	(Volswagen Danum 1600/1)				*(New owner 6.01: current status unknown)*		
G-AYBO	Piper PA-23-250 Aztec D	27-4510	N13874	15. 4.70	Forster and Hales Ltd	Guernsey	7. 5.06
G-AYBP	Jodel D.112	1131	F-PMEK	16. 4.70	G.J.Langston	Bidford	30.10.04P
G-AYBR	Wassmer Jodel D.112	1259	F-BMIG	16. 4.70	I.S.Parker	Damyn's Hall, Upminster	1.11.05P
G-AYBX*	Campbell Cricket				See SECTION 3 Part 1	Alexandria, Egypt	
G-AYCC	Campbell Cricket	CA/336		20. 4.70	D.J.M.Charity	Hinton-in-the-Hedges	16. 5.05P
	(Rotax 582)						
G-AYCE	Scintex CP.301C1 Emeraude	530	F-BJFH	20. 4.70	S.D.Glover	Plymouth	19 . 5.05P
G-AYCF	Reims/Cessna FA150K Aerobat	FA1500055		22. 4.70	E.J.Atkins	Popham	17. 6.06
G-AYCG	SNCAN Stampe SV-4C	59	F-BOHF	24. 4.70	Nancy Bignall	White Waltham	4. 7.07
	(Renault 4P)		F-BBAE, Fr.Mil				
G-AYCJ	Cessna TP206D Turbo Super Skylane	P206-0552	N8752Z	27. 4.70	White Knuckle Airways Ltd	Leeds-Bradford	10. 7.06
	(Regd with c/n T206-0552)						

G-AYCK	AIA Stampe SV-4C	1139	G-BUNT	28. 4.70	The Real Flying Company Ltd	Shoreham	18.10.07T
	(DH Gipsy Major 10) *(Officially regd as built SNCAN)*		G-AYCK, F-BANE				
G-AYCN	Piper J-3C-65 Cub	"13365"	F-BCPO	28. 4.70	W.R. and B.M.Young *(Stored 4.91: current status unknown)*		
	(Frame No.not known: c/n quoted became PH-UCM in 11.46 and p/i is doubtful)					Furze Hill Farm, Rosemarket, Milford Haven	27. 1.89P
G-AYCO	CEA Jodel DR.360 Chevalier	362	F-BRFI	29. 4.70	P.L.Buckley tr Charlie Oscar Club	Hill Farm, Nayland	25. 1.07
G-AYCP	Jodel D.112	67	F-BGKO	30. 4.70	D.J.Nunn	Trevissick Farm, Porthtowan	28. 6.05P
G-AYCT	Reims/Cessna F172H	F17200724		1. 5.70	Haimoss Ltd *(Op Old Sarum Flying Club)*	Old Sarum	9.12.07E
G-AYDI	de Havilland DH.82A Tiger Moth	85910	F-BDOE	7. 5.70	R B and E W Woods and J D M Barr		
			Fr.AF, DF174			Hampstead Norreys	13. 5.06
G-AYDR	SNCAN Stampe SV-4C	307	F-BCLG	13. 5.70	A J.McLuskie	Quebec Farm, Knook, Warminster	27. 3.75
	(Renault 4P)				*(Damaged 16.6.73 and on rebuild 8.93)*		
G-AYDX	Beagle A 61 Terrier 2	B.647	VX121	20. 5.70	R.A Kirby	Spanhoe	10. 4.06
G-AYDY	Phoenix Luton LA-4A Minor	PFA 817		21. 5.70	T.Littlefair	Lymington	15. 8.97P
	(Built L J E Goldfinch - c/n PAL/1302) (Volkswagen 1600)				*(On rebuild 3.00)*		
G-AYDZ	CEA Jodel DR.200	01	F-BLKV	21. 5.70	M.W.Albery	Enstone	29. 3.05
	(Lycoming O-235)		F-WLKV		tr Zero One Group		
G-AYEB	Wassmer Jodel D.112	586	F-BIQR	26. 5.70	R.A.Durance	(Lincoln)	26. 7.05P
G-AYEC	Menavia Piel CP.301A Emeraude	249	F-BIMV	26. 5.70	J.J.Shepherd	Netherthorpe	26. 7.05P
					tr Red Wing Flying Group *"Antoinette"*		
G-AYEE	Piper PA-28-180 Cherokee E	28-5813	N11C	28. 5.70	Halegreen Associates Ltd and Demero Ltd	Oxford	24. 6.05T
G-AYEF	Piper PA-28-180 Cherokee E	28-5815	N11C	28. 5.70	P.A.Coleman and K.Lloyd tr G-AYEF Group	Barton	22. 1.05
G-AYEG	Falconar F-9	PFA 1321		29. 5.70	A.L.Smith	Sackville Lodge Farm, Riseley	4. 8.05P
	(Built G R Gladstone) (Volkswagen 1600)						
G-AYEH	SAN Jodel DR.1050 Ambassadeur	455	F-BLJB	8. 6.70	J.W.Scott	Bidford	17. 6.05
					tr John Scott Jodel Group *"Jemima"*		
G-AYEJ	SAN Jodel DR.1050 Ambassadeur	253	F-BJYG	1. 6.70	J.M.Newbold	Enstone	9. 2.07
G-AYEN	Piper J-3C-65 Cub (L-4H-PI)	12184	F-BGQD	4. 6.70	P.J.Warde and C.F.Morris	Grove Farm, Raveningham	21.10.05P
	(Frame No.12012)		(F-BGQA), Fr.AF, 44-79888				
	(Official identity is c/n 9696/43-835 but fuselages probably exchanged with F-BGQA on conversion in 1952/53)						
G-AYEV	SAN Jodel DR.1050 Ambassadeur	179	F-BERH	10. 6.70	L.G.Evans	Redhill	20. 3.06
			F-OBTH, F-OBRH		tr Echo Victor Group		
G-AYEW	CEA Jodel DR.1050 Sicile	443	F-BLMJ	11. 6.70	J.M.Gale and J.R.Hope *(Noted 6.04)*	Dunkeswell	16.11.03
G-AYFA*	Scottish Aviation Twin Pioneer Mk.3				See SECTION 3 Part 1	Carlisle	
G-AYFC	Druine D.62B Condor	RAE/644		19. 6.70	A R.Chadwick	(Pudsey)	26.12.02P
	(Built Rollason Aircraft and Engines)				*(Fusleage noted 1.04)*		
G-AYFD	Druine D.62B Condor	RAE/645		19. 6.70	B.G.Manning	Little Down Farm, Milson	19. 7.07
	(Built Rollason Aircraft and Engines)				*(Hourds Travel titles) '94'*		
G-AYFE*	Druine D.62C Condor	RAE/646		19. 6.70	P.K.Jenkins	(Redditch)	6.12.01
	(Built Rollason Aircraft and Engines) (Continental O-240-A)				*(New owner 8.04) (Cancelled 29.12.04 by CAA)*		
G-AYFF	Druine D.62B Condor	RAE/647		19. 6.70	D.Ellis and A.W.Maycock		
	(Built Rollason Aircraft and Engines)					Lower Upham Farm, Chiseldon	14.11.05P
G-AYFG	Druine D.62C Condor	RAE/648		19. 6.70	C.Jobling and A.J.Mackay	RAF Barkston Heath	9. 1 06
	(Built Rollason Aircraft and Engines) (Continental O-240-A)						
G-AYFO*	Bücker Bü.133 Jungmeister				See SECTION 3 Part 1	Polk City, Florida, US	
G-AYFP	SAN Jodel D.140 Mousquetaire	18	F-BMSI	24. 6.70	A R.Wood	Audley End	29. 1.08
			F-OBLH, F-WNDO				
G-AYFV	Andreasson BA-4B	PFA 1359		26. 6.70	A R.C.Mathie	RAF Coltishall	5. 7.95P
	(Built Crosby Aviation Ltd - c/n 002) (Lycoming IO-320)				*(Current status unknown)*		
G-AYGA	SAN Jodel D.117	436	F-BHNU	30. 6.70	R.L.E.Horrell	Hawksbridge Farm, Oxenhope	13. 4.05P
G-AYGB*	Cessna 310Q	310Q0111	N7611Q	2. 7.70	Perth College	Perth	23.10.87T
					(Cancelled 23.6.94 by CAA) (Instructional airframe 10.03)		
G-AYGC	Reims/Cessna F150K	F15000556		2. 7.70	S.R.Cooper tr Alpha Aviation Group	Barton	8. 6.07
G-AYGD	CEA Jodel DR.1050 Sicile	515	F-BLRE	3. 7.70	J.P.Morley tr Jodel Flying Group	Netherthorpe	9.12.04
G-AYGE	SNCAN Stampe SV-4C	242	F-BCGM	6. 7.70	I., L.J. and S.Proudfoot	Duxford	4. 6.03
	(Renault 4P)						
G-AYGG	Wassmer Jodel D.120 Paris-Nice	184	F-BJPH	10. 7.70	J.M.Dean	Stoneacre Farm, Farthing Corner	5. 4.05P
G-AYGX	Reims FR172G Rocket	FR17200208		15. 7.70	D.Waterhouse	Preston Court, Ledbury	1.10.06
					tr Reims Rocket Group		
G-AYHA	American AA-1 Yankee Clipper	AA1-0396	N6196L	21. 7.70	S.J.Carr	RAF Leuchars	21. 3.05
G-AYHI*	Campbell Cricket	CA/341		21. 7.70	J.F.MacKay	(North Kessock)	19. 8.86P
	(Volkswagen 1600)				*(Cancelled 2.3.99 by CAA) (Noted 2003)*		
G-AYHX	SAN Jodel D.117A	903	F-BIVE	23. 7.70	L.J.E.Goldfinch	Old Sarum	6. 5.05P
G-AYIA	Hughes 369HS (500)	99-0120S		29. 7.70	G D E Bilton	Sywell	16. 7.88
	(Badly damaged in heavy landing in South.France 1.6.88. to March Helicopters - stored for spares use 8.97: current status unknown)						
G-AYIG	Piper PA-28-140 Cherokee C	28-26878	N11C	31. 7.70	Biggles Ltd	White Waltham	26.11.05T
G-AYII	Piper PA-28R-200 Cherokee Arrow	28R-35736	N11C	4. 8.70	P.W.J.Gove tr Double India Group	Exeter	23. 4.06
G-AYIJ	SNCAN Stampe SV-4B	376	F-BCOM	4. 8.70	T.C.Beadle	Spilsteads Farm, Sedlescombe	23. 6.06
	(DH Gipsy Major 10)						
G-AYIM	Hawker Siddeley HS.748 Series 2A/270	1687	G-11-687	11. 8.70	Emerald Airways Ltd	Liverpool	21.12.04T
			CS-TAG/G-AYIM, G-11-5				
G-AYIT	de Havilland DH.82A Tiger Moth	86343	F-BGEZ	20. 8.70	S.R.Pollitt and H.M.Eassie	Newtownards	15.11.04
	(Built Morris Motors Ltd)		Fr.AF, NL896		tr Ulster Tiger Group		
G-AYJA	SAN Jodel DR.1050 Ambassadeur	150	F-BJJJ	8. 9.70	G.Connell	Weston, Dublin	7. 7.05
G-AYJB	SNCAN Stampe SV-4C	560	F-BDDF	8. 9.70	F.J.M. and J.P.Esson	Bere Farm, Warnford	6. 9.05
	(DH Gipsy Major 10)				*"Odette"*		
G-AYJD	Fournier RF3	11	F-BLXA	8. 9.70	I.O.Bull	Rngmer	1. 9.05P
G-AYJP	Piper PA-28-140 Cherokee C	28-26403	N11C	15. 9.70	RAF Brize Norton Flying Club Ltd	RAF Brize Norton	25. 5.07T
G-AYJR	Piper PA-28-140 Cherokee C	28-26694	N11C	15. 9.70	RAF Brize Norton Flying Club Ltd	RAF Brize Norton	11. 2.07T
G-AYJW	Reims FR172G Rocket	FR17200225		17. 9.70	N.D.Wyndow	Coventry	23.12.06T
					tr Sir W.G.Armstrong-Whitworth Flying Group		
G-AYJY	Isaacs Fury II	PFA 1373		23. 9.70	M.F.Newman	Wreningham	30. 3.04P
	(Built A V Francis) (RR Continental C90)						
G-AYJZ*	Cameron (Ax8) O-84 HAB				See SECTION 3 Part 1	Newbury	
G-AYKD	SAN Jodel DR.1050 Ambassadeur	351	F-BKHR	30. 9.70	M.D.L.Weston *"Isis"*	Popham	20. 6.06

Reg	Type	C/n	Previous identity	Reg date	Owner/operator	Probable base	CofA expiry
G-AYKG*	SOCATA ST-10 Diplomate	117		30. 9.70	Not known	Tulley's Farm, Crawley	
					(Damaged Crowland 4.3.75: cancelled 27.6.75 as WFU).(Noted 10.03)		
G-AYKJ	SAN Jodel D.117A	730	F-BIDX	6.10.70	R.J.Hughes	(Hepworth, Diss)	20. 6.05P
G-AYKK	SAN Jodel D.117	378	F-BHGM	6.10.70	J M Whitham	(Delves Farm, Delves, Huddersfield)	22. 5.85S
					(On rebuild 11.99: new owner 3.04)		
G-AYKL	Reims, Cessna F150L	F15000676		6.10.70	M.A Judge tr Aero Group 78	Netherthorpe	26. 1.06T
G-AYKS	Leopoldoff L.7 Colibri	125	F-PCZX	8.10.70	W.B.Cooper	Walkeridge Farm, Overton	11.11.05P
	(Continental A65)		F-APZQ				
G-AYKT	SAN Jodel D.117	507	F-BGYY	9.10.70	D.I.Walker	Popham	20.10.05P
			F-OAYY				
G-AYKW	Piper PA-28-140 Cherokee C	28-26931	N11C	12.10.70	S.P.Rooney and D.Griffiths	Stapleford	8. 4.06T
G-AYKX	Piper PA-28-140 Cherokee C	28-26933	N11C	12.10.70	B.Malpas tr Robin Flying Group	Liverpool	31. 5.07
G-AYKZ	SAI Kramme KZ-VIII	202	HB-EPB	13.10.70	R.E.Mitchell	RAF Cosford	17. 7.81P
	(DH Gipsy Major 7)		OY-ACB		*(Stored 4.02)*		
G-AYLA	AESL Airtourer T2 (115)	524		12.10.70	D.S.P.Disney	Bristol	30. 6.05
G-AYLC	CEA Jodel DR.1051 Sicile	536	F-BLZG	12.10.70	E.W.B.Trollope	Wing Farm, Longbridge Deverill	4.11.05P
G-AYLF	CEA Jodel DR.1051 Sicile	547	F-BLZQ	14.10.70	R.Twigg and L.Daglish	Rectory Farm, Abbotsley	22. 7.06
					tr Sicile Flying Group		
G-AYLL	CEA Jodel DR.1050 Ambassadeur	11	F-BJHK	27.10.70	C.Joly	Lee-on-Solent	2. 7.06
G-AYLP	American AA-1 Yankee	AA1-0445	EI-AVV	21.10.70	D.Nairn	Haverfordwest	10. 2.02
			G-AYLP		*(Noted 8.04)*		
G-AYLV	Wassmer Jodel D.120 Paris-Nice	300	F-BNCG	27.10.70	M.R.Henham *(Current status unknown)*	(London N2)	13. 9.83P
G-AYLZ	SPP CZL Super Aero 45 Series 04	06-014	9M-AOF	2.11.70	M.J.Cobb	(East Grinstead)	11. 6.76
			F-BILP		*(Damaged Andrewsfield 2.1.76: stored 1997: new owner 3.01)*		
G-AYME	Fournier RF5	5089		6.11.70	R.D.Goodger	Fowle Hall Farm, Laddingford	10. 8.04P
G-AYMK	Piper PA-28-140 Cherokee C	28-26772	N11C	17.11.70	R.Armstrong	Newcastle	8.10.04T
			(PT-DPU)		tr Piper Flying Group		
G-AYMO	Piper PA-23-250 Turbo Aztec C	27-2995	5Y-ACX	18.11.70	A.Wardle	Sheffield City	30. 6.07
			N5845Y, (N5844Y)				
G-AYMP	Phoenix Currie Wot Special	PFA 3014		8.11.70	N.S.Chiitenden	(Herodsfoot, Liskeard)	4.10.94P
	(Built E H Gould)				*(New CofR 11.04)*		
G-AYMR	Lederlin 380L Ladybug	PFA 1513		19.11.70	P.Brayshaw	(Harrogate)	
	(Built J S Brayshaw - c/n EAA/55189) (Continental C90)				*(Last reported under construction 1992: current status unknown)*		
G-AYMU	Wassmer Jodel D.112	1015	F-BJPB	23.11.70	M.R.Baker	(Eastbourne)	5. 6.92P
					(Damaged Hailsham, East Sussex 7.1.92: on rebuild 2004)		
G-AYMV	Western 20 HAB	002		23.11.70	R.F.Turnbull	(Clyro, Hereford)	
					"Tinkerbelle" (New owner 5.02)		
G-AYNA	Phoenix Currie Wot	PFA 3016		25.11.70	I.D.Folland	Bourne Park, Hurstbourne Tarrant	17. 7.05P
	(Built R W Hart) (Continental A65)				*(Noted 11.04)*		
G-AYND	Cessna 310Q	310Q0110	N7610Q	2.12.70	Source Group Ltd	Bournemouth	2. 8.07T
G-AYNF	Piper PA-28-140 Cherokee C	28-26778	N11C	3.12.70	W.S.Bath, M.H.Jones and G.H.Round		
			(PT-DPV)		tr BW Aviation	Wellesbourne Mountford	6. 8.06T
G-AYNJ	Piper PA-28-140 Cherokee C	28-26810	N11C	3.12.70	R.H.Ribbons	Haverfordwest	10. 7.06T
G-AYNN	Cessna 185B Skywagon	185-0518	8R-GCC	11.12.70	Bencray Ltd	Blackpool	7. 3.07T
			VP-GCC, N2518Z		*(Op Blackpool and Fylde Aero Club)*		
G-AYNP*	Westland WS-55 Whirlwind Series 3				See SECTION 3 Part 1	Brandenburg, Germany	
G-AYOW	Cessna 182N Skylane	18260481	N8941G	6. 1.71	D.W.and Susan E.Suttill	Breighton	17. 6.07
G-AYOY	Sikorsky S-61N	61-476		7. 1.71	British International Ltd	Plymouth	21. 4.06T
G-AYOZ	Reims/Cessna FA150L Aerobat	FA1500085		7. 1.71	S.A Hughes	Andrewsfield	21. 4.07
G-AYPE	MBB Bö.209 Monsun 160RV	123	D-EFJA	11. 1.71	Papa Echo Ltd *"Buswells Spirit"*	Biggin Hill	10.11.06
G-AYPG	Reims/Cessna F177RG Cardinal RG	F177RG0007		11. 1.71	D.P.McDermott	Haverfordwest	17. 1 .05
	(Wichita c/n 17700102)						
G-AYPH	Reims/Cessna F177RG Cardinal RG	F177RG0018		11. 1.71	M.R. and K.E.Slack	Cambridge	27. 5.07
	(Wichita c/n 17700146)						
G-AYPJ	Piper PA-28-180 Cherokee E	28-5821	N11C	12. 1.71	R.B.Petrie	RAF Mona	1.10.04T
G-AYPM	Piper PA-18 Super Cub 95	18-1373	ALAT 18-1373	13. 1.71	R.Horner	Trenchard Farm, Eggesford	28. 6.05P
	(L-18C-PI) *(Frame No.18-1282)*		51-15373				
G-AYPO	Piper PA-18 Super Cub 95	18-1615	ALAT 18-1615	13. 1.71	A W.Knowles	Bodmin	24. 6.06
	(L-18C-PI) (RR Continental O-200-A)		51-15615				
	(Rebuilt 1984 using OO-TSJ c/n 18-1398 (Frame No.18-1325) and ex (LN-TSJ)/OO-HMH/51-15398)						
G-AYPR	Piper PA-18 Super Cub 95	18-1631	ALAT 18-1631	13. 1.71	R.G.Manton	Manor Farm, Haddenham	18.12.06T
	(L-18C-PI)		51-15631				
G-AYPS	Piper PA-18 Super Cub 95	18-2092	ALAT 18-2092	13. 1.71	R.J.Hamlett, L.G and D.C.Callow	Andrewsfield	15. 9.05P
	(L-18C-PI)		52-2492				
G-AYPT	Piper PA-18 Super Cub 95	18-1533	(D-EALX)	13. 1.71	B.L.Proctor and T.F.Lyddon	Dunkeswell	27. 5.05
	(L-18C-PI) (RR Continental O-200-A) *(Frame No.18-1508)*		ALAT 18-1533, 51-15533				
G-AYPU	Piper PA-28R-200 Cherokee Arrow B	28R-7135005	N11C	13. 1.71	Monalto Investments Ltd	Jersey	14. 4.05
G-AYPV	Piper PA-28-140 Cherokee D	28-7125039	N11C	13. 1.71	Ashley Gardner Flying Club Ltd	Ronaldsway	6. 9.07T
G-AYPZ	Campbell Cricket	CA/343		13. 1.71	A Melody	Hentsridge	21. 4.04P
	(Volkswagen 1600)						
G-AYRF	Reims/Cessna F150L	F15000665		14. 1.71	D.T.A Rees	Haverfordwest	25.11.00T
G-AYRG	Reims/Cessna F172K	F17200761		14. 1.71	J.H.Mitchell	(Tingley, Wakefield)	8.12.07E
G-AYRH	GEMS MS.892A Rallye Commodore 150	10558	F-BNBX	14. 1.71	S.O'Ceallaigh and J.Barry	(Cork, County Cork)	13. 1.03
G-AYRI	Piper PA-28R-200 Cherokee Arrow B	28R-7135004	N11C	15. 1.71	A E.Thompson and Delta Motor Co (Windsor) Sales Ltd		
						White Waltham	21. 8.05
G-AYRM	Piper PA-28-140 Cherokee D	28-7125049	N11C	19. 1.71	M.J.Saggers	Biggin Hill	7. 8.06T
G-AYRO	Reims/Cessna FA150L Aerobat	FA15000102		21. 1.71	S.M.C.Harvey	Hinton-in-the-Hedges	3. 8.07T
					tr Fat Boys Flying Club		
G-AYRS	Wassmer Jodel D.120A Paris-Nice	255	F-BMAV	22. 1.71	L.R.H.D'Eath	(Diss)	8. 5.05P
G-AYRT	Reims/Cessna F172K	F17200777		22. 1.71	P.E.Crees	Rhosgoch	12. 4.07
G-AYRU	Britten-Norman BN-2A-6 Islander	181	G-51-181	22. 1.71	R.Durie	AAC Netheravon	16. 6.05A
			OH-BNA, G-51-181		tr Army Parachute Association		
G-AYSB	Piper PA-30 Twin Comanche C	30-1916	N8760Y	1. 2.71	N.J.Goff	Biggin Hill	9. 9.05

Reg	Type	C/n	Previous identity	Regd	Owner/operator	Base	CofA
G-AYSD	Slingsby T.61A Falke	1726		4. 2.71	P.W.Hextall	Tatenhill	29. 4.94
					(Stored 1.95: current status unknown)		
G-AYSH	Taylor JT.1 Monoplane	PFA 1413		10. 2.71	C.J.Lodge	Retreat Farm, Little Baddow	16. 5.05P
	(Built C J Lodge)						
G-AYSJ	Bücker Bü.133C Jungmeister	38	D-EHVP	12. 2.71	Patina Ltd	Duxford	10. 7.05P
	(Built AG Fur Dornier-Flugzeuge)		G-AYSJ, HB-MIW, Swiss AF U-91		*(Op The Fighter Collection as "LG+0I" in Luftwaffe c, s)*		
G-AYSK	Phoenix Luton LA-4A Minor	PFA 832		17. 2.71	S.R.Smith	Barton	23. 3.05P
	(Built L Plant) (Continental A65)				tr Luton Minor Group		
G-AYSX	Reims/Cessna F177RG Cardinal RG	F177RG0024		17. 2.71	A P.R.Dean	Hawarden	16. 5.05
	(Wichita c/n 17700175)						
G-AYSY	Reims/Cessna F177RG Cardinal RG	F177RG0026		17. 2.71	S.ATuer	Durham Tees Valley	16. 9.06T
	(Wichita c/n 17700180)						
G-AYTA*	SOCATA MS.880B Rallye Club				See SECTION 3 Part 1	Manchester	
G-AYTR	Menavia Piel CP.301A Emeraude	229	F-BIMD	3. 3.71	R.K.King tr Croft Farm Flying Group	Croft Farm, Defford	9. 9.05P
G-AYTT	Phoenix PM-3 Duet	PFA 841		4. 3.71	H.E.Jenner	Rochester	9.11.04P
	(Built A J Knowles) (Continental C90) *(Officially regd as "Luton Minor III Duet")*						
G-AYTV	Jurca MJ.2D Tempête	PFA 2002		10. 3.71	C W Kirk	Swanborough Farm, Lewes	17. 3.05P
	(Built A Baggallay) (Continental C90)				tr Shoestring Flying Group		
G-AYUA	Auster AOP.9	---	7855M	12. 3.71	P.T.Bolton	Leicester	
	(Officially regd as c/n B5/10-119)		XK416		*(New owner 10.99; frame only)*		
G-AYUB	CEA Jodel DR.253B Régent	185		15. 3.71	D.J.Clark	Sywell	14. 9.05
G-AYUH	Piper PA-28-180 Cherokee F	28-7105042	N11C	17. 3.71	Starpress Ltd	Denham	7. 4.05
G-AYUI*	Piper PA-28-180 Cherokee F	28-7105043	N8557	17. 3.71	Ansair Aviation Ltd	Andrewsfield	5.11.93T
			G-AYUI, N11C		*(Cancelled 27.10.98 by CAA) (Derelict fuselage 1.05)*		
G-AYUJ	Evans VP-1	PFA 1538		17. 3.71	T.N.Howard *"Unforgettable Juliet"*	Woodvale	28. 2.97P
	(Built J A Wills) (Volkswagen 1776)				*(Damaged Ainsdale Beach, Southport 16.6.96: current status unknown)*		
G-AYUK*	Western-Brighton M-B65 HAB				See SECTION 3 Part 1	Munich, Germany	
G-AYUM	Slingsby T.61A Falke	1730		19. 3.71	N A Stone and M H Simms	Shipdham	10. 6.02
					(Unmarked and on rebuild 8.04)		
G-AYUN	Slingsby T.61A Falke	1731		19. 3.71	R.J.Watts tr G-AYUN Group	Rattlesden	15. 5.06
G-AYUP	Slingsby T.61A Falke	1735	XW983	19. 3.71	P.R.Williams	Bicester	15. 7.96
			G-AYUP		*(Stored 2.97: current status unknown)*		
G-AYUR	Slingsby T.61A Falke	1736		19. 3.71	R.Hannigan and R.Lingard	Strubby	7. 4.05
					tr Falke G-AYUR Flying Group		
G-AYUS	Taylor JT.1 Monoplane	PFA 1412		19. 3.71	A M.Sutton	Rayne Hall Farm, Braintree	5. 8.05P
	(Built D G J Barker)						
G-AYUT	SAN Jodel DR.1050 Ambassadeur	479	F-BLJZ	22. 3.71	M.L.Robinson	Kirkbride	16. 6.05
G-AYUV	Reims/Cessna F172H	F17200752		26. 3.71	Justgold Ltd *(Stored 12.01)*	Blackpool	11. 3.00T
G-AYUX*	de Havilland DH.82A Tiger Moth				See SECTION 3 Part 1	Nepi, Rome, Italy	
G-AYVA*	Cameron O-84 HAB				See SECTION 3 Part 1	(Aldershot)	
G-AYVO	Wallis WA-120 Series 1	K/602/X		6. 4.71	K.H.Wallis	Reymerston Hall, Norfolk	31.12.75P
	(RR Continental O-240-B @ 130hp)				*(Stored 8.01)*		
G-AYVP	Aerosport Woody Pusher	PFA 1344		6. 4.71	J.R.Wraight	(Chatham)	
	(Built J.R.Wraight - c/n 181)				*(Stored incomplete: current status unknown)*		
G-AYWA*	Avro 19 Series 2	1361	OO-VIT	14. 4.71	Air Atlantic Historic Flight	(Coventry)	
			OO-DFA, OO-CFA		*(Cancelled 22.8.73 as PWFU) (On long term restoration 3.02)*		
G-AYWD	Cessna 182N Skylane	18260468	N8928G	15. 4.71	S.I.Zorb tr Wild Dreams Group	Leicester	5.12.05T
G-AYWE	Piper PA-28-140 Cherokee C	28-26826	N5910U	16. 4.71	Intelcomm (UK) Ltd	Denham	30. 5.05
G-AYWH	SAN Jodel D.117A	844	F-BIVO	16. 4.71	D.Kynaston	Cambridge	13. 7.04P
G-AYWM	AESL Airtourer T5 (Super 150)	A 534		16. 4.71	H.E.Collett tr The Star Flying Group	Gloucestershire	8. 7.06
G-AYWT	AIA Stampe SV-4C	1111	F-BLEY	21. 4.71	R.A.Palmer	(Weybridge)	20. 3.06T
	(DH Gipsy Major 10)		F-BAGL				
G-AYXP	SAN Jodel D.117A	693	F-BIDD	27. 4.71	G.N.Davies	Enstone	25. 8.04P
G-AYXS	SIAI-Marchetti S.205-18R	4-165	OY-DNG	28. 4.71	P.J.Bloore and Jennifer M.Biles	Croft Farm, Defford	26. 6.07T
G-AYXT*	Westland WS-55 Whirlwind HAS.7 (Series 2)				See SECTION 3 Part 1	Weston-super-Mare	
G-AYXU	Champion 7KCAB Citabria	232-70	N7587F	28. 4.71	Les Wallen Manufacturing Ltd	(Ramsgate)	21. 5.07
G-AYXW*	Evans VP-1	PFA 1544		30. 4.71	M.Howe	Netherthorpe	15. 8.01P
	(Built J S Penny) (Ardem 4C02)				*(Cancelled 7.6.02 by CAA) (Noted 4.04)*		
G-AYYO	CEA Jodel DR.1050/M1 Sicile Record	622	EI-BAI	11. 5.71	D.J.M.White	Boscombe Down	28. 9.05
			G-AYYO, F-BMPZ		tr Bustard Jodel Group		
G-AYYT	CEA Jodel DR.1050/M1 Sicile Record	587	F-BMGU	13. 5.71	W.R.Prescott	Kilkeel, County Down	21. 8.05
					tr Yankee Tango Group		
G-AYYU	Beech C23 Musketeer Custom	M-1353		14. 5.71	D.M.Powell tr G-AYYU Group	Sturgate	17. 5.74
G-AYYX	SOCATA MS.880B Rallye Club	1812		18. 5.71	J.G.MacDonald	Morgansfield, Fishburn	18. 4.05
G-AYZE	Piper PA-39 Twin Comanche C/R	39-92	N8934Y	20. 5.71	J.E.Balmer	Gloucestershire	18. 9.06
G-AYZI	SNCAN Stampe SV-4C	15	(EI-)	24. 5.71	D.M.and Patricia Ann Fenton	Sheffield City	13.11.06
	(Renault 4P)		G-AYZI, F-BBAA, Fr Mil				
G-AYZJ*	Westland WS-55 Whirlwind HAS.7				See SECTION 3 Part 1	Newark	
G-AYZK	CEA Jodel DR.1050/M1 Sicile Record	590	F-BMGY	24. 5.71	D.G.Hesketh	Streethay	8.10.06
G-AYZS	Druine D.62B Condor	RAE/650		4. 6.71	M.N.Thrush	Manor Farm, Inglesham	5.12.05P
	(Built Rollason Aircraft and Engines)						
G-AYZU	Slingsby T.61A Falke	1740		4. 6.71	R.G.Garner	Wellesbourne Mountford	10. 6.07
					tr The Falcon Gliding Group		
G-AYZW	Slingsby T.61A Falke	1743		4. 6.71	I.D.Walton t/a G-ZW Group	(Buckley)	27. 6.07
					tr Portmoak Falke Syndicate		

G-AZAA - G-AZZZ

Reg	Type	C/n	Previous identity	Regd	Owner/operator	Base	CofA
G-AZAB	Piper PA-30 Twin Comanche B	30-1475	5H-MNM	8. 6.71	Bickertons Aerodromes Ltd	Denham	22. 9.07
			5Y-AGB				
G-AZAJ	Piper PA-28R-200 Cherokee Arrow	28R-7135116	N11C	18. 6.71	J.C.McHugh and P.Woulfe	Stapleford	10. 7.06
G-AZAK*	Scottish Aviation Bulldog Series 101 (SK-61)				See SECTION 3 Part 1	Linköping, Sweden	
G-AZAU*	Cierva Rotorcraft CR.LTH.1 Grasshopper III				See SECTION 3 Part 1	Weston-super-Mare	

Reg	Type	C/n	Previous identity	Reg date	Owner	Base	CofA
G-AZAW	Sud-Aviation Gardan GY-80-160 Horizon	104	F-BMUL	24. 6.71	J.W.Foley	(Inverness)	28. 5.05
G-AZAZ*	Bensen B.8M				See SECTION 3 Part 1	RNAS Yeovilton	
G-AZBA	Tipsy T.66 Nipper Series 3B	PFA 1390		30. 6.71	L.A.Brown	Swansea	6. 7.05P
	(Built E Shouler) (Volkswagen 1834)						
G-AZBB	MBB Bö.209 Monsun 160FV	137	D-EFJO	1. 7.71	G.N.Richardson	Shelsley Beauchamp, Worcester	9. 7.04
					t/a GN Richardson Motors		
G-AZBE	AESL Airtourer T5 (Super 150)	A 535		5. 7.71	R.G.Vincent tr BE Flying Group	Gloucestershire	23. 1.06
G-AZBH*	Cameron O-84 HAB				See SECTION 3 Part 1	Newbury	
G-AZBI	SAN Jodel D.150 Mascaret	43	F-BMFB	12. 7.71	F.M.Ward	AAC Dishforth	8. 2.05P
G-AZBL	Jodel D.9 Bébé	PFA 938		12. 7.71	J.Hill	(Dudley)	15.10.85P
	(Built D S Morgans) (Volkswagen 1500)				*(On rebuild 1993 ?: current status unknown)*		
G-AZBN	Noorduyn AT-16-ND Harvard IIB	14A-1431	PH-HON	13. 7.71	Swaygate Ltd	(Steyning)	12. 7.01P
			R.Neth.AF B-97, FT391, 43-13132 *(As "FT391") (Current status unknown)*				
G-AZBU	Auster AOP.9	AUS.183	7862M	15. 7.71	K.Brooks	Tollerton	19. 5.05P
			XR246		tr Auster Nine Group *(As "XR246" in RAE c/s)*		
G-AZBV*	Britten-Norman BN-2A-2 Islander				See SECTION 3 Part 1	Beersheba, Israel	
G-AZCB	SNCAN Stampe SV-4C	140	F-BBCR	21. 7.71	M.L.Martin	Shoreham	12. 8.07
	(DH Gipsy Major 10)						
G-AZCE	Pitts S-1C	PFA/1527		26. 7.71	R.J.Oulton	(Chepstow)	18. 6.76S
	(Built R J Oulton - c/n 373.H) (Lycoming O-235)				*(Crashed Eastbach Farm, Coleford 2.9.75) (Current status unknown)*		
G-AZCK	Beagle B.121 Pup Series 2	B121-153		30. 7.71	D.R.Newell	Newtownards	15. 7.07
G-AZCL	Beagle B.121 Pup Series 2	B121-154		30. 7.71	J.J.Watts and D.Fletcher	Old Sarum	8. 7.07T
G-AZCM*	Beagle B.121 Pup Series 150				See SECTION 3 Part 1	Manston	
G-AZCN	Beagle B.121 Pup Series 2	B121-156	HB-NAY	30. 7.71	D.M.Callaghan, E.J.and Thelma M.Spencer	Egginton	24. 6.07
			G-AZCN				
G-AZCP	Beagle B.121 Pup Series 1	B121-158	(D-EKWA)	30. 7.71	T.J.Watson	Elstree	18. 7.07
			G-AZCP				
G-AZCT	Beagle B.121 Pup Series 1	B121-161		30. 7.71	J.Coleman	Sywell	1. 8.05T
G-AZCU	Beagle B.121 Pup Series 1	B121-162		30. 7.71	A A Harris	Shobdon	31. 8.07
G-AZCV	Beagle B.121 Pup Series 2	B121-163	HB-NAR	30. 7.71	N.R.W.Long	Compton Abbas	14. 8.05
			G-AZCV		*(Great Circle Design titles)*		
G-AZCZ	Beagle B.121 Pup Series 2	B121-167		30. 7.71	L. and J.M.Northover	Cardiff	4. 8.07T
G-AZDA	Beagle B.121 Pup Series 1	B121-168		30. 7.71	B.D.Deubelbeiss	Elstree	29. 1.06
G-AZDD	MBB Bö.209 Monsun 150FF	143	D-EBJC	3. 8.71	J.D.Hall and D.Lawrence	Rochester	20. 3.07
					tr Double Delta Flying Group		
G-AZDE	Piper PA-28R-200 Cherokee Arrow	28R-7135141	N11C	3. 8.71	C.Wilson *(Flyteam Aviation titles)*	Sywell	3.10.05
G-AZDG	Beagle B.121 Pup Series 2	B121-145	(G-BLYM)	17. 6.85	J.R.Heaps	Elstree	28. 4.07
			HB-NAM, (VH-EPT), G-35-145 *(DHL titles)*				
G-AZDJ	Piper PA-32-300 Cherokee Six D	32-7140068	OY-AJK	23. 8.71	Delta Juliet Ltd	Cardiff	25. 8.06T
			G-AZDJ, N5273S				
G-AZDX	Piper PA-28-180 Cherokee F	28-7105186	N11C	25. 8.71	M.Cowan	Hundon, Suffolk	26. 3.05
G-AZDY	de Havilland DH.82A Tiger Moth	86559	F-BGDJ	25. 8.71	J.B.Mills	(Sawbridgeworth)	18. 8.97
	(Built Morris Motors Ltd)		Fr.AF, PG650		*(New CofR 11.04)*		
G-AZDZ*	Cessna 172K Skyhawk	17258501	5N-AIH	25. 8.71	Home Office Fire and Emergency Training Centre		
			N1647C, N84508			Moreton in Marsh	25. 2.83
	(Forced landing Delapre Golf Course, Northants 19.9.81 and extensively damaged: cancelled 5.12.83 as destroyed) (Frame noted 8.03)						
G-AZEE	Morane Saulnier MS.880B Rallye Club	74	F-BKKA	1. 9.71	J.Shelton	Water Leisure Park, Skegness	27. 9.98
	(Composite including fuselage of G-AZNJ c/n 5375 in 1980)				*(Noted 4.04)*		
G-AZEF	Wassmer Jodel D.120 Paris-Nice	321	F-BNZS	1. 9.71	D.Street	Netherthorpe	28. 4.05P
G-AZEG	Piper PA-28-140 Cherokee D	28-7125530	N11C	1. 9.71	The Ashley Gardner Flying Club Ltd	Ronaldsway	3. 7.07T
G-AZER*	Cameron O-42 (Ax5) HAB				See SECTION 3 Part 1	Newbury	
G-AZEV	Beagle B.121 Pup Series 2	B121-131	VH-EPM	15. 9.71	C.J.Partridge	Popham	23. 8.05
			G-35-131				
G-AZEW	Beagle B.121 Pup Series 2	B121-132	VH-EPN	15. 9.71	D.and M.Bonsall	Netherthorpe	18. 5.06
			G-35-132		t/a Dukeries Aviation		
G-AZEY	Beagle B.121 Pup Series 2	B121-136	HB-NAK	15. 9.71	M.E.Reynolds	Popham	5.12.06
			G-AZEY, VH-EPP, G-35-136				
G-AZFA	Beagle B.121 Pup Series 2	B121-143	VH-EPR	15. 9.71	J.Smith	Egginton	23. 7.04
			G-35-143				
G-AZFC	Piper PA-28-140 Cherokee D	28-7125486	N11C	16. 9.71	V.Wyre tr WLS Flying Group	White Waltham	28. 8.05
G-AZFF	Wassmer Jodel D.112	1175	F-BLFI	17. 9.71	T.Mackey	(Tramore, County Waterford)	18. 1.05P
G-AZFI	Piper PA-28R-200 Cherokee Arrow B	28R-7135160	N11C	21. 9.71	GAZFI Ltd	Sherburn-in-Elmet	6. 4.07
G-AZFM	Piper PA-28R-200 Cherokee Arrow B	28R-7135218	N11C	24. 9.71	P.J.Jenness	Bournemouth	21. 9 07
G-AZFR	Cessna 401B	401B0121	N7981Q	30. 9.71	R E Wragg	Blackpool	10. 9.05T
G-AZFW*	Hawker Siddeley HS.121 Trident 2E Series 102				See SECTION 3 Part 1	Guangzou, China	
G-AZGA	Wassmer Jodel D.120 Paris-Nice	144	F-BIXV	30. 9.71	A F.Vizoso	RAF Halton	14.12.04P
G-AZGE	SNCAN Stampe SV-4C	576	F-BDDV	6.10.71	M.R.L.Astor	East Hatley, Tadlow	15. 8.94
	(Renault 4P)				*(Stored 3.97: current status unknown)*		
G-AZGF	Beagle B.121 Pup Series 2	B121-076	PH-KUF	6.10.71	K.Singh	Barton	2. 5.98
			G-35-076		*(Noted 7.04)*		
G-AZGI*	SOCATA MS.880B Rallye Club	1896		7.10.71	B.McIntyre	Mullaghmore	4. 2.05
					(Cancelled 4.3.04 by CAA)		
G-AZGL	SOCATA MS.894A Rallye Minerva 220	119291		7.10.71	The Cambridge Aero Club Ltd	Cambridge	8.12.05T
G-AZGY	Rousseau CP.301B Emeraude	122	F-BRAA	12.10.71	R.H.Braithwaite	RAF Henlow	17. 6.05P
G-AZGZ	de Havilland DH.82A Tiger Moth	86489	F-BGCF	13.10.71	R.J.King	Rush Green	4. 6.05
	(Built Morris Motors Ltd)		Fr.AF, NM181		*(As "NM181")*		
G-AZHB	Robin HR100/200B Royal	118		14.10.71	P.Fenwick	(London SE24)	6. 8.03
G-AZHC	Wassmer Jodel D.112	585	F-BIQQ	18.10.71	A.Galante tr Aerodel Flying Group	Netherthorpe	11. 8.05P
G-AZHD	Slingsby T.61A Falke	1753		18.10.71	Nicola J.Orchard-Armitage	(Deal)	28. 4.07
G-AZHH	K & S Squarecraft SA.102.5 Cavalier	PFA 1393		20.10.71	D.W.Buckle	Morton Carr Farm, Nunthorpe	20. 1.00P
	(Built D Buckle) (Lycoming O-290)						
G-AZHI	AESL Airtourer T5 (Super 150)	A 540		20.10.71	Flying Grasshoppers Ltd	Rochester	23. 4.06T
G-AZHK	Robin HR100/200B Royal	113	G-ILEG	22.10.71	D.J.Sage	(Reigate)	20. 5.05
			G-AZHK				

G-AZHR	Piccard Ax6 HAB	617	N17US	27.10.71	C.Fisher	(Aston, Sheffield)	AC
					tr Halcyon Balloon Group *"Happiness"*		
G-AZHT	AESL Airtourer 115	525		29.10.71	Aviation West Ltd	(Glasgow)	29. 1.89T
	(Continental O-240)				*(Amended CofR 3.03)*		
G-AZHU	Phoenix Luton LA-4A Minor	PFA 839		1.11.71	W.Cawrey	Netherthorpe	4. 1.05P
	(Built A E Morris) (Volkswagen 1834)						
G-AZIB	SOCATA ST-10 Diplomate	141		4.11.71	W.B.Bateson	Blackpool	17. 2.07
G-AZID	Reims/Cessna FA150L Aerobat	FA1500083	N9447	8.11.71	Aerobat Ltd	Wolverhampton	6. 1.06T
G-AZII	SAN Jodel D.117	A 848	F-BNDO	12.11.71	P.J.Brayshaw	Haddock Stone Farm, Markington	11. 4.01P
			F-OBFO				
G-AZIJ	Robin DR360 Chevalier	634		15.11.71	K.J.Fleming	RAF Woodvale	5. 6.06
G-AZIK	Piper PA-34-200 Seneca	34-7250018	N2392T	15.11.71	Walkbury Aviation Ltd	Sibson	30. 4.05T
G-AZIL	Slingsby T.61A Falke	1756		16.11.71	D.W.Savage	Arbroath	20.11.05
G-AZIP	Cameron O-65 HAB	29		24.11.71	P.G.Dunnington	(Hungerford)	5. 5.81A
					tr Dante Balloon Group *"Dante" (Non-airworthy - inflated 4.02)*		
G-AZJC	Fournier RF5	5108		30.11.71	W.S.V.Stoney	(Arezzo, Italy)	1. 5.02P
G-AZJD*	North American AT-6D-NT Harvard IIA				See SECTION 3 Part 1	Yvelines, France	
G-AZJE	Gardan GY-20 Minicab	PFA 1806		1.12.71	J.B.Evans	Ventnor, Isle of Wight	7. 7.82P
	(Built J B Evans - c/n JBE.1 to JB.01 Minicab standard) (Continental C90)				*(Stored 1.98: current status unknown)*		
G-AZJI*	Western O-65 HAB				See SECTION 3 Part 1	Newbury	
G-AZJN	Robin DR300/140 Major	642		6.12.71	Wright Farm Eggs Ltd	Cherry Tree Farm, Monewden	5. 9.05
G-AZJO*	Scottish Aviation Bulldog Series 101 (SK-61)				See SECTION 3 Part 1	Linköping, Sweden	
G-AZJV	Reims/Cessna F172L	F17200810		8.12.71	M.W.Smith tr GAZJV Flying Group	Exeter	13. 3.06
G-AZJY	Reims/Cessna FRA150L Aerobat	FRA1500126		8.12.71	P.J.McCartney	Barton	7. 7.07
G-AZKC	SOCATA MS.880B Rallye Club	1914		8.12.71	L.J.Martin	Bembridge	16.11.06
G-AZKE	SOCATA MS.880B Rallye Club	1950	(LX-SDT)	8.12.71	D.A Thompson and J.D.Headlam	Southend	6. 3.06
G-AZKK	Cameron O-56 HAB	32		13.12.71	P.J.Green and C.Bosley	(Newbury)	23.12.82A
					tr Gemini Balloon Group *"Gemini"*		
G-AZKO	Reims/Cessna F337F Super Skymaster			20.12.71	Willpower Garage Ltd	Wellesbourne Mountford	14. 8.06
	(Wichita c/n 33701380)	F33700041			*"Bird Dog"*		
G-AZKP	SAN Jodel D.117	419	F-BHND	20.12.71	A M. and J.L.Moar	Wick	5. 5.05P
G-AZKR	Piper PA-24-180 Comanche	24-2192	N7044P	23.12.71	J.Van Der Kwast	Rochester	21. 6.04
G-AZKS	American AA-1A Trainer	0334	N6134L	23.12.71	M.D.Henson	Coventry	6..11.06
G-AZKW	Reims/Cessna F172L	F17200836		23.12.71	J.C.C.Wright	Hinton-in-the-Hedges	21. 6.03T
G-AZKZ	Reims/Cessna F172L	F17200814		23.12.71	R.D.Forster *(Op Norfolk Flying Club)*	Beccles	9. 9.07T
G-AZLE	Boeing Stearman E75 (N2S-5) Kaydet	75-8543	CF-XRD	29.12.71	A E.Poulsom	Manor Farm, Tongham	31. 5.07
	(Continental W670)		N5619N, Bu43449		*(As "2" in US Army c/s)*		
G-AZLF	Wassmer Jodel D.120 Paris-Nice	230	F-BLFL	30.12.71	M.S.C.Ball	Garston Farm, Marshfield	20.11.04P
G-AZLH	Reims/Cessna F150L	F15000757		31.12.71	L.Papatheocharis and I.Buck *(Noted 12.04)*	Sywell	30.11.03T
G-AZLM*	Reims/Cessna F172L				See SECTION 3 Part 1	Flixton, Bungay	
G-AZLN	Piper PA-28-180 Cherokee F	28-7105210	N11C	3. 1.72	Liteflite Ltd	Oxford	5. 3.07
G-AZLV	Cessna 172K	17257908	4X-ALM	10. 1.72	P.Williams	Wickenby	27. 5.07E
			N79138		tr G-AZLV Flying Group		
G-AZLY	Reims/Cessna F150L	F15000771		10. 1.72	M.Stewart	Spanhoe	9. 1.06T
G-AZMC	Slingsby T.61A Falke	1757		12. 1.72	B.Molloy tr G-AZMC Group	Challock	27. 6.05
G-AZMD	Slingsby T.61C Falke	1758		12. 1.72	R.A Rice	Wellesbourne Mountford	10. 6.07
G-AZMF	British Aircraft Corporation One-Eleven 530FX		7Q-YKJ	14. 1.72	European Aviation Ltd	Bournemouth	22. 1.04T
		BAC.240	G-AZMF, PT-TYY, G-AZMF		*(European VIP First c/s) "The European Express" (At Aviation Museum 1.05)*		
G-AZMJ	American AA-5 Traveler	0019		27. 1.72	R.T.Love	St.Merryn	1. 5.07
G-AZMN*	AESL Airtourer T5 (Super 150)	A 550		28. 1.72	Not known	Oaksey Park	7. 5.89
					(Crashed near Glasgow 23.6.87: cancelled 14.9.88 by CAA: stored 9.03)		
G-AZMZ	SOCATA MS.893A Rallye Commodore 180	11927		8. 2.72	D.R.Wilcox	Lyveden	14. 8.06
G-AZNA*	Vickers 813 Viscount				See SECTION 3 Part 1	Waarschoot, Belgium	
G-AZNK	SNCAN Stampe SV-4A	290	F-BKXF	15. 2.72	P.D.Jackson and R.A G.Lucas	Redhill	20. 7.07
	(DH Gipsy Major 10)		F-BCGZ		tr November Kilo Group *"Globird"*		
G-AZNL	Piper PA-28R-200 Cherokee Arrow II	28R-7235006	N11C	16. 2.72	B.P.Liversidge	Poplar Hall Farm, Elmsett	15. 1.06T
G-AZNO	Cessna 182P Skylane	18261005	N7365Q	18. 2.72	S.Turton	(Barnstaple)	4. 4.07T
G-AZNT	Cameron O-84 HAB	34		21. 2.72	N.Tasker *"Oberon"*	(Bristol)	5. 6.85
G-AZOA	MBB Bö.209 Monsun 150FF	183	D-EAAY	21. 2.72	M.W.Hurst	Seighford	15. 7.07
G-AZOB	MBB Bö.209 Monsun 150FF	184	D-EAAZ	21. 2.72	G.N.Richardson	Shelsley Beauchamp, Worcester	9. 7.84
					(Crashed Droitwich 21.8.83: stored 8.92: current status unknown)		
G-AZOE	AESL Airtourer T2 (115)	528		21. 2.72	B.J.Edmondson and J.K.Smithson		
					tr G-AZOE 607 Group	Shotton Colliery, Peterlee	28.10.06
G-AZOF	AESL Airtourer T5 (Super 150)	A 549		21. 2.72	R.J.W.Bayliff and A C.Hart	Kirknewton	17. 2.04
					tr Cirrus Flying Group *(On overhaul Perth 9.04)*		
G-AZOG	Piper PA-28R-200 Cherokee Arrow II	28R-7235009	N11C	21. 2.72	Atromin Ltd t/a Southend Flying Club	Southend	18. 8.07
G-AZOL	Piper PA-34-200 Seneca	34-7250075	N4348T	28. 2.72	Stapleford Flying Club Ltd	Stapleford	7. 9.07
G-AZOO	Western O-65 HAB	015		1. 3.72	Semajan Ltd	(Newbury)	6. 6.77S
					"Carousel" (On loan to British Balloon Museum and Library)		
G-AZOT	Piper PA-34-200 Seneca	34-7250073	N4340T	3. 3.72	Northern Aviation Ltd	Durham Tees Valley	13. 9.06T
G-AZOU	SAN Jodel DR.1050 Sicile	354	F-BJYX	7. 3.72	D.Elliott and D.Holl	Wellcross Grange, Slinfold	30. 6.05
					tr Horsham Flying Group		
G-AZOZ	Reims/Cessna FRA150L Aerobat	FRA1500136		7. 3.72	Seawing Flying Club Ltd	Southend	16. 7.05T
					"The Wizard of Oz"		
G-AZPA	Piper PA-25-235 Pawnee C	25-5223	N8797L	7. 3.72	Black Mountains Gliding Club Ltd	Talgarth	17. 2.07
			N9749N				
G-AZPC	Slingsby T.61C Falke	1767		7. 3.72	The Surrey Hills Gliding Club Ltd	Kenley	31. 7.07
G-AZPF	Fournier RF5	5001	D-KOLT	10. 3.72	R.Pye	Blackpool	23. 8.05P
G-AZPH*	Pitts S.1S				See SECTION 3 Part 1	South Kensington, London	
G-AZPV	Phoenix Luton LA-4A Minor	PFA 833		14. 3.72	J.R.Faulkner	(Brize Norton)	18. 9.97P
	(Built J Scott) (Lycoming O-145)				*(Stored 3.01)*		
G-AZPX	Western O-31 HAB	011		20. 3.72	B.L.King	(Coulsdon)	
					tr Eugena Rex Balloon Group *"Eugena Rex"*		
G-AZRA	MBB Bö.209 Monsun 150FF	192	D-EAIH	21. 3.72	Alpha Flying Ltd	Booker	3. 5.07

G-AZRD	Cessna 401B	401B0218	N7999Q	22. 3.72	G.Hatton tr Romeo Delta Group	Blackpool	6. 4.06T
G-AZRH	Piper PA-28-140 Cherokee D	28-7125585	N11C	23. 3.72	H.B.Carter tr Trust Flying Group	Jersey	18. 7.05
G-AZRI	Payne HAB (56,500 cu.ft)	GFP.1		21. 3.72	C.A Butter and J.J.T.Cooke	(Newbury/Southall)	
	(Built G F Payne)				t/a Aardvark Balloon Co *"Shoestring"*		
G-AZRK	Fournier RF5	5112		23. 3.72	A B.Clymo and J.F.Rogers	Shenington	22.12.04P
G-AZRL	Piper PA-18 Super Cub 95	18-1331	OO-SBR	23. 3.72	M.G.Fountain	Leicester	16.10.04
	(L-18C-PI) *(Frame No.18-1213)*		OO-HML, ALAT 18-1331, 51-15331				
G-AZRM	Fournier RF5	5111		24. 3.72	R.Speer and M.Millar	Upper Broyle Farm, Ringmer	2. 5.05P
	(Volkswagen 1834)				tr Romeo Mike Group		
G-AZRN	Cameron O-84 HAB	28		28. 3.72	C.J.Desmet	(Brussels, Belgium)	4. 7.81A
G-AZRP	AESL Airtourer T2 (115)	529		28. 3.72	B.F.Strawford	Shobdon	29.11.07
G-AZRR	Cessna 310Q	310Q0490	N9923F	28. 3.72	M.A Rooney	(Newark)	4. 5.07
G-AZRS	Piper PA-22-150 Tri-Pacer	22-5141	XT-AAH	28. 3.72	R.H.Hulls *"Sandpiper"*	(Broadbury, Okehampton)	29.10.06
			F-OCGZ, ALAT 22-5141, "FMKAC", N10F				
G-AZRV*	Piper PA-28R-200 Cherokee Arrow B	28R-7135191	N2309T	4. 4.72	Not known	Compton Abbas	
	(Crashed on take-off Compton Abbas 30.12.00: cancelled 20.6.01 as destroyed) (Fuselage dumped 2.05)						
G-AZRX*	Sud-Aviation Gardan GY-80-160 Horizon	14	F-BLIJ	4. 4.72	Adventure Island Pleasure Ground	Southend-on-Sea	20. 2.92
	(Damaged Sandtoft 14.8.91: cancelled 21.10.91 by CAA) (On display Crazy Golf Course, Marine Parade 1.05)						
G-AZRZ	Cessna U206F Stationair	U20601803	N9603G	4. 4.72	M.R.Browne and R.G.Wood	Hinton-in-the-Hedges	11. 6.06
					t/a Hinton Skydiving Centre		
G-AZSA	Stampe et Renard Stampe SV-4B	1203	Belgian AF V-61	5. 4.72	J.K.Faulkner	Biggin Hill	6. 9.07P
	(Officially regd with c/n 64)						
G-AZSC	Noorduyn AT-16-ND Harvard IIB	14A-1363	PH-SKK	7. 4.72	Machine Music Ltd	North Weald	8.12.05P
			R.Neth AF B-19, FT323, 43-13064 *(As "43/SC" in USAAF c/s)*				
G-AZSD	Slingsby T.29B Motor Tutor	PFA 1574		7. 4.72	R.G.Boyton	Halstead, Essex	
	(Built R G Boyton - c/n RGB 01/72) (Rebuild of Slingsby c/n 561)				t/a Essex Aviation *(Current status unknown)*		
G-AZSF	Piper PA-28R-200 Cherokee Arrow II	28R-7235048	N11C	10. 4.72	Wellesbourne Flyers Ltd	Wellesbourne Mountford	22. 8.05
					t/a Wellesbourne Aviation		
G-AZSP*	Cameron O-84 HAB				See SECTION 3 Part 1	Newbury	
G-AZSW	Beagle B.121 Pup Series 1	B121-140	PH-VRT	24. 4.72	J.R.Parry	RAF Mona	15. 6.06
			G-35-140				
G-AZSZ	Piper PA-23-250 Aztec D	27-4194	N6851Y	25. 4.72	A A Mattacks *(Noted 1.05)*	Oaksey Park	2. 6.01T
G-AZTA	MBB Bö.209 Monsun 150FF	190	D-EAIF	25. 4.72	K.T.Pierce	North Weald	27.10.07E
G-AZTD*	Piper PA-32-300 Cherokee Six D	32-7140001	N8611N	26. 4.72	(Air Navigation and Trading Co Ltd)	Blackpool	16. 8.98T
					(Cancelled 11.4.01 by CAA) (Noted 10.04)		
G-AZTF	Reims/Cessna F177RG Cardinal RG	F177RG0054		28. 4.72	R.Burgun	Egginton	13. 8.04
G-AZTK	Reims/Cessna F172F	F17200116	PH-CON	27. 4.72	S.O'Ceallaigh	Haverfordwest	20.10.00
			OO-SIR		*(Noted wrecked 8.04)*		
G-AZTS	Reims/Cessna F172L	F17200866		28. 4.72	C.E.Stringer	Bagby	17.12.03T
G-AZTV	Stolp SA.500 Starlet	PFA 1584		19. 5.72	G.G.Rowland	(Christchurch)	19.11.92
	(Built S S Miles - c/n SSM.2) (Continental C90)			*(Damaged Manor Farm, Grateley, Hampshire 4.7.92: current status unknown)*			
G-AZTW	Reims/Cessna F177RG Cardinal RG	F177RG0043		28. 4.72	I.M.Richmond	Panshanger	15. 6.06
G-AZUM	Reims/Cessna F172L	F17200863		11. 5.72	M.S.Hills tr Fowlmere Flyers	Fowlmere	22. 1.07
G-AZUP	Cameron O-65 HAB	36		11. 5.72	R.S.Bailey and A B.Simpson *"Eight of Hearts"*	(Aylesbury)	23.10.77S
G-AZUT	SOCATA MS.893A Rallye Commodore 180	10963	VH-TCH	12. 5.72	J.Palethorpe	Blakedown, Kidderminster	19. 1.06
					tr Rallye Flying Group		
G-AZUV*	Cameron O-65 HAB				See SECTION 3 Part 1	Newbury	
G-AZUW*	Cameron A-140 HAB				See SECTION 3 Part 1	Newbury	
G-AZUY	Cessna 310L	310L0012	SE-FEC	15. 5.72	W.B.Bateson	Blackpool	5.11.05T
			LN-LMH, N2212F				
G-AZUZ	Reims/Cessna FRA150L Aerobat	FRA1500146		16. 5.72	D.J.Parker	Netherthorpe	18.12.06
G-AZVA	MBB Bö.209 Monsun 150FF	177	(D-EAAQ)	16. 5.72	C.Elder	(Cheddar)	30.10.06
G-AZVB	MBB Bö.209 Monsun 150FF	178	(D-EAAS)	16. 5.72	Elizabeth and P.M.L.Cliffe	(Mulbarton, Norwich)	23. 6.06
G-AZVF	SOCATA MS.894A Rallye Minerva 220	11999	(F-OCSR)	16. 5.72	L.Williams tr Minerva Flying Group	Upfield Farm, Usk	9. 3.02T
G-AZVG	American AA-5 Traveler	AA5-0075		16. 5.72	K.M.Whelan tr G-AZVG Group	(Luton)	15. 9.05T
G-AZVH	SOCATA MS.894A Rallye Minerva 220	12017		16. 5.72	P.L.Jubb	Poplar Hall Farm, Elmsett	17. 6.05
G-AZVI	SOCATA MS.892A Rallye Commodore 150	12039		6. 5.72	R.E.Knapton	Turweston	12. 8.04
G-AZVJ	Piper PA-34-200 Seneca	34-7250125	N4529T	16. 5.72	Andrews Professional Colour Laboratories Ltd	Lydd	21. 8.03A
G-AZVL	Jodel D.119	794	F-BILB	19. 5.72	P.T.East	Hill Farm, Nayland	16.10.04P
	(Built Etablissement Valladeau)				tr Forest Flying Group		
G-AZVM	Hughes 369HS	61-0326S	N9091F	19. 5.72	GTS Engineering (Coventry) Ltd	Coventry	20. 5.06
G-AZVP	Reims/Cessna F177RG Cardinal RG	F177RG0057		22. 5.72	Cardinal Flyers Ltd	Denham	25. 6.04
G-AZWB	Piper PA-28-140 Cherokee E	28-7225244	N11C	5. 6.72	B.N.Rides and L.Connor	Kemble	19.12.03
G-AZWD	Piper PA-28-140 Cherokee E	28-7225298	N11C	6. 6.72	County Connections Ltd	Bournemouth	6. 5.05T
					t/a Solent School of Flying		
G-AZWE	Piper PA-28-140 Cherokee E	28-7225303	N11C	6. 6.72	P.M.Tucker tr G-AZWE Flying Group	Dunkeswell	11. 4.05T
G-AZWF	SAN Jodel DR.1050 Ambassadeur	130	F-BJJT	7. 6.72	J.A D. Reedie	Inverness	5.11.04
	(Composite including fuselage of DR.1050M F-BLJX c/n 492)				tr Cawdor Flying Group		
G-AZWO*	Scottish Aviation Bulldog Series 101 (SK-61)				See SECTION 3 Part 1	Linköping, Sweden	
G-AZWS	Piper PA-28R-180 Cherokee Arrow	28R-30749	N4993J	8. 6.72	G.S.Blair tr Arrow 88 Flying Group	Eshott	9. 7.06
G-AZWT	Westland Lysander IIIA	Y1536	RCAF 1582	9. 6.72	Richard Shuttleworth Trustees	Old Warden	15. 8.04P
			V9552		*(As "V9367/MA-B" in 161 Sqdn c/s)*		
G-AZWY	Piper PA-24-260 Comanche C	24-4806	N9310P	16. 6.72	Keymer, Son and Co Ltd	Biggin Hill	27. 5.06
G-AZXB	Cameron O-65 HAB	48		20. 6.72	R.J.Mitchener and P.F.Smart	(Andover)	6. 5.81A
					t/a Balloon Collection *"London Pride II"*		
G-AZXC	Reims/Cessna F150L	F15000793		20. 6.72	D.C.Bonsall	Netherthorpe	28. 4.06T
G-AZXD	Reims/Cessna F172L	F17200878		20. 6.72	Birdlake Ltd *(Op Birdlake Aviation)*	Birmingham	28. 8.06T
G-AZXG*	Piper PA-23-250 Aztec D	27-4328	N6963Y	23. 6.72	Cranfield University	Cranfield	18. 9.94
			(Crashed Little Snoring 25.10.91: cancelled 6.5.93 by CAA) (Iinstructional airframe 8.04)				
G-AZYA	Sud-Aviation Gardan GY-80-160 Horizon	57	F-BLPT	7. 7.72	P.J.Fahie	Draycott Farm, Chiseldon	5. 4.07
G-AZYB*	Bell 47H-1				See SECTION 3 Part 1	Weston-super-Mare	
G-AZYD	GEMS MS.893A Rallye Commodore 180	10645	F-BNSE	30. 6.72	P.Storey	Husbands Bosworth	8. 8.05
					t/a Storey Aviation Services		
G-AZYL*	Portslade School HAB				See SECTION 3 Part 1	Newbury	

G-AZYS	Scintex CP.301C1 Emeraude	568	F-BJAY	7. 7.72	C.G.Ferguson and D.Drew	Jericho Farm, Lambley	10. 5.04P
G-AZYU	Piper PA-23-250 Aztec E	27-4601	N13983	13. 7.72	L.J.Martin	Sandown	16. 5.04
G-AZYY	Slingsby T.61A Falke	1770		12. 7.72	J.A Towers	Yearby	19. 6.05
G-AZYZ	Wassmer WA.51A Pacific	30	F-OCSE	14. 7.72	C.R.Buxton	(Gourvillette, France)	28. 6.04
G-AZZG*	Cessna 188 Agwagon 230	188-0279	OY-AHT N8029V	12. 7.72	N.C.Kensington *(On rebuild 6.00: cancelled 21.9.00 by CAA) (Noted dismantled 10.03)*	Perth	1. 5.81A
G-AZZH	Practavia Pilot Sprite 115 *(Built K G Stewart)*	PFA 1532		13. 7.72	A Moore *(New CofR 9.04)*	(Hutton, Brentwood)	
G-AZZO	Piper PA-28-140 Cherokee	28-22887	N4471J	18. 7.72	R.J.Hind	Stapleford	6. 8.03
G-AZZR	Reims/Cessna F150L	F15000690	LN-LJX	24. 7.72	A.J.Hobbs	Acle Bridge, Norfolk	5. 8.07T
G-AZZV	Reims/Cessna F172L	F15000883		18. 7.72	D.J.Hockings	Rochester	7. 7.05T
G-AZZX*	Reims/Cessna FRA150L Aerobat *(Overturned landing Newtownards 28.2.87: on rebuild 12.93: cancelled 14.12.94 by CAA , restored 28.8.97: cancelled 9.5.01 by CAA) (Stored 5.04)*	FRA1500152		27. 7.72	Not known	Plaistowes Farm, St.Albans	16. 8.88
G-AZZZ	de Havilland DH.82A Tiger Moth *(Built Morris Motors Ltd)*	86311	F-BGJE French AF, NL864	27. 7.72	S.W.McKay	RAF Henlow	21.12.04

G-BAAA - G-BAZZ

G-BAAD	Evans VP-1 *(Built R W Husband)* (Volkswagen 1600)	PFA 1540		27. 7.72	K.Wigglesworth tr Breighton VP-1 Group	Breighton	6.12.05P
G-BAAF	Manning-Flanders MF.1 rep *(Built Personal Plabe Services Ltd)* (Continental C75)	PPS/REP/8		27. 7.72	Bianchi Aviation Film Services Ltd *(No external marks with Flying Aces Movie Aeroplane Collection)*	Compton Abbas	6. 8.96P
G-BAAI	SOCATA MS.893A Rallye Commodore 180	10705	F-BOVG	31. 7.72	R.D.Taylor *(Noted 8.04)*	Thruxton	11. 9.00
G-BAAT	Cessna 182P Skylane	18260835	N399JF G-BAAT, N9295G	10. 8.72	N.D.Johnston t/a British Skysports Paracentre	Grindale	21. 5.06
G-BAAW	Jodel D.119 *(Built Etablissement Valladeau)* (Continental 0-200-A)	366	F-BHMY	11. 8.72	P J Newson tr Alpha Whiskey Flying Group	Cherry Tree Farm, Monewden	20. 7.05P
G-BAAY*	Valtion Lentokonetehdas Viima II				See SECTION 3 Part 1	Wevelghem, Belgium	
G-BABA*	de Havilland DH.82A Tiger Moth				See SECTION 3 Part 1	Nepi, Rome, Italy	
G-BABB	Reims/Cessna F150L	F15000830		15. 8.72	Seawing Flying Club Ltd	Southend	25. 8.06T
G-BABC	Reims/Cessna F150L	F15000831		15. 8.72	G.R.Bright	Little Gransden	14. 3.07T
G-BABD	Reims/Cessna FRA150L Aerobat	FRA1500153		3. 8.72	K.F.Mason and D.Featherby t/a Anglia Flight	Norwich	18. 2.04T
G-BABE	Taylor JT.2 Titch *(Built P E Barker - c/n PEB/01)* (Continental O-200-A) *(Forced landied, overturned Elmton 15.5.04 with serious damage to fuselage, starboard wing and engine)*	PFA 1394		3. 8.72	M.Bonsall	Netherthorpe	3. 2.05P
G-BABG	Piper PA-28-180 Cherokee C	28-2031	PH-APU N7978W	15. 8.72	C.E.Dodge tr Mendip Flying Group	Bristol	16.12.06
G-BABH	Reims/Cessna F150L	F15000820	EI-CCZ G-BABH	15. 8.72	D.B.Ryder and Co Ltd	Cranfield	15. 4.07T
G-BABK	Piper PA-34-200 Seneca	34-7250219	PH-DMN G-BABK, N5203T	18. 8.72	D.F.J.Flashman	Biggin Hill	9. 1.05
G-BACB	Piper PA-34-200 Seneca	34-7250251	N5354T	25. 8.72	A R.Braybrooke t/a Milbrooke Motors	Southend	11. 7.05T
G-BACC	Reims/Cessna FRA150L Aerobat	FRA1500157		16. 8.72	P.J.Dalby and P.Huckle	Sandown	13.12.04
G-BACE	Sportavia-Pützer Fournier RF5	5102	(PT-DVZ) D-KCID	25. 8.72	W.B.Hosie *"Clockwork Mouse"*	(Taunton)	2.12.07E
G-BACJ	Wassmer Jodel D.120 Paris-Nice	315	F-BNZC	1. 9.72	J.M.Allan tr Wearside Flying Association	Eshott	13. 4.05P
G-BACK*	de Havilland DH.82A Tiger Moth				See SECTION 3 Part 1	Santiago, Chile	
G-BACL	SAN Jodel D.150 Mascaret	31	F-BSTY CN-TYY	4. 9.72	G.R.French	Benson's Farm, Laindon	26. 6.05
G-BACN	Reims/Cessna FRA150L Aerobat	FRA1500161		4. 9.72	F.Bundy	Bodmin	10. 7.06T
G-BACO	Reims/Cessna FRA150L Aerobat	FRA1500163		4. 9.72	M.A.Mcloughlin	Turweston	26. 4.07
G-BACP	Reims/Cessna F150L Aerobat *(Built as FRA150L)*	FRA1500164		4. 9.72	M.Markwick	Shoreham	23. 5.07T
G-BADC	Rollason Beta B.2A *(Built H M MacKenzie) (Orig. c/n JJF.1 (J J Feeely): PFA/1384: adopted c/n of Beta G-BETA when cancelled 3.2.87 by CAA as not completed:: probably incorporated*	PFA 02-10140		7. 9.72	D.H.Greenwood *(New owner 10.00)*	Barton	31. 1.85P
G-BADH	Slingsby T.61A Falke	1774		6. 9.72	A P.Askwith	Rhosgoch	7. 8.05
G-BADI*	Piper PA-23-250 Aztec D *(Cancelled 9.7.02 by CAA) (Fuselage noted with fictitious marks "G-ESKY" and" AIR AMBULANCE" titles 1.05)*	27-4235	N6885Y	5. 9.72	West London AeroServices Ltd	North Weald	29.10.92T
G-BADJ	Piper PA-E23-250 Aztec E	27-4841	N14279	11. 9.72	C.Papadakis	(Oxford)	17. 2.05T
G-BADL*	Piper PA-34-200-2 Seneca	34-7250247	N5307T	4. 9.72	Not known *(Damaged on landing Turnhouse 21.10.95: cancelled 15.8.96 as WFU) (Fuselage noted 2.04)*	Willey Park Farm, Caterham	
G-BADM	Druine D.62B Condor *(Built K Worksworth and M.Harris using uncompleted Rollason c/n RAE/653)*	PFA 49-11442		8. 9.72	D.J.Wilson	(Bournemouth)	4.11.05P
G-BADV	Brochet MB.50 Pipistrelle *(Built A Bouriquat)*	78	F-PBRJ	13. 9.72	W.B.Cooper *(New owner 1.05)*	Walkeridge Farm, Overton	9. 5.79P
G-BADW	Pitts S-2A *(Built Aerotek Inc)*	2035		21. 9.72	R.E.Mitchell *(Stored 4.02)*	RAF Cosford	16. 9.95T
G-BADZ*	Pitts S-2A *(Built Aerotek Inc)*	2038		21. 9.72	(D Richardson) *(Cancelled 4.4.02 by CAA) (Reported 1.04)*	(Exeter)	5. 6.00
G-BAEB	Robin DR400/160 Knight	733		19. 9.72	R.Hatton	Jurby, Isle of Man	18. 4.07
G-BAEC	Robin HR100/210 Royal	145	EI-BDG G-BAEC	15. 9.72	Datacorp Enterprises Pty Ltd	Denham	6. 4.03
G-BAEE	CEA Jodel DR.1050/M1 Sicile Record	579	F-BMGN	29. 9.72	R.Little	Jackrells Farm, Southwater	13. 1.07
G-BAEM	Robin DR400/120 Petit Prince	728		25. 9.72	M.A Webb	Turweston	3. 6.06
G-BAEN	Robin DR400/180 Régent	736		25. 9.72	European Soaring Club Ltd	Membury	12.10.06
G-BAEO	Reims/Cessna F172M *(Re-built with original fuselage and remains of G-YTWO)*	F17200911		14. 9.72	L.W.Scattergood	Sherburn-in-Elmet	8.12.05T
G-BAEP	Reims/Cessna F150L Aerobat *(Built as FRA150L)*	FRA1500170		14. 9.72	A M.Lynn t/a Busy Bee *(Op RAF Marham Flying Club)*	RAF Marham	17. 5.07T
G-BAER	LeVier Cosmic Wind *(Built R S Voice - c/n 106)* (Continental O-200-A)	PFA 1571		14. 9.72	R.S.Voice *"Filly"*	Rushett Farm, Chessington	9. 9.05P
G-BAET	Piper J-3C-65 Cub (L-4H-PI) *(Frame No.11430)*	11605	OO-AJI 43-30314	26. 9.72	C.J.Rees	Valley Farm, Winwick	10.11.05P
G-BAEU	Reims/Cessna F150L	F15000873		26. 9.72	L.W.Scattergood	Humberside	1. 7.07T

Reg	Type	C/n	Previous id	Reg date	Owner	Base	CofA
G-BAEV	Reims/Cessna FRA150L Aerobat	FRA1500173		27. 9.72	T.J.Richardson	Popham	25. 3.07
G-BAEY	Reims/Cessna F172M	F17200915		28. 9.72	Skytrax Aviation Ltd	Egginton	9. 3.06T
G-BAEZ	Reims/Cessna FRA150L Aerobat	FRA1500169		28. 9.72	Donair Flying Club Ltd	Nottingham East Midlands	19. 6.03T
G-BAFA	American AA-5 Traveler	AA5-0201	N6136A	6.10.72	C.F.Mackley *(Noted 8.04)*	Sleap	31. 8.01
G-BAFG	de Havilland DH.82A Tiger Moth	85995	F-BGEL	13.10.72	D.Watt	Sibson	18. 8.05
	(Built Morris Motors Ltd)		French.AF, EM778				
G-BAFL	Cessna 182P Skylane	18261469	N21180	15. 8.72	M.Langhammer	Old Sarum	20. 3.06T
G-BAFP	Robin DR400/160 Knight	735		19.10.72	T.A Pugh tr Breidden Flying Group	Sleap	5. 4.07
G-BAFT	Piper PA-18-150 Super Cub	18-5340	(D-E...)	3. 8.72	T.J.Wilkinson	Sackville Lodge Farm, Riseley	2. 5.06
			ALAT 18-5340, N10F				
G-BAFU	Piper PA-28-140 Cherokee	28-20759	PH-NLS	11.10.72	D.Matthews	Humberside	14. 5.06T
G-BAFV	Piper PA-18 Super Cub 95	18-2045	PH-WJK	24.10.72	T.F. and S.J.Thorpe	Coldharbour Farm, Willingham	3.11.07
	(L-18C-PI) *(Frame No.18-2055)*		R.Neth AF R-40, 8A-40, 52-2445				
G-BAFW	Piper PA-28-140 Cherokee	28-21050	PH-NLT	24.10.72	R.S.Chance	(Nueil sur Layon, France)	29 4.06
G-BAFX	Robin DR400/140 Earl	739		30.10.72	R.Foster	(Marston, Oxford)	6. 5.07
G-BAGB	SIAI-Marchetti SF.260	1-07	LN-BIV	20.10.72	British Midland Airways Ltd	Nottingham East Midlands	30. 4.06
G-BAGC	Robin DR400/140 Earl	737		13.10.72	W.P.Nutt	(Scarborough)	15. 6.05
G-BAGF	Jodel D.92 Bébé	59	F-PHFC	13.11.72	E.Evans	(Fobbing)	
	(Built Aero Club Basse-Moselle)				*(Fuselage stored 2.04)*		
G-BAGG (2)	Piper PA-32-300 Cherokee Six	32-7340186	N9562N	7.12.73	Channel Islands Aero Club (Jersey)Ltd	Jersey	19. 2.07
G-BAGL	Westland SA.341G Gazelle 1	1067		26.10.72	Foremans Aviation Ltd	Linley Hill, Leven	9. 6.06T
	(Stuck power cables on take-off Street Farm, Takeley 21.7.03 with damage to cabin, main rotor blades, rotor head and tail boom)						
G-BAGN	Reims/Cessna F177RG Cardinal RG	F177RG0068		24.10.72	R.W.J.Andrews	Wolverhampton	2. 8.04
G-BAGO	Cessna 421B Golden Eagle	421B0356	N7613Q	24.10.72	M.S.Choksey	Coventry	7. 3.07
G-BAGR	Robin DR400/140 Petit Prince	753		30.10.72	J.D.Last	Caernarfon	11. 4.07T
G-BAGS	Robin DR400/100 2+2	760		30.10.72	M Whale and M M A Lunn	Kemble	16. 1.03T
G-BAGT	Helio H.295 Super Courier	1288	CR-LJG	31.10.72	B.J.C.Woodall Ltd	Rushett Farm, Chessington	11.11.07E
G-BAGV	Cessna U206F Stationair	U20601867	N9667G	31.10.72	K.Brady tr The Scottish Parachute Club	Strathallan	14. 5.04
G-BAGX	Piper PA-28-140 Cherokee	28-23633	N3574K	30.10.72	J.R.Clayton tr The Golf X-Ray Group	Conington	28.11.05T
G-BAGY	Cameron O-84 HAB	54		17.10.72	P.G.Dunnington	(Hungerford)	16. 6.81A
					"Beatrice" (Stored 2.97)		
G-BAHB*	de Havilland DH.104 Dove 5				See SECTION 3 Part 1	Port Adlaide, South Australia	
G-BAHD	Cessna 182P Skylane	18261501	N21228	25.10.72	J.W.Hardy	Jericho Farm, Lambley	26. 4.07
					tr Lambley Flying Group		
G-BAHE	Piper PA-28-140 Cherokee C	28-26494	N5696U	30.10.72	A.O.Jones and M.W.Kilvert	Trehelig, Welshpool	8. 6.95
					(Stored 5.96: new owner 10.00)		
G-BAHF	Piper PA-28-140 Flite Liner	28-7125215	N431FL	30.10.72	BJ Services (Midlands) Ltd	Coventry	19. 7.07T
G-BAHH	Wallis WA-121/Mc	K/701/X		7.11.72	K.H.Wallis	Reymerston Hall, Norfolk	27. 5.98P
	(Wallis modified McCulloch)				*(Noted 8.01)*		
G-BAHI	Cessna F150H	F150-0330	PH-EHA	6.11.72	M.Player	Elstree	4. 2.05
	(Built Reims Aviation SA)				t/a MJP Aviation and Sales		
G-BAHJ	Piper PA-24-250 Comanche	24-1863	PH-RED	6.11.72	K.Cooper	Wolverhampton	19.12.07E
			N6735P				
G-BAHL	Robin DR400/160 Knight	704	F-OCSR	8.11.72	B.P.Young tr Robin Group	Dunkeswell	5.10.05T
G-BAHO	Beech C23 Sundowner	M-1456		7.11.72	B.V.Goodman tr G-BAHO Group	Bournemouth	10. 4.06
G-BAHP	Volmer VJ.22 Sportsman	PFA 1313		9.11.72	G.K.Holloway	Aboyne	18.10.93P
	(Built J P Crawford) (Continental C90)				tr Seaplane Group *(Noted engine running 1.01)*		
G-BAHS	Piper PA-28R-200 Cherokee Arrow II	28R-7335017	N15147	9.11.72	A R.N.Morris	Shobdon	26. 6.06
G-BAHX	Cessna 182P Skylane	18261588	N21363	16.11.72	A P.Stone tr Dupost Group	Blackpool	14. 8.06
G-BAIG	Piper PA-34-200 Seneca	34-7250243	OY-BSU	21.11.72	Mid-Anglia Flying Centre Ltd	Cambridge	7.12.06T
			G-BAIG, N5257T		t/a Mid-Anglia School of Flying		
G-BAIH	Piper PA-28R-200 Cherokee Arrow II	28R-7335011	N11C	21.11.72	M.G.West	King's Farm, Thurrock	22. 6.07
G-BAII	Reims/Cessna FRA150L Aerobat	FRA1500178		22.11.72	Cornwall Flying Club Ltd	Bodmin	14. 6.03T
					(Force landed Hendra Farm, Bodmin 9.9.01 and severely damaged) (Wreck noted 4.02)		
G-BAIK	Reims/Cessna F150L	F15000903		22.11.72	Wickenby Aviation Ltd	Wickenby	9. 4.06T
					(Op Lincoln Flight Centre)		
G-BAIN	Reims/Cessna FRA150L Aerobat	FRA1500177		23.11.72	S.J.Windle	Bodmin	4. 8.07T
G-BAIP	Reims/Cessna F150L	F15000898		13.11.72	G. and S.A Jones	Linley Hill, Leven	28. 9.97T
					(Damaged Linley Hill 30.5.95: current status unknown)		
G-BAIS	Reims/Cessna F177RG Cardinal RG	F177RG0069		13.11.72	R.M.Graham and E.P.Howard	Guernsey	22. 8.05
					tr Cardinal Syndicate		
G-BAIW	Reims/Cessna F172M	F17200928		14.11.72	W.J.Greenfield	Humberside	13. 1.07T
G-BAIX	Reims/Cessna F172M	F17200931		14.11.72	R.A Nichols	Elstree	6. 4.06
G-BAIZ	Slingsby T.61A Falke	1776		27.11.72	R.G.Sangster and J.J.Doswell	Hinton-in-the-Hedges	18. 6.06
					tr Falke Syndicate		
G-BAJA	Reims/Cessna F177RG Cardinal RG	F177RG0078		29.11.72	D.W.Ward	Biggin Hill	15. 5.06
G-BAJB	Reims/Cessna F177RG Cardinal RG	F177RG0080		29.11.72	LDJ Ltd	Guernsey	1. 9.06
G-BAJC	Evans VP-1 Series 2	PFA 1548		30.11.72	S.J.Greer	Shenington	10. 6.04P
	(Built J R Clements) (Volkswagen 1834)						
G-BAJE	Cessna 177 Cardinal	17700812	N29322	30.11.72	D.M.Dawson	Blackpool	30. 3.07
G-BAJJ*	Hawker Siddeley HS.121 Trident 2E Series 108				See SECTION 3 Part 1	Beijing, China	
G-BAJN	American AA-5 Traveler	AA5-0259		29.11.72	I.M.Snelson	Blackpool	16. 5.03
G-BAJO	American AA-5 Traveler	AA5-0260		29.11.72	S.Bradshaw	Blackpool	10.10.07E
G-BAJR	Piper PA-28-180 Challenger	28-7305008	N11C	1.12.72	Belfast Flying Club Ltd	Newtownards	11. 5.06
G-BAJY	Robin DR400/180 Régent	758		4.12.72	L.J.Murray	(Surbiton)	28. 8.07T
G-BAJZ	Robin DR400/2+2	759		4.12.72	Weald Air Services Ltd	Headcorn	30. 7.06T
G-BAKD	Piper PA-34-200 Seneca	34-7350013	N1378T	28.11.72	Andrews Professional Colour Laboratories Ltd	Headcorn	7. 5.07A
					(Op Foto Flite)		
G-BAKH	Piper PA-28-140 Cherokee F	28-7325014	N11C	12.12.72	Marham Investments Ltd	Belfast Aldergrove	17.11.06T
					(Op Ulster Flying Club)		
G-BAKJ	Piper PA-30 Twin Comanche B	30-1232	TJ-AAI	13.12.72	G.D.Colover, R.Jones and N. O'Connor	Biggin Hill	19.12.06T
			TJ-ADH, N8122Y				
G-BAKM	Robin DR400/140 Earl	755		15.12.72	D.V.Pieri	Carlisle	9. 3.07

G-BAKN	SNCAN Stampe SV-4C	348	F-BCOY	15.12.72	M.Holloway	Watchford Farm, Yarcombe	29. 7.05
	(Renault 4P)						
G-BAKO*	Cameron O-84 HAB				See SECTION 3 Part 1	(Aldershot)	
G-BAKR	SAN Jodel D.117	814	F-BIOV	27.12.72	R.W.Brown	Stoneacre Farm, Farthing Corner	11. 8.05P
G-BAKV	Piper PA-18-150 Super Cub	18-8993	N9744N	22.12.72	Western Air (Thruxton) Ltd and F.Taylor	Thruxton	17. 8.07T
G-BAKW	Beagle B.121 Pup Series 2	B121-175		15.12.72	H.Beavan	White Waltham	25. 6.06
G-BAKY	Slingsby T.61C Falke	1777		20.12.72	T.J.Wiltshire	Spilsby	7. 8.98
G-BALD	Cameron O-84 HAB	58		2. 1.73	C.A Gould	(Ipswich)	2. 7.78S
					tr Inter-Varsity Balloon Club *"Puffin"*		
					(WFU after severe damage 25.6.78: basket to G-PUFF)		
G-BALF	Robin DR400/140 Earl	772		5. 1.73	G.and D.A Wasey	Kemble	5. 7.06
G-BALG	Robin DR400/180 Régent	771		5. 1.73	R.Jones t/a Southern Sailplanes	Membury	11. 8.07
G-BALH	Robin DR400/140B Earl	766		5. 1.73	C.Johnson tr G-BALH Flying Group	Fenland	21. 6.04
G-BALI	Robin DR400 2+2	764		5. 1.73	A Brinkley	Standalone Farm, Meppershall	3. 9.88
					(On rebuild 3.96: current status unknonwn)		
G-BALJ	Robin DR400/180 Régent	767		5. 1.73	D.A Batt and D.de Lacey-Rowe		
						Fridd Farm, Bethersden, Kent	31. 5.06
G-BALN	Cessna T310Q	310Q0684	N7980Q	8. 1.73	O'Brien Properties Ltd	Shoreham	24. 5.07T
G-BALY	Practavia Pilot Sprite 150	PFA 05-10009		10. 1.73	A L.Young t/a Aly Aviation	(Henstridge)	
	(Built A L Young) (Officially regd as c/n "OS-10009")				*(Project part completed and stored 8.95: current status unknown)*		
G-BALZ	Bell 212	30542	EC-IPD	10. 1.73	Bristow Helicopters Ltd	Redhill	26. 7.07T
			G-BALZ, EC-GCR, EC-931, G-BALZ, 9Y-TIL, G-BALZ, VR-BIB, N99040, G-BALZ, EI-AWK, G-BALZ, VR-BEK, N2961W				
G-BAMB	Slingsby T.61C Falke	1778		9. 1.73	N.J.Clemens tr Flying Group G-BAMB	Eaglescott	21. 1.07
G-BAMC	Reims/Cessna F150L	F15000892		12. 1.73	K.Evans	Welshpool	24. 7.05T
G-BAMF	MBB Bö.105DB	S.36	D-HDAM	10. 1.73	Bond Air Services	Sullom Voe	20. 6.06T
					(Op Sullom Voe Harbour Trust)		
G-BAMH*	Westland S-55 Whirlwind Series 3				See SECTION 3 Part 1	Nottingham East Midlands	
G-BAMJ	Cessna 182P	18261650	N21469	10. 1.73	A E.Kedros	Enstone	29. 5.06
G-BAMK*	Cameron D-96 HA Airship				See SECTION 3 Part 1	Newbury	
G-BAMM	Piper PA-28-235 Cherokee	28-10642	SE-EOA	16. 1.73	D.Clare and P.Moderate tr Group 235	Headcorn	2. 1.05
G-BAMR	Piper PA-16 Clipper	16-392	F-BFMS	12. 1.73	H.Royce	Bradleys Lawn, Heathfield	29. 9.07E
	(Lycoming O-290)		CU-P339				
G-BAMS	Robin DR400/160 Knight	774		15. 1.73	G-BAMS Ltd	Biggin Hill	31. 5.06T
G-BAMT	Robin DR400/160 Knight	775		15. 1.73	S.G.Jones	(Membury)	15. 5.79
					(Crashed Cudham 8.1.78: cancelled 24.4.78 as wfu: wreck stored 1.92) (New CofR 1.05)		
G-BAMU	Robin DR400/160 Knight	778		15. 1.73	J.W.L.Otty tr The Alternative Flying Group	Sywell	22. 4.06
G-BAMV	Robin DR400/180 Régent	777		15. 1.73	K.Jones and E.A Anderson	Booker	3. 5.06
G-BAMY	Piper PA-28R-200 Cherokee Arrow II	28R-7335015	N11C	9. 1.73	S.R.Pool	Lydd	18. 5.07
G-BANA	CEA Jodel DR.221 Dauphin	73	F-BOZR	22. 1.73	G.T.Pryor	Seething	21.10.05
G-BANB	Robin DR400/180 Régent	776		22. 1.73	D.R.L.Jones	Kemble	28. 4.06T
G-BANC	Gardan GY-201 Minicab	A 203	F-PCZV	22. 1.73	C.R.Shipley	Brickhouse Farm, Frogland Cross	31. 5.02P
	(Built M Ducreuzet) (Continental C90)		F-BCZV		*(New owner 9.03)*		
G-BAND*	Cameron O-84 HAB				See SECTION 3 Part 1	(Aldershot)	
G-BANF	Phoenix Luton LA-4A Minor	PFA 838		22. 1.73	W.J.McCollum	Coagh, County Londonderry	5. 6.92P
	(Built D W Bosworth) (Continental A65)				*(Damaged Mullaghmore 27.6.92: noted 11.01)*		
G-BANU	Wassmer Jodel D.120 Paris-Nice	247	F-BLNZ	31. 1.73	W.M. and C.H.Kilner	Shacklewell Farm, Empingham	23. 7.05P
G-BANV	Phoenix Currie Wot	PFA 3010		25. 1.73	K.Knight	(Malvern)	26. 6.84P
	(Built C Turner) (Lycoming O-290)				*(Damaged near Leek, Staffs 15.9.83: current status unknown)*		
G-BANW	CAARP CP.1330 Super Emeraude	941	PH-VRF	30. 1.73	P.S.Milner	Popham	28. 6.05P
	(Lycoming 0-235-C1)						
G-BANX	Reims/Cessna F172M	F17200941		31. 1.73	Oakfleet 2000 Ltd	Biggin Hill	12. 7.06
G-BAOB	Reims/Cessna F172M	F17200949		2. 2.73	R.H.Taylor and S.O.Smith	Andrewsfield	12. 5.07T
G-BAOH	SOCATA MS.880B Rallye Club	2250		6. 2.73	A P.Swain *(Current status unknown)*	Haverfordwest	28. 7.01
G-BAOJ	SOCATA MS.880B Rallye Club	2252		6. 2.73	R.E.Jones *(Noted 4.04)*	Emlyn's Field, Rhuallt	13. 8.01
G-BAOM*	SOCATA MS.880B Rallye Club	2255		6. 2.73	P.J.D.Feehan *(Cancelled 27.11.03)*	Exeter	17. 4.03
G-BAOP	Reims/Cessna FRA150L Aerobat	FRA1500190		5. 2.73	R.D.Forster	Beccles	11. 4.02
					(Noted wrecked Seething 7.02 and dismantled 10.02: new owner 9.03)		
G-BAOS	Reims/Cessna F172M	F17200946		6. 2.73	Wingtask 1995 Ltd	Seething	7. 8.06T
G-BAOU	Grumman AA-5 Traveler	AA5-0298		8. 2.73	R.C.Mark	Shobdon	18. 9.04
G-BAOW*	Cameron O-65 HAB				See SECTION 3 Part 1	(Aldershot)	
G-BAPB	de Havilland DHC-1 Chipmunk 22A	C1/0001	WB549	26. 2.73	G.V.Bunyan *(Current status unknown)*	Bidford	31. 5.98
G-BAPI	Reims/Cessna FRA150L Aerobat	FRA1500195		8. 2.73	Industrial Supplies (Peterborough) Ltd	Sibson	13. 5.07
G-BAPJ	Reims/Cessna FRA150L Aerobat	FRA1500196		8. 2.73	M.D.Page	Manston	10. 6.05
G-BAPL	Piper PA-23-250 Turbo Aztec E	27-7304966	N14377	12. 2.73	Donington Aviation Ltd	Nottingham East Midlands	7.10.07*
G-BAPR	Jodel D.115	295 & PFA 914		14. 2.73	J.P.Liber and J.F.M.Bartlett	Kemble	17. 4.05P
	(Built Crantech Flying Group) (Continental PC60)						
G-BAPS*	Campbell Cougar Gyroplane				See SECTION 3 Part 1	Weston-super-Mare	
G-BAPV	Robin DR400/160 Knight	742	F-OCSR	19. 2.73	J.D. and M.Millne	Newcastle	22. 8.03
G-BAPW	Piper PA-28R-180 Cherokee Arrow	28R-30697	5Y-AIR	21. 2.73	A G.Bourne	Hinton-in-the-Hedges	18.12.06
			N4951J		tr Papa Whisky Flying Group		
G-BAPX	Robin DR400/160 Knight	789		21. 2.73	M.Stanton tr G-BAPX	Group Sywell	11. 6.06
G-BAPY	Robin HR100/210 Royal	153		21. 2.73	D.M.Hansell	Old Buckenham	19. 3.05
G-BARC	Reims FR172J Rocket	FR17200356	(D-EEDK)	5. 3.73	C.H.Porter	Croft Farm, Defford	10. 4.07
					tr Severn Valley Aviation Group		
G-BARF	Wassmer Jodel D.112	1019	F-BJPF	5. 3.73	G.P.Jewell	(Caerphilly)	8. 7.03P
G-BARG	Cessna E310Q	310Q0712	N8237Q	2. 3.73	Tibus Aviation Ltd	Oxford	31. 3.06T
G-BARH	Beech C23 Sundowner	M-1473		2. 3.73	G.Moorby and J.Hinchcliffe	Sherburn-in-Elmet	22. 2.07
G-BARN	Taylor JT.2 Titch	PFA 60-11136		5. 3.73	R.G.W.Newton	Hailsham	23. 6.05P
	(Built R G W Newton) (Continental C90)						
G-BARP	Bell 206B JetRanger II	967	N18092	5. 3.73	South Western Electricity plc	Bristol	9. 5.06T
G-BARS	de Havilland DHC-1 Chipmunk 22	C1/0557	WK520	26. 2.73	J.Beattie	RNAS Yeovilton	7. 8.05
					(As "1377" in Portuguese AF c/s)		
G-BARV	Cessna 310Q	310Q-0774		7. 3.73	Old England Watches Ltd	Elstree	18. 8.07
G-BARZ	Scheibe SF28A Tandem Falke	5724	(D-KAUK)	8. 3.73	K.Kiely	AAC Dishforth	8. 1.06

Reg	Type	C/n	Previous identity	Date	Owner/Operator	Base	CofA
G-BASH	Grumman AA-5 Traveler	AA5-0319	EI-AWV	12. 3.73	G.Jenkins	Popham	30. 9.05T
			G-BASH, N5419L		tr BASH Flying Group		
G-BASJ	Piper PA-28-180 Cherokee Challenger	28-7305136	N11C	13. 3.73	Bristol Aero Club	Filton	7.12.03T
G-BASL	Piper PA-28-140 Cherokee F	28-7325195	N11C	13. 3.73	Justgold Ltd	Blackpool	2. 8.04T
G-BASM	Piper PA-34-200 Seneca	34-7350120	N16272	13. 3.73	M.Gipps and J.R.Whetlor	Denham	21. 2.05
G-BASN	Beech C23 Sundowner	M-1476		13. 3.73	J.Greenwood	Morgansfield, Fishburn	6.11.04
G-BASO	Lake LA-4-180 Amphibian	358	N2025L	16. 3.73	C.J.A Macaulay	Farley Farm, Romsey	19. 6.06
G-BASP	Beagle B.121 Pup Series 1	B121-149	SE-FOC	14. 3.73	B.J.Coutts	Sywell	11. 8.07
			G-35-149				
G-BAST*	Cameron O-84 HAB				See SECTION 3 Part 1	(Aldershot)	
G-BATC	MBB Bö.105DB	S.45	D-HDAW	9. 3.73	Bond Air Services	Stromness	22. 6.06T
	(Originally registered as Bö.105D: rebuilt using new MBB pod 1989 c/n unknown)				*(Op Northern Lighthouse Board with NLB titles)*		
G-BATJ	Jodel D.119	287	F-PIIQ	21. 3.73	D.J.and K.S.Thomas	Fenland	31. 5.045
	(Built Ecole Technique Aéronautique) (Continental C90)						
G-BATN	Piper PA-23-250 Aztec E	27-7304987	N14391	26. 3.73	Marshall of Cambridge Aerospace Ltd	Cambridge	5. 2.06T
G-BATR	Piper PA-34-200 Seneca	34-7250290	9H-ABH	23. 3.73	A.S.Bamrah	Biggin Hill	24. 4.05T
			G-BATR, LN-BDT		t/a Falcon Flying Services		
					(Overran landing Woodchurch 4.8.02: struck hedge and badly damaged)		
G-BATV	Piper PA-28-180 Cherokee F	28-7105022	N5168S	26. 3.73	J.N.Rudsdale tr The Scoreby Flying Group	Full Sutton	11.12.05
G-BATW	Piper PA-28-140 Cherokee Flite Liner	28-7225587	N742FL	26. 3.73	C.D.Sainsbury	Swansea	6. 1.06
G-BAUC	Piper PA-25-235 Pawnee C	25-5243	N8761L	26. 3.73	Southdown Gliding Club Ltd	Parham Park	25. 4.06
G-BAUH	Dormois Jodel D.112	870	F-BILO	29. 3.73	G.A and D.Shepherd tr G-BAUH Flying Group	Seething	9.12.05P
G-BAUI*	Piper PA-23-250 Aztec D	27-4335	LN-RTS	29. 3.73	(Gloucester University)	Gloucestershire	5.12.88
					(Cancelled 26.1.89 by CAA) (Noted as Instructional airframe 11.03)		
G-BAUJ*	Piper PA-23-250 Aztec E	27-7304986	N14390	29. 3.73	S.Bramwell	Cranfield	25. 7.94T
					(Cancelled 31.10.02 by CAA) (Stored 8.04)		
G-BAUR*	Fokker F-27 Friendship 200	10225	PH-FEP	5. 4.73	Not known	Exeter	
			9V-BAP, 9M-AMI, (VR-RCZ), PH-FEP		*(Cancelled 25.1.96 as WFU) (Fuselage on fire dump 9.04).*		
G-BAVB	Reims/Cessna F172M	F17200965		10. 4.73	Taylor Aviation Ltd	Cranfield	26. 5.05T
G-BAVH	de Havilland DHC-1 Chipmunk 22	C1/0841	WP975	10. 4.73	D.C.Murray	Lee-on-Solent	3. 3.06
	(Lycoming O-360 (180hp)				tr Portsmouth Naval Gliding Club		
G-BAVL	Piper PA-23-250 Aztec E	27-4671	N14063	10. 4.73	S.P. and A V.Chilcott	Durham Tees Valley	19. 7.07T
G-BAVN*	Boeing Stearman A75N-1 Kaydet				See SECTION 3 Part 1	Västerås, Sweden	
G-BAVO	Boeing Stearman A75N-1 Kaydet	Not known	4X-AIH	13. 4.73	M.Shaw	Priory Farm, Tibenham	4. 7.07T
	(Continental W670) *(Registered with c/n "3250-1405" which is part number: original identity unknown) (As "26" in US Army c/s)*						
					(Lost power on take-off Old Buckenham 1.8.04 landed in field immediately ahead and substantially damaged)		
G-BAVR	Grumman AA-5 Traveler	AA5-0348		12. 4.73	G.E.Murray	Haverfordwest	25. 9.04
G-BAVU*	Cameron A-105 HAB				See SECTION 3 Part 1	Newbury	
G-BAWG	Piper PA-28R-200 Cherokee Arrow II	28R-7335133	N11C	18. 4.73	Solent Air Ltd	Goodwood	1. 3.07
G-BAWK	Piper PA-28-140 Cherokee Cruiser	28-7325243		24. 4.73	Newcastle Aero Club Ltd	Newcastle	3. 8.04T
G-BAWN*	Piper PA-30-160 Twin Comanche C	30-1948	N8790Y	24. 4.73	Not known		
						Walton New Road Business Park, Bruntingthorpe	3. 5.98
					(Cancelled as WFU 23.6.98: fuselage in scrapyard 8.03)		
G-BAWR	Robin HR100/210 Royal	156		27. 4.73	T.Taylor	Oxford	8. 6.00
G-BAXE	Hughes 269A-1	113-0313	N8931F	2. 5.73	Reeve Newfields Ltd	Sywell	21.12.93S
					(Frame only noted 11.01)		
G-BAXF*	Cameron O-77 HAB				See SECTION 3 Part 1	Newbury	
G-BAXJ	Piper PA-32-300 Cherokee Six B	32-40763	N1362Z	8. 5.73	A G.Knight t/a Airlaunch	Old Buckenham	26. 7.06
			(N59RG), N1362Z, G-BAXJ, 4X-ANY, N5224S				
G-BAXK*	Thunder Ax7-77 HAB				See SECTION 3 Part 1	Newbury	
G-BAXS	Bell 47G-5	7908	5B-CFB	11. 5.73	R.M.Kemp	Fairoaks	17.12.06T
			G-BAXS, N4098G		t/a RK Helicopters		
G-BAXU	Reims/Cessna F150L	F15000959		14. 5.73	M.W.Sheppardson	(Peterborough)	18. 5.07T
G-BAXV	Reims/Cessna F150L	F15000966		14. 5.73	G. and S.A Jones	Linley Hill, Leven	6. 6.07T
G-BAXY	Reims/Cessna F172M	F17200905	N10636	15. 5.73	Eaglesoar Ltd	Humberside	2.11.07E
G-BAXZ	Piper PA-28-140 Cherokee C	28-26760	PH-NLX	15. 5.73	D.Norris and M.Lowen	Turweston	7. 4.07T
			N11C		tr G-BAXZ (87) Syndicate		
G-BAYL*	SNCAN Nord 1203 Norecrin VI	161	F-BEQV	18. 5.73	J.E.Pierce	Ley Farm, Chirk	
					(Cancelled 14.11.91 by CAA) (Fuselage outside 8.04)		
G-BAYO	Cessna 150L	15074435	N19471	18. 5.73	J.A., G.M., D.T.A. and J.A.Rees	Haverfordwest	27. 6.07T
					t/a Messrs Rees of Poyston West		
G-BAYP	Cessna 150L	15074017	N18651	18. 5.73	D.I.Thomas tr Yankee Papa Flying Group	Popham	22. 5.05T
G-BAYR	Robin HR100/210 Royal	164		18. 5.73	P.Chamberlain	Turweston	5. 6.06
G-BAYV*	SNCAN 1101 Noralpha				See SECTION 3 Part 1	Eccleston	
G-BAZC	Robin DR400/160 Knight	824		29. 5.73	R.Jones t/a Southern Sailplanes	Membury	24. 6.88
					(Damaged Crosland Moor 21.5.88: stored 10.01)		
G-BAZJ*	Handley Page HPR.7 Dart Herald 209	209	4X-AHR	30. 5.73	Guernsey Airport Fire Service	Guernsey	24.11.84T
		183	G-8-1		*(Cancelled 4.1.85 as WFU:) (In open storage 3.03)*		
G-BAZM	Jodel D.11	PFA 915		31. 5.73	A F.Simpson	Watchford Farm, Yarcombe	9. 8.05P
	(Built Bingley Flying Group - c/n PAL/1416) (Continental O-200-A) *(Identified as "D.113") "L'Oiseau Jaime"*						
G-BAZS	Reims/Cessna F150L	F15000954		1. 6.73	L.W.Scattergood	Waddington	18.11.04T
G-BAZT	Reims/Cessna F172M	F17200996		1. 6.73	Exeter Flying Club Ltd	Exeter	22. 6.06T

G-BBAA - G-BBZZ

Reg	Type	C/n	Previous identity	Date	Owner/Operator	Base	CofA
G-BBAW	Robin HR100/210 Royal	167		12. 6.73	J.R.Williams	Shoreham	27.10.07E
G-BBAX	Robin DR400/140 Earl	835		12. 6.73	G.J.Bissex and P.H.Garbutt	New Farm, Felton	12. 4.07
G-BBAY	Robin DR400/140 Earl	841		12. 6.73	D.S.Brown and V.H.R.Gray		
					tr Rothwell Group	Rothwell Lodge Farm, Kettering	6. 3.05
G-BBBB	Taylor JT.1 Monoplane	PFA 1422		4. 6.73	S.A.MacConnacher	(Great Houghton, Northampton)	
	(Built S.A MacConnacher - c/n SAM/01)				*(Uncompleted project 2005)*		
G-BBBC	Reims/Cessna F150L	F15000864	N10635	14. 6.73	A.A.Gardner	Humberside	20. 1.05T
G-BBBI	Grumman AA-5 Traveler	AA5-0392		15. 6.73	W.Haddow tr Go Baby Aviation Group	(Glasgow)	2. 4.06T

Registration	Type	C/n	Previous identity	Date	Owner/Operator	Probable base	CofA expiry
G-BBBK	Piper PA-28-140 Cherokee	28-22572	SE-EYF	18. 6.73	Comed Schedule Services Ltd	Blackpool	8. 8.04
					(New owner 1.05)		
G-BBBN	Piper PA-28-180 Cherokee Challenger	28-7305365	N11C	20. 6.73	Estuary Aviation Ltd	Southend	16. 1.06T
G-BBBO	SIPA 903	67	F-BGBQ	16. 1.74	G.K.Brothwood	Liverpool	18. 6.05P
G-BBBV*	Handley Page HP.137 Jetstream				See SECTION 3 Part 1	East Fortune	
G-BBBW	Clutton FRED Series II	PFA 1551		26. 6.73	M.Palfreman	Wathstow Farm, Newby Wiske	2. 4.03P
	(Built D Webster - c/n DLW.1)				*(Current status unknown)*		
G-BBBX	Cessna 310L	310L0134	OY-EGW	28. 6.73	Atlantic Air Transport Ltd	Jersey/Coventry	30.10.05
			N3284X				
G-BBBY	Piper PA-28-140 Cherokee Cruiser	28-7325533	N9501N	28. 6.73	S.J.Mount	Wellesbourne Mountford	25. 6.06
G-BBCA	Bell 206B JetRanger II	1101	N18091	29. 6.73	Heliflight (UK) Ltd	Wolverhampton	21.10.07T
G-BBCB	Western O-65 HAB	018		29. 6.73	G.M.Bulmer *"Cee Bee"*	(Credenhill, Hereford)	19. 5.76S
G-BBCC	Piper PA-23-250 Aztec D	27-4317	N6953Y	29. 6.73	Richard Nash Cars Ltd	Norwich	11. 4.07T
G-BBCH	Robin DR400 2+2	850		4. 7.73	C.D.Palfreyman tr Charlie Hotel Syndicate	Turweston	19. 3.07T
G-BBCI	Cessna 150H	15069282	N50409	4. 7.73	A L and F Alam	Cranfield	9. 8.06T
G-BBCK	Cameron O-77 HAB	76		4. 7.73	W R Teasdale	(Maidenhead)	15. 6.89S
					"Mary Gloster" (Inflated 4.02)		
G-BBCN	Robin HR100/210 Royal	168		11. 7.73	S.J.Goodburn	Gloucestershire	22. 9.06
					tr Gloucestershire Flying Club		
G-BBCS	Robin DR400/140B Earl	851		12. 7.73	A N.Kaschevski	Lydd	12. 5.07T
G-BBCY	Phoenix Luton LA-4A Minor	PFA 825		17. 7.73	J.D.Rooney and G.Turner	Goodwood	7. 3.05P
	(Built C H Difford) (Volkswagen 1600)				tr Wingnuts		
G-BBCZ	Grumman AA-5 Traveler	AA5-0382		18. 7.73	A D Massey	Treddunnock Farm, Llangarron	11. 10.07T
G-BBDC	Piper PA-28-140 Cherokee Cruiser	28-7325437	N11C	18. 7.73	B.Scragg and R.Poole tr G-BBDC Group	Earls Colne	26. 4.04
G-BBDE	Piper PA-28R-200 Cherokee Arrow II	28R-7335250	(EI-...)	18. 7.73	R.L.Coleman, A.Holt and Alison Crozier	Panshanger	25. 9.05T
			G-BBDE, N11C				
G-BBDG*	British Aircraft Corporation-Aérospatiale Concorde 100				See SECTION 3 Part 1	Brooklands	
G-BBDH	Reims/Cessna F172M	F17200990		19. 7.73	J.D.Woodward	Westbury-sub-Mendip	29. 5.05
G-BBDJ	Thunder Ax6-56 HAB	006		20. 7.73	A D.Kent	(Southampton)	5. 8.82A
					tr Balloon Preservation Flying Group *"Jack Tar"*		
G-BBDL	Grumman AA-5 Traveler	AA5-0406		18. 7.73	P.F.Robertshaw	Durham Tees Valley	22.10.06
					tr Delta Lima Flying Group		
G-BBDM	Grumman AA-5 Traveler	AA5-0407		18. 7.73	J.Rees tr Jackeroo Aviation Group	Thruxton	1.11.07E
G-BBDO	Piper PA-23-250 Aztec E	27-7305120	N40361	24. 7.73	J.W.Anstee tr G-BBDO Flying Group	(Filton)	30. 6.06T
G-BBDP	Robin DR400/160 Major	853		25. 7.73	Robin Lance Aviation Associates Ltd	Rochester	9.11.04E
G-BBDS	Piper PA-31 Turbo Navajo B	31-7300956	N97RJ	26. 7.73	Elham Valley Aviation Ltd	Lydd	3. 4.06T
			G-SKKB, G-BBDS, N7565L				
G-BBDT	Cessna 150H	15068839	N23272	26. 7.73	C.I.Beilby tr Delta Tango Group	Full Sutton	18. 6.06
G-BBDV	SIPA 903	7/21	F-BEYY	30. 7.73	W.McAndrew	Sackville Lodge Farm, Riseley	11. 6.04P
	(Continental C90) *(Originally ex F-BEYJ c/n 7 but rebuilt in 1978 from F-BEYY c/n 21)*						
G-BBEA	Phoenix Luton LA-4A Minor	PFA 843		30. 7.73	M.Horner	(Chippenham)	18 8.05P
	(Built G J Hewitt) (Volkswagen 1600)						
G-BBEB	Piper PA-28R-200 Cherokee Arrow II	28R-7335292	N9514N	31. 7.73	R.D.W.Rippingale	Anvil Farm, Hungerford	7. 4.07
					tr Anvils Flying Group		
G-BBEC	Piper PA-28-180 Cherokee Challenger	28-7305478	N11C	30. 7.73	J.B.Conway	Ronaldsway	31. 5.07T
G-BBED	SOCATA MS.894A Rallye Minerva 220	12097		30. 7.73	C.A.Shelley	Alcester	13. 9.87T
					t/a Vista Products *(Stored 9.95: current status unknown)*		
G-BBEF	Piper PA-28-140 Cherokee Cruiser	28-7325527	N9500N	31. 7.73	Liberty Group Assets Ltd	(Heckmondwike)	1. 4.05T
	(Rebuilt using components from damaged G-AVWG by 4.99)						
G-BBEN	Bellanca 7GCBC Citabria	496-73	(D-EAUT)	7. 8.73	C.A G.Schofield	Harpsden, Henley-on-Thames	10. 5.02
			N36416				
G-BBEV	Piper PA-28-140 Cherokee D	28-7125340	LN-MTM	8. 8.73	Comed Schedule Services Ltd	Blackpool	9. 9.01T
					(New owner 1.05)		
G-BBEX	Cessna 185A Skywagon	185-0491	EI-CMC	7. 8.73	V.M.McCarthy	Kildare, County Kildare	1. 1.06T
			G-BBEX, 4X-ALD, N99992, N1691Z				
G-BBEY	Piper PA-23-250 Aztec E	27-7305160	LN-FOE	8. 8.73	D.Nicholls	Bagby	27. 1.05T
			G-BBEY, N40396				
G-BBFC*	Grumman AA-1B Trainer	AA1B-0245	(N9945L)	14. 8.73	Not known	Bournemouth	25.12.96
			(Damaged Perranporth 9.6.96: cancelled as temporarily unregd 14.10.96) (Fuselage noted 11.03)				
G-BBFD	Piper PA-28R-200 Cherokee Arrow II	28R-7335342	N9517N	8. 8.73	C.H.Rose and A R.Annable	White Waltham	17. 4.04T
G-BBFL	Gardan GY-201 Minicab	21	F-BHCQ	17. 8.73	R.Smith	Roughay Farm, Bishops Waltham	23.11.05P
	(Built SRCM) (Continental A65)						
G-BBFS*	Van Den Bemden K-460 (Gas) FB				See SECTION 3 Part 1	Newbury	
G-BBFV	Piper PA-32-260 Cherokee Six	32-778	5Y-ADF	13. 8.73	D.T.Wright tr G-BBFV Syndicate	Wickenby	7. 5.06
G-BBGC	SOCATA MS.893E Rallye 180GT	12215	F-BUCV	16. 8.73	P.M.Nolan	Kilkenny, County Kilkenny	5. 9.07
G-BBGI	Fuji FA.200-160 Aero Subaru	FA200-228		21. 8.73	Tandycel Co Ltd	(Somerton)	9. 3.06T
G-BBGL	Oldfield Baby Lakes	PFA 1593		22. 8.73	F.Ball	Jubilee Farm, Bedford	21. 3.02P
	(Built D S Morgan - c/n 7223-B412-B) (Continental C90)						
G-BBGN*	Cameron A-375 HAB				See SECTION 3 Part 1	Wroughton	
G-BBGR	Cameron O-65 HAB	85		20. 8.73	M.L. and L.P.Willoughby	(Woodcote, Reading)	26. 5.81A
					"Jabberwock"		
G-BBGX	Cessna 182P Skylane	18262350	N58861	30. 8.73	D.I.Sutton tr GX Group	Denham	4. 5.07T
G-BBGZ	Cambridge HA Ballooning Association HAB (42,000 cu.ft)	CHABA 42		31. 8.73	J.L.Hinton, G.Laslett and R.A Laslett	(Newbury)	
					"Phlogiston" (Inflated 4.02)		
G-BBHE	Enstrom F-28A	153	EI-BSD	3. 9.73	Clarke Aviation Ltd	Waterford, County Waterford	21.11.04
			G-BBHE				
G-BBHF	Piper PA-23-250 Aztec E	27-7305166	N40453	5. 9.73	G.J.Williams	Sherburn-in-Elmet	4. 7.05T
G-BBHI	Cessna 177RG Cardinal RG	177RG0225	5Y-ANX	7. 9.73	T.G.W.Bunce	Newtownards	1.12.06
			N1825Q				
G-BBHJ	Piper J-3C-85 Cub	16378	OO-GEC	7. 9.73	J.Stanbridge and R.V.Miller	Wellcross Grange, Slinfold	8. 5.05P
	(Frame No.16037)				tr Wellcross Flying Group		
G-BBHK	Noorduyn AT-16-ND Harvard IIB	14-787	PH-PPS	7. 9.73	R.F.Warner	Kemble	1. 4.04P
			(PH-HTC), R.Neth AF B-158, FH153, 42-12540		t/a Bob Warner Aviation		

Registration	Type	C/n	Previous identity	Regd date	Owner/operator	Probable base	CA expiry
G-BBHL	Sikorsky S-61N Mk.II	61-712	N4032S	7. 9.73	Bristow Helicopters Ltd	Stornoway	4.12.07E
					(Op Marine and Coastguard Agency) "Glamis"		
G-BBHY	Piper PA-28-180 Cherokee Challenger	28-7305474	EI-BBS	7. 9.73	Air Operations Ltd	Guernsey	20. 6.05
			G-BBHY, N9508N				
G-BBIA	Piper PA-28R-200 Cherokee Arrow II	28R-7335287	N11C	7. 9.73	G.H.Kilby	Stapleford	5. 2.07
G-BBIF	Piper PA-23-250 Aztec E	27-7305234	N9736N	10. 9.73	D.M.Davies *"Flying Miss Daisie"*	Tatenhill	14.11.07*
G-BBIH	Enstrom F-28A-UK	026	N4875	12. 9.73	Stephenson Marine Co Ltd *(Noted 1.05)*	Goodwood	28. 6.02T
G-BBII	Fiat G.46-3B	44	I-AEHU	13. 9.73	G-BBII Ltd	(London W1J)	20.12.05P
			MM52-801		*(In Italian markings)*		
G-BBIL	Piper PA-28-140 Cherokee	28-22567	SE-FAR	13. 9.73	John West Consulting Ltd	(South Croydon)	8. 6.07
			N4219J				
G-BBIO	Robin HR100/210 Royal	178		14. 9.73	K.D.Taylor and E.M.Peacock	(Driffield)	27. 9.07E
G-BBIX	Piper PA-28-140 Cherokee E	28-7225442	LN-AEN	17. 9.73	Sterling Aviation Ltd	White Waltham	28. 1.05T
G-BBJI	Isaacs Spitfire	PFA 27-10055		18. 9.73	J.D.Bally	(Painscastle, Builth Wells)	5.11.04P
	(Built J O Isaacs - c/n 2) (Continental O-200-A)				*(As "RN218/N")*		
G-BBJU	Robin DR400/140 Earl	874		19. 9.73	J.C.Lister	Valley Farm, Winwick	13. 6.07
					tr Victor Sierra Aero Club		
G-BBJV	Reims/Cessna F177RG Cardinal RG	F177RG0098		20. 9.73	3GRCOMM Ltd	Shobdon	22.12.06
G-BBJX	Reims/Cessna F150L	F15001017		20. 9.73	L.W.Scattergood	Sherburn-in-Elmet	29. 9.05T
G-BBJY	Reims/Cessna F172M Skyhawk II	F17201075		20. 9.73	R.Windley *(On rebuild 3.04)*	(Mavis Enderby)	26. 5.07
G-BBJZ	Reims/Cessna F172M Skyhawk II	F17201035		20. 9.73	J.K.Green	Gamston	28. 2.04T
					t/a Burks Green and Partners *(Noted 8.04)*		
G-BBKA	Reims/Cessna F150L	F15001029		20. 9.73	W.M.Wilson and R.Campbell	Sandtoft	11. 1.07T
G-BBKB	Reims/Cessna F150L	F15001030		20. 9.73	Justgold Ltd *(Op Blackpool Air Centre)*	Blackpool	10. 1.03T
G-BBKE	Reims/Cessna F150L	F15001026		20. 9.73	Taylor Aviation Ltd	Cranfield	25. 8.07T
G-BBKF	Reims/Cessna FRA150L Aerobat	FRA1500222		20. 9.73	D.W.Mickleburgh	Compton Abbas	13. 6.91T
					(Stored 6.95: current status unknown)		
G-BBKG	Reims FR172J Rocket	FR17200465		20. 9.73	R.Wright	Twycross/Coventry	29. 3.07
G-BBKI	Reims/Cessna F172M Skyhawk II	F17201069		20. 9.73	C.W. and S.A Burman	East Winch	20.12.04
G-BBKL	Menavia Piel CP.301A Emeraude	237	F-BIMK	21. 9.73	R.K.Griggs tr Piel G-BBKL	Perth	13. 6.03P
G-BBKR	Scheibe SF24A Motorspatz	4018	D-KECA	24. 9.73	P.I.Morgans	Furze Hill Farm, Rosemarket, Milford Haven	30. 3.79S
G-BBKU	Reims/Cessna FRA150L Aerobat	FRA1500214		26. 9.73	S.J.Windle	Bodmin	19. 6.04T
G-BBKX	Piper PA-28-180 Cherokee Challenger	28-7305581	N9550N	26. 9.73	RAE Aero Club Ltd	Farnborough	25.11.07E
G-BBKY	Reims/Cessna F150L	F15000991		26. 9.73	Telesonic Ltd	Barton	8. 2.07
G-BBKZ	Cessna 172M	17261495	N20694	27. 9.73	R.S.Thomson tr KZ Flying Group	Exeter	13. 5.06T
G-BBLH	Piper J-3C-65 Cub (L-4B-PI)	10006	F-BFQY	24. 9.73	Shipping and Airlines Ltd	Biggin Hill	9. 5.05T
	(Frame No.9838) (Regd with c/n 10549)		Fr.Mil, 43-1145		*(As "31145/26/G" in 183rd Field Battalion US Army c/s)*		
G-BBLL*	Cameron O-84 HAB				See SECTION 3 Part 1	Newbury	
G-BBLM	SOCATA Rallye 100S	2392		3.10.73	J.R.Rodgers	Wolverhampton	19.12.04
G-BBLS	Grumman AA-5 Traveler	AA5-0440	EI-AYM	8.10.73	A.Grant	Perth	23. 4.05
			G-BBLS				
G-BBLU	Piper PA-34-200 Seneca	34-7350271	N55984	8.10.73	A Phillips	Dunkeswell	28 5.05T
G-BBMB	Robin DR400/180 Regent	848	5Y-ASB	27. 9.73	I.James tr Régent Flying Group	King's Farm, Thurrock	24. 4.07
G-BBMH	EAA Biplane Sport Model P1	PFA 1348		11.10.73	I.S.Parker	Damyn's Hall, Upminster	5.10.05P
	(Built K Dawson) (Continental C90-14F)						
G-BBMI*	Dewoitine D.26				See SECTION 3 Part 1	Polk City, Florida, US	
G-BBMJ	Piper PA-23-250 Aztec E	27-7305150	N40387	12.10.73	Nationwide Caravan Rental Services Ltd	(Holywell)	21. 9.05T
G-BBMN	de Havilland DHC-1 Chipmunk 22	C1/0300	WD359	12.10.73	R.Steiner *"03" (Noted 1.05)*	North Weald	4. 4.04
G-BBMO	de Havilland DHC-1 Chipmunk 22	C1/0550	WK514	12.10.73	D.M.Squires *(As "WK514")*	Wellesbourne Mountford	10. 6.07
G-BBMR	de Havilland DHC-1 Chipmunk 22	C1/0213	WB763	12.10.73	P.J.Wood *(New owner 1.02)*		
						(Twyford, Buckinghamshire)	
G-BBMT	de Havilland DHC-1 Chipmunk 22	C1/0712	WP831	12.10.73	J.Evans and D.Withers tr MT Group	Graveley	17. 5.07
G-BBMV	de Havilland DHC-1 Chipmunk 22	C1/0432	WG348	12.10.73	P.J.Morgan (Aviation) Ltd *(As "WG348")*	Sywell	28. 4.06
G-BBMW	de Havilland DHC-1 Chipmunk 22	C1/0641	WK628	12.10.73	A.Wilson and B.King	Goodwood	15.12.04
					tr Mike Whiskey Group *(As "WK628")*		
G-BBMX	de Havilland DHC-1 Chipmunk 22	C1/0800	WP924	12.10.73	K.A Doornbos	Teuge, The Netherlands	20. 8.05
G-BBMZ	de Havilland DHC-1 Chipmunk 22	C1/0563	WK548	12.10.73	P.C.G.Wyld	Booker	8. 1.07
					tr The Wycombe Gliding School Syndicate		
G-BBNA	de Havilland DHC-1 Chipmunk 22	C1/0491	WG417	12.10.73	Coventry Gliding Club Ltd	Husbands Bosworth	22. 6.06
	(Lycoming O-360)				*"Carrie"*		
G-BBNC*	de Havilland DHC-1 Chipmunk T.10				See SECTION 3 Part 1	Salisbury Hall, London Colney	
G-BBND	de Havilland DHC-1 Chipmunk 22	C1/0225	WD286	12.10.73	W.Norton and D.Fradley	Little Gransden	19. 4.06
					tr Bernoulli Syndicate *(As "WD286/J")*		
G-BBNG	Bell 206B JetRanger II	134	VH-BHX	16.10.73	MB Air Ltd	Winchester Farm, Ouston	27 5.05T
			G-BBNG, VR-BEY, G-BBNG, PK-HBO, N6268N		*(Op Eagle Helicopters)*		
G-BBNH	Piper PA-34-200 Seneca	34-7350339	N56492	16.10.73	A L.Howell, A P.Barrow and M.G.D.Baverstock		
						Bournemouth	4. 4.07
G-BBNI	Piper PA-34-200 Seneca	34-7350312	N56286	16.10.73	Noisy Moose Ltd	(London W5)	22. 9.07T
G-BBNJ	Reims/Cessna F150L	F15001038		16.10.73	Sherburn Aero Club Ltd	Sherburn-in-Elmet	19.11.05T
G-BBNZ	Reims/Cessna F172M Skyhawk II	F17201054		23.10.73	R.E.Nunn	Clipgate Farm, Denton	21. 7.06
G-BBOA	Reims/Cessna F172M Skyhawk II	F17201066		23.10.73	J.D.and A M.Black	Lodge Farm, St.Osyth	18. 7.05
G-BBOC	Cameron O-77 HAB	86		24.10.73	J.A B.Gray	(Cirencester)	6. 1.90A
					t/a Bacchus Balloons *"Bacchus"*		
G-BBOD	Thunder O.5 HAB	013		24.10.73	B.R. and M.Boyle	(Newbury)	
					(On loan to British Balloon Museum and Library) "Little Titch"		
G-BBOE	Robin HR200/100	26		24.10.73	R.J.Powell	(Wickham, Hampshire)	7. 7.02
			(Badly damaged striking hedge and concrete post landing Wells Cross Farm, Horsham 24.6.01: dismantled)				
G-BBOH	Craft-Pitts S-1S	PFA 1570		25.10.73	Techair London Ltd	Popham	8. 9.97P
	(Built Pitts Aviation Enterprises Inc -c/n AJEP-P-S1-S-1)				*(Stored 10.04)*		
G-BBOL	Piper PA-18-150 Super Cub	18-7561	D-EMFE	26.10.73	N.Artt	Aston Down	18. 9.07
			N3821Z				
G-BBOO	Thunder Ax6-56 HAB	012		4.10.73	K.Meehan *"Tiger Jack"*	(Much Wenlock)	31. 5.03A
G-BBOR	Bell 206B JetRanger II	1197	(SE-)	30.10.73	M.J.Easey	Town Farm, Hoxne, Eye	29. 7.05T
			G-BBOR				

Regn	Type	C/n	Previous identity	Regd	Owner/operator	Base	CofA expiry
G-BBOX	Thunder Ax7-77 HAB	011		24.10.73	R.C.Weyda	(Newbury)	23.12.82A
					"Rocinante" (On loan to British Balloon Museum and Library)		
G-BBPN	Enstrom F-28A-UK	166		30.10.73	Smarta Systems Ltd	(Pencader)	1. 4.07T
G-BBPO	Enstrom F-28A	176		30.10.73	Henfield Lodge Aviation Ltd	(Henfield)	1.10.06T
G-BBPS	SAN Jodel D.117	597	F-BHXS	30.10.73	A Appleby	Westfield Farm, Hailsham	26. 5.05P
G-BBPX	Piper PA-34-200 Seneca	34-7250262	N1202T	7.11.73	Richel Investments Ltd	Guernsey	23. 2.06
G-BBPY	Piper PA-28-180 Challenger	28-7305590	N9554N	8.11.73	Sunsaver Ltd	Barton	14. 8.05
G-BBRA	Piper PA-23-250 Aztec E	27-7305197	N40479	12.11.73	R.C.Lough	Stapleford	23. 7.06
G-BBRB	de Havilland DH.82A Tiger Moth	85934	OO-EVB	21.11.73	R.Barham	(Biggin Hill)	
			Belgian AF T-8, ETA-8, DF198		*(Damaged Biggin Hill 16.1.87: current status unknown)*		
G-BBRC	Fuji FA.200-180 Aero Subaru	FA200-235		8.11.73	G-BBRC Ltd	Blackbushe	25. 6.05T
G-BBRI	Bell 47G-5A	25158	N18092	8.11.73	Alan Mann Helicopters Ltd	Fairoaks	28. 7.02T
	(Composite following several major rebuilds)						
G-BBRN	Mitchell-Procter Kittiwake I	PFA 1352	XW784	20.11.73	M.Holliday and H.M.Price	RNAS Yeovilton	8. 9.05P
	(Built Air Engineering HMS Daedalus - c/n 02) (Continental O-200-A)				tr The FAA Squadron Kittiwake Group *(As "XW784/VL")*		
G-BBRV	de Havilland DHC-1 Chipmunk 22	C1/0284	WD347	13.11.73	Yorkshire Vintage Flying Ltd	Sheffield City	15. 5.06T
					(As "WD347" in RAF grey and orange dayglo stripes)		
G-BBRX	SIAI-Marchetti S.205-18F	342	LN-VYH	13.11.73	R.C.West	Popham	21.12.07E
			OO-HAQ				
G-BBRZ	Grumman AA-5 Traveler	AA5-0471	(EI-AYV)	15.11.73	C.P.Osborne	Movenis	30. 4.99
			G-BBRZ		*(Noted 6.04) (Cancelled 10. 2.05 by CAA)*		
G-BBSA	Grumman AA-5 Traveler	AA5-0472		15.11.73	Usworth 84 Flying Associates Ltd	Newcastle	16. 4.05
G-BBSB	Beech C23 Sundowner 180	M-1516		15.11.73	L.J.Welsh	(St. Martins, Perth)	31. 7.05T
G-BBSC*	Beech B24R Sierra 200	MC-217		15.11.73	I.Millar and G.H.Emerson	Belfast Aldergrove	3. 6.99
					(Cancelled 27.7.01 by CAA) (Noted 1.04)		
G-BBSM	Piper PA-32-300 Cherokee Six	32-7440005	N9577N	14.11.73	G.C.Collings	Hardwick	26. 8.06T
G-BBSS	de Havilland DHC-1 Chipmunk 22	C1/0520	WG470	21.11.73	Coventry Gliding Club Ltd	Husbands Bosworth	13. 5.07
	(Lycoming O-360)						
G-BBSW	Pietenpol AirCamper	PFA 1506		21.11.73	J.K.S.Wills	(London SE3)	
	(Built J K S Wills)				*(Current status unknown)*		
G-BBTB	Reims/Cessna FRA150L Aerobat	FRA1500224		26.11.73	BBC Air Ltd *(Op Abbas Air)*	Compton Abbas	2.11.02T
G-BBTG	Reims/Cessna F172M Skyhawk II	F17201097		26.11.73	R.W. and V.P.J.Simpson	Redhill	15. 5.05
					tr Tango Golf Flying Group		
G-BBTH	Reims/Cessna F172M Skyhawk II	F17201089		26.11.73	Tayside Aviation Ltd	Glenrothes	22. 8.05E
G-BBTJ	Piper PA-23-250 Aztec E	27-7305131	N40369	27.11.73	Cooper Aerial Surveys Ltd	Sandtoft	8. 4.047
G-BBTS	Beech V35B Bonanza	D-9551	N3051W	29.11.73	S Wenham t/a Eastern Air	Cannes-Mandelieu, Monaco	5. 6.03
G-BBTY	Beech C23 Sundowner 180	M-1525		29.11.73	A W.Roderick and W.Price tr TY Group	Cardiff	12. 8.07
G-BBTZ	Reims/Cessna F150L	F15001063		30.11.73	Marham Investments Ltd	Cumbernauld	17. 6.07T
					(Op Cumbernauld Flying School)		
G-BBUE	Grumman AA-5 Traveler	0479		6.12.73	G.A Chadfield	(Wokingham)	8. 4.06T
G-BBUF	Grumman AA-5 Traveler	0480		6.12.73	S.and A F.Williams	(Exeter)	22.12.05T
G-BBUG	Piper PA-16 Clipper	16-29	F-BFMC	6.12.73	J.Dolan	Enniskillen, County Fermanagh	19. 7.05
G-BBUJ	Cessna 421B Golden Eagle	421B0335	OY-RYD	7.12.73	Coolflourish Ltd	(Mansfield)	18. 5.00
G-BBUT	Western O-65 HAB	020		11.12.73	R.G.Turnbull	(Clyro, Hereford)	23. 4.97A
					"Christabelle II" (New owner 5.02)		
G-BBUU	Piper J-3C-75 Cub (L-4A-PI)	10529	F-BBSQ	14. 1.74	Cathy Stokes	(Hulcote Farm	16. 5.05P
	(Frame No.10354)		F-OAEZ, Fr.AF, 43-29238		*(New owner 6.04)*		
G-BBVA	Sikorsky S-61N Mk.II	61-718		12. 2.74	Bristow Helicopters Ltd	Lee-on-Solent	24. 2.06T
					(Op Marine and Coastguard Agency) "Vega"		
G-BBVF*	Scottish Aviation Twin Pioneer 3				See SECTION 3 Part 1	East Fortune	
G-BBVJ	Beech B24R Sierra 200	MC-230		21.12.73	E.J.Berek *(Noted 5.04)*	Netherthorpe	7. 6.03
G-BBVO	Isaacs Fury II	PFA 11-10091		20.12.73	J.Moore	Wreningham	23. 6.04P
	(Built D Silsbury) (Lycoming O-320)				*(As Hawker Nimrod "S1579/571" of 408 Flight FAA, HMS Glorious)*		
G-BBVU*	Hawker Siddeley HS.121 Trident 2E Series 109				See SECTION 3 Part 1	Guangzou, China	
G-BBVV*	Hawker Siddeley HS.121 Trident 2E Series 109				See SECTION 3 Part 1	Nanjing, China	
G-BBVW*	Hawker Siddeley HS.121 Trident 2E Series 109				See SECTION 3 Part 1	Guangzou, China	
G-BBVZ*	Hawker Siddeley HS.121 Trident 2E Series 109				See SECTION 3 Part 1	Beijing, China	
G-BBWG*	Hawker Siddeley HS.121 Trident 2E Series 109				See SECTION 3 Part 1	Beijing, China	
G-BBWZ	Grumman AA-1B Trainer	AA1B-0334		4. 1.74	W C Smeaton	Popham	10. 9.06T
G-BBXB	Reims/Cessna FRA150L Aerobat	FRA1500236		16. 1.74	M.G. and Elizabeth A Sweet	(Bridlington)	21.10.06T
G-BBXH	Reims FR172F Rocket	FR17200113	SE-FKG	21. 1.74	D.Ridley	(High Flatts Farm, Chester-le-Street)	8.11.03
G-BBXK	Piper PA-34-200 Seneca	34-7450056	N54366	21. 1.74	J.A Rees	Haverfordwest	25. 5.05T
G-BBXL	Cessna 310Q II	310Q1076	EI-CLX	21. 1.74	Appleton Aviation Ltd	Leeds-Bradford	6. 9.07T
			G-BBXL, (N1223G)				
G-BBXS	Piper J-3C-65 Cub (L-4H-PI)	12214	N9865F	25. 1.74	M.J.Butler	Spanhoe	14. 9.00P
	(Continental C90) *(Frame No.12042)*		G-ALMA, 44-79918		*(Officially regd as c/n "9865") (Noted 11.01)*		
G-BBXW	Piper PA-28-151 Cherokee Warrior	28-7415050	PH-CPL	21. 1.74	Bristol Aero Club	Filton	4.11.07E
			G-BBXW, N9599N				
G-BBXY	Bellanca 7GCBC Citabria	614-74	N57639	1. 2.74	R.R.L.Windus	Truleigh Manor Farm, Edburton	24. 2.06
G-BBXZ	Evans VP-1	PFA 1562		31. 1.74	R.W.Burrows	Swanton Morley	8. 3.96P
	(Built G D Price) (Volkswagen 1600)				*(Noted wingless 4.02)*		
G-BBYB	Piper PA-18 Super Cub 95	18-1627	PH-TMA	4. 2.74	The Tiger Club (1990) Ltd	Headcorn	26. 7.07T
	(L-18C-PI) *(Frame No.18-1628)*		(D-ENCH), ALAT 18-1627, 51-15627				
G-BBYH	Cessna 182P	18262814	N52744	6. 2.74	Ramco (UK) Ltd	Poplar Farm, Croft, Skegness	13. 4.06
G-BBYM*	Handley Page HP.137 Jetstream 200				See SECTION 3 Part 1	RAF Cosford	
G-BBYO*	Fairey Britten-Norman BN-2A Mk.III-1 Trislander	362	ZS-KMH	27. 2.74	Aurigny Air Services Ltd	Guernsey	1. 5.92T
			G-BBYO, G-BBWR		*(WFU 2.92 and cancelled 23.2.95 as WFU) (With Airport Fire Service 3.03)*		
G-BBYP	Piper PA-28-140 Cherokee F	28-7425158	N9620N	19. 2.74	Eurocharter Aviation Ltd	Gamston	6. 7.06T
G-BBYR*	Cameron O-65 HAB				See SECTION 3 Part 1	(Aldershot)	
G-BBYS	Cessna 182P Skylane	18261520	5Y-ATE	14. 2.74	I.M.Jones	Gamston	9. 6.06
			N21256				
G-BBYU*	Cameron O-56 HAB				See SECTION 3 Part 1	Newbury	
G-BBZF	Piper PA-28-140 Cherokee F	28-7425195	N9501N	19. 2.74	J.McGuinness and R.Stamp		
					t/a East Coast Aviation	Waterford, County Waterford	24. 7.06

G-BBZH	Piper PA-28R-200 Cherokee Arrow II	28R-7435102	N9608N	22. 2.74	ZH Flying Ltd	Exeter	13. 6.07
G-BBZJ	Piper PA-34-200-2 Seneca	34-7450088	N40880	26. 2.74	A S.Bamrah t/a Falcon Flying Services	Biggin Hil	5. 7.07T
G-BBZL*	Westland-Bell 47G-3B1				See SECTION 3 Part 1	Stavangar, Norway	
G-BBZN	Fuji FA.200-180 Aero Subaru	FA200-230		26. 2.74	D.Kynaston, J.S.V.Westwood and P.D.Wedd	Cambridge	17.11.05T
G-BBZO	Fuji FA.200-160 Aero Subaru	FA200-238		26. 2.74	M.S.Bird	Pepperbox, Salisbury	26. 8.05
G-BBZV	Piper PA-28R-200 Cherokee Arrow II	28R-7435105	N9609N	11. 3.74	P.B.Mellor	Cambridge	25. 9.05T

G-BCAA - G-BCZZ

G-BCAH	de Havilland DHC-1 Chipmunk 22	C1/0372	WG316	6. 5.74	A W.Eldridge *(As "WG316")*	Leicester	25. 6.05T
G-BCAP*	Cameron O-56 HAB				See SECTION 3 Part 1	(Aldershot)	
G-BCAR*	Thunder Ax7-77 HAB				See SECTION 3 Part 1	Newbury	
G-BCAS*	Thunder Ax7-77 HAB				See SECTION 3 Part 1	(Aldershot)	
G-BCAZ	Piper PA-12 Super Cruiser	12-2312	5Y-KGK	12. 3.74	A D.Williams	Rhos-y-Gilwen Farm, Rhos Hill	21. 2.05
			VP-KGK, ZS-BYJ, ZS-BPH				
G-BCBG	Piper PA-23-250 Aztec E	27-7305224	VP-BBN	13. 3.74	M.J.L Batt	Booker	24.11.07E
			VR-BBN, (VR-BDM), G-BCBG, N40494				
G-BCBH	Fairchild 24R-46A Argus III	975	(VH-AAQ)	13. 3.74	Dreamticket Promotions Ltd	Old Hay, Paddock Wood	28. 6.06
	(UC-61K-FA)		G-BCBH, ZS-AXH, HB737, 43-15011				
G-BCBJ	Piper PA-25-235 Pawnee C	25-2380/R		18. 3.74	Deeside Gliding Club (Aberdeenshire) Ltd	Aboyne	2. 9.07
	(Rebuild of c/n 25-2380/G-ASLA/N6802Z, quoting c/n 25-5544 the new fuselage of G-ASLA !)						
G-BCBL	Fairchild 24R-46A Argus III	989	OO-EKE	19. 3.74	F.J.Cox	Eaglescott	31. 3.96
	(UC-61K-FA)		D-EKEQ, HB-AEC, HB751, 43-15025		*(As "HB751") (Cuurent status unknown)*		
G-BCBM*	Piper PA-23-250 Aztec C	27-3006	N5854Y	19. 3.74	Hatton and Westerman Trawlers	Netherthorpe	12. 5.01
					(Noted derelict 9.03: cancelled 2.6.04 by CAA)		
G-BCBR	Wittman W.8 Tailwind	TW3-380		20. 3.74	D.P.Jones	Top Farm, Croydon, Royston	29. 7.05P
	(Built AJEP Developments)						
G-BCBX	Reims/Cessna F150L	F15001001	F-BUEO	25. 3.74	N.F.O'Neill	Newtownards	30. 3.07T
G-BCBZ	Cessna 337C Super Skymaster	3370942	SE-FKB	28. 3.74	Envirovac 2000 Ltd	(Blackwood)	23. 4.05
	(Robertson STOL conversion)		N2642S				
G-BCCC	Reims/Cessna F150L	F15001041		8. 4.74	D.C.Walker tr Triple Charlie Flying Group	Cranfield	25. 4.07T
G-BCCD	Reims/Cessna F172M Skyhawk II	F17201144		8. 4.74	R.M.Austin t/a Austin Aviation	Elstree	30. 6.07T
G-BCCE	Piper PA-23-250 Aztec E	27-7405282	N40544	3. 4.74	Golf Charlie Echo Ltd *(Op The Flying Hut)*	Shoreham	15. 8.05T
G-BCCF	Piper PA-28-180 Cherokee Archer	28-7405069	N9632N	3. 4.74	Topcat Aviation Ltd	Liverpool	15. 7.06T
G-BCCG	Thunder Ax7-65 HAB	020		4. 4.74	N.H.Ponsford	(Leeds)	7.11.83A
					t/a Rango Balloon and Kite Co *"Zephyr" (Active 1999)*		
G-BCCH*	Thunder Ax6-56A HAB				See SECTION 3 Part 1	(Aldershot)	
G-BCCJ	Grumman AA-5 Traveler	AA5-0546		8. 4.74	T.Needham	Dunkeswell	10. 3.06
G-BCCK	Grumman AA-5 Traveler	AA5-0547		8. 4.74	Prospect Air Ltd	Manchester	19. 9.05
G-BCCR	Piel CP.301A Emeraude	PFA 712		8. 4.74	J.H. and C.J.Waterman		
	(Built Korist Flying Group) (Continental O-200-A)					Armshold Farm, Kingston, Cambridge	24. 4.06P
G-BCCX	de Havilland DHC-1 Chipmunk 22	C1/0531	WG481	17. 4.74	T.Holloway	AAC Dishforth	24. 4.06
	(Lycoming O-360)				tr RAFGSA *(Op Clevelands Gliding Club)*		
G-BCCY	Robin HR200/100 Club	37		18. 4.74	Charlie Yankee Ltd	Filton	21. 4.05
G-BCDJ	Piper PA-28-140 Cherokee	28-24276	PH-NLV	29. 4.74	R.J.Whyham	Blackpool	24. 6.07T
			N1841J		c/o Air Navigation and Trading Ltd		
G-BCDK (2)	Partenavia P68B	32	A6-ALN	4. 7.75	Flyteam Aviation Ltd	(Elstree)	2. 5.05T
			G-BCDK				
G-BCDL	Cameron O-42 HAB	115		24. 4.74	D.P.and B.O.Turner *"Chums"*	(Bath)	13. 7.83A
G-BCDY	Reims/Cessna FRA150L Aerobat	FRA1500237		7. 5.74	I.R.Chaplin	Andrewsfield	18. 8.06T
G-BCEA	Sikorsky S-61N Mk.II	61-721		7. 6.74	British International Ltd	Plymouth	13. 7.06T
G-BCEB	Sikorsky S-61N Mk.II	61-454	N4023S	2.10.74	Veritair Ltd *"The Isles of Scilly"*	Penzance	16.12.05T
G-BCEE	Grumman AA-5 Traveler	AA5-0571		7. 5.74	W.A.L.Mitchell	Dunsfold	5. 6.06
G-BCEF	Grumman AA-5 Traveler	AA5-0572		7. 5.74	J.Lowe tr Enstone Flyers	Enstone	9. 6.05
G-BCEN	Fairey Britten-Norman BN-2A-26 Islander	403	4X-AYG	6. 5.74	Atlantic Air Transport Ltd	Manston	7.11.06A
			SX-BFB, 4X-AYG, N90JA, G-BCEN		*(Op Marine and Coastguard Agency)*		
G-BCEO*	American Aviation AA-5 Traveler	7. 5.74	AA5-0575		W Bateson	Blackpool	
		(Crashed near Selby Farm, Stanton, Morpeth 28.4.02: cancelled 11.07.02 as destroyed) (Wreck noted 1.04)					
G-BCEP	Grumman AA-5 Traveler	AA5-0576		7. 5.74	G.Edelmann	Blackbushe	4. 9.06
G-BCER	Gardan GY-201 Minicab	8	F-BGJP	8. 5.74	D.Beaumont	West Freugh	13. 5.05P
	(Built Con. Aéronautique de Bearn) (Continental A65)						
G-BCEX	Piper PA-23-250 Aztec E	27-7305024	N40225	13. 5.74	DJ Aviation Ltd	(Bulkington, Devizes)	14. 5.05T
G-BCEY	de Havilland DHC-1 Chipmunk 22	C1/0515	WG465	14. 5.74	T.C.B.Dehn and C.A Robey	White Waltham	12. 9.05
					tr Gopher Flying Group *(As "WG465" in RAF c/s)*		
G-BCEZ	Cameron O-84 HAB	107		13. 5.74	P.F.Smart and R.J.Mitchener	(Romsey/Andover)	20. 7.82A
					t/a Balloon Collection *"Stars and Bars"*		
G-BCFD*	West Ax3-15 HAB				See SECTION 3 Part 1	Newbury	
G-BCFE*	Byrne Odyssey 4000 MLB				See SECTION 3 Part 1	Newbury	
G-BCFF	Fuji FA.200-160 Aero Subaru	FA200-237		21. 5.74	C.D.Burleigh	Popham	1. 4.07
G-BCFN	Cameron O-65 HAB	109		23. 5.74	W.G.Johnston and H.M.Savage	(Edinburgh)	15. 5.77S
					"Fireball" (Noted 6.00)		
G-BCFO	Piper PA-18-150 Super Cub	18-5335	(D-EIOZ)	29. 5.74	D.C.Murray	Lee-on-Solent	13. 5.07
			ALAT 18-5335, N10F		tr Portsmouth Naval Gliding Club		
G-BCFR	Reims/Cessna FRA150L Aerobat	FRA1500244		30. 5.74	Bulldog Aviation Ltd and Motorhoods Colchester Ltd		
					(Op Essex Flying School)	Earls Colne	13.12.05T
G-BCFW	SAAB 91D Safir	91-437	PH-RLZ	29. 5.74	D.R.Williams	Peplow	24. 7.06
G-BCFY	Phoenix Luton LA-4A Minor	PFA/824		29. 5.74	G.Capes	(Welton, Brough)	17. 1 92P
	(Built G F M Garner - c/n PAL/1301) (Ardem Mk.6)				*(Stored Sywell 8.92: new owner 10.00)*		
G-BCGB	Bensen B.8P	CL.14		3. 6.74	J.W.Birkett	Chilbolton	10. 6.04P
	(Built P C Lovegrove) (Rotax 503)						
G-BCGC	de Havilland DHC-1 Chipmunk 22	C1/0776	WP903	13. 3.74	J.C.Wright	RAF Henlow	26. 7.04T
					(As "WP903" in Queen's Flight c/s)		
G-BCGH	SNCAN NC.854S	122	F-BAFG	10. 6.74	T.J.N.H.Palmer tr Nord Flying Group	Hill Farm, Nayland	28. 5.05P

Registration	Type	C/n	Previous identity	Regd date	Owner (Operator)	Probable base	CA expiry
G-BCGI	Piper PA-28-140 Cherokee Cruiser	28-7425283	N9573N	10. 6.74	J.C.,T.,T.and H.R.Dodd	Panshanger	11. 6.06T
G-BCGJ	Piper PA-28-140 Cherokee Cruiser	28-7425286	N9574N	10. 6.74	Halegreen Associates Ltd and Demero Ltd		
						Hinton-in-the-Hedges	28. 9.06T
G-BCGM	Wassmer Jodel D.120 Paris-Nice	50	F-BHQM	15. 7.74	J.Pool	Sturgate	14. 4.05P
			F-BHYM				
G-BCGN	Piper PA-28-140 Cherokee F	28-7425323	N9595N	10. 6.74	Golf November Ltd	Oxford	2.. 9.05
G-BCGP*	Gazebo Ax6-65 HAB				See SECTION 3 Part 1	Newbury	
G-BCGS	Piper PA-28R-200 Cherokee Arrow II	28R-7235133	N4893T	13. 6.74	S.Rayne tr Arrow Aviation Group	Cambridge	9. 4.06
G-BCGT	Piper PA-28-140 Cherokee	28-24504	N6779J	17. 6.74	L.Maikowski	Shoreham	6. 8.06T
G-BCGW	Jodel D.11	PFA 912		14. 6.74	G.H.and M.D.Chittenden	Highwood Hall	30. 1.85P
	(Built G.H.and M.D.Chittenden - c/ns CC.001 & EAA/61554) (Lycoming O-290)				*(Current status unknown)*		
G-BCHK	Reims/Cessna F172H	F17200716	9H-AAD	19. 6.74	D.Darby *(Noted 7.04)*	Haverfordwest	23.11.03T
G-BCHL	de Havilland DHC-1 Chipmunk 22A	C1/0680	WP788	20. 6.74	Shropshire Soaring Ltd *(As "WP788")*	Sleap	18.10.04
G-BCHM	Westland SA.341G Gazelle 1	1168	G-17-20	14. 6.74	MW Helicopters Ltd *(New owner 5.02)*	Stapleford	23. 8.99
G-BCHP	Scintex CP.1310-C3 Super Emeraude	902	G-JOSI	24. 6.74	G.Hughes and A G.Just	Earls Colne	6. 8.02P
			G-BCHP, F-BJVQ				
G-BCHT	Schleicher ASK 16	16021	(BGA1996)	25. 6.74	D.E.Cadisch and K.A Lilleywhite	Dunstable	31. 5.07
			D-KAMY		tr Dunstable K16 Group		
G-BCHV	de Havilland DHC-1 Chipmunk 22	C1/0703	WP807	27. 6.74	Katrina.I.Sutherland	(Stevington, Bedford)	20. 6.98
G-BCHX*	Scheibe SF23A Sperling	2013	D-EGIZ	28. 6.74	R.L.McLean t/a DG Powered Sailplanes	Rufforth	29. 6.83P
					(Damaged 7.8.82: frame stored 9.01: cancelled 22.3.02 as WFU)		
G-BCID	Piper PA-34-200 Seneca	34-7250303	N1381T	3. 7.74	Shenley Farms (Aviation) Ltd	Headcorn	7. 3.05T
G-BCIE*	Piper PA-28-151 Cherokee Warrior	28-7415405	N9588N	3. 7.74	Perth College	Perth	19.12.99T
			(Extensively damaged Perth 27.5.99: cancelled 15.9.99 as destroyed) (Dumped 11.03)				
G-BCIH	de Havilland DHC-1 Chipmunk 22	C1/0304	WD363	3. 7.74	J.M.Hosey *(As "WD363")*	Audley End	10. 7.05
G-BCIJ	Grumman AA-5 Traveler	AA5-0603	N6143A	3. 7.74	D.G.Page t/a Arrow Association	Elstree	19. 6.06
G-BCIK*	Grumman AA-5 Traveler	AA5-0604	N6144A	3. 7.74	Trent Aviation Ltd *(Cancelled 3.8.04 by CAA)*	Tatenhill	28. 5.03
G-BCIN	Thunder Ax7-77 HAB	030		5. 7.74	R.A., P.M.G.and N.T.M.Vale	(Kidderminster)	5. 5.84A
					tr Isambard Kingdom Brunel Balloon Group		
G-BCIR	Piper PA-28-151 Cherokee Warrior	28-7415401	N9587N	9. 7.74	P.J.Brennan	Southend	16.11.06
G-BCJH*	Mooney M.20F Executive 21	670126	N9549M	11. 7.74	P.J.Bossard	Bourn	30. 6.91
					(Cancelled 26.9.00 by CAA) (In open store 11.03)		
G-BCJM	Piper PA-28-140 Cherokee F	28-7425321	N9592N	17. 7.74	Topcat Aviation Ltd	Manchester	18. 2.06T
					(Op Manchester School of Flying)		
G-BCJN	Piper PA-28-140 Cherokee Cruiser	28-7425350	N9618N	17. 7.74	Topcat Aviation Ltd	Barton	8. 9.05T
G-BCJO	Piper PA-28R-200 Cherokee Arrow II	28R-7435272	N9640N	17. 7.74	R.Ross	Inverness	19. 9.07
G-BCJP	Piper PA-28-140 Cherokee	28-24187	N1766J	15. 8.74	D.J. and D.Pitman tr Omletair Flying Group	Bournemouth	6. 5.07
G-BCKF*	K & S Squarecraft SA 102.5 Cavalier	PFA 1594		29. 7.74	K.Fairness	(Eyemouth)	
	(Built K Fairness - c/n 71055)				*(No Permit issued and.cancelled 8.7.91 by CAA) (Stored 2001)*		
G-BCKN	de Havilland DHC-1 Chipmunk 22	C1/0707	WP811	5. 8.74	T.Holloway	(RAF Cranwell)	16. 2.07
	(Lycoming O-360)				tr RAFGSA *(Op Cranwell Gliding Club)*		
G-BCKS	Fuji FA.200-180AO Aero Subaru	FA200-250		2. 8.74	J.F.Heath	(Carlisle)	17. 3.07T
G-BCKT	Fuji FA.200-180 Aero Subaru	FA200-251		2. 8.74	P Chilcott tr Kilo Tango Group	Shoreham	20. 5.05
G-BCKU	Reims/Cessna FRA150L Aerobat	FRA1500256		1. 8.74	Stapleford Flying Club Ltd	Stapleford	9.11.07E
G-BCKV	Reims/Cessna FRA150L Aerobat	FRA1500251		1. 8.74	S.N.Bower and P.W.Brown	(Sheffield/Clifton, Bristol)	23 1.06T
					t/a Huck Air		
G-BCLC	Sikorsky S-61N Mk.II	61-737		9. 1.75	Bristow Helicopters Ltd	Sumburgh	12. 1.06T
					(Op Marine and Coastguard Agency) "Craigievar"		
G-BCLD	Sikorsky S-61N Mk.II	61-739		4. 2.75	Bristow Helicopters Ltd *"Slains" (Noted 6.04)*	Aberdeen	2. 2.03T
G-BCLI	Grumman AA-5 Traveler	AA5-0643		12. 8.74	W.D.Smith	Sywell	14. 8.07T
G-BCLL	Piper PA-28-180 Cherokee C	28-2400	SE-EON	13. 8.74	J.Nash and D.F.Amos tr G-BCLL Group	Popham	14.11.07E
G-BCLS	Cessna 170B	20946	N8094A	23. 8.74	N.Simpson	(Lincoln)	27. 1.83
					(Stored 7.99: new owner 12.01)		
G-BCLT	SOCATA MS.894A Rallye Minerva 220	12003	EI-BBW	1. 8.74	K.M.Bowen	Upfield Farm, Whitson	17. 7.05
			G-BCLT, F-BTRL				
G-BCLU	SAN Jodel D.117	506	F-BHXG	28. 8.74	S.J.Wynne	Knettishall	27. 4.05P
G-BCLW	Grumman AA-1B Tr2	AA1B-0463		29. 8.74	J.R.Faulkner	Egginton	22. 8.05T
G-BCMD	Piper PA-18 Super Cub 95	18-2055	OO-SPF	4. 9.74	P.Stephenson	New Farm House, Great Oakley	30. 5.05
	(L-18C-PI) *(Frame No.18-2071)*		R.Neth AF R-70, 52-2455				
G-BCMJ	K & S Squarecraft SA.102.5 Cavalier	PFA 01-1546		9. 9.74	N.F.Andrews	(Oakham)	8. 8.85P
	(Built M Johnson - c/n MJ.1) (Tailwheel u/c)						
G-BCMT	Isaacs Fury II	PFA 1522		9. 9.74	R.W.Burrows	Priory Farm, Tibenham	
	(Built M H Turner) (Continental O-200-A)				*(New owner 2.02)*		
G-BCNC	Gardan GY-201 Minicab	A 202	F-BICF	9. 9.74	J.R.Wraight	(Chatham)	
	(Built Nouvelle Soc Cometal)				*(Current status unknown)*		
G-BCNP	Cameron O-77 HAB	117		16. 9.74	P.Spellward *"Blue Fret"*	(Bristol)	28. 7.00A
G-BCNX	Piper J-3C-65 Cub (L-4H-PI)	11168	F-BEGM	17. 9.74	K.J.Lord	Cherry Tree Farm, Monewden	18.10.05P
	(Frame No.10993)		Fr AF, 43-29877		tr The Grasshopper Flying Group *(As "540" in USAF c/s)*		
G-BCNZ	Fuji FA-200-160 Aero Subaru	FA200-257		16. 9.74	W.Dougan	(Ayr)	23. 1.05
G-BCOB	Piper J-3C-65 Cub (L-4H-PI)	10696	F-BCPV	19. 9.74	R.W. and Mrs.J.Marjoram	Low Farm, South Walsham	17. 6.05P
	(Frame No.10521)		43-29405		*(As "329405/23-A" in USAAC c/s)*		
G-BCOH*	Avro 683 Lancaster Mk.10 AR				See SECTION 3 Part 1	Polk City, Florida, USA	
G-BCOI	de Havilland DHC-1 Chipmunk 22	C1/0759	WP870	24. 9.74	M.J.Diggins	Rayne Hall Farm, Braintree	25. 8.07
					(As "WP870/12" in RAF grey c/s)		
G-BCOJ	Cameron O-56 HAB	124		25. 9.74	T.J.Knott and M.J.Webber	(Rickmansworth)	12. 7.87A
					tr Phoenix Balloon Group *"Red Squirrel"*		
G-BCOL	Reims/Cessna F172M Skyhawk II	F17201233		25. 9.74	A H.Creaser	Old Manor Farm, Anwick	10. 7.06T
G-BCOM	Piper J-3C-90 Cub (L-4A-PI)	10478	F-BDTP	27. 9.74	S.L.McKinnon	Butlers Gyhll, Southwater	8. 7.02P
	(Frame No.10303)		F-BFQP, OO-ADI, 43-29187		tr Dougal Flying Group *"Dougal" (On rebuild 8.03)*		
	(Officially regd as c/n 12040 which is correct identity of G-BGPD: fuselages probably exchanged in France)						
G-BCOO	de Havilland DHC-1 Chipmunk 22	C1/0209	WB760	10.10.74	T.G.Fielding and M.S.Morton	Blackpool	10.11.03
G-BCOR	SOCATA Rallye 100ST	2544	F-OCZK	7. 1.75	P.R.W Goslin and I.M.Speight	Henstridge	27. 9.04
G-BCOU	de Havilland DHC-1 Chipmunk 22	C1/0559	WK522	10.10.74	P.J.Loweth *"Thunderbird 5"*	(High Easter)	30. 3.95
					(As "WK522" in RAF c/s) (Current status unknown)		

G-BCOX	Bede BD-5A	HJC.4523		10.10.74	H.J.Cox and B.L.Robinson *(Noted 7.99)*	Chivenor	27.11.95P
G-BCOY	de Havilland DHC-1 Chipmunk 22	C1/0212	WB762	10.10.74	Coventry Gliding Club Ltd	Husbands Bosworth	23. 1.06
	(Lycoming O-360)						
G-BCPD	Gardan GY-201 Minicab	18	F-BGKN	24.10.74	P.R.Cozens	Hinton-in-the-Hedges	18. 7.05P
	(Built Con. Aéronautique de Bearn) (Continental A65)						
G-BCPG	Piper PA-28R-200 Cherokee Arrow	28R-35705	N4985S	16.10.74	A G.Antoniades tr Roses Flying Group	Barton	17. 8.07
G-BCPH	Piper J-3C-65 Cub (L-4H-PI)	11225	F-BCZA	13.12.74	M.J.Janaway	Siege Cross Farm, Thatcham	27 4.05P
	(Frame No.11050)		Fr.AF, 43-29934		*(As "329934/72-B" in 25th AOP French Armoured Divn of US 3rd Army c/s)*		
G-BCPJ	Piper J-3C-65 Cub (L-4J-PI)	13206	F-BDTJ	5.11.74	S.Hollingsworth	Popham	17. 4.05P
	(Frame No.13036)		45-4466		tr Piper Cub Group		
G-BCPK	Reims/Cessna F172M Skyhawk II	F17201194	(D-ELOB)	21.10.74	D.C.C.Handley	Sywell	12. 1.01T
G-BCPN	Grumman AA-5 Traveler	AA5-0665	N6155A	21.10.74	G.K.Todd	Full Sutton	11. 2.07
G-BCPU	de Havilland DHC-1 Chipmunk 22	C1/0839	WP973	24.10.74	P.Waller	Booker	6. 8.05
G-BCRB	Reims/Cessna F172M Skyhawk II	F17201259		29.10.74	D.E.Lamb	Fenland	7. 5.07
G-BCRE*	Cameron O-77 HAB				See SECTION 3 Part 1	(Aldershot)	
G-BCRI	Cameron O-65 HAB	135		5.11.74	V.J.Thorne *"Joseph"*	Bristol	26. 8.81A
G-BCRK	K & S Squarecraft SA.102.5 Cavalier	PFA 01-10049		5.11.74	P.G.R.Brown	Trenchard Farm, Eggesford	14. 7.00P
	(Built R Y Kendal) (Lycoming O-235)				*(Noted 11.03)*		
G-BCRL	Piper PA-28-151 Cherokee Warrior	28-7415689	N9564N	5.11.74	BCRL Ltd	Humberside	7. 8.06T
G-BCRP	Piper PA-E23-250 Aztec E	27-7305082	N40269	7.11.74	Airlong Charter Ltd *(Op Skydrift Ltd)*	Norwich	10.11.06T
G-BCRR	Grumman AA-5B Tiger	AA5B-0006		7.11.74	N.A Whatling	Spanhoe	12. 1.07
G-BCRT	Reims/Cessna F150M	F15001164		18.11.74	Almat Flying Club Ltd	Coventry	11. 6.05T
G-BCRX	de Havilland DHC-1 Chipmunk 22	C1/0232	WD292	22.11.74	Tuplin Ltd	White Waltham	15. 7.06
					(As "WD292" in RAF c/s)		
G-BCSA	de Havilland DHC-1 Chipmunk 22	C1/0691	WP799	25.11.74	T.Holloway	RAF Halton	14. 2.06
	(Lycoming O-360)				tr RAFGSA		
G-BCSL	de Havilland DHC-1 Chipmunk 22	C1/0524	WG474	26.11.74	Chipmunk Flyers Ltd	Liverpool	26. 3.05
G-BCSM*	Bellanca 8GCBC Scout	108-74		29.11.74	The Furness Gliding Club Propietary Ltd	Eastbach Farm, Coleford	30. 4.04
	(Swung to right landing Walney Island 12.05.02: struck ground causing extensive damage: cancelled 2.12.02) (For potential rebuild 8.03)						
G-BCST	SOCATA MS.893A Rallye Commodore	18010748	F-BPQD	18.11.74	D.R.Wilcox	Sywell	4.12.04
G-BCSX	Thunder Ax7-77 HAB	031		2.12.74	C.Wolstenholme *"Woophski"*	(Macclesfield)	5. 7.86A
G-BCSY	Taylor JT.2 Titch	PFA 1504		5.12.74	I.L.Harding	Sackville Lodge Farm, Riseley	
	(Built P L Mines)				*(Construction abandoned at advanced state: stored 3.97: current status unknown)*		
G-BCTA*	Piper PA-28-151 Cherokee Warrior	28-7515113		6.12.74	T G Aviation Ltd	Fairoaks	28. 6.00T
					(Cancelled 16.7.98 as WFU) (On scrap-heap as "G-OOAT" 4.04)		
G-BCTF	Piper PA-28-151 Cherokee Warrior	28-7515033	N9585N	11.12.74	E.Reed	Durham Tees Valley	20.10.05T
	(Rebuilt 1989/90 using major components from G-BFXZ)				t/a St.George Flight Training		
G-BCTI	Schleicher ASK 16	16029	D-KIWA	23.12.74	A J.Southard	Hinton-in-the-Hedges	15. 8.07
					tr Tango India Syndicate		
G-BCTJ	Cessna 310Q II	310Q1072	N1219G	23.12.74	D.Pearce	Biggin Hill	12. 8.05T
G-BCTK	Reims FR172J Rocket	FR17200546		23.12.74	R.T.Love	Bodmin	26. 3.06
G-BCTT	Evans VP-1	PFA 1543		24.12.74	E.R.G.Ludlow	Pent Farm, Postling	25. 3.05P
	(Built B J Boughton) (Volkswagen 1600)						
G-BCTU*	Reims/Cessna FRA150M Aerobat	FRA1500268		30.12.74	Not known	Willey Park Farm, Caterham	11. 1.00
					(Cancelled 17.1.00 as WFU) (Noted 2.04)		
G-BCUB	Piper J-3C-65 Cub (L-4J-PI)	13370	F-BFBU	13.12.74	A L.Brown	Bourn	13. 6.01P
	(Lippert Reed conversion)		45-4630				
	(Officially regd with c/n 13186 now known to be G-BDOL (qv): airframes possibly switched during conversion in UK)						
G-BCUF	Reims/Cessna F172M Skyhawk II	F17201279		3. 1.75	R.N.Howell	(Mablethorpe)	28. 7.06
					t/a Howell Plant Hire and Construction		
G-BCUH	Reims/Cessna F150M	F15001195		7. 1.75	M.G.Montgomerie tr G-BCUH Group	Elstree	26. 2.07T
G-BCUJ	Reims/Cessna F150M	F15001176		9. 1.75	A.M.Sage	(Swindon)	11. 2.05T
G-BCUL*	SOCATA Rallye 100ST	2545	F-OCZL	27. 1.75	C.A Ussher and Fountain Estates Ltd	Bagby	8. 5.00
					(Cancelled 3.9.03 as wfu)		
G-BCUO	Scottish Aviation Bulldog Series 120/122	BH120/371	Ghana AF G-107	9. 1.75	Cranfield University	Cranfield	3. 3.07T
			G-BCUO				
G-BCUP*	Scottish Aviation Bulldog Series 120/122	BH120/372	Ghana AF G-108	9. 1.75	Aerofab Restorations	Bourne Park, Hurstbourne Tarrant	
			G-BCUP		*(Cancelled 9.6.76 - to Ghana AF as G-108) (Stored as G-108 5.03)*		
G-BCUS	Scottish Aviation Bulldog Series 120/122	BH120/373	Ghana AF G-109	9. 1.75	C.D.Hill	North Weald	29. 4.05
			G-BCUS				
G-BCUV	Scottish Aviation Bulldog Series 120/122	BH120/376	Ghana AF G-112	9. 1.75	Dolphin Property (Management) Ltd	Old Sarum	12. 6.06T
			G-BCUV		*(As "CB733" in RAF c/s)*		
G-BCUY	Reims/Cessna FRA150M Aerobat	FRA1500269		14. 1.75	J.C.Carpenter	Clipgate Farm, Denton	8. 3.07
G-BCVB	Piper PA-17 Vagabond	17-190	F-BFMT	22. 1.75	A T.Nowak	Popham	14. 7.05P
	(Continental A65)		N4890H				
G-BCVC	SOCATA Rallye 100ST	2548	F-OCZO	16. 1.75	W.Haddow	(Glasgow)	25. 1.05
G-BCVE*	Evans VP-2	PFA 7210		16. 1.75	North Western PFA Strut	Barton	
	(Built G Bentley - c/n V2-1015)				*(Cancelled 9.6.93 as TWFU) (Noted 7.04)*		
G-BCVF	Practavia Pilot Sprite 115	PFA 1362		27. 1.75	D.G.Hammersley	Tatenhill	1.10.04P
	(Built G B Castle - c/n GBC.1) (Continental C125)						
G-BCVG	Reims/Cessna FRA150L Aerobat	FRA1500245	(I-AFAD)	16. 1.75	I.G.Cooper tr G-BCVG Flying Group	Compton Abbas	19.12.06
G-BCVH	Reims/Cessna FRA150L Aerobat	FRA1500258		16. 1.75	M.A.James	Perth	18.11.05T
G-BCVJ	Reims/Cessna F172M Skyhawk II	F17201305		16. 1.75	Rothland Ltd	RAF Woodvale	18.12.06
G-BCVY	Piper PA-34-200T Seneca II	34-7570022	N32447	28. 1.75	Oxford Aviation Services Ltd	Oxford	20. 7.06T
G-BCWB	Cessna 182P Skylane II	18263566	N5848J	29. 1.75	M.F.Oliver and A J.Mew	White Waltham	21.12.05T
G-BCWH	Practavia Pilot Sprite 115	PFA 1366		3. 2.75	R.Tasker	Blackpool	12. 6.05P
	(Built K B Parkinson and R.Tasker) (Continental O-240-A)						
G-BCWK	Fournier RF3	24	F-BMDD	7. 2.75	T.J.Hartwell and D.R.Wilkinson	Thurleigh	13. 8.02P
G-BCWO	Fairey Britten-Norman BN-2A-26 Islander	431	SE-LAX	13. 2.75	Cormack (Aircraft Services) Ltd	Cumbernauld	6. 7.05T
			LN-MAC, G-BCWO				
G-BCWR*	Fairey Britten-Norman BN-2A-20 Islander	433	OY-RPZ	13. 2.75	Not known	Picketstone, St.Athan	
			(OY-RPH), G-BCWR				
	(Cancelled 21.4.75 - to (OY-RPH)/OY-RPZ, restored 8.5.86: f/fuselage for technical mock-up as "OY-RPZ" @ 1994: cancelled 26.10.94 as wfu) (Fuselage noted 5.03)						

Reg	Type	C/n	Previous identity	Date	Owner/operator	Probable base	CofA expiry
G-BCXB	SOCATA Rallye 100ST	2546	F-OCZM	7. 2.75	A Smails	Morgansfield, Fishburn	18. 4.07
G-BCXE	Robin DR400 2+2	1015		19. 2.75	Weald Air Services Ltd	Headcorn	11. 6.05T
G-BCXJ	Piper J-3C-65 Cub (L-4J-PI)	13048	F-BFFH	21. 2.75	W.Readman	Old Sarum	17. 9.04P
	(Frame No.12878)		OO-SWA, 44-80752		*(As "480752/39-E" in USAAC c/s)*		
G-BCXN	de Havilland DHC-1 Chipmunk 22	C1/0692	WP800	7. 3.75	G.M.Turner	RAF Halton	8. 5.06
					(As "WP800/2" in Southampton UAS c/s)		
G-BCXO*	MBB Bö.105DD				See SECTION 3 Part 1	Land's End	
G-BCXZ*	Cameron O-56 HAB	154		4. 3.75	Not known	(Firenze, Italy)	
					(Cancelled 19.5.93 by CAA) (Noted inflated 10.03)		
G-BCYH	DAW Privateer Mk.3	PFA 1568	BGA.1158	10. 3.75	A.P.Paskins	Lower Mountpleasant, Chatteris	4 7.05P
	(Built D C Pattison - c/n 2) (Volkswagen 1600)		RAFGSA.264, XA297		*(As "RAFGSA.246")*		
	(Formerly Slingsby T.31B c/n 839 and now officially regd as Cadet III Motor Glider)						
G-BCYK*	Avro (Canada) CF-100 Canuck Mk.IV				See SECTION 3 Part 1	Duxford	
G-BCYM	de Havilland DHC-1 Chipmunk 22	C1/0598	WK577	13. 3.75	C.H.Nicholls tr G-BCYM Group *(As "WK577")*	Kemble	2.10.06
G-BCYR	Reims/Cessna F172M Skyhawk II	F17201288		20. 3.75	Highland Flying School Ltd	Dundee	4.12.04T
G-BCZH	de Havilland DHC-1 Chipmunk 22	C1/0635	WK622	19. 3.75	A C.Byrne	Botany Bay, Horsford	31. 7.87
					(As "WK622" in RAF c/s) (Crashed Pentney, Norfolk 6.9.87: stored 8.93)		
G-BCZI	Thunder Ax7-77 HAB	037		24. 3.75	R.G.Griffin and R.Blackwell	(Newbury)	16. 3.86A
					tr North Hampshire Balloon Group *"Motorway"*		
G-BCZM	Reims/Cessna F172M Skyhawk II	F17201350		3. 4.75	Cornwall Flying Club Ltd	Bodmin	22. 1.07T
G-BCZN	Reims/Cessna F150M	F15001149		27. 3.75	Mona Aviation Ltd	RAF Mona	5. 1.07T
G-BCZO	Cameron O-77 HAB	158		27. 3.75	W.O.T.Holmes *"Leo" (Inflated 6.02)*	Shrewsbury	11.10.86A
G-BCZS*	Fairey Britten-Norman BN-2A-21 Islander				See SECTION 3 Part 1	Stavangar, Norway	

G-BDAA - G-BDZZ

Reg	Type	C/n	Previous identity	Date	Owner/operator	Probable base	CofA expiry
G-BDAD	Taylor JT.1 Monoplane	PFA 1453		2. 4.75	J.Gunson	Brook Farm, Pilling	4.10.05P
	(Built J F Bakewell) (Volkswagen 1700)				tr G-BDAD Group		
G-BDAG	Taylor JT.1 Monoplane	PFA1430		1. 4.75	O.T.Elmer	Horsford	29.11.05P
	(Built R S Basinger)						
G-BDAI	Reims/Cessna FRA150M Aerobat	FRA1500266		21. 4.75	D.F.Ranger	Popham	25. 7.07T
G-BDAK	Rockwell Commander 112A	252	N1252J	10. 4.75	M.C.Wilson	Top Farm, Croydon, Royston	31. 8.06
G-BDAM*	Noorduyn AT-16-ND Harvard IIB				See SECTION 3 Part 1	Niagara Falls, Ontario, Canada	
G-BDAO	SIPA 91	2	F-BEPT	10. 4.75	S.B.Churchill	Eastbach Farm, Coleford	30 3.05P
	(Continental C85)						
G-BDAP	Wittman W.8 Tailwind	PFA 3507		9. 4.75	J.Whiting	Bagby	18. 3.04P
	(Built J and A Whiting - c/n 0387)						
G-BDAR	Evans VP-1 Series 2	PFA 1537 & PFA 62-10461		10. 4.75	R.B.Valler	(Waterlooville)	20. 7.84P
	(Built S C Foggin and M J Dunmore) (Volkswagen 1600)				*(Current status unknown)*		
G-BDAY	Thunder Ax5-42A HAB	042		8. 4.75	T.M.Donnelly *"Meconium"*	(Sprotbrough, Doncaster)	16. 1.93A
G-BDBD	Wittman W.8 Tailwind	133	N1198S	25. 4.75	W.S.Siebert and B.Ohrman	Wellesbourne Mountford	13. 8.04P
	(Built Hamilton Tool Company)				tr Tailwind Group		
G-BDBF	Clutton FRED Series II	PFA 1528		15. 4.75	G.E.and R.E.Collins	Egginton	18. 3.98P
	(Built W T Morrell) (Volkswagen 1600)				*(New owners 10.04)*		
G-BDBH	Bellanca 7GCBC Citabria	758-74	OE-AOL	15. 4.75	C.J.Gray	Finmere	10. 3.07
G-BDBI	Cameron O-77 HAB	162		15. 4.75	C Jones *"Funny Money"*	(Sonning Common, Reading)	11. 7.87A
G-BDBJ	Cessna 182P Skylane II	18263646	N4644K	18. 4.75	H.C.Wilson	Great Ashfield, Suffolk	11. 3.06
G-BDBL*	de Havilland DHC-1 Chipmunk 22				See SECTION 3 Part 1	Ardmore, Auckland, New Zealand	
G-BDBS*	Short SD.3-30 UTT				See SECTION 3 Part 1	Langford Lodge, Belfast	
G-BDBU	Reims/Cessna F150M	F15001174		30. 4.75	Cumbernauld Flying School Ltd	Cumbernauld	20. 8.06T
G-BDBV	Aero Jodel D.11A	V.3	D-EGIB	23. 4.75	G.G.Long	Seething	31.10.05P
	(Buillt Schwabach W Wolfrum) (Continental C90)				tr Seething Jodel Group		
G-BDBZ*	Westland WS-55 Whirlwind 2 (HAR.10)				See SECTION 3 Part 1	Elvington	
G-BDCC*	de Havilland DHC.1 Chipmunk 22	C1/0258	WD321	25. 4.75	Coventry Gliding Club Ltd	Husbands Bosworth	24. 3.02
					(W/o 29.8.99 and cancelled 8.4.04 as wfu) (Fuselage stored 7.04)		
G-BDCD	Piper J-3C-65 Cub (L-4J-PI)	12429	OO-AVS	28. 4.75	D G Pearce	Shoreham	29. 7.05P
	(Continental C90) *(Frame No.12257)*		44-80133		*(As "480133/44-B" in US Army c/s)*		
G-BDCI	Scanor Piel CP.301C Emeraude	503	F-BIRC	25. 4.75	D.L.Sentance	Rothwell Lodge Farm, Kettering	21. 1.05P
G-BDCL	Grumman AA-5 Traveler	AA5-0773	EI-CCI	5. 5.75	J.Crowe	Coventry	29.11.93T
			G-BDCL, EI-BGV, G-BDCL, N1373R *(Stored w/o wings 5.00)*				
G-BDCO	Beagle B.121 Pup Series 1	B121-171		6. 5.75	R.J.Page and M.H.Simms	(Haywards Heath)	28. 7.97
					(New owner 4.01)		
G-BDDD	de Havilland DHC-1 Chipmunk 22	C1/0326	WD387	16. 5.75	RAE Aero Club Ltd	Farnborough	16. 8.02T
G-BDDF	Wassmer Jodel D.120 Paris-Nice	97	F-BIKZ	20. 5.75	J.V.Thompson *(New owner 4.04)*	(Leeds)	10.11.03P
G-BDDG	Dormois Jodel D.112	855	F-BILM	20. 5.75	D.G.Palmer	Sturgate	28. 7.04P
G-BDDS	Piper PA-25-260 Pawnee C	25-4757	CS-AIU	22. 5.75	T.J.Price	Rhigos	16. 5.05
			N10F		tr Vale of Neath Gliding Club		
G-BDDX*	Whittaker MW2B Excalibur				See SECTION 3 Part 1	Trago Mills, Newton Abbot	
G-BDDZ	Menavia Piel CP.301A Emeraude	253	F-BIMZ	30. 5.75	E.C.Mort	(Warrington)	20. 6.84P
					(Damaged Cranwell North 3.6.84: on rebuild 1.01)		
G-BDEC	SOCATA Rallye 100ST	2552	F-OCZS	28. 5.75	M.Mulhall	Kilkenny, County Kilkenny	29. 9.06
G-BDEH	Wassmer Jodel D.120A Paris-Nice	239	F-BLNE	2. 6.75	M.D.Nichol tr EH Group	Oaksey Park	27. 8.05P
G-BDEI	Jodel D.9 Bébé	PFA 936		2. 6.75	R.Q.T.Newns	White Waltham	2.10.05P
	(Built L T Dix - c/n 585) (Volkswagen 1600)				tr The Noddy Group *"Noddy"*		
G-BDEU	de Havilland DHC-1 Chipmunk 22	C1/0704	WP808	17. 6.75	T.E.Earl *(As "WP808")*	Bourne Park, Hurstbourne Tarrant	17.12.06
G-BDEX	Reims/Cessna FRA150M Aerobat	FRA1500279		12. 6.75	R.A Powell	Belle Vue Farm, Yarnscombe	30. 7.06T
G-BDEY	Piper J-3C-65 Cub (L-4J-PI)	12538	OO-AAT	17. 6.75	W.J. and Mrs.J.Morecraft	Highfield Farm, Empingham	17. 6.05P
	(Frame No.12366)		OO-GAC, 44-80242		tr Ducksworth Flying Club		
G-BDEZ	Piper J-3C-65 Cub (L-4J-PI)	12383	OO-SOC	17. 6.75	R.J.M.Turnbull	Rydinghurst Farm, Cranleigh	19. 7.05P
	(Frame No.12211)		OO-EPI, 44-80087				
G-BDFB	Phoenix Currie Wot	PFA 3008		20. 6.75	J.Jennings	Fenland	21.10.04P
	(Built D F Faulkner-Bryant and J Jennings) (Walter Mikron III)						

Reg	Type	C/n	Previous identity	Regd	Owner/Operator	Location	CofA
G-BDFH	Auster AOP.9	B5/10/176	XR240	24. 6.75	R.B.Webber	Trenchard Farm, Eggesford	23.11.04P
	(Frame No.AUS 177 FM)				*(As "XR240" in 'ARMY' camouflage c/s)*		
G-BDFJ	Reims/Cessna F150M	F15001182		25. 6.75	Cassandra J.Hopewell	Sibson	13. 7.02T
G-BDFR	Fuji FA.200-160 Aero Subaru	FA200-262		7. 7.75	P.T.Yates tr Fugi Group	Blackpool	1.11.04
G-BDFS	Fuji FA.200-160 Aero Subaru	FA200-263		7. 7.75	B.Lawrence	Goodwood	24. 9.00T
G-BDFU*	PMPS Dragonfly MPA Mk.1				See SECTION 3 Part 1	Brussels, Belgium	
G-BDFW	Rockwell Commander 112A	308	N1308J	18. 6.75	M.E.and E.G.Reynolds	Blackbushe	22.11.07E
G-BDFX	Taylorcraft J Auster 5	2060	F-BGXG	9. 7.75	J.Eagles	Oaksey Park	3. 6.94T
			TW517		*(Damaged Oaksey Park 10.10.93: on rebuild 4.02)*		
G-BDFY	Grumman AA-5 Traveler	AA5-0806		10. 7.75	G.Robertson	Edinburgh	22. 8.06
					tr The Grumman Group *(Op Edinburgh Flying Club)*		
G-BDFZ	Reims/Cessna F150M	F15001184	(D-EIWB)	14. 7.75	L.W.Scattergood	Sherburn-in-Elmet	19. 6.06T
			(F-BXIH)				
G-BDGB	Gardan GY-20 Minicab	PFA 1819		23. 6.75	D.G.Burden	Armshold Farm, Kingston, Cambridge	12. 6.01P
	(Built D G Burden to JB.01 Minicab standard) (Continental PC-60)						
G-BDGH	Thunder Ax7-77 HAB	049		16. 7.75	R.J.Mitchener and P.F.Smart	(Andover)	30. 8.83A
					t/a Balloon Collection *"London Pride III "*		
G-BDGM	Piper PA-28-151 Cherokee Warrior	28-7415165	N41307	30. 7.75	Comed Schedule Services Ltd	Blackpool	8. 9.07T
	(Abandoned take off Netherthorpe 28.10.04, struck hedge and engine caught fire: damage to engine,both wings and undercarriage)						
G-BDGO*	Thunder Ax7-77 HAB				See SECTION 3 Part 1	(Aldershot)	
G-BDHJ*	Pazmany PL-1	PFA 3604		5. 8.75	Not known	Sleap	5.11.97P
	(Built H Jones and originally regd as Jones-Pazmany PL-1)				*(Cancelled 11.5.01 by CAA) (Noted 3.04)*		
G-BDHK	Piper J-3C-65 Cub (L-4A-PI)	8969	F-PHFZ	24. 7.75	A Liddiard	Eastbach Farm, Coleford	16. 9.05P
	(Frame No.9068)		42-38400		*(As "329417" in USAAC c/s)*		
	(Official c/n quoted as "261" with p/i 42-36414 but this corresponds to c/n 8538/N75366)						
G-BDIE	Rockwell Commander 112A	342	N1342J	14. 8.75	R.J.Adams	RAF Brize Norton	5. 9.07T
G-BDIG	Cessna 182P Skylane II	18263938	N9877E	26. 8.75	P.B.Barrett and A R Bruce	Finningley	15. 8.05
	(Reims-assembled with c/n F18200020)				tr Air Group 6		
G-BDIH	SAN Jodel D.117	812	F-BIOT	22. 8.75	N.D.H.Stokes	Garston Farm, Marshfield	11. 7.05P
G-BDIN	Scottish Aviation Bulldog Series 100/125	BH120/377	RJAF 408	20. 8.75	R.D.Dickson	RAF Wittering	4. 7.76S
			JY-BAI, G-BDIN		tr British Disabled Flying Association *(New CofR 8.04)*		
G-BDIJ	Sikorsky S-61N Mk.II	61-751	9M-AYF	3.10.75	Bristow Helicopters Ltd	Lee-on-Solent	31. 5.07T
	(SAR conversion)		G-BDIJ		*(Op Marine and Coastguard Agency) "Crathes"*		
G-BDIW*	de Havilland DH.106 Comet 4C				See SECTION 3 Part 1	Hermeskeil, Trier, Germany	
G-BDIX*	de Havilland DH.106 Comet 4C				See SECTION 3 Part 1	East Fortune	
G-BDJC	AJEP Wittman W.8 Tailwind	PFA 3508		29. 8.75	M.A.James	(Pitlochry)	5. 1.05P
	(Built A Whiting - c/n 387AW)						
G-BDJD	Jodel D.112	PFA 910		3. 9.75	J.E.Preston	(Ottringham)	29. 5.05P
	(Built J V Derrick) (Continental A65)				*"Marianne"*		
G-BDJG	Phoenix Luton LA-4A Minor	PFA 828		3. 9.75	S.K.Rose	White Waltham	1. 7.04P
	(Built D J Gaskin) (Volkswagen 1835)				tr Very Slow Flying Club		
G-BDJP	Piper J-3C-65 Cub Special	22992	OO-SKZ	11.12.75	S.T.Gilbert	Enstone	18. 5.03T
	(Continental C90) *(Frame No.21017)*		PH-NCV, NC3908K				
G-BDJR	SNCAN NC.858S	2	F-BFIY	30. 9.75	R.F.M.Marson *(On rebuild 9.00)*	(Fleet)	23. 5.92P
G-BDKC	Cessna A185F Skywagon	185-02569	N1854R	30. 9.75	Bridge of Tilt Co Ltd	Blair Atholl	23. 4.07
G-BDKD	Enstrom F-28A	319		30. 9.75	M.A.Crook and A.E.Wright	(Warrington)	29. 5.05
G-BDKH	Menavia Piel CP.301A Emeraude	241	F-BIMN	15.10.75	C.Lobban tr G-BDKH Group	(Edinburgh)	6. 8.05P
G-BDKJ	K & S Squarecraft SA.102.5 Cavalier	PFA 1589		14.10.75	D.A Garner	(Swansea)	5. 6.95P
	(Built H B Yardley c/n 72207) (Continental O-240-A)				*(Damaged Gloucestershire 14.9.97: current status unknown)*		
G-BDKM	SIPA 903	98	F-BGHX	17.11.75	S.W.Markham	Valentine Farm, Odiham	30. 4.05P
G-BDKV*	Piper PA-28R-200 Cherokee Arrow II	28R-7335297	EI-AYE	29.10.75	Bravo Juliet Whiskey Flying Ltd	Fairoaks	8. 4.95
			(G-BBEH), N55837		*(Cancelled 25.5.94 as destroyed) (Stored as "G-OOAT" 4.04)*		
G-BDKW	Rockwell Commander 112A	106	N1277J	3.11.75	J.T.Klaschka	Poplar Hall Farm, Elmsett	4. 9.06T
			ZS-MIB, N1106J				
G-BDLO	Grumman AA-5A Cheetah	AA5A-0026	N6154A	3.11.75	S.and J.Dolan	Elstre	4. 8.07
G-BDLS	Grumman AA-1B	AA1B-0564	N6153A	3.11.75	A.W.Cattermole	Sleap	28.10.04
G-BDLT	Rockwell Commander 112A	363	N1363J	4.11.75	D.L.Churchward	Popham	11. 6.05
G-BDLY	K & S Squarecraft SA.102.5 Cavalier	PFA 01-10011		14.11.75	P.R.Stevens	Thruxton	2. 6.04P
	(Built B S Reeve) (Lycoming O-290)						
G-BDMM	Jodel D.11	PFA/901		5.11.75	P.N.Marshall	Barton	
	(Built D M Metcalf)				*(Cancelled 27.1.97 by CAA) (Noted 11.04)*		
G-BDMO*	Thunder Ax7-77A HAB				See SECTION 3 Part 1	(Aldershot)	
G-BDMS	Piper J-3C-65 Cub (L-4J-PI)	13049	F-BEGZ	4.11.75	A T.H.Martin	Old Sarum	23. 5.54P
			44-80753		*(As "FR886" in RAF c/s)*		
G-BDMW	SAN Jodel DR.100A Ambassadeur	79	F-BIVM	2.12.75	P.R.H.Moore	Yew Tree Farm, Lymm Dam	15. 8.05
					tr G-BDMW Flying Group		
G-BDNC	Taylor JT.1 Monoplane	PFA 1454		8.12.75	D.W.Mathie	(Diss)	25. 4.05P
	(Built N J Cole) (Walter Mikron III)						
G-BDNG	Taylor JT.1 Monoplane	PFA 1405		12.12.75	A C.Beaumont	Shobdon	17. 9.03P
	(Built D J Phillips) (Volkswagen 1834)				*"The Red Sparrow" (Noted 3.04)*		
G-BDNO	Taylor JT.1 Monoplane	PFA 1431		15.12.75	S.D.Glover	Bodmin	4. 9.04P
	(Built A J Gray)						
G-BDNT	Jodel D.92 Bébé	397	F-PINL	2. 1.76	R.F.Morton	Kemble	12. 2.05P
	(Volkswagen 1600)						
G-BDNU	Reims/Cessna F172M Skyhawk II	F17201405		2. 1.76	J.and K.G.McVicar	Elstree	6. 4.06T
G-BDNW	Grumman AA-1B Trainer	AA1B-0588		8. 1.76	P.Mitchell	Humberside	30. 5.06T
G-BDNX	Grumman AA-1B Trainer	AA1B-0590		8. 1.76	R.M.North	Kimbolton	23. 8.07
G-BDOC	Sikorsky S-61N Mk.II	61-765		20. 3.76	Bristow Helicopters Ltd	Sumburgh	28.12.07E
	(SAR conversion)				*(Op Marine and Coastguard Agency) "Tolquhoun"*		
G-BDOD	Reims/Cessna F150M	F15001266		20. 1.76	P.R.Green tr OD Group	(Old Windsor)	2. 7.06
G-BDOE	Reims FR172J Rocket	FR17200559		20. 1.76	D.Sansome	Little Chase Farm, Kenilworth	3. 4.06
G-BDOG	Scottish Aviation Bulldog Series 200	BH200/381		18.12.75	D.C.Bonsall	Netherthorpe	3. 8.05P
					(Phoenix Flying Group titles)		

Reg	Type	C/n	Previous identity	Date	Owner/Operator	Base	CofA expiry
G-BDOL	Piper J-3C-65 Cub (L-4J-PI)	13186	F-BCPC	18.12.75	L.R.Balthazor	Lee-on-Solent	9. 7.05P
	(Frame No.13016)		45-4446				
	(Official c/n 13370 but has c/n and USAAC plates relating to c/n 13370/ex 45-4630 now G-BCUB: airframes switched during UK conversion)						
G-BDON	Thunder Ax7-77A HAB	063		17.12.75	M.J.Smith *"Fred"*	(Westow, York)	24. 6.94A
G-BDOT	Fairey Britten-Norman BN-2A Mk III-2 Trislander	1025	ZK-SFF	21. 1.76	Lyddair Ltd	Lydd	2. 3.06T
			N900TA, N903GD, N3850K, VH-BPB, G-BDOT				
G-BDOW	Reims/Cessna FRA150M Aerobat	FRA1500296		26. 1.76	Joystick Aviation Ltd	(St Albans)	3. 4.07T
G-BDPA	Piper PA-28-151 Cherokee Warrior	28-7615033	N9630N	26. 1.76	Prestwick Flight Centre Ltd	Prestwick	6.11.06T
G-BDPJ	Piper PA-25-235 Pawnee B	25-3665	PH-VBF	2. 2.76	T.M.Holloway	RAF Halton	19.12.06
	(Lycoming O-540A1B5-@ 250hp)		SE-EPZ		tr RAFGSA		
G-BDPK	Cameron O-56 HAB	191		4. 2.76	N.H.Ponsford and A M.Lindsay	(Leeds)	29.12.88A
					t/a Rango Balloon and Kite Co		
G-BDPP*	Fairey Britten-Norman BN-2A-21 Islander				See SECTION 3 Part 1	Grace Hollogne, Belgium	
G-BDPU*	Fairey Britten-Norman BN-2A-21 Islander				See SECTION 3 Part 1	Brussels, Belgium	
G-BDRD	Reims/Cessna FRA150M Aerobat	FRA1500289		9. 2.76	Prestwick Flight Centre Ltd	Prestwick	4. 8.06T
G-BDRG	Taylor JT.2 Titch	PFA 60-10295		19.12.78	D.R.Gray	(Wilmslow)	
	(Built D.R.Gray)				*(Current status unknown)*		
G-BDRJ	de Havilland DHC-1 Chipmunk 22	C1/0742	WP857	19. 2.76	D.Curtis *(As "WP857/24")*	Prestwick	19. 9.05
G-BDRK	Cameron O-65 HAB	205		12. 2.76	D.L.Smith *"Smirk"*	(Eling Hill, Newbury)	20. 6.86A
G-BDRL*	Stits SA-3A Playboy	P-689	N730GF	12. 2.76	O.C.Bradley	Mullaghmore	17. 6.98P
	(Built D R Fisher) (Continental C85)				*(Cancelled 11.5.01 by CAA) (Stored 4.03)*		
G-BDSB	Piper PA-28-181 Cherokee Archer II	28-7690107	N8221C	23. 2.76	Testfair Ltd	Fairoaks	28. 6.07
G-BDSE	Cameron O-77 HAB	210		27. 2.76	British Airways plc *"Concorde"*	(Worplesdon)	31. 3.90A
G-BDSF	Cameron O-56 HAB	209		1. 3.76	J.H.Greensides *"Itzuma"*	(Burton Pidsea, Hull)	24. 5.93A
G-BDSH	Piper PA-28-140 Cherokee Cruiser	28-7625063	N9638N	1. 3.76	D.Jones	Tollerton	26. 8.05
					tr The Wright Brothers Flying Group		
G-BDSK	Cameron O-65 HAB	166		3. 3.76	Semajan Ltd	(France)	9. 9.03A
					tr Southern Balloon Group *"Carousel II"*		
G-BDSL	Reims/Cessna F150M	F15001306		5. 3.76	D.C.Bonsall	Netherthorpe	24. 6.04T
	(Lost power on take-off Netherthorpe 16.2.04: a/c made forced landing, struck runway sign, nosewheel collapsed and overturned: damaged beyond economic repair)						
G-BDSM	Slingsby T.31 Motor Cadet III	PFA 42-10507		5. 3.76	F.C.J.Wevers	(Amersfoort, The Netherlands)	22. 5.02P
	(Built D W Savage - c/n 2464/3B)				*(New owner 4.04)*		
G-BDTB	Evans VP-1 Series 2	PFA 7009		15. 3.76	P.F.Moffatt	Breighton	29.10.04P
	(Built T E Boyes) (Volkswagen 1834)						
G-BDTL	Evans VP-1	PFA 7012		17. 3.76	A K.Lang	(Stoke-sub-Hamdon)	5. 9.85P
	(Built A K.Lang) (Volkswagen 1600)				*(Stored 5.98: current status unknown)*		
G-BDTN	Fairey Britten-Norman BN-2A Mk III-2 Trislander	1026	S7-AAN	16. 3.76	Aurigny Air Services Ltd	Guernsey	10. 6.98T
			VQ-SAN, G-BDTN		*(Stored 3.03)*		
G-BDTO	Fairey Britten-Norman BN-2A Mk.III-2 Trislander	1027	G-RBSI	16. 3.76	Aurigny Air Services Ltd	Guernsey	31. 3.05T
			G-OTSB, G-BDTO, 8P-ASC, G-BDTO, (C-GYOX), G-BDTO *(Merrill Lynch titles) "Nessie"*				
G-BDTU	Van Den Bemden Omega III Gas Balloon (20,000 cu.ft)	VDB-35 & AFB.4		16. 3.76	R.G.Turnbull	Clyro, Hereford	4. 8.99A
					"Omega III"		
G-BDTV	Mooney M.20F Executive	22-1307	N6934V	16. 3.76	S.Redfearn	Gamston	6. 8.06
G-BDTX	Reims/Cessna F150M	F15001275		19. 3.76	F.W.Ellis	Water Leisure Park, Skegness	24. 7.06T
G-BDUI	Cameron V-56 HAB	218		19. 3.76	D.C.Johnson *"True Brit"*	(Farnham)	6. 7.91A
G-BDUL	Evans VP-1	PFA 1557		25. 3.76	C.K.Brown	St Just	19. 5.04P
	(Built C Goodman) (Volkswagen 1834)						
G-BDUM	Reims/Cessna F150M	F15001301	F-BXZB	29. 3.76	S.G.Bishop tr G-BDUM Group	Earls Colne	14. 4.06
G-BDUN	Piper PA-34-200T Seneca II	34-7570163	(EI-BLR)	29. 3.76	Air Medical Fleet Ltd	Oxford	27. 2.07T
			G-BDUN, SE-GIA				
G-BDUO	Reims/Cessna F150M Commuter	F15001304		29. 3.76	D.F.Ranger	Popham	16. 5.07T
G-BDUP*	Bristol 175 Britannia Series 253				See SECTION 3 Part 1	Kemble	
G-BDUY	Robin DR400/140B Major	1120		5. 4.76	J.G.Anderson	Pittrichie Farm, Whiterashes	13. 4.06
G-BDUZ	Cameron V-56 HAB	213		30. 3.76	P.J.Bish	(Newtown Hungerford)	19. 2.00A
					t/a Zebedee Balloon Service *"Hot Lips"*		
G-BDVA	Piper PA-17 Vagabond	17-206	CN-TVY	23. 4.76	I.M.Callier	Liss	13. 8.03P
	(Continental C90)		F-BFFE				
G-BDVB	Piper PA-15 Vagabond	15-229	F-BHHE	23. 4.76	B.P.Gardner	Whittles Farm, Mapledurham	5. 6.05P
	(Continental C90)		SL-AAY, F-BETG				
G-BDVC	Piper PA-17 Vagabond	17-140	F-BFBL	29. 9.76	A R.Caveen	Sandford Hall, Knockin	9. 9.02P
	(Continental C90)						
G-BDVG*	Thunder Ax6-56A HAB				See SECTION 3 Part 1	Newbury	
G-BDVO*	Short SC.7 Skyvan 3-100				See SECTION 3 Part 1	Caracas, Venezuela	
G-BDVS*	Fokker F.27 Friendship 200				See SECTION 3 Part 1	Flixton, Bungay	
G-BDWA	SOCATA Rallye 150ST	2695		20. 4.76	J.T.Wilson *(Noted 4.03)*	Bann Foot, Lough Neagh	7. 6.01
G-BDWE	Flaglor Sky Scooter	PFA 1332		12. 4.76	P.King	Eastbach Farm, Coleford	27. 6.05P
	(Built D W Evernden - c/ns DWE-01 & KF-S-66) (Volkswagen 1600)						
G-BDWH	SOCATA Rallye 150ST	2697		20. 4.76	M.A.Jones	Upper Harford Farm, Bourton-on-the-Water	28.10.07E
G-BDWJ	Replica Plans SE.5A	PFA 20-10034	"C1904"	27. 4.76	D.W.Linney	(Langport)	8. 8.05P
	(Built M L Beach) (Continental C90)		"F8010"		*(As "F8010/Z" in RFC c/s)*		
G-BDWM	Bonsall DB-1 Mustang	PFA 73-10200		3. 5.76	D.C.Bonsall	Netherthorpe	15. 6.98P
	(Built D.C.Bonsall) (Lycoming IO-360)				*(As "FB226/MT-A" in RAF c/s) (Noted on rebuild 5.04)*		
G-BDWO	Howes Ax6 HAB	RBH.2		5. 5.76	R.B. and Mrs C.Howes *"Griffin"*	(Keysoe)	
	(Built R.B.Howes)				*(Complete and extant 11.88 but never certified:)*		
G-BDWP	Piper PA-32R-300 Cherokee Lance	32R-7680176	N8784E	7. 5.76	W.M.Brown	Coventry	3.12.06
G-BDWX	Wassmer Jodel D.120A Paris-Nice	311	F-BNHT	13. 5.76	R.P.Rochester	Wombleton	22.10.04P
G-BDWY	Piper PA-28-140 Cherokee E	28-7225378	PH-NSC	14. 5.76	Comed Schedule Services Ltd	Blackpool	25. 4.07T
			N11C				
G-BDXE	Boeing 747-236B	21350		23. 2.78	European Skybus Ltd	Bournemouth	14. 8.06T
					(Op European Aviation Air Charter)		
G-BDXF	Boeing 747-236B	21351		23. 3.78	European Skybus Ltd	Bournemouth	30. 4.06T
					(Stored minus two engines 1.05)		
G-BDXG	Boeing 747-236B	21536		16. 6.78	European Skybus Ltd	Bournemouth	30. 6.06T
					(Op Air Namibia)		

G-BDXH	Boeing 747-236B	21635		23. 2.79	European Skybus Ltd	Bournemouth	2. 5.04T
					(Stored minus three engines 1.05)		
G-BDXI*	Boeing 747-236B	21830		21. 2.80	British Airways plc	Cardiff	13. 3.04T
					(Cancelled 21.5.03 as wfu) (Stored 5.03)		
G-BDXJ	Boeing 747-236B	21831	N1792B	2. 5.80	Air Atlanta Europe Ltd *(New CofR 12.04)*	Gatwick	7. 5.04T
G-BDXO	Boeing 747-236B	23799	N6055X	22. 4.87	Iceland Ltd *(New owner 10.04)*	(Hamilton, Bermuda)	14. 5.03T
G-BDXX	SNCAN NC.858S	110	F-BEZO	17. 5.76	K.M.Davis	North Weald	3. 7.96P
					(On rebuild 5.04 - newowner 7.04)		
G-BDYA*	CASA 2111				See SECTION 3 Part 1	Mesa, Arizona, US	
G-BDYD	Rockwell Commander 114	14014	N1914J	21. 5.76	M.B.Durkin	Sherburn-in-Elmet	11. 9.06
G-BDYG*	Percival P.56 Provost T.1				See SECTION 3 Part 1	East Fortune	
G-BDYH	Cameron V-56 HAB	233		24. 5.76	B.J.Godding *"Novocastrian"*	(Didcot)	25.11.90A
G-BDZA	Scheibe SF25E Super Falke	4320	(D-KECW)	1. 6.76	D.C.Mason	Crowland	3.10.04
	(Limbach SL-1700)				tr Hereward Flying Group		
G-BDZC	Reims/Cessna F150M	F15001316		1. 6.76	A M.Lynn	Sibson	21. 7.05T
G-BDZD	Reims/Cessna F172M Skyhawk II	F17201478		1. 6.76	J.K.Haig and A L.Jones tr Zephyr Group	South Cerney	20.10.06T
G-BDZU	Cessna 421C Golden Eagle II	421C0094	N98791	14. 6.76	R.Richardson	Nottingham East Midlands	27. 4.06T
					tr Eagle Flying Group		

G-BEAA - G-BEZZ

G-BEAB	CEA Jodel DR.1051 Sicile	228	F-BKGH	18. 8.76	R.C.Hibberd *(Noted 6.04)*	Dunkeswell	9. 3.03
G-BEAC	Piper PA-28-140 Cherokee	28-21963	4X-AND	4. 6.76	C.E.Stringer tr Clipwing Flying Group	Bagby	4. 6.06T
G-BEAG	Piper PA-34-200T Seneca II	34-7670204	N9395K	18. 6.76	Oxford Aviation Services Ltd	Oxford	22.10.03T
G-BEAH	Auster V J/2 Arrow	2366	F-BFUV	28. 6.76	J.G.Parish	Bedwell Hey Farm, Little Thetford, Ely	16. 4.04P
	(Continental C85)		F-BFVV, OO-ABS		tr Bedwell Hey Flying Group *"Llewellyn"*		
G-BEBC*	Westland WS-55 Whirlwind HAR.10				See SECTION 3 Part 1	Norwich	
G-BEBE	Grumman AA-5A Cheetah	AA5A-0154		28. 6.76	Bills Aviation Ltd	Biggin Hill	30. 4.06T
G-BEBG	PZL-Bielsko SZD-45A Ogar	B-655		29. 6.76	D.W.Coultrip tr The Ogar Syndicate	Hinton-in-the-Hedges	18. 6.06
G-BEBN	Cessna 177B Cardinal	17701631	4X-CEW	1. 7.76	R.Turrell and P.Mason	King's Farm, Thurrock	22. 3.06
			N34031				
G-BEBS	Andreasson BA-4B	PFA 38-10157		7. 7.76	N.J.W.Reid	Lee-on-Solent	5. 4.05P
	(Built D M Fenton [t/a Hornet Aviation] - c/n HA/01)		(Continental O-200-A)				
G-BEBU	Rockwell Commander 112A	272	N1272J	8. 7.76	Aeros Engineering Ltd	Gloucestershire	19. 6.04T
G-BEBZ	Piper PA-28-151 Cherokee Warrior	28-7615328	N6193J	14. 7.76	Goodwood Road Racing Co Ltd	Goodwood	19. 4.06T
					(Op Goodwood Flying Club)		
G-BECA	SOCATA Rallye 100ST	2751		14. 7.76	A.C.Stamp	(Dorking)	21.10.06
G-BECB	SOCATA Rallye 100ST	2783		14. 7.76	D.H.Tonkin	Bodmin	24. 5.05
G-BECC	SOCATA Rallye 150ST	2748		14. 7.76	G.R.E.Tapper *(Noted dismantled 5.03)*	Haverfordwest	15. 5.00
G-BECF	Scheibe SF25A Motorfalke	4555	OO-WIZ	14. 7.76	North County Ltd	(Middleton)	1. 3.94P
	(Hirth F23A)		(D-KARA)		*(Current status unknown)*		
G-BECK	Cameron V-56 HAB	136		27. 7.76	M.E.White *"Joyride"*	(Templeogue, Dublin)	21. 3.00A
G-BECL*	CASA 352L				See SECTION 3 Part 1	La Ferté-Alais, Essonne, France	
G-BECN	Piper J-3C-65 Cub (L-4J-PI)	12776	F-BCPS	27. 7.76	G.Denney	Rayne Hall Farm, Braintree	30. 8.05
			HB-OCI (1), 44-80480		*(As "480480/44-E" in USAAC c/s)*		
G-BECS	Thunder Ax6-56A HAB	074		4. 8.76	A.Sieger	(Munster, Germany)	5. 5.05A
G-BECT	CASA 1-131E Jungmann	"3974"	Spanish AF E3B-338	3. 8.76	G.M.S.Scott	Goodwood	1. 9.05P
					tr Alpha 57 Group *(As "A-57" in Swiss AF c/s)*		
G-BECW	CASA 1-131E Jungmann	2037	Spanish AF E3B-423	3. 8.76	R.G.Meredith	Rochester	4.11.05P
	(Incorporating parts of G-BECY ex E3B-459)				*(As "A-10" in Swiss AF c/s)*		
G-BECZ	Avions Mudry CAP.10B	68	F-BXHK	26. 7.76	Avia Special Ltd	White Waltham	2. 6.07T
G-BEDD	SAN Jodel D.117A	915	F-BITY	3. 8.76	M.D.Howlett	Shrove Furlong Farm, Ilmer	7. 6.05P
					tr Dubious Group *(Noted active 9.04)*		
G-BEDF	Boeing B-17G-105-VE Fortress	8693	N17TE	5. 8.76	B-17 Preservation Ltd	Duxford	21. 5.05P
			F-BGSR, 44-85784		*(As "124485/DF-A" in USAAC c/s) "Sally B" (port)/"Memphis Belle" (starboard)*		
G-BEDG	Rockwell Commander 112A	482	N1219J	5. 8.76	Hotels International Ltd	Blackbushe	13.12.05
G-BEDJ	Piper J-3C-65 Cub (L-4J-PI)	12890	F-BDTC	5. 8.76	R.Earl	Roughay Farm, Bishops Waltham	8.10.96P
	(Frame No.12720)		44-80594		*(As "44-805942" in USAAC c/s) (Noted 12.03)*		
G-BEDP	Fairey Britten-Norman BN-2A Mk.III-2 Trislander	1039	ZK-SFG	17. 8.76	Lyddair Ltd	Lydd	28. 5.06T
			N902TA, N1FY, N401JA, G-BEDP				
G-BEDV*	Vickers 668 Varsity T.1				See SECTION 3 Part 1	Duxford	
G-BEEE*	Thunder Ax6-56A HAB				See SECTION 3 Part 1	Newbury	
G-BEEG	Fairey Britten-Norman BN-2A-26 Islander	550	(C-GYUH)	25. 8.76	M.D.Carruthers and S.F.Morris	Cark-in-Cartmel	5. 4.05T
			G-BEEG		t/a North West Parachute Centre		
G-BEEH	Cameron V-56 HAB	250		24. 8.76	Sade Balloons Ltd *"Tywi"*	(Coulsdon)	12. 9.05A
G-BEER	Isaacs Fury II	PFA 1588		31. 8.76	R.S.C.Andrews	Bidford	5. 6.05P
	(Built M J Clark) (Lycoming O-235)				*(As "K2075" in RAF c/s)*		
G-BEEU	Piper PA-28-140 Cherokee F	28-7325247	PH-NSE	9. 9.76	H. and Eve Merkado	Panshanger	6. 3.06T
			N11C				
G-BEEX*	de Havilland DH.106 Comet 4C				See SECTION 3 Part 1	Usworth, Sunderland	
G-BEFA	Piper PA-28-151 Cherokee Warrior	28-7615416	N6978J	8. 9.76	M.A Verran t/a Verran Freight	RAF Benson	6. 4.06
G-BEFF	Piper PA-28-140 Cherokee F	28-7325228	PH-NSF	27. 9.76	H. and Eve Merkado	Panshanger	22. 5.06
		N11C					
G-BEGG	Scheibe SF25E Super Falke	4326	(D-KDFB)	15.10.76	R.M.Murray	Hall Farm, Turweston	21. 5.06
	(Limbach SL1700)				tr G-BEGG Flying Group		
G-BEHH	Piper PA-32R-300 Cherokee Lance	32R-7680323	N6172J	29.10.76	K.Swallow	Sherburn-in-Elmet	5. 9.06
G-BEHU	Piper PA-34-200T Seneca II	34-7670265	N6175J	3.11.76	Pirin Aeronautical Ltd	Stapleford	21. 3.05T
G-BEHV	Reims/Cessna F172N Skyhawk II	F17201541		3.11.76	Edinburgh Air Centre Ltd	Edinburgh	10. 9.05T
G-BEIA	Reims/Cessna FRA150M Aerobat	FRA1500317		8.11.76	Oxford Aviation Services Ltd	Oxford	27.10.03T
					(Noted 7.04)		
G-BEIF	Cameron O-65 HAB	259		17.11.76	C.Vening	(Aldershot)	6.12.04A
					(Op Balloon Preservation Group) "Solitaire"		
G-BEIG	Reims/Cessna F150M	F15001361		18.11.76	T.J.Chapman	(Beccles)	26. 6.05T

Registration	Type	C/n	Previous identity	Date	Owner/Operator	Base	CofA
G-BEII	Piper PA-25-235 Pawnee D	25-7656059	N54918	16.11.76	Burn Gliding Club Ltd	Burn	7. 4.05
G-BEIL	SOCATA Rallye 150T	2653	F-BXDL	1.12.76	J.I.Oakes and R.A Harris	Hill Farm, Nayland	9. 6.07
					tr The Rallye Flying Group		
G-BEIP	Piper PA-28-181 Cherokee Archer II	28-7790158	N6628F	22.11.76	S.Pope	Barton	16. 9.07
G-BEIS	Evans VP-1	PFA 7029		25.11.76	P.J.Hunt	Thruxton	16. 7.90P
	(Built D J Park) (Volkswagen 1600)				*(Stored 2.99: current status unknown)*		
G-BEJB*	Thunder Ax6-56A HAB				See SECTION 3 Part 1	(Aldershot)	
G-BEJD	Avro 748 Series 1/105	1543	LV-HHE	17.12.76	Emerald Airways Ltd	Liverpool	29. 3.06E
			LV-PUF		*(Reed Aviation c/s) "Sisyphus"*		
G-BEJK	Cameron S-31 HAB	256		1.12.76	N.H.Ponsford and A Lindsay	(Leeds)	16. 2.92A
					t/a Rango Balloon and Kite Co *"L'Essence" (or "Esso")*		
G-BEJL	Sikorsky S-61N Mk.II	61-224	EI-BPK	30.12.76	CHC Scotia Ltd	Longside, Peterhead	30. 9.98
			G-BEJL, N4606G		*(Stored 4.04)*		
G-BEJV	Piper PA-34-200T Seneca II	34-7770062	N7657F	31.12.76	Oxford Aviation Services Ltd	Oxford	29. 7.06T
G-BEKL	Bede BD-4E-150	151 & BD4E/2	(G-AYKB)	11. 1.77	F.E.Tofield	(Farnborough)	14.10.80P
	(Built Brookmoor Bede Aircraft)(Lycoming O-320)				*(New owner 4.02)*		
G-BEKM	Evans VP-1	PFA 7025		12. 1.77	G.J.McDill	Westmoor Far, Thirsh	23. 3.95P
	(Built G.J.McDill) (Volkswagen 1834)				*(Stored 7.98: new CofR 1.03)*		
G-BEKN	Reims/Cessna FRA150M Aerobat	FRA1500318		12. 1.77	A.L.Brown	Bourn	8.10.89T
					(Open store without engines 11.01: noted 11.03)		
G-BEKO	Reims/Cessna F182Q Skylane II	F18200037		12. 1.77	G.J. and F.J.Leese	Sherburn-in-Elmet	12. 6.06
G-BELF	Fairey Britten-Norman BN-2A-26 Islander	823	D-IBRA	13. 1.77	Cormack (Aircraft Services) Ltd	Cumbernauld	12. 3.01
	(Built IRMA)		G-BELF		*(New owner 6.04)*		
G-BELP	Piper PA-28-151 Cherokee Warrior	28-7715219	N9543N	18. 1.77	Tatenhill Aviation Ltd	Tatenhill	17. 8.07T
G-BELT	Cessna F150J	F150-0409X		26. 1.77	A.Kumar	Blackpool	22. 8.07T
	(Built Reims Aviation SA) (Mainly rebuild of G-AWUV and parts of G-ATND)						
G-BEMB	Reims/Cessna F172M Skyhawk II	F17201487		27. 1.77	Stocklaunch Ltd	Goodwood	17. 8.07T
G-BEMM	Slingsby Cadet III	1247	BGA942	27. 1.77	B.J.Douglas	Kilrush, County Kildare	10.10.05P
	(Volkswagen 1600) *(Converted from T.31B)*		RAFGSA 289, BGA942				
G-BEMU	Thunder Ax5-42 HAB	097		9. 2.77	M.A Hall *"Chrysophylax"*	(Stoneleigh)	16. 1.99A
G-BEMW	Piper PA-28-181 Cherokee Archer II	28-7790243	N9566N	9. 2.77	Touch and Go Ltd	Lee-on-Solent	30.11.06
G-BEMY	Reims/Cessna FRA150M Aerobat	FRA1500315		9. 2.77	M.J.L.Tondeur and M.M.E.Versyck	(Aalst, Belgium)	13.12.07E
G-BEND	Cameron V-56 HAB	260		14. 2.77	P.J.Bish tr Dante Balloon Group *"Le Billet"*	(Hungerford)	1. 1.94A
G-BENJ	Rockwell Commander 112B	522	N1391J	7. 3.77	D.J.Gibney	Top Farm, Croydon, Royston	7. 8.06
					tr BENJ Flying Group		
G-BENK	Reims/Cessna F172M Skyhawk II	F17201509		2. 3.77	Graham Churchill Plant Ltd	Turweston	4. 6.06
G-BENN	Cameron V-56 HAB	278		4. 3.77	S.J.Hollingsworth and M.K.Bellamy *"English Rose"*		
					(New owners 10.01)	(Bleasby, Nottingham)	15. 3.87A
G-BEOE	Reims/Cessna FRA150M Aerobat	FRA1500322		21. 3.77	W.J.Henderson t/a Air Images	Carlisle	11. 7.06T
G-BEOH	Piper PA-28R-201T Turbo Cherokee Arrow III		N1905H	11. 3.77	K.G.Harper	Gloucestershire	6. 7.07
		28R-7703038			tr Gloucestershire Flying Club		
G-BEOI	Piper PA-18-150 Super Cub	18-7709028	N54976	11. 3.77	Southdown Gliding Club Ltd	Parham Park	17.12.04
	(Lycoming O-360-A4)						
G-BEOK	Reims/Cessna F150M	F15001366		14. 3.77	D.C.Bonsall	Netherthorpe	11. 5.06T
G-BEOL	Short SC.7 Skyvan 3-100	SH.1954	ZS-OIO	16. 3.77	Invicta Aviation Ltd	Manston	27. 3.05T
			JA8803 (2), G-BEOL, G-14-122				
G-BEOX*	Lockheed 414 Hudson IIIA				See SECTION 3 Part 1	Hendon	
G-BEOY	Reims/Cessna FRA150L Aerobat	FRA1500150	F-BTFS	30. 3.77	R.W.Denny *(Op Crowfield Flying Club)*	Crowfield	27. 3.05T
G-BEOZ*	Armstrong-Whitworth AW.650 Argosy 101				See SECTION 3 Part 1	Nottingham East Midlands	
G-BEPC	SNCAN Stampe SV-4C	64	F-BFUM	17.10.77	Papa Charlie's Flying Circus Ltd	Branscombe	24. 3.01T
	(DH Gipsy Major 10)		F-BFZM, Fr.Mil		*(On long termrebuild 1.04)*		
G-BEPF	SNCAN Stampe SV-4C	424	F-BCVD	30. 3.77	L.J.Rice *(Stored 9.04)*	Chilbolton	AC
G-BEPO*	Cameron N-77 HAB				See SECTION 3 Part 1	Newbury	
G-BEPS*	Short SC.5 Belfast C.1	SH.1822	G-52-13	6. 4.77	Heavylift Aviation Holdings Ltd	Southend	31. 8.02T
			XR368		*(Cancelled 8.5.03 by CAA) (Under gradual refurbishment 2.05)*		
G-BEPV	Fokker S.11.1 Instructor	6274	PH-ANK	13. 4.77	L.C.MacKnight	Spanhoe	22. 1.05P
			Dutch Navy 174, Dutch AF E-31 *(As 'K-174' in Dutch Navy c/s)*				
G-BEPY	Rockwell Commander 112B	524	N1399J	20. 4.77	S.A Pigden tr G-BEPY Group	Blackbushe	1. 7.07
G-BEPZ*	Cameron D-96 HA Airship				See SECTION 3 Part 1	Newbury	
G-BERA	SOCATA Rallye 150ST	2821	F-ODEX	13. 4.77	B.Dolby	Exeter	27. 8.06T
G-BERC	SOCATA Rallye 150ST	2858		13. 4.77	R.S.Jones	Welshpool	5. 8.05
					tr The Severn Valley Aero Group		
G-BERD	Thunder Ax6-56A HAB	106		25. 4.77	P.M.Gaines *"Goldfinger"*	(Stockton-on-Tees)	14. 5.04A
G-BERI	Rockwell Commander 114	14234	N4909W	6. 5.77	K.B.Harper	Blackbushe	12. 5.06
G-BERN	Saffery S.330 MLB	4		19. 4.77	B.Martin *"Beeze I"*	(Somersham, Huntingdon)	
G-BERT	Cameron V-56 HAB	273		19. 4.77	Semajan Ltd tr Southern Balloon Group *"Bert"*	(France)	3. 6.04A
G-BERW	Rockwell Commander 114	14214	N4884W	6. 5.77	Romeo Whisky Ltd	Old Buckenham	5. 5.04
G-BERY	Grumman AA-1B Trainer	0193	N9693L	27.10.77	R.H.J.Levi *"79"*	Stapleford	24. 6.07
G-BESY*	British Aircraft Corporation 167 Strikemaster Mk.80A				See SECTION 3 Part 1	Duxford	
G-BETD	Robin HR200/100 Club	20	PH-SRL	28. 4.77	P.I.Acott	Fowlmere	9. 1.06
G-BETE	Rollason Beta B.2A	PFA 02-10169		26. 4.77	T.M.Jones	Egginton	
	(Built T Jones) (Incorporates parts from PFA 1304)				*(Under construction 4.04)*		
G-BETF*	Cameron Champion 35SS HAB				See SECTION 3 Part 1	Newbury	
G-BETG	Cessna 180K Skywagon	180-52873	N64146	17. 5.77	J.A Hart	(Stoke-sub-Hamdon)	17. 7.06
G-BETH*	Thunder Ax6-56A HAB				See SECTION 3 Part 1	Newbury	
G-BETI	Pitts S-1D Special	PFA 09-10156	G-PIII	28. 4.77	N.A. Scully	Leicester	9. 5.01P
	(Built B Bray - c/n 7-0314) (Lycoming O-320)		G-BETI		tr On A Roll Aerobatics Group		
G-BETL	Piper PA-25-235 Pawnee D	25-7656016	N54874	27. 5.77	Cambridge Gliding Club Ltd	Gransden Lodge	22.10.07
G-BETM	Piper PA-25-235 Pawnee D	25-7656066	N54927	5. 5.77	Yorkshire Gliding Club (Pty) Ltd	Sutton Bank	9. 4.07
G-BETO	Morane Saulnier MS.885 Super Rallye	34	F-BKED	18. 5.77	A.J. and A.Hawley	Farley Farm, Romsey	11.12.04
					tr G-BETO Group		
G-BETT	Piper PA-34-200 Seneca	34-7250011	EI-BCD	20. 6.77	D.F.J.Flashman	Biggin Hill	28. 7.99A
			PH-AVM, N1978T		*(New CofR 6.02)*		
					(Made forced landing at Frinstead, near Maidstone 22. 7.04 and extensively damaged)		

G-BETW	Rand Robinson KR-2	PFA 129-10251		26. 4.77	S.C.Solley	Clipgate Farm, Denton	
	(Original c/n allocated as KR-2/TAW.1 (T A Wiffen) but PFA project no issued to VP-1 [62-10251]: re-allocated to S C Solley: current status unknown)						
G-BEUA	Piper PA-18-150 Super Cub	18-8212	D-ECSY	21. 6.77	London Gliding Club Pty Ltd	Dunstable	27. 2.06
	(Lycoming O-360-A4)		N4146Z				
G-BEUD	Robin HR100/285 Tiara	534	F-BXRC	8. 6.77	E.A and L.M.C.Payton	Cranfield	11. 8.05
G-BEUI	Piper J-3C-65 Cub (L-4H-PI)	12174	F-BFEC	19. 5.77	B.W.Webb	(Canterbury)	26. 2.05P
	(Frame No.12002) (Regd as ex 43-29245)		F-OAJF, Fr.AF, 44-79878		tr G-BEUI Group		
G-BEUM	Taylor JT.1 Monoplane	PFA 1438		8. 6.77	J.M.Burgess	St.Just	28. 7.04P
	(Built Speedwell Sailplanes) (Volkswagen 1700)						
G-BEUN	Cassutt Racer IIIM	PFA 34-10241		20. 2.78	R.McNulty	Bourn	7. 7.97P
	(Built M S Crossley) (Continental C90)				*(Noted 1.03)*		
G-BEUP	Robin DR400/180 Régent	1228		19. 5.77	Legal Week Ltd t/a Samuels Aviation	Biggin Hill	13. 6.07T
G-BEUU	Piper PA-18 Super Cub 95	18-1551	F-BOUU	27. 6.77	F.Sharples	Sandown	16. 3.05P
	(L-18C-PI) *(Frame No.should be 18-1523)*		ALAT 18-1551, 51-15551				
G-BEUX	Reims/Cessna F172N Skyhawk II	F17201596		30. 5.77	Multiflight Ltd	Leeds-Bradford	17. 2.07T
G-BEUY	Cameron N-31 HAB	283		31. 5.77	M.L. and L.P.Willoughby	(Woodcote, Reading)	17.10.90A
					(Typhoo Tea titles) (Inflated 4.02)		
G-BEVB	SOCATA Rallye 150ST	2860		2. 6.77	G.G.Hammond	Sorbie Farm, Kingsmuir	16. 5.05
G-BEVC	SOCATA Rallye 150ST	2861		2. 6.77	G-BEVC Ltd	Maypole Farm, Chislet	6. 7.06
G-BEVG	Piper PA-34-200T Seneca II	34-7570060	VQ-SAM	31. 5.77	A G.and J.Wintle	Elstree	3.10.05T
			N32854				
G-BEVI (3)*	Thunder Ax7-77A HAB				See SECTION 3 Part 1	Newbury	
G-BEVO	Fournier RF5	5107	5N-AIX	27. 6.77	D.G.Hey and W.E.R.Jenkins	Little Gransden	14. 7.05P
			D-KAAZ				
G-BEVR*	Fairey Britten-Norman BN-2A Mk.III-2 Trislander	1056	6Y-JQE	10. 6.77	Cormack (Aircraft Services) Ltd	Not known	6. 7.82S
			G-BEVR, XA-THE(2), G-BEVR		*(Cancelled 20.3.03 by CAA) (Noted 5.03)*		
G-BEVS	Taylor JT.1 Monoplane	PFA 1429		8. 6.77	D.Hunter	Kemble	9.12.05P
	(Built D Hunter) (Volkswagen 1835)						
G-BEVT	Fairey Britten-Norman BN-2A Mk.III-2 Trislander	1057		10. 6.77	Aurigny Air Services Ltd	Guernsey	15.11.06T
					(Islands Insurance titles)		
G-BEVV*	Fairey Britten-Norman BN.2A Mk.III-2 Trislander	1059	6Y-JQK	10. 6.77	Cormack (Aero Club Services) Ltd	Cumbernauld	
			G-BNZD, G-BEVV		*(On rebuild but stored by 2.03) (Cancelled 20.3.03 by CAA)*		
G-BEVW	SOCATA Rallye 150ST	2928		2. 6.77	P.G.A Sumner	Town Farm, Woolaston	3. 4.06
G-BEWN	de Havilland DH.82A Tiger Moth	952	VH-WAL	16. 6.77	H.D.Labouchere	Blue Tile Farm, Langham	22. 7.06
	(Built de Havilland Aircraft Pty Ltd, Australia- rebuild c/n T305)		RAAF A17-529				
G-BEWO	Moravan Zlin Z-326 Trener Master	915	CS-ALU	23.11.77	P.A Colman	Luxters Farm, Hambleden, Henley-on-Thames	2. 7.03
G-BEWP*	Reims/Cessna F150M	F15001426		13. 6.77	Perth College	Perth	12. 8.85
					(Crashed Aboyne 4.10.83 and cancelled 5.12.83 as destroyed) (Instructional use 4.03)		
G-BEWR	Reims/Cessna F172N Skyhawk II	F17201613		13. 6.77	Cheshire Air Training Services Ltd	Liverpool	25. 2.07T
G-BEWX	Piper PA-28R-201 Cherokee Arrow III	28R-7737070	N5723V	23. 6.77	A Vickers	North Weald	23. 5.06
G-BEWY	Bell 206B JetRanger II	348	G-CULL	27. 6.77	Polo Aviation Ltd	Bristol	13. 5.07T
			EI-BXQ, G-BEWY, 9Y-TDF				
G-BEXN	Grumman AA-1C Lynx	AA1C-0045	N6147A	7. 9.77	J.S.C.Goodale	Popham	15. 4.06
G-BEXO	Piper PA-23 Apache	23-213	OO-APH	4. 7.77	Aviation Advisory Services Ltd	Stapleford	10. 9.06
			N1176P				
G-BEXW	Piper PA-28-181 Cherokee Archer II	28-7790521	N38122	11. 7.77	J O'Keefe	Mallow, County Cork	12. 5.06T
G-BEXX	Cameron V-56 HAB	274		29. 6.77	K.A Schlussler *"Rupert of Rutland"*	(Bourne, Lincoln)	2. 7.86A
G-BEXZ	Cameron N-56 HAB	294		7. 7.77	D.C.Eager and G.C.Clark *"Valor"*	(Bracknell/Worcester)	13. 4.97A
G-BEYA	Enstrom 280C Shark	1104		15. 8.77	Hovercam Ltd	Staddon Heights, Plymouth	10. 5.07T
G-BEYB*	Fairey Flycatcher rep				See SECTION 3 Part 1	RNAS Yeovilton	
G-BEYF*	Handley Page HPR.7 Dart Herald 401				See SECTION 3 Part 1	Bournemouth	
G-BEYL	Piper PA-28-180 Cherokee Archer	28-7405098	PH-SDW	6. 9.77	D.Midgley and D.Gerrard	Compton Abbas	28. 3.07
			N9518N		tr Yankee Lima Group		
G-BEYN*	Evans VP-2	PFA 63-10271		1. 8.77	H.P.Vox and T.Rayner	East Fortune	
	(Built C D Denham - c/n1)				*(Cancelled 2.9.91 by CAA) (Incomplete airframe stored in hangar roof 6.00)*		
G-BEYO	Piper PA-28-140 Cherokee Cruiser	28-7725215	N9648N	14. 7.77	Eurocharter Aviation Ltd	Gamston	9. 6.07T
G-BEYT	Piper PA-28-140 Cherokee	28-20330	D-EBWO	19. 7.77	G.P. and Frances C.Coleman	RAF Halton	6. 5.06T
			N6280W				
G-BEYV	Cessna T210M Turbo Centurion II	21061583	N732KX	19. 7.77	P.J.W. and N.Austen t/a Austen Aviation	Guernsey	21. 2.07T
G-BEYW	Taylor JT.1 Monoplane	PFA 55-10279		22. 7.77	R.A Abrahams	Barton	28. 5.05P
	(Built R A Abrahams - c/n RJS.100) (Volkswagen 1834)				*"Red Hot"*		
G-BEYZ	CEA Jodel DR.1050/M1 Sicile Record	588	F-BMGV	22. 7.77	M.L.Balding	Biggin Hill	10. 8.06
G-BEZC	Grumman AA-5 Traveler	AA5-0493	F-BUYN	29. 7.77	T.V.Montgomery	Elstree	23. 8.07
			(N7193L)				
G-BEZE	Rutan VariEze	PFA 74-10207		26. 7.77	S.K.Cockburn	Southend	2. 6.04P
	(Built J Berry) (Continental O-200-A)				*(Forced landed 5ms near Rayleigh 17.12.03 and badly damaged: noted 1.05)*		
G-BEZF	Grumman AA-5 Traveler	AA5-0538	F-BVJP	29. 7.77	RAF College Flying Club Ltd	RAF Cranwell	17.12.04T
G-BEZG	Grumman AA-5 Traveler	AA5-0561	F-BVRJ	29. 7.77	M.D.R.Harling	Shobdon	31. 5.05
G-BEZH	Grumman AA-5 Traveler	AA5-0566	F-BVRK	29. 7.77	L. and S.M.Sims	Fenland	6. 3.07
			N9566L				
G-BEZI	Grumman AA-5 Traveler	AA5-0567	F-BVRL	29. 7.77	Heather Matthews	Cranfield	26. 8.07
			N9567L		tr The BEZI Flying Group		
G-BEZK (2)	Cessna F172H	F172-0462	D-EBUD	17. 8.77	S.Jones	Beccles	13. 5.06
	(Built Reims Aviation SA)		D-ENHC, SLN-07, N20462				
G-BEZL	Piper PA-31 Navajo C	31-7712054	SE-GPA	1. 8.77	A Jahanfar *(Op JRB Aviation)*	Southend	20. 7.06T
G-BEZO	Reims/Cessna F172M Skyhawk II	F17201392		24. 8.77	Gloucestershire Flying Services Ltd	Gloucestershire	1. 4.07T
G-BEZP	Piper PA-32-300 Cherokee Six	32-7740087	N38572	19. 8.77	W.D.McNab and T.P.McCormack	White Waltham	19. 5.07
G-BEZR	Reims/Cessna F172M Skyhawk II	F17201395		24. 8.77	Kirmington Aviation Ltd	Sandown	8. 5.07T
G-BEZS*	Reims FR172J Rocket	FR17200562	(I-CCAJ)	11. 8.77	Not known	Bourn	22. 9.79
					(Damaged near Stapleford 15.6.79: front fuselage stored 11.01)		
G-BEZV	Reims/Cessna F172M Skyhawk II	F17201474	(I-CCAY)	24. 8.77	A T.Wilson tr Insch Flying Group	Insch	28. 4.07
G-BEZY	Rutan VariEze	PFA 74-10225		26. 7.77	I.J.Pountney	(Malvern)	18. 5.96P
	(Built R J Jones - c/n 1167) (Continental PC60)				*(Current status unknown)*		
G-BEZZ	Jodel D.112	397	F-BHMC	12. 8.77	M.J.Coles	Barton	11.11.05P
	(Built Passot Aviation)				tr G-BEZZ Jodel Group		

G-BFAA - G-BFZZ

Regn	Type	C/n	Previous identity	Regd	Owner	Location	CofA
G-BFAA	Sud-Aviation Gardan GY-80-160 Horizon	78	F-BLVY	20.10.77	G.R.Williams	(Biddulph, Stoke-on-Trent)	18.11.90
					(New owner 8.04)		
G-BFAB*	Cameron N-56 HAB				See SECTION 3 Part 1	Newbury	
G-BFAF	Aeronca 7BCM Champion	7BCM-11	N797US	15. 8.77	D.C.W.Harper	Finmere	30. 8.01P
	(L-16A-AE)		N2552B, 47-797		*(As "7797" in US Army c/s)*		
G-BFAH	Phoenix Currie Wot	PFA 3017		22. 8.77	R.W.Clarke	(Cheadle)	
	(Continental O-200A) *(Initially allocated with PFA project no 3017 (T Thompson): now being built by N Hamilton-Wright as Replica SE-5A in RFC c/s and regd officially as c/n PFA 58-11376 but probably confused with PFA 101-11376, a Sopwith Pup replica by same former owner/builder)*						
G-BFAI	Rockwell Commander 114	14304	N4984W	17. 8.77	G.Gore-Brown	Sherburn-in-Elmet	8.12.05
					tr Alpha India Flying Group		
G-BFAK	GEMS MS.892A Rallye Commodore 150	10595	F-BNNJ	9. 8.77	J.M.Hedges	Lower Wasing Farm, Brimpton	23. 5.07
G-BFAM*	Piper PA-31P Pressurised Navajo				*(Cancelled 30.10.97 when re-registered as G-SASK qv)*		
G-BFAP	SIAI-Marchetti S.205-20R	4-213	I-ALEN	1. 9.77	A O'Broin	Raby's Farm, Great Stukeley	13. 6.06
G-BFAS	Evans VP-1 Series 2	PFA 7033		15. 8.77	A I.Sutherland	Fearn	18. 3.05P
	(Built A I.Sutherland) (Volkswagen 1834)						
G-BFAW	de Havilland DHC-1 Chipmunk 22	C1/0733	8342M	31. 8.77	R.V.Bowles	Leicester	15. 3.07
			WP848				
G-BFAX	de Havilland DHC-1 Chipmunk 22	C1/0496	8394M	31. 8.77	N.Rushton	Trenchard Farm, Eggesford	23. 6.05
			WG422		*(As "WG422/16")*		
G-BFBA	SAN Jodel DR.100A Ambassadeur	88	F-BIVU	12. 9.77	W.H.Sherlock	Drayton Manor, Drayton St.Leonard	24.10.05
G-BFBB	Piper PA-23-250 Aztec E	27-7405294	SE-GBI	1. 9.77	Air Training Services Ltd	Booker	1. 6.07T
G-BFBC	Taylor JT.1 Monoplane	PFA 55-10280		5. 9.77	G Heins	(Rochdale)	
	(Built A Brooks)				*(Under construction 2.93: new owner 1.02)*		
G-BFBE	Robin HR200/100	12	PH-SRK	9. 9.77	A C.Pearson	Rochester	2. 6.06
G-BFBM	Saffery S.330 MLB	7		1. 9.77	B.Martin *"Beeze II"*	(Somersham, Huntingdon)	
G-BFBP*	Piper PA-25-235 Pawnee D				See SECTION 3 Part 1	Wainfleet	
G-BFBR	Piper PA-28-161 Cherokee Warrior II	28-7716277	N38845	15. 9.77	M.R.Paul t/a Malcolm R Paul Racing	Fairoaks	31. 3.07T
G-BFBU	Partenavia P68B	24	SE-FTM	25. 1.78	Premiair Charter Ltd	Bournemouth	24. 4.06T
G-BFBY	Piper J-3C-65 Cub (L-4H-PI)	10998	F-BDTG	29. 9.77	U.Schuhmacher	Dunkeswell	7. 9.01P
			43-29707		*(Noted 6.04)*		
G-BFCT	Cessna TU206F Turbo Stationair II	U20603202	(LN-TVF)	15. 9.77	D.I.Schellingerhout	(Cottingham)	8. 3.07
			N8341Q				
G-BFCZ*	Sopwith Camel F.1 rep				See SECTION 3 Part 1	Brooklands	
G-BFDC	de Havilland DHC-1 Chipmunk 22	C1/0525	7989M	15.11.77	N.F.O'Neill	Newtownards	4. 9.06
			WG475				
G-BFDE*	Sopwith Tabloid Scout rep				See SECTION 3 Part 1	Hendon	
G-BFDF	SOCATA Rallye 235E	12834	F-GAKT	6.10.77	M.A.Wratten	Bourne Park, Hurstbourne Tarrant	2. 5.05
G-BFDI	Piper PA-28-181 Cherokee Archer II	28-7790382	N2205Q	5.10.77	Truman Aviation Ltd	Tollerton	10.10.07E
G-BFDK	Piper PA-28-161 Warrior II	28-7816010	N40061	23. 9.77	S.T.Gilbert	Kemble	23. 5.07T
G-BFDL	Piper J-3C-65 Cub (L-4J-PI)	13277	HB-OIF	30.11.77	A F.Nicholson	Shempston Farm, Lossiemouth	20. 5.05P
	(Continental O-200-A) *(Frame No.13107)*		45-4537		*(As "454537/04-J" in US Army c/s)*		
G-BFDO	Piper PA-28R-201T Turbo Cherokee Arrow III	28R-7703212	N38396	3.10.77	A.J.Gow	Denham	12.11.05
G-BFDV*	Westland WG.13 Lynx HC.28	WA/028	TAD.013	3.10.77	School of Electrical and Aeronautical Engineering	Princess Marina College, Arborfield	
	(Originally regd as "Lynx 02F")		Qatar AF 1, G-17-20		*(Cancelled 6.78) (Instructional airframe "QP-30" 7.02)*		
G-BFDZ	Taylor JT.1 Monoplane	PFA 55-10185		5.10.77	R.A.Collins	(Chippenham)	7. 8.04P
	(Built D C Barber)						
G-BFEB	SAN Jodel D.150 Mascaret	34	F-BMJR	14.10.77	A.W.Russell	Portmoak	17. 6.05P
			OO-LDY, F-BLDX		tr Jodel Syndicate		
G-BFEF	Agusta-Bell 47G-3B1	1541	XT132	11.10.77	M.P.Wilkinson	Sandtoft	8. 7.05T
G-BFEH	SAN Jodel D.117A	828	F-BITG	5.10.77	J.A.Crabb	Dunkeswell	18.11.05P
G-BFEK	Reims/Cessna F152 II	F15201442		11.10.77	Gloucestershire Flying Services Ltd	Gloucestershire	20. 2.07T
G-BFER	Bell 212	30835	N18099	7.11.77	Bristow Helicopters Ltd	(Kazakhstan)	27.11.05T
G-BFEV	Piper PA-25-235 Pawnee D	25-7756060	N82547	20.10.77	Trent Valley Aerotowing Club Ltd	Kirton-in-Lindsey	18. 4.07
G-BFEW	Piper PA-25-235 Pawnee D	25-7756062	N82553	20.10.77	Cornish Gliding and Flying Club Ltd	Perranporth	26. 4.07
G-BFEY*	Piper PA-25-235 Pawnee D				See SECTION 3 Part 1	Wainfleet	
G-BFFC	Reims/Cessna F152 II	F15201451		27.10.77	Multiflight Ltd	Leeds-Bradford	9.12.07E
G-BFFE	Reims/Cessna F152 II	F15201454		27.10.77	A J.Hastings	Glenrothes	2. 7.06T
G-BFFJ	Sikorsky S-61N Mk.II	61-777	N6231	17. 1.78	Veritair Ltd *"Tresco"*	Penzance	22. 3.06T
G-BFFP	Piper PA-18-150 Super Cub	18-8187	PH-OTC	9.11.77	East Sussex Gliding Club Ltd	Ringmer	7. 5.07
	(Lycoming O-360-A4) *(Frame No.18-8402)*		N10F				
G-BFFT	Cameron V-56 HAB	360		7.11.77	R.I.McKean Kerr and D.C.Boxall	(Bristol)	16. 2.03A
					tr The Red Section Balloon Group *"Red Leader"*		
G-BFFW	Reims/Cessna F152 II	F15201447		14.11.77	Tayside Aviation Ltd	Glenrothes	17. 6.07T
G-BFFY	Reims/Cessna F150M	F15001376		14.11.77	G.A Rodmell	Linley Hill, Leven	6. 7.07T
G-BFGD	Reims/Cessna F172N Skyhawk II	F17201545	F-WZDT	14.11.77	J.T.Armstrong	Fairoaks	18.11.07E
G-BFGG	Reims/Cessna FRA150M Aerobat	FRA1500321	F-WZDS	14.11.77	S.J.Windle	Netherthorpe	5. 4.07T
G-BFGH	Reims/Cessna F337G Skymaster II	F33700081		14.11.77	T.Perkins	Bagby	12. 1.06
	(Wichita c/n 33701754)						
G-BFGK	SAN Jodel D.117	644	F-BIBT	27. 6.78	B.F.J.Hope	Stoneacre Farm, Farthing Corner	23 6.05P
G-BFGL	Reims/Cessna FA152 Aerobat	FA1520339		14.11.77	Multiflight Ltd	Leeds-Bradford	1. 7.07T
G-BFGS	SOCATA MS.893E Rallye 180GT	12571	F-BXYK	31. 8.76	Chiltern Flyers Ltd	Turweston	3.11.06
			Fr.AF 12571 FSCAZ, "41-AZ"				
G-BFGW	Cessna F150H	F150-0370	PH-TGO	24.11.77	C.E.Stringer	Humberside	19.10.95T
	(Built Reims Aviation SA)				*(Current status unknown)*		
G-BFGX	Reims/Cessna FRA150M Aerobat	FRA1500328	F-BUDX	28.11.77	Prestwick Flight Centre Ltd	Prestwick	22. 9.07T
G-BFGZ	Reims/Cessna FRA150M Aerobat	FRA1500329		28.11.77	C.M.Barnes	Garden Piece, Basingstoke	25. 3.06T
G-BFHD*	CASA 352L				See SECTION 3 Part 1	Washington, US	
G-BFHE*	CASA 352L				See SECTION 3 Part 1	Johannesburg, South Africa	
G-BFHG*	CASA 352L				See SECTION 3 Part 1	Polk City, Florida, US	

G-BFHH	de Havilland DH.82A Tiger Moth	85933	F-BDOH	25.11.77	P.Harrison and M.J.Gambrell	Swanborough Farm, Lewes	26. 5.07
	(Built Morris Motors Ltd)		Fr.AF/DF197				
G-BFHI	Piper J-3C-65 Cub (L-4J-PI)	12532	F-BFBT	25.11.77	N.Glass and A.J.Richardson	Bann Foot, Lough Neagh	28. 5.04P
			44-80236				
G-BFHP	Champion 7GCAA Citabria	114	HB-UAX	8.12.77	Poet Pilot Ltd	Liverpool	13.10.05T
G-BFHR	CEA Jodel DR.220 2+2	30	F-BOCX	1.12.77	J.E.Sweetman	Bourne Park, Hurstbourne Tarrant	19. 6.06
G-BFHT	Reims/Cessna F152 II	F15201441		7.12.77	Westward Airways (Lands End) Ltd	St.Just	2. 6.07T
G-BFHU	Reims/Cessna F152 II	F15201461		7.12.77	D.J Cooke and Company Ltd	Hawarden	30.11.7E
G-BFHV	Reims/Cessna F152 II	F15201470		21.12.77	A.S.Bamrah t/a Falcon Flying Services	Rochester	22.11.04T
G-BFHX	Evans VP-1	PFA 62-10283		2.12.77	A.D.Bohanna and D.I.Trussler	Popham	7. 4.99P
	(Built D F Gibson) (Volkswagen 1600)				*(Stored 10.04)*		
G-BFIB	Piper PA-31 Turbo Navajo	31-684	LN-NPE	21.12.77	Richard Hannon Ltd	Lee-on-Solent	29. 1.07T
			OY-DVH, LN-RTJ				
G-BFID	Taylor JT.2 Titch	PFA 60-10311		13.12.77	R.W.Kilham	(Langtoft, Peterborough)	23. 8.99P
	(Built W F Adams) (Continental O-200-A)				*(Damaged Breighton 31.5.99: new owner 5.04)*		
G-BFIE	Reims/Cessna FRA150M Aerobat	FRA1500331		12. 1.78	J.P.A Freeman	Headcorn	17.12.06T
G-BFIG	Reims/Cessna FR172K Hawk XPII	FR17200615		12. 1.78	Tenair Ltd	Barton	27. 1.07
G-BFIJ	Grumman AA-5A Cheetah	AA5A-0486	N6160A	1. 3.78	T.H. and M.G.Weetman	West Freugh/Prestwick	29. 4.07
G-BFIN	Grumman AA-5A Cheetah	AA5A-0520	N6145A	22. 3.78	Prestwick Flight Centre Ltd	Prestwick	2. 5.05T
G-BFIP*	Wallbro Monoplane rep				See SECTION 3 Part 1	Flixton, Bungay	
G-BFIT	Thunder Ax6-56Z HAB	136		20.12.77	J.A G.Tyson *"Folly"*	(Torphins, Banchory)	3. 5.91
G-BFIU	Reims/Cessna FR172K Hawk XP	FR17200591	N96098	12. 1.78	B.M.Jobling	Hinton-in-the-Hedges	20. 8.06
G-BFIV	Reims/Cessna F177RG Cardinal RG	F177RG0161	N96106	12. 1.78	C Fisher	Blackbushe	27. 5.05
G-BFIX	Thunder Ax7-77A HAB	133		9.12.77	R.Owen *"Animal Magic"*	(Wigan)	23. 1.79S
G-BFIY	Reims/Cessna F150M	F15001381	OE-CMT	11. 1.78	R.J.Scott	Blackbushe	15. 4.06T
G-BFJJ	Evans VP-1	PFA 62-10273		30.12.77	N.Clark	Barton Ashes	23. 6.96P
	(Built P R Pykett) (Volkswagen 1800)				*(Noted 10.03)*		
G-BFJR	Reims/Cessna F337G Skymaster II	F33700082	N46297	4. 1.78	Mannix Aviation	Nottingham East Midlands	9. 4.05
	(Wichita c/n 33701761)		(N53658)				
G-BFJZ	Robin DR400/140B Major	1290		20. 1.78	Weald Air Services Ltd	Headcorn	26. 8.07T
G-BFKB	Reims/Cessna F172N Skyhawk II	F17201601	PH-AXO	16. 1.78	R.L.Clarke and D.Tench tr Shropshire Flying Group	Sleap	2. 3.06T
G-BFKF	Reims/Cessna FA152 Aerobat	FA1520337		26. 1.78	Aerolease Ltd	Conington	13. 5.07T
G-BFKH	Reims/Cessna F152 II	F15201464		26. 1.78	TG Aviation Ltd *(Op Thanet Flying Club)*	Manston	26. 3.07T
G-BFKL	Cameron N-56 HAB	369		23. 1.78	Merrythought Ltd	(Telford)	17. 7.92A
G-BFKY	Piper PA-34-200 Seneca	34-7350318	PH-NAZ	22. 2.78	SLH Construction Ltd	Dunsfold	23. 1.05T
			N56332				
G-BFLH	Piper PA-34-200T Seneca II	34-7870065	N2126M	16. 2.78	Air Medical Ltd	Oxford	8. 6.06T
G-BFLI	Piper PA-28R-201T Turbo Arrow III	28R-7803134	N2582M	16. 2.78	J.K.Chudzicki *"Spirit of Rita May"*	Elstree	11. 6.07
G-BFLU	Reims/Cessna F152 II	F15201433		15. 2.78	Atlantic Flight Training Ltd	Coventry	15. 6.07T
					(Atlantic titles on tail and Air Atlantic c/s		
G-BFLX	Grumman AA-5A Cheetah	AA5A-0524	N6147A	14. 3.78	G Force Two Ltd	Blackbushe	23.11.04
G-BFLZ	Beech 95-A55 Baron	TC-220	PH-ILE	16. 3.78	K.A.Graham	Blackpool	23. 8.04
			HB-GOV		t/a Caterite Food Service		
G-BFMG	Piper PA-28-161 Cherokee Warrior II	28-7716160	N3506Q	11. 5.78	Stardial Ltd	Fairoaks	17.10.05T
G-BFMH	Cessna 177B Cardinal	17702034	N34836	18. 4.78	J.C.Owens and A.C.Smith	Newcastle	16. 6.05T
G-BFMK	Reims/Cessna FA152 Aerobat	FA1520344		6. 3.78	RAF Halton Aeroplane Club Ltd	RAF Halton	20. 3.05T
G-BFMR	Piper PA-20 Pacer 125	20-130	N7025K	20. 2.78	J.Knight	Headcorn	6. 3.06
G-BFMX	Reims/Cessna F172N Skyhawk II	F17201732		24. 8.78	A2Z Wholesale Fashion Jewellery Ltd		
						Farley Farm, Romsey	1. 8.03
G-BFMZ	Payne Ax6-62 HAB	GFP.2		1. 3.78	E.G.Woolnough	(Halesworth, Suffolk)	
	(Built G F Payne)						
G-BFNG	Wassmer Jodel D.112	1321	F-BNHI	6. 3.78	A W.Myers and J.MacGregor	Leicester	14. 4.05P
G-BFNI	Piper PA-28-161 Warrior II	28-7816215	N9505N	8. 3.78	P.Elliott	Biggin Hill	26. 7.05
G-BFNJ	Piper PA-28-161 Warrior II	28-7816281	N9520N	8. 3.78	Fleetlands Flying Association Ltd	Lee-on-Solent	6. 6.07T
G-BFNK	Piper PA-28-161 Warrior II	28-7816282	N9527N	8. 3.78	Oxford Aviation Services Ltd	Oxford	9. 1.06T
G-BFNU*	Britten-Norman BN-2B-21 Islander	877		16. 3.78	Isles of Scilly Skybus Ltd	St.Just	18. 8.89T
	(Built IRMA)				*(Cancelled 28.1.94 as WFU) (Fuselage only 8.03)*		
G-BFOD	Reims/Cessna F182Q Skylane II	F18200068		23. 3.78	G.N.Clarke	Alderney	13. 8.05
G-BFOE	Reims/Cessna F152 II	F15201475		23. 3.78	Redhill Air Services Ltd	Rochester	15. 1.06T
G-BFOF	Reims/Cessna F152 II	F15201448		9. 3.78	Gloucestershire Flying Services Ltd	Gloucestershire	21. 6.05T
G-BFOG	Cessna 150M	15076223	N66706	13. 3.78	C.L.Day	(Carmarthen)	12. 5.07T
G-BFOJ	American AA-1 Yankee	AA1-0395	OH-AYB	4. 4.78	N.W.Thomas	Bournemouth	18.11.05
			(LN-KAJ), (N6195L)				
G-BFOM	Piper PA-31-325 Navajo C/R	31-7512017	EI-DMI	17. 3.78	Ashton Air Services Ltd	Oxford	5. 3.05T
			G-BFOM, HB-LHH, N59933				
G-BFOO*	British Aircraft Corporation 167 Strikemaster Mk.80A				See SECTION 3 Part 1	Dharhan, Saudi Arabia	
G-BFOP	Wassmer Jodel D.120 Paris-Nice	32	F-BHTX	23. 3.78	R.J.Wesley and G.D.Western	Sampsons Hall, Kersey	12. 6.04P
					"Jean"		
G-BFOS	Thunder Ax6-56A HAB	147		20. 3.78	N.T.Petty *"Milton Keynes"*	(Sudbury)	25.11.93A
G-BFOU	Taylor JT.1 Monoplane	PFA 55-10333		17. 3.78	G.Bee	(Stockton-on-Tees)	
	(Built I N M Cameron)				*(Current status unknown)*		
G-BFOV	Reims/Cessna F172N Skyhawk II	F17201675		18. 5.78	D.J.Walker	Shoreham	18.10.05
G-BFOZ*	Thunder Ax6-56 Plug HAB				See SECTION 3 Part 1	Newbury	
G-BFPA	Scheibe SF25B Falke	46179	D-KAGM	29. 3.78	R.Gibson and R.Hamilton	(Drumshade)	13. 9.98
	(Volkswagen Danum 1600/1)				*(New owners 3.04)*		
G-BFPB	Grumman AA-5B Tiger	AA5B-0706		7. 4.78	P.Murphy and L.Peake	Coventry	1. 9.05
					tr Papa Bravo Flying Group		
G-BFPH	Reims/Cessna F172K	F17200802	PH-VHN	23. 3.78	M.Pollard tr Linc-Air Flying Group	Gamston	14. 8.05
G-BFPM	Reims/Cessna F172M Skyhawk II	F17201384	PH-MIO	13. 4.78	M.P.Wimsey and J.M.Cope	Strubby	22. 4.06T
G-BFPO	Rockwell Commander 112B	530	N1412J	10. 5.78	J.G.Hale Ltd	Shoreham	16.12.06
G-BFPP	Bell 47J-2 Ranger	2851	F-BJAN	23. 5.78	M.R.Masters	Phoenix Farm, Lower Upham	11.11.99
			TR-LKD, F-OCBU				
G-BFPR	Piper PA-25-235 Pawnee D	25-7856007	SE-KGY	4. 4.78	Booker Gliding Club Ltd	Booker	7. 1.06
			I-TOZU, G-BFPR, N82591				

Reg	Type	C/n	Previous identity	Date	Owner	Base	CofA
G-BFPS	Piper PA-25-235 Pawnee D	25-7856013	N82598	4. 4.78	Kent Gliding Club Ltd	Challock	19. 8.06
G-BFRD	Bowers FlyBaby 1A	PFA 16-10300		27. 1.78	R.A Phillips	(Elgin)	
	(Built F R Donaldson and R A Phillips)				*(Under construction 6.00)*		
G-BFRI	Sikorsky S-61N Mk.II	61-809		26. 5.78	Bristow Helicopters Ltd	Aberdeen	14. 6.04T
G-BFRR	Reims/Cessna FRA150M Aerobat	FRA1500326	LN-ALO	19. 4.78	S.Cosgrove tr Romeo Romeo Flying Group	Tatenhill	16. 9.05
G-BFRS	Reims/Cessna F172N Skyhawk II	F17201555	LN-ALP	19. 4.78	Poplar Hall Ltd	Poplar Hall Farm, Elmsett	8. 5.06T
G-BFRV	Reims/Cessna FA152 Aerobat	FA1520345		17. 4.78	Solo Services Ltd	Shoreham	25. 9.05T
G-BFRY	Piper PA-25-260 Pawnee D	25-7405789	SE-GIB	23. 5.78	Yorkshire Gliding Club (Pty) Ltd	Sutton Bank	30. 6.06
G-BFSA	Reims/Cessna F182Q Skylane II	F18200074	F-WZDG	17. 4.78	Ensiform Aviation Ltd	Elstree	27.11.05
G-BFSC	Piper PA-25-235 Pawnee D	25-7656068	N82302	2. 6.78	Essex Gliding Club Ltd	North Weald	14. 3.07
G-BFSD	Piper PA-25-235 Pawnee D	25-7656084	N82338	2. 6.78	Deeside Gliding Club (Aberdeenshire) Ltd	Aboyne	18. 4.05
G-BFSR	Cessna F150J	F150-0504	OH-CBN	7. 7.78	W.Ali	Stapleford	21. 6.04T
	(Built Reims Aviation SA)						
G-BFSS	Reims FR172G Rocket	FR17200167	OH-CDY	7. 7.78	Albedale Farms Ltd	Grateley, Andover	27 4.06
G-BFSY	Piper PA-28-181 Cherokee Archer II	28-7890200	N9503N	19. 4.78	A.S.Domone t/a Downland Aviation	Goodwood	24. 5.05
G-BFTC	Piper PA-28R-201T Turbo Arrow III	28R-7803197	N3868M	19. 4.78	M.J.Milns	Sherburn-in-Elmet	15. 9.06
G-BFTF	Grumman AA-5B Tiger	AA5B-0879		7. 9.78	F.C.Burrow Ltd	Sherburn-in-Elmet	17. 7.06
G-BFTG	Grumman AA-5B Tiger	AA5B-0777		15. 5.78	D.Hepburn and G.R.Montgomery	Perth	15.10.05
G-BFTH	Reims/Cessna F172N Skyhawk II	F17201671		3. 5.78	J.Birkett	Wickenby	15. 9.05T
G-BFTT	Cessna 421C Golden Eagle II	421C-0462	N6789C	3. 5.78	M.A.Ward	Goodwood	18. 6.05T
G-BFTZ*	SOCATA MS.880B Rallye Club				See SECTION 3 Part 1	Newark	
G-BFUB	Piper PA-32RT-300 Lance II	32R-7885052	N9509C	18. 5.78	Jolida Holdings Ltd	Jersey	3. 4.05
G-BFUD	Scheibe SF25E Super Falke	4313	D-KLDC	19. 5.78	P.A.Lewis	Walney Island	13.12.04
	(Limbach SL1700)				tr The Lakes Libelle Syndicate		
G-BFUZ	Cameron V-77 HAB	398		24. 5.78	A.W.Macdonald	(Leigh-on-Sea)	5.12.04A
					tr Servowarm Balloon Syndicate		
G-BFVG	Piper PA-28-181 Cherokee Archer II	28-7890408	N31746	1. 6.78	M.S.Cornah	Blackpool	22. 6.05
			N9558N		tr G-BFVG Flying Group		
G-BFVH	AirCo DH.2 fsm	WA/4	BAPC.112	1. 6.78	M.J.Kirk	Rugby	20. 7.01P
	(Built Westward Airways) (125 hp Kinner B54)				*(Noted in poor condition as "5964" 9.03)*		
G-BFVS	Grumman AA-5B Tiger	0784	N28736	11. 8.78	S.W.Biroth and T.Chapman	Denham	11. 2.07T
G-BFVU	Cessna 150L Commuter	15074684	N75189	10. 8.78	S.D.Baker	Exeter	23. 7.06T
G-BFWB	Piper PA-28-161 Cherokee Warrior II	28-7816584	N31752	22. 6.78	Mid-Anglia Flight Centre Ltd	Cambridge	17. 7.05T
					t/a Mid-Anglia School of Flying		
G-BFWD	Phoenix Currie Wot	PFA 3009		22. 6.78	D.Silsbury and B.Proctor	Dunkeswell	6.10.96P
	(Built F E Nuthall) (Walter Mikron 3)				*(New owners 3.04)*		
G-BFWE	Piper PA-23-250 Aztec E	27-4583	9M-AQT	13. 7.78	Air Navigation and Trading Co Ltd	Blackpool	15. 2.03T
			9V-BDI, N13968				
G-BFWK*	Piper PA-28-161 Warrior II	28-7816610	N9589N	23. 6.78	Marham Investments Ltd	Belfast Aldergrove	8.12.99T
					(Cancelled 26.5.98 as WFU) (Wrecked fuselage stored 10.01)		
G-BFWL*	Reims/Cessna F150L	F15000971	PH-KDC	4.10.78	P.Maher tr G-BFWL Flying Group	Barton	27. 3.00
					(Cancelled 21.2.00 as WFU) (Fuselage noted 1.04)		
G-BFXF	Andreasson BA-4B	PFA038-10351		10. 7.78	A Brown	Sherburn-in-Elmet	13. 8.03P
	(Built A Brown - c/n AAB-001) (Lycoming O-290-G)						
G-BFXG	Druine D.31 Turbulent	PFA 1663		10. 7.78	E.J.I.Musty and M.J.Whatley	White Waltham	
	(Built S Griffin)				*(Partially complete 6.00)*		
G-BFXK	Piper PA-28-140 Cherokee F	28-7325387	PH-NSK	1. 8.78	D.M.Wheeler	Rochester	9. 6.06
G-BFXL*	Albatros D.Va rep				See SECTION 3 Part 1	RNAS Yeovilton	
G-BFXR	Wassmer Jodel D.112	247	F-BFTM	27. 7.78	J.M.Pearson	Crosland Moor	24.11.05P
G-BFXS	Rockwell Commander 114	14271	N4949W	3. 8.78	Unipak (UK) Ltd	Conington	22. 8.05
G-BFXW	Gulfstream AA-5B Tiger	AA5B-0940		21. 2.79	Campsol Ltd	Sherburn-in-Elmet	3. 7.06
G-BFXX	Gulfstream AA-5B Tiger	AA5B-0917		3.10.78	W.R.Gibson	North Weald	8. 1.07
G-BFYA	MBB Bö.105DB	S.321	D-HJET	31.10.78	Sterling Helicopters Ltd *(Op Norfolk Police)*	Norwich	12. 5.06T
G-BFYC	Piper PA-32RT-300 Lance II	32R-7885200	N36645	31. 7.78	A A Barnes t/a Cyril Silver and Partners	Biggin Hill	22. 5.06
G-BFYI	Westland-Bell 47G-3BI	WA/326	XT167	24. 1.79	B.Walker and Co (Dursley) Ltd	Nympsfield	13. 7.06
G-BFYK	Cameron V-77 HAB	433	EI-BAY	16. 8.78	Louise E.Jones	(Worcester)	31.12.99A
			G-BFYK				
G-BFYL	Evans VP-2	PFA 63-10146		15. 8.78	W.C.Brown	(Camberley)	17.12.98P
	(Built A G Wilford) (Volkswagen 1834)						
G-BFYO*	SPAD XIII rep				See SECTION 3 Part 1	Duxford	
G-BFZA	Fournier RF3	5	F-BLEL	14. 9.78	T.J.Hartwell	Sackville Lodge Farm, Riseley	
G-BFZB	Piper J-3C-65 Cub (L-4J-PI)	13019	D-ECEL	21. 9.78	M.S.Pettit	(Sandbach)	8. 9.05P
	(Continental C85) *(Frame No.12849)*		HB-OSP, 44-80723		*(As "44-80723/E5-J")*		
G-BFZD	Reims/Cessna FR182 Skylane RG II	FR18200010		9.10.78	R.B.Lewis t/a R.B.Lewis and Co	Sleap	1. 2.06
G-BFZH	Piper PA-28R-200 Cherokee Arrow	28R-35307	OY-BDB	25.10.78	A.Mason t/a Mason Aviation	Glasgow	25. 9.06T
G-BFZM	Rockwell Commander 112TC-A	13191	N4661W	9.10.78	J.A.Hart and R.J.Lamplough	Filton	29. 8.07
G-BFZN	Reims/Cessna FA152 Aerobat	FA1520348		20.10.78	A.S.Bamrah t/a Falcon Flying Services	Biggin Hill	29.11.81T
					(Crashed Narborough, Leics 4.10.80: rebuilt 2.95: current status unknown)		
G-BFZO	Gulfstream AA-5A Cheetah	AA5A-0697		1.11.78	J.W.Cross and Andrea E.Kempson	Elstree	18. 6.06
G-BFZR	Gulfstream AA-5B Tiger	AA5B-0979	EI-BJS	3.11.78	P.C.Morrissey	(Bennekerry, County Carlow)	24. 5.07
			G-BFZR				
G-BFZT	Reims/Cessna FA152 Aerobat	FA1520356		4. 7.79	Herefordshire Aero Club Ltd	Shobdon	10. 3.07T
G-BFZU	Reims/Cessna FA152 Aerobat	FA1520355		29. 6.79	C.R.Tilley	Sleap	12. 3.05T
G-BFZV	Reims/Cessna F172M	F17201093	SE-FZR	2.11.78	Wessex Flying Group Ltd	Thruxton	16. 3.06T

G-BGAA - G-BGZZ

Reg	Type	C/n	Previous identity	Date	Owner	Base	CofA
G-BGAA	Cessna 152 II	15281894	N67529	18. 7.78	PJC (Leasing) Ltd	Stapleford	24. 6.07T
G-BGAB	Reims/Cessna F152 II	F15201531		13.10.78	TG Aviation Ltd *(Op Thanet Flying Club)*	Manston	7. 4.03T
G-BGAE	Reims/Cessna F152 II	F15201540		8.11.78	Aerolease Ltd	Conington	7. 5.06T
G-BGAF	Reims/Cessna FA152 Aerobat	FA1520349		13.10.78	P.E.Preston tr G-BGAF Group	Southend	21. 8.06T
G-BGAG	Reims/Cessna F172N Skyhawk II	F17201754	"G-KING"	13.10.78	A.S.Bamrah t/a Falcon Flying Services	Biggin Hill	25. 5.07T
G-BGAJ	Reims/Cessna F182Q Skylane II	F18200096		13.10.78	Ground Airport Services Ltd	Guernsey	4. 5.06

G-BGAS*	Colting Ax8-105A HAB				See SECTION 3 Part 1	Newbury	
G-BGAX	Piper PA-28-140 Cherokee F	28-7325409	PH-NSH	20.10.78	C.D.Brack	Breighton	15. 4.05
G-BGAZ	Cameron V-77 HAB	439		20.10.78	C.J.Madigan and D.H.McGibbon	(Bristol)	E
	(Rebuilt envelope)				*(Cameron Balloons titles) (Noted 8.03)*		
G-BGBA	Robin R2100A Club	133	F-OCBJ	2. 5.78	C and G Property Services Ltd	Gloucestershire	24. 5.06
					(Op Cotswold Aero Club)		
G-BGBE	SAN Jodel DR.1050 Ambassadeur	260	F-BJYT	29.11.78	J.A and B.Mawby	Enstone	11. 6.05
G-BGBF	Druine D.31A Turbulent	PFA 1658		24.10.78	K.Pullen	Eaglescott	25. 3.04P
	(Built L Davies) (Volkswagen 1600)				tr Eaglescott Turbulent Group		
G-BGBG	Piper PA-28-181 Archer II	28-7990012	N39730	2.11.78	Harlow Printing Ltd	Newcastle	11. 6.06
G-BGBI	Reims/Cessna F150L	F15000688	PH-LUA	28.11.78	C.P.Tapp	(Duxford)	25.11.06T
G-BGBK	Piper PA-38-112 Tomahawk	38-78A0433	N9738N	2.11.78	The Sherwood Flying Club Ltd	Tollerton	25. 7.05
G-BGBN	Piper PA-38-112 Tomahawk	38-78A0511	N9657N	29.11.78	Bonus Aviation Ltd	Cranfield	17. 9.06T
G-BGBR	Reims/Cessna F172N Skyhawk II	F17201772		8.11.78	A S.Bamrah t/a Falcon Flying Services	Biggin Hill	16. 2.04T
					(Damaged in freak storm 22.9.03, overturned and on re-build 1.05)		
G-BGBW	Piper PA-38-112 Tomahawk	38-78A0670	N9710N	8.11.78	Truman Aviation Ltd	Tollerton	16. 7.06T
G-BGBY	Piper PA-38-112 Tomahawk	38-78A0711	N9689N	8.11.78	Cheshire Flying Services Ltd	Liverpool	22. 2.07T
G-BGBZ	Rockwell Commander 114	14423	N5878N	9.10.78	R.S.Fenwick	Rochester	9. 8.04
G-BGCF*	Douglas C-47A-90-DL Dakota				See SECTION 3 Part 1	Barksdale, Louisiana, US	
G-BGCM	Gulfstream AA-5A Cheetah	AA5A-0835		23. 3.79	G.and S.A Jones	Linley Hill, Leven	21. 9.07T
G-BGCO	Piper PA-44-180 Seminole	44-7995128	N2103D	20.12.78	J.R.Henderson	Warton	11.11.06
					(Op BAE SYSTEMS (Operations) Ltd)		
G-BGCY	Taylor JT.1 Monoplane	PFA 55-10370		23.11.78	A.T.Lane	Fenland	2. 3.04P
	(Built M T Taylor)						
G-BGEH	Monnett Sonerai II	PFA 15-10254		1.12.78	D.and V.T.Hubbard	(Basingstoke)	16. 8.96P
	(Built R E Finlay- c/n 209) (Volkswagen 2234)				*(Current status unknown)*		
G-BGEI	Oldfield Baby Lakes	PFA 10-10016		1.12.78	M.T.Taylor	Griffins Farm, Temple Bruer	30. 3.05P
	(Built D H Greenwood) (Continental A65) *(Fuselage of PFA 1576 incorporated during construction)*						
G-BGEK	Piper PA-38-112 Tomahawk	38-78A0575	N9662N	13.12.78	Cheshire Flying Services Ltd t/a Ravenair	Liverpool	12. 5.06T
G-BGEW	SNCAN NC.854S	63	F-BFSJ	13.12.78	S.A.Francis	Kemble	7. 9.05P
	(Continental A65)						
G-BGFC	Evans VP-2	PFA 63-10441		15.12.78	S.W.C.Hollins	Llandegla	29. 9.93P
	(Built J A Jones - c/n V2-1278) (Volkswagen 1834)				*(Current status unknown)*		
G-BGFF	Clutton FRED Series II	PFA 29-10261		18.12.78	I.Daniels	Popham	28. 6.05P
	(Built G R G Smith) (Volkswagen 1834)						
G-BGFG	Gulfstream AA-5A Cheetah	AA5A-0687	N6158A	25. 1.79	T.D.Saveker	Bodmin	30. 5.06T
G-BGFH	Reims/Cessna F182Q Skylane II	F18200105		18. 1.79	Rayviation Ltd	Octon Lodge Farm, Thwing	17. 8.07T
	(Rebuilt with fuselage of G-EMMA [F18200099] 1994/95: original fuselage scrapped)						
G-BGFI	Gulfstream AA-5A Cheetah	AA5A-0733	N6142A	5. 3.79	I.J.Hay and A Nayyar tr GFI Group	Biggin Hill	18.10.06
G-BGFJ	Jodel D.9 Bébé	PFA 1324		11.12.78	M.D.Mold	Watchford Farm, Yarcombe	3. 4.05P
	(Built C M Fitton) (Volkswagen 1600)						
G-BGFT	Piper PA-34-200T Seneca II	34-7870218	N9714C	17. 1.79	Oxford Aviation Services Ltd	Oxford	16.12.06T
G-BGFX	Reims/Cessna F152 II	F15201555		28.12.78	A.S.Bamrah	Biggin Hill	23. 6.91T
					t/a Falcon Flying Services *(Last noted 4.01)*		
G-BGGA	Bellanca 7GCBC Citabria 150S	1104-79		5. 2.79	L.A King	North Connel, Oban	2. 3.07
G-BGGB	Bellanca 7GCBC Citabria 150S	1105-79		7. 2.79	G.H.N.Chamberlain	Rattlesden	17. 3.05
G-BGGC	Bellanca 7GCBC Citabria 150S	1106-79		5. 2.79	R.P.Ashfield and J.M.Stone		
						Gorwell Farm, Littlebredy, Dorset	14.10.06
G-BGGD	Bellanca 8GCBC Scout	284-78		5. 2.79	Bristol and Gloucestershire Gliding Club Ltd	Nympsfield	1. 12.07E
G-BGGE	Piper PA-38-112 Tomahawk	38-79A0161	N9673N	10. 1.79	Truman Aviation Ltd	Tollerton	25. 6.06T
G-BGGG	Piper PA-38-112 Tomahawk	38-79A0163	N9675N	10. 1.79	Teesside Flight Centre Ltd	Durham Tees Valley	28. 6.04T
G-BGGI	Piper PA-38-112 Tomahawk	38-79A0165	N9675N	10. 1.79	Truman Aviation Ltd	Tollerton	11. 3.07T
G-BGGL	Piper PA-38-112 Tomahawk	38-79A0169	N9696N	10. 1.79	Grunwick Processing Laboratories Ltd	Cranfield	27. 6.06T
					(Op Bonus Aviation)		
G-BGGM	Piper PA-38-112 Tomahawk	38-79A0170	N9698N	10. 1.79	Grunwick Processing Laboratories Ltd	Cranfield	26.11.06T
					(Op Bonus Aviation)		
G-BGGN	Piper PA-38-112 Tomahawk	38-79A0171	N9706N	10. 1.79	Bell Aviation Ltd	Cranfield	15.12.06T
G-BGGO	Reims/Cessna F152 II	F15201569		8. 3.79	East Midlands Flying School Ltd		
						Nottingham East Midlands	13. 7.06T
G-BGGP	Reims/Cessna F152 II	F15201580		8. 3.79	East Midlands Flying School Ltd		
						Nottingham East Midlands	16.10.06T
G-BGGR*	North American AT-6A Harvard				See SECTION 3 Part 1	Stuttgart, Germany	
G-BGGU	Wallis WA-116 RR	702		28.12.78	K.H.Wallis	Reymerston Hall, Norfolk	AC
	(Subaru EA61)				*(Noted 8.01)*		
G-BGGV	Wallis WA-120 Series 2	703		28.12.78	K.H.Wallis	Reymerston Hall, Norfolk	
					(Not completed)		
G-BGGW	Wallis WA-122 RR	704		28.12.78	K.H.Wallis	Reymerston Hall, Norfolk	2. 6.05P
	(RR Continental O-240-A)						
G-BGHF*	Westland WG.30 Series 100-60				See SECTION 3 Part 1	Weston-super-Mare	
G-BGHI	Reims/Cessna F152 II	F15201560		15. 1.79	V.R.McCready	(Sutton)	27. 5.06T
G-BGHJ	Reims/Cessna F172N Skyhawk II	F17201777	EI-BVF G-BGHJ	15. 1.79	Airplane Ltd	(Middle Rasen, Market Rasen)	18. 8.07T
G-BGHM	Robin R1180T Aiglon	227		19. 2.79	H.Price	Blackpool	12..4.07
G-BGHP	Beech 76 Duchess	ME-190	N60132	16. 1.79	Magenta Ltd *(Op Airways Flight Training)*	Exeter	27. 4.06T
G-BGHS	Cameron N-31 HAB	501		15. 1.79	W.R.Teasdale *"Baby Champion"*	(Newbury)	17. 1.00A
					(On loan to British Balloon Museum and Library) (Inflated 4.02)		
G-BGHT	Falconar F-12	PFA 22-10040		17. 1.79	C.R.Coates	Sneaton Thorpe, Whitby	
	(Built T K Baillie) (Lycoming O-290)				*(Current status unknown)*		
G-BGHU	North American T-6G-NF Texan	182-729	FAP1707 Fr.AF 115042, 51-15042	22. 1.79	C.E.Bellhouse	Headcorn	8. 5.04P
					"Carly" (As "115042/TA-042" in USAF c/s)		
G-BGHV	Cameron V-77 HAB	483		12. 1.79	E.Davies	(Penlan Farm, Llanwrda)	28. 9.03A
					t/a Adeilad Claddings *"Adclad"*		
G-BGHW*	Thunder Ax8-90 HAB	175		30. 1.79	W.G.Johnston	(Edinburgh)	
					(Cancelled 19.5.93 by CAA) (Stored 2001)		

Regn	Type	C/n	Previous identity	Date	Owner/operator	Base	CofA
G-BGHY	Taylor JT.1 Monoplane	PFA 1455		12. 1.79	P.J.Burgess	RAF Cranwell	23. 8.05P
	(Built J Prowse)				*"Shy Talk"*		
G-BGHZ	Clutton FRED Series II	PFA 29-10445		12. 1.79	D.J.Howell	(Stourbridge)	
	(Built A G Edwards)				tr FRED Group *(Under construction 1999: new owner 10.02)*		
G-BGIB	Cessna 152 II	15282161	N68169	3. 7.79	Redhill Air Services Ltd	Shoreham	24. 2.07T
					(Op Airbase Flying Club)		
G-BGIG	Piper PA-38-112 Tomahawk	38-78A0773	N2607A	23. 1.79	Air Claire Ltd *(Op Glasgow Flying Club)*	Glasgow	8. 4.07T
G-BGIO	Montgomerie-Bensen B.8MR	PFA G/01-1259		11. 1.79	R.M.Savage	Carlisle	2211.05P
	(Built C G Johns and R.M.Savage - c/n GJ.1) (Rotax 503)				tr Great Orton Group		
G-BGIP	Colt 56A HAB	038		2. 2.79	J.Foght	(Nyhaven, Copenhagen, Denmark)	23. 7.03A
					(Reserved as OY-FOR for J.F.Rasmussen - Copenhagen 25.2.03R but NTU)		
G-BGIU	Cessna F172H	F172-0620	PH-VIT	26. 2.79	A.G.Arthur	Perranporth	24. 5.07T
	(Built Reims Aviation SA)						
G-BGIX	Helio H.295 Super Courier	1467	(G-BGAO) N68861	17.10.79	Caroline M.Lee	Fanners Farm, Great Waltham	6. 2.05
G-BGIY	Reims/Cessna F172N Skyhawk II	F17201824		31. 1.79	Air Claire Ltd *(Op Glasgow Flying Club)*	Glasgow	7. 9.06T
G-BGJB	Piper PA-44-180 Seminole	44-7995112	G-ISFT EI-CHF, G-BGJB, N3046B	1. 2.79	Ostend Air College NV	Ostend, Belgium	18.12.06T
G-BGJU	Cameron V-65 HAB	499		5. 2.79	Janet A Folkes *"Spoils"*	(Loughborough)	4. 4.93A
G-BGKC	SOCATA Rallye 110ST	3262		25. 4.79	J.H.Cranmer and T.A Timms	Bidford	8. 9.99
					(In open storage 6.04		
G-BGKO	Gardan GY-20 Minicab	PFA 1827		14. 2.79	R.B.Webber	Trenchard Farm, Eggesford	
	(Built R B Webber)				*(Not completed and stored 11.03)*		
G-BGKS	Piper PA-28-161 Warrior II	28-7916221	N9562N	12. 2.79	Marham Investments Ltd	Belfast Aldergrove	14. 4.06T
					(Op Woodgate Executive Air Services)		
G-BGKT	Auster AOP.9	B5-10/137	XN441	28.12.78	A Peters	Trenchard Farm, Eggesford	14. 7.05P
	(Officially regd as AUS/137)				tr KT Group *(As "XN441" in 'ARMY' camouflage c/s)*		
G-BGKU	Piper PA-28R-201 Arrow III	28R-7837237	N31585	8. 3.79	Aerolease Ltd	Conington	19. 2.07T
G-BGKV	Piper PA-28R-201 Cherokee Arrow III	28R-7737156	N44985	21. 5.79	R.Haverson and A K.Lake	Little Snoring	12. 5.07
G-BGKY	Piper PA-38-112 Tomahawk	38-78A0737	N9732N	2. 3.79	Top Cat Aviation Ltd	Manchester	7. 7.06T
G-BGKZ	Auster J/5F Aiglet Trainer	2776	F-BGKZ	15.12.78	R.B.Webber	Trenchard Farm, Eggesford	25. 2.95
					(Noted dismantled 11.04)		
G-BGLA	Piper PA-38-112 Tomahawk	38-78A0741	N9699N	9. 3.79	B.H. and P.M.Illston	Norwich	1. 9.06T
					t/a Norwich School of Flying		
G-BGLB*	Bede BD.5B				See SECTION 3 Part 1	Wroughton	
G-BGLF	Evans VP-1 Series 2	PFA 62-10388		28. 2.79	R.T.Ingram	(Clacton-on-Sea)	8.11.05P
	(Built R A Abrahams) (Volkswagen 1834)				*(New owner 11.04)*		
G-BGLG	Cessna 152 II	15282092	N67909	11. 4.79	L.W.Scattergood	(Selby)	7. 7.07T
G-BGLK	Monnett Sonerai IIL	PFA 15-10304		24. 2.78	J.Bradley	(Coleraine, County Londonderry)	1. 6.05P
	(Built G L Kemp) (Volkswagen 1783)						
G-BGLN	Reims/Cessna FA152 Aerobat	FA1520354		8. 3.79	Bflying Ltd	Bournemouth	18. 9.06T
					(Op Bournemouth Flying Club)		
G-BGLO	Reims/Cessna F172N Skyhawk II	F17201900		8. 3.79	Pulsar Yacht Services Ltd	Southend	29. 1.07T
G-BGLS	Oldfield Super Baby Lakes	PFA 10-10237		11.12.78	J.F.Dowe	Queach Farm, Bury St.Edmunds	18. 6.88P
	(Built D S Morgan) (Lycoming O-235)				*(Noted 5.03)*		
G-BGLW	Piper PA-34-200 Seneca	34-7250132	(G-BFPF) OY-BDZ, SE-FYS	2. 6.78	London Executive Aviation Ltd	Stapleford	28. 8.06T
G-BGLZ	Stits SA-3A Playboy	71-100	N9996	19. 6.79	P.C.Sheard	Strubby	29. 4.05P
	(Built D J Stadler) (Continental C90)						
G-BGME*	SIPA 903	96	G-BCML "G-BCHU", F-BGHU	1. 1.81	M.Emery and C.A Suckling	(Redhill)	17. 6.94P
					(Cancelled 15.11.00 by CAA) (Stored 12.03)		
G-BGMJ	Gardan GY-201 Minicab	12	F-BGMJ	19. 6.78	A.W.Wakefield	Sibson	22. 8.05P
	(Built Con. Aéronautique de Bearn) (Continental A65)				tr G-BGMJ Group		
G-BGMN	Hawker Siddeley HS.748 Series 2A/347	1766	PK-OCH G-BGMN, 9Y-TGH, G-BGMN, 9Y-TGH	9. 3.79	Emerald Airways Ltd	Liverpool	19.11.07E
G-BGMO	Hawker Siddeley HS.748 Series 2A/347	1767	ZK-MCB G-BGMO, 9Y-TGI, V2-LDB, 9Y-TGI, (G-BGMO)	9. 3.79	Emerald Airways Ltd	Liverpool	22. 4.05T
G-BGMP	Cessna F172G	F172-0240	PH-BNV	26. 3.79	R.W.Collings	Hinton-in-the-Hedges	14. 7.07
	(Built Reims Aviation SA)						
G-BGMR	Gardan GY-20 Minicab	PFA 56-10153		12. 3.79	R.A M.Smith	White Waltham	29. 9.04P
	(Built A B Holloway to JB.01 Minicab standard) (Continental C90)				tr Mike Romeo Flying Group		
G-BGMS	Taylor JT.2 Titch	PFA 60-10400		20.10.78	M.A J.Spice	(Middlewich, Cheshire)	
	(Built M.A J.Spice - c/n MS.1)				*(Current status unknown)*		
G-BGMT	SOCATA Rallye 235E	13126		14. 9.78	C.G.Wheeler	Morgansfield, Fishburn	26.12.06
G-BGMV	Scheibe SF25B Falke	4648	D-KEBG	15. 5.79	C.A Bloom and A P Twort	(Kittyhawk Farm, Deanland)	16.11.01
	(Stark-Stamo MS1500)				*(Stored 2004)*		
G-BGND	Reims/Cessna F172N Skyhawk II	F17201576	PH-AYI (F-GAQA)	3. 3.78	A J.M.Freeman	Andrewsfield	7.10.05
G-BGNT	Reims/Cessna F152 II	F15201644		23.10.79	Aerolease Ltd	Conington	15. 3.07T
G-BGNV	Gulfstream GA-7 Cougar	GA7-0078	N790GA	20. 4.79	G.J.Bissex	Kemble	7. 2.07T
G-BGOD	Colt 77A HAB	040		4. 4.79	C.and M.D.Steuer	(London NW1)	18. 6.97A
					"Harvey Wallbanger"		
G-BGOG	Piper PA-28-161 Warrior II	28-7916350	N9639N	8. 6.79	W.D.Moore	Cranfield	1. 4.07
G-BGOI	Cameron O-56 HAB	526		4. 4.79	S.Ellis *"Skymaster" (Active 7.04)*	(Bristol)	13. 5.87A
G-BGOJ	Reims/Cessna F150L	F15000931	G-MABI G-BGOJ, PH-KDA	19. 4.79	D.J.Hockings	Rochester	30. 7.97
					(Current status unknown)		
G-BGOL	Piper PA-28R-201T Turbo Arrow III	28R-7803335	N36705	11. 4.79	R G Jackson	(Winchester)	26. 4.05
G-BGON	Gulfstream GA-7 Cougar	GA7-0095	N9527Z	24. 4.79	J.P.E.Walsh	Elstree	1. 9.06T
					t/a Walsh Aviation *(Op Cabair)*		
G-BGOO*	Colt Flame 56SS HAB				See SECTION 3 Part 1	Newbury	
G-BGOR	North American AT-6D-NT Harvard III	88-14863	FAP1508 SAAF 7504, EX935, 41-33908	28. 3.79	P.Meyrick	Rednal	19. 6.05P
	(Reported as c/n 88-14880)				*(As "14863" in USAAF c/s)*		
G-BGPA	Cessna 182Q Skylane II	18266538	C-GYBW (N94935)	11. 7.79	Tindon Ltd	Little Snoring	24. 4.07

Reg	Type	C/n	Previous identity	Regd	Owner/operator	Base	CofA
G-BGPB	CCF Harvard 4 (T-6J-CCF Texan)	CCF4-538	FAP1747	4. 4.79	J.Romain *(Op Aircraft Restoration Co)*	Duxford	1. 4.06A
			West German AF BF+050, WGAF AA+050, 53-4619 *(As "1747" in Portuguese AF c/s)*				
G-BGPD	Piper J-3C-65 Cub (L-4H-PI)	12040	F-BFQP	18. 4.79	P.R.Whiteman	Marsh Hill Farm, Aylesbury	17. 6.05P
	(Frame No.11867)		F-BDTP, 44-79744		*(As "479744/49-M" in 92nd Armoured FA Btn, US 9th Army c/s)*		
	(Officially regd as c/n 10478 which is ex 43-29187/OO-ADI/F-BFQP: G-BGPD is ex 44-79744/F-BDTP: presumably fuselages exchanged in France - see G-BCOM)						
G-BGPF*	Thunder Ax6-56Z HAB				See SECTION 3 Part 1	Newbury	
G-BGPH	Gulfstream AA-5B Tiger	AA5B-1248	(G-BGRU)	14. 8.79	Shipping and Airlines Ltd	Biggin Hill	2. 1.05T
G-BGPI	Plumb BGP.1 Biplane	PFA 83-10359		26. 6.78	B.G.Plumb	Hinton-in-the-Hedges	29.10.05P
	(Built B.G.Plumb) (Continental O-200A)						
G-BGPJ	Piper PA-28-161 Warrior II	28-7916288	N9602N	24. 4.79	West Lancs Warrior Co Ltd	Woodvale	24. 6.06
G-BGPL	Piper PA-28-161 Warrior II	28-7916289	N9603N	20. 4.79	TG Aviation Ltd *(Op Thanet Flying Club)*	Manston	9. 9.06T
G-BGPN	Piper PA-18-150 Super Cub	18-7909044	N9750N	12. 4.79	D.McHugh and A R.Darke	East Winch	25.11.05T
G-BGPU	Piper PA-28-140 Cherokee F	28-7325282	PH-GNT	25. 4.79	Air Navigation and Trading Co Ltd	Blackpool	3. 9.06T
G-BGRC	Piper PA-28-140 Cherokee B	28-26208	SE-FHF	12. 6.79	Tecair Aviation Ltd and G.F.Haigh	Shipdham	26.10.97T
			N5501U		*(Stored and dismantled 8.04)*		
G-BGRE	Beech 200 Super King Air	BB-568		8. 5.79	Martin-Baker (Engineering) Ltd	Chalgrove	23.10.05T
G-BGRG	Beech 76 Duchess	ME-233		8. 5.79	S.J.Skilton	Bournemouth	7. 4.05T
					t/a Aviation Rentals *(Op Professional Air Training)*		
G-BGRH	Robin DR400 2+2	1411		21. 5.79	C.R.Beard	Grassthorpe Grange, Sutton-on-Trent	5. 4.07
G-BGRI	CEA Jodel DR.1050 Sicile	540	F-BLZJ	27. 4.79	B.J.L.P. and W.J.A.L.de Saar	Shipdham	8.11.07E
					(New owners 9.04)		
G-BGRM	Piper PA-38-112 Tomahawk	38-79A1067	N9673N	1. 8.79	Iris W.Goodger t/a Classair	Biggin Hill	22. 6.06T
G-BGRN*	Piper PA-38-112 Tomahawk	38-79A0897	N9684N	25. 4.79	Goodwood Road Racing Co Ltd	Goodwood	12. 2.00T
					(Cancelled 30.8.01 as WFU) *(Noted @ Fire Section 1.05)*		
G-BGRO	Reims/Cessna F172M Skyhawk II	F17201129	PH-KAB	4. 5.79	A N.Pirie t/a Cammo Aviation	Kirknewton	23. 9.05T
G-BGRR	Piper PA-38-112 Tomahawk	38-78A0336	OO-FLT	8. 5.79	Goodair Leasing Ltd	Cardiff	23. 9.05T
			N9685N				
G-BGRS	Thunder Ax7-77Z HAB	203		21. 5.79	P.M.Gaines	(Stockton-on-Tees)	26. 4.03A
G-BGRT	Steen Skybolt	PFA 64-10171		12. 9.78	O.Meier	(St.Johann In Tirol, Austria)	22.12.04P
	(Built R C Teverson - c/n RCT.001)						
G-BGRX	Piper PA-38-112 Tomahawk	38-79A0609	N9662N	11. 5.79	Bonus Aviation Ltd	Cranfield	13.11.06T
G-BGSA	SOCATA MS.892E Rallye 150GT	12838	F-GAKC	29. 5.79	D.H.Tonkin	Bodmin	14. 6.04
G-BGSB*	Hunting-Percival P.56 Provost T.1				See SECTION 3 Part 1	Bait al Falaj, Muscat, Oman	
G-BGSH	Piper PA-38-112 Tomahawk	38-79A0562	N9719N	11. 5.79	S.Padidar-Nazar	Carlisle	8. 9.05T
G-BGSI	Piper PA-38-112 Tomahawk	38-79A0564	N9720N	18. 5.79	Cheshire Flying Services Ltd	Liverpool	20. 9.03T
					t/a Ravenair *(Had accident and for spares use by 8.01)*		
G-BGSJ	Piper J-3C-65 Cub (L-4A-PI)	8781	F-BGXJ	21. 5.79	A J.Higgins	Dunkeswell	11 11.04P
	(Frame No.8917)		Fr.AF, 42-36657		*(As "236657/72-D" in USAAC c/s)*		
G-BGST*	Thunder Ax7-65 Bolt HAB				See SECTION 3 Part 1	(Aldershot)	
G-BGSV	Reims/Cessna F172N Skyhawk II	F17201830		1. 8.79	Southwell Air Services Ltd	Linley Hill, Leven	18. 3.07
G-BGSW	Beech F33 Bonanza	CD-1253	OH-BDD	30. 5.79	C.Wood	Wellesbourne Mountford	11. 9.06T
					(On rebuild Bournemouth 12.04)		
G-BGSY	Gulfstream GA-7 Cougar	GA7-0096		4. 6.79	Plane Talking Ltd	Biggin Hill	29. 4.05T
G-BGTC	Auster AOP.9	AUS/168	XP282	12.10.79	P.T.Bolton	Leiocester	9. 6.97P
					(Damaged Widmerpool 2.10.96: frame only)		
G-BGTF	Piper PA-44-180 Seminole	44-7995287	N2131Y	20. 6.79	NG Trustees and Nominees Ltd	Jersey	18. 5.06
G-BGTG	Piper PA-23-250 Aztec F	27-7954061	N2454M	23. 5.79	Keen Leasing (IoM) Ltd	Belfast Aldergrove	21.10.06T
G-BGTI	Piper J-3C-65 Cub (L-4J-PI)	12940	F-BFFL	17. 5.79	A P.Broad	Brandy Wharf, Waddingham	7.10.04P
	(Rotax 582) *(Frame No.12770)*		44-80644				
G-BGTJ	Piper PA-28-180 Cherokee Archer	28-7405083	OY-BIO	3. 7.79	Serendipity Aviation Ltd	Gloucestershire	17.12.06
			SE-GAH				
G-BGTT	Cessna 310R II	310R1641	N1AN	13. 7.79	Capital Trading (Aviation) Ltd	Jersey	1. 6.07T
			(N2635D)				
G-BGTX	SAN Jodel D.117	698	F-BIDI	22. 6.79	C.Adams and H.F.Young	Shobdon	10. 3.05P
					tr Madley Flying Group (Cisavia)		
G-BGUB (2)	Piper PA-32-300 Six	32-7940252	N2387U	29.11.79	A J.Diplock	Biggin Hill	17. 3.07
G-BGVB	CEA Jodel DR.315 Petit Prince	308	F-BPOP	20. 7.79	P.J.Leggo	Leicester	24.11.05
G-BGVE	Scintex CP.1310-C3 Super Emeraude	931	F-BMJE	8. 6.79	R.T.L.Arkell	Jackrells Farm, Southwater	9. 3.05P
					tr Victor Echo Group *"Mon Papillon"*		
G-BGVH	Beech 76 Duchess	ME-260		8. 6.79	W.J. and J.C.M.Golden	Bowerchalke, Salisbury	2. 8.07
					t/a Valco Marketing		
G-BGVK	Piper PA-28-161 Cherokee Warrior II	28-7816400	PH-WPT	13. 6.79	R.S.Bristowe	Exeter	29. 5.07
			G-BGVK, N6244C				
G-BGVN	Piper PA-28RT-201 Arrow IV	28R-7918168	N2846U	22. 6.79	C.Smith and S.Carrington tr G-BGVN Syndicate	Fairoaks	15.11.06
G-BGVS	Reims/Cessna F172M	F17200992	PH-HVS	3. 5.79	J.W.Tulloch	Kirkwall	22.12.06T
			(PH-LUK)		tr Kirkwall Flying Club		
G-BGVV	Gulfstream AA-5A Cheetah	AA5A-0750		27. 6.79	A H.McVicar	Carlisle	28. 5.04T
G-BGVY	Gulfstream AA-5B Tiger	AA5B-1080	(G-BGVU)	21. 8.79	R.J.C.Neal-Smith	Old Sarum	1.12.06
			(F-GBOO)				
G-BGVZ	Piper PA-28-181 Archer II	28-7990528	N2886A	12. 7.79	W.Walsh and S.R.Mitchell	Liverpool	3. 7.06T
G-BGWC	Robin DR400/180 Régent	1420		26. 6.79	P.R.Deacon	Headcorn	19. 8.07T
G-BGWH	Piper PA-18-150 Super Cub	18-7605	ST-ABR	18. 6.79	Spectrum Leisure Ltd	Clacton	27. 5.07T
			G-ARSR, N10F				
G-BGWJ	Sikorsky S-61N Mk.II	61-819		20. 8.79	Bristow Helicopters Ltd *"Monadh Mor"*	Faeroe Isles	4. 6.07T
G-BGWK	Sikorsky S-61N Mk.II	61-820	N1346C	10. 9.79	Bristow Helicopters Ltd	Aberdeen	28.11.05T
			G-BGWK		*"Dunrobin"*		
G-BGWM	Piper PA-28-181 Archer II	28-7990458	N2817Y	29. 6.79	Thames Valley Flying Club Ltd	Booker	10. 5.06T
G-BGWN	Piper PA-38-112 Tomahawk	38-79A0918	N9693N	2. 7.79	R.T.Callow	(Sleaford)	1. 7.05
G-BGWO	Jodel D.112	227	F-BHGQ	22. 6.79	R.C.Williams	Breighton	3.12.04P
	(Built Etablissement Valladeau)				tr G-BGWO Group		
G-BGWR	Cessna U206A Super Skywagon	U2060653	G-DISC	6. 7.79	The Parachute Centre Ltd	Tilstock	26. 1.07
			G-BGWR, PH-OTD, N4953F				
G-BGWS	Enstrom 280C Shark	1050		8.11.76	R.L.Heath t/a Whisky Sierra Helicopters	(Ashford)	12. 2.07T
G-BGWU	Piper PA-38-112 Tomahawk	38-79A0788	N9703N	2. 7.79	J.S.and L.M.Markey	Draycott Farm, Chiseldon	29. 3.07

Reg	Type	C/n	Previous id	Date	Owner/operator	Base	CofA
G-BGWV	Aeronca 7AC Champion	7AC-4082	OO-GRI OO-TWR	23. 8.79	J.A Webb tr RFC Flying Group *(Damaged Popham 8.6.86: current status unknown)*	(Alton)	10.10.86P
G-BGWW	Piper PA-23-250 Turbo Aztec E	27-4587	OO-ABH N13971	15. 6.79	Keen Leasing (IoM) Ltd *(New owner 4.02)*	Ronaldsway	28. 9.01T
G-BGWZ*	Eclipse Super Eagle				See SECTION 3 Part 1	RNAS Yeovilton	
G-BGXA	Piper J-3C-65 Cub (L-4H-PI) *(Frame No.10587 - regd with c/n 11170)*	10762	F-BGXA Fr.AF, 43-29471	1. 3.78	E.C. and P.King tr G-XA Group *(As "329471/44-F" in USAAC c/s)*	Eastbach Farm, Coleford	16. 1.05P
G-BGXB	Piper PA-38-112 Tomahawk	38-79A1007	N9728N	2. 7.79	Signtest Ltd	Cardiff	16. 8.04T
G-BGXC	SOCATA TB-10 Tobago	35		19.10.79	D.H.Courtley	Alderney	18. 8.07
G-BGXD	SOCATA TB-10 Tobago	39		19.10.79	D.F.P.Finan	Sturgate	8. 7.07T
G-BGXO	Piper PA-38-112 Tomahawk	38-79A0982	N9703N	5. 7.79	Goodwood Road Racing Co Ltd *(Op Goodwood Flying Club)*	Goodwood	12. 2.07T
G-BGXR	Robin HR200/100	53	F-BVYH	1.10.79	J.R.Cross	Egginton	24. 3.05
G-BGXS	Piper PA-28-236 Dakota	28-7911198	N2836Z	12. 7.79	M Holland tr G-BGXS Group	(Sheffield)	3. 4.07T
G-BGXT	SOCATA TB-10 Tobago	40		3.10.79	I.R.Jones	(Leas Park, Wirral)	31.10.07E
G-BGYN	Piper PA-18-150 Super Cub	18-7709137	N62747	19. 7.79	B.J.Dunford	Long Wood, Morestead	5.12.05
G-BGYT	Embraer EMB-110P1 Bandeirante	110.234	N104VA G-BGYT, PT-SAA	11.10.79	Keen Airways Ltd	Liverpool	12. 1.05T
G-BGZF	Piper PA-38-112 Tomahawk	38-79A1015	N9700N	26. 7.79	Fly Me Ltd *(New owner 11.04)*	Sleap	15. 2.04T
G-BGZW	Piper PA-38-112 Tomahawk	38-79A1068	N9674N	1. 8.79	Cheshire Flying Services Ltd t/a Ravenair	Liverpool	11.12.04T
G-BGZY	Wassmer Jodel D.120 Paris-Nice	118	F-BIQU	17. 8.79	M.Hale	(La Trinite Sur Mer, France)	16. 7.03P
G-BGZZ	Thunder Ax6-56 Bolt HAB	220		10. 8.79	J M.Eaton and K A Wilmore *(New owners 12.01)*	(Milton-under-Wychwood)	16. 7.94A

G-BHAA - G-BHZZ

Reg	Type	C/n	Previous id	Date	Owner/operator	Base	CofA
G-BHAA	Cessna 152 II	15281330	N49809	12. 2.79	Herefordshire Aero Club Ltd	Shobdon	1. 5.06T
G-BHAC	Cessna A152 Aerobat	A1520776	N7595B	12. 2.79	Herefordshire Aero Club Ltd	Shobdon	17. 4.06T
G-BHAD	Cessna A152 Aerobat	A1520807	N7390L	12. 2.79	Shropshire Aero Club Ltd	Sleap	11. 4.06T
G-BHAI	Reims/Cessna F152 II	F15201625	(D-EJAY)	14. 8.79	James D.Peace and Co *(Op Edinburgh Air Centre)*	Edinburgh	18. 2.05T
G-BHAJ	Robin DR400/160 Major 80	1430		22. 8.79	Rowantask Ltd	Rochester	10. 5.07T
G-BHAR	Westland-Bell 47G-3B1	WA/353	XT194	7. 8.79	J.Bird and R.Cove	Cranfield	8. 7.07
G-BHAT*	Thunder Ax7-77 Bolt HAB				See SECTION 3 Part 1	(Aldershot)	
G-BHAV	Reims/Cessna F152 II	F15201633		15. 8.79	T.M.and M.L.Jones *(Op Derby Aero Club)*	Egginton	14. 2.05T
G-BHAW	Reims/Cessna F172N Skyhawk II	F17201858		15. 8.79	C.Wilson	Andrewsfield	28. 6.07T
G-BHAX	Enstrom F-28C-2-UK	486-2	N5689N	22.10.79	J.L.Ferguson	Barton	21. 7.05
G-BHAY	Piper PA-28RT-201 Arrow IV	28R-7918213	N2910N	17. 8.79	Alpha Yankee Ltd	Newcastle	21. 6.07
G-BHBA	Campbell Cricket *(Built S M Irwin)* Rotax 503)	SMI-1		15. 8.79	S.N.McGovern	Henstridge	4.11.05P
G-BHBE	Westland-Bell 47G-3B1 *(Soloy conversion)*	WA/422	XT510	29.10.79	T.R.Smith (Agricultural Machinery) Ltd *(Stored 5.03)*	New Lane Farm, North Elmham	21.12.01
G-BHBF	Sikorsky S-76A II Plus	760022	N4247S	9.11.79	Bristow Helicopters Ltd *"Spirit of Paris"*	(Kazakhstan)	2. 1.07T
G-BHBG	Piper PA-32R-300 Cherokee Lance	32R-7780515	N408RC N9590N	18. 9.79	J.M.Thorpe	Treddunnock Farm, Llangarron	29. 9.07E
G-BHBI	Mooney M.20J Model 201	24-0842	N4764H	24. 9.79	A M.McGlone tr G-BHBI Group	Biggin Hill	10. 4.06
G-BHBT	Marquart MA.5 Charger *(Built R.G.Maidment)* (Lycoming O-320)	PFA 68-10190		3. 9.79	R.G.and C.J.Maidment	Dunsfold	14. 7.05P
G-BHBZ	Partenavia P68B	191		10. 9.79	P.C.Hamer and P.C.W.Landau	Sturgate	17. 7.05
G-BHCC	Cessna 172M Skyhawk II	17266711	(G-BGLY) N80713	26.10.79	D.Wood-Jenkins	Gloucestershire	27. 6.05T
G-BHCE	SAN Jodel D.117A	381	F-BHME	1.10.79	D.M.Parsons tr Parwebb Flying Group	Gloucestershire	30.10.05P
G-BHCM	Cessna F172H *(Built Reims Aviation SA)*	F172-0468	SE-FBD	25. 9.79	J.Dominic	Denham	20. 4.07
G-BHCP	Reims/Cessna F152 II	F15201640		31.10.79	D.Copley *(Current status unknown)*	Sandtoft	12.10.98T
G-BHCZ	Piper PA-38-112 Tomahawk	38-78A0321	N214MD	26. 9.79	Jennifer E.Abbott	Goodwood	15.10.06
G-BHDD	Vickers 668 Varsity T.1	Not known	WL626	18.10.79	G.Vale *(Current CofR 1.05 - in Aeropark Museum as "WL626/P")*	Nottingham East Midlands	AC
G-BHDE	SOCATA TB-10 Tobago	58		2. 1.80	Alpha-Alpha Ltd	Liverpool	7. 3.07
G-BHDK*	Boeing TB-29A-45-BN Superfortress				See SECTION 3 Part 1	Duxford	
G-BHDM	Reims/Cessna F152 II	F15201684		15.10.79	Big Red Kite Ltd	Booker	19. 4.07T
G-BHDP	Reims/Cessna F182Q Skylane II	F18200131		15.10.79	Zone Travel Ltd	Booker	23. 1.06
G-BHDR	Reims/Cessna F152 II *(Rebuilt 2003/4 using parts from G-BPGM)*	F15201680		15.10.79	James D.Peace and Co	Aberdeen	29. 7.07T
G-BHDS	Reims/Cessna F152 II	F15201682		15.10.79	Tayside Aviation Ltd	Glenrothes	14. 7.05T
G-BHDU	Reims/Cessna F152 II	F15201681		15.10.79	A.S.Bamrah t/a Falcon Flying Services	Blackbushe	8. 6.07T
G-BHDV	Cameron V-77 HAB	585		1. 2.80	P.Glydon *"Dorm Ouse"*	(Barnt Green, Birmingham)	8. 7.07A
G-BHDW	Reims/Cessna F152 II	F15201652		15.10.79	Tayside Aviation Ltd	Glenrothes	10. 6.07T
G-BHDX	Reims/Cessna F172N Skyhawk II	F17201889		5.10.79	GDX Ltd	Elstree	8. 7.07
G-BHDZ	Reims/Cessna F172N Skyhawk II	F17201911		3.12.79	Abbey Security Services Ltd	Old Buckenham	15. 5.07T
G-BHEC	Reims/Cessna F152 II	F15201676		3.12.79	Stapleford Flying Club Ltd	Stapleford	22. 7.07T
G-BHED	Reims/Cessna FA152 Aerobat	FA1520359		3.12.79	TG Aviation Ltd *(Op Thanet Flying Club)*	Manston	6. 5.07T
G-BHEG	SAN Jodel D.150 Mascaret	46	PH-ULS OO-SET	3. 7.80	D.M.Griffiths	RAF Mona	25. 6.05P
G-BHEK	Scintex CP.1315-C3 Super Emeraude	923	F-BJMU	11.10.79	D.B.Winstanley	Barton	1. 4.04P
G-BHEL	SAN Jodel D.117	735	F-BIOA	8.10.79	N.Wright and C.M.Kettlewell	Priory Farm, Tibenham	28. 7.04P
G-BHEM	Bensen B.8MV *(Built E Kenny - c/n EK.14)* (Rotax 503)	PFA G/01-1016		8.10.79	G.C.Kerr	(Great Orton)	5.10.00P
G-BHEN	Reims/Cessna FA152 Aerobat	FA1520363		3. 1.80	Leicestershire Aero Club Ltd	Leicester	20. 1.05T
G-BHEU	Thunder Ax7-65 Series 1 HAB	238		16.10.79	D.G.Such *"Polomoche"*	(Redditch)	7. 6.05A
G-BHEV	Piper PA-28R-200 Cherokee Arrow II	28R-7435159	PH-BOY N41244	23.10.79	P.Hardy tr 7-Up Group	Finningley	16. 4.06T

Reg	Type	C/n	Previous identity	Date	Owner/operator	Location	Expiry
G-BHEW*	Sopwith Triplane rep				See SECTION 3 Part 1	Polk City, Florida, US	
G-BHEX	Colt 56A HAB	056		15.10.79	A S.Dear, R.B.Green and W.S.Templeton	(Fordingbridge)	29. 8.04A
					tr Hale Hot Air Balloon Group *"Superwasp"*		
G-BHEZ	SAN Jodel D.150 Mascaret	22	F-BLDO	31. 1.80	A.Shorter tr Air Yorkshire Group	Sherburn-in-Elmet	20. 6.04P
G-BHFC	Reims/Cessna F152 II	F15201436		7. 4.78	Premier Flight Training Ltd	Norwich	1. 8.05T
G-BHFE	Piper PA-44-180 Seminole	44-7995324	Abu Dhabi AF 0052	2.10.79	Grunwick Processing Laboratories Ltd	Cranfield	24. 1.06T
			G-BHFE, N2383U		*(Op Bonus Aviation)*		
G-BHFF	Dormois Jodel D.112	322	F-BEKJ	19.10.79	G.H.Gilmour-White *(New owner 3.03)*	(Exeter)	28. 3.02P
G-BHFG	SNCAN Stampe SV-4C	45	F-BJDN	31.10.79	C.C.Rollings and T.F.J.Hodson	Gloucestershire	20. 3.05T
	(Renault 4P)		Fr.Mil		t/a Tiger Airways		
G-BHFH	Piper PA-34-200T Seneca II	34-7970482	N8075Q	23.10.79	G-WATS Aviation Ltd	Wolverhampton	24. 3.07T
G-BHFI	Reims/Cessna F152 II	F15201685		22.10.79	R.Bilson and D.Turner	Blackpool	11. 5.07T
					tr BAe Warton Flying Club		
G-BHFJ	Piper PA-28RT-201T Turbo Arrow IV	28R-7931298	N8072R	22.10.79	A.D.R.Northeast and S.A.Cook	Booker	26.10.07E
G-BHFK	Piper PA-28-151 Cherokee Warrior	28-7615088	N8325C	12.12.79	Ilkeston Car Sales Ltd	Jericho Farm, Lambley	23. 3.07
G-BHFR	Eiri PIK 20E	20228	(D-KHJR)	8.11.79	J.T.Morgan	Husbands Bosworth	23 4.05
			G-BHFR		*"FR"*		
G-BHFS	Robin DR400/180 Régent	1304		7. 3.78	C.J.Moss	Goodwood	26.11.05
G-BHGC	Piper PA-18-150 Super Cub	18-8793	PH-NKH	3. 4.79	Vectis Gliding Club Ltd	Bembridge	28. 2.06
			N4447Z				
G-BHGF	Cameron V-56 HAB	574		5.11.79	P.Spellward *"Biggles"*	(Bristol)	29. 8.00A
G-BHGJ	Wassmer Jodel D.120 Paris-Nice	336	F-BOYB	15. 1.80	Q.M.B.Oswell	RAF Halton	12. 4.05P
G-BHGO	Piper PA-32-260 Cherokee Six	32-7800007	PH-BGP	16.11.79	I.Parkinson	Eshott	30. 9.04
			N9656C		*(New owner 12.04)*		
G-BHGP	SOCATA TB-10 Tobago	100		17. 1.80	D.Suleyman	Stapleford	29. 5.05
G-BHGY	Piper PA-28R-200-2 Cherokee Arrow II		PH-NSL	23.11.79	R.J.Clark	Elstree	17. 8.07
		28R-7435086	N57365				
G-BHHB	Cameron V-77 HAB	170		26.11.79	R.M.Powell *"Pax"*	(Stockbridge)	13. 5.05T
G-BHHE	CEA Jodel DR.1051/M1 Sicile Record	628	F-BMZC	26. 4.80	P.Bridges	Hamilton Farm, Bilsington	6.12.04
G-BHHG	Reims/Cessna F152 II	F15201725		4. 3.80	TG Aviation Ltd *(Op Thanet Flying Club)*	Manston	24. 6.07T
G-BHHH	Thunder Ax7-65 Bolt HAB	245		5.12.79	C.A Hendley (Essex) Ltd *"Christmas"*	(Loughton)	27. 9.87A
G-BHHK	Cameron N-77 HAB	547		5.12.79	I.S.Bridge *"Shadowfax II"*	(Shrewsbury)	7.12.87A
G-BHHN	Cameron V-77 HAB	549		29.11.79	P.Gooch	(Alresford)	30. 6.05A
					tr The Itchen Valley Balloon Group *"Valley Crusader"*		
G-BHHX	Jodel D.112	223	F-BFAJ	19. 2.80	M.J.Wells	Watchford Farm, Yarcombe	10. 6.05P
	(Built Etablissement Valladeau)				tr G-BHHX Group		
G-BHIB	Reims/Cessna F182Q Skylane II	F18200134		18.12.79	S.N.Chater and B.Payne	Sherburn-in-Elmet	30. 4.06
G-BHIC	Reims/Cessna F182Q Skylane II	F18200135		18.12.79	Oxford Aviation Services Ltd	Oxford	4. 7.05
G-BHIG	Colt 31A Air Chair HAB	060	SE-...	12.12.79	P.A Lindstrand	(Upplands Vasby, Sweden)	13. 3.00A
			G-BHIG		*(Op S.Ericsson)*		
G-BHII	Cameron V-77 HAB	548		10.12.79	R.V.Brown *"Tosca"*	(Maidenhead)	2. 9.96A
G-BHIK	Adam RA.14 Loisir	11-bis	F-PHLK	6. 2.80	L.Lewis	(Redcar)	20. 8.85P
	(Continental A65)				*(Damaged near Lancaster 17.4.85: stored 1.02)*		
G-BHIL	Piper PA-28-161 Warrior II	28-8016069	G-SSFT	17.12.79	A S.Bamrah	Biggin Hill	17. 3.07T
			G-BHIL, N80821		t/a Falcon Flying Services		
		(Crashed in foggy weather conditions near Maidstone Road, Horsmonden, Kent 8.2.05 and badly damaged)					
G-BHIN	Reims/Cessna F152 II	F15201715		28. 1.80	Cristal Air Ltd *(New owner 3.02)*	Rochester	18.11.07E
G-BHIR*	Piper PA-28R-200 Cherokee Arrow	28R-35614	SE-FHP	21. 2.80	Factorcore Ltd	Woodford	6.12.04T
					(Op Manchester School of Flying) (Cancelled 30.1.04 by CAA)		
G-BHIS	Thunder Ax7-65 Bolt HAB	254		26.11.79	J.R.Wilson	(Didcot)	21. 3.96A
					tr The Hedgehoppers Balloon Group *"Yo-Yo"*		
G-BHIT	SOCATA TB-9 Tampico	63		7.12.79	C.J.P.Webster	Biggin Hill	31. 1.01T
					(Reported as noted Mahon, Menorca 6.03 apparently abandoned)		
G-BHIY	Reims/Cessna F150K	F15000627	F-BRXR	18.12.79	G.J.Ball	Old Sarum	17. 4.07
G-BHJF	SOCATA TB-10 Tobago	83		2. 1.80	P.Crutchfield tr Flying Fox Group	Blackbushe	15.10.06
G-BHJI	Mooney M.20J Model 201	24-0925	N3753H	11. 2.80	Hearing Centre Aarhus	(Aarhus, Denmark)	4. 2.05T
G-BHJK	Maule M-5-235C Lunar Rocket	7296C	N56359	25. 2.80	R.I.Gilchrist tr JK Group	Shoreham	30. 5.05T
G-BHJN	Fournier RF4D	4021	F-BORH	3. 1.80	R.F.Wondrak tr RF4 Flying Group	Enstone	23. 3.05P
G-BHJO	Piper PA-28-161 Cherokee Warrior II	28-7816213	OO-FLD	4. 1.80	Anne Sangster, I.Young and T.R.Whittome	Inverness	16. 5.07
			N9507N, N6034H		tr Brackla Flying Group		
G-BHJS	Partenavia P68B	172	I-KLUB	28.12.79	J.J.Watts and D.Fletcher	Bournemouth	15. 7.07T
G-BHJU	Robin DR400 2+2	1288	D-ECDK	9. 1.80	J.Barlow and P.Crow tr Ageless Aeronauts	Lydd	29. 5.07
G-BHKH	Cameron O-65 HAB	592		7. 1.80	P.Donkin	(Caldicot)	30. 5.05A
G-BHKJ	Cessna 421C Golden Eagle III	421C0848	(N26596)	25. 1.80	Totaljet Ltd	Blackpool	6.10.07E
	(Robertson STOL conversion)						
G-BHKN*	Colt 14A Cloudhopper HAB				See SECTION 3 Part 1	Newbury	
G-BHKR*	Colt 14A Cloudhopper HAB				See SECTION 3 Part 1	Newbury	
G-BHKT	Wassmer Jodel D.112	1265	F-BMIQ	10. 1.80	M.G.Davis	(Canterbury)	11.11.05P
G-BHLE	Robin DR400/180 Régent	1466		25. 1.80	B.D.Greenwood	Ronaldsway	16. 5.04
G-BHLH	Robin DR400/180 Régent	1320	F-GBIG	11. 2.80	G.J.Busby tr G-BHLH Group	Booker	7. 7.07
G-BHLJ	Saffery-Rigg S.200 Skyliner MLB	IAR/01		23. 1.80	I.A Rigg	(Manchester)	
	(Built C Saffrey)				*"Skyliner"*		
G-BHLT	de Havilland DH.82A Tiger Moth	84997	ZS-DGA	9. 6.80	P.J. and A J.Borsberry	Kidmore End, Reading	26. 2.90
	(Built Morris Motors Ltd)		SAAF2272, T6697		*(On rebuild 8.90: current status unknown)*		
G-BHLU	Fournier RF3	79	F-BMTN	14. 4.80	Skyview Systems Ltd	(Sudbury)	30.11.05P
G-BHLW	Cessna 120	10210	N73005	24. 3.80	L.W.Scattergood	Sherburn-in-Elmet	8. 9.04P
	(Continental C85)		NC73005		*"Sky Ranger"*		
G-BHLX	Grumman AA-5B Tiger	AA5B-0573	OY-GAR	1. 2.80	M.D.McPherson	Cranfield	22. 7.07T
G-BHMA	SIPA 903	61	OO-FAE	13. 3.80	H.J.Taggart	Ballymoney, County Antrim	18. 5.98P
			F-BGBK				
G-BHMG	Reims/Cessna FA152 Aerobat	FA1520368		10. 6.80	R.D.Smith	Popham	18. 6.05T
G-BHMI	Reims/Cessna F172N Skyhawk II	F17202036	G-WADE	6. 8.80	GMI Aviation Ltd	Blackpool	9. 3.05T
			G-BHMI				

Registration	Type	C/n	Previous identity	Regd	Owner/operator	Location	CofA
G-BHMJ	Avenger T.200-2112 MLB	002		29. 1.80	R.Light	(Stockport)	
	(Built R Light)				*"Lord Anthony I"*		
G-BHMK	Avenger T.200-2112 MLB	003		29. 1.80	P.Kinder	(Stockport)	
	(Built P.Kinder)				*"Lord Anthony II"*		
G-BHMR	Stinson 108-3 Station Wagon	108-4352	F-BABO	12. 2.80	D.G.French	Sandown	23.11.90
	(Built Consolidated Vultee Aircraft)		F-DABO, NC6352M		*(Dismantled 7.04)*		
G-BHMT	Evans VP-1	PFA 62-10473		18. 2.80	P.E.J.Sturgeon	Queach Farm, Bury St Edmunds	8. 7.05P
	(Built P.E.J.Sturgeon) (Volkswagen 1834)						
G-BHMY*	Fokker F.27 Friendship 600				See SECTION 3 Part 1	Norwich	
G-BHNA	Reims/Cessna F152 II	F15201683		12. 2.80	Sheffield Aero Club Ltd	Netherthorpe	19. 2.07T
G-BHNC	Cameron O-65 HAB	588		7. 2.80	D.and C.Bareford	(Kidderminster)	5. 3.94A
					"Hot N'Cold"		
G-BHND	Cameron N-65 HAB	582		7. 2.80	S.M.Wellband	(Frome)	24. 6.89A
G-BHNG*	Piper PA-23-250 Aztec E				See SECTION 3 Part 1	Filching Manor, Wannock	
G-BHNK	Wassmer Jodel D.120A Paris-Nice	243	F-BLNK	26. 3.80	D.A Bates	St.Mary's, Isles of Scilly	16. 4.05P
					tr G-BHNK Flying Group		
G-BHNL	Wassmer Jodel D.112	1206	F-BLNL	30. 1.80	M.D.Mold tr HNL Group	Dunkeswell	15.10.05P
G-BHNO	Piper PA-28-181 Archer II	28-8090211	N81413	7. 2.80	Airfluid Hydraulics and Pneumatics (Wolverhampton) Ltd		
						Sleap	18. 7.07
G-BHNP	Eiri PIK 20E	20253		29. 2.80	D.A Sutton *"NP"*	Sackville Lodge Farm, Riseley	26. 5.05
G-BHNV	Westland-Bell 47G-3B1	WA/700	F-GHNM	11. 3.80	Leyline Helicopters Ltd	Trenholme Farm, Billingham	28. 5.89T
			G-BHNV, XW180		*(Noted 10.02)*		
G-BHNX	SAN Jodel D.117	493	F-BHNX	7. 9.78	A J.Chalkley	(Pwllheli)	12. 1.87P
					(On rebuild 4.91: current status unknown)		
G-BHOA	Robin DR400/160 Major 80	1478		27. 1.80	Goudhurst Service Station Ltd	Goudhurst	1. 9.05
G-BHOG	Sikorsky S-61N Mk.II	61-825	PT-YEK	25. 3.80	Bristow Helicopters Ltd	Aberdeen	
			G-BHOG, (LN-ONK), G-BHOG		*(Stored 6.04)*		
G-BHOH	Sikorsky S-61N Mk.II	61-827		25. 4.80	Bristow Helicopters Ltd	Aberdeen	20. 5.02T
					"Ben Avon" (Stored 6.04)		
G-BHOJ	Colt 12A Cloudhopper HAB	080		27. 2.80	J.A Folkes	(Oswestry)	
	(Originally registered as Colt 14A)						
G-BHOL	CEA Jodel DR.1050 Ambassadeur	35	F-BJQL	6. 2.80	J.E.Sharkey *"Nicolette"*	Inverness	15.12.07
G-BHOM	Piper PA-18 Super Cub 95	18-1391	OO-PIU	7. 3.80	P.Myers	Whitehall Farm, Benington	14. 4.05P
	(L-18C-PI) *(Frame No.18-1272)*		OO-HMT, ALAT 51-15391		tr Oscar Mike Flying Group		
G-BHOO	Yorkshire Air Balloon A66 HAB	001		26. 2.80	D.Livesey and J.M.Purves	(Brampton/York)	
	(Built D.Livesey and J.M.Purves)				*"Scraps"*		
G-BHOR	Piper PA-28-161 Warrior II	28-8016331	N82162	12. 6.80	A J.Harewood	Biggin Hill	26. 6.07
					tr Oscar Romeo Flying Group		
G-BHOT	Cameron V-65 HAB	777		15. 9.81	J.A Baker	Marsh Benham	8. 8.99A
					tr The Dante Balloon Group *"Le Billet Doux"*		
G-BHOZ	SOCATA TB-9 Tampico	84		11. 3.80	G-BHOZ Management Ltd	Kemble	7. 7.07
G-BHPK	Piper J-3C-65 Cub (L-4A-PI)	8979	F-BEPK	26. 2.80	L.B.Smith	Priory Farm, Tibenham	17. 3.05P
	(Frame No.9098: official c/n is 12161/44-79865 [F-BFYU])		Fr.Mil, 42-38410		tr L4 Group *(As "238410/44-A" in USAAF c/s)*		
G-BHPL	CASA 1-131E Jungmann	1058	Spanish AF E3B-350	17. 7.80	R.G.Gray	Thruxton	29. 1.04P
					(As "E3B-350/05-97" in Spanish AF c/s)		
G-BHPM	Piper PA-18 Super Cub 95	18-1501	F-BOUR	10. 4.80	P.I.Morgans	Furze Hill Farm, Rosemarket, Milford Haven	
	(L-18C-PI) *(Frame No.18-1469)*		ALAT 51-15501		*(Stored 5.95: current status unknown)*		
G-BHPN	Colt 14A Cloudhopper HAB	081	(SE-)	6. 3.80	Lindstrand Balloons Ltd	(Upplands Vasby, Sweden)	24. 1.04A
			G-BHPN		*(Op S.Ericsson)*		
G-BHPS	Wassmer Jodel D.120A Paris-Nice	148	F-BIXI	11. 6.80	T.J.Price	Rhigos	7. 8.04P
G-BHPY	Cessna 152 II	15282983	N46009	26. 3.80	Halegreen Associates Ltd	Hinton-in-the-Hedges	4.10.07T
G-BHPZ	Cessna 172N Skyhawk II	17272017	N6411E	26. 3.80	O'Brien Properties Ltd	Shoreham	22. 4.06T
G-BHRB	Reims/Cessna F152 II	F15201707		20. 3.80	LAC (Enterprises) Ltd *(Op Lancashire Aero Club)*	Barton	18. 2.05T
G-BHRC	Piper PA-28-161 Warrior II	28-7916430	N9527N	3. 4.80	The Sherwood Flying Club Ltd	Tollerton	9. 3.07T
G-BHRH	Reims/Cessna FA150K Aerobat	FA1500056	PH-ECB	24. 3.80	Merlin Flying Club Ltd	Hucknall	25. 6.05T
			D-ECBL, (D-EKKW)				
G-BHRM	Reims/Cessna F152 II	F15201718	F-GCHR	8. 4.80	Tatenhill Aviation Ltd	Tatenhill	22. 4.07T
G-BHRN	Reims/Cessna F152 II	F15201728	F-GCHV	8. 4.80	James D.Peace and Co	Edinburgh	31. 7.05T
					(Op Edinburgh Air Centre)		
G-BHRO*	Rockwell Commander 112A	364	N1364J	20. 3.80	J G Lunt *(Cancelled 1.12.04)*	Swansea	2. 9.07
G-BHRP	Piper PA-44-180 Seminole	44-8095021	N81602	1. 4.80	M.S.Farmers	Leicester	20. 5.07T
G-BHRR	Menavia Piel CP.301A Emeraude	270	F-BISK	28. 3.80	T.W.Offen *(Stored 5.01)*	Spilsteads Farm, Sedlescombe	28. 5.87P
G-BHRW	CEA Jodel DR.221 Dauphin	93	F-BPCP	10. 7.80	M.F.Filer and D.H.Williams	Kemble	13. 4.05
G-BHRY	Colt 56A HAB	030		2. 4.80	A S.Davidson *"Turkish Delight"*	(Burton-on-Trent)	29. 4.95A
G-BHSB	Cessna 172N Skyhawk II	17272977	(N1225F)	25. 6.80	SB Aviation Ltd	Leeds-Bradford	5. 2.05T
G-BHSD	Scheibe SF25E Super Falke	4357	D-KDGG	21. 7.80	S.J.Filhol	(Castlefreke, County Cork)	15. 5.05
	(Limbach SL1700)						
G-BHSE	Rockwell Commander 114	14161	N4831W	15. 5.80	604 Squadron Flying Group Ltd	Booker	30. 5.05
			AN-BRL, (N4831W)				
G-BHSN	Cameron N-56 HAB	595		10. 4.80	I.Bentley	(Bath)	9. 7.05A
G-BHSP	Thunder Ax7-77Z HAB	272		15. 4.80	G.A Fisher	(Guildford)	23. 2.94A
	(Originally built as D-TRIER c/n 221)				*"Chicago"*		
G-BHSS	Pitts S-1C	C.1461M	N1704	19. 9.80	C.W.Burkett	(Bedford)	11.12.05P
	(Built C H McClendon) (Lycoming O-320)						
G-BHSY	CEA Jodel DR.1050 Sicile	546	F-BLZO	6. 5.80	T.R.Allebone	Easton Maudit	10.10.04
G-BHTA	Piper PA-28-236 Dakota	28-8011102	N8197H	22. 4.80	Dakota Ltd	Jersey	26. 9.07
G-BHTC	CEA Jodel DR.1051/M1 Sicile Record	581	F-BMGR	1. 5.80	G.Clark	Oaksey Park	20.12.04
G-BHTG	Thunder Ax6-56 Bolt HAB	273		18. 4.80	F.R.and S.H.MacDonald *"Halcyon"*	(Newdigate, Dorking)	18.12.91A
G-BHTH*	North American T-6G-NT Texan	168-176	N2807G	20. 5.80	Northbrook College	Shoreham	11. 5.97T
			49-3072A		*(Damaged Bourne Park, Andover 13.3.95: cancelled 12.4.01) (On rebuild 11.04 as "2807/103")*		
G-BHUB*	Douglas C-47A-85DL Dakota				See SECTION 3 Part 1	Duxford	
G-BHUE	SAN Jodel DR.1050 Ambassadeur	185	F-BERM	21. 4.80	M.J.Harris	(Worcester)	19.10.92
			F-OBRM				
G-BHUG	Cessna 172N Skyhawk II	17272985	N1283F	24. 6.80	F.G.Baulch t/a FGT Aircraft Hire	Dunkeswell	24. 1.05T

G-BHUI	Cessna 152 II	15283144	N46932	27. 5.80	Galair International Ltd	Wellesbourne Mountford	16.12.04T
G-BHUJ	Cessna 172N Skyhawk II	17271932	N5752E	27. 5.80	K.B.Dupuy tr Uniform Juliet Group	Southend	20. 5.05T
G-BHUM	de Havilland DH.82A Tiger Moth	85453	VT-DGA	9. 6.80	S.G.Towers	Beckwithshaw, Harrogate	22. 5.06
			VT-DDN, RIAF, SAAF 4622, DE457				
G-BHUR	Thunder Ax3 Mini Sky Chariot HAB	277		9. 5.80	B.F.G.Ribbans	(Newbury)	16. 5.05A
					"Ben Hur" (On loan to British Balloon Museum and Library)		
G-BHUU	Piper PA-25-260 Pawnee D	25-8056035	N2440Q	28. 5.80	Booker Gliding Club Ltd	Booker	5. 4.07
G-BHUW*	Boeing Stearman A75N1 (N2S-5) Kaydet				See SECTION 3 Part 1	Munich, Germany	
G-BHVB	Piper PA-28-161 Warrior II	28-8016260	N9638N	16. 5.80	Caine Aviation Ltd	Wolverhampton	30. 7.04T
G-BHVF	SAN Jodel D.150A Mascaret	11	F-BLDF	28.10.80	J.D.Walton	Swanborough Farm, Lewes	17. 3.05P
G-BHVP	Cessna 182Q Skylane II	18267071	N97374	15.12.80	R.J.W.Wood	Gregory Farm, Mirfield	7.10.07E
G-BHVR	Cessna 172N Skyhawk II	17270196	N738SG	27. 5.80	M.G.Montgomerie tr G-BHVR Group	Elstree	9. 7.06T
G-BHVV	Piper J-3C-65 Cub (L-4A-PI)	8953	F-BGXF	27. 6.80	C A Ward and Caroline A Cash	Park Farm, Throwley	9. 8.05P
	(Frame No.9048)		Fr Mil, 42-38384		*(Also carries USAAF serial "31430" in yellow on the tai)*		
	(Regd with c/n 10291/43-1430 ex F-BEGF: frames probably exchanged in 1953 rebuild)						
G-BHWA	Reims/Cessna F152 II	F15201775		28. 3.80	Lincoln Aviation Ltd	Wickenby	1. 7.07T
G-BHWB	Reims/Cessna F152 II	F15201776	(G-BHWA)	14. 4.80	Lincoln Aviation Ltd	Wickenby	20. 8.04T
G-BHWH	Weedhopper JC-24A	0074		23. 4.80	G.A Clephane	Basingstoke	Exemption
	(Fuji-Robin EC-34-PM) *(Modified to JC-24C)*				*(As "Bu.126603" in US Navy c/s) "Dream Machine"*		
G-BHWK	SOCATA MS.880B Rallye Club	870	F-BONK	27. 8.80	L.L.Gayther	Shobdon	16.12.04
G-BHWY	Piper PA-28R-200-2 Cherokee Arrow II	28R-7435059	N56904	17. 6.80	P.R.Gould and R.B.Cheek	Sandown	16. 5.05
					tr Kilo Foxtrot Flying Group		
G-BHWZ	Piper PA-28-181 Cherokee Archer II	28-7890299	N3379M	8. 4.80	M A Abbott	(London W9)	16. 8.07T
G-BHXA	Scottish Aviation Bulldog Series 120/1210	BH120/407	Botswana DF OD1	9. 6.80	Air Plan Flight Equipment Ltd	Egginton	3. 6.06T
			G-BHXA		*(Deltair Aviation titles)*		
G-BHXD	Wassmer Jodel D.120 Paris-Nice	258	F-BMIA	3. 7.80	J.M.Fforde and M.Roberts	(Clyro, Hereford)	18. 5.05P
G-BHXK	Piper PA-28-140 Cherokee	28-21106	VR-HGB	14. 7.80	J.Moreland	Bourne Park, Hurstbourne Tarrant	1. 6.06
			9V-BAJ, (9M-AOM)		tr GXK Flying Group		
G-BHXS	Wassmer Jodel D.120 Paris-Nice	133	F-BIXS	27. 8.80	I.R.Willis	Glenrothes	17. 6.05P
G-BHXY	Piper J-3C-65 Cub (L-4H-PI)	11905	D-EAXY	1. 7.80	F.W.Rogers	Dunkeswell	2. 3.05P
	(Frame No.11733)		F-BFQX, 44-79609		*(As "479609/PR-L4" in USAAC c/s) "Heather"*		
G-BHYA	Cessna R182 Skylane RG II	R18200532	N1717R	10. 7.80	B.Davies	(Sandridge)	27. 3.05
G-BHYC	Cessna 172 RG Cutlass II	172RG0404	(N4868V)	24. 6.80	IB Aeroplanes Ltd	City of Derry	1. 5.05
G-BHYD	Cessna R172K Hawk XP II	R1722734	N736RS	11.12.80	Sylmar Aviation and Services Ltd	Lower Wasing Farm, Brimpton	14. 4.06
G-BHYE	Piper PA-34-200T Seneca II	34-8070233	N8225U	27. 6.80	Oxford Aviation Services Ltd	Oxford	10. 6.07T
G-BHYF	Piper PA-34-200T Seneca II	34-8070234	N8225V	27. 6.80	Oxford Aviation Services Ltd	Oxford	14. 1.06T
G-BHYG	Piper PA-34-200T Seneca II	34-8070235	N8225X	30. 6.80	Oxford Aviation Services Ltd	Oxford	27.10.07E
G-BHYI	SNCAN Stampe SV-4A	18	F-BAAF	11. 7.80	G.L.Brown and A M.Plato	Spanhoe	25. 7.07
	(Renault 4P)		Fr.Mil				
G-BHYO	Cameron N-77 HAB	659		30. 6.80	Adventure Balloon Co Ltd	(Aldershot)	8. 5.97A
					"Alcock and Sissons" (Op Balloon Preservation Group) (Inflated 6.02)		
G-BHYP	Reims/Cessna F172M Skyhawk II	F17201108	OY-BFR	30. 6.80	Avior Ltd	Oxford	16. 6.05T
G-BHYR	Reims/Cessna F172M	F17200922	OY-DZH	30. 6.80	R.G.Forster	Stapleford	10. 7.05
			SE-FZH, (OH-CFQ)		tr G-BHYR Group		
G-BHYV	Evans VP-1	PFA 1569		2. 7.80	L.Chiappi	White Waltham	
	(Built L.Chiappi - c/n LC.2) (Volkswagen 1600)				*(Flown 5.89: noted 11.04)*		
G-BHYX	Cessna 152 II	15281832	N67434	4. 7.80	Stapleford Flying Club Ltd	Stapleford	10. 4.05T
G-BHZE	Piper PA-28-181 Cherokee Archer II	28-7890291	OO-FLR	4.11.80	Zegruppe Ltd	White Waltham	22. 1.06T
			(OO-HCM), N3053M				
G-BHZH	Reims/Cessna F152 II	F15201786		25. 7.80	Plymouth School of Flying Ltd	Plymouth	27. 9.07E
G-BHZK	Grumman AA-5B Tiger	AA5B-0743	N28670	8. 9.80	R.G.Seth-Smith tr Zulu Kilo Group	Elstree	29. 5.05
G-BHZO	Gulfstream AA-5A Cheetah	AA5A-0692	N26750	21. 7.80	A H.McVicar *(Op Prestwick Flight Centre)*	Prestwick	23.12.04T
G-BHZR	Scottish Aviation Bulldog Series 120/1210	BH120/410	Botswana DF OD4	23. 7.80	White Knuckle Air Ltd	Haverfordwest	9. 7.06
			G-BHZR				
G-BHZS	Scottish Aviation Bulldog Series 120/1210	BH120/411	Botswana DF OD5	23. 7.80	Air Plan Flight Equipment Ltd	Egginton	20. 2.06T
			G-BHZS				
G-BHZT	Scottish Aviation Bulldog Series 120/1210	BH120/412	Botswana DF OD6	23. 7.80	D.M.Curties	Kemble	14. 3.05T
			G-BHZT				
G-BHZU	Piper J-3C-65 Cub (L-4B-PI)	9775	F-BETO	17. 7.80	J.K.Tomkinson	Brook Farm, Boylestone	28. 5.05P
	(Continental O-200-A)		(F-BFKH), 43-914				
	(Regd with Frame No.9606 fitted to F-BETO in 1961 rebuild replacing c/n 13164 ex 45-4424)						
G-BHZV	Wassmer Jodel D.120A Paris-Nice	278	F-BMON	23. 7.80	K.J.Scott	Stoneacre Farm, Farthing Corner	26. 6.00P
G-BHZX	Thunder Ax7-69A HAB	288		25. 7.80	R.J.and H.M.Beattie *"After Eight"*	(Aylesbury)	10. 6.94A

G-BIAA - G-BIZZ

G-BIAB*	SOCATA TB-9 Tampico	142		17. 7.80	Not known	Willey Park Farm, Caterham	
		(Damaged in forced-landing near Halfpenny Green 6.8.93: cancelled 10.2.94 by CAA) (Fuselage noted 2.04)					
G-BIAC	SOCATA Rallye 235E Gabier	13323		17. 7.80	D.R.Watson and A J.Haigh	Maypole Farm, Chislet	18. 6.05T
G-BIAH	Wassmer Jodel D.112	1218	F-BMAH	20. 8.80	T.K.Duffy	Dunnyvadden, County Antrim	22. 5.05P
G-BIAI	Wallingford WMB.2 Windtracker MLB	008		1. 7.80	I.Chadwick	(Horsham)	
	(Built Wallingford Model Balloons)				tr Unicorn Group *"Amanda I"*		
G-BIAP	Piper PA-16 Clipper	16-732	F-BBGM	25. 6.80	P.J.Bish	Draycott Farm, Chisledon	28. 3.07
	(Frame No.16-733)		F-OAGS				
G-BIAR	Rigg Skyliner II MLB	AKC-59		9. 7.80	I.A Rigg	(Manchester)	
	(Built I.A Rigg - c/n IAR/02)						
G-BIAT*	Sopwith Pup rep				See SECTION 3 Part 1	Christchurch, New Zealand	
G-BIAU*	Sopwith Pup rep				See SECTION 3 Part 1	RNAS Yeovilton	
G-BIAX	Taylor JT.2 Titch	PFA 3228		30. 7.80	J.T.Everest	Popham	11. 5.04P
	(Built J T Everest and G F Rowley - c/n GFR-1) (Continental O-200)				*(Stored during Winter)*		
G-BIAY	Grumman AA-5 Traveler	AA5-0423	OY-GAD	26. 8.80	M D, J H Dupay and S Raven	Southend	30. 7.05
			N7123L				

Registration	Type	C/n	Previous identity	Regd	Owner/Operator	Base	CofA expiry
G-BIAZ*	Cameron AT-165 Helium/HAB				See SECTION 3 Part 1	Newbury	
G-BIBA	SOCATA TB-9 Tampico	149		17. 7.80	TB Aviation Ltd	Denham	1. 5.06
G-BIBB	Mooney M.20C Mark 21	2803	OH-MOD	22. 7.80	Lefay Engineering Ltd	(Havant)	23. 6.05
G-BIBG	Sikorsky S-76A II Plus	760083	5N-BCE	18. 8.80	Bristow Helicopters Ltd	Aberdeen	18. 8.06T
			G-BIBG				
G-BIBJ	Enstrom 280C-UK-2 Shark	1187		13. 8.80	Tindon Ltd	Little Snoring	18. 2.05
G-BIBN	Reims/Cessna FA150K Aerobat	FA1500078	F-BSHN	29.10.80	B.V.Mayo	Maypole Farm, Chislet	7. 1.05
G-BIBO	Cameron V-65 HAB	667		7. 8.80	I.Harris *"Diadem"*	(Devizes)	25. 6.05A
G-BIBS	Cameron P-20 HAB	671		14. 8.80	Cameron Balloons Ltd	(Bristol)	
G-BIBT	Gulfstream AA-5B Tiger	AA5B-1047	N4518V	8. 9.80	Horizon Aviation Ltd	Swansea	3.10.07E
G-BIBW	Reims/Cessna F172N Skyhawk II	F17201756		13.10.78	Drawflight Ltd	Lydd	7. 3.05
G-BIBX	Wallingford WMB.2 Windtracker MLB	9		18. 8.80	I.A Rigg	(Manchester)	
	(Built Wallingford Model Balloons)				*"Bumble"*		
G-BICD	Taylorcraft J Auster 5	735	F-BFXH	20. 8.80	T.R.Parsons	Beeches Farm, South Scarle	3. 8.05P
			MT166				
G-BICE	North American AT-6C-1NT Harvard IIA	88-9755	FAP1545	3. 9.80	C.M.L.Edwards	Cherry Tree Farm, Monewden	6. 9.05P
			SAAF 7084, EX302, 41-33275 *(As "41-33275/CE" in US Army c/s)*				
G-BICG	Reims/Cessna F152 II	F15201796		3. 9.80	A S.Bamrah t/a Falcon Flying Services	Biggin Hill	27. 3.05T
G-BICJ	Monnett Sonerai II	PFA 15-10531		22. 8.80	P.Daukas	(Orton Goldhay, Peterborough)	20.10.90P
	(Built J R Heaton - c/n 726) (Volkswagen 1834)				*(New owner 11.04)*		
G-BICM	Colt 56A HAB	095		1. 9.80	W.S.Templeton and R.B.Green	(Fordingbridge)	20. 4.03A
					tr The Avon Advertiser Balloon Club *"Ladybird"*		
G-BICP	Robin DR360 Chevalier	610	F-BSPH	2.10.80	N.Thorne	Breighton	28. 5.05
G-BICR	Wassmer Jodel D.120A Paris-Nice	135	F-BIXR	5. 9.80	G.L.Perry tr Beehive Flying Group	White Waltham	21.12.05P
G-BICS	Robin R2100A Club	128	F-GBAC	4.12.80	I.Young	Sandown	10. 7.06
G-BICU	Cameron V-56 HAB	680		9. 9.80	S.D.Bather *"Nobby"*	(Melksham)	20. 9.03A
G-BICW	Piper PA-28-161 Warrior II	28-7916309	N2091U	8.10.80	D.Gellhorn tr The Warrior Group	Blackbushe	23. 2.06
G-BICX	Maule M-5-235C Lunar Rocket	7287C	(G-MAUL (1))	2. 2.81	A.T.Jeans and J.F.Clarkson		
			N56352			Compton Chamberlayne, Salisbury	1. 7.05
G-BICY	Piper PA-23-160 Apache	23-1640	PH-ACL	26. 9.80	A M.Lynn *(Op Busy Bee)*	Sibson	15. 7.07T
			N4010P, (PH-ACL), N4010P, N10F				
G-BIDD	Evans VP-1	PFA 62-10974		27.10.78	Jane Hodgkinson	Thruxton	2.12.00P
	(Built J Wedgebury) (Volkswagen 1600) *(Initially regd with c/n PFA 62-10167 and combined with both projects)*						
G-BIDF	Reims/Cessna F172P Skyhawk II	F17202045	(PH-JPO)	18. 9.80	C.J.Chaplin and N.J.C.Howard	Redhill	14. 4.06T
G-BIDG	SAN Jodel D.150A Mascaret	08	F-BLDG	11. 9.80	D.R.Gray	Barton	19.10.05P
G-BIDH	Cessna 152 II	15280546	G-DONA	12. 9.80	Hull Aero Club Ltd	Linley Hill, Leven	16. 7.05T
			G-BIDH, N25234				
G-BIDI	Piper PA-28R-201 Arrow III	28R-7837135	N3759M	11.11.80	Ambrit Ltd	Elstree	9. 5.02
G-BIDJ	Piper PA-18A-150 Super Cub	18-6007	PH-MAY	22. 9.80	Flight Solutions Ltd	Panshanger	14. 7.06T
	(Frame No.18-6089)		N7798D				
G-BIDK	Piper PA-18-150 Super Cub	"18-6591"	PH-MAI	22. 9.80	J.and M.A McCullough	Newtownards	5. 6.05
	(L-21A-PI)		R.Neth AF R-211, 51-15679, N7194K				
	(Composite of PH-MAI originally Frame No.18-6714 [c/n 18-6591] ex LN-TVB/N9285D rebuilt 1976 with Frame No.18-503 [c/n 18-565] ex R.Neth AF R-211)						
G-BIDO	Piel CP.301A Emeraude	327	F-POIO	25. 3.81	A R.Plumb	Hill Farm, Nayland	19. 8.04P
G-BIDV*	Colt 17A Cloudhopper HAB				See SECTION 3 Part 1	Newbury	
G-BIDW*	Sopwith "1½" Strutter rep				See SECTION 3 Part 1	Hendon	
G-BIDX	Dormois Jodel D.112	876	F-BIQY	19. 9.80	P.Turton and H.C.Peake-Jones	Ashcroft Farm, Winsford	5. 9.05P
G-BIEF	Cameron V-77 HAB	679		25. 9.80	D.S.Bush *"Daedalus"*	(Hertingfordbury)	6. 3.94A
G-BIEJ	Sikorsky S-76A II Plus	760097		21.10.80	Bristow Helicopters Ltd *"Glen Lossie"*	Norwich	21. 2.07T
G-BIEN	Wassmer Jodel D.120A Paris-Nice	218	F-BKNK	3. 6.81	C.A J.Van Andel	Booker	11.12.04P
G-BIEO	Wassmer Jodel D.112	1296	F-BMOK	19. 3.82	S.C.Solley tr Clipgate Flyers	Clipgate Farm, Denton	6. 8.05P
G-BIES	Maule M-5-235C Lunar Rocket	7334C	N56394	24. 7.81	W.Procter	Stowe Farm, Tillingham	27. 5.06
					t/a William Procter Farms		
G-BIET	Cameron O-77 HAB	674		30. 9.80	G.M.Westley *"Archimedes"*	(London SW15)	11. 1.02A
G-BIEY	Piper PA-28-151 Cherokee Warrior	28-7715213	PH-KDH	10.11.80	A S.Bamrah	Southend	15. 2.07T
			OO-HCB, N9540N		*(Op.Willowair Flying Club)*		
G-BIFA	Cessna 310R II	310R1606	N36868	29. 1.81	J.S.Lee	Booker	18.12.06
G-BIFB	Piper PA-28-150 Cherokee C	28-1968	4X-AEC	6.10.80	P.Coombs	(Doncaster)	11. 4.05T
G-BIFN	Bensen B.8MR	PFA G/01-1010		7.10.80	B.Gunn	(North Ferriby)	
	(Built K Willows - c/n KW.1)						
G-BIFO	Evans VP-1	PFA 62-10411		29. 9.80	R.Broadhead	Octon Lodge Farm, Thwing	17. 6.05P
	(Built P Raggett) (Volkswagen 1834)						
G-BIFY	Reims/Cessna F150L	F15000829	PH-CEZ	9.10.80	Bonus Aviation Ltd	Cranfield	26. 9.05T
G-BIGJ	Reims/Cessna F172M	F17200936	PH-SKT	2.12.80	V.D.Speck	Clacton	5. 8.02T
G-BIGK	Taylorcraft BC-12D	8302	N96002	29.10.80	N.P.St.J.Ramsay	Barton Ashes	5. 8.05P
	(Continental A65)		NC96002				
G-BIGL	Cameron O-65 HAB	690		22.10.80	P.L.Mossman *"Biggles"*	(Bristol)	12.10.02A
G-BIGP	Bensen B.8M	PFA G/01-1005		14.10.80	R.H.S.Cooper	Cross Houses, Shrewsbury	20.10.97P
	(Built R.H.S.Cooper) (McCulloch Motors O-100-1)				*(Current status unknown)*		
G-BIGR	Avenger T.200-2122 MLB	004		6.10.80	R.Light	(Stockport)	
	(Built R Light)						
G-BIGT*	Colt 77A HAB				See SECTION 3 Part 1	Newbury	
G-BIGU*	Bensen B.8MR	JRM.1 & PFA G/01-1032		5.11.80	I.B.Pitt-Steele	Felthorpe	18. 9.03P
	(Built C G Ponsford) (Rotax 532)				*(Cancelled 19.4.04 as wfu)*		
G-BIGZ	Scheibe SF25B Falke	46142	D-KCAI	22.12.80	P.J.Mcquie	RAF Henlow	22. 7.05
	(Stark-Stamo MS1500)				tr The Big-Z Owners Group		
G-BIHD	Robin DR400/160 Major 80	1510		29.10.80	A J.Fieldman	King's Farm, Thurrock	24. 2.07
G-BIHF	Replica Plans SE.5A	PFA 20-10548		27.10.80	S.H.O'Connell	White Waltham	5. 1.01P
	(Built K J Garrett) (Plans no.079275) (Continental O-200-A)				*"Lady Di" (As "F-943" in 92 Sqdn RFC c/s) (Current status unknown)*		
G-BIHI	Cessna 172M Skyhawk II	17266854	(G-BIHA)	18.11.80	L.R.Haunch	Fenland	15. 5.05T
			N1125U		t/a Fenland Flying School		
G-BIHN*	Airship Industries Skyship 500	1214/02		2.11.80	Airship Heritage Trust	(Huntingon)	1. 4.79
			(Destroyed San Francisco, CA, US 1985: cancelled 31.1.91 by CAA) (Gondola only 2.03)				

Reg	Type	C/n	Previous identity	Regd	Owner/Operator	Base	CofA
G-BIHO	de Havilland DHC-6-310 Twin Otter	738	A6-ADB	9. 1.81	Isles of Scilly Skybus Ltd	St.Just	18. 4.06T
			G-BIHO				
G-BIHP	Van Den Bemden 1000m3 Gas Balloon	VDB-38	OO-VBA	19.12.80	J.J.Harris	London SW6	3. 5.01
	(C/n quoted as "18" on Belgian CofR: believed rebuilt with 600m3 canopy c/n VDB-47) "Belgica"						
G-BIHT	Piper PA-17 Vagabond	17-41	N138N	9. 1.81	Agri Air Services Ltd		
	(Continental A65)		N8N, N4626H, NC4626H			(Newbold on Stour,Stratford-upon-Avon)	4. 7.05P
G-BIHU	Saffery S.200 MLB	25		5.11.80	B.L.King	(Coulsdon)	
	(Built Cupro Sapphire Ltd)						
G-BIHX	Bensen B.8MR	PFA G/01-1003		12.11.80	P.P.Willmott	(Grimsby)	24. 5.05P
	(Built P.P.Willmott) (Rotax 503)						
G-BIIA	Fournier RF3	51	F-BMTA	14.11.80	J.D.Webb and J.D.Bally	Rhosgoch	26. 9.05P
G-BIIB	Reims/Cessna F172M Skyhawk II	F17201110	PH-GRE	18.11.80	Civil Service Flying Club (Biggin Hill) Ltd	Rochester	24. 4.06T
G-BIID	Piper PA-18 Super Cub 95	18-1606	OO-LPA	5. 1.81	D.A Lacey	Cumbernauld	9. 7.05P
	(L-18C-PI) *(Frame No.18-1558)*		OO-HMK, ALAT 18-1606, 51-15606				
G-BIIE	Reims/Cessna F172P Skyhawk II	F17202051		31.12.80	Sterling Helicopters Ltd	Norwich	12. 3.05T
G-BIIG	Thunder Ax6-56Z HAB	307		26.11.80	A Spindler	(Cleish)	19. 8.01A
G-BIIK	SOCATA MS.883 Rallye 115	1552	F-BSAP	28.11.80	K.M.Bowen	Fenland	5. 8.05
G-BIIL	Thunder Ax6-56 Bolt HAB	306		12.11.80	I.M.Ashpole	Ross-on-Wye	3. 9.04A
G-BIIP	Pilatus Britten-Norman BN-2B-26 Islander	2103	6Y-JQJ	1.12.80	Airx Ltd	Alderney	26. 3.07T
			6Y-JKJ, N411JA, G-BIIP		*(Op Rockhoppers)*		
G-BIIT	Piper PA-28-161 Warrior II	28-8116052	N82744	1.12.80	Tayside Aviation Ltd	Glenrothes	1. 4.05T
G-BIIV	Piper PA-28-181 Archer II	28-7990028	N20875	19.12.80	J.Thuret	(Montpellier, France)	21. 8.06T
G-BIIZ	Great Lakes 2T-1A Sport Trainer	57	N603K	1. 4.81	Circa 42 Ltd	(Colchester)	4. 2.99P
	(Warner Super Scarab 165D-5)		NC603K		*(Damaged Upper Harford, Glos 8.8.98: current status unknown)*		
G-BIJB	Piper PA-18-150 Super Cub	18-8009001	N23923	18. 8.80	James Aero Ltd	Stapleford	7. 4.07T
			N2573H				
G-BIJD	Bölkow Bö.208C Junior	636	PH-KAE	9.12.80	P.Singh tr Sikh Syndicate	Leicester	1.10.04
			(PH-DYM), OO-SIS, (D-EGFA)				
G-BIJE	Piper J-3C-65 Cub (L-4A-PI)	8367	F-BIGN	5. 5.81	R.L.Hayward and A G.Scott	Cardiff	
	(Frame No.8504)		Fr.AF, 42-15248		*(On rebuild 4.91: current status unknown)*		
G-BIJS	Phoenix Luton LA-4A Minor	PFA 835		18. 5.78	I.J.Smith	Brook Farm, Boylestone	14.11.95P
	(Built I.J.Smith - c/n PAL/1348) (Volkswagen 1600)						
G-BIJU	Menavia Piel CP.301A Emeraude	221	G-BHTX	10. 6.80	J.R.Large	Stapleford	18. 7.05P
			F-BIJU		tr Eastern Taildraggers Flying Club		
G-BIJV	Reims/Cessna F152 II	F15201813		22.12.80	A.S.Bamrah t/a Falcon Flying Services	Shoreham	21. 3.05T
G-BIJW	Reims/Cessna F152 II	F15201820		22.12.80	A.S.Bamrah t/a Falcon Flying Services	Rochester	26. 2.05T
G-BIKC	Boeing 757-236	22174		31. 1.83	DHL Air Ltd	Brussels, Belgium	9. 2.06T
G-BIKE	Piper PA-28R-200 Cherokee Arrow II	28R-7335173	OY-DVT	18. 4.80	R.V.Webb	Elstree	25.11.07E
			N55047				
G-BIKF (2)	Boeing 757-236	22177	(G-BIKG)	28. 4.83	DHL Air Ltd	Brussels, Belgium	14. 9.06T
G-BIKG (2)	Boeing 757-236	22178	(G-BIKH)	26. 8.83	DHL Air Ltd	Brussels, Belgium	26. 8.06T
G-BIKI (2)	Boeing 757-236	22180	OO-DLO	30.11.83	DHL Air Ltd	Brussels, Belgium	23. 6.05T
			G-BIKI (2)				
G-BIKJ (2)	Boeing 757-236	22181	(G-BIKK)	9. 1.84	DHL Air Ltd	Brussels, Belgium	11. 1.07T
G-BIKK (2)	Boeing 757-236	22182	(G-BIKL)	1. 2.84	DHL Air Ltd	Brussels, Belgium	1. 2.07T
G-BIKM (2)	Boeing 757-236	22184	N8293V	21. 3.84	DHL Air Ltd	Brussels, Belgium	22. 3.07T
			(G-BIKN)				
G-BIKN (2)	Boeing 757-236	22186	(G-BIKP)	23. 1.85	DHL Air Ltd	Brussels, Belgium	22. 9.05T
G-BIKO (2)	Boeing 757-236	22187	(G-BIKR)	14. 2.85	DHL Air Ltd	Brussels, Belgium	18. 2.05T
G-BIKP (2)	Boeing 757-236	22188	(G-BIKS)	11. 3.85	DHL Air Ltd	Brussels, Belgium	14. 3.05T
G-BIKS (2)	Boeing 757-236	22190	(G-BIKU)	31. 5.85	DHL Air Ltd	Brussels, Belgium	2. 6.05T
G-BIKU (2)	Boeing 757-236	23399		7.11.85	DHL Air Ltd	Brussels, Belgium	7.11.05T
G-BIKV	Boeing 757-236	23400		9.12.85	DHL Air Ltd	Brussels, Belgium	11.12.05T
G-BIKZ	Boeing 757-236	23532		15. 5.86	DHL Air Ltd	Brussels, Belgium	30. 9.07T
G-BILB	Wallingford WMB-2 Windtracker MLB	14		22. 1.81	B.L.King	(Coulsdon)	
	(Built Wallingford Model Balloons)						
G-BILE	Scruggs BL-2B MLB	81231		13. 3.81	P.D.Ridout	(Botley)	
	(Built D Morris)						
G-BILG	Scruggs BL-2B MLB	81232		13. 3.81	P.D.Ridout	(Botley)	
	(Built D Morris)						
G-BILI	Piper J-3C-65 Cub (L-4J-PI)	13207	F-BDTB	14. 1.81	S.C.Wilson and J A Goodridge	White Waltham	17. 6.05P
	(Frame No.13044)		45-4467		tr G-BILI Flying Group *(As "454467/44-J" in US Army c/s)*		
G-BILJ	Reims/Cessna FA152 Aerobat	FA1520376		31.12.80	I.R.March	Booker	4. 7.05T
G-BILL	Piper PA-25-235 Pawnee D	25-7856028	N9174T	3. 1.79	A E and W.J.Taylor	East Winch	28. 6.06A
	(Lycoming O-540-G2A5 @ 260hp)				t/a Pawnee Aviation		
G-BILR	Cessna 152 II	15284822	N4822P	19. 3.81	Shropshire Aero Club Ltd	Sleap	8. 5.05T
G-BILS	Cessna 152 II	15284857	N4954P	3. 6.81	Mona Aviation Ltd	RAF Mona	5. 8.07T
					t/a Mona Flying Club		
G-BILU	Cessna 172 RG Cutlass II	172RG0564	N5540V	29. 1.81	R.M.English and Sons Ltd	Full Sutton	13. 7.06T
G-BILZ	Taylor JT.1 Monoplane	PFA 55-10244		15.12.80	A Petherbridge	Sibsey	29. 2.91P
	(Built G Beaumont) (Officially regd as c/n PFA 55-10124)				*(Damaged Ingoldmells 10.6.90: stored 8.00: current status unknown)*		
G-BIMK	Tiger T.200 Series 1 MLB	7/MKB-01		22.12.80	M.K.Baron	(Stockport)	
	(Built M.K.Baron)				*(Current CoR 12.03)*		
G-BIMM	Piper PA-18-150 Super Cub	18-3868	PH-VHO	8. 1.81	Spectrum Leisure Ltd	Clacton	6. 8.04
	(L-21B-PI)*(Frame No.18-3881)*		R.Neth AF R-178, 54-2468				
G-BIMN	Steen Skybolt	PFA 64-10329		31.12.80	R.J.Thomas	(Ashford)	1. 9.05P
	(Built C R Williamson)						
G-BIMO	SNCAN Stampe SV-4C	394	F-BADG	5. 3.81	R.A Robert	Dunsfold	11.11.06
	(DH Gipsy Major 10)		Fr.Mil		*(As "394" in French AF c/s)*		
G-BIMT	Reims/Cessna FA152 Aerobat	FA1520361	N8062L	9. 1.81	Gloucestershire Flying Services Ltd	Gloucestershire	26. 5.05T
G-BIMU	Sikorsky S-61N Mk II	61-752	N8511Z	9. 1.81	Bristow Helicopters Ltd	Stornoway	23.10.05T
	(SAR conversion)		VH-CRU, N4042S		*(Op Marine and Coastguard Agency) "Stac Pollaidh"*		
G-BIMX	Rutan VariEze	PFA 74-10544		6. 1.81	D.G.Crew	Biggin Hill	8. 6.05P
	(Built A S Knowles) (Continental O-200-A)						

G-BIMZ	Beech 76 Duchess	ME-169	N6021K	20. 3.81	R.P.Smith	Gloucestershire	28. 6.07
G-BING	Reims/Cessna F172P Skyhawk II	F17202084		12. 1.81	20th Air Training Group Ltd	(Dunsany, County Meath)	16. 4.05T
G-BINL	Scruggs BL-2B MLB	81216		5. 2.81	P.D.Ridout	(Botley)	
	(Built D Morris)						
G-BINM	Scruggs BL-2B MLB	81217		5. 2.81	P.D.Ridout	(Botley)	
	(Built D Morris)						
G-BINR	Unicorn UE-1A MLB	81004		20. 1.81	I.Chadwick	(Horsham)	
	(Built Unicorn Group)				tr Unicorn Group *"Lady Diana"*		
G-BINS	Unicorn UE-2A MLB	80002		22.12.80	I.Chadwick	(Horsham)	
	(Built Unicorn Group)				tr Unicorn Group *"Caroline"*		
G-BINT	Unicorn UE-1A MLB	80001		22.12.80	D.E.Bint	(Downham Market)	
	(Built Unicorn Group)						
G-BINX	Scruggs BL-2B MLB	81219		5. 2.81	P.D.Ridout	(Botley)	
	(Built D Morris)						
G-BINY	Morton Oriental Air-Bag MLB	OAB-001		22. 1.81	J.L.Morton	(Wokingham)	
	(Built J.L.Morton)						
G-BIOA	Hughes 369D	120-0880D	OO-HFS	9. 2.81	AH Helicopter Services Ltd	Newton Abbot	14. 7.06T
			LX-HLE, OO-HFS, G-BIOA				
G-BIOB	Reims/Cessna F172P Skyhawk II	F17202042		23. 1.81	Simmons Aerofilms Ltd	Elstree	16. 4.05T
G-BIOC	Reims/Cessna F150L	F15000848	F-BUEC	3. 2.81	D.J.Gage and G.Burns tr Southside Flyers	Prestwick	27. 8.05T
G-BIOI	SAN Jodel DR.1050M Excellence	477	F-BLJQ	21. 1.81	R.Pidcock	Fenland	21. 4.05P
G-BIOJ	Rockwell Commander 112TC-A	13192	N4662W	22. 1.82	A T.Dalby	Sywell	13.12.02
G-BIOK	Reims/Cessna F152 II	F15201810		2. 2.81	Tayside Aviation Ltd	Glenrothes	18. 4.05T
G-BIOU	SAN Jodel D.117A	813	F-BIOU	9. 8.78	M.R.Routh and T.de Salis	(Maidenhead)	23. 5.04P
					tr Jemalk Group *(New owners 7.04)*		
G-BIOW	Slingsby T.67A	1988		26. 2.81	A B.Slinger tr Slingsby T67A Group	Sherburn-in-Elmet	25. 8.06
G-BIPA	Grumman AA-5B Tiger	AA5B-0200	OY-GAM	24. 3.81	J.Campbell	Walney Island	14. 7.05
G-BIPH	Scruggs BL-2B MLB	81224		10. 2.81	C.M.Dewsnap	(Camberley)	
	(Built D Morris)						
G-BIPI	Everett Gyroplane	001		30. 4.81	C A Reeves	Apperley, Glos	19. 6.01P
	(Volkswagen 1834)				*(Noted 5.03)*		
G-BIPN	Fournier RF3	35	F-BMDN	26. 2.81	J.C.R.Rogers and I.F.Fairhead	RAF Cranwell	21. 1.04P
					tr G-BIPN Group		
G-BIPO	Mudry CAP.20 LS 200	03	F-GAUB	5. 3.81	A.McClean tr The CAP 20 Group	White Waltham	8. 5.05S
G-BIPT	Wassmer Jodel D.112	1254	F-BMIB	11. 3.81	C.R.Davies	Allensmore, Hereford	14 5.04P
G-BIPV	Gulfstream AA-5B Tiger	AA5B-0981	N28266	10. 3.81	Databridge Services Ltd	Blackbushe	20. 5.05T
G-BIPW	Avenger T.200-2112 MLB	10		24. 2.81	B.L.King	(Coulsdon)	
	(Built R Light)						
G-BIPY	Montgomerie-Bensen B.8MR	PFA G/01-1007		25. 2.81	C.M.Jones	Carlisle	20.12.05P
	(Built A J Wood - c/n AJW.01) (Rotax 532)						
G-BIRB*	SOCATA MS.880B Rallye 100T	2460	F-BVAQ	30. 3.81	Not known Crazy Golf course, Shanklin, Isle of Wight		16. 6.90
					(Cancelled 13.7.92 by CAA) (Noted 9.02)		
G-BIRD	Pitts S-1D	PFA 1596		3.11.77	P.Metcalfe	(Stockton-on-Tees)	19. 5.05P
	(Built R N York - c/n 707-H)						
G-BIRE	Colt Bottle 56SS HAB	323		4. 3.81	K.R.Gafney	(Bracknell)	10. 1.84A
	(Satzenbrau Bottle)				*"Satzenbrau"*		
G-BIRH	Piper PA-18-150 Super Cub (L-21B-PI)	18-3853	PH-LET	19. 3.81	Aquila Gliding Club Ltd	Bicester	19. 6.05
	(Lycoming O-360-A4) *(Frame No.18-3857)*		R Neth AF R-163, 54-2453		*(As 54-2453/R163" in R.Neth AF c/s)*		
G-BIRI	CASA 1-131E Jungmann	1074	Spanish AF E3B-113	14. 4.81	D.Watt	Sibson	9. 9.05P
G-BIRL	Avenger T.200-2112 MLB	008		10. 3.81	R.Light	(Stockport)	
	(Built R Light)						
G-BIRP	Ridout Arena Mk.17 Skyship MLB	01		13. 3.81	Annette S.Viel	(Botley)	
	(Built P D Ridout)						
G-BIRT	Robin R1180TD Aiglon	276		25. 3.81	W.D'A Hall	White Waltham	19.11.05
G-BIRW*	Morane Saulnier MS.505 Criquet				See SECTION 3 Part 1	East Fortune	
G-BIRY*	Cameron V-77 HAB	715		12. 3.81	(P.G.Bish)	Newtown Hungerford	15. 5.99A
					(Cancelled 10.12.02 by CAA) (Extant 1.05)		
G-BIRZ	Zenair CH.250-100 Zenith	PFA 24-10459		10. 3.81	D.Johnston and M.K.McGreavey	Perth	5. 8.04P
	(Built S M Kowalski - kit no.2-454) (Lycoming O-290-G)						
G-BISG	Clutton FRED Series III	PFA 29-10675		13. 3.81	T.Littlefair	Lymington	29.10.86P
	(Built R A Coombe - kit no.RAC 01-224) (Volkswagen 1600)				*"Fuzz Bee" (New owner 10.00)*		
G-BISH	Cameron V-65 HAB	707		16. 3.81	P.J.Bish *"Tsaritsa"*	Newtown Hungerford	14. 7.05A
G-BISL	Scruggs BL-2B MLB	81233		13. 3.81	P.D.Ridout	(Botley)	
	(Built D Morris)						
G-BISM	Scruggs BL-2B MLB	81234		13. 3.81	P.D.Ridout	(Botley)	
	(Built R Light)						
G-BISS	Scruggs BL-2C MLB	81235		13. 3.81	P.D.Ridout	(Botley)	
	(Built R Light)						
G-BIST	Scruggs BL-2C MLB	81236		13. 3.81	P.D.Ridout	(Botley)	
	(Built R Light)						
G-BISX	Colt 56A HAB	324		18. 3.81	C.D.Steel	(St.Boswells)	18. 8.99A
G-BISZ	Sikorsky S-76A II Plus	760156		19. 3.81	Bristow Helicopters Ltd	Redhill	23.10.07E
G-BITA	Piper PA-18-150 Super Cub	18-8109037	N82585	24. 3.81	D.J.Gilmour t/a Intrepid Aviation Co	North Weald	23. 7.05
G-BITE	SOCATA TB-10 Tobago	193		7. 5.81	M.A Smith	Eshott	19.12.05
G-BITF	Reims/Cessna F152 II	F15201822		27. 3.81	S.Webb tr G-BITF Owners Group	(Newburgh, Cupar)	8. 7.06T
G-BITK	Clutton FRED Series II	PFA 29-10369		23. 3.81	D.J.Wood	(Dover)	
	(Built B J Miles) (Volkswagen 1500)						
G-BITM	Reims/Cessna F172P Skyhawk II	F17202046		13. 4.81	Dreamtrade Ltd	Barton	6.11.05T
G-BITO	Wassmer Jodel D.112D	1200	F-BIUO	20. 3.81	A Dunbar *(Noted 7.04)*	Barton	5. 9.02P
G-BITS	Drayton B-56 HAB	MJB-01/81		16. 3.81	M.J.Betts	(Drayton, Norwich)	
					(Op Eastern Region, British Balloon and Airship Club) "Hedger"		
G-BITY	Bell FD.31T Flying Dodo MLB	2604		25. 3.81	A J.Bell	(Luton)	
	(Built A J.Bell)						

G-BIUM	Reims/Cessna F152 II	F15201807		3. 4.81	Sheffield Aero Club Ltd	Netherthorpe	2. 3.06T
G-BIUP	SNCAN NC.854S	54	(G-AMPE)	4. 6.81	J.Greenaway and T.D.Cooper	Popham	10. 7.05P
			G-BIUP, F-BFSC		tr BIUP Flying Group		
G-BIUV	Hawker Siddeley HS.748 Series 2A/275LFD	1701	5W-FAN	11. 5.81	Emerald Airways Ltd	Liverpool	16. 6.05T
			G-AYYH, G-11-8		*"City of Liverpool"*		
G-BIUW	Piper PA-28-161 Warrior II	28-8116128	N9506N	14. 4.81	D.R.Staley	Sheffield City	27. 6.05
G-BIUY	Piper PA-28-181 Archer II	28-8190133	N8318X	3. 4.81	J.S.Develin and Z.Islam	Shoreham	3. 1.05T
G-BIVA	Robin R2112	137	F-GBAZ	6. 5.81	P.A Richardson	Conington	4. 11.05
G-BIVB	Wassmer Jodel D.112	1009	(G-BIVC)	18. 9.81	B.L.Proctor and T.F.Lyddon	Dunkeswell	4. 7.05P
	(Continental A65)		F-BJII				
G-BIVC	Wassmer Jodel D.112	1219	F-BMAI	1. 6.81	M.J.Barnby	Brickhouse Farm, Frogland Cross	13. 7.00P
	(Continental A65)						
G-BIVF	Scintex CP.301C3 Emeraude	594	F-BJVN	4.11.81	R.J.Moore	Sywell	9.12.05P
G-BIVK	Bensen B.8V	PFA G/01-1008		10. 4.81	M.J.Atyeo	(Bognor Regis)	19.11.03P
	(Built J G Toy) (Volkswagen 1834) *(Registered as B.8M)*				*"Skyrider"*		
G-BIVV	Gulfstream AA-5A Cheetah	AA5A-0857	N26979	26. 5.81	R.Afia	North Weald	18. 7.05T
					t/a Robert Afia Consulting Engineer		
					(Hulk dumped 1.05 - fire services trainer as "G-PRAT")		
G-BIWA	Ridout Stevendon Skyreacher MLB	102		8. 6.81	S.D.Barnes	(Botley)	
	(Built P D Ridout)						
G-BIWB	Scruggs RS.5000 MLB	81541		8. 6.81	P.D.Ridout	(Botley)	
	(Built D Morris)						
G-BIWC	Scruggs RS.5000 MLB	81546		26. 6.81	P.D.Ridout	(Botley)	
	(Built D Morris)				*"Waterloo"*		
G-BIWF	Ridout Warren Windcatcher MLB	WW.013		3. 7.81	P.D.Ridout	(Botley)	
	(Built P D Ridout)						
G-BIWG	Ridout Zelenski Mk.2 MLB	Z.401		3. 7.81	P.D.Ridout	(Botley)	
	(Built P D Ridout) (Regd with c/n 2401)						
G-BIWJ	Unicorn UE-1A MLB	81014		14. 7.81	B.L.King	(Coulsdon)	
	(Built Unicorn Group)						
G-BIWK	Cameron V-65 HAB	719		22. 4.81	I.R.Williams and R.G.Bickerdike	(Bedford/Huntingdon)	16. 2.05A
					"Double Fantasy"		
G-BIWL	Piper PA-32-301 Saratoga	32-8106056	N83684	23. 4.81	J D Richardson	Earls Colne	1. 7.05
G-BIWN	Wassmer Jodel D.112	1314	F-BNCN	5. 6.81	C.R.Coates	Sneaton Thorpe, Whitby	8.10.05P
G-BIWR	Mooney M.20F Executive	22-1339	N6972V	1. 6.81	A C.Brink *(Noted WFU 11.03)*	Bourn	19.10.03
G-BIWU	Cameron V-65 HAB	717		15. 5.81	D.J.Groombridge *"Bumble Bee"*	(Bristol)	11. 1.03A
G-BIWW	American AA-5 Traveler	AA5-0263	OY-AYV	2. 6.81	Sandra J.Perkins and D.Dobson	Little Staughton	3. 9.05
G-BIXA	SOCATA TB-9 Tampico	205		7. 5.81	W.Maxwell	Perth	30. 9.05
G-BIXB	SOCATA TB-9 Tampico	208		7. 5.81	Cinque Ports Aviation Ltd	Lydd	3. 4.06T
G-BIXH	Reims/Cessna F152 II	F15201840		30. 4.81	Northern Aviation Ltd	Durham Tees Valley	13. 4.06T
G-BIXI	Cessna 172 RG Cutlass II	172RG0861	N7533B	7. 7.81	L.Gavaghan	Guernsey	19. 6.05
G-BIXL	North American P-51D-20NA Mustang	122-38675	Israel DFAF2343	3. 7.81	R.J.Lamplough *"Miss Helen"*	Manor Farm, East Garston	6. 7.05P
			Fv.26116, 44-72216		*(As "472216/HO-M" in 487th Fighter Sqdn/352nd Fighter Group c/s)*		
G-BIXN	Boeing Stearman A75N1 (PT-17-BW) Kaydet		N51132	15. 6.81	V.S.E.Norman	Rendcomb	3. 8.96
	(Continental W670)	75-2248	41-8689		*(As "FJ777" in RCAF c/s)*		
G-BIXV	Bell 212	30870	N16931	27. 5.81	Bristow Helicopters Ltd	(Kazakhstan)	22. 7.05T
G-BIXW	Colt 56B HAB	348		18. 5.81	N.A P.Bates	(Tunbridge Wells)	17. 8.97A
					"Spam"		
G-BIXX	Pearson Series II MLB	00327		8. 5.81	D.Pearson	(Solihull)	
	(Built D Pearson)						
G-BIXZ	Grob G109	6019	D-KGRO	14. 5.81	D.L.Nind and I.Allum	Booker/Enstone	15. 6.07
G-BIYI	Cameron V-65 HAB	722		21. 5.81	P.F.Smart	(Basingstoke)	16. 4.97A
					tr The Sarnia Balloon Group *"Penny"*		
G-BIYJ	Piper PA-18 Super Cub 95	18-1000	MM51-15303	5. 6.81	S.Russel	Pilmuir Farm, Lundin Links	26. 9.05P
	(L-18C-PI)		I-EIST, MM51-15303, 51-15303				
G-BIYK	Isaacs Fury II	PFA 11-10418		20. 5.81	S.M.Roberts	(Guernsey)	4. 7.05P
	(Built R S Martins) (Continental C90)						
G-BIYP	Piper PA-20 Pacer 125	20-802	CN-TYP	25. 5.83	A W.Hoy and S.W.M.Johnson	(Farnham)	3. 6.05
			F-DACJ, OO-ADP		tr G-BIYP Flying Group		
G-BIYR	Piper PA-18-150 Super Cub	18-3841	(G-BIYB)	26. 5.81	B.H.and M.J.Fairclough	Watchford Farm, Yarcombe	1. 6.07
	(L-21B-PI) *(Frame No. 18-3843)*		PH-GER, R.Neth AF R-151, 5G-96, 54-2441		tr The Delta Foxtrot Flying Group *(As "R-151" in R.Neth AF c/s)*		
G-BIYT	Colt 17A Cloudhopper HAB	344		13. 7.81	J-M Francois	(Salles-Courbatiers, France)	18. 1.05A
G-BIYU	Fokker S.11.1 Instructor	6206	(PH-HOM)	13. 5.81	C.Briggs	Bagby	24.10.01P
			R.Neth AF E-15		*(As "E-15" in R.Neth AF c/s)*		
G-BIYW	Wassmer Jodel D.112	1209	F-BLNR	26. 5.81	K.Balaam	Poplar Hall Farm, Elmsett	9. 6.05P
					tr Pollard/Balaam/Bye Flying Group		
G-BIYX	Piper PA-28-140 Cherokee Cruiser	28-7625064	OY-BLD	19. 6.81	W.B.Bateson	Blackpool	4. 4.05T
G-BIYY	Piper PA-18 Super Cub 95	18-1979	MM52-2379	2. 6.81	A E.and W.J.Taylor	Fenland	6. 3.05T
	(L-18C-PI) *(Frame No.18-1914)*		I-EIGA, MM52-2379, 52-2379				
G-BIZE	SOCATA TB-9 Tampico	209	9H-ABJ	15. 6.81	C.Fordham	Bourn	3. 7.05
			G-BIZE				
G-BIZF	Reims/Cessna F172P Skyhawk II	F17202070		16. 6.81	R.S.Bentley	Bourn	29. 7.07
G-BIZG	Reims/Cessna F152 II	F15201873		16. 6.81	M.A Judge tr Aero Group 78	Netherthorpe	29. 6.06T
G-BIZI	Robin DR400 2+2	1543		29. 5.81	BIZI Club Ltd	RAF Halton	18. 4.06T
G-BIZK	Nord 3202B1	78	N2255E	22.11.85	A I.Milne	Little Snoring	2.12.05P
			ALAT		*(All-yellow French AF c/s)*		
G-BIZM	Nord 3202B	91	N2256K	22.11.85	Global Aviation Ltd	Humberside	31.10.00P
			ALAT		*(New owner 6.02)*		
G-BIZO	Piper PA-28R-200 Cherokee Arrow II	28R-7535339	OY-DLH	16. 6.81	P.J.Mason and J.F.Leather	(Bristol)	23. 2.06E
			N1578X		tr Lemas Air		
G-BIZR	SOCATA TB-9 Tampico	210	G-BSEC	15. 6.81	E.S.Murphy and G.Ward	Fenland	29. 7.07
			G-BIZR		tr Fenland Flying Group		
G-BIZU	Thunder Ax6-56Z HAB	358		15. 6.81	M.J.Loades *"Greenall Whitley"*	(Southampton)	6. 7.03A

Registration	Type	C/n	Previous identity	Regd date	Owner (Operator)	Probable base	CA expiry
G-BIZV	Piper PA-18 Super Cub 95	18-2001	EI-74	12. 6.81	A.W.and Mary E.Jenkins	(Comberton, Cambridge)	27. 6.05P
	(L-18C-PI)		I-EIDE, MM52-2401, 52-2401		*(As "18-2001" in US Army c/s)*		
G-BIZW	Champion 7GCBC Citabria	0157	D-EGPD	16. 7.81	J.C.Read t/a G.Read and Sons	North Reston	30. 8.04
G-BIZY	Wassmer Jodel D.112	1120	F-BKJL	13. 7.81	W.Tunley	Hinton-in-the-Hedges	20. 5.05P
					t/a Wayland Tunley and Associates		

G-BJAA - G-BJZZ

Registration	Type	C/n	Previous identity	Regd date	Owner (Operator)	Probable base	CA expiry
G-BJAD*	Clutton FRED Series II				See SECTION 3 Part 1	Newark	
G-BJAE	Lavadoux Starck AS.80 Holiday	04	F-PGGA	17. 6.81	D.J. and Mrs S.A E.Phillips	(Leamington Spa)	8. 8.92P
	(Continental A65)		F-WGGA		*(Damaged Woburn 17.8.91: current status unknown)*		
G-BJAF	Piper J-3C-65 Cub (L-4A-PI)	8437	D-EJAF	23. 6.81	P.J.Cottle	Craysmarsh Farm, Melksham	25.11.05P
	(Frame No.8540)		HB-OAD, 42-15318				
G-BJAG	Piper PA-28-181 Archer II	28-7990353	PH-LDB	23. 6.81	C.R.Chubb	Manston	18. 9.05T
			(PH-BEG), (OO-FLM), N2244W				
G-BJAJ	Gulfstream AA-5B Tiger	AA5B-1177	N4532V	2. 7.81	A.H.McVicar	Carlisle	22.10.06T
G-BJAL	CASA 1-131E Jungmann	1028	Spanish AF E3B-114	11. 9.78	I.C.Underwood and S.B.J.Chandler	Breighton	18. 7.05P
	(Spanish AF serial no. conflicts with G-BUCC)				tr G-BJAL Group		
G-BJAO	Montgomerie-Bensen B.8MR	PFA G/01-1001		28. 8.81	A P.Lay	Henstridge	2. 4.01P
	(Built A Gault - c/n GLS-01) (Rotax 582) *(Regd as c/n GL5-01)*						
G-BJAP	de Havilland DH.82A Tiger Moth	0482 & PFA 157-12897		15. 6.81	K.Knight	Shobdon	1. 8.05P
	(Built J A Pothecary -composite rebuild)				*(As "K2587" in pre-war 32 Sqdn/CFS c/s)*		
G-BJAS	Rango NA-9 MLB	TL-19		22. 6.81	A Lindsay	(Twickenham)	
G-BJAV	Sud-Aviation Gardan GY-80-160 Horizon	28	OO-AJP	8. 9.81	P.L.Lovegrove	Bournemouth	26.10.06
			F-BLVB				
G-BJAW	Cameron V-65 HAB	745		19. 6.81	G.A McCarthy *"Breezin"*	(Shepton Mallet)	16. 4.86A
G-BJAY	Piper J-3C-65 Cub (L-4H-PI)	12086	F-BFBN	1.11.78	D.W.Finlay	(Lamourache Nord, Aquitaine, France)	7. 6.05P
	(Frame No.11914)		OO-EAC, 44-79790				
G-BJBK	Piper PA-18 Super Cub 95	18-1431	F-BOME	21. 8.81	M.S.Bird	Pepperbox, Salisbury	30. 5.05P
	(L-18C-PI) (Continental O-200-A) *(Frame No.18-1370)*		ALAT, 51-15431				
G-BJBM	Monnett Sonerai I	PFA 15-10022		2. 7.81	T.F.Harrison	(Wolverhampton)	9. 1.97P
	(Built Lyster Aviation Ltd - c/n MEA-117) (Volkswagen 2074)				tr Sonerai G-BJBM Group *(New owner 1.02)*		
G-BJBO	CEA Jodel DR.250/160 Capitaine	40	F-BNJG	24. 8.81	R.C.Thornton tr Wiltshire Flying Group	Oaksey Park	19. 1.07
G-BJBW	Piper PA-28-161 Warrior II	28-8116280	N2913Z	22. 7.81	T.G.Phillips and C.Greenland	Popham	22. 1.06
					tr 152 Group		
G-BJBX	Piper PA-28-161 Warrior II	28-8116269	N8414H	17. 7.81	Haimoss Ltd *(Op Old Sarum Flying Club)*	Old Sarum	2. 4.06T
G-BJCA	Piper PA-28-161 Warrior II	28-7916473	N2846D	30. 7.81	Plane Sailing (South West) Ltd	Plymouth	17. 2.06T
G-BJCF	Scintex CP.1310-C3 Super Emeraude	936	F-BMJH	19.11.81	K.M.Hodson and C.G.H.Gurney	Manor Farm, Binham	7. 1.05P
G-BJCI	Piper PA-18-150 Super Cub	18-6658	N9388D	10. 9.81	The Borders (Milfield) Gliding Club Ltd	Milfield	8. 4.06
	(Lycoming O-360-A4)						
G-BJCL*	Morane Saulnier MS.230 Parasol				See SECTION 3 Part 1	Polk City, Florida, US	
G-BJCW	Piper PA-32R-301 Saratoga SP	32R-8113094	N2866U	6. 8.81	Golf Charlie Whisky Ltd	Fairoaks	27. 5.05
G-BJDE	Reims/Cessna F172M	F17200984	OO-MSS	25. 8.81	S.P.Heathfield	Cranfield	25. 6.06T
			D-EGBR		tr Cranfield Aircraft Partnership		
G-BJDF	SOCATA MS.880B Rallye 100T	3000	F-GAKP	21. 9.81	A J.Wilkinson	Coldharbour Farm,Willingham	11. 3.06
					tr G-BJDF Group		
G-BJDK	Ridout European E.157 MLB	S.2		17. 8.81	E.Osborn	(Southampton)	
	(Built P D Ridout)				t/a Aeroprint Tours		
G-BJDO	Gulfstream AA-5A Cheetah	AA5A-0823	N26936	3. 8.81	J.J.Woodhouse t/a Flying Services	Sandown	15. 5.06T
G-BJDW	Reims/Cessna F172M Skyhawk II	F17201417	PH-JBE	10. 8.81	J.Rae	Earls Colne	16. 1.06T
G-BJEC	Pilatus Britten-Norman BN-2T Turbine Islander	2118	UAE AF 318	28. 7.81	Fly BN Ltd	Bembridge	24. 7.83
	(Originally regd as a BN-2B)		UAE AF 411, G-BJEC				
G-BJEE	Pilatus Britten-Norman BN-2T Turbine Islander	2120	C9-TAH	28. 7.81	Cormack (Aircraft Services) Ltd	Cumbernauld	
	(Originally regd as a BN-2B)		G-BJEE		*(New CofR 8.04)*		
G-BJEF	Pilatus Britten-Norman BN-2T Turbine Islander	2121	C9-TAK	28. 7.81	Cormack (Aircraft Services) Ltd	Cumbernauld	
	(Originally regd as a BN-2B)		G-BJEF		*(New CofR 8.04)*		
G-BJEI	Piper PA-18 Super Cub 95	18-1988	EI-66	27. 7.81	H.J.Cox	Wendover Farm, Sheepwash	3. 8.05P
	(L-18C-PI) *(Frame No.18-1938)*		I-EILO, MM52-2388, 52-2388				
G-BJEJ	Pilatus Britten-Norman BN-2T Turbine Islander	2124	C9-TAJ	28. 7.81	Cormack (Aircraft Services) Ltd	Cumbernauld	
	(Originally regd as a BN-2B)		G-BJEJ		*(New CofR 8.04)*		
G-BJEL	SNCAN NC.854S	113	F-BEZT	7. 8.81	C.A.James *(New owner 1.05)*	Wickwar	5. 9.02P
G-BJEV	Aeronca 11AC Chief	11AC-270	N85897	12. 8.81	R.F.Willcox	Eastbach Farm, Coleford	8. 1.05P
			NC85897		*(As "E/897" in US Navy c/s)*		
G-BJEX	Bölkow Bö.208C Junior	690	F-BRHY	27. 8.81	G.D.H.Crawford	(Henley-on-Thames)	28. 1.88
			D-EEAM				
G-BJFC	Ridout European E.8 MLB	S.1		17. 8.81	P.D.Ridout	(Botley)	
	(Built P D Ridout)						
G-BJFE	Piper PA-18 Super Cub 95	18-2022	EI-91	17. 8.81	P.H.Wilmot-Allistone	Kemble	1. 6.05P
	(L-18C-PI)		I-EISU, MM52-2422, 52-2422				
G-BJFL	Sikorsky S-76A II Plus	760056	N106BH	28. 8.81	Bristow Helicopters Ltd	Khazakstan	17. 9.05T
			N1546T, (G-BHRK)		*"Glen Moray"*		
G-BJFM	Wassmer Jodel D.120 Paris-Nice	227	F-BLFM	8.10.81	J.V.George and P.A Smith	Popham	5. 5.04P
G-BJGK	Cameron V-77 HAB	696		3. 9.81	M E.Orchard	(Bristol)	8. 1.06A
G-BJGM	Unicorn UE-1A MLB	81015		21. 8.81	D.Eaves and P.D.Ridout	(Southampton)	
	(Built Unicorn Group)				*"Capricorn"*		
G-BJGX	Sikorsky S-76A II Plus	760026	N103BH	4. 9.81	Bristow Helicopters Ltd	Norwich	8.10.06T
			N4251S		*"Glen Elgin"*		
G-BJGY	Reims/Cessna F172P Skyhawk II	F17202128		13.10.81	K.and S.Martin	Gunton Hall, Somerton	13. 4.06
G-BJHB	Mooney M.20J Model 201	24-1190	N1145G	23.12.81	Zitair Flying Club Ltd	Booker	23. 5.05T
G-BJHK	EAA Acrosport	PFA 72-10470		20. 3.80	M.R.Holden	Stoneacre Farm, Farthing Corner	31. 8.04P
	(Built J H Kimber) (Lycoming IO-360)						
G-BJHS*	Short S.25 Sandringham				See SECTION 3 Part 1	Polk City, Florida, US	
G-BJHV*	Voisin Scale rep				See SECTION 3 Part 1	Brooklands	

G-BJIA	Allport Aerostatics YUO-1A-1-DA MLB *(Built D J Allport)*	01		2. 9.81	D.J.Allport	(Bourne, Lincoln)	
G-BJIC	Eaves Dodo 1A MLB *(Built D Eaves)*	DD.3		4. 9.81	P.D.Ridout	(Botley)	
G-BJID	Chown Osprey Lizzieliner 1B MLB *(Built A P Chown)*	AKC.28		4. 9.81	P.D.Ridout	(Botley)	
G-BJIG	Slingsby T.67A	1992		16. 9.81	D.Lacy tr G-BJIG Slingsby Syndicate	White Waltham	15. 4.04
G-BJIR	Cessna 550 Citation II	550-0296	N6888C	17. 9.81	Gator Aviation Ltd*(Op Aviation Beauport Ltd)*	Jersey	22. 1.06T
G-BJIV	Piper PA-18-150 Super Cub (Lycoming O-360-A4)	18-8262	N5972Z	17. 9.81	Yorkshire Gliding Club (Pty) Ltd	Sutton Bank	24. 3.06
G-BJKF	SOCATA TB-9 Tampico	240		30. 9.81	P M A Croton	(Charvil, Reading)	27. 4.07
G-BJKW	Wills Aera 2 *(Built J.K.S.Wills)*	A3JKW		1. 3.78	J.K.S.Wills *(Current status unknown)*	(London SE3)	
G-BJKY	Reims/Cessna F152 II	F15201886		22. 9.81	Manx Aero Marine Management Ltd *(Op Westair Flying Services)*	Blackpool	14.10.04T
G-BJLB	SNCAN NC.854S	58	(OO-MVM) F-BFSG	5.11.81	N.F.Hunter *(Crashed nr Newport,Gwent 29.7.84: stored 8.90: current status unknown)*	Wolvesnewton, Chepstow	30. 6.83P
G-BJLC	Monnett Sonerai IIL *(Built J P Whitham - c/n 942L)* (Volkswagen 1835)	PFA 15-10634		18. 9.81	A R.Ansell	AAC Netheravon *"Elsie" (Noted 6.01)*	11. 5.98P
G-BJLF	Unicorn UE-1C MLB *(Built Unicorn Group)*	81018		21. 9.81	I.Chadwick tr Unicorn Group	(Horsham)	
G-BJLG	Unicorn UE-1B MLB *(Built Unicorn Group)*	81017		21. 9.81	I.Chadwick tr Unicorn Group	(Horsham)	
G-BJLX	Cremer Cracker MLB *(Built P A Cremer)*	15.711 PAC		24. 9.81	P.W.May	(Wilmslow)	
G-BJLY	Cremer Cracker MLB *(Built P A Cremer)*	15.709 PAC		24. 9.81	P.Cannon	(Luton)	
G-BJML	Cessna 120 (Continental C90)	10766	N76349 NC76349	5.10.81	D.F.Lawlor	Inverness	7. 5.04P
G-BJMO	Taylor JT.1 Monoplane *(Built R C Mark)*	PFA 55-10612		30. 9.81	R.C.Mark	(Ludlow)	
G-BJMR	Cessna 310R II	310R1624	N2631Z	16. 7.79	J.M.Robinson	Rufforth	14. 4.07
G-BJMW	Thunder Ax8-105 Series 2 HAB	369		14.10.81	G.M.Westley	(London SW15)	11. 1.02A
G-BJMX	Ridout Jarre JR-3 MLB *(Built P D Ridout)*	81601		6.10.81	P.D.Ridout	(Botley)	
G-BJMZ	Ridout European EA-8A MLB *(Built P D Ridout)*	S.5		6.10.81	P.D.Ridout	(Botley)	
G-BJNA	Ridout Arena Mk.117P MLB *(Built P D Ridout)*	202		6.10.81	P.D.Ridout	(Botley)	
G-BJND	Chown Osprey Mk.1E MLB *(Built A P Chown)*	AKC.53		7.10.81	A Billington and D.Whitmore	(Liverpool)	
G-BJNF	Reims/Cessna F152 II	F15201882		21.10.81	D.M.and B.Cloke	Dunkeswell	16. 2.06T
G-BJNG	Slingsby T.67AM	1993		16.10.81	D.F.Hodgkinson *(Noted 6.04)*	Dunkeswell	23. 7.01T
G-BJNH	Chown Osprey Mk.1E MLB *(Built A P Chown)*	AKC.57		8.10.81	D.A Kirk	(Manchester)	
G-BJNN	Piper PA-38-112 Tomahawk	38-80A0064	N9684N	15.10.81	S.Padidar-Nazar	Carlisle	31. 1.05T
G-BJNY	Aeronca 11CC Super Chief	11CC-264	CN-TYZ F-OAEE	28.10.81	P.I.and D.M.Morgans *(Stored 4.91)*	Furze Hill Farm, Rosemarket, Milford Haven	9. 8.90P
G-BJNZ	Piper PA-23-250 Aztec F	27-7954099	G-FANZ N6905A, C-GTJG	5.10.81	Bonus Aviation Ltd	Cranfield	3. 7.05T
G-BJOA	Piper PA-28-181 Archer II	28-8290048	N8453H	29.10.81	Tatenhill Aviation Ltd	Tatenhill	21.11.07E
G-BJOB	SAN Jodel D.140C Mousquetaire III	118	F-BMBD	2.11.81	T.W.M.Beck and M.J.Smith	Monks Gate, Horsham	20. 6.06
G-BJOE	Wassmer Jodel D.120A Paris-Nice	177	F-BJIU	12.11.81	J.F.Govan tr Forth Flying Group	East Fortune	10. 2.05P
G-BJOP	Pilatus Britten-Norman BN-2B-26 Islander	2132		29.10.81	Loganair Ltd	Kirkwall	5. 9.05T
G-BJOT	SAN Jodel D.117	688	F-BJCO CN-TVH, F-DABU	12.11.81	R.H.Ryle tr R.H.Ryle and Partners	High Flatts Farm, Chester-le-Street	13.10.05P
G-BJOV	Reims/Cessna F150K	F15000558	PH-VSD	4. 2.82	J.A Boyd	(Maidstone)	12. 6.06
G-BJPI	Bede BD-5G *(Built M.D.McQueen - c/n 1)* (Hirth 230R)	PFA 14-10218		30.10.81	M.D.McQueen	(Beckenham)	
G-BJPL	Chown Osprey Mk.4A MLB *(Built A P Chown)*	AKC-39		13.10.81	M.Vincent	(Jersey)	
G-BJRA	Chown Osprey Mk.4B MLB *(Built A P Chown)*	AKC.87		23.10.81	E.Osborn	(Southampton)	
G-BJRG	Chown Osprey Mk.4B MLB *(Built A P Chown)*	AKC.95		26.10.81	A de Gruchy	(Jersey)	
G-BJRH	Rango NA-36, Ax3 MLB	NHP-23		4.11.81	N.H.Ponsford t/a Rango Balloon and Kite Co	(Leeds)	
G-BJRP	Cremer Cracker MLB *(Built P A Cremer)*	15.712 PAC		29.10.81	M.D.Williams	(Dunstable)	
G-BJRR	Cremer Cracker MLB *(Built P A Cremer)*	15.715 PAC		29.10.81	M.D.Williams	(Houghton Regis)	
G-BJRV	Cremer Cracker MLB *(Built P A Cremer)*	15.713 PAC		29.10.81	M.D.Williams	(Dunstable)	
G-BJSS	Allport YUO-1B-1-DA Neolithic Invader Superballoon Series 2,0 MLB *(Built D.J.Allport)*	01-8101002		9.11.81	D.J.Allport	(Bourne, Lincoln)	
G-BJST	CCF Harvard 4 (T-6J-CCF Texan)	CCF4-...	MM53-795 SC-66	21.12.81	Classic Aero Engineering Ltd	Thruxton	29. 7.05P
G-BJSV	Piper PA-28-161 Warrior II	28-8016229	PH-VZL (OO-HLM), N35787	25.11.81	Airways Flight Training (Exeter) Ltd	Exeter	30. 9.06T
G-BJSW	Thunder Ax7-65Z HAB	378		16.11.81	Sandicliffe Garage Ltd *"Sandicliffe Ford"*	(Stapleford, Nottingham)	16. 7.04A
G-BJSZ	Piper J-3C-65 Cub (L-4H-PI) *(Regd with c/n 11874)*	12047	D-EHID (D-ECAX), (D-EKAB), PH-NBP, 44-79751	20.11.81	H.Gilbert	Enstone	22. 3.05P

G-BJTB	Cessna A150M Aerobat	A1500627	(G-BIVN) N9818J	28.10.82	V.D.Speck	Clacton	6.11.05T
G-BJTF	Kirk Skyrider Mk.1 MLB *(Built D.A Kirk)*	KSR-01		18.11.81	D.A Kirk	(Manchester)	
G-BJTN	Solent Osprey Mk.4B MLB *(Built Solent Balloon Group)*	ASC-112		23.11.81	M.Vincent	(Jersey)	
G-BJTO	Piper J-3C-65 Cub (L-4H-PI) *(Frame No.11352)*	11527	F-BEGK OO-AAL, 43-30236	1.12.81	K.R.Nunn	Fritton Decoy, Great Yarmouth	4. 7.05P
G-BJTP	Piper PA-18 Super Cub 95 (L-18C-PI)	18-999	EI-51 I-EICO, MM51-15302, 51-15302 *(As "115302/TP" in VMO-6 Sqdn, US Marines c/s) "Sittin' Duck"*	26.11.81	J.T.Parkins	Bidford	20.10.05P
G-BJTY	Solent Osprey Mk.4B MLB *(Built Solent Balloon Group)*	ASC-115		23.11.81	A E.de Gruchy	(Jersey)	
G-BJUB	Wild BVS Special 01 MLB *(Built P.G.Wild)*	VS/PW01		25.11.81	P.G.Wild	(Linley Hill,Leven)	
G-BJUC	Robinson R22HP	0228		13. 1.82	JD Gallagher Estate Agents Ltd	Blackpool	13.11.06T
G-BJUD	Robin DR400/180R Remorqueur *(Rebuilt using new fuselage: original scrapped Membury 11.88)*	870	PH-SRM	27.11.81	Lasham Gliding Society Ltd	Lasham	5. 1.06
G-BJUE	Solent Osprey Mk.4B MLB *(Built Solent Balloon Group)*	ASC-114		23.11.81	M.Vincent	(Jersey)	
G-BJUR	Piper PA-38-112 Tomahawk	38-79A0915	N9722N	5. 2.82	Truman Aviation Ltd *(Op Nottingham School of Flying)*	Tollerton	23.11.03T
G-BJUS	Piper PA-38-112 Tomahawk	38-80A0065	N9690N	10.12.81	Panshanger School of Flying Ltd	High Cross, Ware	12.10.06T
G-BJUU	Solent Osprey Mk.4B MLB *(Built Solent Balloon Group)*	ASC-113		23.11.81	M.Vincent	(Jersey)	
G-BJUV	Cameron V-20 HAB	792		9.12.81	P.Spellward *"Busy Bee"*	(Bristol)	
G-BJVC	Evans VP-2 *(Built R G Fenn)* (Volkswagen 1911)	PFA 63-10599		17. 2.82	C.J.Morris	(Andover)	19. 6.91P
G-BJVH	Reims/Cessna F182Q Skylane II	F18200106	D-EJMO PH-AXU (2)	21.12.81	R.J.D.Cuming	Bolt Head, Salcombe	21. 9.06
G-BJVJ	Reims/Cessna F152 II	F15201906		6. 1.82	Wilkins and Wilkins (Special Auctions) Ltd t/a Henlow Flying Club	RAFHenlow	29. 7.07T
G-BJVK	Grob G109	6074		11. 3.82	B.A Kimberley *(Current status unknown)*	(Banbury)	22. 5.92
G-BJVM	Cessna 172N Skyhawk II	17269374	N737FA	14.12.81	I.C.Maclennan	Gunton Hall, Somerton	1. 8.06T
G-BJVS	Scintex CP.1310-C3 Super Emeraude	903	F-BJVS	5. 1.79	A P.Milton tr BJVS Group	Watchford Farm, Yarcombe	8.10.05P
G-BJVT	Reims/Cessna F152 II	F15201904		12. 1.82	Northern Aviation Ltd	Durham Tees Valley	8. 1.07T
G-BJVU	Thunder Ax6-56 Bolt HAB	397		31.12.81	G.V.Beckwith *"Cooper"*	(York)	26. 4.91A
G-BJVV	Robin R1180TD Aiglon II	79		5.11.81	Medway Flying Group Ltd	Rochester	18. 6.06T
G-BJWH	Reims/Cessna F152 II	F15201919		7. 5.82	Linkcrest Ltd	Elstree	11. 9.06T
G-BJWI	Reims/Cessna F172P Skyhawk II	F17202172		14. 5.82	Bflying Ltd *(Op Bournemouth Flying Club)*	Bournemouth	28. 8.07T
G-BJWJ	Cameron V-65 HAB	802		25. 1.82	R.G.Turnbull and S.G.Farse *"Gawain"*	(Glasbury, Hereford)	24.11.00A
G-BJWO	Fairey Britten-Norman BN-2A-26 Islander	334	4X-AYR SX-BBX, 4X-AYR, G-BAXC	16. 2.82	Stravintower Ltd t/a Falcons Parachute Centre	Hacketstown, County Carlow	5. 5.06A
G-BJWT	Wittman W.10 Tailwind *(Built J.F.Bakewell)* (Lycoming O-290-G)	PFA 31-10688		5. 1.82	J.F.Bakewell tr Tailwind Group	Hucknall	26. 7.05P
G-BJWV	Colt 17A Cloudhopper HAB	391		22. 1.82	D.T.Meyes *(Bryant Homes titles)*	(Leamington Spa)	26. 3.97A
G-BJWW	Reims/Cessna F172P Skyhawk II	F17202148	(D-EFTV)	1. 2.82	Manx Aero Marine Management Ltd *(Op Westair Flying Services)*	Blackpool	15.10.06T
G-BJWX	Piper PA-18 Super Cub 95 (L-18C-PI) (Continental O-200-A)	18-1985	EI-64 I-EIME, MM52-2385, 52-2385	23. 2.82	R A G Lucas tr G-BJWX Syndicate	Sleap	17. 2.05P
G-BJWY*	Sikorsky S-55 (HRS-2) Whirlwind HAR.21				See SECTION 3 Part 1	Carlisle	
G-BJWZ	Piper PA-18 Super Cub 95 (L-18C-PI) *(Frame No.18-1262)*	18-1361	OO-HMO ALAT 18-1361, 51-15361	18. 1.82	R.C.Dean tr G-BJWZ Syndicate	Redhill	30. 3.05P
G-BJXA	Slingsby T.67A	1994		8. 2.82	Liberty Group Assets Ltd	Bagby	20. 1.07T
G-BJXB	Slingsby T.67A	1995		8. 2.82	XRay Bravo Ltd	Barton	21.10.07E
G-BJXK	Fournier RF5	5054	D-KINB	3. 2.82	R.Thompson and S.Jenkins tr G-BJXK Syndicate	Usk	16. 6.07
G-BJXP	Colt 56B HAB	393		29. 3.82	H.J.Anderson *"Bart"*	(Oswestry)	9. 9.00A
G-BJXR	Auster AOP.Mk.9	184	XR267	2. 2.82	I.Churm and J.Hanson *(New CofR 11.03)*	(Belper)	
G-BJXX	Piper PA-23-250 Aztec E	27-4692	F-BTCM N14094	7. 4.82	V.Bojovic *(Noted 3.04)*	Lisicji Jarak, Serbia	23. 6.01
G-BJXZ	Cessna 172N Skyhawk II	17273039	PH-CAA N1949F	24. 3.82	T.M.Jones *(Op Derby Aero Club)*	Egginton	8.11.04T
G-BJYD	Reims/Cessna F152 II	F15201915		25. 3.82	N.J.James	Welshpool	27.10.06T
G-BJYF	Colt 56A HAB	401		1. 3.82	H.Dos Santos *"Fanta"*	(Caxton)	5. 9.04A
G-BJYG	Piper PA-28-161 Warrior II	28-8216053	N8458B	4. 3.82	P.Lodge *(Destroyed when ditched in Liverpool Bay 2nm N of Wallasey 4.7.04)*	Liverpool	20. 6.06T
G-BJYK	Wassmer Jodel D.120A Paris-Nice	185	(G-BJWK) F-BJPK	11. 5.82	T.Fox and D.A Thorpe	Crowland	23. 6.05P
G-BJYN	Piper PA-38-112 Tomahawk	38-79A1076	G-BJTE N24310, N9671N	12. 3.82	Panshanger School of Flying Ltd	High Cross, Ware	6. 3.00T
G-BJZA	Cameron N-65 HAB	820		4. 3.82	A D.Pinner *"Digby"*	(Northampton)	3. 6.97A
G-BJZB	Evans VP-2 *(Built A Graham)* (Volkswagen 1834) *(Scrapped 2002/3 - fuselage noted 4.03: wings and engine to be fitted to a new VP-2 on site and believed to be G-CCEI)*	PFA 63-10633		10. 3.82	I.P.Manley and J.Pearce	Marsh Farm, Sidlesham	15. 1.03P
G-BJZC*	Thunder Ax7-65Z HAB				See SECTION 3 Part 1	(Aldershot)	
G-BJZF	de Havilland DH.82A Tiger Moth *(Built Norfolk Aerial Spraying Ltd from spares and officially redesignated "NAS Tiger Moth")*	NAS-100		8. 3.82	M I Lodge	(Huddersfield)	6. 7.04P
G-BJZN	Slingsby T.67A	1997		31. 3.82	A R.T.Marsland	Breighton	5. 9.07
G-BJZR	Colt 42A HAB	402		18. 3.82	A F.Selby tr Selfish Balloon Group *"Selfish"*	(Loughborough)	24. 7.02A

G-BKAA - G-BKZZ

Registration	Type	C/n	Previous identity	Date	Owner/operator	Probable base	CofA expiry
G-BKAE	Wassmer Jodel D.120 Paris-Nice	200	F-BKCE	5. 5.82	S.J.Harris	RAF Mona	2.12.04P
G-BKAF	Clutton FRED Series II	PFA 29-10337		23. 3.82	J.M.Robinson	(Achill Island, County Mayo)	30. 5.97P
	(Built L G Millen) (Volkswagen 1835)				*(Current status unknown)*		
G-BKAM	Slingsby T.67M-160 Firefly	1999		26. 4.82	R.C.P.Brookhouse	Maypole Farm, Chislet	19.12.05
G-BKAO	Wassmer Jodel D.112	249	F-BFTO	22. 3.82	R.Broadhead	Octon Lodge Farm, Thwing	4. 5.05P
G-BKAS	Piper PA-38-112 Tomahawk	38-79A1075	N24291	16. 4.82	E.Reed	Durham Tees Valley	22. 7.05T
			N9670N		t/a St.George Flight Training		
G-BKAY	Rockwell Commander 114	14411	SE-GSN	28. 9.81	D.L.Bunning tr The Rockwell Group	Dunkeswell	16. 5.04
G-BKAZ	Cessna 152 II	15282832	N89705	27. 4.82	L.W.Scattergood	Sherburn-in-Elmet	24. 5.07A
G-BKBB	Hawker Fury rep	WA/6	OO-HFU	2. 4.82	Brandish Holdings Ltd	Wevelgem, Belgium	3. 6.04P
	(Built Westward Airways) (RR Kestrel 5)		OO-XFU, G-BKBB		*(As "K1930" in 43 Sqdn c/s)*		
G-BKBD	Thunder Ax3 Maxi Sky Chariot HAB	418		5. 4.82	M.J.Casson	(Kendal)	
G-BKBF	SOCATA MS.894A Rallye Minerva 220	11622	F-BSKZ	8. 9.82	K A Hale and L C Clark	Draycott Farm, Chisledon	12. 9.07
G-BKBH*	Hawker Siddeley HS.125 Series 600B	256052	5N-DNL	1. 4.82	Beamalong Ltd	Southampton	
			G-5-698, 5N-DNL, 5N-NBC, G-5-698, G-BKBH, G-5-698, TR-LAU, G-BKBH, G-BDJE, G-5-11				
					(Cancelled 15.7.99 by CAA) (Noted fire dump 4.04)		
G-BKBN	SOCATA TB-10 Tobago	287		4. 6.82	Dateworld Ltd	Bournemouth	25. 8.07
G-BKBO	Colt 17A Cloudhopper HAB	342		1. 9.82	J.Armstrong, M.A Ashworth and H.Davey	(Newquay)	19. 2.04A
					"Captain Courageous"		
G-BKBP	Bellanca 7GCBC Scout	465-73	N8693	1. 6.82	M.G. and J.R.Jefferies	Little Gransden	1.12.06T
					t/a H.G.Jefferies and Son		
					(Damaged Graveley, Herts 23.5.93: stored 9.95: new owner 10.00)		
G-BKBR (2)*	Cameron Chateau 84SS HAB				See SECTION 3 Part 1	Balleroy, Normandy, France	
G-BKBS	Bensen B.8MV	PFA G/01-1027		14. 4.82	C R Gordon	Drummiard Farm, Bonnybank	24. 5.05P
	(Built G Dawe) (Rotax 532)						
G-BKBV	SOCATA TB-10 Tobago	288	F-BNGO	4. 6.82	The Studio People Ltd	Sleap	17. 2.06
G-BKBW	SOCATA TB-10 Tobago	289		4. 6.82	P.J.Bramhall and D.F.Woodhouse	Bristol	26. 5.07
					tr Merlin Aviation		
G-BKCC	Piper PA-28-180 Cherokee Archer	28-7405099	OY-BGY	13. 5.82	DR Flying Club Ltd	Gloucestershire	20. 9.07T
G-BKCE	Reims/Cessna F172P Skyhawk II	F17202135	N9687R	26. 4.82	M.O.Loxton	Parsonage Farm, Eastchurch	4. 6.06T
G-BKCI	Brugger MB.2 Colibri	PFA 43-10692		22. 4.82	M.R.Walters	(Derby)	
	(Built E.R.Newall) (Volkswagen 1600)				*(New owner 7.04)*		
G-BKCJ	Oldfield Baby Lakes	PFA 10-10714		12. 5.82	S.V.Roberts	Sleap	26. 1.99P
	(Built S V Roberts) (Continental O-200-A)						
G-BKCL	Piper PA-30 Twin Comanche C	30-1982	G-AXSP	12. 1.81	R.P.Hodson	Full Sutton	22.11.07E
			N8824Y				
G-BKCN	Phoenix Currie Wot	PFA 3018		27. 4.82	N.A A Pogmore	Kirknewton	10. 9.04P
	(Built S E O Tomlinson) (Continental A65)						
G-BKCR	SOCATA TB-9 Tampico	297		6. 5.82	P.A Little *(Noted 8.04)*	Haverfordwest	3. 8.98T
G-BKCV	EAA Acrosport II	PFA 72A-10776		5. 5.82	M.J.Clark	Dunsfold	15. 9.05P
	(Built M.J.Clark) - c/n 430) (Lycoming 0-360)						
G-BKCW	Wassmer Jodel D.120A Paris-Nice	285	(G-BKCP)	1. 6.82	I.C.Waddell	Perth	19.10.05P
			F-BMYF		tr Dundee Flying Group		
G-BKCX	Avions Mudry CAP.10B	149		28. 7.82	G.P.Gorvett	Swansea	19. 8.07
G-BKCZ	Wassmer Jodel D.120A Paris-Nice	207	F-BKCZ	23. 4.82	J.V.George *(On rebuild off site 11.04)*	(Popham)	AC
G-BKDC	Monnett Sonerai IIL	PFA 15-10597		2. 7.82	K.J.Towell	(Guildford)	18. 6.90P
	(Built J Boobyer - c/n 876) (Volkswagen 1834)				*(Damaged Breighton 7.8.90: current status unknown)*		
G-BKDH	Robin DR400/120 Dauphin 80	1582	PH-CAB	25. 5.82	Dauphin Flying Group Ltd	Draycott Farm, Chisledon	25.11.07E
G-BKDI	Robin DR400/120 Dauphin 80	1583	PH-CAD	25. 5.82	Mistral Aviation Ltd	Goodwood	26. 5.07T
G-BKDJ	Robin DR400/120 Dauphin 80	1584	PH-CAC	25. 5.82	A.Shorter tr Air Yorkshire Flying Group	Sherburn-in-Elmet	26. 5.07
G-BKDK	Thunder Ax7-77Z HAB	428		21. 6.82	A J.Byrne *"Cider Riser"*	(Thatcham)	17. 9.95A
G-BKDP	Clutton FRED Series III	PFA 29-10650		24. 5.82	M.Whittaker	(Wolverhampton)	
	(Built M.Whittaker)						
G-BKDR	Pitts S-1S	PFA 09-10654		14. 6.82	Skyview Systems Ltd	Waits Farm, Belchamp Walter	22. 5.05P
	(Built Maypole Engineering Ltd)						
G-BKDT*	Royal Aircraft Factory SE.5A rep				See SECTION 3 Part 1	Elvington	
G-BKDX	SAN Jodel DR.1050 Ambassadeur	55	F-BITX	1. 6.82	G.J.Slater	Lower Upham Farm, Chiseldon	10. 9.99
					(Noted 6.03)		
G-BKEK	Piper PA-32-300 Cherokee Six	32-7540091	OY-TOP	30. 6.82	P.H.Maynard	Turweston	17. 6.07T
G-BKEP	Reims/Cessna F172M Skyhawk II	F17201095	OY-BFJ	8. 7.82	J.M.Thorpe	Treddunnock Farm, Llangarron	8.10.04
	(Thielert TAE125-01)						
G-BKER	Replica Plans SE.5A	PFA 20-10641		15. 6.82	N.K.Geddes	South Barnbeth Farm, Bridge of Weir	1. 9.04P
	(Built N.K.Geddes) (Continental O-200A)				*(As "F5447/N")*		
G-BKES*	Cameron Bottle 57 SS HAB				See SECTION 3 Part 1	Newbury	
G-BKET	Piper PA-18 Super Cub 95	18-1990	EI-67	17. 6.82	H.M.MacKenzie	Inverness	28. 5.05P
	(L-18C-PI)		I-EIBI, MM52-2390, 52-2390				
G-BKEU	Taylor JT.1 Monoplane	PFA 55-10553		18. 6.82	A.J.Moore	(Turweston, Brackley)	20. 7.95P
	(Built R.J.Whybrow and J.M.Springham)				*lNew owner 1.05)*		
G-BKEV	Reims/Cessna F172M Skyhawk II	F17201443	PH-WLH	8. 7.82	A G.Measey	(Countesthorpe)	17. 3.07T
			OO-CNE		tr Echo Victor Group		
G-BKEW	Bell 206B-3 JetRanger III	3010	D-HDAD	8. 7.82	N.R.Foster t/a Foster Associates	Biggin Hill	17. 7.06T
G-BKEY	Clutton FRED Series III	PFA 29-10208		27. 5.82	G.S.Taylor	(Bewdley)	
	(Built G.S.Taylor) (Volkswagen 1600)						
G-BKFA	Monnett Sonerai IIL	PFA 15-10524		21. 6.82	S.J.N.Robbie	(Eastleigh)	
	(Built R.F.Bridge)				*(New CofR 2.04)*		
G-BKFC	Reims/Cessna F152 II	F15201443	OO-AWB	1. 9.82	C Walton Ltd t/a Sulby Aerial	Sibbertoft	11.11.04T
G-BKFD*	Westland WG.30 Series 100				See SECTION 3 Part 1	Wroughton	
G-BKFF*	Westland WG.30 Series 100				See SECTION 3 Part 1	Wroughton	
G-BKFI	Evans VP-1 Series 2	PFA 62-10491		24. 6.82	P.L.Naylor	Morgansfield, Fishburn	3. 6.05P
	(Built R F A Lavergne) (Volkswagen 1834)						

Reg	Type	C/n	Previous identity	Date	Owner/operator	Base	CofA
G-BKFK	Isaacs Fury II	PFA 11-10038		25. 6.82	G.C.Jones	Waits Farm, Belchamp Walter	27. 8.02P
	(Built G.C.Jones) (Lycoming O-290-D)				*"Cia Cia San" (Persian AF c/s)*		
G-BKFL	Aerosport Scamp	PFA 117-10814		17. 8.82	J.Sherwood	(Barnsley)	
	(Built I D Daniels)				*(Current status unknown)*		
G-BKFM	QAC Quickie 1	PFA 94-10570		28. 6.82	G.E.Meakin	(Ruddington)	29. 6.98P
	(Built R I Davidson and P Cheney) (Rotax 503)				*(Damaged on take off Cranfield 4.7.98: new owner 7.01)*		
G-BKFN	Bell 214ST Super Transport	28109	LZ-CAW	16. 8.82	Bristow Helicopters Ltd	Aberdeen	24.10.07T
			G-BKFN, VH-BEE, VH-LHT, G-BKFN		*"Loch Broome"*		
G-BKFR	Scintex CP.301C Emeraude	519	F-BUUR	30. 6.82	D.G.Burgess	Trenchard Farm, Eggesford	25. 4.05P
			F-BJFF		tr Devonshire Flying Group		
G-BKFW	Percival P.56 Provost T.1	PAC/F/303	XF597	21. 9.82	Sylmar Aviation and Services Ltd	Bossington	19. 7.05P
					(As "XF597/AH" in RAF College c/s)		
G-BKFY	Beech C90-1 King Air	LJ-1028	N213CT	11. 8.82	Blackbrook Aviation LLP	Bournemouth	10.11.07E
			VP-CCT, VR-CCT, N6420H, G-BKFY				
G-BKFZ	Piper PA-28R-200 Cherokee Arrow II	28R-7635127	OY-BLE	17. 8.82	R.S.Watt	Shacklewell Farm, Empingham	9.12.06
					tr Shacklewell Flying Group		
G-BKGA	SOCATA MS.892E Rallye 150GT	13287	F-GBXJ	15. 7.82	P.F.Salter	Wadswick Manor Farm, Corsham	25. 9.06
					tr BJJ Aviation		
G-BKGB	Wassmer Jodel D.120 Paris-Nice	267	F-BMOB	21. 6.82	B.A Ridgway	Rhigos	20. 3.05P
G-BKGC	Maule M-6-235C Super Rocket	7413C	N56465	23. 7.82	B.F.Walker	Gloucestershire	21. 8.06
G-BKGD*	Westland WG.30 Series 100				See SECTION 3 Part 1	Wroughton	
G-BKGL	Beech D18S (3TM)	CA-164	CF-QPD	14. 7.82	A.T.J.Darrah	Duxford	1.11.04
		(Beech c/n A-764)	RCAF 5193, 1564		*(As "1164" in 1942 USAAC c/s)*		
G-BKGM	Beech 3NM (D18S)	CA-203	N5063N	14. 7.82	Skyblue Aviation Ltd	Exeter	10. 7.06
		(Beech c/n A-853)	G-BKGM, CF-SUQ, RCAF 2324		*(As "HB275" in RAF SEAC c/s)*		
G-BKGR	Cameron O-65 HAB	864		6. 8.82	K.Kidner and L.E.More	(Newton Abbot)	8. 5.93P
G-BKGT	SOCATA Rallye 110ST Galopin	3361		23. 7.82	A.G.Morgan	Wellesbourne Mountford	21. 1.07
					tr Long Marston Flying Group		
G-BKGW	Reims/Cessna F152 II	F15201878	N9071N	11. 8.82	Leicestershire Aero Club Ltd	Leicester	15. 5.07T
G-BKHD	Oldfield Baby Lakes	PFA 10-107182		5. 8.82	P.J.Tanulak	Sleap	11. 4.96P
	(Built P.J.Tanulak - c/n 8133-F-802B) (Continental O-200-A)				*(Damaged Shrewsbury 22.10.95: current status unknown)*		
G-BKHG	Piper J-3C-65 Cub (L-4H-PI)	12062	F-BCPT	13. 9.82	K.G.Wakefield	Brickhouse Farm, Frogland Cross	3. 6.05P
			NC79807, 44-79766		*(As "479766/63-D" in HQ 9th Army, USAAC c/s) "Puddle Jumper"*		
G-BKHJ	Cessna 182P Skylane II	18264129	PH-CAT	25. 8.82	Augur Films Ltd	Shipdham	15. 7.05
	(Reims c/n F18200040)		D-EATV, N6223F				
G-BKHR	Luton LA-4A Minor	PFA 51-10228		24. 8.82	C.B.Buscombe and R.Goldsworthy	Bodmin	6. 1.04P
	(Built R J Parkhouse) (Volkswagen 1834)						
G-BKHW	Stoddard-Hamilton Glasair RG	PFA 149-11312		27.8.82	D.Callabritto	Stapleford	22. 3.05P
	(Built N Clayton) - c/n 357) (Lycoming O-320)						
G-BKHY	Taylor JT.1 Monoplane	PFA 1416		8. 9.82	B.C.J.O'Neill	Damyn's Hall, Upminster	11. 6.05P
	(Built J Hall)						
G-BKIB	SOCATA TB-9 Tampico	323		25. 8.82	G.A Vickers	Hawarden	3. 1.05T
G-BKIF	Fournier RF6-100	3	F-GADR	8.10.82	D.J.Taylor and J.T.Flint	Kimbolton	17. 2.07
G-BKII	Reims/Cessna F172M Skyhawk II	F17201370	PH-PLO	8.10.82	Sealand Ap Ltd	Goodwood	4. 3.07T
			(D-EGIA)				
G-BKIJ	Reims/Cessna F172M	F17200920	PH-TGZ	15.10.82	V.D.Speck	Clacton	13. 8.05T
G-BKIK*	Cameron DG-19 Helium Airship				See SECTION 3 Part 1	(Aldershot)	
G-BKIR	SAN Jodel D.117	737	F-BIOC	30. 9.82	R.Shaw and D.M.Hardaker	Birds Edge, Penistone	28. 8.92P
					(On rebuild 3.96: current status unknown)		
G-BKIS	SOCATA TB-10 Tobago	329		22. 9.82	R.A Irwin tr Wessex Flyers Group	Thruxton	18. 6.05
G-BKIT	SOCATA TB-9 Tampico	330		22. 9.82	K.P.Dowling and S.J.Allen	Earls Colne	30. 6.07
G-BKIY*	Thunder Ax3 Sky Chariot HAB				See SECTION 3 Part 1	(Aldershot)	
G-BKJB	Piper PA-18-135 Super Cub	18-574	PH-GAI	1. 8.83	Haimoss Ltd	Old Sarum	26. 9.07T
	(L-21A-PI) *(Frame No.18-522)*		R Neth AF R-204, 51-15657, N1003A				
G-BKJF	SOCATA MS.880B Rallye 100T	2300	F-BULF	16.12.82	Journeyman Aviation Ltd	Sywell	26. 7.07
G-BKJS	Wassmer Jodel D.120A Paris-Nice	191	F-BJPS	4.10.82	J.H.Leigh	Clipgate Farm, Denton	23. 7.05P
					tr Clipgate Flying Group		
G-BKJW	Piper PA-23-250 Aztec E	27-4716	N14153	3.11.78	Alan Williams Entertainments Ltd	Southend	26. 5.05
G-BKKN	Cessna 182R Skylane II	18267801	N6218N	30.11.82	R.A Marven	Coleman Green, Hertford	30. 4.07
					t/a Marvagraphic		
G-BKKO	Cessna 182R Skylane II	18267852	N4907H	30.11.82	Emma L.King and D.S.Lightbown	Crosland Moor	7. 4.05
G-BKKZ	Pitts S-1S	PFA 09-10525	(G-BIVW)	10.11.82	J.A Coutts	Nut Tree Farm, Redenhall	13. 6.05P
	(Built J.A Coutts)						
G-BKLO	Reims/Cessna F172M Skyhawk II	F17201380	PH-BET	22. 3.83	Stapleford Flying Club Ltd	Stapleford	11. 6.07T
			D-EFMS				
G-BKLZ*	Vinten Wallis WA-116MC				See SECTION 3 Part 1	Hermeskeil, Trier, Germany	
G-BKMA	Mooney M.20J Model 201	24-1316	N1170N	13.12.82	C.A White t/a Foxtrot Whisky Aviation	Cambridge	9. 6.07
G-BKMB	Mooney M.20J Model 201	24-1307	N1168P	15.12.82	W.A Cook, B.Pearson and P.Turnbull	Sherburn-in-Elmet	3.12.04
G-BKMG	Handley Page 0/400 rep	TPG-1		8.12.82	M.G.King tr The Paralyser Group	(Wroxham, Norwich)	
	(Built The Paralyser Group)				*(Under construction 1993: current status unknown)*		
G-BKMI	Vickers Supermarine 359 Spitfire HF.VIIIc	6S/583793	A58-671	23.12.82	The Aerial Museum (North Weald) Ltd	Filton	11.11.04P
			MV154		*(As "MT928/ZX-M" in 145 Sqdn c/s)*		
G-BKMR*	Thunder Ax3 Maxi Sky Chariot HAB				See SECTION 3 Part 1	Newbury	
G-BKMT	Piper PA-32R-301 Saratoga SP	32R-8213013	N8005Z	4. 2.83	P.R. and B.N.Lewis	Sleap	20. 4.07
					tr Severn Valley Aviation Group		
G-BKMW*	Short SD.3-30 Sherpa Var.100				See SECTION 3 Part 1	Belfast	
G-BKMX	Short SD.3-60 Variant 100	SH.3608	G-14-3608	13.12.82	BAC Leasing Ltd	Exeter	15. 3.06T
					"City of Bristol" (Op BAC Express)		
G-BKNA	Cessna 421	421-0097	F-BUYB	28. 1.83	Launchapart Ltd	Barton	13. 8.97
			HB-LDZ, N4097L		*(Damaged Penbridge, Hereford 3.8.97)*		
G-BKNB	Cameron V-42 HAB	887		10. 1.83	D.N.Close	(Andover)	17. 7.97A
G-BKNI	Sud-Aviation Gardan GY-80-160D Horizon	249	F-BRJN	28. 1.83	A Hartigan	Bourn	13. 5.02
					tr Blue Horizon Flying Group *"Blue Lady" (Stored 5.03)*		
G-BKNN*	Cameron Minar-E-Pakistan HAB				See SECTION 3 Part 1	Balleroy, Normandy, France	

Registration	Type	C/n	Previous identity	Date	Owner / Operator	Location	CofA
G-BKNO	Monnett Sonerai IIL	PFA 15-10528		11. 3.83	S.Hardy	(Hemel Hempstead)	15. 6.99P
	(Built S Tattersfield and K Bailey - c/n 792) (Volkswagen 1834)				*(Current status unknown)*		
G-BKNP	Cameron V-77 HAB	874		22.12.82	I.Lilja *"Winnie The Pooh"*	(Kvanum, Sweden)	22. 3.04A
G-BKNZ	Menavia Piel CP.301A Emeraude	296	F-BISZ	21. 1.83	C.J.Bellworthy	Finmere	9. 5.05P
G-BKOA	SOCATA MS.893E Rallye 180GT	12432	F-BOFB	2. 3.83	P.Howick	Bodmin	31.10.05
			F-ODAT, F-BVAT				
G-BKOB	Moravan Zlin Z-326 Trener Master	757	F-BKOB	28. 9.81	A L.Rae	(Bourne End)	17. 5.05
G-BKOT	Wassmer WA.81 Piranha	813	F-GAIP	17. 2.87	Barbara N.Rolfe *(Stored 9.01)*	Little Gransden	AC
G-BKOU	Hunting Percival P.84 Jet Provost T.3	PAC/W/13901	XN637	17. 2.83	M.P.Grimshaw tr G-BKOU Group	North Weald	25. 8.05P
					(As "XN637/03" in TWU/79 Sqdn c/s)		
G-BKOW*	Colt 77A HAB				See SECTION 3 Part 1	(Aldershot)	
G-BKPA	Hoffmann H 36 Dimona	3522		16. 6.83	C.I.Roberts and C.D.King	Trenchard Farm, Eggesford	26. 6.05
G-BKPB	Aerosport Scamp	PFA 117-10736		23. 2.83	B.R.Thompson	Leicester	17.12.05P
	(Built R Scroby) (Volkswagen 1834)						
G-BKPC	Cessna A185F AGcarryall	185-03809	N4599E	10. 7.80	The Black Knights Parachute Centre Ltd		
						Bank End Farm, Cockerham	11.10.07E
G-BKPD	Viking Dragonfly	PFA 139-10897		11. 3.83	E.P.Browne and G.J.Sargent	Cambridge	20. 1.00P
	(Built P E J Sturgeon - c/n 302) (Revmaster 2100D)				*(Damaged Cambridge 17.7.99: current status unknown)*		
G-BKPE	CEA Jodel DR.250/160 Capitaine	35	F-BNJD	18. 3.83	J.S. and J.D.Lewer	Dunkeswell	16.12.04
G-BKPG*	Luscombe P3 Rattler Strike				See SECTION 3 Part 1	Usworth, Sunderland	
G-BKPN	Cameron N-77 HAB	923		9. 3.83	R.H.Sanderson *"Do It All"*	(Nuneaton)	21. 5.87A
					(Donated to Balloon Preservation Group 1.03)		
G-BKPS	Grumman AA-5B Tiger	AA5B-0007	OO-SAS	7. 3.83	A E.T.Clarke	Manston	9.10.06
			OO-HAO, (OO-WAY), N1507R				
G-BKPX	Wassmer Jodel D.120A Paris-Nice	240	F-BLNG	19. 1.84	D.M.Garrett and C.A Jones	Defford	19. 5.05P
G-BKPZ	Pitts S-1T	PFA 09-10852		4. 3.83	Mary A Frost	Downland Farm, Redhill	2. 7.05P
	(Built G C Masterton)						
G-BKRA	North American T-6G-NH Texan	188-90	MM53-664	19. 8.83	First Air Ltd	Gloucestershire	24. 2.07T
			RM-9, 51-15227		*(As "51-15227/10" in US Navy c/s)*		
G-BKRF	Piper PA-18 Super Cub 95	18-1525	F-BOUI	7.11.83	K.M.Bishop	(Krefeld, Germany)	29. 9.98P
	(L-18C-PI) *(Frame No.18-1502)*		ALAT, 51-15525				
G-BKRG*	Beech C-45G-BH				See SECTION 3 Part 1	Lelystad, The Netherlands	
G-BKRH	Brugger MB.2 Colibri	PFA 43-10150		15. 3.83	M.R.Benwell	Hinton-in-the-Hedges	9. 8.05P
	(Built M.R.Benwell - c/n 142) (Volkswagen 1835)						
G-BKRK	SNCAN Stampe SV-4C	57	Fr.Navy	30. 3.83	J.R.Bisset	Insch	28. 6.98
	(Renault 4P)				tr Strathgadie Stampe Group *(Current status unknown)*		
G-BKRL*	Chichester-Miles Leopard	001		21. 3.83	Chichester-Miles Consultants Ltd	(Old Sarum)	14.12.91P
	(Noel Penny 301)				*(Cancelled 25.1.99 as WFU) (Stored 5.03)*		
G-BKRN	Beech D.18S	A-675	CF-DTN	14. 4.83	A A Marshall and P.L.Turland	Bruntingthorpe	26. 6.83P
	(Official c/n is CA-75 and suggests Canadian rebuild)		RCAF A675/RCAF 1500		*(Under restoration 3.02)*		
G-BKRS	Cameron V-56 HAB	908		23. 3.83	D.N.and L.J.Close	(Andover)	17. 7.97A
					"Bonkers"		
G-BKRV*	Hovey Beta Bird	PFA 135-10875		30. 3.83	M.J.Aubrey	(Kington, Hereford)	25.6.97P
	(Rotax 503)				*(Cancelled 18.6.98 by CAA) (Noted 2002)* .		
G-BKRZ	Dragon 77 HAB	001		11. 4.83	J.R.Barber *"Rupert"*	(Newbury)	5. 3.94A
					(On loan to British Balloon Museum and Library)		
G-BKSB	Cessna T310Q II	310Q0914	VR-CEM	22. 4.83	D.H.and P.M.Smith	Bagby	10. 6.06
			G-BKSB, HB-LMO, OE-FYL, (N69680)		t/a G.H.Smith and Son		
G-BKSD	Colt 56A HAB	361		11. 4.83	M.J.Casson *"Entwhistle Green"*	(Kendal)	2. 6.96A
G-BKSE	QAC Quickie 1	PFA 94-10748		6. 4.83	M.D.Burns	(Bridge of Weir)	8. 5.89P
	(Built Taylor, M D Burns and Ibbott) (Onan B48M)		*(Regd with c/n PFA 94-107<u>84</u>)*		*(Stored 6.00: current status unknown)*		
G-BKSP	Schleicher ASK 14	14028	D-KOMO	25. 5.83	J.H.Bryson	Bellarena	8. 7.06
G-BKST	Rutan VariEze	12718-001		20. 4.83	R.Towle	(Hexham)	
	(Built R Towle)						
G-BKSX	SNCAN Stampe SV-4C	61	F-BBAF	16. 5.83	C.A Bailey and J.A Carr	(Eggesford)	15. 6.89
	(Renault 4P)		Fr.Mil		*(Stored 11.02)*		
G-BKTA	Piper PA-18 Super Cub 95	18-3223	OO-HBA	10. 5.83	M.J.Dyson and M.T.Clark	Roddige	10. 8.05P
	(L-18C-PI) *(Frame No.18-3246)*		Belg AF OL-L49, L-149, 53-4823				
G-BKTH	Hawker Sea Hurricane IB	CCF/41H/4013	Z7015	24. 5.83	Richard Shuttleworth Trustees	Old Warden	10. 5.05P
	(Built Canadian and Car Foundry Co)				*(As "Z7015/7-L" in 880 Sqdn RN c/s)*		
G-BKTM	PZL-Bielsko SZD-45A Ogar	B-656		31. 5.83	J.T.Pajdak	(London SE16)	13. 5.07
G-BKTR	Cameron V-77 HAB	951		6. 6.83	A Palmer *"Diddlybopper"*	(Tonbridge)	7. 5.05A
G-BKTV	Reims/Cessna F152 II	F15201450	OY-BJB	8. 8.83	A Jahanfar *(Op Seawing Flying Club)*	Southend	11. 8.05T
G-BKTZ	Slingsby T.67M Firefly	2004	G-SFTV	26. 8.83	P.R.Elvidge	Octon Lodge Farm, Thwing	22. 8.05
G-BKUE	SOCATA TB-9 Tampico	369	F-BNGX	31. 5.83	Flying Web Ltd	Gloucestershire	20. 2.06T
G-BKUR	Menavia Piel CP.301A Emeraude	280	(G-BKBX)	19.10.83	R.Wells	Morgansfield, Fishburn	21. 6.05P
			F-BMLX, F-OBLY				
G-BKUU	Thunder Ax7-77 Series 1 HAB	522		3. 8.83	M.A Mould *"Tanglefoot"*	(Winchester)	12. 4.03A
G-BKVA	SOCATA Rallye 180T Galerien	3274	SE-GFS	30. 6.83	Buckminster Gliding Club Ltd	Saltby	15. 8.07T
			F-GBXA				
G-BKVB	SOCATA Rallye 110ST Galopin	3258	OO-PIP	22. 6.83	A and K.Bishop	Newtownards	11.12.05
G-BKVC	SOCATA TB-9 Tampico	372	F-BNGQ	4. 7.83	H.P.Aubin-Parvu	Biggin Hill	26. 3.05
G-BKVF	Clutton FRED Series III	PFA 29-10791		29. 7.83	G.E.and R.E.Collins	Egginton	
	(Built N E Johnson)				*(New owners 10.04)*		
G-BKVG	Scheibe SF25E Super Falke	4362	(D-KNAE)	25. 8.83	G-BKVG Ltd	North Hill	4. 6.05
	(Limbach SL1700)						
G-BKVK	Auster AOP.9	AUS/10/2	WZ662	8. 8.83	M.Robinson	Trenchard Farm, Eggesford	11. 2.05P
					tr Victor Kilo Group *(As "WZ662" in Army c/s)*		
G-BKVL	Robin DR400/160 Major	1625		26. 7.83	Tatenhill Aviation Ltd t/a Tatenhill Aviation	Tatenhill	26. 1.06T
G-BKVM	Piper PA-18-150 Super Cub	18-849	PH-KAZ	26. 8.83	D.G.Caffrey *"Spirit of Goxhill"*	Strubby	13.10.05
	(L-21A-PI) *(Frame No.18-824)*		R Neth AF R-214, 51-15684		*(As "115684/VM" in US Army c/s)*		
G-BKVO	Pietenpol AirCamper	PFA 47-10799		8. 8.83	M.C.Hayes	(Woonton, Hereford)	6.11.04P
	(Built M J Honeychurch) (Continental A65)						

G-BKVP	Pitts S-1D	PFA 09-10800		19. 8.83	S.W.Doyle	Leicester	19. 4.05P
	(Built P J Leggo - c/n 002)						
G-BKVS	Campbell Cricket	PFA G/01-1047		11. 8.83	K.Hughes	(Amlwch, Gwynedd)	10. 8.05P
	(Built V Scott) (Volkswagen 1834)						
G-BKVT	Piper PA-23-250 Aztec F	27-7754002	G-HARV N62760	6. 2.84	BKS Surveys Ltd	Belfast Aldergrove	3. 4.06T
G-BKVW	Airtour AH-56 HAB	AH.003		27. 6.84	L.D.and H.Vaughan *"Lunardi"*	(Tring)	
G-BKVX	Airtour AH-56C HAB	AH.002		27. 6.84	P.Aldridge	(Halesworth, Suffolk)	
					"Featherspin" or "Liebling" ?		
G-BKVY	Airtour B-31 HAB	AH.001		9. 8.83	M.Davies *"Day Dream"*	(Callington, Cornwall)	15. 8.01A
G-BKWD	Taylor JT.2 Titch	PFA 60-10232		17. 8.83	J.F.Sully	(Hemswell Cliff, Gainsborough)	17. 6.05P
	(Built E H Booker) (Continental PC60) *(Originally regd as c/n PFA 60-10143 - presumed absorbed into this project on build)*						
G-BKWR	Cameron V-65 HAB	970		26. 8.83	K.J.Foster *"White Spirit"*	(Coleshill, Birmingham)	6. 7.04A
G-BKWW	Cameron O-77 HAB	984		13. 9.83	A M.Marten *"Kouros"*	(Woking)	18. 1.89A
G-BKWY	Reims/Cessna F152	F15201940		22. 9.83	Northern Aviation Ltd	Durham Tees Valley	11.12.05T
G-BKXA	Robin R2100	114	F-GAOS	24.11.83	M.Wilson *(New owner 8.01)*	Little Gransden	22.10.99
G-BKXD	Aérospatiale SA365N Dauphin 2	6088	F-WMHD	7. 9.83	CHC Scotia Ltd	Blackpool	8.12.07E
G-BKXF	Piper PA-28R-200 Cherokee Arrow II	28R-7335351	OY-DZN N56092	10.11.83	P.L.Brunton	Caernarfon	22. 6.05
G-BKXM	Colt 17A Cloudhopper HAB	531		3.10.83	R.G.Turnbull	(Glasbury, Hereford)	8.10.03A
G-BKXN	ICA IS-28M2A	48		24.10.83	D.C.Wellard	Enstone	14. 5.06
G-BKXO	Rutan LongEz	PFA 74A-10580		24.10.83	M.G.Parsons	(Buckie)	27. 6.00P
	(Built P Wareham) (Continental O-200-A)				*(New owner 7.04)*		
G-BKXP	Auster AOP.6	2830	A-14	12.10.83	B.J.Ellis	Thruxton	
	(Frame No.TAY841BJ)		Belg AF, VT987		*(On rebuild 7.91: new owner 12.01)*		
G-BKXR	Druine D.31A Turbulent	303	OY-AMW	1.11.83	M.B.Hill	Draycott Farm, Chiseldon	27.11.05P
	(Built H Husted) (Volkswagen 1700)						
(G-BKXW)*	North American NA.82 B-25J Mitchell				See SECTION 3 Part 1	North Weald	
G-BKZB	Cameron V-77 HAB	995		11.11.83	K.B.Chapple	(Reading)	6. 7.01A
G-BKZE	Aérospatiale AS332L Super Puma	2102	F-WKQE	30. 9.83	CHC Scotia Ltd	Aberdeen	25. 9.05T
G-BKZF	Cameron V-56 HAB	246	F-BXUK	14.11.83	C.F.Sanger-Davies	(Eldersfield, Gloucester)	18. 3.97A
					(New owner 11.04)		
G-BKZG	Aérospatiale AS332L Super Puma	2106	HB-ZBT G-BKZG	30. 9.83	CHC Scotia Ltd	Aberdeen	25. 8.05T
G-BKZI	Bell 206A JetRanger	118	(5B-CGC or 'D?) G-BKZI, N6238N	7.12.83	Dolphin Property (Management) Ltd	Thruxton	19. 9.07T
G-BKZT	Clutton FRED Series II	PFA 29-10715		20.10.83	U.Chakravorty	(Margate)	2. 7.02P
	(Built A E Morris) (Volkswagen 1834)						
G-BKZV	Bede BD-4	380	ZS-UAB	31. 8.84	G.I.J.Thomson	Little Snoring	28.10.05P
	(Built A L Bergamasgo) (Lycoming O-320)						

G-BLAA - G-BLZZ

G-BLAC	Reims/Cessna FA152 Aerobat	FA1520370		25. 3.80	D.C.C.Handley	Bourn	26. 7.04T
G-BLAF	Stolp SA.900 V-Star	PFA 106-10651		13. 9.83	P.R.Skeels	Lymm Dam	9.12.04P
	(Built J E Malloy) (Continental O-200-A)						
G-BLAG	Pitts S-1D	PFA 09-10195		1.12.83	A.G.Truman	Popham	15. 4.05P
	(Built B Bray)						
G-BLAH	Thunder Ax7-77 Series 1 HAB	526		3.10.83	T.M.Donnelly	(Sprotbrough, Doncaster)	19. 8.01A
					"Blah" (Active Albuquerque , New Mexico 10.03)		
G-BLAI	Monnett Sonerai IIL	PFA 15-10583		6.12.83	T.Simpson	(Kirkcaldy)	12. 1.99P
	(Built T.Simpson) (Officially regd with c/n PFA 15-10584)				*(Noted 12.01)*		
G-BLAM	CEA Jodel DR.360 Chevalier	345	F-BRCM	6. 2.84	D.J.Durell	Rochester	11. 8.05
G-BLAT	SAN Jodel D.150 Mascaret	56	F-BNID	30. 1.84	S.A Smith and D.K.Gliddon	Perth	31.10.05P
					tr G-BLAT Flying Group		
G-BLAX	Reims/Cessna FA152 Aerobat	FA1520385		11.10.83	N C and Michelle L Scanlan	(Sleaford)	28. 5.05T
G-BLCC	Thunder Ax7-77Z HAB	532		7.12.83	J.M.Percival	(Burton-on-the-Wolds, Loughborough)	21. 3.99A
					(New owner 6.04)		
G-BLCG	SOCATA TB-10 Tobago	61	G-BHES	17. 3.80	R.Deery and D.Tunks	Shoreham	10. 8.07
					tr Charlie Golf Flying Group		
G-BLCH	Colt 56D HAB	392		14.11.83	Balloon Flights Club Ltd	(Leicester)	
					"Geronimo"		
G-BLCI	EAA Acrosport P	P-10A	N6AS	29. 2.84	M.R.Holden	Stoneacre Farm, Farthing Corner	16. 6.97P
					"Bluebottle" (Damaged Farthing Corner 1996: current status unknown)		
G-BLCM	SOCATA TB-9 Tampico	194	OO-TCT (OO-TBC)	2.12.83	P.J.Tyler	Sleap	27. 6.05T
					tr Charlie Mike Group		
G-BLCT	CEA Jodel DR.220 2+2	23	F-BOCQ	22.12.83	C.J.Snell	Shoreham	7. 8.05
					tr Christopher Robin Flying Group		
G-BLCU	Scheibe SF25B Falke	4699	D-KECC	30.12.83	J.D.Johnson	Rufforth	11. 9.05
	(Stark-Stamo MS1500)				tr Charlie Uniform Syndicate		
G-BLCV	Hoffmann H 36 Dimona	36113	EI-CJO G-BLCV	21. 3.84	R.and M.Weaver	(Walton-on-Thames)	9. 7.05
G-BLCW	Evans VP-1	PFA 62-10835		19.12.83	M.Flint	Fenland	21. 8.04P
	(Built K D Pearce) (Volkswagen 1600)				*"Le Plank"*		
G-BLCY	Thunder Ax7-65Z HAB	487		13. 1.84	C.M.George *"Warsteiner"*	(Brixton, Plymouth)	26. 2.99A
G-BLDB	Taylor JT.1 Monoplane	PFA 55-10506		28.12.83	C.J.Bush	New Farm House, Great Oakley	23. 5.05P
	(Built C J Bush)						
G-BLDD	Wag-Aero CUBy AcroTrainer	PFA 108-10653		29.12.83	A F.Stafford	Combrook	19. 8.05P
	(Built C A Lacock) (Lycoming O-320)						
G-BLDG	Piper PA-25-260 Pawnee C	25-4501	SE-FLB LN-VYM	9. 1.84	Ouse Gliding Club Ltd	Rufforth	23. 6.05
G-BLDK	Robinson R22	0139	C-GSGU	17. 1.84	I.C.Macdonald	(Arnside, Carnforth)	28. 6.05T
G-BLDL*	Cameron Truck 56 SS HAB				See SECTION 3 Part 1	(Aldershot)	

Reg	Type	C/n	Previous identity	Regd	Owner/operator	Base	CofA
G-BLDN	Rand-Robinson KR-2	PFA 129-10913		12. 1.84	S.C.Solley Mavis	Enderby	7. 7.03P
	(Built R Y Kendal) (Volkswagen 1834)						
G-BLDV	Pilatus Britten-Norman BN-2B-26 Islander	2179	D-INEY	13. 1.84	Loganair Ltd	Kirkwall	18. 7.05T
			G-BLDV				
G-BLEB	Colt 69A HAB	537		20. 1.84	I.R.M.Jacobs	(Padworth Common, Reading)	30. 3.85A
					(CurrentCofR 12.03)		
G-BLEJ	Piper PA-28-161 Cherokee Warrior II	28-7816257	N2194M	8. 2.84	Eglinton Flying Club Ltd	City of Derry	17. 4.05T
G-BLEP	Cameron V-65 HAB	102		27. 2.84	D.Chapman	(Maidstone)	10. 9.96A
					tr The Ground Hogs *"Manor Marquees"*		
G-BLES	Stolp SA.750 Acroduster Too	PFA 89-10428		8.12.83	C.J.Kingswood	RAF Church Fenton	26. 5.05P
	(Built W G Hosie - c/n 197) (Lycoming O-360)						
G-BLET	Thunder Ax7-77 Series 1 HAB	539		16. 2.84	Servatruc Ltd *"Servatruc"*	(Nottingham)	15. 8.97A
G-BLEZ	Aérospatiale SA365N Dauphin 2	6131		24. 1.84	CHC Scotia Ltd	(Forties Oil Field)	27. 8.05T
					(Op First Aim Medevac)		
G-BLFE*	Cameron Sphinx 72SS HAB				See SECTION 3 Part 1	Balleroy, Normandy, France	
G-BLFI	Piper PA-28-181 Archer II	28-8490034	N4333Z	22. 2.84	Bonus Aviation Ltd	Cranfield	27. 7.06T
G-BLFL*	Douglas C-47B-45-DK Dakota				See SECTION 3 Part 1	Berlin, Germany	
G-BLFW	Grumman AA-5 Traveler	AA5-0786	OO-GLW	22. 2.84	D.C.A Milne tr Grumman Club	Draycott Farm, Chiseldon	10. 9.05
G-BLFY	Cameron V-77 HAB	1030		16. 3.84	A N.F.Pertwee *"Groupie"*	(Frinton-on-Sea)	5. 4.92A
G-BLFZ	Piper PA-31 Navajo C	31-7912106	PH-RWS	21. 3.84	London Executive Aviation Ltd	RAF Leeming	1. 9.05T
			(PH-ASV), N3538W				
G-BLGH	Robin DR300/180R Remorqueur	570	D-EAFL	10. 4.84	Booker Gliding Club Ltd	Booker	3. 4.06
G-BLGS	SOCATA Rallye 180T	3206		7. 7.78	A Waters t/a London Light Aircraft	Dunstable	21. 5.99
					(Gutted fuselage noted 6.00)		
G-BLGT	Piper PA-18 Super Cub 95	18-1445	D-EAGT	1. 6.84	Technical Flight Services Ltd	Bournemouth	12. 5.06T
	(L-18C-PI) *(Frame No.18-1399)*		D-EOCC, ALAT 51-15445		*(As "15445" in USAAF c/s)*		
G-BLGV	Bell 206B JetRanger II	982	5B-JSB	2. 5.84	Heliflight (UK) Ltd	Wolverhampton	22. 5.05T
			C-FDYL, CF-DYL				
G-BLGX*	Thunder Ax7-65 HAB	551		16. 4.84	"The 45"	(Uttoxeter)	Exemption
					(Cancelled 19.5.93 by CAA) (Inflated 3.03		
G-BLHH	CEA Jodel DR.315 Petit Prince	324	F-BPRH	3. 7.84	Central Certification Service Ltd	Tower Farm, Wollaston	24. 7.06
G-BLHI	Colt 17A Cloudhopper HAB	506		8. 9.86	Janet A Folkes *"Hopping Mad"*	(Loughborough)	24.11.01A
G-BLHJ	Reims/Cessna F172P Skyhawk II	F17202182		26. 3.84	James D.Peace and Co	Aberdeen	18.12.05T
G-BLHK	Colt 105A HAB	576		19. 6.84	A S.Dear, R.B.Green and W.S.Templeton	(Fordingbridge)	12. 7.97A
					tr Hale HA Balloon Group *"Gloworm"*		
G-BLHM (2)	Piper PA-18 Super Cub 95	18-3120	LX-AIM	23. 7.84	A G.Edwards	(Llandegla)	8. 8.04P
	(L-18C-PI) *(Frame No.18-3088)*		D-EOAB, Belgian AF OL-L46, L-46, 53-4720				
G-BLHN	Robin HR100/285 Tiara	539	F-GABF	20. 2.78	N.P.Finch	Goodwood	18.12.06
G-BLHR	Gulfstream GA-7 Cougar	GA7-0109	OO-RTI	12. 4.84	T.E.Westley	Fowlmere	20. 1.06T
			(OO-HRC), N751G				
G-BLHS	Bellanca 7ECA Citabria 115	1342-80	OO-RTQ	12. 4.84	N.J.F.Campbell and D.J.Lockett	Inverness	29. 4.05
					tr Hotel Sierra Group		
G-BLHW	Varga 2150A Kachina	VAC161-80		17. 7.84	W.D.Garlick	Damyn's Hall, Upminster	10. 6.06
					tr Kachina Hotel Whiskey Group		
G-BLIC*	de Havilland DH.112 Venom FB.54 (Mk.17)				See SECTION 3 Part 1	Ardmore, Auckland, New Zealand	
G-BLID	de Havilland DH.112 Venom FB.50 (FB.1)	815	Swiss AF J-1605	13. 7.84	P.G.Vallance Ltd	Charlwood, Surrey	AC
	(Built F + W)				*(In Gatwick Aviation Museum: as "J-1605" in Swiss AF c/s)*		
G-BLIH	Piper PA-18-135 Super Cub	18-3828	(PH-KNG)	12.11.84	I.R.F.Hammond	Stubbington	AC
	(L-21B-PI)*(Frame No.18-3827)*		R Neth AF R-138, (PH-KNG), (PH-GRC), R-138, 54-2428				
G-BLIK	Wallis WA-116/F/S	K-218X		30. 4.84	K.H.Wallis	Reymerston Hall, Norfolk	9. 4.05P
	(Franklin 2A-120)						
G-BLIO*	Cameron R-42 Gas/HAB				See SECTION 3 Part 1	Newbury	
G-BLIP*	Cameron N-77 HAB				See SECTION 3 Part 1	(Aldershot)	
G-BLIT	Thorp T-18CW	PFA 76-10550		24. 4.84	A P.Tyrwhitt-Drake	White Waltham	6.11.05P
	(Built A J Waller) (Lycoming O-320)						
G-BLIW	Percival P.56 Provost T.51	PAC/F/125	IAC.177	12. 6.85	D.Mould and J.De Uphaugh	Shoreham	16. 5.05P
					tr Provost Flying Group *(As "177" in IAC c/s)*		
G-BLIX	Saro Skeeter AOP.12	S2/5094	PH-HOF	3. 5.84	K.M.Scholes	Wilden	4. 5.05P
			(PH-SRE), XL809		*(As "XL809"in Army c/s)*		
G-BLIY	SOCATA MS.892A Rallye Commodore 150	11639	F-BSCX	9. 5.84	A J.Brasher	Enstone	22. 12.06
G-BLJD	Glaser-Dirks DG-400	4-85		15. 6.84	M.I.Gee	Lasham	8. 6.06
G-BLJF	Cameron O-65 HAB	1041		14. 5.84	M.D.and C.E.C.Hammond	(Aldershot)	16. 8.00A
					(Op Balloon Preservation Group) "Fat Lady"		
G-BLJH	Cameron N-77 HAB	1047		14. 5.84	K A Kent	(Aldershot)	27. 6.89A
					(Op Balloon Preservation Group) "Daydream"		
G-BLJM	Beech 95-B55 Baron	TC-1997	SE-GRT	3. 3.78	R.A Perrot	Guernsey	30.10.06
G-BLJO	Reims/Cessna F152 II	F15201627	OY-BNB	21. 6.84	Redhill School of Flying Ltd	Redhill	28.10.07E
					(Op Redhill Flying Club)		
G-BLKA	de Havilland DH.112 Venom FB.Mk.54				See SECTION 3 Part 1	Salisbury Hall, London Colney	
G-BLKJ	Thunder Ax7-65 HAB				See SECTION 3 Part 1	(Aldershot)	
G-BLKK	Evans VP-1	PFA 62-10642		15. 6.84	N.Wright	Queach Farm, Bury St.Edmunds	16.11.05P
	(Built R W Burrows) (Volkswagen 1834)						
G-BLKM	CEA Jodel DR.1051 Sicile	519	F-BLRO	26. 6.84	F.H.Lissimore tr Kilo Mike Group	Biggin Hill	27. 6.06
G-BLKP	British Aerospace Jetstream Series 3102	634	(G-BLEX)	9. 7.84	Global Aviation Ltd	Humberside	19. 4.03T
			G-31-634		*(Eastern Airways titles)*		
G-BLKU*	Colt Flame 56SS HAB				See SECTION 3 Part 1	Manchester	
G-BLKY	Beech 58 Baron	TH-1440		22. 8.84	R.A.Perrot	Guernsey	30. 7.06
G-BLKZ	Pilatus P.2-05	600-45	Swiss AF U-125	30. 7.84	Robert Hinton Design and Creative Communications		
		(Official c/n is "45")	Swiss AF A-125		*(As "A-125" in Swiss AF c/s)*	Duxford	19. 2.05P
G-BLLA	Bensen B.8M	PFA G/01-1055		27. 6.84	K.T.Donaghey	Henstridge	8. 9.05P
	(Built K.T.Donaghey) (Volkswagen 1834)						
G-BLLB	Bensen B.8MR	PFA G/01A-1059		4. 9.84	D.H.Moss	Henstridge	14. 6.01P
	(Built D.H.Moss) (Rotax 532)				*(Noted 12.02)*		
G-BLLD	Cameron O-77 HAB	1060		16. 7.84	G.Birchall	(Ormskirk)	12. 4.03A

Reg	Type	C/n	Previous identity	Regd	Owner	Base	CofA
G-BLLH	CEA Jodel DR.220A/B 2+2	131	F-BROM	17. 7.84	M.D.Hughes	Pauncefoot, Romsey	30. 6.06
G-BLLJ*	Short SD.3-30 (C-23A) Sherpa				See SECTION 3 Part 1	Davis-Monthan, Tucson, Arizona, US	
G-BLLN	Piper PA-18 Super Cub 95 (L-18C-PI)	18-3447	D-ECLN	27. 6.84	P.L.Pilch and C.G.Fisher	Rochester	1. 3.07T
	(Continental O-200A) *(Frame No.18-3380)*		96+23, PY+901, QZ+011, AC+507, AS+508, 54-747				
G-BLLO*	Piper PA-18-95 Super Cub	18-3099	D-EAUB	11. 7.84	D.G. and M.G.Margetts	Sleap	12.10.96P
	(L-18C-PI)*(Frame No 18-3058)*		Belgian AF OL-L25, L-25, 53-4699 *(Cancelled 15.11.00 by CAA) (Fuselage noted 8.03)*				
G-BLLP*	Slingsby T.67B Firefly	2008		19. 7.84	(Cleveland Flying School Ltd)	Sheffield City	4.12.00T
					(Cancelled 20.8.02 by CAA) (Stored remains dumped 1.05)		
G-BLLR	Slingsby T.67C Firefly	2011		19. 7.84	R.L.Brinklow	Gloucestershire	28.11.04T
	(Lycoming O-320) *(Regd as "T.67B (mod)")*				*(Op Cotswold Aero Club) (www.cotswoldaeroclub.co.uk titles)*		
G-BLLS	Slingsby T.67B Firefly	2013		19. 7.84	Western Air (Thruxton) Ltd	Thruxton	11. 3.06T
G-BLLW	Colt 56B HAB	578		11. 9.84	G.Fordyce, R.Wickens and S.A Sawyer	(Olney)	17. 4.05A
					"Angel Clare"		
G-BLLZ	Rutan LongEz	PFA 74A-10830		16. 7.84	R.S.Stoddart-Stones	Henstridge	22. 6.94P
	(Built G E Relf, D G Machin and E F Braddon) (Lycoming O-235)						
G-BLMA	Moravan Zlin Z-526 Trener Master	922	F-BORS	23. 7.84	G.P.Northcott	Redhill	26. 5.06
(G-BLMC)*	Avro 698 Vulcan B.2A				See SECTION 3 Part 1	Nottingham East Midlands	
G-BLME	Robinson R22HP	0032	N90261	16. 4.85	Heli Air Ltd	Liverpool	24.10.05T
G-BLMG	Grob G109B	6322		27. 9.84	R.W.Littledale tr Mike Golf Syndicate	Enstone	7.11.05
G-BLMI	Piper PA-18 Super Cub 95	18-2066	D-ENWI	5. 6.84	R.Gibson	White Waltham	9. 5.05P
	(L-18C-PI) *(Frame No.18-2086)*		R Neth AF R-55, 52-2466		tr G-BLMI Flying Group *(As "52-2466/R-55" in R Neth AF c/s)*		
G-BLMN	Rutan LongEz	PFA 74A-10643		3. 7.84	S.E.Bowers	Thruxton	24. 8.05P
	(Built Farrington and Farrington) (Lycoming O-235) *(Regd as c/n PFA 74A-10648)* tr G-BLMN Flying Group						
G-BLMP	Piper PA-17 Vagabond	17-193	F-BFMR	15. 5.84	M.Austin	Longwood Farm, Morestead	29. 6.05P
	(Continental A65)		N4893H				
G-BLMR	Piper PA-18-150 Super Cub (L-18C-PI)	18-2057	PH-NLD	29. 5.84	Southern Flight Centre Ltd	Shoreham	22. 7.05T
	(Lycoming O-320) *(Frame No.18-2070)*		R Neth AF R-72, 52-2457				
G-BLMT	Piper PA-18-135 Super Cub	18-2706	D-ELGH	12. 9.84	I.S.Runnalls	Enstone	25. 9.05
	(Frame No.18-2724)		N8558C				
G-BLMW	Nipper T.66 Series 3	PFA 25-11020		31. 8.84	S.L.Millar	Crowland	20.10.05P
	(Built S L Millar) (Ardem 10)						
G-BLMZ	Colt 105A HAB	404		24. 9.84	Mandy D.Dickinson *"Zulu"*	(Bristol)	28. 3.97A
G-BLNJ	Pilatus Britten-Norman BN-2B-26 Islander	2189		3. 9.84	Loganair Ltd	Kirkwall	3.12.05T
G-BLNO	Clutton FRED Series III	PFA 29-10559		17.10.84	L.W.Smith	(Sale, Cheshire)	
	(Built L.W.Smith)						
G-BLOB	Colt 31A Air Chair HAB	599		11. 9.84	Jacques W.Soukup Enterprises Ltd		
						Tyndale, South Dakota, US	5. 6.91A
G-BLOL	SNCAN Stampe SV-4A				*See entry for G-AXRP*		
G-BLOR	Piper PA-30 Twin Comanche	30-59	HB-LAE	19. 7.85	R.L.C.Appleton	Sheepwash	21. 5.05T
			N7097Y, N10F				
G-BLOS	Cessna 185A Skywagon	185-0359	LN-BDS	17. 9.84	Elizabeth Brun	Great Massingham	24. 4.06
			N4159Y				
G-BLOT	Colt 56B HAB	424		11. 9.84	H.J.Anderson *"Pathfinder"*	(Oswestry)	17. 7.96A
G-BLOV	Thunder Ax5-42 Series 1 HAB	590		11. 9.84	A.G.R.Calder *"Puff The Magic Dragon"*	(London SE16)	29.11.02A
G-BLPA	Piper J-3C-65 Cub (L-4H-PI)	11327	OO-AJL	27. 9.84	A C.Frost	Rectory Farm, Abbotsley	26. 8.05P
	(Frame No.11152)		OO-JOE, 43-30036				
G-BLPB	Turner TSW Hot Two Wot	PFA 46-10606		19.10.84	I.R.Hannah	Shoreham	18. 7.05P
	(J R Woolford and K M Thomas) (Lycoming O-320-A)						
G-BLPE	Piper PA-18 Super Cub 95 (L-18C-PI)	18-3084	D-ECBE	28. 9.84	A.A.Haig-Thomas	Thorpe-le-Soken	4. 4.05P
	(Continental O-200-A) *(Also quoted as 18-3083)*		Belgian Army L-10, 53-4684				
G-BLPF	Reims FR172G Rocket	FR17200187	N4594Q	29. 1.85	P.Kohl	Cambridge	2. 5.06T
			D-EEFL				
G-BLPG	Auster J/1N Alpha	3395	G-AZIH	21. 5.82	D.Taylor *(As "16693" in RCAF c/s)*	Clacton	18. 3.04
G-BLPI	Slingsby T.67B Firefly	2016		24. 9.84	RAF Wyton Flying Club Ltd	RAF Wyton	18. 8.06T
G-BLPM	Aerospatiale AS332L Super Puma	2122	LN-ONB	5.10.84	Bristow Helicopters Ltd	Aberdeen	27. 7.06T
			G-BLPM, C-GQCB, G-BLPM				
G-BLPP	Cameron V-77 HAB	432		19. 9.78	L.P.Purfield *"Merlin"*	(Leicester)	30. 4.94A
G-BLPZ*	de Havilland DH.104 Devon C.2				See SECTION 3 Part 1	Stauning, Denmark	
G-BLRA	British Aerospace BAe 146 Series 100	E1017	N117TR	3.10.84	BAE SYSTEMS (Operations) Ltd	Woodford	15.10.06T
			N462AP, CP-2249, N462AP, G-BLRA, G-5-02				
G-BLRC	Piper PA-18-135 Super Cub	18-3602	OO-DKC	27.11.84	S.Hornung tr Supercub Group	Seething	20.12.06
	(L-21B-PI) *(Frame No.18-3790)*		PH-DKC, R NethAF R-112, 54-2402				
G-BLRD	MBB Bö.209 Monsun 150FV	101	D-EBOA	15.10.84	T.A Crone	Bicester	16. 8.04
			(OE-AHM)				
G-BLRF	Slingsby T.67C Firefly	2014		30.11.84	R.C.Nicholls	Wellesbourne Mountford	11. 2.06T
G-BLRG	Slingsby T.67B Firefly	2020		30.11.84	R.L.Brinklow	Turweston	17. 7.00T
G-BLRJ*	CEA Jodel DR.1051 Sicile	502	F-BLRJ	8. 2.78	M.P.Hallam	Deenethorpe	17. 7.00
					(Cancelled 6.3.02 as WFU) (Noted 6.03)		
G-BLRL	Scintex CP.301C1 Emeraude	552	(G-BLNP)	5.11.84	J.A.Macleod and I.M.Macleod	Stornoway	21.12.05P
			F-BJFT				
G-BLRM	Glaser-Dirks DG-400	4-107		5. 2.85	J.A and W.S.Y.Stephen	Aboyne	23. 4.06
G-BLRN	de Havilland DH.104 Dove 8	04266	N531WB	30.10.84	J.F.M.Bleeker	Lelystad, The Netherlands	13. 3.96
			G-BLRN, WB531		*(To Pionier Hangaar Collection: as "WB531" in RAF c/s) (Stored 4.01)*		
G-BLRW*	Cameron Elephant 77SS HAB				See SECTION 3 Part 1	Balleroy, Normandy, France	
G-BLRY	Aérospatiale AS332L Super Puma	2111	LN-ONA	5. 2.85	Bristow Helicopters Ltd	(Egypt)	17. 6.05T
			G-BLRY, LN-ONA, G-BLRY, P2-PHP, VR-BIJ, G-BLRY, C-GQGL, G-BLRY				
G-BLSD*	de Havilland DH.112 Venom FB.54	928	N203DM	20. 5.85	R.J.Lamplough	North Weald	
	(Built F + W)		G-BLSD, Swiss AF J-1758		*(Cancelled 5.6.96 as WFU) (Partially dismantled as "J-1758" 1.05)*		
G-BLSF	Gulfstream AA-5A Cheetah	AA5A-0802	G-BGCK	21. 2.83	Plane Talking Ltd	Elstree	19. 6.06T
G-BLSH*	Cameron V-77 HAB				See SECTION 3 Part 1	(Aldershot)	
G-BLST	Cessna 421C Golden Eagle III	421C0623	N88638	29.11.78	Cecil Aviation Ltd	Cambridge	17.12.04T
G-BLTA	Colt 77A Coil HAB	525		8. 6.84	K.A Schlussler *"James Sadler"*	(Bourne, Lincoln)	7. 8.91A
G-BLTC	Druine D.31A Turbulent	PFA 48-10964		18.12.84	S.J.Butler	(Henlow)	15. 5.05P
	(Built G.P.Smith and A W.Burton) (Volkswagen 1600)						

Registration	Type	C/n	Previous identity	Date	Owner/operator	Location	Expiry
G-BLTF	Robinson R22 Alpha	0428	N8526A	10. 1.85	Brian Seedle Helicopters Ltd	Blackpool	5. 4.04T
G-BLTK	Rockwell Commander 112TC-A	13106	SE-GSD	11.12.84	B.Rogalewski	Blackbushe	26. 5.06
G-BLTM	Robin HR200/100 Club	96	F-GAEC	21.11.84	J.S.Swale tr Barton Robin Group	Barton	15. 1.07
G-BLTN	Thunder Ax7-65 HAB	621		4. 1.85	J.A Liddle *"Frederica"*	(Backwell, Bristol)	3. 9.88A
G-BLTR	Scheibe SF25B Falke	4823	D-KHEC	23. 1.85	V.Mallon	RAF Bruggen	1. 4.94
	(Built Sportavia-Pützer) (Stark-Stamo MS1500)				*(Current status unknown)*		
G-BLTS	Rutan LongEz	PFA 74A-10741		14. 1.85	R.W.Cutler	(Thorverton, Exeter)	
	(Built R.W.Cutler)				*(Current status unknown)*		
G-BLTU	Slingsby T.67B Firefly	2024		16. 1.85	RAF Wyton Flying Club Ltd	RAF Wyton	4. 4.07T
G-BLTW	Slingsby T.67B Firefly	2026		16. 1.85	R.L.Brinklow	Turweston	15. 9.06T
G-BLTY	Westland WG.30 Series 160	019	VT-EKG	14. 1.85	D.Brem-Wilson	Biggin Hill	AC
			G-17-9, G-BLTY, G-17-19		*(Amended CofR 8.04)*		
G-BLUE*	Colting Ax7-77A HAB				See SECTION 3 Part 1	(Aldershot)	
G-BLUI	Thunder Ax7-65 HAB	553		22. 2.85	Susan Johnson *"Rhubarb and Custard"*	(Blackpool)	31. 7.00A
G-BLUL	CEA Jodel DR.1050/M1 Sicile Record	601	F-BMPJ	7. 3.85	J.Owen	Spilsteads Farm, Sedlescombe	24.10.91
					(On overhaul 11.01: current status unknown)		
G-BLUM	Aérospatiale SA365N Dauphin 2	6101		21. 1.85	CHC Scotia Ltd	Humberside	14. 4.05T
G-BLUN	Aérospatiale SA365N Dauphin 2	6114	PH-SSS	21. 1.85	CHC Scotia Ltd	Liverpool	5. 3.05T
			G-BLUN				
G-BLUV	Grob G109B	6336		1. 2.85	R.J.Buckels and S.K.Durso	North Weald	6.12.05
					tr The 109 Flying Group		
G-BLUX	Slingsby T.67M-200 Firefly	2027	G-7-145	31. 1.85	R.L.Brinklow	Rochester	27. 4.07T
			G-BLUX, G-7-113		t/a Richard Brinklow Aviation		
G-BLUZ	de Havilland DH.82B Queen Bee	1435 & SAL.150	LF858	9. 4.85	C.I.Knowles and J.Flynn	RAF Henlow	22. 7.05P
					tr The Bee Keepers Group *(As "LF858")*		
G-BLVA	Airtour AH-31 HAB	AH.004		12. 2.86	A van Wyk *(Noted active 11.02)*	(Caxton, Cambridge)	
G-BLVB	Airtour AH-56 HAB	AH.005		12. 2.86	R.W.Guild *"Bluejay"*	(Vilharino do Bairo, Portugal)	
G-BLVI	Slingsby T.67M Firefly II	2017	(PH-KIF)	1. 2.85	Northern Aviation Ltd	Durham Tees Valley	17. 2.06T
			G-BLVI				
G-BLVK	Avions Mudry CAP.10B	141	JY-GSR	11. 3.85	E.K.Coventry	Childerditch	15. 5.06
G-BLVL	Piper PA-28-161 Warrior II	28-8416109	N43677	11. 2.85	TG Aviation Ltd	Manston	15. 5.06T
G-BLVS	Cessna 150M Commuter	15076869	EI-BLS	19. 2.85	R.Collier	Fenland	9.11.06T
			N45356				
G-BLVW	Cessna F172H	F172-0422	D-ENQU	16. 5.85	R and D Holloway Ltd	Stapleford	10. 7.00
	(Built Reims Aviation SA)						
G-BLVY*	Colt 21A Cloudhopper HAB	634		18. 2.85	Not known	(US)	
			(Cancelled 7.5.91 on sale to Holland) (Active Albuquerque, New Mexico, US 10.03 without marks)				
G-BLWB*	Thunder Ax6-56 Srs 1 HAB	645		22. 2.85	Not known	Not known	10.11.99
					(Cancelled 28.11.01 as wfu) (Noted 4.02)		
G-BLWD	Piper PA-34-200T Seneca II	34-8070334	ZS-KKV	14. 3.85	Bencray Ltd	Blackpool	7. 5.05T
			ZS-XAT, N8253E				
G-BLWF	Robin HR100/210 Safari	183	F-BUSR	8. 3.85	Starguide Ltd	Stapleford	31. 8.06
G-BLWH	Fournier RF6-100	7	F-GADF	3. 4.85	I.R.March *(Noted stored dismantled 4.04)*	Booker	5. 9.03
G-BLWM*	Bristol 20 M.1C rep				See SECTION 3 Part 1	Hendon	
G-BLWP	Piper PA-38-112 Tomahawk	38-78A0367	OY-BTW	7. 6.85	J.C.,T.,T. and H.R Dodd	Panshanger	8. 3.04T
G-BLWT	Evans VP-1 Series 2	PFA 62-10639		27. 3.85	G Malpass	(Copplestone, Crediton)	25. 4.05P
	(Built G B O'Neill) (Volkswagen 1834)						
G-BLWV	Reims/Cessna F152 II	F15201843	EI-BIN	25. 2.85	Redhill Aviation Ltd *(Op Redhill Flying Club)*	Blackbushe	4. 6.06T
G-BLWW*	Aerocar Taylor Mini-Imp Model C	PFA 136-10880		1. 3.85	M.K.Field tr The Brize Group	Sleap	4. 6.87P
	(Built W E Wilks) (Continental O-200-A)				*(Cancelled 13.10.00 by CAA) (Noted 8.03)*		
G-BLWY	Robin R2160	176	F-GCUV	15. 4.85	K.D.Boardman	Perth	9.11.06
			SE-GXE				
G-BLXA	SOCATA TB-20 Trinidad	284	SE-IMO	11. 4.85	Trinidad Flyers Ltd	Blackbushe	28. 9.06
			F-ODOH				
G-BLXG	Colt 21A Cloudhopper HAB	605		2. 5.85	A Walker *"Britannia Park"*	(Richmond, Surrey)	6. 5.98A
G-BLXH	Fournier RF3	39	F-BMDQ	25. 3.85	D.M.Boxell	RAF Colerne	1. 9.05P
(G-BLXI (1))*	Bleriot XI rep				See SECTION 3 Part 1	Cannes, France	
G-BLXI	Scintex CP.1310-C3 Super Emeraude	937	F-BMJI	1. 4.85	R.Howard	Grove Moor Farm, Grassthorpe	11. 4.05P
G-BLXO	SAN Jodel D.150 Mascaret	10	F-BLDB	9. 5.85	P.R.Powell	Allensmore, Hereford	26. 2.05P
G-BLXP	Piper PA-28R-200 Cherokee Arrow II	28R-7235200	N5226T	29. 7.85	M.B.Hamlett	Le Plessis-Belleville, France	5. 8.05
G-BLXR	Aérospatiale AS332L Super Puma	2154		14. 5.85	Bristow Helicopters Ltd *"Cromarty"*	Aberdeen	1. 7.06T
G-BLXT*	Royal Aircraft Factory SE.5A				See SECTION 3 Part 1	Fort Rucker, Alabama, US	
G-BLYA*	Douglas C-53D-DO Skytrooper				See SECTION 3 Part 1	Torp, Norway	
G-BLYD	SOCATA TB-20 Trinidad	518		1. 5.85	Yankee Delta Corporation Ltd	Redhill	23. 3.07T
G-BLYE	SOCATA TB-10 Tobago	521		1. 5.85	G.Hatton	Blackpool	6. 6.07T
G-BLYK	Piper PA-34-220T Seneca III	34-8433083	N4371J	30. 5.85	Trans Euro Air Ltd	(Ash, Canterbury)	31.10.03T
					(New owner 11.04)		
G-BLYP	Robin R3000/120	109		15. 5.85	Weald Air Services Ltd	Headcorn	5. 5.01T
G-BLYT	Airtour AH-77 HAB	AH.008		7. 7.87	I.J.Taylor and R.C Kincaid *"Signal 2"*	(Bristol)	9. 8.03A
G-BLZA	Scheibe SF25B Falke	4684	D-KBAJ	22. 5.85	T.A Lacey	RAF Halton	21. 9.06
	(Stark-Stamo MS1500)				tr Chiltern Gliding Club		
G-BLZB*	Cameron N-65 HAB				See SECTION 3 Part 1	(Aldershot)	
G-BLZE	Reims/Cessna F152 II	F15201579	G-CSSC	3. 5.85	Redhill Aviation Ltd	Redhill	3. 6.07T
			PH-AYF (2)		*(Op Redhill Flying Club)*		
G-BLZF	Thunder Ax7-77 HAB	660		3. 6.85	H.M.Savage *"Hector"*	(Edinburgh)	10. 9.03A
G-BLZH	Reims/Cessna F152 II	F15201965		21. 6.85	G.Price	(Coxheath, Maidstone)	1. 6.07T
					t/a Skytrek Aviation Services		
G-BLZN	Bell 206B JetRanger II	314	ZS-HMV	12. 7.85	GP Helicopters Ltd	(Haywards Heath)	9. 7.04T
			C-GWDH, N1408W		*(New owner 1.05)*		
G-BLZP	Reims/Cessna F152 II	F15201959		10. 7.85	East Midlands Flying School Ltd		
						Nottingham East Midlands	21.12.06T
G-BLZS	Cameron O-77 HAB	479		22. 5.85	C.D.Steel	(St.Boswells, Melrose)	26.11.04A
G-BLZW*	Republic P-47D-30-RA Thunderbolt				See SECTION 3 Part 1	Palm Springs, California, US	

G-BMAA - G-BMZZ

Registration	Type	C/n	Previous identity	Regd	Owner	Base	CofA
G-BMAD	Cameron V-77 HAB	1166		10. 6.85	M.A Stelling *"Nautilus"*	(Bedford)	29. 9.99A
G-BMAL	Sikorsky S-76A II Plus	760120	F-WZSA	27.11.80	CHC Scotia Ltd	Humberside	9. 5.07T
			G-BMAL				
G-BMAO	Taylor JT.1 Monoplane	PFA 1411		29. 7.85	S.J.Alston	Hinton-in-the-Hedges	5. 9.05P
	(Built V A Wordsworth)						
G-BMAV	Aérospatiale AS350B Ecureuil	1089		1. 6.79	PLM Dollar Croup Ltd	Dublin Heliport/Westpoint	4. 8.06T
G-BMAX	Clutton FRED Series II	PFA 29-10322		20.12.78	D.A Arkley	(Chelmsford)	24. 8.99P
	(Built P Cawkwell and D.A Arkley) (Volkswagen 1834)				*(Current status unknown)*		
G-BMAY	Piper PA-18-135 Super Cub	18-3925	OO-LWB	3. 7.85	R.W.Davies	Little Robhurst Farm, Woodchurch	15.11.04T
	(L-21B-PI) *(Frame No.18-3961)*		"EI-229", I-EIJZ, MM54-2525, 54-2525				
G-BMBB	Reims/Cessna F150L	F15001136	OO-LWM	2. 8.85	M.Bonsall	Netherthorpe	21. 1.06T
			PH-GAA				
G-BMBJ	Schempp-Hirth Janus CM	20/209	(G-BLZL)	9. 9.85	J.Hallam tr BJ Flying Group	(Ashby-de-la-Zouch)	30. 6.07
G-BMBS	Colt 105A HAB	704		18. 7.85	H.G.Davies	(Cheltenham)	27. 8.91A
G-BMBW	Bensen B.8MR	PFA G/01-1064		27. 8.85	M.E.Vahdat	Uxbridge	30. 6.93P
	(Built M.E.Vahdat - c/n MV-001) (Rotax 503)						
G-BMBZ	Scheibe SF25E Super Falke	4322	D-KEFQ	17. 7.85	Cornish Gliding and Flying Club Ltd	Perranporth	19. 5.07
	(Limbach SL1700)						
G-BMCC	Thunder Ax7-77 HAB	705		12. 7.85	A K. and C.M.Russell *"Charlie Charlie"*	(Stafford)	23. 2.99A
G-BMCD	Cameron V-65 HAB	1234		26. 6.85	M.C.Drye *"My Second Fantasy"*	(Winkfield)	4. 3.04A
G-BMCG	Grob G109B	6362	(EAF673)	25. 7.85	Lagerholm Finnimport Ltd	Booker	26. 7.07
G-BMCI	Reims/Cessna F172H	F17200683	OO-WID	19. 8.85	A B.Davis *(Op Edinburgh Flying Club)*	Edinburgh	13.11.04T
G-BMCN	Reims/Cessna F152 II	F15201471	D-ELDM	7. 8.85	Eastern Air Centre Ltd	Gamston	9. 3.07T
G-BMCS	Piper PA-22-135 Tri-Pacer	22-1969	5Y-KMH	6. 9.85	P.R.Deacon	Headcorn	15. 7.01
			VP-KMH, ZS-DJI		*(Noted 8.03)*		
G-BMCV	Reims/Cessna F152 II	F15201963		2.10.85	Leicestershire Aero Club Ltd	Leicester	4. 5.07T
G-BMCW	Aérospatiale AS332L Super Puma	2161	F-WYMG	4.10.85	Bristow Helicopters Ltd	(China)	7.11.05T
			G-BMCW		*"Monifieth"*		
G-BMCX	Aérospatiale AS332L Super Puma	2164		7.10.85	Bristow Helicopters Ltd *"Lossiemouth"*	(China)	14.11.07T
G-BMDB	Replica Plans SE.5A	PFA 20-10931		12. 8.85	D.Biggs	Lee-on-Solent	9. 6.05P
	(Built D.Biggs) (Continental O-200-A)				*(As "F235/B" in RFC c/s)*		
G-BMDC	Piper PA-32-301 Saratoga	32-8006075	OO-PAC	13. 8.85	J.D.M.Tickell	Aberdeen	5.10.06T
			OO-HKK, N8242A		t/a MacLaren Aviation		
G-BMDE	Pietenpol AirCamper	PFA 47-10989		12. 8.85	P.B.Childs	New Farm, Felton	18. 8.03P
	(Built D Silsbury) (Continental O-200-A)						
G-BMDJ	Price Ax7-77S HAB	003		1. 8.85	R.A Benham	(Burton-on-Trent)	
	(Built T P Price - c/n TPB.1) (Regd as Price TPB.1)				*"Wings of Phoenix" (New owner 10.01)*		
G-BMDK	Piper PA-34-220T Seneca III	34-8133155	ZS-LOS	16. 9.85	Air Medical Fleet Ltd	Oxford	24.11.04T
			N84209, N9553N				
G-BMDP	Partenavia P64B Oscar 200	08	HB-EPQ	20. 8.85	S.T.G.Lloyd	Cardiff	25. 5.04
G-BMDS	Wassmer Jodel D.120 Paris-Nice	281	F-BMOS	12. 8.85	R.T.Mosforth *(New owner 4.04)*	Netherthorpe	25. 3.05P
G-BMEA	Piper PA-18 Super Cub 95	18-3204	(D-ECZF)	27. 8.85	M.J.Butler	Spanhoe	23.10.05P
	(L-18C-PI)		Belg AF OL-L07, L-130, 53-4804				
	(Frame No. reported as 18-3206 [c/n 18-3194 ex OL-L20/L-120/53-4794]: c/n 18-3204 has Frame No.18-3216)						
G-BMEB	RotorWay Scorpion 145	2896	VR-HJB	10.12.85	P.Trainor	(Newry, County Armagh)	
	(Built Hong Kong Aircraft Engineering Co Ltd)						
G-BMEE	Cameron O-105 HAB	1189		4. 9.85	A G.R.Calder	(Los Angeles, California, US)	8.10.89A
G-BMEG	SOCATA TB-10 Tobago	530		23.10.85	P.Farmer	Luxembourg	17.11.04
G-BMEH	Jodel D.150 Special Super Mascaret	PFA 151-11047		15. 8.85	R.J.and C.J.Lewis	Garston Farm, Marshfield	21. 5.05P
	(Built E J Horsfall) (Lycoming O-235) *(Rebuild of incomplete SAN Jodel D.150 Mascaret c/n 62)*						
G-BMET	Taylor JT.1 Monoplane	PFA 1465		4. 9.85	M.K.A Blyth	Little Gransden	18. 8.05P
	(Built M Blythe)						
G-BMEU	Isaacs Fury II	PFA 11-10179		11. 9.85	I.G. Harrison	Egginton	
	(Built G R G Smith) (Salmson 90hp)				*(90% complete 6.99: new owner 11.02)*		
G-BMEW*	Lockheed 18-56 Lodestar				See SECTION 3 Part 1	Gardermoen, Norway	
G-BMEX	Cessna A150K Aerobat	A1500169	N8469M	18. 9.85	N.A M.and R.A Brain	Netherthorpe	19. 5.05
G-BMEZ*	Cameron DP-70 HA Airship				See SECTION 3 Part 1	Newbury	
G-BMFB*	Douglas Skyraider AEW.1 (AD-4W)				See SECTION 3 Part 1	Tillamook, Oregon, US	
G-BMFD	Piper PA-23-250 Aztec F	27-7954080	G-BGYY	6. 9.79	Gold Air International Ltd	Cambridge	15. 6.07T
			N6834A, N9741N				
G-BMFG	Dornier Do.27A-1	27-1003-342	FAP 3460	23. 9.85	R.F.Warner t/a Sigma Services	(Broughton, Norfolk)	AC
			AC+955		*(On rebuild 2.99: current status unknown)*		
G-BMFI	PZL-Bielsko SZD-45A Ogar	B-657		23. 9.85	S.L.Morrey	Andreas, Isle of Man	29. 4.05
G-BMFN	QAC Quickie Tri-Q 200	PFA 94A1-11062		27. 9.85	A H.Hartog	Thruxton	1. 5.02P
	(Built EMK Aeroplane Ltd - c/n EMK.017) (Continental O-200-A)				*(Noted 8.04)*		
G-BMFP	Piper PA-28-161 Warrior II	28-7916243	N3032L	1.11.85	T.J.Froggatt and C.A Lennard	Blackbushe	15. 7.07
					tr Bravo Mike Fox Papa Group		
G-BMFU	Cameron N-90 HAB	628		1.10.85	J.J.Rudoni	(Rugeley)	19. 5.03T
G-BMFY	Grob G109B	6401		8.10.85	P.J.Shearer	Kirkwall	3. 7.07
G-BMFZ	Reims/Cessna F152 II	F15201953		3.12.85	Cornwall Flying Club Ltd	Bodmin	9. 2.07T
G-BMGB	Piper PA-28R-200 Cherokee Arrow II	28R-7335099	N15864	8.11.85	A.L.Ings t/a Malmesbury Specialist Cars	Kemble	16. 8.04
G-BMGC*	Fairey Swordfish II	-	G-BMGC	23.10.85	Royal Navy Historic Flight	RNAS Yeovilton	
	(Built Blackburn Aircraft)		RCN W5856, RN W5856		*(Cancelled 2.9.91 by CAA)*		
					(As "W5856/A2A" in 810 Sqdn c/s "City of Leeds") (Noted 11.03)		
G-BMGG	Cessna 152 II	15279592	OO-ADB	10.10.85	A S Bamrah	Shoreham	24.10.06T
			PH-ADB, D-EHUG, F-GBLM, N757AT		t/a Falcon Flying Services		
G-BMGR	Grob G109B	6396		27.11.85	M.Clarke tr BMGR Group	Lasham	9. 2.07
G-BMHA	Rutan LongEz	PFA 74A-10973		18.10.85	S.F.Elvins	(Staple Hill, Bristol)	
	(Built S.F.Elvins)				*(Current status unknown)*		
G-BMHC	Cessna U206F Stationair II	U20603427	N10TB	17.11.76	P.Marsden	Sorbie Farm, Kingsmuir	19. 2.07T
			G-BMHC, N8571Q				

Reg	Type	C/n	Previous identity	Reg date	Owner	Base	CofA
G-BMHJ	Thunder Ax7-65 Series 1 HAB	743		2. 1.86	M.G.Robinson *"Kittylog"*	(Great Milton, Oxford)	19. 5.92A
G-BMHL	Wittman W.8 Tailwind *(Built T G Hoult)*	PFA 31-10503		28.11.85	C.R.Nash	Old Sarum	5. 4.05P
G-BMHS	Reims/Cessna F172M	F17200964	PH-WAB	7. 4.86	R.A Hall tr Tango Xray Flying Group	Rayne Hall Farm, Braintree	21. 7.07
G-BMHT	Piper PA-28RT-201T Turbo Arrow IV	28R-8231010	ZS-LCJ N8462Y	18.11.85	G.Lungley	Leeds-Bradford	22. 6.07T
G-BMID	Wassmer Jodel D.120 Paris-Nice	259	F-BMID	18. 8.81	P.E.S.Latham tr G-BMID Flying Group	RAF Shawbury	16. 5.05P
G-BMIG	Cessna 172N Skyhawk II	17272376	ZS-KGI (N48630)	13. 5.86	R.B. Singleton-McGuire tr BMIG Group	Elstree	21. 7.07T
G-BMIM	Rutan LongEz *(Built K A I Christensen)* (Lycoming O-235)	8102/160	OY-CMT OY-8102	12.12.85	R.M.Smith	Biggin Hill	1. 8.05P
G-BMIO	Stoddard-Hamilton Glasair RG *(Buillt A H Carrington)* (Lycoming O-360)	PFA 149-11016		25.11.85	J.M.Ayres and S.C.Ellerton	Gloucestershire	3 6.05P
G-BMIP	Wassmer Jodel D.112	1264	F-BMIP	7.12.78	F.J.E.Brownsill	(Fairford)	3.11.04P
G-BMIR (2)*	Westland Wasp HAS.1	F9670	XT788	24. 1.86	Park Aviation Supply Little Glovers Farm, Charlwood, Surrey *(Cancelled 22.12.95 by CAA) (Exhibited in Flightaid's travelling roadshow as "XT78?/316" in Royal Navy c/s 4.04)*		
G-BMIS	Monnett Sonerai II *(Built B A Bower - c/n 755)* (Revmaster R2100DQ)	PFA 15A-10813	VR-HIS	26. 2.87	S.R Edwards *(New owner 5.04)*	Kemble	26.10.89P
G-BMIV	Piper PA-28R-201T Turbo Cherokee Arrow III	28R-7703154	ZS-JZW N5816V	7. 1.86	Firmbeam Ltd	Booker	5. 7.07
G-BMIW	Piper PA-28-181 Archer II	28-8190093	ZS-KTJ N8301J	6.12.85	Oldbus Ltd	Shoreham	14. 5.07T
G-BMIY	Oldfield Baby Lakes *(Built J B Scott, Parkinson and Brown)* (Continental O-200-A)	PFA 10-10194	G-NOME	3.12.85	J.B.Scott *(Stored 12.01)*	Blackpool	27. 8.87P
G-BMJA	Piper PA-32R-301 Saratoga SP	32R-8113019	ZS-KTH N8309E	23.12.85	H.Merkado	Panshanger	15. 8.05T
G-BMJB*	Cessna 152 II	15280030	N757VD	3. 2.86	Not known *(Cancelled 12.06.00 as wfu) (Noted 11.03).*	Wolverhampton	
G-BMJC	Cessna 152 II	15284989	N623AP	3. 2.86	Northern Aviation Ltd	Durham Tees Valley	22. 7.07T
G-BMJD	Cessna 152 II	15279755	N757HP	21.11.85	Donair Flying Club Ltd	Tatenhill	3 8.07T
G-BMJL	Rockwell Commander 114	14006	A2-JRI ZS-JRI, N1906J	8. 1.86	D.J.and S.M.Hawkins	(Woking)	10. 7.06T
G-BMJM	Evans VP-1 *(Built J A Mawby)* (Volkswagen 1834)	PFA 62-10763		21.11.85	S.E.Clarke	(Coventry)	1. 9.05P
G-BMJN	Cameron O-65 HAB	1212		6.12.85	P.M.Traviss *"F'red"*	(Yarm)	29. 5.05A
G-BMJO	Piper PA-34-220T Seneca III	34-8533036	N6919K N9565N	5.12.85	Deep Cleavage Ltd	Exeter	5. 1.07T
G-BMJR	Cessna T337H Turbo Super Skymaster II	33701895	G-NOVA N1259S	10. 7.84	Eastcote Services Ltd	Cranfield	10.10.05
G-BMJT	Beech 76 Duchess	ME-376	ZS-KMI N3718W	4.12.85	Mike Osborne Properties Ltd	Elstree	1. 4.07
G-BMJX	Wallis WA-116/X Series 1 (Limbach L-2000)	K/219/X		31.12.85	K.H.Wallis *(Stored 8.01)*	Reymerston Hall, Norfolk	1. 4.89P
G-BMJY	SPP Yakovlev Yak C-18A	Not known	(France) Egypt AF 627	21. 1.86	R.J.Lamplough *(As "07" (yellow) in Russian AF c/s)*	Manor Farm, East Garston	27.11.01P
G-BMJZ	Cameron N-90 HAB	1219		16.12.85	P.Spellward *"Uvistat"* tr Bristol University Hot Air Ballooning Society	(Bristol)	31. 3.94A
G-BMKB	Piper PA-18-135 Super Cub (L-21B-PI) *(Frame No.18-3818)*	18-3817	OO-DKB PH-DKB, (PH-GRP), R Neth AF R-127, 54-2417	11.12.85	Cubair Flight Training Ltd	Redhill	3. 7.06T
G-BMKC	Piper J-3C-65 Cub (L-4H-PI) (Continental C90)*(Frame No.10970)*	11145	F-BFBA 43-29854	2. 1.86	J.W.Salter *"Little Rockette Jnr"* *(As "329854/44-R" in USAAC 533rd BS/381st Bomb Group c/s)*	Newtownards	11.10.04P
G-BMKD	Beech C90A King Air	LJ-1069	N223CG N67516	30.12.85	A E.Bristow	Fairoaks	13. 4.06
G-BMKF	CEA Jodel DR.221 Dauphin	96	F-BPCS	3. 2.86	S.T. and L.A Gilbert	Enstone	13. 8.06
G-BMKG	Piper PA-38-112 Tomahawk II	38-82A0050	ZS-LGC N91544	3. 2.86	Glasgow Aviation Ltd	Cumbernauld	15.12.07E
G-BMKI	Colt 21A Cloudhopper HAB	753		30.12.85	A C.Booth	(Bristol)	31.12.04A
G-BMKJ	Cameron V-77 HAB	1235		2. 1.86	R.C.Thursby	(Barry)	17. 5.05A
G-BMKK	Piper PA-28R-200 Cherokee Arrow II	28R-7535265	ZS-JNY N9537N	16. 1.86	Comed Schedule Services Ltd	Blackpool	5. 4.07T
G-BMKP	Cameron V-77 HAB	724	(G-BMFX)	10. 1.86	R.Bayly *"And Baby Makes 10" (Noted active 8.02)*	(Clutton, Bristol)	7. 8.93A
G-BMKR	Piper PA-28-161 Warrior II	28-7916220	G-BGKR N9561N	14. 6.84	D.R.Shrosbee tr Field Flying Group	Goodwood	5. 6.06
G-BMKW	Cameron V-77 HAB	608		29. 1.86	M.H.Redman *(New owner 6.04)*	(Stalbridge, Sturminster Newton)	21. 9.00A
G-BMKX*	Cameron Elephant 77SS HAB				See SECTION 3 Part 1	(Aldershot)	
G-BMKY	Cameron O-65 HAB	1246		4. 3.86	Ann R.Rich *"Orion"*	(Hyde)	11. 4.05A
G-BMLB	Wassmer Jodel D.120A Paris-Nice	295	F-BNCI	20. 1.86	C.A.Croucher	Headcorn	23. 5.05P
G-BMLC	Short SD.3-60 Variant 100	SH.3688	SE-LDA G-BMLC, G-14-3688	18. 2.86	Emerald Airways Ltd	Coventry	24. 5.05T
G-BMLJ	Cameron N-77 HAB	1263		7. 3.86	C.J.Dunkley t/a Wendover Trailers *"Mr Funshine"*	(Aylesbury)	26. 3.03A
G-BMLK	Grob G109B	6424		24. 2.86	J.J.Mawson tr Brams Syndicate	Rufforth	3. 6.07
G-BMLL	Grob G109B	6420		13. 3.86	C.Rupasinha tr G-BMLL Flying Group	Denham	25. 8.07
G-BMLM	Beech 95-58 Baron	TH-405	N111LM G-BMLM, F-GEPV, 3D-ADF, ZS-LOZ, G-BMLM, G-BBJF	2. 7.79	Atlantic Bridge Aviation Ltd	Lydd	5. 3.06T
G-BMLS	Piper PA-28R-201 Cherokee Arrow III	28R-7737167	N47496	11. 2.86	R.M.Shorter	Booker	24. 4.05T
G-BMLT	Pietenpol AirCamper *(Built R A and F Hawke)* (Continental C90)	PFA 47-10949		28. 1.86	W.E.R.Jenkins	Waits Farm, Belchamp Walter	28. 4.05P
G-BMLW	Cameron O-77 HAB	813		6. 2.86	M.L. and L.P.Willoughby *"Stelrad"*	(Woodcote, Reading)	7. 8.95A
G-BMLX	Reims/Cessna F150L	F15000700	PH-VOV	21. 3.86	J.P.A.Freeman	Headcorn	20.12.04T

G-BMMC	Cessna 310Q	310Q0041	YU-BGY	11. 2.86	Airtime Aviation Paint Ltd	Bournemouth	30. 3.02
			N7541Q		*(On slow overhaul 12.04)*		
G-BMMF	Clutton FRED Series II	PFA 29-10296		20. 2.86	E.C.King	Eastbach Farm, Coleford	18. 7.03P
	(Built J M Jones) (Volkswagen 1834)				*"Thankyou Girl"*		
G-BMMI	Pazmany PL-4A	PFA 17-10149		6. 2.86	P.I.Morgans	Haverfordwest	18. 7.03P
	(Built M L Martin) (Continental PC 60)				*(Noted dismantled on trailer 8.03)*		
G-BMMK	Cessna 182P Skylane II	18264117	OO-AVU	24. 3.86	G.G.Weston	Denham	7.10.07E
	(Reims-assembled c/n F18200038)		N6129F				
G-BMML	Piper PA-38-112 Tomahawk	38-80A0079	PH-TMG	2. 4.86	J.C.and C.H.Strong	Thruxton	10. 5.07T
			OO-HKD, N9662N				
G-BMMM	Cessna 152 II	15284793	N4652P	10. 9.86	A S.Bamrah t/a Falcon Flying Services	Biggin Hill	20.10.04T
G-BMMP	Grob G109B	6432		27. 6.86	E.W.Reynolds	Tatenhill	24. 5.05
G-BMMV	ICA IS-28M2A	57		10. 3.86	J.F.Miles	Wing Farm, Longbridge Deverill	24. 8.06
G-BMMW	Thunder Ax7-77 HAB	782		10. 3.86	P.A George *"Ethos"*	(Princes Risborough)	3. 6.96A
G-BMMY	Thunder Ax7-77 HAB	716		11. 3.86	S.M.Wade and Sheila E.Hadley *"Winco"*	(Salisbury)	5. 4.04A
G-BMNL	Piper PA-28R-200 Cherokee Arrow II	28R-7535040	N32280	17. 9.86	Elston Ltd	Elstree	29. 5.05
			(N18MW), N32280		tr Arrow Flying Group		
G-BMNV	SNCAN Stampe SV-4C	108	F-BBNI	14. 3.86	Wessex Aviation and Transport Ltd	Haverfordwest	29. 8.03P
	(Lycoming IO-360)						
G-BMOE	Piper PA-28R-200 Cherokee Arrow II	28R-7635226	PH-PCB	20. 5.86	E.P.C.Rabson	Lee-on-Solent	22.11.05T
			OO-HAS, N9221K				
G-BMOF	Cessna U206G Stationair 6	U20603658	N7427N	17. 4.86	D.M.Penny	Movenis	23. 3.06
					tr Wild Geese Skydiving Centre		
G-BMOG	Thunder Ax7-77 HAB	793		2. 4.86	R.M.Boswell	(Bawburgh, Norwich)	28. 8.95A
G-BMOH	Cameron N-77 HAB	1270		2. 4.86	P.J.Marshall and M.A Clarke *"Ellen Gee"*	(Ruislip)	24.11.03A
G-BMOI	Partenavia P68B	103	I-EEVA	4. 4.86	Simmette Ltd	Exeter	16. 6.07
G-BMOK	ARV Aviation ARV-1 Super 2	011		14. 4.86	R.E.Griffiths	Middle Stoke, Isle of Grain	19. 11.06
G-BMOM	ICA IS-28M2A	50		30. 6.86	R.M.Cust	Sandtoft	5. 9.04
	(Rebuilt 2001 with forward fuselage of G-BKAB)						
G-BMOT	Bensen B.8M	PFA G/01-1066		17. 4.86	Austin Trueman Ltd	Henstridge	13. 8.01P
	(Built R S W Jones) (Volkswagen 1834)				*(Noted 12.02)*		
G-BMOV	Cameron O-105 HAB	1307		11. 4.86	Cheryl Gillott *"Up and Down"*	(Stroud)	1. 7.99A
G-BMPC	Piper PA-28-181 Cherokee Archer II	28-7790436	LN-NAT	23. 4.86	C.J. and R.J.Barnes	Nottingham East Midlands	16. 3.05T
G-BMPD	Cameron V-65 HAB	1200		4. 6.86	D.Triggs	(Alresford)	9. 4.05A
G-BMPL	Optica OA.7 Optica	016		14. 4.86	Aces High Ltd *(Noted 8.04)*	Dunsfold	2. 8.97T
G-BMPP	Cameron N-77 HAB	1303		15. 4.86	P.F.Smart *"Tuppence"*	(Basingstoke)	14. 5.93A
					tr The Sarnia Balloon Group*(Inflated 4.02: new owner 5.02)*		
G-BMPR	Piper PA-28R-201 Arrow III	28R-7837175	ZS-LMF	22. 4.86	B.Edwards	Full Sutton	20. 5.05
			N417GH				
G-BMPS	Strojnik S-2A	045		18. 4.86	G.J.Green	(Matlock)	
	(Built T J Gardiner)						
G-BMPY	de Havilland DH.82A Tiger Moth	"82619"	ZS-CNR	25. 4.86	S.M.F.Eisenstein	Sandford Hall, Knockin	27. 3.03
			SAAF??				
G-BMRA	Boeing 757-236	23710		2. 3.87	DHL Air Ltd	Brussels, Belgium	2. 8.06T
G-BMRB	Boeing 757-236	23975		25. 9.87	DHL Air Ltd	Brussels, Belgium	29. 9.06T
G-BMRC	Boeing 757-236	24072	(N....)	2.12.87	DHL Air Ltd	Brussels, Belgium	26. 1.07T
G-BMRD	Boeing 757-236	24073	(N....)	2.12.87	DHL Air Ltd	Nottingham East Midlands	3. 3.07T
			G-BMRD				
G-BMRE	Boeing 757-236	24074	(N)	2.12.87	DHL Air Ltd	Nottingham East Midlands	28. 4.07T
G-BMRF	Boeing 757-236	24101		13. 5.88	DHL Air Ltd	Brussels, Belgium	17. 5.07T
G-BMRH	Boeing 757-236	24266		21. 2.89	DHL Air Ltd	Brussels, Belgium	28. 2.05T
G-BMRJ	Boeing 757-236	24268		6. 3.89	DHL Air Ltd	Brussels, Belgium	13. 3.05T
G-BMSA	Stinson HW-75 Voyager	7040	G-BCUM	26. 3.86	M.A Thomas	Barton	9. 9.05P
			F-BGQO, NC21189		tr The Stinson Group *"Iron Eagle"*		
G-BMSB	Vickers Supermarine 509 Spitfire T.IX	CBAF.7722	G-ASOZ	3. 5.78	M.S.Bayliss	East Kirkby	25. 4.05P
			IAC158, G-15-171, MJ627		*(As "MJ627/9G-P" in 441 Sqdn c/s)*		
G-BMSC	Evans VP-2	PFA 63-10785		25. 8.82	M.S.Barron	Sittles Farm, Alrewas	29. 4.05P
	(Built Youth Opportunity Project - c/n V2-482MSC) (Volkswagen 1834)						
G-BMSD	Piper PA-28-181 Cherokee Archer II	28-7690070	EC-CVH	2. 7.86	H.Merkado	Panshanger	21. 9.07T
			N9646N				
G-BMSE	Valentin Taifun 17E	1082	D-KHVA(17)	20. 5.86	A.J.Nurse	Kemble	19. 7.02
G-BMSF	Piper PA-38-112 Tomahawk	38-78A0524	N4277E	9. 2.79	B.Catlow	Haverfordwest	30. 6.99
G-BMSG	SAAB 32A Lansen	32028	Fv.32028	22. 7.86	J.E.Wilkie *(Open store in bare-metal finish.8.04)*	Cranfield	AC
G-BMSL	Clutton FRED Series III	PFA 29-11142		19. 5.86	A.C.Coombe	Long Marston	22. 8.05P
	(Built A.C.Coombe)						
G-BMST*	Cameron N-31 HAB				See SECTION 3 Part 1	(Aldershot)	
G-BMSU	Cessna 152 II	15279421	N714TN	29. 8.86	S.Waite tr G-BMSU Group	Gamston	24. 6.06T
G-BMTA	Cessna 152 II	15282864	N89776	27. 8.86	Alarmond Ltd	Perth	16. 6.05T
G-BMTB	Cessna 152 II	15280672	N25457	19. 8.86	Sky Leisure Aviation (Charters) Ltd	Blackbushe	7. 6.06T
G-BMTJ	Cessna 152 II	15285010	N6389P	19. 6.86	The Pilot Centre Ltd	Denham	17. 6.07T
G-BMTN	Cameron O-77 HAB	1305		4. 6.86	Industrial Services (MH) Ltd	(Bristol)	1. 6.97A
					t/a Flete Rental *"Fletie"*		
G-BMTO	Piper PA-38-112 Tomahawk II	38-81A0051	N25679	28.11.86	A.S.Bamrah t/a Falcon Flying Services	Biggin Hill	11. 8.05T
G-BMTR	Piper PA-28-161 Warrior II	28-8116119	N83179	19. 6.86	Aeros Leasing Ltd	Gloucestershire	27. 3.05T
G-BMTS	Cessna 172N Skyhawk II	17270606	N739KP	17. 7.86	A.S.Bamrah t/a Falcon Flying Services	Shoreham	23. 8.07T
G-BMTU	Pitts S-1E	PFA 09-10801		4. 6.86	B.Brown and Eileen Evans	Breighton	28. 9.05P
	(Built O R Howe)						
G-BMTX	Cameron V-77 HAB	733		19. 6.86	J.A Langley	(Stroud)	8. 8.05A
					(Buses for Bristol titles) "Boondoggle"		
G-BMUD	Cessna 182P Skylane	18261786	OY-DVS	6.11.81	Mescal E.Taylor	Turweston	31. 7.03T
			N78847		*(On rebuild 7.04)*		
G-BMUG	Rutan LongEz	PFA 74A-10987		17. 6.86	A G.Sayers	Perth	12. 1.05P
	(Built P Richardson) (Lycoming O-235)						

Registration	Type	C/n	Previous identity	Regd date	Owner/operator	Base	CofA/expiry
G-BMUJ*	Colt Drachenfisch SS HAB *(Futuristic special shape)*	835		3. 6.86	(Virgin Airship and Balloon Co Ltd) *"Drachenfisch/Dragon Fish" (* *Cancelled 4.11.03 as wfu) (Stored in good condition 2004)*	Not known	27. 7.91A
G-BMUK*	Colt UFO SS HAB *(Futuristic special shape)*	836		3. 6.86	(Virgin Airship and Balloon Co Ltd) *"UFO/Dream Station"* *(Cancelled 4.11.03 as wfu) (Stored in good condition 2004)*	Not known	26. 4.95A
G-BMUL*	Colt Kindermond SS HAB *(Futuristic special shape)*	837		3. 6.86	(Virgin Airship and Balloon Co Ltd) *"Kindermond/Childrens' Moon"* *(Cancelled 4.11.03 as wfu) (Stored in good condition 2004)*	Not known	26. 9.91A
G-BMUN*	Cameron Harley 78SS HAB				See SECTION 3 Part 1	Balleroy, Normandy, France	
G-BMUO	Cessna A152 Aerobat	A1520788	4X-ALJ N7328L	4. 6.86	Sky Leisure Aviation (Charters) Ltd	Shoreham	14. 9.07T
G-BMUR*	Cameron Zero 25 HA Airship				See SECTION 3 Part 1	(Aldershot)	
G-BMUT	Piper PA-34-200T Seneca II	34-7570320	EC-CUH N3935X	23. 1.87	G-DAD Air Ltd	Booker	14.11.05T
G-BMUU	Thunder Ax7-77 HAB	827		1. 8.86	A R.Hil *"Fiesta" (New owner 3.03)*	(Cranbrook)	29.10.98A
G-BMUZ	Piper PA-28-161 Warrior II	28-8016329	EC-DMA N9559N	24. 7.86	Newcastle Aero Club Ltd	Newcastle	5. 3.05T
G-BMVA	Scheibe SF25B Falke (Limbach SL1700	46223	RAFGGA.512 D-KAEN	28. 7.86	M.L.Jackson	Headcorn	19. 2.05
G-BMVB	Reims/Cessna F152 II	F15201974		10. 9.86	N.D.Plumb	Hinton-in-the-Hedges	10. 2.00T
G-BMVG	QAC Quickie Q.1 *(Built P.M.Wright)* (Rotax 503)	PFA 94-10749		11. 6.86	P.M.Wright *(Noted 5.03)*	Cambridge	1. 1.02P
G-BMVL	Piper PA-38-112 Tomahawk	38-79A0033	N2391B	5. 9.86	Airways Aero Associations Ltd *(Op British Airways Flying Club)*	Booker	20.12.07E
G-BMVM	Piper PA-38-112 Tomahawk	38-79A0025	N2359B	5. 9.86	Airways Aero Associations Ltd *(Op British Airways Flying Club)*	Booker	17. 3.07T
G-BMVS	Cameron Benihana 70SS HAB *(Aka Chef"s Hat)*	1252		27.10.86	Benihana (UK) Ltd	(London W1)	24. 4.04A
G-BMVT	Thunder Ax7-77A HAB	102	SE-ZYY	15. 7.86	M.L. and L.P.Willoughby *"Trygg Hansa"*	(Woolcote, Reading)	
G-BMVU	Monnett Moni *(Built S R Lee)* (kEF-107)	PFA 142-10948		14. 8.86	R.M.Edworthy *(New owner 3.03)*	(Derby)	20. 9.99P
G-BMVW	Cameron O-65 HAB	1331		27. 6.86	S.P.Richards *"Olau Ferries"*	(Cranbrook)	15. 8.91A
G-BMWA	Hughes 269C	14-0271	N8998F	1. 7.86	R.J.H.Strong	(Yeovil)	19.12.05T
G-BMWE	ARV Aviation ARV-1 Super 2	012		1. 7.86	R.J.N.Noble	Farnborough	13. 5.07
G-BMWF	ARV Aviation ARV-1 Super 2 (Rotax 914 Turbo)	013		1. 7.86	N.R.Beale *(Under re-build 7.96: current status unknown)*	Deppers Bridge, Southam	2. 4.90T
G-BMWM	ARV Aviation ARV-1 Super 2	020		30. 3.87	P.F.Lorriman	Rochester	8. 8.05P
G-BMWN*	Cameron Temple 80SS HAB				See SECTION 3 Part 1	Balleroy, Normandy, France	
G-BMWR	Rockwell Commander 112A	365	N1365J	23. 9.86	M. and J.Edwards	Fairoaks	23. 5.05
G-BMWU	Cameron N-42 HAB	1346		22.12.88	I.Chadwick *"Baby Helix" (Op Balloon Preservation Group) (Active 10.02)*	(Partridge Green)	
G-BMWV	Putzer Elster B	024	D-EEKB 97+14, D-EBGI	5. 8.86	E.A J.Hibbard *(Noted 5.03)*	Hill Farm, Nayland	
G-BMXA	Cessna 152 II	15280125	N757ZC	14. 7.86	I.R.Chaplin	Andrewsfield	28. 9.05T
G-BMXB	Cessna 152 II	15280996	N48840	14. 7.86	Rybec Ltd	Leicester	19.11.06T
G-BMXC	Cessna 152 II	15280416	N24858	14. 7.86	Devon School of Flying Ltd	Dunkeswell	1. 4.05T
G-BMXD	Fokker F.27 Friendship 500	10417	TF-FLR HL5210, (HL5206), PH-FOR	6.10.86	BAC Express Airlines Ltd *"Scottish Trader"*	Gatwick	12.12.05E
G-BMXJ	Reims/Cessna F150L	F15000853	F-BUBA	18. 7.86	R.Harman tr Arrow Aircraft Group	Tatenhill	20. 6.03
G-BMXL	Piper PA-38-112 Tomahawk	38-80A0018	N25060	4. 9.86	Airways Aero Associations Ltd *(Op British Airways Flying Club) (Benyhone Tartan t/s)*	Booker	17. 6.05T
G-BMXM	Colt 180A HAB	838	(C-) G-BMXM	28. 7.86	D.A Michaud	(Sprotborough, Doncaster)	20.12.04A
G-BMXX	Cessna 152 II	15284953	N5469P	10. 9.86	Evensport Ltd	(Rayleigh)	1. 9.02T
G-BMYC	SOCATA TB-10 Tobago	696		1. 9.86	Elizabeth A Grady	Old Buckenham	29. 4.05T
G-BMYD	Beech A36 Bonanza	E-2350		28.11.86	Seabeam Partners Ltd	Wellesbourne Mountford	21. 3.05
G-BMYF	Bensen B.8M *(Built P Entwistle)*	PE-01		18. 8.86	G.Callaghan	Rich Hill, County Armagh	
G-BMYG	Reims/Cessna FA152 Aerobat	FA1520365	OO-JCA (OO-JCC), PH-AXG	23.10.86	Greer Aviation Ltd	Prestwick	13. 6.05T
G-BMYI	Grumman AA-5 Traveler	AA5-0568	EI-BJF F-BVRM, N9568L	1. 9.86	W.C. and S.C.Westran	Shoreham	16. 2.07T
G-BMYJ	Cameron V-65 HAB	726		8. 9.86	S.P.Harrowing *"Skylark II"*	(Port Talbot)	29. 4.05A
G-BMYP	Fairey Gannet AEW.3 *(Built Westland Aircraft Ltd)*	F.9461	8610M XL502	16. 9.86	D.Copley *(As "XL502" in 849 Sqdn/"B" Flight RN c/s: in open store 11.03)*	Sandtoft	29. 9.89P
G-BMYS	Thunder Ax7-77Z HAB	887		3.11.86	J.E.Weidema *(Op Pinkel Balloons)*	(Baambrugge, The Netherlands)	1. 6.01A
G-BMYU	Wassmer Jodel D.120 Paris-Nice	289	F-BMYU	23. 6.78	N.P.Chitty	Drayton Manor, Drayton St.Leonard	4.11.05P
G-BMZB	Cameron N-77 HAB	1370		30.10.86	D.C.Eager *"Dreamland"*	(Bracknell)	30. 4.95A
G-BMZE	SOCATA TB-9 Tampico	708		5.12.86	H.J.Samples	(Radcliffe-on-Trent, Nottingham)	15. 8.05T
G-BMZF*	WSK-Mielec LIM-2				See SECTION 3 Part 1	RNAS Yeovilton	
G-BMZN	Everett Gyroplane Series 1 (Volkswagen 1835)	008		13.11.86	K.Ashford	(Walsall)	2.12.02P
G-BMZP	Everett Gyroplane Series 1 (Volkswagen 1835)	010		14.11.86	D.H.Kirton	(Berkhampsted)	10. 4.02P
G-BMZS	Everett Gyroplane Series 1 (Volkswagen 1835)	012		13.11.86	L.W.Cload	St.Merryn	23.10.03P
G-BMZW	Bensen B.8MR *(Built P.D.Widdicombe)* (Rotax 532)	PFA G/01-1021		16.10.86	P.D.Widdicombe	Huntingdon, York	25. 8.99P
G-BMZX	Wolf W-11 Boredom Fighter *(Built J Penney)* (Continental A65)	PFA 146-11042		31.10.86	J.Nugent	Kilrush, County Kildare	10.12.04P

G-BNAA - G-BNZZ

Reg	Type	C/n	Previous identity	Regd	Owner/Operator	Probable Base	CA Expiry
G-BNAD	Rand Robinson KR-2	PFA 129-11077		10.11.86	M.C.Davies	(Roade, Northants)	27. 2.90P
	(Built M J Field) (Volkswagen 1834)				*(Amended CofR 3.02)*		
G-BNAG	Colt 105A HAB	906		31.10.86	R.W.Batchelor	(Thame)	19.12.89A
G-BNAI	Wolf W-11 Boredom Fighter	PFA 146-11083		31.10.86	C.M.Bunn	Haverfordwest	24. 3.05P
	(Built PJ Gronow) (Continental A65) *(Represents Spad rep)*				*(As "146-11083/5" in AEF France 94th Aero Sqdn c/s)*		
G-BNAJ	Cessna 152 II	15282527	C-GZWF (N69173)	3.11.86	Galair Ltd	Biggin Hill	11. 3.05T
					(Op Surrey and Kent Flying Club)		
G-BNAN	Cameron V-65 HAB	1333		28.10.86	Anne M.Lindsay and N.H.Ponsford	(Leeds)	7. 7.01A
					t/a Rango Balloon and Kite Co *"Actually"*		
G-BNAU	Cameron V-65 HAB	1395		13.11.86	T J Ellenrieder tr 4-Flight Group	(Bristol)	21. 8.04A
G-BNAW	Cameron V-65 HAB	1366		24.10.86	A and P.A Walker	(Richmond, Surrey)	25. 6.95A
					(HMS Recruitment titles) "Hippo-Thermia"		
G-BNBL	Thunder Ax7-77 HAB	910		7. 1.87	G.J.Bell	(Petersfield)	8. 4.05A
G-BNBU	Bensen B.8MV	PFA G/01-1070		1.12.86	B.A Lyford	(St.Merryn)	
	(Built D T Murchie)						
G-BNBV	Thunder Ax7-77 HAB	915		2.12.86	Jennifer M.Robinson *"Layla"*	(Milton-under-Wychwood)	29.11.01A
G-BNBW	Thunder Ax7-77 HAB	914		11.12.86	I.S. and S.W.Watthews *"Mutley"*	(Grange-over-Sands)	9. 9.99A
G-BNBY	Beech 95-B55A Baron	TC-1347	G-AXXR	14. 2.83	J.Butler	(Lisle Sur Tarn, France)	17. 4.07
G-BNBZ	LET L-200D Morava	171329	D-GGDC	16.12.86	C.A Suckling	Rushett Farm, Chessington	15. 5.00
			EI-AOY, (D-GLIN), EI-AOY, OK-SHB				
G-BNCB	Cameron V-77 HAB	1401		2.12.86	C.W.Brown	(Melton Mowbray)	31. 5.04A
G-BNCC	Thunder Ax7-77 HAB	924		11.12.86	Celia J.Burnhope *"Charlie"*	(US)	9.10.99A
G-BNCE*	Grumman G159 Gulfstream I	9	N436M	7. 4.87	Dundee Airport Fire Service	Dundee	9. 4.92T
			N436/N436M, N43M, (N709G)				
			(WFU 10.91 due to corrosion and cannibalised: cancelled 4.5.93 as WFU) (Fuselage dumped 2.04)				
G-BNCJ	Cameron V-77 HAB	815		16.12.86	D.Scott *"Sunshine Desserts"*	(Melksham)	17.10.05A
G-BNCO	Piper PA-38-112 Tomahawk	38-79A0472	N2482F	8. 1.87	Diane K.Walker	Gamston	25.11.04T
G-BNCR	Piper PA-28-161 Warrior II	28-8016111	G-PDMT	10.12.86	Airways Aero Associations Ltd	Booker	16. 5.05T
			ZS-LGW, N8103D		*(Op British Airways Flying Club) (Chelsea Rose t/s)*		
G-BNCS	Cessna 180	30022	OO-SPA	7. 1.87	C.Elwell Transport (Repairs) Ltd	Tatenhill	17. 2.95
			D-ENUX, N2822A				
G-BNCU	Thunder Ax7-77 HAB	928		7. 1.87	W.De Bock	(Peterborough)	14. 7.00A
G-BNCX*	Hawker Hunter T.7				See SECTION 3 Part 1	Brooklands	
G-BNCZ	Rutan LongEz	PFA 74A-10723		8. 1.87	M.C.Davies	Turweston	4.10.94P
	(Built R Bainbridge and P.A Ellway) (Lycoming O-235)				*(Landing accident Sherburn-in-Elmet 2.94: noted dismantled for rebuild 1.04)*		
G-BNDE	Piper PA-38-112 Tomahawk	38-79A0363	EI-BUR	13. 1.87	A.R.Willis	Earls Colne	7. 4.07T
			G-BNDE, N2541D				
G-BNDG	Wallis WA-201/R Series 1	K/220/X		22. 1.87	K.H.Wallis	Reymerston Hall, Norfolk	3. 3.88P
	(Rotax 64hp x 2)				*(Stored 8.01)*		
G-BNDN	Cameron V-77 HAB	1443		8. 1.87	J.A Smith	(Bristol)	22.10.93A
G-BNDO	Cessna 152 II	15284574	N5387M	11. 2.87	Simair Ltd *(Op Essex School of Flying)*	Andrewsfield	12. 9.05T
G-BNDP	Brugger MB.2 Colibri	PFA 43-10956		8. 1.87	J.P.Kynaston	Burcott Lodge Farm, Leighton Buzzard	19. 8.04P
	(Built M Black) (Volkswagen 1834)						
G-BNDR	SOCATA TB-10 Tobago	740		12. 2.87	Delta Fire Ltd	Norwich	4. 3.06T
G-BNDT	Brugger MB.2 Colibri	PFA 43-10981		8. 1.87	H.Haigh	Bagby	14. 7.05P
	(Built A Szep) (Volkswagen 1834)				tr Colibri Flying Group		
G-BNDV	Cameron N-77 HAB	1427		25. 2.87	R.E.Jones *"English Lake Hotels"*	(Lytham St.Annes)	9. 5.93A
G-BNDW	de Havilland DH.82A Tiger Moth	3942	N6638	10.12.86	N.D.Welch *(Noted 5.03)*	Caernarfon	
G-BNDY	Cessna 425 Conquest I	425-0236	N1262T	2. 6.87	Standard Aviation Ltd	Newcastle	14.10.05
G-BNED	Piper PA-22-135 Tri-Pacer	22-1640	OO-JEF	26. 1.87	P.Storey	Sywell	AC
			N3385A		*(Current status unknown)*		
G-BNEE	Piper PA-28R-201 Arrow III	28R-7837084	N630DJ	28. 1.87	J.W.Reid	Turweston	3.11.06
			N9518N		tr Britannic Management Aviation		
G-BNEK	Piper PA-38-112 Tomahawk II	38-82A0081	N9096A	28. 1.87	APB Leasing Ltd	Trehelig, Welshpool	17. 5.00T
G-BNEL	Piper PA-28-161 Warrior II	28-7916314	N2246U	27. 4.87	S.C.Westran	Shoreham	17.10.05T
G-BNEN	Piper PA-34-200T Seneca II	34-8070262	N8232V	18. 2.87	Air Taxis Ltd	Birmingham	12. 5.05T
G-BNEO	Cameron V-77 HAB	1408		9. 2.87	J.G.O'Connell *"Rowtate"*	(Braintree)	2.11.00
G-BNES	Cameron V-77 HAB	1426		19. 2.87	G.Wells	(Congleton)	7. 1.99A
					t/a Northern Counties Photographers		
G-BNET	Cameron O-84 HAB	1368		22. 1.87	C.and A I.Gibson *"Gordon Bennett"*	(Stockport)	1. 6.03A
G-BNEV	Viking Dragonfly	PFA 139-10935		28.11.86	N.W.Eyre	(Wombleton)	
	(Built N.W.Eyre) (Volkswagen 1834)				*(Current status unknown)*		
G-BNEX	Cameron O-120 HAB	1414		3. 4.87	The Balloon Club Ltd	(Bristol)	6. 5.90A
					t/a Bristol Balloons *"Sue Sheppard Employment Agency"*		
G-BNFG	Cameron O-77 HAB	1416		5. 3.87	Capital Balloon Club Ltd *"Dolores"*	(London NW1)	13. 1.94A
G-BNFI	Cessna 150J	15069417	N50588	8. 1.87	A Waters	Enstone	2. 6.06
G-BNFK*	Cameron Egg 89SS HAB				See SECTION 3 Part 1	Balleroy, Normandy, France	
G-BNFM	Colt 21A Cloudhopper HAB	668		5. 3.87	M.E.Dworski	(Vermenton, France)	24. 7.01A
G-BNFN	Cameron N-105 HAB	1442		13. 3.87	P.Glydon	(Barnt Green, Birmingham)	17. 6.97T
G-BNFO	Cameron V-77 HAB	816		5. 3.87	J.King and T.Ellenrieder tr Fox Group *"Funshine"*	(Bristol)	29. 5.04A
G-BNFP	Cameron O-84 HAB	1474		29. 4.87	B.F.G.Ribbans *"Dragonfly"*	(Woodbridge)	16. 5.05A
G-BNFR	Cessna 152 II	15282035	N67817	8. 4.87	Eastern Executive Air Charter Ltd	Southend	29. 4.06T
G-BNFS	Cessna 152 II	15283899	N5545B	10. 4.87	C and S Aviation Ltd	Wolverhampton	10. 7.05T
G-BNFV	Robin DR400/120 Dauphin 80	1767		4. 3.87	J.P.A Freeman	Headcorn	15. 7.02T
G-BNGE	Auster AOP.6	1925	7704M	18. 3.87	M.Pocock	AAC Middle Wallop	15. 4.05P
			TW536		*(Op Military Auster Flight as "TW536/T-SV" in 657 AOP Sqdn c/s)*		
G-BNGJ	Cameron V-77 HAB	1487		18. 3.87	Lathams Ltd *"Latham Timber"*	(High Wycombe)	6. 4.04A
G-BNGN	Cameron N-77 HAB	817		3. 4.87	Catherine B.Leeder *"Falcon"*	(Diss)	21. 4.05A
G-BNGO	Thunder Ax7-77 HAB	971		26. 3.87	J.S.Finlan	(Hamilton, New Zealand)	12. 4.04A
					tr The G-BNGO Group *"Thunderbird" (Philips titles)*		
G-BNGR	Piper PA-38-112 Tomahawk	38-79A0479	N2492F	26. 3.87	Teesside Flight Centre Ltd	Durham Tees Valley	1. 8.03T

Registration	Type	C/n	Previous identity	Reg date	Owner/operator	Base	CofA expiry
G-BNGS	Piper PA-38-112 Tomahawk	38-78A0701	N2463A	26. 3.87	Teesside Flight Centre Ltd	Durham Tees Valley	AC
					(Noted 1.04)		
G-BNGT	Piper PA-28-181 Archer II	28-8590036	N149AV	29. 4.87	Edinburgh Flying Club Ltd	Edinburgh	18. 5.05T
			N9559N				
G-BNGV	ARV Aviation ARV-1 Super 2	021		4. 6.87	N.A Onions	Andrewsfield	28. 4.06
G-BNGW	ARV Aviation ARV-1 Super 2	022		4. 6.87	Southern Gas Turbines Ltd	Manston	8. 7.90T
					(Stored 6.94: current status unknown)		
G-BNGY	ARV Aviation ARV-1 Super 2	019	(G-BMWL)	9. 6.87	S.C.Smith	(Leominster)	28.11.04
G-BNHB	ARV Aviation ARV-1 Super 2	026		13. 7.87	C.J.Challener	Barton	24. 7.04
G-BNHG	Piper PA-38-112 Tomahawk II	38-82A0030	N91435	23. 3.87	D.A Whitmore	icester	31. 5.06T
G-BNHI	Cameron V-77 HAB	1249		26. 3.87	C.J.Nicholls *"Fun-Der-Bird"*	(Warwick)	8. 9.05A
G-BNHJ	Cessna 152 II	15281249	N49418	4. 6.87	The Pilot Centre Ltd	Denham	19.12.05T
G-BNHK	Cessna 152 II	15285355	N80161	30. 3.87	J.P.Slack	Egginton	16. 7.06T
G-BNHL*	Colt Beer Glass 90SS HAB				See SECTION 3 Part 1	(Aldershot)	
G-BNHN*	Colt Ariel Bottle SS HAB				See SECTION 3 Part 1	Newbury	
G-BNHT	Fournier RF3	80	(D-KITX)	13. 4.87	D.G.Hey	Little Gransden	24. 5.05P
			G-BNHT, F-BMTO		tr G-BNHT Group		
G-BNID	Cessna 152 II	15284931	N5378P	24. 4.87	M.J.Ireland	Wellesbourne Mountford	3. 7.06T
G-BNII	Cameron N-90 HAB	1497		15. 4.87	S.Saunders tr Topless Balloon Group	(Farnham)	16. 8.01
G-BNIK	Robin HR200/120 Club	43	LX-AIK	15. 4.87	N.J.Wakeling	Leicester	17. 5.06
			LX-PAA				
G-BNIM	Piper PA-38-112 Tomahawk	38-78A0148	N9631T	18. 6.87	Aurs Aviation Ltd *(Op Glasgow Flying Club)*	Glasgow	30. 3.06T
G-BNIN	Cameron V-77 HAB	1079	G-RRSG (1)	15. 4.87	M.K.Grigson	(Shoreham)	10. 6.02A
			(G-BLRO)		tr Cloud Nine Balloon Group *"Cloud Nine"*		
G-BNIO	Luscombe 8AC Silvaire	2120	N45593	15. 4.87	W.H.Bliss	Eshott	12. 4.05P
	(Continental A75)		NC45593				
G-BNIP	Luscombe 8A Silvaire	3547	N77820	15. 4.87	S.Maric	Cumbernauld	10. 2.93P
	(Continental A65)		NC77820				
G-BNIU	Cameron O-77 HAB	1499		28. 4.87	MC VH SA	(Brussels, Belgium)	3. 4.00A
G-BNIV	Cessna 152 II	15284866	N4972P	24. 4.87	Aerohire Ltd	Wolverhampton	17. 8.03T
					(Op Halfpenny Green Flight Centre)		
G-BNIW	Boeing Stearman A75NI (PT-17) Kaydet	75-1526	N49291	22. 4.87	R.C.Goold	Priory Farm, Tibenham	6. 5.07
	(Pratt and Whitney R985)		41-7967		*(Noted damaged 8.04)*		
G-BNIZ	Fokker F.27 Friendship 600F	10405	OY-SRA	1. 6.87	Dart Group plc	Bournemouth	3.11.06T
			G-BNIZ, 9Q-CLQ, PH-FOD		t/a Channel Express		
G-BNJA*	Wag-Aero Wag-a-Bond	PFA 137-10886		3. 4.87	B.E.Maggs	Willey Park Farm, Caterham	3. 8.03P
	(Built R A Yates) (Continental O-200-A) *(Crashed on landing Staverton/Gloucestershire 28.7.92: cancelled 17.2.04 by CAA) (Fuselage noted 2.04)*						
G-BNJB	Cessna 152 II	15284865	N4970P	27. 4.87	Aerolease Ltd	Conington	20. 7.06T
G-BNJC	Cessna 152 II	15283588	N4705B	27. 4.87	Stapleford Flying Club Ltd	Stapleford	26. 8.05T
G-BNJG	Cameron O-77 HAB	1502		9. 5.89	A M.Figiel	(High Wycombe)	4. 4.97
G-BNJH	Cessna 152 II	15285401	C-GORA	21. 7.87	J.McAuley	Perth	5. 9.03T
			(N93101)		*(Noted 1.04)*		
G-BNJL	Bensen B.8MR	PFA G/01-1020		30. 4.87	S.Ram	(Lowestoft)	29. 3.03P
	(Built C G Ponsford) (Rotax 532)				*(New owner 2.04)*		
G-BNJM	Piper PA-28-161 Warrior II	28-8216078	N8015V	27. 5.87	A S Bamrah	*(Noted 8.04)*	4. 6.90T
					t/a Falcon Flying Services *(Noted 8.04)*		
G-BNJO	QAC Quickie Q.2	2217	N17LM	6.10.87	J.D.McKay	Crowfield	14. 5.93P
	(Revmaster 2100D)						
G-BNJR	Piper PA-28RT-201T Turbo Arrow IV	28R-8031104	N8212U	8. 5.87	D.Crocker	Thruxton	14. 8.05
G-BNJT	Piper PA-28-161 Warrior II	28-8116184	N8360T	11. 6.87	M.Jones tr Hawarden Flying Group	Hawarden	10.12.05T
G-BNJU*	Cameron Bust 80SS HAB	1324		13. 5.87	Ballon Team Bonn GmbH and Co KG	(Meckenheim, Germany)	19. 1.03A
					"Ludwig von Beethoven" (Cancelled 1.5.03 as wfu)		
G-BNJZ	Cassutt Racer IIIM	PFA 34-11228		14. 5.87	J.Cull	(Eastleigh)	16. 7.05P
	(Built Miller Aerial Spraying) (Continental O-200-A)						
G-BNKC	Cessna 152 II	15281036	N48894	26. 5.87	Herefordshire Aero Club Ltd	Shobdon	23. 9.05T
G-BNKD	Cessna 172N Skyhawk II	17272329	N4681D	19. 5.87	Barnes Olson Aero Leasing Ltd	Bristol	12. 2.07T
					(Op Bristol Flying Centre)		
G-BNKE	Cessna 172N Skyhawk II	17273886	N6534J	20. 5.87	T.Jackson tr Kilo Echo Flying Group	Manchester	3. 6.06T
G-BNKF*	Colt AS-56 HA Airship	899		20. 5.87	Formtrack Ltd	(London SE16)	14. 9.98A
					(Cancelled by CAA 20.10.00) (Extant 5.03)		
G-BNKH	Piper PA-38-112 Tomahawk II	38-81A0078	N25874	14. 5.87	Goodwood Road Racing Co Ltd	Goodwood	24. 7.05T
					(Op Goodwood Flying Club)		
G-BNKI	Cessna 152 II	15281765	N67337	19. 5.87	RAF Halton Aeroplane Club Ltd	RAF Halton	6. 7.05T
G-BNKP	Cessna 152 II	15281286	N49460	18. 5.87	Spectrum Leisure Ltd	Clacton	21. 8.06T
G-BNKR	Cessna 152 II	15281284	N49458	18. 5.87	Keen Leasing (IoM) Ltd	Newtownards	10. 7.02T
G-BNKS	Cessna 152 II	15283186	N47202	18. 5.87	Shropshire Aero Club Ltd	Sleap	9. 5.05T
G-BNKT	Cameron O-77 HAB	1356		13. 2.87	British Airways plc *"Katie II"*	(West Drayton)	6. 9.04A
G-BNKV	Cessna 152 II	15283079	N46604	18. 5.87	S.C.Westran	Shoreham	29.10.05T
G-BNLA	Boeing 747-436	23908	N60665	30. 6.89	British Airways plc	Heathrow	29. 6.05T
G-BNLB	Boeing 747-436	23909		31. 7.89	British Airways plc *(Stored 4.02)*	Cardiff	13. 3.05T
G-BNLC	Boeing 747-436	23910		21. 7.89	British Airways plc	Gatwick	26. 7.05T
G-BNLD	Boeing 747-436	23911	N6018N	5. 9.89	British Airways plc	Heathrow	5. 9.05T
G-BNLE	Boeing 747-436	24047		14.11.89	British Airways plc	Heathrow	16.11.05T
G-BNLF	Boeing 747-436	24048		23. 2.90	British Airways plc	Heathrow	11.10.05T
G-BNLG	Boeing 747-436	24049		23. 2.90	British Airways plc	Heathrow	1.10.05T
G-BNLH	Boeing 747-436	24050	VH-NLH	28. 3.90	British Airways plc	Heathrow	13. 5.06T
			G-BNLH				
G-BNLI	Boeing 747-436	24051		19. 4.90	British Airways plc	Heathrow	20. 4.06T
G-BNLJ	Boeing 747-436	24052	N60668	23. 5.90	British Airways plc	Heathrow	24. 5.06T
G-BNLK	Boeing 747-436	24053	N6009F	25. 5.90	British Airways plc	Heathrow	28. 5.06T
G-BNLL	Boeing 747-436	24054		13. 6.90	British Airways plc	Heathrow	13. 6.06T
G-BNLM	Boeing 747-436	24055	N6009F	28. 6.90	British Airways plc	Gatwick	27. 6.06T
G-BNLN	Boeing 747-436	24056		26. 7.90	British Airways plc	Heathrow	26. 7.06T

G-BNLO	Boeing 747-436	24057		25.10.90	British Airways plc	Heathrow	24.10.06T
G-BNLP	Boeing 747-436	24058		17.12.90	British Airways plc	Gatwick	10. 9.06T
G-BNLR	Boeing 747-436	24447	N6005C	15. 1.91	British Airways plc	Heathrow	16. 1.07T
G-BNLS	Boeing 747-436	24629		13. 3.91	British Airways plc	Heathrow	12. 3.07T
G-BNLT	Boeing 747-436	24630		19. 3.91	British Airways plc	Gatwick	14.11.06T
G-BNLU	Boeing 747-436	25406		28. 1.92	British Airways plc	Heathrow	27. 1.08E
G-BNLV	Boeing 747-436	25427		20. 2.92	British Airways plc	Gatwick	19. 2.05T
G-BNLW	Boeing 747-436	25432		4. 3.92	British Airways plc *"City of Norwich"*	Heathrow	4. 3.05T
G-BNLX (2)	Boeing 747-436	25435		1. 4.92	British Airways plc	Heathrow	2. 4.05T
G-BNLY (3)	Boeing 747-436	27090	N60659	10. 2.93	British Airways plc	Heathrow	9. 2.06T
G-BNLZ (3)	Boeing 747-436	27091		4. 3.93	British Airways plc	Heathrow	3. 3.06T
G-BNMA	Cameron O-77 HAB	830		15.12.87	N.Woodham *"Finian" (Active2003)*	(Bristol)	4. 4.03A
G-BNMB	Piper PA-28-151 Cherokee Warrior	28-7615369	N6826J	6.10.87	Britannia Airways Ltd	Liverpool	27. 1.06T
					(Op Britannia Airways Flying Club)		
G-BNMC	Cessna 152 II	15282564	N69218	29. 5.87	M.L.Jones *(Op Derby Aero Club)*	Egginton	10. 8.03T
G-BNMD	Cessna 152 II	15283786	N5170B	28. 5.87	T.M.Jones *(Stored 9.02)*	Egginton	28. 7.01T
G-BNME	Cessna 152 II	15284888	N5159P	25. 9.87	Northamptonshire School of Flying Ltd	Sywell	19.12.05T
G-BNMF	Cessna 152T	15285563	N93858	21. 7.87	Central Aircraft Leasing Ltd	Wolverhampton	8. 7.07T
					(Op Midland Flight Centre)		
G-BNMG	Cameron O-77 HAB	1500		27. 5.87	J.H.Turner	(Bridgnorth)	4. 4.03A
G-BNMH	Pietenpol AirCamper	NH-1-001		2. 6.87	N.M.Hitchman	(Leicester)	
	(Built N M Hitchman)						
G-BNMI*	Colt Black Knight HAB	1096		1. 6. 87	Virgin Airship and Balloon Co Ltd	Not known	26. 9.91A
					(Cancelled 4.11.03 as wfu) (Stored in good condition 2004)		
G-BNMK	Dornier Do.27A-1	271	OE-DGO	14. 8.87	G.Mackie	Belfast Aldergrove	AC
			Luftwaffe 56+04, BD+397, BA+399 *(New Cof 8.03)*				
G-BNML	Rand Robinson KR-2	PFA 129-11240		23. 6.87	R.F.Cresswell	(Alfreton)	17. 8.00P
	(Built R J Smyth) (Volkswagen 1834)						
G-BNMO	Cessna R182 Skylane RG II	R18200956	N738RK	3. 7.87	Kenrye Developments Ltd	Trim, County Meath	10. 7.06
G-BNMT*	Short SD.3-60 Variant.100	SH3723	N160DD	18. 6.87	Not known	Farnborough	
			G-BNMT, G-14-3723				
			(Crashed into Firth of Forth near Granton harbour 27.02.01: cancelled 10.12.02 as destroyed) (Noted 4.04)				
G-BNMU	Short SD.3-60 Variant 100	SH.3724	N161DD	18. 6.87	BAC Express Airlines Ltd	Guernsey	22.11.05T
			G-BNMU, G-14-3724		*(Op Aurigny Air Services Ltd)*		
G-BNMX	Thunder Ax7-77 HAB	1003		15. 6.87	S.A.D.Beard	(Cheltenham)	30. 4.05A
G-BNNA	Stolp SA.300 Starduster Too	1462	N8SD	29. 6.87	M.A.Simpson	Leicester	6.10.05P
	(Built T C Maxwell) (Lycoming O-360)				tr Banana Group *(Noted 5.04)*		
G-BNNE	Cameron N-77 HAB	1413		15. 6.87	Balloon Flights International Ltd *(Active 3.03)*	(Bath)	
G-BNNO	Piper PA-28-161 Warrior II	28-8116099	N8307X	15. 6.87	Tindon Ltd	Norwich	5. 8.05T
G-BNNR	Cessna 152 II	15285146	N40SX	15. 6.87	Sussex Flying Club Ltd	Shoreham	29.10.05T
			N40SU, N6121Q				
G-BNNS	Piper PA-28-161 Warrior II	28-8116061	N8283C	26. 6.87	S.J.French	Turweston	29. 5.05
G-BNNT	Piper PA-28-151 Cherokee Warrior	28-7615056	N7624C	12. 6.87	S.T.Gilbert and D.J.Kirkwood	Hinton-in-the-Hedges	1. 7.07T
G-BNNU	Piper PA-38-112 Tomahawk II	38-81A0037	N25650	12. 6.87	Edinburgh Flying Club Ltd	Edinburgh	2.12.05T
G-BNNX	Piper PA-28R-201T Turbo Cherokee Arrow III	28R-7703009	N9005F	14. 7.87	J.G.Freeden	Bristol	4.12.05T
G-BNNY	Piper PA-28-161 Warrior II	28-8016084	N8092M	1. 9.87	A.S.Bamrah	Biggin Hill	14.12.05T
					t/a Falcon Flying Services *(Op Southern Air)*		
G-BNNZ	Piper PA-28-161 Warrior II	28-8016177	N8135Y	24. 7.87	A S Bamrah	Biggin Hill	21.10.05T
					t/a Falcon Flying Services		
G-BNOB	Wittman W.8 Tailwind	PFA 3502		13. 7.87	M.Robson-Robinson	(Abbots Bromley)	14. 5.02P
	(Built D G Hammersley) - c/n 258/DH1) (Continental.PC60)				*"Imogen"*		
G-BNOE	Piper PA-28-161 Warrior II	2816013	N9121X	26. 6.87	Sherburn Aero Club Ltd	Sherburn-in-Elmet	15. 2.07T
			N9568N				
G-BNOF	Piper PA-28-161 Warrior II	2816014	N9122B	26. 6.87	Tayside Aviation Ltd	Dundee	29. 3.07T
G-BNOG	Piper PA-28-161 Warrior II	2816015	N9122D	26. 6.87	Flight Training Europe SL	Jerez, Cadiz, Spain	24. 9.05T
G-BNOH	Piper PA-28-161 Warrior II	2816016	N9122L	26. 6.87	Sherburn Aero Club Ltd	Sherburn-in-Elmet	25. 1.07T
G-BNOJ	Piper PA-28-161 Warrior II	2816018	N9122R	26. 6.87	R.D.Turner and W.M.Brown	Blackpool	28. 7.06T
					tr BAE (Warton) Flying Club		
G-BNOK	Piper PA-28-161 Warrior II	2816019	N9122U	26. 6.87	Flight Training Europe SL	Jerez, Cadiz, Spain	19. 2.05T
G-BNOM	Piper PA-28-161 Warrior II	2816024		26. 6.87	Air Navigation and Trading Company Ltd	Blackpool	20.10.07E
G-BNON	Piper PA-28-161 Warrior II	2816025		26. 6.87	Tayside Aviation Ltd	Perth	8. 3.04T
G-BNOP	Piper PA-28-161 Warrior II	2816027		26. 6.87	R.D.Turner and F.J.Smith	Blackpool	14. 8.05T
					tr BAE (Warton) Flying Club		
G-BNOT	Piper PA-28-161 Warrior II	2816030		26. 6.87	Flight Training Europe SL	Jerez, Cadiz, Spain	24. 6.05T
G-BNOZ	Cessna 152 II	15281625	EI-CCP	22. 6.87	CSW Flying Hire Ltd	(Willenhall)	21. 3.05T
			G-BNOZ, N65570				
G-BNPE	Cameron N-77 HAB	1519	(G-BNPX)	25. 8.87	Zebedee Balloon Service Ltd	Newtown Hungerford	6. 8.05A
G-BNPF	Slingsby T.31M Cadet III	PFA 42-11122	XA284	3.11.87	S.Luck, P.Norman and D.R.Winder	Audley End	10. 8.00P
	(Built S Luck and Partners) - c/n 826) (Stark Stamo MS.1400A) *(Contains wings from XE791 which became OO-ZDQ) "Noddy"*						
G-BNPG*	Percival P.66 Pembroke C.1				See SECTION 3 Part 1	Västerås, Sweden	
G-BNPH	Percival P.66 Pembroke C.1	P66/41	WV740	30. 6.87	A and G.A Gainsford Dixon	Bournemouth	19. 5.05P
	(Regd with c/n "PAC66/027")				*(As "WV740" in 60 Sqdn RAF c/s)*		
G-BNPL	Piper PA-38-112 Tomahawk	38-79A0524	N2420G	28. 7.87	Cardiff-Wales Flying Club	Gloucestershire	30. 1.03T
					(Noted 11.04 for major surgery with G-BPPE to provide a single flyer)		
G-BNPM	Piper PA-38-112 Tomahawk	38-79A0374	N2561D	28. 7.87	D and Linda K Britten t/a Papa Mike Aviation	Cranfield	10. 5.05T
G-BNPO	Piper PA-28-181 Cherokee Archer II	28-7890123	N47720	28. 7.87	Bonus Aviation Ltd	Cranfield	15. 6.06T
G-BNPV	Bowers Fly Baby 1A	PFA 16-11120		2. 7.87	J.G.Day and R.Gauld-Galliers		
	(Built J.G.Day)					Rushett Farm, Chessington	21.11.05P
					(As "JU.FB. D.I" in German Army Air Service c/s)		
G-BNPY	Cessna 152 II	15280249	N24388	30. 6.87	Eastern Air Centre Ltd	Gamston	2. 1.06T
G-BNPZ	Cessna 152 II	15285134	N6109Q	30. 6.87	Tatenhill Aviation Ltd	Tatenhill	2. 2.06T
G-BNRA	SOCATA TB-10 Tobago	772		15. 7.87	M.Walshe	Tollerton	12. 5.06
					tr Double D Airgroup *"Triple One"*		

Reg	Type	C/n	Previous identity	Regd	Owner/operator	Base	CofA
G-BNRG	Piper PA-28-161 Warrior II	28-8116217	N83810	7. 7.87	RAF Brize Norton Flying Club Ltd	RAF Brize Norton	8. 4.06T
G-BNRK	Cessna 152 II	15284659	N6297M	29. 7.87	Redhill Aviation Ltd	Blackbushe	10. 2.06T
G-BNRL	Cessna 152 II	15284250	N5084L	13. 7.87	Modi Aviation Ltd	Earls Colne	10. 6.06T
G-BNRP	Piper PA-28-181 Cherokee Archer II	28-7790528	N984BT	25.11.87	Bonus Aviation Ltd	Cranfield	9. 6.06T
G-BNRR	Cessna 172P Skyhawk II	17274013	N5213K	13. 7.87	P.H.Archard t/a PHA Aviation	Elstree	29. 1.06T
G-BNRX	Piper PA-34-200T Seneca II	34-7970336	N2898A	25.11.87	Truman Aviation Ltd	Tollerton	31. 3.06T
G-BNRY	Cessna 182Q Skylane II	18265629	N735RR	20. 7.87	Reefly Ltd	Booker	5. 6.06T
G-BNSG	Piper PA-28R-201 Arrow III	28R-7837205	N9516C	30. 7.87	Stronghold Trust Ltd	Standalone Farm, Meppershall	19. 2.06
G-BNSI	Cessna 152 II	15284853	N4945P	6. 8.87	Sky Leisure Aviation (Charters) Ltd	Shoreham	13. 1.06T
G-BNSL	Piper PA-38-112 Tomahawk II	38-81A0086	N25956	21. 7.87	APB Leasing Ltd	Tatenhill	12. 4.07T
G-BNSM	Cessna 152 II	15285342	N68948	23. 7.87	Cornwall Flying Club Ltd	Bodmin	26. 7.06T
G-BNSN	Cessna 152 II	15285776	N94738	21. 7.87	The Pilot Centre Ltd	Denham	8. 5.06T
G-BNSO	Slingsby T.67M Firefly II	2021		20. 8.87	A C.Lees	Sherburn-in-Elmet	13. 3.06T
G-BNSP	Slingsby T.67M Firefly II	2044		20. 8.87	Michelle Susan Roadhouse	Netherthorpe	11. 3.06T
					tr Slingsby Group		
G-BNSR	Slingsby T.67M Firefly II	2047		20. 8.87	R.Harris and N.R.Thorburn	Spamhoe	26. 6.06T
G-BNST	Cessna 172N Skyhawk II	17273661	N4670J	21. 9.87	J.Revill t/a CSG Bodyshop	Netherthorpe	23. 4.06T
G-BNSU	Cessna 152 II	15281245	N49410	2.12.87	A L.Brown	Bourn	13.10.06T
					t/a Channel Aviation *(Op Rural Flying Corps)*		
G-BNSV	Cessna 152 II	15284531	N5322M	4.12.87	A L.Brown	Bourn	12. 7.97T
					t/a Channel Aviation		
G-BNSY	Piper PA-28-161 Warrior II	28-8016017	N4512M	18. 8.87	BCT Aircraft Leasing Ltd	Kemble	12. 3.06T
G-BNSZ	Piper PA-28-161 Warrior II	28-8116315	N8433B	20. 8.87	Carill Aviation Ltd	Lee-on-Solent	8.12.05T
G-BNTC	Piper PA-28RT-201T Turbo Arrow IV	28R-8131081	N83428	4.11.87	Halfpenny Green Flight Centre Ltd	Wolverhampton	20. 6.06T
G-BNTD	Piper PA-28-161 Cherokee Warrior II	28-7716235	N38490	5. 8.87	A M.and F.Alam	Cranfield	7.11.05T
			N9539N				
G-BNTP	Cessna 172N Skyhawk II	17272030	N6531E	4. 9.87	Westnet Ltd	Barton	23. 2.06
G-BNTS	Piper PA-28RT-201T Turbo Arrow IV	28R-8131024	N8296R	6. 8.87	Nasaire Ltd	Liverpool	11. 2.06
G-BNTT	Beech 76 Duchess	ME-228	N54SB	8.10.87	S.J.Skilton	Bournemouth	12. 6.06T
					t/a Aviation Rentals *(Op CTC Aviation)*		
G-BNTW	Cameron V-77 HAB	1574		13. 8.87	P.Goss *"Cecilia"*	(Alton)	6.11.99A
G-BNTZ	Cameron N-77 HAB	1518		27. 8.87	P.M.Watkins t/a Balloon Team	(Chippenham)	26. 8.02A
G-BNUC	Cameron O-77 HAB	1575		18. 8.87	T.J.Bucknall *"Bridges Van Hire II"*	(Hawarden)	
G-BNUL	Cessna 152 II	15284486	N4852M	2.10.87	Big Red Kite Ltd	RAF Bensen	3. 5.06T
G-BNUN	Beech 58PA Baron	TJ-256	N6732Y	19. 8.87	British Midland Airways Ltd	Nottingham East Midlands	8. 5.05T
G-BNUO	Beech 76 Duchess	ME-250	N6635Y	29. 9.87	G.A F.Tilley	Bournemouth	26. 5.05T
G-BNUS	Cessna 152 II	15282166	N68179	26. 8.87	Stapleford Flying Club Ltd	Stapleford	10. 7.06T
G-BNUT	Cessna 152 II	15279458	N714VC	26. 8.87	Stapleford Flying Club Ltd	Stapleford	5. 6.06T
G-BNUV	Piper PA-23-250 Aztec F	27-7854038	N97BB	2.10.87	L.J.Martin	Sandown/Redhill	9. 1.05
			N63894				
G-BNUX	Hoffmann H 36 Dimona	36236		26. 8.87	G.Hill tr Buckminster Dimona Syndicate	Saltby	13. 6.06
G-BNUY	Piper PA-38-112 Tomahawk II	38-81A0093	N26006	10. 9.87	D.J.Whitcombe	Cardiff	17. 8.03T
G-BNVB	Grumman AA-5A Cheetah	AA5A-0758	N26843	28. 8.87	V.R.Coultan	Turweston	23. 2.06T
	(Regd as such but plate indicates Gulfstream American production)				tr Grumman Group		
G-BNVD	Piper PA-38-112 Tomahawk	38-79A0055	N2421B	16.11.87	D.A Whitmore	Carlisle	14. 4.06T
G-BNVE	Piper PA-28-181 Archer II	28-8490046	N4338D	28. 8.87	Solent Flight Ltd	Southampton	10.12.05T
G-BNVT	Piper PA-28R-201T Turbo Cherokee Arrow III	28R-7703157	N5863V	26. 1.88	T.Yeung	Glasgow	18. 2.07T
					tr Victor Tango Group *(Op Glasgow Flying Club)*		
G-BNVZ	Beech 95-B55 Baron	TC-2042	N17720	25. 9.87	W.J.Forrest	White Waltham	14. 6.04
G-BNWA	Boeing 767-336	24333	N6009F	19. 4.90	British Airways plc	Heathrow	24. 4.06T
G-BNWB	Boeing 767-336	24334	N6046P	2. 2.90	British Airways plc	Heathrow	12. 2.06T
G-BNWC	Boeing 767-336	24335		2. 2.90	British Airways plc	Heathrow	21. 2.06T
G-BNWD	Boeing 767-336	24336	N6018N	2. 2.90	British Airways plc	Heathrow	31. 8.06T
G-BNWH	Boeing 767-336	24340	N6005C	31.10.90	British Airways plc	Heathrow	30.10.06T
G-BNWI	Boeing 767-336	24341		18.12.90	British Airways plc	Gatwick	17.12.06T
G-BNWM	Boeing 767-336	25204		24. 6.91	British Airways plc	Manchester	24. 6.07T
G-BNWN	Boeing 767-336	25444		30.10.91	British Airways plc	Heathrow	29.10.07E
G-BNWO	Boeing 767-336	25442		2. 3.92	British Airways plc *"City of Barcelona"*	Gatwick	1. 3.05T
G-BNWR	Boeing 767-336	25732		20. 3.92	British Airways plc	Gatwick	19. 3.05T
G-BNWS	Boeing 767-336	25826	N6018N	19. 2.93	British Airways plc	Heathrow	18. 2.06T
G-BNWT	Boeing 767-336	25828		8. 2.93	British Airways plc	Birmingham	29.11.05T
G-BNWU	Boeing 767-336	25829		16. 3.93	British Airways plc	Gatwick	15. 3.06T
G-BNWV	Boeing 767-336	27140		29. 4.93	British Airways plc	Heathrow	28. 4.06T
G-BNWW	Boeing 767-336	25831		3. 2.94	British Airways plc	Heathrow	18. 3.06T
G-BNWX	Boeing 767-336	25832		1. 3.94	British Airways plc	Cambridge	28. 2.06T
					(Used for Future Strategic Tanker Transport programme)		
G-BNWY	Boeing 767-336	25834	N5005C	22. 4.96	British Airways plc	Birmingham	21. 4.05T
G-BNWZ	Boeing 767-336	25733		25. 2.97	British Airways plc	Heathrow	24. 2.06T
G-BNXC	Cessna 152 II	15285429	N93171	24. 9.87	N.D.Wyndow	Coventry	18. 3.07T
					tr Sir W.G.Armstrong-Whitworth Flying Group		
G-BNXD	Cessna 172N Skyhawk II	17272692	N6285D	25. 9.87	Hecray Co Ltd	Southend	14. 6.07T
					t/a Direct Helicopters *(Op Southend School of Flying)*		
G-BNXE	Piper PA-28-161 Warrior II	28-8116034	N8262D	24. 9.87	M.S.Brown t/a Rugby Autobody Repairs	Leicester	21. 1.06
G-BNXG*	Cameron DP-70 HA Airship	1558		23. 9.87	Not known	(US)	
					(Cancelled 18.6.93 by CAA) (Extant 5.03)		
G-BNXI	Robin DR400/180R Remorqueur	1021	SE-FNI	13.10.87	London Gliding Club Pty Ltd	Dunstable	13. 3.06
G-BNXK	Airship Industries NCA ULD/3 HAB	7 & 1110	(G-BLJN)	23. 9.87	J.R.P.Nott	(US/Bristol)	
			(Hot air envelope stored Twain-Harte, California, US 7.03 - helium inner envelope stored Bristol 1995)				
G-BNXL	Glaser-Dirks DG-400	4-216		2.10.87	C.I.Cowley tr G-BNXL Syndicate	Seighford	14. 4.06
G-BNXM	Piper PA-18 Super Cub 95 (L-21B-PI)	18-4019	MM54-2619	23.11.87	R.Thorp tr G-BNXM Group	Gipsy Wood Farm, Warthill	19. 3.05P
	(Continental O-200-A) *(Italian Frame rebuild No.0006)*		EI-276, I-EIVC, MM54-2619, 54-2619				
G-BNXR	Cameron O-84 HAB	1515		23. 9.87	J.A B.Gray *"Bacchus II"*	(Cirencester)	9.12.04T
G-BNXT	Piper PA-28-161 Cherokee Warrior II	28-7716168	N4047Q	23. 9.87	A S.Bamrah	Biggin Hill	27.10.05T
					t/a Falcon Flying Services *(Op Euroflyers)*		

Registration	Type	C/n	Previous identity	Date	Owner/Operator	Base	CofA expiry
G-BNXU	Piper PA-28-161 Warrior II	28-7916129	N2082C	23. 9.87	D.J.G.Carphin and R.E.Woolsey tr Friendly Warrior Group	Newtownards	19. 5.06
G-BNXV	Piper PA-38-112 Tomahawk	38-79A0826	N2399N	10.12.87	W.B.Bateson *(New owner 10.02)*	Blackpool	5.10.01T
G-BNXX	SOCATA TB-20 Trinidad	664	N20GZ	15. 9.87	D.M.Carr	Wellesbourne Mountford	12. 5.06
G-BNXZ	Thunder Ax7-77 HAB	1105		13.10.87	W.S.Templeton, R.B.Green and A S.Dear tr Hale Hot Air Balloon Group *"Dragonfly"*	(Fordingbridge)	29. 8.04A
G-BNYB	Piper PA-28-201T Turbo Dakota	28-7921040	N2856A N9533N	27. 1.88	G-BNYB Ltd	Goodwood	18. 2.07T
G-BNYD	Bell 206B JetRanger II	1911	N3254P C-GTWM, N49712	1.10.87	Sterling Helicopters Ltd	Norwich	12.11.05T
G-BNYK	Piper PA-38-112 Tomahawk II	38-82A0059	N2376V	23.10.87	APB Leasing Ltd	Sheffield City	17. 5.05T
G-BNYL	Cessna 152 II	15280671	N25454	6.10.87	V.J.Freeman	Headcorn	24.11.06T
G-BNYM	Cessna 172N Skyhawk II	17273854	N6089J	13.11.87	D.J.Skinner tr Kestrel Syndicate	AAC Middle Wallop	16. 4.06
G-BNYN	Cessna 152 II	15285433	N93185	2.10.87	Redhill Aviation Ltd *(Broken in half by car and wrecked by vandals overnight 14/15.4.03)*	Redhill	17. 2.03T
G-BNYO	Beech 76 Duchess	ME-78	N2010P	28.10.87	R.E.Wragg t/a Harding Wragg	Blackpool	5. 7.04T
G-BNYP	Piper PA-28-181 Archer II	28-8490027	N4330K	19.10.87	R.D.Cooper *(Op Sandra's Flying Group)*	Turweston	13. 3.06T
G-BNYS	Boeing 767-204ER	24013	TF-ATO G-BNYS, N6009F	22. 2.88	Air Atlanta Europe Ltd *(Op Excel)*	Gatwick	8. 5.06T
G-BNYV	Piper PA-38-112 Tomahawk	38-78A0073	N9364T	13.11.87	Goodair Leasing Ltd	Cardiff	17. 9.06T
G-BNYX	Denney Kitfox Model 1 *(Built R W Husband)* (Rotax 532)	PFA 172-11285		28.10.87	W.J.Husband	Blackbrook Farm, Sheffield	23. 9.05P
G-BNYZ	SNCAN Stampe SV-4E (Lycoming O-360)	200	F-BFZR Fr Mil *(Also reported as ex N180SV(?) with c/n "200-53" but may be ex Belgian V-53 c/n 1195)*	10.12.87	M.J.Heudebourck and D.E.Starkey	White Waltham	29.12.06
G-BNZB	Piper PA-28-161 Warrior II	28-7916521	N2900U	18.11.87	Falcon Flying Services Ltd	Biggin Hill	24. 1.06T
G-BNZC	de Havilland DHC-1 Chipmunk 22	C1/0778	G-ROYS 7438M, WP905	11.11.87	Richard Shuttleworth Trustees *(As "18671" in RCAF c/s)*	Old Warden	12. 4.07
G-BNZK	Thunder Ax7-77 HAB	1104		10.11.87	T.D.Marsden *"Shropshire Lass"*	(Grimsby)	28. 5.97A
G-BNZL	RotorWay Scorpion 133 *(Built J Evans)*	2839		2.11.87	J.R.Wraight *(Complete but stored 5.95: current status unknown)*	Stoneacre Farm, Farthing Corner	AC
G-BNZM	Cessna T210N Turbo Centurion II	21063640	N4828C	9.11.87	A J.M.Freeman	North Weald	6. 4.06
G-BNZN	Cameron N-56 HAB	1471	SE-ZFA G-BNZN	9.11.87	Balloon Sports HB *(New CofR 3.04)*	(Partille, Sweden)	19. 4.90A
G-BNZO	RotorWay Executive *(Built M J Wiltshire)* (RotorWay RW162)	RW152/3535		9.11.87	D.Collins and R.Ayres *"Bonzo"*	Street Farm, Takeley	8. 4.05P
G-BNZR	Clutton FRED Series II *(Built R.M.Waugh)*	PFA 29-10727		10.11.87	R.M.Waugh *(Current status unknown)*	Newtownards	25. 5.99P
G-BNZV	Piper PA-25-235 Pawnee D	25-7405649	C-GSKU N9548P	22. 2.88	Aeroklub Alpski Letalski Center Lesce	(Lesce, Slovenia)	21. 4.07
G-BNZZ	Piper PA-28-161 Warrior II	28-8216184	N8253Z	17.11.87	Providence Aviation Ltd	Wellesbourne Mountford	8. 3.06T

G-BOAA - G-BOZZ

Registration	Type	C/n	Previous identity	Date	Owner/Operator	Base	CofA expiry
G-BOAA*	British Aircraft Corporation-Aérospatiale Concorde 102				See SECTION 3 Part 1	East Fortune	
G-BOAB*	British Aircraft Corporation-Aérospatiale Concorde 102				See SECTION 3 Part 1	Heathrow	
G-BOAC*	British Aircraft Corporation-Aérospatiale Concorde 102				See SECTION 3 Part 1	Manchester	
G-BOAD*	British Aircraft Corporation-Aérospatiale Concorde 102				See SECTION 3 Part 1	New York, US	
G-BOAE	British Aircraft Corporation-Aérospatiale Concorde 102				See SECTION 3 Part 1	Barbados	
G-BOAF*	British Aircraft Corporation-Aérospatiale Concorde 102				See SECTION 3 Part 1	Filton	
G-BOAG*	British Aircraft Corporation-Aérospatiale Concorde 102				See SECTION 3 Part 1	Washington, US	
G-BOAH	Piper PA-28-161 Warrior II	28-8416030	N43401 N9554N	21. 1.88	Prestwick Flight Centre Ltd	Prestwick	30.11.06T
G-BOAI	Cessna 152 II	15279830	C-GSJH N757LS	8. 1.88	Galair Ltd	Biggin Hill	22. 5.06T
G-BOAL	Cameron V-65 HAB	1600		5.11.87	A Lindsay *"No Name Balloon"*	(Twickenham)	7. 2.02A
G-BOAM	Robinson R22 Beta	0717		10.12.87	Plane Talking Ltd	Cranfield	27. 1.06T
G-BOAO	Thunder Ax7-77 HAB	1162		2.12.87	D.V.Fowler	(Cranbrook)	17.10.02A
G-BOAS	Air Command 503 Commander *(Built R Robinson - c/n 0388)*	PFA G/04-1094		3.12.87	R.Robinson *(Current status unknown)*	(Leighton Buzzard)	
G-BOAU	Cameron V-77 HAB	1606		10.12.87	G.T.Barstow *"Flying Colours/Duster I"*	(Llandrindod Wells)	9.12.96A
G-BOBA	Piper PA-28R-201 Arrow III	28R-7837232	N31249	4. 1.88	Atlantic Flight Training Ltd *(Atlantic Airlines c/s)*	Coventry	4. 7.06T
G-BOBB	Cameron O-120 HAB	1609		24.11.87	Over The Rainbow Balloon Flights Ltd	(Mansfield)	23. 3.05A
G-BOBH	Airtour AH-77B HAB	009		2.12.87	J. and K.Francis *"Gloworm"*	(Southampton)	29. 6.02A
G-BOBL	Piper PA-38-112 Tomahawk II	38-81A0140	N91335	4. 1.88	Cardiff Wales Aviation Services Ltd	Cardiff	28. 9.03T
G-BOBR	Cameron N-77 HAB	1623		10.12.87	M.Morris and P.A Davies	(Oswestry)	1. 4.05A
G-BOBT	Stolp SA.300 Starduster Too *(Built C J Anderson)* (Lycoming O-360)	CJ-01	N690CM	15.12.87	S.C.Lever tr G-BOBT Group	White Waltham	15. 8.05P
G-BOBV	Reims/Cessna F150M	F15001415	EI-BCV	14.12.87	Sheffield Aero Club Ltd	Netherthorpe	12. 5.06T
G-BOBY	Monnett Sonerai II *(Built R G Hallam)* (Volkswagen 2233)	PFA 15-10223		26.10.78	R.G.Hallam *(Damaged near Barton 31.10.82: stored 9.96: current status unknown)*	Netherthorpe	8.11.82P
G-BOBZ	Piper PA-28-181 Archer II	28-8090257	N81671	21.12.87	Trustcomms International Ltd	Goodwood	9. 3.98T
G-BOCB*	Hawker Siddeley HS.125 Series 1B/522				See SECTION 3 Part 1	Doncaster	
G-BOCF*	Colt 77A HAB				See SECTION 3 Part 1	(Aldershot)	
G-BOCG	Piper PA-34-200T Seneca II	34-7870359	N36759	30.12.87	Oxford Aviation Services Ltd	Oxford	13. 1.07T
G-BOCI	Cessna 140A (Continental C90)	15497	N5366C	17.11.87	J.B.Bonnell *"Whitey"*	Thruxton	12.11.05
G-BOCK	Sopwith Triplane rep *(Built Northern Aeroplane Workshops - c/n NAW-1)* (Clerget Rotary 9B @ 130 hp) *(As "N6290" in RNAS 8 Sqdn c/s) "Dixie II"*	153		26. 1.88	Richard Shuttleworth Trustees	Old Warden	31. 7.05P
G-BOCL	Slingsby T.67C Firefly	2035		5. 1.88	Richard Brinklow Aviation Ltd *(Op The Flying Hut)*	Shoreham	26. 2.07T
G-BOCM	Slingsby T.67C Firefly	2036		5. 1.88	Richard Brinklow Aviation Ltd	Hinton-in-the-Hedges	15. 6.06T

Reg	Type	C/n	Previous identity	Date	Owner	Base	CofA
G-BOCN	Robinson R22 Beta	0726	N...	8. 1.88	Auto-Rotation Ltd	(Sproughton, Ipswich)	1. 4.06T
			G-BOCN				
G-BOCS	Piper PA-34-220T Seneca III	3433112		26. 2.88	Flight Training Europe SL	Jerez, Cadiz, Spain	18. 2.05T
G-BODB	Piper PA-28-161 Warrior II	2816042	N9606N	23. 2.88	Sherburn Aero Club Ltd	Sherburn-in-Elmet	13.11.06T
G-BODC	Piper PA-28-161 Warrior II	2816041	N9605N	23. 2.88	Sherburn Aero Club Ltd	Sherburn-in-Elmet	8. 5.06T
G-BODD	Piper PA-28-161 Warrior II	2816040	N9604N	23. 2.88	L.W.Scattergood	Leeds-Bradford	9.10.06T
G-BODE	Piper PA-28-161 Warrior II	2816039	N9603N	23. 2.88	Sherburn Aero Club Ltd	Sherburn-in-Elmet	15. 7.06T
G-BODH	Slingsby Cadet III	PFA 42-10108	BGA.474	5. 1.88	M.M.Bain *"Fochinell"*	Wick	13. 8.02P
	(Built C D Denham) (Volkswagen 1834) *(If p/i is correct then converted ex T.8 Tutor c/n MHL/RT.13 ex G-ALNK/BGA.474) (Noted 5.04)*						
G-BODI	Stoddard-Hamilton Glasair III SH-3R	& 3088	(HB-...)	14. 4.89	H.Arlt	(Viersen, Germany)	16. 3.05P
	(Built Jackson Barr Ltd - c/n EMK.030) (Lycoming O-360)		G-BODI				
G-BODM	Piper PA-28-180 Cherokee Challenger	28-7305519	N56016	2. 2.88	R.Emery	Clutton Hill Farm, Clutton	24.10.05T
G-BODO	Cessna 152 II	15282404	N68923	29. 1.88	Annie R.Sarson	Popham	29. 5.06
G-BODP	Piper PA-38-112 Tomahawk II	38-81A0010	N25616	5. 1.88	M.A.Potter	Eaglescott	18. 1.07T
G-BODR	Piper PA-28-161 Warrior II	28-8116318	N8436B	5. 1.88	Airways Aero Associations Ltd	Booker	29. 9.06T
					(Op British Airways Flying Club) (Waves and Cranes t/s)		
G-BODS	Piper PA-38-112 Tomahawk	38-79A0410	N2379F	3. 2.88	Coulson Flying Services Ltd	Cranfield	11. 9.04T
G-BODT	Jodel D.18	PFA 169-11290		14. 1.88	L.D.McPhillips	Portmoak	22. 5.04P
	(Built R A Jarvis - c/n 173) (Rotax 912-UL)				tr Jodel G-BODT Syndicate		
G-BODU	Scheibe SF25C-2000 Falke	44434	D-KIAA	19. 1.88	Faulkes Flying Foundation Ltd	Dunstable	9. 5.06
	(Limbach L2000)						
G-BODX	Beech 76 Duchess	ME-309	N67094	26. 2.88	S.J.Skilton	Bournemouth	3. 9.06T
					t/a Aviation Rentals *(Op Professional Air Training)*		
G-BODY	Cessna 310R II	310R1503	N4897A	17.12.87	Atlantic Air Transport Ltd	Coventry	23. 2.06T
G-BODZ	Robinson R22 Beta	0729		8. 1.88	Langley Holdings Plc	Gamston	26. 4.07
G-BOEE	Piper PA-28-181 Cherokee Archer II	28-7690359	N6168J	20. 1.88	T.B.Parmenter	Lodge Farm, St.Osyth	13. 6.06
G-BOEG	Short SD.3-60 Variant 100	SH.3733	D-CFXE	27. 1.88	BAC Express Airlines Ltd	CDG, Paris	21.11.05E
			N163DD, N133PC, G-BOEG, G-14-3733		*(Op BAC Express) "City of Paris"*		
G-BOEH	Robin DR340 Major	434	F-BRVN	4. 1.88	G.Bowles	Bradleys Lawn, Heathfield	17. 6.07
					tr Piper Flyers Group		
G-BOEI	Short SD.3-60 Variant 100	SH3735	D-CFLX	27. 1.88	BAC Express Airlines Ltd *(Op Aeroscot)*	Edinburgh	4. 4.05T
	(Originally regd as Variant 300)		VP-BKL, VR-BKL, G-BOEI, (VR-B..), G-BOEI, G-14-3735				
G-BOEK	Cameron V-77 HAB	1658		25. 1.88	R.I.M.Kerr, R.S.McLean and P.McCheyne	(Bristol)	28. 2.05A
					"Secret Leader"		
G-BOEM	Pitts S-2A	2255	N31525	17. 2.88	Margaret Murphy	Spanhoe	21. 7.01
	(Built Aerotek Inc)						
G-BOEN	Cessna 172M Skyhawk	17261325	N20482	12. 2.88	C.Barlow	Fairoaks	27. 6.06T
G-BOER	Piper PA-28-161 Warrior II	28-8116094	N83030	21. 1.88	M.and W.Fraser-Urquhart	Blackpool	18. 4.06
G-BOET	Piper PA-28RT-201 Arrow IV	28R-8018020	G-IBEC	28. 1.88	B.C.Chambers	Jersey	4. 9.06
			G-BOET, N8116V				
G-BOEW	Robinson R22 Beta	0750		27. 1.88	Plane Talking Ltd	Cranfield	13. 4.06T
G-BOEZ	Robinson R22 Beta	0753		27. 1.88	Plane Talking Ltd	Sywell	26. 3.06T
G-BOFC	Beech 76 Duchess	ME-217	N6628M	28. 1.88	Magenta Ltd *(Op Airways Flight Training)*	Exeter	8. 4.06T
G-BOFD	Cessna U206G Stationair 6	U20604181	N756LS	27. 1.88	D.M.Penny	Cark-in-Cartmel	16. 8.06
					(Op Wild Geese Parachute Centre)		
G-BOFE	Piper PA-34-200T Seneca II	34-7870381	N39493	22. 2.88	Alstons Upholstery Ltd	Southend	10. 8.06T
G-BOFF	Cameron N-77 HAB	1666		26. 1.88	R.C.Corcoran *(Systems 80 titles)*	(Bristol)	23. 7.03A
G-BOFL	Cessna 152 II	15284101	N5457H	28. 1.88	Gem Rewinds Ltd	Coventry	8. 5.06T
G-BOFM	Cessna 152 II	15284730	N6445M	28. 1.88	Gem Rewinds Ltd	Coventry	17. 7.06T
G-BOFW	Cessna A150M Aerobat	A1500612	N9803J	15. 2.88	D.F.Donovan	Elstree	28.10.06T
G-BOFX	Cessna A150M Aerobat	A1500678	N9869J	15. 2.88	N.F.O'Neill *(Noted 10.03)*	Newtownards	27.11.00T
G-BOFY	Piper PA-28-140 Cherokee Cruiser	28-7425374	N43521	3. 2.88	BCT Aircraft Leasing Ltd	Kemble	6. 5.07T
G-BOFZ	Piper PA-28-161 Cherokee Warrior II	28-7816255	N2189M	10. 2.88	R.W.Harris *(Op Willowair Flying Club)*	Southend	19. 1.05T
G-BOGC	Cessna 152 II	15284550	N5346M	8. 2.88	Mona Aviation Ltd t/a Mona Flying Club	RAF Mona	29. 8.04T
G-BOGG	Cessna 152 II	15282960	N45956	15. 2.88	A.S.Bamrah	Biggin Hill	26. 9.03T
					t/a Falcon Flying Services *(New owner 4.04)*		
G-BOGI	Robin DR400/180 Régent	1821		15. 2.88	A L.M.Shepherd	Rochester	28. 5.06T
G-BOGK	ARV Aviation ARV-1	PFA 152-11138		10. 2.88	H.N.Stone	(Morpeth)	15. 2.05P
	(Built Monewden Flying Group - kit no K.006)						
G-BOGM	Piper PA-28RT-201T Turbo Arrow IV	28R-8031077	N8173C	10. 2.88	R.J.Pearce t/a RJP Aviation	Wolverhampton	21.12.06T
G-BOGO	Piper PA-32R-301T Turbo Saratoga SP	32R-8029064	N8165W	6. 4.88	A S.Doman	Biggin Hill	28. 7.06T
G-BOGP	Cameron V-77 HAB	896		30. 3.88	T.Gunn *"Dire Straits"*	(Crowborough)	10. 7.00A
G-BOGR*	Colt 180A HAB				See SECTION 3 Part 1	Newbury	
G-BOGT*	Colt 77A HAB				See SECTION 3 Part 1	(Aldershot)	
G-BOGV	Air Command 532 Elite	PFA G/04-1102		10. 3.88	G.M.Hobman	Heworth, York	10. 1.91P
	(Built Deandell Products Ltd - c/n 0399)						
G-BOGY	Cameron V-77 HAB	1650		15. 2.88	R.A Preston *"Bella"*	(Bristol)	29. 5.04A
G-BOHA	Piper PA-28-161 Cherokee Warrior II	28-7816352	N3526M	16. 3.88	C.H.Bough and S.J.Gardiner	Shoreham	12. 6.06T
G-BOHD	Colt 77A HAB	1214		4. 3.88	D.B.Court *"Bluebird"*	Ormskirk	5. 8.02A
G-BOHF	Thunder Ax8-84 HAB	1197		8. 4.88	J.A Harris	(Sturminster Newton)	1. 9.05A
G-BOHH	Cessna 172N Skyhawk II	17273906	N131FR	19. 2.88	T.Scott	Gamston	23. 5.07T
			N7333J				
G-BOHI	Cessna 152 II	15281241	N49406	29. 2.88	V.D.Speck	Clacton	16. 7.03T
G-BOHJ	Cessna 152 II	15280558	N25259	29. 2.88	A.G.Knight t/a Airlaunch	Old Buckenham	5. 3.05T
G-BOHL	Cameron A-120 HAB	1701		11. 3.88	T.J.Bucknall *"Son of City of Bath"*	Hawarden	12.10.02
G-BOHM	Piper PA-28-180 Cherokee Challenger	28-7305287	N55000	18. 2.88	B.F.Keogh and R.A Scott		
						Lockmead Farm, South Marston	8. 5.06
G-BOHO	Piper PA-28-161 Warrior II	28-8016196	N747RH	25. 2.88	H.M.Sherriff and D.G.Buchanan	Duxford	20.10.06T
			N9560N		tr Egressus Flying Group		
G-BOHR	Piper PA-28-151 Cherokee Warrior	28-7515245	C-GNFE	29. 2.88	Karen L.Rivers	(Coleshill, Birmingham)	20. 4.07
G-BOHS	Piper PA-38-112 Tomahawk	38-79A0988	N2418P	26. 2.88	A S.Bamrah t/a Falcon Flying Services	Biggin Hill	15 4.07T
G-BOHT	Piper PA-38-112 Tomahawk	38-79A1079	N25304	14. 4.88	E.Reed	Durham Tees Valley	28. 9.06T
			C-GAYW, N24052		t/a St.George Flight Training		
G-BOHU	Piper PA-38-112 Tomahawk	38-80A0031	N25093	26. 2.88	D.A Whitmore	Norwich	1. 5.06T

G-BOHV	Wittman W.8 Tailwind	PFA 31-11151		3. 3.88	R.A Povall	Yearby	10. 9.05P
	(Built R A Povall - c/n 621)						
G-BOHW	Van's RV-4	PFA 181-11309		16. 6.88	P.J.Robins	Deenethorpe	20. 8.04P
	(Built R W H Cole) (Lycoming O-320)						
G-BOHX	Piper PA-44-180 Seminole	44-7995008	N36814	9. 3.88	Airpart Supply Ltd	Oxford	11. 8.06T
G-BOIA	Cessna 180K Skywagon II	18053121	N2895K	3. 3.88	R.E.,P.E.R.,J.E.R.and R.J.W.Styles	Rush Green	22. 7.07
					tr Old Warden Flying and Parachute Group		
G-BOIB	Wittman W.10 Tailwind	PFA 31-10551		3. 3.88	R.F.Bradshaw	Valley Farm, Winwick	17. 4.05P
	(Built P H Lewis) (Continental O-300-D)						
G-BOIC	Piper PA-28R-201T Turbo Arrow III	28R-7803123	N2336M	7. 4.88	M.J.Pearson	Stapleford	17. 6.06
G-BOID	Bellanca 7ECA Citabria	1092-75	N8676V	3. 3.88	D.Mallinson	Birds Edge, Penistone	28. 4.07
G-BOIG	Piper PA-28-161 Warrior II	28-8516027	N4390B	1. 3.88	D.Vallance-Pell	Netherthorpe	5. 9.03
			N9519N		*(Noted 5.04)*		
G-BOIJ	Thunder Ax7-77 Series 1 HAB	964		11. 3.88	K.Dodman	(Stowmarket)	29. 4.05A
G-BOIK	Air Command 503 Commander	PFA G/04-1087		8. 3.88	F.G.Shepherd	Alston, Cumbria	22. 1.90P
	(Built G R Horner - c/n 0420) (Officially regd as c/n PFA G/04-1090)						
G-BOIL	Cessna 172N Skyhawk II	17271301	N23FL	2. 3.88	Upperstack Ltd	Barton	17. 7.06T
			N23ER, (N2494E)				
G-BOIO	Cessna 152 II	15280260	N24445	7. 3.88	AV Aviation Ltd	Eshott	7. 8.04T
G-BOIP	Cessna 152 II	15283444	N49264	7. 3.88	Stapleford Flying Club Ltd	Stapleford	26. 5.91T
					(Damaged Uckington 11.1.90: stored 5.98: current status unknown)		
G-BOIR	Cessna 152 II	15283272	N48041	7. 3.88	Shropshire Aero Club Ltd	Sleap	13. 6.06T
G-BOIT	SOCATA TB-10 Tobago	810		10. 3.88	M.J.Ryan t/a G-BOIT Flying Group	RNAS Yeovilton	21.11.04T
G-BOIU	SOCATA TB-10 Tobago	811		10. 3.88	R and B Aviation Ltd	Guernsey	20. 4.07
					(Crashed, caught fire attempting to return to Bournemouth 28.8.04 and badly damaged)		
G-BOIV	Cessna 150M Commuter	15078620	N704HH	30. 3.88	M.J.Page tr India Victor Group	Seething	4.12.06
G-BOIX	Cessna 172N Skyhawk II	17271206	C-GMMX	9. 3.88	JR Flying Ltd	Bournemouth	28. 2.07T
			N2253E				
G-BOIY	Cessna 172N Skyhawk II	17267738	N73901	9. 3.88	L.W.Scattergood	Sherburn-in-Elmet	25. 8.06T
G-BOIZ	Piper PA-34-200T Seneca II	34-8070014	N81081	25. 2.88	R.W.Tebby	Bristol	2.10.06T
					t/a S.F.Tebby and Son *(Op Bristol Flying Centre)*		
G-BOJB	Cameron V-77 HAB	1615		11. 3.88	R.M.Trotter	(Bristol)	6. 7.04A
G-BOJD	Cameron N-77 HAB	1653		11. 3.88	R.S.McDonald *"Bluebird"*	(Chesham)	3. 9.05
G-BOJI	Piper PA-28RT-201 Arrow IV	28R-7918221	N2919X	31. 3.88	T.A Stoate and K.D.Head	Blackbushe	23. 4.06
					tr Arrow Two Group		
G-BOJK	Piper PA-34-220T Seneca III	3433020	G-BRUF	11. 3.88	Redhill Aviation Ltd	Blackbushe	3. 7.06T
			N9113D		*(Op Redhill Flying Club)*		
G-BOJM	Piper PA-28-181 Archer II	28-8090244	N8155L	21. 3.88	R.P.Emms	Gamston	12. 5.06
G-BOJR	Cessna 172P Skyhawk II	17275574	N64539	22. 4.88	Exeter Flying Club Ltd	Exeter	13. 5.06T
G-BOJS	Cessna 172P Skyhawk II	17274582	N52699	29. 3.88	Beatrice A.Paul	Denham	19. 6.06T
G-BOJU	Cameron N-77 HAB	1718		21. 3.88	M.A Scholes *"GB Transport"*	(London SE25)	7. 9.97A
G-BOJW	Piper PA-28-161 Cherokee Warrior II	28-7716038	N1668H	28. 3.88	Brewhamfield Farm Ltd	(Wantage)	9. 2.07T
G-BOJZ	Piper PA-28-161 Warrior II	28-7916223	N2113J	28. 3.88	A.S.Bamrah t/a Falcon Flying Services	Rochester	1. 4.07T
G-BOKA	Piper PA-28-201T Turbo Dakota	28-7921076	N2860S	15. 3.88	CBG Aviation Ltd	Fairoaks	14. 5.06
G-BOKB	Piper PA-28-161 Warrior II	28-8216077	N8013Y	29. 3.88	Apollo Aviation Advisory Ltd	Shoreham	20. 8.06T
G-BOKD	Bell 206B-3 JetRanger III	3654	G-ISKY	30. 8.01	Sterling Helicopters Ltd	Norwich	13. 4.06T
			G-PSCI, G-BOKD, N3171A				
G-BOKF	Air Command 532 Elite	PFA G/04-1101		28. 3.88	J.K.Padden	(Morpeth)	22. 9.99P
	(Built N J Brunskill - c/n 0404)				*(Current status unknown)*		
G-BOKH	Whittaker MW7	PFA 171-11281	(G-MTWT)	21. 3.88	I.D.Evans	Chiltern Park, Wallingford	30. 3.05P
	(Built M Whittaker) (Officially regd as PFA 171-11231)						
G-BOKX	Piper PA-28-161 Cherokee Warrior II	28-7816680	N39709	28. 3.88	Shenley Farms (Aviation) Ltd	Headcorn	21. 9.03T
G-BOKY	Cessna 152 II	15285298	N67409	6. 4.88	D.F.F.Poore	Bournemouth	1. 7.07T
G-BOLB	Taylorcraft BC-12-65	3165	N36211	17. 5.88	A D.Pearce, E.C.and P.King		
	(Continental A65)		NC36211		*"Spirit of California"*	Eastbach Farm, Coleford	7. 9.05P
G-BOLC	Fournier RF6-100	1	F-BVKS	28. 3.88	J.D.Cohen	Dunkeswell	19. 1.07
G-BOLD	Piper PA-38-112 Tomahawk	38-78A0180	N9740T	8. 7.88	Harnett Air Services Ltd *(New owner 2.03)*	Eaglescott	21. 1.96T
G-BOLE	Piper PA-38-112 Tomahawk	38-78A0475	N2506E	13. 7.88	J.and G.Stevenson	Tollerton	28. 5.07
G-BOLF	Piper PA-38-112 Tomahawk	38-79A0375	N583P	13. 7.88	Teesside Flight Centre Ltd	Durham Tees Valley	19. 7.04T
			YV-583P, YV-133E, YV-1696P, N9666N				
G-BOLG	Bellanca 7KCAB Citabria	517-75	N8706V	25.11.88	B.R.Pearson t/a Aerotug	Eaglescott	22. 5.04
G-BOLI	Cessna 172P Skyhawk II	17275484	N63794	30. 3.88	W.White tr BOLI Flying Club	Denham	11. 7.06T
G-BOLL	Lake LA-4-200 Skimmer	295	(F-GRMX)	4. 5.88	M.C.Holmes	City of Derry	1. 4.07
			G-BOLL, EI-ANR, N1133L				
G-BOLN	Colt 21A Cloudhopper HAB	1226		4. 5.88	G.Everett	(Maidstone)	19. 5.04A
G-BOLO	Bell 206B JetRanger II	1522	N59409	2.11.87	Hargreaves Leasing Ltd	Goodwood	2. 3.06T
					(Op Blades Helicopters)		
G-BOLP	Colt 21A Cloudhopper HAB	1227		4. 5.88	J.E.Rose	(Abingdon)	1. 5.02A
G-BOLR	Colt 21A Cloudhopper HAB	1228		3. 5.88	C.J.Sanger-Davies	(Uttoxeter)	15. 1.04A
G-BOLS	Clutton FRED Series II	PFA 29-10676		6. 4.88	I.F.Vaughan	(Melton Mowbray)	
	(Built R J Goodburn)				*"The Ruptured Uck" (Current status unknown)*		
G-BOLT	Rockwell Commander 114	14428	N5883N	16.10.78	H. Gafsen	Elstree	4. 7.07T
G-BOLU	Robin R3000/120	106	F-GFAO	14. 4.88	I.W.Goodger	Biggin Hill	18. 7.06T
			SE-IMS		t/a Classair		
G-BOLV	Cessna 152 II	15280492	N24983	8. 4.88	A S.Bamrah t/a Falcon Flying Services	Biggin Hill	16. 1.06T
G-BOLW	Cessna 152 II	15280589	N25316	9. 6.88	JRB Aviation Ltd *(Op Seawing Flying Club)*	Southend	30. 8.06T
G-BOLX	Cessna 172N Skyhawk II	17269099	N734TK	8. 4.88	R.J.Burrough	Lydd	9.12.06
G-BOLY	Cessna 172N Skyhawk II	17269004	N734PJ	31. 3.88	Simair Ltd	Earls Colne	10. 6.07T
G-BOLZ	Rand Robinson KR-2	PFA 129-10866		6. 4.88	B.Normington	Coventry	24. 7.05P
	(Built B.Normington) (Volkswagen 1834)						
G-BOMB	Cassutt Racer IIIM	PFA 34-10386		18.12.78	S.Adams *"Blind Panic"*	RAF Weston on the Green	23. 5.98P
	(Built D Ford) (Continental O-200-A)				*(Damaged Weston Park, Telford 22.6.97: current status unknown)*		
G-BOMG	Pilatus Britten-Norman BN-2B-26 Islander	2205	D-IBNF	6. 4.88	Loganair Ltd	Glasgow	12. 9.05T
			G-BOMG				

Registration	Type	C/n	Previous identity	Regd	Owner/Operator	Base	CofA
G-BOMN	Cessna 150F	15063089	N6489F	25. 4.89	Auburn Air Ltd	Abbeyshrule, County Longford	28. 8.06T
G-BOMO	Piper PA-38-112 Tomahawk II	38-81A0161	N91324	8. 4.88	APB Leasing Ltd	Swansea	28. 6.07T
G-BOMP	Piper PA-28-181 Cherokee Archer II	28-7790249	N8482F	8. 4.88	Dawne Carter *(New owner 8.04)*	Elstree	12. 2.04T
G-BOMS	Cessna 172N Skyhawk II	17269448	N737JG	11. 4.88	Almat Flying Club Ltd and Penchant Ltd	Coventry	25.10.07E
G-BOMU	Piper PA-28-181 Cherokee Archer II	28-7790318	N1631H	8. 4.88	J.Sawyer	Blackbushe	5. 8.06T
G-BOMY	Piper PA-28-161 Warrior II	28-8216049	N8457S	28. 6.88	D.Knight t/a Southern Care Maintenance	Headcorn	16. 2.07T
G-BOMZ	Piper PA-38-112 Tomahawk	38-78A0635	N2315A	30. 6.88	I.C.Barlow and G.W.G.Young tr BOMZ Aviation	Booker	25. 4.06T
G-BONC	Piper PA-28RT-201 Arrow IV	28R-7918007	C-GXYX	13. 5.88	Finglow Ltd	Fowlmere	16. 7.06T
			N3069K				
G-BONG	Enstrom F-28A-UK	154	N9604	22. 4.88	J.G. Dunn	(Eccles)	16. 8.98T
					(Noted 8.04 - new CofR 10.04)		
G-BONK*	Colt 180A HAB				See SECTION 3 Part 1	(Aldershot)	
G-BONO	Cessna 172N Skyhawk II	17270299	C-GSMF	11. 5.88	J.D.McCandless	(Douglas, Isle of Man)	4. 8.07T
			N738WS				
G-BONP	CFM Streak Shadow	PFA 161A-11344		4. 5.88	T.J.Palmer	Prestwick	23.10 04P
	(Built CFM Metal-Fax - c/ns 108 & SS-01P) (Rotax 582)						
G-BONR	Cessna 172N Skyhawk II	17268164	C-GYGK	18. 4.88	D.I.Claik	Biggin Hill	5. 7.06
			(N733BH)				
G-BONS	Cessna 172N Skyhawk II	17268345	C-GIUF	18. 4.88	M.G.Montgomerie tr G-BONS Group	Elstree	11. 9.06
G-BONT	Slingsby T.67M Firefly II	2054		3. 5.88	Babcock Support Services Ltd	AAC Middle Wallop	26.10.06T
					t/a Babcock Defence Services		
G-BONU	Slingsby T.67B Firefly	2037		3. 5.88	R.L.Brinklow *(Noted 4.02)*	Hinton-in-the-Hedges	29. 6.00T
G-BONV*	Colt 17A Cloudhopper HAB				See SECTION 3 Part 1	(Aldershot)	
G-BONW	Cessna 152 II	15280401	OY-CPL	15. 4.88	Lincoln Aero Club Ltd	Sturgate	17. 8.06T
			N24825				
G-BONY	Denney Kitfox Model 1	PFA 172-11351		11. 5.88	J.M.Vinall	Rayne Hall Farm, Braintree	9. 6.05P
	(Built J S Penny - c/n 166) (Rotax 532) *(Inscribed as "Mk.2")*						
G-BONZ	Beech V35B Bonanza	D-10282	N6661D	6. 4.88	P.M.Coulten	Boughton, Norfolk	16. 2.07
G-BOOB	Cameron N-65 HAB	515		12.11.79	J.Rumming *"Cracker"*	(Swindon)	8. 4.90A
G-BOOC	Piper PA-18-150 Super Cub	18-8279	SE-EPC	29. 4.88	S.A C.Whitcombe	Meon, Petersfield	7. 9.06
G-BOOD	Slingsby T.31M Motor Tutor	PFA 42-11264		4. 5.88	K.A Hale	Lotmead Farm, Wanborough, Swindon	29. 4.05P
	(Built P J Titherington) (Fuji-Robin EC-44-2PM) *(Wings ex XE810 c/n 923)*						
G-BOOE	Gulfstream GA-7 Cougar	GA7-0093	N718G	7. 6.88	N.Gardner	Southampton	4. 8.06T
G-BOOF	Piper PA-28-181 Cherokee Archer II	28-7890084	N47510	16. 6.88	H.Merkado	Panshanger	13.11.06T
G-BOOG	Piper PA-28RT-201T Turbo Arrow IV	28R-8331036	N4303K	6. 5.88	Simair Ltd	Earls Colne	4. 4.05
G-BOOH	Jodel D.112	481	F-BHVK	16. 5.88	R.M. MacCormac	RAF Henlow	28. 7.05P
	(Built Etablissement Valladeau)						
G-BOOI	Cessna 152 II	15280751	N25590	22. 8.88	Stapleford Flying Club Ltd	Stapleford	19. 5.07T
G-BOOJ	Air Command 532 Elite II	PFA G/04-1098		4. 5.88	Roger Savage Gyroplanes Ltd		
	(Built H P Barlow - c/n.PB206)					Sorbie Farm, Kingsmuir	6.12.91P
G-BOOL	Cessna 172N Skyhawk II	17272486	C-GJSY	27. 4.88	Surrey and Kent Flying Club Ltd	Biggin Hill	15.12.06T
			N5271D				
G-BOOM*	Hawker Hunter T.7 (T.53)				See SECTION 3 Part 1	Amman, Jordan	
G-BOON*	Piper PA-32RT-300 Lance II	32R-7885253	N361DB	25. 4.88	(A Clark and Sons Ltd)	(Felsted, Essex)	28. 5.00T
			(Damaged landing Connemara 10.10.97: cancelled 3.3.98 by CAA) (Remains to scrap 6.03)				
G-BOOP*	Cameron N-90 HAB				See SECTION 3 Part 1	(Aldershot)	
G-BOOV	Aérospatiale AS355F2 Ecureuil 2	5374		3. 5.88	Atlantic Air Charters Ltd	Lee-on-Solent	27.11.06T
G-BOOW	Aerosport Scamp	PFA 117-10709		10. 5.88	D.A.Weldon	(Dublin)	1. 9.05P
	(Built C Tyers) (Volkswagen 1834)						
G-BOOX	Rutan LongEz	PFA 74A-10844		3. 5.88	I.R.Wilde	Deenethorpe	27. 3.05P
	(Built I R Thomas and I.R.Wilde) (Lycoming O-235)						
G-BOOZ	Cameron N-77 HAB	904	(G-BKSJ)	21. 6.83	J.E.F.Kettlety	(Chippenham)	
	(New home built envelope c 6.98)				*"Bluebell"*		
G-BOPA	Piper PA-28-181 Archer II	28-8490024	N43299	28. 4.88	Flyco Ltd	Denham	21. 8.06
G-BOPB	Boeing 767-204ER	24239	TF-ATP	1.11.88	Air Atlanta Europe Ltd	Gatwick	20. 4.07T
			G-BOPB, N6009F		*(Op Excel)*		
G-BOPC	Piper PA-28-161 Warrior II	28-8216006	N2124X	6. 5.88	Aeros Ltd	Gloucestershire	24. 7.06T
G-BOPD	Bede BD-4	632	N632DH	25. 5.88	S.T.Dauncey	Wellesbourne Mountford	17. 6.05P
	(Lycoming O-320)				*(Noted 11.04)*		
G-BOPH	Cessna TR182 Turbo Skylane RG II	R18201031	N756BJ	11. 5.88	Grandsam Investments Ltd	Cambridge	16. 9.06T
G-BOPO	FLS Aerospace OA.7 Optica Series 301	021	EC-FVM	17. 5.88	Aces High Ltd	North Weald	27. 5.96T
			EC-435, G-BOPO		*(Noted 1.05)*		
G-BOPR	FLS Aerospace OA.7 Optica Series 301	023		17. 5.88	Aces High Ltd *(Noted 1.05)*	North Weald	
G-BOPT	Grob G115	8046		10. 5.88	LAC (Enterprises) Ltd *(Op Lancashire Aero Club)*	Barton	6. 1.07T
G-BOPU	Grob G115	8059		10. 5.88	LAC (Enterprises) Ltd *(Op Lancashire Aero Club)*	Barton	20. 1.06T
G-BOPX	Cessna A152 Aerobat	A1520932	N761BK	11. 5.88	Aerohire Ltd	Bourne Park, Hurstbourne Tarrant	24. 1.98T
					(Stored 10.01 less wings)		
G-BORA*	Colt 77A HAB				See SECTION 3 Part 1	(Aldershot)	
G-BORB	Cameron V-77 HAB	1348		24. 8.88	M.H.Wolff	(Liskeard)	9. 7.02A
G-BORD	Thunder Ax7-77 HAB	1164		26. 5.88	D.D.Owen *"Marvin"*	(Wotton-under-Edge)	11.12.99A
G-BORE	Colt 77A HAB	642		24. 5.88	J.D.Medcalf and C.Wilson	(Enfield)	24. 8.02A
					tr Little Secret Hot Air Balloon Group *"My Little Secret"*		
G-BORG	Campbell Cricket	PFA G/03-1085		8. 6.88	G.Davison and H.Hayes	Carlisle	25. 4.07P
	(Built N G Bailey) (Rotax 503)						
G-BORH	Piper PA-34-200T Seneca II	34-8070352	N8261V	7. 6.88	Aerolease Ltd	Conington	15. 2.07T
G-BORI	Cessna 152 II	15281672	N66936	8. 6.88	S.Copeland	Bakersfield, Northampton	16.10.06T
G-BORJ	Cessna 152 II	15282649	N89148	27. 5.88	Pool Aviation (NW) Ltd	Blackpool	22.12.06T
G-BORK	Piper PA-28-161 Warrior II	28-8116095	G-IIIC		V.R.Coultan	Turweston	21.12.06E
			G-BORK, N83036		t/a The Warrior Group *(Orange Communications titles)*		
G-BORL	Piper PA-28-161 Cherokee Warrior II	28-7816256	N2190M	28. 9.88	Westair Flying School Ltd	Blackpool	26. 1.07T
G-BORM*	Hawker Siddeley HS.748 Series 2B/217	1670	RP-C1043	29. 7.88	Not known	Exeter	
			V2-LAA, VP-LAA, 9Y-TDH		*(Cancelled 18.6.92 by CAA) (On fire dump 9.04)*		
G-BORN	Cameron N-77 HAB	1777		13. 5.88	I.Chadwick *"Ian"*	(Partridge Green, Bolney)	5.12.04A
G-BORO	Cessna 152 II	15283767	N5130B	27. 5.88	Tatenhill Aviation Ltd	Tatenhill	3. 3.07T

Reg	Type	C/n	Prev. id	Regd	Owner/Operator	Base	CofA
G-BORR	Thunder AX8-90 HAB	1256		13. 6.88	W.J.Harris	(Scwarzhofen, Germany)	10. 9.05A
G-BORS	Piper PA-28-181 Archer II	28-8090156	N8127C	31. 5.88	Modern Air (UK) Ltd	Fowlmere	29. 6.06T
G-BORT	Colt 77A HAB	1255		7. 6.88	J.Triquet	St.Gemmes-Le-Robert, France	10. 9.01A
G-BORW	Cessna 172P Skyhawk II	17274301	N51357	23. 8.88	Briter Aviation Ltd	(Coventry)	3.10.07E
G-BORY	Cessna 150L	15072292	N6792G	27. 5.88	C.J.Twelves t/a Alexander Aviation	Fenland	28. 6.05T
G-BOSB	Thunder Ax7-77 HAB	1199		7. 6.88	M.Gallagher	(Consett)	17. 9.99A
	(Regd as c/n 581 but built as above)						
G-BOSD	Piper PA-34-200T Seneca II	34-7570085	N33086	7. 6.88	Barnes Olson Aeroleasing Ltd	Bristol	11. 8.07T
					(Op Bristol Flying Centre)		
G-BOSE	Piper PA-28-181 Archer II	28-8590007	N143AV	17. 5.88	R.Wolf and D.J.Blundell tr G-BOSE Group	White Waltham	4. 6.06T
G-BOSJ	Nord 3400	124	N9048P	26. 5.88	A I.Milne	Etthornes Farm, Swaffham	1.11.94P
			ALAT "MOO"		*(As "124" in Fr.AF c/s) (Damaged Fenland 12.6.94: stored 9.02)*		
G-BOSM	CEA Jodel DR.253B Régent	168	F-BSBH	24. 5.88	A.G.Stevens	(Coalville)	11.12.04
G-BOSN	Aérospatiale AS355F1 Ecureuil 2	5266	N2109L	22. 8.88	L.Smith	Booker	10. 3.06T
			5N-AYL, G-BOSN, 5N-AYL		t/a Helicopter Services		
G-BOSO	Cessna A152 Aerobat	A1520975	N761PD	25. 5.88	J.S.Develin and Z.Islam	Redhill	16. 9.07T
G-BOSR	Piper PA-28-140 Cherokee	28-22092	N7464R	26. 5.88	C R.Guggenheim tr Sierra-Romeo Group	Bournemouth	29.10.06T
G-BOSU	Piper PA-28-140 Cherokee Cruiser	28-7325449	N55635	19. 7.88	R.A Sands	Oxford	8. 6.07
G-BOTD	Cameron O-105 HAB	1611		6. 6.88	P.J.Beglan	(Belves, France)	13. 5.04A
G-BOTE*	Thunder Ax8-90 HAB				See SECTION 3 Part 1	(Aldershot)	
G-BOTF	Piper PA-28-151 Cherokee Warrior	28-7515436	C-GGIF	8. 6.88	P.E.Preston	Southend	8.10.06T
					tr G-BOTF Group *(Op Southend Flying Club)*		
G-BOTG	Cessna 152 II	15283035	N46343	9. 6.88	Donington Aviation Ltd	Nottingham East Midlands	21.10.06T
G-BOTH	Cessna 182Q Skylane II	18267558	N202PS	9. 6.88	A C.Hinton-Lever	Barton	17. 5.07
			N114SP, N5172N		tr G-BOTH Group		
G-BOTI	Piper PA-28-151 Cherokee Warrior	28-7515251	C-GNFF	9. 6.88	A J.Bamrah	Blackbushe	29. 8.06T
	(Converted to PA-28-161 model)				t/a Falcon Flying Services		
G-BOTK	Cameron O-105 HAB	1765		9. 6.88	N.Woodham	(Fawley, Southampton)	23. 5.05T
G-BOTL*	Colt 42A SS HAB				See SECTION 3 Part 1	Newbury	
G-BOTM	Bell 206B-3 JetRanger III	3881	N31940	9. 6.88	Mexsky Ltd	Earls Colne	19. 8.06
G-BOTN	Piper PA-28-161 Warrior II	28-7916261	N2173N	9. 6.88	Apollo Aviation Advisory Ltd	Shoreham	8. 2.07T
G-BOTO	Bellanca 7ECA Citabria	939-73	N57398	9. 6.88	A K.Hulme	Rayne Hall Farm, Braintree	7. 10.04
					tr G-BOTO Group *(Noted 1.05)*		
G-BOTP	Cessna 150J	15070736	N61017	2. 8.88	R.F.Finnis and C.P.Williams	Thruxton	2. 6.07
G-BOTU	Piper J-3C-65 Cub	19045	N98803	8. 7.88	T.L.Giles	Browns Farm, Hitcham	17. 8.04P
	(Continental A75)		NC98803				
G-BOTV	Piper PA-32RT-300 Lance II	32R-7885153	N36039	7. 6.88	Robin Lance Aviation Associates Ltd	Rochester	7. 7.02
G-BOTW	Cameron V-77 HAB	1761		14. 6.88	M.R.Jeynes	(Worcester)	17. 6.03A
G-BOUD	Piper PA-38-112 Tomahawk II	38-82A0017	N91365	26. 7.88	A J.Wiggins	(Longhope, Glos)	11. 7.05T
G-BOUE	Cessna 172N Skyhawk II	17273235	N6535F	8. 8.88	Castleridge Ltd	Leeds-Bradford	24. 3.07T
G-BOUF	Cessna 172N Skyhawk II	17271900	N5605E	24. 6.88	Betty P.and M.I.Sneap	Ripley, Derby	3. 8.07T
G-BOUJ	Cessna 150M Commuter	15076373	N3058V	25. 8.88	R.D.Billins tr UJ Flying Group	Cranfield	15. 1.06T
G-BOUK	Piper PA-34-200T Seneca II	34-7570124	N33476	31. 8.88	C.J.and R.J.Barnes	Nottingham East Midlands	10.10.05T
G-BOUL	Piper PA-34-200T Seneca II	34-7670157	N8936C	28. 6.88	Oxford Aviation Services Ltd	Oxford	18. 4.07T
G-BOUM	Piper PA-34-200T Seneca II	34-7670136	N8401C	3. 8.88	Oxford Aviation Services Ltd	Oxford	7.10.07E
G-BOUN*	Rand-Robinson KR-2	PFA 129-10945		23. 6.88	Not known	Charterhall	27. 7.99P
					(Cancelled 25.10.01 by CAA) (Stored 4.04)		
G-BOUP	Piper PA-28-161 Warrior II	2816059	N9139X	12. 7.88	Aeros Holdings Ltd	Gloucestershire	10. 2.06T
G-BOUR	Piper PA-28-161 Warrior II	2816060	N9139Z	12. 7.88	Oxford Aviation Services Ltd	Oxford	17. 3.07T
G-BOUT	Colomban MC-12 Cri-Cri	12-0135	N120JN	14. 6.88	C.K.Farley	Southampton	
	(Built J A Nelson)				*(Current status unknown)*		
G-BOUV	Montgomerie-Bensen B.8MR	PFA G/01-1092		23. 6.88	G.C.Kerr	Carlisle	13. 6.03P
	(Built P Wilkinson) (Rotax 532)				*(Noted 11.04)*		
G-BOUZ	Cessna 150G	15065606	N2606J	15. 6.88	Atlantic Bridge Aviation Ltd	Lydd	16. 3.04T
G-BOVB	Piper PA-15 Vagabond	15-180	N4396H	23. 6.88	J.R.Kimberley	Andrewsfield	18. 5.05P
	(Lycoming O-145)		NC4396H				
	(Failed to gain height on take-off Whitefields Farm, near South Molton, Devon 16.10.04 and suffered major damage)						
G-BOVG*	Cessna F172H	F172-0627	OO-ANN	2. 8.88	No.1476 Squadron, ATC	Rayleigh	14. 9.91
	(Built Reims Aviation SA)		D-ELTR		*(Damaged Southend 1991: cancelled 26.9.95 as WFU) (Instructional airframe 1.05)*		
G-BOVK	Piper PA-28-161 Warrior II	28-8516061	N69168	7. 9.88	Multiflight Ltd	Leeds-Bradford	14. 5.07T
G-BOVR	Robinson R22HP	0176	N9069D	28. 6.88	J.O'Brien	(Gorey, County Wexford)	19. 8.00T
G-BOVS	Cessna 150M Commuter	15078663	N704KC	21. 7.88	K.J.Marchelak	Swansea	5. 2.07T
G-BOVT	Cessna 150M Commuter	15078032	N8962U	1.12.88	C.J.Hopewell	Fenland	13. 4.07T
G-BOVU	Stoddard-Hamilton Glasair III	3090		16. 9.88	B.R.Chaplin	Deenethorpe	20. 5.05P
	(Buillt A H Carrington) (Lycoming IO-540)						
G-BOVV	Cameron V-77 HAB	1724		26. 9.88	P.Glydon	(Knowle, Bristol)	12.10.02A
G-BOVW	Colt 69A HAB	1286		13. 7.88	V.Hyland	(Nottingham)	6. 4.94A
					"Enderby-Hyland Painting"		
G-BOVX	Hughes 269C	38-0673	N58170	12. 7.88	P.E.Tornberg	Sywell	25. 9.04T
G-BOWB	Cameron V-77 HAB	1767		13. 7.88	R.C.Stone *"Richard's Rainbow"*	(Reading)	27. 5.04A
G-BOWC*	Cessna 150J	15070458	N60626	24.10.88	Missionary Aviation Fellowship	Croft Farm, Defford	
	(Force landed near Barton 10.7.94: cancelled as WFU 16.9.94) (As "G-BMAF" and used for fund-raising events - noted 6.04)						
G-BOWE	Piper PA-34-200T Seneca II	34-7870405	N39668	14. 7.88	Oxford Aviation Services Ltd	Oxford	15 2.07T
G-BOWL	Cameron V-77 HAB	1780		26. 7.88	P.G. and G.R.Hall *"Matrix"*	(Chard)	12. 5.00A
G-BOWM	Cameron V-56 HAB	1781		26. 7.88	C.G.Caldecott and G.Pitt	(Newcastle-under-Lyme)	30. 7.00A
G-BOWN	Piper PA-12 Super Cruiser	12-1912	N3661N	26. 7.88	T.L.Giles	Brown Farm, Hitcham, Ipswich	4.10.07E
	(Lycoming O-235)		NC3661N				
G-BOWO	Cessna R182 Skylane RG II	R18200146	(G-BOTR)	20. 7.88	J.J.Feeney	Elstree	12.11.04
			N2301C				
G-BOWP	Wassmer Jodel D.120A Paris-Nice	319	F-BNZM	26. 7.88	J.M.Pearson	Crosland Moor	21. 4.04P
	(Continental O-200-A)				*(New owner 11.04)*		
G-BOWU	Cameron O-84 HAB	1779		1. 8.88	C.F.Pooley and D.C.Ball	(Gloucester)	7. 8.01A
					tr St.Elmos Fire Syndicate *"Elmo"*		
G-BOWV	Cameron V-65 HAB	1800		24. 8.88	R.A Harris *"Sigmund"*	(Axminster)	19.11.04A

Reg	Type	C/n	Previous identity	Reg date	Owner/operator	Base	CofA expiry
G-BOWY	Piper PA-28RT-201T Turbo Arrow IV	28R-8131114	N404EL	8. 8.88	J.S.Develin and Z.Islam	Blackbushe	18. 8.04T
			N83648				
G-BOWZ	Bensen B.80V	PFA G/01-1060		27. 7.88	M D Cole	Longside, Peterhead	28. 3.05P
	(Built W M Day) (Rotax 532)						
G-BOXA	Piper PA-28-161 Warrior II	2816075	N9149Q	1.11.88	Channel Islands Aero Club (Jersey) Ltd	Jersey	8. 1.07T
G-BOXB*	Piper PA-28-161 Warrior II	2816064	N9142H	12. 8.88	First Class Ltd	Willey Farm, Caterham	6. 2.03T
	(Failed to gain height on take-off from Little Bredy 19.3.03 and badly damaged: cancelled 26. 8.03 by CAA) (Wreck noted 5.03 for scrap)						
G-BOXC	Piper PA-28-161 Warrior II	2816063	N9142D	12. 8.88	Channel Islands Aero Club (Jersey) Ltd	Jersey	18. 5.06T
G-BOXG	Cameron O-77 HAB	1792		26. 8.88	R.A Wicks	(Norwich)	28. 7.05A
G-BOXH	Pitts S-1S	MP4	N8LA	29. 7.88	Pittsco Ltd	Sherburn-in-Elmet	10. 8.05P
	(Built B Halstock and J Mills)						
G-BOXJ	Piper L-4H (L-4H-PI)	12193	OO-ADJ	1. 8.88	J.D.Tseliki	Kittyhawk Farm, Deanland	4. 8.05P
	(Continental C90)*(Frame No.12021)*		44-79897		*(As "479897/JD" in USAF Army Air Corps c/s)*		
G-BOXR	Grumman American GA-7 Cougar	GA7-0059	N772GA	19.10.88	Plane Talking Ltd	Denham	11. 9.06T
	(C/n plate shows manufacturer as Gulfstream American)						
G-BOXT	Hughes 269C	104-0367	SE-HMR	1. 8.88	Jetscape Leisure Ltd	Gloucestershire	17. 3.06T
			PH-JOH, D-HBOL				
G-BOXU	Grumman AA-5B Tiger	AA5B-0026	N1526R	28. 7.88	G.C.Baker tr Marcher Aviation Group	Trehelig, Welshpool	19. 5.07
G-BOXV	Pitts S-1S	7-0433	N27822	8. 8.88	C.Waddington	(Hereford)	28. 9.04P
	(Built J R Castrillo)						
G-BOXW	Cassutt Racer IIIM	PFA 34-11317		11. 8.88	D.I.Johnson	(Leigh-on-Sea)	
	(Built D.I.Johnson)				*(Current status unknown)*		
G-BOYB	Cessna A152 Aerobat	A1520928	N761AW	29. 7.88	Northamptonshire School of Flying Ltd	Sywell	14.12.03T
					(On rebuikt 2.05)		
G-BOYC	Robinson R22 Beta	0837		22. 8.88	M.D.Thorpe t/a Yorkshire Helicopters	Coney Park, Leeds	9. 8.04T
G-BOYF	Sikorsky S-76B	760343		15. 9.88	Darley Stud Management Co Ltd	Cambridge	24.11.05E
					(Op Air Hanson)		
G-BOYH	Piper PA-28-151 Cherokee Warrior	28-7715290	N8795F	8. 8.88	Superpause Ltd	White Waltham	26. 4.07T
	(Converted to 28-161 model)				*(Op West London Aero Club)*		
G-BOYI	Piper PA-28-161 Warrior II	28-7816183	N9032K	8. 8.88	A P.Ware and S.Hoy tr G-BOYI Group	Sleap	3. 6.07
G-BOYL	Cessna 152 II	15284379	N6232L	11. 8.88	Aerohire Ltd	Wolvehampton	26. 2.01T
G-BOYM	Cameron O-84 HAB	1796		25. 8.88	M.P.Ryan *"Frontline"*	(Newbury)	
G-BOYO	Cameron V-20 HAB	1843		27. 9.88	J.M.Willard	(Burgess Hill)	
G-BOYP	Cessna 172N Skyhawk II	17270349	N738YU	22. 8.88	I.D and Delia Brierley	North Weald	19. 5.07
G-BOYR	Reims/Cessna F337G Super Skymaster		RA-04147	9. 9.88	Tri-Star Farms Ltd	Andreas, Isle of Man	27. 8.06T
	(Wichita c/n 33701589)	F33700070	G-BOYR, PH-RPE				
G-BOYS	Cameron N-77 HAB	1759		16. 6.88	Julie King	(Chippenham)	8. 3.05T
G-BOYU	Cessna A150L Aerobat	A1500497	N8121L	31. 8.88	Upperstack Ltd	Barton	26.10.06T
G-BOYV	Piper PA-28R-201T Turbo Cherokee Arrow III		N1143H	1. 9.88	Arrow Air Ltd	Wellesbourne Mountford	14. 6.07T
		28R-7703014					
G-BOYX	Robinson R22 Beta	0862	N90813	25. 8.88	R.Towle	Hexham	28. 9.91T
					(Damaged Teesside 18.7.90: current status unknown)		
G-BOZI	Piper PA-28-161 Warrior II	28-8116120	(G-BOSZ)	14. 7.88	Aerolease Ltd	Conington	11.12.06T
			N8318A				
G-BOZN	Cameron N-77 HAB	1807		1. 9.88	Calarel Developments Ltd	(Chipping Campden)	17. 9.02A
					"Calarel Developments"		
G-BOZO	Gulfstream AA-5B Tiger	AA5B-1282	N4536Q	12. 8.88	Caslon Ltd	Biggin Hill	29. 1.07T
G-BOZR	Cessna 152 II	15284614	N6083M	7. 9.88	Gem Rewinds Ltd	Coventry	1. 4.07T
G-BOZS	Pitts S-1C	221-H	N10EZ	31. 8.88	T.A S.Rayner	Perth	1. 8.05P
	(Built W A Orr) (Lycoming O-320-A2B)						
G-BOZU	Aero Dynamics Sparrow Hawk Mk II	PFA 184-11371		12.12.88	R.V.Phillimore	(Bexhill-on-Sea)	
	(Built R.V.Phillimore)						
G-BOZV	Robin DR340 Major	416	F-BRTS	9. 8.88	C.J.Turner and S.D.Kent	Garston Farm, Marshfield	21.12.06
G-BOZW	Bensen B.8MR	PFA G/01-1096		1. 9.88	M.E.Wills	Lytchett Matravers	13.10.04P
	(Built M.E.Wills) (Rotax 532)						
G-BOZY	Cameron RTW-120 HAB	1770		1. 9.88	Magical Adventures Ltd		
						(West Bloomfield, Michigan, US)	21. 4.97A
G-BOZZ	Gulfstream AA-5B Tiger	AA5B-1155	N4530N	22. 8.88	A W.Matthews tr Solent Tiger Group	Lee-on-Solent	21. 11.06

G-BPAA - G-BPZZ

Reg	Type	C/n	Previous identity	Reg date	Owner/operator	Base	CofA expiry
G-BPAA	Akro Advanced	PFA 200-11528		26. 8.88	Acro Engines and Airframes Ltd	Yearby	27. 6.04P
	(Built B O Smith - c/n AA-001) (Volkswagen Acro 2100)						
G-BPAB	Cessna 150M Commuter	15077244	N63335	21. 9.88	M.J.Diggins	Rayne Hall Farm, Braintree	11. 8.07
G-BPAF	Piper PA-28-161 Cherokee Warrior II	28-7716142	N3199Q	6. 9.88	RAF Brize Norton Flying Club Ltd	RAF Brize Norton	25. 6.06T
G-BPAH*	Colt 69A HAB				See SECTION 3 Part 1	(Aldershot)	
G-BPAI	Bell 47G-3B-1	6528	N8588F	9. 9.88	LRC Leisure Ltd	Barton	29. 3.01
					(Op Manchester Helicopter Centre)		
G-BPAJ	de Havilland DH.82A Tiger Moth	83472	G-AOIX	5.11.80	P.A Jackson	Raby's Farm, Great Stukeley	22. 6.06
	(Built Morris Motors Ltd) (May be composite rebuil d with original G-AMNN ? qv) T7087						
G-BPAL	de Havilland DHC-1 Chipmunk 22	C1/0437	G-BCYE	29.10.86	K.F.and P.Tomsett	Kemble	18. 8.06
			WG350		*(As "WG350")*		
G-BPAO*	Air Command 503 Commander	PFA/G 04-1097		8. 9.88	D.J.A.L.Sagar	Croft Farm, Defford	8. 8.91P
	(Built D J Sagar - c/n 0424)				*(Cancelled 23.2.99 as PWFU) (Noted 1.02)*		
G-BPAS	SOCATA TB-20 Trinidad	283	A2-ADR	9.11.88	Syndicate Clerical Services Ltd	Exeter	21. 6.07T
			F-GDBO				
G-BPAW	Cessna 150M Commuter	15077923	N8348U	5. 9.88	P.D.Sims	Popham	17.10.07E
G-BPAX	Cessna 150M Commuter	15077401	N63571	5. 9.88	M.K.Casely and C.J.Horlock	Shoreham	4. 8.07
					tr The Dirty Dozen		
G-BPAY	Piper PA-28-181 Archer II	28-8090191	N3568X	12. 9.88	Leicestershire Aero Club Ltd	Leicester	13. 5.07T
G-BPBB	Evans VP-2	PFA 63-11261		2. 9.88	A Bleese	Hill Farm, Nayland	9. 6.97P
	(Built J S Penny)				*(Noted 5.03: new CofR 1.04)*		
G-BPBG	Cessna 152 II	15284941	N5418P	16. 9.88	Tatenhill Aviation Ltd	Wellesbourne Mountford	18. 7.07T

Registration	Type	C/n	Previous identity	Date	Owner/operator	Base	CofA
G-BPBJ	Cessna 152 II	15283639	N4793B	9. 9.88	W.Shaw and P.G.Haines	Whaley Farm, New York, Lincoln	28. 1.07
G-BPBK	Cessna 152 II	15283417	N49095	9. 9.88	Atlantic Flight Training Ltd	Coventry	28. 7.07T
G-BPBM	Piper PA-28-161 Warrior II	28-7916272	N3050N	12. 9.88	Halfpenny Green Flight Centre Ltd	Wolverhampton	1.12.07T
G-BPBO	Piper PA-28RT-201T Turbo Arrow IV	28R-8131195	N8431H	28. 9.88	Tile Holdings Ltd	Leicester	21. 6.07
G-BPBP	Brugger MB.2 Colibri Mk.II	PFA 43-10246		6. 2.78	D.A.Preston	(Ulverston)	7. 9.05P
	(Built B Perkins) (Volkswagen 1600)						
G-BPBV	Cameron V-77 HAB	1821		21. 9.88	S.J.Farrant *"Sugar Plumb"*	(Godalmimg)	10. 6.05A
G-BPBW	Cameron O-105 HAB	1841		14.10.88	R.J.Mansfield	(Bowness-on-Windermere)	15. 9.96A
					"October Gold"		
G-BPBY	Cameron V-77 HAB	1818	(G-BPCS)	9.12.88	Louise Hutley *"Brewster's Toy"*	(Guildford)	15. 8.97A
G-BPCA	Pilatus Britten-Norman BN-2B-26 Islander	2198	G-BLNX	28. 1.88	Loganair Ltd	Kirkwall	16. 2.05T
G-BPCF	Piper J-3C-65 Cub	4532	N140DC	12. 5.89	T.I.Williams	Shoreham	17. 9.05P
	(Continental O-200-A)		N28033, NC28033		*(Lippert Reed clipped-wing conversion - s/no.SA811SW)*		
G-BPCG	Colt AS-80 Mk.II HA Airship	1300		14.10.88	N.Charbonnier	(Aosta, Italy)	10.10.96A
	(Built Thunder and Colt Ltd)				*"Greensport/Napapijri"*		
G-BPCI	Cessna R172K Hawk XP II	R1722360	N9976V	3. 1.89	A.J.Flisher	Eshott	24. 5.05T
G-BPCJ*	Cessna 150J	15070797	N61096	26. 9.88	Not known	Amen Corner, Binfield, Bracknell	
	(Badly damaged in gales Compton Abbas 25.1.90: cancelled 4.7.90 by CAA) (Fuselage noted 12.04)						
G-BPCK	Piper PA-28-161 Warrior II	28-8016279	N8529N	26. 9.88	Compton Abbas Airfield Ltd	Compton Abbas	5. 9.07T
			C-GMEI, N9519N				
G-BPCL	Scottish Aviation Bulldog Series 120/128	BH120/393	HKG-6	20. 9.88	Isohigh Ltd	North Weald	5. 8.07T
			G-31-19		tr 121 Group *(As "HKG-6" in Hong Kong DF c/s)*		
G-BPCM	RotorWay Executive	E.3293	N979WP	21. 9.88	R.J.Turner tr Aircare Group	Weavers Loft, Wem	25.11.91P
	(Built W Petrie) (RotorWay RW152)				*(Stored 7.96: current status unknown)*		
G-BPCR	Mooney M.20K Model 231	25-0532	N98433	23. 9.88	T. and R.Harris *"Over The Moony"*	Biggin Hill	15.11.04
G-BPCV	Montgomerie-Bensen B.8MR	PFA G/01-1088		11.10.88	M.A Hayward	(Liskeard)	25. 7.91P
	(Built J Fisher) (Rotax 532)				*(New owner 2.03)*		
G-BPCX	Piper PA-28-236 Dakota	28-8211004	N8441S	25.10.88	D.J.Mountain	Southend	20. 5.07
G-BPDF	Cameron V-77 HAB	1806		6.10.88	The Ballooning Business Ltd	(Northampton)	31. 7.00T
					"Burning Ambition"		
G-BPDG	Cameron V-77 HAB	1839		21.10.88	F.R.Battersby *"Pretty Damn Good"*	(Worsley, Manchester)	16. 5.05A
G-BPDJ	Chris Tena Mini Coupe	275	N13877	4.10.88	J.J.Morrissey	(Teddington)	
	(Volkswagen 1835)				*(To owner's home mid 2002)*		
G-BPDM	CASA 1-131E Jungmann	2058	Spanish AF E3B-369	24.10.88	J.D.Haslam	(Northallerton)	22. 6.96P
					(As "E3B-369/781-32" in Spanish AF c/s)		
G-BPDT	Piper PA-28-161 Warrior II	28-8416004	N4317Z	22.12.88	Channel Islands Aero Club (Jersey) Ltd	Jersey	18.12.07T
G-BPDV	Pitts S-1S	27P	N330VE	15. 9.88	J.Vize	Leicester	22. 6.05P
	(Built C H Pitts)						
G-BPEC	Boeing 757-236	24882		6.11.90	British Airways plc	Heathrow	12.11.06T
G-BPED	Boeing 757-236	25059		30. 4.91	British Airways plc	Heathrow	29. 4.07T
G-BPEE	Boeing 757-236	25060		3. 5.91	British Airways plc	Heathrow	2. 5.07T
G-BPEI	Boeing 757-236	25806	(G-BMRK)	9. 3.94	British Airways plc	Heathrow	8.12.05T
					"Chatham Historic Dockyard"		
G-BPEJ	Boeing 757-236	25807	(G-BMRL)	22. 4.94	British Airways plc	Heathrow	24. 4.06T
G-BPEK	Boeing 757-236	25808	(G-BMRM)	17. 3.95	British Airways plc	Heathrow	16. 3.07T
G-BPEL*	Piper PA-28-151 Cherokee Warrior	28-7415172	C-FEYM	10.10.88	R.W.Harris and A Jahanfar	Southend	8. 2.92T
					(Cancelled 28.2.02 as WFU) (Dismantled wreck stored 1.05)		
G-BPEM	Cessna 150K	15071707	N6207G	24.10.88	R.Strong and R.G.Lindsey	Netherthorpe	17. 6.07
G-BPEO	Cessna 152 II	15283775	C-GQVO	10.10.88	Hecray Co Ltd	Southend	30. 7.05T
			(N5147B)		t/a Direct Helicopters *(Op Southend School of Flying)*		
G-BPES	Piper PA-38-112 Tomahawk II	38-81A0064	N25728	2.11.88	The Sherwood Flying Club Ltd	Tollerton	27. 1.07T
G-BPEX*	Boeing Stearman A75-N1 Kaydet				See SECTION 3 Part 1	La Ferté-Alais, Essonne, France	
G-BPEZ	Colt 77A HAB	1324		14.10.88	J.E.F.Kettlety and W.J.Honey	(Chippenham)	6. 6.01A
G-BPFB	Colt 77A HAB	1334		26.10.88	S.Ingram	(Oldham)	5. 7.04A
G-BPFC	Mooney M.20C Ranger	20-1243	N3606H	21.10.88	D.P.Wring	Dunkeswell	20. 5.05
G-BPFD	Jodel D.112	312	F-PHJT	3.11.88	M.R.Sallows	Damyn's Hall, Upminster	10. 6.05P
G-BPFF	Cameron DP-70 HA Airship	1831		24.10.88	John Aimo Balloons SAS	(Mondovi, Italy)	16.10.04A
G-BPFH	Piper PA-28-161 Warrior II	28-8116201	N83723	3.11.88	Muriel H.Kleiser *(Op Edinburgh Flying Club)*	Edinburgh	19. 1.07T
G-BPFI	Piper PA-28-181 Archer II	28-8090113	N8103G	5. 1.89	F.Teagle	Truro	12. 7.07
G-BPFJ*	Cameron Can 90SS HAB				See SECTION 3 Part 1	(Aldershot)	
G-BPFL	Davis DA-2A	051	N72RJ	27.10.88	B.W.Griffiths	Coventry	17. 1.06P
	(Continental O-200-A)						
G-BPFM	Aeronca 7AC Champion	7AC-4751	N1193E	13.10.88	T.J.Roberts	Rochester	22. 6.05P
			NC1193E				
G-BPFN	Short SD.3-60 Variant 100	SH.3747	N747HH	2.11.88	Aurigny Air Services Ltd	Guernsey	8. 9.05T
			N747SA, G-BPFN, G-14-3747				
G-BPFX	Colt 21A Cloudhopper HAB				See SECTION 3 Part 1	(Aldershot)	
G-BPFZ	Cessna 152 II	15285741	N94594	27.10.88	Devon School of Flying Ltd	Dunkeswell	24. 1.05T
G-BPGC	Speich Air Command Gyroplane	PFA G/04-1108		11.10.88	G.A Speich	Beausdale, Kenilworth	1.12.04P
	(Built G.A Speich - c/n 0440) (Modified Air Command 532 Elite)						
G-BPGD	Cameron V-65 HAB	2000		9. 9.88	Gone With The Wind Ltd	(Bristol)	28. 5.05A
	(New envelope c 8.01 c/n 4969)				*"Silver Lining"*		
G-BPGE	Cessna U206C Super Skywagon	U2061013	N29017	7.11.88	K.Brady tr The Scottish Parachute Club	Strathallan	18.11.07E
G-BPGF	Thunder Ax7-77 HAB	1355		22.11.88	M.Schiavo *"Dovetail"*	(Manchester)	25. 8.95A
G-BPGH	EAA Acrosport II	422	N12JE	14.11.88	G.M.Bradley	Crowfield	6. 9.05P
	(Built J Ellenbaas) (Continental IO-346)						
G-BPGK	Aeronca 7AC Champion	7AC-7187	N4409E	7. 2.89	D.A Crompton	Yeatsall Farm, Abbots Bromley	14. 4.05P
	(Continental A65)						
G-BPGT	Colt AS-80 Mk.II HA Airship	1248	(I -. . . .)	14.11.88	P.Porati	(Milano, Italy)	20. 7.00A
	(Built Thunder and Colt Ltd)		G-BPGT				
G-BPGU	Piper PA-28-181 Archer II	28-8490025	N4330B	26.10.88	G.Underwood	Tollerton	20. 2.07
G-BPGV	Robinson R22 Beta	0887		3.11.88	G.Gazza	(Monte Carlo, Monaco)	24. 5.06T
G-BPGZ	Cessna 150G	15064912	N3612J	14.11.88	J.B.Scott	Blackpool	17. 5.07
G-BPHB	Piper PA-28-161 Warrior II	2816069	N9148G	14.11.88	M.J.Wade	Turweston	8. 3.07T

Registration	Type	C/n	Previous identity	Regd	Owner/operator	Location	CofA
G-BPHD	Cameron N-42 HAB	1863		21. 2.89	P.J.Marshall and M.A Clarke *"Ellen Gee II"*	(Ruislip)	9. 6.02A
G-BPHG	Robin DR400/180 Régent	1887		29.11.88	A Hildreth	Enstone	6. 6.07T
G-BPHH	Cameron V-77 HAB	1840		2.12.88	C.D.Aindow *"Office Angels"*	(Tonbridge)	29. 7.05A
G-BPHI	Piper PA-38-112 Tomahawk	38-79A0002	N2535T	22.11.88	J.S.Develin and Z.Islam	Blackbushe	22. 7.07T
G-BPHJ	Cameron V-77 HAB	1881		23.11.88	C.W.Brown *"Twiggy"*	(Nottingham)	11. 5.02A
G-BPHK	Whittaker MW7	PFA 171-11389		24.11.88	J.S.Shufflebottom	(Chepstow)	23. 6.04P
	(Built J G Beesley)						
G-BPHL	Piper PA-28-161 Warrior II	28-7916315	N555PY	2.12.88	J.D.Swales	(Ruckinge, Ashford)	19. 7.04T
			N2247U		*(New owner 12.04)*		
G-BPHO	Taylorcraft BC-12D	8497	N96197	10. 1.89	A A Alderdice	Woodview, Armagh, County Armagh	16. 4.04P
			NC96197		*"Spirit of Missouri"*		
G-BPHP	Taylorcraft BC-12-65	2799	N33948	12.12.88	J.S.Jackson	(Nottingham)	2.11.99P
	(Continental A65)		NC33948		*(New owner 8.04)*		
G-BPHR	de Havilland DH 82A Tiger Moth	45	N48DH	3. 1.89	N.Parry	Lotmead Farm, Wanborough, Swindon	15. 5.06
	(Built de Havilland Aircraft Pty Ltd, Australia)		VH-BLX, A17-48		tr A17-48 Group *(As "A17-48" in RAAF c/s)*		
G-BPHT	Cessna 152 II	15282401	N961LP	5.12.88	Evensport Ltd	Corporation Farm, Cock Clarks	29. 5.06T
G-BPHU	Thunder Ax7-77 HAB	1365		19.12.88	R.P.Waite	(St.Helens)	20. 4.05A
G-BPHW	Cessna 140	11035	N76595	13. 1.89	L.J.A.Bell	White Waltham	18. 6.05
	(Continental C85)		NC76595				
G-BPHX	Cessna 140	12488	N2252N	2.12.88	M.McChesney	Enniskillen, County Fermanagh	23. 5.93
	(Continental C85)		NC2252N		*(Stored 2.93: re-regd to same owner 10.00)*		
G-BPHZ	Morane Saulnier MS.505 Criquet	53/7	F-BJQC	17. 4.89	Aero Vintage Ltd	Duxford	13. 4.05P
			Fr Mil		*(As "TA+RC" in I/JG54 Luftwaffe c/s)*		
G-BPID	Piper PA-28-161 Warrior II	28-7916325	N2137V	16. 3.89	J.T.Nuttall	Liverpool	3. 6.07T
G-BPIF	Bensen-Parsons Two Place	UK-01		19.12.88	B.J.L.P.de Saar	Shipdham	28. 3.96P
	(Built W Parsons) (Rotax 532)				*(Current status unknown)*		
G-BPII	Denney Kitfox Model 1	PFA 172-11496		15.12.88	P.Etherington	Sandtoft	27. 5.05P
	(Built R Derbyshire - c/n 213) (IAME KFM 112)				tr G-BPII Group		
G-BPIJ	Brantly B.2B	465	N2293U	23. 3.89	I.Davies tr Seething Brantly Group	Seething	3. 4.06
G-BPIK	Piper PA-38-112 Tomahawk II	38-82A0028	N3947M	2.12.88	M.J.White and L.Chadwick	Sleap	29. 4.07T
			ZP-EAP, N91423				
G-BPIL	Cessna 310B	35620	N620GS	16.11.89	A L.Brown	Bourn	28. 4.00T
			OO-SEF, N5420A		*"Fast Lady" (On rebuild 11.03)*		
G-BPIN	Glaser-Dirks DG-400	4-242		14.12.88	J.N.Stevenson	Lasham	11. 4.07
G-BPIO	Reims/Cessna F152 II	F15201556	PH-VSO	23. 1.89	I.D.McClelland	Biggin Hill	23. 9.07T
			PH-AXS				
G-BPIP	Slingsby T.31 Cadet III	PFA 42-10771		14.11.88	J.H.Beard	Bodmin	27. 9.96P
	(Built J.H.Beard) (Volkswagen 1600)				*(Current status unknown)*		
G-BPIR	Scheibe SF25E Super Falke	4332	N25SF	15.12.88	K.E.Ballington	Yeatsall Farm, Abbots Bromley	5. 4.04
	(Limbach SL1700)		(D-KDFX)				
G-BPIT	Robinson R22 Beta	0907	N80011	22.12.88	NA Air Ltd	Hawarden	18. 7.05T
G-BPIU	Piper PA-28-161 Warrior II	28-7916303	N3028T	28.12.88	I.Petersen tr Golf India	Fairoaks	14. 4.07
G-BPIV	Bristol 149 Blenheim IV	---	"Z5722"	152.89	Blenheim (Duxford) Ltd	Duxford	19. 6.04P
	(Built Fairchild Aircraft Ltd as Bollingbroke IVT)		RCAF 10201		*(As "R3281/UX-N") "Spirit of Britain First"*		
					(On rebuild after being badly damaged on landing Duxford 18.8.03)		
G-BPIZ	Gulfstream AA-5B Tiger	AA5B-1154	N4530L	14. 2.89	N.R.F.McNally	Shoreham	21. 8.07
G-BPJB	Schweizer 269C	S.1331	N75065	7.11.88	Elborne Holdings Ltd	Cascais, Portugal	5.11.05
G-BPJD	SOCATA Rallye 110ST	3253	OY-CAV	22.12.88	J.G.Murphy	Morgansfield, Fishburn	22. 6.07
					tr G-BPJD Rallye Group		
G-BPJE	Cameron A-105 HAB	1864		8.11.88	J.S.Eckersley *"Burley Stables"*	(Henley-on-Thames)	28. 3.05A
G-BPJF*	Piper PA-38-112 Tomahawk	38-78A0021	N9312T	5. 4.89	S McNulty	Coventry	4.10.98
			(Crashed on take-off at Egginton 20.6.98: cancelled 2.10.98 by CAA) (For rescue training 5.00)				
G-BPJG	Piper PA-18-150 Super Cub	18-8350	SE-EZG	4. 1.89	M.W.Stein	Oaksey Park	12. 8.07
			N4172Z				
G-BPJH	Piper PA-18 Super Cub 95	18-1980	EI-59	24. 5.83	P.J.Heron	Newtownards	14.12.04P
	(L-18C-PI)		I-EICA, MM52-2380, 52-2380				
G-BPJK	Colt 77A HAB	1362		22.12.88	Saran UK Ltd	(Cheltenham)	10.10.04A
G-BPJO	Piper PA-28-161 Cadet	2841014	N9153Z	15.12.88	J.P.E.Walsh t/a Walsh Aviation	Denham	20.12.04T
					(Op Denham School of Flying)		
G-BPJP	Piper PA-28-161 Cadet	2841015	N9154K	22.12.88	S.J.Skilton	Southampton	27. 6.07T
					t/a Aviation Rentals *(Op Bookajet Operations Ltd)*		
G-BPJR	Piper PA-28-161 Cadet	2841024	N9154X	17. 1.89	J.P.E.Walsh t/a Walsh Aviation	Elstree	21.12.04T
G-BPJU	Piper PA-28-161 Cadet	2841032	N9156Z	11. 1.89	S.J.Skilton t/a Aviation Rentals	Southampton	15. 4.07T
					t/a Aviation Rentals *(Op Bookajet Operations Ltd)*		
G-BPJV	Taylorcraft F-21	F-1005	N2004L	12. 1.89	P.Glennon tr TC Flying Group	Enstone	9. 9.05P
G-BPJW	Cessna A150K Aerobat	A1500127	C-FAJX	4. 1.89	Heald Ltd	(Hornsea)	2. 8.07T
			CF-AJX, N8427M				
G-BPKF	Grob G115	8075		3. 1.89	W R Field tr Dorset Aviation Group	Compton Abbas	27. 1.07
G-BPKK	Denney Kitfox Model 1	PFA 172-11411		19.12.88	D.Moffat	Lochview House, Limerigg	11.10.02P
	(Built R W Holmes - c/n 264) (Rotax 532) *(Laid down as a Model 2)*						
G-BPKM	Piper PA-28-161 Warrior II	28-7916341	PH-CKO	6. 1.89	R.Cass	(Darlington)	27. 6.07T
			N2140X, N9630N				
G-BPKN*	Colt AS-80 Mk.II HA Airship				See SECTION 3 Part 1	Newbury	
G-BPKO	Cessna 140	8936	N89891	12. 1.89	M.J.Patrick	Goodwood	18. 5.06
			NC89891				
G-BPKR	Piper PA-28-151 Cherokee Warrior	28-7515446	N4341X	13. 3.89	Aeros Leasing Ltd	Filton	6. 5.07T
G-BPLD*	Thunder and Colt AS-261 HA Airship				See SECTION 3 Part 1	Newbury	
G-BPLH	CEA Jodel DR.1051 Sicile	401	F-BLAE	27. 2.89	D.W.Tovey	Dunkeswell	1.12.06
	(Potez 4E20A)						
G-BPLM	AIA Stampe SV-4C	1004	F-BHET	8. 2.89	C.J.Jesson	Headcorn	30. 9.05T
	(Officially reg as built SNCAN) (DH Gipsy Major 10)		Fr.Mil, F-BDKC				
G-BPLR	Pilatus Britten-Norman BN-2B-26 Islander	2209	OY-BNT	20. 1.89	Hebridean Air Services Ltd	Cumbernauld	4. 1.05T
			JA5298, G-BPLR				
G-BPLT*	Bristol 20 M.1C rep				See SECTION 3 Part 1	Los Cerillos, Santiago, Chile	

Registration	Type	c/n	Previous identity	Date	Owner/operator	Base	Expiry
G-BPLV	Cameron V-77 HAB	1822		23. 1.89	MC VH SA	(Brussels, Belgium)	1. 2.00A
G-BPLY	Pitts S-2B	5149		25. 1.89	J L Dixon	Sherburn In Elmet	15. 7.07
	(Built Christen Industries Inc) (Lycoming AEIO-540)						
G-BPLZ	Hughes 369HS	91-0342S	N126CM	15. 2.89	R.Kibble	(Burntwood)	27. 9.07E
G-BPMB	Maule M-5-235C Lunar Rocket	7284C	N5635T	13. 8.79	Earth Products Ltd	Crosland Moor	23. 1.07
G-BPME	Cessna 152 II	15285585	N94021	24. 1.89	Eastern Executive Air Charter Ltd	Southend	24. 8.07T
G-BPMF	Piper PA-28-151 Cherokee Warrior	28-7515050	C-GOXL	2. 2.89	A.Hill and P.A.Lewis tr Mike Foxtrot Group	Blackpool	5. 5.07
G-BPML	Cessna 172M Skyhawk II	17267102	N1435U	17.11.89	N.A Bilton	Priory Farm, Tibenham	7. 7.06T
G-BPMM	Champion 7ECA Citabria	498	N5132T	22. 3.89	J.Murray *(Noted 1.04)*	(Ballymena, County Antrim)	25. 2.97P
G-BPMP*	Douglas C-47A-50-DL Dakota				See SECTION 3 Part 1	Aalsmeerderbrug, The Netherlands	
G-BPMR	Piper PA-28-161 Warrior II	28-8416119	N4373S	25. 1.89	Aeros Holdings Ltd	Gloucestershire	17. 5.07T
			N9620N				
G-BPMU	Nord 3202B	70	(G-BIZJ)	26. 1.89	A I.Milne	Little Snoring	19.10.90P
			N22546, ALAT		*"AIX" (Stored 9.97)*		
G-BPMW	QAC Quickie Q.2	PFA 94A-10790	G-OICI	13. 3.89	P.M.Wright	Enstone	17. 8.91P
	(Built R Davidson-Outbridge) (Revmaster R2100DQ)		G-OGKN	*(Damaged near Basingstoke 16.2.91: on repair 9.00: current status unknown)*			
G-BPMX	ARV Aviation ARV-1 Super 2	PFA 152-11128		30. 1.89	T.P.Toth	Enstone	20. 8.04P
	(Built B Houghton - kit no.K.005)						
G-BPNA	Cessna 150L	15073042	N1742Q	10. 2.89	M.P.Whitley and G.W.Todd	(Driffield/Hull)	16. 4.06T
					tr Wolds Flyers Syndicate		
G-BPNI	Robinson R22 Beta	0948		6. 2.89	Heliflight (UK) Ltd	Wolverhampton	12. 4.07T
G-BPNJ	Hawker Siddeley HS.748 Series 2A/263	1680	ZS-ODJ	3. 2.89	Clewer Aviation Ltd	(Maidenhead)	
			F-GHKA, G-BPNJ, 9J-ABW, G-11-4 *(New CofR 1.04)*				
G-BPNN	Montgomerie-Bensen B.8MR	MV-003		3. 2.89	M.E.Vahdat	(Uxbridge)	
	(Built M.E.Vahdat)						
G-BPNO	Moravan Zlin Z-526 Trener Master	930	F-BPNO	18. 2.86	J.A S.Baldry and S.T.Logan	RAF Cranwell	6. 5.07
G-BPNT	British Aerospace BAe 146 Series 300	E3126		4. 1.89	Flightline Ltd	Manchester	31. 5.05T
G-BPNU	Thunder Ax7-77 HAB	1011		9. 2.89	M.J.Barnes *"Firefly"*	(Ivybridge)	16. 8.02T
G-BPOA*	Gloster Meteor T.7	Not known-	WF877	16. 3.89	Not known	Duxford	
					(Cancelled 5.6.96 as WFU) (Noted 10.04)		
G-BPOB	Sopwith Camel F.1 rep	TM-10	N8997	14. 3.89	Bianchi Aviation Film Services Ltd	Compton Abbas	24. 7.05P
	(Built Tallmantz Aviation Inc) (Warner Scarab 165)				*(As "B2458/R" in RFC c/s with Flying Aces Movie Aeroplane Collection)*		
G-BPOL	Pietenpol AirCamper	PFA 47-10941		16. 2.89	G.W.Postance	(Burgess Hill)	
	(Built G.W.Postance)				*(Current status unknown)*		
G-BPOM	Piper PA-28-161 Warrior II	28-8416118	N4373Q	15. 2.89	C.Dale	Humberside	27. 3.05T
			N9619N		tr POM Flying Group		
G-BPON	Piper PA-34-200T Seneca II	34-7570040	N675ES	13. 2.89	Aeros Leasing Ltd	Gloucestershire	15. 6.04T
			N32644				
G-BPOO	Montgomerie-Bensen B.8MR	PFA G/01A-1109		3. 2.89	M.E.Vahdat	(Uxbridge)	
	(Built M.E.Vahdat - c/n MV-002)				*(Believed not constructed)*		
G-BPOS	Cessna 150M	15075905	N66187	21. 2.89	Brooke Park Ltd	(Darlington)	23. 1.06T
G-BPOT	Piper PA-28-181 Cherokee Archer II	28-7790267	N8807F	7. 2.89	Icarus Flyers Ltd	Rochester	17.12.04
G-BPOU	Luscombe 8A Silvaire	4159	N1432K	14. 2.89	P.K.Jordan	(Aylesford)	4.11.02P
	(Continental A65)		NC1432K		tr Luscombe Trio		
G-BPOV*	Cameron Magazine 90SS HAB				See SECTION 3 Part 1	Balleroy, Normandy, France	
G-BPPA	Cameron O-65 HAB	1930		15. 2.89	Rix Petroleum Ltd *"Rix Petroleum"*	(Hull)	26. 9.05A
G-BPPD	Piper PA-38-112 Tomahawk	38-79A0457	N2456F	15. 2.89	S.Snodgrass and M.A Wood	Kemble	26. 9.04T
					t/a AS Belting Products		
G-BPPE	Piper PA-38-112 Tomahawk	38-79A0189	N2445C	15. 2.89	First Air Ltd	Gloucestershire	5. 9.04T
					(Noted 11.04 for major surgery with G-BNPL to provide a single flyer)		
G-BPPF	Piper PA-38-112 Tomahawk	38-79A0578	N2329K	15. 2.89	D.J.Bellamy	Wellesbourne Mountford	5.12.04
					tr Bristol Strut Flying Group		
G-BPPK	Piper PA-28-151 Cherokee Warrior	28-7615054	N7592C	10. 3.89	B and I Ltd	(Lingfield)	30. 6.05T
G-BPPM	Beech B200 Super King Air	BB-1044	N7061T	16. 2.89	Gama Aviation Ltd	Farnborough	17.10.05E
			C-GJJT, N815CE, (N815CF), N815CE, N62895				
G-BPPO	Luscombe 8A Silvaire	2541	N3519M	15. 2.89	M.G.Rummey	Goodwood	28. 6.05P
	(Continental A65)		N71114, NC71114				
G-BPPP	Cameron V-77 HAB	1700		29. 2.88	P.F.Smart	(Basingstoke)	28. 6.97A
					tr The Sarnia Balloon Group *"Thruppence"*		
G-BPPS	Avions Mudry CAP.21	09	F-GDTD	3. 5.85	L.Van Vuuren *(New owner 11.04)*	Durham Tees Valley	14. 3.02S
G-BPPU	Air Command 532 Elite	PFA G/04-1120		22. 2.89	J.Hough	Alresford	18.10.91P
	(Built J Hough - c/n 0438)						
G-BPPY	Hughes 269B	20-0448	N9554F	10. 3.89	D.G.Lewendon	Dunkeswell	16. 4.05T
G-BPPZ	Taylorcraft BC-12D	7988	N28286	22. 3.89	J.Gordon and M.Hart	Sorbie Farm, Kingsmuir	24.11.05P
	(Continental C85)		NC28286		tr Zulu Warriors Flying Group		
G-BPRA	Aeronca 11AC Chief	11AC-1344	N9702E	22. 3.89	P.L.Clements	Beeches Farm, South Scarle	22. 9.05P
			NC9702E				
G-BPRC	Cameron Elephant 77SS HAB	1871		21. 2.89	A Schneider *"Elefant Benjamin"*	(Borken, Germany)	25. 3.05A
G-BPRD	Pitts S-1C	ZZ.1	N10ZZ	21. 2.89	P.Rhodes and R.Trickey	(Aberdeen/Ellon)	6.10.04P
	(Built P C Serkland)				tr Parrot Aerobatic Group		
G-BPRI	Aérospatiale AS355F1 Ecureuil 2	5181	G-TVPA	22. 2.89	MW Helicopters Ltd	Stapleford	5. 11.05T
			G-BPRI, N364E				
G-BPRJ	Aérospatiale AS355F1 Ecureuil 2	5201	N368E	22. 2.89	PLM Dollar Group Ltd	Cumbernauld	14.12.07E
G-BPRL	Aérospatiale AS355F1 Ecureuil 2	5154	N362E	22. 2.89	Gas and Air Ltd *(Op Virgin Helicopters)*	Booker	10. 6.06T
G-BPRM	Reims/Cessna F172L	F17200825	G-AZKG	20. 4.88	BJ Aviation Ltd	Welshpoo)	7. 1.05T
G-BPRN	Piper PA-28-161 Warrior II	28-8116109	N83112	6. 3.89	Air Navigation and Trading Co Ltd	Blackpool	2. 8.04T
G-BPRP*	Cessna 150E	15061269	N3569J	10. 3.89	P.A Griffin	Shoreham	3. 5.98
			(Badly damaged by gales 12.97 Shoreham: cancelled 22.10.99 as WFU) (Noted fire dump 9.02)				
G-BPRR	Rand Robinson KR-2	PFA 129-11105		1. 3.89	P.E.Taylor	(Ferndown)	
	(Built M W Albery)				*(Under construction 6.01: new owner 1.03)*		
G-BPRX	Aeronca 11AC Chief	11AC-94	N86288	3. 3.89	D.J.Dumolo and C.R.Barnes	(Selby)	23. 8.99P
	(Continental A75)		NC86288		*(On rebuild 12.01)*		
G-BPRY	Piper PA-28-161 Warrior II	28-8416120	N4373Y	2. 3.89	R.C.White	Nottingham East Midlands	30. 6.07T
			N9621N		t/a White Wings Aviation *"17"*		

G-BPSH	Cameron V-77 HAB	1837		21. 2.89	P.G.Hossack *"Coconut Ice"*	(Pewsey)	5. 4.97T
G-BPSJ	Thunder Ax6-56 HAB	1479		13. 3.89	V Hyland	(Turnditch, Derby)	5. 6.04A
G-BPSK	Montgomerie-Bensen B.8M	PFA G/01-1100		15. 3.89	P.T.Ambrozik	(Great Orton)	25.11.99P
	(Built P Harrison) (Rotax 532)						
G-BPSL	Cessna 177 Cardinal	17701138	N659SR	3. 3.89	N.P.Bendle tr G-BPSL Group	Dunkeswell	5.12.07E
G-BPSO	Cameron N-90 HAB	1959		10. 3.89	J.Oberprieler	(Mauern, Germany)	25. 7.05A
G-BPSP*	Cameron Ship 90SS HAB				See SECTION 3 Part 1	Balleroy, Normandy, France	
G-BPSR	Cameron V-77 HAB	1962		10. 3.89	K.J.A Maxwell *"Norma Jean"*	(Haywards Heath)	11. 5.05T
G-BPSS	Cameron A-120 HAB	1947		27. 2.89	Anglian Countryside Balloons Ltd	(Burnham-on-Crouch)	26. 3.05T
G-BPSZ*	Cameron N-180 HAB				See SECTION 3 Part 1	(Aldershot)	
G-BPTA	Stinson 108-2 Station Wagon	108-3429	N429C	22. 3.89	M.L.Ryan	Garston Farm, Marshfield	26.10.07
	(Built Consolidate Vultee Aircraft)		NC429C				
G-BPTD	Cameron V-77 HAB	2001		14. 3.89	J.Lippett *"Visions 2001"*	(South Petherton, Somerset)	31. 5.04A
G-BPTE	Piper PA-28-181 Cherokee Archer II	28-7690178	N8553E	9. 3.89	J.S.Develin and Z.Islam	Redhill	25. 8.07T
G-BPTG	Rockwell Commander 112TC	13067	N4577W	31. 3.89	L.G.Watteau	Shoreham	25.10.06
G-BPTI	SOCATA TB-20 Trinidad	414	N41BM	21. 4.89	N.Davis	Blackbushe	3. 7.07
G-BPTL	Cessna 172N Skyhawk II	17268652	N733YJ	22. 3.89	Cleveland Flying School Ltd	Durham Tees Valley	8. 4.05T
G-BPTS	CASA 1-131E Jungmann	Not known	Spanish AF E3B-153	23. 5.89	Aerobatic Displays Ltd	Duxford	23. 8.05P
			"781-75"		*(Op The Old Flying Machine Co as "E3B-153/781-75" in Spanish AF c/s)*		
G-BPTU	Cessna 152 II	15282955	N45946	22. 3.89	A M.Alam	North Weald	21.10.06T
G-BPTV	Bensen B.8	PFA G/01-1058		30. 3.89	C.Munro	(Colne)	
	(Built L Chiappi)						
G-BPTX	Cameron O-120 HAB	1972		29. 3.89	S.J.Colin t/a Skybus Ballooning	(Cranbrook)	2. 6.05T
G-BPTZ	Robinson R22 Beta	0958		22. 3.89	Aero Maintenance Ltd	Walton Wood, Pontefract	16. 11.98
					(New owner 2.04)		
G-BPUA	EAA Biplane Sport	SAAC-02	EI-BBF	30. 3.89	M.D.Gorlov	Elstree	30.10.04P
	(Built B B Feeley) (Lycoming O-235)						
G-BPUB	Cameron V-31 Air Chair HAB	1114		15. 3.89	M.T.Evans	(Peasedown St.John, Bath)	3. 6.94A
					(Noted active 1.04)		
G-BPUE	Air Command 532 Elite	PFA G/04-1136		29. 3.89	A H.Brent	Brough	11. 9.91P
	(Built R A Fazerkerley -0441)				*(Current status unknown)*		
G-BPUF	Thunder Ax6-56Z HAB	270	(G-BHRL)	30. 4.80	R.C.and M.A Trimble *"Buf Puf"*	(Henley-on-Thames)	10. 2.90A
G-BPUG	Air Command 532 Elite	PFA G/04-1157		29. 3.89	T.A Holmes	Melrose Farm, Melbourne	18. 4.91P
	(Built C Slater - c/n 0401)				*(Possibly moved to Spain by 2000: current status unknown)*		
G-BPUJ	Cameron N-90 HAB	1977		17. 4.89	D.Grimshaw	(Preston)	29.12.02T
G-BPUL	Piper PA-18A-150 Super Cub	18-2517	OO-LUL	12. 4.89	C.D.Duthy-James	(Presteigne)	14. 7.05
	(L-18C-PI) *(Frame No thought to be in 18-25xx series)*		PH-NEV				
G-BPUM	Cessna R182 Skylane RG II	R18200915	N738DZ	2. 5.89	R.C.Chapman	Marley Hall, Ledbury	30. 4.04
G-BPUP	Whittaker MW7	PFA 171-11473		2. 8.89	J.H.Beard	(Buckfastleigh, Devon)	
	(Built J.H.Beard)						
G-BPUR	Piper J-3L-65 Cub	4708	N30228	14. 6.89	H.A D.Monro	(New Alresford, Winchester)	
	(Frame No.4764)		NC30228		*(On rebuild 2000)*		
G-BPUU	Cessna 140	13722	N4251N	31. 3.89	D.R.Speight	(Pudsey)	16. 1.06T
			NC4251N				
G-BPUW	Colt 90A HAB	1436		12. 4.89	Gefa-Flug GmbH	(Aachen, Germany)	8. 7.05A
G-BPUY*	Cessna 150K	15071427	N5927G	25. 4.89	Not known	Plaistowes Farm, St.Albans	
					(No UK CofA issued: cancelled 22.12.95 by CAA) (Unmarked fuselage under plastic 5.04)		
G-BPVA	Cessna 172F Skyhawk	17252286	N8386U	13. 4.89	J.Pilkington and P.Makin	Barton	5. 8.06
					tr South Lancashire Flyers Group		
G-BPVC	Cameron V-77 HAB	1302		7. 4.89	B.D.Pettitt	(Bury St.Edmunds)	16. 8.05A
G-BPVE	Bleriot Type XI 1909 rep	1	N1197	20. 6.89	Bianchi Aviation Film Services Ltd	Compton Abbas	29. 6.01P
	(Built R.D.Henry, Texas 1967)				*(As "10" with Flying Aces Movie Aeroplane Collection)*		
G-BPVH	Piper Cub J-3 Prospector	178C	CF-DRY	7. 4.89	D.E.Cooper-Maguire	Washington, West Sussex	18. 8.05P
	(Continental C85)						
G-BPVI	Piper PA-32R-301 Saratoga SP	3213021	N91685	24. 4.89	M.T.Coppen	Goodwood	29. 6.06
G-BPVK	Varga 2150A Kachina	VAC-85-77	N4626V	4. 5.89	H.W.Hall	Southend	12.12.05P
G-BPVM	Cameron V-77 HAB	1970		4. 4.89	J.Dyer	(Farnborough)	22..4.05A
G-BPVN	Piper PA-32R-301T Turbo Saratoga SP	32R-8029073	N8178W	14. 4.89	Y.Leysen	Biggin Hill	22. 4.05T
G-BPVO	Cassutt Racer IIIM	DG.1	N19DD	13. 4.89	A J.Brown	Poplar Hall Farm, Elmsett	15. 9.07P
	(Continental O-200-A)				*"VooDoo"*		
G-BPVU	Thunder Ax7-77 HAB	965		12. 4.89	B.J.Hammond	(Chelmsford)	28. 3.02T
G-BPVW	CASA 1-131E Jungmann	2133	Spanish AF E3B-559	17. 5.89	C. and J.W.Labeij	Goodwood	5. 4.05P
G-BPVY	Cessna 172D Skyhawk	17250568	N2968U	20. 4.89	S.J.Davies	(Doncaster)	13. 6.05
G-BPVZ	Luscombe 8E Silvaire	5565	N2838K	9. 5.89	W.E.Gillham and P.Ryman	Croft Farm, Darlington	3. 6.05P
	(Continental C85)		NC2838K				
G-BPWB	Sikorsky S-61N	61-822	EI-BHO	4. 5.89	Bristow Helicopters Ltd	Portland	10. 7.07T
			G-BPWB, EI-BHO		*(Op Marine and Coastguard Agency) "Portland Castle"*		
G-BPWC	Cameron V-77 HAB	1986		12. 4.89	H.B.Roberts *"Hot Flush"*	(Bristol)	28. 5.05T
G-BPWD	Cessna 120	10026	N72839	14. 4.89	M.W.Albery	Hucknall	27. 8.05P
	(Continental O-240-E)		NC72839		tr Peregrine Flying Group		
G-BPWE	Piper PA-28-161 Warrior II	28-8116143	N8330P	2. 5.89	RPR Associates Ltd	Swansea	22. 6.05T
G-BPWG	Cessna 150M	15076707	(G-BPTK)	10. 4.89	G.Addison	Nanbeck Farm, Wilsford, Grantham	24. 8.07
			N45029		tr G-B Pilots Wilsford Group		
G-BPWI	Bell 206B-3 JetRanger III	3087	9M-BSR	14. 4.89	M.J.Coates	Goodwood	12. 9.07T
			VH-HXZ, ZK-HXX, XC-PFH		t/a Warren Aviation		
G-BPWK	Sportavia-Pützer RF5B Sperber	51036	N56JM	17. 4.89	S.L.Reed	Usk	20.10.05P
			(D-KEAR)				
G-BPWL	Piper PA-25-235 Pawnee	25-2304	N6690Z	14. 4.89	Tecair Aviation Ltd	Shipdham	1. 5.06
			G-BPWL, N6690Z				
G-BPWM	Cessna 150L	15072820	N1520Q	17. 4.89	M.E.Creasey	Crowfield	11.11.05
G-BPWN	Cessna 150L	15074325	N19308	17. 4.89	T.P. Hadley and J.L.Brooks	Compton Abbas	19.11.05T
G-BPWP	Rutan LongEz	PFA 74A-11132		17. 4.89	J.F.O'Hara and A J.Voyle	Biggin Hill	27. 6.05P
	(Built J.F.O'Hara and A J.Voyle) (Continental O-240)						

Registration	Type	c/n	Previous identity	Reg date	Owner/operator	Base	CofA
G-BPWR	Cessna R172K Hawk XPII	R1722953	N758AZ	21. 4.89	J.A., D.T.A., J. and Gwendoline M.Rees t/a Messrs Rees	Haverfordwest	14.11.07
G-BPWS	Cessna 172P Skyhawk II	17274306	N51387	21. 4.89	Chartstone Ltd	Redhill	4. 5.07T
G-BPXA	Piper PA-28-181 Archer II	28-8390064	N4305T	12. 5.89	D.Howdle and D.L.Heighington tr Cherokee Flying Group	Netherthorpe	4. 6.04
G-BPXB	Glaser-Dirks DG-400	4-248		2. 5.89	G.C.Westgate tr Guy Westgate and Syndicate Partners	Parham Park	25. 7.05
G-BPXE	Enstrom 280C Shark	1089	N379KH C-GMLH, N660H	21. 4.89	A Healy	Littlehampden, Bucks	27. 2.05
G-BPXF	Cameron V-65 HAB	2003		21. 4.89	D.Pascall *(Active 2003)*	(Croydon)	
G-BPXH	Colt 17A Cloudhopper HAB	667	OO-BWG	21. 4.89	Sport Promotion SRL	(Belbo, Italy)	8. 9.00A
G-BPXJ	Piper PA-28RT-201T Turbo Arrow IV	28R-8231023	N8061U	21. 4.89	K.M.Hollamby	Biggin Hill	6.10.05
G-BPXX	Piper PA-34-200T Seneca II	34-7970069	N923SM N9556N	21. 4.89	Yorkshire Aviation Ltd	Sherburn-in-Elmet	18. 8.07T
G-BPXY	Aeronca 11AC Chief	11AC-S-50	N3842E	10. 4.89	P.L.Turner	Morgansfield, Fishburn	29.10.05P
G-BPYI	Cameron O-77 HAB	1988		9. 5.89	N.J.Logue	(Pembroke Dock)	9. 7.05A
G-BPYJ	Wittman W.8 Tailwind *(Built J Dixon)* (Continental PC60)	PFA 31-11028		12. 5.89	J.Dixon	Bagby	17. 8.05P
G-BPYK	Thunder Ax7-77 HAB	1166		15. 5.89	A R.Swinnerton *"Yorick"*	(London EC2)	29. 5.93
G-BPYL	Hughes 369D	100-0796D	N65AM G-BPYL, HB-XKT	10. 5.89	Morcorp (BVI) Ltd	Wolverhampton	5. 9.07T
G-BPYN	Piper J-3C-65 Cub (L-4H-PI)	11422	F-BFYN HB-OFN, 43-30131	14. 3.79	D.W.Stubbs tr The Aquila Group	White Waltham	5. 9.05P
G-BPYO	Piper PA-28-181 Archer II	2890114	SE-KIH	22. 5.89	Sherburn Aero Club Ltd	Sherburn-in-Elmet	18. 8.07T
G-BPYR	Piper PA-31 Navajo C	31-7812032	G-ECMA N27493	15. 5.89	West Wales Airport Ltd	Aberporth	11. 5.06T
G-BPYS	Cameron O-77 HAB	2008		9. 5.89	D.J.Goldsmith *"Aqualisa II"*	(Edenbridge)	14.12.99A
G-BPYT	Cameron V-77 HAB	1984		9. 5.89	M.H.Redman	(Sturminster Newton)	
G-BPYV	Cameron V-77 HAB	1992		17. 5.89	R.J.Shortall *(Spa Vehicle Electrics titles)*	(Bath)	29. 3.05A
G-BPYZ	Thunder Ax7-77 HAB	1521		11. 5.89	J.E.Astall *"Axis"* *(Stolen Crewkerne, Somerset 23.10.97)*	(Hinton St.George)	7. 7.96A
G-BPZA	Luscombe 8A Silvaire (Continental A65)	4326	N1599K NC1599K	18. 4.89	M.J.Wright	Rochester	23. 2.04P
G-BPZB	Cessna 120 (Continental C90)	8898	N89853 NC89853	25. 5.89	J.F.Corkin tr Cessna 120 Group	(Horley)	4. 9.05P
G-BPZC	Luscombe 8A Silvaire (Continental A65)	4322	N1595K NC1595K	6. 6.89	C.C.Lovell *(Damaged by gales Cranfield 25.1.90: used for spares 10.96)*	(Winchester)	5. 7.90P
G-BPZD	SNCAN NC.858S (Continental C90) *(Built as NC.854S with Continental C65)*	97	F-BEZD	26. 1.79	S.J.Gaveston, G.Richards and M.S.Regendanz	Headcorn	19.10.05P
G-BPZE	Luscombe 8E Silvaire (Continental C85)	3904	N1177K NC1177K	6. 6.89	B.A Webster tr WFG Luscombe Associates	Airfield Farm, Hardwick	16. 2.05P
G-BPZK	Cameron O-120 HAB	1982		7. 4.89	D.L.Smith *"Hot Stuff"*	(Newbury)	12. 5.97T
G-BPZM	Piper PA-28RT-201 Arrow IV	28R-7918238	G-ROYW G-CRTI, SE-ICY	12. 5.89	Magenta Ltd	Exeter	28.10.07E
G-BPZP	Robin DR400/180R Remorqueur	1471	D-EFZP	4. 5.89	Lasham Gliding Society Ltd	Lasham	23. 5.07
G-BPZS	Colt 105A HAB	1312		25. 5.89	L.E.Giles *(New owner 12.04)*	(Westbury-on-Trym)	12.10.03A
G-BPZU	Scheibe SF25C-2000 Falke (Limbach L2000)	44471	D-KIAV	21. 7.89	Southdown Gliding Club Ltd	Parham Park	12. 8.07
G-BPZY	Pitts S-1C *(Built R N Newbauer)* (Lycoming O-320)	RN-1	N1159	15. 5.89	J.S.Mitchell	White Waltham	27. 3.05P
G-BPZZ	Thunder Ax8-105 HAB	1441		25. 5.89	Capricorn Balloons Ltd	(Loughborough)	25. 5.05T

G-BRAA - G-BRZZ

Registration	Type	c/n	Previous identity	Reg date	Owner/operator	Base	CofA
G-BRAA (2)*	Pitts S-1C *(Built G.R.Miller)*	101-GM	N14T	12. 5.89	Not known *(Stored 4.04)*	(Tranent)	
G-BRAF	Vickers Supermarine 394 Spitfire FR.XVIIIe	6S/663052	Indian AF HS877 SM969	29.12.78	Wizzard Investments Ltd *(As "SM969/D-A" ?)*	North Weald	23. 9.93P
G-BRAK	Cessna 172N Skyhawk II	17273795	C-GBPN (N5438J)	23. 6.88	The Burnett Group Ltd	(Cirencester)	11. 3.07T
G-BRAM*	Mikoyan MiG-21PF				See SECTION 3 Part 1	Farnborough	
G-BRAR	Aeronca 7AC Champion	7AC-6564	N2978E NC2978E	14. 6.89	C.D.Ward	Wombleton	18. 8.05P
G-BRAX	Payne Knight Twister KT-85B *(Built M Anderson and C Sunderland)* (Continental O-200-A)	203	N9792	4. 5.89	R.Earl *(Current status unknown)*	(Brentford)	29. 9.93P
G-BRBA	Piper PA-28-161 Warrior II	28-7916109	N2090B	25. 5.89	Susan H.Pearce	Wolverhampton	29.11.04T
G-BRBB	Piper PA-28-161 Warrior II	28-8116030	N8260W	28. 6.89	Aeros Leasing Ltd	Gloucestershire	3.11.07E
G-BRBC	North American T-6G Texan	182-156	Italian AF MM54-099 51-14470	4. 9.92	A P.Murphy	Audley End	
G-BRBD	Piper PA-28-151 Cherokee Warrior	28-7415315	N41702	28. 6.89	Compton Abbas Airfield Ltd *"Shaftesbury Belle"*	Compton Abbas	10. 4.05
G-BRBE	Piper PA-28-161 Warrior II	28-7916437	N2815D	13. 6.89	Solo Services Ltd *(Op Sussex Flying Club)*	Shoreham	20. 1.05T
G-BRBG	Piper PA-28-180 Cherokee Archer	28-7505248	N3927X	12. 6.89	Surrey Flying Services Ltd and M.D.N.and Andra C.Fisher t/a F and H (Aircraft)	Fenland	22. 8.07
G-BRBH	Cessna 150H	15069283	N50410	13. 6.89	J.Maffia	Panshanger	7. 8.07T
G-BRBI	Cessna 172N Skyhawk II	17269613	N737RJ	7. 7.89	M.D.Harcourt-Brown tr G-BRBI Flying Group	Popham	12. 9.07
G-BRBJ	Cessna 172M Skyhawk II	17267492	N73476	26. 5.89	L.C.Macknight	Elstree	12. 1.02
G-BRBK	Robin DR400/180 Régent	1915		31. 5.89	R.Kemp	Thruxton	25.11.07E
G-BRBL	Robin DR400/180 Régent	1920		5. 7.89	C.A Marren	Upavon	6. 3.07
G-BRBM	Robin DR400/180 Régent	1921		5. 7.89	R.W.Davies	Little Robhurst Farm, Woodchurch	23. 1.05
G-BRBN	Pitts S-1S *(Built W L Garner)*	G.3	N81BG	14. 7.89	D.R.Evans	Gloucestershire	9. 2.05P
G-BRBO	Cameron V-77 HAB	1877		30. 5.89	M.B.Murphy *"Patches"*	(Cheltenham)	15. 5.05A

Registration	Type	C/n	Previous identity	Regd	Owner / operator	Base	CofA
G-BRBP	Cessna 152 II	15284915	N5324P	14. 6.89	Staverton Flying Services Ltd	Gloucestershire	24. 7.07T
G-BRBS	Bensen B.8M *(Built J Simpson)* (Rotax 503)	PFA G/01-1039		30. 5.89	K.T.MacFarlane *(Under construction 6.00)*	(Kilmacolm)	
G-BRBT	Trotter Ax3-20 HAB *(Built R.M.Trotter)*	RMT-001		13. 6.89	R.M.Trotter	(Chew Magna, Bristol)	
G-BRBV	Piper J/4A Cub Coupe	4-1080	N27860	13. 6.89	Janette Schonburg and T.J.Pearson	Exeter	6. 9.05P
G-BRBW	Piper PA-28-140 Cherokee Cruiser	28-7425153	N40737	3. 7.89	Air Navigation and Trading Company Ltd	Blackpool	1.11.04
G-BRBX	Piper PA-28-181 Cherokee Archer II	28-7690185	N8674E	20. 7.89	M.J.Ireland t/a Archer Air	Leicester	25. 3.05T
G-BRBY	Robinson R22 Beta	1027		15. 6.89	D Brown	Cumbernauld	11. 8.07T
G-BRCA	Jodel D.112 *(Built Etablissement Valladeau)*	1203	F-BLIU	11. 7.89	R.C.Jordan	Marsh Hill Farm, Aylesbury	28. 6.05P
G-BRCD	Cessna A152 Aerobat	A1520796	N7377L	8. 6.89	D.E.Simmons tr Charlie Delta Group	Shoreham	14. 9.07
G-BRCE	Pitts S-1C *(Built Davis and Blake)* (Lycoming O-290)	1001	N4611G	22. 6.89	R.D.Rogers *(Op Skylark Aerobatic Co)*	Hulcote Farm, Salford, Bedford	25.11.97P
G-BRCF	Montgomerie-Bensen B.8MR *(Built B E Trinder)* (Rotax 532)	PFA G/01A-1131		12. 6.89	J.S.Walton *(Current status unknown)*	Mold	30.10.91P
G-BRCG	Grob G109	6077	N64BG D-KGRO	15. 6.89	M.P.Flanagan	Gamston	19. 5.07
G-BRCI	Pitts S-1C *(Built J Ballentyne)* (Lycoming O-320)	4668	N351S	6. 7.89	G.L.A Vandormael	Wevelgem, Belgium	18. 7.05P
G-BRCM	Cessna 172L Skyhawk	17259960	N3860Q	19. 6.89	S.G.E.Plessis and D.C.C.Handley *(Op Osprey Flying Club)*	Cranfield	18. 9.05T
G-BRCT	Denney Kitfox Model 2 *(Built M.L.Roberts - c/n 396)*	PFA 172-11521		23. 6.89	M.L.Roberts	Bodmin	5. 4.05P
G-BRCV	Aeronca 7AC Champion (Continental A65)	7AC-282	N81661 NC81661	19. 9.89	M.A.N.Newall	Bagby	26. 8.04P
G-BRCW	Aeronca 11BC Chief (Continental C85)	11AC-366	N85964 NC85964 *(N85964's c/n is 11AC-386)*	16.10.89	R.B.McComish	Bow, Totnes	5. 6.05P
G-BRDB	Zenair CH-701 STOL *(Built D.L.Bowtell) (Regd as "Zenith CH,701 STOL")*	PFA 187-11412		11. 7.89	D.L.Bowtell	Whitehall Farm, Benington	26. 8.05P
G-BRDD	Avions Mudry CAP.10B	224		3. 8.88	R.D.Dickson	Gamston	17. 5.07
G-BRDE	Thunder Ax7-77 HAB	1538		22. 6.89	C.C.Brash *"Veronica"*	(Maidenhead)	21. 7.05A
G-BRDF	Piper PA-28-161 Cherokee Warrior II	28-7716085	N1139Q	26. 6.89	White Waltham Airfield Ltd *(Op West London Aero Services)*	White Waltham	24. 6.05T
G-BRDG	Piper PA-28-161 Cherokee Warrior II	28-7816047	N44934	26. 6.89	White Waltham Airfield Ltd *(Op West London Aero Services)*	White Waltham	17. 1.05T
G-BRDJ	Luscombe 8A Silvaire (Continental A65)	3411	N71984 NC71984	28. 6.89	P.G.Stewart	Popham	8. 1.05P
G-BRDM	Piper PA-28-161 Cherokee Warrior II	28-7716004	N8464F	26. 6.89	White Waltham Airfield Ltd *(Op West London Aero Services)*	White Waltham	4.12.04T
G-BRDN	SOCATA MS.880B Rallye Club	1212	OY-DTV	14. 7.89	A J.Gomes *(Noted 7.04)*	Sandown	27. 4.02
G-BRDO	Cessna 177B Cardinal	17702166	N35030	13. 7.89	I.Jane and A Lidster t/a Cardinal Aviation	Durham Tees Valley	21.12.01
G-BRDT	Cameron DP-70 HA Airship (Konig SD 570)	2029		3. 7.89	Tim Balloon Promotion Airships Ltd	(Ceva, Italy)	4. 8.05A
G-BRDV*	Replica Viking Spitfire prototype				See SECTION 3 Part 1	Southampton	
G-BRDW	Piper PA-24-180 Comanche	24-1733	N6612P	12. 3.90	I.P.Gibson	Southampton	3. 6.06
G-BREA	Bensen B.8MR *(Built R Firth)* (Rotax 503)	PFA G/01-1006		6. 7.89	P.Robichaud	(Bournemouth)	6. 9.05P
G-BREB	Piper J-3C-65 Cub	7705	N41094 NC41094	3. 7.89	J.R.Wraight	Pent Farm, Postling	23. 8.05P
G-BREE	Whittaker MW7 *(Built M J Hayman)* (Rotax 503)	PFA 171-11497		22. 6.89	P.J.Fell	Lower Upham Farm, Chiseldon	4. 7.05P
G-BREH	Cameron V-65 HAB	2049		7. 7.89	S.E. and V.D.Hurst *"Promise"*	(Mansfield)	13. 5.04A
G-BRER	Aeronca 7AC Champion (Continental A65)	7AC-6758	N3157E NC3157E	12. 7.89	I.Sinnett tr Rabbit Flight	Bodmin	10. 7.04P
G-BREU	Montgomerie-Bensen B.8 *(Built M Hayward)* (Rotax 582)	PFA G/01A-1137		20. 7.89	J.S.Firth	Sherburn-in-Elmet	30. 9.04P
G-BREY	Taylorcraft BC-12D	7299	N43640 NC43640	14. 7.89	R.J.Pitts tr BREY Group	Leicester	5. 6.02P
G-BRFB	Rutan LongEz *(Built R A Gardiner)* (Lycoming O-290)	PFA 74A-10646		14. 7.89	R.Young	Perth	17. 4.05P
G-BRFC*	Percival P.57 Sea Prince T.1				See SECTION 3 Part 1	Bournemouth	
G-BRFE	Cameron V-77 HAB	1835		20. 7.89	D.L.C.Nelmes tr Esmerelda Balloon Syndicate *"Esmerelda"*	(Bristol)	23. 4.05A
G-BRFI	Aeronca 7DC Champion (Continental C85)	7AC-4609	N1058E NC1058E	1. 8.89	A C.Lines *(Damaged 1990: on rebuild 4.96)*	Leicester	19. 2.91P
G-BRFJ	Aeronca 11AC Chief (Continental A65)	11AC-796	N9163E NC9163E	28. 7.89	J.M.Mooney *(Stored 2.03)*	Lochview House, Limerigg	11. 9.02P
G-BRFL	Piper PA-38-112 Tomahawk	38-79A0431	N2416F	17. 8.89	Teesside Flight Centre Ltd	Durham Tees Valley	16. 4.06T
G-BRFM	Piper PA-28-161 Warrior II	28-7916279	N2234P	17.10.89	Atlantic Flight Training Ltd *(Atlantic Airlines c/s)*	Coventry	9. 1.05T
G-BRFN*	Piper PA-38-112 Tomahawk	38-79A0397	N2326F	23.10.89	Light Aircraft Leasing (UK) Ltd *(Cancelled 25.9.03 by CAA) (Noted 11.03)*	Exeter	10.12.03T
G-BRFO	Cameron V-77 HAB	2025		6. 7.89	N.J.Bland tr Hedgehoppers Balloon Group *"Lurcher"*	(Oxford)	31. 7.00A
G-BRFR*	Cameron N-105 HAB				See SECTION 3 Part 1	(Aldershot)	
G-BRFW	Montgomerie-Bensen B.8 Two-Seat *(Built J M Montgomerie)* (Rotax 582)	PFA G/01-1073		20. 7.89	A J.Barker	(Dundee)	4. 8.05P
G-BRFX	Pazmany PL-4A *(Built D E Hills)* (Volkswagen 1700)	PFA 17-10079		14. 7.89	D.E.Hills	(Ipswich)	
G-BRGD	Cameron O-84 HAB	2043		20. 7.89	D.J.Phillips *(Watergate Bay Hotel titles) (Active 8.04)*	(Langford, Bristol)	

Regn	Type	C/n	Previous identity	Regd	Owner	Base	CofA
G-BRGF	Luscombe 8E Silvaire	5475	N23FP	20. 7.89	N.Surman tr Luscombe Flying Group	RAF Henlow	30. 6.05P
	(Continental C85)		N944BL, N2748K, NC2748K				
G-BRGG	Luscombe 8A Silvaire	3795	N1068K	20. 7.89	M.A Lamprell	Popham	6.10.05P
	(Continental A65)		NC1068K				
G-BRGI	Piper PA-28-180 Cherokee E	28-5827	N77VG	24. 7.89	Redhill Air Services Ltd	Rochester	20. 6.05
			NIIVG				
G-BRGN	British Aerospace Jetstream Series 3102	637	G-BLHC	20. 3.87	Cranfield University	Coventry	8. 7.07T
			G-31-637		*(Op National Flying Laboratory Centre)*		
G-BRGO	Air Command 532 Elite	PFA G/04-1149		7. 8.89	A McCredie	(Lazonby, Penrith)	13. 2.91P
	(Built D A Wood - c/n 0615)				*(Frame noted Sorbie Farm, Kingsmuir 9.02)*		
G-BRGP*	Colt Flying Stork SS HAB	1409		25. 7.89	Not known *"Great Eggspectations"*	(US)	NE(A)
			(Cancelled 10.3.95 by CAA) (Active Albuquerque, New Mexico, US 10.03)				
G-BRGT	Piper PA-32-260 Cherokee Six	32-658	N3744W	7.11.89	P.Cowley	Nottingham East Midlands	10. 7.05
G-BRGW	Gardan GY-20 Minicab	PFA 1823		13.11.78	R.G.White	Bossington	18. 6.05P
	(Built R G White to JB.01 Minicab standard) (Continental O-200-A)						
G-BRGX	RotorWay Executive	3597		3. 8.89	D.W.J.Lee	South Burlingham, Norwich	8. 12.03P
	(Built D W J Lee) (RotorWay RW 152D)						
G-BRHA	Piper PA-32RT-300 Lance II	32R-7985076	N2093P	27. 7.89	D.J.Chatterton and P.MacKinnon	Earls Colne	8.12.07E
					tr Lance G-BRHA Group		
G-BRHB	Boeing Stearman B75N1 (N2S-3) Kaydet	75-6508	EC-AID	10. 8.89	P.R.Bennett and R.Sage	Priory Farm, Tibenham	AC
			N67955, Bu.05334		*(Noted 8.04)*		
G-BRHG	Colt 90A HAB	1568		11. 9.89	Bath University Students Union *(Badgerline titles)*	(Bath)	17. 8.04A
G-BRHL	Montgomerie-Bensen B.8MR	PFA G/01A-1123		7. 8.89	R.M.Savage and T.M.Jones	Carlisle	26. 8.03P
	(Built N D Marshall) (Rotax 503)						
G-BRHO	Piper PA-34-200 Seneca	34-7350037	N15222	20. 9.89	D.A Lewis	Luton	20.10.07E
G-BRHP	Aeronca O-58B Grasshopper	058B-8533	N58JR	2. 8.89	C.J.Willis	Dunkeswell	22. 2.01P
	(Continental A65)		N46536, 43-1923		*(As "3-1923" in US Army c/s)*		
	(If US Army serial is correct, type should be L-3C-AE)						
G-BRHR	Piper PA-38-112 Tomahawk	38-79A0969	N2377P	21. 8.89	The Royal Artillery Aero Club Ltd	AAC Middle Wallop	28. 9.04T
G-BRHT	Piper PA-38-112 Tomahawk	38-79A0199	N2474C	4. 8.89	P.A Murphy tr Romeo Hotel Tango Group	RAF Mona	9. 8.04T
G-BRHW	de Havilland DH.82A Tiger Moth	85612	7Q-YMY	26. 7.89	P.J.and A J.Borsberry	Kidmore End, Reading	
	(Built Morris Motors Ltd)		VP-YMY, ZS-DLB, SAAF 4606, DE671 *(On rebuild 6.95: current status unknown)*				
G-BRHX	Luscombe 8E Silvaire	5114	N176M	8. 8.89	J.Lakin	Eaglescott	5. 8.05P
	(Continental C90)		N2387K, NC2387K				
G-BRHY	Luscombe 8E Silvaire	5138	N2411K	8. 8.89	A R.W.Taylor	Sleap	29. 4.05P
	(Continental C85)		NC2411K				
G-BRIA	Cessna 310L	310L0010	N2210F	4. 8.89	B.J.Tucker	Kemble	26. 5.06T
G-BRIE	Cameron N-77 HAB	2076		8. 8.89	S.F.Redman	(Sturminster Newton)	1. 9.05A
G-BRIF	Boeing 767-204	24736	(PH-AHM)	10. 3.90	Britannia Airways Ltd	Luton	18.11.05T
			G-BRIF		*"Lord Horatio Nelson"*		
G-BRIG	Boeing 767-204	24757	(PH-AHN)	10. 4.90	Britannia Airways Ltd *(Thomsonfly.com titles)*	Luton	17. 4.06T
			G-BRIG		*"Eglantyne Jebb Founder of the Save the Children Fund"*		
G-BRIH	Taylorcraft BC-12D	7421	N43762	24. 8.89	A D.Duke	Leicester	5. 4.05P
	(Continental A75)		NC43762		tr IH Flying Group		
G-BRII	Zenair CH.600 Zodiac	PFA 162-11392		18. 8.89	A C.Bowdrey	(Hemel Hempstead)	
	(Built A C.Bowdrey)				*(Under build 2000)*		
G-BRIJ	Taylorcraft F-19	F-119	N3863T	23. 8.89	M.W.Olliver *(New owner 3.03)*	Farley Farm, Romsey	12. 6.01P
G-BRIK	Nipper T.66S Series 3	PFA 25-10174		26. 4.77	P.R.Bentley	Roughay Farm, Bishops Waltham	1. 8.02P
	(Built C W R Piper being rebuild of G-AVKH) (Volkswagen 1834)				*(Fuselage on further rebuild 11.01: current status unknown)*		
G-BRIL	Piper J-5A Cub Cruiser	5-572	N35183	2. 8.89	P.L.Jobes and D. J.Bone		
	(Continental A75)		NC35183			Spite Hall Farm, Pinchinthorpe	19.12.05P
G-BRIO	Turner Super T-40A	PFA 104-10636		7. 8.89	S.Bidwell	(Brighton)	15. 8.00P
	(Built D McIntyre) (Continental O-200-A) *(Regd incorrectly as PFA 104-10736)*				*(New owners 8.03)*		
G-BRIR	Cameron V-56 HAB	2056		17. 8.89	H.G.Davies and C.Dowd	(Cheltenham)	6. 9.97A
					(Skyviews Windows titles) "Spirit of Century"		
G-BRIV	SOCATA TB-9 Tampico Club	939		24. 8.89	S.J.Taft	Sturgate	10. 2.06
G-BRIY	Taylorcraft DF-65	6183	N59687	1. 2.90	S.R.Potts	Eshott	10. 7.98P
	(Continental A65) *(Built as TG-6 glider)*		NC59687, 42-58678		*(As "42-58678/IY" in L-2A USAAC c/s) (New owner 1.02)*		
G-BRJA	Luscombe 8A Silvaire	3744	N1017K	12. 9.89	A D.Keen	Dunkeswell	18. 6.03P
	(Continental A65)		NC1017K				
G-BRJB	Zenair CH.600 Zodiac	PFA 162-11573		2. 8.89	C.A.Hasell	Graveley	
	(Built D Collinson - c/n 6-1283)				*(F/f 6.7.04 - new owner 1.05)*		
G-BRJC	Cessna 120	12077	N1833N	21. 8.89	A.L.Hall-Carpenter	Shipdham	21. 3.04P
	(Continental C85)		NC1833N		*(New owner 1.05)*		
G-BRJK	Luscombe 8A Silvaire	4205	N1478K	21. 8.89	C.J.L.Peat and M.Richardson	Popham	11. 4.02P
	(Continental A65)		NC1478K		*(Stored dismantled 9.02)*		
G-BRJL	Piper PA-15 Vagabond	15-157	N4370H	21. 8.89	A.R.Williams	Garston Farm, Marshfield	16. 9.05P
	(Continental C85)		NC4370H				
G-BRJN	Pitts S-1C	1-MA	N6A	23. 8.89	W.Chapel	Sherburn-in-Elmet	20. 5.04P
	(Built M G Acker) (Lycoming O-320)						
G-BRJR	Piper PA-38-112 Tomahawk	38-79A0144	N2598B	31. 8.89	M.McGovern	(Ormskirk)	27. 5.05T
G-BRJT	Cessna 150H	15068426	N44SS	31. 8.89	P.K.Jenkins	Sleap	26. 9.07T
G-BRJV	Piper PA-28-161 Cadet	2841167	N9185G	24. 8.89	Newcastle Aero Club Ltd	Newcastle	30.12.04T
G-BRJW	Bellanca 7GCBC Citabria 150S	1200-80	OO-LPG	7. 4.82	A.J.Sillis and F.A.L.Castleden	Hardwick	8. 9.06
					tr Juliet Whiskey Flying Club		
G-BRJX	Rand-Robinson KR-2	PFA 129-11386		22. 8.89	J.R.Bell	(Parcllyn, Cardigan)	15. 4.97P
	(Built C Willcocks)				*(New CofR 11.03)*		
G-BRJY	Rand-Robinson KR-2	PFA 129-11308		22. 8.89	R.E.Taylor	(Bonar Bridge)	23. 5.96P
	(Built J M Scott) (Revmaster 2100D)				*(Under restoration 6.00)*		
G-BRKC	Auster V J/1 Autocrat	2749	F-BFYT	31. 8.89	J.W.Conlon	High Easter	17.11.04P
G-BRKH	Piper PA-28-236 Dakota	28-7911003	N21444	30. 8.89	A P.H.Hay and C.C.Bennett	Popham	28.12.04T
G-BRKL	Cameron H-34 HAB	2075		29. 8.89	P.L.Harrison	(Rushden, Northampton)	30.12.02A
G-BRKR	Cessna 182R Skylane II	18268468	N9896E	2. 6.89	A R.D.Brooker	Springfield Farm, Ettington	27. 1.05
G-BRKW	Cameron V-77 HAB	2093		1. 9.89	T.J.Parker	(Burnham-on-Crouch)	26. 3.05

Registration	Type	C/n	Previous identity	Reg date	Owner / Operator	Base	CofA
G-BRKY	Viking Dragonfly Mk II	PFA 139-11117		7. 9.89	G.D.Price	Deanland	8. 6.94P
	(Built G D Price) (Volkswagen 2180)				*(Stored less engine 6.04)*		
G-BRLB	Air Command 532 Elite	0622		4. 9.89	F.G.Shepherd	(Great Orton)	
	(Built H R Bethune)						
G-BRLF	Campbell Cricket rep	PFA G/03-1077		6. 9.89	P.G.Rawson	(Huddersfield)	7. 9.05P
	(Built D Wood) (Rotax 503)				*(New owner 6.04)*		
G-BRLG	Piper PA-28RT-201T Turbo Arrow IV	28R-8431027	N4379P	12. 9.89	C.G.Westwood	Welshpool	4. 2.05
			N9600N				
G-BRLI	Piper J-5A Cub Cruiser	5-822	N35951	23. 8.89	Little Bear Ltd	Exeter	19. 7.05P
	(Lycoming O-290)		NC35951				
G-BRLL	Cameron A-105 HAB	2032		7. 9.89	Aerosaurus Balloons Ltd	(Whimple, Exeter)	18. 2.05T
G-BRLO	Piper PA-38-112 Tomahawk	38-78A0621	N2397K	26.10.89	E.Reed	Durham Tees Valley	4. 3.07T
			N9680N		t/a St.George Flight Training		
G-BRLP	Piper PA-38-112 Tomahawk	38-78A0011	N9301T	4.10.89	P D Brooks	Inverness	5. 4.07T
G-BRLR	Cessna 150G	15064822	N4772X	4.10.89	D.Carr and M.R.Muter	Newcastle	5. 7.07
G-BRLS	Thunder Ax7-77 HAB	1603		29. 9.89	Elizabeth C.Meek	Oswestry	19. 5.04A
G-BRLT	Colt 77A HAB	1588		12. 9.89	D.Bareford *"Pro-Sport"*	(Kidderminster)	13. 6.05A
G-BRLV	CCF Harvard 4 (T-6J-CCF Texan)	CCF4-194	N90448	14. 9.89	Extraviation Ltd *"Texan Belle"*	Wevelgem, Belgium	14. 6.05P
			RCAF 20403		*(As "93542/LTA-542" in 6148th TCS USAF c/s)*		
G-BRLX*	Cameron N-77 HAB				See SECTION 3 Part 1	(Aldershot)	
G-BRMA*	Westland-Sikorsky S-51 Dragonfly HR.5				See SECTION 3 Part 1	Weston-super-Mare	
G-BRMB*	Bristol 192 Belvedere HC.1				See SECTION 3 Part 1	Weston-super-Mare	
G-BRMC*	Stampe et Renard SV-4B				See SECTION 3 Part 1	Antwerp, Belgium	
G-BRME	Piper PA-28-181 Cherokee Archer II	28-7790105	OY-BTA	14. 9.89	Keen Leasing Ltd	Belfast Aldergrove	16. 7.06T
G-BRMI	Cameron V-65 HAB	2104		14. 9.89	M.Davies *"Sapphire"*	(Callington, Cornwall)	25. 8.01A
G-BRML	Piper PA-38-112 Tomahawk	38-79A1017	N2510P	3.10.89	P.H.Rogers	Wolverhampton	3. 6.02T
G-BRMS	Piper PA-28RT-201 Arrow IV	28R-8118004	N82708	25. 9.89	Fleetbridge Ltd	White Waltham	11. 6.05
G-BRMT	Cameron V-31 Air Chair HAB	2038		31. 8.89	T.C.Hinton *(Current CoR12.03)*	(Tunbridge Wells)	
G-BRMU	Cameron V-77 HAB	2109		19. 9.89	K.J. and G.R Ibbotson *"Hyperion"*	(Gloucester)	11. 4.04A
G-BRMV	Cameron O-77 HAB	2103		25. 9.89	P.D.Griffiths *"Viscount"*	(Southampton)	10. 7.03A
G-BRMW	Whittaker MW7	PFA 171-11395		25. 9.89	G.S.Parsons	(Coventry)	21. 4.03P
	(Built M Grunwell)						
G-BRNC	Cessna 150M Commuter	15078833	N704SG	29. 9.89	D.C.Bonsall	Netherthorpe	9. 9.05T
G-BRND	Cessna 152 II	15283776	N5148B	7.11.89	T.M.and M.L.Jones *(Op Derby Aero Club)*	Egginton	18. 7.05T
G-BRNE	Cessna 152 II	15284248	N5082L	4.10.89	Redhill Air Services Ltd	Shoreham	20. 3.06T
					(Op Airbase Flying Club)		
G-BRNJ	Piper PA-38-112 Tomahawk	38-79A0415	N2395F	22. 9.89	Cardiff Wales Aviation Services Ltd	Cardiff	15.12.05T
G-BRNK	Cessna 152 II	15280479	N24969	22. 9.89	Sheffield Aero Club Ltd	Netherthorpe	12. 3.05T
G-BRNM*	Chichester-Miles Leopard				See SECTION 3 Part 1	Bournemouth	
G-BRNN	Cessna 152 II	15284735	N6452M	22. 9.89	Sheffield Aero Club Ltd	Netherthorpe	24. 1.05T
G-BRNT	Robin DR400/180 Régent	1935		3.10.89	M.J.Cowham	Top Farm, Croydon, Royston	3. 1.05
G-BRNU	Robin DR400/180 Régent	1937		31.10.89	November Uniform Travel Syndicate Ltd	White Waltham	19. 5.05
G-BRNV	Piper PA-28-181 Cherokee Archer II	28-7790402	N2537Q	7.12.89	B.S.Hobbs	Goodwood	10. 3.05
G-BRNW	Cameron V-77 HAB	2138		2.10.89	G.Smith and N.Robertson	(Bristol/Walton-on-Thames)	5. 8.04A
					"Mr Blue Sky"		
G-BRNX	Piper PA-22-150 Tri-Pacer	22-2945	N2610P	3.10.89	S.N.Askey	(Wheelock, Sandbach)	19.12.05
G-BRNZ	Piper PA-32-300 Cherokee Six B	32-40594	N4229R	7. 2.90	W.Anderson tr Longfellow Flying Group	(Camberley)	20. 6.05T
G-BROB	Cameron V-77 HAB	2073		29. 8.89	J.W.Tomkinson	(Beaconsfield)	17. 4.05A
G-BROE	Cameron N-65 HAB	2098		5.10.89	A.I.Attwood *(New owner 7.04)*	(Aylesbury)	3. 8.97A
G-BROG	Cameron V-65 HAB	2121		6. 9.89	R.Kunert *"The Dodger"*	Finchampstead, Wokingham	19. 7.05A
G-BROH	Cameron O-90 HAB	2120		6.10.89	Patricia A Derbyshire	(Codsall Wood, Wolverhampton)	1. 8.99T
G-BROI	CFM Streak Shadow	PFA 161-11586		16.11.89	T.D.Wolstenholme	Brook Farm, Pilling	6. 8.03P
	(Built G W Rowbotham - kit no K.115-SA) (Rotax 532)				*(New owner 12.03)*		
G-BROJ	Colt 31A HAB	1468		6.10.89	N.J.Langley	(Clapton in Gordano, Bristol)	23. 9.92A
					(Amended CofR 1.05)		
G-BROO	Luscombe 8E Silvaire	6154	N75297	28. 9.89	P.R.Bush	RAF Kinloss	8. 5.00P
	(Continental PC.60)		N1527B, NC1527B		*(Damaged landing Bedwell Hay Farm, Cambridge 13.6.99 and on rebuild 8.02)*		
G-BROP	Van's RV-4	3	N19AT	25.10.89	K.E.Armstrong	Armshold Farm, Kingston, Cambridge	3.11.05P
	(Built AE Tolle) (Lycoming O-360)						
G-BROR	Piper J-3C-65 Cub (L-4H-PI)	10885	F-BHMQ	7.12.89	J.H.Bailey and A P.J.Wiseman	Sturgate	6. 7.05P
			43-29594		tr White Hart Flying Group		
G-BROX	Robinson R22 Beta	1127	N8061V	13.10.89	TLC Handling Ltd	(Doncaster)	9.11.04T
G-BROY	Cameron O-90 HAB	2173		6. 9.89	T.G.S.Dixon *(Dixon Furnace Division tiles)*	(Bromsgrove)	13. 7.05A
G-BROZ	Piper PA-18-150 Super Cub	18-6754	HB-ORC	20. 9.89	P.G.Kynsey	Rushett Farm, Chessington	7. 3.05T
			N9572D				
G-BRPE	Cessna 120	13326	N3068N	11.10.89	W.B.Bateson	Blackpool	7. 7.05P
	(Continental C85)		NC3068N				
G-BRPF	Cessna 120	9902	N72723	11.10.89	A.L.Hall-Carpenter	Shipdham	15. 5.05P
	(Continental C85)		NC72723				
G-BRPG	Cessna 120	9882	N72703	11.10.89	I.C.Lomax	Ottringham	29. 8.94P
	(Continental C85)		NC72703				
G-BRPH	Cessna 120	12137	N1893N	11.10.89	J.A Cook	Pent Farm, Postling	22. 6.05P
	(Continental C85)		NC1893N				
G-BRPJ	Cameron N-90 HAB	2071		11. 9.89	Paul Johnson	(Consett)	10. 3.99T
					t/a Cloud Nine Balloon Co *"Presto"*		
G-BRPK	Piper PA-28-140 Cherokee Cruiser	28-7325070	N15449	17.11.89	G.R.Bright tr G-BRPK Group	Little Gransden	17. 6.05
G-BRPL	Piper PA-28-140 Cherokee Cruiser	28-7325160	N15771	13.10.89	Comed Schedule Services Ltd	Blackpool	6.11.05T
G-BRPM	Nipper T.66 Series 3B	PFA 25-11038		4. 3.85	T.C.Horner	(Barrhead)	
	(Built R Morris)				*(Under construction 6.00)*		
G-BRPP	Gyroflight Brookland Hornet	DC-1		16.10.89	B.J.L.P.and W.J.A L.de Saar	(Great Yarmouth)	19. 8.93P
	(Volkswagen 1776)				*(For rebuild 2000)*		
G-BRPR	Aeronca 0-58B Grasshopper	058B-8823	N49880	17.10.89	C.S.Tolchard	Earls Colne	29. 4.04P
	(Continental A65)		43-1952		*(As "31952" in US Army c/s)*		
	(If US Army serial is correct, type should be L-3C-AE)						

Reg	Type	C/n	Previous identity	Regd	Owner/Operator	Base	CofA
G-BRPS	Cessna 177B Cardinal	17702101	N34935	23.10.89	R.C.Tebbett	Shobdon	19. 3.05
G-BRPT	Rans S-10 Sakota	PFA 194-11554		18.10.89	A R.Hawes	Mendlesham	1. 9.05P
	(Built J G Beesley) (Rotax 532)				*(Noted 6.04)*		
G-BRPU	Beech 76 Duchess	ME-140	N6007Z	17.10.89	Plane Talking Ltd *(Op Cabair)*	Bournemouth	19.10.07E
G-BRPV	Cessna 152 II	15285228	N6311Q	6.11.89	GEM Rewinds Ltd	Coventry	7. 3.05T
G-BRPX	Taylorcraft BC-12D	6462	N39208	12.12.89	R.A C.Lees	Leicester	14. 9.05P
	(Continental A65)		NC39208		tr The BRPX Group		
G-BRPY	Piper PA-15 Vagabond	15-141	N4356H	23.10.89	J.and V.Hobday	Barton	27. 6.05P
	(Continental C85)		NC4356H				
G-BRPZ	Luscombe 8A Silvaire	911	N22089	13.12.89	S.L.and J.P.Waring	Shacklewell Farm, Empingham	30. 5.02P
	(Continental A65)		NC22089				
G-BRRB	Luscombe 8E Silvaire	2611	N71184	23.10.89	J.Nicholls	(Bishops Waltham)	14. 5.00P
	(Continental C85)		NC71184		*(New CofR 5.04)*		
G-BRRD	Scheibe SF25B Falke	4811	D-KBAT	30.10.89	P.J.Gill	Hinton-in-the-Hedges	16. 4.07
	(Built Sportavia-Pützer) (Stark-Stamo MS1500)				tr G-BRRD Group		
G-BRRF	Cameron O-77 HAB	2101		24.10.89	K.P.and G.J.Storey	(Sawbridgeworth)	2. 8.05T
G-BRRG	Glaser-Dirks DG-500M	5E7-M5		7.11.89	D.C.Chaplin tr Glider Syndicate *"492"*	Sutton Bank	15. 4.07
G-BRRJ	Piper PA-28RT-201T Turbo Arrow IV	28R-8431021	N4353T	27.11.89	M.Stower	Elstree	6. 7.05
G-BRRK	Cessna 182Q Skylane II	18266160	N759PW	30.10.89	Werewolf Aviation Ltd	Elstree	7. 5.05
G-BRRL	Piper PA-18 Super Cub 95	18-1615	G-AYPO (1)	17. 9.90	A J.White	Whitehall Farm, Benington	
	(L-18C-PI)		ALAT 18-1615, 51-15615		tr Acebell G-BRRL Syndicate		
	(Regd using paperwork of wrecked D-EMKE [18-2050])				*(On rebuild 4.93: current status unknown)*		
G-BRRN	Piper PA-28-161 Warrior II	28-8216043	N84533	30.10.89	G.Whitlow and I.C.Barlow	Booker	17. 2.05T
G-BRRO	Cameron N-77 HAB	2142		30.10.89	B.Birch *"Newbury Building Society II"*	(Bath)	12. 9.03A
G-BRRR	Cameron V-77 HAB	2070		13.10.89	K.P.and G.J Storey *"Breezy"*	(Sawbridgeworth)	2. 8.05A
G-BRRS	Pitts S-1S	TM-1	N18TM	1.11.89	R.C.Atkinson	Ranksborough Hall Farm, Langham	25. 6.93P
	(Built T D McNamara)				*(Stored 5.95: current status unknown)*		
G-BRRU	Colt 90A HAB	1591		1.11.89	Reach For The Sky Ltd	(Guildford)	29. 7.05T
G-BRRY	Robinson R22 Beta	1193		14.11.89	J.L.Leonard t/a Findon Air Services	Thruxton	7. 1.05T
G-BRSA (2)	Cameron N-56 HAB	2113		8.11.89	C.Wilkinson	(Newcastle)	17.10.92A
G-BRSC (2)	Rans S-10 Sakota	0589.051		8.11.89	J.Lynden	Brook Farm, Pilling	12. 8.97P
	(Built R A Buckley) (Rotax 532)				*(New owner 11.04)*		
G-BRSD (2)	Cameron V-77 HAB	2174		8.11.89	J.E. and T.J.Porter *(New CofR 1.04)*	(Belper)	3. 2.94A
G-BRSE (2)	Piper PA-28-161 Warrior II	28-8016276	N8163R	5.12.89	Meridian Aviation Ltd	Bournemouth	6. 3.05T
	(Engine being converted to Thielert TAE125 diesel @ 1.04)						
G-BRSF (2)	Vickers Supermarine 361 Spitfire HF.IXc	56322	SAAF 5632	2.11.89	M.B.Phillips	Sandown	
			RR232		*(As "RR232") (Noted 8.04)*		
	(Composite including tail/parts ex Mk.VIII/JF629 from W.Australia and wings ex Mk.XIV/R.Thai AF U14-6/93/RAF RM873)						
G-BRSH (2)	CASA 1-131E Jungmann	2156	Spanish AF E3B-540	29.11.89	L.Ness	(Nannestad, Norway)	9. 5.05P
	(C/n also reported as 2140: Spanish AF serial conflicts with F-AZGG)				*(As "781-25" in Spanish AF c/s)*		
G-BRSJ (2)	Piper PA-38-112 Tomahawk II	38-81A0044	N25664	29.12.89	APB Leasing Ltd	Sheffield City	25. 3.05T
G-BRSK	Boeing Stearman B75N1 (N2S-3) Kaydet	75-1180	N5565N	15.11.89	C.R.Lawrence	(Wymondham)	20. 1.97
	(Continental W670)		Bu.3403		t/a Wymondham Engineering *(On rebuild 12.01)*		
G-BRSN	Rand Robinson KR-2	PFA 129-11178		10.11.89	K.W.Darby	(Teignmouth)	
	(Built K.W.Darby) (Volkswagen 1834)						
G-BRSO	CFM Streak Shadow	PFA 161A-11601		16.11.89	B.C.Norris	Trenchard Farm, Eggesford	14.11.03P
	(Built P H Slade - kit no K.133-SA) (Rotax 618)				*(Noted 11.04)*		
G-BRSP	MODAC (Air Command) 503	PFA G04-1158		13.11.89	G.M.Hobman	Askern	21. 1.05P
	(Built D R G Griffith - c/n 0626)						
G-BRSW	Luscombe 8AC Silvaire	3249	N71822	15.11.89	P.H.Needham	Fenland	25. 6.05P
	(Continental A75)		NC71822		tr Bloody Mary Aviation *"Bloody Mary"*		
G-BRSX	Piper PA-15 Vagabond	15-117	N4334H	27.10.89	P.M.Newman	Stoneacre Farm, Farthing Corner	2. 8.05P
	(Continental A65)		NC4334H				
G-BRSY	Hatz CB-1	6	N2257J	15.11.89	C.Knight	Maypole Farm, Chislet	14.10.04P
	(Built M Ondrus) (Lycoming O-290-D)						
G-BRTD	Cessna 152 II	15280023	N757UW	11. 1.90	R.G.Prince, T.G.Phillips and C.Greenland	Popham	21. 6.05
					tr 152 Group		
G-BRTH	Cameron A-180 HAB	2016		21.11.89	The Ballooning Business Ltd	(Northampton)	1. 7.04T
	(Replacement envelope c/n 3199 fitted 1994)				*"Burning Ambition II"*		
G-BRTJ	Cessna 150F	15061749	N8149S	22.11.89	Avon Aviation Ltd	Bristol	24. 7.06T
G-BRTK	Boeing Stearman E75 (PT-13D) Kaydet	75-5949	N16716	29.11.89	Eastern Stearman Ltd	Rendcomb	24. 4.93
	(Continental W670)		42-17786, Bu.38728		*(Donating parts to N52485 [75-4494] @ 12.01 - see SECTION 4)*		
G-BRTL	MD Helicopters Hughes 369E	0356E	(F-GHLF)	5. 1.90	Crewhall Ltd	Leatherhead	31. 3.05
G-BRTM	Piper PA-28-161 Warrior II	28-8416083	N4334L	12.12.89	Oxford Aviation Services Ltd	Oxford	3. 3.05T
G-BRTP	Cessna 152 II	15281275	N49448	28.11.89	K R Emery	Sittles Farm, Alrewas	22. 8.05T
G-BRTT	Schweizer 269C	S.1411		29.11.89	Technical Exponents Ltd	Bennett's Field, Denham	30. 7.05T
G-BRTV	Cameron O-77 HAB	2182		1.12.89	M.C.Gibbons *"Solitaire II"*	(Bristol)	13. 9.05A
G-BRTW	Glaser-Dirks DG-400	4-259		22.12.89	I.J.Carruthers	(Great Orton)	4. 4.05
G-BRTX	Piper PA-28-151 Cherokee Warrior	28-7615085	N8307C	27.12.89	J.Phelan and D.G.Scott	Belfast Aldergrove	16. 5.07T
					tr Spectrum Flying Group		
G-BRUA	Cessna 152 II	15281212	N49267	11. 1.90	BBC Air Ltd *(Op Abbas Air)*	Compton Abbas	5.11.05T
G-BRUB	Piper PA-28-161 Warrior II	28-8116177	N8351Y	27.12.89	Flytrek Ltd	Compton Abbas	3. 2.06T
G-BRUD	Piper PA-28-181 Archer II	28-8390010	N8300S	9. 2.90	Wilkins and Wilkins (Special Auctions) Ltd		
					t/a Henlow Flying Club	RAF Henlow	18. 3.05T
G-BRUG	Luscombe 8E Silvaire	4462	N1735K	15.12.89	N.W.Barratt and K.Reeve	Compton Abbas	21. 3.05P
	(Continental C85)		NC1735K				
G-BRUH	Colt 105A HAB	1650		15.12.89	D.C.Chipping	(Grantham)	29. 7.93T
G-BRUI	Piper PA-44-180 Seminole	44-7995150	N2230E	15.12.89	Marina Griffiths	Tatenhill	4.11.05T
			G-BRUI, N2230E				
G-BRUJ	Boeing Stearman A75N1 (PT-17) Kaydet	75-4299	N55557	6. 4.90	R.M.Hughes	Liverpool	25. 7.07T
	(Continental R670)		42-16136		*(As "16136/205" in USN c/s)*		
G-BRUM	Cessna A152 Aerobat	A1520870	N4693A	12. 3.86	Central Aircraft Leasing Ltd	Wolverhampton	6. 9.04T
G-BRUN	Cessna 120	9294	G-BRDH	29. 8.89	O.C.Brun	Great Massingham	16. 3.05P
	(Continental C85)		N72127, NC72127				

Registration	Type	C/n	Previous identity	Date	Owner/operator	Base	CofA expiry
G-BRUO	Taylor JT.1 Monoplane	PFA 55-10859		15.12.89	R.Hatton	Kirkbride	1. 6.05P
	(Built P C Cardno)						
G-BRUU	EAA Biplane Model P1	1	N41MW	22.12.89	E.C.Murgatroyd	Sackville Lodge Farm, Riseley	17. 6.98P
	(Built L E Lasnier) (Lycoming O-360)		N4775G		*(Badly damaged in accident Sackville 30.9.00) (Current status unknown)*		
G-BRUV	Cameron V-77 HAB	2100		16. 8.89	T.W.and R.F.Benbrook *"biG-BRUVver"*	(Romford)	30. 3.05A
G-BRUX	Piper PA-44-180 Seminole	44-7995151	N2245E	8. 3.79	C J Thomas	Tatenhill	13.12.06
G-BRVB	Stolp SA.300 Starduster Too	409	N33MH	21.12.89	M.N.Petchey	Andrewsfield	13. 9.05P
	(Built M Hoover) (Lycoming O-360)				tr G-VB Group		
G-BRVE	Beech D17S Traveller (UC-43-BH)	6701	N1193V	12. 3.90	P.A Teichman	North Weald	25. 2.05
			NC1193V, Bu.32874, FT475, 44-67724, (Bu.23689)				
G-BRVF	Colt 77A HAB	1651		19.12.89	The Ballooning Business Ltd *(NAPS titles)*	(Northampton)	23. 4.05T
G-BRVG	North American SNJ-7C Texan	88-17676	N830X	24. 1.90	D.J.Gilmour t/a Intrepid Aviation Co	North Weald	6. 6.05
			N4134A, Bu.90678, (42-85895)		*(As "90678/27" in VS-932 Sqdn, USN c/s)*		
G-BRVI	Robinson R22 Beta	1240		27.12.89	M.D.Thorpe t/a Yorkshire Helicopters	Coney Park, Leeds	12. 5.05T
G-BRVJ	Slingsby Cadet III	PFA 42-11382	(BGA3360)	24. 1.90	B.Outhwaite	(Middlesborough)	28. 9.05P
	(Built D F Micklethwait and J R Paskins - c/n 701)		WT906				
	(Volkswagen 1600) *(Modified ex T.31B)*						
G-BRVL	Pitts S-1C	559H	N2NW	10. 1.90	M.F.Pocock	RAF Mona	18. 8.05P
	(Built N Williams) (Lycoming IO-320)						
G-BRVN	Thunder Ax7-77 HAB	1614		28.12.89	D.L.Beckwith	(Northampton)	13. 6.04A
G-BRVO	Aérospatiale AS350B Ecureuil	2315		3. 1.90	Ferns Surfacing Ltd	Sandhurst, Kent	6. 5.05T
G-BRVR	Barnett Rotorcraft J4B-2	216-2		20. 2.90	M.Richardson t/a Ilkeston Contractors	Ilkeston	AC
G-BRVS	Barnett Rotorcraft J4B-2	210-2		20. 2.90	M.Richardson	Ilkeston	AC
	(Built M Richardson)				t/a Ilkeston Contractors		
G-BRVT	Pitts S-2B	5189		6. 4.90	R.Woollard and D.Tilley	Biggin Hill	10. 3.06T
	(Built Christen Industries Inc) (Lycoming AEIO-540)				*"The Tart"*		
G-BRVU	Colt 77A HAB	1652		4. 1.90	J.K.Woods *"Concorde Watches"*	(Chatham)	25. 6.02A
G-BRVY	Thunder Ax8-90 HAB	1676		9. 1.90	G.E.and J.V.Morris *"Golden Gem"*	(Cheltenham)	6. 5.04A
G-BRVZ	SAN Jodel D.117	433	F-BHNR	22.12.89	L.Holland	(Nottingham)	19. 6.05P
G-BRWA	Aeronca 7AC Champion	7AC-351	N81730	20. 3.90	D.D.Smith and J.R.Edwards	Scotland Farm, Hook	23.12.04P
			NC81730				
G-BRWD	Robinson R22 Beta	1231	N8064U	15. 1.90	R.C.Hayward t/a Rotorways Helicopters	Manston	15. 5.05T
G-BRWO	Piper PA-28-140 Cherokee Cruiser	28-7325548	N55985	11. 1.90	G-BRWO Ltd	(Grimsby)	26. 8.06T
G-BRWP	CFM Streak Shadow	PFA 161A-11596		17. 1.90	R.Biffin	Perth	2.10.04P
	(Built D F Gaughan - kit no K.122) (Rotax 532)						
G-BRWR	Aeronca 11AC Chief	11AC-1319	N9676E	17. 1.90	A W.Crutcher	Cardiff	2. 7.05P
	(Continental A65)						
G-BRWT	Scheibe SF25C-2000 Falke	44480	D-KIAY	11. 1.90	Booker Gliding Club Ltd	Booker	5. 6.05
	(Limbach L2000)						
G-BRWU	Phoenix Luton LA-4A Minor	PAL/1141		18. 1.90	R B Webber	Trenchard Farm, Eggesford	18. 7.05P
	(Built R B Webber and P K Pike) (JAP J.99) *(Officially regd as PFA 1141 but correct project no. not known)*						
G-BRWV	Brugger MB.2 Colibri	PFA 43-11027		18. 1.90	M.P.Wakem	Barton	26. 7.05P
	(Built S J McCollom) (Volkswagen 1834)						
G-BRWX	Cessna 172P Skyhawk II	17274729	N53363	17. 1.90	BCT Aircraft Leasing Ltd	Kemble	24. 3.06T
	(Thielert TAE125-01)						
G-BRWZ	Cameron Macaw 90SS HAB	2206		29. 1.90	Forbes Global Inc	(Balleroy, Normandy)	3. 9.00
					"Capitalist Tool" (New CofR 9.03)		
G-BRXA	Cameron O-120 HAB	2217		19. 1.90	R.J.Mansfield	(Bowness-on-Windermere)	22. 2.05T
G-BRXB	Thunder Ax7-77 HAB	1631		18. 1.90	H.Peel	(Worcester)	22. 7.02A
					(Donated to Balloon Preservation Group 2004)		
G-BRXC	Piper PA-28-161 Warrior II	28-8416043	N4339X	19. 2.90	Oxford Aviation Services Ltd	Oxford	10. 4.05T
			N9563N				
G-BRXD	Piper PA-28-181 Archer II	28-8290126	D-EHWN	19. 2.90	D.D.Stone	Wellesbourne Mountford	15. 4.06
			N9690N, N8203E				
G-BRXE	Taylorcraft BC-12D	9459	N95059	25. 1.90	Wendy J.Durrad	Eastbach Farm, Coleford	27.10.05P
	(Continental A65)		NC95059		*"Flying Fishes"*		
G-BRXF	Aeronca 11AC Chief	11AC-1033	N9396E	25. 1.90	C.G.Nice	Andrewsfield	10. 5.05P
	(Continental A65)		NC9396E		tr Aeronca Flying Group		
G-BRXG	Aeronca 7AC Champion	7AC-3910	N85178	1. 3.90	J.D.Webb	Hill Farm, Nayland	15. 9.05P
	(Continental A65)		NC85178		tr X-Ray Golf Flying Group		
G-BRXH	Cessna 120	10462	N76068	25. 1.90	A C.Garside	Headcorn	6. 4.05P
	(Continental C85)		NC76068		tr BRXH Group		
G-BRXL	Aeronca 11AC Chief	11AC-1629	N3254E	31. 1.90	T.Smith	Retreat Farm, Little Baddow	7. 7.05P
	(Continental A65)		NC3254E		*(As "42-78044" in US Army L-3F c/s)*		
G-BRXN	Montgomerie-Bensen B.8MR	PFA G/01-1160		31. 1.90	C.M.Frerk	Spanhoe	22. 1.05P
	(Built J C Aitken) (Rotax 532)						
G-BRXO	Piper PA-34-200T Seneca II	34-7970149	N111ED	12. 4.90	Aviation Services Ltd	Blackpool	3. 9.05
			N9618N				
G-BRXP	SNCAN Stampe SV-4C	678	N33528	2. 2.90	T.Brown	Maypole Farm, Chislet	19. 7.07
	(Gipsy Major 10)		F-BGGU, French AF 678, (F-BDNX)			*(Noted 8.04)*	
G-BRXS	Howard Special T-Minus	REC-1	N2278C	14. 2.90	A Shuttleworth	Barton	10.10.05P
	(Built C Howard) (Lycoming O-290) *(Modified Taylorcraft BC)*						
G-BRXV	Robinson R22 Beta	1246		7. 2.90	Heliflight (UK) Ltd	Wolverhampton	31. 5.05T
G-BRXW	Piper PA-24-260 Comanche	24-4069	N8621P	16. 2.90	P.A Jenkins tr Oak Group	Coventry	11. 2.06
G-BRXY	Pietenpol AirCamper	PFA 47-11416		7. 2.90	P.S.Ganczakowski	Great Eversden	19.10.05P
	(Built A E Morris) (Continental C90)						
G-BRYI	de Havilland DHC-8-311A	256	C-GEOA	26. 3.91	British Airways Citiexpress Ltd	Bristol	27. 3.06T
					(Chelsea Rose t/s) "Northumberland/Drigantes"		
G-BRYJ	de Havilland DHC-8-311A	319	C-GEOA	27. 3.92	British Airways Citiexpress Ltd	Bristol	2. 4.06T
					"Somerset/Gwlad-yr-Haff"		
G-BRYU	de Havilland DHC-8-311A *(Q300)*	458	(9M-PGA)	4. 4.98	British Airways Citiexpress Ltd	Bristol	3. 4.07T
			C-GFEN		*(Benyhone Tartan t/s)*		
G-BRYV	de Havilland DHC-8-311A *(Q300)*	462	(9M-PGD)	10. 4.98	British Airways Citiexpress Ltd	Bristol	9. 4.07T
			C-GFHZ		*(Colum t/s)*		

Regn	Type	C/n	Previous identity	Regd	Owner	Base	CofA
G-BRYW	de Havilland DHC-8-311A *(Q300)*	474	(9M-PG?)	26. 5.98	British Airways Citiexpress Ltd	Bristol	24. 5.07T
			C-GDIU		*(Koguty Lowickie t/s)*		
G-BRYX	de Havilland DHC-8-311A *(Q300)*	508	C-GDOE	25. 9.98	British Airways Citiexpress Ltd	Bristol	27. 9.07T
G-BRYY	de Havilland DHC-8-311A *(Q300)*	519	C-FDHD	11.12.98	British Airways Citiexpress Ltd	Plymouth	10.12.07E
					(Rendezvous t/s)		
G-BRYZ	de Havilland DHC-8-311A *(Q300)*	464	C-FCSG	16.10.98	British Airways Citiexpress Ltd	Plymouth	15.10.07E
G-BRZA	Cameron O-77 HAB	2231		7. 2.90	L. and R.J.Mold *"Breezy"*	(High Wycombe)	5. 4.05A
	(Originally regd with c/n 2237 - see G-BSCA)				*(Phil Dunson/Wycombe Insurance titles)*		
G-BRZB*	Cameron A-105 HAB	2212		7. 2.90	Cornwall Ballooning Adventures Ltd	(Newquay)	21.12.04A
					(Cancelled 8.9.04 by CAA)		
G-BRZC*	Cameron N-90 HAB				See SECTION 3 Part 1	Newbury	
G-BRZD	HAPI Cygnet SF-2A	PFA 182-11443		8. 2.90	C.I.Coghill	Popham	7.10.04P
	(Built L G Millen) (Volkswagen 2078)						
G-BRZE	Thunder Ax7-77 HAB	1633		8. 2.90	G.V.Beckwith and F.Schoeder		
					"Jenlain"	(Zanten/Mulheim Ruhr, Germany)	31. 8.97A
G-BRZG	Enstrom F-28A	169	N9053	8. 2.90	Gillian Elaine Challinor	Barton	6. 6.06T
G-BRZI	Cameron N-180 HAB	2215		8. 2.90	C.E.Wood t/a Eastern Balloon Rides	(Witham)	24. 2.00T
G-BRZK	Stinson 108-2 Voyager	108-2846	N9846K	17. 4.90	P.C.G.Wyld	Booker	1. 7.06
	(Built Consolidated Vultee Aircraft)		NC9846K		tr Voyager G-BRZK Syndicate		
G-BRZL	Pitts S-1D	01	N899RN	26. 2.90	T.R.G.Barnby	Headcorn	2. 8.96P
	(Built R C Nelson)				*(Noted on rebuild 5.00: new CofR 8.03)*		
G-BRZO	Jodel D.18	PFA 169-11275		14. 2.90	J.D.Anson	(Liskeard)	
	(Built J.D.Anson)						
G-BRZS	Cessna 172P Skyhawk II	17275004	N54585	2.10.90	H.Hargreaves and P.F.Hughes	Blackpool	5. 3.06
					tr G-BHYP Flying Group		
G-BRZT	Cameron V-77 HAB	2241		21. 2.90	Beverley Drawbridge *"Hoopla"*	(Cranbrook)	19. 2.01A
G-BRZV	Colt Flying Apple SS HAB	1662		26. 2.90	Obst Vom Bodensee Marketing Gbr		
						(Tettnang-Siggenweiler, Germany)	14. 9.97A
G-BRZW	Rans S-10 Sakota	PFA 194-11932		21. 2.90	D.L.Davies	Emlyn's Field, Rhuallt	6. 8.98P
	(Built D L Davies) (Rotax 532 - *kit no.0789.058)*				*(Noted 4.04)*		
G-BRZX	Pitts S-1S	711-H	N272H	22. 2.90	J.S.Dawson	Sherburn-in-Elmet	30. 4.05P
	(Built M M Lotero) (Lycoming O-320)						
G-BRZZ	CFM Streak Shadow	PFA 161A-11628		22. 2.90	T.Mooney	Sumburgh	5.11.04P
	(Built P R Oakes - K.135) (Rotax 532)				tr Shetland Flying Group		

G-BSAA - G-BSZZ

Regn	Type	C/n	Previous identity	Regd	Owner	Base	CofA
G-BSAI	Stoddard-Hamilton Glasair III	3102		31. 1.90	K.J. and P.J.Whitehead	Booker	7. 7.05P
	(Buillt K J Whitehead) (Lycoming IO-540)						
G-BSAJ	CASA 1-131E Jungmann	2209	Spanish AF E3B-209	23. 1.90	P.G.Kynsey	Headcorn	15. 7.05P
G-BSAK	Colt 21A Sky Chariot HAB	1696		26. 2.90	K.Meehan t/a Northern Flights	(Much Wenlock)	18. 9.05A
G-BSAS	Cameron V-65 HAB	2191		27. 2.90	J.R.Barber	(King's Lynn)	2. 5.04A
G-BSAV	Thunder Ax7-77 HAB	1555		26. 2.90	I.G.Lloyd *"Burnt Savings"*	(Derby)	22. 6.03A
G-BSAW	Piper PA-28-161 Warrior II	28-8216152	N8203C	27. 2.90	Carill Aviation Ltd	Lee-on-Solent	6. 8.05T
			YV-2265P, N8203C				
G-BSAZ	Denney Kitfox Model 2	PFA 172-11664	(G-BRVW)	5. 3.90	A J.Lloyd, D.M.Garrett and J.T.Lane	(Bromyard)	26. 6.97P
	(Built P E Hinkley - c/n 602)						
G-BSBA	Piper PA-28-161 Warrior II	28-8016041	N2574U	1. 3.90	London Transport Flying Club Ltd	Fairoaks	27. 5.06T
G-BSBG	CCF Harvard 4 (T-6J-CCF Texan)	CCF4-483	Moz.PLAF 1753	5. 3.90	A P.St.John	Tatenhill	26. 8.05P
			FAP 1753, BF+053, AA+053, 52-8562		*(As "20310/310" in RCAF c/s)*		
G-BSBI	Cameron O-77 HAB	2245		6. 3.90	D.M.Billing *"Calibre"*	(Uckfield)	30. 6.05A
G-BSBM*	Cameron N-77 HAB				See SECTION 3 Part 1	(Aldershot)	
G-BSBN	Thunder Ax7-77 HAB	1531		6. 3.90	B.Pawson *"Venus"*	(Cambridge)	9.12.93A
G-BSBR	Cameron V-77 HAB	2247		26. 2.90	R.P.Wade *"Honey"*	(Wigan)	11. 4.05A
G-BSBT	Piper J-3C-65 Cub	17712	N70694	9. 3.90	R.W.H.Watson	Grimmet Farm, Maybole	9. 5.05P
			NC70694				
G-BSBV	Rans S-10 Sakota	PFA 194-11769		9. 3.90	J.D.C.Henslow	(Ingrams Green, Midhurst)	16.12.05P
	(Built J Whiting - kit no.1089.064) (Rotax 532)						
G-BSBW	Bell 206B-3 JetRanger III	3664	N43EA	12. 3.90	D.T.Sharpe	Sherburn-in-Elmet	21. 8.05T
			N6498V, 9Y-THC				
G-BSBX	Montgomerie-Bensen B.8MR	PFA G/01A-1135		12. 3.90	W.Toulmin and R.J.Roan	(Peterborough)	26. 5.93P
	(Built B Ibbott) (Rotax 503)				*(New owner 3.03)*		
G-BSBZ	Cessna 150M	15077093	N63086	29. 3.90	D.T.Given t/a DTG Aviation	Newtownards	27. 6.05T
G-BSCA	Cameron N-90 HAB	2237	(9M-)	12. 3.90	J.Steiner	(Kuala Lumpur, Malaysia)	21. 5.05A
	(Originally regd with c/n 2239 - also see G-BRZA)		G-BSCA				
G-BSCB	Air Command 532 Elite	PFA G/04-1172		16. 3.90	P.H.Smith	Tollerton	18. 9.97P
	(Built P H Smith - c/n 0627)						
G-BSCC	Colt 105A HAB	1006		15. 3.90	Capricorn Balloons Ltd	(Loughborough)	11. 3.03T
G-BSCE	Robinson R22 Beta	1245		15. 3.90	H.Sugden	Humberside	17. 3.05T
G-BSCF	Thunder Ax7-77 HAB	1537		14. 3.90	V.P.Gardiner *"Charlie Farley"*	(Stoke-on-Trent)	22.11.02A
G-BSCG	Denney Kitfox Model 2	PFA 172-11620		23. 4.90	S.Burrow *(New CofR 12.04)*	(Pontefract)	30.10.01P
G-BSCH	Denney Kitfox Model 2	PFA 172-11621		16. 3.90	R.B.Wilson	(Kendal)	12.10.05P
	(Built Baldoon Leisure Flying Co Ltd - c/n 510)						
G-BSCI	Colt 77A HAB	1683		16. 3.90	J.L.and S.Wrigglesworth *"Brody"*	(Ilminster)	7. 8.05A
G-BSCK	Cameron H-24 HAB	2263		16. 3.90	J.D.Shapland *"Monacle"*	(Wadebridge)	11. 6.95A
G-BSCM*	Denney Kitfox Model 2	PFA 172-11745		28. 3.90	S.A Hewitt	Sheepcote Farm, Stourbridge	8. 7.03P
	(Built D M Richardson - c/n 638)				*(Cancelled 27.8.03 by CAA)*		
G-BSCN	SOCATA TB-20 Trinidad	1070	D-EGTC	27. 3.90	B.W.Dye	Biggin Hill	1. 6.02
			G-BSCN				
G-BSCO	Thunder Ax7-77 HAB	1635		6. 3.90	F.J.Whalley "Bluebell"	(Cleish)	8. 9.02A
G-BSCP	Cessna 152 II	15283289	N48135	20. 3.90	Moray Flying Club (1990) Ltd	RAF Kinloss	6.10.05T
G-BSCS	Piper PA-28-181 Cherokee Archer II	28-7890064	N47392	3. 4.90	Wingtask 1995 Ltd	Seething	16. 5.05T
G-BSCV	Piper PA-28-161 Cherokee Warrior II	28-7816135	C-GQXW	22. 3.90	R.J.L.Beynon tr Southwood Flying Group	Earls Colne	6. 1.06

Reg	Type	C/n	Previous identity	Regd	Owner/Operator	Location	Expiry
G-BSCW	Taylorcraft BC-65	1798	N24461	22. 3.90	S.Leach	(Honiton)	10. 6.05P
			NC24461		*(Carries "C24461" on fin)*		
G-BSCX	Thunder Ax8-105 HAB	1748		21. 3.90	Balloon Flights Club Ltd *"Balloon Flights"*	(Leicester)	14. 7.99T
G-BSCY	Piper PA-28-151 Cherokee Warrior	28-7515046	C-GOBE	22. 3.90	A.S.Bamrah	Lydd	15. 7.05T
	(Converted to PA-28-161 model)				t/a Falcon Flying Services		
G-BSCZ	Cessna 152 II	15282199	N68226	22. 3.90	The Royal Air Force Halton Aeroplane Club Ltd		
						RAF Halton	8. 8.06T
G-BSDA	Taylorcraft BC-12D	7316	N43657	15.11.90	D.G.Edwards	Ellens Green, West Sussex	8.10.01P
	(Continental A75)		NC43657		*(Noted 7.03 minus engine)*		
G-BSDB	Pitts S-1C Special	01	(N1867)	22. 3.90	J.T.Mielech	Yeatsall Farm, Abbots Bromley	31. 5.99P
	(Built C R Rogers) (Lycoming O-320)		N77R		*(Noted 9.03)*		
G-BSDD	Denney Kitfox Model 2	PFA 172-11797		28. 3.90	D.C.Crawley	(Calverton, Nottingham)	21. 6.96P
	(Built M Richardson and J Cook - c/n 639)				*(New owner 6.02)*		
G-BSDH	Robin DR400/180 Régent	1980		18. 4.90	R.L.Brucciani	Leicester	7. 5.05
G-BSDI*	Corben Junior Ace Model E	3961	N91706	28. 3.90	C.P.Whitwell	Oak Farm, Cowbit	23.10.02P
	(Built Oliver and Clark)				*(Cancelled 23.8.04 by CAA)*		
G-BSDJ	Piper J-4E Cub Coupé	4-1456	N35975	13. 2.91	B.M.Jackson	(Thame)	22. 6.04P
	(Continental C85)		NC35975				
G-BSDK	Piper J-5A Cub Cruiser	5-175	N30337	28. 3.90	J.E.Mead	(Cowbridge)	1. 9.05P
	(Continental A75)		NC30337				
G-BSDL	SOCATA TB-10 Tobago	156		7.10.80	P.Middleton and G.Corbin	Sherburn-in-Elmet	18. 6.06
					tr Delta Lima Group		
G-BSDN	Piper PA-34-200T Seneca II	34-7970335	N2893A	2. 4.90	McCormick Consulting Ltd	Manchester	20. 6.05T
G-BSDO	Cessna 152 II	15281657	N65894	23. 5.90	L.W.Scattergood	Full Sutton	27. 7.06T
G-BSDP	Cessna 152 II	15280268	N24468	11. 6.90	Beatrice A.Paul	Denham	31. 7.06T
G-BSDS	Boeing Stearman E75 (PT-13A) Kaydet	75-118	N57852	6. 4.90	A Basso	Biel, Switzerland	13. 5.06
	(Continental W670)		38-470		*(As "118" in US Army c/s)*		
G-BSDV	Colt 31A HAB	1722		30. 3.90	C.D.Monk	(Bath)	8. 9.03A
G-BSDW	Cessna 182P Skylane II	18264688	N9125M	9. 4.90	Parker Diving Ltd	St.Just	30. 5.05T
G-BSDX	Cameron V-77 HAB	541	(G-BGWA)	30. 3.90	G.P.and Susan J.Allen	(Abingdon)	
	(Officially regd with c/n 2050)		G-SNOW (1)		*(New owners 11.04)*		
G-BSDZ	Enstrom 280FX	2051	OO-MHV	3. 4.90	Avalon Group Ltd	Hawarden	25. 7.05
			(OO-JMH), G-ODSC, G-BSDZ				
G-BSED	Piper PA-22-160 Tri-Pacer	22-6377	N9404D	7. 6.90	Tayflite Ltd	Perth	5. 9.03
	(Hoerner wing-tips: tail-wheel conversion)				*(Noted stripped down 1.03 following accident Stonehaven 21.6.02)*		
G-BSEE	Rans S-9	PFA 196-11635		2. 3.90	R.P.Hothersall	St Michaels	26.11.01P
	(Built P.M.Semler) (Rotax 532)				*(New owner 1.05)*		
G-BSEF	Piper PA-28-180 Cherokee C	28-1846	N7831W	18. 4.90	I.D.Wakeling	Franklyn's Field, Chewton Mendip	2.11.06
G-BSEG	Ken Brock KB-2	PFA G/<u>06</u>-1106		3. 4.90	S.J.M.Ledingham	Carlisle	2. 7.02P
	(Built H Bancroft-Wilson)						
G-BSEJ	Cessna 150M Commuter	15076261	N66767	4. 5.90	I.Shackleton	Wolverhampton	19.11.05T
G-BSEK	Robinson R22	0027	N45AD	10. 4.90	S.J.Strange	(Chorley)	28. 4.05T
			N90193				
G-BSEL	Slingsby T.61G Super Falke	1986		31. 3.80	T.Holloway	RAF Keevil	23. 5.07
					tr RAFGSA *(Op Bannerdown Gliding Club)*		
G-BSEP	Cessna 172	46555	N6455E	12. 4.90	R.J.Watts tr EP Aviation	Biggin Hill	15. 3.05
G-BSER	Piper PA-28-160 Cherokee B	28-790	N5665W	19. 4.90	Yorkair Ltd	Sandtoft	14. 9.06T
G-BSET*	Beagle B.206 Basset CC.1	B.006	XS765	3.12.86	IAE Ltd	Cranfield	28. 7.98
					(As "XS765" in RAF Transport Command c/s) (Cancelled 25.3.04 as wfu).(Stored)		
G-BSEU	Piper PA-28-181 Cherokee Archer II	28-7890108	N47639	1. 5.90	Euro Aviation 91 Ltd	Blackbushe	25. 5.05T
G-BSEV	Cameron O-77 HAB	2271		20. 4.90	P B Kenington	(Devauden, Chepstow)	9. 4.05A
					(New owner 3.04)		
G-BSEY	Beech A36 Bonanza	E-1873	N1809F	17. 5.90	K.Phillips Ltd	Coventry	10. 9.05
G-BSFA	Aero Designs Pulsar	PFA 202-11754		18. 4.90	S.Eddison and R.Minett	Gloucestershire	14. 7.05P
	(Built S A Gill - c/n 176) (Tri-cycle u/c)						
G-BSFB	CASA 1-131E Jungmann Series 2000	2053	Spanish AF E3B-449	27. 4.90	M.L.J.Goff	Old Buckenham	23. 2.05P
					(As "S5+B06" in Luftwaffe c/s)		
G-BSFD	Piper J-3C-65 Cub	16037	N88419	25. 5.90	AJD Engineering Ltd	Milden	19.10.01P
			NC88419				
G-BSFE	Piper PA-38-112 Tomahawk II	38-82A0033	N91452	26. 4.90	D.J.Campbell	Cumbernauld	10. 6.07T
G-BSFF	Robin DR400/180R Remorqueur	1295	D-ELMM	20. 4.90	Lasham Gliding Society Ltd	Lasham	11. 7.05
G-BSFK	Piper PA-28-161 Warrior II	28-8516062	N6918D	1. 5.90	Oxford Aviation Services Ltd	Oxford	10. 7.05T
G-BSFP	Cessna 152T	15285548	N93764	9. 5.90	Walkbury Aviation Ltd	Sibson	5. 8.05T
G-BSFR	Cessna 152 II	15282268	N68341	9. 5.90	Galair Ltd	Biggin Hill	8. 7.05T
G-BSFV	Woods Woody Pusher	201	N16WP	30. 4.90	M.J.Wells	Watchford Farm, Yarcombe	7. 5.05P
	(Continental C85)				*"Woody's Pusher"*		
G-BSFW	Piper PA-15 Vagabond	15-273	N4484H	26. 4.90	J.R.Kimberley	Andrewsfield	21.12.05P
	(Continental A65)		NC4484H				
G-BSFX	Denney Kitfox Model 2	PFA 172-11723		23. 4.90	H.Hedley-Lewis	Croft Farm, Defford	23.12.04P
	(Built D A McFadyean - c/n 506)						
G-BSFY	Denney Kitfox Model 2	PFA 172-11632		16. 3.90	C.I.Bates	Long Marston	8. 6.04P
	(Built J R Howard)						
G-BSGB	Gaertner Ax4 Skyranger HAB	SR.0001		30. 3.90	B.Gaertner	(Biddenden, Ashford)	
	(Built B.Gaertner)						
G-BSGD	Piper PA-28-180 Cherokee E	28-5691	N3463R	4. 5.90	R.J.Cleverley	Draycott Farm, Chiseldon	6. 6.06
G-BSGF	Robinson R22 Beta	1383		1. 5.90	Hecray Co Ltd t/a Direct Helicopters	Southend	24. 6.05T
G-BSGG	Denney Kitfox Model 2	PFA 172-11666		1. 5.90	C.G.Richardson	Fulbeck, Lincoln	17. 6.05P
	(Built C G Richardson) (Jabiru 2200A)						
G-BSGH	Airtour AH-56B HAB	014		1. 5.90	A R.Hardwick	(Shefford)	
					"Battle of Britain" (Active 8.04)		
G-BSGJ	Monnett Sonerai II	300	N34WH	1. 5.90	G.A Brady	Enstone	6. 9.91P
	(Built W Hossink) (Volkswagen 1835)				*(Noted 7.02)*		
G-BSGK	Piper PA-34-200T Seneca II	34-7870331	N36450	22. 5.90	Aeros Holdings Ltd	Gloucestershire	9. 1.06T

G-BSGL	Piper PA-28-161 Warrior II	28-8116041	N82690	10. 5.90	Keywest Air Charter Ltd	Liverpool	4. 9.05T
					(Op Liverpool Flying School) "Liverbird V"		
G-BSGP	Cameron N-65 HAB	2293		1. 5.90	T.D.Gibbs	(Billingshurst)	1. 4.05A
G-BSGR*	Boeing Stearman E75 (PT-17) Kaydet	75-4721	N75864	19. 6.90	A G.Dunkerley	Kemble	
			EC-ATY, N55050, 42-16558		*(Cancelled 10.3.99 by CAA) (Noted unmarked 2.04)*		
	(Composite rebuild of EC-AMD {75-4721} and EC-ATY {75-6714} and also reported as (ex?) N126SE)						
G-BSGS	Rans S-10 Sakota	PFA 194-11724		9. 5.90	M.R.Parr	Kirkbride	15. 9.05P
	(Built R Handley) (Rotax 532 - *kit no.1289.076)*						
G-BSGT	Cessna 210N Turbo Centurion II	21063361	LX-ATL	21. 5.90	E A T Brenninkmeyer	(Kingston upon Thames)	14. 1.06
	(Reims-assembled c/n F2100020)		D-EOGB, N5308A				
G-BSHA	Piper PA-34-200T Seneca II	34-7670216	N9707K	2. 5.90	Justgold Ltd	Blackpool	24. 7.99T
G-BSHC	Colt 69A HAB	1668		8. 5.90	Magical Adventures Ltd		
						(West Bloomfield, Michigan, US)	12.10.98A
G-BSHD	Colt 69A HAB	1736		8. 5.90	D.B.Court *"Jester"*	(Ormskirk)	15. 5.05A
G-BSHH	Luscombe 8E Silvaire	3981	N1254K	11. 5.90	S.L.Lewis	Draycott	29.10.05P
	(Continental C85)		NC1254K				
G-BSHI	Luscombe 8DF Silvaire Trainer	1821	N39060	11. 5.90	Catcott Garage Ltd	Dunkeswell	14. 8.01P
	(Continental C90)		NC39060				
G-BSHK	Denney Kitfox Model 2	PFA 172-11752		11. 5.90	D.Doyle and C.Aherne	Kilrush, County Kildare	25. 8.05P
	(Built A C Cree - c/n 449) (Rotax 532)						
G-BSHO	Cameron V-77 HAB	2313		16. 5.90	D J Duckworth and C Stewart	(Chesham)	1. 7.05A
					(Peugeot/Talbot titles)		
G-BSHP	Piper PA-28-161 Warrior II	28-8616002	N190X	31. 5.90	Plane Talking Ltd	Denham	28.11.07E
			G-BSHP, N9107Y				
G-BSHS	Colt 105A HAB	1674	(D-OCAT)	16. 5.90	I.Novosad	(Planegg, Germany)	9. 9.01A
			G-BSHS				
G-BSHT	Cameron V-77 HAB	2321		30. 5.90	E.C.Moore *"Buckshot Too"*	(Great Missenden)	19.11.03T
G-BSHV	Piper PA-18-135 Super Cub	18-3123	OO-GDG	5. 7.90	G.T Fisher	Northside, Thorney	20. 5.04
	(L-18C-PI)		Belg Army L-49, 53-4723				
G-BSHY	EAA Acrosport	PFA 72-10928		17. 4.90	R.J.Hodder	Eastfield Farm, Manby	20.12.05P
	(Built T Butterworth and R.J.Hodder) (Lycoming O-290)						
G-BSHZ	Enstrom F-28F	427	N51702	16. 5.90	K.G.Ward	(Wrexham)	20. 2.05
G-BSIC	Cameron V-77 HAB	2322		17. 5.90	J.M.and A Cornwall	(Brimington, Chesterfield)	14. 8.05A
G-BSIF	Denney Kitfox Model 2	PFA 172-11889		5. 7.90	P.Annable	Sandtoft	1. 6.05P
	(Built R M Kimbell and M H Wylde - c/n 563)						
G-BSIG	Colt 21A Cloudhopper HAB	1322		18. 5.90	S.J.Humphreys	(Great Missenden)	14. 8.04A
G-BSIH	Rutan LongEz	PFA 74A-11492		31. 5.90	W.S.Allen	(Cheltenham)	
	(Built W S Allen - c/n 1200-1)						
G-BSII	Piper PA-34-200T Seneca II	34-8070336	N8253N	16. 5.90	T.Belso *(Noted 6.04)*	Fowlmere	4. 8.02
G-BSIJ	Cameron V-77 HAB	2164		23. 5.90	A S.Jones	(Wolverhampton)	24. 5.04A
G-BSIK	Denney Kitfox Model 1	51		5. 6.90	S.P.Collins	Hill Farm, Nayland	13. 7.04P
	(Built R L Cunliffe) (Rotax 503)						
G-BSIM	Piper PA-28-181 Archer II	28-8690017	N9092Y	22. 5.90	Halfpenny Green Flight Centre Ltd	Wolverhampton	29. 8.05T
G-BSIO	Cameron Furness House 56SS HAB	2310		25. 5.90	R.E.Jones *"Pinkie"*	(Lytham St.Annes)	19. 3.05A
G-BSIU	Colt 90A HAB	1774		25. 5.90	S.Travaglia	F(irenze, Italy)	12. 7.04A
G-BSIY	Schleicher ASK 14	14005	5Y-AID	4. 6.90	E.V.Goodwin	(Huntingdon)	8. 9.07
			D-KOIC		tr Winwick Flying Group		
G-BSIZ	Piper PA-28-181 Archer II	28-7990377	N2162Y	25. 5.90	A M.L.Maxwell	Alderney	20. 6.05
G-BSJB	Bensen B.8	PFA G/01-1080		5. 6.90	J.W.Limbrick	(Bewdley)	
	(Built J.W.Limbrick)						
G-BSJU	Cessna 150M	15076430	N3230V	14. 6.90	A C.Williamson *(Op Crowfield Flying Club)*	Crowfield	4. 4.07T
G-BSJW	Everett Gyroplane Series 2	020		6. 6.90	R.Sarwan	(Beccles)	25.10.91P
	(Rotax 532)				*(Current status unknown)*		
G-BSJX	Piper PA-28-161 Warrior II	28-8216084	N8036N	30. 5.90	D.A Shields and L.C.Brekkeflat	Denham	23. 7.05T
G-BSJZ	Cessna 150J	15070485	N60661	7. 5.91	BCT Aircraft Leasing Ltd	Kemble	25. 4.05T
G-BSKA	Cessna 150M	15076137	N66588	31. 7.90	R.J.Cushing	Beccles	16. 3.06T
G-BSKD	Cameron V-77 HAB	2336		4. 6.90	M.J.Gunston *"Skulduggery"*	(Camberley)	30. 7.05A
G-BSKE	Cameron O-84 HAB	1604	ZS-HYD	4. 6.90	S.F.Redman	(Sturminster Newton)	19. 8.03A
			G-BSKE				
G-BSKG	Maule MX-7-180 Star Rocket	11072C		7. 6.90	J.R.Surbey	Blockmoor Farm, Barway, Ely	10. 2.06
G-BSKI	Thunder Ax8-90 HAB	1623		18. 5.90	K.P.Barnes and L.A.Pibworth	(Bristol)	12.10.04A
G-BSKK	Piper PA-38-112 Tomahawk	38-79A0671	N2525K	11. 6.90	A.S.Bamrah	Biggin Hill	20.11.05T
					t/a Falcon Flying Services		
G-BSKL	Piper PA-38-112 Tomahawk	38-78A0509	N4252E	11. 6.90	A.S.Bamrah	Biggin Hill	20.11.05T
					t/a Falcon Flying Services		
G-BSKO	Maule MXT-7-180 Star Rocket	14008C		7. 6.90	M.A Ashmole	Perth	12. 5.06
G-BSKP	Vickers Supermarine 379 Spitfire F.XIVe	6S/663417	Belgian AF SG-31	27. 6.90	Historic Flying Ltd	Duxford	12. 6.05P
			RN201		*(As "RN201" in 41 Sqdn silver and red c/s)*		
G-BSKS*	Nieuport 28C-1				See SECTION 3 Part 1	Fort Rucker, Alabama, US	
G-BSKU	Cameron O-84 HAB	2330		8. 6.90	Alfred Bagnall and Sons (West) Ltd *"Bagnalls II"*	(Bristol)	7. 6.02A
G-BSKW	Piper PA-28-181 Archer II	2890138	N91940	1. 6.90	Shropshire Aero Club Ltd	Sleap	9. 4.06T
G-BSLA	Robin DR400/180 Régent	1997		22. 6.90	A B.McCoig tr Robin Lima Alpha Group	Deanland	19. 3.05
G-BSLE	Piper PA-28-161 Warrior II	28-8116028	N8260L	25. 6.90	Oxford Aviation Services Ltd	Oxford	1.10.05T
G-BSLH	CASA 1-131E Jungmann Series 2000	2222	Spanish AF E3B-622	27. 7.90	P.Warden	Biel, Switzerland	20. 8.05P
	(Officially quoted c/n and p/i incorect: built new Bücker Prado SL, Albacete, Spain)						
G-BSLI	Cameron V-77 HAB	2115		15. 6.90	R.C.Corcoran	(Thornbury, Bristol)	18. 7.05T
G-BSLK	Piper PA-28-161 Warrior II	28-7916018	N20849	15. 6.90	R.A Rose	Wellesbourne Mountford	23. 6.06T
G-BSLM	Piper PA-28-160 Cherokee	28-308	N5262W	22. 6.90	R.Fulton	Popham	6. 3.06
G-BSLT	Piper PA-28-161 Warrior II	28-8016303	N81817	19. 6.90	L.W.Scattergood	Sheffield City	25. 1.06T
G-BSLU	Piper PA-28-140 Cherokee	28-24733	OY-PJL	19. 6.90	D.J.Budden Ltd	Shobdon	15. 1.06
			OH-PJL, SE-FFA				
G-BSLV	Enstrom 280FX	2054	D-HHAS	26. 6.90	Keswick Outdoor Clothing Co Ltd	(Keswick)	10. 7.03T
			G-BSLV				
G-BSLW	Champion 7ECA Citabria	431-66	N9696S	16. 7.90	D.W.Mann	Shoreham	6. 9.03
	(Officially regd as "Bellanca 7ECA Citabria")				tr Shoreham Citabria Group		

Registration	Type	C/n	Previous ID	Reg date	Owner	Location	CofA
G-BSLX	WAR Focke-Wulf 190 rep	24	N698WW	19. 6.90	D.Featherby	Norwich	2. 8.02P
					tr FW190 Gruppe *(As "1+4" in Luftwaffe c/s)*		
G-BSMB	Cessna U206E Stationair	U20601659	N9459G	25. 6.90	C.J.Francis	Draycott Farm, Chiseldon	13. 3.06
			C-GUUW, N9459G				
G-BSMD	SNCAN 1101 Noralpha	139	F-GDPQ	26. 6.90	J.W.Hardie	North Weald	4. 5.96P
			F-YEEE, F-YCZK, CAN-11, Fr.Mil *(As "+14" in Luftwaffe c/s) (Noted 1.05)*				
G-BSME	Bölkow Bö.208C Junior	596	D-ECGA	25. 6.90	D.J.Hampson	Fenland	27.10.06
G-BSMG	Montgomerie-Bensen B.8M	PFA G/01-1170		22. 6.90	A C.Timperley	(Aberfeldy)	16. 7.97P
	(Built A C.Timperley) (Rotax 532)				*(Current status unknown)*		
G-BSMK	Cameron O-84 HAB	2328		26. 6.90	D.F.Maine and D.M.Newton	(Redditch)	3. 7.05A
					tr G-BSMK Shareholders		
G-BSML	Schweizer 269C	S.1462	PH-HUH	10.10.90	K.P.Foster and B.I.Winsor	Bodmin	3. 4.06
			N134DM				
G-BSMM	Colt 31A Sky Chariot HAB	1779		27. 6.90	D.V.Fowler	(Cranbrook)	11. 7.04A
G-BSMN	CFM Streak Shadow	PFA 161A-11656		26. 6.90	P.J.Porter	(Wincanton)	17. 9.04P
	(Built K Daniels - kit no K.137-SA) (Rotax 582)						
G-BSMO	Denney Kitfox Model 2	PFA 172-11773		16. 7.90	M.W.Sayers and G.P.Bridgwater		
	(Built F S Beckett)					Sittles Farm, Alrewas	13. 5.05P
G-BSMS	Cameron V-77 HAB	2356		26. 6.90	Sade Balloons Ltd *"Sadie"*	(Coulsdon)	25. 4.05A
G-BSMT	Rans S-10 Sakota	PFA 194-11793		29. 6.90	P.J.Barker	Romney Street, Sevenoaks	8. 6.05P
	(Built N Woodworth) (Rotax 532 - *kit no.1289.077)*						
G-BSMU	Rans S-6-116N Coyote II	PFA 204-11732	G-MWJE	27. 6.90	S.Yelland	(Ferrensby, Knaresborough)	12. 7.05P
	(Built W D Walker - kit no.1089.090)						
G-BSMV	Piper PA-17 Vagabond	17-94	N4696H	29. 6.90	A Cheriton	Wellesbourne Mountford	5. 4.05P
	(Continental C85)		NC4696H		*"Sophie" (Carries 'N4696H' on tail)*		
G-BSMX	Bensen B.8MR	PFA G/01-1171		3. 7.90	J.S.E.R.McGregor	(Birmingham)	
	(Built J.S.E.R.McGregor)						
G-BSND	Air Command 532 Elite	PFA G/04-1180		16. 7.90	B Gunn and W.B.Lumb	Melrose Farm, Melbourne	
	(Built B J Castle)				*(New owner 4.02)*		
G-BSNE	Luscombe 8E Silvaire	5757	N1130B	2.11.90	N.Reynolds and C.Watts	(Guildford)	16. 8.05P
	(Continental C85)		NC1130B				
G-BSNF	Piper J-3C-65 Cub	3070	N23317	17. 8.90	D.A Hammant	Bere Farm, Warnford	17. 9.05P
	(Continental O-200-A) *(Frame No.3116)*		NC23317		*(Lippert Reed conversion)*		
G-BSNG	Cessna 172N Skyhawk II	17270192	N738SB	19. 7.90	A J. and P.C.MacDonald	Edinburgh	4.11.05T
G-BSNJ	Cameron N-90 HAB	2335		6. 7.90	D.P.H.Smith	(Ilkley)	8. 6.04A
G-BSNL	Bensen B.8MR	PFA G/01-1181		16. 7.90	A C.Breane	(Ballybofey, County Donegal)	20. 7.97P
	(Built T R Grief) (Rotax 532)						
G-BSNN	Rans S-10 Sakota	PFA 194-11846		31. 7.90	O.and S.D.Barnard	Leicester	21. 8.05P
	(Built S Adams) (Rotax 532)						
G-BSNP	Piper PA-28R-201T Turbo Cherokee Arrow III	28R-7703236	N38537	18. 7.90	D.F.K.Singleton	(Teck, Germany)	24. 9.05
G-BSNR	British Aerospace BAe 146 Series 300	E3165	EC-FGT	13. 7.90	Trident Aviation Leasing Services (Jersey) Ltd)	Filton	20.11.03T
			EC-807, G-6-165, G-BSNR, N886DV, G-BSNR, (N886DV), G-6-165 *(Stored 6.04)*				
G-BSNS	British Aerospace BAe 146 Series 300	E3169	EI-CTN	13. 7.90	Flightline Ltd	Southend	13. 4.07T
			G-BSNS, EC-FHU, EC-839, G-6-169, G-BSNS, N887DV, G-BSNS, N887DV, G-6-169				
G-BSNT	Luscombe 8A Master	1679	N37018	16. 7.90	P.K.Jordan	Stoneacre Farm, Farthing Corner	17. 5.05P
	(Continental A65) *(Built as Model 8C)*		NC37018		tr Luscombe Quartet		
G-BSNU	Colt 105A HAB	1811		23. 7.90	M.P.Rich tr Gone Ballooning *(Sun Rise titles)*	(Bristol)	7. 5.05A
G-BSNX	Piper PA-28-181 Archer II	28-7990311	N3028S	19. 7.90	Halfpenny Green Flight Centre Ltd	Wolverhampton	21. 8.05T
G-BSNY	Bensen B.8M	PFA G/01-1176		16. 7.90	H.McCartney	Newtownards	6. 9.01P
	(Built A S Deakin) (Arrow GT500R)						
G-BSNZ	Cameron O-105 HAB	2364		16. 7.90	J.Francis *"Firefly"*	(Southampton)	3. 7.05A
G-BSOE	Luscombe 8A Silvaire	4331	N1604K	22. 8.90	S.B.Marsden	Sturgate	
	(C/n would indicate Model 8E)		NC1604K		*(Stored dismantled as "N1604K" 4.02)*		
G-BSOF	Colt 25A Sky Chariot Mk.II HAB	1820		27. 7.90	L.P.Hooper *(Noted 1.02)*	(Bristol)	20.10.04A
G-BSOG	Cessna 172M Skyhawk II	17263636	N1508V	16. 7.90	B.Chapman and A R.Budden	Goodwood	19. 1.06
G-BSOJ	Thunder Ax7-77 HAB	1818	JA-...	31. 7.90	R.J.S.Jones	(Stourbridge)	15. 3.04A
			G-BSOJ				
G-BSOK	Piper PA-28-161 Cherokee Warrior II	28-7816191	N9749K	19. 7.90	Aeros Leasing Ltd	Gloucestershire	4. 2.06T
G-BSOM	Glaser-Dirks DG-400	4-126	LN-GMC	12. 7.90	M.J.Watson	Tibenham	14. 5.06
			D-KGDG		tr G-BSOM Group *"403"*		
G-BSON	Green S-25 HAB	001		7. 6.90	J.J.Green	(Newbury)	
	(Built J.J.Green)						
G-BSOO	Cessna 172F	17252431	N8531U	19. 7.90	P.W.Lawrence tr Double Oscar Flying Group	Seething	6.11.05
G-BSOR	CFM Streak Shadow	PFA 161A-11602		23.10.89	A.Parr	(Lewes, Sussex)	13. 5.05P
	(Built J P Sorensen - kit no.K.131-SA) (Rotax 532)						
G-BSOT	Piper PA-38-112 Tomahawk II	38-81A0053	N25682	23. 7.90	APB Leasing Ltd	Tatenhill	26. 9.05T
G-BSOU	Piper PA-38-112 Tomahawk II	38-81A0130	N23373	23. 7.90	D.J.Campbell *(Noted 5.04)*	Perth	10. 9.00T
G-BSOX	Luscombe 8AE Silvaire	2318	N45791	7. 8.90	R.S.Lanary	Sixpenny Handley, Dorset	19. 5.05P
	(Continental C85)		NC45791		*"Bobby Sox"*		
G-BSOZ	Piper PA-28-161 Warrior II	28-7916080	N30220	14. 8.90	The Moray Flying Club 1990	RAF Kinloss	28.12.05T
G-BSPA	QAC Quickie Q.2	2227	N227T	16. 8.90	G.V.Mckirdy and B.K.Glover	Enstone	21. 8.01P
	(Revmaster R2100DQ)						
G-BSPB	Thunder Ax8-84 HAB	1803		24. 7.90	A.N.F.Pertwee	(Great Holland, Frinton-on-Sea)	22. 9.00T
					(New owner 8.04)		
G-BSPC*	SAN Jodel D.140C Mousquetaire III	150	F-BMFN	2.11.81	Not known	Headcorn	31.10.85
					(Cancelled 15.8.94 by CAA) (Noted 4.03)		
G-BSPE	Reims/Cessna F172P Skyhawk II	F17202073		31.12.80	T.W.Williamson	Bagby	17. 6.05
G-BSPF*	Cessna T303 Crusader	T30300100	OY-SVH	31. 7.90	K.P.Gibben tr G-BSPF Crusader Group	Blackpool	
			N3116C		*(Crashed Burton Joyce, Nottingham 16.7.98: cancelled 25.8.98 as WFU) (Wreck noted 12.01)*		
G-BSPG	Piper PA-34-200T Seneca II	34-8070168	N8176S	8. 8.90	Airtime Aviation Ltd	Bournemouth	23. 1.06
G-BSPI	Piper PA-28-161 Warrior II	28-8116025	N8258V	26. 7.90	Halegreen Associates Ltd	Hinton-in-the-Hedges	22. 4.07T
G-BSPJ	Campell Cricket rep	PFA G/01-1061		3. 8.90	D.Ross	(Londonderry)	8. 1.04P
	(Built P Barlow ,A Scott and C Jones)						

Reg	Type	C/n	Previous identity	Regd	Owner	Base	CofA
G-BSPK	Cessna 195A	7691	N1079D	14. 8.90	A G. and D.L.Bompas	Biggin Hill	25. 4.06
	(Jacobs R-755-9)						
G-BSPL	CFM Streak Shadow	K.140-SA		26. 7.90	G.L.Turner	(Ellon)	6. 8.04P
	(Built CFM Metal-Fax) (Rotax 582)						
G-BSPM	Piper PA-28-161 Warrior II	28-8116046	N82679	27. 7.90	White Waltham Airfield Ltd	White Waltham	12. 1.07T
					(Op West London Aero Services)		
G-BSPN	Piper PA-28R-201T Turbo Cherokee Arrow III	28R-7703171	N5965V	31. 7.90	V.E.H.Taylor	Haverfordwest	19. 2.06
G-BSPW	Avid Speed Wing	PFA 189-11840		17. 7.90	M J Sewell	Blackpool	4. 9.04P
	(Built P D Wheatland)						
G-BSRD	Cameron N-105 HAB	1568	D-ORSD	3. 8.90	A Ockelmann	(Buchholz, Germany)	4. 9.04A
			G-BSRD		t/a Ballon Reisen		
G-BSRH	Pitts S-1C	LS-2	N4111	7. 8.90	J.Glen-Davis Gorman and C.D.Swift	(Bexhill-on-Sea)	21. 7.05P
	(Built L Smith)				*(Carries "N4111" on rudder)*		
G-BSRI	Neico Lancair 235	PFA 191-11467		9. 8.90	G.Lewis	Liverpool	1. 8.05P
	(Built G Lewis) (Lycoming O-235) *(Tri-cycle u/c)*						
G-BSRJ	Colt AA-1050 Gas Balloon	1782		20. 8.90	Trezpark Ltd	(Boulder, Colorado, US)	14. 6.05A
					(Op D Levin) "White Fang"		
	(After landing Crescent, OK 4.10.04 draped across electricity cables and canopy and basket caught fire and burnt out)						
G-BSRK	ARV Aviation ARV-1 Super 2	K.007	ZK-FSQ	8. 8.90	D.M.Blair	RAF Mona	14. 6.05P
G-BSRL	Campbell Cricket Mk.4 rep	PFA G/03-1325		8. 8.90	I.Rosewall	Henstridge	3. 8.05P
	(Built I Rosewall) (Regd as, and rebuilt from, Everett Gyroplane Series 2 c/n 0022 - converted by Peter Lovegrove)						
G-BSRP	RotorWay Executive	3824		15. 8.90	R.J.Baker	Hawarden	7. 4.04P
	(Built J P Dennison) (RotorWay RW 152)						
G-BSRR	Cessna 182Q Skylane II	18266915	N96961	25. 7.90	C.M.Moore	(Buckingham)	10. 7.06
G-BSRT	Denney Kitfox Model 2	PFA 172-11873		9. 8.90	A J.Lloyd	Henstridge	3.10.04P
	(Built L R James - c/n 742)						
G-BSRX	CFM Streak Shadow	PFA 206-11870		15. 8.90	P.Williams	Netherthorpe	22. 5.04P
	(Built C Penman - kit no K.148-SA) (Rotax 618)						
G-BSRZ	Air Command 532 Elite Two-Seat	PFA G/05-1188		15. 8.90	A S.G.Crabb	(Buxton)	
	(Built A S G Crabb)						
G-BSSA	Luscombe 8E Silvaire	4176	N1449K	15. 8.90	K.R.Old	White Waltham	21. 5.05P
	(Continental C85)		NC1449K		tr Luscombe Flying Group		
G-BSSB	Cessna 150L Commuter	15074147	N19076	15. 8.90	D.T.A Rees	Haverfordwest	30. 6.06T
G-BSSC	Piper PA-28-161 Warrior II	28-8216176	N81993	15. 8.90	Faber Developments Ltd	Norwich	6. 3.06T
			N9529N, N8234B				
G-BSSE	Piper PA-28-140 Cherokee Cruiser	28-7525192	N33440	22.10.90	Comed Schedule Services Ltd	Blackpool	19. 5.05T
G-BSSF	Denney Kitfox Model 2	PFA 172-11796		15. 8.90	A M.Hemmings	Sandtoft	20.11.05P
	(Built D M Orrock - c/n 738)						
G-BSSI	Rans S-6-116N Coyote II	PFA 204-11782	(G-MWJA)	17. 8.90	J.Currell	(Bangor, County Down)	16.11.99P
	(Built D AFarnwroth - kit no.0190.112)				*(New owner 9.01)*		
G-BSSK	QAC Quickie Q.200	PFA 94A-11354		5. 9.90	D.G.Greatrex	Enstone	23. 9.99P
	(Built D.G.Greatrex) (Continental O-200-A)				*(Current status unknown)*		
G-BSSO	Cameron O-90 HAB	2255		23. 7.90	R.R. and J.E.Hatton *"Just So"*	(Bristol)	28. 6.04A
G-BSSP	Robin DR400/180R Remorqueur	2015		24. 9.90	Soaring (Oxford) Ltd	RAF Syerston	8. 1.06
					(Op Air Cadets Gliding School)		
G-BSST*	British Aircraft Corporation-Sud Concorde SST				See SECTION 3 Part 1	RNAS Yeovilton	
G-BSSV	CFM Streak Shadow	PFA 206-11657		21. 8.90	R.W.Payne	Eddsfield	5. 5.98P
	(Built A M Green - kit no K.129-SA) (Rotax 532)				*(Current status unknown)*		
G-BSSW	Piper PA-28-161 Cherokee Warrior II	28-7816143	N47850	29. 8.90	R.L.Hayward *(Op Bristol Flying Club)*	Filton	5. 9.05T
G-BSSX	Piper PA-28-161 Warrior II	2816056	N9141H	11. 9.90	Airways Aero Associations Ltd	Booker	14.12.05T
					(Op British Airways Flying Club)		
G-BSSY*	CSS-13 Aeroklubowy				See SECTION 3 Part 1	Old Warden	
G-BSTC	Aeronca 11AC Chief	11AC-1660	N3289E	15.10.90	J Armstrong and D Lamb	(Crook)	26. 6.93P
	(Continental A65)		NC3289E		*(Damaged Henstridge 18.4.93: on rebuild 12.95: new owners 1.02)*		
G-BSTE	Aérospatiale AS355F2 Ecureuil 2	5453		29. 8.90	Oscar Mayer Ltd	Biggin Hill	11. 8.06
G-BSTH	Piper PA-25-235 Pawnee C	25-5009	N8599L	25. 9.90	Scottish Gliding Union Ltd	Portmoak	15. 5.06
G-BSTI	Piper J-3C-65 Cub	19144	N6007H	31. 8.90	G.L.Nunn and J.D.Barwick	North Lopham	31. 5.05P
	(Continental C85) *(Frame No.19073)*		NC6007H				
G-BSTK	Thunder Ax8-90 HAB	1838		17. 9.90	M.Williams	(Wadhurst, East Sussex)	4. 5.95A
G-BSTL	Rand Robinson KR-2	PFA 129-11863		6. 9.90	C.S.Hales and N. Brauns	Shenington	30. 6.05P
	(Built C S Hales) (May incorporate G-BYLP qv)						
G-BSTM	Cessna 172L Skyhawk	17260143	N4243Q	25. 9.90	A H.Windle tr G-BSTM Group	Duxford	18. 3.06
G-BSTO	Cessna 152 II	15282133	N68005	4. 9.90	Plymouth School of Flying Ltd	Plymouth	5.12.05T
G-BSTP	Cessna 152 II	15282925	N89953	4. 9.90	Cobham Leasing Ltd	Bournemouth	18. 3.06T
G-BSTR	Grumman AA-5 Traveler	AA5-0688	OO-ALR	8.10.90	James Allan (Aviation and Engineering) Ltd		
			OO-HAN, (OO-WAZ)			Sorbie Farm, Kingsmuir	22. 1.06
G-BSTT	Rans S-6 Coyote II	PFA 204-11880		5. 9.90	D.G.Palmer	Fetterangus	2.12.02P
	(Built M W Holmes - kit no. 0190.115) (Rotax 582)						
G-BSTV	Piper PA-32-300 Cherokee Six	32-40378	N4069R	13. 9.90	B.C.Hudson *(Open store 5.04)*	Popham	AC
G-BSTX	Luscombe 8A Silvaire	3301	EI-CDZ	10. 9.90	G.R.Nicholson	(Newry, County Armagh)	25 4.05P
			G-BSTX, N71874, NC71874				
G-BSTY	Thunder Ax8-90 HAB	394		12. 9.90	W.D.Mackinnon	(West Clandon, Guildford)	6.12.04A
					tr Shere Ballooning Group *"Beastie"*		
G-BSTZ	Piper PA-28-140 Cherokee Cruiser	28-7725153	N1674H	10.10.90	Air Navigation and Trading Co Ltd	Blackpool	7. 7.07T
G-BSUA	Rans S-6 Coyote II	PFA 204-11910		29.10.90	A J.Todd	Abbey Warren Farm, Bucknall, Lincoln	5. 9.05P
	(Built P S Dopson) (Rotax 582)						
G-BSUB	Colt 77A HAB	1801		30.10.90	M.P.Hill and J. M.Foster	(Bristol)	14. 7.05A
G-BSUD	Luscombe 8A Master	1745	N37084	14. 9.90	I.G.Harrison	Egginton	23.12.04P
	(Continental A65)		NC37084				
G-BSUE	Cessna U206G Stationair 6	U20604334	N756TB	6. 9.90	J.Dyer and I.C.Austin	(Cambridge)	26. 4.07
G-BSUF	Piper PA-32RT-300 Lance II	32R-7885240	N32PL	17. 9.90	S.A.Fell and J.Gibbs	Guernsey	9.10.06
			ZP-PJQ, N9641N				

Regn	Type	C/n	Previous identity	Regd	Owner/operator	Base	CofA
G-BSUH*	Cessna 140	8092	N89088	15.10.90	Not known	Abbeyshrule, County Longford	2. 5.94
	(Continental C85)		NC89088	*(Damaged Gowran Grange 6.93: cancelled 28.4.95 by CAA) (Airframe stored 4.03)*			
G-BSUJ	Brugger MB.2 Colibri	PFA 43-10726		17. 9.90	M.A Farrelly	(Liverpool)	
	(Built M.A and J P Farrelly)						
G-BSUK	Colt 77A HAB	1374		21. 9.90	A J.Moore	(Northwood, Middlesex)	2. 8.94A
G-BSUO	Scheibe SF25C-2000 Falke	44501	D-KIOK	6.12.90	J.G.Smith and J.L.Riley	Portmoak	9. 7.06
	(Limbach L2000)				tr Portmoak Falke Syndicate		
G-BSUT	Rans S-6-ESA Coyote II	PFA 204-11897		2.10.90	J.Bell	Barton	19. 1.05P
	(Built P Clegg) (Rotax 582 - *kit no.0990.138) (Tri-cycle u/c)*						
G-BSUV	Cameron O-77 HAB	2407		26. 9.90	R.Moss	(Banchory)	8. 9.02A
G-BSUW	Piper PA-34-200T Seneca II	34-7870081	N2360M	26. 9.90	NPD Direct Ltd	Fenland	12. 1.06T
G-BSUX	Carlson Sparrow II	PFA 209-11794		5.10.90	J.Stephenson	Wombleton	17. 6.05P
	(Built J Stephnson) (Rotax 532)						
G-BSUZ	Denney Kitfox Model 3	PFA 172-11875		10. 9.90	P.C.Avery	Shipdham	22. 8.05P
	(Built E T Wicks - c/n 745) (Converted from Model 2)						
G-BSVB	Piper PA-28-181 Archer II	2890098	N9155S	10. 9.90	K.A Boost	Stapleford	2. 2.06T
G-BSVE	Binder CP.301S Smaragd	113	HB-SED	27. 9.90	R.E.Perry tr Smaragd Flying Group	Halesland	13. 4.05P
G-BSVG	Piper PA-28-161 Warrior II	28-8516013	C-GZAV	2.10.90	Airways Aero Associations Ltd	Booker	22.12.05T
					(Op British Airways Flying Club) (Colum t/s)		
G-BSVH	Piper J-3C-65 Cub	15360	N87702	2.10.90	C.R.and Karen A.Maher	(Formby, Liverpool)	14. 7.05P
	(Continental A75) *(Frame No.15003)*		NC87702		*(New owners 8.04)*		
G-BSVI	Piper PA-16 Clipper	16-186	N5379H	7.11.90	I.R.Blakemore *"Spirit of St.Petersburg"*	Old Sarum	1.11.05
G-BSVK	Denney Kitfox Model 2	PFA 172-11731		2.10.90	C.M.Looney	(Leatherhead)	5. 4.94P
	(Built K P Wordsworth)						
G-BSVM	Piper PA-28-161 Warrior II	28-8116173	N8351N	7.11.90	EFG Flying Services Ltd	Biggin Hill	2. 2.06T
G-BSVN	Thorp T-18	107	N4881	17. 9.90	J.H.Kirkham	Barton	8. 7.05P
	(Lycoming O-290)						
G-BSVR	Schweizer 269C	S.1236	OO-JWW	14.11.90	Martinair Ltd	Sherburn-in-Elmet	7. 7.07
			D-HLEB				
G-BSVS	Robin DR400/100 Cadet	2017		22.10.90	D.M.Chalmers	Upper Harford	22. 3.06
G-BSVW	Piper PA-38-112 Tomahawk	38-79A0149	N2606B	9.11.90	Cardiff Wales Aviation Services Ltd	Cardiff	25. 4.04T
G-BSVZ	Pietenpol AirCamper	1008	N3265	6.11.90	M.J.Kirk	Haverfordwest	6. 9.93P
	(Built H Challis) (Regd as a Pietenpol/Challis Chaffinch)				*(New owner 10.03)*		
G-BSWB	Rans S-10 Sakota	PFA 194-11560		8.10.90	F.A Hewitt	Garston Farm, Marshfield	25. 8.05P
	(Built F A Hewitt) (Rotax 532 - *kit no.0489.046)*						
G-BSWC	Boeing Stearman E75 (PT-13D) Kaydet	75-5560	N17112	16.11.90	Richard Thwaites Aviation Ltd	Gloucestershire	27. 8.06T
	(Lycoming R-680)		N5021V, 42-17397		*(As "112" in US Army c/s)*		
G-BSWF	Piper PA-16 Clipper	16-475	N5865H	12.10.90	T.M.Storey	Newells Farm, Lower Beeding	24. 2.05
	(Lycoming O-320)						
G-BSWG	Piper PA-17 Vagabond	15-99	N4316H	8.10.90	P.E.J.Sturgeon	Queach Farm, Bury St.Edmunds	24.10.05P
	(Continental A65-8)		NC4316H				
G-BSWH	Cessna 152 II	15281365	N49861	15.10.90	Airspeed Aviation Ltd	Egginton	14. 3.02T
G-BSWL	Slingsby T.61F Venture T.2	1974	EI-CCQ	15.10.90	K.Richards	Talgarth	20. 4.07
			G-BSWL, ZA655				
G-BSWM	Slingsby T.61F Venture T.2	1965	ZA629	12.10.90	P.S.Holmes tr Venture Gliding Group	Bellamena	29. 4.06
G-BSWR	Pilatus Britten-Norman BN-2T Islander	2245		22.10.90	Police Authority for Northern Ireland	Belfast Aldergrove	2. 3.07T
G-BSWV	Cameron N-77 HAB	2369		22.10.90	S.Charlish *"Leicester Mercury"*	(Leicester)	29. 3.05A
G-BSWX	Cameron V-90 HAB	2401		22.10.90	B.J.Burrows *"Beeswax"*	(Bristol)	29. 6.01A
G-BSWY	Cameron N-77 HAB	2428		12.10.90	A.S.Davidson	(Woodville, Swadlincote)	5. 7.05A
					tr Nottingham Hot Air Balloon Club		
G-BSWZ	Cameron A-180 HAB	2419	C-FGWZ	22.10.90	G.C.Ludlow	(Aldershot)	19. 7.99T
			G-BSWZ		*(Op Balloon Preservation Group) "Keep Britain Farming"*		
G-BSXA	Piper PA-28-161 Warrior II	28-8416121	N4373Z	11.12.90	A.S.Bamrah	Biggin Hill	2. 8.06T
			N9622N		t/a Falcon Flying Services		
G-BSXB	Piper PA-28-161 Warrior II	28-8416125	N4374D	4.12.90	Aeros Leasing Ltd	Filton	30. 4.06T
			N9626N				
G-BSXC	Piper PA-28-161 Warrior II	28-8416126	N4374F	4.12.90	L.T.Halpin	Compton Abbas	5.10.07E
			N9627N				
G-BSXD	Soko P-2 Kraguj	030	Yugoslav Army 30146	22.10.90	L.C.MacKnight	Elstree	22. 4.99P
					(As "30146" in Yugoslav Army c/s)		
G-BSXI	Mooney M.20E Chapparal	700056	N6766V	31.10.90	A.N.Pain	Southend	8. 5.06
G-BSXM	Cameron V-77 HAB	2446		5.11.90	C.A Oxby *"Oxby"*	(Doncaster)	31. 8.05A
G-BSXS	Piper PA-28-181 Archer II	28-7990151	N3055C	26.11.90	Horizon Aviation Ltd	Swansea	9.12.05T
G-BSXT	Piper J-5A Cub Cruiser	5-498	N33409	8.11.90	M.J.Kirk	Haverfordwest	26. 7.03P
	(Continental C85)		NC33409				
G-BSXX	Whittaker MW7	PFA 171-11469		16.10.90	H.J.Stanley	(Abingdon)	
	(Built H.J.Stanley)						
G-BSXZ	British Aerospace BAe 146 Series 300	E3174	G-NJIB	14.11.90	Flightline Ltd	Exeter	11. 4.04T
			B-1776, G-BSXZ, G-6-174		*(Stored 9.04)*		
G-BSYA	Jodel D.18	PFA 169-11316		7.11.90	K.Wright	(Douglas, Isle Of Man)	1.10.05P
	(Built S.Harrison) (Volkswagen 1834)						
G-BSYB	Cameron N-120 HAB	2406		7.11.90	M.Buono	(Mondovi, Italy)	24. 7.03A
G-BSYC	Piper PA-32R-300 Cherokee Lance	32R-7780159	N7745T	2. 4.91	Olympia Homes Ltd	Wolverhampton	11. 8.06
			N1435H				
G-BSYD	Cameron A-180 HAB	2426		18.10.90	A A Brown t/a Balloon Company *"Discovery"*	(Guildford)	15. 9.05T
G-BSYF	Luscombe 8A Silvaire	3455	N72028	12.11.90	Atlantic Connexions Ltd	Little Gransden	27. 1.05P
			NC72028		t/a Atlantic Aviation		
G-BSYG	Piper PA-12 Super Cruiser	12-2106	N3228M	12.11.90	E.R.Newall	Breighton	9. 8.05P
	(Lycoming O-235)		NC3228M		tr Fat Cub Group		
G-BSYH	Luscombe 8A Silvaire	2842	N71415	13.11.90	N.R.Osborne	Insch	21.12.04P
	(Continental A65)		NC71415				
G-BSYI	Aérospatiale AS355F1 Ecureuil 2	5197	M-MJI	14.11.90	Premiair Aviation Services Ltd	Denham	28.11.05T
G-BSYL*	Piper PA-38-112 Tomahawk II	38-81A0172	N91333	23. 1.91	Flychoice Ltd	Wolverhampton	
					(No CofA issued: cancelled 10.3.99 by CAA) (Stored 8..03)		

G-BSYM*	Piper PA-38-112 Tomahawk II	38-82A0072	N2507V	30. 1.91	Flychoice Ltd	Wellesbourne Mountford	4. 9.94T
					(Damaged 27.7.94: cancelled 26.10.00 by CAA) (Dumped 7.03)		
G-BSYO	Piper J-3C-65 Cub (L-4B-PI)	12809	(G-BSMJ)	19. 2.91	C.R.Reynolds and J.D.Fuller	Pent Farm, Postling	13. 5.05P
	(Continental O-200-A) *(Frame No.12639)*		(G-BRHE), EC-AIY, HB-ODO, HB-OUA, 44-80513				
	(Officially regd as c/n 10244 which is HB-OVG ex 43-1383/F-BFYF)						
G-BSYU	Robin DR400/180 Régent	2027		26.11.90	P.D.Smoothy	Hinton-in-the-Hedges	15. 4.06
G-BSYV	Cessna 150M	15078371	N9423U	16.11.90	L.R.Haunch t/a Fenland Flying School	Fenland	14. 4.06T
G-BSYW	Cessna 150M	15078446	N9498U	16.11.90	D.R.Calo	(Chipperfield, Kings Langley)	1.12.06T
G-BSYZ	Piper PA-28-161 Warrior II	28-8516051	N6908H	22.11.90	S.W.Cowie	Glasgow	23.10.06T
					tr Yankee Zulu Group *(Op Glasgow Flying Club)*		
G-BSZB	Stolp SA.300 Starduster Too	545	N5495M	3.12.90	D.T.Gethin	Haverfordwest	8.12.04P
	(Built J W Matthews) (Lycoming O-360)						
G-BSZC	Beech C-45H-BH Expeditor	AF-258	N9541Z	14.12.90	A P.H.Walsh	Weston, Dublin	4. 6.06
	(Built as AT-7 42-2490 [4166]: re-manufactured 4.52)			51-11701	*"Southern Comfort" (As "51-11701/AF258/AF313" in USAF c/s)*		
G-BSZD	Robin DR400/180 Régent	2029		21.11.90	R.J.Hitchman and M.Rowland	Draycott Farm, Chiseldon	15. 4.06
G-BSZF	CEA Jodel DR.250/160 Capitaine	32	F-BNJB	29.11.90	J.B.Randle	Biggin Hill	8. 6.06
G-BSZG	Stolp SA.100 Starduster	101	N70P	27.11.90	D.F.Chapman	Headcorn	4. 7.00P
	(Built L Stolp and G Adams) (Lycoming O-320)				*(Noted 4.03)*		
G-BSZH	Thunder Ax7-77 HAB	1848		27.11.90	P.K.Morris	(Cleethorpes)	20. 5.05T
G-BSZI	Cessna 152 II	15285856	N95139	17.12.90	Eglinton Flying Club Ltd	City of Derry	26. 3.06T
G-BSZJ	Piper PA-28-181 Archer II	28-8190216	N8373Z	6.12.90	R.D.Fuller and M.L.A Pudney *"21"*		
					West Newlands Farm, St.Lawrence, Bradwell-on-Sea		3. 5.06
G-BSZM	Montgomerie-Bensen B.8MR	PFA G/01-1193		30.11.90	A McCredie	Carlisle	6. 9.02P
	(Built J H H Turner) (Rotax 582)						
G-BSZN	Bücker Bü.133D-1 Jungmeister	2002	N8103	30.11.90	J Hufner	(Albstadt, Germany)	25. 5.05P
	(Built Bitz Flugzeugbau GmbH) (Siemens Bramo SH14A)		D-ECAY (1)				
G-BSZO	Cessna 152 II	15280221	N24334	30.11.90	Hecray Co Ltd	Southend	17.12.04T
					t/a Direct Helicopters *(Op Southend School of Flying)*		
G-BSZT	Piper PA-28-161 Warrior II	28-8116027	N8260D	31.12.90	Golf Charlie Echo Ltd *(Op The Flying Hut)*	Shoreham	10 4.06T
G-BSZU	Cessna 150F Commuter	15063481	N6881F	3.12.90	J.E.Jones	Bournemouth	18.12.05T
G-BSZV	Cessna 150F	15062304	N3504L	3.12.90	Kirmington Aviation Ltd	Sandown	23. 9.06T
G-BSZW	Cessna 152 II	15281072	N48958	3.12.90	Haimoss Ltd	Old Sarum	9.12.06T

G-BTAA - G-BTZZ

G-BTAB	British Aerospace BAe 125 Series 800B	258088	G-5-563	12. 7.88	Aravco Ltd	Farnborough	6. 5.05T
			G-BOOA, (ZK-RHP), G-5-563				
G-BTAG	Cameron O-77 HAB	2454		12.11.90	R.A.Shapland *"Tag-Along"*	(Petworth)	13. 1.02A
G-BTAH	Bensen B.8M	PFA G/07-1196		13.12.90	C.J.Toner	Abbeyshrule, County Longford	31. 8.98P
	(Built T B Johnson) (Arrow GT500R)				*(Noted 5.00)*		
G-BTAK	EAA Acrosport II	<u>1468</u>	N440X	27.12.90	P.G.Harrison	Leicester	9. 4.05P
	(Built A P Savage) (Lycoming O-320)				*"The Duck"*		
G-BTAL	Reims/Cessna F152 II	F15201444		7. 4.78	TG Aviation Ltd *(Op Thanet Flying Club)*	Manston	16. 3.06T
G-BTAM	Piper PA-28-181 Archer II	2890093	RA-01765	10. 1.91	Tri-Star Farms Ltd	Mount Rule, Isle of Man	19.11.05T
			G-BTAM, N9153D				
G-BTAN	Thunder Ax7-65Z HAB	517		4. 5.83	A S.Newnham	(Southampton)	2. 8.00A
G-BTAP	Piper PA-38-112 Tomahawk	38-78A0141	N9603T	8. 1.91	R.Farrer	(Stevington, Bedford)	19. 8.06T
G-BTAS	Piper PA-38-112 Tomahawk	38-79A0545	F-GTAS	21. 2.91	M.Lowe	Cardiff	17. 2.07T
			G-BTAS, N2492G				
G-BTAT	Denney Kitfox Model 2	PFA 172-11832		6.11.90	M.Lawton	Otherton, Cannock	29. 4.04P
	(Built D G Marwick - c/n 689)						
G-BTAU	Thunder Ax7-77 HAB	1429		13.12.90	S.and G.Gebauer	(Lippstadt, Germany)	14. 4.05A
G-BTAW	Piper PA-28-161 Warrior II	28-8616031	N9259T	14.12.90	A J.Wiggins	Gloucestershire	16. 5.06T
					(Op Gloucester and Cheltenham Flying School)		
G-BTAZ	Evans VP-2	PFA 63-11474		13.12.90	G.S.Poulter	Norwich	
	(Built G.S.Poulter)				*(Noted complete 6.00)*		
G-BTBA	Robinson R22 Beta	1717		18. 3.91	Heliflight (UK) Ltd	Wolverhampton	30. 4.06
G-BTBB	Thunder Ax8-105 Series 2 HAB	1871		23.11.90	W.J.Brogan	(Steiermark, Austria)	21. 9.05T
G-BTBC	Piper PA-28-161 Warrior II	28-7916414	N28755	19.12.90	Wellesbourne Flyers Ltd	Wellesbourne Mountford	24. 4.06T
G-BTBF	Fisher Super Koala	PFA 158-11954	(G-MWOZ)	24.12.90	E.A Taylor	(Southend-on-Sea)	
	(Built E A Taylor - c/n SK.067)				*(Construction Suspended 1.05)*		
G-BTBG	Denney Kitfox Model 2	PFA 172-11845		18.12.90	P.D.Brookes	Long Marston	31. 5.05P
	(Built J Catley)						
G-BTBH	Ryan ST3KR (PT-22-RY)	2063	N854	18. 2.91	P.R.Holloway	(Southam)	26. 4.05P
	(Kinner R56)		N50993, 41-20854		*(As "854" in US Army Air Corp c/s)*		
G-BTBI	WAR P-47 Thunderbolt rep	0054	N47DL	8. 1.91	R.D.Myles	Perth	26. 8.05P
	(Continental O-200-A) *(Marked as "Project No.52685A")*				*(As "85" in USAF c/s)*		
G-BTBJ	Cessna 190	16046	N4461C	2.10.91	P.Camus	(Dijon, France)	9. 3.07
	(Originally regd as Cessna 195B: re-designated 6.01 when fitted with Jacobs Aircraft and Engines R-755-B2)						
G-BTBL	Montgomerie-Bensen B.8MR Merlin	PFA G/01A-1183		21.12.90	N.H.Collins	Melrose Farm, Melbourne	2. 9.01P
	(Built J M Montgomerie) (Rotax 532)				t/a AES Radionic Surveillance Systems *(Noted 4.04)*		
G-BTBN	Denney Kitfox Model 2	PFA 172-11859		31.12.90	R.C.Bowley	Croft Farm, Defford	19. 8.04P
	(Built Valley Avon Flying Group - c/n 686)						
G-BTBP	Cameron N-90 HAB	2464		21.12.90	M.Catalani *(New owner 2.04)*	(Pistoia, Italy)	21. 6.00A
G-BTBU	Piper PA-18-150 Super Cub	18-7509010	N9665P	3. 1.91	A J.White tr G-BTBU Syndicate	Redhill	18. 9.06
G-BTBW	Cessna 120	14220	N2009V	24. 1.91	Melanie J.Willies	Top Farm, Croydon, Royston	19. 8.07
	(Continental C90)		NC2009V				
G-BTBX	Piper J-3C-65 Cub	6334	N35367	29. 1.91	J.B.Hargrave and D.T.C.Collins	RAF Henlow	6.11.06
			NC35367		tr Henlow Taildraggers		
G-BTBY	Piper PA-17 Vagabond	17-195	N4894H	4. 1.91	G.J.Smith	Clipgate Farm, Denton	27. 4.05P
	(Continental C85)						
G-BTCA	Piper PA-32R-300 Cherokee Lance	32R-7780381	N5941V	10. 1.91	R.Page tr Lance Group	Sleap	26. 5.06

Reg	Type	C/n	Previous identity	Reg date	Owner/operator	Base	CofA
G-BTCB	Air Command 582 Sport *(Built G Scurrah - 0634)*	PFA G/04-1198		9. 1.91	G.Scurrah *(Was nearing completion 5.95: current status unknown)*	Millom	
G-BTCC	Grumman F6F-5K Hellcat *(Composite with centre section of Bu.08831 [A-218])*	A-11286	(N10CN) N100T, FN80142, Bu.80141	31.12.90	Patina Ltd *(Op The Fighter Collection as "40467/19" in VF-6 Sqdn/US ex F6F-3 Navy c/s)*	Duxford	6. 7.05P
G-BTCD	North American P-51D-25NA Mustang	122-39608	N51JJ N6340T, RCAF 9568, 44-73149	11. 1.91	Pelham Ltd *"Ferocious Frankie"* *(Op The Old Flying Machine Co as "413704/B7-H" in 374th FG/USAAF c/s)*	Duxford	19. 5.05P
G-BTCE	Cessna 152 II *(Tail-wheel conversion)*	15281376	N49876	10. 1.91	S.T.Gilbert	Enstone	22. 6.07T
G-BTCH	Luscombe 8E Silvaire (Continental C85)	6403	N1976B NC1976B	11. 2.91	J.Grewcock and R.C.Carroll tr G-BTCH Flying Group	Popham	16. 9.05P
G-BTCI	Piper PA-17 Vagabond (Continental A65)	17-136	N4839H NC4839H	11. 1.91	T.R.Whittome	Inverness	9. 8.05P
G-BTCJ	Luscombe 8AE Silvaire (Continental O-200A)	1869	N41908 NC41908	16. 1.91	Mrs J.M.Lovell	Chilbolton	9. 6.05P
G-BTCM	Cameron N-90 HAB	1306	(G-BMPW)	8. 5.86	G.Everett	(Maidstone)	10. 7.05A
G-BTCR	Rans S-10 Sakota *(Built S H Barr)* (Rotax 532)	PFA 194-11877		11. 1.91	B.J.Hewitt	Newtownards	11. 1.02P
G-BTCS	Colt 90A HAB	1895		11. 1.91	R.C.Stone *"Rosie Rags" (Variety Club of GB titles)*	(Reading)	27. 5.04A
G-BTCZ	Cameron Chateau-84SS HAB *(Forbes Chateau de Balleroy shape)*	2246		18. 1.91	Forbes Global Inc *"Chateau II" (New CofR 9.03)*	(London SW11)	7. 7.02
G-BTDA	Slingsby T.61F Venture T.2	1870	XZ550	17. 4.91	T.Holloway tr RAFGSA *(Op Anglia Gliding Club)*	RAF Wattisham	11. 9.06
G-BTDC	Denney Kitfox Model 2 *(Built D Collinson and O.Smith - c/n 405)*	PFA 172-11483		11. 1.91	O.Smith	Croft Farm, Darlington	19. 2.05P
G-BTDD	CFM Streak Shadow *(Built S J Evans - kit no.K.127-SA)* (Rotax 582)	PFA 161A-11622		14. 1.91	N.D.Ewer	Plaistowes Farm, St.Albans	21. 9.05P
G-BTDE	Cessna C-165 Airmaster	551	N21911 NC21911	18. 1.91	G.S.Moss	Popham	17.12.06
G-BTDF	Luscombe 8AF Silvaire (Continental C90)	2205	N45678 NC45678	17. 4.91	R.Harrison tr Delta Foxtrot Group	(Sunderland)	19. 8.93P
G-BTDN	Denney Kitfox Model 2 *(Built A B Butler - c/n 688)*	PFA 172-11826		22. 1.91	S.D.Arnold tr Foxy Flyers Group *(Noted 11.04)*	Wellesbourne Mountford	29. 6.05P
G-BTDP	Grumman TBM-3R Avenger	3381	N3966A Bu.53319	5. 2.91	A.Haig-Thomas *(As "53319/RB-319" in USN c/s)*	North Weald	28. 5.05P
G-BTDR	Aero Designs Pulsar *(Built R M Hughes and T Packe)*	PFA 202-11962		24. 1.91	R.A Blackwell	Eshott	28. 7.05P
G-BTDS	Colt 77A HAB	1897		29. 1.91	CP Witter Ltd *"Witters II"*	(Chester)	26. 3.05A
G-BTDT	CASA 1-131E Jungmann Series 2000	2131	Spanish AF E3B-505	5. 2.91	T.A Reed	Watchford Farm, Yarcombe	22. 5.05P
G-BTDV	Piper PA-28-161 Cherokee Warrior II	28-7816355	N3548M	25. 2.91	Leeds Flying School Ltd	Sherburn-in-Elmet	21. 6.07T
G-BTDW	Cessna 152 II	15279864	N757NC	25. 2.91	J.A Blenkharn	Carlisle	23.10.06T
G-BTDZ	CASA 1-131E Jungmann Series 2000	2104	Spanish AF E3B-524	5. 2.91	R.J.Pickin and I.M.White	Headcorn	17. 6.05P
G-BTEA	Cameron N-105 HAB	284		31. 5.77	M.W.A Shemilt *"Big Red"*	(Henley-on-Thames)	8. 5.99A
G-BTEE	Cameron O-120 HAB	2499		24. 1.91	W.H. and J.P.Morgan *"Y Ddraig Goch/The Red Dragon"*	(Swansea)	17. 5.02T
G-BTEF	Pitts S-1 *(Built D R Brewer)*	515H	N88PR	19. 2.91	C.Davidson *(New CofR 3.02)*	Blackpool	28.10.97P
G-BTEI*	Everett Campbell Cricket Srs.3 *(Built J W Highton)*	023		31. 1.91	R A Jarvis *(Damaged in forced landing nr.Great Orton 15.8.95: stored 8.98) (Canceleld 23.5.01 by CAA) (Reported 2004)*	(Lisbellow, Inniskillen)	21.12.98P
G-BTEL	CFM Streak Shadow *(Built J E Eatwell - kit no K.125-SA)* (Rotax 618)	PFA 206-11667		31. 1.91	J.E.Eatwell	Boscombe Down	9.11.05P
G-BTES	Cessna 150H	15068371	N22575	29. 4.91	R.A Forward	Spilsteads Farm, Sedlescombe	14.10.07
G-BTET	Piper J-3C-65 Cub	18296	N98141 NC98141	5. 2.91	R.M.Jones	Blackpool	19.10.05P
G-BTEU	Aérospatiale SA365N2 Dauphin 2	6392		11. 2.91	CHC Scotia Ltd	Humberside	1. 4.07T
G-BTEW	Cessna 120 (Continental C90)	10238	CF-ELE	29. 4.91	S.D.Pryke and Linda M.Hamblyn	Old Buckenham	29. 6.07
G-BTEX	Piper PA-28-140 Cherokee	28-23773	CF-XXL N3907K	24. 4.91	McAully Flying Group Ltd	Little Snoring	11. 2.07T
G-BTFC	Reims/Cessna F152 II	F15201668		23. 5.79	Tayside Aviation Ltd	Dundee	4. 5.07T
G-BTFE	Parsons Gyroplane Model 1 *(Built I Brewster)* (Rotax 582) *(Tandem Trainer)*	38		13. 2.91	J.R.Goldspink	Haverfordwest	27.10.01P
G-BTFF	Cessna T310R II	310R0718	N1363G	25. 2.91	Clear Prop Ltd	Blackbushe	29. 5.06
G-BTFG	Boeing Stearman A75NI (N2S-4) Kaydet (Continental W670)	75-3441	N4467N Bu.30010	20. 2.91	TG Aviation Ltd *(As "441" in USN c/s)*	Manston	2. 7.05
G-BTFJ	Piper PA-15 Vagabond (Lycoming O-145)	15-159	N4373H NC4373H	13. 2.91	C.W.Thirtle *(Noted 9.04)*	Old Sarum	26. 9.05P
G-BTFK	Taylorcraft BC-12D (Continental A65)	10540	N599SB N5240M	13. 2.91	A O'Rourke *(Noted 7.04)*	Pallas West, Toomyvara, County Tipperary	16. 2.00P
G-BTFL	Aeronca 11AC Chief	11AC-1727	N3403E NC3403E	18. 2.91	J.G.Vaughan tr BTFL Group	Eastbach Farm, Coleford	26.11.05P
G-BTFM	Cameron O-105 HAB	2623		12. 8.91	P.Forster and J.Trehern tr Edinburgh University Hot Air Balloon Club	(Edinburgh)	27. 8.05A
G-BTFO	Piper PA-28-161 Cherokee Warrior II	28-7816580	N31728	12. 3.91	Flyfar Ltd	Blackpool	22. 5.04
G-BTFP	Piper PA-38-112 Tomahawk	38-78A0340	N6201A	17. 4.91	Teesside Flight Centre Ltd *(Noted 2.04)*	Bagby	6. 8.00T
G-BTFS	Cessna A150M Aerobat	A1500719	N20331	20. 2.91	P.A James	Redhill	2. 6.06T
G-BTFT	Beech 58 Baron	TH-979	N2036W	14. 3.91	Fastwing Air Charter Ltd	Thruxton	8. 4.07T
G-BTFU	Cameron N-90 HAB	2391		28. 2.91	J.J.Rudoni and A C.K.Rawson t/a Wickers World Hot Air Balloon Co *"Maltesers II"*	(Stafford)	26. 3.05A
G-BTFV	Whittaker MW7 *(Built S J Luck)*	PFA 171-11722		8. 2.91	S.J.Luck	Tower Farm, Wollaston	12.12.05P
G-BTFW	Montgomerie-Bensen B.8MR *(Built A Mansfield)* (Rotax 532)	PFA G/01A-1141		20. 2.91	J.R.J.Read *(Heavy landing North Coates 16.7.03 and badly damaged: all remains but cockpit pod to Henstridge)*	North Coates	27. 5.05P

G-BTFX	Bell 206B JetRanger II	1648	N400MH	20. 2.91	A.C.Watson	(Wakefield)	12. 6.06T
			N90219				
G-BTGD	Rand Robinson KR-2	PFA 129-11150		22. 2.91	P.A Spurr	(Sanquhar)	23. 5.05P
	(Built D W Mullin) (Volkswagen 1915)						
G-BTGG	Rans S-10 Sakota	PFA 194-11944		20. 2.91	C.A James	Doynton	22. 6.96P
	(Built A R Cameron) (Rotax 582)				*(New owner 12.03)*		
G-BTGH*	Cessna 152 II	15281048	N48919	2. 4.91	C and S Aviation Ltd	Sleap	25. 4.07T
					(Cancelled 2.11.04 by CAA)		
G-BTGI	Rearwin 175 Skyranger	1517	N32308	26. 2.91	N.A Evans	Branscombe	12. 5.05P
	(Continental A75)		NC32308				
G-BTGJ	Smith DSA-1 Miniplane	NM.II	N1471	25. 3.91	G.J.Knowles	Goodwood	31. 3.05P
	(Built S J Malovic) (Continental C90)						
G-BTGL	Avid Speed Wing	PFA 189-11885		27. 2.91	I.Kazi	(Cranbrook)	4. 9.03P
	(Built A J Maxwell)						
G-BTGM	Aeronca 7AC Champion	7AC-3665	N84943	11. 3.91	G.P.Gregg	Shacklewell Farm, Empingham	5. 7.05P
	(Continental A65)		NC84943				
G-BTGO	Piper PA-28-140 Cherokee D	28-7125613	N1998T	20. 2.91	Halegreen Associates Ltd and Demero Ltd	Oxford	6. 8.06T
G-BTGP	Cessna 150M Commuter	15078921	N704WA	28. 2.91	Billins Air Services Ltd *(Op City Air)*	Cranfield	28. 5.06T
G-BTGR	Cessna 152 II	15284447	N6581L	28. 2.91	A J.Gomes	Shoreham	25. 7.00T
					(Crashed Mile Oak, Portslade, Sussex 7.10.99:)		
G-BTGS (2)	Stolp SA.300 Starduster Too	PFA 35-10076	G-AYMA	30. 9.87	G.N.Elliott	Shoreham	3. 7.05P
	(Build initiated T G Solomon: completed P G Leggo) (Lycoming O-320 - *kit no.EAA/50553)*				tr G.N.Elliott and Partners		
G-BTGT	CFM Streak Shadow	PFA 206-11964	(G-MWPY)	1. 3.91	G.D.Bailey	(Winchester)	4.10.05P
	(Built M Allinson) (Rotax 582 - *kit no.K.164-SA)*						
G-BTGU	Piper PA-34-220T Seneca III	34-8233106	N999PW	1. 3.91	Carill Aviation Ltd	Lee-on-Solent	20. 7.06T
			N8160V				
G-BTGV	Piper PA-34-200T Seneca II	34-7970077	N3004H	26. 3.91	Westflight Aviation Ltd	Gloucestershire	3. 8.06T
G-BTGW	Cessna 152 II	15279812	N757KY	5. 3.91	Stapleford Flying Club Ltd	Stapleford	13. 8.06T
G-BTGX	Cessna 152 II	15284950	N5462P	5. 3.91	Stapleford Flying Club Ltd	Stapleford	24. 8.06T
G-BTGY	Piper PA-28-161 Warrior II	28-8216199	N209FT	5. 3.91	Stapleford Flying Club Ltd	Stapleford	23. 6.06T
			N9574N				
G-BTGZ	Piper PA-28-181 Cherokee Archer II	28-7890160	N47956	8. 4.91	Allzones Ltd	Biggin Hill	9.10.06T
G-BTHA	Cessna 182P	18263420	N2932P	22. 3.91	T.P.Hall tr Hotel Alpha Flying Group	Liverpool	7.10.06
G-BTHD	Yakovlev Yak-3U	170101	(France)	7. 3.91	Patina Ltd	Duxford	AC
	(Conversion of LET Yak C.11)		EAF.533		*(Exported to Planes of Fame Air Museum, Chino, California, US 1.05)*		
G-BTHE	Cessna 150L	15075340	N11348	7. 3.91	J.H.Loose and F.P.White	Mount Airey Farm, South Cave	18. 8.06T
					tr Humberside Police Flying Club		
G-BTHF	Cameron V-90 HAB	2543		7. 3.91	N.J. and S.J.Langley	(Bristol)	1. 6.05A
G-BTHH	CEA Jodel DR.100A Ambassadeur	5	F-BJCH	28. 2.91	H.R.Leefe	Bourg-en-Bresse, France	25.11.06
G-BTHI	Robinson R22 Beta	1732		26. 3.91	M.D.Thorpe t/a Yorkshire Helicopters	Coney Park, Leeds	15. 4.07T
G-BTHJ	Evans VP-2	PFA 63-10901		14. 3.91	C.J.Moseley	Bournemouth	
	(Built C.J.Moseley)				*(Under construction 8.92)*		
G-BTHK	Thunder Ax7-77 HAB	1906		11. 3.91	M.S.Trend	(Maidstone)	6. 7.05A
G-BTHM	Thunder Ax8-105 HAB	1925		11. 3.91	J.K.Woods	(Chatham)	1. 8.05A
G-BTHN	Meridian Renegade 912	PFA 188-12005		12. 3.91	F.A Purvis	Eshott	22. 8.03P
	(Built Meridian Ultralights Ltd - c/n 384)				*"Spirit of England II" (Noted 1.05)*		
G-BTHP	Thorp T.211	101		13. 6.91	M.J.Newton	Barton	27.11.04P
G-BTHU	Avid Flyer	PFA 189-11427		14. 3.91	R.C.Bowley	(Earls Croome, Worcester)	
	(Built M Morris)		*(Damaged Field Head Farm, Denholme, Bradford 7.6.92: on rebuild 5.95: current status unknown)*				
G-BTHV	MBB Bö.105DBS-4	S.855	D-HMBV	20. 3.91	Bond Air Services Ltd	Gloucestershire	12. 5.06T
			G-BTHV, D-HFHM		*(Op County Air Ambulance)*		
G-BTHW	Beech F33C Bonanza	CJ-130	PH-BNA	18. 3.91	Robin Lance Aviation Associates Ltd	Rochester	25. 9.06
			N23787				
G-BTHX	Colt 105A HAB	1939		18. 3.91	N.C.Lindsay	(Storrington, Pulborough)	13. 6.05A
					tr Elmer Balloon Group		
G-BTHY	Bell 206B-3 JetRanger III	2290	N6606M	20. 3.91	Sterling Helicopters Ltd	Norwich	19. 5.06T
			VH-BIQ, ZK-HBQ, DQ-FEN, ZK-HLU				
G-BTHZ	Cameron V-56 HAB	486	OO-BBC	20. 3.91	C.N.Marshall	(Nairobi, Kenya)	
					(Noted as "OO-BBC" 9.95)		
G-BTID	Piper PA-28-161 Warrior II	28-8116036	N82647	25. 6.91	Plymouth School of Flying Ltd	Plymouth	24. 5.06T
G-BTIE	SOCATA TB-10 Tobago	187		30. 3.81	Aviation Spirit Ltd	Clee Hill, Ludlow	2. 6.06T
G-BTIF	Denney Kitfox Model 3	PFA 172-11862		27. 2.91	D.A Murchie	(Blackwaterfoot, Arran)	10. 8.01P
	(Built C R Thompson and J Scott -c/n 684) (Converted from Model 2)						
G-BTIG*	Montgomerie-Bensen B.8MR Merlin	PFA G/01-1093		21. 3.91	K.Jarvis	Carlisle	10. 4.04P
	(Built N Beale and D Beevers) (Rotax 532)				*(Cancelled 13.4.04 by CAA)*		
G-BTII	Gulfstream AA-5B Tiger	AA5B-1256	N4560S	5. 6.91	S.A Niechial and P.W.Gillott tr BTII Group	Biggin Hill	13. 5.05T
G-BTIJ	Luscombe 8E Silvaire	5194	N2467K	3. 4.91	S.J.Hornsby	Compton Abbas	21. 9.05P
	(Continental C85)		NC2467K				
G-BTIK	Cessna 152 II	15282993	N46068	26. 3.91	I.R.Chaplin	Andrewsfield	12. 6.05T
G-BTIL	Piper PA-38-112 Tomahawk	38-80A0004	N24730	26. 3.91	B.J.Pearson	Eaglescott	AC
					(Noted as "N24730" dismantled 11.03)		
G-BTIM	Piper PA-28-161 Cadet	2841159	N9185D	24. 8.89	S.J.Skilton	Bournemouth	25.10.04T
			(SE-KIO)		t/a Aviation Rentals		
G-BTIN*	Cessna 150C	15059905	N7805Z	26. 3.91	Perth Technical College	Perth	17. 4.00
			(Overturned by gales Edinburgh 23.12.99 and cancelled 10.5.01 as wfu) (As instructional airframe 10.03)				
G-BTIO	SNCAN Stampe SV-4C	303	N73NS	28. 3.91	M.D.and C.F.Garratt	(Salisbury)	8.10.05
	(Reanult 4P)		F-BCLC				
G-BTIR	Denney Kitfox Model 2	PFA 172-11952		26. 3.91	R.B.Wilson	(Kendal)	12.10.05P
	(Built J N C Shields and D J Millar) (Hewland AE75)						
G-BTIU	SOCATA MS.892A Rallye Commodore 150	10914	F-BPQS	7. 5.91	Cole Aviation Ltd	Spilsteads Farm, Sedlescombe	16. 6.05
G-BTIV	Piper PA-28-161 Warrior II	28-8116044	N82697	10. 5.91	B.R.Pearson tr Warrior Group	Eaglescott	24. 7.06T
G-BTIZ	Cameron A-105 HAB	2546		11. 3.91	Wendy A.Board	(Alicante, Spain)	8. 7.02A
					t/a Glen Board Promotions *(Amended CofR 1.05)*		

Reg	Type	C/n	Previous identity	Date	Owner	Base	CofA
G-BTJA	Luscombe 8E Silvaire	5037	N2310K	4. 4.91	M.W.Rudkin	Blackpool	1. 9.05P
	(Continental C85)		NC2310K				
G-BTJB	Luscombe 8E Silvaire	6194	N1567B	4. 4.91	M.Loxton	Parsonage Farm, Eastchurch	16. 7.04P
	(Continental C85)		NC1567B				
G-BTJC	Luscombe 8F Silvaire	6589	N2162B	4. 4.91	Alison M.Noble	Chilbolton	18.10.99P
	(Lycoming O-290)				*(Damged Glebe Farm, Stockton, Warminster 31.7.99: noted 7.04)*		
G-BTJD	Thunder Ax8-90 Series 2 HAB	1865		28. 3.91	R.E.Vinten *"Beez Neez"*	(Wellingborough)	11. 5.05A
G-BTJF*	Thunder Ax10-180 Series 2 HAB				See SECTION 3 Part 1	(Aldershot)	
G-BTJH	Cameron O-77 HAB	2559		3. 4.91	H.Stringer *"Oriel"*	(Scarborough)	31. 5.05T
G-BTJK	Piper PA-38-112 Tomahawk	38-79A0838	N2427N	3. 4.91	Western Air (Thruxton) Ltd	Thruxton	22.10.06T
G-BTJL	Piper PA-38-112 Tomahawk	38-79A0863	N2477N	3. 4.91	J.S.Develin and Z.Islam	Shoreham	22. 8.04T
G-BTJN	Montgomerie-Bensen B.8MR	PFA G/01-1194		3. 4.91	A Hamilton	Strathaven	9.12.00P
	(Built A Hamilton) (Rotax 532)						
G-BTJO	Thunder Ax9-140 HAB	1948		3. 4.91	G.P.Lane	(Gerrards Cross)	28. 4.92A
G-BTJS	Montgomerie-Bensen B.8MR	PFA G/01-1083		8. 4.91	T.A.Holmes	(Leeds)	18.10.00P
	(Built J M P Annand) (Rotax 532)				*(New owner 11.04)*		
G-BTJU	Cameron V-90 HAB	2554		8. 4.91	C.W.Jones (Floorings) Ltd	Bristol	6. 8.05A
					(C W Jones Carpets titles)		
G-BTJX	Rans S-10 Sakota	PFA 194-12014		9. 4.91	T.Scarborough	(Boston)	15. 9.04P
	(Built M Goacher) (Rotax 582)				*(New owner 7.03)*		
G-BTKA	Piper J-5A Cub Cruiser	5-954	N38403	11. 4.91	Janet M.Lister	Valley Farm, Winwick	9. 5.05P
			NC38403				
G-BTKB	Meridian Renegade 912	PFA 188-11876		11. 4.91	P.and J.Calvert	Rufforth	7. 9.05P
	(Built G S Blundell - kit no.376)						
G-BTKD	Denney Kitfox Model 4	PFA 172-11941		15. 4.91	M.Sansom and R.A.Hills	(Seaton/Sturminster Newton)	16. 8.05P
	(Built J F White - kit no.853 - indicates Model 3 and conflicts with N653CP)						
G-BTKG	Avid Flyer	PFA 189-12037		16. 4.91	I.Holt	Drayton Manor, Drayton St.Leonard	9. 9.05P
	(Built P R Snowden)						
G-BTKL	MBB Bö.105DB-4	S.422	D-HDMU	2. 5.91	Veritair Ltd	Wolverhampton	2. 3.06T
			Swedish Army, D-HDMU		*(Op Central Counties Police Air Operations Unit)*		
G-BTKN	Cameron O-120 HAB	2579	OO-BQQ	24. 4.91	R.H.Etherington	(Siena, Italy)	16. 4.03A
G-BTKP	CFM Streak Shadow	PFA 206-12036		24. 4.91	G.D.Martin	(Cambridge)	20. 8.02P
	(Built K S Woodward - kit no.K.174) (Rotax 582)						
G-BTKT	Piper PA-28-161 Warrior II	28-8216218	N429FT	9. 5.91	Biggin Hill Flying Club Ltd	Shoreham	7. 4.07T
			N9606N				
G-BTKV	Piper PA-22-160 Tri-Pacer	22-7157	N3216Z	25. 4.91	R.A Moore	Rathfriland	22. 8.07
G-BTKW	Cameron O-105 HAB	2566		25. 4.91	P.Spellward	(Bristol)	9. 3.01A
					tr Bristol University Hot Air Ballooning Society		
G-BTKX	Piper PA-28-181 Cherokee Archer II	28-7890146	N47866	14. 5.91	R.M.Pannell	Eaglescott	30. 4.06T
G-BTKZ	Cameron V-77 HAB	2573		26. 4.91	S.P.Richards *"Lancaster Jaguar"*	(Tonbridge)	7. 6.97T
G-BTLB	Wassmer WA.52 Europa	42	F-BTLB	17. 4.89	A.S.Cowan tr Popham Flying Group G-BTLB	Popham	21.10.07E
G-BTLG	Piper PA-28R-200 Cherokee Arrow	28R-35811	N5045S	29. 4.91	W.B.Bateson	Blackpool	15. 3.04
G-BTLL*	Pilatus P.3-03	323-5	A-806	18. 4.91	Not known	Headcorn	23. 6.94P
					(Cancelled 27.10.95 by CAA) (Stored 4.03 as "A-806" in Swiss AF c/s)		
G-BTLM	Piper PA-22-160	22-6162	N9025D	16. 5.91	A C.and M.D.N.Fisher	Fenland	27.11.06
	(Tail-wheel conversion)						
G-BTLP	Grumman AA-1C Lynx	AA1C-0109	N9732U	13. 5.91	Partlease Ltd	Stapleford	2. 2.07
					(Op Stapleford Flying Club) "28"		
G-BTMA	Cessna 172N Skyhawk II	17273711	N5136J	2. 5.91	East of England Flying Group Ltd	North Weald	8.10.06T
G-BTMH	Colt 90A HAB	1963		14. 5.91	European Balloon Corporation	(Espinette, Belgium)	19. 8.01A
G-BTMK	Cessna R172K Hawk XP II	R1722787	N736TZ	10. 6.91	A.C.Barker	Tatenhill	15. 2.07T
G-BTML*	Cameron Rupert Bear 90SS HAB				See SECTION 3 Part 1	(Aldershot)	
G-BTMN	Thunder Ax9-120 Series 2 HAB	2003		17. 5.91	M.E.White *(Inflated 4.02)*	(Dublin)	8. 3 99T
G-BTMO	Colt 69A HAB	2004		20. 5.91	Cameron Balloons Ltd	(Bristol)	
					t/a Thunder and Colt		
G-BTMP	Everett Campbell Cricket	PFA G/03-1226		20. 5.91	P.W.McLaughlin	Henstridge	8. 1.05P
	(Built D G Hill - c/n 024) (Rotax 532)						
G-BTMR	Cessna 172M Skyhawk II	17264985	N64047	20. 5.91	Linley Aviation Ltd	Linley Hill, Leven	6. 7.06T
G-BTMS	Avid Speed Wing	PFA 189-12023	(CS-)	24. 4.91	F.Sayyah	Redhill	22. 3.05P
	(Built D L Docking and M J Kay - c/n 908)		G-BTMS				
G-BTMT	Denney Kitfox Model 1	66		10. 5.91	L.G.Horne	(Ashford, Kent)	5.11.04P
	(Built J Lindzalone) (Rotax 532)						
G-BTMV	Everett Gyroplane Series 2	025		21. 5.91	L.Armes	Basildon	
G-BTMW	Zenair CH-701 STOL	PFA 187-11808		21. 5.91	L.Lewis	Yearby	9. 4.96P
	(Built L Lewis) (Rotax 582)				*(Stored 1.02)*		
G-BTMX	Denney Kitfox Model 3	PFA 172-12079		13. 5.91	V.T.Betts	Otherton, Cannock	29.10.05P
	(Built M E F Reynolds and E A Gilbert - c/n 916)						
G-BTMY	Cameron Train-80 SS HAB	2561	(SE-...)	22. 5.91	Balloon Sports HB	(Partille, Sweden)	9. 2.97A
			G-BTMY				
G-BTNA	Robinson R22 Beta	1800	N40820	23. 5.91	The Air Group Ltd	Booker	13.10.05T
G-BTNC	Aérospatiale SA365N2 Dauphin 2	6409		21. 6.91	CHC Scotia Ltd	Humberside	9.10.07T
G-BTND	Piper PA-38-112 Tomahawk	38-78A0155	N9671T	23. 5.91	R.B.Turner	Gloucestershire	14. 3.05T
G-BTNE	Piper PA-28-161 Warrior II	28-8116212	N8379H	22. 7.91	D.Rowe	Wellesbourne Mountford	8. 7.07T
G-BTNH	Piper PA-28-161 Warrior II	28-8216202	G-DENH	28. 5.91	A S.Bamrah	Biggin Hill	14.10.06T
			G-BTNH, N253FT, N9577N		t/a Falcon Flying Services		
G-BTNI	British Aerospace ATP	2038	EC-GSE	29. 5.91	Trident Aviation Leasing Services (Jersey) Ltd	Woodford	
			EC-GKJ, G-OEDI, G-BTNI, (N238JX), G-BTNI, TC-THU, G-BTNI, (G-SLAM)				
G-BTNN	Colt 21A Cloudhopper HAB	2018		3. 6.91	J.Mason	(Loughton)	11. 4.04A
G-BTNO	Aeronca 7AC Champion	7AC-3132	N84441	31. 5.91	B.J.and B.G.Robe	(Riding Mill)	27. 6.05P
			NC84441				
G-BTNR	Denney Kitfox Model 3	PFA 172-12035		31. 5.91	D.E.Steade	Croft Farm, Defford	14. 9.05P
	(Built J W G Ellis - c/n 921)						
G-BTNT	Piper PA-28-151 Cherokee Warrior	28-7615401	N6929J	31. 5.91	Britannia Airways Ltd	Luton	24.11.06T
					(Op Britannia Airways Flying Club)		

Registration	Type	C/n	Previous identity	Date	Owner	Location	CofA
G-BTNU	British Aerospace BAe 146 Series 300	E3155	EI-CLJ	19. 7.90	Trident Jet (Jersey) Ltd	Filton	
			G-BTHU, (G-BSLS), G-6-155		*(Stored 6.04)*		
G-BTNV	Piper PA-28-161 Cherokee Warrior II	28-7816590	N31878	20. 6.91	G.M.Bauer and A M.Davies	Wickenby	18. 8.06T
G-BTNW	Rans S-6-ESA Coyote II	PFA 204-12077		3. 6.91	R.H.Hughes	(Deeside)	28. 4.03P
	(Built A Barbone) (Rotax 582 - *kit no.0391.174)*						
G-BTOC	Robinson R22 Beta	1801	N23004	10. 6.91	N.Parkhouse	Chelwood Gate, West Sussex	28. 8.06T
G-BTOG	de Havilland DH.82A Tiger Moth	86500	F-BGCJ	5. 9.91	P.T.Szluha	Audley End	
	(Built Morris Motors Ltd)		Fr.AF, NM192		*(As "F-BGCJ") (Stored 7.03)*		
G-BTOL	Denney Kitfox Model 3	PFA 172-12052		26. 6.91	P.J.Gibbs	Truro	28. 7.05P
	(Built C R Phillips - c/n 919)						
G-BTON	Piper PA-28-140 Cherokee Cruiser	28-7425343	N43193	15. 7.91	S.Quinn	Poplar Hall Farm, Elmsett	20. 7.07T
G-BTOO	Pitts S-1C	5215-24A	N37H	12. 6.91	G.H.Matthews	Sandown	
	(Built E Lawrence)				*(On overhaul 5.92 - current status unknown*		
G-BTOP	Cameron V-77 HAB	2484		14. 6.91	J.J.Winter *"Big Top"*	(Cardiff)	
G-BTOS	Cessna 140	8353	N89325	7. 6.91	J.L.Kaiser	Nancy-Essey, France	1. 7.99
	(Continental C85)		NC89325				
G-BTOT	Piper PA-15 Vagabond	15-60	N4176H	22. 5.91	M.S.Rogerson	Morgansfield, Fishburn	11. 5.05P
	(Lycoming O-145)		NC4176H		tr Vagabond Flying Group		
G-BTOU	Cameron O-120 HAB	2606		2. 7.91	J.J.Daly	(Halfway House, County Waterford)	3.11.04T
G-BTOW	SOCATA Rallye 180T Galerien	3360	F-BNGZ	9.11.82	Cambridge Gliding Club Ltd	Gransden Lodge	20. 4.07
G-BTOZ	Thunder Ax9-120 Series 2 HAB	2008		28. 6.91	H.G.Davies	(Cheltenham)	10.10.04T
G-BTPA	British Aerospace ATP	2007	EC-HGC	19. 8.88	Capital Bank Leasing 12 Ltd	Not known	18.11.98T
			G-BTPA, EC-GYE, G-BTPA, (N377AE)				
G-BTPB	Cameron N-105 HAB	1536		6. 7.87	D.J.Farrar	(Collingham, Wetherby)	14. 8.05A
G-BTPC	British Aerospace ATP	2010	EC-HGB	1. 9.88	Capital Bank Leasing 1 Ltd	Not known	29.12.98T
			G-BTPD, EC-GYF, G-BTPC, G-11-10, (N380AE)				
G-BTPD	British Aerospace ATP	2011	EC-HGD	1. 9.88	Seaforth Marime (JARL) Ltd and Flexifly Ltd t/a NWS2		
			G-BTPC, EC-GYR, G-BTPD, (N381AE)			Not known	6. 2.99T
G-BTPE	British Aerospace ATP	2012	EC-HGE	1. 9.88	Capital Bank Leasing 3 Ltd	Not known	12. 3.99T
			G-BTPE, EC-GZH, G-BTPE, (N382AE)				
G-BTPF	British Aerospace ATP	2013	EC-HCY	2. 9.88	Capital Bank Leasing 5 Ltd	Not known	17. 4.99T
			G-BTPF, G-11-013, G-BTPF, (N383AE)				
G-BTPG	British Aerospace ATP	2014	EC-HEH	2. 9.88	Capital Bank Leasing 5 Ltd	Not known	22. 5.99T
			G-BTPG, (N384AE)				
G-BTPH	British Aerospace ATP	2015	EC-HFM	2. 9.88	Capital Bank Leasing 6 Ltd	Zaragoza, Spain	11. 6.99T
			G-BTPH, (N385AE)				
G-BTPJ	British Aerospace ATP	2016	EC-HFR	2. 9.88	Capital Bank Leasing 7 Ltd	Not known	9. 7.99T
			G-BTPJ, (N386AE)				
G-BTPL	British Aerospace ATP	2042	EC-HES	3.10.91	Trident Aviation Leasing Services (Jersey) Ltd	Exeter	21.11.95T
			G-BTPL, EC-GLH, G-BTPL, G-11-042		*(Stored 10.04)*		
G-BTPN	British Aerospace ATP	2044	EC-GSI	19.11.91	Trident Aviation Leasing Services (Jersey) Ltd	Kemble	19.12.98T
			EC-GNJ, G-BTPN, G-11-044		*(Stored 11.04)*		
G-BTPT	Cameron N-77 HAB	2575		10. 6.91	H.J.Andrews *(New owner 1.05)*	Selborne, Alton	15.12.01A
G-BTPV	Colt 90A HAB	1956		14. 6.91	R.S.Kent	(Lancing)	27. 7.05A
					tr Balloon Preservation Flying Group *(Mondial titles)*		
G-BTPX	Thunder Ax8-90 HAB	1873		18. 6.91	E.Cordall	(Chichester)	3. 9.05
G-BTPZ	Isaacs Fury II	PFA 11-11927		1. 7.91	M.A Farrelly	Ormskirk	
	(Built M.A Farrelly)				*(As "85" in Portuguese AF c/s)*		
G-BTRB	Colt Mickey Mouse SS HAB	1959		4. 7.91	Benedikt Haggeney GmbH	(Ennigerloh, Germany)	18. 3.05A
					"Calibre"		
G-BTRC	Avid Speed Wing	PFA 189-12076		2. 7.91	Grangecote Ltd	Goodwood	2.11.04P
	(Built A A Craig - c/n 913) (BMW R100)						
G-BTRF	Aero Designs Pulsar	PFA 202-12051		4. 7.91	C.Smith	Spilsteads Farm, Sedlescombe	19.10.05P
	(Built C Smith)						
G-BTRG	Aeronca 65C Super Chief	C4149	N22466	4. 7.91	A.Welburn	South Cave, Hull	8. 9.05P
	(Continental A65)		NC22466				
G-BTRI	Aeronca 11CC Super Chief	11CC-246	N4540E	4. 7.91	P.A Wensak	Bounds Farm, Ardleigh	18. 3.05P
	(Continental C85)		NC4540E				
G-BTRK	Piper PA-28-161 Warrior II	28-8216206	N297FT	8. 7.91	Stapleford Flying Club Ltd	Stapleford	26.10.06T
			N9594N				
G-BTRL	Cameron N-105 HAB	2622		5. 7.91	J.Lippett *"Harrods"*	(South Petherton, Somerset)	24. 8.02A
G-BTRN	Thunder AX9-120 Series 2 HAB	1983		11. 7.91	P.B.D.Bird	(Bristol)	18. 1.04T
G-BTRO	Thunder Ax8-90 HAB	1872		11. 7.91	Capital Balloon Club Ltd	(London NW1)	17. 4.05A
G-BTRP	MD Helicopters Hughes 369E	0475E	N1607D	11. 7.91	P.C.Shann and P.C.Shann Management and Research Ltd		
						Fulford, York	25. 3.05
G-BTRR	Thunder Ax7-77 HAB	1905		12. 7.91	M.F.Comerford	(Stoke-on-Trent)	6. 3.05A
G-BTRS	Piper PA-28-161 Warrior II	28-8116004	N8248V	12. 7.91	K.D.Taylor and T.Bailey tr Airwise Flying Group	Barton	4. 2.05
G-BTRT	Piper PA-28R-200 Cherokee Arrow II	28R-7535270	N1189X	24. 7.91	C.D.Barden and D.G.Smith	Barton	8. 3.07
					tr Romeo Tango Group		
G-BTRU	Robin DR400/180 Régent	2089		12. 7.91	R.H.Mackay	Easterton	1. 2.07
G-BTRW	Slingsby T.61F Venture T.2	1968	ZA632	5. 7.91	G.B.Monslow tr The Falke Syndicate	Long Marston	3.12.06
G-BTRX*	Cameron V-77 HAB	1143	VH-HIH	12. 7.91	R P Jones and N P Hemsley	(Horsham/Crawley)	16. 4.97A
					(Cancelled 2.5.97 as WFU) (Inflated 6.02)		
G-BTRY	Piper PA-28-161 Warrior II	28-8116190	N8363L	18. 7.91	Oxford Aviation Services Ltd	Oxford	6.12.04T
G-BTRZ	Jodel D.18	PFA 169-11271		16. 7.91	A.P.Aspinall	(Bassingbourn, Royston)	26. 9.04P
	(Built R Collin - c/n 148) (Volkswagen 1834)						
G-BTSB	Corben Baby Ace D	JC-1	N3599	16. 7.91	J.Horovitz and J.W.Macleod	Bourne	27. 5.05P
	(Built J Cole) (Continental A65)						
G-BTSC*	Evans VP-2	PFA 63-10342		20.10.78	G.B.O'Neill	(Upwood, Cambridge)	13. 2.96P
	(Built D J Keam/Truro SChool) (Arrow GT500)				*(Stored 6.00: cancelled 20.1.03 by CAA)*		
G-BTSJ	Piper PA-28-161 Cherokee Warrior II	28-7816473	N9417C	23. 7.91	Plymouth School of Flying Ltd	Plymouth	29. 1.07T
G-BTSL	Cameron Glass 70SS HAB	1627		27. 1.88	M.R.Humphrey and J.R.Clifton	(Brackley)	25. 4.90A
	(Tennent's Lager Glass shape)				*"Tennent's Glass"*		
G-BTSM	Cessna 180A	32678	P2-DEQ	9. 7.91	C.Couston	Manor Farm, Garford	17. 4.05
			VH-DEQ, VH-DEC, N7781A		tr Sierra Mike Group		

G-BTSN	Cessna 150G	15065106	N3806J	30. 8.91	C.P.Whitwell	Fenland	7. 5.05
G-BTSP	Piper J-3C-65 Cub	7647	N41013	30. 8.91	J.A.Walshe and A.Corcoran	Strandhill, County Sligo	25. 4.05P
			NC41013		*(Noted 7.04)*		
G-BTSR	Aeronca 11AC Chief	11AC-785	N9152E	30. 8.91	R.D.and E.G.N.Morris	Perth	4. 7.05P
	(Continental A65)		NC9152E				
G-BTSV	Denney Kitfox Model 3	PFA 172-11920		24. 7.91	M.G.Dovey	Popham	30. 6.05P
	(Built D J Sharland)						
G-BTSW	Colt AS-105GD HA Airship	1999		24. 7.91	Gefa-Flug GmbH	Aachen, Germany	13. 4.05A
	(Built Thunder and Colt Ltd)				*(Adler Modemarkt titles)*		
G-BTSX	Thunder Ax7-77 HAB	2027		24. 7.91	C.Moris-Gallimore	(Sao Bras de Alportel, Portugal)	18. 9.94
G-BTSY*	English Electric Lightning F.6				See SECTION 3 Part 1	Binbrook	
G-BTSZ	Cessna 177A Cardinal	17701198	N30332	30. 7.91	W.J.Peachment	Henstridge	3. 7.06T
G-BTTB	Cameron V-90 HAB	2624		22. 7.91	R.M.Tonkins	(Liverpool)	27. 4.05A
					tr Royal Engineers Balloon Club *"Sapper IV"*		
G-BTTD	Montgomerie-Bensen B.8MR	PFA G/01-1204		31. 7.91	A.J.P.Herculson	(Fakenham)	28. 4.05P
	(Built K J Parker) (Rotax 582)						
G-BTTE	Cessna 150L	15075558	N11602	31. 7.91	C.A.Wilson and W.B.Murray	Hill Farm, Nayland	19. 7.04T
G-BTTL	Cameron V-90 HAB	2649		12. 8.91	A.J.Baird *"Hyde Farm Dairy"*	(Cheltenham)	10. 8.02A
G-BTTO	British Aerospace ATP	2033	EC-HNA	16. 8.91	Trident Aviation Leasing Services (Jersey) Ltd	Woodford	
			EC-GJU, G-BTTO, G-OEDE, G-BTTO, TC-THV, G-BTTO, S2-ACZ, G-11-033 *(For Emerald Airways 2004)*				
G-BTTP	British Aerospace BAe 146 Series 300	E3203	G-6-203	20. 8.91	Trident Aviation Leasing Services (Jersey) Ltd	Filton	11.11.03T
					(Stored 6.04)		
G-BTTR	Pitts S-2A	2208	N38MP	16. 8.91	Yellowbird Adventures Ltd	(College Town, Sandhurst)	12. 4.07T
	(Built Aerotek Inc)						
G-BTTS	Colt 77A HAB	1861		16. 8.91	J.A.Lomas tr Rutland Balloon Club	(Melton Mowbray)	28. 7.05A
G-BTTW	Thunder Ax7-77 HAB	2016		27. 8.91	J.Kenny	(Athlone, County Roscommon)	21. 9.05A
G-BTTY	Denney Kitfox Model 2	PFA 172-11823		29. 7.91	K.J.Fleming	(Liverpool)	
	(Built K.J.Fleming)						
G-BTTZ	Slingsby T.61F Venture T.2	1961	ZA625	30. 7.91	M.W.Olliver	Halesland	3. 9.03
G-BTUA	Slingsby T.61F Venture T.2	1985	ZA666	20. 8.91	C.Edmunds tr Shenington Gliding Club	Shenington	24. 6.07
G-BTUB	LET Yakovlev C.11	172623	(France)	29. 8.91	M.G. and J.R.Jefferies	Little Gransden	27. 8.05P
	(Identity of 039 quoted)		Egyptian AF 543		*(Soviet AF c/s without serial)*		
G-BTUD*	CFM Image	PFA/222-12012	G-MWPV	21. 8.91	Not known	Wick	21. 1.95P
	(Built D G Cook - c/n IM-01)				*(Cancelled 5.2.99 as PWFU) (Noted 4.04)*		
G-BTUC*	Embraer EMB-312 Tucano				See SECTION 3 Part 1	Langford Lodge, Belfast	
G-BTUE	British Aerospace ATP	2039	EC-GSF	5. 9.91	Trident Aviation Leasing Services (Jersey) Ltd	Southend	
			EC-GKI, G-OEDH, (G-OGVA), G-OEDH, G-BTUE, TC-THT, G-11-039, G-BTUE, G-11-039 *(Noted 1.05)*				
G-BTUG	SOCATA Rallye 180T	3208		10. 7.78	Herefordshire Gliding Club Ltd	Shobdon	28. 1.06
G-BTUH	Cameron N-65 HAB	1452		28. 8.91	B.J.Godding *(Zanussi titles)*	(Didcot)	
G-BTUJ	Thunder Ax9-120 HAB	2022		30. 8.91	ECM Construction Ltd	(Great Missenden)	20. 5.02T
G-BTUK	Pitts S-2A	2260	N5300J	2. 9.91	S.H.Elkington	Wickenby	24.10.06T
	(Built Aerotek Inc)						
G-BTUL	Pitts S-2A	2200	N900RS	2. 9.91	J.M.Adams	RAF Syerston	26. 2.07
	(Built Aerotek Inc)						
G-BTUM	Piper J-3C-65 Cub	19516	N6335H	6. 9.91	I.M.Mackay	White Waltham	14. 7.04P
	(Continental C85) *(Frame No.19586)*		NC6335H		tr G-BTUM Syndicate *"Jingle-Belle"*		
G-BTUR	Piper PA-18 Super Cub 95 (L-18C-PI)	18-3205	OO-LVM	11. 9.91	A.P.Meredith	Lasham	25. 3.05T
	(Continental C90) *(Frame No.18-3218)*		Belgian AF OL-L08, L-131, 53-4805				
G-BTUS	Whittaker MW7	PFA 171-11999		5. 9.91	C.T.Bailey	Croft Farm, Defford	27. 6.05P
	(Built J F Bakewell) (Rotax 503)				*(Noted 6.04)*		
G-BTUU	Cameron O-120 HAB	2669		16. 9.91	P Dubois-Dauphin	Montfrim, France	8. 4.03T
G-BTUV	Aeronca 65TAC Defender	C.1661TA	N36816	12. 9.91	M.B.Hamlett and R.E.Coates	Le Plessis-Belleville, France	
			NC36816		*(Noted as "C1661-TA" in USAF c/s 6.03)*		
G-BTUW	Piper PA-28-151 Cherokee Warrior	28-7415066	N54458	12. 9.91	T.S.Kemp	Enstone	22. 9.07T
G-BTUZ	American General AG-5B Tiger	10075	N11939	3.10.91	Grocontinental Ltd	Sleap	26. 2.06
G-BTVA	Thunder Ax7-77 HAB	2009		16. 9.91	A.H.Symonds *"Bertie Bassett"*	(Chelmsford)	14. 6.05A
G-BTVB	Everett Gyroplane Series 3	026		24. 9.91	J.Pumford	Henstridge	18. 4.04P
	(Rotax 532)						
G-BTVC	Denney Kitfox Model 2	PFA 172-11784		23. 9.91	P.Mitchell	Long Marston	18. 3.04P
	(Built R Swinden)				*"Zebedee"*		
G-BTVE	Hawker Demon I	---	2292M	18. 9.91	Demon Displays Ltd	Rotary Farm, Hatch	AC
	(Kestrel V)		K8203		*(As "K8203" in 64 Sqdn c/s) (On rebuild 10.99)*		
	(Composite of ex IAC Hector - front and K8203 - rear)						
G-BTVR	Piper PA-28-140 Cherokee Cruiser	28-7625012	N4328X	16. 9.91	Full Sutton Flying Centre Ltd	Full Sutton	18. 7.07T
G-BTVU	Robinson R22 Beta	1937		26. 9.91	B.Enzo	Bologna, Italy	26. 3.06
G-BTVV	Reims/Cessna F337G Skymaster	F33700058	PH-RPD	25. 9.91	C.Keane	Weston, Dublin	12. 1.03T
	(Wichita c/n 33701476)		N1876M		*(Noted 8.04)*		
G-BTVW	Cessna 152 II	15279631	N757CK	23. 9.91	Halegreen Associates Ltd	Hinton-in-the-Hedges	24. 1.05T
G-BTVX	Cessna 152 II	15283375	N48786	23. 9.91	Eastern Air Centre Ltd	Gamston	19. 9.07T
G-BTWB	Denney Kitfox Model 3	PFA 172-12278	(G-BTTM)	21. 8.91	J.E.Tootell	East Fotrtune	9. 6.05P
	(Built J E Toothill - kit no.920)						
G-BTWC	Slingsby T.61F Venture T.2	1975	ZA656	23. 9.91	T.M.Holloway tr RAFGSA	Upavon	21. 3.05
G-BTWD	Slingsby T.61F Venture T.2	1976	ZA657	23. 9.91	York Gliding Centre Ltd	Rufforth	25. 3.07
					t/a York Gliding Centre		
G-BTWE	Slingsby T.61F Venture T.2	1980	ZA661	23. 9.91	T.M.Holloway	RAF Syerston	19. 3.07
					tr RAFGSA *(Op Four Counties Gliding Club)*		
G-BTWF	de Havilland DHC-1 Chipmunk 22	C1/0564	WK549	30. 9.91	J.A and V.G.Simms *(As "WK549")*	Breighton	20. 5.07
G-BTWI	EAA Acrosport	230	N10JW	2.10.91	S.Alexander and W.M.Coffee	Bidford	3. 6.05P
	(Built J N Wharton) (Lycoming O-290)						
G-BTWJ	Cameron V-77 HAB	2670		3.10.91	S.J. and J.A Bellaby *"Windy Jack"*	(Nottingham)	23. 4.05A
G-BTWL	Wag-Aero CUBy Acro Sport Trainer	PFA 108-10893		3.10.91	I.M.Ashpole	Preston Court, Ledbury	14.10.05P
	(Built Penair) (Lycoming O-235)						
G-BTWM	Cameron V-77 HAB	2163		4.10.91	R.C.Franklin *"Aerolus"*	(Chesham)	13. 9.04A

Registration	Type	C/n	Previous identity	Date	Owner/operator	Base	CofA
G-BTWU*	Piper PA-22-135 Tri-Pacer	22-2135	N3320B	10.10.91	Prestige Air (Engineers) Ltd	Haverfordwest	
					(Noted as "N3320B" 4.01: cancelled 15.5.03 by CAA)		
G-BTWV	Cameron O-90 HAB	2675		10.10.91	S.F.Hancke	(Sunbury-on-Thames)	30. 6.01A
G-BTWX	SOCATA TB-9 Tampico Club	1401		14.10.91	Cinque Ports Aviation Ltd	Lydd	25. 1.07T
G-BTWY	Aero Designs Pulsar	PFA 202-12040		15.10.91	M.Stevenson	Pepperbox, Salisbury	6. 8.03P
	(Built J J Pridal and A K Pirie) (Tail-wheel u/c)						
G-BTWZ	Rans S-10 Sakota	PFA 194-12117		15.10.91	R.V.Barber	Little Gransden	1.12.05P
	(Built D G Hey) (Rotax 582)						
G-BTXB	Colt 77A HAB	2072		16.10.91	A.Derbyshire *"Shellgas"*	(Telford)	1. 8.99T
G-BTXD	Rans S-6-ESA Coyote II	PFA 204-12104		22.10.91	A.I.Sutherland	Fearn	19. 7.05P
	(Built M.Isterling) (Rotax 582 - *kit no.0591.191) (Tail-wheel u/c)*						
G-BTXE*	Vickers Supermarine 394 Spitfire FR.XVIIIe				See SECTION 3 Part 1	Urbana, Illinois, US	
G-BTXF	Cameron V-90 HAB	2692		2.10.91	G.Thompson	(Ambleside)	3. 3.03
G-BTXG	British Aerospace Jetstream Series 3102	719	SE-FVP	23.10.91	Highland Airways Ltd	Inverness	9. 7.06T
			G-BTXG, OK-REJ, G-BTXG, OY-EEC, G-BTXG, N418MX, G-31-719				
G-BTXH	Colt AS-56 HA Airship	2078		23.10.91	L.Kiefer	(March-Flugstetten, Germany)	26. 3.93A
	(Built Thunder and Colt Ltd)						
G-BTXI	Noorduyn AT-16-ND Harvard IIB	14-429	Fv.161052	5.10.91	Patina Ltd	Duxford	29. 3.05P
			RCAF FE695, FE695, 42-892 *(Op The Fighter Collection as "FE695/94")*				
G-BTXK	Thunder Ax7-65 HAB	1910	ZS-HYP	28.10.91	A.F.Selby	(Woodhouse Eaves , Loughborough)	6. 5.04
			G-BTXK				
G-BTXN	British Aerospace BAe 146 Series 300	E3129	EI-CTM	20.11.91	Flightline Ltd	Southend	9. 9.07T
			G-JEAL, G-BTXN, HS-TBM, G-5-129				
G-BTXS	Cameron O-120 HAB	2141		16.10.91	Semajan Ltd tr Southern Balloon Group	(France)	2. 6.03A
G-BTXT	Maule MXT-7-180 Star Rocket	14027C		7.10.91	D.Carr and M.R Muter	Morgansfield, Fishburn	10. 6.07
G-BTXV	Cameron A-210 HAB	2703		30.10.91	The Ballooning Business Ltd	(Northampton)	11. 8.05T
					"Burning Ambition III"		
G-BTXW	Cameron V-77 HAB	2717		31.10.91	P.C.Waterhouse	(Wadhurst, East Sussex)	12. 6.04A
					"Scott's Whisky"		
G-BTXX	Bellanca 8KCAB Decathlon	595-80	OY-CYC	1.10.91	Tatenhill Aviation Ltd	Tatenhill	12. 5.07T
			SE-IEP, N5063G				
G-BTXZ	Zenair CH.250	PFA 113-12170		24.10.91	I.Parris and P.W.J.Hull	Hinton-in-the-Hedges	9.11.05P
	(Built B F Arnall) (Lycoming O-290)						
G-BTYC	Cessna 150L	15075767	N66002	4.11.91	Polestar Aviation Ltd	Jersey	5. 3.05T
G-BTYE	Cameron A-180 HAB	2704		5.11.91	K.J.A.Maxwell and D.S.Messmer	(Haywards Heath)	26. 3.00T
					"Rolling Rock"		
G-BTYF	Thunder Ax10-180 Series 2 HAB	2086		7.11.91	P.Glydon	(Barnt Green, Birmingham)	5. 4.01T
G-BTYH	Pottier P.80S	PFA 160-11121		11.11.91	R.Pickett	Tatenhill	13.10.03P
	(Built R.Pickett) (Volkswagen 1834)						
G-BTYI	Piper PA-28-181 Archer II	28-8190078	N8287T	15.11.91	C.E.Wright	Fenland	28. 2.07
G-BTYK	Cessna 310R II	310R0138	N200VC	21.11.91	Revere Aviation Ltd	Jersey	23. 7.06T
			N5018J				
G-BTYT	Cessna 152 II	15280455	N24931	25.11.91	Cristal Air Ltd *(New CofR 9.04)*	Deanland	20. 3.99T
G-BTYW	Cessna 120	11725	N77283	27.11.91	C.J.Parker	Shacklewell Farm, Empingham	6.10.07
	(Continental C85)		NC77283		tr G-BTYW Group		
G-BTYX*	Cessna 140	11004	N76568	27.11.91	Not known	Rochester	23. 2.98T
			NC76568		*(Cancelled 10.4.01 by CAA) (Stored 4.03)*		
G-BTYY	Curtiss Robin	C-2475	N348K	8.10.91	R.R.L.Windus	Truleigh Manor Farm, Edburton	1. 9.97P
	(Continental W-670)		NC348K		*(Noted 7.04)*		
G-BTYZ	Colt 210A HAB	2083		17.10.91	T.M.Donnelly	(Sprotbrough, Doncaster)	19.10.03T
G-BTZA	Beech F33A Bonanza	CE-957	PH-BNT	22.11.91	H.Mendelssohn tr G-BTZA Group	Kirknewton	19. 5.07
G-BTZB	Yakovlev Yak-50	801810	DOSAAF 77	27.11.91	D.H.Boardman	Lee-on-Solent	25. 9.05P
					(As "69" in Soviet AF c/s) (CAA Mark Exemption states "10")		
G-BTZD	Yakovlev Yak-1 Series 1	8188	1342	10.12.91	Historic Aircraft Collection Ltd	Audley End	AC
	(C/n stamped on engine bearers)		(Soviet AF) *(Salvaged from lake in N.Russia mid 1991 after forced landing c.1942: stored 3.96)*				
G-BTZE	LET Yakovlev C.11	171312	(France)	11. 2.92	Bianchi Aviation Film Services Ltd	Booker	AC
			Egypt AF, OK-JIK		*(On rebuild 6.03)*		
G-BTZG	British Aerospace ATP	2046	PK-MTV	11.12.91	Trident Aviation Leasing Services (Jersey) Ltd	Woodford	
			(PK-MAA), G-BTZG		*(Stored as "PK-MTV" 2.04)*		
G-BTZH	British Aerospace ATP	2047	PK-MTW	11.12.91	Trident Aviation Leasing Services (Jersey) Ltd	Woodford	
			(PK-MAC), G-BTZH		*(Stored 2.04)*		
G-BTZK	British Aerospace ATP	2050	PK-MTZ	11.12.91	Trident Aviation Leasing Services (Jersey) Ltd	Woodford	
			G-BTZK, (PK-MAF), G-BTZK *(Stored 2.04)*				
G-BTZL	Oldfield Baby Lakes	8506-M-28B	N2288B	12.12.91	M.R.Winter	Sleap	9. 9.05P
	(Built H R Swack) (Continental C85)						
G-BTZN	British Aerospace BAe 146 Series 300	E3149	EI-CLY	10. 1.92	Trident Aviation Leasing Services (Jersey) Ltd	Filton	
			G-BTZN, N146PZ, ZP-CCY, N146PZ, G-BTZN, HS-TBN, G-11-149 *(Stored as "EI-CLY" 6.04)*				
G-BTZO	SOCATA TB-20 Trinidad	1409		18.12.91	P.R.Draper	Thruxton	2. 4.07
G-BTZP	SOCATA TB-9 Tampico Club	1421		18.12.91	M.W.Orr	Oxford	21. 8.04T
G-BTZR	Colt 77B HAB	2087		18.12.91	P.J.Fell *"Bullet"*	(Maidenhead)	13. 4.03A
G-BTZS	Colt 77B HAB	2088		18.12.91	P.T.R.Ollivere *"Petal"*	(Sutton)	27. 5.04A
G-BTZU	Cameron Concept 60 HAB	2734		20.12.91	S.A.Simington	(Eccles, Norwich)	1. 5.04A
G-BTZV	Cameron V-77 HAB	2410		20.12.91	D.J.and H.M.Brown *"Vulcan"*	Redditch)	13. 7.05A
G-BTZX	Piper J-3C-65 Cub	18871	N98648	27. 2.92	D.A.Woodhams and J.T.Coulthard	Bidford	14. 2.04P
			NC98648		tr ZX Cub Group		
G-BTZY	Colt 56A HAB	2084		17.10.91	S.J.Wardle	(Kettering)	21. 9.05A
G-BTZZ	CFM Streak Shadow	PFA 206-12155		23.12.91	D.R.Stennett	Mendlesham	21. 9.05P
	(Built D R Stennett - kit no.K.169-SA) (Rotax 582)						

G-BUAA - G-BUZZ

Registration	Type	C/n	Previous identity	Date	Owner/operator	Base	CofA
G-BUAA	Corben Baby Ace	D561	N516DH	19.11.91	M.W.Chamberlain	Eastbach Farm, Coleford	16. 5.02P
	(Built D E Hale) (Continental A65)						

Reg	Type	C/n	Previous identity	Date	Owner	Base	CofA
G-BUAB	Aeronca 11AC Chief	11AC-1759	N3458E	17. 1.92	J.Reed	Craysmarsh Farm, Melksham	30. 6.05P
	(Continental A65)		NC3458E				
G-BUAC	Slingsby Cadet III	PFA 42-12059	(ex??)	17. 1.92	D.A.Wilson and C.R.Partington	Milfield	4.10.94P
	(Built D C Pattison) (Volkswagen 1200) *(P/I unknown, possibly home-built)*				*(Noted 1.04)*		
G-BUAF	Cameron N-77 HAB	2746		2. 1.92	Zebedee Balloon Service Ltd	Szekszárd, Hungary	6. 7.05
	(Rebuilt from 5N-ATT)						
G-BUAG	Jodel D.18	PFA 169-11651		3. 1.92	A.L.Silcox	Bodmin	1. 7.03P
	(Built A L.Silcox) (Volkswagen 1834)						
G-BUAI	Everett Gyroplane Series 3	030		6. 1.92	D.Bateson	(Haslemere)	9. 4.04P
	(Rotax 532)						
G-BUAJ	Cameron N-90 HAB	2735		7. 1.92	J.R. and S.J.Huggins *"Chunnel Plant Hire"*	(Dover)	12.10.05A
G-BUAM	Cameron V-77 HAB	2470		10. 1.92	N.Florence *"J and E Page Flowers"*	(London SW11)	17.12.02T
G-BUAO	Luscombe 8A Silvaire	4089	N1362K	15. 1.92	K.E.Ballington	Yeatsall Farm, Abbots Bromley	11. 8.04P
	(Continental A65)		NC1362K		*(New owner 11.04)*		
G-BUAT	Thunder Ax9-120 HAB	2093		24. 1.92	J.Fenton *"Calor"*	(Preston)	17. 3.00T
G-BUAV	Cameron O-105 HAB	2767		27. 1.92	C.D.Monk	(Leigh upon Mendip, Radstock)	26. 3.97T
					(Noted active 9.03)		
G-BUAX	Rans S-10 Sakota	PFA 194-11848		28. 1.92	G.J.Mcdill	Bagby	29. 6.04P
	(Built J W Topham) (Rotax 582)				*(New owner 11.04)*		
G-BUBL*	Thunder Ax8-105 HAB				See SECTION 3 Part 1	Newbury	
G-BUBN	Pilatus Britten-Norman BN-2B-26 Islander	2270		14. 2.92	Isles of Scilly Skybus Ltd	St.Just	18. 2.06T
G-BUBR*	Cameron A-250 HAB	2779		5. 5. 92	Balloon Flights International Ltd	(Bristol)	2. 9.01T
					(Cancelled 24. 6.03 as wfu) (Stored for sale 2004)		
G-BUBS	Lindstrand LBL 77B HAB	144		10.10.94	Beaulah J.Bower *"Bubbles Balloon"*		
	(Possibly a new envelope c 9.95?)					Middle Wyke Farm, St.Mary Bourne, Andover	4. 4.05A
G-BUBT	Stoddard-Hamilton Glasair IIS RG	PFA 149-11633		6. 2.92	M.D.Evans	Dunkeswell	11. 5.04P
	(Buillt M D Evans - c/n 2026) (Lycoming IO-320)						
G-BUBU	Piper PA-34-220T Seneca III	34-8233060	N8043B	9. 7.87	Brinor (Holdings) Ltd	Poplar Hall Farm, Elmsett	7. 8.06
G-BUBW	Robinson R22 Beta	2048		7. 2.92	Forth Helicopter Services Ltd	Glenrothes	9. 9.07T
G-BUBY	Thunder Ax8-105 Series 2 HAB	2115		3. 2.92	T.M.Donnelly	(Sprotbrough, Doncaster)	10. 6.03T
					"Jorvik Viking Centre"		
G-BUCA	Cessna A150K Aerobat	A1500220	N5920J	14. 6.89	D.Featherby tr BUCA Group	Norwich	17. 3.05T
G-BUCB	Cameron H-34 HAB	2777		11. 2.92	A S.Jones	(Wolverhampton)	14. 8.02A
G-BUCC	CASA 1-131E Jungmann	1109	G-BUEM	11. 9.78	P.L.Gaze	Goodwood	10. 6.05P
	(Spanish AF serial conflicts with G-BJAL)		G-BUCC, Spanish AF E3B-114 *(As "BU+CC" in Luftwaffe c/s)*				
G-BUCF*	Grumman F8F-1B Bearcat				See SECTION 3 Part 1	Chino, California, US	
G-BUCG	Schleicher ASW 20L TOP	20396	BGA.3140	19. 2.92	W.B.Andrews	Davidstow Moor	3. 8.06
	(Konig SD430)		I-FEEL		*"344"*		
G-BUCH	Stinson V-77 (AT-19) Reliant	77-381	N9570H	21. 2.92	Gullwing Trading Ltd	White Waltham	6. 3.06
			FB531(RN)				
G-BUCJ	de Havilland DHC-2 Beaver AL.1	1442	XP772	23. 3.92	Propshop Ltd and G.Warner	Duxford	
					t/a British Aerial Museum *(As "XP772" in Army c/s)*		
G-BUCK	CASA 1-131E Jungmann Series 1000	1113	Spanish AF E3B-322	11. 9.78	R.A Cayless and J.G.Brander	White Waltham	27. 4.05P
					tr Jungmann Flying Group *(As "BU+CK" in Luftwaffe c/s)*		
G-BUCM	Hawker Sea Fury FB.11	-	VX653	26. 2.92	Patina Ltd	Duxford	
					(Op The Fighter Collection as "VX653")		
G-BUCO	Pietenpol AirCamper	PFA 47-11829		10. 2.92	A James	Siege Cross Farm, Thatcham	2. 9.04P
	(Built A James) (Continental C90)						
G-BUCS	Cessna 150F	15062368	N3568L	25. 8.89	Atlantic Bridge Aviation Ltd	Lydd	1. 4.04T
G-BUCT	Cessna 150L	15075326	N11320	14. 6.89	Atlantic Bridge Aviation Ltd	Lydd	5.11.03T
G-BUDA	Slingsby T.61F Venture T.2	1963	ZA627	18. 2.92	T.M.Holloway tr RAFGSA	RAF Cranwell	18. 8.07
G-BUDB	Slingsby T.61F Venture T.2	1964	ZA628	18. 2.92	T.M.Holloway	RAF Marham	5. 9.07
					tr RAFGSA *(Op Fenlands Gliding Club)*		
G-BUDC	Slingsby T.61F Venture T.2	1971	ZA652	18. 2.92	I.P.Litchfield tr T.61 Group	Enstone	15. 6.07
G-BUDE	Piper PA-22-135 Tri-Pacer	22-980	N1144C	9. 4.92	B.A Bower	Branscombe	20.11.06
	(Tail-wheel conversion)						
G-BUDF	Rand-Robinson KR-2	PFA 129-11155		26. 2.92	M.Stott	(Bedford)	3.12.03P
	(Built J B McNab) (HAPI Magnum 75)						
G-BUDI	Aero Designs Pulsar	PFA 202-12185		25. 2.92	R.W.L.Oliver	Popham	5. 8.04P
	(Built R.W.L.Oliver)						
G-BUDK	Thunder Ax7-77 HAB	2076		2. 3.92	W.Evans	(Wrexham)	2. 4.05A
G-BUDL	Taylorcraft E Auster III	458	PH-POL	5. 3.92	M.Pocock *(As "NX534")*	AAC Middle Wallop	
	(Regd with Frame No.TAY 5810)		8A-2, R Neth AF R-17, NX534		*(On rebuild 6.98 for Military Auster Flight)*		
G-BUDN	Cameron Shoe 90SS HAB	2761		6. 3.92	Magical Adventures Ltd *"Converse Allstar Boot"*		
	(Converse Allstar Trainers shape)					(West Bloomfield, Mi.chigan, US)	1. 8.00A
G-BUDO	PZL-110 Koliber 150	03900045	(D-EIVT)	12. 3.92	A S.Vine	Haverfordwest	27. 7.06
G-BUDR	Denney Kitfox Model 3	PFA 172-12107		16. 3.92	N.J.P.Mayled	Dunkeswell	16. 9.05P
	(Built D Silsbury - c/n 1086)						
G-BUDS	Rand Robinson KR-2	PFA 129-10937		31.12.85	D.W.Munday	Popham	
	(Built D W Munday)				*(Noted unmarked nearing completion 9.01)*		
G-BUDT	Slingsby T.61F Venture T.2	1883	XZ563	30. 3.92	R.V.Andrews	Belle Vue Farm, Yarnscombe	4. 8.07P
					tr G-BUDT Group		
G-BUDU	Cameron V-77 HAB	2447		16. 3.92	T.M.G.Amery	(Llandeilo)	18. 9.03A
G-BUDW	Brugger MB.2 Colibri	PFA 43-10644	G-GODS	19. 3.92	D.R.Mickleburgh	(Sutton St James, Spalding)	3.10.00P
	(Built J.M.Hoblyn) (Volkswagen 1600)			*(Crashed Taunton Racecourse 13.8.00: under repair 1.03: new owner 1.04)*			
G-BUEC	Van's RV-6	PFA 181C-11884		17. 3.92	A.H.Harper	Henstridge	9. 2.05P
	(Built D W Richardson and R D Harper - c/n 21015) (Lycoming O-360)						
G-BUED	Slingsby T.61F Venture T.2	1979	ZA660	12. 3.92	F.B.Rutterford tr 617 VGS Flying Group	Waldershare Park	5. 7.07
G-BUEF	Cessna 152 II	15280862	N25928	17. 3.92	A.L.Brown t/a Channel Aviation	Bourn	1. 9.05T
G-BUEG	Cessna 152 II	15280347	N24736	17. 3.92	Plymouth School of Flying Ltd	Plymouth	2.10.07T
G-BUEI*	Thunder Ax8-105 HAB	2172		23. 3.92	Elinore French Ltd	(Morpeth)	13. 4.05T
					t/a Imagination Balloon Flights *(Cancelled 3.11.04 by CAA)*		
G-BUEK	Slingsby T.61F Venture T.2	1879	XZ559	30. 3.92	Norfolk Gliding Club Ltd	Tibenham	12.12.07

G-BUEN	VPM M-14 Scout	VPM14-UK101		19. 3.92	F.G.Shepherd	Carlisle	4.12.96P
	(Arrow GT1000R)				*(Noted 9.02*		
G-BUEP	Maule MXT-7-180 Star Rocket	14023C		24. 3.92	N.J.B.Bennett	(Yeovil)	6. 6.07
G-BUET*	Colt Flying Drinks Can SS HAB				See SECTION 3 Part 1	(Aldershot)	
G-BUEU*	Colt 21A Cloudhopper HAB				See SECTION 3 Part 1	(Aldershot)	
G-BUEV	Cameron O-77 HAB	2810	EI-CFW	31. 3.92	R.R.McCormack	(Belfast)	22. 9.04A
			G-BUEV				
G-BUEW	Rans S-6 Coyote II	PFA 204-12021	G-MWYF	1. 4.92	J.D.Clabon	Weston Zoyland	10.12.03P
	(Built D J O'Gorman - kit no.0190-111) (Rotax 582)		(EI-CEL)				
G-BUEX	Schweizer 269C	S.1412	G-HFLR	14. 4.92	Tout-Saints Hotels Ltd	Thruxton	17. 7.05T
G-BUEZ*	Hawker Hunter F.6A	S4/U/3275	8736M	3. 4.92	The Old Flying Machine (Air Museum)	Spanhoe	
	(Built Armstrong-Whitworth Aircraft)		XF375		*(Cancelled 28.8.01 as WFU) (Noted 4.03)*		
G-BUFA	Cameron R-77 Gas/HAB	2712		19. 3.92	Noble Adventures Ltd *(Stored 1996)*	(The Netherlands)	10. 6.93A
G-BUFC	Cameron R-77 Gas/HAB	2823		19. 3.92	Noble Adventures Ltd *(Stored 1996)*	(The Netherlands)	23. 6.93A
G-BUFE	Cameron R-77 Gas/HAB	2825		19. 3.92	Noble Adventures Ltd *(Stored 1996)*	(The Netherlands)	21. 6.93A
G-BUFF*	Wassmer Jodel D.112	1302	F-BMYD	9. 8.78	(M.Mold)	Watchford Farm, Yarcombe	
				(Cancelled 29.8.91 by	*CAA - no CofA or Permit issued) (Fuselage and wings noted 1.04)*		
G-BUFG	Slingsby T.61F Venture T.2	1977	ZA658	3. 4.92	Halegreen Associates Ltd	Hinton-in-the-Hedges	16. 8.07
G-BUFH	Piper PA-28-161 Warrior II	28-8416076	N43520	15. 4.92	M.P.Rainford and J.E.Slee	Blackpool	24. 6.07T
					tr The Tiger Leisure Group		
G-BUFJ	Cameron V-90 HAB	2809		7. 4.92	S.P.Richards	(Tonbridge)	29. 7.05T
G-BUFN	Slingsby T.61F Venture T.2	1967	ZA631	8. 4.92	S.C.Foggin	Sandhill Farm, Shrivenham	16.11.07
					tr BUFN Group		
G-BUFR	Slingsby T.61F Venture T.2	1880	XZ560	9. 4.92	East Sussex Gliding Club Ltd	Ringmer	16. 12.07
	(Rollason RS Mk.2)						
G-BUFT	Cameron O-120 HAB	2814		9. 4.92	D.Bron	(St.Barthelemy, France)	22. 1.05T
G-BUFV	Avid Speed Wing Mk.4	PFA 189-12192		15. 4.92	M.and B.Gribbin	(Antrim, County Antrim)	16. 9.05P
	(Built S C Ord) (BMW R100)						
G-BUFW	Aérospatiale AS355F1 Ecureuil 2	5112	5N-BAK	21. 4.92	RCR Aviation Ltd	(Horsham)	31. 5.92A
			G-BUFW, N57904				
G-BUFX	Cameron N-90 HAB	2835		22. 4.92	Kerridge Computer Co Ltd *"Kerridge II"*	(Newbury)	7. 7.05A
G-BUFY	Piper PA-28-161 Warrior II	28-8016211	N130CT	14. 4.92	Bickertons Aerodromes Ltd	Denham	1. 7.07T
			N8TS, N3571K		*(Op The Pilots Centre)*		
G-BUGB	Stolp SA.750 Acroduster Too	PFA 89-11942		22. 4.92	R.M.Chaplin	Rochester	7. 9.04P
	(Built D Burnham) (Lycoming O-360)						
G-BUGD	Cameron V-77 HAB	2195		23. 4.92	P.Haslett	(Arcy sur Cure, France)	11. 9.04A
G-BUGE	Bellanca 7GCAA Citabria	339-77	N4165Y	23. 4.92	V.Vaughan and N.O'Brien	Mullinahone, County Tipperary	14. 9.07
G-BUGG	Cessna 150F	15062479	N8379G	24. 3.92	C.P.J.Taylor and D.M.Forshaw	Panshanger	28. 9.05
G-BUGI	Evans VP-2	PFA 7201		16. 4.92	J.A.Rees	Haverfordwest	12. 1.04P
	(Built D Silsbury) (Continental A65-8)						
G-BUGJ	Robin DR400/180 Régent	2137		28. 4.92	W.M.Patterson	Letterkenny, County Donegal	3.11.07E
G-BUGL	Slingsby T.61F Venture T.2	1966	ZA630	29. 4.92	S.Bradford and M.Bean tr VMG Group	Tibenham	1. 6.04
G-BUGM	CFM Streak Shadow	PFA 206-12069		29. 4.92	D.Penn-Smith	Sywell	14. 9.05P
	(Built W J de Gier - kit no K.176-SA) (Rotax 582)				tr The Shadow Group		
G-BUGO	Colt 56B HAB	2143		18. 5.92	Escuela de Aerostacion Mica	(Valencia, Spain)	19. 7.00A
G-BUGP	Cameron V-77 HAB	2278	OO-BEE	10. 3.92	R.Churcher	(Canterbury)	13. 8.05A
G-BUGS	Cameron V-77 HAB	2482		14. 4.92	S.J.Dymond *"Bugs Bunny"*	(Tidworth)	14. 9.01T
G-BUGT	Slingsby T.61F Venture T.2	1871	XZ551	22. 4.92	R.W.Hornsey	Rufforth	5. 8.05
G-BUGV	Slingsby T.61F Venture T.2	1884	XZ564	28. 4.92	Oxfordshire Sportflying Ltd	Enstone	28. 6.07
G-BUGW	Slingsby T.61F Venture T.2	1962	ZA626	22. 4.92	Halegreen Associates Ltd	Hinton-in-the-Hedges	13.10.07
G-BUGY	Cameron V-90 HAB	2800		9. 4.92	I.J.Culley	(Hungerford)	18. 7.05A
					tr Dante Balloon Group *"Florence"*		
G-BUGZ	Slingsby T.61F Venture T.2	1981	ZA662	22. 4.92	R.W.Spiller tr Dishforth Flying Group	AAC Dishforth	24. 5.05
G-BUHA	Slingsby T.61F Venture T.2	1970	ZA634	29. 4.92	K.E. Ballington *(As "ZA634/C")*	Saltby	25. 9.05
G-BUHC	British Aerospace BAe 146 Series 300	E3193	EI-CTO	30. 6.92	Flightline Ltd	Southend	21. 7.07T
			G-BUHC, G-BTMI, (N883DV), G-6-193				
G-BUHM	Cameron V-77 HAB	2481		7. 5.92	L.A Watts *"Blue Horizon"*	(Pangbourne, Reading)	24. 9.05A
G-BUHO	Cessna 140	14402	N2173V	1. 5.92	W.B.Bateson	Blackpool	19.11.04T
	(Continental C90)						
G-BUHR	Slingsby T.61F Venture T.2	1874	XZ554	8. 5.92	W.G.Miller tr Connel Motor Glider Group	Sleap	16.10.05
G-BUHS	Stoddard-Hamilton Glasair SH TD-1	149	C-GYMB	8. 5.92	E.J.Spalding	Wick	20. 3.04P
	(Built F L Binder) (Lycoming O-360)						
G-BUHU	Cameron N-105 HAB	2785		13. 5.92	Unipart Group Ltd tr Unipart Balloon Club	(Cowley)	21.11.96A
G-BUHY	Cameron A-210 HAB	2858		14. 5.92	D.J.Farrar *(New owner 9.04)*	(Collingham, Wetherby)	21. 9.99T
G-BUHZ	Cessna 120	14950	N3676V	1. 5.92	M.R.Houseman tr C140 Group	RAF Henlow	12. 2.05P
G-BUIC	Denney Kitfox Model 2	PFA 172-11802		1. 5.92	C.R.Northrop and B.M.Chilvers	(Huntingdon)	
	(Built C R Northrop)						
G-BUIE	Cameron N-90 HAB	2863		22. 5.92	B.Conway	(Wheatley, Oxford)	24. 1.01A
G-BUIF	Piper PA-28-161 Warrior II	28-7916406	N28375	29. 5.92	Newcastle Aero Club Ltd	Newcastle	8. 9.04T
G-BUIG	Campbell Cricket	PFA G/03-1173		27. 5.92	J.A English	Carlisle	27. 4.05P
	(Built T A Holmes) (Rotax 532)						
G-BUIH	Slingsby T.61F Venture T.2	1876	XZ556	29. 5.92	Yorkshire Gliding Club (Pty) Ltd	Sutton Bank	8. 6.07
G-BUIJ	Piper PA-28-161 Warrior II	28-8116210	N83784	3. 6.92	Tradecliff Ltd	Blackbushe	25. 8.07T
G-BUIK	Piper PA-28-161 Warrior II	28-7916469	N2845P	2. 6.92	A.S.Bamrah t/a Falcon Flying Services	Shoreham	25.10.07E
					(Op The Flying Hut)		
G-BUIL	CFM Streak Shadow	PFA 206-12121		8. 5.92	P.N.Bevan and L.M.Poor	Perth	15. 5.04P
	(Built P.N.Bevan and L.M.Poor - kit no K.182-SA) (Rotax 582)				*"Mr Bounce"*		
G-BUIN	Thunder Ax7-77 HAB	1882		5. 6.92	P.C.Johnson	(Gloucester)	23. 9.05A
G-BUIP	Denney Kitfox Model 2	PFA 172-11874		8. 6.92	Avcomm Developments Ltd	Enstone	16. 9.05P
	(Built G D Lean - c/n 710)						
G-BUIR	Avid Speed Wing Mk.4	PFA 189-12213		9. 6.92	M.C.J.Myers	(Cheltenham)	29. 4.97P
	(Buillt K N Pollard)				*(Damaged near Gainsborough 26.1.97: on rebuild 5.97: new owner 7.02)*		
G-BUIU	Cameron V-90 HAB	2641		11. 6.92	H.Micketeit	(Bielefeld, Germany)	18. 4.03A

Reg	Type	C/n	Previous identity	Date	Owner/operator	Base	CofA
G-BUIZ	Cameron N-90 HAB	2850		12. 6.92	R.S.Kent	(Lancing)	31 8.05A
	(Telecoms Shape)				tr Balloon Preservation Flying Group *"Hutchinson"*		
G-BUJA	Slingsby T.61F Venture T.2	1972	ZA653	22. 5.92	T.M.Holloway	RAF Cosford	14. 1.06
					tr RAFGSA *(Op Wrekin Gliding Club)*		
G-BUJB	Slingsby T.61F Venture T.2	1978	ZA659	21. 5.92	O.F.Vaughan and D.A Fall tr Falke Syndicate	Shobdon	11. 7.07
G-BUJE	Cessna 177B Cardinal	17701920	N34646	10. 6.92	J.Flux tr FG93 Group	Old Sarum	5. 5.07
G-BUJH	Colt 77B HAB	2207		23. 6.92	R.P.Cross and R.Stanley	(Luton/Harpenden)	27. 3.05A
G-BUJI	Slingsby T.61F Venture T.2	1882	XZ562	22. 5.92	Solent Venture Syndicate Ltd	Lee-on-Solent	15. 4.07
G-BUJJ	Avid Speed Wing *(Built M Cox)*	213	N614JD	20.10.92	R.A.Dawson	(Totnes)	30. 9.04P
G-BUJK	Montgomerie-Bensen B.8MR Merlin	PFA G/01-1211		25. 6.92	K.J.Robinson	Manor Farm, Garford	1. 6.05P
	(Built J M Montgomerie) (Rotax 582)						
G-BUJL	Aero Designs Pulsar	PFA 202-11892		16. 6.92	J.J.Lynch	(Dunstable)	
	(Built J.J.Lynch)						
G-BUJM	Cessna 120	11784	N77343	19. 6.92	D.H.Mackay	RNAS Yeovilton	27. 2.06
	(Continental C85)		NC77343		tr Cessna 120 Flying Group		
G-BUJN	Cessna 172N Skyhawk II	17272713	N6315D	19. 6.92	Central Aircraft Leasing Ltd	Wolverhampton	30.11.95T
					(New owner 7.03)		
G-BUJO	Piper PA-28-161 Cherokee Warrior II	28-7716077	N1014Q	19. 6.92	A.S.Bamrah t/a Falcon Flying Services	Shoreham	24. 3.06T
					(Op The Flying Hut)		
G-BUJP	Piper PA-28-161 Warrior II	28-7916047	N21624	19. 6.92	J.M.C.Manson *(Op Ace Aviation)*	Shoreham	16.10.05T
G-BUJR	Cameron A-180 HAB	2821		22. 6.92	Dragon Balloon Company Ltd	(Castleton, Hope Valley)	16.10.00T
G-BUJV	Avid Speed Wing Mk.4	PFA 189-12250		3. 7.92	C.Thomas	(Tamworth)	28. 7.94P
	(Built D N Anderson)				*(Damaged Caernarfon 13.8.93)*		
G-BUJW	Thunder Ax8-90 Series 2 HAB	2208		6. 7.92	R.T.Fagan	(Bath)	6. 8.95T
G-BUJX	Slingsby T.61F Venture T.2	1873	XZ553	7. 7.92	J.R.Chichester-Constable	Burton Constable, Hull	10. 9.06
G-BUJZ	RotorWay Executive 90	5119		9. 7.92	M.P.Swoboda	Street Farm, Takeley	20.10.03P
	(Built T W Aisthorpe and R J D Crick) (RotorWay RI 162)				*(Rolled onto side Street Farm 28.6.03 and extensively damaged: noted 1.05)*		
G-BUKA	Fairchild SA.227AC Metro III	AC-706B	ZK-NSQ	24. 8.88	Atlantic Air Transport Ltd	Coventry	11. 6.07T
			N27185, G-BUKA, N27185		*(Atlantic Express titles)*		
G-BUKB	Rans S-10 Sakota	PFA 194-12078		13. 7.92	M.K.Blatch	RAF Brize Norton	14. 7.05P
	(Built M K Blatch) (Rotax 582)						
G-BUKC*	Cameron A-180 HAB				See SECTION 3 Part 1	(Lincoln)	
G-BUKF	Denney Kitfox Model 4	PFA 172A-12247		2. 6.92	A.G.V.McClintock	East Fortune	20. 6.05P
	(Built M R Crosland)				tr Kilo Foxtrot Group		
G-BUKH	Druine D.31 Turbulent	PFA 48-11419		14. 8.92	R.B.Armitage	Stoneacre Farm, Farthing Corner	26. 8.05P
	(Built J S Smith) (Volkswagen 1600)						
G-BUKI	Thunder Ax7-77 HAB	2239		8. 7.92	Airxcite Ltd t/a Virgin Balloon Flights	(Wembley)	26. 4.05T
G-BUKJ	British Aerospace ATP	2052	EC-HCO	5. 8.92	Trident Aviation Leasing Services (Jersey) Ltd	Exeter	
			G-BUKJ, EC-GLD, G-OEDF, G-BUKJ, TC-THZ, G-BUKJ		*"Rafael Alberti" (Stored 12.04)*		
G-BUKK	Bücker Bü.133D Jungmeister	27	N44DD	15.11.89	E.J.F.McEntee	(Kirdford, Billingshurst)	10. 6.05P
	(Built Dornier-Werke AG)		HB-MKG, Swiss AF U-80		*(As "U-80" in Swiss AF c/s)*		
G-BUKN	Piper PA-15 Vagabond	15-215	N4427H	15. 7.92	Mary A.Goddard	(Southampton	
			NC4427H		*(New owner 1.05)*		
G-BUKO	Cessna 120	13089	N2828N	15. 7.92	Sasha Warrener	(Woodhall Spa)	31. 7.98P
			NC2828N		*(Damaged on take-off Manor Farm, Bishopstone 28.12.97: new CofR 11.04)*		
G-BUKP	Denney Kitfox Model 2	PFA 172-12301		22. 7.92	K.N.Cobb	Weston Zoyland	5. 7.05P
	(Built T D Reid)				*(Noted 11.04)*		
G-BUKR	SOCATA MS.880B Rallye 100T	2923	LN-BIY	27. 7.92	G.R.Russell tr G-BUKR Flying Group	Bridport	23. 2.06
G-BUKS	Colt 77B HAB	2241		6. 7.92	R.and M.Bairstow	(Middlewich, Cheshire)	5. 7.05A
G-BUKT	Luscombe 8E Silvaire	2197	N45670	30. 7.92	R.J.F.Swain	Sleap	18. 1.05P
	(Continental C85)		NC45670				
G-BUKU	Luscombe 8E Silvaire	4720	N1993K	30. 7.92	D.J.Warren	Rochester	11.12.03P
	(Continental C85)		NC1993K		tr Silvaire Flying Group		
G-BUKV	Colt AS-105 Mk.II HA Airship	2212	ZS-HYO	3. 8.92	A.Ockelmann	Buchholz, Germany	17. 4.02A
	(Built Thunder and Colt Ltd)		G-BUKV		t/a Ballon Reisen		
G-BUKX	Piper PA-28-161 Cherokee Warrior II	28-7816674	N231PA	5. 8.92	LNP Ltd	Exeter	12.12.04T
G-BUKZ	Evans VP-2	PFA 63-10761		5. 8.92	P.R.Farnell	Wombleton	
	(Built P.R.Farnell)				*(Extant 7.02)*		
G-BULB	Thunder Ax7-77 HAB	1968		3. 7.92	Richard Nash Cars Ltd	(Norwich)	16. 2.05A
G-BULC	Avid Flyer Mk.4	PFA 189-12202		6. 7.92	C.Nice	Popham	8. 7.05P
	(Built C Nice)						
G-BULD	Cameron N-105 HAB	2136		6. 8.92	R.J.Collins	(Hurtmore, Godalming)	9. 4.05T
G-BULF	Colt 77A HAB	2043		10. 8.92	P.Goss and T.C.Davies *"Nursery"*	(Christchurch)	7. 5.05A
G-BULG	Van's RV-4	JRV4-1	C-FELJ	28. 7.92	M.J.Aldridge	Priory Farm, Tibenham	21. 5.05P
	(Built L Johnson) (Lycoming O-320)						
G-BULH	Cessna 172N Skyhawk II	17269869	N738CJ	2. 7.92	Comed Schedule Services Ltd	Blackpool	16. 1.06T
G-BULJ	CFM Streak Shadow	PFA 206-12199		10. 8.92	C.C.Brown	Wellesbourne Mountford	2. 9.05P
	(Built C.C.Brown - kit no K.191-SA) (Rotax 582)						
G-BULK	Thunder Ax9-120 Series 2 HAB	2237		3. 7.92	S.J.Colin t/a Skybus Ballooning	(Cranbrook)	11.10.04T
G-BULL	Scottish Aviation Bulldog Series 120/128	BH120/392	HKG-5	20. 9.88	Solo Leisure Ltd	Old Sarum	17.10.07T
			G-31-18		*(Also carries "HKG-5" in Hong Kong Guard c/s)*		
G-BULM	Aero Designs Pulsar	PFA 202-12010		11. 8.92	J.Lloyd	(St. Georges, France)	5. 5.54P
	(Built W R Davis-Smith) (Tri-cycle u/c)						
G-BULN	Colt 210A HAB	2265		13. 8.92	H.G.Davies	(Woodmancote, Cheltenham)	19. 7.05T
G-BULO	Luscombe 8F Silvaire	4216	N1489K	13. 8.92	A.F.S.Caldecourt	Popham	17. 9.05P
	(Continental A65)		NC1489K				
G-BULR	Piper PA-28-140 Cherokee B	28-25230	HB-OHP	8. 7.92	G.R.Bright	Little Gransden	13. 5.05T
			N7320F				
G-BULT	Everett Gyroplane Series 1	PFA G/03A-1213		20. 8.92	A T.Pocklington	(Bishops Stortford)	26. 5.05P
	(Built A T.Pocklington) (Originally regd as Campbell Cricket c/n PFA G/03-1213: type and c/n officially amended 12.02)						
G-BULY	Avid Flyer	PFA 189-12309		12. 8.92	J.K.Davies	(Eccleston, Chester)	23. 7.05P
	(Built M O Breen)						
G-BULZ	Denney Kitfox Model 2	PFA 172-11546		31. 7.92	T.G.F.Trenchard		
	(Built D J Dumulo)					Newton Peverill Farm, Sturminster Marshall	22.11.05P

G-BUMP	Piper PA-28-181 Cherokee Archer II	28-7790437	PH-MVA	17. 1.79	Marham Investments Ltd	Belfast Aldergrove	26.11.06T
			OO-HCH, N3105Q				
G-BUNB	Slingsby T.61F Venture T.2	1969	ZA633	25. 8.92	T.M.Holloway tr RAFGSA	Lee-on-Solent	9. 1.05
G-BUNC	PZL-104 Wilga 35A	129444	SP-TWP	2. 9.92	R.F.Goodman	Husbands Bosworth	18. 6.05
G-BUND	Piper PA-28RT-201T Turbo Arrow IV	28R-8031107	N8219V	18. 7.88	Datalake Ltd	Fairoaks	20. 8.04
G-BUNG	Cameron N-77 HAB	2905		2. 9.92	A.Kaye	(Wellingborough)	27. 3.05A
					tr The Bungle Balloon Group *(Aspen titles) "Bungle"*		
G-BUNH	Piper PA-28RT-201T Turbo Arrow IV	28R-8031166	N8255H	26. 8.92	J.H.Sandham t/a JH Sandham Aviation	(Workington)	20. 1.05T
G-BUNJ	K & S Squarecraft SA.102.5 Cavalier	PFA 01-10058		10. 9.92	J.A.Smith	Great Massingham	
	(Built J S Smith)				*(Nearing completion 9.97)*		
G-BUNM	Denney Kitfox Model 3	PFA 172-12111		15. 9.92	P.N.Akass	(Ireland)	5. 5.04P
	(Built P J Carter) (Floatplane)						
G-BUNO	Neico Lancair 320	PFA 191-12332		11. 9.92	J.Softley	(Newbury)	
	(Built J.Softley)				*(On build 2000)*		
G-BUNV	Thunder Ax7-77 HAB	1967		23. 9.92	R.M.Garnett *"Skylark"*	(Eastleigh)	14. 6.05A
G-BUNZ	Thunder Ax10-180 Series 2 HAB	2271		7. 9.92	M.A Scholes	(Haywards Heath)	25. 2.05T
G-BUOA	Whittaker MW6-S Fatboy Flyer Series A			25. 9.92	R.Blackburn	(Raphoe, County Donegal)	18. 6.04P
	(Built D A Izod) (Rotax 582)	PFA 164-11959					
G-BUOB	CFM Streak Shadow	PFA 206-12156		29. 9.92	A. M.Simmons	Belle Vue Farm, Yarnscombe	21.11.05P
	(Built A M Simmons - kit no K.186-SA) (Rotax 582)						
G-BUOD	Replica Plans SE.5A	PFA 20-10474		5.10.92	M.D.Waldron	Kemble	13. 4.05P
	(Built M D Waldron) (Continental C90)				*(As "B595/W" in 56 Sqdn, RFC c/s)*		
G-BUOE	Cameron V-90 HAB	2938		6.10.92	B.and J.Smallwood	(Chippenham)	12. 5.05A
					t/a Dusters and Co *"Flying Colours 2"*		
G-BUOF	Druine D.62B Condor	PFA 49-11236		6.10.92	R.P.Loxton	Wyke Farm, Sherbourne	15. 3.05P
	(Built K Jones)						
G-BUOI	Piper PA-20-135 Pacer	20-571	OY-ALS	18. 9.92	R.A L.Hubbard	Meon	15. 5.05
	(Lycoming O-320) *(Hoerner wing-tips)*		D-EHEN, N7750K		tr Foley Farm Flying Group		
G-BUOJ	Cessna 172N Skyhawk II	17271701	N5064E	8.10.92	Falcon Flying Services Ltd	Biggin Hill	13. 7.07T
G-BUOK	Rans S-6-116 Coyote II	PFA 204A-12317		9.10.92	M.Morris	Fieldhead Farm, Denholme, Bradford	5.10.05P
	(Built M.Morris) (Rotax 912-UL - *kit no.0692.314)*						
G-BUOL	Denney Kitfox Model 3	PFA 172-12142		12.10.92	E.C.King	Eastbach Farm, Coleford	12. 8.04P
	(Built J G D Barbour)				*(New owner 6.04)*		
G-BUON	Avid Aerobat	PFA 189-12160		13.10.92	S.R.Winder	(Bolton)	6.11.01P
	(Built I A J Lappin)						
G-BUOR	CASA 1-131E Jungmann Series 2000	2134	N89542	21.10.92	M.I.M.Schermer Voest	Den Helder, The Netherlands	15. 6.04P
			EC-336, Spanish AF E3B-508				
G-BUOS	Vickers Supermarine 394 Spitfire FR.XVIIIe	6S/672224	Indian AF HS687	19.10.92	Historic Flying Ltd	Duxford	21. 8.05P
			SM845		*(Op Aircraft Restoration Company as "SM845/GZ-J")*		
G-BUOW	Aero Designs Pulsar XP	PFA 202-12206		22.10.92	T.J.Hartwell	Sackville Lodge Farm, Riseley	8. 6.95P
	(Built D F Gaughan)				*(New owner 7.01)*		
G-BUOX	Cameron V-77 HAB	2925		23.10.92	I.G.H.Woodmansey *"High Flyer"*	(Thame)	9. 4.05A
G-BUOZ	Thunder Ax10-180 HAB	1962	(SX-)	29.10.92	Zebedee Balloon Service Ltd	(Newtown, Hungerford)	13. 5.04T
			G-BUOZ				
G-BUPA	Rutan LongEz	750	N72SD	22. 9.92	N.G.Henry	Gloucestershire	13. 8.03P
	(Built D Moore) (Lycoming O-235)						
G-BUPB	Stolp SA.300 Starduster Too	RH.100	N8035E	3.11.92	J.R.Edwards	Popham	10.10.05P
	(Built R Harte) (Lycoming IO-360)				tr Starduster PB Group		
G-BUPC	Rollason Beta B.2	PFA 02-12369		29.10.92	C.A Rolph	Liverpool	3. 6.03P
	(Built C.A Rolph) (Continental C90)						
G-BUPF	Bensen B.8MR	PFA G/01-1209		5.11.92	P.W.Hewitt-Dean	(Wootton Bassett)	1. 8.02P
	(Built G M Hobman) (Rotax 532)						
G-BUPG	Cessna 180J Skywagon	18052490	N52086	15.10.92	T.P.A.Norman	Rendcomb	6.11.05
G-BUPH	Colt 25A Sky Chariot HAB	2023		10.11.92	BAB-Ballonwerbung GmbH	(Hanover, Germany)	11. 2.05A
G-BUPI	Cameron V-77 HAB	1778	G-BOUC	28. 7.88	Sally A.Masey *"Bristol United Press"*	(Bristol)	30. 4.00A
					(Western Daily Press/Evening Post titles)		
G-BUPJ	Fournier RF4D	4119	N7752	10.11.92	M.R.Shelton	Tatenhill	
G-BUPM	VPM M-16 Tandem Trainer	VPM16-UK-102		16.10.92	Roger Savage Gyroplanes Ltd	Carlisle	9. 5.05P
	(Rotax 914)						
G-BUPP	Cameron V-42 HAB	2789		21. 7.92	L.J.Schoeman	(Basildon)	13. 5.04A
G-BUPR	Jodel D.18	PFA 169-11289		23.11.92	R.W.Burrows	Priory Farm, Tibenham	9. 5.05P
	(Built R.W.Burrows) (Limbach L2000)						
G-BUPT	Cameron O-105 HAB	2960		25.11.92	P.M.Simpson	(Hemel Hempstead)	13. 4.05A
G-BUPU	Thunder Ax7-77 HAB	2305		25.11.92	R.C.Barkworth and D.G.Maguire *"Puzzle"*	(Pulborough)	26. 3.01A
G-BUPV	Great Lakes 2T-1A Sport Trainer	126	N865K	26.11.92	R.J.Fray	Sibson	25. 9.04P
	(Gladden Kinner R55)		NC865K				
G-BUPW	Denney Kitfox Model 3	PFA 172-12281		22.10.92	G.M.Park	(Lochwinnoch)	10. 7.05P
	(Built D Sweet) (Rotax 912)				tr Forfoxake Flyers		
G-BURD	Reims/Cessna F172N Skyhawk II	F17201677	PH-AXI	26. 4.78	Tayside Aviation Ltd	Glenrothes	12.12.05T
G-BURE	Jodel D.9 Bébé	PFA 944		30.11.92	Lucy J.Kingsford	Headcorn	
	(Built P B Shilling and C R Kingsford)				*(Noted 4.03)*		
G-BURF*	Rand Robinson KR-2	PFA 129-11345		30.11.92	P.J.H.Moorhouse and B.L.Hewart	Green St., Sunbury	
	(Built P J H Moorhouse and B L Hewart) (Volkswagen 1834)				*(Cancelled 14.8.02 by CAA - no PtoF issued) (Noted 4.04)*		
G-BURG	Colt 77A HAB	2042		12. 1.93	S.T.Humphreys *"Lily"*	(Great Missenden)	13. 8.04A
G-BURH	Cessna 150E	15061225	EI-AOO	2.12.92	Christine A.Davis	Farley Farm, Romsey	4.11.06
			G-BURH, EI-AOO, N2125J				
G-BURI	Enstrom F-28C	433	N51743	11.12.92	R.L.Heath tr India Helicopters Group	Goodwood	26. 5.05T
G-BURL	Colt 105A HAB	2297		18.11.92	J.E.Rose *"Isis"*	(Abingdon)	20. 2.05T
G-BURM*	English Electric Canberra TT.18				See SECTION 3 Part 1		
						Temora, New South Wales, Australia	
G-BURN	Cameron O-120 HAB	2793		18. 2.92	Innovation Ballooning Ltd *"Innovations"*	(Bath)	17. 9.05T
G-BURP	RotorWay Executive 90	5116		8.10.92	N.K.Newman	(Buckingham)	13. 9.96P
	(Built A GA Edwards) (RotorWay RI 162)				*(Stored 7.99)*		

Reg	Type	C/n	Previous identity	Date	Owner/operator	Base	CofA expiry
G-BURS	Sikorsky S-76A II Plus	760040	(HP-)	4. 5.89	Premiair Aviation Services Ltd	Denham	25.10.05T
			G-BURS, G-OHTL				
G-BURT	Piper PA-28-161 Cherokee Warrior II	28-7716105	N2459Q	10. 6.81	B.A Paul	Denham	2. 6.05T
G-BURU	British Aerospace Jetstream Series 3202	974	F-GMVH	14. 1.93	Trident Aviation Leasing Services (Ireland) Ltd	(Dublin)	
	(Originally regd as Series 3206)		G-BURU, (F-OHFT), G-31-974				
G-BURZ	Hawker Nimrod II	41H-59890	K3661	22.12.91	Historic Aircraft Collection Ltd	St.Leonards-on-Sea	AC
					(On rebuild 8.95)		
G-BUSB	Airbus A320-111	0006	(G-BRAA)	30. 3.88	British Airways plc	Heathrow	19. 4.07T
	(Originally flown as G-BRSA)		F-WWDD		*(Koguty Lowickie t/s)*		
G-BUSC	Airbus A320-111	0008	(G-BRAB)	26. 5.88	British Airways plc	Heathrow	1. 6.07T
			F-WWDE				
G-BUSD	Airbus A320-111	0011	(G-BRAC)	21. 7.88	British Airways plc	Heathrow	21. 7.07T
			F-WWDF		*"Isle of Mull"*		
G-BUSE	Airbus A320-111	0017	F-WWDG	1.12.88	British Airways plc	Heathrow	30.11.07E
G-BUSF	Airbus A320-111	0018	F-WWDH	26. 5.89	British Airways plc	Heathrow	25. 5.05T
G-BUSG	Airbus A320-211	0039	F-WWDM	30. 5.89	British Airways plc	Heathrow	30. 5.05T
G-BUSH	Airbus A320-211	0042	F-WWDT	19. 6.89	British Airways plc	Heathrow	18. 6.05T
G-BUSI	Airbus A320-211	0103	F-WWDB	23. 3.90	British Airways plc	Heathrow	21. 3.06T
G-BUSJ	Airbus A320-211	0109	F-WWIC	6. 8.90	British Airways plc	Heathrow	5. 8.06T
G-BUSK	Airbus A320-211	0120	F-WWIN	12.10.90	British Airways plc	Heathrow	11.10.06T
G-BUSN	RotorWay Executive 90	5141		6. 1.93	J.A McGinley	(Dublin)	6. 6.02P
	(Built B Seymour) (RotorWay RI 162)						
G-BUSR	Aero Designs Pulsar	PFA 202-12356		15.12.92	S.S.Bateman and R.A Watts	Cheddington	21. 6.05P
	(Built S.S.Bateman and R.A Watts) (Tail-wheel u/c)						
G-BUSS	Cameron Bus 90SS HAB	1685		11. 3.88	Magical Adventures Ltd *(National Express tiles)*		
						(West Bloomfield, Michigan, US)	31. 1.96A
G-BUSV	Colt 105A HAB	2324		12. 1.93	M.N.J.Kirby	(Northwich)	3. 6.04
G-BUSW	Rockwell Commander 114	14079	N4749W	18. 1.93	J.M.J.Palmer	Biggin Hill	1. 5.06
G-BUSY	Thunder Ax6-56A HAB	111		20. 6.77	M.E.Hooker *"Busy Bodies"*	(Whitchurch)	27. 4.86A
G-BUTA	CASA 1-131E Jungmann Series 2000	1101/A	Spanish AF E3B-336	20. 1.93	Angela Patrice Pearson	Breighton	24. 5.05P
		(Correct c/n not known)					
G-BUTB	CFM Streak Shadow	PFA 206-12243		20. 1.93	S.Vestuti	Swansea	26. 4.05P
	(Built F A H Ashmead - kit no K.190) (Hirth 2706 R05)						
G-BUTD	Van's RV-6	PFA 181-12152		21. 1.93	N.W.Beadle	Airfield Farm, Hardwick	17.11.05P
	(Built N Reddish) (Lycoming O-320)						
G-BUTE	Anderson EA-1 Kingfisher Amphibian	PFA 132-10798	G-BRCK	15. 8.91	T.Crawford	Cumbernauld	15.10.99P
	(Built T.Crawford) (Lycoming O-235)				*(Current status unknown)*		
G-BUTF	Aeronca 11AC Chief	11AC-1578	N3231E	21. 1.93	D Horne	North Weald	15. 2.05P
			NC3231E		tr Fox Flying Group		
G-BUTG	Zenair CH-601HD Zodiac	PFA 162-12225		22. 1.93	I.J.McNally	(Lewes)	7. 6.05P
	(Built J M Scott) (Continental C90)				*(New owner 7.04)*		
G-BUTH	CEA Jodel DR.220 2+2	6	F-BNVK	10. 2.93	T.V.Thorp	Membury	23. 6.06T
G-BUTJ	Cameron O-77 HAB	2991		25. 1.93	A.J.A and Patricia Anne Bubb *"Purple Haze"*		
					(New CofR 11.04)	(Storrington, Pulborough)	29. 6.02A
G-BUTK	Murphy Rebel	PFA 232-12091		25. 1.93	G.S.Claybourn	Walton Wood, Doncaster	17. 6.05P
	(Built D Webb) (Rotax 912-UL)						
G-BUTM	Rans S-6-116 Coyote II	PFA 204A-12414		22. 1.93	N.D.White	Buttermilk Hall Farm, Blisworth	2. 7.05P
	(Built M Rudd) (Rotax 912-UL) *(Tailwheel u/c)*				tr G-BUTM Group		
G-BUTW	British Aerospace Jetstream Series 3202	975	F-GMVI	23. 2.93	Trident Aviation Leasing Services (Ireland) Ltd		
	(Originally regd as Series 3206)		G-BUTW, (F-OHFU), G-31-975			Southampton	
G-BUTX	Bücker Bü.133C Jungmeister	Not known	Spanish AF E1-4	3. 2.93	A J.E.Smith	Leeward Air Ranch, Florida, US	9. 2.05P
	(Warner Super Scarab)		Spanish AF ES.1-4, 35-4		*(Op Real Aeroplane Company)*		
	(Possibly c/n 1010 or a CASA 1.133L)						
G-BUTY	Brugger MB.2 Colibri	PFA 43-12387		30.11.92	R.M.Lawday	(Milford, Derby)	
	(Built R.M.Lawday)						
G-BUTZ	Piper PA-28-180 Cherokee C	28-3107	G-DARL	23. 4.93	M.H.Canning	Leicester	13. 7.06T
			4R-ARL, 4R-ONE, SE-EYD				
G-BUUA	Slingsby T.67M Firefly II	2111		17. 3.93	Babcock Support Services Ltd	AAC Middle Wallop	22. 7.05T
					t/a Babcock Defence Services		
G-BUUB	Slingsby T.67M Firefly II	2112		17. 3.93	Babcock Support Services Ltd	AAC Middle Wallop	1. 6.05T
					t/a Babcock Defence Services		
G-BUUC	Slingsby T.67M Firefly II	2113		17. 3.93	Babcock Support Services Ltd	AAC Middle Wallop	14. 8.05T
					t/a Babcock Defence Services		
G-BUUD	Slingsby T.67M Firefly II	2114		17. 3.93	J.T.Matthews	Old Buckenham	12. 9.05T
G-BUUE	Slingsby T.67M Firefly II	2115		17. 3.93	J.R.Bratty	Tollerton	30. 9.05T
G-BUUF	Slingsby T.67M Firefly II	2116		17. 3.93	Northamptonshire School of Flying Ltd	Sywell	3.12.05T
G-BUUG	Slingsby T.67M Firefly II	2117		17. 3.93	Flight Training Europe SL		
						Jerez De La Frontera,Cadiz, Spain	19. 7.05T
G-BUUI	Slingsby T.67M Firefly II	2119		17. 3.93	P.M.Harrison	Sturgate	14. 1.06T
G-BUUJ	Slingsby T.67M Firefly II	2120		17. 3.93	R.Manning	Netherthorpe	20. 8.05T
G-BUUK	Slingsby T.67M Firefly II	2121		17. 3.93	Babcock Support Services Ltd	AAC Middle Wallop	22. 1.06T
					t/a Babcock Defence Services		
G-BUUL	Slingsby T.67M Firefly II	2122		17. 3.93	Witham (Specialist Vehicles) Ltd		
						(Coltersworth, Grantham)	18. 2.06T
G-BUUM	Piper PA-28RT-201 Arrow IV	28R-7918090	N2145X	14. 1.93	J.Phelan and D.G.Scott	Belfast Aldergrove	3. 4.05
					tr Bluebird Flying Group		
G-BUUO	Cameron N-90 HAB	2994		9. 2.93	M.P.Rich tr Gone Ballooning Group	(Bristol)	19. 6.03A
G-BUUP	British Aerospace ATP	2008	G-MANU	18. 2.93	Trident Aviation Leasing Services (Jersey) Ltd		
			G-BUUP, CS-TGA, G-11-8, (N378AE)			Woodford	24. 3.03T
G-BUUR	British Aerospace ATP	2024	EC-GUX	18. 2.93	Trident Aviation Leasing Services (Jersey) Ltd		
			G-OEDJ, G-BUUR, CS-TGC, G-BUUR, CS-TGC, G-11-024			Woodford	14. 8.98T
G-BUUT	Interavia 70TA HAB	04509-92		21. 1.93	Aero Vintage Ltd	(Rye)	
G-BUUU*	Cameron Bottle 77SS HAB				See SECTION 3 Part 1	Newbury	
G-BUUX	Piper PA-28-180 Cherokee D	28-5128	OY-BCW	17. 2.93	M.A.Judge tr Aero Group 78	Netherthorpe	6.11.05T

G-BUUZ	British Aerospace Jetstream Series 3202 *(Originally regd as Series 3206)*	976	F-GMVJ G-BUUZ, (F-OHFV), G-31-976	10. 3.93	Trident Aviation Leasing Services (Ireland) Ltd	(Dublin)	
G-BUVA	Piper PA-22-135 Tri-Pacer	22-1301	N8626C	12. 2.93	K.W.Thomas tr Oaksey VA Group	Oaksey Park	20. 8.06
G-BUVC	British Aerospace Jetstream Series 3202	970	F-GMVP F-GLPY, G-BUVC, F-GLPY, (F-OHFS), G-BUVC, G-31-970	10. 3.93	Air Kilroe Ltd	Manchester	5. 3.04T
G-BUVD	British Aerospace Jetstream Series 3202	977	F-GMVK G-BUVD, (F-OHFR), (F-OHFW), G-31-977 *(Op Eastern Airways)*	10. 3.93	Air Kilroe Ltd	Manchester	26. 4.07T
G-BUVE	Colt 77B HAB	2376		8. 3.93	G.D.Philpot *"Trident"*	(Hemel Hempstead)	4. 4.05A
G-BUVG	Cameron N-56 HAB	3012		8. 3.93	G.J.Bell	(Wokingham)	6. 8.05A
G-BUVL	Fisher Super Koala *(Built A D.Malcolm)* (Jabiru 2200A)	PFA 228-11399		3. 3.93	A.D.Malcolm *"Spirit of Throwley"*	Park Farm, Throwley, Faversham	30. 9.04P
G-BUVM	CEA Jodel DR.250/160 Capitaine	54	OO-NJR F-BNJR	11. 3.93	M.Lodge tr G-BUVM Group	(Huddersfield)	29.11.04
G-BUVN	CASA 1-131E Jungmann Series 2000	2092	Spanish AF EC-333 Spanish AF E3B-487	12. 3.93	W.Van Egmond *(As "BI-005" in R.Neth AF c/s)*	Hoogeveen, The Netherlands	17. 7.05P
G-BUVO	Reims/Cessna F182P Skylane II	F18200022	G-WTFA PH-VDH, D-EJCL	10. 3.93	P.N.Stapleton tr Romeo Mike Flying Group	Plymouth	1. 6.05
G-BUVR	Christen A-1 Husky	1162		12. 3.93	A.E.Poulsom	Manor Farm, Tongham	26. 4.05
G-BUVS	Colt 77A HAB	2381		12. 3.93	S.J.Chatfield	(Guildford)	11. 5.04A
G-BUVT	Colt 77A HAB	2382		12. 3.93	N.A Carr	(Leicester)	28. 3.05A
G-BUVW	Cameron N-90 HAB	3020		19. 3.93	Bristol Balloon Fiestas Ltd	(Bristol)	7. 2.01A
G-BUVX	CFM Streak Shadow SA *(Built G.K.R.Linney - kit no K.214SA)* (Rotax 582)	PFA 206-12410		22. 3.93	R.Barcis	Brook Farm, Pilling	20.12.05P
G-BUVZ	Thunder Ax10-180 Series 2 HAB	2380		24. 3.93	A.Van Wyk	(Caxton)	4. 9.04T
G-BUWE	Replica Plans SE.5A *(Built D Biggs)* (Continental C90)	PFA 20-11816		25. 3.93	Airpark Flight Centre Ltd *(As "C9533/M" in RFC c/s)*	Coventry	21. 8.05P
G-BUWF	Cameron N-105 HAB	3036		26. 3.93	R.E.Jones *"British Aerospace II"*	(Lytham St.Annes)	17. 3.05T
G-BUWH	Parsons Two-Place Gyroplane *(Built R.V.Brunskill)* (Rotax 532)	PFA G/08-1215		1. 4.93	R.V.Brunskill	Melrose Farm, Melbourne	22. 8.95P
G-BUWI	Lindstrand LBL 77A HAB	023		5. 4.93	Capital Balloon Club Ltd *"Throw Up"*	(London NW1)	30. 7.05T
G-BUWJ	Pitts S-1C *(Built J T Griffins)* (Lycoming O-320)	2002	N110R	25. 3.93	G.Breen tr G-BUWJ Flying Group	Portimão, Faro, Portugal	14. 5.05P
G-BUWK	Rans S-6-116N Coyote II *(Built R.Warriner)* (Rotax 912)	PFA 204A-12448		7. 4.93	R.Warriner	Bradleys Lawn, Heathfield	12. 6.05P
G-BUWL	Piper J/4A Cub Coupé	4-1047	N27828 NC27828	8. 4.93	M L Ryan	Garston Farm, Marshfield	
G-BUWM	British Aerospace ATP	2009	CS-TGB G-BUWM, CS-TGB, G-11-9	19. 4.93	BAE SYSTEMS (Operations) Ltd *(Stored 2.04)*	Woodford	
G-BUWR	CFM Streak Shadow *(Built T.Harvey - kit no K.177-SA)* (Rotax 582)	PFA 206-12068		26. 4.93	T.Harvey	Grove Farm, Raveningham	28.10.05P
G-BUWS	Denney Kitfox Model 2 *(Built J.E.Brewis)*	PFA 172-11831		26. 4.93	J.E.Brewis	(Castletown, Isle of Man)	
G-BUWT	Rand Robinson KR-2 *(Built C M Coombe)*	PFA 129-10952		5. 4.93	Cynthia M.Coombe	(Greenford, Middlesex)	
G-BUWU	Cameron V-77 HAB	3053		27. 4.93	T.R.Dews	(Hill Deverill, Warminster)	21. 4.04A
G-BUWZ	Robin HR200/120B	254		22. 4.93	A.N.Kaschevski	Lydd	13. 2.05T
G-BUXA*	Colt 210A HAB				See SECTION 3 Part 1	(Aldershot)	
G-BUXC	CFM Streak Shadow *(Built T Hosier - kit no K.188)* (Rotax 582)	PFA 206-12177		20. 4.93	J.P.Mimnagh	Guy Lane Farm, Waverton	12. 4.05P
G-BUXD	Maule MXT-7-160 Star Rocket *(Tri-cycle u/c)*	17001C	N9231R	4. 5.93	S.Baigent	East Winch	19.10.07E
G-BUXI	Steen Skybolt *(Built M Frankland)*	PFA 64-10755		16. 3.93	D.Tucker	Kemble	12. 8.05P
G-BUXJ	Slingsby T.61F Venture T.2	1878	XZ558	6. 5.93	D.Mihailovic tr Venture Motor Glider Club	Dunsfold	26. 6.05
G-BUXK	Pietenpol AirCamper *(Built G R G Smith)* (Continental C90)	PFA 47-11901		12. 5.93	B.P.Hogan	Sywell	14. 7.01P
G-BUXL	Taylor JT.1 Monoplane *(Built M W Elliott)*	PFA 55-11819		12. 5.93	M.Fields *(New owner 8.04)*	(Henlow)	23. 5.05P
G-BUXN	Beech C23 Sundowner 180	M-1752	N9256S	13. 5.93	D.G.Tudor tr Private Pilots Syndicate	Bournemouth	1.10.05
G-BUXO	Pober P-9 Pixie *(Built T Moore)*	PFA 105-10647		17. 5.93	J.Mangiapane tr P-9 Flying Group *(Nearing completion 2000)*	(Matlock)	
G-BUXR	Cameron A-250 HAB	3056		13. 5.93	D.S.King t/a Celebration Balloon Flights	(Nottingham)	18. 6.03T
G-BUXS	MBB Bö.105DBS-4 *(Originally c/n S.41: rebuilt 1993)*	S.913	G-PASA G-BGWP, F-ODMZ, G-BGWP, HB-XFD, N153BB, D-HDAS	19. 5.93	Bond Air Services *"Irn Bru"*	Aberdeen	25. 5.05T
G-BUXV	Piper PA-22-160 Tri-Pacer *(Super Pacer Tail-wheel conversion)*	22-6685	N9769D	20. 5.93	J.Mathews tr Romeo Delta Juliet Group *(Noted 7.04 damaged and stored)*	Kilrush, County Kildare	11.10.03
G-BUXW	Thunder Ax8-90 Series 2 HAB	2405		25. 5.93	A.S.Davidson tr Nottingham Hot Air Balloon Club *"Silver Lady"*	(Woodville, Swadlincote)	6. 7.05A
G-BUXX	Piper PA-17 Vagabond (Continental A75)	17-28	N4611H NC4611H	31. 3.93	R.H.Hunt	Old Sarum	10. 6.05P
G-BUXY	Piper PA-25-235 Pawnee	25-2705	C-GZCR N6959Z	18. 3.93	Bath, Wilts and North Dorset Gliding Club Ltd	Kingston Deverill	1.12.07E
G-BUYB	Aero Designs Pulsar *(Built A P.Fenn) (Tail-wheel u/c)*	PFA 202-12193		28. 5.93	A P.Fenn	Shobdon	10.11.05P
G-BUYC	Cameron Concept 80 HAB	3095		28. 5.93	R.P.Cross *(Windrush titles)*	(Luton)	16. 9.05A
G-BUYD	Thunder Ax8-90 HAB	2422		28. 5.93	S.and S.McGuigan	(Draperstown, Magherafelt)	21. 9.05A
G-BUYF	American Aircraft Falcon XP *(Built R W Harris)* (Rotax 503)	600179	N512AA	13. 5.93	M.J.Hadland *(New owner 11.04)*	Tarn Farm, Cockerham	21. 5.04P
G-BUYJ	Lindstrand LBL 105A HAB	039		1. 6.93	D.K.Fish and G.Fordyce	(Swindon or Olney)	14.11.04T
G-BUYK	Denney Kitfox Model 4 *(Built R D L Mayes)* (Rotax 912-UL)	PFA 172A-12214		1. 6.93	M.S.Shelton	Otherton, Cannock	25. 4.05P
G-BUYL	Rotary Air Force RAF 2000 *(Built D A Lafleur: rebuilt by Newtonair using parts from G-TXSE)*	H2-92-361	C-FPFN	2. 6.93	Newtonair Gyroplanes Ltd	Dunkeswell	5. 6.04P
G-BUYN	Cameron O-84 HAB	1214	OE-KZG	4. 6.93	Reach For The Sky Ltd	(Guildford)	4.10.03A

Reg	Type	C/n	Previous identity	Regd	Owner/Operator	Base	CA Expy
G-BUYO	Colt 77A HAB	2398		4. 6.93	S.F.Burden	(Noordwijk, The Netherlands)	17. 5.05A
G-BUYR	Mooney M.20C Mark 21	2650	N1369W	7. 6.93	Charmaine R.Weldon	Haverfordwest	15. 9.00
G-BUYS	Robin DR400/180 Régent	2197		21. 6.93	F.A Spear tr G-BUYS Flying Group	Nuthampstead	7. 4.05
G-BUYT	Ken Brock KB-2	PFA G/06-1214		7. 6.93	J.L.G.McLane and D.G.Chaplin	Melrose Farm, Melbourne	31. 3.05P
	(Built J E Harris)						
G-BUYU	Bowers Fly Baby 1A	PFA 16-12222		7. 6.93	R.Metcalfe	Rushett Farm, Chessington	5. 11.05P
	(Built J A Nugent) (Continental A65)				*(As "CL.1 1803/18" in German Army Air Service c/s to represent Junkers CL1)*		
G-BUYY	Piper PA-28-180 Cherokee B	28-1028	C-FXDP	18. 3.93	A J.Hedges and C.E.Yates	Bristol	4. 7.05T
			CF-XDP, N7214W		tr G-BUYY Group		
G-BUZA	Denney Kitfox Model 3	PFA 172-12547		10. 6.93	J.Thomas	Shefford	2. 4.05P
	(Built R Hill - c/n 1178						
G-BUZB	Aero Designs Pulsar XP	PFA 202-12312		14. 6.93	S.M.Lancashire	Lymm Dam	30. 6.05P
	(Built M J Whatley) (Tail-wheel u/c)						
G-BUZC	Everett Gyroplane Series 3A	034		14. 7.93	M.P.Lhermette	(Faversham)	
					(Damaged 7.94: stored Sproughton 12.95)		
G-BUZD	Aérospatiale AS332L Super Puma	2069	C-GSLJ	11. 2.93	CHC Scotia Ltd	Aberdeen	14.12.05T
			N189EH, C-GSLJ, HC-BNB, C-GSLJ, PT-HRN, C-GSLJ				
G-BUZE	Avid Speed Wing	PFA 189-12047		16. 6.93	H.R.Bell	Perth	24. 3.04P
	(Built N L E Nupee)						
G-BUZF	Colt 77B HAB	1993		16. 6.93	A.E.Austin *"Gee Buzz"*	(Naseby, Northampton)	27. 3.05A
G-BUZG	Zenair CH.601HD Zodiac	PFA 162-12457		17. 6.93	N.C.White	Sorbie Farm, Kingsmuir	4.10.05P
	(Built N C White) (Continental O-200-A)						
G-BUZH	Star-Lite SL-1	119	N4HC	17. 6.93	C.A.McDowall	Farley Farm, Romsey	8. 9.00P
	(Built H M Cottle) (Rotax 447)		*(Damaged Farley Farm 10.8.00: in container 9.00 - fuselage only) (Current status unknown)*				
G-BUZJ	Lindstrand LBL 105A HAB	038		17. 6.93	Eastgate Motor Co Ltd	(Bristol)	19. 7.05A
					t/a Eastgate Mazda *(Mazda titles)*		
G-BUZK	Cameron V-77 HAB	2962		17. 6.93	J.T.Wilkinson	(Blackland, Calne)	8. 6.05A
G-BUZL	VPM M-16 Tandem Trainer	VPM16-UK-105		18. 6.93	R.M.Savage	Carlisle	14. 1.04P
	(Rotax 914)				t/a Roger Savage (Photography)		
G-BUZM	Avid Speed Wing Mk.3	PFA 189-12179		30. 4.93	R.McLuckie and O.G.Jones	RAF Mona	3. 7.04P
	(Built R.McLuckie and O.G.Jones) (Jabiru 2200) *(Officially regd with Rotax 582)*						
G-BUZN	Cessna 172H	17256056	N2856L	24. 6.93	H.Jones	Barton	26.11.05
G-BUZO	Pietenpol AirCamper	PFA 47-12408		28. 6.93	D.A.Jones	(Maidenhead)	
	(Built D.A Jones) (Salmson AD9)						
G-BUZR	Lindstrand LBL 77A HAB	044		29. 6.93	Lindstrand Balloons Ltd	(Oswestry)	3. 6.05A
G-BUZS	Colt Flying Pig SS HAB	2415		2. 7.93	Banco Bilbao Vizcaya	(Bilbao, Spain)	20. 5.96A
G-BUZT	Kolb Twinstar Mk.3	PFA 205-12367		1. 7.93	M.D.Burns, T.C.Horner and D.M.Stewart		
	(Built A C.Goadby - kit no.K0009-0193)					Cumbernauld	25. 6.05P
G-BUZU*	Vickers Supermarine 379 Spitfire FR.XIV				See SECTION 3 Part 1	Wanaka, New Zealand	
G-BUZV	Ken Brock KB-2	PFA G/06-1152		1. 7.93	K.Hughes	(Amlwch, Gwynedd)	
	(Built K Hughes)						
G-BUZY	Cameron A-250 HAB	2936		29. 4.93	P.J.D.Kerr	(Bridgwater)	21. 6.05T
G-BUZZ	Agusta-Bell 206B JetRanger II	8178	F-GAMS	13. 4.78	European Skytime Ltd	Gloucestershire	3. 7.05T
			HB-XGI, OE-DXF				

G-BVAA - G-BVZZ

Reg	Type	C/n	Previous identity	Regd	Owner/Operator	Base	CA Expy
G-BVAB	Zenair CH.601HDS Zodiac	PFA 162-12475		26. 5.93	N.J.Keeling, R.F.Mclachlan and J.A.Charlton	(Ashbourne)	27. 5.04T
	(Built A R Bender) (Rotax 912-UL)						
G-BVAC	Zenair CH.601HD Zodiac	PFA 162-12504		1. 6.93	A.G.Cozens	Goodwood	6. 9.05P
	(Built A G.Cozens) (Rotax 912-UL)						
G-BVAF	Piper J-3C-65 Cub	4645	OO-UBU	14. 6.93	N.M.Hitchman	Leicester	18. 6.05P
	(Continental C85)		N28199, NC28199				
G-BVAG	Lindstrand LBL 90A HAB	022		7. 7.93	S.A.Townley *(New owner 6.04)*	Ellesmere Port	10. 6.03A
					tr The Whitchurch Aviation Training Syndicate		
G-BVAH	Denney Kitfox Model 3	PFA 172-12031		22.10.91	S.Allinson	Enstone	10.11.05P
	(Built V A Hutchinson) (Rotax 912)						
G-BVAI	PZL-110 Koliber 150	03900040	OY-CYJ	7. 7.93	A.R.Howard	Gamston	24. 7.06
G-BVAM	Evans VP-1	PFA 62-12132		7. 7.93	R.F.Selby	(Littlehampton)	
	(Built R F Selby)						
G-BVAN	SOCATA MS.892E Rallye 150GT	12376	F-BVAN	21.11.88	A.J.A.Weal	Shoreham	29.11.07E
G-BVAO	Colt 25A Sky Chariot HAB	2024		9. 7.93	Janice M.Frazer	(Hexham)	7. 5.05A
G-BVAW	Staaken Z-1 Flitzer	PFA 223-12058		12. 7.93	L.R.Williams	(Aberdare)	29. 6.99P
	(Bult L R Williams and D J Evans) (Volkswagen 1834)				t/a Flitzer Sportflugverein *(As "D692") (New owner 4.02)*		
G-BVAX	Colt 77A HAB	1213		30. 3.88	P.H.Porter *"Vax"*	(Tenbury Wells)	5. 8.95A
G-BVAY	Rutan VariEze	RS.8673/345	N5MS	3. 9.93	D.A.Young	(Sunderland)	11.11.02P
	(Built R N Saunders)						
G-BVAZ	Montgomerie-Bensen B.8MR	PFA G/01-1190		12. 7.93	N.Steele	Newtownards	22. 7.04P
	(Built R Patrick) (Rotax 582)						
G-BVBJ*	Colt Flying Coffee Jar 1 SS HAB				See SECTION 3 Part 1	(Aldershot)	
G-BVBK*	Colt Flying Coffee Jar 2 SS HAB				See SECTION 3 Part 1	(Lincoln)	
G-BVBN*	Cameron A-210 HAB	2904		2. 8.93	M.L. and S.M.Gabb	(Alcester)	10.10.01T
					t/a Heart of England Balloons *(Cancelled 13.5.03 by CAA)*		
G-BVBP	Avro 683 Lancaster B.10	Not known	RCAF KB994	4. 8.93	D.Copley	North Weald	
	(Built Victory Aircraft, Canada as B.X)				*(Forward fuselage only hangared 8.01)*		
G-BVBR	Avid Speed Wing	PFA 189-12085		3. 8.93	P.D.Thomas	(Whitchurch)	11. 9.05P
	(Built H R Rowley)						
G-BVBS	Cameron N-77 HAB	3128		4. 8.93	Marley Building Materials Ltd	(Birmingham)	3. 3.05A
G-BVBU	Cameron V-77 HAB	3076	(OO-BYS)	5. 8.93	J.Manclark *(Op Alba Ballooning)*	(Haddington)	15. 3.04A
G-BVBV	Avid Speed Wing	PFA 189-12187		4. 8.93	L.W.M.Summers	Popham	19. 5.05P
	(Built D A Jarvis)						
G-BVBX*	Cameron N-90M HAB				See SECTION 3 Part 1	Newbury	
G-BVCA	Cameron N-105 HAB	3129		11. 8.93	Unipart Group Ltd tr Unipart Balloon Club	(Cowley)	25. 6.00A
G-BVCC	Monnett Sonerai IILT	PFA 15-10547		12. 8.93	J.Eggleston	(Northallerton)	
	(Built J.Eggleston)						

G-BVCG	Van's RV-6	PFA 181-11783		17. 8.93	D.Cook	Shenstone	30. 4.05P
	(Built G J Newby and E M Farquharson) (Lycoming O-320)						
G-BVCL	Rans S-6-116 Coyote II	PFA 204A-12551		25. 8.93	J.Powell	(Fareham)	27. 6.05P
	(Built W E Willetts) (Rotax 912-UL) *(Kit no.0493.486) (Tri-cycle u/c)*						
G-BVCM	Cessna 525 CitationJet	525-0022	N1329N	2. 5.94	Kenmore Aviation Ltd and BLP 2003-19 Ltd	Edinburgh	22. 5.06
					(both) Trustees of The Aircraft Trust		
G-BVCO	Clutton FRED Series II	PFA 29-10947		25. 8.93	I.W.Bremner	(Dornoch, Sutherland)	27. 8.04P
	(Built I.W.Bremner)						
G-BVCP	Piper CP.1 Metisse	PFA 253-12512		24. 6.93	C.W.R.Piper	Hinton-in-the-Hedges	11.11.05P
	(Built C.W.R.Piper) (Revmaster 2200)						
G-BVCS	Aeronca 7AC Champion	7AC-1346	N69BD	1. 9.93	A.C.Lines	Leicester	10. 5.05P
	(Continental A65)		N82702, NC82702				
G-BVCT	Denney Kitfox Model 4-1200	PFA 172A-12456		27. 8.93	A.F.Reid	Comber, County Down	28. 2.05P
	(Built A F.Reid - c/n 1761) (Rotax 912-UL)						
G-BVCX	Sikorsky S-76A	760183	OY-HIW	21. 9.93	CHC Scotia Ltd	Aberdeen	28. 6.99T
			G-BVCX, N951L, N5450M		*(New owner 1.05)*		
G-BVCY	Cameron H-24 HAB	3136		3. 9.93	A.C.K.Rawson and J.J.Rudoni	(Stafford)	19. 1.02A
G-BVDB	Thunder Ax7-77 HAB	2364	G-ORDY	6. 9.93	M.J.Smith and J.Towler	(York)	17.10.99A
G-BVDC	Van's RV-3	PFA 99-12218		12. 7.93	J.A.A.Schofield	(Henley-on-Thames)	27. 5.00P
	(Buit D Calibritto) (Lycoming O-235)				*(On rebuild 6.01)*		
G-BVDD	Colt 69A HAB	2170		6. 9.93	R.M.Cambidge	(Oswestry)	27. 8.02P
					"Delta Dawn Fantasia" (New owner 4.03)		
G-BVDH	Piper PA-28RT-201 Arrow IV	28R-7918030	N2176L	13. 9.93	H.M.John	Cardiff	25.11.06T
G-BVDI	Van's RV-4	2058	N55GJ	13. 9.93	J.Glen-Davis Gorman	Redhill	10. 4.05P
	(Built G P Larson) (Lycoming O-320)						
G-BVDJ	Campbell Cricket rep	PFA G/03-1189		13. 9.93	Shirley Jennings	St.Merryn	5. 9.05P
	(Built C D Julian and S Jennings) (Rotax 582)						
G-BVDM	Cameron Concept 60 HAB	3141		15. 9.93	M.P.Young	(Dover)	31. 5.01A
G-BVDN	Piper PA-34-220T Seneca III	34-8133185	G-IGHA	16. 9.93	Convergence Aviation Ltd	Bournemouth	12.10.07T
			G-IPUT, N8424D				
G-BVDO	Lindstrand LBL 105A HAB	055		16. 9.93	A.E.Still	(Edgcott, Aylesbury)	11. 5.05T
G-BVDP	Sequoia F.8L Falco	PFA 100-10879		17. 9.93	T.G.Painter	Cranfield	25. 1.05P
	(Built T G Painter)						
G-BVDR	Cameron O-77 HAB	2452		21. 9.93	N.J.Logue *(New owner 8.04)*	(Pembroke Dock)	28. 3.05T
G-BVDS	Lindstrand LBL 69A HAB	102		23. 9.93	Lindstrand Hot Air Balloons Ltd *(New CofR 10.03)*	(Oswestry)	26. 6.01A
G-BVDT	CFM Streak Shadow SA-1	PFA 206-12462		23. 9.93	H.J.Bennet	(Helensburgh)	21. 4.04P
	(Built H.J.Bennet - kit no.K.223) (Rotax 582)						
G-BVDW	Thunder Ax8-90 HAB	2507		30. 9.93	S.C.Vora *"Cosmic"*	(Oadby)	30. 6.05
G-BVDX	Cameron V-90 HAB	3159	OO-BMY	30. 9.93	R.K.Scott	(Yeovil)	25. 9.05A
			G-BVDX		*"Merlin"*		
G-BVDY	Cameron Concept 60 HAB	3167		30. 9.93	J.L.Bond	(Warninglid, Haywards Heath)	12. 4.05A
					(New owner 8.04)		
G-BVDZ	Taylorcraft BC-12D	9043	N96743	21. 1.94	P.N.W.England	(Hove)	
			NC96743				
G-BVEA	Nostalgair N.3 Pup	PFA 212-11837	G-MWEA	7. 6.93	N.Lynch	Red Moor Farm, Dufield	9.11.05P
	(Built B D Godden - c/n 01-GB) (Mosler MM-CB35) *(Officially regd as "Mosler Motors N-3 Pup")*						
G-BVEH	Wassmer Jodel D.112	1294	F-BMOH	29.10.93	M.L.Copland	Breighton	6. 7.05P
G-BVEK	Cameron Concept 80 HAB	3133		5.10.93	A.D.Malcolm	(Devizes)	1. 5.05A
G-BVEN	Cameron Concept 80 HAB	3164		6.10.93	Hildon Associates Ltd	(Stockbridge)	13. 5.05A
G-BVEP	Luscombe 8A Master	1468	N28707	8.10.93	B.H.Austen	Oaksey Park	21.10.06
			NC28707				
G-BVER	de Havilland DHC-2 Beaver 1	1648	G-BTDM	13. 8.91	Seaflite Ltd	Cumbernauld	23. 4.95T
			XV268		*(As "XV268" in AAC c/s) (Stored fursela ge 1.05)*		
G-BVES	Cessna 340A II	340A0077	N1378G	8. 9.93	K.P.Gibbin and I.M.Worthington	Tollerton	8. 4.06T
G-BVEU	Cameron O-105 HAB	3145		12.10.93	H.C.Wright	(Kelfield, York)	15. 2.05T
G-BVEV	Piper PA-34-200 Seneca	34-7250316	N1428T	8.10.93	R.W.Harris, M.F.Hatt and JRB Aviation Ltd	Southend	9. 8.06T
			HB-LLN, D-GHSG, N1428T		*(Op Southend Flying Club)*		
G-BVEW	Lindstrand LBL 150A HAB	057		14.10.93	A Van Wyk	(Cambridge)	15. 8.02T
G-BVEY	Denney Kitfox Model 4-1200	PFA 172A-12527		14.10.93	J.H.H.Turner	(Houston)	7. 7.05P
	(Built J S Penny)						
G-BVEZ	Hunting Percival P.84 Jet Provost T.3A	PAC/W/9287	XM479	13.10.93	Newcastle Jet Provost Co Ltd	Newcastle	5. 9.05P
					(As "XM479/54" in RAF c/s)		
G-BVFA	Rans S-10 Sakota	PFA 194-12298		7. 9.93	D.S.Wilkinson	Kemble	27. 3.03P
	(Built D Allam and D Parkinson) (Rotax 582)				*(Noted 1.05)*		
G-BVFB	Cameron N-31 HAB	3175		20.10.93	R.Kunert	Finchampstead, Wokingham	14. 7.05A
G-BVFF	Cameron V-77 HAB	3161		26.10.93	R.J.Kerr and G.P.Allen	(Abingdon)	28. 4.05A
G-BVFM	Rans S-6-116 Coyote II	PFA 204A-12579		2.11.93	Gbvfm Flying Syndicate Ltd	Dunkeswell	17. 5.05P
	(Built P G Walton) (Rotax 912-UL - *kit no.0793.522) (Tri-cycle u/c)*						
G-BVFO	Avid Speed Wing	PFA 189-12053		9. 9.93	P.Chisman	Enstone	20. 4.05P
	(Built P Chisman)						
G-BVFP	Cameron V-90 HAB	3179		2.11.93	D.E. and J.M.Hartland	(Ashbourne)	19. 3.04A
G-BVFR	CFM Streak Shadow	PFA 206-12567		3.11.93	R.W.Chatterton	Griffins Farm, Temple Bruer	16.11.53P
	(Built M G B Stebbing - kit no.K.237-SA) (Rotax 582)						
G-BVFT	Maule M-5-235C Lunar Rocket	7183C	N6180M	5.11.93	S.R.Clark	Turweston	6. 6.06T
					tr Newnham Joint Flying Syndicate		
G-BVFU	Cameron Sphere 105SS HAB	3137		18.11.93	Stichting Phoenix	(Amsterdam, The Netherlands)	24. 6.05A
					(Greenpeace titles)		
G-BVFY	Colt 210A HAB	2493	DQ-BVF	30. 9.93	T.J.Bucknall	(Malpas)	11. 5.00T
			G-BVFY		*(Op Balloon Preservation Group) "Scotair"*		
G-BVFZ	Maule M-5-180C Lunar Rocket	8082C	N5664D	21. 2.94	J.W.Macleod	Beccles	8. 3.06
G-BVGA	Bell 206B-3 JetRanger III	2922	N54AJ	11.11.93	J.L.Leonard	Shoreham	30. 1.06T
			VH-SBC		t/a Findon Air Services		
G-BVGB	Thunder Ax8-105 Series 2 HAB2	408		11.11.93	M.E.Dunstan-Sewell	(Bristol)	5. 7.05A

Reg	Type	C/n	Previous identity	Regd	Owner/operator	Base	CofA
G-BVGE	Westland WS-55 Whirlwind HAR.10	WA/100	8732M	18.11.93	J.F.Kelly	(Mullingar, County Westmeath)	10.10.05P
			XJ729		*(As "XJ729" in RAF Rescue c/s)*		
G-BVGF	Europa Aviation Europa	PFA 247-12565		18.11.93	A.Graham and G.G.Beal	Ewesley Farm, Morpeth	23.10.05P
	(Built A Graham - kit no.034) (Tri-gear u/c)						
G-BVGG	Lindstrand LBL 69A HAB	011		30.11.93	Lindstrand Hot Air Balloons Ltd	(Oswestry)	6. 4.01A
G-BVGH	Hawker Hunter T.7	HABL 004328	XL573	26.11.93	Global Aviation Services Ltd	Humberside	22.10.05P
	(Centre fuselage no.is HABL 003360)				*(As "XL573")*		
G-BVGI	Pereira Osprey 2	PFA 70-10536		29.11.93	A.A.Knight	North Connel, Oban	25.11.03P
	(Built B Weare) (Lycoming O-320)						
G-BVGJ	Cameron Concept 80 HAB	3099		7.12.93	J.M.J and Valerie Frances Roberts	(Epping)	10.12.04A
					(New owners 12.03)		
G-BVGO	Denney Kitfox Model 4-1200	PFA 172A-12362		15.11.93	T.Marriott	Tollerton	4.11.05P
	(Built R K Dunford)						
G-BVGP	Bücker Bü.133C Jungmeister	42	F-AZMN	3.12.93	M.V.Rijkse	Booker	10.10.05P
	(Built Dornier-Werke AG)		G-BVGP, F-AZFQ, N15696, HB-MIE, D-EIII, MB-MIE, Swiss AF U-95				
G-BVGS	Robinson R22 Beta	2389	N2363S	9.12.93	Glendale Helicopter Services Ltd	(Barrhead, Glasgow)	6. 8.06T
G-BVGT	Crofton Auster V J/1A Special	PFA 00-220		19.11.93	P.N.Birch	RAF Coltishall	12. 9.03P
	(Built L A Groves) (Blackburn Cirrus 2) *(Rebuild of unregd Auster J/1 Autocrat frame used as engine test rig)*						
G-BVGW	Luscombe 8A Silvaire	4823	N2096K	18.11.93	J.C.Holland	Chilton Foliat, Hungerford	14. 7.05P
			NC2096K				
G-BVGY	Luscombe 8E Silvaire	4754	N2027K	18.11.93	M.C.Burlock	Siege Cross Farm, Thatcham	8. 6.05P
			NC2027K				
G-BVGZ	Fokker Dr.1 Triplane rep	PFA 238-12654		20.12.93	R.A.Fleming	Breighton	20. 5.05P
	(Built V H Bellamy - c/n VHB-10) (Lycoming AIO-360)				*(As "450/17" in German Army Air Service c/s)*		
G-BVHC	Grob G115D-2 Heron	82005	D-EARG	14.12.93	VT Aerospace Ltd *(Op Royal Navy)*	Plymouth	30. 3.06T
G-BVHD	Grob G115D-2 Heron	82006	D-EARJ	14.12.93	VT Aerospace Ltd *(Op Royal Navy)*	Plymouth	5. 6.06T
G-BVHE	Grob G115D-2 Heron	82008	D-EARQ	14.12.93	VT Aerospace Ltd *(Op Royal Navy)*	Plymouth	27. 3.06T
G-BVHF	Grob G115D-2 Heron	82011	D-EARV	14.12.93	VT Aerospace Ltd *(Op Royal Navy)*	Plymouth	18. 5.06T
G-BVHG	Grob G115D-2 Heron	82012	D-EARX	14.12.93	VT Aerospace Ltd *(Op Royal Navy)*	Plymouth	8. 5.06T
G-BVHI	Rans S-10 Sakota	PFA 194-12608		20.12.93	P.D.Rowley	(Godalming)	2. 6.99P
	(Built P.D.Rowley) (Rotax 582)				*(Stored 6.99)*		
G-BVHK	Cameron V-77 HAB	3209		23.12.93	Ann R.Rich *"Intel Inside"*	(Hyde)	11. 4.05A
G-BVHL	Nicollier HN.700 Menestrel II	PFA 217-12614		24.12.93	W.Goldsmith	(Boldon Colliery)	
	(Built I H R Walker and C Herbert)						
G-BVHM	Piper PA-38-112 Tomahawk	38-79A0313	G-DCAN	14.11.91	A.J.Gomes	Shoreham	18. 8.05T
			N2490D		*(Op Sky Leisure Aviation)*		
G-BVHO	Cameron V-90 HAB	3158		29.12.93	N.W.B.Bews	(Tenbury Wells)	18. 7.03
G-BVHP	Colt 42A HAB	2533		31.12.93	Danny Bertels Ballooning BVBA	(Wommelgem, Belgium)	24. 7.04
G-BVHR	Cameron V-90 HAB	3174		5. 1.94	G.P.Walton	(Bagshot)	17. 7.04T
G-BVHS	Murphy Rebel	PFA 232-12180		5. 1.94	J.R.Malpass	(Coal Aston)	8. 7.04P
	(Built J Brown, B Godden and M Hanley - c/n 050)						
G-BVHT	Avid Speed Wing Mk.4	PFA 189-12226		28.10.93	R.S.Holt	Long Marston	17. 6.04P
	(Built R.S.Holt)						
G-BVHV	Cameron N-105 HAB	3215		6. 1.94	Wye Valley Aviation Ltd *(Rover titles)*	(Ross-on-Wye)	9. 9.05T
G-BVHY*	Pilatus Britten-Norman BN-2T-4R Defender 4000	4004		21. 1.94	B N Group Ltd	Bembridge	
					(Stored 11.00: cancelled 6.4.04 as WFU)		
G-BVIA	Rand Robinson KR-2	PFA 129-11004		14. 1.94	K.Atkinson	(Ulverston)	2. 3.05P
	(Built K.Atkinson)						
G-BVIE	Piper PA-18 Super Cub 95 (L-18C-PI)	18-1549	G-CLIK	26. 1.94	J.C.Best tr C'est La Vie Group	Andrewsfield	19. 3.05P
	(Continental O-200-A) *(Frame No.18-1521)*		(G-BLMB), D-EDRB, ALAT 18-1549, 51-15549 *"C'est La Vie"*				
G-BVIF	Montgomerie-Bensen B.8MR	PFA G/01A-1228		26. 1.94	R.M. and D.Mann	(Brodick, Arran)	21. 8.95P
	(Built R M Mann) (Rotax 582)				*(Noted 4.00)*		
G-BVIG*	Cameron A-250 HAB	3213		26. 1.94	Balloon Flights International Ltd	(Bristol)	2. 9.01T
					(Cancelled 24. 6.03 as wfu) (Stored 2004)		
G-BVIK	Maule MXT-7-180 Star Rocket	14056C		31. 1.94	D.S.Simpson tr Graveley Flying Group	Graveley	17. 8.06
G-BVIL	Maule MXT-7-180 Star Rocket	14059C		31. 1.94	K. and S.C.Knight	Shobdon	5. 8.06
G-BVIN	Rans S-6-ESA Coyote II	PFA 204-12533		25.10.93	T.J.Wilkinson	Sackville Lodge Farm, Riseley	9. 7.05P
	(Built K J Vincent - kit no.1292.406) (Tail-wheel u/c)						
G-BVIO*	Colt Flying Drinks Can SS HAB				See SECTION 3 Part 1	(Aldershot)	
G-BVIR	Lindstrand LBL 69A HAB	079		2. 2.94	Aerial Promotions Ltd (*Vauxhall titles)*	(Cannock)	25. 5.01A
G-BVIS	Brugger MB.2 Colibri	PFA 43-10666		2. 2.94	B.H.Shaw	Spanhoe	8. 1.05P
	*(Built B.H.Shaw) (*Volkswagen 1600)						
G-BVIT	Campbell Cricket rep	PFA G/03-1229		4. 2.94	D.R.Owen	(Blackburn)	24. 7.97P
	(Built A N Nesbit) (Rotax 582)				*(New owner 3.01)*		
G-BVIV	Avid Speed Wing	PFA 189-12034		25.10.93	J.K.Davies	(Eccleston, Chester)	17. 3.00P
	(Built V and J Hobday)				*(New owner 11.04*		
G-BVIW	Piper PA-18-150 Super Cub	18-8277	SE-EPD	4. 2.94	M.J.Medland	(Leicester)	29. 6.06T
G-BVIX	Lindstrand LBL 180A HAB	082		8. 2.94	European Balloon Display Co Ltd	(Great Missenden)	31. 3.01T
					"Drifter"		
G-BVIZ	Europa Aviation Europa	PFA 247-12601		24. 1.94	T.J.Punter and P.G.Jeffers	Booker	21. 3.05P
	(Built T.J.Punter and P.G.Jeffers - kit no.052) (Monowheel u/c)						
G-BVJC	Fokker F.28 Mk.0100	11497	PH-EZJ	2.12.94	British Midland Airways Ltd		
	(Fokker 100)				*(Op bmiRegional)*	Nottingham East Midlands	1.12.06T
G-BVJD	Fokker F.28 Mk.0100	11503	PH-EZO	14.12.94	British Midland Airways Ltd		
	(Fokker 100)				*(Op bmiRegional)*	Nottingham East Midlands	13.12.06T
G-BVJE	Aérospatiale AS350B1 Ecureuil	1991	SE-HRS	3. 2.94	PLM Dollar Group Ltd	Inverness	24. 2.06T
G-BVJF	Montgomerie-Bensen B.8MR	PFA G/01-1082		18. 2.94	D.M.F.Harvey	(Yate, Bristol)	
	(Built D M F Harvey)						
G-BVJG	Cyclone AX3K/582	PFA 245-12663	G-69-14	15. 2.94	T.D.Reid	Tandragee, County Armagh	2.12.05P
	(Built T D Reid - c/n C.3123187)		(G-MYOP)				
G-BVJH	Aero Designs Pulsar	PFA 202-12196		22. 2.94	J.Stringer	Stapleford	24. 5.05P
	(Built J A C Tweedie) (Tri-cycle u/c)						
G-BVJK	Glaser-Dirks DG-800A	8-24-A21		30. 3.94	B.A.Eastwell and J.S.Forster	Ringmer	1. 4.07

Registration	Type	C/n	Previous identity	Reg date	Owner/operator	Location	CofA expiry
G-BVJN	Europa Aviation Europa *(Built N Adam - kit no.066) (Tri-gear u/c)*	PFA 247-12666		2. 3.94	C.C.Burgess tr JN Europa Group *"Better by Redesign"*	White Waltham	19. 5.05P
G-BVJT	Reims/Cessna F406 Caravan II	F406-0073		2. 2.94	M.Evans and A Jay t/a Nor Leasing	Farnborough	29. 3.06
G-BVJU	Evans VP-1 *(Built R Waring and B A Schlussler)*	PFA 62-10691		10. 3.94	B A Schlussler	(Bourne)	
G-BVJX	Marquart MA.5 Charger *(Built M L Martin)* (Lycoming O-360)	PFA 68-11239		12. 1.94	E.Newsham	High Flatts Farm, Chester Le Street	8. 7.05P
G-BVJZ	Piper PA-28-161 Cherokee Warrior II	28-7816248	N2088M	22. 3.94	A.R.Fowkes	Denham	13. 8.06T
G-BVKA	Boeing 737-59D	24694	SE-DNA (SE-DLA)	15. 2.94	British Midland Airways Ltd *(Op bmiBaby)*	Nottingham East Midlands	28. 2.06T
G-BVKB	Boeing 737-59D	27268	SE-DNM	24. 3.94	British Midland Airways Ltd *(Op bmiBaby)*	Nottingham East Midlands	11. 4.06T
G-BVKD	Boeing 737-59D	26421	SE-DNK	25.11.94	British Midland Airways Ltd *"Wales logojet"* *(Op bmiBaby)*	Nottingham East Midlands	15.12.06T
G-BVKF	Europa Aviation Europa *(Built T.R.Sinclair - kit no.050) (Tri-gear u/c)*	PFA 247-12638		11. 3.94	T.R.Sinclair	Lamb Holm Farm, Orkney	1. 6.05P
G-BVKH	Thunder Ax-8-90 HAB	2574		15. 3.94	L.Ashill	(Combe Down, Bath)	12. 3.05
G-BVKJ	Bensen B.8M *(Built J Bagnall)* (Arrow GT500R)	PFA G/01-1221		17. 3.94	A.G.Foster	Grimsby	27. 8.99P
G-BVKK	Slingsby T.61F Venture T.2	1984	ZA665	22. 2.94	Buckminster Gliding Club Ltd	Saltby	27. 3.05
G-BVKL	Cameron A-180 HAB	3255		17. 3.94	Dragon Balloon Company Ltd	(Castleton, Hope Valley)	1. 4.01T
G-BVKM	Rutan VariEze *(Built K H Duncan)* (Continental O-200-A)	1933	N7137G	5. 4.94	J.P.G.Lindquist	(Kilchberg, Switzerland)	18.10.05P
G-BVKR	Sikorsky S-76A	760115	RJAF 734	4. 3.94	Bristow Helicopters Ltd	Aberdeen	8.12.06T
G-BVKU	Slingsby T.61F Venture T.2	1877	XZ557	22. 3.94	N.R.Bowers tr G-BVKU Syndicate	Kingston Deverill	6.12.07
G-BVKX	Colt 14A Cloudhopper HAB	2580		28. 3.94	H.C.J.Williams	(Langford, Bristol)	
G-BVKZ	Thunder Ax9-120 HAB	2547		23. 3.94	D.J.Head	(Newbury)	25. 7.00T
G-BVLD	Campbell Cricket *(Built C Berry)* (Arrow GT500)	PFA G/01A-1163		29. 3.94	C.Berry	Melrose Farm, Melbourne	14. 6.05P
G-BVLE	McCandless M.4 *(Built H Walls)*	PFA G/10-1232		29. 3.94	H.Walls *(Under construction 11.01)*	Victoria Bridge, Strabane	
G-BVLF	CFM Starstreak Shadow SS-D *(Built B.R.Johnson)*	K.250-SSD		4. 3.94	Skyview Systems Ltd	Waits Farm, Belchamp Walter	5. 8.05P
G-BVLG	Aérospatiale AS355F1 Ecureuil 2	5011	N57745	31. 3.94	PLM Dollar Group Ltd	Cumbernauld	6. 4.06T
G-BVLI	Cameron V-77 HAB	5568	N9544G	30. 3.94	J.Lewis-Richardson	(Wakefield, Nelson, New Zealand)	29.12.04A
G-BVLL	Lindstrand LBL 210A HAB	101		9. 3.94	Aerial Promotions Ltd	(Cannock)	19. 7.05T
G-BVLM*	de Havilland DH.115 Vampire T.55				See SECTION 3 Part 1	Amman, Jordan	
G-BVLP	Piper PA-38-112 Tomahawk II	38-82A0002	N91355	8. 4.94	Turweston Aero Club Ltd	Turweston	26. 5.06T
G-BVLR	Van's RV-4 *(Built S.D.Arnold)* (Lycoming 0320-E2A	PFA 181-12306		13. 4.94	S.D.Arnold and S.J.Moody tr RV4 Group *(Under construction 7.99 - current status unknown)*	(Coventry)	
G-BVLT	Bellanca 7GCBC Citabria 150S	1103-79	SE-GHV	6. 4.94	M.D.Hinge	Old Sarum	2. 9.05T
G-BVLU	Druine D.31 Turbulent *(Built C.D.Bancroft)* (Jabiru 2200A)	PFA 1604		18. 4.94	C.D.Bancroft	Litlte Down Farm,Milson	17.11.04P
G-BVLV	Europa Aviation Europa *(Built J.T.Naylor - kit no.039) (Monowheel u/c)*	PFA 247-12585		10. 3.94	J.T.Naylor tr Euro 39 Group	Bidford	18. 4.05P
G-BVLW	Avid Hauler Mk.4 *(Built D M Johnstone)* (Hirth F30)	PFA 189-12577		24. 3.94	D.M.Johnstone	Shobdon	20.10.05P
G-BVLX	Slingsby T.61F Venture T.2	1973	ZA654	19. 4.94	T.M.Holloway tr RAFGSA *(Op Fulmar Gliding Club)*	Easterton	28. 4.07
G-BVLZ	Lindstrand LBL 120A HAB	063		4. 3.94	Balloon Flights Club Ltd	(Kings Norton, Leicester)	6.10.03T
G-BVMA	Beech 200 Super King Air	BB-797	G-VPLC N84B	22. 7.93	Dragonfly Aviation Services LLP	(Rhoose, Barry)	26.10.07E
G-BVMC	Robinson R44 Astro	0060		15. 4.94	B.E.Llewllyn t/a Bell Commercials	Swansea	23. 7.06T
G-BVMD	Luscombe 8E Silvaire	5265	9Q-CGB KAT-?, VP-YRB, ZS-BWC, NC2538K	15. 4.94	P.J.Kirkpatrick	Top Farm, Croydon, Royston	26. 6.05P
G-BVMF	Cameron V-77 HAB	3195		22. 4.94	P.A Meecham	(Milton-under-Wychwood)	1. 9.05A
G-BVMG*	Bensen B.80V *(Built D Moffat)*	PFA G/01-1056		25. 4.94	D Moffat *(Cancelled by CAA 11.12.00 - no PtoF issued) (Stored 2.03)*	Lochview House, Limerigg	
G-BVMH	Wag-Aero Sport Trainer *(Built D M Jagger)* (Continental C90-8)	PFA 108-12647		28. 4.94	J.Mathews *(As "624/D-39" in US Army c/s)*	(Trim, County Meath)	12. 6.05P
G-BVMI	Piper PA-18-150 Super Cub *(Frame No.18-4613)* *(Officially regd with c/n 18-8482 ex OH-PIN/N4262Z but rebuilt from D-EIAC [18-4649] after crash 15.8.95)*	18-4649	D-EIAC (PH-WDP), D-EIAC, D-EKAF, N10F	6. 4.94	S Sampson	Bagby	23. 3.07
G-BVMJ	Cameron Eagle 95SS HAB	3262		28. 4.94	R.D.Sargeant	(Wollerau, Switzerland)	23. 8.05A
G-BVML	Lindstrand LBL 210A HAB	094		29. 4.94	Ballooning Adventures Ltd	(Hexham)	25. 3.03T
G-BVMM	Robin HR200/100 Club	41	F-BVMM	18. 8.80	R.H.Ashforth	Gloucestershire	30. 7.04
G-BVMN	Ken Brock KB-2 *(Built S McCullagh)*	PFA G/06-1218		29. 4.94	S.A.Scally	(Cockermouth)	4. 7.01P
G-BVMP	British Aerospace BAe 146 Series 200	E2210	I-FLRE G-BVMP, G-6-210	20. 5.94	Trident Jet (Jersey) Ltd *(Stored 9.04)*	Kemble	19. 5.95
G-BVMR	Cameron V-90 HAB	3269		28. 3.94	I.R.Comley *"Midnight Rainbow"*	(Churchdown, Gloucester)	30. 5.05A
G-BVMS	British Aerospace BAe 146 Series 200	E2227	I-FLRO G-BVMS, G-6-227	21. 6.94	Trident Jet (Jersey) Ltd *(Stored 9.04)*	Kemble	11.10.94
G-BVMU	Aerostar Yakovlev Yak-52	9411809	YR-013	11. 5.94	A.L.Hall-Carpenter *(As "09" in DOSAAF c/s)*	Shipdham	7.11.05P
G-BVNG	de Havilland Moth Major	Not known	EC-AFK EE1-81, 30-81	17. 5.94	P. and G.Groves *(On rebuild 4.03)*	Lee-on-the-Solent	
G-BVNI	Taylor JT.2 Titch *(Built T.V.Adamson)*	PFA 60-11107		20. 5.94	T.V.Adamson *(Noted 7.01: current status unknown)*	Rufforth	
G-BVNL	Rockwell Commander 114	14118	I-ECCE N4798W	13. 5.94	A.W.Scragg	Leicester	5. 7.06
G-BVNR	Cameron N-105 HAB *(New envelope c/n 4994 @ 1.01)*	3288		24. 5.94	Liquigas SpA	(Milano, Italy)	6. 1.05A

G-BVNS	Piper PA-28-181 Cherokee Archer II	28-7690358	N6163J	13. 4.94	Scottish Airways Flyers (Prestwick) Ltd	Prestwick	27. 8.06T
G-BVNU	FLS Aerospace Sprint Club	004		25. 5.94	Aces High Ltd *(Noted 4.04)*	(Dunsfold)	17.10.98T
G-BVNY	Rans S-7 Courier	PFA 218-11951		24. 5.94	D.M.Byers-Jones	(Alton)	8. 6.05P
	(Built J Whiting) (Rotax 532)				*(New owner 7.04)*		
G-BVOA	Piper PA-28-181 Archer II	28-7990145	N2132C	31. 5.94	M.J. and R.J.Millen	Rochester	20. 7.06T
					t/a Millen Aviation Services *(Cable Consult Ltd titles)*		
G-BVOB	Fokker F.27 Friendship 500	10366	PH-FMN	5. 7.94	BAC Express Airlines Ltd	Exeter	6.10.06T
			PT-LZM, F-BPNA, PH-FMN		*"Euro Trader"*		
G-BVOC	Cameron V-90 HAB	3291		8. 6.94	Sally A.Masey	(Bristol)	14. 6.04A
					(Bristol Evening Post/Western Daily Press titles) "Scoop"		
G-BVOD*	Montgomerie-Parsons Two-Place Gyroplane	G/08-1238		8. 6.94	J.M.Montgomerie	Crosshill	
					(Cancelled 23.11.00 as WFU - no PtoF issued) (Noted 5.04)		
G-BVOH	Campbell Cricket rep	PFA G/03-1220		14. 6.94	G.A.Kitson	(Holyhead)	25. 6.04P
	(Built B F Pearson) (Rotax 532)						
G-BVOI	Rans S-6-116 Coyote II	PFA 204A-12712		14. 6.94	G.J.Chater	Chiltern Park, Wallingford	2. 6.05P
	(Built A P Bacon) (Rotax 582)						
G-BVOK	Aerostar Yakovlev Yak-52	9111505	RA-9111505	14. 6.94	T.Maloney, J.Ormerod and D.Treacher	Shoreham	9. 7.05P
			DOSAAF55		t/a Transair Aviation *(As "55" in DOSAAF c/s)*		
G-BVON	Lindstrand LBL 105A HAB	001	N532LB	16. 6.94	N.D.Hicks	(Blacknest, Alton)	28.12.04
			G-BVON				
G-BVOO	Lindstrand LBL 105A HAB	123		16. 6.94	T.G.Church	(Blackburn)	18. 1.01T
G-BVOP	Cameron N-90 HAB	3317		21. 6.94	October Gold Ballooning Ltd	(Windermere)	22. 2.05T
					t/a Mr.Lazenbys		
G-BVOR	CFM Streak Shadow	PFA 206-12695		31. 3.94	K.Fowler	(Basingstoke)	19. 7.05P
	(Built D Lord - kit no K.238-SA) (Rotax 582)						
G-BVOS	Europa Aviation Europa	PFA 247-12562		11. 4.94	D.A Young	Shotton Colliery, Peterlee	3. 9.05P
	(Built D Collinson and D.A Young- kit no.003) (Mid-West AE100R) *(Monowheel u/c)*				tr Durham Europa Group		
G-BVOU	Hawker Siddeley HS.748 Series 2A/270	1721	CS-TAH	21. 6.94	Emerald Airways Ltd	Exeter	30. 7.07T
			G-11-6		*(Lynx Parcels titles)*		
G-BVOV	Hawker Siddeley HS.748 Series 2A/372	1777	CS-TAO	21. 6.94	Emerald Airways Ltd	Liverpool	11. 5.07T
			G-11-4				
G-BVOW	Europa Aviation Europa	PFA 247-12679		27. 6.94	N.M.Robbins	Sleap	27.10.05P
	(Built M.W.Cater - kit no.084) (Monowheel u/c)						
G-BVOX	Taylorcraft F-22	2208	N221UK	20. 5.94	Jones Samuel Ltd	Leicester	14. 8.06
G-BVOY	RotorWay Executive 90	5238		17. 6.94	Southern Helicopters Ltd	Street Farm, Takeley	AC
	(Built N J Bethell) (RotorWay RI 162)				*(On rebuild 12.03)*		
G-BVOZ	Colt 56A HAB	2595		21. 6.94	Balloon School (International) Ltd	(Petworth)	2. 8.05A
					t/a British School of Ballooning		
G-BVPA	Thunder Ax8-105 Series 2 HAB	2600		24. 6.94	J.Fenton t/a Firefly Balloon Promotions	(Preston)	23. 4.05T
G-BVPD	CASA 1-131E Jungmann	2086	F-AZNG	12. 7.94	D.Bruton	Abbeyshrule, County Longford	2. 2.05P
			Spanish AF E3B-482				
G-BVPK	Cameron O-90 HAB	3313		1. 7.94	D.V.Fowler	(Cranbrook)	5. 7.05T
G-BVPL	Zenair CH.601HD Zodiac	PFA 162-12693		4. 7.94	J.G.Munro	Perth	10.11.05P
	(Built D Harker) (Continental O-200-A)						
G-BVPM	Evans VP-2	PFA 7205		6.11.78	P.Marigold	(Locking, Weston-super-Mare)	31. 5.94P
	(Built P.Marigold - c/n V2-1016) (Continental A65-8)					*(Stored 7.95)*	
G-BVPN	Piper J-3C-65 Cub	6917	G-TAFY	6. 7.94	K.I.Munro	Draycott Farm, Chiseldon	20. 7.05P
			N31073, N38207, N38307, NC38307				
	(Officially regd with c/n 5298 but has Frame No.7002 which was N38207: probably used in rebuild of N31073 in early 1970s)						
G-BVPO*	de Havilland DH.100 Vampire FB.6				See SECTION 3 Part 1	Amman, Jordan	
G-BVPP	Folland Gnat T.1	FL.536	8620M	22. 4.94	Red Gnat Ltd	North Weald	14. 1.05P
			XP534		*(As "XR993" in Red Arrows c/s)*		
G-BVPR	Robinson R22 Beta	1612	G-KNIT	17. 6.94	Helicentre Blackpool Ltd	Blackpool	4. 2.06T
G-BVPS	Jodel D.11	PFA 917		6. 7.94	P.J.Sharp	Rush Green	22. 7.05P
	(Built P J Sharp - plans no.1403) (Continental A65)						
G-BVPV	Lindstrand LBL 77B HAB	119		13. 7.94	A.R.Greensides	(Burton Pidsea, Hull)	20. 6.05A
					"Reverend Leonard"		
G-BVPW	Rans S-6-116 Coyote II	PFA 204A-12737		12. 7.94	J.G.Beesley	Halwell, Totnes	17. 3.05P
	(Built J.G.Beesley - kit no. 0294.587) (Rotax 582) *(Tri-cycle u/c)*						
G-BVPX	Bensen B.8 Tyro Gyro Mk.II	PFA G/11-1237		13. 7.94	A.W.Harvey	Henstridge	22. 1.05P
	(Built P C Lovegrove - kit no.PCL125)						
G-BVPY	CFM Streak Shadow	PFA 206-12375		14. 6.94	R.J.Mitchell	(Scalloway, Shetland)	30. 6.04P
	(Built R.J.Mitchell - kit no.K.204) (Rotax 582)				*(Operates from Tingwall)*		
G-BVRA	Europa Aviation Europa	PFA 247-12635		25. 7.94	N.E.Stokes	Sleap	10.11.05P
	(Built E J J Pels - kit no.008) (Monowheel u/c)				*"Hummingbird"*		
G-BVRH	Taylorcraft BL-65	1657	N23929	15. 7.94	M.J.Smith	Partridge Green	23. 9.06
			G-BVRH, N24322, NC24322				
G-BVRI	Thunder Ax6-56 HAB	2622		2. 8.94	P.C.Bailey *(New owner 6.03)*	(Cambridge)	30.11.02A
G-BVRK	Rans S-6-ESA Coyote II	1193.566	G-MYPK	14. 7.94	J.Secular	(Beckenham)	
	(Built J Secular)						
G-BVRL	Lindstrand LBL 21A HAB	130		3. 8.94	A.M.Holly t/a Exclusive Ballooning	(Berkeley)	17. 1.05A
G-BVRR	Lindstrand LBL 77A HAB	133		9. 8.94	M.Icam	(Carquefou, France)	11. 7.04A
G-BVRU	Lindstrand LBL 105A HAB	131		15. 8.94	R.P.Nash	(Beighton, Norwich)	10. 9.05A
G-BVRV	Van's RV-4	793	N144TH	23. 6.94	A.Troughton	Armagh Field, Woodview	28. 4.05P
	(Built C Thomas Hahn) (Lycoming AEIO-320)						
G-BVRZ	Piper PA-18 Super Cub 95	18-3442	SE-ITP	22.11.94	R.W.Davison	(Wirral)	8. 9.07
	(Regd with Frame No.18-3381)		LN-LJG, D-EDCM, 96+19, QW+901, QZ+001, AC+507, AS+506, 54-752				
G-BVSB	TEAM mini-MAX 91A	PFA 186-12241		1. 7.94	D.G.Palmer	Fetterangus	30. 9.05P
	(Built C Nice) (Rotax 503)						
G-BVSD	Sud-Aviation SE.3130 Alouette II	1897	Swiss AF V-54	8. 9.94	M.J.Cuttell	Gloucestershire	17. 4.05
					(As "V-54" in Swiss AF c/s)		
G-BVSF	Aero Designs Pulsar	PFA 202-12071		1. 7.94	S.N. and R.J.Freestone	Deanland	26. 7.05P
	(Built S.N. and R.J.Freestone) (Tri-cycle u/c)						

Reg	Type	C/n	Previous identity	Reg date	Owner (Operator)	Probable Base	CA Expy
G-BVSM	Rotary Air Force RAF 2000	EW-42		24. 8.94	S.Ram	(Lowestoft)	24. 1.97P
	(Built K Quigley and T M Truesdale) (Subaru EA82)				*(Noted 7.02)*		
G-BVSN	Avid Speed Wing	PFA 189-12088		24. 8.94	S.J.Pemberton	Franklyn's Field, Chewton Mendip	9.11.05P
	(Built D J Park)						
G-BVSO	Cameron A-120 HAB	3339		25. 8.94	A.Kaye	(Wellingborough)	6. 5.05T
					t/a Khaos Ballooning *(Cameron Balloons titles)*		
G-BVSP	Hunting Percival P.84 Jet Provost T.3A	PAC/W/6327	XM370	31. 8.94	H.G.Hodges and Son Ltd	Long Marston	10. 6.04P
G-BVSS	Jodel D.150 Mascaret	PFA 151-11878		22. 8.94	A.P.Burns	RAF Woodvale	19. 5.04P
	(Built A P Burns - c/n 118) (Continental O-200-A)						
G-BVST	Jodel D.150 Mascaret	PFA 235-12198		11. 8.94	A.Shipp	Full Sutton	5.10.04P
	(Built A Shipp - c/n 130) (Continental O-200-A)						
G-BVSX	TEAM mini-MAX 91A	PFA 186-12463		9. 9.94	J.A Clark	Orchard Farm, Sittingbourne	18. 7.03P
	(Built G N Smith) (Mosler MM CB-35)						
G-BVSY	Thunder Ax9-120 HAB	2631		16. 8.94	G.R.Elson	Newtown Hungerford	6. 6.03T
G-BVSZ	Pitts S-1E	PFA 09-11235		9. 9.94	R.C.F.Bailey	Swinmore Farm, Ledbury	24.10.05P
	(Built K Garrett and R P Millinship)						
G-BVTA	Tri-R Kis	PFA 239-12450		26. 8.94	P.J.Webb	Dunkeswell	8. 2.05P
	(Built P.J.Webb) (Continental O-240-E)						
G-BVTC	British Aircraft Corporation BAC.145 Jet Provost T.5A	EEP/JP/997	XW333	7. 9.94	Global Aviation Ltd	Humberside	28. 8.05P
					(As "XW333")		
G-BVTD	CFM Streak Shadow SA	PFA 206-11972		14. 9.94	M.Walton	Old Sarum	26. 5.04P
	(Built M.Walton - kit no K.159-SA) (Rotax 582)						
G-BVTL	Colt 31A Air Chair HAB	2572		5. 7.94	A Lindsay	(Twickenham)	15. 5.97
G-BVTM	Reims/Cessna F152 II	F15201827	G-WACS D-EFGZ	31. 8.94	RAF Halton Aeroplane Club Ltd	RAF Halton	16. 9.07T
G-BVTN	Cameron N-90 HAB	3361		16. 9.94	P.Zulehner	(Peterskirchen, Austria)	27. 9.02A
G-BVTO	Piper PA-28-151 Cherokee Warrior	28-7415253	G-SEWL D-EDOS, N9550N	19. 9.94	A.S.Bamrah t/a Falcon Flying Services	Lydd	10. 2.07T
G-BVTV	RotorWay Executive 90	5243		16. 9.94	M.J.Wasley	Street Farm, Takeley	12. 8.04P
	(Built J J Bull) (RotorWay RI 162)						
G-BVTW	Aero Designs Pulsar	PFA 202-12172		14. 9.94	J.D.Webb	Rhosgoch	14. 9.05P
	(Built J.D.Webb)						
G-BVTX	de Havilland DHC-1 Chipmunk 22A	C1/0705	WP809	2. 8.94	M.W.Cater	Sywell	11.11.04
					tr TX Flying Group *(As "WP809/78" in RN c/s) (Noted 1.05)*		
G-BVUA	Cameron O-105 HAB	3369		27. 9.94	D.C.Eager	(Bracknell)	28. 5.00A
G-BVUC	Colt 56A HAB	2608	G-639	30. 9.94	Cameron Balloons Ltd	(Pickering)	8. 3.05A
	("B" Conditions markings carried as such 9.94)				t/a Thunder and Colt *(Corks and Cans Norton titles)*		
G-BVUG	Betts TB.1	PFA 265-12770		3.10.94	William Tomkins Ltd	Spanhoe	25.10.05P
	(Built T A Betts) (Tigre G IV-A2 @ 120hp) *(Modified AIA Stampe SV.4C c/n1045 ex G-BEUS/F-BKFK/F-DAFK/Fr.Mil)*						
G-BVUH	Thunder Ax6-65B HAB	243	JA-A0075	3.10.94	K.B.Chapple	(Sulham, Reading)	1.11.05E
G-BVUI	Lindstrand LBL 25A Cloudhopper HAB	148		5.10.94	J.W.Hole	Much Wenlock	29. 9.04A
G-BVUJ	Ken Brock KB-2	PFA G/06-1244		10.10.94	R.J.Hutchinson	Kemble	17. 5.99P
	(Built R.J.Hutchinson) (Rotax 503)						
G-BVUK	Cameron V-77 HAB	3372		11.10.94	H.G.Griffiths and W.A.Steel	(Reading)	31. 8.05A
G-BVUM	Rans S-6-116 Coyote II	PFA 204A-12685		11.10.94	M.A.Abbott	Glenrothes	19.10.05P
	(Built J L Donaldson) (Rotax 582)						
G-BVUN	Van's RV-4	PFA 181-12488		11.10.94	A.E.Kay	Parsons Farm, Waterperry Common, Oakley	15. 5.05P
	(Built I G Glenn - kit no.3363UK) (Lycoming O-360)						
G-BVUO	Cameron R-150 Gas/HAB	3365		13.10.94	M.Sevrin	(Court St.Etienne, Belgium)	21.12.95A
G-BVUT	Evans VP-1 Series 2	PFA 62-12092		24.10.94	M.J.Barnett	Shobdon	29. 9.99P
	(Built P.J.Weston) (Volkswagen 1600)				*(Damaged on take off Pepperbox 13.3.99: new owner 11.04)*		
G-BVUU	Cameron C-80 HAB	3383		11.10.94	T M C McCoy	(Peasedown St.John, Bath)	15. 6.05T
					"Ascent" (Op Ascent Balloons)		
G-BVUV	Europa Aviation Europa	PFA 247-12762		23. 9.94	R.J.Mills	Gamston	8. 8.05P
	(Built R.J.Mills - kit no.141) (Monowheel u/c)						
G-BVUZ	Cessna 120	11334	Z-YGH VP-YGH, VP-NAM, VP-YGH	20. 9.94	N.O.Anderson	(Trumpington, Cambridge)	
G-BVVA	IAV-Bacau Yakovlev Yak-52	8776109	LY-ANN DOSAAF 52	24.10.94	T.W.Freeman	Swansea	14. 1.05P
G-BVVB	Carlson Sparrow II	PFA 209-11809		26. 9.94	L.M.McCullen	North Connel, Oban	26. 5.05P
	(Built L M McCullen) (Rotax 532)						
G-BVVE	Wassmer Jodel D.112	1070	F-BKAJ	28.10.94	G.D.Gunby	Crowfield	16. 5.05P
G-BVVG	Nanchang CJ-6A	2751219	(F-....) G-BVVG, Chinese PLAAF	10.10.94	K.E.Wells tr Peeking Duck Group	White Waltham	2.12.05P
	(Yak 18)				*(As "68" in PR China A/F c/s plus "1219" on u/c door)*		
G-BVVH	Europa Aviation Europa	PFA 247-12505		31.10.94	T.G.Hoult	Octon Grange Farm, Foxholes	1. 9.05P
	(Built T.G.Hoult and M P Whitley - kit no.014) (Monowheel u/c)						
G-BVVI	Hawker Audax I	Not known	2015M K5600	3.11.94	Aero Vintage Ltd	St.Leonards-on-Sea	
	(Built Avro Aircraft Ltd)				*(On rebuild 8.95: current status unknown)*		
G-BVVK	de Havilland DHC-6-310 Twin Otter	666	LN-BEZ	21.12.94	Loganair Ltd *"Chatham Historic Dockyard"*	Glasgow	12. 1.05T
G-BVVL	EAA Acrosport II	PFA 72A-10887		11.11.94	G.A.Breen	Portimão, Faro, Portugal	12. 8.05P
	(Built D Park, A J Maxwell and P Price) (Lycoming O-360)						
G-BVVM	Zenair CH.601HD Zodiac	PFA 162-12539		3.10.94	D.Macdonald	Popham	24. 5.05P
	(Buillt J G Small) (Rotax 912-UL)						
G-BVVN	Brugger MB.2 Colibri	PFA 43-10979		12.10.94	T.C.Darters	Valley Farm, Wlinwick	19. 9.05P
	(Built N F Andrews) (Volkswagen 1834)						
G-BVVP	Europa Aviation Europa	PFA 247-12697		20. 9.94	I.Mansfeld	Kemble	21. 4.05P
	(Built J S Melville - kit no.088) (Monowheel u/c)						
G-BVVR	Stits SA-3A Playboy	P-736	N4620S	14.11.94	I.T.James	Enstone	19. 7.05P
	(Built S Goins) (Continental A65)						
G-BVVS	Van's RV-4	PFA 181-12324		15.11.94	E.C.and N.S.C.English	North Weald	14.12.04P
	(Built E.C.English) (Lycoming O-320)						
G-BVVU	Lindstrand LBL Four SS HAB	155	HB-QAP G-BVVU	18.11.94	Magical Adventures Ltd	(West Bloomfield, Michigan, US)	6.12.01P

G-BVVW	IAV-Bacau Yakovlev Yak-52 *(Official c/n suspect as plate shows 833519)*	844605	RA-01361 DOSAAF 15, DOSAAF 95	16.11.94	J.E.Blackman	Andrewsfield	21.12.05P
G-BVVZ	Corby CJ-1 Starlet *(Built A E Morris)* (Volkswagen 1834)	PFA 134-12293		9.11.94	P.V.Flack	Lasham	27. 1.04P
G-BVWB	Thunder Ax8-90 Series 2 HAB	3000		2.12.94	M.A Stelling, K.C.and K.Tanner *"Starship"*	(Barton-Le-Clay)	18. 5.05
G-BVWC	English Electric Canberra B.2 *(Built Avro Aircraft Ltd) (Regd as B.6 but c/n relates to nose section originally fitted to XH568) (As "WK163" in 617 sqdn c/s) (Noted 1.05)*	71399	WK163	2.12.94	Classic Aviation Projects Ltd	Coventry	26. 5.04P
G-BVWG*	Hawker Hunter T.8C				See SECTION 3 Part 1	Cape Town, South Africa	
G-BVWH*	Cameron N-90 Lightbulb SS HAB				See SECTION 3 Part 1	(Aldershot)	
G-BVWI	Cameron Light Bulb 65SS HAB	3405		8.12.94	A D.Kent tr Balloon Preservation Flying Group *"Phillips Energy Saver" (New CofR 6.04)*	(Lancing)	2. 6.97A
G-BVWK*	Air and Space 18A Gyroplane	18-14	SE-HID N6108S	19.12.94	Not known *(Cancelled 18.10.00 as temporarily WFU) (Noted 2004 on road)*	(Kinnetties, Forfar)	
G-BVWL*	Air and Space 18A Gyroplane	18-63	SE-HIE N90588, N6152S	19.12.94	Not known *(Cancelled 18.10.00 as temporarily WFU) (Noted 2004 on road)*	(Kinnetties, Forfar)	
G-BVWM	Europa Aviation Europa *(Built A Aubeelack and C J Hadley - kit no.070) (Monowheel u/c)*	PFA 247-12620		14.12.94	A Aubeelack tr Europa Syndicate	White Waltham	15. 7.05P
G-BVWP	de Havilland DHC-1 Chipmunk 22	C1/0741	WP856	19.12.94	T.W.M.Beck *(As "WP856/904" in RN c/s)*	Monks Gate, Horsham	2. 6.07
G-BVWV*	Hawker Hunter F.6A				See SECTION 3 Part 1	Cape Town, South Africa	
G-BVWW	Lindstrand LBL 90A HAB	169		28.12.94	Drawflight Ltd *"Double Whiskey"*	(Hastings)	27. 3.05A
G-BVWY	Porterfield CP-65 (Continental A65)	720	N27223 NC27223	23.11.94	B.Morris	Manor Farm, Garford	17. 9.05P
G-BVWZ	Piper PA-32-301 Saratoga	3206055	I-TASP N9184N	3. 1.95	Maureen A. Kesteven tr The Saratoga (WZ) Group	(Ryton)	3.12.06
G-BVXA	Cameron N-105 HAB	3441		4. 1.95	R.E.Jones *(Ribby Hall titles)*	(Lytham St.Annes)	19. 3.05T
G-BVXB	Cameron V-77 HAB	3442		4. 1.95	J.A.Lawton *"Pat McLean"*	(Godalming)	20.10.04A
G-BVXC*	English Electric Canberra B(I).8	6649	WT333	9. 1.95	Classic Aviation Projects Ltd *(Cancelled 22.4.03 as PWFU) (As "WT333" 3.04)*	Bruntingthorpe	AC
G-BVXD	Cameron O-84 HAB	3432		5. 1.95	N.J.Langley *(Prudential titles)*	(Bristol)	7. 6.05A
G-BVXE	Steen Skybolt *(Built T C Humphreys and T J Reeve)*	PFA 64-11123	G-LISA	5. 1.95	J.Buglass	Sleap	19. 5.05P
G-BVXF	Cameron O-120 HAB	3400		21. 9.94	Off The Ground Balloon Co Ltd	(Kendal)	7. 3.05T
G-BVXJ	Bücker Bü.133 Jungmeister *(Built CASA) (Official c/n is "E1-9")*	Not known	Spanish AF E1-9 Spanish AF ES1-9, 35-9	11. 1.95	A C Mercer *(As "ES-9/G-BVXJ" in Spanish Air Force c/s) (Noted 1.05)*	Breighton	7.10.04P
G-BVXK	Aerostar Yakovlev Yak-52	9111306	RA-44508 (1) DOSAAF 26	12. 1.95	E.Gavazzi *(As "26" in DOSAAF c/s)*	White Waltham	7. 5.05P
G-BVXM	Aérospatiale AS350B Ecureuil	2013	I-AUDI I-CIOC	10. 1.95	The Berkeley Leisure Group Ltd	Sparkford	4. 3.07T
G-BVXR	de Havilland DH.104 Devon C.2	04436	XA880	13. 1.95	M.Whale and M.W.A.Lunn *(As "XA880" in RAE c/s) (Stored 5.02)*	Kemble	
G-BVXS	Taylorcraft BC-12D (Continental A65)	9284	N96984 NC96984	27. 1.95	D.Riley	(Somerleyton, Lowestoft)	7. 6.05P
G-BVXW	Short SC.7 Skyvan 3A-100	SH.1889	LX-DEF Arg.Coast Guard PA-52, G-14-61	15.11.95	Babcock Support Services Ltd t/a Babcock Defence Services *(Parked outside under polythene sheeting with various parts missing 4.03)*	Oxford	30. 1.03T
G-BVYA	Airbus A320-231	0354	OO-TCF G-BVYA, F-WQAY, (N301SA), F-WWDZ	7. 4.95	Gustav Leasing IV Ltd	(Shannon, County Clare)	20. 1.04T
G-BVYC	Airbus A320-231	0411	OO-TCC G-BVYC, F-WQBA, (N303SA), F-WWDX	26. 4.95	Gustav Leasing IV Ltd	(Shannon, County Clare)	25. 4.04
G-BVYF	Piper PA-31-350 Navajo Chieftain	31-7952102	G-SAVE N3518T	8. 2.95	J.A., G.M, D.T.A. and J.A.Rees t/a Messrs Rees of Poynston West	Haverfordwest	23. 1.06T
G-BVYG	Robin DR300/180R Remorqueur	611	F-BSQB F-BSPI	9. 1.95	Ulster Gliding Club Ltd	Bellarena	9. 4.05
G-BVYK	TEAM mini-MAX 91A *(Built S.B.Churchill)* (DAF) *(Officially regd with Rotax 447)*	PFA 186-12598		13. 2.95	P.C.E.Roberts	Truro)	2.11.05P
G-BVYM	Robin DR300/180R Remorqueur	656	F-BTBL	9.12.94	London Gliding Club Pty Ltd	Dunstable	19. 7.07
G-BVYO	Robin R2160	288		11. 1.95	G.Johnson	Earls Colne	15. 4.07T
G-BVYP	Piper PA-25-235 Pawnee B	25-3481	N7475D OY-CLT, N7475Z	13. 2.95	Bidford Airfield Ltd	Bidford	26. 4.07
G-BVYR	Cameron A-250 HAB	3411		2. 2.95	A van Wyk	(Caxton, Cambridge)	8.10.04T
G-BVYU	Cameron A-140 HAB	3544		17. 2.95	Balloon Flights Club Ltd	(Leicester)	29. 3.05T
G-BVYX	Avid Speed Wing Mk.4 *(Built G J Keen)*	PFA 189-12370		16. 2.95	M E Lloyd *(Noted 11.04)*	Shobdon	1. 7.04P
G-BVYY	Pietenpol AirCamper *(Built J.R.Orchard)*	PFA 47-12559		20. 2.95	T F Harrison tr Pietenpol G-BVYY Group *(New owners 9.04)*	(Perton, Wolverhampton)	28. 6.98P
G-BVYZ	Stemme S 10-V	14-011	D-KGDD	6. 3.95	A D.Gubbay	Bicester	26. 3.05
G-BVZD	Tri-R Kis *(Tri-cycle u/c)* *(Built R T Clegg)* (Canadian Air Motive CAM.100)	PFA 239-12416		21. 2.95	D.R.Morgan	Old Sarum	4.11.05P
G-BVZE	Boeing 737-59D	26422	SE-DNL	7. 3.95	British Midland Airways Ltd *"little costa baby"* *(Op bmiBaby)*	Nottingham East Midlands	22. 3.07T
G-BVZG	Boeing 737-5Q8	25160	SE-DNF	12. 4.95	British Midland Airways Ltd *(Wales logojet)* *(Op bmiBaby)*	Nottingham East Midlands	1. 5.07T
G-BVZH	Boeing 737-5Q8	25166	SE-DNG	25. 4.95	British Midland Airways Ltd *"baby blue Skies"* *(Op bmiBaby)*	Nottingham East Midlands	26. 5.07T
G-BVZI	Boeing 737-5Q8	25167	SE-DNH	15. 5.95	British Midland Airways Ltd *(Op bmiBaby)*	Nottingham East Midlands	11. 6.07T
G-BVZJ	Rand Robinson KR-2 *(Built J.P.McConnell-Wood)* (Revmaster)	PFA 129-11049		21. 2.95	J.P.McConnell-Wood *(Damaged landing Phoenix Farm 15.7.98: current status unknown)*	Phoenix Farm, Upham	
G-BVZN	Cameron C-80 HAB	3546		28. 2.95	Sally J.Langley t/a Sky Fly Balloons	(Bristol)	4. 8.04A
G-BVZO	Rans S-6-116 Coyote II *(Built P.Atkinson kit no.0494.606)* (Rotax 582) *(Tri-cycle u/c)*	PFA 204A-12710		1. 3.95	P.Atkinson	Sandtoft	7. 9.04P
G-BVZR	Zenair CH.601HD Zodiac *(Built J D White)* (Rotax 912-UL)	PFA 162-12417		2. 3.95	J.D.White	Tollerton	27.11.04P

Reg	Type	C/n	Previous identity	Regd	Owner/operator	Base	CofA
G-BVZT	Lindstrand LBL 90A HAB	183		9. 3.95	F.W.Farnsworth Ltd t/a Pork Farms Bowyers *(Bowyers Pork Farms titles)*	(Nottingham)	15. 8.05A
G-BVZV	Rans S-6-116N Coyote II *(Built J Fothergill - kit no.1294.708)* (Rotax 582)	PFA 204A-12832		16. 2.95	A.R.White	Popham	13. 6.05P
G-BVZX	Cameron H-34 HAB	3564		15. 3.95	Julia B.Turnau tr Chianti Balloon Club	(Siena, Italy)	21. 6.96A
G-BVZZ	de Havilland DHC-1 Chipmunk 22	C1/0687	WP795	5. 1.95	D.C.Murray tr Portsmouth Naval Gliding Club *(As "WP795/901" in RN c/s)*	Lee-on-Solent	14. 6.04

G-BWAA - G-BWZZ

Reg	Type	C/n	Previous identity	Regd	Owner/operator	Base	CofA
G-BWAA	Cameron N-133 HAB	3471		9. 3.95	C.and J.Bailey t/a Bailey Balloons *(Brunel Ford titles)*	(Bristol)	14.11.04T
G-BWAB	Jodel D.140 Mousquetaire *(Built W A Braim)* (Lycoming O-360)	PFA 251-12469		25. 1.95	W.A.Braim	(Driffield)	31.10.05P
G-BWAC	Waco YKS-7 (Jacobs R-755)	4693	N50RA N2896D, NC50	19. 8.92	D.N.Peters	Little Gransden	19.10.04
G-BWAD	Rotary Air Force RAF 2000 *(Built J R Legge - c/n 147)* (Subaru EA82)	PFA G/13-1254		27. 2.95	Newtonair Gyroplanes Ltd *(Op A Melody)*	Henstridge	31.10.05P
G-BWAF	Hawker Hunter F.6A *(Built Armstrong-Whitworth Aircraft)*	S4/U/3393	8831M XG160	24. 2.95	RV Aviation Ltd *(As "XG160/U" in Black Arrows c/s in Aviation Museum 2.04)*	Bournemouth	
G-BWAG	Cameron O-120 HAB	3478		3. 2.95	P.M.Skinner *"Joker"*	(Maidstone)	16. 6.05T
G-BWAH	Montgomerie-Bensen B.8MR *(Built S J O Tinn)* (Rotax 582)	PFA G/01-1208		16. 3.95	J.B.Allan	Henstridge	3. 4.05P
G-BWAI	CFM Streak Shadow SA *(Built J M Heath - kit no K.235SA)* (Rotax 582) *(Originally allocated as BMAA/HB/052 c.1990) (Noted 1.05)*	PFA 206-12556		21. 3.95	C.M.James	Kemble	15. 7.04P
G-BWAJ	Cameron V-77 HAB	3579		22. 3.95	R.S. and S.H.Ham *"Robsel"*	(Axbridge)	9. 8.01A
G-BWAN	Cameron N-77 HAB	3499		24. 3.95	I.Chadwick tr Balloon Preservation Flying Group *(BPG tiltes)*	(Horsham)	6.12.04A
G-BWAO	Cameron C-80 HAB	3436		24. 3.95	Airxcite Ltd t/a Virgin Balloon Flights	(Wembley)	16. 4 05T
G-BWAP	Clutton FRED Series III *(Built R J Smyth)*	PFA 29-10959		24. 3.95	G.A.Shepherd	Seething	
G-BWAR	Denney Kitfox Model 3 *(Built C E Brookes)*	PFA 172-12432		16. 3.95	I.Wightman	(Berwick-upon-Tweed)	16.12.04P
G-BWAT	Pietenpol AirCamper *(Built D R Waters)* (Continental C90)	PFA 47-11594		15. 3.95	P.W.Aitchison	Shobdon	19. 2.04P
G-BWAU	Cameron V-90 HAB	3569		27. 3.95	K.M. and A.M.F.Hall	(London N10)	15. 5.05A
G-BWAV	Schweizer 269C	S.1204	SE-JAY LN-OTS, OY-HDW, N41S	28. 2.95	B.Maggs t/a Helihire	Shere, Guildford	31. 8.07T
G-BWAW	Lindstrand LBL 77A HAB	207		28. 3.95	D.Bareford *(Seton Healthcare titles)*	(Kidderminster)	17. 6.03A
G-BWBA	Cameron V-65 HAB	3456		27. 2.95	P.G.Dunnington tr Dante Balloon Group *(British Airways titles)*	(Hungerford)	4. 7.05A
G-BWBB	Lindstrand LBL 14A HAB *(Gas filled)*	222		3. 4.95	Oxford Promotions (UK) Ltd *(Op F Prell)*	(Kentucky, US)	
G-BWBC	Cameron N-90AS HAB	3574		12. 6.95	Wetterauer Montgolfieren EV *"Zeppelin"*	(Budingen, Germany)	18. 3.05A
G-BWBE	Colt Flying Ice Cream Cone SS HAB	3560		3. 4.95	Benedikt Haggeney GmbH	(Ennigerloh, Germany)	5. 5.05A
G-BWBF	Colt Flying Ice Cream Cone SS HAB	3561		3. 4.95	Benedikt Haggeney GmbH	(Ennigerloh, Germany)	17. 4.03A
G-BWBG	Cvjetkovic CA-65 Skyfly *(Built T.White and M.C.Fawkes)*	PFA 1566		6. 4.95	T.White and M.C.Fawkes	Charity Farm, Baxterley	
G-BWBI	Taylorcraft F-22A	2207	N22UK	3. 4.95	P.J.Wallace	Carlisle	31. 7.05
G-BWBJ	Colt 21A Cloudhopper HAB	3532		6. 4.95	U.Schneider	(Giessen, Germany)	12. 3.04A
G-BWBO	Lindstrand LBL 77A HAB	157		10. 4.95	T.J.Orchard, N.J.Glover and S.R.Godfrey	(Aylesbury)	28.12.04A
G-BWBT	Lindstrand LBL 90A HAB	184		3. 4.95	British Telecommunications plc *(BT titles)*	(Newbury)	9. 4.05A
G-BWBY	Schleicher ASH 26E	26076		30. 8.95	J.S.Wand	Bidford	24.11.07E
G-BWBZ	ARV Aviation ARV-1 Super 2 *(Built J N C Shields)* (Mid-West AE.100R) *(Regd as "ARV K1")*	PFA 152-12802		10. 3.95	P.I.Lewis	Sleap	10. 6.05P
G-BWCA	CFM Streak Shadow *(Built R Thompson - kit no K.160)* (Rotax 582)	PFA 206-11985		19. 4.95	I.C.Pearson	(Horsham)	8. 9.04P
G-BWCC	Van Den Bemden 460m3 (Gas) Free Balloon *(C/n may be a corruption of Dutch CofR 622)* *(Dutch records indicate some parts came from OO-BGX which itself became G-BBFS)*	"022"	PH-BOX	5. 4.95	R.W.Batchelor tr Piccard Balloon Group *"Prof A Piccard"*	(Thame)	
G-BWCF*	Yakovlev Yak-50				See SECTION 3 Part 1	Wanaka, Neew Zealand	
G-BWCG	Lindstrand LBL 42A HAB	223		25. 4.95	Oxford Promotions (UK) Ltd *(Op F Prell)*	(Kentucky, US)	10. 1.97A
G-BWCK	Everett Gyroplane Series 3 (Rotax 582)	036		26. 4.95	A C.S.M.Hart	Farley Farm, Romsey	28. 3.05P
G-BWCO	Dornier Do.28D-2 Skyservant	4337	EI-CJU (N5TK), 5N-AOH, D-ILIF	19. 6.95	Wingglider Ltd *(Stored 7.04)*	Hibaldstow	19. 5.99A
G-BWCS	British Aircraft Corporation BAC.145 Jet Provost T.5A	EEP/JP/957	XW293	28. 4.95	R.E.Todd *(As "XW293/Z") (Noted Bournemouth on overhaul 1.04)*	Sandtoft	15. 3.02P
G-BWCT	Tipsy Nipper T.66 Series 1 *(Built Avions Fairey SA)*	11	"OO-NIC" PH-MEC, D-EMEC, OO-NIC	27. 4.95	J.S.Hemmings and C.R.Steer	(Rye/Bexhill-on-Sea)	
G-BWCV	Europa Aviation Europa *(Built M P Chetwynd-Talbot - kit no.041)* (NSI EA-81/100) *(Monowheel u/c)*	PFA 247-12591		4. 5.95	G.V.McKirdy *(Rebuilt by 2004)*	Enstone	14. 4.98P
G-BWCW	Barnett Rotorcraft J4B-2 *(Built S H Kirkby)* (Lycoming)	PFA G/14-1256		5. 5.95	S.H.Kirkby *(Noted 11.02)*	Farley Farm, Romsey	
G-BWCY	Murphy Rebel *(Built A Jones, R Hallam and A Koneczek)*	PFA 232-12135		15. 5.95	M.Stow	Newcastle	5. 7.05P
G-BWCZ	Revolution Helicopters Mini-500 (Rotax 582)	0010		1. 5.95	D.Nieman	(Thame)	AC
G-BWDA	Aérospatiale-Alenia ATR72-202	444	F-W... G-BWDA, F-WWEQ	29. 6.95	Aurigny Air Services Ltd	Guernsey	28. 8.07T
G-BWDB	Aérospatiale-Alenia ATR72-202	449	F-WQNI G-BWDB, F-WWEE	14. 6.95	Aurigny Air Services Ltd	Guernsey	13. 6.07T

Registration	Type	C/n	Previous identity	Regd date	Owner/operator	Probable base	CofA expiry
G-BWDE	Piper PA-31P Pressurised Navajo	31P-7400193	G-HWKN HB-LIR, D-IAIR, N7304L	12. 5.95	Tomkat Aviation Ltd *(Stored 9.02)*	Shoreham	18.12.96T
G-BWDF	PZL-104 Wilga 35A	21950955		17. 5.95	P.G.Marks	Husbands Bosworth	3. 6.07
G-BWDH	Cameron N-105 HAB	3549		22. 5.95	Bridges Van Hire Ltd	(Awsworth, Nottingham)	11. 6.01T
G-BWDO	Sikorsky S-76B	760356	VR-CPN N9HM	2. 6.95	Haughey Air Ltd	(Newry, County Armagh)	8. 6.05T
G-BWDP	Europa Aviation Europa *(Built I Valentine - kit no.062) (Monowheel u/c)*	PFA 247-12637		7. 6.95	W.Hueltz *(Suffered fire refuelling 14.5.00: severe damage deforming fuselage - possible new fuselage/kit no??)*	(Bonn, Germany)	20. 8.05P
G-BWDR	Hunting Percival P.84 Jet Provost T.3A	PAC/W/6603	XM376	6. 6.95	W.O.Bayazid *(As "XM376/27")*	Humberside	24.10.02P
G-BWDS	Hunting Percival P.84 Jet Provost T.3A	"PAC/W/932" *(Correct c/n PAC/W/9231?)*	XM424 (N77506?), XM424	6. 6.95	S.T.G.Lloyd *(As "XM424")*	Swansea	4.10.05P
G-BWDT	Piper PA-34-220T Seneca III	34-8233045	PH-TWI G-BKHS, N8472H	21. 9.88	H.R.Chambers	Redhill	19.11.06T
G-BWDU	Cameron V-90 HAB	3143		19. 6.95	D.M.Roberts	(Llandeilo)	4. 9.04A
G-BWDV	Schweizer 269C	S.1712	N86G	16. 6.95	Helicopter One Ltd	Bournemouth	3. 9.07T
G-BWDX	Europa Aviation Europa *(Built J.B.Crane - kit no.056) (Monowheel u/c)*	PFA 247-12603		13. 6.95	J.B.Crane	Fenland	20. 7.05P
G-BWDZ	Sky 105-24 HAB	002		13. 6.95	Skyride Balloons Ltd	(King's Lynn)	26. 5.04T
G-BWEA	Lindstrand LBL 120A HAB	252		14. 6.95	S.R.Seager *(Parrott and Coales titles)*	(Aylesbury)	8. 7.00T
G-BWEB	British Aircraft Corporation BAC.145 Jet Provost T.5A	EEP/JP/1044	XW422	19. 6.95	S.Patrick *(As "XW422".in red/white/blue RAF c/s)*	Kemble	8. 7.05P
G-BWEE	Cameron V-42 HAB	3480		8. 3.95	J.A Hibberd	Newtown Hungerford	4. 8.05A
G-BWEF	SNCAN Stampe SV-4C(G) (DH Gipsy Major 10)	208	G-BOVL N20SV, F-BHES, F-BBLC	13. 5.93	D A Smith tr Acebell BWEF Syndicate	Redhill	29. 4.07
G-BWEG	Europa Aviation Europa *(Built B.A Selmes and R.J.Marsh - kit no.053) (Conventional u/c)*	PFA 247-12600		4. 4.95	B.A Selmes and R.J.Marsh tr Wessex Europa Group	Dunkeswell	24. 5.05P
G-BWEH	HOAC DV-20 Katana	20123		19. 6.95	Lowlog Ltd	Elstree	19. 7.07T
G-BWEM	Vickers Supermarine 358 Seafire L.III	Not known	IAC.157 RX168	28. 6.95	C.J.Warrilow and S.W.Atkins *(On rebuild 10.01)*	(Exeter)	
G-BWEN	Macair Merlin GT *(Built B.W.Davies - c/n 050194)* (Subaru EA81)	PFA 208A-12859		20. 6.95	D.A.Hill *(New owner 5.04)*	(Terrington St Clement, King's Lynn)	7.12.95P
G-BWEO	Lindstrand LBL 14M HAB (Gas filled)	285		23. 6.95	Lindstrand Balloons Ltd	(Oswestry)	1. 9.04A
G-BWER	Lindstrand LBL 14M HAB (Gas filled)	287		23. 6.95	Lindstrand Balloons Ltd	(Oswestry)	4.11.97A
G-BWEU	Reims/Cessna F152 II	F15201894	EI-BNC N9097Y	15. 6.95	Affair Aircraft Leasing Ltd	Full Sutton	23. 9.07T
G-BWEV	Cessna 152 II	15283182	EI-BVU N47184	28. 6.95	Haimoss Ltd	Old Sarum	9.11.07E
G-BWEW	Cameron N-105 HAB	3637		30. 6.95	Unipart Group Ltd tr Unipart Balloon Club *(Unipart titles)*	(Cowley)	17. 1.00A
G-BWEY	Bensen B.8 *(Built F.G.Shepherd)*	PFA G/01-1197		3. 7.95	F.G.Shepherd	Alston, Cumbria	
G-BWEZ	Piper J-3C-85 Cub	6021	N29050 NC29050	3. 7.95	Rebecca J.Simpson tr PJ L4 Group *(As "436021" in silver c/s)*	Cumbernauld	31.10.05P
G-BWFG	Robin HR200/120	293		20. 7.95	Atlantic Air Transport Ltd	Coventry	7. 2.05T
G-BWFH	Europa Aviation Europa *(B.L.Wratten and R W Baylie - kit no.201) (Monowheel u/c)*	PFA 247-12842		14. 7.95	B.L.Wratten	Shoreham	9. 7.05P
G-BWFI	HOAC DV-20 Katana	20128		17. 7.95	Air Aqua Ltd	Scotland Farm, Hook	22. 8.07T
G-BWFJ	Evans VP-1 *(Built W E Jones)* (Volkswagen 1600)	PFA 62-10349		1. 9.78	P.A West *(Stored 5.94: current status unknown)*	(Old Sarum)	27. 1.93P
G-BWFK	Lindstrand LBL 77A HAB	289		17. 7.95	R.S.Kent tr Balloon Preservation Flying Group *"Mr Orange"*	(Lancing)	22. 6.05A
G-BWFM	Yakovlev Yak-50	781208	NX5224R DDR-WQX, DM-WQX	19. 7.95	Classic Aviation Ltd *(Op The Old Flying Machine Co)*	Warton	30. 11.04P
G-BWFN	HAPI Cygnet SF-2A *(Built T.Crawford)*	PFA 182-11335		19. 7.95	K.Shelton and G.J.Green tr G-BWFN Group *(New owners 8.04)*	(Nottingham/Darley Bridge,Matlock)	
G-BWFO	Colomban MC-15 Cri-Cri *(Built O.G.Jones)* (JPX PUL-212)	PFA 133-11253		19. 7.95	O.G.Jones *(Current status unknown)*	(Llanbedr)	
G-BWFP	IAV-Bacau Yakovlev Yak-52 *(Official c/n suspect as plate shows 855606 - composite?)*	855503	RA-44501 (1) DOSAAF 43, DOSAAF 61 *(blue)*	20. 7.95	M.C.Lee	Blackpool	19. 8.05P
G-BWFR	Hawker Hunter F.58	41H-697398	Swiss AF J-4031	24. 7.95	The Old Flying Machine Air Museum Co Ltd *(Stored as "J-4031" 5.04)*	Scampton	3. 8.99P
G-BWFT	Hawker Hunter T.8M	41H-695332	XL602	24. 7.95	B.R.Pearson tr T8M Group *(As "XL602") (Current status unknown)*	Exeter	23. 7.99P
G-BWFV	HOAC DV-20 Katana	20132		26. 7.95	J.P.E.Walsh t/a Walsh Aviation	Cranfield	10.10.07T
G-BWFX	Europa Aviation Europa *(Built A D.Stewart - kit no.038) (Monowheel u/c)*	PFA 247-12586		26. 7.95	A D.Stewart	Rayne Hall Farm, Braintree	12. 4.05P
G-BWFZ	Murphy Rebel *(Built I E Spencer)*	PFA 232-12536	G-SAVS	19. 7.95	S.Beresford	(Gringley-on-the-Hill, Doncaster)	10. 7.05P
G-BWGA	Lindstrand LBL 105A HAB	295		2. 8.95	I.Chadwick tr Balloon Preservation Flying Group *"Freebird"*	(Horsham)	24. 7.05A
G-BWGF	British Aircraft Corporation BAC.145 Jet Provost T.5A	EEP/JP/989	XW325	10. 8.95	J.W.Cullen tr Specialscope Jet Provost Group *(As "XW325/E")*	Blackpool	5. 8.05P
G-BWGG	Max Holste MH.1521C1 Broussard	20	F-GGKG F-WGKG, Fr mil	10. 7.95	M.J.Burnett Jnr. and R.B.Maalouf *(As "315-SQ/20" in ALAT c/s)*	Kemble	16. 9.06
G-BWGH*	Europa Aviation Europa *(Built M W Handley - kit no.045)*	PFA 247-12589		23. 8.95	Not known *(Crashed on Island of Sark 10.01 and cancelled 4.3.02 by CAA) (Noted 3.03)*	Guernsey	2. 4.02P
G-BWGJ	Chilton DW.1A *(Built T.J.Harrison)* (Lycoming O-145-A2)	PFA 225-12615		11. 8.95	T.J.Harrison *(Complete 5.00)*	Lower Upham, Hampshire	
G-BWGK	Hawker Hunter GA.11 *(Centre fuselage no.is 41HR HABL 003032)*	HABL-003032	XE689	15. 8.95	B.R.Pearson tr GA11 Group *(Open store as "XE689/864/VL" 5.04)*	Exeter	11. 7.01P
G-BWGL	Hawker Hunter T.8C *(Officially regd with c/n 41H-695946)*	HABL-003086	XF357	15. 8.95	Elvington Events Ltd *(As T.11 prototype "XJ615")*	Elvington	1. 9.05P
G-BWGM	Hawker Hunter T.8C *(Officially regd with c/n 41H-695940)*	HABL-003008	XE665	15. 8.95	B.J.Pearson tr The Admirals Barge *(Open store as "XE665/876/VL" 5.04)*	Exeter	24. 6.98P

Registration	Type	C/n	Previous identity	Reg date	Owner/operator	Base	CofA expiry
G-BWGN	Hawker Hunter T.8C	41H-670689	WT722	15. 8.95	B.J.Pearson	Exeter	3. 9.97P
					tr T8C Group *(Open store as "WT722/878/VL" 5.04)*		
G-BWGO	Slingsby T.67M-200 Firefly	2048	SE-LBC	15. 8.95	R.Gray	Fairoaks	9. 6.05
G-BWGP	Cameron C-80 HAB	3631		17. 8.95	D.J.Groombridge	(Bristol)	10.10.04A
					(London Camera Exchange titles)		
G-BWGR	North American TB-25N-NC Mitchell	108-34200	N9494Z	18. 8.95	D Copley	(Belton, Doncaster)	
	(Official c/n 108-30925 - corruption of USAAF serial)			44-30925	*(As "151632/N9494Z") (Current status unknown)*		
G-BWGS	British Aircraft Corporation BAC.145 Jet Provost T.5A	EEP/JP/974	XW310	18. 8.95	G-BWGS Ltd	North Weald	25. 8.05P
G-BWGT	Hunting Percival P.84 Jet Provost T.4	PAC/W/21624	8991M	21. 8.95	R.E.Todd	Sandtoft	9. 7.05P
	(Reported as c/n PAC/W/19992)		XR679		*(Op The Jet Provost Club)*		
G-BWGX	Cameron N-42 HAB	3633		21. 8.95	Newbury Building Society	Newtown Hungerford	19. 1.00A
G-BWGY	HOAC DV.20 Katana	20134		22. 8.95	Stars Fly Ltd	Elstree	12.10.04T
G-BWGZ	HOAC DV.20 Katana	20135		22. 8.95	Plane Talking Ltd *(Op Denham School of Flying)*	Denham	17.11.07E
G-BWHB	Cameron O-65 HAB	2759		24. 8.95	G.Aimo	(Mondovi, Italy)	19. 8.05A
G-BWHC	Cameron N-77 HAB	3647		25. 8.95	R.B.Craik *(Travelsphere Holidays titles)*	(Northampton)	23. 6.04A
G-BWHD	Lindstrand LBL 31A HAB	292		29. 8.95	J.C.E.Price	(Portadown)	12. 1.01A
					tr Army Air Corps Balloon Club		
G-BWHF	Piper PA-31-325 Navajo C/R	31-7612076	F-GECA	7. 9.95	Awyr Cymru Cyf	Shobdon	31.10.05T
			D-IBIS, N59862				
G-BWHG	Cameron N-65 HAB	3619		7. 9.95	M.Stefanini and F.B.Alaoui	(Firenze, Italy)	20. 8.05A
G-BWHI	de Havilland DHC-1 Chipmunk 22	C1/0637	WK624	8. 9.95	N.E.M.Clare	RAF Woodvale	9.12.04T
	(Hulk of WK624/M may have been used in rebuild of G-AOSY c1998/99)				*(As "WK624/M")*		
G-BWHK	Rans S-6-116 Coyote II	PFA 204A-12908		15. 9.95	D.A.Buttress	Charity Farm, Baxterley	12. 5.05P
	(Built N D White - kit no.0695.834) (Rotax 582) *(Tri-cycle u/c)*						
G-BWHM*	Sky 140-24 HAB	006		18. 9.95	C.J.S.Limon	(London NW1)	19. 9.04T
					(Cancelled 10.5.04 as WFU)		
G-BWHP	CASA 1-131E Jungmann	2109	Spanish AF E3B-513	18. 8.95	J.F.Hopkins	Watchford Farm, Yarcombe	14.12.04P
					(As "S4+A07" in Luftwaffe c/s)		
G-BWHR	Tipsy Nipper T.66 Series 1	PFA 25-12843	(OO-KAM)	19. 9.95	L.R.Marnef	(Koningshooikt, Belgium)	
	(Built L.R.Marnef)		OO-69				
	(Composite homebuild of original Fairey build c/n 29 & 71)						
G-BWHS	Rotary Air Force RAF 2000	PFA G/13-1253		25. 9.95	J.M.Cox	(Windsor)	2. 8.05P
	(Built V G Freke) (Subaru EA82)						
G-BWHU	Westland Scout AH.1	F9517	XR595	27. 9.95	N.J.F.Boston	Oreston, Plymouth	16. 2.05P
					(As "XR595/M" in Army c/s)		
G-BWHV	Denney Kitfox Model 2	PFA 172-11857		28. 9.95	A.C.Dove	(Ashtead)	23. 5.05P
	(Built A C.Dove)						
G-BWHW	Cameron A-180 HAB	3634		29. 9.95	Societé Bombard SARL	(Meursanges, Côte-d'Or, France)	24. 9.05A
G-BWHY	Robinson R22	0098	N90366	24. 3.87	E.Warren	(Goley, County Wexford)	11. 6.05T
G-BWIA	Rans S-10 Sakota	PFA 194-12044		15. 9.95	J.B.Lawton	Fenland	1. 8.05P
	(Built P A Beck) (Rotax 582)				*(Noted 6.04)*		
G-BWIB	Scottish Aviation Bulldog Series 120/122	BH120/227	Ghana AF G-103	10.10.95	B.I.Robertson	(Pontiac, Michigan, US)	16. 6.07T
					(As "XX514")		
G-BWID	Druine D.31 Turbulent	201	F-PHFR	16.10.95	A.M.Turney	Cheddington	21. 7.05P
	(Built R Druine and H Gindre) (Volkswagen 1200)						
G-BWII	Cessna 150G	15065308	N4008J	22. 9.95	J.D.G.Hicks	Beeches Farm, South Scarle	25. 2.05
			(G-BSKB), N4008J				
G-BWIJ	Europa Aviation Europa	PFA 247-12513		19.10.95	R.Lloyd	(Hereford)	
	(Built R.Lloyd - kit no. 006				*(New CofR 9.04)*		
G-BWIK	de Havilland DH.82A Tiger Moth	86417	7015M	20.10.95	B.J.Ellis	Little Gransden	
			NL985		*(On rebuild as "NL985")*		
G-BWIL	Rans S-10 Sakota	PFA 194-11770	G-WIEN	4.10.95	J.C.Longmore	Netherthorpe	30. 3.05P
	(Built J.C.Longmore) (Rotax 582 - *kit no.1089.065)*						
G-BWIP	Cameron N-90 HAB	3668		20.10.95	S.H.Fell *(New owner 9.02)*	(Carlisle)	27.10.96A
G-BWIR	Dornier 328-100	3023	D-CDXF	18.10.95	Suckling Airways (Norwich) Ltd	London City	19.10.05E
			N328DA, D-CDHH		t/a Scot Airways		
G-BWIV	Europa Aviation Europa	PFA 247-12871		27.10.95	T.G.Ledbury	White Waltham	9. 9.99P
	(Built J R Lockwood-Goose - kit no.210) (Rotax 912) *(Monowheel u/c)*				*(Current status unknown)*		
G-BWIW	Sky 180-24 HAB	008		1.11.95	J.A.Cooper	(Ivybridge)	4. 4.04T
G-BWIX	Sky 120-24 HAB	009		31.10.95	J.M.Percival *"Mayfly III"*	(Loughborough)	16. 8.05
G-BWJG	Mooney M.20J Model 201MSE	24-3319	N1083P	7.11.95	Samic Ltd	Elstree	27. 5.05
G-BWJH	Europa Aviation Europa	PFA 247-12643		10.11.95	T.P.Cripps	(LLangwm, Haverfordwest)	3. 7.05P
	(Built D, J and A R D Hood - kit no.007) (Tri-gear u/c)						
G-BWJI	Cameron V-90 HAB	3727		13.11.95	Calarel Developments Ltd	(Chipping Camden)	5. 5.05A
G-BWJM	Bristol 20 M.1C rep	NAW-2		23.11.95	Richard Shuttleworth Trustees	Old Warden	7. 5.03P
	(Built Northern Aeroplane Workshops)				*(As "C4918" in 72 Sqdn c/s)*		
G-BWJN	Montgomerie-Bensen B.8MR	PFA G/01-1262		16.11.95	G.C.Kerr	Carlisle	30.11.04P
	(Built M G Mee) (Rotax 582)						
G-BWJP	Cessna 172C	17249424	N1824Y	21.11.95	T.W.R.Case	Moorlands Farm, Farway Common	
					(Noted 5.04)		
G-BWJR	Sky 120-24 HAB	007		22.11.95	W.J.Brogan *"Filzmooser"*	(Steiermark, Austria)	21.12.96
G-BWJT	Yakovlev Yak-50	812003	RA-01385	23.11.95	R.J.Luke	Lee-on-Solent	10.11.05P
			DOSAAF50				
G-BWJW	Westland Scout AH.1	F9705	XV130	29.11.95	S.Dadak and G.Sobell	Bournemouth	23. 6.04P
					(As "XV130/R" in 666 Sqdn c/s)		
G-BWJZ*	de Havilland DHC-1 Chipmunk 22				See SECTION 3 Part 1	Manston	
G-BWJY	de Havilland DHC-1 Chipmunk 22	C1/0519	WG469	5.12.95	K.J.Thompson *(As "WG469")*	Strandhill, County Sligo	31. 7.06
G-BWKA*	Hawker Hunter F.58				See SECTION 3 Part 1	Amman, Jordan	
G-BWKC*	Hawker Hunter F.58				See SECTION 3 Part 1	Amman, Jordan	
G-BWKD	Cameron O-120 HAB	3773		8.12.95	L.J. and Maureen Schoeman	(Basildon)	7. 3.03T
G-BWKE	Cameron AS-105GD HA Airship	3685		8.12.95	W.Arnold	(Kassel, Germany)	4. 2.05A
G-BWKF	Cameron N-105 HAB	3736		8.12.95	R.M.M.Botti	(Grosseto, Italy)	6. 1.05A
G-BWKG	Europa Aviation Europa	PFA 247-12451		28.11.95	E.H.Keppert	(Vienna, Austria)	19. 8.05P
	(Built T C Jackson - kit no.004) (Monowheel u/c)						

Registration	Type	C/n	Previous identity	Reg date	Owner/operator	Base	CofA
G-BWKJ	Rans S-7 Courier	PFA 218-12918		14.12.95	B.Tierney	Kilrush, County Kildare	4.11.04P
	(Built J P Kovacs) (Verner SVS1400) *(Officially recorded as Rotax 582)*				tr Three Point Aviation		
G-BWKK	Auster AOP.9	AUS.166 & B5/10/165	XP279	30. 7.79	C.A Davis and D.R.White	(Winchester)	1. 8.96P
					(As "XP279" in Army c/s)		
G-BWKR	Sky 90-24 HAB	014		18.12.95	Beverley Drawbridge	(Cranbrook)	16. 3.03T
G-BWKT	Laser Z200 Lazer	PFA 123-11421		19.12.95	P.D.Begley	Sywell	21. 3.05P
	(Built P.D.Begley) (Lycoming IO-360)						
G-BWKU	Cameron A-250 HAB	3730		21.12.95	Balloon School (International) Ltd	(Petworth)	13. 7.05T
					t/a British School of Ballooning		
G-BWKV	Cameron V-77 HAB	3780		27.12.95	Poppies (UK) Ltd	(Wootton Fitzpaine, Dorset)	17. 7.05A
G-BWKW	Thunder Ax8-90 HAB	3770		28.12.95	Venice Simplon Orient Express Ltd	(Tutzing, Germany)	18. 7.05A
					"Road to Mandalay"		
G-BWKX	Cameron A-250 HAB	3731		2. 1.96	Balloon School (International) Ltd	(Petworth)	13. 7.05T
					t/a Hot Airlines *(Hot Airlines titles)*		
G-BWKZ	Lindstrand LBL 77A HAB	340		21.12.95	J.H.Dobson	(Reading)	1. 1.05T
G-BWLA	Lindstrand LBL 69A HAB	339		3. 1.96	I.Chadwick tr Balloon Preservation Flying Group	(Horsham)	5. 8.04A
G-BWLD	Cameron O-120 HAB	3774	(I-....)	16. 1.96	D. and P.Pedri and C.Nicolodi		
						(Villa Lagarina, Rovereto, Italy)	4. 8.05A
G-BWLF	Cessna 404 Titan II	404-0414	G-BNXS	26.10.94	M.Evans and A.Jay	Farnborough	19. 3.06
			HKG-4, (N8799K)		t/a Nor Leasing		
G-BWLJ	Taylorcraft DCO-65	O-4331	C-GUSA	16. 1.96	C.Evans	Hill Farm, Nayland	9. 8.05P
			?, 42-35870		*(As "42-35870/129" in USN c/s)*		
G-BWLL	Murphy Rebel	PFA 232-12499		22. 1.96	F.W.Parker	Richmond, North.Yorkshire	19. 7.05P
	(Built F.W.Parker)						
G-BWLN	Cameron O-84 HAB	3737		24. 1.96	Reggiana Riduttori SRL	(San Polo d'Enza, Reggio, Italy)	25. 9.05A
G-BWLP	HOAC DV.20 Katana	20141	OE-UDV	6. 2.96	3business.com Ltd	Cumbernauld	9. 9.07T
G-BWLR	Max Holste MH.1521C1 Broussard	185	F-GGKJ	25. 1.96	Chicory Crops Ltd	Sywell	19. 6.06
			F-WGKJ, French AF		*(As "185/44-CA" in French AF c/s)*		
G-BWLS	HOAC DV.20 Katana	20142	OE-UHK	6. 2.96	M.Reed t/a Shadow Aviation	Blackbushe	23. 7.06T
G-BWLW	Avid Speed Wing Mk.4	PFA 189-12763		26. 1.96	P.C.and Susan A.Creswick	Weston Zoyland	
	(Built P.C.Creswick)						
G-BWLX*	Westland Scout AH.1	F9709	XV134	29.12.95	JM Helicopters Ltd	Draycott Farm, Chiseldon	13. 8.04P
					(As "XV134" in AAC c/s) (Cancelled 7.9.04 by CAA)		
G-BWLY	RotorWay Executive	5142		11. 1.93	P.W.and I.P.Bewley	Ley Farm, Chirk	25. 3.05P
	(Built P W and I P Bewley) (RotorWay RI 162)						
G-BWLZ	Wombat Gyrocopter	PFA G/09-1255		28.12.95	M.R.Harrisson	Guernsey	
	(Built J M Shippen)				*(Stored 3.98)*		
G-BWMA	Colt 105A HAB	1853		31.10.90	L.Lacroix	(St.Paul en Chablais, Haute-Savoie, France)	24. 6.02A
G-BWMB	Wassmer Jodel D.119	77-1492	F-BGMA	17. 2.78	C.Hughes	Finmere	22. 5.05P
	(Orig F-BGMA [77] became F-PHQH and rebuilt as Larrieu JL.2: presumed a rebuild using some components of c/n 77 and new build c/n 1492)						
G-BWMC	Cessna 182P Skylane II	18263117	N5462J	30. 1.96	P.F.N.Burrow and E.N.Skinner	Trenchard Farm, Eggesford	11. 7.06
			G-BWMC, OO-RGM, (OO-RAN), F-BVOU, N7333N		tr Eggesford Eagles Flying Group		
G-BWMF	Gloster Meteor T.7	G5/356460	7917M	15.12.95	M.Jones	Yatesbury	
			WA591		tr Meteor Flight *(Noted 1.04)*		
G-BWMH	Lindstrand LBL 77B HAB	152		7. 2.96	J.W.Hole	(Much Wenlock)	10. 6.03A
G-BWMI	Piper PA-28RT-201T Turbo Arrow IV	28R-8031131	F-GCTG	31. 1.96	O.Cowley	Fairoaks	18.12.05T
			N82482, N9571N				
G-BWMJ	Nieuport Scout 17/23 rep	PFA 121-12351		8. 2.96	R.Gauld-Galliers and Lisa J.Day	Popham	27. 7.05P
	(Built R Gauld-Galliers and L J Day) (Warner Scarab 165)				*(As "B3459/2" in RFC c/s)*		
G-BWMK	de Havilland DH.82A Tiger Moth	84483	T8191	9. 2.96	APB Leasing Ltd	Trehelig, Welshpool	AC
	(Built Morris Motors Ltd)				*(New owner 2.02)*		
G-BWML	Cameron A-275 HAB	3725		12. 2.96	A.J.Street *(Exeter Balloons titles)*	(Exeter)	17. 8.01T
G-BWMN	Rans S-7 Courier	PFA 218-12446		14. 2.96	G.J.Knee	Turweston	9. 4.05P
	(Built T M Turnbull) (Rotax 912-UL)						
G-BWMO	Oldfield Baby Lakes	JAL.3	G-CIII	14. 2.96	A.G.Fowles	Sleap	15. 2.05P
	(Built J A List) (Continental C85)		N11JL				
G-BWMS	de Havilland DH.82A Tiger Moth	82712	OO-EVJ	14. 2.96	Foundation Early Birds	(Nederhorst, The Netherlands)	
			T-29, R4771				
G-BWMU	Cameron Monster Truck 105SS HAB	3607		20. 2.96	Magical Adventures Ltd		
					"Skycrusher"	(West Bloomfield, Michigan, US)	2. 8.01A
G-BWMV	Colt AS-105 Mk.II HA Airship	3775		22. 2.96	D.Stuber	(Bad Krenznach, Germany)	17. 8.04A
	(Built Cameron Balloons Ltd)						
G-BWMX	de Havilland DHC-1 Chipmunk 22	C1/0481	WG407	19. 2.96	K.S.Kelso	Top Farm, Croydon, Royston	2. 4.05
					tr 407th Flying Group *(As "WG407/67")*		
G-BWMY	Cameron Bradford and Bingley 90SS HAB	3808		23. 2.96	Magical Adventures Ltd		
						(West Bloomfield, Michigan, US)	20. 4.00A
G-BWNB	Cessna 152 II	15280051	N757WA	23. 8.96	Galair International Ltd	Wellesbourne Mountford	13.11.05T
G-BWNC	Cessna 152 II	15284415	N6487L	23. 8.96	Galair International Ltd	Wellesbourne Mountford	11. 5.06T
G-BWND	Cessna 152 II	15285905	N95493	23. 8.96	Galair International Ltd, M.R.Galiffe and G.Davis		
						Wellesbourne Mountford	22. 7.06T
G-BWNI	Piper PA-24-180 Comanche	24-136	N5123P	15. 2.96	W A Stewart	North Connel, Oban	30. 7.05
G-BWNJ	Hughes 269C	86-0528	N42LW	29. 2.96	L.R.Fenwick	Long Fosse House, Beelsby, Grimsby	12. 6.05
			N27RD, N7458F				
G-BWNK	de Havilland DHC-1 Chipmunk 22	C1/0317	WD390	4. 3.96	B.Whitworth *(As "WD390")*	Breighton	7. 5.06
G-BWNL*	Europa Aviation Europa	PFA 247-12675		27. 2.96	H.Smith	Morgansfield, Fishburn	
	(Built H Smith - kit no.068) (NSI?) *(Tri-gear u/c)*		*(Damaged Fishburn 14.12.97: cancelled 11.12.00 by CAA - no PtoF issued) (On rebuild 2002)*				
G-BWNM	Piper PA-28R-180 Cherokee Arrow	28R-30435	N934BD	5. 3.96	M. and Rosa C.Ramnial	Gloucestershire	17.10.05
G-BWNO	Cameron O-90 HAB	3716		5. 3.96	T.Knight	(Roydon, Harlow)	2. 9.05A
G-BWNP	Cameron Club-90 SS HAB	1717	EI-BVQ	6. 3.96	C.J.Davies and P.Spellward	(Castleton, Hope Valley)	4. 7.04
	(Club Orange Soft Drink Can shape)						
G-BWNS	Cameron O-90 HAB	3842		6. 3.96	Smithair Ltd	(Billingshurst)	20. 9.05T
					(Self Assessment Tax titles) "Hector"		
G-BWNT	de Havilland DHC-1 Chipmunk 22	C1/0772	WP901	7. 3.96	R.A.Stafford	Nottingham East Midlands	28. 5.06T
					t/a Three Point Aviation *(As "WP901")*		

Registration	Type	C/n	Previous identity	Regd	Owner/operator	Location	CofA
G-BWNU	Piper PA-38-112 Tomahawk	38-78A0334	N9294T	8. 3.96	Kemble Aero Club Ltd	Kemble	24. 9.05
G-BWNY	Aeromot AMT-200 Super Ximango	200055		11. 6.96	A E.Mayhew	Rochester	31. 5.05
G-BWNZ	Agusta A109C	7654		3. 4.96	Anglo Beef Processors Ltd	(Ardlee, County Louth)	13. 4.05T
G-BWOA	Sky 105-24 HAB	027		13. 3.96	Akhter Group plc	(Harlow)	11. 7.02A
G-BWOB	Luscombe 8F Silvaire	6179	N1552B NC1552B	14. 3.96	P.J.Tanulak and H.T.Law	(Shrewsbury)	
G-BWOD	IAV-Bacau Yakovlev Yak-52	833810	LY-ALY DOSAAF 139	14. 3.96	Insurefast Ltd *(As "DOSAAF 139")*	Sywell	3. 5.05P
G-BWOF	British Aircraft Corporation BAC 145 Jet Provost T.5	EEP/JP/955	XW291	18. 3.96	Techair London Ltd	Bournemouth	21. 1.05P
G-BWOI	Piper PA-28-161 Cadet	2841307	N270X G-BWOI, D-EJTM, N9264N, N9208P	18. 3.96	Plane Talking Ltd	Cranfield	12. 8.07T
G-BWOJ	Piper PA-28-161 Cadet	2841331	N630X G-BWOJ, D-ESTM, N92242, (N123ND), N92242	18. 3.96	Plane Talking Ltd	Cranfield	26. 8.07T
G-BWOK	Lindstrand LBL 105G HAB	370		19. 3.96	Lindstrand Balloons Ltd	(Oswestry)	7.10.03A
G-BWOL*	Hawker Sea Fury FB.11	ES.3617 & 61631	D-CACY (2) G-9-66/WG599	18. 3.96	The Old Flying Machine (Air Museum) Co Ltd *(Cancelled 4.1.01 by CAA) (On restoration for K Weeks 8.02)*	Catfield, Norfolk	
G-BWON	Europa Aviation Europa *(Built G T Birks - kit no.112) (Conventional u/c)*	PFA 247-12720		29. 1.96	H.J.Fish	(Ripley)	1. 9.05P
G-BWOR	Piper PA-18-135 Super Cub (L-18C)	18-2547	OO-WIS OO-HMF, ALAT, 52-6229	21. 3.96	C.D.Baird	Roughay Farm, Bishops Waltham	12. 8.05
G-BWOT	Hunting Percival P.84 Jet Provost T.3A	PAC/W/10138 *(C/n reported as PAC/W/949267)*	XN459	25. 3.96	Red Pelicans Ltd *(As "XN459" in all-red Red Pelicans c/s)*	North Weald	28.11.05P
G-BWOU	Hawker Hunter F.58A *(May be a composite: regd with c/n 41H-003067 ex XF306/7776M/G-9-402: this became J-4133)*	HABL.003067	Swiss AF J-4105 G-9-315, A2565, XF303	26. 3.96	The Old Flying Machine (Air Museum) Co Ltd *(As "105")*	Scampton	20. 1.99P
G-BWOV	Enstrom F-28A	222	N690BR G-BWOV, F-BVRG	26. 3.96	P.A.Goss	Draycott Farm, Chiseldon	11. 5.06T
G-BWOW	Cameron N-105 HAB	3805		31. 1.96	S.J.Colin t/a Skybus Ballooning *"Skybus"*	(Cranbrook)	14. 7.03T
G-BWOX	de Havilland DHC-1 Chipmunk 22	C1/0728	WP844	27. 3.96	J.St.Clair-Quentin *(As "WP844")*	Spanhoe	10. 7.00
G-BWOY	Sky 31-24 HAB	029		28. 3.96	C.Wolstenholme	(Bristol)	15.11.04A
G-BWOZ	CFM Streak Shadow SA *(Built H Witt - kit no K.154SA)* (Rotax 582)	PFA 206-12988		1. 4.96	N.P.Harding	Plaistowes Farm, St.Albans	13. 4.04P
G-BWPB	Cameron V-77 HAB	3866		1. 4.96	R.H. and N.K.Calvert t/a The Fair Weather Friends Ballooning Co	(Bristol)	13. 4.05A
G-BWPC	Cameron V-77 HAB	3867		1. 4.96	Helen Vaughan *"Olive"*	(Tring)	20. 1.05A
G-BWPE	Meridian Renegade 912 *(Built G Wilson)*	PFA 188-12791		2. 4.96	J.Hatswell *(Also wears "83-EB" under starboard wing)*	(Menton, France)	23. 5.05P
G-BWPF	Sky 120-24 HAB	028		3. 4.96	Zebedee Balloon Service Ltd *"Whisper"*	(Newtown Hungerford)	16.10.04T
G-BWPH	Piper PA-28-181 Cherokee Archer II	28-7790311	N1408H	4. 4.96	H. and E.Merkado	Panshanger	1. 5.05T
G-BWPJ	Steen Skybolt *(Built W R Penaluna)* (Continental IO-346)	PFA 64-12854		9. 4.96	D.Houghton *(New owner 11.04)*	Croft Farm, Defford	20.10.03P
G-BWPL*	Airtour AH-56 HAB				See SECTION 3 Part 1	(Aldershot)	
G-BWPP	Sky 105-24 HAB	031		9. 4.96	P.F.Smart tr The Sarnia Balloon Group *(Nice Day titles)*	(Basingstoke)	17. 7.00A
G-BWPS	CFM Streak Shadow SA *(Built P G A Sumner - kit no K.275SA)* (Rotax 618)	PFA 206-12954		9. 2.96	P.M.E.D.McNair-Wilson	Old Sarum	12.10.05P
G-BWPT	Cameron N-90 HAB	3838		5. 3.96	G.Everet *(New owner 6.04)*	(Sandway, Maidstone)	18. 5.05A
G-BWPV	Pilatus Britten-Norman BN-2T-4S Defender 4000	4012		24. 4.96	B-N Group Ltd *(To MOD as "ZG997" 2005)*	Bembridge	AC
G-BWPX	Pilatus Britten-Norman BN-2T-4S Defender 4000	4014		24. 4.96	Britten-Norman Aircraft Ltd	Bembridge	
G-BWPY	HOAC DV.20 Katana	20158	OE-UDV	10. 6.96	S.Phillips t/a SAS Flight Services	Gamston	14. 6.04T
G-BWPZ	Cameron N-105 HAB	3889		19. 4.96	D.M.Moffat	(Bristol)	2.10.04A
G-BWRA	Sopwith LC-1T Triplane rep *(Built J S Penny)* (Warner Scarab 165)	PFA 21-10035	G-PENY	19. 4.96	S.M.Truscott and J.M.Hoblyn *(As "N500" in RNAS c/s)*	Watchford Farm, Yarcombe	25. 4.05P
G-BWRC	Avid Hauler Mk.4 *(Built B.Williams)* (Hirth F30)	PFA 189-12979		22. 2.96	B.Williams	Shobdon	17. 6.05P
G-BWRM	Colt 105A HAB	3734		23. 4.96	N.Charbonnier	(Aosta, Italy)	24. 1.05A
G-BWRO	Europa Aviation Europa *(Built E C Clark - kit no.196) (Monowheel u/c)*	PFA 247-12849		22. 4.96	J.G.M.McDiarmid	Bodmin	25. 8.05P
G-BWRP	Beech 58 Baron	TH-1737	VR-BVB N3217H	23. 4.96	Astra Aviation Ltd	Guernsey	8. 8.05
G-BWRR	Cessna 182Q Skylane II	18266660	N95861	29. 3.94	D.O.Halle	Nottingham East Midlands	21.10.06T
G-BWRS	SNCAN Stampe SV-4C	437	(N) F-BCVQ	24. 4.96	G.P.J.M.Valvekens	(Diest, Belgium)	
G-BWRT	Cameron Concept 60 HAB	3078	EI-BYP	22.10.96	W.R.Teasdale *(Inflated 4.02)*	(Maidenhead)	
G-BWRV	Lindstrand LBL 90A HAB	371		23. 4.96	Flying Pictures Ltd *(Audi titles)*	(Chilbolton)	3. 6.04A
G-BWRY	Cameron N-105 HAB	3817		24. 4.96	G.Aimo *"Ferodo"*	(Mondovi, Italy)	27. 5.05A
G-BWRZ	Lindstrand LBL 105A HAB	383		26. 4.96	D.J.Palmer *(New owner 6.03)*	(Bury St. Edmunds)	16. 6.04A
G-BWSB	Lindstrand LBL 105A HAB	384		26. 4.96	R.Calvert-Fisher *(New owner 3.03)*	(Abingdon)	6. 9.05A
G-BWSC	Piper PA-38-112 Tomahawk II	38-81A0125	N23203	29. 4.96	APB Leasing Ltd	Long Marston	17. 9.05T
G-BWSD	Campbell Cricket *(Built R.F.G.Moyle)*	PFA G/03-1216		3. 5.96	R.F.G.Moyle	(Penryn)	
G-BWSG	British Aircraft Corporation BAC 145 Jet Provost T.5	EEP/JP/988	XW324	13. 5.96	R.M.Kay *(As "XW324" in 6FTS c/s)*	Bournemouth	15. 6.05P
G-BWSH	Hunting Percival P.84 Jet Provost T.3A	PAC/W/10159	XN498	13. 5.96	Global Aviation Ltd *(Noted 8.04)*	Humberside	8. 7.03P
G-BWSI	K & S Squarecraft SA.102.5 Cavalier *(Built B.W.Shaw)* (Lycoming O-235)	PFA 01-10624		18. 4.84	B.W.Shaw	Wathstow Farm, Newby Wiske	18. 1.05P
G-BWSJ	Denney Kitfox Model 3 *(Built J.M.Miller)*	PFA 172-12204		15. 5.96	J.M.Miller	Sutton Meadows, Ely	25. 4.05P
G-BWSL	Sky 77-24 HAB	004		16. 5.96	D.Baggley	(Stoke-on-Trent)	28. 2.05A

Regn	Type	C/n	Previous identity	Regd	Owner	Location	CofA
G-BWSN	Denney Kitfox Model 3	PFA 172-12141		16. 5.96	W.J.Forrest	Siege Cross Farm, Thatcham	15. 9.05P
	(Built W.J.Forrest)						
G-BWSO	Cameron Apple-90 HAB	3915		17. 5.96	Flying Pictures Ltd *"Sainsbury's Apple"*	(Chilbolton)	11. 6.02A
G-BWSP	Cameron Carrots-80 HAB	3914		17. 5.96	Flying Pictures Ltd *"Sainsbury's Carrots"*	(Chilbolton)	5. 7.02A
G-BWST	Sky 200-24 HAB	036		20. 5.96	S.A.Townley t/a Sky High Leisure	(Wrexham)	8. 4.05T
G-BWSU	Cameron N-105 HAB	3848		20. 5.96	A.M.Marten *"Wonder Bra"*	(Guildford)	2. 9.03A
G-BWSV	IAV-Bacau Yakovlev Yak-52	877601	DOSAAF 43	20. 5.96	M.W.Fitch	North Weald	2. 2.05P
G-BWSZ	Montgomerie-Bensen B.8MR	PFA G/01-1268		14. 5.96	D.Cawkwell	Goole	6. 1.98P
	(Built D.Cawkwell) (Rotax 582)						
G-BWTA	HOAC DV.20 Katana	20159	OE-UDV	10. 6.96	Lowlog Ltd	Elstree	21.10.05T
G-BWTB	Lindstrand LBL 105A HAB	374		29. 5.96	Servatruc Ltd	(Nottingham)	18. 5.03A
G-BWTC	Moravan Zlin Z-242L	0697		2. 8.96	Oxford Aviation Services Ltd	Oxford	17.11.05T
G-BWTD	Moravan Zlin Z-242L	0698		2. 8.96	Oxford Aviation Services Ltd	Oxford	4.11.05T
G-BWTE	Cameron O-140 HAB	3885		30. 5.96	R.J. and A.J.Mansfield	(Bowness-on-Windermere)	22. 2.05T
G-BWTF	Lindstrand Bear SS HAB	375		3. 6.96	Free Enterprise Balloons Ltd		
					"Mr Biddle"	(East Leroy, Michigan, US)	3. 3.04A
G-BWTG	de Havilland DHC-1 Chipmunk 22	C1/0119	WB671	4. 6.96	P.M.M. de Graaf	Teuge, The Netherlands	24. 9.06
					tr Chipmunk 4 Ever Foundation *(As "WB671/910")*		
G-BWTH	Robinson R22 Beta	1767	HB-XYD	5. 6.96	L.Smith	Booker	27. 6.05T
			N4052R		t/a Helicopter Services		
G-BWTJ	Cameron V-77 HAB	3917		7. 6.96	A.J.Montgomery	(Yeovil)	9. 5.03A
G-BWTK	Rotary Air Force RAF 2000 GTX-SE	PFA G/13-1264		7. 6.96	M.Love	(Camberley)	8. 8.05P
	(Built M P Lehermette)						
G-BWTN	Lindstrand LBL 90A HAB	357		12. 6.96	Clarks Drainage Ltd	(Oakham)	28. 7.05A
					(Clark's Drainage/Polypipe titles)		
G-BWTO	de Havilland DHC-1 Chipmunk 22	C1/0852	WP984	5. 6.96	Skycraft Services Ltd *(As "WP984/H")*	Little Gransden)	21. 6.04
G-BWTR	Slingsby T.61F Venture T.2	1881	XZ561	12. 6.96	P.R.Williams	(Brackley)	
G-BWTW	Mooney M.20C	20-1188	EI-CHI	5. 6.96	R.C.Volkers	RNAS Yeovilton	23.12.05
			N6955V				
G-BWUA	Campbell Cricket rep	PFA G/03-1248		17. 6.96	N J Orchard	(Portishead, Bristol)	13. 1.05P
	(Built R.T.Lancaster)						
G-BWUB	Piper PA-18S-135 Super Cub	18-3986	N786CS	13. 6.96	Caledonian Seaplanes Ltd	Dunkeswell	2. 4.05T
	(L-21C) *(Regd with c/n 18-3786)*		G-BWUB, SX-AHB, EI-263, I-EIUO, MM54-2586, 54-2586				
G-BWUE	Hispano HA.1112-MIL Buchon	223	N9938	14. 6.96	R.A.Fleming	Breighton	AC
	(Reported as c/n 172: C4K-155 was c/n 223)		G-AWHK, C4K-102		*(On rebuild 1.05 as "1" in Luftwaffe c/s)*		
G-BWUH	Piper PA-28-181 Archer III	2843048	N9272E	30. 8.96	B.K.Ambrose	Fowlmere	7.10.05
			(G-BWUH)		tr G-BWUH Flying Group		
G-BWUJ	RotorWay Executive 162F	6153		2. 7.96	Southern Helicopters Ltd	Street Farm, Takeley	16.12.04P
	(*Built Southern Helicopters Ltd*) (RotorWay RI 162F)						
G-BWUK	Sky 160-24 HAB	043		2. 7.96	Cameron Flights Southern Ltd	(Pewsey)	20. 9.05T
G-BWUL	Noorduyn AT-16 Harvard IIB	14A-1415	N16NA	4. 7.96	Aereo Servizi Bresciana SRL	Brescia-Montichiari, Italy	13.10.97
			G-BWUL, FT375, 43-13116		*(Noted as "FT375" 7.04)*		
G-BWUM	Sky 105-24 HAB	038		5. 7.96	P.Stern and F.Kirchberger	(Regen/Lam, Germany)	21. 7.04
					"Wanninger"		
G-BWUN	de Havilland DHC-1 Chipmunk 22	C1/0253	WD310	5. 7.96	T.Henderson *(As "WD310/B")*	Deanland	6. 5.06
G-BWUP	Europa Aviation Europa	PFA 247-12703		3. 7.96	G.A.Haines	Inverness	25. 4.05P
	(Built T J Harrison - kit no.104) (NSI EA-81/100) *(Conventional u/c)*						
G-BWUR	Thunder Ax10-210 Series 2 HAB	3910		11. 7.96	T.J.Bucknall *"Kinetic"*	(Malpas)	7. 4.01T
					(Op Balloon Preservation Group)		
G-BWUS	Sky 65-24 HAB	040		16. 7.96	N.A.P.Bates	(Tunbridge Wells)	17. 8.05A
G-BWUT	de Havilland DHC-1 Chipmunk 22	C1/0918	WZ879	4. 6.96	Aero Vintage Ltd *(As "WZ879/X")*	Duxford	22. 5.06
G-BWUU	Cameron N-90 HAB	3954		17. 7.96	South Western Electricity plc	(Bristol)	27. 6.01A
G-BWUV	de Havilland DHC-1 Chipmunk 22A	C1/0655	WK640	18. 7.96	P.Ray *(As "WK640/C")*	Wombleton	5. 9.07
G-BWUW	British Aircraft Corporation BAC 145 Jet Provost T.5A	EEP/JP/1045	XW423	18. 7.96	S.Jackson	(Connah's Quay, Deeside)	14. 2.02P
					tr Deeside College *(As "XW423/14")*		
G-BWUZ	Campbell Cricket rep	PFA G/03-1267		24. 6.96	K.A.Touhey	(Wells)	18. 9.05P
	(Built M A Concannon) (Rotax 582)						
G-BWVB	Pietenpol AirCamper	PFA 47-11777		24. 7.96	D.Platt	(Keighley)	1.11.03P
	(Built M J Whatley) (Continental O-200-A)						
G-BWVC	Jodel D.18	PFA 169-11331		29. 7.96	R.W.J. Cripps	(Spondon, Derby)	
	(Built R.W.J. Cripps)						
G-BWVH	Robinson R44 Astro	0072	SX-HDE	10. 9.96	Derg Developments Ltd	(Killaloe, County Clare)	9.10.05T
			(D-HBBT)				
G-BWVI	Stern ST.80 Balade	PFA 166-11190		7. 8.96	I.M.Godfrey-Davies	Bourne Park, Hurstbourne Tarrant	6. 8.03P
	(Built P E Parker) (Volkswagen 1834)				*(New owner 7.04)*		
G-BWVL	Cessna 150M	15077229	N50NA	13. 8.96	A.H.Shaw	Gloucestershire	18. 1.03T
			N63286				
G-BWVN	Whittaker MW7	PFA 171-11839		19. 8.96	R.K.Willcox	Wickwar	
	(Built J W May)				*(Noted 1.05)*		
G-BWVR	IAV-Bacau Yakovlev Yak-52	878202	LY-AKQ	27. 8.96	P.Shaw	Sturgate	17. 8.05P
			DOSAAF 134		*"52"*		
G-BWVS	Europa Aviation Europa	PFA 247-12686		28. 8.96	D.R.Bishop	Kemble	9. 4.05P
	(Built D.R.Bishop - kit no.085) (Monowheel u/c)						
G-BWVT	de Havilland DHA.82A Tiger Moth	1039	N1350	27. 8.96	R.Jewitt	(Tunbridge Wells)	20. 6.05
	(Built de Havilland Aircraft Pty Ltd, Australia)		VH-SNZ, A17-604, VH-AIN, A17-604				
G-BWVU	Cameron O-90 HAB	3204		28. 8.96	J.Atkinson	(Dorchester)	26. 1.99A
G-BWVV	Jodel D.18	PFA 169-12699		29. 8.96	D.S.Howarth	Hawksbridge Farm, Oxenhope	17. 2.04P
	(Built P Cooper) (Volkswagen 1834)				*(Noted 7.04 less wings)*		
G-BWVY	de Havilland DHC-1 Chipmunk 22	C1/0766	WP896	3. 9.96	P.W.Portelli *(As "WP896/)*	White Waltham	25. 7.07
G-BWVZ	de Havilland DHC-1 Chipmunk 22	C1/0614	WK590	16. 7.96	D.Campion *(As "WK590/69")*	Grimbergen, Belgium	18.10.05
G-BWWA	Ultravia Pelican Club GS	PFA 165-12242		6. 9.96	D.S.Simpson	Graveley	5. 7.05P
	(Built E F Clapham) (Rotax 912-UL)				tr Pelican Group		
G-BWWB	Europa Aviation Europa	PFA 247-12670		9. 9.96	S.M.O'Reilly	Redhill	28.10.05P
	(Built M.G.Dolphin- kit no.080) (Monowheel u/c)						

Registration	Type	c/n	Previous identity	Reg date	Owner/Operator	Base	CofA expiry
G-BWWC	de Havilland DH.104 Dove 7 *(Wings from G-APSO fitted early 2000)*	04498	XM223	14. 6.96	Air Atlantique Ltd *(As "XM223") (Stored 6.02)*	(Coventry)	
G-BWWE	Lindstrand LBL 90A HAB	410		11. 9.96	B.J.Newman	(Rushden, Northampton)	26. 3.05T
G-BWWF	Cessna 185A Skywagon	185-0240	N4893K G-BWWF, 9J-MCK, 5Y-BBG, ET-ACI, N4040Y	13. 9.96	S.M.Craig Harvey	Hinton-in-the-Hedges	15. 7.07
G-BWWG	SOCATA Rallye 235E Gabier	13121	EI-BIF HB-EYT, N344RA	23.10.96	J.J.Frew	City of Derry	28. 7.06
G-BWWH	Yakovlev Yak-50	853010	LY-ABL LY-XNI, DOSAAF	16. 9.96	R.S.Partridge-Hicks and I C Austin *"853010"*	North Weald	17. 6.05P
G-BWWI	Aérospatiale AS332L Super Puma	2040	OY-HMF (G-TIGT)	11. 9.96	Bristow Helicopters Ltd *"Johnshaven"*	Aberdeen	8.11.05T
G-BWWK	Hawker Nimrod I (RR Kestrel)	41H-43617	S1581	13. 9.96	Patina Ltd *(As "S1581/573" in 802 Sqdn c/s)*	Duxford	24. 6.05P
G-BWWL	Colt Flying Egg SS HAB	1813	JA-A0513	19. 9.96	Magical Adventures Ltd	(West Bloomfield, Michigan, US)	2. 8.01A
G-BWWN	Isaacs Fury II *(Built D H Pattison)* (Lycoming O-235-H2C)	PFA 11-10957		23. 9.96	F.J.Ball *(As "K8303/D")*	Fenland	13. 6.05P
G-BWWP	Rans S-6-116 Coyote II *(Built S A Beddus)* (Rotax 582) *(Tailwheel u/c)*	PFA 204A-12648		2.10.96	P.Lewis	Park Farm, Eaton Bray	15.12.04P
G-BWWS	Rotary Air Force RAF 2000 GTX-SE *(Built G R Williams)*	PFA G/13-1277		7.10.96	A.Gibbon and C.Cruickshank	Sorbie Farm, Kingsmuir	30. 8.05P
G-BWWT	Dornier 328-100	3022	D-CDXO VT-VIG, D-CDHG	12.11.96	Suckling Airways (Luton) Ltd t/a Scot Airways	London City	12.11.05T
G-BWWU	Piper PA-22-150 Tri-Pacer *(Hoerner wing-tips: tail-wheel conversion)*	22-5002	N7139D	9.10.96	J.D.Bally	Gamston	30.10.05
G-BWWW	British Aerospace Jetstream Series 3102	614	G-31-6141	8. 7.83	BAE SYSTEMS (Operations) Ltd	Warton	4.12.06A
G-BWWX	Yakovlev Yak-50	853003	LY-AOI DOSAAF	11.10.96	J.L.Pfundt	Hilversum, The Netherlands	11. 3.05P
G-BWWY	Lindstrand LBL 105A HAB	411		14.10.96	M.J.Smith *(Corks and Cans Norton titles)*	(Westow, York)	8. 3.05T
G-BWWZ	Denney Kitfox Model 3 *(Built K M Allan)* (Rotax 912)	PFA 172-13054		15.10.96	A.I.Eskander	Lymm Dam	21.12.05P
G-BWXA	Slingsby T.67M-260 Firefly	2236		19. 3.96	Babcock Support Services Ltd t/a Babcock Defence Services	RAF Barkston Heath	27. 6.05T
G-BWXB	Slingsby T.67M-260 Firefly	2237		19. 3.96	Babcock Support Services Ltd t/a Babcock Defence Services	RAF Barkston Heath	17. 7.05T
G-BWXC	Slingsby T.67M-260 Firefly	2238		19. 3.96	Babcock Support Services Ltd t/a Babcock Defence Services *(Op DEFTS)*	RAF Barkston Heath	18. 8.05T
G-BWXD	Slingsby T.67M-260 Firefly	2239		19. 3.96	Babcock Support Services Ltd t/a Babcock Defence Services *(Op DEFTS)*	RAF Barkston Heath	18. 8.05T
G-BWXE	Slingsby T.67M-260 Firefly	2240		19. 3.96	Babcock Support Services Ltd t/a Babcock Defence Services	RAF Barkston Heath	28. 8.05T
G-BWXF	Slingsby T.67M-260 Firefly	2241		19. 3.96	Babcock Support Services Ltd t/a Babcock Defence Services *(Op DEFTS)*	RAF Barkston Heath	5. 9.05T
G-BWXG	Slingsby T.67M-260 Firefly	2242		19. 3.96	Babcock Support Services Ltd t/a Babcock Defence Services *(Op DEFTS)*	RAF Barkston Heath	1. 6.06T
G-BWXH	Slingsby T.67M-260 Firefly	2243		19. 3.96	Babcock Support Services Ltd t/a Babcock Defence Services	AAC Middle Wallop	20.10.05T
G-BWXI	Slingsby T.67M-260 Firefly	2244		19. 3.96	Babcock Support Services Ltd t/a Babcock Defence Services *(Op DEFTS)*	RAF Barkston Heath	3.11.05T
G-BWXJ	Slingsby T.67M-260 Firefly	2245		19. 3.96	Babcock Support Services Ltd t/a Babcock Defence Services *(Op DEFTS)*	RAF Barkston Heath	1.12.05T
G-BWXK	Slingsby T.67M-260 Firefly	2246		19. 3.96	Babcock Support Services Ltd t/a Babcock Defence Services	RAF Barkston Heath	13.11.05T
G-BWXL	Slingsby T.67M-260 Firefly	2247		19. 3.96	Babcock Support Services Ltd t/a Babcock Defence Services	RAF Barkston Heath	12.12.05T
G-BWXM	Slingsby T.67M-260 Firefly	2248		19. 3.96	Babcock Support Services Ltd t/a Babcock Defence Services	RAF Barkston Heath	16. 1.06T
G-BWXN	Slingsby T.67M-260 Firefly	2249		19. 3.96	Babcock Support Services Ltd t/a Babcock Defence Services	RAF Barkston Heath	5. 2.06T
G-BWXO	Slingsby T.67M-260 Firefly	2250		19. 3.96	Babcock Support Services Ltd t/a Babcock Defence Services *(Op DEFTS)*	RAF Barkston Heath	19. 2.06T
G-BWXP	Slingsby T.67M-260 Firefly	2251		19. 3.96	D.S.McGregor	Rayne Hall Farm, Braintree	13. 2.06T
G-BWXR	Slingsby T.67M-260 Firefly	2252		19. 3.96	Babcock Support Services Ltd t/a Babcock Defence Services *(Op DEFTS)*	RAF Barkston Heath	16. 1.06T
G-BWXS	Slingsby T.67M-260 Firefly	2253		19. 3.96	Babcock Support Services Ltd t/a Babcock Defence Services *(Op DEFTS) "6"*	RAF Barkston Heath	10. 3.06T
G-BWXT	Slingsby T.67M-260 Firefly	2254		19. 3.96	Babcock Support Services Ltd t/a Babcock Defence Services *(Op DEFTS)*	RAF Barkston Heath	6. 2.06T
G-BWXU	Slingsby T.67M-260 Firefly	2255		19. 3.96	Babcock Support Services Ltd t/a Babcock Defence Services	RAF Barkston Heath	9. 3.06T
G-BWXV	Slingsby T.67M-260 Firefly	2256		19. 3.96	Babcock Support Services Ltd t/a Babcock Defence Services	RAF Barkston Heath	9. 4.06T
G-BWXW	Slingsby T.67M-260 Firefly	2257		19. 3.96	Babcock Support Services Ltd t/a Babcock Defence Services	RAF Barkston Heath	24. 3.06T
G-BWXX	Slingsby T.67M-260 Firefly	2258		19. 3.96	Babcock Support Services Ltd t/a Babcock Defence Services	RAF Barkston Heath	30. 3.06T
G-BWXY	Slingsby T.67M-260 Firefly	2259		19. 3.96	Babcock Support Services Ltd t/a Babcock Defence Services *(Op DEFTS)*	RAF Barkston Heath	6. 4.06T
G-BWXZ	Slingsby T.67M-260 Firefly	2260		19. 3.96	Babcock Support Services Ltd t/a Babcock Defence Services	RAF Barkston Heath	24. 4.06T
G-BWYB	Piper PA-28-160 Cherokee	28-263	N6374A G-BWYB, 6Y-JLO, 6Y-JCH, VP-JCH	16. 9.96	I.M.Latiff	Little Staughton	9. 8.03
G-BWYD	Europa Aviation Europa *(Built H.J.Bendiksen - kit no.072) (Monowheel u/c)*	PFA 247-12621		28. 8.96	F.H.Mycroft	Biggin Hill	10.10.05P

Reg	Type	C/n	Previous identity	Date	Owner	Base	CofA
G-BWYE	Cessna 310R II	310R1654	F-GBPE	6. 9.96	Air Charter Scotland Ltd	Prestwick	8.12.02T
			(N26369)		*(Noted 1.05)*		
G-BWYG	Cessna 310R II	310R1580	F-GBMY	28.10.96	R.F.Jones	Biggin Hill	7. 1.07T
			(N1820E)		t/a Kissair Aviation		
G-BWYH	Cessna 310R II	310R1640	F-GBPC	28.10.96	Air Charter Scotland Ltd	Prestwick	4. 6.06T
			N2634Y				
G-BWYI	Denney Kitfox Model 3	PFA 172-12143		30.10.96	J.Adamson	Beeches Farm, South Scarle	25. 4.05P
	(Built J Adamson) (Rotax 912)						
G-BWYK	Yakovlev Yak-50	812004	RA-01386	9. 8.96	R.A L.Hubbard	Meon	22. 1.05P
			DOSAAF 51		tr Foley Farm Flying Group		
G-BWYM	HOAC DV-20 Katana	20067	D-EWAU	27. 1.97	Plane Talking Ltd *(Op Denham School of Flying)*	Denham	18. 6.06T
G-BWYN	Cameron O-77 HAB	1162	G-ODER	13.11.96	W.H.Morgan *"Hobo"*	(Swansea)	29. 4.01A
G-BWYO	Sequoia F.8L Falco	PFA 100-10920		7.11.96	N.G.Abbott and J.Copeland	Bossington	5.. 7.05P
	(Built S Harper) (Lycoming O-320-E2A)						
G-BWYP	Sky 56-24 HAB	053		8.11.96	S.A Townley t/a Sky High Leisure	(Wrexham)	9.11.01A
G-BWYR	Rans S-6-116 Coyote II	PFA 204A-13058		8.11.96	E.A.Pearson	Weston Zoyland	18.10.05P
	(Built S Palmer) (Rotax 912-UL) *(Tailwheel u/c)*						
G-BWYS	Cameron O-120 HAB	3997		30. 9.96	J.M.Stables t/a Aire Valley Balloons	(Knaresborough)	25. 5.05T
G-BWYU	Sky 120-24 HAB	052		13.11.96	D.J.Tofton *"Bramley Red"*	(Warboys)	13. 7.05A
G-BWZA	Europa Aviation Europa	PFA 247-12626		1.11.96	T.G.Cowlishaw	(Knaresborough)	6. 9.05P
	(Built M.C.Costin - kit no.063) (Monowheel u/c)						
G-BWZD*	Avid Hauler Mk.4	PFA 189-12453		29.11.96	T.S.D.Lyle	(Dorchester)	1. 5.04P
	(Built B Moore) (Officially regd as "Flyer")				*(Cancelled by CAA 12.2.04)*		
G-BWZG	Robin R2160	311	F-WZZZ	6.11.96	Sherburn Aero Club Ltd	Sherburn-in-Elmet	15. 9.06T
G-BWZI	Agusta A109A II	7269	OH-HAD	29.11.96	P.W.Harris	Pendley Farm, Aldbury, Tring	5. 3.06T
			N109AK		t/a Pendley Farm		
G-BWZJ	Cameron A-250 HAB	4021		2.12.96	Balloon School (International) Ltd	(Petworth)	5. 3.06T
					t/a Balloon Club of Great Britain		
G-BWZK	Cameron A-210 HAB	4020		2.12.96	Balloon School (International) Ltd	(Petworth)	13. 7.05T
					t/a Balloon Club of Great Britain		
G-BWZP*	Cameron Home Special 105SS HAB				See SECTION 3 Part 1	(Aldershot)	
G-BWZT	Europa Aviation Europa	PFA 247-12727		9.12.96	G.W.N.Stewart	Crowfield	16. 9.05P
	(Built A M.Smyth, J Lusher and M Farr - kit no.115) (Monowheel u/c)						
G-BWZU	Lindstrand LBL 90B HAB	418		12.12.96	K.D.Pierce	(Cranbrook)	7. 9.03
G-BWZX	Aérospatiale AS332L Super Puma	2120	F-WQDX	12.12.96	Bristow Helicopters Ltd	Aberdeen	5. 5.07T
			G-BWZX, F-WQDX, 5V-MCD, 5V-TAH, LN-OLE		*"Muchalls"*		
G-BWZY	Hughes 269A	95-0378	G-FSDT	4.12.96	Reeve Newfields Ltd	Sywell	19. 6.04
			N269CH, N1336D, 64-18066				

G-BXAA - G-BXZZ

Reg	Type	C/n	Previous identity	Date	Owner	Base	CofA
G-BXAB	Piper PA-28-161 Warrior II	28-8416054	G-BTGK	7.10.96	TG Aviation Ltd	Manston	25. 4.06T
			N4344C				
G-BXAC	Rotary Air Force RAF 2000 GTX-SE	PFA G/13-1279		21.11.96	J.A.Robinson	(Morchard Bishop,Crediton)	13. 6.05P
	(Built D C Fairbrass)						
G-BXAD	Cameron Thunder Ax11-225 Series .2 HAB	4052		18.12.96	M E White	(Templeogue, Dublin)	24. 7.04T
G-BXAF	Pitts S-1D	PFA 09-12258		6.12.96	N.J.Watson	Sandown	19. 5.05P
	(Built F Sharples)						
G-BXAH	Piel CP.301A Emeraude	AB.422	D-EBAH	29.10.96	A.P.Goodwin	Bournemouth	9. 9.05P
	(Continental C90)						
G-BXAI	Cameron Colt 120A HAB	4056		20.12.96	E.F and R.F.Casswell	(Maidstone)	9. 6.00T
G-BXAJ	Lindstrand LBL 14A HAB (Gas filled)	425		23.12.96	Oscair Project AB	(Taby, Sweden)	
G-BXAK	IAV-Bacau Yakovlev Yak-52	811508	LY-ASC	23.12.96	T.N.Jinks	Stoke Golding	30. 6.05P
			DOSAAF				
G-BXAL	Cameron Bertie Bassett 90SS HAB	4034		13. 1.97	Trebor Bassett Ltd	(Aldershot)	27. 1.02A
					(Op Balloon Preservation Group) "Bertie Bassett"		
G-BXAM	Cameron N-90 HAB	4035		13. 1.97	Trebor Bassett Ltd	(Howden)	22. 4.04A
					(Bassett's titles) "Bertie Junior"		
G-BXAN	Scheibe SF25C Falke	44299	D-KDGQ	13. 1.97	E.R.Boyle	Winthorpe	15.10.06
	(Limbach SL1700)				tr C Falke Syndicate		
G-BXAO	Jabiru Aircraft Jabiru SK	PFA 274-13066		14. 1.97	P.J.Thompson	(Gaerwen)	23. 4.99P
	(Built I M Donnelly)				*(Damaged Ledicot near Shobdon 3.5.98: current status unknown)*		
G-BXAR	British Aerospace Avro 146-RJ100	E3298	G-6-298	27. 3.97	British Airways Citiexpress Ltd	Manchester	29. 3.06T
G-BXAS	British Aerospace Avro 146-RJ100	E3301	G-6-301	23. 4.97	British Airways Citiexpress Ltd	Manchester	29. 4.06T
G-BXAU	Pitts S-1	GHG.9	N9GG	22. 1.97	P.J.Tomlinson	Gloucestershire	14. 7.05P
	(Built G Goodrich) (Lycoming O-320)						
G-BXAV	Aerostar Yakovlev Yak-52	9111608	RA-01325	24. 1.97	Skytrace (UK) Ltd	Wolverhampton	15. 6.05P
			DOSAAF 73		tr G-BXAV Group *(As "DOSAAF 72")*		
G-BXAX*	Cameron N-77 HAB				See SECTION 3 Part 1	(Aldershot)	
G-BXAY	Bell 206B-3 JetRanger III	3946	N85EA	24. 1.97	Viewdart Ltd	Conington	18. 8.06T
			N521RC, N3210D				
G-BXBA	Cameron A-210 HAB	4072		10. 1.97	Reach For The Sky Ltd	(Guildford)	1. 8.05T
G-BXBB*	Piper PA-20-135 Pacer	20-959	EC-AOZ	24. 1.97	M.E.R.Coghlan	Farley Farm, Romsey	
			N1133C		*(Noted as "EC-AOZ" 3.03) (Cancelled 22.4.03 by CAA)*		
G-BXBC	Anderson EA-1 Kingfisher Amphibian	PFA 132-11302		28. 1.97	S.Bichan	Swanbister Farm, Orphir, Kirkwall	
	(Built S.Bichan)						
G-BXBD	CASA 1-131 Jungmann	1052	Spanish AF E3B-117	28. 1.97	P.B.Childs and B.L.Robinson	Kemble	6. 7.05P
	(Jungmann N2BU also claims to be ex E3B-117 c/n not confirmed)				*(As "CW+BG/50" in Luftwaffe c/s)*		
G-BXBG	Cameron A-275 HAB	4023		28. 1.97	M.L.Gabb	(Alcester)	20. 3.04T
G-BXBH*	Hunting Percival P.84 Jet Provost T.3A	PAC/W/9241	XM365	29. 1.97	G-BXBH Provost Ltd	Norwich	31. 8.01P
					(Cancelled 10.10.02 by CAA) (Noted 4.03 as "XM365")		
G-BXBI	Hunting Percival P.84 Jet Provost T.3A	PAC/W/11799	XN510	29. 1.97	Global Aviation Ltd	(Binbrook)	
G-BXBK	Avions Mudry CAP.10B	17	N170RC	30. 1.97	S.Skipworth	White Waltham	31. 7.06
			French AF "307-SO"				

Reg	Type	C/n	Previous identity	Date	Owner/Operator	Base	CofA
G-BXBL	Lindstrand LBL 240A HAB	317		31. 1.97	J.Fenton t/a Firefly Balloon Promotions	(Preston)	20. 7.06T
G-BXBM	Cameron O-105 HAB	3990		31. 1.97	P.Spellward *(Beneficial Bank titles)* tr Bristol University Hot Air Ballooning Society	(Bristol)	15. 2.05A
G-BXBN	Rans S-6-116 Coyote II *(Built W S Long)* (Rotax 582) *(Tri-cycle u/c)*	PFA 204A-13062		31. 1.97	A G.Foster *(Crashed on approach c.7.01: stored 12.02)*	(Grimsby)	17. 5.01P
G-BXBP	Denney Kitfox Model 2 *(Built G.S.Adams)*	PFA 172-12149		3. 2.97	G.S.Adams	Enniskillen, County Fermanagh	18.11.05P
G-BXBR	Cameron A-120 HAB	1983	SE-ZDY	4. 2.97	M.G.Barlow	(Skipton)	
G-BXBU	Avions Mudry CAP.10B	103	N173RC French AF	11. 2.97	J.F.Cosgrave and H.R.Pearson	Not known	31. 7.06
G-BXBY	Cameron A-105 HAB	4077		13. 2.97	S.P.Watkins *(Op D Littlewood) (Roman Baths titles)*	(Bath)	29. 3.07T
G-BXBZ	PZL-104 Wilga 80	CF21930941	EC-GDA ZK-PZQ	13. 2.97	P.G.Marks	Dunstable	25. 6.06
G-BXCA	Hapi Cygnet SF-2A *(Built G E Collard)* (Rotax 912-UL)	PFA 182-12921		22. 1.97	J.N.Harley	Popham	5. 9.05P
G-BXCC	Piper PA-28-201T Turbo Dakota	28-7921068	D-EKBM N2855A	19. 2.97	Greer Aviation Ltd	(Prestwick)	29. 9.06T
G-BXCD	TEAM mini-MAX 91A *(Built R.Davies)* (Rotax 503)	PFA 186-12393		18. 2.97	R.Davies	South Cerney	12. 4.05P
G-BXCG	Jodel DR.250 rep *(Built J M Scott - c/n 60)*	PFA 299-13146	D-EHGG	22. 5.97	I.Matthews tr CG Group	Cambridge	8.11.05P
G-BXCH	Europa Aviation Europa *(Built D.M.Stevens - kit no.186) (Monowheel u/c)*	PFA 247-12980		19. 2.97	D.M.Stevens	Haverfordwest	27. 5.05P
G-BXCJ	Campbell Cricket rep *(Built R A Friend)* (Rotax 532)	PFA G/03-1177		24. 2.97	F.Knowles	Henstridge	11. 7.05P
G-BXCK	Cameron Douglas-Lurpak Butterman 110SS HAB	4076		25. 2.97	Flying Pictures Ltd *(Lurpak Douglas titles)*	(Chilbolton)	8.10.02A
G-BXCL	Montgomerie-Bensen B.8MR *(Built A V Francis)* (Rotax 582)	PFA G/01-1287		26. 2.97	M.L.L.Temple	RAF Leconfield	14.12.04P
G-BXCM	Lindstrand LBL 150A HAB	443		26. 2.97	Aerosaurus Balloons Ltd	(Whimple, Exeter)	19. 7.05T
G-BXCN	Sky 105-24 HAB	047		27. 2.97	A.S.Davidson tr Nottingham Hot Air Balloon Club *"Rainbow"*	(Swadlincote)	22. 4.05
G-BXCO	Colt 120A HAB	4086		3. 3.97	T.G.Church	(Blackburn)	22. 4.05T
G-BXCP	de Havilland DHC-1 Chipmunk 22	C1/0744	WP859	27. 2.97	Rebecca E.Tyers tr Echo Flying Group *(As "WP859")*	Spanhoe	13. 3.05
G-BXCS	Cameron N-90 HAB	4122		4. 3.97	Flying Pictures Ltd *(Lurpak titles)*	(Chilbolton)	26. 3.03A
G-BXCT	de Havilland DHC-1 Chipmunk 22	C1/0145	WB697	3. 3.97	Wickenby Aviation Ltd *(As "WB697/95")*	Wickenby	2. 5.03T
G-BXCU	Rans S-6-116 Coyote II *(Built M R McNeil - kit no.1096.1044.0797)* (Rotax 912-UL) *(Tri-cycle u/c)*	PFA 204A-13105		6. 3.97	R.S.Gent	Leicester	23. 4.05P
G-BXCV	de Havilland DHC-1 Chipmunk 22	C1/0807	WP929	3. 3.97	Ocean Flight Holdings Ltd *(As "WP929")*	Duxford	22.10.06T
G-BXCW	Denney Kitfox Model 3 *(Built M.J.Blanchard)*	PFA 172-12619		6. 3.97	M.J.Blanchard *(Noted 1.05)*	Old Sarum	
G-BXDA	de Havilland DHC-1 Chipmunk 22	C1/0747	WP860	7. 3.97	S.R.Cleary *(As "WP860/6")*	Perth	8. 7.06
G-BXDB	Cessna U206F Stationair	U20602233	G-BMNZ F-BVJT, N1519U	18.12.96	D.A.Howard	Colonsay	2. 9.07T
G-BXDD	Rotary Air Force RAF 2000 GTX-SE *(Built R M Savage)*	PFA G/13-1284		9. 1.97	A.Wane *(New CofR 11.04)*	(Bolton)	4. 7.00P
G-BXDE	Rotary Air Force RAF 2000 GTX-SE *(Built A McCredie)*	PFA G/13-1280		14. 1.97	A.McRedie	Carlisle	23. 1.02P
G-BXDF	Beech 95-B55 Baron	TC-2011	SE-IXG OY-ASB	7. 3.97	Chesh-Air Ltd	Liverpool	6. 2.06T
G-BXDG	de Havilland DHC-1 Chipmunk 22	C1/0644	WK630	7. 3.97	Felthorpe Flying Group Ltd *(As "WK630" in RAF c/s)*	Felthorpe	1. 9.07
G-BXDH	de Havilland DHC-1 Chipmunk 22	C1/0270	WD331	10. 3.97	Victory Workwear Ltd *(As "WD331")*	Kemble	17. 1.05
G-BXDI	de Havilland DHC-1 Chipmunk 22	C1/0312	WD373	10. 3.97	Propshop Ltd *(As "WD373/12" in RAF c/s)*	Duxford	22.10.06T
G-BXDL	Hunting Percival P.84 Jet Provost T.3	PAC/W/9286	A8983M XM478	18. 3.97	de Havilland Aviation Ltd *(As "XM478")*	(Bournemouth)	22. 4.05P
G-BXDM	de Havilland DHC-1 Chipmunk 22	C1/0723	WP840	28. 2.97	The RAF Halton Aeroplane Club Ltd *(As "WP840/9")*	RAF Brize Norton	25. 6.06T
G-BXDN	de Havilland DHC-1 Chipmunk 22	C1/0618	WK609	18. 3.97	W.D.Lowe and L.A Edwards *(As "WK609/93")*	Booker	1.11.03
G-BXDO	Rutan Cozy *(Built C R Blackburn)* (Lycoming O-235)	PFA 159-12032		21. 3.97	R.James	Shobdon	4. 7.05P
G-BXDP	de Havilland DHC-1 Chipmunk 22	C1/0659	WK642	27. 2.97	T.A McBennet and J.Kelly *(As "WK642")*	Kilrush, County Kildare	7.10.06
G-BXDR	Lindstrand LBL 77A HAB	441		25. 3.97	British Telecommunications plc *"Bright Future"*	(Thatcham)	10. 9.05A
G-BXDS	Bell 206B-3 JetRanger III	2734	G-OVBJ G-BXDS, OY-HDK, N661PS	19. 2.98	Sterling Helicopters Ltd	Norwich	24. 7.06T
G-BXDT	Robin HR200/120B	315		25. 3.97	Multiflight Ltd	Leeds-Bradford	3. 6.06T
G-BXDU	Aero Designs Pulsar *(Built M.P.Board)*	PFA 202-11991		25. 3.97	M.P.Board	(London E4)	1. 6.05P
G-BXDV	Sky 105-24 HAB	049		26. 3.97	A.Parsons tr Loughborough Students Union Hot Air Balloon Club	(Loughborough)	28. 3.05
G-BXDY	Europa Aviation Europa *(Built D.G.Watts - kit no.229) (Monowheel u/c) (PFA c/n conflicts with G-MYZP)*	PFA 247-12914		27. 3.97	D.G. and S.Watts	Laddingford *"The Rocketeer"*	1. 2.05P
G-BXDZ	Lindstrand LBL 105A HAB	437		4. 4.97	D.J.and A D.Sutcliffe	(Harrogate)	20. 6.05A
G-BXEA	Rotary Air Force RAF 2000 GTX-SE *(Built R Firth)*	PFA G/13-1270		2. 4.97	R.Firth	Netherthorpe	5. 4.05P
G-BXEB	Rotary Air Force RAF 2000 GTX-SE *(Built J S Penny)*	PFA G/13-1285		2. 4.97	Penny Hydraulics Ltd *(Noted 5.04)*	Netherthorpe	10. 4.04P
G-BXEC	de Havilland DHC-1 Chipmunk 22	C1/0647	WK633	3. 4.97	K.P. and D.S.Hunt *(As "WK633/B")*	Redhill	28. 6.07
G-BXEE	Enstrom 280C Shark	1117	OH-HAN N336AT	9. 4.97	R.E.Harvey	Lodge Farm, West Deeping	10. 8.05

Regn	Type	C/n	Previous identity	Regd	Owner/operator	Base	Expiry
G-BXEF	Europa Aviation Europa	PFA 247-12790		7. 4.97	C.and W.P.Busuttil-Reynaud	(Emsworth, Hampshire)	
	(Built C.Busuttil-Reynaud - kit no.159) (Jabiru 3300) *(Monowheel u/c)*						
G-BXEJ	VPM M-16 Tandem Trainer	D-9302	D-MIFF	8. 4.97	N.H.Collins	Cork Farm, Streethay	26. 5.04P
	(Arrow GT 1000)				t/a AES Radionic Surveillance Systems		
G-BXEN	Cameron N-105 HAB	4090		11. 4.97	G.Aimo	(Mondovi, Piedmont, Italy)	28. 5.05A
	(New envelope c/n 10288 @ 3.03)				*"Liquigas"*		
G-BXEP	Lindstrand LBL 14M HAB (Gas filled)	460		14. 4.97	Lindstrand Balloons Ltd	(Oswestry)	19. 5.00A
G-BXES	Hunting Percival P.66 Pembroke C.1	P66/101	N4234C	14. 4.97	Atlantic Air Transport Ltd	Coventry	9. 5.05P
	(Regd with c/n PAC/W/3032)		9042M, XL954		*(As "XL954" in 60 Sqdn c/s)*		
G-BXET	Piper PA-38-112 Tomahawk	38-80A0028	N25089	14. 4.97	APB Leasing Ltd	Tatenhill	10. 8.03T
G-BXEX	Piper PA-28-181 Cherokee Archer II	28-7790463	N3562Q	16. 4.97	R.Mayle	Biggin Hill	11. 6.06T
G-BXEY	Colt AS-105GD HA Airship	3936		15. 4.97	D.Mayer	Felsberg, Germany	24.11.04A
	(Built Thunder and Colt Ltd)						
G-BXEZ	Cessna 182P Skylane II	18264344	OH-CHJ	16. 4.97	Forhawk Ltd	Bodmin	12. 2.06T
	(Reims assembled c/n F18200054)		N1479M				
G-BXFB	Pitts S-1	9543	N77ZZ	16. 4.97	D.Dobson	Little Staughton	5. 2.05P
	(Built B J Dziuba)						
G-BXFC	Jodel D.18	PFA 169-11322		17. 4.97	B.S.Godbold	Little Gransden	26. 7.05P
	(Buillt B.S.Godbold) (Revmaster 2100D)						
G-BXFD	Enstrom 280C Shark	1084	N88MD	18. 4.97	Buckland Newton Hire Ltd	Buckland Newton, Dorset	11. 9.06T
			N632H				
G-BXFE	Avions Mudry CAP.10B	135	N175RC	18. 4.97	Avion Aerobatic Ltd	(London N1)	28. 3.07T
			French AF				
G-BXFG	Europa Aviation Europa	PFA 247-12500		21. 4.97	A Rawicz-Szczerbo	Eaglescott	22. 5.05P
	(Built A Rawicz-Szczerbo - kit no.018) (Monowheel u/c)						
G-BXFI	Hawker Hunter T.7	41H-670815	WV372	24. 4.97	Fox-One Ltd *(As "WV372/R" in 2 Sqdn c/s)*	Kemble	16.10.05P
G-BXFK	CFM Streak Shadow	PFA 206-12329		24. 4.97	S.J.M.French and T.I.Gorell	(Milton Keynes)	18.11.03P
	(Built D Adcock - kit no K.206) (Rotax 582)						
G-BXFN	Cameron Colt 77A HAB	4145		25. 4.97	Cameron Balloons Ltd	(Bristol)	10. 7.04A
G-BXFU	British Aircraft Corporation 167 Strikemaster Mk.83	EEP/JP/???? & PS.158	Botswana DF OJ1	29. 4.97	Global Aviation Ltd	Humberside	6. 8.02P
			ZG805, Kuwait AF 110, G-27-151 *(As "OJ-1")*				
G-BXFV	British Aircraft Corporation 167 Strikemaster Mk.83	EEP/JP/???? & PS.173	Botswana DFOJ8	29. 4.97	Global Aviation Ltd	Humberside	29. 7.04P
			ZG811, Kuwait AF 119, G-27-188 *(As "OJ-8")*				
G-BXFY	Cameron Bierkrug 90 SS HAB	4133	D-OIBP	29. 4.97	Ballooning Bavaria	(Ruhstorf, Germany)	13. 1.04A
			G-BXFY				
G-BXGA	Eurocopter AS350B2 Ecureuil	2493	OO-RCH	30. 4.97	PLM Dollar Group Ltd	Inverness	27. 8.06T
			OO-XCH, F-WZFX				
G-BXGC	Cameron N-105 HAB	4137		6. 5.97	Cliveden Ltd	(Bath)	16. 3.05T
					(Op Ascent Balloons) (Royal Crescent Hotel titles)		
G-BXGD	Sky 90-24 HAB	067		6. 5.97	Servo and Electronic Sales Ltd *(Ocean FM titles)*	(Lydd)	11. 6.05T
G-BXGG	Europa Aviation Europa	PFA 247-12803		29. 4.97	C.J.H and P.A J.Richardson		
	(Built B W Faulkner - kit no.178) (Monowheel u/c)					Bremridge Farm, Shillingford	29. 5.04P
G-BXGH	Diamond DA.20-A1 Katana	10151		20. 5.97	Cumbernauld Flying School Ltd	Cumbernauld	21. 6.07T
G-BXGK	Lindstrand LBL 203M HAB	468		12. 5.97	Lindstrand Balloons Ltd	(Oswestry)	
G-BXGL	de Havilland DHC-1 Chipmunk 22	C1/0924	WZ884	12. 5.97	Airways Aero Associations Ltd	Booker	26.10.06T
					(Op British Airways Flying Club) (BOAC titles)		
G-BXGM	de Havilland DHC-1 Chipmunk 22	C1/0806	WP928	9. 5.97	A.T.Stolton	Shoreham	17.11.06
					tr Chipmunk Golf Mike Group *(As "WP928/D" ARMY)*		
G-BXGO	de Havilland DHC-1 Chipmunk 22	C1/0097	WB654	13. 5.97	I.C.Barlow and T.J.Orchard	Booker	26.10.03
					tr Trees Group *(As "WB654/U")*		
G-BXGP	de Havilland DHC-1 Chipmunk 22	C1/0927	WZ882	12. 5.97	T.K.Pullen	Eaglescott	20. 8.07T
					tr Eaglescott Chipmunk Group *(As "WZ882/K" in "ARMY" c/s)*		
G-BXGS	Rotary Air Force RAF 2000	PFA G/13-1290		14. 5.97	C.R.Gordon	(Cupar)	11.11.03P
	(Built N C White)						
G-BXGT	III Sky Arrow 650 T	PFA 298-13085		7. 5.97	Sky Arrow (Kits) UK Ltd	Old Sarum	7. 7.05P
	(Built Sky Arrow (Kits) UK Ltd)						
G-BXGV	Cessna 172R Skyhawk	17280240	N9300F	7. 1.98	D.Varns tr Skyhawk Group	White Waltham	8. 2.07T
G-BXGW	Robin HR200/120B	317		16. 5.97	Multiflight Ltd	Leeds-Bradford	9.10.06T
					(Op Multiflight Flying Club)		
G-BXGX	de Havilland DHC-1 Chipmunk 22	C1/0609	WK586	19. 5.97	Interflight (Air Charter) Ltd *(As "WK586")*	Blackbushe	3.11.06
G-BXGY	Cameron V-65 HAB	4125		18. 4.97	R.J.Plume tr Dante Balloon Group	(Mondovi, Italy)	19. 8.05A
G-BXGZ	Stemme S-10V	14-023	D-KSTE	18. 8.97	D.Tucker and K.Lloyd	Aston Down	10. 3.07
			EC-GGD, D-KGDF		*"S10"*		
G-BXHA	de Havilland DHC-1 Chipmunk 22	C1/0801	WP925	20. 5.97	F.A de Munck and C.S.Huijers	(Seppe, The Netherlands)	14. 9.06
					(As "WP925/C" ARMY)		
G-BXHD	Beech 76 Duchess	ME-284	OY-ARM	22. 5.97	S.J.Skilton	Bournemouth	7. 8.06T
			N223JC		t/a Aviation Rentals *(Op CTC Aviation)*		
G-BXHE	Lindstrand LBL 105A HAB	459		23. 5.97	L.H.Ellis	(Princes Risborough)	3. 9.05T
G-BXHF	de Havilland DHC-1 Chipmunk 22	C1/0808	WP930	28. 5.97	R.A Wallis	Redhill	5. 7.07
					tr Hotel Fox Sydicate *(As "WP930/J" in ARMY c/s)*		
G-BXHH	Grumman AA-5A Cheetah	AA5A-0105	N9705U	3. 6.97	M.G.Greenslade t/a Oaklands Flying	Biggin Hill	12. 6.06T
G-BXHJ	Hapi Cygnet SF-2A	PFA 182-12159		29. 5.97	I.J.Smith	Brook Farm, Boylestone	
	(Built I.J.Smith) (Volkswagen 1835)						
G-BXHL	Sky 77-24 HAB	055		29. 5.97	R.K.Gyselynck *"Harlequin"*	(Port Erin, Isle of Man)	10. 9.05
G-BXHM*	Lindstrand LBL-25A Cloudhopper HAB				See SECTION 3 Part 1	(Aldershot)	
G-BXHN*	Lindstrand Budweiser Can SS HAB				See SECTION 3 Part 1	(Aldershot)	
G-BXHO	Lindstrand Telewest Sphere SS HAB	474		30. 5.97	Magical Adventures Ltd		
						(West Bloomfield, Michigan, US)	25. 2.03A
G-BXHR	Stemme S 10-V	14-030	D-KSTE	23. 7.97	J.H.Rutherford	Durham Tees Valley	26.11.06
G-BXHT	Bushby-Long Midget Mustang	PFA 168-13077		3. 6.97	P.P.Chapman	Headcorn	
	(Built P.P.Chapman)				*(Noted 11.04)*		
G-BXHU	Campbell Cricket Mk.6	PFA G/16-1293		3. 6.97	P.J.Began	Henstridge	28. 7.05P
	(Built P.C.Lovegrove) (Rotax 503)						

Registration	Type	C/n	Previous identity	Date	Owner / Operator	Base	CofA expiry
G-BXHY	Europa Aviation Europa	PFA 247-12514		6. 6.97	A.L.Thorne and B.Lewis	White Waltham	25. 2.04P
	(Built A L.Thorne - kit no.022) (Monowheel u/c)				tr Jupiter Flying Group		
G-BXIA	de Havilland DHC-1 Chipmunk 22	C1/0056	WB615	9. 6.97	W.Askew and C.Duckett	Blackpool	18. 8.05T
					t/a Dales Aviation *(As "WB615/E")*		
G-BXIC	Cameron A-275 HAB	4162		9. 6.97	Aerosaurus Balloons Ltd	(Whimple, Exeter)	21. 7.05T
G-BXID	IAV-Bacau Yakovlev Yak-52	888802	LY-ALG	10. 6.97	E.S.Ewen	Kemble	31. 3.05P
			DOSAAF 74				
G-BXIE	Cameron Colt 77B HAB	4181		11. 6.97	R.M.Horn	(Hatfield Pevereil, Chelmsford)	12.12.04A
G-BXIF	Piper PA-28-181 Cherokee Archer II	28-7690404	PH-SWM	12. 6.97	Piper Flight Ltd	RAF Brize Norton	9. 7.06T
			OO-HAY, N6827J				
G-BXIG	Zenair CH-701 STOL	PFA 187-12065		16. 6.97	A.J.Perry	Marsh Farm, Bracklesham	22. 8.05P
	(Built A J.Perry) (Rotax 912-UL)						
G-BXIH	Sky 200-24 HAB	076		16. 6.97	Skyview Ballooning Ltd t/a Kent Ballooning	(Ashford)	7. 9.05T
G-BXII	Europa Aviation Europa	PFA 247-12812		30. 4.97	D.A McFadyean	Long Marston	11. 6.05P
	(Built D.A McFadyean - kit no.175) (Tailwheel u/c)G-BX						
G-BXIJ	Europa Aviation Europa	PFA 247-12698		16. 6.97	R.James	Shobdon	17. 4.05P
	(Built D.G. and E.A Bligh - kit no.076) (Monowheel u/c)						
G-BXIM	de Havilland DHC-1 Chipmunk 22	C1/0548	WK512	13. 5.97	P.R.Joshua and A B.Ascroft	RAF Brize Norton	2. 7.06
					(As "WK512/A/ARMY")		
G-BXIO	SAN Jodel DR.1050M Excellance	493	F-BNIO	16. 5.97	D.N.K. and M.A Symon	Perth	14. 7.07
G-BXIT	Zebedee V-31 HAB	Z1/3999		8. 5.97	P.J.Bish	Newtown Hungerford	
	(Built Zebedee Balloon Service)						
G-BXIW	Sky 105-24 HAB	073		24. 6.97	L.A.Watts	(Pangbourne, Reading)	8. 1.05
G-BXIX	VPM M-16 Tandem Trainer	PFA G/12-1292		13. 6.97	D.Beevers	Melrose Farm, Melbourne	17. 8.05P
	(Built D Beevers) (Arrow GT1000R)						
G-BXIY	Blake Bluetit	01	BAPC.37	26. 6.97	J.Bryant	(North Weald)	
	(Built W H C Blake) (Gnat 32hp) *(Pre-war composite from Spartans G-AAGN/G-AAJB and Avro 504K) (On rebuild 8.03)*						
G-BXIZ	Lindstrand LBL 31A HAB	476		3. 7.97	I.Chadwick	(Horsham)	5. 4.05A
					tr Balloon Preservation Flying Group		
G-BXJA	Cessna 402B	402B0356	N5753M	17. 7.97	Air Charter Scotland Ltd	Prestwick	17.10.07T
			XA-RFK, N5753M				
G-BXJB	IAV-Bacau Yakovlev Yak-52	877403	LY-ABR	30. 6.97	A.M.Playford, D.J.Young and N. Willson		
			DOSAAF 15		*"15"*	Poplar Hall Farm, Elmsett	7.11.05P
G-BXJC	Cameron A-210 HAB	419		12. 7.97	Balloon School (International) Ltd	(Petworth)	3. 6.05T
					t/a British School of Ballooning		
G-BXJD	Piper PA-28-180 Cherokee C	28-4215	OY-BBZ	27. 6.97	BCT Aircraft Leasing Ltd	Gamston	3.12.06T
G-BXJG	Lindstrand LBL 105B HAB	478		11. 7.97	C.E.Wood	(Witham)	12. 8.05T
G-BXJH	Cameron N-42 HAB	4194		15. 7.97	B.Conway	(Wheatley, Oxford)	22. 9.01A
G-BXJI*	Tri-R Kis	PFA 239-12573		2. 7.97	R.M.Wakeford	Cumbernauld	20. 5.03P
	(Bult R M Wakeford)	*(Overturned landing Cumbernauld 11.8.02 with substantial damage: cancelled 20.2.03 as temporarily wfu)*					
G-BXJJ	Piper PA-28-161 Cadet	2841200	EC-IGN	26. 6.97	S.J.Skilton	Bournemouth	8.12.06
			G-BXJJ, G-GFCC, N9189N		t/a Aviation Rentals		
G-BXJM	Cessna 152 II	15282380	OO-HOQ	15. 7.97	I.R.Chaplin	Andrewsfield	16. 9.06T
			F-GHOQ, N68797				
G-BXJO	Cameron O-90 HAB	4190		16. 7.97	Dragon Balloon Company Ltd	(Castleton, Hope Valley)	26. 3.05T
G-BXJP	Cameron C-80 HAB	4171		17. 7.97	AR Cobaleno Pasta Fresca SRL	(Perugia, Italy)	23. 7.03A
G-BXJS	Schempp-Hirth Janus CM	35/265	OH-819	7. 7.97	R.A.Hall tr Janus Syndicate	Enstone	12. 3.06
G-BXJT	Sky 90-24 HAB	072		18. 7.97	J.G. O'Connell	(Braintree)	22. 8.05A
G-BXJV	Dimona DA.20-A1 Katana	10152		23. 7.97	Tayside Aviation Ltd	Dundee	30. 7.06T
G-BXJW	Dimona DA.20-A1 Katana	10211	(OE-)	23. 7.97	Tayside Aviation Ltd	Dundee	15. 8.06T
			N811CH				
G-BXJY	Van's RV-6	PFA 181-12447		23. 7.97	D.J.Sharland	Popham	6. 5.05P
	(Built D J Sharland) (Lycoming O-320-D3G)						
G-BXJZ	Cameron C-60 HAB	4168		23. 7.97	R.S.Mohr	(Chippenham)	24.10.04A
G-BXKA	Airbus A320-214	714	N714AW	24.11.97	Thomas Cook Airlines UK Ltd	Manchester	23. 4.05T
			G-BXKA, F-WWIX				
G-BXKB	Airbus A320-214	716	N716AW	10.12.97	Thomas Cook Airlines UK Ltd	Manchester	5. 5.05T
			G-BXKB, F-WWIZ				
G-BXKC	Airbus A320-214	730	F-WWBQ	15.12.97	Thomas Cook Airlines UK Ltd	Manchester	14.12.06T
G-BXKD	Airbus A320-214	735	F-WWBV	17.12.97	Thomas Cook Airlines UK Ltd	Manchester	25.10.06T
G-BXKF	Hawker Hunter T.7	HABL-003314	8676M	28. 7.97	R.F.Harvey	Kemble	20. 9.05P
	(Regd with c/n 41H-003315)		XL577		*(As "XL577/V" in colours of 92 Sqdn "Blue Diamonds" aerobatic team)*		
G-BXKH	Cameron Colt Sparkasse Box 90SS HAB	4161		4. 8.97	Westfalisch-Lippischer Sparkassen und Giroverband		
					"Sparduerschen"	(Münster, Germany)	9. 8.05A
G-BXKJ	Cameron A-275 HAB	4215		4. 8.97	Ballooning Network Ltd	(Bristol)	6. 9.05T
G-BXKL	Bell 206B-3 JetRanger III	3006	N5735Y	8.10.97	Swattons Aviation Ltd	Thruxton	18.11.06T
G-BXKM	Rotary Air Force RAF 2000 GTX-SE	PFA G/13-1291		5. 8.97	J.R.Huggins	Lamberhurst Farm, Faversham	21.10.04P
	(Built J.R.Huggins)						
G-BXKO	Sky 65-24 HAB	083		11. 8.97	J-M.Reck	(Evette-Salbert, France)	23. 1.05
G-BXKU	Colt AS-120 Mk.II HA Airship	4165		15. 8.97	D.C.Chipping	(Grantham)	19. 4.01A
	(Built Cameron Balloons Ltd)						
G-BXKW	Slingsby T.67M-200 Firefly	2061	VR-HZS	15. 8.97	N.A Whatling	Spanhoe	19. 5.07T
			HKG-13, G-7-129		*(As "HKG-13")*		
G-BXKX	Taylorcraft J Auster 5	803	D-EMXA	19. 8.97	A.L.Jubb	Orchard Farm, Sittingbourne	21. 5.04
			HB-EOK, MS938				
G-BXKY*	Cameron DP-90 HA Airship	4198		19. 8.97	Cameron Balloons Ltd	(Rio de Janeiro, Brasil)	
					"Oceanica" (Cancelled 23.1.01) (Noted 5.03)		
G-BXLC	Sky 120-24 HAB	085		20. 8.97	Dragon Balloon Company Ltd	(Castleton, Hope Valley)	18. 9.05A
G-BXLF	Lindstrand LBL 90A HAB	487		3. 9.97	Wendy Rousell and J.Tyrrell	(Wellingborough)	23. 6.05A
					(Variohm Components titles)		
G-BXLG	Cameron C-80 HAB	4250		5. 3.98	D. and L.S.Litchfield *"Borne Again"*	(Reading)	12. 8.03A
G-BXLI	Bell 206B-3 JetRanger III	4041	N206JR	8. 9.97	N.K.Sherborne and R.H.Stevens	Gloucestershire	27.11.06T
			G-JODY		*(Crashed in field Holman Clavel, Culmhead, Somerset 22.1.05 and substantially damaged)*		

Registration	Type	C/n	Previous identity	Date	Owner/Operator	Base	CofA
G-BXLK	Europa Aviation Europa	PFA 247-12613		11. 9.97	R.G.Fairall	Redhill	28. 4.05P
	(Built R.G.Fairall - kit no.071) (Monowheel u/c)						
G-BXLN	Fournier RF4D	4022	F-BORK	15. 9.97	Elizabeth A.Wiseman and N.Thorne	Breighton	7. 5.06
G-BXLO	Hunting Percival P.84 Jet Provost T.4	PAC/W/19986	9032M	14. 8.97	S.J.Davies and S.Eagle	North Weald	19. 9.05P
			XR673		*(As "XR673" in silver RAF c/s and yellow trainer bands)*		
G-BXLP	Sky 90-24 HAB	084		18. 9.97	G.B.Lescott *"Heart of Gold"*	(Oxford)	18.10.04A
G-BXLR	PZL-110 Koliber 160A	04980077	SP-WGF (2)	10. 6.98	Oakmast Systems Ltd	North Weald	31.10.04T
			(SP-WGF (1)), (N150CD)		*(Noted 1.05)*		
G-BXLS	PZL-110 Koliber 160A	04980078	SP-WGG (2)	23. 6.98	P.A Rickells	Gamston	23. 9.07
			SP-PEB(2), (SP-WGG(1), (N150CP)				
G-BXLT	SOCATA TB-200 Tobago XL	1457	F-GRBB	28. 4.97	R.M.Shears	Blackbushe	25. 5.06
			EC-FNX, EC-234, F-GLFP				
G-BXLW	Enstrom F-28F	734	N279SA	11. 9.97	M and P Food Products Ltd	(Sutton Coldfield)	14.10.07E
			G-BXLW, 1712 Thai Government/KASET				
G-BXLY	Piper PA-28-151 Cherokee Warrior	28-7715220	G-WATZ	19. 9.97	Multiflight Ltd	Leeds-Bradford	13. 5.07T
			N7641F				
G-BXMF	Cassutt Racer IIIM	PFA 34-13003		19. 9.97	J.F.Bakewell	Hucknall	25. 4.05P
	(Built J.F.Bakewell)						
G-BXMG	Rotary Air Force RAF 2000	H2-92-3-59	PH-TEN	18. 8.97	J.S.Wright	(Huddersfield)	17.12.03P
	(Built G Hansen)				*(Extensively damaged landing Long Marston 8.2.03: new owner 2.04)*		
G-BXMH	Beech 76 Duchess	ME-168	F-GDMO	19. 9.97	Plane Talking Ltd	Bournemouth	25. 2.07T
			N6021Y		*(Op Cabair)*		
G-BXML	Mooney M.20A	1594	OY-AIZ	26. 9.97	G.Kay	Shipdham	19. 3.05
G-BXMM	Cameron A-180 HAB	4252		28.10.97	B.Conway	(Wheatley, Oxford)	3.11.00A
G-BXMV	Scheibe SF25C Falke	44223	D-KDFV	7. 8.97	J.B.Marett	Sandhill Farm, Shrivenham	23. 6.07
	(Limbach SL1700)				tr Falcon Flying Group		
G-BXMX	Phoenix Currie Wot	PFA 58-13055		23. 9.97	M.J.Hayman	Watchford Farm, Yarcombe	15. 3.05P
	(Built M J Hayman)						
G-BXMY	Hughes 269C	74-0328	N9599F	20.10.97	Oxford Aviation Services Ltd	Oxford	7.12.03T
G-BXMZ	Diamond DA.20-A1 Katana	10236		4.12.97	Tayside Aviation Ltd	Dundee	12. 2.07T
G-BXNA	Avid Flyer	118	N5531J	10.10.97	G.Haynes	(Hemel Hempstead)	21. 1.05P
	(Built L Pickett)						
G-BXNC	Europa Aviation Europa	PFA 247-12970		13.10.97	J.K.Cantwell	(Ashton-under-Lyne)	
	(Built J.K.Cantwell - kit no.122)				*"The Magic Leprechaun"*		
G-BXNG	Beech 58 Baron	TH-874	N18747	13.10.97	Bonanza Flying Club Ltd	Booker	17.12.06
G-BXNH	Piper PA-28-161 Cherokee Warrior II	28-7816314	N2828M	22.10.97	CC Management Associates Ltd	Redhill	21.12.06T
G-BXNN	de Havilland DHC-1 Chipmunk 22	C1/0849	WP983	4. 8.97	J.N.Robinson	Trenchard Farm, Eggesford	27. 7.04
					(As "WP983/B" in RAF c/s)		
G-BXNS	Bell 206B-3 JetRanger III	2385	N16822	3.11.97	Sterling Helicopters Ltd	Norwich	19.12.06T
G-BXNT	Bell 206B-3 JetRanger III	2398	N94CA	11.11.97	Sterling Helicopters Ltd	Norwich	6.11.06T
			N123AL				
G-BXNU*	Jabiru Aircraft Jabiru SK	PFA 274-13218		31.10.97	J.Smith	(Oaklands Farm, Horsham)	14. 7.01P
	(Built J Smith)				*(Crashed 7.00 and written off: fuselage converted into static exhibition: cancelled 23.1.03 as wfu) (Current status unknown)*		
G-BXNV	Colt AS-105 GD HA Airship	4231		19. 2.98	The Sleeping Society	(Edegem, Belgium)	24. 7.04A
	(Built Cameron Balloons Ltd)						
G-BXNX	Lindstrand LBL 210A HAB	318		3.11.97	Balloon School (International) Ltd	(Petworth)	24. 3.05T
G-BXOA	Robinson R22 Beta	1614	N41132	10.11.97	MG Group Ltd	Sywell	4. 2.07
			JA7832				
G-BXOB	Europa Aviation Europa	PFA 247-12892		6.11.97	S.J.Willett	(Maidstone)	
	(Built S.J.Willett - kit no.209) (Tri-gear u/c)				*(Current status unknown)*		
G-BXOC	Evans VP-2	PFA 63-10305		29. 9.97	H.J. and E.M.Cox	(Bideford)	5. 8.05P
	(Built H J Cox)						
G-BXOF	Diamond DA.20-A1 Katana	10256		4.12.97	Cumbernauld Flying School Ltd	Cumbernauld	30. 1.07T
G-BXOI	Cessna 172R Skyhawk	17280145	N9990F	17.11.97	E.J.Watts	Bodmin	4. 2.07T
G-BXOJ	Piper PA-28-161 Warrior III	2842010	N9265G	15.12.97	P.Foster	Southampton	20. 1.07T
G-BXOM	Isaacs Spitfire	PFA 27-12768		25.11.97	J.H.Betton	(Ammanford)	
	(Built J.H.Betton)				*(Current status unknown)*		
G-BXON	Auster AOP.9	AUS/10/60	WZ729	1.12.97	C.J.and D.J.Baker	Carr Farm, Thorney, Newark	
					(On rebuild 1.05)		
G-BXOR	Robin HR200/120B	321		1.12.97	L.Burrow	Leeds-Bradford	22. 4.07T
G-BXOS	Cameron A-200 HAB	4286		19. 2.98	Airborne Balloon Management Ltd	(Beltring, Kent)	22. 5.07T
G-BXOT	Cameron C-70 HAB	4200		21.10.97	R.J.Plume tr Dante Balloon Group	(Hungerford)	1. 5.05A
G-BXOU	CEA Jodel DR.360 Chevalier	312	F-BPOU	6.10.97	S.H.and J.A.Williams	Blackpool	25. 4.07
G-BXOW	Cameron Colt 105A HAB	4228		9. 1.98	The Aerial Display Co Ltd *(Michelin titles)*	(Looe)	28. 9.03A
G-BXOX	Grumman American AA-5A Cheetah	AA5A-0694	F-GBDS	27. 2.98	R.L.Carter and P.J.Large	Turweston	11. 4.07T
G-BXOY	QAC Quickie Q.235	PFA 94-12183		17.11.97	C.C.Clapham	North Weald	3. 9.04P
	(Built C.C.Clapham) (Lycoming O-235)				*(Noted 1.05)*		
G-BXOZ	Piper PA-28-181 Cherokee Archer II	28-7790173	N6927F	14.10.97	Spritetone Ltd	White Waltham	9. 5.07T
G-BXPC	Diamond DA.20-A1 Katana	10258		4.12.97	Cubair Flight Training Ltd	Redhill	4. 3.07T
G-BXPD	Diamond DA.20-A1 Katana	10259		4.12.97	Cubair Flight Training Ltd	Redhill	13. 5.07T
G-BXPE	Diamond DA.20-A1 Katana	10263		4.12.97	Tayside Aviation Ltd	Dundee	5. 3.07T
G-BXPF	Venture Thorp T.211	105	N6524Y	8.12.97	AD Aviation Ltd	Bicester	1. 7.07T
G-BXPI	Van's RV-4	PFA 181-12426		2. 1.98	Cavendish Aviation Ltd	Gamston	31. 8.04P
	(Built E M Marsh) (Lycoming O-360-A1A)						
G-BXPK	Cameron A-250 HAB	4226		2. 2.98	Richard Nash Cars Ltd *(Sign Express titles)*	(Norwich)	16. 2.05T
G-BXPL	Piper PA-28-140 Cherokee	28-24560	N7224J	10.12.97	C.R.Guggenheim	Old Sarum	15. 4.07T
G-BXPM	Beech 58 Baron	TH-1677	N207ZM	10.10.97	Foyle Flyers Ltd	City of Derry	19. 4.07
G-BXPO	Venture Thorp T.211	104	N6524Q	10.12.97	AD Aviation Ltd	Durham Tees Valley	17. 5.07T
G-BXPP	Sky 90-24 HAB	092		17.12.97	S.J.Farrant	(Hydestile, Godalming)	8. 3.05A
G-BXPR	Cameron Colt Can 110SS HAB	4218		2. 2.98	FRB Fleishwarenfabrik Rostock-Bramow	(Rostock, Germany)	25. 2.05A
G-BXPS	Piper PA-23-250 Aztec C	27-3498	G-AYLY	10.12.90	Wendy A.Moore	Redhill	11. 6.05T
			N6258Y				
G-BXPT	Ultramagic H-77 HAB	77/140		22.12.97	G.D.O.Bartram	(Ordino, Andorra)	6. 4.05A

Registration	Type	c/n	Previous identity	Date	Owner/operator	Base	CofA expiry
G-BXPV	Piper PA-34-220T Seneca III	3448035	A7-FCH N9198X	24.12.97	Oxford Aviation Services Ltd	Gloucestershire	7. 2.04T
G-BXPW	Piper PA-34-220T Seneca III	3448034	A7-FCG N9171R	9. 2.98	Oxford Aviation Services Ltd	Gloucestershire	3. 4.04T
G-BXPY	Robinson R44 Astro	0154	OY-HFV	22.12.97	O.Desmet and B.Mornie	Amougies, Belgium	6. 7.07
G-BXRA	Avions Mudry CAP.10B	03	Fr AF 03	12.12.97	B H D H Frere	Branscombe	22. 9.07
			(P/i F-TFVR also quoted but this may be c/n 3)				
G-BXRB	Avions Mudry CAP.10B	100	FrAF 100	12.12.97	T.T.Duhig	Spilsted Farm, Sedlescombe	27. 7.07
G-BXRC	Avions Mudry CAP.10B	134	FrAF 134	12.12.97	I.F.Scott tr Group Alpha	Fenland	20. 3.05
G-BXRD	Enstrom 280FX	2012	PH-JVM N213M	22.12.97	R.P.Bateman	Biggin Hill	8. 2.06
G-BXRF	Scintex CP.1310-C3 Super Emeraude	935	OO-NSF F-BMJG	9. 1.98	D.T.Gethin	Swansea	30. 6.06
G-BXRG	Piper PA-28-181 Archer II	28-7990036	PH-LEC N21173	29. 1.98	Alderney Flying Training Ltd	Alderney	1. 3.07T
G-BXRH	Cessna 185A Skywagon *(Hoerner wing-tips)*	185-0413	HB-CRX N1613Z	10.12.97	R.E.M.Holmes	Ronaldsway	7. 6.07
G-BXRM	Cameron A-210 HAB	4237		23. 4.98	Dragon Balloon Company Ltd	(Castleton, Hope Valley)	3.10.04T
G-BXRO	Cessna U206G Stationair II	U20604217	OH-ULK N756NE	9. 2.98	M.Penny	Strathallan	17. 4.07
G-BXRP	Schweizer 269C	S.1334	OH-HSP N7506U	27. 1.98	AH Helicopter Services Ltd *(Noted 11.04)*	Dunkeswell	21. 3.04T
G-BXRR	Westland Scout AH.1	F9740	XW612	28. 1.98	T.K.Phillips	Draycott Farm, Chiseldon	6.10.05P
G-BXRS	Westland Scout AH.1	F9741	XW613	28. 1.98	B-N Group Ltd	Bembridge	7. 5.05P
G-BXRT	Robin DR400/180 Régent	2382		23. 2.98	R.A Ford	White Waltham	2. 7.07
G-BXRV	Van's RV-4 *(Built B J Oke)* (Lycoming O-320)	PFA 181-12482		12. 1.98	B.J.Oke tr Cleeve Flying Group	Gloucestershire	9. 6.05P
G-BXRY	Bell 206B JetRanger II	208	N4054G	19. 3.98	S.Lee	(Wilden, Bedford)	23. 9.07T
G-BXRZ	Rans S-6-116 Coyote II	PFA 204A-13195		3. 2.98	G.F.M.Garner	Clench Common	23.10.04A
	(Built C M White - kit no.0897.1146.ES) (Rotax 912-UL) *(Tailwheel u/c)*						
G-BXSB*	Cameron PM-80 HAB *(Coca Cola bottle)*	4298		11. 3.98	Flying Pictures Ltd	(US)	13. 5.05A
				(Cancelled 31.1.02 by CAA) (Active Albuquerque, New Mexico, US 10.03)			
G-BXSC	Cameron C-80 HAB	4251		12.12.97	S.J.Coates *"Keepsake"*	Barton-le-Clay	13. 5.05A
G-BXSD	Cessna 172R Skyhawk	17280310	N431ES	12. 3.98	K.K.Freeman	Breighton	29. 3.07
G-BXSE	Cessna 172R Skyhawk	17280352	N9321F	19. 5.98	MK Aero Support Ltd	Dunkeswell	15. 7.07T
G-BXSG	Robinson R22 Beta II	2789		3. 2.98	R.M.Goodenough	Wotton-under-Edge	21. 4.07T
G-BXSH	DG Flugzeugbau DG-800B	8-121B50		5. 2.98	R.O'Conor	Sutton Bank	4. 11.05
G-BXSI	Jabiru Aircraft Jabiru SK *(Built V R Leggott)*	PFA 274-13204		5. 2.98	M.H.Molyneux	Wickenby	6. 7.05P
G-BXSJ	Cameron C-80 HAB	4330		24. 3.98	Balloon School (International) Ltd t/a British School of Ballooning	(Petworth)	26. 7.05T
G-BXSL*	Westland Scout AH.Mk.1				See SECTION 3 Part 1	Hermeskeil, Trier, Germany	
G-BXSM	Cessna 172R Skyhawk	17280320	N432ES	10. 3.98	East Midlands Flying School Ltd	Nottingham East Midlands	22. 4.07T
G-BXSP	Grob G109B	6335	D-KNEA	25. 3.98	J.R.Dransfield tr Deeside Grob Group	Aboyne	13.10.07
G-BXSR	Reims/Cessna F172N Skyhawk II	F17202003	PH-SPY D-EITH	6. 2.98	S.A Parkes	Stapleford	21. 4.07T
G-BXST	Piper PA-25-235 Pawnee C *(Frame No.25-4971)*	25-4952	PH-BAT N8532L	9. 2.98	The Northumbria Gliding Club Ltd	Currock Hill	8. 7.05A
G-BXSU	TEAM mini-MAX 91A *(Built A R Carr)* (Rotax 503)	PFA 186-12357	G-MYGL	20. 2.98	M.R.Overall	Wetherfield, Suffolk	17. 6.05P
G-BXSV	SNCAN Stampe SV-4C	556	N21PM F-BDDB	10.10.02	B.A Bower *(Noted 12.03)*	Roughay Farm, Bishops Waltham	
G-BXSX	Cameron V-77 HAB	4329		6. 4.98	D.R.Medcalf *"All Tech"*	(Bromsgrove)	23. 5.05A
G-BXSY	Robinson R22 Beta II	2778		27. 1.98	N.M.G.Pearson	Bristol	5. 2.07T
G-BXTB	Cessna 152 II	15282516	OH-CMS N69151	25. 2.98	Haimoss Ltd	Old Sarum	25. 4.07T
G-BXTC	Taylor JT.1 Monoplane *(Built R.Holden-Rushworth)*	PFA 55-13142		25. 2.98	R.Holden-Rushworth	(Devizes)	
G-BXTD	Europa Aviation Europa	PFA 247-12772		26. 2.98	P.R.Anderson	Hucknall	14. 5.04P
	(Built P.R.Anderson - kit no.155) (Monowheel u/c)						
G-BXTE	Cameron A-275 HAB	4028		30. 3.98	Global Ballooning Ltd	(Uckfield)	26. 4.05T
G-BXTF	Cameron N-105 HAB	4304		2. 4.98	Flying Pictures Ltd *"Sainsbury's Strawberry"*	(Chilbolton)	8. 5.03A
G-BXTG	Cameron N-42 HAB	4305		2. 4.98	Flying Pictures Ltd *(Sainsbury's titles)*	(Chilbolton)	14. 9.05A
G-BXTH	Westland SA.314D Gazelle HT.3	1120	XW866	13. 3.98	Flightline Ltd *(On rebuild Blackpool 1.05)*	Southend	AC
G-BXTI	Pitts S-1S *(Built N J Pesch)*	NP-1	ZS-VZX N96MM	9. 3.98	A B.Treherne-Pollock	White Waltham	25. 7.05P
G-BXTJ	Cameron N-77 HAB	4332		6. 4.98	J.M.Albury *(Chubb titles)*	(Cirencester)	28. 3.05A
G-BXTL	Schweizer 269C-1	0075		13. 3.98	Oxford Aviation Services Ltd	Oxford	2. 4.07T
G-BXTN	Aérospatiale-Alenia ATR72-202	483	F-WQNR G-BXTN, F-WWEV	24.10.97	Aurigny Air Services Ltd	Guernsey	19. 4.07T
G-BXTO	Hindustan HAL-26 Pushpak (Continental C90-8F)	PK-128	9V-BAI VT-DWM	12. 2.98	A.M.Pepper tr Pushpak Flying Group	Yeatsall Farm, Abbots Bromley	13. 5.05P
G-BXTP	Diamond DA.20-A1 Katana	10306	N636DA	10. 3.98	3business.com Ltd	Cumbernauld	20. 6.07T
G-BXTS	Diamond DA.20-A1 Katana	10308	N638DA	10. 3.98	I.M.Armitage	Bristol	26. 8.07T
G-BXTT	Grumman AA-5B Tiger	AA5B-0749	F-GBDH	27. 2.98	P.Curley and R.Thomas tr G-BXTT Group	Tatenhill	25. 3.07
G-BXTV	Cope Bug	BUG.2		12. 3.98	B.R.Cope	(Bewdley)	
G-BXTW	Piper PA-28-181 Archer III	2843137	N41279 (G-BXTW), N41279	26. 5.98	J.N.Davison t/a Davison Plant Hire	Compton Abbas	17. 6.07
G-BXTY	Piper PA-28-161 Cadet	2841179	PH-LED	11. 3.98	Bflying Ltd *(Op Bournemouth Flying Club)*	Bournemouth	27. 6.07T
G-BXTZ	Piper PA-28-161 Cadet	2841181	PH-LEE	11. 3.98	Bflying Ltd *(Op Bournemouth Flying Club)*	Bournemouth	25. 3.07T
G-BXUA	Campbell Cricket Mk.5 *(Built P C Lovegrrove)* (Rotax 582)	PFA G/03-1272		12. 3.98	R.N.Bodley	Henstridge	1. 7.03P
G-BXUB	Lindstrand Syrup Bottle SS HAB	508		30. 4.98	Free Enterprise Balloons Ltd *"Mrs Butterworth"*	(Mason, Wisconsin, US)	9. 4.04A

Reg	Type	C/n	Previous identity	Regd	Owner	Base	CofA
G-BXUC	Robinson R22 Beta	0908	OY-HFB	17. 3.98	EBG (Helicopters) Ltd	Redhill	15. 4.07T
G-BXUE	Sky 240-24 HAB	098		30. 4.98	G.M.Houston t/a Scotair Balloons	(Lesmahagow)	16. 4.05T
G-BXUF	Agusta-Bell 206B JetRanger II	8633	EC-DUS	12. 5.98	SJ Contracting Services Ltd	Beckley	26. 7.07T
			OE-DXE				
G-BXUG	Lindstrand Baby Bel SS HAB	512		14. 5.98	A.D.Kent tr Balloon Preservation Flying Group	(Lancing)	24. 7.05A
G-BXUH	Lindstrand LBL-31A HAB	513		2. 6.98	A.D.Kent tr Balloon Preservation Flying Group	(Lancing)	23.10.04A
G-BXUI	DG Flugzeugbau DG-800B	8-105B39	BGA.4382	12. 5.98	J.Le Coyte	Sandhill Farm, Shrivenham	4. 5.07
			D-KKLC				
G-BXUK	Robinson R44 Astro	0093	D-HIFF	19. 6.95	Staske Construction Ltd *(Noted 10.04)*	Booker	14. 7.07T
G-BXUL	Vought FG-1D Corsair	3205	N55JP	25. 3.98	J.H.Slade	Duxford	1. 1.05P
	(Built Goodyear Aircraft Corporation)		"NZ5611", NZ5648, Bu.88391		*(As "92844/8" of VF-17 in US Navy c/s)*		
	(Officially regd as c/n 32823 and p/i of Bu.88439 also quoted)						
G-BXUM	Europa Aviation Europa	PFA 247-12611		19. 3.98	D.Bosomworth	Popham	8. 5.05P
	(Built D.Bosomworth - kit no.067) (Monowheel u/c)				*(Based owners' home)*		
G-BXUO	Lindstrand LBL 105A HAB	520		27. 3.98	Lindstrand Balloons Ltd	(Oswestry)	9.11.04A
G-BXUS	Sky 65-24 HAB	111		6. 4.98	PSH Skypower Ltd	(Pewsey)	10. 6.05A
G-BXUU	Cameron V-65 HAB	4362		23. 4.98	M.D.Freeston and Sandra Mitchell	(Hertford)	24. 4.05A
G-BXUW	Cameron Colt 90A HAB	4317		23. 4.98	Zycomm Electronics Ltd	(Ripley)	14. 6.04A
G-BXUX	Brandli Cherry BX-2	PFA 179-12571		4. 4.98	M.F.Fountain	Maypole Farm, Chislet	4.11.05P
	(Built M.F.Fountain) (Continental C90-12F)						
G-BXUY	Cessna 310Q	310Q0231	N137SA	16. 4.98	Massair Ltd	Liverpool	28.11.07E
			D-IHMT, N7731Q				
G-BXVA	SOCATA TB-200 Tobago XL	1325	F-GJXL	15. 4.98	H.R.Palser	Cardiff	18. 8.07
			F-WJXL				
G-BXVB	Cessna 152 II	15282584	N69250	15. 4.98	PJC (Leasing) Ltd	Stapleford	25. 9.07T
G-BXVD	CFM Streak Shadow SA	PFA 206-13304		1. 4.98	Rotech Frabrication Ltd	Inverness	25.11.05P
	(Built CFM Aircraft Ltd - kit no K.301SA) (Rotax 912)						
G-BXVE	Lindstrand LBL 330A HAB	492		6. 5.98	Adventure Balloon Co Ltd	(London W7)	24. 3.03T
					(Adventure Balloons titles)		
G-BXVF	Cameron Thunder Ax11-250 Series 2 HAB	4371		22. 5.98	M.E.White	(Dublin)	24. 7.05T
G-BXVG	Sky 77-24 HAB	99		28. 5.98	M.Wolf	(Wallingford)	27. 5.05
G-BXVH	Sky 25-16 HAB	120		23. 4.98	(Flying Pictures Ltd) *(AXA titles)*	(Chilbolton)	7. 3.02A
G-BXVI	Vickers Supermarine 361 Spitfire LF.XVIe	CBAF.IX.4644	6944M	27.12.84	Wizzard Investments Ltd	North Weald	AC
			"RF114", RW386				
G-BXVJ	Cameron O-120 HAB	2201	PH-VVJ	12. 3.98	Aerosaurus Balloons Ltd	(Whimple, Exeter)	18. 2.05T
			G-IMAX				
G-BXVK	Robin HR200/120B	326		1. 7.98	Northamptonshire School of Flying Ltd	Sywell	23. 6.07T
G-BXVL	Sky 180-24 HAB	113		16. 6.98	S.Stanley t/a Purple Balloons	(Sudbury)	24. 7.05T
G-BXVM	Van's RV-6A	PFA 181-13103		26. 2.98	J.G.Small	RAF Woodvale	22. 5.05P
	(Built J G Small) (Lycoming O-320)						
G-BXVO	Van's RV-6A	PFA 181-12575		28. 4.98	P.J.Hynes and M.E.Holden	Sleap	24. 8.05P
	(Built P J Hynes) (Lycoming O-320-D1A)						
G-BXVP	Sky 31-24 HAB	056		28. 4.98	L.Greaves	(Doulting, Somerset)	18. 5.05A
G-BXVR	Sky 90-24 HAB	061		20. 7.98	P.Hegarty	(Magherafelt, County Londonderry)	22. 9.05
G-BXVS	Brugger MB.2 Colibri	PFA 43-11948		5. 5.98	G.T.Snoddon	Newtownards	19. 5.03P
	(Built G.T.Snoddon) (Volkswagen 1834)						
G-BXVT	Cameron O-77 HAB	1444	PH-MKB	30. 7.98	R.P.Wade	(Wigan)	
G-BXVU	Piper PA-28-161 Cherokee Warrior II	28-7816063	N47372	5. 5.98	London Ashford Airport Ltd	Lydd	24. 8.07T
G-BXVV	Cameron V-90 HAB	4369		5. 5.98	Floating Sensations Ltd	(Thatcham)	14. 5.05A
G-BXVW	Colt Piggy Bank SS HAB	4366		2. 7.98	G.Binder	(Sonnennbuhl, Germany)	2. 8.03A
G-BXVX	Rutan Cozy Classic	PFA 159-12680		6. 5.98	G.E.Murray	Swansea	21.10.03P
	(Built G.E.Murray) (Lycoming O-320)						
G-BXVY	Cessna 152	15279808	N757KU	11. 5.98	Stapleford Flying Club Ltd	Stapleford	13.11.04T
G-BXVZ	WSK-PZL Mielec TS-11 Iskra	3H-1625	SP-DOF	27. 3.98	J.Ziubrzynski *(Noted 1.04)*	Manston	AC
G-BXWA	Beech 76 Duchess	ME-232	OY-CYM	8. 4.98	Plymouth School of Flying Ltd	Plymouth	28. 6.07T
			(SE-IUY), D-GBTD				
G-BXWB	Robin HR100/200B Royal	08	HB-EMT	29. 4.98	W.A.Brunwin	Oaksey Park	14.10.07E
G-BXWC	Cessna 152	15283640	N4794B	11. 5.98	PJC (Leasing) Ltd	Leicester	12. 7.07T
G-BXWE	Fokker F.28 Mk.0100	11327	PH-CFE	6. 7.98	British Midland Airways Ltd	Nottingham East Midlands	30. 8.07T
	(Fokker 100)		F-GJAO, PH-CFE, PH-EZL, (G-FIOX), PH-EZL *(Op bmiRegional)*				
G-BXWF	Fokker F.28 Mk.0100	11328	PH-CFF	13. 7.98	British Midland Airways Ltd	Nottingham East Midlands	31. 8.07T
	(Fokker 100)		F-GKLX, PH-CFF, PH-EZM, (G-FIOY), PH-EZM *(Op bmiRegional)*				
G-BXWG	Sky 120-24 HAB	114		28. 5.98	K.B.Chapple	(Sulham, Reading)	6. 5.05T
G-BXWH	Denney Kitfox Model 4-1200 Speedster			4. 3.98	B.J.Finch	Croft Farm, Defford	23. 5.04P
	(Built B.J.Finch)	PFA 172A-12343					
G-BXWI	Cameron N-120 HAB	4395		12. 6.98	Flying Pictures Ltd	(Chilbolton)	30. 7.02A
	(Rebuilt 1999 with new envelope c/n 4395(2)				*(Energis titles)*		
G-BXWK	Rans S-6-ESA Coyote II	PFA 204-13317		19. 5.98	R.J.Teal	Baxby Manor, Husthwaite	2. 2.05P
	(Built J Whiting) (Rotax 582 - *kit no.0298.1020)* (*Tri-cycle u/c)*						
G-BXWL	Sky 90-24 HAB	117		20. 7.98	I.S.Bridge	(Shrewsbury)	12.10.04A
					t/a The Shropshire Hills Balloon Company		
G-BXWO	Piper PA-28-181 Archer II	28-8190311	D-ENHA (2)	22. 5.98	J.S.Develin and Z.Islam	Biggin Hill	23. 7.04T
			N8431C				
G-BXWP	Piper PA-32-300 Cherokee Six	32-7340088	N8143D	26. 5.98	J.M.Hill and I.Jones	Barton	27. 5.07
			G-BXWP, OE-DRR, N16452		tr Alliance Aviation		
G-BXWR	CFM Streak Shadow SA	PFA 206-13205	G-MZMI	22. 5.98	M.A Hayward	Bodmin	24. 8.05P
	(Built M Hayward - kit no K.289-SA) (Rotax 912)						
G-BXWT	Van's RV-6	PFA 181-12639		19. 7.96	R.C.Owen	Danehill	18. 8.05P
	(Built R C Owen) (Lycoming O-360						
G-BXWU	FLS Aerospace Sprint 160	003	G-70-503	5. 6.98	Aces High Ltd	North Weald	
					(Noted as [c/n] "003" only 1.05)		
G-BXWV	FLS Aerospace Sprint 160	005	G-70-505	5. 6.98	Aces High Ltd	North Weald	
					(Noted as [c/n] "005" only 1.05)		
G-BXWX	Sky 25-16 HAB	082		29. 5.98	Zebedee Balloon Service Ltd	Newtown Hungerford	14. 3.00A

Registration	Type	C/n	Previous identity	Date	Owner	Base	Expiry
G-BXXE	Rand Robinson KR-2S	PFA 129-10927		8. 6.98	N.Rawlinson	(Leek)	
	(Built N.Rawlinson)						
G-BXXG	Cameron N-105 HAB	3662		19. 6.98	Allen Owen Ltd	(Wotton-under-Edge)	2. 4.02A
G-BXXH	Hatz CB-1	PFA 143-12445		9. 6.98	R.F.Shingler	Forest Farm, Welshpool	
	(Built R.F.Shingler)						
G-BXXI	Grob G109B	6400	F-CAQR	9. 6.98	M.N.Martin	Lyveden	23.10.07
			F-WAQR				
G-BXXJ	Colt Flying Yacht SS HAB	1797	JA-A0515	10. 6.98	Magical Adventures Ltd		
						(West Bloomfield, Michigan, US)	7. 9.01A
G-BXXK	Reims/Cessna F172N Skyhawk II	F17201806	D-EOPP	15. 6.98	I.R.Chaplin	Andrewsfield	2. 9.07T
G-BXXL	Cameron N-105 HAB	4408		16. 7.98	Flying Pictures Ltd) *(Blue Peter titles)*	(Chilbolton)	11. 9.05A
G-BXXN	Robinson R22 Beta	0720	N720HH	16. 6.98	L.L.Smith t/a Helicopter Services	Booker	26. 7.07T
G-BXXO	Lindstrand LBL 90B HAB	534		6. 7.98	K.Temple	(Diss)	20. 8.04A
G-BXXP	Sky 77-24 HAB	124		20. 7.98	C.J.James	(Wincanton)	21. 1.05A
G-BXXR	Lovegrove AV-8 Gyroplane	PFA G/15-1263		29. 6.98	P.C.Lovegrove	Didcot	
	(Built P.C.Lovegrove) (Regd. as "Lovegrove BGL Four Runner" c/n PFA G/15-1273)						
G-BXXS	Sky 105-24 HAB	116		30. 7.98	L.D.and H.Vaughan *"Skylark"*	(Tring)	20. 1.05A
G-BXXT	Beech 76 Duchess	ME-212	(N212BE)	17. 7.98	Pridenote Ltd	Sturgate	29. 9.07E
			F-GBOZ				
G-BXXU	Colt 31A HAB	4427		21. 8.98	Sade Balloons Ltd	(Coulsdon)	17.10.03
G-BXXW	Enstrom F-28F	771	G-SCOX	2. 7.98	Falcon Helicopters Ltd	Barton	8.12.07E
			N330SA, G-BXXW, JA7823				
G-BXYA*	CSS-13 Aeroklubowy				See SECTION 3 Part 1	Polk City, Florida, US	
G-BXYC	Schweizer 269C	S.1716	D-HFDZ	8. 7.98	D.W.Wilson and P.Sanders t/a Sycamore Aviation	(Hull)	31. 8.07T
G-BXYD	Eurocopter EC120B Colibri	1006		7. 7.98	Aero Maintenance Ltd	Walton Wood, Pontefract	13.12.07E
G-BXYE	Scintex CP.301-C1 Emeraude	559	F-BTEO	8. 7.98	D.T.Gethin	Swansea	
			F-PTEO, F-WTEO, F-BJFV				
G-BXYF	Colt AS-105GD HA Airship	4433		7. 8.98	LN Flying Ltd	(Germany)	14.11.05E
	(Built Cameron Balloons Ltd)						
G-BXYG	Cessna 310D	39089	HB-LSF	14. 8.98	Equitus SARL	Merville-Calonne, France	28. 1.05T
			F-GEJT, 3A-MCA, F-BBOT, F-OBOT, (N6789T)				
G-BXYJ	SAN Jodel DR.1050 Ambassadeur	143	F-BJNA	28. 7.98	C.Brooke tr G-BXYJ Group	Netherthorpe	15.11.07
G-BXYK	Robinson R22 Beta	1579	N4037B	27. 7.98	D.N.Whittlestone	(Oxenhope)	16. 1.05T
G-BXYM	Piper PA-28-235 Cherokee B	28-10858	SE-FAM	18. 8.98	Ashurst Aviation Ltd	Shoreham	22.12.07E
G-BXYN	Van's RV-6	PFA 181-13265		29. 7.98	J.A.Tooley and R.M.Austin	(Thatcham)	
	(Built J A Tooley and R.M.Austin)						
G-BXYO	Piper PA-28RT-201 Arrow IV	28R-8018046	PH-SDD	18. 8.98	Airways Flight Training (Exeter) Ltd	Exeter	1.12.04T
			N8164M				
G-BXYP	Piper PA-28RT-201 Arrow IV	28R-8018050	PH-SBO	18. 8.98	Westflight Aviation Ltd	Gloucestershire	7.12.07E
			N8168H				
G-BXYR	Piper PA-28RT-201 Arrow IV	28R-8018101	PH-SDA	3. 8.98	A.Dayani	Exeter	26.11.04T
			N8251B				
G-BXYT	Piper PA-28RT-201 Arrow IV	28R-7918198	PH-SBN	3. 8.98	Checkflight Ltd	(Winscombe)	9. 9.07T
			(PH-SBM), OO-HLA, N2878W				
G-BXYU*	Reims/Cessna F152 II	F15201804	OH-CKD	31. 7.98	Exeter Flying Club Ltd	Dunkeswell	24. 8.01T
			SE-IFY		*(Cancelled 16.10.99 as destroyed Whiddon Down, Okehampton 2.8.99: fuselage noted 6.04)*		
G-BXYX	Van's RV-6	22293	N2399C	31. 7.98	A.G.Palmer	Wellesbourne Mountford	30. 3.05P
	(Built A G Palmer) (Lycoming O-320-E2D)						
G-BXZA	Piper PA-38-112 Tomahawk	38-79A0864	N2480N	6. 8.98	P.D.Brooks	Inverness	29.10.07E
G-BXZB	Nanchang CJ-6A	2632019	Chinese AF	18. 9.98	Wingglider Ltd *(As "2632019")*	Hibaldstow	31. 5.02P
G-BXZD	Westland SA.314C Gazelle HT.2	1174	XW895	25. 8.98	S.Atherton *(New owner 5.04)*	(Tadcaster)	12.12.05P
					tr Gazelle Flying Group *(As "XW895/51" in Royal Navy c/s)*		
G-BXZF	Lindstrand LBL 90A HAB	575		8. 1.99	R.G.Carrell	(Havant)	1. 3.05A
G-BXZG	Cameron A-210 HAB	4424		21. 8.98	Societé Bombard SARL	(Meursanges, Côte-d'Or, France)	22. 9.04A
G-BXZH	Cameron A-210 HAB	4423		21. 8.98	Societé Bombard SARL	(Meursanges, Côte-d'Or, France)	22. 9.04A
G-BXZI	Lindstrand LBL 90A HAB	543		14. 8.98	I.Little and J.A.Viner	(Rudgwick, Horsham/Caterham)	17. 6.05A
G-BXZK	MD Helicopters MD 900	900-00057	N9238T	27. 8.98	Dorset Police Air Support Unit	Winfrith	17. 2.05T
			G-76-057				
G-BXZM	Cessna 182S Skylane	18280310	N2683L	8.10.98	AB Integro Aviation Ltd	Booker	18.11.07E
G-BXZN	Advanced Technologies Firebird CH1 ATI	00002	N8186E	25. 8.98	Intora-Firebird plc	(Brentwood)	
G-BXZO	Pietenpol AirCamper	PFA 47-12818		10. 7.98	P.J.Cooke	Westfield Farm, Hailsham	14. 7.05P
	(Built P J Cooke) (Continental A65)						
G-BXZS	Sikorsky S-76A II Plus	760287	N190AL	14. 9.98	Bristow Helicopters Ltd	Redhill	3. 5.05T
			N190AE, N153AE, N7265A				
G-BXZT	Morane Saulnier MS.880B Rallye Club	1733	OO-EDG	2. 9.98	Limerick Flying Club (Coonagh) Ltd		
			D-EBDG, F-BSVL			Coonagh, County Limerick	11.12.04
G-BXZU	Micro Aviation B22.S Bantam	98-015	ZK-JJL	21. 9.98	M.E.Whapham and R.W.Hollamby		
						Corn Wood Farm, Adversane	17. 5.05P
G-BXZY	CFM Shadow Series DD	296-DD		21. 9.98	P.A.James t/a Cloudbase Aviation G-BXZY	Redhill	16.12.05P
G-BXZZ	Sky 160-24 HAB	109		14. 7.98	S.J.Colin t/a Skybus Ballooning	(Cranbrook)	2. 6.05T

G-BYAA - G-BYZZ

Registration	Type	C/n	Previous identity	Date	Owner	Base	Expiry
G-BYAA	Boeing 767-204ER	25058	PH-AHM	23. 4.91	Britannia Airways Ltd	Luton	13.11.05T
			G-BYAA, N60659		*"Sir Matt Busby CBE"*		
G-BYAB	Boeing 767-204ER	25139	(PH-AHN)	11. 6.91	Britannia Airways Ltd	Luton	26. 3.05T
			G-BYAB		*"Brian Johnston CBE MC"*		
G-BYAD	Boeing 757-204ER	26963		6. 5.92	Britannia Airways Ltd *(Thomson c/s)*	Luton	22. 2.05T
G-BYAE	Boeing 757-204ER	26964		12. 5.92	Britannia Airways Ltd *(Thomson c/s)*	Luton	26. 4.07T
G-BYAF	Boeing 757-204ER	26266		13. 1.93	Britannia Airways Ltd *(TUI c/s)*	Luton	19. 1.06T
G-BYAH	Boeing 757-204	26966		5. 2.93	Britannia Airways Ltd *(Thomsonfly.com titles)*	Luton	10. 2.06T
G-BYAI	Boeing 757-204	26967		1. 3.93	Britannia Airways Ltd	Luton	4. 3.06T
G-BYAJ	Boeing 757-204ER	25623		4. 3.93	Britannia Airways Ltd	Luton	23. 1.05T

G-BYAK	Boeing 757-204	26267		6. 4.93	Britannia Airways Ltd *(Thomson c/s)*	Luton	13. 4.06T
G-BYAL	Boeing 757-204	25626		13. 5.93	Britannia Airways Ltd *(TUI c/s)*	Luton	18. 5.06T
G-BYAN	Boeing 757-204	27219		26. 1.94	Britannia Airways Ltd *(Thomson c/s)*	Luton	14. 2.07T
G-BYAO	Boeing 757-204	27235		3. 2.94	Britannia Airways Ltd	Luton	2. 2.06T
					"Eric Morecambe OBE" (Thomson c/s)		
G-BYAP	Boeing 757-204	27236		15. 2.94	Britannia Airways Ltd *"John Lennon" (TUI c/s)*	Luton	14. 2.06T
G-BYAR	Boeing 757-204	27237	PH-AHT	1. 3.94	Britannia Airways Ltd	Luton	11. 6.06T
			G-BYAR				
G-BYAS	Boeing 757-204	27238		9. 3.94	Britannia Airways Ltd (*Thomsonfly.com titles)*	Luton	31. 1.08E
G-BYAT	Boeing 757-204	27208		21. 3.94	Britannia Airways Ltd *(Thomson c/s)*	Luton	24. 3.07T
G-BYAU	Boeing 757-204	27220		18. 5.94	Britannia Airways Ltd *(Thomson c/s)*	Luton	17. 5.06T
G-BYAV	Taylor JT.1 Monoplane	PFA 55-11010		27. 8.98	J.S.Marten-Hale	RAF Henlow	4.10.01P
	(Built C J Pidler)				*(New owner 10.03)*		
G-BYAW	Boeing 757-204	27234		3. 4.95	Britannia Airways Ltd	Luton	2. 4.04T
					("Phil Stanley - Britannia Employee of the Year") (Thomson c/s)		
G-BYAX	Boeing 757-204	28834		24. 2.99	Britannia Airways Ltd (*Thomsonfly.com titles)*	Luton	28. 2.05T
G-BYAY	Boeing 757-204	28836	N1786B	13. 4.99	Britannia Airways Ltd	Luton	12. 4.05T
G-BYAZ	CFM Streak Shadow SA	PFA 206-12656		1. 9.98	A G.Wright	(Camberley)	9. 3.04P
	(Built A G.Wright - kit no K.244) (Rotax 582)						
G-BYBA	Agusta-Bell 206B-3 JetRanger III	8596	G-BHXV	31. 3.98	R.Forests Ltd	White Waltham	5. 9.05T
			G-OWJM, G-BHXV				
G-BYBC	Agusta-Bell 206B JetRanger II	8567	G-BTWW	31. 3.98	Sky Charter UK Ltd	Oxenfoord Castle, Dalkeith	17. 7.06T
			EI-BJV, G-BTWW				
G-BYBD	Cessna F172H	F172-0487	G-OBHX	6. 7.98	D.G.Bell and S.J.Green	Egginton	5. 7.07T
	(Built Reims Aviation SA)		G-AWMU				
G-BYBE	Wassmer Jodel D.120A Paris-Nice	269	OO-FDP	24. 7.98	R.J.Page	Shipdham	15. 7.05
G-BYBF	Robin R2160i	329		1.10.98	D.J.R.Lloyd-Evans	Compton Abbas	2. 5.05T
G-BYBH	Piper PA-34-200T Seneca II	34-8070078	N119SA	9. 6.00	Goldspear (UK) Ltd	(Holyport, Maidenhead)	7. 6.07
			(G-BYBH), N4023K, N3567B				
G-BYBI	Bell 206B-3 JetRanger III	3668	ZS-RGP	19.10.98	Winkburn Air Ltd	Elstree	19. 7.05T
			N5757M				
G-BYBJ	Medway Hybred 44XLR-C	MR156/135-C		22. 1.99	M.Gardner	Rochester	25. 4.01P
	(Rotax 503)						
G-BYBK	Murphy Rebel	260R	N95LD	19. 8.98	R.K..Hyatt	Bodmin	1. 4.03P
	(Built L A Dyer)						
G-BYBL	Sud-Aviation Gardan GY-80-160D Horizon	127	F-BMUY	25. 9.98	R.H.W.Beath	Wadswick Manor Farm, Corsham	23. 6.06
G-BYBM	Jabiru Aircraft Jabiru SK	PFA 274-13377		18. 9.98	P.J.Hatton	(Okehampton)	24.10.05P
	(Built M Rudd)						
G-BYBN	Cameron N-77 HAB	3082	N6004M	30. 9.98	M.G.and R.D.Howard	(Bristol)	20. 3.05A
G-BYBO	Medway EclipseR	155/134		14. 9.98	D.R.Purslow	Pound Green, Buttonoak, Bewdley	1.12.05P
	(Jabiru 2200A)						
G-BYBP	Cessna A185F	185-03804	OO-DCD	15.10.98	G.M.S.Scott	Bradley's Lawn, Heathfield	28. 2 05
			F-GDCD, F-ODIA, N4593E				
G-BYBR	Rans S-6-116 Coyote II	PFA 204A-13081		10. 7.98	J.B.Robinson	Blackpool	24. 8.05P
	(Built J B Robinson - kit no.0996.1042) (Rotax 912-UL) *(Tri-cycle u/c)*						
G-BYBS	Sky 80-16 HAB	136		27.10.98	G.W.Mortimore	(Wantage)	10. 8.04
G-BYBU	Meridian Renegade Spirit UK	PFA 188-13229		12.10.98	L.C.Cook	Sywell	25. 6.04P
	(Built K R Anderson)						
G-BYBV	Mainair Rapier	1183-1198-7 & W986		20.10.98	M.W.Robson	York	1.11.02P
G-BYBW	TEAM mini-MAX 88	PFA 186-12120		19.10.98	R.M.Laver	Shoreham	11.10.05P
	(Built J E Johnson)						
G-BYBX	Slingsby T.67M-260 Firefly	2261		21.10.98	Slingsby Aviation Ltd	Wombleton	
G-BYBY	Thorp T-18C Tiger	492	N77KK	17. 7.98	L.J.Joyce	Liverpool	21.11.03P
G-BYBZ	Jabiru Aircraft Jabiru SK	PFA 274-13290		7. 9.98	R.P.Lewis	(King's Lynn)	26. 2.05P
	(Built A W.Harris)						
G-BYCA	Piper PA-28-140 Cherokee D	28-7125223	PH-VRZ	24. 9.98	A.Reay	Caernarfon	8. 2.05T
			N11C				
G-BYCB	Sky 21-16 HAB	142		28.10.98	S.J.Colin *(New owner 7.03)*	(Rolvenden, Cranbrook)	
G-BYCD	Cessna 140	13744	N4273N	28. 9.98	G.P.James	Oak Farm, Cowbit	17. 4.05
	(Continental O-200-A)	NC4273N					
G-BYCE	Robinson R44 Astro	0520		12.10.98	Jim Davies Civil Engineering Ltd	(Blackwood)	30.10.07T
G-BYCF	Robinson R22 Beta II	2866		12.10.98	Teleology Ltd	(Todmorden)	9.12.04T
G-BYCJ	CFM Shadow Series DD	PFA 161-13258		14.10.98	N.Jones	(Hutton, Brentwood)	9. 2.05P
	(Built J W E Pearson - kit no.K.294-DD)						
G-BYCL	Raj Hamsa X'Air Jabiru	BMAA/HB/088		15.10.98	D.O'Keefe, K.Rutter and A J.Clarke	London Colney	12.11.05P
	(Built G A J Salter - kit no.331)						
G-BYCM	Rans S-6-ES Coyote II	PFA 204-13315		15. 9.98	E.W.McMullan	Dunnyvadden, County Antrim	7.11.00P
	(Built E.W.McMullan)						
G-BYCN	Rans S-6-ES Coyote II	PFA 204-13314		15. 9.98	T.J.Croskery	City of Derry	9. 5.05P
	(Built J.K.and R L Dunseath) (Rotax 582)						
G-BYCO*	Rans S-6-ES Coyote II	PFA 204-13318		17. 9.98	T.J.Croskery	City of Derry	8. 5.02P
	(Built T J Croskery) (Rotax 582)		*(Struck ground in practice forced landing Limavady 23.8.01: cancelled 3.1.02 as WFU) (Noted 4.03)*				
G-BYCP	Beech B200 Super King Air	BB-966	F-GDCS	15.10.98	London Executive Aviation Ltd	Stapleford	11. 2.05
G-BYCS	CEA Jodel DR.1051 Sicile	201	F-BJUJ	28.10.98	Fire Defence plc	Trenchard Farm, Eggesford	24. 5.05
G-BYCT	Aero L-29A Delfin	395142	ES-YLH	29.10.98	Propeller BVBA	North Weald	2. 6.04P
			Estonian AF, Soviet AF		*(Noted 1.05)*		
G-BYCU	Robinson R22 Beta	1094	G-OCGJ	3.11.98	Tiger Helicopters Ltd	Shobdon	15. 9.04T
G-BYCV	Murphy Maverick	PFA 259-12925		24. 9.98	P.Shackleton	Old Sarum	4. 1.05P
	(Built P C Vallance) (Rotax 503)						
G-BYCW	Mainair Blade	1185-1198-7 & W988		5.11.98	P.C Watson	(Hebden Bridge)	23. 5.05P
G-BYCX	Westland Wasp HAS.1	F9754 & WA-B-Z3	ZK-HOX	9.11.98	M P Blokland	Draycott Farm, Chiseldon	15. 5.05P
			South African Navy 92		*(As "92" in South African Navy c/s)*		
G-BYCY	III Sky Arrow 650 T	PFA 298-13332		10.11.98	K.A Daniels	(Caldicote, Newport)	4. 8.05P
	(Built A S Spriglings)						

Reg	Type	C/n	Previous ID	Date	Owner/Operator	Base	CofA
G-BYCZ	Jabiru Aircraft Jabiru SK	PFA 274-13388		16.10.98	R.Scroby	Leicester	12. 9.05P
	(Built C Hewer)						
G-BYDA	McDonnell Douglas DC-10-30	46990	OY-CNO	25. 3.99	MyTravel Airways Ltd	Manchester	29. 3.05T
			XA-SYE, F-GGMZ, C-GFHX, 9V-SDA				
G-BYDB	Grob G115B	8025	VH-JVL	26. 3.99	J.B.Baker	Tatenhill	16. 4.05
			D-EFCG				
G-BYDE	Vickers Supermarine 361 Spitfire IX	Not known	Soviet AF PT879	11.11.98	A H.Soper	(Romford)	
G-BYDF	Sikorsky S-76A	760364	JA6615	9. 1.98	Brecqhou Development Ltd	Guernsey	8. 7.07T
G-BYDG	Beech C24R Sierra	MC-627	OY-AZL	9.11.98	Professional Flight Simulation Ltd	Bournemouth	21. 5.05T
G-BYDI	Cameron A-210 HAB	4495		4. 2.99	N.J.Appleton	(Bristol)	26. 3.05T
					t/a First Flight *(Park Furnishers titles)*		
G-BYDJ	Colt 120A HAB	3527		17.11.98	D.K.Hempleman-Adams	(Corsham)	7. 8.04A
G-BYDK	Stampe SV-4C	55	F-BCXY	20.11.98	Bianchi Aviation Film Services Ltd	Booker	
	(P/i quoted officially as F-BCXV which was c/n 298)						
G-BYDL	Hawker Hurricane IIB	Not known	Soviet AF Z5207	17.11.98	Retro Track and Air (UK) Ltd	(Dursley)	
G-BYDM	Pegasus Quantum 15-912	7488		18.11.98	P.Roberts	Belle Vue Farm, Yarnscombe	11. 7.05P
G-BYDT	Cameron N-90 HAB	4499		28. 1.99	N.J.Langley	(Bristol)	3. 4.04A
G-BYDV	Van's RV-6	PFA 181-13264		3.12.98	R.G.Andrews	Damyn's Hall, Upminster	8. 6.05P
	(Built G.L.Carpenter) (Lycoming O-320)						
G-BYDW	Rotary Air Force RAF 2000 GTX-SE	PFA G/13-1302		4.12.98	R.I.Young	(Nottingham)	19. 1.05P
	(Built M T Byrne)						
G-BYDX	American General AG-5B Tiger	10051	N374SA	25. 3.99	A.J.Watson	Southampton	19. 4.05
			G-BYDX, F-GKBH, N1191Y		tr Bibit Group		
G-BYDY	Beech 58 Baron	TH-1852	C-GBWF	10.11.98	J.F.Britten	Blackbushe	16.12.04
G-BYDZ	Pegasus Quantum 15-912	7493		22.12.98	G.Shaw	Sutton Meadows, Ely	22. 4 05P
G-BYEA	Cessna 172P	17275464	PH-ILL	7.10.98	Plane Talking Ltd	Biggin Hill	21.10.07E
			N63661				
G-BYEB	Cessna 172P	17274634	PH-ILM	7.10.98	Plane Talking Ltd	Redhill	21.10.07E
			N52917				
G-BYEC	DG Flugzeugbau DG-800B	8-102B36)	D-KSDG	13.11.98	P.R.Redshaw "23"	Walney Island	9. 1.05
G-BYEE	Mooney M.20K Model 231	25-0282	N231JZ	20. 7.88	R.J.Baker and W.Woods	Coventry	11. 5.07
					tr Double Echo Flying Group		
G-BYEH	CEA Jodel DR.250/160 Capitaine	15	OO-SOL	6.10.98	Nicholson Decommissioning Ltd	(Kilkeel, Newry)	9. 3.06
			F-BMZL				
G-BYEI	Cameron 90SS Chick	4519		1. 4.99	Bic UK Ltd *(Bic SoftFeel titles)*	Harefield	5.10.00A
G-BYEJ	Scheibe SF28A Tandem Falke	5713	OE-9070	18.12.98	D.Shrimpton	RAF Keevil	16. 9.05
			(D-KDAM)				
G-BYEK	Stoddard-Hamilton Glastar	PFA 295-13087	ZK-NEW	14. 9.98	G.M.New	Bagby	23. 9.05P
	(Built G.M.New) (Continetal IO-240) (Tailwheel u/c)		G-BYEK				
G-BYEL	Van's RV-6	PFA 181-12568		7. 1.99	D.Millar	Bidford	13. 7.05P
	(Built D T Smith) (Lycoming O-320)						
G-BYEM	Cessna R182 Skylane RG II	R18200822	N494	8. 1.99	Wycombe Air Centre Ltd	Booker	30. 1.05T
			D-ELVI, N737FT				
G-BYEO	Zenair CH.601HDS Zodiac	PFA 162-13345		11. 1.99	M.J.Diggins	White Waltham	4. 7.05P
	(Built B S Carpenter and M W Elliott) (Rotax 912-UL) *(Tail-wheel u/c)*				tr Cloudbase Flying Group		
G-BYEP	Lindstrand LBL 90B HAB	560		20.11.98	D.G Macguire	(Pulborough)	26. 3.01A
G-BYER	Cameron C-80 HAB	4513		19.11.98	Cameron Balloons Ltd *"E2"*	(Bristol)	11. 7.03A
G-BYES	Cessna 172P	17274514	PH-ILN	7.10.98	Redhill Air Services Ltd	Rochester	24.10.04T
			N172TP, N52424				
G-BYET	Cessna 172P	17275122	PH-ILP	7.10.98	Redhill Air Services Ltd	Rochester	4.11.07E
			N55158				
G-BYEW	Pegasus Quantum 15-912	7499		15. 1.99	R.J.Murphy	(Edinburgh)	13. 7.05P
G-BYEX	Sky 120-24 HAB	135		21. 1.99	Ballongflyg Upp and Ner AB	(Stockholm, Sweden)	5. 8.05A
G-BYEY	Lindstrand LBL 21 Silver Dream HAB	577		15. 1.99	Oscair Project Ltd	(Taby, Sweden)	
G-BYEZ	Dyn'Aéro MCR-01 Club	PFA 301-13185		25.11.98	J.P.Davies	Leicester	7. 9.05P
	(Built J.P.Davies - kit no.47) (Rotax 912)						
G-BYFA	Reims/Cessna F152 II	F15201968	G-WACA	19.11.98	A J.Gomes	Biggin Hill	5. 5.96
G-BYFC	Jabiru Aircraft Jabiru SK	PFA 274-13344		5. 2.99	A C.N.Freeman	Bournemouth	17. 6.05P
	(Built A C.N.Freeman)						
G-BYFD	Grob G115A	8100	EI-CCN	15. 1.99	M.Kane	Weston, Dublin	28. 8.06T
			G-BSGE		tr Kane Group		
G-BYFE	Pegasus Quantum 15-912	7496		21. 6.99	J.L.Pollard	Knapthorpe Lodge, Caunton	23. 2.05P
					tr G-BYFE Flying Group		
G-BYFF	Pegasus Quantum 15-912	7500		1. 2.99	D.Young tr Kemble Flying Club	Kemble	22. 2.05P
G-BYFG	Europa Aviation Europa XS	PFA 247-13407		22. 1.99	P.R.Brodie	(Guildford)	18.11.05P
	(Built P R Brodie - kit no.396) (Jabiru 3300) *(Tri-gear u/c)*						
G-BYFI	CFM Starstreak Shadow SA	PFA 206-13300		11. 2.99	D.G.Cook	Leiston	
	(Built D.G.Cook)						
G-BYFJ	Cameron N-105 HAB	4545		4. 3.99	R.J.Mercer	(Belfast)	23. 6.05A
G-BYFK*	Cameron Printer 105 SS HAB	4522		4. 3.99	Flying Pictures Ltd	(Chilbolton)	26. 5.03A
					(Samsung Printers titles) (Cancelled 29.9.03 as wfu)		
G-BYFL	Diamond HK 36 TTS Super Dimona	36.623		5. 2.99	C.N.J.Squibb tr Seahawk Gliding Club	RNAS Culdrose	23. 6.05
G-BYFM	Jodel DR.1050-M1 Sicile Record rep	PFA 304-13237		26. 2.99	A.J.Roxburgh	Barton	5. 6.05P
	(Built P.M.Standen and A J.Roxburgh) (Continental O-200-A)						
G-BYFR	Piper PA-32R-301 Saratoga IIHP	3246133	N4135P	13. 4.99	Buckleton Ltd	(Jersey)	5. 7.05T
			G-BYFR, N9515N				
G-BYFT	Pietenpol AirCamper	PFA 47-13057		22.12.98	M.W.Elliott	Shenstone	12. 7.07P
	(Built M W Elliott)						
G-BYFU	Lindstrand LBL 105B HAB	594		9. 3.99	Balloons Lindstrand France	(Curcay-sur-Dive, France)	13 5.04A
G-BYFV	TEAM mini-MAX 91	PFA 186-13431		5. 2.99	W.E.Gillham	Croft Farm, Darlington	16. 7.05P
	(Built W.E.Gillham)						
G-BYFX	Colt 77A HAB	4547		4. 3.99	Flying Pictures Ltd *(Agfa titles)*	(Chilbolton)	5. 5.05A
G-BYFY	Avions Mudry CAP.10B	263	F-GKKD	9. 3.99	R.W.H.Cole	Spilsteads Farm, Sedlescombe	AC
					t/a Cole Aviation *(Noted 11.01)*		

Registration	Type	C/n	Previous identity	Reg date	Owner	Base	CofA expiry
G-BYGA	Boeing 747-436	28855		15.12.98	British Airways plc	Heathrow	13.12.07E
G-BYGB	Boeing 747-436	28856		17. 1.99	British Airways plc	Heathrow	16. 1.07T
G-BYGC	Boeing 747-436	25823		19. 1.99	British Airways plc	Heathrow	23.10.07E
G-BYGD	Boeing 747-436	28857		26. 1.99	British Airways plc *(Rendezvous t/s)*	Heathrow	23.10.07T
G-BYGE	Boeing 747-436	28858		5. 2.99	British Airways plc	Heathrow	4. 2.05T
G-BYGF	Boeing 747-436	25824		17. 2.99	British Airways plc	Heathrow	16. 2.05T
G-BYGG	Boeing 747-436	28859		29. 4.99	British Airways plc	Heathrow	28. 4.05T
G-BYHC	Cameron Z-90 HAB	4555		16. 3.99	S.M.Sherwin *(Darlows titles)*	(Morton, Bourne)	27. 3.05
G-BYHE	Robinson R22 Beta	2023	N82128	14. 1.99	Kent Aviation Ltd	Booker	24. 2.05T
			LV-VAB				
G-BYHG	Dornier 328-100	3098	D-CDAE	7. 4.99	Suckling Aviation (Cambridge) Ltd	London City	6. 4.05T
			D-CDXZ		t/a Scot Airways		
G-BYHH	Piper PA-28-161 Warrior III	2842050	N4126Z	15. 6.99	Stapleford Flying Club Ltd	Stapleford	19. 6.05T
			G-BYHH, N9527N				
G-BYHI	Piper PA-28-161 Warrior II	28-8116084	SE-IDP	4. 1.99	Haimoss Ltd	Old Sarum	25. 2.05T
G-BYHJ	Piper PA-28R-201 Arrow	2844020	N41675	25. 2.00	Bflying Ltd	Bournemouth	22. 4.06T
			G-BYHJ, N41675		*(Op Bournemouth Flying Club)*		
G-BYHK	Piper PA-28-181 Archer II	2843240	N4128V	20. 5.99	T-Air Services Ltd	(Kirk Michael, Isle of Man)	23. 6.05T
			(G-BYHK), N9519N				
G-BYHL	de Havilland DHC-1 Chipmunk 22	C1/0361	WG308	15. 3.99	M.R.and I.D.Higgins	RAF Cranwell	31. 8.06
G-BYHM	British Aerospace BAe 125 Series 800B	258233	VP-BTM	12. 2.99	Corporate Aircraft Leasing Ltd	Southampton	5. 5.05T
			VR-BTM, (VR-BQH), F-WQCD, D-CAVW, G-5-770				
G-BYHN	Mainair Blade 912	1191-0399-7 & W994		9. 4.99	R.Stone	(Stoke-on-Trent)	10. 4.04P
G-BYHO	Mainair Blade 912	1197-0599-7 & W1000		16. 3.99	B.Hall	(Ingleton, Carnforth)	24. 3.05P
G-BYHP	CEA Jodel DR.253B Régent	161	OO-CSK	29. 3.99	C.P.Course, M.Walker and J.C.Harmon		
					t/a The G-BYHP Group	Sywell	25. 8.05T
G-BYHR	Pegasus Quantum 15-912	7518		6. 4.99	I.D.Chantler	Longacres Farm, Sandy	15. 8.05P
G-BYHS	Mainair Blade 912	1187-0299-7 & W990		11. 3.99	C.I.Poole	(Southport)	26. 4.05P
G-BYHT	Robin DR400/180R Remorqueur	811	HB-EUU	9. 4.99	R.C.Wilson tr Deeside Robin Group	Aboyne	1. 9.05
G-BYHU	Cameron N-105 HAB	4567		30. 4.99	Freeup Ltd *(Iveco Ford Truck titles)*	(Bristol)	1. 6.05A
G-BYHV	Raj Hamsa X'Air 582	BMAA/HB/090		25. 3.99	J.S.Mason	(Southwell)	21. 4.05P
	(Built J Bowditch - kit no.381)						
G-BYHX	Cameron A-250 HAB	4565		16. 4.99	Global Ballooning Ltd	(Uckfield)	9. 9.05T
G-BYHY	Cameron V-77 HAB	4493		22. 3.99	P.Spellward *"Biggles"*	(Bristol)	25. 4.05A
G-BYHZ	Sky 160-24 HAB	140		13. 5.99	Breckland Balloons Ltd	(Wendling, Dereham)	26. 5.05T
G-BYIA	Jabiru Aircraft Jabiru SK	PFA 274-13436		10. 2.99	G.M.Geary	Morgansfield, Fishburn	16. 5.05P
	(Built M F Cottam)						
G-BYIB	Rans S-6-ESA Coyote II	PFA 204-13387		26. 3.99	W.Anderson	Polmont	6. 9.05P
	(Built G A Clayton) (Rotax 582) *(Tail-wheel u/c)*						
G-BYIC	Cessna TU206G Turbo Stationair	U20605476	OY-NUA	27. 4.99	D.M.Penney	Shotton Colliery, Peterlee	29. 7.05
			N113RS, N3RS, N6398U				
G-BYID	Rans S-6-ES Coyote II	PFA 204-13348		11. 5.99	J.A E.Bowen	Davidstow Moor	4.11.05P
	(Built D J Brotherhood) (Rotax 582 - *kit no.0498.1218) (Tri-cycle u/c)*						
G-BYIE	Robinson R22 Beta II	2933		22. 4.99	J.W.Ramsbottom t/a Jepar Rotorcraft	(Preston)	22. 5.05T
G-BYII	TEAM mini-MAX 91	PFA 186-11820		22. 1.99	J.Edwards	Barton	11. 3.05P
	(Built J S R Moodie)						
G-BYIJ	CASA 1-131E Jungmann	2110	Spanish AF E3B-514	16. 7.90	P.R.Teager and R.N.Crosland	Deanland	2. 6.05P
G-BYIK	Europa Aviation Europa	PFA 247-12771		2. 2.99	P.M.Davis	Oxford	7.10.05P
	(Built P.M.Davis - kit no.154) (Monowheel u/c)						
G-BYIL	Cameron N-105 HAB	4591		29. 4.99	Oakfield Farm Products Ltd	(Broadway)	12. 4.05A
					(Oakfield Farm Products Ltd titles)		
G-BYIM	Jabiru Aircraft Jabiru UL	PFA 274A-13397		22.12.98	S.D.Miller	(Dry Drayton, Cambridge)	21. 3.05P
	(Built W J Dale and R F Hinton)						
G-BYIN	Rotary Air Force RAF 2000 GTX-SE	PFA G/13-1305		19. 1.99	J.R.Legge	(Rossendale)	30. 9.05P
	(Built J R Legge)						
G-BYIO	Colt 105A HAB	4601		30. 4.99	N.Charbonnier *(Lindt titles)*	(Aosta, Italy)	24. 1.05A
G-BYIP	Pitts S-2A	2244	N109WA	23. 2.99	D.P.Heather-Hayes	Perth	10.10.05T
	(Built Aerotek Inc)		TC-ECN				
G-BYIR	Pitts S-1S	1-0063	N103WA	23. 2.99	Hampshire Aeroplane Co Ltd	(Penzance)	7. 9.06
	(Built Aerotek Inc)		TC-ECP				
G-BYIS	Pegasus Quantum 15-912	7508		25. 2.99	Lillian M.Tidman	North Coates	4. 5.05P
G-BYIT	Robin DR500/200i Président	0010		27. 1.99	P.R.Liddle	Rochester	23. 5.05
	(Registered as DR400/500)						
G-BYIU	Cameron V-90 HAB	4552		6. 4.99	H.Micketeit	(Bielefeld, Germany)	17. 3.05A
G-BYIV	Cameron PM-80 HAB	4595		14. 5.99	A.Schneider	(Borken, Germany)	27. 4.05A
	(Coca Cola bottle)						
G-BYIW*	Cameron PM-80 HAB	4596		14. 5.99	A.Schneider	(Borken, Germany)	27. 4.05A
	(Coca Cola bottle)				*(Cancelled 28.1.05 by CAA)*		
G-BYIX	Cameron PM-80 HAB	4597		14. 5.99	A.Schneider	(Borken, Germany)	28. 4.05A
	(Coca Cola bottle)						
G-BYIY	Lindstrand LBL 105B HAB	601		26. 3.99	J.H.Dobson *(Tutti Frutti titles)*	(Reading)	25. 6.04A
G-BYIZ	Pegasus Quantum 15-912	7504		8. 2.99	J.D.Gray	Eshott	18. 5.05P
G-BYJA	Rotary Air Force RAF 2000 GTX-SE	PFA G/13-1297		6. 4.99	B.Errington-Weddle	Henstridge	18. 7.02P
	(Built B.Errington-Weddle)				*(Active 7.03)*		
G-BYJB	Mainair Blade 912	1192-0499-7 & W995		6. 4.99	J.H.Bradbury	Ince Blundell	14. 4.05P
G-BYJC	Cameron N-90 HAB	4562		30. 4.99	J.M.Percival	(Burton-on-the-Wolds, Loughborough)	31. 7.05A
					(New owner 6.04)		
G-BYJD	Jabiru Aircraft Jabiru UL	PFA 274-13376		16. 4.99	M.W.Knights	Blue Tile Farm, Hindolveston	21. 5.05P
	(Built G Wallis and M W Knights) (PFA c/n prefix should be "274A")						
G-BYJE	TEAM mini-MAX 91	PFA 186-12327		6. 4.99	C.Trollope	Watnall	17. 2.05P
	(Built A W.Austin and M.F.Cottam)						
G-BYJF	Thorp T.211	107	N2545C	20. 5.99	AD Aviation Ltd	Liverpool	25. 7.05T
G-BYJG	Lindstrand LBL 77A HAB	600		16. 4.99	Lindstrand Hot Air Balloons Ltd	(Oswestry)	5. 9.05A
G-BYJH	Grob G109B	6512	D-KFRI	19. 5.99	A.J.Buchanan	Parham Park	26. 6.05

Registration	Type	C/n	Previous identity	Date	Owner	Base	CofA
G-BYJI	Europa Aviation Europa	PFA 247-13010	G-ODTI	19. 4.99	P.S.Jones	Lower Grounds Farm, Sherlowe	1. 9.05P
	(Built Europa Aviation Ltd - kit no."F0004") (Monowheel u/c)						
G-BYJJ	Cameron C-80 HAB	4436	SX-MAX	20. 4.99	Proxim Franchising SRL	(Milano, Italy)	6. 1.05A
G-BYJK	Pegasus Quantum 15-912	7524		7. 5.99	B.S.Smy	East Fortune	2. 6.05P
G-BYJL	Aero Designs Pulsar 3	PFA 202-13311		20. 4.99	F.A H.Ashmead	(Sway)	4. 7.05P
	(Built F.A H.Ashmead)						
G-BYJM	Cyclone AX2000	7523		25. 5.99	A.R.Hood	Knapthorpe Lodge, Caunton	30. 5.05P
					tr Caunton AX2000 Syndicate		
G-BYJN	Lindstrand LBL 105A HAB	605		30. 4.99	B.Meeson	(Pwllheli)	29. 4.00A
G-BYJO	Rans S-6-ES Coyote II	PFA 204-13338		4. 3.99	G.Ferguson	Blue Tile Farm, Hindolveston	9. 6.03P
	(Built G.Ferguson) (Rotax 582 - *kit no.0498.1217) (Tail-wheel u/c)*						
G-BYJP	Pitts S-1S	1-0064	N105WA	16. 3.99	T.Riddle	Eaglescott	1. 4.06
	(Built Aerotek Inc)		TC-ECR		tr Eaglescott Pitts Group		
G-BYJR	Lindstrand LBL 77B HAB	608		30. 4.99	C.D.Duthy-James *(Noted 1.05)*	(Poitiers, France)	18. 1.05A
G-BYJS	SOCATA TB-20 Trinidad	1875	F-OIGE	15. 1.99	A.P.Bedford tr Juliet Sierra Group	Oxford	9. 5.05
G-BYJT	Zenair CH.601HD Zodiac	PFA 162-13130		4. 5.99	J.D.T.Tannock	Tollerton	9. 3.05P
	(Built J.D.T.Tannock) (Rotax 912S)						
G-BYJU	Raj Hamsa X'Air 582	BMAA/HB/098		6. 5.99	G.P.Morling	(Douglas, Isle of Man)	29. 4.05P
	(Built C W Payne - kit no.429)						
G-BYJV	Cameron A-210 HAB	4612		4. 6.99	Societe Bombard SRL	(Beaune, France)	22. 9.04A
G-BYJW	Cameron Sphere 105SS HAB	4585		15. 6.99	Forbes Global Inc	(Far Hills, New Jersey, US)	7. 4.05A
					(New CofR 9.03)		
G-BYJX	Cameron C-70 HAB	4580		30. 4.99	B.Perona	(Torino, Italy)	28. 5.04A
G-BYJZ	Lindstrand LBL 105A HAB	609		27. 5.99	M.A..Webb	(Chard)	26. 7.02A
G-BYKA	Lindstrand LBL 69A HAB	612		7. 5.99	Aerial Promotions Ltd *(Vauxhall titles)*	(Cannock)	6.10.03A
G-BYKB	Rockwell Commander 114	14121	SE-GSM	18. 5.99	A.Walton	Little Staughton	27. 5.05P
			N4801W				
G-BYKC	Mainair Blade 912	1196-0599-7 & W999		7. 5.99	P.J.Burridge	(Liverpool)	23. 6.05P
G-BYKD	Mainair Blade 912	1198-0599-7 & W1001		7. 5.99	D.C.Boyle	(Chorley)	24. 6.05P
G-BYKE	Rans S-6-ESA Coyote II	PFA 204-13327		22. 1.99	C.Townsend	Kemble	6. 3.03P
	(Built C.Townsend) (Rotax 912-UL) *(Tri-cycle u/c)*				*(Rolled to starboard on take-off Enstone 17.3.02, struck runway, caught fire and destroyed)*		
G-BYKF	Enstrom F-28F	725	JA7684	19. 5.99	Battle Helicopters Ltd	(Battle)	25. 2.06T
G-BYKG	Pietenpol AirCamper	PFA 47-12827		17. 3.99	K.B.Hodge	(Mold)	
	(Built K.B.Hodge)				*(Nearing completion 2000)*		
G-BYKI	Cameron N-105 HAB	4635		4. 6.99	J.A. Leahy	(Navan, County Meath)	21. 9.05A
G-BYKJ	Westland Scout AH.1	F9696	XV121	6. 8.99	B.H.Austen t/a Austen Associates	Oaksey Park	21.12.05P
G-BYKK	Robinson R44 Astro	0572		4. 3.99	Banner Helicopters Ltd	(Heywood)	18. 3.05T
G-BYKL	Piper PA-28-181 Archer II	28-8090162	HB-PFB	15. 7.99	MPFC Ltd	Biggin Hil	18. 7.05T
			N8129Y		*(Op Metropolitan Police Flying Club)*		
G-BYKN	Piper PA-28-161 Warrior II	28-7916307	HB-PDO	22. 6.99	Oxford Aviation Services Ltd	Oxford	2. 9.05T
			N2838C, N9613N				
G-BYKO	Piper PA-28-161 Warrior II	28-8516063	HB-PKA	22. 6.99	Oxford Aviation Services Ltd	Gloucestershire	1. 8.05T
			F-GECN, N6920C				
G-BYKP	Piper PA-28RT-201T Turbo Arrow IV	28R-7931029	HB-PDB	22. 6.99	Westflight Aviation Ltd	Gloucestershire	10. 9.05T
			N3010G				
G-BYKR	Piper PA-28-161 Warrior II	2816061	HB-PLM	22. 6.99	Oxford Aviation Services Ltd	Oxford	30. 7.05T
G-BYKS	Leopoldoff L.6 Colibri	129	N10LC	19. 4.99	I.M.Callier	(Basingstoke)	
			F-BGIT, F-WGIT		*(On restoration 10.01)*		
G-BYKT	Pegasus Quantum 15-912	7529		28. 5.99	D.A Bannister and N.J.Howarth	Deenethorpe	7. 6.05P
G-BYKU	BFC Challenger II	PFA 177A-13252		25. 5.99	K.W.Seedhouse	Otherton, Cannock	
	(Built K W Seedhouse)				*(Noted 5.04)*		
G-BYKW	Lindstrand LBL 77B HAB	620		22. 6.99	P-J.Fuseau *(Noted 1.05)*	(Chanteloup, France)	24. 3.05A
G-BYKX	Cameron N-90 HAB	4657		10. 8.99	G.Davis *"Knowledgepool"*	(Reading)	14. 5.05A
G-BYKZ	Sky 140-24 HAB	147		25. 2.99	D.J.Head	(Newbury)	6. 8.05T
G-BYLA	Clutton FRED Series III	PFA 29-10775		11. 5.99	R.Holden-Rushworth	(Devizes)	
	(Built R.Holden-Rushworth)						
G-BYLC	Pegasus Quantum 15-912	7528		25. 6.99	M.Hurn	Hunsdon	7.12.05P
G-BYLD	Pietenpol AirCamper	PFA 47-13392		27. 4.99	S.Bryan	(Banbury)	
	(Built S.Bryan)						
G-BYLE	Piper PA-38-112 Tomahawk II	38-82A0031	N91437	18. 6.99	Surrey and Kent Flying Club Ltd	Biggin Hill	6. 3.06T
G-BYLF	Zenair CH.601HDS	PFA 162-13179		3. 6.99	G.Waters	Swansea	18. 8.05P
	(Built M and J S Thomas and G Waters) (Rotax 912-UL)						
G-BYLG	Robin HR200/120B	336		20. 7.99	Leinster Aero Club Ltd	Weston, Dublin	1. 8.05T
G-BYLH	Robin HR200/120B	335		9. 7.99	Multiflight Ltd	Leeds-Bradford	19. 7.05T
G-BYLI	Nova Vertex 22	14319		9. 4.99	M Hay *(New owner 1.03)*	(Dundee)	
G-BYLJ	Letov LK-2M Sluka	PFA 263-13464		9. 6.99	W.J.McCarroll	(Randalstown, Antrim)	
	(Built N.E.Stokes) (Rotax 447)				*(New owner 6.04)*		
G-BYLL	Sequoia F.8L Falco	PFA 100-10843		6.12.85	N.J.Langrick	Breighton	6. 3.05P
	(Built N.J.Langrick) (Lycoming O-320-A3C)						
G-BYLO	Tipsy Nipper T.66 Series 1	04	OO-NIA	27. 4.99	M.J.A Trudgill	RAF Henlow	24. 6.04P
	(Built Avions Fairey SA) (Volkswagen 1600)						
G-BYLP	Rand Robinson KR-2	PFA 129-11431		19. 4.99	C.S.Hales	(Walsall)	
	(Built C.S.Hales)				*(See G-BSTL)*		
G-BYLR	Cessna 404 Titan	404-0046	OH-CDC	14. 6.99	Air Charter Scotland Ltd	Prestwick	19. 6.06T
			SE-GZH, N5428G				
G-BYLS	Bede BD-4	PFA 37-11288		13.12.90	G.H.Bayliss	Shobdon	21.11.05P
	(Built G.H.Bayliss) (Lycoming O-320-E2F)						
G-BYLT	Raj Hamsa X'Air 582	BMAA/HB/095		8. 6.99	T.W.Phipps and B.G.Simons		
	(Built R J Turner - kit no 411)					Craysmarsh Farm, Melksham	23. 9.05P
G-BYLV	Thunder Ax8-105 S2 HAB	4061		6. 7.99	K B Voli S A S Di Bartolomeo Chiozzi and Cia		
						(Cappella Cantone (CR), Italy)	26. 9.05A
G-BYLW	Lindstrand LBL 77A HAB	615		11. 6.99	Associazione Gran Premio Italiano	(Perugia, Italy)	10. 6.00A
G-BYLX	Lindstrand LBL 105A HAB	614		11. 6.99	Italiana Aeronavi	(Cervignano, Italy)	9. 5.04A
G-BYLY	Cameron V-77 HAB	3375	G-ULIA (2)	16. 7.97	R.Bayly *(Rob Bayly titles) (See G-ULIA)*	(Bristol)	19. 7.05A

Registration	Type	C/n	Previous identity	Regd	Owner/Operator	Base	CofA
G-BYLZ	Rutan Cozy Mk.4	PFA 159-12464		21. 5.99	E.R.Allen	Dunsfold	4.11.05P
	(Built E R Allen)						
G-BYMB	Diamond DA.20-C1 Katana	C0051		9. 7.99	S.C.Brown t/a Enstone Flying Club	Enstone	5. 6.06T
G-BYMC	Piper PA-38-112 Tomahawk II	38-82A0034	N91457	18. 6.99	Central Aircraft Leasing Ltd	Wolverhampton	2. 7.06T
G-BYMD	Piper PA-38-112 Tomahawk II	38-82A0009	N91342	18. 6.99	Surrey and Kent Flying Club Ltd	Shoreham	19. 8.05T
					(Op The Flying Hut)		
G-BYME	Sud-Aviation Gardan GY-80-180 Horizon	207	F-BPAA	24. 5.99	Air Venturas Ltd	Bagby	12.11.05
G-BYMF	Pegasus Quantum 15-912	7540		9. 7.99	G.R.Stockdale	Rufforth	29.11.05P
G-BYMG	Cameron A-210 HAB	4631		17. 9.99	P.Johnson t/a Cloud Nine Balloon Co	(Consett)	12. 2.05T
G-BYMH	Cessna 152	15284980	N6127P	15. 7.99	PJC (Leasing) Ltd	Stapleford	25. 7.05T
G-BYMI	Pegasus Quantum 15	7533		9. 7.99	N.C.Grayson	Rufforth	13. 7.05P
	(Rotax 503)						
G-BYMJ	Cessna 152	15285564	N93865	16. 7.99	PJC (Leasing) Ltd	Stapleford	26.11.05T
G-BYMK	Dornier 328-100	3062	LN-ASK	9. 6.99	Suckling Aviation (Cambridge) Ltd	London City	8. 6.05T
			D-CDXE		t/a Scot Airways		
G-BYML	Dornier 328-100	3069	D-CDUL	27. 7.99	Suckling Aviation (Cambridge) Ltd	London City	14. 8.05T
			LN-ASL, D-CDXT (2)		t/a Scot Airways		
G-BYMM	Raj Hamsa X'Air 582	BMAA/HB/093		29. 4.99	R.W.F Boarder	Field Farm, Oakley	23. 7.05P
	(Built J R Pearce - kit no 417)						
G-BYMN	Rans S-6-ESA Coyote II	PFA 204-13477		16. 6.99	R.L.Barker	Chase Farm, Billericay	20.12.05P
	(Built H Smith) (Rotax 582) *(Tri-cycle u/c)*						
G-BYMO	Campbell Cricket	PFA G/03-1266		16. 7.99	D.G.Hill	(Stockton-on-Tees)	7. 7.03P
	(Built D.G.Hill) (Rotax 532)						
G-BYMP	Campbell Cricket Mk 1	PFA G/03-1265		16. 6.99	J.J.Fitzgerald	(Newtownards)	
	(Built J.J.Fitzgerald)						
G-BYMR	Raj Hamsa X'Air 582	BMAA/HB/094		18. 6.99	W.M McMinn	Slieve Croob , County Down	6.11.05P
	(Built W.M McMinn - kit no.434)						
G-BYMT	Pegasus Quantum 15-912	7549		16. 7.99	C.M.Mackinnon	Cumbernauld	30. 7.05P
G-BYMU	Rans S-6-ESN Coyote II	PFA 204-13424		25. 6.99	I.R.Russell and G.Frogley	Swinford, Rugby	24. 5.03P
	(Built I.R.Russell) (Vernier SVS1400)						
G-BYMV	Rans S-6-ESN Coyote II	PFA 204-13444		25. 6.99	G.A.Squires	Rufforth	25. 7.05P
	(Built G.A Squires) (Rotax 582) *(Tri-cycle u/c)*						
G-BYMW	Boland 52-12 HAB	001		25. 6.99	C.Jones	(Sonning Common, Reading)	
	(Built C.Jones)						
G-BYMX	Cameron A-105 HAB	4629		16. 7.99	H.Reis	(Aachen, Germany)	23.10.03A
G-BYMY	Cameron N-90 HAB	4653		19. 7.99	Cameron Balloons Ltd *(Cameron Balloons titles)*	(Bristol)	15. 8.05A
G-BYNA	Cessna F172H	F172-0626	OO-VDW	15. 1.99	Heliview Ltd	Blackbushe	15. 4.05T
	(Built Reims Aviation SA)		PH-VDW, (G-AWTH), F-WLIT				
G-BYND	Pegasus Quantum 15	7546		16. 7.99	D.G.Baker	Lee-on-Solent	21.10.05P
	(Rotax 582)						
G-BYNE	Pilatus PC-6/B2-H4 Turbo Porter	631	HB-FLW	10. 8.99	D.M.Penny	Le Luc, Cennes, France	1.10.05
			C-FRAV, N631SA, N62148, HS-..., N62148, XW-PFC, XW-PDK, HB-FCR				
G-BYNF	North American NA-64 Yale I	64-2171	N55904	10. 1.00	R.S.Van Dijk	Duxford	AC
			RCAF 3349		*(Under protracted rebuild 4.03)*		
G-BYNH	RotorWay Executive 162F	6323		5. 7.99	R.C.Mackenzie	Street Farm, Takeley	
	(Built R.Mackenzie) (RotorWay RI 162F)				*(Noted 12.03- completion abandoned?)*		
G-BYNI	RotorWay Executive 90	5216		16. 7.99	M.Bunn	Fundenhall, Norfolk	28. 6.05P
	(Built M Bunn) (RotorWay RI 162)						
G-BYNJ	Cameron N-77 HAB	4661		26. 7.99	G.Aimo *(Primagaz titles)*	(Mondovi, Italy)	28. 5.05A
G-BYNK	Robin HR200/160	338		28. 7.99	R.J.Stainer and D.A Healey	Bodmin	30. 9.05T
					tr Penguin Flight Group		
G-BYNL	Jabiru Aircraft Jabiru SK	PFA 274-13328		20. 7.99	GBYNL Ltd	(Tadworth)	25. 4.05P
	(Built R C Daykin)				*(New owner 7.04)*		
G-BYNM	Mainair Blade 912	1204-0799-7 & W1007		20. 7.99	M.W.Holmes	(Ilkeston)	27. 7.05P
G-BYNN	Cameron V-90 HAB	4643		16. 7.99	M.K.Grigson *"Cloud Nine"*	(Shoreham)	3. 7.05
G-BYNO	Pegasus Quantum 15-912	7556		5. 8.99	R.J.Newsham and G.J.Slater	Clench Common	19. 8.05P
G-BYNP	Rans S-6-ES Coyote II	PFA 204-13414		22. 7.99	R.J. Lines	Sandtoft	9. 6.05P
	(Built R J Lines) (Rotax 582)						
G-BYNR	Jabiru Aircraft Jabiru UL	0129	EI-MAT	23. 7.99	M.P.Maughan	(Mirfield)	14. 7.05P
	(Built A Parker)						
G-BYNS	Jabiru Aircraft Jabiru SK	PFA 274-13235		23. 7.99	D.K.Lawry	Tibenham	10. 8.05P
	(Built D.K.Lawry)						
G-BYNT	Raj Hamsa X'Air Victor2	BMAA/HB/107		20. 7.99	G.R.Wallis	Lower Mountpleasant, Chatteris	13.12.02P
	(Built G.R.Wallis - kit no.457)						
G-BYNU	Cameron Thunder AX7-77 HAB	3520		29. 7.99	P.M.Gaines *"Soup Dragon"*	(Sedgefield, Stockton-on-Tees)	29. 6.05A
G-BYNV	Sky 105-24 HAB	165		11. 8.99	Par Rovelli Construzioni SRL	(Mazzini, Italy)	12. 5.05
G-BYNW	Cameron H-34 HAB	4666		27. 7.99	I.M.Ashpole *(New owner 12.03)*	(Bridstow, Ross-on-Wye)	13. 9.05A
G-BYNX	Cameron RX-105 HAB	4656		26. 7.99	Cameron Balloons Ltd	(Bristol)	1.11.00A
G-BYNY	Beech 76 Duchess	ME-247	N247ME	4. 8.99	Magenta Ltd	Shoreham	17.10.05T
			OE-FES, N6635H		*(Op European Flight Training)*		
G-BYOA*	Slingsby T.67M-260	2262	(RA-....)	8. 6.99	Leisair Ltd	(London SW10)	17.10.05T
			G-BYOA		*(Cancelled 26.1.04 by CAA)*		
G-BYOB	Slingsby T.67M-260 Firefly	2263		8. 6.99	Stapleford Flying Club Ltd	Stapleford	6.10.05T
G-BYOD	Slingsby T.67C Firefly	2265		13. 6.00	TDR Aviation Ltd	Newtownards	9.12.04T
G-BYOF	Robin R2160i	337		29. 7.99	Lifeskills Ltd	Wellesbourne Mountford	20.10.05T
G-BYOG	Pegasus Quantum 15-912	7555		15. 9.99	M.D.Hinge	Old Sarum	28.11.04P
G-BYOH	Raj Hamsa X'Air 582	BMAA/HB/101		23. 7.99	P.H.J.Kent	Belle Vue Farm, Yarnscombe	25. 7.05P
	(Built G A J Salter - kit no.443)						
G-BYOI	Sky 80-16 HAB	163		5. 8.99	I.S.and S.W.Watthews	(Cark-in-Cartmel)	21. 8.03
G-BYOJ	Raj Hamsa X'Air 582	BMAA/HB/108		23. 7.99	H.M.Owen	(Llanelli)	24. 3.04P
	(Built R R Hadley - kit no.458)						
G-BYOK	Cameron V-90 HAB	3726		9. 8.99	D.S.Wilson	(Norwich)	4. 4.05A
G-BYOM	Sikorsky S-76C	760464	G-IJCB	25. 8.99	Starspeed Ltd	Blackbushe	26. 3.05T

Reg	Type	C/n	Previous identity	Date	Owner/operator	Location	CofA
G-BYON	Mainair Blade	1199-0599-7 & W1002		4. 8.99	S.Mills and G.M.Hobman	(North Ferriby)	20. 8.05P
	(Rotax 503)						
G-BYOO	CFM Streak Shadow SA	PFA 206-12806		6. 8.99	G.R.Eastwood	Full Sutton	12.12.05P
	(Built C I Chegwen - kit no K.270) (Rotax 912-UL)						
G-BYOR	Raj Hamsa X'Air 582	BMAA/HB/117		11. 8.99	K.J.Draper	Middle Stoke, Isle of Grain	19. 7.05P
	(Built A R Walker) (3-blade Ivoprop - kit no.478)						
G-BYOS	Mainair Blade 912	1209-0899-7 & W1012		6. 8.99	J.L.Guy	Baxby Manor, Husthwaite	15. 8.05P
G-BYOT	Rans S-6-ES Coyote II	PFA 204-13363		29. 7.99	H.F.Blakeman	Arclid Green, Sandbach	9.11.03P
	(Built H.F.Blakeman - kit no.0498.1221) (Tri-cycle u/c)				*(Noted 9.04)*		
G-BYOU	Rans S-6-ES Coyote II	PFA 204-13460		1. 6.99	J.R.Bramley	Wickenby	20.10.05P
	(Built Light Flight Ltd and R Germany) (Rotax 582 - *kit no.1298.1288) (Tri-cycle u/c)*						
G-BYOV	Pegasus Quantum 15-912	7554		17. 8.99	Microlight Hire Ltd	Wickenby	9.11.04P
G-BYOW	Mainair Blade	1207-0899-7 & W1010		9. 8.99	M.Forsyth	(Kelso)	23. 3.05P
	(Rotax 582)						
G-BYOX	Cameron Z-90 HAB	4672		31. 8.99	Airship and Balloon Company Ltd	(Bristol)	3. 4.04A
G-BYOY	Canadair CL-30 (T-33AN) Silver Star Mk.3	T33-231	N36TH	8. 2.00	K.K.Gerstorfer	North Weald	AC
			N333DV, N134AT, N10018, N134AT, RCAF 21231 *(Noted as "N36TH "in USAF c/s 1.05)*				
G-BYOZ	Mainair Rapier	1208-0899-7 & W1011		12. 8.99	M.Morgan	Arclid Green, Sandbach	11. 9.03P
G-BYPA	Aérospatiale AS355F2 Ecureuil 2	5348	G-NWPI	20. 8.99	P. and Judith Carter	(Peterborough)	28. 3.06T
			F-GMAO				
G-BYPB	Pegasus Quantum 15-912	7566		3. 9.99	S.Graham	Redlands	2. 9.05P
G-BYPC	Lindstrand LBL AS2 Gas Balloon	634		17. 8.99	Lindstrand Balloons Ltd	Plano, Texas, US	
					(Super 2 - Superpressure titles)		
G-BYPD	Cameron A-105 HAB	4680		6. 1.00	Headland Hotel Co Ltd	(Newquay)	10. 3.05A
G-BYPE	Sud-Aviation Gardan GY-80-160 Horizon	180	F-BNYD	10. 8.99	H.I.Smith and P.R.Hendry-Smith	Little Snoring	8. 4.06
G-BYPF	Thruster T.600N	9089-T600N-034		17. 8.99	R.J.Oakley	Shobdon	3.10.04P
	(Rotax 582UL)				tr Canary Syndicate		
G-BYPG	Thruster T.600N	9089-T600N-035		17. 8.99	J.I.Greeshields	Dunkeswell	29. 3.04P
	(Rotax 582UL)				tr G-BYPG Syndicate		
G-BYPH	Thruster T.600N	9089-T600N-036		17. 8.99	D.M.Canham	Leicester	19. 5.05P
	(Rotax 582UL) *(Officially regd with incorrect c/n as 9099-T600N-036)*						
G-BYPJ	Pegasus Quantum 15-912	7565		17. 9.99	C.K.Stow	(Thorpe-on-the-Hill, Lincoln)	19.10.05P
G-BYPL	Pegasus Quantum 15-912	7558		9. 9.99	I.T.Carlse	(Fowlmere)	26. 9.04P
G-BYPM	Europa Aviation Europa XS	PFA 247-13418		16.12.98	P.Mileham	(Saffron Walden)	
	(Built P.Mileham - kit no.407) (Tri-gear u/c)						
G-BYPN	SOCATA MS.880B Rallye Club	2043	F-BTPN	23. 7.99	R.and T.C.Edwards	Sturgate	9. 6.06
G-BYPO	Raj Hamsa X'Air 582	BMAA/HB/111		25. 8.99	D.W.Willis	Brook Farm, Pilling	8. 6.05P
	(Built N G Woodhall and A S Leach - kit no.439)				tr X'Air Group		
G-BYPP	Medway Rebel SS	168/146		25.10.99	J.L.Gowens	(Maidstone)	17. 3.01P
	(2 Stroke International 690L70)						
G-BYPR	Zenair CH.601HD Zodiac	PFA 162-12816		25. 8.99	D.Clark	Portmoak	28. 6.05P
	(Built D.Clark) (Lycoming O-235)						
G-BYPT	Rans S-6-ESN Coyote II	PFA 204-13508		27. 8.99	B.S.Keene	Compton Abbas	8.12.05P
	(Built G R Pritchard) (Jabiru 2200A - *kit no.0499.1316)*						
G-BYPU	Piper PA-32R-301 Saratoga IIHP	3246150	N4160K	2.12.99	AM Blatch Electrical Contractors Ltd	East Winch	16. 1.06T
			G-BYPU, N9518N				
G-BYPW	Raj Hamsa X'Air 582	BMAA/HB/113		1. 9.99	A D.Worrall and B.J.Ellis	Tarn Farm, Cockerham	22.12.04P
	(Built P A Mercer - kit no.441)				tr G-BYPW		
G-BYPY	Ryan ST3-KR	1001	F-AZEV	5.10.99	Tracey Curtis-Taylor	Duxford	29. 6.05P
			N18926		*(As "001")*		
G-BYPZ	Rans S-6-116 Super 6	PFA 204A-13448		14. 7.99	R.A Blackbourn	Perth	26. 3.05P
	(Built P G Hayward) (Rotax 912-UL - *kit no.0299.1304) (Tri-cycle u/c)*						
G-BYRA	British Aerospace Jetstream Series 3202	845	OH-JAG	26.10.99	Air Kilroe Ltd	Humberside	21.11.05T
			N845JX/N845AE/G-31-845		*(Op Eastern Airways)*		
G-BYRC	Westland Wessex HC.2	WA539	XT671	23. 9.99	D.Brem-Wilson	Honey Crock Farm, Redhill	AC
G-BYRE*	Rans S-10 Sakota	PFA 194-11729		23. 7.91	R.J. and M.B.Trickey	Longside, Peterhead	
	(Built R J Trickey)				*(Cancelled 8.5.99 by CAA) (Noted 4.04)*		
G-BYRG	Rans S-6-ES Coyote II	PFA 204-13518		9. 9.99	W.H.Mills	Haverfordwest	6.11.05P
	(Built J Whiting) (Rotax 582 - *kit no 1298.1289) (Tri-cycle u/c)*						
G-BYRH	Medway Hybred 44XLR	MR165/143		25.10.99	A.Scotney	Wickenby	5. 9.05P
	(Rotax 503)						
G-BYRJ	Pegasus Quantum 15-912	7548		24. 9.99	D.A.Chamberlain	Long Marston	27. 9.04P
G-BYRK	Cameron V-42 HAB	4662		14. 7.99	R.Kunert	Finchamstead, Wokingham	22. 6.05A
G-BYRM	British Aerospace Jetstream Series 3202	847	OH-JAF	16.12.99	Air Kilroe Ltd *(Op Eastern Airways)*	Humberside	18. 1.05T
			N847JX, N847AE, N332QN, G-31-847				
G-BYRO	Mainair Blade	1210-0899-7 & W1013		20. 8.99	P.W.F.Coleman	Corn Wood Farm, Adversane	19 10.05P
	(Rotax 582)						
G-BYRP	Mainair Blade 912	1075-1295-7 & W877		15. 9.99	M.Bromage	(Haresfield, Stonehouse)	3. 9.04P
	(C/n amended to c/n 1075-0396-7 & W877 by Mainair)				*(New owner 1.05)*		
G-BYRR	Mainair Blade 912	1211-0999-7 & W1015		17. 8.99	G.R.Sharples	(Harrow)	18. 6.04P
	(C/n amended to 1222-0999-7 & W1015 by Mainair)						
G-BYRS	Rans S-6-ES Coyote II	PFA 204-13425		17. 9.99	A.E.Turner	(Wing, Leighton Buzzard)	19. 7.05P
	(Built R Beniston) (Rotax 582 - *kit no.0998.1266) (Tri-cycle u/c)*						
G-BYRU	Pegasus Quantum 15-912	7574		24. 9.99	V.R.March tr The Sarum QTM912 Group	Old Sarum	27. 9.05P
G-BYRV	Raj Hamsa X'Air 582	BMAA/HB/106		10. 9.99	D.R.Darby	Pound Green, Buttonoak, Bewdley	18. 4.05P
	(Built A Hipkin - kit no.387)						
G-BYRX	Westland Scout AH.1	F9640	XT634	5.10.99	Historic Helicopters Ltd *(As "XT634")*	Thruxton	2.11.05P
G-BYRY	Slingsby T.67M-200 Firefly	2042	B-HZQ	28. 9.99	T.R.Pearson	Oxford	21. 3.06T
			VR-HZQ, HKG-11		*(As "HKG-11")*		
G-BYRZ	Lindstrand LBL - 77M HAB	643		28. 9.99	Challenge Transatlantique	(Metz, France)	5.12.00A
	(Reported to be rebuild of G-BXDX)				*"Conseil Régional de Lorraine"*		
G-BYSA	Europa Aviation Europa XS	PFA 247-13199		23. 8.99	B.Allsop	Bentley Farm, Cold Aston	21. 4.05P
	(Built B.Allsop - kit no.360) (Monowheel u/c)				*"Sadie"*		
G-BYSE	Agusta-Bell 206B JetRanger II	8553	G-BFND	3.11.81	Alspath Properties Ltd	(Stratford-upon-Avon)	18. 9.05T

Reg	Type	C/n	Previous identity	Date	Owner/Operator	Location	CofA
G-BYSF	Jabiru Aircraft Jabiru UL	PFA 274A-13356		5.10.99	S.J.Marshall	Sittles Farm, Alrewas	13.11.05P
	(Built M M Smith)						
G-BYSG	Robin HR200/120B	339		22.11.99	Modi Aviation Ltd	Earls Colne	22.12.05T
G-BYSI	PZL-110 Koliber 160A	04990081	SP-WGI	21. 1.00	J.and D.F.Evans	Gamston	15. 3.06
G-BYSJ	de Havilland DHC-1 Chipmunk 22	C1/0021	SE-BON	12.10.99	Silver Victory BVB	Antwerp-Deurne, Belgium	23.10.06T
			WB569		*(As "WB569/R")*		
G-BYSK	Cameron A-275 HAB	4699		23. 2.00	Balloon School (International) Ltd	(Petworth)	16. 7.05T
					(British School of Ballooning titles)		
G-BYSM	Cameron A-210 HAB	4698		12. 4.00	Balloon School (International) Ltd	(Bath)	12. 8.05P
					(Op Heritage Balloons) "Bath Heritage"		
G-BYSN	Rans S-6-ES Coyote II	PFA 204-13459		19.10.99	A.L.and A R.Roberts	RAF St.Athan	7. 7.05P
	(Built A L.Roberts - kit no.1098.1270) (Rotax 582) *(Tri-cycle u/c)*						
G-BYSP	Piper PA-28-181 Archer II	28-8590047	D-EAUL	12.10.99	Central Aircraft Leasing Ltd	Defford	21.11.05T
			N6909D				
G-BYSR	Pegasus Quantum 15-912	7560		7. 3.00	A.C.Stuart	(Parksville, British Columbia, Canada)	10. 5.05P
G-BYSS	Medway Rebel SS	167/145		25.10.99	D.W.Allen	Stoke, Isle of Grain	18. 6.04P
	(2 Stroke International 690L70)						
G-BYSV	Cameron N-120 HAB	4704		15.10.99	S.Simmington	(Eccles, Norwich)	15. 1.05T
G-BYSW	Enstrom 280FX Shark	2026	I-LUST	19. 9.00	D.A.Marks	(Bedford)	10.12.06
			N88CV				
G-BYSX	Pegasus Quantum 15-912	7586		23.11.99	M.E.Howard	RAF Henlow	18.12.04P
					tr RAF Microlight Flying Association (SX)		
G-BYSY	Raj Hamsa X'Air 582	BMAA/HB/109		21.10.99	J.M.Davidson	(Tewkesbury)	24. 5.05P
	(Built J.M.Davidson - kit no.448)						
G-BYTA	Kolb Twinstar Mk.3	PFA 205-13240		2. 9.99	R.E.Gray	(Oxted)	16. 7.04P
	(Built R E Gray)						
G-BYTB	SOCATA TB-20 Trinidad	2002	F-OILE	18. 5.00	Ottoman Empire Ltd	Goodwood	28. 7.06P
G-BYTC	Pegasus Quantum Q.2 Sport 15-912	7571		25.10.99	R.J.Marriott *(Noted 3.04)*	Headon Farm, Retford	1. 8.05P
G-BYTD	Robinson R22 Beta II	3003		25.10.99	J.M.D.Moloney	(Beaumont, Dublin)	6.11.05T
				(Made forced landing between Brafield and Hackleton 13.10.04 and destroyed)			
G-BYTE	Robinson R22 Beta	1250		18. 4.90	Patriot Aviation Ltd	Cranfield	17. 6.02T
G-BYTG	Glaser-Dirks DG-400	4-211	D-KBBP	18.11.99	P.R.Williams and B.Sebestik	Fuentemilanos, Spain	19. 3.06
G-BYTI	Piper PA-24-250 Comanche	24-3489	D-ELOP	9.11.99	P.Marsden	Netherthorpe	5. 5.07
			N8297P, N10F		tr G-BYTI Syndicate		
G-BYTJ	Cameron Concept 80 HAB	4703		19.11.99	M.White *(Rapido titles)*	(Cirencester)	5. 8.05A
G-BYTK	Jabiru Aircraft Jabiru UL-450	PFA 274A-13465		8.11.99	P.J.Reilly	(Burgess Hill)	28. 3.05P
	(Built K A Fagan and S.R.Pike)						
G-BYTL	Mainair Blade 912	1224-0999-7 & W1017		19.10.99	M.E.Keefe	St.Michaels	17.10.03P
G-BYTM	Dyn'Aéro MCR-01 Club	PFA 301-13440		1.10.99	I.Lang	Gloucestershire	19.10.05P
	(Built I.Lang - c/n 83) (Rotax 912-UL)						
G-BYTN	de Havilland DH.82A Tiger Moth	3993	7014M	18.11.99	B.D.Hughes	(Rotary Farm, Hatch)	19. 8.07T
			N6720		*(As "N6720/VX")*		
G-BYTR	Raj Hamsa X'Air 582	BMAA/HB/105		5.10.99	R.Dunn	Dunkeswell	30. 8.05P
	(Built A P.Roberts and R.Dunn - kit no.460)						
G-BYTS	Montgomerie-Bensen B.8MR	MGM-2		22. 9.99	M.G.Mee	Shotton Colliery, Peterlee	18. 7.05P
	(Built M G Mee) (Rotax 912)						
G-BYTT	Raj Hamsa X'Air 582	BMAA/HB/100		22. 9.99	J.L.Pearson	Kilrush, County Kildare	29. 7.05P
	(Built R P Reeves - kit no.402)						
G-BYTU	Mainair Blade 912	1225-1099-7 & W1018		26.11.99	K.Roberts	Caernarfon	18. 4.05P
G-BYTV	Jabiru Aircraft Jabiru UL-450	PFA 274A-13454		3.11.99	A.D.Tomlins	Popham	14. 4.05P
	(Built E Bentley)						
G-BYTW	Cameron O-90 HAB	4747		11. 4.00	Sade Balloons Ltd	(London EC2)	19. 8.07A
G-BYTX	Whittaker MW6-S Fatboy Flyer	PFA 164-12819		2.12.99	J.K.Ewing	Mapperton Farm, Newton Peverill, Dorset	17.12.04P
	(Built J.K.Ewing) (Rotax 532)				*(Active 4.04)*		
G-BYTY	Dornier 328-100	3104	D-CDXJ	2.12.99	Suckling Airways (Cambridge) Ltd	London City	8.12.03T
			5N-BRI		t/a Scot Airways		
G-BYTZ	Raj Hamsa X'Air 582	BMAA/HB/120		26.10.99	K.C.Millar	(Dromara, County Down)	1. 7.05P
	(Built A B Wilson and K.C.Millar - kit no.472)						
G-BYUA	Grob G115E Tutor	82086E	D-EUKB	22. 7.99	VT Aerospace Ltd *(Op Cambridge/London UAS)*	RAF Wyton	5. 8.05T
G-BYUB	Grob G115E Tutor	82087E		22. 7.99	VT Aerospace Ltd	RAF St.Athan	5. 8.05T
					(Op University of Wales UAS)		
G-BYUC	Grob G115E Tutor	82088E		22. 7.99	VT Aerospace Ltd *(Op CFS Tutor Sqdn)*	RAF Cranwell	5. 8.05T
G-BYUD	Grob G115E Tutor	82089E		22. 7.99	VT Aerospace Ltd	Glasgow	5. 8.05T
					(Op Universities of Glasgow and Strathclyde AS)		
G-BYUE	Grob G115E Tutor	82090E		12. 8.99	VT Aerospace Ltd *(Op CFS Tutor Sqdn)*	RAF Cranwell	30. 8.05T
G-BYUF	Grob G115E Tutor	82091E		12. 8.99	VT Aerospace Ltd *(Op Cambridge/London UAS)*	RAF Wyton	30. 8.05T
G-BYUG	Grob G115E Tutor	82092E		22. 9.99	VT Aerospace Ltd	Glasgow	27. 9.05T
					(Op Universities of Glasgow and Strathclyde AS)		
G-BYUH	Grob G115E Tutor	82093E		22. 9.99	VT Aerospace Ltd *(Op Bristol UAS)*	RAF Colerne	27. 9.05T
G-BYUI	Grob G115E Tutor	82094E		24. 9.99	VT Aerospace Ltd	RAF Woodvale	27. 9.05T
					(Op Liverpool/Manchester UAS)		
G-BYUJ	Grob G115E Tutor	82095E		24. 9.99	VT Aerospace Ltd *(Op Southampton UAS)*	Boscombe Down	27. 9.05T
G-BYUK	Grob G115E Tutor	82096E		18.10.99	VT Aerospace Ltd *(Op Cambridge/London UAS)*	RAF Wyton	28.10.05T
G-BYUL	Grob G115E Tutor	82097E		18.10.99	VT Aerospace Ltd *(Op Cambridge/London UAS)*	RAF Wyton	28.10.05T
G-BYUM	Grob G115E Tutor	82098E		18.10.99	VT Aerospace Ltd *(Op Southampton UAS)*	Boscombe Down	28.10.05T
G-BYUN	Grob G115E Tutor	82099E		18.10.99	VT Aerospace Ltd	RAF St.Athan	28.10.05T
					(Op University of Wales UAS)		
G-BYUO	Grob G115E Tutor	82100E		19.11.99	VT Aerospace Ltd *(Op Cambridge/London UAS)*	RAF Wyton	28.11.05T
G-BYUP	Grob G115E Tutor	82101E		19.11.99	VT Aerospace Ltd *(Op Oxford UAS)*	RAF Benson	28.11.05T
G-BYUR	Grob G115E Tutor	82102E		19.11.99	VT Aerospace Ltd	RAF Leuchars	28.11.05T
					(Op Aberdeen, Dundee and St.Andrews UAS)		
G-BYUS	Grob G115E Tutor	82103E		19.11.99	VT Aerospace Ltd *(Op Oxford UAS)*	RAF Benson	28.11.05T
G-BYUT	Grob G115E Tutor	82104E		7.12.99	VT Aerospace Ltd	RAF St.Athan	14.12.05T
					(Op University of Wales UAS)		

G-BYUU	Grob G115E Tutor	82105E	7.12.99	VT Aerospace Ltd	RAF Leuchars	14.12.05T
				(Op Aberdeen, Dundee and St.Andrews UAS)		
G-BYUV	Grob G115E Tutor	82106E	7.12.99	VT Aerospace Ltd *(Op Oxford UAS)*	RAF Benson	14.12.05T
G-BYUW	Grob G115E Tutor	82107E	7.12.99	VT Aerospace Ltd	RAF Leuchars	14.12.05T
				(Op Aberdeen, Dundee and St.Andrews UAS)		
G-BYUX	Grob G115E Tutor	82108E	18. 1.00	VT Aerospace Ltd	RAF Woodvale	16. 12.05T
				(Op Liverpool/Manchester UAS)		
G-BYUY	Grob G115E Tutor	82109E	18. 1.00	VT Aerospace Ltd	RAF Leuchars	31. 1.06T
				(Op Aberdeen, Dundee and St.Andrews UAS)		
G-BYUZ	Grob G115E Tutor	82110E	18. 1.00	VT Aerospace Ltd	RAF Woodvale	31. 1.06T
				(Op Liverpool/Manchester UAS)		
G-BYVA	Grob G115E Tutor	82111E	18. 1.00	VT Aerospace Ltd *(Op CFS Tutor Sqdn)*	RAF Cranwell	31. 1.06T
G-BYVB	Grob G115E Tutor	82112E	17. 2.00	VT Aerospace Ltd	RAF Leuchars	28. 2.06T
				(Op Aberdeen, Dundee and St.Andrews UAS)		
G-BYVC	Grob G115E Tutor	82113E	17. 2.00	VT Aerospace Ltd *(Op Bristol UAS)*	RAF Colerne	2. 3.06T
G-BYVD	Grob G115E Tutor	82114E	17. 2.00	VT Aerospace Ltd *(Op Cambridge/London UAS)*	RAF Wyton	2. 3.06T
G-BYVE	Grob G115E Tutor	82115E	17. 2.00	VT Aerospace Ltd *(Op Southampton UAS)*	Boscombe Down	2. 4.06T
G-BYVF	Grob G115E Tutor	82116E	22. 2.00	VT Aerospace Ltd	Plymouth	28. 2.06T
G-BYVG	Grob G115E Tutor	82117E	22. 3.00	VT Aerospace Ltd *(Op Yorkshire UAS)*	RAF Church Fenton	2. 4.06T
G-BYVH	Grob G115E Tutor	82118E	22. 3.00	VT Aerospace Ltd	RAF Leuchars	4. 4.06T
				(Op Aberdeen, Dundee and St.Andrews UAS)		
G-BYVI	Grob G115E Tutor	82119E	22. 3.00	VT Aerospace Ltd	Glasgow	4. 4.06T
				(Op Universities of Glasgow and Strathclyde AS)		
G-BYVJ	Grob G115E Tutor	82120E	14. 4.00	VT Aerospace Ltd *(Op Birmingham UAS)*	RAF Cosford	25. 4.06T
G-BYVK	Grob G115E Tutor	82121E	14. 4.00	VT Aerospace Ltd *(Op Southampton UAS)*	Boscombe Down	25. 4.06T
G-BYVL	Grob G115E Tutor	82122E	14. 4.00	VT Aerospace Ltd *(Op Oxford UAS)*	RAF Bensen	26. 4.03T
G-BYVM	Grob G115E Tutor	82123E	14. 4.00	VT Aerospace Ltd	Glasgow	26. 4.06T
				(Op Universities of Glasgow and Strathclyde AS)		
G-BYVN	Grob G115E Tutor	82124E	18. 5.00	VT Aerospace Ltd *(Op Bristol UAS)*	RAF Colerne	31. 5.06T
G-BYVO	Grob G115E Tutor	82125E	18. 5.00	VT Aerospace Ltd *(Op Birmingham UAS)*	RAF Cosford	31. 5.06T
G-BYVP	Grob G115E Tutor	82126E	18. 5.00	VT Aerospace Ltd *(Op Oxford UAS)*	RAF Benson	31. 5.06T
G-BYVR	Grob G115E Tutor	82127E	18. 5.00	VT Aerospace Ltd *(Op CFS Tutor Sqdn)*	RAF Cranwell	31. 5.06T
G-BYVS	Grob G115E Tutor	82128E	20. 6.00	VT Aerospace Ltd *(Op Cambridge/London UAS)*	RAF Wyton	29. 6.06T
G-BYVT	Grob G115E Tutor	82129E	20. 6.00	VT Aerospace Ltd *(Op Cambridge/London UAS)*	RAF Wyton	29. 6.06T
G-BYVU	Grob G115E Tutor	82130E	20. 6.00	VT Aerospace Ltd *(Op Oxford UAS)*	RAF Benson	29. 6.06T
G-BYVV	Grob G115E Tutor	82131E	20. 6.00	VT Aerospace Ltd *(Op Yorkshire UAS)*	RAF Church Fenton	29. 6.06T
G-BYVW	Grob G115E Tutor	82132E	21. 7.00	VT Aerospace Ltd	RAF St.Athan	6. 8.06T
				(Op University of Wales UAS)		
G-BYVX	Grob G115E Tutor	82133E	21. 7.00	VT Aerospace Ltd *(Op Yorkshire UAS)*	RAF Church Fenton	6. 8.06T
G-BYVY	Grob G115E Tutor	82134E	21. 7.00	VT Aerospace Ltd *(Op Yorkshire UAS)*	RAF Church Fenton	6. 8.06T
G-BYVZ	Grob G115E Tutor	82135E	21. 7.00	VT Aerospace Ltd *(Op Yorkshire UAS)*	RAF Church Fenton	6. 8.06T
G-BYWA	Grob G115E Tutor	82136E	21. 8.00	VT Aerospace Ltd *(Op Oxford UAS)*	RAF Bensen	30. 8.06T
G-BYWB	Grob G115E Tutor	82137E	21. 8.00	VT Aerospace Ltd *(Op Bristol UAS)*	RAF Colerne	30. 8.06T
G-BYWC	Grob G115E Tutor	82138E	18. 9.00	VT Aerospace Ltd *(Op Bristol UAS)*	RAF Colerne	27. 9.06T
G-BYWD	Grob G115E Tutor	82139E	18. 9.00	VT Aerospace Ltd	RAF Woodvale	27. 9.06T
				(Op Liverpool/Manchester UAS)		
G-BYWE	Grob G115E Tutor	82140E	18. 9.00	VT Aerospace Ltd *(Op Bristol UAS)*	RAF Colerne	27. 9.06T
G-BYWF	Grob G115E Tutor	82141E	18. 9.00	VT Aerospace Ltd *(Op CFS Tutor Sqdn)*	RAF Cranwell	27. 9.06T
G-BYWG	Grob G115E Tutor	82142E	13.10.00	VT Aerospace Ltd *(Op Bristol UAS)*	RAF Colerne	29.10.06T
G-BYWH	Grob G115E Tutor	82143E	13.10.00	VT Aerospace Ltd *(Op Northumbrian UAS)*	RAF Leeming	29.10.06T
G-BYWI	Grob G115E Tutor	82144E	13.10.00	VT Aerospace Ltd *(Op Bristol UAS)*	RAF Colerne	29.10.06T
G-BYWJ	Grob G115E Tutor	82145E	13.10.00	VT Aerospace Ltd	RAF Woodvale	29.10.06T
				(Op Liverpool/Manchester UAS)		
G-BYWK	Grob G115E Tutor	82146E	17.11.00	VT Aerospace Ltd *(Op CFS Tutor Sqdn)*	RAF Cranwell	28.11.06T
G-BYWL	Grob G115E Tutor	82147E	17.11.00	VT Aerospace Ltd	RAF Woodvale	28.11.06T
				(Op Liverpool/Manchester UAS)		
G-BYWM	Grob G115E Tutor	82148E	17.11.00	VT Aerospace Ltd *(Op727 Sqdn)*	Plymouth	28.11.06T
G-BYWN	Grob G115E Tutor	82149E	17.11.00	VT Aerospace Ltd	RAF Woodvale	28.11.06T
				(Op Liverpool/Manchester UAS)		
G-BYWO	Grob G115E Tutor	82150E	7.12.00	VT Aerospace Ltd *(Op Birmingham UAS)*	RAF Cosford	14. 1.07T
G-BYWP	Grob G115E Tutor	82151E	7.12.00	VT Aerospace Ltd *(Op Yorkshire UAS)*	RAF Church Fenton	14. 1.07T
G-BYWR	Grob G115E Tutor	82152E	18. 5.00	VT Aerospace Ltd *(Op Cambridge/London UAS)*	RAF Wyton	28. 1.07T
G-BYWS	Grob G115E Tutor	82153E	7.12.00	VT Aerospace Ltd *(Op Northumbrian UAS)*	RAF Leeming	16. 1.07T
G-BYWT	Grob G115E Tutor	82154E	12.12.00	VT Aerospace Ltd *(Op Northumbrian UAS)*	RAF Leeming	16. 1.07T
G-BYWU	Grob G115E Tutor	82155E	19. 1.01	VT Aerospace Ltd *(Op Cambridge/London UAS)*	RAF Wyton	28. 1.07T
G-BYWV	Grob G115E Tutor	82156E	19. 1.01	VT Aerospace Ltd *(Op Birmingham UAS)*	RAF Cosford	28. 1.07T
G-BYWW	Grob G115E Tutor	82157E	19. 1.01	VT Aerospace Ltd *(Op CFS Tutor Sqdn)*	RAF Cranwell	28. 1.07T
G-BYWX	Grob G115E Tutor	82158E	14. 2.01	VT Aerospace Ltd *(Op Cambridge/London UAS)*	RAF Wyton	25. 2.07T
G-BYWY	Grob G115E Tutor	82159E	14. 2.01	VT Aerospace Ltd *(Op CFS Tutor Sqdn)*	RAF Cranwell	25. 2.07T
G-BYWZ	Grob G115E Tutor	82160E	14. 2.01	VT Aerospace Ltd *(Op CFS Tutor Sqdn)*	RAF Cranwell	4. 3.07T
G-BYXA	Grob G115E Tutor	82161E	14. 2.01	VT Aerospace Ltd	RAF Woodvale	4. 3.07T
				(Op Liverpool/Manchester UAS)		
G-BYXB	Grob G115E Tutor	82162E	19. 3.01	VT Aerospace Ltd *(Op Southampton UAS)*	Boscombe Down	1. 4.07T
G-BYXC	Grob G115E Tutor	82163E	19. 3.01	VT Aerospace Ltd *(Op CFS Tutor Sqdn)*	RAF Cranwell	1. 4.07T
G-BYXD	Grob G115E Tutor	82164E	19. 3.01	VT Aerospace Ltd *(Op CFS Tutor Sqdn)*	RAF Cranwell	1. 4.07T
G-BYXE	Grob G115E Tutor	82165E	19. 3.01	VT Aerospace Ltd *(Op Yorkshire UAS)*	RAF Church Fenton	1. 4.07T
G-BYXF	Grob G115E Tutor	82166E	12. 4.01	VT Aerospace Ltd *(Op Birmingham UAS)*	RAF Cosford	25. 4.07T
G-BYXG	Grob G115E Tutor	82167E	12. 4.01	VT Aerospace Ltd *(Op Birmingham UAS)*	RAF Cosford	25. 4.07T
G-BYXH	Grob G115E Tutor	82168E	12. 4.01	VT Aerospace Ltd *(Op Cambridge/London UAS)*	RAF Wyton	29. 4.07T
G-BYXI	Grob G115E Tutor	82169E	12. 4.01	VT Aerospace Ltd	RAF Woodvale	29. 4.07T
				(Op Liverpool/Manchester UAS)		
G-BYXJ	Grob G115E Tutor	82170E	16. 5.01	VT Aerospace Ltd *(Op Southampton UAS)*	Boscombe Down	28. 5.07T
G-BYXK	Grob G115E Tutor	82171E	16. 5.01	VT Aerospace Ltd	Plymouth	28. 5.07T
G-BYXL	Grob G115E Tutor	82172E	16. 5.01	VT Aerospace Ltd *(Op Birmingham UAS)*	RAF Cosford	28. 5.07T

Regn	Type	C/n	Previous identity	Regd	Owner/Operator	Base	CofA
G-BYXM	Grob G115E Tutor	82173E		16. 5.01	VT Aerospace Ltd *(Op Southampton UAS)*	Boscombe Down	28. 5.07T
G-BYXN	Grob G115E Tutor	82174E		8. 6.01	VT Aerospace Ltd *(Op CFS Tutor Sqdn)*	RAF Cranwell	11. 6.07T
G-BYXO	Grob G115E Tutor	82175E		8. 6.01	VT Aerospace Ltd *(Op Birmingham UAS)*	RAF Cosford	11. 6.04T
G-BYXP	Grob G115E Tutor	82176E		8. 6.01	VT Aerospace Ltd	RAF Wyton	11. 6.07T
					(Op Cambridge/London UAS)		
G-BYXR	Grob G115E Tutor	82177E		8. 6.01	VT Aerospace Ltd *(Op Oxford UAS)*	RAF Benson	11. 6.07T
G-BYXS	Grob G115E Tutor	82178E		18. 7.01	VT Aerospace Ltd *(Op Oxford UAS)*	RAF Benson	29. 7.07T
G-BYXT	Grob G115E Tutor	82179E		18. 7.01	VT Aerospace Ltd *(Op Southampton UAS)*	Boscombe Down	1.11.07E
G-BYXU	Piper PA-28-161 Cherokee Warrior II	28-7716097	EI-BXU	8. 1.99	F.P.McGovern and F.O'Sullivan		
			G-BNUP, N2282Q			Waterford, County Waterford	22. 5.05
G-BYXV	Medway EclipseR	162/140		25.10.99	K.A Christie	Easter Balgillo Farm, Finavon	10. 5.04P
		(C/n not confirmed)					
G-BYXW	Medway EclipseR	166/147		25.10.99	G A Hazell	Redlands	25. 7.05P
	(Officially registered as c/n 166/144)				*(Noted 5.04)*		
G-BYXX	Grob G115E Tutor	82180E		18. 7.01	VT Aerospace Ltd	RAF Woodvale	1.11.07E
					(Op Liverpool/Manchester UAS)		
G-BYXY	Grob G115E Tutor	82181E		18. 7.01	VT Aerospace Ltd *(Op Northumbrian UAS)*	RAF Leeming	30.9.07E
G-BYXZ	Grob G115E Tutor	82182E		15. 8.01	VT Aerospace Ltd *(Op CFS Tutor Sqdn)*	RAF Cranwell	17. 9.07T
G-BYYA	Grob G115E Tutor	82183E		15. 8.01	VT Aerospace Ltd *(Op Northumbrian UAS)*	RAF Leeming	23. 9.07T
G-BYYB	Grob G115E Tutor	82184E		15. 8.01	VT Aerospace Ltd *(Op CFS Tutor Sqdn)*	RAF Cranwell	17. 9.07T
G-BYYC	Hapi Cygnet SF-2A	PFA 182-12311		25.11.99	G.H.Smith	Shenstone Hall Farm, Shenstone	9. 6.05P
	(Built C.D.Hughes and G.H.Smith) (Volkswagen 2180)						
G-BYYD	Cameron A-250 HAB	4712		31. 3.00	C. and J.M.Bailey	(Bristol)	16. 4.05T
G-BYYE	Lindstrand LBL 77A HAB	151		25.11.99	D.J.Cook	(Norwich)	25. 6.05A
G-BYYG	Slingsby T.67C Firefly	2101	PH-SGI	30.11.99	B Dixon and S E Marples	Newcastle	5. 2.06T
G-BYYI	British Aerospace Jetstream Series 3107	620	VH-JSW	1. 3.00	Jetstream Executive Travel Ltd	(Stafford)	
			G-31-620		*(New owner 6.04)*		
G-BYYJ	Lindstrand LBL 25A Cloudhopper HAB	651		10.12.99	A M.Barton	(Coulsdon)	17. 8.05A
G-BYYK	Boeing 737-229C	20916	OO-SDK	11. 1.00	European Aviation Air Charter Ltd	Bournemouth	AC
					(Stored as "OO-SDK" 12.04)		
G-BYYL	Jabiru Aircraft Jabiru UL-450	PFA 274A-13480		10.12.99	K.C.Lye	(Steeple Ashton, Trowbridge)	10. 6.04P
	(Built C.Jackson)						
G-BYYM	Raj Hamsa X'Air 582	BMAA/HB/119		21.10.99	D.J.McCall and B.Pilling	Dunkeswell	1. 8.03P
	(Built J J Cozens - kit no.476)						
G-BYYN	Pegasus Quantum 15-912	7601		6. 1.00	S.E.Robinson and A.Dixon	Tarn Farm, Cockerham	14. 6.05P
G-BYYO	Piper PA-28R-201 Arrow II	2837061	(N182ND)	11. 2.00	Stapleford Flying Club Ltd	Stapleford	26. 4.06T
			N9249C, G-BYYO, N9249C				
G-BYYP	Pegasus Quantum 15	7603		11. 2.00	D.A Linsey-Bloom	Kingston Seymour	14. 3.05P
	(Rotax 582)						
G-BYYR	Raj Hamsa X'Air 582	BMAA/HB/115		23.12.99	T.D.Bawden	Weston Zoyland	11. 5.04P
	(Built T D Bawden - kit no.453)						
G-BYYT	Jabiru Aircraft Jabiru UL-450	PFA 274A-13452		18.11.99	S.C.E.Twiss	Newton Peverill	19. 5.05P
	(Built T.D.Saveker)						
G-BYYW*	de Havilland DHC-1 Chipmunk Mk.20	57	CS-DAR	22 .2.00	R.Farrer	(Bedford)	
	(Built OGMA)		Portuguese AF FAP 1367		*(Stored 2000: cancelled 15.5.03 by CAA)*		
G-BYYX	TEAM mini-MAX 91	PFA 186-13410		6. 1.00	J.Batchelor	(Benfleet)	14. 4.05P
	(Built P L Turner)						
G-BYYY	Pegasus Quantum 15-912	7564		8.12.99	Sarah L.Smith t/a Redlands Airfield	Redlands	24. 1.05P
G-BYYZ	Staaken Z-21A Flitzer	PFA 223-13324		12.11.99	P.K.Jenkins	(Headcorn)	1. 7.05P
	(Built A E.Morris) (Volkswagen 1834)						
G-BYZA	Aérospatiale AS355F2 Ecureuil 2	5518	JA6784	20.12.99	MMAir Ltd	Stapleford	17. 4.06T
			F-OHNK				
G-BYZD	Tri-R Kis Cruiser	PFA 302-13156		22.11.99	R.T.Clegg	Netherthorpe	26. 5.05P
	(Built R.T.Clegg) (Lycoming IO-360)						
G-BYZE	Aérospatiale AS350B2 Ecureuil	2773	F-OGVR	8. 2.00	T.Clark	Alderney	14. 3.06T
			(F-GJIP)				
G-BYZF	Raj Hamsa X'Air Victor1	BMAA/HB/110		7. 1.00	R.P...Davies	Baxby Manor, Husthwaite	
	(Built S W Grainger - kit no.461)				*(Noted 8.04)*		
G-BYZG	Cameron A-275 HAB	4706		23. 2.00	Horizon Ballooning Ltd *(Horizon titles)*	(Alton)	13. 4.05T
G-BYZJ	Boeing 737-3Q8	24962	G-COLE	11. 1.00	British Midland Airways Ltd	Nottingham East Midlands	20.11.07T
			PP-VOX		*(Op bmiBaby)*		
G-BYZL	Cameron GP-65 HAB	4494		6. 4.00	P.Thibo	(Junglinster, Luxembourg)	27. 7.04A
G-BYZM	Piper PA-28-161 Warrior II	28-8116317	HB-PNK	4. 2.00	Goodair Leasing Ltd	Cardiff	30. 6.06T
			N8436A				
G-BYZO	Rans S-6-ES Coyote II	PFA 204-13560		14. 1.00	B.E.J.Badger and J.E.Storer	Long Marston	2. 6.05P
	(Built S C Jackson) (Rotax 582 - *kit no.1298.1287) (Tri-cycle u/c)*						
G-BYZP	Robinson R22 Beta II	3018		9.12.99	Wallis and Son Ltd	Cambridge	8. 1.06T
G-BYZR	III Sky Arrow 650 TC	C001	D-ENGF	24. 1.00	G.H.Hackson and R.Moncrieff	Gamston	12. 5.06P
	(Built Iniziative Industriali Italian)		I-TREI		tr G-BYZR Flying Group		
G-BYZS	Jabiru Aircraft Jabiru UL-450	PFA 274A-13489		25. 1.00	N.Fielding	Ince Blundell	23. 6.05P
	(Built N.Fielding)						
G-BYZT	Nova Vertex 26	13345		21. 1.00	M Hay *(New owner 1.03)*	(Dundee)	
G-BYZU	Pegasus Quantum 15	7613		15. 2.00	N I Clifton	East Fortune	25. 2.05P
	(Rotax 582)						
G-BYZV	Sky 90-24 HAB	174		15. 8.00	P.Farmer	(Wadhurst)	19. 1.01
G-BYZW	Raj Hamsa X'Air 582	BMAA/HB/129		19. 1.00	H.C.Lowther	Kirkbride	24. 6.05P
	(Built P A Gilford - kit no.499)						
G-BYZX	Cameron R-90 HAB	4751		31. 3.00	D.K.Hempleman-Adams	(Corsham)	22. 3.05A
	(Rebuilt with envelope c/n 10369)				*"Britannic Challenge"*		
G-BYZY	Pietenpol AirCamper	PFA 47-12190		2.12.99	D.N.Hanchet	White Waltham	AC
	(Built D.N.Hanchet)						
G-BYZZ	Robinson R22 Beta II	3000		1.12.99	Astra Helicopters Ltd	Kemble	15. 5.06T

G-BZAA - G-BZZZ

Registration	Type	C/n	Previous identity	Date	Owner	Location	CofA
G-BZAA	Mainair Blade 912	1142-0198-7 & W945		22.11.99	J.A.Kentzer	Headon Farm, Retford	23. 2.05P
	(Rotax 462) *(Trike c/n amended to 1142-1299-7 by Mainair Sports Ltd)*						
G-BZAB	Mainair Rapier	1228-1299-7 & W1021		23.12.99	R.H.Stockton	(Christleton, Chester)	9. 1.05P
G-BZAD	Cessna 152	15279563	N303MA	22. 3.00	Cristal Air Ltd	Lydd	18.11.07E
			N714ZN				
G-BZAE	Cessna 152	15281300	N49480	22. 3.00	Horizon Aviation Ltd	Swansea	9 11.06T
G-BZAF	Raj Hamsa X'Air 582	BMAA/HB/130		18. 1.00	Y.A Evans	Rufforth	26.10.05P
	(Built P Hassett - kit no.503)						
G-BZAG	Lindstrand LBL 105A HAB	542		29..2.00	A M.Figiel	(High Wycombe)	18. 8.05A
G-BZAH	Cessna 208B Grand Caravan	208B0811	N5196U	28. 2.00	R.Durie	AAC Netheravon	13. 4.05A
					tr Army Parachute Association		
G-BZAI	Pegasus Quantum 15	7614		9. 2.00	D.Paget	Dunkeswell	4. 4.05P
	(Rotax 503)						
G-BZAJ	PZL-110 Koliber 160A	04990082	SP-WGK	10. 2.00	PZL International Aviation Marketing and Sales plc		
						North Weald	29. 7.06T
G-BZAK	Raj Hamsa X'Air 582	BMAA/HB/114		20. 1.00	R.J.Ripley	Field Farm, Oakley	1. 4.05P
	(Built B W Austen - kit no.477)						
G-BZAL	Mainair Blade 912	1205-0799-7 & W1008		27. 1.00	K.Worthington	Tarn Farm, Cockerham	25. 5.05P
G-BZAM	Europa Aviation Europa	PFA 247-12969		6.12.99	D.Corbett	Shobdon	24. 5.05P
	(Built D.U Corbett - kit no.265) (Monowheel u/c)						
G-BZAO	Rans S-12XL Airaile	PFA 307-13394		1. 2.00	M.L.Robinson	Kirkbride	AC
	(Built M L Robinson) (Rotax 582) *(Lost power on take-off Kirkbride 13.12.01, force landed: u/c legs detached, fuselage distorted: on rebuild with new fuselage 4.03)*						
G-BZAP	Jabiru Aircraft Jabiru UL-450	PFA 274A-13479		13.12.99	S.Derwin	Morgansfield, Fishburn	30. 6.05P
	(Built S.Derwin)						
G-BZAR	Denney Kitfox Model 4-1200 Speedster		G-LEZJ	17. 2.00	C.E.Brookes	Little Battleflats Farm, Ellistown, Coalville	30. 8.03P
	(Build L A James) (Rotax 912-UL)	PFA 172B-12529			*"Ol Red"*		
G-BZAS	Isaacs Fury II	PFA 11-10837		10. 2.00	H.A Brunt and H.Frick	Bournemouth	4. 7.04P
	(Built H.A Brunt and H.Frick) (Canadian Air Motive CAM100)				*(As "K5673") "Spirit of Dunsfold"*		
G-BZAT	British Aerospace Avro 146-RJ100	E3320	G-6-320	18.11.97	British Airways Citiexpress Ltd	Birmingham	8. 1.07T
G-BZAU	British Aerospace Avro 146-RJ100	E3328		25. 4.98	British Airways Citiexpress Ltd	Birmingham	11. 6.07T
G-BZAV	British Aerospace Avro 146-RJ100	E3331		19. 5.98	British Airways Citiexpress Ltd	Birmingham	23. 7.07T
G-BZAW	British Aerospace Avro 146-RJ100	E3354		11. 6.99	British Airways Citiexpress Ltd	Manchester	15. 7.05T
G-BZAX	British Aerospace Avro 146-RJ100	E3356	G-6-356	9. 7.99	British Airways Citiexpress Ltd	Manchester	16. 8.05T
G-BZAY	British Aerospace Avro 146-RJ100	E3368		15. 2.00	British Airways Citiexpress Ltd	Manchester	27. 3.06T
G-BZAZ	British Aerospace Avro 146-RJ100	E3369		15. 2.00	British Airways Citiexpress Ltd	Manchester	13. 4.06T
G-BZBC	Rans S-6-ES Coyote II	PFA 204-13525		2. 2.00	A J.Baldwin	Egginton	7.12.05P
	(Built A J.Baldwin - kit no.0499.1314) (Rotax 582) *(Tri-cycle u/c)*						
G-BZBD*	Westland Scout AH.1	F.9638	XT632	7. 3.00	(Kennet Aviation)	North Weald	10. 7.01P
					(Cancelled 5.10.00 as PWFU) (For spares or restoration 1.05)		
G-BZBE	Cameron A-210 HAB	4708		9. 5.00	Dragon Balloon Company Ltd	(Castleton, Hope Valley)	27. 3.05T
G-BZBF	Cessna 172M	17262258	N126SA	20.12.99	L.W.Scattergood	Breighton	7. 9.06T
	(Lycoming O-360)		G-BZBF, 9H-ACV, N12785				
G-BZBH	Thunder Ax7-65 Bolt HAB	173		28.11.78	R.B. and G.Craik *"Serendipity II"*	(Northampton)	16. 4.05A
G-BZBI	Cameron V-77 HAB	4740		4. 4.00	C.and A I.Gibson *"Flying Colours"*	(Stockport)	30. 5.05A
G-BZBJ	Lindstrand LBL 77A HAB	646		29. 2.00	P.T.R.Ollivere	(Eastbourne)	30. 6.05A
G-BZBL	Lindstrand LBL 120A HAB	676		23. 2.00	C.J.Dunkley	(Aylesbury)	23. 6.05A
G-BZBM	Cameron A-315 HAB	4741		7. 4.00	Listers of Coventry (Motors) Ltd	(Alcester)	18. 3.04T
G-BZBN	Thunder AX9-120 Series 2	4786		21. 2.00	K.Willie	(Maldegem, Belgium)	25. 9.03A
G-BZBO	Stoddard-Hamilton Glasair III	3032		21. 2.00	M.B.Hamlett	Lagny le Sec, France	
	(Built M B Hamlett)						
G-BZBP	Raj Hamsa X'Air 582	BMAA/HB/131		29. 2.00	R.J.Howlett	Middle Stoke, Isle of Grain	21. 7.05P
	(Built D F Hughes - kit no.470)						
G-BZBR	Pegasus Quantum 15	7631		26. 5.00	E.Lewis	Weston Zoyland	28. 9.05P
	(Rotax 503)						
G-BZBS	Piper PA-28-161 Warrior III	2842080	N4180H	10. 5.00	S.J.Skilton	White Waltham	18. 5.06T
			G-BZBS, N9529N		t/a Aviation Rentals		
G-BZBT	Cameron Hopper H-34 HAB	4730		18. 5.00	British Telecommunications plc	(Thatcham)	9. 4.05A
G-BZBU	Robinson R22	0131	OH-HLB	23. 5.00	Helicentre Blackpool Ltd	Blackpool	26. 4.07T
			SE-HOH				
G-BZBW	RotorWay Executive 162F	6415		23. 2.00	Southern Helicopters Ltd	Street Farm, Takeley	AC
	(Built M Gardiner) (RotorWay RI 162F)				*(Noted unmarked 1.05)*		
G-BZBX	Rans S-6-ES Coyote II	PFA 204-13501		26. 1.00	R.Johnstone	Otherton, Cannock	14.11.05P
	(Built R.Johnstone) (Rotax 582) *(Tri-cycle u/c)*						
G-BZBZ	Jodel D.9 Bébé	519	OO-48	29. 2.00	S.Marom	Whitehall Farm, Benington	23. 3.05P
	(Built Etienne de Schrevel, Gent 1970-77) (Volkswagen 1600)						
G-BZDA	Piper PA-28-161 Warrior III	2842087	N41814	29. 6.00	S.J.Skilton	White Waltham	3. 7.06T
			G-BZDA, N41814		t/a Aviation Rentals		
G-BZDB	Thruster T.600T 450 Sprint	0030-T600T-041		7. 3.00	R.M.Raikes	(Bridgend)	19. 5.05P
	(Rotax 582) *(Jabiru to be installed 2005)*						
G-BZDC	Mainair Blade	1232-0100-7 & W1025		13. 3.00	E.J.Wells and P.J.Smith	Over Farm, Gloucester	26. 3.05P
	(Rotax 462)						
G-BZDD	Mainair Blade 912	1238-0200-7 & W1031		21. 1.00	A.S.Facey tr Barton Blade Group	Barton	13. 4.05P
G-BZDE	Lindstrand LBL 210A HAB	665		6. 3.00	Toucan Travel Ltd *(Toucan Travel titles)*	(Basingstoke)	26.10.05T
G-BZDH	Piper PA-28R-200 Cherokee Arrow II	28R-7235028	5B-CJU	8. 3.00	I.R.Chaplin	Andrewsfield	10. 2.07T
			G-BZDH, HB-OHH, N4390T				
G-BZDI	Aero L-39C Albatros	031822	ES-ZLB	7. 6.00	M.Gainza and E.Gavazzi	North Weald	1. 7.03P
			Soviet AF		*(Noted 1.05)*		
G-BZDJ	Cameron Z-105 HAB	4832		27. 6.00	BWS Security Systems Ltd	(Bath)	5. 3.05A
	(Carries incorrect ident tab as "N-105")				*(BWS Security Systems titles)*		
G-BZDK	Raj Hamsa X'Air 582	BMAA/HB/124		8. 2.00	B.Park	Truro	27.11.04P
	(Built B.Park and R Barnes - kit no.447)						

Reg	Type	C/n	Previous identity	Regd	Owner/operator	Base	CofA
G-BZDL	Pegasus Quantum 15-912	7629		18. 4.00	D.M.Merritt-Colman	(Benicolet, Spain)	12. 6.05P
G-BZDM	Stoddard-Hamilton GlaStar	PFA 295-13283		13. 3.00	F.G.Miskelly	(London SW6)	23. 7.05P
	(Built F.G.Miskelly)						
G-BZDN	Cameron N-105 HAB	2840	D-OABB	26. 4.00	J.D.and K.Griffiths	(Bingham)	13. 6.05T
			D-Saxonia (2)				
G-BZDP	Scottish Aviation Bulldog Series 120/121	BH120/244	XX551	31. 3.00	D.J.Rae *(As "XX551/E")*	RAF Colerne	8. 7.07
G-BZDR	Tri-R Kis	9403		8. 3.00	T.J.Johnson	Old Buckenham	26. 8.04P
G-BZDS	Pegasus Quantum 15-912	7633		17. 4.00	P.K.Dale	Bagby	21. 4.05P
G-BZDT	Maule MXT-7-180 Star Rocket	14099C		11. 8.00	Strongcrew Ltd	Exeter	2. 8.06
G-BZDU	de Havilland DHC-1 Chipmunk 22	C1/0714	WP833	31. 3.00	M.R.Clark *(As "WP833" in RAF c/s)*	Newcastle	6. 7.06
G-BZDV	Westland SA.314C Gazelle HT.2	1150	3D-HXL	31. 3.00	MW Helicopters Ltd	Stapleford	AC
			G-BZDV, XW884		*(Stored 10.04)*		
G-BZDX	Cameron Colt Sugarbox 90 SS HAB	4814		17. 5.00	Stratos Ballooning GmbH and Co KG		
						(Ennigerloh, Germany)	6. 5.04A
G-BZDY	Cameron Colt Sugarbox 90 SS HAB	4815		22. 5.00	Stratos Ballooning GmbH and Co KG		
						(Ennigerloh, Germany)	7. 4.04A
G-BZDZ	Jabiru Aircraft Jabiru SP-430	232	ZU-BVB	14. 5.01	R.M.Whiteside	Shoreham	27. 8.04P
	(Built R.M.Whiteside)						
G-BZEA	Cessna A152	A1520824	N7606L	13. 3.00	Sky Leisure Aviation (Charters) Ltd	Blackbushe	19. 2.07T
G-BZEB	Cessna 152	15282772	N89532	31. 1.00	Sky Leisure Aviation (Charters) Ltd	Blackbushe	24. 9.03T
					(Noted 8.04)		
G-BZEC	Cessna 152	15284475	N4655M	21. 1.00	Sky Leisure Aviation (Charters) Ltd	Redhill	14. 8.05T
G-BZED	Pegasus Quantum 15-912	7600		17. 3.00	M.P.Wimsey	Strubby	9. 5.05P
G-BZEE	Agusta-Bell 206B JetRanger II	8554	G-OJCB	22 .2.00	Yateley Helicopters Ltd	Blackbushe	17.11.06T
G-BZEG	Mainair Blade	1239-0200-7 & W1032		3. 3.00	Mainair Microlight School Ltd	Ince Blundell	24. 3.05P
	(Rotax 912-UL)						
G-BZEH	Piper PA-28-235 Cherokee B	28-10838	9M-ARW	31. 3.00	A.D.Wood	Spanhoe	3. 5.07
			RP-C704, PI-C704, N9182W				
G-BZEJ	Raj Hamsa X'Air 582	BMAA/HB/134		31. 3.00	P.J.Perry	Otherton, Cannock	12.10.05P
	(Built H Hall - kit no.500)				tr X'Air Flying Group		
G-BZEK	Cameron C-70 HAB	4860		30. 5.00	Ballooning 50 Degrees Nord	(Fouhren, Luxembourg)	6. 7.05A
G-BZEL	Mainair Blade	1245-0300-7 & W1038		27. 3.00	M.W.Bush	Belle Vue Farm, Yarnscombe	17. 4.05P
	(Rotax 582)						
G-BZEM	Glaser-Dirks DG-800B	8-194-B116	BGA 4887	8. 5.00	I.M.Stromberg	Sutton Bank	19.12.05
			G-BZEM				
G-BZEN	Jabiru Aircraft Jabiru UL	PFA 274-13272		4. 4.00	B.W.Stockil	Bagby	21.10.05P
	(Built B.W.Stockil)						
G-BZEP	Scottish Aviation Bulldog Series 120/121	BH120/257	XX561	4. 4.00	A.J.Amato *(As "XX561/7")*	Biggin Hill	25. 5.05T
G-BZER	Raj Hamsa X'Air BMW R100	BMAA/HB/133		22. 3.00	N.P.Lloyd and H.Lloyd-Hughes	Emlyn's Field, Rhuallt	29. 9.05P
	(Built N.P.Lloyd and H.Lloyd-Hughes - kit no.526)						
G-BZES	RotorWay Executive 90	6191	G-LUFF	25. 4.00	Southern Helicopters Ltd	Street Farm, Takeley	
	(Built D C Luffingham)				*(Noted 2.03)*		
G-BZET	Robin HR200/120B	345	F-GTZG	9. 5.00	Modi Aviation Ltd	Earls Colne	22. 5.06T
G-BZEU	Raj Hamsa X'Air 582	BMAA/HB/140		20. 4.00	M.Bundy	Lower Mountpleasant, Chatteris	10. 4.05P
	(Built J C Harris - kit no.518)						
G-BZEV	Vahdat Semicopter 1 Gyroplane	002		10.10.00	M.E.Vahdat	(Uxbridge)	
	(Built M.E.Vahdat)				*(Current status unknown)*		
G-BZEW	Rans S-6-ES Coyote II	PFA 204-13450		5. 4.00	J.E.Gattrell and A.R.Trace	Sittles Farm, Alrewas	13. 9.05P
	(Built D Kingslake - kit no 0998.1268.0199.ES) (Rotax 582) *(Tri-cycle u/c)*						
G-BZEX	Raj Hamsa X'Air BMW R100	BMAA/HB/135		5. 4.00	J.M.McCullough and R.T.Henry		
	(Built J.M.McCullough and R.T.Henry - kit no.530)					Slieve Croob, County Down	29. 7.05P
G-BZEY	Cameron N-90 HAB	4829		15. 5.00	Northants Auto Parts and Service Ltd	(Northampton)	27. 6.05T
					(*Wrangler Footwear titles)*		
G-BZEZ	CFM Streak Shadow SA	PFA 206-13503		1 .2.00	G.J.Pearce	Jackrells Farm, Southwater	13. 5.05P
	(Built M F Cottam - kit no K.332) (Rotax 582)						
G-BZFB	Robin R2112A Alpha	175	EI-BIU	7. 4.00	M.R.Brown	Bidford	6. 7.06
G-BZFC	Pegasus Quantum 15	7640		14. 4.00	G.Addison	East Fortune	6 7.05P
G-BZFD	Cameron N-90 HAB	2725	OO-BFD	24. 5.00	David Hataway Holdings Ltd *(Active 8.04)*	(Bristol)	
G-BZFF	Raj Hamsa X'Air 582	BMAA/HB/137		6. 4.00	D.L.Brown	(Plymouth)	16. 6.05P
	(Built G Chatwick, A L H Seed and G Statham - kit no.521)						
G-BZFH	Pegasus Quantum 15-912	7660		15. 5.00	J.S.Hamilton t/a Kent Scout Microlights	(Edenbridge)	7. 6.05P
G-BZFI	Jabiru Aircraft Jabiru UL-450	PFA 274A-13497		27. 3.00	A W.J.Findlay	Leicester	13. 4.05P
	(Built A W.J.and A I Findlay)				tr Group Family		
G-BZFJ	Westland SA.314C Gazelle HT.2	1098	XW861	9. 5.00	European Marine Ltd	Goodwood	12. 7.05P
	(Official p/i = c/n 1102: c/n reported as 1096 which corresponds to G-BBHV - while c/n 1098 translates to G-BBHW) (As "XW861/CU-52")						
G-BZFK	TEAM mini-MAX 88	PFA 186-12060		17. 4.00	I.Macleod	Redlands	1. 9.05P
	(Built C Vandenberghe)						
G-BZFN	Scottish Aviation Bulldog Series 120/121	BH120/325	XX667	18. 4.00	Thomas Aviation Ltd *(As "XX667/16")*	Ashbourne	7.10.04T
G-BZFO	Mainair Blade	1235-0100-7 & W1028		29. 3.00	I.Bell and M.Chambers	Crosland Moor	14. 6.05P
	(Rotax 503)						
G-BZFP	de Havilland DHC-6-310 Twin Otter	696	C-GGNF	11. 8.00	Loganair Ltd *"Chatham Historic Dockyard"*	Glasgow	13. 8.05T
			N712PV, N696WJ, F-ODUH, TR-LZN, C-GKIQ				
G-BZFR	Extra EA.300/L	203		26. 6.00	Powerhunt Ltd	Biggin Hill	7. 8.06T
	(Official c/n outside normal EA300L c/n batch: possibly ex D-EDGE [03?])						
G-BZFS	Mainair Blade 912	1243-0300-7 & W1036		23. 3.00	S.P.Stone and F.A Stephens	Baxby Manor, Husthwaite	29. 4.05P
G-BZFT	Murphy Rebel	PFA 232-13224		7. 4.00	N.A.Evans	Branscombe	14. 8.04P
	(Built N.A Evans) (Lycoming O-320)						
G-BZFU	Lindstrand LBL HS-110 HA Airship	671		25. 4.00	PNB Entreprenad AB	Malmo, Sweden	11. 1.03A
G-BZFV	Zenair CH.601UL Zodiac	PFA 162A-13547		14. 4.00	T.R.Sinclair and T.Clyde	Lamb Holm Farm, Orkney	21. 3.05P
	(Built I M Donnelly) (Rotax 912S)						
G-BZGA	de Havilland DHC-1 Chipmunk 22	C1/0608	WK585	31. 3.00	The Real Flying Company Ltd *(As "WK585)*	Shoreham	30. 4.07T
G-BZGB	de Havilland DHC-1 Chipmunk 22	C1/0905	WZ872	31. 3.00	Chipmunk Aviation Ltd	Duxford	18. 8.06
G-BZGC	Aérospatiale AS355F1 Ecureuil 2	5077	G-CCAO	26. 3.99	McAlpine Helicopters Ltd	Warton	21.11.05T
			G-SETA, G-NEAS, G-CMMM, G-BNBJ, C-GLKH *(Police c/s)*				

G-BZGD	Piper PA-18-150 Super Cub	18-8109049	N90943	15. 5.00	M.G.and S.J.White t/a Proline Aviation	Compton Abbas	16. 7.06T
G-BZGE	Medway EclipseR	159/139		6. 5.99	G.Evans	Arclid Green, Sandbach	30. 6.02P
	(Jabiru 2200A) *(Believed to contain original sailwing of G-MZGE)*				*(New owner 9.03)*		
G-BZGF	Rans S-6-ES Coyote II	PFA 204-13594		25. 4.00	D.F.Castle	London Colney	27. 7.05P
	(Built D.F.Castle - kit no 0899.1334) (Rotax 582)						
G-BZGG	Sud-Aviation SE313B Alouette II	1430	G-POSE	22. 2.00	J.T.Meall	(Liverpool)	17. 9.04T
			G-BZGG, EI-CTH, F-GKML, ALAT				
G-BZGH	Reims/Cessna F172N Skyhawk II	F17201789	EI-BGH	1.12.98	D.Behan tr Golf Hotel Group	Weston, Dublin	4. 9.05
G-BZGI	Ultramagic M-145 HAB	145/12		9. 6.00	European Balloon Co Ltd	(Great Missenden)	7. 8.05T
G-BZGJ	Thunder Ax10-180 S2 HAB	3956	LN-CBT	8. 5.00	M.Wady t/a Merlin Balloons	(Hamstreet)	12. 5.05T
G-BZGK	North American OV-10B Bronco	338-17	Luftwaffe 9932	9. 6.00	Invicta Aviation Ltd	Duxford	AC
			D-9561, Bu 158308		*(Op Aircraft Restoration Co as "99+32")*		
G-BZGL	North American OV-10B Bronco	338-11	Luftwaffe 9926	9. 6.00	Invicta Aviation Ltd	Duxford	AC
			D-9555, Bu 158302		*(Op Aircraft Restoration Co as "99+26")*		
G-BZGM	Mainair Blade 912	1247-0400-7 & W1040		14. 4.00	D.Young	North Coates	22. 6.05P
G-BZGN	Raj Hamsa X'Air 582	BMAA/HB/128		3. 5.00	C.S.Warr and P.A Pilkington	North Coates	17. 8.05P
	(Built B G M Chapman - kit no.445)						
G-BZGO	Robinson R44 Astro	0757		14. 4.00	P.Durkin	Blackpool	14. 5.06T
G-BZGP	Thruster T.600N 450	0400-T600N-043		25. 4.00	M.L.Smith	Popham	8. 9.04P
G-BZGR	Rans S-6-ES Coyote II	PFA 204-13595		3. 5.00	J.M.Benton	Hadzor Farm, Worcester	23. 7.05P
	(Built J M Benton and G R Pritchard - kit no 0999.1338.ES) (Jabiru 2200A) *(Tri-cycle u/c)*						
G-BZGS	Mainair Blade 912	1242-0300-7 & W1035		10. 5.00	S.C.Reeve	Ashbourne	3.11.05P
G-BZGT	Jabiru Aircraft Jabiru UL-450	PFA 274A-13539		4. 5.00	P.H.Ronfel	Crosland Moor	
	(Built P.H.Ronfel)				*(Noted 7.02)*		
G-BZGU	Raj Hamsa X'Air 582	BMAA/HB/138		4. 5.00	C.Kiernan	(Mostrim, County Longford)	8.10.05P
	(Built C.Kiernan - kit no.512)						
G-BZGV	Lindstrand LBL 77A HAB	695		9. 5.00	J.H.Dryden *"Skylark"*	(Okehampton)	9. 8.04A
G-BZGW	Mainair Blade	1246-0400-7 & W1039		5. 5.00	C.S.M.Hallam	Barton	5 6.05P
G-BZGX	Raj Hamsa X'Air 582	BMAA/HB/099		2. 6.00	A.Crowe	Newtownards	7 5.05P
	(Built A Crowe - kit no.400)						
G-BZGY	Dyn'Aéro CR100C	21	F-TGCI	7. 6.00	D.Hayes	Spilsteads Farm, Sedlescombe	1. 9.05P
G-BZGZ	Pegasus Quantum 15-912	7674		7. 6.00	W.H.J.Knowles	Weston Zoyland	30. 9.05P
G-BZHA	Boeing 767-336	29230	N60668	22. 5.98	British Airways plc	Heathrow	21. 5.07T
G-BZHB	Boeing 767-336	29231		30. 5.98	British Airways plc	Heathrow	29. 5.07T
G-BZHC	Boeing 767-336	29232		29. 6.98	British Airways plc	Heathrow	28. 6.07T
G-BZHE	Cessna 152	15281303	D-EAOC	20. 4.00	MK Aero Support Ltd	Andrewsfield	24. 7.06T
			N49484				
G-BZHF	Cessna 152	15283986	D-EMJA	20. 4.00	Modi Aviation Ltd	Earsl Colne	11. 7.06T
			N4858H				
G-BZHG	Tecnam P92-EM Echo	PFA 318-13606		24. 5.00	M. and J.Turner	Sturgate	25. 1.05P
	(Built M Rudd)						
G-BZHI	Enstrom F-28A-UK	281	G-BPOZ	14.12.99	Tindon Ltd	Litle Snoring	20. 6.05
			N246Q				
G-BZHJ	Raj Hamsa X'Air 582	BMAA/HB/126		10. 5.00	P.J.Smith	Otherton, Cannock	23.11.03P
	(Built A P Harvey and B Baker - kit no.482)				*(New owner 7.04)*		
G-BZHK	Piper PA-28-181 Archer III	2843347	N41647	14. 7.00	Meldform Metals Ltd	Fowlmere	17. 7.06T
			N9519N				
G-BZHL	North American AT-16 Harvard IIB	14A-1158	FT118	6. 6.00	R.H.Cooper and S.Swallow	Hemswell	
	(Built Noorduyn Aviation, Canada)		43-12859		*(Stored 1.02)*		
	(Officially quoted USAF p/i is "43-12959")						
G-BZHN	Pegasus Quantum 15-912	7677		20. 6.00	P.L.Cummings t/a Eaglescott Microlights	Eaglescott	17. 6.05P
G-BZHO	Pegasus Quantum 15 Super Sport	7658		19. 5.00	R.L.Williams	Roddige	19. 5.05P
	(Rotax 582)						
G-BZHP	Quad City Challenger II	PFA 177-13153		11. 5.00	F.Payne	Plaistowes Farm, St.Albans	21. 6.54P
	(Built F Payne - kit no CH2-0995-CW-1398) (Rotax 582)						
G-BZHR	Jabiru Aircraft Jabiru UL-450	PFA 274A-13493		16. 5.00	G.W.Rowbotham	(Loughborough)	16. 8.05P
	(Built G.W.Rowbotham)						
G-BZHS	Europa Aviation Europa	PFA 247-12865		16. 5.00	P.Waugh	(Llangollen)	21. 8.04P
	(Built P.Waugh - kit no.207) (Monowheel u/c)						
G-BZHT	Piper PA-18A-150 Super Cub	18-5886	ZK-BTF	25. 5.00	The Furness Gliding Club Proprietary	Walney Island	18.12.04
					t/a Lakes Gliding Club		
G-BZHU	Wag-Aero CUBy Sport Trainer	AACA/351	ZK-MPH	25. 5.00	D M Lewington	Gloucestershire	13. 6.05P
G-BZHV	Piper PA-28-181 Archer III	2843382	N41848	17.10.00	M.Basson	Enstone	28. 9.06T
			G-BZHV, N41848				
G-BZHW	Piper PA-28-181 Archer III	2843409	N4184D	16. 2.01	Delta Kilo Services LLP	Bournemouth	8. 3.07T
			G-BZHW, N4184D				
G-BZHX	Thunder Ax11-250 S2 HAB	4880		21. 6.00	T.H.Wilson *"Slim Your Bin"*	(Diss)	17. 8.05T
G-BZHY	Mainair Blade 912	1250-0500-7 & W1043		7. 6.00	M.D.Harris	Earls Barton, Northampton	29. 5.05P
G-BZIA	Raj Hamsa X'Air HK700	BMAA/HB/116		1. 6.00	G.A.J.Salter	(Taunton)	23. 9.04P
	(Built A U.I.Hudson - kit no.475)						
G-BZIC	Lindstrand LBL Sun SS HAB	702		8. 6.00	Ballongaventyr I Skane AB	(Lund, Sweden)	25. 1.05A
G-BZID	Montgomerie-Bensen B.8MR	PFA G/01-1315		31. 5.00	S.C.Gillies	(Buckie)	
	(Built A Gault from cannibalised Air Command Elite G-BOGW)				*(New owner 10.02)*		
G-BZIG	Thruster T.600N	0400-T600N-042		25. 4.00	Ultra Air Ltd	Leicester	8. 5.05P
	(Rotax 582 UL)						
G-BZIH*	Lindstrand LBL 31A HAB				See SECTION 3 Part 1	(Aldershot)	
G-BZII	Extra EA.300/L	119		13. 9.00	R.J.Verrall	Blackbushe	18. 9.06T
G-BZIJ	Robin DR500/200i Président	0023		9. 3.00	Rob Airways Ltd	Guernsey	13. 4.06
	(Registered as DR400/500 but c/n plate denotes type as DR.500/200)						
G-BZIK	Cameron A-250 HAB	4890		27. 6.00	Breckland Balloons Ltd	(Dereham)	26. 5.05T
G-BZIL	Cameron Colt 120A HAB	4876		7. 7.00	S.R.Seager	(Aylesbury)	27. 3.05T
					t/a Champagne Flights *(Parrott and Coles Solicitors titles)*		
G-BZIM	Pegasus Quantum 15-912	7678		20. 6.00	A.Cuthbertson	(Port Talbot)	9. 4.05P
G-BZIN	Robinson R44 Raven	0776		23. 6.00	Helicentre Ltd	Blackpool	11. 7.06T

Reg	Type	C/n	Previous identity	Date	Owner/Operator	Base	CofA
G-BZIO	Piper PA-28-161 Warrior III	2842085	EC-IBJ	29.06.00	S.J. Skilton t/a Aviation Rentals	Exeter	11.12.06T
			G-BZIO, N41796, (VH-PWF), N41796				
G-BZIP	Montgomerie-Bensen B.8MR Merlin	PFA G/01A-1319		11. 5.00	S.J.Boxall	Askern	12. 6.05P
	(Built S.J.Boxall) (Rotax 582)						
G-BZIS	Raj Hamsa X'Air 582	BMAA/HB/142		12. 6.00	M.J.Badham	(Woodbridge)	12.11.04P
	(Built J.Way and R.Bonnett - kit no.520)						
G-BZIT	Beech 95-B55 Baron	TC-564	HB-GBS	12. 6.00	Propellorhead Aviation Ltd	Oxford	17.10.07E
			I-ALGE, HB-GBS, N6845Q				
G-BZIV	Jabiru Aircraft Jabiru SPL-450	PFA 274A-13587		20. 6.00	V.R.Leggott	Coldharbour Farm, Willingham	30. 8.05P
	(Built V.R.Leggott)						
G-BZIW	Pegasus Quantum 15-912	7681		17. 7.00	J.M.Hodgson	Baxby Manor, Husthwaite	23. 8.05P
G-BZIX	Cameron N-90 HAB	4867		3. 8.00	Sport Promotion SRL	(La Morra, Italy)	19. 9.03A
G-BZIY	Raj Hamsa X'Air 582	BMAA/HB/141		19. 6.00	I.K.Hogg	Kirkbride	27. 9.04P
	(Built I.K.Hogg - kit no.488)						
G-BZIZ	Ultramagic H-31 HAB	31/02		12. 6.00	G.D.O.Bartram	(Andorra la Vella)	14.11.04A
G-BZJA	Cameron Fire 90 SS HAB	4757		5. 5.00	Chubb Fire Ltd	(Sunbury-on-Thames)	28. 3.05A
	(Chubb Fire Extinguisher shape)				*(Chubb titles) (New owner 6.04)*		
G-BZJB	Aerostar Yakovlev Yak-52	811601	ZU-YAK	18. 9.00	D.Watt	(Yaxley, Peterborough)	22.10.04P
			ZS-YAK, RA-01356, DOSAAF				
G-BZJC	Thruster T.600N Sprint)	0070-T600N-044		21. 6.00	P.Johns *(Op Solent Microlights)*	Sandown	26. 5.05P
G-BZJD	Thruster T.600T 450 Jab	0070-T600T-045		21. 6.00	P.G.Valentine	(Rochechouart, France)	28. 5.05P
G-BZJF	Pegasus Quantum 15	7696		21. 7.00	R.S.McMaster	Sywell	20. 7.05P
	(Rotax 582)						
G-BZJH	Cameron Z-90 HAB	4920		10. 7.00	Cameron Balloons Ltd *"Greeco"*	(Italy)	2.11.04A
G-BZJI	Nova X-Large 37	18946		28. 6.00	M Hay *(New owner 1.03)*	(Dundee)	
G-BZJJ	Robinson R22 Beta II	3081		12. 6.00	Seiontair Ltd	(Caernarfon)	29. 6.06T
G-BZJL	Mainair Blade 912S	1252-0600-7 & W1046		4. 7.00	D.N.Powell	(Bootle)	20. 2.04P
G-BZJM	VPM M-16 Tandem Trainer	PFA G/12-1301		19. 6.00	J.Musil	Mount Airey Farm, South Cave	
	(Built J.Musil)				*(Noted 10.02)*		
G-BZJN	Mainair Blade 912	1254-0600-7 & W1048		13. 7.00	I.A.Forrest	East Fortune	23. 7.05P
G-BZJO	Pegasus Quantum 15	7699		6. 9.00	J.D.Doran	(Mullingar, County Westmeath)	29. 9.05P
	(Rotax 503)						
G-BZJP	Zenair CH.701 UL	PFA 187-13579		30. 6.00	J.A.Ware	(Maesybont, Llanelli)	4. 8.05P
	(Built D.Jerwood) (Verner SVS1400)						
G-BZJR	Montgomerie-Bensen B.8MR	PFA G/01-1320		11. 7.00	N.H.Collins	Sittles Farm, Alrewas	AC
	(Built N H Collins) (Rotax 582)				t/a AES Radionic Surveillance Systems		
G-BZJS	Taylor JT.2 Titch	PFA 60-13622		12. 7.00	R.W.Clarke	(Warminster)	
	(Built R W Clarke)				*(Wings and tail noted 10.01: construction abandoned temporarily?)*		
G-BZJU	Cameron A-200 HAB	4810		30. 6.00	Leeds Castle Enterprises Ltd	(Leeds Castle, Maidstone)	23. 5.05T
					(Leeds Castle titles)		
G-BZJV	CASA 1-131E Jungmann Series 1000	1075	Spanish AF E3B-367	31. 7.00	J.A Sykes	Stretton	1. 3.05P
G-BZJW	Cessna 150F	15062054	OO-WIH	27. 6.01	R.J.Scott	Popham	14. 3.07T
			OO-SIH, N8754S		*(Horton STOL Craft titles)*		
G-BZJX	Ultramagic N-250 HAB	250/12		4. 7.00	Hot Air Balloons Ltd	(Henley-on-Thames)	10. 6.05T
					(e-homes titles)		
G-BZJZ	Pegasus Quantum 15 Super Sport	7697		2. 8.00	S.Baker	Long Marston	22. 8.05P
	(Rotax 503)						
G-BZKC	Raj Hamsa X'Air 582	BMAA/HB/144		12. 7.00	P.J.Cheyney	New House Farm, Birds Edge	18. 5.05P
	(Built P.J.Cheyney and M C Reed - kit no.502)						
G-BZKD	Stolp SA.300 Starduster Too	1	N70DM	3. 7.00	P.and C.Edmunds	Enstone	29. 4.05P
	(Built R D Merritt) (Lycoming IO-360)						
G-BZKE	Lindstrand LBL 77B HAB	708		17. 7.00	R.M.Cambidge	(Kinnerley, Oswestry)	10. 2.03A
G-BZKF	Rans S-6-ES Coyote II	PFA 204-13610		17. 7.00	P.R.Cowie	Eshott	5. 4.05P
	(Built A W.Hodder) (Rotax 582)						
G-BZKG	Extreme/Silex	E761 01A		17. 7.00	R.M.Hardy	(Baldock)	
G-BZKH	Flylight Airsports Doodle Bug/Target	DB023		17. 7.00	B.Tempest	(Halifax)	
G-BZKI	Flylight Airsports Doodle Bug/Target	DB063		17. 7.00	S.Bond	(Huddersfield)	
G-BZKJ	Flylight Airsports Doodle Bug/Target	DB067		17. 7.00	Flylight Airsports Ltd	Sywell	
G-BZKK	Cameron V-56 HAB	396		2. 8.78	P.J.Green and C.Bosley	(Newbury)	13. 8.96A
					tr Gemini Balloon Group *"Gemini II"*		
G-BZKL	Piper PA-28R-201 Cherokee Arrow III	28R-7737152	D-EFFZ	20. 7.00	I.R.Chaplin	King's Farm, Thurrock	2. 8.06T
			N40000				
G-BZKN	Campbell Cricket Mk.4	PFA G/03-1304		20. 7.00	C.G.Ponsford	(Braintree)	
	(Built C G Hooghkirk)				*(New owner 12.03)*		
G-BZKO	Rans S-6-ES Coyote II	PFA 204-13564		20. 7.00	J.A R.Hartley	Long Marston	21.11.04P
	(Built J.A R.Hartley - c/n 0199.1293 (Rotax 912-UL)		*(Tri-cycle u/c)*		tr The GBZKO Syndicate		
G-BZKR	Cameron Colt Sugarbox 90 SS HAB	4922		4. 8.00	Stratos Ballooning GmbH and Co KG		
						(Enningerloh, Germany)	25.11.04A
G-BZKS	Ercoupe 415CD	4834	EI-CIH	22. 8.00	Adrienne H.Harper	(Langport)	
			OO-AIA, (PH-NDO), N94723, NC94723		*(New owner 11.03)*		
G-BZKT	Cyclone Pegasus Quantum 15	7711		23. 8.00	Rochester Microlights Ltd	Rochester	21. 8.05P
	(Rotax 582)						
G-BZKU	Cameron Z-105 HAB	4931		21. 7.00	N.A.Fishlock *(New owner 1.04)*	(Welland, Malvern)	31. 8.01P
G-BZKV	Cameron Sky 90-24 HAB	4857		5. 9.00	Omega Selection Services Ltd	(Stonehouse)	18. 7.04A
					(Omega titles)		
G-BZKW	Ultramagic M-77 HAB	77/179		25. 7.00	T.G.Church	(Blackburn)	5.12.04T
G-BZKX	Cameron V-90 HAB	4505		19. 7.00	Cameron Balloons Ltd	(Dalien, PRC)	26. 7.01A
G-BZKY	Focke-Wulf FW.189-A1	2100	Luftwaffe V7+1H	9. 8.00	M.T.Pearce-Ware	(Worthing)	
	(Built Aero-Avia)				*(Restoration being undertaken in UK and Germany)*		
G-BZKZ	Lindstrand LBL 25A Cloudhopper HAB	721		14. 8.00	Lindstrand Hot Air Balloons Ltd	(Oswestry)	8. 1.03A
G-BZLA	Aérospatiale SA.341G Gazelle 1	1392	N2TV	31. 7.00	P.J.Brown	Redhill	14.10.04T
			N49534		t/a PJ Brown Civil Engineering and Haulage		
G-BZLC	PZL-110 Koliber 160A	04980084	SP-WGL	13. 9.00	G.F.Smith	Cranfield	29. 4.05T

Reg	Type	C/n	Previous identity	Date	Owner/operator	Base	Expiry
G-BZLE	Rans S-6-ES Coyote II	PFA 204-13608		12. 7.00	R.T.P.Harris	Haddenham, Buckingham	18. 9.04P
	(Built W S Long) (Rotax 582) *(Tricycle u/c)*						
G-BZLF	CFM Shadow Series CD	BMAA/HB/053		31. 7.00	D.W.Stacey	(St.Albans)	
	(Built D.W.Stacey - kit no. K.236)						
G-BZLG	Robin HR200/120B	353		7. 7.00	G.S.McNaughton *(Op Prestwick Flying Club)*	Prestwick	24.11.06T
G-BZLH	Piper PA-28-161 Warrior II	28-8316075	N43069	23. 8.00	S.J.Skilton	Southampton	8. 9.06T
					t/a Aviation Rentals *(Op Solent Flight Centre)*		
G-BZLI	SOCATA TB-21 Trinidad TC	500	F-GENI	29. 9.00	K.B.Hallam	(Woking)	27.10.06T
G-BZLK	Slingsby T.31M Motor Tutor	PFA 42-13629		2. 8.00	I.P.Manley	(Chichester)	
	(Built I.P.Manley) (Formerly T.31B BGA2976/EVA ex WT873 [683])						
G-BZLL	Pegasus Quantum 15-912	7693		9. 8.00	J.L.Merriman	Knapthorpe Lodge, Caunton	14.10.04P
					tr Caunton Graphites Syndicate		
G-BZLM	Mainair Blade	1257-0800-7 & W1051		8. 8.00	R.R.Celentano	Perth	11. 4.05P
	(Rotax 582)						
G-BZLO	Denney Kitfox Model 2	PFA 172-13630		8. 8.00	M.W.Hanley	(Truro)	23. 1.04P
	(Built M W Hanley)						
G-BZLP	Robinson R44 Raven	0814		17. 7.00	R.C.Hayward t/a Rotorways Helicopters	Manston	11. 9.06P
G-BZLS	Cameron Sky 77-24 HAB	4858		17. 8.00	D.W.Young	(Stenhousemuir)	15. 3.04A
G-BZLT	Raj Hamsa X'Air 582	BMAA/HB/125		10. 8.00	G.S Millar	Moygashel, County Tyrone	27. 7.05P
	(Built G.S Millar - kit no.486)						
G-BZLU	Lindstrand LBL 90A HAB	719		9. 8.00	A E.Lusty *"Tetris 1"*	(Bourne, Lincoln)	27. 3.05A
G-BZLV	Jabiru Aircraft Jabiru UL-450	PFA 274A-13537		15. 8.00	G.Dalton	Bodmin	6. 8.04P
	(Built G.Dalton)						
G-BZLX	Pegasus Quantum 15-912	7714		30. 8.00	G.D.Ritchie	(Longniddry)	22. 4.05P
G-BZLY	Grob G109B	6242	D-KLMG	24. 8.00	A.Baker	Lasham	1.12.07E
			G-BZLY, OE-9230				
G-BZLZ	Pegasus Quantum 15-912	7721		13. 9.00	T.Bale and A Martin	(Barnstaple)	5.10.04P
G-BZMB	Piper PA-28R-201 Arrow III	28R-7837144	HB-PBY	20. 4.00	D.S.Seex	King's Farm,Thurrock	17. 4.06T
			N3963M				
G-BZMC	Jabiru Aircraft Jabiru UL-450	PFA 274A-13593		18. 8.00	D.G.Harkness	Newtownards	29. 9.04P
	(Built J.R.Banks)						
G-BZMD	Scottish Aviation Bulldog Series 120/121	BH120/247	XX554	18. 8.00	D.M.Squires *(As "XX554/09")*	Green Farm, Combrook	13.11.04
G-BZME	Scottish Aviation Bulldog Series 120/121	BH120/347	XX698	18. 8.00	B.Whitworth *(As "XX698/9")*	Breighton	28.11.07
G-BZMF	Rutan LongEz	PFA 74-10698		30. 8.00	R.A.Gardiner	Cumbernauld	13. 6.05P
	(Built A McCaughlin)						
G-BZMG	Robinson R44 Raven	0815		16. 8.00	Ramsgill Aviation Ltd	Sherburn-in-Elmet	14. 9.06T
G-BZMH	Scottish Aviation Bulldog Series 120/121	BH120/341	XX692	21. 8.00	M.E.J.Hingley and Co Ltd	Wellesbourne Mountford	16.12.07
					(As "XX692/A")		
G-BZMI	Pegasus Quantum 15-912	7716		22. 9.00	L.M.Bassett *(Noted 1.05)*	Longacres Farm, Sandy	18.10.04P
G-BZMJ	Rans S-6-ES Coyote II	PFA 204-13631		31. 8.00	J.Seddon and F.J.Lloyd	Tarn Farm, Cockerham	21. 7.05P
	(Built J.Seddon, F.J.Lloyd and T I Bull) (Rotax 582) *(Tri-cycle u/c)*				tr Heskin Flying Group		
G-BZML	Scottish Aviation Bulldog Series 120/121	BH120/342	XX693	1. 9.00	I.D.Anderson *(As "XX693/07")*	Poplar Hall Farm, Elmsett	14.11.04
G-BZMM	Robin DR400/180R Remorqueur	918	OE-KIR	17. 7.00	N.A C.Norman	Feshiebridge	5.10.06
			D-EAWR				
G-BZMO	Robinson R22 Beta	1219	N24282	31. 7.00	Sloane Helicopters Ltd	Sywell	4. 9.06T
			JA7814, N8056H				
G-BZMR	Raj Hamsa X'Air 582	BMAA/HB/149		11. 9.00	M.Grime	Brook Farm, Pilling	1. 5.05P
	(Built M.Grime - kit no.480)						
G-BZMS	Mainair Blade	1256-0700-7 & W1050		2. 8.00	A.P.Jackso	(Beccles)	10. 9.04P
	(Rotax 582)				tr Beccles Buzzards		
G-BZMT	Piper PA-28-161 Warrior III	2842107	N4147D	29.11.00	S.J.Skilton	Wellesbourne Mountford	28.11.06T
			G-BZMT, N9519N, N4147D		t/a Aviation Rentals		
G-BZMV	Cameron Concept 80 HAB	4930		26. 9.00	Latteria Soresinese Soc Coop ARL	(Soresina, Italy)	26. 8.05P
G-BZMW	Pegasus Quantum 15-912	7720		26. 9.00	J.I.Greenshields	Dunkeswell	3.10.05P
G-BZMX	Cameron Z-90 HAB	4942		1. 9.00	Cameron Balloons Ltd *(Cameron Balloons titles)*	(Bristol)	11. 4.05A
G-BZMY	SPP Yakovlev Yak C-11	171314	F-AZSF	4.10.00	Griffin Aviation Ltd	(London EC3N)	7. 7.05P
			Egyptian AF		*(As "1" in Soviet AF c/s)*		
G-BZMZ	CFM Streak Shadow	BMAA/HB/051		13. 9.00	J.F.F.Fouche	(London EC1)	
	(Built J.F F.Fouche - kit no.K265-CD)						
G-BZNB	Pegasus Quantum 15	7739		10.11.00	R.C.Whittall	Weston Zoyland	9.11.03P
	(Rotax 503)						
G-BZNC	Pegasus Quantum 15-912	7736		25.10.00	D.E.Wall	Long Marston	26. 4.05P
G-BZND	Sopwith Pup rep	PFA 101-11815		27. 9.00	Mary A.Goddard	(Southampton)	
	(Built B.F.Goddard)				*(New owner 1.05) (As "N5199")*		
G-BZNE	Beech B300 Super King Air	FL-286	N4486V	17.10.00	G.Davies	Blackbushe	16.11.06
	(Aka "King Air 350")						
G-BZNF	Cameron Colt 120A HAB	4866		13.11.00	N.Charbonnier	(Aosta, Italy)	23. 1.05A
G-BZNG	Raj Hamsa X'Air Jab22	BMAA/HB/147		4.10.00	G.L.Craig	Newtownards	10. 8.05T
	(Built G.L.Craig - kit no.571)						
G-BZNH	Rans S-6-ES Coyote II			18.10.00	R.B.Skinner	(Beaworthy)	15. 7.05P
	(Built V Whiting - c/n 0899.1333) (Rotax 582) *(Tri-cycle u/c)*						
G-BZNI	Bell 206B JetRanger II	2142	G-ODIG	4.10.00	Trimax Ltd	Manston	3. 1.04T
			G-NEEP, N777FW, N3CR				
G-BZNJ	Rans S-6-ES Coyote II	0700.1382 & PFA 204-13640		23.10.00	S.P.Read and M.H.Wise	Weston Zoyland	25. 4.05P
	(Built S P Read) (Rotax 582) *(Tailwheel u/c)*						
G-BZNK	Morane Saulnier MS.315E D2	354	F-BCNY	2.11.00	R.H.Cooper and S.Swallow	Gamston	
			French AF		*(Dismantled, in advanced stage of restoration 1.05).*		
G-BZNM	Pegasus Quantum 15 Super Sport	7754		20.11.00	M.Tomlinson	(Burton-on-Trent)	26. 1.05P
	(Rotax 582)						
G-BZNN	Beech 76 Duchess	ME-343	N6133P	25.10.00	S.J.Skilton	Bournemouth	14.12.06T
			F-GHSU, N6722L		t/a Aviation Rentals *(Op Bournemouth Flying Club)*		
G-BZNO	Ercoupe 415C	2118	N99495	9.11.00	D.K.Tregilgas	New Farm House, Great Oakley	
G-BZNP	Thruster T.600N-450	0100-T600N-047		27.10.00	J.D.Gibbons	(Newry)	3.12.05P
	(Rotax 582)						

Reg	Type	C/n	Previous identity	Reg date	Owner/operator	Base	CofA
G-BZNR	British Aerospace BAe 125 Series 800B	258180	G-XRMC	30.10.00	RMC Group Services Ltd	Farnborough	11.12.04T
			G-5-675				
G-BZNS	Mainair Blade	1263-1000-7 & W1057		23.11.00	M.K.B.Molyneux	(Nantwich)	11. 2.05P
	(Rotax 582)						
G-BZNT	Aero L-29 Delfin	893019	ES-YLG	3.11.00	H.P.Odone	North Weald	10.11.04P
			Estonian AF, Soviet AF		*(Noted 1.05)*		
G-BZNU	Cameron A-300 HAB	4960		29.11.00	Balloon School (International) Ltd	(Petworth)	10. 6.05T
G-BZNV	Lindstrand LBL 31A HAB	741		12.12.00	G.R.Down	(Gillingham)	17. 4.05A
G-BZNW	Isaacs Fury II	PFA 11-13402		10.11.00	J.E.D.Rogerson	Morgansfield, Fishburn	17. 6.05P
	(Built J.E.D.Rogerson)				*(As "K2048" in RAF c/s)*		
G-BZNX	SOCATA MS.880B Rallye Club	2113	F-BTVX	17.11.00	R.K.Stewart	Hinton-in-the-Hedges	30. 1.07
G-BZNY	Europa Aviation Europa XS	PFA 247-13355		14.11.00	W.J.Harrison	Duxford	1. 1.06P
	(Built A K.Middlemas - kit no.401) (Tri-gear u/c)						
G-BZNZ	Lindstrand LBL Cake SS HAB	747		21.12.00	Oxford Promotions (UK) Ltd *(Op F Prell)*	(Kentucky, US)	3. 4.03A
G-BZOB	Slepcev Storch	PFA 316-13592		21.11.00	J.E.and A Ashby	(Papworth Everard)	18. 8.05P
	(Built J.E.Ashby)				*(As "6G+ED" in Luftwaffe c/s)*		
G-BZOD	Pegasus Quantum 15-912	7763		18.12.00	S M Wilson	(Port Glasgow)	14. 4.05P
G-BZOE	Pegasus Quantum 15	7723		14. 9.00	B.N.Thresher	Dunkeswell	2.10.04P
	(Rotax 582)						
G-BZOF	Montgomerie-Bensen B.8MR	MGM3 & SJML1		7.11.00	S.J.M.Ledingham	Carlisle	7. 8.05P
	(Built M G Mee and S.J.M.Ledingham) (Rotax 912-UL)						
G-BZOG	Dornier 328-100	3088	D-CDXN(5)	19.12.00	Suckling Airways (Cambridge) Ltd	London City	18.11.05T
			F-GNPR		t/a Scot Airways		
G-BZOI	Nicollier HN700 Menestrel II	PFA 217-12604		27.10.00	S.J.McCollum	Newtownards	22. 5.05P
	(Built S.J.McCollum - c/n 122) (Volkswagen 2180)						
G-BZOL	Robin R3000/140	124	F-GEKZ	20.12.00	S.D.Baker	Exeter	31. 8.07T
G-BZOM	RotorWay Executive 162F	6243	N767SG	27. 3.01	J.A Jackson	Street Farm, Takeley	5. 5.05P
	(Built G and S Waugh)						
G-BZON	Scottish Aviation Bulldog Series 120/121	BH120/214	XX528	19.12.00	Roger Savage Gyroplanes Ltd *(As "XX528/X")*	Carlisle	18. 8.07
G-BZOO	Pegasus Quantum 15-912	7702		15. 8.00	K.Brown	Sywell	16.11.05P
G-BZOP	Robinson R44	0958		11. 1.01	20:20 Logistics Ltd	(Stoke-on-Trent)	1. 2.07T
G-BZOR	TEAM mini-MAX 91	PFA 186-13312		9. 8.00	A.Watt	Insch	13. 5.05P
	(Built A Watt)						
G-BZOU	Pegasus Quantum 15-912	7768		22. 3.01	A J.Gordon	Field Farm, Oakley	3. 4.05P
G-BZOV	Pegasus Quantum 15-912	7769		22. 3.01	D.Turner	(Bicester)	29. 5.03P
G-BZOW	Whittaker MW7	PFA 171-13118		15.12.00	G.W.Peacock	(Doncaster)	
	(Built G.W.Peacock)						
G-BZOX	Cameron Colt 90B HAB	10000		8. 2.01	D.J.Head	(Newbury)	21.1 2.04P
G-BZOY	Beech 76 Duchess	ME-144	EC-ICJ	3.1.01	S.J.Skilton t/a Aviation Rentals	Bournemouth	9. 6.06T
			G-BZOY, F-GFFH, 5T-AOH, F-ODJQ, F-GBLO		*(Op CTC Aviation)*		
G-BZOZ	Van's RV-6	PFA 181A-12455		14. 9.00	V.Edmundson	Blackpool	9. 9.05P
	(Built V Edmundson)						
G-BZPA	Mainair Blade 912S	1264-1100-7 & W1058		13.12.00	J.McGoldrick	Newtownards	28. 1.04P
G-BZPB	Hawker Hunter GA.11	41H-670758	WV256	15. 1.01	B.R.Pearson	Exeter	17. 7.03P
					(As "WB188" Hunter prototype in duck-egg green c/s)		
G-BZPC	Hawker Hunter GA.11	HABL-003061	XF300	15. 1.01	B.R.Pearson	Exeter	AC
					(Open store as "WB188" Hunter prototype in all-red c/s 5.04)		
G-BZPD	Cameron V-65 HAB	4700		10.11.00	P.Spellward	(Heidelburg, Germany)	16. 4.05A
G-BZPE	Lindstrand LBL 310A HAB	746		16. 2.01	Aerosaurus Balloons Ltd	(Whimple, Exeter)	1. 9.05T
G-BZPF	Scheibe SF 24B Motorspatz I	4028	D-KROA	19. 1.01	J.S.Gorrett	(Pontyclun)	17. 2.05
			PH-971, OE-9005, (D-KECO)				
G-BZPG	Beech C24R Sierra	MC-556	N23840	27. 3.01	S.J.Skilton	Bournemouth	30. 4.07T
					t/a Aviation Rentals *(Op Professional Flight Training)*		
G-BZPH	Van's RV-4	PFA 181-12867		6. 9.00	A G.Truman tr G-BZPH RV-4 Group	Kemble	24.11.04P
	(Built A G.Truman) (Lycoming O-360)		*(Bounced landing Lundy Island 4.7.04 and u/c collapsed: damage to u/c, forward fuselage and engine compartment)*				
G-BZPI	SOCATA TB-20 Trindad	1814	SX-ATT	20.12.00	K.M.Brennan	Leicester	15. 4.07
G-BZPJ	Beech 76 Duchess	ME-227	N6630Z	2. 3.01	S.J.Skilton	Bournemouth	23. 3.07T
					t/a Aviation Rentals *(Op Bournemouth Flying Club)*		
G-BZPK	Cameron C-80 HAB	4183		23. 2.01	Horizon Ballooning Ltd	(Alton)	20.11.04T
G-BZPL	Robinson R44 Clipper	0948		10. 1.01	M.K.Shaw	San Bonet, Majorca, Spain	17. 2.07T
G-BZPM	Cessna 172S Skyhawk SP	172S8561	N72760	11. 1.01	Pooler-LMT Ltd	Lower Grounds Farm, Sherlowe	28. 3.07T
G-BZPN	Mainair Blade 912S	1268-0101-7 & W1062		25. 1.01	S.Wing	(Harlow)	22. 4.05P
G-BZPP	Westland Wasp HAS.1	F9675	XT793	15. 1.01	G.P.Hinkley	Moat Farm, Otley, Ipswich	4. 3.05P
G-BZPR	Ultramagic N-210 HAB	210/14		16. 1.01	European Balloon Display Co Ltd	(Great Missenden)	8. 4.05T
G-BZPS	Scottish Aviation Bulldog Series 120/121	BH120/316	XX658	8. 1.01	D.M.Squires	(Wellesbourne Mountford)	AC
					(As "XX658/03") (Noted 1.03)		
G-BZPT	Ultramagic N-210 HAB	210/15		16. 1.01	European Balloon Display Co Ltd	(Great Missenden)	9. 2.05T
G-BZPU	Cameron V-77 HAB	5433	N20726	2. 5.01	J.Vonka	(New Malden)	
G-BZPV	Lindstrand LBL 90B HAB	727		17. 1.01	D.P.Hopkins	(Pidley, Huntingdon)	29. 7.05A
					(LakesideLodge Golf Centre titles)		
G-BZPW	Cameron V-77 HAB	6245	N4463V	2. 2.01	J.Vonka	(New Malden)	11. 8.05A
G-BZPX	Ultramagic S-105 HAB	105/78		12. 2.01	G.M.Houston t/a Scotair Balloons	(Lesmahagow)	16. 4.05T
G-BZPY	Ultramagic H-31 HAB	31/03		12. 2.01	G.M.Houston t/a Scotair Balloons	(Lesmahagow)	16. 4.05A
G-BZPZ	Mainair Blade	1265-1200-7 & W1059		23. 1.01	M.C.W.Robertson	(Leek)	18. 4.05P
	(Rotax 582)						
G-BZRA	Rans S-6-ES Coyote II	PFA 204-13683		16. 1.01	K.J.Warburton	Egginton	28. 3.05P
	(Built A W.Fish) (Rotax 912-UL)						
G-BZRB	Mainair Blade	1270-0201-7 & W1064		7. 3.01	V.P.Haynes	Mill Farm, Shifnal	16. 6.05P
	(Rotax 582)						
G-BZRC	de Havilland DH.115 Vampire T.11	15143	WZ584	26. 3.01	D.Copley	(Belton, Doncaster)	
					(Dismantled as "WZ584/K" - current status unknown)		
G-BZRD	de Havilland DH.115 Vampire T.11	15687	XH313	27. 3.01	D.Copley	(Surrey)	
					(Noted dismantled as "XH313/E" 1.05)		

Reg	Type	C/n	Previous identity	Reg date	Owner/Operator	Base	CofA
G-BZRE	Percival P.56 Provost T.1	PAC/F/234	7688M	15. 5.01	D.Copley	Booker	
			WW421		*(Noted dismantled as "WW421/P" 1.05)*		
G-BZRF	Percival P.56 Provost T.1	PAC/F/062	7698M	15. 5.01	D.Copley	Booker	
			WV499		*(Noted dismantled as "WV499/P-G" 1.05)*		
G-BZRG	Huntwing Avon	BMAA/HB/154		16. 1.01	W.G.Reynolds	(Cromer)	
	(Built W.G.Reynolds - kit no.0009090)				*(Frame only completed by 7.01)*		
G-BZRJ	Pegasus Quantum 15-912	7783		5. 2.01	P.J.Barton tr G-BZRJ Group	Longacre Farm, Sandy	11. 4.05P
G-BZRN	Robinson R44 Raven	0971		1. 2.01	Toriamos Ltd	(Naas, County Kildare)	30. 3.07T
G-BZRO	Piper PA-30 Twin Comanche C	30-1923	SE-IYL	2. 3.01	Comanche Hire Ltd	Gloucestershire	21. 6.07T
			D-GATI, I-KATI, N8767Y				
G-BZRP	Pegasus Quantum 15-912 Super Sport	7758		24. 1.01	M.E.Howard	RAF Henlow	15. 2.05P
					tr RAF Microlight Flying Association (RP)		
G-BZRR	Pegasus Quantum 15-912	7727		4.10.00	F.G.Green and T.Hudson		
					(New owner 10.04)	Knapthorpe Lodge, Caunton	8.10.03P
G-BZRS	Eurocopter EC135T2	0166		22. 3.01	Bond Air Services Ltd	Aberdeen	8. 4.07T
G-BZRT	Beech 76 Duchess	ME-89	EC-HYO	21. 3.01	S.J.Skilton	Bournemouth	12.11.05T
			G-BZRT, F-GHBL, N2074G		t/a Aviation Rentals		
G-BZRU	Cameron V-90 HAB	10053		1. 5.01	J.Yates, P.W.Limpus and I.J.Liddiard		
					t/a Dragon Balloon Group	Crowthorne/Newbury	31. 5.05A
G-BZRV	Van's RV-6	PFA 181A-13573		12.10.00	N.M.Hitchman	Leicester	11. 8.05P
	(Built E.Hicks, P Hicks and N.M.Hitchman) (Lycoming O-320)						
G-BZRW	Mainair Blade 912S	1266-0101-7 & W1060		6. 2.01	N.D.Kube	Leicester	12. 8.04P
G-BZRX	Ultramagic M-105 HAB	105/80		3. 5.01	Specialist Recruitment Group plc	(Huntingdon)	2. 8.04A
					(Interaction Recruitment/Dixon Finance Division titles) "Angel Baby"		
G-BZRY	Rans S-6-ES Coyote II	PFA 204-13666		1. 2.01	C.J.Powell	Shobdon	30. 7.05P
	(Built S Forman - c/n 0600-1375) (Rotax 582)						
G-BZRZ	Thunder Ax11-250 S2 HAB	10013		11.10.01	A.C.K Rawson and J.J.Rudoni	(Stafford)	7.10.05T
G-BZSA	Pegasus Quantum 15	7784		25. 1.01	Cyclone Airsports Ltd	Manton	13.2.05P
	(Rotax 503)				t/a Pegasus Aviation		
G-BZSB	Pitts S-1S	PFA 09-13697		2. 2.01	A.D.Ingold	(Harlow)	
	(Built A D Ingold)				*(On build 2004)*		
G-BZSC	Sopwith Camel F.1 rep	NAW-3		15. 1.01	Richard Shuttleworth Trustees	Old Warden	AC
	(Built Northern Aeroplane Workshops)				*(Under consruction 3.02)*		
G-BZSD	Piper PA-46-350P Mailbu Mirage	4636168	N838DB	14. 2.01	Meridian Aviation Ltd	Bournemouth	23. 3.07T
			N333DB				
G-BZSE	Hawker Hunter T.8C	41H-670788	9096M	6. 2.01	Towerdrive Ltd	Kemble	12. 5.05P
	(Officially.regd as T.8B (c/n 41H-670792) see G-FFOX)		WV322		*(As "WV322" in RAF camouflage with dayglow tail and wingtips)*		
G-BZSF	Hawker Hunter T.8B	HABL-003150	9237M	6. 2.01	Towerdrive Ltd	Kemble	AC
			XF995		*(As "XF995" in RAF c/s 7.04)*		
G-BZSG	Pegasus Quantum 15-912 Super Sport	7766		22. 2.01	K.J.Gay	Newtownards	5. 4.05P
G-BZSH	Ultramagic H-77 HAB	77/191		12. 4.01	J.L.Hutsby	(Oxhill, Warwick)	19. 3.04A
G-BZSI	Pegasus Quantum 15	7787		12. 3.01	M.O.O'Brien	(Flaxton, York)	6. 4.05P
G-BZSL	Sky 25-16 HAB	138		31. 1.01	A.E.Austin	(Naseby)	10.10.03A
G-BZSM	Pegasus Quantum 15	7788		23. 2.01	J.Walker and A.J.Johnson	Deenethorpe	20. 9.05P
	(Rotax 503)						
G-BZSO	Ultramagic M-77C HAB	77/190		22. 3.02	C.C.Duppa-Miller	(Warwick)	23. 5.05A
G-BZSP	Stemme S-10	10-14	HB-2217	10. 5.01	A.Flewelling and L.Bleaken	Aston Down	8. 8.07
			D-KDNE		"626"		
G-BZSR*	Hawker Hunter T.7	41H-693832	A2617	15. 2.01	Stick and Rudder Aviation Ltd	(Meetkerke, Belgium)	
			XL601		*(On rebuild 2002 in red and white livery of 4 FTS: cancelled 14.5.04 as wfu)*		
G-BZSS	Pegasus Quantum 15-912	7770		6. 2.01	T.R.Marsh		
						Brown Shutters Farm, Norton St.Philips, Somerset	18. 4.05P
G-BZST	Jabiru Aircraft Jabiru UL-450	PFA 274A-13616		13. 2.01	G.Hammond	Headcorn	4. 9.05T
	(Built G.Hammond)						
G-BZSU	Cameron A-315 HAB	10009		13. 6.01	Ballooning Network Ltd	(Bristol)	6. 9.05T
					(Bath Building Society titles)		
G-BZSV	Aherne Barracuda	631		20. 2.01	M.J.Aherne	(St.Albans)	AC
	(Built M.J.Aherne)						
G-BZSX	Pegasus Quantum 15-912	7789		23. 2.01	J.B.Greenwood	Rufforth	20. 4.05P
G-BZSY	SAN Stampe SV-4A	677	N12426	12. 3.01	G.P.J.M.Valvekens	Diest, Belgium	AC
			F-BGGT/French AF				
G-BZSZ	Jabiru Aircraft Jabiru UL-450	PFA 274A-13432		16. 2.01	M.P.Gurr and D.R.Burridge	Pent Farm, Postling	22. 8.05P
	(Built R Riley and F Overall)						
G-BZTA	Robinson R44 Raven	0968		20. 2.01	Thurston Helicopters Ltd	Headcorn	23. 3 07T
G-BZTC	TEAM mini-MAX 91	PFA 186-13336		23. 1.01	G.G.Clayton	Roche, Cornwall	23. 2.05P
	(Built G G Clayton)						
G-BZTD	Thruster T.600T 450 Jab	0021-T600T-049		22. 2.01	B.O. and B.C.McCartan	(Banbridge, County Down)	4. 7.05P
G-BZTE	Cameron A-275 HAB	10028		26. 6.01	Richard Nash Cars Ltd *(Richard Nash titles)*	(Norwich)	2. 9.05T
G-BZTF	IAV-Bacau Yakovlev Yak-52	866703	LY-AKE	28. 2.01	A C.Pledger	Duxford	2. 9.05P
			DOSAAF		tr KY Flying Group		
G-BZTG	Piper PA-34-220T Seneca V	3449126	EC-HGK	4. 4.01	L.R.Chiswell	Fowlmere	1. 6.07T
			N4141N				
G-BZTH	Europa Aviation Europa	PFA 247-12494		21.12.00	T.J.Houlihan	Kemble	27. 1.05P
	(Built T J Houlihan - kit no.010) (Monowheel u/c)						
G-BZTI	Europa Aviation Europa XS	PFA 247-13172		30. 3.01	W.Hoolachan	Perranporth	12. 9.02P
	(Built W.Hoolachan - kit no.124) (Rotax 914-UL) *(Tri-gear u/c)*						
G-BZTJ	Bücker Bü.133C Jungmeister	41	Spanish AF ES1-41	7. 3.01	R.A.Seeley	Turweston	20.12.05P
	(Officially recorded as built CASA but c/n dubious and believed to be built Bücker)				*(As "17+TF" in Luftwaffe c/s)*		
G-BZTK	Cameron V-90 HAB	10083		6. 3.01	E.Appollodorus	(London W12)	22. 4.05A
G-BZTL	Cameron Colt Flying Ice Cream Cone SS HAB	10008		16. 5.01	Stratos Ballooning GmbH and Co KG		
						(Ennigerloh, Germany)	14. 4.05A
G-BZTM	Mainair Blade	1273-0201-7 & W1068		12. 2.01	H.N.Barrott	(Gorebridge)	21. 3.05P
G-BZTN	Europa Aviation Europa XS	PFA 247-13715		16. 3.01	S.A Smith	(Dunfermline)	
	(Built W Pringle and J Dewberry - kit no.504) (Tri-gear u/c)				*(New owner 4.03)*		

Registration	Type	C/n	Previous identity	Regd	Owner/operator	Probable base	CofA expiry
G-BZTR	Mainair Blade	1276-0301-7 & W1071		8. 3.01	A Raithby, A Bower and N.McCusker	Rufforth	30. 3.05P
G-BZTS	Cameron Bertie Bassett 90 SS HAB	10050		3. 5.01	Trebor Bassett Ltd	(Rickmansworth)	17. 5.03A
G-BZTT	Cameron A-275 HAB	4953		28. 8.01	Cameron Flights Southern Ltd	(Pewsey)	11. 8.05T
G-BZTU*	Mainair Blade 912	1272-0201 & 7-W1066		8. 2.01	I.Johnson	Beccles	24. 3.05P
					(Cancelled 25.11.04 as destroyed) (Noted 12.04)		
G-BZTV	Mainair Blade 912S	1278-0301-7 & W1073		2. 4.01	D.J.Cook	Guy Lane Farm, Waverton	10. 5.05P
G-BZTW	Huntwing Avon	BMAA/HB/136		17. 1.01	T.S.Walker	(Sandbach)	15. 6.05P
	(Built T.S.Walker) (Rotax 582 - *kit no.9906092)*						
G-BZTX	Mainair Blade 912	1267-0101-7 & W1061		9. 2.01	K.A Ingham	(Norwich)	20. 7.05P
G-BZTY	Jabiru Aircraft Jabiru UL-450	PFA 274A-13533		1. 3.01	R.P.Lewis	White House Farm, Southery	19. 7.05P
	(Built R.P.Lewis)						
G-BZTZ	MD Helicopters MD 600N	RN056	N70412	15. 5.01	Amberley Aviation Ltd	(Jersey)	1. 8.07T
G-BZUB	Mainair Blade	1274-0201-7 & W1069		27. 3.01	S.E.Kearney	(Newry)	13. 6.05P
G-BZUC	Pegasus Quantum 15-912	7796		10. 4.01	G.A Breen	Portimão, Faro, Portugal	10. 4 05P
G-BZUD	Lindstrand LBL 105A HAB	780		27. 3.01	A.Nimmo	(London SW18)	21. 1.05A
G-BZUE	Pegasus Quantum 15	7800		23. 4.01	W.J.Byrd	Long Marston	28. 5.05P
	(Rotax 503)						
G-BZUF	Mainair Rapier	1277-0301-7 & W1072		27. 3.01	S.J.Perry	(Sandbach)	1. 9.04P
G-BZUG	Tiger Cub RL7A XP Sherwood Ranger			23. 3.01	J.G.Boxall	Pittrichie Farm, Whiterashes	6. 9.05P
	(Built S P Sharp) (Rotax 618)	PFA 237-13040			*(In pseudo RAF c/s as fake serial "SR-XP020")*		
G-BZUH	Rans S-6-ES Coyote II	PFA 204-13716		26. 3.01	J.D.Sinclair-Day	Eshott	9. 7.05P
	(Built G M Prowling) (Jabiru 2200A)						
G-BZUI	Pegasus Quantum 15-912	7798		8. 5.01	A.P.Slade	(High Wycombe)	16.12.04P
G-BZUK	Lindstrand LBL 31A HAB	776		7. 3.01	G.R.J.Luckett	(Fort Collins, Colorado, US)	8.10.04A
G-BZUL	Jabiru Aircraft Jabiru UL-450	PFA 274A-13678		28. 3.01	P.Hawkins	Rufforth	22. 8.05P
	(Built P.Hawkins)						
G-BZUM	Mainair Blade 912	1271-0201-7 & W1065		13. 3.01	R.B.Milton	(London E2)	21. 3.02P
G-BZUN	Mainair Blade 912	1279-0301-7 & W1074		18. 4.01	E.Paxton and A Jones	Ince Blundell	2. 6.05P
G-BZUO	Cameron A-340HL HAB	4952		18. 5.01	Anglian Countryside Balloons Ltd	(Burnham-on-Crouch)	29. 3.05T
G-BZUP	Raj Hamsa X'Air 582	BMAA/HB/164		24. 4.01	I.A J.Lappin	Newtownards	2. 8.05P
	(Built I.A J.Lappin - kit no 624)						
G-BZUU	Cameron O-90 HAB	10058		14. 6.01	D.C.Ball and C.F.Pooley *"Elmo 11"*	(London SE1)	6. 8.05A
G-BZUV	Cameron H-24 HAB	2665	LX-JLW	27. 4.01	J.N.Race *"The Gerkin"*	(Lewes)	9. 5.05A
G-BZUX	Pegasus Quantum 15	7819		22. 5.01	K.M.MacRae, J.D.and C.A Capewell	East Fortune	8. 6.05P
G-BZUY	Van's RV-6	PFA 181A-13471		23. 5.01	R.D.Masters	Standalone Farm, Meppershall	15. 1.04P
	(Built D.M.Gale and K.F.Crumplin) (Lycoming O-320)				*(Damaged on take-off Franklyn's Field 28.6.03 and on rebuild 12.03)*		
G-BZUZ	Hunt Avon-Blade BMW R100	BMAA/HB/162		9. 2.01	J.A Hunt	(Abergavenny)	14. 6.04P
	(Built J.A Hunt) (Uses Mainair sailwing c/n W1067)						
G-BZVA	Zenair CH.701 UL	PFA 187-13635		21. 3.01	M.W.Taylor	Insch	30. 3.05P
	(Built M.W.Taylor) (Rotax 912-UL)						
G-BZVB	Reims FR172H Rocket	FR17200327	G-BLMX PH-RPC	29. 8.00	R.and E.M.Brereton	(King's Lynn)	25.11.06T
G-BZVC	Mickleburgh L107	PFA 256-12549		21. 3.01	D.R.Mickleburgh	(Milton Keynes)	
	(Built D.R.Mickleburgh)				*(Exhibited incomplete @ 1996 PFA Rally)*		
G-BZVD	Cameron Colt Forklift-105 SS HAB	10084		15. 6.01	Stratos Ballooning GmbH and Co KG *"JungHeinrich"*	(Ennigerloh, Germany)	18. 3.05A
G-BZVE	Cameron N-133 HAB	10092		20. 6.01	I.M.Ashpole	(Bridstow, Ross-on-Wye)	7.10.04A
					(New owner 10.04)		
G-BZVF	Cessna T182T Turbo Skylane	T18208009	N109LP	12. 6.01	R.Macaire t/a Denston Hall Estate	Appleacre Farm, Stradishall	8. 7.07T
G-BZVG	Eurocopter AS350B3 Ecureuil	3368	F-WQOR	5. 3.01	Windrush Aviation Ltd	Oxford	3. 7.07T
G-BZVH	Raj Hamsa X'Air 582	BMAA/HB/160		20. 4.01	W.Bracken	(Mountmellick, County Laois)	17. 8.05P
	(Built B and D Bergin - kit no.561)						
G-BZVI	Nova Vertex 24	13379		24. 5.01	M Hay *(New owner 1.03)*	(Dundee)	
G-BZVJ	Pegasus Quantum 15	7821		12. 6.01	T.McMahon	Perth	17. 6.05P
G-BZVK	Raj Hamsa X'Air 582	BMAA/HB/152		22. 2.01	A.P.and Jane M.Cadd	(Steeple Aston, Bicester)	7. 1.05P
	(Built K.P.Taylor - kit no.592)						
G-BZVM	Rans S-6-ES Coyote II	PFA 204-13705		1. 3.01	N.N.Ducker	Tatenhill	20. 7.05P
	(Built N N Ducker) (Rotax 912-UL)						
G-BZVN	Van's RV-6	PFA 181-13188		25. 4.01	J.A Booth	Sheffield City	27. 6.05P
	(Built J A Booth) (Lycoming O-360)						
G-BZVO	Cessna TR182 Turbo Skylane RG II	R18200990	D-EPOL N739CX	3. 4.01	Swiftair Ltd	Elstree	21.10.07E
G-BZVR	Raj Hamsa X'Air 582	BMAA/HB/146		13. 3.01	P.Travis tr Hummingbird Club	Davidstow Moor	15.11.04P
	(Built R.P.Sims, E Bowen and P Travis - kit no.566)						
G-BZVS	CASA 1-131E Jungmann Series 2000	2013	D-EHEP (2) D-EDEE, Spanish AF E3B-409	3. 5.01	W.R.M.Beesley	Tollerton	21. 7.04P
	(Avco Lycoming-180hp)						
G-BZVT	III Sky Arrow 650 T	PFA 298-13333		23. 3.01	D.J.Goldsmith	(Edenbridge)	14. 8.05P
	(Built R N W Wright)						
G-BZVU	Cameron Z-105 HAB	10078		9. 8.01	The Mall Balloon Team Ltd	(Bristol)	26. 3.05A
					(The Mall/Cribbs Causeway titles)		
G-BZVV	Pegasus Quantum 15-912	7793		12. 3.01	D.R.Morton	Headon Farm, Retford	18. 3.05P
G-BZVW	Ilyushin Il-2 Stormovik	1870710	Soviet AF 1870710	16. 5.01	S.Swallow and R.H.Cooper	(Stowe, Lincoln)	
G-BZVX	Ilyushin Il-2 Stormovik	1878576	Soviet AF 1878576	16. 5.01	S.Swallow and R.H.Cooper	(Stowe, Lincoln)	
G-BZVZ	Eurocopter AS355N Ecureuil 2	5691		24. 4.01	Ilona My Ltd *(Based on board "MV Ilona")*	(Jersey)	14. 6.07
G-BZWB	Mainair Blade 912	1284-0507-7 & W1079		26. 4.01	L.Cottle	Baxby Manor, Husthwaite	13. 5.05P
G-BZWC	Raj Hamsa X'Air Falcon 912	BMAA/HB/157		9. 5.01	C.McAfee	Nether Glastry, Dunblane	29. 4.04P
	(Built G A J Salter - kit no.587)						
G-BZWF	Colt AS-120 Mk.II HA Airship	10095	(HS-) G-BZWF	2. 5.01	MA Flying Ltd	(Pattaya, Thailand)	20. 6.02A
	(Built Cameron Balloons Ltd)						
G-BZWG	Piper PA-28-140 Cherokee Cruiser	28-7625188	N9656K	17. 5.01	H and E Merkado	Panshanger	12. 7.07T
G-BZWH	Cessna 152	15281339	N49819	17. 5.01	J and H Aviation Services Ltd	North Weald	1. 7.07T
G-BZWI	Medway EclipseR	170/148		3. 5.01	R.A Keene	Over Farm, Gloucester	28. 3.05P

Registration	Type	C/n	Previous identity	Registered	Owner/operator	Base	CofA expiry
G-BZWJ	CFM Streak Shadow SA	PFA 206-13553		8. 5.01	T.A Morgan	Headcorn	17. 8.05P
	(Built T.A Morgan - kit no K.338) (Rotax 582)				*"Harmony Angel"*		
G-BZWK	Jabiru Aircraft Jabiru SK	PFA 274-13292		8. 5.01	N.J.Mines	Kemble	21. 4.05P
	(Built R Thompson)						
G-BZWM	Solar Wings Pegasus XL-Q	7792		18. 5.01	D.T.Evans	(Hereford)	24. 6.04P
	(Built using trike SW-TE-0136 from G-MVKM)						
G-BZWN	Van's RV-8	PFA 303-13692		14. 5.01	A J.Symms and R.D.Harper	(Langport)	9. 2.05P
	(Built A J.Symms and R.D.Harper) (Lycoming IO-360) *(Tailwheel u/c)*						
G-BZWR*	Mainair Rapier	1275-0301-7 & W1070		7. 3.01	R.J.Swann *(Cancelled 22.9.04 by CAA)*	(Preston)	24. 8.04P
G-BZWS	Pegasus Quantum 15-912	7813		26. 4.01	W.Williams tr G-BZWS Syndicate	(Sheffield)	9. 5.05P
G-BZWT	Tecnam P92-EM Echo	PFA 318-13681		17. 5.01	R.F.Cooper	Field Farm, Oakley	22.12.04P
	(Built R.F.Cooper) (Marked as "P92S")						
G-BZWU	Pegasus Quantum 15-912	7831		19. 7.01	P.C.Hogg	Longacres Farm, Sandy	28. 9.04P
G-BZWV	Steen Skybolt	PFA 64-10751		30. 5.01	P.D.and K.Begley	Sywell	3. 4.05P
	(Built P.D.Begley)						
G-BZWX	Whittaker MW5-D Sorcerer	PFA 163-13599		1. 6.01	G.E.Richardson	Sywell	11.12.04P
	(Built P G Depper)						
G-BZWY	CFM Streak Shadow SA	PFA 206-13601		31. 5.01	B.Cartwright	(Craigavon, County Armagh)	17.10.05P
	(Built B.Cartwright)						
G-BZWZ	Van's RV-6	PFA 181A-13419		26. 4.01	J.Shanley	Morgansfield, Fishburn	21. 3.05P
	*(Built J Shanley) (*Lycoming O-320)						
G-BZXA	Raj Hamsa X'Air Victor2	BMAA/HB/148		31. 5.01	D.W.Mullin	Hawarden	27. 6.05P
	(Built D.W.Mullin - kit no.560)						
G-BZXB	Van's RV-6	PFA 181A-13625		4. 6.01	B.J.King-Smith and D.J.Akerman	Goodwood	17. 3.05P
	(Built B.J.King-Smith and D.J.Akerman) (Lycoming O-360)						
G-BZXC	Scottish Aviation Bulldog Series 120/121	BH120/260	XX612	8. 6.01	A R.Oliver *(As "XX612/A03")*	Bourn	17.12.05P
G-BZXD	RotorWay Executive 162F	6494		5. 6.01	P.G.King	(Gravesend)	AC
	*(Built P G King) (*RotorWay RI 162F)				*(Noted 1.05)*		
G-BZXE	de Havilland DHC-1 Chipmunk 22	C1/0722	WP839	5. 6.01	K.Moore	Blackpool	
G-BZXF	Cameron A-210 HAB	4999		5. 2.02	Off The Ground Balloon Co Ltd	(Kendal)	7. 3.05T
G-BZXG	Dyn'Aéro MCR-01 ULC	PFA 301B-13815		26.10.01	J.Rankin	Perth	12. 8.05P
	(Built J M Scott and G J Sargent) (Rotax 912-ULS)						
G-BZXI	Nova Philou 26	11207		16. 5.01	M.Hay *(New owner 1.03)*	(Dundee)	
G-BZXJ	Schweizer 269C-1	0128		20. 6.01	Helicentre Ltd	Liverpool	1. 8.07T
G-BZXK	Robin HR200/120B	286	F-GNNV	12. 6.01	S.J. Skilton t/a Aviation Rentals	Bournemouth	15. 7.07T
G-BZXL	Whittaker MW5-D Sorcerer	PFA 163-13738		7. 6.01	R.Hatton	Kirkbride	2. 7.04P
	(Built K.Wright and C Gale) (Rotax 503)						
G-BZXM	Mainair Blade 912	1283-0501-7 & W1078		20. 4.01	M.E.Fowler	(Botolph Claydon, Buckingham)	1. 8 05P
G-BZXN	Jabiru Aircraft Jabiru UL-450	PFA 274A-13747		7. 6.01	J.Armstrong	Shobdon	19. 1.05P
	(Built A R Silvester)						
G-BZXO	Cameron Z-105 HAB	10125		27. 6.01	Airship and Balloon Company Ltd	(Bristol)	1. 6.05A
					(Innogy/nPower titles)		
G-BZXP	Air Création 582/Kiss 400	BMAA/HB/169		14. 6.01	E.D.Deed	Sywell	15. 4.05P
	(Built P M Dewhurst - kit no.FL001, Trike T00100, Wing A00056-0054)						
G-BZXR	Cameron N-90 HAB	10124		3. 9.01	H.J.Andrews *(New owner 1.05)*	Selborne, Alton	21. 8.02A
G-BZXS	Scottish Aviation Bulldog Series 120/121	BH120/296	XX631	21. 6.01	K.J.Thompson *(As "XX631/W")*	Newtownards	8. 5.06
G-BZXT	Mainair Blade 912	1286-0501-7 & W1081		25. 5.01	S.R.Vinsun tr Barton 912 Flyers	Barton	17. 6.05P
G-BZXV	Pegasus Quantum 15-912	7828		28. 6.01	S.Laws	Rufforth	13. 7.05P
G-BZXW	VPM M-16 Tandem Trainer	PFA G/12-1249	G-NANA	30. 4.01	S.J.Tyler	(Carlisle)	31. 8.00P
	(Built J W P Lewis) (Rotax 912S)						
G-BZXX	Pegasus Quantum 15-912	7812		20. 4.01	C.S.Povey tr G-BZXX Group	Tarn Farm, Cockerham	24. 4 05P
G-BZXY	Robinson R44 Raven	1027		12. 6.01	Heliceair Ltd	North Weald	8. 7.07T
G-BZXZ	Scottish Aviation Bulldog Series 120/121	BH120/294	XX629	1. 6.01	A.G.Fowles *(As "XX629/V")*	Sleap	15.11.07T
G-BZYA	Rans S-6-ES Coyote II	PFA 204-13529		12. 6.01	M.R.Osbourn	(Worcester)	25. 8.05P
	(Built D.J.Clack - kit no.0499.1313) (Rotax 582) *(Tri-cycle u/c)*						
G-BZYB	Westland SA.341D Gazelle HT.3	1272	XX382	14. 6.01	S.Atherton tr Gazelle Flying Group	Breighto	9. 7.05P
G-BZYD	Westland SA.341B Gazelle AH.1	1648	XZ329	14. 6.01	Aerocars Ltd	Filton	3. 6.05P
	(C/n 1652 quoted also)				*(As "XZ329" in Army c/s)*		
G-BZYE	Robinson R22 Beta II	3231		15. 6.01	Plane Talking Ltd	Elstree	8. 7.07T
G-BZYG	DG Flugzeugbau DG-500MB	(5E220B15)		25. 9.01	R.C.Bromwich *"94"*	Kingston Deverill	28.11.07E
G-BZYI	Nova Phocus 123	9748		8. 6.01	M Hay *(New owner 1.03)*	(Dundee)	
G-BZYK	Jabiru Aircraft Jabiru UL-450	PFA 274A-13227		21. 6.01	P.A James	Redhill	28. 8.05P
	(Built A S Forbes)				tr Cloudbase Aviation G-BZYK		
G-BZYL	Rans S-6-ES Coyote II	PFA 204-13718		22. 6.01	C.B.Heslop	Lower Wasing Farm, Brimpton	6. 5.05P
	(Build J D Harris) (Jabiru 2200A) *(Tri-cycle u/c)*						
G-BZYM	Raj Hamsa X'Air HK700	BMAA/HB/172		21. 6.01	G.Fleck	Kirkbride	21. 4.05P
	(Built G.Fleck - kit no.649)						
G-BZYN	Pegasus Quantum 15-912	7835		15. 8.01	A.S.Docherty	(St. Albans)	25.10.05P
G-BZYO	Colt 210A HAB	3523	D-OSPM	19. 7.01	P.M.Forster *(Op Alba Ballooning)*	(Edinburgh)	17. 3.04T
G-BZYP	British Aerospace Jetstream Series 3200	978	F-GMVL	1. 8.01	Trident Aviation Leasing Services (Ireland) Ltd	(Dublin)	
			F-OHFX, G-31-978		*(Exported 1.03)*		
G-BZYR	Cameron N-31 HAB	10137		6. 8.01	N.J.Langley	(Clapton in Gordano, Bristol)	1. 6.05A
G-BZYS	Micro Aviation B.22S Bantam	94-001	ZK-JDO	12. 7.01	D.L.Howel *(Noted 5.04)*	Rochester	12. 9.02P
G-BZYT	Interavia 80TA HAB	04309-92		6. 7.01	J.King *(New owner 10.03)*	(Leighton Buzzard)	
G-BZYU	Whittaker MW6 Merlin	PFA 164-13647		2. 7.01	K.J.Cole	Over Farm, Gloucester	8. 7.04P
	(Built K.J.Cole) (Rotax 582)						
G-BZYV	Snowbird Mk.V 582	BMAA/HB/175		5. 7.01	S.Jones	(Tregaron)	
	(Built S Jones)						
G-BZYW	Cameron N-90 HAB	10134		6. 8.01	C.and J.M.Bailey	(Pill, Bristol)	30. 7.05T
					t/a Bailey Balloons *(SWEB titles)*		
G-BZYX	Raj Hamsa X'Air HK700	BMAA/HB/173		15. 6.01	A G.Marsh	Ingliston	8. 7.05P
	(Built A G.Marsh - kit no.653)						
G-BZYY	Cameron N-90 HAB	10130		30. 8.01	Mason Zimbler Ltd	(Bristol)	12. 8.05A
G-BZZD	Reims/Cessna F172M Skyhawk II	F17201436	G-BDPF	14. 4.98	R.H.M.Richardson-Bunbury	Bodmin	7. 8.05T

G-CAAA - G-CZZZ

Registration	Type	C/n	Previous identity	Date	Owner	Base	CofA
G-CAHA	Piper PA-34-200T Seneca II	34-7770010	N23PL	7. 7.98	H. and R.Marshall	Sandtoft	9. 1.05T
			SE-GPY, (D-IIIC), SE-GPY				
G-CAIN	CFM Shadow Series CD	062	G-MTKU	26. 1.99	S.K.Starling	Suton, Norfolk	27. 9.05P
G-CALL	Piper PA-23-250 Aztec F	27-7754061	N62826	21.12.77	J.D.Moon	Ronaldsway	22. 6.07T
G-CAMB	Aérospatiale AS355F2 Ecureuil 2	5416RAF	N813LP	17.12.96	Cambridge and Essex Air Support Consortium		
						RAF Wyton	22. 5.06T
G-CAMM	Hawker Cygnet rep	PFA 77-10245	(G-ERDB)	30. 5.91	D.M.Cashmore	Old Warden	20. 7.05P
	(Built D M Cashmore) (Mosler MM-CB35)				"6" *(On loan to Richard Shuttleworth Trustees)*		
G-CAMP	Cameron N-105 HAB	4546		24. 3.99	R.D.Parry	(Hong Kong, PRC)	7. 8.05A
					(Op Hong Kong Balloon and Airship Club)		
G-CAMR	BFC Challenger II	PFA 177-12569		26. 3.99	P.R.A Walker	Ringwood	
	(Built P R A Walker) (BFC kit so should carry PFA/177A type prefix)				*(Current status unknown)*		
G-CAPI	Avions Mudry CAP.10B	76	G-BEXR	16. 3.99	I.C.Matthews tr PI Group	(Newmarket)	27.11.06T
G-CAPX	Akrotech Europe CAP.10B	280		21. 9.98	H.J.Pessall	Leicester	26. 3.05
G-CBAB	Scottish Aviation Bulldog Series 120/121	BH120/235	XX543	14. 6.01	Propshop Ltd *(As "XX543/F")*	Duxford	14. 3.05
G-CBAD	Mainair Blade 912	1287-0601-7 & W1082		8. 6.01	P.Lister *(New owner 3.04)*	(Southport)	19. 8.05P
G-CBAE	British Aerospace BAe 146 Series 200	E2057	SE-DRB	30. 4.02	BAE SYSTEMS (Operations) Ltd	Exeter	
			N698AA, N146AC, G-5-057		*(Open storage 9.04 as "SE-DRB")*		
G-CBAF	Neico Lancair 320	PFA 191-13567		11. 6.01	R.W.Fairless	Lee-on-Solent	17. 7.05P
	(Built R.W.Fairless) (Lycoming IO-320)						
G-CBAH	Raj Hamsa X'Air 582	BMAA/HB/174		4. 7.01	D.N.B.Hearn	(Ventnor)	25. 4.05P
	(Built D.N.B.Hearn - kit no.640)						
G-CBAI	Flight Design CT2K	01-02-04-07	G-69-52	4. 7.01	K.D.Taylor	(Driffield)	9. 5.05P
	(Assembled Pegasus Aviation as c/n 7xxx?)						
G-CBAK	Robinson R44 Clipper	1089		2. 8.01	J.Robinson	(Bridlington)	8. 8.07T
G-CBAL	Piper PA-28-161 Warrior II	28-8116087	LN-MAD	25. 3.94	Britannia Airways Ltd	Filton	13. 4.06T
			N83007				
G-CBAN	Scottish Aviation Bulldog Series 120/121	BH120/326	XX668	26. 7.01	C.Hilliker *(As "XX668/I")*	RAF Colerne	4.10.04T
G-CBAO	Not yet allocated						
G-CBAP	Zenair CH.601UL	PFA 162A-13656		12. 7.01	L.J.Lowry	Lower Mountpleasant, Chatteris	2.12.04P
	(Built L.J.Lowry) (Rotax 912S)						
G-CBAR	Stoddard-Hamilton GlaStar	PFA 295-13133		18. 5.01	C.M.Barnes	(Tadley)	10. 5.05P
	(Built C.M.Barnes) (Tricycle u/c)						
G-CBAS	Rans S-6-ES Coyote II	PFA 204-13688		4. 7.01	S.R.Green	Wickwar	27. 1.05P
	(Built S.R.Green) (Rotax 912-UL) *(Tailwheel u/c)*						
G-CBAT	Cameron Z-90 HAB	10099		1. 6.01	British Telecommunications plc	(Thatcham)	9. 4.05A
					(BT Ignite titles)		
G-CBAU	Rand Robinson KR-2	PFA 129-12789		11. 7.01	B.Normington	(Leamington Spa)	
	(Built B.Normington)						
G-CBAV	Raj Hamsa X'Air Victor2	BMAA/HB/127		7. 9.01	D.W.Stamp and G.J.Lampitt		
	(Built D.W.Stamp and G.J.Lampitt - kit no.399)					Pound Green, Buttonoak, Bewdley	3. 9.05P
G-CBAW	Cameron A-300 HAB	10148		16. 4.02	D.K.Hempleman-Adams	(Corsham)	
G-CBAX	Tecnam P92-EM Echo	PFA 318-13698		26. 6.01	R.P.Reeves	Dunkeswell	1.12.05P
	(Built R.P.Reeves)						
G-CBAY	Pegasus Quantum 15-912	7829		4. 6.01	P.R.Jones	Felthorpe	18. 6.03P
					(Damaged Felthorpe in arson attack 18.2.03)		
G-CBAZ	Rans S-6-ES Coyote II	PFA 204-13596		12. 7.01	G.V.Willder	Barton	27.11.04P
	(Built G.V.Willder) (Rotax 582)						
G-CBBA	Robin DR400/180 Régent	2505		27. 7.01	H.P.K.Ferdinand	North Weald	15. 8.07
G-CBBB	Pegasus Quantum 15-912	7827		22. 6.01	A Rose	Knapthorpe Lodge, Caunton	21. 6.05P
					tr Charlie Bravo Group		
G-CBBC	Scottish Aviation Bulldog Series 120/121	BH120/201	XX515	8. 6.01	Bulldog Flyers Ltd *(As "XX515/4")*	Blackbushe	15.11.04T
G-CBBF	Beech 76 Duchess	ME-352	OY-BED	23. 7.01	Liddell Aircraft Ltd	Bournemouth	6.11.07E
			EI-BHS		*(Op CTC Aviation)*		
G-CBBG	Mainair Blade	1291-0601-7 & W1086		23. 7.01	P.J.Donoghue	Warrington	5. 5.05P
G-CBBH	Raj Hamsa X'Air V2	BMAA/HB/143		19. 7.01	R.D.Parkinson	Lower Mountpleasant, Chatteris	13. 8.05P
	(Built W G Colyer - kit no.435)						
G-CBBL	Scottish Aviation Bulldog Series 120/121	BH120/243	XX550	8. 8.01	I.R.Bates *(As "XX550/Z")*	Fenland	16.10.04
G-CBBM	ICP MXP-740 Savannah Jabiru	BMAA/HB/176		10. 8.01	C.E.Passmore	East Barling, Essex	1. 5.05P
	(Built P.J.Wilson and S.Whittaker - kit no.01-03-51-062) (C/n translation is 01 [= year] - 03 [= month] - 51 [= model], 062 [= 62nd ICP unit])						
G-CBBN	Pegasus Quantum 15-912	7844		9. 8.01	N.I.Garland tr G-CBBN Group	Dunkeswell	31. 8.05P
G-CBBO	Whittaker MW5-D Sorcerer	PFA 163-13443		23. 7.01	P.J.Gripton	Baxby Manor, Husthwaite	12. 2.05P
	(Built P.J.Gripton)						
G-CBBP	Pegasus Quantum 15-912	7843		31. 7.01	P.F.Warren	Dunkeswell	30. 7.05P
G-CBBR	Scottish Aviation Bulldog Series 120/121	BH120/290	XX625	8. 8.01	Elite Consultancy Corporation Ltd	Norwich	AC
					(As "XX625/01") (Noted 8.04)		
G-CBBS	Scottish Aviation Bulldog Series 120/121	BH120/343	XX694	8. 8.01	European Light Aviation Ltd	(Newcastle Upon Tyne)	13. 4.06T
					(As "XX694/E")		
G-CBBT	Scottish Aviation Bulldog Series 120/121	BH120/344	XX695	8. 8.01	Newcastle Bulldog Group Ltd	Durham Tees Valley	25. 1.07T
					(As "XX695/3")		
G-CBBU	Scottish Aviation Bulldog Series 120/121	BH120/360	XX711	6. 8.01	Elite Consultancy Corporation Ltd	Egginton	AC
					(As "XX711/X") (In open storage 4.04)		
G-CBBV	Westland SA.341G Gazelle HT.3	1750	XZ940	28. 8.01	A.Cook t/a C3 Consulting	(Hartlepool)	
	(Officially regd with c/n 1757)				*(New owners 6.04)*		
G-CBBW	Scottish Aviation Bulldog Series 120/121	BH120/277	XX619	1. 8.01	S.E.Robottom-Scott *(As "XX619/T")*	Coventry	27. 8.03T
G-CBBX	Lindstrand LBL 69A HAB	805		2. 8.01	J.L.F.Garcia	(Guadalajara, Spain)	5. 8.02A
G-CBBZ	Pegasus Quantum 15-912	7840		14. 8.01	J.R.Horn	(Ashwell, Oakham)	13. 4.05P
G-CBCA	Piper PA-32R-301T Saratoga IITC	3257244	N5338S	4.10.01	W.P.J.Davison	(Fontvielle, Monaco)	7.12.07E
G-CBCB	Scottish Aviation Bulldog Series 120/121	BH120/223	XX537	25. 9.01	The General Aviation Trading Co Ltd	North Weald	28.11.04T
					(As "XX537/C") (Noted 1.05)		
G-CBCC	Not yet allocated						

Reg	Type	C/n	Prev. Id	Regd	Owner/Operator	Base	CofA
G-CBCD	Pegasus Quantum 15	7845		6. 8.01	I.A Lumley	(Penrith)	2.12.04P
G-CBCE	Jungmann rep *(Built E.B.Toulson)*	PFA 242-13771		7. 8.01	E.B.Toulson *(As "A 50" in pre-WW2 Swiss AF dark green/red c/s)*	Sherburn-in-Elmet	27. 5.05P
G-CBCF	Pegasus Quantum 15-912	7846		23. 8.01	S Speake tr G-CBCF Group	Newton Bank, Daresbury	22. 8.04P
G-CBCH	Zenair CH.701 UL *(Built L.G.Millen)*	PFA 187-13568		8. 8.01	L.G.Millen	(Sittingbourne)	
G-CBCI	Raj Hamsa X'Air 582 *(Built P A Gilford - kit no.659)*	BMAA/HB/180		9. 8.01	A J Varga	Rufforth	30.11.04P
G-CBCJ	Rotary Air Force RAF 2000 GTX-SE *(Built J.P.Comerford)*	PFA G/13-1331		13. 8.01	J.P.Comerford	Henstridge	18. 9.05P
G-CBCK	Nipper T.66 Series 3 *(Built C Edwards fromFairey c/n 30)* (Jabiru 2200A)	PFA 25-11051	G-TEDZ	14. 8.01	N.M.Bloom	Abbots Hill Farm, Hemel Hempstead	14.12.05P
G-CBCL	Stoddard-Hamilton GlaStar *(Built C J Norman)*	PFA 295-13089		5. 9.97	A H.Harper	High Ham, Langport	21. 1.05P
G-CBCM	Raj Hamsa X'Air HK 700 *(Built A Hipkin - kit no.656)*	BMAA/HB/177		23. 7.01	G.Firth	(Barnsley)	27. 3.05P
G-CBCN	Schweizer 269C-1	0129	EI-CWS G-CBCN	17. 9.01	Helicentre Liverpool Ltd	Liverpool	29. 9.06T
G-CBCP	Van's RV-6A *(Built AM Smith)*	PFA 181A-13643		6. 8.01	A M.Smith tr G-CBCP Group	Crowfield	27. 5.05P
G-CBCR	Scottish Aviation Bulldog Series 120/121	BH120/351	XX702	5. 9.01	S.C.Smith *(As "XX702")*	Tollerton	1. 7.05T
G-CBCS	British Aerospace Jetstream Series 3201	842	SE-LHA N842JX, N842AE, G-31-842, N332QL, G-31-842	27. 9.01	Air Kilroe Ltd t/a Eastern Airways	Humberside	4.10.05E
G-CBCT	Scottish Aviation Bulldog Series 120/121	BH120/322	XX664	23. 8.01	T.Brun *(Noted 9.01)*	(Paris, France)	
G-CBCU	Hawker Siddeley Harrier GR.3 *(Officially quoted c/n FL/41H-0250295 is forward fuselage no.)*	712229	ZD668	9.11.01	Y.Dumortier Hannants Model Warehouse, Oulton Broad *(Noted 8.02)*		
G-CBCV	Scottish Aviation Bulldog Series 120/121	BH120/348	XX699	30. 8.01	C.A.Patter *(As "XX699/F")*	Newark	8. 7.05
G-CBCX	Pegasus Quantum 15	7848		10. 9.01	D.V.Lawrence	(Stourbridge)	9. 9.05P
G-CBCY	Beech C24R Sierra	MC-491	N881RS PH-HLA	26. 9.01	Plane Talking Ltd *(Op Cabair)*	Bournemouth	3. 3.05T
G-CBCZ	CFM Streak Shadow *(Built J.A Hambleton - kit no K.340-SLA)* (Rotax 582)	PFA 206-13586		13. 9.01	J.A Hambleton *"Alana Rose"*	Otherton, Cannock	5.12.05P
G-CBDA	British Aerospace Jetstream Series 3217	986	JA8590 G-31-986	30.10.01	Eastern Airways (UK) Ltd *(Op Air Kilroe)*	Humberside	16. 4.06T
G-CBDC	Thruster T.600N 450 Jab Sprint	0071-T600N-054		12. 7.01	D.Clarke t/a David Clarke Microlight Aircraft	Great Massingham	14. 7.04P
G-CBDD	Mainair Blade	1293-0701-7 & W1088		1. 8.01	R D Smith	(Urmston, Manchester)	4. 8.05P
G-CBDG	Zenair CH.601HD Zodiac *(Built R.E.Lasnier)*	PFA 162-13375		3. 9.01	R.E.Lasnier	(Moreton, Wirral)	
G-CBDH	Flight Design CT2K *(Assembled Pegasus Aviation as c/n 7849)*	01-07-02-17		4.10.01	K.Tuck	(Dereham)	18.10.04P
G-CBDI	Denney Kitfox Model 2 *(Built J.G.D.Barbour)*	PFA 172-11888		4. 9.01	J.G.D.Barbour	Sherriff Hall Estate, Balgone	25. 3.05P
G-CBDJ	Flight Design CT2K *(Assembled Pegasus Aviation as c/n 7850)*	01-07-01-17		11.10.01	P.J.Walker	Griffins Farm, Temple Bruer	27.10.04P
G-CBDK	Scottish Aviation Bulldog Series 120/121	BH120/259	XX611	26. 9.01	J.N.Randle *(As "XX611")*	Coventry	20.10.05
G-CBDL	Mainair Blade	1292-0701-7 & W1087		1. 8.01	D.Lightwood	(Macclesfield)	14.11.03P
G-CBDM	Tecnam P92-EM Echo *(Built J J Cozens and C J Willy)*	PFA 318-13756		11. 7.01	J.J.Cozens	Dunkeswell	17. 6.05P
G-CBDN	Mainair Blade	1297-0801-7 & W1092		20. 9.01	A and R.W.Osborne	Beccles	26. 9.05P
G-CBDO	Raj Hamsa X'Air 582 *(Built R.T.Henry - kit no.583)*	BMAA/HB/170		12.11.01	A.Campbell	(Dungannon)	28.10.05P
G-CBDP	Mainair Blade 912	1295-0801-7 & W1090		17. 8.01	D.S.Parker	Carlisle	16. 8.05P
G-CBDS	Scottish Aviation Bulldog Series 120/121	BH120/356	XX707	27. 7.01	H R M Tyrell *(As "XX707/4")*	Sleap	6.12.04T
G-CBDT	Zenair CH.601HD Zodiac *(Built D.G.Watt)*	PFA 162-12474		17. 9.01	D.G.Watt	Cark-in-Cartmell	4. 7.05P
G-CBDU	Quad City Challenger II *(Built B A Hiscox)*	PFA 177-13000		14. 9.01	Hiscox Cases Ltd *(Noted 1.04)*	Otherton, Cannock	
G-CBDV	Raj Hamsa X'Air 582 *(Built R.J.Brown and D J Prothero - kit no.616)*	BMAA/HB/161		6. 8.01	R.J.Brown	Davidstow Moor	9. 5.05P
G-CBDW	Raj Hamsa X'Air Jab22 *(Built P.R.Reynolds - kit no.575)*	BMAA/HB/150		18. 9.01	A.W.G.Ambler	Wickenby	5.10.05P
G-CBDX	Pegasus Quantum 15	7857		11.10.01	C.C.Beck	(Croydon)	24.10.04P
G-CBDY	Raj Hamsa X'Air Victor2 *(Built D Mahajan - kit no.588)*	BMAA/HB/155		26. 9.01	P.M.Stoney	Andrewsfield	26. 5.05P
G-CBDZ	Pegasus Quantum 15-912	7852		11. 9.01	C.I.D.H.Garrison	Sutton Meadows, Ely	13. 9.05P
G-CBEB	Air Création 582/Kiss 400 *(Built P J R Bradshaw.and A R.R.Williams - kit no.FL003/135)*	BMAA/HB/184		3.10.01	P J R Bradshaw.and A R.R.Williams	Kingston Seymour	22. 4.05P
G-CBEC	Cameron Z-105 HAB	10105		16.10.01	A L.Ballarino	(Piedimonte Matese, Italy)	9. 2.05A
G-CBED	Cameron Z-90 HAB	10121		15.10.01	John Aimo Balloons SAS	(Mondovi, Italy)	28. 5.05A
G-CBEE	Piper PA-28R-200 Cherokee Arrow II	28R-7635055	N4479X	5.10.01	IHC Aviation Ltd	Biggin Hill	15.11.04T
G-CBEF	Scottish Aviation Bulldog Series 120/121	BH120/286	XX621	3.10.01	M.A Wilkinson *(As "XX621/H")*	Tollerton	31. 1.05
G-CBEG	Robinson R44 Raven	1124		28. 9.01	Abel Developments Ltd	(Thetford)	11.10.04T
G-CBEH	Scottish Aviation Bulldog Series 120/121	BH120/207	XX521	28. 9.01	J.E.Lewis *(As "XX521/H")*	(Murcott, Kidlington)	9. 9.05
G-CBEI	Piper PA-22-108 Colt	22-9136	SE-CZR	5. 6.02	M E Grogan	(Leeds)	9.12.06T
G-CBEJ	Colt 120A HAB	10181		11.10.01	J.A Gray	(Cirencester)	23. 2.05T
G-CBEK	Scottish Aviation Bulldog Series 120/121	BH120/349	XX700	26. 9.01	S.Landregan *(As "XX700/17")*	Blackbushe	23.11.07T
G-CBEL	Hawker Iraqi Fury FB.11	37579	N36SF Iraqi AF 315	6. 8.01	J.A D.Bradshaw *"361/NAVY""*	Kemble	30.10.05P
G-CBEM	Mainair Blade	1294-0801-7 & W1089		17. 8.01	M.Earp	(Macclesfield)	14.11.05P
G-CBEN	Pegasus Quantum 15-912	7855		8.10.01	N J P West	Chilbolton	1.12.04P
G-CBEO	British Aerospace Jetstream Series 3206	979	F-GMVM (F-OHFZ), G-31-979	29.12.04	Trident Aviation Leasing Services (Ireland) Ltd	(Dublin)	

Reg	Type	C/n	Previous identity	Regd	Owner/Operator	Base	CofA
G-CBEP	British Aerospace Jetstream Series 3206	980	F-GMVN G-31-980	23.11.01	Trident Aviation Leasing Services (Ireland) Ltd *(Current CofR 1.05)*	(Dublin)	
G-CBER	British Aerospace Jetstream Series 3206	982	F-GMVO G-31-982	12. 3.02	Trident Aviation Leasing Services (Ireland) Ltd *(Current CofR 1.05)*	(Dublin)	
G-CBES	Europa Aviation Europa *(Built M R Hexley - kit no.061) (Monowheel u/c)*	PFA 247-12691		27. 9.01	M R Hexley	(Penmaenmawr)	20.12.05P
G-CBET	Mainair Blade 912S	1296-0801-7 & W1091		6. 9.01	D F Kenny	Finmere	7. 9.05P
G-CBEU	Pegasus Quantum 15-912	7869		16.10.01	C.Lee	Longacres Farm, Sandy	15.10.05P
G-CBEV	Pegasus Quantum 15-912	7854		16.10.01	B.J.Syson	Longacres Farm, Sandy	15.10.05P
G-CBEW	Flight Design CT2K *(Assembled Pegasus Aviation as c/n 7868)*	01-08-02-23		19.10.01	A J Webb tr Shy Talk Group	Oxford	29.11.04P
G-CBEX	Flight Design CT2K *(Assembled Pegasus Aviation as c/n 7867)*	01-08-01-23		29.10.01	B.W.T.Rood	Sywell	17.12.04P
G-CBEY	Cameron C-80 HAB	10190		31.10.01	D.V.Fowler	(Cranbrook)	26.10.04A
G-CBEZ	Robin DR400/180 Regent	2511		26. 2.02	K.V.Field	Turweston	8. 4.05T
G-CBFA	Diamond DA.40 Star	40063		25.10.01	Lyrastar Ltd	Redhill	5. 3.05T
G-CBFB	Diamond DA.40 Star	40079	OE-VPW	30. 4.02	Phantom Air Ltd	Sleap	1. 7.05T
G-CBFE	Raj Hamsa X'Air Victor2 *(Built S Whittle - kit no.636)*	BMAA/HB/186		19.10.01	M.L.Powell	Brook Farm, Pilling	13. 6.05P
G-CBFF	Cameron O-120 HAB	10167		20.11.01	T M C McCoy *(Op Ascent Balloons) (Ascent titles)*	(Peasedown St.John, Bath)	30.10.04T
G-CBFH	Cameron Thunder AX8-105 S2 HAB	10188		13.11.01	D.V.Fowler and A N.F.Pertwee	(Cranbrook/Frinton-on-Sea)	17. 1.05A
G-CBFJ	Robinson R44 Raven	1131		7.11.01	Scotia Helicopters Ltd	Cumbernauld	8.11.07E
G-CBFK	Murphy Rebel *(Built D.Webb)*	PFA 232-13340		13. 9.01	P.J.Gibbs *(New owner 8.04)*	Truro	13.10.03P
G-CBFM	SOCATA TB-21 Trinidad GT	710	PH-BLM D-EFAK (4)	2. 1.02	Execflight Ltd	Southend	17. 3.05T
G-CBFN	Robin HR100/200B Royal	112	F-BTBP	1. 3.02	A C.Barton tr Foxtrot November Group	Blackbushe	11. 6.05T
G-CBFO	Cessna 172S Skyhawk SP	172S8929	N3520A	22.10.01	Halegreen Associates Ltd	Oxford	29.10.07E
G-CBFP	Scottish Aviation Bulldog Series 120/121	BH120/306	XX636	29.10.01	Taylor Aviation Ltd *(As "XX636/Y")*	Cranfield	23. 1.05P
G-CBFR	Not yet allocated						
G-CBFT	Raj Hamsa X'Air 582 *(Built T Collins - kit no.685)*	BMAA/HB/190		6.11.01	L Reilly	(Dring, County Longford)	25. 7.05P
G-CBFU	Scottish Aviation Bulldog Series 120/121	BH120/293	XX628	12.11.01	J.R.and S.J.Huggins *(As "XX628/9")*	Chalksole Green Farm, Alkham	15. 4.05
G-CBFV	Comco Ikarus C42 FB UK *(Built P A D Chubb)*	PFA 322-13774		5.11.01	P.A D.Chubb	Mergate Hall, Bracon Ash	31. 3.05P
G-CBFW	Bensen B.8 *(Built B.F.Pearson)*	PFA G/01-1312		6.11.01	B.F.Pearson	(Newark)	AC
G-CBFX	Rans S-6-ESN Coyote II *(Built J Whiting)* (Rotax 582)	PFA 204-13820		8.11.01	G.P.Jones	(Stoke-on-Trent)	9. 6.05P
G-CBFY	Cameron Z-250 HAB	10023		9.11.01	M.L.Gabb	(Haselor, Alcester)	5. 3.05A
G-CBFZ	Jabiru Aircraft Jabiru SPL-450 *(Built A H King) (Regd as "UL-450")*	PFA 274A-13617		8.11.01	A H.King	(Orpington)	29. 4.05P
G-CBGA	PZL-110 Koliber 160A	04010086	SP-WGM	14.11.01	Rapidangel Ltd	(Orpington)	17. 1.05T
G-CBGB	Zenair CH.601UL *(Built R.Germany)* (Rotax 912-S)	PFA 162A-13819		12.11.01	J.F.Woodham	Hollow Hill Farm, Granborough	21. 4.05P
G-CBGC	SOCATA TB-10 Tobago	1584	VH-YHB	21. 9.01	Tobago Aviation Ltd	Blackbushe	6.12.05E
G-CBGD	Zenair CH.701 UL *(Built I.S.Walsh)*	PFA 187-13785		13.11.01	I.S.Walsh *(Noted 6.04)*	Dunkeswell	
G-CBGE	Tecnam P92-EM Echo *(Built T.C.Robson)*	PFA 318-13680		9.11.01	I.D.Rutherford	(Great Kingshill, High Wycombe)	17. 8.05P
G-CBGF	Piper PA-31 Navajo 310	31-749	F-BTCP N7227L	2. 1 02	Datalake Ltd	Fairoaks	3. 3.05T
G-CBGG	Pegasus Quantum 15 (Rotax 503)	7874		27.11.01	P.C.Bishop	Dunkeswell	17. 1.05P
G-CBGH	Teverson Bisport *(Built R.C.Teverson)*	PFA 267-12784		7.11.01	R.C.Teverson	Waits Farm, Belchanp Walter	29.11.05P
G-CBGI	CFM Streak Shadow *(Built M.W.W.Clotworthy)*	PFA 206-13559		30.10.01	M.W.W.Clotworthy	(Bath)	
G-CBGJ	Aeroprakt A22 Foxbat *(Built W R Davis-Smith) (Noted marked "002" - could denote 2nd UK built)*	PFA 317-13803		14.11.01	M.J. Whiteman-Haywood	Pound Green, Buttonoak, Bewdley	18. 5.05P
G-CBGK	Hawker Siddeley Harrier GR.3 *(Forward fuselage no.FL/41H-0150252)*	41H-712218	9220M XZ995	13.12.01	Y.Dumortier Hannants Model Warehouse, Oulton Broad *(Noted 8.02)*		
G-CBGL	Max Holste MH.1521M Broussard	19	F-BMJO F-BNEN, Fr AF	3.12.01	A I.Milne tr Broussard Flying Group	Horsford	AC
G-CBGM	Mainair Blade 912	1299-1001-7 & W1094		30.10.01	J.R.Pearce	Chilbolton	7.10.04P
G-CBGN	Van's RV-4 *(Built G.A Nash)* (Lycoming O-320)	PFA 181-12443		16.10.01	G.A Nash *"Spirit Of Youth"*	Kemble	26. 3.05P
G-CBGO	Murphy Maverick 430 *(Built C R Ellis)* (Jabiru 2200A)	PFA 259-13470		24.10.01	C.R.Ellis and E.A Wrathall	(Chapel-en-le-Frith)	7. 7.05P
G-CBGP	Comco Ikarus C42 FB UK *(Built A R.Lloyd)*	PFA 322-13741		22.11.01	G.F.Welby	Deenethorpe	20. 5.05P
G-CBGR	Jabiru Aircraft Jabiru UL-450 *(Built K.R.Emery)*	PFA 274A-13682		21.11.01	K.R.Emery	Sittles Farm, Alrewas	28..3.05P
G-CBGS	Cyclone AX2000 HKS	7866		8. 3.02	P.J.Rooke	Clench Common	1. 6.05P
G-CBGT	Mainair Blade 912	1300-1001-7 & W1095		30.10.01	J.A Cresswell	(Sittles Farm, Alrewas)	3.10.04P
G-CBGU	Thruster T.600N 450	0121-T600N-055		21.11.01	K.Ford and M.Gill	Wickenby	23. 1.05P
G-CBGV	Thruster T.600N 450	0121-T600N-056		21.11.01	B.S.Waycott tr Red Arrow Syndicate	Eastbach Farm, Coleford	9. 3.05P
G-CBGW	Thruster T.600N 450 Sprint	0121-T600N-058		21.11.01	N.J.S.Pitman	Brock Farm, Billericay	18. 3.05P
G-CBGX	Scottish Aviation Bulldog Series 120/121	BH120/287	XX622	26.11.01	G.B.Pearce *(As "XX622/B")*	Washington, West Sussex	7.10.05T
G-CBGY	Mainair Blade 912	1304-1101-7 & W1099		28. 8.02	M.Talbot	Onecote, Leek	20. 8.05P

Registration	Type	C/n	Previous identity	Date	Owner	Base	CofA
G-CBGZ	Westland SA.341C Gazelle HT.2	1923	ZB646	30.10.01	D.Weatherhead Ltd	Cambridge	19. 9.05P
G-CBHA	SOCATA TB-10 Tobago	1583	VH-YHA	6.11.01	Oscar Romeo Aviation Ltd	Redhill	9. 5.05T
G-CBHB	Raj Hamsa X'Air 582	BMAA/HB/189		22.11.01	R.A J.Graham	Kirkbride	12. 1.05P
	(Built G Fleck - kit no.611)						
G-CBHC	Rotary Air Force RAF 2000 GTX-SE	PFA G/13-1326		22.11.01	A J.Thomas	(Sutton Coldfield)	11. 8.05P
	(Built A J.Thomas)						
G-CBHD	Cameron Z-160 HAB	10225		1. 3.02	Ballooning 50 Degrees Nord	(Fouhren, Luxembourg)	9. 4.05A
					(*Nichobetzen Founzen titles.*)		
G-CBHE	Slingsby T.67M-200 Firefly	2050	SE-LBE	28.12.01	R.Swann	Bournemouth	16. 2.06T
			LN-TFE, G-7-125				
G-CBHF	Not yet allocated						
G-CBHG	Mainair Blade 912S	1298-1001-7 & W1093		12.12.01	B S Hope	(Marsh Gibbon, Bicester)	26. 5.05P
G-CBHI	Europa Aviation Europa XS	PFA 247-13245		31.10.01	B.Price	(Southampton)	14.12.04P
	(Built B.Price - kit no.373) (Monowheel u/c)				tr Hotel India Group		
G-CBHJ	Mainair Blade 912	1305-1201-7 & W1100		28. 1.02	B.C.Jones	(Altrincham)	17. 7.04P
G-CBHK	Pegasus Quantum 15	7871		6.12.01	B.Dossett	London Colney	5. 3.05P
	(HKS 700E)						
G-CBHL	Eurocopter AS350B2 Ecureuil	2673	C-GKHS	28. 1.02	Helimac Ltd	Lanark	19. 3.05P
			JA6123				
G-CBHM	Mainair Blade 912	1301-1100-7 & W1096		3.12.01	W.T.Milburn	Barton	4. 3.05P
G-CBHN	Pegasus Quantum 15-912	7872		6.12.01	G.G.Cook	Clench Common	17. 1.05P
G-CBHO	Gloster Gladiator II	Not known	N5719(?)	11.12.01	Retro Track and Air (UK) Ltd	(Dursley)	
G-CBHP	Corby CJ-1 Starlet	PFA 134-12498		12.12.01	D.H.Barker	(London SE10)	
	(Built J D Muldowney)				*(New owner 8.02)*		
G-CBHR	Stephens Lazer Z200	Q056	VH-IAC	31.12.01	Acro Laser Company Ltd	(Heywood)	24. 3.05P
	(Built H Selvey)						
G-CBHT	Dassault Falcon 900EX	48	G-GPWH	3. 1.02	TAG Aviation (UK) Ltd	Farnborough	23.11.05T
			F-WWFP				
G-CBHU	Tiger Cub RL5A LW Sherwood Ranger			12.12.01	M.J.Gooch	Ashcroft Farm, Winsford	
	(Built M.J.Gooch)	PFA 237-12477			*(Noted 4.04)*		
G-CBHV	Raj Hamsa X'Air 582	BMAA/HB/139		12.12.01	N.Hancock and Sheila Mathison	Barton	16.12.04P
	(Built J.D.Buchanan - kit no.525)						
G-CBHW	Cameron Z-105 HAB	10217		16. 1.02	Bristol Chamber of Commerce, Industry and Shipping		
					(*Crea8ive/Business West titles) "Bridg-it"*	(Bristol)	9. 12.04A
G-CBHX	Cameron V-77 HAB	3950		19.12.01	N.A Apsey *"Irene"*	(Hazlemere)	16. 4.05A
G-CBHY	Pegasus Quantum 15-912	7859		7. 1.02	M.W.Abbott	Enstone	1. 3.05P
G-CBHZ	Rotary Air Force RAF 2000 GTX-SE	PFA G/13-1321		2. 1.02	M P Donnelly	(Thurso)	
	(Built M P Donnelly)						
G-CBIB	Flight Design CT2K	01-08-06-23		21. 1.02	J.A Moss	Grove Farm, Needham	6. 3.05P
	(Assembled Pegasus Aviation as c/n 7878)						
G-CBIC	Raj Hamsa X'Air Victor2	BMAA/HB/156		2. 1.02	J T Blackburn and D R Sutton	Brook Farm, Pilling	19. 9.04P
	(Built J T Blackburn and D R Sutton - kit no.608)						
G-CBID	Scottish Aviation Bulldog Series 120/121	BH120/242	XX549	14.12.01	D.A Steven *(As "XX549/6")*	White Waltham	22. 8.05T
G-CBIE	Flight Design CT2K	01.09.01.23		10. 1.02	S.J.Page	Dunkeswell	18. 1.05P
	(Assembled Pegasus Aviation as c/n 7879)						
G-CBIF	Jabiru Aircraft Jabiru UL-450	PFA 274A-13789		3. 1.02	J A Iszard	Parham/Framlingham	17. 7.05P
	(Built J A Iszard)						
G-CBIG	Mainair Blade 912	1303-1101-7 & W1098		29.11.01	T.R.Clark	(New Mills, High Peak)	1.12.04P
G-CBIH	Cameron Z-31 HAB	10243		4. 1.02	Cameron Balloons Ltd	(Bristol)	19. 7.03A
G-CBII	Raj Hamsa X'Air 582	BMAA/HB/185		7. 1.02	A Worthington	Tarn Farm, Cockerham	7. 7.05P
	(Built A Worthington - kit no.676)						
G-CBIJ	Comco Ikarus C42 FB UK	PFA 322-13720		3. 1.02	G.S.Hill and J.D.Arthurs		
	(Built A Jones)					(Uppermill, Oldham/Dobcross, Oldham)	2. 4.05P
G-CBIK	RotorWay Executive 162F	6112		9. 1.02	J.Hodson	(Ashbourne)	AC
	(Built J Hodson) RotorWay RI 162F)						
G-CBIL	Cessna 182K Skylane	18257804	(G-BFZZ)	9.10.78	E Bannister and J R C Spooner		
			D-ENGO, N2604Q			Nottingham East Midlands	18. 3.06T
G-CBIM	Lindstrand LBL 90A HAB	817		28. 1.02	R.K.Parsons	(South Petherton)	19. 8.04A
G-CBIN	TEAM mini-MAX 91	PFA 186-13111		7. 1.02	K J Walton	RAF Halton	30. 6.05P
	(Built D.E.Steade) (Rotax 503) *(Enclosed cockpit)*						
G-CBIO	Thruster T.600N 450 Jab	0022-T600N-062		7. 1.02	G.J.Slater	Clench Common	1. 7.05P
G-CBIP	Thruster T.600N 450	0022-T600N-060		7. 1.02	K.D.Mitchell	Deanland	15. 6.05P
G-CBIR	Thruster T.600N 450	0022-T600N-061		7. 1.02	M.L.Smith	Popham	3. 4.05P
G-CBIS	Raj Hamsa X'Air 582	BMAA/HB/199		15. 1.02	P.T.W.T.Derges	Long Marston	11.10.05P
	(Built P.T.W.T.Derges - kit no.708)						
G-CBIT	Rotary Air Force RAF 2000 GTX-SE	PFA G/13-1340		27.11.01	Terrafirma Services Ltd		
	(Built M P Lehermette)					Lamberhurst Farm, Faversham	
G-CBIU	Cameron Flame 95 SS HAB	10222		6. 2.02	PSH Skypower Ltd	(Pewsey)	26. 8.05A
					(British Gas-Think Energy titles)		
G-CBIV	Best Off Skyranger 912	BMAA/HB/201		25. 1.02	P.M.Dewhurst and S.N.Bond	Sywell	4.10.05P
	(Built P.M.Dewhurst - kit no.SKRxxxxxxx?)						
G-CBIW	Lindstrand LBL 310A HAB	821		24. 1.02	C.E.Wood	(Witham)	2. 3.05P
G-CBIX	Zenair CH.601UL	PFA 162A-13765		24.12.01	R A and Brenda M Roberts		
	(Built M F Cottam) (Rotax 912S)					Griffins Farm, Temple Bruer	3. 6.05P
G-CBIY	Aerotechnik EV-97 Eurostar	PFA 315-13846		23. 1.02	E.M.Middleton	Broadmeadow Farm, Hereford	8. 5.05P
	(Built E.M.Middleton)						
G-CBIZ	Pegasus Quantum 15-912	7870		6.12.01	A.P.Lambert	(Grantham)	11. 1.05P
G-CBJA	Air Création 582/Kiss 400	BMAA/HB/195		11.12.01	M.S.R.Burak	(Dursley)	22. 4.05P
	(Built C W Lark - kit no.FL006, Trike T01101, Wing A01192-1194)						
G-CBJC	Not yet allocated						
G-CBJD	Stoddard-Hamilton GlaStar	PFA 295-13853		23. 1.02	K.F.Farey	(Bourne End)	
	(Built K F Farey)						
G-CBJE	Rotary Air Force RAF 2000 GTX-SE	PFA G/13-1342		23. 1.02	K.F.Farey	(Bourne End)	
	(Built K F Farey)						

Registration	Type	c/n	Previous identity	Reg date	Owner/operator	Base	CofA
G-CBJG	de Havilland DHC-1 Chipmunk T.20 *(Built OGMA)*	63	CS-AZT Portuguese AF FAP 1373	8. 2.02	C.J.Rees	(Oundle)	
G-CBJH	Aeroprakt A22 Foxbat *(Built H Smith)*	PFA 317-13847		30. 1.02	H.Smith	Morgansfield, Fishburn	28. 9.05P
G-CBJI	Cameron N-90 HAB	4169		22. 3.02	C.D.H.Oakland *(Impian Desaru titles)*	(Bristol)	29. 4.04A
G-CBJJ	Scottish Aviation Bulldog Series 120/121	BH120/211	XX525	3.12.01	Elite Consultancy Corporation Ltd *(As "XX525/8") (Noted 5.03)*	Norwich	
G-CBJK	Scottish Aviation Bulldog Series 120/121	BH120/362	XX713	3.12.01	Elite Consultancy Corporation Ltd *(As "XX713/2") (Noted 8.04)*	Norwich	
G-CBJL	Air Création 582/Kiss 400 *(Built R.E.Morris - kit no.FL005, Trike T01099, Wing A01158-1164)*	BMAA/HB/205		8. 2.02	R.E.Morris	(Kidwelly, Dyfed)	30. 7.04P
G-CBJM	Jabiru Aircraft Jabiru SP-470 *(Built A T Moyce)*	PFA 274B-13769		11.12.01	A T Moyce	Newtownards	22. 5.05P
G-CBJN	Rotary Air Force RAF 2000 GTX-SE *(Built R Hall)*	PFA G/13-1335		30. 1.02	R.Hall	(Truro)	
G-CBJO	Pegasus Quantum 15-912	7861		10.12.01	J.H.Bradbury	Arclid Green, Sandbach	10. 3.05P
G-CBJP	Zenair CH.601UL *(Built R.E.Peirse)* (Rotax 912S)	PFA 162A-13590		31. 1.02	R.E.Peirse	(Royston)	31. 8.05P
G-CBJR	Aerotechnik EV-97 Eurostar *(Buillt B.J.Crockett)*	PFA 315-13845		31. 1.02	B.J.Crockett	(Hereford)	4. 4.05P
G-CBJS	Cameron C-60 HAB	10253		17. 4.02	J.M.Stables	(Knaresborough)	16. 4.03A
G-CBJT	Mainair Blade	1302-1101-7 & W1097		12.12.01	T.K.I.Dearden	Chiltern Park, Wallingford	25. 1.05P
G-CBJU	Van's RV-7A *(Built T W Waltham)*	PFA 323-13868		1. 2.02	T.W.Waltham	(Salisbury)	
G-CBJV	RotorWay Executive 162F *(Built Southern Helicopters Ltd)* (RotorWay RI 162F)	6589		13. 2.02	Sacker Potatoes Ltd	(Stamford)	1. 9.05P
G-CBJW	Comco Ikarus C42 FB UK *(Built T J Cale)*	PFA 322-13811		13. 2.02	P.Harper and P.J.Morton	St Michaels	14. 5.05P
G-CBJX	Raj Hamsa X'Air Falcon J 22 *(Built M.R.Coreth - kit no.622)*	BMAA/HB/181		13. 2.02	M.R.Coreth	Henstridge	2.12.04P
G-CBJY	Jabiru Aircraft Jabiru UL-450 *(Built D.L.H.Person)*	PFA 274A-13613		14. 2.02	D.L.H.Person	Great Ashfield, Suffolk	18. 8.05P
G-CBJZ	Westland SA.341G Gazelle HT.3	1734	3D-HGW XZ932	13. 2.02	K.G.Theurer	(Stuttgart, Germany)	9. 3.05P
G-CBKA	Westland SA.341G Gazelle HT.3	1746	XZ937	20. 2.02	MW Helicopters Ltd *(Stored 10.04)*	Stapleford	AC
G-CBKB	Bücker Bü.181C Bestmann	121	F-PCRL F-BCRU	28. 1.02	W.R.and G.D.Snadden	(Alexandria)	
G-CBKC	Westland SA.341D Gazelle HT.3	1104	XW862	20. 2.02	T.E.Westley	Fowlmere	7. 1.05P
G-CBKD	Westland SA.341C Gazelle HT.2	1130	XW868	20. 2.02	Flying Scout Ltd	(Montgomery)	9. 3.05P
G-CBKE	Air Création 582/Kiss 400 *(Built R.J. Howell - kit no.FL010, Trike T02011, Wing A02014-2007)*	BMAA/HB/206		18. 2.02	R.J. Howell	Kingston Seymour	20. 6.05P
G-CBKF	Reality Easy Raider Jab22 *(Built R.J.Creasey - kit no.0003)*	BMAA/HB/202		24. 1.02	R.R.Armstrong	Headcorn	14.11.04P
G-CBKG	Thruster T.600N 450	0022-T600N-059		1. 3.02	G.E Hillyer-Jones	Shobdon	3. 3.05P
G-CBKI	Cameron Z-90 HAB	10236		22. 3.02	Wheatfields Park Ltd *(Roundtrees Garden Centres titles)*	(Winscombe)	19. 3.05A
G-CBKJ	Cameron Z-90 HAB	10251		6. 3.02	Invista (UK) Holdings Ltd	(Brockworth, Gloucester)	13. 5.05A
G-CBKK	Ultramagic S-130 HAB	130/32		19. 3.02	Airborne Adventures Ltd *(Co-Op Bank titles)*	(Skipton)	1. 4.05T
G-CBKL	Raj Hamsa X'Air 582 *(Built J.Garcia - kit no.682)*	BMAA/HB/203		18. 2.02	S.Smith tr Caithness X-Air Group	Wick	29.12.04P
G-CBKM	Mainair Blade 912	1310-0102-7 & W1105		21. 1.02	N.Purdy	Headon Farm, Retford	7. 4.05P
G-CBKN	Mainair Blade 912	1316-0302-7 & W1111		11. 3.02	D.S.Clews	(Morton, Gainsborough)	6. 4.05P
G-CBKO	Mainair Blade 912S	1311-0102-7 & W1106		11. 2.02	B.Jackson	Carlisle	19. 2.05P
G-CBKR	Piper PA-28-161 Warrior III	2842143	N5334N	15. 3.02	Devon School of Flying Ltd	Dunkeswell	21. 3.05T
G-CBKS	Air Création 582/Kiss 400 *(Built S.Kilpin - kit no.FL007, Trike T01113, Wing A01193-1203)*	BMAA/HB/197		28. 1.02	S.Kilpin	(Hackleton, Northants)	25. 9.05P
G-CBKU	Comco Ikarus C42 FB UK *(Built R G Q Clarke and P Walton - kit no.0112-6431)*	PFA 322-13862	(EI-)	4. 3.02	C.Blackburn	Carnowen, County Donegal	4. 6.05P
G-CBKV	Cameron Z-77 HAB	4946		15. 3.02	J.F.Till	(Welburn)	8. 3.05A
G-CBKW	Pegasus Quantum 15-912	7892		25. 3.02	I.W.Trench	East Fortune	12. 6.05P
G-CBKY	Jabiru Aircraft Jabiru SP-470 *(Built P.R.Sistern)*	PFA 274B-13764		6. 3.02	P.R.Sistern	City of Derry	6.10.05P
G-CBLA	Aero Designs Pulsar XP *(Built J.L.Reeves)*	367	N367JR	15. 2.02	J.P.Kynaston	(Luton)	4. 7.05P
G-CBLB	Tecnam P92-EM Echo *(Built M.A Lomas)*	PFA 318-13770		12. 3.02	M.A Lomas	Spanhoe	4. 3.05P
G-CBLD	Mainair Blade 912S	1306-1201-7 & W1101		21. 3.02	N.E.King	St Michaels	20. 4.05P
G-CBLE	Robin R2120U Alpha 120T	364		16. 4.02	Aviation Now Ltd	(Cardiff)	16. 7.05T
G-CBLF	Raj Hamsa X'Air 582 *(Built E.G.Bishop and E N Dunn - kit no.696)*	BMAA/HB/194		18. 3.02	E.G.Bishop	Dunkeswell	29. 4.05P
G-CBLH	Raj Hamsa X'Air 582 *(Built S.Rance - kit no.673)*	BMAA/HB/182		18. 3.02	S.Rance	Mapperton Farm, Newton Peverill	24. 5.05P
G-CBLI	Aerostar Yakovlev Yak-52	867110	LY-ANU DOSAAF 136 *(yellow)*	9. 4.02	R.H.Reeves	Rednal	15. 1.05P
G-CBLJ	IAV-Bacau Yakovlev Yak-52	888615	RA-44472 LY-AHF/DOSAAF 57 *(yellow)*	19. 4.02	G.H.Wilson *"Black/02"*	Audley End	25. 4.05P
G-CBLK	Hawker Hind	41H-82971	Afghan AF L7181	20. 3.02	Aero Vintage Ltd *(Noted 1.05)*	Duxford	
G-CBLL	Pegasus Quantum 15-912	7891		22. 3.02	P R Jones	Felthorpe	30. 4.04P
G-CBLM	Mainair Blade 912	1308-0102-7 & W1103		12. 2.02	J.Dearn tr G-CBLM Flying Group	Barton	17. 2.04P
G-CBLN	Cameron Z-31 HAB	10285		26. 4.02	N.J.Langley	(Clapton in Gordano, Bristol)	1. 6.05A
G-CBLO	Lindstrand LBL 42A HAB	854		3. 4.02	Airship and Balloon Company Ltd *(nPower titles)*	(Bristol)	3. 4.04A
G-CBLP	Raj Hamsa X'Air Falcon Jab22 *(Built M J Kay and S Litchfield - kit no.646)*	BMAA/HB/213		26. 3.02	T.W.Oakley	(Scarborough)	17. 5.05P
G-CBLS	Not yet allocated						

Reg	Type	C/n	Previous identity	Date	Owner/operator	Base	CofA
G-CBLT	Mainair Blade 912	1315-0202-7 & W-1110		25. 4.02	G Edwards	(Andover)	13. 5.05P
G-CBLU	Cameron C-90 HAB	10128		30. 4.02	A G.Martin *"Harlequin"*	(Bristol)	10. 8.05A
G-CBLV	Flight Design CT2K	Not known		11. 4.02	A K.Pickering	(Villefollett, Deux-Sèvres, France)	11. 2.05P
	(Assembled Pegasus Aviation as c/n 7886)						
G-CBLW	Raj Hamsa X'Air Falcon Victor2	BMAA/HB/209		13. 3.02	A R.Cundill	Dunkeswell	15. 8.05P
	(Built R.R.Hadley - kit no.641)						
G-CBLX	Air Création 582/Kiss 400	BMAA/HB/208		3. 4.02	J.H.Hayday	Plaistowes Farm, St.Albans	30.10.54P
	(Built J.H.Hayday - kit no.FL008, Trike T02010, Wing A02013-2003)						
G-CBLY	Grob G109B	6403	D-KITZ (2)	12. 3.02	D.A Smith	(Warminster)	20. 5.05
			(F-WAQS)		tr G-CBLY Syndicate		
G-CBLZ	Rutan LongEz	1046	F-PYYV	5. 6.02	R.P.H.Hancock	Cambridge	9.10.05P
	(Built N W Ruston) (Lycoming O-235)						
G-CBMA	Raj Hamsa X'Air 582	BMAA/HB/204		14. 2.02	B.L.Crick	(Whitstable)	24.11.05P
	(Built K.Angel - kit no.739)						
G-CBMB	Cyclone AX2000 HKS	7894		18. 6.02	York Microlight Centre Ltd	Rufforth	24. 6.05P
G-CBMC	Cameron Z-105 HAB	10274		30. 4.02	The Balloon Co Ltd	(Bristol)	16. 4.05T
					t/a First Flight *(Edward Ware Homes titles)*		
G-CBMD	IAV-Bacau Yakovlev Yak-52	822710	RA-44460	4.11.02	R.J.Hunter	Headcorn	28.11.05P
			LY-AHE, DOSAAF 100 *(yellow)*				
G-CBME	Reims/Cessna F.172M	F17201060	TF-FTV	28. 2.02	Skytrax Aviation Ltd	Blackbushe	2. 4.05T
			TF-POP, SE-FZP				
G-CBMI	IAV-Bacau Yakovlev Yak-52	855907	LY-AOZ	24. 7.02	A Burani *(ETPS c/s)*	Elstree	20. 7.05P
			RA-02050, DOSAAF 107 *(blue)*				
G-CBMJ	Rotary Air Force RAF 2000 GTX-SE	PFAG/13-1336		22. 3.02	C.D.Upsall	Easter Poldar Farm, Thornhill	1.12.04P
	(Built M J Barton)						
G-CBMK	Cameron Z-120 HAB	10293		11. 4.02	Flying Pictures Ltd	(Chilbolton)	16. 4.03A
					(Woolwich Building Society titles)		
G-CBML	de Havilland DHC-6-310 Twin Otter	695	C-FZSP	5. 6.02	Isles of Scilly Skybus Ltd	St.Just	4. 6.05T
			HB-LSN, C-FZSP, TR-LZO, C-GJZK				
G-CBMM	Mainair Blade 912	1312-0202-7 & W1107		9. 9.02	M.R.Mosley	Headon Farm, Retford	4.11.04P
G-CBMO	Piper PA-28-180 Cherokee D	28-4806	ZS-ONK	16. 5.02	C Woodliffe	(South Cockerington, Louth)	23. 7.05T
			9J-RHN, N6391J				
G-CBMP	Cessna R182 Skylane RG II	R18201325	ZS-MWT	9. 4.02	Orman (Carrolls Farm) Ltd	Great Massingham	12. 8.07T
			N38MH, YV-2034P, N2286S				
G-CBMR	Medway EclipseR	172/150		27. 3.02	A Bradfield	Middle Stoke, Isle of Grain	3. 4.05P
G-CBMS	Medway EclipseR	173/151		27. 3.02	R.R.Bagge	Field Farm, Oakley	29. 6.04P
G-CBMT	Robin DR400/180 Régent	2538		3. 5.02	A C.Williamson	Crowfield	26. 6.05T
G-CBMU	Whittaker MW6-S Fatboy Flyer	PFA 164-13339		30. 4.02	F.J.Brown	(Flitwick)	9.11.05P
	(Built F.J.Brown)						
G-CBMV	Pegasus Quantum 15 Super Sport	7893		3. 5.02	B.Hamilton	Long Marston	4. 7.05P
G-CBMW	Zenair CH.701 UL	PFA 187-13788		9. 4.02	C.Long	Broadmeadow Farm, Hereford	
	(Built C Long)				*(Noted 6.04)*		
G-CBMX	Air Création 582/Kiss 400	BMAA/HB/207		28. 3.02	D.L.Turner	(Sidcup)	10. 7.05P
	(Built D.L.Turner - kit no.FL009, Trike T02009, Wing A02012-2004)						
G-CBMY	Not yet allocated						
G-CBMZ	Aerotechnik EV-97 Eurostar	PFA 315-13890		12. 4.02	P.Grenet and J.C.O'Donnell		
	(Built P.Grenet and J.C.O'Donnell)					Church Farm, Shotteswell	10. 7.05P
G-CBNA	Flight Design CT2K	Not known		31. 5.02	D.M.Wood	(Banbury)	16. 6.05P
	(Assembled Pegasus Aviation as c/n 7887)						
G-CBNB	Eurocopter EC120B Colibri	1040		8. 6.99	Arenberg Consultadoria e Servicos LDA	(Madeira, Portugal)	29. 6.05
G-CBNC	Mainair Blade 912	1319-0402-7 & W1114		17. 4.02	A C.Rowlands	(Nether Heyford, Northants)	26. 4.05P
G-CBNF	Rans S-7 Courier	PFA 218-13762		12. 4.02	M.Rockliff	(Leeds)	
	(Built T R Grief)				*(New owner 12.03)*		
G-CBNG	Robin R2112 Alpha	180	PH-ROL	20. 5.02	Solway Flyers Ltd	Carlisle	20. 6.05
			F-GCAF				
G-CBNI	Lindstrand LBL 90A HAB	857		16. 4.02	Cancer Research UK *(Cancer Research UK titles)*	(Bath)	29. 3.05A
G-CBNJ	Raj Hamsa X'Air 912	BMAA/HB/187		23. 4.02	M.K.Slaughter	Farley Farm, Romsey	10. 2.05P
	(Built M.K.Slaughter, Francis and Hunt - kit no.680)				tr 912 X'Air Group		
G-CBNK	Aerotechnik EV-97 Eurostar	PFA 315-13888		17. 4.02	M.R.M.Welch	Goodwood	7. 7.05P
	(Built M.R.M.Welch)						
G-CBNL	Dyn'Aéro MCR-01 Club	PFA 301A-13805		12. 4.02	D.H.Wilson	Tatenhill	17. 8.05P
	(Built D.H.Wilson)						
G-CBNM	North American P-51D Mustang	122-31590	SE-BKG	29. 4.02	Patina Ltd *(Op The Fighter Collection)*	Duxford	9. 7.05P
			4X-AIM, Israeli DFAF 2338, Swed AF Fv26158, 44-63864 *(As ""463864/HL-W")*				
G-CBNN	Not yet allocated						
G-CBNO	CFM Streak Shadow	PFA 206-13809		8. 3.02	D.J.Goldsmith	(Crockham Hill)	9.10.04P
	(Built D.J.Goldsmith)						
G-CBNT	Pegasus Quantum 15-912	7860		14. 5.02	R.K.Watson	Rochester	28. 6.05P
G-CBNU	Vickers Supermarine 361 Spitfire LF.IX	CBAF IX 2115	Turkish AF ML411	27. 8.02	M.Aldridge *(Current CofR4.03)*	(Ashford)	
G-CBNV	Rans S-6-ES Coyote II	PFA 204-13817		23. 4.02	C.W.J.Davis	Roddige	6.10.0P
	(Built C.W.J.Davis - kit no 1000 1393 ES) (Rotax 582) *(Nose-wheel u/c)*						
G-CBNW	Cameron N-105 HAB	10283		16. 5.02	C.and J.M.Bailey	(Bristol)	10. 5.05T
					t/a Bailey Balloons *(Bristol and West Building Society titles)*		
G-CBNX	Montgomerie-Bensen B.8MR	PFA G/01A-1345		26. 4.02	K.Ashford	(Walsall)	14. 9.05P
	(Built C Hewer) (Rotax 912-UL)						
G-CBNY	Air Création 582/Kiss 400	BMAA/HB/218		30. 4.02	R.Redman	(Grantham)	30. 9.05P
	(Built R.Redman - kit no.FL013, Trike T02035, Wing A02051-2047)						
G-CBNZ	TEAM hi-MAX 1700R	PFA 272-13624		30. 4.02	J.J.Penney	(Neath)	
	(Built J J Penny)						
G-CBOA	Auster B.8 Agricola Series 1	AIRP/860	ZK-BXO	22. 4.02	C.J.Baker	Carr Farm, Thorney, Newark	AC
	(Built from spares by Airepair, New Zealand 1966 using parts of ZK-BMN c/n B.106)		ZK-BMN		*(Outer wings removed 1.05)*		
G-CBOC	Raj Hamsa X'Air 582	BMAA/HB/166		1. 5.02	A J.McAleer	(Dungannon)	28. 8.04P
	(Built A J.McAleer - kit no.623)						

Registration	Type	C/n	Previous identity	Reg date	Owner/operator	Base	CofA
G-CBOD	Comco Ikarus C42 FB UK	PFA 322-13854		30. 4.02	B.Hunter	Octon Lodge Farm, Thwing	17. 2.05P
	(Built B Hunter)				*(Crashed near Leucate in south-east France 3.9.04 and badly damaged)*		
G-CBOE	Hawker Hurricane IIB	R30040	RCAF 5487	24. 5.02	Classic Aero Engineering Ltd	Thruxton	
	(Built Canadian and Car Foundry Co)				*(Noted 5.04)*		
G-CBOF	Europa Aviation Europa XS	PFA 247-13462		1. 5.02	I.W.Ligertwood	(Liverpool)	
	(Built I.W.Ligertwood - kit no.431)						
G-CBOG	Mainair Blade 912S	1309-0102-7 & W1104		26. 3.02	J.S.Littler	(Standish)	4. 5 05P
G-CBOK	Rans S-6-ES Coyote II	PFA 204-13864		19. 4.02	S.A Clarehugh	Eshott	28.3.05P
	(Built C.J.Arthur) (Rotax 912-UL)						
G-CBOL	Mainair Blade	1320-0402-7 & W1115		3. 5.02	R.M.Nutt and J.A.Nutt	Carlisle	22. 6.05P
G-CBOM	Mainair Blade 912	1314-0202-7 & W1109		30. 4.02	G.Suckling	(Saffron Walden)	30. 4.05P
G-CBON	Cameron Bull-110 SS HAB	10261		14. 6.02	Stratos Ballooning GmbH and Co KG	(Ennigerloh, Germany)	26. 5.05A
G-CBOO	Mainair Blade 912S	1317-0302-7 & W1112		4. 4.02	T.F.Casey	(Galashiels)	3. 4.05P
G-CBOP	Jabiru Aircraft Jabiru UL-450	PFA 274A-13611		2. 5.02	D.W.Batchelor	Sandtoft	13. 5.05P
	(Built D.W.Batchelor)						
G-CBOR	Reims/Cessna F172N Skyhawk II	F17201656	PH-BOR	28. 5.87	Pauline Seville	Barton	24. 5.06T
			PH-AXG (1)				
G-CBOS	Rans S-6-ES Coyote II	PFA 204-13859		8. 5.02	R.Skene	Rochester	19. 8.04P
	(Built R Skene) (Jabiru 2200A) *(Nosewheel configuration)*						
G-CBOT	Robinson R44 Raven	1194		11. 4.02	S.J.Skilton	Bournemouth	8. 5.05T
					t/a Aviation Rentals *(Op Bournemouth Helicopters)*		
G-CBOU	Bensen-Parsons Two-Place Gyroplane	PFA G/8-1311		8. 5.02	R.Collin and M.S.Sparkes	Templehall Farm, Midlem	
	(Built R Collin)				*(Noted 5.04)*		
G-CBOV	Mainair Blade	1327-0502-7 & W1122		16. 5.02	M J Devane	East Fortune	26. 6.05P
G-CBOW	Cameron Z-120 HAB	10302		7. 8.02	Associated Technologies Ltd *(Torex titles)*	(Banbury)	2. 8.04A
G-CBOX							
G-CBOY	Pegasus Quantum 15-912	7898		17. 4.02	T G Jackson	Compton Abbas	16. 4.05P
G-CBOZ	IAV-Bacau Yakovlev Yak-52	811308	LY-AOC	15.11.02	T.M.Knight	Headcorn	16.12.05P
			DOSAAF 30				
G-CBPA	Not yet allocated						
G-CBPC	Sportavia-Pützer RF5B Sperber	51013	OY-XKC	27. 6.02	J.Bennett tr Lee RF5B Group	Lee-on-Solent	26. 8.05
G-CBPD	Comco Ikarus C42 FB UK	PFA 322-13863		14. 5.02	A.Haslam	Kirkbride	18. 3.05P
	(Built M L Robinson - kit no.0112-6444)				tr Waxwing Group		
G-CBPE	SOCATA TB-10 Tobago	129	HB-EZR	13. 6.02	A F.Welch	(Cambridge)	25. 6.05T
G-CBPF	Not yet allocated						
G-CBPG	The Balloon Works Firefly 7 HAB	FS7-001	N9045C	14. 6.02	I.Chadwick	(Horsham)	
					tr Balloon Preservation Flying Group		
G-CBPH	Lindstrand LBL 105A HAB	850		29. 5.02	Vastano Ivan	(Firenze, Italy)	18. 8.05A
G-CBPI	Piper PA-28R-201 Arrow	2844073	N53496	23. 5.02	Benair Aviation Ltd	(Jersey)	23. 7.05
					(Damaged in UK 8.04)		
G-CBPJ	Not yet allocated						
G-CBPK	Rand-Robinson KR-2	PFA 129-11461		22. 5.02	R.J.McGoldrick	Rochester	
	(Built R.J.McGoldrick)				*(Noted 4.03)*		
G-CBPL	TEAM mini-MAX 93	PFA 186-13100		24. 5.02	K.M.Moores	(Boston)	
	(Built K.M.Moores)						
G-CBPM	Yakovlev Yak-50	812101	LY-ASG	10. 7.02	P.W.Ansell	Audley End	14. 7.05P
			DOSAAF 58 ?		"50"		
G-CBPN	Thruster T.600N 450	0052-T600N-065		23. 5.02	J.S.Webb	Old Sarum	22. 7.05P
G-CBPO	Yakovlev Yak-50	853101	LY-AOT	.05R	(M Jefferies)	(Little Gransden)	
			DOSAAF 59 *(blue)*		*"59" (blue)*		
G-CBPP	Jabiru Aircraft Jabiru UL-450	PFA 274A-13607		23. 4.02	C J Cullen	Dunkeswell	
	(Built J.N.Pearson)				*(Noted 11.04)*		
G-CBPR	Jabiru Aircraft Jabiru UL-450	PFA 274A-13492		16. 5.02	P.L.Riley and F.B.Hall	Dunkeswell	26.10.04P
	(Built P.L.Riley and F.B.Hall)						
G-CBPS	Not yet allocated						
G-CBPT	Robinson R22 Beta II	3329		22. 5.02	Plane Talking Ltd	Blackbushe	10. 7.05T
G-CBPU	Raj Hamsa X'Air BMW R100	BMAA/HB/123		27. 5.02	M.S.McCrudden	Newtownards	2. 9.04P
	(Built M.S.McCrudden and W.P.Byrne - kit no.442)						
G-CBPV	Zenair CH.601UL	PFA 162A-13689		28. 5.02	R.D.Barnard	Ley Farm, Chirk	27. 8.05P
	(Built R.D.Barnard)						
G-CBPW	Lindstrand LBL 105A HAB	863		12. 6.02	Flying Pictures Ltd *(Samsung titles)*	(Chilbolton)	2. 7.04A
G-CBPX	IAV-Bacau Yakovlev Yak-52	8910004	RA-02956	20.12.02	M.Richardson	Sibson	5. 2.05P
			LY-ABV, DOSAAF 106 *(yellow)*				
G-CBPY	IAV-Bacau Yakovlev Yak-52	800708	RA-44474	8. 1.03	Lyttondale Associates Ltd	Sherburn-in-Elmet	18. 1.05P
			LY-AMP, DOSAAF 52				
G-CBPZ	Ultramagic N-300 HAB	300/04		25. 6.02	Skyview Ballooning Ltd t/a Kent Ballooning	(Ashford)	23 6.05T
G-CBRB	Ultramagic S-105 HAB	105/103		19. 6.02	I.S.Bridge	(Shrewsbury)	17.10.04A
G-CBRC	Jodel D.18	PFA 169-11408		31. 5.02	B.W.Shaw	Wathstones Farm, Newby Wiske	
	(Built B.W.Shaw)				*(Noted 5.04)*		
G-CBRD	Jodel D.18	PFA 169-11484		31. 5.02	J.D.Haslam	Wathstones Farm, Newby Wiske	
	(Built J.D.Haslam)				*(Noted 5.04)*		
G-CBRE	Mainair Blade 912	1330-0602-7 & W1125		19. 6.02	R.J.Davey	(Sleaford)	20. 6.05P
G-CBRF	Comco Ikarus C42 FB UK	PFA 322-13900		7. 6.02	T.W.Gale	Trim, County Meath	10.10.05P
	(Built T W Gale)						
G-CBRG	Cessna 560XL Citation Excel	560-5266	N5245D	13. 8.02	Stadium City Ltd	(Brough)	14. 8.07T
G-CBRH	IAV-Bacau Yakovlev Yak-52	844815	LY-ALO	6. 9.02	B.M.Gwynnett	Haverfordwest	27. 4.05P
			DOSAAF 135				
G-CBRJ	Mainair Blade 912	1321-0502-7 & W1116		24. 4.02	R.W.Janion	(Weaverham, Northwich)	10. 7.05P
	(Rotax 582-2V)						
G-CBRK	Ultramagic M-77 HAB	77/212		8. 7.02	R.T.Revel *"Solero"*	(High Wycombe)	5. 6.05P
G-CBRL	IAV-Bacau Yakovlev Yak-52	833708	RA-44468	2.12.02	P.S. Mirams	Biggin Hill	11. 1.05P
			LY-AOX, DOSAAF 122?		tr Norbert Group		
G-CBRM	Mainair Blade	1326-0502-7 & W1121		19. 6.02	M.H.Levy	(Northwich)	19. 7.05P

G-CBRN	Not yet allocated						
G-CBRO	Robinson R44 Raven	1221		17. 6.02	R.D.Jordan	Cranfield	8. 7.05T
G-CBRP	IAV-Bacau Yakovlev Yak-52	822603	RA-02041 LY-AID, DOSAAF 105	22. 9.04	R.J.Pinnock	North Weald	AC
G-CBRR	Aerotechnik EV-97 Eurostar *(Built C M Theakstone)*	PFA 315-13919		18. 6.02	J.E.Timms tr G-CBRR Group	Redhill	1. 8.04P
G-CBRT	Murphy Elite *(Built R W Baylie)*	PFA 232-13461		19. 6.02	R.W.Baylie	(Hailsham)	
G-CBRU	IAV-Bacau Yakovlev Yak-52	888911	RA-02042 DOSAAF 98 *(yellow)*	21. 1.03	S.M.Jackson tr Romeo Alpha 42 Group *(As "42" (White) in Soviet Air Force c/s)*	Rochester	17. 3.05P
G-CBRV	Cameron C-90 HAB	10323		31. 7.02	C.J.Teall	(Salford, Chipping Norton)	23. 4.05
G-CBRW	Aerostar Yakovlev Yak-52	9111415	RA-44464 DOSAAF 50	4. 2.03	M A Gainza *(As "50")*	North Weald	7. 4.05P
G-CBRX	Zenair CH.601UL Zodiac *(Built J.B.Marshall)* (Rotax 912S)	PFA 162A-13833		21. 6.02	J.B.Marshall	Plaistowes Farm, St Albans	25. 3.05P
G-CBRY	Pegasus Quik	7902		24. 6.02	Cyclone Airsports t/a Pegasus Aviation	(Manton)	
G-CBRZ	Air Création 582/Kiss 400 *(Built B.Chantry - kit no.FL015, Trike T02052, Wing A02086-2080)*	BMAA/HB/226		21. 6.02	B.Chantry	Plaistowes Farm, St.Albans	13. 2.05P
G-CBSB	Westland SA.341C Gazelle HT.2	1081	XW857	6. 6.02	Flying Machinery Ltd *(As "XW857/55")*	(Godalming)	
G-CBSC	Westland SA.341C Gazelle HT.2	1148	XW871	6. 6.02	Flying Machinery Ltd *(As "XW871/44")*	(Godalming)	
G-CBSD	Westland SA.341C Gazelle HT.2	1045	XW854	6. 6.02	London Helicopter Centres Ltd *(As "XW854/46")*	Redhill	14.12.04P
G-CBSE	Westland SA.341C Gazelle HT.2	1402	XX436	6. 6.02	Estates UK (Maintenance) Ltd *(As "XW436/39")*	Breighton	12. 8.05P
G-CBSF	Westland SA.341C Gazelle HT.2	1924	ZB647	6. 6.02	Falcon Aviation Ltd *(As "ZB647/40")*	(Reading)	
G-CBSH	Westland SA.341G Gazelle HT.3	1344	XX406	28.10.02	Alltask Ltd *(New owner 5.04) (As "XX406/P")*	(Rochester)	
G-CBSI	Westland SA.341G Gazelle HT.3	1736	XZ934	6. 6.02	P.S.Unwin *(As "XZ934/U")*	Redhill	10. 6.05P
G-CBSK	Westland SA.341G Gazelle HT.3	1914	ZB627	6. 6.02	Knoland Aviation Ltd *(As "ZB627/A")*	Battlesbridge	14.11.05P
G-CBSL	IAV-Bacau Yakovlev Yak-52	822013	RA-44534	13. 1.03	N. and A.D.Barton	Leicester	6. 2.05P
G-CBSM	Mainair Blade 912	1331-0602-7 & W1126		10. 5.02	Mainair Sports Ltd	Rochdale	6. 2.05P
G-CBSN	Aerostar Yakovlev Yak-52	9111413	RA-44514 DOSAAF 48	8. 7.03	P.K.Murtagh	Manston	17. 8.05P
G-CBSO	Piper PA-28-181 Archer II	28-7690376	D-EOFL N9595N	18. 7.02	Archer One Ltd *(Op Lydd Aero Club)*	Lydd	12. 8.05T
G-CBSP	Pegasus Quantum 15-912	7903		9. 7.02	D.S.Carstairs	Perth	14. 7.04P
G-CBSR	IAV-Bacau Yakovlev Yak-52	877913	LY-AQB Ukraine AF 10 (yellow), DOSAAF 100 *(yellow?)*	10. 7.02	L.Olivier tr Pegasus U2W	Wevelgem, Belgium	14. 7.05P
G-CBSS	IAV-Bacau Yakovlev Yak-52	833707	RA-44475 LY-AIJ, DOSAAF 121?	19. 2.03	M Chitty	White Waltham	20 .2.05P
G-CBSU	Jabiru Aircraft Jabiru UL-450 *(Built P.K.Sutton)*	PFA 274A-13812		15. 7.02	P.K.Sutton	(Dudley)	23. 6.05P
G-CBSV	Montgomerie-Bensen B8MR *(Built J A McGill)* (Rotax912-UL)	PFA G/01A-1344		1. 7.02	J.A McGill	Biggin Hil	17. 2.05P
G-CBSW	Not yet allocated						
G-CBSX	Air Création 582/Kiss 400 *(Built N.Hartley - kit no.FL014, Trike T02051, Wing A02085-2079)*	BMAA/HB/225		3. 7.02	N.Hartley	Baxby Manor, Husthwaite	8. 4.05P
G-CBSZ	Mainair Blade 912S	1334-0602-7 & W1129		6. 8.02	D.M.Newton	Easter Balgillo Farm, Finavon	29.11.05P
G-CBTA	Not yet allocated						
G-CBTB	III Sky Arrow 650T *(Built D.A .and J.A S.T.Hood)*	PFA 298-13832		25. 6.02	D.A and J.A S.T.Hood	Wombleton	23. 5.05P
G-CBTC	Not yet allocated						
G-CBTD	Pegasus Quantum 15-912	7904		9. 7.02	D.Baillie	Carlisle	14. 7.05P
G-CBTE	Mainair Blade 912S	1328-0602-7 & W1123		10. 7.02	K J Miles	St Michaels	10. 8.05P
G-CBTF	Not yet allocated						
G-CBTG	Comco Ikarus C42 FB UK *(Built J.A Way and R.Bonnett)*	PFA 322-13849		25. 6.02	J.A Way and R.Bonnett tr Ikarus Group	Maypole Farm, Chislet	10. 2.05P
G-CBTH	Flying Pictures Elson Apoly1 44000 HAB (Gas Filled)	2002/07		1. 8.02	Flying Pictures Ltd	(Chilbolton)	
G-CBTI	Flying Pictures Elson Apoly1 44000 HAB (Gas Filled)	2002/08		1. 8.02	Flying Pictures Ltd	(Chilbolton)	
G-CBTK	Raj Hamsa X'Air 582 *(Built C.D.Wood - kit no.589)*	BMAA/HB/168		9. 7.02	G.V.Rodgers	(Pontyclun)	22. 9.05P
G-CBTL	Monnett Moni *(Built G Dawes)*	PFA 142-11558		8. 7.02	G.Dawes	(Dover)	
G-CBTM	Mainair Blade (Rotax 582)	1322-0502-7 & W1117		2. 7.02	M.J.Ryder	(Lincoln)	1. 6.05P
G-CBTN	Piper PA-31 Navajo C	31-7812073	OO-VLH N27636	7. 8.02	C.and P.Wood t/a Durban Aviation Services	Biggin Hill	10. 9.05T
G-CBTO	Rans S-6-ES Coyote II *(Built B.J.Mould and M Walsh - c/n 1201-1427)* (Rotax 912-UL) *(Tri-cycle u/c)*	PFA 204-13910		16. 7.02	C.G.Deeley	Sittles Farm, Alrewas	26.10.05P
G-CBTR	Lindstrand LBL 120A HAB	733		22. 7.02	R.H.Etherington	(Siena, Italy)	13. 8.05A
G-CBTS	Gloster Gamecock rep *(Built Retro Track and Air (UK) Ltd)*	GA 97		17. 7.02	Retro Track and Air (UK) Ltd	(Dursley)	
G-CBTT	Piper PA-28-181 Archer II	28-7890127	G-BFMM N47735	22. 7.02	Citicourt Aviation Ltd	Denham	4.10.07E
G-CBTU	Cessna 550 Citation II	550-0601	G-OCDB G-ELOT, (N1303M)	12. 8.02	Thames Aviation Ltd	Fairoaks	28. 2.05T
G-CBTV	Tri-R Kis *(Built T.V.Thorp)*	PFA 239-12467		14. 5.02	T.V.Thorp	(Marlborough)	
G-CBTW	Mainair Blade 912	1329-0602-7 & W1124		20. 6.02	D.Hyatt	Middle Stoke, Isle of Grain	27. 7.05P
G-CBTX	Denney Kitfox Model 2 *(Built W M Farrell and G I Doake)*	PFA 172-11721		19. 7.02	G.I.Doake	(Craigavon, County Armagh)	
G-CBTY	Raj Hamsa X'Air Victor2 *(Built K.Quigley - kit no.720)*	BMAA/HB/222		19. 7.02	K.Quigley	Newtownards	4. 4.05P
G-CBTZ	Pegasus Quantum 15-912	7909		29. 7.02	P.M.Connelly	(London W4)	27. 9.05P

Regn	Type	C/n	Previous identity	Regd	Owner	Base	CofA
G-CBUA	Extra EA.230	009	N230KR N286PA	5. 9.02	C.Butler	Netherthorpe	19. 3.05P
G-CBUC	Raj Hamsa X'Air 582 *(Built A P.Fenn and D.R.Lewis - kit no.779)*	BMAA/HB/228		22. 7.02	A P.Fenn and D.R.Lewis	Shobdon	20. 2.05P
G-CBUD	Pegasus Quantum 15-912	7906		30. 7.02	G.N.S.Farrant	Drayton Manor. Drayton St. Leonard	30. 7.05P
G-CBUE	Ultramagic N-250 HAB	250/25		5.12.02	Elinore French Ltd t/a Imagination Balloon Flights	(Morpeth)	13. 4.05T
G-CBUF	Flight Design CT2K *(Assembled Pegasus Aviation as c/n 7901)*	02-04-04-18		26. 7.02	Cyclone Airsports Ltd t/a Pegasus Aviation	Manton	14. 8.04P
G-CBUG	Tecnam P92-EM Echo *(Built R.C.Mincik)* (Rotax 912-ULS) *(Marked as "P92S")*	PFA 318-13662		20. 6.01	R.C.Mincik	Bournemouth	6. 5.05P
G-CBUH	Westland Scout AH.1	F9475	XP849	5. 8.02	C J Marsden	(Petersfield)	21.12.04P
G-CBUI	Westland Wasp HAS.1	F9590	XT420	5. 8.02	Military Helicopters Ltd *(Noted 7.02)*	Yaxham, Dereham	AC
G-CBUJ	Raj Hamsa X'Air 582 *(Built J.T.Laity - kit no.651)*	BMAA/HB/212		1. 8.02	J.T.Laity tr G-CBUJ Flying Group	Kemble	19.10.05P
G-CBUK	Van's RV-6A *(Built P.G.Greenslade)*	PFA 181A-13614		25. 7.02	P.G.Greenslade	(Billingshurst)	
G-CBUL	Not yet allocated						
G-CBUM	Not yet allocated						
G-CBUN	Barker Charade *(Built P.E.Barker)*	PFA 166-13520		31. 7.02	P.E.Barker	(Bedford)	
G-CBUO	Cameron O-90 HAB	3353	CC-PMH	15. 8.02	W.J.Treacy and P.M.Smith	(Trim, County Meath)	22. 9.05A
G-CBUP	VPM M-16 Tandem Trainer *(Built R.W.Husband - c/n SA-M16-10M)*	PFA G/12-1346	ZU-AIH	28. 8.02	J S Firth *(New owner 2.04)*	(Huddersfield)	
G-CBUR	Zenair CH 601UL *(Built R J Kelly)* (Rotax 912S)	PFA 162A-13891		19. 7.02	N.A Jack	(Insch)	18. 9.03P
G-CBUS	Pegasus Quantum 15	7916		29. 8.02	J.Liddiard	Yatesbury	15. 9.04P
G-CBUU	Pegasus Quantum 15-912	7917		27. 8.02	K.B.Woods	Newnham, Baldock	27. 8.05P
G-CBUV	Not yet allocated						
G-CBUW	Cameron Z-133 HAB	10322		29. 8.02	Balloon School (International) Ltd	(Petworth)	31. 8.05T
G-CBUX	Cyclone AX2000 *(Original HKS700E engine replaced by Rotax 582 @ 2003)*	7918		2.10.02	J.Madhvani	Plaistowes Farm, St.Albans	27.10.05P
G-CBUY	Rans S-6-ES Coyote II *(Built S.C.Jackson - c/n 0302-1436)* (Rotax 582) *(Tri-cycle u/c)*	PFA 204-13954		13. 8.02	S.C.Jackson and J.S.Coster	Rufforth	3.11.05P
G-CBUZ	Pegasus Quantum 15 (Rotax 503)	7907		31. 7.02	S.T.Allen	Redlands	2. 8.05P
G-CBVA	Thruster T.600N 450	0082-T600N-068		14. 8.02	G.St.Clair Moseley	(Portrush, County Antrim)	6.12.05P
G-CBVB	Robin R2120U	365		26. 7.02	Aviation Now Ltd	Cardiff	26. 5.06T
G-CBVC	Raj Hamsa X'Air 582 *(Built M.J.Male - kit no.792)*	BMAA/HB/230		15. 8.02	M.J.Male	Dunkeswell	6.10.04P
G-CBVD	Cameron C-60 HAB	10338		31.10.02	Phoenix Balloons Ltd CBVD *"Popeye"*	(Bristol)	10.10.04
G-CBVE	Raj Hamsa X'Air Falcon 912 *(Built D.F.Hughes - kit no.766)*	BMAA/HB/229		19. 8.02	P.K.Bennett	Middle Stoke, Isle of Grain	28. 9.05P
G-CBVF	Murphy Maverick *(Built J.Hopkinson)*	PFA 259-12876		19. 8.02	J.Hopkinson	(Bradford)	
G-CBVG	Mainair Blade 912S	1338-0802-7 & W1133		27. 8.02	A.M.Buchanan	Perth	10.12.04P
G-CBVH	Lindstrand LBL 120A HAB	870		2. 9.02	Line Packaging and Display Ltd	(Gillingham)	17. 4.05A
G-CBVI	Robinson R44 Raven	1259		21. 8.02	R.C.Hields t/a Hields Aviation	Sherburn-in-Elmet	10. 9.05T
G-CBVK	Schroeder Fire Balloons G HAB	408	D-OVHS	30. 9.02	S.Travaglia t/a Idea Balloon	(Fiorentino, Italy)	
G-CBVL	Robinson R22 Beta II	3353	N71650	23. 8.02	Helicopter Training and Hire Ltd	Newtownards	17. 9.05T
G-CBVM	Aerotechnik EV-97 Eurostar *(Built J.Cunliffe and A Costello)*	PFA 315-13932		8. 8.02	J.Cunliffe and A Costello	Brook Farm, Pilling	11.10.05P
G-CBVN	Pegasus Quik	7919		27. 8.02	N.F.Mackenzie	East Fortune	19. 8.05P
G-CBVO	Raj Hamsa X'Air 582 *(Built W.E.Richards - kit no.627)*	BMAA/HB/227		27. 8.02	M D Vearncombe	(Combwich, Bridgwater)	6. 5.05P
G-CBVR	Best Off Skyranger 912 *(Built R.H.J.Jenkins - kit no.SKRxxxx209)*	BMAA/HB/231		6. 9.02	S.H.Lunney	Ince Blundell	18. 5.05P
G-CBVS	Best Off Skyranger 912 *(Built S.C.Cornock - kit no.SKRxxxx215)*	BMAA/HB/234		19. 8.02	S.C.Cornock	(Birmingham)	17. 7.05P
G-CBVT	IAV-Bacau Yakovlev Yak-52	9010305	LY-AGR Ukraine AF 02 *(yellow)*, DOSAAF 02 *(yellow)*	19. 9.02	Lancair Espana SL	(Alicante, Spain)	14.12.04P
G-CBVU	Piper PA-28R-200 Cherokee Arrow II	28R-7135007	ZS-RER N11C	12. 9.02	E.W.Guess (Holdings) Ltd *(Stored as "ZS-RER" 3.03)*	Blue Tile Farm, Langham	AC
G-CBVV	Cameron N-120 HAB	10331		13. 9.02	John Aimo Balloons SAS	(Mondovi, Italy)	19. 8.05A
G-CBVX	Cessna 182P Skylane	18263419	ZS-IYZ N9653G	16.12.02	R.Martin	(Brough)	24. 4.06T
G-CBVY	Comco Ikarus C42 FB UK *(Built M.J.Hendra and R.Gossage - kit no.0112-6436)*	PFA 322-13835		4. 9.02	M.J.Hendra and R.Gossage	Brook Farm, Pilling	23. 2.05P
G-CBVZ	Flight Design CT2K *(Assembled Pegasus Aviation as c/n 7914 but officially regd as 9714)*	02.05.06.04		19. 9.02	Mainair Sports Ltd	Weston Zoyland	12.11.04P
G-CBWA	Flight Design CT2K *(Assembled Pegasus Aviation as c/n 7921)*	02.06.01.04		11.10.02	S R Pike and E O Otun tr Charlie Tango Group	Booker	16.10.05P
G-CBWB	Piper PA-34-200T Seneca II	34-7770188	N2495Q	31.10.02	Fairoaks Airport Ltd	Fairoaks	10.12.05T
G-CBWD	Piper PA-28-161 Warrior III	2842160	N5357G	1.10.02	Plane Talking Ltd	Blackbushe	17.10.05T
G-CBWE	Aerotechnik EV-97 Eurostar *(Built E.Clarke)*	PFA 315-13958		16. 9.02	E.Clarke	Brook Farm, Pilling	15.12.05P
G-CBWF	Europa Aviation Europa XS *(Built A Burrows - kit no.511)* (Rotax 914-UL) *(Tri-gear u/c)*	PFA 247-13879		17. 9.02	Celebrations Ltd	(Ronaldsway)	7. 4.05P
G-CBWG	Aerotechnik EV-97 Eurostar *(Built M.Rhodes - kit no 03-1162)*	PFA 315-13918		17. 9.02	M.Rhodes	Sittles Farm, Alrewas	11.11.05P
G-CBWI	Thruster T.600N 450	0102-T600N-071		20. 9.02	Tina Lee	(Luton)	17.10.05P
G-CBWJ	Thruster T.600N 450 Sprint	0092-T600N-069		20. 9.02	T Harrison-Smith	Middle Stoke, Isle of Grain	26.11.04P
G-CBWK	Ultramagic H-77 HAB	77/218		4.11.02	H.C. Peel	(Worcester)	18. 1.05P

Reg	Type	C/n	Previous identity	Regd	Owner/operator	Base	CofA
G-CBWM	Mainair Blade 912	1339-0802-7 & W1134		22. 8.02	C.Middleton	(Little Hayfield, High Peak)	27.10.05P
	(Rotax 503)*(Rebuilt with new frame and c/n of G-BZUM (1271-0201-7 & W1065)).*						
G-CBWN	Campbell Cricket Mk.6	PFA G/16-1328		24. 9.02	G.J.Layzell	Henstridge	
	(Built G.J.Layzell) (Marked as "Layzell AV-18A")				*(Noted 8.04)*		
G-CBWO	RotorWay Executive 162F	6597		24. 9.02	Handyvalue Ltd		
	(Built S P Tetley) (RotorWay RI 162F)				Thrifty Car Rental/Total Garage, A23, Horley		AC
G-CBWP	Europa Aviation Europa	PFA 247-12930		1.10.02	T.W.Greaves	(Hull)	23. 5.05P
	(Built T.W.Greaves - kit no.233) (Monowheel u/c)						
G-CBWR	Thunder Ax7-77 HAB	2348	N754TC	7.10.02	A Lutz	(Sudbury)	14.10.03A
G-CBWS	Whittaker MW6 Merlin	PFA 164-12863		7.10.02	D.W.McCormack	(Atherstone)	
	(Built D.W.McCormack)						
G-CBWU	RotorWay Executive 162F	6416		4.10.02	F.A Cavaciuti	(Llanbadoc)	12. 6.05P
	(Built F A Cavciuti) (RotorWay RI 162F)				t/a Usk Valley Trout Farm		
G-CBWV	Falconar F-12A Cruiser	PFA 22-13904		7.10.02	A Ackland	(Reading)	
	(Built A Ackland)						
G-CBWW	Best Off Skyranger 912	BMAA/HB/232		30. 8.02	R.L.and S.H.Tosswill	(Workington)	9.11.05P
	(Built R.L.and S.H.Tosswill - kit no.SKRxxxx210)						
G-CBWX	Slingsby T.67M-260 Firefly	2282	G-7-194	11.10.02	Slingsby Aviation Ltd	Wombleton	10.12.05T
G-CBWY	Raj Hamsa X'Air 582	BMAA/HB/244		17.10.02	G.C.Linley	Redlands	27. 1.05P
	(Built T Collins - kit no.775)						
G-CBWZ	Robinson R22 Beta II	3101	N141DC	23.10.02	Plane Talking Ltd	Elstree	13.11.05T
G-CBXA	Raj Hamsa X'Air 582	BMAA/HB/245		18.10.02	N.Stevenson-Guy	(Beaminster)	4. 3.05P
	(Built N.Stevenson-Guy - kit no.790)						
G-CBXC	Comco Ikarus C42 FB UK	PFA 322-13955		23.10.02	A R.Lloyd	Deenethorpe	27. 1.05P
	(Built A R.Lloyd)						
G-CBXD	Bell 206L-3 Long Ranger III	51328	D-HAUA	22.10.02	Whirlybird Charters Ltd Automotive and General Supply Co Ltd		
			N21AH, N21830		*(Op Direct Helicopters)*	Stapleford	8.12.05T
G-CBXE	Reality Easy Raider Jab22	BMAA/HB/198		22. 8.02	A Appleby	(Hailsham)	
	(Built A Appleby - kit no.0006)						
G-CBXF	Reality Easy Raider Jab22	BMAA/HB/196		19.11.02	F.Colman	Eshott	
	(Built F.Colman - kit no.0001)						
G-CBXG	Thruster T.600N 450 Sprint	0112-T600N-073		29.10.02	P.J.Fahie tr The Swallow 1 Group	Compton Abbas	27.11.04P
G-CBXH	Thruster T.600N 450 Sprint	0122-T600N-075		29.10.02	M L Smith	Popham	22. 4.05P
G-CBXI	Not yet allocated						
G-CBXJ	Cessna 172S Skyhawk SP	172S8125	N2391J	30.10.02	Caernarfon Airworld Ltd	Caernarfon	19.12.05T
G-CBXK	Robinson R22 Mariner	2302M	N3052P	4.11.02	County Garage (Cheltenham) Ltd	Gloucestershire	13. 1.06T
			LQ-BLD, N80524				
G-CBXM	Mainair Blade	1335-0802-7 & W1130		19. 8.02	B.A Coombe	(Billingshurst)	15. 9.05P
G-CBXN	Robinson R22 Beta II	3385		11.11.02	N.M.Pearson	Bristol	27.11.05T
G-CBXR	Raj Hamsa X'Air Falcon 582	BMAA/HB/224		11.11.02	A.R.Rhodes	Kirkbride	
	(Built J.F Heath - kit no.612)				*(New owner 8.04)*		
G-CBXS	Best Off Skyranger J2.2	BMAA/HB/248		13.11.02	C.J.Erith	(Reading)	
	(Built C.J.Erith - kit no.SKRxxxx246)						
G-CBXT	Westland SA.341G Gazelle HT.3	1191	XW898	26. 9.02	C.D.Evans and R.Paskey *(As "XW898/G" in Royal Navy c/s)*		
						(Baxterley, Atherstone/Tamworth)	14. 6.05P
G-CBXU	TEAM mini-MAX 91A	PFA 186-13037		13.11.02	C.D.Hatcher	Deenethorpe	27. 8.05P
	(Built T.J.Shaw)						
G-CBXV	Mainair Blade	1343-1002-7 & W1138		4.10.02	C.Turner	Priory Farm, Tibenham	15.12.04P
G-CBXW	Europa Aviation Europa XS	PFA 247-13674		18.11.02	R.G.Fairall	Redhill	
	(Built R.G.Fairall - kit no.494) (Monowheel u/c)						
G-CBXZ	Rans S-6-ESN Coyote II	PFA 204-13988		20.11.02	D.Tole	Long Marston	12. 4.05P
	(Built D.Tole)						
G-CBYB	RotorWay Executive 162F	6623		20.11.02	T.Clark	(Amersham)	AC
	(Built T Clark) (RotorWay RI 162F)				t/a Clark Contracting		
G-CBYC	Cameron Z-275 HAB	10342		20. 3.03	The Balloon Co Ltd	(Bristol)	9. 4.05T
					t/a First Flight *(Park Furnishers titles)*		
G-CBYD	Rans S-6-ESA Coyote II	PFA 204-13871		21.11.02	R.Burland	Perth	9.12.05P
	(Built R.Burland)						
G-CBYE	Pegasus Quik	7933		27. 1.03	A D.Griffin	Enstone	29. 1.05P
G-CBYF	Mainair Blade	1349-1202-7 & W1144		2. 1.03	C.P.Lemon	(Chorley)	19. 5.05P
G-CBYH	Aeroprakt A22 Foxbat	PFA 317-13902		2.12.02	G.C Moore	Otherton, Cannock	28. 3.05P
	(Built G C Moore and P C de-Ville)				tr G-CBYH Foxbat Group		
G-CBYI	Pegasus Quantum 15	7931		2. 1.03	J.M.Hardy	Deenethorpe	5. 2.05P
	(Rotax 503)						
G-CBYJ	Steen Skybolt	PFA 64-13354		2.12.02	F.G Morris	Newtownards	22. 7.05P
	(Built F.G Morris)						
G-CBYM	Mainair Blade	1323-0502-7 & W1118		13. 9.02	A Clarke	(Macclesfield)	22. 9.04P
	(Rotax 582)						
G-CBYN	Europa Aviation Europa XS	PFA 247-13751		5.12.02	A B.Milne	Lower Wasing Farm, Brimpton	25. 8.05P
	(Built A B.Milne - kit no.518 (Tri-gear u/c)						
G-CBYO	Pegasus Quik	7928		5.12.02	L Hogan	Glenrothes	15.12.04P
G-CBYP	Whittaker MW6-S Fatboy Flyer	PFA 164-13131		6.12.02	R.J.Grainger	(Northampton)	
	(Built R.J.Grainger)						
G-CBYS	Lindstrand LBL 21A HAB	156		17.12.02	J.J.C.Bernardin	(Curcay-sur-Dive, France)	18. 1.05A
G-CBYT	Thruster T.600N 450	0102-T600N-072		10.10.02	B.E.Smith	Eshott	8.11.05P
G-CBYU	Piper PA-28-161 Warrior III	2842173	N53606	12. 2.03	Stapleford Flying Club Ltd	Stapleford	19. 2.06T
G-CBYV	Pegasus Quantum 15-912	7920		19. 9.02	N.B.Sanghrajka	(Dunstable)	10. 9.05P
G-CBYW	Hatz CB-1	PFA 143-13710		16. 1.03	T.A Hinton	Doynton	29. 9.05P
	(Built T.A Hinton)						
G-CBYX	Bell 206B JetRanger III	480	HB-ZBX	5. 3.03	RCR Aviation Ltd	(Horsham)	21. 4.06T
			N203WB, C-GRGP, N2502M				
G-CBYY	Robinson R44 Raven	1250	N71837	11. 9.02	Helicopter Training and Hire Ltd	Newtownards	3.10.05T
G-CBYZ	Tecnam P92-EA Echo-Super	PFA 318A-13984		17.12.02	B.Weaver	(Osmington, Weymouth)	12. 8.05P
	(Built M Rudd)						

G-CBZA	Mainair Blade	1344-1002-7 & W1139		28.10.02	S Bedford	(Harrogate)	26. 1.05P
G-CBZB	Mainair Blade	1346-1102-7 & W1141		6.12.02	A Bennion	Arclid Green, Sandbach	21. 1.05P
G-CBZC	Not yet allocated						
G-CBZD	Mainair Blade	1348-1102-7 & W1143		12.12.02	J.Shaw	Leicester	29. 5.05P
G-CBZE	Robinson R44 Raven	1276		12.12.02	Blue Aviation Ltd	(Carnforth)	4. 2.06T
G-CBZF	Robinson R22 Beta II	3393	N71878	6.12.02	MAL Associates Ltd	(Bishopthorpe, York)	5. 2.06T
G-CBZG	Rans S-6ES Coyote II *(Built N.McKenzie)*	PFA 204-13894		9. 1.03	A.D.Thelwall	(Copt Hewick, Ripon)	22. 5.05P
G-CBZH	Pegasus Quik	7934		30. 1.03	M.Bond	(Bias, France)	26. 4.05P
G-CBZI	RotorWay Executive 162F *(Built T Stock) (*RotorWay RI 162F)	6718		3. 1.03	T.D.Stock	(London SE18)	AC
G-CBZJ	Lindstrand LBL 25A Cloudhopper HAB	892		9. 1.03	J.L.and J.Hilditch t/a Pegasus Ballooning	(Southwick, Brighton)	2. 1.05A
G-CBZK	Robin DR400/180 Régent	2543		12. 2.03	R.A.Fleming	Breighton	26. 2.06T
G-CBZL	Westland SA.314G Gazelle HT.3	WA2010	ZB629	17. 1.03	Armstrong Aviation Ltd *(As "ZB629/CU")*	Kirkbride)	25. 8.04P
	(Struck power cables landing Mouswald near Dumfries 22.11.03 and received major damage)						
G-CBZM	Jabiru Aircraft Jabiru SPL-450 *(Built M.E.Ledward)*	PFA 274A-13827		2. 1.03	M.E.Ledward	Wigtown	
G-CBZN	Rans S-6-ES Coyote II *(Built A James) (Nose-wheel u/c)*	PFA 204-13652		6. 1.03	A James	Otherton, Cannock	16. 3.05P
G-CBZP	Hawker Fury 1	41H-67550	SAAF??	2. 4.03	Historic Aircraft Collection Ltd	(Jersey)	AC
G-CBZR	Piper PA-28R-201 Arrow	2837029	EC-IJX N175ND	13. 1.03	S.J.Skilton t/a Aviation Rentals	Bournemouth	24. 2.06T
G-CBZS	Lynden Aurora *(Built J.Lynden)*	PFA 313-13534		13. 1.03	J.Lynden *(Noted 8.04)*	Brook Farm, Pilling	AC
G-CBZT	Pegasus Quik	7936		6. 1.03	M.Brown	Ince Blundell	29. 1.05P
G-CBZU	Lindstrand LBL 180A HAB	877		13. 1.03	Great Escape Ballooning Ltd *"Rainbow"*	(Olney)	11.12.04T
G-CBZV	Ultramagic S-130 HAB	130/36		5. 3.03	J.D.Griffiths	(Bingham)	17. 3.05T
G-CBZW	Zenair CH.701 UL *(Built T.M.Stiles)*	PFA 187-13731		13. 1.03	T.M.Stiles	(Heathfield)	4. 8.05P
G-CBZX	Dyn'Aéro MCR-01 ULC *(Built S.L.Morris)*	PFA 301B-13957		15. 1.03	S.L.Morris	(Salisbury)	
G-CBZY	Flylight Airsports Doodle Bug/Target	DB022		22.11.02	A I.Calderhead-Lea	(Basildon)	
G-CBZZ	Cameron Z-275 HAB	10346		12. 2.03	A C K Rawson and J J Rudoni	(Stafford)	30. 4.05T
G-CCAB	Mainair Blade	1345-1002-7 & W1140		28. 1.03	R.W.Street	(Edinburgh)	20. 2.05P
G-CCAC	Aerotechnik EV-97 Eurostar *(Built P.J Ladd and J.S.Holden)*	PFA 315-13979		26.11.02	J.S.Holden	Wadswick Manor Farm, Corsham	16. 3. 05P
G-CCAD	Pegasus Quik	7924		3.12.02	J.M.Hardstaff	Rufforth	20. 1.05P
G-CCAE	Jabiru Aircraft Jabiru UL-450 *(Built C.E.Daniels)*	PFA 274A-13938		17. 1.03	C.E.Daniels	Felthorpe	22. 7.05P
G-CCAF	Best Off Skyranger 912 *(Built D.W.Squire - kit no.SKRxxxx212)*	BMAA/HB/235		28.11.02	D.W.and M.L.Squire	(St.Austell)	
G-CCAG	Mainair Blade 912	1350-1202-7 & W1145		22.11.02	J.R.North t/a West Lancashire Microlight School	Ince Blundell	26. 1.05P
G-CCAH	Not yet allocated						
G-CCAJ	TEAM hi-MAX 1700R *(Built A P S John)*	PFA 272-13916		16.12.02	A P.S.John *"Chocolat"*	Hartpury	14. 4.05P
G-CCAK	Zenair CH.601 HD Zodiac *(Built A Kinmond)*	PFA 162-13469		11.12.02	A Kinmond	(Blairgowrie)	
G-CCAL	Tecnam P92-EM Echo *(Built D.Cassidy)*	PFA 318-13842		6.12.02	D.Cassidy	Maypole Farm, Chislet	24. 4.05P
G-CCAM	Mainair Blade	1347-1102-7 & W1142		6.12.02	M.D.Peacock	(Leatherhead)	19. 5.05P
G-CCAN	Cessna 182P Skylane	18264069	SE-LON OH-COZ, C-GWXC, (N6052F)	16. 1.03	D.J.Hunter	Priory Farm, Tibenham	12. 5.06T
G-CCAP	Robinson R22 Beta II	3413		11. 2.03	S.G.Simpson t/a HJS Helicopters	Lower Baads Farm, Peterculter	25. 2.06T
G-CCAR	Cameron N-77 HAB	464		5.12.78	D.P.Turner	(Leigh-on-Mendip)	2. 6.05A
	(Rebuilt with envelope c/n 670 @ 8.1980: with c/n 2108 @ 1989 and with c/n 2658 @ 1992) (Mitsubishi Cars titles)						
G-CCAS	Pegasus Quik	7935		11. 2.03	A W Buchan tr Quik Alpha Sierra	Knapthorpe Lodge, Caunton	17. 3.05P
G-CCAT	Gulfstream AA-5A Cheetah	AA5A-0893	G-OAJH G-KILT, G-BJFA, N27169	16. 1.92	Plane Talking Ltd	Cranfield	31.10.05T
G-CCAU	Eurocopter EC135T1	0040	G-79-01	30. 6.98	West Mercia Constabulary *(Op for Central Counties Police)*	Wolverhampton	21. 7.07T
G-CCAV	Piper PA-28-181 Archer II	28-8090353	D-EXRT N8233A	31. 3.03	S.Turner tr Alpha Victor Group	Southend	23. 4.06T
G-CCAW	Mainair Blade 912	1351-0103-7 & W1146		5. 2.03	C A Woodhouse	Oxton	14. 2.05P
G-CCAX	Raj Hamsa X'Air 582 *(Built N.Farrell - kit no.791)*	BMAA/HB/251		20. 1.03	N.Farrell	(Tarmonbarry, County Roscommon)	19. 1.05P
G-CCAY	Cameron Z-42 HAB	10373		27. 2.03	P Stern	(Deggendorf, Germany)	21. 7.04A
G-CCAZ	Pegasus Quik	7927		3.12.02	P.A Bass	Sywell	6. 3.05P
G-CCBA	Best Off Skyranger BMW R100 *(Built R.M.Bremner - kit no.SKR0211277)*	BMAA/HB/256		23. 1.03	R.M.Bremner tr Fourstrokes Group	Popham	4.10.05P
G-CCBB	Cameron N-90 HAB *(Note: G-TEEZ (1) stolen and cut up)*	10085	G-TEEZ (2)	11. 2.03	L E and S C A Craze *"Fresh Air"*	(Berkhamsted)	17. 4.05A
G-CCBC	Thruster T.600N 450	0013-T600N-077		23. 1.03	M.L.Smith	Popham	27. 4.045
G-CCBF	Maule M-5-235C Lunar Rocket	7276C	G-NHVH N5634N	29.11.02	R Windley	(Lincoln)	27. 3.05
G-CCBG	Best Off Skyranger V2P *(Built G.R.Wallis - kit no.SKR0207214)*	BMAA/HB/240		28. 1.03	C.A.Hardy	(Stevenage)	15. 6.05P
G-CCBH	Piper PA-28-235 Cherokee	28-10648	PH-ABL F-BNFY, N9054W	29. 1.03	J.R.Hunt and S.M.Packer	Thruxton	11. 8.06T
G-CCBI	Raj Hamsa X'Air R100 *(Built H Adams - kit no.600)*	BMAA/HB/192		4. 2.03	H Adams	Kirkbride	30. 8.05P

G-CCBJ	Best Off Skyranger 912	BMAA/HB/262		4. 2.03	A T Hayward	Mill Farm, Hughley, Much Wenlock	4. 8.05P
	(Built A T Hayward -kit no.SKRxxxx285)						
G-CCBK	Aerotechnik EV-97 Eurostar	PFA 315-14025		5. 2.03	J A and G R Pritchard	(Hay-on-Wye)	24. 3.05P
	(Built J A and G R Pritchard - kit no.03-1198(?)						
G-CCBL	Agusta-Bell 206B JetRanger III	8732	OO-VCI	5. 3.03	S.W.Adamson	Durham Tees Valley	17. 6.06T
			PH-VCP, OO-VCI, (OO-XCI)		t/a SW Adamson - Haulage		
G-CCBM	Aerotechnik EV-97 Eurostar	PFA 315-14023		5. 2.03	W Graves	Sackville Lodge Farm, Riseley	7. 5.05P
	(Built W Graves)						
G-CCBN	Replica SE5A rep	077246	PH-WWI	2. 4.03	V.C.Lockwood	Thruxton	15. 8.05P
	(Built B.Barra)		N8010S		*(As "80105/19" in US Air Service c/s)*		
G-CCBO	Not yet allocated						
G-CCBP	Lindstrand LBL 60X HAB	908		12. 2.03	Lindstrand Hot Air Balloons Ltd	(Oswestry)	23. 8.05A
G-CCBR	Wassmer Jodel D.120 Paris-Nice	59	OO-JAL	18. 2.03	R R Walters	Poplar Hall Farm, Elmsett	9. 6.05P
			(OO-CMF), F-BHYP				
G-CCBS	Not yet allocated						
G-CCBT	Cameron Z-90 HAB	10340		24. 3.03	British Telecommunications plc *(BT titles)*	(Thatcham)	9. 4.05A
G-CCBU	Raj Hamsa X'Air 582	BMAA/HB/237		19. 2.03	J S Rakkar	Middle Stoke, Isle of Grain	14. 4.05P
	(Built M.L.Newton - kit no.758)						
G-CCBV	Cameron Z-225 HAB	10365		10. 4.03	Compagnie Aéronautique du Grand-Duché de Luxembourg	(Junglister, Luxembourg)	24. 3.05A
G-CCBW	Tiger Cub RL5A LW Sherwood Ranger			18. 2.03	P H Wiltshire	(Southampton)	7.11.05P
	(Built P H Wiltshire)	PFA 237-13002					
G-CCBX	Raj Hamsa X'Air 582	BMAA/HB/286		28. 5.03	A D'Amico	(Northampton)	
	(Built A D'Amico - kit No.745)						
G-CCBY	Jabiru Aircraft Jabiru UL-450	PFA 274A-13528		21. 2.03	D.M.Goodman	(Driffield)	
	(Built D.M.Goodman)						
G-CCBZ	Aero Designs Pulsar	1936	N4075X	17. 2.03	J M Keane	(Brighton)	8.10.04P
	(Built R N Wasserman) (Tricycle u/c)						
G-CCCA	Vickers Supermarine 509 Spitfire T.IX	CBAF.9590	G-TRIX	18. 2.03	Historic Flying Ltd	Duxford	13. 6.00P
			(G-BHGH), IAC 161, G-15-174, PV202 *(As "161" in IAC c/s) (Flew 13.1.05)*				
G-CCCB	Thruster T.600N 450 Sprint	0033-T600N-078		24. 2.03	D.G.Stanley	Leicester	2. 6.05P
G-CCCC	Cessna 172H	17255822	SE-ELU	9. 2.79	Cheshire Flying Services Ltd	Liverpool	12. 5.07T
			N2622L		t/a Ravenair		
G-CCCD	Mainair Pegasus Quantum 15	7929		12. 6.03	R.N.Gamble	Plaistowes Farm, St.Albans	15. 6.05P
	(Rotax 582)						
G-CCCE	Aeroprakt A22 Foxbat	PFA 317-14002		16. 1.03	C.V.Ellingworth	Henstridge	23. 5.05P
	(Built C V Ellingworth)						
G-CCCF	Thruster T.600N 450 Sprint	0033-T600N-081		24. 2.03	C.H.Spragg tr Thruster Group 2004	Shipdham	17. 3.05P
G-CCCG	Mainair Pegasus Quik	7946		16. 4.03	C.J.Gordon	Perth	21. 4.05P
G-CCCH	Thruster T.600N 450 Sprint	0033-T600N-079		24. 2.03	S.K.Maxwell tr G-CCCH Group	Newtownards	8. 4.05P
G-CCCI	Medway EclipseR	174/152		11. 2.03	N.H.Morley	(Ashford, Kent)	28. 2.05P
G-CCCJ	Nicollier HN.700 Menestrel II	PFA 217-13707		26. 2.03	R Y Kendall	Ewesley Farm, Morpeth	20.12.05P
	(Built R Y Kendall)						
G-CCCK	Best Off Skyranger 912	BMAA/HB/265		26. 2.03	P.L.Braniff	Newtownards	1. 8.05P
	(Built J S Liming - kit no.SKRxxxx289)						
G-CCCM	Best Off Skyranger 912	BMAA/HB/263		3. 3.03	J.R.Moore	(Topcliffe)	30. 6.05P
	(Built J.R.Moore - kit no.SKRxxxx292)						
G-CCCN	Robin R.3000/160	167	OE-KOM	5. 3.03	R.W.Denny	Poplar Hall Farm, Elmsett	5. 5.06T
G-CCCO	Aerotechnik EV-97 Eurostar	PFA 315-14006		11. 3.03	D.R.G.Whitelaw	(Connel, Oban)	29. 4.04P
	(Built Connel Flying Club Eurostar Group - kit no.03-1181)				tr Connel Flying Club Eurostar Group		
G-CCCP	IAV-Bacau Yakovlev Yak-52	899404	LY-AKV	30.11.93	A H.Soper	Jenkins Farm, Navestock	5. 8.05P
			DOSAAF16 *(yellow)*				
G-CCCR	Best Off Skyranger 912	BMAA/HB/266		19. 3.03	D M Robbins	Goodwood	23. 6.05P
	(Built T.C.Viner - kit no.SKR0301290)						
G-CCCT	Comco Ikarus C42 FB UK	PFA 322-13975		19. 3.03	G.A Pentelow	(Kettering)	14. 8.05P
	(Built G.A Pentelow - kit no.0211-6504)						
G-CCCU	Thruster T.600N 450	0034-T600N-084		29. 4.03	K.J.Draper t/a Medway Microlights	Middle Stoke, Isle of Grain	13. 5.05P
	(Official c/n incorrect - date of manufacture [April 2003] indicates correct version should be 0043-T.600N-084)						
G-CCCV	Raj Hamsa X'Air Falcon 133	BMAA/HB/252		20. 3.03	A J Fraley	Kingston Seymour	15. 9.05P
	(Built G.A J.Salter - kit no.614)						
G-CCCW	Pereira Osprey 2	PFA 70-13408		10. 4.03	D.J.Southward	(Beckermet)	
	(Built D.J.Southward)						
G-CCCX	Cameron GP-70 HAB	10383		24. 3.03	Cameron Balloons Ltd	(Bristol)	18. 6.04P
G-CCCY	Best Off Skyranger 912	BMAA/HB/260		25. 3.03	D.M.Cottingham and A Grimsley	Branscombe	27. 8.05P
	(Built D.M.Cottingham - kit no.SKRxxxxxxx)						
G-CCCZ	Raj Hamsa X'Air 582	BMAA/HB/200		1. 4.03	R.A.Merrigan	(Clarina, County Limerick)	27. 1.05P
	(Built M.B.Cooke - kit no.718)						
G-CCDB	Mainair Pegasus Quik	7948		8. 4.03	C.J.Van Dyke	Redlands, Swindon	2.11.05P
G-CCDC	Rans S-6-ES Coyote II	PFA 204-13992		28. 1.03	G.N.Smith	Inglenook Farm, Maydensole, Dover	26. 8.04P
	(Built G N Smith) (Tri-cycle u/c)				*(Noted 11.04)*		
G-CCDD	Mainair Pegasus Quik	7951		31. 3.03	M.P.Hadden and M.H.Rollins	Long Marston	16. 4.05P
G-CCDE	Robinson R22 Beta II	3400	N71906	2. 4.03	C.W.B.Wrightson	Sherburn-in-Elmet	24. 4.06T
					t/a Wrightson Aviation and Engineering		
G-CCDF	Mainair Pegasus Quik	7949		23. 4.03	A Featherstone	Knapthorpe Lodge, Caunton	29. 4.05P
					tr G-CCDF Flying Group		
G-CCDG	Best Off Skyranger 912	BMAA/HB/271		1. 4.03	T.H.Filmer	(Bangor)	26. 9.05P
	(Built W.P.Byrne - kit no.SKR0302294)						
G-CCDH	Best Off Skyranger 912	BMAA/HB/233		5. 2.03	P.and Vivien C.Reynolds	Haddington	12. 8.05P
	(Built D M Hepworth - kit no.SKRxxxx211)						
G-CCDJ	Raj Hamsa X'Air Falcon 582	BMAA/HB/214		18. 2.03	J M Spitz	Longacres Farm, Sandy	
	(Built J M Spitz - kit no.692)				*(Noted 11.04)*		
G-CCDK	Pegasus Quantum 15-912 Super Sport	7947		19. 3.03	S.Brock	(Hardwick, Cambridge)	6. 5.05P
G-CCDL	Raj Hamsa X'Air Falcon 582	BMAA/HB/274		1. 4.03	H.Burroughs	Dunkeswell	11. 5.05P
	(Built H.Burroughs - kit no.819)						

Reg	Type	C/n	Previous id	Date	Owner	Base	CofA
G-CCDM	Mainair Blade	1352-0203-7 & W1147		21. 3.03	M K Ashmore	(Dereham, Norfolk)	25. 3.05P
G-CCDN	Piper PA-28-181 Archer III	2843398	HB-PQA	22. 4.03	J.D.Scott	Elstree	8. 4.06T
			N4176W				
G-CCDO	Mainair Pegasus Quik	7944		19. 3.03	A G.Quinn	Chilbolton	16. 4.05P
G-CCDP	Raj Hamsa X'Air BMW R100	BMAA/HB/276		10. 4.03	J.A McKie	Middle Stoke, Isle of Grain	
	(Built J.A McKie - kit no.847)						
G-CCDR	Raj Hamsa X'Air Falcon Jab22	BMAA/HB/253		16. 4.03	P.D.Sibbons	(Duxford)	16. 6.05P
	(Built P.D.Sibbons - kit no.787)						
G-CCDS	Nicollier HN.700 Menestrel II	PFA 217-13915		13. 3.03	B.W.Gowland	(Pwllheli)	
	(Built B.W.Gowland)						
G-CCDT	Rockwell Commander 114	14397	D-EKGD	25. 3.03	D.R.Robertson	(Elsecar, Barnsley)	10. 4.06
			OE-KGD, D-EIBC				
G-CCDU	Tecnam P92-EM Echo	PFA 318-13721		23. 4.03	M.J.Barrett	Eaglescott	9.11.04P
	(Built M.J.Barrett)						
G-CCDV	Thruster T600N 450 Sprint	0034-T600N-082		24. 4.03	D J Whysall	(Ripley)	14. 6.05P
	(Official c/n incorrect - date of manufacture [April 2003] indicates correct version should be 0043-T600N-082)						
G-CCDW	Best Off Skyranger 582	BMAA/HB/268		27. 3.03	P.Reed	(Luton)	17.12.04P
	(Built P.Reed - kit no.SKR0302309)				tr Debts R Us Family Group		
G-CCDX	Aerotechnik EV-97 Eurostar	PFA 315-14013		18. 2.03	J.M.Swash	Sittles Farm, Alrewas	17. 6.05P
	(Built H F Breakwell and R A Morris)						
G-CCDY	Best Off Skyranger 912	BMAA/HB/275		10. 4.03	A V.Dunne and G.S.Gee-Carter		
	(Built A V.Dunne and G.S.Gee-Carter - kit no.SKRxxxx310)					Plaistowes Farm, St Albans	25. 8.05P
G-CCDZ	Mainair Pegasus Quantum 15-912	7952		24. 4.03	K.D.Baldwin	Graveley	27. 4.05P
G-CCEA	Mainair Pegasus Quik	7950		8. 4.03	J.D.Ash	Wickenby	21. 4.05P
G-CCEB	Thruster T.600N 450 Sprint	0035-T600N-085		24. 4.03	V.Goddard	Craysmarsh Farm, Melksham	15. 6.05P
	(Official c/n incorrect - date of manufacture [May 2003] indicates correct version should be 0053-T600N-085)						
G-CCEC	Evans VP-1	PFA 7031	G-ROSE	15. 4.03	C.R.Harrison	(Highbridge, Taunton)	
	(Built W K Rose)						
G-CCED	Zenair CH.601UL	PFA 162A-13946		4. 4.03	R.P.Reynolds	(Walsall)	2.11.05P
	(Built R.P.Reynolds)						
G-CCEE	Piper PA-15 Vagabond	15-248	G-VAGA	31. 3.03	I.M.Callier	Draycott Farm, Chiseldon	18.12.05P
	(Lycoming O-145-B2)		N4458H, NC4458H				
G-CCEF	Europa Aviation Europa	PFA 247-13038		24. 4.03	C.P.Garner	Eaglescott	21. 6.05P
	(Built C.P.Garner - kit no.302)						
G-CCEG	Rans S-6-ES Coyote II	PFA 204-13831		24. 4.03	K.A.Wright and B.Fukes	North Coates	2. 3.05P
	(Built E.O.Bartle)						
G-CCEH	Best Off Skyranger 912	BMAA/HB/267		28. 4.03	A Eastham	(Brook Farm, Pilling)	4.11.04P
	(Built A Eastham - kit no.SKRxxxx291)						
G-CCEI	Evans VP-2	PFA 63-11377		16. 4.03	I.P.Manley and J.Pearce	Marsh Farm, Sidlesham	
	(Built I.P.Manley and J.Pearce) (Volkswagen 1834) *(Incorporates wings and engine from G-BJZB [PFA 63-10633] scrapped 2002/3: new fuselage noted 4.03)*						
G-CCEJ	Aerotechnik EV-97 Eurostar	PFA 315-14011		1. 5.03	C.R.Ashley	Sittles Farm, Alrewas	2.11.05P
	(Built C.R.Ashley)						
G-CCEK	Air Création 582/Kiss 400	BMAA/HB/272		2. 5.03	G.S.Sage	RAF He0nlow	17.12.04P
	(Built G.S.Sage)						
G-CCEL	Jabiru Aircraft Jabiru UL-450	PFA 247A-13976		12. 2.03	R Pyper	Newtownards	27. 4.05P
	(Built R Pyper)						
G-CCEM	Aerotechnik EV-97A Eurostar	PFA 315-13987		19. 2.03	G.K.Kenealy	Barton	10. 8.05P
	(Built E.Atherden - kit no.03-1148)						
G-CCEN	Cameron Z-120 HAB	10399		9. 5.03	R.Hunt	(Ripley)	7. 9.05A
G-CCEO	Thunder Ax10-180 Series 2 HAB	4634	OE-RZH	10. 6.03	P.Heitzeneder	(Desselbrunn, Austria)	14. 7.04A
					t/a 1 Oberosterreichischer and t/a Ballonfahrerverein		
G-CCEP	Raj Hamsa X'Air Falcon Jabiru	BMAA/HB/264		6. 5.03	K Angel	Middle Stoke, Isle of Grain	14. 9.05P
	(Built A McIvor - kit no.716)						
G-CCER	Mainair Gemini/Flash IIA	799-0890-7 & W592		9. 5.03	A Hillyer	(Erding, Germany)	6. 7.05P
	(Originally sold to Club ULM Fontenay, Tresigny, France and may have French ULM p/i, therefore)						
G-CCES	Raj Hamsa X'Air H2706	BMAA/HB/104		8. 5.03	G.V.McCloskey	(Londonderry, County Londonderry)	
	(Built G.V.McCloskey - kit no.401)						
G-CCET	Nova Vertex 28	14296		25. 3.03	M.Hay *(New owner 1.04)*	Dundee	
G-CCEU	Rotary Air Force RAF 2000 GTX-SE	001	N97ZP	24. 6.03	N.G.Dovaston	(Chester Le Street)	27. 9.05P
	(Built J L Rollins)						
G-CCEW	Mainair Pegasus Quik	7966		17. 6.03	C.S.Mackenzie	East Fortune	17. 6.05P
G-CCEX	Cameron Z-90 HAB	10260		28. 4.03	Cameron Balloons Ltd	(Bristo)l	28. 7.05A
G-CCEY	Raj Hamsa X'Air 582	BMAA/HB/258		12. 5.03	P.J.F.Spedding	(Ashford)	11.11.05P
	(Built P.J.F.Spedding - kit no.833)						
G-CCEZ	Reality Easy Raider Jab22	BMAA/HB/220		7. 5.03	S.A Chambers	(Buckfastleigh)	
	(Built S.A Chambers - kit no.0010)						
G-CCFA	Air Création 582/Kiss 400	BMAA/HB/282		16. 5.03	N.Hewitt	Mill Farm, Shifnal	
	(Built N.Hewitt)				*(Noted 5.04)*		
G-CCFB	Mainair Pegasus Quik	7955		16. 6.03	W.T.Davis	Perth	15. 6.05P
G-CCFC	Robinson R44 Raven II	10151		16. 9.03	M Entwistle	Booker	4.11.06T
G-CCFD	Quad City Challenger II	PFA 177-13180		20. 5.03	W.Oswald	(Aberdeen)	
	(Built W.Oswald - c/n CH-0597-UK-1617)						
G-CCFE	Tipsy Nipper T.66 Series 2	37	OO-PLG	12. 6.03	A R.Way	Dunkeswell	1.10.05P
	(Built Avions Fairey SA)						
G-CCFF	Lindstrand LBL 150A HAB	918		30. 6.03	Airborne Adventures Ltd	(Skipton)	28. 7.05T
G-CCFG	Dyn'Aéro MCR-01 Club	PFA 301A-14047		8. 4.03	G.J.Sargent	Bourn	15. 8.05P
	(Built M P Sargent)						
G-CCFH	Not yet allocated						
G-CCFI	Piper PA-32-260 Cherokee Six	32-7400002	OO-PCT	30. 6.03	McManus Truck and Trailer Spares Ltd		
			N56630			Trim, County Meath	20. 8.06P
G-CCFJ	Kolb Twinstar Mk.3 Extra	PFA 205-14014		29. 5.03	D Travers	(Ealand, Scunthorpe)	27. 4.05P
	(Built M.H.Moulai)						
G-CCFK	Europa Aviation Europa	PFA 247-13744		29. 5.03	C.R.Knapton	(Humbleton)	17. 8.05P
	(Built C.R.Knapton - kit no.502)						

Registration	Type	C/n	Previous identity	Reg date	Owner	Location	CofA
G-CCFL	Mainair Pegasus Quik	7960		16. 6.03	S.A Noble	(Bishops Stortford)	15. 6.05P
G-CCFM	Mainair Blade 912	1354-0603-7 & W1149		8. 5.03	J.G.I.Muncey	Rayne Hall Farm, Braintree	8. 7.05P
G-CCFN	Cameron N-105 HAB	10442		5. 8.03	Procter and Gamble (Health and Beauty Care) Ltd		
					(Clairol titles)	(Weybridge)	22. 7.04A
G-CCFO	Pitts S-1S *(Built R.B.Innes)*	001	C-FYXO	28. 5.03	R.J.Anderson	Priory Farm, Tibenham	27. 9.05P
G-CCFP	Diamond DA.40D Star	D4.026		3. 9.03	N P de Gruchy Lambert	Blackbushe	15.10.06T
G-CCFR	Diamond DA.40D Star	D4.032		3. 9.03	B.Wronski	Gloucestershire	12.10.06T
G-CCFS	Diamond DA.40D Star	D4.034		3. 9.03	R.H.Butterfield and A M.Dyson	Gamston	4.12.06T
					t/a Principle Aircraft		
G-CCFT	Mainair Pegasus Quantum 15-912	7961		17. 6.03	D.Tasker	Rufforth	16. 6.04P
G-CCFU	Diamond DA.40D Star	D4.035		3. 9.03	R.H.Butterfield and A M.Dyson	Gamston	27.11.06T
					t/a Principle Aircraft		
G-CCFV	Lindstrand LBL 77A HAB	934		23. 6.03	Alton Aviation Ltd	(Oswestry)	7. 7.05A
G-CCFW	WAR Focke-Wulf Fw 190 rep	PFA 81-12729		27. 5.03	D.B.Conway	Kemble	1. 6.05P
	(Builtb D.B.Conway)				*(As "- + 9")*		
G-CCFX	EAA Acrosport II	PFA 72-11221		23. 6.03	C.D.Ward	(Leyburn)	
	(Built C D Ward)						
G-CCFY	RotorWay Executive 162F	6719		27. 6.03	M.Hawley	Street Farm, Takeley	AC
	(Built M Hawley)				*(Noted unmarked 1.05)*		
G-CCFZ	Comco Ikarus C42 FB UK	PFA 322-14040		2. 5.03	B.W.Drake	Over Farm, Gloucester	18.10.05P
	(Built B.W.Drake)						
G-CCGA	Medway EclipseR	175/153	"G68"	18. 6.03	C.J.Raper	Middle Stoke, Isle of Grain	30. 8.05P
					t/a Medway Microlights		
G-CCGB	TEAM mini-MAX 91	PFA 186-13767		4. 6.03	A D.Pentland	Baxby Manor, Husthwaite	22.12.04P
	(Built A D.Pentland)						
G-CCGC	Mainair Pegasus Quik	7958		27. 5.03	A Crozier	(Bathgate)	1. 6.05P
G-CCGE	Robinson R22 Beta II	3453		25. 6.03	Heli Aitch Be Ltd	(London E3)	9. 7.06T
G-CCGF	Robinson R22 Beta II	3454		25. 6.03	Ground Effect Ltd	Shoreham	13. 7.06T
G-CCGG	Jabiru Aircraft Jabiru J400	PFA 325-14055		16. 6.03	K.D.Pearce	Southery	2. 9.05P
	(Built K.D.Pearce)						
G-CCGH	Super Marine Spitfire Mk.26	PFA 324-14054		17. 6.03	K.D.Pearce	Southery	
	(Built K.D.Pearce - c/n 021) (Jabiru 5100cc @ 180hp)						
G-CCGI	Mainair Pegasus Quik	7967		7. 7.03	M.C.Kerr	Clench Common	21. 7.05P
G-CCGJ*	Raj Hamsa X'Air 582	BMAA/HB/188		30. 6.03	M.C.Rangeley	North Coates	
	(Built M.C.Rangeley - kit no.604)				*(Noted 4.04: cancelled 1.7.04 as WFU - no PtoF issued)*		
G-CCGK	Mainair Blade	1355-0603-7 & W1150		18. 7.03	G.Kerr	East Fortune	19. 7.05P
G-CCGL	SOCATA TB-20 Trinidad	2187	F-OIMN	28. 5.03	Pembroke Motor Services Ltd	Haverfordwest	10. 6.06T
G-CCGM	Air Création 582/Kiss 450	BMAA/HB/277		23. 5.03	A I Lea	(Basildon)	22. 5.05P
	(Built G.P.Masters - c/ns: Trike T03026 & Wing A2049-2044)						
G-CCGN	Bell 206L-1 Long Ranger II	45534	N76LC N17FH, N5752F	15. 7.03	Hughes Helicopter Company Ltd	Biggin Hill	14. 9.06T
					t/a Biggin Hill Helicopters		
G-CCGO	Medway AV8R	176/154		11. 4.03	C.J.Draper	Middle Stoke, Isle of Grain	
	(Rotax 582 - no. 5589987)				t/a Medway Microlights		
G-CCGP	Holman Bristol Type 2000	PFA 270-12858		1. 7.03	R.G.Holman	Kemble	
	(Built R G Holman)				*(Noted on build 1.04)*		
G-CCGR	Raj Hamsa X'Air VM133	BMAA/HB/284		12. 6.03	J.M.Weston	(Preston)	11. 5.05P
	(Built J.M.Weston - kit no.865)						
G-CCGS	Dornier Do.328-100	3101	D-CPRX D-CDXR	18. 3.04	Suckling Airways (Cambridge) Ltd	London City	17. 3.05T
					t/a Scot Airways		
G-CCGT	Cameron Z-425 HAB	10398		16. 6.03	A A Brown	(Guildford)	20. 6.05A
G-CCGU	Van's RV-9A	PFA 320-13798		10. 7.03	B.J.Main, C.W.Hague and A Strachan	Dunkeswell	9. 6.05P
	(Built B.J.Main)						
G-CCGV	Lindstrand LBL 150A HAB	273	HB-BUU	8. 7.03	Lindstrand Hot Air Balloons Ltd	(Oswestry)	24. 9.05T
G-CCGW	Europa Aviation Europa	PFA 247-12548		8. 7.03	G.C.Smith	Cambridge	17. 6.05P
	(Built G.C.Smith - kit no.026)						
G-CCGX	Not yet allocated						
G-CCGY	Cameron Z-105 HAB	10422		8. 7.03	Cameron Balloons Ltd	(Bristol)	6.10.05A
G-CCGZ	Cameron Z-250 HAB	10438		4. 9.03	Ballooning Adventures Ltd	(Hexham)	4. 9.05T
G-CCHA	Diamond DA.40D Star	D4.046		14. 7.03	Cabair College of Air Training Ltd	Cranfield/Elstree	4. 9.06T
G-CCHB	Diamond DA.40D Star	D4.047		16. 7.03	Cabair College of Air Training Ltd	Cranfield/Elstree	7. 9.06T
G-CCHC	Diamond DA.40D Star	D4.050		26. 8.03	Cabair College of Air Training Ltd	Cranfield/Elstree	21. 9.06T
G-CCHD	Diamond DA.40D Star	D4.051		3. 9.03	Cabair College of Air Training Ltd	Cranfield/Elstree	21. 9.06T
G-CCHE	Diamond DA.40D Star	D4.054		29. 8.03	Cabair College of Air Training Ltd	Cranfield/Elstree	21. 9.06T
G-CCHF	Diamond DA.40D Star	D4.055		3. 9.03	Cabair College of Air Training Ltd	Cranfield/Elstree	21. 9.06T
G-CCHG	Diamond DA.40D Star	D4.058		3. 9.03	Plane Talking Ltd	Elstree	30. 9.06T
G-CCHH	Mainair Pegasus Quik	7963		24. 6.03	S.Sebastian	Southend	23. 6.05P
G-CCHI	Mainair Pegasus Quik	7971		29. 7.03	A.L.Bagnall	(Derby)	12. 8.05P
G-CCHJ	Air Création 582/Kiss 400	BMAA/HB/257		28. 4.03	H.C.Jones	(Wolvey)	31. 8.05P
	(Built H.C.Jones - kit no.FL016, Trike T02117, Wing A02184-02179)						
G-CCHK	Diamond DA.40D Star	D4.059		3. 9.03	Plane Talking Ltd	Cranfield	24. 9.06T
G-CCHL	Piper PA-28-181 Archer III	2843176	OY-JAA N9501N	4.08.03	Archer Three Ltd	Lydd	21. 8.06T
G-CCHM	Air Création 582/Kiss 450	BMAA/HB/292		15. 7.03	M.J.Jessup	(Edenbridge)	10. 6.05P
	(Built W.G.Colyer (Kit no.FL022, Trike T03057, Wing A03099-3016)						
G-CCHN	Corby CJ-1 Starlet	PFA 134-12848		15. 7.03	D.C.Mayle	(Reading)	
	(Built D.C.Mayle)						
G-CCHO	Mainair Pegasus Quik	7968		9. 7.03	M.Allan	(Dalkeith)	8. 7.05P
G-CCHP	Cameron Z-31 HAB	10443		20. 8.03	M.H.Redman	(Stalbridge, Sturminster Newton)	
G-CCHR	Reality Easy Raider 503	BMAA/HB/223		24. 6.03	R.B.Hawkins	(Plymouth)	
	(Built R.B.Hawkins - kit no.0008)						
G-CCHS	Raj Hamsa X'Air 582	BMAA/HB/291		4. 7.03	I.Lonsdale	Tarn Farm, Cockerham	13. 7.05P
	(Built I.Lonsdale - kit no.840)						
G-CCHT	Cessna 152	15285176	9H-ACW N6159Q	17. 7.03	J.S.Devlin and Z.Islam	Blackbushe	18. 9.06T

Regn	Type	C/n	Previous identity	Regd	Owner/Operator	Base	CofA
G-CCHU	Flight Design CT2K	03-04-02-15		28. 8.03	A N.D.Arthur	Denhaml	27. 8.05P
	(Assembled Pegasus Aviation as c/n 7976)						
G-CCHV	Mainair Rapier	1353-0403-7 & W1148		15. 9.03	A Butterworth	(Poynton)	29. 9.04P
G-CCHW	Cameron Z-77 HAB	10426		26. 6.03	A Murphy	(Dunshaughlin, County West Meath)	8. 6.05A
G-CCHX	Scheibe SF25C Rotax-Falke	44694		25. 9.03	Lasham Gliding Society Ltd	Lasham	3.11.06
G-CCHY	Bücker Bü.131 Jungmann	21	I-CABI	12.11.03	M.V.Rijkse	(London WC1)	12. 4.05P
	(Built Dornier-Werke AG)		HB-UTZ, Swiss AF A-12		*(As "A-12" in Swiss AF c/s)*		
G-CCHZ	Robinson R22 Beta II	3466		22. 7.03	Helicopter Training and Hire Ltd	Newtownards	2. 9.06T
G-CCIA	Lindstrand LBL 105A HAB	935		28. 7.03	J.J-C.Bernadin	(Curcay-sur-Dive, France)	1. 7.05A
G-CCIC	Thruster T600N 450 Sprint	0036-T600N-086		25. 7.03	M.L.Smith	Popham	2. 9.04P
G-CCID	Jabiru Aircraft Jabiru J400	PFA 325-14059		25. 7.03	J.Bailey	Crowfield	
	(Built J Bailey)				*(Noted 1.05)*		
G-CCIE	Colt 315A HAB	10176		25. 7.03	T.M.Donnelly	(Sprotbrough, Doncaster)	31. 7.05T
G-CCIF	Mainair Blade	1356-0703-7 & W1151		28. 7.03	D.J. Kennedy	(Cleethorpes)	9.11.05P
G-CCIG	Aero Designs Pulsar	PFA 202-12133		15. 7.03	P.Maguire	(Accrington)	
	(Built P.Maguire)						
G-CCIH	Mainair Pegasus Quantum 15	7973		31. 7.03	R.Bennett	(Canterbury)	12. 8.05P
G-CCII	ICP MXP-740 Savannah Jabiru	BMAA/HB/285		6. 6.03	J.R.Livett and D.Chaloner	Redlands	15.10.05P
	(Built M.J.Kaye - kit no.01-04-51-063)						
G-CCIJ	Piper PA-28R-180 Cherokee Arrow	28R-30873	SE-FDZ	14. 7.03	I.R.Chaplin *(Op Southend School of Flying)*	Southend	14. 9.06T
G-CCIK	Best Off Skyranger 912	BMAA/HB/278		30. 7.03	L.E.Cowling and A P.Chapman		
	(Built L.E.Cowling and A P.Chapman - kit no.SKRxxxx279)					Hawksbridge Farm, Oxenhope	20. 1.05P
G-CCIO	Best Off Skyranger 912	BMAA/HB/261		4. 8.03	B.Berry	Crosland Moor	20. 1.05P
	(Built B.Berry - kit no.SKRxxxx284)						
G-CCIR	Van's RV-8	PFA 303-13732		7. 8.03	B F Hill	Shenstone	11. 1.05P
	(Built D.Marsh)						
G-CCIS	Scheibe SF28A Tandem-Falke	5791	OE-9154	15.10.03	P.T.Ross	(Falmouth)	28.10.06
			(D-KDFZ)				
G-CCIT	Zenair CH.701 UL	PFA 187-13911		30. 6.03	I.M.Sinclair	Glenrothes	5.10.05P
	(Built I.M.Sinclair)						
G-CCIU	Cameron N-105 HAB	10485		18. 8.03	Bianchi Aviation Film Services Ltd	(Booker)	14. 9.04A
G-CCIV	Mainair Pegasus Quik	7977		13. 8.03	D.Little	(Crawley)	18. 8.05P
G-CCIW	Raj Hamsa X'Air 582	BMAA/HB/281		11. 8.03	G.Wilkinson	Otherton, Cannock	31. 3.05P
	(Built G.Wilkinson - kiit no.838)						
G-CCIX*	Vickers Supermarine 361 Spitfire LF.IXe				See SECTION 3 Part 1	Polk City, Florida, US	
G-CCIY	Best Off Skyranger 912	BMAA/HB/250		14. 8.03	L.F.Tanner	(Hazlemere, Buckingham)	1. 4.05P
	(Built L.F.Tanner - kit no.SKRxxxx344)						
G-CCIZ	PZL-110 Koliber 160A	04010087	SP-WGN (2)	21. 8.03	Horizon Aviation Ltd	Swansea	26.10.06T
G-CCJA	Best Off Skyranger 912	BMAA/HB/299		6. 8.03	T.R.Southall	Shobdon	20. 1.05P
	(Built T.R.Southall - kit no.SKRxxxx364)						
G-CCJB	Zenair CH.701 STOL	PFA 187-13270		24. 6.03	E.G.Brown	(Luton)	AC
	(Built E.G.Brown - kit no.7-3659)						
G-CCJD	Mainair Pegasus Quantum 15	7974		3. 9.03	P.Clark	(Dorking)	29. 8.05P
G-CCJE	Schweizer 269C-1	0148	N828DK	19. 8.03	Dragon Helicopters Ltd	Sheffield City	3. 9.06T
G-CCJF	Cameron C-90 HAB	10483		14.10.03	Balloon School (International) Ltd	(Petworth)	21. 9.05T
G-CCJG	Cameron A-200 HAB	10484		12. 2.04	J.M.Stables t/a Aire Valley Balloons	(Knaresborough)	26. 1.05T
G-CCJH	Lindstrand LBL 90A HAB	906		5. 8.03	J.R.Hoare	(Plymouth)	27. 8.05A
G-CCJI	Van's RV-6	PFA 181A-13572		4. 7.03	S.Longstaff	Gamston	9. 8.05P
	(Built E M Marsh)						
G-CCJJ	Medway SLA 100 Executive	18803		20. 8.03	C.J.Draper	Middle Stoke, Isle of Grain	
	(Orig regd as Medway Piranah until 10.03)				t/a Medway Microlights		
G-CCJK	Aerostar Yakovlev Yak-52	9612001	RA-02622	3. 3.04	R.K.Howell	White Waltham	18. 3.05P
			LY-AFH				
G-CCJL	Super Marine Spitfire Mk.26	PFA 324-14053		22. 8.03	P.M.Whitaker	(Ilkley)	
	(Built P.M.Whitaker and M Hanley)						
G-CCJM	Mainair Pegasus Quik	7970		24. 7.03	P.Crosby	Ince Blundell	24. 7.05P
G-CCJN	Rans S-6ES Coyote II	PFA 204-13575		28. 8.03	M.G.A Wood	(Tadcaster)	
	(Built M.G.A Wood)						
G-CCJO	ICP MXP-740 Savannah Jabiru	BMAA/HB/295		28. 8.03	R.and I.Fletcher	Sandtoft	12. 8.05P
	(Built R.and I.Fletcher - kit no.03-05-01-213)						
G-CCJS	Reality Easy Raider Jab16	BMAA/HB/293		2. 9.03	K. Wright	(Douglas, Isle of Man)	
	(Built K Wright - kt no.0002)						
G-CCJT	Best Off Skyranger 912	BMAA/HB/300		30. 7.03	J.W.Taylor	Over Farm, Gloucester	
	(Built J.W.Taylor - kit no.SKRxxxx317)				*(Noted 7.04)*		
G-CCJU	ICP MXP-740 Savannah Jabiru	BMAA/HB/294		3. 9.03	K.R.Wootton and A Colverson	Hull	
	(Built K.R.Wootton and A Colverson) (Kit no.03-05-01-214)						
G-CCJV	Aeroprakt A22 Foxbat	PFA 317-14082		3. 9.03	J C Forrester	Otherton, Cannock	6. 5.05P
	(Built M.J.Barrett, A Dace, S McRoberts and J C Forrester)				tr Foxbat UK015 Syndicate		
G-CCJW	Best Off Skyranger 912	BMAA/HB/303		3. 9.03	J.R.Walter	Hunterson Farm, Stair	8. 6.05P
	(Built J.R.Walter - kit no.SKRxxxx366)						
G-CCJX	Europa Aviation Europa XS	PFA 247-13727		9. 9.03	J.S.Baranski	(Maidenhead)	
	(Built J.S.Baranski - kit no.509)						
G-CCJY	Cameron Z-42 HAB	10465		12. 9.03	Cameron Balloons Ltd	(Bristol)	6. 1.05P
G-CCKF	Best Off Skyranger 912	BMAA/HB/289		4. 9.03	Thelma P M Turnbull	Eshott	6. 5.05P
	(Built S.A Owen - kit no.SKRxxxx314)						
G-CCKG	Best Off Skyranger 912	BMAA/HB/302		26. 8.03	J.Hannibal	Pound Green, Buttonoak, Bewdley	27. 7.05P
	(Built J.Hannibal - kit no.SKRxxxx375)						
G-CCKH	Diamond DA.40D Star	D4.039		23.10.03	Diamond Aircraft UK Ltd	Elstree	14. 1.07T
G-CCKI	Diamond DA.40D Star	D4.038		23.10.03	S.C.Horwood	Panshanger	20.11.06T
G-CCKJ	Raj Hamsa X'Air 582	BMAA/HB/306		2.10.03	S.Thompson	(Hull)	
	(Built S.Thompson - kit no.855)						
G-CCKK	Aerotechnik EV-97 Eurostar	PFA 315-14076		10. 7.03	C.Townsend	Kemble	9.11.05P
	(Built C.Townsend)						

Reg	Type	C/n	Previous identity	Reg date	Owner	Base	CofA expiry
G-CCKL	Aerotechnik EV-97 Eurostar	PFA 315-14117		15. 9.03	J.S.Liming and Annette U.I.Hudson		
	(Built J.S.Liming and A U.I.Hudson)					Priory Farm, Tibenham	9.11.05P
G-CCKM	Mainair Pegasus Quik	7985		22. 9.03	J W McCarthy	(Southport)	25. 1.05P
G-CCKN	Nicollier HN.700 Menestrel II	PFA 217-13943		23. 9.03	C.R.Partington	Brunton	
	(Built C.R.Partington)						
G-CCKO	Mainair Pegasus Quik	7982		27. 8.03	M.J.Mawle and C.R.Bunce	Redlands	31. 8.05P
G-CCKP	Robin DR400/120 Dauphin 2+2	2044	F-GKQD	15. 9.04	B.A.Mills t/a Duxford Flying Group	Duxford	AC
G-CCKR	Pietenpol Aircamper	PFA 047-12295		22. 8.03	T.J.Wilson	Dunkeswell	2.11.05P
	(Built T.J.Wilson)						
G-CCKS	Hughes 369E	0303E	N7065C	22.10.03	Storm Aviation Group International Ltd	(Horley)	17. 3.07T
			N314JP				
G-CCKT	HAPI Cygnet SF-2A	PFA 182-13366		15. 7.03	P.W.Abraham	(Bridgend)	
	(Built P.W.Abraham)						
G-CCKU	Canadian Home Rotors Safari	S2113		7.10.03	J.C.Collingwood	(Cranbrook)	AC
	(Built J.C.Collingwood)						
G-CCKV	Isaacs Fury II	PFA 011-13695		10.10.03	S.T.G.Ingram	(Cambourne)	
	(Built S.T.G.Ingram)						
G-CCKW	Piper PA-18-135 Super Cub	18-3535	G-GDAM	16. 9.03	G T Fisher	(Northside, Thorney)	11. 8.91
	(L-21B-PI) *(Frame No.18-3648)*		PH-PVW, (PH-DKE), R.Neth AF R-107, 54-2335 *(New owner 11.03) (Also see G-CUBI)*				
G-CCKX	Lindstrand LBL 210A HAB	931		23. 1.04	Alba Ballooning Ltd	(Edinburgh)	17.11.04T
G-CCKY	Lindstrand LBL 240A HAB	943		21.11.03	Dance With Balloons Ltd	(Witham)	21.10.04T
G-CCKZ	Customcraft A25 HAB	CC005		10.10.03	A Van Wyk	(Caxton, Cambridge)	
G-CCLA	Not yet allocatd						
G-CCLB	Diamond DA.40D Star	D4.074		19. 1.04	R.J.Millen and M.J.Millen	Rochester	17. 5.07T
					t/a The Millen Corporation		
G-CCLC	Diamond DA.40D Star	D4.073		22. 1.04	Diamond Aircraft UK Ltd	Gamston	14. 6.07T
G-CCLE	Aerotechnik EV-97 Eurostar	PFA 315-14074		1. 7.03	W.S.Long	Mayfield Farm, Stevenston	7. 8.05P
	(Built W.S.Long)						
G-CCLF	Best Off Skyranger 912	BMAA/HB/311		16.10.03	J.Fleming	(Cromdale, Grantown-on-Spey)	19. 7.05P
	(Built G.K.R.Linney - kit no.SKRxxxxxxx?)						
G-CCLH	Rans S-6-ES Coyote II	PFA 204-13658		24.10.03	K.R.Browne	Marshland, Wisbech	16. 3.05P
	(Built K.R.Browne)						
G-CCLJ	Piper PA-28-140 Cherokee Cruiser	28-7525049	OY-TOJ	1. 9.03	A M George	Stapleford	13. 1.07T
G-CCLL	Zenair CH601XL Zodiac	PFA 162B-14081		4. 9.03	L.Lewis	Yearby	
	(Built L Lewis)						
G-CCLM	Mainair Pegasus Quik	7986		7.10.03	P.M.Ryder	Newton Bank, Daresbury	6.10.05P
G-CCLO	Ultramagic H-77 HAB	77/244		19. 2.04	J.P.Moore	(Great Missenden)	19. 2.05T
G-CCLP	ICP MXP-740 Savannah Jabiru	BMAA/HB/314		31.10.03	Barbara Fraser	Perth	18. 5.05P
	(Built M.J.Kaye)						
G-CCLR	Schleicher ASH 26E	26209		26.11.03	C.R.Lear	RAF Keevil	26.11.06
G-CCLR	Not yet allocatd						
C-CCLS	Comco Ikarus C42 FB UK	PFA 322-14050		19. 9.03	SLS Computing Services Ltd	(Nuneaton)	
	(Built J Spinks and T Greenhill)						
G-CCLU	Best Off Skyranger 912	BMAA/HB/316		11.11.03	L.Stanton	Stoke Golding	1. 6.05P
	(Built L.Stanton - kit no.SKRxxxxxxx?)						
G-CCLV	Diamond DA.40D Star	D4.052		22.12.03	Plane Talking Ltd	Elstree	23. 3.07T
G-CCLW	Diamond DA.40D Star	D4.068		19.12.03	Plane Talking Ltd	Elstree	4. 2.07T
G-CCLX	Mainair Pegasus Quik	7996		13.11.03	S.D.Pain	Rayne Hall Farm, Braintree	12.11.04P
G-CCLY	Bell 206B-3 JetRanger III	3594	G-TILT	26. 4.95	Mulberry Homes Ltd	(Pontesbury, Shrewsbury)	30.10.04T
			G-BRJO, N2295Z				
G-CCMA	Boeing 747-267B	22872	G-VCAT	12.12.03	ALG-VIR-747RR Ltd	Bournemouth	27.11.06T
			TF-ATK, G-VCAT, B-HIE, VR-HIE *(Stored 1.05 all white and two engines) (To be ZS-PJH)*				
G-CCMC	Jabiru Aircraft Jabiru UL-450	PFA 274A-13775		9. 9.03	J.T.McCormack	Broomhill Farm, West Calder	12. 4.05P
	(Built J.T.McCormack)				*(F/f 4.2.04)*		
G-CCMD	Mainair Pegasus Quik	7991		23.10.03	J.T.McCormack	Broomhill Farm, West Calder	2.11.04P
G-CCME	Mainair Pegasus Quik	7995		7.11.03	R.R.Nichol	Carlisle	6.11.05P
G-CCMF	Diamond DA.40D Star	D4.075		19.12.03	Plane Talking Ltd	Elstree	23. 3.07T
G-CCMG							
G-CCMH	Miles M.2H Hawk Major	172	EC-ABI	20.10.03	J.A Pothecary	(Salisbury)	
			EC-CAS, EC-DDB, EC-W44				
G-CCMI	Scottish Aviation Bulldog Series 120/121	BH120/199	G-KKKK	20.11.03	E.Cummings	Biggin Hill	16. 1.05T
			XX513		tr 617 Syndicate *(As "XX513/10")*		
G-CCMJ	Reality Easy Raider Jab22	BMAA/HB/254		13.11.03	G.F.Clews	(Burton-on-Trent)	
	(Built G.F.Clews - kiit no.0009)						
G-CCMK	Raj Hamsa X'Air Falcon Jab22	BMAA/HB/301		17.11.03	Cambridge Design Partnership Ltd	(Toft)	6. 7.05P
	(Built M A Beadman - kit no.827)						
G-CCML	Mainair Pegasus Quik	7992		14.10.03	P.Ritchie	Perth	23.10.05P
G-CCMM	Dyn'Aéro MCR-01 Club	PFA 301B-13945		8. 9.03	J D Harris	Kingston Seymour	23. 3.05P
	(Built J.P.Davis - c/n.131)						
G-CCMN	Cameron C-90 HAB	10519		27. 1.04	A E.Austin	(Naseby)	27. 1.05A
G-CCMO	Aerotechnik EV-97A Eurostar	PFA 315-14155		11.11.03	E.M.Woods	Lydney-St. Briavels	12. 2.05P
	(Built E.M.Woods)						
G-CCMP	Aerotechnik EV-97A Eurostar	PFA 315-14127		23.10.03	W.K. Wilkie	Newtownards	2. 3.05P
	(Built W.K.Wilkie)						
G-CCMR	Robinson R22 Beta II	3497	N75273	9. 1.04	Mash Enterprises Ltd	Goodwood	14. 1.07
G-CCMS	Mainair Pegasus Quik	7997		1.12.03	A.J.Roche	Elm Farm, Wickford	30.11.04P
G-CCMT	Thruster T600N 450	1031-T600N-092		14.11.03	S.P.McCaffrey	(Abingdon)	11.12.04P
G-CCMU	Rotorway Executive 162F	6720		20.11.03	M.Irving	Street Farm, Takeley	4. 7.05P
	(Built M.Irving)						
G-CCMV*	Vought FG-1D Corsair						
G-CCMW	CFM Shadow Series DD	PFA 161-13869		2. 9.03	M.Wilkinson	(Milton Keynes)	
	(Built M.Wilkinson)						
G-CCMX	Best Off Skyranger 912	BMAA/HB/255		25.11.03	K.J.Cole	(Highnam, Glos)	19.12.05P
	(Built K.J.Cole - kit no.SKRxxxx243)						

Registration	Type	C/n	Previous identity	Regd date	Owner/operator	Probable base	CofA expiry
G-CCMZ	Best Off Skyranger 912	BMAA/HB/288		23.10.03	D.D.Appleford	Kemble	
	(Built D.D.Appleford - kit no.SKRxxxx316)						
G-CCNA	Jodel DR.100A rep	PFA 304-13519		21.11.03	W.R.Davis-Smith and R.Everitt	Ley Farm, Chirk	
	(Built R.Everitt) (Reported as rebuild of G-ATHX)				*(Noted 4.04)*		
G-CCNB	Rans S-6ES Coyote II	PFA 204-14027		2. 9.03	M.S.Lawrence	Mill Farm, Shifnal	7. 6.05P
	(Built D.Bedford)						
G-CCNC	Cameron Z-275 HAB	10504		16. 4.04	D. Ling	(Nuthall, Nottingham)	3. 4.05T
G-CCND	Van's RV-9A	PFA 320-14142		10.12.03	K.S.Woodard	Priory Farm, Tibenham	
	(Built K.S.Woodard)				*(On build 8.04)*		
G-CCNE	Mainair Pegasus Quantum 15	7093	T2-2795	11.12.03	C.D.Waldron	Redlands	17.12.05P
G-CCNF	Raj Hamsa X'Air Falcon Jab22	BMAA/HB/211		9.12.03	M.F.Eddington	(Wincanton)	
	(Built M.F.Eddington - kit no.644)						
G-CCNG	Flight Design CT2K	03-06-02-27		5. 1.04	David Goode Sculpture Ltd	Enstone	4. 1.05P
	(Assembled Pegasus Aviation as c/n 8004)						
G-CCNH	Rans S-6ES Coyote II	PFA 204-14114		11. 9.03	N.C.Harper	Grove Farm, Needham	9.12.05P
	(Built N.C.Harper - c/n 0503.1498-0304ES)				tr Coyote Group		
G-CCNJ	Best Off Skyranger 912	BMAA/HB/330		17.12.03	J.D.Buchanan	Coldharbour Farm, Willingham	
	(Built J.D.Buchanan - kit no.SKRxxxx932)						
G-CCNK	Robinson R44 Raven	1357		13. 1.04	C.W.and K.A Bootman t/a Aircol	(Turvey, Bedford)	5. 2.07T
G-CCNL	Raj Hamsa X'Air Falcon 133	BMAA/HB/326		24.12.03	G.A.J.Salter	(Taunton)	27. 6.05P
	(Built S Rance and A Davis - kit no.909)						
G-CCNM	Mainair Pegasus Quik	8002		2.12.03	G T Snoddon	Newtownards	10.12.04P
G-CCNN	Cameron Z-90 HAB	10512		8. 1.04	J.H.Turner *(Gottex titles)*	(Salisbury)	15.12.04A
G-CCNO	Not yet allocatd						
G-CCNP	Flight Design CT2K	Not known		22. 1.04	M.Clare	(Gayton, Northampton)	10. 3.05P
	(Assembled Pegasus Aviation as c/n 8005)						
G-CCNR	Best Off Skyranger 912	BMAA/HB/315		4.12.03	S.J.Huxtable	Weston Zoyland	20. 5.05P
	(Built S.J.Huxtable - kit no.SKRxxxx381)						
G-CCNS	Best-Off Skyranger 912	BMAA/HB/356		24. 2.04	G.G.Rowley and M.Liptrot	Carlisle	
	(Built G.G.Rowley and M.Liptrot - kit no.SKRxxxx434)						
G-CCNT	Comco Ikarus C42 FB80	0311-6585		19.12.03	Mainair Mircolight School Ltd	Barton	11. 1.05P
	(Rotax 912-UL)						
G-CCNU	Best Off Skyranger J22	BMAA/HB/297		9. 1.04	D.P.Toulson	Redlands	29. 9.05P
	(Built D.P.Toulson and R.L Nyman - kit no.SKRxxxx319)						
G-CCNV	Cameron Z-210 HAB	10505		22. 4.04	J.A.Cooper	(Ugborough, Ivybridge)	18. 4.05T
G-CCNW	Mainair Pegasus Quantum 15	8010		28. 1.04	J. Childs	Kettering	11. 2.05P
G-CCNX	CAB CAP.10B	311		6. 1.04	Arc Input Ltd	Crowfield	31. 3.07T
	(Marked as "CAP.10C")						
G-CCNY	Robinson R44 Raven	1349		18.11.03	Carole M.Evans and J.W.Blaylock	(Kirton, Boston)	15.12.06T
G-CCNZ	Raj Hamsa X'Air VM133	BMAA/HB/308		5.11.03	K.J.Foxall	(Tamworth)	5. 8.05P
	(Built K.J.Foxall - kit no.888)						
G-CCOA*	Scottish Aviation Bulldog Series 120/122	BH120-375	Ghana AF G-111	4. 9.96	Not known	(Isle of Wight)	
			G-BCUU		*(Damaged Cranfield 22.8.01: cancelled 11.6.02 as wfu)*		
			(Fuselage noted 2004 as "G-AXEH" to represent prototype Bulldog for proposed Beagle museum)				
G-CCOB	Aero C.104	247	N2348	21. 1.04	William Tomkins Ltd	Spanhoe	
	(Bücker Bü.131 Jungmann)		LN-BNG, OK-AXV, Czech AF		*(Noted 6.04)*		
G-CCOC	Mainair Pegasus Quantum 15	7999		16.12.03	F.E.J.Moore	(Bexleyheath)	11. 1.05P
G-CCOE	Lindstrand LBL 35A Cloudhopper HAB	971		15. 1.04	Lindstrand Hot Air Balloons Ltd	(Oswestry)	29. 1.05A
G-CCOF	Rans S-6ESA Coyote II	PFA 204-14037		8. 1.04	A J.Wright and M.Govan	(Derby)	30. 6.05P
	(Built A J.Wright and M.Govan)						
G-CCOG	Mainair Pegasus Quik	8001		16.12.03	J.Hood	Eshott	15.12.05P
G-CCOH	Raj Hamsa X'Air Falcon 133	BMAA/HB/338		13. 1.04	A R.Emerson and J.C.Ambrose		
	(Built A R.Emerson - kit no.831)					Priory Farm, Tibenham	4.10.05P
G-CCOI	Lindstrand LBL 90A HAB	945		21. 1.04	M.J.Warne	(Tavistock)	3. 2.05A
G-CCOJ	Not yet allocatd						
G-CCOK	Mainair Pegasus Quik	8000		5. 1.04	E.McCallum	Eshott	4. 1.05P
G-CCOM	Westland Lysander IIIA	Y1363	N3093K	10.12.03	Propshop Ltd	Duxford	
			RCAF V9312				
G-CCOO	Raj Hamsa X'Air VM133	BMAA/HB/320		12.11.03	A Hipkin	Droppingwell Farm, Bewdley	
	(Built A Hipkin - kit no.754)						
G-CCOP	Ultramagic M-105 HAB	105/113		19. 1.04	T.R.Tillson tr Firefly Balloon Team	(Ilkeston)	12. 6.05
G-CCOR	Sequoia F.8L Falco	PFA 100-10588		9.12.03	D.J.and K.S.Thomas	Fenland	
	(Built D.J.Thomas)						
G-CCOS	Cameron Z-350 HAB	10513		8. 3.04	M.L.Gabb	(Haselor, Alcester)	8. 3.05T
G-CCOT	Cameron Z-105 HAB	10517		14. 1.04	Airborne Adventures Ltd *(Invista titles)*	(Skipton)	12. 1.05A
G-CCOU	Mainair Pegasus Quik	8012		21. 1.04	J.R.Pearce	(Andover)	1. 2.05P
G-CCOV	Europa Aviation Europa XS	PFA 247-13998		19. 1.04	G.N.Drake	Tunbridge Wells	22. 9.05P
	(Built G.N.Drake - kit no.5463)						
G-CCOW	Mainair Pegasus Quik	8008		28. 1.04	G.P.Couttie	(Dundee)	27. 1.05P
G-CCOX	Piper J-3C-65 Cub	7278	EI-CCH	21. 1.04	R.P.Marks	(Honiton)	
			N38801, NC38801				
G-CCOY	North American AT-6D Harvard II	88-14555	Portuguese AF 1513	22. 3.04	Classic Aero Services Ltd	(Norwich)	
			SAAF 7426, EX884, 41-33857				
G-CCOZ	Monnett Sonerai II	PFA 15-10107		31. 5.78	P.R.Cozens	Hinton-in-the-Hedges	26. 3.04P
	(Built P.R.Cozens - c/n 0197) (Volkswagen 1900)						
G-CCPA	Air Création 582/Kiss 400	BMAA/HB/334		13. 1.04	C.P Astridge	(Milton Keynes)	
	(Built C.P Astridge - kit no.FL023, Trike T03102, Wing A03183-3172)						
G-CCPB	Mainair Blade 912	1361-0104-7 & W1156		4. 2.04	D.G.Fortune	Kilrush, County Kildare	1. 3.05P
G-CCPC (2)	Mainair Pegasus Quik	7994		26.11.03	P.M.Coppola	East Fortune	27.11.04P
G-CCPD	Campbell Cricket Mk.4	PFA G/03-1333		27. 1.04	N.C.Smith	(Newport, Isle of Wight)	
	(Built N.C.Smith)						
G-CCPE	Steen Skybolt	PFA 064-12830		10.12.03	C.Moore	(Egremont)	
	(Built C.Moore)						

Reg	Type	c/n	Previous identity	Reg date	Owner	Base	CofA
G-CCPF	Best Off Skyranger 912	BMAA/HB/340		26. 1.04	R.K.Willcox, M.Phillips and T.A Willcox	Wickwar	8. 6.05P
	(Built T.A Willcox - kit no.SKRxxxx396)						
G-CCPG	Mainair Pegasus Quik	8016		13. 5.04	M.A.Rhodes	(Congleton)	12. 5.05P
G-CCPH	Aerotechnik EV-97 teamEurostar UK	1814		9. 1.04	A H Woolley	(Hucknall)	27. 1.05P
G-CCPI	Extra EA 300/L	1171		17. 3.04	A Caramella	Biggin Hill	
G-CCPJ	Aerotechnik EV-97 teamEurostar UK	1909		13. 2.04	S.R.Winter	(Broxbourne)	12. 2.05P
G-CCPK	Murphy Rebel	274R	N2283B	20. 1.04	B.A Bridgewater and D.Webb	Shobdon	
	(Built M.C.Sentall)				*(Noted 4.04)*		
G-CCPL	Best Off Skyranger 912	BMAA/HB/342		29. 1.04	John Charles Turner	Tarn Farm, Cockerham	15.12.05P
	(Built P.Openshaw and Partners - kit no.SKR0310385)				tr G-CCPL Group		
G-CCPM	Mainair Blade 912	1360-1203-7 & W1155		12. 1.04	T.D.Thompson	(Knutsford	11. 1.05P
G-CCPN	Dyn'Aéro MCR-01 Club	PFA 301A-14133		28.11.03	P.H.Nelson	(Colyton)	
	(Built P.H.Nelson)						
G-CCPO	Cameron N-77 HAB	3217	(ZS-HPR)	4. 2.04	M.J.Woodcock and A C Woodcock		
			G-MITS			(East Grinstead/Bristol)	15. 4.05A
	(Originally c/n 1115, re-built with unknown new envelope 1994: c/n 3217 is now third envelope)						
G-CCPP	Cameron Concept-70 HAB	10515		16. 3.04	P.F.Smart	(Oakley, Basingstoke)	16. 3.05A
					tr The Sarnia Balloon Group		
G-CCPS	Comco Ikarus C42 FB100 VLA	PFA 322-14138		5. 2.04	H Cullens	Easter Poldar Farm, Thornhilll	11. 8.05P
	(Built H.Cullens)						
G-CCPT	Cameron Z-90 HAB	10534		14. 4.04	Blue Sky Ballooning Ltd	(Liphook)	4. 4.05A
					(Centrepoint titles)		
G-CCPU	Pilatus PC-12/45	549	HB-FQE	6. 5.04	Technical Flight Services Ltd	(Spain)	20. 5.07T
					(Op Norest Air)		
G-CCPV	Jabiru Aircraft Jabiru J400	PFA 325-14058		12. 2.04	J R Lawrence	Nether Huntlywood Farm, Gordon	
	(Built J.R.Lawrence)				*(Under construction 4.04)*		
G-CCPW	British Aerospace Jetstream Series 3102	785	SE-LDI	23.04.04	Keen Leasing (IOM) Ltd	(Castletown, Isle of Man)	1. 9.07T
			C-FHOE, G-31-785				
G-CCPX	Diamond DA.40D Star	D4.092		18. 3.04	R.T.Dickinson	(London SW1	4. 8.07P
G-CCPY	Hughes 369D	20-0674D	N622WA	22. 3.04	London Air Ltd	Biggin Hil	
			N833RW, N58388				
G-CCPZ	Cameron Z-225 HAB	10506		11. 3.04	Horizon Ballooning Ltd	(Blacknest, Alton)	11. 3.05T
G-CCRA	Glaser-Dirks DG-808B	8-308B208		19. 1.04	R Arkle *"RA"*	(Aboyne)	4. 2.07
G-CCRB	Kolb Twinstar Mk.3	PFA 205-13993		9.12.03	R W Burge	(Ilfracombe)	
	(Built R.W.Burge)						
G-CCRC	Cessna TU206G Turbo Stationair 6	U20607001	9A-DLC	24. 2.04	D M Penny	Movenis	6. 4.07
			YU-DLC, N9960R				
G-CCRD	Robinson R44 Raven II	10275		12. 2.04	Warrenpark Ltd	(Tiverton)	10. 3.07T
G-CCRE	Bell 206L-3 Longranger III	51350	N98D	22. 3.04	Hughes Helicopter Co Ltd	Biggin Hill	
			D-HILF, C-FHZE		t/a Biggin Hill Helicopters		
G-CCRF	Mainair Pegasus Quantum 15	8009		3. 3.04	R D Ballard	(Bexhill-on-Sea)	2. 4.05P
G-CCRG	Ultramagic M-77 HAB	77/249		19. 4.04	Aerial Promotions Ltd	Cannock	5. 5.05A
G-CCRH	Cameron Z-315 HAB	10489		19. 3.04	Ballooning Network Ltd	(Bristol)	3. 4.05T
					(Bath Building Society titles)		
G-CCRI	Raj Hamsa X'Air 582	BMAA/HB/354		26. 2.04	R A Wright	(Spilsby)	10.10.05P
	(Built R A Wright - kit no.891)						
G-CCRJ	Europa Aviation Europa	PFA 247-12966		27. 2.04	J F Cliff	(Bracknell)	
	(Built J F Cliff -kit no.259)						
G-CCRK	Luscome 8A Silvaire	3186	N71759	16. 2.04	J R Kimberley	(Manningtree)	
			NC71759				
G-CCRN	Thruster T600N 450 Sprint	1031-T600N-096		25. 2.04	P.Johns	(Sandown, Isle of Wight)	17. 3.05P
G-CCRR	Best Off Skyranger 912	BMAA/HB/329		16. 1.04	J A Hunt	(Abergavenny)	9. 6.05P
	(Built J.A Hunt - kit no.SKR0310393)						
G-CCRS	Lindstrand LBL 210A HAB	981		4. 3.04	Aerosaurus Balloons Ltd	(Whimple, Exeter)	8. 3.05T
G-CCRT	Mainair Pegasus Quantum 15	8014		3. 2.04	C R Whitton	East Fortune	5. 4.05T
G-CCRU	Not yet allocatd						
G-CCRV	Best Off Skyranger 912	BMAA/HB/283		20. 2.04	M R Mosley	Headon Farm, Retford	15. 8.05P
	(Built M R Mosley - kit no.SKRxxxx315)						
G-CCRW	Mainair Pegasus Quik	8003		16. 3.04	M P Jackson	Barton	15. 3.05P
G-CCRX	Jabiru Aircrfaft Jabiru UL-450	PFA 274A-14032		3. 3.04	M Everest	Deanland	
	(Built M Everest)				*(Noted 12.04)*		
G-CCSA	Cameron Z-350 HAB	10490		19. 3.04	Ballooning Network Ltd	Bristol	24. 3.05T
G-CCSB	Reims FR172H Rocket	FR172H0265	OO-RTC	9. 3.04	R J Scott	Popham	
			F-BSHK		*(Noted as "OO-RTC" 5.04)*		
G-CCSD	Mainair Pegasus Quik	8023		19. 3.04	S E Dancaster	Newton Bank Farm, Daresbury	21. 4.05P
G-CCSE	Not yet allocatd						
G-CCSF	Mainair Pegasus Quik 912S	8030		1. 4.04	J.S.Walton	(Mold)	31. 3.05P
G-CCSG	Cameron Z-275 HAB	10518		2. 4.04	M.L.Gabb	(Haselor, Alcester)	10. 9.05T
G-CCSH	Mainair Pegasus Quik	8020		1. 3.04	S D J Harvey	Broadmeadow Farm, Hereford	2. 3.05P
G-CCSI	Cameron Z-42 HAB	10563		30. 3.04	Ikea Ltd	(Eastgate, Bristol)	4. 4.05P
G-CCSJ	Cameron A-275 HAB	10510		7. 5.04	Dragon Balloon Co Ltd	(Castleton, Hope Valley)	22. 4.05T
G-CCSK	Zenair CH.701	PFA 187-14188		11. 3.04	Thomas and Thomas Surveyors Ltd	(Ipswich)	
	(Built S J Thomas)				*(F/f 7.10.04 @ Sibsey)*		
G-CCSL	Mainair Pegasus Quik	8029		26. 4.04	A.J.Harper	(Brackley)	25. 4.05P
G-CCSM	Lindstrand LBL 105A HAB	991		18. 3.04	M A Webb	(Kinnerley, Oswestry)	22. 3.05A
G-CCSN	Cessna U206G Stationair 6	U20604224	F-GECP	26. 3.04	K Brady *(Op Strathallan Parachute Club)*	Strathallan	6. 5.07
			D-EKAX, (OY-ASG), N756NM				
G-CCSO	Raj Hamsa X`Air Falcon VM133	BMAA/HB/364		15. 3.04	P Richardson	(Newark)	28. 7.05P
	(Built P Richardson - kit.no.921)						
G-CCSP	Cameron N-77 HAB	2882	SE-ZFV	17. 3.04	Balloon Sports HB	(Partille, Sweden)	
G-CCSR	Aerotechnik EV-97A Eurostar	PFA 315-14174		18. 3.04	M Lang	(Eddsfield)	20. 6.05P
	(Built M Lang)						
G-CCSS	Lindstrand LBL-90A HAB	973		11. 2.04	British Telecommunications plc	(Thatcham)	7. 4.05A
G-CCST	Piper PA-32R-301 Saratoga IIHP	3246182	N4180T	14. 2.01	G.R.Balls	Biggin Hill	7. 4.07T

Reg	Type	C/n	Previous identity	Reg date	Owner/operator	Location	CofA expiry
G-CCSU	IAV-Bacau Yakovlev Yak-52	888712	LY-APO	26. 4.04	S.Ullrich	(Munich, Germany)	27 .4.05P
			DOSAAF 69 *(yellow)*				
G-CCSV	ICP MXP-740 Savannah Jabiru	BMAA/HB/362		18. 3.04	R D Wood	(Dover)	
	(Built R D Wood - kit no.03-12-51-261)						
G-CCSW	Nott PA HAB	9		24. 3.04	J R P Nott	(London NW3)	
	(Built J R P Nott)						
G-CCSX	Best Off Skyranger 912	BMAA/HB/366		24. 3.04	T Jackson	Wickwar	26. 9.05P
	(Built T Jackson - kit no.SKRxxxx425)						
G-CCSY	Mainair Pegasus Quik	8022		27. 2.04	D Sykes	Rufforth	29. 4.05P
G-CCTA	Zenair CH.601UL Zodiac	PFA 162A-13725		4. 2.04	R E Gray and G T Harris	(Oxted)	22. 9.05P
	(Built R E Gray and G T Harris)						
G-CCTB	British Aerospace Avro 146-RJ100	E3234	TC-THB	29. 3.04	Trident Jet Leasing (Ireland) Ltd	Exeter	
			G-6-234		*(Noted 9.04)*		
G-CCTC	Mainair Pegasus Quik	8021		23. 2.04	D.G.Emery and M.R.Smith	(Oldbury)	22. 2.05P
G-CCTD	Mainair Pegasus Quik	8040		16. 3.04	Mainair Sports Ltd	(Rochdale)	9. 6.05P
G-CCTE	Dyn'Aéro MCR-01 Club	PFA 301-13268		22 .3.04	G J Slater	Clench Common	
	(Built G J Slater)						
G-CCTF	Pitts S-2A	2146	N51ST	26. 3.04	Pitts Aircraft (UK) Ltd	Sherburn-in-Elmet	18. 8.07P
	(Built Aeroteck Inc)						
G-CCTG	Van's RV-3B	PFA 99-10518		9. 3.04	Marjorie and I G Glenn	(Barton, Cambridge)	
	(Built I G Glenn) (C/n is same as Van's RV-3 G-BHXN (c/n EAA 105098) - builder P.Hing: cancelled 2.9.91 by CAA but last known on rebuild Bourn April 1994: presumed to be same aircraft and now on completion)						
G-CCTH	Aerotechnik EV-97 teamEurostar UK	2005		12. 3.04	Fly CB Ltd	Oxford	14. 3.05P
G-CCTI	Aerotechnik EV-97 teamEurostar UK	2009		6. 4.04	Flylight Airsports Ltd	Sywell	26. 4.05P
G-CCTK	Glaser-Dirks DG-800B	8-322B222		21. 7.04	G.W.English *(Noted 11.04)*	Lasham	10. 8..07
G-CCTL	Robinson R44 II	10309		30. 3.04	Auto Corporation Ltd	Liverpool	20. 4.07T
G-CCTM	Mainair Blade	1363-0504-7 & W1158		5. 4.04	J.N.Hanson	Brook Farm, Pilling	31. 5.05P
G-CCTN	Ultramagic T-180 HAB	180/48		5. 7.04	A.Derbyshire	(Woodseaves, Stafford)	18. 6.05T
G-CCTO	Aerotechnik EV-97 Eurostar	PFA 315-14136		17. 3.04	A J Boulton	Sittles Farm, Alrewas	7. 6.05P
	(Built A J Boulton)						
G-CCTP	Aerotechnik EV-97 Eurostar	PFA 315-14185		18. 2.04	G M Yule	Shobdon	23. 5.05P
	(Built G.M.Yule)						
G-CCTR	Best-Off Skyranger 912	BMAA/HB/350		2. 3.04	A H Trapp	(Kidderminster)	
	(Built A H Trapp - kit no.SKRxxxx410)						
G-CCTS	Cameron Z-120 HAB	10570		22. 6.04	F.R.Hart	(Bishops Sutton, Bristol)	21. 6.05A
					(Snap Survey titles)		
G-CCTT	Cessna 172SP	172S8157	N957SP	12. 2.04	A. Reay	Caernarfon	18. 3.07T
G-CCTU	Mainair Pegasus Quik	8024		21. 4.04	B.J.Syson	Chilbolton	29. 4.05P
G-CCTV	Rans S-6-ESA Coyote II	PFA 204-14069		19. 6.03	RIDS Ltd	Wickenby	11. 8.04P
	(Built R M Broom - kit no.0302.1437) (Tricycle u/c)						
G-CCTW	Cessna 152	15279882	N757NW	26. 4.04	R.J.Dempsey	(Newbury)	
G-CCTX	Rans S-6-ES Coyote II	PFA 204-14143		19. 2.04	J.Lynch	(Bacup)	9. 6.05P
	(Built L.M.Leachman)						
G-CCTY	Cameron TR-70 HAB	10437		13. 4.04	Cameron Balloons Ltd	(Bristol)	13. 5.05A
G-CCTZ	Mainair Pegasus Quik 912S	8031		13. 4.04	S.Baker	Long Marston	12. 4.05P
G-CCUA	Mainair Pegasus Quik	8032		27. 4.04	H.M.Manning	Rochester	29. 4.05P
G-CCUB	Piper J-3C-65 Cub	2362A	N33528	2. 4.81	Cormack (Aircraft Services) Ltd	Rothesay	
			NC33528, NX33528		*(On rebuild 2001)*		
G-CCUC	Best Off Skyranger J2.2	BMAA/HB/373		13. 4.04	M.Kerrison	(Ballyclough, County Limerick)	10.10.05P
	(BuiltT M.Kerrison - kit no.SKRxxxx453)						
G-CCUD	Best Off Skyranger J2.2	BMAA/HB/374		13. 4.04	J.Johnston	(Knocklong, County Limerick)	12.10.05P
	(Built J.Johnston - kit no.SKRxxxx454)						
G-CCUE	Ultramagic T-180 HAB	180/45		10. 5.04	Espiritu Balloon Flights Ltd	(Minsterley, Shrewsbury)	16 .5.05T
G-CCUF	Best Off Skyranger 912	BMAA/HB/375		15. 4.04	C.D.Hogbourne and D.J.Parrish		
	(Built C.D.Hogbourne and D.J.Parrish - kit no.SKRxxxx459)					Longacres Farm, Sandy	29. 9.05P
G-CCUG	Bell 206L-1 Long Ranger	45368	N18UG	5.12.03	Hughes Helicopter Co Ltd	Biggin Hill	1. 3.07T
			N18UC, N13UC, N48ZP, N1075T		t/a Biggin Hill Helicopters		
G-CCUH	Rotary Air Force RAF 2000 GTX-SE	PFA G/13-1356		16. 4.04	J.H.Haverhals	(Graffham, Petworth)	28. 6.05P
	(Built D.R.Lazenby)						
G-CCUI	Dyn'Aéro MCR-01	PFA 301-13963		1. 4.04	J.T.Morgan	Husbands Bosworth	15. 9.05P
	(Built J.T.Morgan)						
G-CCUJ	Cameron C-90 HAB	10576		5. 7.04	R.D.Jones	(Easton-in-Gordano, Bristol)	27. 5.05A
					t/a Rudgleigh Inn *(Rudgleigh Inn titles)*		
G-CCUK	Agusta A109 II	7263	RP-C109	5. 4.04	Castle Air Charters Ltd	Liskeard	
			I-SEIE/N109AE		*(Noted 10.04)*		
G-CCUL	Europa Aviation Europa XS	PFA 247-13119		20. 4.04	J.Hellan	Rayne Hall Farm, Braintree	
	(Built I Dole - ki t no.336				tr Europa 6 *(Noted 1.05)*		
G-CCUN	Hughes 369D	117-0220D	N644WA	30. 4.04	London Air Ltd	Biggin Hill	
			N58169				
G-CCUO	Hughes 369D	40-0711D	N655WA	30. 4.04	London Air Ltd	Biggin Hill	
			C-GKHI				
G-CCUP	Westland S.58 Series 2	WA/127	XR502	12.11.04	D.Brem-Wilson and J.Buswell		
						Honey Crock Farm, Redhill	
G-CCUR	Mainair Pegasus Quantum 15-912	8034		30. 4.04	P.S. and N.Bewley	(Bristol)	12. 5.05P
G-CCUS	Diamond DA.40D Star	D4.082		29. 4.04	R.J.and M.J.Millen	Rochester	4. 8.07
					t/a The Millen Corporation		
G-CCUT	Aerotechnik EV-97 Eurostar	PFA 315-14191		9. 3.04	C K Jones	Sywell	17. 8.05P
	(Built C K Jones)						
G-CCUU	Vahdat-Hagh Shiraz Gyroplane	MV-009		15. 3.04	M E Vahdat-Hagh	(Uxbridge)	
	(Built M E Vahdat-Hagh)						
G-CCUV	Piper PA-25-260 Pawnee C	25-5201	VT-EBH	31. 1.05	D.B.Almey	(Spalding)	
			N8745L				
G-CCUW	Piper PA-25-260 Pawnee C	25-5235	VT-EBI	31. 1.05	D.B.Almey	(Spalding)	
			N8801L				

G-CCUX	B-N Group BN-2T Islander	2302		26. 4.04	B-N Group Ltd	Bembridge	
G-CCUY	Europa Aviation Europa *(Built N.Evans)*	PFA 247-13189		14. 4.04	N.Evans	(Wymondham)	
G-CCUZ	Thruster T600N 450	0044-T600N-102		29. 4.04	Fly 365 Ltd	Wickenby	24. 5.05P
G-CCVA	Aerotechnik EV-97 Eurostar *(Built T A.Jones)*	PFA 315-14226		21. 4.04	A.Jones	Headon Farm, Retford	7. 7.05P
G-CCVB	Mainair Pegasus Quik	8033		6. 5.04	L.Chesworth	(Malpas)	5. 5.05P
G-CCVC							
G-CCVD	Cameron Z-105 HAB	10583		25. 5.04	Associazione Sportiva Sorvolvare	(Crevalcore, Italy)	1. 6.05A
G-CCVE	Raj Hamsa X'Air Jabiru *(Built G J Slater - kit.no.885)*	BMAA/HB/361		25. 3.04	G J Slater	Clench Common	
G-CCVF	Lindstrand LBL 105A HAB	953		6. 5.04	S.Villiers and A.W.Patterson t/a Alan Patterson Design	Bangor	5. 5.05A
G-CCVG	Schweizer 269C-1	0164		2. 4.04	Radcliffe Engineering Services Ltd	(Radcliffe, Manchester)	18. 4.07T
G-CCVI	Zenair CH.701 SP *(Built C.R.Hoveman)*	PFA 187-14181		5. 5.04	C.R.Hoveman	(Stratford-upon-Avon)	
G-CCVJ	Raj Hamsa X'Air Falcon 133 *(Built G.A.J.Salter - kit no.916)*	BMAA/HB/381		7. 5.04	A.Davis	(Macclesfield)	
G-CCVK	Aerotechnik EV-97 teamEurostar UK	2016		19. 5.04	S.A.Kirk tr Kent Eurostar Group	Rochester	18. 5.05P
G-CCVL	Zenair CH.601XL Zodiac *(Built A. Y-T.Leung and G.Constantine)*	PFA 162B-14204		22. 4.04	A. Y-T.Leung and G.Constantine	(Sutton Coldfield)	
G-CCVM	Van's RV-7A *(Built J G Small)*	PFA 323-14213		12. 3.04	J G Small	(Southport)	
G-CCVN	Jabiru Aircraft Jabiru SP-470 *(Built J.C.Collingwood)*	PFA 274B-13677		10. 5.04	J.C.Collingwood	(Cranbrook)	
G-CCVO	Bell 206B JetRanger III	4326	N471M JA6150, N20334, C-GLZU	22. 6.04	Carecoast Ltd	(Grimsby)	29. 7.07T
G-CCVP	Beech 58 Baron	TH-1948	PH-ZEM N80VS	13. 5.04	Richard Nash Cars Ltd	Norwich	9. 6.07T
G-CCVR	Best Off Skyranger 912 *(Built M.J.Batchelor - kit no.SKRxxxx407)*	BMAA/HB/353		29. 4.04	M.J.Batchelor	(Bristol)	14. 9.05P
G-CCVS	Van's RV-6A *(Built J Edgeworth)*	PFA 181A-13413	G-CCVC	29. 3.04	J Edgeworth	(Darlington)	
G-CCVT	Zenair CH.601UL Zodiac *(Built D.McCormack)*	PFA 162A-14160		2. 4.04	D.McCormack	Broomhill Farm, West Calder	10. 6.05P
G-CCVU	Robinson R22 Beta II	3600		11. 5.04	M.Horrell	Conington	3. 6.07P
G-CCVV*	Vickers Supermarine 379 Spitfire FR.XIVe				See SECTION 3 Part 1	Polk City, Florida	
G-CCVW	Nicollier HN.700 Menestrel II *(Built B.F.Enock)*	PFA 217-11950		13. 5.04	B.F.Enock	(Leamington Spa)	
G-CCVX	Mainair Tri Flyer 250/Flexiform Striker	AS-001		18. 5.04	J.A.Shufflebotham	(Macclesfield)	
G-CCVY	Robinson R22 Beta	1666	N101SK N4041W, C-GJKD, N4041W	9. 7.04	S.Klinge *(Noted 8.04)*	Prestwick	9. 8.07P
G-CCVZ	Cameron O-120 HAB	10586		27. 7.04	T.M.C.McCoy	(Peasedown St.John, Bath)	18. 7.05T
G-CCWA	Piper PA-28-181 Archer III	2843328	D-ELEM PH-AEG, N41776	16. 6.04	T.P.Gooley	White Waltham	24. 6.07T
G-CCWB	Aero L-39ZA Albatross	132036	N404ZA Romanian AF 136	4. 6.04	Freespirit Charters Ltd	Bristol	
G-CCWC	Best Off Skyranger 912 *(Built C.Hewer - kit no.SKRxxxx422)*	BMAA/HB/367		4. 5.04	C.Hewer	Carlisle	21. 9.05P
G-CCWD	Robinson R44 Raven	1296		10. 3.03	J Henderson	Newtownards	3. 4.06T
G-CCWE	Lindstrand LBL 330A HAB	984		22. 4.04	Adventure Balloons Ltd	(Hartley Wintney, Hook)	27. 4.05A
G-CCWF	Raj Hamsa X'Air 133 *(Built G.A.J.Salter - kit no.675)*	BMAA/HB/331		19. 5.04	N G Middleton	(Barton-le-Clay,Bedford)	6.10.05P
G-CCWG	Whittaker MW6 Merlin *(Built D.E.Williams)*	PFA 164-11998		8. 4.04	D.E.Williams	(Maesteg)	
G-CCWH	Dyn'Aéro MCR-01 Club *(Built M.G.Rasch)*	PFA 301-13949		20. 4.04	M.G.Rasch	(Poole)	
G-CCWI	Robinson R44 Raven II	10362		16. 6.04	Saxon Logistics Ltd	Elstree	2. 9.07T
G-CCWJ	Robinson R44 Raven II	10363		16. 6.04	Saxon Logistics Ltd	Elstree	2. 9.07T
G-CCWK	Aérospatiale AS355F2 Ecureuil 2	5439	N8066G LN-OES, F-GGRS	10.09.04	Helicopter Express Ltd	Booker	13. 9.07T
G-CCWL	Mainair Blade	1364-0504-7 & W1159		19. 5.04	W.D.Joyner	(Preston)	26. 5.05P
G-CCWM	Robin DR400/180 Regent	2457	F-GTZM	3. 6.04	M.R.Clark	Newcastle	21. 6.07
G-CCWN	Mainair Pegasus Quantum 15-912	8045		11. 6.04	S.Jeffrey	Perth	10. 6.05P
G-CCWO	Mainair Pegasus Quantum 15-912	8042		16. 6.04	P.K.Dean	(Brentwood)	16. 6.05P
G-CCWP	Aerotechnik EV-97 teamEurostar UK	2010		9. 6.04	T.R.Murfet	Sutton Meadows, Ely	8. 6.05P
G-CCWR	Mainair Pegasus Quik	8053		1. 6.04	J.A.Robinson	(Kendal)	18. 7.05P
G-CCWS	Balony Kubicek BB30Z HAB	270	OK-0270	18. 5.04	H.C.J and Sara L.G.Williams	(Langford, Bristol)	
G-CCWT	Balony Kubicek BB20GP HAB	298	OK-0298	18. 5.04	H.C.J and Sara L.G.Williams	(Langford, Bristol)	13. 5.05A
G-CCWU	Best Off Skyranger 912 *(Built D.M.Lane - kit no.SKRxxxx461)*	BMAA/HB/386		1. 6.04	D.M.Lane	(Stourbridge)	
G-CCWV	Mainair Pegasus Quik	8043		1. 6.04	R.A.Taylor	Wickenby	31. 5.05P
G-CCWW	Mainair Pegasus Quantum 15-912	8035		4. 5.04	Virginia G.Concannon tr Double Whisky Syndicate	Knapthorpe Lodge, Caunton	3. 5.05P
G-CCWX							
G-CCWY	Pilatus PC-12/45	568		8. 9.04	Meridian Aviation Group Ltd	Bournemouth	10.10.07E
G-CCWZ	Raj Hamsa X'Air Falcon 133 *(Built M.A.Evans - kit no.925)*	BMAA/HB/380		4. 5.04	M.A.Evans	Weston Zoyland	26. 9.05P
G-CCXA	Boeing Stearman A75N1 (N2S-4 Kaydet)	75-3616	N75TL N5148N, Bu.37869	1. 6.04	Skymax (Aviation) Ltd *(As "669" in USAAC c/s)*	Old Buckenham	6. 9.07P
G-CCXB	Boeing Stearman B75N1	75-7854	N1363M	R	Pluto Inc *(Under rebuild 9.04 as "996" in error - to be "699")*	Priory Farm, Tibenham	

Reg	Type	C/n	Previous identity	Regd	Owner/Operator	Base	CofA
G-CCXC	Avions Mudry CAP.10B *(Officially regd as c/n 165)*	65?	N4247M Mexican AF EPC-165	26. 5.04	Skymax (Aviation) Ltd	Damyns Hall, Upminster	6. 9.07T
G-CCXD	Lindstrand LBL 105B HAB	996		14. 6.04	J.H.Dobson	(Streatley, Reading)	13. 6.05T
G-CCXE	Cameron Z-120 HAB	10596		3. 6.04	Cameron Balloons Ltd	(Bristol)	
G-CCXF	Cameron Z-90 HAB	10593		3. 8.04	R.G.March and T.J.Maycock *(Unison titles)*	(Market Harborough)	3. 8.07A
G-CCXG	Replica Plans SE-5A *(Built C.Morris)*	PFA 20-11785		11. 6.04	C.Morris	(Wrexham)	
G-CCXH	Best Off Skyranger J2 *(Built Skyranger UK Ltd - kit no.SKRxxxxxxx?)*	BMAA/HB/377		4. 6.04	Skyranger UK Ltd	Sywell	12. 7.05P
G-CCXI	Thorp T211 *(Built S G R and J.Gilroy)*	PFA 305-13504		22. 9.03	J Gilroy	(Crawley)	
G-CCXJ	Cessna 340A	340A0912	N25PJ HB-LNM, LN-TEA, N27026	13. 7.04	Caernarfon Airworld Ltd	Caernarfon	22. 7.07T
G-CCXK	Pitts S-1S Special *(Built E.C.Roberts)*	AACA/1061	ZK-ECO	14. 6.04	P.G.Bond *(Noted 8.04)*	Shipdham	
G-CCXL	Best Off Skyranger 912 *(Built R.G.Cameron - kit no.SKRxxxx411)*	BMAA/HB/335		18. 6.04	R.G.Cameron	(Broughty Ferry)	
G-CCXM	Best Off Skyranger 912 *(Built C.J.Finnigan - kit no.SKRxxxx394)*	BMAA/HB/337		16. 6.04	C.J.Finnigan *(Noted 11.04)*	Enstone	16.10.05P
G-CCXN	Best Off Skyranger 912 *(Built C.I.Chegwen - kit no.SKRxxxx833)*	BMAA/HB/323		4. 6.04	C.I.Chegwen *(Noted 9.04)*	Mill Farm, Hughley, Much Wenlock	
G-CCXO	Corby CJ-1 Starlet *(Built I.W.L.Aikman)*	PFA 134-13267		21. 6.04	I.W.L.Aikman	(Tadley)	
G-CCXP	ICP Savannah Jabiru *(Built B.J.Harper - kit no. 03-09-51-231)*	BMAA/HB/318		30. 4.04	J.H.Tope and B.J.Harper	(Newton Abbot)	
G-CCXR	Mainair Blade	1367-0604-7 & W1162		5. 7.04	M.Fowler	(Cambourne)	27. 7.05P
G-CCXS	Montgomerie-Bensen B.8MR *(Built S.A.Sharp)*	PFA G/01A-1350		26. 5.04	S.A.Sharp	(Kilkerran)	
G-CCXT	Mainair Pegasus Quik	8046		18. 6.04	G. and P.Verity	Newton Bank, Daresbury	23. 6.05P
G-CCXU	Diamond DA.40D Star	D4.037	ZK-SFH	28. 6.04	McAlpine Aviation Training Ltd	Blackbushe	4. 8.07T
G-CCXV	Thruster T600N 450	0045-T600N-103		18. 6.04	Thruster Air Services Ltd	Ginge, Wantage	4. 7.05P
G-CCXW	Thruster T600N 450	0045-T600N-104		15. 7.04	Joan Walsh *(Op Saxon Microlight School)*	Sheepcoates Farm Little Totham	14. 7.05P
G-CCXX	American General AG-5B Tiger	10160	PH-MLG YL-CAH	19. 4.04	Alexson Ltd *(Noted 12.04)*	Sturgate	AC
G-CCXY	British Aerospace BAe.146 Series 100	E1083	N83HN B-634L, B-2709, G-5-083	21. 7.04	Avtrade Aircraft Leasing Ltd *(In open store as "B-634L" 2.05)*	Southend	
G-CCXZ	Mainair Pegasus Quik	8038		24. 5.04	K.J.Sene	(Warrington)	23. 5.05P
G-CCYA	Jabiru Aircraft Jabiru J450 *(Built D.J.Royce)*	PFA 325-14060		22. 6.04	D.J.Royce	(Ludham)	
G-CCYB	Just/Reality Escapade 912 *(Built B.E.Renehan - kit no.JAESC 0034)*	BMAA/HB/391		24. 6.04	B.E.Renehan	(Alton)	
G-CCYC	Robinson R44 Raven II	10388		24. 6.04	Derg Developments Ltd	(Killaloe, County Clare)	20. 7.07T
G-CCYE	Mainair Pegasus Quik	8050		30. 7.04	J.Lane	Clench Common	29. 7.05P
G-CCYF	Aerophile 5500 Balloon (Gas filled)	6	F-GPRS	30. 6.04	High Point Balloons Ltd *(GWR titles)*	(Almondsbury, Bristol)	13. 7.05T
G-CCYG	Robinson R44 Raven II	10424		9. 7.04	P Durkin t/a Moorland Windows	(Lytham St.Annes)	8. 8.07T
G-CCYH	Embraer EMB-145EP	145.070	SE-DZA PT-SAO	23. 9.04	Air Kilroe Ltd t/a Eastern Airways Lsd *(To be op BMI Regional and re-regd G-RJXX 2005)*	Humberside	22. 9.05T
G-CCYI	Cameron O-105 HAB	10604		30. 7.04	Media Balloons	(Policoro, Italy)	18. 7.05A
aG-CCYJ	Mainair Pegasus Quik	8054		2. 8.04	J.Ellis	Perth	8. 8.05P
G-CCYK	Cessna 180	32470	SE-KEA LN-TSD, N6573A	17. 6.04	K.V.McKinnon *(Noted dismantled and unmarked 9.04)*	East Winch	
G-CCYL	Mainair Pegasus Quantum 15	8055		26. 7.04	M.J.L.Morris	(Swansea)	24.11.05P
G-CCYM	Best Off Skyranger 912 *(Built D.McDonagh - kit no.SKRxxxx412)*	BMAA/HB/390		16. 7.04	D.McDonagh	(Darlington)	
G-CCYN	Cameron C-80 HAB	10133		2. 8.04	D.R.Firkins	(Beckford, Tewkesbury)	27. 7.05A
G-CCYO	Christen Eagle II *(Built R.Hayner)*	HAYNER 0001	N56RJ	13. 9.04	J.R.Pearce	Chilbolton	
G-CCYP	Thunder and Colt 56A HAB	302	SE-ZXG	23. 7.04	L.P.Hooper	(Bristol)	
G-CCYR	Comco Ikarus C42 FB80	0408-6612		20. 9.04	M.L.Smith	Popham	20 .9.05P
G-CCYS	Reims/Cessna F182Q Skylane	F18200126	OY-BNG	31. 8.04	D.J.Colledge	(Watford)	
G-CCYT	Robinson R44 Raven II	10443		28. 7.04	B.E.Llewellyn t/a Bell Commercials	Swansea	13. 9.07T
G-CCYU	Ultramagic S-90 HAB	90/70		17. 8.04	A.R.Craze	(St. Leonards-on-Sea)	17. 8.05T
G-CCYV							
G-CCYW*	Robinson R22 Beta II	3655		28. 7.04	Heli Air Ltd *(Cancelled 18.11.04 by CAA)*	Wellesbourne Mountford	13. 9.07T
G-CCYX	Bell 412	34001	PK-HMI	2. 8.04	RCR Aviation Ltd	Thruxton	
G-CCYY	Piper PA-28-161 Warrior II	2816094	HB-PML	17. 6.04	Flightcontrol Ltd	Fairoaks	7. 7.07T
G-CCYZ	Dornier EKW C3605 *(Built F + W)*	338	N31624 Swiss AF C-558	30. 9.04	William Tomkins Ltd	(Peterborough)	
G-CCZA							
G-CCZB	Mainair Pegasus Quantum 15	8052		21. 7.04	P.Stewart	Thornhill	28. 7.05P
G-CCZC							
G-CCZD	Van's RV-7 *(Built R.T.Clegg)*	PFA 323-14087		28. 5.04	R.T.Clegg	Netherthorpe	22. 9.05P
G-CCZE	Scottish Aviation Bulldog Series 100/125	BH120/417	RJAF 417 RJAF 1138, G-31-48	25. 8.04	R.D.Dickson tr British Disabled Flying Association	RAF Wittering	
G-CCZF	Scottish Aviation Bulldog Series 100/125	BH120/433	RJAF 418 RJAF 1140, G-31-41	25. 8.04	R.D. Dickson tr British Disabled Flying Association	RAF Wittering	
G-CCZG	Robinson R44 Raven II	10474		2. 9.04	Westinbrook Ltd	Shoreham	7.10.07E
G-CCZH	Robinson R44 Raven	1423		6. 9.04	Newtown Aviation Ltd	(Enfield, County Meath)	17.10.07E

Reg	Type	C/n	Previous identity	Regd	Owner/operator	Base	CofA
G-CCZI							
G-CCZJ	Raj Hamsa X'Air Falcon 582	BMAA/HB/401		5. 8.04	A.B.Gridley	(Oakley, Bedford)	
	(Built A.B.Gridley)						
G-CCZK	Zenair CH.601 UL Zodiac	PFA 162A-14270		11. 8.04	R.J.Hopkins	Popham	
	(Built R.J.Hopkins)						
G-CCZL	Comco Ikarus C42 FB80	0410-6620		29.11.04	W.H.J.Knowles	Mill Farm, Shifnal	28.11.05P
G-CCZM	Best Off Skyranger 912S	BMAA/HB/372		1. 7.04	D.Woodward	Latch Farm,Kirknewton	
	(Built D M Hepworth - kit no.SKRxxxx455)				*(Noted 10.04)*		
G-CCZN	Rans S-6ES Coyote II	PFA 204-14275		9. 8.04	M.Taylor	(Withernsea)	
	(Built M.Taylor)						
G-CCZO	Mainair Pegasus Quik	8066		2. 9.04	S.B.Williams	Headcorn	30. 8.05P
G-CCZP	Super Marine Spitfire Mk.26	PFA 324-14062		10. 6.04	J.W.E.Pearson	(St. Albans)	
	(Built J.W.E.Pearson and H Luck - kit no.25)						
G-CCZR	Medway Raven Eclipse 912	177/155		15. 7.04	R.A.Keene	Over Farm, Gloucester	14. 7.05P
G-CCZS	Raj Hamsa X'Air Falcon 582	BMAA/HB/403		16. 8.04	P.J.Sheehy	Barton Ashes	
	(Built P.J.Sheehy)						
G-CCZT	Van's RV-9A	PFA 320-13777		10. 8.04	N.A.Henderson	Bicester	
	(Built N.A.Henderson)				*(Noted 1.05)*		
G-CCZU	Diamond DA40D Star	D4.125	OE-VPU	24. 8.04	Aviation Now Ltd	Cardiff	17.10.07E
G-CCZV	Piper PA-28-151 Cherokee Warrior	28-7715089	OY-CHR	7. 7.04	H.A.Barrs	Aberdeen	5. 8.07
			SE-GNY				
G-CCZW	Mainair Blade	1368-0904-7 & W1163		21.10.04	C.J.Wright	(Kendal)	13.11.05P
G-CCZX	Robin DR400/180 Regent	2127	F-GLKY	15. 9.04	J.B.Mills	Bourn	
G-CCZY	Van's RV-9A	PFA 320-14154		23. 8.04	G.Williams	RAF Mona	
	(Built G.Williams and partners)				tr Mona RV-9 Group		
G-CCZZ	Aerotechnik EV-97 Eurostar	PFA 315-14158		28. 5.04	B.M.Starck and R.Bastin	(Hatfield)	5. 8.05P
	(Built B.M.Starck and R.Bastin)						
G-CDAA	Mainair Pegasus Quantum 15-912	8069		27. 8.04	I.A.Macadam	(London SE13)	26. 8.05P
G-CDAB	Stoddard-Hamilton Glasair IIS RG	PFA 149-13231		7. 7.04	W.L.Hitchins	(Bampton)	
	(Built W.L Hitchins)						
G-CDAC	Aerotechnik EV-97 teamEurostar UK	2116		30. 9.04	Fly CB Ltd	Oxford	29. 9.05P
G-CDAD	Lindstrand LBL 25A Cloudhopper HAB	1003		21. 9.04	G.J.Madelin	(Farnham)	4. 8.05A
G-CDAE	Van's RV-6A	PFA 181-13018		5. 8.04	K.J.Fleming	Woodvale	
	(Built K.J.Fleming)						
G-CDAF	Bell 412	33105	PK-HMS	2. 8.04	RCR Aviation Ltd	Thruxton	
G-CDAG	Mainair Blade	1325-0502-7 & W1120		10. 9.04	M.J.Gerrish	(Sandbach)	
G-CDAH							
G-CDAI	Robin DR400/135	2574		13.12.04	Cole Aviation Ltd	Sedlescombe	
G-CDAK	Zenair CH601UL Zodiac	PFA 162A-14210		21. 7.04	K.Kerr	Sittles Farm, Alrewas	
	(Built K.Kerr)						
G-CDAL	Zenair CH601UL Zodiac	PFA 162A-14195		7. 6.04	D.Cassidy	(Canterbury)	22. 9.05P
	(Built D.Cassidy)						
G-CDAM	Sky 77-24 HAB	057	CS-BAO	16. 8.04	M.Morris and Pauline A.Davies	(Park Hall, Oswestry)	10. 8.05A
G-CDAN*	Vickers Supermarine 361 Spitfire LF.XVIe				See SECTION 3 Part 1	Wanaka, New Zealand	
G-CDAO	Mainair Pegasus Quantum 15-912	8061		17. 8.04	A.M.Dalgetty	Perth	16. 8.05P
G-CDAP	Aerotechnik EV-97 teamEurostar UK	2114		28. 7.04	A.Parker	(Bingley)	27. 7.05P
G-CDAR	Mainair Pegasus Quik	8060		17. 8.04	A.R.Pitcher	(Rye)	16. 8.05P
G-CDAS							
G-CDAT	ICP MXP-740 Savannah Jabiru	BMAA/HB/327		7. 7.04	R.Simpson	Eshott	
	(Built R.Simpson - kit no.03-05-51-211)				*(Noted 1.05)*		
G-CDAW	Robinson R22 Beta	3703		11.10.04	Aeromega Ltd	Stapleford	10.11.07E
G-CDAX	Mainair Pegasus Quik	8068		8. 9.04	M.Winship	Eshott	9. 9.05P
G-CDAY	Best Off Skyranger 912	BMAA/HB/394		7. 7.04	M.E.Furniss	Foston	30.11.05P
	(Built M.E.Furniss - kit no.SKRxxxx473)						
G-CDAZ	Aerotechnik EV-97 Eurostar	PFA 315-14268		13. 8.04	M.C.J.Ludlow	Pent Farm, Postling	10.10.05P
	(Built M.C.J.Ludlow)						
G-CDBA	Best Off Skyranger 912	BMAA/HB/406		21.10.04	P.J.Brennan	(Altrincham)	
G-CDBB	Mainair Pegasus Quik	8062		16. 8.04	J.L.Ker	Eshott	16. 8.05P
G-CDBC	Aviation Enterprises Magnum	001	G-61-2	27. 8.04	Aviation Enterprises Ltd	Membury	AC
G-CDBD	Jabiru Aircraft Jabiru J400	PFA 325-14077		16. 8.04	S.Derwin and E.Bentley	Morgansfield, Fishburn	
	(Built S.Derwin and E.Bentley)						
G-CDBE	Montgomerie-Bensen B.8M	PFA G/01-1360		7. 9.04	P.Harwood	(Laurencekirk)	
	(Built P.Harwood)						
G-CDBF	Robinson R22 Beta II	3681		11.10.04	R.C.Hields t/a Hields Aviation	Sherburn-in-Elmet	31.10.07E
G-CDBG	Robinson R22 Beta II	3682		11.10.04	R.C.Hields t/a Hields Aviation	Sherburn-in-Elmet	28.10.07E
G-CDBH	Pitts S-2B	5151	SE-LVI	28. 1.05	J.W.Sullivan	Maypole Farm, Chislet	
	(Built Christen Industries Inc)		F-GMOV, OO-MOV, N71ZX, N10ZX				
G-CDBI							
G-CDBJ							
G-CDBK	Rotorway Executive 162F	6834		20. 9.04	N.and M.Foreman	(Staplehurst)	
	(Built N.and M.Foreman)				t/a NF Auto Development		
G-CDBL							
G-CDBM	CAB Robin DR400/180 Regent	2573		16.11.04	C.M.Simmonds	(Camborne)	23.11.07E
G-CDBN	Thruster T600N 450	0047-T600N-105		24. 9.04	Thruster Air Services Ltd	Ginge, Wantage	29. 9.05P
G-CDBO	Best Off Skyranger 912	BMAA/HB/370		13. 8.04	A.C.Turnbull	Perth	2.11.05P
	(Built A.M.Dalgetty - kit no. UK/424)						
G-CDBP	Agusta-Bell 206B JetRanger II	8041	R Saudi AF 1207	4.10.04	Eastern Atlantic Helicopters Ltd	Shoreham	
G-CDBR	Stolp SA.300 Starduster Too	PFA 035-13036		15. 9.04	R.J.Warren	(West Drayton)	
	(Built R.J.Warren)						
G-CDBS	MBB Bö.105DBS-4	S.738	D-HDRZ	29. 9.89	Bond Air Services Ltd	RAF St.Mawgan	8.11.07E
			VH-MBK, N970MB, D-HDRZ		*(Op County Air Ambulance)*		
G-CDBT	Agusta-Bell 206B JetRanger II	8038	R Saudi AF 1206	4.10.04	Eastern Atlantic Helicopters Ltd	Shoreham	
G-CDBU	Comco Ikarus C42 FB100	0411-6632		26. 1.05	S.E.Meehan and J.K.Agarwala	(Ormskirk)	

G-CDBV	Best Off Skyranger 912S *(Built K.Hall - kit.no. UK/499)*	BMAA/HB/409		23. 9.04	K.Hall	(Ashford)	
G-CDBX	Europa Aviation Europa XS *(Built R.Marston - kit no.568)*	PFA 247-13971		16. 9.04	R.Marston	(Bridgnorth)	
G-CDBY	Dyn'Aéro MCR-01 ULC *(Built R.Germany)*	PFA 301B-14269		23. 8.04	R.Germany	Knapthorpe Lodge, Caunton	20.12.05P
G-CDBZ	Thruster T600N 450	0047-T600N-106		24. 9.04	Thruster Air Services Ltd	Ginge, Wantage	29. 9.05P
G-CDCA	Robinson R44 Raven II	10551		17.11.04	Heli Air Ltd	Wellesbourne Mountford	19.12.07E
G-CDCB	Robinson R44 Raven II	10532		8.11.04	Heli Air Ltd	Wellesbourne Mountford	9.12.07E
G-CDCC	Aerotechnik EV-97A Eurostar *(Built R.E.and N.G.Nicholson)*	PFA 315A-14262		11. 8.04	R.E.and N.G.Nicholson	(Sidmouth)	29. 9.05
G-CDCD	Van's RV-9A *(Built M.Weaver and S.D Arnold)*	PFA 320-13925		20. 1.04	M.Weaver and S.D Arnold tr RV9ers	(Coleshill)	
G-CDCE	Avions Mudry CAP.10B	39	F-BNDC CN-TBW, F-BUDG	3.11.04	The Tiger Club 1990 Ltd	Headcorn	
G-CDCF	Mainair Pegasus Quik	8076		8.11.04	B.L.Benson	(Malpas)	7.11.05P
G-CDCG	Comco Ikarus C42 FB UK *(Built N.E.Ashton and R.H.J.Jenkins)*	PFA 322-14281		16. 8.04	N.E.Ashton and R.H.J.Jenkins	Ince Blundell	
G-CDCH	Best Off Skyranger 912 *(Built K.Laud - kit no.UK/436)*	BMAA/HB/384		28. 9.04	K.Laud	(Swadlincote)	
G-CDCI	Mainair Pegasus Quik	8077		14. 1.05	S.G. Murray	Rochester	
G-CDCJ	Britten-Norman BN-2B-20 Islander	2303		5.10.04	Britten-Norman Aircraft Ltd	Bembridge	
G-CDCK	Mainair Pegasus Quik	8078		22.10.04	S.G.Ward	(Chatham)	26.10.05P
G-CDCL	Piper PA-26-161 Warrior III	2842217	N3072G	7.10.04	Senate Aviation Ltd	Bournemouth	16.11.07E
G-CDCM	Comco Ikarus C42 FB UK *(Built S.T.Allen)*	PFA 322-14280		26.10.04	S.T.Allen	Redlands	
G-CDCN	BAE SYSTEMS Avro 146-RJ100	E3236	TC-THC G-6-236	13.10.04	Trident Jet Leasing (Ireland) Ltd	Kemble	
G-CDCO	Comco Ikarus C42 FB UK *(Built G.G.Bevis)*	PFA 322-14315		7.10.04	G.G.Bevis	(Southampton)	
G-CDCP	Jabiru Aircraft Jabiru J400 *(Built M.W.T.Wilson)*	PFA 325-14094		7.10.04	M.W.T.Wilson	(Tursdale, Durham)	
G-CDCR	ICP MX-740 Savannah Jabiru *(Built T.Davidson and G.McKinstry - kit no.04-06-51-291)*	BMAA/HB/405		14. 1.05	T.Davidson and G.McKinstry	(Banbridge)	
G-CDCS	Piper PA-12 Super Cruiser	12-2907	N854CC CS-ACC	29. 9.04	Dijana Todorovic	Spanhoe	
G-CDCT	Aerotechnik EV-97 teamEurostar UK	2117		15. 9.04	Cosmik Aviation Ltd	Church Farm, Shotteswell	17.10.05P
G-CDCU	Mainair Blade	1369-1004-7 & W1164		19.10.04	D.K.Jones	(Llanrwst)	15.12.05P
G-CDCV	Robinson R44 Clipper II	10536		28.10.04	A.R.Mcrae	Lanark	29.11.07E
G-CDCW	Just/Reality Escapade 912 *(Built P.Nicholls - kit no. JAESC 0050)*	BMAA/HB/413		4.10.04	P.Nicholls	(Ludlow)	
G-CDCX	Cessna 750 Citation X	750-0194	N194CX N5192E	26. 6.03	P.W.Harris t/a Pendley Farm	Luton	25. 6.06T
G-CDCY	Mainair Pegasus Quantum 15	8065		29.10.04	Gaynor Stocker	Hunsdon	5.12.05P
G-CDCZ	Mainair Pegasus Quantum 15-912	8072		1.10.04	Light Flight Ltd	Knapthorpe Lodge, Caunton	3.10.05P
G-CDDA	SOCATA TB-20 Trinidad	1860	PH-SXE F-OIGL	22.10.04	Oxford Aviation Services Ltd	Oxford	
G-CDDB	Grob/Schemmp-Hirth CS-11 Standard Cirrus	577G	BGA5102/KHJ F-CEMG	13.10.04	K.D.Barber	(St. Nicholas de la Grave, France)	
G-CDDD	Robinson R22 Beta II	3658		24. 8.04	TDR Aviation Ltd	Newtownards	27. 9.07E
G-CDDE	PZL-110 Koliber 160A	04020088	SP-WGO	26.10.04	PZL International Aviation Marketing and Sales plc *(Noted 1.05)*	North Weald	AC
G-CDDF	Mainair Pegasus Quantum 15-912	8079		8.11.04	B.C.Blackburn	Perth	8.11.05P
G-CDDG	Piper PA-26-161 Warrior II	2816065	HB-PLU	4.10.04	A.Oxenham *(Noted 11.04)*	Hinton-in-the-Hedges	
G-CDDH	Raj Hamsa X'Air Falcon Jabiru *(Built B Stanbridge - kit no. 944)*	BMAA/HB/419		26.10.04	B.and Lorna Stanbridge	(Boston)	
G-CDDI	Thruster T600N 450	1040-T600N-109		26.10.04	R.Nayak	South Elkington, Louth	28.11.05P
G-CDDJ							
G-CDDK	Cessna 172M Skyhawk	17265258	TF-SIX N64478	1.12.04	Margaret H and P.R.Kevern	Enstone	
G-CDDL							
G-CDDM	Lindstrand LBL 90A HAB	902	ZS-HAI	14. 1.05	Lindstrand Hot Air Balloons Ltd	(Oswestry)	
G-CDDN	Lindstrand LBL 90A HAB	903	ZS-HAK	14. 1.05	Lindstrand Hot Air Balloons Ltd	(Oswestry)	
G-CDDO	Raj Hamsa X'Air 133 *(Built R.N.Tarrant - kit no. 941)*	BMAA/HB/407		2.11.04	R.N.Tarrant and B.J.Reynolds	(Rushden)	
G-CDDP	Laser Z230 Lazer *(Built F.L.Thomson)*	001	N230RT	16.11.04	A.Smith *(Noted dismantled 1.05)*	Gamston	
G-CDDR	Best Off Skyranger 582 *(Built R.J.Milward - kit no. UK/502)*	BMAA/HB/418		26.10.04	R.J.Milward	(Rushden)	
G-CDDS	Zenair CH.601HD Zodiac *(Built S.Foreman)*	PFA 162-14223		8.10.04	S.Foreman	Priory Farm, Tibenham	
G-CDDT	SOCATA TB-20 Trindad	1858	PH-SXC F-OIGJ	24.11.04	Oxford Aviation Services Ltd	Oxford	
G-CDDU	Best Off Skyranger 912 *(Built R.C.Reynolds - kit no. UK/506)*	BMAA/HB/422		10.11.04	R.C.Reynolds	(Stoke-on-Trent)	
G-CDDV							
G-CDDW	Aeroprakt A22 Foxbat *(Built D.A.A.Wineberg)*	PFA 317-14261		8. 9.04	D.A.A.Wineberg	(Henley-in-Arden)	
G-CDDX	Thruster T600N 450	0049-T600N-107		11.11.04	Linda M.Leachman	(Lincoln)	10.11.05P
G-CDDY	Van's RV-8 *(Built C.S.Ziekle)*	80912	N701CZ	24.11.04	R.W.H.Cole	Spilsted Farm, Sedlescombe	
G-CDDZ	Lindstrand LBL 260A HAB	1018		16.11.04	Lindstrand Hot Air Balloons Ltd	(Oswestry)	
G-CDEA	SAAB-Scania 2000	2000-009	SE-LOX HB-IZF, (D-ADIC), SE-009	20. 1.05	Air Kilroe Ltd t/a Eastern Airways	Humberside	

Registration	Type	C/n	Previous identity	Date	Owner/operator	Base	CofA
G-CDEB	SAAB-Scania 2000	2000-036	SE-036	30.11.04	Air Kilroe Ltd	Humberside	
			HZ-IZT, SE-036		t/a Eastern Airways		
G-CDEC	Mainair Pegasus Quik	8081		27. 1.05	Senga Bradie and A Huyton	(West Calder)	
G-CDED	Robinson R22 Beta II	3747	N74365	19. 1.05	Heli Air Ltd	Wellesbourne Mountford	
G-CDEE							
G-CDEF	Piper PA-28-161 Cadet	2841341	D-ESTD	12.11.04	Western Air (Thruxton) Ltd	Thruxton	
			N9184X, N621FT, (OH-PFB)				
G-CDEG							
G-CDEH	ICP MXP-740 Savannah LS	BMAA/HB/349		18.11.04	S.Whittaker and P.J.Wilson	Sandtoft	
	(Built S.Whittaker and P.J.Wilson - kit no. 03-03-51-200)				t/a Sandtoft Ultralights Partnership		
G-CDEI							
G-CDEJ	Diamond DA.40D Star	D4.142	OE-VPU	1.12.04	Diamond Aircraft UK Ltd	Gamston	
G-CDEK	Diamond DA.40D Star	D4.143	OE-VPW	1.12.04	Diamond Aircraft UK Ltd	Gamston	
G-CDEL	Diamond DA.40D Star	D4.144	OE-VPW	1.12.04	Diamond Aircraft UK Ltd	Gamston	
G-CDEL							
G-CDEM	Raj Hamsa X'Air 133	BMAA/HB/421		17.11.04	R.J.Froud	(Newport, Isle of Wight)	
	(Built R.J.Froud - kit no. 939)						
G-CDEN	Mainair Pegasus Quantum 15-912	8087		30.11.04	R.E.J.Pattenden	Rochester	5.12.05P
G-CDEO							
G-CDEP	Aerotechnik EV-97 teamEurostar UK	2128		6.12.04	S.R.Pike	Booker	5.12.05P
G-CDER							
G-CDES	Bell 206B JetRanger III	3627	N22751	16.12.04	Apple International Inc Ltd	North Creake	
G-CDET	Culver LCA Cadet	129	N29261	10.11.86	H.B.Fox	Booker	11.11.05P
	(Continental O-200-A)		NC29261		*(As "29261" in USAAF c/s)*		
G-CDEU	Lindstrand LBL 90B HAB	1015		7. 2.05	Nadia Florence and P.J.Marshall	(Ruislip)	25.11.05E
G-CDEV	Just/Reality Escapade 912	BMAA/HB/360		26.10.04	M.B.Devenport	(Basingstoke)	
	(Built M.B.Devenport - kit no.JAESC 0024)						
G-CDEW	Mainair Pegasus Quik	8083		13. 1.05	M.M.Chittenden	Lower Road, Hockley	
G-CDEX	Europa Aviation Europa	PFA 247-12507		21.10.04	J.M.Carter	(Trimdon)	
	(Built S Collins -kit no.012)						
G-CDEY	Robinson R44 Raven II	10593		24.12.04	Heli Air Ltd	Wellesbourne Mountford	
G-CDFA	Kolb Twinstar Mk.3 Extra	PFA 205-14274		25.11.04	M.H.Moulai	Sandtoft	
	(Built M.H.Moulai - c/n M3X-04-7-00057)						
G-CDFB	Raj Hamsa X'Air 582	BMAA/HB/411		26.11.04	P.K.Morley	(Malton)	
	(Built P.K.Morley - kit no. 960)						
G-CDFC							
G-CDFD	Scheibe SF25C Falke	44705		1.12.04	T.M.Holloway	RAF Halton	
					tr RAF Gliding and Soaring Association		
G-CDFE	IAV- Bacau Yakovlev Yak-52	855712	LY-APU	29.11.04	D.R.Farley	(Tring)	11. 1.06P
			G-CDFE, N151PA, Krygyz AF, DOSAAF 82 *(blue)*?				
					(To LY-APU 2.12.04 but new UK CofR dated 7.1.05)		
G-CDFF	Aérospatiale-Alenia ATR42-300	331	LN-FAI (5)	30. 3.04	Air Wales Ltd	Swansea	1. 4.07T
			(LN-FAQ), G-BVEF, F-GKNF, F-WWLP				
G-CDFG	Mainair Pegasus Quik	8082		21.12.04	D.Gabbott	(Preston)	
G-CDFH	Just/Reality Escapade 912	BMAA/HB/423		29.11.04	J.Deegan	(Portarlington, County Laios)	
	(Built J.Deegan - kit no. JAESC 0040)						
G-CDFI	Cameron Colt 31A HAB	10655		6.12.04	A.M.Holly t/a Exclusive Ballooning	(Berkeley)	
G-CDFJ	Best Off Skyranger 912	BMAA/HB/424		6.12.04	W.C.Yates	Higher Barn Farm, Houghton	
	(Built W.C.Yates - kit no.UK/490)						
G-CDFK	Jabiru Aircraft Jabiru SPL-450	PFA 274A-14144		2.12.04	H.J.Bradley	(Bridgnorth)	
	(Built H.J.Bradley)						
G-CDFL	Zenair CH.601UL Zodiac	PFA 162A-14309		30.11.04	F.G.Green	Knapthorpe Lodge, Caunton	
	(Built F G Green, R Welch and V Causey)				tr Caunton Zodiac Group		
G-CDFM	Raj Hamsa X'Air 582(	BMAA/HB/417		2.12.04	J.Griffiths	(Bletchley)	
	(Built J.Griffiths - kit no.920)						
G-CDFN							
G-CDFO	Mainair Pegasus Quik	8080		17. 1.05	C.J.Gordon	Perth	
G-CDFP	Best Off Skyranger 912	BMAA/HB/431		10.12.04	J.M.Gammidge	(Thrapston)	
	(Built J.M.Gammidge)						
G-CDFR							
G-CDFS	Embraer EMB-135ER	145.431	EI-ORK	23.12.04	LCY Flight Ltd	Cork, County Cork	
			PT-SUC, (CN-RLG), PT-SUC				
G-CDFT	Gefa-Flug AS105GD HA Airship	0031		15.12.04	Cameron Balloons Ltd	(Bristol)	
G-CDFU	Rans S-6-ES Coyote II	PFA 204-14232		3.11.04	P.W.Taylor	Priory Farm, Tibenham	
	(Built P.W.Taylor - kit no.0304.1560)						
G-CDFV	Aérospatiale AS355F2 Ecureuil 2	5424	RP-C1688	31. 1.05	Lloyd Helicopters Europe Ltd	Redhill	
			JA9964				
G-CDFW							
G-CDFY	Beech B200 Super King Air	BB-1715	N607TA	4. 2.05	BAE Systems Marine Ltd	Walney Island	
G-CDGA	Taylor JT.1 Monoplane	PFA 55-10382		28.12.78	R.M.Larimore	(Spondon, Derby)	
	(Built D G Anderson - kit no.6020/1)				*(Current status unknown)*		
G-CDGB	Rans S-6-116 Coyote II	PFA 204A-13047		21.12.04	S.Penoyre	(Windlesham)	
	(Built S.Penoyre)						
G-CDGC	Mainair Pegasus Quik	8090		26. 1.05	A.T.K. Crozier	(Bathgate)	
G-CDGD	Mainair Pegasus Quik	8086		11.11.04	P.Harper and P.J.Morton	St Michaels	12.12.05P
G-CDGE	AirBorne T912-B/Streak III-B	XT912-028	T2-2253	22.12.04	C.T.Guest	Mill Farm, Shifnal	
G-CDGF							
G-CDGG	Dyn'Aéro MCR-01 Club	PFA 301A-14267		26.10.04	P.Simpson and P.A.B.Morgan	(Potters Bar)	
	(Built P.Simpson and P.A.B.Morgan)						
G-CDGH	Rans S-6-ES Coyote II	PFA 204-14209		22.12.04	K.T.Vinning	(Wilmcote, Stratford-upon-Avon)	
	(Built K.T.Vinning- kit no.457)				tr G-CDGH Group		
G-CDGI	Thruster T600N 450	1041-T600N-108		5. 1.05	Thruster Air Services Ltd	Ginge, Wantage	9. 1.06P
G-CDGJ	American Champion 7ECA Citabria Aurora	1391-2004		14.12.04	T.A.Mann	(Ballasalla, Isle of Man)	

Registration	Type	C/n	Previous identity	Date	Owner	Base	CofA
G-CDGK							
G-CDGL							
G-CDGM	Eurocopter MBB BK-117C-2	9063		22.12.04	McAlpine Helicopters Ltd	Oxford	
G-CDGN							
G-CDGO	Mainair Pegasus Quik	8084		11.11.04	Mainair Sports Ltd	(Rochdale)	
G-CDGP	Zenair CH.601XL Zodiac *(Built T.J.Bax)*	PFA 162B-14313		6. 1.05	T.J.Bax	Henstridge	
G-CDGR	Zenair CH.701UL *(Built M.Morris)*	PFA 187-14327		6. 1.05	M.Morris	(Thornton Cleveleys)	
G-CDGS	American General AG-5B Tiger	10097	PH-BMA N1195Q	7. 2.05	K.Hennessy tr Premier Flying Group	(Clonmel, County Tipperary	
G-CDGT	Montgomerie-Parsons Two Place Gyroplane *(Built A.A.Craig)*	PFA G/08-1361		10. 1.05	A.A.Craig	(Kilwinning)	
G-CDGU	Vickers Supermarine 300 Spitfire Mk.1	6S-75156	X4276	7. 1.05	A.J.E.Smith *(Noted 1.05)*	Breighton	
G-CDGV	Bell 206B JetRanger III	3997	N217PM XC-PFS	11. 1.05	Multiflight Ltd	Leeds-Bradford	
G-CDGW	Piper PA-28-181 Archer II	28-7990402	HB-PDZ N2156Z	13. 1.05	Barry Fielding Aviation Ltd	Ronaldsway	
G-CDGX							
G-CDGY	Vickers Supermarine 349 Spitfire Mk.Vc *(Built Westland Aircraft Ltd)*	WWA3832	ZK-MKV A58-149, EF545	19. 1.05	Aero Vintage Ltd	(Duxford)	
G-CDGZ							
G-CDHA	Best Off Skyranger 912S *(Built K.J.Gay - kit no.UK/508)*	BMAA/HB/428		18. 1.05	K.J.Gay	Newtownards	
G-CDHB	British Aircraft Corporation 167 Strikemaster Mk.80A	EEP/JP/4096	R.Saudi AF 1130 G-27-296	31. 1.05	C.C.Hudson	(Bromley)	
G-CDHC	Slingsby T.67C Firefly	2081	PH-SGC (PH-SBC), G-7-138	20. 1.05	N.J.Morgan	Tatenhill	
G-CDHD							
G-CDHE	Best Off Skyranger 912 *(Built S.Owen)*	BMAA/HB/412		26. 1.05	M.P.Lobb, S.Owen, K. Farrow and P. Burgess	(Cheadle)	
G-CDHF							
G-CDHG							
G-CDHH							
G-CDHI	North American P-51D-25NA Mustang	122-39232	G-SUSY N12066, FAN GN120, 44-72773	31. 1.05	A.J.E.and Ann E.Smith	Breighton	20. 5.01P
G-CDHJ	Lindstrand LBL 90B HAB	1038		4. 2.05	Lindstrand Hot Air Balloons Ltd	(Oswestry)	
G-CDHK	Lindstrand LBL 330A HAB	1022		9. 2.05	Richard Nash Cars Ltd	(Norwich)	
G-CDHL	Lindstrand LBL 330A HAB	1028		9. 2.05	Richard Nash Cars Ltd	(Norwich)	
G-CDHM	Mainair Pegasus Quantum 15	8085		18. 1.05	M.K.Morgan	(Cardiff)	
G-CDHN							
G-CDHO	Raj Hamsa X'Air 133 *(Built W E Corps - kit no:902)*	BMAA/HB/408		8. 2.05	W.E.Corps	Broad Farm, Eastbourne	
G-CDHP	Lindstrand LBL 105A HAB	754	G-OHRH	11.02.05	A.M.Holly	(Berkeley)	12. 4.05T
G-CDHR							
G-CDHS							
G-CDHT	IAV-Bacau Yakovlev Yak-52	867114	RA-01813	11.02.05	G.G.L James	Sleap	
G-CDHU							
G-CDHV							
G-CDHW							
G-CDHX							
G-CDHY							
G-CDHZ							
G-CDIA							
G-CDIB							
G-CDIC							
G-CDID							
G-CDIE							
G-CDIF							
G-CDIG							
G-CDIH							
G-CDII	Schleicher ASW22BLE 50R	22079	BGA4656/JNV	17. 1.05	R.A.Cheetham	Husbands Bosworth	8. 4.05
G-CDIJ							
G-CDIK							
G-CDIL	Mainair Pegasus Quantum 15-912	8093		21.12.04	J.I.Greenshields	Dunkeswell	20.12.05P
G-CDIM							
G-CDIN							
G-CDIO							
G-CDIP	Best Off Skyranger 912S *(Built M.S.McCrudden - kit no.UK/507)*	BMAA/HB/429		11. 1.05	M.S.McCrudden	Newtownards	
G-CDIR							
G-CDIS							
G-CDIT							
G-CDIU							
G-CDIV							
G-CDIW							
G-CDIX							
G-CDIY	Aerotechnik EV-97A Eurostar *(Built G.R.Pritchard)*	PFA 315-14345		30.12.04	G.R.and Judith A.Pritchard	(Hay-on-Wye)	
G-CDIZ	Just/Reality Escapade 912 *(Built E.G.Bishop and E N.Dunn - kit no.JAESC 0033)*	BMAA/HB/393		7. 1.05	E.G.Bishop and Elizabeth N.Dunn	Dunkeswell	
G-CDJA							
G-CDJB	Van's RV-4 *(Built T.A.Rudisill and R.G.Pettyjohn)*	1270	N21RP	21. 1.05	J.M.Keane	(Brighton)	

Reg	Type	C/n	Previous identity	Reg date	Owner/operator	Base	CofA
G-CDJC							
G-CDJD							
G-CDJE							
G-CDJF							
G-CDJG							
G-CDJH							
G-CDJI							
G-CDJJ							
G-CDJK							
G-CDJL	Jabiru Aircraft Jabiru J400	PFA 325-14215		4. 2.05	T.R.Sinclair and T.Clyde	Kirkwall	
	(Built T.R.Sinclair)						
G-CDJM	Zenair CH.601XL Zodiac	PFA 162B-14303		30.12.04	T.J.Adams-Lewis	(Feidrhenffordd, Cardigan)	
	(Built T.J. Adams-Lewis)						
G-CDJN							
G-CDJO							
G-CDJP	Best Off Skyranger 912	BMAA/HB/435		18.01.05	J.S.Potts	(Kilmarnock)	
	(Built J.S.Potts)						
G-CDJR							
G-CDJS							
G-CDJT							
G-CDJU							
G-CDJV							
G-CDJW							
G-CDJX							
G-CDJY							
G-CDJZ							
G-CDKA							
G-CDKB							
G-CDKC							
G-CDKD							
G-CDKE							
G-CDKF							
G-CDKG							
G-CDKH							
G-CDKI							
G-CDKJ							
G-CDKK							
G-CDKL							
G-CDKM	Mainair Pegasus Quik	8091		28. 1.05	P.Barrow	Arclid Green, Sandbach	
G-CDKN							
G-CDKO							
G-CDKP							
G-CDKR							
G-CDKS							
G-CDKT							
G-CDKU							
G-CDKV							
G-CDKW							
G-CDKX							
G-CDKY							
G-CDKZ							
G-CDOG	Lindstrand LBL Dog SS HAB	938		7. 1.04	Airship and Balloon Company Ltd	(Clapton-in-Gordano, Bristol)	8. 3.05A
G-CDON	Piper PA-28-161 Warrior II	28-8216185	N8254D	24. 5.88	East Midlands Flying School Ltd	Nottingham East Midlands	17. 5.06T
G-CDPD	Mainair Pegasus Quik	8051		5. 8.04	P.C.Davis	Weston Zoyland	15. 8.05P
G-CDPL	Aerotechnik EV-97 teamEurostar UK	2207		13. 1.05	C.D.H.Garrison	Sutton Meadows, Ely	12. 1.06P
G-CDPY	Europa Aviation Europa	PFA 247-13029		8. 3.00	A Burrill	(Reading)	
	(Built A Burrill - kit no.303) (Monowheel u/c)						
G-CDRU	CASA 1-131E Jungmann	2321	EC-DRU	19. 1.90	P.Cunniff	White Waltham	30. 6.04P
			Spanish AF E3B-530		*"Yen a Bon"*		
G-CDUO	Boeing 757-236	24792	SE-DUO	20. 2.02	Britannia Airways Ltd *(Thomson c/s)*	Luton	10. 3.05T
			G-BRJI, SX-BBZ, G-BRJI, SX-BBZ, G-BRJI, EC-FMQ, EC-786, EC-EVC, EC-446, G-BRJI				
G-CDUP	Boeing 757-236	24793	SE-DUP	30. 4.02	Britannia Airways Ltd *(Thomson c/s)*	Luton	1. 5.05T
			G-OOOT, G-BRJJ, EC-490, G-BRJJ				
G-CDUX	Piper PA-32-300 Cherokee Six	32-7340074	EC-DUX	31. 7.02	D.J.Mason	(Glan Maye, Isle of Man)	17.12.05T
			F-BSGY, 5T-TJR, N11C				
G-CDXX	Robinson R44 Raven II	10624		31. 1.05	Heli Air Ltd	Wellesbourne Mountford	
G-CEAC	Boeing 737-229	20911	OO-SDE	11. 6.99	European Aviation Air Charter Ltd	Bournemouth	25. 8.05T
			C-GNDX, OO-SDE, C-GNDX, OO-SDE		*(Op by Bmibaby in Palmair European c/s)*		
G-CEAD	Boeing 737-229	21137	OO-SDM	11.10.99	European Aviation Air Charter Ltd	Bournemouth	16.11.05T
G-CEAE	Boeing 737-229	20912	OO-SDF	25. 1.00	European Aviation Air Charter Ltd	Bournemouth	28. 2.06T
G-CEAF	Boeing 737-229	20910	G-BYRI	13. 1.00	European Aviation Air Charter Ltd	Bournemouth	3. 4.06T
			OO-SDD, EC-EEG, OO-SDD		*(Op Palmair European)*		
G-CEAG	Boeing 737-229	21136	OO-SDL	6. 6.00	European Aviation Air Charter Ltd	Bournemouth	14. 6.06T
			(OO-SDM)				
G-CEAH	Boeing 737-229	21135	OO-SDG	1. 8.00	European Aviation Air Charter Ltd	Bournemouth	14.11.06T
G-CEAI	Boeing 737-229	21176	OO-SDN	7. 3.01	European Aviation Air Charter Ltd		
			9M-MBP, OO-SDN, N8277V			(Saltzberg, Germany)	28. 8.07T
G-CEAJ	Boeing 737-229	21177	OO-SDO	5.12.00	European Aviation Air Charter Ltd	Bournemouth	22. 4.07T
					(Op Palmair European)		
G-CEAL	Short SD.3-60 Variant 100	SH.3761	N161CN	11. 9.95	BAC Express Airlines Ltd	Southend	12. 1.05T
			N161SB, G-BPXO		*"City of Belfast" (In open store 2.05.)*		
G-CEEE	Robinson R44 Raven II	10005		26.11.02	R.D.Masters	Standalone Farm, Meppershall	17.12.05T

Reg	Type	C/n	Previous identity	Date	Owner/operator	Base	CofA
G-CEGA	Piper PA-34-200T Seneca II	34-8070367	N8272B	30.12.80	Oxford Aviation Services Ltd	Oxford	22. 8.05T
G-CEGP	Beech 200 Super King Air	BB-726	G-BXMA (N58AJ), G-BXMA, N622JA, N522JA, N222JD	14. 5.01	Cega Aviation Ltd	Goodwood	7. 8.06T
G-CEGR	Beech 200 Super King Air	BB-351	N68CP N351FW, N6666C, N6666K	23. 7.97	Henfield Lodge Aviation Ltd *(Op CEGA)*	Goodwood	18. 8.06T
G-CEJA	Cameron V-77 HAB	2469	G-BTOF	17. 6.91	L. and C.Gray	(Farnborough)	15. 7.04A
G-CELA	Boeing 737-377QC	23663	VH-CZK	15. 8.03	Dart Group plc *(Op Jet 2) (In service 2.05)*	Leeds-Bradford	AC
G-CELB	Boeing 737-377	23664	VH-CZL	3. 7.03	Dart Group plc *(Op Jet 2) (yorkshire Jet2 titles)*	Leeds-Bradford	10. 5.07T
G-CELC	Boeing 737-33A	23831	N190FH VH-CZV, G-OBMA	4. 7.03	Dart Group plc *(Op Jet 2) (prague Jet2 titles)*	Leeds-Bradford	12.10.06T
G-CELD	Boeing 737-33A	23832	N191FH VH-CZW, G-OBMB	4. 7.03	Dart Group plc *(Op Jet 2) (espana Jet2 titles)*	Leeds-Bradford	1. 2.07T
G-CELE	Boeing 737-33A	24029	VH-CZX G-MONN	11. 7.03	Dart Group plc *(Op Jet 2) (belfast Jet2 titles)*	Leeds-Bradford	18. 8.07T
G-CELF	Boeing 737-377	24302	S7-ABB VH-CZM, N113AW, VH-CZM	5.10.04	Dart Group plc *(Op Jet2) (valencia Jet2 titles)*	Leeds-Bradford	7.11.07E
G-CELG	Boeing 737-377	24303	S7-ABD VH-CZN	26.10.04	Dart Group plc *(Op Jet2) (london Jet2 titles)*	Leeds-Bradford	19.12.07E
G-CELH	Boeing 737-330	23525	D-ABXD (PR-GLB), D-ABXD, TF-ABL, D-ABXD	23. 8.04	Dart Group plc *(Op Jet2) (faro Jet2 titles)*	Manchester	23. 9.07T
G-CELI	Boeing 737-330	23526	D-ABXE (PR-GLC), D-ABXE	21. 9.04	Dart Group plc *(Op Jet2) (mancester Jet2 titles)*	Manchester	17.11.07E
G-CELJ	Boeing 737-330	23529	LZ-BOG (PR-GLF) (D-ABXI), LZ-BOG, D-ABXI	23.12.04	Dart Group plc *(Op Jet 2)*	Leeds Bradford	
G-CELK	Boeing 737-330	23530	LZ-BOH D-ABXK	11.02.05	Dart Group plc *(Op Jet 2)*	Leeds-Bradford	
G-CELP	Boeing 737-330QC	23522	TF-ELP D-ABXA, N1786B	27.10.03	Dart Group plc *(Op Channel Express)*	Stansted	26.10.06T
G-CELR	Boeing 737-330QC	23523	TF-ELR D-ABXB	4.11.03	Dart Group plc *(Op Channel Express in Flyglobespan com titles)*	Glasgow	3.11.06T
G-CELS	Boeing 737-377	23660	VH-CZH	17. 5.02	Dart Group plc *(Op Jet2) (leeds-bradford Jet2 titles)*	Leeds-Bradford	19. 6.05T
G-CELU	Boeing 737-377	23657	VH-CZE	13. 6.02	Dart Group plc *(Op Jet2) (barcelona Jet2 titles)*	Leeds-Bradford	30. 7.05T
G-CELV	Boeing 737-377	23661	VH-CZI	2.10.02	Dart Group plc *(Op Jet2) (amsterdam Jet2 titles)*	Leeds-Bradford	10. 2.06T
G-CELW	Boeing 737-377F	23659	N659DG G-CELW, VH-CZG	4. 7.02	Dart Group plc *(Op Channel Express)*	Stansted	14.10.06T
G-CELX	Boeing 737-377	26354	VH-CZB N5573B	5. 3.03	Dart Group plc *(Op Jet 2) (malaga Jet2 titles)*	Leeds-Bradford	15. 4.06T
G-CELY	Boeing 737-377F	23662	N662DG G-CELY, VH-CZJ	29. 4.03	Dart Group plc *(Op Jet2) (ireland Jet2 titles)*	Leeds-Bradford	22. 4.07T
G-CELZ	Boeing 737-377F	23658	VH-CZF	22. 9.03	Dart Group plc *(Op Jet2) (paris Jet2 titles)*	(Leeds-Bradford	2. 8.07T
G-CEPT	SOCATA TB-20 Trinidad	1240	G-BTEK	27. 8.04	P.J.Caiger	Biggin Hill	30. 6.05T
G-CERI	Europa Aviation Europa XS *(Built S.J.M.Shepherd - kit no.541)*	PFA 247-13970		20. 8.03	S.J.M.Shepherd	(Rhyl)	
G-CERT	Mooney M.20K Model 252TSE	25-1134		5.10.87	K.A Hemming	Fowlmere	15. 4.06
G-CEXB	Fokker F.27 Friendship 500RF	10550	N743A PH-EXF	15.11.95	Dart Group plc *(Op Channel Express)*	(Bournemouth)	30. 1.05TC
G-CEXE	Fokker F.27 Friendship 500	10654	SU-GAF PH-EXJ	2. 4.97	Dart Group plc *(Op Channel Express)*	(Bournemouth)	14. 5.06TC
G-CEXG	Fokker F.27 Friendship 500	10459	G-JEAP 9Q-CBI, OY-APF, 9Q-CBI, PH-RUA, VH-EWR, F-BYAH, OY-APF, PH-EXD	13.11.00	Dart Group plc *(Op Channel Express)*	(Bournemouth)	15. 6.07T
G-CEXH	Airbus A300B4-203F	117	D-ASAZ N14966, N966C, F-OGTB, 9V-STA, F-WZER	30. 3.98	Dart Group plc *(Op Channel Express)*	Liege	1. 4.07T
G-CEXI	Airbus A300B4-203	121	D-ASAA N15967, N967C, F-OGTC, 9V-STB, F-WZEK	3. 9.98	Dart Group plc *(Op Channel Express)*	Liege	3. 9.07T
G-CEXJ	Airbus A300B4-203F	147	N300FV F-WQIP, 9M-MHD, F-WZMA	17. 3.00	Channel Express (Air Services) Ltd	(Bournemouth)	19. 3.06T
G-CEXK	Airbus A300B4-103	105	PH-ABF N304FV, F-WQJR, SX-BEF, F-WZED	9. 4.02	Channel Express (Air Services) Ltd	(Bournemouth)	1. 8.05T
G-CEXP*	Handley Page HPR.7 Dart Herald 209	195	I-ZERC G-BFRJ, 4X-AHO	29.10.87	British Airports Authority *(WFU 8.3.96: cancelled 22.3.96 by CAA) (Stored 4.04)*	Gatwick	7.11.96T
G-CEYE	Piper PA-32R-300 Lance	32R-7780533	SE-KCD OH-PAS	24.10.02	Fleetlands Flying Association Ltd	Lee-on-Solent	13. 1.06T
G-CFAA	BAE SYSTEMS Avro 146-RJ100	E3373		9. 5.00	British Airways Citiexpress Ltd	London City	15. 6.06T
G-CFAB	BAE SYSTEMS Avro 146-RJ100	E3377		23. 8.00	British Airways Citiexpress Ltd	London City	1. 8.05T
G-CFAC	BAE SYSTEMS Avro 146-RJ100	E3379		23. 8.00	British Airways Citiexpress Ltd	Manchester	14.12.06T
G-CFAD	BAE SYSTEMS Avro 146-RJ100	E3380		23. 8.00	British Airways Citiexpress Ltd	Manchester	24. 1.07T
G-CFAE	BAE SYSTEMS Avro 146-RJ100	E3381		12. 1.01	British Airways Citiexpress Ltd	London City	22. 2.07T
G-CFAF	BAE SYSTEMS Avro 146-RJ100	E3382		15. 1.01	British Airways Citiexpress Ltd	London City	22. 3.07T
G-CFAH	BAE SYSTEMS Avro 146-RJ100	E3384		15. 1.01	British Airways Citiexpress Ltd	London City	5. 6 07T
G-CFBI	Colt 56A HAB	570		11. 7.84	G.A Fisher tr Out-of-the-Blue *(Op Balloon Preservation Group) "Air O"*	(Aldershot)	24. 7.91A
G-CFGL	Cessna 560XL Citation Excel	560-5361	VP-CGG N361XL, N52397	31. 1.05	St. Merryn Air Ltd	Leeds-Bradford	30. 1.08E
G-CFME	SOCATA TB-10 Tobago	1795	F-GNHU	15. 4.98	Charles Funke Associates Ltd	Goodwood	16. 6.07T
G-CFRY	Zenair CH.601UL Zodiac *(Built C.K.Fry)*	PFA 162A-14302		5. 1.05	C.K.Fry	(Lytchett Minster, Poole)	
G-CFWR	Best Off Skyranger 912 *(Built R.W.Clarke - kit no.UK/525)*	BMAA/HB/426		2.12.04	R.W.Clarke	(Bradford)	

G-CGDH	Europa Aviation Europa XS *(Built G.D.Harding - kit no.457)*	PFA 247-13746		8. 7.03	G.D.Harding	Wombleton	2. 9.05P
G-CGHM	Piper PA-28-140 Cruiser	28-7425143	PH-NSM N9614N	25. 4.79	A.Reay	Caernarfon	10. 2.07T
G-CGOD	Cameron N-77 HAB	2647		5. 9.91	G.P.Lane *"Neptune"*	(Waltham Abbey)	12. 8.05A
G-CHAD	Aeroprakt A22 Foxbat *(Built D Winsper)*	PFA 317-13909		30. 4.02	C.J.Rossiter	Otherton, Cannock	23. 9.05P
G-CHAH	Europa Aviation Europa *(Built T.Higgins - kit no.252)*	PFA 247-12949		14. 6.04	T.Higgins	(Caersws)	
G-CHAM	Cameron Pot 90SS HAB *(Chambourcy Pot shape)*	2912		29. 9.92	B.J.Reeves and C.Walker t/a High Exposure Balloons *"Yogpot"*	(Brighouse)	3. 7.05A
G-CHAP	Robinson R44 Astro	0326		9. 4.97	Brierley Lifting Tackle Co Ltd	Wolverhampton	5. 5.06T
G-CHAR	Grob G109B	6435		21. 5.86	T.Holloway tr RAFGSA	RAF Halton	7. 5.05
G-CHAS	Piper PA-28-181 Archer II	28-8090325	N82228	18. 3.91	C.H.Elliott	Stapleford	12. 6.06
G-CHAV	Europa Aviation Europa *(Built M B Stoner - kit no.117) (Monowheel u/c)*	PFA 247-12769		28.12.94	R.P.Robinson	Gloucestershire	11.11.04P
G-CHCD	Sikorsky S-76A II Plus	760101	OY-HEZ G-CHCD, G-CBJB, N288SP, C-GIMN, YV-326C	16. 1.98	CHC Scotia Ltd	North Denes	28.11.07E
G-CHCF	Eurocopter AS332L2 Super Puma	2567		30.11.01	CHC Scotia Ltd	Aberdeen	15. 1.05T
G-CHCG	Eurocopter AS332L2 Super Puma	2592		1. 7.03	CHC Scotia Ltd	Aberdeen	23. 7.06T
G-CHCH	Eurocopter AS332L2 Super Puma	2601		16.12.03	CHC Scotia Ltd	Aberdeen	26. 1.07T
G-CHEB	Europa Aviation Europa *(Built C.H.P.Bell - kit no.263)* (NSI EA-81/100)	PFA 247-12967		16. 9.96	T.C.Butterworth	Leicester	22. 9.05P
G-CHEL	Colt 77B HAB	4823		18. 5.00	Chelsea Financial Services plc *(Chelsea Financial Service titles)*	(Cirencester)	19. 7.05P
G-CHEM	Piper PA-34-200T Seneca II	34-8170032	N8292Y	26. 8.87	London Executive Aviation Ltd	Stapleford	27. 2.06T
G-CHER	Piper PA-38-112 Tomahawk II	38-82A0004	G-BVBL N91339	19.12.00	C.L.Goodsell	(Kingswinford)	14. 5.06T
G-CHET	Europa Aviation Europa XS *(Built C R Arkle - kit no.376)* (Rotax 914-UL) *(Tri-gear u/c)*	PFA 247-13277		12. 2.98	H.P.Chetwynd-Talbot	Wombleton	19. 4.05P
G-CHEZ	Pilatus Britten-Norman BN-2B-20 Islander	2234	9M-TAM G-BSAG	30. 4.01	The Cheshire Police Authority	Liverpool	11. 8.05T
G-CHGL	Bell 206B JetRanger II	1669	EI-WSN G-CHGL, G-BPNG, G-ORTC, G-BPNG, N20EA, C-GHVB	29. 4.98	Elmridge Ltd	(Dungannon)	27. 9.07E
G-CHIK	Reims/Cessna F152 II	F15201628	G-BHAZ (D-EHLE)	19.10.81	Stapleford Flying Club Ltd	Stapleford	7.12.06T
G-CHIP	Piper PA-28-181 Archer II	28-8290095	N81337	22. 2.82	A.D.Krzeptowski	Shoreham	27. 5.06
G-CHIS	Robinson R22 Beta	1740		5. 4.91	I.R.Chisholm t/a Bradmore Helicopter Leasing	Costock	1. 4.07T
G-CHIX	Robin DR500/200i Président *(Registered as DR400/500)*	0036	F-GXGD F-WQPN	29.11.01	P.A and R.Stephens	Moor Farm, West Haslerton	9. 1.05
G-CHKN	Air Création 582/Kiss 400 *(Built I.Tomkins - kit no FL002/134)*	BMAA/HB/183		18. 9.01	D.A.Edwards	(Spalding)	11. 2.05P
G-CHLL	Lindstrand LBL 90A HAB	941		7. 1.04	Airship and Balloon Company Ltd *(www.churchill.com titles)*	(Clapton-in-Gordano, Bristol)	8. 1.05A
G-CHLT	Stemme S-10	10-30	D-KGCD	3. 7.91	J.Abbess	Tibenham	28. 4.07
G-CHOK	Cameron V-77 HAB	1752		25. 5.88	Amanda J.Moore *"S'il Vous Plait"*	(Great Missenden)	2.11.02A
G-CHOP	Westland-Bell 47G-3B1	WA/380	XT221	19.12.78	Classic Rotors Ltd	(London, WC2)	20. 3.06T
G-CHOX	Europa Aviation Europa XS *(Built P Field - kit no.566)*	PFA 247-13974		2. 4.03	Chocks Away Ltd	(Chertsey)	
G-CHPY	de Havilland DHC-1 Chipmunk 22	C1/0093	WB652	7. 3.97	Devonair Executive Business Transport Ltd	Kemble	14. 3.05T
G-CHSL	Agusta A109A II	7352	EC-HSL HB-XRK, I-AGSC	5. 8.03	Swift Copter Ltd	(London EC3)	22. 7.07T
G-CHSU	Eurocopter EC135T1	0079		4. 2.99	Thames Valley Police Authority *(Op Chiltern Air Support Unit)*	RAF Benson	12. 4.05T
G-CHTA	Grumman AA-5A Cheetah	AA5A-0631	G-BFRC	3. 3.86	Quickspin Ltd *(Op Biggin Hill School of Flying)*	Biggin Hill	20. 3.06T
G-CHTG	RotorWay Executive 90 *(Built Rotorbuild Helicopters Ltd)* (RotorWay RI 162)	5118	G-BVAJ	19.11.99	G.Cooper	Street Farm, Takeley	10. 3.05P
G-CHTT*	Varga 2150A Kachina	VAC-162-80		7. 9.84	H.W.Hall *(Damaged near Hatherleigh, Devon 27.4.86 and cancelled 9.8.94 by CAA) (Wreck stored dismantled 2.05)*	Southend	6. 9.87
G-CHUB*	Colt Cylinder Two N-51 HAB				See SECTION 3 Part 1	Newbury	
G-CHUG	Europa Aviation Europa *(Built C.M.Washington - kit no.260) (Monowheel u/c)*	PFA 247-12960		29. 7.96	C.M.Washington	Sleap	10.3.05P
G-CHUK	Cameron O-77 HAB	2773		6. 3.92	L.C.Taylor	(Burton-on-Trent)	5. 4.93A
G-CHUM	Robinson R44 Raven	0839		2. 8.00	Vitapage Ltd	Elstree	7. 9.06T
G-CHYL	Robinson R22 Beta	1197		28.11.89	Caroline M.Gough-Cooper	(Macclesfield)	16. 1.05T
G-CHZN	Robinson R22 Beta	0884	G-GHZM G-FENI	9. 4.99	Cloudbase Ltd	Welshpool	21. 3.05T
G-CIAO	III Sky Arrow 650 T *(Built J Hosier)*	PFA 298-13095		23. 7.97	G.Arscott	Popham	4. 7.05P
G-CIAS	Pilatus Britten-Norman BN-2B-21 Islander	2162	HC-BNS G-BKJM	1. 5.91	Channel Island Air Search Ltd	Guernsey	11. 3.06
G-CIBO	Cessna 180K Skywagon	18053177	VH-JNS N19029	23. 7.04	CIBO Ops. Ltd	(Hemel Hempstead)	
G-CICI	Cameron R-15 Gas/HAB	673	(N) G-CICI, (G-BIHP)	11.11.80	Noble Adventures Ltd	(Bristol)	5. 6.91P
G-CIDD	Bellanca 7ECA Citabria	1002-74	N86577	29.11.00	A and P.West	(Germany)	12.10.07E
G-CIFR	Piper PA-28-181 Cherokee Archer II	28-7790208	PH-MIT OO-HBB, N7654F	18. 6.97	Shropshire Aero Club Ltd	Sleap	4. 9.06T
G-CIGY	Westland-Bell 47G-3B1	WA/350	G-BGXP XT191	26.10.98	R.A Perrot	Guernsey	9. 8.06
G-CITR	Cameron Z-105 HAB	10278		22. 2.02	Flying Pictures Ltd *(Citroën C3 titles)*	(Chilbolton)	19. 5.04A
G-CITY	Piper PA-31-350 Navajo Chieftain	31-7852136	N27741	12. 9.78	Woodgate Aviation (IoM) Ltd	Ronaldsway	21.12.06T
G-CIVA	Boeing 747-436	27092		19. 3.93	British Airways plc	Heathrow	18. 3.06T

Registration	Type	C/n	Previous identity	Reg date	Owner/operator	Probable base	CA expiry
G-CIVB	Boeing 747-436	25811	(G-BNLY)	15. 2.94	British Airways plc	Heathrow	14. 2.07T
G-CIVC	Boeing 747-436	25812	(G-BNLZ)	26. 2.94	British Airways plc	Heathrow	25. 2.06T
G-CIVD	Boeing 747-436	27349		14.12.94	British Airways plc *(Stored 10.02)*	Cardiff	3. 8.06T
G-CIVE	Boeing 747-436	27350		20.12.94	British Airways plc	Heathrow	22. 8.06T
G-CIVF	Boeing 747-436	25434	(G-BNLY)	29. 3.95	British Airways plc	Gatwick	20. 9.06T
G-CIVG	Boeing 747-436	25813	N6009F	20. 4.95	British Airways plc	Heathrow	18. 4.07T
G-CIVH	Boeing 747-436	25809		23. 4.96	British Airways plc	Gatwick	22. 4.05T
G-CIVI	Boeing 747-436	25814		2. 5.96	British Airways plc	Gatwick	1. 5.05T
G-CIVJ	Boeing 747-436	25817		11. 2.97	British Airways plc	Heathrow	10. 9.05T
G-CIVK	Boeing 747-436	25818		28. 2.97	British Airways plc	Heathrow	30. 8.05T
G-CIVL	Boeing 747-436	27478		28. 3.97	British Airways plc	Heathrow	26.11.05T
G-CIVM	Boeing 747-436	28700		5. 6.97	British Airways plc	Heathrow	4. 6.06T
G-CIVN	Boeing 747-436	28848		29. 9.97	British Airways plc	Gatwick	28. 9.06T
G-CIVO	Boeing 747-436	28849	N6046P	5.12.97	British Airways plc	Heathrow	4.12.06T
G-CIVP	Boeing 747-436	25850		17. 2.98	British Airways plc	Heathrow	16. 2.07T
G-CIVR	Boeing 747-436	25820		2. 3.98	British Airways plc	Gatwick	21. 2.07T
G-CIVS	Boeing 747-436	28851		13. 3.98	British Airways plc	Heathrow	12. 3.07T
G-CIVT	Boeing 747-436	25821	(G-CIVN)	20. 3.98	British Airways plc	Heathrow	9.11.06T
G-CIVU	Boeing 747-436	25810	(G-CIVO)	24. 4.98	British Airways plc	Heathrow	23. 4.07T
G-CIVV	Boeing 747-436	25819	N6009F	23. 5.98	British Airways plc	Heathrow	10. 4.07T
			(G-CIVP)				
G-CIVW	Boeing 747-436	25822	(G-CIVR)	15. 5.98	British Airways plc	Heathrow	14. 5.07T
G-CIVX	Boeing 747-436	28852		3. 9.98	British Airways plc	Heathrow	2. 9.07T
G-CIVY	Boeing 747-436	28853		29. 9.98	British Airways plc	Heathrow	28. 9.07T
G-CIVZ	Boeing 747-436	28854		31.10.98	British Airways plc	Heathrow	30.10.07E
G-CJAA	British Aerospace BAe 125 Series 800B	258240	G-HCFR	28.11.02	Club 328 Ltd	Southampton	28.12.07E
	(Built Corporate Jets Ltd)		HB-VLT, G-SHEA, G-BUWC, G-5-772				
G-CJAD	Cessna 525 CitationJet	525-0435	N525AD	28. 6.02	A B.Davis	Edinburgh	27. 6.06T
			N5244F		t/a Davis Aircraft Operations		
G-CJAE	Cessna 560 Citation V	560-0046	G-CZAR	28.11.02	Club 328 Ltd	Southampton	5. 4.05T
			(N26656)				
G-CJAY	Mainair Pegasus Quik	7979		18. 8.03	J.Madhvani	Plaistowes Farm, St. Albans	17. 8.05P
					t/a Jay Airsports		
G-CJBC	Piper PA-28-180 Cherokee D	28-5470	OY-BDE	28.11.80	J.B.Cave	Wolverhampton	29. 8.05
G-CJCI	Pilatus P.2-06	600-63	Swiss AF U-143	30. 7.84	J.Briscoe and P.G.Bond tr Pilatus P2 Flying Group	Norwich	27. 6.05P
					(As "CC+43"" in Luftwaffe c/s in Arado Ar.96B guise)		
G-CJUD	Denney Kitfox Model 3	PFA 172-11939		17. 1.91	A Thomas	Shobdon	17. 6.05P
	(Built C W Judge - c/n 847)				t/a AV8 Air		
G-CKCK	Enstrom 280FX Shark	2071	OO-PVL	5. 5.95	Farmax Ltd *(On rebuild 7.04*	Goodwood	14. 5.98T
G-CLAC	Piper PA-28-161 Warrior II	28-8116241	N8396U	18. 5.87	M.J.Steadman	Blackbushe	22. 12.05
G-CLAS	Short SD.3-60 Variant 200	SH.3635	EI-BEK	28. 7.93	BAC Express Airlines Ltd	Exeter	20. 7.05T
			G-BLED, G-14-3635		*"City of Cardiff"*		
G-CLAV	Europa Aviation Europa	PFA 247-12641		11.10.02	C.Laverty	(Tobermory, Isle of Mull)	
	(Built C.Laverty - kit no.060) (Monowheel u/c)						
G-CLAX	Jurca MJ.5 Sirocco	PFA 2204	G-AWKB	22. 4.99	G.D.Claxton	(Pontyclun)	
	(Built G.D.Claxton)				*(Current status unknown)*		
G-CLAY	Bell 206B-3 JetRanger III	4409	G-DENN	16. 9.02	Claygate Distribution Ltd	Paynetts Farm, Goudhurst	21. 7.05
			N75486, C-GFNO				
G-CLEA	Piper PA-28-161 Warrior II	28-7916081	N30296	28. 8.80	R.J.Harrison and A R.Carpenter	Oaksey Park	19. 2.05
G-CLEE	Rans S-6-ES Coyote II	PFA 204-13670		29. 6.01	R.Holt	Mill Farm, Shifnal	29. 9.05P
	(Built R Holt) (Rotax 582)						
G-CLEM	Bölkow Bö.208A-2 Junior	561	G-ASWE	22. 9.81	J.J.Donely and K.Herbert	Coventry	10. 6.05P
			D-EFHE		tr Bölkow Group		
G-CLEO	Zenair CH.601HD Zodiac	PFA 162-13500		9. 8.99	K.M.Bowen	Goldcliff	
	(Built K.M.Bowen)				*(Current status unknown)*		
G-CLFC	Mainair Blade	1324-0502-7 & W1119		11. 6.02	G.N.Cliffe and G.Marshall	(Winsford)	16. 6.05P
G-CLHD	British Aerospace BAe 146 Series 200	E2023	G-DEBF	2. 5.00	Flightline Ltd	Aberdeen	26. 9.05T
			N165US, N347PS		*(Integrated Aviation Consortium c/s)*		
G-CLIC	Cameron A-105 HAB	2557		18. 4.91	R.S.Mohr	(Corsham)	26. 9.03A
	(New envelope c/n 3395 @ 4.95)				*"Clic Trust"*		
G-CLKE	Robinson R44 Astro	0185	G-HREH	22. 9.98	J.Clarke	(Burnley)	10. 9.06T
			D-HREH		t/a Clarke Business		
G-CLOE	Sky 90-24 HAB	019		11. 3.96	Zebedee Balloon Service Ltd	Newtown Hungerford	12. 6.05A
G-CLOS	Piper PA-34-200T Seneca II	34-7870361	HB-LKE	17. 6.86	P.S.Kirby	Coventry	3.10.04
			N36783				
G-CLOW	Beech 200 Super King Air	BB-821	N821RC	2.11.99	Clowes Estates Ltd	(Ashbourne)	7.11.05T
			TC-DBY, N144TM, F-GDCB				
G-CLRK	Sky 77-24 HAB	101		3. 3.98	William Clark and Son (Parkgate) Ltd	(Dumfries)	25. 3.04A
G-CLUB	Reims/Cessna FRA150M Aerobat	FRA1500347	OO-AWZ	10. 2.83	D.C.C.Handley	Cranfield	24. 6.05T
			F-WZAZ, (F-WZDZ)				
G-CLUE	Piper PA-34-200T Seneca II	34-7970502	N8089Z	15. 9.92	Flyuk.com Ltd	Gloucestershire	4. 4.05T
G-CLUX	Reims/Cessna F172N Skyhawk II	F17201996	PH-AYG (3)	1. 5.80	J.G.Jackman and K.M.Drewitt	Hawarden	20. 8.07T
					t/a J and K Aviation		
G-CMED	SOCATA TB-9 Tampico Club	1867	F-GSZK	19. 3.01	S.C.Brown t/a Enstone Flying Club	Enstone	15. 4.07T
G-CMGC	Piper PA-25-235 Pawnee D	25-7756042	G-BFEX	19.11.91	Midland Gliding Club Ltd	Long Mynd	24. 6.07
			N82525				
G-CMSN	Robinson R22 Beta	1669	G-MGEE	29. 6.04	Kuki Helicopters Ltd	Gamston	31. 3.07T
			G-PHEL, G-RUMP, N2405T				
G-CNAB	Jabiru Aircraft Jabiru UL-450	PFA 274-13651		27. 9.00	W.A Brighouse	Octon Lodge Farm, Thwing	1.10.05P
	(Built W.A Brighouse) (PFA c/n prefix should be "274A")						
G-COAI	Cranfield A 1-400 Eagle	001	G-BCIT	1. 6.98	Cranfield University	Cranfield	
					(Noted 7.99: current status unknown)		
G-COCO	Reims/Cessna F172M Skyhawk II	F17201373	PH-SMO	27.10.80	P.C.Sheard and R.C.Larder	Strubby	27. 3.05
			OO-ADI				

Registration	Type	C/n	Previous identity	Regd	Owner/Operator	Base	CofA
G-CODE	Bell 206B-3 JetRanger III	3850	N222DM	27. 8.96	C.R.Lear	RAF Keevil	26.11.05
			N84TC				
G-COIN	Bell 206B JetRanger II	897	EI-AWA	11. 3.85	Sandra Pool and J.Woodward		
						(Ramsey, Isle of Man/Stevenage)	12. 9.07
G-COLA	Beech F33C Bonanza	CJ-137	G-BUAZ	31. 3.92	J R C Spooner and P M Scarratt		
			PH-BNH			Nottingham East Midlands	3.11.07
G-COLH	Piper PA-28-140 Cherokee	28-23143	G-AVRT	13.10.00	Full Sutton Flying Centre Ltd	Full Sutton	28. 2.05T
			N11C				
G-COLL	Enstrom 280C-UK-2 Shark	1223		17. 8.81	Taylor Air Services Ltd	(Cockermouth)	7. 7.07
G-COLS	Van's RV-7A	PFA 323-14312		6.10.04	C.Terry	(Truro)	
	(Built C.Terry)						
G-COLR*	Colt 69A HAB				See SECTION 3 Part 1	(Aldershot)	
G-COMB	Piper PA-30 Twin Comanche B	30-1362	G-AVBL	14. 9.84	J.T.Bateson	Blackpool	27.11.04
			N8236Y				
G-COMU	Flight Design CT2K	Not known		16. 6.03	Comunica Industries International Ltd	(Winchester)	10. 6.04P
	(Assembled Pegasus Aviation as c/n 7965)						
G-COMP*	Cameron N-90 HAB				See SECTION 3 Part 1	(Aldershot)	
G-CONB	Robin DR400/180 Régent	2176	G-BUPX	14. 4.93	C.C.Blakey t/a CCB Aviation	Thruxton	24. 2.05T
G-CONC	Cameron N-90 HAB	2139		13.11.89	British Airways plc *"Concorde"*	Heathrow	15. 9.05T
G-CONI*	Lockheed 749A-79 Constellation				See SECTION 3 Part 1	Wroughton	
G-CONL	SOCATA TB-10 Tobago	173	F-GCOR	22.12.98	J.M.Huntington	Full Sutton	29. 6.05
G-CONV	Convair 440-54	484	CS-TML	19. 7.01	Atlantic Air Transport Ltd	Coventry	AC
			N357SA, N28KE, N28KA, N4402				
					(Atlantic c/s with Air Atlantique titles - minus engines - in open store 1.05)		
G-COOT	Aerocar Taylor Coot A	EE-1A		16. 9.81	P.M.Napp	(Newcastle upon Tyne)	
	(Built D A Hood)				*(Current status unknown)*		
G-COPS	Piper J-3C-65 Cub (L-4H-PI)	11911	F-BFYC	17. 7.79	R.W.Sproat	Lenox Plunton Farm, Borgue	17. 2.05P
	(Frame No.11739) (Regd as c/n 36-817 - USAAC Contract No) French AF, 44-79615						
G-COPZ	Van's RV-7	PFA 323-14150		8.12.03	R.S.Horan	(Melrose)	
	(Built R.S.Horan)						
G-CORB	SOCATA TB-20 Trinidad	1178	F-GKUX	12. 4.99	G.D.Corbin	Flamstone Park, Bishopstone	2. 5.05
G-CORD	Nipper T.66 RA.45 Series 3	S.129/1676	G-AVTB	21. 3.88	A V.Lamprell	Charity Farm, Baxterley	3. 7.04P
	(Built Slingsby Sailplanes Ltd from c/n S.105/1565)						
G-CORN	Bell 206B-3 JetRanger III	3035	G-BHTR	4. 6.99	John A Wells Ltd	Costock	21. 4.07T
			N18098				
G-CORP	British Aerospace ATP	2037	G-BTNK	2. 3.98	Trident Aviation Leasing Services (Jersey) Ltd		
			N860AW, G-BTNK, G-11-037		*(In open store 2.05)*	Southend	12. 8.04T
G-CORT	Agusta-Bell 206B-3 JetRanger III	8739		21. 6.96	Helicopter Training and Hire Ltd	Newtownards	28. 7.05T
G-COSY	Lindstrand LBL 56A HAB	017		18. 2.93	D.D.Owen	(Wotton-under-Edge)	2. 4.03A
G-COTT	Cameron Flying Cottage 60SS HAB	687	"G-HOUS"	13. 2.81	Dragon Balloon Company Ltd	(Castleton, Hope Valley)	5. 7.04A
G-COUP	Ercoupe 415C	1903	N99280	27. 5.93	S.M.Gerrard	Goodwood	25. 2.07
	(Continental C75)		NC99280		*"Jenny Lin"*		
G-COVE	Jabiru Aircraft Jabiru UL	PFA 274A-13409		23. 7.99	A A Rowson	Cwm	29. 6.04P
	(Built A A Rowson)						
G-COXS	Aeroprakt A22 Foxbat	PFA 317-14168		9. 3.04	S Cox	(Hinkley)	30. 6.05P
	(Built S Cox)						
G-COXY	Air Création Clipper/Kiss 400-582	BMAA/HB/351		11. 3.04	B G Cox	Hunsdon	27. 6.05P
	(Built B G Cox - kit no FL025, Trike T03107, Wing A03186-3184)						
G-COZI	Rutan Cozy	PFA 159-12162		19. 7.93	D.G.Machin	Lydd	7. 3.05P
	(Built D G Machin) (Lycoming O-320)						
G-CPCD	CEA Jodel DR.221 Dauphin	81	F-BPCD	11.12.90	D.J.Taylor	Enstone	10. 7.07
G-CPDA	de Havilland DH.106 Comet 4C	6473	XS235	10. 8.00	C.Walton Ltd	Bruntingthorpe	
G-CPEL	Boeing 757-236	24398	N602DF	24. 8.92	British Airways plc	Heathrow	26.10.05T
			EC-EOL, EC-597, G-BRJE, EC-EOL, EC-278, G-BRJE				
G-CPEM	Boeing 757-236	28665		28. 3.97	British Airways plc	Heathrow	27. 3.06T
G-CPEN	Boeing 757-236	28666		23. 4.97	British Airways plc	Heathrow	22. 4.06T
G-CPEO	Boeing 757-236	28667		11. 7.97	British Airways plc	Heathrow	10. 7.06T
G-CPEP	Boeing 757-2Y0	25268	C-GTSU	16. 4.97	First Choice Airways Ltd	Manchester	9. 7.06T
			EI-CLP, N400KL, XA-TAE				
G-CPER	Boeing 757-236	29113		29.12.97	British Airways plc	Gatwick	28.12.06T
G-CPES	Boeing 757-236	29114		17. 3.98	British Airways plc	Heathrow	16. 3.07T
G-CPET	Boeing 757-236	29115		12. 5.98	British Airways plc	Heathrow	11. 5.07T
G-CPEU	Boeing 757-236	29941		1. 5.99	First Choice Airways Ltd	Manchester	30. 4.05T
G-CPEV	Boeing 757-236	29943	(G-CPEW)	11. 6.99	First Choice Airways Ltd	Manchester	10. 6.05T
G-CPFC	Reims/Cessna F152 II	F15201430		1.12.77	Willowair Flying Club (1996) Ltd	Southend	20. 7.07T
G-CPMK	de Havilland DHC-1 Chipmunk 22	C1/0866	WZ847	28. 6.96	P.A.Walley *(As "WZ847")*	Elstree	24.10.05T
G-CPMS	SOCATA TB-20 Trinidad	1607	F-GNHA	7. 4.98	Charlotte Park Management Services Ltd	Goodwood	13. 6.07T
G-CPOL	Aérospatiale AS355F1 Ecureuil 2	5007	N5775T	30.11.95	MW Helicopters Ltd	(Stapleford)	30. 1.05T
			C-GJJB, N5775T				
G-CPSF	Cameron N-90 HAB	3747	G-OISK	21. 4.99	S.A Simington and J.D.Rigden	(Norwich)	1. 5.04A
G-CPSH	Eurocopter EC135T1	0209	D-HECJ	8. 4.02	Thames Valley Police Authority	Luton	20. 6.05T
G-CPTM	Piper PA-28-151 Cherokee Warrior	28-7715012	G-BTOE	9. 7.91	T.J.Mackay and C.M.Pollett	Woodford	19. 2.07T
			N4264F				
G-CPTS	Agusta-Bell 206B JetRanger II	8556		1. 6.78	A R.B.Aspinall	Skipton	5. 8.06
G-CPXC	CAB CAP.10C	301		11.12.01	J M Wicks	(Braintree)	29. 7.07
G-CRAB	Best Off Skyranger 912	BMAA/HB/246		1.11.02	M.W.Houghton	Deenethorpe	5. 9.05P
	(Built R.A Bell - kit no.SKRxxxx245)						
G-CRAY	Robinson R22 Beta	0919		12. 1.89	F.C.Owen	(Burnley)	16. 3.06P
G-CRDY	Agusta-Bell 206A JetRanger	8112	G-WHAZ	22.12.03	Cardy Construction Ltd	(Manston)	11.11.06T
			OH-HRE, G-WHAZ, OH-HRE				
G-CRES	Denney Kitfox Model 2	PFA 172-11574		7. 6.90	K.M.James	Higher Barn Farm, Houghton	28. 4.05P
	(Built R J Cresswell) (Rotax 912)						
G-CRIB	Robinson R44 Raven	0980	G-JJWL	28. 3.03	D.G.Williams t/a Cribarth Helicopters	Builth Wells	6. 4.07T

G-CRIC Colomban MC-15 Cri-Cri PFA 133-10915 22. 7.83 R.S.Stoddart-Stones (Woldingham, Caterham) 5. 5.99P
(Built A J Maxwell) (JPX PUL.212) *(Current status unknown)*

G-CRIK Colomban MC-15 Cri-Cri PFA 133-13289 10.11.04 A.R.Robinson (Prestbury, Macclesfield)
(Built A.R.Robinson)

G-CRIL Rockwell Commander 112B 521 N1388J 22. 6.79 J.W.Reynolds tr Rockwell Aviation Group Cardiff 14.12.06

G-CRIS Taylor JT.1 Monoplane PFA 55-10318 5. 6.79 C.R.Steer Spilsteads Farm, Sedlescombe
(Built C J Bragg) *(Bare fuselage noted 5.01: current status unknown)*

G-CRLH Bell 206B-3 JetRanger III 4551 G-RJTT 29. 4.03 R.L.Hartshorn Sywell 20.12.04T
C-GRJE

G-CROB Europa Aviation Europa XS PFA 247-13510 25. 4.02 R.G.Hallam Sleap 17. 8.04P
(Built R.G.Hallam - kit no.442) (Jabiru 3300*)*

G-CROL Maule MXT-7-180 Star Rocket 14032C N9232F 24.11.93 N.G.P.Evans Oaksey Park 4. 3.04

G-CROW Robinson R44 Raven 0754 19. 4.00 Longmoore Ltd *(Op FAST Helicopters Ltd)* Shoreham 22. 5.06T

G-CROY Europa Aviation Europa PFA 247-12896 7. 2.97 M.T.Austin Kirkwall 12. 7.05P
(Built A T Croy - kit no.101) (Monowheel u/c)

G-CRPH Airbus A320-231 0424 F-WQBB 10. 4.95 MyTravel Airways Ltd Manchester 14. 4.07T
F-WWIV

G-CRUM Westland Scout AH.1 F9712 XV137 17. 3.98 D.O.Sears Draycott Farm, Chiseldon 28.11.05P
tr G-CRUM Group *(As "XV137")*

G-CRUZ Cessna T303 Crusader T30300004 N9336T 7.12.90 Bank Farm Ltd Bank Farm, Benwick, Cambs 9. 7.06T

G-CSAV Thruster T.600N 450 0032-T600N-064 14. 3.02 Thruster Air Services Ltd Rayne Hall Farm, Braintree 24. 7.03P
(Noted 1.05)

G-CSBM Reims/Cessna F150M F15201359 PH-AYC 24. 5.78 Halegreen Associates Ltd Hinton-in-the-Hedges 17.12.06T

G-CSCS Reims/Cessna F172N Skyhawk II F17201707 PH-MEM 28.11.86 Cheryl Sullivan Stapleford 10. 6.05T
(PH-WEB), N9899A

G-CSDJ Jabiru Aircraft Jabiru UL PFA 274A-13337 23. 3.99 D W, and J Johnston, C D and S Slater Kemble 22. 5.05P
(Built D W Johnston and C D Slater)

G-CSFC Cessna 150L 15075360 (G-BFLX) 21. 3.78 I.G.McDonald RNAS Culdrose 6. 3.05
N11370 tr Foxtrot Charlie Flying Group

G-CSFD Ultramagic M-90 HAB 90/56 12.12.02 L.A Watts (Pangbourne, Reading) 28.12.04A

G-CSFT* Piper PA-23-250 Aztec D 27-4521 G-AYKU 20. 9.84 Not known North Weald 3.12.94T
N13885 *(Cancelled 5.6.96 as WFU) (Gutted fuselage, wings dumped in grass 1.05)*

G-CSIX Piper PA-32-300 Cherokee Six 32-7840030 ZS-OMX 15. 6.01 G.A Ponsford Goodwood 19. 7.07
Z-WJM, VP-WJM, HB-PCX, ZS-KBR, N9857K *(Op CEGA)*

G-CSMK Aerotechnik EV-97 Eurostar PFA 315-13813 4.12.01 R.Frey Leicester 4. 4.05P
(Built N R Beale)

G-CSNA Cessna 421C Golden Eagle III 421C0677 (D-IOSS) 11. 6.79 Blue Swan Aviation Ltd Lee-on-Solent 18. 8.05T
N26522

G-CSWH Piper PA-28R-180 Cherokee Arrow 28R-30541 N4647J 5. 4.02 The Gould Group International Ltd (Burton-on-Trent) 9. 9.05T
G-CSWH, N4647J

G-CSWL Bell 206L-1 Long Ranger II 45565 G-VOLK 6. 5.97 G.Webb (Stoke-on-Trent) 10.8.06T
G-GBAY, G-CSWL, G-SIRI, G-CSWL, F-GDAD

G-CTAV Aerotechnik EV-97 teamEurostar UK 2129 25.11.04 C.M.Theakstone Sywell 21.12.05P

G-CTCL SOCATA TB-10 Tobago 1107 G-BSIV 16. 7.90 Gift Aviation Ltd Gamston 30. 9.05T

G-CTCT Flight Design CT2K 00-04-04-94 G-69-51 26. 6.00 Cyclone Airsports Ltd Elm Tree, Park, Manton AC
t/a Pegasus Aviation *(Noted 7.04)*

G-CTDH Flight Design CT2K Not known 1. 5.03 D.Haygreen (Rhos-on-Sea) 6. 9.05P
(Assembled Pegasus Aviation as c/n 7939)

G-CTEC Stoddard-Hamilton GlaStar PFA 295-13260 9.11.99 B.N.C.Mogg (Bibberne Farm, Stalbridge)
(Built A J Clarry) *(Current status unknown)*

G-CTEL Cameron N-90 HAB 3933 27. 8.96 M.R.Noyce and A P.Kettle (Alresford) 25. 4.05A
tr G-CTEL Flying Group

G-CTFF Cessna T206H Turbo Stationair T20608150 N24309 29.10.01 Rajair Ltd (Ellesmere) 4 11.07E

G-CTGR Cameron N-77 HAB 1775 G-CCDI 28. 8.97 T.G.Read *(Charles Church titles)* (Knutsford) 13. 9.03T

G-CTIO SOCATA TB-20 Trinidad GT 2174 F-OIMH 7.11.02 Cityiq Ltd Biggin Hill 6.11.05T

G-CTIX Vickers Supermarine 509 Spitfire T.IX Not known N462JC 9. 4.85 A A Hodgson Bryn Gwyn Bach 19. 5.05P
(Major rebuild from parts pre 1994) G-CTIX, IDFAF 2067, 0607, MM4100, PT462 *(As "PT462/SW-A")*

G-CTKL Noorduyn AT-16 Harvard IIB 07-30 (G-BKWZ) 22.11.83 M.R.Simpson North Weald 26. 5.04P
(C/n also quoted as "76-80") MM54-137, RCAF3064 *(As "FE788" in RAF c/s) (Noted 1.05)*

G-CTLA Airbus A321-211 1887 D-AVZC 11. 2.03 Mytravel Airways AS Copenhagen, Denmark 10. 2.06T

G-CTOY Denney Kitfox Model 3 PFA 172-12150 14.10.91 B.McNeilly Newtownards 10. 5.93P
(Built G S Cass and G C Brooke - c/n 1176) *(Current status unknown)*

G-CTRL Robinson R22 Beta II 3601 13. 5.04 Central Helicopters Ltd Tollerton 3. 6.07T

G-CTUG Piper PA-25-235 Pawnee 25-4448 N4713Y 13. 9.04 The Borders (Milfield) Gliding Club Ltd Milfield 27.10.07E

G-CTWW Piper PA-34-200T Seneca II 34-7970191 G-ROYZ 21. 7.93 Centreline Air Charter Ltd Southend 19. 1.06T
G-GALE, N3052X *(Op Fly Now Air Charter)*

G-CTZO SOCATA TB-20 Trinidad GT 2166 F-OIME 7.10.02 Trinidad Hire Ltd Turweston 6.10.05T

G-CUBB Piper PA-18-150 Super Cub (L-18C-PI) 18-3111 PH-WAM 5.12.78 Bidford Airfield Ltd Bidford 18. 4.07
(Lycoming O-360-C2) *(Frame No.18-3009)* Belgian AF OL-L37, 53-4711

G-CUBE Best Off Skyranger 912 BMAA/HB/336 13. 1.04 T.R.Villa Priory Farm, Tibenham 6. 7.05P
(Built T.R.Villa - kit no SKR0311409) *(F/f 6.6.04)*

G-CUBI Piper PA-18-125 Super Cub 18-3181 PH-GAV 26. 2.79 G.T.Fisher (Northside, Thorney) 4.11.94T
(L-18C-PI) PH-VCV, R.Neth AF R-83, Belgian AF L-107, 53-4781
(Official c/n 18-559 related to PH-GAV prior to 1970 rebuild when it incorporated Frame No.18-3170 from PH-VCV: possible link to G-CCKW qv)

G-CUBJ Piper PA-18-150 Super Cub 18-2036 PH-MBF 15.12.82 P B Rice Breighton 9.11.06
(L-18C-PI) *(Frame No.18-2035)* PH-NLF, R.Neth AF R-43, 8A-43, 52-2436 *(As "18-5395/CDG" in French Army c/s)*
(Regd with c/n 18-5395 after 1974 rebuild of PH-NLF: acquired data plate from, and took identity of, PH-MBF - note G-SUPA also carries this c/n)

G-CUBP Piper PA-18-150 Super Cub 18-8482 N1136Z 8. 8.96 D.W.Berger Trenchard Farm, Eggesford 29. 4.05
(Frame No.18-8725) G-BVMI, OH-PIN, N4262Z
(Regd with c/n 18-8823 the "official" identity of N1136Z/D-EIAC: rebuilt 1984/85 with Frame No.18-4613 ex D-EKAF: this frame fitted to G-BVMI following accident on 15.8.95: repaired frame of G-BVMI has now become G-CUBP)

G-CUBS Piper J-3C-65 Cub "17792" G-BHPT 26.10.01 S.M.Rolfe t/a Sunbeam Aviation Willington, Bedford 13. 7.05P
(Frame No.17792) F-BSGQ, LX-AIH, N70688, NC70688
(Quoted p/i is suspect - possibly c/n 18105 ex NC71076/N71076)

Regn	Type	C/n	Previous identity	Regd	Owner	Base	CofA
G-CUBW	Wag-Aero Acro Trainer *(Built B.G, N.D.Plumb and A G. Bourne)*	PFA 108-13581		26.11.02	B.G, N.D.Plumb and A G. Bourne	Hinton-in-the-Hedges	
G-CUBY	Piper J-3C-65 Cub *(Rebuilt with new fuselage 1996/97)*	16317	G-BTZW N88689, NC88689	2. 3.95	Claudine A Bloom	Kittyhawk Farm, Deanland	24. 6.05P
G-CUCU	Colt 180A HAB	3869		22. 4.96	S.R.Seager	(Aylesbury)	2. 5.05T
G-CUIK	QAC Quickie Q200 *(Built C.S.Rayner)*	PFA 094A-11204		15.12.04	C.S.Rayner	(Faringdon)	
G-CUPS	IAV-Bacau Yakovlev Yak-52	9010312	LY-AMD Ukraine AF 09 *(yellow)*, DOSAAF 09 *(yellow)*	13. 6.03	L.R.Haunch t/a Fenland Flying School	Fenland	29. 6.05P
G-CURE*	Colt 77A HAB				See SECTION 3 Part 1	(Aldershot)	
G-CURR	Cessna 172R Skyhawk	17280143	G-BXOH N9989F	27. 5.98	JS Aviation Ltd	Booker	18. 5.07T
G-CURV	Avid Speed Wing *(Built K S Kelso)*	PFA 189-12169		28. 3.00	K.S.Kelso	(Cambridge)	
G-CUTE	Dyn'Aéro MCR-01 Club *(Built E.G.Shimmin - kit no.132)* (Rotax 912-ULS)	PFA 301-13511		7. 9.99	E.G.Shimmin	Shobdon	20.12.05P
G-CUTY	Europa Aviation Europa *(Built D.J. and M.Watson - kit no.224) (Tri-gear u/c)*	PFA 247-12910		20. 8.96	D.J.and M.Watson *(Current status unknown)*	(Selby)	
G-CVBF	Cameron A-210 HAB	3588		2. 6.95	Airxcite Ltd t/a Virgin Balloon Flights	(Wembley)	15. 8.01T
G-CVIP	Bell 206B-3 JetRanger III	3228	SX-HDJ N824C, N824H, N3902L	29. 4.02	Sloane Helicopters Ltd	Sywell	30. 4.05T
G-CVIX	de Havilland DH.110 Sea Vixen D.3 *(Regd as FAW.2 with c/n 10132)*	10125	XP924	26. 2.96	de Havilland Aviation Ltd *(Red Bull titles)*	Bournemouth	29. 5.05P
G-CVLH	Piper PA-34-200T Seneca II	34-8070332	F-GCPK N8252D, N8250H	5. 9.02	Atlantic Aviation Ltd	St Brieuc, France	12. 9.05T
G-CVPM	VPM M-16 Tandem Trainer (Arrow GT1000R)	VPM16-UK-110		26. 3.98	C.S.Teuber	(Hannover, Germany)	2. 6.05P
G-CVST	Jodel D.140 Mousquetaire *(Built A Shipp)*	PFA 251-13384		21. 5.03	A Shipp	(Hull)	
G-CWAG	Sequoia F.8L Falco *(Built C C Wagner)* (Lycoming O-320)	PFA 100-10895		11. 5.92	D.R.Austin	(High Cross, Ware)	5. 4.05P
G-CWAL	Raj Hamsa X'Air 133 *(Built C.Walsh - kit no.777)*	BMAA/HB/339		27. 4.04	C.Walsh	Roche, Cornwall	
G-CWBM	Phoenix Currie Wot *(Built B.V.Mayo)* (Continental C85)	PFA 3020	G-BTVP	28. 3.94	B.V.Mayo	Maypole Farm, Chislet	30. 9.05P
G-CWFA	Piper PA-38-112 Tomahawk	38-78A0120	G-BTGC N9507T	17. 8.99	Cardiff-Wales Flying Club Ltd	Cardiff	1. 6.07T
G-CWFB	Piper PA-38-112 Tomahawk	38-78A0623	G-OAAL N4471E	13. 1.00	P.M.Moyle	Cardiff	8. 2.07T
G-CWFD	Piper PA-38-112 Tomahawk	38-79A0038	G-BSVY N2396B	10. 8.00	M.Lowe and K.Hazelwood	Plymouth	15. 1.07T
G-CWFE	Piper PA-38-112 Tomahawk	38-80A0020	G-BPBR N25082, N9652N	29.11.01	Cardiff Wales Flying Club Ltd	Cardiff	30. 6.01T
G-CWFY	Cessna 152 II	15284639	G-OAMY N6214M	13. 1.00	Cardiff Wales Aviation Services Ltd	Cardiff	14.11.03T
G-CWFZ	Piper PA-28-151 Cherokee Warrior	28-7715131	G-CPCH G-BRGJ, (G-BPGP), N5425F	27.10.99	Cardiff Wales Flying Club Ltd	Cardiff	6. 6.03T
G-CWIC	Mainair Pegasus Quik	8067		21. 9.04	R.Rainey	Newtownards	5.10.05P
G-CWIK	Mainair Pegasus Quik	8018		23. 4.04	L.Kirk	Easter Poldar Farm, Thornhill	29 .4.05P
G-CWOT	Phoenix Currie Wot *(Built D A Lord)* (Walter Mikron 2)	PFA 3019		31. 1.78	D.Doyle and Helena Duggan *(Noted 8.04)*	Kilrush, County Kildare	19. 1.04P
G-CWTD	Aeroprakt A22 Foxbat *(Built J.V.Harris)*	PFA 317-14131		21.10.03	J.V.Harris	Ley Farm, Chirk	17. 2.05P
G-CWVY	Mainair Pegasus Quik	7984		29. 9.03	I.A Gaetan	Enstone	28.10.05P
G-CXCX	Cameron N-90 HAB *(Replacement envelope c/n 3332)*	1242		14. 3.86	Cathay Pacific Airways (London) Ltd *"Cathay Pacific IV"*	(Swindon)	6. 7.03A
G-CXDZ	Cassutt Speed Two *(Build Whinsper, J.A H.Chadwick and Thompson)*	PFA 34-13816		27.12.02	J.A H.Chadwick *(Noted Goodwood 9.04)*	(London N1)	
G-CXHK	Cameron N-77 HAB	4978		22. 2.01	Cathay Pacific Airways (London) Ltd	(London SW1)	6. 3.05A
G-CXIP	Thruster T600N Jab Sprint	1031-T600N-095		17.11.03	R.J.Howells t/a India Papa Syndicate	Shobdon	16.11.04P
G-CYLS	Cessna T303 Crusader	T30300005	N20736 G-BKXI, N303CC, (N9355T)	20.12.90	Gledhill Water Storage Ltd	Blackpool	15. 4.06
G-CYMA	Gulfstream GA-7 Cougar	GA7-0083	G-BKOM N794GA	15. 8.83	Cyma Petroleum (UK) Ltd	Elstree	30. 6.07
G-CYRA	Kolb Twinstar Mk.3 *(Built S J Fox and A P Pickford)* (Rotax 503)	PFA 205-12434	G-MYRA	30. 1.03	S.J.Fox	Popham	10.10.05P
G-CZAC	Zenair CH601XL Zodiac *(Built D Pitt)*	PFA 162B-14113		26. 3.04	D Pitt	(Wickham, Fareham)	
G-CZAG	Sky 90-24 HAB	171		5.10.99	S.McCarthy	(Rothersthorpe, Northampton)	9. 4.05
G-CZBE	CFM Streak Shadow *(Built N J Bushell - kit no K.271)* (Rotax 618)	PFA 206-12905	G-MZBE	17. 6.03	M.I.M.Smith	Charterhall	30.11.05P
G-CZCZ	Avions Mudry CAP.10B	54	OE-AYY F-WZCG, HB-SAK, F-BUDT	28. 7.94	P.R.Moorhead and M.Farmer	Garston Farm, Marshfield	21. 8.06
G-CZMI	Best Off Skyranger 912 *(Built T.W.Thiele - kit no.SKR0308377)*	BMAA/HB/307		18.11.03	T.W.Thiele	Newnham, Baldock	13. 1.05P
G-CZNE	Pilatus Britten-Norman BN-2B-20 Islander	2301	G-BWZF	27. 7.04	G.Davies	Blackbushe	26. 9.05T

G-DAAA - G-DZZZ

Regn	Type	C/n	Previous identity	Regd	Owner	Base	CofA
G-DAAH	Piper PA-28RT-201T Turbo Arrow IV	28R-7931104	N3026U	27. 4.79	R.Peplow	Wolverhampton	24. 5.06
G-DAAM	Robinson R22 Beta	2043		3. 6.92	Hecray Co Ltd t/a Direct Helicopters (Op Bournemouth Helicopters)	Bournemouth	20. 7.07T
G-DAAT	Eurocopter EC135T2	0312		4. 5.04	Bond Air Services Ltd *(Op Devon Air Ambulance)*	Exeter	13. 7.07T

G-DAAZ	Piper PA-28RT-201T Turbo Arrow IV	28R-7931247	N2896B	17. 1.03	Calais Ltd	Guernsey	23. 3.06
G-DABS	Robinson R22 Beta II	3083		15. 5.00	B.Seymour	(Middlesbrough)	1. 6.06
G-DACA	Percival P.57 Sea Prince T.1	P57/12	WF118	6. 5.80	P.G.Vallance Ltd	Charlwood, Surrey	17. 7.81P
					(Gatwick Aviation Museum as "WF118")		
G-DACC	Cessna 401B	401B-0112	N77GR	1. 9.86	Niglon Ltd	Coventry	12. 3.05
			N4488A, G-AYOU, N7972Q				
G-DACF	Cessna 152 II	15281724	G-BURY	13. 6.97	T.M.and M.L.Jones	Egginton	25. 9.06T
			N67285		*(Op Derby Aero Club)*		
G-DADG	Piper PA-18-150 Super Cub	18-5237	N45498	13.10.04	F.J.Cox	Eaglescott	AC
			IDFAF 069				
G-DAEX	Dassault Falcon 900EX	78	F-WWFR	22. 2.01	Triair (Bermuda) Ltd	Farnborough	22. 2.04T
G-DAFY	Beech 58 Baron	TH-1591	N5684C	6.10.93	P.R.Earp	Gloucestershire	5.12.04
G-DAIR	Luscombe 8A Master	1474	G-BURK	3.10.97	D.F.Soul	Standalone Farm, Meppershall	19.10.99P
	(Diesel Air 100hp)		N28713, NC28713				
G-DAIV	Ultramagic H-77 HAB	77/184		2.11.00	D.Harrison-Morris	(Ellesmere)	
G-DAJB	Boeing 757-2T7	23770		26. 2.87	Monarch Airlines Ltd	Luton	13. 5.05T
G-DAJC	Boeing 767-31K	27206		15. 4.94	MyTravel Airways Ltd	Manchester	14. 4.06T
G-DAJW*	K & S Jungster I	PFA 1517		20.11.78	Not known *(Cancelled 2.9.91 by CAA)*	Tranent	
G-DAKK	Douglas C-47A-35DL Skytrain	9798	(G-OFON)	26. 7.94	General Technics Ltd	Lee-on-Solent	23. 5.03T
			F-GEOM, Fr Navy 36, OK-WZB, OK-WDU, 42-23936				
G-DAKO	Piper PA-28-236 Dakota	28-7911187	PH-ARW	29. 7.99	Methods Application Ltd	Denham	5.11.05T
			(PH-MFB), D-EECG, PH-ARW, OO-HCX, N29718				
G-DAMY	Europa Aviation Europa	PFA 247-12781		21.10.94	Ursula A.Schliessler and R.J.Kelly	(London NW11	6. 7.05P
	(Built Hart Aviation Ltd - kit no.105) (Tri-gear u/c)						
G-DANA	Jodel DR.200 rep	PFA 304-13351	G-DAST	2.12.02	F.A Bakir	(Sale)	
	(Built F.A Bakir)						
G-DAND	SOCATA TB-10 Tobago	72		5.12.79	Portway Aviation Ltd	Shobdon	20. 9.04
G-DANT	Rockwell Commander 114	14298	N4978W	9. 7.96	D.P.Tierney	Biggin Hill	10. 3.07
G-DANY	Jabiru Aircraft Jabiru UL	PFA 274A-13588		28.12.00	D.A Crosbie	(Sudbury)	
	(Built D.A Crosbie)						
G-DANZ	Eurocopter AS355N Ecureuil 2	5658		14. 9.98	Frewton Ltd	Oxford	9. 2.05T
G-DAPH	Cessna 180K Skywagon II	18053016	N2620K	29. 1.92	M.R.L.Astor	East Hatley, Tadlow	9. 9.06
G-DARA	Piper PA-34-220T Seneca III	34-8333060	PH-TCT	8.11.88	Sys (Scaffolding Contractors) Ltd	Gamston	3. 4.07
			N83JR, N4297J, N9632N				
G-DARK	CFM Shadow Series DD	PFA 161-13308		13. 7.00	R.W.Hussey	Old Sarum	20. 8.05P
	(Built P M Dewhurst - kit no K.295)						
G-DASH	Rockwell Commander 112A	237	G-BDAJ	31. 3.87	D.and M.Nelson	Bourn	26. 3.06
			N1237J				
G-DASU	Cameron V-77 HAB	2300		6. 4.90	D and L. S Litchfield *"Borne Free"*	(Reading)	11. 8.97A
G-DATE	Agusta A109C	7633	G-RNLD	30. 3.00	Rich and Lucky Aviation Ltd	(Derby)	8. 8.05
			I-ANAG				
G-DATG	Reims/Cessna F182P Skylane II	F18200013	D-EATG	8.11.01	Oxford Aeroplane Co Ltd	Oxford	14. 3.05T
G-DATH	Aerotechnik EV-97 Eurostar	PFA 315-13967		8.10.02	D.N.E.D'Ath	Sackville Lodge Farm, Riseley	21. 3.05P
	(Built D.N.E.D'Ath)						
G-DAVD	Reims/Cessna FR172K Hawk XPII	FR17200632	D-EFJT	23.12.99	D M Driver	Conington	11. 3.06T
			(PH-ADL), PH-AXO				
G-DAVE	Jodel D.112	667	F-BICH	16. 8.78	D.A Porter	Griffins Farm, Temple Bruer	23. 8.05P
	(Built Etablissement Valladeau)						
G-DAVG	Robinson R44 Raven II	10038	G-WOWW	28. 4.03	D.A Gold	(London W8)	25. 2.06T
G-DAVO	Gulfstream AA-5B Tiger	AA5B-1226	G-GAGA	5. 1.96	Douglas Head Consulting Ltd	Elstree	12. 2.07T
			G-BGPG, (G-BGRW)				
G-DAVT	Schleicher ASH 26E	26090		24. 4.96	H.Bill and C.Hansen	(Baekke/Vejen, Denmark)	10. 5.05
G-DAWG	Scottish Aviation Bulldog Series 120/121	BH120/208	XX522	13. 3.02	R.H.Goldstone *(As "XX522/06")*	(Worsley)	2.10.06
G-DAWS	Sikorsky S-61N Mk.II	61-824	G-LAWS	22.11.04	Air Harrods Ltd	Stansted	9. 7.05E
			G-BHOF, LN-ONK, G-BHOF, LN-ONK, G-BHOF				
G-DAYI	Europa Aviation Europa	PFA 247-13027		19. 8.96	A F.Day	(West Wickham)	
	(Built A F.Day - kit no.298) (Monowheel u/c)				*(Current status unknown)*		
G-DAYS	Europa Aviation Europa	PFA 247-12810		9. 5.95	D.J.Bowie	Sleap	9. 8.05P
	(Built S D, A and A J Hall - kit no.177) (Monowheel u/c)						
G-DAYZ	Pietenpol AirCamper	PFA 47-12342		22. 6.01	J.G.Cronk	(Chichester)	
	(Built J G Cronk)						
G-DAZY	Piper PA-34-200T Seneca II	34-7770335	N953A	4. 2.03	Centreline Air Charter Ltd	Bristol	23. 2.06T
			PH-DLM, OE-FGG, N38727				
G-DAZZ	Van's RV-8	PFA 303-14245		20.10.04	D.M.Hartfree-Bright	(Farnham)	
	(Built D.M.Hartfree-Bright)						
G-DBAT	Lindstrand LBL 56A HAB	1001		9. 6.04	G.J.Bell	(Petersfield)	22. 8.05A
G-DBCA	Airbus A319-131	2098	D-AVYV	23. 2.04	British Midland Airways Ltd	Nottingham East Midlands	22. 2.07T
G-DBCB	Airbus A319-131	2188	D-AVYA	23. 4.04	British Midland Airways Ltd	Nottingham East Midlands	22. 4.07T
G-DBCC	Airbus A319-131	2194	D-AVYT	14. 5.04	British Midland Airways Ltd	Nottingham East Midlands	13. 5.07T
G-DBCD	Airbus A319-131	2389	D-AVYJ	9. 2.05	British Midland Airways Ltd	Nottingham East Midlands	
					(Delivered 9.2.05)		
G-DBCE	Airbus A319-131	2429	D-AVWG	.05R	British Midland Airways Ltd	Nottingham East Midlands	
G-DBCF	Airbus A319-131	2488		.05R	British Midland Airways Ltd	Nottingham East Midlands	
G-DBOY	Agusta A109C	7622	N621MM	29.10.04	Maison Air Ltd	Fairoaks	24.11.07E
			HB-ZEE, OE-XSG, N67SH, 9M-SJI, TUDM M38-03, 9M-TMJ				
G-DBUG	Robinson R44 Clipper	1256	G-OBHI	4. 3.04	Pietas Ltd	Cambridge	1.10.05T
G-DBYE	Mooney M.20M	27-0098	N91462	24. 3.98	A J.Thomas	Cranfield	6. 5.07
G-DCAV	Piper PA-32R-301 Saratoga IIHP	3246075	N92864	8. 5.97	Airquest Aviation LLP	Elstree	15. 5.06T
			G-DCAV				
G-DCDB	Bell 407	53137	C-FCDB	19.10.99	Paycourt Ltd	Knocksedan, County Dublin	19.10.05T
			N7238A				
G-DCEA	Piper PA-34-200T Seneca II	34-8070079	N3567D	13. 2.91	Barnes Olson Aero Leasing Ltd	Bristol	16. 7.06T
					(Op Bristol Flying Centre)		

Reg	Type	C/n	Previous identity	Reg date	Owner/Operator	Base	CofA
G-DCKK	Reims/Cessna F172N Skyhawk II	F17201589	PH-GRT PH-AXA	19. 5.80	J.Maffia	North Weald	12. 8.07T
G-DCMI	Mainair Pegasus Quik	7972		7. 8.03	S.J.E.Smith	Eshott	6. 8.05P
G-DCPA	MBB BK-117C-1C	7511	D-HECU D-HXXL, G-LFBA, D-HECU, D-HMBF	16.12.97	Devon and Cornwall Constabulary	Exeter	16. 6.05T
G-DCSE	Robinson R44 Astro	0659		23. 9.99	Foxtrot Golf Helicopters Ltd	Edinburgh	13.10.05T
G-DCXL	SAN Jodel D.140C Mousquetaire III	101	F-BKSM	27. 5.88	C.F.Mugford tr X-Ray Lima Group	Little Gransden	7. 5.06
G-DDAY	Piper PA-28R-201T Turbo Arrow III	28R-7703112	G-BPDO N3496Q	24.11.88	K.E.Hogg tr G-DDAY Group	Tatenhill	15. 4.07
G-DDBD	Europa Aviation Europa XS *(Built B.Davies - kit no.454)*	PFA 247-13569		11. 2.03	B.Davies	(Horsham)	
G-DDMV	North American T-6G-NF Texan	168-313	N3240N Haitian AF 3209, 49-3209	30. 4.90	E.A Morgan *(As "493209" in Californian ANG c/s)*	Gloucestershire	6. 4.06
G-DDOG	Scottish Aviation Bulldog Series 120/121	BH120/210	XX524	18. 6.01	Deltaero Ltd *(As "XX524/04")*	(London SW1)	28. 5.06
G-DEAN	Solar Wings Pegasus XL-Q	SW-TE-0117 & SW-WQ-0123	G-MVJV	30.11.98	A J Hodson	(Sissinghurst, Cranbrook)	17. 8.05P
G-DEBE	British Aerospace BAe 146 Series 200	E2022	N163US N346PS	5. 8.96	Flightline Ltd	Southend	6. 8.05T
G-DEBR	Europa Aviation Europa *(Built A J Calvert and C T Smallwood - kit no.232) (Tri-gear u/c)*	PFA 247-12922		31. 1.01	A J.Calvert and C.T.Smallwood	(Buxton/Ripley)	
G-DEBT	Alpi Pioneer 300 *(Built N.J.T.Tonks)*	PFA 330-14291		5.10.04	N.J.T.Tonks	(Wigmore, Leominster)	
G-DECK	Cessna T210N Turbo Centurion II	21064017	N958MK D-ERDK, N4834Y	29. 2.00	R.J.Howard	Sherburn-in-Elmet	23. 2.06
G-DECO	Dyn'Aéro MCR-01 Club *(Built A.W.Bishop and G.Castelli)*	PFA 301A-14246		15. 9.04	A.W.Bishop and G.Castelli	(Cambridge)	
G-DEER	Robinson R22 Beta II	2827		17. 7.98	Westinbrook Ltd *(Op FAST Helicopters Ltd)*	Shoreham	28. 7.07T
G-DEFM	British Aerospace BAe 146 Series 200	E2016	G-DEBM C-FHAZ	22.10.99	Avtrade Leasing Ltd *(In open store 2.05)*	Southemd	16 .3.05T
G-DEFY	Robinson R22 Beta II	3633	N73750	5. 7.04	P.M.M.P.Silveira	(Lisboa, Portugal)	26. 6.07T
G-DEKA	Cameron Z-90 HAB	10665	16.12.04		Sport Promotion srl	(La Morra, Italy)	
G-DELB*	Robinson R22 Beta				See SECTION 3 Part 1	Doncaster	
G-DELF	Aero L-29A Delfin	194555	ES-YLM Soviet AF 12 (Red)	28. 8.97	B.R.Green *"12"*	Manston	26. 9.01P
G-DELT	Robinson R22 Beta *(Port skid struck ground on lift-off Coventry 16.10.03 and rolled onto left side: damaged beyond economic repair)*	0898		11.11.88	Flightworks (Midlands) Ltd	Coventry	22. 4.05T
G-DEMH	Reims/Cessna F172M Skyhawk II (Lycoming O-360)	F17201137	G-BFLO PH-DMF, (EI-AYO)	18.11.91	M.Hammond	Airfield Farm, Hardwick	15. 6.07
G-DEMM	Eurocopter AS350B2 Ecureuil	3741		6.10.03	Abbeyflight Ltd	(Addlestone)	15.12.06T
G-DENA	Cessna F150G *(Built Reims Aviation SA)*	F150-0204	G-AVEO EI-BOI, G-AVEO	14.12.95	W.M.Wilson and R.Campbell	Sandtoft	20. 1.02T
G-DENB	Cessna F150G *(Built Reims Aviation SA)*	F150-0136	G-ATZZ	14.12.95	D.C.Tissington	Standalone Farm, Meppershall	23. 3.07T
G-DENC	Cessna F150G *(Built Reims Aviation SA)*	F150-0107	G-AVAP	14.12.95	M.Dovey *(www.mdaeroservices.com.titles)*	Manor Farm, Glatton	29. 9.05T
G-DEND	Reims/Cessna F150M	F15001201	G-WAFC G-BDFI, (OH-CGD)	6. 6.97	R.N.Tate	Durham Tees Valley	13.10.07E
G-DENE	Piper PA-28-140 Cherokee	28-21710	G-ATOS N11C	5. 2.98	Avon Aviation Ltd t/a The Bristol and Wessex Aeroplane Club	Bristol	11. 7.05T
G-DENI	Piper PA-32-300 Cherokee Six	32-7340006	G-BAIA N11C	7.12.95	A Bendkowski	Rochester	3. 6.07T
G-DENR	Reims/Cessna F172N Skyhawk II	F17201839	G-BGNR	30. 4.97	I.D.Gumbrell t/a Dorset Flying Club	Bournemouth	9.10.05T
G-DENS	Binder CP.301S Smaragd *(Also carries c/n AB.429 denoting completion as Amateur Build)*	121	D-ENSA	20.11.85	G.E.Roe and I.S.Leader tr Garston Smaragd Group	Garston Farm, Marshfield	16. 9.04P
G-DENT	Cameron N-145 HAB	4135		8. 4.97	Horizon Ballooning Ltd	(Alton)	27. 5.05T
G-DENZ	Piper PA-44-180 Seminole	44-7995327	G-INDE G-BHNM, N8077X	3. 7.97	W.J.Greenfield	Humberside	3. 4.05T
G-DERB	Robinson R22 Beta	1005	G-BPYH	28. 6.95	S Thompson	(Coventry)	22. 9.05T
G-DERI	Piper PA-46-500TP Malibu Meridian	4697078	G-PCAR N51151	9.11.03	Intesa Leasing SpA	(Milano, Italy)	30. 7.07T
G-DERK	Piper PA-46-500TP Malibu Meridian	4697152	N165MA	3. 4.03	D.Priestley	Dunkeswell	6. 5.06T
G-DERV	Cameron Truck 56SS HAB	1719		21. 3.88	J.M.Percival *"Shell UK Truck"*	(Loughborough)	22. 2.00A
G-DESS	Mooney M.20J Model 201	24-1272	N11598	20.10.87	R.M.Hitchin *(Badly damaged landing near Manor Farm, Wadswick 14.11.04)*	Old Sarum	18. 5.06
G-DEST	Mooney M.20J	24-3429		6.11.98	Allegro Aviation Ltd	(Guernsey)	12. 1.05
G-DEVL	Eurocopter EC120B Colibri	1273		7. 6.02	Swift Frame Ltd	Norwich	4. 7.05T
G-DEVN*	de Havilland DH.104 Devon C.2/2				See SECTION 3 Part 1	Merseburg, Germany	
G-DEVS	Piper PA-28-180 Cherokee B	28-830	G-BGVJ D-ENPI, N7066W	5. 3.85	B.J.Hoptroff and J.M.Whiteley tr 180 Group	Blackbushe	23. 1.05
G-DEXP	ARV Aviation ARV-1 Super 2	003 & PFA 152-11154		24. 4.85	W.G.McKinnon	Perth	28. 4.05P
G-DEZC	British Aerospace HS 125 Series 700B	257070	G-BWCR G-5-604, HB-VGG, G-5-604, HB-VGG	28. 5.96	Bunbury Aviation Ltd	Gloucestershire	2. 8.05T
G-DFKI	Westland SA.314C Gazelle HT.2	1216	G-BZOT XW907	12. 2.02	Foremans Aviation Ltd	Full Sutton	1. 1.05P
G-DFLY	Piper PA-38-112 Tomahawk	38-79A0450	N9655N	15. 2.79	P.M.Raggett	Rochester	13. 6.06
G-DGCL	DG Flugzeugbau DG-800B	8-185-B109		27. 3.00	C.J.Lowrie *"CL"*	Parham Park	19. 4.06
G-DGHD	Robinson R44 Raven II	10436		20. 7.04	Ramsgill Aviation Ltd	Sherburn-in-Elmet	24. 8.07T
G-DGHI	Dyn'Aéro MCR-01 Club *(Built D.G.Hall)*	PFA 301A-14128		11.12.03	D.G.Hall	Bourn	18. 8.05P
G-DGIV	DG Flugzeugbau DG-800B	8-145-B69		27.11.98	W.R.McNair	(Holywood, County Down)	12.12.07E
G-DGWW	Rand Robinson KR-2 *(Built W Wilson)* (Hapi Magnum 75)	PFA 129-11044		7. 3.91	W.Wilson	Liverpool	9. 8.05P

G-DHCB	de Havilland DHC-2 Beaver 1 *(Floatplane)*	1450	G-BTDL XP779	20. 6.91	Seaflite Ltd *(Stored 2001)*	(Lochearnhead)	16. 9.97T
G-DHCC	de Havilland DHC-1 Chipmunk 22	C1/0393	WG321	28. 5.97	Eureka Aviation NV *(As "WG321")*	Antwerp, Belgium	24. 9.06T
G-DHDV	de Havilland DH.104 Dove 8	04205	VP981	26.10.98	Air Atlantique Ltd *(As "VP981" in 'Royal Air Force Transport Command' titles)*	Coventry	9. 5.05T
G-DHJH	Airbus A321-211	1238	D-AVZL	7. 6.00	MyTravel Airways Ltd	Manchester	6. 6.06T
G-DHLB	Cameron N-90 HAB	3261		20. 4.94	B.A Bower	(Seaton)	4. 4.05A
G-DHLI	Colt World 90 SS HAB	2603		2. 6.94	A D.Kent tr Balloon Preservation Flying Group *"DHL World" (New CofR 6.04)*	Lancing	17.12.98A
G-DHLZ*	Colt 31A Air Chair HAB				See SECTION 3 Part 1	(Aldershot)	
G-DHPM	de Havilland DHC-1 Chipmunk T.20 *(Built OGMA)*	55	CS-AZS Portuguese AF FAP 1365	28. 3.02	P.Meyrick *(As "1365" in Portuguese AF c/s)*	Sywell	27. 5.05
G-DHSS	de Havilland DH.112 Venom FB.50 (FB.1) *(Built F + W)*	836	Swiss AF J-1626	26. 3.99	D J L Wood *(Stored as "WR360" in white RAF c/s 1.05)*	Bournemouth	22. 4.03P
G-DHTM	de Havilland DH.82A Tiger Moth *(Built L Causer and E.G.Waite-Roberts)*	PFA 157-11095		6. 1.86	E.G.Waite-Roberts *(Current status unknown -believed parts consumed within rebuild of G-APPN qv)*	(Basingstoke)	
G-DHTT	de Havilland DH.112 Venom FB.50 (FB.1) *(Built F + W)*	821	(G-BMOC) Swiss AF J-1611	17.10.96	D.J.Lindsay Wood *(As "WR421" in Aviation Museum 1.05)*	Bournemouth	17. 7.99P
G-DHUU	de Havilland DH.112 Venom FB.50 (FB.1) *(Built F + W)*	749	(G-BMOD) Swiss AF J-1539	26. 2.96	D.J.Lindsay Wood *(Stored as "WR410" in RAF 6 Sqdn RAF 1.05)*	Bournemouth	24. 5.02P
G-DHVM	de Havilland DH.112 Venom FB.50 (FB.1) *(Built F + W)*	752	G-GONE Swiss AF J-1542	26.11.03	Aviation Heritage Ltd *(As "WR470" in RAF 208 Sqdn c/s 1.05)*	Coventry	15. 6.04P
G-DHVV	de Havilland DH.115 Vampire T.55 *(Built F + W) (Reported as built with c/n 974)*	55092	Swiss AF U-1214	5. 9.91	Lindsay Wood Promotions Ltd *(Stored as "XE897" in RAF 54 Sqdn c/s 1.05)*	Bournemouth	5. 6.03P
G-DHWW	de Havilland DH.115 Vampire T.55 *(Built F + W) (Reported as built with c/n 974)*	979	Swiss AF U-1219	5. 9.91	Lindsay Wood Promotions Ltd *(Stored as "XG775" in RN FOFT Yeovilton c/s 1.05)*	Bournemouth	23. 4.03P
G-DHXX	de Havilland DH.100 Vampire FB.6 *(Built F + W)*	682	Swiss AF J-1173	5. 9.91	Lindsay Wood Promotions Ltd *(Stored as "VT871" in RAF 54 Sqdn c/s 1.05)*	Bournemouth	14. 8.02P
G-DHZF	de Havilland DH.82A Tiger Moth	82309	G-BSTJ OO-MEH, OO-GEB, OO-MOR, RNeth AF A-13, PH-UFB, A-13, N9192	7. 7.99	C.A.Parker and M.R.Johnson *(As "N9192" in RAF c/s)*	Sywell	10.10.05
G-DHZZ	de Havilland DH.115 Vampire T.55 *(Built F + W)*	990	Swiss AF U-1230	5. 9.91	Lindsay Wood Promotions Ltd *(Stored as "WZ589" in 54 Sqdn RAF c/s 1.05)*	Bournemouth	4. 8.03P
G-DIAL	Cameron N-90 HAB	1851		7.11.88	A J.Street *"London"*	(Exeter)	11. 5.00A
G-DIAT	Piper PA-28-140 Cherokee Cruiser	28-7425322	G-BCGK N9594N	19. 7.89	Bristol Flying Centre Ltd	Bristol	25. 3.07T
G-DICK	Thunder Ax6-56Z HAB	159		6. 7.78	R.D.Sargeant *"Dandag"*	(Wollerau, Switzerland)	23. 8.05A
G-DIGI	Piper PA-32-300 Cherokee Six	32-7940224	D-EIES N2947M	13.10.98	D.Stokes tr Security UN Ltd Group	Stapleford	18.11.07T
G-DIKY	Murphy Rebel *(Built R.J.P.Herivel)*	PFA 232-13182		13. 2.98	R.J.P.Herivel	Alderney	9. 6.05P
G-DIMB	Boeing 767-31K	28865		28. 4.97	MyTravel Airways Ltd	Manchester	27. 4.06T
G-DIME	Rockwell Commander 114	14123	N49829	9. 3.88	H.B.Richardson	Badminton	9. 9.04
G-DINA	Gulfstream AA-5B Tiger	AA5B-1218	N4555Y	27. 2.81	Portway Aviation Ltd	Shobdon	20. 6.05
G-DING	Colt 77A HAB	1862		28. 6.91	G.J.Bell *"Dingbat"*	(Albuquerque, New Mexico, US)	12.10.04A
G-DINK	Lindstrand Bulb SS HAB	785		28. 6.01	Dinkelacker-Schwaben Brau AG	(Stuttgart, Germany)	26. 8.05A
G-DINO	Pegasus Quantum 15 (Rotax 582)	7225	G-MGMT	15.12.98	R.A Watering	(Bourne)	8. 9.04P
G-DINT	Bristol 156 Beaufighter IF	STAN B1 184604	3858M X7688	17. 6.91	T.E.Moore *(On rebuild from various ex Australian components 10.99)*	Rotary Farm, Hatch	
G-DIPI	Cameron Tub 80SS HAB	1745		6. 5.88	R.A Preston *"KP Choc Dips Tub"*	(Bristol)	2. 5.04A
G-DIPM	Piper PA-46-350P Malibu Mirage	4636325	N5350V	20. 2.02	Intesa Leasing SpA	(Milano, Italy)	25. 2 05T
G-DIPS*	Taylor JT.1 Monoplane *(Built B.J.Halls)* (Volkswagen 1500)	PFA 55-10320		19.12.78	B.J.Halls *(Cancelled 31.3.99 by CAA) (Fuselage stored 8.00)*	Sibsey	
G-DIRK	Glaser-Dirks DG-400	4-124	D-KEKT	18. 9.86	P.D.Barker tr G-DIRK Syndicate *"RK"*	Parham	29. 1.05
G-DISA	Scottish Aviation Bulldog Series 100/125	BH120/435	RJAF 420 RJAF 1142, G-31-44	25. 8.04	R.D.Dickson tr British Disabled Flying Association	RAF Wittering	
G-DISK	Piper PA-24-250 Comanche	24-1197	G-APZG EI-AKW, N10F	9. 8.89	G.A Burtenshaw	Guernsey	30. 5.06
G-DISO	SAN Jodel D.150 Mascaret	24	9Q-CPK OO-APK, F-BLDT	16.12.86	P.F.Craven	Wombleton	10. 6.05P
G-DIVA	Cessna R172K Hawk XPII	R1723071	N758FX	10. 2.86	R.J.Harris	Finningley	10.10.07E
G-DIWY	Piper PA-32-300 Cherokee Six	32-40731	OY-DLW D-EHMW, N8931N	26.11.91	IFS Chemicals Ltd	East Winch	14. 6.07
G-DIXY	Piper PA-28-181 Archer III	2843195	N41284 G-DIXY, N41284	10.12.98	M.G.Bird	Fowlmere	16.12.07E
G-DIZI	Just/Reality Escapade 912 *(Built N Baumber - kit no JAESC 0012)*	BMAA/HB/355		1. 3.04	N Baumber	(Grantham)	
G-DIZO	Wassmer Jodel D.120A Paris-Nice	326	G-EMKM F-BOBG	30. 5.91	D.and E.Aldersea	Breighton	28. 3.05P
G-DIZY	Piper PA-28R-201T Turbo Cherokee Arrow III	28R-7703401	N47570	13.10.88	Calverton Flying Group Ltd	Rochester	2. 9.07T
G-DIZZ	Hughes 369HE	89-0105E	N9029F	19. 2.97	H.J.Pelham	Cleeves Farm, Salisbury	28. 5.06
G-DJAE	Cessna 500 Citation I	500-0339	G-JEAN N300EC, N707US, G-JEAN, (N5339J)	3.11.98	Source Group Ltd	Bournemouth	4. 4.05T
G-DJAY	Jabiru Aircraft Jabiru UL-450 *(Built D.J.Pearce)*	PFA 274A-13633		8. 8.00	D.J.Pearce	(Reading)	13. 5.05P
G-DJCR	Varga 2150A Kachina	VAC 155-80	EI-CFK G-BLWG, OO-HTD, N8360J	11. 4.96	D.J.C.Robertson	Perth	30. 4.99
G-DJEA	Cessna 421C Golden Eagle II	421C0654	TC-AAA N37379, (N24BS), N37379	16. 4.98	Bettany Aircraft Holdings Ltd	Jersey	4. 2.05
G-DJJA	Piper PA-28-181 Archer II	28-8490014	N4326D	14. 9.87	S.M. and A P.Price t/a Choice Aircraft *(Op Modern Air)*	Fowlmere	18.12.05T
G-DJNH	Denney Kitfox Model 3 *(Built D.J.N.Hall - c/n 772)*	PFA 172-11896		20. 9.90	D.J.N.Hall *(Noted 3.04)*	(Downwood, Dorset)	29. 5.01P

Registration	Type	C/n	Previous identity	Reg date	Owner	Base	CofA
G-DJRV	Van's RV-8	PFA 303-13665		15. 9.03	S.G.Hunt	(Goole)	
	(Built D J.Hunt)						
G-DJST	Air Création Clipper/iXess 912	BMAA/HB/416		12.10.04	D.J.Stimpson	Sywell	
	(Buiilt D.J.Stimpson)				*(Noted 1.05)*		
G-DKDP	Grob G109	6100	(G-BMBD)	9. 7.85	D.W.and Janet E.Page	Tibenham	28. 4.07
			D-KAMS		tr Grob 4		
G-DKGF*	Viking Dragonfly Mk.1	PFA 139-10898		16.10.86	P.C.Dowbor	Enstone	
	(Built K G Fathers) (Volkswagen 1834)				*(Cancelled 29.3.01 by CAA) (Dumped less engine 10.03)*		
G-DLCB	Europa Aviation Europa	PFA 247-12652		16.11.95	K.Richards	Talgarth	4. 6.05P
	(Built D J Lockett - kit no.046) (Monowheel u/c)				*(Noted 8.04)*		
G-DLDL	Robinson R22 Beta	1971		2. 1.92	Aeromega Ltd	Stapleford	21. 4.07T
G-DLEE	SOCATA TB-9 Tampico Club	884	G-BPGX	18. 2.04	D.A. Lee	Dunkeswell	3. 7.06
G-DLFN	Aero L-29 Delfin	294872	ES-YLE	28. 5.98	W.Elsen	North Weald	5. 7.05P
			Estonian AF, Soviet AF				
G-DLOM	SOCATA TB-20 Trinidad	1102	N2823Y	13.12.90	J.N.A Adderley	Rochester	2. 9.07
G-DLTR	Piper PA-28-180 Cherokee E	28-5803	G-AYAV	15. 3.96	BCT Aircraft Leasing Ltd	Kemble	18. 7.05T
			N11C				
G-DMAC	Jabiru Aircraft Jabiru UL	PFA 274-13321		15.10.98	C.J.Pratt	Goodwood	6. 8.05P
	(Built B Macfadden) (PFA c/n prefix should be "274A")						
G-DMAH	SOCATA TB-20 Trinidad GT	2039	F-OILY	2. 4.01	C.A Ringrose	Biggin Hill	9. 4.07T
G-DMCA*	McDonnell Douglas DC-10-30				See SECTION 3 Part 1	Manchester	
G-DMCD	Robinson R22 Beta	1201	G-OOLI	14.11.89	Heliair Ltd	Wellesbourne Mountford	8. 6.07T
			G-DMCD				
G-DMCS	Piper PA-28R-200 Cherokee Arrow II	28R-7635284	G-CPAC	29. 5.84	W.G.Ashton and J Bingley	Goodwood	19. 4.05
			PH-SMW, OO-HAU, N75220		t/a Arrow Associates		
G-DMCT	Flight Design CT2K	01-04-02-12		10. 7.01	Buha Zajac	(Gloucester)	15. 3.05P
	(Assembled Pegasus Aviation as c/n 7xxx?)						
G-DMRS	Robinson R44 Raven II	10513		29.10.04	Nottinghamshire Helicopters (2004)Ltd	Tollerton	AC
G-DMSS	Westland SA.341D Gazelle HT.3	1089	XW858	13. 7.01	Woods of York Ltd *(As "XW858")*	Murton, York	21. 7.05P
G-DMWW	CFM Shadow Series DD	304-DD		12.10.98	Microlight Sport Aviation Ltd		
						Lower Mountpleasant, Chatteris	25. 5.04P
G-DNLB*	MBB Bo.105DBS-4	S.60	G-BUDP	10. 4.92	Not known	Farnborough	
	(Rebuilt with new pod S.850 .92)		G-BTBD, VH-LCS, VH-HRM, G-BCDH, EC-DUO, G-BCDH, D-HDBK				
	(Crashed at Brough of Birsay, Orkney 24.5.02 when underslung load became unstable and struck tail rotor: cancelled 25. 7.03 as destoyed) (Noted 4.04)						
G-DNCN	Agusta-Bell 206A JetRanger	8185	9H-AAJ	21.11.97	J.J.Woodhouse	Sandown	2. 3.07T
			Libyan Arab Rep.AF 8185, 5A-BAM		t/a Flying Services		
G-DNCS	Piper PA-28R-201T Turbo Arrow III	28R-7803024	N47841	3. 1.89	BC Arrow Ltd	Liverpool	15. 7.07T
G-DNGA	Balony Kubicek BB20 HAB	235	OK-0235	1. 5.03	G.J.Bell	(Wokingham)	10.10.04A
G-DNGR	Colt 31A HAB	10162		18.10.01	G.J.Bell	(Wokingham)	12.10.04A
G-DNOP	Piper PA-46-350P Malibu Mirage	4636303	N4174A	26. 7.00	Campbell Aviation Ltd	Denham	3. 8.06T
G-DNVT	Gulfstream Aerospace Gulfstream IV	1078	(G-BPJM)	29. 9.89	Shell Aircraft Ltd	Rotterdam, The Netherlands	7.10.06T
			N17589				
G-DOCA	Boeing 737-436	25267		21.10.91	British Airways plc *(Benyhone Tartan t/s)*	Heathrow	20.12.06T
G-DOCB	Boeing 737-436	25304		16.10.91	British Airways plc	Gatwick	15. 2.07T
					(Leased to Air One, Italy .04)		
G-DOCE	Boeing 737-436	25350		20.11.91	British Airways plc *(Blomsterang t/s)*	Heathrow	6. 8.07T
G-DOCF	Boeing 737-436	25407		9.12.91	British Airways plc *(Koguty Lowickie t/s)*	Heathrow	9. 7.07T
G-DOCG	Boeing 737-436	25408		16.12.91	British Airways plc *(Chelsea Rose t/s)*	Gatwick	15. 8.07T
G-DOCH	Boeing 737-436	25428		19.12.91	British Airways plc	Heathrow	18. 8.07T
G-DOCL	Boeing 737-436	25842		2. 3.92	British Airways plc *(Ndebele Martha t/s)*	Heathrow	1. 3.05T
G-DOCN	Boeing 737-436	25848		21.10.92	British Airways plc	Gatwick	20.10.05T
G-DOCO	Boeing 737-436	25849		26.10.92	British Airways plc	Gatwick	25.10.05T
G-DOCP	Boeing 737-436	25850		2.11.92	British Airways plc	Gatwick	1.11.05T
G-DOCR	Boeing 737-436	25851		6.11.92	British Airways plc	Gatwick	5.11.05T
					(Leased to Air One, Italy .04)		
G-DOCS	Boeing 737-436	25852		1.12.92	British Airways plc	Gatwick	30.11.05T
G-DOCT	Boeing 737-436	25853		22.12.92	British Airways plc *(Crossing Borders t/s)*	Gatwick	23.12.05T
G-DOCU	Boeing 737-436	25854		18. 1.93	British Airways plc *(Ndebele Martha t/s)*	Heathrow	19. 1.06T
G-DOCV	Boeing 737-436	25855		25. 1.93	British Airways plc	Heathrow	24. 1.06T
G-DOCW	Boeing 737-436	25856		2. 2.93	British Airways plc *(Rendezvous t/s)*	Heathrow	3. 2.06T
G-DOCX	Boeing 737-436	25857		29. 3.93	British Airways plc	Gatwick	28. 3.06T
G-DOCY	Boeing 737-436	25844	OO-LTQ	17.10.96	British Airways plc	Heathrow	17.10.05T
			G-BVBY, TC-ALS, G-BVBY, (G-DOCY)				
G-DOCZ	Boeing 737-436	25858	EC-FXJ	12.12.94	British Airways plc	Heathrow	11. 1.07T
			EC-657, G-BVBZ, (G-DOCZ)				
G-DODB	Robinson R22 Beta	0911	N8005R	3. 5.96	M.Gallagher	(Ballinamore, County Leitrim)	5.12.05T
G-DODD	Reims/Cessna F172P Skyhawk II	F17202175		5.10.82	K.Watts	Moorlands Farm, Farway Common	28. 8.06
G-DODG	Aerotechnik EV-97A Eurostar	PFA 315-14258		23. 6.04	R.Barton	(Keal Cotes, Spilsby)	
	(Built R.Barton)						
G-DODI	Piper PA-46-350P Malibu Mirage	4636019		26.10.95	CAVOK SRL	(Milano, Italy)	22. 2.07T
G-DODR	Robinson R22 Beta	1325	N80721	5. 6.96	Exmoor Helicopters Ltd	Cranfield	12. 2.06T
G-DOEA	Gulfstream AA-5A Cheetah	AA5A-0895	G-RJMI	30. 4.96	Critical Simulations Ltd	Panshanger	26. 7.06T
			N27170				
G-DOFY	Bell 206B-3 JetRanger III	3637	N2283F	26. 8.87	Cinnamond Ltd	Silver Springs, Denham	17 6.05T
					(Op Cabair Helicopters)		
G-DOGG	Scottish Aviation Bulldog Series 120/121	BH120/308	XX638	3.10.01	P.Sengupta	Bourne Park, Hurstbourne Tarrant	26. 3.05T
					(As "XX638")		
G-DOGZ	Rogerson Horizon 1	PFA 241-13129		10. 8.98	J.E.D.Rogerson	Morgansfield, Fishburn	19. 9.05P
	(Built J.E.D.Rogerson) (Rotax 912-UL) *(Marked as "Fisher Super Koala")*						
G-DOIN	Best Off Skyranger 912	BMAA/HB/379		26. 4.04	C.D.and L.J.Church	(Wool, Wareham)	24.11.05P
	(Built C.D.and L.J.Church - kit no.SKRxxxx460)						
G-DOIT	Aérospatiale AS350B2 Ecureuil	1902	F-GMAZ	10.10.01	C.C.Blakey	Redhill	22.11.07T
			LN-OTA, SE-JAC, LN-OBD, (F-GHYU), LN-OBD, SE-JAC, HB-XPH				

Registration	Type	C/n	Previous identity	Date	Owner/operator	Base	CofA
G-DOLY	Cessna T303 Crusader	T30300107	N303MK	20. 7.94	R.M.Jones	Blackpool	24. 9.06
			G-BJZK, (N3645C)				
G-DOME	Piper PA-28-161 Warrior III	2842062	N4160V	12. 1.00	Haimoss Ltd	Old Sarum	16. 1.06T
G-DOMS	Aerotechnik EV-97A Eurostar	PFA 315-14254		24. 6.04	D.J.Cross	Draycott Farm, Chiseldon	9. 8.05P
	(Built D.J.Cross)						
G-DONI	Gulfstream AA-5B Tiger	AA5B-1029	G-BLLT	20. 7.95	W.P.Moritz	Elstree	16.12.06T
			OO-RTG, (OO-HRS)				
G-DONS	Piper PA-28RT-201T Turbo Arrow IV	28R-8131077	N8336L	22. 4.88	C.E.Griffiths	Blackbushe	26.11.06
G-DONT	Zenair CH.601XL Zodiac	PFA 162B-14172		24. 5.04	N.C.and A.C.J.Butcher	Sturgate	AC
	(Built N.C.Butcher)				*(Noted 12.04)*		
G-DOOZ	Aérospatiale AS355F2 Ecureuil 2	5367	G-BNSX	13. 5.88	Premiair Aviation Services Ltd	Denham)	4. 4.06T
G-DORA	Focke-Wulf Fw 190-D9	211028	Luftwaffe 211028	21. 5.03	G.R.Lacey	(Fairoaks)	
G-DORN	Dornier EKW C-3605	332	HB-RBJ	15. 5.98	R.G.Gray	Bournemouth	11.11.02P
	(Built F + W)		Swiss AF C-552		*(As "C-552")*		
G-DOTT	CFM Streak Shadow	PFA 206-13582		30.11.04	R.J.Bell	(Antrim)	
	(Built R.J.Bell)						
G-DOVE	Cessna 182Q Skylane II	18266724	N96446	26. 6.80	J.Sinclair-Dean	Guernsey	23. 6.07
G-DOWN	Colt 31A Air Chair HAB	1570		3. 8.89	M.Williams *"Up and Down"*	(Wadhurst, Sussex)	8. 6.00A
G-DPPF	Agusta A109E Power	11216		26. 6.03	Dyfed-Powys Police Authority	(Llangunnor, Carmarthen)	18. 2.07T
G-DPST	Phillips ST.2 Speedtwin	PFA 207-12674		10. 5.96	Speedtwin Developments Ltd *(Current status unknown)*		
	(Built P J C Phillips)					Upper Cae Garw Farm, Trelleck, Monmouth	
G-DRAG	Cessna 152 II	15283188	G-REME	27. 4.90	L.A Maynard and M.E.Scouller	Old Sarum	5. 8.05T
	(Tail-wheel conversion)		G-DRAG, G-BRNF, N47217		*(Op Old Sarum Flying Club)*		
G-DRAM	Reims FR172F Rocket *(Floatplane)*	FR17200102	OH-CNS	18. 9.98	T.A Crumpton t/a Clyde River Rats	Loch Earn	11. 4.05T
G-DRAW	Colt 77A HAB	1830		31. 8.90	C.Wolstenholme	(Oswestry)	24. 3.05A
G-DRAY	Taylor JT.1 Monoplane	PFA 1452		13. 7.78	L.J.Dray	(Sidmouth)	
	(Built L J Dray)				*(Current status unknown)*		
G-DRBG	Cessna 172M Skyhawk	17265263	G-MUIL	18. 1.95	Wilkins and Wilkins (Special Auctions)	RAF Henlow	17. 5.07T
			N64486		t/a Henlow Flying Club Ltd		
G-DRCI*	Jabiru Aircraft Jabiru UL	PFA 274A-14301		20. 9.04	D.R.Calo	(Chipperfield, Kings Langley)	
	(Built D.R.Calo)				*(Cancelled 19.1.05 by CAA no PtoF yet issued)*		
G-DREX	Cameron Saturn 110SS HAB	4217		28.10.97	LRC Products Ltd	(Broxbourne, Hertford)	3.11.99A
G-DRFC	Aérospatiale-Alenia ATR42-300	007	OY-CIB	1. 6.04	Atlantic Air Transport Ltd	Coventry	10. 6.07T
			A2-ALJ, OY-CIB, F-WWEC				
G-DRGN	Cameron N-105 HAB	2024		13. 6.91	W.I.Hooker and C.Parker	(Nottingham)	4. 7.01T
G-DRGS	Cessna 182S Skylane	18280375	N2389X	17.11.98	D.R.G.Scott	Edinburgh	30. 1.05
G-DRID	Reims FR172J Rocket	FR1720434	I-ALGB	19. 9.03	D.Ridley	High Flatts Farm, Chester Le Street	22.10.06
G-DRIV	Robinson R44 Raven II	10126	N75233	19. 9.03	Driver Hire Group Services Ltd	Leeds-Bradford	28.10.06T
G-DRKJ	Schweizer 269C	S.1172	G-BPPW	19.10.00	D.R.Kenyon	Shoreham	27. 9.04T
			N3624J		t/a Aviation Bureau		
G-DRMM	Europa Aviation Europa	PFA 247-13201		27. 7.98	M.W.Mason	(Nantwich)	
	(Built M.W.Mason - kit no.362) (Tri-gear u/c)						
G-DRNT	Sikorsky S-76A II Plus	760201	N93WW	5. 4.90	CHC Scotia Ltd	Norwich	1. 5.06T
			N3WQ, N3WL, N3121G				
G-DROP	Cessna U206C Super Skywagon	U2061230	G-UKNO	7. 8.87	Peterborough Parachute Centre Ltd	Sibson	5. 3.06
			G-BAMN, 4X-ALL, N71943				
G-DRSV	Robin DR315X Petit Prince	624	F-ZWRS	7. 6.90	R.S.Voice	Rushett Farm, Chessington	12. 6.05P
	(Officially regd with c/n PFA 210-11765 following major rebuild by R.S.Voice)						
G-DRYI	Cameron N-77 HAB	2046		7. 8.89	C.A Butter	(Marsh Benham)	4. 6.94A
					(New owner 2.01) (Barbour titles)		
G-DRYS	Cameron N-90 HAB	3377		1.12.95	C.A Butter *(Barbour titles)*	(Marsh Benham)	27. 7.03A
G-DRZF	CEA Jodel DR.360 Chevalier	451	F-BRZF	4. 9.91	P.K.Kaufeler	Earls Colne	7.12.06
G-DSFT	Piper PA-28R-200 Cherokee Arrow II	28R-7335157	G-LFSE	22.11.00	J.Jones	Rochester	10. 4.05T
			G-BAXT, N11C				
G-DSGC	Piper PA-25-260 Pawnee C	25-4890	OY-BDA	3. 5.95	Devon and Somerset Gliding Club Ltd	North Hill	24. 3.05
G-DSID	Piper PA-34-220T Seneca III	3447001		21. 7.95	D.S.J.Tait	Fairoaks	28. 6.07
G-DSKI	Aerotechnik EV-97 Eurostar	PFA 315-14088		25. 6.04	D.R.Skill	(Ryton)	
	(Built D.R.Skill)						
G-DSLL	Pegasus Quantum 15-912	7836		5. 7.01	R.G.Jeffery	(Sandbach)	5. 7.05P
G-DSPI	Robinson R44 Astro	0661	G-DPSI	25.10.99	Safe-Sec Group Ltd	Tatenhill	21.11.05T
	(Marked as "Raven")						
G-DSPK	Cameron Z-140 HAB	10640		7. 1.05	Bailey Balloons Ltd	(Pill, Bristol)	17. 1.06E
G-DSPZ	Robinson R44 Raven II	10351		5. 5.04	Focal Point Communications Ltd	(Alresford)	8. 6.07T
G-DTCP	Piper PA-32R-300 Lance	32R-7780255	G-TEEM	26. 1.93	G-DTCP Aviation Ltd	Cranfield	17. 5.07
			N2604Q				
G-DTOO*	Piper PA-38-112 Tomahawk	38-79A0312	N9713N	15. 2.79	Not known	Panshanger	29. 7.94T
				(Damaged Seething 9.7.94: cancelled 31.1.95 as WFU) (Fuselage noted 8.02)			
G-DTOY	Comco Ikarus C42 FB100	0309-6570		20.10.03	C.W.Laskey	Shobdon	23.10.05P
	(Rotax 912-ULS)						
G-DTUG	Wag-Aero Super Sport	PFA 108-14026		13. 5.04	D.A.Bullock	Bicester	
	(Built D.A.Bullock)				*(Noted 1.05)*		
G-DUCK*	Grumman G.44 Widgeon (OA-14)				See SECTION 3 Part 1	Biscarrosse, Landes, France	
G-DUDE	Van's RV-8	PFA 303-13246		16. 7.99	W.M.Hodgkins	Crowfield	20. 6.05P
	(Built W M Hodgkins)				*"Capt Midnight" ("Vans Air Force" & "19" on tail)*		
G-DUDS	CASA 1-131E Jungmann	2108	D-EHDS	27. 6.90	B R Cox	(Bristol)	3. 6.00P
	(Enma Tigre G-1V-B)		Spanish AF E3B-512				
G-DUDZ	Robin DR400/180 Régent	2367	G-BXNK	3.12.97	D.H.Pattison	Lower Upham Farm, Chiseldon	26.12.06
G-DUGE	Comco Ikarus C42 FB UK	PFA 322-13855		30. 7.02	D.Stevenson	Plaistowes Farm, St.Albans	20.10.05P
	(Built D.Stevenson)						
G-DUGI	Lindstrand LBL 90A HAB	562		16. 8.99	D.J.Cook	(Norwich)	25. 6.05A
G-DUGS	Van's RV-9A	PFA 320-13966		21.10.02	D.M.Provost	(Salford)	
	(Built D M Provost)						
G-DUKK	Extra EA.300/L	125	D-EXAC	27.11.00	R.A and K.M.Roberts	Godwood	3. 3.07
					t/a Puddleduck Plane Partnership		

Registration	Type	c/n	Previous identity	Date	Owner/Operator	Location	CofA
G-DUMP	Customcraft A25 HAB	CC003		19. 5.04	P.C.Bailey	(Over, Cambridge)	
G-DUNG	Sky 65-24 HAB	125		20. 7.98	G.J.Bell	(Albuquerque, New Mexico, US)	24. 5.05A
G-DUOA	Canadair CL600-2C10	10028	G-MRSK	22. 7.03	Maersk Aircraft A/S	Copenhagen, Denmark	3.11.05E
	(Built Bombardier Inc) (CRJ 700)		C-GIBG				
G-DUOC	Canadair CL600-2C10	10039	G-MRSI	12. 9.03	Maersk Aircraft A/S	Copenhagen, Denmark	3.11.05E
	(Built Bombardier Inc) (CRJ 700)		C-GICL				
G-DUOD	Canadair CL600-2C10	10048	G-MRSH	22. 9.03	Maersk Aircraft A/S	Copenhagen, Denmark	3.11.05E
	(Built Bombardier Inc) (CRJ 700)		C-GIAI				
G-DUOE	Canadair CL600-2C10	10052	G-MRSG	30. 9.03	Maersk Aircraft A/S	Copenhagen, Denmark	3.11.05E
	(Built Bombardier Inc) (CRJ 700)		C-GIAR				
G-DURO	Europa Aviation Europa	PFA 247-12554		15.11.93	D.J.Sagar	Bidford	30. 4.05P
	(Built R Swinden - kit no.033) (Monowheel u/c)						
G-DURX	Colt 77A HAB	1522		25. 5.89	V.Trimble *(Durex/Avanti titles)*	(Henley-on-Thames)	26. 6.02A
G-DUSK	de Havilland DH.115 Vampire T.11	15596	XE856	1. 2.99	R.M.A Robinson and R.Horsfield	RAF Henlow	
					(On rebuild 4.04)		
G-DUST	Stolp SA.300 Starduster Too	JP-2	N233JP	28. 4.88	J.V.George	(Winchester)	22. 5.90P
	(Built J O Perritt) (Lycoming O-360)			*(Damaged in collision with G-AKTM Badminton 16.7.89: on rebuild 11.04)*			
G-DUVL	Reims/Cessna F172N Skyhawk II	F17201723	G-BFMU (1)	16. 8.78	A J.Simpson	White Waltham	29. 4.07
G-DVBF	Lindstrand LBL 210A HAB	188		6. 3.95	Airxcite Ltd t/a Virgin Balloon Flights	(Wembley)	1. 6.05T
G-DVON	de Havilland DH.104 Devon C.2/2	04201	(G-BLPD)	26.10.84	C.L.Thatcher	Kemble	29. 5.96
			VP955		tr The 955 Preservation Group *(As "VP955": stored 5.02)*		
G-DWEL	SIPA 903	8	G-ASXC	3.12.04	B.L.Procter and T.F.Lyddon	Dunkeswell	
	(Continental C90) *(Wide-track sprung-steel undercarriage)*		F-BEYK		*(On rebuild 11.04)*		
G-DWIA	Chilton DW.1A	PFA 225-12256		25. 1.93	D.Elliott	(Horsham)	
	(Built D.Elliott)				*(Current status unknown)*		
G-DWIB	Chilton DW.1B	PFA 225-12374		22.12.93	J.Jennings	(Bedford)	
	(Built J.Jennings)				*(Current status unknown)*		
G-DWMS	Jabiru Aircraft Jabiru UL-450	PFA 274A-13491		21. 6.00	D.H.S.Williams	Sutton Meadows, Ely	13. 6.05P
	(Built D.H.S.Williams - kit no 0266)						
G-DWPF	Tecnam P92-EM Echo	PFA 318-13838		17. 5.02	P.I.Franklin and D.J.M.Williams	Guernsey	18.10.05P
	(Built P.I.Franklin and D.J.M.Williams)						
G-DWPH	Ultramagic M-77 HAB	77/109		17. 3.95	Jennifer M.Robinson	(Chipping Norton)	15. 7.05
					t/a Ultramagic UK *"Miguel"*		
G-DYCE	Robinson R44 Raven II	10148		3. 9.03	G Walters (Leasing) Ltd	(Aberdare)	29. 9.06T
G-DYKE	Dyke JD-2 Delta	PFA 1331		6. 1.04	M.S.Bird	Pepperbox, Salisbury	
	(Built P Wilson and M.S.Bird)						
G-DYNE	Cessna 414 Chancellor	414-0070	N8170Q	4. 8.87	Commair Aviation Ltd	Tollerton	16.12.06
					t/a Commodore International		
G-DYNG	Colt 105A HAB	1721	G-HSHS	9. 2.98	M.J.Gunston *"High Society"*	(Camberley)	11.10.04A

G-EAAA - G-EZZZ (see SECTION 1, PART 1 for original G-EA.. and G-EB..[1919 to 1928] registrations)

Registration	Type	c/n	Previous identity	Date	Owner/Operator	Location	CofA
G-EAGA (2)	Sopwith Dove rep	"3004/1"	(G-BLOO)	22.11.89	A Wood	Old Warden	16. 5.01P
	(Le Rhone 80hp)				*(On loan to Richard Shuttleworth Trustees)*		
	(Original Dove G-EAGA c/n 3004/1 to Australia and in use as K-157 by 11.12.19: remains of unregistered Dove, thought to have been K-157 and which crashed Essendon, Victoria 9.3.30, brought to UK circa 1987/88, rebuilt as G-BLOO and subsequently re-registered as above)						
G-EAVX (2)	Sopwith Pup	PFA 101-10523	B1807	16. 1.78	K.A.M.Baker	(Winscombe)	
	(Officially regd with c/n "B1807" : claimed as rebuild of original Sopwith a/c cancelled as written-off Hendon 21.7.21) (To carry "B1807/A7" in RFC c/s)						
G-EBJI (2)	Hawker Cygnet rep	PFA 77-10240		9. 8.77	C.J.Essex	(Coventry)	
	(Built C J Essex)				*(Under construction 7.99:)*		
G-EBZN (2)	de Havilland DH.60X Moth	608	VP-NAA	28.10.88	Jane Hodgkinson	(Gravesend)	
	(Cirrus I)		VP-YAA, ZS-AAP, G-UAAP		*(On rebuild from original components)*		
G-ECAH	Fokker F.27 Friendship 500	10669	G-JEAH	14. 4.00	Nordic Aviation Contractor A/s		
			VH-EWY, PH-EXL		*(New owner 7.04)*	Skive , Hojslev, Denmark	14. 2.03T
G-ECAN	de Havilland DH.84 Dragon	2048	VH-DHX	11. 1.01	A J.Norman	Rendcomb	25. 6.06
	(Built de Havilland Aircraft Pty Ltd, Australia)		VH-AQU, RAAF A34-59		tr Norman Aeroplane Trust *(Railway Air Services titles)*		
G-ECAS	Boeing 737-36N	28554		16.12.96	British Midland Airways Ltd *"golden jubilee baby"*		
					(Op bmiBaby)	Nottingham East Midlands	19.12.05T
G-ECAT*	Fokker F-27 Friendship 500	10672	G-JEAI	14. 4.00	Euroceltic Airways Ltd	Strandhill, Sligo	16.12.02T
			VH-EWZ, PH-EXS				
	(Skidded off end of runway landing Sligo 3.11.02: cancelled 17.8.03 by CAA) (Noted 7.04 minus engines, outer wings and tail)						
G-ECBH	Reims/Cessna F150K	F15000577	D-ECBH	16. 5.85	G.Harber tr ECBH Flying Group	Swansea	8. 9.05T
G-ECDX	de Havilland DH.71 Tiger Moth rep	SP.7		1.11.94	M.D.Souch and N.Parkhouse	Hill Farm, Durley	
	(DH Gipsy I)				*(Under build 2.03)*		
G-ECGC	Reims/Cessna F172N Skyhawk II	F17201850		10.10.79	Euroair Flying Club Ltd	Cranfield	26. 7.04T
G-ECGO	Bölkow Bö.208C Junior	599	D-ECGO	24. 8.89	A Flight Aviation Ltd	Prestwick	3. 4.06T
					(Op Prestwick Flying Club)		
G-ECJM	Piper PA-28R-201T Turbo Arrow III	28R-7803178	G-FESL	25. 9.90	Regishire Ltd	Bournemouth	9. 3.07
			G-BNRN, N321EC, N3561M				
G-ECLI	Schweizer 269C	S 1784	N69A	16. 7.99	P.T.Fellows	Biggin Hill	13.11.05
					tr Schweizer 300C Helicopter Group		
G-ECMM	Agusta A109E Power	11115	G-SIVC	7. 6.04	The Meade Corporation Ltd	(Malmesbury)	14. 6.07T
G-ECON	Cessna 172M Skyhawk II	17264490	G-JONE	18. 9.03	Meridian Aviation Group Ltd	Bournemouth	29. 3.06T
	(Thielert TAE 125-01)		N9724V				
G-ECOX	Pietenpol AirCamper GN.1	PFA 47-10356		5.12.78	H.C.Cox	(Bristol)	
	(Built H.C.Cox - c/n WLAW.1)				*(Under construction 2001- current status unknown)*		
G-ECUB	Piper PA-18-150 Super Cub	18-6279	G-CBFI	28.11.03	E.Hopper	Sherburn-in-Elmet	12. 1.07T
			SE-FDY, LN-HHA, SE-CTA, N8675D				
G-ECVB	Pietenpol AirCamper	PFA 47-13014		20. 4.00	K.S.Matcham	Lee-on-Solent	29 .6.05P
	(Built K.S.Matcham) (Continental O-200-A)				*(Noted 9.04)*		
G-EDAV	Scottish Aviation Bulldog Series 120/121	BH120/220	XX534	8. 8.01	Edwalton Aviation Ltd *(As "XX534/B")*	Tollerton	6. 8.05
G-EDEN	SOCATA TB-10 Tobago	66		8. 1.80	N.G.Pistol tr Group Eden	Elstree	17. 4.05
G-EDES	Robinson R44 Raven II	10480		8. 9.04	A.D.Russell	(Cambridge)	10.10.07E

Reg	Type	C/n	Previous identity	Date	Owner/Operator	Base	CofA
G-EDFS	Pietenpol AirCamper	PFA 47-13206		24. 3.98	D.F.Slaughter	(Redhill)	
	(Built D.F.Slaughter)				*(Current status unknown)*		
G-EDGE	Jodel D.150 Mascaret	PFA 151-11223		14. 9.88	A D.Edge	RAF Wyton	25. 1.05
	(Built A D.Edge - c/n 111) (Continental O-200-A)						
G-EDGI	Piper PA-28-161 Warrior II	28-7916565	D-EBGI	19. 1.99	R.A Forster	Cardiff	26..3.05
			N2941R				
G-EDMC	Pegasus Quantum 15-912	7513		11. 3.99	M.W.Riley	Eshott	13. 4.04P
G-EDNA	Piper PA-38-112 Tomahawk	38-78A0364	OY-BRG	4. 9.84	D.J.Clucas	Woodford	17.12.05T
G-EDRV	Van's RV-6A	PFA 181A-13451		20. 8.99	E.A Yates	North Weald	19.10.05P
	(Built E.A Yates)						
G-EDTO	Reims FR172F Rocket	FR17200090	D-EDTQ	21. 3.01	N.G.Hopkinson	Fenland	13. 5.07
G-EDVL	Piper PA-28R-200 Cherokee Arrow II	28R-7235245	G-BXIN	30. 6.97	J.S.Develin and Z.Islam	Shoreham	24. 5.06T
			D-EDVL, N1243T		*(Op Sky Leisure)*		
G-EECO	Lindstrand LBL 25A Cloudhopper HAB	668		1. 2.00	Patricia Anne and A.J.A.Bubb		
						(Storrington, Pulborough)	16. 5.05A
G-EEGL	Christen Eagle II	AES/01/0353	5Y-EGL	14.12.90	M.P.Swoboda	Andrewsfield	1.10.04P
	(Lycoming AEIO-360)				*(Damaged in heavy landing Andrewsfield 16.11.03)*		
G-EEGU	Piper PA-28-161 Warrior II	28-7916457	D-EEGU	7. 5.02	Premier Flight Training Ltd	Norwich	29. 5.05T
			N2831A				
G-EEJE	Piper PA-31 Navajo B	31-825	OH-PNG	18. 5.01	Geeje Ltd	Full Sutton	6. 8.07T
G-EELS	Cessna 208B Caravan I	208B0619		3. 3.97	Glass Eels Ltd	Gloucestershire	29. 7.06T
G-EENA	Piper PA-32R-301 Saratoga SP	32R-8013011	C-GBBU	3.10.97	Gamit Ltd	North Weald	5. 9.07
G-EENI	Europa Aviation Europa	PFA 247-12831		28. 7.98	M.P.Grimshaw	(London W5)	
	(Built M.P.Grimshaw - kit no.199)				*(Current status unknown)*		
G-EENY	Gulfstream GA-7 Cougar	GA7-0094	N721G	21. 6.79	Jade Air plc *(Noted 11.04)*	Not known	20. 7.03T
G-EERH	Ruschmeyer R90-230RG	003	D-EERH	5. 4.01	D.Sadler	Perth	2. 5.07
G-EERV	Van's RV-6	PFA 181-13381	G-NESI	13. 9.01	C.B.Stirling	Damyn's Hall, Upminster	5. 4 05P
	(Built G Ness, P G Stewart and C B Stirling) (Lycoming O-320)						
G-EESA	Europa Aviation Europa	PFA 247-12535	G-HIIL	9. 4.96	S.Collins	RAF Shawbury	4. 6.05P
	(Built C B Stirling - kit no.025) (NSI EA-81/100) *(Monowheel u/c)*						
G-EESE*	Cessna U206G Stationair 6	U20603883	OO-DMA	28. 2.85	Not known	Movenis	1. 4.91
			N7344C				
	(Crashed Magilligan, County Londonderry 31.12.88: cancelled 17.7.90 as destroyed) (Fuselage noted 5.04)						
G-EEST	British Aerospace Jetstream Series 3102	781	SE-LGM	16. 8.00	Eastern Airways (Europe) Ltd	Humberside	23.10.03T
			OY-SVY, C-FASJ, G-31-781				
G-EEUP	SNCAN Stampe SV-4C	451	F-BCXQ	1. 9.78	A M.Wajih	Redhill	11.10.02
	(Renault 4P)						
G-EEVA	Piper PA-23-250 Aztec	27-134	G-ASND	29. 8.03	D.C.Woods	Elstree	24 7.06T
			N4800P				
G-EEYE	Mainair Blade 912	1313-0202-7 & W1108		13. 5.02	B.J.Egerton	(Bootle)	4.11.05
G-EEZA	Robinson R44 Raven II	10071	N71959	29. 4.03	London and Wessex Ltd	Thruxton	28. 5.06T
G-EEZS	Cessna 182P Skylane	18261338	D-EEZS	8.11.99	W.B.Bateson	Blackpool	13. 2.06
			N63054, D-EEZS, (N20981)				
G-EFBP	Reims/Cessna FR172K Hawk XP	FR1720664	D-EFBP	7. 1.04	R.J.Howard	Sherburn-in-Elmet	6. 6.07
			PH-AXF				
G-EFGH	Robinson R22 Beta	1487	G-ROGG	3. 5.01	Foxtrot Golf Helicopters Ltd	Edinburgh	16. 9.05T
G-EFIR	Piper PA-28-181 Archer II	28-8090275	D-EFIR	5. 5.99	Leicestershire Aero Club Ltd	Leicester	8. 6.05T
			N8179R				
G-EFIS*	Westland WG.30 Series 100				See SECTION 3 Part 1	Weston-super- Mare	
G-EFOF	Robinson R22 Beta II	3605	N73323	19. 5.04	N.T.Burton t/a NT Burton Aviation	Costock	17. 6.07T
G-EFPA	Airbus A321-211	1960	D-AVZG	25. 4.03	MyTravel Airways Ltd	Manchester	24. 4.06T
G-EFRY	Avid Aerobat	PFA 189-12096		22. 3.93	P.A Boyden	Burtenshaw Farm, Barcombe	7. 6.05P
	(Built J J Donley)						
G-EFSM	Slingsby T.67M-260 Firefly	2072	G-BPLK	16. 7.92	Pooler-LMT Ltd	Lower Grounds Farm, Sherlowe	17.12.05T
G-EFTE	Bölkow Bö.207	218	D-EFTE	4. 1.90	L.J. and A A Rice	Quebec Farm, Knook, Warminster	7.10.05
G-EFTF	Aérospatiale AS350B Ecureuil	1847	G-CWIZ	1. 4.03	T.J.French	(Cumnock)	6. 4.05T
			CS-HDF, G-DJEM, G-ZBAC, G-SEBI, G-BMCU				
G-EGAL	Christen Eagle II	0042-86	SE-XMU	11. 3.96	J H Penfold	Swanborough Farm, Lewes	20. 5.04P
	(Lycoming AEIO-360)						
G-EGEE	Cessna 310Q	310Q0040	G-AZVY	14.11.83	P.G.Lawrence	Bournemouth	20. 11.06T
			SE-FKV, N7540Q				
G-EGEG	Cessna 172R Skyhawk	17280894	N7262H	4. 7.00	C.D.Lever	Elstree	27. 6.06
G-EGEL	Christen Eagle II	S.308	N388AG	4. 2.91	P.Miny	(Riehen, Switzerland)	15. 7.05P
	(Built MLP Aviation Ltd) (Lycoming AEIO-360)		G-EGEL		tr G-EGEL Flying Group		
G-EGGG*	Lindstrand LBL -90A HAB	269		14. 6.95	S M Edwards	(Houston. Texas, US)	
	(Humpty Dumpty shape)		*(Cancelled 26.2.97 by CAA: to N912HD 5.97) (Active Albuquerque, NM, US 10.03 as "G-EGGG")*				
G-EGGI	Comco Ikarus C42 FB UK	PFA 322-13872		18. 4.02	A G.and G.J.Higgins	Bitteswell	1. 8.05P
	(Built A G.and G.J.Higgins)						
G-EGGS	Robin DR400/180 Régent	1443		15.11.79	R.Foot	Lasham	17. 7.07
G-EGGY	Robinson R22 Beta II	3452	G-CCGD	8.11.04	R.S.and A.G.Higgins	Bitteswell	9. 7.06E
			N75107				
G-EGHB	Ercoupe 415D	1876	N3414G	1. 9.95	P.G.Vallance	Rochester	23. 9.05
	(Continental O-200-A)		N99253, NC99253				
G-EGHH	Hawker Hunter F.58	41H-697450	Swiss AF J-4083	4. 7.95	J.H.Goodwin	Bournemouth	
					(In open store in bare metal 1.05)		
G-EGHR	SOCATA TB-20 Trinidad	795	F-GGIQ	19.12.97	B.M.Prescott	Goodwood	10. 5.07T
G-EGJA	SOCATA TB-20 Trinidad	1101	N2807D	13.12.90	D.A Williamson	Alderney	15.12.06
G-EGLE	Christen Eagle II	F.0053		30. 3.81	S.J.Hampton	Standalone Farm, Meppershall	17.11.05P
	(Built Airmore Aviation) (Lycoming AEIO-360)				tr Eagle Group		
G-EGLS	Piper PA-28-181 Archer III	2843348	N4187C	5. 6.00	D.J.Cooke	Old Sarum	25. 6.06
G-EGLT	Cessna 310R II	310R1874	G-BHTV	9. 9.93	Capital Trading (Aviation) Ltd	Jersey	15. 1.08E
			N1EU, (N3206M)				
G-EGNR	Piper PA-38-112 Tomahawk	38-79A0233	OY-VIG	6.10.97	R.Bestek	(Ballymena)	7. 6.07T
			SE-KNI, N2570C				

Registration	Type	C/n	Previous identity	Date	Owner/Operator	Location	CofA
G-EGTB	Piper PA-28-161 Cherokee Warrior II	28-7816074	G-BPWA	21. 1.04	Airways Aero Association Ltd	Booker	24. 6.07T
			N47450				
G-EGTR	Piper PA-28-161 Cadet	2841281	G-BRSI	25. 4.98	Stars Fly Ltd	Elstree	23. 1.05T
			N92001				
G-EGUL	Christen Eagle II	Argence 0001	G-FRYS	19. 1.93	I.S.Smith	RAF Coltishall	18.11.05P
	(Built Argence EA) (Lycoming AEIO-360)		N66EA		tr G-EGUL Flying Group		
G-EGUR	SAN Jodel D.140B Mousquetaire II	52	D-EGUR	9. 1.04	D.A Hood	(Naxxar, Malta)	22. 9.07P
G-EGUY	Sky 220-24 HAB	103		24. 4.98	Sky Trek Ballooning Ltd	(Hartley, Longfield)	15 4.05T
G-EHBJ	CASA 1-131E Jungmann Series 2000	2150	Spanish AF E3B-550	19. 7.90	E.P.Howard	Airfield Farm, Hardwick	18. 7.05P
					(Also carries "E.3B-550")		
G-EHGF	Piper PA-28-181 Archer II	28-7790188	D-EHGF	23.10.00	E.Stokes and J.Lamb	Barton	29. 3.07T
			N9534N		tr Pegasus Flying Group		
G-EHIC	SAN Jodel D.140B Mousquetaire II	53	D-EHIC	20.10.04	M.Tolson and D.W.Smith	(Kettering)	AC
G-EHIL*	EH Industries EH-101				See SECTION 3 Part 1	Weston-super-Mare	
G-EHLX	Piper PA-28-181 Archer II	28-8090317	D-EHLX	5.11.99	ASG Leasing Ltd	Guernsey	13. 1.06
			N8218S				
G-EHMF	Isaacs Fury II	PFA 011-14109		8.10.03	M.A Farrelly	(Frodsham)	
	(Built M.A Farrelly)						
G-EHMJ	Beech S35 Bonanza	D-7879	D-EHMJ	12. 1.99	A L.Burton and A J.Daley	Gamston	14. 3.05
G-EHMM	Robin DR400/180R Remorqueur	867	D-EHMM (1)	10.12.84	Booker Gliding Club Ltd	Booker	1. 4.03
G-EHMS	MD Helicopters MD 900	900-00068	N3212K	12. 7.00	Virgin HEMS (London) Ltd		
						The Royal London Hospital, London E1	11.10.06T
G-EHUP	Aérospatiale SA.341G Gazelle 1	1407	F-GIJR	3.10.97	M W Helicopters Ltd	Stapleford	22. 3.07T
			N869GT, N869, N49523				
G-EHXP	Rockwell Commander 112A	227	D-EHXP	27. 1.00	A L.Stewart	Durham Tees Valley	15. 4.06T
			N1227J				
G-EIBM	Robinson R22 Beta	1993	G-BUCL	25. 3.94	XL Aviation Ltd	Lower Baads Farm, Peterculter	5. 3.07T
					(Op HJS Helicopters)		
G-EIII	Extra EA.300	057	G-HIII	4.12.00	D Dobson	Little Staughton	19. 6.06T
			D-ETYD				
G-EIKY	Europa Aviation Europa	PFA 247-12634		27. 9.94	J.D.Milbank	Insch	7. 6.05P
	(Built J.D.Milbank - kit no.054) (Monowheel u/c)						
G-EIRE	Cessna T182T Turbo Skylane	T18208049	N3500U	24. 7.01	J.Byrne	Hawley Farm, Tadley	20. 6.07T
G-EISO	SOCATA MS.892A Rallye Commodore 150	10563	D-EISO	23. 1.01	T.E.Harry Simmons	Sandown	30. 6.07T
			F-BNSO		tr G-EISO Group *(Noted 7.04)*		
G-EITE	Luscombe 8A Silvaire	3407	N71980	27. 7.88	S R H Martin *(Noted 9.04)*	Manor Farm, Haddenham	27 2.03P
G-EIWT	Reims/Cessna FR182 Skylane RG II	FR18200052	D-EIWT	28. 1.86	P.P.D.Howard-Johnston	Glenrothes	10. 4.04T
			OO-BLI				
G-EIZO	Eurocopter EC120B Colibri	1120	N20GH	31.12.04	R.M.Bailey	Addiston Mains, Dalmahoy	AC
			D-HSUN				
G-EJAR	Airbus A319-111	2412	D-AVWH	R	EasyJet Airline Co Ltd	Luton	
G-EJEL	Cessna 550 Citation II	550-0643	N747CR	19.12.01	A J.and E.A Elliott	Leeds-Bradford	8. 1.08E
			N643MC, PT-ODW, N13091, (N1259S)				
G-EJGO	Moravan Zlin Z-226T Trener 6HE Spezial	199	D-EJGO	7. 8.85	Aerotation Ltd	Rochester	14. 8.05
			OK-MHB				
G-EJJB	Airbus A319-111	2380	D-AVWV	1. 2.05	EasyJet Airline Co Ltd	Luton	31. 1.08E
G-EJMG	Cessna F150H	F150-0301	D-EJMG	27. 4.98	T.A White	Sheffield City	20.12.07E
	(Built Reims Aviation SA)				tr Bagby Aviation		
G-EJOC	Aérospatiale AS350B Ecureuil	1465	G-GEDS	21.12.94	E.and Susan Vandyk	Redhill	8. 7.05T
			G-HMAN, G-SKIM, G-BIVP		t/a Leisure and Retail Helicopters		
G-EJRS	Piper PA-28-161 Cadet	2841115	D-EJRS	5. 5.04	Small World Aviation Ltd	North Weald	2.12.07E
			N9175X				
G-EJTC	Robinson R44 Clipper II	10623		31. 1.05	Heli Air Ltd	Wellesbourne Mountford	
G-EKIR	Piper PA-28-161 Cadet	2841157	SE-KIR	17. 6.02	Aeros Leasing Ltd	Gloucestershire	6. 8.05T
			(SE-KII)		*(Op Aeros Flying Club)*		
G-EKKC	Reims FR172G Rocket	FR17200185	D-EKKO	23. 2.04	L.B.W. and Fiona H.Hancock	Kemble	14. 4.07
G-EKKL	Piper PA-28-161 Warrior II	28-8416087	D-EKKL	24. 3.99	Apollo Aviation Advisory Ltd	Shoreham	18. 4.05T
			N43588				
G-EKKO	Robinson R44 Raven	0821		18. 7.00	D.I.Pointon	Meriden, Coventry	11. 7.07T
G-EKMN	Moravan Zlin Z-242L	0652	SE-KMN	15. 5.01	R.C.Poolman	Gloucestershire	23. 9.07T
G-EKMW	Mooney M.20J Model 205	24-3213	D-EKMW	13. 5.02	A C.Armstrong	Jersey	16. 7.05T
					(Crashed shortly after takeoff Jersey 16.10.04)		
G-EKOS	Reims/Cessna FR182 Skylane RG II	FR18200017	D-EKOS	15. 7.98	S.Charlton	Sherburn-in-Elmet	22. 9.07
G-EKYD	Robinson R44 Raven II	10081		20. 5.03	EK Aviation Ltd	Cambridge	12. 6.06T
G-ELAM	PA-30 Twin Comanche B	30-1477	N26PJ	14.10.04	Hangar 39 Ltd	North Weald	20.10.07E
			G-BAWU, (G-BAWV), 9J-RFW, ZS-FAM, N8332Y				
G-ELDR	Piper PA-32-260 Cherokee Six	32-7400027	SE-GBK	21. 1.03	Elder Aviation Ltd	Oxford	27.1..06T
G-ELEC*	Westland WG.30 Series 200				See SECTION 3 Part 1	Weston-super-Mare	
G-ELEE	Cameron Z-105 HAB	4882		11. 7.00	D.Eliot	(Aberdeen)	21. 9.03A
G-ELEN	Robin DR400/180 Régent	2363		16. 9.97	N.R. and E.Foster	Cannes, France	12. 4.07T
G-ELIS	Piper PA-34-200T Seneca II	34-8070265	G-BOPV	11. 9.03	Ellis and Co (Restoration and Building) Ltd		
			N82323			(Shepton Mallet)	19. 8.06T
G-ELIT	Bell 206L LongRanger	45091	SE-HTK	28. 7.99	Henfield Lodge Aviation Ltd	Goodwood	18. 8.05T
			N2652				
G-ELIZ	Denney Kitfox Model 2	PFA 172-11835		19. 7.90	A J.Ellis t/a Tiger Helicopters	Sandown	5.11.93P
	(Built A J Ellis - c/n 717)				*(Damaged Brighstone, Isle of Wight 10.5.93: current status unknown)*		
G-ELKA	Christen Eagle II	0001	N121DJ	18.10.94	P.J.Lawton	Thruxton	17. 5.05P
	(Lycoming AEIO-360)				*"Ping Pong"*		
G-ELKS	Avid Speed Wing Mk.4	PFA 189-13109		6. 1.98	H.S.Elkins	Garston Farm, Marshfield	9. 7.05P
	(Built H S Elkins) (Jabiru 2200A)						
G-ELLA	Piper PA-32R-301 Saratoga IIHP	3246050	N9279Q	13. 8.96	C.C.W.Hart	Old Buckenham	27. 4.06
			G-ELLA				
G-ELLE	Cameron N-90 HAB	4498		11. 1.99	N.D.Eliot *(L. E. Electrical titles)*	(London SW19)	17. 5.05A
G-ELLI	Bell 206B-3 JetRanger III	4231	D-HMOF	24. 6.97	RA Fleming Ltd	(Turweston)	6. 7.06T
					(Op Total Air Management Services (TAMS)		

Reg	Type	C/n	Previous identity	Reg date	Owner/operator	Base	CofA
G-ELMH	North American AT-6D-NT Harvard III	88-16336	FAP1662	22. 7.92	M.Hammond	Airfield Farm, Hardwick	27. 5.05P
			EZ341, 42-84555		*"Fools Rush-In" (As "42-84555/EP-H" in USAAC c/s)*		
G-ELMO	Robinson R44 Raven II	10509		7.10.04	Locumlink Associates Ltd	(Ballsbridge, Dublin)	9.11.07E
G-ELNX	Canadair CL600-2B19	7508	VH-KXJ	26. 4.02	Eurolynx Corporation	Farnborough	23. 5.06
	(Built Bombardier Inc)		C-FMNY				
G-ELUN	Robin DR400/180R Remorqueur	1102	D-ELUN	29. 5.02	P.Harper-Little and I.A Lane	Kemble	11. 7.05
			I-ALSA		tr Cotswold DR400 Syndicate		
G-ELUT	Piper PA-28R-200 Cherokee Arrow II	28R-7435009	D-ELUT	21.11.03	Green Arrow Europe Ltd	Goodwood	16. 2.07
			N56514				
G-ELZN	Piper PA-28-161 Warrior II	28-8416078	D-ELZN	20. 7.99	Northamptonshire School of Flying Ltd	Sywell	5. 9.05T
			N9579N				
G-ELZY	Piper PA-28-161 Warrior II	28-8616027	D-ELZY	13. 4.99	Goodwood Road Racing School Ltd	Goodwood	23. 5.05T
			N9095Z, (N163AV), N9641N		*(Op Goodwood Flying Club)*		
G-EMAS	Eurocopter EC135T1	0107		6. 7.99	East Midlands Air Support Unit	Sibbertoft	7.10.05T
G-EMAX	Piper PA-31-350 Navajo Chieftain	31-7952029	N276CT	8.12.98	A M and T Aviation Ltd	Southend	14. 1.05T
			SE-KKP, 54202 Swedish Navy, SE-KKP, LN-PAI *(Op Fly Now Air Charter)*				
G-EMAZ	Piper PA-28-181 Archer II	28-8290088	N8073W	26. 4.90	E.J.Stanley	RAF Woodvale	27.10.05T
			G-EMAZ, N8073W				
G-EMBC	Embraer EMB-145EU	145.024	PT-SYU	1.10.97	British Airways Citiexpress Ltd	Ronaldsway	8.10.06T
G-EMBD	Embraer EMB-145EU	145.039	PT-SZE	7. 1.98	British Airways Citiexpress Ltd	Ronaldsway	11. 1.07T
G-EMBE	Embraer EMB-145EU	145.042	PT-SZH	3. 2.98	British Airways Citiexpress Ltd	Ronaldsway	2. 2.07T
G-EMBF	Embraer EMB-145EU	145.088		10.11.98	British Airways Citiexpress Ltd	Ronaldsway	9.11.07T
G-EMBG	Embraer EMB-145EU	145.094	PT-SBQ	18.11.98	British Airways Citiexpress Ltd	Ronaldsway	17.11.07T
					(Water Dreaming t/s)		
G-EMBH	Embraer EMB-145EU	145.107		20. 1.99	British Airways Citiexpress Ltd	Birmingham	19. 1.05T
					(Blomsterang t/s)		
G-EMBI	Embraer EMB-145EU	145.126	PT-SDD	23. 4.99	British Airways Citiexpress Ltd	Manchester	22. 4.05T
G-EMBJ	Embraer EMB-145EU	145.134	PT-SDL	24. 5.99	British Airways Citiexpress Ltd	Manchester	26. 5.05T
					(Youm-Al-Suq t/s)		
G-EMBK	Embraer EMB-145EU	145.167		26. 8.99	British Airways Citiexpress Ltd	Ronaldsway	25. 8.05T
G-EMBL	Embraer EMB-145EU	145.177		4.10.99	British Airways Citiexpress Ltd	Ronaldsway	3.10.05T
G-EMBM	Embraer EMB-145EU	145.196		22.11.99	British Airways Citiexpress Ltd	Ronaldsway	21.11.05T
G-EMBN	Embraer EMB-145EU	145.201		13. 1.00	British Airways Citiexpress Ltd	Ronaldsway	12. 1.06T
G-EMBO	Embraer EMB-145EU	145.219		14. 3.00	British Airways Citiexpress Ltd	Ronaldsway	13. 3.06T
G-EMBP	Embraer EMB-145EU	145.300	PT-SKR	25. 8.00	British Airways Citiexpress Ltd	Ronaldsway	24. 8.06T
G-EMBS	Embraer EMB-145EU	145.357	PT-SNW	20.12.00	British Airways Citiexpress Ltd	Ronaldsway	19.12.06T
G-EMBT	Embraer EMB-145EU	145.404		22. 3.01	British Airways Citiexpress Ltd	Ronaldsway	21. 3.07T
G-EMBU	Embraer EMB-145EU	145.458	PT-SVD	22. 6.01	British Airways Citiexpress Ltd	Ronaldsway	21. 6.07T
G-EMBV	Embraer EMB-145EU	145.482	PT-SXB	12. 9.01	British Airways Citiexpress Ltd	Ronaldsway	11. 9.07T
G-EMBW	Embraer EMB-145EU	145.546	PT-SZJ	19.12.01	British Airways Citiexpress Ltd	Southampton	18.12.07T
G-EMBX	Embraer EMB-145EU	145.573	PT-SBJ	21. 3.02	British Airways Citiexpress Ltd	Plymouth	20. 3.05T
G-EMBY	Embraer EMB-145EU	145.617	PT-SDF	17. 7.02	British Airways Citiexpress Ltd	Ronaldsway	16. 7.05T
G-EMCA	Commander Aircaft Commander 114B	14661	D-EMCA	23. 7.04	Drive-Tech Ltd *(Noted 9.04)*	Kemble	22.7.07T
G-EMDM	Diamond DA.40-P9 Star	40009	OE-KPO	7.10.02	D.J.Munson	Enstone	20.10.05T
			OE-VPO				
G-EMER	Piper PA-34-200 Seneca	34-7350002	N3081T	29. 7.91	Haimoss Ltd *(Op Old Sarum Flying Club)*	Old Sarum	4. 3.07T
G-EMHH	Aérospatiale AS355F2 Ecureuil 2	5169	G-BYKH	3. 8.99	Hancocks Holdings Ltd	Costock	26. 7.05T
			SX-HNP, VR-CCM, N57967				
G-EMIL	Messerschmitt Bf109E-3	1983	Luftwaffe 1983?	11.12.03	G.R.Lacey	(Fairoaks)	
G-EMIN	Europa Aviation Europa	PFA 247-12673		1. 3.94	S.A Lamb	Rochester	27. 8.05P
	(Built G M Clarke and E W Gladstone - kit no.083) (Monowheel u/c)						
G-EMJA	CASA 1-131E Jungmann Series 2000	PFA 242-12340	(Spanish AF)	2. 9.94	P.J.Brand	High Cross, Ware	20. 9.05P
	(Built P.J.Brand - c/n 013) (Enma Tigre G-IV-B) *(Composite from Spanish spares imported in 1991)*						
G-EMLE	Aerotechnik EV-97 Eurostar	PFA 315-14251		9. 6.04	A.R.White	Scotland Farm, Hook	15. 7.05P
	(Built A.R.White)						
G-EMLY	Pegasus Quantum 15-912	7531		30. 6.99	Sandra J.Reid	Redlands	27. 6.05P
G-EMMI	Robinson R44 Clipper II	10075		6. 6.03	Hub Of The Wheel Ltd	Headcorn	22. 6.06T
G-EMMS	Piper PA-38-112 Tomahawk	38-78A0526	OO-TKT	14. 9.79	Cheshire Flying Services Ltd	Liverpool	13. 1.07T
			N4414E		t/a Ravenair		
G-EMMY	Rutan VariEze	PFA 74-10222		21. 8.78	M.J.Tooze	Biggin Hill	3.11.05P
	(Built M.J.Tooze - cn 577) (Lycoming O-235)						
G-EMSB	Piper PA-22-160 Tri-Pacer	22-7602	G-ARHU	13. 8.03	M.S.Bird	Pepperbox, Salisbury	10.12.98
			N3726Z				
G-EMSI	Europa Aviation Europa	PFA 247-12817		24. 1.95	P.W.L.Thomas	(York)	
	(Built P.W.L.Thomas - kit no.191) (Tri-gear u/c)				*(Current status unknown)*		
G-EMSL	Piper PA-28-161 Warrior II	28-8216117	G-TSFT	20. 2.02	Environmental Maintenance Services Ltd	Biggin Hill	14. 2.05T
			G-BLDJ, N9632N				
G-EMSY	de Havilland DH.82A Tiger Moth	83666	G-ASPZ	27. 6.91	B.E.Micklewright	Bourne Park, Hurstbourne Tarrant	8. 3.07
	(Built Morris Motors Ltd) (Rebuilt with parts ex OO-MOT)		D-EDUM, T7356				
G-ENCE	Partenavia P68B	141	G-OROY	1. 6.84	J.J.H.and A E.Hanna	Exeter	15.10.06T
			G-BFSU		t/a Bicton Aviation		
G-ENEE	CFM Streak Shadow	PFA 206-13628		14. 8.00	T.Green	Wombleton	4. 6.04P
	(Built T.Green - c/n K.280) (Rotax 912-UL)						
G-ENGO	Steen Skybolt	PFA 64-13429		15.11.00	C.Docherty	(Mount Pleasant, Falklands)	
	(Built C.Docherty)						
G-ENIE	Nipper T.66 Series 3	PFA 25-10214		17. 3.78	E.J.Clarke	Seighford	18. 6.05P
	(Built A J Waller) (Volkswagen 1800)						
G-ENII	Reims/Cessna F172M Skyhawk II	F17201352	PH-WAG	18. 1.79	J.Howley	Fenland	4 2.06T
			(D-EDQM)				
G-ENNI	Robin R3000/180	128	F-GGJA	5.10.99	F.R.Traynor	White Waltham	18 12.05T
G-ENNK	Cessna 172S Skyhawk SP	172S8538	N72729	15. 9.00	AK Enterprises Ltd	Booker	10.11.06T
G-ENNY	Cameron V-77 HAB	1399		1.12.86	B.G.Jones *"Crocks of Frome"*	Newtown Hungerford)	27. 8.05A
G-ENOA	Cessna F172F	F172-0138	G-ASZW	2. 9.81	M.K.Acors	King's Farm, Thurrock	3.11.06
	(Built Reims Aviation SA)						

Registration	Type	C/n	Previous identity	Date	Owner/Operator	Base	CofA
G-ENRE	Jabiru Aircraft Jabiru UL-450	PFA 274A-13755		28. 6.01	J.C.Harris	Rochester	10.11.05P
	(Built J.C.Harris)						
G-ENRI	Lindstrand LBL 105A HAB	294		4. 8.95	P.G.Hall	(Meanwood, Chard)	5. 8.04T
					(Henry Numatic Vacuum Cleaners titles)		
G-ENRM	Cessna 182L Skylane	18259220	D-ENRM	12.12.03	P.Murray	Plymouth	6. 1.07
			N808RM, D-EFKH, N42860		tr Romeo Mike Group		
G-ENRY	Cameron N-105 HAB	2096		26. 9.89	P.G.and G.R.Hall	(Aldershot)	7. 7.94T
					"Henry" (To Balloon Preservation Group 2.04)		
G-ENSI	Beech F33A Bonanza	CE-699	D-ENSI	17. 3.78	G.Garnett	Denham	19. 5.05
G-ENTS	Van's RV-9A	PFA 320-13917		6. 1.04	L.G.Johnson	(Washington)	
	(Built L.G.Johnson)						
G-ENTT	Reims/Cessna F152 II	F15201750	G-BHHI	9.11.93	C.and A R.Hyett	Blackbushe	31. 3.05T
			(PH-CBA)				
G-ENTW	Reims/Cessna F152 II	F15201479	G-BFLK	21. 1.93	Firecrest Aviation Ltd, M.Peavan and A Strutt	Elstree	30. 9.07E
G-ENVY	Mainair Blade	1260-1000-7 & W1054		20.12.00	D A Pollitt and P Millership	(Bolton)	17.12.03P
	(Rotax 912-UL)						
G-ENYA	Robinson R44 Raven II	10335		6. 5.04	GT Investigations (International) Ltd	(Sligo, County Sligo)	23. 5.07T
G-EOFM	Reims/Cessna F172N Skyhawk II	F17201988	D-EOFM	2.11.01	20th Air Training Group Ltd	Dublin	7. 3.05T
G-EOFS	Europa Aviation Europa	PFA 247-13033		22. 7.98	G.T.Leedham	Gunby Lea Farm, Overseal	24. 9.05P
	(Built G.T.Leedham - kit no.296) (Rotax 914-UL) *(Tri-gear u/c)*						
G-EOFW	Pegasus Quantum Q.2 Sport 15-912	7582		15.10.99	G.C.Weighell	Enstone	14.10.05P
G-EOHL	Cessna 182L Skylane	18259279	D-EOHL	4. 3.99	G.B.Dale and M.C.Terris	Inverness	10. 4.05
			N70505				
G-EOIN	Zenair CH.701 UL	PFA 187-13490		19.11.99	D.G.Palmer	Longside, Peterhead	30. 9.05P
	(Built I M Donnelly) (Verner SVS1400)						
G-EOLD	Piper PA-28-161 Warrior II	28-8516030	D-EOLD	31. 3.00	Goodwood Road Racing Co Ltd	Goodwood	25. 5.06T
			N4390F, N9531N		*(Op Goodwood Flying Club)*		
G-EOMA	Airbus A330-243	265	F-WWKU	26. 4.99	Monarch Airlines Ltd	Luton	25. 4.05T
G-EORD	Cessna 208B Grand Caravan	208B0935	N5263D	19. 3.02	Air Medical Fleet Ltd	Oxford	18. 4.05T
G-EORG	Piper PA-38-112 Tomahawk	38-78A0427	N9734N	18. 9.78	Airways Aero Associations Ltd	Booker	22. 7.06T
	(Rebuilt with new fuselage: old one stored 9.96)				*(Op British Airways Flying Club) (Whale Rider t/s)*		
G-EORJ	Europa Aviation Europa	PFA 247-13139		23. 7.99	P.E.George	(Sutton Coldfield)	30. 6.05P
	(Built P.E.George - kit no.347) (Monowheel u/c)						
G-EPAR	Robinson R22 Beta II	2781		26. 2.98	J.W.Ramsbottom	Blackpool	11. 6.06T
					t/a Jepar Rotorcraft and J.A Bickerstaffe		
G-EPDI	Cameron N-77 HAB	370		25. 1.78	R.Moss *"Pegasus"*	(Banchory)	29. 6.91A
G-EPED	Piper PA-31-350 Chieftain	31-8252040	G-BMCJ	22. 3.95	Pedley Furniture International Ltd	Oxford	17. 4.06T
			N121CF, N41060				
G-EPIC	Jabiru Aircraft Jabiru UL-450	PFA 274A-14125		4.11.03	T.Chadwick	(Market Weighton)	
	(Built T.Chadwick)						
G-EPMI	Robinson R44 Raven II	10573		6.12.04	Mealey Construction Ltd	(Carlow, County Carlow)	5. 1.08E
G-EPOC	Jabiru Aircraft Jabiru UL-450	PFA 274A-13531		9. 6.04	S.Cope	(Laceby)	
	(Built S.Cope)						
G-EPOL	Aérospatiale AS355F1 Ecureuil 2	5302	G-SASU	13. 1.98	Cambridge and Essex Air Support Unit	Stapleford	12.11.05T
			G-BSSM, G-BMTC, G-BKUK				
G-EPOX	Aero Designs Pulsar XP	PFA 202-12355		27. 4.94	K.F.Farey	(Bourne End)	3. 6.04P
	(Built W A Stewart and K.F.Farey)						
G-EPPO	Robinson R44 Raven	0784	G-JBBS	25. 3.03	North West Service Srl	(Torino Aeritalia Nord, Italy)	23.10.06T
G-EPTR	Piper PA-28R-200 Cherokee Arrow II	28R-7235090	D-EPTR	26. 5.98	Tayflite Ltd	Perth	13. 8.04T
			OH-PTR, (SE-KVF), N4558T				
G-ERAD	Beech C90A King Air	LJ-1565	N213NC	18. 7.01	G.R.Kinally	Fairoaks	27. 7.07
	(Marked as "C90B")				t/a GKL Management Services Ltd		
G-ERBL	Robinson R22 Beta II	2711		26. 6.97	G.V.Maloney	(Cavan, County Cavan)	28. 8.06T
G-ERCO	Ercoupe 415D	3210	N2585H	7. 4.93	A R.and M.V.Tapp	Manston	15. 8.05
	(Continental C85)		NC2585H				
G-ERDA	Staaken Z-21A Flitzer	PFA 223-13947		15. 1.03	J.Cresswell	(Lymington)	
	(Built J.Cresswell)						
G-ERDS	de Havilland DH.82A Tiger Moth	85028	ZS-BCU	27. 7.94	W.A Gerdes	(Lee-on-Solent)	11. 7.07
			SAAF 2267, T6741				
G-ERFS	Piper PA-28-161 Warrior II	28-8216051	D-EPFS	29.11.02	S.Harrison	Lee-on-Solent	23. 2.06T
			N84570				
G-ERIC	Rockwell Commander 112TC	13010	SE-GSA	26. 9.78	Atomchoice Ltd	Cranfield	11. 5.06
G-ERIK	Cameron N-77 HAB	1753		18. 5.88	T.M.Donnelly *"Norsewind"*	(Sprotbrough, Doncaster)	24. 2.00A
G-ERIS	Hughes 369D	11-0871D	G-PJMD	1. 3.96	R.J.Howard	Leeds	13. 9.07
	(Modified to 500E standard)		G-BMJV, N1110S				
G-ERIW	Staaken Z-21 Flitzer	PFA 223-13834		9. 1.04	R.I.Wasey	(Cheltenham)	
	(Built R.I.Wasey)						
G-ERJA	Embraer EMB-145EP	145.229		25. 2.00	Brymon Airways Ltd	Bristol	24. 2.06T
G-ERJB	Embraer EMB-145EP	145.237	PT-SIC	13. 3.00	Brymon Airways Ltd	Bristol	12. 3.06T
G-ERJC	Embraer EMB-145EP	145.253		25. 4.00	Brymon Airways Ltd	Bristol	25. 4.06T
G-ERJD	Embraer EMB-145EP	145.290	PT-SKH	20. 7.00	Brymon Airways Ltd	Bristol	19. 7.06T
G-ERJE	Embraer EMB-145EP	145.315	PT-SMG	15. 9.00	Brymon Airways Ltd	Bristol	14. 9.06T
G-ERJF	Embraer EMB-145EP	145.325		24.10.00	Brymon Airways Ltd	Bristol	23.10.06T
G-ERJG	Embraer EMB-145EP	145.394		8. 3.01	Brymon Airways Ltd	Bristol	7. 3.07T
G-ERMO	ARV Aviation ARV-1 Super 2	018	G-BMWK	7. 1.87	T.Pond	Sandtoft	19. 6.05
G-ERMS	Thunder AS-33 HA Airship	A 1		28.11.78	B.R. and M.Boyle	Newbury	
					"Microbe" (On loan to British Balloon Museum and Library)		
G-ERNI	Piper PA-28-181 Archer II	28-8090146	G-OSSY	9.10.91	H.A Daines	Seething	9. 3.07
			N81215				
G-EROL	Westland SA.341G Gazelle 1	1108	G-NONA	18.10.02	The Coin Group Ltd	Booker	21. 2.05T
			G-FDAV, G-RIFA, G-ORGE, G-BBHU				
G-EROM	Robinson R22 Beta II	3383		19.11.02	Aeromega Ltd	Cambridge	27.11.05T
G-EROS	Cameron H-34 HAB	2296		6. 4.90	A A Brown	(Guildford)	
					t/a Reach For The Sky *(New CofR 5.03)*		

Registration	Type	C/n	Previous identity	Registered	Owner/Operator	Base	CofA expiry
G-ERRI	Lindstrand LBL 77A HAB	811		20. 2.02	K.J.Baxter	(Norton, Worcester)	21. 6.05A
G-ERRY	Grumman AA-5B Tiger	AA5B-0725	G-BFMJ	20. 3.84	M.D.Savage and A F.K.Horne	Shobdon	13. 6.05
					t/a Gemini Aviation		
G-ESCA	Just/Reality Escapade Jabiru	BMAA/HB/280		25. 4.03	W R Davis-Smith	Sleap	17. 9.05P
	(Built T.F.Francis - kit no.JAESC 0002)						
G-ESCC	Just/Reality Escapade 912	BMAA/HB/414		7.10.04	G. and Susan Simons	(Littlehampton)	
	(Built G Simons - kit no.JAESC 0032)						
G-ESCP	Just/Reality Escapade Jabiru	BMAA/HB/313		19. 1.04	R.G.Hughes	(Stanford Bridge, Worcester)	
	(Built R.G.Hughes - kt no.JAESC 0004)						
G-ESEX	Eurocopter EC135T2	0267	D-HECP	9. 4.03	Essex Police Authority	Boreham	18. 6.06T
					(Op Essex Police Air Support Unit)		
G-ESFT	Piper PA-28-161 Warrior II	28-7916060	G-ENNA	16. 5.97	Plane Talking Ltd	Denham	8. 5.06T
			N22065				
G-ESKA	Just/Reality Escapade 912	BMAA/HB/371		13. 5.04	T.F.Francis	(Old Sarum)	
	(Built T.F.Francis - kit no.JAESC 0006)						
G-ESKY	Piper PA-23-250 Aztec D	27-4172	G-BBNN	24.11.95	C.J.Williams	Sandown, Isle of Wight	13. 8.06T
			N6832Y	*(Fuselage @ North Weald 2.04 as "G-ESKY/AIR AMBULANCE" is G-BADI qv)*			
G-ESLH	Agusta A109E	11173		22. 3.04	Euroskylink Ltd	(Coventry)	21. 3.07T
G-ESME	Cessna R182 Skylane RG II	R18201026	G-BNOX	10. 6.03	G.C.Cherrington	Draycott, Chiseldon	7. 4.06
			N756AW				
G-ESSX	Piper PA-28-161 Warrior II	28-8016261	G-BHYY	30. 7.82	S.Harcourt	Cardiff	16. 1.97T
			N9639N		t/a Courtenay Enterprises *(Noted 3.01)*		
G-ESSY	Robinson R44 Raven	1281		17. 1.03	EW Guess (Holdings) Ltd	Sibson	25. 2.06T
G-ESTA	Cessna 550 Citation II	550-0127	G-GAUL	24. 6.98	Executive Aviation Services Ltd	Gloucestershire	18. 8.05T
			N550TJ, (N29TG), N29TC, N2631N				
G-ESTR	Van's RV-6	PFA 181A-13638		11. 9.00	R.M.Johnson	Midlem Farm, Midlem	30. 4.05P
	(Built R M Johnson)				*"Jester"*		
G-ESUS	RotorWay Executive 162F	6169		7.10.96	J.Tickner	Street Farm, Takeley	29. 7.02P
	(Built J Tickner) (RotorWay RI 162F)				*(Noted 1.05)*		
G-ETBY	Piper PA-32-260 Cherokee Six	32-211	G-AWCY	13. 7.89	K.Richards-Green and M.B.Smithson	Oxford	25. 9.05
	(Rebuilt with spare Frame No.32-858S)		N3365W		tr G-ETBY Group		
G-ETCW	Stoddard-Hamilton GlaStar	5627	D-ETCW	12.12.01	P.G.Hayward	Little Snoring	16. 1.05P
	(Built T Wright) (Tri-cycle u/c)						
G-ETDA	Piper PA-28-161 Warrior II	28-8116256	N84051	9. 3.88	J M Thorpe	Treddunnock Farm, Llangarron	13. 4.06T
G-ETDC	Cessna 172P Skyhawk II	17274690	N53133	4. 5.88	Osprey Air Services Ltd	RAF Kinloss	29. 7.06T
					(Op Moray Flying Club)		
G-ETFT*	Colt Financial Times SS HAB				See SECTION 3 Part 1	(Aldershot)	
G-ETHI	IAV-Bacau Yakovlev Yak-52	899714	LY-AOY	23. 4.03	Touchwood and Associates Ltd	Sleap	27. 4.05P
			UR-BFP, Ukraine AF 71 *(yellow)*, DOSAAF 71 *(yellow)*				
G-ETHU	Eurocopter EC135T1	0198		17. 1.02	Helimand Ltd	Jethou, Channel Islands	24. 2.05
G-ETHY	Cessna 208B Caravan	20800293	N1295M	19.10.98	N.A Moore	Movenis	2. 4.06
			G-ETHY				
G-ETIN	Robinson R22 Beta	0853	N9081D	7. 9.88	Getin Helicopters Ltd	(Nazeing)	25. 9.06T
G-ETIV	Robin DR400/180 Régent	2454		12. 7.00	J.Macgilvray	North Connel, Oban	4. 9.06T
G-ETME	Nord 1002 Pingouin	274	N108J	18. 4.00	S.H.O'Connell and J.N.Pittock	White Waltham	18. 9.06
	(Renault 6Q)		F-BFRV, French AF 274		tr 108 Flying Group *(As "10/KG+EM" in WW2 Luftwaffe North Africa c/s)*		
G-ETPS	Hawker Hunter FGA.Mk.9	41H-679959	XE601	15. 9.04	Skyblue Aviation Ltd	Exeter	
G-EUGN	Robinson R44 Raven	0822		19. 7.00	Twinlite Developments Ltd	(Maynooth, County Kildare)	27. 8.06T
G-EUOA	Airbus A319-131	1513	D-AVYE	15. 6.01	British Airways plc	Heathrow	14. 6.07T
G-EUOB	Airbus A319-131	1529	D-AVWH	4. 7.01	British Airways plc	Heathrow	3. 7.07T
G-EUOC	Airbus A319-131	1537	D-AVYP	16. 7.01	British Airways plc	Heathrow	15. 7.07T
G-EUOD	Airbus A319-131	1558	D-AVYJ	16. 8.01	British Airways plc	Heathrow	15. 8.07T
G-EUOE	Airbus A319-131	1574	D-AVWF	5. 9.01	British Airways plc	Heathrow	4. 9.07T
G-EUOF	Airbus A319-131	1590	D-AVYW	23.10.01	British Airways plc	Heathrow	22.10.07T
G-EUOG	Airbus A319-131	1594	D-AVWU	23.10.01	British Airways plc	Heathrow	22.10.07T
G-EUOH	Airbus A319-131	1604	D-AVYM	14.12.01	British Airways plc	Heathrow	13.12.07T
G-EUOI	Airbus A319-131	1606	D-AVYN	13.11.01	British Airways plc	Heathrow	12.11.07T
G-EUOJ	Airbus A319-131			1.08R	British Airways plc	Heathrow	
G-EUOK	Airbus A319-131			2.08R	British Airways plc	Heathrow	
G-EUOL	Airbus A319-131			5.08R	British Airways plc	Heathrow	
G-EUPA	Airbus A319-131	1082	D-AVYK	6.10.99	British Airways plc	Birmingham	5.10.05T
G-EUPB	Airbus A319-131	1115	D-AVYT	9.11.99	British Airways plc	Heathrow	8.11.05T
G-EUPC	Airbus A319-131	1118	D-AVYU	12.11.99	British Airways plc	Birmingham	11.11.05T
G-EUPD	Airbus A319-131	1142	D-AVWG	10.12.99	British Airways plc	Birmingham	9.12.05T
G-EUPE	Airbus A319-131	1193	D-AVYT	27. 3.00	British Airways plc	Birmingham	26. 3.06T
G-EUPF	Airbus A319-131	1197	D-AVWS	30. 3.00	British Airways plc	Birmingham	29. 3.06T
G-EUPG	Airbus A319-131	1222	D-AVYG	25. 5.00	British Airways plc	Heathrow	24. 5.06T
G-EUPH	Airbus A319-131	1225	D-AVYK	23. 5.00	British Airways plc	Birmingham	22. 5.06T
G-EUPJ	Airbus A319-131	1232	D-AVYJ	30. 5.00	British Airways plc	Heathrow	29. 5.06T
G-EUPK	Airbus A319-131	1236	D-AVYO	30. 5.00	British Airways plc	Birmingham	29. 5.06T
G-EUPL	Airbus A319-131	1239	D-AVYP	8. 6.00	British Airways plc	Heathrow	7. 6.06T
G-EUPM	Airbus A319-131	1258	D-AVYR	30. 6.00	British Airways plc	Heathrow	29. 6.06T
G-EUPN	Airbus A319-131	1261	D-AVWA	10. 7.00	British Airways plc	Heathrow	9. 7.06T
G-EUPO	Airbus A319-131	1279	D-AVYU	1. 8.00	British Airways plc	Heathrow	31. 7.06T
G-EUPP	Airbus A319-131	1295	D-AVWU	14. 8.00	British Airways plc	Heathrow	13. 8.06T
G-EUPR	Airbus A319-131	1329	D-AVYH	9.10.00	British Airways plc	Heathrow	8.10.06T
G-EUPS	Airbus A319-131	1338	D-AVYM	23.10.00	British Airways plc	Heathrow	22.10.06T
G-EUPT	Airbus A319-131	1380	D-AVWH	5.12.00	British Airways plc	Heathrow	4.12.06T
G-EUPU	Airbus A319-131	1384	D-AVWP	14.12.00	British Airways plc	Heathrow	13.12.06T
G-EUPV	Airbus A319-131	1423	D-AVYE	13. 2.01	British Airways plc	Heathrow	12. 2.04T
G-EUPW	Airbus A319-131	1440	D-AVYP	6. 3.01	British Airways plc	Birmingham	5. 3.07T
G-EUPX	Airbus A319-131	1445	D-AVWB	14.12.01	British Airways plc	Heathrow	13.12.07E
G-EUPY	Airbus A319-131	1466	D-AVYK	12. 4.01	British Airways plc	Heathrow	11. 4.07T
G-EUPZ	Airbus A319-131	1510	D-AVYY	7. 6.01	British Airways plc	Heathrow	6. 6.07T

G-EURA	Agusta-Bell 47J-2 Ranger	2061	G-ASNV	21. 7.83	L.Goddard	Thornicombe, Dorset	28. 3.07
G-EURX	Europa Aviation Europa XS	PFA 247-13661		15.12.00	C C Napier	Newtownards	
	(Built C C Napier - kit no.482) (Tri-gear u/c)						
G-EUSO	Robin DR400/140 Major	904	F-BUSO	18. 9.03	Weald Air Services Ltd	Headcorn	14.10.06T
G-EUUA	Airbus A320-232	1661	F-WWIH	31. 1.02	British Airways plc	Heathrow	30. 1.05T
G-EUUB	Airbus A320-232	1689	F-WWBE	14. 2.02	British Airways plc	Heathrow	13. 2.05T
G-EUUC	Airbus A320-232	1696	F-WWIO	28. 2.02	British Airways plc	Heathrow	27. 2.05T
G-EUUD	Airbus A320-232	1760	F-WWBN	29. 4.02	British Airways plc	Heathrow	28. 4.05T
G-EUUE	Airbus A320-232	1782	F-WWDO	30. 5.02	British Airways plc	Heathrow	29. 4 05T
G-EUUF	Airbus A320-232	1814	F-WWIY	29. 7.02	British Airways plc	Heathrow	28. 7.05T
G-EUUG	Airbus A320-232	1829	F-WWIU	30. 8.02	British Airways plc	Heathrow	29. 8.05T
G-EUUH	Airbus A320-232	1665	F-WWIG	25.10.02	British Airways plc	Heathrow	24.10.05T
G-EUUI	Airbus A320-232	1871	F-WWBI	22.11.02	British Airways plc	Heathrow	21.11.05T
G-EUUJ	Airbus A320-232	1883	F-WWBQ	25.11.02	British Airways plc	Heathrow	24.11.05T
G-EUUK	Airbus A320-232	1899	F-WWDO	20.12.02	British Airways plc	Heathrow	19.12.05T
G-EUUL	Airbus A320-232	1708	F-WWIV	20.12.02	British Airways plc	Heathrow	19.12.05T
G-EUUM	Airbus A320-232	1907	F-WWDN	23.12.02	British Airways plc	Heathrow	22.12.05T
G-EUUN	Airbus A320-232	1910	F-WWDP	31. 1.03	British Airways plc	Heathrow	30. 1.06T
G-EUUO	Airbus A320-232	1958	F-WWIT	11. 4.03	British Airways plc	Heathrow	10. 4.06T
G-EUUP	Airbus A320-232	2038	F-WWDB	27. 6.03	British Airways plc	Heathrow	26. 6.06T
G-EUUR	Airbus A320-232	2040	F-WWID	29. 7.03	British Airways plc	Heathrow	28. 7.06T
G-EUUS	Airbus A320-232			11.07R	British Airways plc	Heathrow	
G-EUUT	Airbus A320-232			12.07R	British Airways plc	Heathrow	
G-EUUU	Airbus A320-232			1.08R	British Airways plc	Heathrow	
G-EUXC	Airbus A321-231	2305	D-AVZE	15.10.04	British Airways plc	Heathrow	14.10.07E
G-EUXD	Airbus A321-231	2320	D-AVZO	28.10.04	British Airways plc	Heathrow	27.10.07E
G-EUXE	Airbus A321-231	2323	D-AVZP	29.10.04	British Airways plc	Heathrow	AC
G-EUXF	Airbus A321-231	2324	D-AVZQ	4.11.04	British Airways plc	Heathrow	3.11.07E
G-EUXG	Airbus A321-231	2351	D-AVZU	2.12.04	British Airways plc	Heathrow	1.12.07E
G-EUXH	Airbus A321-231	2363	D-AVZW	17.12.04	British Airways plc	Heathrow	16.12.07E
G-EUXI	Airbus A321-231			5.05R	British Airways plc	Heathrow	
G-EUXJ	Airbus A321-231			8.05R	British Airways plc	Heathrow	
G-EUXK	Airbus A321-231			9.05R	British Airways plc	Heathrow	
G-EUXL	Airbus A321-231			10.05R	British Airways plc	Heathrow	
G-EVET	Cameron Concept 80 HAB	3703		30.10.95	K.J.Foster	(Coleshill, Birmingham)	18. 8.05A
G-EVEY	Thruster T.600N 450 Sprint	0121-T600N-057		22.11.01	K J Crompton	Newtownards	16.12.05P
G-EVIE	Piper PA-28-161 Warrior II	28-8316043	G-ZULU	31. 3.04	Lorraine Richardson	Compton Abbas	16. 9.06T
			N4292X				
G-EVLE	Rearwin 8125 Cloudster	803	G-BVLK	10. 4.03	M.C.Hiscock	Popham	
			N25403, NC25403		*(Noted 10.04)*		
G-EVLN	Gulfstream Aerospace Gulfstream IV	1175	N18WF	3. 6.02	Metropix Ltd	(London EC4)	5. 9.03T
			VH-CCA, (N1175B), HB-ITJ, N17588				
G-EVPI	Evans VP-1 Series 2	PFA 62-13136		10. 4.03	C.P.Martyr	(Haywards Heath)	2. 9.05P
	(Built C.P.Martyr)						
G-EVRO	Aerotechnik EV-97 Eurostar	PFA 315-14137		30. 1.04	R.I., D.Blain and D.Breen	Newtownards	16. 3.05P
	(Built R.I. and D.Blain)				t/a Movie-Go *(Noted 2.04)*		
G-EWAN	Protech PT-2C-160 Prostar	PFA 249-12425		23. 6.93	C.G.Shaw	Truleigh Manor Farm, Edburton	29. 1.05P
	(Built C.G.Shaw) (Lycoming O-320-B2B)						
G-EWAW	Bell 206B-3 JetRanger III	3955	G-DORB	25. 7.03	A J Renham	Durham Tees Valley	19. 2.06T
			SE-HTI, TC-HBN				
G-EWBC	Jabiru Aircraft Jabiru SK	PFA 274-13457		3.11.00	E.W.B.Comber	Fenland	21. 9.05P
	(Built E.W.B.Comber)						
G-EWES	Alpi Pioneer 300	PFA 330-14322		24.11.04	R.Y.Kendal and D.A.Ions	(Ewesley)	
	(Built R.Y.Kendal and D.A.Ions)						
G-EWHT	Robin R2112 Alpha	371		4. 5.04	Ewan Ltd *(Op Cotswold Aero Club)*	Gloucestershire	2. 6.07T
G-EWIZ	Pitts S-2S	S18	VH-EHQ	12.11.82	P.A Soper	Rands Farm, Layham, Ipswich	11. 5.04P
	(Built H M Shelvey) (Lycoming AEIO-540)						
G-EWME	Piper PA-28-235 Cherokee Charger	28-7310156	D-EECN	24. 9.04	C.J.Mewis and E.S.Ewen	Oaksey Park	29. 9.07E
			N55766				
G-EWRT	Eurocopter EC135T2	0347	D-HECF	6.12.04	McAlpine Helicopters Ltd	Oxford	
G-EXEA	Extra EA.300/L	082		9. 3.99	S.A W.Becker t/a Stewart Becker Aviation	Goodwood	29.10.05T
G-EXEC	Piper PA-34-200 Seneca	34-7450072	(G-EXXC)	11. 5.78	Sky Air Travel Ltd	Stapleford	31. 3.06T
			OY-BGU				
G-EXES	Europa Aviation Europa XS	PFA 247-13574		25. 4.03	D.Barraclough	Brunton	
	(Built D.Barraclough - kit no.578)						
G-EXEX	Cessna 404 Titan II	404-0037	SE-GZF	3. 5.79	Atlantic Air Transport Ltd	Aberdeen	29. 7.06A
			(N5418G)		*(COASTGUARD titles, red/white c/s with MCA logo on tail)*		
G-EXIT	SOCATA MS.893E Rallye 180GT	12979	F-GARX	22. 9.78	M.A Baldwin	Rochester	31. 5.04
G-EXLL	Zenair CH601XL Zodiac	PFA 162B-14205		4. 3.04	B McFadden	Mount Airey Farm, South Cave	
	(Built B McFadden, B Gardner and R Fox)						
G-EXON	Piper PA-28-161 Cadet	2841283	G-EGLD	9.12.03	Plane Talking Ltd	Elstree	6. 1.05T
			N92007				
G-EXPD	Stemme S 10-VT	11-063		5. 7.01	Global Gliding Expeditions Ltd		
						Lane Farm, Painscastle, Builth Wells	8. 9.07
G-EXPL	American Champion 7GCBC Citabria Explorer	1220-96		9. 5.96	E.J.F.McEntee	Goodwood	12. 6.06
G-EXTR	Extra EA.260	004	D-EDID	10. 8.92	S.J.Carver	Netherthorpe	21. 4.05P
G-EXXO	Piper PA-28-161 Cadet	2841210	G-CBXP	22. 1.04	Plane Talking Ltd	Biggin Hill	2.12.05T
			N117ND				
G-EYAK	Yakovlev Yak-50	801804	RA-01193	19. 2.03	P N A Whitehead	Leicster	13. 3.05P
			DOSAAF?				
G-EYAS	Denney Kitfox Model 2	PFA 172-11858		3. 3.93	K.Hamnett	Long Marston	24. 5.05P
	(Built E J Young)						
G-EYCO	Robin DR400/180 Régent	1949		12. 3.90	Charlie Oscar Ltd	Perth	25. 4.05

G-EYES	Cessna 402C II	402C0008	SE-IRU G-BLCE, N4648N	16. 7.90	Atlantic Air Transport Ltd	Coventry	15. 8.05T
G-EYET	Robinson R44 Astro	0052	G-JPAD	30.11.98	Warwickshire Flight Training Centre Ltd	Coventry	12. 6.06T
G-EYLE	Bell 206L-1 LongRanger II	45232	G-OCRP V4-AAB, G-OCRP, G-BWCU, N2758A, C-FPET, N2758A, JA9234	20.11.01	Eyles Construction Ltd	Manston	20.11.06T
G-EYNL	MBB Bö.105DBS-5	S.382	LN-OTJ D-HDLR, EC-DSO, D-HDLR	19. 8.96	Sterling Helicopters Ltd	Norwich	11.12.05T
G-EYOR	Van's RV-6 *(Built S I Fraser)* (Lycoming O-320)	PFA 181A-13259		15.10.99	S.I.Fraser	Henstridge	29. 1.05P
G-EYRE	Bell 206L-1 LongRanger II	45229	G-STVI N60MA, N5019K	12.11.90	Hideroute Ltd	Manston	14. 7.06T
G-EZAM	Airbus A319-111	2037	HB-JZA G-CCKA, D-AVYS	14. 9.04	EasyJet Airline Co Ltd	Gatwick	13. 9.07T
G-EZAR	Pegasus Quik	7942		18. 3.03	M.W.Houghton	Deenethorpe	27. 4.05P
G-EZBS	Airbus A319-111	2387	D-AVYF (G-EZCL)	7. 2.05	EasyJet Airline Co Ltd	Luton	
G-EZDC	Airbus A319-111	2043	HB-JZB G-CCKB, D-AVYU	20. 9.04	EasyJet Airline Co Ltd	Gatwick	19. 9.07T
G-EZEA	Airbus A319-111	2119	D-AVWZ	18. 2.04	EasyJet Airline Co Ltd	Gatwick	17. 2.07T
G-EZEB	Airbus A319-111	2120	D-AVYK	25. 3.04	EasyJet Airline Co Ltd	Gatwick	24. 3.07T
G-EZEC	Airbus A319-111	2129	D-AVWR	19. 3.04	EasyJet Airline Co Ltd	Gatwick	18. 3.07T
G-EZED	Airbus A319-111	2170	D-AVWT	7. 4.04	EasyJet Airline Co Ltd	Gatwick	6. 4.07T
G-EZEF	Airbus A319-111	2176	D-AVYS	19. 3.04	EasyJet Airline Co Ltd	Gatwick	15. 3.07T
G-EZEG	Airbus A319 111	2181	D-AVWF	1. 4.04	EasyJet Airline Co Ltd	Gatwick	31 .3.07T
G-EZEJ	Airbus A319-111	2214	D-AVYO	5. 5.04	EasyJet Airline Co Ltd	Gatwick	4. 5.07T
G-EZEK	Airbus A319-111	2224	D-AVYZ	6. 5.04	EasyJet Airline Co Ltd	Gatwick	5. 5.07T
G-EZEL	Westland SA.341G Gazelle 1	1073	(F-GIVQ) I-ATOM, F-BXPG, G-BAZL	1.12.00	W R Pitcher/Regal Group UK	Leatherhead	1. 4.07T
G-EZEO	Airbus A319-111	2249	D-AVYN	17. 6.04	EasyJet Airline Co Ltd	Gatwick	16. 6.07T
G-EZEP	Airbus A319-111	2251	D-AVYQ	1. 7.04	EasyJet Airline Co Ltd	Gatwick	30. 6.07T
G-EZER	Cameron H-34 HAB	2366	LX-ROM	31.10.02	D.D.Maltby	(Bristol)	13. 8.05A
G-EZET	Airbus A319-111	2271	D-AVWY	11. 8.04	EasyJet Airline Co Ltd	Gatwick	10. 8.07T
G-EZEU	Airbus A319-111	2283	D-AVYP	5.10.04	EasyJet Airline Co Ltd	Gatwick	4.10.07E
G-EZEV	Airbus A319-111	2289	D-AVYV	9. 9.04	EasyJet Airline Co Ltd	Gatwick	8. 9.07T
G-EZEW	Airbus A319-111	2300	D-AVWH	15.10.04	EasyJet Airline Co Ltd	Gatwick	14.10.07E
G-EZEX	Airbus A319-111	2319	D-AVYL	4.11.04	EasyJet Airline Co Ltd	Gatwick	3.11.07E
G-EZEY	Airbus A319 111	2353	D-AVYM	14. 1.05	EasyJet Airline Co Ltd	Gatwick	13. 1.08E
G-EZEZ	Airbus A319-111	2360	D-AVWP	9.12.04	EasyJet Airline Co Ltd	Gatwick	8.12.07E
G-EZIA	Airbus A319-111	2420	D-AVYL	R	EasyJet Airline Co Ltd	Gatwick	
G-EZIB	Airbus A319-111	2427	D-AVWD	R	EasyJet Airline Co Ltd	Gatwick	
G-EZIC	Airbus A319-111	2436	D-AVWC	R	EasyJet Airline Co Ltd	Gatwick	
G-EZID	Airbus A319-111	2442		R	EasyJet Airline Co Ltd	Gatwick	
G-EZIE	Airbus A319-111	2446		R	EasyJet Airline Co Ltd	Gatwick	
G-EZIF	Airbus A319-111	2450		R	EasyJet Airline Co Ltd	Gatwick	
G-EZIG	Airbus A319-111	2460		R	EasyJet Airline Co Ltd	Gatwick	
G-EZIH	Airbus A319-111	2463		R	EasyJet Airline Co Ltd	Gatwick	
G-EZII	Airbus A319-111	2471		R	EasyJet Airline Co Ltd	Gatwick	
G-EZIJ	Airbus A319-111	2477		R	EasyJet Airline Co Ltd	Gatwick	
G-EZIK	Airbus A319-111	2481		R	EasyJet Airline Co Ltd	Gatwick	
G-EZIL	Airbus A319-111	2492		R	EasyJet Airline Co Ltd	Gatwick	
G-EZIM	Airbus A319-111	2495		R	EasyJet Airline Co Ltd	Gatwick	
G-EZIN	Airbus A319-111	2503		R	EasyJet Airline Co Ltd	Gatwick	
G-EZJA	Boeing 737-73V	30235		13.10.00	EasyJet Airline Co Ltd	Luton	12.10.06T
G-EZJB	Boeing 737-73V	30236	N1787B	22.11.00	EasyJet Airline Co Ltd	Luton	21.11.06T
G-EZJC	Boeing 737-73V	30237		15.12.00	EasyJet Airline Co Ltd	Luton	13.12.06T
G-EZJD	Boeing 737-73V	30242		13. 7.01	EasyJet Airline Co Ltd	Luton	12. 7.07T
G-EZJE	Boeing 737-73V	30238		10. 8.01	EasyJet Airline Co Ltd	Luton	9. 8.07T
G-EZJF	Boeing 737-73V	30243		15. 8.01	EasyJet Airline Co Ltd	Luton	14. 8.07T
G-EZJG	Boeing 737-73V	30239		28. 9.01	EasyJet Airline Co Ltd	Luton	27. 9.07T
G-EZJH	Boeing 737-73V	30240		15.10.01	EasyJet Airline Co Ltd	Luton	14.10.07T
G-EZJI	Boeing 737-73V	30241		20.12.01	EasyJet Airline Co Ltd	Luton	19.12.07E
G-EZJJ	Boeing 737-73V	30245		30. 1.02	EasyJet Airline Co Ltd	Luton	29.1.05T
G-EZJK	Boeing 737-73V	30246		7. 2.02	EasyJet Airline Co Ltd	Luton	6. 2.05T
G-EZJL	Boeing 737-73V	30247		12. 2.02	EasyJet Airline Co Ltd	Luton	11. 2.05T
G-EZJM	Boeing 737-73V	30248		24. 4.02	EasyJet Airline Co Ltd	Luton	23. 4.05T
G-EZJN	Boeing 737-73V	30249		8. 5.02	EasyJet Airline Co Ltd	Luton	7. 5.05T
G-EZJO	Boeing 737-73V	30244		6. 6.02	EasyJet Airline Co Ltd	Luton	5. 6.05T
G-EZJP	Boeing 737-73V	32412		11. 6.02	EasyJet Airline Co Ltd	Luton	10. 6.05T
G-EZJR	Boeing 737-73V	32413		20. 8.02	EasyJet Airline Co Ltd	Luton	19. 8.05T
G-EZJS	Boeing 737-73V	32414		23. 9.02	EasyJet Airline Co Ltd	Luton	20. 9.05T
G-EZJT	Boeing 737-73V	32415		19.12.02	EasyJet Airline Co Ltd	Luton	18.12.05T
G-EZJU	Boeing 737-73V	32416	N6046P	21.12.02	EasyJet Airline Co Ltd	Luton	20. 2.06T
G-EZJV	Boeing 737-73V	32417		3. 3.03	EasyJet Airline Co Ltd	Luton	2. 3.06T
G-EZJW	Boeing 737-73V	32418		28. 3.03	EasyJet Airline Co Ltd	Luton	27. 3.06T
G-EZJX	Boeing 737-73V	32419	N1787B	12. 5.03	EasyJet Airline Co Ltd	Luton	11. 5.06T
G-EZJY	Boeing 737-73V	32420		27. 6.03	EazyJet Airline Co Ltd	Luton	26. 6.06T
G-EZJZ	Boeing 737-73V	32421		31. 7.03	EasyJet Airline Co Ltd	Luton	30. 7.06T
G-EZKA	Boeing 737-73V	32422		12. 8.03	EasyJet Airline Co Ltd	Luton	12. 8.06T
G-EZKB	Boeing 737-73V	32423		21. 1.04	EasyJet Airline Co Ltd	Luton	20. 1.07T
G-EZKC	Boeing 737-73V	32424		12. 2.04	EasyJet Airline Co Ltd	Luton	11.2.07T
G-EZKD	Boeing 737-73V	32425	N1787B	13. 2.04	EasyJet Airline Co Ltd	Luton	12. 2.07T
G-EZKE	Boeing 737-73V	32426		30. 3.04	EasyJet Airline Co Ltd	Luton	29. 3.07T
G-EZKF	Boeing 737-73V	32427		22. 4.04	EasyJet Airline Co Ltd	Luton	21. 4.07T
G-EZKG	Boeing 737-73V	32428		27. 5.04	EasyJet Airline Co Ltd	Luton	26. 5.07T

G-EZMH	Airbus A319-111	2053	HB-JZD	24. 9.04	EasyJet Airline Co Ltd	Luton	23. 9.07T
			G-CCKD, D-AVYB				
G-EZMK	Airbus A319-111	2370	D-AVWE	13. 1.05	EasyJet Airline Co Ltd *(Delivered 13.1.05)*	Stansted	
G-EZMS	Airbus A319-111	2378	D-AVWS	21. 1.05	EasyJet Airline Co Ltd	Gatwick	20. 1.08E
G-EZNC	Airbus A319-111	2050	HB-JZC	22. 9.04	EasyJet Airline Co Ltd	Luton	21. 9.07T
			G-CCKC, D-AVWF				
G-EZNM	Airbus A319-111	2402	D-AVYY	R	EasyJet Airline Co Ltd	Luton	
G-EZOS	Rutan VariEze	PFA 74-10221		10. 7.78	C.Moffat	Blackbushe	18.10.03P
	(Built O Smith - c/n 002) (Continental O-200-A)						
G-EZPG	Airbus A319-111	2385	D-AVYD	15. 2.05	EasyJet Airline Co Ltd	Luton	AC
G-EZSM	Airbus A319-111	2062	HB-JZE	8.10.04	EasyJet Airline Co Ltd	Gatwick	7.10.07E
			G-CCKE, D-AVYD				
G-EZUB	Zenair CH.601HD Zodiac	PFA 162-12765		13. 9.04	R.A.C.Stephens	(Billingshurst)	
	(Built R.A.C.Stephens)						
G-EZVS	Colt 77B HAB	063	SE-ZVS	6. 7.04	A.J.Lovell	(Goteborg, Sweden)	3. 7.05A
G-EZXO	Colt 56A HAB	421	SE-ZXO	6. 7.04	A.J.Lovell	(Goteborg, Sweden)	3. 7.05A
G-EZYD	Boeing 737-3M8	24022	N798BB	5. 2.97	EasyJet Airline Co Ltd	Luton	10. 2.06T
			I-TEAE, OO-LTC, (OO-BTC)				
G-EZYH	Boeing 737-33V	29332		17. 9.98	EasyJet Airline Co Ltd	Luton	17. 9.04T
G-EZYI	Boeing 737-33V	29333	N1787B	24.11.98	EasyJet Airline Co Ltd	Luton	22.11.04T
G-EZYJ	Boeing 737-33V	29334		18.12.98	EasyJet Airline Co Ltd	Luton	17.12.04T
G-EZYK	Boeing 737-33V	29335		31. 1.99	EasyJet Airline Co Ltd	Luton	30. 1.05T
G-EZYL	Boeing 737-33V	29336	N1787B	12. 3.99	EasyJet Airline Co Ltd	Luton	11. 3.05T
G-EZYM	Boeing 737-33V	29337	HB-IIK	23. 6.99	EasyJet Airline Co Ltd	Luton	3. 9.06T
			G-EZYM				
G-EZYN	Boeing 737-33V	29338	HB-III	8. 7.99	EasyJet Airline Co Ltd	Luton	17.12.06T
			G-EZYN				
G-EZYO	Boeing 737-33V	29339	HB-IIT	23. 8.99	EasyJet Airline Co Ltd	Stansted	15. 9.06T
			G-EZYO				
G-EZYP	Boeing 737-33V	29340		17. 9.99	EasyJet Airline Co Ltd	Stansted	16. 9.05T
G-EZYR	Boeing 737-33V	29341	N1787B	20.10.99	EasyJet Airline Co Ltd	Stansted	18.10.05T
G-EZYS	Boeing 737-33V	29342	HB-IIJ	9.11.99	EasyJet Airline Co Ltd	Luton	25. 8.06T
			G-EZYS				
G-EZYT	Boeing 737-3Q8	26307	HB-IIE	28. 6.00	EasyJet Airline Co Ltd	Luton	27. 6.06T
			N721LF, (HB-IIE)		*(To become G-TOYD 3.05)*		
G-EZYU	Piper PA-34-200-2 Seneca	34-7450110	G-BCDB	4. 7.01	P.A S.Dyke	Elstree	30. 7.05T
			N41346				
G-EZZA	Europa Aviation Europa XS	PFA 247-13841		10. 5.02	J.C.R.Davey	(Bicester)	
	(Built J.C.R.Davey - kit no.537) (Rotax 914) *(Monowheel u/c)*						

G-FAAA - G-FZZZ

G-FABB	Cameron V-77 HAB	822	LX-FAB	13.12.89	P.Trumper	(Ashford)	8. 5.03T
G-FABI	Robinson R44 Astro	0325		5. 4.97	J.Froggatt Ltd	(Dukinfield)	12. 5.06T
G-FABM	Beech 95-B55A Baron	TC-2259	G-JOND	22. 2.91	F.B.Miles	Gloucestershire	18. 8.07
			G-BMVC, N66456				
G-FABS	Thunder Ax9-120 S2 HAB	2399		8. 6.93	R.C.Corrall	(Cretingham, Woodbridge)	19.10.02T
					(New owner 4.04)		
G-FACE	Cessna 172S Skyhawk SP	172S9194	N52733	24.10.02	M.O.Loxton	Parsonage Farm, Eastchurch	28.10.05T
G-FAIR	SOCATA TB-10 Tobago	241		13.10.81	Fairwings Ltd	Rochester	27. 7.07T
G-FALC	Aeromere F.8L Falco	3224	G-AROT	19. 2.81	P.J.Jones	Oxford	28. 6.04
G-FALO	Sequoia F.8L Falco	1401		10. 5.02	M.J.and S.E.Aherne	Top Farm, Croydon, Royston	AC
G-FAME	CFM Starstreak Shadow SA-II	PFA 206A-12973		23. 5.96	T.J.Palmer	Prestwick/Oban	20.11.04P
	(Built T.J.Palmer - kit no K.273SA) (Jabiru 2200)						
G-FAMH	Zenair CH.701 STOL	PFA 187-13301		26. 6.98	G.T.Neale	Sandown	11. 2.04P
	(Built A M Harrhy) (Jabiru 2200A)						
G-FANC*	Fairchild 24R-46 Argus	R46-347	N77647	16.10.89	A T.Fines	Priory Farm, Tibenham	26. 5.03T
			NC77647				
			(Cancelled 30.7.03 by CAA after arson attack Felthorpe 18.2.03: remnants only 8.04)				
G-FANL	Cessna R172K Hawk XPII	R1722873	N736XQ	7. 6.79	J.A Rees	Haverfordwest	6. 7.06T
G-FANN*	Hawker Siddeley HS 125 Series 600B	256019	HZ-AA1	13. 2.89	British Airways Aircraft Recovery Unit	Dunsfold	
			G-BARR	*(No CofA issued: cancelled 29.3.93 as WFU: on fire dump 3.00 as "HZ-AA1")*			
G-FARE	Robinson R44 Raven II	10454		16. 8.04	Toriamos Ltd	(Harolds Cross, Dublin)	7.10.07E
G-FARL	Pitts S-1E	1	N333AB	22.10.03	F.L.McGee	Andrewsfield	23. 3.05P
	(Built S.C. Burgess)				*(Noted also as "N333AB" 1.05)*		
G-FARM	SOCATA Rallye 235E	12832	F-GARF	10.10.78	Bristol Cars Ltd	White Waltham	27. 6.04
G-FARO	Star-Lite SL-1	PFA 175-11359		19. 6.89	M.K.Faro	Henstridge	6.10.03P
	(Built M.K.Faro) (Rotax 447)						
G-FARR	SAN Jodel D.150 Mascaret	58	F-BNIN	21. 7.81	G.H.Farr	Dairy House Farm, Worleston	19. 5.05P
G-FARY	QAC Quickie Tri-Q	PFA 94A-10951		2. 4.02	F.Sayyah	Enstone	17.10.03P
	(Built J C Simpson and F.Sayyah) (Limbach L2000)						
G-FATB	Commander Aircraft Commander 114B	14624	N6037Y	3. 7.96	James D.Peace and Co	(Kirkwall)	19. 8.05
G-FAUX	Cessna 182S Skylane	18280190	D-EWEI	4.12.03	R.S.Faux	Southend	18. 3.07T
G-FAVC	de Havilland DH.80A Puss Moth	DHC.225	CF-AVC	21.11.03	R.A Seeley	(London SW13)	
	(Built de Havilland Canada)						
G-FAYE	Reims/Cessna F150M	F15001252	PH-VSK	24. 1.80	Cheshire Air Training Services Ltd	Liverpool	29. 7.07T
G-FBAT	Aeroprakt A22 Foxbat	PFA 317-13591		16. 5.00	The Small Aeroplane Co Ltd	Otherton, Cannock	4. 5.05P
	(Built G Faulkner) (Rotax 912-S)						
G-FBFI	Canadair CL600-2B16 Challenger	5152	N601FB	24. 5.03	Gama Aviation Ltd	Farnborough	20. 1.05T
	(Built Bombardier Inc)		G-FBFI, N388PG, (N933PG), N18RF, N605BA, VP-COJ, VR-COJ, N777XX, C-GLWX				
G-FBII	Comco Ikarus C42 FB100	0310-6574		18.12.03	F.Beeson	(Nantwich)	17.12.04P
G-FBIX	de Havilland DH.100 Vampire FB.9	22100	7705M	24. 7.91	D.G.Jones	(St.Mary Hill, Bridgend)	
			WL505		*(Stored as "WL505" 7.04)*		

Reg	Type	C/n	Previous identity	Regn date	Owner (Operator)	Probable base	CA expiry
G-FBMW	Cameron N-90 HAB	3019		23. 4.93	K-J.Schwer	(Erbach-Donaurieden, Germany)	10. 9.05A
G-FBPI	Air Navigation and Engineering Co ANEC IV Missel Thrush			19. 1.99	R.Trickett	(Wolverhampton)	
	(Built R Trickett)	PFA 312-13417			*(Noted incomplete 3.02)*		
G-FBRN	Piper PA-28-181 Archer II	28-8290166	D-ERBN	3. 8.98	Herefordshire Aero Club Ltd	Shobdon	28.10.07E
			N82628				
G-FBWH	Piper PA-28R-180 Cherokee Arrow	28R-30368	SE-FCV	23. 8.78	F.T.Short	Whaley Farm, New York, Lincoln	6. 5.07
G-FCDB	Cessna 550 Citation Bravo	550-0985	N5269J	10. 9.01	Eurojet Aviation Ltd	(Belfast)	9. 9.05T
G-FCED	Piper PA-31T2 Cheyenne IIXL	31T-8166013	C-FCED	27. 9.04	Air Medical Fleet Ltd	Oxford	AC
			N2501Y				
G-FCLA	Boeing 757-28A	27621	N1789B	26. 2.97	Thomas Cook Airlines UK Ltd	Manchester	25. 2.06T
G-FCLB	Boeing 757-28A	28164	N751NA	25. 3.97	Thomas Cook Airlines UK Ltd	Manchester	29. 4.07T
			G-FCLB				
G-FCLC	Boeing 757-28A	28166		9. 5.97	Thomas Cook Airlines UK Ltd	Manchester	8. 5.06T
G-FCLD	Boeing 757-25F	28718		25. 4.97	Thomas Cook Airlines UK Ltd	Manchester	24. 4.06T
G-FCLE	Boeing 757-28A	28171		24. 5.98	Thomas Cook Airlines UK Ltd	Manchester	23. 5.07T
G-FCLF	Boeing 757-28A	28835		24. 3.99	Thomas Cook Airlines UK Ltd	Manchester	22. 3.05T
G-FCLG	Boeing 757-28A	24367	N701LF	18.12.98	Thomas Cook Airlines UK Ltd	Manchester	2. 4.05T
			EI-CLM, N381LF, N240LA, C-GTSK, C-GNXI, G-GAWB				
G-FCLH	Boeing 757-28A	26274	N751LF	17. 2.99	Thomas Cook Airlines UK Ltd	Manchester	12. 5.05T
			EI-CLU, N161LF				
G-FCLI	Boeing 757-28A	26275	N651LF	17. 3.99	Thomas Cook Airlines UK Ltd	Manchester	1. 6.05T
			EI-CLV, N151LF				
G-FCLJ	Boeing 757-2Y0	26160	N160GE	26. 4.99	Thomas Cook Airlines UK Ltd	Manchester	25. 4.05T
			EI-CJX, N3519M, N1786B, (B-2830)		*(Apple Vacations titles - Apple t/s)*		
G-FCLK	Boeing 757-2Y0	26161	N161GE	6. 4.99	JMC Airlines Ltd	Manchester	5. 4.05T
			EI-CJY, N3521N				
G-FCSP	Robin DR400/180 Régent	2022		24.10.90	F.C.Smith t/a FCS Photochemicals	Biggin Hill	20. 3.06
G-FCUK	Pitts S-1C	02	OH-XPB	9. 8.02	M.O'Hearne	Sherburn-in-Elmet	21. 5.05P
	(Built A Ronnberg)						
G-FDPS	Pitts S-2C	6066	N130PS	11.02.05	Flights and Dreams Ltd	(Bedford)	
	(Built Aviat Aircraft Inc)						
G-FEAB	Piper PA-28-181 Archer III	2843567	N53690	7. 7.04	Feabrex Ltd	Rochester	18. 7.07T
G-FEBE	Cessna 340A II	340A-0345	N405LS	12. 7.88	Terence and Gabriela Montague-Moore	Denham	10. 8.07
			(N37320)				
G-FEBY	Robinson R22 Beta II	3179		23. 4.01	Astra Helicopters Ltd	Bristol	12. 5.07T
G-FEDA	Eurocopter EC120B Colibri	1129	F-WQOD	2. 8.00	Federal Aviation Ltd	(Bury St.Edmunds)	22. 8.06T
G-FEES	Eurocopter EC135T2	0311		3.12.03	Cairnsilver Ltd c/o Marshall Wace LLP	(London, SW1)	27. 5.07T
G-FEFE	Scheibe SF25B Falke	46126	EI-BVZ	11. 4.94	D.G.Roberts	Aston Down	8. 6.06
	(Stark-Stamo MS1500)		D-KADB		tr G-FEFE Syndicate		
G-FELL	Europa Aviation Europa	PFA 247-13208		17. 3.98	R.A.Blackwell	North Weald	5.10.05P
	(Built J A Fell - kit no.372) (Tri-gear u/c)						
G-FELT	Cameron N-77 HAB	1174		19. 7.85	Allan Industries Ltd *"Fuzzy Felt"*	(Chinnor)	27. 3.04A
G-FERN	Mainair Blade 912	1342-1002-7-W1137		18.10.02	M.H.Moulai *(Op Silver Fern Microlights)*	Sandtoft	27.10.05P
G-FEWG	Fuji FA.200-160 Aero Subaru	FA200-232	G-BBNV	15.10.04	Caseright Ltd	Turweston	25. 4.99
G-FEZZ	Agusta-Bell 206B JetRanger II	8317	SU-YAD	16. 9.98	R.J.Myram	Booker	1.11.07E
			YU-HAT				
G-FFAB	Cameron N-105 HAB	4067		20. 2.97	B J Hammond *(Forever Friends titles)*	(Chelmsford)	2. 9.04A
G-FFEN	Reims/Cessna F150M	F15001204	PH-VGL	25. 8.78	D.Thrower	Poplar Hall Farm, Elmsett	19. 3.06T
G-FFFT	Lindstrand LBL 31A HAB	705		30. 5.00	The Aerial Display Co Ltd *(FT titles)*	(Looe)	23.12.03A
G-FFOX	Hawker Hunter T.7B	41H-670788	WV318	10. 1.96	Delta Engineering Aviation Ltd	Kemble	14. 5.05P
	(Composite including components of WV322- see G-BZSE)				*(As "WV318/D" in all-black c/s)*		
G-FFRA	Dassault Falcon 20DC	132	N902FR	28. 5.92	Cobham Leasing Ltd	Bournemouth	20.10.05A
			(N23FR), (N149FE), N2FE, N560L, N4348F, F-WMKG				
G-FFRI	Aérospatiale AS355F1 Ecureuil 2	5120	G-GLOW	15. 4.93	ATC Trading Ltd	Lasham	6. 5.06T
			G-PAPA, G-CNET, G-MCAH				
G-FFTI	SOCATA TB-20 Trinidad	1065		23. 2.90	R Lenk	Shipdham	19. 8.05T
G-FFTT	Lindstrand LBL Newspaper SS HAB	673		10. 7.00	The Aerial Display Co Ltd *(FT titles)*	(Looe)	19. 1.05A
G-FFUN	Pegasus Quantum 15	6655	G-MYMD	9. 6.99	P R Mailer	Longacres Farm, Sandy	12.12.04P
	(Rotax 503)						
G-FFWD	Cessna 310R II	310R0579	G-TVKE	20. 2.90	T.S.Courtman	(Gaulby, Leicester)	7. 4.07
			G-EURO, N87468				
G-FGID	Vought FG-1D Corsair	3111	N8297	1.11.91	Patina Ltd	Duxford	10. 9.05P
	(Built Goodyear Aircraft Corporation)		N9154Z, Bu.88297		*(Op The Fighter Collection as "KD345/130" in 1850 Sqdn RN c/s)*		
G-FGSK	Cameron Beer Crate-120 SS HAB	10417		6. 4.04	Ballon-Sport und Luftwerbung Dresden GmbH		
						(Dresden, Germany)	12. 5.05A
G-FHAJ	Airbus A320-231	0444	D-ACAF	14.11.01	MyTravel Airways Ltd	Manchester	15.11.04T
			N444RX, TC-ONF, N444RX, F-WWBY		*(Op MyTravel Lite)*		
G-FHAS	Scheibe SF25E Super Falke	4359	(D-KOOG)	14. 5.81	Burn Gliding Club Ltd	Burn	30.12.05
	(Limbach SL1700)						
G-FIAT	Piper PA-28-140 Cherokee F	28-7425162	G-BBYW	19. 7.89	Halegreen Associates Ltd and Demero Ltd		
			N9622N			Hinton-in-the-Hedges	31. 7.05T
G-FIBS	Aérospatiale AS350BA Ecureuil	2074	JA9732	14. 6.94	Pristheath Ltd	Denham	8. 8.06T
G-FIFE	Reims/Cessna FA152 Aerobat	FA1520351	G-BFYN	15. 2.95	Tayside Aviation Ltd	Glenrothes	4. 2.06T
G-FIFI	SOCATA TB-20 Trinidad	688	G-BMWS	16. 1.87	F.A Saker	Denham	23. 3.06
G-FIFT	Comco Ikarus C42 FB100	0409-6623		2. 8.04	Fly Buy Ultralights Ltd	Mill Farm, Shifnal	17.10.05P
G-FIGA	Cessna 152 II	15284644	N6243M	3. 6.87	Central Aircraft Leasing Ltd	Wolverhampton	10. 7.06T
					(Op Midland Flight Centre)		
G-FIGB	Cessna 152 II	15285925	N95561	16.11.87	Aerohire Ltd	Wolverhampton	12. 2.00T
G-FII	Extra EA.300/L	091	G-RGEE	6.12.04	G.G.Ferriman	Jericho Farm, Lambley	23. 7.04T
			D-ESEW		*(New CofR 12.04)*		
G-FIJJ	Reims/Cessna F177RG Cardinal RG	F177RG0031	G-AZFP	29. 4.99	D.R.Vale	Egginton	15. 6.06
	(Wichita c/n 17700194)						
G-FIJR	Lockheed L188PF Electra	1138	(EI-HCF)	12. 9.91	Atlantic Airlines Ltd	Coventry	12. 9.07T
			G-FIJR, C-FIJR, CF-IJR, N134US				

G-FIJV	Lockheed L188C Electra	1129	EI-HCE	29. 8.91	Atlantic Airlines Ltd	Coventry	27. 9.07T
			G-FIJV, C-FIJV, CF-IJV, N7143C				
G-FILE	Piper PA-34-200T Seneca II	34-8070108	N8140Z	23. 7.87	Alyson Jane Warren	Bristol	23.11.05T
G-FILL	Piper PA-31 Navajo C	31-7912069	OO-EJM	28. 6.96	P.V.Naylor-Leyland	Milton, Peterborough	23. 8.05
			N3521C7				
G-FINA	Reims/Cessna F150L	F15000826	G-BIFT	12.10.93	D.Norris	Cranfield	8.12.05T
			PH-CEW				
G-FIND	Reims/Cessna F406 Caravan II	F406-0045	OY-PEU	16. 8.90	Atlantic Air Transport Ltd	Coventry	5. 5.07T
			5Y-LAN, G-FIND, PH-ALV, F-WZDT				
G-FINZ	III Sky Arrow 650 T	PFA 298-13824		8. 1.03	A G.Counsell	(Banchory)	27. 5.05P
	(Built A G.Counsell)						
G-FIRE*	Vickers Supermarine 379 Spitfire FR.XIVc				See SECTION 3 Part 1	Palm Springs, California, US	
G-FIRM	Cessna 550 Citation Bravo	550-0940	N5263S	29. 9.00	Marshall of Cambridge Aerospace Ltd	Cambridge	2.10.07T
G-FIRS	Robinson R22 Beta II	2807		15. 4.98	M. and S.Chantler	(Crewe)	13. 5.07
G-FIRZ	Meridian Renegade 912	PFA 188-13494		10.12.99	P.J.Houtman	Park Farm, Eaton Bray	5. 4.05P
	(Built D M Wood and M Hanley)						
G-FISH	Cessna 310R II	310R1845	N2740Y	8. 5.81	Air Charter Scotland Ltd	Prestwick	28.10.05T
G-FISK*	Pazmany PL-4A	PFA 17-10129		14.12.88	K.S.Woodard	Little Snoring	11. 4.96P
	(Built K.S.Woodard) (Volkswagen 1834)				*(Cancelled 8.11.00 by CAA) (Noted damaged 3.03)*		
G-FIST*	Fieseler Fi.156C-3 Storch				See SECTION 3 Part 1	Vigna di Valle, Italy	
G-FITZ	Cessna 335	335-0044	G-RIND	20. 4.95	J.R.Naylor and D.Hughes	Leeds-Bradford	14. 3.05
			N2710L				
G-FIZU	Lockheed L188CF Electra	2014	EI-CHY	6. 4.93	Atlantic Airlines Ltd	Coventry	3. 1.08E
			G-FIZU, SE-IZU, (N857ST), N857U, PH-LLG				
G-FIZY	Europa Aviation Europa XS	PFA 247-13291	G-DDSC	16.12.99	G.N Holland	(Shoscombe, Bath)	
	(Built G.N Holland - kit no.384) (Jabiru 3300) *(Tri-gear u/c)*				*(Current status unknown)*		
G-FIZZ	Piper PA-28-161 Cherokee Warrior II	28-7816301	N2721M	1.12.78	Tecair Aviation Ltd	Shipdham	10. 3.07T
G-FJCE	Thruster T600T	9128-T600T-032		25.11.98	F.Cameron	(Craigavon, Belfast)	31. 7.04P
	(Rotax 532)						
G-FJEA	Boeing 757-23A	24636	G-LCRC	16. 5.03	Flyjet Ltd	Gatwick	9. 5.05T
			G-IEAB				
G-FJEB	Boeing 757-23A	24290	N290AN	21. 7.03	Flyjet Ltd	Gatwick	27. 7.06T
			G-OOOJ, N510FP, EC-EMU, EC-248				
G-FJET	Cessna 550 Citation II	550-0419	G-DCFR	7. 7.97	London Executive Aviation Ltd	London City	17. 1.05T
			G-WYLX, VH-JVS, G-JETD, N1217N				
G-FJMS	Partenavia P68B	113	G-SVHA	7. 9.92	J.B.Randle	Church Farm, Piltdown	13. 6.05T
			OY-AJH				
G-FJTH	Aeroprakt A22 Foxbat	PFA 317-13928		16. 7.03	F.J.T.Hancock	Shobdon	19.12.05P
	(Built F.J.T.Hancock)						
G-FKNH	Piper PA-15 Vagabond	15-291	CF-KNH	19. 3.97	M.J.Mothershaw	RAF Woodvale	15. 5.06
	(Continental C85)		N4517H, NC4517H				
G-FLAG	Colt 77A HAB	2000		20. 9.90	B.A Williams	(Maidstone)	10. 6.97T
G-FLAK	Beech E55 Baron	TE-1128	N4771M	26. 9.89	D.Clark *"Red Baron"*	Great Massingham	29. 7.05T
G-FLAP	Cessna A152 Aerobat	A1520856	G-BHJB	14. 6.02	Walkbury Aviation Ltd	Sibson	2. 9.05T
			N4662A				
G-FLAV	Piper PA-28-161 Warrior II	28-8016283	N8171X	7. 4.94	S.W.Parker tr The Crew Flying Group	Tollerton	23. 5.06
G-FLBI	Robinson R44 Raven II	10158		10. 9.03	Heli Air Ltd	Wellesbourne Mountford	16.10.06T
G-FLCA	Fleet 80 Canuck	068	CS-ACQ	18. 7.90	E.C.Taylor	(Balsall Common)	
			CF-DQP		*(On rebuild 3.01)*		
G-FLCT	Hallam Fleche	PFA 309-13389		21.10.98	R G Hallam	(Macclesfield)	
	(Built R G Hallam)				*(Current status unknown)*		
G-FLDG	Best Off Skyranger 912	BMAA/HB/328		21. 4.04	A.J.Gay	Doynton	1. 9.05P
	(Built A.J.Gay - kit no.SKRxxxx390)						
G-FLEA	SOCATA TB-10 Tobago	235	PH-TTP	31. 7.81	R.Kilburn	Leicester	26. 7.05
			G-FLEA		tr TB Group		
G-FLEW	Lindstrand LBL 90A HAB	586		21. 1.99	Lindstrand Hot Air Balloons Ltd	(Oswestry)	17. 6.05A
G-FLEX	Mainair Pegasus Quik	7953		15. 5.03	A J.Tyler	Beccles	24. 5.05P
G-FLGT	Lindstrand LBL 105A HAB	888		5.12.02	Ballongaventyr I Skane AB	(Kavlinge, Sweden)	26. 1.05A
G-FLII	Grumman GA-7 Cougar	GA7-0003	G-GRAC	18.12.91	Plane Talking Ltd	(Elstree)	23.10.04T
			C-GRAC, (N1367R), N730GA				
G-FLIK	Pitts S-1S	PFA 09-10513		7. 1.81	R.P.Millinship	Leicester	22. 5.05P
	(Built R P Millinship) (Lycoming O-320)						
G-FLIP	Reims/Cessna FA152 Aerobat	FA1520375	G-BOES	29.12.80	Walkbury Aviation Ltd	Sibson	2.10.06T
			G-FLIP				
G-FLIT	RotorWay Executive 162F	6324		22.12.98	R S Snell	Street Farm, Takeley	4. 7.05P
	(Built R F Rhodes) (RotorWay RI 162F)						
G-FLIZ	Staaken Z-21 Flitzer	PFA 223-13115		24. 3.97	M.A Wood	Shempston Farm, Lossiemouth	6. 7.05P
	(Built G L Brown - c/n 006)				*(As "D-694")*		
G-FLKE	Scheibe SF25C Rotax-Falke	44673		5.10.01	Faulkes Flying Foundation Ltd	Lasham	5.12.07E
	(Rotax 912S) *(Officially regd as "SF25C Falke")*						
G-FLKS	Scheibe SF25C Rotax-Falke	44662	D-KIEQ	16.10.00	Faulkes Flying Foundation Ltd	Dunstable	27.10.06
	(Rotax 912S) *(Officially regd as "SF25C Falke")*						
G-FLOA	Cameron O-120 HAB	4006		4.10.96	Floating Sensations Ltd	(Thatcham)	2.10.03T
G-FLOR	Europa Aviation Europa	PFA 247-12793		11.11.98	A F.C.Van Eldik	Pent Farm, Postling	1.10.05P
	(Built A F.C.Van Eldik - kit no.171) (Monowheel u/c)						
G-FLOW	Cessna 172S Skyhawk	172S9677	N6127S	19. 8.04	M P Dolan	(Eglinton, Londonderry)	19. 9.07T
G-FLOX	Europa Aviation Europa	PFA 247-12732		28. 6.95	T.W.Eaton	Fowle Hall Farm, Laddingford	13. 8.04P
	(Built P S Buchan, T W Eaton and B Lewer - kit no.129) (Jabiru 2200A) *(Monowheel u/c)*				tr DPT Group		
G-FLPI	Rockwell Commander 112A	205	SE-FLP	16. 3.79	Freshfield Lane Brickworks Ltd	Gamston	15. 4.06T
			(N1205J)				
G-FLRT	Europa Aviation Europa	PFA 247-12926		10. 3.04	K G Atkinson and D B Southworth	(York)	
	(Built K G Atkinson and D B Southworth)						
G-FLSH	IAV- Bacau Yakovlev Yak-52	877409	RA-44550	9. 6.03	Flashair Ltd	North Weald	17. 8.05P
			LY-AKF, DOSAAF 21 *(yellow)*				

G-FLSI	FLS AerospaceSprint 160	001		20. 8.93	Aces High Ltd *(Noted 1.05)*	North Weald	AC
G-FLTA	British Aerospace BAe 146 Series 200	E2048	N189US	25. 2.98	Flightline Ltd	Aberdeen	26. 2.07T
			N365PS		*(Op IAC)*		
G-FLTC	British Aerospace BAe 146 Series 300	E3205	G-JEBH	15.12.04	Flightline Ltd	Heathrow	18.10.06T
			G-BTVO, G-NJID, B-1777, G-BTVO, G-6-205 *(Op QANTAS)*				
G-FLTG	Cameron A-140 HAB	4506		3.11.00	Floating Sensations Ltd	(Thatcham)	14. 9.05T
G-FLTY	Embraer EMB-110P1 Bandeirante	110.215	G-ZUSS	28. 8.92	Skydrift Ltd	Glasgow	5. 8.05T
			G-REGA, N711NH, PT-GMH		*(Op Air Caledonian/ClasAir)*		
G-FLTZ	Beech 58 Baron	TH-1154	G-PSVS	21. 9.93	Flightline Ltd	Southend	21.10.07E
			N5824T, YV-266P				
G-FLUF	Lindstrand Bunny SS HAB	002		7. 4.93	Lindstrand Balloons Ltd *(Not built)*	(Oswestry)	
G-FLUX	Piper PA-28-181 Archer III	2843484	N5339U	13. 3.02	TEC Air Hire Ltd	Goodwood	21. 3.05T
G-FLVU	Cessna 501 Citation I	501-0178	N83ND	11. 6.98	AD Aviation Ltd	Liverpool	23. 6.07T
			N4246A, LV-PML, N67749				
G-FLYA	Mooney M.20J Model 201SE	24-3124		8. 6.89	BRF Aviation Ltd	Full Sutton	18. 2.05
G-FLYB	Comco Ikarus C42 FB100	0309-6572		15. 9.03	Ultrasport Aviation Ltd	Old Sarum	23.10.05P
G-FLYE	Cameron A-210 HAB	4216		12.12.97	A M.Holly t/a Exclusive Ballooning	(Berkeley)	8. 7.05T
G-FLYG	Slingsby T.67C	2074	PH-SGA	23. 8.02	G.Laden	(North Ferriby)	20. 2.06T
			(PH-SBA)				
G-FLYH	Robinson R22 Beta	1932	CS-HEQ	4.10.02	C.D.Cochrane	(Manston)	3.10.05T
			G-BXMR, N923FM, N2306E		t/a Cyclone Helicopters		
G-FLYI	Piper PA-34-200 Seneca	34-7250144	G-BHVO	1. 9.81	S Papi	Southend	10. 8.06T
			SE-FYY		*(Op Willowair Flying Club)*		
G-FLYP	Beagle B.206 Srs 2	B.058	N40CJ	15.10.98	Key Publishing Ltd	Cranfield	28. 1.06T
			N97JH, G-AVHO, VQ-LAY, G-AVHO				
G-FLYS	Robinson R44 Astro	0347		5. 6.97	Newmarket Plant Hire Ltd	Cambridge	27. 7.06T
G-FLYT	Europa Aviation Europa	PFA 247-12653		15. 5.95	K.F.and R.Richardson	Wellesbourne Mountford	6. 5.00P
	(Built D W Adams - kit no.057) (NSI EA-81/100) *(Conventional u/c)*						
G-FLYY	British Aircraft Corporation 167 Strikemaster Mk.80A	EEP/JP/163	R.Saudi AF 1112	3. 9.01	D.T.Barber	City of Derry	10. 3.05
			G-27-31				
G-FLZR	Staaken Z-21 Flitzer	PFA 223-13219		21. 9.01	J.F.Govan	East Linton	
	(Built J.F.Govan)				*(Current status unknown)*		
G-FMAH	Fokker F.28 Mk.0100	11286	N856US	26.8.03	Gazelle Ltd	Farnborough	AC
	(Fokker 100)		PH-EZU				
G-FMAM	Piper PA-28-151 Cherokee Warrior	28-7415056	G-BBXV	7. 6.90	P B Anderson	Southend	12. 3.06T
			N9603N		tr Lima Tango Flying Group		
G-FMGG	Maule M-5-235C Lunar Rocket	7260C	G-RAGG	30. 4.02	S.Bierbaum	Bodmin	18. 4.05
			N5632M				
G-FMKA	Diamond HK 36 TC Super Dimona	36.672		26. 4.00	D.J.M.Williams	(Guernsey)	11. 6.06T
G-FMSG	Reims/Cessna FA150K Aerobat	FA1500081	G-POTS	4. 1.95	G.Owen	Humberside	21. 1.07T
			G-AYUY				
G-FNEY	Reims/Cessna F177RG Cardinal RG	F177RG0059	F-BTFQ	24. 2.04	F.Ney	Egginton	12 .8.07
G-FNLD	Cessna 172N Skyhawk II	17270596	(G-BOUG)	3. 8.88	D.Wright and R.C.Laming	Fenland	26. 2.07
			N739KD		tr Papa Hotel Flying Group		
G-FNLY	Reims/Cessna F172M	F17200910	G-WACX	20. 3.89	C.F.Dukes	Exeter	10. 8.06T
			G-BAEX				
G-FNPT	Piper PA-28-161 Warrior III	2842163	N5346Y	2.10.02	Plane Talking Ltd	Elstree	17.10.05T
G-FOFO	Robinson R44 Raven II	10320		6. 4.04	P.A.Williams t/a Towers Aviation	Gamston	10. 5.07T
G-FOGG	Cameron N-90 HAB	1365		21.11.86	J.P.E.Money-Kyrle *"Phileas Fogg"*	(Chippenham)	25. 9.96A
G-FOGI	Europa Aviation Europa XS	PFA 247-13313		29.10.04	B.Fogg	(Crewe)	
	(Built B.Fogg)						
G-FOGY	Robinson R22 Beta	1020	N62991	5. 7.99	M.N. Cowley	Red House Farm, Preston Capes	25. 7.05T
			F-GGAI		t/a Dragonfly Aviation		
G-FOLI	Robinson R22 Beta II	2813		25. 4.98	K.Duckworth	Wolverhampton	13. 6.07
G-FOLY	Pitts S-2A	2213	N31477	26. 7.89	D.G.Gilmour	(Bearsden, Glasgow)	17. 8.06
	(Built Aerotek Inc)						
G-FONZ	Best Off Skyranger 912	BMAA/HB/304		15. 9.03	A A Pacitti	(Strathaven)	28. 7.05P
	(Built A A Pacitti - kit no.SKRxxxx369)						
G-FOPP	Neico Lancair 320	PFA 191-12319		14. 8.92	Airsport (UK) Ltd	Cranfield	17. 4.05P
	(Built M A Fopp) (Lycoming IO-320)						
G-FORC	SNCAN Stampe SV-4C	665	(G-BLTJ)	6. 6.85	C.C.Rollings	Gloucestershire	12. 6.06T
	(Renault 4P)		F-BDNJ				
G-FORD	SNCAN Stampe SV-4C	129	F-BBNS	7. 2.78	P.H.Meeson	Chilbolton	31. 7.98
	(DH Gipsy Major 10)				*(Damaged East Tytherley 16.7.96: stored 9.04)*		
G-FORN	Learjet Inc Learjet 45	45-2019	N50111	28. 1.05	Broomco 3598 Ltd	(Sheffield)	
G-FORR	Piper PA-28-181 Archer III	2843336	N4160Z	20. 4.00	B.and A E.Galt	Dundee	23. 4.06T
			G-FORR, N4160Z		t/a Buchanan Partnership		
G-FORS	Slingsby T.67C-3 Firefly	2082	PH-SGD	17.11.99	V.R.Coultan, M.J.Golding, E.P.Dablin and M.Glazer		
			(PH-SBD)		t/a Open Skies Partnership	Turweston	18.12.05T
G-FORT*	Boeing 299-O (B-17G-95-DL) Fortress				See SECTION 3 Part 1	Galveston, New Mexico, US	
G-FORZ	Pitts S-1S	PFA 09-13393		3.11.98	N.W.Parkinson	Sywell	
	(Built N Parkinson)				*(Noted 1.05)*		
G-FOSY	SOCATA MS.880B Rallye Club	1304	G-AXAK	7.12.00	A G.Foster	Humberside	17. 5.04
G-FOTO	Piper PA-E23-250 Aztec F	27-7654089	G-BJDH	27. 2.79	Simmons Aerofilms Ltd	Sywell	10. 3.06A
			G-BDXV, N62614				
G-FOWL	Colt 90A HAB	1198		11. 3.88	The Packhouse Ltd *(New owner 1.05)*	(Farnham)	6. 8.03A
G-FOWS	Cameron N-105 HAB	3995		11.12.96	F R Hart *(New owner 2.04)*	(Bishops Sutton, Bristol)	5.12.02A
G-FOXA	Piper PA-28-161 Cadet	2841240	N9192B	17.11.89	Leicestershire Aero Club Ltd	(Leicester)	8. 5.06T
G-FOXB	Aeroprakt A22 Foxbat	PFA 317-13878		15. 3.02	M.Raflewski	(Omagh, County Tyrone)	23. 8.05P
	(Buillt M Raflewski)						
G-FOXC	Denney Kitfox Model 3	PFA 172-11900		8. 1.91	G.Hawkins	Newton Peverill Farm, Sturminster Marshall	10. 6.05P
	(Built B W Davis - c/n 773)				*"Foxe Lady"*		
G-FOXD	Denney Kitfox Model 2	PFA 172-11618		22.11.89	P.P.Trangmar	(Hailsham)	13. 8.04P
	(Built D Hanley)						

Registration	Type	c/n	Previous identity	Reg date	Owner/operator	Base	CofA expiry
G-FOXF	Denney Kitfox Model 4 *(Built M.S.Goodwin)* (Rotax 912-UL)	PFA 172-12399		24. 3.00	M.S.Goodwin	Bridge of Weir	8. 7.05P
G-FOXG	Denney Kitfox Model 2 *(Built S M Jackson - c/n 452)* (Rotax 532)	PFA 172-11886		15. 8.90	A C.Newman tr Kitfox Group	Rochester	14.12.04P
G-FOXI	Denney Kitfox Model 2 *(Built I N Jennison)* (Rotax 532)	PFA 172-11508		21. 9.89	B.Johns	Combrook	21. 8.04P
G-FOXM	Bell 206B JetRanger II	1514	G-STAK G-BNIS, N35HF, N135VG	5. 2.93	R.P.Maydon t/a Milton Keynes City Air *(Op CSE Helicopters - Fox FM Radio)*	Oxford	3. 6.06T
G-FOXS	Denney Kitfox Model 2 *(Built S.P.Watkins and C.C.Rea - c/n 458)*	PFA 172-11571		15. 8.90	S.P.Watkins and C.C.Rea	Sheepcote Farm, Stourbridge	28. 4.05P
G-FOXX	Denney Kitfox *(Built R O F Harper)*	PFA/172-11509		1.11.89	R O F Harper *(Noted 9.04)*	Barton	
G-FOXZ	Denney Kitfox *(Built M Smalley and J C Whittle)*	PFA 172-11834		4.12.90	S.C.Goozee	(Wimborne)	30.11.05P
G-FPIG	Piper PA-28-151 Cherokee Warrior	28-7615001	G-BSSR N1190X	22. 3.00	G.F.Strain	Bournemouth	3. 4.06
G-FPLA	Beech B200 Super King Air	BB-944	N31WL HB-GHZ, HL5260, N1824V	3.12.97	FR Aviation Ltd *(Op Flight Precision)*	Durham Tees Valley	9. 3.07T
G-FPLB	Beech B200 Super King Air	BB-1048	N739MG N223MD, 9Y-TGY	3.12.97	FR Aviation Ltd *(Op Flight Precision)*	Durham Tees Valley	11. 1.07T
G-FPLC	Cessna 441 Conquest II	441-0207	G-FRAX G-BMTZ, N27280	14. 1.98	FR Aviation Ltd *(Op Flight Precision)*	Durham Tees Valley	29. 3.06T
G-FPLD	Beech B200 Super King Air	BB-1433	N43CE N43AJ, C-GMEV, C-GMEH, N8043K	2.11.01	Flight Precision Ltd	Durham Tees Valley	19.11.07T
G-FPSA	Piper PA-28-161 Warrior II	28-8616038	G-RSFT G-WARI, N9276Y	28.2.03	Deep Cleavage Ltd	Exeter	10. 3.07T
G-FRAF	Dassault Falcon 20E	295/500	N911FR I-EDIM, F-WRQQ	1. 9.87	Cobham Leasing Ltd *(Op FR Aviation)*	Bournemouth	2. 7.06A
G-FRAG	Piper PA-32-300 Six	32-7940284	N3566L	21. 1.80	T.A Houghton	Rochester	21. 5.04
G-FRAH	Dassault Falcon 20DC	223	G-60-01 N900FR, (N904FR), N22FE, N4407F, F-WPUX	31. 5.90	Cobham Leasing Ltd *(Op FR Aviation)*	Bournemouth	7.10.05A
G-FRAI	Dassault Falcon 20E	270	N901FR N37FE, N4435F, F-WPUZ	17.10.90	Cobham Leasing Ltd *(Op FR Aviation)*	Bournemouth	3. 6.06A
G-FRAJ	Dassault Falcon 20DC	20	N903FR (N25FR), N5FE, (N146FE), N5FE, N367GA, N367, N842F, F-WMKJ	30. 4.91	Cobham Leasing Ltd *(Op FR Aviation)*	Bournemouth	12.12.05A
G-FRAK	Dassault Falcon 20DC	213	N905FR N32FE, N4390F, F-WJMM	9.10.91	Cobham Leasing Ltd *(Op FR Aviation)*	Bournemouth	13. 4.06A
G-FRAL	Dassault Falcon 20DC	151	N904FR (N24FR), N3FE, (N148FE), N3FE, N810PA, N810F, N4360F, F-WMK	17. 3.93	Cobham Leasing Ltd	Durham Tees Valley	22.12.05A
G-FRAM	Dassault Falcon 20DC	224	N907FR N23FE, N4408F, F-WPUY	13. 5.93	Cobham Leasing Ltd *(Op FR Aviation)*	Bournemouth	26. 5.05A
G-FRAN	Piper J-3C-65 Cub (L-4J-PI) (Continental C90) *(Frame No.12447)*	12617	G-BIXY F-BDTZ, 44-80321	14. 7.86	I.Dole tr Essex L-4 Group *(As "480321/44-H" in USAAC c/s)*	Rayne Hall Farm, Braintree	2.11.05P
G-FRAO	Dassault Falcon 20DC	214	N906FR N33FE, N4400F, F-WNGO	23.10.92	Cobham Leasing Ltd *(Op FR Aviation)*	Bournemouth	28. 1.06A
G-FRAP	Dassault Falcon 20DC	207	N908FR N27FE, N4395F, F-WMKF	12. 7.93	Cobham Leasing Ltd *(Op FR Aviation)*	Bournemouth	19.10.05A
G-FRAR	Dassault Falcon 20DC	209	N909FR N28FE, N4396F, F-WLCX	2.12.93	Cobham Leasing Ltd *(Op FR Aviation)*	Durham Tees Valley	15. 2.06A
G-FRAS	Dassault Falcon 20C	82/418	CAF117501 20501, F-WJMM	31. 7.90	Cobham Leasing Ltd *(Op FR Aviation)*	Bournemouth	1.12.05A
G-FRAT	Dassault Falcon 20C	87/424	CAF117502 20502, F-WJMJ	31. 7.90	Cobham Leasing Ltd *(Op FR Aviation)*	Bournemouth	21. 2.06A
G-FRAU	Dassault Falcon 20C	97/422	CAF117504 20504, F-WJMJ	31. 7.90	Cobham Leasing Ltd *(Op FR Aviation)*	Bournemouth	15.12.05A
G-FRAW	Dassault Falcon 20C	114/420	CAF117507 20507, F-WJMM	31. 7.90	Cobham Leasing Ltd *(Op FR Aviation)*	Durham Tees Valley	9. 4.05A
G-FRAY	Cassutt Racer IIIM *(Built C.I.Fray)*	PFA 34-11211		24.10.90	C.I.Fray *(Current status unknown)*	(Macclesfield)	
G-FRBA	Dassault Falcon 20C	178/459	OH-FFA F-WPXF	16. 7.96	FR Finances Ltd *(Op FR Aviation)*	Bournemouth	16. 5.06A
G-FRCE	Folland Gnat T.1	FL.598	8604M XS104	28.11.89	Airborne Innovations Ltd	North Weald	3. 5.05P
G-FRGN	Piper PA-28-236 Dakota	2811046	N9244N	8. 2.96	Fregon Aviation Ltd	Church Farm, North Moreton	11. 3.05T
G-FRJB*	Britten SA-1 Sheriff				See SECTION 3 Part 1	Nottingham East Midlands	
G-FRNK	Best Off Skyranger 912 *(Built F.Tumelty)*	BMAA/HB/439		26. 1.05	F.Tumelty	(Castlewellan)	
G-FROH	Eurocopter AS350B2 Ecureuil	9024		16. 5.00	Specialist Helicopters Ltd	Nairn	28. 1.07T
G-FROM	Comco Ikarus C42 FB100	0309-6554		15. 9.03	Thames Valley Airsports Ltd	Chiltern Park,Wallingford	14.10.05P
G-FRYI	Beech 200 Super King Air	BB-210	G-OAVX G-IBCA, G-BMCA, N5657N	15. 3.96	London Executive Aviation Ltd	Stapleford	26. 3.04T
G-FRYL	Raytheon 390 Premier 1	RB-97	N6197F	11. 8.04	Houston Jet Services Ltd t/a Gregg Air	Oxford	17. 8.05T
G-FSHA	Denney Kitfox Model 2 *(Built S J Alston)*	PFA 172-11906		20. 9.99	P.P.Trangmar *(Current status unknown)*	(Hailsham)	
G-FSIX*	English Electric Lightning F.6				See SECTION 3 Part 1	Cape Town, South Africa	
G-FTAX*	Cessna 421C Golden Eagle II	421C0308	N8363G G-BFFM, N8363G	23. 8.84	Gold Air International Ltd *(Cancelled 8. 7.03 as WFU:) (Noted in open store minus various parts 9.03)*	Cambridge	16. 5.01T
G-FTFT*	Colt Financial Times 90SS HAB				See SECTION 3 Part 1	Newbury	
G-FTIL	Robin DR400/180 Régent	1825		10. 3.88	RAF Wyton Flying Club Ltd	RAF Wyton	20. 7.07T
G-FTIM	Robin DR400/100 Cadet	1829		6. 5.88	C.McGee tr Madley Flying Group	Shobdon	5.12.07E
G-FTIN	Robin DR400/100 Cadet	1830		6. 5.88	G.D.Clark and M.J.D.Theobald tr YP Flying Group	Blackpool	10.10.06
G-FTSE	Fairey Britten-Norman BN-2A Mk.III-2 Trislander	1053	G-BEPI	23. 5.00	Aurigny Air Services Ltd *(Hambros Banking titles)*	Guernsey	18.12.05T

Reg	Type	C/n	Previous identity	Regd	Owner/Operator	Base	CofA
G-FTSL	Canadair CL600-2B16 Challenger	5416	N161MD	8.12.04	Farglobe Transport Services Ltd		
			(G-), N161MN, N161MM, N604MG, C-GLXG			(Hamilton, Bermuda)	AC
G-FTUO	Van's RV-4	926	C-FTUQ	23.12.97	J.E. Singleton	Hinton-in-the-Hedges	7. 3.05P
	(Built T Martin) (Lycoming IO-360-B4A)				tr G-FTUO Flying Group *"Raven"*		
G-FTWO	Aérospatiale AS355F2 Ecureuil 2	5347	G-OJOR	27. 1.87	McAlpine Helicopters Ltd	Oxford	18. 3.05T
			G-FTWO, G-BMUS				
G-FUEL	Robin DR400/180 Régent	1537		15. 5.81	R.Darch	East Chinnock, Yeovil	18. 9.06
G-FUKM	Westland SA.314B Gazelle AH.1	1799	ZA730	18. 8.03	Falcon Aviation Ltd	(Reading)	
G-FULL	Piper PA-28R-200 Cherokee Arrow II	28R-7435248	G-HWAY	26.11.84	Stapleford Flying Club Ltd	Stapleford	17.12.05T
			G-JULI, (G-BKDC), OY-POV, CS-AQF, N43128				
G-FUNK	Yakovlev Yak-50	852908	RA852908	27. 3.98	Transair (UK) Ltd	Shoreham	4. 6.05P
			DOSAAF (46 blue ?)				
G-FUNN	Plumb BGP-1 Biplane	PFA 83-12744		16.10.95	J.D.Anson	(Liskeard)	
	(Built J.D.Anson)				*(Current status unknown)*		
G-FUSE	Cameron N-105 HAB	10639		30.11.04	LE Electrical Ltd	(Norwich)	
G-FUZY	Cameron N-77 HAB	1751		6. 5.88	Allan Industries Ltd *"Fuzzy Felt II"*	(Chinnor)	5.11.96A
G-FUZZ	Piper PA-18 Super Cub 95	18-1016	(OO-HMY)	11. 9.80	G.W.Cline	Gipsy Wood Farm, Warthill	17. 3.05P
	(L-18C-PI) *(Frame No.18-1086)*		ALAT-FMBIT, 51-15319		*(As "51-15319/319-A" in yellow US Army c/s)*		
G-FVBF	Lindstrand LBL 210A HAB	311		6.12.95	Virgin Airship and Balloon Co Ltd	(Hoccum, Bridgnorth)	14. 7.03T
					"Red November" (New owner 7.04)		
G-FVEL	Cameron Z-90 HAB	10580		24. 8.04	Fort Vale Engineering Ltd	(Nelson)	24. 8.05A
G-FWAY	Lindstrand LBL 90A HAB	967		7. 5.04	Harding and Sons Ltd t/a Fairway Furniture	(Plymouth)	19. 5.05A
G-FWPW	Piper PA-28-236 Dakota	2811018	N9145L	10.10.88	P.A and F.C.Winters	Oxford	29.10.06
G-FXBT	Aeroprakt A22 Foxbat	PFA 317-13787		7. 2.02	R.H.Jago	Mapperton Farm, Newton Peverill	28.10.05P
	(Built R Jago)						
G-FXII*	Vickers Supermarine 366 Spitfire F.XIIE	6S/197707	EN224	4.12.89	P.R.Arnold	(Newport Pagnell)	
	(On rebuild from components 3.00 as "EN224")				t/a Peter R.Arnold Collection *(Cancelled 9.5.02 as temporarily WFU)*		
G-FXIV*	Vickers Supermarine 379 Spitfire FR.XIVc				See SECTION 3 Part 1	Hannover, Germany	
G-FYAN	Williams Westwind MLB	MDW-1		6. 1.82	M.D.Williams	(Dunstable)	
	(Built M D Williams)						
G-FYAO	Williams Westwind MLB	MDW-001		6. 1.82	M.D.Williams	(Dunstable)	
	(Built M D Williams)						
G-FYAU	Williams Westwind Mk.Two MLB	MDW-002		6. 1.82	M.D.Williams	(Dunstable)	
	(Built M D Williams)						
G-FYAV	Osprey Mk.4E2 MLB	ASC-247		12. 1.82	C.D.Egan and C.Stiles	(Hounslow)	
	(Built Solent Balloon Group)						
G-FYBD	Osprey Mk.1E MLB	ASC-136		20. 1.82	M.Vincent	(Jersey)	
	(Built Solent Balloon Group)						
G-FYBE	Osprey Mk.4D MLB	ASC-128		20. 1.82	M.Vincent	(Jersey)	
	(Built Solent Balloon Group)						
G-FYBF	Osprey Mk.5 MLB	ASC-218		20. 1.82	M.Vincent	(Jersey)	
	(Built Solent Balloon Group)						
G-FYBG	Osprey Mk.4G2 MLB	ASC-204		20. 1.82	M.Vincent	(Jersey)	
	(Built Solent Balloon Group)						
G-FYBH	Osprey Mk.4G MLB	ASC-214		20. 1.82	M.Vincent	(Jersey)	
	(Built Solent Balloon Group)						
G-FYBI	Osprey Mk.4H MLB	ASC-234		20. 1.82	M.Vincent	(Jersey)	
	(Built Solent Balloon Group)						
G-FYCL	Osprey Mk.4G MLB	ASC-213		9. 2.82	P.J.Rogers	(Banbury)	
	(Built Solent Balloon Group)						
G-FYCV	Osprey Mk.4D MLB	ASK-276		19. 2.82	M.Thomson	(London SW11)	
	(Built Solent Balloon Group)						
G-FYCZ	Osprey Mk.4D2 MLB	ASC-244		24. 2.82	P.Middleton	(Colchester)	
	(Built Solent Balloon Group)						
G-FYDF	Osprey Mk.4D MLB	ASK-278		22. 3.82	K.A Jones	(Thornton Heath)	
	(Built Solent Balloon Group)						
G-FYDI	Williams Westwind Two MLB	MDW-005		29. 3.82	M.D.Williams	(Dunstable)	
	(Built M.D.Williams)						
G-FYDN	European 8C MLB	DD34/S.22		5. 4.82	P.D.Ridout	(Botley)	
	(Built D Eaves)						
G-FYDO	Osprey Mk.4D MLB	ASK-262		15. 4.82	N.L.Scallan	(Hayes)	
	(Built Solent Balloon Group)						
G-FYDP	Williams Westwind Three MLB	MDW-006		29. 3.82	M.D.Williams	(Dunstable)	
	(Built M.D.Williams)						
G-FYDS	Osprey Mk.4D MLB	ASK-261		15. 4.82	M.E.Scallan	(Hayes)	
	(Built Solent Balloon Group)						
G-FYEK	Unicorn UE-1C MLB	82024		2. 7.82	D. and D.Eaves	(Southampton)	
	(Built Unicorn Group)						
G-FYEO	Scallan Eagle Mk.1A MLB	001		20. 7.82	M.E.Scallan	(Hayes)	
	(Built M.E.Scallan)						
G-FYEV	Osprey Mk.1C MLB	ASK-294		10. 8.82	M.E.Scallan	(Hayes)	
	(Built Solent Balloon Group)						
G-FYEZ	Scallan Firefly Mk.1 MLB	MNS-748		22. 9.82	M.E.and N.L.Scallan	(Hayes)	
	(Built M.E.Scallan)						
G-FYFI	European E.84PS MLB	S.29		1.12.82	M.A Stelling	(Barton-le-Clay)	
	(Built D Eaves)						
G-FYFJ	Williams Westwind Two MLB	MDW-010		14.12.82	M.D.Williams	(Dunstable)	
	(Built M.D.Williams)						
G-FYFN	Osprey Saturn 2 DC3 MLB	ATC-250/MJS-11		17. 2.83	J.Woods and M.Woods	(Bracknell)	
	(Built Solent Balloon Group)						
G-FYFW	Rango NA-55 MLB	NHP-40		8.10.84	N.H.Ponsford and A M.Lindsay	(Leeds)	
	(Radio controlled)				t/a Rango Kite and Balloon Co *"Vaughan Williams"*		
G-FYFY	Rango NA-55RC MLB	AL-43		28. 2.85	A M.Lindsay	(Leeds)	
	(Radio controlled)				*"Fifi"*		

Reg	Type	c/n	Previous ID	Date	Owner	Base	CofA
G-FYGI	Rango NA-55RC MLB	NHP-54		26. 6.90	D.K.Fish	(Bedford)	
	(Radio controlled)						
G-FYGJ	Wells Airspeed-300 MLB	001		8.10.91	N.Wells	(Tunbridge Wells)	
	(Built N Wells)						
G-FYGM	Saffery/Smith Princess MLB	551		24.11.97	A and N.Smith	(Goole)	
	(Built C Saffery and N Smith)						
G-FZZA	General Avia F22-A	018		13. 8.98	APB Leasing Ltd	(Netherthorpe)	15.10.04T
		(Forced landed Welshpool 28.11.01: nose u/c collapsed causing damage to propeller and nose: noted 10.02)					
G-FZZI	Cameron H-34 HAB	2105		30.10.89	Magical Adventures Ltd		
						(West Bloomfield, Michigan, US)	30. 7.96A
G-FZZY*	Colt 69A HAB				See SECTION 3 Part 1	(Aldershot)	
G-FZZZ*	Colt 56A HAB				See SECTION 3 Part 1	Newbury	

G-GAAA - G-GZZZ

Reg	Type	c/n	Previous ID	Date	Owner	Base	CofA
G-GABD	Gulfstream GA-7 Cougar	GA7-0043	D-GABD	13. 4.82	C.B.Stewart *(Op Prestwick Flight Centre)*	Prestwick	18. 3.06T
G-GACA	Percival P.57 Sea Prince T.1	P57/58	WP308	2. 9.80	P.G.Vallance Ltd	Charlwood, Surrey	4.11.80P
					(Gatwick Aviation Museum as "WP308/572")		
G-GACB	Robinson R44 Raven II	10243		9. 1.04	A C Barker	Tatenhill	12. 2.07T
G-GAFA	Piper PA-34-200T Seneca II	34-7970218	D-GAFA	12.10.99	Oxford Aviation Services Ltd	Oxford	6.12.05T
			N2247Z				
G-GAFT	Piper PA-44-180 Seminole	4496162	N5324Q	24. 1.03	Atlantic Flight Training Ltd	Coventry	28. 1.06T
G-GAJB	Gulfstream AA-5B Tiger	AA5B-1179	G-BHZN	6. 4.87	G.A J.Bowles	Carlisle	23. 1.05T
			N37519				
G-GAJW	Bell 407	53186	N52245	15. 6.01	Titan Airways Ltd	Stansted	15. 6.04T
G-GALA	Piper PA-28-180 Cherokee E	28-5794	G-AYAP	31. 7.89	Flyteam Aviation Ltd	Elstree	10. 5.05T
			N11C				
G-GALB	Piper PA-28-161 Warrior II	28-8616021	D-EHMP	1. 9.00	Goodair Leasing Ltd	Cardiff	15. 6.07T
			N9097E, (N157AV), N9635N				
G-GALL	Piper PA-38-112 Tomahawk	38-78A0025	G-BTEV	1. 6.00	M.Lowe and K.Hazelwood	Cardiff	29. 1.05T
			N9315T				
G-GAME	Cessna T303 Crusader	T30300098	(F-GDFN)	25. 2.83	P.Heffron	Swansea	25. 8.07
			N2693C				
G-GAND	Agusta-Bell 206B JetRanger II	8073	G-AWMK	11. 1.00	Scotia Helicopters Ltd	Cumbernauld	1. 6.06T
			9Y-TFC, G-AWMK, (VR-BCV), G-AWMK				
	(Officially regd with c/n 8073 airframe exchanged with 5N-AQJ [8051] on rebuild 1999: original 5N-AQJ remains as c/n 8051 and became VH-JEF 1.00)						
G-GANE	Sequoia F.8L Falco	PFA 100-11100		25. 9.85	S.J.Gane	Kemble	15. 6.05P
	(Built S.J.Gane - c/n 906) (Lycoming IO-320)						
G-GASC	Hughes 369HS	110-0270S	G-WELD	11. 7.85	Aerolease Ltd	Conington	11. 5.05T
			G-FROG, OO-KAR				
G-GASP	Piper PA-28-181 Cherokee Archer II	28-7790013	N4328F	15.10.90	D.J.Turner tr G-GASP Flying Group	Fairoaks	2. 2.06
G-GASS	Thunder Ax7-77 HAB	1746		19. 4.90	M.W.Axon	(Brentwood)	4. 7.03A
					tr Servowarm Balloon Syndicate *"Travel Gas III"*		
G-GATE	Robinson R44 Raven II	10448		28. 7.04	J.W.Gate	Durham Tees Valley	16. 9.07T
G-GATT	Robinson R44 Raven II	10531		15.11.04	N R Gatt	(Tarporley)	22.11.07E
G-GAWA*	Cessna 140	9619	G-BRSM	17. 9.91	D.B.Almey	(Spalding)	11. 6.05
	(Continental C85)		N72454, NC72454		*(Cancelled 12.2.03 as wfu)*		
G-GAZA	Aérospatiale SA.341G Gazelle 1	1187	G-RALE	19. 6.92	The Auster Aircraft Co Ltd	Waltham, Leics	13. 7.05
			G-SFTG, N87712				
G-GAZI	Aérospatiale SA.341G Gazelle 1	1136	G-BKLU	29. 6.90	Sharpness Dock Ltd	(Plymouth)	13. 6.05T
			N32PA, N341VH, N90957		*(New owner 5.02)*		
G-GAZL	Westland SA.341C Gazelle HT.2	1078	G-CBBY	3. 4.02	R.M.Bailey	Addison Mains, Dalmahoy	6. 2.05P
			XW856				
G-GAZZ	Aérospatiale SA.341G Gazelle 1	1271	F-GFHD	14. 3.90	Stratton Motor Co (Norfolk) Ltd	Long Stratton	19. 6.06T
			YV-242CP, HB-XGA, F-WMHC		*(Op Cheqair)*		
G-GBAO	Robin R1180TD Aiglon	277	F-GBAO	9. 9.81	J.Kay-Movat	Slinfold	25.11.07E
	(Rebuild of R.1180 prototype F-WVKU c/n 01)						
G-GBEE	Mainair Pegasus Quik	8039		21. 5.04	L.G.White	Finmere	20. 5.05P
G-GBFF	Reims/Cessna F172N Skyhawk II	F17201565	F-GBFF	16. 6.99	Meridian Aviation Group Ltd	Bournemouth	17. 6.05T
	(Thielert TAE125-01)						
G-GBFR	Reims/Cessna F177RG Cardinal RG	F177RG0172	F-GBFR	7. 4.04	Airspeed Aviation Ltd	Egginton	
G-GBGA	Scheibe SF25C Rotax-Falke	44683	D-KIEJ	28. 8.02	British Gliding Association Ltd	Bicester	25. 9.05
	(Rotax 912S) *(Officially regd as "SF25C Falke")*						
G-GBGB	Ultramagic M-105 HAB	105/126		30.12.04	Universal Car Services Ltd	(Aldershot)	
G-GBHI	SOCATA TB-10 Tobago	19	F-GBHI	12.11.97	Robert Purvis Plant Hire Ltd	Glenrothes	21.12.06T
G-GBJP	Mainair Pegasus Quantum 15	8036		16. 8.04	R.G.Mulford	(Chatham)	15. 8.05P
G-GBLP	Reims/Cessna F172M Skyhawk II	F17201042	G-GWEN	9.11.84	Aviate Scotland Ltd	Edinburgh	20.11.06T
			G-GBLP, N14496		*(Op Edinburgh Air Centre)*		
G-GBLR	Reims/Cessna F150L	F15001109	N961L	30. 4.85	Almat Flying Club Ltd	Coventry	11. 7.05T
			(D-EDJE)				
G-GBRB	Piper PA-28-180 Cherokee C	28-2583	N8381W	2. 2.00	Border Air Training Ltd	Carlisle	6. 3.06T
G-GBSL	Beech 76 Duchess	ME-265	G-BGVG	27. 3.81	M.H.Cundey	Redhill/Alderney	29. 5.05
G-GBTA	Boeing 737-436	25859	G-BVHA	7. 2.94	British Airways plc	Gatwick	31.10.05T
			(G-GBTA)		*(Youm-Al-Suq t/s)*		
G-GBTB	Boeing 737-436	25860	OO-LTS	23.10.96	Aurigny Air Services Ltd	Guernsey	28.10.05T
			G-BVHB, OO-LTS, G-BVHB, (G-GBTB)				
G-GBUE	Robin DR400/120A Petit Prince	1354	G-BPXD	11. 5.89	J.A Kane	Bagby	5.10.07E
			F-GBUE		tr G-GBUE Group		
G-GBUN	Cessna 182T Skylane	18281280	N2157P	11.12.03	G.M.Bunn	Goodwood	2. 2.07
G-GBXS	Europa Aviation Europa XS	PFA 247-13196	"G-2000"	1. 4.98	A.J.Draper	Wombleton	5. 6.05P
	(Built Europa Aviation Ltd) (Rotax 914-UL) *(Monowheel u/c)*		G-GBXS				
G-GCAC	Europa Aviation Europa XS	PFA 247-13940		21. 8.02	J L Gunn	(Norwich)	
	(Built G.J.Cattermole - kit no.559) (Tri-gear u/c)				*(New owner 3.04)*		

G-GCAT	Piper PA-28-140 Cherokee B	28-26032	G-BFRH	22.10.81	P.F.Jude	Sturgate	5. 9.05
			OH-PCA		tr Group CAT		
G-GCCL	Beech 76 Duchess	ME-322	(G-BNRF)	5. 8.87	Aerolease Ltd	Conington	10.12.05T
			N6714U				
G-GCJL*	British Aerospace Jetstream Series 4100	41001		5. 2.91	BAE SYSTEMS (Operations) Ltd	Woodford	29. 4.95S
					(Cancelled 15.11.02 as wfu) (Stored 2.04)		
G-GCKI	Mooney M.20K Model 231	25-0401	N4062H	15. 8.80	B.Barr	Seething	15. 8.04
G-GCNZ*	Cessna 150M Commuter	15075933	C-GCNZ	8.11.88	Not known	Elstree	27. 3.98T
					(Cancelled 8.6.99 as destroyed) (Wreck noted 2.01)		
G-GCYC	Reims/Cessna F182Q Skylane II	F18200157	F-GCYC	11. 2.00	G-GCYC Ltd	Barton	12. 4.06
G-GDER	Robin R1180TD Aiglon II	280	F-GDER	15. 5.97	Berkshire Aviation Services Ltd	Fairoaks	24. 7.06
G-GDEZ	British Aerospace BAe 125 Series 1000B	259026	N9026	30.10.95	Frewton Ltd	Jersey	20.11.05
			G-5-743, ZS-ACT, ZS-CCT, G-5-743				
G-GDMW	Beech 76 Duchess	ME-316	D-GDMW	29.10.04	Apollo Aviation Advisory Ltd	Shoreham	AC
			LX-DRS, F-GCGB				
G-GDOG	Piper PA-28R-200 Cherokee Arrow II	28R-7635227	G-BDXW	17. 4.89	B Alidina	(Thornton Heath)	7.11.05
			N9235K				
G-GDRV	Van's RV-6	21367	C-GDRV	26.11.01	M.A Jardim de Queiroz	Gloucestershire	7. 4.03P
	(Built D.Piper) (Lycoming O-320)				tr G-GDRV Group *(Noted dismantled 9.04)*		
G-GDTU	Avions Mudry CAP.10B	193	F-GDTU	27. 5.99	Sherburn Aero Club Ltd	Sherburn-in-Elmet	15. 9.05T
			(N.....), F-GDTK, F-WZCI				
G-GEDY	Dassault Falcon 2000	208	F-WWVV	6. 5.04	Victoria Aviation Ltd	(Geneva, Switzerland)	5. 5.07
G-GEEP	Robin R1180TD Aiglon	266		9. 4.80	C.J.P.Green	Booker	26. 9.07
G-GEES	Cameron N-77 HAB	357		8.11.77	N.A Carr	(Leicester)	31. 5.00A
G-GEEZ	Cameron N-77 HAB	1159		3. 5.85	Charnwood Forest Turf Accountants Ltd	(Leicester)	7. 4.96A
					"Tic Tac"		
G-GEGE*	Robinson R-22 Beta	2994		19.10.99	Not known	Wellesbourne Mountford	28.10.02T
					(Cancelled 2.7.01 by CAA) (Stored in damaged condition 6.04		
G-GEHL	Cessna 172S Skyhawk SP	172S8324	N163RA	18. 6.03	Ebryl Ltd	White Waltham	30. 7.06T
G-GEHP	Piper PA-28RT-201 Arrow IV	28R-8218014	F-GEHP	24. 4.98	Aeros Leasing Ltd	Gloucestershire	11. 7.07T
			N82023				
G-GEMS	Thunder Ax8-90 Series 2 HAB	2287	G-BUNP	6.11.92	B.Sevenich, B.and S.Harren and W.Christoph		
					(Stolen 7.6.00)	(Aachen, Germany)	2. 4.01T
G-GENN	Gulfstream GA-7 Cougar	GA7-0114	G-BNAB	2.12.94	Plane Talking Ltd	Elstree	23. 2.07T
			G-BGYP				
G-GEOF	Pereira Osprey 2	PFA 70-10384		7. 9.78	G.Crossley	(Blackpool)	
	(Built G.Crossley)				*(Current status unknown)*		
G-GERT	Van's RV-7	PFA 323-13836		22. 3.04	M Castle-Smith	(Milton Abbas, Blandford Forum)	
	(M Castle-Smith, B West and A Burroughs)				tr Barnstormers		
G-GERY	Stoddard-Hamilton GlaStar	PFA 295-13475		6. 7.01	G.E.Collard	Popham	3. 7.05P
	(Built G.E.Collard) (Tailwheel u/c)						
G-GEUP*	Cameron N-77 HAB				See SECTION 3 Part 1	(Aldershot)	
G-GFAB	Cameron N-105 HAB	2048		4. 8.89	L Greaves	(Shepton Mallett)	7. 8.05A
G-GFCA	Piper PA-28-161 Cadet	2841100	N9174X	24. 4.89	Aeros Leasing Ltd	Gloucestershire	4.10.07T
G-GFCB	Piper PA-28-161 Cadet	2841101	N9175F	24. 4.89	AM and T Aviation Ltd	Bristol	15. 7.07T
G-GFCD	Piper PA-34-220T Seneca III	34-8133073	G-KIDS	31. 5.90	Stonehurst Aviation Ltd	Shoreham	7. 4.05T
			N83745				
G-GFEY	Piper PA-34-200T Seneca II	34-7870343	D-GFEY	13. 5.98	West Wales Airport Ltd	Aberporth	15. 1.05
			D-IFEY, N36599				
G-GFFA	Boeing 737-59D	25038	G-BVZF	10. 2.00	British Airways plc	Gatwick	2. 5.07T
			SE-DND, (SE-DNC)				
G-GFFB	Boeing 737-505	25789	LN-BRT	15. 2.00	British Airways plc	Gatwick	8. 5.06T
G-GFFC	Boeing 737-505	24272	LN-BRG	23. 3.00	British Airways plc	Manchester	8. 6.06T
G-GFFD	Boeing 737-59D	26419	LY-BFV	3. 7.00	British Airways plc	Manchester	13. 8.06T
			OY-SEG, G-OBMY, SE-DNI				
G-GFFE	Boeing 737-528	27424	LX-LGR	16. 6.00	British Airways plc	Gatwick	10. 7.06T
			(F-GJNP)				
G-GFFF	Boeing 737-53A	24754	G-OBMZ	2. 1.01	British Airways plc	Manchester	20. 9.05T
			SE-DNC				
G-GFFG	Boeing 737-505	24650	LN-BRC	20. 9.00	British Airways plc	Manchester	29.10.06T
			N5573K				
G-GFFH	Boeing 737-5H6	27354	VT-JAW	24.10.00	British Airways plc	Gatwick	23. 1.04T
			9M-MFG				
G-GFFI	Boeing 737-528	27425	LX-LGS	9.11.00	British Airways plc	Manchester	23. 1.07T
			(F-GJNQ)				
G-GFFJ	Boeing 737-5H6	27355	VT-JAZ	19. 1.01	British Airways plc	Manchester	12. 3.07T
			9M-MFH				
G-GFKY	Zenair CH.250	34	C-GFKY	23. 4.93	J.J.Beal	Earls Colne	22. 7.04P
	(Built D Koch) (Lycoming O-235)						
G-GFLY	Reims/Cessna F150L	F15000822	PH-CES	28. 8.80	Leagate Ltd	Norwich	1.12.04
G-GFMT	Cessna 172S Skyhawk	172S8258	C-GFMT	2.11.04	Upperstack Ltd	Barton	23.11.07E
			N341SP				
G-GFRD	Robin ATL	53	F-GFRD	2. 2.05	R H Ashforth tr Gloster Aero Group	Gloucestershire	
G-GFTA	Piper PA-28-161 Warrior III	2842047	N4132L	1. 4.99	One Zero Three Ltd	Guernsey	31. 3.05T
			G-GFTA, N9525N				
G-GFTB	Piper PA-28-161 Warrior III	2842048	N4120V	7. 5.99	One Zero Three Ltd	Guernsey	6. 5.05T
			G-GFTB, N4120V				
G-GGCT	Flight Design CT2K	Not known		18. 2.03	G.R.Graham	Kirkbride	23. 2.05P
	(Assembled Pegasus Aviation as c/n 7938)				*(Noted 5.03)*		
G-GGGG	Thunder Ax7-77 HAB	162		2. 8.78	T.A.Gilmour tr Flying G Group *"Flying G"*	(Stockbridge)	17. 8.99A
G-GGHZ	Robin ATL	123	F-GGHZ	10. 2.05	P.L.Bowman *(Noted 10.04)*	Headcorn	AC
G-GGLE	Piper PA-22-108 Colt	22-8914	N5234Z	13. 5.93	M.and S.Leonard	Crowfield	2.10.05T
	(Frame No.108-915) (Tail-wheel conversion incorporating parts from G-AROM c/n 22-8805)						

G-GGOW	Colt 77A HAB	1542		19. 6.89	G.Everett *"Charles Rennie Mackintosh"*	(Dartford)	26. 4.05A
G-GGRR	Scottish Aviation Bulldog Series 120/121	BH120/272	G-CBAM	11. 7.01	F.P.Corbett	White Waltham	28. 9.07
			XX614		*(As "XX614/V")*		
G-GGTT	Agusta-Bell 47G-4A	2538	F-GGTT	21. 8.97	Face and Fragrance Ltd	(Manchester)	8. 7.07
			I-ANDO				
G-GHEE	Aerotechnik EV-97 Eurostar	PFA 315-13840		14.12.01	C.J.Ball	(Cheltenham)	5. 7.05P
	(Built C.J.Ball)						
G-GHIA	Cameron N-120 HAB	2442		13.11.90	J.A Marshall	(Billingshurst)	25. 8.04T
G-GHIN	Thunder Ax7-77 HAB	1802		16. 7.90	N.T.Parry *"Pegasus"*	(Binfield)	4. 9.00A
G-GHKX	Piper PA-28-161 Warrior II	28-8416005	N380X	10. 6.99	Plane Talking Ltd	Elstree	28.11.07E
			G-GHKX, F-GHKX, N4318X		*(New CofR 10.04)*		
G-GHOW	Reims/Cessna F182Q Skylane II	F18200151	OO-MCD	20. 2.01	G.How	Top Farm, Croydon, Royston	15. 4.07T
			F-BJCE				
G-GHPG	Cessna 550 Citation II	550-0897	EI-GHP	22. 2.02	MCP Aviation (Charter) Ltd	Farnborough	18. 3.05T
			N5079V				
G-GHRW	Piper PA-28RT-201 Arrow IV	28R-7918140	G-ONAB	8.12.83	Bonus Aviation Ltd	Cranfield	14. 1.07T
			G-BHAK, N29555				
G-GHSI	Piper PA-44-180T Turbo Seminole	44-8107026	SX-ATA	2.12.94	M.G.Roberts	Bournemouth	1.12.97
			N8278Z		*(On long term rebuild by Airtime Aviation 12.04)*		
G-GHZJ	SOCATA TB-9 Tampico	941	F-GHZJ	4. 3.98	M.Haller	Little Snoring	25. 5.07
G-GIDY	Europa Aviation Europa XS	PFA 247-13467		18. 2.04	I.N.Robson tr Gidy Group	(Repton)	
G-GIGI	SOCATA MS.893A Rallye Commodore 180	11637	G-AYVX	28. 9.81	D.J.Moore	(Aston Down)	13. 4.00
			F-BSFJ		*(Noted 3.03)*		
G-GILI	Robinson R44 Raven	1436		5. 1.05	Twylight Management Ltd	(Douglas, Isle of Man)	
G-GILT	Cessna 421C Golden Eagle III	421C0515	G-BMZC	3. 7.97	M C Daniels t/a Skymaster Air Services	Lee-on-Solent	28. 3.07T
			N555WV, N555WW, N885WW, N885EC, N88541				
(G-GINO)	Auster V J/1 Autocrat				*See entry for G-AJEM*		
G-GIRA	British Aerospace HS 125 Series 700B	257103	YL-VIR	16. 1.02	EAS Aeroserviza SAS	(Venice, Italy)	25. 3.06T
			YL-VIP, VP-BOJ, VR-BOJ, G-LTEC, G-BHSU, G-5-12				
G-GIRY	American General AG-5B Tiger	10146	F-GIRY	5. 2.99	F.Neefs tr Romeo Yankee Flying Group	Elstree	25. 4.05T
G-GIWT	Europa Aviation Europa XS	PFA 247-13623		29. 3.01	A.Twigg	(Wootton Bassett)	
	(Built A Twigg - kit no.463) (Monowheel u/c)						
G-GJCD	Robinson R22 Beta	0966		22. 2.89	J.C.Lane	Wolverhampton	21.10.07E
G-GJKK	Mooney M.20K Model 252TSE	25-1227	F-GJKK	26.11.93	Pergola Ltd	Weston, Dublin	15. 3.06
G-GKAT	Enstrom 280C Shark	1200	F-GKAT	26. 8.97	C.D.Lacey	Compton Abbas	12. 7.07
			N5694Y				
G-GKFC	Tiger Cub RL5A LW Sherwood Ranger		G-MYZI	24.11.98	K.F.Crumplin	Franklyn's Field, Chewton Mendip	18.12.04P
	(Built K.F.Crumplin) (Jabiru 2200A)	PFA 237-12947					
G-GKRG	Cessna 172RG Cutlass RG	172RG0500	F-GKRG	3.12.03	Skytrax Aviation Ltd	Egginton	27.10.07E
			N5297V				
G-GLAD	Gloster Gladiator II	Not known	N5903	5. 1.95	Patina Ltd	Duxford	
					(Op The Fighter Collection as "N2276": on rebuild 9.02)		
G-GLAW	Cameron N-90 HAB	1808		10.10.88	R.A Vale	(Hurcott, Kidderminster)	5.12.04A
G-GLED	Cessna 150M	15076673	C-GLED	6. 1.89	Firecrest Aviation Ltd	Elstree	26.11.07E
G-GLHI	Best Off Skyranger 912S	BMAA/HB/392		30. 6.04	G.L.Higgins	Bitteswell	
	(Built G.L.Higgins - kit no.SKRxxxx468)						
G-GLIB	Robinson R44 Raven	1226		12. 6.02	Helisport UK Ltd	Earls Colne	11. 7.05T
G-GLST	Great Lakes Sport Trainer	PFA 321-13646		21. 7.03	D.A Graham	(Farnborough)	
	(Built D.A Graham)						
G-GLSU	Bucker Bu.181B-1 Bestmann	25071	D-EDUB	21. 7.04	G.R.Lacey	Fairoaks	29. 9.07
	(Built Hagglund and Soner)		Swedish AF Fv25071				
G-GLTT	Piper PA-31-350 Chieftain	31-8452004	N27JV	19. 9.97	Airtime Aviation Ltd	Bournemouth	26. 5.07T
			XA-SVW, XA-SGZ, N606SM, N4115D				
G-GLUC	Van's RV-6	20153	C-GLUC	15.10.99	Speedfreak Ltd	Crosland Moor	16. 8.05P
	(Built L De Sadeleer) (Lycoming O-320)						
G-GLUE	Cameron N-65 HAB	390		17. 3.81	L.J.M.Muir and G.D.Hallett	(East Molesey)	17. 7.90A
					(Mobile Windscreens titles) "Tacky Jack/Jack of Herts"		
G-GLUG	Piper PA-31-350 Chieftain	31-8052077	N2287J	1. 9.94	Champagne-Air Ltd	Newcastle	13. 4.06T
			G-BLOE, G-NITE, N3559A				
G-GMAA	Learjet Inc Learjet 45	45-167	N5012V	1. 5.02	Gama Aviation Ltd	Farnborough	1. 5.06T
G-GMAB	British Aerospace BAe 125 Series 1000B	259034	N81HH	21.11.01	Gama Aviation Ltd	Farnborough	11. 3.07T
			N290H, G-BUWX, G-5-761				
G-GMAC	Gulfstream Aerospace Gulfstream G-IVSP	1058	VP-BME	23. 7.04	Gama Aviation Ltd	Farnborough	24. 7.07T
			VP-BSF, N70PS, N458GA				
			(Extensively damaged on landing Teterboro, New Jersey, US 1.12.04 when aircraft went off right side of runway and struck trees)				
G-GMAX	SNCAN Stampe SV-4C	141	G-BXNW	19. 6.87	Glidegold Ltd	Booker	29. 8.93T
	(Renault 4P)		F-BBPB		*(Damaged in crash Booker 3.6.91: on rebuild 5.96: current status unknown)*		
G-GMPA	Aérospatiale AS355F2 Ecureuil 2	5409	G-BPOI	26. 9.89	Greater Manchester Police Authority	Barton	9. 1.05T
G-GMPB	Pilatus Britten-Norman BN-2T-4S Defender 4000	4011	G-BWPU	5. 4.02	B-N Group Ltd	Bembridge	1. 7.06T
			(9M-TPD), G-BWPU		*(Op Greater Manchester Police)*		
G-GMPS	MD Helicopters MD 900	900-00081	N7033K	8. 1.01	Greater Manchester Police Authority	Barton	12. 2.07T
G-GMSI	SOCATA TB-9 Tampico	145		18. 9.80	M.L.Rhodes	Gloucestershire	4. 6.06T
G-GNAA	MDH MD.900 Explorer	900-00079	PH-RVD	15.12.04	Police Aviation Services Ltd	Durham Tees Valley	14.12.07E
			N70279		*(Op Great Northern Air Ambulance)*		
G-GNJW	Comco Ikarus C42 FB100 VLA	PFA 322-13717		21. 8.01	I.R.Westrope	(Haverhill)	18. 7.05P
	(Built I.R.Westrope)						
G-GNMG	Cessna U206F Stationair	U2062173	F-GNMG	26. 5.04	P.Marsden	Dunkeswell	28. 7.07T
			F-MJAE, F-BRGK, N7325Q				
G-GNTB	SAAB-Scania SF.340A	340A-082	HB-AHL	30. 9.91	Loganair Ltd	Glasgow	13. 3.05T
			SE-E82				
G-GNTC	SAAB-Fairchild SF.340A	340A-020	HB-AHE	25. 9.92	Aurigny Air Services Ltd	Guernsey	24. 9.05T
			OK-RGS, HB-AHE, SE-E20				
G-GNTZ	British Aerospace BAe 146 Series 200	E2036	G-CLHB	31. 3.00	British Airways Citiexpress Ltd	Aberdeen	25.11.06T
			G-GNTZ, HB-IXB, N175US, N355PS				

Reg	Type	C/n	Previous identity	Date	Owner	Base	CofA
G-GOAC	Piper PA-34-200T Seneca II	34-7770007	D-GOAC N5329F	9.12.03	Oxford Aviation Services Ltd	Oxford	15. 2.07T
G-GOAL	Lindstrand LBL 105A HAB	420		18.11.96	I.Chadwick tr Balloon Preservation Flying Group	(Horsham)	3. 7.05A
G-GOBT	Colt 77A HAB	1815		13. 2.91	British Telecommunications plc *"Sky Piper"*	(Thatcham)	18. 3.00A
G-GOCX	Cameron N-90 HAB	2619		7. 8.91	R.D.Parry	(Hong Kong, PRC)	9. 4.05A
G-GOGB	Lindstrand LBL 90A HAB	1011	G-CDFX	20. 1.05	J.Dyer	(Farnborough)	
G-GOGS	Piper PA-34-200T Seneca II	34-7570228	N1172X	23. 7.03	A Semple	Shoreham	4. 8.06
G-GOGW	Cameron N-90 HAB	3304		31. 8.94	S.E.Carrol *(Great Western titles)*	(Caversham, Reading)	17. 5.05A
G-GOJP	Piper PA-46-350P Malibu Mirage	4636309	G-CREW N41865	17.03.03	Plato Management Ltd	Oxford	22.10.06
G-GOLF	SOCATA TB-10 Tobago	250		21.12.81	A C.Scamell tr Golf Golf Group	Biggin Hill	30. 7.06
G-GONN	Eurocopter AS355N Ecureuil 2	5557	HB-XIQ VR-BQM, G-BVNW, (D-HWPC)	20. 3.02	Taggart Homes Ltd	(Londonderry)	7. 4.05T
G-GOOD	SOCATA TB-20 Trinidad	1657	F-GNHJ	4.11.94	T.M.Sloan and M.P.Bowcock	Goodwood	12. 2.07T
G-GOON	MD Helicopters MD 600N	RN030	VP-CCW N9230Q	20. 2.04	Cumbrian Seafoods Ltd	Maryport	25. 3.07T
G-GORE	CFM Streak Shadow *(Built D N Gore - kit no K.138-SA)* (Rotax 532) *(PFA project no. duplicates TEAM mini-MAX G-MWFD)*	PFA 206-11646		12. 4.90	M.S.Clinton	Old Sarum	21. 8.04P
G-GORF	Robin HR200/120B	291	F-GORF	14. 1.00	J.A Ingram	Tollerton	24. 4.06T
G-GOSL	Robin DR400/180 Régent	1974	G-BSDG	14. 1.02	R.M.Gosling	Stones Farm, Wickham St.Pauls, Essex	23. 5.05
G-GOTC	Gulfstream GA-7 Cougar	GA7-0074	G-BMDY OO-LCR, OO-HRA	25. 6.97	Wakelite Ltd	Elstree	10. 3.05T
G-GOTH	Piper PA-28-161 Warrior III	2842208	N3088U	21. 6.04	J.Gosling	(Stockport)	24. 6.07T
G-GOTO	Piper PA-32R-301T Saratoga IITC	3257026	N92965 G-GOTO, N92965	8. 1.98	J.A Varndell	Blackbushe	27. 1.07
G-GOUP	Robinson R22 Beta	1663	G-DIRE	9. 1.01	Electric Scribe 2000 Ltd	Panshanger	27. 3.06T
G-GPAG	Van's RV-6 *(Built P A Green)* (Lycoming O-320)	PFA 181A-13306		18. 5.01	P.A Green	Old Sarum	9. 1.05P
G-GPAS	Jabiru Aircraft Jabiru UL-450 *(Built G.D.Allen)*	PFA 274A-13823		15. 1.02	G.D.Allen	Priory Farm, Tibenham	18. 8.05P
G-GPEG	Cameron Sky 90-24 HAB	4849		31. 5.00	N.T.Parrry "*Pegasus*"	(Bracknell)	6. 9.05A
G-GPFI	Boeing 737-229	20907	F-GVAC OO-SDA, LX-LGN, OO-SDA	31. 7.03	European Skybus Ltd *(WFU 9.04 and stored engineless 12.04: current CofR 1.05)*	Bournemouth	15.11.06T
G-GPMW	Piper PA-28RT-201T Turbo Arrow IV	28R-8031041	N3576V	3. 7.89	Calverton Flying Group Ltd	(London W4)	14. 5.07T
G-GPST	Phillips ST.1 Speedtwin *(Built P J C Phillips - c/n 1)* (Continental O-200-A) *(PFA project no. duplicates Kolb Twinstar G-MWWM)*	PFA 207-11645		21. 6.90	Speedtwin Developments Ltd	Upper Cae Garw Farm, Trelleck, Monmouth	3.12.03P
G-GRAY*	Cessna 172N Skyhawk II	17272375	N4859D	3.12.79	Trueman Aviation Ltd *(Damaged ditching Firth of Forth, Mussleburgh 2.4.93: cancelled 27.9.00 as WFU: noted 6.04)*	Tollerton	13. 2.95
G-GREY	Piper PA-46-350P Malibu Mirage	4636155	OY-LAR N1280K, N4129D	28.11.03	S.A and K.J.Williams	Gloucestershire	10.12.06T
G-GRID	Aérospatiale AS355F1 Ecureuil 2	5012	TG-BOS	28. 3.89	National Grid Co plc	Sheffield City	18. 6.07T
G-GRIN	Van's RV-6 *(Built A Phillips)* (Lycoming O-320)	PFA 181-12409		8. 1.98	A Phillips	Boarhunt Farm, Fareham	1. 6.05P
G-GRIP	Cameron Colt Bibendum 110SS HAB	4224		5. 1.98	The Aerial Display Co Ltd *(Michelin titles)* *(Donated to Balloon Preservation Group 2004)*	(Looe)	25. 1.02A
G-GROE	Grob G.115A	8054	I-GROE (D-EGVV)	22. 6.04	H.and Eve Merkado	Panshanger	30. 9.07
G-GROL	Maule MXT-7-180 Star Rocket	14091C		16. 6.98	D.C., C.and C.Croll	Southend	16.12.04
G-GRPA	Comco Ikarus C42 FB100	0407-6609		26. 8.04	G.R.Page	Taunton)	27 .8.05P
G-GRRC	Piper PA-28-161 Warrior II	2816076	G-BXJX HB-POM, D-EJTB, N9149X	9. 3.98	Goodwood Road Racing Co Ltd *(Op Goodwood Flying Club)*	Goodwood	30.11.06T
G-GRRR	Scottish Aviation Bulldog Series 120/122	BH120/229	G-BXGU Ghana AF G-105	19.10.98	Horizons Europe Ltd	Old Sarum	25. 5.05T
G-GRWW	Robinson R44 Raven II	10382	G-HEEL	6.12.04	G.R.Williams	(Easton, Winchester)	17. 6.07E
G-GRYZ	Beech F33A Bonanza	CE-1668	F-GRYZ D-ESNE, N80011, (OY-GEN), N80011	4.10.99	J.Kawadri and M.Kaveh	Booker	13.11.05
G-GSCV	Comco Ikarus C42 FB UK *(Built G.Sipson)*	PFA 322-13939		5. 9.02	G.Sipson	(Coventry)	1. 3.05P
G-GSJH	Bell 206B-3 JetRanger III	3958	G-PENT G-IIRB, N903CA	15. 3.02	S.J.Hanson t/a Interheli	Farleton, Lancaster	10. 6.05T
G-GSPG	Hughes 369HS	45-0738S	G-GEEE G-BDOY	17. 1.03	S.P.Giddings t/a S.Giddings Aviation	Gowles Farm, Sherrington, Buckingham	8.10.04
G-GSPN	Boeing 737-31S	29267	(G-SPAN) D-ADBW, N60436, N1787B	9. 2.04	Globespan Airways Ltd t/a Flyglobespan.com	Glasgow	21. 3.07T
G-GSSA	Boeing 747-47UF	29256	N495MC (N496MC)	23. 1.02	Global Supply Systems Ltd *(British Airways c/s)*	Stansted	27. 1.05TC
G-GSSB	Boeing 747-47UF	29252	N491MC	17. 1.03	Global Supply Systems Ltd	Stansted	20. 1.06TC
G-GSSC	Boeing 747-47UF	29255	N494MC OO-TJA, N494MC	27. 8.03	Global Supply Systems Ltd	Stansted	28. 8.06TC
G-GTHM	Piper PA-38-112 Tomahawk II	38-81A0171	C-GTHM N91338	17.11.86	Turweston Aero Club Ltd	Turweston	16.11.07E
G-GUAY	Enstrom 480	5036		1.12.98	Testactual Ltd t/a Heliway Aviation	(Fareham)	30.12.04T
G-GUCK	Beech C23 Sundowner 180	M-2221	G-BPYG N6638R	9. 4.92	J.T.Francis	Headcorn	28.10.07E
G-GUFO	Cameron Saucer 80SS HAB	1641	C-GUFO G-BOUB	10. 6.98	Magical Adventures Ltd	(West Bloomfield, Michigan, US)	14. 8.05A
G-GULF	Lindstrand LBL 105A HAB	320		3.11.95	M.A Webb	(Meifod, Shrewbury)	17. 7.02A
G-GULP	III Sky Arrow 650 T *(Built H R Rotherwick - kit no K130)* (Rotax 914)	PFA 298-13664		4.12.00	S.Marriott	Old Sarum	9.12.05P
G-GUMS	Cessna 182P Skylane	18261643	G-CBMN ZS-KJS, N21458	11.11.02	L.W.Scattergood	Breighton	24. 6.05T

Reg	Type	C/n	Previous identity	Regd	Owner	Base	CofA
G-GUNS	Cameron V-77 HAB	2221		9. 5.90	Royal School of Artillery Hot Air Balloon Club		
					"Guns"	Newtwon Hungerford	23. 7.04A
G-GURL*	Cameron A-210 HAB				See SECTION 3 Part 1	(Aldershot)	
G-GURN	Piper PA-31 Turbo Navajo C	31-7912117	G-BHGA	20. 6.01	Neric Ltd	Fowlmere	23.11.06
	(Winglets)		N3539M				
G-GURU	Piper PA-28-161 Warrior II	28-8316018	PH-SVJ	12. 2.02	Leeds Flying School Ltd	Leeds-Bradford	25. 2.05T
			N83085				
G-GUSS	Piper PA-28-151 Cherokee Warrior	28-7415497	G-BJRY	16. 8.95	M.J.Cleaver and J.M.Newman	Southend	23. 6.06T
			N43453				
G-GUST	Agusta-Bell 206B JetRanger II	8192	G-CBHH	30. 8.96	Gatehouse Estates Ltd	Sywell	27. 5.06T
			F-GALU, G-AYBE				
G-GUYS	Piper PA-34-200T Seneca II	34-7870283	G-BMWT	14. 7.87	R.J.and J.M.Z.Keel	Sturgate	3. 5.05
			N31984				
G-GVPI	Evans VP-1	PFA 62-10668		9. 8.02	G.Martin	(Hinckley)	
	(Built P A Schafle and G.Martin)						
G-GWIZ	Colt Clown SS HAB	1369	(G-BPWU)	25. 4.89	Magical Adventures Ltd		
						(West Bloomfield, Michigan, US)	13. 4.99A
G-GWYN	Reims/Cessna F172M Skyhawk II	F17201217	PH-TWN	5. 3.81	C.G.Tandy tr Magic Carpet Flying Company	Denham	27. 4.05
G-GYAK	Yakovlev Yak-50	852905	RA-02246	9.12.02	M.V.Rijske and M.W.Levy	North Weald	23.12.05P
			DOSAAF (43 *blue* ?)		*"46"*		
G-GYAT	Sud-Aviation Gardan GY-80-160 Horizon	136	D-EAZZ	13.12.02	J. Luck	Rochester	13. 3.06T
			HB-DCL, F-BMUU		tr Rochester GYAT Flying Group Club		
G-GYAV	Cessna 172N Skyhawk II	17271362	C-GYAV	26. 8.87	Southport and Merseyside Aero Club (1979) Ltd	Liverpool	15. 5.06T
G-GYBO	Sud-Aviation Gardan GY-80-180 Horizon	228	OY-DTN	4. 8.98	A L.Fogg	(Warwick)	28. 8.06T
			SE-FGL, OY-DTN				
G-GYMM	Piper PA-28R-200 Cherokee Arrow B	28R-7135049	G-AYWW	22. 2.90	L.E.Vandervort	(Bedford)	16.10.07E
			N11C				
G-GYRO	Campbell Cricket rep	PFA G/03-1046		26. 2.82	J.W.Pavitt	St.Merryn	20. 6.05P
	(Built Howell, Pitcher and J W Pavitt) (Rotax 532) *(Originally registered as Bensen B.8 (c/n 01 & PFA G/01-1046)*						
G-GYTO	Piper PA-28-161 Warrior III	2842082	N160FT	11. 5.00	Wellesbourne Flyers Ltd	Wellesbourne Mountford	1. 6.06T
			N9511N		t/a Wellesbourne Aviation		
G-GZDO	Cessna 172N Skyhawk II	17271826	C-GZDO	11.10.88	G.Cambridge and G.W.J.Hall	Elstree	16. 6.07T
			(N5299E)		t/a Cambridge Hall Aviation *(Op Firecrest Aviation)*		
G-GZLE	Aérospatiale SA.341G Gazelle 1	1145	G-PYOB	8. 5.01	R.G.Fairall	Redhill	22.12.06
			G-WELA, G-SFTD, G-RIFC, G-SFTD, N641HM, N341BB, F-WKQH				

G-HAAA - G-HZZZ

Reg	Type	C/n	Previous identity	Regd	Owner	Base	CofA
G-HACK	Piper PA-18-150 Super Cub	18-7168	SE-CSA	20.11.97	S.J.Harris	Kemble	4.12.06
			N10F				
G-HADA	Enstrom 480	5017		17. 9.96	W.B.Steele	Whitchurch, Shropshire	9.10.05
G-HAEC	Commonwealth CAC-18 Mustang 22	CACM-192-1517	VR-HIU	1. 5.85	R.W.Davies	Little Robhurst Farm, Woodchurch	13. 6.05P
			(RP-C651), PI-C651, VH-FCB, A68-192		*(Op The Old Flying Machine Ltd)*		
	(Composite rebuilt 1974-76 using major components ex Philippine AF P-51D 44-72917) (As "472218/WZ-I" in 78th FG USAAF c/s) "Big Beautiful Doll"						
G-HAFG	Cessna 340A	340A1806	JY-AFG	5. 8.04	Haller and Sons (Dereham) Ltd	Little Snoring	AC
			N1230V				
G-HAIB	Aviat A-1B Husky	2255	N53HY	16.11.04	Aviat Aircraft (UK) Ltd	Lower Grounds Farm, Sherlowe	AC
G-HAIG	Rutan LongEz	PFA 74A-11149		20. 5.86	L.J.Bayliss	Gloucestershire	29.11.05P
	(Built P N Haig c/n 1983-L) (Lycoming O-235)						
G-HAIR	Robin DR400/180 Régent	2479		7.12.00	Arden Ridge Developments Ltd	(Southham)	25. 1.07T
					t/a Racoon International		
G-HAJJ	Glaser-Dirks DG-400	4-225		15. 2.88	W.G.Upton and J.G.Kosak	RNAS Culdrose	21. 4.06
G-HALC	Piper PA-28R-200 Cherokee Arrow II	28R-7335042	N91253	26.11.90	Halcyon Aviation Ltd	Barton	3. 9.06
			C-FFQO, CF-FQO				
G-HALE	Robinson R44 Astro	0492		6. 8.98	Barhale Surveying Ltd	Elstree	16. 9.07T
G-HALJ	Cessna 140	8336	N89308	30. 4.96	H.A Lloyd-Jennings	Scotland Farm, Hook	8.10.05
	(Continental C85)		NC89308				
G-HALL	Piper PA-22-160 Tri-Pacer	22-7423	G-ARAH	8.11.79	F.P.Hall	Clipgate Farm, Denton	17. 8.06
			N10F				
G-HALP	SOCATA TB-10 Tobago	192	G-BITD	19. 8.81	D.Halpern	Booker	25. 9.06T
G-HALT	Mainair Pegasus Quik	8063		1. 9.04	J.A.Horn	Shotton Colliery, Peterlee	31. 8.05P
G-HAMA	Beech 200 Super King Air	BB-30	N244JB	16.11.84	Gama Aviation Ltd	Farnborough	19.11.05T
	(To Model B200 status + 4-blade propellers 1999)		N211JB, N3090C, N3030C, N200CA				
G-HAMI	Fuji FA.200-180 Aero Subaru	FA200-188	G-OISF	31. 1.92	K.G.Cameron, M.P.Antoniak and Renzacci (UK) plc		
			G-BAPT			White Waltham	31. 3.05T
G-HAMM	Yakovlev Yak-50	832409	LY-ANG	15.10.02	A D.Hammond	Little Gransden	31.10.05P
			DOSAAF 81				
G-HAMP	Bellanca 7DCA Champ	30-72	N9173L	8. 8.88	K.Macdonald	Rushett Farm, Chessington	14. 9.05P
G-HANS-	Robin DR400 2+2	1384		2. 3.79	J.S.Russell	(Cleghorn, Lanark)	23.11.06T
G-HANY	Agusta-Bell 206B-3 JetRanger III	8598	G-JEKP	5. 1.01	BAC Management (IoM) Ltd	(Douglas, Isle Of Man)	13. 4.06T
			D-HMSF, G-ESAL, G-BHXW				
G-HAPI	Lindstrand LBL 105A HAB	669		21. 3.00	Adventure Balloons Ltd	(Hartley Wintney, Hook)	21. 5.05T
G-HAPR	Bristol 171 Sycamore HC.14	13387	8010M	15. 6.78	E.D.ap Rees	Weston-super-Mare	
			XG547		t/a The Helicopter Museum *(As "XG547/T-S" in CFS c/s)*		
G-HAPY	de Havilland DHC-1 Chipmunk 22	C1/0697	WP803	3. 7.96	Astrojet Ltd *(As "WP803")*	Booker	1.10.05
G-HARE	Cameron N-77 HAB	1467		12. 3.87	D.H.Sheryn and Catherine A.Buck		
						(London SE16/Widmer End, High Wycombe)	28. 6.04A
G-HARF	Gulfstream Aerospace Gulfstream IV	1117	N1761J	9.10.91	Fayair (Jersey) Co Ltd *(Op Harrods)*	Stansted	20.12.05T
G-HARH	Sikorsky S-76B	760391	N7600U	30. 9.91	Air Harrods Ltd	Stansted	17. 1.05T
G-HARI	Raj Hamsa X'Air Victor 2	BMAA/HB/103		11. 6.99	A.Chidlow	Brook Farm, Pilling	24 10.04P
	(Built D Mahajan - kit no.455 but conflicts with G-HITM [BMAA/HB/112])						
G-HARN	Piper PA-28-181 Archer II	28-8290108	G-DENK	3. 2.00	J.P.and T.M.Jones	Denham	24. 3.07T
			G-BXRJ, HB-PGO				

G-HARR	Robinson R22 Beta II	3514	N75301	14. 6.04	Tananmera Properties Ltd	Newtownards	1. 7.07T
G-HART	Cessna 152 II	15279734	(G-BPBF)	2. 2.89	Atlantic Air Transport Ltd	Coventry	29. 6.07T
	(Tail-wheel u/c conversion)		N757GS		*(Air Atlantic c/s)*		
G-HARY	Alon A-2 Aircoupe	A 188	G-ATWP	15. 3.93	R.E.Dagless	Holly Hill Farm, Guist	15. 7.07
G-HASI	Cessna 421B Golden Eagle	421B0654	G-BTDK	17. 2.98	Chester Air Maintenance Ltd	Hawarden	7. 3.02
			OY-BFA, N1558G		*(New owner 12.03)*		
G-HASO	Diamond DA.40D Star	D4.070	G-CCLZ	21. 4.04	Hasso Enterprises Ltd	Gamston	8. 3.07T
G-HATF	Thorp T-18CW	PFA 76-11481		6.12.01	A T.Fraser	(Crowthorne)	
	(Built A T.Fraser and GHill)				*(Current status unknown)*		
G-HATZ	Hatz CB-1	17	N54623	11. 5.89	S.P.Rollason	Long Marston	8.11.05P
	(Lycoming O-320)				*(Carries 'N54623' on tail)*		
G-HAUL*	Westland WG.30-300				See SECTION 3 Part 1	Weston-super-Mare	
G-HAUS	Hughes 369HM	52-0214M	G-KBOT	20. 7.99	J Pulford	Sywell	21.10.05
			G-RAMM, EI-AVN, N9037F		t/a Pulford Aviation		
G-HAZE	Thunder Ax8-90 HAB	989		3. 8.88	T.G.Church	(Blackburn)	23. 6.97T
G-HBBC	de Havilland DH.104 Dove 8	04211	G-ALFM	24. 1.96	BBC Air Ltd	Compton Abbas	4. 7.04
			VP961, G-ALFM, VP961				
G-HBMW	Robinson R22	0170	G-BOFA	7. 7.94	Northumbria Helicopters Ltd	Kintore	6. 9.07T
			N9068D				
G-HBOS	Scheibe SF25C Falke	44574	D-KTIN	26. 7.01	Coventry Gliding Club Ltd	Husbands Bosworth	29. 9.07E
	(Rotax 912-A)						
G-HBUG	Cameron N-90 HAB	1991	G-BRCN	21. 6.89	R.T.and H.Revel	(High Wycombe)	30.11.04A
					(Thorn-EMI Computeraid titles)		
G-HCSL	Piper PA-34-220T Seneca III	34-8133237	N84375	9. 5.91	Southern Flight Centre Ltd	Shoreham	11.12.06T
G-HDAE	de Havilland DHC-1 Chipmunk 22	C1/0280	CS-DAE	4. 6.03	G.P.J.M.Valvekens and R.F.T.Baetens	Diest, Belgium	13. 7.06
			Portuguese AF 1304				
G-HDEW	Piper PA-32R-301 Saratoga SP	3213026	G-BRGZ	4.12.89	Plantation Stud Ltd	(Newmarket)	31. 3.07
			N91787				
G-HDIX	Enstrom 280FX	2076	N506DH	19. 2.98	D.H.Brown	(Burnley)	3. 5.04T
			D-HDIX				
G-HDTV	Agusta A109A II	7266	G-BXWD	16. 9.04	Castle Air Charters Ltd	Liskeard	6. 7.07T
			N565RJ, I-URIA, D-HEMZ, N109BD				
G-HEBE	Bell 206B-3 JetRanger III	3745	CS-HDN	5. 2.97	M and E Building and Civil Engineering Contractors Ltd		
			N3179A			(Deeside)	24. 6.06T
G-HELA	SOCATA TB-10 Tobago	135	F-GCOF	31.12.03	S.J. Heller tr Group TB-10	Panshanger	27. 1.07
G-HELE	Bell 206B-3 JetRanger III	3789	G-OJFR	21. 2.91	B.E.E.Smith	White Waltham	12. 5.06T
			N18095				
G-HELN	Piper PA-18 Super Cub 95	"18-3400"	G-BKDG	10. 1.86	J.J.Anziani tr Helen Group	Booker	12. 6.04P
	(L-21B-PI) *(Frame No.18-3400)*		MM52-2392, EI-69, EI-141, I-EIWB, MM53-7765, 53-7765				
G-HELP	Colt 17A Cloudhopper HAB	902		16. 2.87	A D.Kent	(Lancing)	20. 1.05A
					tr Balloon Preservation Flying Group *"Mondial Cloudhopper"*		
G-HELI*	Saro Skeeter AOP.12				See SECTION 3 Part 1	Gatow, Berlin, Germany	
G-HELV	de Havilland DH.115 Vampire T.55	975	Swiss AF U-1215	17. 9.91	Aviation Heritage Ltd	Coventry	24. 5.05P
	(Built F + W)				*(Op Air Atlantique as "XJ771" in RAF c/s)*		
G-HENS*	Cameron N-65 HAB				See SECTION 3 Part 1	(Aldershot)	
G-HENT	SOCATA Rallye 110ST Galopin	3210	OO-MBV	28.11.01	R.J.Patton	Strandhill, County Sligo	25. 2.05T
G-HENY	Cameron V-77 HAB	2486		9. 1.91	R.S.D'Alton *"Henny"*	(Newbury)	1. 1.05A
G-HEPY	Robinson R44 Astro	0695		11. 1.00	Beoley Mill Software Ltd	Wellesbourne Mountford	5. 2.06T
G-HERB	Piper PA-28R-201 Arrow III	28R-7837118	ZS-LAG	5. 6.86	Consort Aviation Ltd	Leeds-Bradford	19.10.07
			N3504M				
G-HERC	Cessna 172S Skyhawk SP	172S8985	N5113P	10.12.01	The Cambridge Aero Club Ltd	Cambridge	15. 1.05T
G-HERD	Lindstrand LBL 77B HAB	707		31. 7.00	S.W.Herd	(Mold)	7.11.04A
G-HERO*	Piper PA-32RT-300 Lance II	32R-7885086	G-BOGN	26. 4.88	Air Alize Communication	Biarritz, France	9. 7.97
			N33LV, N30573		*(Noted 10.00) (Cancelled 10.4.02 by CAA)*		
G-HEVN	SOCATA TB-200 Tobago XL	2013	D-EVHN	29.10.03	I.K.Maclean	(Banbury)	6.11.06T
G-HEWI	Piper J-3C-65 Cub (L-4J-PI)	12566	G-BLEN	20. 7.84	R.Preston	Denham	8. 5.06
	(Continental C90) *(Frame No.12396)*		D-EBEN, HB-OFZ, 44-80270		tr Denham Grasshopper Flying Group		
G-HEXE	Colt 17A HAB	2221		24. 2.04	A.Dunnington	(Tutzing, Germany)	
	(Built by Thunder and Colt Ltd 1994)				*(Stored 2004)*		
G-HEYY	Cameron Bear 72SS HAB	1244		21. 1.86	Magical Adventures Ltd *(Hofmeister Lager Bear)*		
					"George"	(West Bloomfield, Michigan, US)	30.11.98A
G-HFBM	Curtiss Robin C-2	352	LV-FBM	24. 4.90	D.M.Forshaw	High Cross, Ware	9.12.05P
	(Continental W-670)		NC9279				
G-HFCA	Cessna A150L Aerobat	A1500381	N6081J	30. 8.91	T.H.Scott	Rayne Hall Farm, Braintree	8.11.07E
	(Texas tail-wheel u/c conversion)						
G-HFCB	Reims/Cessna F150L	F15000798	G-AZVR	10. 2.87	P.Grace tr G-HFCB Group	(Great Tey, Colchester)	1. 5.06T
G-HFCI	Reims/Cessna F150L	F15000823	PH-CET	11. 9.80	A Modi	Earls Colne	14. 2.05T
G-HFCL	Reims/Cessna F152 II	F15201663	G-BGLR	11.10.88	Modi Aviation Ltd	Earls Colne	16. 7.06T
G-HFCT	Reims/Cessna F152 II	F15201861		27. 1.81	Stapleford Flying Club Ltd	Stapleford	17. 6.05T
G-HFLA	Schweizer 269C	S.1428		8.12.89	Sterling Helicopters Ltd	Norwich	12. 3.05T
G-HFTG	Piper PA-23-250 Aztec E	27-7405378	G-BSOB	30. 4.87	A Mannheim	(Mulheim-Karlich, Germany)	6. 4.05T
			G-BCJR, N54040				
G-HGPI	SOCATA TB-20 Trinidad	851		4. 8.88	M.J.Jackson	Bournemouth	5. 2.07
G-HHAA	Hawker Siddfeley Buccaneer S.2B	B3-01-73	9225M	6.12.02	Hawker Hunter Aviation Ltd	Scampton	AC
	(C/n officially quoted as B3-R-50-67)		XX885				
G-HHAB	Hawker Hunter F.58	41H-697439	Swiss AF J-4072	13. 1.03	Hawker Hunter Aviation Ltd	Scampton	
					(Stored as "J-4072" 5.04)		
G-HHAC	Hawker Hunter F.58	41H-691770	G-BWIU	10.12.02	Hawker Hunter Aviation Ltd	Scampton	14. 5.05P
			Swiss AF J-4021		*(As "J-4021")*		
G-HHAD	Hawker Hunter F.58	41H-697425	G-BWFS	10.12.02	Hawker Hunter Aviation Ltd	Scampton	25. 6.05P
			Swiss AF J-4058		*(A "J-4058")*		
G-HHAE	Hawker Hunter F.58	41H-697433	G-BXNZ	10.12.02	Hawker Hunter Aviation Ltd	Scampton	
	(C/n officially quoted as 41H-28364)		Swiss AF J-4066		*(Stored as "J-4066" 5.04)*		

G-HHAF	Hawker Hunter F.58	41H-697448	G-BWKB	13. 1.03	Hawker Hunter Aviation Ltd	Scampton	
			Swiss AF J-4081		*(Stored as "J-4081" 5.04)*		
G-HHAV	SOCATA MS.894A Rallye Minerva 220	11620	G-AYDG	9.10.02	R.E.Dagless	Holly Hill Farm, Guist	4. 8.07
G-HHOG	Robinson R44 Raven II	10584		23.12.04	Heli Air Ltd	Wellesbourne Mountford	AC
G-HIBM	Cameron N-145 HAB	3197		8. 2.94	Alba Ballooning Ltd	(Edinburgh)	25. 2.05T
G-HIEL	Robinson R22 Beta	1120		28. 9.89	Naylors Timber Recovery Ltd	Sherburn-in-Elmet	18. 2.07T
G-HIJK	Cessna 421C Golden Eagle III	421C-0218	G-OSAL	25. 2.00	Oxford Aviation Services Ltd	Oxford	22.11.05E
			G-HIJK, OY-BEC, SE-GZI, N5471G				
G-HIJN	Comco Ikarus C42 FB100	0403-6597		19. 5.04	J.R.North	Ince Blundell	18. 5.05P
G-HILO	Rockwell Commander 114	14224	N4894W	6. 2.98	F.H.Parkes	Stapleford	6. 5.07
G-HILS	Cessna F172H	F172-0522	G-AWCH	20.12.88	B.F.W.Lowdon	Blackbushe	28. 2.07
	(Built Reims Aviation SA)				tr Lowdon Aviation Group		
G-HILT	SOCATA TB-10 Tobago	298	(G-BMYB)	13. 5.82	Cheshire Aircraft Leasing Ltd	Hawarden	12.11.05T
			EI-BOF, G-HILT				
G-HIND	Maule MT-7-235 Star Rocket	18037C		26. 3.98	R.G.Humphries	Bramshill Farm, Hatchgate	29. 4.07T
G-HINZ	Jabiru Aircraft Jabiru SK	PFA 274-13441		1. 2.00	B.Faupel	Bourn	1. 8.04P
	(Built B.Faupel)						
G-HIPE	Sorrell SNS-7 Hyperbipe	209	N18RS	6. 4.93	J K Cook	Hill Farm, Nayland	19. 8.05P
	(Built R Stephen) (Lycoming IO-360)						
G-HIPO	Robinson R22 Beta	1719	G-BTGB	11. 9.92	Patriot Aviation Ltd	Cranfield	24. 4.06T
G-HIRE	Gulfstream GA-7 Cougar	GA7-0091	G-BGSZ	10.12.81	London Aerial Tours Ltd	Rochester	26. 5.06T
			N704G				
G-HISS	Pitts S-2A	2137	G-BLVU	17. 3.92	L.V.Adams and J.Maffia	Panshanger	28. 8.05T
	(Built Aerotek Inc)		SE-GTX		*"Always Dangerous"*		
G-HITM	Raj Hamsa X'Air Jab22	BMAA/HB/112		23. 2.00	J.A C.Cockfield	RNAS Culdrose	16. 1.05P
	(Built D J Hickey - kit no.455 but see G-HARI)				tr G-HITM Flying Group		
G-HITS	Piper PA-46-310P Malibu	46-8508063	G-BMBE	24.10.00	Law 2000 Ltd	Bournemouth	24. 5.07T
			N6908W				
G-HIUP	Cameron A-250 HAB	4464		16. 4.99	Bridges Van Hire Ltd	(Nottingham)	19. 8.04T
G-HIVA	Cessna 337A Super Skymaster	33700429	G-BAES	28. 3.88	G.J.Banfield	Gloucestershire	7. 1.07
			SE-CWW, N5329S				
G-HIVE	Reims/Cessna F150M	F15001186	G-BCXT	19. 4.85	M.P.Lynn	(Fenland)	10. 5.07T
G-HIZZ	Robinson R22 Beta II	2677	G-CNDY	2. 8.04	M.J.Gee	Gamston	8. 6.06T
			G-BXEW				
G-HJSM	Schempp-Hirth Nimbus 4DM	22/32	G-ROAM	19. 2.01	R.Jones tr 60 Group *"60"*	Lasham	16. 4.06
G-HJSS	AIA Stampe SV-4C	1101	G-AZNF	7. 9.92	H.J.Smith	Shoreham	16. 9.05
	(Renault 4P)		F-BGJM, Fr Mil				
G-HKHM	Hughes 369D	71-1019D	B-HHM	8. 4.99	Heli Air Ltd	Denham	25. 5.05
			VR-HHM, N50605				
G-HLCF	CFM Starstreak Shadow SA	PFA 206-12796		10. 5.96	A B Atkinson	Wombleton	3.11.03P
	(Built S M E Solomon - kit no K.256) (Rotax 618)						
G-HLIX*	Cameron Helix Oilcan 61SS HAB				See SECTION 3 Part 1	(Aldershot)	
G-HMBJ	Commander Aircraft Commander 114B	14636	N6036F	30. 6.97	Bravo Juliet Aviation Ltd	Guernsey	13. 8.06
G-HMCC	Airbus A319-111	2398	D-AVYS	R	EasyJet Airline Co Ltd	Luton	
G-HMED	Piper PA-28-161 Warrior III	2842020	LX-III	21. 7.97	H.Faizal	Little Gransden	18.12.06T
G-HMEI	Dassault Falcon 900	1	F-HOCI	2. 7.04	Executive Jet Group Ltd	RAF Northolt	15. 7.07T
			F-GIDE, F-WIDE				
G-HMJB	Piper PA-34-220T Seneca III	34-8133040	N8356R	12. 7.89	Cross Atlantic Ventures Ltd	Blackpool	10.10.04
G-HMMV	Cessna 525 CitationJet	525-0358	N51564	16. 2.00	D.Sears t/a Stellar Aviation	Biggin Hill	12. 3.05T
G-HMPF	Robinson R44 Astro	0730		8. 3.00	Mightycraft Ltd	White Waltham	20. 3.06T
G-HMPH	Bell 206B JetRanger II	1232	G-BBUY	20. 6.88	Sturmer Ltd	(Tring)	31. 3.05T
			N18090				
G-HMPT	Agusta-Bell 206B JetRanger II	8168	D-HARO	7.11.91	Helicopter Express Ltd	Beckley	13. 7.07T
G-HMSS	Bell 206B JetRanger II	1010	ZS-HMS	7. 5.02	A D Nevin	Goodwood	30.10.05T
			C-GXOI, N58008		t/a Jemaal Management Services		
G-HNTR*	Hawker Hunter T.7				See SECTION 3 Part 1	Elvington	
G-HOBO	Denney Kitfox Model 4	PFA 172A-12140		10. 9.92	J P Donovan	(Milton Keynes)	9.11.05P
	(W M Hodgkins)				*"Navy Baby"*		
G-HOBZ	Westland SA.341G Gazelle HT.3	1792	G-CBSJ	27. 5.03	SE Hobbs (UK) Ltd	(Staines)	10. 6.05P
			ZA802		*(As "ZA802/W")*		
G-HOCK	Piper PA-28-180 Cherokee D	28-4395	G-AVSH	15. 5.86	J.I.Simper	Goodwood	12. 8.07T
			N11C		tr G-HOCK Flying Group		
G-HOFC	Europa Aviation Europa	PFA 247-12736		25. 9.95	W.R.Mills	Upfield Farm, Whitson	7. 6.05P
	(Built J W Lang - kit no.119) (Monowheel u/c)				*(Resides in personal trailer)*		
G-HOFM	Cameron N-56 HAB	1245		21. 1.86	Magical Adventures Ltd	(Aldershot)	30.11.98A
					(Op Balloon Preservation Group)		
G-HOGS	Cameron Pig 90SS HAB	4121		7. 4.97	Magical Adventures Ltd *"Britannia Piggy Bank"*		
						(West Bloomfield, Michigan, US)	1. 7.99A
G-HOHO	Colt Santa Claus SS HAB	1671		21.12.89	Oxford Promotions (UK) Ltd	(Kentucky, US)	3. 4.03A
G-HOLY	SOCATA ST-10 Diplomate	108	F-BSCZ	31. 1.90	M.K.Barsham	Booker	26. 9.05
G-HOME	Colt 77A HAB	032		26. 2.79	Anglia Balloon School Ltd *"Tardis"*	(Newbury)	27. 5.86A
					(On loan to British Balloon Museum and Library)		
G-HONG	Slingsby T.67M-200 Firefly	2060	VR-HZR	24. 3.94	Jewel Aviation Ltd	Blackbushe	22. 2.07T
			HKG-12, G-7-128				
G-HONI	Robinson R22 Beta	0871	G-SEGO	27. 1.00	Patriot Aviation Ltd	Cranfield	14.12.06T
			N9081N				
G-HONK	Cameron O-105 HAB	1813		30. 9.88	T.F.W.Dixon and Son Ltd *"Dixons"*	(Bromsgrove)	14. 9.97A
G-HONY	Lilliput Type 1 Series A MLB	L-01		31. 7.98	A E. and D.E.Thomas	(Honiton)	
G-HOOD	SOCATA TB-20 Trinidad GT	2008	F-OILJ	25. 7.00	M.J.Hoodless	Blackbushe	13. 8.06
G-HOOT	Aérospatiale AS355F2 Ecureuil 2	5346	G-SCOW	27.11.02	Squirrel Helicopter Hire Ltd	(Weybridge)	28. 1.06T
			ZS-HSW, G-POON, G-MCAL				
G-HOOV	Cameron N-56 HAB	388		2. 3.78	Heather R.Evans *"Hoover"*	(Ross-on-Wye)	26. 5.89A
G-HOPA	Lindstrand LBL 35A Cloudhopper HAB	972		16. 1.04	S.F.Burden	(Noordwijk, The Netherlands)	18. 9.05A
G-HOPE	Beech F33A Bonanza	CE-805	N2024Z	27. 2.79	Hope Aviation	Bournemouth	24. 5.07

Reg	Type	C/n	Previous identity	Date	Owner/Operator	Base	CofA
G-HOPI	Cameron N-42 HAB	2724		5.12.91	Ballonverbung Hamburg GmbH	(Kiel, Germany)	30. 4.05A
G-HOPR	Lindstrand LBL 25A Cloudhopper HAB	999		21. 6.04	K.C.Tanner	(Thame)	18. 7.05A
G-HOPY	Van's RV-6A	PFA 181-12742		4.12.95	R.C.Hopkinson	Manor Farm, Garford	13. 4.05P
	(Built R C Hopkinson) (Lycoming O-320-B2B)						
G-HORN	Cameron V-77 HAB	570		29.11.79	S.Herd	(Mold)	11.12.98A
G-HORS	Cameron Flying Dala Horse SS HAB	2954		20. 1.93	Flygande Dalahasten AB	Partille, Sweden	14. 2.04A
G-HOTI	Colt 77A HAB	750		13. 7.87	R.Ollier *"Horace Hot One"*	(Northwich)	30. 9.90A
G-HOTT	Cameron O-120 HAB	2581		30. 4.91	D.L.Smith *"Floating Sensations"*	(Newbury)	17. 5.97T
G-HOTZ	Colt 77B HAB	2218		16. 6.92	C.J. and S.M.Davies	(Castleton, Sheffield)	21. 9.05A
G-HOUS	Colt 31A Air Chair HAB	099		7.10.80	The British Balloon Museum and Library	(Newbury)	3. 5.90A
					(Barratts titles) "K9"		
G-HOWE	Thunder Ax7-77 HAB	1340		10. 4.89	M.F.Howe *"Howie/Howzat"*	(Linley Hill, Leven)	15. 8.95A
G-HOWL	Rotary Air Force RAF 2000 GTX-SE	H2-95-6-164	N4994U	2. 7.01	C.J.Watkinson	Charity Farm, Baxterley	25. 4.05P
	(Built M Urbanczyk)						
G-HPAD	Bell 206B JetRanger III	1997	G-CITZ	2. 9.02	Helipad Ltd	(Turweston)	29. 7.05T
			G-BRTB, N9936K		*(Op Total Air Management Services (TAMS)*		
G-HPOL	MD Helicopters MD 900	900-00082	N70082	24. 1.01	Humberside Police Authority	Leconfield	3. 9.07T
G-HPSB	Commander Aircraft Commander 114B	14678	N6118R	24.10.01	Propwash Investments Ltd	Swansea	23.10.04T
G-HPSE	Commander Aircraft Commander 114B	14638	N6038V	26. 8.97	Al Nisr Ltd	Guernsey	28. 9.06A
G-HPSF	Commander Aircraft Commander 114B	14590	N6003F	16.11.04	J.Barnett tr Three Foxtrot Group	Guernsey	15.11.07E
G-HPSL	Commander Aircraft Commander 114B	14682	N115KL	26. 8.04	Halliard (Gsy) Ltd	Guernsey	12.10.07E
G-HPUX	Hawker Hunter T.7	41H-693455	8807M	12. 3.99	Hawker Hunter Aviation Ltd	Scampton	
			XL587		*(Stored as "XL587" 5.04)*		
G-HRAK	Aérospatiale AS350B Ecureuil	1749	F-GMPA	28. 7.04	R.A.Kingston	(Stapleford)	11.10.07E
			D-HSAN, SE-HUV, D-HCHL				
G-HRBS	Robinson R22 Beta II	3537	N75353	11. 3.04	Universal Energy Ltd	Booker	17. 3.07T
G-HRDS	Gulfstream V-SP *(Gulfstream 550)*	5032	N932GA	14.12.04	Fayair (Jersey) Co Ltd *(Noted 12.04)*	Stansted	AC
G-HRHE	Robinson R22 Beta	1950	G-BTWP	24. 1.97	Mandarin Aviation Ltd	Redhill	7.11.07E
G-HRHI	Beagle B.206 Basset CC.1	B.014	XS770	6. 7.89	M.D.Lewis	(Royston)	20.10.06
					(As "XS770" in Queens Flight c/s)		
G-HRHS	Robinson R44 Astro	0323		15. 4.97	Stratus Aviation Ltd	North Weald	16. 4.06
G-HRIO	Robin HR100/210 Safari	149	F-BTZR	22. 1.87	C.D.B.Cope	(Ramsey, Isle of Man)	17.12.07E
G-HRLI	Hawker Hurricane I	41H-136172	V7497	25. 4.02	Hawker Restorations Ltd	Milden	
G-HRLK	SAAB 91D/2 Safir	91376	G-BRZY	6. 3.90	Sylmar Aviation and Services Ltd		
			PH-RLK			Lower Wasing Farm, Brimpton	8. 6.07
G-HRLM	Brugger MB.2 Colibri	PFA 43-10118		28.12.78	D.G.Reid	Morgansfield, Fishburn	30. 3.05P
	(Built R A Harris) (Volkswagen 1834)				*"Titch"*		
G-HRNT	Cessna 182S Skylane	18280395	N2369H	29.1.99	Dingle Star Ltd	Denham	6. 3.05
G-HROI	Rockwell Commander 112A	326	N1326J	19. 6.89	Intereuropean Aviation Ltd	Jersey	28. 4.07
G-HRON*	de Havilland DH.114 Heron 2B	14102	XR442	4. 4.91	M.E.R.Coghlan	Gloucestershire	
			G-AORH		*(Cancelled 10.4.02 by CAA) (Stored unmarked 12.03)*		
G-HRPN	Robinson R44 Raven II	10007		26.11.02	Harpin Ltd	(York)	22.12.05T
G-HRVD	CCF Harvard 4 (T-6J-CCF Texan)	CCF4-548	G-BSBC	8.12.92	K.F.Mason and D.Featherby t/a Anglia Flight	Norwich	
			Moz PLAF 1741, FAP 1741, BF+055, AA+055, 53-4629				
	(Possibly a composite with rear fuselage of Moz PLAF/FAP 1780/AA+614/53-4622) (On rebuild 1999: new owners 10.01)						
G-HRZN	Colt 77A HAB	536		14.12.83	A J.Spindler *"Tequila Sunrise"*	(Cleish)	4. 5.88A
G-HSDW	Bell 206B JetRanger II	1789	ZS-HFC	16.12.85	Greatsearch Ltd	Rossendale	20. 2.05
G-HSLA	Robinson R22 Beta	1130	G-BRTI	22.11.01	D.I.Pointon	(Coventry)	17. 6.06T
			EI-CDW, (EI-CFJ), G-BRTI, N8044U				
G-HSLB	Agusta-Bell 206B JetRanger II	8690	F-GUJR	4. 4.02	B3 Aviation Services Ltd	(Laxey, Isle of Man)	29. 5.05T
			SX-HEN, F-GRCY, I-ELEP				
G-HSOO	Hughes 369HE	109-0208E	G-BFYJ	3.11.93	Edwards Aviation Ltd	(Wilmslow)	27. 9.03T
			F-BRSY				
G-HSTH*	Lindstrand LBL HS-110 HA Airship	546		20. 8.98	Not known	(Germany)	17.12.00
					(Thomapyrin titles) (Cancelled 22.8.02 by CAA) (Noted active 1.03)		
G-HTEL	Robinson R44 Raven	1155	N70319	25. 1.02	Forestdale Hotels Ltd	Burley, Ringwood	7. 2.05T
G-HTRL	Piper PA-34-220T Seneca III	34-8333061	G-BXXY	8. 2.00	Air Medical Fleet Ltd	Oxford	18. 2.07T
			PH-TLN, N4295X				
G-HUBB	Partenavia P68B	194	OY-BJH	27. 5.83	G-HUBB Ltd	Denham	29. 7.04
			SE-GXL				
G-HUCH	Cameron Carrots 80SS HAB	2258	G-BYPS	13. 3.91	Magical Adventures Ltd		
					"Magic Carrots"	(West Bloomfield, Michigan, US)	2. 8.01A
G-HUEW	Europa Aviation Europa XS	PFA 247-14156		22. 7.04	C.R.Wright	(Yoxall, Burton-on-Trent)	
	(Built C.R.Wright)						
G-HUEY	Bell UH-1H-BF Iroquois	13560	AE-413	23. 7.85	Argonauts Holdings Ltd	North Weald	12. 4.00P
			(Argentine Army), 73-22077		*(Noted 1.05)*		
G-HUFF	Cessna 182P Skylane II	18264076	PH-CAS	31.10.78	A E.G.Cousins	Southend	20. 6.06T
	(Reims-assembled with c/n F18200033)		N6059F				
G-HUGO	Colt 260A HAB	2559		20. 1.94	P.G.Hall t/a Adventure Ballooning	(Meanwood, Chard)	5. 8.04T
G-HUGS	Robinson R22 Beta	1455	G-BYHD	27. 2.02	C.G.P.Holden	Gamston	19. 3.05T
			N900AB				
G-HUKA	MD Helicopters Hughes 369E	0298E	G-OSOO	12. 2.02	B.P.Stein	(London WC2)	1. 9.07T
G-HULL	Reims/Cessna F150M	F15001255	PH-TGR	19. 1.79	Hull Aero Club Ltd	Linley Hill, Leven	19.10.07T
G-HUNI	Bellanca 7GCBC Scout	541-73	OO-IME	21.10.96	T.I.M.Paul	Denham	14.10.07E
	(Officially regd as "Scout" although 7GCBC = Citabria)		D-EIME				
G-HUNK	Lindstrand LBL 77A HAB	551		9. 9.98	Lindstrand Balloons Ltd	(Oswestry)	23. 9.00A
G-HUNN*	Hispano HA.1112-MIL Buchon				See SECTION 3 Part 1	Tico, Florida, US	
G-HUNT*	Hawker Hunter F.51				See SECTION 3 Part 1	Oshkosh, Wisconsin, US	
G-HUPW	Hawker Hurricane I	G5-92301	R4118	21. 8.01	P.J.and P.M.A Vacher	Oxford	AC
	(Built Gloster Aircraft Co Ltd)				t/a Minmere Farm Partnership *(F/f 23.12.04 as "R4118/UP-W")*		
G-HURI	Hawker Hurricane XIIA (IIB)	72036	RCAF 5711	9. 6.83	Historic Aircraft Collection Ltd	Duxford	6. 5.05P
	(Built Canadian and Car Foundry Co)				*(Op The Fighter Collection) (As "Z5140/HA-C" in 126 Sqdn RAF c/s)*		
	(Composite - probably includes parts from c/n 44019/RCAF 5424, RCAF 5625 and RCAF 5547)						
G-HURN	Robinson R22 Beta	1441		18. 7.90	Sloane Helicopters Ltd	Sywell	16. 8.07T

Registration	Type	C/n	Previous identity	Date	Owner/operator	Probable base	CofA expiry
G-HURR	Hawker Hurricane XII (IIB) *(Built Canadian and Car Foundry Co)*	52024	RCAF 5589	30. 7.90	R.A Fleming *(As "52024LK-A")*	Breighton	21. 4.05P
G-HURY	Hawker Hurricane IV *(RAF p/i unlikely as KZ321 was written off 23.5.43)*	---	(Israel) Yugoslav AF, KZ321	31. 3.89	Patina Ltd *(Op The Fighter Collection as "KZ321/JV-N")*	Duxford	9. 7.05P
G-HUSK	Aviat A-1B Husky	2214		2. 1.03	P.H.Yarrow and A.T.Duke	Wisbridge Farm, Reed, Royston	24. 4.06T
G-HUTT	Denney Kitfox Model 2 *(Built M A J Hutt and B Davies - c/n 509)*	PFA 172-11634		24. 1.90	P.C.E.Roberts	Truro	23. 2.03P
G-HVAN	Tiger Cub RL5A LW Sherwood Ranger *(Built H.T.H.Van Neck)* (BMW R100)	PFA 237-13074l		10.12.98	H.T.H.Van Neck	Ince Blundell	
G-HVBF	Lindstrand LBL 210A HAB	372		23. 5.96	Airxcite Ltd t/a Virgin Balloon Flights	(Wembley)	5. 4.05T
G-HVDM*	Vickers Supermarine 361 Spitfire LF.IXc				See SECTION 3 Part 1	Glize-Rijen, The Netherlands	
G-HVIP	Hawker Hunter T.68	HABL-003215	Swiss AF J-4208 G-9-415, Fv.34080, G-9-56	7. 7.95	K.G.Theurer	(Stuttgart, Germany)	2. 7.05P
G-HVRD	Piper PA-31-350 Navajo Chieftain	31-7305052	G-BEZU SE-GDP, N74920, N9666N	11. 6.87	N.Singh	Prestwick	13. 5.07T
G-HXTD	Robin DR400/180 Régent	2510		24.10.01	Hayley Aviation Ltd	(Banstead)	12.12.04T
G-HYAK	IAV-Bacau Yakovlev Yak-52	9011107	LY-ALU DOSAAF 124	27. 8.02	Goodridge (UK) Ltd	Exeter	19. 9.05P
G-HYLT	Piper PA-32R-301 Saratoga SP	32R-8213001	N84588	23. 4.86	G.G.L.James	Sleap	31. 1.05
G-HYST	Enstrom 280FX Shark	2082		9. 7.98	Patten Helicopter Services Ltd	Barton	8.10.04

G-IAAA - G-IZZZ

Registration	Type	C/n	Previous identity	Date	Owner/operator	Probable base	CofA expiry
G-IAFT	Cessna 152 II	15285123	EI-BVW N6093Q	20. 6.95	Marham Investments Ltd *(Op Woodgate Executive Air Services)*	Newtownards	5. 9.07T
G-IAGD	Robinson R22 Beta	0918	N2018Y G-DRAI, N8808V	16.11.99	M.Kenyon and A Ingham t/a A and M Engineering	Blackpool	26. 1.06T
G-IAMP	Cameron H-34 HAB	2541		11. 3.91	R.S.Kent tr Balloon Preservation Flying Group *(BPG titles)*	(Lancing)	3. 5.05A
G-IAMS	Cessna 560XL Citation Excel	560-5183	G-CFRA N5090V	15. 4.03	Amsair Ltd *(Op London Executive Aviation)*	Stansted	5. 9.06T
G-IANB	DG Flugzeugbau DG-808B	8-246B159		12. 3.02	I.S.Bullous	Sutton Bank	25. 9.05
G-IANC	SOCATA TB-10 Tobago	150	G-BIAK	15.12.04	I.Corbin *(New CofR 12.04)*	Biggin Hill	13. 8.03
G-IANH	SOCATA TB-10 Tobago	1843	F-OILI	13. 3.00	XD Flight Management Ltd	Goodwood	26. 3.06T
G-IANI	Europa Aviation Europa XS *(Built I.F.Rickard and I.A Watson - kit no.505)* (Rotax 914)	PFA 247-13714		20. 4.01	I.F.Rickard and I.A Watson	Dunsfold	20 .6.05P
G-IANJ	Reims/Cessna F150K	F15000548	G-AXVW	19. 5.98	J.A.,G.M.,D.T.A. and J.A.Rees t/a Messrs Rees of Poyston West	Swansea	20. 9.07T
G-IANN	Kolb Twinstar Mk.3 Extra *(Built I.Newman)*	PFA 205-14259		7.10.04	I.Newman	(Cleethorpes)	
G-IANW	Eurocopter AS350B3 Ecureuil	3447	F-WQPU	18. 9.01	Milford Aviation Services Ltd	(Harrow)	20. 1.08E
G-IARC	Stoddard-Hamilton GlaStar *(Built A A Craig) (Tri-cycle u/c)*	PFA 295-13261		9.11.99	A A Craig	Prestwick	4. 7.05P
G-IASL	Beech 60 Duke	P-21	G-SING D-IDTA, SE-EXT	18. 4.97	Applied Sweepers Ltd	Perth	13. 4.04
G-IATU	Cessna 182P Skylane	18261436	G-BIRS G-BBBS, N21131	8. 1.03	Auto Corporation Ltd	Liverpool	21. 4.07T
G-IAWE	Reims/Cessna F150L	F15000877	EI-AWE	9. 2.04	P.P.Maguire *(On rebuild 3.04)*	Sleap	
G-IBAZ	Comco Ikarus C42 FB100	0409-6622		30. 9.04	B.R.Underwood	Swinford, Rugby	10.10.05P
G-IBBC	Cameron Sphere 105SS HAB	4082		2. 4.97	R.S.Kent tr Balloon Preservation Group	(Lancing)	17. 7.03A
G-IBBS	Europa Aviation Europa *(Built R.H.Gibbs - kit no.118) (Monowheel u/c)*	PFA 247-12745		8. 9.94	R.H.Gibbs	Popham	14. 4.05P
G-IBED	Robinson R22 Alpha	0500	G-BMHN	7. 9.93	B.C.Seedle t/a Brian Seedle Helicopters	Blackpool	30. 9.94
G-IBET	Cameron Can 70SS HAB	1625		25. 1.88	M.R.Humphrey and J.R.Clifton *"Carling Black Label"*	(Brackley)	25. 8.02A
G-IBEV	Cameron C-90 HAB	10375		10. 4.03	B.Drawbridge	(Cranbrook)	9. 4.05A
G-IBFC	BFC Challenger II Long Wing *(Built K N Dickinson - kit no.CH2-0898-UK-1774)*	PFA 177B-13369		9.11.98	K.N.Dickinson	(Lytham St.Annes)	1. 7.05P
G-IBFW	Piper PA-28R-201 Arrow III	28R-7837235	N31534	22. 1.79	D A Triplett	Sleap	16.10.06T
G-IBHH	Hughes 269C	74-0327	G-BSCD PH-HSH, SE-HFG	20. 8.99	The Hughes Helicopter Co Ltd t/a Biggin Hill Helicopters	Biggin Hill	21. 6.07T
G-IBIG	Bell 206B JetRanger III	2202	G-BORV C-GVTY, N16763	20. 3.02	Big Heli-Charter Ltd	Earls Colne	5. 5.05T
G-IBLU	Cameron Z-90 HAB	4913		4. 8.00	John Aimo Balloons SAS	(Mondovi, Italy)	27. 5.05A
G-IBRI	Eurocopter EC120B Colibri	1073	LX-HCR	7. 2.02	Colibri Aviation Ltd	Welshpool	19. 3.05T
G-IBRO	Reims/Cessna F152 II	F15201957	EI-BRO	11.10.95	Leicestershire Aero Club Ltd	Tollerton	14. 3.05T
G-IBZS	Cessna 182S Skylane	18280529	N7269A	11.12.99	Patrick Eddery Ltd	Oxford	9.12.05
G-ICAB	Robinson R44 Astro	0086		28.11.94	Northumbria Helicopters Ltd	Newcastle	4. 3.07T
G-ICAS	Pitts S-2B *(Built Aviat Inc)*	5344	N511P	19. 6.97	J.C.Smith	Sherburn-in-Elmet	10. 7.06T
G-ICBM	Stoddard-Hamilton Glasair III Turbine *(Built G V Waters)* (Allison 250-B17B)	3337		18.12.00	G V Waters and D N Brown *(Current status unknown)*	Attleborough	AC
G-ICCL	Robinson R22 Beta	1608	G-ORZZ	25.11.93	JK Aviation Services Ltd	Headcorn	13. 3.06T
G-ICES	Thunder Ax6-56 SP.1 HAB *(Ice Cream special shape)*	283		3. 7.80	British Balloon Museum and Library Ltd *"Ashfords"*	(Newbury)	3. 6.94A
G-ICEY	Lindstrand LBL 77A HAB	043		11. 8.93	G.C.Elson t/a Lindstrand Balloon School	(Ronda, Spain)	6. 6.03A
G-ICKY	Lindstrand LBL 77A HAB	029		19. 5.93	M.J.Green	(Shrewsbury)	8. 9.04T
G-ICOI	Lindstrand LBL 105A HAB	564	(D-O...) G-ICOI	3.11.98	F.Schroeder	(Mulheim Ruhr, Germany)	20. 3.03A
G-ICOM	Reims/Cessna F172M Skyhawk II	F17201212	G-BFXI PH-ABA, D-EEVC	25. 4.94	C G Elesmore	Lydd	23. 6.06T

G-ICON	Rutan LongEz	PFA 74A-11104		29.11.00	S.J.and M.A Carradice	Gamston	
	(Built S.J.Carradice)				*(Noted 10.04)*		
G-ICRS	Comco Ikarus C42 FB UK	PFA 322-13873		11. 3.02	Ikarus Flying Group Ltd	(Aylesbury)	6. 4.05P
	(Built A J Whitlock)						
G-ICSG	Aérospatiale AS355F1 Ecureuil 2	5104	G-PAMI	6. 4.93	Stratton Motor Company (Norfolk) Ltd	Long Stratton	7. 2.05T
			G-BUSA				
G-ICWT	Pegasus Quantum 15-912	7632		7. 4.00	C.W.Taylor	Mill Farm, Shifnal	7. 6.05P
G-IDAB	Cessna 550 Citation Bravo	550-0917	EI-DAB	16. 3.04	Errigal Aviation Ltd	(Blackpool)	24. 3.05T
			N5100J				
G-IDAY	Skyfox CA-25N Gazelle	CA25N-028	VH-RCR	29. 4.96	G.G.Johnstone	Glenrothes	26. 6.06T
	(Rotax 912)						
G-IDDI	Cameron N-77 HAB	2383		21. 8.90	PSH Skypower Ltd	(Pewsey)	
					(Allen and Harris - Royal Sun Alliance titles)		
G-IDII	Dan Rihn DR.107 One Design	PFA 264 12953		16. 6.99	C.Darlow	Jericho Farm, Lambley	4.10.05P
	(Built C.Darlow) (Lycoming O-360)						
G-IDPH	Piper PA-28-181 Archer III	2843585	N3054D	3. 2.04	D.Holland	Cambridge	2. 2.07T
G-IDSL	Flight Design CT2K	Not known		28.10.02	D.S.Luke	Kemble	27.10.04P
	(Assembled Pegasus Aviation as c/n 7922)						
G-IDUP	Enstrom 280C Shark	1163	G-BRZF	11. 5.92	Antique Buildings Ltd	Hunterswood Farm, Dunsfold	21. 7.07
			N5687D				
G-IDWR	Hughes 369HS	69-0101S	G-AXEJ	26. 5.81	M.A and M.Gradwell	Barton	16.12.04
G-IEIO	Piper PA-34-200T Seneca II	34-7670274	EI-EIO	6.12.02	Jade Air plc	Shoreham	14. 1.06T
			N6257J				
G-IEJH	SAN Jodel D.150A Mascaret	02	G-BPAM	28. 2.95	A Turner and D.Worth	Crowfield	28. 4.05P
			F-BLDA, F-WLDA				
G-IEYE	Robin DR400/180 Régent	2123		29. 1.92	G.Wood	Sherburn-in-Elmet	28. 6.07
G-IFAB	Reims/Cessna F182Q Skylane II	F18200127	N61AN	6. 1.98	Manda Construction Ltd	Inverness	21.1.06T
			G-IFAB, OO-ELM, (OO-HNU)				
G-IFBP	Eurocopter AS350B2 Ecureuil	9051	F-GTKR	27.10.03	F.Bird t/a Frank Bird Aviation	(Penrith)	19.11.06
			(F-GYBR)				
G-IFDM	Robinson R44 Astro	0707		24. 1.00	M.Paffey	Goodwood	16. 3.06T
G-IFFR	Piper PA-32-300 Cherokee Six	32-7340123	G-BWVO	1. 4.97	D.J.D and G.D.Ritchie and J.C.Gilbert	RAF Henlow	20. 3.06
			OO-JPC, N55520				
G-IFIT	Piper PA-31-350 Chieftain	31-8052078	G-NABI	31.12.85	Dart Group plc	Bournemouth	4.11.05T
			G-MARG, N3580C		*(Op Channel Express)*		
G-IFLE	Aerotechnik EV-97 teamEurostar UK	2113		1. 7.04	M.Strom t/a Euravia Flight	Rochester	4. 7.05P
G-IFLI	Gulfstream AA-5A Cheetah	AA5A-0831	N26948	7. 7.82	C M Petherbridge tr I Fly Group	Linley Hill, Leven	22.12.06
G-IFLP	Piper PA-34-200T Seneca II	34-8070029	N81WS	4. 1.88	Tayflite Ltd	Perth	16. 7.06T
			N81149				
G-IFTE	British Aerospace HS 125 Series 700B	257037	G-BFVI	16. 5.96	Albion Aviation Management Ltd	Biggin Hill	9. 9.05T
			G-5-18				
G-IFTS	Robinson R44 Astro	0366		16. 9.97	Frank Bird (Poultry) Ltd	(Penrith)	11.11.06T
G-IGEL*	Cameron N-90 HAB				See SECTION 3 Part 1	(Aldershot)	
G-IGGL	SOCATA TB-10 Tobago	146	G-BYDC	26. 3.99	M.P.Perkin	White Waltham	3. 1.05T
			F-GCOL		tr G-IGGL Flying Group		
G-IGHH	Enstrom 480	5034		1.12.98	H.Hyndman	(Ramsey, Huntingdon)	16. 1.05T
G-IGIE	SIAI Marchetti SF.260	2-42	D-EHGB	13. 3.02	D.Fletcher and J.J.Watts	Bournemouth	12. 5.06T
G-IGII	Europa Aviation Europa	PFA 247-12506		9. 4.02	C.D.Peacock	Sywell	17. 6.05P
	(Built W.C.Walters - kit no.011) (NSI EA/81-100) *(Conventional u/c)*						
G-IGLA	Colt 240A HAB	2228		3. 7.92	M.L. and S.M.Gabb	(Alcester)	29. 8.03T
					t/a Heart of England Balloons *(Barclaycard titles)*		
G-IGLE	Cameron V-90 HAB	2609		11. 6.91	A A Laing *"Giggle"*	(Aberdeen)	4. 7.04A
G-IGLS	American Champion 7GCAA Citabria Adventure	477-2003		25. 9.03	Blue Yonder Aviation Ltd	Earsl Colne	23.11.07E
G-IGLZ	American Champion 8KCAB	914-2003		18. 2.03	Woodgate Aviation (IoM) Ltd	Belfast	1. 6.06T
G-IGOB	Boeing 737-36Q	28660	EC-GNU	17.12.01	Easyjet Airline Company Ltd	Luton	24. 1.05T
G-IGOG	Boeing 737-3Y0	23927	F-GLLE	3. 9.98	Easyjet Airline Company Ltd	Luton	3. 9.07T
			PT-TEK				
G-IGOH	Boeing 737-3Y0	23926	F-GLLD	6.11.98	Easyjet Airline Company Ltd	Luton	12.12.07T
			PT-TEJ				
G-IGOI	Boeing 737-33A	24092	G-OBMD	30.12.98	Easyjet Airline Company Ltd	Luton	13. 2.05T
G-IGOJ	Boeing 737-36N	28872	N1795B	11.11.98	Easyjet Airline Company Ltd	Luton	20.11.07T
G-IGOK	Boeing 737-36N	28594		24. 4.99	Easyjet Airline Company Ltd	Bristol	23. 4.05T
G-IGOL	Boeing 737-36N	28596	N1015X	26. 6.99	Easyjet Airline Company Ltd	Luton	25. 6.05T
G-IGOM	Boeing 737-36N	28599		13. 7.99	Easyjet Airline Company Ltd	Luton	12. 7.05T
G-IGOO	Boeing 737-36N	28557	G-SMDB	15. 3.02	Easyjet Airline Company Ltd	Luton	20. 3.06T
G-IGOP	Boeing 737-36N	28602		12. 8.99	Easyjet Airline Company Ltd	Bristol	11. 8.05T
G-IGOR	Boeing 737-36N	28606		22.10.99	Easyjet Airline Company Ltd	Luton	20.10.05T
G-IGOS	Boeing 737-3L9	27336	D-ADBH	21. 2 01	Easyjet Airline Company Ltd	Luton	9. 3.07T
G-IGOT	Boeing 737-3L9	24571	N2393W	12. 7.01	Easyjet Airline Company Ltd	Luton	11. 7.07T
			TC-IAE, D-ADBF, OY-MMF				
G-IGOU	Boeing 737-3L9	27337	D-ADBI	9. 5.01	Easyjet Airline Company Ltd	Luton	14. 5.07T
			OY-MAP				
G-IGOV	Boeing 737-3M8	25017	N250GE	13. 9.01	Easyjet Airline Company Ltd	Luton	13. 9.07T
			LZ-BOF, N250GE, 9V-TRD, N760BE, N35030, (OO-LTH)				
G-IGOW	Boeing 737-3YO	23923	N923AP	24. 8.01	Easyjet Airline Company Ltd	Luton	23. 8.07T
			(G-OBWW), N923AP, LZ-BOE, EC-FJZ, EC-898, EI-BZP, EI-CEE, G-TEAB, (N117AW), EI-BZP, (LN-AEQ), EI-BZP, EC-EIA, EC-152				
G-IGOY	Boeing 737-36N	28570	CS-TGQ	21. 3.02	Easyjet Airline Company Ltd	Luton	10. 4.05T
G-IGOZ	Boeing 737-3Q8	24699	G-OBWZ	13. 3.02	Easyjet Airline Company Ltd	Luton	27. 3.06T
			N699PU, PK-GWG				
G-IGPW	Eurocopter EC120B Colibri	1027	G-CBRI	31. 7 99	J.Havakin	(Darlington)	30. 5.05T
G-IHOT	Aerotechnik EV-97 teamEurostar UK	2007		1. 4.04	A.J.Turner	Andrewsfield	31. 3.05P
G-IHSB	Robinson R22 Beta	0982		16. 3.89	Patriot Aviation Ltd	Cranfield	10. 3.07T
G-IIAC	Aeronca 11AC Chief	11AC-169	(G-BTPY)	2. 7.91	G.R.Moore	Black Spring Farm, Castle Bytham	16. 5.05P
	(Continental A65)		N86359, NC86359				

Reg	Type	C/n	Previous identity	Date	Owner/Operator	Base	CofA
G-IIAN	Aero Designs Pulsar *(Built I.G.Harrison)*	PFA 202-12123		10. 9.91	I.G.Harrison *(Under construction 2000)*	(Ripley)	
G-IICI	Pitts S-2C *(Built Aviat Inc)* (Lycoming AEIO-540)	6017	N113PS	20. 5.02	D.G.Cowen tr Charlie India Group	Redhill	15. 7.05
G-IICM	Extra EA.300/L	100		19.11.99	Phonetiques Ltd *"Charlie Macaw II"*	Denham	27.11.05T
G-IIDI	Extra EA.300/L *(Tail-wheel u/c)*	047	G-XTRS D-EXJH	5.10.01	Power Aerobatics Ltd *(Op Xtreme Team)*	Membury	13. 3.05T
G-IIDY	Pitts S-2B *(Built Aerotek Inc)*	5000	G-BPVP N5302M	11.11.02	R.P.Millinship tr The S-2B Group	Leicester	29.10.06
G-IIFR	Robinson R22 Beta II	2841		2. 9.98	C.W.B.Wrightson t/a Wrightson Aviation and Engineering	Sherburn-in-Elmet	22.11.04T
G-IIGI	Van's RV-4 *(Built R.F.and C.Palmer)*	381	N44BZ	7. 4.04	A.Darlington tr G-IIGI Flying Club *(Noted 11.04)*	Lasham	
G-IIID	Dan Rihn DR.107 One Design *(Built M A N and A J Newall)*	PFA 264-12766		6. 7.00	V.Millard *(New owner 1.04)*	Crowfield	
G-IIIE	Pitts S-2B *(Built Aerotek Inc)*	5017	N9WQ N9WR	8. 3.04	D Dobson	Little Staughton	18. 8.07T
G-IIIG	Boeing Stearman A75N1 (PT-17) Kaydet (Continental W670)	75-4354	G-BSDR N61827, 42-16191	25. 3.91	F.and S.Vormezeele *"Annie"*	(Brasschaat, Belgium)	21. 7.06T
G-IIII	Pitts S-2B *(Built Christen Industries Inc)* (Lycoming AEIO-540)	5010	N5330G	6. 1.89	Four Eyes Aerobatics Ltd	Barton	28. 3.07T
G-IIIL	Pitts S-1T *(Built J L Edwardson)*	008	OH-XPT G-IIIL, N15JE	15. 2.89	Empyreal Airways Ltd	Leicester	14. 9.05P
G-IIIO	Schempp-Hirth Ventus 2cM	41/73	PH-1110 D-KBBF	10.08.04	M.A.V.Gatehouse	(Plymouth)	23. 9.07
G-IIIR	Pitts S-1S *(Built Milam)*	604	N27M	21. 1.93	R.O.Rogers	Hulcote Farm, Salford, Berkshire	18. 6 04P
G-IIIS	Sukhoi Su-26M2	06-07	55 *(Black)* RA-0607, N626RM, RA-01321	7.10.03	S.Jones t/a Airtime Aerobatics	Kemble	25. 5.05P
G-IIIT	Pitts S-2A *(Built Aerotek Inc)*	2222	N7YT	16. 1.89	Aerobatic Displays Ltd	Booker	19. 9.07A
G-IIIV	Pitts Super Stinker 11-260 *(Built A N R Houghton)* (Lycoming IO-540) *(Marked as a "Pitts S1.11B")*	PFA 273-13005		4. 2.97	G G Ferriman	Breighton	16. 3.05P
G-IIIX	Pitts S-1S *(Built J Tarascio)*	AJT	G-LBAT G-UCCI, G-BIYN, N455T	22. 5.89	D.A Kean	Gloucestershire	18.11.05P
G-IIIZ	Sukhoi Su-26M	04-05	RA-44444 DOSAAF 35 *(black)*	14. 5.03	P.M.M.Bonhomme	White Waltham	26. 5.05P
G-IIMI	Extra EA.300/L	141	D-EXLE	2. 5.01	A.Birch	(Winchester)	15. 6.07T
G-IIMT	Bushby-Long Midget Mustang *(P/I formerly c/n PFA/1327)*	PFA 168-1327	G-BDGA	13.11.03	M.J.A Trudgill	RAF Henlow	
G-IINI	Van's RV-9A *(Built S.Sampson)*	PFA 320-13781		6. 8.04	S.Sampson	Bagby	
G-IIPM	Aérospatiale AS350B Ecureuil	1790	G-GWIL	18.12.96	Helicopter Ltd	(West Derby, Liverpool)	28.11.05T
G-IIPT	Robinson R22 Beta	2506	G-FUSI N83306	10. 5.01	Patriot Aviation Ltd	Cranfield	30. 6.07T
G-IIRG	Stoddard-Hamilton Glasair IIS RG *(Built D S Watson)* (Lycoming IO-360)	PFA 149-11937		29. 6.93	A C.Lang	(Ottery St.Mary)	27. 6.05P
G-IITI	Extra EA.300	018	D-EFRR	12. 5.92	Aerobatic Displays Ltd *(International Watch Company titles)*	Booker	24. 6.05A
G-IIUI	Extra EA.300/S	004	G-CCBD OK-XTA, D-EBEW	26.11.03	M.G, J.R Jeffries and C. Scrope	Fullers Hill Farm, Little Gransden	5. 6.06
G-IIVI	Mudry CAP 232	21	F-GUJM	20. 1.05	Skylane Aviation Ltd	(Halifax)	
G-IIXI	Extra EA.300/L	134	YR-EWG	6.08.03	R.Jones t/a Southern Sailplanes and Pelham Ltd	Membury	4.12.06T
G-IIXX	Parsons Two Place Gyroplane *(Built J.M.Montgomerie)* (Rotax 912)	PFA G/8-1225		13.10.93	J.M.Montgomerie *(Stored 5.04)*	Crosshill	
G-IIYK	Yakovlev Yak-50	842706	LY-AFZ DOSAAF 24	15.10.02	D.A Hammant	Bere Farm, Warnford	22.10.05P
G-IIZI	Extra EA.300	037	JY-RNB D-ETXA	12.12.96	S.G.Jones and Power Aerobatics Ltd *(Op Extreme Team)*	Kemble	12. 2.07T
G-IJAC	Avid Speed Wing Mk.4 *(Built I J A Charlton)*	PFA 189-12095		31.12.92	I.J.A Charlton *(Current status unknown)*	(Petworth)	
G-IJBB	Enstrom 480	5010	G-LIVA N900SA, G-PBTT, JA6169	17. 9.99	G.Kidger	Sheffield City	11.11.07E
G-IJMC	VPM M-16 Tandem Trainer (Arrow GT1000R)	VPM16-UK-106	G-POSA G-BVJM	10. 6.98	D C Fairbrass *(New owner 9.04)*	Willingale	3. 6.02P
G-IJOE	Piper PA-28RT-201T Turbo Arrow IV	28R-8031178	N8265X N9599N	14. 8.90	J.H.Bailey tr G-IJOE Group	Sturgate	22. 7.05
G-IJYS	British Aerospace Jetstream Series 3102	715	G-BTZT N416MX, G-31-715	5.10.92	Air Kilroe Ltd *(Op Eastern Airways) "Flying Scotsman"*	Humberside	18.11.06T
G-IKAP	Cessna T303 Crusader	T30300182	N63SA D-IKAP, N9518C	4. 3.99	T.M.Beresford	Cambridge	29. 4.05T
G-IKAT	Diamond DA.20-C1 Katana	C0096	N966CT	14.12.04	Diamond Aircraft UK Ltd	Gamston	AC
G-IKBP	Piper PA-28-161 Warrior II	28-8216132	N81762	16. 7.90	K.B.Page	Goodwood	24.10.05
G-IKEA	Cameron IKEA-120 SS HAB	10562		14. 6.04	IKEA Ltd	(Bristol)	22. 6.05A
G-IKEV	Jabiru Aircraft Jabiru UL-450 *(Built K.J.Bream)*	PFA 274A-14075		29. 6.04	K.J.Bream *(On build 12.04)*	Priory Farm, Tibenham	
G-IKOS	Cessna 550 Citation Bravo	550-0957	N957PH N51780	27. 5.04	London Executive Aviation Ltd	Stansted	9. 6.05T
G-IKRK	Europa Aviation Europa *(Built K.R.Kesterton - kit no.202) (Monowheel u/c)*	PFA 247-12903		16. 4.02	K.R.Kesterton *(Based at home)*	Andrewsfield	24. 6.05P
G-IKRS	Comco Ikarus C42 FB UK *(Built P.G.Walton)*	PFA 322-13719		1. 8.01	A.W.Fish	Trehelig, Welshpool	23. 2.05P

G-IKUS	Comco Ikarus C42 FB UK *(Built C.I.Law)*	PFA 322-14130		10.10.03	C.I.Law	Wickenby	12. 4.05P
G-ILDA	Vickers Supermarine 361 Spitfire HF.IX	CBAF.10164	G-BXHZ SAAF???, SM520	11. 7.02	P.W.Portelli *(On rebuild 5.04)*	Thruxton	AC
G-ILEA	Piper PA-31 Navajo C	31-7812117	(8P-) G-ILEA, D-ILEA, N27775	7. 7.97	I.G.Fletcher *(Ditched 54 miles W of Barbados 18.5.03 and lost without trace)*	(London SW1)	7. 8.03
G-ILEE	Colt 56A Duo Chariot HAB	2624		29. 7.94	G.I.Lindsay *"Gillie"*	(Storrington, Pulborough)	15. 6.03A
G-ILLE	Boeing Stearman E75 (PT-13D) Kaydet (Continental W670)	75-5028	N68979 42-16865, Bu.60906	7. 3.90	M.Minkler *(Also carries "379" in USAAC c/s)*	Priory Farm, Tibenham	22. 8.05T
G-ILLY	Piper PA-28-181 Cherokee Archer II	28-7690193	SE-GND	21. 2.80	R.A and G.M. Spiers *(New owners 6.03)*	Hinton-in-the-Hedges	19.12.93
G-ILRS	Comco Ikarus C42 FB UK *(Built L R Smith)*	PFA 322-13927		19. 6.02	Knitsley Mill Leisure Ltd	(Consett)	17.12.04P
G-ILSE	Corby CJ-1 Starlet *(Built S.Stride)* (HAPI Magnum 75)	PFA 134-10818		9. 1.84	S.Stride	Wolverhampton	31. 3.05P
G-ILTS	Piper PA-32-300 Six	32-7940217	G-CVOK OE-DOH, N2941C	28. 3.90	Foremans Aviation Ltd	Full Sutton	24. 2.05T
G-ILUM	Europa Aviation Europa XS *(Built A R.Haynes - kit no.467) (Tri-gear u/c)*	PFA 247-13565		15. 6.00	A R.Haynes	(Stevenage)	
G-IMAG*	Colt 77A HAB				See SECTION 3 Part 1	(Aldershot)	
G-IMAN	Colt 31A Sky Chariot HAB	2605		23. 6.94	Benedikt Haggeney GmbH	(Ennigerloh, Germany)	17. 3.05A
G-IMBI	QAC Quickie 1 (Rotax 503)	484	G-BWIT N4482Z	7.10.02	J.D.King	Biggin Hill	20. 1.05P
G-IMBY	Pietenpol AirCamper *(Built P F Bockh)*	PFA 47-12402		22.12.93	P.F.Bockh *(Current status unknown)*	(Horsham)	
G-IMCD	Van's RV-7 *(Built I.G.McDowell)*	PFA 323-13965		14. 4.04	I.G.McDowell	(Horstead, Norwich)	
G-IMGL	Beech B200 Super King Air	BB-1564	VP-CMA N205JT	9. 8.99	IM Aviation Ltd	Coventry	6. 9.05T
G-IMIC	IAV-Bacau Yakovlev Yak-52	8910001	RA-02149 DOSAAF 103 *(Yellow)*	21. 8.02	C.Vogelgesang and R.Hockey	Booker	27. 8.05P
G-IMLI	Cessna 310Q	310Q0491	G-AZYK N4182Q	3. 4.86	Leisure Park Management Ltd	(Ealand, Scunthorpe)	17. 8.06T
G-IMME	Zenair CH.701 STOL *(Built M.Spearman)*	PFA 187-14080		29. 8.03	M.Spearman	(Greenhithe)	
G-IMNY	Just/Reality Escapade 912 *(Built D.S.Bremner - kit no.JAESC 0022)*	BMAA/HB/358		25. 5.04	D.S.Bremner	(Bury)	
G-IMOK	HOAC HK 36R Super Dimona	36317	I-NELI OE-9352	31. 7.97	A L.Garfield	Dunstable	6. 8.06
G-IMPX	Rockwell Commander 112B	512	N1304J	25.10.90	J.C.Stewart	(Calne)	1. 6.06
G-IMPY	Avid Flyer C *(Built T R C Griffin)* (Rotax 532)	PFA 189-11439		10. 4.89	T.R.C.Griffin	Haverfordwest	15. 9.05P
G-INAV	Aviation Composites Europa *(Built Aviation Composites Co.Ltd)*	AC.001		23. 2.87	I.Shaw	(York)	AC
G-INCA	Glaser-Dirks DG-400	4-199		22. 1.87	K.D.Hook *"320"*	Portmoak	17. 3.05
G-INCE	Best Off Skyranger 912 *(Built N.P.Sleigh - kit no.SKR0302293)*	BMAA/HB/270		12. 3.03	N.P.Sleigh	Ashcroft Farm, Winsford	28. 7.04P
G-INDC	Cessna T303 Crusader	T30300122	G-BKFH N4766C	28. 6.83	Crusader Aviation Ltd	Turwewston	10. 4.05
G-INDX	Robinson R44 Clipper II	10491		20.10.04	Kinetic Computers Ltd *(Noted 11.04)*	Elstree	8.11.07E
G-INGA	Thunder Ax8-84 HAB	2149		16. 6.92	M.L.J.Ritchie	(Weybridge)	30. 9.94A
G-INGE	Thruster T.600N Sprint	9039-T600N-033		23. 2.99	Thruster Air Services Ltd	Ginge, Wantage	7.10.05P
G-INIS	Robinson R22 Beta	1982	G-UPMW	31. 5.00	M.Fowkes t/a Rotorfun Aviation	(Derby)	13. 3.04T
G-INIT	SOCATA TB-9 Tampico Club	1384	I-IAFS	19. 1.05	Air Touring Ltd	Biggin Hill	
G-INJA	Comco Ikarus C42 FB100 VLA *(Built J.W.G.Andrews)*	PFA 322-14044		24. 4.03	J.W.G.Andrews	(Welwyn)	22.11.05P
G-INNI	Wassmer Jodel D.112	540	F-BHPU	30. 8.94	V.E.Murphy	(Fethard, County Tipperary)	5.12.03P
G-INNY	Replica Plans SE.5A *(Built R M Ordish)* (Continental C90)	PFA 20-10439		18.12.78	K.S.Matcham *(As "F5459/Y" in RFC c/s)*	Barton Ashes	15. 5.05P
G-INOW	Monnett Moni *(Built ARV Aviation Ltd - c/n 223)* (KEF 107)	PFA 142-10953		30. 3.84	W.C.Brown *(Stored 8.97: current status unknown)*	Fairoaks	20. 8.88P
G-INSR	Cameron N-90 HAB	4320		23. 4.98	The Smith and Pinching Group Ltd and P.Phillips	(Norwich/Wymondham)	29. 7.05A
G-INTL	Boeing 747-245F	20826	N640FE (N631FE), N811FT, N701SW	8.12.00	AFX Capital Ltd	Stansted	14. 3.04T
G-INVU	Agusta-Bell 206B JetRanger II	8530	G-XXII G-GGCC, G-BEHG	1. 3.95	Elmtree Estates Ltd	Henstridge	8. 5.06T
G-IOCO	Beech 58 Baron	TH-1783		6. 6.96	Arenberg Consultadoria E Servicos LDA	(Madeira, Portugal)	20. 6.05
G-IOIA	III Sky Arrow 650 T *(Built P.J.Lynch, P.G.Ward and N.J.C.Ray - c/n P/98/024)*	PFA 298-14008		20. 3.03	P.J.Lynch, P.G.Ward and N.J.C.Ray	Old Sarum	4.11.05P
G-IOIT*	Lockheed L1011-385-1 Tristar 200	193N-1145	G-CEAP SE-DPM, G-BEAL	6. 5.98	Classic Airways *(Cancelled 1.10.98 by CAA) (Stored 9.03)*	Stansted	
G-IONA	Aérospatiale-Alenia ATR42-300	017	N971NA F-WWER	19.12.02	Highland Airways Ltd	Inverness	19.12.05T
G-IOOI	Robin DR400/160 Major 80	1700		31. 5.85	N.B.Mason	Rendcomb	15. 1.05
G-IOOX	Learjet Learjet 45	45-243	N4004Q	8. 6.04	Hundred Percent Aviation Ltd	Coventry	7. 6.07T
G-IOPT	Cessna 182P Skylane	18261731	N182EE D-ECVM, N21585	9. 6.98	A J.Marks and D.Madden tr Indy Oscar Group	Elstree	26. 9.04
G-IORG	Robinson R22 Beta	1679	OH-HRU G-ZAND	28. 1.00	G.M.Richardson t/a Commission-Air	(Market Deeping)	19. 4.06T
G-IOSI	CEA Jodel DR.1050 Sicile	526	F-BLRS	6.10.80	G.A Saxby tr Sicile Flying Group	Bidford	9. 7.05
G-IOSO	CEA Jodel DR.1050 Ambassadeur	46	OO-VDV F-BJUE	13. 7.00	A E.Jackson	Duxford	2.12.06

G-IOWE	Europa Aviation Europa XS	PFA 247-13303		30. 7.99	P.A Lowe	Wolverhampton	30. 3.05P
	(Built P.A Lowe - kit no.388) (Tri-gear u/c)						
G-IPAL	Cessna 550 Citation Bravo	550-0935	EI-PAL	21. 1.04	Pacific Aviation Ltd	Belfast Aldergrove	22. 1.06T
			N5264A				
G-IPAT	Jabiru Aircraft Jabiru SP-470	PFA 274B-14227		14. 4.04	M.G.Thatcher	Sleap	4.11.05P
	(Built M.G.Thatcher)						
G-IPAX	Cessna 560XL Citation Excel	560-5228	EI-PAX	13. 2.04	Pacific Aviation Ltd	Belfast	19. 2.05T
G-IPKA	Alpi Pioneer 300	PFA 330-14355		9. 2.05	I.P.King	(Wetherby)	
G-IPSI (2)	Grob G109B	6425	G-BMLO	29. 5.86	D.G.Margetts	Vaynor Farm, Llanidloes, Powys	12. 6.05
G-IPSY	Rutan VariEze	PFA 74-10284	(G-IPSI)	19. 6.78	R.A Fairclough	Biggin Hill	8. 7.05P
	(Built R.A Fairclough - c/n 1512) (Continental PC60)						
G-IPUP	Beagle B.121 Pup Series 2	B121-036	HB-NAC	17. 7.95	R.G.Hayes and S.Tvietan	North Weald	23.11.07T
			G-35-036		tr Skyway Group		
G-IRAF	Rotary Air Force RAF 2000 GTX-SE	PFA G/13-1278		17. 6.96	M.S.R.Allen	(Oakham)	16.10.03P
	(Built C D Julian)						
G-IRAL	Thruster T.600N 450	0035-T600N-083		24. 4.03	J.Giraldez	(Burgh Le Marsh, Skegness)	29. 5.04P
	(Official c/n incorrect - date of manufacture [May 2003] indicates correct version should be 0053-T600N-083) (New owner 11.04)						
G-IRAN	Cessna 152 II	15283907	OH-CKM	19. 8.97	I.R.Chaplin	King's Farm, Thurrock	28. 4.07T
			C-GBJY, (N6150B)				
G-IRIS	Gulfstream AA-5B Tiger	AA5B-1184	G-BIXU	14.12.87	A H.McVicar	Carlisle	2. 6.06T
			N4533N		*(Op Carlisle Flight Centre)*		
G-IRJX*	BAE SYSTEMS Avro 146-RJX100				See SECTION 3 Part 1	Manchester	
G-IRKB	Piper PA-28R-201 Cherokee Arrow III	28R-7737071	D-EJDS	7. 3.00	R.K.Brierley	Earls Colne	31. 5.06
			N5814V				
G-IRLY	Colt 90A HAB	1620		28.12.89	C.E.R.Smart *(New owner 8.04)*	(Waterlooville)	6. 8.05A
G-IRON	Europa Aviation Europa XS	PFA 247-14235		4. 5.04	T.M.Clark	(Guildford)	
	(Built T.M.Clark)						
G-IROW	VPM M-16 Tandem Trainer	PFA G/12-1239	G-DBDB	1. 4.04	B.Jones	(Boroughbridge)	7. 7.04P
	(Built D.R.Bolsover) (Rotax 914-UL)						
G-IRPC	Cessna 182Q Skylane II	18266039	G-BSKM	15. 5.91	J.W.Halfpenny	Cambridge	23. 7.05T
			N559CT, N759JV				
G-IRTH	Lindstrand LBL 150A HAB	772	G-BZTO	20. 6.02	A M.Holly *(Cowlin Construction titles)*	(Berkeley)	11 8.04T
G-ISAX	Piper PA-28-181 Archer III	2843453	N5325G	28. 6.01	Airpark Flight Centre Ltd	Coventry	27. 6.07T
G-ISCA	Piper PA-28RT-201 Arrow IV	28R-8118012	N8288Y	12. 2.91	D.J. and P.Pay	Exeter	25. 5.06
			N9608N				
G-ISDB	Piper PA-28-161 Cherokee Warrior II	28-7716074	G-BWET	19. 2.96	Action Air Services Ltd	White Waltham	19. 5.07T
			SX-ALX, D-EFFQ, N9612N				
G-ISDN	Boeing Stearman B75N1 (N2S-3) Kaydet	75-1263	N4197X	6. 2.95	D.R.L.Jones	Rendcomb	18. 6.05
	(Officially regd as "A75N1")		XB-WOV, Bu.3486		*(As "14" in US Army c/s)*		
G-ISEH	Cessna 182R Skylane II	18267843	G-BIWS	9.11.90	C.M.White and R.MacFarlane	Perth	29. 5.06
			N6601N				
G-ISEL	Best Off Skyranger 912	BMAA/HB/312		7.11.03	P.A Robertson	(Royston)	
	(Built P.A Robertson - kit no.SKRxxxxxxx?)						
G-ISFC	Piper PA-31 Turbo Navajo B	31-7300970	G-BNEF	23. 3.94	I.M.Latiff	Little Staughton	16. 5.03
			N7574L				
G-ISHA	Piper PA-28-161 Warrior III	2842211	N3092D	21. 7.04	Clever Clogs (Middleton) Ltd	Barton	28. 7.07T
G-ISKA	WSK-PZL Mielec TS-11 Iskra	1H1018	Polish AF 1018	11. 5.00	P.C.Harper *(Noted 3.02)*	Bruntingthorpe	
G-ISLA	Britten-Norman BN-2A-26 Islander	206	PH-PAR	7. 5.97	Cormack (Aircraft Services) Ltd	Cumbernauld	21. 6.04T
			G-BNEA, SE-FTA, G-51-206				
G-ISMO	Robinson R22 Beta	0870	OH-HOR	14.10.88	Moy Motorsport Ltd	Sywell	7. 8.05T
			G-ISMO, N8214T				
G-ISSY	Eurocopter EC120B Colibri	1236	G-CBCG	11.10.01	D R Williams	Stapleford	3.12.04T
			F-WQPT				
G-ISTT	Thunder Ax8-84 HAB	1787		12. 6.90	RAF Halton Hot Air Balloon Club	(RAF Halton)	5.12.04A
					"RAF Halton"		
G-ITII	Pitts S-2A	2223	I-VLAT	5. 7.95	Aerobatic Displays Ltd	Booker	21. 1.05A
	(Built Aerotek Inc)						
G-ITOI	Cameron N-90 HAB	4785		14. 1.00	A E.Lusty	(Bourne)	17..8.05A
G-ITON	Maule MX-7-235 Star Rocket	10050C	N5670R	11. 9.96	J.R.S.Heaton	Hawksbridge Farm, Oxenhope	14. 1.06
G-ITUG	Piper PA-28-180 Cherokee C	28-4121	G-AVNR	14. 8.02	S.I.Tugwell	Rayne Hall Farm, Braintree	25.10.07E
			N11C				
G-ITVM	Lindstrand LBL 105A HAB	1017		8.11.04	N.C.Lindsay	(Storrington, Pulborough)	5.12.05E
G-ITWB	de Havilland DHC-1 Chipmunk 22	48	CS-AZO	16. 9.04	I.T.Whitaker-Bethel	(Bury St. Edmunds)	
	(Built OGMA) (Officially regd as c/n 1348)		Portuguese AF 1358				
G-IUAN	Cessna 525 CitationJet	525-0324	N5163C	30. 6.99	RF Celada SpA	(Milano, Italy)	23. 3.05
			(N428PC)				
G-IVAC	Airtour AH-77B HAB	012		28.11.89	T.D.Gibbs	(Billingshurst)	1. 4.05A
G-IVAL	CAB CAP 10B	307		8. 4.03	I.Valentine	Kilrush, County Kildare	6. 7.06
G-IVAN	Shaw Twin-Eze	PFA 74-10502		11. 9.78	A M.Aldridge	Ostend, Belgium	5.10.90P
	(Built I Shaw - c/n 39) (Norton-Wankel)				*"Mistress" (Noted 2000)*		
G-IVAR	Yakovlev Yak-50	791504	D-EIVI	24. 2.89	A H.Soper	North Weald	19. 4.05P
			(N5219K), DDR-WQT, DM-WQT				
G-IVDM	Schempp-Hirth Nimbus-4DM	39/55	D-KABV	18.10.02	G.W.Lynch	Rattlesden	14. 1.06
G-IVEL	Fournier RF4D	4029	G-AVNY	29. 6.95	V.S.E.Norman *(St.Ivel/Utterly Butterly titles)*	Rendcomb	14. 4.01A
G-IVEN	Robinson R44 Raven II	10442		28. 7.04	OKR Group	(Dublin)	13. 9.07T
G-IVER	Europa Aviation Europa XS	PFA 247-13632		14. 8.00	I.Phillips	(Orpington)	
	(Built I.Phillips - kit no.486) (Convertible u/c)						
G-IVET	Europa Aviation Europa	PFA 247-12511		23. 5.97	K.J.Fraser	(Abingdon)	
	(Built K.J.Fraser - kit no.020) (Conventional u/c)				*(Current status unknown)*		
G-IVII	Van's RV-7	PFA 323-14222		31. 8.04	M.A.N.Newall	Bagby	
	(Built M.A.N.Newall)						
G-IVIV	Robinson R44 Astro	0016	(N803EH)	2. 8.93	Helitrain Ltd	(Bristol)	23. 7.05T
G-IVOR	Aeronca 11AC Chief	11AC-1035	EI-BKB	18. 6.82	P.R.White and C.P.Matthews	Bodmin	3. 6.05P
			G-IVOR, EI-BKB, N9397E		tr South Western Aeronca Group		

G-IVYS	Parsons Two Place Gyroplane	PFA G/8-1275		11. 1.00	R.M.Harris	(Nottingham)	
	(Built R.M.Harris) (Mazda RX-7)						
G-IWON	Cameron V-90 HAB	2504	G-BTCV	17. 2.92	D.P.P.Jenkinson *(Cameron - 21 years titles)*	(Tring)	17. 4.05A
G-IWRC	Eurocopter EC135T2	0241	D-HECA	9. 9.02	Hundred Percent Aviation Ltd	(Banbury)	13.11.05T
G-IXES	Air Création Clipper/iXess 912	BMAA/HB/357		3. 3.04	Flylight Airsports Ltd	Sywell	14. 7.05P
	(Built R Grimwood - kit no FL027, Trike T03027, Wing A04019-4014) (Thought to be Trike unit from French 07-JF with new wing)						
G-IXII	Christen Eagle II	T0001	G-BPZI	9. 1.03	R P Marks	Dunkeswell	4.12.04P
	(Built J Trent and R Eicher) (Lycoming IO-360)		N48BB		tr Eagle Flying Group		
G-IXIX	III Sky Arrow 650 T	PFA 298-13257		24.10.00	W.J.De Gier	North Weald	23. 5.05P
	(Built W.J.De Gier)						
G-IXTI	Extra EA.300/L	121		15. 9.00	J.C.Merry	(East Coker, Yeovil)	22.10.06T
G-IYAK	SPP Yakovlev Yak C-11	171103	OK-JIM	12. 1.94	G.G.L.James	Sleap	23. 9.05P
			(French AF, Egypt AF)				
G-IYCO	Robin DR500/200i Président	0031		23. 2.01	Timgee Holdings Ltd	Jersey	21. 3.07
	(Registered as DR400/500)						
G-IZIT	Rans S-6-116 Coyote II	PFA 204A-12965		7. 3.96	J.R.Caylow	Headon Farm, Retford	17. 6.05P
	(Built D A Crompton) (Rotax 912-UL) *(Tri-cycle u/c)*						
G-IZOD	Jabiru Aircraft Jabiru UL-450	PFA 274A-13541		26. 5.00	D.A Izod	Damyn's Hall, Upminster	16. 9.05P
	(Built D.A Izod)						
G-IZZI	Cessna T182T Turbo Skylane	T18208100	N51197	19. 3.02	T.J. and P.S.Nicholson	Oxford	7. 5.05T
G-IZZS	Cessna 172S Skyhawk SP	172S8152	N952SP	1. 7.99	Walkbury Aviation Ltd	Sibson	26. 9.05T
G-IZZY	Cessna 172R Skyhawk	17280419	G-BXSF	7. 9.99	R.Parsons	Haverfordwest	6. 6.07T
			N9967F				
G-IZZZ	American Champion 8KCAB	939-2003		29. 1.04	A M.Read	Goodwood	9. 6.07T

G-JAAA - G-JZZZ

G-JABB	Jabiru Aircraft Jabiru UL-450	PFA 274A-13555		27. 4.00	D.J.Abbott	Plaistowes Farm, St. Albans	13. 4.05P
	(Built D.J.Royce)						
G-JABI	Jabiru Aircraft Jabiru J400	PFA 325-14098		3.10.03	R.A Shaw Aviation Ltd	(Thornton-Cleveleys)	
	(Built R.A Shaw)						
G-JABJ	Jabiru Aircraft Jabiru J400	PFA 325-14126		17.11.03	Hamsard 2668 Ltd	(Luton)	1.11.05P
	(Built P G Leonard)						
G-JABO	WAR Focke-Wulf FW190-A3 rep	PFA 81-11786		23. 8.01	S.P.Taylor	Clench Common	
	(Built S.P.Taylor)				*(Noted 8.04)*		
G-JABS	Jabiru Aircraft Jabiru UL-450	PFA 274A-13704		27. 6.02	P.E.Todd	Chilbolton	21. 1.05P
	(Built I R Cook and P.E.Todd)				tr Jabiru Flyer Group		
G-JABY	Jabiru Aircraft Jabiru SPL-450	PFA 274A-13672		2. 2.01	J.T.Grant	(Norwich)	28. 8.04P
	(Built J.T.Grant)						
G-JACA	Piper PA-28-161 Warrior III	2842139	N5328Q	28. 2.02	Channel Islands Aero Club (Jersey) Ltd	Jersey	6. 3.05T
G-JACB	Piper PA-28-181 Archer III	2843278	G-PNNI	23. 7.02	Channel Islands Aero Club (Jersey) Ltd	Jersey	24.10.05T
			N41651				
G-JACC	Piper PA-28-181 Archer III	2843222	G-GIFT	23.12.02	Channel Islands Aero Club (Jersey) Ltd	Jersey	6.11.05T
			G-IMVA, SE-KIH, N9524N, N4166F				
G-JACK	Cessna 421C Golden Eagle III	421C1411	N421GQ	29. 4.97	JCT 600 Ltd	Leeds-Bradford	3. 5.07
			N125RS, N12028				
G-JACO	Jabiru Aircraft Jabiru UL	PFA 274A-13371		14. 4.99	C.D.Matthews	(Kilmurry Bray, County Wicklow)	21.10.04P
	(Built S Jackson)						
G-JACS	Piper PA-28-181 Archer III	2843078	N9287J	15. 4.97	Vector Air Ltd	Fowlmere	2. 6.06T
			(G-JACS)				
G-JADJ	Piper PA-28-181 Archer III	2843009	N49TP	27. 7.99	S.J.Skilton	Bournemouth	28. 7.05T
	(Originally intended as c/n 2890240)		N92552		t/a Aviation Rentals *(Op Solent Flight Training)*		
G-JAEE	Van's RV-6A	PFA 181A-13571		16. 9.02	J.A E.Edser	(Luton)	
	(Built J A E Edser)						
G-JAES	Bell 206B JetRanger II	1513	G-STOX	13. 1.04	Heli Charter Wales Ltd	Haverfordwest	4. 7.05T
			G-BNIR, N59615				
G-JAGS	Reims/Cessna FRA150L Aerobat	FRA1500167	G-BAUY	24.10.01	S Brecken	RAF Coltishall	19. 3.05T
			N10633		tr RAF Coltishall Flying Club		
G-JAHL	Bell 206B-3 JetRanger III	3565	N666ST	2. 1.98	D.T.Gittins t/a Jet Air Helicopters	Shobdon	26. 2.07T
G-JAIR	Mainair Blade	1249-0500-7 & W1042		11. 7.00	J.Loughran	Newton Bank, Daresbury	9. 5.05P
	(Rotax 582)				*"Joe Loughran"*		
G-JAJB	Grumman American AA-5A Cheetah	AA5A-0590	OY-CJE	30. 4.02	J.Bradley	Thruxton	29. 5.05
			N26434				
G-JAJK	Piper PA-31-350 Chieftain	31-8152014	G-OLDB	16.12.99	Keen Leasing (IoM) Ltd	Belfast Aldergrove	24. 7.06T
			OY-SKY, G-DIXI, N40717				
G-JAJP	Jabiru Aircraft Jabiru UL-450	PFA 274A-13627		1.12.00	J.W.E.Pearson and J.Anderson		
	(Built J.W.E.Pearson)					Plaistowes Farm, St.Albans	18.10.05P
G-JAKI	Mooney M.20R Ovation	29-0030		7. 2.95	J.M.Moss and D.M.Abrahamson	Dublin	12. 5.07
G-JAKS	Piper PA-28-160 Cherokee	28-339	G-ARVS	2. 7.99	K.Harper	Stapleford	29. 1.07
G-JALC	Boeing 757-225	22194	N504EA	6. 3.95	MyTravel Airways Ltd	Manchester	22. 4.07T
G-JAMA	Schweizer 269C-1	0165		2. 4.04	JWL Helicopters Ltd	Biggin Hill	20. 4.07T
G-JAME	Zenair CH.601UL Zodiac	PFA 162A-14279	G-CDFZ	14. 1.05	J.P.Harris	(Ossett)	
	(Built J P Harris, B Yoxall, K Yoxall and N Barnes)						
G-JAMP	Piper PA-28-151 Cherokee Warrior	28-7515026	G-BRJU	3. 4.95	Lapwing Flying Group Ltd	Denham	22. 7.07T
			N44762				
G-JAMY	Europa Aviation Europa XS	PFA 247-13557		5. 1.01	J P Sharp	Rayne Hall Fam, Braintree	13. 6.05P
	(Built J P Sharp - kit no.449)						
G-JANA	Piper PA-28-181 Archer II	28-7990483	N2838X	12. 2.87	C.Dashfield t/a Croaker Aviation	Stapleford	3. 6.05
G-JANB*	Colt Flying Bottle SS HAB				See SECTION 3 Part 1	(Aldershot)	
G-JANG	Piper PA-28RT-201T Turbo Arrow IV		C-GGDU	2. 6.03	Abertawe Aviation Ltd	Swansea	29. 6.06T
		28R-8531002	N4389E, N9648N				
G-JANN	Piper PA-34-220T Seneca III	3433133	N9154W	23. 6.89	MBC Aviation Ltd	Fairoaks	30. 8.07T

Reg	Type	C/n	Previous identity	Date	Owner	Base	CofA
G-JANO	Piper PA-28RT-201 Arrow IV	28R-7918091	SE-IZR	14. 5.98	Blackpool Aviators Ltd	Blackpool	9. 9.07T
			N2146X				
G-JANS	Reims FR172J Rocket	FR17200414	PH-GJO	11. 8.78	I.G.Aizlewood	Rush Green	29.10.04
			D-EGJO				
G-JANT	Piper PA-28-181 Archer II	28-8390075	N4297J	23. 2.87	Janair Aviation Ltd	Denham	1. 4.05T
	(Originally built as c/n 28-8290117/N81992/YV-2234P: not delivered and re-manufactured as c/n stated)						
G-JAPS	Europa Aviation Europa XS	PFA 247-13980		22.10.03	P.E.Tait	(Stockton-on-Tees)	
	(Built P.E.Tait - kit no.551)						
G-JARA	Robinson R22 Beta	1837		11. 6.91	Northumbria Helicopters Ltd	Newcastle	4. 9.06T
G-JARV	Aérospatiale AS355F1 Ecureuil 2	5164	G-OGHL	15.10.01	PLM Dollar Group Ltd	Gloucestershire	27. 5.06T
			N5796S				
G-JASE	Piper PA-28-161 Warrior II	28-8216056	N8461R	13. 2.91	Mid-Anglia Flight Centre Ltd	Cambridge	3. 8.07T
					t/a Mid-Anglia School of Flying		
G-JAST	Mooney M.20J Model 201	24-1010	OO-RYL	1.10.04	S.J.Tillotson	Elstree	6.12.07E
			(N4004H)				
G-JATD	Robinson R22 Beta	0534	G-HUMF	10. 7.03	Cycamore Ltd	Elstree	24. 1.05T
			N23743				
G-JAVO	Piper PA-28-161 Warrior II	28-8016130	G-BSXW	17. 9.97	Victor Oscar Ltd	Wellesbourne Mountford	7. 7.06T
			N8119S				
G-JAWC	Pegasus Quantum 15-912	7692		21. 7.00	M.H.Husey	(Tunbridge Wells)	7. 9.05P
G-JAWZ	Pitts S-1S	PFA 09-12846		6.11.95	A R.Harding	Leicester	22.12.05P
	(Built S Howes)						
G-JAXS	Jabiru Aircraft Jabiru UL-450	PFA 274A-13548		10.12.99	J P Pullin	Weston Zoyland	29. 9.05P
	(Built C A Palmer)						
G-JAYI	Auster V J/1 Autocrat	2030	OY-ALU	5. 2.93	Bravo Aviation Ltd	Coventry	28. 5.06
			D-EGYK, OO-ABF				
G-JAYS	Best Off Skyranger 912S	BMAA/HB/433		14.12.04	J.Williams	(Mansfield)	
	(Built J.Williams)						
G-JAZZ	Gulfstream AA-5A Cheetah	AA5A-0819	N26932	30. 3.82	A J.Radford	Tatenhill	7.10.05
G-JBAS	Neico Lancair 200	PFA 191-11465		21.11.03	B.A Slater	(Wellington)	
G-JBBZ	Eurocopter AS350B3 Ecureuil	3580	F-WQDE	13. 1.03	Powersense Ltd	Elstree	16. 2.06T
			F-WQPV				
G-JBDB	Agusta-Bell 206B JetRanger II	8238	G-OOPS	11. 4.96	Dicksons Van World Ltd	Newcastle	20. 1.06T
			G-BNRD, Oman AF 602				
G-JBDH	Robin DR400/180 Régent	1901		17. 3.89	W.A Clark	Netherthorpe	2. 6.07
G-JBEN	Mainair Blade 912	1337-0802-7 & W1132		13. 9.02	G.J.Bentley	(Chester)	12. 9.04P
G-JBHH	Bell 206B JetRanger II	1129	G-SCOO	26. 4.04	Hughes Helicopter Co Ltd	Biggin Hill	12. 7.06T
			G-CORC, G-CJHI, G-BBFB, N18094		t/a Biggin Hill Helicopters		
G-JBII	Robinson R22 Beta	1368	G-BXLA	13. 3.03	Jet Black 11 Ltd	Shoreham	27. 5.06T
			SE-HVX, N4014G		*(Op FAST Helicopters Ltd)*		
G-JBJB	Colt 69A HAB	1274		26. 7.88	Justerini and Brooks Ltd *"J and B Jeremy"*	(London SW1)	19. 5.02A
G-JBKA	Robinson R44 Raven	1175		12. 3.02	Deeside Helicopters and Promotions	Glenrothes	26. 3.05T
G-JBMC	SOCATA TB-10 Tobago	1230	F-GKVA	10. 5.04	J.McCloskey	(Maghera)	12. 8.07
G-JBRN	Cessna 182S Skylane	18280029	N432V	11. 6.99	Parallel Flooring Accessories Ltd	Wickenby	26. 5.07
			G-RITZ, N9872F				
G-JBSP	Jabiru Aircraft Jabiru SP-470	PFA 274B-13486		12.10.99	C.R.James	Ludham	26. 6.05P
	(Built C.R.James - c/n 289)						
G-JCAP	Robinson R22 Beta II	3415	EI-EWM	25. 6.04	Italian Clothes Ltd	(Birmingham)	8. 7.07T
G-JCAR	Piper PA-46-350P Malibu Mirage	4636223	N4148N	17.12.99	Aquarelle Investments Ltd	Jersey	21.12.05T
G-JCAS	Piper PA-28-181 Archer II	28-8690036	N9093N	12. 6.89	Charlie Alpha Ltd	Jersey	19. 6.07T
			(N170AV), N9648N				
G-JCBA	Sikorsky S-76B	760352	N95UT	25.11.99	J C Bamford Excavators Ltd	Nottingham East Midlands	2. 3.05T
			N95LT, N120PP, N120PM				
G-JCBJ	Sikorsky S-76C	760502		9. 7.99	J C Bamford Excavators Ltd	Nottingham East Midlands	21. 7.05T
G-JCBV	Gulfstream Aerospace Gulfstream V	682	(VP-BFD)	15. 7.03	J C Bamford Excavators Ltd	Nottingham East Midlands	14. 7.07
			N682GA				
G-JCKT	Stemme S 10-VT	11-004	D-KSTE	8. 4.98	J.C.Taylor	(Castletown, Isle of Man)	13. 5.07
G-JCMW	Rand Robinson KR-2	PFA 129-11064		3. 2.99	M.Wildish and J.Cook	(Gainsborough)	
	(Built M.Wildish)				*(Current status unknown)*		
G-JCUB	Piper PA-18-135 Super Cub	18-3531	PH-VCH	21. 1.82	N.Cummins and S.Bennett	Weston, Dublin	28. 7.05T
	(L-21B-PI) *(Frame No.18-3630)*		R.Neth AF R-103 , 54-2331				
G-JDBC	Piper PA-34-200T Seneca II	34-7570150	G-BDEF	9.10.02	Bowdon Aviation Ltd	(Cheadle)	24.10.05T
			N33695				
G-JDEE	SOCATA TB-20 Trinidad	333	G-BKLA	1. 5.84	A W Eldridge and J A Heard	Leicester	12. 6.05
			F-BNGX				
G-JDEL	Jodel D.150 Mascaret	PFA 151-11276	G-JDLI	19. 9.95	K.F.and R.Richardson	(Solihull)	
	(Built K.F.Richardson - c/n 112)				*(Current status unknown)*		
G-JDIX	Mooney M.20B Mark 21	1866	G-ARTB	28.11.85	A L.Hall-Carpenter *(In open storage 7.02)*	Shipdham	16. 1.00
G-JDJM	Piper PA-28-140 Cherokee C	28-26877	(G-HSJM)	11.10.00	R.Jackson-Moore and D.J.Street	Booker	22. 9.07
			G-AYIF, N11C		tr The Hare Flying Group		
G-JEAD	Fokker F.27 Friendship 500	10627	VH-EWU	14.11.90	BAC Group Ltd	Stansted	21.11.05T
			PH-EXL		*(Op BAC Express) "Midland Trader"*		
G-JEAJ	British Aerospace BAe 146 Series 200	E2099	G-OLCA	20. 9.93	Jersey European Airways (UK) Ltd	Gatwick	17. 7.05T
			G-5-099		*"Pride of Guernsey" (flybe.com titles)*		
G-JEAK	British Aerospace BAe 146 Series 200	E2103	G-OLCB	18. 3.93	Jersey European Airways (UK) Ltd	Gatwick	20. 6.05T
			G-5-103		*"Pride of Birmingham" (Op Air France Express)*		
G-JEAM	British Aerospace BAe 146 Series 300	E3128	G-BTJT	24. 5.93	Jersey European Airways (UK) Ltd	Heathrow	23. 5.06T
			HS-TBK, G-11-128		*"Pride of Jersey" (Op Air France Express)*		
G-JEAO	British Aerospace BAe 146 Series 100	E1010	G-UKPC	19. 9.94	Trident Aviation Leasing Services (Jersey) Ltd	Filton	4. 6.05T
			C-GNVX, N802RW, G-5-512, PT-LEP, G-BKXZ, PT-LEP		*(Stored 6.04)*		
G-JEAS	British Aerospace BAe 146 Series 200	E2020	G-OLHB	13. 2.96	Jersey European Airways (UK) Ltd	Heathrow	13. 7.06T
			G-BSRV, G-OSUN, C-FEXN, N604AW		*(Op Air France Express)*		
G-JEAT*	British Aerospace BAe 146 Series 100	E1071	N171TR	11.10.96	Jersey European Airways (UK) Ltd	Exeter	23.10.05T
			J8-VBB, G-BVUY, B-2706, G-5-071		*(Cancelled 4.8.04 as WFU) (On fire dump 9.04)*		

Registration	Type	C/n	Previous identity	Date	Owner/Operator	Base	CofA expiry
G-JEAU	British Aerospace BAe 146 Series 100	E1035	N135TR	30.12.96	Jersey European Airways (UK) Ltd	Exeter	24. 1.06T
			J8-VBC, G-BVUW, B-584L, B-2704, G-5-035		*(Stored 9.04)*		
G-JEAV	British Aerospace BAe 146 Series 200	E2064	N764BA	17. 6.97	Jersey European Airways (UK) Ltd	Exeter	19. 6.06T
			CC-CEN, N414XV, G-5-064, N404XV		*(Stored 1.05)*		
G-JEAW	British Aerospace BAe 146 Series 200	E2059	(N759BA)	21. 7.97	Jersey European Airways (UK) Ltd	Gatwick	21. 8.06T
			CC-CEJ, N401XV, G-5-059, N401XV, G-5-059 *(flybe.com titles)*				
G-JEAX	British Aerospace BAe 146 Series 200	E2136	N136JV	16. 2.98	Jersey European Airways (UK) Ltd	Heathrow	19. 2.07T
			C-FHAP, N136TR, N882DV, (N719TA), N882DV, G-5-136 *(Op Air France Express)*				
G-JEAY	British Aerospace BAe 146 Series 200	E2138	SE-DRL	27. 3.01	Jersey European Airways (UK) Ltd	Gatwick	26. 3.07T
			N138JV, C-FHAA, N138TR, (N719TA), N883DV, G-5-138				
G-JEBA	British Aerospace BAe 146 Series 300	E3181	HS-TBL	16. 6.98	Jersey European Airways (UK) Ltd	Heathrow	27. 7.07T
			G-6-181, G-BSYR, G-6-181		*(Op Air France Express)*		
G-JEBB	British Aerospace BAe 146 Series 300	E3185	HS-TBK	26. 6.98	Jersey European Airways (UK) Ltd	Heathrow	1.11.07T
			G-6-185		*(Op Air France Express)*		
G-JEBC	British Aerospace BAe 146 Series 300	E3189	HS-TBO	4. 6.98	Jersey European Airways (UK) Ltd	Gatwick	1. 7.07T
			G-6-189				
G-JEBD	British Aerospace BAe 146 Series 300	E3191	HS-TBJ	14. 7.98	Jersey European Airways (UK) Ltd	Gatwick	17. 9.07T
			G-6-191				
G-JEBE	British Aerospace BAe 146 Series 300	E3206	HS-TBM	28. 5.98	Jersey European Airways (UK) Ltd	Gatwick	25. 6.07T
			G-6-206				
G-JEBF	British Aerospace BAe 146 Series 300	E3202	G-BTUY	25. 6.04	Jersey European Airways (UK) Ltd	Gatwick	25. 6.07T
			G-NJIC, B-17811, B-1781, G-BTUY, G-6-202				
G-JEBG	British Aerospace BAe 146 Series 300	E3209	G-BVCE	20. 7.04	Jersey European Airways (UK) Ltd	Gatwick	13. 7.07T
			G-NJIE, B-1778, G-BVCE, G-6-209				
G-JECE	de Havilland DHC-8-402	4094	C-FDHU	26. 8.04	Jersey European Airways (UK) Ltd	Gatwick	2. 9.07T
G-JECF	de Havilland DHC-8-402	4095	C-FDHV	7.10.04	Jersey European Airways (UK) Ltd	Gatwick	12.10.07E
G-JEDE	de Havilland DHC-8-311A *(Q300)*	534	C-GERL	25.11.99	Jersey European Airways (UK) Ltd	Gatwick	2.12.05T
G-JEDF	de Havilland DHC-8-311A *(Q300)*	548	C-GDIW	10. 7.00	Jersey European Airways (UK) Ltd	Gatwick	9. 7.06T
G-JECG	de Havilland DHC-8-402	4098	C-FAQH	7. 1.05	Jersey European Airways (UK) Ltd	Gatwick	
					(Delivered 12.1.05)		
G-JEDH	Robin DR400/180 Régent	2343		3. 2.97	J.B.Hoolahan	(Sevenoaks)	8. 7.06
G-JEDI	de Havilland DHC-8-402Q *(Q400)*	4052	C-GFOD	25.10.01	Jersey European Airways (UK) Ltd	Gatwick	24.10.07T
G-JEDJ	de Havilland DHC-8-402Q *(Q400)*	4058	C-FDHZ	23. 1.02	Jersey European Airways (UK) Ltd	Gatwick	4. 2.05T
G-JEDK	de Havilland DHC-8-402Q *(Q400)*	4065	C-GEMU	23. 4.02	Jersey European Airways (UK) Ltd	Gatwick	30. 4.05T
					(flybe.com titles)		
G-JEDL	de Havilland DHC-8-402Q *(Q400)*	4067	C-GEOZ	17. 6.02	Jersey European Airways (UK) Ltd	Gatwick	30. 6.05T
G-JEDM	de Havilland DHC-8-402	4077	C-FGNP	18. 7.03	Jersey European Airways (UK) Ltd	Exeter	22. 7.06T
G-JEDN	de Havilland DHC-8-402	4078	C-FNGB	31. 7.03	Jersey European Airways (UK) Ltd	Exeter	7. 8.06T
G-JEDO	de Havilland DHC-8-402	4079	C-GDFT	1. 8.03	Jersey European Airways (UK) Ltd	Exeter	13. 8.06T
G-JEDP	de Havilland DHC-8-402	4085	C-FDHO	30. 1.04	Jersey European Airways (UK) Ltd	Exeter	4. 3.07T
G-JEDR	de Havilland DHC-8-402	4087	C-FDHI	5. 3.04	Jersey European Airways (UK) Ltd	Exeter	15. 3.07T
G-JEDS	Andreasson BA-4B	PFA 38-10158	G-BEBT	17.12.02	S.B.Jedburgh	White Waltham	28. 9.05P
	(Built A Horsfall - c/n HA/02 [Hornet Aviation]) (Lycoming O-235-C)						
G-JEDT	de Havilland DHC-8-402	4088	C-FDHP	19. 3.04	Jersey European Airways (UK) Ltd	Exeter	23. 3.07T
G-JEDU	de Havilland DHC-8-402	4089	C-GEMU	7. 4.04	Jersey European Airways (UK) Ltd	Belfast	13. 4.07T
G-JEDV	de Havilland DHC-8-402	4090	C-FDHX	7. 5.04	Jersey European Airways (UK) Ltd	Exeter	18. 5.07T
G-JEDW	de Havilland DHC-8-402	4093	C-GFBW	27. 7.04	Jersey European Airways (UK) Ltd	Exeter	2. 8.07T
G-JEET	Reims/Cessna FA152 Aerobat	FA1520369	G-BHMF	10.12.87	Willowair Flying Club (1996) Ltd	Southend	6. 9.07E
G-JEFA	Robinson R44 Astro	0710		7. 2.00	Simlot Ltd	Denham	11. 8.06T
G-JEJE	Rotary Air Force RAF 2000 GTX-SE	PFA G/13-1352		21. 1.03	J.W.Erswell	(South Brent)	18. 8.04P
	(Built J.W.Erswell)						
G-JEMA	British Aerospace ATP	2028	N854AW	24. 2.04	Emerald Airways Ltd	Blackpool	6. 5.07T
			G-11-028, N5000R		*(Flyjem c/s)*		
G-JEMB	British Aerospace ATP	2029	N855AW	29. 3.04	Emerald Airways Ltd	Blackpool	11. 7.07T
			G-11-029		*(Flyjem c/s)*		
G-JEMC	British Aerospace ATP	2032	N856AW	29.12.03	Emerald Airways Ltd	Blackpool	15. 4.07T
			G-11-032		*(Flyjem c/s)*		
G-JEMD	British Aerospace ATP	2026	S2-ACX	3. 2.04	Emerald Airways Ltd	Blackpool	
			(SE-LHX), S2-ACX, G-11-026				
G-JEME	British Aerospace ATP	2027	S2-ACY	3. 2.04	Emerald Airways Ltd	Blackpool	
			(SE-LHY), S2-ACY, G-11-027				
G-JEMX	Short SD.3-60 Variant 100	SH3715	G-SSWX	10.03.04	Emerald Airways Ltd	Exeter	2.12.05E
			N711PM, G-BNDL, G-14-3715				
G-JEMY	Lindstrand LBL 90A HAB	742		22.11.00	J.A Lawton	(Godalming)	9.12.04A
G-JENA	Mooney M.20J Model 201	24-1304	N1168D	5. 7.82	P.Leverkuehn	Antwerp-Deurne, Belgium	16. 9.07
					t/a Mooney Partnership		
G-JENI	Cessna R182 Skylane RG II	R18200267	N3284C	17. 9.87	R.A Bentley	Stapleford	15. 5.06
G-JENN	Gulfstream AA-5B Tiger	AA5B-1187	N4533T	7.12.81	M. Reed t/a Shadow Aviation	Cranfield	23. 3.06T
G-JENO	Lindstrand LBL 105A HAB	916		28. 4.03	S.F.Redman	(Sturminster Newton)	23. 5.05A
G-JERL	Agusta A109E Power	11118		29. 5.01	Perment Ltd	(Clitheroe)	6. 6.07T
G-JERO	Europa Aviation Europa XS	PFA 247-13691		13. 6.02	B.Robshaw and P.Jenkinson	Wombleton	23. 5.05P
	(Built B.Robshaw and P.Jenkinson - kit no.492) (Rotax 914) *(Tri-gear u/c)*						
G-JERS	Robinson R22 Beta	1610		21.12.90	Sloane Helicopters Ltd	Sywell	7. 6.06T
G-JESA	Mainair 582 Gemini/Southdown Raven X	664-688-6 & SN2232/0117	G-MNLB	14. 4.04	Jesa UK Ltd	Redlands	25. 6.05P
			(Officially regd as "Southdown Raven X (modified Gemini F2A trike")				
G-JESI	Aérospatiale AS350B Ecureuil	1205	G-JOSS	16.12.03	Staske Construction Ltd	(Milton Keynes)	29.10.05T
			F-WQJY, 3A-..., G-WILX, G-RAHM, G-UNIC, G-COLN, G-BHIV				
G-JESS	Piper PA-28R-201T Turbo Arrow III	28R-7803334	G-REIS	18. 9.95	R.S.Tomlinson	White Waltham	23. 9.06
			N36689				
G-JETC	Cessna 550 Citation II	550-0282	G-JCFR	28. 5.81	Ability Air Ltd	London City	24. 3.05T
			G-JETC, N68644T				
G-JETH	Hawker Sea Hawk FGA.6	AW-6385	"XE364"	10. 8.83	P.G.Vallance Ltd	Charlwood, Surrey	
	(Buillt Armstrong-Whitworth Aircraft)		XE489		*(Gatwick Aviation Museum as "XE489")*		
	(Composite with WM983/A2511)						

Registration	Type	C/n	Previous identity	Reg date	Owner	Base	CofA expiry
G-JETI	British Aerospace BAe 125 Series 800B	258056	G-5-509	9. 7.86	Ford Motor Co Ltd	Stansted	16. 3.07T
G-JETJ	Cessna 550 Citation II	550-0154	G-EJET	9. 2.93	G-JETJ Ltd	Liverpool	26. 8.05T
			G-DJBE, (N8887N)				
G-JETM	Gloster Meteor T.7	-	VZ638	10. 8.83	P.G.Vallance Ltd	Charlwood, Surrey	
					(Gatwick Aviation Museum as "VZ638" in RN/FRU c/s)		
G-JETU	Aérospatiale AS355F2 Ecureuil 2	5450	VR-CET	18. 4.96	Marlborough Aviation Ltd	Oxford	22. 5.05T
			JA6623				
G-JETX	Bell 206B-3 JetRanger III	3208	N3898L	9. 2.88	AA Consultants Ltd	Shoreham	17. 5.06T
G-JETZ	MD Helicopters Hughes 369E	0450E	VR-HJI	26. 3.97	John Matchett Ltd	Sywell	26. 6.06
G-JEZZ	Best Off Skyranger 582	BMAA/HB/368		6. 4.04	J.W.Barwick and P.J.Harris	(Weedon, Northampton)	
	(Built J.W.Barwick - kit no.SKRxxxx456)						
G-JFMK	Zenair CH.701 SP	PFA 187-14264		24. 9.04	J.D.Pearson	(Glasgow)	
	(Built J.D.Pearson)						
G-JFRV	Van's RV-7A	PFA 323-13851		8.10.03	J.H.Fisher	Haverfordwest	2. 9.05P
	(Built J.H.Fisher)				*(Noted 9.04)*		
G-JFWI	Reims/Cessna F172N Skyhawk II	F17201622	PH-DPA	1. 9.80	Staryear Ltd	Barton	23. 2.06T
			PH-AXY				
G-JGBI	Bell 206L-4 LongRanger IV	52257	N91285	13. 8.01	Dorbcrest Homes Ltd	(Wigan)	31. 5.07
			C-GBUP				
G-JGMN	CASA 1-131E Jungmann Series 2000	2011	Spanish AF E3B-407	17. 4.91	P.D.Scandrett	Rendcomb	15. 1.05P
	(C/n as officially regd but carries c/n plate 2104 in rear cockpit: c/n 2011 is regd as N65522)						
G-JGSI	Pegasus Quantum 15-912	7515		19. 4.99	P.Thompson	Deenethorpe	31. 5.05P
G-JHAC	Reims/Cessna FRA150L Aerobat	FRA1500160	EI-BRX	16. 9.02	J.H.A Clarke	(Bromham, Chippenham)	7. 7.06T
			G-BACM, EI-BRX, G-BACM				
G-JHEW	Robinson R22 Beta	0672	N23677	20. 7.87	Burbage Farms Ltd	Hinckley	8.12.05
G-JHKP	Europa Aviation Europa XS	PFA 247-13828		5.11.03	J.D.Heykoop	(Pulborough)	
	(Built J.D.Heykoop - kit no.536)						
G-JHNY	Cameron A-210 HAB	10487		17. 3.04	Floating Sensations Ltd	(Llandeilo)	17. 3.05P
G-JHYS	Europa Aviation Europa	PFA 247-13307		6. 3.01	J.D.Boyce and G.E.Walker	Burnham-on-Crouch	2. 8.05P
	(Built J.D.Boyce and G.E.Walker - kit no.314) (Tri-gear u/c)						
G-JIGS*	Lindstrand LBL 90A HAB	656		9. 3.00	Jigsaw Connections Ltd	(Slough)	15. 4.04A
					(Cancelled 9.3.04 by CAA)		
G-JIII	Stolp SA.300 Starduster Too	2-3-12	N9043	27. 5.93	J.G.McTaggart	Archerfield Estate, Dirleton	8. 7.05P
	(Built C S Johnson) (Lycoming IO-360)				t/a VTIO Company		
G-JILL	Rockwell Commander 112TC-A	13304	(OO-HPB)	25. 7.80	D.Carlton	Full Sutton	19. 2.05T
			G-JILL, N8070R, HB-NCW				
G-JILY	Robinson R44 Raven	0959		5. 1.01	AG Aviation Ltd	Naas, County Kildare	27. 2.07
G-JIMB	Beagle B.121 Pup Series 1	B121-033	G-AWWF	7. 4.94	P.G.Fowler	Enstone	26. 5.07T
G-JIMM	Europa Aviation Europa XS	PFA 247-14071		13. 7.04	J.Riley	(Billericay)	
	(Built J.Riley)						
G-JIVE	MD Helicopters Hughes 369E	0486E	G-DRAR	24. 5.01	Sleekform Ltd	(Sowerby Bridge)	8.11.04T
			N101LH, N1608Z				
G-JJAN	Piper PA-28-181 Archer II	2890007	N9105Z	28. 3.88	Redhill Aviation Ltd t/a Redhill Flying Club	Blackbushe	15. 5.06T
G-JJEN	Piper PA-28-181 Archer III	2843370	N4190D	25. 8.00	J.E.Jenkins	Jersey	24. 8.06
G-JJMX	Dassault Falcon 900EX	112	F-WWFK	17.10.02	J.Hargreaves t/a J-Max Air Services	(Preston)	16.10.06T
G-JJSI	British Aerospace BAe 125 Series 800B	258058	G-OMGG	16. 4.04	Gama Aviation Ltd	Farnborough	24.11.07T
			N125JW, G-5-637, N125JW, VH-NMR, ZK-EUI, (ZK-EUR), G-5-510				
G-JKMF	Diamond DA.40D Star	D4.033		3. 9.03	A D and Carole Realff t/a ADR Aviation	Manston	17. 2.07T
G-JLAT	Aerotechnik EV-97 Eurostar	PFA 315-14068		14. 5.03	J.Latimer	Barton	8. 7.05P
	(Built J.Latimer)						
G-JLCA	Piper PA-34-200T Seneca II	34-7870428	G-BOKE	3. 9.97	C.A S.Atha	Durham Tees Valley	12 2.07T
			N21030				
G-JLEE	Agusta-Bell 206B-3 JetRanger III	8588	G-JOKE	10. 2.88	J S Lee	Booker	23.10.06
			G-CSKY, G-TALY				
G-JLHS	Beech A36 Bonanza	E-2571	N8046U	30.11.90	I.G.Meredith	Lydd	9. 3.06
G-JLMW	Cameron V-77 HAB	1768		23. 6.88	J.L.M.Watkins	(Ivybridge)	26. 2.99T
G-JLRW	Beech 76 Duchess	ME-165	N60206	4.11.87	Magenta Ltd *(Op Airways Flight Training)*	Exeter	19. 1.06T
G-JMAA	Boeing 757-3CQ	32241		24. 4.01	Thomas Cook Airlines UK Ltd	Manchester	23. 4.07T
G-JMAB	Boeing 757-3CQ	32242		14. 5.01	Thomas Cook Airlines UK Ltd	Manchester	13. 5.07T
G-JMAC*	British Aerospace Jetstream Series 4100				See SECTION 3 Part 1	Liverpool	
G-JMAN	Mainair Blade 912S	1290-0601-7 & W1085		12. 7.01	J.Manuel	(Southport)	15. 7.02P
G-JMAX	Hawker 800XP	258456	N41762	13.10.04	J.Hargreaves	Blackpool	AC
			N800EM		t/a J-Max Air Services		
G-JMCD	Boeing 757-25F	30757	N1795B	26. 5.00	Thomas Cook Airlines UK Ltd	Manchester	26. 5.06T
					(Op Ryan International)		
G-JMCE	Boeing 757-25F	30758		24. 6.00	Thomas Cook Airlines UK Ltd	Manchester	22. 6.06T
					(Apple Vacations titles)		
G-JMCF	Boeing 757-28A	24369	C-FOOE	20. 5.00	Thomas Cook Airlines UK Ltd	Manchester	26. 5.06T
G-JMCG	Boeing 757-2G5	26278	SX-BLV	27. 4.00	Thomas Cook Airlines UK Ltd	Manchester	1. 6.06T
			G-JMCG, D-AMUQ				
G-JMDI	Schweizer 269C	S.1398	G-FLAT	24. 9.91	J.J.Potter	Sherburn-in-Elmet	25. 4.05T
G-JMDW	Cessna 550 Citation II	550-0183	HB-VGS	16. 2.04	MAS Airways Ltd	(Horley)	24. 3.07T
			(XC-DUF), N98630				
G-JMJR	Cameron Z-90 HAB	10611		17. 1.05	J-M.Reck	(Evette-Salbert, France)	
G-JMKE	Cessna 172S Skyhawk SP	172S9248	N53012	17.12.02	115CR (146) Ltd	Wellesbourne Mountford	17.12.05T
G-JMTS	Robin DR400/180 Régent	2045		29.11.90	C.J.M.de Verenne	Dunkeswel	20. 6.06
G-JMTT	Piper PA-28R-201T Turbo Arrow III	28R-7803190	G-BMHM	8. 7.86	Jonathan Dunn	Cranfield	9. 5.05T
			N3735M				
G-JMXA	Agusta A109E Power Elite	11156		31. 5.02	J.J.Hargreaves (J-Max Air Services)	(Preston)	30. 5.05T
G-JNAS	Grumman American AA-5A Cheetah	AA5A-0604	SE-GEI	28.11.00	J.R.Nutter and A L.Shore	Bembridge	9. 3.07T
			LN-KLE				
G-JNNB	Colt 90A HAB	2063		20.12.91	N.A P.Godfrey	(Aylesbury)	17. 4.04A
G-JODI	Agusta A109A II	7265	G-BVCJ	27. 2.04	Foxdale Consulting Ltd	Ronaldsway	21.11.06T
			G-CLRL, G-EJCB				

Regn	Type	C/n	Previous identity	Date	Owner/operator	Location	CofA expiry
G-JODL	SAN Jodel DR.1050M Excellence	99	F-BJJC	28. 4.86	D.Silsbury *(Noted 6.04)*	Dunkeswell	26.11.99
G-JOEL	Bensen B.8M	PFA G/03-1300		6. 7.99	G.C.Young	Swansea	
	(Built G C Young) (Converted from Air Command)				*(Current status unknown)*		
G-JOEM	Airbus A320-231	0449	G-OUZO	17. 4.00	MyTravel Airways Ltd *(Op MyTravel Lite)*	Manchester	7.11.07E
			EI-VIR, N449RX, SX-BSV, N449RX, F-WWIG				
G-JOEY	Fairey Britten-Norman BN-2A Mk.III-2 Trislander	1016	G-BDGG	27.11.81	Aurigny Air Services Ltd	Guernsey	18. 4.07T
			C-GSAA, G-BDGG		*"Joey"*		
G-JOJO	Cameron A-210 HAB	2674		20. 9.91	A C.Rawson and J.J.Rudoni	(Little Haywood, Stafford)	29. 4.03T
G-JOLY	Cessna 120	13872	OO-ACE	3. 9.81	B.V.Meade	Garston Farm, Marshfield	13. 6.05P
	(Continental C85)						
G-JONB	Robinson R22 Beta II	2593		29. 4.96	J.Bignall	Mistletoe Farm, Pinner	27. 5.05
G-JONG	Rotorway Executive 162F	6168	N630GH	27. 4.04	J.V.George	(Winchester)	AC
	(Built S A Foster)						
G-JONH	Robinson R22 Beta	2170		3. 6.93	Productivity Computer Solutions Ltd	Sherburn-in-Elmet	22. 8.05T
G-JONI	Reims/Cessna FA152 Aerobat	FA1520346	G-BFTU	6. 7.84	R.F.and J.S.Pooler	Sleap	4. 9.06
G-JONO	Colt 77A HAB	1086		22. 6.87	The Sandcliffe Motor Group Ltd	(Stapleford, Nottingham)	17. 9.95A
					"Sandcliffe Ford"		
G-JONY	Cyclone AX2000 HKS	7503		12. 3.99	K.R.Matheson *(USAF c/s)*	Sandtoft	11. 3.05P
G-JONZ	Cessna 172P Skyhawk II	17276233	N97835	28. 9.89	Truman Aviation Ltd	Tollerton	26. 5.05T
G-JOOL	Mainair Blade 912	1262-1000-7 & W1056		8.12.00	J R Gibson	Ince Blundell	13.12.04P
G-JOON	Cessna 182D	18253067	(N....)	9. 6.81	B.Walsh	(Kildare Town, County Kildare)	14.10.05T
			G-JOON, OO-ACD, N9967T		tr Go Skydive Group		
G-JOPF*	Smyth Model S Sidewinder	PFA 92-12313		19. 4.01	Not known	Gedney Hill	
	(Built J Furby)				*(Cancelled 6.12.02 as wfu, no PtoF issued) (Noted 4.04)*		
G-JOSH	Cameron N-105 HAB	1319		13. 8.86	M.White	(Cirencester)	16. 8.96T
G-JOST	Europa Aviation Europa	PFA 247-12916		17. 6.98	J.A Austin	(Bangor)	
	(Built J.A Austin - kit no.234) (Tri-gear u/c)				*(Current status unknown)*		
G-JOYT	Piper PA-28-181 Archer II	28-7990132	G-BOVO	13. 2.90	John K.Cathcart Ltd	City of Derry	30. 3.06T
			N2239B				
G-JOYZ	Piper PA-28-181 Archer III	2843018	N9262R	19. 1.96	S.W. and Joy E.Taylor	Biggin Hill	7. 2.05
			(G-JOYZ)				
G-JPAL	Eurocopter AS355N Ecureuil II	5692	F-GSJP	9.10.01	JPM Ltd	(Horsham)	6.11.07E
G-JPAT	Robin HR200/100 Club	76	G-BDJN	13. 9.00	L.Girardier and A J.McCulloch	Sherburn-in-Elmet	31. 5.07
G-JPMA	Jabiru Aircraft Jabiru UL	PFA 274A-13399		24. 5.99	J.P.Metcalfe	Lydd	12.11.05P
	(Built J.P.Metcalfe)				*"Sheila"*		
G-JPOT	Piper PA-32R-301 Saratoga SP	32R-8113065	G-BIYM	1. 8.94	S.W.Turley	Wickenby	29. 8.05T
			N8385X				
G-JPRO	British Aircraft Corporation BAC 145 Jet Provost T.5A	EEP/JP/1055	XW433	10. 8.95	Edwalton Aviation Ltd	Humberside	9. 4.05P
					(As "XW433 in CFS c/s)		
G-JPSX	Dassault Falcon 900EX	132	F-WWFJ	17. 2.04	Sorven Aviation Ltd	Gloucestershire	16. 2.05T
G-JPTT	Enstrom 480	5032	G-PPAH	10. 4.02	P.G.Lawrence	Gloucestershire	10. 6.07T
G-JPTV	British Aircraft Corporation BAC 145 Jet Provost T.5A		XW355	2. 5.96	S.J.Davies	(Doncaster)	21. 8.04P
	(C/n '...1002' reported)	EEP/JP/1005					
G-JPVA	British Aircraft Corporation BAC 145 Jet Provost T.5A	EEP/JP/953	G-BVXT	22. 2.95	T.J.Manna	North Weald	27. 5.05P
			XW289		t/a Kennet Aviation *(As "XW289/73" in 1FTS c/s)*		
G-JREE	Maule MX-7-180 Star Rocket	11096C	N99MX	13. 4.01	J.M.P.Ree	Draycott Farm, Chiseldon	19. 4.07A
			N30051				
G-JRME	Jodel D.140E	PFA 251-13155		13.11.02	J.E.and L.L.Rex	(Goole)	
	(Built J.E.and L.L.Rex)						
G-JSAK	Robinson R22 Beta II	2959		30. 6.99	S.M. and J.W.F.Tuke t/a Tukair Aircraft Charter	Headcorn	7. 7.05T
G-JSAR	Eurocopter AS332L2 Super Puma	2576	F-WQRE	3. 9.02	Bristow Helicopters Ltd	Den Helder, The Netherlands	18.12.05T
G-JSAT	Pilatus Britten-Norman BN-2T Islander	2277	G-BVFK	5. 2.98	P.Moore	Sennelager, Germany	5. 3.05A
					tr Rhine Army Parachute Association		
G-JSCL*	Rans S-10 Sakota	PFA 194-11781		12. 4.90	Not known	Emlyns Field, Rhuallt	
	(Kit no.1289-075)				*(Crashed Emlyns Field 16.7.91: cancelled 16.12.97 as WFU) (Remains noted 4.04)*		
G-JSON	Cameron N-105 HAB	2933		21. 5.92	Up and Away Ballooning Ltd *"Jason"*	(High Wycombe)	7. 9.04A
G-JSPC	Pilatus Britten-Norman BN-2T Islander	2264	G-BUBG	21.12.94	P.Moore	Sennelager, Germany	13. 2.06A
					tr Rhine Army Parachute Association		
G-JSPL	Jabiru Aircraft Jabiru SPL-450	PFA 274A-13604		27.12.00	J A Lord	Knettishall	31. 5.05P
	(Built J A Lord)						
G-JSSD*	Handley Page HP.137 Jetstream 1				See SECTION 3 Part 1	East Kilbride	
G-JTCA	Piper PA-23-250 Aztec E	27-7305112	G-BBCU	29.12.80	J.D.Tighe	Sturgate	8.11.06T
			N40297		t/a Eastern Air Executive		
G-JTEM	Van's RV-7	PFA 323-14237		30. 4.04	J.C.Bacon	(Gwehelog, Usk)	
	(Built J.C.Bacon)						
G-JTNC	Cessna 500 Citation	500-0264	G-OEJA	9. 1.04	Eurojet Aviation Ltd	Birmingham	8.12.05T
			G-BWFL, F-GLJA, N205FM, N5264J				
G-JTPC	Aeromot AMT-200 Super Ximango	200067		28. 5.97	J.T.Potter and P.G.Cowling	Rufforth	22. 6.06
					tr G-JTPC Falcon 3 Group		
G-JTWO	Taylor J-2 Cub	1754	G-BPZR	23.10.89	C.C.Silk	Bericote Farm, Blackdown, Leamington Spa	20.12.05P
	(Built Taylor Aircraft Co Inc) (Continental A65)		N19554, NC19554		*(Carries "NC19554" on tail)*		
G-JTYE	Aeronca 7BM Champion	7AC-4185	N85445	26. 9.91	G.D.Horn	(Old Sarum)	17. 6.99P
	(Continental C85) *(Modified ex 7AS standard)*		NC85445		*(Damaged Longwood Farm, Southampton 2.8.98: current status unknown)*		
G-JUDD	Jabiru Aircraft Jabiru UL-450	PFA 274A-13570		9. 8.00	C.Judd	Lark Engine Farmhouse, Prickwillow, Ely	16. 9.05P
	(Built C.Judd)						
G-JUDE	Robin DR400/180 Régent	1869		14.10.88	Bravo India Flying Group Ltd	RAF Woodvale	15. 2.07
G-JUDI	North American AT-6D-NT Harvard III	88-14722	FAP 1502	17.11.78	A A Hodgson	Bryn Gwyn Bach	10. 8.05P
	(Regd as c/n "EX915-326165")		SAAF7439, EX915, 41-33888		*(As "FX301/FD-NQ")*		
G-JUDY	Grumman AA-5A Cheetah	AA5A-0620	(G-BFWM)	31. 8.78	Plane Talking Ltd	Biggin Hill	26.11.05T
			N26480				
G-JUGE	Aerotechnik EV-97 teamEurostar UK	1709		7.10.03	L.J. Appleby	Leicester	12.10.05P
G-JUIN	Cessna T303 Crusader	T30300014	OO-PEN	29. 2.88	M.J. and J M Newman	Denham	28. 4.06
			N9401T				
G-JULL	Stemme S-10VT	11-039		10. 2.00	J.P.C.Fuchs	Rufforth	14. 5.06

Reg	Type	C/n	Previous identity	Date	Owner/operator	Base	CofA
G-JULU	Cameron V-90 HAB	3611		7. 7.95	N.J.Appleton	(Bristol)	12. 4.05A
G-JULZ	Europa Aviation Europa	PFA 247-13045		8.10.96	M.Parkin	Sandtoft	25. 4.05P
	(Built M.Parkin - kit no.312) (Rotax 914) *(Monowheel u/c)*						
G-JUNG	CASA 1-131E Jungmann	1121	Spanish AF E3B-143	23.11.88	K.H.Wilson	Kemble	30. 6.05P
G-JUPP	Piper PA-32RT-300 Lance II	32R-7885098	G-BNJF	3.10.02	Jupp Air LLP	Wolverhampton	5.12.05T
			N31539				
G-JURA	British Aerospace Jetstream Series 3102	772	SE-LDH	21. 5.01	Highland Airways Ltd *"City of Inverness"*	Inverness	11. 7.07T
			OY-SVK, C-FAMJ, G-31-772				
G-JURE	SOCATA TB-10 Tobago	597	N106U	6.11.92	P.M.Ireland	South Lodge Farm, Widmerpool	14. 1.05
G-JURG	Rockwell Commander 114A GT	14516	N4752W	19. 9.79	N A Southern	Fairoaks	4. 5.07
		(Laid-down as c/n 14449)					
G-JUST	Beech F33A Bonanza	CE-1165	N334CW	11.10.00	Budge It Aviation Ltd	Elstree	1. 4.07
G-JVBF	Lindstrand LBL 210A HAB	265		5. 6.95	Airxcite Ltd t/a Virgin Balloon Flights	(Wembley)	23. 7.05T
G-JWBI	Agusta-Bell 206B JetRanger II	8435	G-RODS	3. 4.96	J.W.Bonser	Walsall	26. 9.05T
			G-NOEL, G-BCWN				
G-JWCM	Scottish Aviation Bulldog Series 120/1210	BH120/408	G-BHXB	19.10.99	M.L.J.Goff	Old Buckenham	15.10.06T
			Botswana DF OD2, G-BHXB				
G-JWDS	Cessna F150G	F150-0216	G-AVNB	15.12.88	G Sayer	(Caerphilly)	29. 9.94T
	(Built Reims Aviation SA)				*(New owner 2.04)*		
G-JWEB	Robinson R44 Raven	1334		3. 9.03	R.C.Hields t/a Hields Aviation	Sherburn-in-Elmet	1.10.06T
G-JWFT	Robinson R22 Beta	0989		16. 3.89	J.P.O'Brien	(Gorey, County Wexford)	10. 5.04
G-JWIV	CEA Jodel DR.1051 Sicile	431	F-BLMD	6. 9.78	C.M.Fitton	Watchford Farm, Yarcombe	12. 1.05P
G-JWJW	CASA 1-131E Jungmann Series 2000	2000/419	PH-MRK	15. 5.03	J.W.and J.T.Whicher	(York)	
		(Official c/n is dubious)	(PH-MRN), D-EDWC, Spanish AF E3B-419				
G-JYAK	Yakovlev Yak-50	853001	RA-01493	26.11.02	J.W.Stow	North Weald	4.12.05P
			DOSAAF 49 (*blue*) ?)		*"R-93"*		

G-KAAA - G-KZZZ

Reg	Type	C/n	Previous identity	Date	Owner/operator	Base	CofA
G-KAAT	MD Helicopters MD 900	900-00056	G-PASS	22. 2.00	Police Aviation Services Ltd	Marden	19. 4.05T
			N9234P		*(Op Kent Air Ambulance Trust)*		
G-KAEW	Fairey Gannet AEW.Mk.3	F9459	XL500	9. 1.04	T.J. Manna	North Weald	
	(Built Westland Aircraft Ltd)		A2701, XL500		*(Noted for rebuild to flying condition 1.05)*		
G-KAFE	Cameron N-65 HAB	1505		18. 5.87	J.R.Rivers-Scott	(Loughborough)	11. 5.02A
G-KAIR	Piper PA-28-181 Archer II	28-7990176	N3075D	28.12.78	Keen Leasing (IoM) Ltd	Cumbernauld	20.10.06T
					(Op Cumbernauld Flying School)		
G-KAMM	Hawker Hurricane XIIA	CCF/R32007	BW881	23. 2.95	Alpine Deer Group Ltd	Wanaka, New Zealand	AC
	(Built Canadian and Car Foundry Co)				*(Rebuilt for American Flying Heritage Collection, Seattle, WA 11.99)*		
G-KAMP	Piper PA-18-135 Super Cub	18-3451	D-EDPM	9. 5.97	J.R.G.Furnell	Perth	26. 8.07T
	(L-18C)		96+27, NL+104, AC+502, AS+501, 54-751				
G-KAOM	Scheibe SF25C Falke	4417	D-KAOM	3. 2.98	W.T.Barnard, G.Mckay and J.Murdoch	Strathaven	25. 3.05
	(Limbach SL1700)				tr Falke G-KAOM Syndicate		
G-KAOS	Van's RV-7	PFA 323-13956		20. 5.03	A E.N.Nicholas and D.F.McGarvey	(Sevenoaks)	
	(Built A E.N.Nicholas and D.F.McGarvey)						
G-KAPW	Percival P.56 Provost T.1	PAC/F/311	XF603	22. 9.97	Richard Shuttleworth Trustees *(As "XF603/H")*	Old Warden	28. 8.04P
G-KARA	Brugger MB.2 Colibri	PFA 43-10980	G-BMUI	1. 6.95	Cara L.Reddish	Netherthorpe	21. 8.04P
	(Built Carlton Flying Group) (Volkswagen 1834)						
G-KARI	Fuji FA.200-160 Aero Subaru	FA200-236	G-BBRE	19.12.84	The Scottish Civil Service Flying Club Ltd	Perth	11. 4.06T
G-KARK	Dyn'Aéro MCR-01 Club	PFA 301A-14010		29.12.03	R.Bailes-Brown	Tatenhill	
	(Built R.Bailes-Brown)						
G-KART	Piper PA-28-161 Warrior II	28-8016088	N8097B	10. 7.91	Newcastle Aero Club Ltd	Newcastle	24. 1.04T
G-KASX	Vickers Supermarine 384 Seafire F.XVII	FLWA.25488	G-BRMG	30.10.03	T.J.Manna	North Weald	
	(Built Westland Aircraft Ltd)		A2055, SX336		*(Noted 1.05)*		
G-KATI	Rans S-7 Courier	PFA 218-12917		5. 3.96	S.M and K.E.Hall	Netherthorpe	21. 3.05P
	(Built S M Hall - c/n 0795.151) (Jabiru 2200A)						
G-KATS	Piper PA-28-140 Cherokee Cruiser	28-7325022	G-BIRC	26. 8.83	A G.Knight	Old Buckenham	17.11.02T
			OY-BGE		t/a Airlaunch *(Noted 1.05)*		
G-KATT	Cessna 152 II	15285661	G-BMTK	10. 6.93	Central Aircraft Leasing Ltd	Wolverhampton	7. 8.05T
			N94387				
G-KAWA	Denney Kitfox Model 2	PFA 172-11822		11. 3.91	D G Burrows	Shobdon	16. 5.05P
	(Built T W C Maton)						
G-KAWW	Westland Wasp HAS.1	F9663	NZ3908	29. 3.99	S.J.Davies	Sandtoft	2. 9.05P
			XT781		*(As "XT781/426")*		
G-KAXF	Hawker Hunter F.6A	S4/U/3361	8830M	20.12.95	T.J.Manna	North Weald	15. 9.05P
	(Built Armstrong-Whitworth Aircraft)		XF515		t/a Kennet Aviation *(As "XF515/R")*		
G-KAXL	Westland Scout AH.1	F9715	XV140	16.11.95	T.J.Manna	North Weald	15. 7.04P
	(Regd with c/n F8-7976)				t/a Kennet Aviation *(As "XV140/K")* *(Noted 1.05)*		
G-KAXT	Westland Wasp HAS.1	F9669	NZ3905	5. 3.02	T.J.Manna	North Weald	17. 5.05P
			XT787		*(As "XT787" - HMS Endurance)*		
G-KAYH	Extra EA.300/L	144		9. 4.02	Integrated Management Practices Ltd		
						Standalone Farm, Meppershall)	16. 5.05T
G-KBKB	Thunder Ax8-90 Series 2 HAB	2089		30.10.91	G.Boulden *"KB Cars"*	(Aldershot)	14. 9.03A
G-KBPI	Piper PA-28-161 Cherokee Warrior II	28-7816468	G-BFSZ	21. 5.81	Goodwood Road Racing Co Ltd	Goodwood	1. 9.05T
			N9556N		*(Op Goodwood Flying Club)*		
G-KCIG	Sportavia-Pützer RF5B Sperber	51005	D-KCIG	19. 6.80	J.R.Bisset tr Deeside Fournier Group	Aboyne	6. 1.05P
G-KDET	Piper PA-28-161 Cadet	2841158	(SE-KIR)	8. 8.89	Rapidspin Ltd	Biggin Hill	26. 1.07T
			N9184Z		*(Op Biggin Hill School of Flying)*		
G-KDEY	Scheibe SF25E Super Falke	4325	D-KDEY	8. 1.99	J.French	Aston Down	14. 8.05
	(Limbach SL1700)				tr Falke Syndicate		
G-KDIX	Jodel D.9 Bébé	PFA 54-10293		23.11.78	P.M.Bowden	(Stockport)	11. 6.05P
	(Built K Barlow) (Volkswagen 1600)						
G-KDLN	LET Zlin Z.37A-2 Cmelak	19-05	OK-DLN	14. 8.95	J.Richards	Henstridge	16. 7.06
G-KDMA	Cessna 560 Citation Ultra	560-0553	N5145V	4. 4.01	Forest Aviation Ltd	Gamston	3. 4.05T

Registration	Type	C/n	Previous identity	Reg date	Owner/operator	Probable base	CofA expiry
G-KDOG	Scottish Aviation Bulldog Series 120/121	BH120/289	XX624	18. 6.01	Gamit Ltd *(As "XX624/E")*	North Weald	6. 7.06
G-KEAB*	Beech 65-B80 Queen Air	LD-344	G-BSSL	3. 8.88	(N Franklin)	Bruntingthorpe	27. 9.87T
			G-BFEP, F-BRNR, OO-VDE		*(Cancelled 24.5.91 as WFU) (Fuselage noted 3.04)*		
G-KEAC*	Beech 65-A80 Queen Air	LD-176	G-REXY	3. 8.88	(E.A Prentice)	Little Gransden	18. 9.89T
			G-AVNG, D-ILBO		*(Cancelled by CAA 3.4.01) (Stored 10.03)*		
G-KEAM	Schleicher ASH 26E	26116	D-KEAM	3.03.04	D T Reilly	(Taunton)	AC
G-KEEF	Commander Aircraft Commander 114B	14610	N828DL	17. 6.04	K.D.Pearse	(Tadworth)	18. 7.07T
			VT-PVA, (F-GSDV), VT-PVA, N6025M				
G-KEEN	Stolp SA.300 Starduster Too	800	PH-HAB	19. 7.78	H.Sharp tr Sharp Aerobatics		
	(Built R E Ellenbest) (Lycoming IO-540)		(PH-PET), G-KEEN, N800RE			(Ballymena, County Antrim)	27. 4.04P
G-KEES	Piper PA-28-180 Cherokee Archer	28-7505025	OO-AJV	29. 5.97	C.N.Ellerbrook	Wicklewood	15. 9.06
			OO-HAC, N32102				
G-KEJY	Aerotechnik EV-97 teamEurostar UK	2017		23. 6.04	D.Young tr Kemble Eurostar 1	Kemble	22. 6.05P
G-KELL	Van's RV-6	PFA 181-12845		16. 5.95	I.R.Thomas	(Pulborough)	6. 3.05P
	(Built J D Kelsall) (Lycoming O-320)						
G-KELS	Van's RV-7	PFA 323-13801		22. 2.02	J.D.Kelsall	Netherthorpe	6. 7.05P
	(Built J D Kelsall)				*"Lady Lucy"*		
G-KEMC	Grob G109	6024	D-KEMC	19.10.84	Norfolk Gliding Club Ltd	Tibenham	19. 11.06
G-KEMI	Piper PA-28-181 Archer III	2843180	N41493	28.10.98	R.B.Kempster	Cambridge	16.11.07T
G-KEMY	Cessna 182T Skylane	18281206	N53397	27. 8.03	Allen Aircraft Rental Ltd	Cambridge	11. 9.06T
G-KENB	Air Command 503 Commander	PFA G/4-1153		7.11.89	K.Brogden	(Heywood, Lancashire)	24. 9.93P
	(Built K.Brogden)				*(Current status unknown)*		
G-KENI	RotorWay Executive 152	3599		14. 3.89	A J.Wheatley	Street Farm, Takeley	29. 6.05P
	(Built K Hassall) (RotorWay RW 152)						
G-KENM	Luscombe 8EF Silvaire	2908	N21NK	9. 1.91	M.G.Waters	Compton Abbas	3.11.05P
	(Continental C90)		N71481, NC71481				
G-KENW	Robin DR400/500 Président	39		20. 2.03	K.J.White	(Crowhurst)	28. 5.06P
G-KENZ	Rutan VariEze	PFA 74-10960	G-BNUI	13. 8.04	K.M.McConnell	Belfast Aldergrove	26. 4.05P
	(Built T N F Snead) (Continental O-200-A)						
G-KEPP	Rans S-6-ES Coyote II	PFA 204-14308		19.10.04	S.Munday	(St. Neots)	
	(Built S.Munday)						
G-KEST	Steen Skybolt	1	G-BNKG	11. 6.91	B.Tempest	Leicester	22.11.05P
	(Built A Todd)		G-RATS, G-RHFI, N443AT		tr G-KEST Syndicate		
G-KETH	Agusta-Bell 206B JetRanger II	8418	OO-HOP	14.10.03	DAC Leasing Ltd	(Mannington, Norwich)	2. 6.07T
			PH-HAP, SX-HAP, (HB-XEX)				
G-KEVB	Piper PA-28-181 Archer III	2843098	N9289E	29. 8.97	Palmair Ltd	Elstree	19.10.06T
G-KEVI	Jabiru Aircraft Jabiru J400	PFA 325-14321		19.10.04	K.A.Allen	(Upton, Norwich)	
	(Built K.A.Allen)						
G-KEWT	Ultramagic M-90 HAB	90/66		27. 5.04	Kew Technik Ltd	(Basingstoke)	7. 6.05A
G-KEYS	Piper PA-23-250 Aztec F	27-7854052	N63909	6.10.78	R.E.Myson	Hardings Farm, Ingatestone	13.11.06T
G-KEYY	Cameron N-77 HAB	1748	G-BORZ	14. 6.88	B.N.Trowbridge	(Allestree, Derby)	14. 8.05A
G-KFAN	Scheibe SF25B Falke	46301	D-KFAN	14. 5.96	R.G and J.A Boyes	Eaglescott	29. 5.99
	(Stark-Stamo MS1500)				*(Current status unknown)*		
G-KFOX	Denney Kitfox Model 2	PFA 172-11447		11.10.88	I.R.Lawrence	Eaglescott	25. 6.03P
	(Built J Hannibal - c/n 298)						
G-KFRA	Piper PA-32-300 Six	32-7840182	G-BGII	9. 9.97	M.Drake and W.Rankin	Weston, Dublin	22. 7.06
			N20879		tr West India Flying Group		
G-KFZI	Williams KFZ-1 Tigerfalck	PFA 153-11054		2. 2.89	L.R.Williams	(Aberdare)	
	(Built L.R.Williams) (Continental C90) *(Originally laid-down as Kestrel Sport c/n PFA 1530)*						
G-KGAO	Scheibe SF25C Falke	44386	D-KGAG	30. 7.99	C.R.Ellis	Long Mynd	8. 8.05
	(Limbach L2000)				tr Falke 2000 Group		
G-KGED	Campbell Cricket Mk.4	PFA G/03-1337		27. 2.04	K.G.Edwards	(Bridgwater)	
G-KHOM	Aeromot AMT200 Super Ximango	200091		5. 5.98	W.R.Morris	(Tunbridge Wells)	28. 6.07
G-KHRE	SOCATA Rallye 150SV Garnement	2931	F-GAYR	25. 3.82	D.M.Gale and K.F.Crumplin		
						Franklyn's Field, Chewton Mendip	19. 5.07
G-KICK	Pegasus Quantum 15-912	7679		28. 6.00	Graham van der Gaag	Lower Mountpleasant, Chatteris	6. 7.05P
G-KIMB	Robin DR300/140 Major	470	F-BPXX	23. 3.90	R M Kimbell	Sywell	2. 6.06
			F-WPXX				
G-KIMK	Partenavia P68B	27	G-BCPO	23. 2.01	M Konstantinovic	King's Farm, Thurrock	24. 4.07T
G-KIMM	Europa Aviation Europa XS	PFA 247-13404		20. 7.99	P A D.Clarke	Wadswick Manor Farm, Corsham	28.11.05P
	(Built P.A D.Clarke - kit no.404) (Monowheel u/c)						
G-KIMY	Robin DR400/140B Major	1401	PH-SRX	7. 6.00	D.C.Writer	Rochester	4. 7.03T
G-KINE	Gulfstream AA-5A Cheetah	AA5A-0896	N27173	20. 7.82	J.P.E.Walsh t/a Walsh Aviation	Blackbushe	27. 7.06T
G-KIPP	Thruster T600N 450	1031-T600N-094		19.12.03	R.C.Kelly and S.Munday	Lower Mountpleasant, Chatteris	18.12.04P
G-KIRK	Piper J-3C-65 Cub	10536	F-BBQC	28. 2.79	M.J.Kirk	Wanaka, New Zealand	17. 5.01P
	(Frame No.12490)		Fr AF, 43-29245		*"Liberty Girl" (Noted 4.04)*		
G-KISS	Rand-Robinson KR-2	PFA 129-10899		2. 8.83	E.A Rooney	(Whitstable)	
	(Built A C Waller) (Volkswagen 1835)						
G-KITE	Piper PA-28-181 Archer II	28-8490053	N4338X	12. 4.88	Dateworld Ltd	Bournemouth	18. 4.06T
G-KITF	Denney Kitfox	1156	N156BH	10. 5.89	P Smith	Long Marston	2. 6.05P
	(Built J B Hartline) (Rotax 532)						
G-KITI	Pitts S-2E	002	N36BM	21. 6.90	B.R.Cornes	RAF Colerne	17. 4.01P
	(Built R Jones)				*"Super Turkey II"*		
G-KITS	Europa Aviation Europa XS	PFA 247-12844		13. 6.94	J.R.Evernden	Bidford	28. 3.05P
	(Built Europa Aviation Ltd - kit no.468) (Mid-West AE.100R) *(Tri-gear u/c)*						
G-KITT	Curtiss TP-40M Kittyhawk	27490	F-AZPJ	4. 3.98	Patina Ltd *(Op The Fighter Collection)*	Duxford	21. 6.05P
	(Officially c/n quoted as "31423")		N1009N, N1233N, RCAF 840, 43-5802		*(As "49/Bengal Tiger" in US Army c/s)*		
	(C/n 31423 was P-40N 43-23484/RCAF 877/N1009N (1) which was scrapped in 1965 when identity adopted by RCAF 840)						
G-KITY	Denney Kitfox Model 2	PFA 172-11565		18. 8.89	J.P.Jenkins	South Lodge Farm, Widmerpool	11. 2.05P
	(Built T Ringshaw - c/n 456) (IAME KFM 112)				tr Kitfox KFM Group		
G-KIZZ	Air Création Buggy 582/Kiss 450	BMAA/HB/388		24. 6.04	D.C.P.Cardey and P.David		
	(Built P.David - trike T04028 & sailwing A04068-4969)				*(Noted 6.04)*	Broadmeadow Farm, Hereford	
G-KKCW	Flight Design CT2K	Not known		17. 6.03	K.C.Wigley and Co Ltd	(Belper)	30. 6.05P
	(Assembled Pegasus Aviation as c/n 7964)						

Reg	Type	C/n	Previous identity	Date	Owner/operator	Base	CofA
G-KKER	Jabiru Aircraft Jabiru UL-450	PFA 274A-13474		1.10.99	W.K.Evans	Swansea	29. 4.05P
	(Built K Kerr)						
G-KKES	SOCATA TB-20 Trinidad	1316	G-BTLH	2. 3.92	Island Brokers Ltd	Andreas, isle of Man	18. 5.07T
G-KMRV	Van's RV-9A	PFA 320-14093		17.11.04	G.K.Mutch	(Mold)	
	(Built G.K.Mutch)						
G-KNAP	Piper PA-28-161 Warrior II	28-8116129	G-BIUX	15. 2.90	Keen Leasing (IoM) Ltd	Belfast Aldergrove	28. 4.02T
			N9507N		*(Crashed on take off Stevensons Field, Letterkenny, County Donegal 13.7.99: wreck stored 2.01)*		
G-KNEE	Ultramagic M-77C HAB	77/234		20. 6.03	M.A Green	(Rednal)	13. 6.05A
G-KNEK	Grob G109B	6437	D-KNEK	22. 5.00	R.A Winley tr Syndicate 109	Currock Hill	8. 6.06
G-KNIB	Robinson R22 Beta II	3145		30.10.00	C.G.Knibb	Sywell	4.12.06T
G-KNNY	Aérospatiale-Alenia ATR42-300	346	G-ORFH	1. 4.03	Air Wales Ltd	Swansea	3. 4.06T
			F-WWEI				
G-KNOB	Lindstrand LBL 180A HAB	065		20.12.93	Wye Valley Aviation Ltd	(Ross-on-Wye)	16. 4.01T
G-KNOT	Hunting Percival P.84 Jet Provost T.3A	PAC/W/13893	G-BVEG	9. 6.99	R.S.Partridge-Hicks	North Weald	6.10.05P
			XN629		*(As "XN629/49" in RAF c/s)*		
G-KNOW	Piper PA-32-300 Six	32-7840111	N9694C	21. 9.88	B.R.and G.E.Mullaly	Stapleford	22. 7.07
G-KNOX	Robinson R22 Beta II	3603		5. 5.04	TA Knox Shopfitters Ltd	(Stockport)	3. 6.07T
G-KNYT	Robinson R44 Astro	0723		13. 3.00	C.Bootman t/a Aircol	Cranfield	20. 3.06T
G-KODA	Cameron O-77 HAB	1448		26. 3.87	P J Garrad and J E Nolan	(Bracknell)	27. 7.05A
G-KOFM	Glaser-Dirks DG-600/18M	6-66M16	D-KOFM	13. 7.99	A Mossman	Feshiebridge	28. 7.05
G-KOHF	Schleicher ASK 14	14033	D-KOHF	4. 9.01	J. Houlihan	Gowran Grange, Dublin	6.11.04
G-KOKL	Hoffmann H 36 Dimona	36276	D-KOKL	4. 3.98	R.Smith and R.Stembrowicz	Rufforth	17. 4.07
G-KOLB	Kolb Twinstar Mk.3A	PFA 205-12228		30. 6.93	J.L.Moar	Wick	29. 9.03P
	(Built P A Akines) (Rotax 912-UL)						
G-KOLI	PZL-110 Koliber 150	03900038		23. 7.90	J.R.Powell	Sackville Lodge Farm, Riseley	1. 2.05
G-KONG	Slingsby T.67M-200 Firefly	2041	VR-HZP	24. 3.94	R.C.Morton	North Weald	15. 9.06T
			HKG-10, G-7-119				
G-KOOL	de Havilland DH.104 Devon C.2/2	04220	"G-DOVE"	12. 1.82	K.P.Hunt c/o 135 Sqdn (Reigate and Redhill) ATC	Redhill	AC
			VP967		*(Noted 12.02)*		
G-KORN	Cameron Berentzen Bottle 70SS HAB	1655		10. 5.88	A D and R.S.Kent, I M Martin and I Chadwick *"Berentzen"*		
					tr Balloon Preservation Flying Group	(Aldershot)	23. 6.00A
G-KOTA	Piper PA-28-236 Dakota	28-8011044	N8130R	23.12.88	D.J.Fravigar	Clough Farm, Croft, Skegness	14. 3.05
					t/a JF Packaging		
G-KPAO	Robinson R44 Astro	0382	G-SSSS	19.11.98	Avonair Ltd	Bristol	26.11.06T
G-KPTT	SOCATA TB-20 Trinidad	1821	F-GRBI	13. 6.01	IAE Ltd	Cranfield	19. 9.07T
G-KRES	Stoddard-Hamilton Glasair Super II-SRG			12. 6.96	A.D.Murray	(Horley)	29. 4.05P
	(Built G Kresfelder) (Lycoming IO-360) PFA 149-12984				*(New owner 7.04)*		
G-KRII	Rand Robinson KR-2	PFA 129-10934		4. 8.89	M.R.Cleveley	(Halesworth, Suffolk)	
	(Built M.R.Cleveley)				*(Current status unknown)*		
G-KRIS	Maule M-5-235C Lunar Rocket	7357C	N56420	21. 4.81	A C.Vermeer	(Antrim, County Antrim)	5. 4.07
G-KRNW	Eurocopter EC135T2	0175		9. 7.01	Bond Air Services Ltd	RAF St.Mawgan	11. 7.07T
					(Op Cornwall Air Ambulance)		
G-KSIR	Stoddard-Hamilton Glasair IIS RG	PFA 149-12137		15. 4.94	G T Grimward	Booker	12. 6.05P
	(Built R.Cayzer - kit no.2151) (Lycoming IO-360)						
G-KSKS	Cameron N-105 HAB	4963		21. 3.01	A M Holly	(Berkeley)	25. 2.05T
					t/a Exclusive Ballooning *(Kwik Save titles)*		
G-KSKY	Sky 77-24 HAB	170		15.10.99	J.R.Howard	(Poulton-le-Fylde)	13. 6.05A
G-KSVB	Piper PA-24-260 Comanche B	24-4657	G-ENIU	8.11.91	S.Juggler	Stapleford	28. 5.07
			G-AVJU, N9199P, N10F				
G-KTCC	Schempp-Hirth Ventus-2cM	33/57	D-KTCC	29. 4.02	D.Heslop *"2C"*	Wormingford	29. 5.05
G-KTEE	Cameron V-77 HAB	2177		28.12.89	D.C.and N.P.Bull	Princes Risborough	10.10.04A
					tr Katie Group *"Katie"*		
G-KTKT	Sky 260-24 HAB	110		19. 5.98	T.M.Donnelly *"Kit Kat"*	(Sprotbrough, Doncaster)	28. 4.05T
G-KTOL	Robinson R44 Clipper	0780	G-DCOM	10.10.02	K.N.Tolley t/a JNK 2000	(Bromyard)	16. 7.06T
G-KUBB	SOCATA TB-20 Trinidad GT	2026	F-OILS	1.12.00	Offshore Marine Consultants Ltd	Gamston	29.12.06
G-KUIK	Mainair Pegasus Quik	7990		17.10.03	G.R.Hall and P.R.Brooker	(Canterbury)	23.10.05P
G-KUKI	Robinson R22 Beta	1802	G-BTNB	15. 8.02	C.C.Brook	Gamston	30. 7.06T
			N23006				
G-KULA	Best Off Skyranger 912	BMAA/HB/344		26. 1.04	C.R Mason	Sywell	
	(Built C.R Mason - kit no.SKRxxxx364)				*(Noted 1.05)*		
G-KUTU	QAC Quickie Q.2	PFA 94A-10758		8. 3.82	J.Parkinson and R.Nash	Booker	29. 4.86P
	(Built Quick Construction Group) (Limbach L2000)				*(Damaged Cranfield 18.5.85: stored 4.99)*		
G-KVBF	Cameron A-340HL HAB	4313		6. 4.98	Airxcite Ltd t/a Virgin Balloon Flights	(Wembley)	27. 7.05T
G-KVIP	Beech 200 Super King Air	BB-487	G-CBFS	17. 5.02	Capital Trading (Aviation) Ltd	Filton	29. 5.07T
			G-PLAT, N8PY, VH-PIL, N198SC, PT-OYR, N40QN, VH-NIC, N40QN, N400N, N243KA				
G-KWAK	Scheibe SF25C Rotax-Falke	44581	D-KWAK	8. 1.03	Mendip Gliding Club Ltd	Halesland	6. 2.06T
	(Rotax 912A) *(Officially regd as "SF25C Falke")*						
G-KWAX	Cessna 182E Skylane	18253808	N9902	18. 5.78	D.Shaw	Egginton	16. 4.06T
			YV-T-PTS, N2808Y				
G-KWIC	Mainair Pegasus Quik	7962		25. 6.03	T.Southwell	(Spalding)	19. 7.04P
G-KWIK	Partenavia P68B	152		27. 9.78	ACD Cidra NV	Wevelgem, Belgium	3. 7.06T
G-KWKI	QAC Quickie Q.200	PFA 94-12158		22.10.91	B.M.Jackson	Enstone	17. 9.04P
	(Built D G Greatrex and B.M.Jackson) (Continental O-200-A)						
G-KWLI	Cessna 421C Golden Eagle II	421C0168	G-DARR	13.11.98	Langley Holdings plc	Gamston	6. 2.05T
			G-BNEZ, N87386				
G-KYAK	SPP Yakovlev Yak C-11	171101	F-AZQI	21.12.78	M.Gainza *(As "36" (white))*	North Weald	3.12.05P
			G-KYAK, F-AZHQ, G-KYAK, Israeli DFAF, Egyptian AF 590, Czech AF				
G-KYDD	Robinson R44 Astro	0106	N2123E	16. 9.99	Heli Air Ltd	Wellesbourne Mountford	3.10.05T
			D-HDLW				

G-LAAA - G-LZZZ

Reg	Type	C/n	Previous identity	Date	Owner/operator	Base	CofA
G-LABS	Europa Aviation Europa	PFA 247-12595		1. 3.94	C.T.H.Pattinson	Turweston	15.12.04P
	(Built C.T.H.Pattinson - kit no.049) (Monowheel u/c)				*(Noted 3.04)*		

Reg	Type	C/n	Previous identity	Regd	Owner/Operator	Probable base	CA expiry
G-LACA	Piper PA-28-161 Cherokee Warrior II	28-7816036	N44883	22. 6.90	LAC (Enterprises) Ltd *(Op Lancashire Aero Club)*	Barton	18. 3.05T
G-LACB	Piper PA-28-161 Warrior II	28-8216035	N8450A	12. 6.90	LAC (Enterprises) Ltd *(Op Lancashire Aero Club)*	Barton	9. 7.05T
G-LACD	Piper PA-28-181 Archer III	2843157	G-BYBG	11.11.98	D.H.Brown	(Burnley)	25.11.07E
			N47BK		t/a David Brown Aviation		
G-LACE	Europa Aviation Europa	PFA 247-12962		15. 4.96	J.H.Phillingham	(Wallingford)	
	(Built J.H.Phillingham - kit no.256) (Monowheel u/c)				*(Current status unknown)*		
G-LACR	Denney Kitfox	PFA 172-11945		4.12.90	C.M.Rose	(Edinburgh)	
	(Built C M Rose)				*(Under construction 6.00)*		
G-LADD	Enstrom 480	5037		20. 5.99	Combi-Lift Ltd	(Clontibret, County Monaghan)	9. 7.05T
G-LADI	Piper PA-30 Twin Comanche	30-334	G-ASOO	8. 4.94	S.H.Eastwood	Blackbushe	3.10.05T
			N10F				
G-LADS	Rockwell Commander 114	14314	N4994W	6.12.90	D.F.Soul	Emberton, Olney	2. 2.06
			(N114XT), N4994W				
G-LAGR	Cameron N-90 HAB	1628		25. 1.88	J.R.Clifton	(Brackley)	11.10.03A
G-LAIN	Robinson R22 Beta	1992		7. 2.92	Patriot Aviation Ltd	(Cranfield)	4. 6.07T
G-LAIR	Stoddard-Hamilton Glasair IIS	2106		12. 9.91	A I.O'Broin and S.T.Raby	Grange Farm, Woodwalton	
	(Built D L Swallow and S T Raby)				*(Noted 10.02)*		
G-LAKE	Lake LA-250 Renegade	70	(EI-PJM)	12. 7.88	P.J.McGoldrick	Lough Derg Marina, Killaloe	10. 6.05
	(Built Aerofab Inc)		G-LAKE, N8415B				
G-LAKI	CEA Jodel DR.1050 Sicile	534	G-JWBB	12.11.79	V Panteli	(Canterbury)	30. 6.05
			G-LAKI, F-BLZD				
G-LAMA	Aérospatiale SA315B Lama	2348	SE-HET	17. 3.98	PLM Dollar Group Ltd	Cumbernauld	19. 3.07T
G-LAMM	Europa Aviation Europa	PFA 247-12941		20.11.95	S.A Lamb	(Paddock Wood)	
	(Built S.A Lamb - kit no.244) (Monowheel u/c)				*(Current status unknown)*		
G-LAMP	Cameron Lightbulb 110 SS HAB	4899		21. 7.00	LE Electrical Ltd	(Norwich)	17. 5.05A
G-LAMS	Reims/Cessna F152 II	F15201431	N54558	23. 6.88	Horizon Aviation Ltd	Swansea	30.10.06T
G-LANC*	Avro 683 Lancaster B.X				See SECTION 3 Part 1	Duxford	
G-LANE	Reims/Cessna F172N Skyhawk II	F17201853		27. 6.79	G.C.Bantin	Sproatley	29. 6.06
G-LAOK	IAV-Bacau Yakovlev Yak-52	877404	LY-AOK	22. 1.03	I.F.Vaughan and J.P.Armitage	Tollerton	20. 2.04P
			DOSAAF 16 *(yellow)*				
G-LAOL	Piper PA-28RT-201 Arrow IV	28R-7918211	D-EAOL	6.10.99	G.P.Aviation Ltd	Goodwood	13. 1.06T
			N2903Y		*"19"*		
G-LAOR	Raytehon Hawker 800XP	258384	N955MC	2. 3.04	Select Plant Hire Co. Ltd	Southend	4. 3.07T
			N23455, TC-MDC, N23455				
G-LAPN	Avid Aerobat	PFA 189-12146		4. 3.93	I A P Harper	(Stourbridge)	21. 5.04P
	(Built R M Shorter)				*(New owner 9.04)*		
G-LARA	Robin DR400/180 Régent	2050		14. 2.91	K.D.and C.A Brackwell	Goodwood	23. 4.06
G-LARE	Piper PA-39 Twin Comanche C/R	39-16	N8861Y	20. 2.91	Glareways (Neasden) Ltd	Biggin Hill	23. 4.06
G-LARK	Helton Lark 95	9517	N5017J	3.12.85	J.Fox	Booker	26. 3.05P
G-LARS	Dyn'Aéro MCR-01 Club	PFA 301-14271		31. 8.04	L.A.Oyno	(Hovik, Norway)	
	(Built L.A.Oyno)						
G-LARY	Robinson R44 Raven II	10255	G-CCRZ	10. 2.04	L Behan and Sons Ltd	(Rathcoole, County Dublin)	19. 2.07T
G-LASN	Best Off Skyranger J2.2	BMAA/HB/396		19. 7.04	L.C.F.Lasne	(Newry)	
	(Built L.C.F.Lasne - kit no.SKRxxxxxxx?)						
G-LASR	Stoddard-Hamilton Glasair II	2027		8. 1.90	G.Lewis	(Heswall, Wirral)	
	(Built P Taylor)				*(Current status unknown)*		
G-LASS	Rutan VariEze	PFA 74-10209		20. 9.78	J.Mellor	Sleap	14. 6.05P
	(Built Calvert, Foreman and O'Hara) (Continental O-200-A)						
G-LASU	Eurocopter EC135T2	0228	D-HTSH	3. 9.02	Lancashire Constabulary Air Support Unit	Warton	15.10.05T
G-LAVE	Cessna 172R Skyhawk	17280663	G-BYEV	10. 3.99	M.L.Roland	Hill Farm, Nayland	17. 4.05
			N2377J, N41297				
G-LAZA	Laser Z200 Lazer	PFA 123-12682		15. 6.95	D G Jenkins	(Stanton, Bury St. Edmunds)	8.12.05P
	(Built M.Hammond) (Lycoming AEIO-360)						
G-LAZL	Piper PA-28-161 Warrior II	28-8116216	D-EAZL	9. 6.99	S.Bagley and K.J.Amies	Luton	25. 7.05T
			N9536N		t/a Hawk Aero Leasing		
G-LAZR	Cameron O-77 HAB	2240		6. 3.90	Laser Holdings (UK) Ltd *(New CofR 8.04)*	(Worcester)	10. 6.97A
G-LAZY	Lindstrand LBL Armchair SS HAB	129		18. 9.94	The Air Chair Co Ltd *"The Chair"*	(Westville, Indiana, US)	27. 4.03A
G-LAZZ	Stoddard-Hamilton GlaStar	PFA 295-13059		31.10.96	A N.Evans	Ashcroft Fam, Winsford	22.11.05P
	(Built G K Brunwin and A N Evans) (Tricycle u/c)						
G-LBLI	Lindstrand LBL 69A HAB	010		4.11.92	N.M.Gabriel	(Kimberley, Nottingham)	2. 7.05A
G-LBMM	Piper PA-28-161 Cherokee Warrior II	28-7816440	N6940C	28.11.89	Flexi-Soft Ltd	Wellesbourne Mountford	8. 5.05T
G-LBRC	Piper PA-28RT-201 Arrow IV	28R-7918051	N2245P	20. 7.88	D.J.V.Morgan	Wolverhampton	28. 3.07
G-LBUK	Lindstrand LBL 77A HAB	922		15. 5.03	Lindstrand Hot Air Balloons Ltd	(Oswestry)	22. 7.05A
G-LCGL	Comper CLA.7 Swift rep	PFA 103-11089		1.7.92	J.M.Greenland	Blackacre Farm, Holt, Trowbridge	25. 4.05P
	(Built J.M.Greenland) (Pobjoy Niagara 1A)						
G-LCIO*	Colt 240A HAB				See SECTION 3 Part 1	Newbury	
G-LCOC	Britten-Norman BN-2A Mk.III-1 Trislander	366	G-BCCU	30. 7.01	AirX Ltd	Alderney	6. 1.07T
			4X-CCK, G-BCCU, 9L-LAR, G-BCCU, (LN-VIV)		t/a Lecocqs.com *(Op Rockhoppers)*		
G-LCYA	Dassault Falcon 900EX	105	F-WWFC	5. 8.02	London City Airport Jet Centre Ltd	London City	4. 8.05T
G-LDAH	Best Off Skyranger 912	BMAA/HB/241		8.10.02	A S.Haslam and L.Dickinson	Long Marston	11. 9.05P
	(Built A S.Haslam and L.Dickinson - kit no.SKRxxxx216)						
G-LDWS	SAN Jodel D150 Mascaret	48	G-BKSS	13.2.04	D.H.Wilson-Spratt	Ronaldsway	
			F-BMFC				
G-LDYS	Thunder Ax6-56Z HAB	347		18. 5.81	M.J.Myddelton	(Bristol)	27. 3.00A
	(Originally regd as Colt 56A)				*"Gladys"*		
G-LEAF	Reims/Cessna F406 Caravan II	F406-0018	EI-CKY	7. 3.96	Atlantic Air Transport Ltd	Inverness	20. 5.07T
			PH-ALN, OO-TIW, F-WZDX		*(Op Highland Airways) (Atlantic Airlines titles)*		
G-LEAM	Piper PA-28-236 Dakota	28-8011061	G-BHLS	1. 7.80	C.S.Doherty	Gamston	8. 7.07
			N35650				
G-LEAP	Pilatus Britten-Norman BN-2T Islander	2183	G-BLND	19. 8.87	R.Durie tr Army Parachute Association	AAC Netheravon	18. 4.05A
G-LEAS	Sky 90-24 HAB	158		4. 5.99	C.I.Humphrey	(Tilehurst, Reading)	8. 6.05A
					(The Leasing Co titles)		
G-LEAU	Cameron N-31 HAB	761		5. 8.81	P.L.Mossman	(Trellech, Monmouth)	24. 2.97A
					"Perrier" (Inflated 4.02)		

Reg	Type	C/n	Previous identity	Reg date	Owner/operator	Base	CofA
G-LEBE	Europa Aviation Europa *(Built P.Atkinson - kit no.237)* (Wilksch WAM-120) *(Monowheel u/c)*	PFA 247-12927		17. 5.01	P.Atkinson	(Carnforth)	
G-LECA	Aérospatiale AS355F1 Ecureuil 2	5043	G-BNBK C-GBKH	6. 2.87	South Western Electricity plc	Bristol	24. 7.05T
G-LEED	Denney Kitfox Model 2 *(Built G T Leedham - c/n 450)*	PFA 172-11577		24. 4.91	M.J.Beding *(Swung to left off runway Compton Abbas 16.9.02 and substantially damaged: noted 4.04)*	Bodmin	9.12.05P
G-LEEE	Jabiru Aircraft Jabiru UL-450 *(Built L E G Fekete)*	PFA 274A-13516		18. 1.00	J.P.Mimnagh and L.E.G.Fekete	(Neston)	21.11.05P
G-LEEN	Aero Designs Pulsar XP *(Built D F Gaughan)*	PFA 202-12147	G-BZMP G-DESI	16. 7.01	R.B.Hemsworth	Eaglescott	6.12.05P
G-LEES	Glaser-Dirks DG-400	4-238		4.10.88	J Bradley	RAF Upavon	14. 3.07
G-LEEZ	Bell 206L-1 LongRanger II	45761	G-BPCT D-HDBB, N3175G	22. 1.92	Pennine Helicopters Ltd	(Oldham)	27. 1.07T
G-LEGG	Reims/Cessna F182Q Skylane II	F18200145	G-GOOS	26. 6.96	P.J.Clegg	Barton	5. 2.06
G-LEGO	Cameron O-77 HAB	1975		14. 4.89	P.M.Traviss *"Jigsaw II"*	(Yarm)	29. 5.05A
G-LEIC	Reims/Cessna FA152 Aerobat	FA1520416		16. 9.86	Leicestershire Aero Club Ltd	Leicester	8. 8.05T
G-LEKT	Robin DR400/180 Régent	1181	D-EEKT	1. 2.05	Exavia Ltd	Exeter	
G-LELE	Lindstrand LBL 31A HAB	806		16. 8.01	L.E.Electrical Ltd	(Norwich)	17. 5.05A
G-LENA	IAV-Bacau Yakovlev Yak-52	833901	LY-AMU DOSAAF 42 *(red)*	5.11.02	Yak-52 Ltd	Compton Abbas	26.10.05P
G-LEND*	Cameron N-77 HAB	2012		25. 5.89	Southern Flight Co Ltd *"Southern Finance Co/Glenda" (Cancelled 8.10.01 as WFU)*	(Southampton)	12. 9.96T
G-LENF	Mainair Blade 912S	1362-0104-7 & W1157		18. 2.04	G.D.Fuller	(Grimsby)	4. 3.05P
G-LENI	Aérospatiale AS355F1 Ecureuil 2	5311	G-ZFDB G-BLEV	9. 8.95	Grid Aviation Ltd	Denham	21. 4.06T
G-LENN	Cameron V-56 HAB	1833		29. 9.88	M.Jesper	(Schwäbisch Hall, Germany)	28. 3.05A
G-LENS	Thunder Ax7-77Z HAB	168		3.11.78	R S Breakwell	(Bridgnorth)	23. 7.05A
G-LENX	Cessna 172N Skyhawk II	17272232	G-BMVJ N9347E	15. 2.02	M.W.Glencross	(Luton)	17. 6.07T
G-LENY	Piper PA-34-220T Seneca III	34-8233205	N111PS (OK-MKN), PH-SMS, (PH-CCC), D-GAPN, N82396	26. 7.00	Air Medical Fleet Ltd	Oxford	27. 8.06T
G-LEOS	Robin DR400/120 Dauphin 2+2	1884		29.11.88	R J O Walker	Gamston	20. 6.07
G-LESJ	Denney Kitfox Model 3 *(Built L A James)*	PFA 172-12001		4.10.94	S.E.Spencer	(Bath)	7. 7.05P
G-LESZ	Skystar Kitfox Model 5 *(Built L A James)* (Rotec)	PFA 172C-12822		25.10.02	L.A James	Wharf Farm, Market Bosworth	4.11.04P
G-LEVI	Aeronca 7AC Champion	7AC-4001	N85266 NC85266	17. 4.90	Jean P.A Pumphrey tr G-LEVI Group *(Carries "NC85266" on fin)*	White Waltham	2.11.04P
G-LEXI	Cameron N-77 HAB	438		26.10.78	T.Gilbert *(Rolls Royce titles)*	(Weston-super-Mare)	9. 7.05A
G-LEXX	Van's RV-8 *(Built A A Wordsworth)*	PFA 303-13896		11. 4.02	A A Wordsworth	(Sutton-in-Ashfield)	27. 9.05P
G-LEZE*	Rutan LongEz *(Built K.G.M.Loyal)* (Wilksch diesel	PFA 74A-10702		31. 3.82	K.G.M.Loyal, A J.Draper, J.R.J.Giesler and C.McGeachy *(Cancelled 22.5.03 by CAA) (Noted 12.04*	Nympsfield	5.11.01P
G-LEZZ	Stoddard-Hamilton GlaStar *(Built L A James) (Tri-cycle u/c)*	PFA 295-13241	G-BYCR	4.11.98	L.A James	Wharf Farm, Market Bosworth	27. 6.05P
G-LFBW	Robinson R44 Astro	0722	G-ODES	4.10.04	L.Ferdinand	Elstree	3. 3.06E
G-LFIX	Vickers Supermarine 509 Spitfire T.IX	CBAF.8463 *(C/n is firewall plate no)*	IAC162 G-15-175, ML407	1. 2.80	Carolyn S.Grace *"Nicholson Leslie"* *(As "ML407/OU-V" (stbd)" in 485 Sqdn c/s and "ML407/"NL-D" (port) in 341 Sqdn c/s)*	Duxford	16. 4.05P
G-LFSA	Piper PA-38-112 Tomahawk	38-78A0430	G-BSFC N9739N	22.10.90	Liverpool Flying School Ltd	Liverpool	8. 5.06T
G-LFSB	Piper PA-38-112 Tomahawk	38-78A0072	G-BLYC D-ELID, N9715N	20.10.94	Spencer Davies Engineering Ltd	Swansea	30. 6.07T
G-LFSC	Piper PA-28-140 Cherokee Cruiser	28-7425005	G-BGTR OY-BGO, SE-GDS	4. 9.95	P.Dion	(Lahr, Germany)	27.10.04T
G-LFSD	Piper PA-38-112 Tomahawk II	38-82A0046	G-BNPT G-LFSD, N91522	21.10.96	Liverpool Flying School Ltd	Liverpoo	8. 7.06T
G-LFSG	Piper PA-28-180 Cherokee E	28-5799	G-AYAA N11C	19. 6.00	Liverpool Flying School Ltd	Liverpool	10.11.07E
G-LFSH	Piper PA-38-112 Tomahawk	38-78A0352	G-BOZM N6247A	16. 7.01	Liverpool Flying School Ltd	Liverpool	19. 6.07T
G-LFSI	Piper PA-28-140 Cherokee C	28-26850	G-AYKV N11C	14. 7.89	M.J.Green, P.S.Hoyle and S.Merriman	Humberside	13. 3.05T
G-LFSJ	Piper PA-28-161 Warrior II	28-7916536	G-BPHE N2911D	4.11.02	Leeds Flying School Ltd	Leeds-Bradford	26. 3.07T
G-LFSK	Piper PA-28-161 Cherokee Warrior II	28-7816599	SE-IAD	23. 4.04	Leeds Flying School Ltd	Leeds-Bradford	9. 6.07T
G-LFSM	Piper PA-38-112 Tomahawk	38-78A0449	G-BWNR N2361E	21. 9.04	Liverpool Flying School Ltd	Liverpool	6. 5.05T
G-LFVB	Vickers Supermarine 349 Spitfire LF.V	CBAF.2403	8070M 5377M, EP120	9. 5.94	Patina Ltd *"City of Winnipeg"* *(Op The Fighter Collection as "EP120/AE-A" in 402 Sqdn c/s)*	Duxford	9. 4.05P
G-LFVC	Vickers Supermarine 349 Spitfire L.Vc	Not known	ZK-MKV A58-178, JG891	28. 9.99	Historic Flying Ltd	(Duxford)	AC
G-LGCA	Robin DR400/180R Remorquer	1686	HB-KAP	17. 2.04	London Gliding Club Proprietary Ltd	Dunstable	10. 3.07
G-LGNA	SAAB-Scania SF.340B	340B-199	N592MA SE-F99	11. 6.99	Loganair Ltd	Glasgow	13. 6.05T
G-LGNB	SAAB-Scania SF.340B	340B-216	N595MA SE-G16	8. 7.99	Loganair Ltd	Glasgow	8. 7.05T
G-LGNC	SAAB-Scania SF.340B	340B-318	SE-KXC F-GTSF, EC-GMI, F-GMVZ, SE-KXC, SE-C18	9. 6.00	Loganair Ltd *"Chatham Historic Dockyard"*	Glasgow	11. 6.05T
G-LGND	SAAB-Scania SF.340B	340B-169	G-GNTH N588MA, SE-F69	7. 9.01	Loganair Ltd	Glasgow	4. 2.05T
G-LGNE	SAAB-Scania SF.340B	340B-172	G-GNTI N589MA, SE-F72	31. 8.01	Loganair Ltd	Glasgow	5. 2.05T
G-LGNF	SAAB-Scania SF.340B	340B-192	N192JE G-GNTJ, N591MA, SE-F92	8. 8.02	Loganair Ltd	Glasgow	7. 8.05T

G-LGNG	SAAB-Scania SF.340B	340B-327	SE-C27	16.12.02	Loganair Ltd	Glasgow	16.12.05E
			VH-CMH, SE-C27				
G-LGNH	SAAB-Scania SF.340B	340B-333	SE-C33	28. 5.04	Loganair Ltd	Glasgow	30. 5.07T
			VH-XDA, F-GMVX, SE-C33				
G-LGTE	Boeing 737-3Y0	24908	TC-SUP	25. 1.01	British Airways plc	Gatwick	26. 3.07T
G-LGTF	Boeing 737-382	24450	N115GB	7. 3.01	British Airways plc	Gatwick	30. 4.07T
			TC-IAC, CS-TIE				
G-LGTG	Boeing 737-3Q8	24470	N696BJ	4. 4.01	British Airways plc	Gatwick	14. 6.07T
			SX-BFT, N470KB, PK-GWD				
G-LGTH	Boeing 737-3Y0	23924	OO-LTV	4. 4.01	British Airways plc	Gatwick	8. 6.07T
			XA-SEM, G-BNGL				
G-LGTI	Boeing 737-3Y0	23925	OO-LTY	2. 4.01	British Airways plc	Gatwick	25. 7.07T
			XA-SEO, G-BNGM				
G-LHCA	Robinson R22 Beta II	2947	N299FA	28.10.02	Rotorcraft Ltd	Redhill	28.11.05T
G-LHCB	Robinson R22 Beta II	3241	G-SIVX	14. 6.04	London Helicopter Centres Ltd	Redhill	30. 7.07T
G-LHEL	Aérospatiale AS355F2 Ecureuil 2	5462	N42AT	29. 3.04	Lloyd Helicopters Europe Ltd	Redhill	12. 5.07T
			N70PB				
G-LIBB	Cameron V-77 HAB	2463		21. 6.91	R.J.Mercer	(Belfast)	21. 6.04A
G-LIBS	Hughes 369HS	43-0469S	N9147F	20. 8.85	R.J.H.Strong	(Yeovil)	28. 6.07T
G-LICK	Cessna 172N Skyhawk II	17270631	N172AG	17. 7.02	Sky Back Ltd	North Weald	16. 9.05T
			G-LICK, G-BNTR, N739LQ				
G-LIDA	HOAC HK 36R Super Dimona	36355		15. 4.92	Bidford Airfield Ltd	Bidford	6.12.04
G-LIDE	Piper PA-31-350 Navajo Chieftain	31-7852156	(G-VIDE)	26.10.78	Keen Leasing (IoM) Ltd	Ronaldsway	27.10.07T
			N27800				
G-LIFE	Thunder Ax6-56Z HAB	135		11. 1.78	D.P.Hopkins	(Pidley, Huntingdon)	7.10.05A
					t/a Lakeside Lodge Golf Centre *"Golden Delicious"*		
G-LILP	Europa Aviation Europa XS	PFA 247-13802		22. 5.02	G.L.Jennings	(Shoreham-by-Sea)	
	(Built G.L.Jennings - kit no.487) (Monowheel u/c)						
G-LILY	Bell 206B-3 JetRanger III	4107	G-NTBI	14. 3.95	T.S.Brown	Goodwood	11. 4.05T
			C-FIJD				
G-LIMO	Bell 206L-1 Long Ranger II	45476	N5742H	12. 6.03	Heliplayer Ltd	Sheffield City	3. 9.06T
			G-LIMO, N5742H				
G-LIMP	Cameron C-80 HAB	10391		4. 6.03	T.and B.Chamberlain	(Melborne, York)	28. 6.05A
G-LINC	Hughes 369HS	43-0467S	C-FDUZ	14. 5.87	Wavendon Social Housing Ltd	Sywell	28. 2.07T
			CF-DUZ				
G-LINE	Eurocopter AS355N Ecureuil 2	5566		22. 3.94	National Grid Co plc	Oxford	12. 5.06T
G-LINN	Europa Aviation Europa XS	PFA 247-14118		20. 8.04	T.Pond	Sandtoft	
	(Built T.Pond)						
G-LION	Piper PA-18-135 Super Cub	18-3857	PH-KLB	29. 9.80	J.G.Jones t/a JG Jones Haulage	Caernarfon	13. 1.06
	(L-21B-PI) *(Frame No.18-3841)*		(PH-DKG), R.Neth AF R-167,		54-2457 *"Grin'n Bare It" (As "R-167" in R.Neth AF c/s)*		
G-LIOA*	Lockheed 10A Electra				See SECTION 3 Part 1	Wroughton	
G-LIOT	Cameron O-77 HAB	2378		7. 8.90	N.D.Eliot	(London SW19)	27. 5.05A
G-LIPA	Cessna U206G Stationair 6	U20604128	HZ-SAM	19. 2.04	Navigando Air SpA	(Lipari, Italy)	19. 5.07T
	(Built Reims c/n 0002)		N756KM				
G-LIPE	Robinson R22 Beta	1882	G-BTXJ	23. 1.92	Highland Helicopter Leasing Ltd	Edinburgh	4. 4.07T
G-LIPS	Cameron Lips 90 SS HAB	4846	G-BZBV	15.11.00	Reach For The Sky Ltd	(Worplesdon, Guildford)	30. 7.02A
					(New owner 3.04)		
G-LISE	Robin DR500/200i Président	0001		27. 7.98	J.Marks	Goodwood	28.11.07E
	(Registered as DR400/500)						
G-LISO	SIAI Marchetti SM.1019	045	Italian Army MM57-237	1. 7.04	Castiglioni Daliso	(Legnago, Verona, Italy)	
G-LITE	Rockwell Commander 112A	291	OY-RPP	13. 6.80	R.B.Kay tr TR Air	Gloucestershire	27. 1.07
G-LITZ	Pitts S-1E	PFA 09-11131		3. 3.92	P.J.L Caruth	Henstridge	14. 6.05P
	(Built K Eld and J Hughes)				*"Glitz"*		
G-LIVH	Piper J-3C-65 Cub (L-4H-PI)	11529	OO-JAN	31. 3.94	M.D.Cowburn	Welshpool	1. 7.06
	(Frame No.11354)		OO-AAT, OO-PAX, 43-30238		*(As "330238/24-A" in US Army c/s)*		
G-LIZA	Cessna 340A II	340A1021	G-BMDM	15. 2.90	Tayflite Ltd	Gamston	12. 6.05T
			ZS-KRH, N4620N				
G-LIZI	Piper PA-28-160 Cherokee	28-52	G-ARRP	26. 1.89	N.F.Andrews and A.J.Kingston	Netherthorpe	1. 8.05
			N5050W		tr G-LIZI Group		
G-LIZY*	Westland Lysander III				See SECTION 3 Part 1	Duxford	
G-LIZZ	Piper PA-E23-250 Aztec E	27-7405268	G-BBWM	26. 7.93	T.J.Nathan	Fairoaks	12. 12.06
			N40532				
G-LJCC	Murphy Rebel	PFA 232-13335		8. 7.98	P.H.Hyde	(Biggleswade)	
	(Built J Clarke)				*(Current status unknown)*		
G-LKET	Cameron Kindernet Dog-100 SS HAB	4765	OO-BKN	17.11.03	G.R.J.Luckett	(Fort Collins, Colorado, US)	4. 12.04A
G-LKTB	Piper PA-28-181 Archer III	2843496	N5339X	18.12.01	Top Cat Aviation Ltd	Manchester	19.12.04T
G-LLAI*	Colt 21A Cloudhopper HAB				See SECTION 3 Part 1	(Aldershot)	
G-LLAN	Grob G109B	6398	OH-747	19.11.04	J.D.Scott	(Crickhowell)	AC
G-LLEW	Aeromot AMT-200S Super Ximango	200126		15.11.00	N.J.Watt	Glenrothes	6. 5.07
					tr Echo Whiskey Ximango Syndicate		
G-LMAX	Sequoia F.8L Falco	PFA 100-13423		28.10.02	J.Maxwell	(Ascot)	
	(Built J.Maxwell)						
G-LMLV	Dyn'Aéro MCR-01 Club	PFA 301A-13524		25.10.99	L.and M.La Vecchia	Cambridge	21.11.05P
	(Built L.La Vecchia) (Rotax 912 ULS)						
G-LNAA	MD Helicopters MD 900	900-00074	G-76-074	26. 9.00	Police Aviation Services Ltd	RAF Waddington	27.11.06T
			G-LNAA, N7030B		*(Op Lincs and Notts Air Ambulance)*		
G-LNIC	Robinson R22 Beta II	3524		5. 1.04	Linic Consultants Ltd	(Sutton)	5. 2.07T
G-LNTY	Aérospatiale AS355F1 Ecureuil 2	5300	G-ECOS	29. 9.03	LNT Aviation Ltd	Coney Park, Leeds	24.11.06T
			G-DOLR, G-BPVB, OH-HAJ,		D-HEHN		
G-LNYS	Reims/Cessna F177RG Cardinal RG	F177RG0120	G-BDCM	30.11.92	Heliview Ltd	Blackbushe	28. 4.06T
			OY-BIP				
G-LOAD	Dan Rihn DR.107 One Design	PFA 264-13776		7. 6.02	M.J.Clark	(Horsham)	
	(Built M.J.Clark)						
G-LOAG*	Cameron N-77 HAB				See SECTION 3 Part 1	Newbury	

Reg	Type	C/n	Previous identity	Date	Owner	Location	Date
G-LOAN	Cameron N-77 HAB	1434		9. 1.87	P.Lawman	(Northampton)	8. 5.01A
					(Newbury Building Society titles)		
G-LOBL	Bombardier BD-700-1A10 Global Express	9038	G-52-24	22. 2.02	1427 Ltd	Jersey	21. 2.07T
			C-GFJR				
G-LOBO	Cameron O-120 HAB	3389		3. 1.95	C.A Butler t/a Solo Aerostatics	(Newbury)	26. 7.03A
G-LOCH	Piper J-3C-90 Cub (L-4J-PI)	12687	HB-OCH	10.12.84	J.M.Greenland	Blackacre Farm, Holt, Trowbridge	5.11.05P
	(Frame No.12517)		44-80391				
G-LOFA*	Lockheed L188CF Electra	2002	N359Q	10. 2.94	Atlantic Air Transport Ltd	(Coventry)	9. 2.00T
			F-OGST, N359AC, TI-LRM, N359AC, HC-AVX, N359AC, VH-ECA				
					(Cancelled 29.7.98 as WFU) (With Fire Section 11.02)		
G-LOFB	Lockheed L188CF Electra	1131	N667F	28. 6.94	Atlantic Airlines Ltd	Coventry	8. 2.07T
			N133AJ, CF-IJW, N131US				
G-LOFC	Lockheed L188CF Electra	1100	N665F	15. 6.95	Atlantic Airlines Ltd	Coventry	10. 7.07T
			N289AC, N6123A				
G-LOFD	Lockheed L188CF Electra	1143	LN-FOG	12. 6.97	Atlantic Airlines Ltd	Coventry	15. 6.06T
			LN-MOD, N9745C, (CF-IJC), N9745C				
G-LOFE	Lockheed L188CF Electra	1144	EI-CET	5. 1.99	Atlantic Airlines Ltd *(Atlantic Airlines titles)*	Coventry	19. 3.05T
			(G-FIGF), N668Q, N668F, N24AF, N138US				
G-LOFF	Lockheed L188C Electra	1128	LN-FON (2)	21. 6.00	Atlantic Airlines Ltd	Coventry	AC
			N342HA, N417MA, OB-R-1138, HP-684, N417MA, CF-ZST, N7142C				
					(As "..N" in DHL titles: Fred Olsen c/s on tail - airframe - in open store 1.05)		
G-LOFG*	Lockheed L188C Electra	1116	LN-FOL (2)	21. 6.00	Atlantic Air Transport Ltd	Coventry	
			N669F, N404GN, N6126A		*(Cancelled 16.6.04 as destroyed)*		
			(On fire dump as "LN-FOL" 1.05: DHL titles painted out - airframe, minus starboard outer wing)				
G-LOFM	Maule MX-7-180A Sportplane	20027C	N31110	19. 7.95	Atlantic Air Transport Ltd	Coventry	10. 9.04T
G-LOFT	Cessna 500 Citation I	500-0331	LN-NAT	12. 1.95	Atlantic Air Transport Ltd	Coventry	25. 3.06T
			EC-FUM, EC-500, LN-NAT, N40AC, N96RE, N86RE, N331CC, (N5331J)				
					(Atlantic Executive Aviation c/s)		
G-LOGO	MD Helicopters Hughes 369E	0454E	G-BWLC	4.10.96	Eastern Atlantic Helicopters Ltd	Shoreham	25. 7.05T
			HB-XIJ, SE-JAM		*(Crashed Kensworth, Bedford 2.12.03 and badly damaged)*		
G-LOIS	Jabiru Aircraft Jabiru UL	SAAC-68	EI-JAK	14. 9.00	D W Newman	(Rushden)	19. 5.05P
	(Built S Walshe - kit no 0144) (C/n shown officially as "PFA 274A-0144")						
G-LOKM	PZL-110 Koliber 160A	04990080	G-BYSH	26.11.99	PZL International Aviation Marketing and Sales plc		
			SP-WGH			Earls Colne	16. 1.07T
G-LOKO	Cameron Locomotive 105SS HAB	3680	HB-QBN	19. 9.95	Warsteiner Brauerei Haus Cramer KG		9. 8.05A
			G-LOKO			(Warstein, Germany)	
G-LOLA	Beech A36 Bonanza	E-2116	N67501	18. 2.02	J.H. and L.F.Strutt	Earsl Colne	11. 4.05P
G-LOLL	Cameron V-77 HAB	2964		4.12.92	C.N.Rawnson	(Stockbridge)	13. 8.05A
					tr Test Valley Balloon Group		
G-LONE	Bell 206L-1 Long Ranger	45729	G-CDAJ	5.10.04	C.W.and Keeley A.M.Bootman	Cranfield	27.10.07E
			N20AP, N3174W		t/a Aircol		
G-LOOP	Pitts S-1C	850	5Y-AOX	11. 5.78	D.Shutter	Leicester	28. 6.05P
	(Built D Mallinson) (Lycoming O-320) *(Marked as "S-1D")*						
G-LORC	Piper PA-28-161 Cadet	2841339	D-ESTC	12. 1.99	Sherburn Aero Club Ltd	Sherburn-in-Elmet	12. 3.05T
			N9184W, (N620FT), (SE-KMP)				
G-LORD	Piper PA-34-200T Seneca II	34-7970347	N2908W	6. 5.88	Carill Aviation Ltd and R.P.Thomas	Lee-on-Solent	8. 5.06T
G-LORN	Avions Mudry CAP.10B	282		4. 3.99	J.D.Gailey	Old Sarum	10. 5.05
G-LORR	Piper PA-28-181 Archer III	2843037	N9268X	19. 4.96	VA Technology Ltd	Wolverhampton	13. 6.05
			G-LORR				
G-LORT	Avid Speed Wing Mk.4	PFA 189-12219		12. 2.92	G.E.Laucht	Long Marston	5. 7.05P
	(Built G.E.Laucht - c/n 1124)						
G-LORY	Thunder Ax4-31Z HAB	171		28.11.78	A J.Moore *"Glory"*	(Northwood)	
G-LOSI	Cameron Z-105 HAB	10011		15. 1.01	Aeropubblicita Vicenza SRL	(Caldogno, Italy)	19. 8.05A
G-LOSM	Gloster Meteor NF.11	S4/U/2342	WM167	8. 6.84	Aviation Heritage Ltd	Coventry	2. 9.05P
	(Built Armstrong-Whitworth Aircraft)				*(As "WM167" in 151 Sqdn RAF camouflage c/s: also carries "G-LOSM")*		
G-LOST	Denney Kitfox Model 3	PFA 172-12055		10. 8.95	J.H.S.Booth	Perth	6. 8.01P
	(Built R Baily and H Balfour-Paul) (Rotax 618) *(Floatplane)*				*(Stored 10.03)*		
G-LOSY	Aerotechnik EV-97 Eurostar	PFA 315-14161		22.12.03	J.A Shufflebotham	(Macclesfield)	16. 3.05P
	(Built J.A Shufflebotham)						
G-LOTA	Robinson R44 Raven	1232		8. 7.02	Rahtol Ltd	Redhill	10. 7.05
G-LOTI	Bleriot Type XI rep	PFA 88-10410		21.12.78	Brooklands Museum Trust Ltd	Brooklands	19. 7.82P
	(Built M L Beach) (ABC Scorpion II)						
G-LOVB	British Aerospace Jetstream Series 3102	622	VH-HSW	12. 8.99	London Flight Centre (Stansted) Ltd	Inverness	26. 1.04T
			G-31-622, G-BLCB, G-31-622		*(Stored 11.04)*		
G-LOWS	Sky 77-24 HAB	025		19. 3.96	A J.Byme and D.J.Bellinger *"Dawn Treader"*	(Thatcham)	25. 4.05
G-LOYA	Reims FR172J Rocket	FR17200352	G-BLVT	4. 8.89	T.R.Scorer	Earsl Colne	31. 5.06
			PH-EDI/D-EEDI				
G-LOYD	Aérospatiale SA.341G Gazelle 1	1289	G-SFTC	19. 6.85	I.G.Lloyd	Ripley, Derbyshire	26 6.06
	(Rebuilt 1990 with major components of N6957 [c/n 1060])		N47298				
G-LPAD	Lindstrand LBL 105A HAB	632		5. 8.99	Line Packaging and Display Ltd	(Gillingham, Kent)	17. 4.05A
G-LRBW	Lindstrand HS-110 HA Airship	253		2. 8.95	Croymark Ltd *(New CofR 12.02)*	(Ottawa, Canada)	30. 3.05A
G-LRGE	Lindstrand LBL 330A HAB	929		31. 7.03	Adventure Balloons Ltd	(Hartley Wintney, Hook)	31. 3.05T
G-LRSN	Robinson R44 Raven	0984		28. 3.01	Larsen Manufacturing Ltd	Newtownards	4. 4.07T
G-LSCM	Cessna 172S Skyhawk	172S8445	N612TG	7. 6.04	G.A.Luscombe	Exeter	15. 6.07T
			N165ME				
G-LSFI	Gulfstream AA-5A Cheetah	AA5A-0770	G-BGSK	13. 2.84	A D.Prothero tr G-LSFI Group	North Moor, Scunthorpe	9. 7.06T
G-LSFT	Piper PA-28-161 Warrior II	28-8516008	G-BXTX	10.11.99	Biggin Hill Flying Club Ltd	Biggin Hill	1. 4.07T
			PH-LEH, N130AV, N43682				
G-LSHI	Colt 77A HAB	1264		20. 7.88	J H Dobson	(Streatley, Berkshire)	12. 7.95A
					(Lambert Smith and Hampton titles)		
G-LSMI	Reims/Cessna F152 II	F15201710		1. 2.80	A S.Bamrah t/a Falcon Flying Services	(Blackbushe)	4. 4.05T
G-LSTR	Stoddard-Hamilton GlaStar	PFA 295-13093		20. 4.98	R.Y.Kendal	Ewesley Farm, Morpeth	27. 6.05P
	(Built R Y Kendal) (Tail-wheel u/c)						

Registration	Type	C/n	Previous identity	Regd date	Owner/operator	Probable base	CA expiry
G-LTFB	Piper PA-28-140 Cherokee	28-23343	G-AVLU	28. 2.97	C.W.J.Cunningham	Popham	16. 5.07T
			N11C				
G-LTFC	Piper PA-28-140 Cherokee B	28-26259	G-AXTI	8. 6.94	London Transport Flying Club Ltd	Fairoaks	16.10.06T
			N11C				
G-LTNG*	English Electric Lightning T.5				See SECTION 3 Part 1	Cape Town, South Africa	
G-LTRF	Fournier RF7	7001	G-EHAP	10.12.97	Skyview Systems Ltd	Waits Farm, Belchamp Walter	22. 5.05P
			(G-BGVC), D-EHAP, F-WPXV				
G-LTSB	Cameron LTSB 90SS HAB	4483		15. 1.99	Airship and Balloon Company Ltd	(Bristol)	1. 6.05A
					(Lloyds TSB titles)		
G-LUBE	Cameron N-77 HAB	1127		25. 2.85	A C.Rawson *"Lubey Loo"*	(Stafford)	27. 7.05A
G-LUCK	Reims/Cessna F150M	F15001238	PH-LEO	13.12.79	Taylor Aviation Ltd	Cranfield	23. 6.07T
			D-EHRA				
G-LUED	Aero Designs Pulsar	PFA 202-12122		9. 3.92	J.C.Anderson	Sturgate	24. 6.04P
	(Built J.C.Anderson)						
G-LUFT	Putzer Elster C	011	G-BOPY	31. 3.92	A and E.A Wiseman	(Selby)	
			D-EDEZ		*(On rebuild 2003)*		
G-LUKE	Rutan LongEz	PFA 74A-10978		4. 7.84	R.A.Pearson	(Knowle, Solihull)	14. 8.03P
	(Built S G Busby) (Lycoming O-235)				*(New owner 7.04)*		
G-LUKI	Robinson R44 Raven	0818	G-BZLN	20.10.00	Marcella Air Ltd	Panshanger	27.11.06T
G-LUKY	Robinson R44 Astro	0357		10. 7.97	Melissa A.Hack	(Llantrisant, Pontyclun)	11. 9.06T
					t/a Hack Aviation		
G-LULU	Grob G109	6137		6. 9.82	A P.Bowden	Enstone	15. 5.07
G-LUMA	Jabiru Aircraft Jabiru SP-430	PFA 274B-13458		11. 5.99	B.Luyckx	Keuheuvel, Belgium	14. 7.03P
	(Built B.Luyckx)						
G-LUNA	Piper PA-32RT-300T Turbo Lance II	32R-7987108	N2246Q	19. 3.79	Lance Aviation Ltd	Humberside	19. 4.06T
G-LUND	Cessna 340 II	340-0305	G-LAST	27. 3.03	Prospect Developments (Northern) Ltd	Manchester	17. 9.06T
			G-UNDY, G-BBNR, N69452				
G-LUNE	Mainair Pegasus Quik	8017		18.2.04	D.Muir	St. Michaels	1. 3.05P
G-LUSC	Luscombe 8E Silvaire	3975	D-EFYR	1.11.84	M.Fowler	Bruntingthorpe	
			LN-PAT, (NC1248K)		*(On rebuild 9.97: current status unknown)*		
G-LUSH	Piper PA-28-151 Cherokee Warrior	28-7515201	OH-PAB	25. 7.01	A.Jahanfar *(Op Willowair Flying Club)*	Southend	4.11.04*
G-LUSI	Temco Luscombe 8F Silvaire	6770	N838B	3.10.89	J.P.Hunt and D.M.Robinson		
	(Continental C85)					Bourne Park, Hurstbourne Tarrant	12. 5.05P
G-LUST	Luscombe 8E Silvaire	6492	N2065B	9.11.89	M.Griffiths	Chilbolton	14. 7.05P
	(Continental C85)		NC2065B				
G-LUVY	Aérospatiale AS355F1 Ecureuil 2	5134	N358E	25. 2.00	DNH Helicopters Ltd	Biggin Hill	31. 8.07T
			ZS-HUA, (G-BPDP), D-HOCH, N358E, N5792M				
G-LUXE	British Aerospace BAe 146 Series 301	E3001	G-5-300	9. 4.87	BAE SYSTEMS (Operations) Ltd	Woodford	5. 5.07T
			G-SSSH, (G-BIAD)				
G-LVBF	Lindstrand LBL 330A HAB	936		30. 1.04	Airxcite Ltd t/a Virgin Balloon Flights	(Wembley)	17.8.05T
G-LVES	Cessna 182S Skylane	18280741	G-ELIE	19. 8.02	R.W.and A M.Glaves	Nottingham East Midlands	31. 8.06T
			N23754				
G-LVLV	Canadair CL604 Challenger	5372	N314FX	27. 4.04	Gama Aviation Ltd	Farnborough	27. 4.07T
			(N413LV), N314FX, C-GLWR				
G-LVPL	AirBorne Edge XT912-B/Streak III-B	XT912-035		21.12.04	C.D.Connor	Mill Farm, Shifnal	
G-LWAY	Robinson R44 Raven	1244	N71822	22. 8.02	Glenkerrin Aviation Ltd	(Maynooth, County Kildare	15. 9.05T
					(Op Lantway Properties Ltd)		
G-LWNG	Aero Designs Pulsar	PFA 202-11866	G-OMKF	14.10.02	C.Moffat	Eaglescott	15. 9.05P
	(Built M K Faro) (Tri-cycle u/c)						
G-LWUK	Robinson R44 Raven	1343	N75283	6.11.03	Wyberton Developments Ltd	(Boston)	30.11.06T
G-LYAK	IAV-Bacau Yakovlev Yak-52	822113	LY-AGN	18.12.02	Lee 52 Ltd *(Poke Software titles)*	Lee-on-Solent	22.12.05P
			Ukraine AF 140 *(yellow)*, DOSAAF 40 *(yellow)*				
G-LYDA	Hoffmann H 36 Dimona	3515	OE-9213	5. 4.94	J.W.Hagley	Booker	9.10.06
					tr G-LYDA Flying Group		
G-LYDD*	Piper PA-31 Turbo Navajo	31-537	G-BBDU	8. 5.89	Not known	Blackpool	12. 5.89T
			N6796L		*(Damaged Lydd 17.7.91: cancelled 30.3.93 as WFU) (Fuselage on fire dump 1.04)*		
G-LYFA	IAV-Bacau Yakovlev Yak-52	822608	LY-AFA	31. 3.03	M.I.Boyd	Barton	26. 5.05P
			DOSAAF 110		tr Fox Alpha Flying Group *"110"*		
G-LYNC	Robinson R22 Beta II	3069		5. 5.00	Traffic Management Services Ltd	Gamston	25. 5.06T
G-LYND	Piper PA-25-235 Pawnee D	"25-6309"	SE-IXU	8. 9.93	York Gliding Centre Ltd	Rufforth	18.11.05
	(Rebuild of G-ASFZ [25-2246] with new frame)		G-BSFZ, N6672Z				
G-LYNK	CFM Shadow Series DD	303-DD		12.10.98	J.Walton	Morgansfield, Fishburn	1.12.04P
G-LYNX*	Westland WG.13 Lynx 800				See SECTION 3 Part 1	Weston-super-Mare	
G-LYPG	Jabiru Aircraft Jabiru UL-450	PFA 274A-13466		6. 7.99	P.G.Gale	Fitzroy Farm, Bratton, Wiltshire	1. 7.05P
	(Built P.G.Gale)						
G-LYTE	Thunder Ax7-77 HAB	1113		29. 9.87	G.M.Bulmer *"Crispen"*	(Hereford)	19. 5.91A
G-LZZY	Piper PA-28RT-201T Turbo Arrow IV	28R-8031001	G-BMHZ	8. 5.01	A.C.Gradidge	(West Wellow, Romsey)	28. 5.05
			ZS-KII, N8096D				

G-MAAA - G-MZZZ

Registration	Type	C/n	Previous identity	Regd date	Owner/operator	Probable base	CA expiry
G-MAAC*	Advanced Airship Corporation ANR-1				See SECTION 3 Part 1	(Lincoln)	
G-MAAH	British Aircraft Corporation One-Eleven 488GH	BAC.259	VP-CDA	6.10.98	Gazelle Ltd	*(Stored 1.05)* Bournemouth	21. 9.05T
			G-MAAH, PK-TAL, G-BWES, PK-TAL, G-BWES, 5N-UDE, LX-MAM, HZ-MAM				
G-MAAN	Europa Aviation Europa XS	PFA 247-14009		7. 1.03	P.S.Maan	(Desborough)	
	(Built P.S.Maan - kit no.567) (Tri-gear u/c)						
G-MABE	Reims/Cessna F150L	F15001119	G-BLJP	20. 6.97	A.C.Saunders	Shobdon	14. 9.06T
			N962L				
G-MABH	Fokker F.28 Mk.0100	11291	N858US	20. 5.03	Gazelle Ltd	Farnborough	14. 7.05T
	(Fokker 100)		PH-EZH				
G-MABR	British Aerospace BAe 146 Series 100	E1015	G-DEBN	13. 1.00	British Airways Citiexpress Ltd	Ronaldsway	22.12.07T
			EC-GEP, EC-971, N568BA, XA-RST, N461AP, G-5-01				
G-MACE	Hughes 369E	0015E	HA-MSA	3. 2.05	Arc Helicopters Ltd	(Taunton)	
			SE-HNA				

G-MACH	SIAI-Marchetti SF.260	1-14	F-BUVY	29.10.80	Cheyne Motors Ltd	Old Sarum	29. 5.05
			OO-AHR, OO-HAZ, (OO-RAB)				
G-MACK	Piper PA-28R-200 Cherokee Arrow II	28R-7635449	N5213F	18. 8.78	Haimoss Ltd	Old Sarum	18.12.04T
G-MAFA	Reims/Cessna F406 Caravan II	F406-0036	G-DFLT	2. 6.98	Directflight Ltd	Exeter	6. 6.047
			F-WZDZ		*(Op DEFRA)*		
G-MAFB	Reims/Cessna F406 Caravan II	F406-0080	F-WWSR	27. 5.98	Directflight Ltd	Exeter	28. 9.07T
					(Op DEFRA - carries Fisheries Patrol titles on nose)		
G-MAFE	Dornier 228-202K	8009	G-OALF	21.12.92	FR Aviation Ltd	Bournemouth	4.11.05T
			G-MLDO, PH-SDO, D-IDON		*(Op DEFRA)*		
G-MAFF	Pilatus Britten-Norman BN-2T Islander	2119	G-BJED	20. 4.82	Cobham Leasing Ltd	Durham Tees Valley	25. 9.05T
(					*Op DEFRA)*		
G-MAFI	Dornier 228-202K	8115	D-CAAE	16. 2.87	Cobham Leasing Ltd *(Op DEFRA)*	Bournemouth	15. 7.05T
G-MAGC	Cameron Grand Illusion SS HAB	4000		19. 1.95	Magical Adventures Ltd	(West Bloomfield, Michigan, US)	16. 8.03A
G-MAGG	Pitts S-1SE	PFA 09-10873		17. 3.83	C.A Boardman	Little Gransden	2. 4.05
	(Built G C Masterton)						
G-MAGL	Sky 77-24 HAB	164		14. 7.99	RCM SARL	(Stuppicht, Luxembourg)	6. 6.05
G-MAIE	Piper PA-32R-301T Saratoga IITC	3257046	N47BK	1.12.00	B R Sennett	(Murcia, Spain)	9.11.06
			N41283				
G-MAIK	Piper PA-34-220T Seneca IV	3448078	N73BS	17.11.97	TEL (IoM) Ltd	Ronaldsway	23. 1.07
G-MAIN	Mainair Blade 912	1202-0699-7 & W1005		16. 6.99	W.Dawson	Baxby Manor, Husthwaite	4. 7.05P
G-MAIR	Piper PA-34-200T Seneca II	34-7970140	N3029R	15. 2.89	Alyson Jane Warren	Bristol	26. 4.07T
					(Op Bristol Flying Centre)		
G-MAJA	British Aerospace Jetstream Series 4100	41032	G-4-032	22. 4.94	Air Kilroe Ltd t/a Eastern Airways	Humberside	24. 5.05T
G-MAJB	British Aerospace Jetstream Series 4100	41018	G-BVKT	1. 6.94	Air Kilroe Ltd t/a Eastern Airways	Humberside	8. 6.06T
			N140MA, G-4-018				
G-MAJC	British Aerospace Jetstream Series 4100	41005	G-LOGJ	12. 9.94	Air Kilroe Ltd t/a Eastern Airways	Humberside	20.12.06T
G-MAJD	British Aerospace Jetstream Series 4100	41006	G-WAWR	27. 3.95	Air Kilroe Ltd t/a Eastern Airways	Humberside	2. 3.07T
G-MAJE	British Aerospace Jetstream Series 4100	41007	G-LOGK	12. 9.94	Air Kilroe Ltd t/a Eastern Airways	Humberside	24. 2.06T
G-MAJF	British Aerospace Jetstream Series 4100	41008	G-WAWL	6. 2.95	Air Kilroe Ltd t/a Eastern Airways	Humberside	18. 3.05T
G-MAJG	British Aerospace Jetstream Series 4100	41009	G-LOGL	16. 8.94	Air Kilroe Ltd t/a Eastern Airways	Humberside	30. 3.05T
G-MAJH	British Aerospace Jetstream Series 4100	41010	G-WAYR	4. 4.95	Air Kilroe Ltd t/a Eastern Airways	Humberside	13. 4.06T
G-MAJI	British Aerospace Jetstream Series 4100	41011	G-WAND	20. 3.95	Air Kilroe Ltd t/a Eastern Airways	Humberside	27. 4.07T
G-MAJJ	British Aerospace Jetstream Series 4100	41024	G-WAFT	27. 2.95	Air Kilroe Ltd t/a Eastern Airways	Humberside	28.10.05T
			G-4-024				
G-MAJK	British Aerospace Jetstream Series 4100	41070	G-4-070	27. 7.95	Air Kilroe Ltd t/a Eastern Airways	Humberside	2. 9.06T
G-MAJL	British Aerospace Jetstream Series 4100	41087	G-4-087	1. 4.96	Eastern Airways (UK) Ltd *"R J Mitchell"*	Humberside	16. 5.06T
G-MAJM	British Aerospace Jetstream Series 4100	41096	G-4-096	23. 9.96	Air Kilroe Ltd t/a Eastern Airways	Humberside	29.10.05T
G-MAJN	British Aerospace Jetstream Series 4100	41014	OY-SVS	8. 9.04	Air Kilroe Ltd t/a Eastern Airways	Humberside	8. 9.07T
			G-4-014				
G-MAJR	de Havilland DHC-1 Chipmunk 22	C1/0699	WP805	25. 9.96	C.Adams	Lee-on-Solent	22.10.06
					tr Chipmunk Shareholders *(As "WP805" in RAF c/s)*		
G-MAJS	Airbus A300B4-605R	604	F-WWAX	26. 4.91	Monarch Airlines Ltd	Luton	25. 4.05T
G-MALA	Piper PA-28-181 Archer II	28-8190055	G-BIIU	6. 3.81	D.C. and M.E.Dowell	Kemble	16. 4.05T
			N82748		t/a M and D Aviation		
G-MALC	Grumman AA-5 Traveler	AA5-0664	G-BCPM	19.11.79	B.P.Hogan	Sywell	7. 6.06
			N6170A				
G-MALK*	Reims/Cessna F172N Skyhawk II	F17201886	PH-SVS	1. 7.81	Edinburgh Airport Fire Service	Edinburgh	
			PH-AXF (3)				
			(Crashed near Lochgilphead 23.7.97: cancelled 23.12.97 as destroyed) (Fuselage for instructional use 2.04)				
G-MALS	Mooney M.20K Model 231	25-0573	N1061T	16. 8.84	J.Houlberg tr G-MALS Group	Blackbushe	25. 4.05
G-MALT	Colt Flying Hop SS HAB	1447		14. 4.89	P.J.Stapley	(London Colney)	11. 9.97A
					"Hoppie" (CofR restored 20.11.01)		
G-MAMC	RotorWay Executive 90	5057		24. 5.94	J.R.Carmichael	(Inverary)	19. 2.99P
	(Built J Carmichael) (RotorWay RI 162)				*(Damaged landing Cumbernauld 22.9.98: current status unknown)*		
G-MAMD	Beech B200 Super King Air	BB-1549	N1069S	16. 7.99	Forest Aviation Ltd	Gamston	16. 7.05T
G-MAMH	Fokker F.28 Mk.0100	11293	PK-TWI	26. 3.03	Gazelle Ltd	Farnborough	31.10.06
	(Fokker 100)		G-MAMH, N859US, PH-EZY				
G-MAMK	Robinson R44 Clipper	1201		17. 4.02	M.J.Hayward	Coventry	13. 5.05T
	(Marked as "Raven")				t/a M and M Aviation		
G-MAMO	Cameron V-77 HAB	1616		17.11.87	The Marble Mosaic Co Ltd *"Osprey"*	(Portishead)	27. 8.03A
G-MANA	British Aerospace ATP	2056	G-LOGH	21. 2.94	Trident Aviation Leasing Services (Jersey) Ltd		
			G-11-056		*(For freight-door conversion 12.04)*	(Bucharest, Romania)	21. 3.04T
G-MANC	British Aerospace ATP	2054	G-LOGF	7.11.94	Trident Aviation Leasing Services (Jersey) Ltd		
			G-11-054		*(In open store 2.05)*	Southend	20.10.03T
G-MANE	British Aerospace ATP	2045	G-LOGB	7. 6.94	British Airways Citiexpress Ltd	Southend	26. 2.07T
			G-11-045		*(In open store 2.05)*		
G-MANF	British Aerospace ATP	2040	G-LOGA	19. 9.94	British Airways Citiexpress Ltd	Southend	5.11.07E
					(In open store 2.05)		
G-MANG	British Aerospace ATP	2018	G-LOGD	22. 8.94	BAE Systems (Operations) Ltd	Southend	28. 9.07T
					(In open store 2.05)		
G-MANH	British Aerospace ATP	2017	G-LOGC	16.11.94	BAE Systems (Operations) Ltd	Ronaldsway	14. 8.05T
					(Stored 4.04)		
G-MANI	Cameron V-90 HAB	3038		8. 3.93	M.P.G.Papworth	(Ilkley)	3. 7.01A
G-MANJ	British Aerospace ATP	2004	G-LOGE	6. 9.94	British Airways Citiexpress Ltd	Glasgow	14. 4.07T
					(Op Inflite Aviation		
G-MANL	British Aerospace ATP	2003	G-ERIN	3.10.94	British Airways Citiexpress Ltd	Glasgow	25. 5.05T
			G-BMYK		*(Op Inflite Aviation*		
G-MANN	Aérospatiale SA.341G Gazelle 1	1295	G-BKLW	14. 4.86	N.E.R.Brunt	(Knebworth)	31. 5.07T
			N4DQ, N4QQ, N444JJ, N47316, F-WKQH				
G-MANO	British Aerospace ATP	2006	OK-TFN	28.11.94	BAE Systems (Operations) Ltd	Southend	18. 1.05T
			G-MANO, G-UIET, G-11-5, (N376AE) *(In open store 2.05)*				
G-MANP	British Aerospace ATP	2023	OK-VFO	28.10.94	Manx Airlines Ltd	Southend	25.10.06T
			G-MANP, G-PEEL		*(In open store 2.05)*		

Reg	Type	C/n	Previous identity	Date	Owner/Operator	Base	CofA
G-MANS	British Aerospace BAe 146 Series 200	E2088	G-CLHC	22. 5.00	British Airways Citiexpress Ltd	Manchester	25. 4.06T
			G-MANS, G-CHSR, G-5-088				
G-MANT*	Cessna 210L Centurion II	21060970	G-MAXY	22. 5.85	Sea-Front Crazee Golf	Great Yarmouth	2.10.94
			N550SV		*(Damaged near Oxford 16.2.92: cancelled 3.4.92 by CAA) (Noted 9.02)*		
G-MANW	Tri-R Kis	PFA 239-12628		12. 9.96	M.T.Manwaring	(Barking)	
	(Built M.T.Manwaring)				*(Current status unknown)*		
G-MANX	Clutton FRED Series II	PFA 29-10327		31. 5.78	S.Styles	(Birmingham)	17. 8.82P
	(Built P Williamson - c/n PW.2) (Ardem 4C02)				*(Crashed near Ronaldsway 30.10.81: on rebuild Wellesbourne Mountford 7.90)*		
G-MAPL	Robinson R44 Raven	0929	G-BZVP	7. 6.02	M.P.Lafuente	(Monte Carlo, Monaco)	27. 5.07T
G-MAPP	Cessna 402B	402B0583	D-INRH	16. 4.99	Simmons Mapping (UK) Ltd	Cranfield	5.10.05T
			N1445G				
G-MAPR	Beech A36 Bonanza	E-2713	N55916	17. 9.92	Moderandum Ltd	(Bournemouth/Guernsey)	2. 9.07
G-MAPS*	Sky Flying Map SS HAB				See SECTION 3 Part 1	(Aldershot)	
G-MARA	Airbus A321-231	0983	D-AVZB	31. 3.99	Monarch Airlines Ltd	Luton	30. 3.05T
G-MARE	Schweizer 269C	S-1320		12. 8.88	The Earl of Caledon	Caledon Castle, County Tyrone	30. 6.07
G-MARO	Best Off Skyranger J2.2	BMAA/HB/348		22.12.04	E.Daleki	(Chinnor)	
	(Built E.Daleki - kit no.UK/318)						
G-MARX	Van's RV-4	2394-1211	SE-XUU	16.11.04	M.W.Albery	Enstone	
	(Built T.L.Berry)		N42BN				
G-MARZ	Thruster T600N 450	1031-T600N-093		27. 1.04	D.P.Tassart	Scotland Farm, Hook	26. 1.05P
G-MASC	SAN Jodel D.150A Mascaret	37	F-BLDZ	1. 2.91	K.F. and R.Richardson	Wellesbourne Mountford	24.10.05P
G-MASF	Piper PA-28-181 Cherokee Archer II	28-7790191	OY-EPT	24. 6.97	Mid-Anglia Flight Centre Ltd	Cambridge	17. 8.06T
			LN-NAP		t/a Mid-Anglia School of Flying		
G-MASH	Westland-Bell 47G-4A	WA/725	G-AXKU	3.11.89	Defence Products Ltd	Redhill	21. 5.05
			G-17-10		*(US Army c/s)*		
G-MASS	Cessna 152 II	15281605	G-BSHN	6. 3.95	MK Aero Support Ltd	Denham	2. 5.05T
			N65541		*(Op The Pilot Centre)*		
G-MASX	Masquito Masquito M.80	03		19. 6.98	Masquito Aircraft NV	(Roosdaal, Belgium)	
G-MASY	Masquito Masquito M.80	02		19. 6.98	Masquito Aircraft NV	(Roosdaal, Belgium)	
G-MASZ	Masquito Masquito M.58	01		29. 4.97	Masquito Aircraft NV	(Roosdaal, Belgium)	AC
G-MATE	Moravan Zlin Z-50LX	0068		26.10.90	J.H.Askew	(Holmfirth)	5. 7.07
G-MATS	Colt GA-42 Gas Airship	738	JA1009	11. 6.87	P.A Lindstrand	Oswestry	23. 5.90A
			G-MATS		*(New owner 6.01)*		
G-MATT	Robin R2160	97	G-BKRC	7. 5.85	P.White	(Fethard, County Tipperary)	22. 5.06
			F-BZAC, F-WZAC				
G-MATY	Robinson R22 Beta II	3686		28.10.04	MT Aviation Ltd	Cambridge	24.11.07E
G-MATZ	Piper PA-28-140 Cherokee Cruiser	28-7325200	G-BASI	11.12.90	R.B.Walker,	Coventry	20.12.06T
			N11C		t/a Midland Air Training School		
G-MAUD	British Aerospace ATP	2002	(G-MANK)	14.12.93	British Airways Citiexpress Ltd	Glasgow	13. 6.07T
			G-MAUD, G-BMYM		*(Blue Poole t/s) (Op Loganair)*		
G-MAUK	Colt 77A HAB	901		16. 2.87	B.Meeson *"Mondial Assistance"*	(Walsall)	4. 6.92A
G-MAVI	Robinson R22 Beta	0960		7. 2.89	Northumbria Helicopters Ltd	Newcastle	26. 4.07T
G-MAXG	Pitts S-1S	PFA 09-13233		27. 4.01	Jenks Air Ltd	(Berkhampsted)	11. 8.05P
	(Built T.P.Jenkinson)						
G-MAXI	Piper PA-34-200T Seneca II	34-7670150	N8658C	11. 2.81	Draycott Seneca Syndicate Ltd	Kemble	2. 6.06T
G-MAXV	Van's RV-4	PFA 181-13266		20. 1.00	R.S.Partridge-Hicks	(Bury St.Edmunds)	3.11.05P
	(Built T P Jenkinson) (Lycoming IO-360)						
G-MAYB	Robinson R44 Raven	1429		7.10.04	Highmark Aviation Ltd	(Pilgrims Hatch,Brentwood)	31.10.07T
G-MAYO	Piper PA-28-161 Cherokee Warrior II	28-7716278	G-BFBG	20. 2.81	M.P.Catto	(Zeals, Warminster)	20. 5.07T
			N38848		t/a Jermyk Engineering		
G-MBAA	Hiway Skytrike II/Excalibur	01		23. 4.81	M.J.Aubrey	Kington, Hereford	
	(Hiro Delta 22)				*(Noted 2002)*		
G-MBAB	Hovey Whing Ding II	PFA 116-10706		26. 5.81	M.J.Aubrey	Kington, Hereford	1. 2.98P
	(Built R F Morton - c/n MA-59) (Konig SC340)				*(Noted 2002)*		
G-MBAW	Pterodactyl Ptraveler	017		14. 7.81	J.C.K.Scardifield	(Lymington)	Exemption
	(Cuyana 430R)				*(Current status unknown)*		
G-MBBB	Skycraft Scout II	0388W		3. 8.81	A J.and B.Chalkley	(Pwllheli)	
	(Pixie 173)				*(Current status unknown)*		
G-MBBM	Eipper Quicksilver MX	10960		11. 9.81	J.Brown	(Markfield, Leics)	Exemption
	(Cuyana 430R)				*(In storage: current status unknown)*		
G-MBBZ*	Volmer VJ-24W				See SECTION 3 Part 1	Newark	
G-MBCJ	Mainair Tri-Flyer/Solar Wings Typhoon S			30. 9.81	R.A Smith	(Doncaster)	Exemption
		JRN-1 & T881-225	*(May have replacement wing T382-390L)*			*(Current status unknown)*	
G-MBCK	Eipper Quicksilver MX	GWR-10962		30. 9.81	P.Rowbotham	(Loughborough)	Exemption
	(Rotax 503)				*(Current status unknown)*		
G-MBCL	Hiway Skytrike 160/Solar Wings Typhoon			30. 9.81	P.J.Callis	Halwell, Totnes	Exemption
		2332 & T1181-307			*(New CofR 4.04)*		
G-MBCU	American Aerolights Eagle Amphibian	3181		5.10.81	J.L.May	(Portsmouth)	25.10.04P
	(Rotax 377)						
G-MBCX	Hornet 250/Airwave Nimrod 165	H090 & 0090 LJH		12.10.81	M.Maylor	(Louth)	Exemption
	(Built Airwave Gliders Ltd) (Fuji-Robin EC-25-PS)				*(Current status unknown)*		
G-MBDG	Eurowing Goldwing	E.20		19.10.81	B.Fussell	(Llanelli)	14.12.94P
	(Konig SC 430)				*(Current status unknown)*		
G-MBDL*	Striplin (AES) Lone Ranger				See SECTION 3 Part 1	Usworth, Sunderland	
G-MBDM	Southdown Trike/Southdown Sigma	SST/001		26.10.81	A R.Prentice	(Dartford)	Exemption
	(Fuji-Robin EC-25-PS)				*(Current status unknown)*		
G-MBEP*	American Aerolights Eagle 215B				See SECTION 3 Part 1	Caernarfon	
G-MBET	Micro Engineering (Aviation) Mistral	MEA.103		10.11.81	B.H.Stephens	Old Sarum	27. 9.98P
	(Fuji-Robin EC-44-PM)				*(Noted in trailer 2002)*		
G-MBEU	Chargus T.250/Hiway Demon	T.250/06		10.11.81	R.C.Smith	(Clacton-on-Sea)	Exemption
	(Fuji-Robin EC-25-PS)				*(Current status unknown)*		
G-MBFK	Hiway Skytrike 250/Hiway Demon 175	LR17D		16.11.81	D.W.Stamp	(Kidderminster)	Exemption
	(Fuji-Robin EC-25-PS)				*(Current status unknown)*		

Reg	Type	C/n	Regd	Owner/Operator	Location	CofA
G-MBFO	Eipper Quicksilver MX (Cuyuna 430R)	MLD-01	17.11.81	J.C.Larkin *(Current status unknown)*	(Maryport, Cumbria)	20. 8.93P
G-MBFS*	American Aerolights Eagle (Fuji-Robin EC-25-PS)	RF-01	19.11.81	M.J.Aubrey *(Cancelled 24.5.90 as WFU) (Noted 2002)*	Kington, Hereford	
G-MBFZ	Eurowing Goldwing (Fuji-Robin EC-34-PM)	MSS-01	25.11.81	D.G.Palmer *(Under active rebuild 2001)*	Fetterangus	5. 9.00P
G-MBGA	Mainair Tri-Flyer/Flexiform Solo Sealander *(Originally regd as Mainair Tri-Flyer/Solar Wings Typhoon with same c/n)*	001	25.11.81	D.A Caig *(Current status unknown)*	(Southport)	14. 9.97P
G-MBGF	Twamley Trike/Birdman Cherokee *(Built Birdman Enterprises Ltd)*	RWT-01	26.11.81	T.B.Woolley *(Current status unknown)*	(Leicester)	
G-MBGS	Rotec Rally 2B	PCB-1	2.12.81	P.C.Bell *(Current status unknown)*	(Yalding, Kent)	
G-MBGX	Southdown Lightning DS (Sachs-Dolmar 340?) *(Believed fitted with Ultralight Aviation Systems Storm Buggy trike ex G-MBKD)*	RBDB-1	7.12.81	T.Knight *(Current status unknown)*	(Newton Abbot)	Exemption
G-MBHE	American Aerolights Eagle (Cuyuna 430R)	4210	18.12.81	R.J.Osborne *(Current status unknown)*	Long Marston	12.10.96P
G-MBHK	Mainair Tri-Flyer 330/Flexiform Solo Striker (Fuji-Robin EC-34-PM) *(Original Tri-Flyer 250 trike c/n 036 replaced by Tri-Flyer 330 c/n 060-382 in 1982)*	EB-1 & 036-241181	30.12.81	K.T.Vinning *(Current status unknown)*	(Stratford-upon-Avon)	11. 8.98P
G-MBHZ	Pterodactyl Ptraveler (Cuyuna 430R)	TD-01	6. 1.82	J.C.K.Scardifield *(Current status unknown)*	(Lymington)	Exemption
G-MBIA	Hiway Skytrike/Flexiform Sealander (Fuji-Robin EC-34-PM)	6172349/336	6. 1.82	I.P.Cook *(Current status unknown)*	(Oldham)	Exemption
G-MBIT	Hiway Skytrike/Demon (Fuji-Robin EC-25-PS)	2501	18. 1.82	K.S.Hodgson *(Reported as taken to Canada 9.89 but new owner 6.00!)*	(Yarm)	Exemption
G-MBIY	Ultrasports Tri-Pacer/Southdown Lightning Phase II (Fuji-Robin EC-34-PM) *(Wing c/n L170-439)*	330	19. 1.82	J.W.Burton *(Stored 2.03)*	Tarn Farm, Cockerham	18. 4.99P
G-MBIZ	Mainair Tri-Flyer 250/Hiway Vulcan (Fuji-Robin EC-25PS)	039-251181 & SD9V	20. 1.82	E.F.Clapham, W.B.S.Dobi, S.P.Slade and D.M.A Templeman *(Current status unknown)*	(Bristol)	
G-MBJD	American Aerolights Eagle 215B (Zenoah G25B1)	4169	21. 1.82	R.W.F.Boarder *(Current status unknown)*	(Tring)	Exemption
G-MBJF	Hiway Skytrike II/Vulcan C (Fuji-Robin EC-25-PS) *(C/n is engine serial no)*	80-00099	22. 1.82	C.H.Bestwick *(Current status unknown)*	(Nottingham)	Exemption
G-MBJG	Chargus T.250/Airwave Nimrod UP (Fuji-Robin EC-25-PS)	CMT165045	25. 1.82	D.H.George	Sandown	5. 7.05P
G-MBJK	American Aerolights Eagle (Chrysler 820)	2742	16. 1.82	B.W.Olley *(Stored 2000)*	(Ely)	
G-MBJL	Hornet 250/Airwave Nimrod (Fuji-Robin EC-25-PS)	JSRM-01	26. 1.82	A G.Lowe *(Noted at owner's home 4.02)*	(Aberdeen)	20.10.96P
G-MBJM	Striplin Lone Ranger *(Built C K Brown)* (Fuji-Robin) *(C/n 81-00138 is engine serial no)*	LR-81-00138	26. 1.82	C.K.Brown *(Current status unknown)*	(Loughborough)	
G-MBJX*	Hiway Skytrike I/Hiway Super Scorpion			See SECTION 3 Part 1	East Fortune	
G-MBKY	American Aerolights Eagle 215B	BF-01	12. 2.82	M.J.Aubrey *(Noted 2002)*	(Kington, Hereford)	
G-MBKZ	Hiway Skytrike/Super Scorpion (Fuji-Robin EC-25-PS) *(C/n is corruption of engine type)*	EC25P8-04	12. 2.82	S.I.Harding *(Current status unknown)*	(Camberley)	
G-MBLK*	Ultrasports Tri-Pacer/Southdown Lightning DS (Fuji-Robin EC-44)	DS-390	18. 2.82	M.J.Aubrey *(Cancelled 23.6.97 as WFU) (Noted 2002).*	(Kington, Hereford)	
G-MBLN	MEA Pterodactyl Ptraveller 4300	HCM-01	19. 2.82	P.H.and K.H.Risdale *(New owners 7.04)*	Tower Farm, Wollaston	2. 4.94P
G-MBLU	Ultrasports Tri-Pacer/Southdown Lightning L195 (Fuji-Robin EC-25-PS)	L195/191	26. 2.82	C.R.Franklin *(Current status unknown)*	(Barnstaple)	Exemption
G-MBMG	Rotec Rally 2B	RJP-01	3. 3.82	J.R.Pyper *(Current status unknown)*	(Craigavon, County Armagh)	
G-MBMT	Mainair Tri-Flyer/Southdown Lightning 195 (Fuji-Robin EC-25-PS) *(Wing c/n L195-195?)*	TRY-01	8. 3.82	A G.Rodenburg and T.Abro *(Noted 4.04)*	Hillfoots Nurseries and Golf Driving Range, Tillicoultry, Stirling	Exemption
G-MBOF	Pakes Jackdaw *(Built L.G.Pakes)*	LGP-01	26. 3.82	L.G.Pakes *(Current status unknown)*	(Ryde, Isle of Wight)	
G-MBOH	Micro Engineering (Aviation) Mistral (Fuji-Robin EC-44-PM)	008	29. 3.82	N.A Bell *(Current status unknown)*	(Fordingbridge)	Exemption
G-MBPB (2)	Pterodactyl Ptraveller *(Built P E Bailey)*	PEB-01	7. 4.82	N.A Bell *(For rebuild 12.01)*	(Fordingbridge)	
G-MBPG	Mainair Tri-Flyer/Solar Wings Typhoon (Fuji-Robin EC-25-PS) *(Original trike was c/n 067-582 and may have been used for G-MMGT)*	189-1983 & T381-105	13. 4.82	S.D.Thorpe	Otherton, Cannock	14. 6.01P
G-MBPJ	Centrair Moto-Delta G.11 *(Built Moto Delta)*	001	14. 5.82	J.B.Jackson *(Current status unknown)*	(Chester)	
G-MBPM*	Eurowing Goldwing			See SECTION 3 Part 1	East Fortune	
G-MBPU	Hiway Skytrike 250/Demon	DSS-01	21. 4.82	D.Hines	(Crewe)	22. 1.04P
G-MBPX	Eurowing Goldwing SP (Konig SC 430)	EW-42	21. 4.82	A R.Channon *(Current status unknown)*	(Sawston, Cambridge)	6.11.96P
G-MBPY	Ultrasports Tri-Pacer 330/Wasp Gryphon II *(Built Wasp)* (Fuji-Robin EC-34-PM)	RKP-01	21. 4.82	D.Hawkes and C.Poundes	(Milton Keynes)	12. 4.03P
G-MBRB	Electraflyer Eagle Mk.I	E.2229	9.12.81	R.C.Bott *(Current status unknown)*	(Tywyn)	
G-MBRD	American Aerolights Eagle 215B (Fuji-Robin EC-25-PS)	E.2635	20. 4.82	R.J.Osborne *(Current status unknown)*	(Tiverton)	Exemption
G-MBRH	Ultraflight Mirage II	83-009 & RALH-01	20. 4.82	R.W.F.Boarder	Field Farm, Oakley	8. 1.01P
G-MBRS	American Aerolights Eagle 215B (Zenoah G25B1)	RWC.1	23. 4.82	W.J.Phillips *(Stored 6.90: current status unknown)*	Haverfordwest	Exemption
G-MBST	Mainair Gemini/Puma Sprint (Fuji-Robin EC-44-PM) *(Fitted with Trike from G-MJXA)*	141-29383	10. 4.84	G.J.Bowen	(Llanelli)	25. 5.03P
G-MBSX	Ultraflight Mirage II	240	14. 6.82	P.J.Careless and P.Samal	Sandy	30. 5.03P
G-MBTF	Mainair Gemini/Southdown Sprint (Fuji-Robin EC-44-PM)	168-30683	26. 4.82	D.E.J.McVicar	(Antrim, County Antrim)	26. 3.00P
G-MBTG*	Mainair Gemini/Southdown Sprint *(Originally regd as Mainair Tri-Flyer Dual)*	064-19482 & P.431	26. 4.82	Not known *(Cancelled 31.3.00 by CAA) (Noted 1.04)*	Roddige	15.10.94P

Registration	Type	C/n	Previous identity	Regd date	Owner	Base	CofA expiry
G-MBTH	Whittaker MW4	T1081-262L	(G-MBPB (1))	6. 4.82	L.Greenfield and M.Whittaker	Otherton, Cannock	12. 7.04P
	(Built M Whittaker - c/n 001) (Fuji-Robin EC-34-PM)				tr The MW4 Flying Group		
G-MBTJ	Ultrasports Tri-Pacer/Solar Wings Typhoon	CSRS-01		2. 4.82	H.A Comber	(Poole)	6. 7.05P
	(Fuji-Robin EC-25-PS) *(Wing c/n may be T1081-286L)*						
G-MBTW	Aerodyne Vector 600	1188		10. 5.82	W.I.Fuller	Cambridge	Exemption
	(Zenoah G25B1)				*(Current status unknown)*		
G-MBUA	Hiway Skytrike/Demon	RJN-01		30. 4.82	R.J.Nicholson *(Current status unknown)*	(Lightwater)	
G-MBUE*	MBA Tiger Cub 440				See SECTION 3 Part 1	Newark	
G-MBUS*	MEA Mistral Trainer	FGJ-01		7. 5.82	(N.A Bell)	(Fordingbridge)	
					(Cancelled 25.10.88 as destroyed) (For spares 12.01)		
G-MBUZ	Skycraft Scout II	0366		4. 5.82	A C.Thorne *(Current status unknown)*	(Yelverton)	
G-MBVW	Skyhook TR2/Cutlass	TR2/23		14. 5.82	M.Jobling	(Harrogate)	Exemption
	(Solo 210 x 2)				*(Current status unknown)*		
G-MBWE*	American Aerolights Eagle	2937		18. 5.82	M.J.Aubrey	(Kington, Hereford)	
	(Fuji-Robin EC-25-PS)				*(Cancelled 24.3.99 by CAA) (Noted 2002)*		
G-MBWG	Huntair Pathfinder Mk.1	006		19. 5.82	T.Mahmood	(Aberdeen)	14. 7.99P
	(Fuji-Robin EC-34-PM)				*(New owner 6.02)*		
G-MBWH*	Jordan Duet Series 1	D82001		20. 5.82	Designability Ltd	Kemble	
					(Cancelled 22.3.02 as WFU) (Noted 2004)		
G-MBWI*	Lafayette Hi-Nuski Mk.1	30680		8. 6.82	M.J.Aubrey	(Kington, Hereford)	
					(Cancelled 13.6.90 by CAA) (Noted 11.03)		
G-MBXX	Ultraflight Mirage II	111		21. 1.82	E.J.Girling	St.Just	Exemption
	(Built Breen Aviation Ltd) (Kawasaki TA440)				*(Stored 5.94: current status unknown)*		
G-MBYH*	Maxair Hummer	001		4. 6.82	Not known	Doynton	31.12.87P
					(Cancelled 19.5.97 by CAA) (Noted derelict 2.03)		
G-MBYI	Ultraflight Lazair IIIE	A464/001		4. 6.82	S.Barr	Perth	8. 8.05P
	(Built AMF Microflight Ltd) (Rotax 185 x 2) *(Originally kit no A522 and amended during rebuild c.1982)*						
G-MBYK*	Huntair Pathfinder Mk.1	012		4. 6.82	Not known	Letterkenny, County Donegal	
					(Cancelled 30.5.01 by CAA) (Noted 4.04)		
G-MBYL	Huntair Pathfinder Mk.1	009		4. 6.82	A R.Hobbins	(Limavady, County Londonderry)	17. 2.02P
	(Fuji-Robin EC-44-PM)						
G-MBYM	Eipper Quicksilver MX	JW-01		4. 6.82	M.P.Harper and L.L.Perry	Priory Farm, Tibenham	21. 9.96P
	(Cuyuna 430R)				*(Current status unknown)*		
G-MBZH	Eurowing Goldwing	EW-50		14. 6.82	J.Spavins	Longacres Farm, Sandy	31. 3.03P
	(Fuji-Robin EC-34-PM)						
G-MBZJ	Southdown Puma/Lightning	L170-415		14. 6.82	A K.Webster	(Wallingford)	1. 8.98P
	(Fuji-Robin EC-34-PM)				*(Stolen 24.8.97 from Chiltern Park: current status unknown)*		
G-MBZO	Mainair Tri-Flyer/Flexiform Medium Striker			15. 6.82	A N.Burrows	(Kirk Michael, Isle of Man)	15. 4.98P
	(Fuji-Robin EC-34-PM)	GRH-01 & 021-101081			*(Current status unknown)*		
G-MBZV	American Aerolights Eagle 215B	4227-Z		16. 6.82	M.J.Aubrey *(New owner 7.03)*	Kington, Hereford	Exemption
G-MCAI	Robinson R44 Raven II	10423		9. 7.04	M.C.Allen	(London W4)	8. 8.07T
G-MCAP	Cameron C-80 HAB	10186		30. 7.02	L.D.Thurgar *(Mencap titles)*	(Bristol)	20. 8.05A
G-MCCF	Thruster T.600N Sprint	0100-T600N-048		25. 4.01	C.C.F.Fuller	Craysmarsh Farm, Melksham	26. 5.05P
G-MCCY	IAV-Bacau Yakovlev Yak-52	9011112	LY-AQF	11.11.04	D.P.McCoy	(Wicklow, County Wicklow)	11.11.05P
			UR-BBP, Ukraine AF 129, DOSAAF 129				
G-MCEA	Boeing 757-225	22200	N510EA	6. 2.95	MyTravel Airways Ltd	Manchester	23. 3.07T
G-MCEL	Pegasus Quantum 15-912 Super Sport	7858		10.10.01	F.Hodgson	Sywell	21.10.05P
G-MCJL	Pegasus Quantum 15-912	7497		16. 3.99	A Gillett	Southend	14. 5.05P
G-MCMS	Aero Designs Pulsar	PFA 202-11982		3. 2.93	B.R.Hunter	Perth	21.12.04P
	(Built M C Manning)						
G-MCOX	Fuji FA.200-180AO Aero Subaru	FA200-296	(G-BIMS)	29.12.81	West Surrey Engineering Ltd	Fairoaks	15. 6.06
G-MCOY	Flight Design CT2K	01-04-01-12		25. 7.01	D.Young	Kemble	24. 7.05P
					t/a Pegasus Flight Training (Cotswolds)		
G-MCPI	Bell 206B-3 JetRanger III	3191	G-ONTB	4. 4.90	D.A C.Pipe	Liskeard	3. 4.06T
			N3896C				
G-MCXV	Colomban MC-15 Cri-Cri	371	F-PYVA	1. 3.00	H.A Leek	(Scalford, Melton Mowbray)	
	(Built J.P.Lorre)				*(Current status unknown)*		
G-MDAC	Piper PA-28-181 Archer II	28-8290154	N8242T	6.11.87	B.R.McKay	Compton Abbas	15. 6.06
					tr Alpha Charlie Flying Group		
G-MDAY	Cessna 170B	26350	N2807C	2. 5.03	M.Day	Bourne Park, Hurstbourne Tarrant	16. 6.06
					(Carries "N2807C" on tail)		
G-MDBC	Pegasus Quantum 15-912	7814		4. 5.01	D.B.Caiden	East Fortune	26. 8.05P
G-MDBD	Airbus A330-243	266	F-WWKG	24. 6.99	MyTravel Airways Ltd	(Manchester	24. 6.05T
G-MDCA	Piper PA-34-220T Seneca V	3449273	N53643	25. 6.04	MDC-Aviation LLP	(Wetherby)	24. 6.07T
G-MDGE	Robinson R22 Beta	1475	G-OGOG	25. 6.04	Mandarin Aviation Ltd	Redhill	9.10.06T
			G-TILL				
G-MDJN	Beech 95-B55 Baron	TC-1574	G-SUZI	3. 8.04	D.J.Nock	(Stourbridge)	30.11.07E
			G-BAXR				
G-MDKD	Robinson R22 Beta	1247		18. 4.90	B.C.Seedle t/a Brian Seedle Helicopters	Blackpool	20. 5.02T
G-MDPI	Agusta A109A II	7393	G-PERI	11. 8.04	Castle Air Charters Ltd	Liskeard	3. 7.06T
			G-EXEK, G-SLNE, G-EEVS, G-OTSL				
G-MEAH	Piper PA-28R-200 Cherokee Arrow II	28R-7435104	G-BSNM	14. 6.91	Stapleford Flying Club Ltd	Stapleford	10. 4.06T
			N46PR, G-BSNM, N46PR, N54439				
G-MEDA	Airbus A320-231	480	N480RX	12.10.94	British Mediterranean Airways Ltd	Heathrow	11.10.06T
			F-WWDU		t/a BMED *(Whale Rider t/s)*		
G-MEDE	Airbus A320-232	1194	F-WWDY	25. 4.00	British Mediterranean Airways Ltd	Heathrow	24. 4.06T
					t/a BMED		
G-MEDF	Airbus A321-231	1690	D-AVZX	28. 2.02	British Mediterranean Airways Ltd	Heathrow	27. 2.05T
					t/a BMED		
G-MEDG	Airbus A321-231	1711	D-AVZK	5. 4.02	British Mediterranean Airways Ltd	Heathrow	4. 4.05T
					t/a BMED		
G-MEDH	Airbus A320-232	1922	F-WWBX	6. 3.03	British Mediterranean Airways Ltd	Heathrow	5. 3.06T
					t/a BMED		
G-MEDJ	Airbus A321-231	2190	D-AVZD	8. 4.04	British Mediterranean Airways Ltd	Heathrow	7. 4.07T
					t/a BMED		

Registration	Type	C/n	Previous identity	Date	Owner/Operator	Location	CofA
G-MEDK	Airbus A321-231	2441		5.05R	British Mediterranean Airways Ltd		
					t/a BMED		
G-MEGA	Piper PA-28R-201T Turbo Arrow III	28R-7803303	N999JG	13. 2.86	A.W.Bean	(Goole)	7. 1.05T
G-MEGG	Europa Aviation Europa XS	PFA 247-13202		14. 6.00	M.E.Mavers	(Macclesfield)	8. 6.05P
	(Built M.E.Mavers - kit no.358) (Monowheel u/c)						
G-MELT	Cessna F172H	F172-0580	G-AWTI	23. 9.83	J.H.Haycock and J.G.Baggott	Goodwood	10. 3.06
	(Built Reims Aviation SA)						
G-MELV	SOCATA Rallye 235E Gabier	13328	G-BIND	21. 5.86	J.W.Busby	Grove Fields Farm, Wasperton	3 .2.06
G-MEME	Piper PA-28R-201 Arrow	2837051	N9219N	17. 8.90	Henry J.Clare Ltd	Bodmin	6.10.05
G-MEOW	CFM Streak Shadow	PFA 206-12025		23. 4.93	G.J.Moor	Craysmarsh Farm, Melksham	11. 6.04P
	(Built S D Hicks - kit no K.172) (Rotax 582)						
G-MERC	Colt 56A HAB	842		11. 6.86	A F.and C.D.Selby	(Loughborough)	4. 4.04A
G-MERE	Lindstrand LBL 77A HAB	092		7. 4.94	R.D.Baker	(Canterbury)	
G-MERF	Grob G115A	8091	EI-CAB	24. 7.95	G.Wylie tr G-MERF Group	White Waltham	24. 5.05
G-MERI	Piper PA-28-181 Archer II	28-8090267	N8175J	17. 7.80	S.Padidar-Nazar	(Rosley, Wigton)	27.10.05T
G-MERL	Piper PA-28RT-201 Arrow IV	28R-7918036	N2116N	27. 6.86	M.Giles	Cardiff	22. 8.04
G-MESS	SNCAN Nord 1101 Noralpha	87	F-BEEV	21. 5.03	G.R.Lacey	(Fairoaks)	
			F-WZBI, French AF 87				
G-MEUP	Cameron A-120 HAB	2117		5.10.89	Innovation Ballooning Ltd	(Bath)	30. 7.05T
					(Sopwith Aviation Co titles)		
G-MFAC	Cessna F172H	F172-0387	G-AVBZ	23. 8.01	Cheshire Flying Services Ltd	Liverpool	30. 6.06T
	(Built Reims Aviation SA)				t/a Ravenair		
G-MFEF	Reims FR172J Rocket	FR17200426	D-EGJQ	19.10.00	M.and E.N.Ford	Butlers Gyhll, Southwater	26.11.06
G-MFHI	Europa Aviation Europa	PFA 247-12841		14.11.97	P.Rees	(Shorne, Gravesend)	5. 1.05P
	(Built M.F.Howe - kit no.202) (Monowheel u/c)				tr Hi Fliers		
G-MFHT	Robinson R22 Beta II	2601	N8334H	20. 6.96	MFH Helicopters Ltd	Blackpool	3. 7.05T
G-MFLI	Cameron V-90 HAB	2650		14. 8.91	J.M.Percival *"Mayfly" (Mouldform titles)*	(Loughborough)	31. 7.05A
G-MFLY	Mainair Rapier	1359-1103-7-W1154		30. 3.04	J J Tierney	Chiltern Park, Wallingford	29. 3.05P
G-MFMF	Bell 206B-3 JetRanger III	3569	G-BJNJ	4. 6.84	South Western Electricity plc	Bristol	3.12.06T
G-MFMM	Scheibe SF25C Falke	4412	(G-MBMM)	20. 4.82	J.E.Selman	(Ardagh, County Limerick)	11. 8.06T
	(Limbach SL1700)		D-KAEU				
G-MGAA	BFC Challenger II	PFA 177A-13124		18. 8.97	S.G.Beeson	(Stoke-on-Trent)	13. 4.05P
	(Built G A Archer and J W E Pearson - c/n CH2-0297-1568) (Rotax 582)						
G-MGAG	Aviasud Mistral 532GB	BMAA/HB/009		20. 6.89	M.Raj	Otherton, Cannock	27. 6.00P
		(Kit no.0587-045)					
G-MGAN	Robinson R44 Astro	0588		10. 5.99	Carnbeg Golf Ltd	Weston, Dublin	9. 7.05T
G-MGCA	Jabiru Aircraft Jabiru UL	PFA 274-13228		8. 5.98	P.A James	Redhill	26. 3.04P
	(Built P A James) (C/n prefix should be "PFA 274A")				t/a Cloudbase Aviation		
G-MGCB	Solar Wings Pegasus XL-Q	SW-TE-0344 & 7267	*(Trike ex G-MWUT)*	16.10.96	M.G.Gomez	Headon Farm, Retford	5. 9.05P
G-MGCK	Whittaker MW6 Merlin	PFA 164-11262		30. 3.93	M.W.J.Whittaker and L.R.Orriss	Askern	
	(Built M.W.J.Whittaker)				*(New CofR 8.04)*		
G-MGDL	Pegasus Quantum 15	7400		17. 2.98	I.Fernihough	Bradley	23. 5.05P
	(Rotax 582)						
G-MGEC	Rans S-6-ESD Coyote II XL	PFA 204-13209		13.10.97	G.Clipston	Sywell	19.10.04P
	(Built E Carter) (Tri-cycle u/c)						
G-MGEF	Pegasus Quantum 15-912 Super Sport	7261		18. 9.96	G.D.Castell	Longacres Farm, Sandy	29. 1.05P
G-MGFK	Pegasus Quantum 15-912	7396		2. 2.98	F.A A Kay	London Colney	13. 9.05P
G-MGFO*	Pegasus Quantum 15	7410		24. 3.98	Not known	Clench Common	19. 3.00P
					(Cancelled 10.5.99 by CAA) (Stored 1.04)		
G-MGGG	Pegasus Quantum 15-912	7377		3.11.97	R.A Beauchamp	Shenstone Hall Farm, Shenstone	17. 5.05P
G-MGGT	CFM Streak Shadow SA-M	PFA 206-12723		3. 6.94	B E Trinder	(Rushden, Northampton)	18.10.05P
	(Built J W V Edmonds - kit no K.252) (Rotax 618)						
G-MGGV	Pegasus Quantum 15-912	7484		12.10.98	S.M.Green	Clench Common	6. 6.05P
G-MGMC	Pegasus Quantum 15-912	7430		28. 4.98	G.J.Slater	Clench Common	26. 6.05P
G-MGMM	Piper PA-18-150 Super Cub	18-7909189	D-EBRG	1. 6.04	M.J.Martin	Challock	1. 7.07T
			N9750N				
G-MGND	Rans S-6-ESD Coyote II XL	PFA 204-13152		27. 6.97	P.Vallis	(Alfreton)	28. 9.05P
	(Built N N Ducker)						
G-MGOD	Medway Raven X	MRB110/106		6. 7.93	A Wherrett, N.R.Andrew and D.J.Millward	Doynton	1. 5.00P
					(Noted 1.05)		
G-MGOO	Meridian Renegade Spirit UK	PFA 188-11580		14.11.89	P.J.Dale	(Chesham)	12. 7.05P
	(Built A R Max - kit no.301)						
G-MGPA	Comco Ikarus C42 FB100	0412-6635		25. 1.05	P.D.Ashley	(Brixham)	AC
G-MGPD	Solar Wings Pegasus XL-R	6905		9. 1.95	A.Armsby	Weston Zoyland	15. 2.05P
	(Rotax 462)						
G-MGPH	CFM Streak Shadow SA-M	PFA 206-13166	G-RSPH	27.11.97	R.S.Partridge-Hicks	(Bury St.Edmunds)	29. 7.00P
	(Built CFM Aircraft Ltd - kit no.K.286) (Rotax 582)		*(Force landed Cockfield, Suffolk 29.7.01: damage to propeller, nose u/c and port flap)*				
G-MGRH	Quad City Challenger II	CH2-1189-0482		20. 2.90	R.A and B.M.Roberts	Griffins Farm, Temple Bruer	16. 2.00P
	(Built R T Hall) (Hirth 2705.R06)						
G-MGTG	Pegasus Quantum 15-912	7369A	G-MZIO	19.12.97	R.B.Milton	Plaistowes Farm, St.Albans	19. 9.05P
	(Original c/n 7369 amended after rebuild 11.98)						
G-MGTR	Huntwing/Experience	BMAA/HB/067		24. 7.97	A C.Ryall	(Cardiff)	
	(Built A C Rydall) (Listed as "Huntwing Avon" in BMAA's records)				*(Current status unknown)*		
G-MGTV	Thruster T.600N 450 Sprint	0052-T600N-070		14. 3.02	R.I.Blain	Newtownards	3.10.05P
G-MGTW	CFM Shadow Series DD	K.287 & 287-DD		23. 1.98	G.T.Webster	Easter Poldar Farm, Thornhill	29.11.05P
G-MGUN	Cyclone AX2000	7284		18.12.96	I.Lonsdale	Tarn Farm, Cockerham	3. 5.04P
G-MGUY	CFM Shadow Series CD	078		23.11.87	F.J.Luckhurst and R.G.M.Proost	(Old Sarum)	16. 8.91P
	(Rotax 447)		*(Crashed Home Farm, Pontisbury, Shrewsbury 20.7.91: current status unknown)*				
G-MGWH	Thruster T.300	9013-T300-507		8.12.92	S.Bell and D.J.Flower	Hooton Hall, Hooton	21. 4.05P
	(Built Tempest Aviation Ltd) (Rotax 582)						
G-MGWI	Robinson R44 Astro	0663	G-BZEF	4. 5.00	N.Currie t/a Arran Heli-Tours	(Shiskine, Isle of Arran)	22. 5.06T
G-MHBD*	Cameron O-105 HAB				See SECTION 3 Part 1	(Aldershot)	
G-MHCB	Enstrom 280C Shark	1031	N892PT	11.10.95	Springbank Aviation Ltd	(Castletown, Isle of Man)	23. 7.06T

G-MHCD	Enstrom 280C-UK Shark	1112	(SX-...)	12. 7.96	Dayrise EPE	(Thessalonki, Greece)	25. 9.04T
			G-MHCD, G-SHGG, N627H				
G-MHCE	Enstrom F-28A	150	G-BBHD	22. 8.96	Wyke Commercial Services Ltd	Barton	3. 7.05T
G-MHCF	Enstrom 280C-UK Shark	1149	G-GSML	19. 9.96	K., H.K. and D.Collier	Barton	11. 8.07T
			G-BNNV, SE-HIY		t/a HKC Helicopter Services		
G-MHCG	Enstrom 280C-UK Shark	1155	G-HAYN	7. 3.97	E.Drinkwater	Barton	5. 9.03
			G-BPOX, N51776				
G-MHCI	Enstrom 280C Shark	1152	N100WZ	20. 5.97	B and B Helicopters Ltd	Barton	17. 9.03T
G-MHCJ	Enstrom F-28C-UK	453	G-CTRN	30. 3.98	P.E.Toleman t/a Paradise Helicopters	Hawarden	21. 6.04T
G-MHCK	Enstrom 280FX Shark	2006	G-BXXB	5. 6.98	Manchester Helicopter Centre Ltd	Barton	16. 9.04T
			ZK-HHN, JA7702				
G-MHCL	Enstrom 280C Shark	1144	N51740	30. 6.98	J A Newton *(New owner 3.04)*	(Knutsford)	24.11.01T
G-MHGS	Stoddard-Hamilton Glastar	PFA 295-13473		30. 7.03	M.Henderson	Aberdeen	
	(Built M.Henderson)						
G-MHRV	Van's RV-6A	PFA 181A-13422		28. 7.04	M.R.Harris	(Luton)	
	(Built M.R.Harris)						
G-MICH	Robinson R22 Beta	0647	G-BNKY	3. 9.87	Tiger Helicopters Ltd	Shobdon	10.10.05T
			(Tail rotor struck ground Shobdon 10.6.03, rolled onto side and substantially damaged)				
G-MICI	Cessna 182S Skylane	18280546	G-WARF	14. 6.01	DI Aviation LLP	Booker	25. 7.05T
			N7089F				
G-MICK	Reims/Cessna F172N Skyhawk II	F17201592	PH-JRA	9. 1.80	S.J.Gronow	Blackpool	23. 8.07
			PH-AXB		tr G-MICK Flying Group		
G-MICY	Everett Gyroplane Series 1	018	(G-BOVF)	26. 2.90	D.M.Hughes	St.Merryn	2. 5.92P
	(Volkswagen 1835)				*(Current status unknown)*		
G-MIDA	Airbus A321-231	806	D-AVZQ	31. 3.98	British Midland Airways Ltd	Nottingham East Midlands	30. 3.07T
G-MIDC	Airbus A321-231	835	D-AVZZ	12. 6.98	British Midland Airways Ltd	Nottingham East Midlands	11. 6.07T
G-MIDD	Piper PA-28-140 Cherokee Cruiser	28-7325444	G-BBDD	20. 1.97	R.B.Walker	Coventry	25. 5.07T
			N11C		t/a Midland Air Training School		
G-MIDE	Airbus A321-231	864	D-AVZB	14. 8.98	British Midland Airways Ltd	Nottingham East Midlands	13. 8.07T
G-MIDF	Airbus A321-231	810	D-AVZS	24. 4.98	British Midland Airways Ltd	Nottingham East Midlands	23. 4.07T
G-MIDG	Bushby-Long Midget Mustang	385	N11DE	14. 3.90	C.E.Bellhouse	Headcorn	9.12.05P
	(Built T L Owens) (Lycoming O-320)						
G-MIDH	Airbus A321-231	968	D-AVXZ	22. 3.99	British Midland Airways Ltd	Nottingham East Midlands	21. 1.05T
G-MIDI	Airbus A321-231	974	D-AVZA	26. 3.99	British Midland Airways Ltd	Nottingham East Midlands	25. 3.05T
G-MIDJ	Airbus A321-231	1045	D-AVZO	16. 7.99	British Midland Airways Ltd	Nottingham East Midlands	15. 7.05T
G-MIDK	Airbus A321-231	1153	D-AVZF	12. 1.00	British Midland Airways Ltd	Nottingham East Midlands	11. 1.06T
G-MIDL	Airbus A321-231	1174	D-AVZH	22. 2.00	British Midland Airways Ltd	Nottingham East Midlands	21. 2.06T
					(Star Alliance titles)		
G-MIDM	Airbus A321-231	1207	D-AVZR	18. 4.00	British Midland Airways Ltd	Nottingham East Midlands	17. 4.06T
G-MIDO	Airbus A321-232	1987	F-WWIR	29. 4.03	British Midland Airways Ltd	Nottingham East Midlands	28. 4.06T
G-MIDP	Airbus A320-232	1732	F-WWBK	24. 5.02	British Midland Airways Ltd	Nottingham East Midlands	23. 5.05T
G-MIDR	Airbus A320-232	1697	F-WWIQ	22. 4.02	British Midland Airways Ltd	Nottingham East Midlands	21. 4.05T
G-MIDS	Airbus A320-232	1424	F-WWBO	21. 3.01	British Midland Airways Ltd	Nottingham East Midlands	20. 3.07T
G-MIDT	Airbus A320-232	1418	F-WWBI	14. 3.01	British Midland Airways Ltd	Nottingham East Midlands	13. 3.07T
G-MIDU	Airbus A320-232	1407	F-WWDC	27. 2.01	British Midland Airways Ltd	Nottingham East Midlands	26. 2.07T
G-MIDV	Airbus A320-232	1383	F-WWIQ	30. 1.01	British Midland Airways Ltd	Nottingham East Midlands	29. 1.07T
G-MIDW	Airbus A320-232	1183	F-WWDT	29. 3.00	British Midland Airways Ltd	Nottingham East Midlands	28. 3.06T
					(Star Alliance titles)		
G-MIDX	Airbus A320-232	1177	F-WWDP	21. 3.00	British Midland Airways Ltd	Nottingham East Midlands	20. 3.06T
					(Star Alliance titles)		
G-MIDY	Airbus A320-232	1014	F-WWDQ	28. 6.99	British Midland Airways Ltd	Nottingham East Midlands	27. 6.05T
G-MIDZ	Airbus A320-232	934	F-WWII	19. 1.99	British Midland Airways Ltd	Nottingham East Midlands	18. 1.05T
G-MIFF	Robin DR400/180 Régent	2076		31. 5.91	J.C.Harvey	Spilsteads Farm, Sedlescombe	3. 8.06
					tr Westfield Flying Group		
G-MIGG	WSK PZL-Mielec Lim-5	1C1211	G-BWUF	17. 1.03	D.Miles	Bournemouth	AC
			Polish AF 1211		*(At Aviation Museum 1.05 in North Vietnamese c/s)*		
G-MIII	Extra EA.300/L	013	D-EXFI	5. 9.95	R.C.Berger	Bicester	27. 9.07E
G-MIKE	Gyroflight Brookland Hornet	MG.1		15. 5.78	M.H.J.Goldring	St.Merryn	25. 9.92P
	(Volkswagen 1830)				*(Current status unknown)*		
G-MIKG	Robinson R22 Mariner	3332M		5. 6.02	Sloane Helicopters Ltd	Sywell	26. 6.05T
G-MIKI	Rans S-6-ESA Coyote II	PFA 204-13094		28. 2.97	S.P.Slade	Wickwar	16. 6.05P
	(Built N R Beale - c/n 0996.1040) (Rotax 912-UL) *(Tri-cycle u/c)*						
G-MIKS	Robinson R44 Clipper II	10314		15. 4.04	Direct Timber Ltd	(Coalville)	6. 5.07T
G-MILA	Reims/Cessna F172N Skyhawk II	F17201686	D-EGHC (2)	9. 6.98	P.J.Miller	Cuckoo Tye Farm, Long Melford	16. 9.07A
			PH-AYJ				
G-MILE	Cameron Z-90 HAB	2411		26. 9.90	Miles Air Ltd	(Bristol)	14. 7.03A
	(Originally regd as Cameron N-77 and now N-90 - new envelope fitted 2004)				*(Miles Architectural Ironmongery Ltd titles)*		
G-MILI	Bell 206B-3 JetRanger III	2275	C-GGAR	5.10.94	Shropshire Aviation Ltd	(Telford)	1.10.04T
			5H-MPV				
G-MILN	Cessna 182Q Skylane	18265770	N735XQ	9. 7.99	Meon Hill Farms (Stockbridge) Ltd	Thruxton	22. 8.05T
G-MILY	Grumman American AA-5A Cheetah	AA5A-0672	G-BFXY	2. 9.96	Plane Talking Ltd	Elstree	17.10.05T
	(C/n plate shows maufacturer as Gulfstream American)						
G-MIMA	British Aerospace BAe 146 Series 200	E2079	G-CNMF	3. 3.93	Manx Airlines Ltd	Ronaldsway	25.11.07T
			G-5-079		*(Op Manx Airlines)*		
G-MIME	Europa Aviation Europa	PFA 247-12850		26. 9.97	N.W.Charles	Kemble	12.10.05P
	(Built N.W.Charles - kit no.203) (Monowheel u/c)						
G-MIND	Cessna 404 Titan II	404-0004	G-SKKC	27. 4.93	Atlantic Air Transport Ltd	Inverness	13. 2.06T
			G-OHUB, SE-GMX, (N3932C)		*(Environment Agency titles + logo on tail)*		
G-MINN	Lindstrand LBL 90A HAB	883		30.10.02	S.M.and D.Johnson	(Bromley)	14. 4.05A
G-MINS	Nicollier HN.700 Menestrel II	PFA 217-12354		23.10.92	R.Fenion	Kirkbride	12.10.03P
	(Built R.Fenion) (Volkswagen 1900)						
G-MINT	Pitts S-1S	PFA 09-10292		7. 2.83	T.G.Sanderson	Leicester	17. 8.05P
	(Built T.G.Sanderson)				*(Noted 5.04)*		

Regn	Type	C/n	Previous identity	Regd date	Owner/operator	Base	CofA expiry
G-MIOO	Miles M.100 Student 2	M1008	G-APLK G-MIOO, G-APLK, XS941, G-APLK, G-35-4	26.10.84	Aces High Ltd *(On rebuild as "G-APLK" 3.02: current status unknown)*	(Woking)	6. 5.86P
G-MISH	Cessna 182R Skylane II	18267888	G-RFAB G-BIXT, N6397H	16. 6.95	L.H.Robinson	Full Sutton	15. 4.06
G-MISS	Taylor JT.2 Titch *(Built A Brenen)*	PFA 3234		18.12.78	P L.Brenen *(Noted 10.02)*	RAF Halton	
G-MITE	Raj Hamsa X'Air Falcon Jabiru *(Built T Jestico - kit no 830)*	BMAA/HB/296		3. 3.04	T Jestico *(Noted 5.04)*	Popham	
G-MITT	Jabiru Aircraft Jabiru SK *(Built N.C.Mitton)*	PFA 274-13427		29. 2.00	K.J.Betteley	Sandown	28.10.05P
G-MIWS	Cessna 310R II	310R1585	G-ODNP N19TP, N2DD, N1836E	1. 2.96	Wilcott Sport and Construction Ltd	Welshpool	3.10.05P
G-MJAE	American Aerolights Eagle	1021 *(C/n not confirmed)*		12. 7.82	T.B.Woolley *(Current status unknown)*	(Leicester)	
G-MJAJ	Eurowing Goldwing (Fuji-Robin EC-44-PM)	EW-36		18. 6.82	R.D.J Brixton tr Canard Flyers Group *(Noted 4.04)*	Leebotwood	6. 8.03P
G-MJAL*	Wheeler Scout Mk.III/3/R *(Built Ron Wheeler Aircraft Sales Pty)*	0433 R/3		18. 6.82	G W Wickington *(Cancelled 22.8.00 by CAA) (Displayed Popham 5.04)*	(Hamble, Southampton)	
G-MJAM	Eipper Quicksilver MX (Cuyuna 430)	JCL-01		18. 6.82	J.C.Larkin *(Current status unknown)*	Maryport, Cumbria	20. 8.93P
G-MJAN	Hiway Skytrike I/Flexiform Hilander (Valmet)	RPFD-01 & 21U9		21. 6.82	G.M.Sutcliffe *(Current status unknown)*	(Stockport)	Exemption
G-MJAV	Hiway Skytrike II/Demon 175 (Fuji-Robin 250)	817003		23. 6.82	J.N.J.Roberts *(Current status unknown)*	Longacres Farm, Sandy	Exemption
G-MJAY	Eurowing Goldwing (Fuji-Robin EC-34-PM)	EW-58		23. 6.82	M.Anthony *(Current status unknown)*	(Alfreton)	Exemption
G-MJAZ	Aerodyne Vector 627SR Ultravector (Konig SC430) *(Originally regd as Vector 610 but converted 4.88 when PH-1J1)*	1251	PH-1J1 G-MJAZ	23. 6.82	B.Fussell *(Dismantled 11.04)*	Swansea	Exemption
G-MJBK	Swallow AeroPlane Swallow B (Rotax 447)	582007-2		18.11.83	M.A Newbould *(Current status unknown)*	(Harrogate)	Exemption
G-MJBL	American Aerolights Eagle 215B (Chrysler 820)	2892		25. 6.82	B.W.Olley	(Ely)	19. 9.04P
G-MJBS	Ultralight Aviation Systems Storm Buggy/Solar	JL814S		29. 6.82	G.I.Sargeant *(BMAA records as damaged in 1982: current status unknown)*	(Bridgwater)	
G-MJBT*	Eipper Quicksilver MX II (Cuyana 430R)	DJ/NBII & 3662		30. 6.82	Not known *(Cancelled 24.1.95 by CAA) (Noted 4.04)*	Letterkenny, County Donegal	Exemption
G-MJBV	American Aerolights Eagle 215B (Fuji-Robin EC-25-PS)	RSP-001		1. 7.82	B.H.Stephens *(Current status unknown)*	(Southampton)	11. 8.96P
G-MJBZ	Huntair Pathfinder Mk.1 (Fuji-Robin EC-34-PM)	PK-17		2. 7.82	J.C.Rose *(Noted 9.03)*	Eastbach Farm, Coleford	28.12.93P
G-MJCE	Ultrasports Puma/Southdown Sprint X (Fuji-Robin EC-44-PM) *(Designation amended by BMAA 1990)*	RGC-01		5. 7.82	L.I.Bateup	(Salisbury)	25. 8.01P
G-MJCF*	Hill (Maxair) Hummer (Fuji-Robin EC-25-PS)	SMC-01		5. 7.82	Not known *(Cancelled 24.1.95 by CAA) (Noted derelict 2.03)*	Doynton	Exemption
G-MJCU	Tarjani/Solar Wings Typhoon (Fuji-Robin EC-25-PS)	SCG-01 & T982-610		7. 7.82	J.K.Ewing *(Current status unknown)*	(Old Sarum)	1. 9.94P
G-MJDE	Huntair Pathfinder Mk.1 (Fuji-Robin EC-34-PM)	020		9. 7.82	P.Rayson	(Holsworthy, Devon)	16. 9.05P
G-MJDH*	Huntair Pathfinder Mk.1 (Fuji-Robin EC-44-PM)	015		9. 7.82	T.Mahmood *(Cancelled 10.6.02 as WFU) (Noted 5.04)*	Maryculter, Aberdeen	12. 8.01P
G-MJDJ	Hiway Skytrike/Demon	VW17D		9. 7.82	A J.Cowan *(Current status unknown)*	(Billingham)	
G-MJDP	Eurowing Goldwing (Fuji-Robin EC-34-PM)	GW-001		12. 7.82	B.L.Keeping tr G-MJDP Flying Group *(New owner 1.03)*	Davidstow Moor	15.11.92P
G-MJDR	Hiway Skytrike/Demon	PJB-01		14. 7.82	D.R.Redmile *(Current status unknown)*	(Leicester)	
G-MJDU	Eipper Quicksilver MXII (Rotax 503)	14002		15. 7.82	J.Brown *(New owner 5.00)*	Markfield, Leics	Exemption
G-MJDW	Eipper Quicksilver MXII (Cuyuna 430)	RI-01 *(C/n noted as 3506)*		15. 7.82	J.A Brumpton	Abbey Warren Farm, Bucknall, Lincoln	11. 4.05P
G-MJEB	Southdown Puma Sprint (Rotax 447)	SN1231/0041		18. 4.85	R.J.Shelswell *(Current status unknown)*	(Warwick)	1. 5.96P
G-MJEE	Mainair Tri-Flyer 250/Solar Wings Typhoon (Fuji-Robin EC-25-PS)	038-251181		20. 7.82	M.F.Eddington	(Wincanton)	11.11.00P
G-MJEO	American Aerolights Eagle 215B (Zenoah G25B1)	4562		26. 7.82	A M.Shaw *(Current status unknown)*	(Stoke-on-Trent)	Exemption
G-MJER	Ultrasports Tri-Pacer/Flexiform Solo Striker (Rotax 447)	DSD-01		23. 7.82	D.S.Simpson	Radwell, Letchworth	26.12.00P
G-MJEY*	Mainair Tri-Flyer 440/Southdown Lightning DS (Fuji-Robin EC-44-PM)	085-26782 & PMC-01		27. 7.82	M.McKenzie *(Cancelled 12.5.03 by CAA) (Trike only stored 5.03)*	Insch	7. 6.96P
G-MJFB	Ultrasports Tri-Pacer/Flexiform Solo Striker (Fuji-Robin EC-34-PM)	AJK-01		27. 7.82	B.Tetley	(Cowes)	27.10.04P
G-MJFM	Huntair Pathfinder Mk.1 (Fuji-Robin EC-34-PM)	ML-0		12. 9.82	R.Gillespie and S.P.Girr *(Current status unknown)*	Mullaghmore	23. 7.99P
G-MJFX	Skyhook TR1/Sabre (Hunting HS.525A)	TR1/38		2. 8.82	M.R.Dean *(Current status unknown)*	(Hebden Bridge)	Exemption
G-MJFZ	Hiway Skytrike/Demon	JAL-01		29. 7.82	A.W.Lowrie *(New owner 4.04)*	(West Rainton, Houghton Le Spring)	
G-MJHC	Ultrasports Tri-Pacer 330/Southdown Lightning Mk II (Fuji-Robin EC-34)	82-00044 *(C/n is engine serial no)*		9. 8.82	E.J.Allen *(Current status unknown)*	(Cambridge)	Exemption
G-MJHR	Mainair Dual Tri-Flyer/Southdown Lightning	GNS-01		12. 8.82	B.R.Barnes *(Current status unknown)*	(Bristol)	
G-MJHV	Hiway Skytrike II/Demon	AG-17		13. 8.82	A G.Griffiths *(Current status unknown)*	(Avenchurch, Birmingham)	

G-MJHX	Eipper Quicksilver MXII (Rotax 503)	1033		13. 8.82	P.D.Lucas *(Noted 12.04)*	Grove Farm, Needham	14. 5.95P
G-MJIA	Ultrasports Tri-Pacer/Flexiform Solo Striker (Rotax 377)	SE-007		13. 8.82	D.G.Ellis *(Noted 1.04)*	Otherton, Cannock	20. 9.96P
G-MJIC	Ultrasports Tri-Pacer/Flexiform Solo Striker (Fuji-Robin EC-34-PM)	82-00043		13. 8.82	J.Curran *(Current status unknown)*	(Newry, County Armagh)	15.10.94P
G-MJIF	Mainair Tri-Flyer/Flexiform Striker (Fuji-Robin EC-34-PL) *(C/n was original engine type)*	"E-1 EC25PS-04"		16. 8.82	R.J.Payne *(Current status unknown)*	(Newmarket)	Exemption
G-MJIR	Eipper Quicksilver MXII (Rotax 503)	1392		18. 8.82	H.Feeney *(Stored 8.96: current status unknown)*	Long Marston	26. 1.95P
G-MJIY	Ultrasports Puma/Southdown Sprint (Fuji-Robin EC-34-PM) *(Originally regd as Ultrasports Tri-Pacer/Flexiform Striker)*	X002 CSRS		23. 8.82	M.I.McClelland t/a McClelland Aviation	Old Sarum	10. 7.00P
G-MJJA	Huntair Pathfinder Mk.1	031		23. 8.82	R.D.Bateman and J.M.Watkins	Davidstow Moor	25. 8.02P
G-MJJK	Eipper Quicksilver MXII (Rotax 503)	3397		25. 8.82	Janet A.Brumpton *(New owner 8.04)*	Abbey Warren Farm, Bucknall, Lincoln	13. 8.05P
G-MJKB	Striplin Sky Ranger *(Officially quoted as c/n SRI-6-I)*	ST 161		2. 9.82	A P.Booth *(Current status unknown)*	(Newbury)	
G-MJKF	Hiway Demon	WGR-01		2. 9.82	S.D.Hill *(Current status unknown)*	(Henley-on-Thames)	
G-MJKO	Hiway Skytrike/Gold Marque Gyr 188 (Fuji-Robin EC-25-PS) *(Assembled from spares by Windsports)*	90030P		7. 9.82	M.J.Barry *(Current status unknown)*	(Bridgwater)	Exemption
G-MJKP*	Hiway Skystrike/Super Scorpion				See SECTION 3 Part 1	Doncaster	
G-MJKX	Skyrider Airsports Phantom (Fuji-Robin EC-50)	PH.82005		14. 9.82	A P.Love	Long Marston	19. 6.04P
G-MJMB*	Weedhopper JC-24 (Chotia 460)	846		23. 9.82	M.J.Aubrey *(Cancelled 7.9.94 by CAA) (Noted 2002)*	(Kington, Hereford)	
G-MJMD	Hiway Skytrike II/Demon 175 (Fuji-Robin EC-34-PM)	OE17D		27. 9.82	T.A N.Brierley	Baxby Manor, Husthwaite	14. 6.04P
G-MJMN	Mainair Tri-Flyer/Flexiform Striker (Fuji-Robin EC-34-PM)	087-04882		29. 9.82	K.Medd	(Manchester)	22. 7.05P
G-MJMR	Mainair Tri-Flyer 250/Solar Wings Typhoon	DR-01 & 048-5182		30. 9.82	J.C.S.Jones *(Stored 12.97: current status unknown)*	Emlyn's Field, Rhuallt	
G-MJMS	Hiway Skytrike II/Demon 175	EEW-01		30. 9.82	D.E.Peace *(Current status unknown)*	(Rawdon, Leeds)	
G-MJMU	Hiway Skytrike II/Demon (Fuji-Robin EC-25-PS) *(C/n duplicates several a/c incl G-MJOI and PH-1B2 and is suspect!) (Current status unknown)*	817003		1.10.82	P.Hunt	(Bishop Auckland)	
G-MJNK	Hiway Skytrike II/Demon 175 (Fuji-Robin EC-34-PM)	EA17D		14.10.82	S.W.Barker *(Noted 8.04)*	Baxby Manor, Husthwaite	28.10.96P
G-MJNM	American Aerolights Eagle 430B (Cuyuna 430R)	702		25.11.82	B.H.Stephens *(Current status unknown)*	(Southampton)	19. 9.93P
G-MJNO	American Aerolights Eagle Amphibian (Rotax 447)	703		24.11.82	R.S.Martin	(Gosport)	23. 6.05P
G-MJNU	Skyhook TR1/Cutlass	TR1/17		19.10.82	R.W.Taylor *(Current status unknown)*	(Sheffield)	
G-MJNY	Skyhook TR1/Sabre	TR1/35		3.11.82	P.Ratcliffe *(Current status unknown)*	(Sheffield)	
G-MJOC	Huntair Pathfinder (Fuji-Robin EC-34-PM)	048		25.10.82	A J.Glynn *(Current status unknown)*	Gerpins Lane, Upminster	31. 7.99P
G-MJOE	Eurowing Goldwing (Rotax 377)	EW-55		29.10.82	R.J.Osborne *(Current status unknown)*	(Tiverton)	Exemption
G-MJOI*	Ultrasports Tri-Pacer/Hiway Demon	817003		1.11.82	Not known *(Cancelled 1.11.89 as WFU) (Noted 6.03)*	Biggin Hill	
G-MJPA	Rotec Rally 2B	AT-01		5. 1.83	R.Boyd *(Current status unknown)*	(Armagh, County Armagh)	
G-MJPB*	Manuel Ladybird				See SECTION 3 Part 1	Brooklands	
G-MJPE	Mainair Tri-Flyer 330/Demon 175 (Fuji-Robin EC-34-PM)	117-151282 & OG17D		10.11.82	E.G.Astin *(Current status unknown)*	(Whitby)	7. 8.96P
G-MJPV	Eipper Quicksilver MX (Cuyuna 430R)	JBW-01		30.11.82	F.W.Ellis	Water Leisure Park, Skegness	17. 8.04P
G-MJRA*	Mainair Tri-Flyer 250/Hiway Demon				See SECTION 3 Part 1	Elvington	
G-MJRL	Eurowing Goldwing (Rotax 377)	EW-79 & SWA-5K		30.12.82	M.Daniels	(Heanor)	15. 6.00P
G-MJRO	Eurowing Goldwing (Rotax 447)	EW-77 & SWA-04		31.12.82	H.P.Welch *(Current status unknown)*	(Taunton)	22. 9.99P
G-MJRR	Reece SkyRanger Series 1	JR-3		26. 4.82	J.R.Reece *(Current status unknown)*	(Formby)	
G-MJRS	Eurowing Goldwing (Rotax 377)	EW-80 & SWA-6K		5. 1.83	R M Newlands *(New owner 3.04)*	(East Cowes)	12.10.01P
G-MJRU	MBA Tiger Cub 440	SO.86		6. 1.83	D.J.Short *(Current status unknown)*	(Nailsea, Bristol)	Exemption
G-MJSE	Skyrider Airsports Phantom (Fuji-Robin EC-40-PL)	SF-101		24. 1.83	K H A Negal *(Noted 5.04)*	Red House Farm, Preston Capes	20. 5.02P
G-MJSF	Skyrider Airsports Phantom (Rotax 462)	SF-105	SE-... G-MJSF	24. 1.83	B.J.Towers *(On rebuild 5.00)*	(Pershore)	
G-MJSL	Dragon Light Aircraft Dragon 200 (Rotax 503)	0018		24. 2.83	G.Kingston *(New owner 1.02)*	Long Marston	22. 9.99P
G-MJSO	Hiway Skytrike III/Demon 175 (Hiro 22)	SA17D		1. 2.83	D.C.Read *(Current status unknown)*	(Ledbury)	Exemption
G-MJSP	Romain MBA Super Tiger Cub Special 440 *(Tri-cycle u/c)*	S0.54		7. 2.83	A R.Sunley *(On rebuild 3.03: new owner 4.03)*	(Chelmsford)	Exemption
G-MJST	MEA Pterodactyl Ptraveler (Fuji-Robin EC-34-PM)	GCS-01		2.12.81	C.H.J.Goodwin *(Current status unknown)*	(Bedford)	7. 5.99P
G-MJSU*	MBA Tiger Cub 440				See SECTION 3 Part 1	Flixton, Bungay	
G-MJSV*	MBA Tiger Cub 440 *(Officially regd with c/n SO.287)*	SO.87/2		2. 2.83	Not known *(Cancelled 9.11.89 by CAA) (Stored 2001)*	(Kinloss)	
G-MJSY	Eurowing Goldwing (Rotax 377)	EW-63		8. 2.83	A J.Rex	(Wrexham)	5. 1.01P
G-MJSZ	Harker DH Wasp *(Built D Harker)* (Rotax 447)	HA.5		10. 2.83	J.J.Hill *(On overhaul Stokesley 2004)*	Baxby Manor, Husthwaite	24. 3.01P
G-MJTC	Ultrasports Tri-Pacer/Typhoon Medium	T1282-677		14. 2.83	V.C.Redhead *(Current status unknown)*	(Saxmundham)	

Reg	Type	C/n	Prev. Id	Regd	Owner/Operator	Location	CofA
G-MJTD	Gardner T-M Scout	83/001		14. 2.83	D.Gardner	(Rugby)	AC
	(Built D.Gardner) (Thomas-Morse S4 Scout 2/3rd rep) (Possibly c/n PFA 111-10664) (As "41386" in US Army Signal Corps c/s)						
G-MJTE	Skyrider Airsports Phantom	SF-106		15. 2.83	L.Zivanovic	Droitwich	27. 4.05P
	(Fuji-Robin EC-44-PM)						
G-MJTM	Southdown Aerostructure Pipistrelle P2B			21. 2.83	M.A Collins and G.J.Jones	(St. Neots)	27. 4.05P
	(KFM-107ER)	019 & SAL/P2B/002					
G-MJTP	Mainair Tri-Flyer/Flexiform Dual Sealander			25. 2.83	P.Milton	(Bedford)	22. 8.00P
	(Fuji-Robin EC-44-PM)	AJDH-01 & 139-7383			*(Possibly fitted with Dual Striker wing after accident 29.10.87)*		
G-MJTR	Southdown Puma DS Mk.1	H362		9. 3.83	A G.Rodenburg and T.Abro *(Noted wrecked 4.04)*		
	(Fuji-Robin EC-44-PM)					Hillfoots Nurseries , Tillicoultry, Stirling	15. 7.96P
G-MJTX	Skyrider Aviation Phantom	SF-110		1. 3.83	P.D.Coppin	Lee-on-Solent	22. 4.96P
	(Fuji-Robin EC-44-PM)				*(Noted 9.04)*		
G-MJTZ	Skyrider Aviation Phantom	MBS-01		29. 4.83	B.J.Towers	(Pershore)	Exemption
	(Fuji-Robin EC-44-PM) *(Engine No.82-00119)*				*(Current status unknown)*		
G--MJUC	MBA Tiger Cub 440	RRH-01 & PFA/140-10908		7. 3.83	P A Avery *(New CofR 3.04)*	Shipdham	Exemption
G-MJUF*	MBA Super Tiger Cub 440	MCT-01		8. 3.83	D G Palmer	Fetterangus	
					(Cancelled 27.4.90 by CAA) (Stored 7.02)		
G-MJUO*	Eipper Quicksilver MXII	104C		22. 3.83	A Hamilton	Strathaven	
					(Cancelled 24.1.95 by CAA: noted 8.00)		
G-MJUR*	Skyrider Aviation Phantom	SF-108		5. 4.83	A L Lewis	Droppingwell Farm, Bewdley	20. 9.98P
	(Fuji-Robin EC-44-PM)				*(Cancelled 31.5.01 by CAA) (Noted 2.03)*		
G-MJUU	Eurowing Goldwing	EW-70		28. 3.83	E.F.Clapham	(Oldbury-on-Severn)	3. 5.97P
	(Fuji-Robin EC-34-PM)				*(Current status unknown)*		
G-MJUV	Huntair Pathfinder Mk.1	045		18. 5.83	S.J.Overton	(Colchester)	31. 7.99P
	(Fuji-Robin EC-44-PM)						
G-MJUW	MBA Tiger Cub 440	SO.69		29. 3.83	D.G.Palmer *(Noted 4.04)*	Longside, Peterhead	7. 6.02P
G-MJUX	Skyrider Aviation Phantom	RFF-01 & PH00094		29. 2.84	K.H.A Negal	Sittles Farm, Alrewas	10. 3.02P
	(Fuji-Robin EC-44-PM)						
G-MJVE	Medway Hybred 44XL/Solar Wings Typhoon XLII			19. 4.83	T.A Clark	(Rheda-Wiedenbrueck, Germany)	5. 6.00P
		4483/1 & T483-761XL	*(Original wing c/n T283-703XL)*				
G-MJVF	CFM Shadow Series CD	002		12. 4.83	J.A Cook	(Thorpeness)	24. 6.04P
G-MJVN	Ultrasports Tri-Pacer/Flexiform Striker	82-00030-PR1		18. 4.83	R.McGookin	(West Kilbride)	5.10.93P
	(Fuji-Robin EC-44-PM) *(Original Trike and engine fitted in G-MJRP)*				*(Current status unknown)*		
G-MJVP	Eipper Quicksilver MXII	1149		19. 4.83	G.J.Ward	(Dorchester)	10. 7.96P
	(Rotax 503) *(Original c/n 1124 became G-MTDO?)*				*(Current status unknown)*		
G-MJVR*	Ultrasports Panther/Flexiform Dual Striker	LAH		20. 4.83	D J Evans	Longacres Farm, Sandy	11. 6.93P
	(Fuji-Robin EC-44-PM) *(Originally regd as a Tripacer)*				*(Cancelled 28.11.95 by CAA) (Trike noted 7.03)*		
G-MJVU	Eipper Quicksilver MXII	1118		3. 4.84	F.J.Griffith	(Denbigh)	5. 7.05P
	(Rotax 503)						
G-MJVX	Skyrider Aviation Phantom	JAG-01 & SF-102		27. 4.83	J.R.Harris	Droppingwell Farm, Bewdley	1. 9.05P
	(Fuji-Robin EC-44-PM)						
G-MJVY	Dragon Light Aircraft Dragon 150	D.150/013		4. 5.83	J.C.Craddock	(Freshwater, Isle of Wight)	26.10.04P
	(Rotax 503)						
G-MJWB	Eurowing Goldwing	EW-59		24. 5.83	D.G.Palmer	Fetterangus	25. 8.93P
	(Fuji-Robin EC-34-PM)				*(Noted 7.01)*		
G-MJWF	MBA Tiger Cub 440	BRH-001 & SO.79		4. 5.83	T and R L Maycock *(New owners 6.02)*	(Glasgow)	
G-MJWH*	Chargus Vortex 120				See SECTION 3 Part 1	Coventry	
G-MJWJ	MBA Tiger Cub 440	013/191		9. 5.83	J.W.Barratt *(Current status unknown)*	(Langport)	18. 3.96P
G-MJWK	Huntair Pathfinder 1	JWK-01		1.10.82	D.Young	Kemble	3. 7.05P
	(Rotax 447)				tr Kemble Flying Club		
G-MJWS*	Eurowing Goldwing				See SECTION 3 Part 1	Langford Lodge, Belfast	
G-MJWW	MBA Super Tiger Cub 440	MU-001		11. 5.83	J.J.Littler and T.J.Gayton-Polley	(Chichester)	23. 5.98P
					(New CofR 11.03)		
G-MJWZ	Solar Wings Panther XL-S	T583-781XL		9. 9.85	A L.Davies	(Holywell)	27. 1.01P
	(Originally regd as XL)				*(New owner 10.02)*		
G-MJXD*	MBA Tiger Cub 440	011/061		16. 5.83	Not known	Halwell, Totnes	
					(Cancelled 3.4.02 as wfu) (Noted engineless on trailer 7.03)		
G-MJXE*	Mainair Tri-Flyer 330/Hiway Demon 175				See SECTION 3 Part 1	Manchester	
G-MJXF*	MBA Tiger Cub 440	EJH-01		1. 6.83	Not known	Jackrells Farm, Southwater	
					(Cancelled 5.9.94 by CAA) (Stored 8.03)		
G-MJXY	Hiway Skytrike 330/Demon 175	KQ17D		31. 5.83	H.C.Lowther	(Penrith)	25. 7.00P
	(Fuji-Robin EC-34-PM)						
G-MJYD	MBA Tiger Cub 440	SO.179		1. 6.83	R.A Budd *(New CofR 6.02)*	(Ashbourne)	30. 7.92P
G-MJYP	Mainair Gemini/Flexiform Dual Striker	167-13683		7. 6.83	M.S.Whitehouse	(Solihull)	23. 7.02P
	(Fuji-Robin EC-44-PM)						
G-MJYV	Mainair Rapier 1+1/Flexiform Solo Striker			23.11.83	L.H.Phillips	(Solihull)	6.12.04P
	(Fuji-Robin EC-34-PM)	175-19783					
G-MJYW	Lancashire Micro-Trike Dual 330/Wasp Gryphon III			28. 6.83	P.D.Lawrence	(Munlochy, Ross-shire)	
		2/330PM/PGK/6.83/K			*(Dismantled and Trike used on G-MMPL: parts noted 7.01)*		
G-MJYX	Mainair Tri-Flyer/Hiway Demon	108-251182		9. 6.83	K.A Wright	North Coates	7. 9.05P
	(Fuji-Robin EC-33-PM)						
G-MJZD	Mainair Gemini/Flash	311-585-3 & W50		18. 4.85	A R.Gaivoto	Popham	14. 8.03P
	(Fuji-Robin EC-44-PM)						
G-MJZE	MBA Tiger Cub 440	SO.168		14. 6.83	J.E.D.Rogerson	Morgansfield, Fishburn	Exemption
					tr Fishburn Flying Tigers *(Current status unknown)*		
G-MJZK (2)	Southdown Puma Sprint	SN1111/0081		3. 3.86	R.J.Osborne	(Tiverton)	18.10.91P
	(Fuji-Robin EC-44-PM)				*(Current status unknown)*		
G-MJZL	Eipper Quicksilver MXII	EEW-01		15. 6.83	P.D.Lucas	Grove Farm, Needham	6. 9.03P
	(Rotax 503)				*(Noted 11.04)*		
G-MJZU	Mainair Gemini/Flexiform Dual Striker			21. 6.83	C.G.Chambers	Swinford, Rugby	7. 8.04P
	(Fuji-Robin EC-44-PM)	214-41183 & JDR-02	*(Trike fitted ex G-MMVX (1))*		*(New owner 11.04)*		
G-MKAK	Colt 77A HAB	2039		15. 8.91	M.Kendrick	(Bridgnorth)	18. 7.03A
G-MKAS	Piper PA-28-140 Cherokee Cruiser	28-7425338	G-BKVR	30. 4.98	MK Aero Support Ltd	Andrewsfield	16.12.07E
			OY-BGV				

Reg	Type	C/n	Previous identity	Date	Owner	Location	CofA
G-MKIA	Vickers Supermarine 300 Spitfire I	6S-30565	P9374	16.11.00	S.J.Marsh	(Castelcucco, Italy)	
G-MKIV*	Bristol 149 Bolingbroke IVT	-	(G-BLHM)	26. 3.82	G.A Warner *(As "V6028/GB-D" in 105 Sqdn c/s)*	Duxford	28. 5.88P
			RCAF 10038		*(Crashed Denham 21.6.87: cancelled 1.11.88 as destroyed) (For spares 1.02)*		
G-MKSS	British Aerospace HS 125 Series 700B	257175	VP-BEK	29. 3.01	Markoss Aviation Ltd	Biggin Hill	4. 4.05T
			VP-CEK, N770TJ, C9-TAC, (C9-TTA)				
G-MKVB	Vickers Supermarine 349 Spitfire LF.Vb	CBAF.2461	5718M	2. 5.89	Historic Aircraft Collection Ltd	Duxford	25. 3.05P
			BM597		*(As "BM597/JH-C" in 317 Sqdn c/s)*		
G-MKVI	de Havilland DH.100 Vampire FB.6	676	Swiss AF J-1167	2. 6.92	T.C.Topen	Hemswell Cliff	14. 9.95P
	(Built F + W)				*(Noted 7.04 as "J-1167")*		
G-MKXI	Vickers Supermarine 365 Spitfire PR.Mk.XI	6S-504719	N965RF	13.11.89	P.A.Teichman	North Weald	29. 6.05P
			G-MKXI, R.Netherlands AF, PL965 *(As "PL965/R")*				
G-MLFF	Piper PA-23-250 Aztec E	27-7305194	G-WEBB	31. 1.90	W.C.Cullinane	Waterford, County Waterford	9. 1.06
			G-BJBU, N40476				
G-MLGL	Colt 21A Cloudhopper HAB	527		3. 4.84	H.C.J.Williams	(Langford, Bristol)	
G-MLHI	Maule MX-7-180 Star Rocket	11073C	G-BTMJ	20. 4.04	T.J.Westcott tr Maulehigh	White Waltham	4. 9.07
G-MLJL	Airbus A330-243	254	F-WWKT	15. 6.99	MyTravel Airways Ltd	Manchester	14. 6.05T
G-MLSN	MDH Hughes 369E	0357E	G-HMAC	12.12.03	Molson Holdings Ltd	Gloucestershire	29. 5.05T
			HB-XUO				
G-MLTY	Aérospatiale AS365N2 Dauphin	6431	N365EL	4. 6.99	Multiflight Ltd	Leeds-Bradford	6. 6.06
			JA6673				
G-MLWI	Thunder Ax7-77 HAB	1000		3. 9.86	M.L.and L.P.Willoughby	(Woodcote, Reading)	12. 8.03A
					"Mr Blue Sky"		
G-MMAC	Dragon Light Aircraft Dragon Series 200	003	OY-...	14. 7.82	J.F.Ashton and J.Kirwan	(Liverpool)	Exemption
	(Fuji-Robin EC-44-PM)		G-MMAC		*(Current status unknown)*		
G-MMAE	Dragon Light Aircraft Dragon Series 200	005		7. 9.82	P.J.Sheehy and K.S.Matcham	Barton Ashes	1. 5.05P
	(Fuji-Robin EC-44-PM)				*(Noted 3.02)*		
G-MMAG	MBA Tiger Cub 440	SO.47		22. 6.83	M.J.Aubrey *(Noted 2002)*	(Kington, Hereford)	14. 9.93P
G-MMAH*	Eipper Quicksilver MXII	TM.1016		23. 6.83	Not known	Doynton	
	(Engine No.14805)				*(Cancelled 26.10.95 by CAA) (Derelict 2.03)*		
G-MMAI	Dragon Light Aircraft Dragon Series 150	0032		1. 7.83	G.S.Richardson	(Cleethorpes)	13. 7.97P
	(Fuji-Robin EC-44-PM)		*(Dismantled and parts split between North Coates and owner's home) (New CofR 6.01)*				
G-MMAO*	Southdown Puma Sprint X	HS.549		28.12.83	P A Kershaw	Ince Blundell	14. 3.00P
					(Cancelled 31.5.01 by CAA) (Stored 6.03)		
G-MMAR	Mainair Gemini/Puma Sprint MS	195-11083-2		23. 9.83	A R. and J.Fawkes	(Newbury)	17. 9.98P
	(Fuji-Robin EC-44-PM)				*(Current status unknown)*		
G-MMAX*	Garland Trike/Flexiform Dual Striker	0011		5. 8.93	M T Wells	(Newcastle-under-Lyme, Stafford)	18. 2.02P
	(Fuji-Robin EC-44-PM)				*(Cancelled 4.5.04 by CAA)*		
G-MMAZ	Southdown Puma Sprint X	MAPB-01		5. 8.83	A R.Smith	(Chelmsford)	22. 7.96P
	(Fuji-Robin EC-44-PM)				*(Current status unknown)*		
G-MMBE*	MBA Tiger Cub 440	SO.74		30. 6.83	Not known	Wick	
					(Cancelled 4.2.92 by CAA) (Noted 4.04)		
G-MMBL	Ultrasports Puma/Southdown Lightning DS	80-00083		4. 7.83	B.J.Farrell	(Preston)	Exemption
	(Fuji-Robin EC-44-PM) *(C/n is engine serial no.)*				*(Current status unknown)*		
G-MMBN	Eurowing Goldwing	EW-89		28. 6.83	E.H.Jenkins	(Newcastle upon Tyne)	Exemption
	(Rotax 447)				*(Current status unknown)*		
G-MMBT	MBA Tiger Cub 440	SO.131 & TA.01		19. 7.83	B.Chamberlain	(Otley, Ipswich)	Exemption
	(Probably either c/n PFA 140-10924 or 10990)				*(Stored 1.91: current status unknown)*		
G-MMBU	Eipper Quicksilver MXII	CAL-222		8. 7.83	D.A Norwood	Ashcroft Farm, Winsford	11. 6.05P
	(Rotax 503)						
G-MMBV	Huntair Pathfinder	044		8. 7.83	P.J.Bishop	Tarn Farm, Cockerham	1. 1.05P
	(Fuji-Robin EC-44-PM) *(New sailwing 1999)*						
G-MMBY	Solar Wings Panther XL	T483-759XL		20. 7.83	R.M.Sheppard and P.Huddleston	(Wantage/Marlborough)	3. 8.03P
G-MMBZ	Solar Wings Typhoon P	T981-5217		20. 7.83	S.C.Mann	(Kirbymoorside)	28. 4.96P
	(Fuji-Robin EC-34-PM)				*(New owner 10.00)*		
	(Originally thougth to have sailwing c/n T781-217 - 5217: possible corruption of S217 for Typhoon Small: rebuilt as c/n T981-228)						
G-MMCB*	Huntair Pathfinder II				See SECTION 3 Part 1	Wroughton	
G-MMCI	Ultrasports Puma Sprint X	DMP-01 & P.421		28. 9.83	R.J.Webb	Long Marston	24. 6.03P
	(Fuji-Robin EC-44-PM)						
G-MMCN	Hiway Skytrike 250/Solar Wings Storm	SMB.8069		19. 7.83	P.J.Ramsay	Farnham	Exemption
G-MMCV	Hiway Skytrike II/Solar Wings Typhoon	T583-783		27. 7.83	G.Addison	(Kinross)	8. 6.97P
	(Fuji-Robin EC-34-PM)				*(Current status unknown)*		
G-MMCX	MBA Super Tiger Cub 440	MU.002		8. 8.83	D.Harkin	(Johnstone, Renfrew)	
					(Current status unknown)		
G-MMCZ	Mainair Tri-Flyer/Flexiform Dual Striker	TE-01		10. 8.83	T.D.Adamson	Wombleton	5. 8.04P
	(Fuji-Robin EC-44-PM) *(Mainair Trike c/n 180-6883)*						
G-MMDE	Mainair Tri-Flyer 250/Solar Wings Typhoon S	DES-1 & 025-211081-6		12. 8.83	D.J.Moore	(Oakington)	11. 6.01P
G-MMDF	Southdown Wild Cat Mk.II/Lightning Phase II	007		24. 8.83	J.C.Haigh	(Tonbridge)	4.11.03P
	(Fuji-Robin EC-34-PM)						
G-MMDK	Mainair Merlin/Striker	181-16883		7. 9.83	P.E.Blyth	(Rotherham)	30. 5.99P
	(Fuji-Robin EC-34-PM)				*(Current status unknown)*		
G-MMDN	Mainair Tri-Flyer 330/Flexiform Dual Striker	197-983 & RPO.12	*(Mainair c/n not confirmed)*	30. 9.83	M.G.Griffiths	(Monmouth)	Exemption
					(Current status unknown)		
G-MMDP	Mainair Gemini/Sprint X	183-22883		20. 9.83	G.V. Cowle	(Port St.Mary, Isle of Man)	25. 1.95P
	(Fuji-Robin EC-44-PM)				*(New owner 6.02)*		
G-MMDR	Huntair Pathfinder Mk.II	137		30. 8.83	C.Dolling	(United Arab Emirates)	
	(Rotax 377)				*(Current status unknown)*		
G-MMDY	Ultrasports Panther Sprint I	S.064		7. 9.83	C.Duffin	(Portlaoise, County Laois)	Exemption
	(Fuji-Robin EC-44-PM)				*(Current status unknown)*		
G-MMEJ	Mainair Tri-Flyer/Flexiform Striker	215-41183 & FF/LAI/83/JDR/03		15. 9.83	R.B.Tweedie	(Stoke-on-Trent)	9. 11.97P
					(Current status unknown)		
G-MMEK	Medway Hybred 44XL/Solar Wings Typhoon XL2	12983/6		16. 9.83	M.G.J.Bridges	(Exeter)	28. 8.00P
	(Typhoon sailwing c/n either T883-884XL or '887XL - both originally supplied to Medway for G-MMEK and G-MMEN)						

Regn	Type	C/n	Regd	Owner	Location	CofA
G-MMFD	Mainair Tri-Flyer 440/Flexiform Dual Striker		20. 9.83	M.E.and W.L.Chapman	(Oldham)	6.12.93P
		210-31082-2 & FF/LAI/83/JDR/12	*(Trike unit is believed to be c/n 210-31083-2)*			
G-MMFE	Mainair Tri-Flyer/Flexiform Striker		20. 9.83	W.Camm	(Barnsley)	16. 6.94P
	(Fuji-Robin EC-44-PM)	FF/LAI/83/JDR/13		*(Current status unknown)*		
	(Trike unit replaced by c/n 256-784-2 and probably now podded to 440 Gemini standard)					
G-MMFG	Lancashire Micro-Trike/Flexiform Dual Striker		20. 9.83	M.G.Dean and M.J.Hadland	Tarn Farm, Cockerham	Exemption
	(Fuji-Robin EC-44-PM)	FF/LAI/83/JDR/15		*(Current status unknown)*		
G-MMFS	MBA Tiger Cub 440	SO.64	1.11.83	G.S.Taylor	Otherton, Cannock	27. 7.01P
				"Black Adder" (Noted 8.04)		
G-MMFV	Mainair Tri-Flyer 440/Flexiform Dual Striker		8.12.83	R.A Walton	(Slough)	26. 4.97P
	(Fuji-Robin EC-44-PM)	83-00130 & 212-271083		*(New CofR 7.03)*		
G-MMFY	Cliff Sims Aztec trike/Dual Striker	AZT001CS	14.12.83	K.R.M.Adair and S.R.Browne	(Bognor/Pulborough)	Exemption
				(Amended CofR 1.05)		
G-MMFZ*	Striplin (AES) Sky Ranger	HAW-01	18.11.83	M.J.Aubrey	(Kington, Hereford)	
	(Cuyana 430)			*(Cancelled 3.10.01.by CAA) (Noted 2002)*		
G-MMGF	MBA Tiger Cub 440	SO.124	18.11.83	J.G.Boxall	Pittrichie Farm, Whiterashes	22. 8.02P
G-MMGL	MBA Tiger Cub 440	SO.148 & BMAA/HB/050	23.11.83	H.E.Dunning	Baxby Manor, Husthwaite	20. 4.05P
	(Built H.E.Dunning)					
G-MMGS	Solar Wings Panther XL	T1283-939XL	28.12.83	R.J.Hood *(New owner 11.04)*	London Colney	12. 8.98P
G-MMGT	Huntwing/Pegasus Classic	JAH-7	28.11.83	H.Cook	(Newport, Gwent)	25. 8.05P
	(Built J A Hunt) (BMW R100) *(Currently with Trike c/n SW-TB-1228 ex G-MTOH)*					
G-MMGU	SMD Gazelle/Flexiform Sealander	30-4883	1.12.83	A D.Cranfield	(Wincanton)	Exemption
	(Fuji-Robin EC-44-PM)			*(Current status unknown)*		
G-MMGV	Whittaker MW5 Sorcerer Series A	001	2.12.83	G.N.Haffey and M.W.J.Whittaker	Askern	1. 9.05P
	(Built Microknight Aviation Ltd) (Fuji-Robin EC-34-M)					
G-MMHG*	Hiway Skytrike 250/Solar Wings Storm	DRB-01 1	3.12.83	M.J.Aubrey	(Kington, Hereford)	
	(Fuji-Robin EC-25-PM)			*(Cancelled 22.9.93 by CAA) (Noted 2002)*		
G-MMHK	Hiway Skytrike/Super Scorpion	KSC83	19.12.83	S.Davison	(Newcastle upon Tyne)	Exemption
	(Fuji-Robin EC-25-PS)			*(New CofR 3.03)*		
G-MMHL	Hiway Skytrike II/Super Scorpion	KSC84	19.12.83	E.J.Blyth	(Pickering)	Exemption
	(Fuji-Robin EC-44)			*(Current status unknown)*		
G-MMHN	MBA Tiger Cub 440	SO.136	19.12.83	M.J.Aubrey *(Noted 2002)*	(Kington, Hereford)	
G-MMHS	SMD Gazelle/Flexiform Dual Striker	104-11283	21.12.83	C.J.Meadows	(Shepton Mallet)	
				(Current status unknown)		
G-MMIE	MBA Tiger Cub 440	G7-7	3. 1.84	B.W.Olliver *(Current status unknown)*	(Telford)	Exemption
G-MMIH	MBA Tiger Cub 440	SO.130	25. 4.84	R.A Davis *(Current status unknown)*	(Gloucester)	19. 8.93P
G-MMIR	Mainair Gemini/Puma Sprint	051-20182	25. 1.84	J.P.Wilson	Long Marston	15. 8.97P
	(Fuji-Robin EC-44-PM)			*(Stored 10.00)*		
	(Regd with original Trike c/n ex G-MBKX then G-MJDO: rebuilt with Trike 314-585-3 ex G-MMZK: wing ex G-MMTI)					
G-MMIW	Southdown Puma Sprint	590	9. 2.84	J.Ryland	(Swanley)	26. 9.04P
	(Fuji-Robin EC-44-PM)					
G-MMIX	MBA Tiger Cub 440	MBCB-01	14. 2.84	N.J.McKain *(To Dumfries & Galloway Museum)*	(Dumfries)	Exemption
G-MMIZ	Southdown Lightning Mk.II	CB-01	24. 2.84	R.G.Earp	(Peterborough)	25. 4.05P
G-MMJD	Southdown Puma Sprint	SP/1001	28. 6.83	C.A Sargent	Wickhambrook, Newmarket	22. 4.05P
	(Fuji-Robin EC-44-PM)					
G-MMJF	Solar Wings Panther Dual XL-S		27. 2.84	K.J.Hoare	Sywell	13. 9.04P
		PXL842-150 & T284-988XL		*(Noted 1.05)*		
G-MMJG	Mainair Tri-Flyer/Flexiform Dual Striker	185-1983	31. 9.83	A Strang	(Ashgill, Larkhall)	10. 9.03P
	(Fuji-Robin EC-44-PM)			*(Under repair 2004)*		
G-MMJM	Southdown Puma Sprint 440	PD.500 & SN1111/001	27. 2.84	R.J.Sanger *(New owner 7.01)*	(Wickford)	31. 5.97P
G-MMJT	Mainair Gemini/Sprint X	JBT-01	20.12.83	W.F.Murray	Swinford, Rugby	1. 9.04P
	(Fuji-Robin EC-44-PM) *(No Mainair identity and probably plans-built by J B Tate)*					
G-MMJV	MBA Tiger Cub 440	SO.195 & PFA 140-1090	25. 3.84	D.G.Palmer	Fetterangus	9. 5.93P
	(Built K Bannister)			*(Noted 7.01)*		
G-MMJX	Teman Mono-Fly	01	6. 3.84	M.Ingleton	Cripps Barn, Eastchurch	17. 7.04P
	(Built B F J Hope) (Rotax 377)					
G-MMKA	Solar Wings Panther Dual XL	T284-986XL	8. 3.84	R.S.Wood	(Wallacestone, Falkirk)	Exemption
				(Current status unknown)		
G-MMKE	Birdman WT-11 Chinook	01817	2. 4.84	D.M.Jackson	(Belper)	Exemption
	(Rotax 277)			*(Current status unknown)*		
G-MMKG	Medway Hybred 44XL/Solar Wings Typhoon XL2	22284/7	9. 3.84	G.P.Lane	(Bristol)	18. 7.97P
	(Typhoon sailwing c/n T-?84-1035XL - either '384 or '484) (Reported with wing marked "G-MNYX" 8.96) (Current status unknown)					
G-MMKH	Medway Hybred 44XL/Solar Wings Typhoon XL2	22284/8	9. 3.84	C.Richardson	(Ilkeston)	16.10.01P
	(Typhoon sailwing c/n T.?84-1047XL - either '384 or '484)					
G-MMKL	Mainair Gemini/Flash	238-384-2-W11	12. 3.84	D.W.Cox	(Kenilworth)	29. 9.93P
	(Fuji-Robin EC-44-PM)			*(Current status unknown)*		
G-MMKM	Mainair Gemini/Flexiform Dual Striker	221-184-2	12. 3.84	S.W.Hutchinson	(Northallerton)	11. 6.99P
	(Fuji-Robin EC-44-PM) *(Regd/stamped with c/n 221-0184-0002)*					
	(Originally fitted with Mainair 440 Tri-Flyer trike [210-1083] and part- exchanged for 440 Gemini as fitted: rebuild of trike originally exported to US and re-imported)					
G-MMKP	MBA Tiger Cub 440	SO.203	13. 3.84	J.W.Beaty *(Current status unknown)*	(Kettering)	
G-MMKR	Mainair Tri-Flyer/Southdown Lightning DS		14. 3.84	C.R.Madden	(Great Orton)	22. 7.05P
	(Fuji-Robin EC-44-PM)	209-171083 & CM-01	*(Regd as G-MNDK in error and then restored as G-MMKR)*			
G-MMKV	Southdown Puma Sprint X	P.521	24. 4.84	A L.Flude	(Saxmundham)	28. 8.02P
	(Fuji-Robin EC-44-PM)					
G-MMKX	Skyrider Aviation Phantom 330	PH-107R	18. 3.85	C A James	Doynton	17. 6.01P
	(Fuji-Robin EC-34-PL-02)			*(Noted 2.03)*		
G-MMKY*	Jordan Duet Series 1	CHS-01	19. 3.84	Not known	Field Farm, Oakley	
	(Rotax 503)		*(Cancelled by CAA 1.9.95: composite airframe being assembled 7.02 from this and G-MNIN)*			
G-MMLE	Eurowing Goldwing SP	EW-81	21. 3.84	B.K.Harrison *(Current status unknown)*	(Glasgow)	
G-MMLH	Hiway Skytrike II 330/Demon	PMH-01 & DJL-01	28. 3.84	P.M.Hendry and D.J.Lukey	(Folkestone)	
				(Current status unknown)		
G-MMLI*	Mainair Tri-Flyer 250/Solar Wings Typhoon S			See SECTION 3 Part 1	East Fortune	
G-MMMB	Mainair Tri-Flyer/Sprint	CR-01/170 & 170-16583	5. 4.84	K.Birkett	(Southampton)	22. 9.02P
	(Fuji-Robin EC-44-PM)*(Trike unit ex G-MJYU)*					

Regn	Type / Engine	C/n	Previous identity	Regd	Owner (Notes)	Base	CofA / Permit
G-MMMG	Eipper Quicksilver MXL (Rotax 447)	1383		5. 6.84	J.G.Campbell	Sandtoft	27. 9.04P
G-MMMH	Hadland Willow/Flexiform Striker (BMW R80/7)	MJH 383		9.12.83	M.J.Hadland	(Wigan)	4. 6.05P
G-MMML	Dragon Light Aircraft Dragon Series 150 (Fuji-Robin EC-44-PM)	D150/002	OY-... G-MMML	28. 6.83	R.G.Huntley *(Noted in poor state 5.03)*	South Wraxall, Bradford-on-Avon	6. 8.00P
G-MMMN	Solar Wings Panther Dual XL-S	PXL 843-150 & T484-105?XL	*(Probably '1059)*	4. 4.84	C.Downton	(Newton Abbot)	16. 7.04P
G-MMNA	Eipper Quicksilver MXII *(C/n conflicts with Quicksilver G-MMIL)*	1046		30. 3.84	J.W.Dodson	Leicester	2. 1.05P
G-MMNB	Eipper Quicksilver MX (Cuyuna 430R)	4286		30. 3.84	J.M Lindop *(New owner 6.01)*	Long Marston	12.10.97P
G-MMNC	Eipper Quicksilver MX	4276		30. 3.84	W.S.Toulmin *(New CofR 2.04)*	(Great Gidding, Huntingdon)	31. 5.96P
G-MMNH	Dragon Light Aircraft Dragon Series 150 (Fuji-Robin EC-44-PM)	D150/42		27. 7.83	T.J.Barlow *(Current status unknown)*	Dromore, County Down	Exemption
G-MMNN	Sherry Buzzard *(Built E.W.Sherry c/n 1)*	PFA 190-10430		6. 4.84	E.W.Sherry *(Current status unknown)*	(Stoke-on-Trent)	
G-MMNS	Mitchell U-2 Super Wing *(Built C Baldwin)*	PFA 114-10690		11. 4.84	C.Baldwin and J.C.Lister *(Current status unknown)*	Valley Farm, Winwick	
G-MMNT	Flexiform trike/Flexiform Solo Striker (Rotax 277)	SSL-1		16. 4.84	C.R.Thorne *(Current status unknown)*	(Lyndhurst, Hampshire)	Exemption
G-MMOB	Mainair Gemini/Sprint (Fuji-Robin EC-44-PM) *(C/n 'K' denotes kit built)*	244-584-2(K) & EM-01		11. 5.84	D.Woolcock	St.Michaels	27. 5.05P
G-MMOG*	Huntair Pathfinder Mk.I	011		9. 5.84	Not known *(Cancelled 26.7.95 on sale to Ireland) (Noted 4.04)*	Letterkenny, County Donegal	5. 5.95P
G-MMOH	Solar Wings Pegasus XL-R *(Trike fitted replacing one formerly on G-MBTT: new Trike now fitted ex G-MYGA)*	SW-TB-1450 & T484-1054XL		4. 5.84	T.H.Scott *(Current status unknown)*	Rayne Hall Farm, Braintree	
G-MMOK	Solar Wings Panther XL-S	PXL844-157 & T584-1066XL		9. 5.84	R.F. and A J.Foster	(Woodbridge)	19.10.05P
G-MMPG	Southdown Puma Sprint/Tripacer/Lightning Mk.II (Fuji-Robin EC-34-PM)	NEA-01		8. 6.84	T.J.Hector	(Royston)	15. 4.01P
G-MMPH	Southdown Puma Sprint (Fuji-Robin EC-44-PM)	P.545		20. 6.84	D.A.Frank	(Reading)	27. 4.05P
G-MMPI*	Pterodactyl Ptraveler (Fuji-Robin EC-25)	108		23. 5.84	M.J.Aubrey *(Cancelled 24.8.94 as WFU) (Noted 2002)*	(Kington, Hereford)	
G-MMPJ*	Mainair Gemini/Southdown Sprint *(Orig regd as Tri-Flyer 440)*	MSB-01, 264-884-2 & P.567		10. 8.84	Mainair Ltd *(Cancelled 24.4.92 by CAA).(Stored 12.04)*	Rochdale	
G-MMPL	Lancashire Micro-Trike 440/Flexiform Dual Striker (Fuji-Robin EC-44-PM) *(Trike unit from G-MJYW: maybe flown with exchangeable sailwings)*	PDL-02 & 2/330PM/PGK/683/K		5.12.83	P.D.Lawrence	Insch	21. 7.04P
G-MMPO	Mainair Gemini/Flash (Fuji-Robin EC-44-PM)	325-785-3 & W65		18. 4.85	F.H.Cook	(Whitchurch, Hampshire)	2. 6.05P
G-MMPR*	Dragon Light Aircraft Dragon Series 150	0011		18. 4.83	Not known *(Cancelled 8.10.93 by CAA) (Noted 4.04)*	Letterkenny, County Donegal	Exemption
G-MMPU	R J Heming Trike/Typhoon S4 (Fuji-Robin EC-34-PM)	RJH-01 & T782-553L		5. 6.84	J.T.Halford *(Current status unknown)*	(Holt, Norfolk)	22. 5.96P
G-MMPZ	Teman Mono-Fly (Rotax 447)	JWH-01		2. 7.84	P.B.Kylo	(Consett)	6. 5.05P
G-MMRK	Ultrasports/Solar Wings Panther XL-S	PXL846-175 & T684-1107XL		9. 7.84	J.A Churchill *(Current status unknown)*	(Worthing)	28. 9.95P
G-MMRL	Ultrasports/Solar Wings Pegasus XL-S	PXL846-174 & T684-1102XL		17. 7.84	R.J.Hood	London Colney	3. 9.05P
G-MMRN	Southdown Puma Sprint (Fuji-Robin EC-44-PM)	P.544		16. 7.84	D.C.Read	(Ledbury)	18. 4.01P
G-MMRP	Mainair Gemini/Sprint (Fuji-Robin EC-44-PM)	259-884-2-P.561		7. 2.85	J.C.S.Jones	Emlyn's Field, Rhuallt	5 .9.05P
G-MMRW	Mainair Gemini 440/Flexiform Dual Striker	LAI/DS/25 & 216-71283		5. 1.84	M.D.Hinge *(Current status unknown)*	Salisbury	Exemption
G-MMSA	Solar Wings Panther XL-S	PXL847-189 and T184-1142XL	*(C/n probably T784-1142XL)*	9. 8.84	T.W.Thiele and G.Savage *(Current status unknown)*	(Baldock)	27. 5.98P
G-MMSG	Solar Wings Panther XL-S *(Regd with c/n 8841/65XC)*	T884-1165XL		6. 9.85	R.W.McKee *(Amended CofR 5.03)*	(Deeside)	4. 6.01P
G-MMSH	Solar Wings Panther XL-S	PXL847-192 & T884-1163XL		28. 5.85	I.J.Drake *(Current status unknown)*	(Billericay)	7. 5.90P
G-MMSO	Mainair Gemini/Sprint (Fuji-Robin EC-44-PM)	255-784-2-P.539		14. 1.86	K.A Maughan *(Noted 4.04)*	Askern	26. 7.99P
G-MMSP	Mainair Gemini/Flash (Fuji-Robin EC-44-PM) *(Original sailwing c/n W03 later sold to G-MNGF 1998: current sailwing identity not yet known)*	265-984-2		17. 8.84	J.Whiteford	East Fortune	24. 4.01P
G-MMTA	Solar Wings Panther XL-R (Rotax 462HP)	PXL848-194 & T884-1164XL		25.10.84	P.A McMahon	(Dun Laoghaire, County Dublin)	29. 6.03P
G-MMTC	Solar Wings Pegasus XL-R	SW-TB-1037 & T684-1101XL		28. 9.84	T.L.Moses *(Original Trike was Ultrasports c/n PXL847-170 and later fitted to G-MNHH) (Dismantled 2.03)*	LlanSt., Carmarthen	8. 2.02P
G-MMTD	Mainair Tri-Flyer/Hiway Demon 175 (Fuji-Robin EC-34-PM)	150-30583 & EIA-01	*(Trike originally exported to Denmark)*	16. 8.84	W.E.Teare	(Ramsey, Isle of Man)	10. 9.03P
G-MMTG	Mainair Gemini/Southdown Sprint	267-984-2 & P.577		21. 8.84	J.C.F.Dalton *(Originally regd as Mainair Tri-Flyer with c/n RPWJ-01; type amended on 12.10.84) (New CofR 1.03)*	(St.Neots)	13.8.94P
G-MMTI	Southdown Puma Sprint (Fuji-Robin EC-44-PM) *(C/n duplicates ZS-VLZ)*	SN1221/0005		13. 9.84	S.A Jackson *(See G-MMIR - possibly fitted with new wing)*	(Polegate)	26. 5.02P
G-MMTJ	Southdown Puma Sprint (Fuji-Robin EC-44-PM)	SN1221/0006		17. 1.85	P J Kirwan	(Geashill, County Offaly)	16. 4.00P
G-MMTL	Mainair Gemini/Sprint (Fuji-Robin EC-44-PM)	268-1084-2-P.576		3.10.84	K.Birkett	Lee-on-Solent	3. 8.05P

G-MMTR Solar Wings Pegasus XL-R KND-03 27. 9.84 P.M.Kelsey (Rufforth) 11.12.04P
(Originally fitted with Ultrasports trike/Typhoon wing c/n T984-1211XL: trike replaced by Solar Wings XL c/n SW-TB-1092 circa 8.86)
G-MMTS Solar Wings Panther XL T784-1157XL 18. 9.84 A E.James tr Slow Thrusters Redlands 29. 7.05P
G-MMTT Ultrasports/Solar Wings Panther XL-S T684-1165XL 12.12.84 C.T.H.Tenison *(Current status unknown)* (Abergavenny) 7.11.97P
(C/n possibly T884-1165XL but duplicates G-MMSG: Solar Wings records sailwing as G-MMTT when returned for repair: G-MMSG possibly had a replacement wing)
G-MMTV American Aerolights Eagle 215B Seaplane SGP-1 25. 5.84 P.J.Scott (Seaview, Isle of Wight) 21.11.96P
(Fuji-Robin EC-25-PS) *(Current status unknown)*
G-MMTX Mainair Gemini/Sprint 275-1284-2-P.590 25. 3.85 A.Worthington Tarn Farm, Cockerham 13. 8.05P
(Fuji-Robin EC-44-PM) *(Original fitted with wing P.577 now from G-MMTG)* *(New owner 8.04)*
G-MMTY Fisher FP.202U 2140 28. 9.84 B.E.Maggs Brickhouse Farm, Frogland Cross
(Stored 4.96: current status unknown)
G-MMTZ Eurowing Goldwing EW-60 & SWA-7 28. 9.84 R.B.D.Baker (Torquay) 15. 7.03P
(Rotax 447)
G-MMUA Southdown Puma Sprint SN1221/0007 21.12.84 M.R.Crowhurst (Ramsey, Isle of Man) 21. 7.05P
(Fuji-Robin EC-44-PM)
G-MMUH Mainair Tri-Flyer/Sprint 270-1084-2-P.579 8.11.84 J.P.Nicklin (Hayling Island) 20. 9.04P
(Fuji-Robin EC-44-PM)
G-MMUL* Ward Elf E-47 16.10.84 N H Ponsford Breighton
(Cancelled 12.4.89 by CAA) (Noted 1.05)
G-MMUM MBA Tiger Cub 440 SO.019 8. 3.83 Coulson Flying Services Ltd (Skegness)
(Current status unknown)
G-MMUO Mainair Gemini/Flash 272-1084-2 & W08 29.10.84 B.D.Bastin and D.R.Howells Long Marston 3.11.05P
(Fuji-Robin EC-44-PM) *(Noted 8.03)*
G-MMUR Hiway Skytrike II/Solar Wings Storm SLI.80180 28.12.84 R.J.Ripley Field Farm, Oakley
(Fuji-Robin EC-25) *(Stored @ owner's house 1998)*
G-MMUT* Mainair Gemini/Flash II 235-484-2* &-W04 5.10.84 S.C.Briggs Lauder 5. 7.01P
(Fuji-Robin EC-44-PM) *(Original c/n 62-884-2 & W04: fitted with new Trike* first used on G-MMFC (3) and wing W73 @ 6.98- also see G-MNAC)*
(Cancelled as PWFU 14.1.05) (Stored 1.05)
G-MMUV Southdown Puma Sprint SN1121/0010 7.11.84 D.C.Read (Ledbury) 2.11.89P
(Fuji-Robin EC-44-PM) *(Current status unknown)*
G-MMUW Mainair Gemini/Flash II 60-784-2 & W13 17. 1.85 J.C.K.Scardifield (Lymington) 23. 3.87P
(Fuji-Robin EC-44-PM) *(Current status unknown)*
G-MMUX Mainair Gemini/Sprint 285-185-3 & P587 28.12.84 G.A Harper Priory Farm, Tibenham 24.11.05P
(Fuji-Robin EC-44-PM) *(Trike c/n confirmed as "284-185-3")*
G-MMVA Southdown Puma Sprint SN1121/0011 & P.588 7.11.84 C.H.Tomkins (Kettering) 26. 3.92P
(Fuji-Robin EC-44-PM) *(Current status unknown)*
G-MMVH Southdown Raven X SN2122/0015 10. 1.85 G.W. and K.M.Carwardine (Isle of Grain) 29. 4.01P
G-MMVI Southdown Puma Sprint SN1121/0012 28.11.84 G.R.Williams (Haverfordwest) 2.11.97P
(Fuji-Robin EC-44-PM) *(Current status unknown)*
G-MMVO Southdown Puma Sprint SN1232/0017 20. 3.85 D.M.Pearson Chilton Park, Wallingford 17.10.05P
(Rotax 447)
G-MMVP* Mainair Gemini/Flash II 76-1284-2 & W12 17.12.84 S.C.McGowan (Rufforth) 24. 1.05P
(Fuji-Robin EC-44-PM) *(Cancelled 7.9.04 by CAA)*
G-MMVS Skyhook TR1 Pixie/Zeus TR1/52 28. 2.85 B.W.Olley (Ely) Exemption
(Solo 210) *(Current status unknown)*
G-MMVX Southdown Puma Sprint 41183 & P.452 29.11.83 M.P.Jones Haverfordwest 5. 4.03P
(Fuji-Robin EC-44-PM)
G-MMVZ Southdown Puma Sprint SN1121/0016 15. 1.85 C Colclough (Celbridge, County Kildare) 2. 8.04P
(Fuji-Robin EC-44-PM)
G-MMWA Mainair Gemini/Flash II 271-1184-1 & W07 22.11.84 N Roberts Askern 12.10.04P
(Fuji-Robin EC-44-PM) *(Trike c/n stamped as "KR271-1184-2")*
G-MMWC Eipper Quicksilver MXII 1041 22.10.84 J.S.Harris and M.Holmes Old Sarum 27. 7.03P
(Rotax 503)
G-MMWG P Greenslade Trike/Flexiform Solo Striker 17.12.84 C.R.Green (Redruth) 26. 6.99P
(Rotax 377) FF/LAI/83/JDR/11 *(Trike originally fitted to G-MJGN: wing no. duplicates G-MMFC) (Current status unknown)*
G-MMWL Eurowing Goldwing SWA-09 & EW-91 9. 4.85 P.J.Brookman Knapthorpe Lodge, Caunton 24. 5.05P
(Rotax 447)
G-MMWS Ultrasports Tri-Pacer/Flexiform Solo Striker 983.SH 21.11.84 P.H.Risdale Tower Farm, Wollaston 5. 7.05P
(Rotax 377) *(Originally fitted with Mainair trike) (Original owners of G-MMW<u>N</u> and 'MMW<u>S</u> were Nigel and Sally Huxtable: believed trikes inter-changed)*
G-MMWX Southdown Puma Sprint SN1121/0047 10. 4.85 S H Leahy Longacres Farm, Sandy 28. 5.04P
(Fuji-Robin EC-44-PM)
G-MMXD Mainair Gemini/Flash II 282-185-3 & W20 28.12.84 W A Bibby Brook Farm, Pilling 13. 8.05P
(Rotax 447)
G-MMXG Mainair Gemini/Flash II 288-485-1 & W32 17. 1.85 T.Birch (Wolverhampton) 15. 6.01P
(Fuji-Robin EC-44-PM) *(Damaged c6.00 - trike used to rebuilt G-MNBD and rest to store)*
G-MMXJ Mainair Gemini/Flash II 289-185-3 & W22 17. 1.85 R.Meredith- Hardy Radwell, Letchworth 6. 8.96P
(Rotax 447) *(Current status unknown)*
G-MMXL Mainair Gemini/Flash II 292-385-3 & W36 17. 1.85 J.M.Marshall (Urmston) 16. 5.97P
(Fuji-Robin EC-44-PM) *(Current status unknown)*
G-MMXO Southdown Puma Sprint SN1121/0018 23. 1.85 D.J.Tasker Swinford, Rugby 1. 3.05P
(Fuji-Robin EC-44-PM)
G-MMXU Mainair Gemini/Flash II 254-784-2 & W21 29. 1.85 T.J.Franklin Graveley 14. 7.01P
(Fuji-Robin EC-44-PM) *(Stored 7.03)*
G-MMXV Mainair Gemini/Flash II 298-385-3 & W37 29. 1.85 K.C.Beattie Cumbernauld 20. 4.05P
(Rotax 503)_
G-MMXW Mainair Gemini/Sprint 286-185-3-P.597 23. 1.85 A Hodgson (Milton Keynes) 4. 6.02P
(Fuji-Robin EC-44-PM)
G-MMYA Solar Wings Pegasus XL-R/Se *(Originally regd as XL)* 30. 1.85 J North (Felmersham, Bedford) 5. 8.05P
XL-P Proto & T784-1151XL
G-MMYF Southdown Puma Sprint SN1121/0026 28. 3.85 M Campbell (Halifax) 14. 3.05P
(Fuji-Robin EC-44-PM)
G-MMYL Cyclone 70/Aerial Arts 130SX CH.01 8. 3.85 E.W.P.Van Zeller (Ashford, Kent) 6. 5.05P
G-MMYN Solar Wings Panther XL-R T784-1158XL 27. 2.85 H.J.Long (Enniscorthy, County Wexford) 24. 4.05P

Reg	Type / Engine	C/n	Prev id	Reg date	Owner / Remarks	Base	CofA
G-MMYO	Southdown Puma Sprint (Fuji-Robin EC-44-PM) *(Fitted with rainbow Medway sailwing c3.96 after accident 20.9.95)*	SN1121/0037		11. 4.85	P.R.Whitehouse	Otherton, Cannock	29. 8.00P
G-MMYR	Eipper Quicksilver MXII (Rotax 503)	3345		27. 2.85	P.A Pilkington	North Coates	17. 6.01P
G-MMYT	Southdown Puma Sprint (Fuji-Robin EC-44-PM)	SN1121/0046 & T569/P621		15. 4.85	J.K.Divall *(Current status unknown)*	(Chichester)	25. 3.94P
G-MMYU	Southdown Puma Sprint (Rotax 447)	SN1231/0045		11. 6.85	M.V.Hearns	Glenrothes	21. 4.02P
G-MMYV	John Webb trike/Flexiform Striker (Rotax 277)	JW-2		22. 3.85	S.B.Herbert *(Current status unknown)*	(Presteigne)	20.12.95P
G-MMYY	Southdown Puma Sprint (Rotax 447)	SN1231/0042		18. 7.85	D.J.Whittle	(Liverpool)	12. 7.04P
G-MMYZ*	Southdown Puma Sprint (Rotax 447)	SN1231/0034		28. 2.85	M.Bodill *(Damaged in gales Roddige 1.98: cancelled 31.5.01 by CAA) (Trike noted in poor state 1.04)*	Roddige	19. 2.99P
G-MMZA	Mainair Gemini/Flash II (Fuji-Robin EC-44-PM)	266-984-3 & W60		4. 3.85	G.T.Johnston	(Craigavon, Co Armagh)	30. 6.00P
G-MMZB	Mainair Gemini/Flash (Fuji-Robin EC-44-PM)	319-685-3 & W58		4. 3.85	M.A Nolan	(Great Orton)	23. 5.02P
G-MMZF	Mainair Gemini/Flash II (Fuji-Robin EC-44-PM)	299-485-3 & W38		4. 3.85	J.Tait *(New owner 10.04)*	(Houghton le Spring)	13. 9.03P
G-MMZG	Solar Wings Panther XL-S	SW-TA-1008 & SW-WA-1022		12. 8.85	K.A Sutton	Enstone	24. 5.05P
G-MMZI	Medway Half Pint/Aerial Arts 130SX	2385/1 & 130SX-057		6. 3.85	J.Messenger *(Current status unknown)*	(Workington)	Exemption
G-MMZJ	Mainair Gemini/Flash (Rotax 462)	312-585-3 & W51		18. 3.85	P.J.Glover	North Coates	15. 3.05P
G-MMZK	Mainair Gemini/Flash (Fuji-Robin EC-44-PM) *(Trike ex G-MMEZ: originally regd with trike c/n 314-585-3: to G-MMIR) (Current status unknown)*	326-785-3 & W53		18. 3.85	G.Jones and B.Lee	(Warrington)	3.11.99P
G-MMZM	Mainair Gemini/Flash (Fuji-Robin EC-44-PM)	304-585-3 & W44		18. 3.85	H.Brown	(Dunbar)	4. 1.04P
G-MMZN	Mainair Gemini/Flash II (Fuji-Robin EC-44-PM)	283-185-3 & W23		18. 3.85	W.K.Dalus *(Current status unknown)*	(Keyworth)	28. 9.93P
G-MMZP	Solar Wings Panther XL *(Built H Phipps) (Possibly original trike from G-MJWZ)*	HP-01		14. 3.85	B.Richardson *(Current status unknown)*	(Sunderland)	12. 1.94P
G-MMZR	Southdown Puma Sprint (Fuji-Robin EC-44-PM)	SN1121/0039 & T560/P622		4. 7.85	J.E.Hicks tr International Animal Rescue *(Current status unknown)*	Dunkeswell	6.12.93P
G-MMZV	Mainair Gemini/Flash (Rotax 447)	313-585-3 & W52		18. 4.85	P.R.M.Spengler	(Bracknell)	12. 5.02P
G-MMZW	Southdown Puma Sprint (Fuji-Robin EC-44-PM)	SN1121/0043 & T566/P620		28. 3.85	M.G.Ashbee *(Damaged c.8.00)*	(Cranbrook)	30. 9.00P
G-MMZZ*	Maxair Hummer (Fuji-Robin EC-25-)	0010		8. 4.82	M.J.Aubrey *(Cancelled 12.6.00 by CAA) (Noted 2002)*	(Kington, Hereford)	
G-MNAC	Mainair Gemini/Flash *(C/n now verified as 262-884-2 and W04 ex G-MMUT qv)*	335-885-3 & W72		18. 4.85	C Bayliss	(Liverpool)	30. 7.05P
G-MNAE	Mainair Gemini/Flash (Rotax 447)	343-885-3 & W77		18. 4.85	G.C.Luddington	(Bletsoe)	29. 7.00P
G-MNAH	Solar Wings Panther XL-S	SW-TA-1002 & SW-WA-1002		24. 4.85	J.H.Button and G.A Harman *(Current status unknown)*	(Sandy)	18. 9.99P
G-MNAI	Solar Wings Panther XL-S	SW-TA-1003 & SW-WA-1003		15. 5.85	R.G.Cameron *(Noted stored 8.04)*	Errol	23. 6.98P
G-MNAK*	Solar Wings Panther XL-S	SW-TA-1005 & SW-WA-1005		15. 5.85	F.J.McVey *(Cancelled 16.1.04 as WFU)*	Insch	15. 8.04P
G-MNAR	Solar Wings Pegasus XL-R	SW-TB-0014 & SW-WA-1011		6. 8.85	D.A Cansdale	(Harlow)	3. 3.03P
G-MNAV	Southdown Puma Sprint (Fuji-Robin EC-44-PM)	SN1121/0033		28. 2.85	G.P.Morling	(Douglas, Isle of Man)	15.11.04P
G-MNAW	Solar Wings Pegasus XL-R	SW-TB-1010 & SW-WA-1014		16. 8.85	D.J.Harber	(Henley-on-Thames)	9. 2.05P
G-MNAX	Solar Wings Pegasus XL-R	SW-TB-1011 & SW-WA-1015		16. 8.85	B.J.Phillips *(Current status unknown)*	(Newbury)	21. 7.96P
G-MNAY	Solar Wings Pegasus XL-R	SW-TB-1015 & SW-WA-1016		6. 8.85	A Seaton *(New owner 5.02)*	(Sleaford)	11. 9.99P
G-MNAZ	Solar Wings Pegasus XL-R	SW-TB-1016 & SW-WA-1017		6. 8.85	R.W.Houldsworth	(Rochford)	2. 8.04P
G-MNBA	Solar Wings Pegasus XL-R	SW-TB-1024 & SW-WA-1018		6. 9.85	V.F.Clemmens	Graveley	19. 8.05P
G-MNBB	Solar Wings Pegasus XL-R	SW-TB-1020 & SW-WA-1019		20. 9.85	R.Piper	Ince Blundell	3. 8.04P
G-MNBC	Solar Wings Pegasus XL-R (Rotax 503)	SW-TB-1026 & SW-WA-1020		11.10.85	N Kelly	(Carmarthen)	9. 9.04P
G-MNBD	Mainair Gemini/Flash (Fuji-Robin EC-44-PM) *(Originally built as Mainair 440 Tri-Flyer c/n 341-585-3 and W42: unsold and.reworked by Mainair as c/n 162-683 and fitted to G-MMSN. This was podded to become a Gemini and used in rebuild of G-MNBD after late 1996 accident)*	162-683 & W42	G-MMSN	6. 1.86	P.Woodcock	Sittles Farm, Alrewas	30.12.03P
G-MNBE	Southdown Puma Sprint (Rotax 447)	SN1121/0050		17. 5.85	J.Liversuch and C.Hershaw	Doynton	22. 8.04P
G-MNBF	Mainair Gemini/Flash (Fuji-Robin EC-44-PM)	306-585-3 & W46		2. 5.85	P.Mokryk and S.King	(Derby)	25. 5.05P
G-MNBG	Mainair Gemini/Flash (Rotax 447)	347-585-3 & W66		9. 5.85	T.Barnett	Baxby Manor, Husthwaite	11. 9.05P
G-MNBI	Solar Wings Panther XL-S	PXL884-178 & T884-1161XL	G-MMVF?	3. 5.85	Ann-Marie Whelan *(New owner 8.04)*	(Athy, County Kildare	29. 4.97P
G-MNBM	Southdown Puma Sprint (Rotax 447)	SN1231/0058		25. 6.85	D.A Hopewell *(New owner 6.02)*	(Newcastle-under-Lyme, Staffs)	7.10.01P
G-MNBN	Mainair Gemini/Flash (Fuji-Robin EC-44-PM)	303-485-3 & W43		11. 6.85	I.H.Gates	Long Marston	28. 5.05P

Reg	Type	C/n	Regd	Owner/Operator	Base	CofA
G-MNBP	Mainair Gemini/Flash	338-885-3 & W75	15. 5.85	N.L.Zaman	(London Colney)	29. 3.04P
	(Fuji-Robin EC-44-PM)					
G-MNBR*	Mainair Gemini/Flash	345-985-3 & W79	15. 5.85	N.A P.Gregory	Long Marston	5. 2.94P
	(Rotax 447)			*(Cancelled 31.5.00 by CAA: stored 8.03)*		
G-MNBS	Mainair Gemini/Flash	308-585-3 & W48	15. 5.85	P.A Comins	(Nottingham)	20. 6.94P
	(Fuji-Robin EC-44-PM)			*(Current status unknown)*		
G-MNBT	Mainair Gemini/Flash	322-685-3 & W62	15. 5.85	T.H.Parr	(Carnforth)	6. 3.05P
G-MNBV	Mainair Gemini/Flash	333-685-3 & W70	15. 5.85	J.Walshe	(Newtownards)	21. 8.04P
	(Rotax 447)					
G-MNBW*	Mainair Gemini/Flash	332-685-3 & W69	15. 5.85	G.A Brown and N.S.Brotherton	Weston Zoyland	10. 9.03P
	(Rotax 447) *(C/n now SW-WF-0005 & W95 ex G-MNJI)*			*(Cancelled 21.6.04 by CAA)*		
G-MNCA	Hunt Avon Sky-Trike/Hiway Demon 175	DA-01	28. 5.85	M.A.Sirant	Monkswell Farm	Exemption
	(Built Hiway Hang Gliders Ltd and originally regd as Adams Trike)			*(New CofR 10.04)*		
G-MNCF	Mainair Gemini/Flash	321-685-3 & W61	3. 6.85	M.Atkinson	(Blackpool)	26. 4.05P
	(Rotax 447)					
G-MNCG	Mainair Gemini/Flash	320-685-3 & W59	3. 6.85	J E F Fletcher	Tarn Farm, Cockerham	20. 9.05P
	(Rebuilt c2000)					
G-MNCI	Southdown Puma Sprint	SN1231/0059	7. 6.85	R.M.Wait and N.Hewitt	Mill Farm, Shifnal	25. 5.05P
	(Rotax 447)					
G-MNCJ	Mainair Gemini/Flash	351-785-3 & W83	3. 6.85	R.S.McLeister	(Accrington)	16.11.93P
	(Fuji-Robin EC-44)	*(Original trike stolen, new one c/n 282-1284-2 ex G-MMXF fitted c.12.89) (Current status unknown)*				
G-MNCM	CFM Shadow Series C	006	31. 5.85	K.G.D.Macrae	Drummiard Farm, Bonnybank	12. 8.05P
G-MNCO	Eipper Quicksilver MXII	1045	3. 6.85	S.Lawton *(CofR restored 4.02)*	(Barnoldswick)	
G-MNCP	Southdown Puma Sprint	SN1231/0071	24. 6.85	S.Baker and D.M.Lane	(Barton under Needwood)	10. 4.00P
	(Rotax 447)			t/a Freedom Sports Aviation		
G-MNCS	Skyrider Aviation Phantom	PH.00098	2. 1.86	S.P.Allen	(Kettering)	25. 7.03P
	(Fuji-Robin EC-44-PM)					
G-MNCU	Medway Hybred/Solar Wings Typhoon 44XL	26485/10 & SW-WA-1029	13. 6.85	A Thornley	(Louth)	24. 6.05P
G-MNCV	Medway Hybred/Solar Wings Typhoon 44XL		13. 6.85	P.D.Mickleburgh	Swinford, Rugby	29.11.04P
	(Fuji-Robin EC-44-PM)	26485/11 & SW-WA-1030 *(Pegasus XL-R wing)*				
G-MNDC	Mainair Gemini/Flash	336-885-3 & W73	12. 6.85	M.Medlock	(Guildford)	9. 6.05P
G-MNDD	Mainair Scorcher	358-885-1 & W85	12. 6.85	L.Hurman	(Oxford)	9. 6.05P
G-MNDE	Medway Half Pint/Aerial Arts 130SX	3/8685	19. 6.85	C.D.Wills	(Andover)	3.10.03P
	(Wing ex G-MNBZ)					
G-MNDF	Mainair Gemini/Flash	327-785-3 & W67	25. 6.85	R.Bowden	Dunkeswell	17. 6.05P
	(Rotax 447)					
G-MNDG	Southdown Puma Sprint	SN1121/0057	18. 7.85	P.J.Kirwan	(Geashill, County Offaly)	14. 6.99P
	(Fuji-Robin EC-44-PM)			*(Current status unknown)*		
G-MNDO	Solar Wings Pegasus Flash	SW-WF-0001	2. 7.85	R.H.Cooke	Lee-on-Solent	26. 7.05P
	(Trike is c/n SW-TB-1012 and Mainair sailwing c/n W86)			*(Noted 9.04)*		
G-MNDU	Midland Ultralights Sirocco 377GB	MU-011	22. 7.85	M.A Collins	Longacres Farm, Sandy	14. 2.04P
	(Rotax 377)					
G-MNDY	Southdown Puma Sprint	DY-01 & P.536	2. 5.84	A M.Marshall	(Oswestry)	19. 6.03P
	(Fuji-Robin EC-44-PM) *(Trike rebuilt c4.99)*					
G-MNEF	Mainair Gemini/Flash	344-885-3 & W78	8. 7.85	P.G.Richards	(Dunfermline)	15. 7.04P
	(Rotax 447)					
G-MNEG	Mainair Gemini/Flash	360-885-3 & W92	8. 7.85	A.Sexton	(Nurney, County Kildare	18.10.99P
	(Rotax 447)			*(New owner 8.04)*		
G-MNEH	Mainair Gemini/Flash	361-885-3 & W90	8. 7.85	I.Rawson	St.Michaels	21. 7.05P
G-MNEI	Medway Hybred/Solar Wings Typhoon/XL-R		9. 7.85	L.G.Thompson	Long Marston	26. 7.93P
	(Fuji-Robin EC-44-PM)	8785/12 & SW-WA-1035		*(Damaged 28.11.92 and stored 8.96: current status unknown)*		
G-MNEK	Medway Half Pint/Aerial Arts 130S	4/8785	12. 7.85	M.I.Dougall	(Maidstone)	25. 9.94P
				(Damaged Stoke 6.7.93: current status unknown)		
G-MNER	CFM Shadow Series CD	008	15. 7.85	F.C.Claydon	Wickhambrook, Newmarket	14. 6.05P
	(Rotax 462)					
G-MNET	Mainair Gemini/Flash	349-885-3 & W81	23. 7.85	I P Stubbins	North Coates	6. 8.04P
	(Fuji-Robin EC-44-PM)					
G-MNEV	Mainair Gemini/Flash	362-1085-3 & W108	23. 7.85	C.A Denver	St.Michaels	30. 9.04P
	(Rotax 447)					
G-MNEY	Mainair Gemini/Flash	365-1085-3 & W94	23. 7.85	D.A Spiers	East Fortune	17. 9.04P
	(Rotax 447)					
G-MNFA*	Mainair Tri-Flyer/Solar Wings Typhoon		29.12.83	Not known	Mill Farm, Hughley, Much Wenlock	10.10.92P
	(Trike ex G-MJFA)	DRJ-01 & GWW-01		*(Cancelled 4.5.01 by CAA: noted 9.04)*		
G-MNFB	Southdown Puma Sprint	SN1231/0077	22. 7.85	C.Lawrence	Weston Zoyland	16. 8.05P
	(Rotax 447)					
G-MNFE	Mainair Gemini/Flash	350-885-3 & W82	29. 7.85	D.R.Kennedy	East Fortune	20.10.01P
	(Fuji-Robin EC-44-PM)					
G-MNFF	Mainair Gemini/Flash	371-1185-3 & W110	29. 7.85	R.P.Cook and C.H.Spencer	St.Michaels	24. 5.02P
	(Rotax 447)					
G-MNFG	Southdown Puma Sprint	SN1231/0078	31. 7.85	A C.Hing	Longacres Farm, Sandy	12. 6.03P
	(Rotax 447)					
G-MNFH	Mainair Gemini/Flash	364-1085-3 & W93	6. 8.85	K.Glynn	(Loughrea, County Galway)	30. 6.95P
	(Rotax 447)			*(Current status unknown)*		
G-MNFL	AMF Microflight Chevvron 2-32A	CH.002	19. 8.85	P.W.Wright	Saltby	13.12.00P
G-MNFM	Mainair Gemini/Flash	366-1085-3 & W98	10.10.85	P.M.Fidell	Wombleton	1. 9.05P
	(Rotax 447)					
G-MNFN	Mainair Gemini/Flash	367-1085-3 & W99	6.11.85	J.R.Martin	(Bedale)	13. 8.04P
	(Rotax 447)					
G-MNFP	Mainair Gemini/Flash	368-1085-3 & W100	23.10.85	S.Farnsworth and P.Howarth	Tarn Farm, Cockerham	7. 6.05P
	(Rotax 447)					
G-MNFW	Medway Hybred 44XL	10885/13	15. 8.85	A T.Palmer *(Current status unknown)*	(Plymouth)	15. 8.99P
G-MNFX	Southdown Puma Sprint	SN1231/0079	14. 8.85	A M.Shaw	(Stoke-on-Trent)	6. 9.04P
	(Rotax 447)	*(Nosewheel detached landing Arclid Industrial Estate, Hemmingshaw Lane, Sandbach 15.8.04 and seriously damaged)*				

Regn	Type	C/n	Date	Owner	Base	CofA
G-MNGD	Ultrasports Tri-Pacer/Solar Wings Medium Typhoon		13. 8.85	F.H.Cook	(Whitchurch, Hampshire)	15. 5.05P
	(Fuji-Robin EC-34-PM)	012 & T681-171				
G-MNGF	Solar Wings Pegasus Flash		21. 8.85	R.G.Smith	Oak Farm, Woodton	2. 8.05P
		W-TB-1022 & SW-WF-0003	*(Correct trike c/n is SW-TB-1022 plus Mainair sailwing c/n W87)*			
G-MNGG	Solar Wings Pegasus XL-R	T784-1159XL	21. 8.85	T.Peckham	(Faversham)	10. 6.04P
		(Trike c/n is US.TPR.0002)				
G-MNGK	Mainair Gemini/Flash	374-1085-3 & W112	5. 9.85	J.Pulford	(Attleborough)	2. 9.05P
	(Rotax 447)					
G-MNGM	Mainair Gemini/Flash	394-1285-3 & W109	5. 9.85	J.E.Caffull and D.R.Beale	Over Farm, Gloucester	2. 8.05P
	(Rotax 447) *(Originally supplied with Mainair trike c/n 377: this, and sailwing ex G-MNIO, stolen from Popham 15/16.3.86 and, subsequently, trike ex G-MNIO fitted with sailwing ex G-MNGM)*					
G-MNGS	Southdown Puma/Lightning 195	GJS-02	8. 5.84	R.J.Turner	(Spalding)	14. 8.02P
	(Fuji-Robin EC-34-PM) *(Tripacer Trike from G-MJRF)*					
G-MNGT	Mainair Gemini/Flash	372-1085-3 & W106	30. 9.85	J.W.Biegus	Arclid Green, Sandbach	7. 6.02P
	(Rotax 447)					
G-MNGU	Mainair Gemini/Flash	373-1085-3 & W111	30. 9.85	G.Macpherson-Irvine	(Perth)	25. 6.04P
G-MNGW	Mainair Gemini/Flash	386-1185-3 & W121	30. 9.85	F R Stephens	(Worthing)	23. 4.03P
	(Rotax 447)					
G-MNGX	Southdown Puma Sprint	SN1231/0088	26. 9.85	A.J.Morris	Sutton Meadows, Ely	26.11.04P
	(Rotax 447)			*(New owner 1.05)*		
G-MNHB	Solar Wings Pegasus XL-R/Se		1.11.85	G.Charles-Jones	Eshott	13. 4.05P
		SW-TB-1031 & SW-WA-1045				
G-MNHC	Solar Wings Pegasus XL-R		31.10.85	C.Thomas	Haverfordwest	12. 3.02P
		SW-TB-1032 & SW-WA-1046/2				
	(Original sailwing [SW-WA-1046] damaged: replaced by SW-WA-1065, probably so marked on the sailwing, but re-numbered as SW-WA-1046/2)					
G-MNHD	Solar Wings Pegasus XL-R		5.11.85	P.D.Stiles	(Ashley Down, Bristol)	22. 6.05P
		SW-TB-1033 & SW-WA-1047				
G-MNHE	Solar Wings Pegasus XL-R/Se		11.12.85	J W Coventry	Davidstow Moor	17.10.04P
		SW-TB-1036 & SW-WA-1048				
G-MNHF	Solar Wings Pegasus XL-R		29.11.85	J E Cox	Shobdon	29. 9.05P
		SW-TB-1038 & SW-WA-1049				
G-MNHH	Solar Wings Pegasus XL-S	SW-WA-1051	22. 1.86	F.J.Williams	(Shefford, Bedford)	24. 6.01P
	(Trike is an Ultrasports unit c/n PXL847-170)					
G-MNHI	Solar Wings Pegasus XL-R		8. 1.86	M.Lewis	(Canterbury)	13. 7.95P
		SW-TB-1042 & SW-WA-1052		*(New owner 5.03)*		
G-MNHJ	Solar Wings Pegasus XL-R		11. 3.86	S.J.Woodd	(Oxford)	26. 6.93P
		SW-TB-1056 & SW-WA-1053		*(Current status unknown)*		
G-MNHK	Solar Wings Pegasus XL-R		9. 7.86	R.D.Proctor	(Stamford)	13. 6.92P
	(Rotax 462)	SW-TE-0005 & SW-WA-1054		*(Current status unknown)*		
G-MNHL	Solar Wings Pegasus XL-R/Se		9. 7.86	The Microlight School Ltd	Roddige	27. 5.05P
	(Rotax 503)	SW-TB-1077 & SW-WA-1055				
G-MNHM	Solar Wings Pegasus XL-R		11. 7.86	A C Bell	(Nottingham)	20. 4.05P
		SW-TB-1078 & SW-WA-1056				
G-MNHN	Solar Wings Pegasus XL-R		11. 8.86	P.K.Peppard	Sandtoft	20. 3.05P
		SW-TB-1079 & SW-WA-1057				
G-MNHR	Solar Wings Pegasus XL-R		7. 8.86	B.D.Jackson	(Wincanton)	22. 3.05P
		SW-TB-1081 & SW-WA-1060				
G-MNHS	Solar Wings Pegasus XL-R		21. 8.86	M.D.Packer	Weston Zoyland	12.10.03P
		SW-TB-1082 & SW-WA-1061		*(Noted 7.04)*		
G-MNHT	Solar Wings Pegasus XL-R		4. 8.86	J.W.Coventry	Davidstow Moor	19. 3.04P
		SW-TB-1084 & SW-WA-1062				
G-MNHV	Solar Wings Pegasus XL-R		18. 8.86	E.Jenkins	(Crymych, Dyfed)	31. 7.99P
		SW-TB-1095 & SW-WA-1064		*(Current status unknown)*		
G-MNHZ	Mainair Gemini/Flash	310-585-3 & W118	15.10.85	I.O.S.Ross	(Cowie)	4. 7.05P
	(Fuji-Robin EC-44-PM)					
G-MNIA	Mainair Gemini/Flash	370-1185-3 & W105	10.10.85	A E.Dix	Long Marston	10. 4.89P
	(Rotax 447)			*(Noted wrecked 1990)*		
G-MNID	Mainair Gemini/Flash	369-1185-5 & W104	7. 2.86	D.Sykes	Rufforth	23. 4.98P
	(Rotax 447)			*(New owner 1.05)*		
G-MNIE	Mainair Gemini/Flash	388-1185-3 & W123	21.11.85	G.M.Hewer	(Cheltenham)	8. 7.02P
	(Rotax 447)					
G-MNIF	Mainair Gemini/Flash	403-286-4 & W147	7. 1.86	D.G.Bowden	(Wilpshire, Blackburn)	22.11.04P
	(Rotax 447)			*(New owner 1.05)*		
G-MNIG	Mainair Gemini/Flash	391-1285-3 & W129	9. 1.86	I.S.Everett	(Astwood)	22. 5.05P
	(Rotax 447)					
G-MNIH	Mainair Gemini/Flash	379-1185-3 & W116	10.12.85	A R.Richardson	Askern	6. 5.05P
	(Rotax 447)					
G-MNII	Mainair Gemini/Flash	390-1285-3 & W128	6.11.85	R.F.Finnis	(Guildford)	6. 9.91P
	(Rotax 447)			*(Trike reported at St.Michaels 9.96: current status unknown)*		
G-MNIK	Solar Wings Pegasus Photon		29.10.85	J Grotrian	Wing Farm, Longbridge Deverill	26. 9.04P
		SW-TP-0002 & SW-WP-0002				
G-MNIL	Southdown Puma Sprint	SN1231/0094	4.11.85	A Bishop	Ince Blundell	31. 8.05P
	(Rotax 447)					
G-MNIM	Maxair Hummer	PJB-01	29.10.85	K.Wood *(Current status unknown)*	(Leicester)	
G-MNIS	CFM Shadow Series C	014	11.11.85	R.W.Payne *(Current status unknown)*	(Peterborough)	25. 4.92P
G-MNIT	Aerial Arts Alpha Mk.II/130SX	130SX/176	27. 2.86	M.J.Edmett	(London N3)	15. 8.99
	(Originally regd as Hiway Skytrike II with same c/n)			*(New owner 5.02)*		
G-MNIU	Solar Wings Pegasus Photon		27.11.85	S.Ferguson	(Bearsden)	Exemption
	(Fuji-Robin EC-34)	SW-TP-0003 & SW-WP-0003		*(Damaged and stored 3.90: new owner 10.02)*		
G-MNIX	Mainair Gemini/Flash	395-1285-3 & W136	29.11.85	S.Farnworth	(Kempston, Bedford)	11. 7.98P
	(Rotax 447)			*(Current status unknown)*		
G-MNIZ	Mainair Gemini/Flash	392-1285-3 & W130	26. 2.86	A G.Power	(Darwen)	7. 6.05P
	(Rotax 447)					
G-MNJB	Southdown Raven X	SN2232/0098	10.12.85	T.A Simpson	Longacres Farm, Sandy	26. 5.03P

Registration	Type	C/n	Previous identity	Date	Owner / Notes	Base	CofA
G-MNJC	MBA Tiger Cub 440	SO.215		8. 6.84	J.G.Carpenter *(Current status unknown)*	(Romsey)	Exemption
G-MNJD	Mainair Tri-Flyer 440/Sprint (Fuji-Robin EC-44-PM)	243-10484-2-P.537		2. 4.84	M.E.Smith	(Verwood)	7. 5.05P
G-MNJF	Dragon Light Aircraft Dragon Series 150 (Fuji-Robin EC-44-PM)	0068	(OY) 9-17	2. 1.86	B.W.Langley	South Wraxall, Bradford-on-Avon	2. 8.05P
G-MNJG	Mainair Gemini/Puma Sprint MS (Fuji-Robin EC-44-PM)	SA.2030 & 251-684-2-P.593		29. 9.83	P.Batchelor	(Crawley)	24. 8.05P
G-MNJH	Solar Wings Pegasus Flash	SW-TB-1023 & SW-WF-0004	*(Mainair sailwing c/n W89)*	22.10.85	C.P.Course	Church Farm, Wellingborough	18. 8.02P
G-MNJJ	Solar Wings Pegasus Flash	SW-TB-1029 & SW-WF-0006	*(Mainair sailwing c/n W96)*	22.10.85	P.A Shelley *(Current status unknown)*	Sutton Meadows, Ely	26.11.96P
G-MNJL	Solar Wings Pegasus Flash	SW-TB-1028 & SW-WF-0008	*(Mainair sailwing c/n W101)*	21.10.85	S.D.Thomas *(Current status unknown)*	(Bilston)	11.11.94P
G-MNJN	Solar Wings Pegasus Flash	SW-TB-1034 & SW-WF-0010	*(Mainair sailwing c/n W103)*	19.11.85	D.Thorn	Davidstow Moor	5. 7.05P
G-MNJO	Solar Wings Pegasus Flash	SW-TB-1035 & SW-WF-0011	*(Mainair sailwing c/n W126)*	19.11.85	S.Clarke	Swinford, Rugby	17. 5.05P
G-MNJR	Solar Wings Pegasus Flash	SW-TB-1041 & SW-WF-0013	*(Mainair sailwing c/n W133)*	30.12.85	M.G.Ashbee	(Cranbrook)	18. 4.05P
G-MNJS	Southdown Puma Sprint (Rotax 447)	SN1231/0085		18. 9.85	E.A.Frost *(New owner 12.04)*	Sutton Meadows, Ely	20. 6.03P
G-MNJT	Southdown Raven X	SN2232/0087		20. 9.85	P.A Harris *(Forced landing East of Exford, Devon 11.7.03 causing extensive damage to aircraft)*	RAF Henlow	7. 3.04P
G-MNJU	Mainair Gemini/Flash (Rotax 447)	384-1185-3 & W119		20. 9.85	E.J.Wells *(Noted 11.04)*	Finmere	3. 6.03P
G-MNJX	Medway Hybred 44XL	15885/14		9.12.85	H.A Stewart *(Current status unknown)*	(Sittingbourne)	23. 7.98P
G-MNKB	Solar Wings Pegasus Photon	SW-TP-0005 & SW-WP-0005		14. 1.86	M.E.Gilbert	Drummaird Farm, Bonnybank	3. 5.05P
G-MNKC	Solar Wings Pegasus Photon	SW-TP-0006 & SW-WP-0006		14. 1.86	C Murphy	Ince Blundell	17. 3.04P
G-MNKD	Solar Wings Pegasus Photon *(Originally allocated trike c/n SW-TP-0007 but believed exported: current trike is possibly c/n SW-TP-0016)*	SW-WP-0007		14. 1.86	D.Glasper	(Darlington)	8. 4.05P
G-MNKE	Solar Wings Pegasus Photon	SW-TP-0008 & SW-WP-0008		14. 1.86	M.J.Olsen	Wombleton	2. 8.04P
G-MNKG	Solar Wings Pegasus Photon	SW-TP-0010 & SW-WP-0010		28. 1.86	T.W.Thompson *(Trike stored 9.97: current status unknown)*	Eshott	11. 6.95P
G-MNKK	Solar Wings Pegasus Photon (Fuji-Robin EC-34-PM)	SW-TP-0014 & SW-WP-0014	*(To be fitted with Zanzottera 340cc engine) (Current status unknown)*	28. 1.86	M.E.Gilbert	(Inverkeithing)	7. 5.95P
G-MNKM	MBA Tiger Cub 440	SO.213		30.12.85	A.R.Sunley *(New CofR 11.04)*	(Chelmsford)	17. 2.04P
G-MNKN*	Wheeler Scout Mk.III/3/R *(Built Skycraft (UK) Ltd)* (Fuji-Robin EC-25)	410		6. 1.86	M.J.Aubrey *(Cancelled 19.2.99 as WFU).(Noted 2002)*	(Kington, Hereford)	
G-MNKO	Solar Wings Pegasus XL-Q (Rotax 447)	SW-TB-1158 & SW-WX-0001		2. 1.86	G.Sharp	Eshott	13. 7.03P
G-MNKP	Solar Wings Pegasus Flash	SW-TB-1043 & SW-WF-0014	*(Mainair sailwing c/n W131)*	9. 1.86	I.N.Miller	Hunsdon	30.11.05P
G-MNKR	Solar Wings Pegasus Flash	SW-TB-1045 & SW-WF-0015		14. 1.86	T.R.Murfet	Sutton Meadows, Ely	3.11.04P
G-MNKS*	Solar Wings Pegasus Flash	SW-TB-1044 & SW-WF-0016	*(Mainair sailwing c/n W132)*	9. 1.86	P.L.Dowd *(Cancelled 6.8.04 by CAA)*	(Wirral)	6. 6.04P
G-MNKU	Southdown Puma Sprint (Rotax 447)	SN1231/0100		29. 1.86	S.P.O'Hannrachain	(Coolaney, County Sligo)	30. 8.03P
G-MNKV	Solar Wings Pegasus Flash	SW-TB-1047 & SW-WF-0017	*(Mainair sailwing c/n W137)*	15. 1.86	G.Wakerley	Dunkeswell	25. 4.05P
G-MNKW	Solar Wings Pegasus Flash	SW-TB-1049 & SW-WF-0018	*(Mainair sailwing c/n W140)*	28. 1.86	S.P.Halford	Mill Farm, Hughley, Much Wenlock	29. 6.05P
G-MNKX	Solar Wings Pegasus Flash	SW-TB-1054 & SW-WF-0019	*(Mainair sailwing c/n W139)*	28. 2.86	P.Samal	(Sandy)	3. 8.05P
G-MNKZ	Southdown Raven X	SN2232/0102		4. 2.86	G.B.Gratton	Chilbolton	13. 7.03P
G-MNLE	Southdown Raven X	SN2232/0128		30. 4.86	I.D. and P.G.Cresswell *(Current status unknown)*	(Rochester)	6.10.98P
G-MNLH	Romain Cobra Biplane (Midwest AE50R)	001		23. 1.86	J.W.E.Romain	(Welwyn)	27.10.04P
G-MNLI	Mainair Gemini/Flash II	407-286-4 & W152		28. 1.86	P.M.Fessi	Swinford, Rugby	9. 5.04P
G-MNLM	Southdown Raven X	SN2232/0110		6. 2.86	A P.White *(Current status unknown)*	(Exmouth)	9. 6.93P
G-MNLN	Southdown Raven X	SN2232/0111		6. 2.86	A S.Windley	(Matlock)	27.12.00P
G-MNLT	Southdown Raven X	SN2232/0115		6. 2.86	J.L.Stachini	Middle Stoke, Isle of Grain	12. 8.01P
G-MNLV	Southdown Raven X	SN2232/0118		6. 2.86	J.Murphy	(Tonbridge)	26. 5.02P
G-MNLY	Mainair Gemini/Flash	406-386-4 & W151		14. 2.86	P D Parry	(Ruthin)	26. 7.05P
G-MNLZ	Southdown Raven X	SN2232/0123		6. 2.86	E.L.Jenkins	(Bexleyheath)	12. 6.02P
G-MNMC	Mainair Gemini/Southdown Puma Sprint MS *(Originally regd with c/n MLC-01/KR/226184-2, then with c/n 226-184-2 - which is a corruption on the original c/n)*	222-284-2 & P.524		20. 3.84	A.J Lockley	(Crediton)	23. 4.05P
G-MNMD	Southdown Raven X *(Originally regd with c/n SN2232/0121) (SN2000 sailwing c/n prefix indicates sold without trikeand suggests this may have changed also)*	SN2000/0121		10. 2.86	P.G.Overall	(Crawley)	31. 5.05P
G-MNME*	Hiway Skytrike/Demon (Rotax 377)	WTP-01 & 3535009		12. 2.86	M.J.Aubrey *(Cancelled 28.4.00 by CAA) (Noted 2002)* .	(Kington, Hereford)	10. 6.93
G-MNMG	Mainair Gemini/Flash II (Rotax 447)	419-386-4 & W177		11. 2.86	N.A M.Beyer-Kay *(Current status unknown)*	(Southport)	20. 8.94P
G-MNMI	Mainair Gemini/Flash II (Fuji-Robin EC-44) *(Trike and engine ex G-MMZL following accident 8.9.91)*	317-685-3 & W178		11. 2.86	A D.Bales *(Under repair 8.04)*	Priory Farm, Tibenham	21. 4.05P
G-MNMK	Solar Wings Pegasus XL-R	SW-TB-1021 & SW-WA-1038		19. 8.85	A F.Smallacombe	(Okehampton)	2. 7.00P
G-MNML	Southdown Puma Sprint (Fuji-Robin EC-44-PM)	SN1111/0065		4. 8.83	R.C.Carr *(Current status unknown)*	(Launceston)	14. 7.97P

Registration	Type	C/n	Previous identity	Regd	Owner/Operator	Base	CofA
G-MNMM	Aerotech MW.5(K) Sorcerer	5K-0001-02		11. 2.86	S.F.N.Warnell	(Staines)	17. 8.99P
	(Originally regd as c/n SR101-R4008-01 - now officially regd. as c/n 5K-0001-01)				*(New CofR 9.04)*		
G-MNMN	Medway Hybred 44XLR	8286/16		7. 3.86	D.S Blofeld	Middle Stoke, Isle of Grain	4. 7.05P
G-MNMU	Southdown Puma Raven	SN2232/0127		17. 2.86	M.J.Curley	(London Colney)	20. 5.01P
G-MNMV	Mainair Gemini/Flash (Rotax 447)	375-1085-3 & W113		3. 3.86	B.Light	Tarn Farm, Cockerham	6. 6.05P
G-MNMW	Whittaker MW6-1-1 Merlin *(Built E.F.Clapham)* (Rotax 582)	PFA 164-11144		16. 4.86	E.F.Clapham tr G-MNMW Flying Group	Otherton, Cannock	23. 7.02P
G-MNMY	Cyclone 70/Aerial Arts 110SX	CH-02		6. 3.86	N.R.Beale	Deppers Bridge, Southam	15. 7.05P
G-MNNA	Southdown Raven X	SN2232/0129		4. 3.86	D. and G.D.Palfrey *(Current status unknown)*	(Tiverton)	20. 7.88P
G-MNNB	Southdown Raven (Fuji-Robin EC-44-PM)	SN2122/0130		4. 3.86	J.F.Horn	(Yelverton)	3. 6.03P
G-MNNC	Southdown Raven X	SN2232/0131		4. 3.86	S.A Sacker	Deenethorpe	5. 8.00P
G-MNNF	Mainair Gemini/Flash II (Rotax 447)	402-286-4 & W148		28. 2.86	W.J.Gunn *(Stored 1.98: current status unknown)*	Long Marston	8. 4.97P
G-MNNG	Squires Lightfly/Solar Wings Photon	SW-WP-0019		25. 2.86	C.C.Bilham	Huntingdon	4. 8.02P
	(Rotax 277) *(Trike may be Mainair Tri-Flyer c/n 032-221181 ex G-MJKY?)*						
G-MNNI	Mainair Gemini/Flash II	427-486-4 & W170		28. 2.86	J.C.Miller *(Amended CofR 3.02)*	(Edinburgh)	2. 6.98P
G-MNNJ	Mainair Gemini/Flash II *(ID Plate incorrectly marked as "G-MNNZ")*	405-286-4 & W150		28. 2.86	H.D.Lynch	(Fermoy, County Cork)	6. 5.05P
G-MNNL	Mainair Gemini/Flash II	429-486-4 & W186		28. 2.86	D.Wilson	(Nottingham)	14. 9.04P
G-MNNM	Mainair Scorcher Solo	424-486-1 & W182	(G-MNPE)	20. 3.86	S.R.Leeper and L.L.Perry *(Noted 11.04)*	Grove Farm, Needham	8. 9.91P
G-MNNO	Southdown Raven X	SN2232/0133		26. 3.86	M.J.Robbins	(Tunbridge Wells)	16.12.01P
G-MNNR	Mainair Gemini/Flash II *(Wing originally quoted as c/n W157)*	430-586-4 & W188		6. 3.86	W.A B.Hill *(Noted 5.03)*	Davidstow Moor	9. 6.02P
G-MNNS	Eurowing Goldwing (Rotax 377)	EW-74		8. 4.86	J.S.R.Moodie *(Current status unknown)*	Rovie Farm, Rogart	
G-MNNV	Mainair Gemini/Flash II	431-586-4 & W187		10. 3.86	M.J.Lea	Tarn Farm, Cockerham	11. 4.02P
G-MNNY	Solar Wings Pegasus Flash	SW-TB-1059 & SW-WF-0023	*(Mainair sailwing c/n W161)*	14. 3.86	C.W.Payne	Croft Farm, Defford	20. 3.04P
G-MNNZ	Solar Wings Pegasus Flash II	SW-TB-1060 & SW-WF-0101	*(Mainair sailwing c/n W162)*	24. 4.86	R.D.A Henderson *(Current status unknown)*	(Exeter)	1. 4.98P
G-MNPA	Solar Wings Pegasus Flash II (Rotax 462)	SW-TB-1061 & SW-WF-0102		18. 4.86	N.T.Murphy *(New owner 9.01)*	(Rathongon, County Kildare)	30. 5.98P
	(Originally Mainair sailwing c/n W174 but now acquired c/n W210 from G-MNZA)						
G-MNPC	Mainair Gemini/Flash II (Rotax 462)	423-586-4 & W181		17. 3.86	M.S.McGimpsey	Newtownards	29. 7.04P
					(Crashed on football pitch 1.04 and remains hangared 7.04)		
G-MNPG	Mainair Gemini/Flash II (Rotax 447)	437-686-4 & W204		20. 3.86	P.Kirton	Easter Poldar Farm, Thornhill	4. 9.05P
G-MNPV	Mainair Scorcher Solo	432-586-1 & W189		24. 3.86	A W.Fish	(Telford)	21.10.05P
G-MNPY	Mainair Scorcher Solo	452-886-1 & W229		25. 3.86	R.N.O.Kingsbury	(Tunbridge Wells)	26. 8.04P
G-MNPZ	Mainair Scorcher Solo (Rotax 503) *(3-Blade Propeller Test a/c)*	449-886-1 & W226		25. 3.86	S.Stevens *(Current status unknown)*	(North Shields)	4. 9.93P
G-MNRD	Ultraflight Lazair IIIE	81		17. 6.83	F.P.Welsh tr Sywell Lazair Group	Sywell	12. 4.05P
G-MNRE	Mainair Scorcher Solo	453-886-1 & W230		25. 3.86	A P.Pearce	Wickhambrook, Newmarket	26. 3.05P
G-MNRI	Hornet Dual Trainer/Southdown Raven	HRWA 0051 & SN2000/0119		26. 3.86	R.H.Goll *(New owner 1.04)*	(Llansawel, Llandeilo)	2. 8.02P
G-MNRK	Hornet Dual Trainer/Southdown Raven (Rotax 447)	HRWA 0053 & SN2000/0183		26. 3.86	M.A H.Milne *(New owner 3.03)*	(Huntly)	30. 7.95P
G-MNRM	Hornet Dual Trainer/Southdown Raven (Rotax 447)	HRWA 0055 & SN2000/0214		26. 3.86	R.I.Cannan	(Ramsey, Isle of Man)	26.12.04P
G-MNRP	Southdown Raven X	SN2232/0135		7. 4.86	C.Moore *(Current status unknown)*	(Egremont)	5. 7.95P
G-MNRS	Southdown Raven X	SN2232/0137		7. 4.86	M.C.Newman	(St.Leonards-on-Sea)	29. 7.04P
G-MNRT	Midland Ultralights Sirocco 377GB	MU-016		1. 4.86	R.F.Hinton	(Mansfield)	18. 8.01P
G-MNRW	Mainair Gemini/Flash II (Rotax 462)	411-486-4 & W156		7. 4.86	D.Buckthorpe *(Noted 11.04)*	Clench Common	12.10.04P
G-MNRX	Mainair Gemini/Flash II	434-686-4 & W220		8. 4.86	M.Garside	(Fleetwood)	12. 7.05P
G-MNRZ	Mainair Scorcher Solo	426-586-1 & W184		4. 4.86	R Pattrick	Barton	11. 4.04P
G-MNSA	Mainair Gemini/Flash II	442-786-4 & W219		18. 4.86	W.F.G.Panayiotiou	(Llanelli)	23. 8.04P
G-MNSB	Southdown Puma Sprint (Fuji-Robin EC-44-PM)	539 & SN1121/0066		15. 6.83	T.D.Gibson	(Ledbury)	17. 7.00P
G-MNSD	Ultrasports Tri-Pacer 250/Solar Wings Typhoon S4 (Hunting HS.260A)	T182-341L		23. 4.86	A Strydom *(New owner 10.02)*	(London WC1)	Exemption
G-MNSH	Solar Wings Pegasus Flash II (Rotax 447)	SW-TB-1063 & SW-WF-0104	*(Mainair sailwing c/n W163)*	14. 4.86	D.Lee	Newton Bank, Daresbury	14. 5.05P
G-MNSI	Mainair Gemini/Flash II (Rotax 462)	445-786-4 & W213		9. 4.86	E.J.I.Dejaegher *(New owner 12.04)*	(Retford)	9. 6.03P
G-MNSJ	Mainair Gemini/Flash II	443-886-4 & W223		11. 4.86	G.J.Cadden	Smithboro, County Monaghan	9. 5.05P
G-MNSL	Southdown Raven X	SN2232/0145		17. 4.86	P.B.Robinson *(New CofR 5.02)*	(Ely)	11. 8.00P
G-MNSN	Solar Wings Pegasus Flash II (Rotax 447)	SW-TB-1066 & SW-WF-0105	*(Mainair sailwing c/n W173)*	25. 4.86	F.R. and V.L.Higgins	Weston Zoyland	19. 4.97P
					(Stored in wrecked condition 5.98: current status unknown)		
G-MNSR*	Mainair Gemini/Flash II	399-486-4 & W144	(G-MNLJ)	17. 4.86	A M.Bell	Rufforth	22. 5.00P
					(Cancelled 12.4.02 by CAA) (Trike noted 5.03)		
G-MNSX	Southdown Raven X	SN2232/0148		30. 4.86	S.F.Chave	(Honiton)	18. 7.03P
G-MNSY	Southdown Raven X	SN2232/0149		30. 4.86	L.A Hosegood	(Swindon)	8. 3.03P
G-MNTC	Southdown Raven X	SN2232/0150		30. 4.86	D.S.Bancalari *(New owner 11.01)*	(Norwich)	12.10.92P
G-MNTD	Aerial Arts Chaser/110SX	110SX/255 *(C/n duplicates G-MTSF)*		24. 4.86	B.Richardson *(Current status unknown)*	(Sunderland)	
G-MNTE	Southdown Raven X	SN2232/0151		30. 4.86	E Foster	St.Michaels	27. 6.04P
G-MNTI	Mainair Gemini/Flash II	447-886-4 & W231		8. 5.86	R.T.Strathie *(Noted 5.04)*	Nether Huntlywood Farm, Gordon	19. 8.01P
G-MNTK	CFM Shadow Series CD	024		8. 5.86	A B.Potts	Eshott	17. 8.04P
G-MNTM	Southdown Raven X	SN2232/0154		19. 5.86	D.M.Garland	(Atherstone)	24. 7.01P

G-MNTN	Southdown Raven X	SN2232/0155		2. 6.86	J.Hall	(Wolverhampton)	3. 1.05P
G-MNTP	CFM Shadow Series C (Rotax 462)	K.022		19. 5.86	E.G.White	Lower Upham Farm, Chiseldon	23. 9.04P
G-MNTS*	Mainair Gemini/Flash II (Rotax 462)	450-886-4 & W227		3. 4.86	J.A Colley *(Cancelled 9.6.04 by CAA) (Noted 7.04)*	Over Farm, Gloucester	16. 1.02P
G-MNTT	Medway Half Pint/Aerial Arts 130SX (Rotax 462)	12/1486		7. 4.86	P.J.Burrow	(Crediton)	20. 6.03P
G-MNTU	Mainair Gemini/Flash II (Rotax 462?)	460-886-4 & W233		9. 7.86	S.Cogger	(Wickford)	15. 7.05P
G-MNTV	Mainair Gemini/Flash II (Rotax 462)	455-886-4 & W241		9. 7.86	D.R.Coles	Davidstow Moor	17.10.04P
G-MNTX*	Mainair Gemini/Flash II	415-486-4 & W166		20. 5.86	S.S.Thornton *(Cancelled 17.10.03 as WFU)*	St.Michaels	19. 9.03P
G-MNTY	Southdown Raven X	SN2232/0157		29. 5.86	S.Phillips	(Snodland)	16. 5.05P
G-MNTZ	Mainair Gemini/Flash II	457-886-4 & W243		3. 6.86	D.E.Milner	(Leeds)	6. 9.04P
G-MNUA	Mainair Gemini/Flash II (Rotax 462)	458-886-4 & W235		29. 5.86	P.Hughes and S.Beggan *(New owners 11.04)*	Newtownards	26. 9.04P
G-MNUD	Solar Wings Pegasus Flash II (Rotax 462)	SW-TE-0003 & SW-WF-0110	*(Mainair sailwing c/n W195)*	10. 6.86	P.G.H.Milbank	Sutton Meadows, Ely	20. 9.03P
G-MNUE	Solar Wings Pegasus Flash II (Rotax 462)	SW-TE-0002 & SW-WF-0108	*(Originally fitted with Mainair sailwing c/n W193 but now with W209 ex original G-MNYA)*	10. 6.86	P.M.Rogers	(Rochdale)	21.12.04P
G-MNUF	Mainair Gemini/Flash II	472-786-4 & W252		13. 6.86	C.Hannaby	Guy Lane Farm, Waverton	9. 8.04P
G-MNUG	Mainair Gemini/Flash II (Rotax 462)	465-986-4 & W245		13. 6.86	A S Nader	(Liverpool	19.10.05P
G-MNUI	Mainair Tri-Flyer/Skyhook Cutlass (Fuji-Robin EC-44-PM)	MH-01		21. 5.86	M.Holling *(Current status unknown)*	(Goole)	Exemption
G-MNUO	Mainair Gemini/Flash II (Rotax 462)	421-586-4 & W179		9. 7.86	P.S.Taylor	(Weybridge)	11. 5.02P
G-MNUR	Mainair Gemini/Flash II	470-986-4 & W250		14. 8.86	J.C.Greves *(Current status unknown)*	(Cobham)	30. 3.90P
G-MNUU	Southdown Raven X	SN2232/0162		26. 6.86	P.N.Jackson	Davidstow Moor	10. 9.02P
G-MNUW	Southdown Raven X	SN2232/0163		17. 6.86	B A McDonald	(Cambridge)	19.12.96P
G-MNUX	Solar Wings Pegasus XL-R	SW-TB-1072 & SW-WA-1076		24. 6.86	A M.Smith	Shotton Colliery, Peterlee	3. 5.03P
G-MNUY*	Mainair Gemini/Flash II	422-586-4 & W180		23. 6.86	M.Nymark *(Cancelled 14.12.04 by CAA)*	Newtownards	14.10.05P
G-MNVB	Solar Wings Pegasus XL-R	SW-TB-1073 & SW-WA-1077		7. 7.86	M.J.Melvin	(Spalding)	13. 8.05P
G-MNVC	Solar Wings Pegasus XL-R	SW-TB-1074 & SW-WA-1078		7. 7.86	M.N.C.Ward	Shobdon	11. 6.00P
G-MNVE	Solar Wings Pegasus XL-R	SW-TB-1075 & SW-WA-1079		19. 6.86	M.P.Aris	(Welwyn)	11. 8.00P
G-MNVG	Solar Wings Pegasus Flash II (Rotax 447)	SW-TB-1069 & SW-WF-0109	*(Mainair sailwing c/n W194)*	11. 6.86	D.J.Ward	Low Farm, South Walsham	6. 8.05P
G-MNVH	Solar Wings Pegasus Flash II (Rotax 462)	SW-TE-0001 & SW-WF-0122	*(Mainair sailwing c/n W260)*	23. 6.86	J.A Clarke and C.Hall *(Current status unknown)*	(London N22/E8)	9. 4.97P
G-MNVI	CFM Shadow Series C	026		17. 6.86	D.R.C.Pugh	(Caersws, Powys)	17. 9.04P
G-MNVJ	CFM Shadow Series CD	028		17. 6.86	V.C.Readhead	(Saxmundham)	4. 5.02P
G-MNVK	CFM Shadow Series CD	029		17. 6.86	M.Cheetham *(Noted 6.03)*	Plaistowes Farm, St Albans	31. 7.02P
G-MNVL	Medway Half Pint/Aerial Arts 130SX	3/21585 & 130SX-100	G-MNBZ	22. 9.86	B.W.Austin	Croft Farm, Defford	27. 6.04P
G-MNVN	Southdown Puma Raven (Fuji-Robin EC-34-PM)	SN2132/0165		27. 6.86	R.J.Styles	(Worcester)	5.10.05P
G-MNVO	Hovey Whing-Ding II	CW-01		14. 8.86	C.Wilson *(New CofR 5.02)*	(Basildon)	
G-MNVT	Mainair Gemini/Flash II	477-786-4 & W258		27. 6.86	A C.Barker, t/a ACB Hydraulics *(Stored 4.90: current status unknown)*	Hinton-in-the-Hedges	28. 7.87P
G-MNVU	Mainair Gemini/Flash II	468-986-4 & W248		26. 6.86	W.R.Marsh *(Current status unknown)*	Newhouse Farm, Hardwicke, Hereford	28. 6.99P
G-MNVV	Mainair Gemini/Flash II	467-986-4 & W247		26. 6.86	R.P.Hothersall *(Op Northern Microlight School)*	St.Michaels	15. 9.05P
G-MNVW	Mainair Gemini/Flash II	466-986-4 & W246		26. 6.86	J.C.Munro-Hunt *(Current status unknown)*	Little Down Farm, Milson	20. 9.98P
G-MNVZ	Solar Wings Pegasus Photon	SW-TP-0021 & SW-WP-0021		27. 6.86	J.J.Russ *(Current status unknown)*	Eshott	27. 6.94P
G-MNWD	Mainair Gemini/Flash II (Rotax 462)	474-986-4 & W254		27. 6.86	M.B.Rutherford	Swinford, Rugby	7. 7.01P
G-MNWG	Southdown Raven X	SN2232/0170		4. 8.86	D.Murray	(Clevedon)	9. 5.05P
G-MNWI	Mainair Gemini/Flash II	478-986-4 & W264		9. 7.86	P.Bayliss	Blackpool	10. 6.05P
G-MNWK	CFM Shadow Series C	030		9. 7.86	J.E.Hunt *(Current status unknown)*	(Welling, Kent)	19. 8.98P
G-MNWL	Arbiter Services Trike/Aerial Arts 130S	130SX/333		23. 7.86	E.H.Snook *(Current status unknown)*	(Newport Pagnell)	
G-MNWP	Solar Wings Pegasus Flash II	SW-TB-1083 & SW-WF-0113	*(Mainair sailwing c/n W198)*	4. 8.86	P.J.Harrison *(Cancelled 28.8.02 by CAA)*	(Beccles)	19. 3.03P
G-MNWU	Solar Wings Pegasus Flash II (Rotax 462)	SW-TE-0006 & SW-WF-0111	*(Mainair sailwing c/n W196)*	4. 8.86	S.P.Wass	(Springfield, Chelmsford)	26. 4.05P
G-MNWV	Solar Wings Pegasus Flash II	SW-TB-1090 & SW-WF-0121	*(Mainair sailwing c/n W206)*	4. 8.86	A T.Palmer tr Pegasus Group	Davidstow Moor	29. 8.04P
G-MNWW	Solar Wings Pegasus XL Tug (Rotax 462)	SW-TE-0008 & SW-WA-1085		8.10.86	N.P.Chitty tr Chiltern Flyers Aero Tow Group	Ginge, Wantage	21. 9.05P
G-MNWY	CFM Shadow Series C	K.021 & PFA 161-11130		28. 7.86	N.E.Gormley	(Dublin)	19. 8.03P
G-MNWZ	Mainair Gemini/Flash II	436-686-4 & W203	(G-MNXV)	19. 8.86	W.T.Hume *(Current status unknown)*	(Newmilns)	16. 6.98P
G-MNXA	Southdown Raven X	SN2232/0180		5. 8.86	B.D.Acres	(Maidstone)	28. 6.02P
G-MNXB	Mainair Tri-Flyer/Solar Wings Photon (Fuji-Robin EC-34-PM)	016-29981 & SW-WP-0022		29. 7.86	G.W.Carwardine *(Current status unknown)*	(Uckfield)	16. 6.98P
G-MNXE	Southdown Raven X	SN2232/0202		7. 8.86	A E.Silvey	Wilburton, Ely	29. 1.05P
G-MNXF	Southdown Puma Raven	SN2132/0176		2. 9.86	D.E.Gwenin	(Tring)	13. 5.99P
G-MNXG	Southdown Raven X	SN2232/0181		3. 9.86	M.A Williams	(Tonbridge)	22. 7.02P

G-MNXI	Southdown Raven X	SN2232/0179		19. 8.86	A M.Yates *(Current status unknown)*	(Wisbech)	13. 7.96P
G-MNXO	Medway Hybred 44XLR	29786/19		3. 9.86	D.L.Turner	(Chatham)	6. 7.02P
G-MNXP	Solar Wings Pegasus Flash II (Rotax 447)	SW-TB-1094 & SW-WF-0117	*(Mainair sailwing c/n W207)*	16. 9.86	I K Priestley *(New owner 3.04)*	(Thurleigh, Bedford)	6. 8.96P
G-MNXS	Mainair Gemini/Flash II (Rotax 462)	480-986-4 & W267		8. 9.86	F.T.Rawlings *(Believed exported to Portugal c1988?)*	(Hereford)	16. 3.89P
G-MNXU	Mainair Gemini/Flash II	482-1086-4 & W272		18. 8.86	J.M.Hucker *(Current status unknown)*	(Abertillery)	10. 3.98P
G-MNXX	CFM Shadow Series CD	K.027		13. 8.86	P.J.Mogg	Henstridge	29.11.05P
G-MNXZ	Whittaker MW5 Sorcerer *(Built P.J.Cheyney)* (Fuji-Robin EC-34-PM)	PFA 163-11156		13. 8.86	A.J.Glynn	Gerpins Farm, Upminster	1. 9.05P
G-MNYA	Solar Wings Pegasus Flash II (Rotax 447)	SW-TB-1098 & SW-WF-0119	*(Originally laid down with Mainair sailwing c/n W209 but changed to W259 - see G-MNUE)*	3. 9.86	C.Trollope	Watnall	25. 3.05P
G-MNYB	Solar Wings Pegasus XL-R	SW-TB-1096 & SW-WA-1089		8. 9.86	G.Hanna *(Noted 4.04)*	Newtownards	23. 6.05P
G-MNYC	Solar Wings Pegasus XL-R	SW-TB-1097 & SW-WA-1090		3. 9.86	A N.Papworth	Sutton Meadows. Ely	16. 7.05P
G-MNYD	Aerial Arts Chaser/110SX (Rotax 377)	110SX/320		19. 8.86	B.Richardson	(Sunderland)	17.10.04P
G-MNYE	Aerial Arts Chaser/110SX (Rotax 337)	110SX/321		19. 8.86	R.J.Ripley *(New owner 6.00)*	(Oakley, Bedford)	18.11.99P
G-MNYF	Aerial Arts Chaser/110SX (Rotax 377)	110SX/322		19. 8.86	B.Richardson	(Sunderland)	31. 8.04P
G-MNYG	Southdown Puma Raven	SN2122/0172		19. 8.86	K.Clifford	(Stanmore)	3. 7.00P
G-MNYJ	Mainair Gemini/Flash II (Rotax 462)	485-1086-4 & W275		8. 9.86	G.B.Jones	Otherton, Cannock	13. 7.03P
G-MNYK	Mainair Gemini/Flash II (Rotax 582)	494-1086-4 & W296		11. 9.86	J.J.Ryan *(Current status unknown)*	(Enniscorthy, County Wexford)	4.10.95P
G-MNYL	Southdown Raven X	SN2232/0195		2. 9.86	A D.F.Clifford *(Current status unknown)*	Broadmeadow Farm, Hereford	9. 6.98P
G-MNYM	Southdown Raven X	SN2232/0196		2. 9.86	R.L.Davis	Dunkeswell	11. 6.04P
G-MNYP	Southdown Raven X	SN2232/0207		3. 9.86	A G.Davies	(Bristol)	14. 5.01P
G-MNYS	Southdown Raven X	SN2232/0208		8. 9.86	B.Ward *(Current status unknown)*	(Sheerness)	9. 1.99P
G-MNYU	Solar Wings Pegasus XL-R/Se	SW-TB-1100 & SW-WA-1092		16. 9.86	Jean-Bernard Weber *(Trike noted Mill Farm, Shifnal 5.04)*	(Carshalton)	22. 4.05P
G-MNYW	Solar Wings Pegasus XL-R	SW-TB-1102 & SW-WA-1094		11. 9.86	M.P.Waldock *(Current status unknown)*	(Selsdon, Surrey)	7. 8.98P
G-MNYX	Solar Wings Pegasus XL-R (Rotax 462)	SW-TE-0009 & SW-WA-1095		19. 9.86	P.Mayes and J.P.Widdowson *(See G-MMKG)*	(Bridgnorth)	19. 7.05P
G-MNYZ	Solar Wings Pegasus Flash II (Rotax 462)	SW-TE-0010 & SW-WF-0114	*(Mainair sailwing c/n W199)*	11. 9.86	A C.Bartolozzi	(Ely)	7. 8.04P
G-MNZB	Mainair Gemini/Flash II	483-1086-4 & W273		8. 9.86	P.A Ryde	(Knebworth)	13. 2.05P
G-MNZC	Mainair Gemini/Flash II	484-1086-4 & W274		6. 9.86	C.J.Whittaker *(New CofR 11.02)*	(Ledbury)	19. 1.89P
G-MNZD	Mainair Gemini/Flash II	493-1086-4 & W295		8. 9.86	N.D.Carter *(Stored 9.96: current status unknown)*	Little Gransden	4. 4.96P
G-MNZF	Mainair Gemini/Flash II	496-1186-4 & W291		8. 9.86	A L.Wright	Swinford, Rugby	1. 8.04P
G-MNZJ	CFM Shadow Series CD	033		19. 9.86	T.E.P.Eves tr G-MNZJ Shadow Group	Baxby Manor, Husthwaite	13. 6.05P
G-MNZK	Solar Wings Pegasus XL-R/Se	SW-WA-1096		24. 9.86	P.J.Appleby	(Carrick-on-Shannon, County Leitrim)	29. 3.04P
G-MNZP	CFM Shadow Series BD	K.039 & PFA 161-11206		19. 9.86	J.G.Wakeford	Deanland	19. 8.04P
G-MNZR	CFM Shadow Series BD	040		19. 9.86	P.J.Watson	(King's Lynn)	24. 7.04P
G-MNZS	Aerial Arts Alpha/130SX (Rotax 277)	130SX/376		23. 9.86	N.R.Beale	Deppers Bridge, Southam	1. 8.00P
G-MNZU	Eurowing Goldwing (Fuji-Robin EC-34-PM)	EW-88		24. 9.86	P.D.Coppin and P.R.Millen	Colemore Common, Hampshire	5.11.05P
G-MNZW	Southdown Raven X	SN2232/0220		17.10.86	T A Willcox *(New owner 11.03)*	(Bristol)	7. 7.02P
G-MNZX	Southdown Raven X	SN2232/0221		10.10.86	B.F.Hole	Middle Stoke, Isle of Grain	30. 8.05P
G-MNZZ	CFM Shadow Series CD	036		19. 9.86	P.B.Merritt	(Kingsclere)	31. 8.05P
G-MOAC	Beech F33A Bonanza	CE-1349	N1563N	25. 5.89	R.L.Camrass	Le Touquet	30. 5.07
G-MOAN	Aeromot AMT-200S Super Ximango	200133		29. 3.04	A J Leigh	Gamston	20. 4.07
G-MODE	Eurocopter EC120B Colibri	1295	F-WQPU	19. 8.02	N.J.Ferris t/a Brilliant	Gloucestershire	17.11.05T
G-MOFB	Cameron O-120 HAB	4275		13. 1.98	D.M.Moffat	(Chateaux d'Oex, Switzerland)	4. 1.04A
G-MOFF	Cameron O-77 HAB	2040		27. 7.89	D.M.Moffat *"Moff" (Current status unknown)*	(Alveston, Bristol)	7. 9.95A
G-MOFZ	Cameron O-90 HAB	3350		7. 9.94	D.M.Moffat	(Alveston, Bristol)	14. 5.05A
G-MOGI	Grumman AA-5A Cheetah	AA5A-0630	G-BFMU	1. 5.86	J.G.Stewart tr MOGI Flying Group	Cranfield	16.12.05
G-MOGY	Robinson R22 Beta	0899		23.11.88	Northumbria Helicopters Ltd	Newcastle	10. 2.07T
G-MOHS	Piper PA-31-350 Chieftain	31-8152115	G-BWOC, N40898, CP-1665	29. 4.96	Sky Air Travel Ltd	Stapleford	19.10.06T
G-MOKE	Cameron V-77 HAB	3686		4.10.95	D.D.Owen	(Wotton-under-Edge)	13.12.03A
G-MOLE	Taylor JT.2 Titch *(Buillt S.R.Mowle)*	PFA 60-10725		20. 1.87	K.R.H.Wingate *(Noted 6.04)*	Dunkeswell	AC
G-MOLI(1))	Cameron A-250 HAB	3429(1)		26. 1.95	See SECTION 3 Part 1	(Aldershot)	
G-MOLI(2))	Cameron A-250 HAB *(New envelope fitted - date unknown)*	3429(2)		26. 1.95	J.J.Rudoni and A.C.K.Rawson t/a Wickers World Air Balloon Company *"Stan Robertson Transport"*	Stafford	9.10.04T
G-MOLL	Piper PA-32-301T Turbo Saratoga	32-8024040	N82535	25. 3.91	N.A M.and R.A Brain	Netherthorpe	3. 5.06T
G-MOLY	Piper PA-23-160 Apache	23-1686	EI-BAW, G-APFV, EI-ALK, N10F	7. 6.79	J L Thorogood	Longside, Peterhead-	28. 2.05
G-MOMA	Thruster T.600N 450 Sprint	0036-T600N-088	G-CCIB	22. 8.03	Turley Farms Ltd	(Dunholme, Lincoln)	25. 8.05P
G-MOMO	Agusta A109E Power Elite	11154		30. 4.02	Air Harrods Ltd	Stansted	30. 4.05T
G-MONB	Boeing 757-2T7	22780		7. 3.83	Monarch Airlines Ltd	Luton	1. 2.06T
G-MONC	Boeing 757-2T7	22781	PH-AHO, D-ABNY, G-MONC, EC-211, G-MONC	15. 4.83	Monarch Airlines Ltd	Luton	29. 4.05T
G-MOND	Boeing 757-2T7	22960	D-ABNZ, G-MOND	28. 4.83	Monarch Airlines Ltd	Luton	13. 5.05T

G-MONE	Boeing 757-2T7	23293		27. 2.85	Monarch Airlines Ltd *(Renaissance Cruise titles)*	Luton	25. 2.06T
G-MONI	Monnett Moni *(Built ARV Aviation Ltd)* (IAME KFM.107)	PFA 142-10925		12. 1.84	R.M.Edworthy	(Littleover)	16. 4.02P
G-MONJ	Boeing 757-2T7	24104		26. 2.88	Monarch Airlines Ltd	Luton	23. 1.06T
G-MONK	Boeing 757-2T7	24105		26. 2.88	Monarch Airlines Ltd	Luton	31. 5.05T
G-MONR	Airbus A300B4-605R	540	VH-YMJ G-MONR, F-WWAT	15. 3.90	Monarch Airlines Ltd	Luton	2. 4.05T
G-MONS	Airbus A300B4-605R	556	VH-YMK G-MONS, F-WWAY	17. 4.90	Monarch Airlines Ltd	Luton	23. 3.06T
G-MONX	Airbus A320-212	392	F-WWDR	19. 3.93	Monarch Airlines Ltd	Luton	17. 3.06T
G-MOOR	SOCATA TB-10 Tobago	82	G-MILK	23. 7.91	M.Watkin	Sheffield City	4.11.07E
G-MOOS	Hunting Percival P.56 Provost T.1	PAC/F/335	G-BGKA 8041M/XF690	5. 4.91	T.J.Manna t/a Kennet Aviation *(As "XF690" in RAF c/s)*	North Weald	8. 7.05P
G-MOPB	Diamond DA.40 Star	40067		19.11.01	Papa Bravo Aviation Ltd	Bagby	5. 3.05T
G-MOSI*	de Havilland DH.98 Mosquito TT.35				See SECTION 3 Part 1	Dayton, Ohio, US	
G-MOSS	Beech D55 Baron	TE-548	G-AWAD	12. 6.95	A G.E.Camisa	Elstree	5. 6.06
G-MOSY	Cameron O-84 HAB	2315	EI-CAO	17. 4.96	P.L.Mossman	(Trellech, Monmouth)	19. 9.04A
G-MOTA	Bell 206B-3 JetRanger III	4494	N81521	20.10.98	J W Sandle	Runcton Holme, King's Lynn	1.10.07E
G-MOTH	de Havilland DH.82A Tiger Moth *(Built Morris Motors Ltd)*	85340	7035M DE306	31. 1.78	M.C.Russell *(As "K-2567") (Composite rebuild to DH.82 standard)*	Audley End	3. 4.05
G-MOTI	Robin DR500/200i Président *(Registered as DR400/500)*	0006		23.11.98	The Lord Saville of Newdigate tr The Tango India Flying Group	Biggin Hill	11. 2.05
G-MOTO	Piper PA-24-180 Comanche	24-3239	G-EDHE N51867, G-ASFH, EI-AMM, N7998P	24. 3.87	L.T.and S.Evans	Sandown	11.11.05
G-MOUL	Maule M-6-235C Super Rocket	7518C		1. 5.90	M.Klinge	Prestwick	2. 6.06
G-MOUN	Beech B200 Super King Air	BB-1734	N123NA JA200N, N123NA	27. 9.02	G.and H.L.Mountain	Leeds-Bradford	7.11.05T
G-MOUR	Folland Gnat T.1	FL.596	8624M XS102	16. 5.90	R.F.Harvey and M.J.Gadsby tr Yellowjack Group *(As "XR991" in Yellowjacks c/s)*	Kemble	16. 4.05P
G-MOUT	Cessna 182T Skylane	18281315	N2104H	23. 3.04	G Mountain	Leeds-Bradford	3. 5.07T
G-MOVE	Piper Aerostar 601P	61P-0593-7963263	OO-PKB G-MOVE, N8144J	5. 1.79	A Kazaz	Leicester	6. 8.05
G-MOVI	Piper PA-32R-301 Saratoga SP	32R-8313029	G-MARI N8248H	6. 2.89	G-BOON Ltd	Booker	17. 3.06T
G-MOZZ	Avions Mudry CAP.10B	256		30.10.90	N.Skipworth and J.R.W.Luxton	Shrove Furlong Farm, Ilmer	19. 4.06
G-MPAC	Ultravia Pelican PL *(Built M.J.Craven)* (Rotax 912-UL)	PFA 165-12944		6. 4.00	M.J.Craven	Rayne Hall Farm, Braintree	9. 8.05P
G-MPBH	Reims/Cessna FA152 Aerobat	FA1520374	G-FLIC G-BILV	8.12.88	The Moray Flying Club (1990)	RAF Kinloss	18. 8.05T
G-MPBI	Cessna 310R II	310R0584	F-GEBB HB-LMD, N87473	21. 7.97	M.P.Bolshaw	Elstree	23. 9.06
G-MPCD	Airbus A320-212	379	C-GZCD G-MPCD, C-FTDU, G-MPCD, C-FTDU, G-MPCD, C-FTDU, G-MPCD, C-FTDU, G-MPCD, F-WWDY	14. 3.94	Monarch Airlines Ltd	Luton	1. 5.07T
G-MPRL	Cessna 210M Centurion	21061892	EC-GKD N732YY	5. 8.02	Myriad Public Relations Ltd	Top Farm, Croydon, Royston	17.10.05T
G-MPWI	Robin HR100/210 Royal	163	F-GBTY F-ODFA, F-BUPD	3. 3.80	P.G.Clarkson and Sue King	Bournemouth	9. 5.05
G-MPWT	Piper PA-34-220T Seneca III *(Originally built as c/n 34-8233163)*	34-8333068	N4294X N888DB, N4294X, N9539N, N8218K	26. 9.88	Modern Air (UK) Ltd *"Duke 2"*	Fowlmere	19. 5.07T
G-MRAF	Aeroprakt A22 Foxbat *(Built.M Raflewski)*	PFA 317-14370		7. 2.05	M.Raflewski	(Dungannon)	
G-MRAJ	MD Helicopters Hughes 369E	0010E	N51946	19. 3.98	A Jardine	Dundee	10. 6.05
G-MRAM	Mignet HM-1000 Balerit	134		15.11.99	R.A Marven	Tower Hill Lane, Coleman Green, Herts	25. 2.05P
G-MRED	Elmwood CA-05 Christavia Mk.1 *(Built E.Hewett)*	PFA 185-12935		2. 8.96	E.Hewett *(Current status unknown)*	(Fareham)	
G-MRJJ	Mainair Pegasus Quik	7940		16. 4.03	J.H.Sparks	Kingston Semour	21. 4.04P
G-MRKI	Extra EA.300/200	05	N694M	15.12.04	Extra 200 Ltd	(London W1)	AC
G-MRKS	Robinson R44 Raven	0771	G-RAYC	20. 5.03	TJD Trade Ltd	Cranfield	22. 6.06T
G-MRKT	Lindstrand LBL 90A HAB	037		7. 6.93	Marketplace Public Relations (London) Ltd *"Kaytee"*	(Crowthorne)	9. 7.05A
G-MRLN	Sky 240-24 HAB	161		4. 8.99	M.Wady t/a Merlin Balloons	(Hamstreet)	1. 4.05T
G-MRMR	Piper PA-31-350 Navajo Chieftain	31-7952092	OH-PRE G-WROX, G-BNZI, N3517T	21. 8.97	I.D. and P.J.Margetson-Rushmor t/a MRMR Flight Services	Stapleford	22.12.04T
G-MROC	Pegasus Quantum 15-912	7498		22. 1.99	M.Convine	Tower Farm, Wollaston	28. 9.05P
G-MROY	Comco Ikarus C42 FB UK *(Built R Beckham)*	PFA 322-13758		9.10.01	D.M.Jobbins and K.R.Rowland	(Northampton)	9. 1.03P
G-MRSN	Robinson R22 Beta	1654		21. 1.91	M D Thorpe t/a Yorkshire Helicopters	Coney Park, Leeds	11. 6.06T
G-MRST	Piper PA-28RT-201 Arrow IV	28R-7918068	9H-AAU 5B-CEC, N3019U	27.11.86	Calverton Flying Group Ltd	Rochester	8. 4.05T
G-MRTN	SOCATA TB-10 Tobago	62	G-BHET	9. 7.98	Underwood Kitchens Ltd	Turweston	31. 5.07
G-MRTY	Cameron N-77 HAB	1008		24. 4.84	R.A, P.M.G. and N.T.M. Vale *"Marty" (New owners 7.01)*	(Kidderminster)	19. 5.96A
G-MSAL	Morane Saulnier MS.733 Alcyon	143	F-BLXV Fr.Mil	16. 6.93	North Weald Flying Services Ltd *(As "143" in Aéronavale c/s) (Dismantled 1.05)*	North Weald	
G-MSFC	Piper PA-38-112 Tomahawk II	38-81A0067	N25735	11. 5.90	The Sherwood Flying Club Ltd	Tollerton	28.10.05T
G-MSFT	Piper PA-28-161 Warrior II	28-8416093	G-MUMS N118AV	2. 4.97	Western Air (Thruxton) Ltd	Thruxton	20. 5.06T
G-MSIX	DG Flugzeugbau DG-800B *(New fuselage c/n 274 fitted c8.02)* *(Struck hedge making forced landing Cropthorne near Bidford 15.6.02 due to engine failure: substantial damage to fuselage and tailplane)*	8-156B80		21. 4.99	E.Coles tr G-MSIX Group *"M6"*	Dunstable	29. 5.05
G-MSKY	Comco Ikarus C42 FB UK *(Built C K Jones)*	PFA 322-13722		3.10.01	M.W.Fitch tr G-MSKY Group	Plaistowes Farm, St Albans	6. 9.05P
G-MSPY	Pegasus Quantum 15-912	7625		17. 3.00	J.Madhvani and R.K.Green	Plaistowes Farm, St.Albans	15. 3.05P

G-MSTC	Gulfstream AA-5A Cheetah	AA5A-0833	G-BIJT N26950	30. 1.95	J.Crook tr Association Of Manx Pilots	Andreas, Isle of Man	19. 2.05T
G-MSTG	North American P-51D-25-NT Mustang	124-48271	NZ2427 45-11518	2. 9.97	M.Hammond *"Janie"* *(As "414419/LH-F" in USAAF c/s of 350th Fighter Sqdn/353rd Fighter Group)*	Airfield Farm, Hardwick	19. 8.05P
G-MSTR	Cameron Monster 110SS HAB	4957	G-OJOB	18. 7.01	Airship and Balloon Company Ltd	(Bristol)	10. 8.05A
G-MTAA	Solar Wings Pegasus XL-R	SW-TB-1108 & SW-WA-1102		15.10.86	R.Scott	London Colney	30. 5.02P
G-MTAB	Mainair Gemini/Flash II	492-1086-4 & W290		8.10.86	C.Thompson	(Wolverhampton)	22. 5.05P
G-MTAE	Mainair Gemini/Flash II	500-1186-4 & W302		15.10.86	C.E.Hannigan	Perth	26. 6.05P
G-MTAF	Mainair Gemini/Flash II	499-1186-4 & W301		5.10.86	P.A Long	St.Michaels	21.12.04P
G-MTAG	Mainair Gemini/Flash II	487-1086-4 & W281		15.10.86	M.J.Cowie and J.P.Hardy	Ince Blundell	28. 5.04P
G-MTAH	Mainair Gemini/Flash II	488-1086-4 & W282		16.10.86	T.G.Elmhirst	St.Michaels	15. 9.05P
G-MTAI	Solar Wings Pegasus XL-R (Rotax 503)	SW-TB-1109 & SW-WA-1103		14.10.86	D.Ruston	Watnall	11. 5.05P
G-MTAJ	Solar Wings Pegasus XL-R	SW-TB-1110 & SW-WA-1104		16.10.86	G.A and S.D.Batchelor	Deenethorpe	14.12.05P
G-MTAL	Solar Wings Pegasus Photon *(Rotax 277?)*	SW-TP-0023 & SW-WP-0023		15.10.86	R.P.Wilkinson *(Current status unknown)*	(Batheaston, Bath)	29.10.95P
G-MTAO	Solar Wings Pegasus XL-R	SW-TB-1107 & SW-WA-1107		21.10.86	S.P.Disney *(New CofR 8.03)*	Swinford, Rugby	26. 6.01P
G-MTAP	Southdown Raven X	SN2232/0225		15.10.86	M.C.Newman *(New owner 9.02)*	(St.Leonards-on-Sea)	13. 6.98P
G-MTAR	Mainair Gemini Flash II (Rotax 462)	504-1286-4-W307		16.10.86	J.B.Woolley	(Madrid, Spain)	9. 6.05P
G-MTAS	Whittaker MW5 Sorcerer *(Built E A Henman)* (Rotax 503) *(May be Model MW5C?)*	PFA 163-11166		14.10.86	R.L.Nyman	Redlands	12. 8.05P
G-MTAV	Solar Wings Pegasus XL-R	SW-TB-1115 & SW-WA-1110		21.10.86	Susan Fairweather and Carolyn L.Harris	(Nottingham/Warrington)	13. 6.05P
G-MTAW	Solar Wings Pegasus XL-R	SW-TB-1116 & SW-WA-1111		21.10.86	M.G.Ralph	Weston Zoyland	13. 7.05P
G-MTAX	Solar Wings Pegasus XL-R	SW-TB-1117 & SW-WA-1115		27.10.86	G Hawes	Deenethorpe	17.10.04P
G-MTAY	Solar Wings Pegasus XL-R	SW-TB-1118 & SW-WA-1113		27.10.86	S.A McLatchie	Enstone	20. 8.04P
G-MTAZ	Solar Wings Pegasus XL-R	SW-TB-1119 & SW-WA-1114		28.10.86	T.L.Moses *(New owner 5.03)*	Llansalnt, Carmarthen	25. 3.01P
G-MTBA	Solar Wings Pegasus XL-R	SW-TB-1120 & SW-WA-1115		27.10.86	R.J.W.Franklin and M.C.Buffery *(Wrecked 5.97: current status unknown)*	(Cheltenham)	24. 6.93P
G-MTBB	Southdown Raven X	SN2232/0226		16.10.86	A Miller	(Woking)	15.10.02P
G-MTBD	Mainair Gemini/Flash II *(Wing regd as W229)*	498-1186-4 & W299		16.10.86	J.Williams	Oxton	24. 8.05P
G-MTBE	CFM Shadow Series CD (Rotax 462HP)	K.035		16.10.86	S.K.Brown	Old Sarum	18. 3.04P
G-MTBH	Mainair Gemini/Flash II (Rotax 462)	524-187-5 & W327		28.10.86	T.and P.Sludds	Newtownards	20. 7.05P
G-MTBJ	Mainair Gemini/Flash II	509-1286-4 & W312		27.10.86	R.M. and P.J.Perry *(Op Staffordshire Aero Club)*	Otherton, Cannock	29.10.05P
G-MTBK	Southdown Raven X (Rotax 503) *(Officially regd with Rotax 447)*	SN2232/0230		28.10.86	I.Garwood *(New owner 12.03)*	(Northampton)	27. 6.99P
G-MTBL	Solar Wings Pegasus XL-R	SW-TB-1121 & SW-WA-1117		6.11.86	R.N.Whiting	Lower Mountpleasant, Chatteris	16. 7.04P
G-MTBN	Southdown Raven X	SN2232/0227		28.10.86	A J. and S.E.Crosby-Jones	Hailsham	5. 6.05P
G-MTBO	Southdown Raven X	SN2232/0233		28.10.86	J.Liversuch	Doynton	24. 6.05P
G-MTBP	Aerotech MW-5B Sorcerer (Fuji-Robin EC-44-PM)	SR102-R440B-02		28.10.86	G.Bennett *(Current status unknown)*	(Caister-on-Sea)	21. 9.94P
G-MTBR	Aerotech MW-5B Sorcerer (Fuji-Robin EC-44-PM)	SR102-R440B-03		20. 1.87	P.W.Hastings	Long Marston	31.10.02P
G-MTBS	Aerotech MW-5B Sorcerer (Fuji-Robin EC-44-PM)	SR102-R440B-04		27.10.86	T B Fowler	(Newent)	27. 9.04P
G-MTBU	Solar Wings Pegasus XL-R	SW-TB-1122 & SW-WA-1118		13.11.86	R.P.R.Staveley	(Alfreton)	7. 3.05P
G-MTBV	Solar Wings Pegasus XL-R	SW-WA-1119		6.11.86	T.H.Scott *(New owner 6.03)*	(Coggeshall)	
G-MTBX	Mainair Gemini/Flash II (Rotax 447)	510-1286-4 & W313		6.11.86	A F Grimwood	(Wigan)	24. 3.05P
G-MTBY	Mainair Gemini/Flash II (Rotax 447)	507-1286-4-W310		6.11.86	D Pearson	(Heywood)	22. 7.05P
G-MTBZ	Southdown Raven X	SN2232/0232		10.11.86	K.W.E.Brunnenkant	(Lincoln)	6. 3.05P
G-MTCA	CFM Shadow Series C	K.011		6.11.86	J.R.L.Murray	East Fortune	4. 8.04P
G-MTCC*	Mainair Gemini/Flash II	497-1186-4 & W298		13.11.86	J.Madhvani *(Damaged trike noted 9.00: cancelled 4.10.00 as WFU) (Noted 7.03)*	Plaistowes Farm, St.Albans	21.10.96P
G-MTCE	Mainair Gemini/Flash II (Rotax 462)	511-1286-4 & W314		2.12.86	H.Shaw	(Woodseaves, Stafford)	9. 8.04P
G-MTCG	Solar Wings Pegasus XL-R/Se	SW-TB-1125 & SW-WA-1123		16.12.86	M K Nicholson	Eshott	11.12.02P
G-MTCK	Solar Wings Pegasus Flash II (Rotax 447)	SW-TB-1127 & SW-WF-0127	*(Mainair sailwing c/n W263)*	11.12.86	J J Littler	(Aldingbourne, Chichester)	21. 9.05P
G-MTCM	Southdown Raven X	SN2232/0239		11.12.86	J.C Rose *(New owner 9.01)*	Field Farm, Oakley	2. 7.97P
G-MTCN	Solar Wings Pegasus XL-R	SW-TB-1128 & SW-WA-1126		16.12.86	S.R.Hughes	Over Farm, Gloucester	11. 7.05P
G-MTCO	Solar Wings Pegasus XL-R	SW-TB-1129 & SW-WA-1127		7. 1.87	R.Johnson	(Bury St.Edmunds)	3. 4.04P
G-MTCP	Aerial Arts Chaser/110SX (Rotax 377)	110SX/476		16.12.86	B.Richardson	(Sunderland)	28. 6.00P
G-MTCR	Solar Wings Pegasus XL-R	SW-TB-1130 & SW-WA-1128		16.12.86	G.M.Cruise-Smith	Roddige	25. 3.05P

G-MTCT	CFM Shadow Series CD	042	16.12.86	F.W.McCann	Cumbernauld	19. 9.05P
G-MTCU	Mainair Gemini/Flash IIA	451-1286-4 & W228	5. 1.87	T.J.Philip	Ashcroft Farm, Winsford	12. 9.05P
G-MTCW	Mainair Gemini/Flash II	502-1186-4 & W304	5. 1.87	S.B.Walters	Deenethorpe	20. 1.05P
	(Rotax 462)					
G-MTDD	Aerial Arts Chaser/110SX	110SX/437	26. 1.87	B.Richardson	(Sunderland)	4. 7.00P
	(Rotax 377)					
G-MTDE	Aerial Arts Chaser/110SX	110SX/438	5. 1.87	M.N.Hudson	North Coates	25. 4.05P
	(Rotax 377) *(May now have Rotax 330)*					
G-MTDF	Mainair Gemini/Flash II	515-287-5 & W319	5. 1.87	P G Barnes *(New owner 2.04)*	(Harwich)	1. 5.03P
G-MTDG	Solar Wings Pegasus XL-R/Se	SW-TB-1132 & SW-WA-1130	20. 1.87	E.W.Laidlaw *(Noted 8.00)*	(Turriff)	7. 7.01P
G-MTDH	Solar Wings Pegasus XL-R	SW-TB-1133 & SW-WA-1131	22. 1.87	F.J.McVey	Insch	15. 5.05P
G-MTDI	Solar Wings Pegasus XL-R/Se	SW-TB-1134 & SW-WA-1132	22. 1.87	D Allan	Eshott	10. 3.05P
G-MTDK	Aerotech MW-5B Sorcerer	SR102-R440B-06	22. 1.87	C.C.Wright *(Noted 5.04)*	Easter Balgillo Farm, Finavon	23. 6.00P
	(Fuji-Robin EC-44-PM) *(To be converted to Rotax 447)*					
G-MTDN	Ultraflight Lazair IIIE	A465/002	22. 1.87	M.J.Broom *(New owner 6.01)*	Long Marston	27. 6.97P
	(Rotax 185)					
G-MTDO	Eipper Quicksilver MXII	1124	27. 2.87	D.L.Ham *(Current status unknown)*	(Honiton)	Exemption
	(Rotax 503)					
G-MTDR	Mainair Gemini/Flash II	516-287-5 & W276	26. 1.87	J.W.and C.Richardson	Church Farm, Askern	11. 6.04P
G-MTDU	CFM Shadow Series CD	K.037	26. 1.87	R.C.Osler	Over Farm, Gloucester	31. 8.05P
G-MTDW	Mainair Gemini/Flash II	517-387-5 & W212	2. 2.87	S.R.Leeper	Priory Farm, Tibenham	4.12.04P
G-MTDX*	CFM Shadow Series BD	K.043	10. 2.87	L.Fekete *(Cancelled 12.4.02 by CAA)*	(Ellesmere Port)	4. 6.00P
	(Rotax 503)					
G-MTDY	Mainair Gemini/Flash II	513-187-5 & W317	11. 2.87	S.Penoyre	(Windlesham)	13.10.00P
	(Rotax 462)					
G-MTEC	Solar Wings Pegasus XL-R	SW-TB-1142 & SW-WA-1140	9. 2.87	R.W.Glover *(Trike noted 2000)*	Kemble	11. 6.94P
G-MTED*	Solar Wings Pegasus XL-R	SW-TB-1143 & SW-WA-1141	9. 2.87	D.Marsh *(Cancelled 19.5.03 by CAA)*	Charminster, Bournemouth	31. 8.01P
G-MTEE	Solar Wings Pegasus XL-R	SW-TB-1144 & SW-WA-1142	13. 2.87	M J Moulton	Roddige	17. 9.04P
	(C/n plate incorrectly shows SW-WA-1144 & SW-WA-1142) (New wing ? - see G-MTLG)					
G-MTEJ	Mainair Gemini/Flash II	522-387-5 & W277	18. 2.87	G.J.Moore	Ince Blundell	29. 5.03P
	(Rotax 462)					
G-MTEK	Mainair Gemini/Flash II	523-387-5 & W279	3. 3.87	M.O'Hearne and G.M.Wrigley	Rufforth	17. 6.03P
G-MTEN	Mainair Gemini/Flash II	527-487-5 & W285	25. 2.87	B.Bennison	(Brough)	18.10.05P
G-MTER	Solar Wings Pegasus XL-R/Se	SW-TB-1146 & SW-WA-1144	19. 2.87	I.Stratford	(Stoke-on-Trent)	7. 4.02P
G-MTES	Solar Wings Pegasus XL-R	SW-TB-1147 & SW-WA-1145	19. 2.87	N.P.Read	Davidstow Moor	29. 8.04P
G-MTET	Solar Wings Pegasus XL-R	SW-TB-1148 & SW-WA-1146	19. 2.87	P.A S.Talbot	(Camborne)	2. 5.00P
G-MTEU	Solar Wings Pegasus XL-R/Se	SW-TB-1149 & SW-WA-1147	19. 2.87	B.Harris	(Northwich)	9. 4.01P
G-MTEW	Solar Wings Pegasus XL-R/Se	SW-TB-1151 & SW-WA-1149	19. 2.87	R.W. and P.J.Holley	Mill Farm, Shifnal	4. 5.05P
G-MTEX	Solar Wings Pegasus XL-R	SW-TB-1152 & SW-WA-1150	19. 2.87	C.M.and K.M.Bradford	(Marlborough)	4. 5.04P
G-MTEY	Mainair Gemini/Flash II	518-387-5 & W217	20. 2.87	A Wells *(Current status unknown)*	Thornton Watlass	30. 5.05P
G-MTFA	Solar Wings Pegasus XL-R	SW-TB-1158 & SW-WA-1156	24. 2.87	I.Armitstead *(Noted 8.04)*	Tarn Farm, Cockerham	14. 3.94P
	(Rotax 462)					
G-MTFB	Solar Wings Pegasus XL-R	SW-TE-0015 & SW-WA-1157	24. 2.87	S.J.M.Morling	(Taunton)	13. 9.05P
	(Rotax 462)					
G-MTFC	Medway Hybred 44XLR	22087/24	23. 3.87	J.K.Masters *(Current status unknown)*	(Chigwell)	25. 7.97P
G-MTFE	Solar Wings Pegasus XL-R	SW-TB-1157 & SW-WA-1155	6. 3.87	D.A Eastough	(Chellaston)	17. 5.02P
	(New sailwing fitted 1999)					
G-MTFF	Mainair Gemini/Flash II	528-487-5 & W286	12. 3.87	T.N.Taylor *(Current status unknown)*	(Sidcup)	19. 4.96P
G-MTFG	AMF Microflight Chevvron 2-32C	CH.004	9. 3.87	R.Gardner	(Stratford-upon-Avon)	19. 8.04P
G-MTFI	Mainair Gemini/Flash II	531-487-5 & W289	12. 3.87	M.Carolan	Annaghmore, County Tyrone	27. 6.05P
G-MTFJ*	Mainair Gemini/Flash II	532-487-5 & W320	12. 3.87	G.Souch and M.D.Peacock *(Cancelled 2.6.03 by CAA)*	(Leatherhead)	3. 7.03P
G-MTFK*	Moult Trike/Flexiform Striker			See SECTION 3 Part 1	Flixton, Bungay	
G-MTFL	Ultraflight Lazair IIIE	A466/003	12. 3.87	P.J.Turrell *(Current status unknown)*	(Halesowen)	26. 9.89P
	(Built AMF Microflight Ltd) (Rotax 185 x 2)					
G-MTFM	Solar Wings Pegasus XL-R	SW-TE-0016 & SW-WA-1158	13. 3.87	P.R.G.Morley	Newnham, Baldock	26. 4.05P
	(Rotax 462)					
G-MTFN	Whittaker MW5 Sorcerer	PFA 163-11207	13. 3.87	S.M.King	Errol	21. 2.05P
	(Built K Southam and D C Britton) (Fuji-Robin EC-44-PM) *(May be Model MW5B)*					
G-MTFO	Solar Wings Pegasus XL-R/Se	SW-TB-1159 & SW-WA-1159	18. 3.87	A Gonzalez and W.Highton	Carlisle	1. 5.02P
G-MTFP	Solar Wings Pegasus XL-R	SW-TB-1160 & SW-WA-1160	18. 3.87	C.Rickards	(Swansea)	25. 4.04P
G-MTFR	Solar Wings Pegasus XL-R/Se	SW-TB-1161 & SW-WA-1161	18. 3.87	S.Ballantyne *(Current status unknown)*	(Blanefield)	19. 9.99P
G-MTFT	Solar Wings Pegasus XL-R	SW-TB-1163 & SW-WA-1163	18. 3.87	A T.Smith	Hughley, Much Wenlock	30. 7.00P
G-MTFZ	CFM Shadow Series CD	053	24. 3.87	R.P.Stonor	Long Marston	7.12.05P
G-MTGA	Mainair Gemini/Flash II	535-587-5 & W293	26. 3.87	B.S.Ogden	Tarn Farm, Cockerham	25. 5.04P
G-MTGB	Thruster TST Mk.1	837-TST-011	10. 4.87	G.Arthur	Tarn Farm, Cockerham	11. 9.00P
G-MTGC	Thruster TST Mk.1	837-TST-012	10. 4.87	B.Foster and P.Smith	Gerpins Lane, Upminster	31. 5.04P
G-MTGD	Thruster TST Mk.1	837-TST-013	10. 4.87	W.J.Lister	Strathaven	23.10.04P

Reg	Type	C/n	Date	Owner	Location	CofA
G-MTGE	Thruster TST Mk.1	837-TST-014	10. 4.87	G.W.R.Swift *(Current status unknown)*	(Hartfield)	17.10.99P
G-MTGF	Thruster TST Mk.1	837-TST-015	10. 4.87	B.Swindon	(Chesham)	16. 9.04P
G-MTGH	Mainair Gemini/Flash II (Rotax 462)	536-587-5 & W294	31. 3.87	J.R.Gillies	Graveley	27. 5.05P
G-MTGJ	Solar Wings Pegasus XL-R	SW-TB-1165 & SW-WA-1165	1. 4.87	M.S.Taylor	(Gillingham)	17. 2.02P
G-MTGK	Solar Wings Pegasus XL-R	SW-TB-1166 & SW-WA-1166	1. 4.87	I.A Smith *(Current status unknown)*	(Canterbury)	1. 8.91P
G-MTGL	Solar Wings Pegasus XL-R	SW-TB-1167 & SW-WA-1167	1. 4.87	P.J.and R.Openshaw	Ince Blundell	31. 5.05P
G-MTGM	Solar Wings Pegasus XL-R/Se	SW-TB-1168 & SW-WA-1168	1. 4.87	D.J.Barnes tr TGM Syndicate *(Original trike destroyed in gales Roddige@ 1.98 : new trike fitted ex G-MNYT (c/n SW-TB-1099)*	Roddige	1. 9.04P
G-MTGN	CFM Shadow Series BD	K.041	31. 3.87	N.G.Price	Bricket Wood, Radlett	15. 6.03P
G-MTGO	Mainair Gemini/Flash IIA (Rotax 462)	550-587-5 & W336	10. 4.87	P.Jephcott	Long Marston	22. 7.05P
G-MTGP*	Thruster TST Mk.1	847-TST-016	10. 4.87	L.A Hosegood *(Cancelled 30.12.02 by CAA) (Noted wrecked 5.03)*	South Wraxall, Bradford-on-Avon	15. 3.03P
G-MTGR	Thruster TST Mk.1	847-TST-017	10. 4.87	M.R.Grunwell	Gerpins Farm, Upminster	9.12.04P
G-MTGS	Thruster TST Mk.1	847-TST-018	10. 4.87	R.Dennett	Barton	21. 8.04P
G-MTGT	Thruster TST Mk.1	847-TST-019	10. 4.87	A W.Paterson and P.McVay	Strathaven	16. 1.05P
G-MTGU	Thruster TST Mk.1	847-TST-020	10. 4.87	J.Jordan *(New owner 1.05)*	(Stockland Green, Birmingham)	30. 9.04P
G-MTGV	CFM Shadow Series CD	052	8. 4.87	V.R.Riley	(Amlwch, Gywnedd)	11. 3.01P
G-MTGW	CFM Shadow Series CD	054	8. 4.87	J.O.Kane	Rayne Hall Farm, Braintree	6. 9.05P
G-MTGX	Hornet Dual Trainer/Southdown Raven	HRWA 0061 & SN2000/0270	13. 4.87	M.A Pantling	Weston Zoyland	30. 7.05P
G-MTHB	Aerotech MW-5B Sorcerer (Fuji-Robin EC-44-PM)	SR102-R440B-08	10. 4.87	F.R.Wilson	Roddige	16. 7.05P
G-MTHC	Raven Aircraft Raven X	SN2232/0257	15. 4.87	E.Bayliss	Ince Blundell	27. 2.05P
G-MTHG	Solar Wings Pegasus XL-R	SW-TB-1170 & SW-WA-1171	13. 4.87	H.E.Paterson	(Poynton)	12. 3.03P
G-MTHH	Solar Wings Pegasus XL-R	SW-TB-1171 & SW-WA-1172	13. 4.87	J.Palmer	(Winkleigh)	28.12.98P
G-MTHI	Solar Wings Pegasus XL-R	SW-TB-1172 & SW-WA-1173	13. 4.87	Thames Valley Airsports Ltd	Chiltern Park, Wallingford	26. 9.04P
G-MTHJ	Solar Wings Pegasus XL-R	SW-TB-1173 & SW-WA-1174	13. 4.87	S.A Watson	Longacres Farm, Sandy	26. 4.05P
G-MTHN	Solar Wings Pegasus XL-R	SW-TB-1177 & SW-WA-1178	13. 4.87	M.T.Seal	(Cilcennin, Lampeter)	15. 6.05P
G-MTHT	CFM Shadow Series CD	058	22. 4.87	C.A.S.Powell	(Cheadle)	24. 5.04P
G-MTHV	CFM Shadow Series BD	K.049	7. 5.87	K.R.Bircher *(Noted 7.04)*	Over Farm, Gloucester	24. 7.00P
G-MTHW	Mainair Gemini/Flash II (Rotax 462)	540-587-5 & W325	14. 5.87	M.D.Kirby	Chase Farm, Billericay	24. 1.05P
G-MTHZ	Mainair Gemini/Flash IIA	541-587-5 & W329	14. 5.87	A I.Kinnear	East Fortune	12. 7.05P
G-MTIA	Mainair Gemini/Flash IIA	544-687-5 & W332	14. 5.87	T.J.Burrow	(Penwortham, Preston)	29.11.04P
G-MTIB	Mainair Gemini/Flash IIA	545-687-5 & W333	14. 5.87	K.P.Hayes	St.Michaels	7.12.04P
G-MTIE	Solar Wings Pegasus XL-R (Rotax 462)	SW-TE-0019 & SW-WA-1183	18. 5.87	P.Wibberley	(Chesterfield)	29. 4.05P
G-MTIH	Solar Wings Pegasus XL-R	SW-TB-1183 & SW-WA-1186	18. 5.87	B Chapman	(Chelmsford)	20. 6.05P
G-MTIJ	Solar Wings Pegasus XL-R/Se	SW-TB-1185 & SW-WA-1188	18. 5.87	M.J.F.Gilbody *(Current status unknown)*	(Urmston, Manchester)	1. 4.98P
G-MTIK	Raven Aircraft Raven X	SN2232/0272	19. 5.87	G.A Oldershaw	Sutton Meadows, Ely	14. 2.05P
G-MTIL	Mainair Gemini/Flash IIA (Rotax 462)	549-687-5 & W338	21. 5.87	P.G.Nolan	Ince Blundell	16. 7.05P
G-MTIM	Mainair Gemini/Flash IIA	553-687-5 & W341	21. 5.87	W.M.Swan	East Fortune	1. 5.05P
G-MTIN	Mainair Gemini/Flash IIA	547-687-5 & W335	1. 6.87	S.J.Firth	(Dallerie, Crieff)	3. 5.05P
G-MTIO	Solar Wings Pegasus XL-R	SW-TB-1187 & SW-WA-1190	26. 5.87	J W Mount	(Lowestoft	6. 8.05P
G-MTIP	Solar Wings Pegasus XL-R	SW-TB-1188 & SW-WA-1191	26. 5.87	M.P.Williams	Redlands	8.12.05P
G-MTIR	Solar Wings Pegasus XL-R/Se	SW-TB-1189 & SW-WA-1192	26. 5.87	D.Raybould	(Chesterfield)	17. 7.05P
G-MTIS	Solar Wings Pegasus XL-R	SW-TB-1190 & SW-WA-1193	26. 5.87	N.P.Power	(Eastbourne)	25. 7.05P
G-MTIU	Solar Wings Pegasus XL-R	SW-TB-1191 & SW-WA-1194	26. 5.87	D E Pedder	(Hampton)	6. 8.05P
G-MTIV	Solar Wings Pegasus XL-R	SW-TB-1192 & SW-WA-1195	26. 5.87	P.J.Culverhouse, tr Syndicate IV	Sittles Farm, Alrewas	19.10.04P
G-MTIW	Solar Wings Pegasus XL-R	SW-TB-1193 & SW-WA-1196	26. 5.87	G.S.Francis	(Bristol)	12. 9.05P
G-MTIX	Solar Wings Pegasus XL-R	SW-TB-1194 & SW-WA-1197	26. 5.87	S.Pickering	Sutton Meadows, Ely	15. 1.01P
G-MTIY	Solar Wings Pegasus XL-R	SW-TB-1195 & SW-WA-1198	26. 5.87	P.J.Tanner	Weston Zoyland	27. 5.04P
G-MTIZ	Solar Wings Pegasus XL-R	SW-TB-1196 & SW-WA-1199	26. 5.87	S.L.Blount	Sutton Meadows. Ely	22.10.03P
G-MTJA	Mainair Gemini/Flash IIA	551-687-5 & W339	15. 6.87	K.A.Brunton	Finmere	3. 4.05P
G-MTJB	Mainair Gemini/Flash IIA (Rotax 462)	554-687-5 & W343	2. 6.87	B.Skidmore	Tarn Farm, Cockerham	26. 5.05P
G-MTJC	Mainair Gemini/Flash IIA (Honda BF52 @ 808cc)	555-687-5 & W344	1. 6.87	T.A Dockrell	Kingston Seymour	8. 7.05P
G-MTJD	Mainair Gemini/Flash IIA (Rotax 462)	552-687-5 & W340	5. 6.87	D.J.Richards	(Nivillac, France)	19.10.02P

Reg	Type	C/n	Previous identity	Regd	Owner/operator	Probable base	CofA expiry
G-MTJE	Mainair Gemini/Flash IIA	556-687-5 & W345		24. 6.87	I.O.S.Ross	(Stirling)	26. 9.05P
G-MTJG	Medway Hybred 44XLR	22587/25		16. 6.87	M A Trodden *(Current status unknown)*	Tupton, Chesterfield	24. 2.99P
G-MTJH	Solar Wings Pegasus Flash	SW-TB-1050 & W342-687-3	*(Trike previously fitted to G-MMUF)*	17. 6.87	C.G.Ludgate	(Norwich)	28. 3.04P
G-MTJL	Mainair Gemini/Flash IIA	548-687-5 & W337		17. 6.87	J.Murphy	(Washington)	26. 7.05P
G-MTJM	Mainair Gemini/Flash IIA (Rotax 462)	560-787-5 & W349		24. 6.87	K.J.Regan	(Teddington)	6. 8.00P
G-MTJN	Midland Ultralights Sirocco 377GB (Rotax 377)	MU-020		23. 6.87	S.Armstrong *(New owner 7.01)*	(Canterbury)	19. 3.94P
G-MTJP	Medway Hybred 44XLR	25687/27		6. 7.87	I.J.Alexander and P.Fitzsimmons	Plaistowes Farm, St.Albans	8.12.04P
G-MTJS	Solar Wings Pegasus XL-Q	SW-TE-0022 & SW-WX-0013		6. 7.87	R.J.H.Hayward	Broadmeadow Farm, Hereford	8. 4.05P
G-MTJT	Mainair Gemini/Flash IIA (Rotax 462)	558-787-5 & W347		16. 7.87	D.T.A Rees	Haverfordwest	26. 4.05P
G-MTJV	Mainair Gemini/Flash IIA	562-787-5 & W351		16. 7.87	N.Charles	Finmere	5. 7.05P
G-MTJW	Mainair Gemini/Flash IIA	563-787-5 & W352		16. 7.87	J.F.Ashton *(Current status unknown)*	(Liverpool)	4.10.95P
G-MTJX	Hornet Dual Trainer/Southdown Raven	HRWA 0063 & SN2000/0279		5. 8.87	J.P.Kirwan *(Noted 6.03*	Ince Blundell	31. 3.99P
G-MTJZ	Mainair Gemini/Flash IIA (Rotax 462)	561-787-5 & W350		16. 7.87	J.D.Harriman	(Dudley)	4. 7.05P
G-MTKA	Thruster TST Mk.1	867-TST-021		21. 7.87	C.M.Bradford and D.Marsh	Clench Common	19. 8.05P
G-MTKB	Thruster TST Mk.1	867-TST-022		21. 7.87	M.Hanna	Rathfriland, County Down	16. 2.05P
G-MTKD	Thruster TST Mk.1	867-TST-024		21. 7.87	Enda Spain	Kilrushn, County Kildare	31.10.04P
G-MTKE	Thruster TST Mk.1	867-TST-025		21. 7.87	M.R.Jones	Wing Farm, Longbridge Deverill	27. 8.03P
G-MTKG	Solar Wings Pegasus XL-R/Se	SW-TB-1199 & SW-WA-1201		13. 7.87	W.J.Hodgins	Deenethorpe	19. 6.04P
G-MTKH	Solar Wings Pegasus XL-R	SW-TB-1200 & SW-WA-1202		13. 7.87	K.Brooker	(Horsham)	17. 8.05P
G-MTKI	Solar Wings Pegasus XL-R	SW-TB-1201 & SW-WA-1203		13. 7.87	M.Wady	Hamstreet, Ashford	17. 8.05P
G-MTKN	Mainair Gemini/Flash IIA	566-887-5 & W355		15. 7.87	Antonia Jane Altori	(Colne)	4. 6.03P
G-MTKR	CFM Shadow Series CD	067	9H-ABL G-MTKR	20. 7.87	P.A James t/a Cloudbase Aviation	Redhill	28. 5.04P
G-MTKV	Mainair Gemini/Flash IIA	565-887-5 & W354		26. 8.87	L.A Davidson	Sandtoft	22. 5.02P
G-MTKW	Mainair Gemini/Flash IIA	569-887-5 & W358		13. 7.87	J.H.McIvor	Newtownards	20. 6.05P
G-MTKX	Mainair Gemini/Flash IIA	568-887-5 & W357		13. 7.87	G.E.Jones	(Chorley)	27. 8.00P
G-MTKZ	Mainair Gemini/Flash IIA	571-887-5 & W360		31. 7.87	W J F Mclean and I S Mcneill	(Dalkeith/Longniddry)	14. 8.05P
G-MTLB	Mainair Gemini/Flash IIA	573-887-5 & W362		31. 7.87	D.N.Bacon	Hucknall	27. 6.04P
G-MTLC	Mainair Gemini/Flash IIA	574-887-5 & W363		31. 7.87	R.J.Alston	(Cromer)	13. 7.02P
G-MTLD	Mainair Gemini/Flash IIA	575-887-5 & W364		31. 7.87	C.Kearney	Ince Blundell	13. 4.05P
G-MTLG	Solar Wings Pegasus XL-R	SW-TB-1207 & SW-WA-1211		31. 7.87	G.J.Simoni	Kemble	23. 5.05P
G-MTLI	Solar Wings Pegasus XL-R	SW-TB-1209 & SW-WA-1213		31. 7.87	M.McKay	(Robertsbridge)	3. 9.04P
G-MTLJ	Solar Wings Pegasus XL-R/Se	SW-TB-1210 & SW-WA-1214		31. 7.87	S.Wild	(Rotherham)	26. 5.05P
G-MTLL	Mainair Gemini/Flash IIA	578-987-5 & W367		14. 8.87	M.S.Lawrence	Mill Farm, Shifnal	14. 7.05P
G-MTLM	Thruster TST Mk.1	887-TST-027		5. 8.87	E.F.Howells tr Chloe's Flying Group *"Chloe"*	Leicester	24. 6.05P
G-MTLN	Thruster TST Mk.1	887-TST-028		5. 8.87	A.R.Brew	Roddige	20. 5.05P
G-MTLT	Solar Wings Pegasus XL-R	SW-TB-1212 & SW-WA-1216		12. 8.87	K.M.Mayling	Plaistowes Farm, St Albans	10.10.03P
G-MTLU	Solar Wings Pegasus XL-R/Se	SW-TB-1213 & SW-WA-1217		12. 8.87	M.W.Riley	(Morpeth)	22. 9.02P
G-MTLV	Solar Wings Pegasus XL-R	SW-TB-1214 & SW-WA-1218		12. 8.87	D.E.Watson	Long Marston	22. 1.04P
G-MTLX	Medway Hybred 44XLR	20687/26		14. 8.87	D.A Coupland	RAF Wyton	28. 6.04P
G-MTLY	Solar Wings Pegasus XL-R (Rotax 462)	SW-TE-0026 & SW-WA-1220		12. 8.87	I.Johnston *(Current status unknown)*	(Bolton)	5. 7.92P
G-MTLZ	Whittaker MW5 Sorcerer *(Built E H Gould)* (Rotax 377)	PFA 163-11241		13. 8.87	M.J.Davenport	Weston Zoyland	17. 8.04P
G-MTMA	Mainair Gemini/Flash IIA	579-987-5 & W368		14. 8.87	D.Bussell	St.Michaels	5. 9.05P
G-MTMC	Mainair Gemini/Flash IIA	581-987-5 & W370		14. 8.87	A R.Johnson	Brenzett, Kent	30. 6.05P
G-MTME	Solar Wings Pegasus XL-R	SW-TB-1216 & SW-WA-1221		18. 8.87	M.T.Finch	Sutton Meadows, Ely	23. 7.05P
G-MTMF	Solar Wings Pegasus XL-R	SW-TB-1217 & SW-WA-1222		18. 8.87	J.T.W.Smith	(Mallaig)	4. 8.05P
G-MTMG	Solar Wings Pegasus XL-R	SW-TB-1218 & SW-WA-1223		18. 8.87	C.W. and P.E.F.Suckling	(Rushden, Northampton)	8.11.04P
G-MTMI	Solar Wings Pegasus XL-R/Se	SW-TB-1220 & SW-WA-1225		18. 8.87	D.Crozier	Eshott	23. 4.05P
G-MTMK	Raven Aircraft Raven X	SN2000/0289		2. 9.87	D.W.Thomas	Long Marston	1. 8.02P
G-MTML	Mainair Gemini/Flash IIA (Rotax 462)	582-1087-5 & W371		27. 8.87	J.F.Ashton *(Trike only noted 5.03)*	Perth	30. 7.00P
G-MTMO	Raven Aircraft Raven X	SN2232/0278	(G-MTKL)	11. 9.87	H.Tuvey	(South Ockendon)	1. 9.04P
G-MTMP	Hornet Dual Trainer/Southdown Raven (Rotax 462)	HRWA 0064 & SN2000/0288		28. 8.87	P.G.Owen *(Current status unknown)*	Baxby Manor, Husthwaite	6. 8.99P
G-MTMR	Hornet Dual Trainer/Southdown Raven (Rotax 462)	HRWA 0065 & SN2000/0297		28. 8.87	D.J.Smith	Hucknall	4. 8.04P
G-MTMT	Mainair Gemini/Flash IIA (Rotax 462)	583-1087-5 & W372		3. 9.87	C.Pickvance	Tarn Farm, Cockerham	6. 9.04P
G-MTMV	Mainair Gemini/Flash IIA	585-1087-5 & W374		3. 9.87	M.F.Botha	(Gnosall, Stafford)	23. 4.05P

Reg	Type	C/n	Date	Owner	Location	Date
G-MTMW	Mainair Gemini/Flash IIA	587-1087-5 & W376	9. 9.87	F Lees	(Walsall)	1. 9.05P
G-MTMX	CFM Shadow Series CD	070	4. 9.87	D.R.White	Plaistowes Farm, St Albans	27. 9.045
G-MTMY	CFM Shadow Series CD	071	4. 9.87	A.J.Harpley	(Hawes)	1. 2.05P
G-MTNC	Mainair Gemini/Flash IIA	588-1087-5 & W377	15. 9.87	D.J.Kelly and M Titmus	Shobdon	5. 9.05P
G-MTNE	Medway Hybred 44XLR	7987/32	12.10.87	A G.Rodenburg	Latch Farm, Kirknewton	8. 8.05P
	(Fitted with new trike as original was transferred to G-MVDC in 1988)					
G-MTNF	Medway Hybred 44XLR	1987/31	12.10.87	P.A Bedford	(Tewkesbury)	1. 9.04P
G-MTNG	Mainair Gemini/Flash IIA	590-1087-5 & W379	21. 9.87	A N.Bellis	Shobdon	25. 7.04P
G-MTNH	Mainair Gemini/Flash IIA (Rotax 462)	589-1087-5 & W378	17. 9.87	J.R.Smart	Over Farm, Gloucester	7. 9.05P
G-MTNI	Mainair Gemini/Flash IIA	595-1187-5 & W384	18. 9.87	D.Gatland	Rufforth	25.11.02P
G-MTNJ	Mainair Gemini/Flash IIA (Rotax 462)	593-1187-5 & W382	17. 9.87	S.F.Kennedy	Swinford, Rugby	26. 4.05P
G-MTNK	Weedhopper JC-24B	1936	28. 9.87	S R Davis	(Frampton Mansell, Stroud)	Exemption
	(Fuji-Robin EC-34-PM) *(Test flown under "B" Conditions 29.6.00 as "G-???")*					
G-MTNL	Mainair Gemini/Flash IIA	591-1187-5 & W380	21. 9.87	R.A Matthews	Otherton, Cannock	14. 1.03P
G-MTNM	Mainair Gemini/Flash IIA	592-1187-5 & W381	22. 9.87	C.J.Janson	Shobdon	12. 8.05P
G-MTNO	Solar Wings Pegasus XL-Q (Rotax 447)	SW-TB-1252 & SW-WQ-0001	23. 9.87	A F.Batchelor *(Noted 1.05)*	Rayne Hall Farm, Braintree	6.11.04P
G-MTNP	Solar Wings Pegasus XL-Q (Rotax 447)	SW-TB-1253 & SW-WQ-0002	23. 9.87	G.G.Roberts	Rayne Hall Farm, Braintree	17. 8.03P
G-MTNR	Thruster TST Mk.1	897-TST-032	1.10.87	A M Sirant	Monkswell Farm, Horrabridge	6. 5.05P
G-MTNS	Thruster TST Mk.1	897-TST-033	1.10.87	G.and B.W.Evans	Arclid Green, Sandbach	25. 5.02P
G-MTNT	Thruster TST Mk.1	897-TST-034	1.10.87	M J Clifford *(Noted 6.04)*	Mendlesham	11. 3.01P
G-MTNU	Thruster TST Mk.1	897-TST-035	1.10.87	T.H.Brearley	Wickwar	15. 9.05P
G-MTNV	Thruster TST Mk.1	897-TST-036	1.10.87	J.B.Russell *(Current status unknown)*	Larne, County Antrim	11.10.88P
G-MTNX	Mainair Gemini/Flash IIA	606-1187-5 & W393	29. 9.87	C.Evans	RAF Wyton	24. 6.02P
G-MTNY	Mainair Gemini/Flash IIA	594-1187-5 & W383	2.10.87	R.C.Granger	(Burnham-on-Crouch)	8. 8.03P
G-MTOA	Solar Wings Pegasus XL-R	SW-TB-1221 & SW-WA-1226	15. 9.87	R.A Bird *(Amended CofR 5.03)*	East Hunsbury, Northampton	8. 8.01P
G-MTOB	Solar Wings Pegasus XL-R	SW-TB-1222 & SW-WA-1227	15. 9.87	P.S.Lemm *(Current status unknown)*	Otherton, Cannock	1.10.97P
G-MTOD	Solar Wings Pegasus XL-R	SW-TB-1224 & SW-WA-1229	15. 9.87	T A Gordon *(New CofR 3.03)*	(Liskeard)	3. 9.00P
G-MTOE	Solar Wings Pegasus XL-R	SW-TB-1225 & SW-WA-1230	15. 9.87	K.J.Bright	Old Sarum	13. 7.03P
G-MTOG	Solar Wings Pegasus XL-R	SW-TB-1227 & SW-WA-1232	15. 9.87	D.S.F.McNair	(Lochgilphead)	9. 4.05P
G-MTOH	Solar Wings Pegasus XL-R	SW-TB-1228 & SW-WA-1233	15. 9.87	H.Cook	(Pontypool)	2. 3.02P
G-MTOI	Solar Wings Pegasus XL-R	SW-TB-1229 & SW-WA-1234	15. 9.87	G.Taylor	(Horsford, Norwich)	7. 6.04P
G-MTOJ	Solar Wings Pegasus XL-R/Se	SW-TB-1230 & SW-WA-1235	15. 9.87	S.Jelley	(Chichester)	16.10.04P
G-MTOK	Solar Wings Pegasus XL-R	SW-TB-1231 & SW-WA-1236	2.10.87	W.S.Davis	Oxton, Nottingham	19.11.04P
G-MTOM	Solar Wings Pegasus XL-R/Se	SW-TB-1233 & SW-WA-1238	2.10.87	R.J.Hood	London Colney	27. 7.01P
	(Original sailwing now written off and shares with that from G-MMRL)					
G-MTON	Solar Wings Pegasus XL-R	SW-TB-1234 & SW-WA-1239	2.10.87	D.J.Willett	(Malpas)	11. 3.05P
G-MTOO	Solar Wings Pegasus XL-R	SW-TB-1235 & SW-WA-1240	2.10.87	G.W.Bulmer *(Noted 1.05)*	Wickwar	25. 3.04P
G-MTOP	Solar Wings Pegasus XL-R/Se	SW-TB-1236 & SW-WA-1241	2.10.87	P.D.Larkin	Field Farm, Oakley	18.10.05P
G-MTOR	Solar Wings Pegasus XL-R	SW-TB-1237 & SW-WA-1242	9.10.87	W.F.G.Panayiotiou	(Llanelli)	28. 7.03P
G-MTOS	Solar Wings Pegasus XL-R	SW-TB-1238 & SW-WA-1243	9.10.87	C.McKay	Perth	16.10.04P
G-MTOT	Solar Wings Pegasus XL-R	SW-TB-1239 & SW-WA-1244	9.10.87	A J Lloyd *(New owner 9.04)*	Henstridge	25. 3.04P
G-MTOU	Solar Wings Pegasus XL-R/Se	SW-TB-1240 & SW-WA-1245	9.10.87	D.T.Smith	(Thornaby)	14. 4.04P
G-MTOY	Solar Wings Pegasus XL-R	SW-TB-1244 & SW-WA-1249	19.10.87	C.M.Bradford tr G-MTOY Group	Yatesbury	20. 1.05P
G-MTOZ	Solar Wings Pegasus XL-R	SW-TB-1245 & SW-WA-1250	19.10.87	M.A Furber	(Hanslope, Milton Keynes)	20. 1.05P
G-MTPA	Mainair Gemini/Flash IIA (Rotax 462)	598-1187-5 & W394	13.10.87	P.G.Eastlake	(Harlow)	14. 8.05P
G-MTPC	Raven Aircraft Raven X	SN2232/0309	15.10.87	G.W.Carwardine *(Current status unknown)*	(Uckfield)	3.11.90P
	(Rotax 582) *(Modified to "Phillips Swphift" standard 1999)*					
G-MTPE	Solar Wings Pegasus XL-R (Rotax 503)	SW-TB-1258 & SW-WA-1260	21.10.87	J.Basset	Brown Shutters Farm, Norton St.Philips, Somerset	25. 4.05P
G-MTPF	Solar Wings Pegasus XL-R	SW-TB-1259 & SW-WA-1261	21.10.87	P.M.Watts and A.S.Mitchel *(New owners 12.04)*	Halwell, Totnes	27. 9.04P
G-MTPG	Solar Wings Pegasus XL-R	SW-TB-1260 & SW-WA-1262	21.10.87	J.Sullivan	Davidstow Moor	16. 7.03P
G-MTPH	Solar Wings Pegasus XL-R	SW-TB-1261 & SW-WA-1263	30.10.87	L.M.Sams	Long Marston	9. 6.05P
G-MTPI	Solar Wings Pegasus XL-R/Se	SW-TB-1262 & SW-WA-1264	30.10.87	R.J.Bullock	Long Marston	1. 8.05P
G-MTPJ	Solar Wings Pegasus XL-R	SW-TB-1263 & SW-WA-1265	30.10.87	D.Lockwood	(Birmingham)	14.10.05P
G-MTPK	Solar Wings Pegasus XL-R	SW-TB-1264 & SW-WA-1266	30.10.87	S.H.James	Deenethorpe	21.10.01P

Registration	Type / c/n	Date	Owner	Location	CofA
G-MTPL	Solar Wings Pegasus XL-R SW-TB-1265 & SW-WA-1267	30.10.87	C.J.Jones	(Bath)	13. 7.05P
G-MTPM	Solar Wings Pegasus XL-R SW-TB-1266 & SW-WA-1268	30.10.87	D.K.Seal	Roddige	4. 8.04P
G-MTPN	Solar Wings Pegasus XL-Q (Rotax 447) SW-TB-1267 & SW-WQ-0004	21.10.87	H N Graham	(Enniskillen)	24. 1.05P
G-MTPP	Solar Wings Pegasus XL-R SW-TB-1257 & SW-WA-1259	21.10.87	P Molyneux	(Southport)	8. 7.05P
G-MTPR	Solar Wings Pegasus XL-R SW-TB-1256 & SW-WA-1257	21.10.87	T.Kenny *(Current status unknown)*	(Ballygar)	16. 6.96P
G-MTPS	Solar Wings Pegasus XL-Q SW-TE-0021 & SW-WX-0011	23.10.87	G.Tyler	(Cambridge)	28. 7.05P
G-MTPT	Thruster TST Mk.1 8107-TST-038	23.10.87	G.B.Gratton tr Chilbolton Thruster Group	Chilbolton	12. 5.05P
G-MTPU	Thruster TST Mk.1 8107-TST-039	23.10.87	D.R.Sims	Halwell, Totnes	8. 8.05P
G-MTPW	Thruster TST Mk.1 8107-TST-041	23.10.87	K.Hawthorne	(Armagh)	18. 8.05P
G-MTPX	Thruster TST Mk.1 8107-TST-042	23.10.87	T.Snook *(Current status unknown)*	Long Marston	2. 5.93P
G-MTPY	Thruster TST Mk.1 8107-TST-043	23.10.87	C.M.Bradford	Clench Common	1. 8.04P
G-MTRA	Mainair Gemini/Flash IIA 605-1187-5 & W395	28.10.87	E.N.Alms *"Yellow Bird"*	Guy Lane Farm, Waverton	2. 2.05P
G-MTRC	Midland Ultralights Sirocco 377GB MU-021	2.11.87	D.Thorpe	Grantham	13. 4.03P
G-MTRL	Hornet Dual Trainer/Southdown Raven HRWA 0068 & SN2000/0326	4.11.87	J.McAlpine	(Largs)	27. 4.05P
G-MTRM	Solar Wings Pegasus XL-R (Rotax 462) SW-TE-0030 & SW-WA-1276	10.11.87	M.S.Ahmadu	(Kettering)	26. 6.05P
G-MTRN	Solar Wings Pegasus XL-R SW-TB-1270 & SW-WA-1269	2.12.87	K.McCoubrey *(Noted 5.04)*	Roddige	3. 8.01P
G-MTRO	Solar Wings Pegasus XL-R/Se SW-TB-1271 & SW-WA-1270	2.12.87	H.Lloyd-Hughes	Emlyn's Field, Rhuallt	26. 4.05P
G-MTRS	Solar Wings Pegasus XL-R SW-TB-1274 & SW-WA-1273	2.12.87	J.J.R.Tickle	Llanerchymedd, Gwynedd	13. 6.01P
G-MTRT	Raven Aircraft Raven X SN2232/0325	12.11.87	G.S.Stokes	Pound Green, Buttonoak, Bewdley	14. 9.04P
G-MTRU	Solar Wings Pegasus XL-Q (Rotax 477) SW-TB-1275 & SW-WQ-0009	10.11.87	A L.S.Routledge *(Noted wrecked 7.01)*	Rufforth	15.10.00P
G-MTRV	Solar Wings Pegasus XL-Q (Rotax 477) SW-TB-1276 & SW-WX-0010	10.11.87	R.M.Adams	(Kettering)	3. 6.04P
G-MTRW	Raven Aircraft Raven X SN2232/0328	12.11.87	P.K.J.Chun	Rochester	8.12.04P
G-MTRX	Whittaker MW5 Sorcerer PFA 163-11202 *(Built W.Turner)*	11.11.87	W.Turner *(Stored 5.04)*	Otherton, Cannock	13. 2.95P
G-MTRZ	Mainair Gemini/Flash IIA 611-1287-5 & W400	17.11.87	D.F.G.Barlow	St Michaels	4.10.04P
G-MTSC	Mainair Gemini/Flash IIA 618-188-5 & W407	17.11.87	K.Wilson *(New owner 12.04)*	(Llanddulas, Abergele)	4. 6.02P
G-MTSG	CFM Shadow Series CD 079	24.11.87	C.A Purvis	Plaistowes Farm, St.Albans	26. 4.05P
G-MTSH	Thruster TST Mk.1 8117-TST-044	3.12.87	R R Orr	Dromore, County Down	1. 6.03P
G-MTSJ	Thruster TST Mk.1 8117-TST-046	3.12.87	P.J.Mogg	Sturminster Newton	5. 4.05P
G-MTSK	Thruster TST Mk.1 8117-TST-047	3.12.87	J.S.Pyke	Westfield Farm, Hailsham	24.10.05P
G-MTSM	Thruster TST Mk.1 8117-TST-049 *(Modified to T300 standard)*	3.12.87	J J Hill	Baxby Manor, Husthwaite	2. 9.05P
G-MTSN	Solar Wings Pegasus XL-R SW-TB-1278 & SW-WA-1280	14.12.87	G.P.Lane	Doynton	15. 4.03P
G-MTSO*	Solar Wings Pegasus XL-R/Se SW-TB-1279 & SW-WA-1281	14.12.87	P.Wibberley *(Cancelled 4.5.04 as wfu)*	(Chesterfield)	20. 5.03P
G-MTSP	Solar Wings Pegasus XL-R SW-TB-1280 & SW-WA-1282	14.12.87	R.J.Nelson	Swinford, Rugby	11.10.05P
G-MTSR	Solar Wings Pegasus XL-R SW-TB-1281 & SW-WA-1283	14.12.87	J.Norman	Longacres Farm, Sandy	25. 5.04P
G-MTSS	Solar Wings Pegasus XL-R (Rotax 462) SW-TE-0031 & SW-WA-1284	14.12.87	R.J.Turner	(Spalding)	8. 5.05P
G-MTSU	Solar Wings Pegasus XL-R SW-TB-1289 & SW-WA-1285	4. 1.88	L.Earls	Newtownards	26.12.04P
G-MTSY	Solar Wings Pegasus XL-R/Se SW-TB-1283 & SW-WA-1289	14. 1.88	N.F.Waldron *(Current status unknown)*	Swinford, Rugby	24. 5.99P
G-MTSZ	Solar Wings Pegasus XL-R/Se SW-TB-1284 & SW-WA-1290	14. 1.88	J.R.Appleton	(Colne)	28. 4.05P
G-MTTA	Solar Wings Pegasus XL-R (Rotax 462) SW-TE-0035 & SW-WA-1291	14. 1.88	J.J.McMennum *(Noted 7.04)*	Morgansfield, Fishburn	4. 9.00P
G-MTTB	Solar Wings Pegasus XL-R (Rotax 447) SW-TB-1285 & SW-WA-1292	14. 1.88	P.J.Soukup	(Winkleigh)	29. 4.04P
G-MTTD	Solar Wings Pegasus XL-Q (Rotax 447) SW-TB-1286 & SW-WQ-0011	15. 1.88	J.P.Dilley	Hunsdon	11.10.05P
G-MTTE	Solar Wings Pegasus XL-Q SW-TB-1287 & SW-WQ-0012	15. 1.88	P.Sinkler	Rufforth	8. 2.05P
G-MTTF	Whittaker MW6 Merlin PFA 164-11273 *(Built V E Booth)* (Rotax 532)	14.12.87	P.Cotton *(Current status unknown)*	Long Marston	29. 3.95P
G-MTTH	CFM Shadow Series BD K.061	15.12.87	G.F.Hill and A Y-T.Leung	(Shenstone)	6. 9.05P
G-MTTI	Mainair Gemini/Flash IIA 620-188-5 & W409	14.12.87	S.M.Savage *(Current status unknown)*	(Guildford)	19. 7.96P
G-MTTL*	Ultrasports Tri-Pacer 330/Excalibur EXS-872 (Fuji-Robin EC-34-PM)	23. 3.88	M.J.Aubrey *(Cancelled 20.11.95 by CAA) (Noted 2002)* .	(Kington, Hereford)	
G-MTTM	Mainair Gemini/Flash IIA 609-1287-5 & W398	5. 1.88	M.Anderson	East Fortune	11.10.05P
G-MTTN	Skyrider Aviation Phantom PH.00100 *(Officially registered as Ultralight Flight Phantom)*	22. 1.88	C.G.Benham	(Towcester	4. 8.05P
G-MTTP	Mainair Gemini/Flash IIA 612-188-5 & W401 (Rotax 462)	18. 1.88	A Ormson	St.Michaels	23. 6.05P
G-MTTR	Mainair Gemini/Flash IIA 614-188-5 & W403 (Rotax 462)	27. 1.88	A Westoby	Hucknall	22. 7.00P
G-MTTU	Solar Wings Pegasus XL-R SW-TB-1332 & SW-WA-1294	25. 2.88	A Friend *(New owner 9.04)*	(Gurney Slade, Radstock)	25. 7.04P

G-MTTW	Mainair Gemini/Flash IIA (Rotax 462)	622-188-5 & W411	15. 1.88	A F Glover	(Woolston, Warrington)	4. 6.05P
G-MTTX	Solar Wings Pegasus XL-Q (Rotax 447)	SW-TB-1293 & SW-WQ-0013	15. 2.88	M.J.Sunter	(Leyburn)	16. 8.05P
G-MTTY	Solar Wings Pegasus XL-Q	SW-WQ-0014	21. 1.88	G.A.Tegg	Clench Common	3. 9.05P
G-MTTZ	Solar Wings Pegasus XL-Q	SW-TE-0039 & SW-WQ-0015	21. 1.88	J.Haskett	(King's Lynn)	17.10.05P
G-MTUA	Solar Wings Pegasus XL-R/Se	SW-TB-1294 & SW-WA-1295	15. 1.88	M.D.Reardon	(Leeds)	6. 8.05P
G-MTUB	Thruster TST Mk.1	8018-TST-050	15. 1.88	M.Curtin	(Clonmel, County Tipperary)	4. 9.04P
G-MTUC	Thruster TST Mk.1	8018-TST-051	15. 1.88	E.J.Girling	Davidstow Moor	21. 3.04P
G-MTUD	Thruster TST Mk.1 *(Modified to T300 standard)*	8018-TST-052	15. 1.88	A J.Best	Baxby Manor, Husthwaite	5.10.05P
G-MTUE*	Thruster TST Mk.1	8018-TST-053	15. 1.88	J.P.McVitty *(Cancelled 23.5.00 by CAA) (Noted 7.04)*	Gort, Derrykee	11. 6.94P
G-MTUF	Thruster TST Mk.1	8018-TST-054	15. 1.88	P.Stark *(Noted 3.04)*	Strathaven	16. 1.05P
G-MTUI	Solar Wings Pegasus XL-R/Se	SW-TB-1296 & SW-WA-1296	21. 1.88	R.Green	(Shirley, Solihull)	15. 2.05P
G-MTUJ	Solar Wings Pegasus XL-R	SW-TB-1297 & SW-WA-1297	21. 1.88	R.W.Pincombe *(Current status unknown)*	(Chumleigh)	31. 5.94P
G-MTUK	Solar Wings Pegasus XL-R	SW-TB-1298 & SW-WA-1298	21. 1.88	G.L.Hall	Rufforth	30. 9.04P
G-MTUL	Solar Wings Pegasus XL-R/Se	SW-TB-1299 & SW-WA-1299	21. 1.88	M.Worthington	Roddige	25. 4.05P
G-MTUN	Solar Wings Pegasus XL-Q (Rotax 447)	SW-TB-1301 & SW-WQ-0016 *(Fitted with Wing from G-MVUK?)*	20. 1.88	P.Boardman	(Sale)	26. 4.05P
G-MTUP	Solar Wings Pegasus XL-Q (Rotax 447)	SW-TB-1303 & SW-WA-0018	20. 1.88	S.J.Allen	Blisworth, Northampton	25. 8.03P
G-MTUR	Solar Wings Pegasus XL-Q (Rotax 447)	SW-TB-1304 & SW-WQ-0019	20. 1.88	G.Ball	(Tewkesbury)	26. 3.04P
G-MTUS	Solar Wings Pegasus XL-Q (Rotax 447)	SW-TB-1305 & SW-WQ-0020	20. 1.88	A I.McPherson	Pratis Farm, Leven	9.11.05P
G-MTUT	Solar Wings Pegasus XL-Q	SW-TE-0040 & SW-WQ-0021	21. 1.88	F.A Dimmock	(Corby)	5. 5.05P
G-MTUU	Mainair Gemini/Flash IIA	623-288-5 & W412	10. 2.88	M.Harris	Eshott	10. 2.04P
G-MTUV	Mainair Gemini/Flash IIA (Rotax 462)	624-288-5 & W413	28. 1.88	J.F.Bolton	Plaistowes Farm, St.Albans	6. 8.05P
G-MTUX	Medway Hybred 44XLR	241287/33	2. 2.88	P.A R.Wilson *(Current status unknown)*	(Cloughton, Scarborough)	29. 8.99P
G-MTUY	Solar Wings Pegasus XL-Q	SW-TE-0041 & SW-WQ-0022	28. 1.88	H.C.Lowther	(Penrith)	1. 4.01P
G-MTVB	Solar Wings Pegasus XL-R	SW-TB-1307 & SW-WA-1302	28. 1.88	J.Williams	(Worcester)	15.11.03P
G-MTVG	Mainair Gemini/Flash IIA	628-388-6 & W417	12. 2.88	D.A Whitworth and L.L.Perry	Grove Farm, Needham	4.12.04P
G-MTVH	Mainair Gemini/Flash IIA	626-288-6 & W415	17. 2.88	D.K.May	Tarn Farm, Cockerham	22. 5.05P
G-MTVI	Mainair Gemini/Flash IIA	629-388-6 & W416	12. 2.88	R.A McDowell *(Current status unknown)*	(Slough)	10. 5.92P
G-MTVJ	Mainair Gemini/Flash IIA	627-388-6 & W418	12. 2.88	D.W.Buck	(Chesterfield)	27. 5.05P
G-MTVK	Solar Wings Pegasus XL-R	SW-TB-1311 & SW-WA-1306	15. 2.88	J D MacNamara *(Current status unknown)*	(Crediton)	17. 3.98P
G-MTVL	Solar Wings Pegasus XL-R/Se	SW-TB-1312 & SW-WA-1307	15. 2.88	K.P.Roper	Weston Zoyland	1. 8.05P
G-MTVM	Solar Wings Pegasus XL-R	SW-TB-1313 & SW-WA-1308	15. 2.88	M.D.Howard	(Great Yarmouth)	23. 9.03P
G-MTVN	Solar Wings Pegasus XL-R	SW-TB-1314 & SW-WA-1309	15. 2.88	A I.Crighton	Lower Mountpleasant, Chatteris	23. 4.05P
G-MTVO	Solar Wings Pegasus XL-R	SW-TB-1315 & SW-WA-1310	15. 2.88	D A Payne	Long Marston	19. 4.05P
G-MTVP	Thruster TST Mk.1 *(C/n plate marked incorrectly as 8208-TST-056)*	8028-TST-056	10. 2.88	J.M.Evans	Manor Farm, Garford	5.10.05P
G-MTVR	Thruster TST Mk.1	8028-TST-057	10. 2.88	G.Hawes	Deenethorpe	2.10.04P
G-MTVS	Thruster TST Mk.1	8028-TST-058	10. 2.88	J.G.McMinn	(Craigavon)	20. 6.05P
G-MTVT	Thruster TST Mk.1	8028-TST-059	10. 2.88	M.L.Walsh and A T.Farmer	Mill Farm, Shifnal	27. 6.05P
G-MTVV	Thruster TST Mk.1	8028-TST-061	10. 2.88	Carol Jones *(New owner 2.04)*	(Westbury)	22.10.02P
G-MTVX	Solar Wings Pegasus XL-Q	SW-TE-0042 & SW-WQ-0025	3. 3.88	D.A.Foster	(Melton Mowbray)	7. 8.05P
G-MTWA	Solar Wings Pegasus XL-R	SW-TB-1317 & SW-WA-1311	25. 2.88	J.D.J.Spragg and R.A.Hawkes	(Tamworth)	14. 6.05P
G-MTWB	Solar Wings Pegasus XL-R *(Originally fitted with trike c/n SW-TB-1318: latter damaged, repaired and resold with sailwing c/n SW-WA-1330 as SE-YOK)*	SW-TB-1342 & SW-WA-1312	25. 2.88	M.W.A Shemilt	(Henley-on-Thames)	22. 4.04P
G-MTWC	Solar Wings Pegasus XL-R	SW-TB-1321 & SW-WA-1313	25. 2.88	E.K.Perchard	Newton Bank, Daresbury	8.10.05P
G-MTWD	Solar Wings Pegasus XL-R	SW-TB-1320 & SW-WA-1314	25. 2.88	J.C.Rawlings	Sywell	6.12.05P
G-MTWF	Mainair Gemini/Flash IIA	630-388-6 & W419	25. 2.88	W.Porter	Knapthorpe Lodge, Caunton	23. 4.05P
G-MTWG	Mainair Gemini/Flash IIA	631-288-6 & W420	25. 2.88	N.Mackenzie and P.S.Bunting	(Southport)	28. 7.00P
G-MTWH	CFM Shadow Series CD	K.064	25. 2.88	A A Ross	Knockbain Farm, Dingwall	17. 8.04P
G-MTWK	CFM Shadow Series CD	073	25. 2.88	P Dass tr Bagby Shadow Whiskey Kilo Group	Bagby	21.10.05P
G-MTWL	CFM Shadow Series BD	076	25. 2.88	M.J.Gray	Manor Farm, Croughton	26.12.03P
G-MTWR	Mainair Gemini/Flash IIA	632-388-6 & W421	3. 3.88	J.B.Hodson	Arclid Green, Sandbach	6. 5.05P
G-MTWS	Mainair Gemini/Flash IIA	633-388-6 & W422	3. 3.88	K W Roberts	Askern	3. 6.05P
G-MTWX	Mainair Gemini/Flash IIA	634-488-6 & W423	11. 3.88	M Rushworth	(Leyland)	24. 3.05P
G-MTWY	Thruster TST Mk.1	8038-TST-062	15. 3.88	M.F.Eddington	(Wincanton)	5. 4.05P

Registration	Type	C/n	Date	Owner	Location	Expiry
G-MTWZ	Thruster TST Mk.1	8038-TST-063	15. 3.88	T.A Colman	Mapperton Farm, Newton Peverill	8. 9.05P
G-MTXA	Thruster TST Mk.1	8038-TST-064	15. 3.88	J.Upex	(Harrogate)	23. 5.05P
G-MTXB	Thruster TST Mk.1	8038-TST-065	15. 3.88	J.J.Hill	Baxby Manor, Husthwaite	30. 8.03P
G-MTXC	Thruster TST Mk.1	8038-TST-066	15. 3.88	Joan A Huntley	South Wraxall, Bradford-on-Avon	8. 6.04P
G-MTXD	Thruster TST Mk.1 *(Modified to T300 standard)*	8038-TST-067	15. 3.88	D.J.Flower	Baxby Manor, Husthwaite	28. 3.05P
G-MTXE	Hornet Dual Trainer/Southdown Raven	HRWA 0070 & SN2000/0332	11. 3.88	F.J.Marton t/a Charter Systems	Long Marston	22. 5.00P
G-MTXH	Solar Wings Pegasus XL-Q (Rotax 447)	SW-TB-1328 & SW-WQ-0030	11. 3.88	J.Rhodes *(Current status unknown)*	(Pontefract)	21. 7.97P
G-MTXI	Solar Wings Pegasus XL-Q (Rotax 447)	SW-TB-1329 & SW-WQ-0031	11. 3.88	J.J.Brutnell	Sutton Meadows, Ely	23. 1.05P
G-MTXJ	Solar Wings Pegasus XL-Q (Rotax 447)	SW-TB-1330 & SW-WQ-0032	11. 3.88	J.E.Wright	Biggin Hill	4. 9.04P
G-MTXK	Solar Wings Pegasus XL-Q (Rotax 447)	SW-TB-1331 & SW-WQ-0033	11. 3.88	M.J.McManamon *(Noted 5.04)*	Insch	4. 8.02P
G-MTXL	Noble Hardman Snowbird Mk.IV	SB-006	4. 5.88	P.J.Collins	Kilrush, County Kildare	9. 8.05P
G-MTXM	Mainair Gemini/Flash IIA	636-488-6 & W425	10. 5.88	W.Archibald	East Fortune	1.11.04P
G-MTXO	Whittaker MW6 Merlin *(Built N A Bailes)* (Rotax 503)	PFA 164-11326	11. 3.88	S.J.Whyatt *(Noted 5.04)*	Mill Farm, Shifnal	4. 8.98P
G-MTXP	Mainair Gemini/Flash IIA	637-488-6 & W426	23. 3.88	G.S.Duerden	St Michaels	31. 1.04P
G-MTXR	CFM Shadow Series CD	K.038	23. 3.88	M.E.H.Quick	Old Sarum	9. 8.04P
G-MTXS	Mainair Gemini/Flash IIA	638-488-6 & W427	23. 3.88	J.Kennedy	Mill Farm, Shifnal	16. 5.05P
G-MTXU	Noble Hardman Snowbird Mk.IV	SB-007	3. 5.88	J.A Rees *(Current status unknown)*	Haverfordwest	16. 5.89P
G-MTXZ	Mainair Gemini/Flash IIA	641-588-6 & W430	10. 5.88	P.Cave	(Caerphilly)	3. 8.05P
G-MTYA	Solar Wings Pegasus XL-Q	SW-TE-0047 & SW-WQ-0037	29. 3.88	I.Clarkson	Long Marston	22. 9.04P
G-MTYC	Solar Wings Pegasus XL-Q	SW-TE-0049 & SW-WQ-0039	30. 3.88	C.I.D.H.Garrison	Sutton Meadows, Ely	20. 9.05P
G-MTYD	Solar Wings Pegasus XL-Q	SW-TE-0050 & SW-WQ-0040	29. 3.88	D.Young	Kemble	13. 6.05P
G-MTYE	Solar Wings Pegasus XL-Q	SW-TE-0051 & SW-WQ-0041	29. 3.88	K.L.Chorley and A Cook	Enstone	26. 5.05P
G-MTYF	Solar Wings Pegasus XL-Q	SW-TE-0052 & SW-WQ-0042	29. 3.88	J.Hyde	(Spalding)	11. 8.04P
G-MTYH	Solar Wings Pegasus XL-Q	SW-TE-0054 & SW-WQ-0044	7.11.88	P.R.Hanman	Over Farm, Gloucester	8. 8.05P
G-MTYI	Solar Wings Pegasus XL-Q	SW-TE-0055 & SW-WQ-0045	30. 3.88	M.J.Hall	(Hollywood, Birmingham)	22.11.05P
G-MTYL	Solar Wings Pegasus XL-Q	SW-TE-0058 & SW-WQ-0048	30. 3.88	E.T.H.Cox *(Original sailwing c/n SW-WQ-0048 replaced by c/n 6412)*	(Church Stretton)	20.10.02P
G-MTYP	Solar Wings Pegasus XL-Q	SW-TE-0062 & SW-WQ-0052	30. 3.88	J.L.Ker	Eshott	12. 7.05P
G-MTYR	Solar Wings Pegasus XL-Q	SW-TE-0063 & SW-WQ-0053	30. 3.88	M.E.Grafton *(Noted 6.04)*	Broadmeadow Farm, Hereford	30. 4.99P
G-MTYS	Solar Wings Pegasus XL-Q	SW-TE-0064 & SW-WQ-0054	30. 3.88	R.G.Wall	Caerleon	4.10.05P
G-MTYT	Solar Wings Pegasus XL-Q	SW-TE-0065 & SW-WQ-0055	30. 3.88	M.G.Walsh	Rufforth	13. 9.99P
G-MTYU	Solar Wings Pegasus XL-Q	SW-TE-0066 & SW-WQ-0056	30. 3.88	S.East	Baxby Manor, Husthwaite	8. 9.05P
G-MTYW	Raven Aircraft Raven X	SN2232/0344	8. 4.88	R.Solomans	Middle Stoke, Isle of Grain	12. 7.05P
G-MTYX	Raven Aircraft Raven X	SN2232/0345	8. 4.88	C.Rean	(Petersfield)	19. 9.04P
G-MTYY	Solar Wings Pegasus XL-R	SW-WA-1326	6. 5.88	G.A Tegg	Clench Common	24. 1.05P
G-MTZA	Thruster TST Mk.1	8048-TST-068	13. 4.88	R.Bingham	(Craigavon, Co Armagh)	13. 5.05P
G-MTZB	Thruster TST Mk.1	8048-TST-069	13. 4.88	S.J.O.Tinn	Popham	4. 7.05P
G-MTZC	Thruster TST Mk.1	8048-TST-070	13. 4.88	R.W.Marshall	(Armagh)	24. 8.05P
G-MTZD	Thruster TST Mk.1	8048-TST-071	13. 4.88	A.Spence	Popham	21. 9.05P
G-MTZF	Thruster TST Mk.1	8048-TST-073	13. 4.88	D C Marsh	Doynton	28. 2.05P
G-MTZG	Mainair Gemini/Flash IIA	642-588-6 & W431	10. 5.88	P.J.Bent	(Hinckley)	12. 7.05P
G-MTZH	Mainair Gemini/Flash IIA (Rotax 462)	643-588-6 & W433	9. 6.88	D.C.Hughes	St.Michaels	17.12.01P
G-MTZJ	Solar Wings Pegasus XL-R	SW-TB-1335 & SW-WA-1328	6. 5.88	C.Gogarty	Long Acres Farm, Sandy	15.10.05P
G-MTZK	Solar Wings Pegasus XL-R	SW-TB-1336 & SW-WA-1329	6. 5.88	G F Jones	Mill Farm, Shifnal	8. 1.05P
G-MTZL	Mainair Gemini/Flash IIA	645-588-6 & W435	10. 5.88	N.S.Brayn	Popham	31. 7.03P
G-MTZM	Mainair Gemini/Flash IIA	646-588-6 & W436	3. 5.88	K.L.Smith	(Leicester)	7. 7.05P
G-MTZO	Mainair Gemini/Flash IIA (Rotax 462)	649-688-6 & W439	6. 5.88	R.C.Hinds	(Newnham, Glos)	23. 6.05P
G-MTZP	Solar Wings Pegasus XL-Q (Rotax 447)	SW-TB-1337 & SW-WQ-0059	6. 5.88	M.J.Newman	Sandown	23. 6.05P
G-MTZR	Solar Wings Pegasus XL-Q (Rotax 447)	SW-TB-1338 & SW-WQ-0060	6. 5.88	P.J.Hatchett *(Current status unknown)*	Emlyn's Field, Rhuallt	19. 8.98P
G-MTZS	Solar Wings Pegasus XL-Q (Rotax 447)	SW-TB-1339 & SW-WQ-0061	6. 5.88	P.A Darling *(Current status unknown)*	(Wilmslow)	15. 7.93P
G-MTZT	Solar Wings Pegasus XL-Q (Rotax 447)	SW-TB-1340 & SW-WQ-0062	6. 5.88	M.Y.Brown	Eshott	14. 5.05P
G-MTZV	Mainair Gemini/Flash IIA	650-688-6 & W440	6. 5.88	G.J.Donnellon	Barton	7. 9.04P
G-MTZW	Mainair Gemini/Flash IIA	651-688-6 & W441	25. 5.88	L.McIntyre	Ince Blundell	7. 7.05P
G-MTZX	Mainair Gemini/Flash IIA	652-688-6 & W442	23. 6.88	R.G.Cuckow and J.C.Thompson	Rufforth	2. 3.04P
G-MTZY	Mainair Gemini/Flash IIA	653-688-6 & W443	24. 5.88	T C Palmer	Baxby Manor, Husthwaite	26. 7.05P
G-MTZZ	Mainair Gemini/Flash IIA	654-688-6 & W444	14. 6.88	P.J.Litchfield *(Noted 2.03)*	Tarn Farm, Cockerham	14. 4.01P
G-MUCK	Lindstrand LBL 77A HAB	982	25. 1.05	C.J.Wootton	(Ormskirk)	

Registration	Type	C/n	Previous identity	Date	Owner/Operator	Base	CofA
G-MUFY	Robinson R22 Beta	1248	D-HICH	13.12.96	Helicentre Liverpool Ltd	Liverpool	9. 1.06T
G-MUIR	Cameron V-65 HAB	2037		23. 6.89	Lindsay J.M.Muir *"Muriel"*	(East Molesey)	9. 4.04A
G-MULT	Beech 76 Duchess	ME-396	N810Y	26.10.04	Folada Aero and Technical Services Ltd	Southampton	AC
G-MUMY	Van's RV-4	PFA 181-13401		11. 1.05	S.D.Howes	(Darlington)	
	(Built S.D.Howes)						
G-MUNI	Mooney M.20J Model 201SE	24-3118		12. 5.89	H.A Daines	Seething	1. 9.07
G-MURP	Aérospatiale AS350B Ecureuil	2386	RP-C2388	15.12.04	M.Murphy	(Armagh)	AC
			N82632, JA6054, N49GA				
G-MURR	Whittaker MW6 Merlin	PFA 164-12502		16. 4.99	D.Murray	Charmy Down, Bath	
	(Built M Whittaker)				*(Noted 1.05)*		
G-MURY*	Robinson R44 Astro				See SECTION 3 Part 1	Washington, US	
G-MUSH	Robinson R44 Raven II	10278		18. 2.04	Flightpath Ltd	(Loughborough)	10. 3.07T
G-MUSO	Rutan LongEz	PFA 74A-10590		11. 6.83	P.A Willis	RAF Coningsby	19. 1.05P
	(Built G B Castle) (Lycoming O-235)						
G-MUST*	Commonwealth CAC-18 (P-51D) Mustang Mk.22				See SECTION 3 Part 1		
						Raymond Terrace, New South Wales, Australia	
G-MUTE	Colt 31A Air Chair HAB	2099		2.12.91	K.Temple	(Eye)	11.11.99A
G-MUTZ	Jabiru Aircraft Jabiru J400	PFA 325-14171		29.12.03	N.C.Dean	(Raunds, Wellingborough)	
	(Built N.C.Dean)						
G-MUVG	Cessna 421C Golden Eagle III	421C1064	N421DD	13. 1.97	Martin Collins Aviation Ltd	Nottingham East Midlands	12. 3.05T
			N6865P				
G-MVAA	Mainair Gemini/Flash IIA	655-688-6 & W445		8. 6.88	R.M.Wigman	Oxton	5.10.05P
G-MVAB	Mainair Gemini/Flash IIA	656-688-6 & W446		10. 5.88	B.Hindley	Newton Bank Farm, Daresbury	1. 7.05P
G-MVAC	CFM Shadow Series CD	K.077		12. 5.88	M.Field	(Burnley)	15. 9.04P
G-MVAD	Mainair Gemini/Flash IIA	657-688-6 & W447		10. 5.88	N.D.Fox and C.I.Hemmingway	Brook Farm, Pilling	21. 1.05P
G-MVAF	Southdown Puma Sprint	P.455	G-MBAF	24. 6.87	J.F.Horn	Ezenridge Farm, Bere Alston, Yelverton	17. 5.05P
	(Fuji-Robin EC-44-2PM)						
G-MVAG	Thruster TST Mk.1	8058-TST-074		18. 5.88	W.A Stephenson	(Newry, County Armagh)	11. 4.04P
G-MVAH	Thruster TST Mk.1	8058-TST-075		18. 5.88	M.W.H.Henton *"Times Four"*	Popham	18. 8.05P
G-MVAI	Thruster TST Mk.1	8058-TST-076		18. 5.88	D.J.Townsend	Mendlesham	26. 7.05P
G-MVAJ	Thruster TST Mk.1	8058-TST-077		18. 5.88	A T.Harvey *(Noted 8.03)*	Long Marston	16. 6.02P
G-MVAK	Thruster TST Mk.1	8058-TST-078		18. 5.88	L.A Hosegood *(Noted 11.04)*	Clench Common	24. 8.04P
G-MVAL	Thruster TST Mk.1	8058-TST-079		18. 5.88	G.C.Brooke *(Current status unknown)*	(Colchester)	7. 8.96P
G-MVAM	CFM Shadow Series CD	082		18. 5.88	C.P.Barber	Brook Farm, Pilling	14. 5.04P
G-MVAN	CFM Shadow Series CD	K.048 & PFA 161-11219		18. 5.88	I.Brewster	Longacres Farm, Sandy	17.10.05P
G-MVAO	Mainair Gemini/Flash IIA	658-688-6 & W448		24. 5.88	S.J.Robson	Brook Farm, Pilling	24. 7.05P
					(Op Brook Farm Microlight Centre)		
G-MVAP	Mainair Gemini/Flash IIA	659-688-6 & W449		24. 5.88	R.J.Miller	Long Marston	21. 4.05P
G-MVAR	Solar Wings Pegasus XL-R			24. 5.88	A J Thomas	Deenethorpe	14. 5.04P
		SW-TB-1343 & SW-WA-1331					
G-MVAT	Solar Wings Pegasus XL-R			24. 5.88	R.Hickman	(Hucknall, Nottingham)	2.12.05P
		SW-TB-1345 & SW-WA-1333					
G-MVAV	Solar Wings Pegasus XL-R			24. 5.88	D.J.Utting	(Bungay)	16. 2.03P
		SW-TB-1347 & SW-WA-1335					
G-MVAW	Solar Wings Pegasus XL-Q			24. 5.88	G.Sharman	Long Marston	9.11.05P
	(Rotax 447)	SW-TB-1348 & SW-WQ-0064					
G-MVAX	Solar Wings Pegasus XL-Q			24. 5.88	P.M.Golden	(Reading)	2. 6.05P
	(Rotax 447)	SW-TB-1349 & SW-WQ-0065					
G-MVAY	Solar Wings Pegasus XL-Q			24. 5.88	V.O.Morris	(Swansea)	16. 4.97P
	(Rotax 447)	SW-TB-1350 & SW-WQ-0066			*(Current status unknown)*		
G-MVBB	CFM Shadow Series BD	K.051		24. 5.88	R.Garrod *(Stored 7.04)*	Mendlesham	8.10.00P
G-MVBC	Mainair Tri-Flyer/Aerial Arts 130SX	130SX-616		24. 5.88	D.Beer	(Ilfracombe)	
	(Believed to be using Mainair Tri-Flyer 250 trike from G-MJIX)				*(Current status unknown)*		
G-MVBD	Mainair Gemini/Flash IIA	660-688-6 & W450		8. 6.88	J.Batchelor	(Benfleet)	17.12.03P
	(Rotax 462)						
G-MVBE	Mainair Scorcher	661-688-6 & W451		28. 7.88	W.R.C.Williams-Wynne	Talybont, Gwynedd	29. 1.05P
G-MVBF	Mainair Gemini/Flash IIA	662-688-6 & W452		14. 6.88	D.P.Stacey	Deenethorpe	15. 9.05P
	(Rotax 462) *(Original trike now fitted to G-JESA qv)*						
G-MVBG	Mainair Gemini/Flash IIA	663-688-6 & W453		25. 5.88	M.P.Edwards	Shifnal	27. 7.05P
G-MVBI	Mainair Gemini/Flash IIA	665-788-6 & W455		7. 6.88	P.Thomas	(Barrow-in-Furness)	26. 9.05P
G-MVBK	Mainair Gemini/Flash IIA	666-788-6 & W456		7. 6.88	M.E.Lawes	(Ormskirk)	14. 5.05P
	(Rotax 462)						
G-MVBL	Mainair Gemini/Flash IIA	669-788-6 & W459		7. 6.88	P.M.Wright	Higher Barn Farm, Houghton	30.12.03P
G-MVBM	Mainair Gemini/Flash IIA	667-788-6 & W457		7. 6.88	A M.Wood	(Chelmsford)	9. 8.03P
G-MVBN	Mainair Gemini/Flash IIA	668-788-6 & W458		8. 6.88	M.Frankcom *(Current status unknown)*	(Darwen)	2. 6.99P
G-MVBO	Mainair Gemini/Flash IIA	671-788-6 & W461		8. 6.88	R.Brasher	(Rugeley)	11.12.05P
G-MVBP	Thruster TST Mk.1	8068-TST-080		14. 6.88	K.J.Crompton	Newtownards	12. 3.05P
G-MVBT	Thruster TST Mk.1	8068-TST-083		14. 6.88	E.L.Everitt	Ley Farm, Chirk	23.11.04P
	(BMW R100)						
G-MVBY	Solar Wings Pegasus XL-R			17. 6.88	J.E.Harman	(Napton, Southam)	9. 9.03P
		SW-TB-1357 & SW-WA-1344			tr Pigs R Us Flying Group		
G-MVBZ	Solar Wings Pegasus XL-R			17. 6.88	A G.Butler	Shenstone Hall Farm, Shenstone	17. 9.05P
		SW-TB-1358 & SW-WA-1345					
G-MVCA	Solar Wings Pegasus XL-R			17. 6.88	R.Walker	Sutton Meadows, Ely	3.10.05P
		SW-TB-1359 & SW-WA-1346					
G-MVCB	Solar Wings Pegasus XL-R			17. 6.88	G.T.Clipstone	(Ipswich)	23. 5.05P
		SW-TB-1360 & SW-WA-1347					
G-MVCC	CFM Shadow Series CD	K.045		17. 6.88	A Buzuk	(Rodmell, Lewes)	9. 5.05P
G-MVCD	Medway Hybred 44XLR	MR001/34		14. 6.88	J.Thompson	Longacres Farm, Sandy	27. 7.03P
	(Marked as "Raven") (Original sailwing transferred to G-MVOS: new wing c/n not yet known)						
G-MVCE	Mainair Gemini/Flash IIA	672-788-6 & W462		23. 6.88	J.D.Berry *(Current status unknown)*	Ince Blundell	5. 4.99P
G-MVCF	Mainair Gemini/Flash IIA	673-788-6 & W463		14. 7.88	J.S.Harris	Old Sarum	1. 6.05P
	(Rotax 462)						

G-MVCI	Noble Hardman Snowbird Mk.IV	SB-011	11.10.88	W.L.Chapman *(Current status unknown)*	Tarn Farm, Cockerham	13. 4.95P
G-MVCJ	Noble Hardman Snowbird Mk.IV (Rotax 582) *(Officially regd with Rotax 532)*	SB-012	11.10.88	C.W.Buxton	Barton	14. 9.04P
G-MVCK	Cosmos Trike/La Mouette Profil 19	SDA-01	19. 7.88	S.D.Alsop *(Current status unknown)*	(Bath)	
G-MVCL	Solar Wings Pegasus XL-Q	SW-TE-0069 & SW-WQ-0075	27. 6.88	T.E.Robinson	Insch	20. 9.04P
G-MVCM	Solar Wings Pegasus XL-Q	SW-TE-0070 & SW-WQ-0076	27. 6.88	P.J.Croney	Retreat Farm, Little Baddow	7. 4.05P
G-MVCN	Solar Wings Pegasus XL-Q	SW-TE-0071 & SW-WQ-0077	27. 6.88	S.R.S.Evans	(Chelmsford)	20. 3.01P
G-MVCP	Solar Wings Pegasus XL-Q	SW-TE-0073 & SW-WQ-0079	27. 6.88	J.R.Fulcher	Deenethorpe	27. 6.03P
G-MVCR	Solar Wings Pegasus XL-Q	SW-TE-0069 & SW-WQ-0080	27. 6.88	P.Hoeft	(Stickney, Boston)	18. 6.05P
G-MVCS	Solar Wings Pegasus XL-Q	SW-TE-0075 & SW-WQ-0081	27. 6.88	J.J.Sparrow	Sywell	19. 7.02P
G-MVCT	Solar Wings Pegasus XL-Q	SW-TE-0076 & SW-WQ-0082	27. 6.88	G.J.Lampitt	Pound Green, Buttonoak, Bewdley	19. 9.04P
G-MVCV	Solar Wings Pegasus XL-Q *(Original trike c/n SW-TE-0078 damaged and replaced by SW-TE-0108: later repaired and fitted with sailwing SW-WQ-0105 and regd G-MVHP)*	SW-WQ-0084	27. 6.88	G.E.and B.T.Nunn	Longacres Farm, Sandy	20. 4.02P
G-MVCW	CFM Shadow Series BD	084	28. 6.88	W.A Douthwaite	Baxby Manor, Husthwaite	1. 8.05P
G-MVCY	Mainair Gemini/Flash IIA	674-788-6 & W464	14. 7.88	A M.Smith	Otherton, Cannock	11.10.05P
G-MVCZ	Mainair Gemini/Flash IIA	675-788-6 & W465	26. 8.88	I.Simpson	Tarn Farm, Cockerham	8. 2.05P
G-MVDA	Mainair Gemini/Flash IIA (Rotax 462)	676-788-6 & W466	13. 7.88	C.Tweedley	(Great Orton)	15. 7.04P
G-MVDD	Thruster TST Mk.1	8078-TST-086	12. 7.88	D.J.Love *(Current status unknown)*	(Witton, Norwich)	9.11.99P
G-MVDE	Thruster TST Mk.1	8078-TST-087	12. 7.88	R.H.Davis *(Derelict 2.03)*	Doynton	26. 8.99P
G-MVDF	Thruster TST Mk.1	8078-TST-088	12. 7.88	J.Walsh and A R.Sunley tr G-MVDF Syndicate	Rayne Hall Farm, Braintree	2. 5.05P
G-MVDG	Thruster TST Mk.1	8078-TST-089	12. 7.88	D.G.,P.M. and A B.Smith	Popham	26. 7.00P
G-MVDH	Thruster TST Mk.1	8078-TST-090	12. 7.88	P.E.Terrell	(Plymouth)	22. 9.04P
G-MVDJ	Medway Hybred 44XLR	MR010/38	20. 7.88	W.D.Hutchings	(Nottingham)	8. 9.05P
G-MVDK	Aerial Arts Chaser S	CH.702	5. 8.88	S.Adams *(Noted 9.01)*	Leicester	29.11.98P
G-MVDL	Aerial Arts Chaser S (Rotax 462) *(Officially regd with Rotax 377)*	CH.701	11. 8.88	J.R.Hall *"112"*	(Bibberne Farm, Stalbridge)	26. 4.05P
G-MVDP	Aerial Arts Chaser S 447	CH.706	11. 8.88	P.Corke *(Noted 7.03)*	Longacres Farm, Sandy	9.11.02P
G-MVDR	Aerial Arts Chaser S 447 *(Officially regd with Rotax 377)*	CH.708	11. 8.88	A M.Sutton	(Stourbridge)	8. 3.05P
G-MVDS*	Hiway Skytrike/Demon 175	PB-01	30. 6.88	Not known *(Cancelled 24.1.95 by CAA) (Stored 6.04)*	Broadmeadow Farm, Hereford	
G-MVDT	Mainair Gemini/Flash IIA	670-788-6 & W460	20. 7.88	D.C.Stephens	(Coleford)	26. 5.00P
G-MVDU	Solar Wings Pegasus XL-R	SW-TB-1361 & SW-WA-1348	13. 7.88	R.J.Parkman	Weston Zoyland	25.10.04P
G-MVDV	Solar Wings Pegasus XL-R	SW-TB-1362 & SW-WA-1349	13. 7.88	I.Hutchinson	Wombleton	17.12.04P
G-MVDW	Solar Wings Pegasus XL-R	SW-TB-1363 & SW-WA-1350	13. 7.88	R.P.Brown *(Noted wrecked 7.03)*	Longacres Farm, Sandy	20. 7.97P
G-MVDX	Solar Wings Pegasus XL-R	SW-TB-1364 & SW-WA-1351	13. 7.88	C.Kett *(Current status unknown)*	Weston Zoyland	8. 8.98P
G-MVDY	Solar Wings Pegasus XL-R	SW-TB-1365 & SW-WA-1352	13. 7.88	C.G.Murphy *(Current status unknown)*	Biggin Hill	1. 6.92P
G-MVDZ	Solar Wings Pegasus XL-R	SW-TB-1366 & SW-WA-1353	12. 7.88	A K.Pickering	(Robertsbridge)	19. 5.00P
G-MVEC	Solar Wings Pegasus XL-R	SW-TB-1369 & SW-WA-1356	20. 7.88	J.A Jarvis	Davidstow Moor	18. 4.03P
G-MVED	Solar Wings Pegasus XL-R/Se	SW-TB-1370 & SW-WA-1357	20. 7.88	P.A Sleightholme	Baxby Manor, Husthwaite	29. 5.05P
G-MVEE	Medway Hybred 44XLR *(Trike c/n same as G-MYMJ and suggests this has a replacement unit)*	MR004/35	22. 7.88	D.S.L.Evans	(Gravesend)	4.10.05P
G-MVEF	Solar Wings Pegasus XL-R (Rotax 462)	SW-TE-0079 & SW-WA-1358	19. 7.88	E.J.Blyth *(Current status unknown)*	(Pickering)	15.11.93P
G-MVEG	Solar Wings Pegasus XL-R (Rotax 462)	SW-TE-0080 & SW-WA-1359	19. 7.88	A W.Leadley *(Current status unknown)*	(Strabane, County Tyrone)	25. 7.05P
G-MVEH	Mainair Gemini/Flash IIA	677-788-6 & W468	26. 8.88	K Bailey	(Oldham)	18.10.05P
G-MVEI	CFM Shadow Series CD	085	26. 7.88	R.H.Faux	Kirkbride	10.11.05P
G-MVEJ	Mainair Gemini/Flash IIA (Rotax 462)	678-888-6 & W469	27. 7.88	M.Thornburn and S.Mair	(Moffat/Lockerbie)	28. 5.05P
G-MVEK	Mainair Gemini/Flash IIA	679-888-6 & W470	27. 7.88	R.M.Rea	(Lubenham, Market Harborough)	28. 9.05P
G-MVEL	Mainair Gemini/Flash IIA	680-888-6 & W471	27. 7.88	M.R.Starling	Swanton Morley	25. 7.03P
G-MVEN	CFM Shadow Series CD	K.047	26. 7.88	N.J.Mepham	(Crowborough)	4. 8.04P
G-MVEO	Mainair Gemini/Flash IIA	682-888-6 & W472	28. 7.88	S.Macmillan	(Dumbarton)	20. 2.05P
G-MVER	Mainair Gemini/Flash IIA	684-888-6 & W474	28. 7.88	J.R.Davis	(Cheltenham)	16. 5.05P
G-MVES	Mainair Gemini/Flash IIA	685-888-6 & W475	5. 8.88	F.W.McLean and S.H.Mitchell	East Fortune	4. 4.05P
G-MVET	Mainair Gemini/Flash IIA	686-888-6 & W476	19. 8.88	T.Bailey	Otherton, Cannock	16. 2.05P
G-MVEV	Mainair Gemini/Flash IIA	687-888-6 & W477	5. 8.88	C.Allen	(Alderley Edge)	8. 7.01P
G-MVEW	Mainair Gemini/Flash IIA	688-988-6 & W478	16. 9.88	N.A Dye *(Current status unknown)*	Swanton Morley	27. 7.98P
G-MVEX	Solar Wings Pegasus XL-Q	SW-TE-0082 & SW-WQ-0088	5. 8.88	D.Maher	(Nenagh, CountyTipperary)	5. 3.04P
G-MVEZ	Solar Wings Pegasus XL-Q	SW-TE-0084 & SW-WQ-0090	9. 8.88	P.W.Millar *(Current status unknown)*	(Newnham)	13. 6.99P
G-MVFA	Solar Wings Pegasus XL-Q	SW-TE-0085 & SW-WQ-0091	9. 8.88	A Johnson	Deenethorpe	25. 6.05P
G-MVFB	Solar Wings Pegasus XL-Q	SW-TE-0086 & SW-WQ-0092	9. 8.88	M.O.Bloy	(King's Lynn)	4.10.05P

Reg	Type	C/n	Prev.Id	Regd	Owner	Location	CofA
G-MVFC	Solar Wings Pegasus XL-Q	SW-TE-0087 & SW-WQ-0093		9. 8.88	D.R.Joint	(Wimborne)	3. 8.05P
G-MVFD	Solar Wings Pegasus XL-Q	SW-TE-0088 & SW-WQ-0094		9. 8.88	C.D.Humphries	Long Marston	28.11.05P
G-MVFE	Solar Wings Pegasus XL-Q	SW-TE-0089 & SW-WQ-0095		9. 8.88	S.J.Weeks	Kemble	30. 4.00P
G-MVFF	Solar Wings Pegasus XL-Q	SW-TE-0090 & SW-WQ-0096		9. 8.88	A Makepeace	(Guildford)	31. 8.05P
G-MVFH	CFM Shadow Series CD (Rotax 447)	086		9. 8.88	G.R.Read	(Mendlesham)	4. 5.05P
G-MVFJ	Thruster TST Mk.1	8088-TST-092		11. 8.88	B.E.Renehan tr Kestrel Flying Group	Popham	2. 4.05P
G-MVFK	Thruster TST Mk.1	8088-TST-093		11. 8.88	A Evans	Longside, Peterhead	13.10.05P
G-MVFL	Thruster TST Mk.1	8088-TST-094		11. 8.88	C.Scoble tr G-MVFL Group	(Poole)	22. 5.02P
G-MVFM	Thruster TST Mk.1	8088-TST-095		11. 8.88	G.J.Boyer	Weston Zoyland	1. 9.05P
G-MVFN	Thruster TST Mk.1	8088-TST-096		11. 8.88	A G.Ward	Longacres Farm, Sandy	29. 5.04P
G-MVFO	Thruster TST Mk.1	8088-TST-097		11. 8.88	A L.Higgins and D.H.King tr G-MVFO Group	(Newport Pagnell)	20. 4.05P
G-MVFP	Solar Wings Pegasus XL-R	SW-TB-1371 & SW-WA-1365		9. 8.88	D J Brixton tr Shropshire Tow Group	Bishops Castle, Shropshire	24. 3.03P
G-MVFR	Solar Wings Pegasus XL-R	SW-TB-1372 & SW-WA-1366		9. 8.88	P.Newton *(Current status unknown)*	(Macclesfield)	21.11.99P
G-MVFS	Solar Wings Pegasus XL-R/Se	SW-TB-1373 & SW-WA-1367		9. 8.88	A Cordes	Deenethorpe	25. 6.05P
G-MVFT	Solar Wings Pegasus XL-R	SW-TB-1374 & SW-WA-1368		9. 8.88	S.J.Whalley	Roddige	24.10.05P
G-MVFV	Solar Wings Pegasus XL-R	SW-TB-1376 & SW-WA-1370		9. 8.88	L.R.M.Grigg	Deenethorpe	25. 6.05P
G-MVFW	Solar Wings Pegasus XL-R	SW-TB-1377 & SW-WA-1371		9. 8.88	S.F.Chaplin *(Noted 9.04)*	Haverfordwest	28. 8.00P
G-MVFY	Solar Wings Pegasus XL-R	SW-TB-1379 & SW-WA-1373		9. 8.88	T.D.Bawden	Weston Zoyland	19. 6.04P
G-MVFZ	Solar Wings Pegasus XL-R	SW-TB-1380 & SW-WA-1374		9. 8.88	R.K.Johnson	Popham	29. 5.05P
G-MVGA	Aerial Arts Chaser S 508	CH.859 *(Officially regd with CH.707)*		11. 8.88	N.R.Beale	Deppers Bridge, Southam	2. 7.04P
G-MVGB	Medway Hybred 44XLR	MR011/39		1. 9.88	R.Graham	Rochester	28. 8.02P
G-MVGC	AMF Microflight Chevvron 2-32C	010		2. 9.88	W.Fletcher	(Usk)	14. 7.05P
G-MVGD	AMF Microflight Chevvron 2-32C	011		5. 9.88	T.R.James	(Southam)	2. 9.05P
G-MVGE	AMF Microflight Chevvron 2-32C	012		26. 9.88	J.P.Bennett	Old Sarum	13. 6.05P
G-MVGF	Aerial Arts Chaser S	CH.720		2. 9.88	P.J.Higgins "The Dingbat" *(New owner 4.03)*	Gedney Dyke, Lutton	11.11.00P
G-MVGG	Aerial Arts Chaser S 508	CH.721		2. 9.88	J J Bowen	Newton Bank, Daresbury	17. 6.05P
G-MVGH	Aerial Arts Chaser S 447	CH.722		2. 9.88	J.E.Orbell	North Connel, Oban	14.11.03P
G-MVGI*	Aerial Arts Chaser S	CH.723		1. 9.88	Not known *(Cancelled 6.1.03 as sold to Ireland but noted stored as "G-MVGi" 9.04)*	Mill Farm, Hughley, Much Wenlock	
G-MVGL	Medway Hybred 44XLR	MR012/40		1. 9.88	W.Stacey	(Chelmsford)	6. 9.05P
G-MVGM	Mainair Gemini/Flash IIA	691-988-6 & W481		25. 8.88	A R.Pitcher	(Cranbrook)	27.10.04P
G-MVGN	Solar Wings Pegasus XL-R/Se	SW-TB-1381 & SW-WA-1377		23. 8.88	M.J.Smith	(Taunton)	24. 6.05P
G-MVGO	Solar Wings Pegasus XL-R	SW-TB-1382 & SW-WA-1378		23. 8.88	J.B.Peacock	Lower Mountpleasant, Chatteris	25. 7.05P
G-MVGP	Solar Wings Pegasus XL-R	SW-TB-1383 & SW-WA-1379	(EC-) G-MVGP	23. 8.88	J.P.Cox	(Kettering)	9. 6.00P
G-MVGU	Solar Wings Pegasus XL-Q	SW-TB-0092 & SW-WQ-0100		23. 8.88	T.D.Turner	Redlands	21. 6.04P
G-MVGW	Solar Wings Pegasus XL-Q	SW-TE-0095 & SW-WQ-0102		23. 8.88	M.J.L.de Carvalho and V.V.P.Pedro *(Current status unknown)* tr G-MVGW Group	Lagos, Algarve, Portugal	8. 2.92P
G-MVGY	Medway Hybred 44XLR	MR015/41		31. 8.88	D.G.Baker	Lee-on-Solent	24. 8.05P
G-MVGZ	Ultraflight Lazair IIIE (Rotax 185 x 2)	A338	C-?	21.10.88	D.M.Broom	(Towcester)	25. 3.03P
G-MVHA	Aerial Arts Chaser S-1000 (Mosler MM CB-38)	CH.729		24. 8.88	R.Meredith-Hardy	Radwell Lodge, Baldock	9. 5.04P
G-MVHB	Powerchute Raider	80105		26. 8.88	A E.Askew	(Melton Mowbray)	18. 5.05P
G-MVHC	Powerchute Raider	80106		26. 8.88	N.and S.A Melrose	(Ripley)	19. 7.03P
G-MVHD	CFM Shadow Series CD	088		8. 9.88	M.J.Day and D.F.Randall	Beccles	8.12.05P
G-MVHE	Mainair Gemini/Flash IIA	692-988-6 & W482		4.10.88	B.R.Thomas	Oak Farm, Woodton	21.11.05P
G-MVHF	Mainair Gemini/Flash IIA	693-988-6 & W483		4.10.88	M.G.Nicholson	St Michaels	28. 5.05P
G-MVHG	Mainair Gemini/Flash IIA	694-988-6 & W484		14.10.88	C.A J.Elder	(Bo'ness)	4. 4.05P
G-MVHH	Mainair Gemini/Flash IIA *(Original trike 695.-.. replaced by 607-... ex G-MTSA 1995)*	607-1187-5 & W485		24.10.88	I.J.Cleland and M.D.Calder	(Glasgow)	15.11.05P
G-MVHI	Thruster TST Mk.1	8098-TST-100		26. 9.88	R.A.Samulis	Enstone	28. 2.05P
G-MVHJ	Thruster TST Mk.1	8098-TST-101		26. 9.88	S.P.Macdonald	Lower Mountpleasant, Chatteris	3. 4.05P
G-MVHK	Thruster TST Mk.1	8098-TST-102		27. 9.88	D.J.Gordon	Roche, Cornwall	24. 6.05P
G-MVHL	Thruster TST Mk.1	8098-TST-103		27. 9.88	G.Jones	(Llanfairfechan)	16. 7.02P
G-MVHO*	Solar Wings Pegasus XL-Q	SW-TE-0097 & SW-WQ-0104		23. 9.88	S.J.Barkworth *(Cancelled 12.7.04 by CAA)*	Rufforth	8. 5.05P
G-MVHP	Solar Wings Pegasus XL-Q	SW-TE-0078 & SW-WQ-0105		23. 9.88	J.B.Gasson *(Damaged trike from G-MVCV repaired and fitted to sailwing)*	Lower Mountpleasant, Chatteris	13. 2.05P
G-MVHR	Solar Wings Pegasus XL-Q	SW-TE-0099 & SW-WQ-0106		23. 9.88	J.M.Hucker	Broadmeadow Farm, Hereford	26. 5.04P
G-MVHS	Solar Wings Pegasus XL-Q	SW-TE-0100 & SW-WQ-0107		23. 9.88	C.L.Lebeter	(Derby)	14. 7.05P
G-MVHW	Solar Wings Pegasus XL-Q	SW-TE-0101 & SW-WQ-0111		23. 9.88	Ultralight Training Ltd	Roddige	6. 2.05P

Regn	Type	C/n	Prev.Id	Regd	Owner	Base	CofA
G-MVHX	Solar Wings Pegasus XL-Q	SW-TE-0105 & SW-WQ-0112		23. 9.88	E.J.Carass	(Norwich)	31. 5.04P
G-MVHY	Solar Wings Pegasus XL-Q	SW-TE-0106 & SW-WQ-0113		23. 9.88	R.P.Paine	Headon Farm, Retford	17. 8.05P
G-MVHZ	Hornet Dual Trainer/Southdown Raven	HRWA 0076 & MHR-101		26. 9.88	J.M.Addison *(New owner 10.04)*	(Monymusk, Inverurie)	3. 8.03P
G-MVIA	Solar Wings Pegasus XL-R (Rotax 462)	SW-TE-0107 & SW-WA-1375		4.10.88	K.Parkyn *(Unmarked and stored 8.03)*	St.Just	14. 6.01P
G-MVIB	Mainair Gemini/Flash IIA	700-1088-4 & W490		14.10.88	S, A and L Rosser t/a LSA Systems	Arclid Green, Sandbach	5.12.04P
G-MVIE	Aerial Arts Chaser S	CH.732		14.10.88	T.M.Stiles *(Noted 8.03)*	Perth	6. 6.97P
G-MVIF	Medway Raven X *(Originally regd as Hybred 44XLR)*	MR020/43		4.10.88	J.R.Harrison	(Bolsover)	16. 5.05P
G-MVIG	CFM Shadow Series B (Rotax 447)	K.044		5.10.88	M.P.and P.A G.Harper *(Damaged 1993: stored 8.93: current status unknown)*	Priory Farm, Tibenham	20. 1.94P
G-MVIH	Mainair Gemini/Flash IIA	697-1088-6 & W487		14.10.88	T.M.Gilsenan	(Eaton Bray)	12. 6.05P
G-MVIL	Noble Hardman Snowbird Mk.IV (Rotax 582)	SB-014		6. 2.89	Marine Power (Scotland) Ltd	Kirkbride	27. 4.05P
G-MVIM	Noble Hardman Snowbird Mk.IV	SB-015		6. 2.89	W.G.Goodall *(New owner 10.03)*	(Goole)	26. 8.91P
G-MVIN	Noble Hardman Snowbird Mk.IV *(Rebuilt to Mk.V standard with Rotax 582)*	SB-016		6. 2.89	C.P.Dawes	Darley Moor	7.10.05P
G-MVIO	Noble Hardman Snowbird Mk.IV	SB-017		12. 4.89	B.Mason-Baker tr Mobility Advice Line	Mill Farm, Hughley, Much Wenlock	3. 5.04P
G-MVIP	AMF Microflight Chevvron 2-32C	008		11. 5.88	P C Avery	Lower Mountpleasant, Chaterris	26. 1.05P
G-MVIR	Thruster TST Mk.1	8108-TST-104 *(C/n plate marked as 8<u>11</u>8-TST-104)*		21.10.88	T D Gardner	Popham	7. 9.05P
G-MVIT	Thruster TST Mk.1	8108-TST-106	(C-) G-MVIT	21.10.88	A P.Trumper	(Grantham)	10.10.03P
G-MVIU	Thruster TST Mk.1 *(Rebuilt to part T.600 standard)*	8108-TST-107		21.10.88	E.Bentley	Morgansfield, Fishburn	28. 3.05P
G-MVIV	Thruster TST Mk.1	8108-TST-108		21.10.88	P.J.Sears	(Ivybridge)	24. 6.01P
G-MVIW	Thruster TST Mk.1 (Rotax 532)	8108-TST-109		21.10.88	J.M.Nicholson	(Denham)	4. 6.01P
G-MVIX	Mainair Gemini/Flash IIA	702-1088-6 & W492		14.10.88	R.S.T.MacEwen	East Fortune	19.10.05P
G-MVIY	Mainair Gemini/Flash IIA	701-1088-6 & W491		14.10.88	J.J.Valentine	Broadmeadow Farm, Hereford	23. 8.05P
G-MVIZ	Mainair Gemini/Flash IIA	703-1088-6 & W493		14.10.88	P.R.Hutty	(Hessle)	22. 7.05P
G-MVJA	Mainair Gemini/Flash IIA	696-988-6 & W486		5.12.88	J.R.Harrison	(Wisbech)	17. 6.05P
G-MVJC	Mainair Gemini/Flash IIA	705-1088-6 & W495		24.10.88	B.Temple *(Engine failure on take off Priory Farm, Tibenham 15.8.04 and seriously damaged)*	Broadmeadow Farm, Hereford	25. 4.05P
G-MVJD	Solar Wings Pegasus XL-R (Rotax 462)	SW-TE-0109 & SW-WA-1386		24.10.88	R.S.Finlay	Kemble	2. 5.05P
G-MVJE	Mainair Gemini/Flash IIA	706-1188-6 & W496		21.10.88	S D Morris	(Wrexham)	1. 4.05P
G-MVJF	Aerial Arts Chaser S	CH.743		21.11.88	Victoria S.Rudham	Dunkeswell	2. 8.05P
G-MVJG	Aerial Arts Chaser S	CH.749		22.11.88	T.H.Scott *(Noted 1.05)*	Rayne Hall Farm, Braintree	13. 6.03P
G-MVJH	Aerial Arts Chaser S	CH.751		14.11.88	M.Van Rompaey	(Scunthorpe)	1. 9.03P
G-MVJI	Aerial Arts Chaser S	CH.752		17.11.88	T.Beckham	(Newcastle upon Tyne)	1.11.03P
G-MVJJ	Aerial Arts Chaser S 508	CH.753		14.11.88	C.W.Potts	(Newcastle upon Tyne)	24. 8.03P
G-MVJK	Aerial Arts Chaser S	CH.754		14.11.88	K.J.Samuels	(Loughton)	9. 4.05P
G-MVJL	Mainair Gemini/Flash IIA	698-1188-6 & W488		21.10.88	F.Huxley	(Morpeth)	20. 5.01P
G-MVJM	Microflight Spectrum	007		21.10.88	S.E.Matthews tr Poppy Syndicate	Otherton, Cannock	9. 8.05P
G-MVJN	Solar Wings Pegasus XL-Q	SW-TE-0110 & SW-WQ-0116		26.10.88	R.A.Paintain	Chilbolton	13.11.04P
G-MVJO	Solar Wings Pegasus XL-Q	SW-TE-0111 & SW-WQ-0117		26.10.88	C M Wilkes	(Fareham)	22. 9.05P
G-MVJP	Solar Wings Pegasus XL-Q	SW-TE-0112 & SW-WQ-0118		26.10.88	S.H.Bakowski	Rochester	18. 4.05P
G-MVJR	Solar Wings Pegasus XL-Q	SW-TE-0113 & SW-WQ-0119		26.10.88	B Goldsmith	Chiltern Park, Wallingford	9. 8.05P
G-MVJS	Solar Wings Pegasus XL-Q	SW-TE-0114 & SW-WQ-0120		26.10.88	S.D.Morley	Rayne Hall Farm, Braintree	10. 8.05P
G-MVJT	Solar Wings Pegasus XL-Q	SW-TE-0115 & SW-WQ-0121		26.10.88	L A Hosegood	Redlands	18. 7.04P
G-MVJU*	Solar Wings Pegasus XL-Q	SW-TE-0116 & SW-WQ-0122		26.10.88	G.B.Hutchison *(Cancelled 9.2.05 by CAA)*	Sandtoft	26. 8.04P
G-MVJW	Solar Wings Pegasus XL-Q	SE-TE-0118 & SW-WQ-0124		26.10.88	R.Dainty and D.W.Stamp	Pound Green, Buttonoak, Bewdley	18. 4.05P
G-MVKB	Medway Hybred 44XLR	MR023/45		11.11.88	J.Newby	Sandtoft	11. 9.00P
G-MVKC	Mainair Gemini/Flash IIA	709-1188-6 & W499		16.11.88	R.L.Bladon	Sittles Farm, Alrewas	21. 4.05P
G-MVKF	Solar Wings Pegasus XL-R	SW-TB-1389 & SW-WA-1392		14.11.88	B.Shaw	(Northampton)	6. 7.05P
G-MVKH	Solar Wings Pegasus XL-R	SW-TB-1393 & SW-WA-1396		14.11.88	K.M.Elson	Roddige	14.11.04P
G-MVKJ	Solar Wings Pegasus XL-R (Rotax 462)	SW-TE-0132 & SW-WA-1398	*(Officially regd with Rotax 447)*	14.11.88	G.V.Warner	Croughton	7. 6.02P
G-MVKK	Solar Wings Pegasus XL-R (Rotax 462)	SW-TE-0131 & SW-WA-1397		14.11.88	P.G.Sayers	Graveley	15. 8.05P
G-MVKL	Solar Wings Pegasus XL-R	SW-TB-1391 & SW-WA-1394	*(Although pod marked as "XL-Q" remains a XL-R model) (Current status unknown)*	14.11.88	J.Powell-Tuck	(Pontypool)	6. 6.91P
G-MVKM	Solar Wings Pegasus XL-R	SW-TE-0136 & SW-WA-1399	*(<u>Now uses Trike c/n SW-TB-1152 - see G-BZWM</u>*	14.11.88	A.E.Dobson	Broadmeadow Farm, Hereford	12. 2.05P
G-MVKN	Solar Wings Pegasus XL-Q	SW-TE-0120 & SW-WQ-0126		14.11.88	R.A and C.A Allen	Siege Cross Farm, Thatcham	17. 6.05P
G-MVKO	Solar Wings Pegasus XL-Q	SW-TE-0121 & SW-WQ-0127		14.11.88	C.R.Bunce	Yatesbury	9. 1.05P

Regn	Type	C/n	Date	Owner/Operator	Base	CofA
G-MVKP	Solar Wings Pegasus XL-Q	SW-TE-0122 & SW-WQ-0128	14.11.88	P.Mokryk and S.King	(Derby)	19. 9.04P
G-MVKS	Solar Wings Pegasus XL-Q	SW-TE-0124 & SW-WQ-0130	14.11.88	K.S.Wright *(Stored 8.95: current status unknown)*	Long Marston	13. 5.94P
G-MVKT	Solar Wings Pegasus XL-Q	SW-TE-0125 & SW-WQ-0131	14.11.88	N.C.Williams	Enstone	10. 6.05P
G-MVKU	Solar Wings Pegasus XL-Q	SW-TE-0126 & SW-WQ-0132	14.11.88	J.R.F.Shepherd	Longacres Farm, Sandy	4.10.05P
G-MVKV	Solar Wings Pegasus XL-Q	SW-TE-0127 & SW-WQ-0152	14.11.88	D.M.Taylor	(Thornton-Cleveleys)	26. 3.05P
	(Original sailwing c/n SW-WQ-0133 damaged 14.8.91 and replaced by '0152)					
G-MVKW	Solar Wings Pegasus XL-Q	SW-TE-0128 & SW-WQ-0134	14.11.88	A T.Scott	(London SW17)	10. 2.05P
G-MVKY	Aerial Arts Chaser S	CH.755	5.12.88	R.W.Whitehead	Swinford, Rugby	3. 7.02P
G-MVKZ	Aerial Arts Chaser S	CH.756	5.12.88	T J Barley	(Harlow)	29. 7.04P
G-MVLA	Aerial Arts Chaser S	CH.762	12.12.88	T.Birch	Sittles Farm, Alrewas	21. 4.05P
G-MVLB	Aerial Arts Chaser S	CH.763	5.12.88	R.P.Wilkinson *(New owner 8.04)*	Charmy Down, Bath	13. 9.97P
G-MVLC	Aerial Arts Chaser S 477 *(Officially regd with Rotax 377)*	CH.764	22.11.88	B.R.Barnes	(Bristol)	10. 5.03P
G-MVLD	Aerial Arts Chaser S	CH.765	22.11.88	A.W.Leadley *(New owner 5.04)*	(Letterkenny, County Donegal)	4. 6.05P
G-MVLE	Aerial Arts Chaser S	CH.766	5.12.88	R.G.Hooker	Brunton	22. 9.05P
G-MVLF	Aerial Arts Chaser S 508	CH.767	11. 1.89	I B Smith	Deenethorpe	1. 8.05P
G-MVLG	Aerial Arts Chaser S	CH.768	14.11.88	A Strang	East Fortune	26.11.04P
G-MVLJ	CFM Shadow Series CD	092	11.11.88	R.S.Cochrane	Sutton Meadows, Ely	22. 1.05P
G-MVLL	Mainair Gemini/Flash IIA *(Sailwing c/n now W396 ex G-MTSA)*	708-1188-6 & W498	23.11.88	M.J.A.New *(New owner 8.04)*	Mill Farm, Hughley, Much Wenlock	27.10.02P
G-MVLP	CFM Shadow Series C (Rotax 447) *(Officially regd with Rotax 503)*	095	22.11.88	D.Bridgland and D.T.Moran	Old Sarum	29. 7.04P
G-MVLR	Mainair Gemini/Flash IIA	713-1288-6 & W503	30.11.88	K.B.A Judson	(Colchester)	18.11.00P
G-MVLS	Aerial Arts Chaser S 477	CH.773	21. 2.89	T.C.Brown	(London SW20)	20. 4.05P
G-MVLT	Aerial Arts Chaser S	CH.774	5.12.88	B.D.Searle	(Portsmouth)	19. 7.05P
G-MVLW	Aerial Arts Chaser S	CH.778	28.12.88	E.W.P.van Zeller *(Current status unknown)*	(Ashford)	5. 9.99P
G-MVLX	Solar Wings Pegasus XL-Q	SW-TE-0133 & SW-WQ-0114	30.11.88	J.F.Smith	Enstone	25. 6.05P
G-MVLY	Solar Wings Pegasus XL-Q	SW-TE-0137 & SW-WQ-0142	5.12.88	I.B.Osborn	Manston	28. 9.05P
G-MVMA	Solar Wings Pegasus XL-Q	SW-TE-0139 & SW-WQ-0144	5.12.88	T.J.Gayton-Polley	(Billingshurst)	10. 7.05P
G-MVMC	Solar Wings Pegasus XL-Q	SW-TE-0141 & SW-WQ-0146	5.12.88	P.Smith and I.W.Barlow	(Ilkeston)	6. 3.05P
G-MVMD	Powerchute Raider	80924	15.12.88	S.M.Paulin *(Current status unknown)*	(Reading)	13. 7.90P
G-MVME	Thruster TST Mk.1	8128-TST-110	12.12.88	S J Payne	Shoreham	19. 1.05P
G-MVMG	Thruster TST Mk.1	8128-TST-112	12.12.88	A.D.McCaldin	(Banbridge)	12. 6.05P
G-MVMI	Thruster TST Mk.1	8128-TST-114	12.12.88	G.J.Johnson	North Coates	11. 2.05P
G-MVMK	Medway Hybred 44XLR	MR022/46	12.12.88	D J Lewis (*Current status unknown)*	Perth	5. 2.94P
G-MVML	Aerial Arts Chaser S	CH.781	28.12.88	G C Luddington	Wilden, Bedford	29. 7.00P
G-MVMM	Aerial Arts Chaser S	CH.797	21. 2.89	D.Margereson	(Chesterfield)	30. 7.04P
G-MVMO	Mainair Gemini/Flash IIA	715-1288-6 & W507	12.12.88	K.Austwick	Carlisle	16. 9.04P
G-MVMR	Mainair Gemini/Flash IIA	717-1288-6 & W509	9. 1.89	P.W.Ramage *(Current status unknown)*	(Rufforth)	20. 9.96P
G-MVMT	Mainair Gemini/Flash IIA	718-189-6 & W510	22.12.88	R.F.Sanders t/a Independent Financial Advisory Service *(Current status unknown)*	Otherton, Cannock	25. 9.98P
G-MVMU	Mainair Gemini/Flash IIA	719-189-6 & W511	22.12.88	M.J.A New and A Clift *"Icarus"*	Mill Farm, Hughley, Much Wenlock	26. 8.03P
G-MVMV	Mainair Gemini/Flash IIA	720-189-6 & W512	22.12.88	M.J.A New and A Clift	Mill Farm, Hughley, Much Wenlock	24. 3.05P
G-MVMW	Mainair Gemini/Flash IIA	710-1188-6 & W500	11.11.88	K.Downes and B.Nock	(Wolverhampton)	31. 7.03P
G-MVMX	Mainair Gemini/Flash IIA (Rotax 462) *(Trike stamped incorrectly as "W512")*	721-189-6 & W513	23.12.88	N.M.Corr	Newtownards	9. 4.05P
G-MVMY	Mainair Gemini/Flash IIA	722-189-6 & W514	22.12.88	N.G.Leteney	(Congleton)	21.12.02P
G-MVMZ	Mainair Gemini/Flash IIA	723-189-6 & W515	22.12.88	S.Richards	Otherton, Cannock	24. 5.02P
G-MVNA	Powerchute Raider	81230	12. 7.89	J.McGoldrick	Newtownards	6. 7.04P
G-MVNB	Powerchute Raider	81231	12. 7.89	K.J.Crompton	(Bangor, County Down)	23. 5.05P
G-MVNC	Powerchute Raider	81232	12. 7.89	W.R.Hanley	(Edinburgh)	25. 7.00P
G-MVNI	Powerchute Raider	90625	12. 7.89	N.J.Staib	Kemble	13. 7.02P
G-MVNK	Powerchute Raider	90623	12. 7.89	J.Cunliffe *(Current status unknown)*	(Stoke-on-Trent)	16. 7.95P
G-MVNL	Powerchute Raider)	90624	12. 7.89	S.Penoyre	(Windlesham)	17. 3.01P
G-MVNM	Mainair Gemini/Flash IIA	725-189-6 & W517	6. 1.89	M.Castle and T.Hartwig	(Shrewsbury)	8. 6.00P
G-MVNN	Whittaker MW-5K Sorcerer *(Built K N Dando - kit no.5K-0003-02)*	BMAA/HB/022	28. 3.90	J.A T.Merino	(Malaga, Spain)	21. 5.04P
G-MVNO	Whittaker MW-5K Sorcerer *(Built Aerotech International Ltd)*	5K-0004-02	4. 5.89	R.L.Wadley	Middle Stoke, Isle of Grain	31. 5.04P
G-MVNP	Whittaker MW-5K Sorcerer *(Built Aerotech International Ltd)*	5K-0005-02	13. 7.89	A M Edwards	(Wokingham)	30. 6.05P
G-MVNR	Whittaker MW-5K Sorcerer *(Built Aerotech International Ltd)*	5K-0006-02	4. 5.89	E.I.Rowlands-Jones *(New owner 1.04)*	(Mochdre, Newtown)	23. 8.01P
G-MVNS	Whittaker MW-5K Sorcerer *(Built Aerotech International Ltd)*	5K-0007-02	19. 7.89	R.D.Chiles	Shenstone Hall Farm, Shenstone	25. 7.04P
G-MVNT	Whittaker MW-5K Sorcerer *(Built Aerotech International Ltd)*	5K-0008-02	28. 3.90	P.E.Blyth	Wombleton	5 8.05P
G-MVNU	Whittaker MW-5K Sorcerer *(Built Aerotech International Ltd)*	5K-0009-02	4. 5.89	J.C.Rose	Field Farm, Oakley	30. 6.05P
G-MVNW	Mainair Gemini/Flash IIA	726-189-6 & W518	25. 1.89	A Weatherall	(Preston)	14. 5.05P
G-MVNX	Mainair Gemini/Flash IIA	727-289-6 & W519	10. 1.89	I.Sidebotham	Barton	13. 5.05P
G-MVNY	Mainair Gemini/Flash IIA (Rotax 462)	724-189-6 & W516	11. 1.89	M.K.Buckland	(Daventry)	25. 7.05P

Regn	Type	C/n	Prev.id	Date	Owner	Location	CofA
G-MVNZ	Mainair Gemini/Flash IIA	728-289-6 & W520		11. 1.89	J.Howarth	(Mansfield)	8.11.04P
G-MVOB	Mainair Gemini/Flash IIA	729-289-6 & W521		16. 1.89	B.J.Bader	(Taunton)	8. 8.04P
G-MVOD	Aerial Arts Chaser/110SX (Rotax 377)	110SX/653		16. 1.89	M.A Hodgson	Baxby Manor, Husthwaite	9. 4.05P
G-MVOF	Mainair Gemini/Flash IIA	730-289-6 & W522		31. 1.89	P.J.Nolan	(Coventry)	1.12.04P
G-MVOH	CFM Shadow Series CD	K.090		23. 1.89	D.I.Farmer	Dunkeswell	2. 9.02P
G-MVOJ	Noble Hardman Snowbird Mk.IV	SB-019		26. 7.89	T.D.Thwaites tr The HFC Group *(Current status unknown)*	(Penrith)	28. 7.99P
G-MVOL	Noble Hardman Snowbird Mk.IV	SB-021		29. 8.89	E.J.Lewis tr Swansea Snowbird Fliers	(Swansea)	26. 1.02P
G-MVON	Mainair Gemini/Flash IIA	731-289-6 & W523		30. 1.89	D.W.Beech *(New owner 10.03)*	Ince Blundell	14. 7.02P
G-MVOO	AMF Microflight Chevvron 2-32C	014		10. 1.89	I.R.F.Hammond	Lee-on-Solent	8. 4.05P
G-MVOP	Aerial Arts Chaser S	CH.787		21. 2.89	D.Thorpe	(Grantham)	4. 5.03P
G-MVOR	Mainair Gemini/Flash IIA (Rotax 462)	732-289-6 & W524	(EC-) G-MVOR	6. 2.89	P.T. and R.M.Jenkins	Dunkeswell	5.10.03P
G-MVOT	Thruster TST Mk.1	8029-TST-116		17. 2.89	B.L.R.J.Keeping	Davidstow Moor	10.10.04P
G-MVOU	Thruster TST Mk.1	8029-TST-117		17. 2.89	G.E.Norton	(Lincoln)	5 11.05P
G-MVOV	Thruster TST Mk.1	8029-TST-118		17. 2.89	D.A Duthie	Otherton, Cannock	19. 9.05P
G-MVOW	Thruster TST Mk.1	8029-TST-119		17. 2.89	J.Short and B.J.Merret	Dunkeswell	17. 7.00P
G-MVOX	Thruster TST Mk.1	8029-TST-120		17. 2.89	J.E.Davies	Haverfordwest	1. 9.05P
G-MVOY	Thruster TST Mk.1	8029-TST-121		17. 2.89	G.R.Breaden	Tarn Farm, Cockerham	14.12.04P
G-MVPA	Mainair Gemini/Flash IIA	735-289-7 & W527		29. 3.89	J.E.Milburn *(Current status unknown)*	Eshott	30. 8.95P
G-MVPB	Mainair Gemini/Flash IIA	736-389-7 & W528		29. 3.89	O.Carter	Baxby Manor, Husthwaite	30. 7.05P
G-MVPC	Mainair Gemini/Flash IIA *(Mis-stamped with c/n plate for 740-389-7 & W532 which is G-MVPI)*	737-389-7 & W529		7. 2.89	N.C.Marciano	Blackpool	26. 6.05P
G-MVPD	Mainair Gemini/Flash IIA	738-389-7 & W530		7. 2.89	A S.Bates	(Ashton-under-Lyne)	4.12.04P
G-MVPE	Mainair Gemini/Flash IIA	739-389-7 & W531		7. 2.89	A.Croucher	Newton Bank Farm, Daresbury	10. 2.05P
G-MVPF	Medway Hybred 44XLR	MR036/52		27. 2.89	G H Crick	Plaistowes Farm, St.Albans	20. 6.05P
G-MVPG	Medway Hybred 44XLR	MR026/53		15. 2.89	M.A Jones *(Current status unknown)*	(Wigan)	30.12.98P
G-MVPH	Whittaker MW6-S Fatboy Flyer *(Built E A Henman)* (Rotax 503)	PFA 164-11404		7. 2.89	A A Rowson *(Amended CofR 8.02)*	Emlyn's Field, Rhuallt	23. 8.99P
G-MVPI	Mainair Gemini/Flash IIA	740-389-7 & W532		9. 2.89	R.J.Bowden	Ince Blundell	27. 5.05P
G-MVPJ	Rans S-5 Coyote *(Built J Whiting)*	88.083 & PFA 193-11470		15. 2.89	J.E.D.Rogerson *(New owner 3.02)*	Morgansfield, Fishburn	2. 8.99P
G-MVPK	CFM Shadow Series CD (Rotax 447) *(Officially regd with Rotax 503)*	K.091		15. 2.89	P.Sarfas	Margaretting	23. 4.04P
G-MVPL	Medway Hybred 44XLR	MR034/50		1. 3.89	J.N.J.Roberts *(Current status unknown)*	Longacres Farm, Sandy	30. 4.98P
G-MVPM	Whittaker MW6 Merlin *(Built S J Field)* (Rotax 503) *(Reported as Type MW6-T)*	PFA 164-11272		21. 2.89	K.W.Curry *(New owner 1.05)*	(Nantmel, Llandrindod Wells)	30. 4.03P
G-MVPN	Whittaker MW6 Merlin *(Built A M.Field)* (Rotax 503)	PFA 164-11280		21. 2.89	A M.Field *(Current status unknown)*	(Glastonbury)	18. 5.93P
G-MVPR	Solar Wings Pegasus XL-Q	SW-TE-0149 & SW-WQ-0163		14. 3.89	R.S.Swift	Finmere	13. 6.05P
G-MVPS	Solar Wings Pegasus XL-Q	SW-TE-0143 & SW-WQ-0140		14. 3.89	B.R.Chamberlain	London Colney	27 10.04P
G-MVPW	Solar Wings Pegasus XL-R (Rotax 462)	SW-TE-0177 & SW-WA-1411		28. 3.89	C.A Mitchell *(Current status unknown)*	(Newport, Gwent)	24.10.98P
G-MVPX	Solar Wings Pegasus XL-Q	SW-TE-0144 & SW-WQ-0158		28. 3.89	M.M.P.Evans	Plaistowes Farm, St.Albans	26. 5.05P
G-MVPY	Solar Wings Pegasus XL-Q	SW-TE-0178 & SW-WQ-0188		28. 3.89	G.H.Dawson	(Swavesey, Cambridge)	27. 3.05P
G-MVRA	Mainair Gemini/Flash IIA	743-489-7-& W535		10. 4.89	L.E.Donaldson	Brunton	6.11.04P
G-MVRB	Mainair Gemini/Flash	747-489-7 & W539		29. 3.89	J Walshe	(Dungannon	4. 3.05P
G-MVRC	Mainair Gemini/Flash IIA	748-489-7 & W540		29. 3.89	M.O'Connell	Rufforth	16. 5.03P
G-MVRD	Mainair Gemini/Flash IIA	749-489-7 & W541		9. 5.89	A R.Helm	St Michaels	5. 9.05P
G-MVRE*	CFM Shadow Series CD	K.087		10. 4.89	J.Madhvani *(Cancelled 13.8.02 by CAA) (Noted 7.03)*	Plaistowes Farm, St.Albans	21.10.02P
G-MVRF	Rotec Rally 2B	AIE-01		28. 4.89	A I.Edwards *(Current status unknown)*	(Stafford)	
G-MVRG	Aerial Arts Chaser S	CH.798		14. 4.89	J.P.Kynaston *(Current status unknown)*	(Luton)	31. 8.99P
G-MVRH	Solar Wings Pegasus XL-Q	SW-TE-0160 & SW-WQ-0177		10. 4.89	K.Farr	Swinford, Rugby	29.11.04P
G-MVRI	Solar Wings Pegasus XL-Q	SW-TE-0145 & SW-WQ-0159		10. 4.89	P.Martin	(Stevenage)	12. 4.05P
G-MVRJ	Solar Wings Pegasus XL-Q	SW-TE-0172 & SW-WQ-0165		10. 4.89	J.Goldsmith-Ryan	Eaglescott	3.10.05P
G-MVRL	Aerial Arts Chaser S 447	CH.801		18. 4.89	C.N.Beale	Mill Farm, Shifnall	23.10.03P
G-MVRM	Mainair Gemini/Flash IIA (Rotax 462)	752-489-7 & W545		12. 4.89	G G Wood	(Melrose)	11. 2.05P
G-MVRO	CFM Shadow Series CD	K.105		3. 4.89	K.H.Creed	(Langar)	10.11.05P
G-MVRP	CFM Shadow Series CD	097		7. 4.89	B.Barrass	Sywell	27. 7.04P
G-MVRR	CFM Shadow Series CD	098		7. 4.89	S.Fairweather and S.P.Christian	Hougham, Lincoln	5. 9.04P
G-MVRS*	CFM Shadow Series BD				See SECTION 3 Part 1	Doncaster	
G-MVRT	CFM Shadow Series BD	104		7. 4.89	A J.Thomas	(Nottingham)	26.11.04P
G-MVRU	Solar Wings Pegasus XL-Q	SW-TE-0166 & SW-WQ-0183		12. 4.89	P.Copping	(Olney)	12. 9.04P
G-MVRV	Powerchute Kestrel	90210		28. 4.89	G.M.Fletcher *(Current status unknown)*	(Chesterfield)	3. 2.97P
G-MVRW	Solar Wings Pegasus XL-Q	SW-TE-0161 & SW-WQ-0178		12. 4.89	L Harland *(Rebuilt 1999 including new factory supplied sailwing)*	Rochester	13. 9.04P
G-MVRX	Solar Wings Pegasus XL-Q	SW-TE-0151 & SW-WQ-0165		12. 4.89	M.Everest	(Hailsham)	18. 6.05P
G-MVRY	Medway Hybred 44XLR	MR049/56		12. 4.89	K.Dodman *(Current status unknown)*	(Cambridge)	4. 3.99P
G-MVRZ	Medway Hybred 44XLR (Rotax 503)	MR043/57		9. 5.89	I.Oswald	(London SE9)	13.11.01P
G-MVSB	Solar Wings Pegasus XL-Q			18. 4.89	M.Jennings and D.Forde	Rufforth	23.11.02P

	SW-TE-0184 & SW-WQ-0193			*(New owners 3.03)*		
G-MVSD	Solar Wings Pegasus XL-Q SW-TE-0186 & SW-WQ-0195		18. 4.89	M.T.Aplin tr G-MVSD Group	Dunkeswell	5. 7.05P
G-MVSE	Solar Wings Pegasus XL-Q SW-TE-0187 & SW-WQ-0196		18. 4.89	L.B.Richardson	Bagby	9.11.05P
G-MVSG	Aerial Arts Chaser S CH.804		24. 4.89	M.Roberts	(Melksham)	24. 7.05P
G-MVSI	Medway Hybred 44XLR MR040/58		18. 4.89	G.Cousins	(Maidstone)	26. 4.05P
G-MVSJ	Aviasud Mistral 532GB 072 & BMAA/HB/013		18. 4.89	A J.Record	(Selby)	25. 9.03P
G-MVSM	Midland Ultralights Sirocco 377GB MU-023		21. 4.89	C A Greenwood	(Port Talbot)	15. 8.04P
G-MVSN	Mainair Gemini/Flash IIA 754-589-7 & W547		28. 4.89	D W Watson *(Noted 4.04)*	Eshott	16.11.02P
G-MVSO	Mainair Gemini/Flash IIA 755-589-7 & W548		27. 4.89	A W.Hallam	Beccles	13. 8.04P
G-MVSP	Mainair Gemini/Flash IIA 756-589-7 & W549		27. 4.89	D.R.Buchanan	Pulborough	4. 4.04P
G-MVST	Mainair Gemini/Flash IIA 750-589-7 & W543 (Rotax 462)		12. 6.89	D.Curtis	Rufforth	2. 6.05P
G-MVSU*	Microfligt Spectrum 008		4. 5.89	M W Shepherd *(Cancelled 22.4.04 as destroyed) (Noted 8.04)*	Otherton, Cannock	21. 5.04P
G-MVSV	Mainair Gemini/Flash IIA 757-589-7 & W550		11. 5.89	P.Shelton	St.Michaels	26. 9.04P
G-MVSW	Solar Wings Pegasus XL-Q SW-TE-0189 & SW-WQ-0198		17. 5.89	G.F.Ryland	Oxton, Nottingham	2. 8.04P
G-MVSX	Solar Wings Pegasus XL-Q SW-TE-0190 & SW-WQ-0199		11. 5.89	D.J.Ackroyd	(Halifax)	2. 9.05P
G-MVSY	Solar Wings Pegasus XL-Q SW-TE-0191 & SW-WQ-0200		11. 5.89	G.P.Turnbull	Weston Zoyland	12. 9.04P
G-MVSZ	Solar Wings Pegasus XL-Q SW-TE-0192 & SW-WQ-0201		11. 5.89	D.M.Goldsmith and R.M.Gill	Rufforth	17. 6.05P
G-MVTA	Solar Wings Pegasus XL-Q SW-TE-0193 & SW-WQ-0202		11. 5.89	P.Hanby	(Maidstone)	3.11.05P
G-MVTC	Mainair Gemini/Flash IIA 759-689-7 & W552		30. 5.89	R.A Neal	Otherton, Cannock	31. 5.05P
G-MVTD	Whittaker MW6 Merlin PFA 164-11367 *(Built J S Yates)* (Rotax 503)		11. 5.89	G.J.Green	(Matlock)	27. 3.04P
G-MVTF	Aerial Arts Chaser S CH808		30. 5.89	P.Mundy *(New owner 12.04)*	(Prestwich, Manchester)	15. 7.92P
G-MVTI	Solar Wings Pegasus XL-Q SW-TE-0217 & SW-WQ-0206		25. 5.89	D. Burdett	(Chatteris)	24. 9.04P
G-MVTJ	Solar Wings Pegasus XL-Q SW-TE-0197 & SW-WQ-0207		25. 5.89	P.D.Rowe	Dunkeswell	7. 6.05P
G-MVTK	Solar Wings Pegasus XL-Q SW-TE-0198 & SW-WQ-0208		25. 5.89	I P Sissons	(St. Martins, Oswestry)	12. 6.05P
G-MVTL	Aerial Arts Chaser S CH.809		13. 6.89	N.D.Meer	Roddige	26. 5.05P
G-MVTM	Aerial Arts Chaser S 447 CH.810 *(Officially regd with Rotax 377)*		13. 6.89	E.W.Laidlaw	Longside, Peterhead	26. 9.04P
G-MVUA	Mainair Gemini/Flash IIA 760-689-7 & W553 (Rotax 462)		14. 6.89	T.V.Almond and R.J.Boydell	Ince Blundell	27. 6.04P
G-MVUB	Thruster T.300 089-T300-373 (Rotax 532)		13. 6.89	S.Silk	Middle Stoke, Isle of Grain	26. 6.03P
G-MVUD	Medway Hybred 44XLR MR037/55 (Rotax 503)		19. 6.89	T.W.Nelson	Carlisle	18. 3.04P
G-MVUF	Solar Wings Pegasus XL-Q SW-TE-0203 & SW-WQ-0213		13. 6.89	G. and Susan Simons	(Littlehampton)	23. 4.05P
G-MVUG	Solar Wings Pegasus XL-Q SW-TE-0204 & SW-WQ-0214		13. 6.89	A R.Hughes	Yatesbury	14. 9.04P
G-MVUI	Solar Wings Pegasus XL-Q SW-TE-0206 & SW-WQ-0216	*(Wing marked incorrectly as c/n SW-TE-0216)*	13. 6.89	J.K.Edgecombe	(Coalville)	7. 6.05P
G-MVUJ	Solar Wings Pegasus XL-Q SW-TE-0207 & SW-WQ-0217		13. 6.89	J.H.Cooper	Enstone	6. 9.05P
G-MVUK	Solar Wings Pegasus XL-Q SW-TE-0208 & SW-WQ-0218		13. 6.89	D.Greenslade	(Nailsea)	27.10.05P
G-MVUL	Solar Wings Pegasus XL-Q SW-TE-0209 & SW-WQ-0219		13. 6.89	D.Hamilton-Brown	(Pevensey)	26. 4.05P
G-MVUM	Solar Wings Pegasus XL-Q SW-TE-0210 & SW-WQ-0220		13. 6.89	M.A.Howson	Ince Blundell	21.12.05P
G-MVUO	AMF Microflight Chevvron 2-32C 015		14. 6.89	E.Maguire	(Cavan, County Cavan)	4. 3.04P
G-MVUP	Aviasud Mistral 532GB1087-48 & BMAA/HB/003	83-CQ	10. 8.89	C.J.and B.W.Foulds	Ashbourne	1. 8.02P
G-MVUR	Hornet RS-ZA HRWA-0050 & ZA107 (Rotax 532) *(Originally regd with Trike c/n HRWA-0076 but now confirmed with HRWA-0050 [ex G-MVLK] (Noted 5.04)*		3. 7.89	M.J.Moulton	Roddige	30.11.03P
G-MVUS	Aerial Arts Chaser S CH.813		3. 7.89	H.Poyzer	Eshott	16.12.01P
G-MVUT	Aerial Arts Chaser S CH.814		4. 7.89	A J.Tyler *(Damaged 2003?)*	(Beccles)	21. 4.02P
G-MVUU	Hornet R-ZA HRWB0061 & ZA110		13. 7.89	K.W.Warn *(New CofR 7.03)*	(Newton Abbot)	11. 8.92P
G-MVVH	Medway Hybred 44XLR MR047/63		11. 7.89	M.S.Henson	(Portsmouth)	12. 2.05P
G-MVVI	Medway Hybred 44XLR MR050/64		12. 7.89	D.W.Allen	Middle Stoke, Isle of Grain	8.12.04P
G-MVVK	Solar Wings Pegasus XL-R SW-TB-1414 & SW-WA-1423		11. 7.89	A J.Weir	(Bath)	3.10.05P
G-MVVM	Solar Wings Pegasus XL-R SW-TB-1416 & SW-WA-1425		12. 7.89	Alexandra F.Cunningham	(Sandbach)	16. 6.05P
G-MVVN	Solar Wings Pegasus XL-Q SW-TE-0214 & SW-WQ-0226		11. 7.89	D.J.Swanwick	Deenethorpe	5.11.04P
G-MVVO	Solar Wings Pegasus XL-Q SW-TE-0215 & SW-WQ-0227		11. 7.89	A L.Scarlett	Clench Common	29. 7.04P
G-MVVP	Solar Wings Pegasus XL-Q SW-TE-0216 & SW-WQ-0228		11. 7.89	M.P.Wimsey	(Louth)	2. 5.05P
G-MVVR	Medway Hybred 44XLR MR058/66 (Rotax 503)		20. 7.89	C.J.Meadows	Franklyn's Field, Chewton Mendip	4. 6.05P
G-MVVT	CFM Shadow Series CD K.101 & PFA 161-11569		26. 7.89	I.Macleod *(Noted 5.04)*	Redlands	30. 5.04P
G-MVVU	Aerial Arts Chaser S CH.816 (Rotax 462)		19. 7.89	S.Bradie	East Fortune	29. 5.05P

Reg	Type	C/n	Previous id	Date	Owner	Base	CofA
G-MVVV	AMF Microflight Chevvron 2-32C	016	PH-1W9 G-MVVV	11. 5.89	P.R.Turton	Old Sarum	16. 2.05P
G-MVVZ	Powerchute Raider	90628		25. 7.89	J.H.Cadman	(Melton Mowbray)	13. 7.02P
G-MVWJ	Powerchute Raider	90738		25. 7.89	N.J.Doubek	(Stanford-le-Hope)	24. 5.03P
G-MVWN	Thruster T.300 (Rotax 503)	089-T300-374		26. 7.89	T.B.Reakes tr Whisky November Group	Franklyn's Field, Chewton Mendip	6. 6.05P
G-MVWR	Thruster T.300 (Rotax 503)	089-T300-377		26. 7.89	A Allan	North Connel, Oban	26.10.05P
G-MVWS	Thruster T.300 (Rotax 503)	089-T300-378		26. 7.89	R.J.Humphries *(Current status unknown)*	(Southampton)	15. 8.95P
G-MVWV	Medway Hybred 44XLR (Rotax 447)	MR060/69		24. 7.89	K.Smith	(Rainham)	17. 8.03P
G-MVWW	Aviasud Mistral 532GB	0389-81 & BMAA/HB/005		25. 7.89	P.S.Balmer and B.H.D.Minto	Tarn Farm, Cockerham	1. 1.05P
G-MVWZ	Aviasud Mistral 532GB	1288-70 & BMAA/HB/008		2. 8.89	G Gates tr Chilbolton Mistral Group	Chilbolton	18. 5.05P
G-MVXA	Whittaker MW6 Merlin *(Built I.Brewster)* (Fuji-Robin EC-44-PM)	PFA 164-11337		17. 8.89	I.Brewster	Little Gransden	24. 4.05P
G-MVXB	Mainair Gemini/Flash IIA (Rotax 462)	762-789-7 & W555		3. 8.89	S.P.Elliot	Baxby Manor, Husthwaite	11. 7.05P
G-MVXC	Mainair Gemini/Flash IIA	763-889-7 & W556		4. 8.89	D.Wood	Arclid Green, Sandbach	27.10.04P
G-MVXD	Medway Hybred 44XLR (Rotax 503) *(Marked as "Raven")*	MR061/70		3. 8.89	P.R.Millen	Lee-on-Solent	6. 5.04P
G-MVXE	Medway Hybred 44XLR (Rotax 447)	MR063/71		23. 8.89	A M.Brittle	Sittles Farm, Alrewas	31. 7.00P
G-MVXG*	Aerial Arts Chaser S	CH.820		5. 9.89	R J Cook *(Cancelled 15.10.01 by CAA) (Noted 2004?)*	(Bearsden, Glasgow)	18. 7.97P
G-MVXI	Medway Hybred 44XLR (Rotax 447)	MR064/72		9. 8.89	T de Landro	Middle Stoke, Isle of Grain	11. 7.05P
G-MVXJ	Medway Hybred 44XLR (Rotax 447)	MR065/73		25. 8.89	P.J.Wilks *(Current status unknown)*	(Edenbridge)	26. 9.90P
G-MVXL	Thruster TST Mk.1	8089-TST-122		18. 8.89	A J.Smith	(Cardiff)	30. 8.00P
G-MVXM	Medway Hybred 44XLR (Rotax 503) *(Reported as Medway Raven)*	MR055/75		17. 8.89	P J Short	(Worcester)	2. 8.04P
G-MVXN	Aviasud Mistra 532GB	065 & BMAA/HB/002		18. 8.89	N.C. and Annette C.J.Butcher	Abbey Warren Farm, Bucknall, Lincoln	25. 8.05P
G-MVXR	Mainair Gemini/Flash IIA (Rotax 462)	764-889-7 & W557		22. 8.89	D M Bayne	East Fortune	29. 7.05P
G-MVXS	Mainair Gemini/Flash IIA	766-889-7 & W559		22. 8.89	J.W.Wood	Tarn Farm, Cockerham	25. 7.02P
G-MVXV	Aviasud Mistral 532GB	092 & BMAA/HB/004		22. 8.89	R.Bilson tr Mistral G-MVXV Group	Tarn Farm, Cockerham	5. 1.03P
G-MVXW	Rans S-4 Coyote *(Built D Hedley-Goddard - c.n 89.098)*	PFA 193-11545		22. 8.89	J.Kilpatrick	(Lifford, County Donegal	11. 8.05P
G-MVXX	AMF Microflight Chevvron 2-32	018		27. 7.89	C.K.Brown	(Loughborough)	27. 6.05P
G-MVYC	Solar Wings Pegasus XL-Q	SW-TE-0224 & SW-WQ-0239		8. 9.89	P.E.L.Street	(Lincoln)	1.10.03P
G-MVYD	Solar Wings Pegasus XL-Q	SW-TE-0225 & SW-WQ-0240		8. 9.89	T.M.Wakeley *(Noted 11.04)*	Clench Common	30. 7.04P
G-MVYE	Thruster TST Mk.1	8089-TST-123		13. 9.89	G.Elwes *(As "GM-VYE")*	Graveley	24. 6.03P
G-MVYK	Hornet R-ZA	HRWB-0076 & ZA117		22. 9.89	P.Asbridge *(Current status unknown)*	Emlyn's Field, Rhuallt	22. 7.99P
G-MVYL	Hornet R-ZA	HRWB-0077 & ZA115		22. 9.89	J.L.Thomas	Kingston Seymour	28. 3.04P
G-MVYN	Hornet R-ZA	HRWB-0079 & ZA136		22. 9.89	W.M.Studley	Weston Zoyland	25. 5.05P
G-MVYP	Medway Hybred 44XLR (Rotax 447)	MR071/77		19. 9.89	T.Almond	Middle Stoke, Isle of Grain	22. 5.05P
G-MVYR	Medway Hybred 44XLR (Rotax 447)	MR068/76		19. 9.89	K.J.Clarke	(Hockley)	10. 5.05P
G-MVYS	Mainair Gemini/Flash IIA	770-989-7 & W563		19. 9.89	S.D. and E.C.Perkins	Tarn Farm, Cockerham	14. 2.05P
G-MVYT	Noble Hardman Snowbird Mk.IV	SB-022		26. 9.89	D.T.A Rees	Haverfordwest	26. 4.05P
G-MVYU	Noble Hardman Snowbird Mk.IV	SB-023		7.11.89	B.Foster and P.Meah *(Noted 10.04)*	Gerpins Farm, Upminster	26. 6.99P
G-MVYV	Noble Hardman Snowbird Mk.IV	SB-024		21. 8.90	D.W.Hayden	Swansea	25. 7.05P
G-MVYW	Noble Hardman Snowbird Mk.IV	SB-025		22.10.90	T.J.Harrison	(Bristol)	25. 7.05P
G-MVYX	Noble Hardman Snowbird Mk.IV	SB-026		25.11.91	R.McBlain	Kilkerran	23. 1.04P
G-MVYY	Aerial Arts Chaser S 508	CH.824		26. 9.89	C.J.Gordon and R.H.Bird	Perth	21.10.05P
G-MVYZ	CFM Shadow Series BD	121		25. 9.89	A.P.Worbey *(Noted 1.05)*	Longacres Farm, Sandy	19.10.04P
G-MVZA	Thruster T.300 (Rotax 582)	089-T300-379		26. 9.89	C.C.Belcher	Popham	21. 9.02P
G-MVZC	Thruster T.300 (Rotax 532)	089-T300-381		26. 9.89	R.A Knight	Chilbolton	17. 6.05P
G-MVZD	Thruster T.300 (Rotax 532)	089-T300-382		26. 9.89	T.Pearce tr G-MVZD Syndicate	(Twickenham)	26.11.04P
G-MVZE	Thruster T.300 (Rotax 532)	089-T300-383		26. 9.89	T.L.Davis	(Graiguenamanagh, County Kilkenny)	9. 7.02P
G-MVZG	Thruster T.300 (Rotax 532)	089-T300-385		26. 9.89	R.Lewis-Evans *(Noted 3.04)*	Mapperton Farm, Newton Peverill, Dorset	24. 1.05P
G-MVZI	Thruster T.300 (Rotax 503)	089-T300-387		26. 9.89	R.R.R.Whittern	South Wraxall, Bradford-on-Avon	8. 8.05P
G-MVZJ	Solar Wings Pegasus XL-Q	SW-TE-0226 & SW-WQ-0241		26. 8.89	P.Mansfield	(Peterborough)	19. 9.05P
G-MVZK	Quad City Challenger II UK *(Built K B Tolley)* (BMW R.100)	PFA 177-11498		28. 9.89	D.K.Maclennan tr G-MVZK Group	(Glasgow)	23. 9.04P
G-MVZL	Solar Wings Pegasus XL-Q	SW-TE-0227 & SW-WQ-0242		4.10.89	P.R.Dobson	(Brentwood)	16. 6.05P
G-MVZM	Aerial Arts Chaser S 447	CH.825		2.11.89	J L Parker	(Maidstone)	3. 5.05P
G-MVZO	Medway Hybred 44XLR (Rotax 503)	MR072/78		25.10.89	D.L.Wright	Sywell	11. 5.05P

Regn	Type	C/n	Regd	Owner	Location	CofA
G-MVZP	Meridian Renegade Spirit UK	PFA 188-11630	17.10.89	H.M.Doyle	Lower Mountpleasant, Chaterris	21. 1.05P
	(Built G S Hollingsworth - kit no.256)					
G-MVZR	Aviasud Mistral 532GB	090 & BMAA/HB/011	9.10.89	S.E.and J.A Robinson	Crosland Moor	7. 6.01P
G-MVZS	Mainair Gemini/Flash IIA	771-1089-7 & W564	17.10.89	R.L.Beese	(Tarporley)	22. 5.05P
G-MVZT	Solar Wings Pegasus XL-Q		6.10.89	C.J.Meadows	Franklyn's Field, Chewton Mendip	25. 8.02P
		SW-TE-0228 & SW-WQ-0243				
G-MVZU	Solar Wings Pegasus XL-Q		6.10.89	R.D.Proctor	RAF Wittering	27.10.04P
		SW-TE-0229 & SW-WQ-0244				
G-MVZV	Solar Wings Pegasus XL-Q		6.10.89	I.M.Gibson	(Horsham)	17. 8.05P
		SW-TE-0230 & SW-WQ-0245				
G-MVZW	Hornet R-ZA	HRWB-0063 & ZA142	27.10.89	K.W.Warn	Popham	9. 8.02P
G-MVZX	Meridian Renegade Spirit UK	PFA 188-11590	18.10.89	G.Holmes	(Pickering)	27. 6.05P
	(Built G Holmes)					
G-MVZZ	AMF Microflight Chevvron 2-32	019	27. 7.89	D.Patrick	Kirkbride	22. 5.05P
G-MWAB	Mainair Gemini/Flash IIA	772-1089-7 & W565	24.10.89	J.E.Buckley	(Sandbach)	28. 7.05P
G-MWAC	Solar Wings Pegasus XL-Q		25.10.89	P.A Tabberer	Cwm	2.10.03P
		SW-TE-0236 & SW-WQ-0260				
G-MWAD	Solar Wings Pegasus XL-Q		25.10.89	J.G.McNally	Sutton Meadows, Ely	14. 4.05P
		SW-TE-0237 & SW-WQ-0261				
G-MWAE	CFM Shadow Series CD	130	24.10.89	D.J.Adams	North Coates	29. 5.04P
G-MWAF	Solar Wings Pegasus XL-R		30.10.89	I.W.Skeldon	(Crewe)	6. 7.05P
		SW-TB-1422 & SW-WA-1441				
G-MWAG	Solar Wings Pegasus XL-R		30.10.89	X.Norman	(Thame)	22. 7.05P
		SW-TB-1423 & SW-WA-1442				
G-MWAJ	Meridian Renegade Spirit UK	PFA 188-11438	1.11.89	M.Mailey	(Antrim)	6. 4.04P
	(Built J Hall) (BMW R.100RS)					
G-MWAL	Solar Wings Pegasus XL-Q		2.11.89	A W.Hill	Bluntisham	24. 6.05P
		SW-TE-0240 & SW-WQ-0263				
G-MWAN	Thruster T.300	089-T300-389	14.11.89	R.B.Hawkins	(Plymouth)	25. 6.05P
	(Rotax 532)					
G-MWAP	Thruster T.300	089-T300-391	14.11.89	S.F.Chave and A G.Spurway	(Honiton)	9. 8.04P
	(Rotax 503)			*"Wanda"*		
G-MWAR	Thruster T.300	089-T300-392	14.11.89	S.M.Birbeck	Popham	8.12.02P
	(Rotax 532)					
G-MWAT	Solar Wings Pegasus XL-Q		13.11.89	D.G.Seymour	Yatesbury	7. 7.05P
		SW-TE-0241 & SW-WQ-0265				
G-MWAU	Mainair Gemini/Flash IIA	773-1189-7 & W566	7.12.89	L.Roberts	(Ammanford)	22. 6.05P
	(Rotax 582)					
G-MWAV	Solar Wings Pegasus XL-R		13.11.89	S.P.Waine	(Fordingbridge)	6. 9.03P
		SW-TB-1424 & SW-WA-1444		*(See G-MWBL)*		
G-MWAW	Whittaker MW6 Merlin	PFA 164-11460	10.11.89	J.A Hindley	Brook Farm, Pilling	3. 4.04P
	(Built P Palmer) (Rotax 503)					
G-MWBJ	Medway Puma Sprint	MS003/1	21.11.89	C.C.Strong	(Bures, Cornwall)	14. 7.00P
	(Rotax 447)					
G-MWBK	Solar Wings Pegasus XL-Q		16.11.89	P.J.Harrison	Bracknell	29. 3.04P
		SW-TE-0248 & SW-WQ-0271				
G-MWBL	Solar Wings Pegasus XL-R/Se		16.11.89	J.A Valentine	Eshott	3. 7.04P
		SW-TB-1424 & SW-WA-1446	*(Trike c/n appears to have been duplicated with G-MWAV so two '1424s exist)*			
G-MWBO	Rans S-4 Coyote	PFA 193-11583	29.11.89	A R Thomson and B M Tibenham	Longside, Peterhead	12.10.04P
	(Built L R H d'Ath c/n 89.097)			tr G-MWBO Group		
G-MWBP	Hornet R-ZA	HRWB-0083 & ZA144	29.11.89	S.Brader	Tarn Farm, Cockerham	17. 8.05P
G-MWBR	Hornet RS-ZA	HRWB-0084 & ZA145	29.11.89	I.A Clark	North Coates	8. 8.05P
G-MWBS	Hornet R-ZA	HRWB-0085 & ZA146	29.11.89	P.D.Jaques	Sandtoft	22.11.04P
	(BMW R100)					
G-MWBU	Hornet R-ZA	HRWB-0087 & ZA148	29.11.89	J.D.Nelson	Plaistowes Farm, St Albans	12. 9.04P
G-MWBW	Hornet R-ZA	HRWB-0089 & ZA150	29.11.89	C.G.Bentley	(Chesterfield)	15. 5.00P
G-MWBY	Hornet R-ZA	HRWB-0091 & ZA152	29.11.89	G.P.Austin	Mill Farm, Hughley, Much Wenlock	19. 6.04P
G-MWCB	Solar Wings Pegasus XL-Q		1.12.89	M.Foreman	(Telford)	23. 8.05P
		SW-TE-0250 & SW-WQ-0273				
G-MWCC	Solar Wings Pegasus XL-R/Se		1.12.89	L.Robinson	(Milton Keynes)	26. 3.05P
		SW-TB-1387 & SW-WA-1447	*(Trike ex G-MVKD when latter's sailwing sold)*			
G-MWCE	Mainair Gemini/Flash IIA	775-1289-7 & W568	19.12.89	B.A Tooze	Shobdon	6. 9.05P
G-MWCF	Solar Wings Pegasus XL-Q		13.12.89	J.D.Amos	Redlands	20. 9.05P
		SW-TE-0252 & SW-WQ-0276		tr G-MWCF Group		
G-MWCG	Microflight Spectrum	011	15.12.89	M.W.Shepherd	Otherton, Cannock	2. 6.05P
G-MWCH	Rans S-6-ESD Coyote II	PFA 204-11632	15.12.89	J G Burns and T Briton	Morgansfield, Fishburn	15. 9.05P
	(Built J Whiting - 0989.067) (PFA c/n duplicates Kitfox G-BSFY)			tr G-MWCH Group		
G-MWCI	Powerchute Kestrel	91245	3. 1.90	E.G.Bray	Clacton	2. 7.05P
G-MWCJ	Powerchute Kestrel	91246	3. 1.90	B.A Dowland	(Thorney)	13. 7.02P
G-MWCK	Powerchute Kestrel	91247	3. 1.90	F W Downham	(Kendal)	28.12.04P
G-MWCM	Powerchute Kestrel	91249	3. 1.90	G.E.Lockyer *(New owner 6.03)*	(Stoke-on-Trent)	17. 6.96P
G-MWCN	Powerchute Kestrel	91250	3. 1.90	H.J.Goddard	(Fleet)	16. 8.04P
G-MWCO	Powerchute Kestrel	91251	3. 1.90	T.F.Bakker *(Current status unknown)*	(Fairford)	17. 5.93P
G-MWCP	Powerchute Kestrel	91252	3. 1.90	I Fraser	(Corgarff, Strathdon)	20.10.01P
G-MWCR	Southdown Puma Sprint	P.516 & SN1121/0070	24. 2.84	S.R.Hall	(Stroud)	2. 8.02P
	(Fuji-Robin EC-44-PM)					
G-MWCS	Powerchute Kestrel	91253	3. 1.90	R S McFadyen	(Tamworth)	5. 7.05P
G-MWCU	Solar Wings Pegasus XL-R		27.12.89	T.P.Noonan	Newtownards	5. 8.05P
		SW-TB-1412 & SW-WA-1449				
G-MWCW	Mainair Gemini/Flash IIA	776-0190-7 & W569	29.12.89	A J Thomas	(Nottingham)	11. 8.05P
	(Rotax 462)					
G-MWCX	Medway Hybred 44XLR	MR076/80	8. 1.90	P.A Harris	(Petersfield)	31. 3.96P
	(Rotax 503)			*(Current status unknown)*		

Reg	Type	C/n	Previous identity	Date	Owner	Location	Expiry
G-MWCY	Medway Hybred 44XLR	MR077/81		15. 1.90	J.K.Masters	(Chigwell)	10. 9.04P
	(Rotax 503)						
G-MWCZ	Medway Hybred 44XLR	MR078/82		10. 1.90	A Titcombe	(Aylesford)	22. 6.04P
	(Rotax 503)						
G-MWDB	CFM Shadow Series CD	100		3. 7.89	M.D.Meade	(St.Albans)	10. 7.04P
G-MWDC	Solar Wings Pegasus XL-R/Se			5. 1.90	A N.Edwards	(Great Orton)	27. 7.05P
	(Rotax 462)	SW-TE-0255 & SW-WA-1450					
G-MWDD	Solar Wings Pegasus XL-Q			15. 1.90	D.J.Billham	Enstone	20. 5.05P
		SW-TE-0258 & SW-WQ-0280					
G-MWDE	Hornet RS-ZA	HRWB-0094 & ZA126		10. 1.90	H.G.Reid	Roddige	13. 6.98P
	(Rotax 532)				*(Trike noted 1.04)*		
G-MWDI	Hornet RS-ZA	HRWB-0098 & ZA158		10. 1.90	R.J.Perrin	Brook Farm, Pilling	29. 6.05P
	(Rotax 532)						
G-MWDJ	Mainair Gemini/Flash IIA	777-0190-7 & W570		17. 1.90	M.Gardiner	Crosland Moor	27. 6.05P
G-MWDK	Solar Wings Pegasus XL-Q			17. 1.90	T.Wicks	(Devizes)	12.10.04P
		SW-TE-0259 & SW-WQ-0281					
G-MWDL	Solar Wings Pegasus XL-Q			17. 1.90	P.K.Dean	Sutton Meadows, Ely	13. 4.05P
		SW-TE-0260 & SW-WQ-0282					
G-MWDM	Meridian Renegade Spirit UK	PFA 188A-11628		18. 1.90	A Haugh tr Doctor and The Medics	Long Marston	13. 7.05P
	(Built M L Smith - c/n 319) (Jabiru 2200A) *(PFA c/n duplicates Streak Shadow G-BRZZ)*						
G-MWDN	CFM Shadow Series CD	K.102		17. 1.90	A A Duffus	RAF Halton	1. 8.04P
G-MWDP	Thruster TST Mk.1	8129-TST-124		30. 1.90	J.Walker	(Ballymena, County Antrim)	5. 5.95P
					(Current status unknown)		
G-MWDS	Thruster T.300	089-T300-395		30. 1.90	C.N.Nairn	Nether Huntlywood Farm, Gordon	20. 1.05P
	(Rotax 532)						
G-MWDZ	Eipper Quicksilver MXL II	022		29. 1.90	R.G.Cook	Cranfield	6. 7.05P
	(Rotax 503) *(Originally c/n PFA214-11869)*						
G-MWEE	Solar Wings Pegasus XL-Q			12.12.88	R.J.Sharp	Rochester	2. 9.01P
		SW-TE-0175 & SW-WQ-0147					
G-MWEF*	Solar Wings Pegasus XL-Q			30. 1.90	N.R.Williams	Long Marston	3. 7.03P
		SE-TE-0261 & SW-WQ-0283			*(Cancelled 21.1.04 by CAA)*		
G-MWEG	Solar Wings Pegasus XL-Q			30. 1.90	S.P.Michlig	Long Marston	28.12.04P
		SW-TE-0262 & SW-WQ-0284					
G-MWEH	Solar Wings Pegasus XL-Q			7. 2.90	K.A Davidson	(Leuchars)	1. 6.05P
		SW-TE-0264 & SW-WQ-0286					
G-MWEK	Whittaker MW5 Sorcerer	PFA 163-11284		20. 2.90	D.W. and M.L.Squire	(St.Austell)	12. 7.05P
	(Built J T Francis)						
G-MWEL	Mainair Gemini/Flash IIA	780-0290-7 & W573		13. 2.90	B.L.Benson	(Malpas)	20 6.05P
G-MWEN	CFM Shadow Series CD	K.113		20. 2.90	C.Dawn *(New owner 7.03)*	(Market Rasen)	8. 8.01P
G-MWEO	Whittaker MW5 Sorcerer	PFA 163-11263		21. 2.90	P.Stewart	(Portrush, County Antrim)	9. 9.04P
	(Built C D Wills) (Fuji-Robin EC-34-PM)						
G-MWEP	Rans S-4 Coyote	PFA 193-11616		21. 2.90	E.J.Wallington	Inglenook Farm, Maydensole, Dover	16.11.05P
	(Built K E Wedl - c/n 89.096)						
G-MWER	Solar Wings Pegasus XL-Q			1. 3.90	S.V.Stojanovic	(Swansea)	21. 8.02P
		SW-TE-0265 & SW-WQ-0287					
G-MWES	Rans S-4 Coyote	PFA 193-11737		1. 2.90	G.Scott	Lower Mountpleasant, Chatteris	18. 8.04P
	(Built I Fleming and R W Sage - c/n 89.099)				*(Carries "N89099" on tail which matches c/n but is not a p/i)*		
G-MWEU	Hornet RS-ZA	HRWB-0100 & ZA160		21. 2.90	K.H.Hicks	(March)	11.11.01P
	(Rotax 532)						
G-MWEY	Hornet R-ZA	HRWB-0104 & ZA135		21. 2.90	J.Kidd	Tarn Farm, Cockerham	24. 9.00P
G-MWEZ	CFM Shadow Series CD	136		22. 2.90	T.D.Dawson tr G-MWEZ Group	Plaistowes Farm, St.Albans	6. 6.04P
G-MWFA	Solar Wings Pegasus XL-R			27. 2.90	A W.Edwards	Davidstow Moor	23. 6.02P
		SW-TB-1406 & SW-WA-1454					
G-MWFB	CFM Shadow Series CD	K.119		1. 3.90	K.W.E.Brunnenkant	(Lincoln)	21.10.04P
G-MWFC	TEAM mini-MAX 88	PFA 186-11648	G-BTXC	1. 3.90	M.Bradley	North Coates	13. 4.05P
	(Built M H D Soltau - c/n 294)		G-MWFC				
G-MWFD	TEAM mini-MAX 88	PFA 186-11646		1. 3.90	J.Flanagan	(Hadfield, Glossop)	7. 7.05P
	(Built J Riley - c/n 293) (PFA c/n duplicates Shadow G-GORE)						
G-MWFF	Rans S-5 Coyote	PFA 193-11639		10. 1.90	J S Sweetingham	(Helsby, Frodsham)	27. 1.05P
	(Built M W Holmes - c/n 89.106)						
G-MWFG	Powerchute Kestrel	00358		20. 3.90	Rosemary I.Simpson	Rochester	18. 5.05P
G-MWFI	Powerchute Kestrel	00360		20. 3.90	R.R.O'Neill *(New CofR 8.03)*	(Dungannon, Belfast)	28. 8.02P
G-MWFL	Powerchute Kestrel	00363		20. 3.90	A Vincent	Fenland	31. 3.02P
G-MWFS	Solar Wings Pegasus XL-Q			14. 3.90	C.P.Hughes	(Cwm)	23. 5.05P
		SW-TE-0267 & SW-WQ-0289					
G-MWFT	MBA Tiger Cub 440	WFT-02		24.11.83	J.R.Ravenhill	Kemble	20. 4.05P
G-MWFU	Quad City Challenger II UK	PFA 177-11654		16. 3.90	M.Ellis	Sandtoft	10.10.05P
	(Built K N Dickinson)						
G-MWFV	Quad City Challenger II UK	PFA 177-11655		16. 3.90	P.Bowers	(Carlisle)	26. 5.05P
	(Built E G Astin)						
G-MWFW	Rans S-4 Coyote	PFA 193-11662		16. 3.90	M P Hallam	(Horsham)	21. 4.05P
	(Built G R Hillary - c/n 89.107)						
G-MWFX	Quad City Challenger II UK	PFA 177-11706		20. 3.90	I.M.Walton	Wellesbourne Mountford	9. 6.04P
	(Built I M Walton - c/n CH2-1189-UK-0485) (Rotax 462)						
G-MWFY	Quad City Challenger II UK	PFA 177-11668		20. 3.90	P.J.Ladd	Craysmarsh Farm, Melksham	14. 7.05P
	(Built P J Ladd)						
G-MWFZ	Quad City Challenger II UK	PFA 177-11707		20. 3.90	A Slade	(Enfield)	
	(Built A Slade - c/n CH2-0190-UK-0506)				*(Current status unknown)*		
G-MWGA	Rans S-5 Coyote	PFA 193-11810		20. 3.90	G.N.Smith	Inglenook Farm, Maydensole, Dover	23. 9.04P
	(Built M A C Stevenson - c/n 89.092)						
G-MWGC	Medway Hybred 44XLR	MR087/85		26. 3.90	C.Spalding	Brock Farm, Billericay	26. 5.05P
	(Rotax 503)						
G-MWGG	Mainair Gemini/Flash IIA	785-0390-7 & W578		26. 3.90	D.Lopez	Ashcroft Farm, Winsford	26. 4.05P
	(Rotax 462)						

Registration	Type / C/n	Previous identity	Regd	Owner	Base	CofA
G-MWGI	Whittaker MW5-K Sorcerer 5K-0012-02 *(Built Aerotech International Ltd) (Wings to G-MTBT by 7.96)*		28. 3.90	B.Barrass	Sywell	13. 2.04P
G-MWGJ	Whittaker MW-5K Sorcerer 5K-0014-02 *(Built Aerotech International Ltd)*		6. 9.90	V.J.Morris	(Truro)	28. 5.05P
G-MWGK	Whittaker MW-5K Sorcerer 5K-0015-02 *(Built Aerotech International Ltd)*	(G-MWLV)	19. 9.90	R.M.Thomas	Wombleton	29. 7.05P
G-MWGL	Solar Wings Pegasus XL-Q SW-TE-0270 & SW-WQ-0293		28. 3.90	G.D.Haimes and A.R.Campbell	Newton Bank Farm, Daresbury	26. 3.05P
G-MWGM	Solar Wings Pegasus XL-Q SW-TE-0271 & SW-WQ-0294		28. 3.90	G C Christopher	Longacres Farm, Sandy	25. 8.05P
G-MWGN	Rans S-4 Coyote PFA 193-11709 *(Built B H Ashman - c/n 89.113)*		26. 3.90	V.Hallam	(Torquay)	16. 5.05P
G-MWGO	Aerial Arts 110SX/Chaser 110SX/566 (Rotax 377)		28. 3.90	B.Nicolson *(Current status unknown)*	(Middlesbrough)	28. 4.97P
G-MWGR	Solar Wings Pegasus XL-Q SW-TE-0272 & SW-WQ-0296		6. 4.90	A.Maskell	(Wolverhampton)	10. 8.05P
G-MWGU	Powerchute Kestrel 00368	(9H-) G-MWGU	26. 4.90	M.Pandolfino *(Current status unknown)*	Luqa, Malta	19. 7.91P
G-MWGW	Powerchute Kestrel 00370		26. 4.90	S.P.Tomlinson *(Current status unknown)*	(Leominster)	23. 5.05P
G-MWGZ	Powerchute Kestrel 00373		26. 4.90	L.J.Lynch *(Current status unknown)*	(Ventnor)	27. 5.97P
G-MWHC	Solar Wings Pegasus XL-Q SW-TE-0274 & SW-WQ-0304		24. 4.90	P.J.Lowery *(Stored 7.03)*	Longacres Farm, Sandy	14. 5.99P
G-MWHF	Solar Wings Pegasus XL-Q SW-TE-0275 & SW-WQ-0305		24. 4.90	N.J.Troke	Swinford, Rugby	26. 3.05P
G-MWHG	Solar Wings Pegasus XL-Q SW-TE-0276 & SW-WQ-0306		24. 4.90	I.A Lumley	(Great Orton)	4. 2.03P
G-MWHH	TEAM mini-MAX 88 PFA 186-11814 *(Built B.F.Crick - c/n 326)*		23. 4.90	B.F.Crick *(Current status unknown)*	Desborough	4. 8.00P
G-MWHI	Mainair Gemini/Flash IIA 784-0390-5 & W577 (Rotax 503)		26. 4.90	P.Harwood	(Chesterfield)	6. 7.05P
G-MWHL	Solar Wings Pegasus XL-Q SW-TE-0278 & SW-WQ-0308		1. 5.90	G J Chapman	Swinford, Rugby	4. 7.05P
G-MWHM	Whittaker MW6-S Fatboy Flyer PFA 164-11463 *(Built D W Squire)* (Rotax 532)		18. 5.90	G.H.Davies	(Walsall)	16. 6.05P
G-MWHO	Mainair Gemini/Flash IIA 778-0190-5 & W571		10. 5.90	B.Epps	Arclid Green, Sandbach	31.10.04P
G-MWHP	Rans S-6-ESD Coyote II PFA 204-11768 *(Built J.F.Bickerstaffe - c/n 1089.093)* (Rotax 532) *(Tri-cycle u/c)*		8. 5.90	J.F.Bickerstaffe	Higher Barn Farm, Houghton	27.11.04P
G-MWHR	Mainair Gemini/Flash IIA 787-0590-7 & W580		16. 5.90	B.Brazier	Brook Farm, Pilling	31.10.05P
G-MWHT	Solar Wings Pegasus Quasar TC SW-TQ-0005 & SW-WQQ-0314		15. 5.90	E.H.Gatehouse	Pound Green, Buttonoak, Bewdley	13. 8.05P
G-MWHU	Solar Wings Pegasus Quasar SW-TQ-0006 & SW-WQQ-0315		15. 5.90	A F.Frost and S.J.Park	Sywell	13.11.02P
G-MWHX	Solar Wings Pegasus XL-Q SW-TE-0280 & SW-WQ-0318		15. 5.90	N.P.Kelly	Trim, County Meath	15. 7.05P
G-MWIA	Mainair Gemini/Flash IIA 789-0690-7 & W582		21. 5.90	M.Raj	Otherton, Cannock	26. 9.04P
G-MWIB	Aviasud Mistral 532GB 094 & BMAA/HB/010		16. 5.90	N.W.Finn-Kelcey *"Weston Belle"*	Weston Underwood, Olney	14. 6.05P
G-MWIC	Whittaker MW5-C Sorcerer PFA 163-11224 *(Built I P Croft)*		20. 2.90	A.M.Witt	(Milton Keynes)	29. 4.05P
G-MWIE	Solar Wings Pegasus XL-Q SW-TE-0282 & SW-WQ-0325		30. 5.90	R.Mercer	Long Marston	22. 7.05P
G-MWIF	Rans S-6-ESD Coyote II PFA 204-11749 *(Built M G K Prout - c/n 1089.095) (Tri-cycle u/c)*		30. 5.90	J.H.Cooling	Fenland	30. 4.05P
G-MWIG	Mainair Gemini/Flash IIA 790-0690-7 & W583 (Rotax 462)		4. 6.90	A P.Purbrick	(King's Lynn)	12. 9.05P
G-MWIH	Mainair Gemini/Flash IIA 791-0690-5 & W584		4. 6.90	J P Norton	Headon Farm, Retford	27. 3.05P
G-MWIL	Medway Hybred 44XLR MR096/90 (Rotax 447)		8. 6.90	P.J.Bosworth	Mill Farm, Hughley, Much Wenlock	7.10.05P
G-MWIM	Solar Wings Pegasus Quasar TC SW-TQ-0008 & SW-WQQ-0326		11. 6.90	P.J.Bates and T.S.Smith	Long Marston	18. 4.05P
G-MWIO	Rans S-4 Coyote PFA 193-11774 *(Built G Ferguson - c/n 90.117)*		11. 6.90	P.D.Lucas *(Noted 11.04)*	Grove Farm, Needham	29.10.02P
G-MWIP	Whittaker MW6 Merlin PFA 164-11360 *(Built D.Beer and B.J.Merrett)* (Rotax 582)		7. 6.90	D.Beer and B.J.Merrett	(Ilfracombe)	12. 8.05P
G-MWIR	Solar Wings Pegasus XL-Q SW-TE-0283 & SW-WQ-0330		8. 6.90	C.E.Dagless	Yaxham, Dereham	14. 7.03P
G-MWIS	Solar Wings Pegasus XL-Q SW-TE-0284 & SW-WQ-0331		8. 6.90	C.J.Hunt	(Ilkeston)	28.10.05P
G-MWIU	Solar Wings Pegasus Quasar TC SW-TQ-0010 & SW-WQQ-0333		8. 6.90	D.G.Bond	RAF Halton	29. 3.05P
G-MWIV	Mainair Gemini/Flash 792-0690-5 & W585		15. 6.90	P.and J.Calvert	(Pickering)	14. 7.03P
G-MWIW	Solar Wings Pegasus Quasar SW-TQ-0011 & SW-WQQ-0334		18. 6.90	T.Yates	(Alfreton)	25. 6.04P
G-MWIX	Solar Wings Pegasus Quasar TC SW-TQ-0012 & SW-WQQ-0335		18. 6.90	T.D.Neal	Shobdon	14. 2.04P
G-MWIY	Solar Wings Pegasus Quasar TC SW-TQ-0014 & SW-WQQ-0336		22. 6.90	R J Coppin	Shobdon	6. 3.05P
G-MWIZ	CFM Shadow Series CD 096 (Rotax 462)		22.11.88	T.P.Ryan	Plaistowes Farm, St.Albans	3. 1.05P
G-MWJD	Solar Wings Pegasus Quasar SW-TQ-0016 & SW-WQ-0339		22. 6.90	A J.Blackwell *(Trike stored 8.03)*	Long Marston	30. 9.99P
G-MWJF	CFM Shadow Series CD K.123 (Rotax 447) *(Officially regd as "Series BD")*		26. 6.90	S.N.White	(Basingstoke)	4.11.05P

Registration	Type / c/n	Date	Owner	Location	Date
G-MWJG	Solar Wings Pegasus XL-R SW-TB-1415 & SW-WA-1472 *(Uses trike ex G-MVVL)*	26. 6.90	M.J.Piggott	Little Gransden	18. 6.05P
G-MWJH	Solar Wings Pegasus Quasar SW-TQ-0017 & SW-WQQ-0340	29. 6.90	N.J.Braund	Broadmeadow Farm, Hereford	25. 8.04P
G-MWJI	Solar Wings Pegasus Quasar SW-TQ-0018 & SW-WQQ-0341	29. 6.90	L.Luscombe	Weston Zoyland	30. 5.05P
G-MWJJ	Solar Wings Pegasus Quasar SW-TQ-0019 & SW-WQQ-0342	29. 6.90	R.Langham	Oxton	31. 7.05P
G-MWJK	Solar Wings Pegasus Quasar SW-TQ-0020 & SW-WQQ-0343	29. 6.90	M.Richardson	(Swansea)	8. 4.03P
G-MWJN	Solar Wings Pegasus XL-Q SW-TE-0288 & SW-WQ-0344	29. 6.90	J.C.Corrall	Lower Mountpleasant, Chatteris	11.10.05P
G-MWJO*	Solar Wings Pegasus XL-Q SW-TE-0289 & SW-WQ-0345	29. 6.90	S.J.Canon *(Cancelled 26.11.04 by CAA)*	(London E17)	25. 2.05P
G-MWJP	Medway Hybred 44XLR MR097/91 (Rotax 503)	29. 6.90	C D Simmons	(Edenbridge)	29. 3.05P
G-MWJR	Medway Hybred 44XLR MR098/92 (Rotax 503)	28. 6.90	T G Almond	(Bishops Itchington, Southam)	9. 9.05P
G-MWJS	Solar Wings Pegasus Quasar TC SW-TQ-0021 & SW-WQQ-0349	6. 7.90	R.J.Milward	Sywell	14. 3.05P
G-MWJT	Solar Wings Pegasus Quasar TC SW-TQ-0022 & SW-WQQ-0350	16. 7.90	K.V.Rands-Allen	Sywell	30. 3.05P
G-MWJV	Solar Wings Pegasus Quasar SW-TQ-0024 & SW-WQQ-0352	6. 7.90	A Davies *(Noted 7.04)*	Weston Zoyland	4.10.03P
G-MWJW	Whittaker MW5 Sorcerer PFA 163-11186 *(Built J D Webb - c/n JDW-02)* (Fuji-Robin EC-44-PM)	11. 5.90	S.Badby	Tibenham	25. 9.02P
G-MWJX	Medway Puma Sprint MS009/3 (Rotax 447)	17. 7.90	R.Potter	Pound Green, Buttonoak, Bewdley	22. 9.05P
G-MWJY	Mainair Gemini/Flash IIA 797-0790-7 & W590	16. 7.90	M.D.Walton	(Ware)	7.10.05P
G-MWJZ*	CFM Shadow Series CD K.132 *(Severely damaged Chatteris during 1999: cancelled 24.12.01 as wfu) (Engine and wings removed and stored 9.03)*	19. 7.90	D.Mahajan	Lower Mountpleasant, Chatteris	23. 7.99P
G-MWKA	Meridian Renegade Spirit UK PFA 188-11864 *(Built Downlands Flying Group) (Incorporates project PFA 188-11690)*	26. 7.90	C.E.Neill tr Downlands Flying Group *"Spirit of Lewes"*	Deanland	8. 4.01P
G-MWKE	Hornet RS-ZA HRWB-0108 & ZA167 (Rotax 532) *(Trike c/n overstamped on HRWB-0107)*	30. 7.90	D.R.Stapleton	Tarn Farm, Cockerham	14.11.04P
G-MWKO	Solar Wings Pegasus XL-Q SW-TE-0290 & SW-WQ-0357	31. 7.90	J.Boath	Carlisle	18. 9.05P
G-MWKP	Solar Wings Pegasus XL-Q SW-TE-0291 & SW-WQ-0358	31. 7.90	K.H.Smalley	Wickenby	20. 7.05P
G-MWKW	Microflight Spectrum 015	3. 8.90	P L Stribling	(Great Cornard, Sudbury)	29. 1.05P
G-MWKX	Microflight Spectrum 016	3. 8.90	C.R.Ions	Eshott	14. 5.04P
G-MWKY	Solar Wings Pegasus XL-Q SW-TE-0292 & SW-WQ-0362	3. 8.90	I.D.Edwards	Roddige	25. 8.04P
G-MWKZ	Solar Wings Pegasus XL-Q SW-TE-0293 & SW-WQ-0363 *(Another unit noted with same c/n probably G-MWEG with manufacturer's incorrect I/D plate)*	3. 8.90	C.Lamb and P.J.Cragg	(Ashford)	9.11.05P
G-MWLA	Rans S-4 Coyote PFA 193-11787 *(Built S H Williams - c/n 89.114)*	3. 8.90	D.C.Lees *(New owner 1.04)*	Andrewsfield	27. 8.03P
G-MWLB	Medway Hybred 44XLR MR104/93 (Rotax 503)	15. 8.90	M.W.Harmer	Longacres Farm, Sandy	2.11.05P
G-MWLD	CFM Shadow Series CD 106	9. 5.89	P.C.Avery	Shipdham	1.11.01P
G-MWLE	Solar Wings Pegasus XL-R SW-TB-1425 & SW-WA-1474	9. 8.90	D.Stevenson	Plaistowes Farm, St.Albans	10. 5.05P
G-MWLF	Solar Wings Pegasus XL-R SW-TB-1426 & SW-WA-1475	9. 8.90	G.Rainey	Weston Zoyland	22. 4.05P
G-MWLG	Solar Wings Pegasus XL-R SW-TB-1427 & SW-WA-1476	9. 8.90	A.S.Docherty	(St. Albans)	14. 7.05P
G-MWLH	Solar Wings Pegasus Quasar SW-TQ-0030 & SW-WQQ-0364	9. 8.90	G.A Gamblin	Old Sarum	3. 7.03P
G-MWLJ	Solar Wings Pegasus Quasar SW-TQ-0032 & SW-WQQ-0366	9. 8.90	C.G.Rouse	Chandlers Ford	18. 9.05P
G-MWLK	Solar Wings Pegasus Quasar TC SW-TQ-0033 & SW-WQQ-0367	9. 8.90	D.J.Shippen	Newton Bank, Daresbury	3. 1.05P
G-MWLL	Solar Wings Pegasus XL-Q SW-TE-0287 & SW-WQ-0338	16. 8.90	J.Bacon	Botany Bay, Horsford	13. 9.05P
G-MWLM	Solar Wings Pegasus XL-Q SW-TE-0286 & SW-WQ-0322	17. 8.90	A.A.Judge	Hunsdon	26. 5.05P
G-MWLN	Whittaker MW6-S Fatboy Flyer PFA 164-11844 *(Built S J Field)* (Rotax 503)	16. 8.90	S.J.Field *"Red Lips" (Current status unknown)*	(Bridgwater)	5. 6.92P
G-MWLO	Whittaker MW6 Merlin PFA 164-11373 *(Built G W Peacock)* (Rotax 503)	21. 8.90	S.P.Ganecki tr G-MWLO Flying Group *(New owner 4.03)*	Otherton, Cannock	6. 6.02P
G-MWLP	Mainair Gemini/Flash 801-0990-5 & W594	24. 8.90	J.S.Potts *(Current status unknown)*	(Kilmarnock)	24. 8.99P
G-MWLS	Medway Hybred 44XLR MR081/95 (Rotax 503)	29. 8.90	M A Oliver	Carlisle	4. 8.05P
G-MWLT	Mainair Gemini/Flash IIA 804-0990-7 & W597	31. 8.90	S.A Sacker	Deenethorpe	18. 7.03P
G-MWLU	Solar Wings Pegasus XL-R/Se (Rotax 462) SW-TE-0294 & SW-WA-1478 *(Trike c/n corrected from '0304 which is G-MWNC) (Stored 9.97: current status unknown)*	6. 9.90	T.P.G.Ward	(Great Orton)	14.10.91P
G-MWLW	TEAM mini-MAX PFA 186-11717 *(Built W T Kirk)* (Rotax 377)	14. 9.90	L.G.Horne *(New owner 12.04)*	(Willesborough, Ashford)	28. 7.04P
G-MWLX	Mainair Gemini/Flash IIA 805-0990-7 & W598	5.10.90	G Good and Elizabeth Jillian Douglas	East Fortune	24. 1.05P
G-MWLZ	Rans S-4 Coyote PFA 193-11887 *(Built T E G Buckett c/n 90.116)*	8.10.90	B.O.McCartan	(Banbridge, County Down)	11. 3.05P
G-MWMB	Powerchute Kestrel 00399	7.11.90	D.J.Whysall	Ripley, Derbyshire	17. 5.05P
G-MWMC	Powerchute Kestrel 00400	7.11.90	R.A Stewart tr Talgarreg Flying Club	(Llandysul)	3.10.05P
G-MWMD	Powerchute Kestrel 00401	7.11.90	D.J.Jackson *(Current status unknown)*	(Melton Constable)	20.11.91P

G-MWMG	Powerchute Kestrel	00404	7.11.90	M.D.Walton	(Tregaron)	13. 7.05P
G-MWMH	Powerchute Kestrel	00405	7.11.90	E.W.Potts	(Crymych, Dyfed)	12. 6.04P
G-MWMI	Solar Wings Pegasus Quasar	SW-TQ-0043 & SW-WQQ-0383	21. 9.90	A.R.Winton	Brock Farm, Billericay	25. 7.05P
G-MWMJ	Solar Wings Pegasus Quasar	SW-TQ-0044 & SW-WQQ-0384	21. 9.90	M.Booth	Kemble	25. 9.04P
G-MWMK	Solar Wings Pegasus Quasar	SW-TQ-0045 & SW-WQQ-0385	21. 9.90	P.Adams	(Bristol)	13. 6.05P
G-MWML	Solar Wings Pegasus Quasar	SW-TQ-0046 & SW-WQQ-0386	21. 9.90	S.C.Key	Deopham Green	28. 9.04P
G-MWMM	Mainair Gemini/Flash IIA (Rotax 462)	800-0890-7 & W593	24. 8.90	R.H.Church	Croft Farm, Defford	11. 8.05P
G-MWMN	Solar Wings Pegasus XL-Q	SW-TE-0297 & SW-WQ-0387	2.10.90	N.A Rathbone and P.A Arnold	Swinford, Rugby	28. 3.05P
G-MWMO	Solar Wings Pegasus XL-Q	SW-TE-0298 & SW-WQ-0388	2.10.90	D.S.F.McNair	(Lochgilphead)	2. 8.04P
G-MWMP	Solar Wings Pegasus XL-Q	SW-TE-0299 & SW-WQ-0389	2.10.90	F.E.Hall	(Rushden)	20.10.05P
G-MWMR	Solar Wings Pegasus XL-R (Rotax 462)	SW-TE-0300 & SW-WA-1483	2.10.90	M.I. Stone *(New owner 4.03)*	(Barnstaple)	13. 3.00P
G-MWMS	Mainair Gemini/Flash	807-1090-5 & W600	3.10.90	A.Gannon	East Fortune	30. 4.05P
G-MWMT	Mainair Gemini/Flash IIA	808-1090-7 & W601	3.10.90	R Findlay	Mill Farm, Shifnal	6. 3.05P
G-MWMU	CFM Shadow Series CD *(Manufacturer's records show as c/n 142 - with c/n 150 sold to Namibia)*	150	2.10.90	A J.Burton	Wickenby	15. 8.05P
G-MWMV	Solar Wings Pegasus XL-R (Rotax 462)	SW-TE-0307 & SW-WA-1484	5.10.90	M A Oakley	Kemble	21. 7.04P
G-MWMW	Meridian Renegade Spirit UK *(Built M W Hanley - c/n 254)*	PFA 188-11544	21. 8.89	H.Feeney *"Spirit of Cornwall" (Noted 12.04)*	Priory Farm, Tibenham	26. 7.02P
G-MWMX	Mainair Gemini/Flash IIA (Rotax 462)	810-1090-7 & W603	17.10.90	L.Campbell	Newtownards	14. 2.05P
G-MWMY	Mainair Gemini/Flash IIA (Rotax 462)	809-1090-7 & W602	17.10.90	P J Harrison	Priory Farm, Tibenham	19. 9.05P
G-MWMZ	Solar Wings Pegasus XL-Q	SW-TE-0301 & SW-WQ-0393	8.10.90	M.Bastin	Redlands	10. 8.05P
G-MWNA	Solar Wings Pegasus XL-Q	SW-TE-0302 & SW-WQ-0394	8.10.90	S.Dixon	Eshott	10. 1.05P
G-MWNB	Solar Wings Pegasus XL-Q	SW-TE-0303 & SW-WQ-0395	8.10.90	P.F.J.Rogers	(London SW17)	23. 7.05P
G-MWNC	Solar Wings Pegasus XL-Q	SW-TE-0304 & SW-WQ-0396	8.10.90	M.E.Oakman	(Watton-at-Stone, Hertford)	14. 8.05P
G-MWND	Tiger Cub RL5A LW Sherwood Ranger *(Built D A Pike - c/n 001)* (Rotax 532)	PFA 237-12229	9.10.90	D.A Pike	Brook Farm, Pilling	13. 6.03P
G-MWNE	Mainair Gemini/Flash IIA	803-1090-7 & W596	17.10.90	T.C.Edwards	(Ware)	1. 4.05P
G-MWNF	Meridian Renegade Spirit UK *(Built D J White)* (BMW R100)	PFA 188-11853	15.10.90	D.J.White	Bodmin	21. 9.04P
G-MWNG	Solar Wings Pegasus XL-Q	SW-TE-0305 & SW -WQ-0399	17.10.90	H.C.Thomson	Easter Balgillo Farm, Finavon	3.11.03P
G-MWNK	Solar Wings Pegasus Quasar TC	SW-TQA-0054 & SW-WQQ-0403	1.11.90	G.S.Lyon	RAF Wyton	1. 8.05P
G-MWNL	Solar Wings Pegasus Quasar	SW-TQ-0055 & SW-WQQ-0404	1.11.90	R J Humphries *(New owner 9.04)*	(Eastleigh)	30. 7.00P
G-MWNO	AMF Microflight Chevvron 2-32C	025	12.11.90	I.K.Hogg	Kirkbride	30. 4.05P
G-MWNP	AMF Microflight Chevvron 2-32C	026	31.10.90	C R Rosby	(Belford)	18. 7.03P
G-MWNR	Meridian Renegade Spirit UK *(Built J.J.Lancaster)*	PFA 188-11926	12.11.90	J.J.Lancaster	Cublington	14. 9.05P
G-MWNS	Mainair Gemini/Flash IIA	811-1190-7 & W604	6.11.90	G.M.Baker	Deenethorpe	25. 6.05P
G-MWNT	Mainair Gemini/Flash IIA (Rotax 582)	812-1190-7 & W605	6.11.90	K.Forsyth	(Lempitlaw, Kelso)	3. 6.05P
G-MWNU	Mainair Gemini/Flash IIA	813-1190-5 & W606	6.11.90	C.C.Muir	Doynton	21. 7.05P
G-MWNV	Powerchute Kestrel	00406	12.11.90	K.N.Byrne *(Current status unknown)*	(Colonsay)	13. 3.92P
G-MWNX	Powerchute Kestrel	00408	12.11.90	J.H.Greenroyd	(Hebden Bridge)	24. 9.02P
G-MWOC	Powerchute Kestrel	00413	12.11.90	H.G.Davies and V.E.Booth	(Woodmancote, Cheltenham/Prestbury, Cheltenham)	28. 7.05P
G-MWOD	Powerchute Kestrel	00414	12.11.90	T.Morgan	(Kidderminster)	4.10.00P
G-MWOE	Powerchute Kestrel	00415	12.11.90	E.G.Woolnough *(New owner 4.04)*	(Halesworth)	21. 1.02P
G-MWOF	Microflight Spectrum	018	13.11.90	P.Williams	Otherton, Cannock	16. 5.05P
G-MWOH	Solar Wings Pegasus XL-R/Se	SW-TB-1429 & SW-WA-1485	28.11.90	P.G.Mallon	Redlands	2. 6.05P
G-MWOI	Solar Wings Pegasus XL-R	SW-TB-1430 & SW-WA-1486	29.11.90	B.T.Geoghegan	Brook Farm, Pilling	17. 6.05P
G-MWOJ	Mainair Gemini/Flash IIA	814-1290-7 & W608	6.12.90	C.J.Pryce	Newton Bank, Daresbury	24. 6.04P
G-MWOK	Mainair Gemini/Flash IIA (Rotax 462)	815-1290-7 & W609	6.12.90	J.C.Miller *(Noted 8.04)*	Errol	19. 8.02P
G-MWOM	Solar Wings Pegasus Quasar TC	SW-TQ-0060 & SW-WQQ-0412	1. 3.91	T.J.Williams	(Tuam, County Galway)	1. 9.05P
G-MWON	CFM Shadow Series CD	K.128	18.12.90	R.E.M.Gibson-Bevan	Wickenby	14. 7.03P
G-MWOO	Meridian Renegade Spirit UK *(Built A Hipkin - c/n 318)*	PFA 188-11811	14. 9.90	R.C.Wood	Lower Mountpleasant, Chatteris	26. 7.05P
G-MWOP	Solar Wings Pegasus Quasar TC	SW-TQC-0059 & SW-WQQ-0410	31.12.90	A Baynes	Sywell	12. 4.05P
G-MWOR	Solar Wings Pegasus XL-Q	SW-TE-0308 & SW-WQ-0411	21.12.90	M.J.Saich	Longacres Farm, Sandy	25. 3.05P
G-MWOV	Whittaker MW6 Merlin *(Built C R Melhuish)* (Rotax 503)	PFA 164-11301	9. 1.91	L.R.Hodgson	Kirkbride	27. 7.05P

G-MWOY	Solar Wings Pegasus XL-Q			7. 1.91	R.V.Barber	Hunsdon	11.10.02P
		SW-TE-0310 & SW-WQ-0414			*(Noted 1.05)*		
G-MWPB	Mainair Gemini/Flash IIA	823-0191-7 & W617		3. 1.91	J.Fenton	St.Michaels	11. 7.05P
G-MWPC	Mainair Gemini/Flash IIA	826-0191-7 & W620		3. 1.91	M.Johnson	(Sale)	26. 4.04P
G-MWPD	Mainair Gemini/Flash IIA	824-0191-7 & W618		9. 1.91	C.J.Roper and P.F.Mayo	East Fortune	19. 2.05P
G-MWPE	Solar Wings Pegasus XL-Q			9. 1.91	E.C.R.Hudson	Upper Stow, Weedon	26. 5.05P
		SW-TE-0096 & SW-WQ-0416	*(Trike ex G-MVGX)*				
G-MWPF	Mainair Gemini/Flash IIA	825-0191-7 & W619		11. 1.91	L.H.Black	(Carrickfergus)	3. 6.05P
G-MWPG	Corbett Farms Spectrum	019		9. 1.91	D.Payn tr G-MWPG Group	Eshott	23. 1.05P
G-MWPH	Corbett Farms Spectrum	020		9. 1.91	S.Rickett and M.J.Deacon	Draycott Farm, Chiseldon	15. 8.05P
G-MWPJ	Solar Wings Pegasus XL-Q			17. 1.91	D.S.Parker	Carlisle	14. 5.03P
		SW-TE-0312 & SW-WQ-0418					
G-MWPK	Solar Wings Pegasus XL-Q			17. 1.91	M.Harris	Redlands	1. 6.05P
		SW-TE-0313 & SW-WQ-0419					
G-MWPN	CFM Shadow Series CD	K.147		22. 1.91	W.R.H.Thomas *(Current status unknown)*	(Swansea)	11. 6.99P
G-MWPO	Mainair Gemini/Flash IIA	827-0191-7 & W621		29. 1.91	R.P.McGann	Headon Farm, Retford	2.11.05P
G-MWPP	CFM Streak Shadow M	PFA 206-11992	G-BTEM	14. 2.91	A.J.Thomas	(Nottingham)	26. 8.04P
	(Built C A Mortlock - c/n K.166-SA) (Rotax 582)						
G-MWPR	Whittaker MW6 Merlin	PFA 164-11260		16.10.90	S.F.N.Warnell	(Staines)	
	(Built P J S Ritchie)				*(New owner 10.01)*		
G-MWPS	Meridian Renegade Spirit UK	PFA 188-11931		18. 2.91	M.D.Stewart	(Leicester)	1. 7.98P
	(Built A R Broughton-Tompkins)				*(New owner 6.01)*		
G-MWPT	Hunt Avon/Huntwing	BMAA/HB/015	EI-CKF	18. 2.91	M.Leyden and S.Cronin	(Ennis, County Clare)	10. 5.97P
	(Built J A Hunt - c/n JAH-8) (Fuji-Robin EC-44-PM)		G-MWPT		*(Current status unknown)*		
G-MWPU	Solar Wings Pegasus Quasar TC			20. 2.91	R.L.Flowerday	(Exeter)	6. 7.04P
		SW-TQC-0062 & SW-WQQ-0426					
G-MWPW	AMF Microflight Chevvron 2-32C	027		26.11.90	E.L.T.Westman	Broadford, Isle of Skye	13.10.05P
G-MWPX	Solar Wings Pegasus XL-R			27. 2.91	R.J.Wheeler	Redlands	1. 9.04P
	(Rotax 462)	SW-TE-0315 & SW-WA-1488					
G-MWPZ	Meridian Renegade Spirit UK	PFA 188-11631		18. 3.91	J.Ievers	Rhosgoch	24. 2.99P
	(Built J Ievers)				*(Noted 2.04)*		
G-MWRB	Mainair Gemini/Flash IIA	819-0191-7 & W613		5. 2.91	A T.Farmer	Roddige	31. 5.05P
G-MWRC	Mainair Gemini/Flash IIA	820-0191-7 & W614		5. 2.91	D.R.Talbot	Chiltern Park, Wallingford	29. 4.05P
G-MWRD	Mainair Gemini/Flash IIA	821-0191-7 & W615		5. 2.91	S.R.McKiernan	(Bangor)	29. 6.04P
	(Motavia?)				*(New owner 12.04)*		
G-MWRE	Mainair Gemini/Flash IIA	822-0191-7 & W616		5. 2.91	A D.Dias	Otherton, Cannock	4. 8.05P
G-MWRF	Mainair Gemini/Flash IIA	829-0191-7 & W623		4. 2.91	N.Hay	Perth	13. 6.05P
G-MWRG	Mainair Gemini/Flash IIA	830-0191-7 & W624		5. 2.91	Fiona J. Clarehugh	Eshott	1.12.05P
G-MWRH	Mainair Gemini/Flash IIA	831-0191-7 & W625		5. 2.91	E.G.Astin	Eshott	24. 8.05P
G-MWRJ	Mainair Gemini/Flash IIA	832-0291-7 & W626		28. 2.91	R.J.Lindley	(Crewe)	26. 3.05P
G-MWRK*	Rans S-6 Coyote II	PFA 204-11930		13. 2.91	R.H.Bambury	Felixkirk	10. 5.99P
	(Built R M Bambury - c/n 0191.154) (Tri-cycle u/c)			*(Damaged Easingwold 4.7.99: cancelled 19.5.00 by CAA) (Wreck noted 2003)*			
G-MWRL	CFM Shadow Series CD	K.152		13. 2.91	R.A.and C.A.Allen	Clench Common	18. 8.01P
					(New owners 4.04)		
G-MWRM	Medway Hybred 44XLR	MR086/94/91/S	G-MWLC	26. 2.91	M.A Jones	(Wigan)	5. 3.05P
	(Rotax 503)						
G-MWRN	Solar Wings Pegasus XL-R			5. 3.91	D.T.MacKenzie	Easter Poldar Farm, Thornhill	28. 3.05P
	(Rotax 462)	SW-TE-0316 & SW-WA-1489					
G-MWRP	Solar Wings Pegasus XL-R			1. 3.91	A R.Hughes	(Calne)	5. 6.04P
	(Rotax 462)	SW-TE-0318 & SW-WA-1491					
G-MWRR	Mainair Gemini/Flash IIA	834-0391-7 & W628		7. 3.91	C P Davis	(Burntwood)	10. 4.05P
G-MWRS	Ultravia Super Pelican	E001-201		9. 5.84	T.B.Woolley	(Narborough, Leicester)	9. 9.87P
					(Current status unknown)		
G-MWRT	Solar Wings Pegasus XL-R			15. 3.91	G.L.Gunnell	Sywell	10. 9.03P
		SW-TB-1431 & SW-WA-1492					
G-MWRU	Solar Wings Pegasus XL-R			15. 3.91	J.McIver	(West Kilbride)	25. 8.96P
		SW-TB-1432 & SW-WA-1493			*(Current status unknown)*		
G-MWRV	Solar Wings Pegasus XL-R			15. 3.91	M.S.Adams	Roddige	20. 4.05P
		SW-TB-1433 & SW-WA-1494					
G-MWRW	Solar Wings Pegasus XL-Q			25. 3.91	M Peters	Wing Farm, Longbridge Deverill	31. 8.05P
		SW-TE-0320 & SW-WQ-0431					
G-MWRX	Solar Wings Pegasus XL-Q			25. 3.91	W Parkes	Long Marston	17. 9.04P
		SW-TE-0321 & SW-WQ-0432					
G-MWRY	CFM Shadow Series CD	K.162		26. 3.91	A W.Hodder	Belle Vue Farm, Yarnscombe	12. 9.04P
G-MWRZ	AMF Microflight Chevvron 2-32C	028		10. 4.91	P.V.Prowse	Davidstow Moor	4. 7.04P
G-MWSA	TEAM mini-MAX 88	PFA 186-11855		8. 4.91	S.Hutchinson	RAF Scampton	15. 4.04P
	(Built A N Baumber) (Rotax 377)						
G-MWSB	Mainair Gemini/Flash IIA	837-0591-7 & W631		30. 4.91	G.R.Whittemore	(Harvington, Evesham)	6. 9.05P
	(Rotax 582)						
G-MWSC	Rans S-6-ESD Coyote II	PFA 204-12019		13. 5.91	E.M.Lear	(Langport)	30. 6.96P
	(Built B E Francis) (Tri-cycle u/c)				*(Current status unknown)*		
G-MWSD	Solar Wings Pegasus XL-Q			6. 3.91	A M.Harley	Sutton Meadows, Ely	18. 7.05P
		SW-TE-0319 & SW-WQ-0430					
G-MWSE	Solar Wings Pegasus XL-R			10. 4.91	Ultra Light Training Ltd	Roddige	28. 2.05P
	(Rotax 462)	SW-TE-0323 & SW-WA-1496	*(Fitted with trike unit from G-MTJR)*				
G-MWSF	Solar Wings Pegasus XL-R			10. 4.91	N.A.and F.W.Milne	(Chedburgh, Bury St.Edmunds)	3. 7.05P
	(Rotax 462)	SW-TE-0324 & SW-WA-1497					
G-MWSH	Solar Wings Pegasus Quasar TC			30. 4.91	G Jones	(Llanfairfechan)	15. 5.05P
		SW-TQC-0064 & SW-WQ-0435					
G-MWSI	Solar Wings Pegasus Quasar TC			23. 5.91	J.A Ganderton	Sywell	18.10.03P
		SW-TQC-0065 & SW-WQ-0436					
G-MWSJ	Solar Wings Pegasus XL-Q			12. 4.91	R.A Barrett	Sutton Meadows, Ely	17.10.04P
		SW-TE-0326 & SW-WQ-0437					

Regn	Type	C/n	Prev.Id	Regd	Owner/Operator	Probable Base	CA Expy
G-MWSK	Solar Wings Pegasus XL-Q	SW-TE-0327 & SW-WQ-0438		12. 4.91	J.Doogan t/a Scottglass	(Galashiels)	26. 5.02P
G-MWSL	Mainair Gemini/Flash IIA	835-0491-7 & W629		16. 4.91	C.W.Frost *(Current status unknown)*	Rufforth	11. 6.98P
G-MWSM	Mainair Gemini/Flash IIA	836-0491-7 & W630		16. 4.91	R.M.Wall	(Longacres Farm, Sandy)	13. 4.04P
G-MWSO	Solar Wings Pegasus XL-R (Rotax 462)	SW-TE-0329 & SW-WA-1503		25. 4.91	M.A Clayton	(New Romney)	11. 6.05P
G-MWSP	Solar Wings Pegasus XL-R (Rotax 462)	SW-TE-0330 & SW-WA-1504		25. 4.91	P.J.Barratt	Headon Farm, Retford	1.10.04P
G-MWSR	Solar Wings Pegasus XL-R (Rotax 462)	SW-TE-0331 & SW-WA-1505		25. 4.91	G.P.J.Davies	Mill Farm, Shifnal	11.10.05P
G-MWSS	Medway Hybred 44XLR (Rotax 503)	MR117/97		7. 5.91	F.S.Ogden	West Hoathly, Haywards Heath	20. 9.03P
G-MWST	Medway Hybred 44XLR (Rotax 503)	MR118/98		8. 5.91	A Ferguson	(Sconser, Skye)	3. 8.05P
G-MWSU	Medway Hybred 44XLR (Rotax 503)	MR119/99		1. 5.92	K.J.Draper t/a Medway Microlights	Middle Stoke, Isle of Grain	6. 5.05P
G-MWSW	Whittaker MW6 Merlin *(Built S.N.F.Warnell)*	PFA 164-11328		15. 2.91	S.N.F.Warnell *(Current status unknown)*	Staines	
G-MWSX	Whittaker MW5 Sorcerer *(Built A T Armstrong)*	PFA 163-11549		3. 5.91	D.R.Drewett	(Worthing)	8. 6.05P
G-MWSY	Whittaker MW5 Sorcerer *(Built J.E.Holloway)*	PFA 163-11218		3. 5.91	J.E.Holloway	Saltash	21.12.04P
G-MWSZ	CFM Shadow Series CD	K.158	(G-MWRY)	4. 4.91	P.G.Bibbey	Old Sarum	17.10.05P
G-MWTA	Solar Wings Pegasus XL-Q	SW-TE-0332 & SW-WQ-0444		8. 5.91	C D Arnold	Craysmarsh Farm, Melksham	21. 5.05P
G-MWTB	Solar Wings Pegasus XL-Q	SW-TE-0333 & SW-WQ-0445		8. 5.91	G.S.Highley	(Corby)	17. 9.04P
G-MWTC	Solar Wings Pegasus XL-Q	SW-TE-0334 & SW-WQ-0446		8. 5.91	P.Nicholson	(London SE18)	14.10.05P
G-MWTD	Corbett Farms Spectrum	022		13. 5.91	J.V.Harris tr Group Delta	Ashbourne	18. 4.00P
G-MWTE	Corbett Farms Spectrum	023		13. 5.91	T.H.Evans	(Ammanford)	20. 6.04P
G-MWTG	Mainair Gemini/Flash IIA (Rotax 582)	838-0591-7 & W632		16. 5.91	G.Keenan	Ince Blundell	27. 5.04P
G-MWTH	Mainair Gemini/Flash IIA	839-0591-7 & W633		21. 5.91	A.Strang	East Fortune	24. 5.05P
G-MWTI	Solar Wings Pegasus XL-Q	SW-TE-0251 & SW-WQ-0274		23. 5.91	O.G.Johns	(Headington, Oxford)	9. 5.05P
G-MWTJ	CFM Shadow Series CD	K.167		16. 5.91	T.D.Wolstenholme	Brook Farm, Pilling	24.11.05P
G-MWTK	Solar Wings Pegasus XL-R/Se (Rotax 462)	SW-TE-0335 & SW-WA-1507		28. 5.91	G.Munro	Watnall	23. 4.05P
G-MWTL	Solar Wings Pegasus XL-R (Rotax 462)	SW-TE-0336 & SW-WA-1508		28. 5.91	B.Lindsay	(Chipping Sodbury)	4. 6.05P
G-MWTM	Solar Wings Pegasus XL-R (Rotax 462)	SW-TE-0337 & SW-WA-1509		28. 5.91	I.R.F.King	(Tunbridge Wells)	31. 3.02P
G-MWTN	CFM Shadow Series CD	K.153		23. 5.91	M.J.Broom	Long Marston	15. 4.05P
G-MWTO	Mainair Gemini/Flash IIA	840-0591-7 & W634		28. 5.91	J.Greenhalgh	St.Michaels	22. 6.05P
G-MWTP	CFM Shadow Series CD	K.107		23. 5.91	R.E.M.Gibson-Bevan	Wickenby	27. 6.05P
G-MWTR	Mainair Gemini/Flash IIA (Rotax 582)	842-0591-7 & W636		31. 5.91	A A Howland	(Battle)	26.11.04P
G-MWTT	Rans S-6-ESD Coyote II *(Built I K Radcliffe - c/n 20391.175) (Tri-cycle u/c)*	PFA 204-12016		30. 4.91	L.E.Duffin *"Warrior 2"*	Insch	8. 1.05P
G-MWTU	Solar Wings Pegasus XL-R	SW-TB-1435 & SW-WA-1501		21. 6.91	S.Woods	(Banagher, County Offaly)	2. 9.01P
G-MWTY	Mainair Gemini/Flash IIA	843-0691-7 & W637		12. 6.91	A McGing and J.C.Townsend	Ince Blundell	28. 4.04P
G-MWTZ	Mainair Gemini/Flash IIA	844-0691-7 & W638		12. 6.91	C.W.R.Felce	Riseley, Bedford	23. 6.04P
G-MWUA	CFM Shadow Series CD	K.161		10. 6.91	P.A James t/a Cloudbase Aviation G-MWUA	Redhill	9. 8.05P
G-MWUB	Solar Wings Pegasus XL-R (Rotax 462)	SW-TE-0338 & SW-WA-1510		12. 6.91	T.R.L.Bayley	(Edenbridge)	25. 6.05P
G-MWUC	Solar Wings Pegasus XL-R (Rotax 462)	SW-TE-0339 & SW-WA-1511		12. 6.91	M.A.Hicks	Wickenby	27. 6.05P
G-MWUD	Solar Wings Pegasus XL Tug (Rotax 462)	SW-TE-0340 & SW-WA-1512		12. 6.91	M.J.Taggart	Clench Common	17. 5.05P
G-MWUF*	Solar Wings Pegasus XL-R	SW-TB-1439 & SW-WA-1514		13. 6.91	J.G.Jackson *(Cancelled 21.1.05 as WFU)*	Enstone	27. 7.05P
G-MWUH	Meridian Renegade Spirit UK *(Built A I Grant)*	343		12. 6.91	A I.Grant	Washington, Sussex	18. 6.04P
G-MWUI	AMF Microflight Chevvron 2-32C	029		2. 7.91	N.D.A Graham	(Lochgilphead)	9. 4.05P
G-MWUK	Rans S-6-ESD Coyote II *(Built G K Hoult - c/n 0491.187) (Tri-cycle u/c)*	PFA 204-12090		1. 7.91	G.K.Hoult	Long Marston	1.12.05P
G-MWUL	Rans S-6-ESD Coyote II *(Built K J Lywood - c/n 0391.172) (Tri-cycle u/c)*	PFA 204-12054		10. 6.91	C.K.Fry	(Lychett Minster)	17. 8.05P
G-MWUN	Rans S-6-ESD Coyote II *(Built S Eland - c/n 0695.841) (Tri-cycle u/c) (Original c/n 0391-170 (?) and rebuilt with new airframe)*	PFA 204-12075		10. 6.91	M.L.Robinson tr Coyote Flying Group	Kirkbride	5. 5.05P
G-MWUO	Solar Wings Pegasus XL-Q	SW-TE-0296 & SW-WQ-0379	(ZS-...?)	26. 6.91	A P.Slade	Long Marston	7. 9.03P
G-MWUP	Solar Wings Pegasus XL-R (Rotax 462)	SW-TE-0341 & SW-WA-1517		21. 6.91	G.R.Hall and P.R.Brooker	(Petham. Canterbury/Smarden, Ashford)	22. 7.05P
G-MWUR	Solar Wings Pegasus XL Tug (Rotax 462)	SW-TE-0342 & SW-WA-1518	*(Officially regd as "XL-R")*	21. 6.91	G T Wirdnam and A J Gilbert tr Nottingham Aerotow Club	Knapthorpe Lodge, Caunton	19. 8.05P
G-MWUS	Solar Wings Pegasus XL-R (Rotax 462)	SW-TE-0343 & SW-WA-1519		21. 6.91	H.R.Loxton	Weston Zoyland	29. 8.00P
G-MWUU	Solar Wings Pegasus XL-R (Rotax 462)	SW-TE-0346 & SW-WA-1521		28. 6.91	B.R.Underwood	Swinford, Rugby	21. 7.05P
G-MWUV	Solar Wings Pegasus XL-R (Rotax 462)	SW-TE-0347 & SW-WA-1522		28. 6.91	C.D.Baines	(Crewe)	27. 1.05P

Reg	Type	C/n	Previous id	Date	Owner	Location	Expiry
G-MWUX	Solar Wings Pegasus XL-Q	SW-WQ-0454		28. 6.91	B.D.Attwell	Caerphilly	27. 3.03P
	(Originally supplied as a sailwing only - trike origin unknown)						
G-MWUY	Solar Wings Pegasus XL-Q	SW-TE-0345 & SW-WQ-0455		28. 6.91	M.J.Sharp	(Kilmarnock)	11. 1.04P
G-MWUZ	Solar Wings Pegasus XL-Q	SW-TE-0350 & SW-WQ-0456		28. 6.91	S.R.Nanson	(Sittingbourne)	31. 5.05P
G-MWVA	Solar Wings Pegasus XL-Q	SW-TE-0351 & SW-WQ-0457		28. 6.91	P.C.Hancox	Croft Farm, Defford	19. 2.05P
G-MWVE	Solar Wings Pegasus XL-R	SW-TB-1441 & SW-WA-1524		18. 7.91	W.A Keel-Stocker	Long Marston	7. 6.05P
G-MWVF	Solar Wings Pegasus XL-R/Se	SW-TB-1442 & SW-WA-1525		18. 7.91	J.B.Wright	Roddige	7.11.05P
G-MWVG	CFM Shadow Series CD	151		5. 8.91	Shadow Aviation Ltd	Old Sarum	16.10.05P
G-MWVH	CFM Shadow Series CD	181		5. 8.91	J.Rochead and J.McAvoy tr Connel VH Group	North Connel, Oban	15. 7.05P
G-MWVK	Mainair Mercury	849-0891-5 & W643		13. 8.91	S.J.Robertson	(Newton Abbot)	27. 7.05P
G-MWVL	Rans S-6 ESD Coyote II	PFA 204-12118		13. 8.91	J.T., A., A and O.D.Lewis	Rufforth	17. 6.05P
	(Built J D Hall - c/n 0892.341) (Tri-cycle u/c) (Originally built with frame c/n 0491-186: damaged, repaired and fitted as a replacement frame to G-MZAH)						
G-MWVM	Solar Wings Pegasus Quasar IITC	SW-TQ-0031 & SW-WX-0020	G-65-8 *(Trike c/n duplicates G-MWLI)*	2. 9.91	J.D.Jones and A A Edmonds	Mill Farm, Shifnal	28. 3.05P
G-MWVN	Mainair Gemini/Flash IIA	850-0891-7 & W644		19. 8.91	J.McCafferty	Enstone	20.11.04P
G-MWVO	Mainair Gemini/Flash IIA (Rotax 582)	852-0891-7 & W646		27. 8.91	P.Webb	(Bristol)	25. 7.05P
G-MWVP	Meridian Renegade Spirit UK *(Built I E Spencer - kit no.345)*	PFA 188-11735		22. 8.91	P.D.Mickleburgh *(Under restoration 4.04)*	Swinford, Rugby	29. 4.94P
G-MWVR	Mainair Gemini/Flash IIA	855-0991-7 & W650		30. 8.91	G.Cartwright	Northampton	29. 4.05P
G-MWVS	Mainair Gemini/Flash IIA	856-0991-7 & W651		30. 8.91	S.J.J.Griffiths	(Leighton, Crewe)	24. 1.05P
G-MWVT	Mainair Gemini/Flash IIA	860-1091-7 & W655		2. 9.91	J.Barlow and C.Osiejuk	Oxton, Nottingham	29. 4.05P
G-MWVU	Medway Hybred 44XLR (Rotax 503)	MR123/102		18. 9.91	H.M.Manning	Rochester	24. 1.02P
G-MWVW	Mainair Gemini/Flash IIA	853-0891-7 & W647		9. 9.91	W.O'Brien	Arclid Green, Sandbach	29. 7.02P
G-MWVY	Mainair Gemini/Flash IIA	854-0991-7 & W649		4. 9.91	J.D.Hinton	(Tunbridge Wells)	4. 5.02P
G-MWWB	Mainair Gemini/Flash IIA	864-1091-7 & W659		18. 9.91	J.H.Bradbury	Arclid Green, Sandbach	19. 2.05P
G-MWWC	Mainair Gemini/Flash IIA (Rotax 582)	868-1191-7 & W663		23. 9.91	A and D.Margereson	(Chesterfield)	30. 7.04P
G-MWWD	Meridian Renegade Spirit UK *(Built A M Smyth and J M Walter - c/n 344)*	PFA 188-11719		23. 9.91	F.Overall *"Winning Spirit"*	Wethersfield	20. 6.05P
G-MWWE	TEAM mini-MAX 88 *(Built J C Longmore)*	PFA 186-11925		1.10.91	J.Entwistle *(Dismantled and stored 9.03)*	Tarn Farm, Cockerham	23. 7.97P
G-MWWG	Solar Wings Pegasus XL-Q	SW-TE-0355 & SW-WQ-0468		3.10.91	A W.Guerri	Rufforth	17. 8.05P
G-MWWH	Solar Wings Pegasus XL-Q	SW-TE-0356 & SW-WQ-0469		3.10.91	M.R.Dunnett	Ludham	17. 7.03P
G-MWWI	Mainair Gemini/Flash IIA	870-1291-7 & W665		11.10.91	R.J.Vaughan	(Northwich)	2.10.03P
G-MWWJ	Mainair Gemini/Flash IIA	865-1191-7 & W660		22.10.91	S.B.Walsh and S.Lysaght	Tarn Farm, Cockerham	5.11.04P
G-MWWK	Mainair Gemini/Flash IIA (Rotax 582)	866-1191-7 & W661		22.10.91	J.C.Boyd tr JDS Group	Davidstow Moor	28. 2.05P
G-MWWL	Rans S-6-ESD Coyote II *(Built J E Carr) (Tri-cycle u/c)*	PFA 204-11849	(G-BTXD)	17.10.91	J.C.Field	North Coates	12. 6.05P
G-MWWM	Kolb Twinstar Mk.2	PFA 205-11645	(G-BTXC)	17.10.91	D.Jordan	RAF Brize Norton	21. 6.04P
	(Built D Jordan) (Rotax 503) *(PFA Plans no duplicates G-GPST)*						
G-MWWN	Mainair Gemini/Flash IIA	872-1291-7 & W667		22.10.91	F.Watts	(Bournemouth)	27. 6.05P
G-MWWP	Rans S-4 Coyote *(Built R H Braihtwaite)*	90.115 & PFA 193-12073		21.10.91	R.P.Cross *(New owner 11.04)*	(Welton, Lincoln)	1. 8.00P
G-MWWR	Corbett Farms Spectrum	024		23.10.91	M.S.J.Bateman	(Hull)	24. 5.05P
G-MWWS	Thruster T.300 *(Built Tempest Aviation Ltd)* (Rotax 532)	089-T300-370	EI-BYW	4.11.91	S.P.McCaffrey	Ginge, Wantage	2. 6.05P
G-MWWV	Solar Wings Pegasus XL-Q	SW-TE-0357 & SW-WQ-0470		30.10.91	R.W.Livingstone	Enniskillen, County Fermanagh	28. 4.05P
G-MWWZ	Cyclone Chaser S 447	CH.829		29.10.91	J.F.Willoughby	Eshott	4. 4.04P
G-MWXB	Mainair Gemini/Flash IIA	869-1191-7 & W664		6.11.91	N.W.Barnett	Sittles Farm, Alrewas	3. 1.04P
G-MWXC	Mainair Gemini/Flash IIA	874-0192-7 & W669		6.11.91	G.Dufton-Kelly	(Wirral)	25. 7.05P
G-MWXF	Mainair Mercury	867-1191-5 & W662		12.11.91	P.E.Jackson	(Ruthin)	23. 3.05P
G-MWXG	Solar Wings Pegasus Quasar IITC	SW-TQC-0074 & SW-WQT-0471		7.11.91	J.E.Moseley	Saffron Walden	14. 9.05P
G-MWXH	Solar Wings Pegasus Quasar IITC	SW-TQC-0075 & SW-WQT-0472		7.11.91	R.P.Wilkinson	Charmy Down, Bath	10. 6.05P
G-MWXJ	Mainair Mercury	861-1091-5 & W656		15.11.91	P.J.Taylor	Sandtoft	13. 3.05P
G-MWXK	Mainair Mercury	862-1191-5 & W657		15.11.91	M.P.Wilkinson *(Current status unknown)*	Sandtoft	18. 7.96P
G-MWXL	Mainair Gemini/Flash IIA (Rotax 582)	859-1091-7 & W654		12.12.91	S.N.Catchpole	Thurton	15. 9.04P
G-MWXN	Mainair Gemini/Flash IIA (Rotax 582)	878-0192-7 & W673		20.11.91	P.I.Miles	(Chesterfield)	30. 5.05P
G-MWXO	Mainair Gemini/Flash IIA	880-0192-7 & W675		25.11.91	T.P.Wright *(Noted 1.04)*	Roddige	19. 9.02P
G-MWXP	Solar Wings Pegasus XL-Q	SW-TE-0359 & SW-WQ-0475		26.11.91	A P.Attfield *(Current status unknown)*	Sutton Meadows, Ely	18. 8.99P
G-MWXR	Solar Wings Pegasus XL-Q	SW-TE-0360 & SW-WQ-0476		26.11.91	G.W.Craig	Insch	15. 6.04P
G-MWXU	Mainair Gemini/Flash IIA (Rotax 582)	882-0192-7 & W677		9.12.91	C.M.Mackinnon	Cumbernauld	23. 1.03P
G-MWXV	Mainair Gemini/Flash IIA (Rotax 582)	879-1291-7 & W674		9.12.91	A.W.Lowrie	(Houghton Le Spring)	25. 5.05P
G-MWXW	Cyclone Chaser S (Rotax 377)	CH.830		9.12.91	K.C.Dodd	Roddige	13. 5.04P

Registration	Type	c/n	Previous identity	Reg date	Owner	Location	CofA
G-MWXX	Cyclone Chaser S 447	CH.831	(G-MWEB) (G-MWCD)	9.12.91	P I Frost	(Guilsborough, Northampton)	21. 7.05P
G-MWXY	Cyclone Chaser S 447	CH.832	(G-MWEC)	19.12.91	C A Benjamin	Longacres Farm, Sandy	16. 8.04P
G-MWXZ	Cyclone Chaser S 508	CH.836		31.12.91	N.R.Beale *"Daedalus"*	Deppers Bridge, Southam	27. 4.01P
G-MWYA	Mainair Gemini/Flash IIA (Rotax 462)	886-0292-7 & W681		3. 1.92	R.F.Hunt	St.Michaels	22.10.05P
G-MWYB	Solar Wings Pegasus XL-Q	SW-TE-0364 & SW-WQ-0485		15. 1.92	R Revill	(Bristol)	5.12.05P
G-MWYC	Solar Wings Pegasus XL-Q	SW-TE-0365 & SW-WQ-0486		15. 1.92	N.J.Duckworth	(Bovington)	25. 7.05P
G-MWYD	CFM Shadow Series C	K.179		8. 1.92	J.Anderson	Plaistowes Farm, St.Albans	21. 7.05P
G-MWYE	Rans S-6-ESD Coyote II *(Built G A Squires - c/n 0591.189) (Tri-cycle u/c)*	PFA 204-12223		10. 1.92	G A M Moffat	(Manchester)	28. 8.05P
G-MWYG	Mainair Gemini/Flash IIA (Rotax 582)	884-0292-7 & W679		15. 1.92	M J Burns	(Castlewellan, County Down)	14. 7.05P
G-MWYH	Mainair Gemini/Flash IIA	887-0292-7 & W682		15. 1.92	A G.Carter and D.C.Jackson	(Giltbrook, Nottingham)	30.10.05P
G-MWYI	Solar Wings Pegasus Quasar IITC	SW-TQC-0083 & SW-WQT-0488		30. 1.92	T.S.Chadfield	Graveley	9. 4.04P
G-MWYJ	Solar Wings Pegasus Quasar IITC	SW-TQC-0084 & SW-WQT-0489		24. 1.92	J.W.Edwards	Finmere	26. 4.05P
G-MWYL	Mainair Gemini/Flash IIA	877-0192-7 & W672		17. 1.92	A J.Hinks	East Fortune	4. 7.05P
G-MWYM	Cyclone Chaser S 1000 (Mosler MM-CB35) *(Reported as rebuild of G-MVJI - perhaps trike only?)*	CH.838		21. 1.92	C J Meadows	(Shepton Mallet)	9. 8.05P
G-MWYN	Rans S-6-ESD Coyote II *(Built W R Tull - c/n 0491.185) (Tri-cycle u/c)*	PFA 204-12168		22. 1.92	W.R.Tull	(Chipping Norton)	12. 9.05P
G-MWYS	CGS Arrow Flight Hawk I Arrow *(Built Arrowflight Ltd - c/n H-T-470-R447)* (Rotax 447)	BMAA/HB/020		17. 2.93	D.W.Hermiston-Hooper t/a Civilair *(Current status unknown)*	(Ryde, Isle of Wight)	
G-MWYT	Mainair Gemini/Flash IIA	881-0392-7 & W676		3. 2.92	M.A Hodgson	Baxby Manor, Husthwaite	17. 9.05P
G-MWYU	Solar Wings Pegasus XL-Q	SW-TE-0364 & SW-WQ-0491		30. 1.92	N.Hammerton	(Oxted)	12. 5.01P
G-MWYV	Mainair Gemini/Flash IIA (Rotax 582)	896-0392-7 & W691		3. 2.92	J.N.Whitworth	Oxton	10. 9.04P
G-MWYY	Solar Wings Pegasus XL-Q	SW-TE-0365 & SW-WQ-0492		17. 2.92	R.D.Allard	Deenethorpe	31. 7.03P
G-MWYZ	Solar Wings Pegasus XL-Q	SW-TE-0358 & SW-WQ-0474		20.11.91	A Boston	Sywell	17. 6.05P
G-MWZA	Mainair Mercury	888-0292-5 & W683		7. 2.92	A J.Malham	Rufforth	9. 2.05P
G-MWZB	AMF Microflight Chevvron 2-32C	033		10. 2.92	A J.Pickup	Membury	7.12.05P
G-MWZC	Mainair Gemini/Flash IIA	899-0492-7 & W694		7. 2.92	L.W.Fowler	(Rochdale)	19. 4.05P
G-MWZD	Solar Wings Pegasus Quasar IITC	SW-TQC-0086 & SW-WQT-0494		2. 3.92	J.R.Burton	Redlands	16. 5.05P
G-MWZE	Solar Wings Pegasus Quasar IITC	SW-TQC-0087 & SW-WQT-0495		17. 2.92	H.Lorimer	(Mauchline)	18. 7.01P
G-MWZF	Solar Wings Pegasus Quasar IITC (Rotax 582)	SW-TQD-0108 & SW-WQT-0496	*(Trike c/n duplicates G-MYEK)*	17. 2.92	R.G.T.Corney	Clench Common	8. 8.05P
G-MWZG	Mainair Gemini/Flash IIA (Rotax 582)	889-0392-7 & W684		7. 2.92	C.J.O'Sullivan	(Thurles, County Tipperary	9. 9.04P
G-MWZH	Solar Wings Pegasus XL-R (Rotax 462)	SW-TE-0366 & SW-WA-1532		17. 2.92	P.A Ord	(Redcar)	17. 4.00P
G-MWZI	Solar Wings Pegasus XL-R (Rotax 462)	SW-TE-0367 & SW-WA-1533		17. 2.92	K.Slater and S.Reader	Roddige	29.10.05P
G-MWZJ	Solar Wings Pegasus XL-R/Se (Rotax 462)	SW-TE-0368 & SW-WA-1534		17. 2.92	P.Kitchen	Shotton Colliery, Peterlee	30. 8.05P
G-MWZL	Mainair Gemini/Flash IIA (Rotax 582)	900-0492-7 & W695		17. 2.92	D.Renton	East Fortune	5. 7.04P
G-MWZM	TEAM mini-MAX 91 *(Built M A J Hutt)* (Mosler MM CB-40)	PFA 186-12211	G-BUDD G-MWZM	18. 2.92	C.Leighton-Thomas *"My Buddy" (Noted 1.05)*	Charmy Down, Bath	29. 7.02P
G-MWZN	Mainair Gemini/Flash IIA (Rotax 582)	902-0492-7 & W697		25. 2.92	K.J.Rexter	(Cumbernauld)	3. 3.05P
G-MWZO	Solar Wings Pegasus Quasar IITC	SW-TQC-0089 & SW-WQT-0498		26. 2.92	R.Oseland	Roddige	27. 3.05P
G-MWZP	Solar Wings Pegasus Quasar IITC	SW-TQC-0090 & SW-WQT-0499		26. 2.92	C.M.Lewis and M.J.Wilson	Longacres Farm, Sandy	12. 6.05P
G-MWZR	Solar Wings Pegasus Quasar IITC	SW-TQC-0091 & SW-WQT-0500		26. 2.92	P.K.Appleton	Weston Zoyland	21. 9.05P
G-MWZS	Solar Wings Pegasus Quasar IITC	SW-TQC-0092 & SW-WQT-0501	EI-CIP G-MWZS	26. 2.92	G.Bennett	Beccles	8. 9.05P
G-MWZT	Solar Wings Pegasus XL-R (Rotax 462)	SW-TE-0370 & SW-WA-1535		26. 2.92	J.R.Lowman	Sutton Meadows, Ely	3. 9.04P
G-MWZU	Solar Wings Pegasus XL-R (Rotax 462)	SW-TE-0371 & SW-WA-1536		26. 2.92	D.W.Palmer	(Bexhill)	8. 6.05P
G-MWZV	Solar Wings Pegasus XL-R (Rotax 462)	SW-TE-0372 & SW-WA-1537		26. 2.92	D.J.Newby	Clench Common	14. 2.05P
G-MWZW	Solar Wings Pegasus XL-R (Rotax 462)	SW-TE-0373 & SW-WA-1538		26. 2.92	G.R.Puffett	Yatesbury	23. 1.05P
G-MWZY	Solar Wings Pegasus XL-R (Rotax 462)	SW-TE-0375 & SW-WA-1540		26. 2.92	S.J.Barkworth	Rufforth	20. 6.05P
G-MWZZ	Solar Wings Pegasus XL-R (Rotax 503)	SW-TE-0376 & SW-WA-1541		26. 2.92	M.P.Shea	Roddige	18. 6.04P
G-MXIV*	Vickers Supermarine 379 Spitfire FR.XIVc				See SECTION 3 Part 1	(Santa Monica, California, US)	
G-MXVI	Vickers Supermarine 361 Spitfire LF.XVIe	CBAF.IX.4394	6850M TE184	17. 2.89	De Cadenet Motor Racing Ltd *(As "TE184/D" in Free French and 328 Squadron RAF c/s) (Noted 7.03)*	RAF Halton	30. 5.02P
G-MYAB	Solar Wings Pegasus XL-R/Se (Rotax 462)	SW-TE-0377 & SW-WA-1542		26. 2.92	A N.F.Stewart	Long Marston	16. 6.04P

Registration	Type	c/n	Previous identity	Regd date	Owner	Location	Expiry
G-MYAC	Solar Wings Pegasus XL-Q			26. 2.92	M.A Garner	Thetford	25. 2.04P
		SW-TE-0378 & SW-WQ-0502					
G-MYAE	Solar Wings Pegasus XL-Q			26. 2.92	R.J.Waller	Redlands	17. 1.05P
		SW-TE-0380 & SW-WQ-0504					
G-MYAF	Solar Wings Pegasus XL-Q			26. 2.92	K.N.Rigley	(Newark)	27. 8.05P
		SW-TE-0381 & SW-WQ-0505					
G-MYAG	Quad City Challenger II	PFA 177-12167		25. 2.92	R H Hughes	Old Sarum	27. 6.05P
	(Built F Payne)						
G-MYAH	Whittaker MW5 Sorcerer	PFA 163-11233		2. 3.92	K.R.Emery	Sittles Farm, Alrewas	1.10.04P
	(Built T Knight)						
G-MYAI	Mainair Mercury	892-0392-5 & W687		11. 3.92	J.Ellerton	(Hazel Grove, Stockport)	6. 5.03P
G-MYAJ	Rans S-6-ESD Coyote II	PFA 204-12227		3. 3.92	K R Haskell	(Blandford Forum)	18. 8.04P
	(Built N J Willmott - c/n 1291.248) (Tail-wheel u/c)						
G-MYAK	Solar Wings Pegasus Quasar IITC		D-M…?	5. 3.92	I.E.Brunning	Ince Blundell	1. 9.03P
		SW-TQC-0093 & SW-WQT-0506	G-MYAK		*(Sailwing only 7.03)*		
G-MYAM	Meridian Renegade Spirit UK	PFA 188-11907		6. 3.92	A F.Reid	Newtownards	14. 9.01P
	(Built S R Groves)						
G-MYAN	Whittaker MW-5K Sorcerer	5K-0017-02	(G-MWNI)	24. 3.92	J.Hollings	Comber, County Down	1. 7.03P
	(Built Aerotech International Ltd) (Full Lotus floats)						
G-MYAO	Mainair Gemini/Flash IIA	894-0392-7 & W689		11. 3.92	K Fowler *(New owner 9.04)*	(Edinburgh)	9. 8.05P
G-MYAP	Thruster T.300	9022-T300-501		12. 3.92	G.Bennett	Blofield	29. 3.05P
	(Built Tempest Aviation Ltd) (Rotax 582)						
G-MYAR	Thruster T.300	9022-T300-502		12. 3.92	H.G.Denton	Knapthorpe Lodge, Caunton	2. 9.05P
	(Built Tempest Aviation Ltd) (Rotax 503)						
G-MYAS	Mainair Gemini/Flash IIA	895-0392-7 & W690		11. 3.92	A N.Duncanson	Redlands	12. 6.04P
G-MYAT	TEAM mini-MAX 88	PFA 186-12017		6. 3.92	M.A Perry	Elm Farm, Wickford	19. 5.04P
	(Built D M Couling)				*(Noted wrecked 10.03)*		
G-MYAU	Mainair Gemini/Flash IIA	890-0392-7 & W685		25. 3.92	P.P.Allen	(Ely)	22. 4.05P
	(Rotax 462)						
G-MYAV	Mainair Mercury	893-0392-5 & W688		23. 3.92	J.Lynch	East Fortune	26. 5.04P
G-MYAY	Corbett Farms Spectrum	027		13. 3.92	P F Craggs *(Noted 1.05)*	Eshott	21.12.00P
G-MYAZ	Meridian Renegade Spirit UK	PFA 188-12027		16. 3.92	R.Smith	Kilkerran	10.10.03P
	(Built R.Smith)						
G-MYBA	Rans S-6-ESD Coyote II	PFA 204-12210		12. 3.92	M.R.Cann	Dunkeswell	29.10.05P
	(Built S M Vickers) *(Tail-wheel u/c)*				tr Climsland Climber Society		
G-MYBB	Maxair Drifter	BMAA/HB/014		10. 4.92	M.Ingleton	(Sheerness)	12. 6.92P
	(Built M.Ingleton - c/n MD.001) (Rotax 503)		*(Initially imported by Medway Microlights: on rebuild under BMAA number as "Ingleton Skylark")*				
G-MYBC	CFM Shadow Series CD	BMAA/HB/047		18. 3.92	M.E.Gilbert	Drummaird Farm, Bonnybank	24. 4.04P
	(Originally regd with c/ns K.195 & PFA 206-12221 - PFA project no. indicates Streak Shadow)						
G-MYBD	Solar Wings Pegasus Quasar IITC			26. 3.92	A.Gunn	Abbey Warren Farm, Bucknall, Lincoln	5. 7.07P
		SW-TQC-0094 & SW-WQT-0511					
G-MYBE	Solar Wings Pegasus Quasar IITC			26. 3.92	D.Lumsdon	Eshott	25. 8.03P
		SW-TQC-0095 & SW-WQT-0512			*(New owner 1.04)*		
G-MYBF	Solar Wings Pegasus XL-Q			26. 3.92	K.H.Pead	(Ipswich)	25. 2.04P
		SW-TE-0384 & SW-WQ-0513					
G-MYBG	Solar Wings Pegasus XL-Q			26. 3.92	P.A Henretty and M.Aylett	(Northampton)	9. 7.03P
		SW-TE-0385 & SW-WQ-0514					
G-MYBI	Rans S-6-ESD Coyote II	PFA 204-12186		26. 3.92	N.C.Tambiah	Otherton, Cannock	25 7.05P
	(Built G A Archer and J A Soilleux - c/n 1291.249) (Tri-cycle u/c)						
G-MYBJ	Mainair Gemini/Flash IIA	908-0593-7 & W706		2. 4.92	G.C.Bowers	(Shipmeadow, Beccles)	24. 5.05P
	(Rotax 462)						
G-MYBL	CFM Shadow Series CD	K.194		2. 4.92	J.P.Pullin *(Noted 7.04)*	Weston Zoyland	12. 3.04P
G-MYBM	TEAM mini-MAX 91	PFA 186-12212		3. 4.92	B Hunter	Brook Farm, Pilling	18. 2.05P
	(Built M K Dring) (Mosler MM CB-35)						
G-MYBN	Hiway Skytrike II/Demon 175	BRL-01		14. 4.92	B.R.Lamming *(Current status unknown)*	Seaton, Hull	
G-MYBO	Solar Wings Pegasus XL-R			16. 4.92	D Gledhill	(London SW19)	27. 1.03P
		SW-TB-1445 & SW-WA-1545					
G-MYBP	Solar Wings Pegasus XL-R/Se			16. 4.92	S.H.Williams	(Kidderminster)	2. 8.03P
		SW-TB-1446 & SW-WA-1546					
G-MYBS	Solar Wings Pegasus XL-Q			16. 4.92	T.Smith	(Challock, Ashford)	28. 7.05P
		SW-TE-0387 & SW-WQ-0518					
G-MYBT	Solar Wings Pegasus Quasar IITC			16. 4.92	G.A Rainbow-Ockwell	Redlands	4. 9.05P
		SW-TQC-0097 & SW-WQT-0519					
G-MYBU	Cyclone Chaser S 447	CH.837	G-69-15	28. 4.92	R.L.Arscott	Longacres Farm, Sandy	21. 1.00P
			G-MYBU		*(Noted 7.03)*		
G-MYBV	Solar Wings Pegasus XL-Q			5. 5.92	P.M.Langdon	(Coniston)	21. 5.05P
		SW-TE-0393 & SW-WQ-0522					
G-MYBW	Solar Wings Pegasus XL-Q			5. 5.92	J.S.Chapman	Baxby Manor, Husthwaite	15. 9.05P
		SW-TE-0394 & SW-WQ-0523					
G-MYBY	Solar Wings Pegasus XL-Q			5. 5.92	I.D.A Spanton	(Malvern)	18. 4.05P
		SW-TE-0396 & SW-WQ-0525			*(New owner 1.04)*		
G-MYBZ	Solar Wings Pegasus XL-Q			5. 5.92	A.J.Blackwell	Long Marston	27. 9.97P
		SW-TE-0397 & SW-WQ-0526			*(New owner 8.04)*		
G-MYCA	Whittaker MW6-T Merlin	PFA 164-11821		14. 5.92	R.A L.Harris	(Andover)	25. 3.05P
	(Built N B Morley and E Barfoot) (Rotax 532)						
G-MYCB	Cyclone Chaser S 447	CH.839		18. 5.92	P.Sykes	Newton Peverill	14. 5.05P
G-MYCE	Solar Wings Pegasus Quasar IITC			14. 5.92	S.W.Barker	(Scarborough)	27. 7.04P
		SW-TQC-0098 & SW-WQT-0527					
G-MYCJ	Mainair Mercury	906-0592-5 & W704		19. 5.92	J.Agnew	East Fortune	28.11.04P
G-MYCK	Mainair Gemini/Flash IIA	909-0592-7 & W707		19. 5.92	J.P.Hanlon and A C.McAllister	Ince Blundell	5.12.04P
	(Rotax 462)						
G-MYCL	Mainair Mercury	910-0592-5 & W708		19. 5.92	Palladium Leisure Ltd	RAF Wyton	13. 9.05P
G-MYCM	CFM Shadow Series CD	196		20. 5.92	T.Jones *(Current status unknown)*	Redhill	20. 5.99P
G-MYCN	Mainair Mercury	901-0492-5 & W696		22. 5.92	P Lowham	Newtownards	14. 9.05P

Reg	Type	C/n	Previous identity	Regd	Owner/operator	Probable base	CA expiry
G-MYCO	Meridian Renegade Spirit UK *(Built C Slater)*	PFA 188-12020		28. 5.92	M.J.Downes	Pool Quay, Breidden	1. 6.05P
G-MYCP	Whittaker MW6 Merlin *(Built R M Clarke)* (Rotax 532)	PFA 164-11505		2. 6.92	A CJones	Otherton, Cannock	4.11.05P
G-MYCR	Mainair Gemini/Flash IIA	875-0192-7 & W670		10. 6.92	A P.King	Long Marston	14. 2.05P
G-MYCS	Mainair Gemini/Flash IIA	911-0592-7 & W710		12. 6.92	G.Penson tr Husthwaite Alpha Group	Baxby Manor, Husthwaite	13.10.05P
G-MYCT	TEAM mini-MAX 91 *(Built M A Curant)*	PFA 186-12163		30. 3.92	D.D.Rayment	(Horsham)	17. 6.04P
G-MYCU	Whittaker MW6 Merlin *(Built P L Lonsdale and D Shackleton)* (Rotax 532) *(PFA Plans No.duplicates Streak Shadow G-ORAF)*	PFA 164-11627		9. 6.92	R.D.Thomasson	London Colney	11. 6.04P
G-MYCV	Mainair Mercury	913-0792-5 & W712		12. 6.92	D.P.Creedy	Guy Lane Farm, Waverton	8. 6.05P
G-MYCW	Powerchute Kestrel	00420		15. 6.92	C.D.Treffers	(Basildon)	30. 4.02P
G-MYCX	Powerchute Kestrel	00421		15. 6.92	S.J.Pugh-Jones	(Kidwelly)	14. 5.05P
G-MYCY	Powerchute Kestrel)	00422		15. 6.92	D.R.M.Powell	(Llandysul)	23. 5.05P
G-MYDA	Powerchute Kestrel	00424		15. 6.92	K.J.Greatrix	(Sleaford)	14. 5.05P
G-MYDC	Mainair Mercury	916-0792-5 & W715		23. 6.92	D.J.Boylan and D.Gordon	Rufforth	18. 4.05P
G-MYDD	CFM Shadow Series CD	K.197		22. 6.92	C.H.Gem *(Current status unknown)*	(Marbella, Spain)	22.11.95P
G-MYDE	CFM Shadow Series CD	K.187		24. 6.92	D.N.L.Howell	Upper Colwall)	19. 9.05P
G-MYDF	TEAM mini-MAX 91 *(Built L G Horne)*	PFA 186-12129		24. 6.92	A.M.Hughes	(Esher)	6. 5.05P
G-MYDI	Solar Wings Pegasus XL Tug (Rotax 462HP)	SW-TE-0402 & SW-WA-1557	*(Officially reg as XL-R)*	26. 6.92	M.D.Howard *(New owner 7.03)*	(Great Yarmouth)	3. 3.03P
G-MYDJ	Solar Wings Pegasus XL Tug (Rotax 462)	SW-TE-0403 & SW-WA-1558	*(Officially reg as XL-R)*	1. 7.92	R C Wood & D C Richardson tr Cambridgeshire Aerotow Club	Sutton Meadows, Ely	24. 4.05P
G-MYDK	Rans S-6-ESD Coyote II *(Built D N Kershaw - c/n 0392.276) (Tri-cycle u/c)*	PFA 204-12239		21. 4.92	J.W.Caush	Eshott	26. 8.05P
G-MYDM	Whittaker MW6-S Fatboy Flyer (Rotax 582)	PFA 164-12105		26. 6.92	K.Gregan	Kilrush, County Kildare	4. 7.05P
G-MYDN	Quad City Challenger II UK *(Built T C Hooks - c/n CH2-1091-UK-0736)* (Rotax 462)	PFA 177-12245		30. 6.92	T.C.Hooks	Newtownards	9. 6.05P
G-MYDO	Rans S-5 Coyote *(Built W A Stevens - c/n 89.110)*	PFA 193-12274		6. 7.92	B.J.Benton	Long Marston	24. 8.05P
G-MYDP	Kolb Twinstar Mk.3 *(Built P M Standen - kit no K0002-1291)* (Rotax 503)	PFA 205-12231		15. 7.92	J.W.Bryan *(New owner 7.03)*	(Rickmansworth)	6. 8.02P
G-MYDR	Thruster T.300 *(Built Tempest Aviation Ltd)* (Rotax 582)	9072-T300-505		21. 7.92	H.G.Soper	Chiddingley, East Sussex	9. 8.05P
G-MYDS	Quad City Challenger II UK *(Built D D Smith - c/n CH2-1289-UK-0500)*	PFA 177-11716		6. 3.90	L.R.Graham *(Damaged in forced-landing following engine failure after take-off Southend 19.1.05)*	Southend	27. 8.05P
G-MYDU	Thruster T.300 *(Built Tempest Aviation Ltd)* (Rotax 582)	9072-T300-504		21. 7.92	S.Collins	(Drogheda, County Louth)	23. 5.05P
G-MYDV	Mainair Gemini/Flash IIA (Rotax 462)	917-0892-7 & W716		29. 7.92	M.Griffiths	(Blackpool)	5.12.05P
G-MYDW	Whittaker MW6 Merlin *(Built W R G West)* (Rotax 503)	PFA 164-12184		27. 7.92	A Chidlow	Askern	2. 9.04P
G-MYDX	Rans S-6-ESD Coyote II *(Built R.J.Goodburn) (Tri-cycle u/c)*	PFA 204-12238		27. 7.92	R.J.Goodburn *"The Ruptured Duck"*	Spanhoe	8. 6.05P
G-MYDZ	Mignet HM-1000 Balerit	66		3. 8.92	D S Simpson	Graveley	8. 8.05P
G-MYEA	Solar Wings Pegasus XL-Q	SW-TE-0404 & SW-WQ-0537		28. 7.92	A M.Taylor	Long Marston	10.11.05P
G-MYEC	Solar Wings Pegasus XL-Q	SW-TE-0406 & SW-WQ-0539		28. 7.92	J.I.King	(Bath)	2. 6.05P
G-MYED	Solar Wings Pegasus XL-R (Rotax 462HP)	SW-TE-0407 & SW-WA-1559		28. 7.92	P.K.Dale	Bagby	28. 3.05P
G-MYEG	Solar Wings Pegasus XL-R	SW-TB-1447 & SW-WA-1560		4. 8.92	D.R.Mortlock	(Maldon)	11. 8.04P
G-MYEH	Solar Wings Pegasus XL-R	SW-TB-1448 & SW-WA-1561		4. 8.92	P.G.Strangward tr G-MYEH Flying Group	Roddige	18. 6.05P
G-MYEI	Cyclone Chaser S 447	CH.841		18. 8.92	A C Parsons	Kingston Seymour	12. 8.04P
G-MYEJ	Cyclone Chaser S 447	CH.842		18. 8.92	D.A Cochrane	Newnham, Baldock	3. 4.00P
G-MYEK	Solar Wings Pegasus Quasar IITC (Rotax 582)	SW-TQD-0108 & SW-WQT-0540	*(See G-MWZF)*	7. 8.92	B.A McWilliams	Long Marston	11.11.04P
G-MYEM	Solar Wings Pegasus Quasar IITC (Rotax 582)	SW-TQD-0101 & SW-WQT-0542		7. 8.92	D.J.Moore	Oakington	22. 4.05P
G-MYEN	Solar Wings Pegasus Quasar IITC (Rotax 582)	SW-TQD-0105 & SW-WQT-0543		7. 8.92	J.C.Higham	(Willenhall)	27. 3.05P
G-MYEO	Solar Wings Pegasus Quasar IITC (Rotax 582)	SW-TQD-0106 & SW-WQT-0544		7. 8.92	A G.Curtis	Deenethorpe	21. 7.05P
G-MYEP	CFM Shadow Series CD	K.205		13. 8.92	J.S.Seddon-Harvey	Broadmeadow Farm, Hereford	8. 6.04P
G-MYER	Cyclone AX2000	B.1052901 & CA.001	G-69-27 G-MYER, G-69-5, 59-GF	19. 8.92	B G de Wert tr Caithness Flying Group	Wick	21. 7.05P
G-MYES	Rans S-6-ESD Coyote II *(Built G R Pritchard - c/n 0392.283) (Tri-cycle u/c)*	PFA 204-12254		3. 7.92	F.J.Percival tr Dairy House Flyers	Dairy House Farm, Worleston	9. 7.05P
G-MYET	Whittaker MW6 Merlin *(Built M B Haine)* (Rotax 503)	PFA 164-12318		19. 8.92	G.Campbell	Newton Peverill Farm, Sturminster Marshall	8. 9.04P
G-MYEU	Mainair Gemini/Flash IIA	918-0892-7 & W718		1. 9.92	G.J.Webster	Mill Farm, Shifnal	27. 5.05P
G-MYEV	Whittaker MW6 Merlin *(Built M M Ruck)*	PFA 164-11250		25. 8.92	M.J.Batchelor *(Noted 2.03)*	Wickwar	
G-MYEW	Powerchute Kestrel *(Fitted with Harley Ram-air parachute)*	00417		28. 8.92	J.Weston	(Grantham)	13. 7.02P
G-MYEX	Powerchute Kestrel	00426		28. 8.92	R.S.McFadyen	(Tamworth)	17. 5.05P
G-MYFA	Powerchute Kestrel	00429		28. 8.95	D.A Gardner *(Noted 12.03)*	Perth	23. 3.98P
G-MYFH	Quad City Challenger II UK *(Built C J and R J Lines - c/n CH2-0292-0798)*	PFA 177-12282		9. 9.92	P.S.Fossey	(Bristol)	21.11.05P

G-MYFI	Cyclone AX3/503	C.3093159 & CA.002		9. 9.92	C.Childs	Middle Stoke, Isle of Grain	11. 5.05P
					t/a Medway Airsports		
G-MYFK	Solar Wings Pegasus Quasar IITC			11. 9.92	R.L.Harris	(Ilford)	8.11.04P
	(Rotax 582)	SW-TQD-0113 & SW-WQT-0553					
G-MYFL	Solar Wings Pegasus Quasar IITC			11. 9.92	S.B.Wilkes	Roddige	13. 6.05P
	(Rotax 582)	SW-TQD-0103 & SW-WQT-0541/A	*(Originally regd as c/n SW-WQT-0554: replacement wing fitted to trike G-MYEL after wing stolen 1.1.93)*				
G-MYFM	Meridian Renegade Spirit UK	PFA 188-12249		9. 9.92	A C.Cale	Marley Hall, Ledbury	7. 1.05P
	(Built J M Walsh)						
G-MYFN	Rans S-5 Coyote	PFA 193-12273		16. 9.92	P.Doran	(Monaghan, County Monaghan)	7. 8.03P
	(Built L Kellner - c/n 89.112)						
G-MYFO	Cyclone Chaser S	CH.843		22. 9.92	D.M.Broom	(Towcester)	29. 5.02P
	(Rotax 377)						
G-MYFP	Mainair Gemini/Flash IIA	920-0992-7 & W719		2.10.92	J.S.Hill and G.H.S.Skilton	(Stone)	7. 9.05P
G-MYFR	Mainair Gemini/Flash IIA	921-0992-7 & W720		30. 9.92	P.G.Bright	(Hull)	17. 7.05P
G-MYFS	Solar Wings Pegasus XL-R			30. 9.92	D.J.Middleton	(Ilkeston)	26. 9.05P
		SW-TB-1453 & SW-WA-1564					
G-MYFT	Mainair Scorcher	922-0992-3 & W234		30. 9.92	N.Crowther-Wilton	Enstone	22. 1.05P
	(Rotax 503)						
G-MYFU	Mainair Gemini/Flash IIA	924-1092-7 & W722		7.10.92	J.Payne	(Bretton, Peterborough)	23. 8.05P
	(Rotax 462)						
G-MYFV	Cyclone AX3/503	C.2083050		6.10.92	J.K.Sargent	Middle Stoke, Isle of Grain	28.10.05P
G-MYFW	Cyclone AX3/503	C.2083051		13.10.92	The Microlight School (Lichfield) Ltd	Roddige	19. 1.05P
G-MYFX	Solar Wings Pegasus XL-Q			25. 6.93	J R Bluett	(Ware)	24. 8.04P
		SW-TE-0295 & SW-WQ-0378					
G-MYFY	Cyclone AX3/503	C.2083047		1.10.92	T.A Simpson	Longacres Farm, Sandy	15. 7.03P
G-MYFZ	Cyclone AX3/503	C.2083048		20.10.92	M.L.Smith tr Buzzard Flying Group	Popham	6.12.04P
G-MYGD	Cyclone AX3/503	C.2083049		21.10.92	D.Young tr Kemble Flying Club	Kemble	3.12.05P
G-MYGE	Whittaker MW6 Merlin	PFA 164-11650		20.10.92	M.D. and S.M.North	Manor Farm, Croughton	24. 6.97P
	(Built M D North) (Rotax 532)				*(Current status unknown)*		
G-MYGF	TEAM mini-MAX 91	PFA 186-12175		22.10.92	R.D.Barnard	Ley Farm, Chirk	31. 7.01P
	(Built R.D.Barnard)						
G-MYGH	Rans S-6-ESD Coyote II	PFA 204-12335		30.10.92	L.J.Field	(Peacehaven)	24. 5.05P
	(Built S M Hall - c/n 0692.318)						
G-MYGJ	Mainair Mercury	923-0992-7 & W721		5.10.92	J.R.Harnett	(Whitmore, Newcastle)	13. 1.05P
G-MYGK	Cyclone Chaser S 508	CH.846		3.11.92	P.C.Collins *(Current status unknown)*	(Bath)	14.11.95P
G-MYGM	Quad City Challenger II UK	PFA 177-12261		6.11.92	J.White and G.J.Williams	Mill Farm, Shifnall	21. 9.04P
	(Built R Holt - c/n CH2-0391-UK-0662)						
G-MYGN	AMF Microflight Chevvron 2-32C	034		29.12.92	A C Barber	(Newmarket)	10. 5.05P
G-MYGO	CFM Shadow Series CD	K.114		28. 7.92	R.C.S.Mason	Sywell	6. 9.04P
G-MYGP	Rans S-6-ESD Coyote II	PFA 204-12368		10.11.92	A N.Hughes	Weston Zoyland	22. 8.05P
	(Built J S Melville - c/n 0992.349) (Tail-wheel u/c)						
G-MYGR	Rans S-6-ESD Coyote II	PFA 204-12378		16.11.92	E.B.Atalay	Longacres Farm, Sandy	9. 5.04P
	(Built D K Haughton)				*(New owner 12.04)*		
G-MYGT	Solar Wings Pegasus XL Tug			13.11.92	J.J.Hoer	Dunkeswell	29. 3.04P
	(Rotax 462)	SW-TE-0413 & SW-WA-1569	*(Officially reg as XL-R)*		tr Condors Aerotow Syndicate		
G-MYGU	Solar Wings Pegasus XL-R			13.11.92	D.R.Western	Weston Zoyland	16. 5.05P
	(Rotax 462)	SW-TE-0414 & SW-WA-1570					
G-MYGV	Solar Wings Pegasus XL Tug			13.11.92	D.J.Brixton	Leebotwood	4. 8.05P
	(Rotax 462HP)	SW-TE-0415 & SW-WA-1571	*(Officially reg as XL-R)*		tr Shropshire Tow Group		
G-MYGZ	Mainair Gemini/Flash IIA	928-1192-7 & W726		18.11.92	P.M.Reddington	(Ormskirk)	7. 6.05P
	(Rotax 582)						
G-MYHF	Mainair Gemini/Flash IIA	929-1092-7 & W727		25.11.92	P.J.Bloor	(Knutsford)	6.10.03P
G-MYHG	Cyclone AX3/503	C.2103070		27.11.92	I.McDiarmid tr G-MYHG Flying Group	Strathaven	2.12.05P
G-MYHH	Cyclone AX3/503	C.2103069 & CA.006		30.11.92	M.L.Smith	Popham	23. 6.05P
G-MYHI	Rans S-6-ESD Coyote II	PFA 204-12279		8.12.92	D.K.Anderson	Otherton, Cannock	9.11.05P
	(Built L N Anderson c/n 0692.312) (Tailwheel u/c)						
G-MYHJ	Cyclone AX3/503	C.2103073		11.12.92	J.G.Campbell	(Barnsley)	17. 6.05P
	(Reported as keel tube c/n C.3093157 - see G-MYME)						
G-MYHK	Rans S-6-ESD Coyote II	PFA 204-12349		3.12.92	M.R.Williamson	Sutton Meadows, Ely	5. 8.04P
	(Built J M Longley - c/n 0692.311) (Tri-cycle u/c)						
G-MYHL	Mainair Gemini/Flash IIA	932-0193-7 & W730		21.12.92	P.J.Lomax and J.A Robinson	St.Michaels	14. 2.05P
G-MYHM	Cyclone AX3/503	C.2103068 & CA.007		18.12.92	A J.Bergman	Popham	16. 8.05P
G-MYHN	Mainair Gemini/Flash IIA	933-0193-7 & W731		29.12.92	T.J.Widdison	(King's Lynn)	26. 9.05P
	(Rotax 582)						
G-MYHP	Rans S-6-ESD Coyote II	PFA 204-12406		8. 1.93	S F Winter	Redlands	11.11.05P
	(Built D A Crompton - c/n 0892.315) (Tri-cycle u/c)				*"Grass Stripper"*		
G-MYHR	Cyclone AX3/503	C.2103071	G-68-8	15. 1.93	G.Humphrey	Long Marston	27. 6.05P
			G-MYHR		tr G-MYHR Flying Group		
G-MYHS	Powerchute Kestrel	00433		26. 1.93	R.Kent	(Newark)	6. 4.01P
	(Frame No.00433/Parachute No.931013/Engine No.4104716)						
G-MYHX	Mainair Gemini/Flash IIA	930-1292-7 & W728		2.12.92	C.P.Simmons	(London N10)	23. 4.01P
	(Rotax 582)						
G-MYIA	Quad City Challenger II UK	PFA 177-12400		21. 1.93	I.J.Arkieson	Ley Farm, Chirk	3.10.00P
	(Built I.J.Arkieson)						
G-MYIE	Whittaker MW6-S Fatboy Flyer	PFA 164-11800		26. 1.93	K.Gair and J.G.E.Lane	Brock Farm, Billericay	29. 7.05P
	(Built P A Mercer) (Rotax 532)						
G-MYIF	CFM Shadow Series CD	217		2. 2.93	L.Walsh	(Huntingdon)	18. 4.04P
G-MYIH	Mainair Gemini/Flash IIA	937-0293-7 & W734		9. 3.93	W.Purcell	Weston Zoyland	27. 5.04P
	(Rotax 582)				*(Noted 10.04)*		
G-MYII	TEAM mini-MAX 91	PFA 186-12119		10.11.92	P.A Gasson	(Woking)	8. 3.05P
	(Built G W Peacock) (Mosler MM CB-40)						
G-MYIJ	Cyclone AX3/503	C.2103072		8. 2.93	Ultralight Training Ltd	(Coventry)	28. 2.05P
G-MYIK	Kolb Twinstar Mk.3	PFA 205-12220		13. 1.93	J.Latimer	Barton	10. 8.05P
	(Built R P Smith)						

Registration	Type	C/n	Previous identity	Regd	Owner/operator	Base	CofA
G-MYIL	Cyclone Chaser S 508	CH.849		3. 3.93	R.A Rawes *"Fricky"*	Over Farm, Gloucester	14. 2.05P
G-MYIM	Solar Wings Pegasus Quasar IITC		(EI-...)	22. 2.93	D.Forde	Clare Galway, County Galway	31. 8.05P
	(Rotax 582)	SW-TQD-0122 & SW-WQT-0579	G-MYIM				
G-MYIN	Solar Wings Pegasus Quasar IITC			22. 2.93	W.P.Hughes	RAF Henlow	1. 8.05P
	(Rotax 582)	SW-TQD-0123 & SW-WQT-0580					
G-MYIO	Solar Wings Pegasus Quasar IITC			22. 2.93	E Foster and J H Peet	(Leyland/Preston)	19. 8.05P
	(Rotax 582)	SW-TQD-0124 & SW-WQT-0581					
G-MYIP	CFM Shadow Series CD	K.198		16. 3.93	S.Marshall and A Halsall	(Southport)	22.11.04P
G-MYIR	Rans S-6-ESD Coyote II	PFA 204-12458		17. 3.93	P.Vergette	North Coates	17.11.05P
	(Built J Simpson - c/n 0892.344) (Tri-cycle u/c)						
G-MYIS	Rans S-6-ESD Coyote II	PFA 204-12382		31.12.92	J.Bolton and R.Shewan	Longside, Peterhead-	13. 4.05P
	(Built A J Wyatt) (Tri-cycle u/c)						
G-MYIT	Cyclone Chaser S 508	CH.850		19. 3.93	R.Barringer	Ravensthorpe, Northampton	28. 3.99P
					(Current status unknown)		
G-MYIU	Cyclone AX3/503	C.3013084		22. 3.93	G.R.Hill	Mullaghmore	8. 7.05P
	(Zanzoterra Z-202) *(Officially regd with Rotax 503)*						
G-MYIV	Mainair Gemini/Flash IIA	938-0393-7 & W735		30. 3.93	P.S.Nicholls	Finmere	15. 6.05P
	(Rotax 582)						
G-MYIX	Quad City Challenger II UK	PFA 177-12260		5. 1.93	A Studley	(Crewkerne)	3. 6.04P
	(Built I M Walton - c/n CH2-0191-UK-0615)						
G-MYIY	Mainair Gemini/Flash IIA	942-0493-7 & W737		1. 4.93	I.C.Macbeth	Arclid Green, Sandbach	22. 6.00P
G-MYIZ	TEAM mini-MAX 91	PFA 186-12347		31. 3.93	S.E.Richardson	Escrick, York	26. 5.03P
	(Built J C Longmore)						
G-MYJB	Mainair Gemini/Flash IIA	943-0593-7 & W738		7. 4.93	S.J.P.Gibbs	(Aberaeron)	6 6.05P
G-MYJC	Mainair Gemini/Flash IIA	944-0593-7 & W739		7. 4.93	R G Hearsey	(Rye)	14. 4.04P
	(Rotax 462)						
G-MYJD	Rans S-6-ESD Coyote II	PFA 204-12360		23. 4.93	D.R.Collier	(Staines)	5. 8.05P
	(Built D J Dimmer and B Robins - c/n 0792.324) (Tail-wheel u/c)						
G-MYJF	Thruster T.300	9013-T300-509		14. 4.93	B.McConville	(Craigavon, County Armagh)	31. 8.05P
	(Built Tempest Aviation Ltd) (Rotax 582)						
G-MYJG	Thruster Super T.300	9043-ST300-510		14. 4.93	S.P.Read and Marjorie H.Wise	Weston Zoyland	17. 9.05P
	(Built Tempest Aviation Ltd) (Rotax 582) *(Single-seat conversion)*						
G-MYJJ	Solar Wings Pegasus Quasar IITC			27. 4.93	D.Murray	(Clevedon)	1. 9.04P
	(Rotax 582)	SW-TQD-0131 & SW-WQT-0591					
G-MYJK	Solar Wings Pegasus Quasar IITC			27. 4.93	B Hall	Tarn Farm, Cockerham	28. 4.05P
	(Rotax 582)	SW-TQD-0132 & SW-WQT-0592	*(Built with Cyclone c/n 6752)*				
G-MYJM	Mainair Gemini/Flash IIA	945-0593-7 & W740		29. 4.93	S.McMaster	Newtownards	28.12.04P
	(Rotax 582)						
G-MYJO	Cyclone Chaser S 508	CH.851		30. 4.93	S.B.Brady	(Sandbach)	5. 7.05P
G-MYJP*	Murphy Renegade Spirit UK	PFA 188-12045		3. 4.91	E.J.Pels	Swinford, Rugby	21. 2.03P
	(Built J W E Pearson - c/n 357)				*(Cancelled 10.6.03 as sold to France but noted 9.04)*		
G-MYJR	Mainair Mercury	947-0593-7 & W742		12. 5.93	C.J.Johnson	(Hornchurch)	15. 7.05P
G-MYJS	Solar Wings Pegasus Quasar IITC	6581		19. 5.93	P.R.Saunders	Longacres Farm, Sandy	1. 1.05P
	(Rotax 582)						
G-MYJT	Solar Wings Pegasus Quasar IITC	6582		19. 5.93	I.M.Bunce	Perth	4. 6.05P
	(Rotax 582)						
G-MYJU	Solar Wings Pegasus Quasar IITC	6573		19. 5.93	P.G.Penhaligan	Rochester	30. 5.05P
	(Rotax 582)						
G-MYJW	Cyclone Chaser S 508	CH.856		19. 5.93	A R.Mikolajczyk	Headon Farm, Retford	22. 5.05P
G-MYJX*	Whittaker MW.8	PFA/243-12345		24. 5.93	Not known	Askern	
	(Built M W Whittaker - c/n 001) (Rotax 508)			*(Cancelled 31.7.01 by CAA - no PtoF believed issued) (Stored in a trailer 4.04)*			
G-MYJY	Rans S-6-ESD Coyote II	PFA 204-12346		24. 5.93	F.N.Pearson	Linley Hill, Leven	28. 7.03P
	(Build G A Clayton - c/n 0692.317) (Tri-cycle u/c)						
G-MYJZ	Whittaker MW5-D Sorcerer	PFA 163-12385		22. 4.93	P.A Aston	(Newton Abbot)	2.12.99P
	(Built J G Beesley)				*(Current status unknown)*		
G-MYKA	Cyclone AX3/503	C.3013086		25. 5.93	Rosemary Nicklin	Othertton, Cannock	14. 2.05P
G-MYKB	Kolb Twinstar Mk.3	PFA 205-12398		31. 3.93	D.Young	Eastbach Farm, Coleford	1. 9.00P
	(Built J D Holt - kit no K0007-0193)						
G-MYKC	Mainair Gemini/Flash IIA	948-0593-7 & W743		26. 5.93	R.W.Lenthall	Headon Farm, Retford	26.12.04P
	(Rotax 582)						
G-MYKD	Cyclone Chaser S 508	CH.857		26. 5.93	J.V.Clewer	(Ashford)	30.10.05P
G-MYKE	CFM Shadow Series BD	K.031		14. 1.88	M.Hughes	Emlyn's Field, Rhuallt	26.10.96P
					t/a MKH Engineering *(Noted 4.04)*		
G-MYKF*	Cyclone AX3/503	C.3013083		8. 6.93	P.Jones	Brook Farm, Pilling	10. 7.01P
	(Rotax 503)				*(Cancelled 1.4.03 by CAA) (Dismantled and stored 9.03:*		
G-MYKG	Mainair Gemini/Flash IIA	950-0693-7 & W745		21. 6.93	P.G.Angus	Higher Barn Farm, Houghton	30.12.04P
	(Rotax 582)						
G-MYKH	Mainair Gemini/Flash IIA	951-0693-7 & W746		21. 6.93	G.F.Atkinson	Rufforth	9.11.04P
	(Rotax 582)						
G-MYKI	Mainair Mercury	953-0693-7 & W748		21. 6.93	R.J.Pattinson and R.C.Matthews	(Portsmouth)	27. 7.05P
					tr Three Counties Microlight Group		
G-MYKJ	TEAM mini-MAX	PFA 186-12215		10. 6.93	P.I.Frost	Guilsborough, Northampton	18. 6.04P
	(Built M Hill) (Rotax 508)						
G-MYKL	Medway Raven X	MRB116/104		6. 7.93	S.Hutchinson	RAF Wyton	24. 1.05P
G-MYKN	Rans S-6-ESD Coyote II	PFA 204-12361		23. 6.93	G.C.Alderson	Lower Mountpleasant, Chatteris	7. 9.05P
	(Built S E Hartles - kit no.0892.338) (Tri-cycle u/c)						
G-MYKO	Whittaker MW6-S Fatboy Flyer	PFA 164-11919		25. 6.93	K.R.Challis and C.S.Andersson	(Sutton)	23. 8.05P
	(Hirth 2706)						
G-MYKP	Solar Wings Pegasus Quasar IITC	6627		7. 7.93	R F Dye and G S B Airth	Perth	24.10.04P
	(Rotax 582)						
G-MYKR	Solar Wings Pegasus Quasar IITC	6635		7. 7.93	C.Stallard	Larkins Farm, Laindon	19. 9.04P
	(Rotax 582)						
G-MYKS	Solar Wings Pegasus Quasar IITC	6636		7. 7.93	N Groome	Hunsdon	12. 4.05P
	(Rotax 582) *(Now fitted with new trike c/n 6780)*						

Reg	Type	C/n	Previous identity	Date	Owner	Location	Expiry
G-MYKT	Cyclone AX3/503	C.3013082		5. 7.93	P.J.Hepburn	Middle Stoke, Isle of Grain	13. 9.05P
G-MYKV	Mainair Gemini/Flash IIA	954-0793-7 & W749		13. 7.93	P.J.Gulliver	Mill Farm, Shifnal	29. 9.05P
G-MYKW	Mainair Mercury	960-0893-7 & W755		9. 7.93	C.Foster *(New owner 6.04)*	Eshott	5. 8.05P
G-MYKX	Mainair Mercury	961-0893-7 & W756		3. 9.93	B.W.Hunter	(Penicuik)	4. 7.05P
G-MYKY	Mainair Mercury	962-0893-7 & W757		6. 8.93	R.P.Jewit	(York)	9.11.03P
G-MYKZ	TEAM mini-MAX 91	PFA 186-11841	G-BVAV	26. 7.93	G H Hills	(Medstead, Alton)	29. 7.05P
	(Built P A Ellis) (Rotax 503)						
G-MYLB	TEAM mini-MAX 91	PFA 186-12419		2. 8.93	J.G.Burns	Morgansfield, Fishburn	10.12.04P
	(Built P Harvey) (Rotax 532)						
G-MYLC	Pegasus Quantum 15	6634		9. 8.93	G.C.Weighell	Enstone	24. 7.05P
	(Rotax 503)				tr G-MYLC Microlight Group		
G-MYLD	Rans S-6-ESD Coyote II	PFA 204-12394		1. 3.93	E.Griffin	(Limerick, County Limerick)	27.10.05P
	(Built L R H d'Eath) (Tail-wheel u/c)						
G-MYLE	Pegasus Quantum 15	6609		9. 8.93	Susan E.Powell	Enstone	21. 9.05P
	(Rotax 503)						
G-MYLF	Rans S-6-ESD Coyote II	PFA 204-12544		4. 8.93	A J Spencer	(Swayfield, Grantham)	22. 7.05P
	(Built G R and J A Pritchard - c/n 0493.483) (Tri-cycle u/c)				*"Low Flyer"*		
G-MYLG	Mainair Gemini/Flash IIA	959-0893-7 & W754		6. 8.93	J.D.and N.G.Philp	Redlands	7. 6.04P
G-MYLH	Pegasus Quantum 15	6632		27. 8.93	J.W.Atkin	(Stoke-on-Trent)	22. 7.05P
	(Rotax 503)						
G-MYLI	Pegasus Quantum 15	6645		11. 8.93	A M.Keyte	(West Wickham)	26. 5.05P
	(Rotax 503)						
G-MYLJ	Cyclone Chaser S 447	CH.858		24. 8.93	A Gunn	North Coates	13. 7.04P
G-MYLK	Pegasus Quantum 15	6602		27. 8.93	C.L.Minter	Deenethorpe	24. 9.04P
	(Rotax 503)				tr G-MYLK Group		
G-MYLL	Pegasus Quantum 15	6650		31. 8.93	C.M.Wilkinson and T.E.Tomlinson	Wickenby	23. 5.05P
	(Rotax 462HP)						
G-MYLM	Pegasus Quantum 15	6651	(EC-)	31. 8.93	P.A.Ashton	(Lincoln)	10.11.05P
	(Rotax 582)		G-MYLM				
G-MYLN	Kolb Twinstar Mk.3	PFA 205-12430		3. 9.93	C.D.Hatcher	Deenethorpe	27. 7.04P
	(Built G E Collard - kit no K0010-0193)						
G-MYLO	Rans S-6-ESD Coyote II	PFA 204-12334		9. 9.93	A Thornton	Ince Blundell	13. 5.04P
	(Built J G Dance and P E Lewis - c/n 0692.313)(Tri-cycle u/c)						
G-MYLP	Kolb Twinstar Mk.3	PFA 205-12391	(G-BVCR)	9. 9.93	R.Thompson	(Bristol)	27. 5.99P
	(Built D M Stevens - kit no K0005-0992)				*(Current status unknown)*		
G-MYLR	Mainair Gemini/Flash IIA	964-0993-7 & W759		17. 9.93	A.V.Cosser	(Rugby)	21.11.05P
	(Rotax 582)						
G-MYLS	Mainair Mercury	966-0993-7 & W761		5.10.93	D.Burnell-Higgs	Shobdon	6. 9.05P
G-MYLT	Mainair Blade	967-1093-7 & W762		23. 9.93	A R.Walsh	Ince Blundell	29. 5.04P
	(Rotax 912)						
G-MYLV	CFM Shadow Series CD	220		24. 9.93	G.Gilhead and R.G.M.Proost	Old Sarum	24. 3.05P
					tr Aviation for Paraplegics and Tetraplegics Trust		
G-MYLW	Rans S-6-ESD Coyote II	PFA 204-12560		4. 8.93	M.J.Phillips	Priory Farm, Tibenham	29. 9.05P
	(Built J R Worswick - c/n 1292.401) (Tri-cycle u/c)						
G-MYLX	Medway Raven X	MRB113/109		6.10.93	T.M.Knight	(Luton)	12. 2.00P
	(Sailwing c/n also quoted for G-MYVV)						
G-MYLY	Medway Raven X	MRB001/108		23. 9.93	C.R.Smith	(Stanford-le-Hope)	3.10.94P
	(Sailwing c/n also quoted for G-MYVU)				*(Current status unknown)*		
G-MYLZ	Pegasus Quantum 15	6672		6.10.93	W.G.McPherson	Perth	30.11.05P
	(Rotax 462)						
G-MYMB	Pegasus Quantum 15	6674		6.10.93	C.A Green	Lee-on-Solent	26. 4.05P
	(Rotax 582)				*"Firebird"*		
G-MYMC	Pegasus Quantum 15	6675		6.10.93	B.J.Topham	Rufforth	14. 5.05P
	(Rotax 582)						
G-MYME	Cyclone AX3/503	C.3093157		13.10.93	M.L.Smith	Broomhill Farm, West Calder	5.12.04P
	(Officially regd wiith c/n as such but keel tube C.3093157 noted as fitted to G-MYHJ)						
G-MYMF*	Cyclone AX3/503	C.3093158		18.10.93	G.Gates	Old Sarum	15.11.04P
					tr Old Sarum Ax3 Group *(Cancelled 23.8.04 by CAA)*		
G-MYMH	Rans S-6-ESD Coyote II	PFA 204-12576		20.10.93	A R.Cattell	Lower Wasing Farm, Brimpton	1. 7.05P
	(Built D J Thompsett - c/n 0793.520) (Tri-cycle u/c)						
G-MYMI	Kolb Twinstar Mk.3	PFA 205-12537		21.10.93	C.Trollope	Watnall	5. 9.05P
	(Built R T P Harris - kit no K0016-0693)						
G-MYMJ	Medway Raven X	MRB004/110		28.10.93	N.Brigginshaw	RAF Wyton	13. 7.05P
	(Sailwing c/n also quoted for G-MYVX)						
G-MYMK	Mainair Gemini/Flash IIA	968-1193-7 & W763		29.10.93	A Britton	(Rickmansworth)	16. 5.05P
	(Rotax 582)						
G-MYML	Mainair Mercury	969-1193-7 & W765		29.10.93	D.J.Dalley	(Weymouth)	7. 6.01P
G-MYMM	Air Création 503/Fun 18S GT bis	93/001		30. 9.93	W.H.Greenwood	Swanborough Farm, Lewes	23. 8.01P
					(New owner 8.02)		
G-MYMN	Whittaker MW6 Merlin	PFA 164-12124		29.10.93	K.J.Cole	Over Farm, Gloucester	25.11.03P
	(Built K.J.Cole) (Rotax 582)						
G-MYMO	Mainair Gemini/Flash IIA	955-0793-7 & W750		24. 6.93	S.P.Moores	(Biddulph, Stoke-on-Trent)	7. 3.05P
G-MYMP	Rans S-6-ESD Coyote II	PFA 204-12436	(G-CHAZ)	5.11.93	F.M.Pearce	(Exmouth)	8. 8.05P
	(Built C H Middleton - c/n 1291.250) (Tri-cycle u/c)						
G-MYMR	Rans S-6-ESD Coyote II	PFA 204-12580		17.11.93	J.Neilands	Letterkenny, County Donegal	3. 6.05P
	(Built M J Kay) (Tri-cycle u/c)						
G-MYMS	Rans S-6-ESD Coyote II	PFA 204-12581		17.11.93	M.R.Johnson and P.G.Briscoe	Long Marston	12. 7.05P
	(Built G C Moore - c/n 0893.526) (Tri-cycle u/c)						
G-MYMT	Mainair Mercury	970-1193-7 & W766		19.11.93	W. and C.A Bradshaw	St.Michaels	17.12.04P
G-MYMV	Mainair Gemini/Flash IIA	971-1193-7 & W767		26.11.93	A Szczepanek	Guy Lane Farm, Waverton	8. 4.05P
G-MYMW	Cyclone AX3/503	C.3093156		23.11.93	L.J.Perring	Field Farm, Oakley	1. 5.05P
G-MYMX	Pegasus Quantum 15	6705		1.12.93	N.F.McKenzie	Insch	23. 6.05P
	(Rotax 582)						
G-MYMY	Cyclone Chaser S 508	CH.860		7. 9.93	D.L.Hadley	(Canterbury)	29. 5.05P

Registration	Type	C/n	Date	Owner	Base	CofA
G-MYMZ	Cyclone AX3/503	C.3093154	7.12.93	The Microlight School (Lichfield) Ltd	Roddige	20. 1.05P
G-MYNA	CFM Shadow Series C	K.023	10. 2.88	D. Gorman	Plaistowes Farm, St Albans	24. 4.05P
G-MYNB	Pegasus Quantum 15	6719	14.12.93	D.M.Savage and J.P.Briggs	Rufforth	24. 1.05P
	(Rotax 582)					
G-MYNC	Mainair Mercury	K973-1293-7 & W769	17.12.93	A Brotheridge	Rufforth	21. 8.03P
G-MYND	Mainair Gemini/Flash IIA	841-0591-7 & W635	28. 5.91	S.Wild	(Bramley, Rotherham)	23. 4.05P
G-MYNE	Rans S-6-ESD Coyote II	PFA 204-12497	25. 6.93	J N W Moss	Enstone	6. 10.04P
	(Built G Ferguson) (Tail-wheel u/c)					
G-MYNF	Mainair Mercury	974-1293-7 & W770	17. 1.94	B.Walker tr G-MYNF Group	Eshott	24. 1.05P
G-MYNH	Rans S-6-ESD Coyote II	PFA 204-12616	30.12.93	E.F.and V.M.Clapham	Oldbury-on-Severn	29. 3.01P
	(Built E F Clapham - kit no.0493.487) (Rotax 912-UL) *(Tail-wheel u/c)*			*(Current status unknown)*		
G-MYNI	TEAM mini-MAX 91	PFA 186-12314	22. 2.93	J.J.Penney	Rhygos, Glamorgan	17.11.99P
	(Built R Barton) (Mosler MM CB-35)			*(Current status unknown)*		
G-MYNJ	Mainair Mercury	K972-1293-7 & W768	14. 1.94	S.M.Buchan	Deenethorpe	10. 8.04P
G-MYNK	Pegasus Quantum 15	6614	17.11.93	N.D.Azevedo	(London N5)	12. 6.04P
	(Rotax 582)					
G-MYNL	Pegasus Quantum 15	6648	17.11.93	C.Hodgkiss	(Burton-on-Trent)	29.11.05P
	(Rotax 582)					
G-MYNN	Pegasus Quantum 15	6679	17.11.93	A C.Thomson	(Belper)	22.12.04P
	(Rotax 582)					
G-MYNO	Pegasus Quantum 15	6724	10. 1.94	L C Stockman	Plaistowes Farm, St Albans	20. 4.05P
	(Rotax 582)					
G-MYNP	Pegasus Quantum 15	6688	17.11.93	M.E.Howard	RAF Henlow	7. 6.05P
	(Rotax 582)			tr RAF Microlight Flying Association (NP)		
G-MYNR	Pegasus Quantum 15 Super Sport	6692	17.11.93	A.S.Wason	Redlands	18. 4.05P
	(Rotax 582)					
G-MYNS	Pegasus Quantum 15	6694	17.11.93	E.G.Cartwright	(Belper)	16. 5.05P
	(Rotax 582)					
G-MYNT	Pegasus Quantum 15	6693	17.11.93	P.A Vernon	Craysmarsh Farm, Melksham	8. 5.05P
	(Rotax 582)					
G-MYNV	Pegasus Quantum 15	6725	10. 1.94	M.D.Gregory	Eaglescott	7. 6.05P
	(Rotax 582)					
G-MYNX	CFM Streak Shadow SA-M	PFA 206-12268	15. 6.92	T.J.and M.D.Palmer	Mill Farm, Shifnal	23.10.04P
	(Built P A White - kit no.K.193-SA-M) (Rotax 618)					
G-MYNY	Kolb Twinstar Mk.3	PFA 205-12478	22.11.93	B.Alexander	Swinford, Rugby	25. 8.98P
	(Built W R C Williams-Wynne - kit no K0014-0693)				*(Current status unknown)*	
G-MYNZ	Pegasus Quantum 15	6709	18. 1.94	P.W.Rogers	(Shawforth, Rochdale)	27. 1.05P
	(Rotax 582)					
G-MYOA	Rans S-6-ESD Coyote II	PFA 204-12578	23.11.93	M.D.Bayliss	Otherton, Cannock	17. 6.05P
	(Built H Lang - c/n 0793.523) (Tri-cycle u/c)			tr Orcas Syndicate		
G-MYOB	Mainair Mercury	976-1293-7 & W772	8.12.93	J.C.and B.E.Barnes	(Wisbech)	19. 7.05P
G-MYOF	Mainair Mercury	975-1293-7 & W771	3.12.93	G.Bonnar	(Leicester)	16. 2.05P
G-MYOG	Kolb Twinstar Mk.3	PFA 205-12449	19. 1.94	A P.de Legh	Redhill	17. 8.05P
	(Built A P.de Legh - kit no K0011-0193) (Hirth 2706)					
G-MYOH	CFM Shadow Series CD	K.201	27. 1.94	D.R.Sutton	(Euxton, Chorley)	15. 8.05P
G-MYOI	Rans S-6-ESD Coyote II	PFA 204-12503	3. 2.94	R.M.Moulton	(Shackerstone, Nuneaton)	18.11.05P
	(Built P F Hill - c/n 1292.409) (Tailwheel u/c)					
G-MYOL	Air Création/Fun 18S GT bis	94/001	7. 2.94	S.N.Bond and D.E.Lord	Crosland Moor	22. 7.05P
	(Rotax 447)					
G-MYOM	Mainair Gemini/Flash IIA	981-0294-7 & W777	14. 2.94	M.A Haughey	Newtownards	14.10.05P
	(Rotax 582)					
G-MYON	CFM Shadow Series CD	240	12. 1.94	D.W.and S.E.Suttill	(Goole)	3.12.04P
G-MYOO	Kolb Twinstar Mk.3M	PFA 205-12200	11. 5.92	P.D.Coppin	Lee-on-Solent	16.11.05P
	(Built P.D.Coppin and P Watmough - kit no K0004-0192)					
G-MYOR	Kolb Twinstar Mk.3	PFA 205-12602	16. 2.94	J.J.Littler	Chichester	5.10.04P
	(Built O J and J H Stodhert) (Hirth 2705 RO6)					
G-MYOS	CFM Shadow Series CD	246	18. 2.94	E.J. and C.A Bowles	Craysmarsh Farm, Melksham	25.10.05P
G-MYOT*	Rans S-6-ESD Coyote II	PFA 204-12668	21. 2.94	D.E.Wilson	Davidstow Moor	15. 7.03P
	(Built R T Mosforth - c/n 0893.525) (Tail-wheel u/c)			*(Cancelled 10.3.04 by CAA)*		
G-MYOU	Pegasus Quantum 15	6726	1. 3.94	M.Botten and O.Kent	Plaistowes Farm, St.Albans	11.10.05P
	(Rotax 582)					
G-MYOV	Mainair Mercury	K979-0294-7 & W775	1. 3.94	P.Newton	(Buxton)	14.12.03P
G-MYOW	Mainair Gemini/Flash IIA	983-0294-7 & W779	16. 3.94	A J.A Fowler	Corn Wood Farm, Adversane	25. 6.04P
G-MYOX	Mainair Mercury	K984-0294-7 & W780	23. 2.94	K.Driver	(Rotherham)	2. 9.05P
G-MYOY	Cyclone AX3/503	C.3123191	23. 2.94	R.Nicklin	Otherton, Cannock	10. 6.05P
G-MYOZ	BFC Challenger II	PFA 177A-12640	24. 2.94	J J Littler	(Aldingbourne, Chichester)	2.11.05P
	(Built P F Bockh - c/n UK CH2-1093-1045) (Rotax 503)					
G-MYPA	Rans S-6-ESD Coyote II	PFA 204-12678	24. 2.94	L.J.Dutch	Tarn Farm, Cockerham	14.12.04P
	(Built E J Garner - c/n 0893.527) (Tail-wheel u/c)					
G-MYPC	Kolb Twinstar Mk.3	PFA 205-12437	2. 3.94	J.Young and S.Hussain	Otherton, Cannock	17.12.05P
	(Built R Pattrick - c/n K0012-0199)					
G-MYPD	Mainair Mercury	982-0294-7 & W778	11. 3.94	R.D.McManus	(Stoke-on-Trent)	3.11.04P
	(Rotax 462)					
G-MYPE	Mainair Gemini/Flash IIA	985-0394-7 & W781	11. 3.94	R.Cant	(Musselburgh)	24. 5.05P
	(Rotax 582)					
G-MYPG	Solar Wings Pegasus XL-Q	SW-WQ-0176	29. 3.89	I.A.G.Hull	(Epping Green, Epping)	19. 6.05P
G-MYPH	Pegasus Quantum 15	6764	11. 3.94	P.M.J.White	Wombleton	29. 6.05P
	(Rotax 582)					
G-MYPI	Pegasus Quantum 15	6767	11. 3.94	P.L.Jarvis	London Colney	31. 8.05P
	(Rotax 582)					
G-MYPJ	Rans S-6-ESD Coyote II	PFA 204-12692	18. 3.94	K A Eden	Brook Farm, Pilling	9. 9.05P
	(Built A W Fish - c/n 1293.569) (Tri-cycle u/c)					
G-MYPL	CFM Shadow Series CD	K.213 & BMAA/HB/080	14. 2.94	G.I.Madden	(Milton Keynes)	27. 2.05P
G-MYPM	Cyclone AX3/503	C.3123188	23. 3.94	Microflight Ireland Ltd	Mullaghmore	26. 6.03P

Registration	Type	C/n	Previous identity	Date	Owner	Base	CofA
G-MYPN	Pegasus Quantum 15 (Rotax 582)	6727		12. 4.94	P J S Albon	(Bedford)	2. 8.05P
G-MYPP	Whittaker MW6-S Fatboy Flyer *(Built D.S.L.Evans)*	PFA 164-12413		11. 4.94	D.S.L.Evans *(Current status unknown)*	(Ashford)	
G-MYPR	Cyclone AX3/503	C.3123190		13. 4.94	N.E.Ashton	Ince Blundell	17. 4.04P
G-MYPS	Whittaker MW6 Merlin *(Built I.S.Bishop)* (Rotax 503)	PFA 164-11585		19. 4.94	I.S.Bishop	Bicester	27.11.04P
G-MYPT	CFM Shadow Series CD	K.212		22. 4.94	M.G. and S.A Collins	(Oldbury-on-Severn)	13.10.05P
G-MYPV	Mainair Mercury (Rotax 582)	986-0394-7 & W782		18. 3.94	J.J.Brex *(Noted 1.05)*	Sywell	26.10.04P
G-MYPW	Mainair Gemini/Flash IIA (Rotax 582)	991-0494-7 & W787		3. 5.94	R.E.Parker	Hunsdon	24. 5.05P
G-MYPX	Pegasus Quantum 15 (Rotax 582) *(Believed to have used "B Conditions" marks "G-69-29" during trials)*	6785		28. 4.94	P.J.Callis and M.Aylett	Swinford, Rugby	13. 5.05P
G-MYPY	Pegasus Quantum 15 (Rotax 582)	6786		12. 5.94	G.and G.Trudgill	Shotton Colliery, Peterlee	30. 8.05P
G-MYPZ	BFC Challenger II UK *(Built E G Astin - c/n CH2-1093-UK-1046)* (Hirth 2706) *(Regd incorrectly as CH2-0194-UK-1046)*	PFA 177A-12689		2. 3.94	E.G.Astin	Whitby	14.10.05P
G-MYRB	Whittaker MW5 Sorcerer *(Built P.J.Careless)*	PFA 163-11543		14. 4.94	P.J.Careless *(Current status unknown)*	(Sandy)	
G-MYRC	Mainair Blade (Rotax 462)	988-0594-7 & W784		1. 6.94	T.C.Brown and J.Murphy	Mill Farm, Shifnal	14. 8.05P
G-MYRD	Mainair Blade (Rotax 582)	989-0594-7 & W785		20. 5.94	W.J.Walker	Tarn Farm, Cockerham	21. 2.05P
G-MYRE	Cyclone Chaser S (Rotax 377)	CH863		10. 5.94	B.Richardson	Eshott	9. 5.05P
G-MYRF	Pegasus Quantum 15 (Rotax 462HP)	6795		13. 5.94	A O.Sutherland	Latch Farm, Kirknewton	23. 1.05P
G-MYRG	TEAM mini-MAX 88 *(Built D.G.Burrows)*	PFA 186-11891		17. 5.94	D.G.Burrows	Shobdon	16. 9.04P
G-MYRH	BFC Challenger II UK *(Built R T Hall - kit no.CH2-1093-1044)* (Rotax 582)	PFA 177A-12690		10. 3.94	C.M.Gray	Lee-on-Solent	27. 4.05P
G-MYRI	Medway Hybred 44XLR (Rotax 503)	MR180/841		23. 5.94	B.D.Acres	(Maidstone)	1. 9.00P
G-MYRJ	BFC Challenger II UK *(Built H F Breakwell and P Woodcock - c/n CH2-1093-1042)* (Rotax 582)	PFA 177A-12658		28. 3.94	A Watson	Siege Cross Farm, Thatcham	12.10.05P
G-MYRK	Meridian Renegade Spirit UK *(Built J Brown - c/n 215)*	PFA 188-11425		3.10.89	P.Crowhurst	Sywell	7.12.05P
G-MYRL	TEAM mini-MAX 91 *(Built W W Vinten)*	PFA 186-11967		17. 5.94	J.N.Hanson	Brook Farm, Pilling	3.11.05P
G-MYRM	Pegasus Quantum 15 (Rotax 582)	6800		26. 5.94	T.Read	Old Sarum	15. 8.05P
G-MYRN	Pegasus Quantum 15 (Rotax 582)	6801		26. 5.94	B.Robertson	Perth	30.12.05P
G-MYRO	Cyclone AX3/503	C.4043211		6. 6.94	Rosemary I.Simpson	Rochester	22. 8.05P
G-MYRP	Letov LK-2M Sluka (Rotax 447)	829409x09? & PFA 263-12725		6. 6.94	R.M.C.Hunter	Blue Tile Farm, Hindolveston	7.11.04P
G-MYRR	Letov LK-2M Sluka (Rotax 447)	829409x05?		10. 6.94	B.C.McCartan	(Banbridge, County Down)	31. 8.05P
G-MYRS	Pegasus Quantum 15 (Rotax 582)	6803		13. 6.94	R.M.Summers	Insch	28. 7.05P
G-MYRT	Pegasus Quantum 15 (Rotax 582)	6732		1. 3.94	M.C.Taylor	(Coleford)	17. 5.05P
G-MYRU	Cyclone AX3/503	C.4043210		7. 6.94	S.Fraser	(Newcastle upon Tyne)	10. 5.05P
G-MYRV	Cyclone AX3/503	C.4043209		8. 6.94	M.Gardiner	Crosland Moor	4. 7.05P
G-MYRW	Mainair Mercury	999-0694-7 & W795		17. 6.94	G.C.Hobson *(Op Northern Microlight School)*	St.Michaels	5. 7.05P
G-MYRY	Pegasus Quantum Lite (Rotax 582)	6813		15. 6.94	J.Needham	Roddige	19.10.05P
G-MYRZ	Pegasus Quantum 15 (Rotax 582)	6812		15. 6.94	G.D.Black	(Perth)	13. 7.05P
G-MYSA	Cyclone Chaser S 508	CH.864		15. 6.94	P.Nicholls	Shobdon	13. 4.05P
G-MYSB	Pegasus Quantum 15 (Rotax 582)	6809		22. 6.94	P.H.Woodward	Roddige	25. 3.05P
G-MYSC	Pegasus Quantum 15 (Rotax 582)	6811		22. 6.94	K.R.White	Dunkeswell	2. 9.05P
G-MYSD	BFC Challenger II *(Built C E Bell - c/n CH2-1093-1043)*	PFA 177A-12688		23. 6.94	C.E.Bell	Spanhoe	2.11.05P
G-MYSG	Mainair Mercury (Rotax 582)	K993-0694-7 & W790		12. 7.94	M.Donnelly	Guy Lane Farm, Waverton	17.10.05P
G-MYSI	Mignet HM.14/93 *(Built A R.D.Seaman)*	PFA 255-12700		18. 7.94	A R.D.Seaman *(Current status unknown)*	(Dagenham)	
G-MYSJ	Mainair Gemini/Flash IIA (Rotax 503)	1001-0894-7 & W797		2. 8.94	E M Christoffersen	Baxby Manor, Husthwaite	12. 3.05P
G-MYSK	TEAM mini-MAX 91 *(Built K Worthington)*	PFA 186-12203		25. 7.94	A D.Bolshaw *(Op Brook Farm Microlight Centre)*	Brook Farm, Pilling	1. 8.02P
G-MYSL	Aviasud Mistral 582GB	066 & BMAA/HB/007	83-DE	27. 2.92	P.C.Piggott and M.E.Hughes	Little Battleflats Farm, Ellistown, Coalville	12. 8.05P
G-MYSM	CFM Shadow Series CD	K.243 & BMAA/HB/049		22. 3.94	L.W.Stevens	(Grantham)	10.10.03P
G-MYSO	Cyclone AX3/503	C.4043215		1. 8.94	M.L.Smith	Popham	25. 3.05P
G-MYSP	Rans S-6-ESD Coyote II *(Built S Palmer - c/n 0392-284)* (Rotax 582) *(Tri-cycle u/c)*	PFA 204-12265		26. 5.92	A J.Alexander, K.G.Diamond and B.Knight	Redhill	24. 8.05P
G-MYSR	Pegasus Quantum 15 (Rotax 582)	6837		3. 8.94	W G Craig	(Dunfermline)	6. 9.05P
G-MYSU	Rans S-6-ESD Coyote II *(Built I Whyte)*	PFA 204-12753		5. 8.94	K.W.Allan	Drummaird Farm, Bonnybank	20. 7.05P

Reg	Type	C/n	Previous identity	Date	Owner	Base	CofA
G-MYSV	Aerial Arts Chaser S (Rotax 377)	CH.812	(ex Korea)	24. 8.94	R J Sims and I G Reason	(Salisbury)	6. 5.05P
G-MYSW	Pegasus Quantum 15 (Rotax 582)	6834		13. 7.94	D J Cornelius	Lee-on-Solent	25. 2.05P
G-MYSX	Pegasus Quantum 15 (Rotax 503)	6832		13. 7.94	J.L.Treves	Longacres Farm, Sandy	29. 7.04P
G-MYSY	Pegasus Quantum 15 (Rotax 582)	6864		15. 8.94	Premier Aviation (UK) Ltd	(Faygate, Horsham)	10. 4.05P
G-MYSZ	Mainair Mercury *(C/n confirmed but see G-MYYY)*	1006-0894-7 & W802		2. 9.94	R.G.McCron	Shobdon	8.11.05P
G-MYTB	Mainair Mercury (Rotax 582)	1004-0894-7 & W800		19. 8.94	P.J.Higgin	(Red House Farm, Gedney Dyke, Lutton)	16. 9.05P
G-MYTC	Solar Wings Pegasus XL-Q	SW-WQ-0246	(ex...)	28. 9.94	M.J.Edmett *(New owner 5.02)*	(London N3)	
G-MYTD	Mainair Blade (Rotax 582)	1002-0894-7 & W798		18. 8.94	M.P.Law and B.E.Warburton	St.Michaels	10. 5.05P
G-MYTE	Rans S-6-ESD Coyote II *(Built AJ Bourner) (Tail-wheel u/c)*	PFA 204-12718		22. 7.94	G Mills tr The Rans Flying Group	Crosland Moor	17.12.05P
G-MYTG	Mainair Blade (Rotax 582)	1008-0994-7 & W804		16. 9.94	O.P.Farrell	(Drogheda, County Louth)	14. 2.05P
G-MYTH	CFM Shadow Series CD	089		7.11.88	H.A Leek	(Melton Mowbray)	20.12.04P
G-MYTI	Pegasus Quantum 15 (Rotax 582)	6874		6.10.94	J.Madhvani	Plaistowes Farm, St.Albans	23. 2.05P
G-MYTJ	Pegasus Quantum 15 (Rotax 582)	6877		29. 9.94	K.Laud	Roddige	4. 6.05P
G-MYTK	Mainair Mercury	1009-1094-7 & W805		29. 9.94	D.A Holroyd	(London W14)	13. 6.04P
G-MYTL	Mainair Blade (Rotax 582)	1010-1094-7 & W807		4.10.94	S.Ostrowski	Davidstow Moor	26. 7.05P
G-MYTM	Cyclone AX3/503	C.3123189		13. 4.94	J P Gardiner	(Farnworth)	1. 9.05P
G-MYTN	Pegasus Quantum 15 (Rotax 503)	6878		30. 9.94	M.Hoggett and M.F.Ambrose	Over Farm, Gloucester	27. 5.05P
G-MYTO	Quad City Challenger II UK *(Built K B Tolley and D M Cottingham)* (Hirth 2705.R06)	PFA 177-12583		22. 7.94	R.W.Sage *(Noted 8.04)*	Priory Farm, Tibenham	16. 4.01P
G-MYTP	CGS Arrow Flight Hawk II *(Built M Whittaker - c/ns 215 & H-CGS-489-P but believed c/n 215 taken from p/i)* (Rotax 503)	PFA 266-12801	N215	6.10.94	R.J.Turner	Otherton, Cannock	12. 7.05P
G-MYTR	Solar Wings Pegasus Quasar IITC (Rotax 582)	6880		11.10.94	M.E.Grafton	(Hay-on-Wye)	22. 5.02P
G-MYTT	Quad City Challenger II *(Built P L Fisk - c/n CH2-0394-UK-111)*	PFA 177-12761		11.10.94	R.J.Shave tr Challenger G-MYTT	Dunkeswell	26. 7.05P
G-MYTU	Mainair Blade (Rotax 582)	1011-1094-7 & W808		21.10.94	C.J.Barker	(Bagthorpe, Nottingham)	20. 6.05P
G-MYTV	Huntwing Avon *(Built M P Hadden - kit no.9204010)* (Rotax 503)	BMAA/HB/029		13.10.94	P.J.Sutton	(Hereford)	13.12.05P
G-MYTX	Mainair Mercury	K1003-0894-7 & W799		23. 9.94	R.Steel	Rufforth	31. 8.05P
G-MYTY	CFM Streak Shadow M *(Built N R Beale - kit no K.242)* (Rotax 912-UL)	PFA 206-12607		11. 7.94	K.H.A Negal	Enstone	6. 6.02P
G-MYTZ	Air Création 503/Fun 18S GT bis	94/003		7.11.94	S.N.Bond and D.E.Lord	Crosland Moor	12. 7.05P
G-MYUA	Air Création 503/Fun 18S GT bis	94/002		8.11.94	J.Leden	Darley Moor	9. 7.04P
G-MYUB	Mainair Mercury	1014-1194-7 & W812		14.12.94	T.A Ross	Arclid Green, Sandbach	7.10.05P
G-MYUC	Mainair Blade (Rotax 462)	1015-1294-7 & W813		16.11.94	A D.Clayton	St.Michaels	4.11.05P
G-MYUD	Mainair Mercury (Rotax 582)	1016-1294-7 & W814		24.11.94	P.W.Margetson	Deenethorpe	17. 5.05P
G-MYUE	Mainair Mercury (Rotax 582)	1017-1294-7 & W815		22.11.94	D.W.Power	(Ammanford)	20. 4.05P
G-MYUF	Meridian Renegade Spirit *(Built C J Dale)* (Jabiru 2200A)	PFA 188-12795		16.11.94	M.A Pantling	Weston Zoyland	7.12.05P
G-MYUH	Solar Wings Pegasus XL-Q *(C/n 6810 refers to new wing - trike unit is ex-G-MVKR (SW-TE-0123)*	6810		28.11.94	K.S.Daniels	(London Colney)	4. 6.05P
G-MYUI	Cyclone AX3/503 *(C/n carried is C.102822 and probably results from a changed monopole)*	C.4043213		13.12.94	R.and M.Bailey	Plaistowes Farm, St.Albans	10. 5.05P
G-MYUK	Mainair Mercury (Rotax 462)	1020-0195-7 & W818		12.12.94	S.Lear	Plaistowes Farm, St.Albans	6. 9.04P
G-MYUL	Quad City Challenger II UK *(Built A J Easson - c/n CH2-1293-UK-1063)*	PFA 177-12687		10. 1.95	P.Knott	(Bradford)	24. 5.02P
G-MYUN	Mainair Blade (Rotax 582)	1019-0195-7 & W817		5.12.94	G.A Barratt	(Preston)	7. 2.05P
G-MYUO	Pegasus Quantum 15 (Rotax 582)	6911		23. 1.95	S M Neil	(Bury, Huntingdon)	17.12.04P
G-MYUP	Letov LK-2M Sluka *(Built F Overall)* (Rotax 447)	829409x24, UK.2 & PFA 263-12785		20.12.94	C J Meadows	Weston Zoyland	18. 6.05P
G-MYUR	Huntwing Avon *(Built S.D.Pain - kit no.9409030)* (Rotax 582)	BMAA/HB/034		24. 1.95	T.C.Saltmarsh	Rayne Hall Farm, Braintree	30. 3.05P
G-MYUS	CFM Shadow Series CD	257		26. 1.95	G.Gilhead and R.G.M.Proost tr Aviation for Paraplegics and Tetraplegics Trust	Old Sarum	15. 7.05P
G-MYUU	Pegasus Quantum 15 (Rotax 462)	6917		30. 1.95	J A Slocombe	(Backfields, Rochester)	9. 7.05P
G-MYUV	Pegasus Quantum 15 (Rotax 582)	6918		6. 2.95	D.W.Wilson	(Jacksdale, Nottingham)	6. 8.05P
G-MYUW	Mainair Mercury	1024-0295-7 & W822		7. 2.95	G.C.Hobson	St Michaels	21.11.04P
G-MYUZ	Rans S-6-ESD Coyote II *(Built D K Ross and B Davies - c/n 1293.568)* (Rotax 582) *(Tri-cycle u/c)*	PFA 204-12741		5. 1.95	B.Davies	Sittles Farm, Alrewas	29. 4.05P
G-MYVA	Kolb Twinstar Mk.3 *(Built S P Read)*	PFA 205-12756		13. 2.95	A S.Milner	(Southport)	23. 3.04P
G-MYVB	Mainair Blade (Rotax 582)	1021-0195-7 & W819		15.12.94	M & Paula L Eardley	Tarn Farm, Cockerham	24. 4.05P

G-MYVC	Pegasus Quantum 15 (Rotax 582)	6904		13. 2.95	R.Howes and O.Lloyd	Roddige	27. 3.05P
G-MYVE	Mainair Blade (Rotax 582)	1027-0295-7 & W825		8. 2.95	R.D.Serle and P.M.Jennings	Shobdon	26. 5.05P
G-MYVG	Letov LK-2M Sluka *(Built L W M Summers)* (Rotax 447)	829409x26 & PFA 263-12786		15. 2.95	K.Cox	(Wirral)	28. 9.05P
G-MYVH	Mainair Blade (Rotax 582)	1028-0295-7 & W826		21. 2.95	S.E.Wilks	(Tollerton, Nottingham)	27. 3.05P
G-MYVI	Air Création 503/Fun 18S GT bis	94/004		17. 2.95	P.Osborne tr Northampton Aerotow Club	(Northampton)	25. 3.04P
G-MYVJ	Pegasus Quantum 15 (Rotax 582)	6974		24. 2.95	P.W.Davidson and A I.McPherson	Pratis Farm, Leven	27. 4.05P
G-MYVK	Pegasus Quantum 15 (Rotax 582)	6970		27. 2.95	J.Hammond	Weston Zoyland	10. 6.05P
G-MYVL	Mainair Mercury (Rotax 462)	1030-0395-7 & W828		1. 3.95	P.J.Judge	Davidstow Moor	8. 7.05P
G-MYVM	Pegasus Quantum 15 (Rotax 582)	6893	G-69-17 G-MYVM	9. 3.95	T.P.Hunt *(New owner 8.04)*	(Wadebridge)	27. 4.97P
G-MYVN	Cyclone AX3/503	C.4043212		16. 3.95	F.Watt	Insch	5.10.03P
G-MYVO	Mainair Blade (Rotax 582)	1013-1194-7 & W811		8.11.94	S.S.Raines	Shobdon	13. 4.05P
G-MYVP	Rans S-6-ESD Coyote II *(Built J S Liming - c/n 0294.593) (Tri-cycle u/c)*	PFA 204-12828		27. 3.95	C.E.Hormaeche	Eshott	23. 7.05P
G-MYVR	Pegasus Quantum 15 (Rotax 582)	6980		21. 3.95	J.M.Webster	(Nottingham)	26. 3.05P
G-MYVS	Mainair Mercury (Rotax 462)	1037-0495-7 & W835		12. 4.95	P.S.Flynn	Sandtoft	25. 4.04P
G-MYVT	Letov LK-2M Sluka *(Built J Hannibal)* (Rotax 447)	829409x25 & PFA 263-12835		17. 3.95	Jean P.Gardiner *(Noted 11.04)*	Newton Bank Farm, Daresbury	11.11.03P
G-MYVV	Medway Hybred 44XLR (Rotax 503) *(Sailwing c/n also quoted for G-MYLX)*	MR127/109		3. 4.95	S.Perity	(Wisbech)	4. 6.04P
G-MYVW	Medway Raven X *(Sailwing c/n also quoted for G-MYMJ)*	MRB128/110		15. 5.95	J.C.Woolgrove *(Current status unknown)*	(Beckenham	5. 6.97P
G-MYVX	Medway Hybred 44XLR (Rotax 503)	MR129/111		3. 4.95	A R.Fricker	(South Ockendon)	4. 6.02P
G-MYVY	Mainair Blade (Rotax 582)	1033-0495-7 & W831		29. 3.95	S.P.Maxwell	North Coates	21.10.04P
G-MYVZ	Mainair Blade (Rotax 582)	1034-0495-7 & W832		31. 3.95	R.Llewellyn	Chiltern Park, Wallingford	7. 6.05P
G-MYWA	Mainair Mercury	1035-0495-7 & W833		30. 3.95	D.James	(Neath)	28. 6.05P
G-MYWC	Huntwing Avon *(Built F.J.C.Binks)* (Rotax 503 - *kit no.9409038)*	BMAA/HB/043		3. 4.95	M.A.Coffin	(Wrotham, Sevenoaks)	24. 6.05P
G-MYWD	Thruster T.600N (Rotax 582)	9035-T600-511	(G-MYOJ)	18. 4.95	M D Kirby	Chase Farm, Billericay	16. 8.05P
G-MYWE	Thruster T.600T (Rotax 503)	9035-T600-512	(G-MYOK)	18. 4.95	W A Stephenson	(Newry, County Armagh)	4. 4.04P
G-MYWF	CFM Shadow Series CD	K.248 & BMAA/HB/068		18. 4.95	M.A Newman	(Saxmundham)	18. 4.05P
G-MYWG	Pegasus Quantum 15 (Rotax 582)	6998		20. 4.95	M.R.Bishop	(Dunstable)	18.10.03P
G-MYWH	Huntwing Experience *(Built G.N.Hatchett - kit no.9409025)*	BMAA/HB/037		20.12.94	R.S.Sanby	(Eastwood, Nottingham)	
G-MYWI	Pegasus Quantum 15 (Rotax 582)	7006		1. 5.95	J.R.Fulcher	Deenethorpe	20. 2.05P
G-MYWJ	Pegasus Quantum 15 (Rotax 582)	6919		24. 1.95	P.A Banks	Longacres Farm, Sandy	3.10.05P
G-MYWK	Pegasus Quantum 15 (Rotax 582)	7011		1. 5.95	G.Hanna	(Drogheda, County Louth)	25. 9.03P
G-MYWL	Pegasus Quantum 15 (Rotax 582)	6995		2. 5.95	E Smith	Swinford, Rugby	22. 5.05P
G-MYWM	CFM Shadow Series CD	K.227 & BMAA/HB/056		9. 5.95	N.J.Mckinley	Plaistowes Farm, St.Albans	4. 8.04P
G-MYWN	Cyclone Chaser S 508 *(Conceived as Aerial Arts Chaser with c/n CH865, this was built under Cyclone c/n 7016) (New owner 11.04)*	CH.865		9. 5.95	N.R.Beale	Deppers Bridge, Southam	1. 9.03P
G-MYWO	Cyclone Pegasus Quantum 15 (Rotax 582)	6932		9. 5.95	J.W.Cope	Strubby	21.10.05P
G-MYWP	Kolb Twinstar Mk.3 *(Built B Albiston - kit no K0017-0993)*	PFA 205-12561		7. 3.95	P.R.Day	(Southampton)	19. 8.04P
G-MYWR	Cyclone Pegasus Quantum 15 (Rotax 582)	7002		10. 5.95	R.Horton	Roddige	11. 8.04P
G-MYWS	Cyclone Chaser S 447 *(Conceived as Aerial Arts Chaser with c/n CH866, this was built under Cyclone c/n 6946)*	6946 & CH.866		17. 5.95	M.H.Broadbent	Westfield Farm, Hailsham	23. 5.03P
G-MYWT	Pegasus Quantum 15 (Rotax 582)	6997		19. 5.95	J.Dyer	Sywell	7.11.05P
G-MYWU	Pegasus Quantum 15 (Rotax 582)	7024		25. 5.95	J.R.Buttle	Dunkeswell	14. 7.04P
G-MYWV	Rans S-4C Coyote *(Built A H Trapp - c/n 093.212)*	PFA 193-12826		30. 5.95	D.Cassidy	Maypole Farm, Chislet	27. 9.05P
G-MYWW	Pegasus Quantum 15 (Rotax 503)	7021		30. 5.95	C.W.Bailie	Newtownards	24. 8.05P
G-MYWX	Pegasus Quantum 15 (Rotax 582)	7019		6. 6.95	D.J.Revell	Lower Mountpleasant, Chatteris	24.10.05P
G-MYWY	Pegasus Quantum 15 (Rotax 582)	6982		20. 3.95	Annabel Czajka	Redlands	29. 9.04P
G-MYWZ	Thruster TST Mk.1 (Rotax 503)	8128-TST-115	G-MVMJ	22. 2.93	W.H.J.KNowles	Yundum/Banjul, Gambia	29. 9.04P
G-MYXA	TEAM mini-MAX 91 *(Built D S Worman)*	PFA 186-12266		13. 6.95	L.H.S.Stephens	Tinnel, Landulph, Saltash	27. 7.04P

Registration	Type	C/n	Date	Owner	Location	Expiry
G-MYXB	Rans S-6-ESD Coyote II *(Built A Aldridge - c/n 1293.567) (Tri-cycle u/c)*	PFA 204-12787	20. 6.95	V.G.J.Davies and D.A.Hall	Lee-on-Solent	21. 9.04P
G-MYXC	Quad City Challenger II UK *(Built K N Dickinson)* (Hirth H2706)	CH2-0294-UK-1099	16. 5.95	K.N.Dickinson *(Current status unknown)*	Higher Barn Farm, Houghton	
G-MYXD	Solar Wings Pegasus Quasar IITC (Rotax 582)	7029	21. 6.95	A Cochrane	Longacres Farm, Sandy	5. 9.05P
G-MYXE	Pegasus Quantum 15 (Rotax 582)	7061	23. 6.95	W.Bowen	(Luton)	1.12.05P
G-MYXF	Air Création 503/Fun 18S GT bis	94/005	23. 6.95	T.A Morgan	Popham	15. 1.01P
G-MYXG	Rans S-6-ESD Coyote II *(Built G H Lee) (Tri-cycle u/c)*	PFA 204-12879	29. 6.95	G.H.Lee *(New CofR 4.03)*	Higher Barn Farm, Houghton	20. 6.01P
G-MYXH	Cyclone AX3/503	7028	3. 7.95	G.M.Brown	Craysmarsh Farm, Melksham	2.10.04P
G-MYXI	Cook Aries 1 *(Built H.Cook) (Design awaiting finalisation 10.01- planned engine fit is BMW R80)*	BMAA/HB/048	4. 7.95	H.Cook	(Newport, Gwent)	
G-MYXJ	Mainair Blade (Rotax 582)	1048-0795-7 & W846	17. 7.95	L Seddon	Eshott	9. 1.05P
G-MYXK	BFC Challenger II *(Built E G Astin - c/n CH2-1194-1254)* (Rotax 503)	PFA 177A-12877	11. 7.95	V.Vaughan	(Mullinahone, County Tipperary)	31. 7.04P
G-MYXL	Mignet HM-1000 Balerit	112	11. 7.95	R.W.Hollamby	Bardown, Wadhurst	17. 7.05P
G-MYXM	Mainair Blade (Rotax 582)	1047-0795-7 & W845	19. 7.95	S.C.Hodgson	(Chesterfield)	30. 7.04P
G-MYXN	Mainair Blade (Rotax 582)	1046-0795-7 & W844	27. 7.95	M.R.Sands	Shotton Colliery, Peterlee	5. 9.05P
G-MYXO	Letov LK-2M Sluka *(Built K C Rutland)* (Rotax 447)	8295s001 & PFA 263-12873	27. 7.95	G.W.Allport	(Kingswinford)	1. 7.05P
G-MYXP	Rans S-6-ESD Coyote II *(Built K J Lywood - c/n 0494-605) (Tail-wheel u/c)*	PFA 204-12886	31. 7.95	R S Amor	Weston Zoyland	1. 9.05P
G-MYXR	Meridian Renegade Spirit UK *(Built S Hooker)*	PFA 188-12755	2. 8.95	S.Hooker *(Current status unknown)*	Ashford	
G-MYXS	Kolb Twinstar Mk.3 *(Built M A Smith - kit no K0015-0693)*	PFA 205-12528	4. 5.94	R.Coar	Higher Barn Farm, Houghton	17. 8.05P
G-MYXT	Pegasus Quantum 15 (Rotax 582)	7073	4. 8.95	A Bloomfield and Allison Underwood	Headon Farm, Retford	10.12.04P
G-MYXU	Thruster T.300 *(Built Tempest Aviation Ltd)* (Rotax 582)	9024-T300-513	16. 8.95	D.W.Wilson	(Collone, County Armagh)	14. 5.05P
G-MYXV	Quad City Challenger II UK *(Built A Hipkin)*	CH2-1194-UK-1243	19. 7.95	M.L.Sumner	(Market Drayton)	19. 5.05P
G-MYXW	Pegasus Quantum 15 (Rotax 582)	7090	24. 8.95	D.Martin	Perth	17. 7.04P
G-MYXX	Pegasus Quantum 15 (Rotax 582)	7081	25. 8.95	J.H.Arnold	Milverton, Taunton	21.12.03P
G-MYXY	CFM Shadow Series CD	K.245 & BMAA/HB/059	29. 8.95	N.H.Townsend	Old Sarum	21. 7.04P
G-MYXZ	Pegasus Quantum 15 (Rotax 582)	7023	21. 6.95	T.P.Wright	Roddige	11.10.05P
G-MYYA	Mainair Blade (Rotax 462)	1052-0995-7 & W850	1. 9.95	D.E.Bassett	(Marple Bridge)	18.10.04P
G-MYYB	Pegasus Quantum 15 Lite (Rotax 582)	7079	4. 9.95	A L.Johnson	Longacres Farm, Sandy	9. 8.05P
G-MYYC	Pegasus Quantum 15 (Rotax 582)	7094	12. 9.95	G.A.Dennett *(New owner 10.04)*	(Congleton)	9. 1.04P
G-MYYD	Cyclone Chaser S 447	CH.7099	15. 9.95	K A Armstrong	Linley Hill, Leven	13. 9.04P
G-MYYE	Huntwing Avon 462 *(Built P J Dickinson - kit no.9409035)*	BMAA/HB/041	21. 9.95	N.S.Payne	Broadmeadow Farm, Hereford	23. 2.05P
G-MYYF	Quad City Challenger II UK *(Built G Ferries)*	PFA 177-12811	27. 9.95	G.Ferries	Insch	16.11.05P
G-MYYG	Mainair Blade (Rotax 462) *(Believed supplied as Kit, if so c/n should be K1054...)*	1054-0995-7 & W852	4.10.95	Angelika Corson	(Beccles)	15.11.03P
G-MYYH	Mainair Blade (Rotax 582)	1056-1095-7 & W854	3.10.95	D.Travers	(Scunthorpe)	19. 5.05P
G-MYYI	Pegasus Quantum 15 (Rotax 582)	7101	28. 9.95	M R Rowlands	(Sutton-in-Ashfield)	17.11.04P
G-MYYJ	Huntwing Avon *(Built M.J.Slatter - kit no.9409033)* (Rotax 503)	BMAA/HB/033	29. 9.95	M.J.Slatter *(Completed 5.95 and stored 5.97: current status unknown)*	(Marlborough)	
G-MYYK	Pegasus Quantum 15 (Rotax 582)	7100	2.10.95	R.Noble	Redlands	16. 6.05P
G-MYYL	Cyclone AX3/503	7110	4.10.95	Flylight Airsports Ltd	Sywell	14.11.05P
G-MYYN	Pegasus Quantum 15 (Rotax 582)	7022	3.10.95	J.Darby *(Noted 5.04)*	Perth	28.12.04P
G-MYYO	Medway Raven X	MRB134/114	5.10.95	J.R.Harrison	(Chesterfield)	30. 5.01P
G-MYYP	AMF Microflight Chevvron 2-32C	036	31.10.95	I.D.Smith	Perth	27.11.04P
G-MYYR	TEAM mini-MAX 91 *(Built D Palmer)*	PFA 186-12724	31.10.95	J.J.James	East Kirkby	7.11.05P
G-MYYS	TEAM mini-MAX *(Built J.R.Hopkinson*	PFA 186-11989	7.11.95	J.R.Hopkinson *(Current status unknown)*	(Chesterfield)	
G-MYYU	Mainair Mercury	1062-1295-7 & W862	17.11.95	J.T.and A C.Swannick	Ince Blundell	10. 2.05P
G-MYYV	Rans S-6-ESD Coyote IIXL *(Built J Whiting - c/n.0896.1026XL) (Tri-cycle u/c)*	PFA 204-12943	17.11.95	M.B.Buttle	Tarn Farm, Cockerham	27. 3.05P
G-MYYW	Mainair Blade (Rotax 582)	1051-0895-7 & W849	8. 8.95	A.Morris-Jones	Swansea	17.11.04P
G-MYYX	Pegasus Quantum 15 (Rotax 582)	7126	17.11.95	I.H.Calder	(Edinburgh)	14. 4.05P
G-MYYY	Mainair Blade (Rotax 582)	1031-0495-7 & W829	15. 3.95	E.D.Locke	Barton	31. 3.05P
G-MYYZ	Medway Raven X	MRB135/116	10. 1.96	J W Leaper	Croft Farm, Defford	27. 6.04P

G-MYZA	Whittaker MW6 Merlin *(Built D.C.Davies)* (Rotax 582)	PFA 164-11396	17. 7.95	D.C.Davies	Over Farm, Gloucester	22.11.04P
G-MYZB	Pegasus Quantum 15 (Rotax 582)	7124	22.11.95	N G Barbour	(Sleaford)	1. 4.05P
G-MYZC	Cyclone AX3/503	7125	5.12.95	P.E.Owen	Eaglescott	2. 6.04P
G-MYZE	TEAM mini-MAX 91 *(Built E H Gould)* (Global GMT-35)	PFA 186-12570	28. 9.95	R.B.M.Etherington	Halwell, Totnes	6. 6.02P
G-MYZF	Cyclone AX3/503	7133	11.12.95	Microflight (Ireland) Ltd	Mullaghmore	8. 7.05P
G-MYZG	Cyclone AX3/503	7137	11. 1.96	R.A Johns	Weston Zoyland	16. 1.05P
G-MYZH*	Chargus Titan 38	JPA-1	16. 1.96	P.A James *(Trike noted 6.03) (Cancelled 24.02.04 as temporarily.wfu)*	Redhill	
G-MYZJ	Pegasus Quantum 15 (Rotax 582)	7150	24. 1.96	P.Millar	Latch Farm, Kirknewton	28. 4.03P
G-MYZK	Pegasus Quantum 15 (Rotax 582)	7157	5. 2.96	J.Douglas	East Fortune	26. 5.04P
G-MYZL	Pegasus Quantum 15 (Rotax 582)	7158	5. 2.96	T.J.Gayton-Polley	(Billingshurst)	28. 7.05P
G-MYZM	Pegasus Quantum 15 (Rotax 582)	7159	5. 2.96	D.Hope	(Uckfield)	29. 4.05P
G-MYZN	Whittaker MW6-S-LW Fatboy Flyer *(Built M.K.Shaw)* (Rotax 582)	PFA 164-12431	31. 1.96	M.K.Shaw	RAF Halton	20. 5.05P
G-MYZO	Medway Raven X	MRB136/115	12. 2.96	M.C.Arnold	Rochester	26. 7.05P
G-MYZP	CFM Shadow Series DD *(Built D G Cook - c/n 249) (PFA c/n conflicts with G-BXDY)*	PFA 161-12914	7. 2.96	R.M.Davies and P.I.Hodgson	(Amersham)	27. 8.04P
G-MYZR	Rans S-6-ESD Coyote II XL *(Built V R Leggott) (Tri-cycle u/c)*	PFA 204-12958	9. 2.96	S.E.J.McLaughlin	Lark Engine Farmhouse, Prickwillow, Ely	27. 7.05P
G-MYZV	Rans S-6-ESD Coyote II XL *(Built H Lammers - c/n 0795.849) (Tri-cycle u/c)*	PFA 204-12946	26. 2.96	B.W.Savory	Long Marston	13. 8.05P
G-MYZY	Pegasus Quantum 15 (Rotax 582)	7156	8. 2.96	D.D.Appleford	Kemble	25. 6.05P
G-MZAA	Mainair Blade (Rotax 462)	1059-1195-7 & W857	24.10.95	P.R.Proost	(Barnsley)	22. 4.05P
G-MZAB	Mainair Blade (Rotax 582)	1043-0695-7 & W841	26. 5.95	A Meadley	Baxby Manor, Husthwaite	7. 6.05P
G-MZAC	BFC Challenger II *(Built M N Calhaem - c/n CH2-0294-1100)* (Rotax 503)	PFA 177A-12716	21. 7.95	M.N.Calhaem	Fradswell, Stafford	26. 3.04P
G-MZAE	Mainair Blade (Rotax 582)	1063-1295-7 & W863	4.12.95	D.J.Guild and M.R.Revelle	(Colchester)	25. 5.05P
G-MZAF	Mainair Blade (Rotax 582)	1045-0795-7 & W843	1.12.95	G.C.Brown	Barton	25. 3.05P
G-MZAG	Mainair Blade (Rotax 582)	1042-0695-7 & W840	26. 5.95	R.D.Kay	St.Michaels	10. 5.05P
G-MZAH	Rans S-6-ESD Coyote II *(Built A Hipkin- c/n 0393.470) (Tri-cycle u/c) (Originally built as tail-wheel u/c: repaired with frame c/n 0491-186 [G-MWVL])*	PFA 204-12553	3. 9.93	C.J.Collett	Long Marston	18. 8.04P
G-MZAJ	Mainair Blade (Rotax 582)	1067-0196-7 & W869	20.12.95	M.P.Daley	Arclid Green, Sandbach	24. 6.02P
G-MZAK	Mainair Mercury	1070-0296-7 & W872	15. 1.96	A Munro	(Bishop's Stortford)	16. 5.05P
G-MZAM	Mainair Blade (Rotax 582)	1044-0695-7 & W842	31. 5.95	B.M.Marsh and N.Cox	Shobdon	4. 8.05P
G-MZAN	Pegasus Quantum 15 (Rotax 582)	7188	7. 3.96	C.G.Veitch tr Zanco Syndicate	Dunkeswell	20. 6.05P
G-MZAP	Mainair Blade 912	1036-0495-7 & W834	31. 3.95	K.D.Adams	Ince Blundell	5. 4.05P
G-MZAR	Mainair Blade (Rotax 582)	1072-0296-7 & W874	13. 2.96	D A Valentine	Deenethorpe	20. 3.05P
G-MZAS	Mainair Blade (Rotax 582)	1049-0895-7 & W847	15. 8.95	T.Carter	Pound Green, Buttonoak, Bewdley	18.10.04P
G-MZAT	Mainair Blade (Rotax 582)	1060-1195-7 & W860	29.11.95	K Horrobin	Ince Blundell	30. 8.04P
G-MZAU	Mainair Blade *(Cancelled when stolen 1999: sailwing located in 2004 and new trike built by P&M from spare parts retaining original build numbers: replacement Rotax 582 fitted)*	1064-0196-7 & W864	29.11.95	B.Light *(Noted 8.04)*	Tarn Farm, Cockerham	13. 8.05P
G-MZAV	Mainair Blade (Rotax 462)	1078-0396-7 & W881	11. 3.96	B.B.Boniface	St.Michaels	5. 5.04P
G-MZAW	Pegasus Quantum 15 (Rotax 503)	7160	14. 2.96	K.A.O'Neill	Plaistows Farm, St.Albans	2. 3.05P
G-MZAX	Pegasus Quantum 15 (Rotax 582)	7152	11. 3.96	D.W.Beach	(Harlow)	30. 3.05P
G-MZAY	Mainair Blade (Rotax 462)	1077-0396-7 & W880	15. 3.96	R.A.Ginders	(Digby, Lincoln)	22. 4.05P
G-MZAZ	Mainair Blade (Rotax 462)	1040-0595-7 & W838	26. 5.95	T Porter and D Whiteley	(Leyland)	19. 7.05P
G-MZBA	Mainair Blade (Rotax 912-UL)	1068-0296-7 & W870	15. 3.96	S.Stone	Kemble	23. 3.05P
G-MZBB	Pegasus Quantum 15 (Rotax 582)	7139	13. 3.96	T.W.Pelan	Perth	27. 8.05P
G-MZBC	Pegasus Quantum 15 (Rotax 582)	7077	15. 8.95	B.M.Quinn	Barlow, Sheffield	6. 7.05P
G-MZBD	Rans S-6-ESD Coyote II XL *(Built H W Foster - c/n 0795.850XL) (Tri-cycle u/c)*	PFA 204-12957	15. 3.96	S P Yardley	Otherton, Cannock	28. 4.05P
G-MZBF	Letov LK-2M Sluka *(Built C R Stockdale)* (Rotax 447)	PFA 263-12881	18. 3.96	T.D.Reid	Newtownards	21.12.05P
G-MZBG	Whittaker MW6-S Fatboy Flyer *(Built A W Hodder)* (Rotax 503)	PFA 164-12891	20. 3.96	M.W.Kilvert and I.Rowlands-Jones	(Newtown, Powys)	1. 6.01P
G-MZBH	Rans S-6-ESD Coyote II *(Built D.Sutherland) (Tri-cycle u/c)*	PFA 204-12244	21. 3.96	D.Sutherland	Breighton	26.10.05P

G-MZBI	Pegasus Quantum 15 (Rotax 582)	7189		21. 3.96	A.F.Halliday	(Arrochar)	29. 6.05P
G-MZBK	Letov LK-2M Sluka *(Built R Painter)* (Rotax 447)	8295s002 & PFA 263-12872		26. 3.96	R.J.Porter	Insch	26. 7.05P
G-MZBL	Mainair Blade (Rotax 582)	1080-0496-7 & W883		11. 4.96	G.Tomlinson *(New owner 12.04)*	Eshott	5. 7.04P
G-MZBM	Pegasus Quantum 15	7196		12. 4.96	W.Gray	Drummiard Farm, Bonnybank	15. 6.05P
G-MZBN	CFM Shadow Series CD *(BMAA c/n issued for rebuild of Shadow BD G-MTWP)*	K069 & BMAA/HB/073		22. 4.96	P.A James t/a Cloudbase Aviation G-MZBN	Redhill	9. 6.04P
G-MZBO	Pegasus Quantum 15 (Rotax 582)	7218		3. 5.96	J.E.Davis	(Broughty Ferry, Dundee)	24. 7.04P
G-MZBR	Southdown Raven X	SN2232/0082		24. 5.96	D.M.Lane *(Current status unknown)*	(Stourbridge)	
G-MZBS	CFM Shadow Series D *(Built P A White - kit no K.274)*	PFA 161-13008		14. 5.96	S.K.Ryan	Plaistowes Farm, St.Albans	3. 7.04P
G-MZBT	Pegasus Quantum 15-912	7224		22. 5.96	G.and Susan Simons	(Littlehampton)	12. 7.05P
G-MZBU	Rans S-6-ESD Coyote II XL *(Built J B Marshall)*	PFA 204-12992		30. 5.96	C.Clark and R.S.Marriott	Otherton, Cannock	25. 7.04P
G-MZBV	Rans S-6-ESD Coyote II XL *(Built P F Hill)- c/n 0396.950XL)* (Rotax 582) *(Tri-cycle u/c)*	PFA 204-13009		30. 5.96	L.C.Barham and R.I.Cannan	Andreas, Isle of Man	1. 6.05P
G-MZBW	Quad City Challenger II UK *(Built C Bird - c/n CH2-0795-UK-1367)* (Rotax 582)	PFA 177-12971		19. 2.96	R.T.L.Chaloner	Dunkeswell	9. 6.05P
G-MZBX	Whittaker MW6-S-LW Fatboy Flyer *(Built S.Rose & P.Tearall)* (Rotax 503)	PFA 164-12563		16. 5.96	A.A.Comper	(Welwyn Garden City)	22. 6.05P
G-MZBY	Pegasus Quantum 15 (Rotax 582)	7227		30. 5.96	L G Wray	(Knaresborough)	16. 6.05P
G-MZBZ	Quad City Challenger II UK *(Built J Flisher - c/n CH2 0695 UK 1360)* (Hirth 2706)	PFA 177-12928		11. 3.96	T.R.Gregory	Dunkeswell	22. 6.05P
G-MZCA	Rans S-6-ESD Coyote II XL *(Built S.J.Everett, K.Kettles and F.Williams - c/n 0396.953XL) (Tri-cycle u/c)*	PFA 204-12997		31. 5.96	G.R.Inston	Overseal, Derby	11. 8.05P
G-MZCB	Cyclone Chaser S 447	7220		4. 6.96	G.P.Hodgson	Rufforth	14. 9.05P
G-MZCC	Mainair Blade (Rotax 912-UL)	1086-0696-7 & W889		7. 6.96	D.E.McGauley	Ince Blundell	18. 5.05P
G-MZCD	Mainair Blade (Rotax 582)	1087-0696-7 & W890		10. 6.96	G.Evans	Swinford, Rugby	20.10.05P
G-MZCE	Mainair Blade (Rotax 462)	K1088-0696-7 & W891		17. 6.96	C.T.Halliday	(Southport)	31. 3.05P
G-MZCF	Mainair Blade (Rotax 462)	1089-0696-7 & W892		30. 8.96	C.Hannaby	(Wrexham)	9. 8.04P
G-MZCG	Mainair Blade (Rotax 462)	1090-0696-7 & W893		17. 6.96	J.F.Bennett	(Upton, Retford)	12. 9.05P
G-MZCH	Whittaker MW6-S Fatboy Flyer *(Built E J Blake and B.G.King)* (Rotax 503)	PFA 164-12131		7. 6.96	B.G.King	Dunkeswell	5. 4.05P
G-MZCI	Pegasus Quantum 15 (Rotax 582)	7231		10. 6.96	P.H.Risdale	Tower Farm, Wollaston	20. 8.05P
G-MZCJ	Pegasus Quantum 15 (Rotax 582)	7233		14. 6.96	A W.Hay	Insch	10. 7.05P
G-MZCK	AMF Microflight Chevvron 2-32C	038		11. 7.96	T.K.Lane	(Symonds Yat, Ross-on-Wye)	22.11.05P
G-MZCM	Pegasus Quantum 15 (Rotax 582) *(Unfaired trike)*	7219		3. 5.96	M.J.Newcombe	Sywell	22. 6.05P
G-MZCN	Mainair Blade (Rotax 582)	1079-0396-7 & W882		27. 6.96	K.Cockersol and K.J.Ratcliffe	Ince Blundell	21. 6.04P
G-MZCO	Mainair Mercury (Rotax 462)	1091-0796-7 & W894		26. 6.96	E.Rush	(Congleton)	27. 3.05P
G-MZCP	Solar Wings Pegasus XL-Q	SW-TE-0434 & SW-WQ-0576		11. 2.93	C.Hamblin	Clench Common	13. 9.05P
G-MZCR	Pegasus Quantum 15 (Rotax 503)	7234		28. 6.96	J.E.P.Stubberfield	(Kenley)	11. 6.05P
G-MZCS	TEAM mini-MAX 91 *(Built C S Cox)* (Rotax 377)	PFA 186-12646		20.12.95	D.T.J.Stanley	Kemble	10. 5.05P
G-MZCT	CFM Shadow Series CD	277		11. 7.96	W.G.Gill	Plaistowes Farm, St.Albans	2. 8.05P
G-MZCU	Mainair Blade (Rotax 462)	1082-0496-7 & W885		1. 5.96	C.E.Pearce	Thurton	23. 6.05P
G-MZCV	Pegasus Quantum 15 (Rotax 503)	7235		11. 7.96	B.S.Toole	Guy Lane Farm, Waverton	24. 9.05P
G-MZCX	Huntwing Avon Skytrike *(Built C Harrison - kit no.9510055)* (Rotax 503)	BMAA/HB/072		17. 7.96	G.R.Coghill and W.Mccarthy tr Huntwing Group	(Wick)	19. 9.05P
G-MZCY	Pegasus Quantum 15 (Rotax 582)	7236		19. 7.96	J F Northey	Weston Zoyland	1. 9.05P
G-MZDA	Rans S-6-ESD Coyote II XL *(Built J Dent and W.C.Lombard - c/n 0396.951)* (Rotax 582) *(Tri-cycle u/c)*	PFA 204-13019		29. 7.96	W.C.Lombard	Newby Wiske	16. 8.04P
G-MZDB	Pegasus Quantum 15-912	7237		31. 7.96	D.T.Mackenzie tr Scottish Aerotow Club	Easter Poldar Farm, Thornhill	29. 5.05P
G-MZDC	Pegasus Quantum 15 (Rotax 582)	7246		2. 8.96	M.T.Jones	Enstone	24. 1.05P
G-MZDD	Pegasus Quantum 15 (Rotax 503)	7114	G-69-23	11. 7.96	C.D.Reeves	Weston Zoyland	29.11.04P
G-MZDE	Pegasus Quantum 15 (Rotax 582)	7238		12. 7.96	I.D.Town	(Bedford)	19. 3.05P
G-MZDF	Mainair Blade (Rotax 462)	1093-0896-7 & W896		15. 8.96	M.Liptrot	Carlisle	24. 4.05P
G-MZDG	Rans S-6-ESD Coyote II XL *(Built R Rhodes - c/n 0696.1002.1100) (Tri-cycle u/c)*	PFA 204-13030		7. 8.96	M.J.Rhodes tr Barton Heritage Flying Group	Barton	9. 5.05P
G-MZDH	Pegasus Quantum 15-912	7248		12. 8.96	B.J.Palfreyman	(Newthorpe, Nottingham)	11.11.05P

Registration	Type	C/n	Previous identity	Regd	Owner / Operator	Base	CofA
G-MZDI	Whittaker MW6-S Fatboy Flyer Series A *(Built M Grunwell)* (Rotax 503)	PFA 164-11929	G-BUNN	15. 8.96	J.M.Brooks	(Bromsgrove)	19. 2.02P
G-MZDJ	Medway Raven X	MRB138/119		19. 8.96	R.Bryan *(Noted 1.05)*	Doynton	11. 8.04P
G-MZDK	Mainair Blade (Rotax 582) *(C/n reported as 1084-0696-7)*	1084-0596-7 & W887		9. 5.96	B L Cook	Sandtoft	26.10.04P
G-MZDL	Whittaker MW6-S Fatboy Flyer *(Built C D Wills)* (Rotax 582)	PFA 164-12412		19. 8.96	S.M.Pink	(Oxted)	18.11.05P
G-MZDM	Rans S-6-ESD Coyote II XL *(Built M E Nicholas - c/n 0396.954XL)* Rotax 503) *(Tri-cycle u/c)*	PFA 204-13022		2. 9.96	M.E.Nicholas *(Noted 7.03)*	Kemble	11. 4.01P
G-MZDN	Pegasus Quantum 15 (Rotax 582)	7255		5. 9.96	P.G.Ford	(Ely)	28. 9.03P
G-MZDO	Cyclone AX3/503	7252		11. 9.96	W.H.J.Knowles	Yundum/Banjul, Gambia	29. 9.04P
G-MZDP	AMF Microflight Chevvron 2-32C	020		3. 4.90	F.Overall *(Damaged mid 1995: stored 2001: new owner 6.02)*	Whitehall Farm, Wethersfield	16. 3.96P
G-MZDR	Rans S-6-ESD Coyote II XL *(Built R Pyper and P McGill)*	PFA 204-13012		8. 8.96	J.D.Gibbons	(Newry, County Armagh	21. 3.04P
G-MZDS	Cyclone AX3/503	7253		16. 9.96	D S Parker	Carlisle	14.12.05P
G-MZDT	Mainair Blade (Rotax 582)	1096-0996-7 & W899		19. 9.96	G.P.Wiley	Otherton, Cannock	8. 7.05P
G-MZDU	Pegasus Quantum 15-912	7260		19. 9.96	G.A Breen	Portimão, Faro, Portugal	26.10.05P
G-MZDV	Pegasus Quantum 15 (Rotax 582)	7199		9. 4.96	P.M.Wilkinson	(Great Orton)	5. 6.01P
G-MZDX	Letov LK-2M Sluka *(Built T J T Dorricott)* (Rotax 447 1-V)	PFA 263-12882		30. 9.96	A.L.Brown	(Redditch)	11.11.05P
G-MZDY	Pegasus Quantum 15 (Rotax 462HP)	7263		2.10.96	R.Bailey	Sutton Meadows, Ely	7.10.04P
G-MZDZ	Hunt Avon/Wing *(Built E.W.Laidlaw - kit no.9501042)*	BMAA/HB/045		23.10.96	E.W.Laidlaw *(Under construction 2001)*	(Turriff)	
G-MZEA	Quad City Challenger II *(Built G S Cridland - c/n CH2-0294-1101)* (Hirth 2706)	PFA 177A-12728		22. 4.96	G.S.Cridland	Wymeswold	2. 4.04P
G-MZEB	Mainair Blade (Rotax 462)	1074-0396-7 & W876		22. 7.96	K P Smith	(Dunmow)	13. 7.05P
G-MZEC	Pegasus Quantum 15 Super Sport (Rotax 582)	7278		24.10.96	A B.Godber	Bradley Ashbourne, Derby	19.11.04P
G-MZED	Mainair Blade (Rotax 582)	1092-0796-7 & W895		3. 7.96	C.W.Potts	Eshott	19. 4.05P
G-MZEE	Pegasus Quantum 15 (Rotax 582)	7245		9. 8.96	M.Hassan	Plaistowes Farm, St.Albans	5. 9.05P
G-MZEG	Mainair Blade (Rotax 582)	1095-0896-7 & W898		8. 8.96	R.and A.Soltysik	Otherton, Cannock	12.10.05P
G-MZEH	Pegasus Quantum 15 (Rotax 582)	7259		19. 9.96	P.S.Hall	Sywell	16.10.05P
G-MZEI	Whittaker MW5-D Sorcerer *(Built W.G.Reynolds)*	PFA 163-12011		28.10.96	S.A.Gill *(US Navy c/s)*	(Winkleigh)	1.10.04P
G-MZEJ	Mainair Blade (Rotax 462)	1097-0996-7 & W900		8.10.96	P.M.Hopewell	(Loughborough)	29. 4.05P
G-MZEK	Mainair Mercury (Rotax 462)	1098-1096-7 & W901		14.10.96	M.Whiteman-Heywood	Arclid Green, Sandbach	15.11.04P
G-MZEL	Cyclone AX3/503	7250		30.10.96	T.I.Bull	Tarn Farm, Cockerham	18. 4.04P
G-MZEM	Pegasus Quantum 15-912	7277		8.11.96	D E J McVicker	(Toomebridge, Antrim)	9.10.05P
G-MZEN	Rans S-6-ESD Coyote II *(Built P Bottomley - c/n 1294.705) (Tri-cycle u/c)*	PFA 204-12823		9. 7.96	C.Slater	Bentley Farm, Coal Aston)	7.10.04P
G-MZEO	Rans S-6-ESD Coyote II XL *(Built J A A Dungey)*	PFA 204-13046		19.11.96	J.R.Dobson tr G-MZEO Group *(Noted 1.05)*	Eshott	26. 8.03P
G-MZEP	Mainair Rapier	1103-1296-7 & W906		13.12.96	B O'Connor	Eddsfield	8. 7.04P
G-MZER	Cyclone AX2000	7251	G-69-28 G-MZER	4.12.96	J H Keep	Henstridge	5. 9.05P
G-MZES	Letov LK-2M Sluka *(Built C Parkinson - kit no 8296K10)* (Rotax 447)	PFA 263-13064		5.12.96	J.L.Self *(Noted 8.04)*	Priory Farm, Tibenham	14. 4.05P
G-MZEU	Rans S-6-ESD Coyote II XL *(Built J E Holloway) (Tri-cycle u/c)*	PFA 204-13023		23.12.96	D.W.Pearce	Enstone	18. 3.04P
G-MZEV	Mainair Rapier	1101-1296-7 & W904		7. 1.97	M.Jones tr G-MZEV Syndicate	Barton	19. 7.05P
G-MZEW	Mainair Blade (Rotax 462)	1105-0197-7 & W908		13. 1.97	S.J.Meehan	Sittles Farm, Alrewas	10. 5.05P
G-MZEX	Pegasus Quantum 15 (Rotax 582)	7292		19.11.96	M.P.Duckett	(Bradford)	22. 4.05P
G-MZEY	Micro Aviation B.22S Bantam	96-002	ZK-TII	7. 1.97	K.T.Bettington and D.Harris	(Brierley Hill/Stourbridge)	15. 6.05P
G-MZEZ	Pegasus Quantum 15-912	7285		8.11.96	M.G.Evans	Finmere	17. 8.05P
G-MZFA	Cyclone AX2000	7301		17.12.96	R.S.Mcmaster	Sackville Lodge Farm, Riseley	2. 3.05P
G-MZFB	Mainair Blade (Rotax 462)	1108-0197-7 & W911		7. 1.97	A.J.Plant	(Manchester)	5. 5.05P
G-MZFC	Letov LK-2M Sluka *(Built G Johnson - kit no 8296K009)* (Rotax 447)	PFA 263-13063		7. 1.97	J.E.Tulett	(Burmarsh, Romney Marsh)	24. 8.04P
G-MZFD	Mainair Rapier (Rotax 462)	1109-0197-7 & W912		24. 1.97	R.Gill	Knapthorpe Lodge, Caunton	17. 8.05P
G-MZFE	Huntwing Avon *(Built G.J.Latham)* (Rotax 503 - *kit no.9507049)*	BMAA/HB/061		16. 1.97	G.J.Latham	Sittles Farm, Alrewas	15. 6.05P
G-MZFF	Huntwing Avon 503 *(Built B.J.Adamson - kit no.9604058)*	BMAA/HB/074		22. 1.97	B.J.Adamson	Ince Blundell	16. 1.03P
G-MZFG	Pegasus Quantum 15 (Rotax 582)	7305		21. 1.97	A M.Prentice	Yundum/Banjul, Gambia	23. 3.05P
G-MZFH	AMF Microflight Chevvron 2-32C	039		27. 3.97	D R Gooby	Henstridge	15. 9.05P
G-MZFI	Lorimer Iolaire *(Built H.Lorimer)* (BMW)	BMAA/HB/035		30. 1.97	H.Lorimer *"Iolaire" (Stored 10.03)*	Hunterston Farm, Stair	

G-MZFK	Whittaker MW6 Merlin *(Built K Worthington)* (Rotax 532)	PFA 164-11626	10. 2.97	G.J.Chadwick, tr G-MZFK Flying Group	Tarn Farm, Cockerham	9.12.02P
G-MZFL	Rans S-6-ESD Coyote II XL *(Built G A Clayton - c/n.0696.999XL) (Tri-cycle u/c)*	PFA 204-13041	12. 2.97	T.W.Stewart tr G-MZFL Flying Group	Eshott	26. 8.05P
G-MZFM	Pegasus Quantum 15 (Rotax 582)	7310	21. 2.97	M.J.A.New *(New owner 11.04)*	Mill Farm, Hughley, Much Wenlock	28. 9.03P
G-MZFN	Rans S-6-ESD Coyote II *(Built C R Wallbank)*	PFA 204-12977	26. 2.97	C.J. and W.R.Wallbank	Ley Farm, Chirk	25. 4.05P
G-MZFO	Thruster T.600N (Rotax 503)	9037-T600N-001	4. 3.97	J.Berry	Barton	7. 6.05P
G-MZFP*	Thruster T.600N (Rotax 503)	9047-T600T-002	4. 3.97	R.Nayak *(Cancelled 27.10.04 by CAA)*	North Coates	12. 4.04P
G-MZFR	Thruster T.600N (Rotax 503)	9047-T600N-003	4. 3.97	M.Firman tr Blue Bird Syndicate	Shobdon	28. 3.04P
G-MZFS	Mainair Blade (Rotax 582) *(Officially regd with trike c/n 1010-0297-7)*	1110-0297-7 & W913	8. 1.97	P.Green *(Noted 4.04)*	North Coates	30. 8.05P
G-MZFT	Pegasus Quantum 15-912	7264	2.10.96	C R Cawley *(New owner 9.04)*	(Chelmsford)	8. 5.04P
G-MZFU	Thruster T.600N 450 Jab	9047-T600N-004	4. 3.97	G.J.Slater	Clench Common	9. 2.05P
G-MZFV	Pegasus Quantum 15-912	7324	13. 3.97	A J.Geary	Baxby Manor, Husthwaite	22. 4.05P
G-MZFX	Cyclone AX2000	7322	14. 3.97	Flylight Airsports Ltd	Sywell	7. 6.05P
G-MZFY	Rans S-6-ESD Coyote II XL *(Built L.G.Tserkezos - c/n.0696.1003) (Tri-cycle u/c)*	PFA 204-13043	17. 3.97	L.G.Tserkezos	Popham	23. 3.05P
G-MZFZ	Mainair Blade (Rotax 582)	1119-0497-7 & W922	2. 4.97	D.C.Keeble	Oak Farm, Woodton	21. 2.05P
G-MZGA	Cyclone AX2000	7303	17.12.96	R.J.Gilbert	Mullaghmore	28. 7.05P
G-MZGB	Cyclone AX2000	7302	28. 1.97	P.Hegarty	(Magherafelt, County Londonderry)	29. 7.05P
G-MZGC	Cyclone AX2000	7304	20.12.96	Carol E.Walls	Mullaghmore	23. 5.05P
G-MZGD	Rans S-5 Coyote *(Built A G Headford - c/n 89.095)*	PFA 193-13096	1. 4.97	M.J.Olsen	Eshott	31. 8.04P
G-MZGF	Letov LK-2M Sluka *(Built R J Cook) - kit no 8296K008)* (Rotax 447)	PFA 263-13073	8. 4.97	G.Lombardi	RAF Wyton	17. 4.05P
G-MZGG	Pegasus Quantum 15 (Rotax 503)	7327	10. 4.97	R.W.Partington	Sywell	17. 4.05P
G-MZGH	Huntwing Avon 462 *(Built G C Horner - kit no.9406021)*	BMAA/HB/070	20.10.96	J.H.Cole	(Rugeley)	1. 8.04P
G-MZGI	Mainair Blade (Rotax 912-UL)	1117-0397-7 & W920	11. 4.97	H.M.Roberts	St.Michaels	11.10.05P
G-MZGJ	Kolb Twinstar Mk.3 *(Built P.Coppock - kit no K0008-0193)* (Hirth 2705 R06)	PFA 205-12421	16. 4.97	P.Coppock	Kemble	28. 7.05P
G-MZGK	Pegasus Quantum 15 (Rotax 582)	7331	30. 4.97	A J.Wells	Sywell	8.11.05P
G-MZGL	Mainair Rapier	1104-0197-7 & W907	18.12.96	D.Savatovich	(Stoke-on-Trent)	8. 6.04P
G-MZGM	Cyclone AX2000	7334	1. 5.97	W.G.Dunn	Winkleigh, Devon	20. 5.05P
G-MZGN	Pegasus Quantum 15 (Rotax 503)	7332	2. 5.97	G Taylor	(Norwich)	27.10.05P
G-MZGO	Pegasus Quantum 15 (Rotax 582)	7320	20. 3.97	S.F.G.Allen	Long Marston	2. 8.04P
G-MZGP	Cyclone AX2000	7333	7. 5.97	D.G.Palmer tr Buchan Light Aeroplane Club	Fetterangus	12. 6.05P
G-MZGS	CFM Shadow Series DD *(Built M J McChrystal - kit no K.284)* (Rotax 447)	PFA 161-13050	8. 5.97	E Fogarty	(Tadworth)	28. 9.05P
G-MZGT	Roger Hardy RH7B Tiger Light *(Built J.B.McNab)*	PFA 230-13013	10. 3.97	J.B.McNab	Dunkeswell	21. 1.05P
G-MZGU	Arrowflight Hawk II (UK) *(Built Arrowflight Aviation Ltd)* (Rotax 503)	PFA 266-13075	8. 5.97	J.N.Holden *(New owner 12.03)*	(Enniskillen)	3. 5.02P
G-MZGV	Pegasus Quantum 15 (Rotax 582)	7339	12. 6.97	R.E.Kilby tr G-MZGV Syndicate	Dunkeswell	22. 6.05P
G-MZGW	Mainair Blade (Rotax 462)	1112-0297-7 & W915	19. 2.97	R.Almond	Wickhambrook, Newmarket	11.10.05P
G-MZGX	Thruster T.600N (Rotax 503)	9057-T600N-005	28. 4.97	I.Shaw tr G-MZGX Group	Arclid Green, Sandbach	10.12.05P
G-MZGY	Thruster T.600N (Rotax 503)	9057-T600N-006	28. 4.97	M.J.and A R.Wolldridge	Siege Cross Farm, Thatcham	16. 9.05P
G-MZGZ	Thruster T.600N (Rotax 503)	9057-T600N-007	28. 4.97	B.E.J.Badger	(Warwick)	29. 9.05P
G-MZHA	Thruster T.600T (Rotax 503)	9057-T600T-008	28. 4.97	R.V.Buxton	Feshiebridge	10. 8.05P
G-MZHB	Mainair Blade (Rotax 462)	1114-0297-7 & W917	19. 2.97	R.J.Butler	Guy Lane Farm, Waverton	27. 4.05P
G-MZHC	Thruster T.600T (HKS 700E) *(Officially regd with Rotax 582)*	9067-T600T-009	13. 5.97	S.W.Tallamy	Davidstow Moor	14. 6.04P
G-MZHD	Thruster T.600T (Rotax 503)	9067-T600T-010	13. 5.97	B.E.Foster	(Tain)	11. 4.05P
G-MZHE	Thruster T.600N (Rotax 503)	9067-T600N-011	13. 5.97	R.S.O'Carroll	Tandragee, Craigavon, County Armagh	18. 5.05P
G-MZHF	Thruster T.600N (Rotax 582)	9067-T600N-012	13. 5.97	M.F.Cottam	(Lincoln)	2. 8.04P
G-MZHG	Whittaker MW6-T *(Built M G Speers)* (Rotax 532)	PFA 164-11420	16. 6.97	M.G.Speers	(Douglas, Isle of Man)	24. 6.05P
G-MZHI	Pegasus Quantum 15 (Rotax 582)	7337	27. 5.97	C.Surman	(Cranleigh)	11. 7.05P
G-MZHJ	Mainair Rapier (Rotax 462)	1123-0697-7 & W926	17. 6.97	D.J.King	(Blackburn)	22. 6.05P
G-MZHK	Pegasus Quantum 15 (Rotax 582)	7352	24. 6.97	O.Goodwin	(York)	22. 8.05P

Reg	Type	C/n	Prev id	Regd	Owner	Base	CofA
G-MZHL	Mainair Rapier	1126-0797-7 & W929		30. 6.97	K Mallin	Pound Green, Buttonoak, Bewdley	1. 9.05P
G-MZHM	TEAM hi-MAX 1700R	PFA 272-12912		8. 1.97	M.H.McKeown	(Gorey, County Wexford)	11. 8.05P
	(Built M.H.McKeown) (Robin 440) *(Officially regd with Rotax 447)*						
G-MZHN	Pegasus Quantum 15	7351		27. 6.97	F.Omarie-Hamdanie	Hunsdon	12.10.05P
	(Rotax 462HP)						
G-MZHO	Quad City Challenger II	PFA 177-12936		15. 7.97	J.Pavelin	West Barling, Essex	27. 7.05P
	(Built J.Pavelin)						
G-MZHP	Pegasus Quantum 15	7353		15. 7.97	A S.Findley	Longacres Farm, Sandy	20. 8.04P
	(Rotax 582)						
G-MZHR	Cyclone AX2000	7307		7. 3.97	J.Leden and C.P.Dawes	Darley Moor	27. 6.05P
G-MZHS	Thruster T.600T	9077-T600T-013		4. 7.97	D.Mahajan	Lower Mountpleasant, Chatteris	20.10.05P
	(Rotax 582) *(Officially regd with Rotax 503)*						
G-MZHT	Whittaker MW6 Merlin	PFA 164-11244		12. 6.97	G J Chadwick	Tarn Farm, Cockerham	18. 8.05P
	(Built P Mogg) (Hirth 2706)						
G-MZHU	Thruster T.600T	9077-T600T-019		4. 7.97	P.A Tarplee	Otherton, Cannock	27. 8.02P
	(Rotax 503)				*(Noted 1.04)*		
G-MZHV	Thruster T.600T	9077-T600T-018		4. 7.97	L.G.M.Maddick	Leicester	4. 1.05P
	(Rotax 503) *(Officially regd as "T.600N")*				tr Hotel Victor Group		
G-MZHW	Thruster T.600N	9077-T600N-017		4. 7.97	Tracey J.Mellor	(Congresbury, Bristol)	31. 3.05P
	(Rotax 503)						
G-MZHY	Thruster T.600N	9077-T600N-015		4. 7.97	J.P.Beeley	Ince Blundell	26. 6.04P
	(Rotax 503)						
G-MZHZ*	Thruster T.600N	9077-T600N-014		4. 7.97	B.S.Waycott tr Red Arrow Syndicate	(Lydney)	29. 6.05P
	(Rotax 582,				*(Cancelled 8.10.04 as destroyed) (Noted 11.04)*		
G-MZIA	TEAM hi-MAX 1700R	PFA 272-13020		25. 4.97	I.J.Arkieson	(Meols, Wirral)	
	(Built I.J.Arkieson)				*(Current status unknown)*		
G-MZIB	Pegasus Quantum 15	7354		15. 7.97	S.Murphy	Trim, County Meath	18. 8.05P
	(Rotax 582)						
G-MZIC	Pegasus Quantum 15	7348		24. 6.97	Helen M.Squire and C.F.Two	(Swansea)	6. 8.03P
	(Rotax 503)				t/a Swansea Airsports Services		
G-MZID	Whittaker MW6 Merlin	PFA 164-11383		15. 7.97	M.G.A Wood	(Tadcaster)	8. 6.05P
	(Built M.G.A Wood) (Rotax 503)						
G-MZIE	Pegasus Quantum 15	7359		6. 8.97	Flylight Airsports Ltd	Sywell	11. 2.05P
	(Rotax 582)						
G-MZIF	Pegasus Quantum 15	7355		16. 7.97	V.Grayson	(Sittingbourne)	27. 7.05P
	(Rotax 503)						
G-MZIH*	Mainair Blade	1128-0797-7 & W931		16. 7.97	E.Scarisbrick	Baxby Manor, Husthwaite	12. 9.01P
	(Rotax 462)				*(Cancelled 19.6.03 by CAA) (Noted 8.04)*		
G-MZII	TEAM mini-MAX 88	PFA 186-11842		19. 3.97	M.J.Kirk	Haverfordwest	10. 4.04P
	(Built G.F.M.Garner)						
G-MZIJ	Pegasus Quantum 15	7362		14. 8.97	C.M.Saysell	Plaistowes Farm, St.Albans	6.12.05P
	(Rotax 582)						
G-MZIK	Pegasus Quantum 15	7368		8. 9.97	L.A Read	Eaglescott	8.10.05P
	(Rotax 582)						
G-MZIL	Mainair Rapier	1132-0897-7 & W935		1. 9.97	G S Highley	Deenethorpe	7.11.05P
	(Rotax 462)						
G-MZIM	Mainair Rapier	1124-0697-7 & W927		9. 6.97	M.J.McKegney	Newtownards	1. 9.05P
	(Rotax 462)						
G-MZIR	Mainair Blade	1134-0997-7 & W937		18. 9.97	S.W.Tallamy	Davidstow Moor	3. 8.05P
	(Rotax 582)						
G-MZIS	Mainair Blade	1115-0397-7 & W918		17. 2.97	K.R.McCartney	Baxby Manor, Husthwaite	23. 4.05P
	(Rotax 462)						
G-MZIT	Mainair Blade	1129-0897-7 & W932		16. 7.97	P.M.Horn	Shotton Colliery, Peterlee	26. 8.05P
	(Rotax 912-UL)						
G-MZIU	Pegasus Quantum 15	7371		15.10.97	M.J.Stainthorpe	(Charfield, Wotton-under-Edge)	27. 2.05P
	(Rotax 582)						
G-MZIV	Cyclone AX2000	7372		21.10.97	C.J.Tomlin	Knapthorpe Lodge, Caunton	31. 1.05P
G-MZIW	Mainair Blade	1127-0797-7 & W930		16. 7.97	N.E.J.Hayes	(Chorley)	12.10.05P
	(Rotax 462)						
G-MZIX	Mignet HM-1000 Balerit	130		23. 9.97	P.E.H.Scott	(Stockbridge)	2. 6.05P
G-MZIY	Rans S-6-ESD Coyote II XL	PFA 204-13184		29. 9.97	P.A Bell	Barton	11.11.05P
	(Built P A Bell - c/n.1096.1050XL) (Tri-cycle u/c) (Rebuilt with new fuselage frame c.1998)						
G-MZIZ	Meridian Renegade Spirit UK	PFA 188-11701	G-MWGP	21.10.92	R J Collins	Belle Vue Farm, Yarnscombe	15. 4.05P
	(Built B Bayley - kit no.257)						
G-MZJA	Mainair Blade	1135-0997-7 & W938		30. 9.97	R.C.McArthur	Ince Blundell	12. 4.05P
	(Rotax 582)						
G-MZJB	Aviasud Mistral	047	(ex ?)	30. 9.97	J M Whitham	(Delves Farm, Delves, Huddersfield)	
					(New owner 3.04)		
G-MZJD	Mainair Blade	1130-0897-7 & W933		7. 8.97	A R.Vincent and R.W.Neal	(Lancaster/Preston)	26. 9.05P
	(Rotax 503)						
G-MZJE	Mainair Rapier	1136-1097-7 & W939		17.10.97	J.E.Davies	(Southport)	5. 8.04P
G-MZJF	Cyclone AX2000	7378		2.12.97	P.W.Hastings	Long Marston	21.12.04P
G-MZJG	Pegasus Quantum 15	7335		2. 5.97	P.D.Myer	Kemble	21. 6.05P
	(Rotax 462)						
G-MZJH	Pegasus Quantum 15	7350		25. 6.97	D.W.Ormond	Deenethorpe	27.11.04P
	(Rotax 503)						
G-MZJI	Rans S-6-ESD Coyote II XL	PFA 204-13221		3.11.97	M A Newbould and C Topp	Baxby Manor, Husthwaite	16. 8.05P
	(Built J Whiting - c/n.1096.1046XL) (Tri-cycle u/c)						
G-MZJJ	Murphy Maverick	PFA 259-13016		5.11.97	P.J.Wood	Roddige	18. 7.05P
	(Built M F Cottam) (Jabiru 2200A)						
G-MZJK	Mainair Blade	1100-1196-7 & W903		19.11.96	A H.Kershaw	(Bury)	15. 7.01P
	(Rotax 582)						
G-MZJL	Cyclone AX2000	7363		11. 8.97	M.H.Owen *(Noted 4.04)*	Dunkeswell	28.10.00P
G-MZJM	Rans S-6-ESD Coyote II XL	PFA 204-13215		19.11.97	K A Hastie	Popham	25. 4.045
	(Built R J Hopkins - kit no.1096.1049XL)						

G-MZJN	Pegasus Quantum 15	7376		11.11.97	J.Nelson	Knapthorpe Lodge, Caunton	9. 2.05P
	(Rotax 582)				*(Op Derbyshire and Nottingham Microlight Club)*		
G-MZJO	Pegasus Quantum 15	7338		17. 6.97	D.J.Cook	Eaglescott	30. 9.04P
	(Rotax 582)						
G-MZJP	Whittaker MW6-S Fatboy Flyer	PFA 164-13049		21.10.97	D.J.Burton and C.A J.Funnell	(Brighton)	
	(Built D.J.Burton and C.A J.Funnell)				*(Current status unknown)*		
G-MZJR	Cyclone AX2000	7385		11.11.97	N.A Martin	Clench Common	17. 1.05P
	(HKS 700E)				tr Marlborough Aerotow Group		
G-MZJS	Murphy Maverick 430	PFA 259-13017		12.12.97	R.B.M.Etherington	Halwell, Totnes	27. 6.05P
	(Built R D Bernard) (Jabiru 2200A)						
G-MZJT	Pegasus Quantum 15-912	7399		23.12.97	M.A McClelland	Old Sarum	26.12.04P
G-MZJV	Mainair Blade	1141-0198-7 & W944		7. 1.98	M.A Roberts	West Malling	13. 6.04P
G-MZJW	Pegasus Quantum 15-912	7390		27. 1.98	W.H.J.Knowles	Yundum/Banjul, Gambia	8.10.04P
G-MZJX	Mainair Blade	1139-0198-7 & W942		9. 1.98	A D Taylor	(Penrith)	25. 5.05P
	(Rotax 503)						
G-MZJY	Pegasus Quantum 15-912	7394	(EI-) G-MZJY	23.12.97	Esther J.Childs	(Clara, County Offaly)	30. 6.05P
G-MZJZ	Mainair Blade	1121-0597-7 & W924		23. 6.97	P.McParlin	Ince Blundell	16. 7.05P
G-MZKA	Pegasus Quantum 15	7380		1.12.97	A S.R.McSherry	West Kilbride	25. 4.05P
G-MZKC	Cyclone AX2000	7398		22. 1.98	D.Cioffi tr Broad Farm Flyers	Broad Farm, Eastbourne	20. 2.05P
G-MZKD	Pegasus Quantum 15	7404		19. 3.98	S.J.E.Smith	Eshott	9. 8.05P
G-MZKE	Rans S-6-ESD Coyote II XL	PFA 204-13248		19. 1.98	I.Findlay	Morgansfield, Fishburn	30. 5.05P
	(Build I Findlay)						
G-MZKF	Pegasus Quantum 15	7407		21. 1.98	T.A Howe	(Margate)	27.10.04P
G-MZKG	Mainair Blade	1145-0198-7 & W948		23. 1.98	N.S.Rigby	Ince Blundell	21.12.04P
	(Rotax 582)						
G-MZKH	CFM Shadow Series DD	292-DD		23. 1.98	S.P.H.Calvert	Deanland	26. 9.05P
G-MZKI	Mainair Rapier	1147-0298-7 & W950		12. 2.98	C.K.Richardson	East Fortune	24. 5.05P
G-MZKJ	Mainair Blade	1039-0595-7 & W837		19. 5.95	L.G.M.Maddick	Leicester	1. 5.05P
	(Rotax 582)						
G-MZKK	Mainair Blade	1140-0198-7 & W943		12. 2.98	D.I.Lee	Finmere	7. 3.05P
G-MZKL	Pegasus Quantum 15	7360		18. 8.97	G.Williams	(Ilkeston)	31. 8.04P
	(Rotax 582)						
G-MZKM	Mainair Blade	1133-0897-7 & W936		15. 8.97	G.F.J.Field	(Hucknall, Nottingham)	11. 8.05P
G-MZKN	Mainair Rapier	1138-1297-7 & W941		12.12.97	G.Craig	Newtownards	21. 6.04P
G-MZKO	Mainair Blade	1131-0897-7 & W934		5. 8.97	A M.Durose	Headon Farm, Retford	28. 9.05P
	(Rotax 503)						
G-MZKR	Thruster T.600N	9038-T600N-021		27. 1.98	R.J.Arnett	(Albuferia, Portugal)	1 .3.03P
	(Rotax 582UL)						
G-MZKS	Thruster T.600N	9038-T600N-022		27. 1.98	S.Jeffrey	Ginge Farm, Wantage	12.10.05P
	(HKS 700E) *(Officially regd with Rotax 582)*						
G-MZKT	Thruster T.600T	9038-T600T-023		27. 1.98	M.J.O'Connor	(Purley , Surrey)	18.12.05P
	(Rotax 582UL)						
G-MZKU	Thruster T.600T	9038-T600T-024		27. 1.98	A S.Day	Ginge, Wantage	24. 9.04P
	(Rotax 582UL) *(Officially regd with Rotax 503)*				*(Noted 1.05)*		
G-MZKV	Mainair Blade	1144-0198-7 & W947		28. 1.98	M.P.J.Moore	(Stoke-on-Trent)	23.10.04P
G-MZKW	Quad City Challenger II	PFA 177-12518		22. 3.94	K.W.Warn	Siege Cross Farm, Thatcham	16.10.05P
	(Built K.W.Warn) (Hirth 2705 R06)						
G-MZKX	Pegasus Quantum 15	7395		15. 1.98	D B Jones	Cranfield	14. 2.05P
	(Rotax 582)						
G-MZKY	Pegasus Quantum 15	7403		16. 1.98	P.S.Constable	Chiltern Park, Wallingford	24. 5.05P
	(HKS 700E)						
G-MZKZ	Mainair Blade	K1137-0298-7 & W940		18. 2.98	R.P.Wolstenholme	Arclid Green, Sandbach	7. 8.04P
	(Rotax 582)						
G-MZLA	Pegasus Quantum 15	7415		27. 2.98	D.A Morgan	Trenchard Farm, Eggesford	21. 4.04P
	(Rotax 582)						
G-MZLB	Huntwing Experience	BMAA/HB/058		25. 2.98	M.Ffrench	(New Ross, County Wexford)	
	(Built M.Ffrench)				*(Current status unknown)*		
G-MZLC	Mainair Blade	1146-0298-7 & W949		26. 2.98	M J Rummery	Ince Blundell	6. 4.05P
G-MZLD	Pegasus Quantum 15-912	7416		24. 3.98	S.G.Mclean	(Blairgowrie)	14. 7.05P
G-MZLE	Murphy Maverick 430	PFA 259-12955	G-BXSZ	27. 2.98	J.Smith	Otherton, Cannock	22. 6.05P
	(Built A A Plumridge) (Jabiru 2200A)						
G-MZLF	Pegasus Quantum 15	7417		30. 3.98	S.Seymour	(Corby)	13. 5.04P
	(Rotax 503)						
G-MZLG	Rans S-6-ESD Coyote II XL	PFA 204-13192		3. 3.98	J.Edwards	(Radcliffe)	6. 8.04P
	(Built R H J Jenkins - c/n 0897.1143XL) (Tri-cycle u/c)				tr G-MZLG Group		
G-MZLH	Pegasus Quantum 15	7426		1. 4.98	G J Howley	(Coleford)	2. 9.05P
	(Rotax 582)						
G-MZLI	Mignet HM-1000 Balerit	133		5. 3.98	A G.Barr	Otherton, Cannock	17.10.04P
G-MZLJ	Pegasus Quantum 15	7421		20. 3.98	M.H.Colin	Otherton, Cannock	29. 3.05P
	(Rotax 503)						
G-MZLK	Ultrasports Tri-Pacer/Solar Wings Typhoon	T285-1471		9. 3.98	J.A Jones	(Winchester)	18. 5.05P
	(Fuji-Robin EC-34-PM) *(Trike unit ex G-MJEC and sailwing is Typhoon S4+ (ex-hanglider) s/n T785-1471M)*						
G-MZLL	Rans S-6-ESD Coyote II	PFA 204-13067		23. 9.97	J.A Willats and G.W.Champion		
	(Built J.A Willats and G.W.Champion)					Maypole Farm, Chislet	20.12.05P
G-MZLM	Cyclone AX2000	7425		22. 4.98	P.E.Hadley	Swinford, Rugby	4. 7.05P
	(Modified to tug version)						
G-MZLN	Pegasus Quantum 15	7431		14. 4.98	C.J.Kew	Abbey Warren Farm, Bucknall, Lincoln	24. 4.03P
	(Rotax 503)				*(Noted 11.04)*		
G-MZLP	CFM Shadow D Series SS	K.299-D		1. 4.98	C.S.Robinson	Newtownards	31. 8.04P
	(Rotax 912-UL)						
G-MZLR	Solar Wings Pegasus XL-Q	7441		28. 5.98	I W Barlow	(Ilkeston)	28. 6.05P
	(Trike c/n SW-TB-1040 ex G-MNJP fitted with new sailwing c/n 7441)						

G-MZLS	Cyclone AX2000 (HKS 700E V3)	7428		6. 7.98	M.J.A New	Mill Farm, Hughley, Much Wenlock	2.12.04P
G-MZLT	Pegasus Quantum 15-912	7438		24. 4.98	C.S.Bourne	(Stone)	22. 4.05P
G-MZLU	Cyclone AX2000 (HKS 700E V3)	7439		28. 7.98	E.Pashley	Popham	26.10.04P
G-MZLV	Pegasus Quantum 15 (Rotax 503)	7437		29. 4.98	A L.Rudge	Weston Zoyland	2. 8.05P
G-MZLW	Pegasus Quantam 15 (Rotax 582)	7440		28. 4.98	R.W.R.Crevel	Sywell	6. 7.05P
G-MZLX	Micro Aviation B.22S Bantam	97-013	ZK-JIV	9.12.97	D.L.Howell	Longacres Farm, Sandy	26. 9.05P
G-MZLY	Letov LK-2M Sluka *(Built B G M Chapman)* (Rotax 447 1V)	PFA 263-13065		20. 4.98	W.McCarthy	Wick	4.11.05P
G-MZLZ	Mainair Blade	1154-0498-7 & W957		21. 4.98	S.R.Winter	Hunsdon	22. 7.05P
G-MZMA	Solar Wings Pegasus Quasar IITC (Rotax 582)	6611		1. 9.93	A C.Barnes and D.J.Parsons	Eshott	18. 5.05P
G-MZMB	Mainair Blade (Rotax 462)	1149-0398-7 & W952		5. 3.98	J T Hearle	(Blackburn)	20. 6.05P
G-MZMC	Pegasus Quantum 15-912	7206		10. 5.96	J.J.Baker	Deenethorpe	20. 5.05P
G-MZMD	Mainair Blade	1148-0398-7 & W951		5. 3.98	T.Gate	(Clitheroe)	22. 4.05P
G-MZME	Medway EclipseR (Jabiru 2200A)	151/129E	G-582	8. 4.98	T.Bowles	Ince Blundell	1.11.03P
G-MZMF	Pegasus Quantum 15 (HKS)	7387		30. 4.98	A J.Tranter	Perth	17.10.05P
G-MZMG	Pegasus Quantum 15 Super Sport (Rotax 503)	7446		27. 5.98	J M Pattison	Weston Zoyland	27. 9.05P
G-MZMH	Pegasus Quantum 15-912	7402		27. 1.98	M.Hurtubise	(Leamington Spa)	29. 3.04P
G-MZMJ	Mainair Blade	1155-0598-7 & W958		8. 5.98	T.F.R.Calladine *(Noted 7.04)*	Oxton	1. 8.02P
G-MZMK	AMF Microflight Chevvron 2-32C	040		19. 5.98	K.D.Calvert	Park Farm, Eaton Bray	20. 9.03P
G-MZML	Mainair Blade	1158-0698-7 & W961		19. 5.98	T.Williams	(Glossop)	16. 8.05P
G-MZMM	Mainair Blade (Rotax 462)	1162-0698-7 & W965		19. 5.98	J.F.Shaw	Baxby Manor, Husthwaite	30. 7.04P
G-MZMN	Pegasus Quantum 15-912	7445		21. 5.98	L.A Hosegood	Redlands	17. 1.05P
G-MZMO	TEAM mini-MAX 91 *(Built I.M.Ross)*	PFA 186-12951		20. 5.98	I.M.Ross	Insch	9. 5.05P
G-MZMP	Mainair Blade (Rotax 582)	1160-0698-7 & W963		20. 5.98	M.A.Hicks	Wickenby	3. 1.05P
G-MZMS	Rans S-6-ES Coyote II *(Built J G Dungey - c/n.1298.1203ES) (Tri-cycle u/c)*	PFA 204-13294		26. 5.98	D.G.Matthews	Oak Farm, Woodton	29. 9.05P
G-MZMT	Pegasus Quantum 15 (Rotax 582)	7449		18. 6.98	B.J.Kitson	Sutton Meadows, Ely	7. 8.04P
G-MZMU	Rans S-6-ESD-XL Coyote II *(Built S Cox)*	PFA 204-13242		5. 6.98	S.Bishop	Mill Farm, Shifnal	2.10.05P
G-MZMV	Mainair Blade (Rotax 462)	1152-0496-7 & W955		30. 3.98	J.Mayer	Otherton, Cannock	22.12.05P
G-MZMW	Mignet HM-1000 Balerit	125		2.10.96	M.E.Whapham	Corn Wood Farm, Adversane	6. 9.05P
G-MZMX	Cyclone AX2000 (HKS 700E V3)	7451		8. 9.98	P.A Tarplee	Otherton, Cannock	18.10.04P
G-MZMY	Mainair Blade (Rotax 462)	1153-0498-7 & W956		16. 3.98	C.J.Millership	St.Michaels	27. 5.05P
G-MZNA	Quad City Challenger II UK *(Built M Tormey)*	CH2-0894-UK-1193	EI-CLE	19. 3.98	S.Hennessy	(Dublin)	14. 4.05P
G-MZNB	Pegasus Quantum 15-912	7456		17. 7.98	F.Gorse	(Caernarfon)	2. 8.05P
G-MZNC	Mainair Blade	1161-0698-7 & W964		22. 6.98	A Costello	Brook Farm, Pilling	10. 6.05P
G-MZND	Mainair Rapier	1170-0898-7 & W973		24. 6.98	S.D.Hutchinson	St.Michaels	20 7.05P
G-MZNE	Whittaker MW6-S Fatboy Flyer *(Built V E Booth)* (Rotax 582)	PFA 164-13120		26. 6.98	D.A.Perkins *(Noted 8.04)*	Tarn Farm, Cockerham	21. 6.02P
G-MZNG	Pegasus Quantum 15-912	7457		11. 8.98	M.J.Hawkins	(Lower Ansty, Dorchester)	28. 1.05P
G-MZNH	CFM Shadow Series DD	K.297-DD		30. 6.98	P.A James	Redhill	1. 7.05P
G-MZNI	Mainair Blade	1163-0698-7 & W966		3. 7.98	V.D.Carmichael	Newtownards	5. 4.05P
G-MZNJ	Mainair Blade (Rotax 462)	1168-0798-7 & W971		6. 7.98	G.E.Cole	Over Farm, Gloucester	16. 8.05P
G-MZNK	Mainair Blade	1164-0798-7 & W967		6. 7.98	R.P.Taylor	(Folkestone)	3.10.05P
G-MZNL	Mainair Blade	1165-0798-7 & W968		6. 7.98	M.A.Williams	Rochester	1. 7.05P
G-MZNM	TEAM mini-MAX 91 *(Built N.P.Thomson)* (Fuji-Robin EC-44) *(Open cockpit)*	PFA 186-12304		10. 7.98	N.P.Thomson	Latch Farm, Kirknewton	27.10.05P
G-MZNN	TEAM mini-MAX 91 *(Built D.M.Dronsfield)*	PFA 186-13125		10. 7.98	D.M.Dronsfield *(Op Brook Farm Microlight Centre)*	Brook Farm, Pilling	17. 6.02P
G-MZNO	Mainair Blade (Rotax 462)	1167-0798-7 & W970		9. 6.98	R.C.Colclough	(Stoke-on-Trent)	24. 4.05P
G-MZNP	Pegasus Quantum 15-912	7466	G-69-55	22. 7.98	C.A Hasell	Graveley Farm, Hertford	13. 9.04P
G-MZNR	Pegasus Quantum 15 (Rotax 503)	7465		17. 8.98	E.S.Wills	(Paignton)	24. 9.04P
G-MZNS	Pegasus Quantum 15-912 Super Sport	7473		31. 7.98	M J Robbins	Rochester	3.11.05P
G-MZNT	Pegasus Quantum 15-912 Super Sport	7470		25. 9.98	J Rodgers	Barton	25.10.05P
G-MZNU	Mainair Rapier	174-0898-7 & W977		5. 8.98	D.N.Carnegie	(St.Bees)	26. 9.04P
G-MZNV	Rans S-6-ESD Coyote II *(Built D E Rubery - c/n 1294.704) (Tri-cycle u/c)*	PFA 204-12884		7. 8.98	A P.Thomas	(Reading)	6. 4.05P
G-MZNX	Thruster T.600N (Rotax 503)	9098-T600N-026		10. 8.98	M.H.Moulai *(Op Silver Fern Microlights) (Noted 4.04)*	Sandtoft	4.11.02P
G-MZNY	Thruster T.600N (Rotax 582)	9098-T600N-027		10. 8.98	L.O.Partington and G.Price	(Warrington)	26. 6.05P
G-MZNZ	Letov LK-2M Sluka *(Built K.T.Vinning - c/n 8295s015)* (Rotax 447)	PFA 263-13274		21. 4.98	K.T.Vinning	Long Marston	25.11.05P
G-MZOC	Mainair Blade	1172-0898-7 & W975		10. 8.98	R.A Carr	Brunton	5.12.05P

Registration	Type	C/n	Previous identity	Regd	Owner/Operator	Probable Base	CA Expy
G-MZOD	Pegasus Quantum 15-912	7435		28. 4.98	J.W.Mann	Enstone	26. 4.05P
G-MZOE	Cyclone AX2000	7472		17. 9.98	York Microlight Centre Ltd	Rufforth	19.10.05P
	(HKS 700E V3)						
G-MZOF	Mainair Blade	1122-0697-7 & W925		5. 6.97	R.M.Ellis	(Tewkesbury)	28. 7.05P
	(Rotax 462)						
G-MZOG	Pegasus Quantum 15	7471		12.10.98	J.Urwin	Eshott	14. 6.05P
	(Rotax 503)						
G-MZOH	Whittaker MW5-D Sorcerer	PFA 163-13060		14. 8.98	D.M.Precious	Davidstow Moor	2. 8.05P
	(Built D.M.Precious) (Fuji-Robin EC-44) *(Officially recorded as Rotax 377)*						
G-MZOI	Letov LK-2M Sluka	PFA 263-13238		17. 8.98	B.S.P.Finch	Redlands, Swindon	1. 7.05P
	(Built K P Taylor - c/n 8296s012*)* (Rotax 447 1V)						
G-MZOJ	Pegasus Quantum 15	7478		9.11.98	A C Lane	Longacres Farm, Sandy	15.12.04P
	(Rotax 582)						
G-MZOK	Whittaker MW6 Merlin	PFA 164-11568		24. 8.97	R.E.Arnold	Otherton, Cannock	1. 9.05P
	(Built R.K.Willcox) (Rotax 582)				tr G-MZOK Syndicate		
G-MZOM	CFM Shadow Series DD	302-DD		8. 9.98	P.S.Winteron and P.Tidd	Lower Mountpleasant, Chatteris	28. 9.05P
					tr Side-Stick Syndicate		
G-MZON	Mainair Rapier	1180-1098-7 & W983		11. 9.98	C.King	(Newcastle Upon Tyne)	1.11.04P
G-MZOP	Mainair Blade	1178-0998-7 & W981		11. 9.98	P.Barrow	Arclid Green, Sandbach	23.10.05P
G-MZOR	Mainair Blade	1173-0898-7 & W976		21. 9.98	D.L.Foxley	Ince Blundell	6.11.04P
G-MZOS	Pegasus Quantum 15-912	7458		6.10.98	R.J.Field	(Gunville West, Newport)	21. 2.05P
G-MZOT	Letov LK-2M Sluka	PFA 263-13346		21. 9.98	J.Bolton	(Aberdeen)	13. 8.05P
	(Built J R Walter) (Rotax 447)						
G-MZOV	Pegasus Quantum 15	7512		9. 3.99	C.S.Garrett	Enstone	13. 6.05P
	(Rotax 503)				tr Pegasus XL Group		
G-MZOW	Pegasus Quantum 15-912	7502		9. 3.99	C.S.Taylor	(Rayleigh)	23. 5.05P
G-MZOX	Letov LK-2M Sluka	PFA 263-13415		15. 2.99	C.M.James	Maypole Farm, Chislet	
	(Built C M James) (Rotax 447)				*(Noted 4.04)*		
G-MZOY	TEAM mini-MAX 91	PFA 186-12526		29. 3.99	E.F.Smith	(Egremont)	
	(Built E.F.Smith)				*(Current status unknown)*		
G-MZOZ	Rans S-6-ESD Coyote II XL	PFA 204-13168		20. 5.98	D.C.and S.G.Emmons	(Reading)	21. 9.05P
	(Built D.C.and S.G.Emmons - c/n 1096.1052XL) (Rotax 912 *-officially recorded with Rotax 503) (Tri-cycle u/c)*						
G-MZPB	Mignet HM-1000 Balerit	124		4.10.96	J.K.Evans	(Lamport)	29. 9.05P
G-MZPD	Pegasus Quantum 15	7013		9. 5.95	P.M.Dewhurst	Sywell	24. 1.05P
	(Rotax 582)						
G-MZPH	Mainair Blade	1177-0998-7 & W980		26. 8.98	J.D.Hoyland	Chilbolton	9.11.05P
	(Rotax 582)						
G-MZPJ	TEAM mini-MAX 91	PFA 186-12277		23.11.92	P.R.Jenson	Sittles Farm, Alrewas	7. 7.05P
	(Built P.R.Jenson) (Rotax 503)						
G-MZPW	Solar Wings Pegasus Quasar IITC	6892		26.10.94	D.R.Griffiths	Weston Zoyland	3. 6.05P
	(Rotax 582)						
G-MZRC	Pegasus Quantum 15	7482		25.11.98	M.Hopkins	Rufforth	23. 2.03P
	(Rotax 582)						
G-MZRH	Pegasus Quantum 15	7269		11.10.96	R.J.Ware	Roddige	8. 4.05P
	(Rotax 582)						
G-MZRM	Pegasus Quantum 15-912	7455		10. 7.98	D.Morrison	Nether Huntlywood Farm, Gordon	11.10.05P
G-MZRS	CFM Shadow Series CD	141		4. 4.90	M.R.Lovegrove	Croft Farm, Defford	29. 4.02P
G-MZSC	Pegasus Quantum 15	7370		3.10.97	R.J.Greaves	Sywell	11.10.05P
	(Rotax 503)						
G-MZSD	Mainair Blade	1179-0998-7 & W978		21. 8.98	D.Sampson	East Fortune	24. 3.05P
G-MZSM	Mainair Blade	1000-0794-7 & W796		15. 7.94	P.R.Anderson	Oxton, Nottingham	10. 9.05P
	(Rotax 582)						
G-MZTA	Mignet HM-1000 Balerit	120		14. 5.96	A Fusco tr Sky Light Group	(Burwash)	8. 5.01P
G-MZTS	Aerial Arts Chaser S 447	CH703	G-MVDM	19. 3.96	D.G.Ellis	(Tamworth)	15.11.04P
G-MZUB	Rans S-6-ESD Coyote II XL	PFA 204-13244		30. 4.98	B.O.Dowsett	Astwood	4. 8.05P
	(Built B.O.Dowsett) (Tri-cycle u/c)						
G-MZZT	Kolb Twinstar Mk.3	PFA 205-12596		1. 5.98	D.E.Martin	Plaistowes Farm, St.Albans	12. 6.05P
	(Built P I Morgans - kit no K0006-0992)						
G-MZZY	Mainair Blade	1050-0895-7 & W848		13.11.95	A Mucznik	Oxton	13. 2.05P

G-NAAA - G-NZZZ

Registration	Type	C/n	Previous identity	Regd	Owner/Operator	Probable Base	CA Expy
G-NAAA	MBB Bö.105DBS-4	S.34/912	G-BUTN	6. 4.99	Bond Air Services Ltd	Blackpool	21. 2.05T
	(Rebuilt with new pod S.912 c.1993)		G-AZTI, EI-BTE, G-AZTI, EC-DRY, G-AZTI, D-HDAN		*(Op Lancashire Air Ambulance)*		
G-NAAB	MBB Bö.105DBS-4	S.416	D-HDMO	23. 3.99	Bond Air Services Ltd	Henstridge	8. 4.05T
			D-HSTP, D-HDMO		*(Dorset & Somerset Air Ambulance)*		
G-NAAS	Aérospatiale AS355F1 Ecureuil 2	5203	G-BPRG	23. 3.90	Police Aviation Services Ltd	Gloucestershire	17. 7.05T
			G-NWPA, G-NAAS, G-BPRG, N370E				
G-NAAT*	Folland Gnat T.1	FL.507	XM697	27.11.89	Hunter Flying Club	Exeter	
					(Cancelled 10.4.95 as WFU) (For rebuild to static display condition 2003)		
G-NACA	Norman NAC-2 Freelance 180	2001		23.11.87	NDN Aircraft Ltd	Not known	AC
					(Current status unknown)		
G-NACI	Norman NAC-1 Freelance 180	NAC.001	G-AXFB	20. 6.84	L.J.Martin	Sandown	16. 5.05P
G-NADS	TEAM mini-MAX 91	PFA 186-12995		8. 2.99	S.Stockill	RAF Halton	25. 4.05P
	(Built G Evans and P M Spencer)						
G-NAPO	Pegasus Quantum 15-912	7799		6. 4.01	A Gulliver	Bagby	26. 5.05P
G-NAPP	Van's RV-7	PFA 323-14115		3. 9.03	R.J.Napp	Brock Farm, Billericay	
	(Built R.J.Napp)				*(Noted under construction 7.04)*		
G-NARO	Cassutt Racer	M.14372	G-BTXR	14. 4.98	D.A Wirdnam	Redhill	7.10.00P
	(Continental O-200-A)		N68PM		*(Aka Musso Racer Original)*		
G-NATT	Rockwell Commander 114A	14538	N5921N	14. 1.80	Northgleam Ltd	Hawarden	2.11.07E
G-NATX	Cameron O-65 HAB	1681		3. 3.88	A G.E.Faulkner	(Willenhall)	5. 5.91T
					(National Express Rapide titles)		

Reg	Type	C/n	Previous identity	Date	Owner/Operator	Base	CofA
G-NATY*	Folland Gnat T.1				See SECTION 3 Part 1	Bournemouth	
G-NBDD	Robin DR400/180 Régent	1103	F-BXVN	26. 9.88	J.N.Binks and I.H.Taylor	Sherburn-in-Elmet	11. 2.07
G-NCFC	Piper PA-38-112 Tomahawk	38-81A0107	N737V	14. 1.99	S.J.Elvery	Trenchard Farm, Eggesford	6. 7.06T
			G-BNOA, N23272				
G-NCFE	Piper PA-38-112 Tomahawk	38-80A0081	G-BKMK	1. 7.99	R.M.Browes	Norwich	19. 8.07T
			OO-GME, (OO-HKD), N9676N				
G-NCUB	Piper J-3C-65 Cub (L-4H-PI)	11599	G-BGXV	6. 7.84	L.W.Usherwood	Headcorn	6.10.05P
			F-BFQT, AO-GAB, 43-30308				
G-NDGC	Grob G109	6150		7. 4.83	J.E.Bedford and M.Mathieson	Tibenham	28. 8.05
G-NDOL	Europa Aviation Europa	PFA 247-12594		30.11.93	S.Longstaff	(Sheffield)	27.10.05P
	(Built G K Brunwin - kit no.044) (NSI EA-81) *(Monowheel u/c)*						
G-NDOT	Thruster T.600N 450	0052-T600N-066		18. 6.02	P.C.Bailey	(Over, Cambridge)	1. 7.05P
G-NEAL	Piper PA-32-260 Cherokee Six	32-1048	G-BFPY	7.11.83	V.Walker	Wolverhampton	2.. 8.07
			N5588J		tr VSD Group		
G-NEAT	Europa Aviation Europa	PFA 247-12642		28. 6.94	M.Burton	Sleap	21. 5.05P
	(Built M.Burton - kit no.065) (Tricycle u/c)						
G-NEAU	Eurocopter EC135T2	0333	D-HECB	13. 9.04	McAlpine Helicopters Ltd	Oxford	AC
G-NEEL	RotorWay Executive 90	5002		7. 8.90	C.Bedford	(Skegness)	5. 5.05P
	(Built P N Haigh) (RotorWay RI 162)						
G-NEGG	EAA Acrosport II	844	N715RJ	16. 1.04	D.K.Keays and R.S.Goodwin	(Alcester)	18. 7.05P
	(Built R.S.Challis)						
G-NEGS	Thunder Ax7-77 HAB	1059		18. 3.87	M.Rowlands *"Hot-Shot"*	(Ashton-in-Makerfield)	29 5.04A
G-NEIL	Thunder AX3/503 Maxi Sky Chariot HAB	379		2.12.81	N.A Robertson *"Neil" (Op A Moore)*	(Great Missenden)	4. 9.03A
G-NELI	Piper PA-28R-180 Cherokee Arrow	28R-31011	OH-PWW	9. 2.01	The Newcastle Aero Club Ltd	Newcastle	8. 4.07T
			D-EMWE, N7693J				
G-NEMO	Raj Hamsa X'Air Jabiru	BMAA/HB/158		11. 3.04	D G Smith	Rochester	
	(Built D G Smith - kit no. 602)				*(Noted 4.04)*		
G-NEON	Piper PA-32-300B Cherokee Six	32-40683	D-EMKW	7. 4.00	S.C.A Lever	Fairoaks	24. 5.06T
			N4246R				
G-NERC	Piper PA-31-350 Navajo Chieftain	31-7405402	G-BBXX	26. 4.94	Natural Environment Research Council	Coventry	5. 7.07T
			N66869		*(Op Air Atlantique)*		
G-NESA	Europa Aviation Europa XS	PFA 247-13544		17. 4.01	K.G.and V.E.Summerhill	(Boston)	
	(Built K.G.and V.E.Summerhill - kit no.450) (Tri-gear u/c)						
G-NESU	Pilatus Britten-Norman BN-2B-20 Islander	2260	G-BTVN	30. 5.95	Northumbria Police Authority	Durham Tees Valley	20. 2.06T
					(Op North East Police)		
G-NESV	Eurocopter EC135T1	0067		4. 2.99	Northumbria Police Authority	Newcastle	30. 3.05T
					(Op North East Air Support Unit)		
G-NESW	Piper PA-34-220T Seneca III	34-8233072	D-GAMO	13.12.02	Scot Wings Ltd	RAF Kinloss	2. 2.06T
			N8064M				
G-NESY	Piper PA-18 Super Cub 95	18-7482	N124SA	18. 8.00	V.Fisher	North Side, Thorney	30.10.06
			SE-CUG				
G-NETA	Cessna 560XL Citation Excel	560-5230	N5085E	19. 2.02	Houston Air Taxis Ltd	Oxford	25. 2.06T
G-NETY	Piper PA-18-150 Super Cub	1809108	N4159K	8. 9.95	N.B.Mason	Rendcomb	27. 3.05
G-NEUF	Bell 206L-1 LongRanger II	45548	G-BVVV	20.11.98	Yendle Roberts Ltd	Booker	7.10.07E
			D-HUGO, OE-KXT, C-GLMM				
G-NEWR	Piper PA-31-350 Navajo Chieftain	31-7952129	N35251	23. 8.79	MAS Airways Ltd	Biggin Hill	3. 8.05T
G-NEWS	Bell 206B-3 JetRanger III	2547	N18098	29.11.78	Lanthwaite Aviation Ltd	(Royston)	3. 4.06T
G-NEWT	Beech 35 Bonanza	D-1168	G-APVW	28. 2.90	J.S.Allison	RAF Halton	16. 7.05
	(Continental E-185 = C35 status)		EI-BIL, G-APVW, N9866F, 4X-ACI, IDFAF 0604, ZS-BTE				
G-NEWZ	Bell 206B-3 JetRanger III	4475	C-GBVZ	28. 1.98	Peter Press Ltd	Blackbushe	1. 4.07
G-NFLC*	Handley Page HP137 Jetstream 1	222	G-AXUI	12.12.95	Air Service Training	Perth	3. 6.04T
			G-8-9		*(Now instructional airframe 7.04) (Cancelled 3.8.04 as WFU)*		
G-NFNF	Robin DR400/180 Régent	2047	VP-BNU	15.11.02	N.French	Lower Wasing Farm, Brimpton	3.12.05
			VR-BNU, G-BTDU				
G-NGRM	Spezio DAL-1 Tuholer	134	N6RM	14. 8.90	S.H.Crook	Roughay Farm, Bishops Waltham	7. 2.00P
	(Lycoming O-290-G)				*(Crashed near Le Touquet 24.7.99 following engine failure: noted 12.03)*		
G-NHRH	Piper PA-28-140 Cherokee	28-22807	OY-BIC	19. 5.82	J.E.Parkinson	Newcastle	22. 5.07
			SE-EZP				
G-NHRJ	Europa Aviation Europa XS	PFA 247-13112		30. 9.99	D.A Lowe	Lower Grounds Farm, Sherlowe	
	(Built D.A Lowe - kit no.333) (Tri-gear u/c)				*(Noted 3.03)*		
G-NICC	Aerotechnik EV-97 teamEurostar UK	1913		1. 3.04	Pickup and Son Property Maintenance Ltd	Leicester	28. 2.05P
G-NIDG	Aerotechnik EV-97 Eurostar	PFA 315-13580		29. 2.00	Skydrive Ltd	Church Farm, Shotteswell	30.11.05P
	(Built N R Beale) (Identified as Aerotechnik 99 Eurostar c/n 990609)						
G-NIGC	Jabiru Aircraft Jabiru UL-450	PFA 274A-13703		3. 5.01	N.Creeney	Brook Farm, Pilling	9. 8.04P
	(Built N.Creeney)						
G-NIGE	Luscombe 8E Silvaire	3525	G-BSHG	6. 6.90	Gardan Party Ltd	Popham	9. 3.05P
	(Continental C85)		N72098, NC72098				
G-NIGL	Europa Aviation Europa	PFA 247-12775		6. 7.95	N.M.Graham	(Southampton)	
	(Built N.M.Graham - kit no.147) (Conventional u/c)				*(Current status unknown)*		
G-NIGS	Thunder Ax7-65 HAB	1663		30. 1.90	A N.F.Pertwee *"Bang Sai"*	(Frinton-on-Sea)	22. 9.00A
G-NIJM	Piper PA-28R-180 Cherokee Arrow	28R-30644	D-EDTM	27. 8.04	Nissr Nicola Nijim	(Solihull)	13.10.07E
			N4923J				
G-NIKE	Piper PA-28-181 Archer II	28-8390086	N4315N	4. 7.89	Key Properties Ltd	White Waltham	26.10.07E
G-NIKO	Airbus A321-211	1250	D-AVZA	21. 6.00	MyTravel Airways Ltd	Manchester	20. 6.06T
G-NINA	Piper PA-28-161 Cherokee Warrior II	28-7716162	G-BEUC	29. 7.88	A.P.Gorrod	Old Buckenham	7. 1.07T
			N3507Q				
G-NINB	Piper PA-28-180 Cherokee Challenger	28-7305234	SE-KHR	16. 7.99	P.A Layzell	Old Buckenham	21. 8.05T
			OY-DLR, CS-AHY, N11C				
G-NINC	Piper PA-28-180 Cherokee G	28-7205016	SE-KVH	2. 2.00	P.A Layzell	Old Buckenham	6. 4.06T
			N2166T				
G-NINE	Meridian Renegade 912	PFA 188-12191		16. 6.93	R.F.Bond	Garston Farm, Marshfield	29. 5.05P
	(Built R.F.Bond - c/n 448)						
G-NIOG	Robinson R44 Clipper II	10471		1. 9.04	Farm Aviation Ltd	(High Wycombe)	27. 9.07E

Registration	Type	C/n	Previous identity	Date	Owner	Base	CofA
G-NIOS	Piper PA-32R-301 Saratoga SP	32R-8513004	N4381Z	28. 9.90	D.J.Everett and R.R.Alderslade	Stapleford	19. 5.05
			N105DX, N4381Z		t/a Plant Aviaton		
G-NIPA	Nipper T.66 RA.45 Series 3	S.120/1627	G-AWDD	7. 6.96	R.J.O.Walker	North Lopham	13. 4.04P
	(Built Slingsby Aircraft Co Ltd) (Volkswagen 1834 (Acro))						
G-NIPP	Nipper T.66 RA.45 Series 3	S.103/1587	G-AVKJ	17. 1.00	T.Dale	(Dunnington, York)	21. 8.97P
	(Built Slingsby Aircraft Co Ltd fromTipsy c/n 32) (Volkswagen 1834)				*(On rebuild 12.02)*		
G-NITA	Piper PA-28-180 Cherokee C	28-2909	G-AVVG	16. 1.84	T.Clifford	Cranfield	17.11.97T
	(Used spare Frame No.28-3807S)		N7517W		*(Wreck noted 8.04)*		
G-NJAG	Cessna 207 Skywagon	20700093	D-EMDN	2. 8.78	G.H.Nolan	Biggin Hill	17. 6.06T
			(N91152)				
G-NJSH	Robinson R22 Beta	0780		19. 4.88	A J.Hawes	Sywell	29. 6.06
G-NLEE	Cessna 182Q Skylane II	18265934	G-TLTD	1.12.93	G.Hall	Bournemouth	12. 6.06
			N759EL				
G-NLYB	Cameron N-105 HAB	10012		19. 4.01	P.H.E.Van Overwalle	(Nazareth, Belgium)	8. 3.05A
	(Pink Elephant Head Shape)						
G-NMID	Eurocopter EC135T2	0300		29. 9.03	Derbyshire Constabulary	(Butterley Hall, Ripley)	29. 3.07T
					(Op North Midlands Police)		
G-NMOS	Cameron C-80 HAB	4966		5. 1.01	C J Thomas and M C East	(Godalming/Alton)	30. 7.05A
					(Riversoft titles)		
G-NNAC	Piper PA-18-135 Super Cub	18-3820	PH-PSW	19. 5.81	PAW Flying Services Ltd	Bagby	6. 5.07T
	(L-21B-PI) *(Frame No.18-3820)*		R.Neth AF R-130, 54-2420				
G-NNON	Mainair Blade	1318-0302-7 & W1113		24. 4.02	A Gannon	East Fortune	23. 5.04P
G-NOBI	Spezio HES-1 Tuholer Sport	162	N1603	28.11.90	M.G.Parsons	(Buckie)	30.10.04P
	(Continental C125)						
G-NOCK	Reims/Cessna FR182 Skylane RG II	FR18200036	G-BGTK	18. 1.94	F.J.Whidbourne	Park Farm, East Worldham, Alton	6. 3.04
			(D-EHZB)				
G-NODE	Gulfstream AA-5B Tiger	AA5B-1182	N4533L	22. 5.81	Strategic Telecom Networks Ltd	Blackbushe	5. 7.05T
G-NODY	American General AG-5B Tiger	10076	N1194C	3.10.91	Curd and Green Ltd	Sywell	12. 3.07T
G-NOIR	Bell 222	47031	G-OJLC	9. 8.91	Arlington Property Developments (2003)	Blackbushe	31. 5.05T
			G-OSEB, G-BNDA, A40-CG				
G-NOIZ	Yakovlev Yak-55M	910104	RA-44537	5. 2.04	S.C.Cattlin	White Waltham	17. 3.05P
			HA-JAM, DOSAAF 04 *(blue)*				
G-NOMO	Cameron O-31 HAB	241		31.10.00	Tim Balloon Promotion Airships Ltd	(Ceva, Italy)	4. 8.05A
G-NONE	Dyn'Aéro MCR-01 ULC	PFA 301B-14238		2. 7.04	J.Flisher	Dunkeswell	
	(Built J.Flisher)						
G-NONI	Grumman AA-5 Traveler	AA5-0383	G-BBDA	1. 8.88	P.T.Harmsworth	Exeter	25. 5.07T
			(EI-AYL), G-BBDA		tr November India Flying Group		
G-NOOK	Mainair Blade 912S	1281-0401-7 & W1076		11. 6.01	P.M.Knight	Elm Farm, Wickford	13. 9.05P
G-NOOR	Commander Aircraft Commander 114B	14656		6. 2.98	As-Al Ltd	Zell-am-See, Austria	9. 6.07
G-NORD	SNCAN NC.854	7	F-BFIS	20.10.78	W.J.McCollum	Coagh, County Londonderry	27. 5.82P
					(Remains noted 11.01 - current CofR 2.05)		
G-NORT	Robinson R22	3404		7. 1.03	Plane Talking Ltd	Elstree	12. 2.06T
G-NOSE	Cessna 402B	402B0823	N98AR	23. 4.96	Atlantic Air Transport Ltd	Coventry	7. 1.05T
			G-MPCU, SE-IRL, OO-TAT, (OO-SEL), N3946C				
G-NOSY	Robinson R44 Astro	0064	G-LATK	6. 3.03	Cotswold Helicopters Ltd	Gloucestershire	3. 9.06T
			G-BVMK				
G-NOTE	Piper PA-28-181 Archer III	2843082	D-ESPI	19. 9.97	J.Beach	Elstree	9.10.06T
			N9282N				
G-NOTR	MD Helicopters MD 500N	LN018	N520MD	26. 2.01	Eastern Atlantic Helicopters Ltd	Shoreham	27. 3.07T
G-NOTT	Nott ULD2 HAB	06		11. 6.86	J.R.P.Nott	(London NW3)	
	(Built J.R.P.Nott)				*(Current status unknown)*		
G-NOTY	Westland Scout AH.1	F9630	XT624	5.11.97	R.P.Coplestone	Draycott Farm, Chiseldon	30. 3.04P
G-NOVO	Colt AS-56 HA Airship	1067		20. 5.87	J.R.Huggins	Dover	29. 4.97A
	(Built Thunder and Colt Ltd) (Konig SC430)				*(Extant 2004)*		
G-NOWW	Mainair Blade 912	1227-1299-7 & W1020		10.12.99	C Bodill	Oxton	23. 2.05P
G-NPKJ	Van's RV-6	PFA 181-13138		12. 2.98	H M Darlington	High Easter	25. 3.05P
	(Built K Jones) (Lycoming IO-360)						
G-NPNP*	Cameron N-105 HAB				See SECTION 3 Part 1	(Aldershot)	
G-NPPL	Comco Ikarus C42 FB100	0306-6543		2. 9.03	D.C.Jarman	(Chilcompton, Radstock)	14.10.05P
	(Officially regd with incorrect c/n 0307-.6543)				tr Papa Lima Group		
G-NPWR*	Cameron RX-100 HAB				See SECTION 3 Part 1	(Lincoln)	
G-NRDC*	Norman NDN-6 Fieldmaster	004		8. 6.81	Not known	Sandown	17.10.87P
					(Cancelled 3.2.95 by CAA) (Wreck noted 7.02)		
G-NROY	Piper PA-32RT-300 Lance II	32R-7985070	G-LYNN	26.11.93	R.L.West	Norwich	28. 2.05T
			G-BGNY, N3024L		t/a Roy West Cars		
G-NRRA	SIAI-Marchetti SF.260W	116	F-GOBF	29.11.00	G.Boot	Lydd	27. 4.05P
			Philippines AF BF8431, OO-SMB				
					(As "BF8431/13" in Burkina Faso Defence Force c/s)		
G-NRSC	Piper PA-23-250 Aztec E	27-7305142	N250MC	23. 6.00	Infoterra Ltd	Leicester	11. 9.06A
			(N244AR), N250MC, EI-BXP, G-BSFL, PH-NOA, 9M-AUS, PH-NOA, N40378				
G-NRYL	Mooney M.20R	29-0303	N10391	10. 3.04	Deltamood Ltd	(Jersey)	7. 4.07T
G-NSBB	Comco Ikarus C42 FB100 VLA	PFA 322-14162		15. 1.04	B.Bayes and N.E.Sams	(Cranfield)	27. 5.05P
	(Built B.Bayes and N.E.Sams)						
G-NSEW	Robinson R44 Astro	0615		6. 7.99	Captive Audience (UK) Ltd	Denham	11. 7.05T
G-NSOF	Robin HR200/120B	334		4. 6.99	Northamptonshire School of Flying Ltd	Sywell	20. 6.05T
G-NSTG	Cessna F150F	F150-0058	G-ATNI	16. 8.89	Westair Flying Services Ltd	Sleap	26. 8.05
	(Built Reims Aviation SA) (Wichita c/n 15063499)		*(Tail-wheel conversion)*				
G-NSUK	Piper PA-34-220T Seneca V	3449256	N126RB (2	27. 2.03	Genus Plc	Blackbushe	4. 3.06T
G-NUKA	Piper PA-28-181 Archer II	28-8290134	OY-CJI	9. 1.04	N.Ibrahim	Panshanger	14. 3.07T
			N8209A				
G-NULA	Flight Design CT2K	02-05-05-04		17.10.02	R.C.Skidmore	Sywell	20.12.05P
	(Assembled Pegasus Aviation as c/n 7913)				tr G-NULA Flying Group		
G-NUTS	Cameron Mr Peanut 35SS HAB	711		18. 2.81	Balloon Flights International Ltd	(Bristol)	7. 4.86A
					"Mr Peanut II" (New Cof R 3.02)		

Registration	Type	C/n	Previous identity	Regd date	Owner/operator	Base	CofA expiry
G-NUTY	Aérospatiale AS350B Ecureuil	1490	G-BXKT	20. 7.98	Arena Aviation Ltd	Redhill	16.10.06T
			F-GXRT, N333FH, N5797V				
G-NVBF	Lindstrand LBL 210A HAB	249		19. 5.95	Airxcite Ltd t/a Virgin Balloon Flights	(Wembley)	4. 9.04T
G-NVSA	de Havilland DHC-8-311A (Q300)	451	C-GDNG	20.11.98	Brymon Airways Ltd	Plymouth	20.11.07E
G-NVSB	de Havilland DHC-8-311A (Q300)	517	C-GHRI	14. 1.99	Brymon Airways Ltd	Plymouth	13. 1.08E
G-NWPB*	Thunder Ax7-77Z HAB				See SECTION 3 Part 1	(Aldershot)	
G-NWPR	Cameron N-77 HAB	1181		15. 8.85	D.B.Court	(Ormskirk)	30. 6.93A
	(Rebuilt with new envelope c/n 1667)				*(CofR restored 4.02)*		
G-NWPS	Eurocopter EC135T1	0063		15.10.98	North Wales Police Authority	Boddelwydden	11. 2.05T
G-NYMF	Piper PA-25-235 Pawnee D	25-7556112	OO-PAL	8. 2.02	The Bristol Gliding Club Pty Ltd	Nympsfield	3. 3.05
			N267JW, N9799P				
G-NYZS	Cessna 182G Skylane	18255135	G-ASRR	8. 4.04	P.Ragg	(Weerberg, Austria)	5. 7.05
			(G-CBIL), EI-ATF, G-ASRR, N3735U				
G-NZGL	Cameron O-105 HAB	1361		3. 9.86	R.A., P.M.G.and N.T.M.Vale *"Nazgul"*	(Kidderminster)	28. 5.00A
G-NZSS	Boeing Stearman E75 (N2S-5) Kaydet	75-8611	N4325	31. 1.89	D.L.H.Barrell	Duxford	27. 6.05T
	(Lycoming R-680)		Bu.43517, 42-109578		*(As "343251/27" in USAAC c/s)*		

G-OAAA - G-OZZZ

Registration	Type	C/n	Previous identity	Regd date	Owner/operator	Base	CofA expiry
G-OAAA	Piper PA-28-161 Warrior II	2816107	N9142N	8. 9.93	Halfpenny Green Flight Centre Ltd	Wolverhampton	17. 9.05T
G-OAAC	Airtour AH-77B HAB	010		13. 9.88	Director, Army Air Corps, Historic Aircraft Board of Management		
					"Go AAC"	(AAC Middle Wallop)	10. 1.00A
G-OABB	SAN Jodel D.150 Mascaret	01	F-BJST	21. 1.97	K.Manley	Swanborough Farm, Lewes	19. 3.06
			F-WJST				
G-OABC	Colt 69A HAB	1159		17.11.87	P.A C.Stuart-Kregor	(Newbury)	26. 6.00A
G-OABO	Enstrom F-28A	097	G-BAIB	10. 7.98	ABO Ltd *(Stored 9.04)*	Shoreham	13.11.04T
G-OABR	American General AG-5B Tiger	10124	C-GZLA	15. 4.98	Vulcan House Management UK Ltd	Biggin Hill	26. 4.04T
			N256ER				
G-OACA	Piper PA-44-180 Seminole	44-7995202	G-GSFT	31.07.02	Plane Talking Ltd	Cranfield	19.12.04T
			EI-BYZ, N2193K				
G-OACC	Piper PA-44-180 Seminole	44-7995190	G-FSFT	15. 1.03	Plane Talking Ltd	Elstree	20.12.07E
			EI-CCO, N2135G		*(Op Cabair)*		
G-OACE	Valentin Taifun 17E	1017	D-KCBA	22. 1.87	D.R.Piercy	Gallows Hill, Bovington Camp	27. 4.02
					(New owner 8.04)		
G-OACF	CAB Robin DR.400/180	2534		30.10.03	A C.Fletcher	Sherburn-in-Elmet	22.12.06
G-OACG	Piper PA-34-200T Seneca II	34-7870177	G-BUNR	10. 3.94	Cega Aviation Ltd	Goodwood	9.12.07E
			EI-CFI, N9245C				
G-OACI	SOCATA MS.893E Rallye 180GT	13086	G-DOOR	5. 5.98	A M.Quayle	Alderney	6. 4.07
			EI-BHD, F-GBCF				
G-OACP	de Havilland DHC-1 Chipmunk 22	35	(CS-DAO)	20. 8.96	Aeroclub de Portugal	(Lisbon, Portugal)	14. 3.03
	(Built OGMA) (Lycoming O-360)		Portuguese AF FAP 1345				
G-OADY	Beech 76 Duchess	ME-56	N5022M	27.10.86	Multiflight Ltd	Leeds-Bradford	31. 1.05T
G-OAER	Lindstrand LBL 105A HAB	359		4. 3.96	T.M.Donnelly *"Aero"*	(Sprotbrough, Doncaster)	25. 6.01A
G-OAFT	Cessna 152 II	15285177	G-BNKM	19. 4.88	Evensport Ltd	Corporation Farm, Cock Clarks	18.11.02T
			N6161Q		*(Noted 4.04)*		
G-OAHC	Beech F33C Bonanza	CJ-133	G-BTTF	2. 9.91	V.D.Speck	Clacton	5. 8.07
			PH-BND				
G-OAJB	Cyclone AX2000	7281	G-MZFJ	16. 2.99	R.S.McMaster	Sackville Lodge Farm, Riseley	21. 5.05P
G-OAJC	Robinson R44 Raven	1381		12. 5.04	A.J.Cain	(Leeds)	17. 5.07T
G-OAJL	Comco Ikarus C42 FB100	0403-6589		18. 5.04	A.J.Longbottom t/a AJL Driver Training	Dunkeswell	17. 5.05P
G-OAJS	Piper PA-39 Twin Comanche C/R	39-15	G-BCIO	9. 3.94	Go-AJs Ltd	Sherburn-in-Elmet	21. 3.07
			N49JA, N57RG, G-BCIO, N8860Y				
G-OAKJ	British Aerospace Jetstream Series 3202	795	G-BOTJ	20. 7.89	Air Kilroe Ltd	Manchester	31. 8.06T
			G-OAKJ, G-BOTJ, G-31-795				
G-OAKR	Cessna 172S Skyhawk	172S9643	N21738	10. 6.04	A.K.Robson	Oaksey Park	23. 6.07
G-OALB	Aero L-39C Albatros	931523	ES-ZLD	27. 6.00	Starindale Ltd	(Manston)	17.11.05P
			Soviet Air Force				
G-OALD	SOCATA TB-20 Trinidad	490	N54TB	17. 3.88	D.A Grief	Biggin Hill	24. 5.06
			F-GBLL		t/a Gold Aviation		
G-OALH	Tecnam P92-EA Echo	PFA 318-13675		12. 6.01	L.Hill	(Ulverston)	20. 3.05P
	(Built L Hill)						
G-OAMF	Pegasus Quantum 15-912	7764		20.12.00	I.B.Smith	Deenethorpe	16. 3.05P
G-OAMG	Bell 206B-3 JetRanger III	2901	G-COAL	25. 2.86	Alan Mann Helicopters Ltd	Fairoaks	9. 6.07T
G-OAMI	Bell 206B JetRanger II	464	G-BAUN	15. 3.01	Techno Solutions Ltd	Goodwood	10. 2.05T
			5N-BAY, G-BAUN, 5N-AOU, VR-BIA, G-BAUN, N2261W				
G-OAML	Cameron AML-105 HAB	3881		4.12.96	Stratton Motor Co (Norfolk) Ltd	(Long Stratton)	4. 6.04A
G-OAMP	Reims/Cessna F177RG Cardinal RG	F177RG0006	G-AYPF	30.11.93	G.Hamilton and R.Sheldon	Liverpool	9. 9.06
	(Wichita c/n 17700098)				tr Vale Aero Group		
G-OANI	Piper PA-28-161 Warrior II	28-8416091	N43570	8. 1.91	J.F.Mitchell	Oxford	8. 9.97
					(Damaged Upton Farm, Dover 16.6.96: wreck noted 9.96: current status unknown)		
G-OANN	Zenair CH.601HDS Zodiac	PFA 162-12932		2. 2.96	E.W.Chapman		
	(Built P Noden) (Rotax 912-UL)					Low Hill Farm, Messingham, Scunthorpe	8. 6.05P
G-OAPE	Cessna T303 Crusader	T30300245	N303MF	3. 2.99	C.Twiston-Davies and P.L.Drew	Jersey	25. 2.05T
			D-INKA, N9960C, M303HW, N9960C				
G-OAPR	Brantly B-2B	446	(G-BPST)	21. 4.89	E.D.ap Rees	Weston-super-Mare	1. 7.07
			N2280U		t/a Helicopter International Magazine		
G-OAPW	Glaser-Dirks DG-400	4-268		17. 4.90	D.Bonucchi	(Watford)	10. 6.05
G-OARA	Piper PA-28R-201 Arrow	2837002	N802ND	28.10.98	S.Evans	White Waltham	30. 5.07T
			N9622N		t/a Airsure		
G-OARC	Piper PA-28RT-201 Arrow IV	28R-7918009	EC-HXO	17. 8.99	S.J.Skilton t/a Aviation Rentals	Bournemouth	4.11.07E
			G-OARC, G-BMVE, N3071K				
G-OARG	Cameron C-80 HAB	3379		20.10.94	G.and R.Madelin *"Argent"*	(Farnham/London SW15)	28. 9.04A

G-OARI	Piper PA-28R-201 Arrow	2837005	N170ND	14.10.02	Plane Talking Ltd	Denham	12.11.05T
					(Op Denham School of Flying)		
G-OARO	Piper PA-28R-201 Arrow	2837006	N171ND	30.10.01	Plane Talking Ltd	Sywell	5.11.07T
G-OART	Piper PA-23-250 Aztec D	27-4293	G-AXKD	26.11.93	Tindon Ltd	Little Snoring	20. 3.06T
			N6936Y				
G-OARU	Piper PA-28R-201 Arrow	2837026	N174ND	24. 5.02	Mind Power Consultancy Ltd	Biggin Hill	17. 7.05T
G-OARV	ARV Aviation ARV-1 Super 2	PFA 152-11060		18. 6.84	N.R.Beale	Sproughton	12.10.87P
	(Originally kit no 001: rebuilt with kit no.008 @ 1986)				*(Stored 1.91: current status unknown)*		
G-OARW	Piper PA-32R-301 Saratoga IIHP	3246193	EC-IJT	27. 8.04	A.R.Ward	Southend	29. 9.07E
			N5339Z				
G-OASH	Robinson R22 Beta	0761	N2627Z	13. 6.88	J.C.Lane *(Op Heliflight)*	Wolverhampton	22. 6.06T
G-OASJ	Thruster T600N 450 Sprint	0037-T600N-090		13.10.03	A S.Johnson	Mapperton Farm, Newton Peverill	14.10.05P
G-OASP	Aérospatiale AS355F2 Ecureuil 2	5479	F-GJAJ	3. 8.95	Helicopter Services Ltd	Booker	14.10.04T
			F-WYMH				
G-OATE	Mainair Pegasus Quantum 15-912	8064		27. 8.04	S.J.Goate	(Wellingborough)	26. 8.05P
G-OATS	Piper PA-38-112 Tomahawk	38-78A0007	N9659N	14. 3.78	Truman Aviation Ltd	Tollerton	15.10.06T
G-OATG	Advanced Technologies AT-10 (Gas) Airship	1001		29.11.01	Advanced Technologies Group Ltd	Cardington	3. 9.04S
G-OATV	Cameron V-77 HAB	2149		14. 2.90	W.G.Andrews	(Plymouth)	23.10.93A
G-OAVA	Robinson R22 Beta II	3303		8. 3.02	B.W.Faulkner	(Petersfield)	2. 4.05T
G-OAVB	Boeing 757-23A	24289	N289AN	21.11.03	Astraeus Ltd	Gatwick	12. 7.07T
			G-OOOI, N510SK, EC-EMV, EC-247				
G-OAWS	Cameron Colt 77A HAB	4340		23. 4.98	E.A and H A Evans	(Chesterfield)	6. 7.05A
G-OBAK	Piper PA-28R-201T Turbo Cherokee Arrow III	28R-7703054	D-EKOR	27. 8.02	D R Freeth	Blackbushe	22. 9.05
			N1146Q		t/a DP Group Aviation		
G-OBAL	Mooney M.20J Model 201LM	24-1601	N56569	27.11.86	Britannia Airways Ltd *(Op Britannia Flying Club)*	Luton	3. 4.05T
G-OBAM	Bell 206B-3 JetRanger III	4511	N6379U	25. 5.99	Cherwell Tobacco Ltd	(Whitchurch, Shropshire)	4. 7.05T
G-OBAN	SAN Jodel D.140B Mousquetaire II	80	G-ATSU	20. 2.92	S.R.Cameron	North Connel, Oban	13. 6.07
			F-BKSA				
G-OBAX	Thruster T.600N 450 Sprint Jab	0051-T600N-053		12. 7.01	J.Northage and M.E.Hutchinson		
					(New owners 1.05)	Baxby Manor, Husthwaite	10. 7.04P
G-OBAZ	Best Off Skyranger 912	BMAA/HB/322		17.11.03	B.J.Marsh	Plaistowes Farm, St.Albans	30. 6.05P
	(Built B.J.Marsh - kit no.SKRxxxxxxx?)						
G-OBBC	Colt 90A HAB	1358		11. 5.89	R.A and M.A Riley	(Bromsgrove)	12.10.02A
					(BBC in the Midlands titles) "Beeb"		
G-OBBJ	Boeing 737-8DR	32777	N379BJ	29.11.01	Multiflight Jet Charter LLP	Leeds-Bradford	26. 6.06T
G-OBBO	Cessna 182S Skylane	18280534	N7274Z	8. 6.99	F.Friedenberg	Goodwood	21. 6.05
G-OBBY	Robinson R44	0939		4.12.00	P.C.and J.A Twigg	Wellesbourne Mountford	2. 1.07T
G-OBDA	Diamond DA.20-A1 Katana	10260		2. 7.98	Oscar Papa Ltd	Wolverhampton	30. 7.07T
G-OBDM	Europa Aviation Europa XS	PFA 247-14048		16.12.03	B.D.McHugh	(Orpington)	
	(Built B.D.McHugh - kit no.572)						
G-OBDN	Piper PA-28-161 Warrior III	2842177	N53586	10. 7.03	Barn Air Ltd	Redhill	15. 7.06T
G-OBEI	SOCATA TB-200 Tobago XL	2096	F-OIUX	26. 6.02	Air Touring Ltd	Biggin Hill	26. 6.05T
	(Carries Tobago GT titles)						
G-OBEN	Cessna 152 II	15281856	G-NALI	16. 8.93	Airbase Aircraft Ltd	Shoreham	29. 3.06T
			G-BHVM, N67477		*(Op Airbase Flying Club)*		
G-OBET	Sky 77-24 HAB	178		22. 2.00	Patricia M.Watkins and Susan M.Carden	(Chippenham)	20. 7.04A
G-OBEV	Europa Aviation Europa	PFA 247-12813		3. 2.98	M.B.Hill and N.I.Wingfield	(Dursley)	
	(Built M.B.Hill - kit no.188) (Monowheel u/c)				*(Current status unknown)*		
G-OBFC	Piper PA-28-161 Warrior III	2816118	N9252X	15. 7.96	Bflying Ltd	Bournemouth	4. 8.05T
					(Op Bournemouth Flying Club)		
G-OBFE	Sky 120-24 HAB	167	D-OBFE	28. 4.03	H.Schmidt	(Siegen, Germany)	15. 6.05
G-OBFS	Piper PA-28-161 Warrior III	2842039	N41274	4.12.98	Plane Talking Ltd	Elstree	3.12.07E
G-OBGC	SOCATA TB-20 Trinidad	1898		13. 5.99	Bidford Airfield Ltd	Bidford	29. 5.05T
G-OBHD	Short SD.3-60 Variant 100	SH.3714	G-BNDK	20. 1.87	Emerald Airways Ltd	Coventry	5. 3.05T
			G-OBHD, G-BNDK, G-14-3714		*(No titles - white overall)*		
G-OBIB	Colt 120A HAB	4229		9. 1.98	The Aerial Display Co Ltd *(Michelin titles)*	(Looe)	4. 4.05A
G-OBIL	Robinson R22 Beta	0792		10. 5.88	C.A Rosenberg	Miskin Manor Hotel, Pontyclun	24. 8.06T
G-OBIO	Robinson R22 Beta	1402	N7724M	29. 6.98	Heli Air Ltd	Wellesbourne Mountford	25. 8.07
G-OBJB	Lindstrand LBL 90A HAB	640		12.11.99	B.J.Bower	(Andover)	4. 4.05A
G-OBJP	Pegasus Quantum 15-912	7847		29. 8.01	S.J.Baker	Sutton Meadows, Ely	28. 8.05P
G-OBJT	Europa Aviation Europa	PFA 247-12623	G-MUZO	16.11.00	B.J.Tarmar	(Fordingbridge)	
	(Built J T Grant - kit no.055) (Wilksch WAM-120) *(Tri-gear u/c)*				*(Current status unknown)*		
G-OBLC	Beech 76 Duchess	ME-249	N6635R	3. 6.87	Pridenote Ltd	Sturgate	12.12.05T
G-OBLN	de Havilland DH.115 Vampire T.11	15664	XE956	14. 9.95	de Havilland Aviation Ltd	(St.Mary Hill, Bridgend)	
	(Regd with Nacelle No.DHP.48700)				*(As "XE956") (On rebuild 1.01)*		
G-OBLU	Cameron H-34 HAB	4914		4. 8.00	John Aimo Balloons SAS	(Mondovi, Italy)	28. 5.03A
G-OBMI	Mainair Blade	1289-0601-7 & W1084		19. 6.01	I.G.Webster and P.Clark	Ashcroft Farm, Winsford	26. 7.05P
G-OBMP	Boeing 737-3Q8	24963		8. 1.92	British Midland Airways Ltd *"baby ET"*		
					(Op bmiBaby)	Nottingham East Midlands	19. 3.05T
G-OBMS	Reims/Cessna F172N Skyhawk II	F17201584	OO-BWA	16. 4.84	D.Beverley, A J.and A P.Ransome	Sherburn-in-Elmet	3. 7.05
			(OO-HWA), D-EBYX				
G-OBMW	Grumman AA-5 Traveler	AA5-0805	G-BDFV	4. 7.79	Fretcourt Ltd	Sherburn-in-Elmet	26. 5.06
G-OBNA	Piper PA-34-220T Seneca V	3449002	N9281D	25. 5.00	Palmair Ltd	Elstree	31. 5.06
			(N338DB)				
G-OBNC	Britten-Norman BN-2B-20 Islander	3000		9. 2.05	Britten-Norman Aircraft Ltd	Bembridge	
G-OBNW	Piper PA-31-350 Navajo Chieftain	31-7305118	OY-EBE	4. 4.03	Liberty Group Assets Ltd	Blackpool	5. 5.07T
			EI-BYE, G-BFDA, SE-GDR, N9684N				
G-OBRI	Medway EclipseR	171/149		25.10.01	B.D.Campbell	(Maidstone)	24.10.02P
G-OBRY	Cameron N-180 HAB	3010		1. 3.93	A C.K.Rawson and J.J.Rudoni	(Stafford)	30. 3.05T
					(Bryant Homes titles)		
G-OBTS	Cameron C-80 HAB	3589		18. 4.95	C.F.Cushion	(Petersfield)	9. 7.05A
G-OBUD*	Colt 69A HAB				See SECTION 3 Part 1	Newbury	
G-OBUN	Cameron A-250 HAB	4711		29. 2.00	A C.K.Rawson and J.J.Rudoni	(Stafford)	5 .11.04T

Registration	Type	c/n	Previous identity	Reg date	Owner / Operator	Base	CofA expiry
G-OBWD*	British Aircraft Corporation One-Eleven 518FG	BAC.203	G-BDAE G-AXMI	14. 1.93	British World Airlines	Southend	14. 4.02T
	(Cancelled as sold Sierra Leone 15.5.02 - allocated marks 9L-LDK but not taken up to date) (In open store 2.05 with Fresh titles)						
G-OBWP	British Aerospace ATP	2051	G-BTPO G-5-051	8.10.99	Trident Aviation Leasing Services (Jersey) Ltd *(Stored 9.03)*	Exeter	17.10.02T
G-OBWR	British Aerospace ATP	2053	G-BUWP G-11-053	8.10.99	Trident Jet (Jersey) Ltd *(In open store 2.05)*	Southend	20. 4.03T
G-OBYB	Boeing 767-304ER	28040		17. 5.96	Britannia Airways Ltd *"Bobby Moore OBE" (Thomson c/s)*	Luton	16. 5.05T
G-OBYC	Boeing 767-304ER	28041	D-AGYC G-OBYC, D-AGYC, G-OBYC	21. 5.96	Britannia Airways Ltd *"Roy Castle OBE"*	Luton	7.11.06T
G-OBYD	Boeing 767-304ER	28042	SE-DZG G-OBYD	4. 3.97	Britannia Airways Ltd *(TUI c/s)*	Luton	2. 5.07T
G-OBYE	Boeing 767-304ER	28979	D-AGYE G-OBYE	26. 2.98	Britannia Airways Ltd *"Bill Travers" (Garuda titles and tail logo)*	Luton	28.10.05T
G-OBYF	Boeing 767-304ER	28208	D-AGYF G-OBYF	8. 6.98	Britannia Airways Ltd *(Garuda titles and tail logo)*	Luton	30. 4.07T
G-OBYG	Boeing 767-304ER	29137		13. 1.99	Britannia Airways Ltd *(Garuda titles and tail logo)*	Luton	12. 1.05T
G-OBYH	Boeing 767-304ER	28883	SE-DZO D-AGYH, G-OBYH	4. 2.99	Britannia Airways Ltd	Luton	24. 4.05T
G-OBYI	Boeing 767-304ER	29138		1. 2.00	Britannia Airways Ltd *(TUI c/s)*	Luton	31. 1.06T
G-OBYJ	Boeing 767-304ER	29384		20. 2.00	Britannia Airways Ltd	Luton	18. 2.06T
G-OBYT	Agusta-Bell 206A JetRanger	8237	G-BNRC Oman AF 601	30. 1.95	R.J.Everett	(Sproughton)	12. 7.03T
G-OCAD	Sequoia F.8L Falco *(Built C.W.Garrard)* (Lycoming IO-320)	PFA 100-12114		8. 6.92	I.R.Court tr Falco Flying Group	Leicester	21.12.05P
G-OCAM	Gulfstream AA-5A Cheetah	AA5A-0741	G-BLHO OO-RTJ, OO-HRN	24. 3.94	Plane Talking Ltd *(Op Cabair)*	Cranfield	21.10.06T
G-OCAR	Colt 77A HAB	1099		6. 8.87	S.C.J.Derham *"Toyota"*	(Bridgnorth)	25. 8.00A
G-OCBA	Hawker Siddeley HS.125 Series 3B	25132	EI-WDC G-OCBA, EI-WDC, G-OCBA, G-MRFB, G-AZVS, OY-DKP	2. 5.89	R J Everett	Bentwaters	13.10.06T
G-OCBI	Schweizer 269C-1	0139	N86G	14. 8.02	JWL Helicopters Ltd	Biggin Hill	26. 9.05T
G-OCBS	Lindstrand LBL 210A HAB	602		21. 7.99	G.Binder	(Sonnenbuhl, Germany)	15. 7.05A
G-OCDS	Aviamilano F.8L Falco Series II	114	G-VEGL OO-MEN, I-VEGL	6. 9.85	C.O.P.Barth	(The Netherlands)	21.11.05
G-OCDW	Jabiru Aircraft Jabiru UL-450 *(Built C D Wood)*	PFA 274A-14122		31. 3.04	C D Wood	Dunkeswell	19.10.05P
G-OCEA	Short SD.3-60 Variant 100	SH.3762	N162CN N162SB, G-BRMX	26.10.95	BAC Express Airlines Ltd *(In open store 2.05)*	Southend	26. 3.04T
)G-OCFC	Robin R2160	374		21. 6.02	Cornwall Flying Club Ltd	Bodmin	11. 7.05T
G-OCFD	Bell 206B JetRanger III	3165	G-WGAL G-OICS, N678TM, N678TW	10. 6.04	Cranfield Helicopters Ltd	Cranfield	29. 4.07T
G-OCFM	Piper PA-34-200 Seneca	34-7350021	G-ELBC G-BANS, N15110	20. 4.04	Stapleford Flying Club Ltd *(Op Capital Radio)*	Stapleford	27.12.06T
G-OCHM	Robinson R44 Raven	1055		4. 5.01	Westleigh Developments Ltd	Whetstone, Leics	30. 5.07T
G-OCIT	Cessna 208B Grand Caravan	208B1041	N51564	21.11.03	Fly CI Ltd	Guernsey	13. 1.07Y
G-OCJK	Schweizer 269C	S.1294	N69A	10.12.87	P Crawley *(New CofR 5.02)*	Shipley	27. 5.00
G-OCMM	Agusta A109A II	7347	G-BXCB F-GJSH, G-ISEB, G-IADT, G-HBCA	20. 3.01	Maison Air Ltd	Fairoaks	6. 8.06T
G-OCMT	Aerotechnik EV-97 teamEurostar UK	1701		14. 7.03	P.Crowhurst	Sywell	30. 7.05P
G-OCND*	Cameron O-77 HAB				See SECTION 3 Part 1	(Aldershot)	
G-OCOV	Robinson R22 Beta II	3217		23. 5.01	Flight Training Ltd	Coventry	17. 6.74T
G-OCPC	Reims/Cessna FA152 Aerobat	FA1520343		20. 1.78	Westward Airways (Lands End) Ltd	St.Just	12. 8.05T
G-OCRI	Colomban MC-15 Cri-Cri *(Built M.J.J.Dunning - c/n 524)*	PFA 133-12288		24. 6.92	M.J.J.Dunning *(Amended CofR 8.02)*	(Bredbury, Stockport)	
G-OCST	Agusta-Bell 206B-3 JetRanger III	8694	N39AH VR-CDG, G-BMKM	14.12.94	Lift West Ltd	(Seavington, Ilminster)	18. 1.07T
G-OCTI	Piper PA-32-260 Cherokee Six	32-288	G-BGZX 9XR-MP, 5Y-ADH, N3427W	26. 7.88	D.G.Williams	Blackbushe	19. 8.07T
G-OCUB	Piper J-3C-90 Cub (L-4J-PI) *(Frame No.13078)*	13248	OO-JOZ PH-NKC, PH-UCH, 45-4508	21. 4.81	C.A Foss and P.A Brook tr Florence Flying Group *(Op Zebedee Flying Group) "Florence"*	Shoreham	14. 2.05P
	(Official regd with c/n 13215 but this was 45-4475/PH-UCW and rebuilt as PH-UCH)						
G-ODAC	Reims/Cessna F152 II *(Rebuilt with cockpit/front fuselage of G-BITG 6.96)*	F15201824	G-BITG	19.12.96	T.M.and M.L.Jones *(Op Derby Aero Club)*	Egginton	15. 9.07T
G-ODAD	Colt 77A HAB	2001		20. 2.91	K.Meehan *"Odyssey"*	(Much Wenlock)	13. 5.05A
G-ODAK	Piper PA-28-236 Dakota	28-7911162	D-EXMA OH-SMO, N386WT, N22328	29. 2.00	Airways Aero Associations Ltd	Booker	8. 4.06T
G-ODAT	Aero L-29 Delfin	194227	ES-YLV Estonian AF, Soviet AF	28. 7.99	Graniteweb Ltd	North Weald	19. 1.05P
G-ODAV	Aerotechnik EV-97 Eurostar *(Built B.R.Davies)*	PFA 315-14299		30. 9.04	B.R.Davies	(Cannock)	9..11.05P
G-ODAY	Cameron N-56 HAB	551		16. 7.79	The British Balloon Museum and Library Ltd	(Southampton)	3. 10.82A
G-ODBN	Lindstrand Flowers SS HAB	389		22. 5.96	Magical Adventures Ltd *"Sainsbury's Flowers"*	(West Bloomfield, Michigan, US)	12.10.03A
G-ODCS	Robinson R22 Beta II	2828		19. 5.98	V.E.Nalbantis	Panshanger	29. 9.07T
G-ODDY	Lindstrand LBL 105A HAB	042		15. 7.93	P.and T.Huckle	(Oakwood)	8. 6.05A
G-ODEE	Van's RV-6 *(Built D Powell)* (Lycoming O-320)	PFA 181-13173		14. 4.00	D.Powell	Shenstone	10. 6.05P
G-ODEN	Piper PA-28-161 Cadet	2841282	N92004	22.11.89	Atrium Aviation Ltd	Denham	18.12.04T
G-ODGS	Jabiru Aircraft Jabiru UL-450 *(Built D.G.Salt)*	PFA 274A-13472		2. 8.99	D.G.Salt	(Ashbourne)	9.11.04P
G-ODHG	Robinson R44 Raven	1024		19. 4.01	Sloane Helicopters Ltd	Sywell	13. 5.07T

Registration	Type	c/n	Previous identity	Date	Owner	Base	CofA
G-ODIN	Avions Mudry CAP.10B	192	F-GDTH	16.12.93	R.P.W.Steele	Sandown	18. 7.07T
G-ODJB	Robinson R22 Beta II	3463		10. 7.03	N.T.Burton	Costock	20. 8.06P
G-ODJD	Raj Hamsa X'Air 582 *(Built D J Davis - kit no.559)*	BMAA/HB/151		25. 4.01	K.P.Roper	Weston Zoyland	16. 8.05P
G-ODJG	Europa Aviation Europa *(Built D.J.Goldsmith - kit no.167) (Monowheel u/c)*	PFA 247-12889		3. 5.96	D.J.Goldsmith	(Edenbridge)	28. 9.05P
G-ODJH	Mooney M.20C Ranger	690083	G-BMLH N9293V	19. 1.93	A.P.Howells	(Pontypridd)	15. 9.05
G-ODLY	Cessna 310J	310J0077	G-TUBY G-ASZZ, N3077L	21. 3.88	Card Tech Ltd	Denham	18. 6.06
G-ODMC	Aérospatiale AS350B1 Ecureuil	2200	G-BPVF	17.10.89	D.M.Coombs t/a DM Leasing Co	Denham	26.10.07E
G-ODNH	Schweizer 269C-1	0112	N41S	5. 9.00	DNH Helicopters Ltd	Leeds-Bradford	25. 9.06T
G-ODOC	Robinson R44 Astro	0372		27. 8.97	Gas and Air Ltd	Booker	21. 1.07T
G-ODOD	MD Helicopters MD 600N	RN052	N3204S	5.10.00	HPM Investments Ltd	Bournemouth	10.11.06T
G-ODOG	Piper PA-28R-200 Cherokee Arrow II	28R-7235197	EI-BPB G-BAAR, N11C	2. 8.96	Advanced Investments Ltd	Sibson	25. 2.06
G-ODOT	Robinson R22 Beta II	2779		23. 1.98	Ribbands Ltd t/a Ribbands Explosives	(Norwich)	31. 3.07T
G-ODPJ	VPM M-16 Tandem Trainer *(Built M L Smith)* (Arrow GT1000R)	VPM16-UK-111	G-BVWX	4. 4.03	A.P.Wilkinson *(New owner 10.04)*	(Hedon, Hull)	17. 6.02P
G-ODSK	Boeing 737-37Q	28537		23. 7.97	British Midland Airways Ltd *"baby dragon fly"* *(Op bmiBaby)*	Nottingham East Midlands	27. 7.06T
G-ODTW	Europa Aviation Europa *(Built D.T.Walters - kit no.215) (Monowheel u/c)*	PFA 247-12890		7. 9.95	D.T.Walters *(Current status unknown)*	(Longfield, Kent)	
G-ODUD	Piper PA-28-181 Cherokee Archer II	28-7790107	G-IBBO D-EPCA, N5389F	15. 3.04	D Rogg	Panshanger	31. 1.05
G-ODUS	Boeing 737-36Q	28659	D-ADBX	17. 3.98	Easyjet Airline Company Ltd	Luton	15. 4.07T
G-ODVB	CFM Shadow Series DD	300-DD	G-MGDB	3.11.98	D.V.Brunt	Plaistowes Farm, St.Albans	24. 4.05P
G-OEAC	Mooney M.20J Model 201	24-1636	N57656	16. 6.88	S.Lovatt	Tollerton	5. 4.06
G-OEAT	Robinson R22 Beta	0650	G-RACH	8. 1.98	C.Y.O.Seeds Ltd	(Didcot)	24. 1.05T
G-OECH	Gulfstream AA-5A Cheetah	AA5A-0836	G-BKBE (G-BJVN), N26952	24. 1.89	Plane Talking Ltd *(Op London School of Flying)*	Cranfield	10. 5.06T
G-OECM	Commander Aircraft Commander 114B	14627	N6107Y	4. 3.04	ECM (Vehicle Delivery Service) Ltd	Carlisle	11. 5.07
G-OEDB	Piper PA-38-112 Tomahawk	38-79A0167	G-BGGJ N9694N	9. 5.89	G.G.L.James *(New owner 7.04)*	Sleap	15. 6.03T
G-OEDP	Cameron N-77 HAB	2189		28.12.89	M.J.Betts *"Eastern Counties Press"*	(Norwich)	12. 6.01A
G-OEGL	Christen Eagle II (Lycoming IO-360)	001	N46JH	12. 1.98	R.Dauncey, tr The Eagle Flight Syndicate	Shoreham	9. 5.05P
G-OELD	Pegasus Quantum 15-912	7765		20.12.00	K M Sullivan	(Ahoghill, Ballymena)	16. 3.05P
G-OERR	Lindstrand LBL 60A HAB	469		30. 6.97	P C Gooch	(Alresford)	12. 4.03
G-OERX	Cameron O-65 HAB	4004		23. 1.96	R.Roehsler *(Current status unknown)*	(Vienna, Austria)	27. 2.97A
G-OESL	Piaggio P.180	1080		27. 5.04	Euroskylink Ltd	Coventry	1. 6.07T
G-OESY	Reality Easy Raider Jab22 *(Built T F Francis - kit no.0005)*	BMAA/HB/193		16.11.01	G C Long	Plaistowes Farm, St.Albans	16. 6.05P
G-OETI	Bell 206B-3 JetRanger III	2533	G-RMIE G-BPIE, N327WM	23. 7.02	Elec-Track Installations Ltd	(Hythe)	5.12.04T
G-OETV	Piper PA-31-350 Navajo Chieftain	31-7852073	N27597	16. 6.04	European Executive Ltd	Shoreham	30. 6.07T
G-OEVA	Piper PA-32-260 Cherokee Six *(Rebuilt using spare Frame No.32-860S)*	32-219	G-FLJA G-AVTJ, N3373W	13. 3.03	M.G.Cookson	North Weald	29. 1.05T
G-OEWA	de Havilland DH.104 Dove 8	04528	G-DDCD G-ARUM	10. 6.98	D.C.Hunter *(On rebuild for East West Airlines 2002)*	Kemble	AC
G-OEYE	Rans S-10 Sakota *(Built J D Haslam)* (Rotax 582)	PFA 194-11955		25. 4.91	I.M.J.Mitchell	Otherton, Cannock	8.10.04P
G-OEZI	Reality Easy Raider Jab22 *(Built M.A Claydon - kit no.0007) (Replacement fuselage fitted as original noted Westfield Farm, Hailsham 2004)*	BMAA/HB/216		31. 5.02	M.A Claydon	Frensham, Wishanger	11. 5.05P
G-OEZY	Europa Aviation Europa *(Built A W.Wakefield - kit no.042) (Monowheel u/c)*	PFA 247-12590		8. 8.95	A W.Wakefield	Conington	22. 6.05P
G-OFAS	Robinson R22 Beta	0559		17. 6.86	J.L.Leonard t/a Findon Air Services	Shoreham	25. 4.05T
G-OFBJ	Thunder Ax7-77A HAB	2050		2. 9.91	N.D.Hicks *"Blue Horizon"*	(Alton)	25. 9.99A
G-OFBU	Comco Ikarus C42 FB UK *(Built Fly Buy Ultralights Ltd - kit no.0301-6328)*	PFA 322-13653		28. 8.01	J Pearce tr Old Sarum C42 Group	Old Sarum	10.12.05P
G-OFCH	Agusta-Bell 206B JetRanger II	8337	HB-XUI G-BKDA, LN-OQX	15. 5.00	Fleet Coast Helicopters Ltd *(Rolled over, struck ground Mom Farm, Chickerell 4.1.02 and badly damaged)*	Shoreham	11. 7.03T
G-OFCM	Reims/Cessna F172L	F17200839	G-AZUN (OO-FCB)	21.10.81	FCM Aviation Ltd	Guernsey	15. 4.06
G-OFER	Piper PA-18-150 Super Cub	18-7709058	N83509	29.12.89	Mary S.W.Meagher	Shenington	29. 5.06
G-OFFA	Pietenpol AirCamper *(Built Offa Group)*	PFA 47-13181		3.11.98	D.J.Street tr Offa Group *(Noted 1.05)*	Bicester	
G-OFIL	Robinson R44 Astro	0555		15. 1.99	B J North t/a North Helicopters	Redhill	10. 4.05T
G-OFIT	SOCATA TB-10 Tobago	938	G-BRIU	11. 9.89	G.M.Richards tr GFI Aviation Group	White Waltham	20. 3.05T
G-OFIZ*	Cameron Can 80SS HAB				See SECTION 3 Part 1	Newbury	
G-OFLG	SOCATA TB-10 Tobago	11	G-JMWT F-GBHF	11.12.91	MRR Aviation Ltd	St. Just	29. 5.06T
G-OFLY	Cessna 210M Centurion II	21061600	(D-EBYM) N732LQ	13.10.79	A P.Mothew	Southend	3. 6.07E
G-OFMB	Rand Robinson KR-2 *(Built M.A Shepard)*	7808	N5337X	29. 4.97	F.M.and S.I.Burden *(Noted 1.05)*	Gloucestershire	
G-OFOA	British Aerospace BAe 146 Series 100	E1006	G-BKMN EI-COF, SE-DRH, G-BKMN, G-ODAN	3. 3.98	Formula One Adminstration Ltd	Biggin Hill	14. 7.05
G-OFOM	British Aerospace BAe 146 Series 100	E1144	N3206T PK-DTA, G-BSLP, (PK-DTA), G-6-144, G-11-144, (G-BRLM)	16. 3.00	Formula One Management Ltd	Biggin Hill	6.10.05
G-OFOX	Denney Kitfox *(Built P.R.Skeels)*	PFA 172-11523		1.11.89	P.R.Skeels *(Current status unknown)*	Barton	
G-OFRA	Boeing 737-36Q	29327		5. 5.98	Easyjet Airline Company Ltd	Luton	17. 5.06T

Registration	Type	C/n	Previous identity	Date	Owner / Operator	Base	CofA
G-OFRB	Everett Gyroplane Series 2	006	(G-BLSR)	7. 8.85	M.P.James	Carlisle	14.10.03P
	(Rotax 582)						
G-OFRT*	Lockheed L188CF Electra	1075	N347HA	29.10.91	Dart Group plc	Coventry	28.10.01T
			N423MA, N23AF, N64405, SE-FGC, N5537 *(Cancelled 12.5.03 by CAA) (Stored 5.03)*				
G-OFRY	Cessna 152 II	15281420	G-BPHS	8. 2.93	Devon School of Flying Ltd	Dunkeswell	8. 1.05T
			N49971				
G-OFTI	Piper PA-28-140 Cherokee Cruiser	28-7325201	G-BRKU	11. 6.90	G.S.A Spencer	Andrewsfield	8.10.05T
			N15926				
G-OGAN	Europa Aviation Europa	PFA 247-12734		28. 7.94	B.W.Rendall	North Coates	18. 9.05P
	(Built M A Jackson, M P Gogan and R S Cullum - kit no.100) (Tri-gear u/c)				tr G-OGAN Group		
G-OGAR	PZL-Bielsko SZD-45A Ogar	B-601	SP-0004	29. 1.90	N.C.Grayson	Boscombe Down	9. 6.00
G-OGAZ	Aérospatiale SA.341G Gazelle 1	1274	G-OCJR	12. 1.94	I.M.and S.M.Graham t/a Killochries Fold	Edinburgh	10. 3.07T
			G-BRGS, F-GEQA, N341SG, (N341P), N341SG, N47295 *(Op Forth Helicopters Ltd)*				
G-OGBD	Boeing 737-3L9	27833	OY-MAR	16. 3.98	British Midland Airways Ltd		
			D-ADBJ, OY-MAR		*(Op bmiBaby)*	Nottingham East Midlands	12. 3.07T
G-OGBE	Boeing 737-3L9	27834	OY-MAS	24.11.98	British Midland Airways Ltd *"Derby's baby pride"*		
					(Op bmiBaby)	Nottingham East Midlands	17.12.07E
G-OGEM	Piper PA-28-181 Archer II	28-8190226	N83816	10. 3.88	GEM Rewinds Ltd	Coventry	19. 6.06T
G-OGEO	Aérospatiale SA.341G Gazelle 1	1417	G-BXJK	28. 1.02	MW Helicopters Ltd	Stapleford	16. 9.06T
			F-GEHC, N341AT, N49536				
G-OGES	Enstrom 280FX Shark	2078	G-CBYL	15.11.02	G.N.Ratcliffe	(Bolton)	10.11.05T
			HB-XAJ				
G-OGET	Piper PA-39 Twin Comanche C/R	39-87	G-AYXY	14. 3.83	D.G. Lewendon	Farley Farm, Romsey	12. 9.05
			N8930Y				
G-OGGY	Aviat A-1B	NF0005	N144HP	27. 2.04	Chris Irvine Aviation Ltd	(Mylor, Falmouth)	24. 8.07T
G-OGIL*	Short SD.3-30 Var.100				See SECTION 3 Part 1	Usworth, Sunderland	
G-OGJM	Cameron C-80 HAB	4869		21.11.00	G.J.Madelin	(Farnham)	20.10.04A
G-OGJP	Hughes 369E	0512E	N685F	23. 1.01	Motortrak Ltd	(Thames Ditton)	30. 1.04T
			N5223X				
G-OGJS	Rutan Cozy	PFA 159-11169		27. 1.89	G.J.Stamper	(Carlisle)	14. 9.98P
	(Built G.J.Stamper) (Lycoming O-360)				*(Stored 2003)*		
G-OGOA	Aérospatiale AS350B Ecureuil	1745	G-PLMD	16. 1.90	Lomas Brothers Ltd	Lake, Bideford	20. 6.05T
			G-NIAL		t/a Lomas Helicopters		
G-OGOB	Schweizer 269C	S.1315	G-GLEE	2.10.90	Kingfisher Helicopters Ltd	Longdown	11. 4.05T
			G-BRUW, N86G				
G-OGOH	Robinson R22 Beta II	2738	G-IPDM	25.11.02	Lomas Brothers Ltd	(Bideford)	27. 5.07T
			G-OMSG		t/a Lomas Helicopters		
G-OGOS	Everett Gyroplane	004	7Q-YES	30. 7.84	N.A Seymour	(Norwich)	12. 9.90P
	(Volkswagen 1834)		G-OGOS		*(Current status unknown)*		
G-OGRG	Cessna 560 Citation Ultra	560-0506	G-RIBV	26.3.03	Aeropublic Sl	(Alzira, Valencia, Spain)	20. 3.06T
			N50820				
G-OGSA	Jabiru Aircraft Jabiru UL-450	PFA 274A-13540		10. 2.00	A Knape	Lower Wasing Farm, Brimpton	10. 6.05P
	(Built G J Slater - kit no.300?)						
G-OGSS	Lindstrand LBL 120A HAB	683		19. 5.00	R.Klarer	(Erbach, Germany)	12. 8.04A
G-OGTS	Air Command 532 Elite	PFA G/104-1125		19.12.88	GTS Engineering (Coventry) Ltd	Coventry	1.10.90P
	(Built G E Heritage - c/n 0432)				t/a GTS Cars *(Current status unknown)*		
G-OHAC	Reims/Cessna F182Q Skylane II	F18200048	D-ENCM	11. 7.01	The RAF Halton Aeroplane Club	RAF Halton	12. 7.07T
G-OHAJ	Boeing 737-36Q	29141		2. 6.98	Easyjet Airline Company Ltd	Luton	30. 8.06T
G-OHAL	Pietenpol AirCamper	PFA 47-12840		25.11.96	H.C.Danby	(Sudbury)	
	(Built H.C.Danby)				*(Current status unknown)*		
G-OHCP	Aérospatiale AS355F1 Ecureuil 2	5249	G-BTVS	14. 3.94	AJJ Developments Ltd	Sheffield City	20. 2.07T
			G-STVE, G-TOFF, G-BKJX				
G-OHDC*	Colt Film Cassette SS HAB				See SECTION 3 Part 1	(Aldershot)	
G-OHEA*	Hawker Siddeley HS 125 Series 3B/RA	25144	G-AVRG	25.11.86	Cranfield University	Cranfield	7. 8.92T
			G-5-12		*(Cancelled 23.6.94 as WFU) (Fuselage dumped as "G-DHEA" 8.04)*		
G-OHFT	Robinson R22 Beta	1040	G-TYPO	18.12.01	Heliflight (UK) Ltd	Wolverhampton	14. 3.05T
			G-JBWI				
G-OHGC	Scheibe SF25C Falke	44695		26. 7.04	D.J.Marpole tr Heron Gliding Club	RNAS Yeovilton	5. 8.07
G-OHHI	Bell 206L-1 LongRanger II	45552	G-BWYJ	30. 4.98	Big Heli-Charter Ltd	Manston	30. 4.06T
			D-HOBD, D-HGAD				
G-OHKS	Pegasus Quantum 15(HKS)	7505		24. 3.99	York Microlight Centre Ltd	Rufforth	8. 5.05T
	(HKS 700E s/n 99030A)						
G-OHMS	Aérospatiale AS355F1 Ecureuil 2	5194	N367E	15. 6.90	South Western Electricity plc	Farnborough	23. 6.05T
					(WPD Electricity titles)		
G-OHOV	Rotorway Executive 162F	6885		14. 9.04	M.G.Bird	(Royston)	AC
	(Built M.G.Bird)						
G-OHSA	Cameron N-77 HAB	4269		2. 2.98	D.N. and L.J.Close *(HSA Healthcare titles)*	(Andover)	15. 4.01A
G-OHSL	Robinson R22 Beta	0967	G-BPNF	4. 7.01	Astons of Kempsey (Holdings) Ltd	Shobdon	24. 8.07T
			N8029Y				
G-OHVA	Mainair Blade 912	1189-0199-7 & W992		6.11.98	M C Metatidj	(La Baule, France)	27. 3.04P
G-OHWV	Raj Hamsa X'Air 582	BMAA/HB/121		18.11.99	H.W.Vasey	Roche, Cornwall	8. 6.05P
	(Built H.W.Vasey - kit no.474)						
G-OHYE	Thruster T600N 450 Sprint	0042-T600N-098	G-CCRO	9. 3.04	P J Read tr G-OHYE Group	Enstone	17. 3.05P
G-OIBM	Rockwell Commander 114	14295	G-BLVZ	14.10.88	E.J.Percival	Blackbushe	22. 7.06
			SX-AJO, N4957W				
G-OIBO	Piper PA-28-180 Cherokee C	28-3794	G-AVAZ	21. 1.87	Britannia Airways Ltd	Wellesbourne Mountford	14. 4.06T
			N11C				
G-OIDW	Cessna F150G	F150-0188	N70163	24. 4.90	T.S.Sheridan-McGinnitty	Wolverhampton	25. 6.06
	(Built Reims Aviation SA)		D-EGTI		*(As "00195700" in pseudo-USAF c/s)*		
G-OIFM	Cameron Dude 90SS HAB	2841		18. 6.92	Magical Adventures Ltd		
	(Radio One FM DJ's Head and Earphones)				*"Cool Dude"*	(West Bloomfield, Michigan, US)	29. 5.99A
G-OIIO	Robinson R22 Beta	2444	G-ULAB	27. 3.02	Un Pied Sur Terre Ltd	Welshpool	17. 7.06T
			N8311Z		t/a Whizzard Helicopters		

Reg	Type	C/n	Previous identity	Date	Owner	Location	Expiry
G-OIMC	Cessna 152 II	15285506	N93521	15. 5.87	East Midlands Flying School Ltd	Nottingham East Midlands	28. 6.05T
G-OINK	Piper J-3C-65 Cub (L-4J-PI)	12613	G-BILD	22. 3.83	A R.Harding	Newton Farm, Sudbury	19. 7.99P
	(Frame no.12443)		G-KERK, F-BBQD, 44-80317				
G-OINV	British Aerospace BAe 146 Series 300	E3171	VH-EWI	17. 2.00	British Airways Citiexpress Ltd	Inverness	15. 5.06T
			G-6-171, VH-EWI, G-6-171		*"Chatham Historic Dockyard"*		
G-OIOI*	EH Industries EH-101 Heliliner				See SECTION 3 Part 1	Hendon	
G-OIOZ	Thunder AX9-120 Series 2 HAB	4434		17.11.98	C.M.Hodges	(Old Sarum, Salisbury)	23.10.03T
G-OIPB	IAV-Bacau Yakovlev Yak-52	877902	RA-44455	3. 4.03	Mastercraft Ltd	(Castletown, Isle of Man)	14. 5.05P
			DOSAAF 89 *(yellow?)*				
G-OISO	Reims/Cessna FA150 Aerobat	FRA1500213	G-BBJW	3. 4.90	Linda A and B A Mills	(Cambridge)	26. 8.05T
	(Built as FRA150L)						
G-OITN	Aérospatiale AS355F1 Ecureuil 2	5088	N400HH	3.10.89	Independent Television News Ltd	Redhill	13.12.04T
			N5788B				
G-OITV	Enstrom 280C Shark	1038	G-HRVY	9. 4.96	C.W.Brierley Jones	(Warrington)	24. 9.04T
			G-DUGY, G-BEEL				
G-OIZI*	Europa Aviation Europa XS	PFA 247-13615		9.10.00	K.S.Duddy	(Malvern)	
	(Built N Hayes - kit no.474)				*(Cancelled 21.9.04 by CAA - no PtoF issued)*		
G-OJAB	Jabiru Aircraft Jabiru SK	PFA 274-13031		19. 9.96	P.A Brigstock	Leicester	9. 5.05P
	(Built K D Pearce)						
G-OJAC	Mooney M.20J Model 201	24-1490	N5767E	20. 8.90	Hornet Engineering Ltd	Biggin Hill	16. 2.06T
G-OJAE	Hughes 269C	90-0966	N1101W	12. 2.90	J.A and C.M.Wilson	Slaithwaite, Huddersfield	20.10.05
G-OJAN	Robinson R22 Beta	2012	G-SANS	22. 5.01	Heliflight (UK) Ltd	Gloucestershire	5. 9.07T
			G-BUHX				
G-OJAS	Auster J/1U Workmaster	3501	F-BJAS	21. 3.00	K.P.and D.S.Hunt	Redhill	
			F-WJAS, (F-OBHT)		*(Noted in "Wings'" Museum 1.04)*		
G-OJAV	Fairey Britten-Norman BN-2A Mk.III-2 Trislander	1024	G-BDOS	6. 6.90	Lyddair Ltd	Lydd	29. 1.07T
			(4X-CCI), G-BDOS				
G-OJBB	Enstrom 280FX	2084		14. 6.99	Pendragon (Design and Build) Ltd	Gloucestershire	1. 7.05T
G-OJBM	Cameron N-90 HAB	2899		28. 9.92	P.Spinlove	(Chalfont St.Giles)	23. 9.93A
G-OJBS	Cameron N-105 HAB	4733		8. 3.00	Up and Away Ballooning Ltd	(High Wycombe)	17. 4.05T
G-OJBW	Lindstrand J & B Bottle SS HAB	436		26. 8.97	N.A P.Godfrey *(New owner 3.03)*	(Aylesbury)	20. 5.02A
G-OJCW	Piper PA-32RT-300 Lance II	32R-7985062	N3016K	9. 1.80	P.G.Dobson tr CW Group	Blackbushe	11. 8.07
G-OJDA	EAA Acrosport II	PFA 72-11067		1. 4.98	D.B.Almey	(Spalding)	2. 6.05P
	(Built D.B.Almey) (Lycoming O-360-A4A) *(C/n should be PFA 72A-...)*						
G-OJDC	Thunder Ax7-77 HAB	875		9. 1.89	Julia Crosby	(Brighton)	1. 8.02A
G-OJDR	Yakovlev Yak-50	792006	LY-JDR	24. 7.02	A M.Holman-West	Wellesbourne Mountford	20. 8.05P
			DOSAAF		*(As "JD-R/J-DR" in US c/s)*		
	(C/n as officially regd but a/c either side (G-BWYK/G-CBPM) are 1981 build so c/n should be 812006?)						
G-OJDS	Comco Ikarus C42 FB80	0411-6633		26.11.04	J.D.Smith	Baxby Manor, Husthwaite	
G-OJEG	Airbus A321-231	1015	D-AVZN	14. 5.99	Monarch Airlines Ltd	Luton	13. 5.05T
G-OJEH	Piper PA-28-181 Archer II	28-8690051	D-EDPA	17.12.02	Pamela C.and M.A.Greenaway	(Sevenoaks)	29. 1.06T
			N9125Y				
G-OJEN	Cameron V-77 HAB	3302		26. 5.94	D.J.Geddes	(High Wycombe)	30. 3.05A
G-OJGT	Maule M-5-235C Lunar Rocket	7285C	LN-AEL	30. 6.98	J.G.Townsend	Lower Upham Farm, Chiseldon	8. 8.07
			(LN-BEK), N5635V				
G-OJHB	Colt Flying Ice Cream Cone SS HAB	2591		23. 6.94	Benedikt Haggeney GmbH	(Ennigerloh, Germany)	16. 4.04A
G-OJHL	Europa Aviation Europa	PFA 247-13039		12. 5.97	J.H.Lace	Prestwick	3. 9.05P
	(Built J.H.Lace - kit no.311) (Monowheel u/c)				*"Lady Lace"*		
G-OJIL	Piper PA-31-350 Navajo Chieftain	31-7652175	OY-BTP	28. 5.97	Redhill Aviation Ltd *(Op Redhill Charters)*	Blackbushe	23.2.07T
G-OJIM	Piper PA-28R-201T Turbo Cherokee Arrow III	28R-7703200	N38299	4. 8.86	Grey Fox Investigations Ltd	Biggin Hill	5. 2.05
G-OJJB	Mooney M.20K Model 252TSE	25-1161		12. 8.88	G Italiano	Roma-Urbe, Italy	7. 7.07
G-OJJF	Druine D.31 Turbulent	378 & 31	OO-30	6. 1.97	J.J.Ferguson	Belle Vue Farm,Yarnscombe	
	(Built J.J.Ferguson) (Volkswagen 1300)				*(Noted 4.04)*		
G-OJKM	Rans S-7 Courier	PFA 218-12982		5. 3.01	M.Jackson	Southend	24.10.05P
	(Built M Jackson)						
G-OJLH	TEAM mini-MAX 91	PFA 186-12164	G-MYAW	12.12.01	J.L.Hamer	Hartpury	13. 4.05P
	(Built J.L.Hamer)						
G-OJMB	Airbus A330-243	427	F-WWYH	8.11.01	Thomas Cook Airlines UK Ltd	Manchester	8.11.07E
G-OJMC	Airbus A330-243	456	F-WWKI	5. 3.02	Thomas Cook Airlines UK Ltd	Manchester	4. 3.05T
G-OJMF	Enstrom 280FX	2086	G-DDOD	12. 6.01	JMF Ltd	(Ballymoney, County Antrim)	9 .12.02
G-OJMR	Airbus A300B4-605R	605	F-WWAY	3. 5.91	Monarch Airlines Ltd	Luton	2. 5.05T
G-OJNB	Lindstrand LBL 21A HAB	085		14. 2.94	Justerini and Brooks Ltd	(London SW1)	17. 4.04A
G-OJOD	Jodel D.18	PFA 169-12774		20. 6.02	D.Hawkes and C.Poundes	(Milton Keynes)	
	(Built D.Hawkes)						
G-OJON	Taylor JT.2 Titch	PFA 3208		6.10.78	A.Donald	Netherthorpe	3. 5.05P
	(Built J H Fell) (Continental C90)						
G-OJRH	Robinson R44 Astro	0321		11. 4.97	Holgate Construction Ltd	Emley Moor, Huddersfield	24. 4.06
G-OJRM	Cessna T182T Turbo Skylane	T18208007	N72778	19. 7.01	SPD Ltd	Oxford	31. 7.04T
G-OJSH	Thruster T.600N 450	0061-T600N-052		29. 5.01	S.A Lewis tr November Whiskey Flying Club	Shobdon	23. 9.05P
G-OJTA	Stemme S-10V	14-018	D-KGDA	18. 9.95	O.J.Truelove t/a OJT Associates	(RAF St. Mawgan)	5. 6.05
G-OJTW	Boeing 737-36N	28558	(G-JTWF)	26. 4.97	British Midland Airways Ltd *"rock-a-bye baby"*		
					(Op bmiBaby)	Nottingham East Midlands	1. 5.06T
G-OJVA	Van's RV-6	PFA 181-12292		6. 9.96	J.A Village	Moorgreen Farm, Barlow	6.11.05P
	(Built J A Village) (Lycoming O-320)						
G-OJVH	Cessna F150H	F150-0356	G-AWJZ	27. 3.81	A W.Cairns	RAF Brize Norton	23. 5.07T
	(Built Reims Aviation SA)						
G-OJVL	Van's RV-6	PFA 181-12441		28.10.02	S.E.Tomlinson	(Bournemouth)	
	(Built S E Tomlinson)						
G-OJWS	Piper PA-28-161 Cherokee Warrior II	28-7816415	N6377C	13. 7.88	P.J.Ward	Denham	9. 9.06
G-OKAG	Piper PA-28R-180 Cherokee Arrow	28R-30075	N3764T	15. 4.88	B.R.Green	Stapleford	13. 4.06T
G-OKAY	Pitts S-1E	12358	N35WH	27. 5.80	D S T Eggleton	Waits Farm, Belchamp Walter	10. 4.04P
	(Built W D Henline)						

Reg	Type	C/n	Previous identity	Regd	Owner/operator	Base	CofA
G-OKBT	Colt 25A Sky Chariot Mk.II HAB	2301		10.11.92	British Telecommunications plc *"Skypiper II"*	(Thatcham)	18. 4.03A
G-OKCC	Cameron N-90 HAB	1741		6. 5.88	D.J.Head	(Newbury)	25. 7.00A
G-OKED	Cessna 150L	15074250	N19223	29.1.93	L.J.Pluck	Clipgate Farm, Denton	12. 3.06T
G-OKEM	Maianair Pegasus Quik	8047		23. 7.04	D.Young tr Kemble Quik 1	Kemble	22. 7.05P
G-OKEN	Piper PA-28R-201T Turbo Cherokee Arrow III	28R-7703390	N47518	20.10.87	W.B.Bateson	Blackpool	10. 6.07T
G-OKER	Van's RV-7 *(Built R.M.Johnson)*	PFA 323-14233		11. 5.04	R.M.Johnson	(Selkirk)	
G-OKES	Robinson R44 Astro	0053		16. 3.94	Hecray Co Ltd t/a Direct Helicopters	Southend	25. 5.06T
G-OKEV	Europa Aviation Europa *(Built K.A Pilcher - kit no.328) (Tri-gear u/c)*	PFA 247-13091		11. 6.97	K.A Pilcher *"Freedom"*	Wolverhampton	1 11.05P
G-OKEY	Robinson R22 Beta	2004		14. 1.92	Key Properties Ltd	Denham	25. 6.05
G-OKGB	IAV-Bacau Yakovlev Yak-52	9010407	LY-AGU Ukraine AF 19 *(yellow)*, DOSAAF 19 *(yellow)*	16.10.02	W.Hanekom	(Bedford)	16.10.05P
G-OKIM	Best Off Skyranger 912 *(Built K.P.Taylor - kit no.SKRxxxx395)*	BMAA/HB/333		23.12.03	K.P.Taylor *(Under construction 4.04)*	RAF Henlow	
G-OKIS	Tri-R Kis *(Built B W Davies)* (Canadian Air Motive CAM.100)	PFA 239-12248		15. 6.92	M.R.Cleveley	Tibenham	31. 5.05P
G-OKMA	Tri-R Kis *(Built K.Miller - tricycle u/c)*	PFA 239-12808		22.11.95	K.Miller	(Coventry)	30. 9.05P
G-OKPW	Tri-R Kis *(Built K.P.Wordsworth - tricycle u/c)* (Continental O-200-A)	PFA 239-12359		17. 8.93	K.P.Wordsworth *(Current CofR 1.05)*	(Deanland)	7. 6.03P
G-OKYA	Cameron V-77 HAB *(Replacement envelope c/n 3331)*	1259		4. 3.87	D.J.B.Woodd tr Army Balloon Club *"Fly Army II"*	(BFPO.17, Germany)	
G-OKYM	Piper PA-28-140 Cherokee	28-23303	G-AVLS N11C	10. 5.88	Hi-Fliers Aviation Ltd	Humberside	11. 3.07
G-OLAU	Robinson R22 Beta	1119		5. 9.89	Thistle Aviation Ltd *(Op Direct Helicopters)*	Southend	17. 7.05T
G-OLAW	Lindstrand LBL 25A Cloudhopper HAB	170		9.12.94	George Law Plant Ltd *"Law Hopper"*	(Kidderminster)	26. 6.04A
G-OLCP	Eurocopter AS355N Ecureuil 2	5580	G-CLIP	18. 2.02	Charterstyle Ltd	Blackbushe	11. 4.07T
G-OLDC	Learjet Inc Learjet 45	45-156	N3017F	12.10.01	Gold Air International Ltd	Biggin Hill	11.10.07ET
G-OLDD	British Aerospace BAe 125 Series 800B	258106	PK-RGM PK-WSJ, G-5-580	11. 3.99	Gold Air International Ltd	Stansted	19. 8.05T
G-OLDF	Learjet Inc Learjet 45	45-055	G-JRJR N45LR, N63MJ	3. 1.03	Gold Air International Ltd	Biggin Hill	23. 1.06T
G-OLDG	Cessna T182T Turbo Skylane	T18208127	G-CBTJ N5170R	17.10.02	Gold Air International Ltd	Stansted	22. 8.05T
G-OLDH	Aerospatiale SA.341G Gazelle 1	1307	G-UTZY G-BKLV, N341SC	10. 3.04	Gold Air International Ltd	Stansted	26. 3.05T
G-OLDJ	Learjet Inc Learjet 45	45-138	N5018G	24. 5.01	Gold Air International Ltd	Biggin Hill	23. 5.05T
G-OLDL	Learjet Inc Learjet 45	45-124	N4003Q	19. 2.01	Gold Air International Ltd	Stansted	18. 2.05T
G-OLDM	Pegasus Quantum 15-912	7589		10.12.99	A P.Watkins	Roddige	3. 4.05P
G-OLDN	Bell 206L LongRanger	45077	G-TBCA G-BFAL, N64689, A6-BCL	2.10.84	Von Essen Aviation Ltd	Thruxton	30. 7.06T
G-OLDP	Mainair Pegasus Quik	7957		28. 5.03	M J Wilson and G Lace	(Childwall, Liverpool)	6. 6.05P
G-OLDR	Learjet Inc Learjet 45	45-161	N3000S	18. 1.02	Gold Air International Ltd	Stansted	17. 1.05T
G-OLDV*	Colt 90A HAB				See SECTION 3 Part 1	(Aldershot)	
G-OLEE	Reims/Cessna F152 II	F15201797		11. 9.80	Redhill Air Services Ltd *(Op Airbase Flying Club)*	Shoreham	8. 4.06T
G-OLEL	American Blimp Corp A-60+ Airship	016	N606LG	9. 3.01	Keelex 195 Ltd	(Chilbolton, Stockbridge)	30. 6.07T
G-OLEM	Jodel D.18 *(Built R T Pratt)* (Revmaster R2100)	PFA 169-11613	G-BSBP	11. 2.02	D.G.H.Oswald *(Noted 1.04)*	Perth	
G-OLEO	Thunder Ax10-210 Series 2 HAB	3974		9. 1.97	P.J.Waller	(Norwich)	11. 6.03T
G-OLEZ	Piper J-3C-65 Cub	18432	G-BSAX N98260, NC98260	8. 8.01	L.Powell *(For restoration)*	(Canterbury)	
G-OLFB	Pegasus Quantum 15-912	7767		2. 3.01	A J.Boyd	Newtownards	5. 4.05P
G-OLFC	Piper PA-38-112 Tomahawk	38-79A0995	G-BGZG N9658N	6.12.85	M.W.Glencross	Cranfield	29. 5.07T
G-OLFO	Robinson R44 Raven	1305		6. 6.03	Crinstown Aviation Ltd	Clane, County Kildare	14. 7.06T
G-OLFT	Rockwell Commander 114	14274	G-WJMN N4954W	28. 3.85	D.A Tubby	(Warrington)	16. 5.05
G-OLGA	CFM Starstreak Shadow SA.II *(Built N.F.Smith - kit no K288)* (Rotax 618)	PFA 206-13164		15.10.97	N.F.Smith	Halstead, Essex	4. 8.05P
G-OLIZ	Robinson R22 Beta	0779		29. 9.88	Just Plane Trading Ltd	Standalone Farm, Meppershall	16. 8.07T
G-OLJT	Mainair Gemini/Flash IIA	570-887-5 & W359	G-MTKY	16. 9.98	A Wraith	Sandtoft	14. 3.04P
G-OLLI	Cameron O-31 HAB *(Golly Special shape)*	196		11. 5.76	N.A Robertson *"Golly III"* *(Loaned British Balloon Museum and Library 2002)*	(Newbury)	17. 7.97A
G-OLLS	Cessna T206H Turbo Stationair 6 *(Floatplane)*	T20608401	N5361L	8. 3.04	Loch Lomond Seaplanes Ltd	Luss, Loch Lomond	8. 3.07T
G-OLMA	Partenavia P68B	159	G-BGBT	15. 4.85	C.M.Evans	Plymouth	16.10.05T
G-OLOW	Robinson R44 Astro	0100		3.10.94	R.C.Hields t/a Hields Aviation	Sherburn-in-Elmet	14. 6.07T
G-OLPG*	Colt 77A HAB	2568		11. 3.94	D.J.Farrar *(Cancelled 9.3.04 by CAA)*	(Collingham, Wetherby)	16. 2.03
G-OLRT	Robinson R22 Beta	1378	N4014R	21. 5.90	S.Farmer t/a First Degree Air	Tatenhill	12. 8.05T
G-OLSF	Piper PA-28-161 Cadet	2841284	G-OTYJ G-OLSF, N92008	23.11.89	Bflying Ltd *(Op Bournemouth Flying Club)*	Bournemouth	23. 1.05T
G-OMAC	Reims FR172E Rocket	FR17200022	PH-HAI (PH-KRC), D-EDDC	3. 7.84	S.G.Shilling	Maypole Farm, Chislet	15.11.04T
G-OMAF	Dornier 228-202K	8112	D-CAAD	16. 2.87	Cobham Leasing Ltd *(Op DEFRA/Fisheries Patrol)*	Bournemouth	22. 6.05T
G-OMAK	Airbus A319-132	0913	F-WWIF G-OMAK, F-WWIF, G-OMAK, D-AVYL	7. 1.99	Twinjet Aircraft Sales Ltd	Luton	7. 1.06T
G-OMAL	Thruster T.600N 450	0061-T600N-050		16. 5.01	M Howland	Wickenby	4.12.04P
G-OMAP	Rockwell Commander 685	12036	F-GIRX F-OCGX, F-ZBBU, N6525V	4.11.94	Cooper Aerial Surveys Ltd *(Ordnance Survey titles)*	Gamston	14. 5.05A

Registration	Type	C/n	Previous identity	Reg date	Owner/operator	Base	CofA
G-OMAT	Piper PA-28-140 Cherokee D	28-7125139	G-JIMY G-AYUG, N11C	27. 8.87	R.B.Walker t/a Midland Air Training School	Leicester	16.11.06T
G-OMAX	Brantly B.2B	473	G-AVJN	7. 8.87	P.D.Benmax	(London W3)	16.11.03
G-OMCC	Aérospatiale AS350B Ecureuil	1836	G-JTCM G-HLEN, G-LOLY, JA9897, N5805T, HP-1084P, HP-1084, N5805T	26.03.03	Michael Car Centres Ltd	(Kirk Michael, Isle of Man)	29. 5.05T
G-OMCD	Robinson R44 Clipper II	10249		21. 1.04	J G M and H D McDiarmid t/a Mcdiarmid Partnership	(Callington, Plymouth)	12. 2.07T
G-OMDB	Van's RV-6A *(Built D A G Roseblade)*	25735		14. 8.02	D.A Roseblade	(Dubai, United Arab Emirates)	
G-OMDD	Cameron Thunder AX8-90 S2 HAB	4345		2. 4.98	M.D.Dickinson	(Bristol)	17. 3.04T
G-OMDG	Hoffmann H 36 Dimona	3510	OE-9215	19.11.98	D. Coulson tr Ards Dimona Group	Newtownards	9. 1.05
G-OMDH	MD Helicopters Hughes 369E	0293E		14.11.88	Stiltgate Ltd	Booker	15. 2.04T
G-OMDR	Agusta-Bell 206B-3 JetRanger III	8610	G-HRAY G-VANG, G-BIZA	8.12.97	Atlas Helicopters Ltd *(Op Total Air Management Services (TAMS)*	(Turweston)	3. 3.07T
G-OMEL	Robinson R44 Astro	0073	G-BVPB	30. 9.96	Coolen-Huybregts Vof	(Heythuysen, The Netherlands)	21.12.06T
G-OMEN	Cameron Z-90 HAB	10614		25. 6.04	MRC Howard Ltd	(Timperley, Altrincham)	24. 8.05T
G-OMEX	Zenair CH.701 STOL *(Built S.J.Perry)*	PFA 187-13556		11.12.01	S.J.Perry *(Noted 4.04)*	(Bucknall)	
G-OMEZ	Zenair CH.601HDS Zodiac *(Built C.J.Gow)*	PFA 162-13552		16. 7.01	C.J.Gow	Perth	6.11.04P
G-OMFG	Cameron A-120 HAB	4965		7. 2.01	M.F.Glue	(Hertford)	9. 4.05T
G-OMGH	Robinson R44 Clipper II	10259		29. 1.04	Universal Energy Ltd	Booker	26. 2.07T
G-OMHC	Piper PA-28RT-201 Arrow IV	28R-7918105	N3072Y	10. 2.81	Tatenhill Aviation Ltd	Tatenhill	27. 5.05T
G-OMHI	Mills MH-1 *(Built J.P.Mills)*	MH.001		8.10.97	J.P.Mills *(Current status unknown)*	(Stockport)	
G-OMHP	Jabiru Aircraft Jabiru UL *(Built M.H.Player)*	PFA 274A-13584		23. 5.00	M.H.Player	(Upton Noble, Shepton Mallet)	25. 8.05P
G-OMIA	SOCATA MS.893A Rallye Commodore 180	12074	D-ENME F-BUGE, (D-ENMH)	21. 7.98	P.W.Portelli	Elstree	19.12.04
G-OMIG*	WSK SBLim-2A				See SECTION 3 Part 1	San Carlos, Brazil	
G-OMIK	Europa Aviation Europa *(Built M.J.Clews - kit no.270)* (Rotax 914-UL) *(Monowheel u/c)*	PFA 247-12991		12. 1.98	M.J.Clews	White Waltham	29. 9.05P
G-OMJC	Raytheon 390 Premier 1	RB-88	N4488F	17. 6.04	Manhattan Jet Charter Ltd	Farnborough	17. 6.07T
G-OMJT	Rutan LongEz *(Built M.J.Timmons - c/n 968)* (Lycoming O-235)	PFA 74A-10703		14.10.92	M.J.Timmons	Prestwick	7. 9.05P
G-OMMG	Robinson R22 Beta	1041	G-BPYX	25. 2.94	Preston Associates Ltd	Yearby	5. 2.07T
G-OMMM	Colt 90A HAB	2328		20. 1.93	V.Trimble *"Red September"*	(Henley-on-Thames)	22. 4.05A
G-OMMT	Robinson R44 Astro	0315	G-IBKA G-USTE	10. 6.02	Morrison Motors (Turriff)	Lower Baads Farm, Peterculter	15. 4.06T
G-OMNH	Beech 200 Super King Air	BB-108	N108BM RP-C1979, TR-LWC	19. 8.98	Newborne Ltd	Norwich	20. 8.06T
G-OMNI	Piper PA-28R-200 Cherokee Arrow II	28R-7335130	G-BAWA N11C	3. 1.84	Air Gloster Ltd	Gloucestershire	31. 7.06T
G-OMOG*	Gulfstream AA-5A Cheetah	AA5A-0793	G-BHWR N26892	4. 3.88	Solent Flight Aircraft Ltd *(Cancelled 23.7.01 by CAA) (Fuselage noted 8.04)*	Phoenix Farm, Lower Upham	15. 4.02T
G-OMOL	Maule MX-7-180C Star Rocket	28012C		15. 8.00	Aeromarine Ltd	Owlesbury	18. 7.07
G-OMPW	Mainair Pegasus Quik	8088		12. 1.05	MPW Decorators Ltd	(Kidgate, Louth)	
G-OMRB	Cameron V-77 HAB	2184		29. 8.90	I.J.Jevons *"Harlequin"*	(Bristol)	5. 9.05A
G-OMRG	Hoffmann H 36 Dimona	36132	G-BLHG	15.11.88	M.R.Grimwood	Kemble	21. 1.06
G-OMSS	Best Off Skyranger 912 *(Built M.S.Schofield - kit no.UK/523)*	BMAA/HB/425		6. 1.05	M.S.Schofield	(Candlesby, Spilsby)	
G-OMST	Piper PA-28-161 Warrior III	2842121	G-BZUA N53363	1. 8.01	Mid-Sussex Timber Co Ltd	Biggin Hill	11. 6.07T
G-OMUC	Boeing 737-36Q	29405		29. 6.98	Easyjet Airline Company Ltd	Luton	22.11.06T
G-OMUM	Rockwell Commander 114	14067	PH-JJJ (PH-MMM), N4737W	24. 1.97	C.E.Campbell	Blackbushe	18. 3.06
G-OMWE	Zenair CH.601HD Zodiac *(Built P J Roy)*	PFA 162-12740	(N) G-OMWE, G-BVXU	21. 3.97	G.Cockburn *(New CofR 12.03)*	(Hawick)	14. 7.01P
G-OMYT	Airbus A330-243	301	G-MOJO F-WWYE	14. 5.03	MyTravel Airways Ltd	Manchester	7.11 05T
G-ONAF	Naval Aircraft Factory N3N-3 (Wright Whirlwind R.760)	Not known	N45192 Bu.4406	31. 1.89	R.P.W.Steele and J.D.Hutchinson	Sandown	30. 8.06T
G-ONAV	Piper PA-31 Navajo C	31-7812004	G-IGAR D-IGAR, N27378	29. 1.93	Panther Aviation Ltd	Elstree	31. 5.06T
G-ONCB	Lindstrand LBL 31A HAB	393		4. 6.96	R.J.Mold	(High Wycombe)	5.4 ..05A
G-ONCL	Colt 77A HAB	1637		4. 4.90	D.R.Pearce	(Slimbridge)	3. 8.02A
G-ONCM	Partenavia P68C	217	I-CITT G-TELE, G-DORE, OY-CAD	5.12.01	M.Arnell	Pretswick	13. 1.05T
G-ONEB	Westland Scout AH.1	F9761	G-BXOE XW798	21. 1.98	E.R.Meredith and E M Smith	Draycott Farm, Chiseldon	7. 7.05P
G-ONES	Slingsby T.67M-200 FIrefly	2046	SE-LBB LN-TFB, G-7-122	12.11.01	L.J.Jones	Exeter	26. 6.06T
G-ONET	Piper PA-28-180 Cherokee E	28-5802	G-AYAU N11C	3. 6.98	K.Tomlin tr Hatfield Flying Club	Elstree	22. 8.05T
G-ONFL	Murphy Maverick *(Built K Godfrey and G Lockwood - c/n 402)* (Rotax 503)	PFA 259-12750	G-MYUJ	27.11 98	R Foster	Otherton, Cannock	20.12.05P
G-ONGA	Robinson R44 Raven II	10479		5.10.04	Mash Enterprises Ltd *(Noted 11.04)*	Bushey Heath	AC
G-ONGC	Robin DR400/180R Remorqueur	1385	EI-CKA SE-GHM	11.11.98	Norfolk Gliding Club Ltd	Tibenham	26. 3.05
G-ONHH	Forney F-1A Aircoupe	5725	G-ARHA N3030G	13.12.89	R.D.I.Tarry *"Easy Rider"*	Pytchley Grange	27. 4.07
G-ONIG	Murphy Elite *(Built N Smith)*	PFA 232-14042		29. 4.03	N.S.Smith	(Derby)	

G-ONIX	Cameron C-80 HAB	4411		12. 8.98	Tracy Louise Gorman	(Weston-super-Mare)	3. 6.05A
G-ONKA	Aeronca K	K283	N19780	21.10.91	N.J.R.Minchin	Hill Top Farm, Hambledon	12.10.04P
	(Lycoming O-145)		NC19780		*"Aggnes"*		
G-ONMT	Robinson R22 Beta II	2963		20. 7.99	Redcourt Enterprises Ltd	Kintore	2. 8.05T
G-ONON	Rotary Air Force RAF 2000 GTX-SE	PFA G/13-1313		13. 8.99	M.P.Lhermette	(Faversham)	
	(Built M S R Allen)				*(New owner 6.03)*		
G-ONOW	Bell 206A JetRanger	605	G-AYMX	8. 8.88	J.Lucketti	(Rochdale)	27. 4.00T
G-ONPA (2)	Piper PA-31-350 Navajo Chieftain	31-7952110	N89PA	6. 5.98	West Wales Airport Ltd	(Ross-on-Wye)	15.10.07T
			N35225				
G-ONSF	Piper PA-28R-201 Cherokee Arrow III	28R-773708	G-EMAK	17. 1.01	Northamptonshire School of Flying Ltd	Sywell	16. 5.07T
			D-EMAK, N38180				
G-ONTV	Agusta-Bell 206B-3 JetRanger III	8733	D-HUNT	1. 4.98	Castle Air Charters Ltd	Liskeard	19. 4.07T
			TC-HKJ, (D-HSAV), I-GPFP, I-PIEF				
G-ONUN	Van's RV-6A	PFA 181-12976		20. 2.96	R.E.Nunn	Clipgate Farm, Denton	23. 5.05P
	(Built R E Nunn) (Lycoming O-360)						
G-ONUP	Enstrom F-28C	348	G-MHCA	18. 1.00	S Brophy	(Widnes)	20. 6.02
			G-SHWW, G-SMUJ, G-BHTF		*(New owner 2.04)*		
G-ONYX	Bell 206B-3 JetRanger III	4160	G-BXPN	22. 1.98	Kenrye Developments Ltd	Newtownards	16. 3.07T
			N18EA, D-HOBA, (D-HOBE)				
G-ONZO	Cameron N-77 HAB	1089		13.11.84	K.Temple *"Gonzo"*	(Eye)	19. 7.99A
G-OOAE	Airbus A321-211	852	(G-UNIF)	14. 7.98	First Choice Airways Ltd	Manchester	13. 7.07T
			D-AVZG				
G-OOAF	Airbus A321-211	677	G-UNID	4.12.98	First Choice Airways Ltdv	Manchester	6. 5.06T
			G-UKLO, D-AVZO				
G-OOAH	Airbus A321-211	781	G-UNIE	4. 1.99	First Choice Airways Ltd	Manchester	2. 3.07T
			D-AVZK				
G-OOAL	Boeing 767-38A	29617		29. 3.99	First Choice Airways Ltd *"Sunrise"*	Manchester	29. 3.05T
G-OOAN	Boeing 767-39H	26256	G-UKLH	26. 1.99	First Choice Airways Ltd *"Caribbean Star"*	Manchester	4. 4.06T
G-OOAP	Airbus A320-214	1306	F-WWBY	23.10.00	First Choice Airways Ltd	Manchester	22.10.06T
G-OOAR	Airbus A320-214	1320	F-WWDT	3.11.00	First Choice Airways Ltd	Manchester	2.11.06T
G-OOAU	Airbus A320-214	1637	F-WWDM	10. 1.02	First Choice Airways Ltd	Manchester	9. 1.08E
G-OOAV	Airbus A321-211	1720	D-AVXA	29. 4.02	First Choice Airways Ltd	Manchester	28. 4.05T
G-OOAW	Airbus A320-214	1777	F-WWDM	27. 5.02	First Choice Airways Ltd	Manchester	26. 5.05T
G-OOAX	Airbus A321-231	2180	F-WWDY	7. 4.04	First Choice Airways Ltd	Manchester	6. 4.07T
G-OOBA	Boeing 757-28A	32446	C-GUBA	9. 2.01	First Choice Airways Ltd	Manchester	20. 4.07T
			G-OOBA, N446GE, (N558NA)				
G-OOBC	Boeing 757-28A	33098		28. 3.03	First Choice Airways Ltd	Manchester	27. 3.06T
G-OOBD	Boeing 757-28A	33099		31. 3.03	First Choice Airways Ltd	Manchester	30. 3.06T
G-OOBE	Boeing 757-28A	33100		19. 5.03	First Choice Airways Ltd	Manchester	18. 5.06T
G-OOBF	Boeing 757-28A	33101		19. 4.04	First Choice Airways Ltd	Manchester	18. 4.07T
G-OOBI	Boeing 757-2B7	27146	N615AU	29. 6.04	First Choice Airways Ltd	Manchester	9.10.07E
G-OOBJ	Boeing 757-2B7	27147	N616AU	21. 5.04	First Choice Airways Ltd	Manchester	5.12.07E
G-OOBK	Boeing 767-324ER	27392	VN-A762	18.11.04	First Choice Airways Ltd	Manchester	AC
			S7-RGV, EI-CMD, N1785B, (N48901)				
G-OOCS	MD Helicopters Hughes 369E	0204E	G-OTDB	19.12.03	R and S Fire and Security Ltd	(Bulcote, Nottingham)	23.1.05T
			G-BXUR, HA-MSC				
G-OODE	SNCAN Stampe SV-4C	500	G-AZNN	9. 5.77	A R.Radford	Redhill	5. 9.05T
	(DH Gipsy Major 10)		F-BDGI				
G-OODI	Pitts S-1D	KH.1	G-BBBU	23.12.80	R.M.Buchan	Leicester	7. 5.05P
	(Built Etheridge and Lincs Aerial)				*"Little Bumble" (Noted 9.04)*		
G-OODW	Piper PA-28-181 Archer II	28-8490031	N4332C	14. 7.87	Goodwood Road Racing Co Ltd	Goodwood	19.12.05T
					(Op Goodwood Flying Club)		
G-OOER	Lindstrand LBL 25A Cloudhopper HAB	125		15. 8.94	Airborne Adventures Ltd	(Skipton)	6. 5.05P
G-OOFE	Thruster T600N 450 Sprint	0036-T600N-087		8. 7.03	Rochester Microlights Ltd	Rochester	8. 7.05P
G-OOFT	Piper PA-28-161 Warrior III	2842083	N170FT	25. 5.00	Lyrical Computing Ltd	Denham	22. 6.06T
					(Op Denham School of Flying)		
G-OOGA	Gulfstream GA-7 Cougar	GA7-0111	SE-IEA	3. 2.86	B.Robinson	Denham	25.11.07E
	(C/n correct but duplicates that for YV-1334P)		N758G				
G-OOGI	Gulfstream GA-7 Cougar	GA7-0077	G-PLAS	16. 1.95	Plane Talking Ltd	Elstree	13.10.06T
			G-BGHL, N789GA				
G-OOGO	Grumman GA-7 Cougar	GA7-0049	N762GA	12.11.97	Leonard F.Jollye (Brookmans Park) Ltd	Elstree	13. 1.07T
G-OOGS	Gulfstream GA-7 Cougar	GA7-0105	G-BGJW	19. 6.98	Leeds Flying School Ltd	Leeds	3. 4.06T
			N737G				
G-OOIO	Eurocopter AS350B3 Ecureuil	3463		17.10.01	Hovering Ltd	Elstree	19.11.04T
G-OOJC	Bensen B.8MR	PFA G/101-1303		4.12.98	J.R.Cooper	Henstridge	
	(Built J R Cooper) (Converted ex Air Command)				*(Noted 11.02)*		
G-OOJP	Commander Aircraft Commander 114B	14567	N92JT	24.12.99	Speedsport Ltd	(Oxford)	13. 2.06
			D-EYCA				
G-OOLE	Cessna 172M Skyhawk II	17266712	G-BOSI	25. 8.89	P.S.Eccersley	Humberside	15. 3.07
			N80714				
G-OOMW	Comco Ikarus C42 FB UK	PFA 322-14056		27. 5.03	R.O'Malley-White	Dunkeswell	20. 9.05P
	(Built R.O'Malley-White)						
G-OONE	Mooney M.20J Model 205	24-3039		31. 7.87	Go One Aviation Ltd	Welshpool	23. 7.06
G-OONI	Thunder Ax7-77 HAB	1534		9. 3.90	Fivedata Ltd *"Bridesnightie"*	(Todmorden)	31. 3.01A
G-OONY	Piper PA-28-161 Warrior II	28-8316015	N83071	26. 7.89	D.A Field and P.B.Jenkins	Compton Abbas	4.11.07T
G-OOOB	Boeing 757-28A	23822	C-FOOB	19. 2.87	Astraeus Ltd	Gatwick	28. 4.07T
	G-OOOB, C-FOOB, G-OOOB, C-FOOB, G-OOOB, C-FOOB, G-OOOB, C-FOOB, G-OOOB, C-FOOB, G-OOOB, C-FOOB, G-OOOB, C-FOOB, G-OOOB, C-FOOB, G-OOOB						
G-OOOG	Boeing 757-23A	24292	C-FOOG	29. 3.89	First Choice Airways Ltd	Manchester	1. 4.07T
			G-OOOG, C-FOOG, G-OOOG, C-FOOG, G-OOOG, C-FOOG, G-OOOG, C-FOOG, G-OOOG, C-FOOG, G-OOOG				
G-OOON	Piper PA-34-220T Seneca III	34-8533024	N822CB	8. 1.03	Goon Aviation Ltd	Fairoaks	20. 2.06T
			ZS-LWI, N2431Q, N9513N				
G-OOOX	Boeing 757-2Y0	26158		24. 2.93	First Choice Airways Ltd	Manchester	22. 3.06T
G-OOOY	Boeing 757-28A	28203		21. 5.98	First Choice Airways Ltd	Manchester	20. 5.04T

G-OOSE	Rutan VariEze	PFA 74-10326		7.12.78	B.O.Smith and J.A Towers	Yearby	
	(Built J A Towers -c/n 1536)				*(Stored dismantled 1.02)*		
G-OOSI	Cessna 404 Titan	404-0855	VT-DAT	31. 1.03	Cooper Aerial Surveys Ltd	Gamston	1. 5.06T
			N404N, F-WQFV, F-ZBDB, F-BRGN, N68104				
G-OOSY	de Havilland DH.82A Tiger Moth	85831	F-BGFI	6. 9.94	M.Goosey	Eccleshall, Stafford	
	(Composite rebuild)		Fr AF, DE971		*(On rebuild 9.94: current status unknown)*		
G-OOTB	SOCATA TB-20 Trinidad GT	2180	D-EADS	11. 5.04	Select Helicopters Ltd	Aberdeen	16. 5.07T
G-OOTC	Piper PA-28R-201T Turbo Cherokee Arrow III	28R-7703086	G-CLIV	18. 1.94	D.G.and Caroline M.King	Turweston	12. 5.06
			N3011Q				
G-OOTW	Cameron Z-275 HAB	10380		6. 6.03	Airborne Balloon Management Ltd	(Tonbridge)	29. 7.05T
G-OOUT	Colt Flying Shuttlecock SS HAB	1938		16. 5.91	Shiplake Investments Ltd	(St. Peter Port, Guernsey)	18.11.00A
					"Shuttlecock"		
G-OOXP	Aero Designs Pulsar XP	PFA 202-11915		25.10.90	T.D.Baker	Corby	18. 4.96P
	(Built G W Associates Ltd)				*(Current status unknown)*		
G-OPAG	Piper PA-34-200 Seneca	34-7250348	N506DM	16.10.90	A H.Lavender	Biggin Hill	10. 4.06
			G-BNGB, F-BTQT, F-BTMT				
G-OPAL	Robinson R22 Beta	0535	N23750	11. 2.86	Heli Air Ltd	Leasowes Farm, Oxhill, Warwick	12. 4.07T
					(Op The Leamington Hobby Centre Ltd)		
G-OPAM	Reims/Cessna F152 II	F15201536	G-BFZS	5. 9.86	PJC (Leasing) Ltd *"Little Red Rooster"*	Stapleford	28. 6.06T
G-OPAS*	Vickers 806 Viscount				See SECTION 3 Part 1	Duxford	
G-OPAT	Beech 76 Duchess	ME-304	G-BHAO	6.12.82	R.D.J.Axford	Booker	18. 3.06
G-OPAZ	Pazmany PL-2	PFA 69-10673		20. 3.98	K.Morris	Thruxton	18.12.05P
	(Built K Morris) (Lycoming O-235)				*"Y Myddryg Bach Melyn"*		
G-OPCG	Cessna 182T Skylane	18280948	N2451Y	18. 2.02	Pye Consulting Group Ltd	Blackpool	25. 2.05T
G-OPCS	Hughes 369E	0333E	CS-HBN	31. 1.01	Productivity Computer Solutions Ltd	(Ossett)	3. 5.07T
			N500AH				
G-OPDS	Denney Kitfox Model 4	PFA 172A-12259		8. 1.93	D.A Lord	Swanborough Farm, Lewes	27. 8.05P
	(Built P D Sparling)						
G-OPEN	Bell 206B JetRanger III	4300	N743BT	20. 1.05	Gazelle Aviation LLP	(Wetherby)	
			N206AJ, N2155K, C-GFNP				
G-OPEP	Piper PA-28RT-201T Turbo Arrow IV	28R-7931070	OY-PEP	3.12.97	S A F Elliott	Cranfield	31. 8.06T
			N2217Q		t/a Sam Aviation		
G-OPET	Piper PA-28-181 Cherokee Archer II	28-7690067	OH-PET	3. 1.02	Cambrian Flying Group Ltd	Cardiff	26. 2.05T
			OY-BLC				
G-OPFA	Alpi Pioneer 300	PFA 330-14298		23.11.04	S.Eddison and Rachel Minett	Gloucestershire	
	(Built S Eddison and R Minett)						
G-OPFT*	Cessna 172R Skyhawk	17280316	N9491F	11. 3.98	Rankart Ltd	Oxford	19. 3.01T
	(Overran runway landing Newtownards 14.11.2000: damage to undercarriage: cancelled 20.11.02 as temporarily wfu) (Noted 4.03)						
G-OPFW	Hawker Siddeley HS.748 Series 2A/266	1714	G-BMFT	1. 7.98	Emerald Airways Ltd *(Parcel Force titles)*	Liverpool	16. 2.07T
			VP-BFT, VR-BFT, G-BMFT, 5W-FAO, G11-10				
G-OPHR	Diamond DA.40 Star	40066		8.11.01	MC Air Ltd	Wellesbourne Mountford	5. 3.05T
G-OPHT	Schleicher ASH 26E	26105		6. 2.97	P Turner *"T1" (Noted 12.04)*	Nympsfield	21. 6.04
G-OPIB*	English Electric Lightning F.6				See SECTION 3 Part 1	Cape Town, South Africa	
G-OPIC	Reims/Cessna FRA150L Aerobat	FRA1500234	G-BGNZ	20. 6.95	S.J.Burke	Bodmin	28. 9.06T
			PH-GAB, D-EIQE		t/a Peak Aviation Photography		
G-OPIK	Eiri PIK 20E	20233	PH-651	27. 1.82	A J.McWilliam	Newtownards	1.10.06
G-OPIT	CFM Streak Shadow	PFA 161A-11624		22.11.89	I.Sinnett	Bodmin	8. 7.03P
	(Built L W Opit - kit no K.126-SA) (Rotax 532)						
G-OPJC	Cessna 152 II	15282280	N68354	7. 6.88	PJC (Leasing) Ltd	Stapleford	17.10.06T
G-OPJD	Piper PA-28RT-201T Turbo Arrow IV	28R-8231028	N8097V	2.10.89	J M McMillan	Thruxton	16.12.04T
G-OPJH	Druine D.62B Condor	RAE/619	G-AVDW	15. 4.97	P.J.Hall	Oaksey Park	29. 4.06
	(Built Rollason Aircraft and Engines)						
G-OPJK	Europa Aviation Europa	PFA 247-12487		29. 4.93	F.D.Hollinshead	Sleap	24. 4.06P
	(Built P.J.Kember - kit no.017) (Monowheel u/c)						
G-OPJM	Bell 206B JetRanger II	4259	D-HABA	25. 2.04	PJM Helicopters LLP	Nottingham East Midlands	10. 3.07T
			C-FOFG		*(Op East Midland Helicopters)*		
G-OPJS	Pietenpol AirCamper	PFA 47-12834		10.11.00	P.J.Shenton	Sywell	
	(Built P.J.and J W Shenton)				*(Noted 1.05)*		
G-OPKF	Cameron Bowler 90 SS HAB	2314		12. 6.90	D.K.Fish	(Burbage)	2. 8.03A
G-OPLB	Cessna 340A II	340A0486	G-FCHJ	11. 7.95	Just Plane Trading Ltd	Newcastle	27. 5.06
			G-BJLS, (N6315X)				
G-OPLC	de Havilland DH.104 Dove 8	04212	G-BLRB	10. 1.91	W.G.T.Pritchard	Redhill	9. 5.05T
			VP962		*(Op Mayfair Dove)*		
G-OPME	Piper PA-23-250 Aztec D	27-4099	G-ODIR	31. 3.94	Portway Aviation Ltd	Shobdon	14.10.07T
			G-AZGB, N878SH, N9...N				
G-OPMT	Lindstrand LBL 105A HAB	052		30. 9.93	Pace Micro Technology plc *"Pace"*	(Shipley)	31. 7.99A
G-OPNH	Stoddard-Hamilton Glasair Super II-SRG		G-CINY	14.10.98	J.L.Mangelschots	(Balen, Belgium)	26. 5.05P
	(Built P N Haigh) (Lycoming IO-360)	PFA 149-13011					
G-OPPL	Gulfstream AA-5A Cheetah	AA5A-0867	G-BGNN	11.10.85	Plane Talking Ltd	Elstree	8. 8.06T
G-OPRC	Europa Aviation Europa XS	PFA 247-13281		22. 6.01	M.J.Ashby-Arnold	Wombleton	7. 3.05P
	(Built I Chaplin - kit no.378) (Tri-gear u/c)						
G-OPSF	Piper PA-38-112 Tomahawk	38-79A0998	EI-BLT	13.10.82	Panshanger School of Flying Ltd	High Cross, Ware	24. 6.06T
			G-BGZI, N9664N				
G-OPSL	Piper PA-32R-301 Saratoga SP	32R-8013085	G-IMPW	4. 1.99	P.R.Tomkins	(Robertsbridge)	24. 6.06
			N8186A				
G-OPSS	Cirrus SR20-G2	1458	N410CD	28.10.04	Public Sector Software	(Warwick)	13.12.07E
G-OPST	Cessna 182R Skylane II	18267932	OO-HFF	16. 6.88	Lota Ltd	Shoreham	2. 6.06T
			N9317H				
G-OPTF	Robinson R44 Raven II	10235		9. 1.04	Franks Helicopter Leasing Ltd	Booker	1. 2.07T
G-OPUB	Slingsby T.67M-160 Firefly	2002	G-DLTA	18.10.96	P.M.Barker	Leeds-Bradford	9. 9.07T
			G-SFTX				
G-OPUP	Beagle B.121 Pup Series 2	B121-062	G-AXEU	31.10.84	A Brinkley	Standalone Farm, Meppershall	2. 6.07
			(5N-AJC)				

Reg	Type	C/n	Previous identity	Date	Owner/Operator	Base	CofA
G-OPUS	Jabiru Aircraft Jabiru SK	PFA 274-13343		16. 7.98	H.H.R.Lagache	Leicester	29. 7.05P
	(Built S Percy)						
G-OPWK	Grumman AA-5A Cheetah	AA5A-0663	G-OAEL	26. 5.92	A H.McVicar	Carlisle	6. 9.02T
			N26706		*(Noted 11.04)*		
G-OPWS	Mooney M.20K Model 231	25-0663	N1162W	12. 4.91	A R.Mills	Fowlmere	13. 8.06
G-OPYE	Cessna 172S Skyhawk SP	172S8059	N653SP	19. 2.99	Far North Aviation	Wick	25. 2.05T
G-ORAC	Cameron Van 110SS HAB	4577		22. 6.99	A.G.Kennedy *(RAC titles)*	(Nelson)	21. 5.02A
G-ORAE	Van's RV-7	PFA 323-14016		20. 3.03	R.W.Eaton	(Chesterfield)	
	(Built R W Eaton)				tr G-ORAE Group		
G-ORAF	CFM Streak Shadow	PFA 161A-11627		18. 5.90	A P.Hunn	Swanton Morley	1.11.00P
	(Built G A Taylor - c/n K.134-SA) (Rotax 532) *(PFA c/n duplicates MW6 G-MYCU)*				*(Dismantled 5.00)*		
G-ORAL	Hawker Siddeley HS.748 Series 2A/334	1756	G-BPDA	13. 8.99	Emerald Airways Ltd	Liverpool	12.11.05T
			G-GLAS, 9Y-TFS, G-11-8		*(Reed Aviation titles) "The Paper Plane"*		
G-ORAR	Piper PA-28-181 Archer III	2890224	N9255G	6. 6.95	P.N and S.M.Thornton	Goodwood	8. 7.07
G-ORAS	Clutton FRED Series II	PFA 29-11002		14. 6.01	A I.Sutherland	(Edderton)	17.11.05P
	(Built A I.Sutherland)						
G-ORAY	Reims/Cessna F182Q Skylane II	F18200132	G-BHDN	18. 3.94	Yorkshire Estates Ltd	(Douglas, Isle of Man)	9.10.04
G-ORBD	Van's RV-6	PFA 181-12677	G-BVRE	23. 7.01	O.R.B.Dixon	Barton	6.11.05P
	(Built J J Martyn) (Lycoming O-320)						
G-ORBK	Robinson R44 Raven II	10213	G-CCNO	28.11.03	GTC (UK) Ltd	Booker	22.12.06T
G-ORBS	Mainair Blade	1336-0802-7 & W1131		19. 8.02	J.W.Dodson	Leicester	11.10.05P
	(Rotax 582)						
G-ORCA	Van's RV-4	PFA 181-12924		25.11.04	M.R.H.Wishart	(Scalloway, Shetland)	
	(Built M.R.H.Wishart)						
G-ORCP	Hawker Siddeley HS.748 Series 2A/242	1647	(G-CLEW)	2. 1.03	Emerald Airways Ltd	Liverpool	18. 2.06T
			ZS-OCF/ZK-CWJ				
G-ORDB	Cessna 550 Citation Bravo	550-1042	N51869	11.12.02	Equipe Air Ltd	Gamston	12.12.05T
G-ORDS	Thruster T600N 450	0042-T600N-100		14. 1.04	Thruster Air Services Ltd	Ginge, Wantage	
G-ORED	Pilatus Britten-Norman BN-2T Islander	2142	G-BJYW	10. 1.85	Fly BN Ltd	(London SE16)	26. 2.07A
G-OREV	Revolution Helicopters Mini-500	0112		8. 8.96	R.H.Everett *(Current status unknown)*	Thruxton	AC
G-ORFC	Jurca MJ.5 Sirocco	PFA 2210		16. 5.85	D.J.Phillips	Lasham	3. 7.04P
	(Built P Phillips) (Lycoming O-290)						
G-ORGI*	Hawker Hurricane IIB				See SECTION Part 1 Niagara Falls, Ontario, Canada		
G-ORGY	Cameron Z-210 HAB	10320		9. 7.02	A M.Holly *(Go Ballooning.co.uk titles)*	(Berkeley)	8. 7.05T
G-ORHE	Cessna 500 Citation	500-0220	(N619EA)	25. 3.96	Weston Ltd	Weston, County Kildare	22. 5.06T
			G-OBEL, G-BOGA, N932HA, N93WD, N5220J				
G-ORIG	Glaser-Dirks DG-800A	8-39-A29	BGA4972, KBY	5. 4.94	I.Godfrey	Lasham	17. 4.07
			G-ORIG		*"386"*		
G-ORIX	ARV K1 Super 2	PFA 152-12424	G-BUXH	16. 9.93	T.M.Lyons	Egginton	1. 7.05P
	(Built P M Harrison - kit no 034) (Norton AE.100R)		(G-BNVK)				
G-ORJA	Beech B200 Super King Air	BB-1570	N1120Z	5.06.03	Airwest Ltd	Gamston	4. 6.06T
			N50PM, N1120Z				
G-ORJW	Laverda F.8L Falco Series 4	403	(PH-...)	2.12.85	W.R.M.Sutton	(Hilversum, The Netherlands)	31. 8.07
			G-ORJW, D-ELDV, D-ELDY				
G-ORJX*	BAE SYSTEMS Avro 146-RJX85	E2376		16. 2.00	BAE SYSTEMS (Operations) Ltd	Woodford	
					(Cancelled 12.12.02 as wfu) (Stored 2.04)		
G-ORMA	Aérospatiale AS355F1 Ecureuil 2	5192	G-SITE	9.11.98	MW Helicopters Ltd	Stapleford	15. 6.07T
			G-BPHC, N365E				
G-ORMB	Robinson R22 Beta	1607		14.12.90	CHC Scotia Ltd	Cumbernauld	19. 4.06T
G-ORMG	Cessna 172R Skyhawk	17280344	N9518F	25. 9.98	J.R.T.Royle	Andrewsfield	19.10.07E
G-OROB	Robinson R22 Beta	0965	G-TBFC	11. 6.90	R.Culff t/a Corniche Helicopters	Redhill	25. 6.95T
			N80287		*(Spares use 9.97: current status unknown)*		
G-OROD	Piper PA-18-150 Super Cub	18-7856	SE-CRD	27. 6.89	B.W.Faulkner	(Petersfield)	10. 3.05
G-ORON	Colt 77A HAB	1149		8. 3.88	J.Charley tr Orion Hot Air Balloon Group	(Wymeswold)	1.10.00A
G-ORPC	Europa Aviation Europa XS	PFA 247-13521		5. 2.04	P.W.Churms	(Farnborough)	
	(Built P.W.Churms)						
G-ORPR	Cameron O-77 HAB	2341		26. 6.90	T.Strauss and A Sheehan *"Batman"*	(London SW1)	21. 3.04A
G-ORRR	Hughes 369HS	114-0673S	G-STEF	20. 6.01	The Lower Mill Estate Ltd	Kemble	12. 4.07
			G-BKTK, OY-HCL, OO-JGR				
G-ORTH	Beech E90 King Air	LW-136	G-DEXY	12.11.03	Kilo Aviation Ltd	Liverpool	2. 2.05T
			N750DC, N30CW, N84GA, N328TB, TR-LTT				
G-ORTM	Glaser-Dirks DG-400	4-209		6. 3.87	A.R.Garcia	(Granada, Spain)	16. 7.06
G-ORUG	Thruster T600N 450 Sprint	0033-T600N-080		2. 9.03	I.Shulver	Baxby Manor, Husthwaite	12. 9.05P
G-ORVB	McCulloch J.2	039	(G-BLGI)	2. 8.89	R.V.Bowles	(Rugby)	AC
			(G-BKKL), Bahrain Public Security BPS-3, N4329G *(Under restoration 6.03)*				
G-ORVG	Van's RV-6	PFA 181A-13509		2. 1.01	R.J.Fray	Sibson	1. 3.05P
	(Built R J Fry) (Lycoming O-360)						
G-ORVR	Partenavia P68	115	G-BFBD	2.10.95	Cheshire Flying Services Ltd	Liverpool	1. 4.05T
	(Officially regd as "P68B")				t/a Ravenair		
G-OSAT	Cameron Z-105 HAB	10564		18. 6.04	Lotus Balloons Ltd *"Astra"*	(Broughton, Stockbridge)	12. 7.05A
G-OSAW	QAC Quickie Q.2	2443	G-BVYT	17. 9.04	S.A.Wilson	Newtownards	2. 9.00P
	(Built N A Evans)		N3797S				
G-OSCC	Piper PA-32-300 Cherokee Six	32-7540020	G-BGFD	27.11.84	BG and G Airlines Ltd	Jersey	17. 4.05
			D-EOSH, N32186				
G-OSCH	Cessna 421C Golden Eagle III	421C0706	G-SALI	13. 9.95	Northern Aviation Ltd	Durham Tees Valley	8.12.05T
			N26552				
G-OSCO	TEAM mini-MAX 91	PFA 186-12878		24.12.96	G.E.Norton	Cadwell Park, Louth	21. 9.05P
	(Built PJ Schofield)						
G-OSDI	Beech 58 Baron	TH-1111	G-BHFY	27. 7.84	A.W.Eldridge	Leicester	18.11.05
G-OSEA	Pilatus Britten-Norman BN-2B-26 Islander	2175	G-BKOL	27. 8.85	W.T.Johnson and Sons (Huddersfield) Ltd	Crosland Moor	23. 3.07
G-OSEE	Robinson R22 Beta	0917		11. 1.89	Aero-Charter Ltd	Manston	22. 1.06T
G-OSEP	Mainair Blade 912	1340-0902-7-W1135		29.10.02	J.D.Smith	Baxby Manor, Husthwaite	19.10.05P
G-OSFA	Diamond HK 36TC Super Dimona	36.649		15. 6.99	Oxfordshire Sportflying Ltd	Enstone	23. 7.05T
G-OSFC	Reims/Cessna F152 II	F15201872	G-BIVJ	31. 1.86	Stapleford Flying Club Ltd	Stapleford	12. 6.06T

Reg	Type	C/n	Previous identity	Date	Owner/Operator	Base	CofA
G-OSFS	Reims/Cessna F.177RG Cardinal RG	F177RG0082	F-BUMP	26. 1.04	Cardinal Sin Ltd	Gloucestershire	26. 4.07T
G-OSGB	Piper PA-31-350 Navajo Chieftain	31-7952155	G-YSKY	25. 1.99	Aerial Support Services Ltd	Fairoaks	20. 6.07T
			N3529D				
G-OSHL	Robinson R22 Beta	1000		19. 4.89	Sloane Helicopters Ltd	Sywell	6. 9.04T
G-OSIC	Pitts S-1C	1921-77	G-BUAW	7.10.02	J.A Dodd	Booker	28. 4.05P
	(Built R Hendry) (Lycoming O-320)		N29DH				
G-OSII	Cessna 172N Skyhawk II	17267768	G-BIVY	17.10.95	K.J.Abrams	Andrewsfield	14. 3.05T
			N73973				
G-OSIP	Robinson R22 Beta II	2916		9. 2.99	Heli Air Ltd	Tatenhill	3 .3.02T
G-OSIS	Pitts S-1S	PFA 09-12043		19. 9.94	C.Butler	Netherthorpe	
	(Built C Butler)				*(Current status unknown)*		
G-OSIT	Pitts S-1T	1023	N96JD	7.12.01	P.Shaw	Sturgate	9.12.04
	(Built Pitts Aerobatics)						
G-OSIX	Piper PA-32-260 Cherokee Six	32-499	G-AZMO	5. 8.86	J T Le Bon	Lee on Solent	29. 5.05T
			SE-EYN				
G-OSJN	Europa Aviation Europa XS	PFA 247-13687		3. 6.03	S.J.Nash	(Harlow)	
	(Built S J Nash - kit no.495)						
G-OSKP	Enstrom 480	5002	F-GSOT	6. 6.94	Churchill Stairlifts Ltd	Hawarden	9. 8.04T
			G-OSKP, N480EN				
G-OSKR	Best Off Skyranger 912	BMAA/HB/249		14. 1.03	Skyranger UK Ltd	Chiltern Park, Wallingford	15. 6.05P
	(Built Skyranger UK Ltd - kit no.SKR0102162)						
G-OSKY	Cessna 172M Skyhawk II	17267389	A6-KCB	27. 2.79	Skyhawk Leasing Ltd	Wellesbourne Mountford	8. 7.06T
			N73343				
G-OSLD	Europa Aviation Europa XS	PFA 247-13641		23. 8.00	Opus Software Ltd	Black Spring Farm, Castle Bytham	7. 3.05P
	(Built S C Percy) (Rotax 914-UL) *(Tri-gear u/c - kit no.485)*						
G-OSLO	Schweizer 269C	S.1360	N7507L	15. 3.89	AH Helicopter Services Ltd	Dunkeswell	16. 3.07T
G-OSMD	Bell 206B JetRanger II	2034	G-LTEK	12. 2.99	Stuart Aviation Ltd	White Waltham	6. 5.07T
			G-BMIB, ZS-HGH				
G-OSMS	Robinson R22 Beta	1528	G-BXYW	22. 2.99	Heliflight (UK) Ltd	Wolverhampton	13. 5.05T
			HA-MIU, N528SH				
G-OSND	Reims/Cessna FRA150M Aerobat	FRA1500272	G-BDOU	16.10.84	Wilkins and Wilkins (Special Auctions) Ltd		
					t/a Henlow Flying Club	RAF Henlow	2. 3.06T
G-OSNI	Piper PA-23-250 Aztec C	27-3852	G-AWER	2. 7.98	Marham Investments Ltd	Belfast Aldergrove	22. 5.04T
			N6556Y				
G-OSOE	Hawker Siddeley HS.748 Series 2A/275	1697	G-AYYG	17.11.97	Emerald Airways Ltd	Liverpool	10.11.05T
			ZK-MCF, C-GRCU, ZK-MCF, G-AYYG, ZK-MCF, G-AYYG, ZK-MCF, G-AYYG, G-11-9		*(Securicor Omega Express titles)*		
G-OSPD	Aerotechnik EV-97 teamEurostar UK	1708		3. 9.03	V.C.Garwood	Rochester	7. 9.05P
G-OSPG	British Aerospace BAe 125 Series 800B	258130	D-CPAS	1.12.03	Houston Jet Services Ltd	Oxford	15. 2.05T
			G-ETOM, G-BVFC, G-TPHK, G-FDSL, G-5-620				
G-OSPS	Piper PA-18 Super Cub 95	18-1555	OO-SPS	9. 7.92	J.P.Morrissey	Weston, Dublin	24. 7.06
	(L-18C-PI) *(Frame No.18-1527)*		G-AWRH, OO-HMI, ?ALAT 51-15555				
G-OSSA	Cessna TU206B Super Skywagon B	U2060824	4X-CHT	17.11.03	Skydive St. Andrews Ltd	Sorbie Farm, Kingsmuir	
			C-GDTO, N139LA, (N3824G)		*(On overhaul Perth 9.04)*		
G-OSSF	Gulfstream AA-5A Cheetah	AA5A-0863	G-MELD	1. 2.00	Hecray Co Ltd	Southend	6. 4.07T
			G-BHCB		t/a Direct Helicopters		
G-OSSI	Robinson R44 Raven II	10470		27. 8.04	Goss Air Ltd	(Bradford)	22. 9.07P
G-OSST	Colt 77A HAB	737		28.10.85	British Airways plc *"Concorde II"*	(West Drayton)	10.10.96A
G-OSTC	Gulfstream AA-5A Cheetah	AA5A-0848	N26967	22. 4.91	5th Generation Designs Ltd	White Waltham	10. 2.07T
G-OSTU	Gulfstream AA-5A Cheetah	AA5A-0807	G-BGCL	18. 4.95	Hecray Co Ltd	Southend	3. 7.03T
					t/a Direct Helicopters *(Noted 1.05)*		
G-OSTY	Cessna F150G	F150-0129	G-AVCU	21. 3.97	R.F.Newman	(Ingatestone)	15. 5.06T
	(Built Reims Aviation SA)						
G-OSUP	Lindstrand LBL 90A HAB	098		17. 3.94	T.J.Orchard	(Booker)	9. 3.05T
					tr British Airways Balloon Club (*British Airways Clubs titles)*		
G-OSUS	Mooney M.20K Model 231	25-0429	OY-SUS	7.11.94	J.B.and M.O.King	Goodwood	11. 2.07
			(N3597H)				
G-OSVY*	Sky 31-24 HAB				See SECTION 3 Part 1	(Aldershot)	
G-OSZB	Pitts S-2B	5200	G-OGEE	10. 2.04	P.M.Ambrose	Popham	28. 6.07T
	(Built Christen Industries Inc) (Lycoming AEIO-540)		OH-SKY				
G-OTAL	ARV Aviation ARV-1 Super 2	024	G-BNGZ	10. 9.87	N.R.Beale	Church Farm, Shotteswelll	26. 2.05P
	(Rotax 914-UL)						
G-OTAM	Cessna 172M Skyhawk II	17264098	N29060	13. 2.89	G.V.White	Norwich	6.12.04T
G-OTAN	Piper PA-18-135 Super Cub	18-3845	OO-TAN	28.10.96	S.D.Turner	Andrewsfield	12. 6.06
	(L-21B-PI) *(Frame No.18-3850)*		(OO-DPD), R.Neth AF R-155, 54-2445				
G-OTBA	Hawker Siddeley HS.748 Series 2A/242	1712	A3-MCA	14. 3.01	Emerald Airways Ltd	Liverpool	3. 5.07T
			ZK-MCA, G-11-7				
G-OTBY	Piper PA-32-300 Six	32-7940219	N2932G	14. 2.91	M.J.Willing	Jersey	5. 4.06
G-OTCH	CFM Streak Shadow	PFA 206-12401		28.10.93	H.E.Gotch	Redhill	3. 9.02P
	(Built H.E.Gotch - kitno K.207) (Rotax 582)						
G-OTCV	Best Off Skyranger 912S	BMAA/HB/436		13.12.04	T.C.Viner	Long Marston	
	(Built T.C.Viner)						
G-OTDA	Boeing 737-31S	29266	D-ADBV	5. 2.04	Globespan Airways Ltd	Glasgow	8. 3.07T
			N1786B		t/a Flyglobespan.com		
G-OTDI	Diamond DA.40D Star	D4.031		10. 9.03	Diamond Aircraft UK Ltd	Gamston	8.10.06T
G-OTEL	Thunder Ax8-90 HAB	1790		13. 6.90	D.N.Belton	(Chard)	31. 7.03A
G-OTFT	Piper PA-38-112 Tomahawk	38-78A0311	G-BNKW	14. 3.97	P.Tribble	Panshanger	11. 8.06T
			N9274T				
G-OTGA	Piper PA-28R-201 Cherokee Arrow III	28R-7837281	ZS-KFI	21. 2.01	J.G.Gleeson t/a The Ellie *(New owner 11.04)*	Manston	29. 3.04T
G-OTHE	Enstrom 280C-UK Shark	1226	G-OPJT	22. 9.87	National Technologies Ltd	(Oldham)	24. 11.05
			G-BKCO				
G-OTHL*	Robinson R22 Beta				See SECTION 3 Part 1	Hendon	
G-OTIB	Robin DR400/180R Remorqueur	1545	D-EGIA	26. 4.00	Norfolk Gliding Club Ltd	Tibenham	29. 5.06
G-OTIG	Gulfstream AA-5B Tiger	AA5B-0996	G-PENN	28. 7.00	D H Green	Elstree	30. 9.07E
			(I-TIGR), N3756L				

G-OTIM	Bensen B.8MV	PFA G/101-1084		5. 6.90	T.J.Deane	(Tilehurst, Reading)	
	(Built T.J.Deane)				*(Current status unknown)*		
G-OTJB	Robinson R44 Raven	0813		4. 8.00	T J Burke	(Montgomery Hill, Wirral)	16. 8.06T
G-OTJH	Pegasus Quantum 15-912	7791		20. 3.01	T.J.Hector	(Royston)	8. 4.06P
G-OTOE	Aeronca 7AC Champion	7AC-4621	G-BRWW	2. 4.90	J.M.Gale	Coombe Farm, Spreyton, Crediton	10. 5.95P
			N1070E, NC1070E		*(Damaged Coombe Farm 31.5.95)*		
G-OTOO	Stolp SA.300 Starduster Too	PFA 35-13352		26. 8.98	I.M.Castle	(Market Harborough)	
	(Built I M Castle)						
G-OTOY	Robinson R22 Beta	0888	G-BPEW	5. 9.97	Tickstop Ltd	Kimpton Park, Hitchin	19. 6.06T
G-OTRG	Cessna TR182 Turbo-Skylane RG II	R18200766	(N736SU)	14. 3.79	P.Mather	Cambridge	16.12.07E
G-OTRV	Van's RV-6	PFA 181-13302		27. 5.98	W.R.C.Williams-Wynne	Talybont, Gwynedd	18. 6.04P
	(Built W.R.C.Williams-Wynne) (Lycoming O-360)						
G-OTSP	Aérospatiale AS355F1 Ecureuil 2	5177	G-XPOL	31. 3.98	Anglia Aviation plc	Stapleford	20. 3.06T
			G-BPRF, N363E		*(Op Essex Police Air Support Unit)*		
G-OTTI	Cameron OTTI 34SS HAB	3490		23. 3.95	Ballonverbung Hamburg GmbH	(Kiel, Germany)	30. 4.05A
G-OTTO	Cameron Katalog 82SS HAB	2843		15. 6.92	Ballonverbung Hamburg GmbH	(Kiel, Germany)	6. 7.03A
	(New envelope 1999 - c/n 4382)				*"Otto Versand Katalog"*		
G-OTUG	Piper PA-18-150 Super Cub	18-5352	(G-BKNM)	17. 2.83	B.F.Walker	Nympsfield	14. 3.07T
	(Frame No.18-5424)		PH-MBA, ALAT 18-5352, N10F				
G-OTUI	SOCATA TB-20 Trinidad	1096	G-KKDL	7. 3.03	P.F.Rothwell	Luton	21.8.05T
			G-BSHU				
G-OTUN	Aerotechnik EV-97 Eurostar	PFA 315-13865		15. 5.02	E.O.Otun	White Waltham	7. 7.05P
	(Built E.O.Otun)						
G-OTUP	Lindstrand LBL 180A HAB	111		28. 3.94	Airborne Adventures Ltd	(Skipton)	1. 4.05T
G-OTWO	Rutan Defiant	114		24. 6.87	B.Wronski	Gloucestershire	14.10.03P
	(Built D G Foreman) (Lycoming O-320)				*(New owner 11.04)*		
G-OTYE	Aerotechnik EV-97 Eurostar	PFA 315-13858		15. 4.02	A B.Godber and J.Tye	(Ashbourne)	29 5.05P
	(Built A B.Godber and J.Tye)						
G-OTYP	Piper PA-28-180 Cherokee Challenger	28-7305166	F-BTYP	13. 1.04	I.R.Chaplin	Andrewsfield	6. 5.07T
			N11C		*(Noted 5.04)*		
G-OUCH	Cameron N-105 HAB	4830		3. 5.00	Flying Pictures Ltd *(Elastoplast titles)*	(Chilbolton)	26. 3.03A
G-OUHI	Europa Aviation Europa XS	PFA 247-13684		7. 6.01	Airplan Flight Equipment Ltd	Barton	
	(Built D R Philpott) (Tri-gear u/c - kit no.488)				*(New owner 8.04)*		
G-OUIK	Mainair Pegasus Quik	7983		22. 8.03	N.A Harwood	(Littlehampton)	25. 8.05P
G-OUMC	Lindstrand LBL 105A HAB	724		14. 9.00	A Holly t/a Executive Ballooning	(Berkeley)	18. 9.04T
					(Uphill Motor Company titles)		
G-OURB	British Aerospace BAe 125 Series 700B	257054	G-NCFR	28.11.02	Club 328 Ltd	Southampton	20. 5.05T
			G-BVJY, RA-02802, G-BVJY, C6-BET				
G-OURO	Europa Aviation Europa	PFA 247-12522		13.12.93	M.Crunden	Kemble	2. 4.04P
	(Buillt D Dufton - kit no.016) (NSI EA-81/100) *(Tri-gear u/c)*				*(Noted 9.04)*		
G-OURS	Sky 120-24 HAB	168		22.12.99	M P A Sevrin	(Albuquerque, New Mexico, US)	27. 8.04
	(Bear's Head shape)				*"Victor"*		
G-OUVI	Cameron O-105 HAB	1766		4. 5.89	P.Spellward	(Bristol)	31. 3.94A
					tr Bristol University Hot Air Ballooning Society *"Uvistat II"*		
G-OVAA	Colt Jumbo SS HAB	1426		11. 5.89	I.Chadwick	(Horsham)	24. 7.05A
	(Conventional HAB with nose/wings/tail of Virgin Boeing 747)				tr Balloon Preservation Flying Group		
G-OVAG	Tipsy Nipper T.66 Series 1	15	OO-VAG	7. 4.04	L.D.Johnston	(Bridge of Allan, Stirling)	20. 6.05P
	(Built Avions Fairey SA)		OO-LYS				
G-OVAL	Comco Ikarus C42 FB100	0407-6608		18. 8.04	J.I.Greenshields	Dunkeswell	24. 8.05P
G-OVAX	Colt AS-105GD Mk II HA Airship	1501		3. 7.89	Gefa-Flug GmbH	(Aachen, Germany)	23 1.04A
	(Built Cameron Balloons Ltd				*"Vax Airship"*		
G-OVBF	Cameron A-250 HAB	3494		1. 3.95	Airxcite Ltd	(Wembley)	27. 9.01T
					t/a Virgin Balloon Flights *"Virgin Oscar"*		
G-OVET	Cameron O-56 HAB	3939		25. 6.96	E.J.A Macholc	(Saltburn-by-the-Sea)	10. 7.05A
G-OVFM	Cessna 120	14720	N2119V	29. 4.88	A Sutherland and A P Bacon	Wiick	16. 2.05P
	(Continental O-200-A)		NC2119V				
G-OVFR	Reims/Cessna F172N Skyhawk II	F17201892		23. 5.79	Western Air (Thruxton) Ltd	Thruxton	13. 7.07T
G-OVIA	Lindstrand LBL 105A HAB	1002		9. 7.04	N.C.Lindsay	Pulborough	30. 6.05A
G-OVIC	Cameron A-250 HAB	4409	SE-ZKA	29.10.04	M.E.White	(Templeogue, Dublin)	26. 8.05E
G-OVID	Avid Flyer	NMFC.11760	N879UP	31. 5.91	G.G.Ansell	RAF Henlow	29. 4.05P
	(Built J Pelafigue) (Rotax 532)				*(New owner 5.04)*		
G-OVII	Van's RV-7	PFA 323-14100		30. 9.04	T.J.Richardson	Popham	
	(Built T.J.Richardson)						
G-OVIN	Rockwell Commander 112TC	13090	OY-DVN	19.11.04	G-OVIN Aviation Ltd	(Ilford)	AC
			D-EIXN, N4585W				
G-OVLA	Comco Ikarus C42 FB UK	PFA 322-14028		4. 2.03	B.C.and Patricia A Webb	Dunkeswell	5.10.05P
	(Built N Sams and B Bayes - kit no.0303-6550)				t/a Webb Plant Sales		
G-OVMC	Reims/Cessna F152 II	F15201667		29. 5.79	S.C.Moss	Gloucestershire	19. 8.07T
G-OVNE*	Cessna 401A	401A-0036	N401XX	11. 3.88	Not known	Norwich	8.10.92.
			(N171SF), N71SF, N6236Q		*(Cancelled 8.2.94 by CAA) (On fire dump 4.04)*		
G-OVNR	Robinson R22 Beta	1634		24.12.90	Helicopter Training and Hire Ltd	Newtownards	15. 9.07T
G-OWAC	Reims/Cessna F152 II	F15201678	G-BHEB	25. 2.80	Aviation South West Ltd	Exeter	10. 4.07T
			(OO-HNW)				
G-OWAK	Reims/Cessna F152 II	F15201677	G-BHEA	25. 2.80	A S.Bamrah t/a Falcon Flying Services	Rochester	23.11.04T
G-OWAL	Piper PA-34-220T Seneca III	3448030	D-GAPN	7. 7.98	R.G.and W.Allison	Gamston	9.11.07E
			N9163K				
G-OWAR	Piper PA-28-161 Warrior II	28-8616054	TF-OBO	18. 2.88	Bickertons Aerodromes Ltd	Denham	27. 3.06T
			N9521N		*(Op The Pilot Centre)*		
G-OWAX	Beech 200 Super King Air	BB-302	N86Y	4. 1.00	Context GB Ltd	Blackpool	2. 3.06T
			N300BW, N600CP				
G-OWAZ	Pitts S-1C	43JM	G-BRPI	22.11.94	P.E.S.Latham	Sleap	10. 3.05P
	(Built J Magueri) (Lycoming O-320)		N199M		*"Tiny Dancer"*		
G-OWCS	Cessna 182J Skylane	18257009	D-EFSA	25.11.02	P.Ragg	(Weerberg, Austria)	15. 1.06
			N2909F				

Reg	Type	C/n	Previous identity	Regd	Owner/Operator	Base	CofA
G-OWDB	Hawker Siddeley HS 125 Series 700B	257040	G-BYFO	18. 2.99	Bizair Ltd	(Jersey)	14. 9.05T
			HB-VMD, VP-BPE, VR-BPE, N47TJ, EC-ETI, EC-375, G-OWEB, HZ-RC1				
G-OWEL	Colt 105A HAB	1773		18. 5.90	S.R.Seager	(Aylesbury)	16. 3.98T
G-OWEN	K & S Jungster 1	PFA 44-10124		13.11.78	R.C.Owen	Danehill	
	(Built R.C.Owen) (Continental C90)				*(Current status unknown)*		
G-OWET	Thurston TSC-1A2 Teal	037	C-FNOR	28. 9.94	D.Nieman	Andrewsfield	18. 7.07
			(N1342W)				
G-OWFS	Cessna A152 Aerobat	A1520805	G-DESY	21. 5.02	MAMM Ltd	Blackpool	19. 4.06T
			G-BNJE, N7386L		*(Op Westair Flying Services)*		
G-OWGC	Slingsby T.61F Venture T.2	1875	XZ555	14. 8.91	Wolds Gliding Club Ltd	Pocklington	21. 1.07
G-OWLC	Piper PA-31 Turbo Navajo	31-679	G-AYFZ	13. 6.91	Channel Airways Ltd	Guernsey	10. 2.07T
			N6771L				
G-OWMC	Thruster T.600N-450	0122-T600N-076		5. 3.03	A R.Hughes t/a Wiltshire Microlight Centre	Yatesbury	19. 4.05P
G-OWND	Robinson R44 Astro	0644		26. 8.99	W.N.Dore	Wellesbourne Mountford	7. 9.05T
G-OWOW	Cessna 152 II	15283199	G-BMSZ	10. 5.95	A S.Bamrah	Biggin Hill	15.11.04T
			N47254		t/a Falcon Flying Services		
G-OWRT	Cessna 182G Skylane	18255077	G-ASUL	24. 8.00	Blackpool and Fylde Aero Club Ltd	Blackpool	17. 5.04
			N3677U				
G-OWWW	Europa Aviation Europa XS	PFA 247-12683		9. 6.94	R.F.W.Holder	High Cross, Ware	25. 5.05P
	(Built W R C Williams-Wynne and R.F.W.Holder - kit no.051) (Tri-gear u/c)				tr Whisky Group		
G-OWYE	Lindstrand LBL 240A HAB	645		27. 4.00	Wye Valley Aviation Ltd	(Ross-on-Wye)	3. 9.04T
G-OWYN	Aviamilano F.14 Nibbio	208	HB-EVZ	2. 2.87	R.Nash	Shoreham	11. 5.05P
			I-SERE				
G-OXBC	Cameron A-140 HAB	4981		2. 2.01	J.E.Rose *"Oxford Balloon Co"*	(Abingdon)	16. 2.05T
G-OXBY	Cameron N-90 HAB	1993	PH-DUM	9. 6.94	C.A Oxby *"The Zit"*	(Doncaster)	
G-OXKB	Cameron Jaguar XK8 Sports Car 110SS HAB	3941		9. 7.96	D.M.Moffat *"Jaguar XK8"*	(Bristol)	15. 5.05A
G-OXOM	Piper PA-28-161 Cadet	2841285	G-BRSG (2)	9.12.03	Plane Talking Ltd	Denham	4. 1.05T
			N92011		*(Cabair and Oil Elite 20W/50 inscriptions)*		
G-OXRG*	Colt Film Can SS HAB				See SECTION 3 Part 1	(Aldershot)	
G-OXTC	Piper PA-23-250 Aztec D	27-4344	G-AZOD	31. 5.89	A S.Bamrah	Biggin Hill	15. 6.98T
			N697RC, N6976Y		t/a Falcon Flying Services		
G-OXVI	Vickers Supermarine 361 Spitfire LF.XVIe	CBAF.IX.4262	7246M	22. 8.89	Silver Victory BVBA	Duxford	7. 7.05P
			TD248		*(As "TD248/D" in 41 Sqdn c/s)*		
G-OXXL	Cameron A-300 HAB	4944		25. 3.03	Exclusive Ballooning Ltd	(Berkeley)	1. 8.05T
G-OYAK	SPP Yakovlev Yak C-11	171205	EAF 705	25. 2.88	A H.Soper	North Weald	4. 8.05P
	(C/n quoted as 1701139 and/or 690120)		OK-KIH		*(As "27" in Soviet AF c/s)*		
G-OYES	Mainair Blade 912	1186-1198-7 & W989		12.11.98	J.Crowe	East Fortune	13. 4.05P
G-OYST	Agusta-Bell 206B JetRanger III	8440	G-JIMW	9.10.02	Oyster Leasing Ltd	Peldon, Mersea Island	29. 4.07T
			G-UNIK, G-TPPH, G-BCYP				
G-OYTE	Rans S-6ES Coyote II	PFA 204-14263		21. 7.04	I.M.Vass	(Hartlepool)	
	(Built I.M.Vass)						
G-OZAR	Enstrom 480	5007	G-BWFF	31. 7.95	Lancroft Air Ltd	(Shrewsbury)	1.11.04T
G-OZBB	Airbus A 320-212	389	C-GZUM	21. 3.94	Monarch Airlines Ltd	Luton	29. 4.07T
	G-OZBB, C-GZUM, G-OZBB, C-GXBB, G-OZBB, C-FTDW, G-OZBB, C-FTDW, G-OZBB, C-FTDW, G-OZBB, C-FTDW, G-OZBB, C-FTDW, G-OZBB, F-WWDI						
G-OZBE	Airbus A321-231	1707	D-AVZH	27. 3.02	Monarch Airlines Ltd	Luton	26. 3.05T
G-OZBF	Airbus A321-231	1763	D-AVZB	20. 6.02	Monarch Airlines Ltd	Luton	19. 6.05T
G-OZBG	Airbus A321-231	1941	D-AVXC	20. 3.03	Monarch Airlines Ltd	Luton	19. 3.06T
G-OZBH	Airbus A321-231	2105	D-AVXB	17. 3.04	Monarch Airlines Ltd	Luton	16. 3.07T
G-OZBI	Airbus A321-231	2234	D-AVZV	4. 6.04	Monarch Airlines Ltd	Luton	3. 6.07T
G-OZEE	Avid Speed Wing Mk.4	PFA 189-12308		18. 4.94	S.C.Goozee	Newton Peverill	25.10.05P
	(Built S.C.Goozee)						
G-OZEF	Europa Aviation Europa XS	PFA 247-14041		23.12.03	Z.M.Ahmad	(Southall)	
	(Built Z.M.Ahmad - kit no.538)						
G-OZOI	Cessna R182 Skylane RG II	R18201950	G-ROBK	31. 5.85	J.R. and F.L.Gibson Fleming	Ranston, Blandford Forum	28. 6.07T
					t/a Ranston Farms		
G-OZOO	Cessna 172N Skyhawk II	17267663	G-BWEI	17.11.99	Atlantic Air Bridge Ltd	Lydd	27. 8.07T
			N73767				
G-OZRH	British Aerospace BAe 146 Series 200	E2047	N188US	29. 1.96	Flightline Ltd	London City	1. 2.05T
			N364PS		*(Op Air France CityJet)*		
G-OZZI	Jabiru Aircraft Jabiru SK	PFA 274-13176		15. 8.97	A H.Godfrey	(Weston-super-Mare)	22. 6.05P
	(Built A H.Godfrey and E J Stradling)						
G-OZZY	Robinson R22 Beta II	2982	EI-RZZ	2. 9.03	G.T.Kozlowski	Redhill	2. 9.06T
			G-PWEL				

G-PAAA - G-PZZZ

Reg	Type	C/n	Previous identity	Regd	Owner/Operator	Base	CofA
G-PACE	Robin R1180T Aiglon	218		16.10.78	Millicron Instruments Ltd	(Denham)	7. 4.07
G-PACL	Robinson R22 Beta	1893	N2314S	17.12.91	R.Wharam	(Rotherham)	3.10.07E
G-PACT	Piper PA-28-181 Archer III	2843546	N5368F	25. 3.03	Burscombe Consulting Ltd	Headcorn	25. 3.06
	(Officially regd as ex N3568F)						
G-PADD	Gulfstream AA-5A Cheetah	AA5A-0780	G-ESTE	27.10.03	BPAD Ltd	Elstree	8.12.04T
			G-GHNC, N26877				
G-PADE	Just/Reality Escapade 912	BMAA/HB/369		2. 6.04	C.L.G.Innocent	(Worthing)	
	(Built C.L.G.Innocent - kit no.JAESC 0027)						
G-PADI	Cameron V-77 HAB	1809		18. 8.88	R.F.Penney *"Padiwac"*	(Watford)	17. 4.05A
G-PAGS	Aérospatiale SA.341G Gazelle 1	1155	G-OAFY	11. 3.96	P.A G.Seers	Willingale	25. 6.06T
			G-SFTH, G-BLAP, N62406				
			(Made forced landing in corn field at High Roding 14.8.04: damage to skids, fuselage and tail boom)				
G-PAIZ	Piper PA-12 Super Cruiser	12-2018	N3215M	11. 4.94	B.R.Pearson	Eaglescott	16. 8.07T
			NC3215M		*(Carries "NC3215M" on tail)*		
G-PALS	Enstrom 280C-UK-2 Shark	1191	N5688M	17. 7.80	J.A Sullivan	(Stockport)	17. 2.06
G-PAPS	Piper PA-32R-301T Turbo Saratoga SP	32R-8529005	F-GELX	8. 7.97	N.G.W.Cragg and C.S.Walker	Gamston	25. 9.06
			N4385D		t/a Nicol Aviation		

G-PARG	Pitts S-1S	19528-1	N18FW	30. 6.03	R.C.Pargeter	RAF Linton-on-Ouse	21.11.05P
	(Built F.G.Weaver)						
G-PARI	Cessna 172RG Cutlass II	172RG0010	N4685R	19.11.79	Applied Signs Ltd	Tatenhill	7. 4.05
G-PARR*	Colt Bottle 90SS HAB				See SECTION 3 Part 1	Newbury	
G-PART	Partenavia P68	62	F-GMPT	19.12.84	Cheshire Flying Services Ltd	Liverpool	21. 1.08E
	(Officially regd as "P68B")		G-PART, OY-CEY, D-GATE, PH-EEO, (N718R) t/a Ravenair				
G-PASB*	MBB Bö.105D				See SECTION 3 Part 1	Weston-super-Mare	
G-PASF	Aérospatiale AS355F1 Ecureuil 2	5033	G-SCHU	7. 3.91	M.F.Sheardown	Manston	30. 1.05T
			N915EG, N5777H				
G-PASG	MBB Bö.105DBS-4	S.819	G-MHSL	7.12.92	Police Aviation Services Ltd	Leeds-Bradford	13. 5.05T
			D-HFCC		*(Op Yorkshire Air Ambulance)*		
G-PASH	Aérospatiale AS355F1 Ecureuil 2	5040	F-GHLI	17. 5.96	Police Aviation Services Ltd	(Llangunnor, Carmarthen)	12. 4.07T
			LX-HUG, F-GHLI, N356E		*(Op for Dyfed-Powys Police)*		
G-PASV	Pilatus Britten-Norman BN-2B-21 Islander	2157	G-BKJH	26. 2.92	Police Aviation Services Ltd	Gloucestershire	18. 7.06T
			HC-BNR, G-BKJH				
G-PASX	MBB Bö.105DBS-4	S.814	D-HDZX	20.12.89	Police Aviation Services Ltd	Shoreham	13. 2.05T
G-PATF	Europa Aviation Europa	PFA 247-12757		5. 1.99	E P Farrell	(Beaconsfield)	
	(Built E P Farrell - kit no.107) (Monowheel u/c)				*(Current status unknown)*		
G-PATG	Cameron O-90 HAB	3856		13. 3.96	Bath University Students Union	(Claverton Down, Bath)	5. 7.04A
G-PATI	Reims/Cessna F172M Skyhawk II	F17201311	G-WACZ	20. 4.00	Nigel Kenny Aviation Ltd	Barton	8. 5.05T
			G-BCUK				
G-PATN	SOCATA TB-10 Tobago	307	G-LUAR	25. 3.97	N Robson	Bagby	21.10.07T
G-PATP	Lindstrand LBL 77A HAB	471		8. 7.97	P.Pruchnickyj	(Weston Turville, Buckingham)	31. 7.05
G-PATS	Europa Aviation Europa	PFA 247-12888		19. 7.95	D.J.G.Kesterton	(Milton Keynes)	
	(Built N Surman - kit no.216) (Monowheel u/c)				*(Current status unknown)*		
G-PATX	Lindstrand LBL 90A HAB	778		19. 6.01	P.A Bubb *"Purple Flame"*	(Guildford)	6. 9.05A
G-PATZ	Europa Aviation Europa	PFA 247-12625		2. 6.98	H.P.H.Griffin	Denham	1. 8.05T
	(Built H.P.H.Griffin - kit no.069) (Monowheel u/c)				tr G-PATZ Group		
G-PAVL	Robin R3000/120	170		22.11.96	S.Baker	White Waltham	2. 3.06T
G-PAWL	Piper PA-28-140 Cherokee	28-24456	G-AWEU	8. 9.82	A E.Davies	Barton	7. 9.07
			N11C		tr G-PAWL Group		
G-PAWN	Piper PA-25-260 Pawnee C	25-5207	G-BEHS	12. 3.01	A P.Meredith	Lasham	25. 6.93A
			OE-AFX, N8755L		*(New owner 3.01)*		
G-PAWS	Gulfstream AA-5A Cheetah	AA5A-0806	N2623Q	8. 2.82	Hecray Co Ltd t/a Direct Helicopters	Southend	11.10.07E
G-PAXX	Piper PA-20-135 Pacer	20-1107	N135XX	20. 5.83	D.W.Grace	(Horsham)	17. 9.05
			G-PAXX, (G-ARCE), F-BLLA, CN-TDJ, F-DADR				
G-PAYD	Robin DR400/180 Régent	847	D-EAYD	14. 1.03	A Head	Bicester	27. 3.06T
G-PAZY	Pazmany PL-4A	PFA 17-10378	G-BLAJ	20.11.89	C.R.Nash	(Fordingbridge)	3.10.95P
	(Built J D Le Pine) (Continental A65)				*(Current status unknown)*		
G-PBEE	Robinson R44 Clipper	0829		11. 9.00	P.Barnard	Guernsey	21. 9.06T
G-PBEK	Agusta A109A	7135	G-BXIV	20.12.04	P.Beck t/a Bek Helicopters	(Knutsford)	22.10.04T
			F-GERU, HB-XOK, D-HFZF				
G-PBEL	CFM Shadow Series DD	305-DD		27.10.98	S.J.Joseph	(Cheshunt)	3.10.05P
G-PBUS	Jabiru Aircraft Jabiru SK	PFA 274-13269		18. 8.98	G.R.Pybus	Morgansfield, Fishburn	7. 5.04P
	(Built G.R.Pybus)						
G-PBYA	Consolidated PBY-5A Catalina	CV-283	C-FNJF	19.11.04	Catalina Aircraft Ltd	Duxford	AC
	(Built Canadian Vickers Ltd)		CF-NJF, F-ZBBD, CF-NJF, F-ZBAY, CF-NJF, RCAF 11005				
G-PBYY	Enstrom 280FX Shark	2077	G-BXKV	15. 8.97	Hogan (Holdings) Ltd	Anglesey	9. 4.04T
			D-HHML				
G-PCAF	Pietenpol AirCamper	PFA 47-12433		1. 6.94	C.C. and F.M.Barley	(Farnborough)	
	(Built C.C.and F.M.Barley)				*(Under construction 2000)*		
G-PCAM	Fairey Britten-Norman BN-2A Mk.III-2 Trislander	1052	G-BEPH	26. 9.01	Aurigny Air Services Ltd	Guernsey	14. 6.04T
			S7-AAG, G-BEPH		*(ABN AMRO Bank titles)*		
G-PCAT	SOCATA TB-10 Tobago	60	G-BHER	17. 7.03	Air Touring Ltd	Biggin Hil	21. 4.07T
			4X-AKK, G-BHER				
G-PCCC	Alpi Pioneer 300	PFA 330-14220		31. 3.04	Pioneer Aviation UK Ltd	(Hardwick, Abergavenny)	24. 5.05P
	(Built F A Civaciuti - c/n 112)						
G-PCDP	Moravan Zlin Z-526F Trener	1163	SP-CDP	24.10.94	J.Mann	Brock Farm, Brentwood	23. 4.05
	(Walter M137A)				*"Ticker"*		
G-PDGE	Eurocopter EC120B Colibri	1211	F-WQPD	20. 7.01	Cadenza Helicopters Ltd	(London W1)	10. 9.07
G-PDGG	Aeromere F.8L Falco Series 3	208	OO-TOS	6. 1.98	P.D.G.Grist	Sibson	17. 5.07
			I-BLIZ				
G-PDGN	Aérospatiale SA365N Dauphin 2	6074	PH-SSU	5. 4.01	PLM Dollar Group Ltd	Inverness	19. 7.07
			5N-ATX, PH-SSU, (G-BLDR), G-TRAF, G-BLDR				
G-PDHJ	Cessna T182R Turbo Skylane II	T18268092	N6888H	3. 1.85	P.G.Vallance Ltd	Redhill	1. 8.07
G-PDOC	Piper PA-44-180 Seminole	44-7995090	G-PVAF	17.12.85	T.White	Newcastle	22.10.07E
			N2242A		t/a Medicare		
G-PDOG	Cessna O-1E Bird Dog	24550	F-GKGP	25. 9.98	J D Needham	Old Manor Farm, Anwick	7. 5.05
	(Regd as Cessna 305C)		ALAT		*(As "24550/GP" in South Vietnamese AF c/s)*		
G-PDSI	Cessna 172N Skyhawk II	17270420	N739BU	4. 1.88	A J.Clements and C.I.Bateman	Frensham, Wishanger	18. 3.07T
					tr DA Flying Group		
G-PDWI	Revolution Helicopters Mini-500	0248		14. 2.97	P.Waterhouse	(Stockport)	
G-PEAK	Agusta-Bell 206B JetRanger II	8242	G-BLJE	7. 3.94	Techanimation Ltd	(Stansted)	4. 6.06T
			SE-HBW				
G-PECK	Piper PA-32-300 Cherokee Six D	32-7140008	G-ETAV	22. 4.03	H.Peck	Gamston	26. 4.05T
			G-MCAR, G-LADA, G-AYWK, N8616N				
G-PEGA	Pegasus Quantum 15-912	7700		14. 8.00	B A Showell	Maypole Farm, Chislet	9. 8.05P
G-PEGG	Colt 90A HAB	1550		28. 6.89	Ballon Vole Association	(Fontaine les Dijon, France)	24. 1.04A
G-PEGI	Piper PA-34-200T Seneca II	34-7970339	N2907A	27.11.89	Tayflite Ltd	Perth	17. 8.07T
G-PEGY	Europa Aviation Europa	PFA 247-12713		16. 5.00	M.T.Dawson	Leeds-Bradford	14. 3.05P
	(Built M.T.Dawson - kit no.096) (Rotax 914-UL) *(Tri-gear u/c)*						
G-PEJM	Piper PA-28-181 Archer III	2843355	N41860	28. 6.00	D.A.Earle	(Tiverton)	9. 7.06E
G-PEKT	SOCATA TB-20 Trinidad	532	N24AS	28. 7.89	A J.Dales	Mount Airey Farm, South Cave	7. 3.05

Regn	Type	C/n	Previous identity	Date	Owner	Base	CofA
G-PEPA	Cessna 206H Stationair	20608181	G-MGMG	15. 6.04	R.D.Lygo	Farnborough	28.10.05T
			N5076D				
G-PEPL	MD Helicopters MD 600N	RN047	N3047L	5. 2.01	Blue Anchor Leisure Ltd	Gamston	26. 4.07T
G-PERC	Cameron N-90 HAB	10127		29. 8.01	P.A Foot and I.R.Warrington	(Stamford)	27. 3.05A
					(Stanton Marris titles)		
G-PERE	Robinson R22 Beta II	3382	N70881	24. 2.03	Central Helicopters Ltd	Tollerton	6. 3.06T
G-PERR*	Cameron Bottle 60SS HAB				See SECTION 3 Part 1	Newbury	
G-PERZ	Bell 206B-3 JetRanger III	4411	N6272T	7. 1.97	C.P.Lockyer	Coventry	31. 3.06T
G-PEST	Hawker Tempest II	12202	HA604	9.10.89	Tempest Two Ltd	Hemswell	AC
	(Built Bristol Aeroplane Co Ltd) (Regd with c/n "1181")		Indian AF, MW401		*(Rebuild nearing completion 7.04)*		
G-PETH	Piper PA-24-260 Comanche C	24-4979	N9469P	15.10.04	S.H.Petherbridge	(Worksop)	AC
G-PETR	Piper PA-28-140 Cherokee Cruiser	28-7425320	G-BCJL	23. 9.85	A A Gardner	(Port St.Mary, Isle of Man)	22. 1.06T
			N9591N				
G-PFAA	EAA Biplane Model P2	PFA 1338		19. 9.78	R J Marshall	Watchford Farm, Yarcombe	4. 8.05P
	(Built P E Barker c/n PEB/03) (Continental PC90)						
G-PFAF	Clutton FRED Series II	PFA 29-10310		30.10.78	M.S.Perkins	Stoke Golding	5. 8.04P
	(Built K Fern and M.S.Perkins)						
G-PFAG	Evans VP-1	PFA 7022		13.11.78	J.A Hatch	Netherthorpe	30. 6.89P
	(Built N S Giles-Townsend) (Volkswagen 1600)				*(Current status unknown)*		
G-PFAH	Evans VP-1	PFA 7004		23.11.78	J.A Scott	Chestnut Farm, Tipps End	26. 7.05P
	(Built J.A Scott) (Volkswagen 1834)						
G-PFAL	Clutton FRED Series II	PFA 29-10243		7.12.78	J.M.Robinson	Bann Foot, Lough Neagh	27. 7.88P
	(Built H Pugh) (Volkswagen 1600)				*(Noted 1.04)*		
G-PFAO	Evans VP-1	PFA 7008		12.12.78	P.W.Price	(Cheadle)	
	(Built P.W.Price)				*(Current status unknown)*		
G-PFAP	Phoenix Currie Wot	PFA 58-10315		12.12.78	J.H.Seed	Black Spring Farm, Castle Bytham	17.12.96P
	(Built P G Abbey) (Continental O-200-A) *(Built as an SE-5A rep)*				*(As "C1904/Z" in RFC c/s)*		
G-PFAR	Isaacs Fury II	PFA 11-10220		18.12.78	J.W.Hale and R.Cooper	Netherthorpe	9. 6.05P
	(Built C J Repik) (Continental O-200-A)				*(As "K2059" in 25 Sqdn RAF c/s)*		
G-PFAT	Monnett Sonerai II	PFA 15-10312		26.10.78	H.B.Carter	(St.Clement, Jersey)	24.10.92P
	(Built H.B.Carter) (Volkswagen 1834)				*(Stored Newcastle 5.93: current status unknown)*		
G-PFAW	Evans VP-1	PFA 62-10183		18.12.78	R.F.Shingler	Forest Farm, Welshpool	15.10.04P
	(Built R.F.Shingler) (Volkswagen 1834)						
G-PFAY	EAA Biplane	PFA 1525		18.12.78	A K.Lang and A L.Young	(Stoke-sub-Hamdon)	
	(Built A K Lang and A L Young) (Officially regd with c/n "1525")				*(Project abandoned 5.98)*		
G-PFCL	Cessna 172S Skyhawk <u>SP</u>	172S9330	N53287	19. 3.03	Prestwick Flight Centre Ltd	Prestwick	2. 4.06T
G-PFFN	Beech 200 Super King Air	BB-456	N456CD	7. 4.00	The Puffin Club Ltd	Leicester	17. 4.07T
			N861D, N124BB, C6-BFP, C6-CAA, N80NF, N80NE, N100FB				
G-PFML	Robinson R44 Astro	0082		9. 9.94	D.J.Parker	Hanover Farm, Addington, Buckingham	6.11.06T
					t/a Skyscraper Aviation		
G-PFSL	Reims/Cessna F152	F15201746	PH-TWF	30. 8.00	P.A Simon	Headcorn	25. 3.07T
			D-ENAX				
G-PGAC	Dyn'Aéro MCR-01 Club	PFA 301-13186		27. 1.99	D.T.S.Walsh and G.A Coatesworth	Cambridge	26. 7.05P
	(Built G.A Coatesworth - c/n 48) (Rotax 912-UL)						
G-PGFG	Tecnam P92-EM Echo	PFA 318-13772		30.10.01	P.G.Fitzgerald	Franklyn's Field, Chewton Mendip	30. 5.05P
	(Built P.G.Fitzgerald)				*"Charlies Angel"*		
G-PGHM	Air Création Buggy 582/Kiss 450	BMAA/HB/341		4. 2.04	P.G.H.Millbank	Sutton Meadows, Ely	21. 7.05P
	(Built P.G.H.Millbank - kit no FL024, Trike T03111)						
G-PGSA	Thruster T.600N	0080-T600N-046		11. 8.00	A J.A Hitchcock	Redlands	7. 1.05T
	(Rotax 582)						
G-PGSI	Robin R2160 Alpha Sport	309	F-GSAF	9. 3.00	M.A Spencer	North Weald	29. 6.06
G-PGUY	Sky 70-16 HAB	131	G-BXZJ	13.12.99	J L Guy t/a Black Sheep Balloons	(Skipton)	17. 6.04
G-PHAA	Reims/Cessna F150M	F15001159	G-BCPE	19. 6.97	P.H.Archard t/a PHA Aviation	Elstree	27. 5.07T
G-PHIL	Gyroflight Brookland Hornet	17		7. 7.78	A J.Philpotts	St.Merryn	11. 8.89P
	(Volkswagen 1600)				*(Stored 5.90: current status unknown)*		
G-PHLB	RAF 2000 GTX-SE	PFA G/13-1359		11. 3.04	P R Bell	Henstridge	23. 5.05P
G-PHSI	Colt 90A HAB	2181		12. 5.92	P.H.Strickland *(New owner 5.03)*	(Bedford)	21. 7.01A
G-PHTG	SOCATA TB-10 Tobago	1008		15.11.89	A J.Baggarley	Shoreham	14.10.05
G-PHXS	Europa Aviation Europa XS	PFA 247-13876		22. 7.02	P.Handford	(Wellingborough)	
	(Built P.Handford - kit no.523) (Tri-gear u/c)						
G-PHYL	Denney Kitfox Model 4	PFA 172A-12189		14. 9.98	J.Dunn	Siege Cross Farm, Thatcham	28. 6.05P
	(Built J Dunn)						
G-PHYS	Jabiru Aircraft Jabiru SP-470	PFA 274B-13926		19. 2.03	P C Knight	(Stafford)	26.11.05P
	(Built P C Knight)						
G-PIAF	Thunder Ax7-65 HAB	1885		19.11.90	L.Battersey *"No Regrets/La Vie en Rose"*	(Newbury)	24. 3.94A
G-PIDG	Robinson R44 Astro	0678		23.11.99	P.J.Rogers	(Lichfield)	11.12.05T
G-PIDS	Boeing 757-225	22195	N505EA	9. 1.95	MyTravel Airways Ltd	Manchester	23. 2.07T
G-PIEL	Menavia Piel CP.301A Emeraude	218	G-BARY	17.11.88	P.R.Thorne	Cublington	4.10.05P
			F-BIJR				
G-PIET	Pietenpol AirCamper	PFA 47-12267		1. 4.93	N.D.Marshall	RAF Halton	14. 5.05P
	(Built N.D.Marshall) (Continental C90)		*(Hit bump and bounced on take-off RAF Halton 31.8.04 and substantially damaged)*				
G-PIGG	Lindstrand Flying Pig SS HAB	473		18. 8.97	Iris Heidenreich	(Remscheid, Germany)	25. 3.05A
G-PIGS	SOCATA Rallye 150ST	2696	G-BDWB	13. 6.88	D.Hodgson tr Boonhill Flying Group	Wombleton	18. 8.06
G-PIGY	Short SC.7 Skyvan 3A-100	SH.1943	LX-JUL	21.12.95	Babcock Support Services Ltd	Oxford	27.11.06T
			5T-MAM, (G-14-111)		t/a Babcock Defence Services		
G-PIIX	Cessna P210N Pressurised Centurion II	P21000130	G-KATH	12. 6.95	J.R.Colthurst	(Winding Wood, Hungerford)	21. 3.05
			(N4898P)				
G-PIKK	Piper PA-28-140 Cherokee	28-22932	G-AVLA	19. 8.88	S J Woodfield	Coventry	10.11.07E
			N11C, (N9509W)		tr Coventry Aviators Flying Group		
G-PILE	RotorWay Executive 90	5143		27. 7.93	J.B.Russell	Magheramorne, County Antrim	5.11.98P
	(Built J B Russell) (RotorWay RI 162)						
G-PILL	Avid Flyer Mk.4	PFA 189-12333		12. 8.97	D.R.Meston	Old Sarum	16. 4.05P
	(Built D.R.Meston) (Rotax 912-UL)						
G-PINC	Cameron Z-90 HAB	10441		23. 9.03	M.Cowling	(Dubai, United Arab Emirates)	1. 9.04A

G-PING	Gulfstream AA-5A Cheetah	AA5A-0878	G-OCWC	6.12.95	Plane Talking Ltd	Cranfield	18. 6.06T
			G-WULL, N27153		*(Op London School of Flying)*		
G-PINT	Cameron Barrel 60 SS HAB	794		4. 1.82	D.K.Fish	(Bedford)	13. 2.98A
	(Wells Brewery Beer Barrel shape)				*"Charles Wells"*		
G-PINX	Lindstrand Pink Panther SS HAB	032		23. 4.93	Magical Adventures Ltd		
						(West Bloomfield, Mi.chigan, US)	30. 5.99A
G-PIPR	Piper PA-18 Super Cub 95	18-826	G-BCDC	11.10.96	D.S.Sweet	Dunkeswell	19. 9.07E
	(Frame No.18-832)		4X-ANQ, IDFAF/4X-ADE				
G-PIPS	Van's RV-4	PFA 181-11836		3. 8.90	C.J.Marsh	Newnham Fam, Whitwell	23. 5.05P
	(Built C J Marsh) (Lycoming O-320-D1A)						
G-PIPY	Cameron Scottish Piper 105SS HAB	3815		30. 1.96	Cameron Balloons Ltd	(Almondsbury)	9.12.04A
					(Op M.Moffat) "Pipy"		
G-PITS	Pitts S-2AE	PFA 09-11001		4. 7.85	P.F.van Lonkhuyzen and E.Goggins	Kilrush, County Kildare	5. 8.04P
	(Built B Bray)				tr The Eitlean Group		
					(Noted 7.04 damaged after landing mishap - crumpled starboard wings and tailplane)		
G-PITZ	Pitts S-2A	100ER	N183ER	2.10.87	J.A.Coutts	Nut Tree Farm, Redenhall	28. 9.05P
	(Built Razorback Air Services)						
G-PIXE	Colt 31A HAB	4883		11. 7.00	N.D.Eliot	(London SW19)	22. 5.05A
G-PIXI	Pegasus Quantum 15-912	7557		27. 8.99	G.R.Craig	Insch	19. 9.05P
G-PIXS	Cessna 336 Skymaster	336-0130	N86648	9. 9.88	Atlantic Bridge Aviation Ltd *(Stored 10.01)*	Lydd	29. 1.95T
G-PIXX	Robinson R44 Raven II	10263		16. 4.04	Flying TV Ltd	(London W4)	10. 5.07T
G-PIZZ	Lindstrand LBL 105A HAB	629		27. 7.99	HD Bargain SRL	(Firenze, Italy)	15. 8.05A
G-PJCC	Piper PA-28-161 Warrior II	2816043	OY-ODN	30. 3.04	PJC (Leasing) Ltd	Stapleford	20. 4.07T
			SE-IUI				
G-PJMT	Neico Lancair 320	PFA 191-12348		8. 5.98	V.Hatton and P.Gilroy *(New owners 1.05)*		
	(Built P J and M.T.Holland) (Lycoming IO-320-D) *(Tri-cycle u/c)*					(Churston Ferrers, Brixham /Lindridge, Teignmouth)	10. 6.04P
G-PJNZ	Commander Aircraft Commander 114B	14618	N6033Z	7. 9.04	P.D.Jackson	(Edenbridge)	12. 9.07
G-PJSY	Van's RV-6	PFA 181-13107		19. 7.04	P.J.York	Leicester	
	(Built P.J.York)				*(Noted 9.04)*		
G-PJTM	Reims/Cessna FR172K Hawk XPII	FR17200611	EI-CHJ	13.10.98	P.J.McNamara	Haverfordwest	23.11.07E
			G-BFIF		t/a Jane Air		
G-PKPK	Schweizer 269C	S.1454	EI-CAR	3. 8.93	C.H.Dobson	(Louth)	26. 9.05T
			N69A				
G-PLAC	Piper PA-31-350 Chieftain	31-8052038	G-OLDA	23.12.98	D.B.Harper	Biggin Hill	26. 2.05T
			G-BNDS, N131PP, N3550N				
G-PLAD	Kolb Twinstar Mk.3 Extra	"PFA 205-23833"		25. 1.05	P.J.Ladd	Wadsworth Manor Farm, Corsham	
	(Built P J Ladd - PFA c/n incorrect)						
G-PLAH	British Aerospace Jetstream Series 3102	640	G-LOVA	1.11.99	Jetstream Executive Travel Ltd	Coventry	26. 7.02T
			G-OAKA, G-BUFM, G-LAKH, G-BUFM, N410MX, G-31-640		*(New owner 6.04)*		
G-PLAJ	British Aerospace Jetstream Series 3102	738	N2274C	30. 3.00	Jetstream Executive Travel Ltd	Coventry	13.10.05E
			C-GJPH, N331QB, G-31-738		*(Platinum Air 2000 titles)*		
G-PLAN	Reims/Cessna F150L	F15001066	PH-SPR	11. 8.78	D.A Johnson tr G-PLAN Flying Group	Barton	23.12.05
G-PLAY	Robin R2112	170	F-ODIT	1. 8.79	D.R.Austin	High Cross, Ware	28. 8.07
G-PLAZ	Rockwell Commander 112A	345	G-RDCI	15. 4.04	Simat Marketing Ltd	(Llantwit Major)	16.11.06T
			G-BFWG, ZS-JRX, N1345J				
G-PLBI	Cessna 172S Skyhawk SP	172S8822	N35368	8. 5.01	Grandfort Properties Ltd	Booker	30. 6.07T
G-PLEE	Cessna 182Q Skylane II	18266570	N95538	4.12.87	Sunderland Parachute Centre Ltd	Shotton Colliery, Peterlee	1. 5.06
					t/a Peterlee Parachute Centre		
G-PLIV	Pazmany PL-4A	PFA 17-10155		19.12.78	B.P.North	RAF Halton	2. 7.05P
	(Built B.P.North) (Continental A65-8)						
G-PLMB	Aérospatiale AS350B Ecureuil	1207	G-BMMB	26. 3.86	PLM Dollar Group Ltd	Inverness	15. 2.07T
			C-GBEW, (N36033)				
G-PLMH	Eurocopter AS350B2 Ecureuil	2156	F-WQDJ	9. 1.95	PLM Dollar Group Ltd	Inverness	25. 2.07T
			G-PLMH, HB-XTE, F-WQPK, HB-XTE				
G-PLMI	Aérospatiale SA365C1 Dauphin 2	5001	F-GFYH	19. 6.95	PLM Dollar Group Ltd	Cumbernauld	8. 7.07T
			F-WZAE				
G-PLOD	Tecnam P92-EM Echo	PFA 318-14152		23. 6.04	G.M.Jupp and S.P.Pearson		
	(Built S.P.Pearson)					Lower Mountpleasant, Chatteris	18. 8.05P
G-PLOW	Hughes 269B	67-0317	G-AVUM	13. 9.83	C Walton Ltd	Sibbertoft	29.11.92
					t/a Sulby Aerial Surveys *(New owner 8.04)*		
G-PLPC	Schweizer 269C	S.1558	G-JMAT	14. 4.97	Power Lines, Piper and Cables Ltd	(Carluke)	19. 9.07
G-PLPM	Europa Aviation Europa XS	PFA 247-13287		17. 5.00	P.L.P.Mansfield	(Hartley Wintney)	
	(Built P.L.P.Mansfield - kit no.383) (Monowheel u/c)						
G-PLSA	Aero Designs Pulsar XP	PFA 202-12283	G-NEVS	20.12.04	C.A.Yardley	(Mansfield)	
	(Built N Warrener)						
G-PLUG*	Colt 105A HAB				See SECTION 3 Part 1	Newbury	
G-PLXI	British Aerospace ATP	2001	G-MATP	26. 8.94	BAE SYSTEMS (Operations) Ltd	Woodford	2.12.92P
	(Development a/c with PW 127D engines)		(G-OATP)		*(Stored 2.04)*		
G-PMAM	Cameron V-65 HAB	1155		29. 5.85	P.A Meecham *"Tempus Fugit"*	(Milton-under-Wychwood)	28. 9.03A
G-PMAX	Piper PA-31-350 Navajo Chieftain	31-7305006	G-GRAM	7. 7.99	Liberty Group Assets Ltd	Blackpool	17. 8.07T
			G-BRHF, N7679L				
G-PMNF	Vickers Supermarine 361 Spitfire HF.IX	CBAF.10372	SAAF??	29. 4.96	P.R.Monk	(Maidstone)	
			TA805		*(On rebuild 1995 as "TA805": current status unknown)*		
G-PNEU	Cameron Colt Bibendum 110SS HAB				See SECTION 3 Part 1	(Aldershot)	
G-PNIX	Reims/Cessna FRA150L Aerobat	FRA1500205	G-BBEO	2.11.04	D.C.Bonsall t/a Dukeries Aviation	Netherthorpe	21. 6.07E
G-POCO	Cessna 152	15283956	N6592B	8.10.04	K.M.Watts	(Ledbury)	
G-POGO	Flight Design CT2K	01-06-02-12		30. 7.01	P.A and M.W.Aston	Exeter	29. 7.05P
G-POLL	Best-Off Skyranger 912	BMAA/HB/290		26. 2.04	Thorne Engineering Ltd	(Bolton)	
	(Built D L Pollitt - kit no.SKRxxxx313)						
G-POLY	Cameron N-77 HAB	428		13. 7.78	D.M.Barnes, N.F.Biggs, J.L.Hinton, M.A C.Life and D.J.Thornley		
					tr The Empty Wallets Balloon Group *"Polywallets"*	(Bristol)	4. 8.00A
G-POND	Oldfield Baby Lakes	01	N87ED	2.10.90	U.Reichert	(Fehrbellin, Germany)	30. 6.05P
	(Built G E Davis) (Continental A80)						
G-PONY*	Colt 31A Air Chair HAB				See SECTION 3 Part 1	(Aldershot)	

Registration	Type	C/n	Previous identity	Date	Owner/operator	Base	CofA
G-POOH	Piper J-3C-65 Cub	6932	F-BEGY	17.10.79	P and H Robinson		
	(Frame No.7015)		NC38324			Upper Harford Farm, Bourton-on-the-Water	9. 8.07
G-POOL	ARV Aviation ARV-1 Super 2	025	G-BNHA	28. 8.87	P.A Dawson *(Fuselage stored 1.01)*	RAF Keevil	9. 9.90T
G-POOP	Dyn'Aéro MCR-01 Club	PFA 301-13190		5.11.97	K.and E.Nicholson,	Leicester	9.10.05P
	(Built P Bondar - c/n 81) (Rotax 912 ULS)				t/a Eurodata Computer Supplies		
G-POPA	Beech A36 Bonanza	E-2177	N7007F	20. 5.92	C.J.O'Sullivan	Southend	28. 1.05
			N7204R				
G-POPE	Eiri PIK 20E	20257		5. 3.80	P Rees *(New owner 9.04)*	(Wolverley, Kidderminster)	6. 6.04
G-POPI	SOCATA TB-10 Tobago	315	G-BKEN	20. 4.90	C.J. Earle	Seething	27. 3.07
			(G-BKEL)		tr G-POPI Flying Group		
G-POPP*	Colt 105A HAB				See SECTION 3 Part 1	(Aldershot)	
G-POPS	Piper PA-34-220T Seneca III	34-8133150	N8407H	11. 6.90	Alpine Ltd	Jersey	12. 5.05
G-POPW	Cessna 182S Skylane	18280204	N9451F	10. 7.98	D.L. Price	Little Staughton	18. 8.07
G-PORK	Grumman AA-5B Tiger	AA5B-0625	EI-BMT	28. 2.84	C.M.M.Grange and D.Thomas	Bournemouth	3. 2.05T
			G-BFHS				
G-PORT	Bell 206B-3 JetRanger III	2784	N37AH	23. 8.89	J.Poole	East Wellow, Romsey	12. 8.07T
			N39TV, N397TV, N2774R				
G-POSH	Colt 56A HAB	822	G-BMPT	10. 6.86	B.K.Rippon	(Didcot)	22. 4.05A
G-POTT	Robinson R44 Astro	0383		21.11.97	S.J.A Brown	(Le Touquet, France)	15. 1.07T
G-POWL	Cessna 182R Skylane II	18267813	N9070G	11.11.82	Hillhouse Estates Ltd	Scotland Farm, Hook	19. 5.07
			D-EOMF, N6265N				
G-POZA	Just/Reality Escapade Jabiru	BMAA/HB/347		5. 2.04	M R Jones	Wing Farm, Longbridge Deverill	
	(Built M R Jones - kit no.JAESC 0014)						
G-PPLL	Van's RV-7A	PFA 323-14240		28. 6.04	P.G.Leonard	(Luton)	
	(Built P.G.Leonard)						
G-PPPP	Denney Kitfox Model 3	PFA 172-11830		9. 1.91	R.Powers	Otherton, Cannock	10. 7.05P
	(Built P Eastwood - c/n 771)						
G-PPTS	Robinson R44 Clipper *(Float equipped)*	0664		14.10.99	Supablast Nationwide Ltd	Coleshill, Birmingham	14. 5.05T
G-PRAG	Brugger MB.2 Colibri	PFA 43-10362		29.11.78	D.Frankland	RAF Mona	26.2.04P
	(Built P Russell) (Volkswagen 1835)				tr Colibri Flying Group		
G-PRAH	Flight Design CT2K	01-06-01-12		31. 7.01	P.R.A Hammond	London Colney	1. 9.04P
G-PRET	Robinson R44 Astro	0381		8.10.97	J.A and C.M. Wilson	Folly Farm, Cop Hill, Slaithwaite	29.10.06T
G-PREY	Pereira Osprey 2	PFA 70-10193	G-BEPB	28. 9.99	D.W.Gibson	(Doseley)	8. 6.98P
	(Built J J and A J C Zwetsloot - c/n 88) (Lycoming IO-320)				*(Current status unknown)*		
G-PREZ	Robin DR500/200i Président	0038		26. 7.02	M.A Wilkinson	Spanhoe	22. 8.05P
	(Registered as DR400/500)						
G-PRII	Hawker Hunter PR.11	41H-670690	N723WT	14. 7.99	Stick and Rudder Aviation Ltd	Exeter	5. 9.03P
			A2616, WT723		*(As "WT723" in RN c/s) (Noted 11.03)*		
G-PRIM	Piper PA-38-112 Tomahawk	38-78A0669	N2398A	28. 1.87	Braddock Ltd *(Noted derelict 11.04)*	Chilbolton	25.12.01T
G-PRIT	Cameron N-90 HAB	1375	G-HTVI	6.11.86	R.D.Stagg	(Spixworth, Norwich)	15. 3.04A
			G-PRIT				
G-PRLY	Jabiru Aircraft Jabiru SK	PFA 274-13385	G-BYKY	11. 3.02	N.C.Cowell	Stoke Golding	8. 3.05P
	(Built N J Bond)						
G-PRNT	Cameron V-90 HAB	2819		23. 3.92	E.K.Gray	(Droitwich)	17. 6.03A
G-PROB	Eurocopter AS350B2 Ecureuil	2825	G-PROD	25. 6.01	Irvine Aviation Ltd	Denham	26. 4.07T
G-PROF	Lindstrand LBL 90A HAB	740		14. 2.01	S.J.Wardle	(Thrapston, Kettering)	29. 7.05A
					(Professional Financial Services titles)		
G-PROM	Aérospatiale AS350B Ecureuil	1486	G-MAGY	11.10.96	Peadar Hughes	Dungannon, County Tyrone	23.10.05T
			G-BIYC		t/a General Cabins and Engineering		
G-PROP	Gulfstream AA-5A Cheetah	AA5A-0845	G-BHKU	16. 2.84	Fortune Technology Ltd	Panshanger	28. 5.01T
			(OO-HTF)				
G-PROV	Hunting Percival P.84 Jet Provost T.52A (T.4)		Singapore AF 352	13.12.83	Hollytree Management Ltd tr Provost Group		
		PAC/W/23905	South Yemen AF 104, G-27-7, XS228			North Weald	24. 5.05P
G-PROW	Aerotechnik EV-97 Eurostar	PFA 315-13968		30.10.02	G.M.Prowling	Baxby Manor, Husthwaite	12. 1.05P
	(Built G.M.Prowling)						
G-PRSI	Pegasus Quantum 15-912	7492		17.12.98	J C Kitchen	Plaistowes Farm, St Albans	16.12.04P
G-PRTT	Cameron N-31 HAB	1374		6.11.86	J.M.Albury *"Baby Pritt"*	(Cirencester)	13.11.00A
G-PRXI	Vickers Supermarine 365 Spitfire PR.XI	6S/583003	PL983	6. 6.83	Propshop Ltd	Duxford	11. 6.01P
			G-15-109, N74138, PL983		*(As "PL983/JV-F" in 4 Sqdn, 2 TAF c/s) (New owner 2.03)*		
G-PSAX	Lindstrand LBL 77B HAB	960		8.10.03	P.A Sachs	(Pulborough)	13.10.04A
G-PSGC	Piper PA-25-260 Pawnee C	25-5324	G-BDDT	29. 4.04	Peterborough and Spalding Gliding Club Ltd		
			CS-AIX, N8820L			Crowland	
G-PSIC	North American P-51C-10 Mustang	103-26778	N51PR	16. 4.98	Patina Ltd *"Princess Elizabeth"*	Duxford	AC
	(Composite of P-51D IDF/AF 13 major components)		43-25147		*(Op The Fighter Collection) (Under rebuild 6.00)*		
G-PSKY	Best Off Skyranger 912S	BMAA/HB/430		3. 2.05	G.Mills	(Huddersfield)	
	(Built S Ivell)				tr Skyranger Flying Group G-PSKY		
G-PSNI	Eurocopter EC135T2	0337		26. 7.04	McAlpine Helicopters Ltd	Oxford	
G-PSON	Colt Cylinder One SS HAB	1780	PH-SON	14. 3.95	R.S.and A D.Kent	(Aldershot)	24. 4.05A
	(Panasonic Battery shape)				tr Balloon Preservation Group *(Panasonic Battery titles)*		
G-PSRT	Piper PA-28-151 Cherokee Warrior	28-7615225	G-BSGN	18. 3.99	P.A S.Dyke	Little Gransden	30. 9.05E
			N9657K				
G-PSST	Hawker Hunter F.58A	HABL-003115	Swiss AF J-4104	12. 2.97	Heritage Aviation Developments Ltd	Kemble	30.10.05P
			G-9-317, A2568, XF947		*"Miss Demeanour"*		
G-PSUE	CFM Shadow Series CD	K.139	G-MYAA	1. 4.99	D A Crosbie *(New owner 3.04)*	(Sudbury)	19. 5.03P
G-PSUK	Thruster T600N 450	0044-T600N-101		26. 5.04	A.J.Dunlop	Longacres Farm, Sandy	25. 5.05P
G-PTAG	Europa Aviation Europa	PFA 247-13121		14.12.98	R.C.Harrison	Wickenby	14. 4.05P
	(Built R.C.Harrison - kit no.337) (Jabiru 3300A) *(Tri-gear u/c)*						
G-PTRE	SOCATA TB-20 Trinidad	762	G-BNKU	14. 6.88	Trantshore Ltd	Rochester	27. 5.07
G-PTTS	Pitts S-2A	2179	N555JR	9. 5.03	D.C.Avery	Booker	5. 6.06T
	(Built Aerotek Inc)		N32TP, N31450				
G-PTWB	Cessna T303 Crusader	T30300306	G-BYNG	6.12.84	F.Kratky	Denham	16. 4.06
			G-PTWB, N6312V		t/a FK Global Aviation		
G-PTWO	Pilatus P.2-05	600-30	Swiss AF U-110	26. 2.81	Bulldog Aviation Ltd	Earls Colne	29.10.05P
			Swiss AF A-110		*(As "U-110" in Swiss AF c/s)*		

Reg	Type	C/n	Previous identity	Regd	Owner	Base	CofA
G-PTYE	Europa Aviation Europa	PFA 247-12496		22. 1.96	P.J.Carnes t/a Hitech International	Egginton	1. 9.03P
	(Built J Tye - kit no.001) (Monowheel u/c)	*(Failed to gain height on take-off Carltonmoor 16.3.03 and struck stone wall: substantially damaged)*					
G-PUBS*	Colt Beer Glass 56SS HAB				See SECTION 3 Part 1	Newbury	
G-PUDL	Piper PA-18-150 Super Cub	18-7292	SE-CSE	24. 2.98	R.A Roberts	Sparr Farm, Wisborough Green	16. 6.07
G-PUDS	Europa Aviation Europa	PFA 247-12999		9.10.97	I.Milner	Carlisle	6. 4.05P
	(Built I.Milner - kit no.253) (Rotax 914-UL) *(Tri-gear u/c)*						
G-PUFF	Thunder Ax7-77 Bolt HAB	165		17.11.78	C.A Gould	(Ipswich)	20. 8.99A
					tr Intervarsity Balloon Club *"Puffin II"*		
G-PUFN	Cessna 340A II	340A0114	N532KG	4.12.96	G.R.Case	Guernsey	19. 2.06T
			N532KC, N5477J				
G-PUGS	Cessna 182H Skylane	18256480	SE-ESM	15. 5.00	N.C.and M.F.Shaw	East Winch	2. 6.06T
			N8380S				
G-PUKA	Jabiru Aircraft Jabiru J400	PFA 325-14120		11. 9.03	D.P.Harris	(South Croydon)	2. 9.05P
	(Built D.P.Harris)						
G-PUMA	Aérospatiale AS332L Super Puma	2038	F-WMHB	31. 1.83	CHC Scotia Ltd	Aberdeen	12. 4.06T
G-PUMB	Aérospatiale AS332L Super Puma	2075		31. 1.83	CHC Scotia Ltd	Aberdeen	15. 5.06T
G-PUMD	Aérospatiale AS332L Super Puma	2077	F-WXFD	31. 1.83	CHC Scotia Ltd	Aberdeen	23. 8.05T
G-PUME	Aérospatiale AS332L Super Puma	2091		3. 8.83	CHC Scotia Ltd	Aberdeen	6. 9.06T
G-PUMH	Aérospatiale AS332L Super Puma	2101		3. 8.83	Bristow Helicopters Ltd	Aberdeen	22. 5.07
G-PUML	Aérospatiale AS332L Super Puma	2073	LN-ODA	20. 7.90	CHC Scotia Ltd	Aberdeen	4. 7.06T
			G-PUML, LN-OMG				
G-PUMN	Eurocopter AS332L2 Super Puma	2484	LN-OHF	16. 7.99	CHC Scotia Ltd	Aberdeen	26. 7.06T
G-PUMO	Eurocopter AS332L2 Super Puma	2467		30. 9.98	CHC Scotia Ltd	Aberdeen	25.10.05T
G-PUMS	Eurocopter AS332L2 Super Puma	2504		18. 8.00	CHC Scotia Ltd	Aberdeen	30. 1.05T
G-PUNK	Thunder AX8-105 HAB	1719		28. 3.90	S.C.Kinsey *(CofR restored 4.02)*	(Amersham)	15. 5.99T
G-PUPP	Beagle B.121 Pup Series 2	B121-174	G-BASD	23.11.93	M.D.O'Brien	(Horsham)	21. 5.05T
			(SE-FOG), G-BASD				
G-PUPY	Europa Aviation Europa XS	PFA 247-13694		10. 9.02	P.G.Johnson	(Leeds)	
	(Built P.G.Johnson - kit no.499) (Monowheel u/c)						
G-PURE*	Cameron Can 70SS HAB				See SECTION 3 Part 1	(Aldershot)	
G-PURR	Gulfstream AA-5A Cheetah	AA5A-0794	G-BJDN	22. 2.82	N.Bass	Elstree	2. 9.05T
			N26893		t/a Nabco Retail Display		
G-PURS	RotorWay Executive	3827		19. 1.90	J.E.Houseman	Clitheroe	5. 6.96P
	(Built J E Houseman) (RotorWay RW 152)						
G-PUSH	Rutan LongEz	PFA 74A-10740		11. 7.83	E.G.Peterson	(Nottingham)	
	(Built E.G.Peterson)				*(Current status unknown)*		
G-PUSI	Cessna T303 Crusader	T30300273	N3479V	26. 7.88	Crusader Aviation Ltd	Oxford	18. 6.06T
G-PUSK	Piper PA-32R-301 Satatoga IIHP	3246143	N237TB	24. 8.01	H.Nathanson	Elstree	21.10.07E
G-PUSS	Cameron N-77 HAB	1577		6.10.87	L.D.Thurgar *"Dick Whittington"*	(Bristol)	18. 6.01A
G-PUSY	Tiger Cub RL-5A LW Sherwood Ranger		G-MZNF	25. 6.99	S.C.Briggs	Charterhall	11. 3.05P
	(Built B.J.Chester-Master) (Rotax 582)	PFA 237-12964					
G-PUTT	Cameron Golfball 76SS HAB	2060	LX-KIK	8. 8.95	D.P.Hopkins	(Pidley, Huntingdon)	
					t/a Lakeside Lodge Golf Centre		
G-PVBF	Lindstrand LBL 260S HAB	504		7. 4.98	Virgin Balloon Flights Ltd	(London SE16)	26. 2.05T
G-PVET	de Havilland DHC-1 Chipmunk 22	C1/0017	WB565	23. 5.97	Connect Properties Ltd	Kemble	12.11.06T
					(As "WB565/X" in Army c/s)		
G-PVIP	Cessna 421C Golden Eagle II	421C0118	G-RLMC	30. 6.04	Passion 4 Health International Ltd	(Chertsey)	13.12.04*
		PH-SBI, D-IMAZ, I-CCNN, N3849C					
G-PVST	Thruster T.600N 450 Sprint	0122-T600N-074		29.10.02	P.V.Stevens	(Uxbridge)	12. 2.05T
G-PWBE	de Havilland DH.82A Tiger Moth	LES.1	VH-KRW	23. 7.99	P.W.Beales	White Waltham	11. 8.03P
	(Built Lawrence Engineering and Sales Pty Ltd, Camden, NSW, Australia ex-RAAF spares)						
G-PWIT	Bell 206L-1 LongRanger II	45193	D-HHSW	18. 5.00	Anne R.King	Gloucestershire	13. 7.06T
			G-DWMI, N18092				
G-PWUL	Van's RV-6	PFA 181-12773		3. 7.02	P.C.Woolley	(Woodhall Spa)	
	(Built P C Woolley)						
G-PYLN*	Cameron Pylon 80SS HAB				See SECTION 3 Part 1	(Aldershot)	
G-PYNE	Thruster T.600N 450	0072-T600N-067		27. 8.02	R.Dereham	(Bungay)	2. 9.05P
G-PYRO	Cameron N-65 HAB	567		8. 1.80	A C.Booth *"Pyromania"*	(Bristol)	30.12.04A
G-PZAZ	Piper PA-31-350 Navajo Chieftain	31-7405214	G-VTAX	18. 1.95	Air Medical Fleet Ltd	Oxford	22. 5.06T
			(G-UTAX), N54266				
G-PZIZ	Piper PA-31-350 Navajo Chieftain	31-7405429	G-CAFZ	30.10.98	Air Medical Fleet Ltd	Oxford	16. 5.05T
			G-BPPT, N54297, N9655N				

G-RAAA - G-RZZZ

Reg	Type	C/n	Previous identity	Regd	Owner	Base	CofA
G-RABA	Reims FR172H Rocket	FR17200292	D-ECSE	14.12.04	Air Ads Ltd	Blackpool	AC
G-RACA*	Percival P.57 Sea Prince T.1	P57/49	WM735	2. 9.80	Not known	Long Marston	4.11.80P
					(Cancelled 28.11.95 by CAA) (Derelict 6.03)		
G-RACI	Beech C90 King Air	LJ-819	G-SHAM	10. 4.03	King Air Ltd	(Douglas, Isle of Man)	3. 4.05T
			N2063A				
G-RACO	Piper PA-28R-200 Cherokee Arrow II	28R-7535300	N1498X	12. 9.91	Graco Group Ltd	Barton	23. 5.07
G-RACY	Cessna 182S Skylane	18280588	N7273Y	19.10.99	N.J.and P.D.Fuller	Cambridge	29.12.05
G-RADA	Soko P-2 Kraguj	024	Yugoslav AF 30140	25. 9.96	M G Roberts	Biggin Hill	5. 9.05P
					t/a Flight Consultancy Services		
G-RADI	Piper PA-28-181 Archer II	28-8690002	N2582X	6. 5.98	G.S. and D.V.Foster	Tatenhill	16. 6.07
			N9608N				
G-RADR	Douglas AD-4NA Skyraider	7722	G-RAID	30.10.03	T.J.Manna	North Weald	23. 3.05P
	(SFERMA c/n 42)		F-AZED, TR-K.., Fr AF 42, Bu.126922 *(As "126922/AK 402" of VA-176 Sqdn/USS Intrepid in USN c/s)*				
G-RAEM	Rutan LongEz	PFA 74A-10638		15. 3.82	G.F.H.Singleton	(Matlock)	18. 6.93P
	(Built G.F.H.Singleton - c/n 557) (Lycoming O-235)				tr Easy Group *(Current status unknown)*		
G-RAES	Boeing 777-236	27491	(G-ZZZN)	10. 6.97	British Airways plc	Heathrow	9. 6.06T
G-RAFA	Grob G115A	8081	D-EGVV	2. 3.89	RAF College Flying Club Ltd	RAF Cranwell	9. 6.07T
G-RAFB	Grob G115A	8079	D-EGVV	2. 3.89	RAF College Flying Club Ltd	RAF Cranwell	19. 4.07T

Reg	Type	C/n	Previous identity	Date	Owner/Operator	Base	CofA
G-RAFC	Robin R2112 Alpha	192		19. 5.80	J.E.Churchill tr RAF Charlie Group	Conington	12. 7.07
G-RAFE	Thunder Ax7-77 Bolt HAB	176		18.12.78	L.P.Hooper	(Bristol)	7. 9.02A
					tr Giraffe Balloon Syndicate *"Giraffe"*		
G-RAFG	Slingsby T.67C Firefly	2076		2.11.89	Arrow Flying Ltd	Popham	18. 4.05T
G-RAFH	Thruster T.600N 450	0032-T600N-063		10. 4.02	M.E.Howard	RAF Halton	9. 4.05P
					tr RAF Microlight Flying Association (FH)		
G-RAFI	Hunting Percival P.84 Jet Provost T.4	PAC/W/17641	8458M	18.12.92	R.J.Everett	North Weald	11. 3.00P
			XP672		*(As "XP672/03") (Noted 1.05)*		
G-RAFJ	Beech B200 Super King Air	BB-1829	N6129N	12.12.03	Serco Ltd	RAF Cranwell	11.12.07T
G-RAFK	Beech B200 Super King Air	BB-1830	N50130	12.12.03	Serco Ltd	RAF Cranwell	11.12.07T
G-RAFL	Beech B200 Super King Air	BB-1832	N5032K	19. 3.04	Serco Ltd	RAF Cranwell	23. 3.05T
G-RAFM	Beech B200 Super King Air	BB-1833	N51283	21. 1.04	Serco Ltd	RAF Cranwell	1.12.05T
G-RAFN	Beech B200 Super King Air	BB-1835	N60275	23. 1.04	Serco Ltd	RAF Cranwell	27. 1.05T
G-RAFO	Beech B200 Super King Air	BB-1836	N60476	11. 3.04	Serco Ltd	RAF Cranwell	18. 3.05T
G-RAFP	Beech B200 Super King Air	BB-1837	N61037	11. 3.04	Serco Ltd	RAF Cranwell	18. 3.05T
G-RAFR	Best Off Skyranger J2.2	BMAA/HB/410		8.10.04	M.E.Howard	RAF Halton	
	(Built P Waters - kit no.UK/487)				tr RAF Microlight Flying Association (FR)		
G-RAFS	Thruster T600N 450	0041-T600N-097		5. 4.04	M.E.Howard	RAF Halton	4. 4.05P
					tr RAF Microlight Flying Association (FS)		
G-RAFT	Rutan LongEz	PFA 74A-10734		9. 8.82	B.Wronski	Gloucestershire	12. 7.05P
	(Built D G Foreman) (Continental O-240-A)				*"A Craft of Graft"*		
G-RAFV	Avid Speed Wing	PFA 189-11738	G-MOTT	28. 7.04	A.F.Vizoso	RAF Halton	26. 2.99P
	(Built M D Ott)						
G-RAFW	Mooney M.20E Super 21	805	G-ATHW	14.11.84	Vinola (Knitwear) Manufacturing Co Ltd	Leicester	4.10.04
			N5881Q				
G-RAFZ	Rotary Air Force RAF 2000 GTX-SE	PFA G/13-1295		7. 5.02	John Pavitt (Engineers) Ltd	(Torrington)	
	(Built J W Pavitt)						
G-RAGS	Pietenpol AirCamper	PFA 47-11551		8. 6.94	R.F.Billington	(Kenilworth)	
	(Built R.F.Billington)				*(Current status unknown)*		
G-RAIG	Scottish Aviation Bulldog Series 100/101	BH100/146	SE-LLI	12. 9.03	C.S.Beevers	Kemble	5. 7 .07T
			Fv61037, G-AZMR				
G-RAIL	Colt 105A HAB	1434		31. 3.89	Ballooning World Ltd *"Railfreight"*	(London NW1)	4. 1.04A
G-RAIN	Maule M-5-235C Lunar Rocket	7262C	N5632J	26. 7.79	D.S.McKay and J.A Rayment	Gloucestershire	2.10.07E
G-RAIX	CCF Harvard 4 (T-6J-CCF Texan)	CCF4...	G-BIWX	16. 2.98	M.R.Paul	Lee-on-Solent	28. 6.05P
	(Possibly c/n CCF4-409 ex 51-17227)		MM53-846, RM-22/51-17		*(As "KF584/'RAI-X'")*		
G-RAJA	Raj Hamsa X'Air 582	BMAA/HB/118		13. 9.99	M.Quarterman	Lower Mountpleasant, Chaterris	12.10.05P
	(Built S R Roberts - kit no.456)						
G-RALD	Robinson R22HP	0218	G-CHIL	25. 1.96	Heli Air Ltd	Wellesbourne Mountford	1. 1.06T
			(G-BMXI), N9074K				
G-RAMI	Bell 206B-3 JetRanger III	2955	N1080N	18.10.90	M.D.Thorpe t/a Yorkshire Helicopters	Coney Park, Leeds	11. 9.06T
G-RAMP	Piper J-3C-65 Cub	6658	N35941	5. 7.90	R N Whittall	Brickhouse Farm, Frogland Cross	31. 8.01P
			NC35941		*(New owner 3.04)*		
G-RAMS	Piper PA-32R-301 Saratoga SP	32R-8013134	N8271Z	17.10.80	Air Tobago Ltd	Gamston	26. 6.05
G-RAMY	Bell 206B JetRanger II	1401	N59554	22. 9.95	Lincair Ltd	Humberside	21.11.07E
G-RANI	Eurocopter AS355N Ecureuil 2	5577	G-CCIN	30.10.03	Helicopter Training And Hire Ltd	Newtownards	23.10.06T
			N625LH, RP-C3688, F-OHNF				
G-RANS	Rans S-10 Sakota	PFA 194-11537		17. 8.89	J.D.Weller	(Sutton Coldfield)	23. 6.00P
	(Built J D Weller) (Rotax 532)				*(Current status unknown)*		
G-RANZ	Rans S-10 Sakota	PFA 194-11536		2.11.89	O M C Dismore	Popham	27. 6.03P
	(Built B A Philips) (Rotax 532)				*(Noted 10.04)*		
G-RAPH	Cameron O-77 HAB	1673		21. 3.88	M.E.Mason *"Walsal Litho"*	(Bristol)	28. 5.00T
G-RAPI	Lindstrand LBL 105A HAB	998		16. 7.04	M.White t/a Rapido Balloons	(Cirencester)	20. 7.05A
G-RAPP	Cameron H-34 HAB	2380		16. 8.90	Cameron Balloons Ltd	(St.Louis, US)	12. 5.04A
G-RARB	Cessna 172N Skyhawk II	17272334	G-BOII	4. 6.96	Cristal Air Ltd	Shoreham	16. 7.06T
			N4702D		*(Op Ace Aviation)*		
G-RARE*	Thunder Ax5-42 SS HAB				See SECTION 3 Part 1	(Aldershot)	
G-RASC*	Evans VP-2	PFA 63-10422		14.12.78	Not known	Breighton	
	(Built R A Codling - c/n V2-1178) (Continental C90-8)				*(Crashed on take-off from Bagby 9.7.95: cancelled 19.9.95 as TWFU) (On rebuild 1.05)*		
G-RASH	Grob G.109B	6217	OH-686	24. 6.04	C.Kaminski tr G-RASH Syndicate	Eaglescott	18. 8.07
G-RATE	Gulfstream AA-5A Cheetah	AA5A-0781	G-BIFF	11. 6.84	Plane Talking Ltd	(Elstree)	25.11.07E
			(G-BIBR), N26879				
G-RATH	Rotorway Executive 162F	6886		12.10.04	M.S.Cole	(Chelmsford)	AC
	(Built M.S.Cole)						
G-RATZ	Europa Aviation Europa	PFA 247-12582		16. 6.95	W.Goldsmith	Morgansfield, Fishburn	23. 8.05P
	(Built R Muller - kit no.037) (Monowheel u/c)						
G-RAVE	Mercury 582 Gemini/Southdown Raven X	538-0487 & SN2232/0219	G-MNZV	22.12.98	M.J.Robbins	Rochester	20. 3.04P
			(Sailwing is ex G-MNCV [SN2000/0219])				
G-RAVN	Robinson R44 Raven	1022		23. 3.01	Heli Air Ltd	Wellesbourne Mountford	1. 4.07T
G-RAWS	RotorWay Executive 162F	6492		14.11.00	Raw Sports Ltd	Street Farm, Takeley	16. 8.05P
	(Built B W Grindle)						
G-RAYA	Denney Kitfox Model 4	PFA 172A-12403		14.12.92	A K.Ray	(Stoke-on-Trent)	
	(Built A K Ray)				*(Amended CofR 1.05)*		
G-RAYE*	Piper PA-32-260 Cherokee Six	32-460	G-ATTY	30. 5.96	G.R.Silver	Willey Park Farm, Caterham	8. 8.03
			N11C		*(Fusleage noted 2.04)*		
					(Damaged landing Panshanger 1.3.03: cancelled 17.6.03 as wfu)		
G-RAYH	Zenair CH.701 UL	PFA 187-13583		7. 7.03	R.Horner	(Richmond, Yorkshire)	
	(Built R.Horner)						
G-RAYO	Lindstrand LBL 90A HAB	949		13.10.03	R.Owen	(Standish, Wigan)	21.10.04A
G-RAYS	Zenair CH.250 Zenith	PFA 24-10460		26.10.78	M.J.Malbon	Egginton	
	(Built R E Delves - c/n RED.001) (Lycoming O-235)				*(Noted 8.04)*		
G-RAZY	Piper PA-28-181 Archer II	28-8090102	G-REXS	11. 2.04	R.W.Cooper	(Rugeley)	16. 6.07T
			N8093Y				
G-RAZZ	Maule MX-7-180 Super Rocket	11050C	N266MM	10.11.04	N.M.Robinson	Compton Abbas	24.11.07E
			D-EOLW, N6118L		tr Compton Maule Syndicate		

Registration	Type	c/n	Previous identity	Date	Owner	Base	CofA
G-RBBB	Europa Aviation Europa *(Built W M Goodman and I H McCleod - kit no.073) (Monowheel u/c)*	PFA 247-12664		6. 5.94	T.J.Hartwell	Sackville Lodge Farm, Riseley	12. 8.05P
G-RBCI	Fairey Britten-Norman BN-2A Mk.III-2 Trislander	1035	G-BDWV 8P-ASF, G-BDWV	16. 3.01	Aurigny Air Services Ltd *(Royal Bank of Canada titles)*	Guernsey	15. 4.05T
G-RBJW	Europa Aviation Europa XS *(Built J.Worthington and R.J.Bull)*	PFA 247-13600		15. 7.04	J.Worthington and R.J.Bull	(Nairn)	
G-RBMV	Cameron O-31 HAB	4658		27. 7.99	P.D.Griffiths	(Romsey)	3. 7.05A
G-RBOS*	Colt AS-105 HA Airship				See SECTION 3 Part 1	Wroughton	
G-RBOW	Thunder Ax7-65 HAB	1439		24. 4.89	R.S.Mcdonald *"Rain-Beau-Lune"*	(Chesham)	14. 6.054A
G-RBSG	Dassault Falcon 900EX	113	F-WWFL	21.10.02	The Royal Bank of Scotland plc	Le Bourget, Paris	22.10.06T
G-RBSN	Comco Ikarus C42 FB80	0407-6610		23. 8.04	P.B.and Mary Robinson	Sutton Meadows, Ely	22. 8.05P
G-RCAF*	North American AT-6C Harvard IIA				See SECTION 3 Part 1	Breckenridge, Texas, US	
G-RCED	Rockwell Commander 114	14241	VR-CED N4917W	19. 6.92	D.G.Welch	Tollerton	16. 5.07
G-RCEJ	British Aerospace BAe 125 Series 800B	258021	VR-CEJ G-GEIL, G-5-15	15. 6.95	Albion Aviation Management Ltd	Biggin Hill	14. 6.05T
G-RCHY	Aerotechnik EV-97 Eurostar *(Built N McKenzie)*	PFA 315-14187		30. 3.04	N McKenzie	Kirkbride	6. 7.05P
G-RCKT	Harmon Rocket II *(Built K.E.Armstrong)*	PFA 314-13536		10.10.03	K.E.Armstrong	Armshold Farm, Kington	
G-RCMC	Meridian Renegade 912 *(Built B D Godden - c/n 485)*	PFA 188-12483		1. 2.93	R.C.M.Collisson	Bicester	30. 6.05P
G-RCMF	Cameron V-77 HAB	1618		23.11.87	J.M.Percival *(New owner 2.02)*	(Burton-on-the-Wolds)	19. 8.97A
G-RCML	Sky 77-24 HAB	148		9. 3.99	RCM SARL	(Stuppicht, Luxembourg)	6.12.04
G-RCMS	Agusta A109E Power	11056	G-BZEI	3. 4.03	Stolkin Helicopters Ltd	Denham	11. 6.06T
G-RCNB	Eurocopter EC120B Colibri	1333	F-WQPX	20. 3.03	JR Clark Ltd	Culverthorpe, Grantham	28. 4.06T
G-RCOM	Bell 206L-3 Long Ranger III	51599	TC-HZT	24.10.02	3GRComm Ltd	(Hereford)	4. 2.06T
G-RDBS	Cessna 550 Citation II	550-0094	G-JETA (N26630)	7. 5.99	Albion Aviation Management Ltd	Biggin Hill	21. 6.05T
G-RDCO	Jabiru Aircraft Jabiru J400 *(Built J M Record)*	PFA 325-14052		15. 4.03	RDCO (International) LLP	(Knutsford)	AC
G-RDEL	Robinson R44 Raven	1071		5. 6.01	J A R Allwright t/a Jara Aviation	Jaggards House, Weeley Heath	27. 6.07T
G-RDHS	Europa Aviation Europa XS *(Built R.D.H.Spencer - kit no.549) (Tri-gear u/c)*	PFA 247-13887		31. 5.02	R.D.H.Spencer	(Braintree)	
G-RDNS	Rans S-6-S Super Coyote *(Built G.J.McDill)*	PFA 204-14307		2.11.04	G.J.McDill	Bagby	
G-READ	Colt 77A HAB	1158	EI-BYI G-READ	16.11.87	J.Keena	(Athlone, County Westmeath)	23. 9.04A
G-REAL	Eurocopter AS350B2 Ecureuil	3032	G-DRHL	26. 8.04	Imagine Leisure Ltd	Dunsfold	29. 4.07T
G-REAP	Pitts S-1S *(Built S D Howes)*	PFA 09-11557		7. 2.90	R.Dixon *"The Grim Reaper"*	Netherthorpe	23.10.04P
G-REAR	Lindstrand LBL-69X HAB	977		12. 2.04	Exclusive Ballooning Ltd *(Sloggi titles)*	(Berkeley)	26. 4.05A
G-REAS	Van's RV-6A *(Built D W Reast)* (Lycoming O-320)	PFA 181-12188		16. 8.94	T.J.Smith	Sleap	22. 8.05P
G-REAT	Grumman GA-7 Cougar	GA7-0033	N29699	6.10.78	Goodtechnique Ltd	Leeds-Bradford	9.10.06T
G-REBA	Rotary Air Force RAF 2000 GTX-SE *(Built D J Pearce)*	PFA G/13-1334		5.10.01	D.J.Pearce	Watchford Farm, Yarcombe	22. 3.05P
G-REBL	Hughes 269B	67-0318	N9493F	25. 7.89	Farmax Ltd *(Stored 11.03)*	Shoreham	9.10.95
G-RECE	Cameron C-80 HAB	10435		16. 6.03	M.Kotsageridis	(Thessaloniki, Greece)	16. 6.04A
G-RECK	Piper PA-28-140 Cherokee B	28-25656	G-AXJW N11C	17. 3.88	R.J.Grantham and D.Boatswain	Clutton Hill Farm, Clutton	26. 8.07
G-RECO	Jurca MJ.5-L2 Sirocco	96	F-PYYD F-WYYD	30. 9.91	J D Tseliki *(Stored 6.97: current status unknown)*	Kittyhawk Farm, Deanland	
G-RECS	Piper PA-38-112 Tomahawk	38-81A0118	N5824H D-EFFX, N23138	23. 4.02	S.H.and C.L.Maynard	(Middlesbrough)	2. 5.05T
G-REDB	Cessna 310Q	310Q0811	G-BBIC N69600	17. 6.93	Red Baron Haulage Ltd	Full Sutton	16. 8.07T
G-REDC	Pegasus Quantum 15-912	7572		30. 9.99	R.F.Richardson	(Aldborough, Norwich)	28.10.05P
G-REDD	Cessna 310R II	310R1833	G-BMGT ZS-KSY, (N2738X)	2.10.96	G.Wightman	Blackpool	25. 3.05
G-REDI	Robinson R44 Clipper	0817		2. 8.00	Redeye.com Ltd	Sheffield City	27. 8.06T
G-REDJ	Eurocopter AS332L2 Super Puma	2608	F-WWOJ	19. 5.04	International Aviation Leasing Ltd	Aberdeen	20. 5.07T
G-REDK	Eurocopter AS332L2 Super Puma	2610	F-WWOM	2. 6.04	International Aviation Leasing Ltd	Aberdeen	6. 6.07T
G-REDL	Eurocopter AS332L2 Super Puma	2612	F-WWOD	30. 6.04	International Aviation Leasing Ltd	Aberdeen	5. 7.07T
G-REDM	Eurocopter AS332L2 Super Puma	2614	F-WWOF	26. 7.04	International Aviation Leasing Ltd	Aberdeen	29.7.07T
G-REDN	Eurocopter AS332L2 Super Puma	2616	F-WQDH	20. 8.04	International Aviation Leasing Ltd	Aberdeen	22. 8.07T
G-REDS	Cessna 560XL Citation Excel	560-5167	N250SM N5188N	10.10.02	Ferron Trading Ltd	Jersey	9.10.05T
G-REDX	Experimental Aviation Berkut *(Built G V Waters - c/n 002)* (Lycoming O-360-A1A)	PFA 252-12481		27. 1.95	G.V.Waters	RAF Coltishall	23. 6.05P
G-REDY	Robinson R22 Beta II	3402	G-CBXO N71909	28. 7.03	Plane Talking Ltd	Elstree	1. 4.06T
G-REDZ	Thruster T.600N 450 Sprint	0037-T600N-091		5. 8.03	S.L.and W.J.Smith t/a Redlands Airfield	Redlands	4. 8.05P
G-REEC	Sequoia F.8L Falco (Lycoming IO-320)	654	LN-LCA	2. 7.96	J.D.Tseliki	Kittyhawk Farm, Deanland	22. 7.05P
G-REED	Mainair Blade 912S	1282-0501-7 & W1077		11. 6.01	D.Jessop *(New owner 8.04)*	Abbey Warren Farm, Bucknall, Lincoln	26. 4.04P
G-REEF	Mainair Blade 912S	1285-0501-7 & W1080		15. 6.01	G.Mowll	(Caernarfon)	9. 7.05P
G-REEK	Grumman AA-5A Cheetah	AA5A-0429		12. 9.77	J.and A Pearson *(Stored at docks 11.02)*	Dundee	10.12.01
G-REEM	Aérospatiale AS355F1 Ecureuil 2	5175	G-EMAN G-WEKR, G-CHLA, N818RL, C-FLXH, N818RL, N818R, N5798U	9. 3.98	Heliking Ltd	Denham	1. 4.07T

Reg	Type	C/n	Previous identity	Regd	Owner/operator	Base	CofA
G-REEN	Cessna 340	340-0063	G-AZYR N5893M	2. 2.84	Beaumont Cornish Securities Ltd	North Weald	6.11.05
G-REES	SAN Jodel D.140C Mousquetaire III	156	F-BMFR	23. 4.80	C.C.Rea tr G-REES Flying Group	Sheepcote Farm, Stourbridge	20. 6.07
G-REGI*	Cyclone Chaser S 508	7165	G-MYZW	16. 5.03	G.S.Stokes *(Cancelled 22.9.03 by CAA)*	Pound Green, Buttonoak, Bewdley	18. 6.04P
G-REKO	Solar Wings Pegasus Quasar IITC	SW-TQC-0073 & SW-WQT-0467	G-MWWA	14.11.01	M.Sims	Haverfordwest	22. 4.05P
G-RENE	Meridian Renegade 912 *(Built D Evans)*	PFA 188-12030		6.11.91	P.M.Whitaker	(Ilkley)	30. 6.04P
G-RENO	SOCATA TB-10 Tobago	249		10.12.81	Lamond Ltd	Birmingham	21. 5.07T
G-RENT*	Robinson R22 Beta				See SECTION 3 Part 1	Langford Lodge, Belfast	
G-REPH	Pegasus Quantum 15-912	7785		6. 2.01	R.S.Partridge-Hicks	(Bury St.Edmunds)	3. 5.03P
G-RESG	Dyn'Aéro MCR-01 Club *(Built R.E.S.Greenwood)*	PFA 301A-13994		10. 4.03	R.E.S.Greenwood	(Newmarket)	30. 9.05P
G-REST	Beech P35 Bonanza	D-7171	G-ASFJ	14.12.82	C.R.E.S.Taylor	North Weald	8.10.05
G-RETA	CASA 1-131E Jungmann Series 2000	2197	Spanish AF E3B-305	24. 3.80	Richard Shuttleworth Trustees *(As "4477/GD+EG" in Luftwaffe WW2 c/s)*	Old Warden	6. 9.05P
G-REVO	Best-Off Skyranger 912 *(Built R.T.Henry - kit no.SKRxxxx408)*	BMAA/HB/346		2. 2.04	R.T.Henry	Newtownards	13. 7.05P
G-REYS	Canadair CL604 Challenger *(Built Bombardier Inc)*	5467	N467RD C-GLWX	17. 9.01	Greyscape Ltd	Farnborough	16. 9.07T
G-RFDS	Agusta A109A II	7411	N1YU VP-CLA, VR-CLA, G-BOLA, VR-CMP, G-BOLA	24. 5.99	Clifford Kent Ltd	Liskeard	22. 7.05T
G-RFIO	Aeromot AMT-200 Super Ximango	200048		6. 3.95	M.D.Evens	Enstone	30. 6.07
G-RFSB	Sportavia-Pützer RF5B Sperber	51045	N55HC D-KEAO	2.12.88	J F Mcaulay and A R Jury	(Lincoln/Oakham)	29. 4.07
G-RFUN	Robinson R44 Raven	1239		17. 7.02	D.Woodgates t/a Bdw Fuels	(Woodlands St.Mary, Hungerford)	29. 7.05T
G-RGEN	Cessna T337D Turbo Super Skymaster	3371062	G-EDOT G-BJIY, 9Q-CPF, PH-JWL, N86056	24. 5.96	Legoprint SpA	(Lavis, Italy)	11. 6.06
G-RGNT	Robinson R44 Raven II	10514	G-DMCG	26. 1.05	P.R.Nott t/a Regent Aviation	Rochester	
G-RGUS	Fairchild 24R-46A Argus III (UC-61K-FA)	1145	(PH-) G-RGUS, ZS-UJZ, ZS-BAY, KK527, 44-83184	16. 9.86	Fenlands Ltd *(As "44-83184/7" in USAAC c/s)*	Sturgate	19. 2.05
G-RHCB	Schweizer 269C-1	0036	N201WL	20. 3.98	Helicopter One Ltd *(Op Bournemouth Helicopters)*	Bournemouth	9. 4.07T
G-RHHT	Piper PA-32RT-300 Lance II	32R-7885190	N36476	3. 7.78	G.R.Bright	Little Gransden	30. 3.06
G-RHOP	Fairey Britten-Norman BN-2A Mk.III-2 Trislander	1042	G-WEAC 5H-AZD, G-BEFP, (4X-CCL), G-BEFP, N30WA, JA6401, G-BEFP	11. 3.04	Airx Ltd	Bournemouth	18. 3.05T
G-RHYM	Piper PA-31 Turbo Navajo B	31-815	G-BJLO F-BTQG, (F-BTDV), N7428L	24. 4.02	ATC Trading Ltd	Lasham	30. 1.05T
G-RHYS	RotorWay Exec 90 *(Built B Williams)* (RotorWay RI 162)	5140		8.11.93	A K.Voase and K.Matthews	(Hornsea)	21. 7.04P
G-RIAN	Agusta-Bell 206A JetRanger	8056	G-SOOR G-FMAL, G-RIAN, G-BHSG, PH-FSW	16. 9.87	B.J.Green	Draycott	17. 2.06
G-RIAT	Robinson R22 Beta II	2684		27. 5.97	R.Cove and J.P.Gordon t/a RMJ Helicopters	Cranfield	29. 6.06T
G-RIBS	Diamond DA.20-A1 Katana	10143	G-BWWM	7. 7.97	Diana Margaret Green	Wolverhampton	28.11.05T
G-RIBZ	Enstrom 480B	5055		14. 8.03	Premiair Aviation Group Ltd	Blackbushe	28. 8.06T
G-RICC	Aérospatiale AS350B2 Ecureuil	2559	G-BTXA	30.10.91	Specialist Helicopters Ltd	Nairn	23. 3.07T
G-RICE	Robinson R22 Beta	2509	N93MK	14. 3.97	P.Newton t/a Heli-Air Wales	Swansea	3. 4.06T
G-RICK	Beech 95-B55 Baron	TC-1472	G-BAAG	23. 5.84	James Jack Lifting Services Ltd	Wick	8.10.06T
G-RICO	American-General AG-5B Tiger	10162	N130U	14. 5.99	I.J.Ward	(Ripley, Woking)	30. 5.05T
G-RICS	Europa Aviation Europa *(Built R.G.Allen - kit no.125)* (NSI EA81) *(Conventional u/c)*	PFA 247-12747		19. 3.96	R.G.Allen t/a The Flying Property Doctor	Kemble	29. 1.05P
G-RIDD	Robinson R22 Beta II	3685		14.10.04	Essex Match Co Ltd	(Woodford Green)	28.10.07E
G-RIDE	Stephens Akro *(Built N Mardis)* (Lycoming AIO-360)	111	N81AC N55NM	10. 8.78	R.Mitchell t/a Mitchell Aviation *(PSA c/s) (Stored 4.02)*	RAF Cosford	13. 8.92P
G-RIDL	Robinson R22 Beta II	3194		30. 3.01	Peterborough Helicopter Hire Ltd	Conington	23. 4.07T
G-RIET	Hoffmann H 36 Dimona	36224	I-RIET	6. 8.02	L.J.McKelvie tr Dimona Gliding Group	Bellarena	16.12.05
G-RIFB	Hughes 269C	116-0562	N7428F	17. 5.90	J.McHugh and Son (Civil Engineering) *(Current status unknown)*	(Romford)	6. 3.03
G-RIFN	Avions Mudry CAP.10B	276		6. 6.96	R.A J.Spurrell	White Waltham	17. 2.06T
G-RIGB	Thunder Ax7-77 HAB	1201		16. 3.88	N.J.Bettin	(Farnham)	15. 9.05A
G-RIGH	Piper PA-32R-301 Saratoga	3246123	N41272 G-RIGH, N41272	23.12.98	Right Aviation Ltd	Fowlmere	22.12.04T
G-RIGS	Piper Aerostar 601P	61P-0621-7963281	N8220J	18. 5.79	G G Caravatti and P G Penati *"Marilyn"*	Milan-Bresso, Italy	27.10.06
G-RIHN	Dan Rihn DR.107 One Design *(Built J.P.Brown)*	PFA 264-14201		18. 5.04	J.P.Brown	(Marlow)	
G-RIIN	PZL-104M Wilga 2000	00010010	SP-WEI	27. 6.01	E.A M.Austin	Oaksey Park	22. 7.04T
G-RIKI	Mainair Blade 912	1280-0401-7 & W1075		29. 8.01	R.Cook	East Fortune	11.10.05P
G-RIKS	Europa Aviation Europa XS *(Built R.Morris - kit no.393) (Tri-gear u/c)*	PFA 247-13329		18.10.01	R.Morris	Cambridge	14. 7.05P
G-RIKY	Mainair Pegasus Quik	8007		17.12.03	R.J.Cook	Easter Poldar Farm, Thornhill	16.12.05P
G-RIMB	Lindstrand LBL 105A HAB	827		15. 3.02	D.Grimshaw	(Leyland)	15. 4.05T
G-RIME	Lindstrand LBL 25A Cloudhopper HAB	954		9.12.03	Poppies (UK) Ltd	(Bridport)	17. 7.05A
G-RIMM	Westland Wasp HAS.1	F9605	NZ3907 XT435	11. 3.99	M.P.Grimshaw and T.Martin *(As "XT435/430")*	North Weald	19. 8.05P
G-RINN	Mainair Blade (Rotax 582)	1261-1000-7 & W1055		2. 1.01	J.P.Lang	Guy Lane Farm, Waverton	21. 1.05P
G-RINO	Thunder Ax7-77 HAB	975		24. 6.87	D.J.Head *"Cerous"*	(Newbury)	5. 3.94T
G-RINS	Rans S-6-ESD Coyote II *(Built D G Watts)* (Rotax 582)	PFA 204-13361		15. 3.99	D.Watt *(Noted 7.04)*	Ladthwaite Farm, Kirkby Steven	18. 5.05P

Reg	Type	C/n	Previous identity	Date	Owner	Base	CofA
G-RINT	CFM Streak Shadow	PFA 206-12251		7.12.93	D.and J.S.Grint	Shoreham	24.11.05P
	(Built D Grint - kit no)(Rotax 582)						
G-RIPS*	Cameron Action Man/Parachutist 110SS HAB				See SECTION 3 Part 1	(Aldershot)	
G-RISE	Cameron V-90 HAB	2395		21. 9.90	D.L.Smith *"Rise N' Shine"*	(Newbury)	18. 9.05T
G-RIST	Cessna 310R II	310R1294	G-DATS	28. 4.81	F.B.Spriggs	Bournemouth	5. 2.05
			(N6128X)				
G-RIVE	Jodel D.153 *(Built P.Fines)*	PFA 235-12856		14. 7.04	P.Fines	(Benniworth, Market Rasen)	
G-RIVR	Thruster T.600F	9029-T600N-031		3.12.99	Thruster Air Services Ltd	Ginge, Wantage	7.12.00P
	(Hirth H2706) *(Officially regd as "T.600N" with Rotax 582)*				*(Noted 1.05 in float configuration)*		
G-RIVT	Van's RV-6	PFA 181-12743		31. 7.95	N.Reddish	Netherthorpe	26. 3.05P
	(Built N Reddish) (Lycoming O-320)						
G-RIXS	Europa Aviation Europa XS	PFA 247-13822		2. 7.02	R.Iddon	(Leyland)	23. 4.05P
	(Built R.Iddon - kit no.533) (Tri-gear u/c)						
G-RIZE	Cameron O-90 HAB	3163		13.12.93	S.F.Burden	(Noordwijk, The Netherlands)	21. 9.04A
G-RIZZ	Piper PA-28-161 Cherokee Warrior II	28-7816494	D-EMFW	11. 2.99	Northamptonshire School of Flying Ltd	Sywell	21. 3.05T
			N9563N				
G-RJAH	Boeing Stearman D75N1 (PT-27BW) Kaydet	75-4041	N75957	6. 4.90	R.J.Horne	Kemble	11. 4.04
	(Continental W670)		RCAF FJ991, 42-15852		*(As "44" in US Army c/s)*		
G-RJAM	Sequoia F.8L Falco	PFA 100-11665		26. 7.00	R.J.Marks	Dunkeswell	
	(Built R.J.Marks)				*(Noted 11.04)*		
G-RJGR	Boeing 757-225	22197	N701MG	22.11.94	MyTravel Airways Ltd	Manchester	1. 2.07T
			N507EA				
G-RJMS	Piper PA-28R-201 Arrow III	28R-7837059	N6223H	19. 1.88	M.G.Hill	Crosland Moor	29. 5.06
G-RJWW	Maule M-5-235C Lunar Rocket	7250C	G-BRWG	6.10.87	PAW Flying Services Ltd	Bagby	21.10.06T
			N5632H				
G-RJWX	Europa Aviation Europa XS	PFA 247-13197		11. 9.00	J.R.Jones	Sleap	20. 1.05P
	(Built J.R.Jones - kit no.359) (Monowheel u/c)						
G-RJXA	Embraer EMB-145EP	145.136	PT-SDN	18. 6.99	British Midland Airways Ltd	Nottingham East Midlands	17. 6.05T
					(Op bmi Regional)		
G-RJXB	Embraer EMB-145EP	145.142	PT-SDS	23. 6.99	British Midland Airways Ltd	Nottingham East Midlands	27. 6.05T
					(Op bmi Regional)		
G-RJXC	Embraer EMB-145EP	145.153	PT-SEE	15. 7.99	British Midland Airways Ltd	Nottingham East Midlands	14. 7.05T
					(Op bmi Regional)		
G-RJXD	Embraer EMB-145EP	145.207		4. 2.00	British Midland Airways Ltd	Nottingham East Midlands	3. 2.06T
					(Op bmi Regional)		
G-RJXE	Embraer EMB-145EP	145.245	PT-SIJ	10. 4.00	British Midland Airways Ltd	Nottingham East Midlands	9. 4.06T
					(Op bmi Regional)		
G-RJXF	Embraer EMB-145EP	145.280	PT-SJW	29. 6.00	British Midland Airways Ltd	Nottingham East Midlands	28. 6.06T
					(Op bmi Regional)		
G-RJXG	Embraer EMB-145EP	145.390		20. 2.01	British Midland Airways Ltd	Nottingham East Midlands	19. 2.07T
					(Op bmi Regional)		
G-RJXH	Embraer EMB-145EP	145.442	PT-SVD	1. 6.01	British Midland Airways Ltd	Nottingham East Midlands	31. 5.07T
					(Op bmi Regional) (Star Alliance titles)		
G-RJXI	Embraer EMB-145EP	145.454	PT-SVD	22. 6.01	British Midland Airways Ltd	Nottingham East Midlands	21. 6.07T
					(Op bmi Regional) (Star Alliance titles)		
G-RJXJ	Embraer EMB-145ER	145.473		23. 7.01	British Midland Airways Ltd	Nottingham East Midlands	22. 7.07T
					(Op bmi Regional)		
G-RJXK	Embraer EMB-145ER	145.494	PT-SXN	14. 9.01	British Midland Airways Ltd	Nottingham East Midlands	13. 9.07T
					(Op bmi Regional) (Star Alliance titles)		
G-RJXL	Embraer EMB-135ER	145.376	PT-SQA	20.12.04	British Midland Airways Ltd	Nottingham East Midlands	AC
			(EI-LCY), PT-SQA, (CN-RLF), PT-SQA				
G-RJXM	Embraer EMB-145ER	145.xxx		R	British Midland Airways Ltd	Nottingham East Midlands	
G-RKEL	Agusta-Bell 206B-3 JetRanger III	8617	HB-XPR	2. 8.01	Nunkeeling Ltd	(Brough)	26. 3.05T
			F-GCVE				
G-RKET	Taylor JT.2 Titch	PFA 3223	G-BIBK	25. 8.99	P.A Dunley	RAF Valley	
	(Built P A Dunley)				*(Current status unknown)*		
G-RKJT	Piper PA-46-500TP Mailbu Meridian	4697111	N338DB	25. 6.03	Harpin Ltd	Leeds-Bradford	26. 6.06T
G-RLFI	Reims/Cessna FA152 Aerobat	FA1520340	G-DFTS	17. 1.90	Tayside Aviation Ltd	Dundee	5. 2.05T
G-RLON	Fairey Britten-Norman BN-2A Mk.III-2 Trislander	1008	G-ITEX	26. 4.02	Aurigny Air Services Ltd	Guernsey	16 12.07E
			G-OCTA, VR-CAA, (G-OLPL), VR-CAA, DQ-FCF, G-BCXW				
					(Royal London Asset Management titles)		
G-RMAC	Europa Aviation Europa	PFA 247-12717		3. 7.97	P.J.Lawless	Kemble	16. 6.05P
	(Built P.J.Lawless - kit no.109) (Monowheel u/c)						
G-RMAN	Aero Designs Pulsar	PFA 202-13071		6. 6.97	M.B.Redman	Old Sarum	22. 6.05P
	(Built M.B.Redman)						
G-RMAX	Cameron C-80 HAB	4705		6.12.99	M.Quinn and D.Curtain	(Dublin)	5. 7.05A
G-RMIT	Van's RV-4	PFA 181-12207		4. 9.96	J.P.Kloos	Truleigh Manor Farm, Edburton	26. 5.05P
	(Built J.P.Kloos) (Lycoming O-320)						
G-RMMT	Europa Aviation Europa XS	A260	N929N	28. 1.05	N.Schmitt	(Nottingham)	
	(Built N.Schmitt) (Rotax 914) *(Tri-cycle u/c)*						
G-RMPY	Aerotechnik EV-97 Eurostar	PFA 315-14139		4. 2.04	N.R.Beale	Deppers Bridge, Southam	27. 6.05P
	(Built N.R.Beale)						
G-RMUG	Cameron Nescafe Mug 90SS HAB	3450		3. 5.95	Nestle UK Ltd *"Nescafe"*	(Croydon)	12. 7.03A
G-RNAC	IAV-Bacau Yakovlev Yak-52	888912	RA-44463	25. 7.03	RNAEC Ltd	RNAS Yeovilton	31.10.05P
			DOSAAF 99				
G-RNAS*	de Havilland DH.104 Sea Devon C.20	04473	XK896	16.11.82	Not known	Filton	3. 7.84
					(Cancelled 17.4.97 by CAA) (For spares 9.01)		
G-RNBW	Bell 206B JetRanger II	2270	F-GQFH	9. 1.98	Rainbow Helicopters Ltd	Whimple	24. 3.07T
			F-WQFH, HB-XUF, F-GFBP, N900JJ, N16UC				
G-RNDD	Robin DR400/500 Président	37		2. 5.03	Sterna Aviation Ltd	(Castletown, Isle of Man)	27. 5.06T
G-RNGO	Robinson R22 Beta II	3035		19. 1.00	Proflight Helicopters Ltd	Swansea	4. 3.06T
G-RNIE	Cameron Ball 70SS HAB	2333		3. 8.90	N.J.Bland *"Schwarzenegger"*	(Didcot)	7. 6.05A
G-RNLI	Vickers Supermarine 236 Walrus 1	S2/5591	W2718	13.12.90	R.E.Melton	Repps, Martham	
					(On rebuild 4.03 as "W2718/AA5Y" in 751 Sqdn RN c/s)		

G-RNRM	Cessna A185F Skywagon	185-02541	N1826R	20. 1.87	Skydive St.Andrews Ltd *"Thunderchild"*	Sorbie Farm, Kingsmuir	3. 3.06
G-RNRS	Scottish Aviation Bulldog Series 100/101	BH100/132	SE-LLF Fv61026, G-AZIT	12. 9.03	Power Aerobatics Ltd	Kemble	20. 5.07T
G-ROAR	Cessna 401	401-0166	G-BZFL G-AWSF, N4066Q	8. 3.82	Special Scope Ltd	Blackpool	6. 3.06
G-ROBD	Europa Aviation Europa *(Built R.D.Davies - kit no.078) (Monowheel u/c)*	PFA 247-12671		23. 2.94	R.D.Davies *(Current status unknown)*	(Cowbridge)	
G-ROBN	Robin R1180T Aiglon	220		16. 8.78	Bustard Flying Group Ltd	Boscombe Down	27. 5.06T
G-ROBT	Hawker Hurricane I	---	P2902	19. 9.94	R.A Roberts	Moat Farm, Milden	
	(Built Gloster Aircraft Co.Ltd) (On rebuild by Hawker Restorations Ltd from remains salvaged 1988 at Dunkirk Beach: to be "P2902/DX-X")						
G-ROCH	Cessna T303 Crusader	T30300129	N4962C	29. 3.90	R.S.Bentley	Cambridge	7. 6.05
G-ROCK	Thunder Ax7-77 HAB	781		25. 2.86	M.A Green *"Rocky"*	(Rednal)	28. 7.05A
G-ROCR	Schweizer 269C	S.1336	N219MS	14. 6.90	C.J.Williams	Sandown	13. 3.06T
G-RODC	Steen Skybolt *(Built R.H. Williams)*	4568	N10624	20. 2.02	J.W.Teesdale and S.Yelland	(Thornton Le Clay, York/Ferrensby, Knaresborough	2.11.05P
G-RODD	Cessna 310R II	310R0544	G-TEDD G-MADI, N87396, G-MADI, N87396	2.10.89	R.J.Herbert Engineering Ltd	Marshland, Wisbech	27.11.06
G-RODG	Jabiru Aircraft Jabiru UL *(Built I M Donnelly)*	PFA 274A-13379		14. 4.99	P.C.Appleton	Davidstow Moor	9. 5.04P
G-RODI	Isaacs Fury *(Built D C J Summerfield)* (Lycoming O-290)	PFA 11-10130		22.12.78	M.R.Baker *(As "K3731" in 43 Sqdn c/s)*	Westfield Farm, Hailsham	9.12.05P
G-ROGY	Cameron Concept 60 HAB	3055		11. 5.93	S.A Laing	(Banchory)	29. 8.04A
G-ROKT	Reims Cessna FR172E Rocket	FR1720046	N261SA D-ECLY	1.05.03	Sylmar Aviation and Services Ltd	Lower Wasing Farm, Brimpton	22. 9.06
G-ROLF	Piper PA-32R-301 Saratoga SP	32R-8113018	N83052	7. 1.81	P.F.Larkins	High Cross, Ware	23. 4.05
G-ROLL	Pitts S-2A *(Built Aerotek Inc)*	2175	N31444	20. 2.80	N.Lamb t/a Aerial and Aerobatic Services	Booker	17. 7.07A
G-ROLY	Reims/Cessna F172N Skyhawk II	F17201945	G-BHIH	1.12.04	R.G.Froggatt	Gamston	12. 9.07E
G-ROME	III Sky Arrow 650 TC *(Built Iniziative Industriali Italiane)*	C011		26. 5.99	Sky Arrow (Kits) UK Ltd	Old Sarum	16. 6.05T
G-ROMP	Extra EA.230H *(Built W Hawickhorst)*	001	S5-MBP OO-JVD, D-EIWH	13. 1.05	J S Allison *(Noted 1.05)*	Gamston	
G-ROMS	Lindstrand LBL 105G HAB	401		13. 9.96	T.D.Donnelly tr Gromit Balloon Group *"Gromit"*	(Sprotbrough, Doncaster)	13. 9.00A
G-ROMW	Cyclone AX2000 (HKS 700E V3)	7486		4. 2.99	L.P.Taylor	Clench Common	29. 5.05P
G-RONA	Europa Aviation Europa *(Built C.M.Noakes - kit no.043) (Monowheel u/c)*	PFA 247-12588		17. 1.95	C.M.Noakes *"Mr Jake"*	Shenstone Hall Farm, Shenstone	17. 6.05P
G-ROND	Short SD.3-60 Variant 100	SH.3604	EI-CWG G-OLAH, G-BPCO, G-RMSS, G-BKKU	1.11.01	Emerald Airways Ltd	Liverpool	26.11.05T
G-RONG	Piper PA-28R-200 Cherokee Arrow II	28R-7335148	N16451	14. 6.90	E.Tang	Elstree	7.11.05
G-RONI	Cameron V-77 HAB	2349		27. 7.90	R.E.Simpson *"Roni"*	(Great Missenden)	15. 8.02A
G-RONN	Robinson R44 Astro	0267	N770SC G-RONN, D-HIRR	8. 1.98	R Hallam and S E Watts	Leicester	17. 2.07
G-RONS	Robin DR400/180 Régent	2088		17. 7.91	R.and K.Baker	Swansea	21. 9.06
G-RONW	Clutton FRED Series II *(Built P Gronow)*	PFA 29-10121		18.12.78	V Magee *(Noted 6.04)*	Dunkeswell	29. 3.04P
G-ROOK	Reims/Cessna F172P Skyhawk II	F17202081	PH-TGY G-ROOK	12. 1.81	Rolim Ltd *(Op Bon Accord Flying Group)*	Aberdeen	18.11.05T
G-ROOV	Europa Aviation Europa XS *(Built D Richardson - kit no.354)* (Rotax 914-UL) *(Tri-gear u/c)*	PFA 247-13214		16. 7.98	E.Sheridan and P.W.Hawkins	Biggin Hill	27. 5.05P
G-RORI	Folland Gnat T.1	FL.549	8621M XR538	18.10.93	M.P.Grimshaw *(As "XR538/01")*	North Weald	10. 6.05P
G-RORY	Focke-Wulf Piaggio FWP.149D *(Piaggio c/n 338)*	014	G-TOWN D-EFFY, 90+06, BB+394 7	2. 8.88	Bushfire Investments Ltd	Booker	24. 7.06
G-ROSI	Thunder Ax7-77 HAB	1284		29. 6.88	J.E.Rose *"Rosi"*	(Abingdon)	21. 9.96A
G-ROSS	Practavia Pilot Sprite *(Built F M T Ross - plans no.132)*	PFA 05-10404		28. 2.80	A.D.Janaway *(New owner 6.04*	Exeter	
G-ROTI	Luscombe 8A (Continental A65)	2117	N45590 NC45590	18. 4.89	R.Ludgate and A L.Chapman	Old Hay,Paddock Wood	9.10.97P
G-ROTR	Brantly B.2B	403	N2192U	9.12.91	P.G.R.Brown	(Crediton)	17.11.02
G-ROTS	CFM Streak Shadow *(Buillt H R Cayzer - kit no K.120-SA)* (Rotax 582)	PFA 161A-11603		21.12.89	A.G.Vallis and C.J.Kendal *(New owner 8.04)*	(Littleborough/Stockport)	30.11.05P
G-ROUP	Reims/Cessna F172M Skyhawk II	F17201451	G-BDPH	23. 5.84	Stapleford Flying Club Ltd	Stapleford	1. 5.06T
G-ROUS	Piper PA-34-200T Seneca II	34-7870187	(G-BFTB) N9412C	26. 4.78	Oxford Aviation Services Ltd	Oxford	3. 3.06T
G-ROUT	Robinson R22 Beta	1241	N8068U	23. 1.90	Preston Associates Ltd	(Guisborough)	19.8.07T
G-ROVE	Piper PA-18-135 Super Cub (L-21B-PI) *(Frame No.18-3853)*	18-3846	PH-VLO (PH-DKF), R Neth AF R-156, 54-2446	6. 5.82	S.J.Gaveston *(As "54-2446/R-156" in R Neth AF c/s)*	Redhill	13. 8.04T
G-ROVY	Robinson R22 Beta II	2957		9. 7.99	Plane Talking Ltd	Elstree	23. 9.05T
G-ROWE	Reims/Cessna F182P Skylane II	F18200007	OO-CNG	18.12.95	D.Rowe	St.Just	4. 3.05
G-ROWI	Europa Aviation Europa XS *(Built R.M.Carson - kit no.435)* (Wilksch WAM-120) *(Monowheel u/c)*	PFA 247-13482		16. 6.99	R.M.Carson *(Current status unknown)*	(Cheltenham)	
G-ROWL	Grumman AA-5B Tiger	AA5B-0595	(N28410)	26.10.77	G.Shapps t/a Airhouse Corporation	Elstree	17. 5.07T
G-ROWN	Beech 200 Super King Air	BB-684	G-BHLC N27L, N8511L, G-BHLC	13.10.87	Valentia Air Ltd	RAF Brize Norton	10. 4.05T
G-ROWR	Robinson R44 Raven	1036		17. 4.01	R.A Oldworth	(Petworth)	24. 6.07T
G-ROWS	Piper PA-28-151 Cherokee Warrior	28-7715296	N8949F	15. 9.78	N.J.Amey	(Royston)	27. 3.06
G-ROXY	Skystar Kitfox Mk.7 *(Built P.N.Akass)*	PFA 172D-14024		23.10.03	P.N.Akass	Inverness	15. 7.05P
G-ROYC	Jabiru Aircraft Jabiru UL-450 *(Built R.Clark)*	PFA 274A-13990		24. 4.03	R.Clark	Rufforth	29. 9.05P
G-ROZI	Robinson R44 Astro	0252		26. 3.96	Walker Plant Services Ltd	(Retford)	29. 4.05T

G-ROZY	Cameron R-36 Gas/HAB	1141		20. 5.85	Jacques W.Soukup Enterprises Ltd	(Florida, US)	18. 9.96A
G-ROZZ	Comco Ikarus C42 FB80	0407-6607		19. 8.04	A.J.Blackwell	Long Marston	18. 8.05P
G-RPAF	Europa Aviation Europa XS *(Built R.P.Frost)*	PFA 247-14202		26. 1.05	R.P.Frost	(Woolley, Wakefield)	
G-RPBM	Cameron Z-210 HAB	10230		6. 3.02	The Balloon Co Ltd t/a First Flight *(Robert Price Builders Merchants titles)*	(Bristol)	31. 3.05T
G-RPEZ	Rutan LongEz *(Built B A Fairston and D Richardson)*	PFA 74A-10746		3. 4.84	D.G.Foreman *(Stored uncomplete 5.00 - new owner 7.03)*	(Swanley)	
G-RPRV	Van's RV-9A *(Built G.R.Pybus)*	PFA 320-13936		17.10.03	G.R.Pybus	(Durham)	
G-RRCU	CEA Jodel DR.221B Dauphin	129	F-BRCU	9.12.99	Merlin Flying Club Ltd	Hucknall	12.11.06T
G-RRFC	SOCATA TB-20 Trinidad GT	2053	F-OILV	9. 5.01	C.A.Hawkins	Blackbushe	8. 6.07
G-RRGN	Vickers Supermarine 390 Spitfire PR.XIX	6S/594677	G-MXIX PS853	23.12.96	Rolls-Royce plc *(As "PS853/C" in 2nd TAF/PRU c/s)*	Filton	3. 9.05P
G-RROB	Robinson R44 Raven II	10011		6.12.02	Ranc Helicopters Ltd	Stapleford	5. 2.06T
G-RROD	Piper PA-30 Twin Comanche B	30-1221	G-SHAW LN-BWS, N10F	20. 6.00	R.P.Coplestone	Thruxton	14. 6.07T
G-RISY	Van's RV-7A *(Built A.J.A Weal)*	PFA 323-14320		10. 2.05	A.J.A Weal	(Worthing)	
G-RSKR	Piper PA-28-161 Warrior II	28-7916181	G-BOJY N3030G	27. 4.95	R.Sherwin-Smith tr Krown Group	Shoreham	4.12.06T
G-RSKY	Best Off Skyranger 912 *(Built C.G. Benham - kit no. UK/452)*	BMAA/HB/382		12.10.04	C.G.Benham	Red House Farm, Preston Capes	
G-RSSF	Denney Kitfox Model 2 *(Built R W Somerville)*	PFA 172-12125		9.10.92	R.W.Somerville *(Noted 10.03)*	Comber, County Down	15. 5.97P
G-RSVP	Robinson R22 Beta II	2788		5. 2.98	Plane Talking Ltd	Elstree	8. 3.07T
G-RSWO	Cessna 172R Skyhawk	17280206	N9401F	25. 2.98	AC Management Associates Ltd	Kemble	28. 3.07T
G-RSWW	Robinson R22 Beta	1775	N40815	16. 5.91	R.S.Weston-Woods t/a Woodstock Enterprises	Brands Hatch, Kent	11. 8.06T
G-RTBI	Thunder Ax6-56 HAB	2584		19. 4.94	P.J.Waller	(Norwich)	8. 7.02A
G-RTMS	Rans S-6ES Coyote II *(Built C.J.Arthur)*	PFA 204-14149		19. 8.04	C.J.Arthur	Eshott	
G-RTWO	Robinson R44 Raven II	10618		21. 1.05	Heli Air Ltd	Wellesbourne Mountford	
G-RTWW	Robinson R44 Astro	0438		20. 3.98	R.Woods t/a Rotorvation	(Longfield, Kent)	7. 5.07T
G-RUBB	Gulfstream AA-5B Tiger	AA5B-0928	(G-BKVI) OO-NAS, (OO-HRC)	20. 9.83	D.E.Gee	Blackbushe	23.11.07E
G-RUBI	Thunder Ax7-77 HAB	1051		27. 2.87	G.Warren t/a Warren and Johnson *"Rubicon Computer Systems"*	(Norwich)	20.11.93A
G-RUBY	Piper PA-28RT-201T Turbo Arrow IV	28R-8331037	G-BROU N4306K	5. 1.90	R.Harman tr Arrow Aircraft Group	Tatenhill	26. 6.05
G-RUDD	Cameron V-65 HAB	844		19. 5.82	N.A Apsey *(Kodak titles) "Smilie"*	(High Wycombe)	20. 5.00A
G-RUES	Robin HR100/210 Safari	185	F-BVCH	31. 7.02	R.H.R.Rue	Oxford	23. 9.05
G-RUFF	Mainair Blade 912	1203-0799-7 & W1006		18. 6.99	J.C.Townsend	Ince Blundell	17. 6.05P
G-RUFS	Jabiru Aircraft Jabiru UL *(Built J.W.Holland)*	PFA 274A-13359		19.11.99	S.Richens	Clench Common	1. 7.05P
G-RUGS	Campbell Cricket Mk.4 *(Built J.L.G.Mclane)*	PFA G/103-1307		11. 2.99	J.L.G.Mclane *(Current status unknown)*	(York)	
G-RUIA	Reims/Cessna F172N Skyhawk II	F17201856	PH-AXA (3)	4.10.79	Knockin Flying Club Ltd	Knockin, Shropshire	26. 7.07
G-RUMI	Noble Hardman Snowbird Mk.IV	SB-018	G-MVOI	9. 9.02	G.Crossley *(Noted 8.04)*	(Anglesey)	13. 6.02P
G-RUMM	Grumman F8F-2P Bearcat	D.1088	NX700HL NX700H, N1YY, N4995V, Bu.121714	20. 3.98	Patina Ltd *(Op The Fighter Collection as "21714/201B" in USN c/s)*	Duxford	5. 7.05P
G-RUMN	American Aviation AA-1A Trainer	AA1A-0086	N87599 D-EAFB, (N9386L)	30. 5.80	M.T.Manwaring	RAF Halton	26. 5.06
G-RUMT	Grumman F7F-3P Tigercat	C.167	N7235C BuA.80425	6. 4.98	Patina Ltd *(Op The Fighter Collection as "80425/4-WT" in US Marines c/s)*	Duxford	5. 7.05P
G-RUMW	Grumman FM-2 Wildcat	5765	N4845V BuA.86711	15. 4.98	Patina Ltd *(Op The Fighter Collection as "F" in FAA c/s)*	Duxford	28. 6.05P
G-RUNG	SAAB-Scania SF.340A	340A-086	F-GGBV SE-E86	3. 6.97	Aurigny Air Services Ltd	Guernsey	5. 6.06T
G-RUNT	Cassutt Racer IIIM *(Built N A Brendish - c/n 161149)* (Lycoming O-235)	PFA 34-10860		12. 4.83	N.A Scully	Leicester	13. 4.05P
G-RUSA	Pegasus Quantum 15-912	7517		7. 4.99	A D.Stewart	Perth	14. 4.04P
G-RUSL	Van's RV-6A *(Built G Russell)*	PFA 181-13522		22.10.01	G.R.Russell	(Crewkerne)	
G-RUVI	Zenair CH.601UL *(Built P.G.Depper)*	PFA 162A-13933		8.11.02	P.G.Depper *"Indulgence"*	(Kidderminster)	31.10.05P
G-RUVY	Van's RV-9A *(Built R D Taylor)*	PFA 320-13807		4. 1.02	R Taylor *(Noted 6.04)*	Dunkeswell	17. 8.05P
G-RUZZ	Robinson R44 Raven II	10082		20. 5.03	Russell Harrison plc	(Chipping Norton)	12. 6.06T
G-RVAB	Van's RV-7 *(Built I.M.Belmore)*	PFA 323-14005		20. 9.04	I.M.Belmore and A.T.Banks	(Horsham)	
G-RVAL	Van's RV-8 *(Built R N York)*	PFA 303-13532		23. 7.01	R.N.York	Dunsfold	18.11.05P
G-RVAN	Van's RV-6 *(Built D Broom)* (Lycoming IO-320)	PFA 181-12657		25. 4.97	D.Broom	Benington	22. 4.05P
G-RVAW	Van's RV-6 *(Built A A Wordsworth)* (Lycoming IO-320-A1A)	PFA 181-13234		24.11.97	P.E.Bates tr High Flatts RV Group	High Flatts Farm, Chester-le-Street	21. 5.05P
G-RVBA	Van's RV-8A *(Buillt S.Hawksworth)*	PFA 303-13309		26.10.99	S.Hawksworth *(Under construction 2000)*	(Nuneaton)	
G-RVBC	Van's RV-6A *(Built T G Gibbs)*	PFA 181-12618		16. 2.00	T G Gibbs	(Radstock)	
G-RVBF	Cameron A-340 HAB	10493		23. 2.04	Airxcite Ltd t/a Virgin Balloon Flights	Wembley	27. 7.05T
G-RVCE	Van's RV-6A *(Built M.D.Barnard and C.Voelger)*	PFA 181-13372		28. 6.01	M.D.Barnard and C.Voelger	(Welwyn)	

Registration	Type / Builder	C/n	Previous identity	Reg date	Owner	Location	Date
G-RVCG	Van's RV-6A *(Built C J Griffin)*	PFA 181A-13602		26. 4.01	C.J.Griffin	(Stratford-upon-Avon)	
G-RVCL	Van's RV-6 *(Built C Lamb)*	PFA 181A-13439		18. 2.99	C.T.Lamb	(Stamford)	
G-RVDG	Van's RV-9A *(Built D.M.Gill)*	PFA 320-14310		6. 1.05	D.M.Gill	(Aylesbury)	
G-RVDJ	Van's RV-6 *(Built J D Hewitt)* (Lycoming O-360)	PFA 181-12938		8. 2.99	J.D.Jewitt	(Selby)	3.10.05P
G-RVDP	Van's RV-4 *(Built D.H.Pattison)*	PFA 181-13416		10. 5.00	D.H.Pattison	Lower Upham Farm, Chiseldon	
G-RVDR	Van's RV-6A *(Built D E Reast)* (Lycoming IO-320)	PFA 181-13098		15. 5.00	T.M.Norman	Tollerton	25. 4.04P
G-RVDS	Van's RV-4 *(Built D.F.Sargant)*	PFA 181-12270		28.10.02	D.F.Sargant	Ludham	24. 4.05P
G-RVEE	Van's RV-6A *(Built J C A Wheeler)* (Lycoming O-360)	PFA 181-12262		16. 2.93	J.C.A Wheeler	Perth	9. 4.05P
G-RVET	Van's RV-6 *(Built D R Coleman)* (Lycoming O-300)	PFA 181-12852		9. 3.98	D.R.Coleman	Rochester	11. 2.02P
G-RVGA	Van's RV-6A *(Built D P Dawson)* (Lycoming IO-320)	PFA 181-13079		11. 5.98	D.P.Dawson	RAF Henlow	2. 5.05P
G-RVIA	Van's RV-6A *(Built A J Rose)* (Lycoming O-320)	PFA 181-12289		13. 8.97	K R Emery	Sittles Farm, Alrewas	3.11.05P
G-RVIB	Van's RV-6 *(Built I M Belmore)* (Lycoming O-320)	PFA 181-13220		22. 6.99	K.Martin and P.Gorman	Kilrush, County Kildare	13. 6.05P
G-RVIC	Van's RV-6A *(Built I.T.Corse)*	PFA 181-13319		11. 6.04	I.T.Corse	(Laurencekirk)	
G-RVII	Van's RV-7 *(Built P H Hall) (Project conceived originally as a RV-6, hence the '181A prefix)*	PFA 181A-13576		13. 9.01	P.H.C.Hall	(Swindon)	
G-RVIN	Van's RV-6 *(Built N Reddish)* (Lycoming O-320)	PFA 181-13236		28.11.97	R.G.Jones	Rednal	22. 6.05P
G-RVIS	Van's RV-8 *(Built I.V.Sharman)*	PFA 303-14031		17. 6.03	I.V.Sharman	(Horley)	
G-RVIT	Van's RV-6 *(Built K F Crumplin)* (Lycoming O-360)	PFA 181-12422		1. 5.95	P.J.Shotbolt	Ingthorpe Farm, Great Casterton, Lincoln	7. 9.05P
G-RVIV	Van's RV-4 *(Built G.S.Scott)* (Lycoming O-320)	PFA 181-12366		31.12.97	G.S.Scott	Truleigh Manor Farm, Edburton	9. 9.05P
G-RVIX	Van's RV-9A *(Built R.E.Garforth)*	PFA 320-13779		11. 9.01	R.E.Garforth	Southend	18.11.05P
G-RVJM	Van's RV-6 *(Built M D Challoner)*	PFA 181A-13861		4.12.02	M.D.Challoner	(Sturminster Newton)	
G-RVMC	Van's RV-7 *(Built M.R.McNeil)*	PFA 323-13897		9. 5.03	M.R.McNeil	(Selby)	22. 9.05P
G-RVMJ	Van's RV-4 *(Built M.J.de Ruiter)*	PFA 181-13433		16. 2.99	M.J.de Ruiter	(Craigavon, County Armagh)	
G-RVMT	Van's RV-6 *(Built M R Tingle)* (Lycoming O-360)	PFA 181A-13644		30. 1.01	M R Tingle	Norwich	8. 5.05P
G-RVMZ	Van's RV-8 *(Built M.W.Zipfell)* (Lycoming O-360)	PFA 303-13395		12.11.99	M.W.Zipfell	Hollow Lane Farm, Thurston	19. 1.06P
G-RVPH	Van's RV-8 *(Built J.C.P.Herbert)*	PFA 303-13906		25. 5.04	J.C.P.Herbert	(Saffron Walden)	
G-RVPL	Van's RV-8 *(Built A.P.Lawton)*	PFA 303-13885		6. 8.04	A.P.Lawton	(Great Massingham)	
G-RVPW	Van's RV-6A *(Built P Waldron)*	PFA 181A-13481		9. 6.03	P.Waldron *(Noted 9.04)*	Netherthorpe	8.11.05P
G-RVRA	Piper PA-28-140 Cherokee Cruiser	28-7625038	G-OWVA N4459X	14. 1.97	Mona Aviation Ltd t/a Mona Flying Club	RAF Mona	19. 4.06T
G-RVRB	Piper PA-34-200T Seneca II	34-7970440	G-BTAJ N22MJ, N45113	24. 2.97	Cheshire Flying Services Ltd t/a Ravenair	Manchester	26. 7.07T
G-RVRC	Piper PA-23-250 Aztec E	27-7405336	G-BNPD N101VH, N40591	14.10.97	West-Tec Ltd c/o Ravenair	(Liverpool	26. 1.07T
G-RVRD	Piper PA-23-250 Aztec E	27-4634	G-BRAV G-BBCM, N14021	16. 3.98	Cheshire Flying Services Ltd t/a Ravenair	Manchester	17.12.05T
G-RVRE	Partenavia P.68B	57	D-GIFR (N4412H), D-GIFR, LN-LMS	8.12.03	Cheshire Flying Services Ltd t/a Ravenair	Liverpool	1. 2.07T
G-RVRF	Piper PA-38-112 Tomahawk	38-78A0714	G-BGEL N9723N	21.11.97	Cheshire Flying Services Ltd t/a Ravenair	Liverpool	5. 6.06T
G-RVRG	Piper PA-38-112 Tomahawk	38-79A1092	G-BHAF N9703N	3. 8.98	Cheshire Flying Services Ltd t/a Ravenair	Manchester	15. 8.05T
G-RVRH	Van's RV-3B *(Built R Hodgson)*	PFA 99-10821		17. 2.03	R.Hodgson	(Guildford)	
G-RVRJ	Piper PA-E23-250 Aztec E	27-7305004	G-BBGB N40206	12.10.04	Cheshire Flying Services Ltd t/a Ravenair	Liverpool	13. 3.06E
G-RVRP	Van's RV-7 *(Built R.C.Parris)*	PFA 323-14085		16. 7.03	R.C.Parris	(Chalgrove)	
G-RVRV	Van's RV-4 *(Built P Jenkins)*	PFA 181-13024		29. 9.98	P Jenkins *(Amended CofR 8.02)*	(Inverness)	
G-RVRW	Piper PA-23-250 Aztec E	27-7305045	G-BAVZ N40241	17.12.04	Cheshire Flying Services Ltd t/a Ravenair	Liverpool	28. 4.07E
G-RVSA	Van's RV-6A *(Buil W H Knott)*	PFA 181A-12574		19. 5.99	W.H.Knott *(F/f 31.7.04)*	Dornoch	9. 8.05P
G-RVSG	Van's RV-9A *(Built S.Gerrish)*	PFA 320-14265		10.11.04	S.Gerrish	(Ferndown)	
G-RVSH	Van's RV-6A *(Built S J D Hall)*	PFA 181A-13026		20. 9.02	S.J.D.Hall	Blackbushe	23. 5.05P

Registration	Type	C/n	Previous identity	Reg date	Owner/operator	Base	CofA
G-RVSX	Van's RV-6	PFA 181-13090		18. 9.97	R.L.and V.A West	Shoreham	9. 8.05P
	(Built R L West)						
G-RVVI	Van's RV-6	PFA 181-12418		26. 1.93	J.E.Alsford and J.N.Parr	Sibson	18. 3.05P
	(Built J.E.Alsford and J.N.Parr) (Lycoming AEIO-360)						
G-RWAY	Rotorway Executive 162F	6414	G-URCH	18.11.04	S.Andrews	(Addlestone)	
	(Built D L Urch now S.Andrews) (RotorWay RI 162F)						
G-RWHC	Cameron A-180 HAB	2700		16. 4.92	J.J.Rudoni and A C.K.Rawson	(Stafford)	13. 4.00T
					t/a Wickers World Hot Air Balloon Co		
G-RWIN	Rearwin 175 Skyranger	1522	N32391	12. 9.90	A B Bourne and N D Battye	RAF Henlow	18. 7.05P
	(Continental A75)		NC32391				
G-RWLY	Europa Aviation Europa XS	PFA 247-13701		22. 3.01	C.R.Arkle	(Ascot)	
	(Built C.R.Arkle - kit no.469) (Tri-gear u/c)						
G-RWRW	Ultramagic M-77 HAB	77/221		11.11.02	Flying Pictures Ltd	(Chilbolton)	
G-RWSS	Denney Kitfox Model 2	PFA 172-12008		16. 4.91	R.W.Somerville	Comber, County Down	14. 6.93P
	(Built R.W.Somerville)				*(Current status unknown)*		
G-RWWW*	Westland WS-55 Whirlwind HCC.12				See SECTION 3 Part 1	Weston-super-Mare	
G-RXUK	Lindstrand LBL 105A HAB	232		29. 3.95	P.A Hames *"Rank Xerox"*	(Reading)	19. 5.02A
G-RYAL	Jabiru Aircraft Jabiru UL	PFA 274A-13365		6. 7.99	A C.Ryall	Cardiff	11.8.05P
	(Built A C.Ryall)						
G-RYPH	Mainair Blade 912	1248-0500-7 & W1041		8. 6.00	I.A Cunningham	Perth	29. 7.05P

G-SAAA - G-SZZZ

Registration	Type	C/n	Previous identity	Reg date	Owner/operator	Base	CofA
G-SAAB	Rockwell Commander 112TC	13002	G-BEFS	5.12.79	J B Barbour	Gamston	7. 3.07T
			N1502J				
G-SAAM	Cessna T182R Turbo Skylane II	T18268200	G-TAGL	23. 5.84	M.D.Harvey, M.A Tokley and J.R.Partner	Earls Colne	22.11.04
			G-SAAM, N2399E				
G-SABA	Piper PA-28R-201T Turbo Cherokee Arrow III	28R-7703268	G-BFEN	22. 8.79	C.A Burton	Sherburn-in-Elmet	9. 5.07
			N38745				
G-SABR	North American F-86A-5NA Sabre	151-43547	N178	6.11.91	Golden Apple Operations Ltd	Duxford	20. 6.05P
	(Regd with c/n 151-083)		N68388, 48-178		*(Op The Old Flying Machine Co as "8178/FU-178" in 4th Fighter Wing USAF c/s)*		
G-SACB	Reims/Cessna F152 II	F15201501	G-BFRB	7. 3.84	P.Wilson	(Mossley, Ashton-under-Lyne)	20. 5.06T
G-SACD	Cessna F172H	F172-0385	G-AVCD	13. 6.83	Northbrook College of Design and Technology		
	(Built Reims Aviation SA)				*(Op Sky Leisure Aviation)*	Shoreham	27. 7.00T
G-SACF*	Cessna 152 II	15283175	G-BHSZ	21. 3.85	Derby Aero Club Ltd	Egginton	8. 6.98T
			N47125		*(Damaged landing Egginton 21.3.97: cancelled 11.8.97 by CAA) (Wreck dumped 5.03)*		
G-SACH	Stoddard-Hamilton GlaStar	PFA 295-13088		27. 8.99	R.S.Holt	(Evesham)	29.10.05P
	(Built R S Holt) (Tailwheel u/c)						
G-SACI	Piper PA-28-161 Warrior II	28-8216123	N81535	26. 7.89	PJC (Leasing) Ltd	Stapleford	2. 5.05T
G-SACK	Robin R2160	316		2. 5.97	Sherburn Aero Club Ltd	Sherburn-in-Elmet	19. 6.06T
G-SACO	Piper PA-28-161 Warrior II	28-8416085	N4358Z	1. 6.89	D.C.and M.Brooks t/a The Barn Gallery	Oxford	7. 9.07
G-SACR	Piper PA-28-161 Cadet	2841046	N91618	6. 2.89	Sherburn Aero Club Ltd	Sherburn-in-Elmet	20. 2.07T
G-SACS	Piper PA-28-161 Cadet	2841047	N91619	6. 2.89	Sherburn Aero Club Ltd	Sherburn-in-Elmet	20. 2.07T
G-SACT	Piper PA-28-161 Cadet	2841048	N9162D	6. 2.89	Sherburn Aero Club Ltd	Sherburn-in-Elmet	25. 2.07T
G-SACZ	Piper PA-28-161 Warrior II	28-7916258	N2098N	26. 7.89	J.R.Santamaria	Guernsey	21. 4.05T
G-SADE	Reims/Cessna F150L	F15000752	G-AZJW	28. 5.91	N.E.Sams *(Stored engineless 8.04)*	Cranfield	21. 9.97T
G-SAFE	Cameron N-77 HAB	511		14. 2.79	P.J.Waller *"The High Flyer"*	(Norwich)	21. 4.91A
G-SAFI	Piel CP.1320	PFA 183-12103		23. 7.01	C.S.Carleton-Smith	(Great Missenden)	
	(Built C.S.Carleton-Smith)						
G-SAFR	SAAB 91D Safir	91-382	PH-RLR	10.10.95	Sylmar Aviation and Services Ltd		
						Lower Wasing Farm, Brimpton	AC
G-SAGA	Grob G109B	6364	OE-9254	28. 6.90	G-GROB Ltd	Booker	16. 7.05
G-SAGE	Luscombe 8A Silvaire	2581	G-AKTL	15. 8.90	C.Howell and J.O'Brien	(Kingsbridge/Launceston)	14. 4.05P
	(Continental A65)		N71154, NC71154				
G-SAHI	FLS Aerospace Sprint 160	001		21.10.80	Aces High Ltd	North Weald	30. 4.94P
	(Lycoming O-235) *(Built Trago Mills Ltd as SAH-1)*				*(Noted 1.05)*		
G-SAIX	Cameron N-77 HAB	626	N386CB	14. 1.99	C.Walther, B.Sevenich, B. and S.Harren	(Aachen, Germany)	21. 2.00A
G-SALA	Piper PA-32-300 Six	32-7940106	(G-BHEJ)	17.10.79	Stonebold Ltd	Booker	16. 2.07
			N2184Z				
G-SALL	Reims/Cessna F150L	F15000682	PH-LTY	19. 1.79	D.and P.A Hailey	Lower Wasing Farm, Brimpton	10. 9.06
			D-ECPH				
G-SAMG	Grob G109B	6278		16. 5.84	T.Holloway tr RAFGSA	RAF Halton	16. 4.05
G-SAMI	Cameron N-90 Sainsbury Strawberry SS HAB	3907	G-BWSE	21. 8.96	Flying Pictures Ltd	(Chilbolton)	15. 7.02A
G-SAMJ	Partenavia P68B	101	D-GERA	27. 4.01	S.M.Jack	Sherburn-in-Elmet	7. 7.07T
	(C/n indicates P68 model)		CS-AYB, D-GERA		tr G-SAMJ Group		
G-SAMM	Cessna 340A II	340A0742	N37TJ	7. 3.88	Calverton Flying Group Ltd	Cranfield	25. 4.07T
	(RAM-conversion)		N2671A				
G-SAMY	Europa Aviation Europa	PFA 247-12901		17. 8.95	P.Vallis	(Alfreton)	
	(Built K.R.Tallent and P.Vallis - kit no.221) (Tri-gear u/c)				*(New ownner 4.04)*		
G-SAMZ	Cessna 150D	15060536	G-ASSO	19. 4.84	Bonus Aviation Ltd	Cranfield	12.12.05P
			N4536U				
G-SAPM	SOCATA TB-20 Trindad	1009	G-EWFN	8.12.04	Trinidair Ltd	Filton	7. 4.05E
			G-BRTY				
G-SARA	Piper PA-28-181 Archer II	28-7990039	N21270	6. 4.81	Airflo Aviation Ltd	Shoreham	13. 5.07T
G-SARH	Piper PA-28-161 Warrior II	28-8216173	N8232Q	18. 2.91	Sussex Flying Club Ltd	Shoreham	18. 2.07T
G-SARK	British Aircraft Corporation 167 Strikemaster Mk.84	EEP/JP/1931	N2146S	13. 1.95	Tubetime Ltd	North Weald	27. 6.05P
			Singapore AF 311, G-27-140				
G-SARO	Saro Skeeter AOP.12	S2/5097	XL812	17. 7.78	B.Chamberlain *(As "XL812")*	Otley, Ipswich	1. 8.01P
G-SARV	Van's RV-4	PFA 181-12606		2.10.00	S.N.Aston	Bicester	4. 6.05P
	(Built S N Aston) (Lycoming O-320)						
G-SASA	Eurocopter EC135T1	0147		12.10.00	Bond Air Services Ltd	Glasgow City Heliport	22.10.06T
					(Op Scottish Ambulance Service)		

Reg	Type	C/n	Previous identity	Reg date	Owner / Operator	Base	CofA
G-SASB	Eurocopter EC135T2	0151		29. 9.00	Bond Air Services Ltd	Glasgow City Heliport	5.10.06T
					(Op Scottish Ambulance Service)		
G-SASK*	Piper PA-31P Pressurised Navajo	31P-39	G-BFAM	30.10.97	Middle East Business Club Ltd	(Guernsey)	30.08.91
			SE-GLV, OH-PNF		*(Noted as "G-BFAM" on repair Biggin Hill 12.00: cancelled 19.6.03 by CAA)*		
G-SATL	Cameron Sphere 105SS HAB	2696		5.12.91	Ballonverbung Hamburg GmbH	(Kiel, Germany)	29. 4.97A
G-SAUF	Colt 90A HAB	1497		25. 5.89	K.H.Medau	(Baden, Germany)	21. 6.04A
	(New envelope c/n 2492 c.1990/1)						
G-SAUK	Rans S-6-ES Coyote II	PFA 204-14346		5. 1.05	D.A.Smith and E.Robshaw	Rufforth	
	(Built D.A.Smith and E.Robshaw)						
G-SAWI	Piper PA-32RT-300T Turbo Lance II	32R-7887069	OY-CJJ	23. 6.99	S.T.Day	(Old Derry Hill, Calne)	25. 4.05E
			N36719				
G-SAXO	Cameron N-105 HAB	3864		1. 4.96	(Flying Pictures Ltd) *(Citroën Saxo titles)*	Chilbolton	25. 5.00A
G-SAYS	Rotary Air Force RAF 2000 GTX-SE	PFA G/13-1322		4. 9.00	The Aziz Corporation Ltd	Popham	9. 7.05P
	(Built K Aziz)				*(Based at owner's home)*		
G-SAZY	Jabiru Aircraft Jabiru J400	PFA 325-14057		16. 4.03	N.J.Bond	Dunkeswell	2.12.05P
	(Built N J Bond)				t/a JC Aviation		
G-SAZZ	Piel CP.328 Super Emeraude	PFA 216-11940		4. 7.01	D.J.Long	(Gloucestershire)	18.10.05P
	(Built D J Long)						
G-SBAE	Reims/Cessna F172P Skyhawk II	F17202200	D-EOCD (3)	3. 6.98	BAE SYSTEMS (Operations) Ltd	Blackpool	27. 7.07T
G-SBHH	Schweizer 269C	S.1314	G-XALP	7. 5.02	Hughes Helicopter Co Ltd	Biggin Hill	8.10.06T
			N41S		t/a Biggin Hill Helicopters		
G-SBIZ	Cameron Z-90 HAB	10348		12.12.02	Snow Business International Ltd	(Stroud)	2.12.04A
G-SBKR	SOCATA TB-10 Tobago	1077	D-EAGG	10. 3.04	S C M Bagley	Luton	28. 4.07T
G-SBLT	Steen Skybolt	MH-01		14. 4.92	S.D.Arnold	Coventry	
	(Built M A McCallum and N Workman)				tr Skybolt Group *(Current status unknown)*		
G-SBMM	Piper PA-28R-180 Cherokee Arrow	28R-30877	G-BBEL	8. 2.02	K.S.Kalsi	Cambridge	27. 5.07T
			SE-FDX				
G-SBMO	Robin R2160i	116	EI-BMO	12. 2.99	D.Henderson, U.Simpson and M.Mannion		
			SE-GSZ			Weston, Dublin	24. 3.06T
G-SBRA	Robinson R44 Raven II	10233		17.12.03	Sabretooth Aviation Ltd	(London SW4)	1. 2.07T
G-SBUS	Britten-Norman BN-2A-26 Islander	3013	G-BMMH	31.10.86	Isles of Scilly Skybus Ltd	St.Just	17. 4.06T
	(Built PADC)		RP-C578				
G-SBUT	Robinson R22 Beta II	2739	G-BXMT	18. 5.98	Heli Air Ltd	Wellesbourne Mountford	12.11.06T
G-SCAH*	Cameron V-77 HAB				See SECTION 3 Part 1	(Aldershot)	
G-SCAN	Vinten Wallis WA-116 Series 100/R	001		5. 7.82	K.H.Wallis	Reymerston Hall, Norfolk	10. 7.91P
	(Rotax 532)				*(Stored 8.01)*		
G-SCAT	Cessna F150F *(Tail-wheel u/c)*	F150-0054	G-ATRN	15. 9.86	Westward Airways (Lands End) Ltd	St.Just	10. 4.05T
	(Built Reims Aviation SA) (Wichita c/n 15063455)		(G-ATMN)				
G-SCBI	SOCATA TB-20 Trinidad	1908	F-OIGV	10. 8.99	S.C.Brown t/a Ace Services	Enstone	17. 8.05T
G-SCFO	Cameron O-77 HAB	1131		3. 5.85	M.K.Grigson	(Aldershot)	24. 5.95A
					(Op Balloon Preservation Group) "Southern Counties"		
G-SCHI	Eurocopter AS350B2 Ecureuil	3337	F-WQOQ	5. 2.01	Patriot Aviation Ltd	(Birmingham)	23. 3.07T
G-SCIP	SOCATA TB-20 Trinidad GT	2014	F-OILO	19. 9.00	J.C.White	Oxford	1.10.06
G-SCLX	FLS Aerospace Sprint 160	002	G-PLYM	14. 7.94	Aces High Ltd	Fairoaks	25. 9.06
G-SCOI	Agusta A109E Power	11051	G-HPWH	16. 8.02	Trustair Ltd	(Euxton, Chorley)	23. 6.05T
			G-HWPH				
G-SCPD	Just/Reality Escapade 912	BMAA/HB/319		30. 1.04	R.Gibson	(Meols, Wirral)	
	(Built R.Gibson - kit no JAESC 0015)						
G-SCPL	Piper PA-28-140 Cherokee Cruiser	28-7725160	G-BPVL	4. 5.89	Aeros Leasing Ltd	Gloucestershire	23. 9.07T
			N1785H				
G-SCRU	Cameron A-250 HAB	3935	G-BWWO	30. 9.96	Societé Bombard SARL	(Meursanges, Côte-d'Or, France)	24. 9.05A
G-SCTA	Westland Scout AH.1	F9701	XV126	18.12.95	G.R.Harrison	(Guildford)	14. 3.05P
G-SCUB	Piper PA-18-135 Super Cub	18-3847	PH-GAX	13.12.78	Mrs C.L.Needham	Old Manor Farm, Anwick	7. 3.07
	(L-21B-PI) *(Frame No.18-3849)*		R.Neth AF R-157, 54-2447		*(As "54-2447" in US Army c/s)*		
G-SCUD	Montgomerie-Bensen B.8MR	PFA G/101-1294		18. 8.97	D.Taylor	Belper	
	(Built D Taylor)				*(Current status unknown)*		
G-SCUL	Rutan Cozy	PFA 159-13212		28. 5.98	K.R.W.Scull	(Usk)	
	(Built K.R.W.Scull)				*(Current status unknown)*		
G-SCUR	Eurocopter EC120B Colibri	1090		1. 3.00	JS Aviation Ltd	Luton	18. 5.06T
G-SDCI	Bell 206B JetRanger II	925	G-GHCL	24. 2.00	S.D.Coomes	(Auldhouse, East Kilbride)	3. 6.05T
			G-SHVV, N72GM, N83106				
G-SDEV	de Havilland DH.104 Sea Devon C.20	04472	XK895	29. 3.90	Wyndeham Press Group plc	Kemble	17. 9.01
					(As "XK895/CU19" in 771 Sqdn RN c/s)		
G-SDFM	Aerotechnik EV-97 Eurostar	PFA 315-13884		23. 8.02	D.F.Randall	Priory Farm, Tibenham	9.12.05P
	(Built A K.Paterson)						
G-SDLW	Cameron O-105 HAB	2460		11. 3.91	P.J.Smart	(Bath)	4. 6.05A
G-SDOI	Aeroprakt A22 Foxbat	PFA 317-14064		26. 6.03	S.A.Owen	Latch Farm, Kirknewton	5. 8.05P
	(Built S P S Dornan)						
G-SDOZ	Tecnam P92-EA Echo-Super	PFA 318A-14287		9. 9.04	S.P.S.Dornan	Moss Side Farm, Carluke	
	(Built S P S Dornan)						
G-SEAI	Cessna U206G Stationair II	U20604059	N756FQ	20. 3.92	K.O'Connor	Weston, Dublin	16. 9.06T
	(Amphibian)						
G-SEDO	Cameron N-105 HAB	10388		28. 3.03	Flying Pictures Ltd *(Agfa Film titles)*	(Chilbolton)	5. 5.05A
G-SEED	Piper J-3C-90 Cub (L-4H-PI)	11098	EI-BAP	28. 1.80	J.H.Seed	Black Spring Farm, Castle Bytham	7. 8.04P
	(Frame No.10932)		F-BFBZ, 44-80203, 43-29807				
	(Official identity is c/n 12499/44-80203 and probably rebuilt 1945)						
G-SEEK	Cessna T210N Turbo Centurion II	21064579	N9721Y	14.10.83	A Hopper	Little Shelford	21. 3.05
G-SEFI	Robinson R44 Raven II	10147	N75271	2.10.03	Kermann Avionics Sales Ltd	(London NW3)	27.11.06T
G-SEGA	Cameron Sonic 90SS HAB	2896		16. 9.92	A D Kent	Lancing	29. 6.00A
	(Sonic The Hedgehog shape)				tr Balloon Preservation Flying Group *"Sonic" (New CofR 6.04)*		
G-SEJW	Piper PA-28-161 Cherokee Warrior II	28-7816469	N9557N	19. 4.78	Keen Leasing Ltd	Belfast Aldergrove	12. 6.06T
G-SELF	Europa Aviation Europa	PFA 247-12996		10. 8.01	N.D.Crisp, A H.Lames and E.J.Hatcher		
	(Built N.D.Crisp, A H.Lames and E.J.Hatcher - kit no.279) (Jabiru 3300) *(Monowheel u/c)*					(Leigh-on-Sea)	
G-SELL	Robin DR400/180 Régent	1153	D-EEMT	7. 3.85	C.Morris tr G-SELL Régent Group	Bidford	10. 4.06

Reg	Type	C/n	Previous identity	Reg date	Owner/operator	Base	CofA
G-SELY	Agusta-Bell 206B-3 JetRanger III	8740		26. 7.96	CT Rental Ltd	Cumbernauld	15. 9.05T
G-SEMI	Piper PA-44-180 Seminole	44-7995052	G-DENW	23. 2.99	Halfpenny Green Flight Centre Ltd	Wolverhampton	22.12.02T
			N21439		*(New owner 10.04)*		
G-SENA	Rutan LongEz	1325	F-PZSQ	11.11.96	G.Bennett	(Great Yarmouth)	
	(Built R Bazin)		F-WZSQ		*(Current status unknown)*		
G-SEND	Colt 90A HAB	2100		2.12.91	J-P.Barre	(Brezolles, France)	21. 1.05T
G-SENE	Piper PA-34-200T Seneca II	34-8170069	N797WA	20.11.03	R.Clarke	Wolverhampton	26.11.06T
			N8314P				
G-SENX	Piper PA-34-200T Seneca II	34-7870356	G-DARE	15. 5.95	Katotech Ltd	Cardiff	15. 7.07T
			G-WOTS, G-SEVL, N36742				
G-SEPA	Eurocopter AS355N Ecureuil 2	5525	G-METD	25. 7.96	Metropolitan Police Authority	Lippits Hill	4. 8.05T
			G-BUJF, F-WYMF				
G-SEPB	Eurocopter AS355N Ecureuil 2	5574	G-BVSE	1. 2.95	Metropolitan Police Authority	Lippits Hill	1. 3.04T
G-SEPC	Eurocopter AS355N Ecureuil 2	5596	G-BWGV	29.11.95	Metropolitan Police Authority	Lippits Hill	20. 3.05T
G-SEPT	Cameron N-105 HAB	1880		22.11.88	P.Gooch *(Septodont - Dentist's Supplies titles)*	(Alresford)	28. 8.05A
G-SERA	Enstrom F-28A-UK	103	G-BAHU	14. 3.91	W.R.Pitcher	Leatherhead	13. 7.06T
			EI-BDF, G-BAHU		t/a Enstrom Associates		
G-SERL	SOCATA TB-10 Tobago	109	G-LANA	28. 5.92	R.J.Searle	Rochester	19. 4.06
			EI-BIH				
G-SERV	Cameron N-105 HAB	10382		16. 4.03	PSH Skypower Ltd *(Servo Connectors titles)*	(Pewsey)	16. 4.05A
G-SETI	Cameron Sky 80-16 HAB	4853		25. 9.00	R.P.Allan	(Chinnor)	4. 5.05A
G-SEUK*	Cameron TV 80SS HAB				See SECTION 3 Part 1	(Aldershot)	
G-SEVA	Replica Plans SE.5A	PFA 20-10955		19. 6.85	I.D.Gregory	Boscombe Down	2. 1.06P
	(Built I D Gregory) (Continental C90)				*(As "F-141/G" in 141 Sqdn RFC c/s)*		
G-SEVE	Cessna 172N Skyhawk II	17269970	N738GR	10. 1.90	MK Aero Support Ltd	Netherthorpe	13. 2.05T
					(Op Sheffield Aero Club)		
G-SEVN	Van's RV-7	PFA 323-13795		13. 9.01	N.Reddish	Netherthorpe	28. 8.05P
	(Built N Reddish)						
G-SEWP	Aérospatiale AS355F2 Ecureuil 2	5480	G-OFIN	14. 8.00	Veritair Ltd	Cardiff Heliport	14. 6.05T
			G-DANS, G-BTNM				
G-SEXE	Scheibe SF.25C Falke	44396	N716SF	31. 7.03	Repulor Ltd	Sleap	21. 1.07
			(D-KNII)				
G-SEXI	Cessna 172M Skyhawk II	17263806	N1964V	21. 4.92	Willowair Flying Club (1996) Ltd	Not known	5. 9.04T
		(Bounced landing Nayland 2.2.02 and overran: struck hedge and came to rest inverted: badly damaged)					
G-SEXX	Piper PA-28-161 Warrior II	28-7816196	SE-GVD	12.05.03	A M.Blatch Electrical Contractors Ltd	Headcorn	12. 5.06T
G-SFCJ	Cessna 525 CitationJet	525-0245	N33CJ	21. 7.04	Sureflight Aviation Ltd	(Sutton Coldfield)	29. 7.05T
			N5124J				
G-SFHR	Piper PA-23-250 Aztec F	27-8054041	G-BHSO	24. 6.82	Comed Aviation Ltd	Blackpool	22.11.01T
			N2527Z		*(Current status unknown)*		
G-SFLY	Diamond DA.40 Star	40362		31. 3.04	F.Pilkington and L.Turner	Sleap	29. 6.07T
G-SFOX	RotorWay Exec 90	5059	G-BUAH	11.10.93	Magpie Technology Ltd	Crabtree Farm, Crowborough	15.12.04P
	(Built I L Griffith) (RotorWay RI 162)				*(New CofR 9.04)*		
G-SFPA	Reims/Cessna F406 Caravan II	F406-0064		11.11.91	Secretary of State for Scotland/Dept of Agriculture and Fisheries		
					(Op Direct Flight for Fisheries Protection Agency)	Inverness	12. 3.06T
G-SFPB	Reims/Cessna F406 Caravan II	F406-0065		11.11.91	Secretary of State for Scotland/Dept of Agriculture and Fisheries		
					(Op Direct Flight for Fisheries Protection Agency)	Inverness	26. 4.06T
G-SFRY	Thunder Ax7-77 HAB	1667		23. 1.90	M.Rowlands	(Wigan)	15. 5.05A
G-SFSG	Beech E90 King Air	LW-239	N24SM	29. 7.03	Premiair Charter Ltd	Thruxton	20. 8.06T
			N89FN, N6EA				
G-SFSL	Cameron Z-105 HAB	10308		31. 7.02	A M.Holly	(Berkeley)	22. 6.05T
					t/a Exclusive Ballooning *(Somerfield Supermarkets titles)*		
G-SFTA*	Westland SA.341G Gazelle 1				See SECTION 3 Part 1	Usworth, Sunderland	
G-SFTZ	Slingsby T.67M-160 Firefly	2000		7. 2.83	Western Air (Thruxton) Ltd	Thruxton	7. 2.05T
G-SGAS	Colt 77A HAB	2073		31.10.91	A Derbyshire	(Tunstall, Woodseaves, Stafford)	26. 4.01A
					(Shelll Gas titles) (Donated to Balloon Preservation Group 2004)		
G-SGEC	Beech B200 Super King Air	BB-1747	N214FW	19. 8.03	Bridgtown Plant Ltd	Gamston	26. 8.06T
G-SGEN	Comco Ikarus C42 FB80	0407-6611		27. 8.04	Fly Buy Ultralights Ltd	Mill Farm, Shifnal	
G-SGSE	Piper PA-28-181 Archer II	28-7890332	G-BOJX	2.12.96	D.Masson	King's Farm, Thurrock	4. 6.06T
			N3774M				
G-SHAA	Enstrom 280C-UK Shark	1011	N280Q	8. 7.88	C.J.and D.Whitehead t/a ELT Radio Telephones	(Burnley)	9. 4.05T
G-SHAH	Reims/Cessna F152 II	F15201839	OH-IHA	7. 2.97	I.R.Chaplin	Andrewsfield	29. 5.06T
			SE-IHA				
G-SHAN	Robinson R44 Clipper II	10617		28. 1.05	Heli Air Ltd	Wellesbourne Mountford	
G-SHAY	Piper PA-28R-201T Turbo Arrow III	28R-7703365	G-JEFS	17. 9.01	R.Rudderham	Earls Colne	27.10.07E
			G-BFDG, N47381		tr Alpha Yankee		
G-SHCB	Schweizer 269C-1	0038	N41S	28. 6.96	Oxford Aviation Services Ltd	Oxford	27.10.05T
G-SHED	Piper PA-28-181 Cherokee Archer II	28-7890068	G-BRAU	12. 6.89	D.R.Allard	Gloucestershire	8. 7.07
			N47411		tr G-SHED Flying Group		
G-SHEZ	Mainair Pegasus Quik	7993		28.10.03	A Anderson	Carlisle	4.11.05P
G-SHIM	CFM Streak Shadow	PFA 206-12501		19. 5.93	J.A.Weston	Shobdon	11.10.05P
	(Built E G Shimmin - kit no K.228-SA) (Rotax 582) *(C/n duplicates G-MURR)*						
G-SHOG	Colomban MC-15 Cri-Cri	001	G-PFAB	3.10.96	V.S.E.Norman	Rendcomb	24. 6.02P
	(Built G Nappez) (JPX PUL-212)		F-PYPU		*(Mitsubishi Shogun titles)*		
G-SHPP	Hughes 269A (TH-55A)	36-0481	N80559	24. 7.89	Helirouge Ltd	Redhill	10. 4.06T
			64-18169				
G-SHRK	Enstrom 280C-UK Shark	1173	N373SA	6. 1.97	D.R.Kenyon t/a Aviation Bureau	Goodwood	26. 9.02T
			G-SHRK, G-BGMX, EI-CCS, G-SHXX, G-BGMX, EI-BHR, G-BGMX, (F-GBOS) *(Noted 10.04)*				
G-SHRT	Robinson R44 Raven II	10473		14. 9.04	Overby Ltd	(Ascot)	16. 9.07T
G-SHSH	Europa Aviation Europa	PFA 247-12722		7. 4.98	D.G.Hillam	(Birkenhead)	23. 6.05P
	(Built D.G.Hillam - kit no.113) (Monowheel u/c)						
G-SHSP	Cessna 172S Skyhawk SP	172S8079	N6535P	25. 3.99	Shropshire Aero Club Ltd	Sleap	15. 4.05T
			N9552Q				
G-SHUF	Mainair Blade	1241-0200-7 & W1034		10. 3.00	R.G.Bradley	Newton Bank Farm, Daresbury	27. 3.05P
	(Rotax 582)						

Reg	Type	c/n	Previous identity	Reg date	Owner/operator	Base	CofA
G-SHUG	Piper PA-28R-201T Turbo Cherokee Arrow III	28R-7703048	N1026Q	17. 5.88	G-SHUG Ltd	Booker	10. 7.06T
G-SHUU	Enstrom 280C-UK-2 Shark	1221	G-OMCP	16.10.89	D.Ellis	Hawarden	20. 2.05
			G-KENY, G-BJFG, N8617N				
G-SHUV	Aerosport Woody Pusher	PFA 07A-13960		20. 9.02	J.R.Wraight	(Chatham)	
	(Built J.R.Wraight)						
G-SHWK	Cessna 172S Skyhawk	172S9642	N21733	2. 6.04	The Cambridge Aero Club Ltd	Cambridge	8. 7.07T
G-SIAI	SIAI-Marchetti SF.260W	361/31-005	F-GVAB (2)	15. 1.01	D.Gage	Booker	30. 4.05P
			OO-XCP, FAB-184		*(As "FAB-184" in Bolivian AF c/s)*		
G-SIAL	Hawker Hunter F.58	41H-697457	Swiss AF J-4090	2.10.95	Classic Aviation Ltd	Scampton	21. 3.01P
					(Stored as "J4090" 5.04)		
G-SIAM	Cameron V-90 HAB	4096	G-BXBS	7. 3.01	D Tuck *"Warners"*	(Bangkok, Thailand)	25.11.04A
G-SIAX	Cameron Z-210 HAB	10281		4. 5.04	Societé Bombard SARL	(Meursanges, Côte-d'Or, France)	
G-SIGN	Piper PA-39 Twin Comanche C/R	39-8	OY-TOO	9. 2.78	D.Buttle	Blackbushe	10. 3.06
			N8853Y				
G-SIIA	Pitts S-2A	2127	D-ECKC	14.11.01	A P.Crumpholt	(Baldock)	9.12.04
	(Built Aerotek Inc)		N8073				
G-SIIB	Pitts S-2B	5218	G-BUVY	24. 3.93	G.Ferriman	Horsford	30. 4.05
	(Built Aviat Inc) (Lycoming AEIO-540)		N6073U				
G-SIID	Sukhoi Su-26M2	01-04	RA-44531	17.11.03	Gold Air International Ltd	Biggin Hill	18.11.04P
			PK-SDM		*(Honda titles)*		
G-SIIE	Pitts S-2B	5057	G-SKYD	6. 2.04	Technoforce Ltd	Biggin Hill	17. 6.05
	(Built Christen Industries Inc) (Lycoming AEIO-540)		N5331N		*(Honda titles)*		
G-SIII	Extra EA.300	058	D-ETYE	10. 1.95	Fun Flight Ltd	(Berkhamsted)	21.12.07E
G-SIIS	Pitts S-1S	PFA 09-13485	G-RIPE	3. 7.02	I.H.Searson	(Mansfield)	
	(Built J A Harris)						
G-SIJJ	North American P-51D-NA Mustang	122-31894	F-AZMU	20.03.03	P.A Teichman *"Jumpin' Jacques"*	North Weald	4. 7.05P
			N5306M, HK-2812P, HK-2812X, N5411V, 44-72035		*(As "472035" in USAF c/s)*		
G-SIJW	Scottish Aviation Bulldog Series 120/121	BH120/295	XX630	31. 3.00	M.Miles *(As "XX630/5")*	Shenington	2. 9.04
G-SILS	Pietenpol AirCamper	PFA 47-13331		29. 6.98	D.Silsbury	Dunkeswell	
	(Built D.Silsbury)				*(On build 8.02)*		
G-SILY	Mainair Pegasus Quantum 15	8074		17.12.04	L.Harland	Rochester	16.12.05P
G-SIMI	Cameron A-315 HAB	3391		10. 3.95	Balloon School (International) Ltd	(Petworth)	16. 7.05T
					t/a Balloon Safaris		
G-SIMM	Comco Ikarus C42 FB100 VLA	PFA 322-14286		6. 9.04	D.Simmons	(Newmarket)	
	(Built D.Simmons)						
G-SIMN	Robinson R22 Beta II	2769		10.12.97	Heli Air Ltd	Cambridge	18. 1.07T
G-SIMP	Jabiru Aircraft Jabiru SP-470	PFA 274B-13794		4. 1.02	J C Simpson	Wellcross Farm, Slinfold	26. 8.05P
	(Built J C Simpson)						
G-SIMS	Robinson R22 Beta	1596	N7800R	14. 6.04	HS (Holdings) Ltd	Durham Tees Valley	20. 7.07T
			LV-RBZ		t/a Heli-One		
G-SIMY	Piper PA-32-300 Cherokee Six	32-7640082	G-OCPF	22. 3.04	I Simpson	Kirkbride	10. 9.06T
			G-BOCH, N9292K				
G-SIPA	SIPA 903	63	G-BGBM	31. 5.83	A C Leak and J H Dilland	(Southampton)	14. 2.89P
			F-BGBM		*(New owners 3.04)*		
G-SIRS	Cessna 560XL Citation Excel	560-5185	N51042	1. 8.01	Amsair Ltd *(Op London Executive Aviation)*	Stansted	1. 8.05T
G-SITA	Pegasus Quantum 15-912	7797		18. 6.01	A R.Oliver	Dunkeswell	17. 6.05P
G-SIVJ	Westland SA.341C Gazelle HT.2	2012	G-CBSG	26. 6.02	Eastern Atlantic Helicopters Ltd	Shoreham	19. 9.04P
			ZB649				
G-SIVN	MDHelicopters MD.500N	LN089	N9RU	2.11.04	Mandarin Aviation Ltd.	Redhill	AC
			G-SIVN, (HB-ZBS), (XA-), N3234D				
	(Initially regd to Eastern Atlantic Helicopters Ltd, then cancelled 19.11.04 as sold as N9RU, restored on 3.12.04)						
G-SIVR	MD Helicopters MD-900	900-00102	N7002S	20. 8.02	Mandarin Aviation Ltd	Redhill	5. 9.05T
G-SIVW	Lake LA-250 Renegade	233	N8553T	6. 2.03	C.J.Siva-Jothy	Redhill	24. 4.06
	(Built Aerofab Inc)						
G-SIXC	Douglas DC-6A/B	45550	N93459	20. 3.87	Atlantic Air Transport Ltd	Coventry	4. 4.05T
			N90645, B-1006, XW-PFZ, B-1006		*(Atlantic c/s with Air Atlantique titles)*		
G-SIXD	Piper PA-32-300 Cherokee Six D	32-7140007	HB-OMH	25. 3.98	M.B.Payne and I.Gordon	King's Farm, Thurrock	9.10.04
			N8615N				
G-SIXS	Whittaker MW-6S Fat Boy Flyer	PFA 164-12521		27. 8.03	R.H.Braithwaite	RAF Henlow	
	(Built R.H.Braithwaite)				*(Noted 4.04)*		
G-SIXX	Colt 77A HAB	1327		21.10.88	M.Dear and M.Taylor	(Marlow)	30. 3.05A
G-SIXY	Van's RV-6	PFA 181-13368		9. 3.99	C.J.Hall and C.R.P.Hamlett	(Cambridge)	
	(Built C.J.Hall and C.R.P.Hamlett)						
G-SJCH	Pilatus Britten-Norman BN-2T-4S Islander	4006	G-BWPK	18.11.99	Hampshire Police Authority	Lee-on-Solent	26. 2.05T
					"Sir John Charles Hoddinott"		
G-SJDI	Robinson R44 Astro	0626		16. 7.99	HJV Ltd	Redhill	7. 8.05T
G-SJEN	Comco Ikarus C42 FB80	0405-6602		1. 7.04	M.C.Henry	Blackbushe	26. 7.05P
	(Built M.C.Henry)				t/a Charles Henry Services		
G-SJKR	Lindstrand LBL 90A HAB	756		26. 1.01	S.J.Roake *"Smarthouse"*	(Frimley, Camberley)	24. 8.05A
G-SJMC	Boeing 767-31K	27205	N6038E	16. 3.94	MyTravel Airways Ltd	Manchester	15. 3.06T
G-SKAN	Reims/Cessna F172M Skyhawk II	F17201120	G-BFKT	8. 7.85	Bustard Flying Club Ltd	Boscombe Down	6. 4.07T
			F-BVBJ				
G-SKCI	Rutan VariEze	PFA 74-12081		30. 3.01	S.K.Cockburn	(Stanford-le-Hope)	
	(Built S.K.Cockburn)		*(Landed short of runway Biggin Hill 16.2.02 on first flight: substantial damage to nose u/c and a/c nose)*				
G-SKEW	Mudry CAP.232	11	F-GXRB	28.11.03	J.H.Askew	Gamston	27.11.06
			F-GRRG, F-GKCK, ALAT				
G-SKIE	Steen Skybolt	AACA/357	ZK-DEN	29. 8.97	S.Gray	Redhill	12. 9.02P
	(Built D Axe)						
G-SKII	Agusta-Bell 206B-3 JetRanger III	8562	EI-BKT	22. 4.03	C and M Coldstores		
			D-HAFD, HB-XIC			(Carrickmacross, County Monaghan	13. 5.06T
G-SKNT	Pitts S-2A	2048	G-PEAL	31. 3.03	T.G.Lloyd	Little Gransden	21. 2.92T
	(Built Aerotek Inc)		N81LF, N48KA		*(Damaged nr Kidderminster 28.6.91: uncovered airframe complete 2.01)*		
G-SKOT	Cameron V-42 HAB	4813		27. 6.00	S.A Laing	(Banchory)	29. 8.04A

G-SKPG	Best Off Skyranger 912S	BMAA/HB/400		11.11.04	P.Gibbs	(Croxley Green)	
	(Built P.Gibbs - kit no. UK/483)						
G-SKRG	Best Off Skyranger 912	BMAA/HB/298		2. 9.03	R.W.Goddin	Longacres Farm, Sandy	21. 1.05P
	(Built R.W.Goddin - kit no.SKR0307352)						
G-SKUL	Aérospatiale SA.341G Gazelle	1491	F-GFEB	25. 7.03	Flaming Skull Aviation Ltd	Stapleford	3. 9.06T
			N9002L				
G-SKYC	Slingsby T.67M Firefly	2009	G-BLDP	13. 6.97	T.W.Cassells	Bagby	7.10.05T
G-SKYE	Cessna TU206G Turbo Stationair 6 II	U20604568	(G-DROP)	1. 8.79	Sarah M.C.Harvey	Hinton-in-the-Hedges	19. 8.07
			N9783M				
G-SKYF	SOCATA TB-10 Tobago	1589	VH-YHG	1. 5.01	D.P.Boyle	Goodwood	24. 5.07T
G-SKYG	III Sky Arrow 650 TC	C008		15.12.98	R.Jones	Membury	20. 2.05
	(Built Iniziative Industriali Italian)						
G-SKYH*	Cessna 172N Skyhawk 100	17268098	A6-GRM	20. 2.79	Elgor Hire Purchase and Credit Ltd		
			N76034			Abbeyshrule, County Longford	9. 8.91
			(Crashed landing Connaught 21.7.91: cancelled 6.2.92 by CAA) (Wrecked fuselage noted 4.03)				
G-SKYK	Cameron A-275 HAB	4879		31. 7.00	Cameron Flights Southern Ltd	(Pewsey)	28. 6.05T
					(Cameron Balloons titles)		
G-SKYL	Cessna 182S Skylane	18280176	N4104D	19. 6.98	Skylane Aviation Ltd	(Halifax)	18. 8.07
G-SKYN	Aérospatiale AS355F1 Ecureuil 2	5185	G-OGRK	21.11.03	Arena Aviation Ltd *(Op Sky News)*	Redhill	24. 4.06T
			G-BWZC, (G-MOBZ), N107KF, N5799R				
G-SKYO	Slingsby T.67M-200 Firefly	2264		20. 9.00	E.D.Fern	Truro	11. 1.07T
G-SKYR	Cameron A-180 HAB	2826		31. 3.92	Cameron Flights Southern Ltd *"Candy Floss"*	(Pewsey)	29. 4.00T
G-SKYT	III Sky Arrow 650 TC	C004		6. 9.96	W.M.Bell and Susan J.Brooks	Bicester	23. 6.06
	(Built Iniziative Industriali Italian)						
G-SKYU	Cameron A-210 HAB	10129		28. 8.01	Cameron Flights Southern Ltd	(Pewsey)	29. 7.05T
					(Evening Advertiser titles)		
G-SKYV	Piper PA-28RT-201T Turbo Arrow IV	28R-8031132	G-BNZG	20. 9.04	Skyviews Pictures.com Ltd	(Selby)	18. 3.02
			N82376				
G-SKYW	Aérospatiale AS355F1 Ecureuil 2	5261	G-BTIS	17. 1.05	Skywalker Aviation Ltd	Fairoaks	8. 7.07E
			G-TALI				
G-SKYX	Cameron A-210 HAB	4613		22. 6.99	Cameron Flights Southern Ltd	(Pewsey)	14. 4.05T
					(Whitely Village titles)		
G-SKYY	Cameron A-250 HAB	3402		9. 3.95	Cameron Flights Southern Ltd	(Pewsey)	9. 3.01T
					"City of Southampton"		
G-SLCE	Cameron C-80 HAB	4022		24. 2.97	A M.Holly	(Berkeley)	11. 6.04T
G-SLEA	Avions Mudry CAP.10B	124		19.12.80	M J M Jenkins	RAF Wittering	11. 8.06
G-SLII	Cameron O-90 HAB	2388		20. 9.90	R.B. and A M.Harris *"Mad Dash"*	(Huntingdon)	17. 8.03A
G-SLIP	Reality Easy Raider BMW R100	BMAA/HB/215		21. 5.02	J.S.Harris	(Porton)	
	(Built J.S.Harris - kit no.0004)						
G-SLMG	Diamond HK 36 TTC Super Dimona	36.727	N267JP	5. 8.04	P.R.Thody and A.R.Morley	Nympsfield	AC
					tr G-SLMG Syndicate *(New owners 11.04)*		
G-SLOW	Pietenpol AirCamper	PFA 47-13488		8.10.99	C.Newton	(Brackley)	
	(Built C.Newton)				*(Current status unknown)*		
G-SLTN	SOCATA TB-20 Trinidad	763	HB-KBR	6. 8.99	Oceana Air Ltd	Elstree	27. 8.05T
G-SLYN	Piper PA-28-161 Warrior II	28-8116204	N161WA	12. 4.89	Haimoss Ltd	Old Sarum	8. 6.07T
			N8373K				
G-SMAN	Airbus A330-243	261	F-WWKR	26. 3.99	Monarch Airlines Ltd	Luton	25. 3.05T
G-SMBM	Pegasus Quantum 15-912 Super Sport	7602		24. 1.00	M.C.Watson	Deenethorpe	27. 2.05P
G-SMDH	Europa Aviation Europa XS	PFA 247-13367		8.10.98	S.W.Pitt	(Petersfield)	
	(Built S.W.Pitt - kit no.403) (Tri-gear u/c)				*(Current status unknown)*		
G-SMDJ	Eurocopter AS350B2 Ecureuil	3187		21. 4.99	M.Ziani de Ferranti	(Llanfairfechan)	12. 8.05
G-SMIG	Cameron O-65 HAB	922		6. 6.83	R.D.Parry *(San Miguel titles) (Active 1.04)*	(Stroud)	28. 7.87A
G-SMIT*	Messerschmitt Bf.109G-10/U-4				See SECTION 3 Part 1	McMinnville, Oregon, US	
G-SMJJ	Cessna 414A Chancellor II	414A0425	N2694H	24. 3.81	Gull Air Ltd	Guernsey	31. 5.06
G-SMTC	Colt Flying Hut SS HAB	1828		7. 1.91	Shiplake Investments Ltd	(St. Peter Port, Guernsey)	18.11.00A
G-SMTH	Piper PA-28-140 Cherokee C	28-26916	G-AYJS	28. 9.90	Feedair Express Ltd	Southend	13. 2.05
			N11C		t/a Aerosen		
G-SMTJ	Airbus A321-211	1972	D-AVXG	15. 5.03	MyTravel Airways Ltd	Manchester	14. 5.06T
G-SNAK	Lindstrand LBL 105A HAB	404		23. 9.96	Ballooning Adventures Ltd	(Hexham)	26.12.04T
G-SNAP	Cameron V-77 HAB	1217		29.11.85	C.J.S.Limon *"Snapshot"*	(Great Missenden)	19. 9.04A
G-SNEV	CFM Streak Shadow SA	PFA 206-13042		17. 9.96	G.Gilhead and R.G.M-J Proostt	Old Sarum	1. 4.04P
	(Built N.G.Smart - kit no K.283) (Rotax 582) *(Official c/n duplicates G-BZNW qv)*				tr Aviation for Paraplegics and Tetraplegics Trust		
G-SNOG	Air Création 582/Kiss 400	BMAA/HB/219		2. 5.02	P.S.Wesley	(Slough)	11. 3.05P
	(Built B.H.Ashman - kit no.FL011, Trike T02033, Wing A02048-2045)						
G-SNOW (2)	Cameron V-77 HAB	2050		21. 6.79	M.J.Ball	(Clitheroe)	10. 7.05A
	(Officially regd with c/n 541 but ftted with replacement envelope 1989)						
G-SNOZ	Europa Aviation Europa	PFA 247-12545	G-DONZ	7.10.04	M.P.Wiseman	(Brough)	
	(Built D.J.Smith and D.McNicholl - kit no.032)						
G-SNUZ	Piper PA-28-161 Warrior II	28-8416021	G-PSFT	19.12.01	J.C.O.and C.A Adams	Biggin Hill	10.12.06T
			G-BPDS, N4328P				
G-SOBI	Piper PA-28-181 Archer II	28-7690212	D-EAQL	3. 5.00	Northern Aviation Ltd	Durham Tees Valley	8. 8.06T
G-SOCK	Mainair Pegasus Quik	8041		25. 5.04	P.W.Lupton	Arclid Green, Sandbach	27. 5.05P
G-SOCT	Yakovlev Yak-50	842804	LY-XCD	17. 3.04	C R Turton	Headcorn	21. 3.05P
			DOSAAF 32		*(Also carries "CT304/RB-A" in pseudo-RAF scheme)*		
G-SOEI	Hawker Siddeley HS.748 Series 2A/242	1689	ZK-DES	25. 2.98	Emerald Airways Ltd *(Mount Cook Airlines c/s)*	Liverpool	17. 4.07T
G-SOFT	Thunder Ax7-77 HAB	1339		5.12.88	A J.Bowen *"Enterprise Software"*	(Edinburgh)	11. 9.99A
G-SOHO	Diamond DA.40D Star	D4.079		12. 2.04	Soho Aviation Ltd	Biggin Hill	29. 3.07T
G-SOKO	Soko P-2 Kraguj	033	G-BRXK	6. 1.94	A G.and G.A G.Dixon	Bournemouth	21. 5.05P
			Yugoslav Army 30149				
G-SOLA	Star-Lite SL-1	PFA 175-11311		9. 6.88	J.P.Roberts-Lethaby	Belle Vue Farm, Yarnscombe	31. 3.93P
	(Built P Clifton and A Clarke - c/n 203TG) (Rotax 447)				*"A Star Is Born" (Noted 4.04)*		
G-SOLH	Bell 47G-5	2639	G-AZMB	5. 3.97	Sol Helicopters Ltd	Elstree	16. 4.06T
			CF-NJW				
G-SOLO	Pitts S-2S	AA/1/1980		30. 5.80	H Staltmeir	(Altenau, Germany)	6. 4.96P
	(Built Anvil-Aviation Ltd) (Lycoming AEIO-540)				*(New owner 2.04)*		

Registration	Type	C/n	Previous identity	Date	Owner	Base	Expiry
G-SONA	SOCATA TB-10 Tobago	151	G-BIBI	24.10.80	M.Kelly	Sherburn-in-Elmet	25. 9.05
G-SOOC	Hughes 369HS	111-0354S	G-BRRX	6.10.93	R.J.H.Strong	(Yeovil)	23. 2.06
			N9083F				
G-SOOE	Hughes 369E	0227E		27. 4.87	R.W.Nash	Rochester	26. 5.05
G-SOOM	Glaser-Dirks DG-500M	5E42M20	JZF, BGA4907	14. 5.92	G.W.Kirton	Saltby	5.10.06
			G-SOOM				
G-SOOS	Colt 21A Cloudhopper HAB	1263		7. 6.88	P.J.Stapley *(Current status unknown)*	(Redcar)	25. 3.95A
G-SOOT	Piper PA-28-180 Cherokee C	28-4033	G-AVNM	19. 8.88	A G.Branch	Exeter	16. 9.07T
			N11C				
G-SOOZ	Rans S-6-ESN Coyote II	PFA 204-13543		27. 4.01	A Batters	(Ilkley)	9. 8.05P
	(Built A Batters - kit no.0899.1335) (Rotax 582) *(Tri-cycle u/c)*						
G-SOPH	Best Off Skyranger 912	BMAA/HB/259		25. 2.03	N.A Read	Nunnington, York	1. 9.05P
	(Built N A Read - kit no.SKR0212286)						
G-SOPP	Enstrom 280FX	2024	G-OSAB	23.10.97	F.P., M.Sopp and L.A Moore	Jefferies Farm, Billingshurst	5.11.04T
			N86259				
G-SORT	Cameron N-90 HAB	2878		13. 7.92	A Brown *"Streamline"*	(Bristol)	8. 7.04A
G-SOUL	Cessna 310R II	310R0140	N5020J	27. 6.88	Atlantic Air Transport Ltd	Coventry	27. 6.07T
G-SPAL	Robinson R44 Raven II	10319		14. 4.04	Productive Investments Ltd	Guildford	10. 5.07T
G-SPAM	Avid Aerobat	PFA 189-12074		9. 5.91	J.Lee	Full Sutton	19. 7.02P
	(Built C M Hicks - c/n 829)						
G-SPAT	Aero AT-3 R100	AT3.008	SP-EAR	20.10.04	S2T Aero Ltd	North Weald	25.10.07E
			(SP-ERM)				
G-SPDR	de Havilland DH.115 Sea Vampire T.35	15641	VH-RAN	19. 5.00	M.J.Cobb	(East Grinstead)	
			Royal Australian Navy N6-766, XG766			*(Noted as "N6-766" 10.03)*	
G-SPEE	Robinson R22 Beta	0939	G-BPJC	20. 7.94	Verve Systems Ltd	Shobdon	22. 12.06
G-SPEL	Sky 220-24 HAB	045		26. 7.96	T.G.Church	(Blackburn)	29. 3.05T
					t/a Pendle Balloon Co *(Pendle titles)*		
G-SPEY	Agusta-Bell 206B-3 JetRanger III	8608	G-BIGO	1. 4.81	G.Elliott	(Luton)	15. 9.05T
G-SPFX	Rutan Cozy	PFA 159-13113		30. 4.97	B.D.Tutty	(Gillingham, Kent)	
	(Built B.D.Tutty)				*(Current status unknown)*		
G-SPHU	Eurocopter EC135T2	0245	D-HKBA	12.11.02	Bond Air Services Ltd	Glasgow City Heliport	27. 2.06T
					(Op Strathclyde Police)		
G-SPIN	Pitts S-2A	2110	N5CQ	13. 3.80	N.M.R.Richards	(London W1)	22. 4.02T
	(Built Aerotek Inc)						
G-SPIT	Vickers Supermarine 379 Spitfire FR.XIVe	6S/649205	(G-BGHB)	2. 3.79	Patina Ltd	Duxford	18. 6.05P
			Indian AF T-20, MV293		*(Op The Fighter Collection as "MV268/JE-J")*		
G-SPOG	San Jodel DR.1050 Ambassadeur	155	G-AXVS	25. 9.95	A C.Frost	(Ware)	13. 6.77S
			F-BJNL		*(Damaged Stonacre Farm, Bredhurst 17.2.91: on rebuild 1995)*		
G-SPOR	Beech B200 Super King Air	BB-1557	N57TL	3. 9.99	Select Plant Hire Co Ltd	Southend	19. 9.05T
			N57TS				
G-SPUR	Cessna 550 Citation II	550-0714	N593EM	27.10.98	Banecorp Ltd	Stansted	15.11.07E
			N12035		*(Op London Executive Aviation)*		
G-SPYI	Bell 206B-3 JetRanger III	3689	G-BVRC	9. 5.96	Heli-bott Ltd	Blackpool	18. 7.05T
			G-BSJC, N3175S				
G-SRII	Reality Easy Raider 503	BMAA/HB/163		2. 3.01	R.J.Creasey	Maypole Farm, Chislet	12.11.04P
	(Built T F Francis - kit no.10: originally regd as Sky Raider II 503 until 8.01)				tr Sierra Romeo India India Group		
G-SROE	Westland Scout AH.1	F9508	XP907	26.10.95	Bolenda Engineering Ltd *(As "XP907")*	Ipswich	31.10.01P
G-SRVO	Cameron N-90 HAB	3551		10. 4.95	Servo and Electronic Sales Ltd *(Servo titles)*	(Lydd)	29. 7.04A
G-SRWN	Piper PA-28-161 Warrior II	28-8116284	G-MAND	30. 7.02	S.Smith	(Alton)	3.12.01T
			G-BRKT, N8082Z				
G-SRYY	Europa Aviation Europa XS	PFA 247-13806		19. 9.02	S.R.Young	(Seaton)	
	(Built S.R.Young - kit no.530) (Tri-gear u/c)						
G-SSAS	Airbus A320-231	0230	C-GTDO	4. 4.02	MyTravel Airways Ltd	Birmingham	3. 4.05T
			D-AFRO, G-BYFS, D-AFRO, A4O-MA, N230RX, SX-BSJ, N230RX, F-WWDI *(Op MyTravel Lite)*				
G-SSCL	MD Helicopters Hughes 369E	0491E	N684F	25. 4.98	Shaun Stevens Contractors Ltd	Rochester	9. 4.07
G-SSEA	Aérospatiale-Alenia ATR42-300	196	OY-CIT	29. 8.03	Air Wales Ltd	Swansea	7.10.06T
			C-FZVZ, C-GITI, F-WWEK				
G-SSIX	Rans S-6-116 Coyote II	PFA 204A-12749		5. 9.94	R.I.Kelly	Wellesbourne Mountford	14. 7.05P
	(Built J V Squires) (Rotax 582) *(Tailwheel u/c)*						
G-SSJP	Robinson R44 Clipper II	10574		7. 1.05	P.Cripps	(Compton Basset, Calne)	23. 1.08E
G-SSKY	Pilatus Britten-Norman BN-2B-26 Islander	2247	G-BSWT	11. 5.92	Isles of Scilly Skybus Ltd	St.Just	31. 3.06T
G-SSLF	Lindstrand LBL 210A HAB	649		29. 2.00	A M.Holly	(Berkeley)	11. 8.05T
					t/a Exclusive Ballooning *(Somerfield titles)*		
G-SSPP	Sky Science Powerhawk L70/500	SS001		18. 7.00	Sky Science Powered Parachutes Ltd	(Tidworth)	
G-SSSC	Sikorsky S-76C	760408		26.10.93	CHC Scotia Ltd	Aberdeen	13. 1.07T
G-SSSD	Sikorsky S-76C	760415		26.10.93	CHC Scotia Ltd	Aberdeen	22.12.05T
G-SSSE	Sikorsky S-76C	760417		23.11.93	CHC Scotia Ltd	Aberdeen	2. 2.06T
G-SSTI	Cameron N-105 HAB	3238		30. 3.94	British Airways plc *"Concorde"*	(West Drayton)	10. 1.05T
G-SSWA	Short SD.3-30 Variant 100	SH.3042	D-CTAG	15.10.99	Emerald Airways Ltd	Exeter	14.12.06T
			G-BHHU, OY-MUC, G-BHHU, N181AP, N332MV, G-BHHU, G-14-3042				
G-SSWB	Short SD.3-60 Variant 100	SH.3690	C6-BFT	17. 8.00	Emerald Airways Ltd *(Op Streamline Aviation)*	Exeter	25. 9.05T
			N690PC, G-BMLE, G-14-3690				
G-SSWC	Short SD.3-60 Variant 100	SH.3686	SE-LGE	2.11.00	Emerald Airways Ltd	Exeter	14.11.05E
			G-BMHX, G-14-3686				
G-SSWE	Short SD.3-60 Variant 100	SH.3705	SE-IXE	15. 7.02	Emerald Airways Ltd *"Laura"*	Exeter	22. 8.05T
			(G-BNBA), G-14-3705				
G-SSWM	Short SD.3-60 Variant 100	SH.3648	SE-KCI	28. 9.01	Emerald Airways Ltd *(No titles)*	Coventry	14.10.05E
			G-OAAS, OY-MMB, G-BLIL, G-14-3648				
G-SSWO	Short SD.3-60 Variant 100	SH.3609	SE-KLO	8.10.01	Emerald Airways Ltd	Exeter	4.12.04T
			N343MV, (G-BKMY), G-14-3609				
G-SSWR	Short SD.3-60 Variant 100	SH.3670	SE-KGV	2.10.01	Emerald Airways Ltd	Exeter	13.11.05E
			HR-IAQ, N108PS, B-3603, G-BLWJ, G-14-3670				
G-SSWV	Sportavia-Pützer RF5B Sperber	51032	N55WV	31. 5.90	N.Fisher and D.Athey	Camphill	14.12.05P
			D-KEAI				

Registration	Type	C/n	Previous identity	Reg date	Owner/Operator	Base	CofA expiry
G-SSXX	Eurocopter EC135T2	0270	G-SSSX	31. 3.03	Bond Air Services Ltd	Boreham	27. 5.06T
					(Op Essex Air Ambulance)		
G-STAF	Van's RV-7A	PFA 323-13875		15. 2.02	A F.Stafford	(Melbourne, Derbyshire)	
	(Built A F Stafford)						
G-STAT	Cessna U206F Stationair II	U20603485	A6-MAM	20. 2.79	K.Brady	Grindale	17.10.05
			N8732Q		tr Scottish Parachute Club *(Force landed 8.03 and wrecked 2004)*		
G-STAY	Reims/Cessna FR172K Hawk XP II	FR17200620	D-EOVX	15.12.00	G.A.Owston	(Haverfordwest)	4. 3.07
			OE-DVX				
G-STCH	Fies<u>e</u>ler Fi 156A-1 Storch	2088	Luftwaffe 2088	21. 5.03	G.R.Lacey	(Fairoaks)	
G-STEA	Piper PA-28R-200 Cherokee Arrow II	28R-7235096	HB-OIH	18. 6.02	D.J.Brown	Redhill	26. 6.05
			N4569T				
G-STEM	Stemme S 10-V	14-027	D-KSTE	2. 7.97	J Abbess tr G-STEM Group	Tibenham	11. 2.07
G-STEN	Stemme S 10	10-32	D-KGCH	9. 1.92	J.P.Lyell tr G-STEN Syndicate *"4"*	Lasham	24. 5.07
G-STEP	Schweizer 269C	S.1494		1.10.90	M.Johnson	Neath	8.12.06T
G-STER	Bell 206B-3 JetRanger III	4116	OO-EGA	23. 3.94	Maintopic Ltd	Sherburn-in-Elmet	23. 6.06T
G-STEV	CEA Jodel DR.221 Dauphin	61	F-BOZD	9. 3.82	S.W.Talbot	Long Marston	17. 4.05
G-STIG	Focke Wulf FW44J Stieglitz	183	OO-JKT (2)	16. 2.04	G.R.Lacey	Fairoaks	
			D-EHDH (2), LV-YYX				
G-STMP	SAN Stampe SV-4A	241	F-BCKB	11. 3.83	A C.Thorne	(Yelverton)	
					(On overhaul Ivybridge 5.93: current status unknown)		
G-STOK	Cameron Colt 77B HAB	4791		4. 5.00	M.H.Read and J.E.Wetters	(Timperley, Altrincham)	11. 4.05A
					(Christows titles) "Hollybush"		
G-STOO	Stolp SA.300 Starduster Too	PFA 35-13870		30. 1.03	K.F.Crumplin	Franklyn's Field, Chewton Mendip	
	(Built K F Crumplin)						
G-STOR	Fieseler Fi 156D-0 Storch	110451	Luftwaffe 110451	21. 5.03	G.R.Lacey	(Fairoaks)	
G-STOT	Robinson R44 Raven II	10212		20.11.03	Howard Stott Demolition Ltd	Rossendale	21.12.06T
G-STOW	Cameron Wine Box 90 SS HAB	4420		2.10.98	I.Martin and D.Groombridge	(Bristol)	21. 7.04A
					t/a Flying Enterprises Partnership *(Stowells of Chelsea titles)*		
G-STPI	Cameron A-250 HAB	4102		26. 2.97	A D.Pinner *(Central Auto Supplies titles)*	(Northampton)	13. 7.03T
G-STRA	Boeing 737-3S3	24059	G-OBWY	25. 3.02	Astraeus Ltd	Gatwick	9. 5.06T
			N202KG, G-DEBZ, RP-C4006, EC-FGG, EC-711, G-BNPB, C-FGHT, G-BNPB				
G-STRB	Boeing 737-3Y0	24255	G-OBWX	27. 3.02	Astraeus Ltd	Kuala Lumpur Intl, Malaysia	13. 6.06T
			SE-DUS, HB-IID, EI-CFQ, OO-IID, XA-RJP, G-MONL			*(Op AirAsia)*	
G-STRE	Boeing 737-36N	28572	G-XBHX	27. 3.03	Astraeus Ltd	Gatwick	5. 9.06T
G-STRF	Boeing 737-76N	29885	EI-CXD	1. 4.04	Astraeus Ltd	Gatwick	31. 3.07T
G-STRG	Cyclone AX2000	7837		24. 7.01	D.Young	Kemble	29. 9.05P
	(HKS 700E V3)				tr Pegasus Flight Training (Cotswolds)		
G-STRH	Boeing 737-36N	32737	EI-CXE	17. 5.04	Astraeus Ltd	Gatwick	16. 5.07T
G-STRK	CFM Streak Shadow SA	PFA 161-11762		4. 4.90	E.J.Hadley	(Arch, Switzerland)	21. 4.04P
	(Built M E Dodd - c/n K.143-SA) (Rotax 582) *(PFA c/n should be "PFA161<u>A</u>")*						
G-STRL	Eurocopter AS355N Ecureuil 2	5733		14.12.04	McAlpine Helicopters Ltd	Oxford	
G-STRM	Cameron N-90 HAB	3568		3. 7.95	B.G.Jones t/a High Profile Balloons	(Devizes)	19. 7.02T
G-STUA	Pitts S-2A	2164	N13GT	6. 3.91	A.C.Cassidy	White Waltham	21. 3.06T
	(Built Aerotek Inc)				tr Rollquick Group		
G-STUB	Pitts S-2B	5163	N260Y	5. 5.94	P.T.Borchert	Old Sarum	13. 8.06T
	(Built Christen Industries Inc) (Lycoming AEIO-540)						
G-STUK	Junkers Ju 87/R4	6234	Luftwaffe 6234	19.12.03	G.R.Lacey	(Fairoaks)	
G-STUY	Robinson R44 Raven II	10508		8.10.04	S.Mayers	(Worplesdon, Guildford)	16.11.07E
G-STWO	ARV Aviation ARV-1 Super 2	002 & PFA 152-11048		24. 4.85	G.E.Morris	Bolt Head, Salcombe	29. 6.05P
G-STYL	Pitts S-1S	GJSN-1P	N665JG	26. 1.88	P.D.Albrow	Rochester	16.10.04P
	(Built G Harben and G Smith) (Lycoming-O-320)						
G-STYX	Pegasus Quik	7932		28. 1.03	T.A E.M.Stewart tr G-STYX Flying Group	Rochester	29. 1.05P
				(Crashed into cliff Eastchurch, Isle of Sheppey 21.8.04 and badly damaged)			
G-SUCH	Cameron N-77 HAB	676	G-BIGD	3. 9.01	D.G.Such	(Redditch)	4. 1.84A
G-SUCK	Cameron Z-105 HAB	10280		16. 5.02	Airship and Balloon Company Ltd	(Bristol)	26. 5.05A
G-SUED	Thunder Ax8-90 HAB	1546	G-PINE	22.10.02	E.C.Lubbock and S.A Kidd	(Billericay)	9. 3.05A
G-SUES*	North American AT-6D Harvard III				See SECTION 3 Part 1	Torp, Sandefjord, Norway	
G-SUEW	Airbus A320-214	1961	F-WWIX	3. 4.03	MyTravel Airways Ltd	Manchester	2. 4.06T
G-SUEY	Bell 206L-1 Long Ranger	45612	C-GCET	5. 3.04	Aerospeed Ltd	Southampton	15.6.07T
			N300CS, N3901Q				
G-SUEZ	Agusta-Bell 206B JetRanger II	8319	SU-YAE	16. 9.98	Aerospeed Ltd	Manston	3. 7.05T
			YU-HAZ				
G-SUFF	Eurocopter EC135T1	0118		1. 2.00	Suffolk Constabulary Air Support Unit	Beccles	23. 8.06T
G-SUKI	Piper PA-38-112 Tomahawk	38-79A0260	G-BPNV	22. 5.91	Western Air (Thruxton) Ltd	Thruxton	5. 6.05T
			N2313D				
G-SUMT	Robinson R22 Beta	2147	G-BUKD	24. 9.92	Aero Maintenance Ltd	Walton Wood	23. 9.04T
			N23381				
G-SUMX	Robinson R22 Beta II	3274		1.11.01	Frankham Brothers Ltd	(Bruntingthorpe)	16.11.07
G-SUMZ	Robinson R44 Raven II	10490		11.11.04	Frankham Brothers Ltd	(Bruntingthorpe)	11.11.07E
G-SUNN	Robinson R44 Clipper	1367	N7531L	16. 7.04	Helicentre Ltd	(Palma, Majorca)	15. 8.07T
G-SUPA	Piper PA-18-150 Super Cub	18-5395	PH-BAJ	13.12.78	D.Sutton tr G-SUPA Owners Group	Headcorn	2.12.04
	(Frame No.18-5512)		PH-MBF, ALAT 18-5395, N10F				
G-SURG	Piper PA-30 Twin Comanche B	30-1424	G-VIST	18. 6.90	A R.Taylor	Turweston	26. 3.05T
			G-AVHZ, N8287Y				
G-SURY	Eurocopter EC135T2	0283		17. 6.03	Surrey Police Authority	Fairoaks	12. 2.07T
G-SUSE	Europa Aviation Europa XS	PFA 247-13905		25. 6.02	P.R.Tunney	(Sale)	
	(Built P.R.Tunney - kit no.554) (BMW1100RS)						
G-SUSI	Cameron V-77 HAB	1133		22. 7.85	J.H.Dyden *"Susi"*	(Okehampton)	9. 8.04A
G-SUSX	MD Helicopters MD 900	900-00065	N3065W	19. 1.00	Sussex Police Authority	Shoreham	18. 2.07T
G-SUTN	III Sky Arrow 650 TC	C007		27. 8.98	G.C.Sutton and Margaret A.Coltman	Headcorn	4.11.04
	(Built Iniziative Industriali Italian)						
G-SUZN	Piper PA-28-161 Warrior II	28-8016187	N3573C	16. 1.91	E.Reed	Durham Tees Valley	16. 4.06T
			N9540N		t/a St.George Flight Training		

G-SUZY	Taylor JT.1 Monoplane	PFA 55-10395		1.12.78	N.Gregson	Ashcroft Farm, Winsford	27. 1.04P
	(Built S A Kaniok)						
G-SVEA	Piper PA-28-161 Warrior II	28-7916082	N30299	16.12.98	Emma-Claire V.Dunning	Coventry	9.12.06E
G-SVET	Yakovlev Yak-50	822210	RA-44459	16. 9.03	Yak-52 Ltd	Compton Abbas	23.11.05P
			LY-AGG, DOSAAF 107		*"Svetlana"*		
G-SWIF*	Vickers Supermarine 552 Swift F.7				See SECTION 3 Part 1	Southampton	
G-SVIP	Cessna 421B Golden Eagle II	421B0820	G-BNYJ	12. 3.97	Tracey Stone-Brown	(Henfield)	26.12.03T
			N4686Q, D-IMVB, N1590G		*(New owner 1.05)*		
G-SVIV	SNCAN Stampe SV-4C	475	N65214	7. 8.90	R.Taylor	Vendee Air Park, France	12. 6.05
	(DH Gipsy Major 10)		F-BDBL				
G-SWAT	Robinson R44 Raven II	10041	N75097	24. 2.03	J.S.Corr	Newtownards	20. 3.06T
G-SWEB	Cameron N-90 HAB	2413		1.10.90	South Western Electricity plc *"SWEB"*	(Bristol)	1. 8.01T
G-SWEE	Beech 95-B55 Baron	TC-1406	G-AZDK	16. 4.03	Mirage Aircraft Leasing Ltd	Poplar Hall Farm, Elmsett	12. 5.07
G-SWEL	Hughes 369HS	61-0328S	G-RBUT	18. 7.96	M.A Crook and A E.Wright	Barton	28. 5.06
			C-FTXZ, CF-TXZ				
G-SWIF*	Vickers Supermarine 552 Swift F.7				See SECTION 3 Part 1	Southampton	
G-SWOT	Phoenix Currie Super Wot	PFA 3011		10. 9.80	P.M.Flint	Sibson	30. 9.03P
	(Built G Chittenden) (Continental O-200-A)				*(As "C3011/S" in SE.5A guise)*		
G-SWPR	Cameron N-56 HAB	829		16. 3.82	A Brown *"Post Code"*	(Bristol)	5. 7.95A
G-SWUN	Pitts S-1M	338-H	G-BSXH	18. 4.95	J.E.Rands	(Sudbrooke, Lincoln)	8. 6.05P
	(Built R Merrick) (Lycoming O-320)		N14RM				
G-SWWM	Westland SA.341C Gazelle HT.2	1033	XW853	6. 5.03	M.S.Beaton	Babcary, Somerset	28. 1.05P
G-SXVI*	Vickers Supermarine 361 Spitfire LF.XVIe				See SECTION 3 Part 1	McMinnville, Oregon, US	
G-SYCO	Europa Aviation Europa	PFA 247-12540		27.11.95	G.W.Huggins	(Rayne Hall Farm, Braintree)	24. 6.04P
	(Built J W E de Fraysinnet - kit no.031) (NSI EA-81/118) *(Conventional u/c)*				tr G-SYCO Syndicate		
G-SYDD	Piper PA-28-181 Archer III	2843593	N5365M	21. 6.04	Sherborne Aviation Ltd	Bournemouth	24. 6.07T
G-SYFW	WAR Focke-Wulf 190 rep	PFA 81-10584		28. 2.83	R.P. Cross	(Lincoln)	10.11.05P
	(Built M R Parr - kit no.269) (Continental O-200-A)				*(As "W/no.7334/2+1" in Luftwaffe c/s)*		
G-SYPA	Aérospatiale AS355F2 Ecureuil 2	5193	LV-WHC	25. 9.96	Veritair Ltd	(Sherborne)	2. 4.06T
			F-WYMS, G-BPRE, N366E		t/a British International		
G-SYPS	MD Helicopters MD 900	900-00104	G-76-104	3. 7.03	South Yorkshire Police Authority	Sheffield City	3. 7.07T
			N7034X				
G-SYTN	Robinson R44 Raven	1156	N70575	11. 2.02	Silverstar Components Ltd	Swansea	3. 3.05T

G-TAAA - G-TZZZ

G-TAAL	Cessna 172R Skyhawk	17280733	N9535G	11. 8.99	Standard Aviation Ltd	Newcastle	18. 9.05T
G-TABS	Embraer EMB-110P1 Bandeirante	110.212	G-PBAC	18. 8.98	Skydrift Ltd *(Op Keenair)*	Liverpool	26.10.05T
			F-GCLA, F-OGME, F-GCLA, PT-GME				
G-TACK	Grob G109B	6279		30. 5.84	A P.Mayne	Exeter	22. 5.05
G-TADC	Aeroprakt A22 Foxbat	PFA 317-13883		16. 4.02	R.J.Sharp	Rochester	26. 4.05P
	(Built R J Sharp)						
G-TAFC	Maule M.7-235C Super Rocket	23062C	N210SA	11. 1.05	The Amphibious Flying Club Ltd	Husbands Bosworth	
			N9164M, C-GFVX				
G-TAFF	CASA 1-131E Jungmann	1129	G-BFNE	7. 9.84	A J.E.Smith	Breighton	22. 7.05P
			Spanish AF E3B-148				
G-TAFI	Bücker Bü.133C Jungmeister	24	N2210	27. 1.93	R.J.Lamplough	Manor Farm, East Garston	5. 7.01P
	(Built Dornier-Werke AG)		HB-MIF, Swiss AF U-77		*(Current status unknown)*		
G-TAGG	Eurocopter EC135T2	0341		3. 9.04	Taggart Homes Ltd	(Londonderry)	1.12.07E
G-TAGR	Europa Aviation Europa XS	PFA 247-13061		23. 8.02	A G.Rackstraw	Tollerton	
	(Built A G.Rackstraw - kit no.317)				*(Noted 6.04)*		
G-TAGS	Piper PA-28-161 Warrior II	28-8416026	N4329D	6. 5.88	Oxford Aviation Services Ltd	Oxford	12.10.06T
G-TAIL	Cessna 150J	15070152	N60220	21. 4.89	L.I.D.Denham-Brown	Blackpool	25. 6.06T
G-TAIR	Piper PA-34-200T Seneca II	34-7970055	N3059H	17.11.87	Nigel Kenny Aviation Ltd	Barton	16. 6.06T
G-TAIT	Cessna 172R Skyhawk	17280781	G-DREY	23. 7.02	Centenary Flying Group Ltd	Cork, County Cork	2.12.05T
			N23726				
G-TAJF	Lindstrand LBL 77A HAB	905		28. 4.03	T.A J.Fowles	(Chester)	12. 6.05A
G-TAME	Schweizer 269D	0035A	N2119S	10. 1.05	Total Air Management Services Ltd	Sheffield City	24. 1.08E
G-TAMR	Cessna 172S Skyhawk SP	172S8480	N2458J	7. 6.00	C.Durbidge t/a Tamair Leasing	Oxford	29. 7.06T
G-TAMS	Beech A23-24 Musketeer Super	MA-190	OY-DKF	30. 6.00	Aerograde Ltd	Old Buckenham	23.11.06T
G-TAMY	Cessna 421B Golden Eagle	421B0512	SE-FNS	14.11.77	Charniere Ltd	Biggin Hill	17.11.05E
			N2BH, N69865				
G-TAND	Robinson R44 Astro	0478		12. 6.98	Global Air Charter Ltd	(Ascot)	14. 7.07T
G-TANI	Gulfstream GA-7 Cougar	GA7-0107	G-VJAI	18. 5.95	S.Spier	Cranfield	6. 2.05T
			G-OCAB, G-BICF, N8500H, N29707				
G-TANJ	Raj Hamsa X'Air 582	BMAA/HB/171		21. 6.01	R.Thorman	Perth	8.10.05P
	(Built R Thorman - kit no.629)						
G-TANK	Cameron N-90 HAB	3625		20. 6.95	Hoyers (UK) Ltd *(DFDS/Hoyer titles)*	(Huddersfield)	14. 6.03A
G-TANS	SOCATA TB-20 Trinidad	1870	F-GRBX	25. 9.98	K.P.Threlfall	Wolverhampton	1.12.07E
G-TANY	EAA Acrosport 2	PFA 072A-13821		19. 2.04	P.J.Tanulak	Sleap	29.11.05P
	(Built P.J.Tanulak)						
G-TAPE	Piper PA-23-250 Aztec D	27-4054	G-AWVW	7.10.83	D.J.Hare	Fairoaks	5. 4.06T
			OY-RPF, G-AWVW, N6799Y, N9654N *(Op Merlix Air)*				
G-TAPS	Piper PA-28RT-201T Turbo Arrow IV	28R-8131080	HB-PLV	2. 6.04	P.G.Doble	Fairoaks	23. 6.07
			N83423				
G-TARN	Pietenpol AirCamper	PFA 47-13349		3. 8.98	P.J.Heilbron	(Guildford)	
	(Built P.J.Heilbron)				*(Current status unknown)*		
G-TART	Piper PA-28-236 Dakota	28-7911261	N2945C	18.12.90	Prescot Planes Ltd	Goodwood	18. 6.06T
G-TARV	ARV Aviation ARV-1 Super 2	PFA 152-12627		1. 6.94	M.F.Filer	Dunkeswell	19. 7.05P
	(Built B Childs and M.F.Filer from ARV fuselage c/n 008 and remnants of G-OARV c/n 001 which crashed 1986)						
G-TASH	Cessna 172N	17270531	PH-KOS	4.11.98	A Ashpitel	Popham	30.11.07E
			N739GL				

G-TASK	Cessna 404 Titan II	404-0829	PH-MPC	10. 3.93	Bravo Aviation Ltd	Coventry	8. 7.06T
			SE-IHL, N6806Q		(*COASTGUARD titles, red/white c/s with MCA logo on tail*)		
G-TASS	Schweizer 269C	S.1600	PH-HPL	14. 8.02	A Tasker	Leeds-Bradford	10. 9.05T
			N69A				
	(Engine failed and force landed Bowscale Tarn, Cumbria 10.5.04, tail rotor struck ground, rolled onto side and destroyed)						
G-TATS	Aérospatiale AS350BA Ecureuil	1905	F-GHSN	14. 5.01	Milford Aviation Services Ltd	(London SW1E)	7. 6.07T
			N37AW				
G-TATT	Gardan GY-20 Minicab	PFA 56-10347		30.11.78	P.W.Tattershall	(Clitheroe)	
	(Built L Tattershall)				tr Tatt's Group *(Current status unknown)*		
G-TATY	Robinson R44 Astro	0627		27. 7.99	W.R.Walker	Denham	22. 8.05T
G-TAWE	Aérospatiale-Alenia ATR42-300	371	G-BVJP	17. 3.03	Air Wales Ltd	Swansea	6. 3.06T
			F-WWLN				
G-TAXI	Piper PA-23-250 Aztec E	27-7305085	N40270	6. 4.78	M.L.Levi, S.Waite and R.Murgatroyd	Blackpool	17.12.04T
					t/a SWL Leasing		
G-TAYI	Grob G115	8008	(D-ENFT)	12. 9.90	K.P.Widdowson and K.Hackshall	Sandtoft	1. 7.07
			G-TAYI, G-DODO, D-ENFT				
G-TAYS	Reims/Cessna F152 II	F15201697	G-LFCA	28.10.91	Tayside Aviation Ltd	Glenrothes	8. 7.07T
G-TBAE	British Aerospace BAe 146 Series 200	E2018	G-JEAR	6. 1.03	BAE SYSTEMS (Corporate Air Travel) Ltd	Warton	7. 4.07T
			G-HWPB, G-6-018, G-BSRU, G-OSKI, N603AW				
G-TBAG	Meridian Renegade 912	PFA 188-11912		11.12.90	M.R.Tetley	Newton-on-Rawcliffe, Yorkshire	19.10.05P
	(Built M Tetley)						
G-TBAH	Bell 206B JetRanger II	2051	G-OMJB	10.12.01	RB Helicopters Ltd	Booker	20.11.04T
			N315JP, N712WG, N712WC, N9989K				
G-TBBC	Pegasus Quantum 15-912	7583		6.12.99	J.Horn	Eshott	27.10.05P
G-TBEE	Dyn'Aéro MCR-01 Club	PFA 301-13514		30.11.99	A D.S.Baker	Shoreham	26. 4.05P
	(Built A D.S.Baker) (Rotax 912-ULS)						
G-TBGL	Agusta A109A II	7412	G-VJCB	6. 1.99	Bulford Holdings Ltd	(Jersey)	22. 3.07T
			G-BOUA				
G-TBGT	SOCATA TB-20 Trinidad GT	2027	F-OILF	1.12.00	A J Maitland-Robinson	(Jersey)	10.12.06T
G-TBIC	British Aerospace BAe 146 Series 200	E2025	N167US	15. 1.97	Flightline Ltd	Aberdeen	16. 1.06T
			N349PS		*(Op IAC)*		
G-TBIO	SOCATA TB-10 Tobago	340	F-BNGZ	10. 2.83	Delta Bird Aviation Ltd	Liverpoo	9. 5.05T
G-TBJP	Mainair Pegasus Quik	8071		29. 9.04	B.J.Partridge	Sutton Meadows, Ely	29. 9.05P
G-TBLY	Eurocopter EC120B Colibri	1192	F-WQOV	12. 3.01	A D.Bly Aircraft Leasing Ltd	(Knebworth)	28. 6.04T
G-TBMW	Meridian Renegade Spirit	PFA 188A-11725	(G-MYIG)	20.10.98	S J and M J Spavins	Longacres Farm, Sandy	
	(Built J R Peters)				*(Noted 1.05)*		
G-TBOK	SOCATA TB-10 Tobago	1111	SX-ABF	26. 6.02	TB10 Ltd	Dunkeswell	24. 9.05T
			F-GKUA				
G-TBRD	Canadair CL-30 (T-33AN) Silver Star Mk.3	T33-261	N33VC	18.12.96	Golden Apple Operations Ltd	Duxford	25. 3.05P
			G-JETT, G-OAHB, CF-IHB, CAF 133261, RCAF 21261				
					(Op The Old Flying Machine Co as "21261" in RCAF c/s)		
G-TBTN	SOCATA TB-10 Tobago	322	G-BKIA	7. 8.03	J.S.Chaggar	Wellesbourne Mountford	20.10.07E
					(Op Avon Flying School)		
G-TBXX	SOCATA TB-20 Trinidad	276		16. 3.82	D.A Phillips	Headcorn	20. 4.06
G-TBZI	SOCATA TB-21 Trinidad TC	871	N21HR	25. 7.96	Skypartners UK Ltd	Coventry	21.11.05T
G-TBZO	SOCATA TB-20 Trinidad	444		8. 8.84	R.P.Lewis and D.L.Clarke	Shoreham	27. 4.06
G-TCAN	Colt 69A HAB	1996		19. 7.91	H.C.J.Williams *"Toucan"*	(Bristol)	13. 6.04A
G-TCAP	British Aerospace BAe 125 Series 800B	258115	G-5-599	24. 4.96	BAE SYSTEMS (Operations) Ltd	Warton	28. 8.06
			104 RSAF, G-5-665, RSAF 104, G-BPGR, G-5-599				
G-TCAS	Cameron Z-275 HAB	10343		28. 2.03	The Ballooning Business Ltd	(Walcote, Alcester)	27. 3.05T
					(Central Auto Supplies titles)		
G-TCKE	Airbus A320-214	1968	F-WWBD	16. 4.03	Thomas Cook Airlines UK Ltd	Manchester	15. 4.06T
G-TCMM	Agusta-Bell 206B-3 JetRanger III	8560	EI-BXX	24. 3.04	Transair (UK) Ltd	Shoreham	30. 3.07T
			G-JMVB, G-OIML				
G-TCNM	Tecnam P92-EA Echo	PFA 318-13922		12. 8.02	J.Quaife	Headcorn	14. 4.05P
	(Built J.Quaife)						
G-TCNY	Mainair Pegasus Quik	8037		13. 5.04	T.Butler	Enstone	12. 5.05P
G-TCOM	Piper PA-30 Twin Comanche C	30-1967	N555JC	29. 1.96	C.A C.Burrough	Jersey	11. 4.05
			N8810Y				
G-TCTC	Piper PA-28RT-201T Turbo Arrow IV	2831001	N9130B	1.12.89	T.Haigh	Wellesbourne Mountford	26. 3.05
	(Originally built as N9524N [28R-8631006])						
G-TCUB	Piper J-3C-65 Cub	13970	N9039Q	31. 7.87	C.Kirk	(Lincoln)	8. 7.07
	(Frame No.13805)		N67666, NC67666, Bu.29684, 45-55204				
G-TDFS	IMCO Callair A-9	1200	G-AVZA	8.10.86	P.Stephenson	Egginton	5. 7.07A
			PH-ABI, SE-EUA, N26D				
G-TDOG	Scottish Aviation Bulldog Series 120/121	BH120/230	XX538	17. 9.01	G.S Taylor *(As "XX538/O")*	Shobdon	25. 4.05
G-TDVB	Dyn'Aéro MCR-01 ULC	PFA 301B-14015		23. 1.03	D.V.Brunt	Plaistowes Farm, St.Albans	2. 9.05P
	(Built D Brunt - kit no.242)						
G-TEBZ	Piper PA-28R-201 Cherokee Arrow III	28R-7737050	N105CC	7. 1.00	R.W.Tebby	Bristol	21. 8.06T
					t/a S.F.Tebby and Son *(Op Bristol Flying Centre)*		
G-TECC	Aeronca 7AC Champion	7AC-5269	N1704E	26. 6.91	Nicola Joy Orchard-Armitage	Waldershare Park	20.11.04P
			NC1704E				
G-TECH	Rockwell Commander 114	14074	G-BEDH	8. 8.85	P.A Reed	Elstree	14. 8.06
			N4744W				
G-TECK	Cameron V-77 HAB	625		21. 3.86	M.W.A Shemilt *(New owner 7.03)*	(Henley-on-Thames)	5. 8.02A
G-TECM	Tecnam P92-EM Echo	PFA 318-13667		1.12.00	D.A Lawrence	Draycott Farm, Chiseldon	5. 9.05P
	(Built D A Lawrence)						
G-TECS	Tecnam P2002 EA Sierra	PFA 333-14325		16. 6.04	D.A.Lawrence	Kemble	
	(Built D A Lawrence) (Originally regd with as PFA 318A-14250)				*(Noted 1.05)*		
G-TEDB	Reims/Cessna F150L	F15000772	G-AZLZ	20. 1.05	E.L.Bamford	Haverfordwest	16. 7.00T
G-TEDF	Cameron N-90 HAB	2634		8. 8.91	Fort Vale Engineering Ltd	(Nelson)	10. 7.05A
G-TEDI	Best Off Skyranger J2.2	BMAA/HB/243		1. 5.03	Karen Lorenzen	(London SW19	25.11.04P
	(Built D.A Smith and E.Robshaw - kit no.SKR0207217)						
G-TEDS	SOCATA TB-10 Tobago	57	G-BHCO	29. 3.83	M.Camp tr G-TEDS Group Aviation	Sleap	28. 5.05

G-TEDW	Air Création 582/Kiss 400	BMAA/HB/343		21. 1.04	D.J.Wood	(Cleckheaton)	26. 5.05P
	(Built D.J.Wood - kit no.FL026, Trike T03103, Wing A03184-3179)						
G-TEDY	Evans VP-1	PFA 62-10383	G-BHGN	4.10.90	N.K.Marston	(Harrow)	1. 7.97P
	(Built A Cameron) (Volkswagen 1834)				*"The Plank"*		
G-TEFC	Piper PA-28-140 Cherokee F	28-7325088	OY-PRC N15530	18. 6.80	P.M.Havard	Andrewsfield	18. 4.05
G-TEHL	CFM Streak Shadow M	185	G-MYJE	20.11.98	A K.Paterson	Sleaford	13. 8.04P
	(Built A K Paterson) (Rotax 503)						
G-TELY	Agusta A109A II	7326	N1HQ N200SH	10. 3.89	Castle Air Charters Ltd	Liskeard	24. 7.05T
G-TEMP	Piper PA-28-180 Cherokee E	28-5806	G-AYBK N11C	15. 5.89	M.J.Groome tr Bev Piper Group	Andrewsfield	26. 8.07T
G-TEMT	Hawker Tempest II	420	HA586 (RIAF), MW763	9.10.89	Tempest Two Ltd *(On rebuild 1.05 to be "MW763/HF-A" in 183 Sqdn c/s)*	Gamston	AC
G-TENG	Extra EA.300/L	172		8.12.03	10G Aerobatics Ltd	North Weald	29.12.06T
G-TENS	HOAC DV.20 Katana	20148	G-BXBW D-ESHM	28. 2.01	Ewan Ltd	Gamston	18. 6.06T
G-TENT	Auster J/1N Alpha	2058	G-AKJU TW513	1. 2.90	R.C.Callaway-Lewis	Turweston)	26. 8.05
G-TERN	Europa Aviation Europa	PFA 247-12780		18. 7.97	J Smith	(York)	10. 3.05P
	(Built J.E.G.Lundesjo - kit no.106) (NSI EA-81/100) *(Monowheel u/c)*						
G-TERR	Pegasus Quik	7925		6. 1.03	T.R.Thomas	Kemble	6. 1.05P
G-TERY	Piper PA-28-181 Archer II	28-7990078	G-BOXZ N22402	13. 1.89	J.R.Bratherton	Durham Tees Valley	26. 2.05T
G-TEST	Piper PA-34-200 Seneca	34-7450116	OO-RPW G-BLCD, PH-PLZ, N41409	28. 7.89	Stapleford Flying Club Ltd	Stapleford	23.12.04T
G-TETI	Cameron N-90 HAB	2877	D-OBMW	9. 2.00	Teti SpA	(Firenze, Italy)	
G-TEWS	Piper PA-28-140 Cherokee B	28-25128	G-KEAN G-AWTM, N11C	23. 5.88	R.A Bentley and P.A Connolly tr G-TEWS Flying Group	Liverpool	4.11.07E
G-TEXS	Van's RV-6	23830	N996SF	28.10.04	W.H.Greenwood	Swamborough Farm, Lewes	21.12.05P
	(Built S.Formhals)						
G-TEXT	Robinson R44 Raven II	10577		7.12.04	Heli Air Ltd	Wellesbourne Mountford	AC
G-TFCI	Reims/Cessna FA152 Aerobat	FA1520358		25.10.79	Tayside Aviation Ltd	Perth	25. 2.07T
G-TFIN	Piper PA-32RT-300T Turbo Lance II	32R-7887012	N221RT D-ELAL	23.4.03	M.D.Parker	Bourn	1. 5.06T
G-TFIX	Mainair Pegasus Quantum 15-912	8048		21. 7.04	T.G.Jones	(Caernarfon)	20. 7.05P
G-TFOX	Denney Kitfox Model 2	PFA 172-11817		3. 6.91	F.A Bakir	Lymm Dam	14.12.04P
	(Built F A Roberts)						
G-TFRB*	Air Command 532 Elite Sport				See SECTION 3 Part 1	Elvington	
G-TFSA	Reims/Cessna F152 II	F15201825	G-BITH	2.10.03	S.J.George	Egginton	31. 8.06T
G-TFUN	Valentin Taifun 17E	1011	D-KIHP	28.12.83	G.F.Wynn tr North West Taifun Group	Blackpool	10. 5.04
G-TFYN	Piper PA-32RT-300 Lance II	32R-7885128	N5HG D-ELAE, N31740	28. 4.00	R C Poolman	(Gloucestershire)	5. 7.03T
G-TGAS	Cameron O-160 HAB	1315		12. 8.87	Zebedee Balloon Service Ltd	Newtown Hungerford	26. 5.00T
G-TGER	Gulfstream AA-5B Tiger	AA5B-0952	G-BFZP	20. 2.86	P.J.Haldenby	Rochester	9. 2.07T
G-TGGR	Eurocopter EC120B Colibri	1224	SE-JMF	22. 7.04	Blue Five Aviation Ltd	(Reigate)	1. 8.07T
G-TGRA	Agusta A109A	7201	D-HEED N3983N, HB-XNF, I-PATZ	15. 2.01	Tiger Helicopters Ltd	Shobdon	2. 3.07T
G-TGRD	Robinson R22 Beta II	2712	G-OPTS	16. 6.04	Tiger Helicopters Ltd	Shobdon	17. 8.06T
G-TGRE	Robinson R22 Alpha	0471	G-SOLD N8559X	11. 9.03	Tiger Helicopters Ltd	Shobdon	17. 6.06T
G-TGRR	Robinson R22 Beta	1235	G-BSZS N8058J	4.12.02	Tiger Helicopters Ltd	Shobdon	3. 4.06T
G-TGRS	Robinson R22 Beta	1069	G-DELL N80466	5.11.97	Tiger Helicopters Ltd *(Rolled over landing Duxford 18.11.01 and damaged)*	Shobdon	9. 10.04T
G-TGRZ	Bell 206B JetRanger II	2288	G-BXZX N27EA, N286CA, N93AT, N16873	15. 6.00	Tiger Helicopters Ltd	Shobdon	21.11.04T
G-THAI	CFM Shadow Series E 912	336 & BMAA/HB/239	(G-PIXY) G-85-26	28. 8.02	D.L.Hendry *(Noted 1.05)*	Sywell	
	(Officially regd as "Series D")						
G-THAT	Raj Hamsa X'Air Falcon 912	BMAA/HB/221		27. 5.02	J.Mccall	Middle Stoke, Isle of Grain	5. 7.05P
	(Built M.G.Thatcher - kit no.613)						
G-THEA	Boeing Stearman E75 (N2S-5) Kaydet	75-5736A	(EI-RYR) N1733B, USN Bu.38122	18. 3.81	C.M.Ryan *(As "33" in Navy c/s)*	(Leixlip, County Kildare)	5.12.05
	(Lycoming R-680)						
G-THEL	Robinson R44 Astro	0159	G-OCCB G-STMM	2. 9.98	N.Parkhouse	Elstree	3. 5.07T
G-THEO	TEAM mini-MAX 91	PFA 186-13099		9. 2.99	C.Fletcher	Dunkeswell	1. 7.05P
	(Buil;t T Willford) (Built up rear fuselage)						
G-THIN	Reims FR172E Rocket	FR17200016	G-BXYY OY-AHO, F-WLIP	4.12.02	C.A Ussher	Bagby	24. 6.07T
G-THLA	Robinson R22 Beta II	3462		18. 7.03	Thurston Helicopters Ltd	Headcorn	3. 8.06T
G-THOA	Boeing 737-5L9	24859	N859CT OY-MAC, G-MSKA, OY-MAC, (OY-MMZ)	10. 3.04	Britannia Airways Ltd *(Op Thomsonfly)*	Coventry	9. 3.07T
G-THOB	Boeing 737-5L9	24928	N928CT OY-MAD, G-MSKB, OY-MAD, (OY-MMO)	19. 3.04	Britannia Airways Ltd *(Op Thomsonfly)*	Coventry	23. 3.07T
G-THOC	Boeing 737-59D	24694	G-BVKA SE-DNA, (SE-DLA)	28. 4.04	Britannia Airways Ltd *(Op Thomsonfly)*	Coventry	28. 2.06T
G-THOD	Boeing 737-59D	24695	G-BVKC SE-DNB, (SE-DLB)	30. 6.04	Britannia Aurways Ltd *(Op Thomsonfly)*	Coventry	15. 5.06T
G-THOE	Boeing 737-3Q8	26313	G-BZZH N14384	9.12.04	Britannia Airways Ltd *(Op Thomsonfly)*	Coventry	19. 3.06E
G-THOF	Boeing 737-3Q8	26314	G-BZZI N73385	9.12.04	Britannia Airways Ltd *(Op Thomsonfly)*	Coventry	27. 3.06E
G-THOG	Boeing 737-3		G-	R	Britannia Airways Ltd *(Op Thomsonfly)*	Coventry	
G-THOH	Boeing 737-3		G-	R	Britannia Airways Ltd *(Op Thomsonfly)*	Coventry	

Registration	Type	c/n	Previous identity	Reg date	Owner	Base	CofA
G-THOH	Boeing 737-3		G-	R	Britannia Airways Ltd *(Op Thomsonfly)*	Coventry	
G-THOJ	Boeing 737-3		D-	R	Britannia Airways Ltd *(Op Thomsonfly)*	Coventry	
G-THOK	Boeing 737-8		D-	R	Britannia Airways Ltd *(Op Thomsonfly)*	Coventry	
G-THOM	Thunder Ax6-56 HAB	366		14. 7.81	T.H.Wilson *"Macavity"*	Diss	17. 8.05A
G-THOS	Thunder Ax7-77 HAB	769		20. 2.86	C.E.A Breton	(Bristol)	14. 3.01A
G-THOT	Jabiru Aircraft Jabiru SK	PFA 274-13159		16. 9.97	D.J. and S.C.Reed	Booker	19. 5.05P
	(Built N V Cook)						
G-THRE	Cessna 182S Skylane	18280454	N2391A	6. 5.99	S.J.G.Mole	Wolverhampton	5. 5.05
G-THSL	Piper PA-28R-201 Arrow III	28R-7837278	N36396	11. 9.78	D.M.Markscheffel	Southend	3. 4.06
G-THUN	Republic P-47D Thunderbolt	"39955731"	N47DD	18. 6.99	Patina Ltd	Duxford	7. 7.05P
					(Op The Fighter Collection as "226671/MH-X/LH-X")		
	(Composite re-build from wreck of original N47DD plus new P-47N fuselage identification unknown: "original" N47DD is c/n 399-55731 ex Peruvian AF 119/Peruvian AF 545/45-49192 and is static exhibit in American Air Museum, Duxford)						
G-THZL	SOCATA TB-20 Trinidad	534	F-GJDR	9. 5.96	Thistle Aviation Ltd	Oxford	16. 4.05
			N65TB				
G-TICH	Taylor JT.2 Titch	PFA 60-3213		12. 2.01	A J.House, C.J.Wheeler and R.Davitt	(Reading)	
	(Built A J.House, C.J.Wheeler and R.Davitt) (Project no. originally allocated as PFA 3213)				*(40% complete in 1973! : current status unknown)*		
G-TIDS	SAN Jodel D.150 Mascaret	44	OO-GAN	15. 4.86	M.R.Parker	Sywell	20.11.05P
G-TIGA	de Havilland DH.82A Tiger Moth	83547	G-AOEG	5. 6.85	D.E.Leatherland	Tollerton	28. 2.07T
	(Built Morris Motors Ltd)		T7120				
G-TIGB	Aérospatiale AS332L Super Puma	2023	G-BJXC	31. 3.82	Bristow Helicopters Ltd	Aberdeen	27. 4.07T
			F-WTNM		*"City of Aberdeen"*		
G-TIGC	Aérospatiale AS332L Super Puma	2024	G-BJYH	14. 4.82	Bristow Helicopters Ltd	Aberdeen	17. 5.05T
			F-WTNJ		*"Royal Burgh of Montrose"*		
G-TIGE	Aérospatiale AS332L Super Puma	2028	G-BJYJ	15. 4.82	Bristow Helicopters Ltd	Aberdeen	7. 6.07T
			F-WTNM		*"City of Dundee"*		
G-TIGF	Aérospatiale AS332L Super Puma	2030	F-WKQJ	15. 4.82	Bristow Helicopters Ltd *"Peterhead"*	Aberdeen	27. 6.06T
G-TIGG	Aérospatiale AS332L Super Puma	2032	F-WXFT	15. 4.82	Bristow Helicopters Ltd *"Macduff"*	Aberdeen	1. 8.07T
G-TIGH*	Aérospatiale AS332L Super Puma	2034	F-WXFL	15. 4.82	Bristow Helicopters Ltd	Aberdeen	24. 8.92T
			(Damaged 100m NE of Shetland Isles 14.3.92: cancelled 3.8.92 as destroyed) (Instruction use 2.04)				
G-TIGI	Aérospatiale AS332L Super Puma	2036	F-WTNP	15. 4.82	Bristow Helicopters Ltd *"Fraserburgh"*	Shenzhen, China	5. 9.05T
G-TIGJ	Aérospatiale AS332L Super Puma	2042	VH-BHT	15. 4.82	Bristow Helicopters Ltd	(Mauretania)	29. 6.05T
			G-TIGJ		*"Rosehearty"*		
G-TIGO	Aérospatiale AS332L Super Puma	2061	PP-MIM	18. 2.83	Bristow Helicopters Ltd	Aberdeen	22. 8.07T
			G-TIGO, F-WMHH		*"Royal Burgh of Arbroath"*		
G-TIGP	Aérospatiale AS332L Super Puma	2064		11. 3.83	Bristow Helicopters Ltd *"Carnoustie"*	Shenzhen, China	8. 5.08T
G-TIGR	Aérospatiale AS332L Super Puma	2071	F-WTNW	11. 3.83	Bristow Helicopters Ltd *"Stonehaven"*	Aberdeen	19. 5.05T
G-TIGS	Aérospatiale AS332L Super Puma	2086		6. 5.83	Bristow Helicopters Ltd *"Findochty"*	Aberdeen	27. 6.05T
G-TIGT	Aérospatiale AS332L Super Puma	2078		6. 5.83	Bristow Helicopters Ltd *"Portknockie"*	Aberdeen	2. 5.07T
G-TIGV	Aérospatiale AS332L Super Puma	2099	LN-ONC	12. 1.84	Bristow Helicopters Ltd *"Burghead"*	Aberdeen	25. 6.07T
			G-TIGV, LN-ONC, G-TIGV, LN-OPF, G-TIGV				
G-TIGZ	Aérospatiale AS332L Super Puma	2115	C-GQKK	8. 8.84	CHC Scotia Ltd	Aberdeen	14.10.06T
			G-TIGZ				
G-TIII	Pitts S-2A	2196	G-BGSE	27. 2.89	S.B.Janvrin	Redhill	1. 5.07
	(Built Aerotek Inc)		N947		tr Treble India Group		
G-TIKO	Hatz CB-1	PFA 143-13396		9. 7.99	K.Robb	(Yeovil)	
	(Built K Robb)				t/a Tiko Architecture *(Current status unknown)*		
G-TILE	Robinson R22 Beta	1100		4. 8.89	Fenland Helicopters Ltd	Cambridge	12. 8.05T
G-TILI	Bell 206B JetRanger II	2061	F-GHFN	6. 3.96	TLC Handling Ltd	Finningley	20. 1.07T
			N7037A, XC-BOQ				
G-TIMB	Rutan VariEze	PFA 74-10795	G-BKXJ	11. 6.85	T.M.Bailey	Shoreham	22. 7.05P
	(Built B Wronski) (Continental O-200-A)				*"Kitty"*		
G-TIME	Piper PA-61P Aerostar 601P	61P-0541-230	N8058J	21. 7.78	T and G Engineering Co Ltd	(West Byfleet)	10. 3.06
G-TIMG	Beagle Terrier 3	"PFA 00-318"		7. 3.01	T.J.Goodwin	(Manningtree)	
	(Built T.J.Goodwin) (To comprise an Auster 6 fuselage frame and one wing of VF505 not used in the build of Terrier G-ASOM - noted 8.03)						
G-TIMK	Piper PA-28-181 Archer II	28-8090214	OO-TRT	25. 8.81	T.Baker	Wolverhampton	16. 6.04
			PH-EAS, OO-HLN, N8142H				
G-TIMM	Folland Gnat T.1	FL.519	8618M	19. 2.92	T.J.Manna	North Weald	2. 3.05P
			XP504		t/a Kennet Aviation *(As "XS111")*		
G-TIMP	Aeronca 7BCM Champion	7AC-3392	N84681	14. 8.92	M.G.Rumney	Goodwood	9. 9.05P
	(Continental C85)		NC84681		*"Nancy"*		
G-TIMS	Falconar F-12A Cruiser	PFA 22-12134		1.10.91	T.Sheridan	Wellingborough	
	(Built T Sheridan)				*(Current status unknown)*		
G-TIMY	Sud-Aviation Gardan GY-80-160 Horizon	36	I-TIKI	17. 1.00	R.G.Whyte	Turweston	28. 7.07
G-TINA	SOCATA TB-10 Tobago	67		30.10.79	A Lister	Shipdham	13. 9.04
G-TING	Cameron O-120 HAB	4007		4.10.96	Floating Sensations Ltd	(Thatcham)	2.11.04T
G-TINK	Robinson R22 Beta	0937	G-NICH	22. 5.01	Heli Air Ltd	Wellesbourne Mountford	5. 5.04T
					(New owner 12.04)		
G-TINS	Cameron N-90 HAB	1626		27. 1.88	J.R.Clifton *(Carling Black Label titles)*	(Brackley)	11.10.03A
G-TINY	Moravan Zlin Z-526F Trener	1257	OK-CMD	10. 5.95	D.Evans	Little Gransden	17. 8.98
			G-TINY, YR-ZAD		*(Current status unknown)*		
G-TIPS	Nipper T.66 Series 3	PFA 25-12696	OO-VAL	27. 3.95	R.F.L.Cuypers and F.V.Neefs	Grimbergen, Belgium	10. 8.05P
	(Built R.F.L.Cuypers from Fairey c/n 50) (Jabiru 2200A)		9Q-CYJ, 9O-CYJ, (OO-CYJ), (OO-CCD)				
G-TIVS	Rans S-6ES Coyote II	PFA 204-14236		10. 5.04	S.Hoyle	(Bagby)	20. 6.05P
	(Built S.Hoyle - kit no.1203.1536) (Nosewheel configuration)						
G-TJAL	Jabiru Aircraft Jabiru SPL-430	PFA 274A-13360		21. 2.03	D W Cross	(Lincoln)	21. 1.05P
	(Built T.J.Adams-Lewis)						
G-TJAV	Mainair Pegasus Quik	8070		23. 9.04	Red Communications Ltd	Sutton Meadows, Ely	27. 9.05P
G-TJAY	Piper PA-22-135 Tri-Pacer	22-730	N730TJ	11. 5.93	D.D.Saint	Garston Farm, Marshfield	28. 8.05
			N2353A				
G-TKAY	Europa Aviation Europa	PFA 247-12804		2. 6.99	A M.Kay	Nuthampstead	1. 8.05P
	(Built A M.Kay - kit no.179) (Monowheel u/c)						
G-TKGR	Lindstrand Racing Car SS HAB	380		28. 8.96	Brown and Williamson Tobacco Corporation (Export) Ltd		
					"Team Green"	(Louisville, Kentucky, US)	20. 8.99A

Reg	Type	C/n	Previous identity	Reg date	Owner/operator	Base	CofA expiry
G-TKIS	Tri-R Kis	PFA 239-12358		23.12.93	J.L.Bone	Biggin Hill	5. 4.05P
	(Built J L Bone - c/n 029) (Lycoming O-290-D2) *(Tail-wheel variant)*						
G-TKPZ	Cessna 310R II	310R1225	G-BRAH	19. 3.90	Air Charter Scotland Ltd	Prestwick	1. 4.05T
			N1909G				
G-TLBC	SOCATA MS.892A Rallye Commodore 150	10283	OO-LBC	21.10.04	Film Funding Inc	(Salisbury)	
			F-BLBC, (D-ECKY), F-BLBC				
G-TLDK	Piper PA-22-150 Tri-Pacer	22-4726	N6072D	27. 1.97	A M.Thomson	Phoenix Farm, Lower Upham	
					(Noted 12.99: current status unknown)		
G-TLEL	American Blimp Corp A-60+ Airship	003	I-TIRE	17. 5.02	Keelex 195 Ltd	Chilbolton	26. 5.05T
			N2017A				
G-TLET	Piper PA-28-161 Cadet	2841259	G-GFCF	25. 8.04	Meridian Aviation Sales Ltd	Bournemouth	12. 2.05T
			G-RHBH, N9193Z				
G-TMCB	Best Off Skyranger 912	BMAA/HB/310		23.10.03	A H.McBreen	(Rugby)	
	(Built A H.McBreen - kit no.SKRxxxxxxx?)						
G-TMCC	Cameron N-90 HAB	4327		30. 3.98	Prudential Assurance Co Ltd	(Bristol)	7. 6.02A
					(The Mall/Cribbs Causeway titles)		
G-TMKI	Percival P.56 Provost T.1	PAC/F/268	WW453	1. 7.92	B.L.Robinson	(Clevedon)	
					(As "WW453/W-S" in RAF c/s)		
G-TMOL	SOCATA TB-20 Trinidad	2103	F-OJBQ	24.12.01	West Wales Airport Ltd	Gloucestershire	13. 1.05T
G-TNTN	Thunder Ax6-56 HAB	1991		25. 4.91	H.M.Savage and J.F.Trehern	(Edinburgh)	10 9.03A
G-TOAD	SAN Jodel D.140 Mousquetaire	27	F-BIZG	27. 9.88	J.H.Stevens	Headcorn	20. 7.06
G-TOAK	SOCATA TB-20 Trinidad	468	N83AV	5.12.89	C.Wade and A Young tr Phoenix Group	Belfast Aldergrove	11. 1.05
G-TOBA	SOCATA TB-10 Tobago	625	N600N	4. 4.91	E.J.Downing	Lee-on-Solent	19.10.06
G-TOBI	Reims/Cessna F172K	F17200792	G-AYVB	5. 1.84	A I Bird	Leicester	14. 7.05
G-TOBY*	Cessna 172B	47852	G-ARCM	8. 4.81	Northbrook College	Shoreham	28. 4.85
			N6952X	*(Damaged Sandown 15.10.83: cancelled 27.2.90 by CAA) (In open store 5.03)*			
G-TODE	Ruschmeyer R90-230RG	016	D-EEAX	20. 6.94	A I.D.Rich	Elstree	1. 7.06
G-TOFT	Thunder and Colt 90A HAB	1693		8. 3.90	C.S.Perceval *"Bumble"*	(Great Missenden)	20. 5.03A
G-TOGO	Van's RV-6A	PFA 181A-13447		6. 4.99	Stephanie G.Schwetz	(Lisciano Niccone, Perugia, Italy)	12. 2.05P
	(Built G.Schwetz)						
G-TOHS	Cameron V-31 HAB	10267		4.11.02	J.P.Moore	(Great Missenden)	19. 9.03P
G-TOLL	Piper PA-28R-201 Arrow III	28R-7837025	N52HV	12.10.00	Plymouth School of Flying Ltd	Plymouth	1.12.06T
			D-ECIW, N9007K				
G-TOLY	Robinson R22 Beta II	2809	G-NSHR	8. 2.01	HelicopterServices Ltd	Booker	20. 3.07T
G-TOMC	North American AT-6D Texan	88-14602	French AF 114700	22. 4.02	A A Marshall	(Bruntingthorpe)	
			42-44514		tr Texan Restoration		
G-TOMJ	Flight Design CT2K	03.01.04.14		28. 7.03	P.T.Knight	Sittles Farm, Alrewas	21. 8.05P
	(Assembled Pegasus Aviation as c/n 7975)						
G-TOMM	Robinson R22 Beta II	3384		5.11.02	Airfleet Aircraft Leasing Ltd	Earls Colne	2.12.05T
G-TOMS	Piper PA-38-112 Tomahawk	38-79A0453	N9658N	22. 1.79	N.King tr G-TOMS Group	Swansea	18. 8.07T
G-TOMZ	Denney Kitfox Model 2	PFA 172-11977		15.11.00	S.Austen	Old Sarum	17. 6.05P
	(Built P T Knight and L James) (Rotax 912)						
G-TONN	Mainair Pegasus Quik	7954		28. 5.03	The Windmill Kennels and Catery Ltd	Bourn	8. 6.05P
G-TONS	Slingsby T.67M-200 Firefly	2045	LN-TFA	22. 7.03	D.I.Stanbridge	(Arnhem, The Netherlands)	17.12.06
			SE-LBA, LN-TFA, G-7-121				
G-TOOL	Thunder Ax8-105 HAB	1670		29. 3.90	D.V.Howard *(Toolmaster Hire titles)*	(Bath)	29. 3.05A
G-TOOT	Dyn'Aéro MCR-01 Club	PFA 301-13542		1. 3.01	E.K.Griffin	Bicester	1.11.05P
	(Built E K Griffin)						
G-TOPC	Aérospatiale AS355F1 Ecureuil 2	5313	I-LGOG	29. 7.97	Bridge Street Nominees Ltd	Denham	6.11.06T
			3A-MCS, D-HOSY, OE-BXV, D-HOSY				
G-TOPK	Europa Aviation Europa	PFA 247-14193		16. 3.04	P J Kember	Fowle Hall Farm, Laddingford	
	(Built P J Kember - kit no.1000)						
G-TOPS	Aérospatiale AS355F1 Ecureuil 2	5151	G-BPRH	7. 5.91	Sterling Helicopters Ltd	Norwich	24. 1.05T
			N360E, N5794F				
G-TORC	Piper PA-28R-200 Cherokee Arrow II	28R-7535036	OE-DIU	18. 7.03	Solo Leisure Ltd	Old Sarum	25. 9.06T
			N32236				
G-TORE*	Hunting Percival P.84 Jet Provost T.3A	PAC/W/9212	XM405	14. 6.91	City University	Islington, London NW	5. 5.95P
					(Cancelled 24.2.04 by CAA) (Instructional airframe 9.03)		
G-TORS	Robinson R22 Beta II	3021		4. 1.00	IW Aviation Ltd	(Warwick)	6. 3.06T
G-TOSH	Robinson R22 Beta	0933	N2629S	14. 3.97	Heli Air Ltd	Leicester	3.10.07E
			LV-RBD, N8012T				
G-TOTN	Cessna 210M Centurion II	21061674	G-BVZM	15. 7.04	Just Plane Trading Ltd	Standalone Farm, Meppershall	16. 5.07
			OO-CNJ, N732PV				
G-TOTO	Reims/Cessna F177RG Cardinal RG	F177RG0049	G-OADE	29. 8.89	Horizon Flyers Ltd	Denham	30. 9.07E
			G-AZKH				
G-TOUR	Robin R2112	187		9.10.79	Mardenair Ltd *(Op Plessey Flying Group)*	Goodwood	23. 3.07T
G-TOWS	Piper PA-25-260 Pawnee C	25-4853	PH-VBT	17. 7.91	Lasham Gliding Society Ltd	Lasham	25. 3.07
	(Hoffman 4 -blade propeller)		D-EAVI, N4370Y, N9722N				
G-TOYA	Boeing 737-3Q8	26310	G-BZZE	13.12.04	British Midland Airways Ltd *"Brummie Baby"*		
			N14381		*(Op BMI Baby)*	Nottingham East Midlands	17.12.05E
G-TOYB	Boeing 737-3Q8	26311	G-BZZF	24.11.04	British Midland Airways Ltd		
			N19382		*(Op BMI Baby)*	Nottingham East Midlands	21. 7.06E
G-TOYC	Boeing 737-3Q8	26312	G-BZZG	13.12.04	British Midland Airways Ltd	Birmingham	3. 3.06E
			N14383		*(Op BMI Baby)*		
G-TOYD	Boeing 737-3Q8	26307	G-EZYT	3.05R	British Midland Airways Ltd		
			HB-IIE, N721LF, (HB-IIE)		*(Op BMI Baby)*	Nottingham East Midlands	
G-TOYZ	Bell 206B-3 JetRanger III	3949	G-RGER	21.11.96	A R.Pocock	(Glasgow)	24.10.03T
			N75EA, JA9452, N32018				
G-TPSL	Cessna 182S Skylane	18280398	N23700	11.12.98	A N.Purslow	Blackbushe	15. 1.05T
G-TRAC	Robinson R44 Astro	0598		10. 5.99	C.Sharples	Newbury	22. 5.05T
G-TRAM	Pegasus Quantum 15-912 Super Sport	7552		29. 7.99	I.W.Barlow	(Ilkeston)	19. 8.05P
G-TRAN	Beech 76 Duchess	ME-408	G-NIFR	15. 3.93	J.A Shufflebotham	Leeds-Bradford	25. 9.07T
			N1808A				

Registration	Type	C/n	Previous identity	Reg date	Owner (operator)	Base	CofA expiry
G-TRAV	Cameron A-210 HAB	3181	PH-GET G-TRAV	6.12.93	A M.Holly	(Berkeley)	17. 1.05T
G-TRCY	Robinson R44 Astro	0668		22.10.99	T.Taylor	(Staddiscombe, Plymouth)	29. 1.06T
G-TREC	Cessna 421C Golden Eagle III	421C0838	G-TLOL (N2659K)	2. 7.96	Sovereign Business Integration plc	Booker	6. 3.05T
G-TRED	Cameron Colt Bibendum 110SS HAB	4222		12.12.97	The Aerial Display Co Ltd	(Looe)	11. 2.02A
G-TREE	Bell 206B-3 JetRanger III	2826	N2779U	15. 6.87	R.B.Eaton t/a Bush Woodlands	(Bridgnorth)	18.12.05T
G-TREK	Jodel D.18 *(Built R.H.Mole - c/n 182)* (JPX 4TX @ 65hp)	PFA 169-11265		1. 5.92	R.H.Mole	Leicester	15. 8.05P
G-TREX	Alpi Pioneer 300 *(Built R.K.King)*	PFA 330-14305		6. 1.05	R.K.King	Croft Farm, Defford	
G-TRIB	Lindstrand HS-110 HA Airship (Rotax 582)	174	(N....)	23. 1.95	J.Addison	Melton Mowbray	17. 8.05A
G-TRIC	de Havilland DHC-1 Chipmunk 22A	C1/0080	G-AOSZ WB635	18.12.89	D.M.Barnett tr Landpro *(As "18013/013" in RCAF c/s)*	Spanhoe	2.11.06
G-TRIG	Cameron Z-90 HAB	10446		28. 7.03	Trigger Concepts Ltd *(Intel Inside/Centrino titles)*	(Swallowfield, Reading)	12. 7.05A
G-TRIM	Monnett Moni *(Built Monnett Aircraft- c/n 00258T)*	PFA 142-11012		16. 2.84	E.A.Brotherton-Ratcliffe *(New owner 1.05)*	(London SW6)	
G-TRIN	SOCATA TB-20 Trinidad	1131		25. 6.90	TL Aviation Ltd	Jersey	9. 3.06
G-TRIO	Cessna 172M Skyhawk II	17266271	G-BNXY N9621H	30. 7.91	C.M.B.Reid	Rochester	23. 1.06T
G-TRNT	Robinson R44 Raven II	10293	OE-XHW	17. 9.04	Charles Trent Ltd	Bournemouth	23. 9.07T
G-TROP	Cessna T310R II	310R1381	N4250C	31.12.86	D.E.Carpenter	Shoreham	29. 4.05T
G-TROY	North American T-28A Fennec	142/174-545	F-AZFV FrAF No 142, 51-7692	21. 4.99	S.G.Howell and S.Tilling *(As "51-7692")*	Duxford	4. 7.05P
G-TRUD	Enstrom 480	5022	XT-BOK	27. 2.01	Sussex Aviation Ltd	Shoreham	4. 3.07T
G-TRUE	MD Helicopters Hughes 369E	0490E	N6TK ZK-HFP	12. 9.94	Bailey Employment Services Ltd	(Sywell)	13. 5.07T
G-TRUK	Stoddard-Hamilton Glasair RG *(Built M P Jackson - 575R)* (Lycoming O-320)	PFA 149-11015		23. 7.84	M.P.Jackson	Fairoaks	26. 5.05P
G-TRUX	Colt 77A HAB	1860		13.11.90	M J Forster	(Humshaugh, Hexham)	6. 5.05A
G-TRYG	Robinson R44	0960		4. 1.01	Enable International Ltd	(Stratford-upon-Avon)	26. 2.07T
G-TRYK	Air Création 582/Kiss 400 *(Built S.Elsbury - kit no.FL004, Trike T01098, Wing A01157-1163)*	BMAA/HB/191		31.10.01	S.Elsbury	(Brentwood)	25. 6.05P
G-TSGJ	Piper PA-28-181 Archer II	28-8090109	N8097W	12. 9.88	A D.S.Peat and G.White tr Golf Juliet Flying Club	Durham Tees Valley	1. 2.07
G-TSIX	North American AT-6C-1NT Harvard IIA	88-9725	FAP1535 SAAF7183, EX289, 41-33262	19. 3.79	S.J.Davies *(As "111836/JZ/6" in USN c/s)*	Sandtoft	15. 9.05P
G-TSKD	Raj Hamsa X'Air Jab22 *(Built T.Sexton and K.B.Dupuy - kit no.633)*	BMAA/HB/165		8. 5.01	T.Sexton and K.B.Dupuy	Southend	20.10.05P
G-TSKY	Beagle B.121 Pup Series 2	B121-010	OE-CFM HB-NAA, G-AWDY, HB-NAA, G-AWDY	6. 4.98	R.G.Hayes	North Weald	7. 5.07T
G-TSOB	Rans S-6ES Coyote II *(Built S.C Luck)*	PFA 204-14066		28. 6.04	S.Luck *"The Spirit of Brooklands"*	Audley End	
G-TSOL	EAA Acrosport *(Built J Sykes)* (Lycoming O-320)	PFA 72-11391	G-BPKI	18. 7.00	A G.Fowles	(Shrewsbury)	31. 3.00P
G-TTAC	SOCATA TB-20 Trinidad GT	2121	F-OIMD (N212GT)	8. 5.02	AC Aviation Ltd	Shoreham	23. 5.05T
G-TTDD	Zenair CH.701 STOL *(Built B E Trinder and D.B.Dainton)* (Jabiru 2200A)	PFA 187-13106		1. 9.97	D.B.Dainton and V.D.Asque	Sackville Lodge Farm, Riseley	16. 1.05T
G-TTFN	Cessna 560 Citation V	560-0537	N5181V	19.11.99	Corporate Administration Management Ltd	Fairoaks	12.12.05T
G-TTHC	Robinson R22 Beta	1196		21.12.89	Multiflight Ltd	Leeds-Bradford	28. 5.05T
G-TTIA	Airbus A321-231	1428	D-AVZA	19. 2.01	GB Airways Ltd *"Mons Teide"*	Gatwick	18. 2.07T
G-TTIB	Airbus A321-231	1433	D-AVZC	27. 2.01	GB Airways Ltd *"Mons Toubkal"*	Gatwick	26. 2.07T
G-TTIC	Airbus A321-231	1869	D-AVZZ	12.12.02	GB Airways Ltd	Gatwick	11.12.05T
G-TTID	Airbus A321-231	2462	D-AV.. (G-EUXI)	5.05R	GB Airways Ltd	Gatwick	
G-TTIE	Airbus A321-231		(G-EUXJ)	3.06R	GB Airways Ltd	Gatwick	
G-TTIF	Airbus A321-231		(G-EUXL)	3.07R	GB Airways Ltd	Gatwick	
G-TTMB	Bell 206B Jet Ranger III	947	G-RNME G-CBDF, N211KR, JA9119, N58064	26. 6.03	Helirentals Ltd	Earls Colne	7. 3.05T
G-TTOA	Airbus A320-232	1215	F-WWDB	18. 5.00	GB Airways Ltd *"Mons Blanc"*	Nottingham East Midlands	17. 5.06T
G-TTOB	Airbus A320-232	1687	F-WWIM	11. 2.02	GB Airways Ltd	Gatwick	13. 2.05T
G-TTOC	Airbus A320-232	1715	F-WWDB	6. 3.02	GB Airways Ltd	Gatwick	5. 3.05T
G-TTOD	Airbus A320-232	1723	F-WWBH	14. 3.02	GB Airways Ltd	Gatwick	13. 3.05T
G-TTOE	Airbus A320-232	1754	F-WWDH	11. 4.02	GB Airways Ltd	Gatwick	10. 4.05T
G-TTOF	Airbus A320-232	1918	F-WWIS	13. 2.03	GB Airways Ltd	Gatwick	12. 2.06T
G-TTOG	Airbus A320-232	1969	F-WWDZ	10. 4.03	GB Airways Ltd	Gatwick	9. 4.06T
G-TTOH	Airbus A320-232	1993	F-WWDO	2. 5.03	GB Airways Ltd	Gatwick	1. 5.06T
G-TTOI	Airbus A320-232	2137	F-WWBN	11.12.03	GB Airways Ltd	Gatwick	10.12.06T
G-TTOJ	Airbus A320-232	2157		30. 3.04	GB Airways Ltd	Gatwick	29. 3.07T
G-TTOY	CFM Streak Shadow SA *(Built D A Payne - kit no K.233)* (Rotax 618)	PFA 206-12805		15. 4.96	D. and B.D.C.Barnard	(Farnborough)	27.10.05P
G-TTWO*	Colt 56A HAB				See SECTION 3 Part 1	(Aldershot)	
G-TUBB	Jabiru Aircraft Jabiru UL-450 *(Built A H.Bower and A Silvester)*	PFA 274A-13484		1.10.99	A H.Bower	Kemble	17. 2.04P
G-TUCK	Van's RV-8 *(Built M.A Tuck)*	PFA 303-13706		25. 9.03	M.A Tuck	(Rushden, Northampton)	
G-TUDR	Cameron V-77 HAB	1135		20. 5.85	Jacques W.Soukup Enterprises Ltd *"Tudor Rose/HVIIIR"*	(US)	29. 9.04A
G-TUGG	Piper PA-18-150 Super Cub (Lycoming O-360-A3) *(Frame No.18-8497)*	18-8274	PH-MAH N5451Y	10. 1.83	Ulster Gliding Club Ltd	Bellarena	9. 4.07

Reg	Type	C/n	Previous identity	Regd date	Owner/Operator	Base	CofA expiry
G-TUGY	Robin DR400/180 Régent	2052	D-EPAR	27. 4.98	J.M.Airey	Tibenham	19. 5.07T
G-TULP	Lindstrand LBL Tulips SS HAB	662	(PH-AJT)	16.10.00	Oxford Promotions (UK) Ltd	(Kentucky, US)	3. 4.03A
			(PH-TLP), PH-ORA)		*(Op F Prell)*		
G-TUNE	Robinson R22 Beta	0818	N60661	12. 1.99	Ecurie Ecosse (Scotland) Ltd	Cumbernauld	6. 2.05T
			G-OJVI, (G-OJVJ)		*(Op Scotia Helicopters)*		
G-TURF	Reims/Cessna F406 Caravan II	F406-0020	PH-FWF	17.10.96	Atlantic Air Transport Ltd	Coventry	20. 1.05T
			(EI-CND), PH-FWF, F-WZDS				
					(COASTGUARD titles, red/white c/s with MCA logo on tail)		
G-TURK*	Cameron Sultan 80SS HAB				See SECTION 3 Part 1	Balleroy, Normandy, France	
G-TURN	Steen Skybolt	PFA 64-11349		14. 7.88	K.W.Hadley and A M.Chester	Turweston	1. 6.04P
	(Built M Hammond - c/n 003)				tr G-TURN Flying Group		
G-TURP*	Aérospatiale SA341G Gazelle 1				See SECTION 3 Part 1	Charlwood	
G-TUSA	Pegasus Quantum 15-912	7841		9. 8.01	N.J.Holt	Weston Zoyland	22. 3.05P
G-TUSK	Bell 206B-3 JetRanger III	4406	G-BWZH	13. 1.97	Heli Aviation Ltd	(Rotherwick, Hook)	23. 2.06T
			N53114				
G-TUTU	Cameron O-105 HAB	10659		17. 1.05	A.C.K.Rawson and J.J.Rudoni	(Stafford)	4. 1.06E
G-TVAM	MBB Bö.105DBS-4	S.392	G-SPOL	26. 6.03	Bond Air Services Ltd	White Waltham	5. 6.05T
			VR-BGV, D-HDLH		(Op *Thames Valley Air Ambulance)*		
G-TVBF	Lindstrand LBL 310A HAB	439		2. 4.97	Airxcite Ltd t/a Virgin Balloon Flights	(Wembley)	20 .8.05T
G-TVEE	Hughes 369HS	14-0557S	N45457	26. 1.05	M.J.Gee	Gamston	AC
			ZK-HCM, N22352, C-GCXK, N500AH, N500WH				
G-TVII	Hawker Hunter T.7	41H-693834	XX467	8.12.97	G.R.Montgomery *(As "XX467" in TWU c/s)*	Kemble	AC
			R Jordanian AF 836, RSAF 70-617, G-9-214, XL605				
G-TVIJ	CCF Harvard 4 (T-6J-CCF Texan)	CCF4-442	G-BSBE	10.12.93	R.W.Davies	Little Robhurst Farm, Woodchurch	17. 6.05P
			Moz PLAF 1730, FAP 1730, AA+652, 52-8521		*(As "28521/TA-521" in USAF c/s)*		
G-TVIP	Cessna 404 Titan Courier II	404-0644	G-KIWI	16. 8.00	Capital Trading (Aviation) Ltd	Filton	5. 2.05T
			G-BHNI, LN-LGM, SE-IFV, G-BHNI, (N5302J)				
G-TVTV	Cameron TV 90SS HAB	2357		14. 9.90	J.Krebs	(Erfstadt, Germany)	19. 8.02A
G-TWEL	Piper PA-28-181 Archer II	28-8090290	N81963	12. 6.80	IAE Ltd	Cranfield	15. 5.05T
G-TWEY	Colt 69A HAB	700		24. 7.85	N.Bland	(Didcot)	12. 1.02A
G-TWIG	Reims/Cessna F406 Caravan II	F406-0014	PH-FWD	21.10.98	Bravo Aviation Ltd *"Wee Dram"*	Coventry	22.10.07E
			F-WZDS		*(Crashed Meall Feith na Stataich, near Ullapool 22.10.04 and badly damaged)*		
G-TWIN	Piper PA-44-180 Seminole	44-7995072	N30267	6.11.78	Bonus Aviation Ltd	Cranfield	22. 5.06T
G-TWIZ	Rockwell Commander 114	14375	SE-GSP	9. 5.90	B.C.and P M Cox	Biggin Hill	17. 6.05
			N5808N				
G-TWST	Silence Twister	PFA 329-14211		7. 9.04	Zulu Glasstek Ltd	Baileys Farm, Long Crendon	
	(Built P M Wells)						
G-TWTW	Denny Kitfox Model 2	PFA 172-11730		24. 3.04	T Willford	Spetisbury, Blandford Forum	
	(Built T Willford)						
G-TXSE	Rotary Air Force RAF 2000 GTX-SE	PFA G/113-1271		1. 3.96	G.J.Layzell	(Quedgeley)	1. 1.98P
	(Built P Green)				*(New owner 2.02)*		
G-TYAK	IAV-Bacau Yakovlev Yak-52	899907	RA-01038	23.12.02	S.J.Ducker	Breighton	10. 8.05P
			LY-AIE, DOSAAF 94 *(yellow)*				
G-TYCN	Agusta A109E Power	11123	G-VMCO	30. 7.03	A J. Walter (Aviation) Ltd	(Horsham)	30. 7.07T
G-TYER	Robin DR500/200i Président	0021	F-GTZB	25. 4.00	Alfred Graham Ltd	Southend	14. 5.06
	(Registered as DR400/500)						
G-TYGA	Gulfstream AA-5B Tiger	AA5B-1161	G-BHNZ	22. 2.82	D H and R J Carman	Biggin Hill	29. 1.07T
			(D-EGDS), N4547L				
G-TYGR	Best Off Skyranger 912S	BMAA/HB/420		8.11.04	M.J.Poole	(Barnsley)	
	(Built M.J.Poole - kit no. UK/498)						
G-TYKE	Jabiru Aircraft Jabiru UL-450	PFA 274A-13739		8. 6.01	A Parker	Rufforth	9. 7.05P
	(Built A Parker)						
G-TYNE	SOCATA TB-20 Trindad	1523	F-GRBM	6.11.97	D.T.Watkins	Newcastle	26.11.06T
			F-WWRW, CS-AZH, F-OHDE				
G-TYRE	Reims/Cessna F172M Skyhawk II	F17201222	OY-BIA	16. 2.79	J.A Lyons	Willington Court, Sandhurst	2. 9.06T
					t/a Staverton Flying School		
G-TZEE	SOCATA TB-10 Tobago	727	F-GFQG	9. 1.03	Zytech Ltd	Earls Colne	4. 2.06T
G-TZII	Thorp T.211B	PFA 305-13285		2. 6.99	AD Aviation Ltd	Barton	AC
	(Built AD Aerospace Ltd)				*(Noted 7.04)*		

G-UAAA - G-UZZZ

Reg	Type	C/n	Previous identity	Regd date	Owner/Operator	Base	CofA expiry
G-UACA	Best Off Skyranger R100	BMAA/HB/324		25. 6.04	R.G.Openshaw	(Faversham)	
	(Built R.Openshaw - kit no.SKRxxxx384)						
G-UAKE	North American P-51D-5-NA Mustang	109-27587	44-13954	17. 2.04	The Mustang Restoration Co Ltd	(Coventry)	
G-UANT	Piper PA-28-140 Cherokee F	28-7325568	OO-MYR	12. 4.02	Air Navigation and Trading Co Ltd	Blackpool	18.12.05
			N56084				
G-UAPA	Robin DR400/140B Major	2213	F-GMXC	11. 1.95	Carlos Saraiva Lda	(Alges, Portugal)	20.10.07E
G-UAPO	Ruschmeyer R90-230RG	019	D-EECT	2. 3.95	P.Randall	Sturgate	12. 7.07
G-UAVA	Piper PA-30 Twin Comanche B	30-1413	D-GLDU	16. 6.03	M.D.Northwood	Enstone	AC
			HB-LDU, N8279Y		*(Noted 10.03)*		
G-UCCC	Cameron Sign 90SS HAB	3918		5. 7.96	B.Conway	(Wheatley, Oxford)	6. 9.99A
G-UDAY	Robinson R22 Beta	1101		4. 8.89	D.J.Fowler	Jenkins Farm, Navestock	29. 5.05T
G-UDGE	Thruster T.600N Sprint	9099-T600N-037	G-BYPI	17. 9.99	A.P.Scott	Shobdon	3.12.04P
	(Rotax 503UL) *(Offically regd with Rotax 582)*				tr G-UDGE Syndicate		
G-UDOG	Scottish Aviation Bulldog Series 120/121	BH.120/204	XX518	24. 1.02	Gamit Ltd *(As "XX518/S")*	Sleap	21. 4.06T
G-UEST	Bell 206B JetRanger II	1484	G-RYOB	8. 9.89	Cairnsilver Ltd	(London SW1H)	1. 2.06T
			G-BLWU, ZS-PAW				
G-UESY	Robinson R22 Beta II	2801		13. 3.98	Plane Talking Ltd *(Op London Helicopters)*	Redhill	2. 4.04T
G-UFAW	Raj Hamsa X'Air 582	BMAA/HB/167		24. 7.01	T.J.Butler	(Beaminster)	8. 7.04P
	(Built J.H.Goddard - kit no.582)						
G-UFCB	Cessna 172S Skyhawk SP	172S8318	N455SP	25. 1.00	The Cambridge Aero Club Ltd	Cambridge	13 .3.06T
G-UFCC	Cessna 172S Skyhawk SP	172S8611	N2466X	8. 1.01	Skypix Aviation Ltd	Old Sarum	31. 3.07T

Reg	Type	c/n	Previous identity	Date	Owner	Base	CofA
G-UFCD	Cessna 172S Skyhawk SP	172S8443	G-OYZK N7262C	4. 1.01	E-Options Ltd	(Huntingdon)	11. 9.06T
G-UFCE	Cessna 172S Skyhawk SP	172S9305	N5318Y	20. 2.03	Ulster Flying Club (1961) Ltd	Newtownards	6. 3.06T
G-UFCF	Cessna 172S Skyhawk SP	172S9306	N5320Y	20. 2.03	Ulster Flying Club (1961) Ltd	Newtownards	4. 3.06T
G-UFCG	Cessna 172S Skyhawk SP	172S9450	N2154T	28. 7.03	Ulster Flying Club (1961) Ltd	Newtownards	7. 8.06T
G-UFCH	Cessna 172S Skyhawk SP	172S9507	N53551	4.11.03	Ulster Flying Club (1961) Ltd	Newtownards	12.11.06T
G-UFLY	Cessna F150H *(Built Reims Aviation SA)*	F150-0264	G-AVVY	29. 9.89	Westair Flying Services Ltd	Blackpool	11. 3.05T
G-UGLY	Sud-Aviation SE.313O Alouette II	1500	G-BSFN XP967	7. 6.00	L.Smith t/a Helicopter Services	Booker	16. 7.04T
G-UILD	Grob G109B	6419		28. 1.86	Runnymede Consultants Ltd	Blackbushe	26. 5.07
G-UILE	Neico Lancair 320 *(Built R.J.Martin)*	PFA 191-12538		17. 1.94	R.J.Martin *(Stored mid 2004)*	Shoreham	
G-UILT	Cessna T303 Crusader	T30300280	G-EDRY N4817V	3. 7.00	W.J.Forrest tr G-UILT Group	Barton	10. 4.05
G-UINN	Stolp SA.300 Starduster Too *(Built Haakon Baaken)* (Lycoming O-360)	HB.1980-1	EI-CDQ C-GTLJ	16. 3.98	J.D.H.Gordon *(Also carries "EI-CDQ") (Noted 4.04)*	Charterhall	9.12.05P
G-UIST	British Aerospace Jetstream Series 3102	750	N190PC (N331QH), N840JS, G-31-750	15. 3.02	Highland Airways Ltd	Inverness	27. 5.05T
G-UJAB	Jabiru Aircraft Jabiru UL *(Built C.A Thomas)*	PFA 274A-13373		27. 1.99	C.A Thomas	Top Farm, Croydon, Royston	26. 8.04P
G-UJGK	Jabiru Aircraft Jabiru UL-450 *(Built W.G.Upton and J.G.Kosak)*	PFA 274A-13558		17. 4.00	W.G.Upton and J.G.Kosak	RNAS Culdrose	16. 5.05P
G-UKAG	British Aerospace BAe 146 Series 300	E3162	G-6-162	28.11.90	KLM UK Ltd *(Stored 4.04)*	Norwich	11.12.04T
G-UKAT	Aero AT-3 *(Built T.Archer)*	PFA 327-14107		14. 4.04	T.Archer tr G-UKAT Group	North Weald	
G-UKOZ	Jabiru Aircraft Jabiru SK *(Built D.J.Burnett)*	PFA 274-13310		16. 6.99	D.J.Burnett	White House Farm, Southery	7. 9.05P
G-UKRC	British Aerospace BAe 146 Series 300	E3158	G-BSMR G-6-158	14. 2.91	KLM UK Ltd	Norwich	24. 2.05T
G-UKSC	British Aerospace BAe 146 Series 300	E3125	G-5-125	26.10.88	Trident Jet (Jersey) Ltd *(Stored 4.04)*	Filton	9. 3.05T
G-UKTA	Fokker F.27 Mk 050 *(Fokker 50)*	20246	PH-KXF	22. 2.95	KLM UK Ltd *"City of Norwich" (Stored 4.04)*	Norwich	21. 2.04T
G-UKTB	Fokker F.27 Mk.050 *(Fokker 50)*	20247	PH-KXG	21. 3.95	KLM UK Ltd *"City of Aberdeen" (Stored 4.04)*	Norwich	21. 3.04T
G-UKTD	Fokker F.27 Mk.050 *(Fokker 50)*	20256	PH-KXT	20. 1.95	KLM UK Ltd *"City of Leeds" (Stored 4.04)*	Norwich	19. 1.04T
G-UKUK	Head Ax8-105 HAB	248	N8303U	1. 9.97	P.A George *"Union Jack"*	(Princes Risborough)	14. 8.03A
G-ULAS	de Havilland DHC-1 Chipmunk 22	C1/0554	WK517	14. 6.96	ULAS Flying Club Ltd *(As "WK517")*	(London N3)	11. 9.05
G-ULES	Aérospatiale AS355F2 Ecureuil 2	5364	G-OBHL G-HARO, G-DAFT, G-BNNN	6. 3.03	Select Plant Hire Company Ltd	Southend	20. 1.06T
G-ULHI	Scottish Aviation Bulldog Series 100/101	BH100/148	G-OPOD SE-LLK, Fv61038, G-AZMS	30. 9.03	Power Aerobatics Ltd	Kemble	10.10.07T
G-ULIA	Cameron V-77 HAB	2860		20. 5.92	J.M.Dean	(Oswestry)	15. 6.04A
G-ULLS	Lindstrand LBL 90A HAB	434		18. 2.97	J.R.Clifton	(Brackley)	11.10.03A
G-ULLY	Thruster T600N 450 Sprint	0043-T600N-099	G-CCRP	17. 3.04	P.J.Fahie tr The Swallow 2 Group *(Op Swallow Aviation)*	Compton Abbas	16. 3.05P
G-ULPS	Everett Gyroplane Series 1 (Volkswagen 1835)	007	G-BMNY	13. 7.93	C.J.Watkinson	(Goole)	10. 7.01P
G-ULSY	Comco Ikarus C42 FB80	0405-6603		26. 7.04	P.J.Fahie tr Ikarus 1 Flying Group *(Op Swallow Aviation)*	Compton Abbas	25. 7.05P
G-ULTR	Cameron A-105 HAB	4100		24. 2.97	P.Glydon *(Ultrafilter titles)*	(Bristol)	6. 7.05T
G-UMMI	Piper PA-31 Navajo C	31-7912060	G-BGSO N3519F	11. 8.92	J.A, G.M, D.T.A and J.A Rees t/a Messrs Rees of Poynston West	Haverfordwest	14. 8.06T
G-UNDD	Piper PA-23-250 Aztec E	27-4832	G-BATX N14271	22. 3.00	G.J.and D.P.Deadman	Goodwood	27. 9.07E
G-UNER	Lindstrand LBL 90A HAB	895		16. 4.03	S.T.Armstrong tr Royal Artillery Display Troop	(London SE18)	14. 4.04A
G-UNGE	Lindstrand LBL 90A HAB	122	G-BVPJ	6.12.96	M.T.Stevens tr Silver Ghost Balloon Club	(Warwick)	13. 7.05A
G-UNGO	Pietenpol AirCamper *(Built A R.Wyatt and P Thody)*	PFA 47-13951		16. 9.02	A R.Wyatt	(Buntingford)	
G-UNIP*	Cameron Oil Container SS HAB				See SECTION 3 Part 1	(Aldershot)	
G-UNIV	Montgomerie-Parsons Two-Place Gyroplane *(Built J M Montgomerie)* (Rotax 618)	PFA G/8-1276	G-BWTP	3. 8.99	Department of Aerospace Engineering, University of Glasgow	(Glasgow)	18. 1.05P
G-UNYT	Robinson R22 Beta	0985	G-BWZV G-LIAN	17.11.97	D.I.Pointon	(Coventry)	16.11.06T
G-UPFS	Waco UPF-7	5660	N32029	27. 8.04	D.N.Peters and N.R.Finlayson	Little Gransden	11.11.07
G-UPHL	Cameron Concept 80 HAB	3002		23. 2.93	CSM (Weston) Ltd t/a Uphill Motor Co *(Uphill Motors titles)*	(Weston-super-Mare)	12. 4.05T
G-UPPI	British Aircraft Corporation 167 Strikemaster Mk.80A	EEP/JP/159	G-CBPB R.Saudi AF 1108, G-27-27	20. 5.03	Gower Jets Ltd *(As "FAE 259/T59" in Ecuadorian AF c/s)*	Swansea	19. 5.04P
G-UPPP	Colt 77A HAB	852		4. 8.86	M.Williams *"Nugget"*	(Wadhurst)	21. 2.04A
G-UPPY	Cameron DP-80 HA Airship	2274		29. 3.90	Jacques W.Soukup Enterprises Ltd *"Jacques Soukup"*	(Beaulieu Court, Wilts/Great Missenden)	27. 8.94A
G-UPUP	Cameron V-77 HAB	1828		21. 7.89	S.R.Burden *"Fantasia"*	(Noordwijk, The Netherlands)	9. 5.04T
G-UPUZ	Lindstrand LBL 120A HAB	969		26. 1.04	C.J.Sanger-Davies	(Oswestry)	27. 1.05A
G-UROP	Beech 95-B55 Baron	TC-2452	N64311	17. 9.90	Pooler International Ltd	Sleap	4. 9.06T
G-URRR	Air Command 582 Sport *(Built L Armes - c/n 0630)*	PFA G/4-1200		13. 6.90	L.Armes *(Current status unknown)*	(Basildon)	
G-URUS	Maule MX-7-180B Super Rocket	22014C	N611BH	28. 6.04	Broomco Ltd	(Camberley)	AC
G-USAM (2)	Cameron Uncle Sam SS HAB *(Uncle Sam head shape) (New envelope c/n 4526 c.3.99)*	1120		20. 5.85	Corn Palace Balloon Club Ltd *(Active Albuquerque, New Mexico, US 10.03)*	(Tyndale, South Dakota, US)	27. 6.00A
G-USIL	Thunder Ax7-77 HAB	1587		22. 8.89	Window on the World Ltd *"Mantis"*	(London SE1)	30. 4.05A
G-USKY	Aviat A-1B Husky	2261		8. 4.04	A D and Carole Realff t/a ADR Aviation	Shoreham	18. 6.07T
G-USMC	Cameron Chestie 90SS HAB *(US Marine Corps Bulldog shape)*	1251		24. 4.86	Jacques W.Soukup Enterprises Ltd *(Active Albuquerque, New Mexico, US 10.03)*	(Tyndale, South Dakota, US)	15. 6.00A

G-USRV	Van's RV-6 *(Built F.M.Carter)*	23771	N200HC	9. 7.04	D.S.Watt	(Sydenham, Chinnor)	6.10.05P
G-USSI	Stoddard-Hamilton Glasair III *(Built H R Rotherwick)*	3380		30.12.03	Lord Rotherwick	Oxford	AC
G-USSR	Cameron Doll 90SS HAB *(Russian Doll shape)*	2273		29. 3.90	Corn Palace Balloon Club Ltd *"Matrioshka" (Active Albuquerque, New Mexico, US 10.03)*	(Tyndale, South Dakota, US)	9. 6.00A
G-USSY	Piper PA-28-181 Archer II	28-8290011	N8439R	7.11.88	Western Air (Thruxton) Ltd	Thruxton	18. 2.07T
			(Damaged Bedlam Street, Hurstpierpoint 27.3.99: cancelled 5.8.99 as WFU) (Stored w/o boom 12.00)				
G-USTB	Agusta A109A	7163	D-HEEG (D-HEEF), VR-CKN, HB-XKM	9. 6.97	Newton Aviation Ltd	Redhill	19. 7.06T
G-USTC	Agusta A109C	7649	JA6695 (G-LAXO)	16.12.04	MW Helicopters Ltd	Stapleford	AC
G-USTS	Agusta A109A II	7275	G-MKSF N18SF, F-GDPR	24.11.03	MB Air Ltd t/a Eagle Helicopters	Newcastle City Heliport	30. 1.05T
G-USTV*	Messerschmitt Bf.109G-2/Trop				See SECTION 3 Part 1	Hendon	
G-USTY	Clutton FRED Series III *(Built S Styles)*	PFA 29-10390		11.10.78	R.G.Hallam *(New owner 7.04)*	Netherthorpe	8. 12.05P
G-USUK*	Colt 2500A HAB				See SECTION 3 Part 1	Duxford	
G-UTSI	Rand Robinson KR-2 *(Built K.B.Gutridge)*	KBG-01		2.10.89	K.B.Gutridge *(Stored 6.03)*	Biggin Hill	
G-UTSY	Piper PA-28R-201 Cherokee Arrow III	28R-7737052	N3346Q	29. 8.86	Arrow Aviation Ltd	Southend	8. 2.05
G-UTTS	Robinson R44 Raven	0865	G-ROAP	20.10.00	Heli Hire Ltd	Gamston	4.12.06T
G-UTZI	Robinson R44 Raven II	10590		24.12.04	Heli Air Ltd	Wellesbourne Mountford	AC
G-UVIP	Cessna 421C Golden Eagle III	421C0603	G-BSKH N88600	23.11.98	Capital Trading Aviation Ltd	Filton	2.11.05T
G-UVNA	Piper PA-24-260 Comanche B	24-4306	G-BAHG 5Y-AFX, N8831P	30. 7.03	D.G.Sheppard	New Farm House, Great Oakley	14. 9.06
G-UVNR	British Aircraft Corporation 167 Strikemaster Mk.87 *(Or PS.174?)*	EEP/JP/2876 & PS.168	G-BXFS Botswana DF OJ10, Kenyan AF 605, G-27-195	4. 5.01	Global Aviation Services Ltd	Humberside	23.12.04P
G-UZEL	Aérospatiale SA.341G Gazelle 1	1413	G-BRNH YU-HBO	21.11.89	Fairalls of Godstone Ltd	Redhill	9. 6.07
G-UZLE	Colt 77A HAB	2021		1. 8.91	Flying Pictures Ltd *"John Courage"*	(Chilbolton)	25. 5.00A
G-UZZY	Enstrom 480	5013	G-BWMD (F-GOTA), G-BWMD	29.10.04	A.A.Wood t/a Shoreham Helicopters	Shoreham	14. 7.07E

G-VAAA - G-VZZZ

G-VAEL	Airbus A340-311	015	F-WWJG	15.12.93	Virgin Atlantic Airways Ltd *"Maiden Toulouse" (For disposal 2004)*	Gatwick	14.12.05T
G-VAIR	Airbus A340-313	164	F-WWJA	21. 4.97	Virgin Atlantic Airways Ltd *"Maiden Tokyo"*	Gatwick	20. 4.06T
G-VALE*	North American AT-6C Harvard IIA				See SECTION 3 Part 1	Bloomingdale, Illinois, US	
G-VALS	Pietenpol AirCamper *(Built J.R.D.Bygraves)*	PFA 47-13157		30. 7.97	J.R.D.Bygraves *(New owner and builder 11.04)*	Little Gransden	
G-VALV	Robinson R44 Raven	1421		9. 9.04	Valve Train Components Ltd	(Matlock)	4.10.07E
G-VALY	SOCATA TB-21 Trinidad	2081	N246SS (N717TB)	30. 3.04	Valley Flying Co Ltd	Valley Farm, Stafford	27. 4.07
G-VALZ	Cameron N-120 HAB	4998		9. 1.01	D Ling	(Nottingham)	10 11.04T
G-VANN	Van's RV-7A *(Built D.N. and J A.Carnegie)*	PFA 323-14034		5. 7.04	D.N.and Julia A.Carnegie	(St. Bees)	
G-VANS	Van's RV-4 *(Built T Saylor)* (Lycoming O-320)	355	N16TS	7. 9.92	M.Swanborough and D.Jones *(Noted 1.05)*	Breighton	17. 3.04P
G-VANZ	Van's RV-6A *(Built S J Baxter)*	PFA 181-12531		15. 7.93	S.J.Baxter *(Under construction 2000)*	(Macclesfield)	
G-VARG	Varga 2150A Kachina	VAC 157-80	OO-RTY N80716	14. 5.84	R.A.Denton	Sturgate	23. 6.05
G-VART	Rotorway Executive 90 *(Built N J Bethell)* (RotorWay RI 162)	5003	G-BSUR	11. 8.03	G.Varty	(Whitley Bay)	1.12.93P
G-VASA	Piper PA-34-200 Seneca	34-7350080	G-BNNB (N...), G-BNNB, N15625	29. 3.96	H.Hafez	(Westhampnett, Chichester)	10. 7.06T
G-VAST	Boeing 747-41R	28757		17. 6.97	Virgin Atlantic Airways Ltd *"Ladybird"*	Heathrow	16. 6.06T
G-VATL	Airbus A340-642	376	F-WWCC	31.10.03	Virgin Atlantic Airways Ltd *"Miss Kitty"*	Heathrow	30.10.06T
G-VBAC	Short SD.3-60 Variant 100	SH.3736	VH-MJU G-BOEJ, G-14-3736	15. 9.97	BAC Leasing Ltd *(Op Aerocondor)*	Glasgow	31.10.05E
G-VBEE	Boeing 747-219B	22723	TF-ATN G-VBEE, ZK-NZW	5. 4.99	Air Atlanta Europe Ltd	Gatwick	6. 5.07T
G-VBIG	Boeing 747-4Q8	26255		10. 6.96	Virgin Atlantic Airways Ltd *"Tinkerbelle"*	Gatwick	9. 6.05T
G-VBUS	Airbus A340-311	013	F-WWJE	26.11.93	Virgin Atlantic Airways Ltd *"Lady in Red"*	Gatwick	25.11.06T
G-VCED	Airbus A320-231	0193	OY-CNI F-WWIX	21. 1.97	MyTravel Airways Ltd	Birmingham	30. 1.06T
G-VCIO	EAA Acrosport II *(Built F Sharples and R F Bond)* (Lycoming O-360)	PFA 72A-12388		9.10.97	V Millard	Crowfield	27. 5.05P
G-VCML	Beech 58 Baron	TH-1346	N2289R	31.10.97	St.Angelo Aviation Ltd	Dunkeswell	14. 3.07T
G-VDIR	Cessna 310R II	310R0211	N5091J	31. 1.91	J.Driver	Elstree	21. 6.07T
G-VECD	Robin R1180T Aiglon II	234	F-GCAD	22. 6.00	Mistral Aviation Ltd	Goodwood	20. 5.07T
G-VECE	Robin R2120U	355		11. 5.01	Mistral Aviation Ltd	Goodwood	8. 7.07T
G-VECG	Robin R2160 Alpha Sport	322	F-GSRD	6. 2.02	Mistral Aviation Ltd	Goodwood	17. 2.05T
G-VEEE	Robinson R22 Beta II	3172	G-REDA	13.10.04	Veee Helicopters Ltd	(Bolney, Haywards Heath)	31. 3.07E
G-VEIL	Airbus A340-642	575	F-WWCK	8. 4.04	Virgin Atlantic Airways Ltd *"Dancing Girl"*	Heathrow	7. 4.07T
G-VEIT	Robinson R44 Raven II	10091		11. 6.03	Field Marshall Helicopters Ltd	(Barkisland, Halifax)	27. 7.06T
G-VELA	SIAI-Marchetti S.208A Waco Vela *(Officially regd as "S.205-22R")*	4-149	N949W	30.10.89	Broadland Flyers Ltd	Norwich	29. 4.05
G-VELD	Airbus A340-313	214	F-WWJY	16. 3.98	Virgin Atlantic Airways Ltd *"African Queen"*	Gatwick	15. 3.07T

G-VENI	de Havilland DH.112 Venom FB.50 (FB.1) *(Built F + W)*	733	Swiss AF J-1523	8. 6.84	Lindsay Wood Promotions Ltd *(Stored as "VV612" in silver RAF c/s 1.05)*	Bournemouth	25. 7.01P
G-VENM	de Havilland DH.112 Venom FB.50 (FB.1) *(Built F + W)*	824	G-BLIE Swiss AF J-1614	16. 6.99	T.J.Manna *(On loan de Havilland Aircraft Heritage Centre as "WK436" in 11 Sqdn c/s)*	Salisbury Hall, London Colney	23. 3.05P
G-VENT	Schempp-Hirth Ventus-2cM	3/17	(BGA 4918) D-KBTL	25. 9.01	D.Rance *"TL"* *(Noted 12.04)*	Nympsfield	5.11.04
G-VERA	Gardan GY-201 Minicab *(Built D K Shipton)*	PFA 56-12236		7. 6.94	D.K.Shipton *(Current status unknown)*	(Peterborough)	
G-VERN	Piper PA-32R-300 Cherokee Lance	32R-7680151	G-BVBG N19BP, N8363C	16. 4.03	P.M.Moyle	Haverfordwest	14. 9.07
G-VETA	Hawker Hunter T.7	41H-693751	G-BVWN A2729, XL600	2. 7.96	Gower Jets Ltd	North Weald	15. 6.05P
G-VETS	Enstrom 280C-UK Shark	1015	G-FSDC G-BKTG, OY-HBP	11. 9.95	A.J.Warburton *(Noted 9.04)*	Barton	28. 7.02
G-VEYE	Robinson R22	0140	G-BPTP N9056H	8. 2.00	RK Transport Services Ltd *(New owners 11.04)*	(Willenhall)	2. 6.01T
G-VEZE	Rutan VariEze *(Built S D Brown, S Evans and M Roper)*	PFA 74-10285		2. 9.77	S.D.Brown, S.Evans and M.Roper	Biggin Hill	3.10.05P
G-VFAB	Boeing 747-4Q8	24958		28. 4.94	Virgin Atlantic Airways Ltd *"Lady Penelope"*	Gatwick	27. 4.06T
G-VFAR	Airbus A340-313X	225	(G-VPOW) F-WWJZ	12. 6.98	Virgin Atlantic Airways Ltd *"Diana"*	Gatwick	11. 6.04T
G-VFLY	Airbus A340-311	058	F-WWJE	24.10.94	Virgin Atlantic Airways Ltd *"Dragon Lady"*	Gatwick	23.10.06T
G-VFOX	Airbus A340-642	449	F-WWCM	23.12.02	Virgin Atlantic Airways Ltd *"Silver Lady"*	Heathrow	22.12.05T
G-VGAL	Boeing 747-443	32337	(EI-CVH)	26. 4.01	Virgin Atlantic Airways Ltd *"Jersey Girl"*	Gatwick	25. 4.07T
G-VGAS	Airbus A340-642	639	F-WW..	5.05R	Virgin Atlantic Airways Ltd *"Varga Girl" (On order 5.05)*		
G-VGMC	Eurocopter AS355N Ecureuil 2	5693	G-HEMH F-WQPV	2. 3.04	Finlay (Holdings) Ltd	(Augher, County Tyro)	20.12.07E
G-VGOA	Airbus A340-642	371	F-WWCB	30. 8.03	Virgin Atlantic Airways Ltd *"Indian Princess"*	Heathrow	29. 8.06T
G-VHOL	Airbus A340-311	002	F-WWAS	30. 5.97	Virgin Atlantic Airways Ltd *"Jetstreamer"*	Gatwick	29. 5.06T
G-VHOT	Boeing 747-4Q8	26326		12.10.94	Virgin Atlantic Airways Ltd *"Tubular Belle"*	Gatwick	11.10.06T
G-VIBA	Cameron DP-80 HA Airship	1729		28. 5.91	Jacques W.Soukup Enterprises Ltd	Beaulieu Court, Wiltshire/Great Missenden	3. 2.99A
G-VIBE	Boeing 747-219B	22791	ZK-NZZ 9M-MHH, ZK-NZZ, N6108N	24. 9.99	Virgin Atlantic Airways Ltd *"Spirit of New York"* *(Stored 12.04)*	Mojave-Kern County, California, US	23. 9.02T
G-VICC	Piper PA-28-161 Warrior II	28-7916317	G-JFHL N2249U	3. 3.92	T.Wenham tr Charlie Charlie Syndicate	Hinton-in-the-Hedges	5. 8.07T
G-VICE	MD Helicopters Hughes 369E	0365E	D-HLIS	16. 5.95	B.T.Andersen	Sywell	2.12.07E
G-VICI	de Havilland DH.112 Venom FB.50 (FB.1) *(Built F + W)*	783	HB-RVB (G-BMOB), Swiss AF J-1573	6. 2.95	Lindsay Wood Promotions Ltd *(Stored as "J-1573" in Swiss AF c/s 1.05)*	Bournemouth	24.11.99P
G-VICM	Beech F33C Bonanza	CJ-136	PH-BNG	3. 7.91	Velocity Engineering Ltd	Elstree	20. 5.06
G-VICS	Commander Aircraft Commander 114B	14655	N655V	3. 2.98	Millennium Aviation Ltd	Guernsey	17. 4.07
G-VICT	Piper PA-31 Turbo Navajo B	31-7401211	G-BBZI N7590L	10. 9.99	Heliquick Ltd	Bournemouth	13. 8.05T
G-VIEW	Vinten Wallis WA-116/L (Limbach L2000)	002		5. 7.82	K.H.Wallis *(Stored 8.01)*	Reymerston Hall, Norfolk	6.10.85P
G-VIIA	Boeing 777-236	27483	N5022E (G-ZZZF)	3. 7.97	British Airways plc	Gatwick	2. 7.06T
G-VIIB	Boeing 777-236	27484	N5023Q (G-ZZZG)	23. 5.97	British Airways plc	Heathrow	30. 1.06T
G-VIIC	Boeing 777-236	27485	N5016R (G-ZZZH)	6. 2.97	British Airways plc	Heathrow	20. 8.05T
G-VIID	Boeing 777-236	27486	(G-ZZZI)	18. 2.97	British Airways plc	Heathrow	15. 9.05T
G-VIIE	Boeing 777-236	27487	(G-ZZZJ)	27. 2.97	British Airways plc	Heathrow	23. 9.05T
G-VIIF	Boeing 777-236	27488	(G-ZZZK)	19. 3.97	British Airways plc	Heathrow	1.11.05T
G-VIIG	Boeing 777-236	27489	(G-ZZZL)	9. 4.97	British Airways plc	Heathrow	8. 4.06T
G-VIIH	Boeing 777-236	27490	(G-ZZZM)	7. 5.97	British Airways plc	Heathrow	6. 5.06T
G-VIII*	Vickers Supermarine 359 Spitfire LF.VIII				See SECTION 3 Part 1	Dallas, Texas, US	
G-VIIJ	Boeing 777-236	27492	(G-ZZZP)	29.12.97	British Airways plc	Gatwick	21.12.06T
G-VIIK	Boeing 777-236	28840		3. 2.98	British Airways plc	Gatwick	2. 2.07T
G-VIIL	Boeing 777-236	27493		13. 3.98	British Airways plc	Heathrow	12. 3.07T
G-VIIM	Boeing 777-236	28841		26. 3.98	British Airways plc	Gatwick	13. 9.06T
G-VIIN	Boeing 777-236	29319		21. 8.98	British Airways plc	Heathrow	20. 8.07T
G-VIIO	Boeing 777-236	29320		26. 1.99	British Airways plc8 *(Chelsea Rose t/s)*	Gatwick	25. 1.05T
G-VIIP	Boeing 777-236	29321		9. 2.99	British Airways plc	Heathrow	8. 2.05T
G-VIIR	Boeing 777-236	29322		18. 3.99	British Airways plc	Gatwick	17. 3.05T
G-VIIS	Boeing 777-236	29323		1. 4.99	British Airways plc *(Chelsea Rose t/s)*	Heathrow	31. 3.05T
G-VIIT	Boeing 777-236	29962		26. 5.99	British Airways plc	Heathrow	25. 5.05T
G-VIIU	Boeing 777-236	29963		28. 5.99	British Airways plc	Heathrow	27. 5.05T
G-VIIV	Boeing 777-236	29964		29. 6.99	British Airways plc	Gatwick	28. 6.05T
G-VIIW	Boeing 777-236	29965		30. 7.99	British Airways plc	Heathrow	29. 7.05T
G-VIIX	Boeing 777-236	29966		11. 8.99	British Airways plc	Gatwick	10. 8.05T
G-VIIY	Boeing 777-236	29967		22.10.99	British Airways plc	Heathrow	21.10.05T
G-VIKE	Bellanca 17-30A Super Viking 300A	79-30911	N302CB	8. 7.80	W.G.Prout	Lee-on-Solent	16. 6.05
G-VIKY	Cameron A-120 HAB	3068		27. 4.93	P.J.Stapley *(Active 1.04)*	(London Colney)	26. 8.05A
G-VILA	Jabiru Aircraft Jabiru UL *(Built D Cassidy)*	PFA 274A-13364	G-BYIF	18. 7.02	R.W.Sage and T.R.Villa	Priory Farm, Tibenham	23. 6.05P
G-VILL	Laser Z200 Lazer *(Built M G Jefferies)* (Lycoming AEIO-360)	10	G-BOYZ	10. 6.96	M.G.Jefferies *(Global Village titles)*	Little Gransden	25. 8.05P
G-VINO	Sky 90-24 HAB	102		25. 2.98	Fivedata Ltd *(Lambrini Bianco titles)*	(Todmorden)	20. 5.04A
G-VIPA	Cessna 182S Skylane	18280720	N148ME	13. 9.00	Rollright Aviation Ltd	(Chipping Norton)	29.10.06T
G-VIPH	Agusta A109C	7643	EI-CUV G-BVNH, G-LAXO	21. 9.01	Cheqair Ltd	(Norwich)	23. 9.07T
G-VIPI	British Aerospace BAe 125 Series 800B	258222	G-5-745	27. 7.92	Yeates of Leicester Ltd	Farnborough	28. 9.06T

Registration	Type	C/n	Previous identity	Reg date	Owner/operator	Base	CofA expiry
G-VIPP	Piper PA-31-350 Navajo Chieftain	31-7952244	G-OGRV	6. 8.93	Capital Trading Aviation Ltd	Filton	28. 8.05T
			G-BMPX, N3543D				
G-VIPY	Piper PA-31-350 Navajo Chieftain	31-7852143	EI-JTC	10.10.97	Capital Trading Aviation Ltd	Filton	12.10.05T
			G-POLO, (EI-...), G-POLO, N27750				
G-VITE	Robin R1180T Aiglon	219		16.10.78	D.C.Perrett and D.T.Scrutton	Stapleford	18. 9.06
					tr G-VITE Flying Group		
G-VITL	Lindstrand LBL 105A HAB	720		24. 8.00	Actionstrength Ltd	(Manchester)	5. 1.05A
					t/a Vital Resources *(Vital Resources titles)*		
G-VITO	British Aerospace Jetstream Series 3102	829	SX-BNJ	6. 8.04	Aceline Air Ltd *(Op Quest Airlines)*	Coventry	20. 1.06E
			C-GMDJ, G-BSIW, G-OEDL, G-OAKK, G-BSIW, HB-AED, G-31-829, C-FCPG, G-31-829				
G-VIVA	Thunder Ax7-65 Bolt HAB	190		28.11.78	R.J.Mitchener *(Inflated 4.02)*	(Andover)	18. 3.99A
G-VIVI	Taylor JT.2 Titch	PFA 60-12405		4.11.96	D.G.Tucker	Hill Farm, Nayland	27. 1.05P
	(Built D.G.Tucker)						
G-VIVM	Hunting Percival P.84 Jet Provost T.5	PAC/W/23907	G-BVWF	25. 3.96	The Skys The Ltd	North Weald	7.12.05P
			XS230		(International Test Pilots School titles)		
G-VIXN	de Havilland DH.110 Sea Vixen FAW.2 (TT)	10145	8828M	5. 8.85	P.G.Vallance Ltd	Charlwood, Surrey	AC
			XS587		*(Gatwick Aviation Museum as "XS587 in RN c/s)*		
G-VIZZ	Sportavia RS.180 Sportsman	6018	D-EFBK	25.10.79	J.D.Howard and S.J.Morris	Exeter	23. 7.07
					tr Exeter Fournier Group		
G-VJAB	Jabiru Aircraft Jabiru UL	PFA 274-13322		25. 6.98	A S.R.Milner	Ince Blundell	27. 4.05P
	(Built S T Aviation Ltd) (PFA c/n prefix should be "274A-")						
G-VJET	Avro 698 Vulcan B.2	-	XL426	7. 7.87	R.J.Clarkson	Southend	
					tr The Vulcan Restoration Trust *(Noted as "XL426/G-VJET" 2.05)*		
G-VJIM	Thunder and Colt Jumbo SS HAB	1298	(G-BPJI)	7. 8.89	Magical Adventures Ltd *(Virgin Atlantic titles)*		
	(Registered as Colt Jumbo-2)				*"Jumbo Jim"*	(West Bloomfield, Michigan, US)	14.10.02A
G-VKIT	Europa Aviation Europa	PFA 247-12783		11. 6.01	T.H.Crow	Bicester	
	(Built T.H.Crow - kit no.163) (Monowheel u/c)				*(Noted 1.05)*		
G-VKNG	Boeing 767-3Z9ER	23765	OE-LAU	10. 9.04	Air Atlanta Europe Ltd	Gatwick	25.11.07E
			N6009N, (OE-LAA), N767PW		*(Operated by Air Atlanta for Excel Airways)*		
G-VLAD	Yakovlev Yak-50	791502	D-EIVR	14.11.88	M.B.Smith	Top Farm, Croydon, Royston	11. 9.02P
			N51980, DDR-WQR, DM-WQR				
G-VLCN	Avro 698 Vulcan B.2		XH558	6. 2.95	C.Walton Ltd	Bruntingthorpe	AC
					(As "XH558") (Noted 12.03)		
G-VLIP	Boeing 747-443	32338	(EI-CVI)	15. 5.01	Virgin Atlantic Airways Ltd *"Hot Lips"*	Gatwick	14. 5.07T
G-VMCG	Piper PA-38-112 Tomahawk	38-79A0950	G-BSVX	12. 9.03	J.McGarry	Shoreham	29. 6.06T
			N2336P				
G-VMDE (2)	Cessna P210N Pressurized Centurion II	P21000088	(N4717P)	20. 7.78	P.L.Goldberg	Standalone Farm, Meppershall	25. 5.07
G-VMEG	Airbus A340-642	391	F-WWCK	5.10.02	Virgin Atlantic Airways Ltd *"Mystic Maiden"*	Heathrow	3.10.05T
G-VMJM	SOCATA TB-10 Tobago	1361	G-BTOK	21. 4.92	Cardonstar Ltd	Enstone	7. 6.07T
G-VMPR	de Havilland DH.115 Vampire T.11	15621	8196M	13. 3.95	de Havilland Aviation Ltd	Bournemouth	2. 2.05P
			XE920		*(As "XE920/A" in 603 (County of Edinburgh) Sqdn c/s)*		
G-VMSL	Robinson R22 Alpha	0483	G-KILY	5. 2.98	L.L.F.Smith	Booker	26.12.03T
			N8561M		*(Force landed and rolled over near Turweston 4.6.01)*		
G-VNAP	Airbus A340-642	615	F-WWCE	2.05R	Virgin Atlantic Airways Ltd *"Sleeping Beauty" (On order 2.05)*		
G-VNOM	de Havilland DH.112 Venom FB.50 (FB.1)	842	Swiss AF J-1632	13. 7.84	T.J.Manna	Salisbury Hall, London Colney	AC
	(Built F + W)				*(Unmarked pod only 5.04) (On loan to de Havilland Heritage Museum)*		
G-VNUS	Hughes 269C	122-0175	G-BATT	20. 9.00	Enable International Ltd	(Stratford-upon-Avon)	14. 1.07T
G-VOAR	Piper PA-28-181 Archer III	2843011	N9256Q	3.11.95	Solent Flight Ltd	Southampton	7.1.05T
G-VODA	Cameron N-77 HAB	2208		8. 2.90	Vodafone Group plc	(Newbury)	30. 7.05A
	(New envelope c/n 4164 @ 12.97)				*(Vodafone titles)*		
G-VOGE	Airbus A340-642	416	F-WWCF	29.11.02	Virgin Atlantic Airways Ltd *"Cover Girl"*	Heathrow	28.11.05T
G-VOID	Piper PA-28RT-201 Arrow IV	28R-8118049	ZS-KTM	17. 8.87	Newbus Aviation Ltd	Shoreham	17. 2.06T
			(G-GCAA), ZS-KTM, N83232				
G-VOLT*	Cameron N-77 HAB				See SECTION 3 Part 1	(Aldershot)	
G-VONA	Sikorsky S-76A	760086	G-BUXB	15. 4.03	Von Essen Aviation Ltd	Blackbushe	16. 8.05T
			(F-GSJG), G-BUXB, VR-CCZ, N399BB, N39RP				
G-VONB	Sikorsky S-76B	760399	G-POAH	8.10.03	Von Essen Aviation Ltd	Blackbushe	17. 5.05T
G-VOND	Bell 222	47041	G-OWCG	17.11.03	Von Essen Aviation Ltd	Blackbushe	12. 3.05T
			G-VERT, G-JLBZ, G-BNDB, A40-CH				
G-VONE	Eurocopter AS.355N Ecureuil 2	5572	G-LCON	11. 3.04	Von Essen Aviation Ltd	Denham	27. 6.06T
G-VONF	Aérospatiale AS.355F1 Ecureuil 2	5262	G-BXBT	3. 2.05	Premiair Aviation Services Ltd	Denham	28. 9.07E
			G-TMMC, G-JLCO				
G-VONS	Piper PA-32R-301T Saratoga II TC	3257155	N602MA	28. 7.03	W.S.Stanley	Gloucestershire	20.11.06T
G-VOOM	Pitts S-1S	PFA 09-12989		8. 4.03	P.G.Roberts	(Maidenhead)	
	(Built P.G.Roberts)						
G-VOTE	Ultramagic M-77 HAB	77/164		10. 3.99	Flying Pictures Ltd	Chilbolton	1. 3.05A
G-VPAT	Evans VP-1 Srs 2	PFA 62-13907		11. 2.04	A P.Twort	Kittyhawk Farm, Deanland	
	(Built A.P.Twort)				*(Under construction 2004)*		
G-VPCB	Evans VP-1 Srs 2	PFA 62-13901		28. 2.03	Claudine A Bloom	Kittyhawk Farm, Deanland	
	(Built C Bloom)				*(Under construction 2004)*		
G-VPSJ	Europa Aviation Europa	PFA 247-12520		29. 7.93	J.D.Bean	(Oxford)	
	(Built J.D.Bean - kit no.023) (Monowheel u/c)				*(Current status unknown)*		
G-VPUF	Boeing 747-219B	22725	ZK-NZY	21. 3.00	Virgin Atlantic Airways Ltd *"High as a Kite"*		
			N6005C		*(Stored 12.04)*	Mojave-Kern County, California, US	21 .3.03T
G-VROC	Boeing 747-41R	32746		22.10.03	Virgin Atlantic Airways Ltd *"Mustang Sally"*	Heathrow	21.10.06T
G-VROD	Aeroprakt A-22 Foxbat	PFA 317-13991		18. 3.03	P.A Sanders	Otherton, Cannock	28. 7.05P
	(Built P A Sanders)						
G-VROE	Avro 652A Anson T.21	3634	G-BFIR	3. 3.98	Air Atlantique Ltd	Coventry	30. 6.05P
			7881M, WD413		*(As "WD413" in standard RAF silver with yellow "T" bands)*		
G-VROM	Boeing 747-443	32339	(EI-CVJ)	29. 5.01	Virgin Atlantic Airways Ltd *"Barbarella"*	Gatwick	28. 5.07T
G-VROS	Boeing 747-443	30885	(EI-CVG)	22. 3.01	Virgin Atlantic Airways Ltd *"English Rose"*	Heathrow	21. 3.07T
G-VROY	Boeing 747-443	32340	(EI-CVK)	18. 6.01	Virgin Atlantic Airways Ltd *"Pretty Woman"*	Gatwick	17. 6.07T
G-VRST	Piper PA-46-350P Malibu Mirage	4636189		7.12.98	Winchfield Development Ltd	Fairoaks	24. 2.05

G-VRTX	Enstrom 280FX Shark	2044	G-CBNH Chilean Army H-180	8. 7.02	Bladerunner Aviation Ltd	(Altrincham)	11.12.05T
G-VRVI	Cameron O-90 HAB	2522		27. 2.91	SNT Property Ltd	(Bristol)	5. 8.05A
G-VSBC	Beech B200 Super King Air	BB-1290	N3185C JA8859, N3185C	17. 6.93	BAE Systems Marine Ltd	Walney Island	21. 6.06
G-VSEA	Airbus A340-311	003	F-WWDA	7. 7.97	Virgin Atlantic Airways Ltd *"Plain Sailing"*	Gatwick	6. 7.06T
G-VSGE	Cameron O-105 HAB	2382	I-VSGE G-BSSD	14. 8.02	G.Aimo	(Roma, Italy)	25. 7.04A
G-VSHY	Airbus A340-642	383	F-WWCD	26. 7.02	Virgin Atlantic Airways Ltd *"Madam Butterfly"*	Heathrow	25. 7.05T
G-VSPN	Piper PA-28-181 Archer III	2843554	OY-PHC	31. 1.05	Caspian Air Sevices Ltd	(Wigton)	
G-VSSH	Airbus A340-642	615	F-WWCZ.	31. 1.05	Virgin Atlantic Airways Ltd *"Sweet Dreamer"*	Heathrow	
G-VSSS	Boeing 747-219B	22724	TF-ATW G-VSSS, ZK-NZX, 9M-MHG, ZK-NZX	22. 5.00	Air Atlanta Europe Ltd	Gatwick	4. 2.07T
G-VSUN	Airbus A340-313	114	F-WWJI (F-GLZJ)	30. 4.96	Virgin Atlantic Airways Ltd *"Rainbow Lady"*	Gatwick	29. 4.05T
G-VTAL	Beech V35 Bonanza	D-7978	HB-EJB D-EFTH	27. 2.03	R.Chamberlain t/r Wellesbourne Bonanza Group	Wellesbourne Mountford	9. 4.06
G-VTEN*	Vinten-Wallis WA-117 Venom (Continental O-200-B)	UMA-01 & 003		22. 4.85	K.H.Wallis *(Cancelled 12.6.89 as WFU) (Stored unmarked 8.03)*	Reymerston Hall, Norfolk	3.12.85P
G-VTII	de Havilland DH.115 Vampire T.11	15127	WZ507	9. 1.80	Vampire Preservation Ltd *(As "WZ507")*	Bournemouth	29. 7.05P
G-VTOL*	Hawker Siddeley Harrier T.52				See SECTION 3 Part 1	Brooklands	
G-VTOP	Boeing 747-4Q8	28194		28. 1.97	Virgin Atlantic Airways Ltd *"Virginia Plain"*	Gatwick	17. 3.06T
G-VUEA	Cessna 550 Citation II	550-0671	G-BWOM N671EA, 9M-TAA, (N6761L)	20. 6.02	AD Aviation Ltd	(Warrington)	18. 4.07T
G-VULC*	Avro 698 Vulcan B.2A	Not known	N655AV G-VULC, XM655	27. 2.84	Radarmoor Ltd *(Cancelled 25.3.02 as WFU) (Noted 3.04 as "XM655")*	Wellesbourne Mountford	
G-VVBF	Colt 315A HAB	4058		3. 3.97	Airxcite Ltd t/a Virgin Balloon Flights	(Wembley)	17. 8.05T
G-VVBK	Piper PA-34-200T Seneca II	34-7570303	G-BSBS G-BDRI, SE-GLG	26. 1.89	Cheshire Flying Services Ltd t/a Ravenair	Liverpool	1. 9.07T
G-VVIP	Cessna 421C Golden Eagle III	421C0699	G-BMWB N2655L	7. 7.92	J.A Robson t/a Air Deluxe	Gloucestershire	30. 4.07T
G-VVVV	Best Off Skyranger 912 *(Built J.Thomas and J.B.Hobbs - kit no.UK/510)*	BMAA/HB/427		15.12.04	J.Thomas and J.B.Hobbs	(St. Neots)	
G-VVWW	Enstrom 280C Shark	2056	N7802J JA7822	22.10.03	P.J.Odendaal	(Shenfield, Brentwood)	3.12.06T
G-VWKD	Airbus A340-642	706	F-WW..	11.05R	Virgin Atlantic Airways Ltd *"Miss Behavin"*		
G-VWOW	Boeing 747-41R	32745		31.10.01	Virgin Atlantic Airways Ltd *"Cosmic Girl"*	Heathrow	12.12.07E
G-VXLG	Boeing 747-41R	29406		30. 9.98	Virgin Atlantic Airways Ltd *"Ruby Tuesday"*	Heathrow	29. 9.07T
G-VYGR	Colt 120A HAB	2479		24. 9.93	A van Wyk	(Caxton)	23. 6.02T
G-VZZZ	Boeing 747-219B	22722	ZK-NZV	7. 7.99	Virgin Atlantic Airways Ltd *"Morning Glory"* *(Stored 12.04)*	Mojave-Kem County, California, US	6. 7.02T

G-WAAA - G-WZZZ

G-WAAC	Cameron N-56 HAB	492		14. 2.79	N.P.Hemsley tr Whacko Balloon Group *"Whacko"*	(Crawley)	26. 6.97A
G-WAAN	MBB Bö.105DB	S.20	G-AZOR EC-DOE, G-AZOR, D-HDAC	14.11.03	PLM Dollar Group Ltd *(Op Welsh Air Ambulance)*	(Swansea)	31. 5.05T
G-WAAS	MBB Bö.105DBS-4 *(Remanufactured with new pod c/n S.911 c.1992)*	S.138/911	G-ESAM G-BUIB, G-BDYZ, D-HDEF	26. 6.03	Bond Air Services Ltd *(Op Welsh Air Ambulance)*	Swansea	24. 6.07T
G-WABH	Cessna 172S Skyhawk	172S8163	N961SP	10. 5.04	P.Green	Booker	3. 6.07T
G-WACB	Reims/Cessna F152 II	F15201972		16. 9.86	Wycombe Air Centre Ltd	Booker	24. 2.05T
G-WACE	Reims/Cessna F152 II	F15201978		16. 9.86	Wycombe Air Centre Ltd	Booker	23. 4.05T
G-WACF	Cessna 152 II	15284852	N628GH (LV-PMB), N628GH	20. 1.87	Wycombe Air Centre Ltd	Booker	24.11.06T
G-WACG	Cessna 152 II	15285536	ZS-KXY (N93699)	4.11.86	Wycombe Air Centre Ltd	Booker	5. 4.07T
G-WACH	Reims/Cessna FA152 Aerobat	FA1520425		18. 6.87	Wycombe Air Centre Ltd	Booker	4. 8.05T
G-WACI	Beech 76 Duchess	ME-289	N6703Y	26. 7.88	Wycombe Air Centre Ltd	Booker	20.11.06T
G-WACJ	Beech 76 Duchess	ME-278	N6700Y	3. 1.89	Wycombe Air Centre Ltd	Booker	7. 5.05T
G-WACL	Reims/Cessna F172N Skyhawk II	F17201912	G-BHGG	19. 6.89	A G Arthur	Perranporth	22. 4.07T
G-WACM	Cessna 172S Skyhawk SP	172S9005	N35526	21.12.01	Wycombe Air Centre Ltd	Booker	7. 2.05T
G-WACO	Waco UPF-7	5400	N29903 NC29903	28. 1.87	R.G.Vincent t/a RGV (Aircraft Services) and Co *(Damaged Liverpool 15.4.89: stored 4.01)*	Gloucestershire	13. 5.90
G-WACP	Piper PA-28-180 Cherokee Archer	28-7405007	G-BBPP N9559N	5. 4.89	Big Red Kite Ltd	Booker	28. 7.07T
G-WACR	Piper PA-28-180 Cherokee Archer	28-7505090	G-BCZF N9517N	18.12.86	Lees Avionics Ltd	Booker	9. 7.06T
G-WACT	Reims/Cessna F152 II	F15201908	G-BKFT	24. 6.86	Lynda A.Flisher	Eshott	20.10.06T
G-WACU	Reims/Cessna FA152 Aerobat	FA1520380	G-BJZU	10. 7.86	Wycombe Air Centre Ltd	Booker	9. 6.06T
G-WACW	Cessna 172P Skyhawk II	17274057	N5307K	16. 5.88	Wycombe Air Centre Ltd	Booker	20.10.06T
G-WACY	Reims/Cessna F172P Skyhawk II	F17202217	F-GDOZ	3.10.86	Wycombe Air Centre Ltd	Booker	21. 1.05T
G-WADI	Piper PA-46-350P Malibu Mirage	4636205		8. 5.99	H.J.D.S.Baioes	Cranfield	2. 6.05T
G-WADS	Robinson R22 Beta	1224	G-NICO	25. 4.96	Un Pied Sur Terre Ltd t/a Whizzard Helicopters	Welshpool	3. 3.05T
G-WAFU	Robinson R44 Raven	1364		25. 2.04	Magic Aviation Ltd *(Fly Magic titles)*	(Thatcham)	4. 3.07T
G-WAGG	Robinson R22 Beta II	2960		7. 7.99	J.B.Wagstaff t/a N.J.Wagstaff Leasing	Costock	28. 7.05T
G-WAHL	QAC Quickie *(Built A A M.Wahlberg)*	PFA 94-10619		20. 9.00	A A M.Wahlberg *(Noted 9.04)*	Lee-on-Solent	
G-WAIR	Piper PA-32-301 Saratoga	32-8506010	N2607X N9577N	14. 1.91	P.H.Burtwhistle t/a Thorne Aviation	Thorne, Doncaster	13. 5.06

G-WAIT	Cameron V-77 HAB	2390		20.11.90	C.P.Brown	(Ely)	24. 7.99A
G-WAKE	Mainair Blade 912	1244-0300-7 & W1037		6. 3.00	B.W.Webster	(Deeside)	6. 3.05P
G-WAKY	Cyclone AX2000 (HKS 700E V3)	7890		5. 4.02	York Microlight Centre Ltd	Rufforth	14. 4.05P
G-WALY	Maule MX-7-180 Star Rocket	11028C	N5668H	23. 1.03	A J.West	Sywell	14. 7.06T
G-WAMS	Piper PA-28R-201 Arrow	2844050	N491A N5328Q	29. 4.04	Amsair Ltd	Stapleford	20. 5.07T
G-WARB	Piper PA-28-161 Warrior III	2842034	N41286 (G-WARB), N41286	4. 9.98	Muller Aircraft Leasing Ltd	Biggin Hill	9. 9.07T
G-WARC	Piper PA-28-161 Warrior III	2842035	N41244 (G-WARC), N41244	11. 9.98	Plane Talking Ltd	Elstree	14. 9.07T
G-WARD	Taylor JT.1 Monoplane *(Built G Ward - c/n WB.VI)* (Volkswagen 1834)	PFA 1407		1.12.80	R.P.J.Hunter *(Damaged Redhill 17.9.99: current status unknown)*	Redhill	22. 2.00P
G-WARE	Piper PA-28-161 Warrior II	28-8416080	N4357L ("N4354Z")	21. 7.89	W.B.Ware	Filton	27. 3.05
G-WARH	Piper PA-28-161 Warrior III	2842063	N4177Y G-WARH	4. 2.00	Newcastle Aero Club Ltd	Newcastle	17. 2.06T
G-WARK	Schweizer 269C	S.1354		13.11.89	K.Sutcliffe	(Halifax)	15. 4.05
G-WARP*	Cessna 182F	18254633	G-ASHB N3233U	6. 6.95	G.Burton tr Army Parachute Association *(Cancelled 23.11.04 as wfu)*	Haverfordwest	16. 8.04
G-WARR	Piper PA-28-161 Warrior II	28-7916321	N3074U	15. 9.88	T.J. and G.M.Laundy *(Op RAF Halton Aeroplane Club)*	RAF Halton	19. 2.07T
G-WARS	Piper PA-28-161 Warrior III	2842022	N9281X (G-WARS), N9281X	7.11.97	Blaneby Ltd	Biggin Hill	12.11.06T
G-WARV	Piper PA-28-161 Warrior III	2842036	N41247 (G-WARV), N41247	9.10.98	Plane Talking Ltd	Elstree	21.10.07T
G-WARW	Piper PA-28-161 Warrior III	2842037	N41254 (G-WARW), N41254	17.11.98	C.J.Simmonds	St.Just	1. 1.05
G-WARX	Piper PA-28-161 Warrior III	2842038	N4126D (G-WARX), N4126D	15.12.98	C.M.A Clark	Wellesbourne Mountford	15. 1.05
G-WARY	Piper PA-28-161 Warrior III	2842024	N9287X (G-WARY), N9287X	13.11.97	Armstrong Aviation Ltd	Blackpool	4.12.06T
G-WARZ	Piper PA-28-161 Warrior III	2842025	EC-IBI G-WARZ, N92944	26.11.97	S.J.Skilton t/a Aviation Rentals *(Op Exeter Flying School)*	Exeter	18. 1.07T
G-WATR	Christen A-1 Husky *(Floatplane)*	1040	N2941W	2. 4.03	S.N.Gregory	Loch Earn	17. 6.06T
G-WATT*	Cameron Cooling Tower SS HAB				See SECTION 3 Part 1	(Aldershot)	
G-WAVA	Robin HR200/120B	352		10. 7.00	Wellesbourne Flyers Ltd t/a Wellesbourne Aviation	Wellesbourne Mountford	3. 8.06T
G-WAVE (2)	Grob G109B	6381		1. 8.85	C.G.Wray	Park Farm, Eaton Bray	3. 5.07
G-WAVI	Robin HR200/120B	346	G-BZDG	8. 5.01	Wellesbourne Flyers Ltd	Wellesbourne Mountford	12. 4.06T
G-WAVN	Robin HR200/120B	344	G-VECA	2. 5.02	Wellesbourne Flyers Ltd t/a Wellesbourne Aviation	Wellesbourne Mountford	9. 3.06T
G-WAVT	Robin R2160I	375	G-CBLG	7.12.04	Wellesbourne Flyers Ltd t/a Wellesbourne Aviation	Wellesbourne Mountford	15. 5.05E
G-WAZP	Best Off Skyranger 912 *(Built K.H.A.Negal - kit no.SKRxxxx288)*	BMAA/HB/273		10. 6.04	K.H.A.Negal	Red House Farm, Preston Capes	26. 9.05P
G-WAZZ	Pitts S-1S *(Built T H Decarlo)*	7-0332	G-BRRP N3TD	17. 6.94	D.T.Knight	White Waltham	19. 8.04P
G-WBAT	Wombat Gyrocopter *(Built C D Julian)* (Rotax 532)	CJ-001	G-BSID	31. 5.90	M.R.Harrison	(Guernsey)	
G-WBEV	Cameron N-77 HAB	4376	G-PVCU	15.12.04	T.J.and M.Turner	(Southampton/Wellingborough)	
G-WBLY	Mainair Pegasus Quik	8057		30. 7.04	A.J.Lindsay *(Noted 8.04)*	Newtownards	29. 7.05P
G-WBMG	Cameron N Ele 90SS HAB	3086	G-BUYV	5. 7.93	M.Sevrin	(Court St.Etienne, Belgium)	22. 6.02A
G-WBTS	Falconar F-11W-200 *(Built A J Watson)* (Continental O-200-A)	PFA 32-10070	G-BDPL	22.10.90	W.C.Brown	Chilbolton	15. 7.05P
G-WBVS	Diamond DA.40D Star	D4.060		23.10.03	G.W.Beavis	Durham Tees Valley)	14. 1.07T
G-WCAO	Eurocopter EC135T2	0204	D-HECU	8. 4.02	Avon and Somerset Constabulary and Gloucestershire Constabulary *(Op Western Counties Air Operations Unit)*	Filton	13. 6 05T
G-WCAT	Colt Flying Mitt SS HAB	1744		30. 5.90	I.Chadwick tr Balloon Preservation Flying Group *"Washcat"*	(Aldershot)	28.11.03A
G-WCEI	SOCATA MS.894E Rallye 220GT	12141	G-BAOC	28. 5.85	R.A L.Lucas	Netherthorpe	18.10.05E
G-WCIL	Agusta A109E Power	11079	S5-HCN D-HARY	19. 1.05	Castle Air Charters Ltd	Liskeard	
G-WCRD	Aérospatiale SA.341G Gazelle	1390	F-GEHD N6KT, N49527	25.10.02	Wickford Aviation Services Ltd	Wickford	10.12.05T
G-WCUB	Piper PA-18-150 Super Cub	18-8278	HB-OLR N5514Y	11. 5.01	P.A Walley	Croft Farm, Defford	16. 8.07T
G-WDEB	Thunder Ax7-77 HAB	1606		26. 9.89	A Heginbottom	(Cheadle)	20. 7.05A
G-WDEV	Westland SA.341G Gazelle 1	1098	G-IZEL G-BBHW	30. 9.98	Mentorvale Construction Ltd	(Balgriffin, Dublin	20. 1.07T
G-WEEM	Europa Aviation Europa XS *(Built N.A Harrison - kit no.584)*	PFA 247-14079		22. 7.03	N.A Harrison	(Douglas, Isle of Man)	
G-WEGO	Robinson R44 Raven II	10325		28. 4.04	J.D.Forbes-Nixon and N.H.Taylor t/a Clifton Helicopter Hire	Bristol	13. 5.07T
G-WELI	Cameron N-77 HAB	1078		26. 9.84	M.A Shannon *"Wellie"*	(Southampton)	5. 7.04A
G-WELL	Beech E90 King Air	LW-198	N202CC (N7PB), N202CC	18. 7.85	Cega Aviation Ltd	Goodwood	5. 6.06T
G-WELS	Cameron N-65 HAB	1297		7. 4.86	K.J.Vickery *"Talisman"*	(Billingshurst)	26. 6.92A
G-WENA	Aérospatiale AS355F2 Ecureuil 2	5260	G-MOBI G-MUFF, G-CORR	19. 4.04	Kensington and Chelsea Aviation Ltd	Redhill	18. 4.06T
G-WEND	Piper PA-28RT-201 Arrow IV	28R-8118026	PH-SYL N8296L	8.11.82	Tayside Aviation Ltd	Perth	20. 5.05T
G-WERY	SOCATA TB-20 Trinidad	305		2. 4.82	Fastour Aviation Ltd	Sherburn-in-Elmet	19. 5.06
G-WEST	Agusta A109A	7213		21. 1.81	Westland Helicopters Ltd	Yeovil	28. 3.05

Registration	Type	C/n	Previous identity	Reg date	Owner/operator	Base	CofA
G-WESX	CFM Streak Shadow	PFA 161A-11561		2. 2.90	M Catania	(Mold)	9. 7.04P
	(Built N Ramsey - kit no K.116-SA) (Rotax 582)						
G-WETI	Cameron N-31 HAB	449		27.11.78	C.A Butter and J.J.T.Cooke	(Marsh Benham)	11. 9.00A
					(New CofR 7.03)		
G-WFFW	Piper PA-28-161 Warrior II	28-8116161	N8342A	26.10.93	N.F.Duke	Bournemouth	30. 1.06
G-WFLY	Mainair Pegasus Quik	8073		8.10.04	D.E.Lord	Crosland Moor	7.10.05P
G-WFOX	Robinson R22 Beta II	2826		2. 6.98	D.W.Fox t/a Fox Air	Wolverhampton	8. 7.07T
G-WGCS	Piper PA-18 Super Cub 95	18-1528	(G-BLSV)	21.12.84	S.C.Thompson	Newells Farm, Bolney	20. 7.05P
	(L-18C-PI) *(Frame No.18-1500)*		ALAT F-MBCH, 51-15528				
G-WGHB	Canadair CL-30 (T-33AN) Silver Star Mk.3	T33-640	CF-EHB	9. 5.74	R.H.and G.C Cooper	Booker	13. 6.77P
			CAF 133640, RCAF 21640		*(Noted 3.04)*		
G-WGSC	Pilatus PC-6/B2-H4 Turbo-Porter	848	(G-BRVM)	2. 1.90	D.M.Penny	Movenis	29. 3.07
			OE-ECS		*(Op Wild Geese Parachute Centre)*		
G-WHAM	Eurocopter AS350B3 Ecureuil	3494		18. 1.02	B.M.Christie t/a Horizon Helicopter Hire	Goodwood	21. 4.05T
G-WHAT	Colt 77A HAB	1911		15. 3.91	M.A Scholes *"Chad"*	(London SE25)	9. 7.04T
G-WHDP*	Cessna 182S Skylane II	18280178	N178TC	12. 5.98	Not known	Farnborough	
			(Crashed on landing St. Mawgan 23.6.01: cancelled 31.12.01 as destroyed) (Noted 4.04)				
G-WHEE	Pegasus Quantum 15-912	7510		26. 3.99	Airways Airsports Ltd	(Darley Moor, Ashbourne)	4 .3.05P
G-WHEN	Tecnam P92-EM Echo	PFA 318-13679		7. 2.01	E.Windle	Old Sarum	20.12.05P
	(Built C.D.Marsh)						
G-WHIM	Colt 77A HAB	1476		10. 4.89	D.L.Morgan	(Ilford)	28. 7.04A
G-WHOG	CFM Streak Shadow	PFA 206-12776		21. 9.94	B.R.Cannell	Old Sarum	10. 8.04P
	(Built B.R.Cannell - K.253-SA) (Rotax 618)				*"Wart Hog"*		
G-WHOO	RotorWay Executive 162F	6495		5. 6.01	C.A Saul	Street Farm, Takeley	5. 7.05P
	(Built C A Saull) (RotorWay RI 162F)						
G-WHRL	Schweizer 269C	S1453	EC-GGX	19. 4.90	M.Gardiner	Henstridge	26. 8.05T
			CS-HDG, G-WHRL, N41S				
G-WHST	Eurocopter AS350B2 Ecureuil	2915	G-BWYA	9. 8.96	Hawkrise Ltd	Wishaw, Warwick	26. 9.05T
G-WIBB	Jodel D.18	PFA 169-11640		18. 6.96	D.Dobson	Little Staughton	22. 9.05P
	(Built J Wibberley) (Subaru EA81)						
G-WIBS	CASA 1-131E Jungmann Series 2000	2005	Spanish AF E3B-401	25. 3.99	C Willoughby	(Ashford)	
G-WIFE	Cessna R182 Skylane RG II	R18200244	G-BGVT	11.12.01	J.Brennan	Strandhill, County Sligo	29. 1.05
			N3162C		tr Wife Group		
G-WIFI	Cameron Z-90 HAB	10624		9. 9.04	Trigger Concepts Ltd	(Swallowfield, Reading)	1. 9.05A
G-WIIZ	Agusta-Bell 206B JetRanger II	8111	G-DBHH	28.10.03	Action Vehicles Ltd	(Dorking)	22. 6.07T
			G-AWVO, VH-BHI, PK-HCA, G-AWVO, 9Y-TDN, PK-HBG, G-AWVO				
G-WIKY	Cessna 208B Grand Caravan	208B1024	N5117U	17. 4.03	Provident Partners Ltd	(Stroud)	14. 5.06T
G-WILD	Pitts S-1T	1017	ZS-LMM	6.12.85	A McClean	White Waltham	15. 6.07
	(Built Pitts Aerobatics)						
G-WILG	WSK PZL-104 Wilga 35A	62153	G-AZYJ	15. 4.97	M.H.Bletsoe-Brown	Sywell	20. 9.07
G-WILS	Piper PA-28RT-201T Turbo Arrow IV	28R-8431005	PH-DPD	16. 1.96	B. Walker and Co (Dursley) Ltd	Gloucestershire	14. 4.05T
			N4330W				
G-WILY	Rutan LongEz	PFA 74A-10724		8. 6.83	W.S.Allen	Vannes, France	19. 5.05P
	(Built W SAllen - c/n 1200) (Lycoming O-320)				*"Time Flies"*		
G-WIMP	Colt 56A HAB	755		13. 2.86	T.and B.Chamberlain	(York)	23.11.03A
G-WINA	Cessna 560XL Citation XL	560-5343	N5145V	17.12.03	Ability Air Ltd	Stansted	21.12.07E
G-WINE*	Thunder AX7-77Z HAB				See SECTION 3 Part 1	(Aldershot)	
G-WINI	Scottish Aviation Bulldog Series 120/121	BH120/238	G-CBCO	23.09.03	B.Robinson	Blackbushe	15. 6.06
			XX546		*(As XX546/03")*		
G-WINK	Grumman AA-5B Tiger	AA5B-0327	N74658	14.12.90	B.S.Cooke	Elstree	1. 5.06
G-WINS	Piper PA-32-300 Cherokee Six	32-7640065	N8476C	24. 4.91	Cheyenne Ltd	Jersey	14. 3.06
G-WIRE	Aérospatiale AS355F1 Ecureuil 2	5312	G-CEGB	22. 1.90	National Grid Co plc	Oxford	18. 6.06T
			G-BLJL				
G-WIRL	Robinson R22 Beta	0671		27. 7.87	Rivermead Aviation Ltd	Goodwood	20. 6.05T
G-WISH	Lindstrand Cake SS HAB	006		14.12.92	Oxford Promotions (UK) Ltd	(Kentucky, US)	3. 4.03A
	(Birthday Cake shape)				*(Op F Prell)*		
G-WIXI	Akrotech Europe CAP.10B	279		27. 1.98	Meridian Aviation Group Ltd	Bournemouth	2. 8.07
G-WIZA	Robinson R22 Beta	0861	G-PERL	16.11.94	Patriot Aviation Ltd	Gloucestershire	14.11.05T
			N90815				
G-WIZD	Lindstrand LBL 180A HAB	066		12.11.93	Wizard Balloons Cambridge Ltd		
						(Barningham, Bury St.Edmunds)	17. 8.05T
G-WIZI	Enstrom 280FX	2040	Chilean Army H-177	8. 7.02	Cynthia A V.Witheridge	(Newport, Isle of Wight)	2.12.05T
G-WIZO	Piper PA-34-220T Seneca III	34-8133171	N8413U	16.12.86	B.J.Booty	Bristol	26. 4.07T
G-WIZR	Robinson R22 Beta II	2799		9. 3.98	Longmint Properties Ltd	(Croydon)	29. 4.07T
G-WIZS	Mainair Pegasus Quik	8019		17. 3.04	G R Barker *(Noted 5.04)*	Bourn	30. 3.05P
G-WIZY	Robinson R22 Beta	0566	G-BMWX	26. 8.97	Heli Air Ltd	Wellesbourne Mountford	18. 6.06T
			N24196				
G-WIZZ	Agusta-Bell 206B JetRanger II	8540		7.12.77	Rivermead Aviation Ltd	(Reading)	1. 5.06T
G-WKRD	Eurocopter AS350B2 Ecureuil	2668	G-BUJG	16. 3.99	S.A.Jackson and Elaine A.Carson		
			G-HEAR, G-BUJG		t/a Eassda Aviation	(Templepatrick, Ballyclare)	26.10.07E
G-WLAC	Piper PA-18-150 Super Cub	18-8899	G-HAHA	2. 6.98	White Waltham Airfield Ltd	White Waltham	12. 8.07T
			G-BSWE, N9194P				
G-WLGA	WSK PZL-104 Wilga 80	CF21910932	EC-FYY	8.11.96	W.E.Willets	Shobdon	21.12.06
			F-GMLR		tr G-WLGA Group		
G-WLLY	Bell 206B JetRanger II	405	G-OBHH	24. 3.93	G.R. and Noelle M.Sighe t/a G and N Aviation		
			G-WLLY, G-RODY, G-ROGR, G-AXMM, N1469W			(Aberdeen)	8. 8.05T
G-WLMS	Mainair Blade 912	1223-0999-7 & W1016		23. 9.99	D.Smart and C.J.Coggins	Finmere	8 8.05P
G-WLSH	Aérospatiale-Alenia ATR42-300	329	PT-MFX	21. 7.03	Air Wales Ltd	Swansea	11. 8.06T
			G-BXEG, ZS-NKY, F-WQAB, F-GKNE, F-WWLO				
G-WMAA	MBB Bö.105DBS-4	S.135/914	G-PASB	8. 9.94	Bond Air Services Ltd	Exeter	29. 9.06T
	(Rebuilt with new airframe S.914 1994 - see G-PASB)		VH-LSA, G-BDMC, D-HDEC			*(Op Devon Air Ambulance)*	
G-WMAN	Aérospatiale SA.341G Gazelle 1	1277	ZS-HUR	4. 8.99	J.Wightman	(Ballynahinch, County Down)	2.12.07E
			N4491R, YV-54CP				

Registration	Type	C/n	Previous identity	Reg date	Owner/operator	Base	CofA expiry
G-WMAS	Eurocopter EC135T1	0174		18. 6.01	Bond Air Services Ltd	RAF Cosford	26. 7.07T
					(Op County Air Ambulance)		
G-WMBT	Robinson R44 Raven II	10353		5. 5.04	P.Winslow	(Courteenhall, Northampton)	10. 6.07T
G-WMID	MD Helicopters MD-900	900-00062	N3063T	12.10.99	West Midlands Police Authority	Birmingham	27. 1.06T
					"Miss Molly Collins"		
G-WMLT	Cessna 182Q Skylane II	18266689	G-BOPG	23. 4.02	G.Wimlett	Blackpool	1. 4.07T
			N95962				
G-WMPA	Aérospatiale AS355F2 Ecureuil 2	5401		7. 2.89	Police Aviation Services Ltd	Gloucestershire	25. 6.07T
G-WMTM	Gulfstream AA-5B Tiger	AA5B-1035	N4517V	8. 1.91	G.Hance	(Bridge of Don, Aberdeen)	10. 8.05T
					tr Falcon Flying Group		
G-WMWM	Robinson R44 Raven	0767		27. 4.00	K.Cummins	Cambridge	12. 6.06T
G-WNAA	Agusta A109E Power	11090	G-TVAC	28. 5.03	Sloane Helicopters Ltd	Sywell	12.11.06T
					(Op Warwickshire and Northampton Air Ambulance)		
G-WNGS	Cameron N-105 HAB	4385		15. 7.98	R.M.Horn	(Chelmsford)	12.12.04A
G-WOCO	Waco YMF-F5C	F5C091	N770MM	10. 1.05	Airpark Flight Centre Ltd	Coventry	AC
	(Built Waco Classic Aircraft Corpn @ 2000)						
G-WOLF	Piper PA-28-140 Cherokee Cruiser	28-7425439	OY-TOD	20. 3.80	Aircraft Management Services Ltd	Elstree	12. 2.05T
G-WOOD	Beech 95-B55A Baron	TC-1283	SE-GRC	17. 9.79	T.D.Broadhurst	Sleap	7. 1.05
			G-AYID, SE-EXK		t/a Baron Aviation		
G-WOOF	Enstrom 480	5027		3. 3.98	Netcopter.co.uk Ltd and Curvature Ltd	Barton	11. 5.07T
G-WOOL	Colt 77A HAB	2044		23. 2.93	T.G.and C.L.Pembrey and N.P.Helmsley	(Steyning)	14. 6.04
					tr Whacko Balloon Group		
G-WORK*	Thunder Ax10-180 Series 2 HAB				See SECTION 3 Part 1	(Lincoln)	
G-WORM	Thruster T.600N	9109-T600N-039		5.10.99	R.Best	North Coates	5. 3.05P
	(Rotax 582 UL)						
G-WOTG	Pilatus Britten-Norman BN-2T Islander	2139	(ZF444)	10.11.83	I.C.Atkinson	RAF Weston on the Green	16. 2.06
			G-WOTG, G-BJYT		tr RAF Sport Parachute Association		
G-WOWA	de Havilland DHC-8-311	296	C-GZOF	23.10.03	Air Southwest Ltd	Plymouth	22.10.07T
			G-BRYS, PH-SDG, D-DKIS, C-GFQL				
G-WOWB	de Havilland DHC-8-311	334	C-GZOU	29.10.03	Air Southwest Ltd	Plymouth	28.10.07T
			G-BRYT, D-BKIR, C-GFEN				
G-WOWC	de Havilland DHC-8-311	311	N784BC	29.10.04	Air South West Ltd	Plymouth	11.11.07E
			G-BRYO, N434AW, C-GEVP				
G-WPAS	MD Helicopters MD-900	900-00053		1. 7.98	Police Aviation Services Ltd	Devizes	9.11.07T
					(Op Wiltshire Police/Ambulance Authority)		
G-WREN	Pitts S-2A	2229	N9472	8. 1.81	Northamptonshire School of Flying Ltd	Sywell	10. 4.05T
	(Built Aerotek Inc)						
G-WRFM	Enstrom 280C-UK Shark	1202	G-CTSI	21. 4.89	W.F.Blake	Goodwood	25. 8.07
			G-BKIO, (G-BKHN), SE-HLB				
G-WRIT	Colt 77A HAB	1328		15. 9.88	G.Pusey *"Legal Eagle"*	(Seville, Spain)	11. 7.04A
G-WRWR	Robinson R22 Beta II	2964		20. 7.99	AJS Helicopters Ltd	Conington	22. 8.05T
G-WSEC	Enstrom F-28C	398	G-BONF	19.12.88	AJD Engineering Ltd	Weston, Dublin	3. 2.05
			N51661		*(Op Helicopter Aviation Sales)*		
G-WSKY	Enstrom 280C-UK-2 Shark	1037	G-BEEK	25. 7.83	M.I.Edwards	Brandon, Suffolk	14. 7.07
G-WUFF	Europa Aviation Europa	PFA 247-12942		19. 1.99	M.A Barker	Breighton	2. 6.05P
	(Built M.A Barker - kit no.235) (Monowheel u/c)						
G-WULF	WAR Focke-Wulf 190	PFA 81-10328		24. 2.78	A Howe	(Birmingham)	22. 6.01P
	(Built SBV Aeroservices Ltd - c/n 204) (Continental O-200-A)				*(As "8+-" in Luftwaffe c/s)*		
G-WUSH	Eurocopter EC120B Colibri	1290		15. 5.02	S Farmer t/a First Degree Air	Tatenhill	15. 8.05T
G-WVBF	Lindstrand LBL 210A HAB	312		6.12.95	Airxcite Ltd t/a Virgin Balloon Flights	(Wembley)	22. 4.04T
G-WVIP	Beech B200 Super King Air	BB-625	N869AM	18.08.04	Capital Trading (Aviation) Ltd	Filton	19. 8.07T
			N8SZ, N8SP, N18BH, N302EC, N6682U				
G-WWAL	Piper PA-28R-180 Cherokee Arrow	28R-30461	G-AZSH	23.10.98	C.and G.Clarke	White Waltham	19. 9.05T
			N4612J				
G-WWAS	Piper PA-34-220T Seneca III	34-8133222	G-BPPB	2. 3.95	D.Intzevidis	Athens, Greece	6. 3.05
			N83270, (N707WF), N83270, N9579N				
G-WWAY	Piper PA-28-181 Archer II	28-8690031	D-ELCX	30. 3.04	R A Witchell	Andrewsfield	6. 5.07T
			N165AV, N9643N				
G-WWBB	Airbus A330-243	404	F-WWKP	30. 5.01	British Midland Airways Ltd	Johannesburg, RSA	29. 5.07T
					(Op SAA)		
G-WWBC	Airbus A330-243	455		R	British Midland Airways Ltd	Manchester	
G-WWBD	Airbus A330-243	401	F-WWKN	9. 5.01	British Midland Airways Ltd	Johannesburg, RSA	8. 5.07T
					(Op SAA)		
G-WWBM	Airbus A330-243	398	F-WWKL	27. 4.01	British Midland Airways Ltd	Manchester	26. 4.07T
G-WWIZ	Beech 58 Baron	TH-429	G-GAMA	18.10.96	Scenestage Ltd	Bournemouth	23. 6.05T
			G-BBSD				
G-WWWG	Europa Aviation Europa	PFA 247-12597	"G-DSEL"	31. 7.95	Chloe F.Williams-Wynne	Bicester	10.11.98P
	(Built W R C Williams-Wynne - kit no.040) (Wilksch WAM120) *(Monowheel u/c)*				*(Noted 1.05)*		
G-WYAT	CFM Streak Shadow SA	PFA 206-12993		9. 6.97	M.G.Whyatt	Brook Farm, Pilling	21.11.04P
	(Built M.G.Whyatt - c/n K.279) (Rotax 618)						
G-WYCH	Cameron Witch 90SS HAB	1330		30. 9.86	Corn Palace Balloon Club Ltd		
					"Hilda"	(Tyndale, South Dakota, US)	13. 7.99A
					(Active Albuquerque, New Mexico, US 10.03)		
G-WYLE	Rans S-6-ES Coyote II	PFA 204-14330		26.11.04	Anita and R.W.Osborne	Priory Farm, Tibenham	
	(Built A and R.W.Osborne)						
G-WYND	Wittman W.8 Tailwind	PFA 31-12407		2. 8.99	R.S.Marriott and C.Clark	(Scunthorpe)	
	(Built R.S.Marriott and C.Clark)				tr Forge Group *(Current status unknown)*		
G-WYNS	Aero Designs Pulsar XP	PFA 202-11976		22. 2.91	S.L.Bauza	(Palma de Mallorca)	27. 4.98P
	(Built G Griffiths)				*(Current status unknown)*		
G-WYNT	Cameron N-56 HAB	1038		3. 4.84	S.L.G.Williams	(Bristol)	14. 4.02A
					"Gwyntoedd Dros Cymru/Winds over Wales"		
G-WYPA	MBB Bö.105DBS-4	S.815	D-HDZY	27.10.89	Police Aviation Services Ltd	Gloucestershire	9. 1.05T
G-WYSP	Robinson R44 Astro	0657		17. 9.99	Calderbrook Estates Ltd	(Sowerby Bridge)	28. 9.05T

Reg	Type	C/n	Previous identity	Regd	Owner/Operator	Base	CofA
G-WZOL	Tiger Cub RL5B LWS Sherwood Ranger		G-MZOL	20. 1.99	G.W.F.Webb	Coldharbour Farm, Willingham	13.12.02P
	(Built G.W.F.Webb) (Jabiru 2200A)	PFA 237-12887					
G-WZZZ	Colt AS-42 HA Airship	459		10.12.82	Lindstrand Balloons Ltd	(Oswestry)	4. 9.01A
	(Built Colt Balloons Ltd) (Rebuilt 1984/85 using new AS-56 envelope c/n 607)				*"Kit Kat"*		

G-XAAA - G-XZZZ

Reg	Type	C/n	Previous identity	Regd	Owner/Operator	Base	CofA
G-XARV	ARV Aviation ARV-1 Super 2	010	G-OPIG	8.11.95	D.J.Burton	Shoreham	5. 4.05P
	(Rotax 912-S)		G-BMSJ				
G-XATS	Pitts S-2A	2147	CS-AZE	29. 3.01	Air Training Services Ltd	Booker	17.11.07E
	(Built Pitts Aerobatics)		N338BD				
G-XAXA	Fairey Britten-Norman BN-2A-26 Islander	530	G-LOTO	22. 8.00	AirX Ltd *(Op Rockhoppers)*	Bournemouth	28.11.06T
			G-BDWG, (N90255), (C-GYUF), G-BDWG				
G-XAYR	Raj Hamsa X'Air 582	BMAA/HB/122		4. 1.00	D.L.Connolly and R.V.Barber	Little Gransden	17. 7.04P
	(Built M J Kaye - kit no.471)						
G-XBCI	Bell 206B JetRanger III	4466	N206EE	17. 6.04	BCI Helicopter Charters Ltd	(Guildford)	15. 7.07T
			RP-C1778, N80706, C-GAJH				
G-XBOX	Bell 206B JetRanger III	3370	G-OOHO	31. 1.05	Mainstream Digital Ltd	(Cirencester)	26. 6.04T
			G-OCHC, G-KLEE, G-SIZL, G-BOSW, N2063T				
G-XCCC	Extra EA.300/L	142		20. 8.01	P.T.Fellows	Rochester	20. 9.07T
G-XCIT	Alpi Pioneer 300	PFA 330-14296		16. 9.04	A.Thomas	Shobdon	
	(Built A.Thomas)				*(Noted 12.04)*		
G-XCUB	Piper PA-18-150 Super Cub	18-8109036	N9348T	1. 5.81	M.C.Barraclough	(Selborne, Alton)	23. 5.07
G-XENA	Piper PA-28-161 Cherokee Warrior II	28-7716158	N3486Q	29. 6.98	Braddock Ltd	Blackbushe	16. 5.05T
G-XFLY	Lambert Aircraft Mission M212-100	PFA 306-13380		3. 2.00	Lambert Aircraft Engineering BVBA	(Kortrijk, Belgium)	AC
G-XIII	Van's RV-7	PFA 323-14165		20. 2.04	G.Wright	(Selby)	
	(Built G.Wright)				tr G-XIII Group		
G-XINE	Piper PA-28-161 Cherokee Warrior II	28-7716112	G-BPAC	14.10.03	P.Tee	Denham	6. 5.07T
			N2567Q				
G-XIOO	Raj Hamsa X'Air V133	BMAA/HB/247		27. 1.03	J.Campbell	(Eastbourne)	4. 7.05P
	(Built R.Paton and A Start - kit no.681)						
G-XKEN	Piper PA-34-200T Seneca II	34-7970003	N3036A	5. 9.01	Choicecircle Ltd	Coventry	16. 9.07E
G-XLAA	Boeing 737-8Q8	28226	G-OKDN	13. 3.01	Excel Airways Ltd	Gatwick	26. 7.07T
G-XLAB	Boeing 737-8Q8	28218	G-OJSW	14. 5.01	Excel Airways Ltd	Gatwick	10.12.07E
G-XLAE	Boeing 737-8Q8	30637	D-ABAA	9.11.01	Excel Airways Ltd	Gatwick	8.11.07T
			G-OKJW, N1787B		*(Sunwing.com titles)*		
G-XLAF	Boeing 737-86N	29883		15. 3.02	Excel Airways Ltd	Gatwick	14. 3.05T
G-XLAG	Boeing 737-86N	33003		29. 4.02	Excel Airways Ltd	Gatwick	28. 4.05T
G-XLAH	Boeing 737-8Q8	29351		25. 3.04	Excel Airways Ltd	Gatwick	24. 3.07T
G-XLIV	Robinson R44 Raven	0810		11. 7.00	Rotorcraft Ltd	Redhill	25. 7.06T
G-XLMB	Cessna 560XL Citation Excel	560-5259	N52526	25. 6.02	Aviation Beauport Ltd	Jersey	25. 6.05T
G-XLNT	Zenair CH601XL Zodiac	PFA 162B-14182		27. 1.04	P.H.Ronfell	Blackpool	AC
	(Built P.H.and S J Ronfell)				tr Zenair G-XLNT Group *(Noted 11.04)*		
G-XLTG	Cessna 182S Skylane	18280234	N9571L	17. 7.98	D.H.Morgan	Denham	30. 7.07
G-XLXL	Robin DR400/160 Knight	813	G-BAUD	3. 1.92	L.R.Marchant	Rochester	29. 7.06
G-XMEN	Eurocopter AS350B3 Ecureuil	3362	G-ZWRC	7. 6.02	Corporate Estates Ltd	Booker	4. 7.07T
			F-GPNE				
G-XMGO	Aeromot AMT-200S Super Ximango	200127		18. 4.01	R.P.Beck and G.McLean	Rufforth	9. 5.07
G-XMII	Eurocopter EC135T2	0215		15. 4.02	Merseyside Police Authority	RAF Woodvale	6. 8.05T
G-XOIL	Eurocopter AS355N Ecureuil 2	5627	G-LOUN	5. 3.03	Firstearl Marine and Aviation Ltd	Denham	15. 6.06T
G-XPBI	Letov LK-2M Sluka	PFA 263-13341		4.12.98	K.Harness	North Coates	26. 4.05P
	(Built P Bishop) (Rotax 447)						
G-XPRS	Bombardier BD-700-1A10 Global Express	9133	C-FZXE	23.12.04	Leferson Holdings Ltd	Farnborough	AC
G-XPSS	Short SD.3-60 Variant 100	SH.3713	EI-CPR	2. 5.01	BAC Express Airlines Ltd	Edinburgh	7. 5.05T
			G-OBOH, G-BNDJ, G-14-3713 *"City of Derby"*				
G-XPXP	Aero Designs Pulsar XP	PFA 202-11958		30. 3.92	B.J.Edwards	Belle Vue Farm, Yarnscombe	27. 6.05P
	(Built B.J.Edwards - c/n 218) (Tail-wheel u/c)						
G-XRAF	Raj Hamsa X'Air 582	BMAA/HB/132		7. 4.00	R.G.Kirkland	RAF Henlow	30. 7.04P
	(Built M E Howard and S.Stockill - kit no.513)						
G-XRAY	Rand Robinson KR-2	PFA 129-11227		30. 4.87	R.S.Smith	Barthol Chapel	
	(Built R S Smith)				*(Under construction 2001)*		
G-XRLD	Cameron A-250 HAB	4820		25. 4.00	Red Letter Days Ltd	(London N10)	22. 3.05T
					(Red Letter Days titles)		
G-XRXR	Raj Hamsa X-Air 582	BMAA/HB/102		13. 9.99	R J Philpotts	Pound Green, Buttonoak, Bewdley	29. 9.05P
	(Built I S Walsh - kit no.431)						
G-XSAM	Van's RV-9A	PFA 320-13797		18. 9.02	D.G.Lucas	Plymouth	
	(Built D.G.Lucas)				*(Under construction 11.03)*		
G-XSDJ	Europa Aviation Europa XS	PFA 247-13378		3. 2.99	D.N.Joyce	(Berkeley)	14. 6.045
	(Built D.N.Joyce - kit no.402) (Rotax 914-UL) *(Monowheel u/c)*						
G-XSFT	Piper PA-23-250 Aztec F	27-7754103	G-CPPC	18. 6.86	T.L.B.Dykes	Carlisle	1. 6.06T
			G-BGBH, N63773				
G-XSKY	Cameron N-77 HAB	2508		26. 3.91	D Hempleman-Adams	(Corsham)	11. 8.00A
G-XTEE	AirBorne XT912-B/Streak III-B	XT912-026	T2-2252	25. 8.04	G.J. Webster	Mill Farm, Shifnal	AC
					t/a Airborne Australia In UK		
G-XTEK	Robinson R44 Astro	0647		11. 8.99	PLM Properties plc	Bromley Hall, Kings Bromley	29. 9.05T
G-XTOR	Fairey Britten-Norman BN-2A Mk.III-2 Trislander	359	G-BAXD	1. 4.96	Aurigny Air Services Ltd	Guernsey	5. 7.07T
	(Fuselage ex N3266G [1065] fitted 2.96)						
G-XTUN	Westland-Bell 47G-3B1	WA/382	G-BGZK	11. 5.99	R.C.Hields	Sherburn-in-Elmet	10. 9.06T
			XT223		t/a Hields Aviation *(As "XT223" in Army Air Corps c/s)*		
G-XVBF	Lindstrand LBL 330A HAB	966		30. 1.04	Airxcite Ltd t/a Virgin Balloon Flights	(Wembley)	8. 9.05T
G-XVIB*	Vickers Supermarine 361 Spitfire LF.XVIe				See SECTION 3 Part 1	Polk City, Florida, US	
G-XVOM	Van's RV-6	PFA 181-12894		6. 4.01	A Baker-Munton	(Leicester)	3. 7.05P
	(Built A Baker-Munton)						

G-XXEA	Sikorsky S-76C	760492		21.12.98	T.C.Hewlett, Director of Royal Travel	Blackbushe	4. 1.06T
					(Op Queen's Flight)		
G-XXIV	Agusta-Bell 206B-3 JetRanger III	8717		27. 4.89	Bart Fifty Nine Ltd	(Portbury, Bristol)	4. 7.07T
G-XXTR	Extra EA.300/L	126	G-ECCC	13. 8.02	Airpark Flight Centre Ltd	Coventry	7. 4.07T
			D-EDGE				
G-XXVI	Sukhoi Su-26M	04-10	RA-0410	2. 4.93	A N.Onn *(As "39")*	Headcorn	5. 4.05P
G-XYAK	IAV-Bacau Yakovlev Yak-52	899413	RA-44469	6. 3.03	M.G.F.di Prima	Denham	16. 5.05P
			LY-AFX, DOSAAF 69 *(blue)*				
G-XYJY	Best Off Skyranger 912	BMAA/HB/309		20.10.03	A V.Francis	(Dunstable)	
	(Built A V.Francis - kit no.SKRxxxxxxx?)						

G-YAAA - G-YZZZ

G-YACB	Robinson R22 Beta II	3092	G-VOSL	24. 1.02	Heli Air Ltd	Wellesbourne Mountford	27. 6.06A
G-YAKA	Yakovlev Yak-50	822303	LY-ANJ	10.11.94	M.Chapman	Stewton, Louth	1.12.05P
			DOSAAF 80				
G-YAKB	Aerostar Yakovlev Yak-52	9211517	RA-44491	25.11.02	Kemble Air Services Ltd	Kemble	9. 2.05P
			LY-AOB				
G-YAKC	IAV-Bacau Yakovlev Yak-52	867212	LY-AKC	25. 6.02	T.J.Wilson	Andrewsfield	4. 7.05P
			DOSAAF 153 *(yellow)?*				
G-YAKH	IAV-Bacau Yakovlev Yak-52	899915	RA-01948	24.12.02	Plus7Minus5 Ltd	RAF Halton	15.12.04P
			LY-AFV, DOSAAF 102 *(yellow)*				
G-YAKI	IAV-Bacau Yakovlev Yak-52	866904	LY-ANM	20. 9.94	Yak One Ltd	Popham	30.12.04P
			DOSAAF 100		*"100" (DOSAAF c/s)*		
G-YAKK	Yakovlev Yak-50	853104	RA-01293	5.11.02	K.J.Pilling	North Weald	16. 5.05P
G-YAKM	Yakovlev Yak-50	842710	RA-44461	6. 2.04	Airborne Services Ltd	Compton Abbas	23. 2.05P
			Ukraine 28 *(blue)*		*(As "61" in Soviet Air Force c/s)*		
G-YAKN	IAV-Bacau Yakovlev Yak-52	855905	RA-44466	6. 2.04	Airborne Services Ltd	Compton Abbas	25. 3.05P
			DOSAAF 105 *(blue)*		*(As "66" in Soviet Air Force c/s)*		
G-YAKO	IAV-Bacau Yakovlev Yak-52	822203	RA-01493 (1)	8. 5.99	M.K.Shaw	Norwich	25. 8.05P
G-YAKR	IAV-Bacau Yakovlev Yak-52	899803	LY-AOV	15.11.02	A S.Nottage and R.A Alexander	North Weald	19.11.05P
			Ukraine AF 75 *(yellow)*, DOSAAF 75 *(yellow) "03"*				
G-YAKS	Aerostar Yakovlev Yak-52	9311708		16.12.93	Two Bees Associates Ltd *"2"*	(Chelmsford)	29. 5.05P
G-YAKT	IAV-Bacau Yakovlev Yak-52	8910302	RA-01564	21. 1.03	S.J.Thomas	White Waltham	19. 2.05P
			DOSAAF 149 *(yellow)*		tr G-YAKT Group		
G-YAKU	Yakovlev Yak-50	822305	RA-44549	6.11.03	D.J.Hopkinson	Compton Abbas	2. 2.05P
			G-BXNO, LY-ASD, DOSAAF 82			*(As "49" in Soviet Airforce c/s)*	
G-YAKV	Aerostar Yakovlev Yak-52	9111311	RA-02209	8. 7.03	P.D.Scandrett	Rendcomb	21. 7.04P
			RA-9111311, DOSAAF 31				
G-YAKX	Aerostar Yakovlev Yak-52	9111307	RA-44473	13. 3.96	The X-Fliers Ltd	Popham	16.12.04P
			G-YAKX, RA-9111307, DOSAAF 27		*(As "27" (Red) in Soviet Air Force c/s)*		
G-YAKY	IAV-Bacau Yakovlev Yak-52	844109	LY-AKX	26. 2.96	W.T.Marriott	(Market Rasen)	7. 3.02P
			DOSAAF 24 *(red)*				
G-YAKZ	Yakovlev Yak-50	853206	RA-44533	7.11.03	Greenhouse New Media Ltd	Compton Abbas	28. 1.05P
G-YAMS	IAV-Bacau Yakovlev Yak-52	844306	LY-AMS	23. 7.03	N.Gooderham	Southend	6.10.05P
			DOSAAF 51 *(red)*		*(Force landed in field near Canewdon, Essex 27.12.04 and badly damaged)*		
G-YANK	Piper PA-28-181 Archer II	28-8090163	N81314	19. 3.93	Janet A Millar-Craig tr G-YANK Flying Group	Tatenhill	30. 5.05
G-YARR	Mainair Rapier	1255-0700-7 & W1049		14. 8.00	D.Yarr	St.Michaels	22. 9.05P
G-YARV	ARV Aviation ARV-1 Super 2	PFA 152-11127	G-BMDO	15.10.01	P.R.Snowden	(Bury St.Edmunds)	12. 6.97P
	(Built Hornet Aviation Ltd - c/n K.004)				*(Current status unknown)*		
G-YAWW	Piper PA-28RT-201T Turbo Arrow IV	28R-8031024	N2929Y	15.11.90	Barton Aviation Ltd	Barton	9. 7.06
G-YBAA	Reims FR172J Rocket	FR17200579	5Y-BAA	15.11.84	A Evans	Bourn	21. 6.06
G-YCII	LET Yakovlev C.11	25III/08	F-AZPA	13. 1.00	R.W.Davies	(Woodchuch, Kent)	14. 6.05P
			Egyptian AF		*(As "11 (yellow) in Soviet Airforce c/s)*		
G-YCUB	Piper PA-18-150 Super Cub	1809077	N4993X	23. 8.96	F.W.Rogers Garage (Saltash) Ltd	Bodmin	23. 3.06
			N4157T				
G-YEAH	Robinson R44 Raven II	10453		3. 8.04	Turboprop Leasing Llp	(Manor Park, Runcorn)	13. 9.07T
G-YELL	Murphy Rebel	PFA 232-12381		1. 5.95	A D.Keen	Bodmin	
	(Built A D.Keen)				*(Noted 10.02)*		
G-YEOM	Piper PA-31-350 Chieftain	31-8352022	N41108	3. 1.89	Foster Yeoman Ltd	Bristol	21. 3.05
G-YEWS	RotorWay Executive	3850		22. 6.89	R.Turrell and P.Mason	(Wickford)	17. 6.93P
	(Built D G Pollard - c/n DGP-1) (RotorWay RW 152)						
G-YFLY	VPM M-16 Tandem Trainer	VPM16-UK114	G-BWGI	14.10.96	A J.Unwin	Kemble	22. 9.04P
	(Arrow GT1000R)						
G-YFUT	IAV-Bacau Yakovlev Yak-52	888410	LY-FUT	6. 2.03	R.Oliver	Swansea	24. 3.05P
			Ukraine AF 22 *(yellow)*, DOSAAF 22				
G-YFZT	Cessna 172S Skyhawk	172S9587	N20974	24. 2.04	AB Integro Ltd	(Reigate)	29. 3.07T
G-YIII	Reims/Cessna F150L	F15000827	PH-CEX	5. 6.80	Sherburn Aero Club Ltd	Sherburn-in-Elmet	18. 9.06T
G-YIIK	Robinson R44 Astro	0640		9. 8.99	The Websiteshop (UK) Ltd	Denham	4. 9.05T
G-YIPI	Reims Cessna FR172K	FR1720616	OY-IPI	9. 1.03	A.J.G.Davis	(St. Marys)	2. 3.06
			D-EIPI				
G-YJET	Montgomerie-Bensen B.8MR	PFA G/01-1072	G-BMUH	25. 9.96	A Shuttleworth	Barton	28. 6.05P
	(Built J M Montgomerie) (Rotax 582)						
G-YKCT	Aerostar Yakovlev Yak-52	9010307	LY-ATI	29. 5.02	C.R.Turton	Headcorn	18. 6.05P
			Ukraine AF 04, DOSAAF 04				
G-YKSO	Yakovlev Yak-50	791506	LY-APT	8. 4.02	Classic Display (Scotland) Ltd	(Little Gransden)	22. 5.05P
G-YKSS	Yakovlev Yak-55	901103	RA-44525	1. 7.03	I.D.Trask	Headcorn	7. 7.05P
			DOSAAF 96 *(blue)*				
G-YKSZ	Aerostar Yakovlev Yak-52	9311709		16.12.93	N.Rhind tr Tzarina Group	(Wokingham)	25. 9.05P
					(As "01" in Soviet AF c/s)		
G-YKYK	Aerostar Yakovlev Yak-52	9812106	LY-AHB	16. 6.03	K.J.Pilling	North Weald	29. 6.05P
G-YLYB	Cameron N-105 HAB	4482		15. 1.99	Airship and Balloon Company Ltd	(Bristol)	14.11.02A
					(Lloyds TSB titles)		

Registration	Type	C/n	Previous identity	Reg date	Owner	Base	CofA
G-YMBO	Robinson R22 Mariner	2054M	OY-HFR	21. 8.95	Helicentre Blackpool Ltd	Blackpool	20. 9.07T
G-YMFC	Waco YMF	F5033	N90B	6. 4.04	S.J.Brenchley	Booker	24. 8.07
	(Built Classic Aircraft Corporation @ 1990)						
G-YMMA	Boeing 777-236	30302	N5017Q	7. 1.00	British Airways plc	Heathrow	6. 1.06T
G-YMMB	Boeing 777-236	30303		18. 1.00	British Airways plc	Heathrow	17. 1.06T
G-YMMC	Boeing 777-236	30304		4. 2.00	British Airways plc	Heathrow	3. 2.06T
G-YMMD	Boeing 777-236	30305		19. 2.00	British Airways plc	Heathrow	17. 2.06T
G-YMME	Boeing 777-236	30306		16. 4.00	British Airways plc	Heathrow	14. 4.06T
G-YMMF	Boeing 777-236	30307		17. 5.00	British Airways plc	Heathrow	16. 5.06T
G-YMMG	Boeing 777-236	30308		28. 9.00	British Airways plc *"City of Chicago"*	Heathrow	28. 7.06T
G-YMMH	Boeing 777-236	30309		14.10.00	British Airways plc	Heathrow	13.10.06T
G-YMMI	Boeing 777-236	30310		2.11.00	British Airways plc	Heathrow	1.11.06T
G-YMMJ	Boeing 777-236	30311		8.12.00	British Airways plc	Heathrow	7.12.06T
G-YMMK	Boeing 777-236	30312		8.12.00	British Airways plc	Heathrow	2. 10.06T
G-YMML	Boeing 777-236	30313		10. 4.01	British Airways plc	Heathrow	13. 4.07T
G-YMMM	Boeing 777-236	30314		31. 5.01	British Airways plc	Heathrow	30. 5.07T
G-YMMN	Boeing 777-236	30316		15. 6.01	British Airways plc	Heathrow	14. 6.07T
G-YMMO	Boeing 777-236	30317		17. 9.01	British Airways plc	Heathrow	13. 9.07T
G-YMMP	Boeing 777-236	30315		30.10.01	British Airways plc	Heathrow	29.10.07T
G-YNOT	Druine D.62B Condor	RAE/649	G-AYFH	10.11.83	A Littlefair	Lymington	4. 9.03P
	(Built Rollason Aircraft and Engines)						
G-YOGI	Robin DR400/140B Major	1090	G-BDME	1.10.86	M.M.Pepper	Sibson	19. 4.07
G-YORK	Reims/Cessna F172M Skyhawk II	F17201354	PH-LUY	14.12.78	H-R.A E.Waetjen	(Athboy, County Meath)	19.2.07
			F-WLIT				
G-YOTS	Aerostar Yakovlev Yak-52	9010308	LY-AOW	4. 5.03	J.D.F.Barke	Southend	16. 5.05P
			Ukraine AF 105, DOSAAF 05 *(yellow)*				
G-YOYO	Pitts S-1E	PFA 09-10885	G-OTSW	22. 5.96	J.D.L.Richardson	Exeter	11. 2.04P
	(Built W R Penaluna)		G-BLHE				
G-YPOL	MD Helicopters MD 900	900-00078	N7038S	4.10.00	West Yorkshire Police Authority	Wakefield	25. 1.07T
G-YPSY	Andreasson BA-4B	PFA 38-10352		7. 6.78	R.W.Hinton	Poplar Hall Farm, Elmsett	15. 4.04P
	(Built H P Burrill) (Continental O-200-A)						
G-YRAF	Rotary Air Force RAF 2000 GTX-SE	PFA G/13-1289		1. 6.01	C.V.King	Henstridge	23. 8.04P
	(Built C V King and J R Cooper)						
G-YRIL	Luscombe 8E Silvaire	5945	N1318B	3. 2.92	C.Potter	North Weald	12.10.05P
	(Continental O-200-A)		NC1318B				
G-YROI	Air Command 532 Elite	0002	N532CG	3. 9.87	W.B.Lumb	(Manchester)	17.12.90P
	(Built Air Command Manufacturing Inc)				*(Current status unknown)*		
G-YROJ	Rotary Air Force RAF 2000 GTX-SE	PFA G/13-1343		26.11.04	J.R.Mercer	(Formby, Liverpool)	
	(Built J.R.Mercer)						
G-YROO	Rotary Air Force RAF 2000 GTX-SE	PFA G/13-1341		27.11.01	K.D.Rhodes and C.S.Oakes	(Wimborne)	22.10.03P
	(Built K.D.Rhodes and C.S.Oakes)						
G-YROS	Montgomerie-Bensen B.8M	PFA G/101-1004		29. 1.81	Flight Acadamy (Gyrocopters) Ltd	Barton	6. 6.97P
	(Built J M Montgomerie) (HAPI 60-6M)				*(New owner 11.04)*		
G-YROW	VPM M-16 Tandem Trainer	PFA G/12-1351	ZU-AOA	16.10.02	B.Jones	Haverfordwest	11. 2.04P
	(Built B Jones - kit no M16-90)						
	(Struck roof of building attempting to re-land on parade ground Pirbright Barracks 24.4.03: rotor damaged, struck ground and rolled onto side)						
G-YROY	Montgomerie-Bensen B.8MR Merlin	PFA G/101A-1145		12. 9.89	S.Brennan	Carlisle	13. 5.05P
	(Built R D Armishaw) (Rotax 532)						
G-YRUS	Jodel D.140E	PFA 251-14090	G-YRNS	8. 9.03	W.E.Massam	(Blagdon, Bristol)	
	(Built W.E.Massam)						
G-YSMO	Mainair Pegasus Quik	8049		12. 7.04	M.P.and Rachel A. Wells	(Earls Croome, Worcester)	11. 7.05P
G-YSPY	Cessna 172Q Cutlass II	17275932	N917AT	4. 2.03	J.Henderson	Deanland	2. 7.06T
			N917ER, (N65957)				
G-YSTT	Piper PA-32R-301 Saratoga IIHP	3246056	N848T	4. 8.97	A W.Kendrick	Wolverhampton	19. 2.06
			N9282D				
G-YTUK	Cameron A-210 HAB	4640		30. 9.99	Societe Bombard SRL	(Beaune, France)	24. 9.05A
G-YULL	Piper PA-28-180 Cherokee E	28-5603	G-BEAJ	30. 3.79	G. Watkinson-Yull	Guernsey	23.10.06
			9H-AAC, N2390R				
G-YUMM	Cameron N-90 HAB	2723		12.12.91	Wunderbar Ltd *"Boulevard"*	(York)	29. 9.054A
G-YUPI	Cameron N-90 HAB	1602		12. 1.88	MCVH SA	(Brussels, Belgium)	22.11.98A
G-YURO*	Europa Aviation Europa	PFA 220-11981		6. 4.92	Europa Management (International) Ltd	Wombleton	9. 6.95P
	(Built Europa Aviation Ltd - kit no.001)				*(Cancelled 22.4.98 as WFU) (Noted incomplete 8.04)*		
G-YVBF	Lindstrand LBL 317S HAB	505		2. 4.98	Airxcite Ltd t/a Virgin Balloon Fights	(Wembley)	16. 4.05T
G-YVES	Alpi Pioneer 300	PFA 330-14290		7.12.04	M.C.Birchall	(Cardiff)	
	(Built M.C.Birchall)						
G-YVET	Cameron V-90 HAB	3182		11.10.93	K.J.Foster	(Coleshill, Birmingham)	6. 7.04A
G-YVFS	de Havilland DH.82A Tiger Moth	85957	G-ANDE	19.10.04	Yorkshire Vintage Flying Ltd	Sheffield City	2.10.06T
	(Built Morris Motors Ltd)		EM726				
G-YYAK	IAV-Bacau Yakovlev Yak-52	878101	LY-AOM	18. 4.02	J Armstrong and D W Lamb	Durham Tees Valley	21. 6.05P
			DOSAAF 118				
G-YYYY	Max Holste MH.1521C1 Broussard	208	F-GDPZ	10. 3.00	P.F. Burrow	Trenchard Farm, Eggesford	31. 7.06
			French Air Force		tr Eggesford Heritage Flight *("St.Ivel Shape' c/s)*		
G-YZMO	American Champion 8KCAB	942-2003		10.12.03	Blue Yonder Aviation Ltd	North Weald	6. 5.07
G-YZYZ	Mainair Blade 912	1357-0803-7 & W1152		14. 8.03	R.Chadwick	Barton	16. 8.05P

G-ZAAA - G-ZZZZ

Registration	Type	C/n	Previous identity	Reg date	Owner	Base	CofA
G-ZAAZ	Van's RV-8	PFA 303-13279		2. 7.02	P.A Soper	Ipswich	9. 6.05P
	(Built P A Soper)						
G-ZABC	Sky 90-24 HAB	062		10. 4.97	P.Donnelly	(Maghera, Belfast)	23. 9.05A
G-ZACE	Cessna 172S Skyhawk SP	172S8808	F-HAMC	3. 9.02	M.C.Tonsbeek	Manston	3. 9.05T
			N3527P				
G-ZACH	Robin DR400/100 Cadet	1831	G-FTIO	20.10.92	A P.Wellings	Sandown	12.10.07E

Registration	Type	C/n	Previous identity	Reg date	Owner	Base	CofA expiry
G-ZAIR	Zenair CH.601HD Zodiac	PFA 162-12194		21. 2.92	J.R.Standring	Crosland Moor	18. 1.05P
	(Built B E Shaw) (Rotax 912-UL)						
G-ZANG	Piper PA-28-140 Cherokee E	28-7225178	SE-FYT	29. 7.04	Angela J.Gale	East Winch	5. 8.07T
G-ZANY	Diamond DA.40D Star	D4.040		23.10.03	Altair Aviation Ltd	Stapleford	4.12.06T
G-ZAPH	Bell 206B-3 JetRanger III	4401	G-DBMW	6. 2.01	Northern Flights Ltd	(Farnham)	3. 7.05T
			C-GAJH				
G-ZAPK	British Aerospace BAe 146 Series 200QC	E2148	G-BTIA	25. 4.96	Titan Airways Ltd	Stansted	17. 4.06T
			ZS-NCB, G-BTIA, G-6-148, G-PRIN				
G-ZAPM	Boeing 737-33A	27285	DQ-FJD	2. 6.99	Titan Airways Ltd	Stansted	2. 6.05T
			N102AN, CS-TKG				
G-ZAPN	British Aerospace BAe 146 Series 200QC	E2119	ZK-NZC	20. 9.99	Titan Airways Ltd	Stansted	15.11.05T
			G-BPBT				
G-ZAPO	British Aerospace BAe 146 Series 200QC	E2176	F-GMMP	28. 7.00	Titan Airways Ltd	Stansted	3. 8.06T
			G-BWLG, VH-NJQ, G-PRCS				
G-ZAPR	British Aerospace BAe 146 Series 200QT	E2114	VH-JJZ	19.12.03	Titan Airways Ltd	Stansted	15. 1.07T
			G-BOXE				
G-ZAPT	Beech B200C Super King Air	BL-141	N200KA	24. 9.01	Titan Airways plc	Stansted	23. 9.05T
			N5141Y				
G-ZAPU	Boeing 757-2YO	26151	EI-MON	22. 4.03	Titan Airways Ltd	Stansted	29. 4.06T
			4X-BAY, SE-DUL, SX-BBY, XA-KWK, XA-SCB				
G-ZAPV	Boeing 737-3YO	24546	G-IGOC	27. 2.04	Hagondale Ltd	Stansted	9. 5.07T
			EI-BZH		*(Op Titan Airways in Royal Mail c/s)*		
G-ZAPW	Boeing 737-3L9	24219	G-IGOX	11.02.05	Titan Airways Ltd	Stansted	20. 3.05E
			N219TY, PH-TSW, OY-MMO, G-BOZB, OY-MMO, N1786B				
G-ZAPY	Robinson R22 Beta	0788	G-INGB	8. 7.98	Heli Air Ltd	Wellesbourne Mountford	15. 1.07T
G-ZARI	Grumman AA-5B Tiger	AA5B-0845	G-BHVY	7. 3.86	ZARI Aviation Ltd	Biggin Hill	23. 2.07
			N28835				
G-ZARV	ARV Aviation ARV-1 Super 2	PFA 152-13035		26. 2.97	P.R.Snowden	Cambridge	21. 6.05P
	(Built P.R.Snowden) (Rotax 914-UL)						
G-ZAZA	Piper PA-18 Super Cub 95	18-2041	D-ENAS	1. 5.84	Airborne Taxi Services Ltd	Chilbolton	14. 4.05P
	(L-18C-PI)		R.Neth AF R-66, 52-2441		*(Op Adrian Swire)*		
G-ZBED	Robinson R22 Beta	1684	N63993	18.11.99	P.D.Spinks	Stream Farm, Sherburn-in-Elmet	22.11.05T
			F-GHHM				
G-ZBLT	Cessna 182S Skylane	18280910	N72764	6. 7.01	Entee Global Services Ltd	(Abingdon)	26. 8.07T
G-ZEBO	Thunder Ax8-105 Series 2 HAB	2197		22. 5.92	S.M.Waterton *"Gazebo"*	(Borehamwood)	2. 4.05T
G-ZEBY	Piper PA-28-140 Cherokee F	28-7325240	G-BFBF	7.04.04	M.Marshall	(Middlestown, Wakefield)	9. 5.05T
			EI-BMG, G-BFBF, PH-SRF				
G-ZEIN	Slingsby T.67M-260 Firefly	2234		19. 7.95	R.C.P.Brookhouse	Manston	22. 8.05T
G-ZELE	Westland SA.341C Gazelle HT.2	1007	G-CBSA	27. 5.03	A Cook	(Hartlepool)	11.10.05P
			XW845		t/a C3 Consulting *(As "XW845/47")*		
G-ZENA	Zenair CH.701 UL	PFA 187-13637		16.10.00	A N.Aston	(Wolverhampton)	
	(Built A N.Aston)						
G-ZEPI	Colt GA-42 Gas Airship	878	G-ISPY	9. 4.92	P A Lindstrand	Oswestry	12. 5.93A
	(RR Continental O-200B)		(G-BPRB)				
G-ZERO	Grumman AA-5B Tiger	AA5B-0051	OO-PEC	3. 9.80	D.M.Ashford tr G-ZERO Syndicate	Bournemouth	26. 3.05T
G-ZETA	Lindstrand LBL 105A HAB	952		27. 1.04	S.Travaglia	Firenze, Italy	26. 9.05A
G-ZHKF	Just/Reality Escapade 912	BMAA/HB/415		26.11.04	C.D.and Claire M.Wills	Chilbolton	
	(Built C D Wills - kit no.JAESC 0045)						
G-ZHWH	RotorWay Executive 162F	6596		19.11.01	B.Alexander	(Canterbury)	AC
	(Built B Alexander) (RotorWay RI 162F)						
G-ZIGI	Robin DR400/180 Régent	2107		19.11.91	R J Dix	(La Roche sur Foron, France)	23. 3.07
G-ZINT	Cameron Z-77 HAB	10488		17. 9.03	Film Production Consultants SRL	(Roma, Italy)	19. 8.05A
G-ZIPA	Rockwell Commander 114A	14505	G-BHRA	3. 9.98	A C Lees	Full Sutton	9. 6.07T
	(Laid down as c/n 14436)		N5891N				
G-ZIPI	Robin DR400/180 Régent	1557		22. 2.82	H.U.and D.C.Stahlberg	Rochester	13. 6.07
G-ZIPY	Wittman W.8 Tailwind	PFA 031-11339		29. 5.91	J.C.Metcalf	Spanhoe	4.11.05P
	(Built M J Butler) (Lycoming O-235)						
G-ZITZ	Aérospatiale AS355F2 Ecureuil 2	5135	N596SJ	26. 4.04	Heli Aviation Ltd	Biggin Hill	20. 9.07T
			9M-BDA, F-GIFR, F-WZKZ				
G-ZIZI	Cessna 525 CitationJet	525-0345	N5185V	10.11.99	Ortac Air Ltd	Guernsey	17.11.05T
G-ZLIN	Moravan Zlin Z-526 Trener Master	916	G-BBCR	30. 6.81	N.J.Arthur	Bicester	6.10.01
	(Modified from Z-326 standard)		OH-TZF		*(Noted 1.05)*		
	(C/n confirmed but duplicates I-ETRM)						
G-ZLLE	Aérospatiale SA.341G Gazelle 1	1012	N504KH	4.10.01	G-ZLLE Ltd	Stapleford	8.11.04T
			JA9098				
G-ZLOJ	Beech A36 Bonanza	E-1677	ZS-LOJ	11. 9.98	W.D.Gray	Bournemouth	20.12.04
			N6748J				
G-ZLYN	Moravan Zlin Z-526F Trener	1255	OK-CMC	4. 8.95	H.G.Philippart	(London N1)	12.10.07
	(Walter M137A)		YR-ZAB				
G-ZMAM	Piper PA-28-181 Archer II	28-7890059	G-BNPN	3.11.00	Z.Mahmood	Elstree	23. 9.06T
			N47379				
G-ZODI	Zenair CH.601UL Zodiac	PFA 162A-13585		13. 3.00	M.J.Dolby	(London W4)	12.10.04P
	(Built B.McFadden) (Rotax 912-UL)				tr ZODI Group		
G-ZODY	Zenair CH.601UL Zodiac	PFA 162A-14239		26. 5.04	B.H.Stephens	(Old Sarum)	
	(Built B.H.Stephens and Partners - kit no 6-2354?)				tr Sarum AX2000 Group		
G-ZOOL	Reims/Cessna FA152 Aerobat	FA1520357	G-BGXZ	11.11.94	G.G.Hammond	Biggin Hill	9. 5.07T
G-ZORO	Europa Aviation Europa	PFA 247-12672		20. 6.95	N.T.Read	(Gillingham, Kent)	
	(Built N.T.Read - kit no.074) (Monowheel u/c)				*(Current status unknown)*		
G-ZTED	Europa Aviation Europa	PFA 247-12492		30. 4.96	J.J.Kennedy and E.W.Gladstone	(Edinburgh)	
	(Built J.J.Kennedy and E.W.Gladstone - kit no.015) (Monowheel u/c)				*(Part built 7.01)*		
G-ZUMI	Van's RV-8	PFA 303-13527		6. 3.02	P.M.Wells	(Aylesbury)	19. 8.05P
	(Built P M Wells)						
G-ZUMP*	Cameron N-77 HAB				See SECTION 3 Part 1	Newbury	
G-ZVBF	Cameron A-400 HAB	4280		21. 1.98	Airxcite Ltd t/a Virgin Balloon Flights	(Wembley)	9. 1.05T

G-ZWAR	Eurocopter EC120B Colibri	1024	D-HVIP	14. 4.00	Hedgeton Trading Ltd	Lower Baads Farm, Peterculter	3. 8.06
G-ZXZX	Learjet Inc Learjet 45	45-005	N455LJ	16. 1.04	Gama Aviation Ltd	Farnborough	18. 1.05T
G-ZYAK	IAV-Bacau Yakovlev Yak-52	877415	LY-AFK	14. 2.03	R.C Amer	Lydd	31. 3.05P
			DOSAAF 27				
G-ZZAP	American Champion 8KCAB Super Decathlon		N900JF	28.7.04	L.Maikowski, Elise Mason, S.Hipwell and J.Pothecary		
		871-2000				Shoreham	10.10.07E
G-ZZEL	Westland Gazelle AH.Mk.1	1152	G-BZYJ	25.11.02	Military Helicopters Ltd	RAF Shawbury	12. 2.05P
			XW885				
G-ZZOE	Eurocopter EC120B Colibri	1196	F-WQOX	21. 3.01	J.F.H.James	Banbury	9. 7.07T
G-ZZWW	Enstrom 280FX	2052	G-BSIE	22. 2.00	J.W.Gough and F.N.James	Wolverhampton	12. 4.07T
			HA-MIN, G-BSIE				
G-ZZZA	Boeing 777-236	27105	N77779	20. 5.96	British Airways plc	Heathrow	19. 5.05T
G-ZZZB	Boeing 777-236	27106	N77771	28. 3.97	British Airways plc	Heathrow	3.12.05T
G-ZZZC	Boeing 777-236	27107	N5014K	11.11.95	British Airways plc	Heathrow	10.11.07E

PART 3 – AIRCRAFT REGISTERED AND CANCELLED IN 2004 AND 2005

Registration	Type	Construction Number	Previous Identity	Date	Owner/Operator	Cancellation details	Date
G-AWGR	Reims/Cessna F.172H Skyhawk	F17200484	N525DB	9. 4.68	R.G.Hallam	To N....	14. 1.05
			G-AWGR		*(Restored 7.1.05)*		
G-BGLJ	Bell 212	30548	ZJ969	5. 3.79	FB Leasing Ltd	See below	
	G-BGLJ, ZJ969, G-BGLJ, EC-HCZ, (EC-HCP), G-BGLJ, (EC-GHP), EC-295, G-BGLJ, 9Y-TIJ, G-BGLJ, 5N-AJX, G-BGLJ, EP-HBZ, VR-BEJ, N2956W						
	(Cancelled 20.10.04 to MoD, restored 9.12.04 and cancelled 10.2.05 to MoD as ZJ969)						
G-BSPU	Pilatus Britten-Norman BN-2B-26 Islander	2241	N200BN	3. 8.90	Fly BN Ltd	To B-....	21.12.04
	(Originally regd as a BN-2B-26)		D-IFLN (2), G-BSPU		*(New CofR 23.11.04)*		
G-BVYB	Airbus A320-231	0357	OO-TCB	20. 4.95	Thomas Cook Airlines UK Ltd	To XA-UCZ	7.12.04
			G-BVYB, F-WQAZ, (N302SA), F-WWBH *(New CofR 7.10.04)*				
G-CCLD	Embraer EMB-145EP	145.169	SE-DZC	12. 2.04	Air Kilroe Ltd t/a Eastern Airways	To N978RP	27. 9.04
G-CCOD	Cameron Nudie-90 SS HAB	10522		13. 1.04	Cameron Balloons Ltd	To VH-NUD	8. 9.04
G-CCPR	ICP MXP-740 Savannah J22	BMAA/HB/345		3. 2.04	N Farrell	To EI-DGI	28. 4.04
	(Built N Farrell - kit no 03-10-51-236)						
G-CCRL	Mainair Pegasus Quik	8015		16. 2.04	B L Benson	Cancelled as destroyed with Permit to 15.2.05	29. 9.04
G-CCRM	Mainair Pegasus Quik	8011		9. 1.04	Mainair Sports Ltd	To SE-VFT	6. 9.04
G-CCRO	Thruster T600N 450	0042-T600N-098		25. 2.04	Thruster Air Services Ltd	To G-OHYE	9. 3.04
G-CCRP	Thruster T600N 450	0043-T600N-099		25. 2.04	Thruster Air Services Ltd	To G-ULLY	17. 3.04
G-CCRY	Piper PA-31 Turbo Navajo	31-480	F-BTMM	4. 3.04	D C Hanss	To N449TA	6. 5.04
			N449TA				
G-CCSZ	Agusta A119 Koala	14013	I-RALY	16. 4.04	M Sport Ltd	To ZS-RSL	29. 7.04
G-CCTJ	SAAB-Scania 2000	2000-007	SE-007	27.10.04	Air Kilroe Ltd	To SE-...	18. 1.05
			HB-IZD, (D-ADIA)		t/a Eastern Airways		
G-CCUM	Robinson R44 Raven	1377	N7536G	20. 4.04	Heli Air Ltd.	To EI-DFW	28. 5.04
G-CCVC	Van's RV-6A	PFA 181A-13413		24. 3.04	J Edgeworth	To G-CCVS	29. 3.04
G-CCVH	Curtiss H-75A-1	12881	French AF 82	25. 5.04	Patina Ltd	To NX80FR	6.10.04
G-CCYD	Robinson R44 Raven II	10410		23. 6.04	Heli Air Ltd	To OY-HPH	28. 9.04
G-CDAJ	Bell 206L-1 Long Ranger	45729	N20AP	10. 9.04	Patriot Aviation Ltd	To G-LONE	5.10.04
			N3174W				
G-CDAU	Robinson R44 Raven II	10507		22.10.04	Heli Air Ltd	To EI-MCC	1.12.04
G-CDBW	IAV-Bacau Yakovlev Yak-52	866915	RA-02705	1.10.04	R.W.H.Cole	To N521YK	24.11.04
			LY-ABG, DOSAAF 111 *(yellow)*				
G-CDEZ	Robinson R44 Raven II	10614	N999RL	19. 1.05	Heli Air Ltd	To N....	24. 1.05
G-CDFX	Lindstrand LBL 90A HAB	1011		21.12.04	Lindstrand Hot Air Balloons Ltd	To G-GOGB	20. 1.05
G-CDFZ	Zenair CH.601UL	PFA 162A-14279		29.10.04	B.Yoxall and J.P.Harris	To G-JAME	14. 1.05
	(Built B.Yoxall, J.P.Harris and Partners)						
G-DMCG	Robinson R44 Raven II	10514		13.10.04	Heli Air Ltd	To G-RGNT	26. 1.05
G-EJBI	Bölkow Bö.207	242	D-EJBI	9. 3.04	J O'Donnell	To D-EJBI	23. 4.04
G-EZEH	Airbus A319-111	2184	D-AVWO	16. 4.04	EasyJet Airline Co Ltd	To HB-JZF	13. 9.04
G-EZEI	Airbus A319-111	2196	D-AVYC	22. 4.04	EasyJet Airline Co Ltd	To HB-JZG	17. 9.04
G-EZEM	Airbus A319-111	2230	D-AVWD	18. 5.04	EasyJet Airline Co Ltd	To HB-JZH	21. 9.04
G-EZEN	Airbus A319-111	2245	D-AVYH	11. 6.04	EasyJet Airline Co Ltd	To HB-JZI	23. 9.04
G-EZES	Airbus A319-111	2265	D-AVWR	4.10.04	EasyJet Airline Co Ltd	To HB-JZJ	7.10.04
			F-WWBD, D-AVWR				
G-HEEL	Robinson R44 Raven II	10382		27. 5.04	River House Properties Ltd	To G-GRWW	6.12.04
G-OESL	Piaggio P.180 Avanti	1080		27. 5.04	Euroskylink Ltd	To N23RF	30. 9.04
G-OILX	Aerospatiale AS.355F1 Ecureuil 2	5327	ZA141	1. 3.04	Premiair Aviation Services Ltd	To MOD as ZH141	1. 3.04
			G-OILX, G-RMGN, G-BMCY				
G-OMLS	Bell 206B JetRanger II	1957	D-HAFN	2. 2.04	M.L.Scott	To N80367	29. 4.04
			N9909K				
G-OOBG	Boeing 757-236	29942	N544NA	11. 3.04	First Choice Airways Ltd	To C-FUBG	6.12.04
			N1795B, (G-CPEV)				
G-OOBH	Boeing 757-236	29944	N545NA	11. 3.04	First Choice Airways Ltd	To C-FUBG	6.12.04
			(G-CPEX)				
G-OXLA	Boeing 737-81Q	30619	N733MA	25. 5.04	Excel Airways Ltd	To N733MA	8.11.04
			G-OXLA, N733MA, G-OXLA, N733MA				
G-PPLA	Robinson R22 Beta II	3684	N74327	20.10.04	Heli Air Ltd	To EI-DIZ	15.12.04
G-SAJP	Canadair CL-601-3A Challenger	5022	VP-CJP	21. 1.05	Jetlinks (CI) Ltd	To VR-C..	1. 2.05
			N449MC, N449ML, C-GLYK				
G-UKHP	British Aerospace BAe 146 Series 300	E3123	G-5-123	26.10.88	KLM UK Ltd	To EI-DEV	10. 3.04
					(Sold Ireland x.3.04, restored 8.3.04)		
G-YHPV	Cessna E310N	310N0054	N510PS	1.12.04	V.E.Young and P.O.Hayes	To N.....	12. 1.05
			G-AWTA, EI-ATB, N4154Q				

SECTION 2

PART 1 – REPUBLIC OF IRELAND REGISTER

Pete Hornfeck has again compiled the Irish Register and we extend our thanks to him once again. The basic updating information comes from Ian Burnett"s monthly Overseas Register section published in Air-Britain News. Other infortmation has been supplied by Miike Cain, Ian Callier, Antoin Daltun, Ken Hearns, Paul Hewins, Ken Parfitt and Tony Pither. Our thanks to one and all.

No official C of A data is available so it is difficult to determine the status of aircraft and we are grateful to the numerous reports which appear in the "Round and About" Section of Air-Britain News. This information is used to update the status notation system which indicates if an aircraft has been observed as being either active ie flying (A) or merely noted ie apparently inactive (N). However, aircraft still listed on the official Irish Aviation Authority's Website Register are retained even if there have not been any recent positive sightings. In several cases their status appears to indicate that several entries should have been removed long ago from the official Register.

Registration	Type	Construction No	Previous Identity	Date	Registered Owner*(Operator)*	(Unconfirmed) Base	Status
EI-ABI (2)	de Havilland DH.84 Dragon 2	6105	EI-AFK	12. 8.85	Aer Lingus plc	Dublin	A2004
			G-AECZ, AV982, G-AECZ		*"Iolar"*		
EI-ADV	Piper PA-12 Super Cruiser	12-3459	NC4031H	11. 5.48	R.E.Levis	Weston	N 7.99
	(Lycoming O-235)		*(Badly damaged in force landing Maynooth, Weston 8.7.99: current status unknown)*				
EI-AFE	Piper J-3C-90 Cub	16687	OO-COR	11. 3.49	J.Conlon	Kildare	N 4.96
			D-ELAB, N9954F, EI-AFE, NC79076 *(Current status unknown)*				
EI-AFF	BA L.25C Swallow II	406	G-ADMF	18. 5.49	J.Molloy, J.J.Sullivan and B.Donoghue	Ashbourne	N 4.96
	(Pobjoy Cataract II)				*(Damaged Coonagh 24.10.61: current status unknown)*		
EI-AGD	Taylorcraft Plus D	108	G-AFUB	26. 5.53	B.and K.O'Sullivan	Abbeyshrule	N 4.96
			HL534, G-AFUB		*(Current status unknown)*		
EI-AGJ	Auster V J/1 Autocrat	2208	G-AIPZ	3.11.53	T.G.Rafter *(Current status unknown)*	(Ballyboughal)	
EI-AHI (2)	de Havilland DH.82A Tiger Moth	85347	G-APRA	17. 9.93	High Fidelity Flyers	Birr	A 8.02
			DE313				
EI-AKM	Piper J-3C-65 Cub	15810	N88194	17.11.58	Setanta Flying Group	Kilmoon	
			NC88194		*(Stored: current status unknown)*		
EI-ALH	Taylorcraft Plus D	106	G-AHLJ	5. 5.60	N.Reilly	Ballyjamesduff	
			HH987, G-AFTZ		*(Current status unknown)*		
EI-ALP	Avro 643 Cadet	848	G-ADIE	12. 9.60	J.C.O'Loughlin	(Ballybougal)	N2002
	(Genet Major)				*(Engine seizure 12.6.77: awaiting spares and stored)*		
EI-AMF*	Taylorcraft Plus D	157	G-ARRK	26. 4.62	C J Baker	Carr Farm,Thorney, Newark	N 1.05
			G-AHUM, LB286		*(Cancelled 3.4.70 as WFU) (Fuselage partly restored)*		
EI-AMK	Auster V J/1 Autocrat	1838	G-AGTV	19. 9.62	Irish Aero Club	Newcastle, Dublin	N1998
			(WFU after engine failure 5.79: sold 4.95: stored for Air Corps Museum) (Current status unknown)				
EI-AMY	Auster J/1N Alpha	2634	G-AJUW	9. 4.63	T.Lennon	Maynooth, County.Kildare	N 4.92
					(Current status unknown)		
EI-ANT	Champion 7ECA Citabria	7ECA-38		13. 1.65	T.Croke, H.Sydner, D.Foley and E.Lennon	Gorey	A 6.01
EI-ANY	Piper PA-18 Super Cub 95	18-7152	G-AREU	18.11.64	The Bogavia Group	Weston	N 1.03
			N3096Z				
EI-AOB	Piper PA-28-140 Cherokee	28-20667		28. 4.65	J.Surdival, L.Moran, J.Kilcoyne and J.Cowell	Waterford	A 9.04
EI-AOK (2)	Cessna F172G	F172-0208		14. 3.66	D.Bruton	Abbeyshrule	N 7.01
	(Built Reims Aviation SA)				*(Stripped hulk noted)*		
EI-AOS	Cessna 310B	35578	G-ARIG	1.11.65	Joyce Aviation Ltd	Kildimo	
			EI-AOS, G-ARIG, N5378A		*(WFU and to scrapyard) (Current status unknown)*		
EI-APF	Cessna F150G	F150-0112		6. 3.66	Sligo Aero Club Ltd	Kirknewton	N 3.04
	(Built Reims Aviation SA)						
EI-APS (2)	Schleicher ASK14	14008	(EI-114)	24.11.69	SLG Group	Gowran Grange	
			G-AWVV, D-KOBB		*(Current status unknown)*		
EI-ARH (2)	Slingsby T.56 SE5 rep	1590	G-AVOT	22. 6.67	L.Garrison	Flabob, California, US	N 5.96
	(Lycoming O-235)				*(Current status unknown)*		
EI-ARM	Slingsby T.56 SE5 rep	1594	G-AVOX	22. 6.67	M L Putman	Sanger, Texas. US	N10.99
	(Lycoming O-235) *(Regd with c/n 1595 ex G-AVOY)*				*(Registered in US as N912AC)*		
EI-ARW	Jodel DR.1050	118		14. 8.67	Joseph Davy *(New owner 2.03)*	Moyne	
EI-ASR (2)	McCandless M.4 Gyroplane	M.4/5	G-AXHZ	29. 9.69	G.J.J.Fasenfeld	Sion Mills, Strabane	N 4.96
	(Volkswagen) *(C/n M4/4 quoted also)*				*(Sold to R.McGregor: stored) (Current status unknown)*		
EI-AST	Cessna F150H	F150-0273		30. 1.68	Ormand Flying Group	Birr	A 9.04
	(Built Reims Aviation SA)						
EI-ASU*	Beagle A.61 Terrier 2	B.633	G-ASRG	10. 1.68	C Lebroda and Partners	Trim	N 1.03
			WE599		*(Cancelled 15.6.77 as "sold abroad" but stored in container)*		
EI-ATJ	Beagle B.121 Pup Series 2	B121-029	G-35-029	10. 2.69	L.O'Leary	Waterford	N 1.04
EI-ATK	Piper PA-28-140 Cherokee	28-24120	G-AVUP	18.10.68	Mayo Flying Club	Abbeyshrule	N 1.03
			N11C		*(Damaged Connaught 14.2.87: stripped hulk noted)*		
EI-ATL	Aeronca 7AC Champion	7AC-4674	N1119E	22. 9.69	Kildare Flying Club	Abbeyshrule	
			(Damaged Weston 26.11.75: used for spares in restoration of EI-AVB) (Current status unknown)				
EI-ATS	SOCATA MS.880B Rallye Club	1582		20. 4.70	ATS Group	Abbeyshrule	N 4.96
					(Stored: current status unknown)		
EI-AUC	Reims Cessna FA150K Aerobat	FA1500040		10. 4.70	Garda Aviation Club Ltd	Weston	
			(Badly damaged in force landing north of Weston 15.7.99) (Current status unknown)				
EI-AUE	SOCATA MS.880B Rallye Club	1359	G-AXHU	1. 4.70	Kilkenny Flying Club Ltd	Kilkenny	A 9.04
EI-AUG	SOCATA MS.894A Rallye Minerva 220	11080		17. 6.70	K.O'Leary	Rathcoole	N 1.03
EI-AUJ	SOCATA MS.880B Rallye Club	1370	G-AXHF	12. 6.70	Ormond Flying Club Ltd	Abbeyshrule	N 8.98
			F-BNGV		*(Stored: current status unknown)*		
EI-AUM	Auster V J/1 Autocrat	2612	G-AJRN	11. 9.70	T.G.Rafter (*Current status unknown)*	(Ballyboughal)	N 6.96
EI-AUO	Reims/Cessna FA150K Aerobat	FA1500074		2. 3.70	Kerry Aero Club Ltd	Waterford	A10.02
EI-AUS	Auster J/5F Aiglet Trainer	2779	G-AMRL	17.11.70	T.Stevens and T.Lennon	Powerscourt	N 4.95
					(Current status unknown)		

EI-AUT	Forney F-1A Aircoupe	5731	G-ARXS	21.12.70	Joyce Aviation Ltd	Bann Foot, Lough Neagh	N 8.97
			D-EBSA, N3037G		*(To N.Glass and A.Richardson?) (Current status unknown)*		
EI-AUY	Morane-Saulnier MS.502 Criquet	338	F-BCDG	30.11.70	G Warner	Duxford	
	(Argus AS.10)		Fr.Mil		*(As "CF+HF" in Luftwaffe c/s)*		
EI-AVB	Aeronca 7AC Champion	7AC-1790	7P-AXK	14. 6.71	J D Cooper	Thonotosassa, Florida, US	N10.99
	(Continental A65)		ZS-AXK		*(Regd in US as N151JC [7AC-71790]*		
EI-AVM	Reims/Cessna F150L	F15000745		3. 3.72	John Logan and Tony Bradford	Ballynacarigy	A 9.04
EI-AWD	Piper PA-22-160 Tri-Pacer	22-6411	G-APXV	17. 1.73	J.P.Montcalm	Carrigtwo hil, County.Cork	N1989
			N9437D		*(Blown over in gales Cork 12.81) (Stored: current status unknown)*		
EI-AWH	Cessna 210J Centurion	21059067	G-AZCC	19. 1.73	Rathcoole Flying Club Ltd	Rathcoole	
			(EI-AWH), G-AZCC, 5N-AIE, N1734C, (N6167F)		*(Current status unknown)*		
EI-AWP	de Havilland DH.82A Tiger Moth	85931	F-BGCL	4. 7.72	Anne.P.Bruton	Abbeyshrule	N 1.03
	(Regd with c/n 19577)		Fr.AF, DF195				
EI-AWR	Malmo MFI-9 Junior	010	LN-HAG	12. 6.73	M.Whyte and J.Brennen	Galway	N 1.03
			(SE-EBW)				
EI-AWU	SOCATA MS.880B Rallye Club	880	G-AVIM	12. 1.74	Longford Aviation Ltd	Milford, County.Donegal	N 9.03
					(Fuselage to scrapyard)		
EI-AYA	SOCATA MS.880B Rallye Club	2256	G-BAON	27. 7.73	Limerick Flying Club (Coonagh) Ltd	Coonagh	A 8.02
EI-AYB	Gardan GY-80-180 Horizon	156	F-BNQP	5.10.73	J.B.Smith	Abbeyshrule	N 1.03
EI-AYD	Grumman-American AA-5 Traveler	0380	G-BAZE	9. 7.73	P.Howick, H.Martini and V.O'Rourke	Powerscourt	
			N5480L		*(Current stattus unknown)*		
EI-AYF	Reims/Cessna FRA150L Aerobat	FRA1500218		26. 3.74	Limerick Flying Club (Coonagh) Ltd	Coonagh	N 1.03
EI-AYI	Morane MS.880B Rallye Club	189	F-OBXE	21.11.73	J.McNamara	Cloncameel	A 9.02
EI-AYK	Reims/Cessna F172M Skyhawk II	F17201092		25. 3.74	D.Gallagher	Trim	A 9.04
EI-AYN	IRMA BN-2A-8 Islander	704	G-BBFJ	26. 3.74	Galway Aviation Services Ltd	Galway	A2004
					(Op Aer Arann Islands) "Inis-Mor"		
EI-AYR	Schleicher ASK 16	16022	(EI-119)	5. 4.74	Brian O'Broin	Kilrush	N 1.03
EI-AYT (2)	SOCATA MS.894A Rallye Minerva 220	11065	G-AXIU	6. 8.74	K.A.O'Connor	Abbeyshrule	N 5.00
					(Damaged Palklasmore 12.11.89: wreck only)		
EI-BAJ	SNCAN Stampe SV-4C	171	F-BBPN	17.10.74	Dublin Tiger Group	Trim	N 1.03
					(Being rebuilt at Abbeyshrule)		
EI-BAR	Thunder Ax8-105 HAB	014	G-BCAM	26. 2.75	J.Burke and V.Hourihane	Cahir	
					"Rockwell" (WFU: current status unknown)		
EI-BAT	Reims/Cessna F150M	F15001196		2. 5.75	K.A.O'Connor	Weston	N 1.03
EI-BAV	Piper PA-22-108 Colt	22-8347	G-ARKO	30. 4.75	Edmond Finnamore and Jerry Deegen	Birr	N 1.03
EI-BBC	Piper PA-28-180 Cherokee B	28-1049	G-ASEJ	18. 6.75	Vero Beach Ltd *(New owner 11.01)*	Strandhill	A 9.04
EI-BBD	Evans VP-1	VP-1-No.2 & SAAC-02		13. 8.76	The Volksplane Group	Celbridge	N 1.03
	(Volkswagen 1600)				*(Damaged 12.9.81: on rebuild)*		
EI-BBE	Champion 7FC Tri-Traveler	7FC-393	G-APZW	7. 9.75	Randal McNally and Cormac Carey	Carnmore, Galway	N 2.03
	(Tail-wheel conversion to 7EC Traveler status)						
EI-BBG	SOCATA Rallye 100ST	2592		27.10.75	Weston Ltd *(Dismantled fuselage)*	Weston	N 4.01
EI-BBI	SOCATA Rallye 150ST	2663		13.10.75	Kilkenny Airport Ltd	Kilkenny	A 8.02
EI-BBJ	SOCATA MS.880B Rallye 100S	2361	F-BUVX	7.11.75	Weston Ltd	Weston	N 1.03
EI-BBO	SOCATA MS.893E Rallye 180GT	12522	F-BVNM	8. 3.76	G.P.Moorhead	Hacketstown	N 1.03
EI-BBV	Piper J-3C-65 Cub (L-4J-PI)	13058	D-ELWY	14. 6.76	F.Cronin	Weston	N 1.03
	(Frame No.12888)		F-BEGB, 44-80762		*(As "480762" in USAAF c/s)*		
EI-BCE	Britten Norman BN-2A-26 Islander	519	G-BDUV	14. 9.76	Galway Aviation Services Ltd	Galway	A2004
					(Op Aer Arann Express) "Inis-Meain"		
EI-BCF	Bensen B-8M Gyrocopter	47941	N....	24. 8.76	P.Flanagan	(Kilrush)	N1997
	(McCulloch.AF O-100)				*(Stored: current status unknown)*		
EI-BCJ (2)	Aeromere F.8L Falco 3	204	G-ATAK	19. 1.77	D.Kelly	Abbeyshrule	N 1.03
			D-ENYB		(On rebuild)		
EI-BCK	Reims/Cessna F172N Skyhawk II	F17201543		22.11.76	K.A.O'Connor	Weston	A11.04
EI-BCL	Cessna 182P Skylane II	18264300	N1366M	22.11.76	L.Burke	Newcastle	A12.01
	(Reims assembled with c/n F1820045)						
EI-BCM	Piper J-3C-65 Cub (L-4H-PI)	12155	11983	26.11.76	Kilmoon Flying Group	Trim	N 1.03
			44-79859, OO-RAL, N9857F, F-BNAV				
EI-BCN	Piper J-3C-65 Cub (L-4H-PI)	12335	F-BFQE	26.11.76	H.Diver	Kilrush	A 1.00
			OO-PIE, 44-80039				
EI-BCO	Piper J-3C-65 Cub	"1"	F-BBIV	26.11.76	J.Molloy	Kilmoon	
					(Not converted and remains stored: current status unknown)		
EI-BCP	Rollason Druine D.62B Condor	RAE/618	G-AVCZ	27. 1.77	A.Delaney	Dolla	A 7.02
EI-BCS	SOCATA MS.880B Rallye 100T	2550	F-BVZV	4. 2.77	Organic Fruit and Vegetables of Ireland Ltd	Kilkenny	A 4.02
EI-BCU	SOCATA MS.880B Rallye 100T	2595	F-BXTH	10. 2.77	Weston Ltd *(Derelict)*	Weston	N 1.03
EI-BCW	SOCATA MS.880B Rallye Club	1783	G-AYKE	18. 4.77	Kilkenny Flying Club	Abbeyshrule	N 6.97
					(Stored: current status unknown)		
EI-BDH	SOCATA MS.880B Rallye Club	1270	G-AWOB	18. 7.77	Munster Wings Ltd	Abbeyshrule	
					(Damaged Cork 5.12.78: current status unknown - believed scrapped)		
EI-BDK	SOCATA MS.880B Rallye 100T	2561	F-BXMZ	10. 8.77	Limerick Flying Club (Coonagh) Ltd	Abbeyshrule	N 9.99
					(Airframe stored: current status unknown)		
EI-BDL	Evans VP-2	V2-2101, PFA 7213 and SAAC-04		7. 9.77	P.Buggle	Kildare	N 1.03
	(Volkswagen)						
EI-BDM	Piper PA-23-250 Aztec D	27-4166	G-AXIV	10.10.77	G.A.Costello	Waterford	N 4.96
			N6826Y		t/a Executive Air Services *(Current status unknown)*		
EI-BDR	Piper PA-28-180 Cherokee C	28-3980	G-BAAO	8.12.77	Cherokee Group	Farranfore	A10.02
			LN-AEL, SE-FAG				
EI-BEA	SOCATA Rallye 100ST	3007		28. 2.78	Weston Ltd *(Dismantled fuselage hangared)*	Weston	N 4.01
EI-BEN	Piper J-3C-65 Cub (L-4J-PI)	12546	G-BCUC	28. 4.78	J.J.O'Sullivan	Weston	A 8.00
	(Frame No.12376)		F-BFMN, 44-80250				
EI-BEP	SOCATA MS.892A Rallye Commodore 150	11947	F-BTJT	14. 4.78	H.Lynch and J.O'Leary	Abbeyshrule	N 1.03
					(Stripped hulk noted)		
EI-BFE	Cessna F150G	F150-0158	G-AVGM	3. 8.78	Joyce Aviation Ltd	Waterford	N 1.02
	(Built Reims Aviation SA)				*(Stored dismantled)*		
EI-BFF	Beech A23-24 Musketeer Super III	MA-352	G-AXCJ	20. 8.78	P.McCoole	Coonagh	N 1.03

Regn	Type	C/n	Previous identity	Date	Owner/operator	Base	Status
EI-BFI	SOCATA Rallye 100ST	2618		10. 8.78	J O'Neill	Abbeyshrule	N 1.03
					(Crashed 14.12.85: stripped hulk noted)		
EI-BFO	Piper J-3C-90 Cub (L-4J-PI)	12701	F-BFQJ	11. 9.78	D.Gordon	Weston	
	(Frame No.12531) (Regd as c/n 8911)		N79856, NC79856, 44-80405			*(Current status unknown)*	
EI-BFP	SOCATA Rallye 100ST	2942	F-GARR	6.10.78	Limerick Flying Club (Coonagh) Ltd	Coonagh	A 1.03
EI-BFR	SOCATA Rallye 100ST	2429	F-OCVK	9.11.78	Wexford Flying Group	Waterford	A 1.02
EI-BFV	SOCATA MS.880B Rallye 100T	2415	F-BVAH	2. 2.79	Ormond Flying Club Ltd		
					(Stag Park, Cloughjordan, County Tipperary)		
					(Stored 8.94: current status unknown but believed scrapped 1976)		
EI-BGA	SOCATA Rallye 100ST	2549	G-BCXC	23.11.78	J.J.Frew	Mullaghmore	N 7.01
			F-OCZQ				
EI-BGB	SOCATA MS.880B Rallye Club	1913	G-AZKB	22. 1.79	Limerick Flying Club (Coonagh) Ltd	Abbeyshrule	N 3.98
					(Stored: current status unknown)		
EI-BGC	SOCATA MS.880B Rallye Club	1265	F-BRDC	22.12.78	P.Moran	Roscommon	
					(WFU and cannibalised: current status unknown))		
EI-BGD	SOCATA MS.880B Rallye Club	2287	F-BUJI	18.12.78	N.Kavanagh *(Stripped hulk noted)*	Abbeyshrule	N 1.03
EI-BGF	Piper PA-28R-180 Cherokee Arrow	28R-30121	SE-FAS	30. 1.79	Arrow Group	(Dublin)	
			(Crashed Mynydd Prescelly near Haverfordwest, Dyfed 6.10.83: current status unknown)				
EI-BGG	SOCATA MS.892E Rallye 150GT	12824	F-GAFS	30. 1.79	J.Dowling and M.Martin	Abbeyshrule	N 1.04
EI-BGJ	Reims/Cessna F152 II	F15201664		14. 5.79	Sligo Aeronautical Club Ltd	Strandhill	A10.02
EI-BGS	SOCATA MS.893E Rallye 180GT	12675	F-BXTY	25. 4.79	M.Farrelly	Abbeyshrule	N 4.96
					(Damaged Claive 3.91: wreck stored: current status unknown)		
EI-BGT	Colt 77A HAB	041		14. 5.79	M.J.Mills	Navan	
	(New envelope c/n 1092 - original to EI-BBM)				*"Spirit of Ireland" (Ryan Air titles) (New owner 9.01)*		
EI-BGU	SOCATA MS.880B Rallye Club	875	F-BONM	9. 5.79	M.F.Neary	Abbeyshrule	N 9.99
					(Wreck stored: current status unknown)		
EI-BHC	Reims/Cessna F177RG Cardinal	F177RG0010	G-AYTG	11. 7.79	B.J.Palfrey and Partners	Dublin	A 5.01
	(Wichita c/n 17700117)				*"Hot Chocolate/90"*		
EI-BHF	SOCATA MS.892A Rallye Commodore 150	10742	F-BPBP	10. 7.79	B.Mullen	Strandhill	
					(WFU Strandhill '87: engine to EI-BYL: scrapped c.1989)		
EI-BHI	Bell 206B JetRanger II	906	G-BAKX	14. 8.79	H.S.S.Ltd *(New owner 9.01)*	Rathcoole	
EI-BHK*	Socata MS.880B Rallye Club	1307	F-BRJE	20. 8.79	Not known	Kilkenny	N 4.02
					(Cancelled 8.9.00 as WFU).(Derelict)		
EI-BHM	Reims Cessna F337E Super Skymaster	F33700004	OO-PDC	1.11.79	City of Dublin VEC	Bolton St, Dublin	N 9.01
	(Wichita c/n 33701217)		OO-PDG		*(At College of Technology: instructional airframe)*		
EI-BHN	SOCATA MS.893A Rallye Commodore 180	11422	F-BRRO	11.10.79	T.Garvan *(On overhaul)*	Hacketstown	N 1.03
EI-BHP	SOCATA MS.893A Rallye Commodore 180	11459	F-BSAA	12.10.79	Spanish Point Flying Club	Spanish Point	
					(Current status unknown)		
EI-BHT	Beech 77 Skipper	WA-77		17.10.79	Waterford Aero Club Ltd	Waterford	A 8.04
EI-BHV	Aeronca 7EC Traveler	7EC-739	G-AVDU	30.10.79	E.P.O'Donnell and Partners	Clonmel	N 1.03
			N9837Y				
EI-BHW	Cessna F150F	F150-0013	G-ATMK	22.11.79	R.Sharpe	Weston	
	(Built Reims Aviation SA) (Wichita c/n 15062671)				*(Current status un known)*		
EI-BHY	SOCATA Rallye 150ST	2929	F-GARL	19.11.79	Limerick Flying Club (Coonagh) Ltd	Coonagh	N 1.03
EI-BIB	Reims/Cessna F152 II	F15201724		30.11.79	Galway Flying Club Ltd	Carnmore, Galway	A 9.04
EI-BIC	Reims/Cessna F172N Skyhawk II	F17201965	(OO-HNZ)	15. 2.80	Oriel Flying Group Ltd	Abbeyshrule	N 1.03
					(Wreck in open storage)		
EI-BID	Piper PA-18 Super Cub 95	18-1524	D-EAES	30.11.79	S.Coghlan and P.Ryan	Carnmore, Galway	N 2.03
	(L-18C-PI)		ALAT 18-1524, 51-15524				
EI-BIG	Moravan Zlin 526 Trener Master	1086	D-EBUP	7.12.79	P A Colman *(Damaged 9.91 and acquired for spares?)*		
			OO-BUT		Luxters Farm, Hambleden, Henley-on-Thames		N2002
EI-BIJ	Agusta-Bell 206B JetRanger II	8432	G-BCVZ	29. 1.80	Medavia Properties Ltd	Dublin Heliport	A12.01
					(Op Celtic Helicopters Ltd)		
EI-BIK	Piper PA-18-150 Super Cub	18-7909088	N82276	1. 2.80	Dublin Gliding Club Ltd	Gowran Grange	N10.01
	(Modified to 180hp)						
EI-BIM	Morane MS.880B Rallye Club	305	F-BKYJ	28. 3.80	D.Millar *(Stored: current status unknown)*	Abbeyshrule	N 6.97
EI-BIO	Piper J-3C-65 Cub (L-4J-PI)	12657	F-BGXP	27. 5.80	Monasterevin Flying Group		
			OO-GAE, 44-80361			Harristown Nurney, Monasterevin	A 8.00
EI-BIR	Reims/Cessna F172M Skyhawk II	F17201225	F-BVXI	24. 3.80	B.Harrison, K.Brereton, P.Rogers and F.Maher		
						Clonbullogue	A11.00
EI-BIS	Robin R1180TD Aiglon	268		14. 5.80	The Robin Aiglon Group	Abbeyshrule	N 1.03
EI-BIT	SOCATA MS.887 Rallye 125	2169	F-BULQ	18. 3.80	Spanish Point Flying Club	Spanish Point, Co.Clare	A 1.00
EI-BIV	Bellanca 8KCAB Super Decathlon	464-79	N5032Q	3. 6.80	Aerocrats Flying Group Ltd	Abbeyshrule	N 7.01
EI-BIW	SOCATA MS.880B Rallye Club	1144	F-BPGB	19. 5.80	E.J.Barr	Buncrana	
					(Crashed on take off Rosnakil 10.8.86 and scrapped 1990)		
EI-BJB	Aeronca 7DC Champion	7AC-925	G-BKKM	16. 4.80	W.Kennedy	Killenaule	N12.00
	(Continental C85)		EI-BJB, N82296, NC82296		*(Stored incomplete)*		
EI-BJC	Aeronca 7AC Champion	7AC-4927	N1366E	2. 4.80	E.Griffin	Blackwater	N 8.04
	(Continental A65)		NC1366E, SE-FBW, OY-DKN		*(Crashed Edenderry 9.82)*		
EI-BJI	Reims FR172E Rocket	FR17200040	G-BAAS	23. 5.80	Irish Parachute Club Ltd	Dublin	
			SE-FBW, OY-DKN		*(Crashed Edenderry 9.82: probably scrapped pre 1990)*		
EI-BJJ	Aeronca 15AC Sedan	15AC-226	(G-BHXP)	6. 6.80	O.Bruton	Abbeyshrule	N 3.98
			EI-BJJ, N1214H		*(Stored: current status unknown)*		
EI-BJK	SOCATA Rallye 110ST	3226	F-GBKY	8. 7.80	Malachy Keenan	Weston	A 1.03
EI-BJM	Cessna A152 Aerobat	A1520936	N761CC	18. 9.80	Leinster Aero Club Ltd	Abbeyshrule	A11.04
EI-BJO	Cessna R172K Hawk XP II	R1723340	N758TD	6. 8.80	P.Hogan and others	Carnmore, Galway	N 2.03
EI-BJT	Piper PA-38-112 Tomahawk	38-78A0818	G-BGEU	16.10.80	S.Corrigan and W.Lennon	Abbeyshrule	N 1.03
			N9650N				
EI-BKC	Aeronca 15AC Sedan	15AC-467	N1394H	5.11.80	G.Hendrick, M.Farrell and J.Keating	Birr	N 1.03
EI-BKE*	Morane MS.885 Super Rallye	278	F-BKUN	9. 2.81	Not known	Abbeyshrule	N 1.03
			F-WKUN		*(Crashed Ballyclumack, Wexford 5.4.81: stripped hulk noted)*		
EI-BKF	Cessna F172H	F172-0476	G-AVUX	4.12.80	D.Darby	Weston	N 7.01
	(Built Reims Aviation SA)						

Reg	Type	C/n	Previous id	Date	Owner	Base	Status
EI-BKK	Taylor JT.1 Monoplane	PFA 1421	G-AYYC	2. 2.81	Waterford Aero Club	Waterford	N 1.02
	(Volkswagen 1500)				*(Stored dismantled)*		
EI-BKN	SOCATA Rallye 100ST	3035	F-GBCK	18. 2.81	Weston Ltd (Stored)	Weston	N 8.02
EI-BKU	SOCATA MS.892A Rallye Commodore 150	10990	F-BRLG	21. 5.81	Limerick Flying Club (Coonagh) Ltd	Abbeyshrule	N 5.99
					(Open stored: current status un known)		
EI-BLB	SNCAN Stampe SV-4C	323	F-BCTE	27. 7.81	J.E.Hutchinson and R.A.Stafford	Abbeyshrule	
					(Crashed Drumsna, Carrick-on-Shannon 1.6.97: current status unknown)		
EI-BLD	MBB Bö.105DB	S.381	D-HDLQ	21. 7.81	Irish Helicopters Ltd	Dublin	A 1.03
EI-BLE	Eipper Quicksilver	IMA-003		20. 8.81	R.P.St.George-Smith	Kilkenny	
	(Yamaha KT100SP)				*(Believed scrapped following accident date unknown)*		
EI-BLN	Eipper Quicksilver MX	MX.01		26. 8.81	O.J.Conway and B.Daffy	Ennis	
	(Cuyana 340)				*(Believed scrapped)*		
EI-BMA	SOCATA MS.880B Rallye Club	1965	F-BTJR	26. 1.82	W.Rankin and M.Kelleher	Abbeyshrule	N 1.03
					(Wings only noted)		
EI-BMB	SOCATA MS.880B Rallye 100T	2505	G-BJCO	5. 1.82	Glyde Court Developments Ltd	Weston	N 1.03
			F-BVLB				
EI-BMF	Laverda F.8L Super Falco Srs.IV	416	G-AWSU	28. 1.82	M.Slazenger and H.McCann	Powerscourt	A 8.02
EI-BMI	SOCATA TB-9 Tampico	203	F-GCOV	12. 5.82	Ashford Flying Group	Weston	A11.04
EI-BMJ	SOCATA MS.880B Rallye 100T	2594	F-BXTG	10. 3.82	Limerick Flying Club (Coonagh)Ltd	Coonagh	N 9.02
EI-BMM	Reims Cessna F152 II	F15201899		10. 3.82	P.Redmond *(Current status unknown)*	Weston	
EI-BMN	Reims/Cessna F152 II	F15201912		10. 3.82	Sligo Light Aviation Club Ltd	Strandhill	A 9.04
EI-BMU	Monnett Sonerai IIL	01224		19. 5.82	A.Fenton	Ballyshannon	N 7.01
	(Volkswagen 2100)						
EI-BMV	American Aviation AA-5 Traveler	AA5-0200	G-BAEJ	28. 7.82	E.Tierney and K.A.Harold	Abbeyshrule	N 1.03
					(Damaged Brittas Bay 3.93: stripped hulk noted)		
EI-BMW	Skytrike/Hiway Vulcan	LM-100		1. 6.82	L.Maddock	Carlow	
	(Built L Maddock) (Fuji-Robin)				*(Current status unknown)*		
EI-BNF	Eurowing Goldwing	Not known		22. 9.82	N Irwin	Cork	
	(Fuji-Robin)				*(WFU & scrapped 1985)*		
EI-BNH	Hiway Skytrike	AS.09		18.10.82	M.Martin	Tullamore	
	(Fuji-Robin EC-25-PS)				*(Current status unknown)*		
EI-BNJ	Evans VP-2	Not known		24. 1.83	G.A.Cashman	Bartlemy	
	(VW 2000)				*(WFU & believed scrapped 1996)*		
EI-BNK	Cessna U206F Stationair	U20601706	G-HILL	23.12.82	Irish Parachute Club Ltd	Clonbulloge	A 1.03
			PH-ADN, D-EEXY, N9506G				
EI-BNL	Rand Robinson KR-2	Not known		13. 1.83	K.Hayes	Birr	N 1.03
	(Volkswagen 2000)				*(Under construction)*		
EI-BNP	Rotorway Executive 145	Not known		1. 3.83	R.L.Renfroe *(Not completed 1989)*	Letterkenny	
EI-BNT	Cvjetkovic CA-65	Not known		23. 3.83	B.Tobin and P.G.Ryan	(Tallaght)	
					(Current status unknown)		
EI-BNU	SOCATA MS.880B Rallye Club	1204	F-BPQV	7. 4.83	P.A.Doyle	Coonagh	N 1.03
EI-BOA	Pterodactyl	Not known		3. 5.83	A.Murphy	Athenry	
EI-BOE	SOCATA TB-10 Tobago	301	F-GDBL	12. 9.83	Tobago Group c/o Denis A.Cody	Dun Laoghaire	N 1.03
EI-BOH	Eipper Quicksilver	Not known		8. 9.83	J.Leech	Waterford	
	(Yamaha 970cc)				*(Current status unknown - believed dismantled)*		
EI-BOV	Rand Robinson KR-2	SAAC-11		7. 5.84	G.O'Hara and G.Callan *"Kitty Hawk"*	(Strandhill)	
	(Volkswagen 1835)				*(Damaged Cammore 3.91 on re-build 1999: current status unknown)*		
EI-BOX	Box Duet	Not known		12.10.84	Dr.K.Riccius	(Newcastle)	
	(Rotax 503)				*(Current status unknown)*		
EI-BPE	Viking Dragonfly	SAAC-16		15.10.84	G.G.Bracken	Castlebar	N 1.01
	(Volkswagen 1835)				*(Not completed and stored)*		
EI-BPL	Reims/Cessna F172K	F17200758	G-AYSG	28. 3.85	Phoenix Flying Ltd	Shannon	A 9.04
EI-BPN	Flexiform Striker	Not known		12. 3.85	P.H.Collins	Dunlaoghaire	
	(Fuji Robin)				*(Current status unknown)*		
EI-BPO	Southdown Puma	1923		12. 3.85	A.Channing	Clane	
	(Fuji-Robin EC-44-PM - E/No.82-00108) *(Also quoted as Southdown Sailwing c/n 1924) (Current status unknown)*						
EI-BPP	Eipper Quicksilver MX	3207		12. 3.85	J.A.Smith	Abbeyshrule	N 1.03
	(Cuyana 430)				*(Stored)*		
EI-BPT	Skyhook Sabre	Not known		26. 3.85	T.McGrath	Glounthane	
	(Solo 210)				*(Current status unknown: thought dismantled)*		
EI-BPU	Hiway Demon	Not known		26. 3.85	A.Channing	Abbeyshrule	N 5.00
	(Fuji-Robin EC-25-PS)						
EI-BRK	Flexiform Trike	LM.102		17. 6.85	L.Maddock	Hacketstown	
	(Fuji Robin)				*(WFU and scrapped 1994)*		
EI-BRS	Cessna P172D	P17257173	G-WPUI	2. 9.85	Paul Mathews	Weston	N 1.03
			G-AXPI, 9M-AMR, N11B, (N8573X) *(In poor condition)*				
EI-BRU	Evans VP-1	V-12-84-CQ and SAAC-18		5.11.85	Home Bru Flying Group	Weston	N 6.02
	(Volkswagen 1600)						
EI-BRV	Hiway Demon/Skytrike	Not known		5.11.85	M.Garvey and C.Tully	Kells	
	(Fuji-Robin EC-25-PS)						
EI-BRW	Hovey Delta Bird	Not known		5.11.85	A and E Aerosport	Curraglass	
	(Volkswagen 1300) *(Originally regd as Ultra-Lite Deltabird but is a Bimax Osprey: was built to re-enact a scene nr Fermoy 1986 from the "Blue Max" film: dismantled and scrapped)*						
EI-BSB	Wassmer Jodel D.112	1067	G-AWIG	23. 6.87	Estartit Ltd	Kilrush	A 1.03
			F-BKAA				
EI-BSC	Reims/Cessna F172N Skyhawk II	F17201651	G-NIUS	10.12.85	S.Phelan	Weston	N 1.03
EI-BSF*	Avro 748 Srs.1/105	1544	EC-DTP	28. 5.86	Ryanair Ltd *"Spirit of Tipperary"*	Dublin	N 4.01
			G-BEKD, LV-HHF, LV-PUM		*(Fuselage used as cabin trainer for fire training in all-white c/s)*		
EI-BSG	Bensen B-80 Gyrocopter	HB		30. 1.86	J.Todd	(Riverstick)	N 3.90
	(McCullough 4318)				*(Stored: current status unknown)*		
EI-BSK	SOCATA TB-9 Tampico	618		9. 4.86	Weston Ltd	Weston	N 1.03
EI-BSL	Piper PA-34-220T Seneca III	34-8233041	N8468X	27. 6.86	P Greenan	Weston	A 6.02
EI-BSN	Cameron O-65 HAB	1278		14. 4.86	Carol O'Neill and Tracy Hooper	Cavan	A 10.02
					"Erin-Go-Bragh"		

Reg	Type	C/n	Previous identity	Date	Owner/Operator	Base	Status
EI-BSO	Piper PA-28-140 Cherokee B	28-25449	C-GOBL	16. 4.86	H.M.Hanley	Waterford	N 1 04
			N8241N				
EI-BSV	SOCATA TB-20 Trinidad	579	G-BMIX	15. 8.86	J.Condron	Abbeyshrule	A 9.04
EI-BSW	Solar Wings Pegasus XL-R			22. 6.87	E.Fitzgerald	Waterford	A 10.02
	(Rotax 447) SW-TB-1124 and SW-WA-1122						
EI-BSX	Piper J-3C-65 Cub	8912	G-ICUB	25. 3.86	J. and T.O'Dwyer	Gowran Grange	
	(Frame No.8999)		F-BEGT, NC79805, 45-4515, 42-36788		*(Current status unknown)*		
	(Official c/n 13255 is incorrect as a/c probably rebuilt c.1945)						
EI-BTX	McDonnell-Douglas MD-82	49660	(N59842)	23. 3.88	Airplanes Holdings Ltd		
					(Op Aeromexico)	Mexico City-Benito Juarez, Mexico	A2004
EI-BTY (2)	McDonnell-Douglas MD-82	49667	(N12844)	6. 5.88	Airplanes Holdings Ltd		
					(Op Aeromexico)	Mexico City-Benito Juarez, Mexico	A2004
EI-BUA	Cessna 172M Skyhawk II	17265451	N5458H	8. 8.86	Skyhawks Flying Club	Weston	N 1.03
EI-BUC	Jodel D.9 Bébé	PFA 929	G-BASY	20. 1.87	Bernard Lyons and Michael Blake	Thurles	A 7.01
	(Volkswagen 1500)				*(Current status unknown)*		
EI-BUF	Cessna 210N Centurion II	21063070	G-MCDS	18.12.86	210 Group	Abbeyshrule	A10.04
			G-BHNB, N6496N				
EI-BUG	SOCATA ST-10 Diplomate	125	G-STIO	4. 2.87	J.Cooke	Weston	N 1.03
			OH-SAB		*(Derelict)*		
EI-BUH	Lake LA-4-200 Buccaneer	543	G-PARK	27. 5.87	P.Redden	Weston	N.1.03
			G-BBGK, N39779				
EI-BUJ	SOCATA MS.892A Rallye Commodore 150	10737	G-FOAM	27. 2.87	T.Cunniffe	Abbeyshrule	N 3.98
			G-AVPL		*(Damaged pre 1992: stored: current status unknown)*		
EI-BUL	Whittaker MW.5 Sorcerer	1		4. 3.87	J.Culleton	Mountmellick, County Laois	
	(Citroen 602cc)				*(Current status unknown)*		
EI-BUN	Beech 76 Duchess	ME-371	(EI-BUO)	26. 6.87	K.A.O'Connor	Weston	A11.04
			N37001				
EI-BUT	GEMS MS.893A Rallye Commodore 180	10559	SE-IMV	30. 7.87	T.Keating	Weston	A 9.02
			F-BNBU		*(Galerien c/s)*		
EI-BVB	Whittaker MW.6 Merlin	1		14. 9.87	R.England	Watergrasshill	N 9.00
	(Rotax)				*(Current status unknown)*		
EI-BVJ (2)	AMF Chevvron 2-32	009		16. 2.88	S.J.Dunne	Bolybeg, Ballymore Eustace	
	(Konig SD570)				*(Current status unknown)*		
EI-BVK	Piper PA-38-112 Tomahawk	38-79A0966	OO-FLG	2. 3.88	M.Martin	Trim	N 1.03
			OO-HLG, N9705N				
EI-BVT	Evans VP-2 V2-2129, PFA 7221 and SAAC-20		G-BEIE	29. 4.88	P.Morrison	(Cobh)	N 1.01
	(Volkswagen 1834)				*(Under construction: current status unknown))*		
EI-BVY	Heintz Zenith CH.200AA-RW	2-582		7. 6.88	J.Matthews, M.Skelly and T.Coleman	Abbeyshrule	A 7.01
	(Lycoming O-320)				*(Current status unknown)*		
EI-BWH	Partenavia P.68C	212	G-BHJP	11.12.87	K.Buckley	Cork	A10.02
EI-BXI	Boeing 737-448	25052		29. 4.91	Aer Lingus Ltd *"St.Finnian"*	Dublin	A2004
EI-BXK	Boeing 737-448	25736		14. 4.92	Aer Lingus Ltd *"St.Caimin"*	Dublin	A2004
EI-BXO	Fouga (Valmet) CM-170 Magister	213	N18FM	21.11.88	G.W.Connolly	Saggart, Dublin	N 4.96
	(C/n FM-28 also quoted)				*(Stored: current status unknown))*		
EI-BXT	Rollason Druine D.62B Condor	RAE/626	G-AVZE	24. 8.88	The Condor Group	Abbeyshrule	A 1.03
EI-BYA	Thruster TST Mk.1	8504	G-MNDA	1. 2.89	E.Fagan	Ballyheelan	
					(Scrapped following storm damage Abbeyshrule 1992)		
EI-BYF	Cessna 150M Commuter	15076654	N3924V	20.11.89	Damien Cashin	Abbeyshrule	A10.02
EI-BYG	SOCATA TB-9 Tampico Club	928		23. 8.89	Weston Ltd	Weston	A 9.02
EI-BYJ	Bell 206B JetRanger II	1897	N49725	23. 6.89	Medeva Properties Ltd	Dublin Heliport	N 1.03
EI-BYL	Heintz Zenith CH-250	MS/FAS 2866	(EI-BYD)	14. 6.89	M.McLoughlin	Kilrush	N 1.03
	(Lycoming O-320) *(C/n quoted as c/n A2-866)*						
EI-BYO	Aérospatiale-Alenia ATR42-310	161	OY-CIS	20.12.02	GPA-ATR Ltd.	Dublin	A 1.03
			EI-BYO, F-WWEH		*(Op Aer Arann Express)*		
EI-BYR	Bell 206L-3 LongRanger III	51284	(EI-LMG)	15. 8.89	H.S.S.Ltd	Rathcoole	A 8.01
			EI-BYR, D-HBAD		*(New owner 9.01)*		
EI-BYX	Champion 7GCAA Citabria	7GCAA-40	N546DS	4. 4.90	P.J.Gallagher	Abbeyshrule	N 1.03
EI-BYY	Piper J-3C-85 Cub	12494	EC-AQZ	12. 4.90	The Cub Club	Galway	A12.01
	(Frame No.12322)		HB-OSG, 44-80198				
	(Regd with c/n 22288 and officially ex G-AKTJ, N3595K, NC3595K)						
EI-BZE	Boeing 737-3Y0	24464		2. 8.89	Paloma Developments Ltd		
					(Op Philippine Airlines)	Manila-Nino Aquino, Philippines	A2004
EI-BZF	Boeing 737-3Y0	24465		7. 8.89	Pergola Ltd	Manila-Nino Aquino, Philippines	A2004
					(Op Philippine Airlines)		
EI-BZJ	Boeing 737-3Y0	24677		29. 3.90	Pergola Ltd	Manila-Nino Aquino, Philippines	A2004
					(Op Philippine Airlines)		
EI-BZN	Boeing 737-3Y0	24770		30.10.90	Airplanes Finance Ltd	Subic Bay, Philippines	A2004
					(Leased Air Philippines)		
EI-CAA*	Reims FR172J Rocket	FR17200486	G-BHTW	17. 8.89	O.Bruton	Abbeysrule	N 9.00
			5Y-ATO		*(Damaged 1993/94: cancelled 27.11.98 as WFU: open store)*		
EI-CAC	Grob G-115A	8092		22.10.89	G.Tracey	Weston	A 1.03
EI-CAD	Grob G-115A	8104	G-WIZB	22. 8.03	Flightwise Training Ltd	Weston	A 9.04
			EI-CAD				
EI-CAE	Grob G-115A	8105		5. 4.90	Kieran A.O'Connor	Weston	A 1.02
EI-CAP	Cessna R182 Skylane RGII	R18200056	G-BMUF	27. 4.90	M.J.Hanlon	Weston	A 8.02
			N7342W				
EI-CAN	Aerotech MW.5(K) Sorcerer	5K-0011-02	(G-MWGH)	15. 6.90	V.Vaughan	Mulinahone	
	(Rotax 447)				*(Current status unknown)*		
EI-CAW	Bell 206B JetRanger II	780	N2947W	11. 7.90	Celtic Helicopters (Maintenance Services) Ltd		
					(Dismantled)	Dublin Heliport	N 5.00
EI-CAX	Cessna P210N Pressurized Centurion II		(EI-CAS)	9. 7.90	Kieran A.O'Connor	Weston	N 1.03
		P21000215	G-OPMB, N4553K				
EI-CAY	Mooney M.20C Ranger	690074	N9272V	14.11.90	Ranger Flights Ltd *(Stored dismantled)*	Hacketstown	N 1.03
EI-CBK	Aérospatiale-Alenia ATR42-310	199	F-WWEM	25. 7.90	GPA-ATR Ltd *(Op Aer Arann Express)*	Connemara	A 1.03

EI-CBR	McDonnell-Douglas MD-83	49939		3.12.90	Airplanes 111 Ltd *"Ciudad de Bucaramanga"* *(Op Avianca)*	Bogota-Eldorado, Colombia	A2004
EI-CBS	McDonnell-Douglas MD-83	49942		10.12.90	GECAS Technical Services Ltd *(Op Sociedad Aeronautica de Medellin Consolida)*	Medellin-Olaya Herrara, Colombia	A2004
EI-CBY	McDonnell-Douglas MD-83	49944		30. 7.91	GECAS Technical Services Ltd *"Ciudad de Barranquilla"* *(Op Avianca)*	Bogota-Eldorado, Colombia	A2004
EI-CBZ	McDonnell-Douglas MD-83	49945		13. 8.91	GECAS Technical Services Ltd *"Ciudad de Santiago de Cali"* *(Op Avianca)*	Bogota-Eldorado, Colombia	A2004
EI-CCC	McDonnell-Douglas MD-83	49946		27. 9.91	Airplanes 111 Ltd *"Ciudad de Pereira"* *(Op Avianca)*	Bogota-Eldorado, Colombia	A2004 A2004
EI-CCD	Grob G-115A	8108	D-EIUD or D-EIWD ?	15. 8.90	Kal Aviation Ltd	Cork	A 1.03
EI-CCE (2)	McDonnell-Douglas MD-83	49947		19. 9.91	GECAS Technical Services Ltd *"Ciudad de Medelin"* *(Op Avianca)*	Bogota-Eldorado, Colombia	A2004
EI-CCF	Aeronca 11AC Chief (Continental A65)	11AC-S-40	N3826E NC3826E	10. 1.91	L.Murray and Partners	Trim	A 1.03
EI-CCJ	Cessna 152 II	15280174	N24251	9.10.90	M.P.Cahill *(Stored: current status unknown)*	Dublin	N 2.95
EI-CCK	Cessna 152 II	15279610	N757BM	9.10.90	M.P.Cahill *(Damaged pre 1995: current status unknown)*	Newcastle	
EI-CCL	Cessna 152 II	15280382	N24791	9.10.90	M.P.Cahil *(Damaged Bray Head, County.Wicklow 4.5.93: current status unknown))*	Dublin	
EI-CCM	Cessna 152 II	15282320	N68679	9.10.90	E Hopkins *(Current status unknown)*	Newcastle	
EI-CCV	Cessna R172K Hawk XPII	R1723039	N758EP	2. 3.91	Kerry Aero Club Ltd *(Current status unknown)*	Farranfore	
EI-CDC	Boeing 737-548	24968	EI-BXG	19. 6.91	Aer Lingus Ltd *"St.Munchen/Maincin"* *(Delivered to Air Baltic 1.2.05 as YLBBH)*	Dublin	
EI-CDD	Boeing 737-548	24989	EI-BXH	3. 7.91	Aer Lingus Ltd *"St.Macartan/Macarthain"*	Dublin	A2004
EI-CDE	Boeing 737-548	25115	PT-SLM EI-CDE, (EI-BXJ)	21. 5.91	Aer Lingus Ltd *"St.Jarlath/Iarflaith"*	Dublin	A2004
EI-CDF	Boeing 737-548	25737		23. 3.92	Aer Lingus Ltd *"St.Cronan"*	Dublin	A2004
EI-CDG	Boeing 737-548	25738		7. 4.92	Aer Lingus Ltd *"St.Moling"*	Dublin	A2004
EI-CDH	Boeing 737-548	25739		14. 4.92	Aer Lingus Ltd *"St.Ronan"*	Dublin	A2004
EI-CDP	Cessna 182L	18258955	G-FALL OY-AHS, N4230S	20. 5.91	Irish Parachute Club Ltd	Clonbulloge	N 1.03
EI-CDV	Cessna 150G	15066677	N2777S	17. 7.91	K.A.O'Connor	Weston	A.1.03
EI-CDX	Cessna 210K Centurion	21059329	G-AYGN N9429M	14. 8.91	Falcon Aviation Ltd	Waterford	A 12.03
EI-CDY	McDonnell-Douglas MD-83	49948		27. 9.91	GECAS Technical Services Ltd *"Ciudad de Santa Maria"* *(Op Sociedad Aeronautica de Medellin Consolida)*	Medellin-Olaya Herrara, Colombia	A2004
EI-CEG	SOCATA MS.893E Rallye 180GT	13083	SE-GTS	31.10.91	M.Farrelly *(Current status unknown)*	Powerscourt	
EI-CEK	McDonnell-Douglas MD-83	49631	EC-FMY EC-113, EI-CEK, EC-EPM, EC-261	13.12.91	Airplanes IAL Finance Ltd *(Op Nouvelair Tunisie)*	Monastir, Tunisia	A2004
EI-CEN	Thruster T.300 (Rotax 582)	9012-T300-500		2. 3.92	P.J.Murphy *(Current status unknown)*	Macroom	
EI-CEP	McDonnell-Douglas MD-83	53122		14. 4.92	GECAS Technical Services Ltd *"San Andres Isla"* *(Op Avianca)*	Bogota-Eldorado, Colombia	A2004
EI-CEQ	McDonnell-Douglas MD-83	53123		14. 4.92	GECAS Technical Services Ltd *"Ciudad de Leticia"* *(Op Avianca)*	Bogota-Eldorado, Colombia	A2004
EI-CER	McDonnell-Douglas MD-83	53125	N9017P	20. 5.92	Airplanes 111 Ltd *"Ciudad de Monteria"* *(Op Avianca)*	Bogota-Eldorado, Colombia	A2004
EI-CES	Taylorcraft BC-65	2231	G-BTEG N27590, NC27590	25. 3.92	N.O'Brien	Kilkenny	N 9.00
EI-CEX	Lake LA-4-200 Buccaneer	1115	N8VG N3VC, N8544Z	18. 5.92	C.L.Cargill	Killaloe	A10.01
EI-CEY	Boeing 757-2Y0	26152		10. 8.92	Pergola Ltd *(Op Avianca)*	Bogota-Eldorado, Colombia	A2004
EI-CEZ	Boeing 757-2Y0	26154		18. 9.92	Airplanes Holdings Ltd *(Op Avianca)*	Bogota-Eldorado, Colombia	A2004
EI-CFE	Robinson R22 Beta	1709	G-BTHG	15. 5.91	ILH Enterprises Ltd	Limerick	A 9.02
EI-CFF	Piper PA-12 Super Cruiser (Lycoming O-235)	12-3928	N78544 NC78544	23. 5.91	J.O'Dwyer and J.Molloy	Gowran Grange	N 1.03
EI-CFG	Rousseau Piel CP.301B Emeraude	112	G-ARIW F-BIRQ	1. 6.91	F.Doyle *(Stored complete)*	Murntown, County Wexford	N 1.02
EI-CFH	Piper PA-12 Super Cruiser (Lycoming O-320)	12-3110	(EI-CCE) N4214M, NC4214M	1. 6.91	G.Treacy *(Current status unknown)*	Shinrone	A10.99
EI-CFN	Cessna 172P Skyhawk II	17274113	N5446K JA4172, N5446K	10. 5.92	B.Fitzmaurice and G.O'Connell	Weston	N 1.03
EI-CFO	Piper J-3C-65 Cub (L-4H-PI)	11947	OO-RAZ OO-RAF, 44-79651	13. 5.92	J.Mathews and Partners *(USAAF c/s) (Current status unknown)*	Trim	N 8.98
EI-CFP	Cessna 172P Skyhawk II	17274428	N52178	15. 7.91	K A O'Connor	Weston	N 1.03
EI-CFV*	SOCATA MS.880B Rallye Club	1850	G-OLFS G-AYYZ	13. 5.92	Not known *(Cancelled 15.11.00 as scrapped) (Stripped hulk noted)*	Abbeyshrule	N.1.03
EI-CFX	Robinson R22 Beta	0793	G-OSPI	16. 6.92	Ballaugh Motors Ltd	Galway	A 8.02
EI-CFY	Cessna 172N Skyhawk II	17268902	N734JZ	18. 6.92	K.A.O'Connor	Weston	N 1.03
EI-CFZ	McDonnell-Douglas MD-83	53120	N6206F	29. 7.92	Airplanes 111 Ltd *"Ciudad de San Juan de Pasto"* *(Op Avianca)*	Bogota-Eldorado, Colombia	A2004
EI-CGB	TEAM miniMAX	SAAC-36		20. 8.92	M.Garvey *(Current status unknown)*	Abbeyshrule	
EI-CGC	Stinson 108-3 Station Wagon	108-5243	OO-IAC OO-JAC, N3B	17. 7.92	Anne P.Bruton	Kildare	N 1.03
EI-CGD	Cessna 172M Skyhawk II	17262309	OO-BMT N12846	30. 7.92	J Murray	Weston	A10.02

EI-CGF	Phoenix Luton LA-5 Major	PAL-1124, PFA 1208 and SAAC-19	G-BENH	31. 7.92	F.Doyle and J.Duggan	(Newlands)	A 9.04
EI-CGG	Ercoupe 415C (Continental C75)	3147	N2522H NC2522H	10. 9.92	Irish Ercoupe Group *(Derelict)*	Weston	N 1.03
EI-CGH	Cessna 210N Centurion II	21063524	N6374A	16.11.92	J.Smith	Abbeyshrule	A10.04
EI-CGJ	Solar Wings Pegasus XL-R (Rotax 447)	SW-WA-1506	G-MWTV	5. 4.93	P.Hearty *(Crashed Portarlington late 1995: current status unknown)*	Portarlington	
EI-CGM	Solar Wings Pegasus XL-R (Rotax 447)	SW-WA-1502	G-MWVC	14.11.92	Microflight Ltd *(Current status unknown)*	Ballyfore	
EI-CGN	Solar Wings Pegasus XL-R (Rotax 447)	SW-WA-1529	G-MWXM	14.11.92	V.Power	Donamore, New Ross	A 8.00
EI-CGP	Piper PA-28-140 Cherokee C	28-26928	G-MLUA G-AYJT, N11C	25.11.92	G.Cashman *(Op Euroair Training)*	Cork	A 9.04
EI-CGQ	Aérospatiale AS350B Ecureuil	2076	G-BUPK JA9740	21. 1.93	Blue Star Helicopters Ltd	Dublin	A10.04
EI-CGT	Cessna 152 II	15282331	G-BPBL N16SU, N68715	10.12.92	J.J.Dunne *(Current status unknown)*	Stamullen	
EI-CGV	Piper J-5A Cub Cruiser	5-624	G-BPKT N35372, NC35372	11.12.92	J5 Grp	(Trim)	A 9.04
EI-CHH (2)	Boeing 737-317	23177	(EI-CGX) PT-WBG, PP-SNU, C-FCPL	15. 1.93	Airplanes Finance Ltd *(Cancelled 2.97 - to N302AL but still shown as current by IAA)*	(Shannon, County Clare)	
EI-CHK	Piper J-3C-65 Cub Special	23019	C-FHNS CF-HNS, N1492N, NC1492N	10. 3.93	N.Higgins *(Current status unknown)*	Longwood	A 7.99
EI-CHM	Cessna 150M Commuter	15079288	G-BSZX N714MU	2. 3.93	K.A.O'Connor	Weston	N 1.03
EI-CHR	CFM Shadow BD (Rotax 447)	063	G-MTKT	20. 5.93	Fergus Maughan	Ardclough, Straffan	A 1.03
EI-CHS	Cessna 172M Skyhawk II	17266742	G-BREZ N80775	26. 4.93	Kerry Aero Club Ltd	Farranfore	A 9.04
EI-CHT	Solar Wings Pegasus XL-R	SW-WA-1568	G-BZXU	27. 3.02	Enda Spain	Monasterevin	A 1.03
EI-CHV	Agusta A109A II	7149	VR-BMM HB-XTJ, D-HASV	10. 6.93	Celtic Helicopters Ltd	Dublin Heliport	N 1.03
EI-CIA	SOCATA MS.880B Rallye Club	1218	G-MONA G-AWJK	26. 4.93	G.Hackett and C.Mason	Thurles	A 8.02
EI-CIF	Piper PA-28-180 Cherokee C *(Rebuilt 1967 with spare frame c/n 28-3808S)*	28-2853	G-AVVV N8880J	12. 6.93	AA Flying Group	Weston	N 1.03
EI-CIG	Piper PA-18-150 Super Cub *(Frame No.18-7360)*	18-7203	G-BGWF ST-AFJ, ST-ABN	12. 6.93	K.A.O'Connor	Weston	A 1.03
EI-CIJ	Cessna 340	3400304	G-BBVE N69451	2. 7.93	Airlink Airways Ltd	Sligo	A10.02
EI-CIM	Avid Flyer Model IV	1125D		17. 8.93	P.Swan	Weston	A 1.03
EI-CIN	Cessna 150K	15071728	G-OCIN EI-CIN, G-BSXG, N6228G	6. 9.93	K.O'Connor	Weston	A 1.03
EI-CIR (2)	Cessna 551 Citation II *(Built as Cessna 550 EI-CIR(1) c/n 550-0128)*	551-0174	N60AR EI-CIR(1), F-WLEF, 9A-BPU, RC-BPU, YU-BPU, N220LA, N536M, N2631V	29.11.93	Air Group Finance Ltd	Dinard, France	A12.01
EI-CIV	Piper PA-28-140 Cherokee Cruiser	28-7725232	G-BEXY N9648N	20.11.93	G.Cashman and E.Callanan	Cork	A10.02
EI-CIW	McDonnell-Douglas MD-83	49785	HL-7271	30.12.93	Carotene Ltd *(Op Meridiana) "Isola Tremiti"*	Olbia, Italy	A2004
EI-CIZ	Steen Skybolt (Lycoming IO-360)	001	G-BSAO N303BC	12.12.93	J.Keane	Coonagh	N 1.03
EI-CJC	Boeing 737-204ADV	22640	G-BJCV CS-TMA, G-BJCV, C-GCAU, G-BJCV, C-GXCP, G-BJCV	25. 1.94	Ryanair Ltd *(Hertz Car Rental titles)*	Dublin	A2004
EI-CJD*	Boeing 737-204ADV	22966	G-BKHE (G-BKGU)	18. 2.94	Dublin Airport Authority *(Cancelled 29.10.04 as WFU)* (Used for tug practice and other duties)	Dublin	N 1.05
EI-CJE*	Boeing 737-204ADV	22639	G-BJCU EC-DVE, G-BJCU	10. 3.94	(Ryanair Ltd) *(Jaguar titles)* *(Cancelled 16.11.04 and WFU (Stored engineless)*	Prestwick	N10.04
EI-CJF*	Boeing 737-204ADV	22967	G-BTZF G-BKHF, (G-BKGV)	24. 3.94	(Ryanair Ltd) *(Cancelled 25.11.04 as WFU) (Stored)*	Prestwick	N10.04
EI-CJG	Boeing 737-204ADV	22058	G-BGYK PP-SRW, G-BGYK, (G-BGRV)	25. 3.94	Ryanair Ltd	Dublin	A2004
EI-CJH*	Boeing 737-204ADV	22057	G-BGYJ (G-BGRU), N8278V	30. 3.94	(Ryanair Ltd) *(Cancelled 1.12.04 as WFU).(Stored)*	Prestwick	N10.04
EI-CJI	Boeing 737-2E7	22875	G-BMDF (PK-RI.), G-BMDF, 4X-BAB, N4570B *(Stored)*	8. 7.94	Ryanair Ltd	Prestwick	N10.04
EI-CJR	SNCAN Stampe SV-4A	318	G-BKBK OO-CLR, F-BCLR	28. 2.94	P.McKenna	Carnmore	N 1.03
EI-CJS	Jodel Wassmer D.120A Paris-Nice	339	F-BOYF	28. 2.94	Anthony Flood	Birr	N 1.03
EI-CJT	Slingsby Cadet III (Volkswagen 1835)	830 & PCW-001	G-BPCW XA288	25. 2.94	J.Tarrant *(Current status unknown)*	Rathcoole	
EI-CJV	Moskito 2 (Rotax 582)	004	D-MBGM	12. 3.94	Messrs Peril, Kingston, Hanly and Fitzgerald *(Dismantled)*	Coonagh	N 1.03
EI-CJZ	Whittaker MW-6S Fatboy Flyer (Rotax 503)	PFA 164-11493	G-MWTW	24. 3.94	M.McCarthy	Watergrasshill	N 9.00
EI-CKG	Hunt Avon (Rotax 447)	92009013		2. 7.94	B.Kenny *(Current status unknown)*	Clara	
EI-CKH	Piper PA-18 Super Cub 95	18-7248	G-APZK N10F	3. 6.94	G.Brady and C.Keenan	Weston	N 1.03
EI-CKI	Thruster TST Mk.1 (Rotax 503)	8078-TST-091	G-MVDI	3. 6.94	N.Furlong	Stradbally, County Laois	A 8.02
EI-CKJ	Cameron N-77 HAB	3305		6. 7.94	F.Meldon *"Goodfellas" (Current status unknown)*	Blackrock	
EI-CKM	McDonnell-Douglas MD-83	49792	TC-INC EI-CKM, (D-ALLW), EI-CKM, XA-RPH, EC-FFP, EC-733, XA-RPH	10. 8.94	Airplanes Finance Ltd *"Isola dell'Asinara"*	Olbia, Italy *(Op Meridiana)*	A2004

EI-CKN	Whittaker MW-6S Fatboy Flyer (Rotax 462)	BCA.8942		29. 7.94	F.Byrne and M.O'Carroll	Kilrush	
EI-CKP	Boeing 737-2K2ADV	22296	PH-TVS PP-SRV, PH-TVS, LV-RBH, PH-TVS, LV-RAO, PH-TVS, EC-DVN, PH-TVS	7.10.94	Ryanair Ltd *(WFU)*	Dublin	N2004
EI-CKS*	Boeing 737-2T5ADV	22023	PH-TVX OE-ILE, PH-TVX, G-BGTW	1. 6.95	Ryanair Ltd *(Cancelled 1.12.04 as WFU) (Stored)*	Prestwick	N10.04
EI-CKT	Mainair Gemini/Flash (Fuji-Robin EC-44-PM)	307-585-3 & W47	G-MNCB	27. 9.94	C.Burke *(Current status unknown)*	Bartlemy	
EI-CKU	Solar Wings Pegasus XL-R (Rotax 447)	SW-TB-1434 & SW-WA-1500	G-MWVB	14.10.94	M.O'Regan *(Current status unknown)*	Edenderry	
EI-CKZ	Jodel D.18 (Volkswagen 1834)	229		5. 4.95	J.O'Brien *(Current status unknown)*	Glen of Imal	
EI-CLA	HOAC DV-20 Katana	20106		24. 3.95	Weston Ltd	Weston	N 1.03
EI-CLB	Aérospatiale-Alenia ATR72-212	423	F-WWEB	23. 2.95	Tarquin Ltd *(Op Alitalia Express) "Lago di Bracciano"*	Rome-Fiumicino, Italy	A2004
EI-CLC	Aérospatiale-Alenia ATR72-212	428	F-WWEF	24. 2.95	Tarquin Ltd *(Op Alitalia Express) "Fiume Simeto"*	Rome-Fiumicino, Italy	A2004
EI-CLD	Aérospatiale-Alenia ATR72-212	432	F-WWEL	3. 3.95	Tarquin Ltd *(Op Alitalia Express) "Fiume Piave"*	Rome-Fiumicino, Italy	A2004
EI-CLG	British Aerospace BAe 146 Srs.300	E3131	G-BRAB HS-TBL, G-BRAB, G-11-131	7. 6.95	Aer Lingus Ltd *"St.Finbar/Fionnbar" (Stored)*	Mojave, California, US	N 6.04
EI-CLH	British Aerospace BAe 146 Srs.300	E3146	G-BOJJ I-ATSC, G-BOJJ, G-6-146	2. 6.95	Aer Lingus Ltd *"St.Aoife" (Stored)*	Mojave, California, US	N 6.04
EI-CLI	British Aerospace BAe 146 Srs.300	E3159	G-BVSA I-ATSD, G-6-159, G-5-159	19. 4.95	Aer Lingus Ltd *"St.Eithne" (Stored)*	Mojave, California, US	N 6.04
EI-CLL	Whittaker MW-6S Fatboy Flyer (Rotax 503)	1069		2. 4.95	F.Stack *(Current status unknown)*	Midleton, County Cork	
EI-CLQ	Reims/Cessna F172N Skyhawk II	F17201653	G-BFLV	26. 5.95	K.Dardis and Partners	Abbeyshrule	N 1.03
EI-CLW	Boeing 737-3Y0	25187	XA-SAB	10. 6.95	Airplanes Finance Ltd *(Op Air One)*	Pescara, Italy	A2004
EI-CLZ	Boeing 737-3Y0	25179	XA-RJR N3521N	27. 7.95	Airplanes Finance Ltd *(Op Air One)*	Pescara, Italy	A2004
EI-CMB	Piper PA-28-140 Cherokee Cruiser	28-7725094	G-BELR N9541N	5. 9.95	Dublin Flyers Ltd	Dublin	A 8.02
EI-CMJ	Aérospatiale-Alenia ATR72-212	467	F-WWLU	21.12.95	Tarquin Ltd *(Op Alitalia Express) "Fiume Volturno"*	Rome-Fiumicino, Italy	A2004
EI-CMK	Eurowing Goldwing ST (Fuji-Robin EC-PM-34)	76 & SAAC-57		22.12.95	M.Garrigan *(Current status unknown)*	Clondara, Longford	
EI-CML	Cessna 150M	15076786	G-BNSS N45207	5. 1.96	K.A.O'Connor	Weston	A 9.04
EI-CMN	Piper PA-12 Super Cruiser (Lycoming O-235)	12-1617	N2363M NC2363M	26. 1.96	D.Graham and Partners	Birr	N 1.03
EI-CMR	Rutan LongEz (Lycoming O-235)	1716		2. 5.96	F.and C.O'Caoimh	Waterford	N12.03
EI-CMS	British Aerospace BAe 146 Srs.200A	E2044	N184US N361PS	24. 4.96	Cityjet Ltd *(Op Air France)*	Paris, France	A2004
EI-CMT	Piper PA-34-200T Seneca II	34-7870088	G-BNER N2590M	23. 4.96	Atlantic Flight Training Ltd	Cork	A 9.04
EI-CMU	Mainair Mercury (Rotax 462)	1071-0296-7 and W873		3. 5.96	L.Langan and L.Laffan *(Restored)*	Wexford/Waterford	N 8.01
EI-CMV	Cessna 150L	150-72747	G-MSES N1447Q	17. 5.96	K.A.O'Connor	Weston	N 1.03
EI-CMW	Rotorway Executive (Rotorway RW 162D)	3550		13. 5.96	B.McNamee *(Current status unknown)*	Dunboyne	
EI-CMY	British Aerospace BAe 146 Srs.200A	E2039	N177US N356PS	19. 6.96	Cityjet Ltd *(Op Air France)*	Paris, France	A2004
EI-CNA	Letov LK-2M Sluka (Rotax 447)	8295S005		28. 6.96	G.Doody	Portlaoise	A 7.00
EI-CNB	British Aerospace BAe 146 Srs.200A	E2046	(EI-CMZ) N187US, N363PS	3. 8.96	Cityjet Ltd *(Op Air France)*	Paris, France	A2004
EI-CNC	Team miniMax 1600 (Rotax 447)	514		10. 9.96	A.M.S.Allen	Enniskillen	A 7.01
EI-CNG	Air and Space 18-A Gyroplane	18-75	G-BALB N6170S	10. 9.96	P.Joyce	Waterford	A 1.02
EI-CNI	British Aerospace BAe 146 Srs.200 *(Avro RJ85)*	E2299	G-6-299	26.11.96	Peregrine Aviation Leasing Co Ltd *"Lombardia" (Stored)*	Exeter	N12.03
EI-CNJ	British Aerospace BAe 146 Srs.200 *(Avro RJ85)*	E2300	G-6-300	2.12.96	Peregrine Aviation Leasing Co Ltd *"Piemonte" (Stored)*	Exeter	N12.03
EI-CNL	Sikorsky S-61N Mk.II	61746	G-BDDA ZS-RBU, G-BDDA, N91201, G-BDDA	19.12.96	CHC Ireland Ltd *"IMES Rescue"*	Cork	A2004
EI-CNM	Piper PA-31-350 Navajo Chieftain	31-7305107	N1201H G-BBNT, N74958	16.12.96	M.Goss *(Op Air Atlantic)*	Dublin	A 1.03
EI-CNN	Lockheed L.1011-385-1 Tristar	1024	VR-HHV G-BAAA	30. 1.97	Aer Turas Teoranta t/a Irish Cargo Airlines *(Stored)*	Dublin	A2002
EI-CNQ	British Aerospace BAe 146 Srs.200	E2031	G-OWLD N173US, N353PS	2. 7.97	Cityjet Ltd *(Op Air France)*	Paris, France	A2002
EI-CNT	Boeing 737-230ADV	22115	D-ABFC	5.12.96	Ryanair Ltd *(Vodaphone titles)*	Dublin	A2003
EI-CNU	Pegasus Quantum 15-912	7326		10. 4.97	M.Ffrench	Donamore, New Ross	A 6.01
EI-CNV	Boeing 737-230ADV	22128	D-ABFX (D-ABFW)	26. 3.97	Ryanair Ltd	Dublin	A2004
EI-CNW	Boeing 737-230ADV	22133	D-ABHC (B-), D-ABHC, (D-ABHB)	31. 5.97	Ryanair Ltd	Dublin	A2004
EI-CNX	Boeing 737-230ADV	22127	D-ABFW N5573K, (D-ABFU)	4. 7.97	Ryanair Ltd	Dublin	A2004
EI-CNZ	Boeing 737-230ADV	22126	D-ABFU (D-ABFT)	5.11.97	Ryanair Ltd	Dublin	A2004

Reg	Type	C/n	Previous identities	Date	Owner/Operator	Base	CofA
EI-COA	Boeing 737-230ADV	22637	CS-TES	16.12.97	Ryanair Ltd	Prestwick	N10.04
			D-ABHX		*(In open storage)*		
EI-COB	Boeing 737-230ADV	22124	D-ABFR	16. 1.98	Ryanair Ltd	Dublin	A2004
EI-COE	Europa Aviation Europa	286		29. 5.97	F.Flynn	(Urlanmore)	N 1.01
	(Jabiru 2200) *(Monowheel u/c)*				*(Under construction)*		
EI-COG	Gyroscopic Rotorcraft Gyroplane	G.120		11. 3.98	R.C.Fidler and D D.Bracken	Letterkenny	A 8.98
	(Subaru AE81) *(Official c/n G.120 may be type designation)*				*(Current status unknown)*		
EI-COH	Boeing 737-430	27001	D-ABKB	6. 6.97	Flightlease (Ireland) Ltd	Pescara, Italy	A2004
			(VT-S), D-ABKB		*(Op Air One)*		
EI-COI	Boeing 737-430	27002	D-ABKC	13.11.97	Challey Ltd *(Op Air One)*	Pescara, Italy	A2004
EI-COJ	Boeing 737-430	27005	D-ABKK	13.11.97	Challey Ltd	Pescara, Italy	A2004
			(D-ABKF)		*(Op Air One)*		
EI-COK	Boeing 737-430	27003	F-GRNZ	22. 4.02	Flightleases (Ireland) Ltd	Milan, Italy	A2004
			EI-COK, D-ABKD		*(Op Air One)*		
EI-COM	Whittaker MW-6S Fatboy Flyer	1		10.10.97	M.Watson	Clonbullogue	N 1.01
	(Rotax 582)				*(Under construction)*		
EI-CON	Boeing 737-2T5	22396	PK-RIW	21. 7.97	Ryanair Ltd	Prestwick	N 1.05
			EI-CON, PK-RIW, VT-EWF, A40-BM, C-GVRE, (EI-B), G-BHVH				
					(To be N396AD 2005)		
EI-COO	Carlson Sparrow II	302		13. 8.97	D.Logue	Weston	
	(Rotax 532)				*(Current status unknown)*		
EI-COP	Reims/Cessna F150L	F15001058	G-BCBY	26. 6.97	High Kings Flying Group Ltd	Greigs, Navan	A 5.01
			PH-TGI, (G-BCBY)				
EI-COQ	British Aerospace BAe 146 Srs.100	E1254	9H-ACM	17.10.97	Peregrine Aviation Leasing Co Ltd	Bergamo, Italy	N12.04
	(Avro RJ70)		(9H-ABW), G-BVRJ, G-6-254			*(To Sweden 12.04)*	
EI-COT	Reims/Cessna F172N Skyhawk II	F17201884	D-EIEF	24.11.97	Kawasaki Distributors (Ireland) Ltd	Newcastle	N 1.03
EI-COX	Boeing 737-230ADV	22123	D-ABFP	9. 1.98	Ryanair Ltd	Dublin	A2003
EI-COY	Piper J-3C-65 Cub Special	22519	N3319N	5.11.97	D.Bruton and W.Flood	Abbeyshrule	N 1.03
			NC3319N				
EI-COZ	Piper PA-28-140 Cherokee C	28-26796	G-AYMZ	5.11.97	G Cashman	Cork	A10.02
			N11C				
EI-CPC	Airbus A321-211	815	D-AVZT	8. 5.98	ILFC *(Op Aer Lingus) "St.Fergus/Faergus"*	Dublin	A2004
EI-CPD	Airbus A321-211	841	D-AVZA	19. 6.98	ILFC *(Op Aer Lingus) "St.Davnet/Damhnat"*	Dublin	A2004
EI-CPE	Airbus A321-211	926	D-AVZQ	11.12.98	ILFC *(Op Aer Lingus) "St.Enda/Eanna"*	Dublin	A2004
EI-CPF	Airbus A321-211	991	D-AVZE	9. 4.99	Aer Lingus Ltd *"St.Ida/Ide"*	Dublin	A2004
EI-CPG	Airbus A321-211	1023	D-AVZR	28. 5.99	Aer Lingus Ltd *"St.Aidan/Aodhan"*	Dublin	A2004
EI-CPH	Airbus A321-211	1094	F-WWDD	22.11.99	Aer Lingus Ltd *"St.Dervilla/Dearbhile"*	Dublin	A2004
			D-AVZA				
EI-CPI	Rutan LongEz	17		18.12.97	D.J.Ryan *"Lady Elizabeth"*	Waterford	A12.03
	(Lycoming O-235)						
EI-CPJ	British Aerospace BAe 146 Srs.100	E1258	9H-ACN	27. 3.98	Peregrine Aviation Leasing Co Ltd	Exeter	N12.03
	(Avro RJ70)		(9H-ABX), G-6-258		*(Stored)*		
EI-CPK	British Aerospace BAe 146 Srs.100	E1260	9H-ACO	27. 3.98	Peregrine Aviation Leasing Co Ltd	Exeter	N12.03
	(Avro RJ70)		(9H-ABY), G-6-261		*(Stored)*		
EI-CPL	British Aerospace BAe 146 Srs.100	E1267	9H-ACP	31. 3.98	Peregrine Aviation Leasing Co Ltd	Exeter	N12.03
	(Avro RJ70)		(9H-ABZ), G-6-267		*(Stored)*		
EI-CPN	Auster J/4	2073	G-AIJR	1. 4.98	E.Fagan	Abbeyshrule	N12.00
EI-CPO	Robinson R.22B2 Beta	2775	G-BXUJ	23. 9.98	Eirecopter Helicopters Ltd	Weston	A 4.04
EI-CPP	Piper J-3C-65 Cub (L-4H-PI)	12052	G-BIGH	23. 3.98	E.Fitzgerald	Newcastle	A10.02
	(Rebuilt Glasthule, Dublin 1994/1998)		F-BFQV, OO-GAS, OO-GAZ, 44-79756				
EI-CPT	Aérospatiale-Alenia ATR42-300	191	C-GIQS	12. 6.98	GPA-ATR Ltd *(Op Aer Arann Express)*	Connemara	A2004
			(ZS-NYP), C-GIQS, F-WWEA				
EI-CPX	III Sky Arrow 650T	K.122 and SAAC-67		24. 6.98	Martin McCarthy	Watergrasshill	N.8.00
EI-CRB	Lindstrand LBL-90A HAB	550		23. 9.98	J.and C.Concannon *(Current status unknown)*	Tuam	
EI-CRD	Boeing 767-31BER	26259	B-2565	29.10.98	ILFC Ireland Ltd *(Op Alitalia)*	Rome-Fiumicino, Italy	A2004
EI-CRE	McDonnell-Douglas MD-83	49854	D-ALLL	11.12.98	AAR Ireland Ltd	Olbia, Italy	A2004
					(Op Meridiana) "Tavolara-Punta Coda Cavallo"		
EI-CRF	Boeing 767-31B	25170	B-2566	4.12.98	ILFC Ireland Ltd *(Op Alitalia)*	Rome-Fiumicino, Italy	A2004
EI-CRG	Robin DR400-180R	2021	D-EHEC	11.12.98	D and B Lodge	Waterford	A 1.04
EI-CRH	McDonnell-Douglas MD-83	49935	HB-IKM	10. 2.99	Airplanes 111 Ltd	Olbia, Italy	A2004
			G-DCAC, N3004C		*(Op Meridiana) "Torre Guaceto"*		
EI-CRJ	McDonnell-Douglas MD-83	53013	D-ALLP	27. 1.99	C A Aviation Ltd	Olbia, Italy	A2004
					(Op Meridiana) "Isole Egadi"		
EI-CRK	Airbus A330-301	070	(EI-NYC)	18.11.94	Aer Lingus Ltd	Dublin	A2004
			F-WWKV		*"St Brigid/Brighid"*		
EI-CRL	Boeing 767-343ER	30008	(I-DEIB)	22. 3.99	GECAS Technical Services Ltd	Rome-Fiumicino, Italy	A2004
					(Op Alitalia) "Leonardo da Vinci"		
EI-CRM	Boeing 767-343	30009		8. 4.99	GECAS Technical Services Ltd	Rome-Fiumicino, Italy	A2004
					(Op Alitalia) "Amerigo Vespucci"		
EI-CRO	Boeing 767-3Q8ER	29383		16. 4.99	ILFC Ireland Ltd	Rome-Fiumicino, Italy	A2004
					(Op Alitalia) "Francesco de Pinedo"		
EI-CRR	Aeronca 11AC Chief	11AC-1605	OO-ESM	13. 4.99	L.Maddock and Partners	Killamaster	N 1.03
			(OO-DEL), OO-ESM				
EI-CRU	Cessna 152	15285621	G-BNSW	21. 9.99	W.Reilly	Inis Mor	
			N94213		*(Current status unknown)*		
EI-CRV	Hoffman H.36 Dimona	3674	OE-9319	2. 6.99	The Dimona Group c/o Victor Young	Dublin	N 1.04
			HB-2081				
EI-CRW	McDonnell-Douglas MD-83	49951	HB-IKN	8. 4.99	Airplanes IAL Ltd	Olbia, Italy	A2004
			G-GMJM, N13627		*(Op Meridiana)*		
EI-CRX	SOCATA TB-9 Tampico	1170	F-GKUL	21. 5.99	Hotel Bravo Flying Club Ltd	Weston	A 9.04
EI-CRY	Medway EclipseR	160/138		2. 6.99	G.A.Murphy *(Current status unknown)*	Rathcoole	
EI-CRZ	Boeing 737-36E	26322	EC-GGE	14. 4.99	ILFC Ireland Ltd	Pescara, Italy	A2004
			EC-798		*(Op Air One)*		

EI-CSA	Boeing 737-8AS	29916	N5537L	12. 3.99	Ryanair Ltd	Dublin	A2004
			N1786B				
EI-CSB	Boeing 737-8AS	29917	N1786B	16. 6.99	Ryanair Ltd	Dublin	A2004
EI-CSC	Boeing 737-8AS	29918	N1786B	25. 6.99	Ryanair Ltd	Dublin	A2004
EI-CSD	Boeing 737-8AS	29919	N1786B	9. 8.99	Ryanair Ltd *"Ryanair.com" titles*	Dublin	A2004
EI-CSE	Boeing 737-8AS	29920	N1786B	31. 8.99	Ryanair Ltd	Dublin	A2004
EI-CSF	Boeing 737-8AS	29921	N1786B	24. 5.00	Ryanair Ltd	Dublin	A2004
					(Carries "Dites Non a la Surcharge Kerosene D'Air France" titles)		
EI-CSG	Boeing 737-8AS	29922	N1786B	31. 5.00	Ryanair Ltd	Dublin	A2004
EI-CSH	Boeing 737-8AS	29923	N1787B	9. 6.00	Ryanair Ltd	Dublin	A2004
					(Carries "Auf Weidersehen Lufthansa" titles)		
EI-CSI	Boeing 737-8AS	29924	N1796B	12. 6.00	Ryanair Ltd	Dublin	A2004
EI-CSJ	Boeing 737-8AS	29925	N1786B	20. 6.00	Ryanair Ltd	Dublin	A2004
EI-CSK	British Aerospace BAe 146 Srs.200A	E2062	N810AS	3. 4.98	Cityjet Ltd *(Op Air France)*	Paris, France	A2004
			N880DV, G-5-062, N406XV, (G-BNDR), G-5-062				
EI-CSL	British Aerospace BAe 146 Srs.200A	E2074	N812AS	8. 5.98	Cityjet Ltd *(Op Air France)*	Paris, France	A2004
			N881DV, G-5-074, G-BNND, HS-TBQ, G-BNND, N146SB, N192US, N368PS, (G-BNND), G-5-074				
EI-CSM	Boeing 737-8AS	29926	N...	7.12.00	Ryanair Ltd	Dublin	A2004
EI-CSN	Boeing 737-8AS	29927	N...	11.12.00	Ryanair Ltd	Dublin	A2004
EI-CSO	Boeing 737-8AS	29928	N...	11. 1.01	Ryanair Ltd	Dublin	A2004
EI-CSP	Boeing 737-8AS	29929	N...	25. 1.01	Ryanair Ltd	Dublin	A2004
EI-CSQ	Boeing 737-8AS	29930	N...	26. 1.01	Ryanair Ltd	Dublin	A2004
EI-CSR	Boeing 737-8AS	29931	N...	5.12.01	Ryanair Ltd	Dublin	A2004
EI-CSS	Boeing 737-8AS	29932	N...	14.12.01	Ryanair Ltd	Dublin	A2004
EI-CST	Boeing 737-8AS	29933	N...	19.12.01	Ryanair Ltd	Dublin	A2004
EI-CSU	Boeing 737-36E	27626	EC-GGZ	14. 4.99	ILFC Ireland Ltd	Pescara, Italy	A2004
			EC-799		*(Op Air One)*		
EI-CSV	Boeing 737-8AS	29934		18. 1.02	Ryanair Ltd	Dublin	A2004
					"City of Nyköping" (Carries"www.Nyköping.se" titles)		
EI-CSW	Boeing 737-8AS	29935		4. 2.02	Ryanair Ltd *"Catalunya"*	Dublin	A2004
EI-CSX	Boeing 737-8AS	32778		21. 5.02	Ryanair Ltd	Dublin	A2004
EI-CSY	Boeing 737-8AS	32779		25. 6.02	Ryanair Ltd	Dublin	A2004
EI-CSZ	Boeing 737-8AS	32780		15. 7.02	Ryanair Ltd	Dublin	A2004
					(Carries "Arrivederci Alitalia" titles)		
EI-CTA	Boeing 737-8AS	29936		19.11.02	Ryanair Ltd	Dublin	A2004
EI-CTB	Boeing 737-8AS	29937		19.11.02	Ryanair Ltd	Dublin	A2004
EI-CTC	Medway EclipseR	158/137		2. 6.99	P.A.McMahon	Dunlaoghaire	N 1.03
EI-CTD	Airbus A320-211	0085	F-GJVZ	6. 5.99	Aerco Ireland Ltd	Verona, Italy	A2004
			F-WWDF		*(Op VolareAirlines)*		
EI-CTG	Stoddard-Hamilton SH-2R Glasair RG		N721WR	3. 6.99	K.Higgins	Carnmore, Galway	A 7.04
EI-CTI	Reims/Cessna FRA150L	FRA1500261	G-BCRN	29. 4.99	John Logan and Tony Bradford	Ballynacarigy	N 1.03
EI-CTL	Aerotech MW-5B Sorcerer	SR102-R440B-07	G-MTFH	21. 5.99	M.Wade	Kilrush	N 1.03
	(Fuji-Robin EC-44-PM)						
EI-CTT	Piper PA-28-161 Cherokee Warrior II	28-7716305	N38974	14. 7.99	Conair Group	Knock	A12.01
EI-CUA	Boeing 737-4K5	24901	D-AHLR	29. 9.99	Gustav Leasing XI Ltd	Rome-Fiumicino, Italy	A2004
					(Op Blue Panorama Airlines)		
EI-CUD	Boeing 737-4Q8	26298	TC-JEI	13. 3.00	Castle 2003-2 Ireland Ltd	Rome-Fiumicino, Italy	A2004
					(Op Blue Panorama Airlines)		
EI-CUE	Cameron N-105 HAB	4683		16. 9.99	Eircom Ltd *'Eircom'*	Celbridge	A 12.03
EI-CUG	Bell 206B JetRanger II	4177	N248BC	21.10.99	Avatar Aviation	Dublin	A2003
			N118GC				
EI-CUJ	Cessna 172N Skyhawk II	17271985	G-BJGO	19.11.99	M.Nally	Abbeyshrule	A 9.04
			N6038E				
EI-CUN	Boeing 737-4K5	27074	D-AHLS	13. 4.00	Gustav Leasing XI Ltd	Rome-Fiumicino, Italy	A2004
			(D-AHLG)		*(Op Blue Panorama Airlines) (Juventus c/s)*		
EI-CUP	Cessna 335	335-0018	N2706X	5. 5.00	J.Greany	Kerry	A11.02
EI-CUQ	Airbus A320-214	1259	F-WWIZ	24. 7.00	Chamonix Aircraft Leasing Ltd	Verona, Italy	A2004
					(Op Volare Airlines)		
EI-CUS	Agusta-Bell 206B-3 JetRanger III	8721	G-BZKA	24. 8.00	Doherty Quarries and Waste Management.	Slane	A 7.01
			(EI-...), G-OONS, G-LIND, G-OONS				
EI-CUT	Maule MX 7-180A	21080C		6. 4.01	Cosair Ltd	Trim	N 1.03
	(Nosewheel u/c)						
EI-CUW	Pilatus BN-2B-20 Islander	2293	G-BWYW	8.11.00	Galway Aviation Services Ltd	Galway	A2004
					(Op Aer Arann Express)		
EI-CVA	Airbus A320-214	1242	F-WWIT	22. 6.00	Aer Lingus Ltd *"St Schira/Scire"*	Dublin	A2004
EI-CVB	Airbus A320-214	1394	F-WWIV	8. 2.01	Aer Lingus Ltd *"St Mobhi"*	Dublin	A2004
EI-CVC	Airbus A320-214	1443	F-WWBG	6. 4.01	Aer Lingus Ltd *"St Kealin/Caoilfhionn"*	Dublin	A2004
EI-CVD	Airbus A320-214	1467	F-WWDK	10. 5.01	Aer Lingus Ltd *"St Kevin/Caoimhin"*	Dublin	A2004
EI-CVL	Ercoupe 415CD	4754	G-ASNF	Not known	Bernard Lyons and Jerry Hackett	Thurles	N10.01
			PH-NCF, NC94647				
EI-CVM	Schweizer Hughes 269C	S1328	G-GIRO	7.11.00	W.Moloney	Tralee, County Kerry	A 1.03
			N41S				
EI-CVN	Boeing 737-4YO	24684	TC-AFK	21.11.00	Airplanes Finance Ltd	Manila-Nino Aquino, Philippines	A2004
					(Op Philippine Airlines)		
EI-CVO	Boeing 737-4S3	25594	SP-LLH	28.10.00	Aerco Ireland Ltd	Manila-Nino Aquino, Philippines	A2004
			N2423N, TC-AVA, 9M-MLJ		*(Op Philippine Airlines)*		
EI-CVP	Boeing 737-4YO	26081	TC-AFU	22.12.00	Airplanes Finance Ltd	Manila-Nino Aquino, Philippines	A2004
					(Op Philippine Airlines)		
EI-CVR	Aérospatiale-Alenia ATR42-310	022	F-GGLK	17. 1.01	Comhfhorbairt (Gaiiimh) Teo	Dublin	A2004
			OH-LTB, F-WWEI		*(Op Aer Arann Express)*		
EI-CVS	Aérospatiale-Alenia ATR42-310	033	F-GIRC	16. 3.01	Comhfhorbairt (Gaiiimh) Teo	Dublin	A2004
			F-WIAF, OH-LTC, F-WWEO		*(Op Aer Arann Express)*		
EI-CVY	Brock KB-2 Gyrocopter	074		7. 8.03	G.Smyth	Wexford	

Reg	Type	C/n	Previous identity	Date	Owner/Operator	Base	Notes
EI-CWA	British Aerospace BAe 146 Srs.200	E2058	G-ECAL	7.11.02	Cityjet Ltd	Dublin	
			N699AA, G-ECAL, (G-BMXE), N148AC, G-5-058				
EI-CWB	British Aerospace BAe 146 Srs.200	E2051	SE-DRE	29. 3.01	Cityjet Ltd *(Op Air France)*	Paris, France	A2004
			N694AA, N141AC, G-5-003, N141AC, G-5-003				
EI-CWC	British Aerospace BAe 146 Srs.200	E2053	SE-DRC	27. 4.01	Cityjet Ltd *(Op Air France)*	Paris, France	A2004
			N695AA, N142AC, G-5-053, N142AC, G-5-053				
EI-CWD	British Aerospace BAe 146 Srs.200	E2108	SE-DRK	13. 6.01	Cityjet Ltd *(Op Air France)*	Paris, France	A2004
			N295UE, G-5-108				
EI-CWE	Boeing 737-42C	24232	N941PG	18. 5.01	Rockshaw Ltd *(Op Air One)*	Milan, Italy	A2004
			PH-BPE, G-UKLD				
EI-CWF	Boeing 737-42C	24814	PH-BPG	16. 5.01	Rockshaw Ltd *(Op Air One)*	Milan, Italy	A2004
			G-UKLG				
EI-CWH	Agusta A109E Power	11106		17. 7.01	Lochbrea Aircraft Ltd	Carnmore, Galway	A12.04
EI-CWL	Robinson R22 Beta	0885	G-BXCX	19. 9.01	J.McLoughlin	Dunboyne	
			G-MFHL				
EI-CWP	Robinson R22 Beta	3233	G-CBBK	23.10.01	D.J.Crowley	Dublin	A 10.04
EI-CWR	Robinson R22 Beta	3234	G-CBDB	2.11.01	Coates Aviation Ltd	Kilcoole, Co.Wicklow	A 10.04
EI-CWW	Boeing 737-4YO	24906	EC-GAZ	19.12.01	Airplanes Holdings Ltd *(Op Air One)*	Milan, Italy	A2004
			EC-850, 9M-MJO				
EI-CWX	Boeing 737-4YO	24912	EC-GBN	6.12.01	Airplanes Holdings Ltd *(Op Air One)*	Milan, Italy	A2004
			EC-851, 9M-MJQ				
EI-CXC	Raj Hamsa X'Air 502T	333	44SU)	6. 9.02	B.Moore	Kernan Valley	
EI-CXI	Boeing 737-46Q	28661	EC-GPI	5. 4.02	Bellevue Aircraft Leasing Ltd *(Op Air One)*	Milan, Italy	A2004
EI-CXJ	Boeing 737-4Q8	25164	G-BUHJ	22. 3.02	Castle 2003-1 Ireland Ltd *(Op Air One)*	Milan, Italy	A2004
			N164LF				
EI-CXK	Boeing 737-4S3	25596	G-OGBA	9. 4.02	Bravo Aircraft Management Ltd (*Op Transaero Airlines)*		
			G-OBMK			Moscow-Domodedovo, Russia	A2004
EI-CXL	Boeing 737-46N	28723	G-SFBH	1. 5.02	Monroe Aircraft Ireland Ltd *(Op Air One)*	Milan, Italy	A2004
EI-CXM	Boeing 737-4Q8	26302	VH-VOZ	17. 5.02	ILFC Ireland Ltd *(Op Air One)*	Milan, Italy	A2004
			TC-JEM				
EI-CXN	Boeing 737-329	23772	OO-SDW	1. 5.02	Embarcadero Aircraft Securitization Trust Ireland Ltd		
			N506GX, OO-SDW		*(Op Transaero Airlines)*	Moscow-Domodedovo, Russia	A2004
EI-CXO	Boeing 767-3GS	28111	N581LF	12. 4.02	ILFC Ireland Ltd	Milan, Italy	A2004
			D-AMUJ		*(Op Blue Panorama Airlines)*		
EI-CXR	Boeing 737-329	24355	OO-SYA	31. 5.02	Embarcadero Aircraft Securitization Trust Ireland Ltd		
			(OO-SQA)		*(Op Transaero Airlines)*	Moscow-Domodedovo, Russia	A2004
EI-CXS	Sikorsky S.61N	61816	IAC257	14.10.04	CHC Ireland Ltd *"IMES Rescue"*	Cork	A2004
			EI-CXS, C-GBKZ, LN-OQU				
EI-CXV	Boeing 737-8CX *(winglets)*	32364		3. 7.02	Jackson Leasing Ireland Ltd	Ulaanbataar , Mongolia	A2004
					"Khubelai Khaan" (Op Mongolian Airlines (Mongol Air))		
EI-CXY	Evektor EV-97 Eurostar	(20000701)	OK-FUR	31. 10.02	G Doody, E McEvoy and S.Pallister	Co.Kildare	A 1.03
EI-CXZ	Boeing 767-216ER	24973	N502GX	25. 7.02	Embarcadero Aircraft Securitization Trust Ireland Ltd		
			VH-RMM, N483GX, CC-CEF		*(Op Transaero Airlines)*	Moscow-Domodedovo, Russia	A2004
EI-CZA	ATEC Zephyr 2000	Z580602A		17. 7.03	M.Higgins	Carnmore, Galway	N 2.03
EI-CZC	CFM Streak Shadow	K269SA11	G-BWHJ	16. 7.02	M.Culhane and D.Burrows	Rathcoole	
EI-CZD	Boeing 767-216ER	23623	N762TA	2. 9.02	Embarcadero Aircraft Securitization Trust Ireland Ltd		
			CC-CJU, N4529T		*(Op Transaero Airlines)*	Moscow-Domodedovo, Russia	A2004
EI-CZG	Boeing 737-4Q8	25740	VH-VGB	8.10.02	ILFC Ireland Ltd *(Op Air One)*	Milan, Italy	A2004
			N257BR, SU-PTA, EC-HAN, TC-JED				
EI-CZH	Boeing 767-3G5ER	29435	D-AMUO	9. 8.02	ILTU Ireland Ltd	Milan, Italy	A2004
					(Op Blue Panorama Airlines)		
EI-CZK	Boeing 737-4YO	24519	N519AP	10. 1.03	Carotene Ltd	Moscow-Domodedovo, Russia	A2004
			TC-ACA, VR-CAB		*(Op Transaero Airlines)*		
EI-CZL	Schweizer 269C-1	0147	N86G	19.12.02	Venturecopters Ltd	Ballycorous/Weston	
EI-CZM	Robinson R.44Raven II	10054		24. 4.03	Wellingford Construction Ltd	Galway	A10.03
EI-CZN	Sikorsky S.61N	61740	G-CBWC	15. 4.03	CHC Ireland Ltd *"IMES Rescue"*	Cork	A2004
			OY-HDO, LN-OSU				
EI-CZO	British Aerospace BAe 146 Srs.200	E2024	G-CLHA	3.03R	CityJet Ltd *(Op Air France)*	Paris, France	A2004
			(G-GNTX), G-DEBC, N168US, N348PS				
EI-CZP	Schweizer 269C-1	0149		2. 5.03	Ng Kam Tim	Weston	
EI-DAA	Airbus A330-202	397	F-WWKX	17. 4.01	Aer Lingus Ltd *"St Keeva/Caoimhe"*	Dublin	A2004
EI-DAC	Boeing 737-8AS	29938		2.12.02	Ryanair Ltd	Dublin	A2004
EI-DAD	Boeing 737-8AS	33544		3.12.02	Ryanair Ltd	Dublin	A2004
					(Carries "Nein Zum Lufthansa Kerosinzuschlag" titles)		
EI-DAE	Boeing 737-8AS	33545		9.12.02	Ryanair Ltd	Dublin	A2004
EI-DAF	Boeing 737-8AS	29939		9. 1.03	Ryanair Ltd	Dublin	A2004
EI-DAG	Boeing 737-8AS	29940		17. 1.03	Ryanair Ltd	Dublin	A2004
EI-DAH	Boeing 737-8AS	33546		22. 1.03	Ryanair Ltd	Dublin	A2004
EI-DAI	Boeing 737-8AS	33547		3. 2.03	Ryanait Ltd	Dublin	A2004
EI-DAJ	Boeing 737-8AS	33548		4. 2.03	Ryanair Ltd	Dublin	A2004
EI-DAK	Boeing 737-8AS	33717		17. 4.03	Ryanair Ltd	Dublin	A2004
EI-DAL	Boeing 737-8AS	33718		22. 4.03	Ryanair Ltd	Dublin	A2004
EI-DAM	Boeing 737-8AS	33719		23. 4.03	Ryanair Ltd	Dublin	A2004
EI-DAN	Boeing 737-8AS	33549		2. 9.03	Ryanair Ltd	Dublin	A2004
EI-DAO	Boeing 737-8AS	33550	N1800B	5. 9.03	Ryanair Ltd (*Pride of Scotland titles)*	Dublin	A2004
EI-DAP	Boeing 737-8AS	33551	N6066U	18. 9.03	Ryanair Ltd	Dublin	A2004
EI-DAR	Boeing 737-8AS	33552	(EI-DAQ)	11. 9.03	Ryanair Ltd	Dublin	A2004
EI-DAS	Boeing 737-8AS	33553	(EI-DAR)	12. 9.03	Ryanair Ltd	Dublin	A2004
EI-DAT	Boeing 737-8AS	33554		5.12.03	Ryanair Ltd	Dublin	A2004
EI-DAV	Boeing 737-8AS	33555		9. 1.04	Ryanair Ltd	Dublin	A2004
EI-DAW	Boeing 737-8AS	33556		8. 1.04	Ryanair Ltd	Dublin	A2004
EI-DAX	Boeing 737-8AS	33557		23. 1.04	Ryanair Ltd	Dublin	A2004
EI-DAY	Boeing 737-8AS	33558		2. 2.04	Ryanair Ltd	Dublin	A2004
EI-DAZ	Boeing 737-8AS	33559		3. 2.04	Ryanair Ltd	Dublin	A2004

EI-DBE	Fokker 28-0100	11329	G-BYDN	27. 6.03	EU-Jetops Ltd *(Op EUJet)*	Shannon	A10.04
	(Fokker 100)		N130ML, SE-DUF, PH-CFE, PH-EZV, (G-FIOZ), PH-EZV				
EI-DBF	Boeing 767-3Q8	24745	F-GHGF	23. 4.03	ACG Acquisition Ireland Ltd (*Op Transaero Airlines)*	Moscow-Domodedovo, Russia	A2004
EI-DBG	Boeing 767-3Q8	24746	F-GHGG	14 .5.03	Charlie Airvraft Management Ltd *(Op Transaero Airlines)*	Moscow-Domodedovo, Russia	A2004
EI-DBH	CFM Streak Shadow SA-11	K278-SA11		14. 4.03	Michael O'Mahony	(County Cork)	
EI-DBI	Raj Hamsa X'Air Mk.2 Falcon	671		16. 4.03	Hamilton Elliot	Kilrush	
EI-DBJ	Huntwing Pegasus XL Classic	BMAA/HB/039	G-MZCZ	19. 5.03	P.A.McMahon	Clonbullogue	
EI-DBK	Boeing 777-243ER	32783		10. 10.03	GECAS Technical Services Ltd (*Op.Alitalia) "Ostuni"*	Rome-Fiumicino, Italy	A2004
EI-DBL	Boeing 777-243ER	32781		14. 11.03	GECAS Technical Services Ltd *(Op Alitalia) "Sestriere"*	Rome-Fiumicino, Italy	A2004
EI-DBM	Boeing 777-243ER	32782		12.12.03	GECAS Technical Services Ltd (*Op Alitalia) "Argentario"*	Rome-Fiumicino, Italy	A2004
EI-DBN	Bell 407	53331	N8268T	16. 5.03	Broadbourne Ltd	Kilkenny	
EI-DBO	Air Création Kiss 400	A03034-3033		13. 5.03	Enda Spain	Monasterevin	
EI-DBP	Boeing 767-35H	26389	C-GGBJ	23. 5.03	CIT Leasing Ltd	Rome-Fiumicino, Italy	A2004
			(VH-BZN), ZK-NCM, N800CZ, N60659		*(Op Alitalia) "Duca degli Abruzzi"*		
EI-DBR	Fokker 28-0100	11323	G-BYDO	30. 5.03	EU-Jetops Ltd *(Op EUJet)*	Shannon	A2004
	(Fokker 100)		N131ML, SE-DUB, PH-CFA, (PH-LNP), PH-EZC, (G-FIOT), PH-EZC				
EI-DBT	Farrington Twinstarr	TS98-018	N2409E	15. 5.03	C.O'Shea	Abbeyshrule	
EI-DBU	Boeing 767-37E	25077	F-GHGH	8. 7.03	Pegasus Aviation Ireland Ltd (*Op Transaero Airlines)*	Moscow, Russia	A2004
EI-DBV	Rand Kar X'Air 602T	516	44-AEE	13. 8.03	S.Scanlon	Tralee	
EI-DBW	Boeing 767-201	23899	N647US N607P	10. 9.03	BA Finance (Ireland) Ltd *(Op Transaero Airlines)*	Moscow-Domodedovo, Russia	A2004
EI-DBX	Magni M-18 Spartan	18-032181		3. 7.03	M.Concannon	Lough Rey	
EI-DCA	Raj Hamsa X'Air	742		18. 7.03	M.O'Connell	Granard, County.Longford	
EI-DCB	Boeing 737-8AS	33560		10. 2.04	Ryanair Ltd	Dublin	A2004
EI-DCC	Boeing 737-8AS	33561		11. 3.04	Ryanair Ltd	Dublin	A2004
EI-DCD	Boeing 737-8AS	33562		12. 3.04	Ryanair Ltd	Dublin	A2004
EI-DCE	Boeing 737-8AS	33563		25. 3.04	Ryanair Ltd	Dublin	A2004
EI-DCF	Boeing 737-8AS	33804		1. 7.04	Ryanair Ltd	Dublin	A2004
EI-DCG	Boeing 737-8AS	33805		2. 7.04	Ryanair Ltd	Dublin	A2004
EI-DCH	Boeing 737-8AS	33566		3. 8.04	Ryaniar Ltd	Dublin	A2004
EI-DCI	Boeing 737-8AS	33567		3. 8.04	Ryanair Ltd	Dublin	A2004
EI-DCJ	Boeing 737-8AS	33564		1. 9.04	Ryanair Ltd	Dublin	A2004
EI-DCK	Boeing 737-8AS	33565		1. 9.04	Ryanair Ltd	Dublin	A2004
EI-DCL	Boeing 737-8AS	33806		1.10.04	Ryanair Ltd *("Dreamliner" c/s)*	Dublin	A2004
EI-DCM	Boeing 737-8AS	33807		1.10.04	Ryanair Ltd	Dublin	A2004
EI-DCN	Boeing 737-8AS	33808	N60436	1.11.04	Ryanair Ltd	Dublin	A2004
EI-DCO	Boeing 737-8AS	33809		1.11.04	Ryanair Ltd	Dublin	A2004
EI-DCP	Boeing 737-8AS	33810		1.11.04	Ryanair Ltd	Dublin	A2004
EI-DCR	Boeing 737-8AS	33811		2.12.04	Ryanair Ltd	Dublin	A2004
EI-DCS	Boeing 737-8AS	33812		3.12.04	Ryanair Ltd	Dublin	A2004
EI-DCT	Boeing 737-8AS	33813		11.12.04	Ryanair Ltd	Dublin	A2004
EI-DCV	Boeing 737-8AS	33814		14.12.04	Ryanair Ltd	Dublin	A2004
EI-DCW	Boeing 737-8AS	33568		1.05	Ryanair Ltd *(Delivered 14.1.05)*	Dublin	
EI-DCX	Boeing 737-8AS	33569		1.05	Ryanair Ltd *(Delivered 22.1.05)*	Dublin	
EI-DCY	Boeing 737-8AS	33570		1.05	Ryanair Ltd *(Delivered 27.1.05)*	Dublin	
EI-DCZ	Boeing 737-8AS	33815		1.05	Ryanair Ltd *(Delivered 27.1.05)*	Dublin	
EI-DDA	Robinson R44 Raven II	10105	G-CCIP	26. 9.03	J.O'R Security Ltd	Craighnamanagh, County.Kilkenny	
EI-DDB	Eurocopter EC-120 Colibri	1341	G-CCIL	30.12.03	J.Cuddy	Weston	
EI-DDC	Reims Cessna F.172M	1082	G-BCEC	15.10.03	Trim Flying Club Ltd	Trim	
EI-DDD	Aeronca 7AC Champion	7AC-2895	G-BTRH N84204, NC84204	27. 4.04	J.Sullivan and M.Quinn	Coonagh	
EI-DDE	British Aerospace BAe 146 Srs.200	E2060	G-CCJC	30.10.03	Cityjet Ltd *(Op Air France)*	Dublin	A2004
			D-AZUR, N352BA, CP-2260, N352BA, XA-RMO, N402XV, G-5-060				
EI-DDF	British Aerospace BAe 146 Srs 200	E2047	G-OZRH N188US, N364PS	3.04	Cityjet Ltd *(Op Air France)*	Dublin	A2004
EI-DDH	Boeing 777-243ER	32784		15. 5.04	GECAS Technical Services Ltd *(Op Alitalia) "Tropea"*	Rome-Fiumicino, Italy	A2004
EI-DDI	Schweizer S.269C-1	0156	OO-SAC N86G	6.11.03	B.Hade	Weston	N.1.04
EI-DDJ	Raj Hamsa X'Air 582	863		24. 9.03	Pat John McHugh	Donegal Town	
EI-DDK	Boeing 737-4S3	24165	N758BC	3.12.03	Boeing Capital Leasing Ltd *(Op Transaero Airlines)*	Moscow-Domodedovo, Russia	A2004
			VT-SIH, VT-JAI, N690MA, G-BPKC				
EI-DDN	CFM Shadow CD	K.034	(G-MTFU)	2. 3. 04	F.Lynch	Fermoy, County Cork	
EI-DDO	Montgomerie Merlin	0072		21. 5.04	C.Condell	Celbridge	
EI-DDP	Southdown Puma Sprint	1121/0031	G-MMYJ	27. 5.04	P.O'Reilly	Kilrush	
EI-DDW	Boeing 767-3S1ER	26608	N979PG C-GGBI, (N769TA)	9. 1.04	Pegasus Aviation *(Op Alitalia)*	Rome-Fiumicino, Italy	
EI-DDX	Cessna 172S	172S8313	G-UFCA N2461P	19. 2.04	Flight Training Ltd	Cork	N.2.04
EI-DDY	Boeing 737-4YO	24904	HA-LEV TC-JDE	6. 5.04	Aerco Ireland Ltd *(Op Transaero Airlines)*	Moscow-Domodedovo, Russia	A2004
EI-DDZ	Piper PA-28-181Cherokee	28-7690211	PH-PDW OO-HAT, N8882E	4. 6.04	Ardnar Ltd	Weston	
EI-DEA	Airbus A320-214	2191	F-WWBX	30. 4.04	ILFC (Aer Lingus Ltd) *"St Fidelma/Fiedeilime"*	Dublin	A2004
EI-DEB	Airbus A320-214	2206	F-WWBP	19. 5.04	ILFC (Aer Lingus Ltd) *"St Nathy"*	Dublin	A2004
EI-DEC	Airbus A320-214	2217	F-WWBH	4. 6.04	ILFC (Aer Lingus Ltd) *"St Fergal/Fearghal"*	Dublin	A2004
EI-DEE	Airbus A320-214	2250	F-WWBE	27. 8.04	Aer Lingus Ltd *"St Fintain/Fionntain"*	Dublin	A2004

EI-DEF	Airbus A320-214	2256	F-WWBK	2. 9.04	Aer Lingus Ltd *"St Declan/Deaglan"*	Dublin	A2004
EI-DEG	Airbus A320-214	2272	F-WWIB	10. 9.04	Aer Lingus Ltd *"St Fachtna"*	Dublin	A2004
EI-DEH	Airbus A320-214	2294	F-WWDF	20.10.04	Aer Lingus Ltd *"St Malachy/Maolmhaoghog"*	Dublin	A2004
EI-DEI	Airbus A320-214	2374	F-WWDI	2.05R	Aer Lingus Ltd *"St Kilian/Cillian"* *(For delivery 2.05)*	Dublin	
EI-DEJ	Airbus A320-214	2364	F-WWDU	2.05	ILFC (Aer Lingus Ltd) *"St Oliver Plunkett"* *(Delivered 3.2.05)*	Dublin	
EI-DEK	Airbus A320-214	2399	F-WWIZ	3.05R	ILFC (Aer Lingus Ltd) *"St Eunan"* *(For delivery 3.05)*	Dublin	
EI-DEL	Airbus A320-214	2409	F-WWDE	4.05R	Aer Lingus Ltd *"St Ibar/Ibhar"* *(For delivery 4.05)*	Dublin	
EI-DEM	Airbus A320-214	2411	F-WWDG	4.05R	ILFC (Aer Lingus Ltd) *"St Canice/Cainneach"* *(For delivery 4.05)*	Dublin	
EI-DEN	Airbus A320-214	2432	F-WWBK	5.05R	ILFC (Aer Lingus Ltd) *"St Kieran/Ciaran"* *(For delivery 5.05)*	Dublin	
EI-DEO	Airbus A320-214	2486		7.05R	Aer Lingus Ltd *"St Senan/Seanan"* *(For delivery 7.05)*	Dublin	
EI-DEP	Airbus A320-214	2542		10.05R	ILFC (Aer Lingus Ltd) *"St Eugene/Eoghan"* *(For delivery 10.05)*	Dublin	
EI-DER	Airbus A320-214	2583		12.05R	ILFC (Aer Lingus Ltd) *"St Mel"* *(For delivery 12.05)*	Dublin	
EI-DES	Airbus A320-214	2608		1.06R	ILFC (Aer Lingus Ltd) *"St Pappin/Paipan"* *(For delivery 1.06)*	Dublin	
EI-DEV	British Aerospace BAe 146 Srs 300	E3123	G-UKHP G-5-123	10. 3.04	Cityjet Ltd	Dublin	
EI-DEW	British Aerospace BAe 146 Srs 300	E3142	G-UKAC G-5-142	4.04R	Cityjet Ltd	Dublin	
EI-DEX	British Aerospace BAe 146 Srs 300	E3157	G-UKID G-6-157	11. 6.04	Cityjet Ltd	Dublin	
EI-DEY	Airbus A319-112	1102	D-ANDA F-WQQE, OO-SSD, D-AVYI	2. 04R	Permeke Aircraft Leasing Co Ltd *(Op Meridiana) "Capo Rizzulo"*	Olbia, Italy	A2004
EI-DEZ	Airbus A319-112	1283	D-ANDE F-WQQF, OO-SSI, D-AVYI	25. 2.04	Permeke Aircraft Leasing Co Ltd *(Op Meridiana) "Capo Gallo"*	Olbia, Italy	A2004
EI-DFA	Airbus A319-112	1305	D-ANDI F-WQQG, OO-SSJ, D-AVWX	30. 3.04	Permeke Aircraft Leasing Co Ltd *(Op Meridiana) "Capo Carbonara"*	Olbia, Italy	A2004
EI-DFB	Fokker F.28-0100 *(Fokker 100)*	11290	F-WQUG PT-MRZ, PH-TAB, F-OLGA, (811150), PH-TAB, PH-EZP	22. 4.04	EU-Jet Ops Ltd *(Op EUJet)*	Naples, Italy	N11.04
EI-DFC	Fokker F.28-0100 *(Fokker 100)*	11296	PT-MQA PH-TAC, F-OLGB, (B-11152), PH-TAC, PH-EZV	2.04R	EU-Jet Ops Ltd *(Op EUJet)*	Naples, Italy	N11.04
EI-DFD	Boeing 737-4S3	24163	G-BVNM G-BPKE, 9M-MJJ, G-BPKA	4. 8.04	Orix Aircraft Management Ltd *(Op Air One)*	Naples, Italy	A2004
EI-DFE	Boeing 737-4S3	24164	G-BVNN G-BPKB, 9M-MLA, G-BPKB	27. 4.04	Orix Aircraft Management Ltd *(Op Air One)*	Naples, Italy	A2004
EI-DFF	Boeing 737-4S3	24167	G-BVNO G-BPKE, 9M-MLB, G-BPKE	.04R	Orix Aircraft Management Ltd *(Op Air One)*	Naples, Italy	
EI-DFG	Embraer 170-100LR	17000008	(I-EMCX) PT-SKA	24. 3.04	GECAS Technical Services Ltd *(Op Alitalia Express) "Via Appia"*	Rome-Fiumicino, Italy	A2004
EI-DFH	Embraer 170-100LR	17000009	PT-SKB	20. 4.04	GECAS Technical Services Ltd *(Op Alitalia Express) "Via Aurélia"*	Rome-Fiumicino, Italy	A2004
EI-DFI	Embraer 170-100LR	17000010	PT-SKC	04	GECAS Technical Services Ltd *(Op Alitalia Express) "Via Cassia"*	Rome-Fiumicino, Italy	A2004
EI-DFJ	Embraer 170-100LR	17000011	PT-SKD	30. 6.04	GECAS Technical Services Ltd *(Op Alitalia Express) "Via Flaminia"*	Rome-Fiumicino, Italy	A2004
EI-DFK	Embraer 170-100LR	17000032	PT-SUA	30. 6.04	GECAS Technical Services Ltd *(Op Alitalia Express) "Via Salaria"*	Rome-Fiumicino, Italy	A2004
EI-DFL	Embraer 170-100LR	17000036	PT-SUF	16. 7.04	GECAS Technical Services Ltd *(Op Alitalia Express) "Via Tiburtina Valeria"*	Rome-Fiumicino, Italy	A2004
EI-DFM	Evektor EV-97 Eurostar	2003-1706		8. 3.04	Gerard Doody	Portlaoise	
EI-DFO	Airbus A.320-211	0371	A6-ABX C-FTDD, SU-LBA, TC-OND, N531LF, C-FLSJ, F-WWIQ	20. 4.04	Triton Aviation Ireland Ltd *(Op Windjet)*	Catania, Italy	A2004
EI-DFP	Airbus A.319-112	1048	F-OHJV F-WIHE, OO-SSA, D-AVYT	4.04R	Permeke Aircraft Leasing Co.Ltd *(Op Meridiana) "Capo Caccia"*	Olbia, Italy	A2004
EI-DFS	Boeing 767-33A	25346	ET-AKW V8-RBE	31. 5.04	Jeritt Ltd *(Op Transaero Airlines)*	Moscow-Domodedovo, Russia	A2004
EI-DFW	Robinson R44 Raven	1377	G-CCUM N7536G	3. 6.04	Blue Star Helicopters Ltd	Cork	
EI-DFX	Air Creation Kiss 400	A04007-4007		13. 5.04	L.Daly	Tullamore	
EI-DFY	Raj Hamsa X'Air R100(2)	430	44-VE	13. 5.04	P.McGirr and R.Gillespie	Killygordon	
EI-DFZ	Fokker F28-0100 *(Fokker 100)*	11265	PT-MQQ PH-SEM, PK-JGC, (N403PA), (N208BN), PH-EZM	24. 6.04	EU-Jet Ops Ltd *(Op EUJet)*	Manston	A2004
EI-DGA	Urban Air Lambada UFM-11UK	16/11		16. 4.04	Dr P and D Dunkin	Abbeyshrule	
EI-DGD	Boeing 737-430	27000	OO-VEF D-ABKA	27.5 04	Challey Ltd *(Op Air One)*	Pescara, Italy	A2004
EI-DGE	Fokker F28-0100 *(Fokker 100)*	11351	PT-MRF PH-LNO	.05R	EU-Jet Ops Ltd *(Op EUJet)*	Manston	
EI-DGG	Raj Hamsa X'Air 133	899		9. 6.04	N.Geh	Garrabeg, County Galway	
EI-DGH	Raj Hamsa X'Air 582	861		9. 6.04	M.Garvey and T.McGowan	Kells, CountyMeath	
EI-DGI	ICP MXP740 Savannah Jab *(Built N Farrell - kit no 03-10-51-236)*	BMAA/HB/345	G-CCPR	7. 9.04	N.Farrell	Ballytooney, County Roscommon	
EI-DGJ	Raj Hamsa X'Air 582 *(Built R.Morelli - kit no.707)*	BMAA/HB/210	G-CCEV	9. 6.04	R.Morelli	Dublin	
EI-DGK	Raj Hamsa X'Air 133	856		9. 6.04	B.Chambers	Castlefinn, County Donegal	
EI-DGL	Boeing 737-46J	27171	SX-BMA D-ABAE	6. 7.04	Gelston Ltd *(Op Air One)*	Pescara, Italy	A2004

EI-DGM	Boeing 737-4C9	26437	LX-LGG(2)	15. 7.04	Lux Aircraft Leasing Ltd *(Op Blue Panorama Airlines)*	Milan, Italy	A2004
EI-DGN	Boeing 737-4C9	25429	LX-LGF(2)	15. 7.04	Lux Aircraft Leasing Ltd *(Op Blue Panorama Airlines)*	Milan, Italy	A2004
EI-DGP	Urban Air Lambada UFM-11UK	15/11	OK-IUA-68	24.11.04	M.Tormey	Abbeyshrule	N 1.05
EI-DGR	Urban Air Lambada UFM-11UK	17/11		21. 7.04	M.Tormey	Abbeyshrule	N 1.05
EI-DGS	ATEC Zephyr 2000	861003A		20.10.04	K.Higgins	Carnmore	
EI-DGT	Urban Air Lambada UFM-11UK	14/11	OK-FUA-09	12. 8.04	A and P Aviation Ltd	Carnmore	A12.04
EI-DGV	ATEC Zephyr 2000			.05R	Not known	Carnmore	
EI-DGW	Cameron Z-90	10607		15. 9.04	J.Leahy	Navan, County Meath	N 9.04
EI-DGX	Cessna 152	15281296	G-BPJL N49473	19.10.04	K.O'Connor	Weston	
EI-DGY	Urban Air UFM-11 Lambada	10/11	OK-EUU-55	15.10.04	J.Keena	Abbeyshrule	
EI-DGZ	Boeing 737-86N	28624	EC-HMK N1786B	27.10.04	GE Capital Aviation Services Ltd *(Op Ryan International Airlines)*	Wichita-Mid Continent, Kansas, US	
EI-DHA	Boeing 737-8AS	33571		2.05	Ryanair Ltd *(Delivered 2.2.05)*	Dublin	
EI-DHB	Boeing 737-8AS	33572		.05R	Ryanair Ltd	Dublin	
EI-DHC	Boeing 737-8AS	33573		.05R	Ryanair Ltd	Dublin	
EI-DHD	Boeing 737-8AS	33816		.05R	Ryanair Ltd	Dublin	
EI-DHE	Boeing 737-8AS	33574		.05R	Ryanair Ltd	Dublin	
EI-DHF	Boeing 737-8AS	33575		.05R	Ryanair Ltd	Dublin	
EI-DHG	Boeing 737-8AS	33576		.05R	Ryanair Ltd	Dublin	
EI-DHI	Boeing 737-8AS	33817		.05R	Ryanair Ltd	Dublin	
EI-DHJ	Boeing 737-8AS	33818		.05R	Ryanair Ltd	Dublin	
EI-DHK	Boeing 737-8AS	33819		.05R	Ryanair Ltd	Dublin	
EI-DHM	Boeing 737-8AS	33820		.05R	Ryanair Ltd	Dublin	
EI-DIA	Solar Wings Pegasus XL-Q	SW-WQ-0503	G-MYAD	15. 9.04	P.Byrne	Hacketstown, County.Carlow	
EI-DIB	Air Creation Kiss 400	A04117-4123		10. 9.04	E.Redmond	Ferns, County.Wexford	
EI-DIG	Airbus A.320-214	1597	EC-IAG F-WWBY	29.10.04	GE Capital Aviation Services Ltd *(Op Ryan International Airlines) (Funjet Vacations titles)*	Wichita-Mid Continent, Kansas, US	
EI-DIH	Airbus A.320-214	1657	EC-ICK F-WWID	13.12.04	GE Capital Aviation Services Ltd *(Op Ryan International Airlines) (Transglobal titles)*	Wichita-Mid Continent, Kansas, US	
EI-DIJ	Airbus A.320-212	391	G-MONW F-WWDO	22.12.04	Eirjet Ltd	Shannon	A11.04
EI-DIP	Airbus A.330-202	339	I-VLEF F-WQQO, N339LF, C-GGWD, F-WWYZ	28.10.04	Calliope Ltd	Shannon	
EI-DIR	Airbus A.330-202	272	I-VLEE F-WQQL, C-GWWC, F-WWKE	18.11.04	Calliope Ltd	Shannon	
EI-DIS	Boeing 737-86N	28610	EC-HHH N1786B	10.12.04	Futura GAEL *(Op Ryan International Airlines)*	Wichita-Mid Continent, Kansas, US	
EI-DIT	Boeing 737-86N	28621	EC-HMJ N1786B	10.12.04	Futura GAEL *(Op Ryan International Airlines)*	Wichita-Mid Continent, Kansas, US	
EI-DIU	Airbus A.320-232	990	I-PEKU N990SE, D-ALAJ, F-WWBC	2.12.04	S.A.L.E.Ireland Ltd *(Op FlyNiki)*	Vienna, Austria	
EI-DIV	Airbus A.320-232	1856	I-PEKS F-WQSK, F-WWDI	1.12.04	S.A.L.E.Ireland Ltd	Dublin	
EI-DIW	Airbus A.320-232	1909	I-PEKT F-WQSL, F-WWDS	7.12.04	S.A.L.E.Ireland Ltd	Dublin	
EI-DIX	Airbus A.320-232	1996	I-PEKV N996SE, D-ALAT, F-WWIM	2.12.04	S.A.L.E.Ireland Ltd	Dublin	
EI-DIZ	Robinson R22 Beta	3684	G-PPLA N74327	23.12.04	Blue Star Helicopters Ltd	Cork	
EI-DJH	Airbus A.320-232	814	I-PEKW N471LF, SU-LBB, F-WWII	20.12.04	ILFC Ireland Ltd *(Op Myway.com)*	Bergamo-Orio al Serio, Italy	
EI-DJI	Airbus A.320-232	1757	I-PEKQ F-WWIR	20.12.04	ILFC Ireland Ltd *(Op Myway.com)*	Bergamo-Orio al Serio, Italy	
EI-DJK	Boeing 737-382	24365	9H-ADM CS-TIB	9. 1.05	Not known *(Op Kaliningradavia KLD*	Kaliningrad-Khrabovo, Russia	
EI-DJM	Piper PA-28-161 Warrior II	28-8316106	HB-POV N4314K	2.05	Not known	Waterford	A 2.05
EI-DLP	Agusta A109C	7657	N611VA N97CN, N97CH, N7CH, N1VN	16. 7.03	DLP Helicopters Ltd	Galway	
EI-DMG	Cessna 441 Conquest	441-0165	N140MP N27214	4. 7.01	Dawn Meats Group	Waterford	A 1.03
EI-DOC	Robinson R44 Raven	1400		3. 9.04	Donville Heli's Ltd	Oranmore	
EI-DUB	Airbus A330-301ER	055	F-WWKP	6. 5.94	Aer Lingus Ltd *"St.Patrick/Padraig"*	Dublin	A2004
EI-DUN	Agusta A.109E	11176		26. 8.04	Barkisland (Developments) Ltd	Dublin	
EI-EBJ	Robinson R44 Raven	1358		18. 5.04	Billy Jet Ltd	Oranmore	A12.04
EI-ECA	Agusta A109A II	7387	N109RP JA9662	28. 2.97	Backdrive Ltd *(Op Ace Helicopters Ltd)*	Drogheda	A12.01
EI-EDR	Piper PA-28R-200 Cherokee Arrow II	28R-7435265	G-BCGD N9628N	19.11.87	Kestrel Flying Group Ltd	Dublin	A 9.04
EI-EGG	Robinson R44 Raven	1344	G-CCLI	10.12.03	In-Flight Aviation Ltd	Weston	N.1.04
EI-EHB	Robinson R22B2 Beta	3569		18. 5.04	Blue Star Helicopters Ltd	Cork	
EI-EHC	Robinson R22B2 Beta	3442	N71850	24. 6.03	Executive Helicopters (Cork)Ltd	Oranmore	
EI-EHD	Robinson R22B2 Beta	3653		14. 9.04	South Coast Helicopters Ltd	Cork	
EI-EHE	Robinson R22B2 Beta	3654		1.10.04	Blue Star Helicopters Ltd	Cork	
EI-EHG	Robinson R22 Beta	3509	N75302	16. 4.04	Executive Helicopters Maintenance Ltd	Oranmore	
EI-ELL	Medway EclipseR	157/136		2. 6.99	Microflex Ltd	Kilrush	N 1.03
EI-EUR	Eurocopter EC.120B Colibri	1138	G-BZMK F-WQOE	14.12.00	Atlantic Helicopters Ltd	Dublin	N 1.03

Regn	Type	C/n	Previous identity	Date	Owner/Operator	Base	CofA
EI-EWR	Airbus A330-202	330	F-WWKV	9. 5.00	Aer Lingus Ltd	Dublin	A2004
					"Laurence O'Toole/Lorcan O'Tuathail"		
EI-EXC	Robinson R44 Raven	1312		24. 6.03	Executive Helicopters(Cork) Ltd	Oranmore	
EI-EXG	Robinson R22B2 Beta	3698	N7337F	18.11.04	21st Century Aviation Ltd	Oranmore, Deerpark	
EI-FBG	Reims/Cessna F182Q Skylane	F18200032	D-EFBG	4. 7.00	Messrs Tunney, Helly and Spelman	Weston	A 1.03
			(F-GAGU)				
EI-GAA	Boeing 767-266ER	23179	N567KM	25. 5.04	Arbor Finance *(Op.Krasnoyarsk Airlines (Kras Air))*		
			ZS-SRB, N573SW, SU-GAI, N1788B			Krasnoyarsk-Yemilianovo, Russia	A2004
EI-GAN	Bell 407	53551	N20446	13. 6.03	Rosewall Property Ltd	Dublin	
			C-GFNR				
EI-GAV	Robinson R-22 Beta	3485	N75264	17.10.03	Airo Helicopters Ltd	Weston	
EI-GBA	Boeing 767-266ER	23180	N573JW	25. 5.04	Arbor Finance *(Op.Krasnoyarsk Airlines (Kras Air))*		
			ZS-SRC, N575SW, SU-GAJ, N1789B			Krasnoyarsk-Yemilianovo, Russia	A2004
EI-GER	Maule MX-7-180A Star Rocket	20006C		7. 1.94	P.J.L.Ryan	Trim	N 1.03
	(Tail-wheel u/c)						
EI-GFC	SOCATA TB-9 Tampico	141	G-BIAA	9.10.93	B.McGrath, J.Ryan and D.O'Neill	Waterford	N 12.03
EI-GKL	Robinson R22B2 Beta	3570		19. 5.04	Gerry Keyes Ltd	Oranmore	
EI-GPT	Robinson R22B2 Beta	3317	N70637	8.11.04R	Treaty Plant and Tool (Hire and Sales) Ltd	Limerick	
EI-GPZ	Robinson R44 Raven	1388		5. 8.04	G and P Transport	Castlebar	
EI-GSE	Cessna F.172M	1105	D-EDXO	12. 4.02	Frank Doherty	Donegal	
EI-GSM	Cessna 182S	18280188	N9541Q	17. 6.98	Westpoint Flying Group	Dublin	N 1.03
EI-GWY	Cessna 172R Skyhawk	17280162	N9497F	31.12.97	Galway Flying Club Ltd	Galway	A 1.03
EI-HAM	Light-Aero Avid Flyer	1072-90		18.11.96	H.Goulding	Bray	
	(Rotax 582)				*(Current status unknown)*		
EI-HCS	Grob G-109B	6414	G-BMHR	18. 8.95	H.Sydner	Boleybeg, Ballymore Castle	A 9.01
EI-HER	Bell 206B JetRanger III	3408	G-HIER	1. 7.94	Irish Helicopters Ltd	Dublin	A10.02
			G-BRFD, N2069N				
EI-HXM	Bell 206B JetRanger II	4105	ZS-HXM	28. 7.00	Premier Star Equipment Ltd	Weston	N 1.03
			N7131J				
EI-IAW	Learjet Inc Learjet 60	218	N8084J	14. 6.01	Voltage Plus Ltd	Shannon	A11.04
			N50157				
EI-IHL	Aérospatiale AS350B1 Ecureuil	1963	G-BWFY	27. 5.04	Irish Helicopters Ltd	Dublin	
			N518R				
EI-IPC	Fairey Britten-Norman BN-2A-26 Islander	2011	G-CHES	24. 1.02	Irish Parachute Club	Clonbulloge	A 1.03
			G-PASY, G-BPCB, G-BEXA, G-MALI, (ZB503), G-DIVE, G-BEXA				
EI-IRE	Canadair CL.600-2B16 Challenger	5515	N515DM	20. 8.02	Starair (Ireland)Ltd	Dublin	A 1.03
			C-GLXF				
EI-IRV	Aérospatiale AS.350B Ecureuil	1713	D-HENY	14.10.03	Harrcops Ltd	Galway	
EI-IZO	Eurocopter EC120B Colibri	1191	G-BZUS	26. 7.01	Cloud Nine Helicopters Ltd	Carlow	A12.01
			F-WQOU				
EI-JAL	Robinson R44 Raven II	10329	N7530N	5. 8.04	Heliwest Ltd	Galway	
EI-JBC	Agusta A109A	7126	F-GATN	24. 7.97	Medeva Properties Ltd	Dublin Heliport	N 5.00
EI-JFD	Robinson R44	0969		13. 3.01	New World Plant Ltd	Galway	A10.03
EI-JFK	Airbus A330-301	086	F-GMDE	11. 7.95	Aer Lingus Ltd *"St.Colmcille"*	Dublin	A2002
EI-JIV	Lockheed L.382G Hercules	4673	ZS-JIV	15.11.02	Air Contractors (Ireland) Ltd	Dublin	A9. 04
			D2-THE, ZS-JIV		*(ORSL titles)*		
EI-JWM	Robinson R.22 Beta	1386	G-BSLB	21.11.92	C.Shiel	Weston	A10.00
EI-KEV	Raj Hamsa X'Air 133	567	G-BZLD	21. 5.04	K.Glynn	Birr	
EI-LAF	Bell 206B JetRanger	4090	N47LM	13. 4.04	Shamrock Helicopters Ltd	Shannon	
			N62AJ, G-DPPA, C-FHPA				
EI-LAX	Airbus A330-202	269	F-WWKV	29. 4.99	Aer Lingus Ltd *"St.Mella/Mella"*	Dublin	A2004
EI-LHD	Bell 206L-3 LongRanger	51206	D-HKLW	10.11.04	Quarry and Mining Equipment Ltd	Dundalk	
			N3205J				
EI-LIT	MBB Bö.105S	S.434	A6-DBH	20. 2.96	Irish Helicopters Ltd	Cork	A 9.01
			Dubai 105, D-HDMH				
EI-LKS	Eurocopter EC130B4 Ecureuil	3643	F-WQDQ	20. 1.03	WIGAF Leasing Co.Ltd	Shannon	N 1.03
					(Op Links Helicopters)		
EI-LNX	Eurocopter EC130B4 Ecureuil	3498	N460AE	10. 6.02	WIGAF Leasing Co.Ltd	Shannon	N 1.03
					(Op Links Helicopters)		
EI-LTE	Airbus A.320-214	1775	EC-IEP	7. 12.04	Celestial Aviation Trading Ltd		
			F-WWDL		*(Op Myway.com)*	Bergamo-Orio al Serio, Italy	
EI-MAG	Robinson R22 Beta	2592	G-DHGS	3. 8.01	Airo Helicopters Ltd	Bagnelstown, Carlow	A 1.03
EI-MAX	Learjet 31A	31A-233	N233BX	26. 4.04	Airlink Airways	Westport, County Mayo	
			LX-PAT, N5005X				
EI-MCC	Robinson R44 Raven II	10507	G-CDAU	1.12.04	Coates Aviation Ltd	Dublin	
EI-MCF	Cessna 172R Skyhawk	172080799	N2469D	20. 1.00	Galway Flying Club	Carnmore, Galway	A 2.03
EI-MEL	Agusta A109C	7672	LV-WXA	20. 6.00	Kildare Aviation Ltd	Dublin Heliport	N 1.03
			N27ET, LV-WXA, N4NM				
EI-MER	Bell 206B JetRanger	4513	N60507	28. 9.99	Gaelic Helicopters Ltd	Westpoint	A10.02
EI-MES	Sikorsky S-61N	61776	G-BXAE	27. 3.97	CHC Ireland Ltd *"IMES Rescue"*	Dublin	A2004
			LN-OQO				
EI-MET	Eurocopter EC.130B4 Ecureuil	3810	SE-JHY	20. 9.04	Skyheli Ltd	Cabinteeley	
EI-MIK	Eurocopter EC.120B Colibri	1104	G-BZIU	22. 6.01	Bachir Ltd	Oranmore	A10.03
EI-MIP	Aérospatiale SA.365N Dauphin 2	6119	G-BLEY	20. 3.96	CHC Ireland Ltd	Cork	A2004
			F-WTNM				
EI-MIT	Agusta A109E Power Elite	11162		17. 1.03	Mercury Engineering Ltd	Dublin	A 1.03
EI-MJR	Robinson R44 Raven 11	1391	N72603	5. 8.04	M.Melville	Oranmore	
EI-MMO	Robinson R44 Raven	1389	(EI-EHF)	3. 9.04	Morrissey Fencing Ltd	Portlaoise	
EI-MOR	Robinson R44 Raven	1392		5. 8.04	Ultimate Flight Ops Ltd	Blarney, County Cork	
EI-MUL	Robinson R44 Raven	1074		29. 8.01	Cotton Box Design Group Ltd	Galway	
EI-MUR	Robinson R22B2 Beta	3464		30. 7.03	B.Murphy	Weston	
EI-NPG	Agusta A109E Power Elite	11169		4.12.03	Anson Logue and William Moffet	Dublin	
EI-NVL	Jora	C129		25 7.03	N.Van Lonkhuyzen	Abbeyshrule	N 8.04
EI-NZO	Eurocopter EC.120B Colibri	1257	G-CBJF	6. 2.04	Executive Helicopters Maintenance	Oranmore	

EI-OBJ	Robinson R-22 Beta	3418		17. 4.03	Billy Jet Ltd	Weston	N.1.04
EI-ORD	Airbus A330-301	059	(EI-USA) F-GMDD	6. 6.97	Aer Lingus plc *"St.Maeve/Maedbh"*	Dublin	A2004
EI-OZA	Airbus A300B4-103F	148	F-GOZA SX-BEG	5. 4.02	Air Contractors (Ireland) Ltd	Dublin	A 6.04
EI-OZB	Airbus A300B4-103F	184	F-GOZB SX-BEH	5. 4.02	Air Contractors (Ireland) Ltd	Dublin	A10.04
EI-OZC	Airbus A300B4-103F	189	F-GOZC SX-BEI	8. 2.02	Air Contractors (Ireland) Ltd	Dublin	
EI-PAT	British Aerospace BAe.146 Srs.200	E2030	G-ZAPL G-WLCY, N172US, N352US	11.10.99	Brimstage Ltd *(Op Cityjet and Air France)*	Paris, France	A2004
EI-PCI	Bell 206B Jet Ranger	4072	N208M JA9850, C-GAJN	6. 2.03	Malcove Ltd	Dublin	
EI-PEC	Robinson R44 Raven 11	10354		2. 6.04	P.Sexton	Mullingar, County Westmeath	
EI-PJD	Aerospatial AS.350B Ecureuil	3594	SE-JGY	15.10.03	New World Plant Ltd	Oranmore	N10.03
EI-PJW	Eurocopter EC.120B Colibri	1111	D-HUAD F-WQDT	23.12.03	Paddy White	Oranmore	
EI-PKS	Bell 206B Jetranger	4480	OE-XAC D-HIFIS	7.11.03	Mountainway Builders Ltd	Clontara, County Clare	
EI-PMI	Agusta-Bell 206B-3 JetRanger III	8614	EI-BLG G-BIGS	19. 9.96	Ping Golf Equipment Ltd *(Current status unknown)*	Dublin	A 8.99
EI-POD	Cessna 177B	17702729	N1444C	3. 8.95	Trim Flying Club Ltd	Trim	N 1.03
EI-PRI	Bell 206B Jet Ranger	4523	N6389V C-GLZM	29. 2.00	Brentwood Properties Ltd	Castleknock	A10.02
EI-RCG	Sikorsky S.61N	61807	G-BZSN LN-OQB	25. 9.01	CHC Ireland Ltd *"IMES Rescue"*	Shannon	A2004
EI-REA	Aérospatiale-Alenia ATR72-201	441	F-WQNC G-BWTL, F-WWLG	30. 5.02	Comhfhorbairt (Gaillimh) Teo *(Op Aer Arann Express)*	Dublin	A2004
EI-REB	Aérospatiale-Alenia ATR72-201	470	F-WONH F-WQOL, F-WQOF, G-BWTM, F-WWED *(Op Aer Arann Express)*	17. 5.02	Comhfhorbairt (Gaillimh) Teo	Dublin	A2004
EI-RED	Aérospatiale-Alenia ATR72-202	373	F-GJRX HS-PGA, F-WQAK, F-GKOL, F-WWEU *(Op Aer Arann Express)*	5. 6.03	Comhfhorbairt (Gaillimh) Teo	Dublin	A2004
EI-REE	Aérospatiale-Alenia ATR72-201	342	G-BVTJ F-WWEV, F-GKOI, F-WWLX	10.12.03	Comhfhorbairt (Gaillimh) Teo *(Op Aer Arann Express)*	Dublin	A2004
EI-REF	Aérospatiale-Alenia ATR72-202	201	F-GKOA	23. 6.04	Comhfhorbairt (Gaillimh) Teo *(Op Aer Arann Express)*	Galway	A2004
EI-RMC	Bell 206B Jet Ranger	488	G-BWLO N2290W	16.12.99	Westair Aviation Ltd	Shannon	A 7.01
EI-SAC	Cessna 172P Skyhawk	17276263	N98149	22. 9.00	Sligo Aeronautical Club	Strandhill	A 8.02
EI-SAM	Extra EA.300/200	031	(D-EDGE (5))	19. 7.01	D.Bruton	Abbeyshrule	N 1.03
EI-SAR	Sikorsky S-61N *(Mitsubishi c/n M61-001)*	61-143	G-AYOM N4585, JA9506, N94565	26. 6.98	CHC Ireland Ltd *"IMES Rescue"*	Waterford	A2004
EI-SAT	Steen Skybolt	1	N52DH	22.10.99	Capt B.O'Sullivan	Trim	N 1.03
EI-SBM	Agusta A.109E	11174		1. 5.04	Ballymore Management Services Ltd	Dublin	
EI-SBP	Cessna T.206H	T20608159	N2354M N4234H	16. 8.00	P.Morrissey	Dublin	N 1.03
EI-SGF	Robinson R44 Raven	1401	N74108	1.10.04	M.Reilly and S.Filan	Carraroe, County Sligo	
EI-SKS	Robin R.2160	307	OO-OBC	13. 7.04	Shemburn Ltd	Weston	
EI-SKT	Piper PA-44-180 Seminole	44-7995004	G-BGSG N36538	27.11.02	Shemburn Ltd	Waterford	N 1.03
EI-SKU	Piper PA28RT-201 Arrow IV	28R-7918145	G-BXYS PH-SBS, N29561	14. 2.03	Shemburn Ltd	Weston	N 12.03
EI-SKV	Robin R.2160D	171	PH-BLO	28. 3.03	Shemburn Ltd	Weston	
EI-SKW	Piper PA-28-161 Warrior II	28-8216115	D-EIBV N9630N]	18. 2.04	Shemburn Ltd	Weston	A 6.04
EI-SLB	Aérospatiale-Alenia ATR42-300	079	OY-CID D-BATA, F-WWEE	27. 8.04	Air Contractors (Ireland) Ltd *(Operated by ACL for FedEx)*	(US)	A2004
EI-SLC	Aérospatiale-Alenia ATR42-300	082	OY-CIE D-BATB, F-WWEH	27. 8.04	Air Contractors (Ireland) Ltd *(Operated by ACL for FedEx)*	(US)	A2004
EI-SLD	Aérospatiale-Alenia ATR42-310	005	F-GHPZ OY-CIA, F-GHPZ, F-WZGH, YU-ALM, F-GEDZ, OY-CIA *(Operated by ACL for FedEx)*	4. 5.04	Air Contractors (Ireland) Ltd	Hamburg, Germany	A2004
EI-SLE	Aérospatiale-Alenia ATR42-310	024	F-WQNE SP-EEE, F-WQJL, OY-CIC, F-WWEC *(Operated by ACL for FedEx)*	23. 6.03	Air Contractors Ltd	Dublin	A2004
EI-SLF	Aérospatiale-Alenia ATR72-202	210	OY-RUA B-22703, F-WWEH	26.11.02	Air Contractors (Ireland) Ltd *(Operated by ACL for FedEx)*	Dublin	A 6.04
EI-SMA	Shorts SD.3-60	SH-3712	G-OBLK G-BNDI, G-OBLK, G-BNDI, G-14-3712	11. 6.04	Air Contractors (Ireland) Ltd	Dublin	A 9.04
EI-SMB	Shorts SD.3-60	SH-3661	G-EXPS TC-AOA, G-BLRT, SE-KRV, G-BLRT, G-14-3661	15.11.04	Air Contractors (Ireland) Ltd	Dublin	A11.04
EI-SMF	Fokker F.27 Friendship 500	10633	G-JEAE VH-EWV, PH-FSO, PH-EXC	19. 7.04	Air Contractors (Ireland) Ltd	Dublin	N11.04
EI-SMK	Zenith CH.701	7-3551		15.10.03	Seamus King	(County Kildare)	
EI-SNJ	Bell 407	53442	N407J PP-MSJ, N6096D, C-GFNR	17.12.03	Thornridge Services Ltd	Castleknock	
EI-SQG	Agusta A109E Power	11084		1. 8.00	Quinn Group Ltd *(Slieve Russel Hotel titles)*	Dublin	A12.01
EI-STR	Bell 407	53282	N44504	19. 5.00	G and H Homes Ltd	Listowel	N 1.03
EI-STT	Cessna 172M	17266228	D-EVBB N9557H	30 .8.00	Garda Aviation Club Ltd	Weston	N 1.03
EI-TAB	Airbus A320-233	1624	(N485TA) F-WWIZ	27. 6.02	CIT Ireland Leasing Ltd *(Op TACA International Airlines) "Mensajero de Esperanza"*	San Salvador-Comalapa Intl, El Salvador	
EI-TAC	Airbus A320-233	1676	F-WWBX (N486TA), F-WWBX	18.10.02	CIT Ireland Leasing Ltd *(Op TACA International Airlines)*	San Salvador-Comalapa Intl, El Salvador	

EI-TAD	Airbus A320-233	2301	F-WWDF	.05R	CIT Ireland Leasing Ltd *(Op TACA International Airlines)* San Salvador-Comalapa Intl, El Salvador		
EI-TAE	Airbus A320-233	874	N455TA F-WWBV	27. 8.04	Pegasus Aviation Ireland *(Op TACA International Airlines)* San Salvador-Comalapa Intl, El Salvador		
EI-TBM	SOCATA TBM.700	232		3. 7.02	Folens Management Services Ltd	Weston	N 1.03
EI-TIP	Bell 430	49074	N430MK N9151Z, C-GAHJ	12. 6.02	Starair (Ireland) Ltd	Cloughran	
EI-TKI	Robinson R22 Beta	1195	G-OBIP	22. 8.91	J.McDaid	Weston	A10.00
EI-TMH	Robinson R44 Raven	1402		26. 8.04	T.Maybury	Weston	
EI-TOY	Robinson R44	1294		2. 4.03	Metroheli Ltd	Weston	
EI-UFO	Piper PA-22-150 Tri-Pacer *(Tail-wheel conversion)*	22-4942	G-BRZR N7045D	12. 2.94	W.Treacy	Trim	N 9.00
EI-VNE	Eurocopter EC.120B Colibri	1253	G-CBHS	17. 5.02	Seafield Demesne Management Ltd	Weston	A11.02
EI-WAC	Piper PA-23-250 Aztec E	27-4683	G-AZBK N14077	26. 5.95	Westair Aviation Ltd	Shannon	A 7.01
EI-WAV	Bell 430	49028	N4213V	24.12.97	Westair Aviation Ltd	Shannon	A12.01
EI-WGV	Gulfstream G.1159 Gulfstream V	505	N505GV	21.11.97	Westair Aviation Ltd *"Born Free"*	Shannon	A12.04
EI-WJN	Hawker Siddeley HS.125 Srs.700A	257062	N416RD N26EA, RA02809, G-5-708, RA02809, (G-BWJX), G-5-708, N7062B, HB-VGF, G-5-708, HB-VGF, G-5-16	30. 5.00	Westair Aviation Ltd	Shannon	A12.01
EI-WMN	Piper PA-23-250 Aztec F	27-7954063	G-ZSFT G-SALT, G-BGTH, N2551M, N9731N	12.10.00	Westair Aviation Ltd	Shannon	A 9.01
EI-WRN	Piper PA-28-151 Cherokee Warrior	28-7615212	G-BDZX N9559N	5.10.99	Waterford Aero Club Ltd	Waterford	A 12.03

Registrations awaited

EI-...	Aerial Arts Chaser S	CH.723	G-MVGI	1.03	Not yet known
	(Officially transferred from the UK to Republic of Ireland 6.1.03 but noted stored as "G-MVGI" @ Mill Farm, Hughley, Much Wenlock 9.04)				
EI--	CFM Streak Shadow SA *(Built J.W.Beckett - kit no K.241)* (Rotax 582)	PFA 206-12609	G-BZDF		Not yet known *(Officially transferred from the UK to Republic of Ireland 21.1.05)*

PART 2 – AIRCRAFT REGISTERED AND CANCELLED IN 2004

Registration	Type	Construction No	Previous Identity	Date	Registerd Owner*(Operator)* Cancellation Details
EI-DFN	Airbus A.320-211	0204	F-GJVC	.04R	(Debis Celtavia 5 Ltd) *(for ops by Windjet)* *(EI- marks carried but then returned to former marks as F-GJVC)*
EI-DFU	Airbus A.320-233	1892	F-WQTJ N892VX, F-WWBU	28.5.04	Debis Celtavia 5 Ltd *(For Wizz Air KfT as HA-LPE and cancelled 9.7.04)*
EI-DGB	Airbus A.320-233	1902	F-WQTK N902VX, F-WWDG	21.5.04	Debis Celtavia 4 Ltd *(For Wizz Air KfT as HA-LPD and cancelled 18.7.04)*
EI-DGC	Airbus A.320-233	1834	F-WQTI N834VX, F-WWBF	28.5.04	Debis Celtavia 6 Ltd *(For Wizz Air KfT as HA-LPF and cancelled 28.7.04)*

SECTION 3

PART 1 – MUSEUMS AND PRIVATE COLLECTIONS

The sole qualification for inclusion in this Section is that all entries were originally allocated either United Kingdom and Ireland registrations, BGA and BAPC allocations and remain in existence. All registrations are now officially cancelled and details are shown. Entries should be usually accessible to the public, in some cases by prior permission, being Gate Guardians or located within Museums and private collections. Some are held in Museum stores and may not be available for viewing: these are annotated. Some Museums and private collections also hold specimens which retain current Certificates of Registration and remain pertinent to SECTION 1, these are annotated.

The Section has been expanded this year to cater for the numerous and formerly registered United Kingdom and Ireland aircraft still to be found elsewhere in Museums and collections world-wide but which have not featured in this annual Registers for many years, if at all. As such, entries are located under their original G- and EI- registration and details of subsequent markings now carried. In extending this Section I am grateful to Bob Ogden's series of "Aircraft Museums and Collections of the World" books, his Air-Britain News Section and Peter Hornfecks's "70 Years of the Irish Civil Aircraft Register". Other updating details have been taken from Museum websites Credit 2005 Bryan Foster

Regn	Type	C/n	P/I	Date	Remarks	CA Expiry
ENGLAND						
BEDFORDSHIRE						
Stondon Transport Museum, Lower Stondon SG16 JN (www.transportmuseum.co.uk)						
G-AXOM	Penn-Smith Gyroplane (Volkswagen 1600)	DJPS.1		26. 9.69	Cancelled 11.10.74 as WFU	24. 2.71P
BAPC.77	Mignet HM.14 Pou-Du-Ciel (Citroën 425cc) *(Modern reproduction)*	---			*(As "G-ADRG")*	
Richard Shuttleworth Trustees, Old Warden SG18 9EP (www.shuttleworth.org)						
BAPC.1	Roe Triplane IV reconstruction	---			See G-ARSG in SECTION 1 Part 2	
BAPC.2	Bristol Boxkite reconstruction	---			See G-ASPP in SECTION 1,Part 2	
BAPC.3	Bleriot Type XI	---			See G-AANG (2) in SECTION 1 Part 2	
BAPC.4	Deperdussin Monoplane	---			See G-AANH (2) in SECTION 1 Part 2	
BAPC.5	Blackburn Monoplane	---			See G-AANI (2) in SECTION 1 Part 2	
BAPC.8	Dixon Ornithopter reconstruction	---				
BAPC.11	English Electric Wren composite	---			See G-EBNV in SECTION 1 Part 2	
BAPC.271	Messerschmitt Me 163B Komet fsm (Walter HWK 509A-2 rocket motor)	---			*(As "191454") (Wingless)*	
G-BSSY	CSS-13 Aeroklubowy *(Licence built Polikarpov Po-2?) (The quoted c/n is suspect)*	0094	YU-CLJ	6.11.90	Cancelled 23.6.94 by CAA - sold as N588NB 7.94 *(As fictitious "ZK-POZ" 7.03)*	
BERKSHIRE						
British Balloon Museum and Library, Newbury						
G-ATGN	Thorn K-800 Coal Gas Balloon	2		12. 7.65	Cancelled 23.6.81 as WFU *"Eccles"*	
G-ATXR	Abingdon Gas/HAB	A.F.B.1		22. 7.66	Cancelled 14.7.86 by CAA *(Basket only)*	1. 9.76
G-AVTL	Brighton Ax7-65 HAB *(Originally regd as Hot-Air Group ¼ FB with c/n 1)*	01		17. 8.67	Cancelled 11.9.81 as WFU	
G-AWCR	Piccard Ax6 HAB	6204		29. 1.68	Cancelled 24.5.78 as WFU *"London Pride 1"*	
G-AWJB	Brighton MAB-65 HAB	MAB-3		3. 5.68	Cancelled 4.12.70 on sale to Italy *(To HB-BOU (1) 2.73) (As "HB-BOU")*	
G-AWMO	Omega O-84 HAB	01		31. 7.68	Cancelled 13.5.69 *(To OY-BOB 5.69) "Blue Strike"*	
G-AWOK	Sussex Gas (Free) Balloon	SARD.1		7. 8.68	Cancelled 29.2.84 as WFU *(Withdrawn 1970) "Sardinia"*	
G-AXMD	Omega O-20 HAB *(Acquired second envelope c/n 07 c.1969/70 but not known which one BBML holds)*	06		7. 8.69	Cancelled 7.12.89 as WFU *"Nimble"*	
G-AXVU	Omega 84 HAB	09		7. 1.70	Cancelled 22.8.89 as WFU *"Henry VIII"*	28. 4.77
G-AXXP	Bradshaw HAB-76 (Ax7) HAB	RB.001		20. 2.70	Cancelled 9.9.81 *(WFU 2.77) "Ignis Volens"*	
G-AYAJ	Cameron O-84 HAB	11		31. 3.70	Cancelled 1.2.90 as WFU *"Flaming Pearl"*	
G-AYAL	Omega 56 HAB	10		2. 4.70	Cancelled 18.10.84 by CAA *"Nimble II"*	25. 8.76
G-AYJZ	Cameron (Ax8) O-84 HAB *(Original canopy replaced by c/n 433)*	16		30. 9.70	Cancelled 21.5.75 - to EI-BAY 5.75 *"Godolphin" (As "EI-BAY" (1))*	
G-AZBH	Cameron O-84 HAB	23		8. 7.71	Cancelled 30.8.85 as WFU *"Serendipity"*	10. 5.81
G-AZER	Cameron O-42 (Ax5) HAB	26		9. 9.71	Cancelled 25.3.92 by CAA *"Shy Tot"*	15. 5.81A
G-AZJI	Western O-65 HAB	007		2.12.71	Cancelled 19.5.93 by CAA *"Peek-A-Boo"*	
G-AZSP	Cameron O-84 HAB	43		18. 4.72	Cancelled 11.1.82 as WFU *"Esso"*	22. 3.82
G-AZUV	Cameron O-65 HAB	41		12. 5.72	Cancelled 6.1.82 *"Icarus" (Damaged and WFU Rendham Green, Suffolk)*	23. 6.83
G-AZUW	Cameron A-140 HAB	45		12. 5.72	Cancelled 7.6.73 - to F-WTVO 6.73 *"Cumulonimbus" (Subsequently F-BTVO, 5Y-SIL)*	
G-AZYL	Portslade School HAB	MK17		10. 7.72	Cancelled 25.4.85 as WFU	
G-BAMK	Cameron D-96 HA Airship	72		11. 1.73	Cancelled 16.8.00 by CAA *"Isibidbi"*	24. 4.90A
G-BAVU	Cameron A-105 HAB	66		11. 4.73	Cancelled 6.12.01 by CAA	5.10.84A
G-BAXF	Cameron O-77 HAB	74		3. 5.73	Cancelled 5.9.95 by CAA *"Granna"*	
G-BAXK	Thunder Ax7-77 HAB	005		9. 5.73	Cancelled 7.9.01 as WFU *"Jack O'Newbury"*	2. 7.91A
G-BBFS	Van Den Bemden K-460 (Gas) FB	VDB-16	OO-BGX	10. 8.73	Cancelled 19.5.93 by CAA *"Le Tomate"*	
G-BBLL	Cameron O-84 HAB	84		2.10.73	Cancelled 19.5.93 by CAA *"Boadicea"*	25. 5.81A
G-BBYU	Cameron O-56 HAB	96		19. 2.74	Cancelled 9.8.89 as WFU *"Chieftain"*	28. 2.82A
G-BCAR	Thunder Ax7-77 HAB	019		5. 3.74	Cancelled 2.4.92 by CAA *"Marie Antoinette"*	
G-BCFD	West Ax3-15 HAB	JW.1		16. 5.74	Cancelled 30.1.87 by CAA *"Hellfire"*	
G-BCFE	Byrne Odyssey 4000 MLB	AJB-2		20. 5.74	Cancelled 19.9.85 as WFU *"Odyssey"*	
G-BCGP	Gazebo Ax6-65 HAB	1		13. 6.74	Cancelled 18.12.79 as WFU *"Aries"*	
G-BDVG	Thunder Ax6-56A HAB	067		2. 4.76	Cancelled 3.4.92 by CAA *"Argonaut"*	
G-BEEE	Thunder Ax6-56A HAB	070		20. 8.76	Cancelled 19.5.93 by CAA *"Avia"*	11. 5.84A
G-BEPO	Cameron N-77 HAB	279		1. 4.77	Cancelled 14.5.98 as WFU *"Sungas"*	

Reg	Type	C/n	Prev id	Regd	Fate / Remarks	CofA
G-BEPZ	Cameron D-96 HA Airship	300		13. 4.77	Cancelled 28.4.94 as WFU *"Zanussi"*	12. 2.90A
					(Damaged Warren Farm, Savernake Forest 8.1.94 and DBR during recovery)	
G-BETF	Cameron Champion 35SS HAB	280		17. 5.77	Cancelled 24.1.92 as WFU	6. 4.84A
	(Champion Spark Plug shape)				*"Champion"*	
G-BETH	Thunder Ax6-56A HAB	113		27. 5.77	Cancelled 11.5.93 as WFU *"Debenhams"*	31. 5.78
G-BEVI (3)	Thunder Ax7-77A HAB	125		30. 5.77	Cancelled 8.1.92 as WFU *"Prime Bang"*	
G-BFAB	Cameron N-56 HAB	297		15. 8.77	Cancelled 21.4.92 by CAA *"Phonogram" (On loan from A.Gibson)*	
G-BFOZ	Thunder Ax6-56 Plug HAB	144		20. 3.78	Cancelled 16.4.92 by CAA *"Motorway" (Current status unknown)*	
G-BGAS	Colting Ax8-105A HAB	001		27. 6.78	Cancelled *(Destroyed Flims, Switzerland 20.9.80) (Basket only)*	
G-BGOO	Colt Flame 56SS HAB	039		27. 4.79	Cancelled 19.5.93 by CAA	
	(Smiling Flame shape)				*"Mr Gas"*	
G-BGPF	Thunder Ax6-56Z HAB	206		13. 7.79	Cancelled 21.11.89 as WFU	27. 6.82A
					"Pepsi" (On loan from P.J.Bish)	
G-BHKN	Colt 14A Cloudhopper HAB	068		17. 1.80	Cancelled 5.12.89 as WFU	
	(Officially regd as Colt 12A)				*"Green Ice 2"*	
G-BHKR	Colt 14A Cloudhopper HAB	071		17. 1.80	Cancelled 5.12.89 as WFU	
	(Officially regd as Colt 12A)				*"Green Ice 5"*	
G-BIAZ	Cameron AT-165 (Helium/Hot-Air) FB	400		7. 2.78	Cancelled 27.5.80 *"Zanussi"*	31.10.78
	(Used for 1978 Atlantic attempt)				*(Hot Air envelope destroyed Trubenbuch, Austria 14.1.80) (Inner helium cell envelope only)*	
G-BIDV	Colt 17A Cloudhopper HAB	789		29. 1.79	Cancelled 20.5.93 by CAA	
	(Originally. was Colt 14A c/n 034)				*"Smirnoff*	
G-BIGT	Colt 77A HAB	078		28. 2.80	Cancelled 4.2.87 by CAA	20. 2.83A
					"Big T" (Damaged Belton Hall, Grantham 23.8.81)	
G-BKES	Cameron Bottle 57 SS HAB	846		25. 6.82	Cancelled 1.5.90 by CAA	
	(Robinsons Barley Water Bottle)				*"Robinsons Barley Water*	
G-BKMR	Thunder Ax3 Maxi Sky Chariot HAB	497		12. 1.83	Cancelled 23.4.98 as WFU *"The Weasel"*	
G-BLIO	Cameron R-42 Gas/HAB	1015		17. 4.84	Cancelled 24.1.90 as destroyed	17. 5.84P
G-BMEZ	Cameron DP-70 HA Airship	1130		18. 9.85	Cancelled 20.6.91	4. 5.89A
	(Originally regd as D-50)				*(Sold as EC-FUS) (Envelope only)*	
G-BNHN	Colt Ariel Bottle SS HAB	1045		30. 3.87	Cancelled 24.1.92 as WFU *"Ariel"*	
G-BOGR	Colt 180A HAB	1183		11. 5.88	Cancelled 28.4.97 as WFU *"Britannia"*	13. 3.92T
G-BOTL	Colt 42A SS HAB	466		23.11.82	Cancelled 21.11.89 as WFU *"Bottle"*	
G-BPKN	Colt AS-80 Mk.II HA Airship	1297		11. 1.89	Cancelled 7.1.91 by CAA *"Fuji"*	14. 3.91A
G-BPLD	Thunder and Colt AS-261 HA Airship	1380		25. 1.89	Cancelled13.6.91 - to F-WGGM / F-GHRI 5.91R *(As F-WGGM)*	
G-BRZC	Cameron N-90 HAB	2227		8. 2.90	Cancelled 29.4.97 as WFU *"Unipart II"*	2.12.92A
G-BUBL	Thunder Ax8-105 HAB	1147		10.12.87	Cancelled 16.6.98 as WFU *"Mercier / l'Espit D'Adventure"*	
G-BUUU	Cameron Bottle 77SS HAB	2980		11. 2.93	Cancelled 22.10.01 by CAA	4. 3.94A
	(Bells Whisky Bottle shape)				*"Bells Whisky"*	
G-BVBX	Cameron N-90M HAB	3102		10. 8.93	Cancelled 10.2.97 as temporary WFU *"Mercury"*	*27. 9.95A*
G-CHUB	Colt Cylinder Two N-51 HAB	1720		11. 4.90	Cancelled 12.12.01 as WFU	19.12.95A
	(Fire Extinguisher shape)				*"Chubb Fire Extinguisher"*	
G-FTFT	Colt Financial Times 90SS HAB	1163		14. 1.88	Cancelled 13.5.98 as WFU	5. 6.95A
					"Financial Times" (On loan from Financial Times Ltd)	
G-FZZZ	Colt 56A HAB	507		23. 2.83	Cancelled 29.4.97 as WFU *"Alka Seltzer 1"*	
G-LCIO	Colt 240A HAB	1381		23. 1.89	Cancelled 25.5.94 as WFU *"Star Flyer 2"*	
					(Damaged landing after first overflight Mt Everest by HAB 21.10.91)	
G-LOAG	Cameron N-77 HAB	359		10.11.77	Cancelled 31.3.93 as destroyed	6. 4.84A
					"Famous Grouse" (Envelope only)	
G-OBUD	Colt 69A HAB	698		26. 6.85	Cancelled 29.4.97 as WFU *"Budweiser"*	1. 2.90A
G-OFIZ	Cameron Can 80SS HAB	2106		30.10.89	Cancelled 10.2.97 as temporary WFU *"Andrews Can"*	2.12.91A
G-OLLI	Cameron O-31 HAB				See SECTION 1, Part 2	
G-PARR	Colt Bottle 90SS HAB	1953		15. 3.91	Cancelled 10.2.97 as temporary WFU	29. 9.94A
	(Old Parr Whisky bottle shape)				*"Old Parr"*	
G-PERR	Cameron Bottle 60SS HAB	699		28. 1.81	Cancelled 24.1.92 as WFU *"Perrier"*	3. 6.84A
G-PLUG	Colt 105A HAB	1958		17. 4.91	Cancelled 23.7.96 by CAA	14. 8.95T
G-PUBS	Colt Beer Glass 56SS HAB	037		7. 6.79	Cancelled 1.12.95 by CAA	
G-ZUMP	Cameron N-77 HAB	377		18. 1.78	Cancelled 8.4.98 as WFU	
	(Rebuilt 1985 with new canopy c/n 1107)				*"Gazump"*	
BAPC.258	Adams RFD-GQ Balloon (5,000 cu.ft)	Not known				
	(Built RFD-GQ Parachutes)					

Museum of Berkshire Aviation/Royal Berkshire Aviation Society and The Herald Society (+) ,Woodley RG5 4UF

Reg	Type	C/n	Prev id	Regd	Fate / Remarks	CofA
G-AJJP	Fairey FB.2 Jet Gyrodyne	F.9420 & FB.2		1. 3.47	Cancelled 9.11.50 - to RAF as XD759 @ 11.50	
					(As "XJ389")	
G-AKKY	Miles M.14A Hawk Trainer 3	2078	T9841	23. 6.48	Cancelled 12.4.73 as WFU	6.11.64
	(Also allocated BAPC.44 to reflect rebuild status from various parts)				*(WFU 11.60) (As "L6906")*	
G-APLK	Miles M.100 Student 2	100/1008		11. 3.58	Cancelled 31.8.84 to G-MIOO - see SECTION 1 Part 2	
G-APWA	Handley Page HPR.7 Dart Herald 100 **(+)**	149	PP-SDM	28. 9.59	Cancelled 29.1.87 as WFU	6. 4.82T
			G-APWA, PP-SDM, PP-ASV, G-APWA *(BEA titles)*			
BAPC.233	Broburn Wanderlust sailplane	---			*Built 1946*	
BAPC.248	McBroom Hang Glider	---			*Built 1974*	

BRISTOL

City Museum and Art Gallery, Clifton BS8 1RL (www.bristol-city.gov.uk/museums)

Reg	Type	C/n	Prev id	Regd	Fate / Remarks	CofA
BAPC.40	Bristol Boxkite reconstruction (Gnome)	BOX 3 & BM.7281			*(Built for "Those Magnificent Men in Their Flying Machines" film)*	

Bristol Aero Collection, Filton ***(also see Gloucestershire/Kemble)***

Reg	Type	C/n	Prev id	Regd	Fate / Remarks	CofA
G-BOAF	British Aircraft Corporation-Aérospatiale Concorde 102				Cancelled 4.5.04 as WFU	11. 6.04T

CAMBRIDGESHIRE

Imperial War Museum, Duxford CB2 4QR (www.iwm.org.uk)

Reg	Type	C/n	Previous identity	Regd	Fate	
G-ACUU	Cierva C.30A	726	(G-AIXE)	26. 6.34	Cancelled 14.11.88 as WFU	30. 4.60
	(Avro 671) (AS Civet)		HM580, G-ACUU		*(WFU 4.60) (As "HM580/KX-K")*	
G-AFBS	Miles M.14A Hawk Trainer 3	539	(G-AKKU)	17. 9.37	Cancelled 22.12.95 by CAA	25. 2.63
			BB661, G-AFBS		*(Dismantled and unmarked 4.03)*	
G-AHTW	Airspeed AS.40 Oxford 1	3083	V3388	6. 6.46	Cancelled 3.4.89 by CAA *(As "V3388")*	15.12.60
G-ALCK	Percival P.34A Proctor III	H.536	LZ766	18. 6.48	WFU *(As "LZ766")*	19. 6.63
G-AMDA	Avro 652A Anson 1	---	N4877	20. 7.50	Cancelled 9.9.81 by CAA *(As "N4877/MK-V")*	14.12.62
G-ASKC	de Havilland DH.98 Mosquito TT.35	---	TA719	8. 7.63	Crashed 27.7.64 *(As "TA719/6 T")*	18. 1.64
G-BCYK	Avro (Canada) CF-100 Canuck Mk.IV	---	RCAF 18393	18. 3.75	Cancelled 15.9.81 as WFU	
					(As "18393" in RCAF c/s)	
G-BEDV	Vickers 668 Varsity T.1	---	WJ945	26. 7.76	Cancelled 15.6.89 by CAA *(As "WJ945/21")*	15.10.87P
G-BESY	British Aircraft Corporation 167 Strikemaster Mk.80A		G-27-299	26. 4.77	Cancelled 7.77	
	(Officially regd as Mk.88)	PS.364	Saudi AF 1133, G-27-299		*(As "1133" in Saudi c/s)*	
G-LANC	Avro 683 Lancaster B.X	---	RCAF KB889	31. 1.85	Cancelled 2.9.91 by CAA	
	(Built Victory Aircraft, Canada)				*(As "KB889/NA-I" in 428 Sqdn c/s)*	
G-LIZY	Westland Lysander III	"504/39"	RCAF 1558	20. 6.86	Cancelled 18.4.89 as WFU	
	(C/n also quoted as "Y1351")		V9300		*(As "V9673/MA-J" in 161 Sqdn c/s)*	
G-USUK	Colt 2500A HAB	1100		1. 6.87	Cancelled 21.8.90 as WFU *"Virgin Atlantic Flyer"*	19. 8.87P
					(On loan from Virgin Atlantic Airways Ltd) (Gondola displayed - remainder stored)	
EI-AUY	Morane-Saulnier MS.502 Criquet	338	F-BCDG	30.11.70	Cancellation details not known	
	(Argus AS.10)		Fr.Mil		*(On loan from G Warner) (As "CF+HF" in Luftwaffe c/s)*	
BAPC.90	Colditz Cock rep	---			*(Built for BBC "The Colditz Story" film)*	
BAPC.93	Fieseler Fi 103 (V-1)	---			*(BAPC identity unconfirmed)*	
BAPC.209	Supermarine Spitfire LF.IXC fsm	---			*(Built for "Piece of Cake" TV series)(As "MJ751/DU-V" in 321 Sqdn c/s)*	
BAPC.267	Hawker Hurricane fsm	---			*(As "R4115/LE-X" in 243 Sqdn c/s)*	

American Air Museum, Duxford

Reg	Type	C/n	Previous identity	Regd	Fate	
G-BFYO	SPAD XIII rep	0035	D-EOWM	16.11.78	Cancelled 14.10.86 as WFU	21. 6.82P
	(Built Williams Flugzeugbau) (Lycoming AIO-360)				*(As "1 4513 S" in 3rd Escadrille French AF c/s)*	
G-BHDK	Boeing TB-29A-45-BN Superfortress	11225	44-61748	27. 9.79	Cancelled 29.2.84 as WFU *"Hawg Wild"*	
					(As "461748/Y" in USAF c/s)	
G-BHUB	Douglas C-47A-85DL Dakota	19975	"G-AGIV"	30. 4.80	Cancelled 19.10.81 as WFU	
			Spanish AF T3-29, N51V, N9985F, SE-BBH, 43-15509 *(As "315509/W7-S" in USAAF c/s)*			
BAPC.255	North American P-51D Mustang fsm	---			*(As "463209/WZ-S" in 78th FG c/s)*	
	(Built Rialto, CA, USA 1990)					

Duxford Aviation Society, Duxford

Reg	Type	C/n	Previous identity	Regd	Fate	
G-ALDG	Handley Page HP.81 Hermes IV	HP.81/8		27.10.49	WFU 9.62	9. 1.63
					(BOAC titles) "Horsa" (Unmarked fuselage only)	
G-ALFU	de Havilland DH.104 Dove 6	04234		14.12.48	Cancelled 14.11.72 as WFU	4. 6.71
G-ALWF	Vickers 701 Viscount	5		2. 1.50	Cancelled 18.4.72 as WFU	16. 4.72
G-ALZO (2)	Airspeed AS.57 Ambassador 2	5226	R Jordan AF 108	5. 4.50	Cancelled 10.9.81 as WFU	14. 5.71
			G-ALZO, (G-AMAD)		*(Unmarked 4.03)*	
G-ANTK	Avro 685 York C.1	Not known	MW232	23. 7.54	WFU Lasham 30.4.64 *(Dan Air titles)*	29.10.64T
G-AOVT	Bristol 175 Britannia 312	13427		23. 6.58	Cancelled 21.9.81 as WFU *(Monarch titles)*	11. 3.75T
G-APDB	de Havilland DH.106 Comet 4	6403	9M-AOB	2. 5.57	Cancelled 18.2.74 as WFU *(Dan-Air titles)*	7.10.74
			G-APDB			
G-APWJ	Handley Page HPR.7 Dart Herald 201	158		28. 9.59	Cancelled 10.7.85 as WFU *(Air UK titles)*	21.12.85
G-ASGC	Vickers Super VC-10 Srs.1151	853		11. 4.63	WFU 15.4.80 *(BOAC-Cunard titles)*	20. 4.80
G-AVFB	Hawker Siddeley HS.121 Trident 2E	2141	5B-DAC	1. 2.67	Cancelled 9.7.82	30. 9.82
			G-AVFB		*(WFU 27.3.82) (BEA titles)*	
G-AVMU	British Aircraft Corporation One-Eleven 510ED	BAC.148		11. 5.67	Cancelled 12.7.93 as WFU	8. 1.95T
					(British Airways titles) "County of Dorset"	
G-AXDN	British Aircraft Corporation-Aérospatiale Concorde	13522 & 01		16. 4.69	Cancelled 10.11.86 as WFU	30. 9.77
G-OPAS	Vickers 806 Viscount	263	G-AOYN	5.10.94	Cancelled 28.7.97 as destroyed	26. 3.97T
					(WFU 6.96 Southend and broken up) (Parcelforce titles) (Nose only)	

CHESHIRE

Hooton Park Trust and Griffin Trust (&), Hooton Park L65 1BQ

Reg	Type	C/n	Previous identity	Regd	Fate	
G-AGPG	Avro 652A Anson 19 Srs.2	1212		15. 6.45	Cancelled 5.11.75	13. 2.71
	(Originally regd as Anson XII, to Anson XIX - 1.47 and to Avro 19 Series.2 - 5.52)				*(On loan from The Aeroplane Collection)*	
G-AJEB	Auster J/1N Alpha	2325		14. 3.47	Cancelled 9.6.81 as WFU	27. 3.69
					(On loan from The Aeroplane Collection)	
BAPC.68	Hawker Hurricane fsm (**&**)	---	"P3975"		*(Built for "Battle of Britain" film)*	
BAPC.204	McBroom Hang Glider	---				

CORNWALL

Land's End Theme Park, Land's End TR19 7AA

Reg	Type	C/n	Previous identity	Regd	Fate	
G-BCXO	MBB Bö.105DD	S.80	D-HDCE	27. 2.75	Cancelled 4.3.92 as WFU	
	(C/n S.80 is the original pod, replaced @1992 and subsequently rebuilt as display piece "G-CDBS")					
BAPC.137	Sopwith Baby Floatplane rep				*(ex Leisure Sport)*	
	(Built FEM Displays Ltd 1978)					

CUMBRIA

Solway Aviation Museum and Edward Haughey Aviation Heritage Centre, Carlisle CA6 4NW (www.solway-aviation-museum.org.uk)

Reg	Type	C/n	Previous identity	Regd	Fate	
G-APLG	Auster J/5L Aiglet Trainer	3148		4. 3.58	Cancelled 11.2.99 by CAA	26.10.68
G-AYFA	Scottish Aviation Twin Pioneer Mk.3	538	G-31-15	15. 6.70	Cancelled 16.5.91 as WFU	24. 5.82
	(Originally regd as a CC.2)		XM285		*(Nose only)*	
G-BJWY	Sikorsky S-55 (HRS-2) Whirlwind HAR.21	55???	A2576	25. 1.82	Cancelled 23.2.94 by CAA	
			WV198, Bu.130191		*(As "WV198/K") (On loan from D.Charles)*	

RAF Millom Museum, Haverigg LA18 4NA

BAPC.231 Mignet HM.14 Pou-Du-Ciel --- *(On loan from South Copeland Aviation Group)*
(Thought originally built @ Ulverston 1936 with Anzani engine) *(As "G-ADRX")*
BAPC.260 Mignet HM.280 ---

Windermere Steamboat Centre, Windermere LA23 1BN (www.steamboat.co.uk)

BGA.266 Slingsby T.1 Falcon 1 Waterglider 237A 29. 5.36

DEVON

Trago Mills Shopping Mall, Newton Abbot

G-BDDX Whittaker MW2B Excalibur PFA 041-10106 28. 5.75 WFU 1976
(Built M W Whittaker - c/n 001) (Volkswagen 1500)

DORSET

Bournemouth Aviation Museum, Bournemouth BH23 6SE (www.aviation-museum.co.uk)

G-AVMN British Aircraft Corporation One-Eleven 510ED See SECTION 1 Part 2
G-BEYF Handley Page HPR.7 Dart Herald 401 175 FM1022 13. 7.77 Cancelled 18.11.99 as WFU *(On loan from Dart Group plc)* 11. 3.01T
G-BRFC Percival P.57 Sea Prince T.1 P57/71 WP321 10. 9.80 Cancelled 12.11.99 - to N7SY 11.99
G-BRNM Chichester-Miles Leopard 002 17.10.89 Cancelled 31.1.05 as WFU
G-BWAF Hawker Hunter F.6A See SECTION 1 Part 2
G-EGHH Hawker Hunter F.58 See SECTION 1 Part 2
G-HELV de Havilland DH.115 Vampire T.55 See SECTION 1 Part 2
G-NATY Folland Gnat T.1 FL.548 8642M 19. 6.90 Cancelled 23.8.02 by CAA - no UK PtoF issued
XR537 *(As "XR537/T")*

ESSEX

Aces High Flying Museum, North Weald CM16 6AA

G-AMSN Douglas C-47B-35DK Dakota IV 16631/33379 N3455 28. 4.52 Cancelled 25.1.00 as WFU 3. 1.68
G-AMSN, EI-BSI, SU-BFZ, G-AMSN, KN673, 44-77047 *(Dismantled 9.02)*
(G-BKXW) North American NA.82 B-25J Mitchell 108-35186 "HD368" 83R NTU and remained as N9089Z *"Bedsheet Bomber"*
N9089Z, "N908", N9089Z, 44-30861 *(As "430861" in USAAF c/s)*

GLOUCESTERSHIRE

Jet Age Museum, Gloucestershire

BAPC.72 Hawker Hurricane fsm *(Built for "Battle of Britain" film)*
(As "V6799/SD-X" of 501 RAAF Sqdn c/s)
BAPC.259 Gloster Gamecock *(Under construction 2.04)*

Bristol Aero Collection, Kemble GL7 6BA(www.bristolaero.com) ***(To move to Bristol/Filton 2005)***

G-ALBN Bristol 173 Mk.1 12871 7648M 22. 7.48 Cancelled - to RAF as XF785 in 1953
XF785
G-ALRX Bristol 175 Britannia Srs.101 12874 (WB473) 25. 6.51 Cancelled 5.4.54 as Withdrawn
(VX447) *(DBR landing Littleton-upon-Severn 4.2.54)*
(Nose only - on loan from Britannia Aircraft Preservation Trust)
G-ANCF Bristol 175 Britannia Srs.308F 12922 5Y-AZP 3. 1.58 Cancelled 21.2.84 as WFU 12. 1.81
(Originally regd as Srs.305) G-ANCF, LV-GJB, LV-PPJ, (G-ANCF), G-14-1, G-18-4, G-ANCF, (N6597C)
(On loan from Britannia Aircraft Preservation Trust) (Fuselage only)
G-ARRM Beagle B.206X B.001 23. 6.61 WFU 1965? and cancelled 9.4.74 as PWFU 23.12.64
(Originally regd as Beagle B.2 Srs.1 [B2/1010]: re-designated 9.61)
G-ATDD Beagle B.206 Srs.1 B.013 (VH-...) 27. 4.65 U/c collapsed Sherburn 6.73 and cancelled
(Originally regd as Beagle B.206R - to Srs.1 1966) G-ATDD *(Nose only)*
BAPC.87 Bristol 30/46 Babe III reconstruction 1 *(Built W.Sneesby) (As "G-EASQ")*

Britannia Aircraft Preservation Trust, Kemble

G-BDUP Bristol 175 Britannia Srs.253 13508 EL-WXA 31. 3.76 Cancelled 9.8.84 to CU-T120 @ 8.84).
CU-T120, G-BDUP, XM496 *(As "XM496" in RAF c/s)*

HAMPSHIRE

Farnborough Air Sciences Trust (FAST Museum) Farnborough GU16 6DH

G-AWZI* Hawker Siddeley HS.121 Trident 3B Srs.101 2310 14. 1.69 Cancelled 9.7.87 as destroyed 5. 8.85T
(WFU 1.5.85: cockpit only)
G-BRAM Mikoyan MiG-21PF Not known Hungarian AF 503 22. 5.89 Cancelled 16.4.99 by CAA
(As "503" in Russian AF c/s)

Whittle Memorial, Ively, Farnborough

BAPC.285 Gloster E28/39 fsm ---
(Built Sir Frank Whittle Commemorative Trust)

Prince's Mead Shopping Centre, Farnborough

BAPC.208 Royal Aircraft Factory SE.5a --- *(As "D276/A")*
(Built AJD Engineering)

Second World War Aircraft Preservation Society, Lasham

G-APXX de Havilland DHA.3 Drover 2 5014 VH-EAS 15.12.59 Cancelled 26.11.73 as WFU - *regn* not taken up
VH-EAZ *(As "VH-FDT")*
G-APIT Percival P.40 Prentice T.1 PAC/016 VR192 28.11.57 Cancelled 8.11.79 as WFU *(As "VR192")* 7. 9.67

Museum of Army Flying, AAC Middle Wallop SO20 8DY (www.flying-museum.org.uk)

G-ABOX (2) Sopwith Pup See SECTION 1 Part 2
G-AKKR Miles M.14A Hawk Trainer 3 1995 "T9967" 23. 6.48 Cancelled 16.7.69 as PWFU 10. 4.65
(May be T9967 [2160] from 1943 rebuild) 8378M, G-AKKR, T9708 *(As "T9707")*
G-AKOW Taylorcraft J Auster 5 1579 PH-NAD (2) 23.12.47 Cancelled 5.8.87 as WFU 26. 6.82
(Regd as c/n TJ569A after rebuild in Holland) PH-NEG, G-AKOW, TJ569 *(As "TJ569")*

Reg	Type	c/n	Previous identity	Regd	Fate / Notes	Date
G-APXW	Lancashire Aircraft EP-9 Prospector	43		22.12.59	Cancelled 20.5.82	22. 5.76
					(Composite rebuild ex G-APWZ and others) (As "XM819" in Army c/s)	
G-ARYD	Auster AOP.6	---	WJ358	8. 3.62	Cancelled 5.8.87 as WFU *(Conversion abandoned 9.63) (As "WJ358")*	
G-AXKS	Westland-Bell 47G-4A	WA.723	G-17-8	22. 7.69	Cancelled 22.4.82 as WFU	21. 9.82
BAPC.80	Airspeed AS.58 Horsa II	---			*(Composite from LH208, TL659 and 8569M) (As "KJ351")*	
BAPC.163	AFEE 10/42 Rotachute Rotabuggy reconstruction				*(On loan from Wessex Aviation Society) (As "B-415")*	
BAPC.185	WACO CG-4A Hadrian Glider	---			*(Fuselage only) (As "243809")*	
BAPC.261	General Aircraft Hotspur replica				*(As "HH379")*	
	(Composite from anonymous cockpit of Mk.1 and rear of Mk.II, HH379)					

Solent Sky, Southampton SO1 1FR (www.spitfireonline.co.uk)

Reg	Type	c/n	Previous identity	Regd	Fate / Notes	Date
G-ADWO	de Havilland DH.82A Tiger Moth	3455	BB807	9.12.35	Cancelled 15.9.58 as destroyed	
			G-ADWO		*(As "BB807")*	
	(Restored @ 3.51 and overhauled with fuselage of BB860 (ex G-ADXT): damaged landing Christchurch 31.7.58 and WFU: fuselage/parts ex G-AOAC and parts ex G-AOJJ used in composite rebuild 1987/90: completed to static condition 1990)					
G-ALZE	Britten-Norman BN-1F	1		16. 3.50	Cancelled 8.6.89 as WFU	
G-BRDV	Replica Viking Spitfire prototype	PFA 130-10796		3. 7.89	Cancelled 19.5.00 as WFU	18. 2.95P
	(Built Viking Wood Products - c/n HD36/001) (Jaguar V-12 350hp)				*(As "K5054" in RAF c/s) (On loan from Replica Spitfire Ltd)*	
G-SWIF	Supermarine 552 Swift F.7	VA.9597	XF114	1. 6.90	Cancelled 19.7.04 as WFU	
BAPC.7	Southampton University Man Powered Aircraft (SUMPAC)					
BAPC.164	Wight Quadraplane Type 1 fsm	---			*(As "N546")*	
BAPC.210	Avro 504J	---			*Built AJD Engineering*	
	(Gnome Monosoupape 100hp)				*(As "C4451")*	
BAPC.215	Airwave Hang Glide	---				
BAPC.253	Mignet HM.14 Pou-Du-Ciel rep	---			*(Built 1990s) (On loan from H.Shore) (As "G-ADZW")*	

HERTFORDSHIRE

Galleria Shopping Mall, Hatfield

Reg	Type	c/n	Previous identity	Regd	Fate / Notes	Date
BAPC.257	de Havilland DH.88 Comet fsm				*(As "G-ACSS") "Grosvenor House"*	

de Havilland Aircraft Heritage Centre, Salisbury Hall, London Colney AL2 1BU (www.dehavillandmuseum.co.uk)

Reg	Type	c/n	Previous identity	Regd	Fate / Notes	Date
G-ABLM	Cierva C.24	710		22. 4.31	Cancelled as WFU 12.34	16. 1.35
	(DH Gipsy III)				*(On loan from Science Museum)*	
G-ADOT	de Havilland DH.87B Hornet Moth	8027	X9326	?.11.35	Cancelled as WFU	15.10.59
			G-ADOT			
G-AMXR	de Havilland DH.104 Dove 6	04379	D-CFSB	21. 1.53	Cancelled 22.7.54 - to D-CFSB 7.54	
			G-AMXR, N4280V		*(Subsequently and as D-IFSB (1))*	
G-ANRX	de Havilland DH.82A Tiger Moth	3863	N6550	25. 5.54	WFU 20.6.61 *"Border City"*	20. 6.61
G-AOJT	de Havilland DH.106 Comet IXB	06020	F-BGNX	11. 5.56	Cancelled 9.7.56 as WFU	5. 7.56
					(Fuselage only) (As "F-BGNX" in Air France titles)	
G-AOTI	de Havilland DH.114 Heron 2D	14107	G-5-19	25. 7.56	Cancelled 17.10.95 as WFU *(Unmarked)*	24. 6.87T
G-AREA	de Havilland DH.104 Dove 8	04520		3. 8.60	Cancelled 19.9.00 by CAA	18. 9.87
G-ARYC	de Havilland DH.125 Srs.1	25003		1. 3.62	Cancelled 31.3.76 as WFU *(WFU 1.8.73)*	1. 8.73
G-AVFH	Hawker Siddeley HS.121 Trident 2E	2147		1. 2.67	Cancelled 12.5.82	18. 5.83T
					(WFU 24.10.81) (Forward fuselage only)	
G-AWJV	de Havilland DH.98 Mosquito TT.35	---	TA634	21. 5.68	Cancelled 19.10.70 as WFU *(As "TA634/8K-K" in 571 Sqdn c/s)*	
G-BBNC	de Havilland DHC-1 Chipmunk T.10	C1/0682	WP790	12.10.73	Cancelled 23.9.74 as WFU *(As "WP790/T")*	
G-BLKA	de Havilland DH.112 Venom FB.Mk.54 (FB.4)	960	(G-VENM (1))	13. 7.84	Cancelled 13.10.00 by CAA	14. 7.95P
	(Built F + W) (Officially regd as c/n 431)		Swiss AF J-1790		*(Unmarked pod only)*	
BAPC.186	de Havilland DH.82B Queen Bee composite	---	"K3584"		*(Original p/i not known) (As "LF789/R2-K")*	
BAPC.216	de Havilland DH.88 Comet fsm	---			*(As "G-ACSS")*	
BAPC.232	Airspeed AS.58 Horsa I/II Glider composite	---			*(Composite airframe from unidentified components)*	

KENT

Brenzett Aeronautical Museum Trust, Brenzett TN29 0EE

Reg	Type	c/n	Previous identity	Regd	Fate / Notes	Date
G-AMSM	Douglas C-47B-20-DK Dakota	15764/27209	KN274	28. 4.52	Cancelled 11.9.78 as WFU	
			43-49948		*(Ground-looped on take-off Lydd 17.8.78) (Nose only)*	

Dover Museum, Dover (www.dovermuseum.co.uk)

Reg	Type	c/n	Previous identity	Regd	Fate / Notes	Date
BAPC.290	Fieseler Fi.104 flying-bomb fsm					

National Battle of Britain Memorial, Capel Le Ferne, Folkestone (www.spitfire-museum.com)

Reg	Type	c/n	Previous identity	Regd	Fate / Notes	Date
BAPC.291	Hawker Hurricane I fsm				*(As "P2970/US-X" in 56 Sqdn c/s) "Little Willie"*	
	(Built GB Replicas)					
BAPC.299	Supermarine Spitfire I fsm				*(As "P9338" in 72 Sqdn c/s)*	
	(Built GB Replicas)					

Kent Battle of Britain Museum, Hawkinge CT18 7AG (www.kbobm.org)

Reg	Type	c/n	Previous identity	Regd	Fate / Notes	Date
BAPC.36	Fieseler Fi 103 (V-1) fsm	---			*(Built for "Operation Crossbow" film)*	
BAPC.63	Hawker Hurricane fsm	---	"L1592"		*(Built for "Battle of Britain" film) (As "P3208/SD-T" in 501 Sqdn c/s)*	
BAPC.64	Hawker Hurricane fsm	---			*(Built for "Battle of Britain" film) (As "P3059/SD-N" in 501 Sqdn c/s)*	
BAPC.65	Supermarine Spitfire fsm	---			*(Built for "Battle of Britain" film) (As "N3289/DW-K" in 610 Sqdn c/s)*	
BAPC.66	Messerschmitt Bf109 fsm	1480			*(Built for "Battle of Britain" film)*	
BAPC.67	Messerschmitt Bf109 fsm	---			*(Built for "Battle of Britain" film) (As "14" in JG52 c/s)*	
BAPC.69	Supermarine Spitfire fsm	---			*(Built for "Battle of Britain" film) (As "N3313/KL-B" in 54 Sqdn c/s)*	
BAPC.74	Messerschmitt Bf109 fsm	6357			*(Built for "Battle of Britain" film)*	
BAPC.133	Fokker Dr.1 fsm	---			*(As "425/17")*	
BAPC.272	Hawker Hurricane fsm	---			*(As "N2532/GZ-H" in 32 Sqdn c/s)*	
BAPC.273	Hawker Hurricane fsm	---			*(As "P2921/GZ-L" in 32 Sqdn c/s)*	
BAPC.278	Hawker Hurricane fsm	---			*(As "P3679/GZ-K" in 32 Sqdn c/s)*	
BAPC.297	Supermarine Spitfire reproduction	---				
BAPC.298	Supermarine Spitfire IX fsm	---				

Lashenden Air Warfare Museum, Headcorn TN27 9HX

Regn	Type	C/n	Previous identity	Regd	Fate/Status	CofA
BAPC.91	Fieseler Fi.103R-IV Reichenberg				*(Under restoration 4.03)*	

RAF Manston History Museum, Manston CT12 5DF (www.raf-manston.fsnet.co.uk)

Regn	Type	C/n	Previous identity	Regd	Fate/Status	CofA
G-AZCM	Beagle B.121 Pup Series .150	B121-155	G-35-155	30. 7.71	Cancelled 4.5.72 as sold abroad *(To HB-NAV 5.72)*	
G-BWJZ	de Havilland DHC-1Chipmunk 22	C1/0653	WK638	23.11.95	Cancelled as WFU 4.4.00	

Medway Aircraft Preservation Society, Rochester ME5 9TX

Regn	Type	C/n	Previous identity	Regd	Fate/Status	CofA
(G-ALSP)	Bristol 171 Sycamore 3	12900	17.11.50		Regn ntu and cancelled 26.3.52 - to RAF as WV783 @ 4.52 *(As "WV783")*	
G-36-1	Short SB.4 Sherpa *(Short & Harland Experimental & Research Aircraft)*	SH.1604	G-14-1		Cancelled - WFU 5.66 *(Fuselage only)*	

LANCASHIRE

Botany Bay Village, Chorley PR6 9AF

Regn	Type	C/n	Previous identity	Regd	Fate/Status	CofA
BAPC.176	Royal Aircraft Factory SE.5a scale model *(Built Slingsby Sailplanes from Currie Wot basic airframe)*	---			*(Built for "The Blue Max" film) (As "A4850")*	

Bygone Times Antique Warehouse, Eccleston PR7 5PD

Regn	Type	C/n	Previous identity	Regd	Fate/Status	CofA
G-BAYV	SNCAN 1101 Noralpha	193	F-BLTN French AF	22. 5.73	Cancelled 28.4.83 as WFU *(Crashed Longbridge Deverill 23.2.74) (On loan from P.Smith) (As "F-OTAN-6")*	2. 8.75

LEICESTERSHIRE

Snibston Discovery Park, Coalville LE67 3LN

Regn	Type	C/n	Previous identity	Regd	Fate/Status	CofA
G-AFTN	Taylorcraft Plus C2	102	HL535 G-AFTN	2. 5.39	Cancelled 13.1.99 by CAA	1.11.57
G-AGOS	Reid and Sigrist RS.4 Desford Trainer	3	VZ728 G-AGOS	?. 5.45	Cancelled 9.11.81 as WFU *(As "VZ728")*	28.11.80P
G-AIJK	Auster V J/4	2067		13.11.46	CofA expired and WFU	24. 8.68

Charnwood Museum, Loughboough LE11 3QU

Regn	Type	C/n	Previous identity	Regd	Fate/Status	CofA
G-AJRH	Auster J/1N Alpha	2606		12. 5.47	Cancelled 18.1.99 by CAA	5. 6.69

Whittle Memorial A426, Lutterworth

Regn	Type	C/n	Previous identity	Regd	Fate/Status	CofA
BAPC.284	Gloster E28/39 fsm *(Built Sir Frank Whittle Commemorative Trust)*	---				

East Midlands Aeropark, Nottingham East Midlands Airport

Regn	Type	C/n	Previous identity	Regd	Fate/Status	CofA
G-APES	Vickers 953C Vanguard Merchantman	721		9. 9.57	Cancelled 28.2.97 as WFU *(Nose only)* *"Swiftsure"*	2.10.95T
G-BAMH	Westland S-55 Whirlwind Srs.3	WA.83	VR-BEP G-BAMH, XG588	10. 1.73	Cancelled 31.10.73 - to VR-BEP 10.73 *(As "XG588 in SAR c/s)*	
G-BEOZ	Armstrong-Whitworth AW.650 Argosy 101	6660	N895U N6502R, G-1-7	28. 3.77	Cancelled 19.11.87 as WFU *(Elan titles) "Fat Albert"*	28. 5.86T
(G-BLMC)	Avro 698 Vulcan B.2A	---	XM575	R	Reservation @ 8.84 not taken up *(As "XM575")*	
G-FRJB	Britten SA-1 Sheriff	0001		18. 5.81	Cancelled 6.2.87 by CAA *(Not completed: unfinished airframe without marks)*	

Stanford Hall and Percy Pilcher Museum, Stanford Hall, Stanford LE17 6DH (www.stanfordhall.co.uk)

Regn	Type	C/n	Previous identity	Regd	Fate/Status	CofA
BAPC.45	Pilcher Hawk Glider reconstruction *(Built Armstrong-Whitworth Aviation apprentices 1957/58)*	---				

Thorpe Camp Visitor Centre, Woodhall Spa (www.thorpecamp.org.uk)

Regn	Type	C/n	Previous identity	Regd	Fate/Status	CofA
BAPC.294	Fairchild Argus					

LINCOLNSHIRE

Lightning Association, Binbrook LN8 6DR

Regn	Type	C/n	Previous identity	Regd	Fate/Status	CofA
G-BTSY	English Electric Lightning F.6	95207	XR724	25. 7.91	Cancelled 26.5.92 as TWFU - no Permit issued *(As "XR724")*	

Battle of Britain Memorial Flight, RAF Coningsby LN4 4SY

Regn	Type	C/n	Previous identity	Regd	Fate/Status	CofA
G-AISU	Vickers Supermarine 349 Spitfire LF.VB	CBAF.1061	AB910	25.10.46	Cancelled as transferred to Military Marks *(As "AB910/ZD-C" in 222 Sqdn c/s)*	
G-AMAU	Hawker Hurricane IIc *(12,780th and final Hurricane built)*	---	PZ865	1. 5.50	Cancelled 19.12.72 as transferred to Military Marks *(As "PZ865/Q" in RAFSEAC c/s)*	
G-AWIJ	Vickers Supermarine 329 Spitfire IIA	CBAF.1	P7350	25. 4.68	Cancelled 29.2.84 to MOD *"Blue Peter" (Returned to RAF) (As "P7350/XT-D" in 603 Sqdn c/s)*	

Royal Air Force Cranwell

Regn	Type	C/n	Previous identity	Regd	Fate/Status	CofA
BAPC.225	Supermarine Spitfire IX fsm	---			*(As "P8448/"UM-D" in 52 Sqdn c/s)*	

Royal Air Force Digby

Regn	Type	C/n	Previous identity	Regd	Fate/Status	CofA
BAPC.229	Supermarine Spitfire IX fsm	---	"L1096"		*(As "MJ832/DN-Y" in 416 Sqdn c/s) "City of Oshawa"*	

Lincolnshire Aviation Heritage Centre, East Kirkby (www.lincsaviation.co.uk)

Regn	Type	C/n	Previous identity	Regd	Fate/Status	CofA
G-ASXX	Avro 683 Lancaster B.VII *(As "NX611/LE-C" in 630 Sqdn c/s "City of Sheffield" [starboard] and "NX611/DX-C" in 57 Sqdn c/s "Just Jane" [port])*	---	(8375M) French Navy WU-15, NX611	22.10.64	Cancelled 16.2.79 as WFU	
BAPC.90	Colditz Cock rep	---			*(Built for BBC "The Colditz Story" film)*	

Bomber County Aviation Musem, Hemswell (www.lineone.net/~bcam)

Regn	Type	C/n	Previous identity	Regd	Fate/Status	CofA
G-AEJZ	Mignet HM.14 Pou-Du-Ciel *(Built T L Crosland)*	TLC.1	G-AEJZ	9. 6.36	Cancelled 31.12.38 in census *(Allocated "BAPC.120") (As "G-AEJZ") (Stored 3.04)*	

Aerial Application Collection, Wainfleet

G-BFBP	Piper PA-25-235 Pawnee D	25-7756033		7. 9.77	Cancelled 21.6.78 as destroyed	
					(Crashed near Comberton, Cambs 11.5.78) (Cockpit only)	
G-BFEY	Piper PA-25-235 Pawnee D	25-7756039		20.10.77	Cancelled 17.7.90 as WFU *(Fuselage frame only)*	19.1.87

GREATER LONDON

Royal Air Force Bentley Priory, HQ 11/18 Groups

BAPC.217	Supermarine Spitfire I fsm	---			*(As "K9926/JH-C" in 317 Sqdn c/s)*	
BAPC.218	Hawker Hurricane IIc fsm	---	"P3386"		*(As "BN230/FT-A"in 43 Sqdn c/s)*	

Royal Air Force Memorial Chapel, Biggin Hill

BAPC.219	Hawker Hurricane I fsm	---			*(As "L1710/AL-D" in 79 Sqdn c/s)*	
BAPC.220	Supermarine Spitfire I fsm	---			*(As "N3194/GR-Z" in 92 Sqdn c/s)*	

Croydon Airport Visitor Centre, Croydon (www.croydon.gov.uk/airport-soc/)

G-ANKV	de Havilland DH.82A Tiger Moth	84166	T7793	30.12.53	Cancelled 9.56 - not converted	
	(Provenance uncertain)				*(As "T7793" in RAF c/s)*	
G-ANUO	de Havilland DH.114 Heron 2D	14062		27. 9.54	Cancelled 9.8.96 as WFU	12. 9.86T
					(As "G-AOXL" in Morton Air Services c/s) (Note original G-AOXL below)	

British Airports Authority, Heathrow

G-BOAB	British Aircraft Corporation-Aérospatiale Concorde 102				Cancelled 4.5.04 as WFU	19. 9.01T

Royal Air Force Museum, Hendon NW9 5LL (www.rafmuseum.org.uk)

G-EBIC	Royal Aircraft Factory.SE.5A	688/2404	"B4563"	26. 9.23	Cancelled 31.12.38	3. 9.30
	(Wolseley Viper 200hp) *(Regd with c/n 687/2404)*		9208M, G-EBIC, F937		*(WFU 9.30) (As "F938")*	
G-EBJE	Avro 504K	927	(9205M)	??. 7.24	Cancelled ?.12.34	29. 9.34
	(Includes components of Avro 548A G-EBKN ex E449)				*(As "E449")*	
G-AAMX (2)	Moth Aircraft Corporation DH.60GM Moth	125	NC926M	11. 9.86	Cancelled 19.8.95 as WFU	7. 5.94P
	(DH Gipsy II)					
G-AANJ (2)	Luft-Verkehrs Gesellschaft C.VI	4503	9239M	29.10.81	Cancelled 11.12.03 as WFU	16. 5.03P
	(Benz @ 230hp)		C7198, 18, "1594", C7198, 18		*(As "7198/18" in German Air Force c/s)*	
	(Composite aircraft including parts from LVG 1594: captured 1916/17 and allotted RFC serial "XG7")					
G-ABBB	Bristol 105A Bulldog IIA	7446	"K2227"	12. 6.30	Cancelled 22.9.61 as PWFU	
			G-ABBB, R-11, G-ABBB		*(As "K2227")*	
G-ABMR	Hawker Hart	H.H-1	"J9933"	28. 5.31	Cancelled 2.2.59 - to "J9933" and later "J9941")	
			G-ABMR		*(As "J9941" in 57 Sqdn c/s)* 11. 6.57	
G-AETA	Caudron G.III	7487	OO-ELA	29. 1.37	Sold to RAF 1972	
	(Anzani 90hp) *(Also reported as c/n 5019 or 5021)*		O-BELA, (9203M)		*(As "3066" in RNAS c/s)*	
G-AFDX	Hanriot HD.1	"HD.1"	N75	4. 5.38	Cancellation details not known	
			OO-APJ, Belgian AF H-1, Belgian AF N75 *(As "HD-75")*			
	(DBR landing Old Warden 17.6.39: wings destroyed in air raid Brooklands 1940, fuselage survived and rebuilt 1968)					
G-AGYX	Douglas C-47A-10DK Dakota 3	12472	5N-ATA	15. 1.46	Cancelled 1.7.65 - to PH-MAG 7.65	
			PH-MAG, G-AGYX, (OD-...), G-AGYX, KG437, 42-9264 *(Subsequently N9050T) (Nose only)*			
G-AITB	Airspeed AS.40 Oxford 1	---	MP425	1.11.46	WFU *(As "MP425" in 1536 BATF c/s)*	24. 5.61
G-AIXA	Taylorcraft Plus D	134	LB264	13. 1.47	Cancelled 13.12.02 by CAA *(As "LB264")*	21. 2.02P
G-APUP	Sopwith Pup rep	B.5292 & PFA 1582	9213M	13. 2.59	Cancelled 4.10.84 by CAA	28. 6.78
	(Le Rhone)		G-APUP, N5182		*(As "N5182")*	
G-ATVP	Vickers FB.5 Gunbus rep	VAFA-01 & FB.5		31. 5.66	Cancelled 27.2.69 as WFU	6. 5.69
	(Gnome Monosoupape 100 hp)				*(As "2345" in RFC c/s)*	
G-AWAU	Vickers FB.27A Vimy rep	VAFA-02	"H651"	8. 1.68	Cancelled 19.7.73 as WFU *"Triple First" (As "F8614")*	4. 8.69
G-BEOX	Lockheed 414 Hudson IIIA	414-6464	VH-AGJ	25. 3.77	Cancelled 22.12.81 as WFU	
	(A-29A-LO)		VH-SMM, R.Australian AF A16-199, FH174, 41-36975 *(As "A16-199/SF-R")*			
G-BFDE	Sopwith Tabloid Scout rep	168 & PFA 067-10186		22. 9.77	Cancelled 8.12.86 as WFU	4. 6.83P
	(Continental PC.60)				*(As "168" in RNAS c/s)*	
G-BIDW	Sopwith "1½" Strutter rep	WA/5	"9382"	24. 9.80	Cancelled 4.2.87 by CAA	29.12.80P
	(Built Westward Airways) (Le Clerget)				*(As "A8226" in 45 Sqdn RFC c/s)*	
G-BLWM	Bristol 20 M.1C rep	PFA 112-10892	"C4912"	12. 3.85	Cancelled 12.5.88 by CAA	12. 8.87P
	(110 hp Gnome)				*(As "C4994" in RFC c/s)*	
G-OIOI	EH Industries EH-101 Heliliner	50008		23.11.88	Cancelled 1.4.96 - to MOD	5. 5.94P
	(Airframe No.PP8)				*(To RAF as "ZJ116")*	
G-OTHL	Robinson R22 Beta	0738	G-DSGN	28.11.94	Cancelled 8.2.00 as WFU *(As "G-RAFM")*	27. 4.03T
G-USTV	Messerschmitt Bf.109G-2/Trop	----	8478M	26.10.90	Cancelled as PWFU 24.9.98	30. 5.98P
	(Built Erla Maschinenwerk GmbH)	10639	RN228, Luftwaffe		*(As "10639/6" in Luftwaffe III/JG77 c/s)*	
BAPC.83	Kawasaki Type 5 Model 1b (Ki 100)	---	8476M		*(As "24")*	
BAPC.92	Fieseler Fi 103 (V-1)	---				
BAPC.100	Clarke TWK	---			*(On loan from Science Museum)*	
BAPC.106	Bleriot Type XI	---	9209M		*(As "164")*	
	(1910 original) (Anzani 40hp)					
BAPC.107	Bleriot Type XXVII	---	9202M			
	(1911 original)					
BAPC.165	Bristol F.2b Fighter	---			*(As "E2466" in 22 Sqdn c/s)*	
	(RR Falcon rep)					
BAPC.181	Royal Aircraft Factory BE.2b reconstruction	---			*(As "687")*	
	(Renault V8) *(Restoration from original components)*					
BAPC.205	Hawker Hurricane IIc fsm	---			*(As "BE421/XP-G" in 174 Sqdn c/s)*	
BAPC.206	Supermarine Spitfire IX fsm	---			*(As "MH486/FF-A" in 132 Sqdn c/s)*	
BAPC.292	Eurofighter Typhoon fsm					
BAPC.293	Supermarine Spitfire fsm					
	(Built Concepts and Innovations)					
BAPC.296	Army Balloon Factory Nulli reproduction					

RAF Northolt

Registration	Type	c/n	Previous identity	Regd	Remarks	Date
BAPC.221	Supermarine Spitfire LF.IX fsm	---			*(As "EN526/SZ-G" in 316 Sqdn c/s)*	

Science Museum, South Kensington, London SW7 2DD (www.sciencemuseum.org.uk)

Registration	Type	c/n	Previous identity	Regd	Remarks	Date
G-EBIB	Royal Aircraft Factory SE.5A *(Regd with c/n 688/2404)*	687/2404	"F939" G-EBIB, F937	26. 9.23	Cancelled 1.12.46 *(WFU)*	8. 8.35
G-AAAH	de Havilland DH.60G Moth *(Original but note two BAPC.reproductions depicted as "G-AAAH" exist –see below) "Jason"*	804		30. 8.28	Cancelled 12.31	23.12.30
G-ACWP	Cierva C.30A (Avro 671)	728	AP507 G-ACWP	24. 7.34	Cancelled 1.6.40 on sale *(Impressment)* *(As "AP507/KX-P" in 529 Sqdn c/s)*	6. 3.41
G-ASSM	Hawker Siddeley HS.125 Srs.1/522	25010	5N-AMK G-ASSM	5. 5.64	Cancelled 28.5.80 as sold in Nigeria	
G-AWAW	Reims/Cessna F150F *(Wichita c/n 15063167)*	F150-0037	OY-DKJ	5. 1.68	Cancelled 16.5.90 as WFU	8. 6.92T
G-AZPH	Pitts S.1S (Lycoming IO-360)	S1S-001-C	N11CB	13. 3.72	Cancelled 8.1.97 as WFU *"Neil Williams" (Ground-looped landing Little Snoring 10.5.91)*	4. 9.91P
G-LIOA	Lockheed 10A Electra	1037	N5171N NC243, NC14959	6. 5.83	Cancelled 26.4.02 as WFU *(As "NC5171N")*	
BAPC.50	Roe Triplane Type I *(1909 original)* (JAP 9hp)	---				
BAPC.51	Vickers FB.27 Vimy IV *(Rebuild of 1919 original)* (RR Eagle VIII 360hp)	13				
BAPC.53	Wright Flyer reconstruction *(Built Hatfield)*	---				
BAPC.54	JAP/Harding Monoplane *(Built J.A.Prestwich & Co 1910) (Modified Bleriot XI* (JAP Anzani 45hp)	---				
BAPC.55	Levasseur-Antoinette Developed Type VII Monoplane *(1909 original)* (Antoinette V8 50hp)	---				
BAPC.56	Fokker E.III (Oberusal 100hp)	---			*(Captured @ Somme, France 4.1916)* *(As "210/16") (Skeletal airframe)*	
BAPC.62	Cody Type V Biplane *(1912 original)* (Austro-Daimler 120hp)	---			*(As "304")*	
BAPC.124	Lilienthal Glider Type XI reconstruction	---			*(Display reproduction of BAPC.52 qv)*	
BAPC.199	Fieseler Fi 103 (V-1)	---			*(As "442795")*	
BGA.2091	Schempp-HirthHS.4 Standard Cirrus	396	AGA...		*(As "DFY")*	9. 5.02

Imperial War Museum, South Lambeth, London SE1 6HZ (www.iwm.org.uk)

Registration	Type	c/n	Previous identity	Regd	Remarks	Date
BAPC.198	Fieseler Fi 103 (V-1)	---			*Transferred to IWM Duxford 2002/3?*	

RAF Uxbridge

Registration	Type	c/n	Previous identity	Regd	Remarks	Date
BAPC.222	Supermarine Spitfire IX fsm	---			*(As "BR600/SH-V" in 64 Sqdn c/s)*	

GREATER MANCHESTER

Museum of Science and Industry in Manchester, Castlefield, Manchester M3 4JP (www.msim.org.uk)

Registration	Type	c/n	Previous identity	Regd	Remarks	Date
G-EBZM	Avro 594A Avian IIIA (ADC Cirrus) *(Fitted with parts from G-ABEE)*	R3/CN/160		??.7.28	Cancelled 1.12.46 by Secretary of State *(On loan from The Aeroplane Collection)*	20. 1.38
G-ABAA	Avro 504K	---	9244M "H2311", G-ABAA	11. 9.30	Cancelled 1939?	11. 4.39
G-ADAH	de Havilland DH.89 Dragon Rapide	6278		30. 1.35	WFU 1969 *"Pioneer"* *(Allied Airways (Gandar Dower) titles) (On loan from The Aeroplane Collection)*	9. 6.47
G-APUD	Bensen B-7Mc	1		11. 5.59	Cancelled 27.2.70 as WFU *(On loan from The Aeroplane Collection)*	27. 9.60
G-AYTA	SOCATA MS.880B Rallye Club	1789		19. 2.71	Cancelled 12.5.93 as WFU	7.11.88
G-AWZP	Hawker Siddeley HS.121 Trident 3B Srs.101	2317		14. 1.69	Cancelled 27.6.86 as destroyed *(Nose section only)*	14. 3.86T
G-BLKU	Colt Flame 56SS HAB	572		17. 7.84	Cancelled 1.5.92 as WFU *"Mr.Wonderfuel II"*	
G-MJXE	Mainair Tri-Flyer 330/Hiway Demon 175	102-131082 & HS-001		17. 5.83	Cancelled 19.10.00 as TWFU *(On loan from The Aeroplane Collection)*	21. 3.95P
BAPC.6	Roe Triplane Type I (JAP @ 9hp)	---			*(As "14") "Bullseye Avroplane"*	
BAPC.12	Mignet HM.14 Pou-Du-Ciel (Scott A2S)	---				
BAPC.98	Yokosuka MXY-7 Ohka II		8485M		*(As "997")*	
BAPC.175	Volmer VJ-23 Swingwing Powered Hang Glider (McCulloch @ 9hp)	---				
BAPC.182	Wood Ornithopter	---			*(Stored 3.04)*	
BAPC.251	Hiway Spectrum Hang-Glider *(Built 1980)*	---			*(Stored 3.04)*	
BAPC.252	Flexiform Wing Hang-Glider *(Built 1982)*	---			*(Stored 3.04)*	
BGA.1156	EoN AP.10 460 Srs.1	EoN/S/007	BGA.2666 AGA.6, BGA.1156	26. 1.64	*(On loan from J.H.May)* *(As "BQT")*	18. 4.97

Aviation Viewing Park, Manchester International Airport

Registration	Type	c/n	Previous identity	Regd	Remarks	Date
G-AWZK	Hawker Siddeley HS.121 Trident 3B Srs.101	2312		14. 1.69	Cancelled 29.5.90 as WFU *(WFU 1.11.85) (BEA "Quarter Union Jack" titles) (On loan from Trident Preservation Society)*	14.10.86T
G-BOAC	British Aircraft Corporation- Aérospatiale Concorde	102		3.4.74	Cancelled 4.5.04 as WFU	16. 5.05T
G-DMCA	McDonnell Douglas DC-10-30	48266	N3016Z	12. 3.96	Cancelled 3.11.03 as destroyed *(Forward 60 feet of fuselage, including flight deck and 70 seats, retained for use an education classroom)*	11. 3.03T
G-IRJX	BAE SYSTEMS Avro 146-RJX100	E3378		24. 5.00	Cancelled 20.2.03 as wfu)	

MERSEYSIDE

Wirral Aviation Society, Liverpool Marriott Hotel South, Liverpool-John Lennon Airport (www.jetstream-club.org)

G-JMAC	British Aerospace Jetstream Srs.4100	41004	G-JAMD	12. 6.92	Cancelled 21.5.03 by CAA	6.10.97A
			G-JXLI			
BAPC.280	de Havilland DH.89 Dragon Rapide fsm	----			*(As "G-AEAJ" in Railway Air Services c/s) "Neptune"*	

WEST MIDLANDS

Boulton Paul Aircraft Heritage Project, Bilbrook, Wolverhampton WV8 1EU

G-AVVR	Avro 652A Anson 19 Srs.2	"34530"	VP519	6.10.67	Cancelled by CAA 16.9.72 - not converted	
					(On loan from The Aeroplane Collection) (Cockpit only as "VP519")	
G-FBPI	Air Navigation and Engineering Co ANEC IV Missel Thrush				See SECTION 1 Part 2	
BAPC.274	Boulton Paul P.6 fsm	---			*(As "X-25")*	
BAPC.281	Boulton Paul Defiant recreation	---			*(As "L7005/PS-B" in 264 Sqdn c/s)*	
BGA.1759	Slingsby T.8 Tutor	---	RAFGSA.178	10.72		
BGA.1992	Hirth Go IV Goevier 3	557	D-5233	13. 7.74	*(As "DBU")*	19. 7.87

NORFOLK

Royal Air Force Coltishall

BAPC.223	Hawker Hurricane I fsm	---			*(As "V7467/LE-D"in 242 Sqdn c/s)*	

City of Norwich Aviation Museum, Norwich Airport NR10 3JE

G-ASKK	Handley Page HPR.7 Dart Herald 211	161	PP-ASU	17. 7.63	Cancelled 29.4.85 as WFU	19. 5.85T
			G-ASKK, PI-C910, CF-MCK			
G-AWON	English Electric Lightning F.53	95291	G-27-56	9. 8.68	Cancelled 9.68 - to R.Saudi AF as 53-686 4.69	
					(Subsequently RSAF 201, 203 and 1305) (As "ZF592"	
G-BEBC	Westland WS-55 Whirlwind HAR.10	WA.371	8463M	25. 6.76	Cancelled 5.12.83 as WFU	
			XP355		*(Not converted) (As "XP355/A")*	
G-BHMY	Fokker F.27 Friendship 600	10196	F-GBDK (2)	6. 5.80	Cancelled as PWFU 22.3.03	22. 5.99T
			(F-GBRV), PK-PFS, JA8606, PH-FDL		*(Donated by KLM (UK) Ltd - less engines)*	

NORTHUMBERLAND and TYNESIDE

Military Vehicle Museum, Newcastle upon Tyne NE2 4PZ (www.military-museum.org.uk)

G-ANFU	Taylorcraft J Auster 5	1748	TW385	31.10.53	Cancelled 3.8.76 as WFU	17. 2.71
					(On rebuild with frame of un-identified Auster 5.93) (As "NJ719" with starboard wing ex G-AKPH)	

North East Aircraft Museum, Usworth, Sunderland SR5 3HZ

G-APTW	Westland WS-51/2 Widgeon	WA/H/150		27. 4.59	Cancelled 24.8.77 as WFU	26. 9.75
G-ARAD	Phoenix Luton LA-5A Major	PAL/1204 & PFA 836		29. 4.60	Cancelled 16.10.02 as WFU - completed but not flown	
G-ASOL	Bell 47D-1	4	N146B	31. 1.64	Cancelled 5.12.83 as WFU	6. 9.71
G-AWRS	Avro 652A Anson C.19 Srs.2	"33785"	TX213	14.10.68	Cancelled 30.5.84 as PWFU *(WFU 5.2.73)*	10. 8.73
G-BEEX	de Havilland DH.106 Comet 4C	6458	SU-ALM	10. 9.76	Cancelled 19.5.83	
					(Not converted and broken up Lasham 8.77: nose section only)	
G-MBDL	Striplin (AES) Lone Ranger	109		21.10.81	Cancelled 13.6.90 by CAA	
G-OGIL	Short SD.3-30 Var.100	SH.3068	G-BITV	23. 1.89	Cancelled 12.11.92 as WFU	21. 4.93T
			G-14-3068		*(Damaged Newcastle 1.7.92)*	
G-SFTA	Westland SA.341G Gazelle 1	1039	G-SFTA	10. 9.82	Cancelled 21.5.86 as WFU	24. 2.86
			HB-XIL, G-BAGJ, (XW858)		*(Crashed near Alston, Cumbria 7.3.84) (As "G-BAGJ")*	
BAPC.96	Brown Helicopter	---				
BAPC.97	Luton LA.4 Minor (JAP J99)	---			*(As "G-AFUG")*	
BAPC.119	Bensen B.7 Gyroglider	---				
BAPC.211	Mignet HM.14 Pou-Du-Ciel	---			*(As "G-ADVU")*	
	(Built Ken Fern/Vintage and Rotary Wing Collection 1993)					
BAPC.228	Olympus Hang Glider	---				

NOTTINGHAMSHIRE

Wonderland Pleasure Park, Farnsfield, Mansfield

BAPC.288	Hawker Hurricane fsm				*(As "V7467/LE-D" in 242 Sqdn c/s)*	

Newark Air Museum, Winthorpe, Newark NG24 2NY (www.newarkairmuseum.co.uk)

G-AHRI	de Havilland DH.104 Dove 1B	04008	4X-ARI	11. 7.46	Cancelled 18.5.72 as WFU	
			G-AHRI			
G-ANXB	de Havilland DH.114 Heron 1B	14048	G-5-14	3.12.54	Cancelled 2.11.81 as PWFU	25. 3.79
					(BEA Scottish Airways titles) "Sir James Young Simpson"	
G-APIY	Percival P.40 Prentice T.1	PAC/075	VR249	28.11.57	Cancelled 19.4.73	18. 3.67
					(WFU 18.3.67) (As "VR249/FA-EL" in RAFC c/s)	
G-APNJ*	Cessna 310	35335	EI-AJY	2. 6.58	Cancelled 5.12.83 as WFU Newark	28.11.74
			N3635D			
G-APVV	Mooney M.20A	1474	N8164E	30. 7.59	Cancelled 3.4.89 by CAA *(Crashed at Barton 11.1.81)*	19. 9.81
G-AVVO	Avro 652A Anson 19 Srs.2	34219	VL348		Cancelled 16.9.72 by CAA *(As "VL348")*	6.10.67
G-AYZJ	Westland WS-55 Whirlwind HAS.7	WA.263	XM685	24. 5.71	Cancelled 29.12.80 as WFU	
	(Also c/n WAG/34)				*(As "XM685/PO-513")*	
G-BFTZ	SOCATA MS.880B Rallye Club	1269	F-BPAX	2. 6.78	Cancelled 14.11.91 by CAA	19. 9.81
					(On loan from The Aeroplane Collection)	
G-BJAD	Clutton FRED Srs.II	CA.1 & PFA/29-10586		11. 6.81	Cancelled 9.9.97 by CAA	
					(No Permit issued, probably not completed)	
G-BKPG	Luscombe P3 Rattler Strike	003		7. 3.83	Cancelled 31.7.91 by CAA *(No Permit issued)*	
G-BKPY	SAAB 91B/2 Safir	91321	R NorAF 56321	23. 3.83	Cancelled 8.2.02 as WFU *(As "56321" in R.NorAF c/s)*	
G-MBBZ	Volmer VJ-24W	7		23. 9.81	Cancelled 29.11.95 as WFU	
G-MBUE	MBA Tiger Cub 440	MBA-001		29. 4.82	Cancelled 6.9.94 by CAA *(Originally regd as Micro-Bipe c/n 001)*	
BAPC.43	Mignet HM.14 Pou-Du-Ciel (Scott A2S)	---				
BAPC.101	Mignet HM.14 Pou-Du-Ciel	---				
BAPC.183	Zurowski ZP-1 Helicopter	---			*(Polish AF c/s)*	
	(Panhard 850cc)					

OXFORDSHIRE

Royal Air Force, RAF Benson

BAPC.226	Supermarine Spitfire XI fsm	---			*(As "EN343" in PRU c/s)*	

SHROPSHIRE

Royal Air Force Museum including Michael Beetham Restoration Centre ($), RAF Cosford TF11 8UP (www.rafmuseum.com)

G-EBMB	Hawker Cygnet I **($)**	1	No.14 (Lympne 1924)	29. 7.25	Cancelled 30.11.61	30.11.61
	(Bristol Cherub III)					
G-AEEH	Mignet HM.14 Pou-Du-Ciel	EGD.1		13. 3.36	Cancelled 8.46 in census - WFU 15.5.38	15. 5.38
	(Built E G Davis)					
G-AEKW	Miles M.12 Mohawk **($)**	298	HM503	14. 7.36	Crashed Spain 1.1.50	1. 3.50
			G-AEKW, "G-AEKN"			
G-AGNV	Avro 685 York C.1	1223	"MW100"	20. 8.45	WFU 9.10.64	6. 3.65
			"LV633", G-AGNV, TS798			
G-AIZE	Fairchild F.24W-41A Argus II	565	N9996F	18.12.46	Cancelled 6.4.73	6. 8.66
	(UC-61A-FA)		G-AIZE, 43-14601		*(WFU) (As "FS628")*	
G-AMOG (2)	Vickers 701 Viscount	7	(G-AMNZ)	23. 5.52	Cancelled 17.5.76 as WFU	14. 6.77
					(BEA titles) "RMA Robert Falcon Scott"	
G-AOVF	Bristol 175 Britannia Srs.312F	13237	9Q-CAZ	13. 2.57	Cancelled 21.11.84 as WFU	
			G-AOVF		*(BOAC titles)*	
G-APAS	de Havilland DH.106 Comet 1A	06022	8351M	23. 5.57	Cancelled 22.10.58	
			XM823, G-APAS, G-5-23, F-BGNZ		*(BOAC titles) (As "XM823" @ 30.1.58)*	
G-APFJ	Boeing 707-436	17711		7. 8.59	WFU 12.6.81 *(British Airtours titles)*	16. 2.82T
G-ARPH	de Havilland DH.121 Trident 1C	2108		13. 4.61	WFU 26.3.82 *(British Airways titles)*	8. 9.82T
G-ARVM (2)	Vickers VC-10 Srs.1101	815	(G-ARVJ)	16. 1.63	WFU 22.10.79 *(British Airways titles)*	5. 8.80
G-AVMO	British Aircraft Corporation One-Eleven 510ED	BAC.143		11. 5.67	Cancelled 12.7.93 as WFU	3. 2.95T
					(British Airways titles) "Lothian Region"	
G-BBYM	Handley Page HP.137 Jetstream 200	243	G-AYWR	13. 2.74	Cancelled 7.6.00 as WFU	20. 9.98A
			G-8-13			
BAPC.82	Hawker Afghan Hind	41H/81899			*(R.Afghan AF c/s)*	
	(RR Kestrel)					
BAPC.84	Mitsubishi Ki 46 III (Dinah)	---	8484M		*(As "5439")*	
BAPC.94	Fieseler Fi 103 (V-1)	---	8483M			
BAPC.99	Yokosuka MXY-7 Ohka II	---	8486M			
BGA.572	Slingsby T.21B Sedbergh TX.1 **($)**	539	8884M		*(As "VX275")*	
			VX275, BGA.572			

Assault Glider Association, RAF Shawbury (www.assaultgliderproject.co.uk)

G-AMHJ*	Douglas C-47A-25DK Dakota 6	13468	SU-AZI	6. 2.51	Cancelled 23.1.03 as wfu	5.12.00A
			G-AMHJ, ZS-BRW, KG651, 42-108962			
BAPC.279	Airspeed AS.58 Horsa reconstruction					

Wartime Aircraft Recovery Group Aviation Museum, Sleap (www.wargroup.homestead.com)

BAPC.148	Hawker Fury II fsm	---			*(As "K7271" in 1 Sqdn c/s)*	
BAPC.234	Vickers FB.5 Gunbus fsm	---			*(Built for "Gunbus" film) (As "GBH-7")*	
	(Built 1985)					

SOMERSET

The Helicopter Museum, Weston-super-Mare BS24 8PP (www.helicoptermuseum.co.uk)

G-ACWM	Cierva C.30A	715	(G-AHMK)	24. 7.34	Cancelled 17.3.59 as WFU	13. 7.40
	(Avro 671)		AP506, G-ACWM		*(On loan from E.D.ap Rees) (No external marks)*	
G-ALSX	Bristol 171 Sycamore 3	12892	G-48-1	17.11.50	Cancelled as WFU	24. 9.65
			G-ALSX, VR-TBS, G-ALSX		*(On loan from E.D.ap Rees)*	
G-ANFH	Westland WS.55 Whirlwind Srs.1	WA.15		27.10.53	Cancelled 2.9.77 as WFU	17. 7.71
					(On loan from E.D.ap Rees) (No external marks)	
G-ANJV	Westland WS-55 Whirlwind Srs.3	WA.24	VR-BET	14.12.53	Cancelled 8.1.74 on sale in Bermuda	
			G-ANJV		*(On loan from E.D.ap Rees) (No external marks) (Stored 3.02)*	
G-AODA	Westland WS-55 Whirlwind Srs.3	WA.113	9Y-TDA	13. 5.55	Cancelled 23.9.93 by CAA	23. 8.91A
			EP-HAC, G-AODA		*(Bristow Helicopters c/s) "Dorado"*	
G-AOUJ	Fairey Ultralight Helicopter	F.9424	XJ928	1. 8.56	Cancelled 26.2.69 as Destroyed	29. 3.59
					(WFU) (On loan from E.D.ap Rees)	
G-AOZE	Westland WS-51/2 Widgeon	WA/H/141	5N-ABW	11. 1.57	Cancelled 20.6.62 on sale in Nigeria	
			G-AOZE		*(On loan from E.D.ap Rees)*	
G-ARVN (2)	Servotec Rotorcraft Grasshopper 1	1		16. 2.63	Cancelled 14.3.77 as WFU	18. 5.63
					(On loan from E.D.ap Rees)	
G-ASCT	Bensen B.7Mc	DC.3		14. 8.62	Cancelled 20.9.73 as PWFU	11.11.66P
	(McCulloch 4318E)				*(Built D.Campbell) (Stored 3.02)*	
G-ASHD	Brantly B.2A	314		2. 4.63	Cancelled 22.6.67 as Destroyed	5. 6.67
					(Crashed into River Colne, Brightlingsea, Essex 19.2.67) (Stored 3.02)	
G-ASTP	Hiller UH-12C	1045	N9750C	4. 6.64	Cancelled 24.1.90 by CAA	3. 7.82
G-ATBZ	Westland WS-58 Wessex 60 Srs.1	WA/461		22. 3.65	Cancelled as TWFU 23.11.82	15.12.81
					(WFU 5.12.81 - to G-17-4) (Stored 3.02)	
G-ATKV	Westland WS-55 Whirlwind 3	WA/493	VR-BEU	11.11.65	Cancelled 8.1.74	
			G-ATKV, EP-HAN (1), G-ATKV		*(To VR-BEU 1.74) (Stored 3.02)*	
G-AVKE	Gadfly HDW-1	HDW-1		19. 4.67	Cancelled 12.10.81 as WFU	
	(Continental IO-340A)				*(On loan from E.D.ap Rees)*	
G-AVNE	Westland WS-58 Wessex 60 Srs.1	WA.561	G-17-3	15. 5.67	Cancelled 23.11.82 as TWFU	7. 2.83
			G-AVNE, 5N-AJL, G-AVNE, 9M-ASS, VH-BHC, PK-HBQ, G-AVNE, (G-AVMC)		*(As "G-AVNE" & "G-17-3")*	
G-AWRP	Cierva Rotorcraft CR.LTH.1 Grasshopper III	GB.1		14.10.68	Cancelled 5.12.83 as WFU	12. 5.72P
G-AXFM	Cierva Rotorcraft CR.LTH.1 Grasshopper III	GB.2		19. 5.69	Cancelled 5.12.83 as WFU	
					(Completed as Ground-Running Rig) (Stored 3.02)	
G-AYXT	Westland WS-55 Whirlwind HAS.7 (Srs.2)	WA.167	XK940	28. 4.71	Cancelled 8.8.00 by CAA *(As "XK940")*	4. 2.99P
G-AZAU	Cierva Rotorcraft CR.LTH.1 Grasshopper III	GB.3		21. 6.71	Cancelled 5.12.83 as WFU *(Incomplete Rig) (Stored 3.02)*	

G-AZYB	Bell 47H-1	1538	LN-OQG SE-HBE, OO-SHW	4. 7.72	Cancelled 22.4.85 as destroyed *(Crashed St.Mary Bourne, Thruxton 21.4.84)* *(On loan from E.D.ap Rees) (As "OO-SHW")*	8. 9.84
G-BAPS	Campbell Cougar Gyroplane (Continental O-240-A)	CA/6000		14. 2.73	Cancelled 21.1.87 by CAA *(On loan from A.M.W.Curzon-Howe-Herrick)*	20. 5.74
G-BGHF	Westland WG.30 Srs.100-60	WA.001.P		4. 1.79	Cancelled 29.3.89 as WFU	1. 8.86
G-BKFD	Westland WG.30 Srs.100	004	G-17-28	22. 6.82	Cancelled 6.12.82 *(To N5820T) (Stored 3.02)*	
G-BKFF	Westland WG.30 Srs.100	006	G-17-30	22. 6.82	Cancelled 6.12.82 *(To N5840T) (Stored 3.02)*	
G-BKGD	Westland WG.30 Srs.100	002	(G-BKBJ)	15. 7.82	Cancelled 15.4.93 as WFU	6. 7.93
G-BRMA	Westland-Sikorsky S-51 Dragonfly HR.5	WA/H/50	WG719	15. 6.78	Cancelled 30.3.89 as WFU *(On loan from E.D.ap Rees) (As "WG719")*	
G-BRMB	Bristol 192 Belvedere HC.1	13347	7997M XG452	15. 6.78	Cancelled 3.7.96 as WFU *(On loan from E.D.ap Rees) (As "XG452")*	
G-EFIS	Westland WG.30 Srs.100	014	G-17-18	24. 6.84	Cancelled 27.11.84 *(To N114WG 11.84) (As "N114WG")*	
G-EHIL	EH Industries EH-101 *(Airframe No.PP3)*	50003	ZH647	9. 7.87	Cancelled 28.4.99 as PWFU *(To MoD as ZH647 9.93 and restored 8.98)*	9. 7.87
G-ELEC	Westland WG.30 Srs.200	007	G-BKNV	17. 6.83	Cancelled 27.2.98 as WFU	28. 6.95P
G-HAUL	Westland WG.30-300	020	G-17-22	3. 7.86	Cancelled 22.4.92 as WFU	27.10.86P
G-LYNX	Westland WG.13 Lynx 800	WA/102	(ZA500) G-LYNX, ZB500	6.11.78	Cancelled 27.2.98 as WFU *(As "ZB500")*	
G-PASB	MBB Bö.105D	S.135	VH-LSA G-BDMC, D-HDEC	2. 3.89	Cancelled 9.8.94 as WFU *(Original pod from 1994 rebuild)*	
G-RWWW	Westland WS-55 Whirlwind HCC.12	WA/418	8727M XR486	21. 6.90	Cancelled 10.7.00 as WFU *(As "XR486" in Queens Flight c/s)*	25. 8.96P
BAPC.10	Hafner R-11 Revoplane (Salmson 45hp)	---				
BAPC.60	Murray M.1 Helicopter (JAPJ99 @ 36hp)	---				
BAPC.128	Watkinson CG-4 Cyclogyroplane Man Powered Gyroplane Mark IV					
BAPC.153	Westland WG-33 Mock-up	---			*(Engineering mock-up)*	
BAPC.212	Bensen B.6 Gyrocopter	---			*(Stored 3.04)*	
BAPC.213	Cranfield Vertigo Man Powered Helicopter	---			*(Stored 3.04)*	
BAPC.264	Bensen B.8M	---			*(Built 1984)*	
BAPC.289	Gyro-Boat					

Fleet Air Arm Museum/FAAM Cobham Hall ($), RNAS Yeovilton BA22 8HT (www.fleetairarm.com)

G-AIBE	Fairey Fulmar 2	F.3707	N1854 G-AIBE, N1854	29. 7.46	Cancelled 30.4.59 - to N1854 *(As "N1854")*	6. 7.59
G-AIZG	Supermarine VS.236 Walrus 1	6S/21840	EI-ACC IAC N-18, L2301	20.12.46	Cancelled 1963 *(As "L2301")*	
G-AOXG	de Havilland DH.82A Tiger Moth **($)**	83805	XL717 G-AOXG, T7291	3.10.56	Cancelled as sold as XL717 @ 10.56 *(As "G-ABUL")*	
G-APNV	Saunders-Roe P.531-1	S2/5268		24. 6.58	Cancelled 1.10.59 - to XN332 @ 10.59) *(As "XN332/759")*	
G-ASTL	Fairey Firefly 1	F.5607	SE-BRD Z2033	1. 6.64	WFU 3.2.82 *(As "Z2033/275/N" in 1771 Sqdn RN c/s)*	
G-AWYY	Slingsby T.57 Sopwith Camel F.1 rep (Clerget)	1701	"C1701" N1917H, G-AWYY	14. 2.69	Cancelled 25.11.91 as WFU *(As "B6401")*	1. 9.85P
G-AZAZ	Bensen B.8M ($)	RNEC.1		2. 7.71	Cancelled 19.9.75 as WFU	
G-BEYB	Fairey Flycatcher rep *(Built Westward Airways)*	WA/3		11. 7.77	Cancelled 12.7.96 as WFU	
G-BFXL	Albatros D.Va rep *(Built Williams Flugzeugbau)* (Ranger 6-440-C5)	0034	D-EGKO	24. 8.78	Cancelled 10.3.97 as WFU *(As "D5397" in German c/s)*	5.11.91P
G-BGWZ	Eclipse Super Eagle **($)**	ESE.007		29. 6.79	Cancelled 5.12.83 as WFU *(No external marks)*	
G-BIAU	Sopwith Pup rep (Le Rhone 80hp)	EMK/002		4. 1.83	Cancelled 10.3.97 as WFU *(As "N6452" in RNAS c/s)*	13. 9.89P
G-BMZF	WSK-Mielec LIM-2 (MiG-15bis)	1B-01420	Polish AF 01420	18.12.86	Cancelled 23.2.90 as WFU *(As "01420" in North Korean c/s)*	
G-BSST	British Aircraft Corporation-Sud Concorde SST	13520 & 002		6. 5.68	Cancelled 21.1.87 as WFU *(WFU 4.3.76) (On loan from Science Museum)*	31.10.74P
BAPC.58	Yokosuka MXY-7 Ohka II	---			*(As "15-1585") (On loan from Science Museum)*	
BAPC.88	Fokker Dr.1 5/8th scale model *(Modified Lawrence Parasol airframe)*	---			*(As "102/17")*	
BAPC.111	Sopwith Triplane fsm	---			*(As "N5492") "Black Maria"*	
BAPC.149	Short S.27 rep **($)**	---				

STAFFORDSHIRE

Royal Air Force Museum Reserve Collection, RAF Stafford

G-ADMW	Miles M.2H Hawk Major	177	DG590 8379M, G-ADMW	30. 7.35	Cancelled 16.9.86 by CAA	30. 7.65
G-AHED	de Havilland DH.89A Dragon Rapide 6	6944	RL962	27. 2.46	Cancelled 3.3.69 *(WFU)*	17. 4.68
BAPC.108	Fairey Swordfish IV				*(As "HS503")*	
BAPC.194	Santos Dumont Type 20 Demoiselle *(Built Personal Plane Services)* (ABC Scorpion 30hp)	PPS/DEM/1	24 bis		*(Built "for Those Magnificent Men in Their Flying Machines" film)*	
BAPC.237	Fieseler Fi 103 (V-1)					

SUFFOLK

Norfolk and Suffolk Aviation Museum, Flixton, Bungay NR35 1NZ (www.aviationmuseum.net)

G-ANLW	Westland WS-51 Srs.2 Widgeon	WA/H/133	"MD497" G-ANLW	23. 3.54	Cancelled 15.11.02 as WFU *(On loan from Sloane Helicopters Ltd)*	27. 5.81A
G-ASRF	Gowland GWG.2 Jenny Wren *(Built G W Gowland - c/n GWG.2 using modified Luton Minor wings once fitted to G-AGEP)*	PFA 1300		18. 3.64	*(Cancelled 11.12.96 by CAA)*	4. 6.71P
G-AWSA	Avro 652A Anson C.19/2	"293483"	(N5054) G-AWSA, VL349	21.10.68	Cancelled 18.8.69 as sold in USA *(Not delivered) (As "VL349/V7-Q")*	

G-AZLM	Reims/Cessna F172L	F17200842		31.12.71	Cancelled 15.4.91 as destroyed *(Crashed on take off Badminton 23.3.91) (Fuselage only)*	
G-BDVS	Fokker F.27 Friendship 200	10232	S2-ABK PH-FEX, PH-EXC, 9M-AMM, PH-FEX	20. 4.76	Cancelled 19.12.96 as WFU *(Scrapped 12.96 - nose only) "Eric Gandar Dower*	
G-BFIP	Wallbro Monoplane rep (McCulloch/Wallis)	WA-1		16.12.77	Cancelled 28.3.01 as TWFU *(On loan from K.H.Wallis) (No external marks)*	22. 4.82P
G-MJSU	MBA Tiger Cub 440	SO.75/1		2. 2.83	Cancelled 23.6.93 by CAA	31. 1.86E
G-MTFK	Moult Trike/Flexiform Striker *(Officially regd with c/n SO.175)*	DIM-01		23. 3.87	Cancelled 13.6.90 by CAA	
BAPC.71	Supermarine Spitfire fsm	---	"P9390" "N3317"		*(Built for "Battle of Britain" film)* *(As "P8140/ZP-K" in 74 Sqdn c/s) "Nuflier"*	
BAPC.115	Mignet HM.14 Pou-Du-Ciel (Douglas 500cc)	---			*(On loan from I Hancock)*	
BAPC.147	Bensen B.7 Gyroglider	---			*(As "LHS-1")*	
BAPC.239	Fokker D.VIII 5/8th scale model	---			*(As "694")*	
BGA.1461	EoN AP.7 Primary	Not known			*(As "CDN")*	
BGA.4757	Colditz Cock rep *(Built Southdown Aero Services and J.Lee)*	---	"JTA"	1.00		

Wings Of Liberty Memorial Park, USAF Lakenheath

BAPC.269	Supermarine Spitfire V fsm	---			*(As "BM631/XR-C" in 71 Sqdn c/s)*	

SURREY

Brooklands Museum, Brooklands, Weybridge KT13 0QN (www.motor-software.co.uk/brooklands)

G-AEKV	Kronfeld (BAC) Drone de luxe	30		13. 1.37	Cancelled 14.1.99 as WFU *(On loan from M.L.Beach) (Allocated BGA.2510 5.79)*	6.10.60P
G-AGRU	Vickers 657 Viking 1	112	VP-TAX G-AGRU	8. 5.46	WFU 9.63 *"Vagrant" (BEA titles)*	9. 1.64
G-APEJ	Vickers 953C Vanguard Merchantman	713		9. 9.57	Cancelled 15.11.96 as WFU *"Ajax" (WFU 24.12.92: broken up 1.6.95: nose section only)*	
G-APEP	Vickers 953C Vanguard Merchantman	719		9. 9.57	Cancelled 28.2.97 as WFU *"Superb" (Hunting Cargo Airlines titles)*	1.10.98T
G-APIM	Vickers 806 Viscount	412		19.11.57	Cancelled *"Viscount Stephen Piercey"* *(DBR struck by Short SD.3-30 G-BHWT on 11.1.88) (British Air Ferries titles)*	19. 7.88T
G-ASIX	Vickers VC-10 Srs.1103	820		29. 5.63	Cancelled 10.10.74 - to A40-AB @ 10.74) *(As "A40-AB")*	
G-ASYD	British Aircraft Corporation One-Eleven 475AM *(Originally regd as Srs.400AM: converted to prototype Srs.500 1967: to Srs.475EM 1970)*	BAC.053		9.11.64	Cancelled 25.7.94 as WFU	13. 7.94
G-BFCZ	Sopwith Camel F.1 rep *(Built Westward Airways)* (Clerget 9B)	WA/2		12.10.77	Cancelled 23. 1.03 as WFU *(As "B7270")*	23. 2.89P
G-BJHV	Voisin Scale rep	MPS-1		1. 9.81	Cancelled 4.7.91 by CAA *(On loan from M.P.Sayer)*	
G-BNCX	Hawker Hunter T.7	41H/695454	XL621	9. 1.87	Cancelled 1.3.93 as WFU *(On loan from J Hallett) (As "XL621")*	28. 3.87P
G-MJPB	Manuel Ladybird	WLM-14		9.11.82	Cancelled 13.6.90 by CAA *(On loan from Estate of W.L.Manuel)*	
G-VTOL	Hawker Siddeley Harrier T.52	B3/41H/735795	ZA250 G-VTOL, (XW273)	27. 7.70	Cancelled 3.90 by CAA	2.11.86S
BAPC.29	Mignet HM.14 Pou-Du-Ciel *(Built P.D.Roberts, Swansea 1960/78)* (Anzani "V")	---			*(As "G-ADRY")*	
BAPC.114	Vickers Type 60 Viking IV reconstruction	---	"R4"		*(Built for "The Land Time Forgot" film) (As "G-EBED")*	
BAPC.177	Avro 504K fsm (Clerget 130hp)	---	"G1381"		*(As "G-AACA" in Brooklands School of Flying c/s)*	
BAPC.187	Roe Type I Biplane reconstruction *(Built M.L.Beach)* (ABC 24hp)	---				
BAPC.249	Hawker Fury I fsm *(Built Brooklands 1990s)*	---			*(As "K5673" in 1 Sqdn 'A' Flight c/s)*	
BAPC.250	Royal Aircraft Factory SE.5a replica *(Built Brooklands 1990s)*	---			*(As "F5475/A") "1st Battalion Honourable Artillery Company"*	
BAPC.256	Santos Dumont Type 20 Demoiselle reconstruction- *(Built J Aubot 1996/97)*					
BGA.162	Manuel Willow Wren	---		9.34	*"The Willow Wren"*	
BGA.643	Slingsby T.15 Gull III	364A	TJ711	11.49	*((On loan from M L Beach) (As "ATH")*	20. 6.04
BGA.3922	Abbott-Baynes Scud I rep.	001		R	*(As "HFZ")*	

Gatwick Aviation Museum, Charlwood, Surrey RH6 0BT (www.gatwick-aviation-museum.co.uk)

G-TURP	Aérospatiale SA341G Gazelle 1	1445	G-BKLS N17MT, N14MT, N49549	21. 1.88	Cancelled 17.8.92 as WFU *(No external marks)* *(Damaged Stanford-le-Hope 9.9.91: restored as G-BKLS 5.11.91 for rebuild but cancelled)*	2.12.91T

SUSSEX

Gatwick Hilton Hotel, Gatwick Airport

BAPC.168	DH.60G Moth reconstruction	8058			*"Jason" (As "G-AAAH")*	

Balloon Preservation Group, Lancing (www.balloons.flyer.co.uk/bpg1.htm)

G-AYVA	Cameron O-84 HAB	17		30. 3.71	Cancelled 19.5.93 by CAA *"April Fool" (Located Alfriston)*	6. 9.76
G-BAKO	Cameron O-84 HAB	57		18.12.72	Cancelled 19.5.93 by CAA *"Pied Piper"*	12. 7.76
G-BAND	Cameron O-84 HAB	52		22. 1.73	Cancelled 17.4.98 as WFU *"Clover" (Located Southampton)*	
G-BAOW	Cameron O-65 HAB	59		6. 2.73	Cancelled 15.10.01 by CAA *"Winslow Boy"* *(Located Southampton)*	9. 5.74S
G-BAST	Cameron O-84 HAB	70		15. 3.73	Cancelled 19.5.93 by CAA *"Honey"*	2. 5.84A
G-BBYR	Cameron O-65 HAB	97		14. 2.74	Cancelled 30.1.87 by CAA *"Phoenix"*	15. 7.81
G-BCAP	Cameron O-56 HAB	92		5. 3.74	Cancelled 30.3.93 as WFU *"Honey Child"*	
G-BCAS	Thunder Ax7-77 HAB	018		5. 3.74	Cancelled 30.11.01 by CAA *"Drifter" (Located Southampton)*	9. 4.91A
G-BCCH	Thunder Ax6-56A HAB	024	G-BCCH	4. 4.74	Cancelled 15.11.82 as sold Belgium but NTU *"Wrangler"*	

G-BCRE	Cameron O-77 HAB	128		30.10.74	Cancelled 19.5.93 by CAA	6.10.83A
					"Snapdragon" (Located Aylesbury)	
G-BDGO	Thunder Ax7-77 HAB	048		16. 7.75	Cancelled 12.6.02 by CAA *"J & B"*	2. 2.82A
G-BDMO	Thunder Ax7-77 HAB	053	(EC-...), G-BDMO	25.11.75	Cancelled 8.3.95 as WFU *"Flash Harry"*	
G-BEIF	Cameron O-65 HAB				See SECTION 1 Part 2	
G-BEJB	Thunder Ax6-56A HAB	096		31.12.76	Cancelled 10.5.02 by CAA	21. 5.87A
	(Original canopy destroyed by fire @ Latimer 4.9.77: replacement c, n not known)				*"Baby J & B"*	
G-BGST	Thunder Ax7-65 Bolt HAB	217		14. 5.79	Cancelled 7.12.01 by CAA *"Black Fred"*	23. 3.91A
G-BHAT	Thunder Ax7-77 Bolt HAB	250		28. 1.80	Cancelled 29.4.93 as WFU *"Witter"*	6. 2.83A
G-BJZC	Thunder Ax7-65Z HAB	416		5. 3.82	Cancelled 8.7.98 as WFU *"Greenpeace Trinity"*	17. 6.94A
G-BKIK	Cameron DG-19 Helium Airship	776		23. 8.82	Cancelled 5.9.00 as WFU	4. 9.88A
	(Rotax 400)				*"B & Q"*	
G-BKIY	Thunder Ax3 Sky Chariot HAB	464		7.10.82	Cancelled 15.11.01 as WFU *"Michaelangelo"*	
G-BKOW	Colt 77A HAB	505		6. 9.84	Cancelled 29.4.97 as WFU *(Elle titles) "Lady Di"*	14. 2.88A
G-BLDL	Cameron Truck 56 SS HAB	990		10. 1.84	Cancelled as WFU *"Europa"*	21.10.96
G-BLIP	Cameron N-77 HAB	1031		17. 4.84	Cancelled 23.6.98 as WFU *"Systems 80"*	26. 3.94A
G-BLJF	Cameron O-65 HAB				See SECTION 1 Part 2	
G-BLJH	Cameron N-77 HAB				See SECTION 1 Part 2	
G-BLKJ	Thunder Ax7-65 HAB	580		18. 7.84	Cancelled 22.5.97 as PWFU *"Up & Coming"*	3. 2.96A
G-BLSH	Cameron V-77 HAB	1085		7.12.84	Cancelled 30.3.98 as WFU *"Compass Rose"*	14. 1.95A
G-BLUE	Colting Ax7-77A HAB	77A-011		2. 5.78	Cancelled 30.11.01 by CAA	20. 9.99
	(Regd as Colt 77A c/n 11)					
G-BLZB	Cameron N-65 HAB	1164		21. 5.85	Cancelled 22.11.01 as WFU *"Pro-Sport"*	25. 4.90A
G-BMKX	Cameron Elephant 77SS HAB	1196		6. 2.86	Cancelled 21.10.96 as WFU *"Benjamin I"*	19. 2.89A
G-BMST	Cameron N-31 HAB	1317		4. 6.86	Cancelled 1.5.92 as WFU	
					"B&Q" (Badly damaged and for spares 12.01)	
G-BMUR	Cameron Zero 25 HA Airship	1169		11. 6.86	Cancelled 21.10.96 as WFU *"Zero 25"*	
G-BMWU	Cameron N-42 HAB				See SECTION 1 Part 2	
G-BNHL	Colt Beer Glass 90SS HAB	1042		24 .3.87	Cancelled 22.6.98 as WFU *"Gatzweiler"*	4. 3.97A
G-BOCF	Colt 77A HAB	1178		4. 1.88	Cancelled 6.10.03 as WFU	25. 7.94T
G-BOGT	Colt 77A HAB	1212		21. 3.88	Cancelled 9.5.97 as WFU *"British Gas"*	2.12.94A
G-BONK	Colt 180A HAB	1167		14.12.87	Cancelled 28.11.01 as WFU *"Bonkette"*	2.11.94T
G-BONV	Colt 17A Cloudhopper HAB	1238		3. 5.88	Cancelled 22.11.01 as WFU *"Bryant Group"*	1. 4.93A
G-BOOP	Cameron N-90 HAB	1702	(G-BOMX)	11. 5.88	Cancelled 31.10.95 by CAA	28. 9.95A
					(Unipart titles) "Betty Boop" (Located Aylesbury)	
G-BORA	Colt 77A HAB	1233		19. 5.88	Cancelled 17.9.98 as WFU *"Carla"*	24. 8.94A
G-BOTE	Thunder Ax8-90 HAB	555		14. 6.88	Cancelled 12.12.95 as WFU *"Barge Fox"*	16. 2.95T
G-BPAH	Colt 69A HAB	512		2. 6.83	Cancelled 13.5.02 by CAA *"J & B Phil"*	15. 8.88A
G-BPDF	Cameron V-77 HAB				See SECTION 1 Part 2	
G-BPFJ	Cameron Can 90SS HAB	1834		14.11.88	Cancelled 9.5.97 as WFU	10.12.93A
	(Budweiser Beer Can shape)				*"Budweiser Can I"*	
G-BPFX	Colt 21A Cloudhopper HAB	1348		7.11.88	Cancelled 23.12.98 as WFU *"Budweiser Hopper"*	
G-BPSZ	Cameron N-180 HAB	1911		14. 3.89	Cancelled 9.5.01 as WFU *"Park Furnishers"*	30. 4.01T
G-BRFR	Cameron N-105 HAB	2042		14. 7.89	Cancelled 9.5.97 as WFU	6.12.93A
					"Rover" (Badly damaged, spares use only)	
G-BRLX	Cameron N-77 HAB	2095		13. 9.89	Cancelled 23.10.01 by CAA *"National Power"*	1. 6.96A
G-BSBM	Cameron N-77 HAB	2229		8. 3.90	Cancelled 15.7.98 as WFU *"Nuclear Electric 1"*	21.11.96A
G-BSWZ	Cameron A-180 HAB				See SECTION 1 Part 2	
G-BTJF	Thunder Ax10-180 Srs.2 HAB	1952		28. 3.91	Cancelled 4.8.03 as destroyed *"Yorkshire Lad"*	4. 5.01T
G-BTML	Cameron Rupert Bear 90SS HAB	2533		16. 5.91	Cancelled 29.4.97 as WFU *"Rupert Bear"*	31.12.94A
G-BTPV	Colt 90A HAB				See SECTION 1 Part 2	
G-BUET	Colt Flying Drinks Can SS HAB	2162		30. 3.92	Cancelled 29.4.97 as WFU *"Bud King Can"*	10.12.93A
	(Budweiser Can shape)					
G-BUEU	Colt 21A Cloudhopper HAB	2163		30. 3.92	Cancelled 29.4.97 as WFU *"Bud King Hopper"*	2.12.94A
G-BUKC	Cameron A-180 HAB	2870		3. 7.92	Cancelled 11.10.99 as WFU *"Cloud Nine" (Located Lincoln)*	
G-BUXA	Colt 210A HAB	2400		28. 4.93	Cancelled 20.12.01 as WFU *"Buxam"*	7. 4.99T
G-BVBJ	Colt Flying Coffee Jar 1 SS HAB	2427		27. 7.93	Cancelled 29.4.97 as WFU *"Maxwell House I"*	21.11.96A
	(Maxwell House Jar)					
G-BVBK	Colt Flying Coffee Jar 2 SS HAB	2428		27. 7.93	Cancelled 29.4.97 as WFU	14. 2.97A
	(Maxwell House Jar)				*"Maxwell House II" (Located Lincoln)*	
G-BVFY	Colt 210A HAB				See SECTION 1 Part 2	
G-BVIO	Colt Flying Drinks Can SS HAB	2538		4. 2.94	Cancelled 6.11.01 as WFU	9. 6.00A
	(Budweiser Can shape)				*"Budweiser Can II"*	
G-BVWH	Cameron N-90 Lightbulb SS HAB	3404		8.12.94	Cancelled 17.12.01 as WFU *"Phillips Light Bulb"*	2. 9.98A
G-BVWI	Cameron Light Bulb 65SS HAB				See SECTION 1 Part 2	
G-BWAN	Cameron N-77 HAB				See SECTION 1 Part 2	
G-BWPL	Airtour AH-56 HAB	011	G-OAFC	19. 3.96	Cancelled 17.9.01 as WFU	
					(As "G-OAFC" - qv) "Paul J Donnellan" (Located Wokingham)	
G-BWUR	Thunder Ax10-210 Srs.2 HAB				See SECTION 1 Part 2	
G-BWZP	Cameron Home Special 105SS HAB	4051		6.12.96	Cancelled 15.1.04 as WFU *(Barclays Mortgages titles)*	10. 4.02A
G-BXAL	Cameron Bertie Bassett 90SS HAB				See SECTION 1 Part 2	
G-BXAX	Cameron N-77 HAB	2010		25. 5.89	Cancelled 31.1.02 as WFU *"Citroen"*	21.11.96A
G-BXHM	Lindstrand LBL-25A Cloudhopper HAB	466		30. 5.97	Cancelled 6.11.01 as WFU *"Bud Ice/Michelob"*	7. 5.00A
G-BXHN	Lindstrand Budweiser Can SS HAB	465		30. 5.97	Cancelled 6.11.01 as WFU *"Budweiser Can III"*	8. 3.01A
G-BXND	Cameron Thomas The Tank Engine 110SS HAB				See SECTION 1 Part 2	
G-BZIH	Lindstrand LBL 31A HAB	700		7. 6.00	Cancelled 6.11.01 as WFU *"Budweiser""Budweiser"*	11. 6.01A
G-CBPG	The Balloon Works Firefly 7 HAB				See SECTION 1 Part 2	
G-COLR	Colt 69A HAB	780		8. 4.86	Cancelled 21.5.93 as WFU *"Bubble"*	
G-COMP	Cameron N-90 HAB	1564		24. 9.87	Cancelled 18.12.01 by CAA *"Computacentre I"*	20. 5.97
G-CURE	Colt 77A HAB	1424		3. 7.89	Cancelled 29.4.97 as WFU	21.11.96A
	(Standard shape plus Tablet blisters)				*"Alka Seltzer III"*	
G-DHLI	Colt World 90SS HAB				See SECTION 1 Part 2	
G-DHLZ	Colt 31A Air Chair HAB	2604		2. 6.94	Cancelled 9.4.99 as WFU *"DHL Parcel"*	23. 7.99A
G-ENRY	Cameron N-105 HAB				See SECTION 1	

G-ETFT	Colt Financial Times SS HAB	1792	G-BSGZ	11. 1.91	Cancelled 10.12.02 by CAA *"Financial Times II"*	14.10.00A
G-FZZY	Colt 69A HAB	779		19. 2.86	Cancelled 29.4.97 as WFU *"Alka-Seltzer II"*	16. 2.90A
G-GEUP	Cameron N-77 HAB	880		8.12.82	Cancelled 9.5.01 as WFU *"Gee-Up"*	19. 7.96
G-GURL	Cameron A-210 HAB	2387		3. 9.90	Cancelled 2.12.98 as WFU *"Hot Airlines" (Located Austria)*	9. 8.96T
G-HELP	Colt 17A Cloudhopper HAB				See SECTION 1 Part 2	
G-HENS	Cameron N-65 HAB	740		8. 7.81	Cancelled 8.4.93 by CAA *"Free Range"*	
G-HLIX	Cameron Helix Oilcan 61SS HAB	1192		20. 9.85	Cancelled 29.4.97 as WFU *"Helix Oil Can"*	25. 4.90A
	(Originally regd as 80SS)					
G-HOFM	Cameron N-56 HAB				See SECTION 1 Part 2	
G-IAMP	Cameron H-34 HAB				See SECTION 1 Part 2	
G-IGEL	Cameron N-90 HAB	2726		7. 4.92	Cancelled 18.12.01 by CAA *"Computacentre II"*	12. 5.97A
G-IMAG	Colt 77A HAB	1718		9. 3.90	Cancelled 31.1.02 as WFU	19. 1.00A
	(Second envelope c/n 2254 as original dbf 6.92)				*"Agfa"*	
G-JANB	Colt Flying Bottle SS HAB	1643		16. 2.90	Cancelled 20.12.01 by CAA	30. 9.96A
	(J & B Whisky Bottle shape)				*"J & B Bottle"*	
G-KORN	Cameron Berentzen Bottle 70SS HAB				See SECTION 1 Part 2	
G-LLAI	Colt 21A Cloudhopper HAB	519	(G-BKTX)	18. 7.83	Cancelled 16.7.90 by CAA	
					(Lowndes Laing Insurance titles) "Llama"	
G-MAAC	Advanced Airship Corporation ANR-1	01		16. 1.89	Cancelled 15.11.00 by CAA *"ANR-1" (Located Lincoln)*	
G-MAPS	Sky Flying Map SS HAB	105		20. 7.98	Cancelled 31.7.01 as WFU *"OS Map"*	28. 2.01A
					(On loan from The Balloon Advertising Co Ltd)	
G-MHBD	Cameron O-105 HAB	1021		23. 2.84	Cancelled as WFU 17.4.98 *(Dawsons Toys titles)*	
G-MOLI (1)	Cameron A-250 HAB				See SECTION 1 Part 2	
G-NPNP	Cameron N-105 HAB	2959	G-BURX	18. 1.93	Cancelled 6.11.01 as WFU *"National Power III"*	18. 8.98A
G-NPWR	Cameron RX-100 HAB	2849		13. 7.92	Cancelled 15.7.98 as WFU	21.11.96A
					"Nuclear Rozier" (Located Lincoln)	
G-NWPB	Thunder Ax7-77Z HAB	278		13. 5.80	Cancelled 27.4.90 by CAA *"Post Office"*	
G-OAFC	Airtour AH-56 HAB			15. 6.89	Re-registered G-BWPL - qv	
G-OCND	Cameron O-77 HAB	1020		6. 2.84	Cancelled 19.5.95 by CAA *"CND Airborne"*	
G-OHDC	Colt Film Cassette SS HAB	2633		8. 8.94	Cancelled 31.1.02 as WFU	26. 8.99A
	(Agfa Film Can shape)				*"Agfa HDC Can"*	
G-OLDV	Colt 90A HAB	2592		5. 5.94	Cancelled 29.6.99 as WFU *"LDV"*	10.11.98A
G-OSVY	Sky 31-24 HAB	104		28. 5.98	Cancelled 31.7.01 as WFU *"OS Hopper"*	11. 3.00A
G-OXRG	Colt Film Can SS HAB	2138		17. 1.92	Cancelled 29.4.97 as WFU	
	(Agfacolor Film Can shape)				*"Agfa XRG Can"*	
G-PNEU	Cameron Colt Bibendum 110SS HAB	4223		5. 1.98	Cancelled 13.08.04 as WFU	26. 6.02A
					(Op Balloon Preservation Group) (Michelin titles)	
G-PONY	Colt 31A Air Chair HAB	434		23. 8.82	Cancelled 19.5.95 by CAA *"Neddie"*	
G-POPP	Colt 105A HAB	1776		1. 3.91	Cancelled 5.2.99 as WFU *"Mercier"*	21.11.96A
G-PSON	Colt Cylinder One SS HAB				See SECTION 1 Part 2	
G-PURE	Cameron Can 70SS HAB	1913		18. 1.89	Cancelled 29.4.97 as WFU *"Guinness Can"*	
	(Guinness Can)					
G-PYLN	Cameron Pylon 80SS HAB	2958	G-BUSO	18. 1.93	Cancelled 6.11.01 as WFU	25. 4.97A
	(Electricity Pylon shape)				*"Essex Girl"*	
G-RARE	Thunder Ax5-42 SS HAB	266		20. 2.80	Cancelled 14.5.02 by CAA	7. 4.95A
	(J & B Rare Whisky Bottle shape)				*"J & B Hamish"*	
G-RIPS	Cameron Action Man/Parachutist 110SS HAB	4092		29. 4.97	Cancelled 23.7.02 as WFU	25. 5.00A
					"Action Man" (Located Stockport)	
G-SCAH	Cameron V-77 HAB	788		18. 1.82	Cancelled 30.11.01 by CAA	24. 7.87A
					"Orpheus" (Located Southampton)	
G-SCFO	Cameron O-77 HAB				See SECTION 1 Part 2	
G-SEGA	Cameron Sonic 90SS HAB				See SECTION 1 Part 2	
G-SEUK	Cameron TV 80SS HAB	3810		12. 4.96	Cancelled 31.1.02 as WFU	24. 3.00A
	(Samsung Computer shape)				*"Samsung Monitor"*	
G-TTWO	Colt 56A HAB	087		14. 5.80	Cancelled 14.11.95 as WFU *"Tea 4 Two"*	1. 9.87A
G-UNIP	Cameron Oil Container SS HAB	2532		15. 3.91	Cancelled 31.1.02 as WFU	7.11.96A
	(Unipart Sureflow Oil Can)				*"Unipart Can"*	
G-VOLT	Cameron N-77 HAB	2157		8.11.89	Cancelled 23.10.01 as WFU *"National Power II"*	25. 4.97A
G-WATT	Cameron Cooling Tower SS HAB	2158		8.11.89	Cancelled 23.10.01 as WFU *"Cooling Tower"*	22. 8.96A
G-WCAT	Colt Flying Mitt SS HAB				See SECTION 1 Part 2	
G-WINE	Thunder AX7-77Z HAB	472		25.11.82	Cancelled 10.4.02 by CAA *"Gemini"*	17. 6.97A
G-WORK	Thunder Ax10-180 Srs.2 HAB	2396	DQ-PBF	12. 5.93	Cancelled 3.11.97 *(Sold as DQ-HBF 11.97)*	
			G-WORK		*(Paradise Balloon Flights titles) (As "DQ-PBF") (Located Lincoln)*	

Visitor Centre and Shoreham Airport Historical Association, Shoreham Airport BN43 5FF (www.thearchivesshoreham.co.uk)

BAPC.20	Lee-Richards Annular Biplane rep	---	*(Built for "Those Magnificent Men in Their Flying Machines" film)*
BAPC.277	Mignet HM.14 Pou-Du-Ciel		
BAPC.300	Piffard Hummingbird reproduction		

Tangmere Military Aviation Museum, Tangmere PO20 6ES (www.tangmere-museum.org.uk)

BAPC.214	Supermarine Spitfire Prototype fsm	---	*(As "K5054")*
BAPC.241	Hawker Hurricane I fsm	---	*(As "L1679/JX-G" in 1 Sqdn c/s)*
	(Built Aerofab 1994)		
BAPC.242	Supermarine Spitfire VB rep	---	*(As "BL924/AZ-G" in 234 Sqdn c/s) "Valde Maar Atterdag"*
	(Built TDL Reps 1994)		

Foulkes-Halbard Collection, Filching Manor, Wannock BN26 5QA

G-BHNG	Piper PA-23-250 Aztec E	27-7405432	N54125	13. 5.80	Cancelled 12.12.86 by CAA
		(Badly damaged following collision with Cessna F152 G-BFRB on take-off Shoreham 19.12.81) (Fuselage only)			
BAPC.127	Halton Man Powered Aircraft Group Jupiter				
IAHC.2	Aldritt Monoplane rep				

WARWICKSHIRE

Midland Air Museum, Coventry CV8 3AZ (www.midlandairmuseum.org.uk)

G-EBJG	Parnall Pixie III	---	No.17/18 (Lympne 1924?)	. 9.24	Cancelled 1.12.46 by Sec of State *(Components only 2.05)*	2.10.36
G-ABOI	Wheeler Slymph	AHW.1		17. 7.31	Cancelled 1.12.46 by Sec of State *(On loan from A.H.Wheeler) (Dismantled components only 2.05)*	
G-AEGV	Mignet HM.14 Pou-Du-Ciel *(Built East Midlands Aviation Company)*	EMAC.1		22. 4.36	Cancelled 12.37 *(WFU 26.5.37)*	26. 5.37
G-ALCU	de Havilland DH.104 Dove 2B	04022	VT-CEH	3. 8.48	Cancelled 8.9.78 as WFU *(Displayed as "G-ALVD" in 'Dunlop Aviation Division' c/s)*	16. 3.73
G-AOKZ	Percival P.40 Prentice 1	PAC/238	VS623	20. 4.56	Not converted *(Became instructional airframe)* *((As "VS623" with wings from G-AONB - No wheel spats)*	
G-APJJ (2)	Fairey Ultralight Helicopter	F.9428		4.12.57	Cancelled 2.3.73 as WFU *(Royal Navy' c/s)*	1. 4.59
G-APRL	Armstrong-Whitworth 650 Argosy Srs.101	AW.6652	N890U N602Z, N6507R, G-APRL	2. 1.59	Cancelled 19.11.87 as WFU *(Elan titles) "Edna"*	23. 3.87T
G-APWN	Westland WS-55 Whirlwind 3	WA.298	VR-BER G-APWN, 5N-AGI, G-APWN	8. 9.59	Cancelled 25.6.81 as WFU *(Bristow Helicopters titles) "Skerries"*	17. 5.78
G-ARYB	de Havilland DH.125 Srs.1	25002		1. 3.62	Cancelled 4.3.69 *(Mounted wingless on plinths)*	22. 1.68
G-MJWH	Chargus Vortex 120 *(Regn reserved 1983 for Chargus T.250 and engine: fitted to 1974 Vortex hang glider: abandoned and only wing on display)*			R		
BAPC.9	Humber-Bleriot XI Monoplane reconstruction (Humber) *(Converted to 1911 Humber Bleriot Monoplane)*	---			*(No markings)*	
BAPC.32	Crossley Tom Thumb	---			*(Not completed Banbury 1937) (Stored 2.04)*	
BAPC.126	Rollason-Druine D.31 Turbulent	---			*(Static airframe) (Yellow c/s, no markings)*	
BAPC.179	Sopwith Pup fsm	---			*(As "A7317")*	
"BGA.804"	Slingsby Cadet TX.1				*(As "BAA")* (Also see SECTION 5 Part 1)	

WILTSHIRE

Royal Air Force Lyneham

G-AMPO	Douglas C-47B-30DK Dakota 3 *(Regd as c/n 16438/33186)*	16437/33185	LN-RTO G-AMPO, KN566, 44-76853	25. 2.52	Cancelled 18.10.01 as WFU *(As "FZ625")*	29. 3.97A

Science Museum Air Transport Collection and Storage Facility, Wroughton SN4 9NS (www.sciencemuseum.org.uk/wroughton

G-AACN	Handley Page HP.39 Gugnunc	1	K1908 G-AACN	2.11.28	Cancelled 12.30 on transfer to RAF	19.9.30
G-ACIT	de Havilland DH84 Dragon 1	6039		24. 7.33	Cancelled 26.4.02 as WFU	25. 5.74
G-AEHM	Mignet HM.14 Pou-Du-Ciel *(Built H J Dolman)* (ABC Scorpion @ 35 hp)	HJD.1		30. 4.36	Cancelled in 3.39 census *"Blue Finch"*	
G-ALXT	de Havilland DH.89A Dragon Rapide	6736	4R-AAI CY-AAI, G-ALXT, NF865	24. 1.50	Cancelled 5.7.51 - to CY-AAI *(Railway Air Service titles) "Star of Scotia"*	
G-ANAV	de Havilland DH.106 Comet 1A	06013	CF-CUM	15. 8.53	Cancelled 1.7.55 as WFU *(Broken up RAE Farnborough 1955: nose section)*	
G-APWY	Piaggio P.166	362		16.12.59	Cancelled 20.10.00 by CAA	14. 3.81
G-APYD	de Havilland DH.106 Comet 4B	6438	SX-DAL G-APYD	21. 1.60	Cancelled 23.11.79 as WFU *(Dan-Air titles)*	3. 8.79T
G-ATTN	Piccard HAB (62,000 cu ft)	15 & 1352		27. 4.66	Cancelled 5.12.77 as PWFU *(Envelope/basket stored 6.94) "The Red Dragon"*	
G-AVZB	LET Z-37 Cmelak	04-08	OK-WKQ	30.11.67	Cancelled 21.12.88 as WFU	5. 4.84A
G-AWZM	Hawker Siddeley HS.121 Trident 3B Srs.101	2314		14. 1.69	Cancelled 18.3.86 as WFU *(WFU 13.12.85) (British Airways titles)*	13.12.85T
G-BBGN	Cameron A-375 HAB	90		23. 8.73	Cancelled 22.8.89 as WFU	
G-BGLB	Bede BD.5B (Hirth 230R)	3796 & PFA 14-10085		2. 3.79	Cancelled 21.11.91 by CAA	4.8.81P
G-CONI	Lockheed 749A-79 Constellation	2553	N7777G N173X), N7777G, TI-1045P, PH-LDT, PH-TET	12. 5.82	Cancelled 13.6.84 as WFU *(As "N7777G" in TWA c/s)*	
G-MMCB	Huntair Pathfinder II	136		13. 7.83	Cancelled 23.11.88 as WFU	
G-RBOS	Colt AS-105 HA Airship	390		9. 2.82	Cancelled 3.4.97 by CAA	6. 3.87A
EI-AYO (2)	Douglas DC-3A-197	1911	N655GP N65556, N255JB, N8695E, N333H, NC16071	5.3.76		
BAPC.52	Lilienthal Glider Type XI *(1895 original)*	Not known				
BAPC.162	Goodhart Man Powered Aircraft Newbury Manflier					
BAPC.172	Chargus Midas Super E Hang Glider	---				
BAPC.173	Birdman Promotions Grasshopper	---				
BAPC.174	Bensen B.7 Gyroglider	---				
BAPC.188	McBroom Cobra 88 Hang Glider	---				
BAPC.276	Hartman Ornithopter	---				

YORKSHIRE

Museum of Army Transport, Beverley HU17 0NG (www.museum/armytransport.co.uk)

G-AOAI	Blackburn Beverley C.1	1002	XB259	15. 3.55	Cancelled 30.3.55 *(Restored as XB259)* *(To be transferred to Yorkshire Air Museum, Elvington 2005?)*	

AeroVenture, Doncaster DN4 5EP

G-ALYB	Taylorcraft J Auster 5	1173	RT520	3. 2.50	Cancelled 29.2.84 by CAA *(Fuselage only)*	26. 5.63
G-AOKO	Percival P.40 Prentice 1	PAC/234	VS621	13. 4.56	Cancelled 9.10.84 as WFU *(On loan from Atlantic Air Transport Ltd)*	23.10.72
G-APMY	Piper PA-23-160 Apache	23-1258	EI-AJT	15. 5.58	WFU 1.11.81 *(On loan from W.Fern)*	1.11.81
G-ARGI (2)	Auster 6A Tugmaster	2299	VF530	8.12.60	WFU at Heathfield in 7.73 *(Fuselage stored)*	4.7.76
G-ARHX	de Havilland DH.104 Dove 8	04513		11. 1.61	WFU 8.9.78	8. 9.78
G-AVAA	Cessna F150G *(Built Reims Aviation SA)*	F150-0164		14.10.66	Cancelled 16.4.96 by CAA *(Fuselage only)*	5. 7.96

G-BOCB	Hawker Siddeley HS.125 Srs.1B/522	25106	G-OMCA	14. 9.87	Cancelled as WFU 22.2.95	16.10.90
			G-DJMJ, G-AWUF, 5N-ALY, G-AWUF, HZ-BIN *(WFU Luton 1994 for spares) (Cockpit only)*			
G-DELB	Robinson R22 Beta	0799	N26461	18. 5.88	Cancelled 20.4.95 by CAA *(DBR Sherburn-in-Elmet 27.12.94)*	
G-MJKP	Hiway Skystrike/Super Scorpion	PEB-01		7. 9.82	Cancelled 9.12.94 as WFU	
	(Fuji-Robin EC-25)					
G-MVRS	CFM Shadow Series BD	099		7. 4.89	Cancelled 5.7.90 as destroyed	4. 6.90P
					(Engineless fuselage shell in Storage Shed 4.04)	
BAPC.207	Austin Whippet fsm	---			*(On loan from D Charles) (As "K.158")*	
	(Built Ken Fern/Vintage and Rotary Wing Collection c.1993)					
BGA.2517	Slingsby T.30B Prefect TX.1	577	WE987	6.79	*(No external marks)*	
BGA.3239	Slingsby T.31B	708	WT913	26.10.86	*(On loan from J.M.Brookes and Partners) (As "WT913")*	21. 7.96

Museum and Art Gallery, Doncaster DN1 2AE (www.doncaster.gov.uk)

G-AEKR	Mignet HM.14 Pou-Du-Ciel	CAC.1		26. 6.36	WFU and cancelled 31.7.38	22. 6.37
	(Buit E Claybourne & Co)		*(Stored Doncaster 1938/1960: dbf RAF Finningley 4.9.70: rebuilt and and allocated "G-AEKR"/BAPC.121)*			
BAPC.275	Bensen B.7	--				
	(Built S J R Wood, Warmsworth) (Volkswagen 1600)					

Yorkshire Air Museum, Elvington YO41 4AU (www.yorkshireairmuseum.gov.uk)

G-AJOZ	Fairchild 24W-41A Argus 1	347	FK338	21. 4.47	Cancelled	15.12.63
	(UC-61-FA)		42-32142		*(Crashed Rennes, France 16.8.62)*	
G-AMYJ	Douglas C-47B-25DK Dakota 6	15968/32716	SU-AZF	23. 2.53	Cancelled 12.12.01 as WFU	4. 4.97A
			G-AMYJ, XF747, G-AMYJ, KN353, 44-76384			
G-ASCD	Beagle A.61 Terrier 2	B.615	PH-SFT	23. 7.62	Cancelled 5.10.89 as WFU	26. 9.71
			(PH-SCD), G-ASCD, VW993		*(As "TJ704/JA")*	
G-AVPN	Handley Page HPR.7 Dart Herald 213	176	I-TIVB	22. 6.67	Cancelled 8.12.97 as WFU	14.12.99T
			G-AVPN, D-BIBI, (HB-AAK)		*(Channel Express titles)*	
G-BDBZ	Westland WS-55 Whirlwind 2 (HAR.10)	WA.62	XJ398	23. 4.75	Cancelled 28.3.85 by CAA *(Not converted) (As "XJ398")*	
	(Regd with c/n WA.386)		(XD768)		*(On loan from Yorkshire Helicopter Preservation Group*	
G-BKDT	Royal Aircraft Factory SE.5A rep	PFA 080-10325		26. 5.82	Cancelled 11.7.91 by CAA	
	(Built J A Tetley - kit no.278)				*(No Permit issued) (As "F943" - note carried by G-BIHF also)*	
G-HNTR	Hawker Hunter T.7	HABL-003311	8834M	7. 7.89	Cancelled 11.10.91 as WFU	
			XL572		*(As "XL571/V" in "Blue Diamonds" c/s)*	
G-MJRA	Mainair Tri-Flyer 250/Hiway Demon	PRJM-01		21.12.82	Cancelled 24.1.95 by CAA	28. 6.91
G-TFRB	Air Command 532 Elite Sport	PFA G/04-1167		26. 4.90	Cancelled 7.6.01 by CAA	6. 8.98P
	(Kit no.0628)					
BAPC.28	Wright Flyer fsm	---				
BAPC.41	Royal Aircraft Factory BE.2c fsm	---	"6232"			
BAPC.42	Avro 504K fsm	---			*(As "H1968")*	
BAPC.76	Mignet HM.14 Pou-Du-Ciel	---			*(As "G-AFFI")*	
	(Modern reproduction) (Scott)					
BAPC.89	Cayley Glider fsm	---				
BAPC.130	Blackburn (1911) Monoplane fsm	---			*"Mercury" (Built for TV Series."The Flambards")*	
BAPC.157	WACO CG-4A Hadrian Glider	---			*(As "237123") (Fuselage frame section only and tail pieces ex 456476)*	
BAPC.240	Messerschmitt Bf.109G fsm	---				
	(Built D.Thorton 1994)					
BAPC.254	Supermarine Spitfire 1 fsm	---			*(As "R6690/PR-A" in 609 Sqdn c/s)*	
BAPC.265	Hawker Hurricane fsm	---			*(As "P3873/YO-H" in 1 Sqdn/RCAF c/s)*	
BAPC.270	de Havilland DH.60 Moth fsm	---			*(As "G-AAAH") "Jason"*	

Street Life Museum, High Street Hull

BAPC.287	Blackburn Lincock fsm				*(As "G-EBVO")I*	
	(Built BAE, Brough c 2002)					

Eden Camp Modern History Theme Museum, Malton YO17 6RT (www.edencamp.co.uk)

BAPC.230	Supermarine Spitfire fsm	---	"AA908"		*(As "AB550/GE-P" in 349 Sqdn c/s)*	
	(Built TDL Replicas 1993)					
BAPC.235	Fieseler Fi 103 (V-1) fsm)	---				
	(Built TDL Replicas 1993)					
BAPC.236	Hawker Hurricane fsm	---			*(As "P2793/SD-M" in 501 Sqdn c/s)*	
	(Built G B Moulders, Norfolk 1993)					

SCOTLAND

Dumfries and Galloway Aviation Museum, Dumfries DG2 9PS (www.members.xoom.com/dgamuseum)

G-AHAT	Auster J/1N Alpha	1849	(HB-EOK)	11. 2.46	Cancelled 7.7.75 as WFU *(Crashed 31.8.74) (Frame only)*	6. 2.75
G-AWZJ	Hawker Siddeley HS.121 Trident 3B Srs.101	2311		14. 1.69	Cancelled 7.3.86 as WFU *(Forward fuselage only)*	12. 9.86T

National Museums of Scotland - Museum of Flight, East Fortune EH39 5LF (www.nms.ac.uk/flight)

G-ABDW	de Havilland DH.80A Puss Moth	2051	VH-UQB	23. 8.30	Cancelled 12.33 WFU and 21.1.82 PWFU	
			G-ABDW		*(To VH-UQB 27.5.31 and restored 25.3.77) (As "VH-UQB")*	
G-ACVA	Kay Gyroplane 33/1	1002		26. 6.34	Cancelled 9.58 *(On loan from Glasgow Museum of Transport)*	
	(Pobjoy R @ 75hp)					
G-ACYK	Spartan Cruiser III	101		2. 5.35	Crashed Largs, Ayrshire 14.1.38 *(Remains recovered 7.73)*	2. 6.38
G-AFJU	Miles M.17 Monarch	789	X9306	25. 8.38	Cancelled 18.11.74 as PWFU	18. 5.64
			G-AFJU		*(On loan from Aircraft Preservation Society of Scotland)*	
G-AGBN	General Aircraft GAL.42 Cygnet 2	111	ES915	4.10.40	Cancelled 15.11.88 as WFU	28.11.80P
			G-AGBN			
G-AHKY	Miles M.18 Srs.2	4426	HM545	26. 4.46	Cancelled 19.3.92 as WFU	20. 9.89P
			U-0224, U-8			
G-ANOV	de Havilland DH.104 Dove 6	04445	G-5-16	11. 3.54	Cancelled 6.7.81 as WFU *(Civil Aviation Authority titles)*	31. 5.75
G-AOEL	de Havilland DH.82A Tiger Moth	82537	N9510	27. 9.55	WFU 18.7.72	18. 7.72
G-APHV	Avro 652A Anson C.19 Srs.2	—	VM360	19. 9.57	Cancelled 21.1.82 as PWFU *As "VM360")*	15. 6.73

G-ARCX	Gloster Meteor NF.14	AW.2163	WM261	8. 9.60	Cancelled 25.10.73 as WFU	20. 2.69S
	(Built Armstrong-Whitworth Aircraft)				*(WFU 2.69)*	
G-ASUG	Beech E18S-9700	BA-111	N575C	3. 7.64	WFU 12.5.75	23. 7.75
			N555CB, N24R		*(Loganair c/s)*	
G-ATFG	Brantly B.2B	448		16. 6.65	Cancelled 25.9.87 as WFU	25. 3.85
	(Composite with parts from G-ASLO/G-AXSR)				*(On loan from Aircraft Preservation Society of Scotland)*	
G-ATOY	Piper PA-24-260 Comanche B	24-4346	N8893P	7. 2.66	Crashed near Elstree 6.3.79 *"Myth Too" (Fuselage only)*	
G-AXEH	Beagle B.125 Bulldog 1	B.125-001		25. 4.69	Cancelled 15.1.77 as WFU	15. 1.77
G-BBBV	Handley Page HP.137 Jetstream	234	N14234	26. 6.73	Cancelled 21.8.74 *(To N200SE 8.74)*	
	(Fuselage used by BAe as Jetstream 31 mock-up)		N102SC, N1BE, (N200SE), G-BBBV, G-8-12		*(As "N14234")*	
G-BBVF	Scottish Aviation Twin Pioneer 3	558	7978M	17.12.73	Cancelled 8.8.83	14. 5.82
			XM961			
G-BDFU	PMPS Dragonfly MPA Mk.1	01		14. 7.75	Cancelled 12.83 as WFU *(On loan from R.J.Hardy and R.Churcher)*	
G-BDIX	de Havilland DH.106 Comet 4C	6471	XR399	1. 9.75	Cancelled 2.9.91 by CAA *(Dan-Air titles)*	11.10.81T
G-BDYG	Percival P.56 Provost T.1	PAC/F/056	7696M	25. 5.76	Cancelled 4.11.91 by CAA	28.11.80P
			WV493		*(As "WV493/29/A-P")*	
G-BIRW	Morane-Saulnier MS.505 Criquet	695/28	OO-FIS	10. 4.81	Cancelled 15.11.88 as WFU	3. 6.83P
			F-BDQS, French AF 695		*(As "FI+S" in Luftwaffe c/s)*	
G-BOAA	British Aircraft Corporation-Aérospatiale Concorde 102				Cancelled 4.5.04 as WFU	24. 2.01T
G-JSSD	Handley Page HP.137 Jetstream 1	227	N510F	14. 6.79	Cancelled 4.1.96 by CAA	9.10.90
	(Conv to BAe Jetstream Srs.3001 prototype.1979/80)		N510E, N12227, G-AXJZ			
G-MBJX	Hiway Skytrike I/Hiway Super Scorpion	MM-01		2. 2.82	Cancelled 13.6.90 by CAA	
	(Valmet SM160 s/n 15108)					
G-MBPM	Eurowing Goldwing	EW-21		14. 4.82	Cancelled 30.8.00 as WFU	21. 8.98P
	(Fuji-Robin EC-34-PM)					
G-MMLI	Mainair Tri-Flyer 250/Solar Wings Typhoon S	RPAT-01 & T484-423L	BAPC.244	26. 3.84	Cancelled 7.9.94 by CAA	
					(Originally regd as Hiway Skytrike Mk.II 250)	
BAPC.49	Pilcher Hawk Glider	---				
	(1896 original) (Rebuilt after fatal crash Stanford Hall, Leicester 30.9.1899)					
BAPC.59	Sopwith F1 Camel fsm	---	"D3419"		*(As "B5577/W")*	
			"F1921"			
BAPC.70	Auster AOP.5	TAY/33153	"GALES"		*(As "TJ398") (On loan from Aircraft Preservation Society of Scotland)*	
BAPC.85	Weir W-2	---			*(As "W-2") (On loan)*	
	(Weir Dryad II @ 50hp)					
BAPC.160	Chargus 18/50 Hang Glider	---				
BAPC.195	Birdman Sports Moonraker 77 Hang Glider	---				
	(Built c.1977)					
BAPC.196	Southdown Sailwings Sigma 2m Hang Glider	---				
	(Built c.1980)					
BAPC.197	Scotkites Cirrus III Hang Glider	---				
	(Built 1977)					
BAPC.245	Electraflyer Floater Hang Glider	---			*(Wing only)*	
	(Built 1979)					
BAPC.246	Hiway Cloudbase Hang Glider	---				
	(Built 1978)					
BAPC.247	Albatros ASG.21 Hang Glider	---				
	(Built 1977)					
BAPC.262	Catto CP-16	---				
BGA.852	Slingsby T.8 Tutor	---	TS291	2. 7.58	*(As "TS291")*	
BGA.902	Slingsby T.12 Gull I	---		15. 5.59	*(As "BED") "G-ALPHA"*	
BGA.1014	Slingsby T.21B	556	SE-SHK	1.62	*(As "BJV")*	

City of Edinburgh Council, Edinburgh Airport

BAPC.227	Supermarine Spitfire IA fsm	---			*(As "L1070/XT-A" in 603 Sqdn c/s)*	

Museum of Transport, Kelvin Hall, Glasgow G3 8DP

BAPC.48	Pilcher Hawk glider rep	---			*(Stored 3.04)*	
	(Built No.2175 Sqdn ATC, Glasgow 1966)					

WALES

Caernarfon Air Museum, Caernarfon, Gwynedd LL54 5TP (www.caeairparc.com)

G-ALFT	de Havilland DH.104 Dove 6	04233		14.12.48	Cancelled 11.2.77 as WFU	13. 6.73
G-AMLZ	Percival P.50 Prince 6E	P.46	(VR-TBN)	23.11.51	Cancelled 9.10.84 as WFU	18. 6.71
G-AWUK	Cessna F150H	F150-0344		25.11.68	Cancelled 13.4.73 as WFU	3. 9.73
	(Built Reims Aviation SA)				*(Crashed Shoreham 4.9.71) (Cockpit only)*	
G-MBEP	American Aerolights Eagle 215B	2877		9.11.81	Cancelled 16.5.96 as WFU	8. 4.96E
	(Chrysler 820)					
BAPC.201	Mignet HM.14 Pou-Du-Ciel	---				
BAPC.286	Mignet HM.14 Flea				See SECTION 3 Part 1	Cardiff
	(Scott A2S)					

National Industrial Museum, Swansea

BAPC.47	Watkins CHW Monoplane	---			Due in 2005	
	(Watkins 40hp)					

Maes Artro Village, Llanbedr, Harlech

BAPC.202	Supermarine Spitfire V fsm				*(Built for "Piece of Cake" TV series)*	
					(As "MAV467/"RO")	

IRELAND

Ulster Folk and Transport Museum, Holywood, Belfast BT18 0EU (www.nidex.com/uftm)

G-ACUX	Short S.16 Scion 1	S.776	VH-UUP	26. 6.34	Cancelled 2.38 and 7.81	
			G-ACUX		*(To VH-UUP 2.10.35 and restored 25.3.77) (Stripped frame)*	

G-AJOC	Miles M.38 Messenger 2A	6370		23. 4.47	Cancelled 5.1.82 as WFU *(Rear fuselage)*	18. 5.72
G-AKEL	Miles M.65 Gemini 1A	6484		8. 9.47	Cancelled 30.5.84 as WFU	29. 4.72
	(Centre section) (But see entry for ***South East Aviation Enthusiasts Group*** *below)*					
G-AKGE	Miles M.65 Gemini 3C	6488	EI-ALM G-AKGE	18.10.47	Cancelled 30.5.84	7. 6.74
G-AKLW	Short SA.6 Sealand 1	SH.1571	(USA)	26.11.47	Sold abroad 8.51	
	R Saudi AF, SU-AHY, G-AKLW *(Stored dismantled 2.04)*					
G-AOUR	de Havilland DH.82A Tiger Moth	86341	NL898	14. 8.56	Crashed Newtownards 6.6.65	19.11.66
G-ARTZ (1)	McCandless M.2 (Triumph 650cc)	M2/1		?.10.61	Replaced by G-ARTZ (2) – see SECTION 1 Part 2	
G-ATXX	McCandless M.4 (Volkswagen 1600)	M4/3		27. 7.66	Cancelled 9.9.70 as WFU	
G-BKMW	Short SD.3-30 Sherpa Var.100	SH3094	G-14-3094	13.12.82	Cancelled 14.11.96 as WFU *(Broken up 3.96) (Cockpit only)*	14. 9.90
BGA.470	Short Nimbus	S.1312		.47	*(As "ALA") (Stored 6.97*	8.75
IAHC.6	Ferguson monoplane rep *(Built Capt J.Kelly Rogers 1974)* (Original engine)	---				
IAHC.9	Ferguson monoplane rep *(Built L.Hannah 1980)*	---			*(Stored 1.04)*	

Ulster Aviation Heritage, Langford Lodge, Belfast BT16 1WQ (www.ulsteraviationsociety.co.uk)

G-BDBS	Short SD.3-30 UTT *(Airframe originally laid down as SC.7 Skyvan c/n SH.1935)*	SH.1935 & SH.3001	G-14-3001	21. 4.75	Cancelled 1.7.93 as WFU	28. 9.92S
G-BTUC	Embraer EMB-312 Tucano	312.007	G-14-007 PP-ZTC	19. 6.86	Cancelled 20.12.96 as WFU	11. 9.93
G-MJWS	Eurowing Goldwing (Fuji-Robin EC-34-PM)	EW-22		16. 5.83	Cancelled 23.6.97 by CAA	
G-RENT	Robinson R22 Beta	0758	N2635M	17. 3.88	Cancelled 11.12.03 as WFU *(Damaged Newtownards 30.9.92)*	12. 6.94T
EI-BAG	Cessna 172A	47571	G-ARAV N9771T	7. 8.74	*(Damaged 3.10.76)*	26. 6.79
EI-BUO	Aero Composites Sea Hawker *(Aka Glass S.005E)*	80		25. 8.87		
BAPC.263	Chargus Cyclone *(Built 1979)*	---				
BAPC.266	Rogallo Hang Glider	---				

Nutgrove Shopping Centre, Churchtown, Dublin

EI-124	Grob G.102 Astir Standard CS 77	1761	D-...	.80		

Meath Aero Museum, Ashbourne, Co.Meath

IGA.6	Slingsby T.8 Tutor	---	IAC.6 VM657	.56		

South East Aviation Enthusiasts Group c/o Cavan and Leitrim Railway, Dromod, Leitrim, County Leitrim

G-ALCS (2)	Miles M.65 Gemini 3C *(Originally regd as Srs.3A with c/n 6534)*	WAL/C/1001		7.11.49	Cancelled 30.5.84 by CAA *(WFU 1983)*	
	(Cockpit only remains and the provenance was suspect but now thought to be from Miles M.65 Gemini 1A G-AKEL when transferred to Ulster Folk & Transport Museum: components only noted 4.96 and was for rebuild with G-AKGE)					
G-AMDD	de Havilland DH.104 Dove 6 *(Originally regd as Srs.2, then 2B)*	04292		8. 8.50	Cancelled 26.9.68 *(As "IAC 176")* *(To VQ-ZJC 9.68, 3D-AAI, VP-YKF and IAC 176)*	
G-AOGA	Miles M.75 Aries 1	75/1007	EI-ANB G-AOGA	9.11.55	Cancelled 30.5.84 *(To EI-ANB 18.5.63, restored 10.9.65: damaged Cork 8.8.69)*	10.10.69
G-AOIE	Douglas DC-7C	45115	PH-SAX G-AOIE	27. 8.56	Cancelled 31.3.70 as WFU *(To PH-SAX 10.4.67, restored 18.11.69)* *(Scrapped 10.97: fuselage only)*	
G-AYAG	Boeing 707-321	18085	N759PA	26. 3.70	Cancelled 8.12.72 *(As "VP-BDF") (Nose only)*	
			(To G-41-2-72, (CS-BDG) (N435MA) and VP-BDF 12.72)			
EI-BDM	Piper PA-23-250 Aztec D	27-4166	G-AXIV N6826Y	10.10.77	*(WFU and scrapped Shannon 4.85) (Fuselage only)*	
IAHC.1	Mignet HM.14 Pou-Du-Ciel	---			*"St.Patrick" (Noted 4.04)*	

ARGENTINA

Museo Nacional de Aeronatica, Buenos Aires

G-AICH	Bristol 170 Freighter Mk.1A	12751		26. 8.46	Cancelled 19.12.46 - to LV-XIM 2.47 *(Subsequently LV-AEY, Argentine AF T-30 and TC-330)*	

AUSTRALIA

Australian War Memorial, Canberra, Australian Capital Territory

G-EAQM	Airco DH.9 (AS Puma)	---	F1278	31.12.19	Cancelled 8.1.20 *(NTU and to Australia 1920 as "G-EAQM")*	1. 1.21

National Museum of Australia, Canberra, Australian Capital Territory

G-AERD	Percival P.3 Gull Six	D.65	HB-OFU	16. 9.77	Cancelled 28.11.86 on sale to Australia	
G-AHTN	Percival P.28 Proctor I	K.279	P6245	29. 5.46	Cancelled 22.6.48 - to VH-BLC 6.48 *(Subsequently VH-FEP)*	

Camden Museum of Aviation, Kogarah, New South Wales

G-AGTB	Percival P.44 Proctor V *(C/n reported as changed to Ae.9)*	Ae.8		8.12.45	Cancelled 5.11.47 - to VH-BCM 11.47 *(Subsequently VH-SST and VH-BCM) (As "NP336"*	
G-AMWI	Bristol 171 Sycamore 4	13070		1.12.52	Cancelled 1958 - to RN as "XN635" *(Subsequently VH-BAW) (As "XR592")*	

Donald and Robert Bunn Collection, Albury, New South Wales

G-AFOW	de Havilland DH.94 Moth Minor	94047		16. 5.39	Cancelled 16.5.39 - to R.Australia AF 10.39 but NTU *(Subsequently VH-ACS 23.1.40)*	

Australian Aviation Museum, Bankstown, New South Wales

G-ASFI	de Havilland DH.114 Heron 2D *(Converted to Riley Heron)*	14108	(N4661T) G-ASFI, West German AF CA+001, G-5-15	4. 3.63	Cancelled 15.9.64 - to CR-GAT *(Subsequently VH-CLW, T3-ATA, DQ-FDY, VH-CLW and DQ-FDY)*	

Greg and Nick Challinor Collection, Murwillumbah Airfield, Murwillumbah, New South Wales

G-ANHM	Taylorcraft G Auster 4	846	MT137	5.12.53	Cancelled 12.10.54 - to VH-AZO 11.54 *(Subsequently VH-ILS and VH-HPM)*	

Fighter World, RAAF Williamtown, Raymond Terrace, New South Wales

G-MUST	Commonwealth CAC-18 (P-51D) Mustang Mk.22	1524	VH-BOZ RAAF A68-199	20.12.79	Cancelled 31.7.91 by CAA - not imported *(Restored 15.6.93 but cancelled 3.3.97 by CAA)*	

Pay's Flying Museum, Scone Aerodrome, Scone, New South Wales

G-AOSP	de Havilland DHC-1 Chipmunk 22	C1/0174	WB722	26. 6.56	Cancelled - to VH-BTL *(Subsequently VH-BWF and VH-AMV)*	

Temora Aviation Museum, Temora, New South Wales

G-BURM	English Electric Canberra TT.18 *(Built Handley Page Ltd)*	---	WJ680	11.12.92	Cancelled 3.7.02	2.12.02P

Central Australian Aviation Museum, Alice Springs, Northern Territory

G-AHVG	Percival P.30 Proctor II	H.224	BV658	17. 6.46	Cancelled 25.3.57 - to VH-AVG 5.58	

Queensland Air Museum, Caloundra Airfield, Queensland

G-ASVC	de Havilland DH.114 Heron 2D *(Converted to Riley Heron)*	14123	EC-AOF	31. 7.64	Cancelled 1.10.64 - to VQ-FAF *(Subsequently VH-KAM)*	

Queensland Museum, South Brisbane, Queensland

G-EACQ	Avro 534 Baby	534/1	VH-UCQ G-AUCQ, G-EACQ, K-131	29. 5.19	Sold 6.21 - to G-AUCQ 12.7.21 *(Subsequently VH-UCQ 10.30) (As "G-EACQ")*	
G-EBOV	Avro 581E Avian *(Originally regd as Avro 581, to 581A in 1927 and modified to 581E)*	5116	No 9 (Lympne 1926)	7. 7.26	Cancelled 14.1.30 as sold in Australia	30. 1.29
G-ABLK	Avro 616 Avian V	R3/CN/523	VH-UQG G-ABLK	4.31	Cancelled - to VH-UQG 16.9.31 *(Restored 16.8.32 and crashed south Reggane, Sahara 12.4.33)*	

Caboolture Warplane Museum Trust, Caboolture Airfield, Caboolture, Queensland

G-AFOR	de Havilland DH.94 Moth Minor	9404		29. 8.39	Cancelled - to VH-AGL 1939 *(Subsequently R.Australian AF A21-14 @ 2.40 and VH-AGO) (As "A21-14")*	
G-AJVL	Miles M.38 Messenger 2A	6372		30. 5.47	Cancelled 1.12.49 - to VH-BJM 11.49 *(Subsequently VH-BJH)*	

Mackay Tiger Moth Museum, Mackay, Queensland

G-AMPM	de Havilland DH.82A Tiger Moth	86128	EM945	23. 2.52	Cancelled 31.3.52 - to ZK-BBF 4.52 *(Subsequently VH-IVN)*	

Classic Jets Fighter Museum, Parafield Airport, Adelaide, South Australia

G-APTS	de Havilland DHC-1 Chipmunk 22A *(Originally regd as Mk.22)*	C1/0683	WP791	21. 4.59	Cancelled 1.7.93 - to VH-ZIZ 7.94.	

Sir Ross and Sir Keith Smith War Memorial, Adelaide,South Australia

G-EAOU	Vickers FB.27A Vimy IV	---	(A5-1) G-EAOU/F8630	23.10.19	Cancelled 1920 *(As "G-EAOU")*	31.10.20

Lincoln Nitschke's Military and Historical Aircraft Collection, Greenock, South Australia

G-APAD	Edgar Percival EP.9	27	(VH-SSW) G-43-6	18. 3.57	Cancelled 1.8.58 - to VH-SSX 8.9.58	

South Australian Aviation Museum, Port Adlaide, South Australia

G-BAHB	de Havilland DH.104 Dove 5	04107	CS-TAC	31.10.72	Cancelled 14.3.74 - to VH-CLD (2) 3.74	

Geoff Davis Collection, Salisbury, South Australia

G-ABLF	Avro 616 Avian Sport	R3/CN/522		. 4.31	Cancelled 8.36 - to VH-UVX 2.36 *(Rebuilt with parts of, and flew as, VH-UQE) (As "VH-UVX")*	

Queen Victoria Museum, Launceston, Tasmania

G-ASCX	de Havilland DH.114 Heron 2D *(Converted to Riley Heron)*	14124	West German AF CA+002	30. 8.62	Cancelled 7.5.70 -to VH-CLV	

Ballarat Aviation Museum, Ballarat, Victoria

G-ATGJ	Riley Dove 400 *(Originally registered as DH.104 Dove 5B)*	04113	R.Jordanian AF D-101 TJ-ACB, YI-ABL	12. 7.65	Cancelled 1.11.65 as TWFU *(Restored 9.10.69: cancelled 20.3.74 - sold as VH-ABK)*	

Moorabbin Air Museum, Melbourne, Victoria

G-AHDI	Percival P.28 Proctor I	K.253	P6194	26. 2.46	Cancelled 1.6.51 - to VH-AUC 6.51 *(As "VH-AUC")*	
G-AJKG (2)	Miles M.38 Messenger 2A	6373		30. 5.47	Cancelled 17.8.53 - to VH-AVQ *(Stored)*	
G-ANPV (2)	de Havilland DH.114 Heron 2D *(Converted to Riley Heron)*	14098	G-5-24 (G-ANPV)	20.11.56	Cancelled 5.1.71 - to VH-CLX.	

Clyde North Aeronautical Preservation Group, Mount Waverley, Victoria

"G-ACSS"	de Havilland DH.88 Comet rep	---	"G-ACSP"			

Royal Australian Air Force Museum, Point Cook, Victoria

Regn	Type	C/n	Previous identity	Regd	Fate	Date
G-AIMI	Bristol 170 Freighter 21E	12799		3.12.46	Cancelled 24.1.49 - to RAF as WB482 30.3.49	
					(Subsequently R.Australian AF A81-1 on 14.4.49 and VH-SJG) (As "A81-1")	
G-AJHW	Sikorsky S-51	51-17	WB220 G-AJHW	27. 2.47	Cancelled 22.5.57 - to CF-JTO 6.57 *(Subsequently R.Australian AF A80-374)*	
G-AJKF (2)	de Havilland DH.84 Dragon III *(Built de Havilland Aircraft Pty Ltd, Australia)*	2081	RAAF A34-92	28. 3.47	Cancelled 19.8.48 as NTU and regd as VH-BDS 28.4.48 *(Subsequently VH-AML) (As "VH-AML")*	
G-AOHD (2)	Hunting Percival P.84 Jet Provost T.2	P.84/12	RAAF A99-001 G-AOHD	26. 3.56	Cancelled - to RAAF as A99-001 8.61 *(Restored to UK but subsequently A99-001)*	

Tyabb Airport Collection, Tyabb Airfield, Tyabb, Victoria

Regn	Type	C/n	Previous identity	Regd	Fate	Date
G-AMPN	de Havilland DH.82A Tiger Moth	85829	DE969	23. 2.52	Cancelled 31.3.52 - to ZK-BBG 7.52 *(Subsequently VH-BYB)*	

Airworld, Wangaratta, Victoria

Regn	Type	C/n	Previous identity	Regd	Fate	Date
G-EBUB	Westland Widgeon III	WA.1695		12. 3.28	Cancelled 8.28 - to G-AUHU 30.7.28 *(Subsequently VH-UHU)*	
G-ACUP	Percival P.3 Gull Six	D.46		. 7.34	Cancelled 5.39 - to VH-ACM 31.7.39 *(Subsequently VH-CCM)*	
G-ALBP	Miles M.38 Messenger 4A *(Originally regd as Messenger 1)*	---	RH376	18. 6.48	Cancelled 7.7.55 - to VH-WYN 4.55	

Nelson Wilson Collection, Yering, Victoria

Regn	Type	C/n	Previous identity	Regd	Fate	Date
G-ANZS	de Havilland DH.82A Tiger Moth	85082	T6813	4. 3.55	Cancelled 18.6.59 - to VH-DCH 2.60.	

RAAF Association Aviation Heritage Museum, Bull Creek, Western Australia

Regn	Type	C/n	Previous identity	Regd	Fate	Date
G-ALIS	Percival P.34A Proctor III	K.392	Z7203	21. 2.49	Cancelled 21.2.52 - to VH-BQR 2.52 *(Subsequently VH-GAS)*	
G-ALZL	de Havilland DH.114 Heron Srs.1	10903	LN-BDH G-ALZL	30. 3.50	Cancelled 25.4.54 - to LN-BDH	
					(Restored 21.5.54: cancelled 2.12.66 - to OY-DGS and subsequently VH-CJS)	
BGA.334	Slingsby T.12 Gull I	293A		4.38	Cancelled - to VH-GHL	

AUSTRIA

McDonald's Restaurant, Vienna-Schwechat Airport, Vienna

Regn	Type	C/n	Previous identity	Regd	Fate	Date
G-AGRW	Vickers 639 Viking 1 *(Originally registered as 498 Viking 1A)*	115	XF640 G-AGRW	8. 5.46	WFU 8.64 *(Noted 9.01)*	9. 7.68

BARBADOS

Grantly Adams Airport

Regn	Type	C/n	Previous identity	Regd	Fate	Date
G-BOAE	British Aircraft Corporation-Aérospatiale Concorde 102				Cancelled 4.5.04 as WFU	18. 7.05T

BELGIUM

Stampe and Vertongen Museum, Antwerp

Regn	Type	C/n	Previous identity	Regd	Fate	Date
G-BRMC	Stampe et Renard SV-4B	1160	SLN-03 Belgium AF V-18	17. 3.78	Cancelled 2.9.93 - to OO-GWD	

Koninklijk Leger Museum / Musée Royal de l'Armée, Brussels

Regn	Type	C/n	Previous identity	Regd	Fate	Date
G-ACGR	Percival P.1C Gull Four IIA (DH Gipsy Major I) *(Originally regd as P.1B)*	D.29		11. 5.33	Cancelled 12.34 - crashed Waterloo, Belgium 12.34	20. 6.35
G-AFJR	Tipsy Trainer 1 *(Converted to Belfair)*	2		20. 8.38	Cancelled 12.4.89 as TWFU *(Stored for static rebuild with remains of G-AFRV)*	10. 9.64
G-AFRV	Tipsy Trainer I	10		15. 7.39	Cancelled 10.2.87 by CAA	
					(Damaged landing Herrings Farm,Cross-in-Hand, Sussex 15.9.79) (Used for spares for rebuild of G-AFJR)	
G-AFVH	Tipsy S.2	29	OO-ASB	7. 6.39	Cancelled 27.7.49 - to OO-TIP 7.49 *(As "OO-TIP")*	
G-AHZY	Percival P.44 Proctor V	Ae.84		14. 8.46	Cancelled 25.7.57 - to OO-ARM 9.57	
G-AKIS	Miles M.38 Messenger 2A	6725		19. 9.47	Cancelled 24.2.70 as WFU	5. 8.70
G-AKNV	de Havilland DH.89A Dragon Rapide	6458	G-AKNV EI-AGK, G-AKNV, R5922	2.12.47	Cancelled 27.9.55 - to OO-AFG 9.55 *(Subsequently OO-CNP 4.64))*	
G-AMJD	de Havilland DH.82A Tiger Moth	83728	T7238	9. 4.51	Cancelled 11.11.52 - to OO-SOI 10.52 *(As "T-24/UR-!")*	
G-AOJX	de Havilland DH.82A Tiger Moth	3272	K4276	18. 4.56	Cancelled 5.6.56 - to OO-EVS 7.56 *(As "OO-EVS")*	
G-AOPO	Percival P.40 Prentice T.1	PAC/215	VS613	30. 5.56	Cancelled 11.9.58 - to OO-OPO 4.58	
G-APPT	de Havilland DH.82A Tiger Moth	84567	T6100	24.10.58	Cancelled 2.1.59 - to OO-SOK 1.59 *(Subsequently OO-SOW) (As "OO-SOW")*	
G-BDPU	Fairey Britten-Norman BN-2A-21 Islander *(Originally regd as BN-2A-26)*	510		5. 2.76	Cancelled 19.11.76 - to Belgium Army as B-06/OTA-LF 8.76 *(As "B-06")*	
BAPC.19	Bristol F2b Fighter fuselage frame *(Rebuilt to static condition by Skysport Engineering 6.89 with parts from J8264)*	---			*(As "66" in Belgian AF c/s)*	

Musée de la Base de Bierset, Grace Hollogne

Regn	Type	C/n	Previous identity	Regd	Fate	Date
G-BDPP	Fairey Britten-Norman BN-2A-21 Islander	501		5. 2.76	Cancelled 19.11.76 - to Belgium Army as B-05/OTA-LE 7.76 *(As "B-05")*	

Tipsy Flying Museum, Keiheuvel

Regn	Type	C/n	Previous identity	Regd	Fate	Date
G-AISB	Tipsy B Srs.1	18		24. 4.47	Cancelled 21.11.89 - to OO-EOT	

Disco N9, Waarschoot

Regn	Type	C/n	Previous identity	Regd	Fate	Date
G-AZNA	Vickers 813 Viscount	350	(G-AZLU) ZS-CDX, (ZS-SBX), ZS-CDX	8. 2.72	Cancelled 17.6.92 by CAA - WFU *(On display 12.01)*	

Vliegtuigverzameling Robert Landuyt, Wevelghem

Regn	Type	C/n	Previous identity	Regd	Fate	Date
G-ADCG	de Havilland DH.82 Tiger Moth	3318	A2126 A728, BB731, G-ADCG	11. 2.35	Cancelled 2.6.93 - to OO-TGM 6.93	
					(Originally cancelled 30.10.40 - to RAF as BB731, later A2126: restored 14.12.84: cancelled 10.12.91 as TWFU: restored 28.4.92)	
G-BAAY	Valtion Lentokonetehdas Viima II	VI-3	OH-VIG Finnish AF VI-3	12. 8.72	Cancelled 28.11.89 - to OO-EBL	

Sabena Old Timer Foundation, Zaventem Airport, Zaventem

G-AMTL	de Havilland DH.82A Tiger Moth	82592	W6420	24. 6.52	Cancelled 28.7.52 - to OO-SOF 7.52
			G-AFWF		
G-AMTP	de Havilland DH.82A Tiger Moth	84875	T6534	17. 7.52	Cancelled 29.8.52 - to OO-ETP 7.53

BRAZIL

Museu de Armas, Veiculos Motorizados e Avioes Antigos "Eduardo Andreia Matarazzo", Bebeduoro

G-AMHB	Westland-Sikorsky S-51 Dragonfly Mk.1B	WA/H/030		18. 1.51	Cancelled 12.5.51 - to OO-CWA 5.51
					(Subsequently XB-JUQ and PT-HAL)
G-AMOI	Vickers 701 Viscount	22		23. 5.52	Cancelled 23.3.63 - to (PP-SRK), PP-SRL 10.62
G-ANHD	Vickers 701C Viscount	64		12.12.53	Cancelled 23.4.63 - to PP-SRO 8.62

TAM Museum, San Carlos

G-OMIG	WSK SBLim-2A	622047	PLW-6247	10.11.92	Cancelled 1. 8.00 by CAA as sold to Brazil
	(MiG-15UTI)		(Polish AF)		*(Stored dismantled 10.02 as "PLW-6247")*
	(Built Aero Vodochody as S.103/MiG 15bis; later rebuilt in Poland)				

Fundacao Museu de Technologia de Sao Paulo, Sao Paulo

G-AOFX (2)	Vickers 701C Viscount	182		20.12.55	Cancelled 25.7.63 - to PP-SRS 8.63

CANADA

Reynolds Heritage Preservation Foundation, Wetaskiwin, Alberta

G-ACPP	de Havilland DH.89 Dragon Rapide	6254		20. 2.35	Cancelled - sold as CF-PTK 6.61.
					(Subsequently "CF-BND")
G-AFGK	Miles M.11A Whitney Straight	509	U-1	6. 4.38	Cancelled 13.9.77 - to N72511 9.77
					(Subsequently CF-FGK)
G-AOYU	de Havilland DH.82A Tiger Moth	82270	N9151	27.12.56	Cancelled 14.7.71 on sale to USA.
					(Subsequently C-GABB 9.78)

Canadian Museum of Flight and Transportation, Langley, British Columbia

G-APRK	de Havilland DH.114 Heron 2D	14050	5N-AGM(2)	12. 1.59	Cancelled 19.1.68 - to OY-DGK 1.68
	(Converted to Saunders ST-27 [003])		G-APRK, OY-DGK, G-APRK, OY-DPN, PH-VLA, G-APRK, SA-R-5 *(Subsequently HK-1299)*		

Quesnel HeritageAircraft Museum, Richbar Store, Quesnel, British Columbia

G-APHY	Scottish Aviation Twin Pioneer Srs.1	508	9K-ACC	2.10.57	Cancelled 16.8.74 - to C-FSTX 8.74
			G-APHY, VR-OAF		*(Subsequently C-GSTX) (Front fuselage only)*

Commonwealth Air Training Plan Museum, Brandon Municipal Airport, Brandon, Manitoba

G-ANOS	de Havilland DH.82A Tiger Moth	85461	DE465	4. 3.54	Cancelled 31.8.72 - to CF-JNF 8.72 *(As "4188"*

Western Canada Aviation Museum, Winnipeg, Manitoba

G-AHLO	de Havilland DH.80A Puss Moth	2187	HM534	1. 5.46	Cancelled 26.9.69 - to CF-PEI 10.69
			(DR630), Bu.A8877		
G-APXG	de Havilland DH.114 Heron 2D	14137		20.11.59	Cancelled 25.6.60 - to Kuwait AF 303 1.60
	(Converted to Saunders ST-27 [008])				*(Subsequently 9K-BAA and CF-CNX) (As C-FCNX)*

Tiger Boys Collection, Kitchener, Ontario

G-APHZ	Thruxton Jackaroo	82168	N6924	3.10.57	Cancelled 25.12.70 - to CF-QOT 5.71 *(Subsequently C-FPHZ)*

International Vintage Aircraft, Markham, Ontario

G-ACYZ	Miles M.2H Hawk Major	123		16.10.34	Cancelled - to VH-ACC 9.12.38
					(Subsequently CF-AUV and C-FAUV)

Ed Russell Museum, Niagara Falls, Ontario

G-BDAM	Noorduyn AT-16-ND Harvard IIB	14-726	LN-MAA	10. 4.75	Cancelled 30.9.03 - to C-GFLR
			Swedish AF Fv.16047, FE992, 42-12479		
G-BRRA	Vickers Supermarine 361 Spitfire LF.IXc	CBAF.IX.1875	Belgian AF SM29	10.10.89	Cancelled 16.2.04 - to C-... 19. 9.04P
	(Regd as c/n CBAF.8185)		R.Netherlands AF H.59, H.119, Fokker B-1, MK912 *(As "MK912/SH-L" of 350 (Belgian) Squadron)*		
G-ORGI	Hawker Hurricane IIB	60372	RCAF 5481	20.11.89	Cancelled 31.1.92 - to N678DP 2.92
	(Built Canadian Car and Foundary Co)				*(As "P2970")*

Canadian Bushplane Heritage and Forest Fire Educational Centre, Sault Sainte Marie, Ontario

G-AKGV	de Havilland DH.89A Dragon Rapide	6796	F-BFPU	9.10.47	Cancelled 25.7.50- sold as F-BFPU
			G-AKGV, NR697		*(Restored 7.11.75: cancelled 21.6.76 - sold as (C-GXFJ) - subsequently C-FAYE)*
G-AOGW	de Havilland DH.114 Heron 2E	14095		2. 2.56	Cancelled 14.2.73 - to C-GCML 9.72
	(Converted to Saunders ST-27 [009])				

CHILE

Museo Nacional deAeronautica de Chile, Los Cerillos, Santiago

G-BACK	de Havilland DH.82A Tiger Moth	85879	F-BDOB	4. 9.72	Cancelled 25.1.89 asold to Chilean AF Museum in 12.87
			French AF, DF130		*(Subsquently CC-DMC)*
G-BPLT	Bristol 20 M.1C rep	AJD-1		18. 1.89	Cancelled by CAA 22.6.89 - to CC-DMA *(As "C4988")*
	(Built AJD Engieering Ltd)				

PEOPLES REPUBLIC OF CHINA

Military Museum of China, Beijing

G-BBVZ	Hawker Siddeley HS.121 Trident 2E Srs.109	2182		3. 1.74	Cancelled 30.8.77 - to China as B-294 @ 8.77 *(Subsequently B-2207)*

China Aviation Museum, Chiangping, Beijing

Reg	Type	c/n	Previous identity	Date	Fate	
G-ASDS	Vickers 843 Viscount	453		8.11.62	Cancelled 19.9.63 - to China as 84303 8.63 *(Subsequently 406 and B-406)* (As "50258")	
G-BAJJ	Hawker Siddeley HS.121 Trident 2E Srs.109	2173		28.11.72	Cancelled 1.6.76 - to China as 276 @ 6.76 *(Subsequently B-276 and B-2213) (As "50031"*	
G-BBWG	Hawker Siddeley HS.121 Trident 2E Srs.109	2188		3. 1.74	Cancelled 22.5.78 - to China as 269 @ 4.78 *(Subsequently B-269) (As "50055")*	

Guangzou Technical Institute Museum, Guangzou

Reg	Type	c/n	Previous identity	Date	Fate	
G-AZFW	Hawker Siddeley HS.121 Trident 2E Srs.102	2160		29. 9.71	Cancelled 6.9.73 - to China as 246 @ 9.73 *(Subsequently "B-2219")*	
G-BBVU	Hawker Siddeley HS.121 Trident 2E Srs.109	2177		3. 1.74	Cancelled 7.2.77 - to China as 284 @ 2.77 *(Subsequently B-284 and B-2116)*	
G-BBVW	Hawker Siddeley HS.121 Trident 2E Srs.109	2179		3. 1.74	Cancelled 1.4.77 - to China as 288 @ 4.77 *(Subsequently B-288 and B-2117)*	

Hong Kong Historic Aviation Association, Chek Lap Kok Airport, Hong Kong

Reg	Type	c/n	Previous identity	Date	Fate	
G-AYAZ	Britten-Norman BN-2A-7 Islander	615	9V-BDW G-AYAZ	6. 4.70	Cancelled - to R.Hong Kong AuxAF as HKG-7 @ 12.71	

Nanjing Hangkong Xueyan, Nanjing

Reg	Type	c/n	Previous identity	Date	Fate	
G-BBVV	Hawker Siddeley HS.121 Trident 2E Srs.109	2178		3. 1.74	Cancelled 7.2.77 - to China as B-286 @ 2.77 *(Subsequently B-2210)*	

CZECH REPUBLIC

Historicky Ustav Armady Ceske Republicky - Letecke Muzeum Kbely, Prague

Reg	Type	c/n	Previous identity	Date	Fate	
G-ACGO	Saunders-Roe A.19 Cloud	A.19/5		8. 5.33	Cancelled - to OK-BAK 7.34	

DENMARK

Danmarks Flyvemuseum, Kongelunden, Billund

Reg	Type	c/n	Previous identity	Date	Fate	
G-AEYF	General Aircraft Monospar ST.25 Ambulance	GAL/ST25/95		31. 5.37	Cancelled - to Denmark as OY-DAZ 3.39	
G-AIWY	de Havilland DH.89A Dragon Rapide	6775	NR676	22.11.46	Cancelled 7.1.47 - to OY-AAO	
G-AKDK 3.70	Miles M.65 Gemini 1A	6469		22. 8.47	Cancelled 5.11.73 as WFU *(For rebuild with parts from G-AJWA c/n 6290)*	27.
G-AOJG	Hunting-Percival P.66 President I *(Originally regd as Prince 5) (Also c/n P66/79)*	HPAL/PEM/79		27. 3.56	Cancelled 4.7.59 - to Danish AF as 69-697 @ 7.59 *(Subequently Danish AF M-697 and OY-AVA)*	
G-AOUF	de Havilland DH.104 Dove 6	04476		19. 7.56	Cancelled 23.4.68 -to D-IBYW *(Subsequently OY-DHZ)*	
G-APJP	de Havilland DH.82A Tiger Moth	82869	(G-AOCV) R4961	30. 7.57	Cancelled 30.10.79 - to SE-GXO 7.80 *(As "SE-GXO")*	
G-AROI	de Havilland DH.104 Dove 5	04474	XK897	24. 5.61	Cancelled - to OY-AJR 6.80	

Egeskov Veteranmuseum, Egeskov

Reg	Type	c/n	Previous identity	Date	Fate	
G-AOFR	de Havilland DH.82A Tiger Moth	86425	NL913	27.10.55	Cancelled 20.11.60 - to SE-COX 9.61 *(Subsequently OY-BAK and "S-11") (As "NL913")*	

Dansk Veteranflysamlung, Stauning

Reg	Type	c/n	Previous identity	Date	Fate	
G-AFWN	Auster J/1 Autocrat *(Originally regd as Taylorcraft Plus D, converted in 1945)*	124		1. 8.39	Cancelled 23.7.56 - to D-EKOM 9.56 *(As "OY-...)*	
G-AHAA	Miles M.28 Mercury 6	6268		26. 1.46	Cancelled 31.8.56 to D-EHAB 9.56 *(Subsequently OY-ALW)*	
G-AHKO	Taylorcraft Plus D	228	LB381	24. 4.46	Cancelled 10.3.56 - to D-ECOD 3.56 *(Subsequently OY-DSH) (As "LB381")*	
G-AJHM	SAI KZ.VII UA Laerke	148	OY-AAN	24. 4.47	Cancelled 15.7.49 - to F-BFXA 7.49 *(Subsequently OY-AAN)*	
G-AMMA	de Havilland DHC-1 Chipmunk 21	C1/0470		21. 9.51	Cancelled 23.4.69 - to OY-DHJ 7.69.	
G-AMZO	de Havilland DH.87B Hornet Moth	8040	SE-ALD OY-DEZ, VR-RAI	15. 5.53	Cancelled 14.2.74 - restored as OY-DEZ 5.74	
G-ANCY	de Havilland DH.82A Tiger Moth	85234	DE164	12. 9.53	Cancelled 1.3.55 - to OO-DLA 5.55	
G-BLPZ	de Havilland DH.104 Devon C.2	04270	WB534	18.10.84	Cancelled 12.3.86 - to OY-BHZ *(As "OY-BHZ") (Stored 5.01)*	

EGYPT

Military Museum, Alexandria

Reg	Type	c/n	Previous identity	Date	Fate	
G-AYBX	Campbell Cricket	CA/331		20. 4.70	Cancelled 1.9.70 on sale to Kuwait *(As "G-AYBX")*	

FINLAND

Keski-Suomen Ilmailumuseo, Tikkakoski, Jyväskylä

Reg	Type	c/n	Previous identity	Date	Fate	
G-EBNU	Avro 504K	---	E448	19. 3.26	Cancelled - to Finnish AF as 1H49 @ 9.26 *(As "AV-57")*	

FRANCE

Musée des Ballons, Chateau de Balleroy, Balleroy, 14 Calvados, Normandy

Reg	Type	c/n	Previous identity	Date	Fate	
G-BKBR (2)	Cameron Chateau 84SS HAB *(Forbes Chateau de Balleroy shape)*	743		11. 5.82	Cancelled 29.4.93 as WFU	
G-BKNN	Cameron Minar-E-Pakistan HAB *(240ft Moslem National Monument shape)*	900		7. 2.82	Cancelled 29.4.93 as WFU	
G-BLFE	Cameron Sphinx 72SS HAB *(Egyptian Sphinx shape)*	1011		22. 2.84	Cancelled 29.4.93 as WFU	
G-BLRW	Cameron Elephant 77SS HAB	1074		14.12.84	Cancelled 14.11.02 by CAA *"Great Sky Elephant"*	1.10.00A
G-BMUN	Cameron Harley 78SS HAB *(Harley Davidson Motorcycle shape)*	1188		10. 6.86	Cancelled 12.11.01 as WFU	23. 5.99

G-BMWN	Cameron Temple 80SS HAB	1211		9. 7.86	Cancelled 14.11.02 by CAA *"Temple"*	17. 6.96A
G-BNFK	Cameron Egg 89SS HAB	1436		20. 2.87	Cancelled 14.11.02 by CAA	15. 7.02A
	(Faberge Rosebud Egg shape)				*"Faberge Easter Egg"*	
G-BPOV	Cameron Magazine 90SS HAB	1890		10. 3.89	Cancelled 14.11.02 by CAA	5. 7.01A
	(Forbes Magazine shape)				*"Forbes Capitalist Tool"*	
G-BPSP	Cameron Ship 90SS HAB	1848		10. 3.89	Cancelled 14.11.02 by CAA	17. 6.94
	(Columbus Santa Maria shape)				*"Santa Maria"*	
G-TURK	Cameron Sultan 80SS HAB	1711		12. 4.88	Cancelled 28.8.02 by CAA	18. 6.00A

Musée Aéronautique Presq'île Côte D'Amour, La Baule-Escoublac Aerodrome, La Baule 44 Loire-Atlantique

G-AVXV	Bleriot XI	225		2.11.67	Cancelled 16.7.92 - to F-AZIN	

Musée Historique de l'Hydraviation, Biscarrosse, 40 Landes

G-DUCK	Grumman G.44 Widgeon (OA-14)	1218	N3103Q	15.11.88	Cancelled 24.3.93 on sale to France	
			N58337, 42-38217, NC28679		*(As "G-DUCK")*	

Concorde Supermarket, Lempdes, Clermont-Ferrand, 63 Puy-de-Dôme

G-ARER	Vickers 708 Viscount	12	F-BGNM	19. 9.60	Cancelled 27.6.66. To F-BOEA in 8.66	
					(Crashed 28.12.71) (As "F-BOEA" 9.97)	

Amicale Aéronautique de Cerny, La Ferté-Alais, 91 Essonne (ww.jbs.com)

G-BECL	CASA 352L	24	Spanish AF T2B-212	27. 7.76	Cancelled 19.6.90 - sold as F-AZJU	
	(Junkers Ju.52/3M) *(C/n also reported as 103)*				*(As "N9+AA")*	

Amicale Jean Salis, La Ferté-Alais, 91 Essonne

G-AFNJ	de Havilland DH.94 Moth Minor	94038	AW113	15. 5.39	Impressed as AW113 9.6.40, restored 6.4.46	
			G-AFNJ		Cancelled 17.7.54 - sold as F-BAOG 7.54 *(Subsequently F-PAOG)*	
G-ALZF	de Havilland DH.89A Dragon Rapide	6541	X7381	24. 3.50	Cancelled - sold as F-BGON 9.52	
					(Subsequently F-AZCA)	
G-ANJG	de Havilland DH.82A Tiger Moth	83875	T7349	12.12.53	Cancelled 26.5.54 - sold as F-BGZT 6.54.	
G-AOAO	de Havilland DH.89A Dragon Rapide	6844	NR756	22. 3.55	Cancelled 4.1.56 - sold as F-BHGR	
G-AWHG	Hispano HA.1112-MIL Buchon	139	Spanish AF C4K-75	14. 5.68	Cancelled as WFU 13.8.70, restored ??	
	(Messerschmitt Bf.109G)				Cancelled 25.10.74 - sold as N3109G *(As "N3109")*	
G-BPEX	Boeing-Stearman A75-N1 Kaydet	75-589	N65D	24.10.88	Cancelled 14.4.89	
	(PT-18-BW)		N61304, 40-2032		*(To F-AZGM 8.89)*	

Association des Cages à Poules d'Acquitaine, Liorac-sur-Louyre, 24 Dordogne

G-ANXP	Piper J3C-65 Cub (L-4H-PI)	12192	N79819	13.12.54	Cancelled 14.10.55 - sold as D-EGUL 10.55.	
			44-79896		*(Subsequently PH-CMS/(OO-LSD)/OO-GMS - to F-GLMS 12.91)*	

Airborne Troops Museum, Sainte-Mère-Eglise, 50 Manche

EI-ALR	Douglas C-47-DL Dakota	4579	EI-ACG	2. 1.61	Cancelled 14.1.61 - sold to French Navy as 8487/87	
			41-18487		*(Call sign F-YFGC later F-YGGB)*	

Musée National de l'Automobile, Mulhouse, 68 Haut-Rhin

G-AIVG	Vickers 610 Viking 1B	220		18.11.46	Crashed Le Bourget 12.8.53 *(Airframe only)*	12. 2.54

Antic Air, Les Mureaux, 78 Yvelines

G-AZJD	North American AT-6D-NT Harvard IIA	88-14948	F-BJBF	30.11.71	Cancelled 9.10.84 - sold as F-WZDU	
			Belgium AF H-9, SAAF 7509, EX959, 41-33932		*(To F-AZDU 6.85)*	

Musée de l'Air et de l'Espace, Le Bourget, 93 Seine-Saint-Denis (www.mae.org)

G-EBYY	Cierva C.8L Mk.2	---		21. 6.28	Sold 4.30 abroad?	13. 7.29
	(Avro 617) (AS Lynx @ 180hp)					
G-AKCO	Short S.25 Sandringham 7	SH57C	JM719	29. 7.47	Cancelled 10.5.55	
	(Sunderland GR.3 c/n SB2022 conversion)				*(To VH-APG 10.54 subsequently F-OBIP)*	
G-ALGB	de Havilland DH.89A Dragon Rapide	6706	HG721	17.12.48	Cancelled - sold as F-BHCD 10.54	
EI-ALT	Douglas C-47A-10-DK Dakota	12471	EI-ACT	2. 1.61	Cancelled in 2.1.61 - to French Navy as 12471/71	
			KG436, 42-92647			

Association France DC-3, Orly, 94 Val-de-Marne

G-AGZF	Douglas C-47A-1-DL Dakota	9172	WZ984	11. 1.46	Cancelled 5.10.52 - to French AF as 42-23310/F-RAFC 10.52	
			G-AGZF, 42-23310		*(Subsequently F-BRGN, TJ-ABB, TL-AAX, (F-ODQE) and F-GDPP)*	

Musée de L'Automoiliste, Aire de Breguieres, Mougins, Cannes, 06 Alpes-Maritimes

(G-BLXI (1))	Bleriot Type XI	PFA 88-10864		3.85R	NTU - to BAPC.132	

(Anzani 25hp) *(Built L.D.Goldsmith 1976 from original components: rebuilt again by EMK @ 1982 - c/n EMK010: possibly same a/c as BAPC.189)*

Ailes Anciennes Toulouse, Blagnac, Toulouse, 31 Haute-Garonne

G-ALWC	Douglas C-47A-25-DK Dakota	13590	KG723	10. 1.50	To F-GBOL in 11.82 but NTU	6.2.83
			42-93654		Cancelled 29.2.84 by CAA, restored 1.5.84, cancelled 3.4.89 by CAA	

GERMANY

Deutches Tecknikmuseum, Berlin

G-BLFL	Douglas C-47B-45-DK Dakota	16954, 34214	N951CA	14. 3.84	Cancelled 12.8.85 - to N951CA - restored 5.11.85	
			G-BLFL, Spanish AF T3-54, N73856, 45-951		Cancelled 27.8.86 - to N951CA *(As "45-0951")*	

Luftwaffen Museum, Gatow, Berlin

G-HELI	Saro Skeeter AOP.12	S2/5110?		15. 6.78	Cancelled 22.3.95 on sale to Germany.	
	(Composite of cabin 7870M/XM556 and boom 7979M/XM529 [S2/5105})				*(As "XM556")*	

Aero Park, Brandenburg

G-AYNP	Westland WS-55 Whirlwind Srs.3	WA/71	ZS-HCY G-AYNP, XG576	14.12.70	Cancelled 22.2.94 by CAA *(To Hubschrauber Museum, Germany 6.95) (As "G-AYNP")*	27.10.85T

Luftbruckengedenkanlage, Rhein-Main Air Force Base, Frankfurt

EI-ARS (2)	Douglas C-54E-5-DO Skymaster	27289	N88887	9.12.69	Cancelled 22.6.77 - restored as N88887 @ 8.77	

EL-ALP, N88887, ZS-LMH, N88887, FAR-91, N88887, EI-ARS, LN-TUR, EI-ARS, HB-ILU, N88887, NC88887, 44-9063
(As "44-9063")

Luftfahrt Museum, Laatzen- Hannover, Hannover

G-FXIV	Vickers Supermarine 379 Spitfire FR.XIVc	----	Indian AF T44 Indian AF HS..., MV370	11. 4.80	Cancelled 5.2.85 as WFU *(As "MV370")*	

Flugausstellung L und P Junior, Hermeskeil, Trier (www.flugaustellung.de)

G-AMXX	de Havilland DH.104 Dove 2A	04406		22. 1.53	Cancelled 21.10.54 - to RN as Sea Devon C.20 XJ348 10.54. Rrestored as G-NAVY 6.1.82: cancelled 2.7.91 as WFU *(As "XJ348")*	
G-ARVF	Vickers VC-10-1101	808		16. 1.63	Cancelled as WFU 11.4.83 *(As "G-ARVF" in UAE titlees)*	23. 7.81
G-BDIW	de Havilland DH.106 Comet 4C	6470	XR398	1. 9.75	Cancelled 23.2.81 to Germany *(As "G-BDIW" in Dan-Air titles)*	8.6.81
G-BKLZ	Vinten Wallis WA-116MC *(Aka Vinten VJ-22 Autogyro)*	UMA-01		8.12.82	Cancelled 8.6.89 as destroyed *(As "G-55-2")*	16.12.83P
G-BXSL	Westland Scout AH.Mk.1	F.9762	XW799	17. 2.98	Cancelled 7.5.02 by CAA *(As "XW799")*	19. 8.02P

Luftfahrt Museum Merseburg, Merseburg Sud

G-DEVN	de Havilland DH.104 Devon C.2/2	04269	WB533	26.10.84	Cancelled 16.11.90 by CAA *(As "WB533")*	4.2.85P

Deutsches Museum, Oberschleisseim, Munich (www.deutsches-museum.de)

G-ALUA	Winter LF-1 Zaunkönig	V-2	VX190 D-YBAR	28. 6.49	Cancelled 16.4.74 - to EI-AYU 5.74 *(Subsequently D-EBCQ)*	
G-AWHA	CASA 2111D (He.111H-16)	025	Spanish AF B2I-77	14. 5.68	Cancelled 15.8.70 - to D-CAGI 8.70	
G-AYUK	Western-Brighton M-B65 HAB	003		17. 3.71	Cancelled 10.12.73 - to D-Westfalen II @ 12.73	
G-BHUW	Boeing-Stearman A75N1 (N2S-5) Kaydet	75-3475	N474 N64639, BuA.30038	16. 5.80	Cancelled 7.12.83 - to D-EFTX 4..92	

Schwäbisches Bauern und Tecknikmuseum, Seifertschofen, Schwäbisch-Gmund

G-AJAZ	Douglas C-47A-50-DL Dakota	10100	FL517 42-24238	13. 1.47	Cancelled 3.12.48 - to EC-ADR 12.48 *(Subsequently Spanish AF T3-61 and N8041B) (As "N569R")*	

Auto und Technik Museum, Schleissheim (www.tecknik.museum.de)

G-AKLL	Douglas C-47A-30-DK Dakota	14005/25450	KG773 43-48189	18.11.47	Cancelled 6.6.50 - to EC-AEU *(Subsequently Spanish AF as T.3-62, N8041A and "D-CORA")* *(As "D-CADE")*	
G-ARUE (2)	de Havilland DH.104 Dove 7	04530	Irish Air Corps 194 (G-ARUE)	7.10.80	Sold as D-IKER 10.83 and cancelled 17.7.86 by CAA	
G-AWHS	Hispano HA.1112-MIL Buchon (Messerschmitt Bf.109G) (Daimler-Benz.605D)	228	Spanish AF C4K-170	14. 5.68	Cancelled 17.2.69 on sale to Spain *(Subsequently N170BG) (As "4+-" in Luftwaffe c/s)*	

Albatros Flugmuseum, Stuttgart Airport, Stuttgart

G-AVHE	Vickers 812 Viscount	363	(G-AVGY) N251V	20. 2.67	Cancelled 14.2.73 as destroyed *(WFU 30.3.70 and broken up 8.72) (Forward fuselage only)*	
G-BGGR	North American AT-6A Harvard	77-4176	Portuguese AF 1608, 41-217	17. 1.79	Cancelled 20.4.79 *(D-FOBY reserved 4.79) (As "D-FOBY")*	

Deutsches Segelfligzeugmuseum, Wassererkuppe

BGA.1711	Aachen FVA.10B Rheinland		CPZ RAFGGA521	4.72	*(As "D-12-354")*	

ICELAND

Islenska Flugogufelagid, Reykjavik

G-AIBA	Douglas C-47A-40-DL Dakota	9860	FD939 42-23998	. 6.46	Cancelled - restored to RAF as FD939 @ 7.46 Restored as G-AKSM 4. 2.48 : cancelled 7.3.51 - to TF-ISB 3.51	

ISRAEL

Israeli Air Force Museum, Hatzerim Air Force Base, Beersheba

G-AHAY	Auster V J/1 Autocrat	1956		12. 3.46	Cancelled - to Israel 10.3.83 for preservation *(Subsequently "VQ-PAS" and 1948) (As "13")*	
G-AIYT	Douglas C-47A-10-DK Dakota	12486	KG451 42-92661	20.12.46	Cancelled 12.3.47 - to ZS-BCJ 4.47	
G-AJMC	Bristol 156 Beaufighter TF.X	---	RD448	10. 4.47	Cancelled 28.5.49 - to Israeli DF/AF *(As "171"*	
G-AKRS	de Havilland DH.89A Dragon Rapide	6952	RL981	21. 1.48	Cancelled 29.2.84 - to Israeli DF/AF Museum as 4X-970/002 @ 5.78 *(As "VQ-PAR/002")*	
G-AZBV	Britten-Norman BN-2A-2 Islander	285	(EI-AVO) G-AZBV, G-51-285	16. 7.71	Cancelled 16.3.72 - to 4X-AYK 5.72 *(Subsequently 4X-FNP) (As 004/4X-FMD)*	

ITALY

Museo Nazionale della Scienza e della Tecnica, Milan

G-ACXA	Cierva C.30A (Avro 671)	753		4. 4.35	Cancelled - to I-CIER 8.35	

Rusty Angels, Nepi, Rome

G-APGM	de Havilland DH.82A Tiger Moth	86550	EI-AHM	9. 1.86	Cancelled 13.5.88 - to I-JJOY 7.88
			(EI-AHH), PG641		
G-APLR	de Havilland DH.82A Tiger Moth	84682	T6256	26. 3.58	Cancelled 6.8.58 - to I-JENA 1.59
G-AYUX	de Havilland DH.82A Tiger Moth	86560	F-BDOQ	26. 3.71	Cancelled 6.5.88 - to I-EDAI
			French AF, PG651		
G-BABA	de Havilland DH.82A Tiger Moth	86584	F-BGDT	11. 8.72	Cancelled 13.6.96 by CAA
			French AF, PG687		*(Components used in rebuild of G-AMHF 1985) (Reported on rebuild in Italy as I-BABA)*

Museo dell'Aeronautica Gianni Caproni, Trento Airport, Trento

G-ACTH	de Havilland DH.85 Leopard Moth	7074		12. 7.34	Cancelled - to I-ACIH 2.36 *(Front fuselage only)*

Museo Storico dell'Aeronautica Militaire Italiana, Vigna di Valle

G-ALMB	Westland-Sikorsky S-51 Dragonfly Mk.1A	WA/H/006		29. 3.49	Cancelled 13.4.51 - to I-MCOM 4.51
G-FIST	Fieseler Fi.156C-3 Storch	156-5802	D-EDEC	23.11.83	Cancelled 4.12.90 by CAA
			I-FAGG, MM12822		*(Restored 25.2.91 and cancelled 24.4.95 on sale to Italy) (As "MM12822/20")*

JAPAN

Tokyo Metropolitan College of Aeronautical Engineering "Fame " Gallery, Tokyo

G-AKVD	Chrislea CH.3 Srs.2 Super Ace	112	(VH-BRP)	8. 3.48	Cancelled 10.2.53 - to JA3062 2.53
			G-AKVD		

JORDAN

Royal Jordanian Air Force Historic Flight, King Abdulla Air Base, Amman

G-AIPW	Taylorcraft Auster 5A Srs.160	2204		9. 1.47	Cancelled 13.11.95 on sale to Jordan
	(Originally regd as J/1 Autocrat)				*(As "A-410" in Arab Legion AF c/s)*
G-ATGK	Riley Dove 400	04288	F-BORJ	12. 7.65	Cancelled 7.3.67 - sold as F-BORJ
	(Originally regd as DH.104 Dove 5B)		G-ATGK, R.Jordanian AF D-102, TJ-ACC, (TJ-ABG) *(As "JY-AEU")*		
					(Restored 21.2.75: cancelled 19.12.75 - sold as JY-AEU)
G-BVLM	de Havilland DH.115 Vampire T.55	976	ZH563	6. 4.94	Cancelled 23.6.97 - to Jordan..1997
	(Built F + W)		Swiss AF U-1216		*(As Jordan AF "209")*
G-BVPO	de Havilland DH.100 Vampire FB.6	615	HB-RVO	11. 7.94	Cancelled 23.6.97 on sale to Jordan.
	(Built F + W)		Swiss AF J-1106		*(As Jordan AF "109")*
G-BWKA	Hawker Hunter F.58	41H-697442	Swiss AF J-4075	12.10.95	Cancelled 23.6.97 on sale to Jordan.
					(As Jordan AF "843")
G-BWKC	Hawker Hunter F.58	41H-697394	Swiss AF J-4025	12.10.95	Cancelled 29.9.99 on sale to Jordan
					(As Jordan AF "712/E")
G-BOOM	Hawker Hunter T.7 (T.53)	41H/693749	G-9-432	6.10.80	Cancelled 23.6.97 on sale to Jordan.
			R.Danish AF ET-274, R.Netherlands AF N-307		*(As Jordan AF "800")*

KUWAIT

Kuwait Air Force Collection, Kuwait International Airport

G-ANNW	Auster J/5F Aiglet Trainer	3120		15. 2.54	Cancelled 25.6.60 - to K-AAAE 8.60
					(Subsequently 9K-AAE) (As "K-AAAE ")
G-APXA	Westland-Sikorsky S-55 Whirlwind Srs.2	WA/318		28.10.59	Cancelled 25.6.60 - to 9K-BHA 5.60
	(Originally regd as Srs.1)				*(Subsequently Kuwait AF 312)*
G-APXB	Westland-Sikorsky S-55 Whirlwind Srs.2	WA/319		28.10.59	Cancelled 25.6.60 - to 9K-BHB 6.60
	(Originally regd as Srs.1)				*(Subsequently Kuwait AF 313)*

Museum of Science and Industry, Kuwait City

G-AXEE	English Electric Lightning F.53	95311	G-27-86	24. 4.69	Cancelled 10.6.69 - to Kuwaiti AF as 418 *(As "53-418")*

MALTA

Malta Aviation Museum, Ta'Qali

G-ANFW	de Havilland DH.82A Tiger Moth	85660	DE730	5.11.53	Cancelled 10.3.00 by CAA	21. 7.01
	(Built Morris Motors) (Regd with Fuselage No.3737)					

MALAYSIA

Muzium Tentera Udara Diraja Malaysia, Sungai Besi Airfield, Kuala Lumpur

G-ANEJ	de Havilland DH.82A Tiger Moth	85592	DE638	1.10.53	Cancelled as WFU 10.9.73 - DBR landing Owstwich, Yorks 15.5.65 *(Sold R.Malaysian AF 2.89)* *(As "T7245")*
G-APJT	Scottish Aviation Twin Pioneer Srs.1	529		18.12.57	Cancelled 18.4.58 - to Malaysian AF as FM1001 4.58

The NETHERLANDS

Aviation Shop, Aalsmeerderbrug

G-BPMP	Douglas C-47A-50-DL Dakota	10073	N54607	2. 2.89	Cancelled 18.4.95 as WFU
			(N9842A), N54607, Morocco AF 20669, CN-CCL, F-BEFA, 42-24211 *(Cockpit only 3.02)*		

Museum Bevrijdende Vleugels/Wings of Liberation Museum, Best, Noord-Brabant (www.wingsofliberation.nl/)

G-AMPP	Douglas C-47B-15-DK Dakota 3	15272/26717	"G-AMSU"	4. 3.52	Cancelled 7.2.71	7. 2.71
			XF756, G-AMPP, KK136, 43-49456			

Duke of Brabant Air Force, Eindhoven

G-AMBY	Beech D-17S (YC-43) Traveler	295	NC91397	24. 5.50	Cancelled 15.9.51 - to VP-YIV 7.51
			39-0139, DR628, 39-0139		*(Subsequently ZS-PWD and N295BS)*

Dutch Spitfire Flight, Glize-Rijen

G-HVDM	Vickers Supermarine 361 Spitfire LF.IXc	CBAF.IX.1732	8633M	18. 1.91	Cancelled - to PH-OUQ 3.10.00	
			R.Netherlands AF 3W-17, H-25, MK732 *(As "MK732, '3W-17' in 322 Sqn. c, s)*			

Glize-Rijen Historic Flight, Glize-Rijen

G-APCU	de Havilland DH.82A Tiger Moth	82535	N9588	18. 6.57	Cancelled 22.6.76 - to PH-TGR (1) but NTU: restored 18.8.76 Cancelled 16.5.85 by CAA - to PH-TGR (2) 10.5.84. but NTU *(Subsequently PH-TYG 5.86) (As "A-12")*	

Aviodrome Museum, Lelystad Airfield, Lelystad

G-AMCA	Douglas C-47B-30DK Dakota 3	16218/32966	KN487 44-76634	1. 6.50	Cancelled 16.10.03 as WFU *(As "PH-ALR" in KLM orange prewar c/s Holland titles)*	10.12.00A
G-BKRG	Beech C-45G-BH *(Regd as C-45H)*	AF-222	N75WB N9072Z/51-11665	5. 5.83	Cancelled 27.4.98 as WFU	
G-BVOL*	Douglas C-47B-40-DL Dakota	9836	ZS-NJE (2) SAAF 6867, FD938, 42-23974	14. 6.94	Cancelled 16.5.96 on sale to the Netherlands for spares *(As "PH-TCB" in postwar KLM c/s)*	

Stichting Vroege Vogels, Lelystad Airfield, Lelystad

G-EAFN	BAT FK.23 Bantam	FK.23, 18	K-155 F1657	22. 7.19	Cancelled .7.19 - NTU and no CofA issued. *(Originally regd as K-155 @ 24. 6.19) (As "K-155")*	

Vliegend Museum Seppe, Seppe Airfield, Rosendaal

G-AIPE	Taylorcraft J Auster 5	1416	TJ347	17.12.46	Cancelled 22.5.52 - to PH-NET (2) 6.52	

Aviodome Museum, Schiphol Centrum, Schiphol Airport

BAPC.22	Mignet HM.14 Pou-Du-Ciel (Scott A2S)	WM.1			*(As "G-AE0F")*	

Stichting Koolhoven Vliegtuigen, Schiphol Centrum, Schiphol Airport

G-EACN	BAT FK.23 Bantam 1 (ABC Wasp)	FK23/15	K-123 F1654	22. 7.19	Cancelled .7.19 - NTU and no CofA issued *(Originally regd as K-123 @ 29. 5.19) (As "K-123")*	

NEW ZEALAND

Stan Smith Collection, Albany, Auckland

G-AETL	Miles M.14A Hawk Trainer III	332		15. 2.37	Cancelled -- to ZK-AEY 2.38	
G-AFON	de Havilland DH.94 Moth Minor	94012		24. 4.39	Cancelled 21.2.40 - to Australia 2.40 - NTU - to ZK-AHK 9.41	
G-AIBV	Auster J/1N Alpha *(Originally regd as J/1 Autocrat)*	2157		2. 9.46	Cancelled 9.8.49 - to ZK-ATS 9.49	
G-AKHW	Miles M.65 Gemini 1A	6524		21.10.47	Cancelled 9.12.94 - reserved as ZK-KHW.16.5.91	

New Zealand Warbirds Association, Ardmore Airfield, Papakura, Auckland

G-AJHR	de Havilland DH.82A Tiger Moth	85349	DE315	12. 2.47	Cancelled 28.7.50 - to ZK-AUZ 8.50	
G-ALAD (2)	de Havilland DH.82A Tiger Moth	84711	"T6296"??	14. 6.51	Cancelled - to ZK-BAW 11.52 *(Subsequently ZK-CDU)*	
G-AOSW	de Havilland DHC-1 Chipmunk 22	C1/0221	WD283	28. 6.56	Cancelled 19.10.59 - to ZK-BSV 12.59	
G-ARTR	de Havilland DHC-2 Beaver 1	25	ZS-DCG	5.10.61	Cancelled 7.8.64 - to ZK-CKH 12.64 *(As New Zealand AF "NZ6001)*	
G-BDBL	de Havilland DHC-1 Chipmunk 22	C1/0633	WK621	12. 5.75	Cancelled 30.5.84 as destroyed	
			(Crashed @ Luxter Farm, Hambleden 21.1.84: restored 24.12.85, cancelled 5.9.94 - to ZK-UAS 11.94)			
G-BLIC	de Havilland DH.112 Venom FB.54 (Mk.17) *(Built F + W)*	969	N502DM (G-BLIC), Swiss AF J-1799	13. 7.84	Cancelled - to N502DM 7.6.85 *(Subsequently ZK-VNM (2)*	
G-BUZU	Vickers Supermarine 379 Spitfire FR.XIV	---	Indian AF NH799	1. 7.93	Cancelled 4.3.94 - to ZK-XIV *(On rebuild 9.04)*	

Jean Batten Memorial, Terminal Building, Auckland International Airport, Auckland

G-ADPR	Percival P.3 Gull Six	D.55	AX866 G-ADPR	29. 8.35	Cancelled 14.3.95 - to ZK-DPR	1. 8.95P

Museum of Transport, Technolgy and Social History, Point Chevalier, Auckland

G-AGCN	Lockheed 18-56 Lodestar II (C-56D-LO) *(Originally regd as 18-08 Lodestar I; converted 1944)*	2020	AX756 42-53504, NC25630	29. 9.41	Cancelled 19.11.47 - restored as AX756 @ 11.47 *(Subsequently EC-A.. 12.48, Spanish AF as T.4-? @ 1.49,N9933F and ZK-BVE)*	
G-AHJR (2)	Short S.25 Sunderland MR.5	SH1552	SZ584	27. 6.46	Cancelled 15.5.48 - restored as SZ584 @ 4.48 *(As "NZ4115")*	
G-AMMC	Miles M.14A Hawk Trainer III	779	L8353	15. 9.51	Cancelled 1.10.53 - to ZK-AYW 11.53 *(As "L8353")*	
G-AMRM	de Havilland DH.82A Tiger Moth	83513	T7106	18. 4.52	Cancelled 3.6.52 - to ZK-BBI 8.52	

Air Force World - Royal New Zealand Air Force Museum, RNZAF Wigram, Christchurch

G-AIKR	Airspeed AS.65 Consul	4338	PK286	25. 9.46	WFU *(As "G-AIKR")*	14. 5.65
G-AINT	Bristol 170 Freighter 31MNZ	12834	G-18-100	27. 1.47	Cancelled 17.4.52 - to R.New Zealand AF as NZ5903 @ 2.52	
G-BIAT	Sopwith Pup rep	EMK001		3.12.82	Cancelled 9.8.89 - to Australia - NTU -to New Zealand as "N6460"	

Croydon Aviation Heritage Trust, Mandeville, Gore

G-AANF (2)	American Moth Corporation DH.60GMW Moth	49	N298M N237K, NC237K	3. 2.87	Cancelled 19.9.90 as TWFU following crash near Popham 8.8.89 *(As "G-AANF")*	
G-AATC	de Havilland DH.80A Puss Moth	2001		23.12.29	Cancelled 12.30 - to VH-UON 11.8.30 *(Subsequently ZK-ADU)*	
G-AAXG	de Havilland DH.60M Moth	1542	F-AJZB G-AAXG	4.30	Cancelled - to (ZK-ADF) and ZK-AEJ 9.35	
G-ABHC	de Havilland DH.80A Puss Moth	2125		.11.30	To New Zealand Permanent Air Force as 2125 @ 8, 31 *(Subsequently ZK-AEV and NZ593) (As ZK-AEV)*	
G-ABYN	Spartan 3-Seater II	102	EI-ABU G-ABYN	. 7.32	Cancelled - to EI-ABU 10.38, restored 10.4.92, cancelled 31.3.99 by CAA *(Reserved as ZK-ARH.but as "EI-ABU")*	
G-ADHA (2)	de Havilland DH.83 Fox Moth	4097	N83DH ZK-ASP, RNew Zealand AF NZ566, ZK-ADI	3.12.84	Cancelled 3.2.97 - to ZK-ADI 2.97	
G-ADSK	de Havilland DH.87B Hornet Moth *(Rebuilt from components incl from G-ADKL)*	8091	D-EJOM AP-AES, G-ADSK, AV952, G-ADSK	9. 4.36	Cancelled by CAA 15.9.86 -reduced to spares 1974 *(As "G-ADSK")*	14.3.70

G-AEDT	de Havilland DH.90 Dragonfly	7508	N2034	9. 5.36	Cancelled 23.3.98 - to ZK-AYR 4.98
			G-AEDT, VH-AAD, G-AEDT, VH-ABM), G-AEDT		
G-AFPR	de Havilland DH.94 Moth Minor	94031	X5122	16. 5.39	Cancelled 19.3.99 - to ZK-AJN 1998
	((Rebuilt from components)		G-AFPR		
G-AFUU	de Havilland DH.94 Moth Minor	94084		. 7.39R	Cancelled - NTU to R.Australian AF as A21-20 @ 3.40
					(Subsequently ZK-AJR)
G-ALZA	de Havilland DH.82A Tiger Moth	83589	ZK-BAH	19. 1.53	Cancelled - restored as ZK-BAH 8.53
			T5853		
G-AMMK	de Havilland DH.82A Tiger Moth	82521	N9494	2.10.51	Cancelled 12.10.51 - to ZK-AYY 3.52
G-ATFU	de Havilland DH.85 Leopard Moth	7007	HB-OTA	30. 6.65	Cancelled 7.11.96 by CAA - to.ZK-ARG 1998
			CH-366		

New Zealand Vintage Aero Club, Hamilton Airport, Hamilton East

G-AHHE	Auster V J/1 Autocrat	1994		9. 4.46	Cancelled as NTU - to G-AERO 5.46 *(Subsequently ZK-AWX)*
BGA.787	Slingsby T.42B Eagle	1091			Cancelled - to ZK-GBG

Stuart Tantrum Collection, Foxton Pines Airfield, Levin

G-AMEX	de Havilland DH.82A Tiger Moth	83346	T5639	20.10.50	Cancelled 3.1.51 - to ZK-AYA 5.51

Argosy Trust, Woodborne, Woodbourne Airfield, Marlborough

G-ASXM	Armstrong-Whitworth 650 Argosy Srs.222	AW.6801		14.10.64	Cancelled 22.6.70 - sold as CF-TAG
					(Subsequently EI-AVJ, CF-TAG and ZK-SAF) (Fuselage only)
G-ASXN	Armstrong-Whitworth 650 Argosy Srs.222	AW.6802		14.10.64	Cancelled 31.1.70- sold as CF-TAJ.
					(Subsequently ZK-SAE)

New Zealand Sport and Vintage Aviation Society, Hood Airport, Masterton

G-AFSH	de Havilland DH.82A Tiger Moth	82139	X5106	20. 4.39	Cancelled - to X5106 @ 5.1.40, restored 4.1.52
			G-AFSH		Cancelled 23.3.52 - to ZK-BAT 3.52
G-ANSJ	de Havilland DH.82A Tiger Moth	85071	T6802	10. 6.54	Cancelled 7.10.56 - to ZK-BLV 10.55
G-ANSU	de Havilland DH.82A Tiger Moth	85768	DE883	29. 6.54	Cancelled 15.8.54 - to ZK-BGY 11.54 *(Subsequently ZK-BVN)*
G-AOAF	de Havilland DH.82A Tiger Moth	82812	R4895	14. 3.55	Cancelled 16.5.55 - to ZK-BLK 8.55

Queenstown Museum, Queenstown

G-ANGV	AusterV J/1B Aiglet	3122		30.11.53	Cancelled 23.3.54 - to ZK-BDX 3.54

New Zealand Fighter Pilots Museum - Alpine Fighter Collection, Wanaka Airport, Wanaka

G-AMUH	de Havilland DHC-1 Chipmunk 21	C1/0834		2. 9.52	Cancelled 9.1.87 - to ZK-MUH 7.87 *(As "WB568")*
G-BWCF	Yakovlev Yak-50	852904	LY-ANQ	25. 4.95	Cancelled 21.12.95 - to ZK-YAC 3.96
			DOSAAF		
G-CDAN	Vickers Supermarine 361 Spitfire LF.XVIe	CBAF/10895	TB863	30.11.82	Cancelled 16.1.89 - to ZK-XVI 1.89.
		(C/n is firewall no.)			

Wanaka Transport Museum, Wanaka Airport, Wanaka

G-AJEC	Auster V J/1 Autocrat	2327		14. 3.47	Cancelled 18.11.54 - to ZK-BJL 1.55
G-AOMF	Percival P.40 Prentice T.1	PAC/252	(VH-...)	3. 5.56	Cancelled - to ZK-DJC 6.72
	(Officially regd with c/n 5820/1)		G-AOMF, VS316		

NORWAY

Forsvarsmuseet Flysamlingen, Gardermoen

G-ASCF	Beagle A.61 Terrier 2	B.617	WE548	23. 7.62	Cancelled 10.4.67 - to SE-ELO 9.67
G-BMEW	Lockheed 18-56 Lodestar	18-2444	OH-SIR	30. 9.85	Cancelled 15.7.86 on sale to "Canada"
	(C-60A-5-LO) *(Gulfstar conversion c.4.59)*		(N283M), OH-MAP, N283M, N9223R, N105G, N69898 ,NC69898, 42-55983 *(As "G-AGIH")*		

Flyhistorick Museum Sola, Stavangar-Lufthavn

G-AKKA	Miles M.65 Gemini 1A	6528		21.10.47	Cancelled 27.4.48 - to LN-TAH
G-AOXL	de Havilland DH.114 Heron 1B	14015	(LN-BFY)	5. 4.57	Cancelled 13.9.71 - to LN-BFY: NTU and restored 21.9.71:
			G-AOXL, PK-GHB		Cancelled 11.10.71 - to LN-BFY 2.72
					(As "LN-PSG" in Braathens/SAFE c/s))
G-BBZL	Westland-Bell 47G-3B1	WA/583	S Yemen AF 404	26. 2.74	Cancelled 2.6.82- to SE-HME 6.82
					(Subsequently "LN-ORB") (As "G-BBZL" 11.02)
G-BCZS	Fairey Britten-Norman BN-2A-21 Islander	441		1. 4.75	Cancelled 19.8.77 - to LN-MAF 5..57 *(Stored 11.02)*

Dakota Norway, Sandefjord, Torp

G-BLYA	Douglas C-53D-DO Skytrooper	11750	Finnish AF DO-9	19. 4.85	Cancelled 25.4.85 -NTU - to N59NA 485,
			OH-LCG, 42-68823		*(Subsequently LN-WND 5.85)*

Warbirds of Norway, Kjeller, Oslo and Torp, Sandefjord (*)

G-ANSE	de Havilland DH.82A Tiger Moth	85738	DE840	29. 5.54	Cancelled 24.6.54 - to LN-BDO 6.54 *(Subsequently LN-MAX)*.
G-SUES	North American AT-6D Harvard III **(*)**	88-14552	FAP1506	18. 1.79	Cancelled 10.2.84. - to (LN-LFW)/LN-WNH 7.86
			South African AF 7424, EX881, 41-33854		

OMAN

Sultanate of Oman Armed Forces Museum, Bait al Falaj Airfield, Muscat

G-BGSB	Hunting-Percival P.56 Provost T.1	PAC/F/057	7922M	21. 5.79	Cancelled 11.12.87 by CAA - to Oman AF 1982
	(C/n officially quoted as 886391) (Also c/n P56/57)			WV494	*(As "XF868")*

PAPUA - NEW GUINEA

Air Nuigini Collection, Jackson Airport, Port Moresby

G-AMZH	Douglas C-47B-20-DK Dakota	15665/27110	KN421	2. 5.53	Cancelled 20.5.65 - to VH-SBW 22.5.65
			43-49849		*(Subsequently P2-SBW 1.6.74 and P2-ANQ 3.12.75)*.

PORTUGAL

Museu do Ar, Alverca (www.emfa.pt/museu)

G-ASCB	Beagle A.109 Airedale	B.527		23. 7.62	Crashed into River Douro, Barqueiros do Douro, Portugal 26.7.64: rebuilt as CS-AIB 6.68 and cancelled	
BGA.619	Slingsby T.21B	551		10.48	Sold as CS-PAI 1.51	

Museu Industrial de Aeronáutica das OGMA, Alverca

G-AKBT	Piper J3C-85 Cub		21984	29. 9.47	Cancelled 27.11.48 - to CS-AAP 2.49	

Museo da Transportes Aéreos Portugueses, Aeroporto de Lisboa, Lisbon

EI-ACK	Douglas C-47A-80-DL Dakota	19503	43-15037	29. 6.46	Cancelled 9.2.60 - sold as 4X-AOC 2.60 *(Subsequently CS-AGA @ 27.3.63) (As "CS-TDA")*	

SAUDI ARABIA

Dhahran Air Force Base Collection, King Abdul-al-Azix Air Base, Dhahran

G-BFOO	British Aircraft Corporation 167 Strikemaster Mk.80A	PS.366	G-27-312	22. 3.78	Cancelled - to R.Saudi AF as 1135 @ 5.78	

SERBIA

Muzej Yugoslovenskog Vazduhplovsta, Surcin Airport, Belgrade

G-AHLX	Douglas C-47A-30-DK Dakota	14035/25480	KG803 43-48219	8. 5.46	Cancelled 23.12.47 - to YU-ABG 12.47	
G-AKLS	Short SA.6 Sealand 1	SH1567	G-14-2 G-AKLS, (VP-TBC), G-AKLS	26.11.47	Cancelled 7.3.51 - To YU-CFK 9.51	

SINGAPORE

Singapore Science Centre, Singapore

G-AJLR	Airspeed AS.65 Consul	5136	R6029	25. 5.47	Cancelled 26.2.73 as WFU *(As "VR-SCO") (Noted wrecked 2.03)*	23. 4.63

REPUBLIC of SOUTH AFRICA

Classic Jets South Africa, Cape Town International Airport, Cape Town

G-BVWG	Hawker Hunter T.8C *(Officially regd with c/n 41H-695320)*	41H-693836	XL598	8.12.94	Cancelled 12.6.95 - to ZU-ATH	
G-BVWV	Hawker Hunter F.6A *(Officially regd with c/n 41H-674112)*	41H-679991	8829M XE653	22.12.94	Cancelled 30.8.95 - to ZU-AUJ 9.95	
G-FSIX	English Electric Lightning F.6	95116	XP693	31.12.92	Cancelled 13.2.97 - to ZU-BEY 2.97	
G-LTNG	English Electric Lightning T.5	B1/95011	8503M XS451	8.11.89	Cancelled 13.2.97 - to ZU-BEX 2.97	
G-OPIB	English Electric Lightning F.6	95238	XR773	31.12.92	Cancelled 13.2.97 - to ZU-BEW 2.97	

South African Airways Museum and Historic Flight, Jan Smuts International Airport, Johannesburg (www.historicflight.co.za)

G-AHOT	Vickers 498 Viking 1A	121	XD635 G-AHOT	27. 6.46	Cancelled 26.9.54 - to ZS-DKH 10.54	
G-AWFM	de Havilland DH.104 Dove 6	04079	9J-RHX VP-YLX, VP-RCL, ZS-BCC, (G-AJOU)	27. 3.68	Cancelled 26.8.81 - to ZS-BCC 10.78	
G-BFHE	CASA 352L (Junkers Ju 52/3m)	164	Spanish AF T2B-273	23.11.77	Cancelled 12.5.81 - to ZS-UYU 8.81 *(Subsequently ZS-AFA)*	

Caesar's Place Casino, Jan Smuts International Airport, Johannesburg

G-AMZW	Douglas C-47B-20-DK Dakota	15654/27099	KN231 43-49838	29. 5.53	Cancelled 1.7.53 - to SN-AAH 7.53 *(Subsequently ST-AAH, Sudan AF 424, ST-AAH) (As SAAF 6850 and/or ZS-BMF)*	

Wings and Wheels Museum , Jack Taylor Airfield, Krugersdorp

G-ASRL	Beagle A.61 Terrier 2	B.631	WE609	18. 3.64	Cancelled as WFU 18.9.69 - crashed Kota, Malawi 18.4.69 *(Repaired as VP-WDN 1970)*	

South African Air Force Museum, Pretoria

G-EAML	Airco DH.6	---	C9449	8. 9.19	Cancelled 19.9.19 - to South Africa *(Components only preserved as "G-EAML")*	18. 9.20
G-AITF	Airspeed AS.40 Oxford 1	---	ED290	1.11.46	Cancelled 31.10.61 as PWFU *(As "G-AITF")*	8. 6.60
G-AJTI	Miles M.65 Gemini 1A	6444		2. 6.47	Cancelled 8.10.47 - to ZS-BRV 10.47	
G-AOPL	Percival P.40 Prentice T.1	PAC/207	VS609	30. 5.56	Cancelled 24.4.67 - to ZS-EUS 5.67	

SPAIN

Asociacion Aeroclasica, Villafria, Burgos

G-AHCF	Auster V J/1 Autocrat	1960		11. 3.46	Cancelled by CAA 1948 - to EC-DAZ 10.48: restored.?? and cancelled 1.1.51 - to EI-ACJ 4.54	
G-AIGH	Auster V J/1 Autocrat	2190		5.11.46	Cancelled 9.9.53 - to EC-AIS 11.54	
G-AJAL	Auster V J/1 Autocrat	2216		13. 2.47	Cancelled 26.6.48 - to EC-ADG 6.48	

Col-Lecccio d'Automobils de Salvador Claret, Sils-Girona

G-AGXS	Auster V J/1 Autocrat	1967		24. 1.46	Cancelled 12.5.55 -to EC-ALD 8.55. .	

Fundación Infante de Orleans, Cuatro-Vientos, Madrid

G-ABUU	Comper CLA.7 Swift	S.32/5		7. 3.32	Cancelled 18.1.99 - to EC-HAM 2.99 *(As "EC-AAT")*	
G-ADLS	Miles M.3C Falcon Six *(Provenance not confirmed)*	231		15. 7.35	Cancelled 8.36 - to EC-ACB	
G-AKAA	Piper J3C-65 Cub (L-4H-PI)	10780	43-29489	23. 6.47	Cancelled 31.10.96 - to EC-GQE 10.96	

Museo del Aire, Cuatro-Vientos, Madrid

G-ABXH	Cierva C.19 Mk.IVP (Avro 620)	5158		1. 6.32	Cancelled - to EC-W13 and EC-ATT 12.32 *(Subsequently Spanish AF 30-62, EC-CAB and EC-AIM)*	
G-ACYR	de Havilland DH.89 Dragon Rapide	6261		15.10.34	WFU *(Olley Air Services titles)*	23. 8.47
G-AERN	de Havilland DH.89A Dragon Rapide	6345		13. 1.37	Cancelled - sold as EC-AKO 12.54	
BGA.1402	Slingsby HP-14C	1637			To EC-BOL 5.68	

Hotel las Americas, Tenerife

G-AOYM	Vickers 806 Viscount	262		20.12.56	Cancelled 29.10.85 - to EC-DYC 10.85 *(Nose only stored 12.99)*	

SRI LANKA

Katunayake-Negombe Base Collection, Negombo

G-APCN	Boulton Paul P.108 Balliol T.2	BPA.10C *(Second issue of c/n)*	WG224	12. 5.57	Cancelled 13.8.57 - to Ceylon AF as CA310 @ 8.57	

Sri Lanka Air Force Musem, Ratamalana Air Force Base, Ratamalana

G-AKEE	de Havilland DH.82A Tiger Moth	---	"T7179"	18. 8.47	Cancelled 25.9.47 - to VP-CAW 9.47 *(Subsequently CY-AAW) (As "CX-123")*	
G-AMJY	Douglas C-47B-40-DK Dakota	16808/33556	KP254 44-77224	2. 6.51	Cancelled 11.11.59 - to 4R-ACI 11.59 *(As "CR-822")*	

SWEDEN

Eskiltuna Flygklubb, Ekeby, Eskiltuna

BGA.3242	Slingsby T.21B Sedbergh TX.1	663	WG496	. 8.86	*(Code "FGD") (Subsequently BGA.3309/FJY and SE-SME)*	

Svedinos Bil Och Flygmuseum, Slöinge, Halmstad

G-EBNO	de Havilland DH.60 Moth (Cirrus I)	261		19. 2.26	Cancelled 26.7.28 - to S-AABS 7.28 *(Subsequently SE-ABS) (As "Fv5555")*	
G-AIZW	Auster V J/1 Autocrat	2230		31. 1.47	Cancelled 14.8.58 - to SE-CGR 8.58	
G-AKAO	Miles M.38 Messenger 2A	6703		27. 6.47	Cancelled 7.9.53 - to SE-BYY 9.53 *(As "L-H")*	
G-ANSO	Gloster Meteor T.7	G5/1525	G-7-1	12. 6.54	Cancelled 11.8.59 - to SE-DCC 8.59 *(As "WS774/4")*	

(Originally built as Meteor F.8 G-AMCJ {G5/1210} regd 19. 6.50: to R.Danish AF as 490 then Egyptian AF as 1424 then rebuilt as G-ANSO)

Ljungbyheds Aeronautiska Sallstap, Ljungbyhed Airfield, Helsingborg

G-ADLV	de Havilland DH.82A Tiger Moth	3364		6. 8.35	Cancelled 30.10.40 - to RAF as BB750 @ 10.40 Restored as G-AORA 17. 5.56: cancelled 21.6.63 - to SE-CWG 8.63	
G-ALCT	Taylorcraft J Auster 5	1532	TJ513	28. 7.48	Cancelled 22.9.53 - to SE-BZB 10.53	
G-APLI	de Havilland DH.82A Tiger Moth	85593	DE639	7. 3.58	Cancelled 15.6.60 - to SE-COG 9.60	
G-APNN	Auster 5 Alpha	3410		10. 6.58	Cancelled 20.8.59 - to SE-CME 10.59	

High Chapperal Park, Hillerstorp

G-AVJB	Vickers 815 Viscount	375	(LX-LGD) G-AVJB, AP-AJF	21. 3.67	Cancelled 28.10.86 - to SE-IVY *"Big Airland"*	

Foreningen Veteranfly Kungsangen Airport, Kungsangen

G-ANTV	de Havilland DH.82A Tiger Moth	84589	T6122	26. 7.54	Cancelled 24.10.55 - to D-EKUR 11.55	

Flyvapenmuseum, Malmslatt, Linköping

G-ANVU	de Havilland DH.104 Dove 1B *(Originally regd as Srs.1)*	04082	VR-NAP	12.11.54	Cancelled 20.6.85 by CAA *(WFU 1977: restored 15.4.86: cancelled 16.9.86 on sale to Sweden: stored 5.03)*	14. 9.77
G-AZAK	Scottish Aviation Bulldog Srs.101 (SK-61)	BH.100/106		18. 6.71	Cancelled 17.9.71 - to Swedish Army as Fv61006 9.71 *(Stored 5.03)*	
G-AZJO	Scottish Aviation Bulldog Srs.101 (SK-61)	BH.100/137		8.12.71	Cancelled 7.2.72 - to Swedish Army as Fv61030 2.72 *(Stored 5.03)*	
G-AZWO	Scottish Aviation Bulldog Srs.101 (SK-61)	BH.100/186		8. 6.72	Cancelled 19.10.72 - to Swedish Army as Fv61068 10.72 *(Stored 5.03)*	

Stiftelsen Aerospace, Arlanda Airport, Stockholm

G-AAHD	Avro 594 Avian IV	R3/CN/318		5.29	Cancelled - to Sweden as SE-ADT 8.33	
G-AGIJ	Lockheed 18-56 Lodestar II (C-60A-5-LO)	2593	43-16433	4. 3.44	Cancelled 9.7.45 - to R.Norwegian AF as 2593/T-AE 7.45 *(Subsequently OH-VKP and SE-BZE)*	
G-ANIU	Taylorcraft J Auster 5	841	MS977	5.12.53	Cancelled 12.10.55 - to LN-BDU 2.55.	

Swedish Veteran Wing Museum, Hässlo Airfield, Västerås

G-BNPG	Percival P.66 Pembroke C.1	PAC/66/082 *(Also c/n K66/045)*	XK884	30. 6.87	Cancelled 4.5.88 - to SE-BKH	

Västerås Flygande, Hässlo Airfield, Västerås

G-ANEF	de Havilland DH.82A Tiger Moth	83226	T5493	28. 9.53	Cancelled by CAA 10.9.96 - damaged on take-off Cranwell 17.9.88 Sold Sweden 1991 for rebuild - to SE-AMM 4.98	
G-APOU	de Havilland DH.82A Tiger Moth	85867	DF118	15. 8.58	Cancelled 20.5.59 - to SE-CHG 9.59	
G-BAVN	Boeing-Stearman A75N-1 Kaydet	75-5659	4X-AMT N5367N, 42-17496	13. 4.73	Cancelled 22.10.84 - to SE-AMT 8.84	

(Regd with c/n "3250-2606", which is a part number)

SWITZERLAND

Technorama Museum, Winterthur

G-AHPB	Vickers 639 Viking 1 *(Originally regd as Type 614)*	132		4. 9.46	Cancelled?? *(To XF638 ?? and restored ??) (As "D-BABY")* *(Also reported as broken up between 1988 and 1993: current status unknown)*	20.5.68

THAILAND

Foundation for the Preservation and the Development of Thai Aircraft, Don Muang Air Force Base, Don Muang, Bangkok

G-AMOU	de Havilland DH.82A Tiger Moth	84695	N200D 9M-ALJ, VR-RBZ, G-AMOU, T6269 *(As "21")*	5. 2.52	Cancelled 18.11.93 on sale to Thailand.	

Royal Thai Air Force Museum, Don Muang Air Force Base, Don Muang, Bangkok
G-AMJW Westland-Sikorsky S-51 Mk.1A WA/H/120 G-17-2 9. 1.52 Cancelled 22.5.53 - to R.Thailand AF as 305-53 @ 5.53

Golden Jubilee Museum of Agriculture, Khlong Luang
G-AWID Britten-Norman BN-2A Islander 26 17. 4.68 Cancelled - to R.Thailand AF as 501 @ 10.68

UNITED ARAB EMIRATES

Al Mahata Museum, The Sharjah Aviation Museum, Sharjah
G-ARDE de Havilland DH.104 Dove 6 04469 I-TONY 15.11.60 Cancelled 30.5.01 by CAA *(As "G-AJPR" in Gulf Air c/s)* 25. 8.91

UNITED STATES

United States Army Aviation Museum, Fort Rucker, Alabama
G-BLXT Royal Aircraft Factory SE.5A ---- N4488, USAAS 22-296 2.10.85 Cancelled 28.9.89 by CAA *(As "18-8010")*
G-BSKS Nieuport 28C-1 6531 "N5246", US Navy 27. 6.90 Cancelled 1.4.93 on sale to USA *(As "6531/5" in 94 Aero Sqn AEF c/s)*

Alaska Aviation Heritage Museum, Anchorage, Alaska
G-AGZI Consolidated-Vultee CV.32-3 (LB-30) Liberator II 55 AL557 11. 1.46 Cancelled 24.2.48 - to SX-DAA 2.48 *(Subsequently N9981F, N68735 and N92MK)*

Aerospace Maintenance and Regeneration Center "Celebrity Row", Davis-Monthan Air Force Base, Tucson, Arizona
G-BLLJ Short SD.3-30 (C-23A) Sherpa SH3100 G-14-3100 26. 7.84 Cancelled 2.11.84 - to USAF as 83-0512 11.84.

Champlin Fighter Museum, Mesa, Arizona
G-AVAV Vickers Supermarine 509 Spitfire Trainer IX CBAF/7269 Irish Air Corps 159, G-15-172, MJ772 8.11.66 Cancelled 18.5.75 - to N8R 5.75 *(Subsequently N8R)*
G-AWHL Hispano HA.1112-MIL Buchon (Messerschmitt Bf.109G) 186 Spanish AF C4K-122 14. 5.68 Cancelled 17.2.69 on sale to Spain. *(Subsequently NX109J) (As"J-392")*

Confederate Air Force (Arizona Wing), Mesa, Arizona
G-AWHE Hispano HA.1112-MIL Buchon 67 Spanish AF C4K-31 14. 5.68 Cancelled 20.2.69 - to N109ME 2.69
(Messerschmitt Bf.109G) *(Regd incorrectly as c/n 64)*
G-BDYA CASA 2111 (Heinkel 111H-16) ---- Spanish AF T8B-124 21. 5.76 Cancelled 2.8.77 - to N72615 8.77

American Aeronautical Foundation Museum, Camarillo, California
G-AOLP Percival P.40 Prentice T.1 5840/7 VS385 25. 4.56 Cancelled 5.3.79 - to N1041P 3.79

Arango Collection, Paso Robles, California
EI-APU (2) Fokker D.VII/65 rep *(Built Rousseau Aviation)* <u>01</u> F-BNDG .68 Cancelled 28.5.85 - to N902AC 5.85
EI-ARC Pfalz D.III rep *(Built Personal Plane Services)* PPS/PFLZ/1 G-ATIF 29. 5.67 Cancelled 28.5.85 - to N906AC 5.85
BAPC.105 Bleriot Type XI 54
(Anzani "V" 25hp) *(Composite from original components including c/n 54: built by L.D.Goldsmith in 1976 @ RAF Colerne)*

Fighter Jets and Air Racing Museum, Chino, California
BAPC.141 Macchi M.39 fsm (Gipsy Queen) *(As "5")*

Planes of Fame Air Museum, Chino, California
G-BTHD Yakovlev Yak-3U See SECTION 1, Part 2
BAPC.110 Fokker D.VIIF fsm *(As "5125/18")*
BAPC.136 Deperdussin 1913 Monoplane fsm *(As "<u>19</u>")*
BAPC.140 Curtiss 42A R3C2 fsm *(As "3" in US Army c/s)*

Yanks Air Museum, Chino, California
G-BUCF Grumman F8F-1B Bearcat D.779 R.Thailand AF 122095, BuA.122095 18. 2.92 Cancelled 12.11.92 - to N2209 8.93

Museum of Flying, Santa Monica, California
G-MXIV Vickers Supermarine 379 Spitfire FR.XIVc 6S/583887 Indian AF T3, NH749 11. 4.80 Cancelled 15.5.85 to NX749DP 5.85

Palm Springs Air Museum, Palm Springs, California
G-BLZW Republic P-47D-30-RA Thunderbolt 399-55744 N47DE 15. 7.85 Cancelled 4.11.85 - to NX47RP 3.86. .
Peruvian AF 122, Peruvian AF 547, 45-49205 *(Subsequently N47DE 6.86) (As "228473"")*
G-FIRE Vickers Supermarine 379 Spitfire FR.XIVc 6S/648206 Belgium AF SG128, G-15-1, NH904 21. 3.79 Cancelled 13.2.89 -to N8118J 2.89. . *(Subsequently N114BP)*

Western Aerospace Museum, Oakland, California
G-AKNP Short S.45 Solent 3 S.1295 NJ203 2.12.47 Cancelled 20.3.51 - to VH-TOB 1.51 *(Subsequently N9946F) (As "NJ203")*

National Air and Space Museum, Washington, DC (www.nasm.sl.edu)
G-AARO (2) Arrow Sport A2-60 341 N932S, NC932S 17. 9.79 Cancelled in 6.83 - to N280AS *(Subsequently N9325) (As "G-AARO" @ Paul E Garber Facility)*
G-BFHD CASA 352L (Junkers Ju 52/3m) 146 Spanish AF T2B-255, Spanish AF "721-8" 23.11.77 Cancelled 21.1.88 on sale to West Germany *(Subsequently to NASM) (As "N8+AA" in Lufthansa c/s - displayed Dulles international Airport, Washington)*
G-MURY Robinson R44 Astro 0201 19. 7.95 Cancelled as sold to USA 9.10.03 17. 5.04T

Miami International Airport, Miami, Florida
G-ASCY Phoenix Luton LA-4A Minor PAL/1124 5. 9.62 Cancelled 8.69 - to EI-ATP 29.8.69 *(Subsequently N924GB)*
(Built Cornelius Bros) *(Displayed Terminal as "EI-ATP/N924GB" @ 11.02)*

Weeks Air Museum, Tamiami, Miami, Florida
G-ASKB de Havilland DH.98 Mosquito TT.35 ---- N35MK 8. 7.63 Cancelled 17.6.83 - to N35MK and restored 16.3.87
G-ASKB, N35MK, G-ASKB, RS712 Cancelled 20.10.87 - to N35MK 3.88
(Used in film "Mosquito Squadron" 6.68 as "HJ690/HT-N")
(On loan to Experimental Aircraft Association Air Adventure Museum, Witttman Field, Oshkosh, Wisconsin)

Fantasy of Flight, Polk City, Florida (www.fantasyofflight.com)
G-AHMJ Cierva C.30A R3/CA/43 K4235 8. 5.46 Cancelled 9.2.50 as WFU
(Avro 671) (Rota I) *(Official c/n quoted incorrectly as 774 - which was a Danish Avro 621)* *(Rrestored 8.4.93: cancelled 12.11.98 - to USA) (As "K4235")*
G-AYAK Yakovlev Yak C.11 172701 OK-KIE 31. 3.70 Cancelled 7.6.84 - to N11YK *(As "N11YK")*
G-AYFO Bucker Bu.133 Jungmeister 4 HB-MIO 24. 6.70 Cancelled 28.4.71 - to N40BJ
(Built Dornier) Swiss AF U-57
G-BBMI Dewoitine D.26 10853 HB-RAA 11.10.73 Cancelled 30.5.84 - to N282DW
(Built EKW) Swiss AF 282
G-BCOH Avro 683 Lancaster Mk.10 AR 277 CF-TQC 24. 9.74 Cancelled 23.2.93
(Built Victory Aircraft, Canada) RCAF KB976 *(As "G-BCOH")*
G-BFHG CASA 352L 153 Spanish AF T2B-262 23.11.77 Cancelled 27.9.94 - to USA.
(Junkers Ju 52/3m) *(C/n reported as 155)* *(Subsequently "D2+60" and "D-TABX") (As "VK-NZ")*
G-BHEW Sopwith Triplane repPPS/REP/9 & PFA 95-10485 "N5430" 11.10.79 Cancelled 14.1.86 - to N5460
(Built Personal Plane Services) (Originally c/n PFA 1539)
G-BJCL Morane Saulnier MS.230 Parasol 1049 EI-ARG 22. 7.81 Cancelled 27.1.88 - to N230MS
F-BGMR, French Mil
G-BJHS Short S.25 Sandringham SH.55C (EI-BYI) 11. 9.81 Cancelled 12.8.93 - to N814ML
(Sunderland GR.3 c/n SH974 conversion) G-BJHS, N158J, VH-BRF, R.NZ AF NZ4108, ML814 *(As "N814ML/G-BJHS/ML814")*
G-BXYA CSS-13 Aeroklubowy 0365 SP-ACP (3) 3. 7.98 Cancelled 13.11.98 - to USA
(Licence built Polikarpov Po-2) (SP-FCN), SP-ACN (3), PLW-... *(As "10" in Soviet c/s)*
G-CCIX Vickers Supermarine 361 Spitfire LF.IXe CBAF.IX.558 G-BIXP 9. 4.85 Cancelled 6.1.93 as TWFU
(C/n is Firewall No.) Israel DF 2046, Czech AF, TE517 *(As "G-CCIX")*
G-CCVV Vickers Supermarine 379 Spitfire FR.XIVe 6S/649186 Indian AF "42" 18. 5.88 Cancelled 6.1.93 as TWFU
MV262 *(As "G-CCVV")*
G-XVIB Vickers Supermarine 361 Spitfire LF.XVIeCBAF.IX.4610 N476TE 3. 7.89 Cancelled 1.2.90 - to N476TE
G-XVIB, 8071M, 7451M, TE476 *(Restored 3.5.94: cancelled 21.9.95 and restored 1.96 as N476TE)*

Florida Military Aviation Museum, St.Petersburg-Clearwater Airport, Florida
G-AGIP Douglas C-47A-1-DK Dakota 11903 FL544 20.12.43 Cancelled 12.8.63 - To CN-ALI 10.63
42-92136 *(Subsequently Maroc AF 29136, N54605 and (N9845A))*

Valiant Air Command Museum, Tico, Florida (www.vacwarbirds.org)
G-AGWE Avro 19 Srs.2 1286 TX201 28.12.45 Cancelled 17.5.73 - to USA *(As "G-AGWE")* 5. 3.73
G-HUNN Hispano HA.1112-M1L Buchon 235 G-BJZZ 29. 4.87 Cancelled 9.10.91 - to N109GU
(Messerschmitt Bf.109G) *(P/i also reportedly as C4K-235)* N48157, Spanish AF C4K-172

Frasca Air Museum, Urbana, Illinois
G-BTXE Vickers Supermarine 394 Spitfire FR.XVIIIe N6 IAF 23.10.91 Cancelled 2.12.92 - to N280TP 2.93
6S/676372-165 Indian AF HS654, TP280

Columbus Flying Corps Museum, Columbus, Indiana
G-ANKN de Havilland DH.82A Tiger Moth 82700 R4759 24.12.53 Cancelled 5.8.57 - to CF-JJI 8.57
(Subsequently "N4808")

Confederate Air Force (Great Lakes Wing), Bloomingdale, Illinois
G-VALE North American AT-6C Harvard IIA 88-10677 (G-BHXF) 17. 9.80 Cancelled 12.11.85 - to N36CA 11.85
G-RBAC, FAP 1522, South African AF 7244, EX584, 41-33557

Eighth Air Force Museum, Barksdale Air Force Base, Louisiana
G-BGCF Douglas C-47A-90-DL Dakota 20596 -10 20.11.78 Cancelled 30.1.80 - to N3753C 4.80
N86453, 43-16130 *(As "43-16130")*

Kalamazoo Aviation History Museum, Kalamazoo, Michigan
G-AWHJ Hispano HA.1112-MIL Buchon 171 Spanish AF C4K-100 14. 5.68 Cancelled 20.2.69 - to N90605 2.69
(Messerschmitt Bf.109G)

Yankee Air Museum, Belleville, Michigan
G-AOZZ Armstrong-Whitworth 650 ArgosySrs.100AW.6651 12. 3.57 Cancelled 1.10.68 - to G-11-1 11.68 and N896U 12.68.

Confederate Air Force (New Mexico Wing), Albuquerque, New Mexico
G-AKZY Messerschmitt Bf.108D-1 Taifun 3059 Luftwaffe 7. 6.48 Cancelled - to HB-DUB 1.50
D-ERPN *(Subsequently N2231)*

War Eagles Museum, Santa Teresa, New Mexico
G-ANNC de Havilland DH.82A Tiger Moth 84569 T6102 22. 1.54 Cancelled 9.4.58 - to OO-SOM (2) 4.58
(Subsequently N7158N)

Lone Star Flight Museum, Galveston Airport, New Mexico
G-FORT Boeing 299-O (B-17G-95-DL) Fortress 8627 F-BEEC 11. 4.84 Cancelled 2.7.87 - to N900RW
ZS-EEC, F-BEEC, 44-85718 *(As "238050")*

Intrepid Air and Space Museum, Manhattan, New York
G-BOAD British Aircraft Corporation-Aérospatiale Concorde 102 Cancelled 4.5.04 as WFU 3.12.04T

Empire State Aerosciences Museum, Schenectady County Airport, Scotia, New York

G-AMJX	Douglas C-47B-20-DK Dakota	15635/27080	KN214 43-49819	2. 6.51	Cancelled 12.8.63 - to Moroccan AF as 49819/CN-ALJ 8.63 *(As "316250")*

Rhinebeck Aerodrome Museum, Rhinebeck, New York

G-ABIH	de Havilland DH.80A Puss Moth	2140		17. 2.31	Cancelled 11.31 - to USA as NC770N 11.31 *(Subsequently N770N and NC770N) (As "N770N")*
G-ATXL	Avro 504K rep *(Built Hampshire Aero Club)*	HAC-1		19. 7.66	Cancelled 6.8.71 - to N2929 8.71 *(As "E2939")*
G-AWYI	Royal Aircraft Factory BE.2c rep *(Built C Boddington)*	001		5. 2.69	Cancelled 17.6.71 - to N1914B

United States Air Force Museum, Wright Patterson Air Force Base, Dayton, Ohio

G-MOSI	de Havilland DH.98 Mosquito TT.35	Not known	N98DH N9797, G-ASKA, RS709	10.11.81	Cancelled 21.1.87 by CAA *(Sold to USAF Museum 2.85) (As "NS519")*	17.12.84P

Captain Michael King Smith Evergreen Aviation Educational Center, McMinnville, Oregon

G-SMIT	Messerschmitt Bf.109G-10/U-4	610937	T2-124 FE-124, Luftwaffe 610937	10.12.79	To N109MS 4.4.90 *(Subsequently N109EV)*
G-SXVI	Vickers Supermarine 361 Spitfire LF.XVIe	CBAF-11470	7001M 6709M, TE356	25. 2.87	Cancelled 15.1.90 - to N356V *(Subsequently N356TE)*

Tillamook Naval Air Station Museum, Tillamook, Oregon

G-AIDN	Vickers Supermarine 502 Spitfire T.8	6S/729058	N32 MT818	22. 8.46	Cancelled 3.7.86 - to N58JE 8.86
G-AWHN	Hispano HA.1112-MIL Buchon (Messerschmitt Bf.109G)	193	Spanish AF C4K-130	14. 5.68	Cancelled 20.2.69 - to N90602 2.69
G-BMFB	Douglas Skyraider AEW.1 (AD-4W)	7850	SE-EBK G-31-12, WV181, BuA.126867	24. 9.85	Cancelled 1.5.90 - to N4277N 5.91

Breckenridge Aviation Museum, Stephens County Airport, Breckenridge, Texas

G-RCAF	North American AT-6C Harvard IIA	88-9723	FAP1560 South African AF 7168, EX287, 41-33260	6. 3.79	Cancelled 16.4.80 - to N42BA 7.80

Cavanaugh Flight Museum, Addison Airport, Dallas, Texas

G-CCMV	Vought FG-1D Corsair *(Built Goodyear Aircraft Corporation)*	3660	N448AG N4717C, Bu.92399	21.11.00	Cancelled 5.9.02 - to N451FG
G-VIII	Vickers Supermarine 359 Spitfire LF.VIII	6S/479770	I-SPIT Indian AF T17, MT719	27. 4.89	Cancelled 9.7.93 - to N719MT 8.93

Museum of Flight, Seattle, Washington

G-AOVU	de Havilland DH.106 Comet 4C	6424	(G-APMD) (G-APDN)	22.10.59	Cancelled 25.3.60 - to XA-NAR 3.60 *(Subsequently N888WA)*
G-BOAG	British Aircraft Corporation-Aérospatiale Concorde 102				See SECTION 1

Experimental Aircraft Association Air Adventure Museum, Witttman Field, Oshkosh, Wisconsin

G-AHXW	de Havilland DH.89A Dragon Rapide	6782	NR683	11. 7.46	Cancelled 16.3.71 - to N683DH 3.71.
G-ASKB	de Havilland DH.98 Mosquito TT.35	----	N35MK	8. 7.63	On loan - see Weeks Air Museum, Tamiami, Miami, Florida
G-AWHO	Hispano HA.1112-MIL Buchon (Messerschmitt Bf.109G)	199	Spanish AF C4K-127	14. 5.68	Cancelled 20.2.69 - to N90601 2.69 *(Subsequently N109BF)*
G-AXTY	de Havilland DH.82A Tiger Moth	85970	F-BDMP EM739	3.12.69	Cancelled 12.12.72 - to N16645
G-HUNT	Hawker Hunter F.51	41H-680277	G-9-440 Danish AF E-418, Danish AF 35-418	5. 7.78	Cancelled 10.12.87 - to N50972 1.88 *(Subsequently N611JR) (As "WB188")*

URUGUAY

Museo Aeronautico, San Gabriel, Montevideo

G-AMNL	Douglas C-47B-35-DK Dakota	16644/33392	XF767 G-AMNL, KN682, 44-77060	30.11.51	Cancelled 23.11.61 - to I-TAVO 1.62 *(Subsequently CX-BDB)*

VENEZUELA

Museo del Transporte, Caracas

G-BDVO	Short SC.7 Skyvan 3-100	SH1949	G-14-117	20. 4.76	Cancelled - to YV-O-MC-9 @ 8.76 *(Subsequently YV-O-MTC-9)*

Escaudron Legendario, Base Area el Libertador, Palo Negro, Aragua, Maracay

G-AJAV	Douglas C-47A-5-DK Dakota	12386	KG377 42-92571	9. 1.47	Cancelled 6.9.50 - to N19E 9.50. *(Subsquently N70, N70F N40G, YV-P-EPO and YV-T-RTC)*

Museo Aeronautica de la Fuerza Aerea Venezolana, Maracay

G-AMXS	de Havilland DH.104 Dove 2A	04382	(N4281C)	21. 1.53	Cancelled 28.8.53 - to YV-T-FTQ *(Subsequently 3C-R1) (As "2531")*

ZIMBABWE

Zimbabwe Military Museum, Gweru

G-AWTD	Percival P.56 Provost T.1	P56/285 (Also c/n PAC/F/285)	(D-....) XF554	8.11.68	Cancelled - to Rhodesian AF as 3614 1973

PART 2 – BRITISH AVIATION PRESERVATION COUNCIL REGISTER

The British Aviation Preservation Council (BAPC) was formed in 1967 and is the national body for the preservation of aviation related items. It is a voluntary staffed body which undertakes a representation, co-ordination and enabling role. BAPC membership includes national, local authority, independent and service museums, private collections, voluntary groups and other organisations involved in the advancement of aviation preservation in the UK. A number of overseas aircraft preservation organisations have affiliated membership.

The BAPC Register of Anonymous Airframes was started in the 1980s as a way of flagging up aircraft which had not managed to be given a formal method of identification, for example a civilian registration, a military serial or a construction number for one reason or another. Such examples include "pioneer" aircraft built and flown before registration systems were devised, unfinished projects, deliberate omissions and Hang gliders and similar devices. Additionally, the register allows other airframes and smilar items which would not normally need an formal identity such as man-powered aircraft, full scale models for use as "gate guardians" or other display purposes and non flying reproductions intended only for display purposes. Most exhibits held in Museums are usually on display.

This year the section has received a detailed overhaul thanks to a considerable amount of new information contained in BAPC's own web-site and Ken Ellis' 19th edition of "Wrecks & Relics" published in May 2004 and, also, from personal observations.

Register No	Type	C/n	P/I	Remarks	Location
BAPC.1	Roe Triplane IV reconstruction			See SECTION 3 Part 1	Old Warden
BAPC.2	Bristol Boxkite reconstruction	BOX.1 & BM.7279		See SECTION 3 Part 1 *(As "12A")*	Old Warden
BAPC.3	Bleriot Type XI			See SECTION 3 Part 1	Old Warden
BAPC.4	Deperdussin Monoplane			See SECTION 3 Part 1	Old Warden
BAPC.5	Blackburn Monoplane			See SECTION 3 Part 1	Old Warden
BAPC.6	Roe Triplane Type I			See SECTION 3 Part 1 *(As "14")*	Manchester
BAPC.7	Southampton University Man Powered Aircraft (SUMPAC)			See SECTION 3 Part 1	Southampton
BAPC.8	Dixon Ornithopter reconstruction			See SECTION 3 Part 1	Old Warden
BAPC.9	Humber-Bleriot XI Monoplane reconstruction			See SECTION 3 Part 1	Coventry
BAPC.10	Hafner R-11 Revoplane			See SECTION 3 Part 1	Weston-super-Mare
BAPC.11	English Electric Wren composite			See SECTION 3 Part 1 *(As "4")*	Old Warden
BAPC.12	Mignet HM.14 Pou-Du-Ciel			See SECTION 3 Part 1	Manchester
BAPC.13	Mignet HM.14 Pou-Du-Ciel (Douglas 600cc)			Brimpex Metal Treatments Ltd *(Under restoration 3.98)*	Sheffield
BAPC.14	Addyman Standard Training Glider			Ponsford Collection *(Stored 2.04)*	Selby
BAPC.15	Addyman Standard Training Glider *(Rebuilt Yorkshire Aeroplanes)*	YA2		Ponsford Collection *(Stored 2.04)*	Wigan
BAPC.16	Addyman Ultralight			Ponsford Collection *(Stored incomplete 2.04)*	Selby
BAPC.17	Woodhams Sprite			BB Aviation *(Stored incomplete 2.04)*	Bossingham, Canterbury
BAPC.18	Killick Man Powered Gyroplane			Ponsford Collection *(Stored 2.04)*	Selby
BAPC.19	Bristol F2b Fighter fuselage frame			See SECTION 3 Part 1	Brussels, Belgium
BAPC.20	Lee-Richards Annular Bi-plane reconstruction			See SECTION 3 Part 1	Shoreham
BAPC.21	Thruxton Jackaroo *(Used as spares in rebuild of G-APAL)*			M.J.Brett *(Conversion abandoned)*	-----
BAPC.22	Mignet HM.14 Pou-Du-Ciel			See SECTION 3 Part 1	Schiphol, The Netherlands
BAPC.23	*Allocated in error – originally used by 1/2th scale SE.5 replica at Newark Air Museum*				
BAPC.24	*Allocated in error – originally used by 2/3rd scale Currie Wot replica at Newark Air Museum*				
BAPC.25	Nyborg TGN.III Sailplane			P Williams *(Stored 1.92)*	Warwick
BAPC.26	Auster AOP.9			*Fuselage frame only - scrapped Swansea*	
BAPC.27	Mignet HM.14 Pou-Du-Ciel reconstruction *(Under construction 1988 - presumably abandoned)*			M J Abbey	-----
BAPC.28	Wright Flyer fsm			See SECTION 3 Part 1	Elvington
BAPC.29	Mignet HM.14 Pou-Du-Ciel			See SECTION 3 Part 1 *(As "G-ADRY")*	Brooklands
BAPC.30	DFS Grunau Baby			*Destroyed by fire Swansea 1969*	
BAPC.31	Slingsby T.7 Tutor			*Believed scrapped Swansea*	
BAPC.32	Crossley Tom Thumb			See SECTION 3 Part 1	Coventry
BAPC.33	DFS 108-49 Grunau Baby IIB		BGA.2400 VN148, LN+ST	*(To Denmark for rebuild 2003)*	
BAPC.34	DFS 108-49 Grunau Baby IIB	030892	RAFGSA281 RAFGGA GK.4/LZ+AR	D.Elsdon *(Originally on rebuild as BGA.2362: possibly used for spares)*	Hazlemere, Buckingham
BAPC.35	EoN AP.7 Primary	EoN/P/063		ex Russavia Collection *(On rebuild as BGA.2493 @ 8.89)*	Pocklington
BAPC.36	Fieseler Fi 103 V1 fsm			See SECTION 3 Part 1	Hawkinge
BAPC.37	Blake Bluetit			See G-BXIY in SECTION 1, Part 2	
BAPC.38	Bristol Scout D fsm (Gnome 80 hp)			K Williams and M Thorn *(As "A1742") (On rebuild 9.03)*	Solihull
BAPC.39	Addyman Zephyr Sailplane			Ponsford Collection *(Parts held for eventual rebuild 2.04)*	Selby
BAPC.40	Bristol Boxkite reconstruction	BOX 3 & BM.7281		See SECTION 3 Part 1	Bristol
BAPC.41	Royal Aircraft Factory BE.2C fsm			See SECTION 3 Part 1	Elvington
BAPC.42	Avro 504K fsm			See SECTION 3 Part 1 *(As "H1968")*	Elvington
BAPC.43	Mignet HM.14 Pou-Du-Ciel			See SECTION 3 Part 1	Newark
BAPC.44	Miles M.14A Magister			See SECTION 3 Part 1 *(As "L6906")*	Woodley
BAPC.45	Pilcher Hawk Glider reconstruction			See SECTION 3 Part 1	Stanford Hall
BAPC.46	Mignet HM.14 Pou-Du-Ciel			*Probably scrapped*	
BAPC.47	Watkins CHW monoplane			See SECTION 3 Part 1	Swansea
BAPC.48	Pilcher Hawk Glider reconstruction			See SECTION 3 Part 1	Glasgow
BAPC.49	Pilcher Hawk Glider			See SECTION 3 Part 1	East Fortune
BAPC.50	Roe Triplane Type I			See SECTION 3 Part 1	South Kensington, London
BAPC.51	Vickers FB.27 Vimy IV			See SECTION 3 Part 1	South Kensington, London
BAPC.52	Lilienthal Glider Type XI			See SECTION 3 Part 1	Wroughton
BAPC.53	Wright Flyer reconstruction			See SECTION 3 Part 1	South Kensington, London
BAPC.54	JAP/Harding Monoplane			See SECTION 3 Part 1	South Kensington, London
BAPC.55	Levasseur-Antoinette Developed Type VII Monoplane			See SECTION 3 Part 1	South Kensington, London

Reg	Type	C/n	Previous identity	Details	Location
BAPC.56	Fokker E.III			See SECTION 3 Part 1 *(As "210/16")*	South Kensington, London
BAPC.57	Pilcher Hawk Glider reconstruction *(Built Martin and Miller, Edinburgh 1930)*			E Littledike *(Under restoration 11.01*	St Albans
BAPC.58	Yokosuka MXY-7 Ohka II			See SECTION 3 Part 1*(As "15-1585")*	RNAS Yeovilton
BAPC.59	Sopwith F1 Camel fsm			See SECTION 3 Part 1*(As "B5577/W")*	East Fortune
BAPC.60	Murray M.1 Helicopter			See SECTION 3 Part 1	Weston-super-Mare
BAPC.61	Stewart Man Powered Ornithopter			*"Bellbird II" (Noted 10.03)*	Salford
BAPC.62	Cody Type V Biplane			See SECTION 3 Part 1 *(As "304")*	South Kensington, London
BAPC.63	Hawker Hurricane fsm			See SECTION 3 Part 1 *(As "P3208/SD-T" in 501 Sqdn c/s)*	Hawkinge
BAPC.64	Hawker Hurricane fsm			See SECTION 3 Part 1 *(As "P3059/SD-N" in 501 Sqdn c/s)*	Hawkinge
BAPC.65	Supermarine Spitfire fsm			See SECTION 3 Part 1 *(As "N3289/DW-K" in 610 Sqdn c/s)*	Hawkinge
BAPC.66	Messerschmitt Bf109 fsm			See SECTION 3 Part 1	Hawkinge
BAPC.67	Messerschmitt Bf109 fsm			See SECTION 3 Part 1 *(As "14" in JG52 c/s)*	Hawkinge
BAPC.68	Hawker Hurricane fsm			See SECTION 3 Part 1	Hooton Park
BAPC.69	Supermarine Spitfire fsm			See SECTION 3 Part 1 *(As "N3313/KL-B" in 54 Sqdn c/s)*	Hawkinge
BAPC.70	Auster AOP.5			See SECTION 3 Part 1 *(As "TJ398")*	East Fortune
BAPC.71	Supermarine Spitfire fsm			See SECTION 3 Part 1 *(As "P8140/ZP-K" in 74 Sqdn c/s)*	Flixton
BAPC.72	Hawker Hurricane fsm			See SECTION 3 Part 1 *(As "V6799/SD-X" of 501 RAAF Sqdn c/s)*	(Gloucestershire)
BAPC.73	Hawker Hurricane fsm			*Displayed "Queens Head" Public House, Bishops Stortford* *(Current status unknown)*	
BAPC.74	Messerschmitt Bf109 fsm			See SECTION 3 Part 1	Hawkinge
BAPC.75	Mignet HM.14 Pou-Du-Ciel			See G-AEFG in SECTION 1 Part 2	Selby
BAPC.76	Mignet HM.14 Pou-Du-Ciel			See SECTION 3 Part 1 *(As "G-AFFI")*	Elvington
BAPC.77	Mignet HM.14 Pou-Du-Ciel			See SECTION 3 Part 1 *(As "G-ADRG")*	Stondon
BAPC.78	Hawker Afghan Hind			See G-AENP in SECTION 1 Part 2	
BAPC.79	Fiat G.46-4b	71	FHE MM53211	*(Stored as "MM53211/ZI-4" 9.00)*	La Ferte Alais, France
BAPC.80	Airspeed AS.58 Horsa II			See SECTION 3 Part 1 *(Fuselage only: as "KJ351")*	AAC Middle Wallop
BAPC.81	Hawkridge Nacelle Dagling	10471	BGA.493	Russavia Collection *(On rebuild)*	Hemel Hempstead
BAPC.82	Hawker Afghan Hind			See SECTION 3 Part 1	RAF Cosford
BAPC.83	Kawasaki Type 5 Model 1b (Ki 100)			See SECTION 3 Part 1	Hendon
BAPC.84	Mitsubishi Ki 46 III (Dinah)			See SECTION 3 Part 1	RAF Cosford
BAPC.85	Weir W-2			See SECTION 3 Part 1	East Fortune
BAPC.86	DH.82A Tiger Moth			*(Current status unknown)*	
BAPC.87	Bristol 30/46 Babe III reconstruction			See SECTION 3 Part 1 *(As "G-EASQ")*	Kemble
BAPC.88	Fokker Dr.1 5/8th scale model			See SECTION 3 Part 1 *(As "102/17")*	RNAS Yeovilton
BAPC.89	Cayley Glider fsm			See SECTION 3 Part 1	Elvington
BAPC.90	Colditz Cock Glider reconstruction			See SECTION 3 Part 1	East Kirkby
BAPC.91	Fieseler Fi 103R-IV (V-1)			See SECTION 3 Part 1	Lashenden
BAPC.92	Fieseler Fi 103 (V-1)			See SECTION 3 Part 1	Hendon
BAPC.93	Fieseler Fi 103 (V-1)			See SECTION 3 Part 1	Duxford
BAPC.94	Fieseler Fi 103 (V-1)			See SECTION 3 Part 1	RAF Cosford
BAPC.95	Gizmer Autogyro			F.Fewsdale *(Current status unknown)*	Darlington
BAPC.96	Brown Helicopter			See SECTION 3 Part 1	Sunderland
BAPC.97	Luton LA.4 Minor			See SECTION 3 Part 1 *(As "G-AFUG")*	Sunderland
BAPC.98	Yokosuka MXY-7 Ohka II			See SECTION 3 Part 1 *(As "997")*	Manchester
BAPC.99	Yokosuka MXY-7 Ohka II			See SECTION 3 Part 1	RAF Cosford
BAPC.100	Clarke TWK			See SECTION 3 Part 1	Hendon
BAPC.101	Mignet HM.14 Pou-Du-Ciel			See SECTION 3 Part 1	Newark
BAPC.102	Mignet HM.14 Pou-Du-Ciel			*Not constructed - parts to BAPC.75*	
BAPC.103	Pilcher Hawk reconstruction *(Built E.A.S.Hulton, London 1969)*			Personal Plane Services Ltd	Booker
BAPC.104	Bleriot Type XI		G-AVXV	Sold as F-AZIN 1992	
BAPC.105	Bleriot Type XI		54	See SECTION 3 Part 1	Paso Robles, Ca, USA
BAPC.106	Bleriot Type XI			See SECTION 3 Part 1	RAF Cosford
BAPC.107	Bleriot Type XXVII			See SECTION 3 Part 1	RAF Cosford
BAPC.108	Fairey Swordfish IV			See SECTION 3 Part 1 *(As "HS503")*	RAF Stafford
BAPC.109	Slingsby T.7 Cadet TX.1	28	8599M BGA.679	*Current status unknown*	
BAPC.110	Fokker D.VIIF fsm			See SECTION 3 Part 1	Chino, CA, USA
BAPC.111	Sopwith Triplane fsm			See SECTION 3 Part 1*(As "N5492")*	RNAS Yeovilton
BAPC.112	AirCo DH.2 fsm			See G-BFVH in SECTION 1 Part 2 *(As "5964")*	
BAPC.113	Royal Aircraft Factory SE.5a fsm			*(ex Leisure Sport) (As "B4863") (Current status unknown)*	
BAPC.114	Vickers Type 60 Viking IV reconstruction			See SECTION 3 Part 1 *(As "G-EBED")*	Brooklands
BAPC.115	Mignet HM.14 Pou-Du-Ciel			See SECTION 3 Part 1	Flixton
BAPC.116	Santos-Dumont Demoiselle XX reconstruction (JAP J99)			*(ex Flambards Theme Park)* *(Current status unknown)*	
BAPC.117	Royal Aircraft Factory BE.2c fsm *(Built Ackland and Shaw 1976)* (Gipsy Major)			P Smith *(Built for BBC TV "Wings") (Stored 3.02)*	(Hawkinge)
BAPC.118	Albatros D.Va fsm			*(As "C19/15") (Current status unknown)*	
BAPC.119	Bensen B.7 Gyroglider			See SECTION 3 Part 1	Sunderland
BAPC.120	Mignet HM.14 Pou-Du-Ciel			See SECTION 3 Part 1 *(As G-AEJZ)*	Hemswell
BAPC.121	Mignet HM.14 Pou-Du-Ciel			See SECTION 3 Part 1 *(As "G-AEKR")*	Doncaster

BAPC.122	Avro 504 fsm *(Built Personal Plane Services 1976)* (Ford 1300)			*(Built for BBC TV "Wings")* *(As "1881") (Current status unknown)*	
BAPC.123	Vickers FB.5 Gunbus fsm *(Built IES Projects Ltd 1975)*	1186/2	ZS-UHN	A.Topen *(As "P641") (Built for "Shout at the Devil" film)* *(Small components only remain and stored 3.90)*	Cranfield
BAPC.124	Lilienthal Glider Type XI reconstruction			See SECTION 3 Part 1	South Kensington, London
BAPC.125	Clay Cherub ground trainer				(Coventry)
BAPC.126	Rollason-Druine D.31 Turbulent			See SECTION 3 Part 1	Coventry
BAPC.127	Halton Man Powered Aircraft Group Jupiter			See SECTION 3 Part 1	Filching Manor, Wannock
BAPC.128	Watkinson CG-4 Cyclogyroplane Man Powered Gyroplane Mark IV			See SECTION 3 Part 1	Weston-super-Mare
BAPC.129	Blackburn (1911) Monoplane fsm			*(Built for TV Series."The Flambards")* *"Mercury" (Sold 1993: current status unknown)*	
BAPC.130	Blackburn (1911) Monoplane fsm			See SECTION 3 Part 1	Elvington
BAPC.131	Pilcher Hawk Glider reconstruction *(Built C.Paton 1972 for film)*			C.Paton *(Current status unknown - probably stored)*	London E
BAPC.132	Bleriot Type XI	PFA 88- 10864		See SECTION 3 Part 1	Mougins, Cannes, France
BAPC.133	Fokker Dr.1 fsm			See SECTION 3 Part 1	Hawkinge
BAPC.134	Aerotek Pitts S.2A		"G-RKSF"	Toyota Cars See "G-CARS" in SECTION 4	Northampton
BAPC.135	Bristol 20 M.1C Monoplane fsm			*(ex Leisure Sport)* *(As "C4912") (Sold 10.87: current status unknown)*	
BAPC.136	Deperdussin 1913 Monoplane fsm			See SECTION 3 Part 1	Chino, CA, USA
BAPC.137	Sopwith Baby Floatplane fsm *(Built FEM Displays Ltd 1978)*			*(ex Leisure Sport)*	Lands End
BAPC.138	Hansa Brandenberg W.29 fsm (Ford 1300)			*(ex Leisure Sport) (As "2292")* *(Sold prior to 10.87: current status unknown)*	
BAPC.139	Fokker Dr.1 Triplane fsm			*(ex Leisure Sport) (As "DR1/17")* *(Sold 10.87: current status unknown)*	
BAPC.140	Curtiss 42A R3C2 fsm			See SECTION 3 Part 1	Chino, CA, USA
BAPC.141	Macchi M.39 fsm			See SECTION 3 Part 1	Paso Robles, CA, USA
BAPC.142	Royal Aircraft Factory SE.5a fsm			*(As "F5459/Y")* *(Sold 1.5.93: current status unknown)*	Switzerland
BAPC.143	Paxton Man Powered Aircraft			R.A.Paxton *(Current status unknown: presumed stored)*	Gloucestershire
BAPC.144	Weybridge Man Powered Aircraft Group Mercury *(Previously "Dumbo" rebuilt)*			*"Mercury"* *(Current status unknown)*	Cranwell
BAPC.145	Oliver Man Powered Aircraft			*(Current status unknown - possibly scrapped)*	Warton
BAPC.146	Pedals Aeronauts Man Powered Aircraft Toucan			*"Toucan" (Current status unknown)* *(Centre section/power train only departed London Colney 1995)*	
BAPC.147	Bensen B.7 Gyroglider			See SECTION 3 Part 1 *As "LHS-1")*	Flixton
BAPC.148	Hawker Fury II fsm			See SECTION 3 Part 1 *(As "K7271" in 1 Sqdn c/s)*	Sleap
BAPC.149	Short S.27 fsm			See SECTION 3 Part 1	RNAS Yeovilton
BAPC.150	Sepecat Jaguar GR.1 fsm		"XX718" "XX732"	RAF Exhibition Production and Transportation Team *(As "XX725/GU" in 54 Sqdn c/s)*	Oman
BAPC.151	Sepecat Jaguar GR.1A fsm		"XX824"	RAF Exhibition Production and Transportation Team *(As"XZ363/A")*	RAF Cranwell
BAPC.152	BAe Hawk T.1A fsm		"XX262" "XX162"	RAF Exhibition Production and Transportation Team *(As "XX226/74" in 74 Sqdn c/s)*	RAF Cranwell
BAPC.153	Westland WG-33 Mock-up			See SECTION 3 Part 1	Weston-super-Mare
BAPC.154	Druine D.31 Turbulent		PFA 1654	Lincolnshire Aviation Society *(Unfinished: stored 3.96)*	East Kirkby
BAPC.155	Panavia Tornado GR.1 fsm		"ZA368" "ZA446/ZA600/ZA322"	RAF Exhibition Production and Transportation Team *(As "ZA556")*	RAF Cranwell
BAPC.156	Supermarine S.6B fsm			National Air Race Museum *(As "S1595")*	Sparks, Nevada, USA
BAPC.157	WACO CG-4A Hadrian Glider			See SECTION 3 Part 1	Elvington
BAPC.158	Fieseler Fi 103 (V1)			Defence Explosives Ordnance Disposal School	Chattenden
BAPC.159	Yokosuka MXY-7 Ohka II			Defence Explosives Ordnance Disposal School	Chattenden
BAPC.160	Chargus 18/50 Hang Glider			See SECTION 3 Part 1	East Fortune
BAPC.161	Stewart Man Powered Ornithopter *(Built A Stewart)*			*"Coppelia" (Stored 8.98: current status unknown)*	(Louth)
BAPC.162	Goodhart Man Powered Aircraft Newbury Manflier			See SECTION 3 Part 1	Wroughton
BAPC.163	AFEE 10/42 Rotachute Rotabuggy reconstruction			See SECTION 3 Part 1 *(As "B-415")*	AAC Middle Wallop
BAPC.164	Wight Quadraplane Type 1 fsm			See SECTION 3 Part 1 *(As "N248")*	Southampton
BAPC.165	Bristol F.2b Fighter			See SECTION 3 Part 1 *(As "E2466" in 22 Sqdn c/s)*	Hendon
BAPC.166	Bristol F.2b Fighter composite			See G-AANM in SECTION 1 Part 2	
BAPC.167	Royal Aircraft Factory SE.5a fsm *(Built TDL Replicas Ltd)*			*(Exported 12.97)*	(USA)
BAPC.168	DH.60G Moth reconstruction	8058		See SECTION 3 Part 1 *(As "G-AAAH")*	Croydon
BAPC.169	Sepecat Jaguar GR.1 fsm *(Engine systems static demonstration airframe)*			RAF/No.1 School of Technical Training *(As "XX110")*	RAF Cosford
BAPC.170	Pilcher Hawk Glider reconstruction *(Built A.Gourlay 1983)*			A Gourlay *(Built for "Kings Royal" BBC film) (Stored 3.93)*	Strathallan
BAPC.171	BAe Hawk T.1 fsm		"XX297" "XX262"	RAF Exhibition Production and Transportation Team *(As "XX253")*	RAF Cranwell
BAPC.172	Chargus Midas Super E Hang Glider			See SECTION 3 Part 1	Wroughton
BAPC.173	Birdman Promotions Grasshopper			See SECTION 3 Part 1	Wroughton
BAPC.174	Bensen B.7 Gyroglider			See SECTION 3 Part 1	Wroughton
BAPC.175	Volmer VJ-23 Swingwing Powered Hang Glider			See SECTION 3 Part 1	Manchester
BAPC.176	Royal Aircraft Factory SE.5a scale model			See SECTION 3 Part 1	Chorley

BAPC.177	Avro 504K fsm		See SECTION 3 Part 1 *(As "G-AACA" in Brooklands School of Flying c/s)*	Brooklands
BAPC.178	Avro 504K fsm	"E373"	By-gone Times Antique Warehouse *(German c/s)*	Eccleston, Lancashire
BAPC.179	Sopwith Pup fsm		See SECTION 3 Part 1 *(As "A7317")*	Coventry
BAPC.180	McCurdy Silver Dart reconstruction		Reynolds Pioneer Museum (?) *(Delivered 4.94)*	Wetaskiwin, Alberta, Canada
BAPC.181	Royal Aircraft Factory BE.2b reconstruction		See SECTION 3 Part 1 *(As "687")*	Hendon
BAPC.182	Wood Ornithopter		See SECTION 3 Part 1	Manchester
BAPC.183	Zurowski ZP-1 Helicopter		See SECTION 3 Part 1	Newark
BAPC.184	Supermarine Spitfire IX fsm *(Built Specialised Mouldings Ltd 1985)*		R.J.Lamplough/Fighter Wing Display Team *(As "EN398)*	North Weald
BAPC.185	WACO CG-4A Hadrian Glider		See SECTION 3 Part 1 *(As "243809")*	AAC Middle Wallop
BAPC.186	de Havilland DH.82B Queen Bee composite		See SECTION 3 Part 1 *(As "LF789/R2-K")*	London Colney
BAPC.187	Roe Type I Biplane reconstruction		See SECTION 3 Part 1	Brooklands
BAPC.188	McBroom Cobra 88 Hang Glider		See SECTION 3 Part 1	Wroughton
BAPC.189	Bleriot Type XI rep (Anzani) *(Some original parts ex Goldsmith Trust)*		See BAPC.132 *(Sold @ Christies 31.10.86, probably to France) (Current status unknown)*	
BAPC.190	Supermarine Spitfire fsm	"K5054"	P.Smith *(Noted 3.02)*	(Hawkinge)
BAPC.191	BAe Harrier GR.7 fsm	"ZD472"	RAF Exhibition Production and Transportation Team *(As "ZH139/01")*	RAF Cranwell
BAPC.192	Weedhopper JC24		M J Aubrey	Kington, Hertford
BAPC.193	Hovey WDII Whing Ding		M J Aubrey	Kington, Hertford
BAPC.194	Santos Dumont Type 20 Demoiselle		See SECTION 3 Part 1	RAF Stafford
BAPC.195	Birdman Sports Moonraker 77 Hang Glider		See SECTION 3 Part 1	East Fortune
BAPC.196	Southdown Sailwings Sigma 2m Hang Glider		See SECTION 3 Part 1	East Fortune
BAPC.197	Scotkites Cirrus III Hang Glider		See SECTION 3 Part 1	East Fortune
BAPC.198	Fieseler Fi 103 (V-1)		See SECTION 3 Part 1	South Lambeth, London
BAPC.199	Fieseler Fi 103 (V-1)		See SECTION 3 Part 1 *(As "442795")*	South Kensington, London
BAPC.200	Bensen B.7 Gyroglider *(Composite of three airframes)*		*(Stored 11.93)* *(Current status unknown)*	Leeds
BAPC.201	Mignet HM.14 Pou-Du-Ciel		See SECTION 3 Part 1	Caernarfon
BAPC.202	Supermarine Spitfire V fsm		See SECTION 3 Part 1 *(As "MAV467/"RO")*	Llanbedr
BAPC.203	Chrislea LC.1 Airguard rep		The Aeroplane Collection *(As "G-AFIN") (Current status unknown: burnt 1998?)*	Wigan
BAPC.204	McBroom Hang Glider		See SECTION 3 Part 1	Hooton Park
BAPC.205	Hawker Hurricane IIc fsm		See SECTION 3 Part 1 *(As "BE421/XP-G" in 174 Sqdn c/s)*	Hendon
BAPC.206	Supermarine Spitfire IX fsm		See SECTION 3 Part 1 *(As "MH486/FF-A" in 132 Sqdn c/s)*	Hendon
BAPC.207	Austin Whippet fsm		See SECTION 3 Part 1 *(As "K.158")*	Doncaster
BAPC.208	Royal Aircraft Factory SE.5a		See SECTION 3 Part 1	Farnborough
BAPC.209	Supermarine Spitfire LF.IXC fsm		See SECTION 3 Part 1 *(As "MH415/FU-N") (Built for "Piece of Cake" TV series)*	Duxford
BAPC.210	Avro 504J *(Built AJD Engineering)*		See SECTION 3 Part 1 *(As "C4451")*	Southampton
BAPC.211	Mignet HM.14 Pou-Du-Ciel		See SECTION 3 Part 1 *(As "G-ADVU")*	Sunderland
BAPC.212	Bensen B.6 Gyrocopter		See SECTION 3 Part 1	Weston-super-Mare
BAPC.213	Cranfield Vertigo Man Powered Helicopter		See SECTION 3 Part 1	Weston-super-Mare
BAPC.214	Supermarine Spitfire Prototype fsm		See SECTION 3 Part 1 *(As "K5054")*	Tangmere
BAPC.215	Airwave Hang Glider		See SECTION 3 Part 1	Southampton
BAPC.216	de Havilland DH.88 Comet fsm		See SECTION 3 Part 1	London Colney
BAPC.217	Supermarine Spitfire I fsm		See SECTION 3 Part 1 *(As "K9926/JH-C" in 317 Sqdn c/s)*	RAF Bentley Priory
BAPC.218	Hawker Hurricane IIc fsm		See SECTION 3 Part 1 *(As "BN230/FT-A"in 43 Sqdn c/s)*	RAF Bentley Priory
BAPC.219	Hawker Hurricane I fsm		See SECTION 3 Part 1 *(As "L1710/AL-D" in 79 Sqdn c/s)*	Biggin Hill
BAPC.220	Supermarine Spitfire I fsm		See SECTION 3 Part 1 *(As "N3194/GR-Z" in 92 Sqdn c/s)*	Biggin Hill
BAPC.221	Supermarine Spitfire LF.IX fsm		See SECTION 3 Part 1 *(As "EN526/SZ-G" in 316 Sqdn c/s)*	RAF Northolt
BAPC.222	Supermarine Spitfire IX fsm		See SECTION 3 Part 1 *(As "BR600/SH-V" in 64 Sqdn c/s)*	RAF Uxbridge
BAPC.223	Hawker Hurricane I fsm		See SECTION 3 Part 1 *(As "V7467/LE-D"in 242 Sqdn c/s) (See BAPC.288)*	RAF Coltishall
BAPC.224	Supermarine Spitfire V fsm *(Built TDL Replicas)*		AJD Engineering/Hawker Restorations Ltd *(As "BR600")*	Sudbury
BAPC.225	Supermarine Spitfire IX fsm		See SECTION 3 Part 1 *(As "P8448/"UM-D" in 52 Sqdn c/s)*	RAF Cranwell
BAPC.226	Supermarine Spitfire XI fsm		See SECTION 3 Part 1 *(As "EN343" in PRU c/s)*	RAF Benson
BAPC.227	Supermarine Spitfire IA fsm		See SECTION 3 Part 1 *(As "L1070/XT-A" in 603 Sqdn c/s)*	Edinburgh
BAPC.228	Olympus Hang Glider		See SECTION 3 Part 1	Sunderland
BAPC.229	Supermarine Spitfire IX fsm		See SECTION 3 Part 1 *(As "MJ832/DN-Y" in 416 Sqdn c/s)*	RAF Digby
BAPC.230	Supermarine Spitfire fsm		See SECTION 3 Part 1 *(As "AB550/GE-P" in 349 Sqdn c/s)*	Malton

BAPC.231	Mignet HM.14 Pou-Du-Ciel		See SECTION 3 Part 1 *(As "G-ADRX")*	RAF Millom
BAPC.232	Airspeed AS.58 Horsa I/II Glider composite		See SECTION 3 Part 1	London Colney
BAPC.233	Broburn Wanderlust sailplane		See SECTION 3 Part 1	Woodley
BAPC.234	Vickers FB.5 Gunbus fsm		See SECTION 3 Part 1 *(As "GBH-7")*	Sleap
BAPC.235	Fieseler Fi 103 (V-1 fsm		See SECTION 3 Part 1	Malton
BAPC.236	Hawker Hurricane fsm		See SECTION 3 Part 1	Malton
			(As "P2793/SD-M" in 501 Sqdn c/s)	
BAPC.237	Fieseler Fi 103 (V-1)		See SECTION 3 Part 1	RAF Stafford
BAPC.238	Waxflatter Ornithopter replica		Personal Plane Services Ltd	Compton Abbas
	(Built Personal Plane Services)		*(Built for "Young Sherlock Holmes")*	
BAPC.239	Fokker D.VIII 5/8th scale model		See SECTION 3 Part 1 *(As "694")*	Flixton
BAPC.240	Messerschmitt Bf.109G fsm		See SECTION 3 Part 1	Elvington
BAPC.241	Hawker Hurricane I fsm		See SECTION 3 Part 1	Tangmere
			(As "L1679/JX-G" in 1 Sqdn c/s)	
BAPC.242	Supermarine Spitfire VB fsm		See SECTION 3 Part 1	Tangmere
			(As "BL924/AZ-G" in 234 Sqdn c/s)	
BAPC.243	Mignet HM.14 Pou-Du-Ciel	"A-FLEA"	P.Ward	Malvern Wells
	(Built Bill Francis) (Scott A2S)		*(As "G-ADYV") (Stored 8.03)*	
BAPC.244	Mainair Tri-Flyer 250/Solar Wings Typhoon S		See SECTION 3 Part 1 *(As "G-MMLI)"*	East Fortune
BAPC.245	Electraflyer Floater Hang Glider		See SECTION 3 Part 1	East Fortune
BAPC.246	Hiway Cloudbase Hang Glider		See SECTION 3 Part 1	East Fortune
BAPC.247	Albatros ASG.21 Hang Glider		See SECTION 3 Part 1	East Fortune
BAPC.248	McBroom Hang Glider		See SECTION 3 Part 1	Woodley
BAPC.249	Hawker Fury I fsm		See SECTION 3 Part 1	Brooklands
			(As "K5673" in 1 Sqdn 'A' Flight c/s)	
BAPC.250	Royal Aircraft Factory SE.5a replica		See SECTION 3 Part 1*(As "F5475/A")*	Brooklands
BAPC.251	Hiway Spectrum Hang Glider		See SECTION 3 Part 1	Manchester
BAPC.252	Flexiform Wing Hang Glider		See SECTION 3 Part 1	Manchester
BAPC.253	Mignet HM.14 Pou-Du-Ciel rep		See SECTION 3 Part 1 *(As "G-ADZW")*	Southampton
BAPC.254	Supermarine Spitfire 1 fsm		See SECTION 3 Part 1	Elvington
			(As "R6690/PR-A" in 609 Sqdn c/s)	
BAPC.255	North American P-51D Mustang fsm		See SECTION 3 Part 1	Duxford
BAPC.256	Santos Dumont Type 20 Demoiselle reconstruction		See SECTION 3 Part 1	Brooklands
BAPC.257	de Havilland DH.88 Comet model		See SECTION 3 Part 1	Hatfield
BAPC.258	Adams Balloon (15,000 cu.ft)		See SECTION 3 Part 1	Newbury
BAPC.259	Gloster Gamecock		See SECTION 3 Part 1	(Gloucestershire)
BAPC.260	Mignet HM.280		See SECTION 3 Part 1	RAF Millom
BAPC.261	General Aircraft Hotspur replica		See SECTION 3 Part 1	AAC Middle Wallop
			(As "HH268")	
BAPC.262	Catto CP-16		See SECTION 3 Part 1	East Fortune
BAPC.263	Chargus Cyclone		See SECTION 3 Part 1	Langford Lodge, Belfast
BAPC.264	Bensen B8M		See SECTION 3 Part 1	Weston-super-Mare
BAPC.265	Hawker Hurricane fsm		See SECTION 3 Part 1	Elvington
			(As "P3873/YO-H" in 1 Sqdn/RCAF c/s)	
BAPC.266	Rogallo Hang Glider		See SECTION 3 Part 1	Langford Lodge, Belfast
BAPC.267	Hawker Hurricane fsm		See SECTION 3 Part 1 *(As "R4115/LE-X")*	Duxford
BAPC.268	Supermarine Spitfire fsm		B Wallond	St Mawgan
			(Built for "Dark Blue World") (As "N3317/AI-A")	
BAPC.269	Supermarine Spitfire V fsm		See SECTION 3 Part 1	USAF Lakenheath
			(As "BM631/XR-C" in 71 Sqdn c/s)	
BAPC.270	de Havilland DH.60 Moth fsm		See SECTION 3 Part 1 *(As "G-AAAH")*	Elvington
BAPC.271	Messerschmitt Me 163B Komet fsm		See SECTION 3 Part 1 *(As "191454")*	Old Warden
BAPC.272	Hawker Hurricane fsm		See SECTION 3 Part 1	Hawkinge
BAPC.273	Hawker Hurricane fsm		See SECTION 3 Part 1	Hawkinge
BAPC.274	Boulton & Paul P.6 fsm		See SECTION 3 Part 1 *(As "X-25")*	Wolverhampton
BAPC.275	Bensen B.7		See SECTION 3 Part 1	Doncaster
BAPC.276	Hartman Ornithopter		See SECTION 3 Part 1	Wroughton
BAPC.277	Mignet HM.14 Pou-Du-Ciel		See SECTION 3 Part 1	Shoreham
BAPC.278	Hawker Hurricane fsm		See SECTION 3 Part 1	Hawkinge
BAPC.279	Airspeed AS.58 Horsa reconstruction		See SECTION 3 Part 1	RAF Shawbury
BAPC.280	de Havilland DH.89 Dragon Rapide fsm		See SECTION 3 Part 1	Liverpool
			(As "G-AEAJ" in Railway Air Services c/s)	
BAPC.281	Boulton Paul Defiant fsm		See SECTION 3 Part 1	Wolverhampton
			(As "L7005/PS-B" in 264 Sqdn c/s)	
BAPC.282	Manx Eider Duck		Not known *(Noted 7.01)*	Ronaldsway
BAPC.283	Supermarine Spitfire fsm		A Saunders	Jurby
	(Built F Brown)		*(Built for "Piece of Cake" TV series) (Noted 2.04)*	
BAPC.284	Gloster E28/39 fsm		See SECTION 3 Part 1	Lutterworth
BAPC.285	Gloster E28/39 fsm		See SECTION 3 Part 1	Farnborough
BAPC.286	Mignet HM.14 Flea		See SECTION 3 Part 1	Cardiff
	(Scott A2S)			
BAPC.287	Blackburn F2 Lincock fsm		See SECTION 3 Part 1 *(As "G-EBVO")*	Hull
BAPC.288	Hawker Hurricane fsm		See SECTION 3 Part 1	Farnsfield
			(As "V7467/LE-D" in 242 Sqdn c/s) (see BAPC.223)	
BAPC.289	Gyro-Boat		See SECTION 3 Part 1	Weston-super-Mare
BAPC.290	Fieseler Fi.104 flying-bomb fsm		See SECTION 3 Part 1	Dover
BAPC.291	Hawker Hurricane I fsm		See SECTION 3 Part 1	Capel Le Ferne, Folkestone
			(As "P2970/US-X" in 56 Sqdn c/s)	
BAPC.292	Eurofighter Typhoon fsm		See SECTION 3 Part 1	Hendon
BAPC.293	Supermarine Spitfire fsm		See SECTION 3 Part 1	Hendon
BAPC.294	Fairchild Argus		See SECTION 3 Part 1	Woodhall Spa
BAPC.295	Leonardo Da Vinci hang-glider reproduction		Skysport Engineering *(Stored 3.04)*	Hatch
	(Built Skysport Engineering)		*(Built for Channel 4 TV Documentary "Leonardo's Dream Machines")*	
BAPC.296	Army Balloon Factory Nulli reproduction			Hendon

BAPC.297	Supermarine Spitfire reproduction	See SECTION 3 Part 1	Hawkinge
BAPC.298	Supermarine Spitfire IX fsm	See SECTION 3 Part 1	Hawkinge
BAPC.299	Supermarine Spitfire I fsm	See SECTION 3 Part 1 *(As "P9338" in 72 Sqdn c/s)*	Capel le Ferne .Folkestone
BAPC.300	Piffard Hummingbird reproduction	See SECTION 3 Part 1	Shoreham

PART 3 – IRISH AVIATION HISTORICAL COUNCIL REGISTER

The IAHC Register came into existence with similar objectives to the BAPC.

Register No.	Type	C/n	P/I	Remarks	Location
IAHC.1	Mignet HM.14 Pou-Du-Ciel			See SECTION 3 Part 1	
IAHC.2	Aldritt Monoplane rep			See SECTION 3 Part 1	
IAHC.3	Mignet HM.14 Pou-Du-Ciel *(Built 1937 but unflown)*			M.Donohoe *(Last noted 4.96)*	Delgany
IAHC.4	Hawker Hector		IAAC....	D.McCarthy *(Believed components on rebuild Florida, USA)*	Not known
IAHC.5	Not known			Not known	
IAHC.6.	Ferguson Monoplane rep			See SECTION 3 Part 1	
IAHC.7	Sligo Concept			G.O'Hara *(Was stored incomplete 8.91 - current status unknown)*	Sligo
IAHC.8	O'Hara Autogyro			G.O'Hara *(Was stored incomplete 8.91 - current status unknown)*	Sligo
IAHC.9	Ferguson Monoplane rep			See SECTION 3 Part 1	

SECTION 4

PART 1 – OVERSEAS CIVIL REGISTERED AIRCRAFT LOCATED IN UK AND IRELAND

From this edition last noted sighting dates have been moved to a discrete column to provide a degree of consistency with other Sections in this book. This move has highlighted a number of entries which are rather long in the tooth and we would welcome clarification of their current status. All overseas civil registered aircraft in museums and private collections in the UK are now to be found in Part 2 with the transfer of those with a UK provenance.

Leased and stored airliners continue to provide difficulties in establishing how long they will remain based in the UK and Ireland and, as last year, they have been ignored unless there is a likelihood they may still be present into the Summer of 2005.

This Section could not have been produced without the assistance of the following to whom due credit is acknowledged: Ray Barber, Peter Budden, Nigel Burch (Biggin Hill), Mike Cain (Southend), Richard Cawsey, Terry Dann, Bryan Foster, Dave Haines (Gloucestershire), Peter Hughes, Paul Jackson, Phil and Nigel Kemp (North Weald), Andy Mac (Stapleford), Alistair Ness (Scotland), Nigel Ponsford (Breighton), Stephen Reglar, Trevor Spedding, Mike Stroud, Bill Teasdale, Ken Tilley (White Waltham) and Barrie Towey, plus the numerous contributors to the AB-Sightings, Airfields and Civil Spotters mailing lists.

Registration	Type	Construction No	Previous Identity	Owner*(Operator)*	(Unconfirmed) Base	Last noted
UNITED ARAB EMIRATES						
A6-ESH	Airbus A319-133X	910		Ruler of Sharjah	Farnborough/Sharjah	9.04
A6-HHH	Gulfstream Gulfstream IV	1011	(A6-DLF) N17581	Government of Dubai	Farnborough/Dubai	12.04
A6-HRS	Boeing 737-7EO	29251		Dubai Royal Flight	Farnborough/Dubai	12.04
A6-MRM	Boeing 737-8EC	32450		Dubai Air Wing	Farnborough/Dubai	10.04
A6-SHK*	British Aerospace BAe 146 Srs.100	E1091	G-BOMA G-5-091	Air Salvage International *(Fuselage only)*	Alton	4.02
MUSCAT and OMAN						
A4O-CT*	Britten Norman BN-2T Islander	2201	G-51-2201 G-BOMC	Britten-Norman Group *(Original fuselage derelict)*	Bembridge	1.02
CANADA						
CF-EPV*	Aviation Traders ATL.98 Carvair	10448/8	EI-AMR N88819, 42-72343	Old Airfield Estate *(Cockpit section only)*	Halesworth	6.03
C-FQIP	Lake LA-4-200 Buccaneer	679	N1068L	P J Molloy	Elstree	1.05
C-GOLZ	Van's RV-6A	1951		John and Sandra Brennan	Waterford	7.04
C-GWJO*	Boeing 737-2A3	20299	HR-SHO HR-TNR, CX-BHM, N1797B, N1787B	Newcastle Aviation Academy *(Instructional use)*	Newcastle	9.04
PORTUGAL						
CS-AZY	de Havilland DHC.1 Chipmunk T.20 *(Built OGMA)*	40	FAP1350	Richard Farrer *(Current status unknown)*	Bourn	6.01
CS-HBK*	Hughes 369E	0165E	N5233N	March Helicopters *(Wreck)*	Sywell	9.02
CS-HBL*	Hughes 369E	0377E		March Helicopters *(Wreck)*	Sywell	9.02
CS-HCE*	Hughes 369D	120-0856D	G-JIMI N1109T	March Helicopters *(Dismantled: current status unknown)*	Sywell	12.01
CS-TMY	Short SD.3-60	SH.3632	G-BLCP OY-MMA, (SE-KSU), OY-MMA, EI-BYU, OY-MMA, G-BLCP	Aerocondor *(In open store engineless)*	Southend	2.05
GERMANY						
D-CALM	Dornier Do 228-101	7051		Deutschen Zentrum fur Luft und Raumfahrt	Oxford	10.04
D-EADD*	Reims/Cessna FR172K Hawk XP	FR17200607		Small World Aviation	North Weald	1.05
D-EAGC	Cessna F172H	F172-0637	(D-EBEB)	Not known	Rochester	1.05
D-EALX	Cessna F150L	F15000766	OE-ALX	S A Young	Hill Farm, Nayland	5.02
D-EAWD*	Reims/Cessna F150M	F15001259		Not known *(Noted w/o engine, damaged fin, wings & rear fuselage)*	Andrewsfield	1.05
D-EAWT	Robin HR200/100 Club	18		Not known	Oxford	1.04
D-EAWW	Piper PA-28R-201 Arrow III	28R-7837199	N9469C	Not known	Blackbushe	12.04
D-EAXX*	SEEMS MS.885 Super Rallye	260	F-BKUI	P Garcia *(Current status unknown)*	King's Farm, Thurrock	2.00
D-EBLI	Bölkow Bö.207	223		Not known	Thruxton	5.04
D-EBLO*	Bölkow Bö.207	224		Not known	Gamston	10.04
D-EBWE*	Piper PA-28-235 Cherokee	28-10431	N8874W	Not known	Manston	2.05
D-EBXR	Reims/Cessna FR172K Hawk XP	FR17200597	HB-CXO	Not known	Cork	10.03
D-ECDL	Navion Rangemaster H	NAV-4-2522	N2520T	G Spooner *(Dismantled)* *(Acquired as spares source for long-term restoration of Navion N3864 qv)*	(Kelvedon)	12.02
D-ECDU*	Reims/Cessna F172E	F172-0068		M Dunn *(Carries fictitious marks "G-ASOK": current status unknown)*	Longside, Peterhead	6.00
D-ECGI	Bölkow Bö.208C Junior	598		Not known	Bidford	1.05
D-EDEL	Piper PA-32-300 Cherokee Six D	32-7140009	N8617N	Not known	Fairoaks	1.05
D-EDEQ	Beech B24R Sierra 200	MC-239		C Jones	Shoreham	9.04
D-EDYQ	Piper PA-32-260 Cherokee Six	32-415	N3529W N11C	Not known	Tatenhill	8.04
D-EEAH	Bölkow Bö.208C Junior	658	(D-EJMH)	J Webb	Bourne Park, Hurstbourne Tarrant	8.04
D-EEHW	Cessna P210N Pressurized Centurion II	P21000455	N731FX	Not known	Exeter	11.04
D-EEPI	Wassmer WA.54 Atlantic	151		R Hunter	Redhill	11.03
D-EEVY	Cessna 170A	19537	D-ELYC N5503C	Not known	Andrewsfield	1.05
D-EFFA	Ruschmeyer RG90-230RG	018	D-ELVY (2) (D-EEBY (2))	K Cropp	Southend	1.05
D-EFJD	Bölkow Bö.209 Monsun 160RV	126		W Williams-Wynne	Old Sarum	1.03

Reg	Type	C/n	Previous identity	Owner/operator	Location	Date
D-EFJG	Bölkow Bö.209 Monsun 160RV	129		R.Truesdale	Newtownards	11.04
D-EFNO*	Bölkow Bö.208A-1 Junior	604		Aero Engines & Airframes	Yearby	3.02
				(For composite rebuild with G-ASFR)		
D-EFQE	Bölkow Bö.207	266		J Webb	Popham	10.04
D-EFQR	Robin DR.400/180 Régent	1369		M Costin	Husbands Bosworth	8.04
D-EFTI	Bölkow Bö.207	219		Mark Hayles	Turweston	1.05
D-EFVS	Wassmer WA.52 Europa	22		M Hales	Little Staughton	8.04
D-EFZC	SIAI-Marchetti S.208	2-18		Not known	RAF Benson	11.04
D-EFZO	Reims Cessna F172F	F172-0156		Not known	Stapleford	1.05
D-EGEU*	Piper PA-22-108 Colt	22-9055	EL-AEU	Not known	Farley Farm, Romsey	1.05
			5N-AEH	*(Badly damaged by storms 27.10.02))*		
D-EGLW	Piper PA-38-112 Tomahawk	38-80A0105	N9694N	(Fire/Rescue trainer)	Old Warden	5.04
D-EHJL	Piaggio FWP.149	045	90+31	C A Tyers/Windmill Aviation	Spanhoe	1.05
			AC+441, AS+441, GA+394, D-EGEW, GA+394			
D-EHKY	Bölkow Bö 207	272		Not known	Haverfordwest	4.04
D-EHLA	Bölkow Bö.207	273		J Webb	Lodge Farm, St Osyth	9.04
D-EHUQ	Bölkow Bö.207	207		J Bally	Nuthampstead	7.04
D-EHYX	Bölkow Bö.207	209		Not known	Haverfordwest	7.04
D-EIAL	Piper PA-28-161 Warrior II	28-8116076	N8291D	Not known	Alderney	1.05
D-EIAR	CEA DR.250/160 Capitaine	98		D G Holmann	Leicester	11.04
D-EIIA	Robin R3000/160	164		T Harrison	Shennington	10.04
D-EJBI	Bölkow Bö.207	242	G-EJBI	John O'Donnell	Biggin Hill	9.04
			D-EJBI			
D-EKDN	Beech A36 Bonanza	E-2353	N7241Y	Not known	Elstree	1.05
D-EKJD	Reims FR172J Rocket	FR17200582		Not known	Bourn	1.05
D-EMFU	Bölkow Bö.208C Junior	574		A Court	Franklyn's Field, Chewton Mendip	8.04
D-EMLS	Cessna T210L Turbo Centurion	21060094	(G-BCJJ)	Not known	Top Farm, Croydon, Royston	12.04
			N59107			
D-EMUH	Bölkow Bö.208C Junior	623		N Beavins	Rayne Hall Farm, Rayne	1.05
D-EMZC	Reims FR.172G Rocket	FR17200154		Not known	Cork	11.03
D-ENCA	Beech V35B Bonanza	D-10134		Not known	Elstree	1.05
D-ETTO	Extra EA.300/L	1174		Not known	Panshanger	1.05
D-EXGC	Extra EA.200	027		Not known	Andrewsfield	1.05
D-FBPS	Cessna 208B Grand Caravan	208B0494	LV-YJC	Not known	Langar	6.04
			N208BA, N1219G	*(Op British Parachute School)*		
D-FLOH	Cessna 208B Grand Caravan	208B0576	N1041F	Not known *(Op British Parachute School)*	Langar	6.04
D-GPEZ	Piper PA-30 Twin Comanche C	30-1871	N8798Y	Not known	Farley Farm, Romsey	1.05
			N9703N			
D-HCKV	Agusta A109A-II	7345	N109HC	Police Aviation Services	Gloucestershire	1.05
			N2GN	*(Damaged near Newby Bridge, Cumbria 2.1.00: hulk marked "G-OPAS")*		
D-IBPN	Beech 58P Baron	TJ-424	N6526S	Small World Aviation	North Weald	1.05
D-KIFF	SFS-31 Milan	6604		N Grayson	Boscombe Down	7.04
D-KMDP	Fournier RF-3	37	F-BMDP	Not known	Carnmore,Galway	8.04
D-MVMM	WDL Fascination			Not known	Ledbury	2.03
D-9004	Schleicher ASW 28-18	28502		R Thirkell *"B3"*	Lasham	.04

SPAIN

Reg	Type	C/n	Previous identity	Owner/operator	Location	Date
EC-AMD*	Boeing Stearman A75N1	75-4721	N55050	Not known	Kemble	9.04
	(P/I/provenance uncertain -previously quoted as N126SE)		42-16558			
EC-AOY*	Aero-Difusion Jodel D.1190-S Compostela	E.56		G.Janney	Sibsey	2001
				(Water damaged remains - thought unuseable: current status unknown)		
EC-CFA*	Boeing 727-256	20811	N907RF	Not known	Shannon	12.04
			EC-CFA	*(Stored)*		
EC-CFI*	Boeing 727-256	20819		Iberia *(Stored)*	Bournemouth	1.05
EC-DDX*	Boeing 727-256A	21779		Iberia *(Stored)*	Bournemouth	1.05
EC-FQI*	Aérospatiale SA.316B Alouette III	1926	I-BYCS	Not known	Breighton	1.05
			F-BYCS, RAN-12	*(Stored unmarked for spares)*		
EC-GQZ	McDonnell Douglas DC-9-81	49571	HB-INY	Not known	Southend	2.05I
				(In open store with no markings)		

LIBERIA (old series)

Reg	Type	C/n	Previous identity	Owner/operator	Location	Date
EL-AKJ*	Boeing 707-321C	19375	(N2NF)	Omega Air *(In open store)*	Southend	2.05
			EL-AKJ, (PP-BRR), EL-AKJ, 9Q-CSW, 5N-TAS, N864BX, OB-R-1243, HK-2473, HK-2473X, N473RN, N473PA			
EL-AKL*	Boeing 707-351C	18922	EL-AKF	Omega Air *(Open store: all white c/s)*	Shannon	7.03
			HR-AME, 5N-JIL, 5N-ASY, N82TF, VR-HGP, (VR-HGQ), N362US			

ESTONIA

Reg	Type	C/n	Previous identity	Owner/operator	Location	Date
ES-YLK*	Aero L-29A Delfin	194521	Est.AF	R Patton	Cork, Countty.Cork	4.03
			Sov AF			

FRANCE

Reg	Type	C/n	Previous identity	Owner/operator	Location	Date
F-BBGH*	Brochet MB.100	01		Not known *(Frame stored: current status unknown)*	Sibsey	2001
F-BBSO*	Taylorcraft Auster 5	1792	G-AMJM	D.J.Baker	Carr Farm, Thorney, Newark	1.05
			TW452	*(Dismantled frame)*		
F-BCJJ	Piper PA-32-260 Cherokee Six	32-226	F-OCJJ	Eurodeal SARL	Popham	10.04
			N3800W, (ZK-CNS)	*(Op P.Marie)*		
F-BGNR*	Vickers 708 Viscount	35	(OY-AFO)	Skysport Engineering	Rotary Farm, Hatch	8.04
			(OY-AFN), F-BGNR	*(Stored)*		
F-BMCY*	Potez 840	02	N840HP	Highlands & Islands Airports Ltd	Sumburgh	2002
			F-BJSU, F-WJSU	*(Damaged Sumburgh 29.3.81: for Fire Service use)*		
F-BMHM	Piper J-3C-65 Cub	11907	Fr Military	Not known	Turweston	1.05
			44-79611			

F-BOJP	Mooney M.20F Executive	670216	N9639M	Not known	Bournemouth	12.04
F-BRHN*	Bölkow Bö.208C Junior	688	D-EEAK	Not known *(Stored)*	Farley Farm, Romsey	2.04
F-BRII	Cessna U206D Skywagon	U206-1318	N72204	Not known	Sorbie Farm, Kingsmuir	1.03
F-BTKO	Robin HR.100/210 Safari	142		Fly Ltd *(Stored dismantled)*	Turweston	1.05
F-BTLO*	Wassmer WA.52 Europa	66		E.Wiseman	Breighton	1.05
F-CARF*	Fournier RF-9	02	F-WARF	Not known	Nympsfield	8.02
			(DBR in aborted take off Nympsfield 11.8.02: cancelled 14.1.03 as destroyed: current status unknown)			
F-GAOM	Robin DR.400/2+2 Tricycle	1220		Not known	Exeter	11.04
F-GBVX	Robin DR.400/120A Petit Prince	1419		Not known	Duxford	12.04
F-GEHA*	Aérospatiale SA.341G Gazelle	1064	N7721Y	MW Helicopters Ltd	(Stapleford)	8.01
			N6952, F-WMHG	*(Spares use: current status unknown)*		
F-GFDG*	Aérospatiale SA.342 Gazelle	1204	TG-KOV	P Holder *(Op MW Helicopters)*	Blackpool	10.04
F-GFGH	SOCATA Rallye 235E Gabier	13337	F-ODNQ	Ian Watts	Bagby	10.04
F-GFLD*	Beech C90 King Air	LJ-741	HB-GGW	RFS Aircraft Engineering	Southend	12.04
			I-AZIO	*(Stored in false marks "N33FL")*		
F-GFNO	Robin ATL	16	F-WFNO	A Orchard	RAF Mona	10.04
F-GFOR	Robin ATL	42		M Godsell	Haverfordwest	9.04
F-GFRO	Robin ATL	64		B Sharpen	North Weald	1.05
F-GGJK	Robin DR.400/140B Major 80	1805		Not known	Headcorn	1.05
F-GGTJ	Aérospatiale SA.342J Gazelle	1473	C-GVWC	M W Helicopters Ltd	Bristol	9.04
	(Converted from SA.341G to SA.342J @ 3.92)		F-WXFX			
F-WGTX	Heli Atlas	01		Intora Firebird plc *(Stored unconfirmed)*	(Brentwood)	2003
F-WGTY	Heli Atlas	02		Intora Firebird plc *(Stored unconfirmed)*	(Brentwood)	2003
F-GIBU	Aérospatiale SA.342J Gazelle	1470	HB-XMU	Global Aviation Services Ltd	Hawarden	4.04
			N9000A			
F-GJJI	Fouga CM.170 Magister	500	AdlA	M. Murphy	Cranfield	1.05
F-GJQI	Robin ATL L	133		C Fox	Wing Farm, Longbridge Deverill	6.04
F-GJSL	Aérospatiale SA.342J Gazelle	1052	C-GPGO	MW Helicopters Ltd	Stapleford	7.04
	(Converted from SA.341G)		N8350			
F-GKKI	Avions Mudry CAP.231EX	02	(G-BVXL)	D Kaberry	Barton	9.04
			F-GKKF, F-WGZC			
F-GKMZ	Mudry CAP.232	09		Not known	Headcorn	9.04
F-GLAO	SOCATA TB-9 Tampico Club	1106		Not known	Bournemouth	1.05
F-GMHH	Robin HR.100/210 Safari	155	3A-MUZ	Fly Ltd	Turweston	1.05
			F-BUHH	*(Stored dismantled)*		
F-GODZ	Pilatus PC-6/340 Porter	340	(N340N)	Not known	Cockerham	7.04
			ST-AFR, HB-FAR	*(Op Black Knights Parachute Club)*		
F-GOTC	Mudry CAP.232	15		T Cassells	Bagby	10.04
F-GOXD	Robin DR400/180RP Remorquer	1817	OO-CXD	G Richardson	Bridge Farm, Carr	12.04
			HB-KBU, D-EAJD			
F-GUJM	Mudry CAP.232	21		Not known	Sherburn-in-Elmet	10.04
F-WWMX	Aerotech Europe CAP.222	CO3		A Cassidy	White Waltham	9.04
F-GXDB	Mudry CAP.232	33		Diana Britten *"Diana"*	Fairoaks	10.04
F-GYRO	Mudry CAP.232	25		A Cassidy *(Securicor titles)*	White Waltham	9.04
F-PAGD	Auster V J/1 Autocrat	2218	G-AJID	J Guerin *(Op Al Mathie)*	(Burgate, Diss)	8.02
F-PFUG*	Adam RA-14	11		Not known *(Stored: current status unknown)*	Sibsey	2001
F-PYOY	Heintz Zenith 100	52		B L Featherstone *(Stored)*	Southend	1.05
31 WI	Canada Benjamin Co Twinstarr Gyrocopter	NK		Woody de Saar	Shipdham	8.04
50 BH	Fisher FP-202 Super Koala	Not known		K.Riches	(Guernsey)	12.01
				t/a MUL International *(Stored: current status unknown)*		
51 HO	Ultralair Weedhopper Ax-3	C3103163		Not known	Belle Vue Farm, Yarnscombe	9.04
95 MR	Power Assist Swift	Not known		Not known	Easter Poldar Farm, Thornhill	2004

HUNGARY

HA-ACL	Dornier Do.28D-2 Skyservant	4125	D-IDRC	Trener Kft	RAF Weston on the Green	12.04
	(Walter M-601 turbo conversion)		58+50	*(Op Wingglider Ltd)*		
HA-ACO	Dornier Do.28D-2 Skyservant	4335	G-BWCN	Trener Kft	RAF Weston on the Green	7.04
	(Walter M-601 turbo conversion)		5N-AYE, D-ILID, 9V-BKL, D-ILID	*(Op Wingglider Ltd)*		
HA-JAB	Yakovlev Yak-18T	22202023842	FLARF-02160	Kobo-Coop 96 Kft	Spilstead Farm, Hastings	12.04
			CCCP-44420			
HA-JAC	Yakovlev Yak-18T	22202034139	RA-44506	Kobo-Coop 96 Kft	(Sleap)	6.04
			HA-JAC, FLA-02159			
HA-LAQ	LET L-410UVP-E4	841332	HAF-332	Farnair Hungary Kft	Hinton-in-the-Hedges	12.04
			HA-YFB			
HA-LFM*	Aérospatiale SA.341G Gazelle	1301	G-OCMJ	S Atherton	Breighton	11.04
			G-HTPS, G-BRNI, YU-HBI	*(Stored)*		
HA-LFZ	Sud SA318C Alouette II	2043	ALAT	Hidroplan Nord Kft	Brierley	11.04
HA-MKE	WSK-PZL Antonov An-2R	1G158-34	UR-07714	Trener Kft	Popham	11.04
			CCCP-07714			
HA-MKF	Antonov An-2	1G233-43	OM-248	Trener Kft	Popham	11.04
			OM-UIN, OK-UIN			
HA-PPY	SOKO SO341 Gazelle	021	HA-LFR	J R Saul	Brierley, South Yorkshire	9.03
	(Aérospatiale c/n 1118)		HA-VLA, YU-HDN, JRV			
HA-VOC	Dornier Do.28D-2 Skyservant	4331	TC-FBC	Trener Kft	Hibaldstow	7.04
	(Turbo conversion)		D-IDWM, CN-..., D-IDWM	*(Op Wingglider Ltd)*		
HA-YAB	Yakovlev Yak-18T	12-35	RA44777	Kobo-Coop 96 Kft	Rendcomb	8.04
HA-YAD	Yakovlev Yak-18T	22202054812	RA81584	Kobo-Coop 96 Kft	Elstree	1.05
HA-YAE	Yakovlev Yak-18T	11-35	RA44000	Kobo-Coop 96 Kft	White Waltham	12.04
HA-YAF	Yakovlev Yak-18T	08-34	RA44480	Kobo-Coop 96 Kft	Dunsfold	1.05
HA-YAG	Yakovlev Yak-18T	05-36	RA-01555	Kobo-Coop 96 Kft	Compton Abbas	2.05
HA-YAH	Technoavia Yak-18T	18-33	RA-44470	Kobo-Coop 96 Kft	White Waltham	12.04
HA-YAI	Yakovlev Yak-18T	22202052122	RA44538	Kobo-Coop 96 Kft	(Barton)	4.04

HA-YAJ	Yakovlev Yak-18T	01-33	LY-APP LY-AOG	Kobo-Coop 96 Kft	Booker	1.05
HA-YAL	Yakovlev Yak-18T	7201513	CCCP-44312	Kobo-Coop 96 Kft	White Waltham	1.05
HA-YAM	Yakovlev Yak-18T	22202047812	RA-44504 LY-AMJ , DOSAAF, CCCP-81558	Kobo-Coop 96 Kft	Earls Colne	8.04
HA-YAN	Yakovlev Yak-18T	10-34	RA-44465 LY-AOL	Kobo-Coop 96 Kft	Oaksey Park	10.04
HA-YAP	Yakovlev Yak-18T	22202034023	RA-44545 LY-AIH, ES-FYE	Kobo-Coop 96 Kft	White Waltham	1.05
HA-YAR	Sukhoi Su-29	75-03	RA-01609 RA7503	Kobo-Coop 96 Kft	White Waltham	10.04
HA-YAU	Yakovlev Yak-18T	15-35	RA-44527	Kobo-Coop 96 Kft	White Waltham	1.05
HA-YAV	Yakovlev Yak-18T	22202047817	RA-01153	Kobo-Coop 96 Kft *(Op Martin Robinson)*	Exeter	1.05
HA-YDF	Technoavia SMG-92 Finist	01-0005		G-92 Kereskedelmi.Kft *(Op Wingglider Ltd)*	Hibaldstow	7.04
HA-YFC	LET L-410-FG	851528		Farnair Hungary	South Cerney	11.04
SWITZERLAND						
HB-CII*	Cessna 210D Centurion	21058321	D-EDEG OE-DEG, N3821Y	Not known	Tatenhill	1.05
HB-DFT	Mooney M.20J	24-0837		Air Link AG	Andrewsfield	1.05
HB-IIY	Gulfstream Gulfstream V	638	N638GA	Aviation G5 Ltd *(Op Jet Aviation)*	Biggin Hill	1.05
HB-IVR	Canadair CL604 Challenger	5318	HB-IKQ (TC-DHE), C-FYYH, C-GLXO	Interline SA *(Op Execujet Charter AG)*	Luton	1.05
HB-IZY*	SAAB 2000	047	SE-047	Air Salvage International *(Fuselage only)*	Alton	2.04
HB-NCN	Rockwell Commander 112TCA	13151	N4620W	R Symmonds	Southend	2.05
SAUDI ARABIA						
HZ-123	Boeing 707-138B	17696	"17696" HZ-123, N138MJ, N220M, N138TA, (N112TA), C-FPWV, CF-PWV, VH-EBA, N31239	Not known *(In open store)*	Southend	2.05
HZ-AB3	Boeing 727-2U5AR	22362	V8-BG1 V8-HM2, V8-HM1, V8-UB1, V8-HM1, JY-HNH	Al Anwa Establishment	Lasham	1.05
HZ-KAA	Gulfstream Gulfstream IV/SP	1294	N416GA HZ-MAL, N416GA	Mawarid Ltd	Farnborough	1.05
HZ-OFC4	Dassault Falcon 900EX	31	F-GSAI F-WWFC	Olayan Finance Co	Luton	1.05
HZ-SJP3	Canadair CL604 Challenger	5346	N604JP C-GLXS	Jouannou and Parskevaides	Farnborough	1.05
ITALY						
I-EIXM*	Piper PA-18-135 Super Cub	18-3572	MM54-2372 54-2372	Not known *(Open store as "EI-184")*	Kesgrave, Ipswich	2.03
I-FXRF	Piaggio P.180 Avanti	1060		Euroskylink	Coventry	7.04
I-IAFS	SOCATA TB-9 Tampico Club	1384		Not known	Biggin Hill	12.04
I-LELF	SIAI-Marchetti SF.260C	568/41-004		Not known	Elstree	1.05
I-TOMI*	Nardi FN.305D	Not known		Not known *(Stored)*	Booker	11.04
I-VFAN*	Agusta Bell AB.206B3	8670	VF-26	Not known	Redhill	8.02
I-6052	Jabiru Jabiru UL	-		Not known	Lower Mountpleasant, Chatteris	3.03
NORWAY						
LN-AMY	North American AT-6D Harvard	88-16849	(LN-LCS) (LN-LCN), N10595, 42-85068	The Old Flying Machine Co *"Amy" "8084" (Breitling Fighter Team titles)*	Duxford	9.04
LN-FOI	Lockheed L-188C Electra	2005	(LN-MOF) N31231, ZK-TEA, (ZK-BMP), N9724C *(Stored derelict in DHL c/s))*	Not known	Coventry	4.04
LN-KKA*	Fokker F.27-050	20117	PH-DLT D-AFKY, PH-DLT, OE-LFZ, (VH-FNL), PH-EXA *(Fuselage only)*	Air Salvage International	Alton	1.05
ARGENTINA						
LV-RIE	Nord 1002 Pingouin	240		R.J.Lamplough *(Stored)*	(East Garston)	2.04
LV-WTY	McDonnell-Douglas MD-81	48011	LV-PMJ N532MD, HB-INM	Not known *(Stored unmarked)*	Shannon	12.04
LUXEMBOURG						
LX-FTA	Dassault Falcon 900C	201	N210FJ F-WWFA	Silver Arrows SA	Farnborough	1.05
LX-HEC	Eurocopter EC.155B Dauphin	6600	F-WQPX	Not known *(Op Roman Abramovich)*	Blackbushe	1.05
LX-PRE	Raytheon 390 Premier 1	RB-60	N6160D	Not known	Blackbushe	1.05
LX-TLB	Douglas DC-8-62F	45925	N822BV CX-BQN, CX-BQN-F, C-GMXR, N922CL, HB-IDG *(Stored)*	Cargo Lion	Manston	1.05
LITHUANIA						
LY-AFB	Yakovlev Yak-52	822610	DOSAAF 112	Termikas Co *"112"*	Little Gransden	8.03
LY-AFO	Antonov An-2R	1G-211-42	LY-ADL CCCP-32683	Not known	Cork	4.03
LY-AHD	Yakovlev Yak-12	30119	SP-CXW PLW....	Not known	Little Gransden	12.04
LY-ALJ*	Yakovlev Yak-52	8910115	DOSAAF 132	D.Hawkins *(Crashed landing Bagsea Farm, Blandford 23.7.97: wreck)*	Little Gransden	9.03
LY-ALS*	Yakovlev Yak-52	855509	DOSAAF 69 DOSAAF 49	M.Jefferies *"69" "Once a Knight" (to become G-KOMI ?)*	Little Gransden	1.05
LY-ALT	Yakovlev Yak-52	822704	DOSAAF 121	Titan Airways Ltd	Elmsett	8.04

Reg	Type	C/n	Previous identity	Owner/Operator	Location	Date
LY-AOT*	Yakovlev Yak-50	853101		R Cayless	White Waltham	11.01
				(To become G-CBPO ?: current status unknown)		
LY-ARH	Yakovlev Yak-18T	22202040114	CCCP-44560	Nerka Ltd	Turweston	1.05
LY-BIG	Antonov An-2T	1G236-23	Ukraine AF 48 *(red)*	Air Unique *"Baltic Bear"*	Tatenhill	1.05
LY-IOO	Yakovlev Yak-50	Not known		Yak UK *"100" (Current status unknown)*	Little Gransden	11.01
LY-MHC	Antonov An-2R	1G215-33	LY-AVK	Lietuvos Svedijos UAB	Little Gransden	1.05
			CCCP40896	*"103"*		

UNITED STATES

Reg	Type	C/n	Previous identity	Owner/Operator	Location	Date
N1FD	SOCATA TB-200 XL Tobago	1614		Siek Aviation Inc	Blackbushe	12.04
N1FY	Cessna 421C Golden Eagle I	421C1067	N345TG	LME Aviation	Kemble	1.05
N2CL	Piper PA-28RT-201T Turbo Arrow IV	28R-8131054	N8333S	Southern Aircraft Consultancy Inc	Elstree	8.04
			N9649N			
N2FU	Learjet Inc. Learjet 31	31-027	N30LJ	Wilmington Trust Company	Biggin Hill	7.04
			N91201	*(Op Formula One Administration)*		
N2MD	Piper J3C-65 Cub	17521	N70515	Merlin Aire Limited	Rendcomb	7.03
			NC70515	*(Op Gypsy Flyers Ltd) (Kia Cars titles)*		
N2NR	Agusta A109A-II	7350	N800AH	N2NR Inc *(Op Sunseeker Leisure)*	Bournemouth	7.04
N2RK	Lockheed L.188PF Electra	2010	N178RV	Electra Aero Inc	Coventry	1.05
			C-GNWC, N178RV, N63AJ, CF-NAX, N33506, ZK-TEB, (ZK-BMQ)			
				(Reeve Illusion titles in open store)		
N3HK	Cessna 340 II	340-0538	G-VAUN	Southern Aircraft Consultancy Inc	North Weald	1.05
			D-IOFW, N5148J			
N4HG	Lockheed L.188PF Electra	1140	G-LOFH	Reeve Aleutian	Coventry	1.05
			N9744C	*(As "44C" on nose wheel door*		
N5LL	Piper PA-31 Navajo C	31-7812041	N27495	Southern Aircraft Consultancy Inc	Shoreham	2.05
N6FL	Latulip LM-3X	LM-3X-1001		M J Aubrey	(Kington, Hereford)	.02
	(Rotax 377) *(Aeronca 7AC scale rep)*			*(Stored)*		
N6NE	Lockheed Jetstar 731	5006/40	(VR-CCC)	Aerospace Finance Leasing Inc	Southampton	9.04
			N6NE, N222Y, N731JS, N227K, N12R, N9280R			
				(Damaged Southampton 27.11.92: on fire dump)		
N7AG	Agusta A109A Mk.II	7436	I-AXLE	Helijet Inc	Stapleford	6.04
N7UK	Cirrus Design SR22	0200		Southern Aircraft Consultancy Inc	Goodwood	11.04
N8MZ	Piper PA-30 Twin Comanche B	30-1648	G-ORDO	Francis Aviation	Lee-on-Solent	10.4
			N8485Y			
N8YG	Piper PA-32R-301T Saratoga II TC	3257151	OY-LAA	Normac Aviation	Lydd	8.04
N9AY	Cessna 421C Golden Eagle III	421C0844	G-NSGI	Sooty Aviation Inc	Jersey	1.05
			N421EL, XA-RAE, N421EB, (N21MW), N421EB, N2659Z			
N9VL	Agusta A109A-II	7325	OO-AHE	Murtagh Aviation	Liskeard	12.04
			OO-XHE, N109LA	*("The Mortgage Group" titles)*		
N10MC	Cirrus Design SR22	1084		La Luisa Aviation	Jersey	1.05
N11FV	Cessna T303 Crusader	T30300133	G-BXRI	Auster Aviation	Guernsey	1.05
			HB-LNI, (N5143C)			
N11ZP	American Blimp Corp A-60+ Airship	011		Virgin Lightships Inc *(Lotto titles)*	Wolverhampton	6.02
N12ZP	American Blimp Corp A-60+ Airship	012		Virgin Lightships Inc	Wolverhampton	5.02
				"Spirit of Europe 2" (Lotto titles) (Marks "N612LG2 requested)		
N14AF	Rockwell Commander 112TC-A	13258	G-GRIF	Southern Aircraft Consultancy Inc	Gamston	9.04
			G-BHXC, N1005C			
N14EP	SOCATA TB-20 Trinidad GT	2061	G-SHEP	Shephard Aviation	Goodwood	9.04
			F-OILU, F-WWRB			
N14HF	Maule MT-7-235 Star Rocket	18084C		Hamilton-Fairley Aviation	Bramshill	1.05
N14MT	Cessna TR182 Skylane RG	R18201227		Andrew Wvensche	Elstree	1.05
N15CK	Maule MX-7-235 Star Rocket	10012C		Rossendale Air	East Winch	9.04
N18V	Beech UC-43-BH Traveler	6869	NC18	R.J.Lamplough	East Garston	2.04
			Bu 32898, FT507, 44-67761	*(As "DR828/PB1")*		
N19F	Cessna 337A Super Skymaster	33700289	N6289F	Southern Aircraft Consultancy Inc	Fakenham	4.01
	(Robertson STOL conversion)			*(Current status unknown)*		
N19GL	Brantly B.2B	2004		Southern Aircraft Consultancy Inc		
					Hill Top Farm, Hambledon	8.03
N20AG	SOCATA TB-20 Trinidad	2003	N29KF	Archway International Inc	Jersey	1.05
N20UK	Mooney M.20F Executive	22-1380	N9155J	E-Plane Inc	Biggin Hill	12.04
			G-BDVU			
N21UH	United Helicopter Corp UH-12C	UH2001		Southern Aircraft Consultancy Inc	Kilrush	8.04
N21VC	Cessna 525 CitationJet	525-0106	N5211A	Mistral Aviation	Guernsey	1.05
N22CG	Cessna 441 Conquest II	441-0119		Jubilee Airways Inc *(Op M.Klinge)*	Prestwick	10.04
N23AM	Cirrus Design SR22	0937		Southern Aircraft Consultancy Inc	Membury	1.05
				(Damaged landing 22.6.04)		
N25KB	Piper PA-24-250 Comanche	24-3034	F-BKRK	Southern Aircraft Consultancy Inc		
			N7814P	*(Op K Bettoney)*	Farley Farm, Romsey	1.05
N25PR	Piper PA-30-160 Twin Comanche B	30-1511	G-AVPR	PSL Aviation	Gloucestershire	12.04
			N8395Y			
N25XZ	Cessna 182G	18255388	HB-CST	Southern Aircraft Consultancy Inc		
		OE-DDM , N2188R			Lower Botrea Farm, Sancreed	11.03
N26HE	Cessna 421C Golden Eagle II	421C0687		Wells Fargo Bank Northwest NA	Biggin Hill	1.05
				(Op Springhurst Ltd)		
N27BG	Cessna 340A	340A0656		Traca Inc *(Op B Gregory)*	Cranfield	1.05
N27MW	Beech B58 Baron	TH-995		B58 Aviation Inc	Fairoaks	12.04
N28TE	Raytheon 58 Baron	TH-1951		Ecosse Aviation Inc	Blackbushe	12.04
N29MR	Cessna 525A CitationJet CJ2	525A0206		Falco Aviation	Bournemouth	12.04
N30NW	Piper PA-30-160 Twin Comanche	30-312	G-ASON	R S Barnett	Norwich	4.02
			N7273Y			
N31NB	Piper PA-31 Turbo Navajo B	31-7401239	G-OSFT	Navajo Aviation Inc	Old Buckenham	11.04
			G-MDAS, 5N-AEP, G-BJCZ, N61427 *(Op N.Brown)*			

Reg	Type	C/n	Previous identity	Owner/operator	Base	Date
N31RB	Grumman-American AA-5B Tiger	AA5B-0156		Southern Aircraft Consultancy Inc *(Op T Cromber t/a Echo Echo Group)*	Bournemouth	1.05
N32LE	Piper PA-32R-301T Turbo Saratoga SP	32R-8329016		Southern Aircraft Consultancy *(Op Light into Europe)*	Great Oakley	7.04
N33EW	Mitsubishi MU-2B-60	1519SA	N331W N33TW, N434MA	Florida Express Corp *(Op King Aviation)*	Southend	12.04
N33NW	SOCATA TB-20 Trinidad	1073	N666HM OO-PDV, F-GLAC	Southern Aircraft Consultancy Inc	Cranfield	1.05
N34FA	SOCATA TB-20 Trinidad	866	G-BPFG	Southern Aircraft Consultancy Inc	Elstree	1.05
N35AD	Piper PA-30 Twin Comanche B	30-1121	HB-LDE N8014Y	Southern Aircraft Consultancy Inc	Jersey	1.05
N35AL	Piper PA-34-220T Seneca IV	3447014	D-GLPE	Able Liston Aviation	Jersey	10.04
N35ZS	Beech V35B Bonanza	D-10098		Mattik Aviation	Guernsey	1.05
N36NB	Beech A36 Bonanza	E-2274	F-GKTZ N7249H	Air Bickerton Inc	Biggin Hill	12.04
N37EL	SOCATA TB-20 Trinidad	378	G-GDGR F-GDGR	Southern Aircraft Consultancy Inc	Stapleford	1.05
N37US	Piper PA-34-200T Seneca II	34-8070111	G-PLUS N81406	Southern Aircraft Consultancy Inc *(Op Skycabs)*	Jersey	1.05
N37VB	Cessna 421C	421C-0418	EC-IFT G-FWRP, N3919C	Lowndes Aviation	Bournemouth	12.04
N39TA	Beech C24R Sierra	MC-639		Southern Aircraft Consultancy Inc	Sturgate	5.04
N40D	Stolp SA-100 Starduster 1	4258549		Southern Aircraft Consultancy Inc	Rochester	5.01
N40GD	Cirrus Design SR22	0473		Aircraft Guaranty Trust	Sherburn-in-Elmet	1.05
N41AK	Beech F90 King Air	LA-188	N41CK N6429M	I and S Graham Aviation	Glasgow	12.04
N41FT	PiperPA-39 Twin Comanche C/R	39-59	G-BZLW ZS-NLF, ZS-MRH, ZS-IKG, N8904Y	Southern Aircraft Consultancy Inc	Biggin Hill	7.04
N41TT	Cessna 414	414-0263		Green Lattice Inc	North Weald	1.05
N42FW	Beech E33 Bonanza	CD-1199	N7682N	Southern Aircraft Consultancy Inc *(Op Feroz Wadia)*	Kirknewton	8.04
N43GG	Piper PA-34-200T Seneca II	34-7670066	G-ROLA N4537X, G-ROLA, N4537X	TT Aviation Rentals Inc	Humberside	12.04
N43SV	Boeing-Stearman PT-13D Kaydet	75-5541		A de Cadenet *"796"*	Oxford	5.04
N44DN	Piper PA-46-350P Malibu Mirage	4622116		Convergence Aviation *(Wreck stored)*	Bournemouth	1.05
N45AW	Piper PA-28RT-201T Turbo Arrow IV	28R-8431003	N43230 N9548N	Andair Inc	Wolverhampton	11.04
N45CD	Piper PA-28-161 Warrior II	28-7916467	PH-AND N2841J	Hill Air Inc	Sigwells, Somerset	9.04
N45YM	Piper PA-46-350P Malibu Mirage	4636217	G-BYLM	Southern Aircraft Consultancy Inc	Alderney	1.05
N46PL	Piper PA-46-500TP Malibu Meridian	4697054	D-FKAI N53263, (D-FKAI), N53263, (PH-EPS)	AvCorp Leasing	Gloucestershire/Jersey	12.04
N47DG*	Republic P-47G Thunderbolt	21962	N42354 42-25068	Flying A Services *(Cancelled 4.00 by FAA) (Stored in container)*	(Greenham Common)	2.04
N48HB	Piper PA-32R-301T Saratoga IITC	3257310		HBC Aviation	(Thurrock	1.05
N48NS	Cessna 550 Citation Bravo	550-0939	VP-BNS (N939BB), N5076K	Tower House Investments	Jersey	1.05
N51AH	Piper PA-32R-301 Saratoga SP	32R-8413017	G-REAH G-CELL, (G-BLRI), N4361D	Southern Aircraft Consultancy Inc	Blackbushe	12.04
N51ER	American Champion 7GCAA	484-2004		Southern Aircraft Consultancy Inc	Abbeyshrule	7.04
N55AE	Beech 95-C55 Baron	TE-84	OY-EBF OH-BBF, SE-EUT	Sam Argo Aviation Inc	Bruntingthorpe	12.02
N55BN	Beech 95-B55 Baron	TC-1572	G-KCAS G-KCEA, N2840W	Snowadam Inc *(Op C.Butler)*	Standalone Farm, Meppershall	8.04
N55CJ	Cessna 525 CitationJet	525-0298	G-RSCJ	Pektron Aviation	Gamston	12.04
N55EN	Beech E55 Baron	TE-942		Monckton Byng Inc	Elstree	12.04
N56PZ	Reims/Cessna F177RG Cardinal	F177RG0079	G-BFPZ (OO-DVE), G-BFPZ, PH-AUK, D-EGBM	Atlantic Bridge Group	Exeter	11.04
N57CR	Hiller UH-12C	909		Pulse Helicopter Corp	Sywell	1.05
N57MT	Cessna T303 Crusader	T30300211	D-IEEG N9748C	Flying Dog Inc	Guernsey	1.05
N58GT	Beech B58 Baron (winglets)	TH-1090	HB-GIK	Swiftair Inc	Elstree	1.05
N58HK	Cessna 550 Citation Bravo	550-1086	N52446	David McLean Inc	Hawarden	1.05
N58YD	Beech 58 Baron	TH-1427	G-OLYD N7255H, ZS-LYC, N7255H	Southern Aircraft Consultancy Inc	Gamston	8.04
N59VT	Beech K35 Bonanza	D-5897	D-EMEF	Southern Aircraft Consultancy Inc	Kemble	1.05
N60B*	Rockwell 690A Turbo Commander	11172	TF-ERR (D-IIGII), TF-ERR, C-GERR, TF-ERR, N9164N	Not known *(Cancelled 5.00 by FAA)*	Gamston	1.05
N60GM	Cessna 421C Golden Eagle III	421C0828		Southern Aircraft Consultancy Inc	Gamston	11.04
N60NB	Mitsubishi MU-2B-60 Marquise	1528SA	5Y-VIZ	Dogfox Airways Inc	Dublin	11.04
N61DE	Piper PA-32-300 Six	32-7940030	G-LADE N3008L	Southern Aircraft Consultancy Inc	Denham	12.04
N61HB	Piper PA-34-220T Seneca V	3449217	G-CBAA N53445	HBC Aviation Inc	Thurrock/Guernsey	1.05
N61MF	Mooney M.20J	24-0847	G-BYDD D-EIWM	Michael Flynn	Fairoaks	1.05
N64GG	Raytheon B300 King Air	FL-274		Specsavers Aviation	Guernsey	1.05
N64JG	Bell 206B-2 Jet Ranger	1507	N589LB	Celtic Aircraft Inc	(Ronaldsway)	12.04
N64VB	Beech 58 Baron	TH-305	N273TB	Rogers Aviation	Leeds-Bradford	1.05
N65JF	Piper PA-28-181 Archer II	28-7990140	N2087C	Southern Aircraft Consultancy Inc	Nottingham	12.04
N65MJ	Beech 58P Baron	TJ-487		Southern Aircraft Consultancy Inc	Donegal	11.04
N66SG	Learjet Inc Learjet 45	45-073	N65U	Woolsington Wunderbus Inc *(Op Sagesoft)*	Luton	1.05
N66SW*	Cessna 340	340-0011	N5035Q	North East Wales Institute *(Cancelled 5.03 by FAA as sold in UK)*	Connah's Quay	11.04

N67TC	Rockwell Turbo Commander 690A	11233	N9192N HR-AAJ, N9192N	Aircraft Guaranty Title Corp	North Weald/Southend	1.05
N69LJ	Learjet Inc. Learjet 60	60-027		Wilmington Trust Co	Biggin Hill	12.04
N70AA	Beech 70 Queen Air	LB-35	G-KEAA G-REXP, G-AYPC	Metals and Alloys International *(In open store)*	Southend	2.05
N70VB	Ted Smith Aerostar 600A	60-0446-150	C-GVHQ N9805Q	Southern Aircraft Consultancy Inc	Blackbushe	12.04
N70XA	Cessna 550 Citation II	550-0008	N70X N550JF, (N108AJ), OE-GIW, N575W, N98840	Wells Fargo Bank Northwest NA	Liverpool	12.04
N71VE	Rockwell Commander 690A	11043	N71VT N2VQ, N2VA	Airbourne Inc	Gamston	1.05
N73MW	Beech B200 Super King Air	BB-22	N7300R	Royal Palm Air Lease	Plymouth	5.04
N74DC	Pitts S-2A Special	2228	I-ALAT	H J Seery *(Op D.Cockburn)*	Rush Green	7.04
N74PM	Agusta A109C	7636	D-HOFP I-SEIN	Ortac Inc *(Op Huktra UK Ltd)*	Hawarden	11.04
N76HN	British Aerospace BAe146 Srs.100	E1076	B-632L B-2707, G-5-076	Aircraft Holdings Network inc *(In open store as "B-632L")*	Southend	2.05
N76JN	Piper PA-31-350 Navajo Chieftain	31-7652176		Aviation Leasing Inc	Nottingham East Midlands	8.03
N77XB	Piper PA-31 Turbo Navajo *(Colemill Panther conversion)*	31-583	G-AXXB N7XB, N6645L, G-AXXB, N6645L	Aircraft Guaranty Title and Trust Llc	Oxford	12.03
N77YY	Piper PA-32R-301T Saratoga II TC	3257120	G-LLYY N4165C	Flying Start Aviation Inc *(Op M J Start)*	Guernsey	1.05
N78HB	Aviat A-1B Husky	2066	N115BB G-FOFF, N115BB	HBC Aviation Inc *(Op T Holding)*	King's Farm, Thurrock	2.05
N79AP	Beech 58P Baron	TJ-206	VH-ORP ZK-TML, N6648Z	Aircraft Guaranty Title LLC *(Op R & B Services Ltd)*	Enstone	1.05
N79EL	Beech 400A Beechjet	RK-214		Edra Lauren Leasing Corp *(Op DFS Furniture)*	Nottingham East Midlands	9.04
N79YK	Yakovlev Yak-50	791602	SE-LBR DOSAAF	Windjammer Aviation *(Still marked "SE-LBR")*	Rochester	8.03
N80HC	Beech 58 Baron	TH-672		Brian Richardson Aviation	Wellcross Grange, Slinfold	12.04
N80JN	Mitsubishi MU-2J	626	EC-GLU OY-ATZ, SE-GHY, N476MA	Aircraft Guaranty Title Corp	Waterford	7.04
N80RF	Beech 60 Duke	P-17	(G-BMSO) I-DUKA, F-BRAX, HB-GDO	Goldwing Aviation Inc *(Op MLP Aviation/E.Lundquist) (Current status unknown)*	Fairoaks	12.00
N82GK	Hawker 800XP	258618	N618XP N896QS, (N895QS)	AGK82 Llc *(Op Signature Aviation/Portland Management Services)*	Luton	1.05
N89WC	Sikorsky S-76B	760311		Icarus Helicopters	(Northern Ireland)	1.05
N90BE	Mooney M.20K	25-1143	G-MOON N252BT	Southern Aircraft Consultancy Inc	Wellesbourne Mountford	9.04
N90SA	Reims Cessna F172M	F17201402	PH-TWS OY-BUL	H K de Carlucci	Denham	1.05
N90TK	Mooney M.20C Mk 21	2620	EI-CIK G-BFXC, 9H-ABD, G-BFXC, OH-MOA, N1349W	N90TK Inc	Elstree	1.05
N90U	Piper PA-46-350P Malibu Mirage *(DLX Jet Prop conversion)*	4622106		Speedair Inc	Gloucestershire	12.04
N91ME	SOCATA TB-20 Trinidad	2152		Komfort Aviation	Denham	1.05
N94SA	Citabria 7ECA Champion	227	OY-AUG D-EFLO	Southern Aircraft Consultancy Inc	Kilkeel, County Down	7.04
N95D	Piper PA-34-220T Seneca V	3449060	N9506N	Zeta Aviation Inc	Welshpool	1.05
N95TA	Piper PA-31 Turbo Navajo B	31-7300971	N7576L	High Flyers Aviation	Durham Tees Valley	2.05
N97GP	SOCATA TB-20 Trinidad	1837	G-WASI D-EVHV	GP Aviation *(Op G Pope)*	Goodwood	11.04
N99ET	SOCATA TB-10 Tobago	226	G-BJDG F-BNGR	E.A.Terris	Oxford	9.04
N100JS	Cessna 525A CitationJet CJ2	525A-0176		Jato Aviation	Northolt	12.03
N101DW	Piper PA-32R-300 Cherokee Lance	32R-7680399		Southern Aircraft Consultancy Inc	Panshanger	12.04
N101UK	Mooney M.20K	25-0631		Southern Aircraft Consultancy Inc	Blackpool	11.04
N103ZZ	Piper PA-31-350 Navajo Chieftain	31-7405444	Z-ICO N103ZZ, N61400, (N2LK), N61400	Ronald Hope	Cranfield	12.04
N105LF	Dassault Falcon 2000	105	N220EJ F-WWVZ	Wells Fargo Bank Northwest NA *(Op Krystel Air Charter)*	Cranfield	11.02
N109AB	Agusta A109E Power	11015		Airabco Inc	Rhyader	12.04
N109AN	Agusta A109A-II	7348		Flytrue Aircraft Inc	Elstree	1.05
N109AR	Agusta A109A	7390	8P-BHA G-BXPX, I-DVRE, (N109LL)	Adrian Raymond Aviation Inc *(Op Castle Air Charters)*	Liskeard	12.04
N109GR	Agusta A109E Power	11043	TC-HCU	Castle Helicopters Inc *(Op Castle Air Charters)*	Liskeard	9.04
N109KH	Agusta A109C	7609	N133H N1NQ	Redwood (IOW) Inc	(Calbourne)	10.04
N109TF	Agusta A109A-II	7328	VH-NWD VH-DMR, (VH-MRS)	Chestham Park Inc *(Op Castle Helicopters Inc)*	Goodwood	1.05
N109TK	Agusta A 109C	7650	N109TW D-HCKM	Botany Aviation	Blackpool	1.05
N109UK	Agusta A109A-II	7304	F-GKGV N109PS, (N109FS), N109FM	SMC Aviation	Stapleford	1.05
N109WF	Agusta A109A-II	7298	N109BC	Agusta 109 LLC *(Op Lenham Racing)*	Elstree	1.05
N111SX	Piper PA-46-350P Malibu Mirage	4636286	EC-HPP	Saxon Aviation	Elstree	1.05
N112JA	Rockwell Commander 112TC-A	13182	5Y-MBK	Southern Aircraft Consultancy Inc	Sandown	11.04
N114ED	Commander Aircraft Commander 114B	14637	G-PADS N60987	Southern Aircraft Consultancy Inc	Guernsey	1.05
N115RT	Piper PA-28-181 Archer III	2843010		Southern Aircraft Consultancy Inc	Old Sarum	8.04
N116HS	Bell UH-1L	6172	Bu154949	Yorkshire Helicopters USA Inc *(Stored dismantled)*	Beckley, Oxford	3.04

Reg	Type	C/n	Previous identity	Owner/operator	Base	Date
N116WG	Westland WG-30-100	016	(G-BLLG)	Cogent plc	Montrose	7.03
				(Op Oil Petroleum Training Industry Board)		
N119BM	Agusta A119 Koala	14016		HSS (USA) Ltd	(Eastleigh)	11.04
N121EL*	Learjet Inc. Learjet 25	25-010	(N121GL)	Kingston University	Roehampton, Surrey	9.03
			(N82UH), (N10BF), N102PS, N671WM, N846HC, N846GA	*(Instructional use)*		
N121HT	Cirrus Design SR22	0794		Aircraft Guaranty Management Llc	Gloucestershire	1.05
N121MT	Britten Norman BN-2T Turbine Islander	880	N200LQ	Swiftair Inc	Elstree	12.04
	(Built IRMA)		USAF 88-0916, N5097R, N73413, (YV-2173P), N413JA, G-BFNX			
N122MG	Cirrus Design SR22	1250		Caseright Inc	Turweston	1.05
N123AX	Piper PA-32R-301 Saratoga IIHP	3246060	G-LLTT	Axis Aircraft Leasing Inc	Gloucestershire	12.04
			N9283P			
N123DU	Piper PA-28-161 Cherokee Warrior II	28-7716195	G-BPDU	Fletcher Aviation	Guernsey	2.05
			N5672V	*(Southern Flight Centre titles)*		
N123SA	Piper PA-18-150 Super Cub	18-1372	ALAT	Southern Aircraft Consultancy Inc	Glenforsa	8.03
			51-15372	*(Op B Walsh)*		
N123UK	Mooney M.20J	24-3167	G-ZZIP	Southern Aircraft Consultancy Inc	Southend	2.05
			N1086N	*(Op H T El-Kasaby and other)*		
N125GP	Learjet Inc. Learjet 31A	31A-162	N162LJ	Trans Air Inc	Dublin	1.05
			N525GP			
N125GW	Learjet Inc. Learjet 45	45-236		Tyneside Thunderbolt Inc	Newcastle	1.05
N125XX	Hawker Siddeley HS.125 Srs.700A	257075 & NA0254	N124AR	Surewings Inc *(Op Ambrion Aviation)*	Luton	12.04
			N125TR, N125AM, (G-BHKF), G-5-13			
N125ZZ	Hawker 800XP	258630	N630XP	Wells Fargo Bank Northwest NA	Luton	1.05
N129SC	Piper PA-32-300 Cherokee Six	32-7440057		Manx Orthopaedic Services	Ronaldsway	8.04
N132CK	Cessna 421A	421A0038	EI-TCK	Southern Aircraft Consultancy Inc	Weston, Dublin	8.04
			G-AXAW, (EI-TCK), G-AXAW, N2238Q			
N132LE	Piper PA-32-300 Cherokee Six	32-40038	G-AVFS	Southern Aircraft Consultancy	Great Oakley	1.05
				(Op Light into Europe)		
N134TT	Cessna 305C Bird Dog	24541	F-GFVE	Southern Aircraft Consultancy Inc		
			F-WFVE, ALAT		Belle Vue Farm, Yarnscombe	9.04
				(As "24541" in US Marines c/s)		
N136SA	American General AG-5B	10164	G-RICA	Southern Aircraft Consultancy Inc	Popham	1.05
N139DB	Piper PA-23-250 Aztec E	27-4611	G-AYUL	Pyramis Inc	White Waltham	12.04
			N13992	*(Op Earlsfield Investments)*		
N142TW	Beech 58 Baron		TH-1841	Specialized Aircraft Services Inc	Fairoaks	1.05
N145AV	Piper PA-32R-301 Saratoga SP	32R-8013132	EC-HHM	Sav Air	Sleap	1.05
			G-TRIP, G-HOSK, PH-WET, OO-HKN, N8261X			
N145DF	Cessna S550 Citation II	S550-0018	N1AF	Star Aviation Ltd	Luton	1.05
			N814CC, N501NB, (N1259K)			
N145DR	Piper PA-34-220T Seneca	3449240		Cleevewood Aviation Inc	Gloucestershire	12.04
N146FL	Beech F90 King Air	LA-59	G-FLTI	Keep Holdings	Guernsey/Southend	2.05
			N7P	*(Operated Flightline)*		
N147CD	Cirrus Design SR20	1043		Plane Holdings Inc *(Op FreeFlight Aviation)*	Denham	12.04
N147DC	Douglas C-47A-75-DL Dakota	19347	G-DAKS	Aces High US Inc	North Weald	1.05
			TS423, "108841, "KG374", "G-AGHY", TS423, 42-100884 *(As "07")*			
N147GT	Cirrus Design SR22-G2	1069		FreeFlight Aviation	Denham	2.05
N147VC	Cirrus Design SR22	0689		FreeFlight Aviation	Denham	1.05
N150JC	Beech A35 Bonanza	D-2084	N8674A	R.M.Hornblower	Southend	1.05
				(Open store, dismantled on trailer)		
N151CG	Cirrus Design SR22	0344		Central Chiswick Surveying Inc	White Waltham	9.04
N153H	Bell 222B	47138		Spansky Aviation Inc	Castleknock, Dublin	3.04
N154DJ	Cessna T303 Crusader	T30300230	F-GGLJ	N154DJ Inc	Denham	2.05
			N9891C			
N156LG	American Blimp Corp A-1-50 Airship	106		Virgin Lightships Inc	Cardington	4.01
				(Current status unknown)		
N160TR	Piper PA-31T Cheyenne II	31T-7920036	N81918	BHS Direct	Sleap	12.04
			EC-FOT, N525CA, C-GEAS, N222, N796SW			
N163J	Dassault Falcon 2000	163	F-WWVW	Wells Fargo Bank Northwest NA	Biggin Hill	1.05
N170AZ	Cessna 170A	19674	HB-CAZ	Southern Aircraft Consultancy Inc *(Op A Gregori: damaged)*		
			N5720C		Nethershields Farm, Chapelton	9.03
N171JB	Piper PA-28R-180 Cherokee Arrow	28R-30756	N7414J	Berry Air Flying Club	Kirknewton	5.04
N172AM	Cessna 172M Skyhawk II	17264993	G-BXHG	Southern Aircraft Consultancy Inc		
			N64057	*(Op Pacnet)*	Brittas House, Limerick	10.04
N173RG	Velocity 173RG	3RE052		Aircraft Guaranty Title and Trust LLC	North Weald	1.05
N176AF	Cessna 650 Citation III	650-0176		General Electric Capital Corp	Coventry	1.05
				(Op Ilmor Engineering)		
N177MA	Piper PA-46-350P Malibu Mirage	4622177		Southern Aircraft Consultancy Inc	Weston, Dublin	7.04
N177SA	Reims/Cessna F177RG Cardinal RG	F177RG0171	F-GBFI	Southern Aircraft Consultancy Inc	Norwood, Clacton	1.05
N180BB	Cessna 180K	18053103		Southern Aircraft Consultancy Inc	Humberside	9.04
N180FN	Cessna 180K	18053201		Rivet Inc	Fordham, Newmarket	7.04
N180LK	Piper PA-28-180 Cherokee F	28-7105121		Southern Aircraft Consultancy Inc	Henstridge	1.05
N181PC	SOCATA TBM-700	261		Longslow Food Group	Norwich	11.04
N181WW	Beagle B.206 Srs.1	B.018	G-BCJF	Southern Aircraft Consultancy Inc	Biggin Hill	12.04
			N181WW, G-BCJF, XS773			
N182PN	Cessna 182R Skylane II	18268316	G-OCJW	Southern Aircraft Consultancy Inc		
			G-SJGM, N357WC	*(Op Pacific Network Air)*	Brittas House, Limerick	1.05
N184CD	Cirrus Design SR20	1087		R F Aviation	Cherry Tree Farm, Monewden	2.05
N185UK	Cessna A185F Skywagon II	18504367	SE-KOC	Jet Blades & Engineering Inc	Lochearnhead	6.03
	(Floatplane)		N96DS, N9903N	*(Op Seaflite)*		
N187SA	Piper PA-28R Cherokee Arrow II	28R-7235139	G-BOJH	Southern Aircraft Consultancy Inc	Glasgow	11.04
			N2821T	*"Knight of the Thistle"*		
N188S	Agusta A109A-II	7349		JSJ Aviation	Leeds-Bradford	11.04
N189SA	Piper PA-31-325 Navajo C/R	31-7512045	G-BMGH	Southern Aircraft Consultancy Inc	Southend	2.05
			ZS-LEU, N8493, A2-CAT	*(Op J Jacques)*		

N191ME	Cessna T206H	T20608188		Southern Aircraft Consultancy Inc	Blackpool	1.05
N200PR	Raytheon RB390 Premier	RB-79		Wells Fargo Bank Northwest NA	Farnborough	1.05
				(Crashed landing Blackbushe 7.4.04 - wreck to AAIB)		
N200UP	Dassault Falcon 50	55	N96UH	Wells Fargo Bank Northwest NA	Farnborough	1.05
			N300CR, N625CR, N332MQ, N332MC, N1CN, (N30N), N839F, N73FJ, F-WZHU			
N201YK	Mooney M.20J	24-0518		Conmacair Inc *(Op C McAfee)*	Cumbernauld	2.04
N202AA	Cessna 421C Golden Eagle	421C1015		Simply Living Ltd	North Weald	1.05
				(Nosewheel collapsed landing North Weald 9.4.04)		
N202MC	Mitsubishi MU-2B-26A	369SA	N755MA	Southern Aircraft Consultancy Inc	Bournemouth	11.04
N202NK	Cirrus Design SR22	1230		Heathfield Aviation	Turweston	1.05
N203CD	Cirrus Design SR20	1451		Hughston Aircraft Corporation	Manchester	1.05
N203SA	Piper AE-1 Cub Cruiser	5-1477	G-BWUG	Southern Aircraft Consultancy Inc	Henstridge	1.05
			(ZK-USN), N62073, NC62073, Bu30274	*(As "Bu30274" in US Navy c/s)*		
N206GF	Bell 206B-3 Jet Ranger	3806	JA9448	Southern Aircraft Consultancy Inc	Bournemouth	8.03
			N206JG			
N206HE	Bell 206B Jet Ranger	2880	N316JP	Southern Aircraft Consultancy Inc	Bournemouth	9.04
N208KP	Cessna 208 Caravan	20800367		Chartfleet Inc	(Fenland)	1.05
N208NJ	Cessna 208B Grand Caravan	208B1051		Dolphin Aviation Group	Coventry	1.05
N208ST	Cessna 208B Grand Caravan	208B1023		Aircraft Guaranty Title and Trust Llc	Strathallan	9.04
				(Op B Munro)		
N209SA	Piper PA-22-108 Colt	22-8448	EI-AYS	Southern Aircraft Consultancy Inc	Abbeyshrule	8.04
			G-ARKT			
N210AD	Cessna 210G Centurion	21058835	OE-DES	Uniplane Inc	Guernsey	1.05
N210CP	Cessna 210M Centurion	21062034		Southern Aircraft Consultancy Inc	Southend	2.05
				(Op Quick Flights Ltd)		
N210NM	Cessna 210K Centurion	21059255	D-ECAL	Alpine Air Leasing	Denham	1.05
			N8255M			
N216GC	Piper PA-28R-200 Cherokee Arrow B	28R-7135151	G-EVVA	American Flight Academy	Elstree	2.05
			G-BAZU, EI-AVH, N11C			
N218BA	Boeing 747-245F	20827	G-GAFX	Wilmington Trust Co	Manston	1.05
			N641FE, VP-BXP, N641FE, (N632FE), N812FT, N702SW	*(Stored)*		
N218SA	Piper PA-24-250 Comanche	24-1877	G-OJOK	Southern Aircraft Consultancy Inc		
			PH-DZE, D-EIEI, N6749P		Boonhill Farm, Fadmoor	11.04
N218Y	Cessna 310Q	310Q0507	G-AZYM	Southern Aircraft Consultancy Inc	Guernsey	1.05
			N218Y, G-AZYM, N5893M, N4592L			
N222SW	Cirrus Design SR22-G2	0977		Staywhite UK Ltd	Rochester	2.05
N222WX	Bell 222A	47021	EI-BOR	Shannon Helicopters Inc	Shannon	10.04
			LN-OSB			
N228CX	SOCATA TBM-700	084		Turbine Aviation Inc *(Op B.Holmes)*	Southend	1.05
N228TM	Raytheon Hawker 800XP	258458		Wells Fargo Bank Northwest NA	Cork, County.Cork	1.05
				(Op EMC Corporation)		
N230MJ	Piper PA-30 Twin Comanche B	30-1302	G-AVCX	Hughston Aircraft Corp	Exeter	11.04
			N8185Y			
N231CM	Piper PA-46-500TP Malibu Meridian	4697146		PB One Aviation	Farnborough	1.05
N232N	Beech F33A Bonanza	CE-971	G-BYRT	Single Beech Inc	Top Farm, Croydon, Royston	11.04
			ZS-LFB, N18384			
N234RG	Pilatus PC-12/45	520		Wells Fargo Bank Northwest NA	Belfast City	1.05
N234SA	Cessna T310R	310R1805	F-GGGG	Southern Aircraft Consultancy Inc	Nottingham	11.04
			N310AF, N2642B			
N235PF	Piper PA-28-235 Pathfinder	28-7410083	OO-DDC	Southern Aircraft Consultancy Inc	Southend	2.05
				(Op Pathfinder Group)		
N242ML	Cessna 525 CitationJet	525-0506		Branksome Aviation	Bournemouth	1.05
N245CB	Piper PA-34-220T Seneca III	34-8333007		Universal Aviation Corp	Biggin Hill	12.04
N249SP	Cessna 210L Centurion	21060990	4X-CGU	Stellios Papi Inc *(Op Willowair Flying Club)*	Southend	2.05
N250AC	Piper PA-31 Navajo C	31-7612040	G-NWAC	North West Air Inc	Liverpool	1.05
			G-BDUJ, N59814			
N250BW	Piper PA-23-250 Aztec C	27-3799	N6602W	Southern Aircraft Consultancy Inc	Hardwick	7.04
			G-AYSA, N6509Y			
N250JF	Neico Lancair 360	644-320-389FB		Clark D Baker	North Weald	1.05
N250MD	Piper PA-31-310 Navajo	31-742	D-ICHY	Oilsearch Aviation	Gloucestershire	12.04
			F-BTCK, N7222L			
N250TB	Piper PA-23-250 Aztec D	27-4577	G-VHFA	Motor City Aviation LLC	Prestwick	7.04
			G-BZFE, G-AZFE, EI-BPA, G-AZFE, N13962	c/o Computaplane *(Stored)*		
N250TM	Beech 200 Super King Air	BB-822	F-GIND	Richard Lewis Aviation	Cranfield	12.04
			N3FH, N3844E			
N250TP	Beech A36TP Bonanza	E-2408	N416HC	Minster Enterprises Inc	Tatenhill	2.03
	(Allison 250-B17)		N600TT, N3107K			
N257SA	Piper PA-32-300 Cherokee Six	32-40755	OY-PCF	Southern Aircraft Consultancy Inc		
			OH-PCF		King's Farm, Thurrock.	2.05
N259SA	Cessna F172G	F172-0278	EI-BAO	Southern Aircraft Consultancy Inc	North Coates	9.04
	(Built Reims Aviation SA)		G-ATNH	*(On rebuild)*		
N260AP	SIAI-Marchetti SF.260D	839	I-ISAK	Pauls Airplane Inc	Old Sarum	12.04
N262J	SOCATA TBM-700	292		Sales Force Management Inc	Southend	2.05
N276SA	Brantly B.2B	474	G-AXSR	Southern Aircraft Consultancy Inc	(Stevenage)	12.04
			G-ROOF, G-AXSR, N2237U			
N277CD	Cessna 210L Centurion	21059663	SE-IGY	Bonner-Davies Aviation Inc	Headcorn	1.05
			N1163Q			
N277SA	Piper PA-28-140 Cherokee	28-21661	SE-EYG	Southern Aircraft Consultancy Inc	Blackpool	7.04
N278SA	Cessna 177RG Cardinal RG	177RG0571	OO-ALT	Southern Aircraft Consultancy Inc	Kemble	1.05
			N2171Q			
N280SA	Maule MX-7-180 Star Rocket	11070C	G-BSKT	Southern Aircraft Consultancy Inc	Barton	5.04
N282CJ	Cessna 525A CitationJet CJ2	525A-0082		CJ Airways *(Op C I Automobiles Ltd)*	Guernsey	1.05
N295S	Piper PA-46-350P Malibu Mirage	4636174	N295SS	Convergence Aviation	Biggin Hill	1.05
	(Jetprop DLX conversion)					

N297CJ	SNCASE SE.313B Alouette II	1847	F-GLPI	Southern Aircraft Consultancy Inc	Redhill	4.04
			(FAP9214), 77+00			
N297GT	SOCATA TB-21 Trinidad	2197		Blackbrook Aviation	Dunkeswell	2.05
N305RD	Mooney M20K	25-0844		Southern Aircraft Consultancy Inc	Exeter	11.04
N305SE	Mooney M.20K	25-0377	N231RH	Navajo Aviation	Deenethorpe	7.04
N309LJ*	Learjet Inc. Learjet 25	25-034		Not known *(Ground Instruction)*	Gloucestershire	1.05
N310QQ	Cessna 310Q	310Q0695	G-BAUE	Veryord Inc	Elstree	1.05
			N8048Q	*(Op H Gold)*		
N310WT	Cessna 310R II	310R1257	G-BGXK	Motor City Aviation	Perth	11.04
			N6070X			
N312CJ	Cessna 525A CitationJet CJ2	525A0031		JCT Aviation Inc	Ronaldsway	1.05
N320MR	Piper PA-30 Twin Comanche C	30-1917	G-CALV (2)	N320MR Inc	Elstree	12.04
	(Modified to PA-39 C/R status)		G-AZFO, N8761Y			
N321KL	Mooney M.20J (201)	24-1102	G-BPKL	Southern Aircraft Consultancy Inc	Stapleford	1.05
			N1008K			
N322MC	MD Helicopters MD 369E	0224E		AAA Flight Inc *(Op Jepar Rotorcraft)*	Blackpool	12.04
N322RJ	Beech 60 Duke	P-322		Aircraft Guaranty Title Corp	Leeds-Bradford	8.04
N324JC	Cessna 500 Citation I	500-0324	N52TC	Foxdale Aviation	Ronaldsway	12.04
			N324C, (N5324J)			
N324JS	SOCATA TBM-700	230		Flamingo 700 Inc	Luton	12.04
N328BX	Canadair CL-604 Challenger	5328	(A6-EJB)	Aircraft Guaranty Trust LLC	Biggin Hill	12.04
			ZS-AVL, N712DG, C-GLXH	*(Op London Executive Aviation)*		
N333UC	Schweizer S269D	0006	JA6140	CSE Aviation	Oxford	2.04
N337UK	Reims/Cessns F337G Skymaster	F33700084	G-BOWD	E-Plane Inc	Biggin Hill	2.05
			N337BC, G-BLSB, EI-BET, D-INAI, (N53697)			
N338DB	Piper PA-46-500TP Malibu Meridian	4697155	N53677	Oakfield Aviation	Jersey	1.05
N340DW	Cessna 340A-II	340A0497	G-BISJ	Southern Aircraft Consultancy Inc	Coventry	1.05
			OO-LFK, N6328X			
N340GJ	Cessna 340A	340A0637		Bee Bee Aviation	Elstree	1.05
N340SC	Cessna 340	340-0363		Hastingwood Management Inc	North Weald	1.05
N340YP	Cessna 340A II	340A0990	VR-CHR	ILEA Inc	Biggin Hill	1.05
			G-OCAN, D-ICIC, (N3970C)			
N345SF	Beech A36 Bonanza	E-2788		Southern Aircraft Consultancy Inc	Blackpool	1.05
N345TB	SOCATA TB-20 Trinidad	1914		Monty 345TB Llc	Biggin Hill	1.05
N346X	Maule M5-210C Strata Rocket	6156C		Emmanuel S Miserey	Biggin Hill	12.04
N350PB	Piper PA-31-350 Chieftain	31-8252028	HP-1309	PFB Self Drive Inc	Coventry	1.05
	(Panther II conversion with winglets)		N3548S			
N350UK	Aérospatiale AS350B Ecureuil	1244	F-GJYG	M W Helicopters Inc	(Brentwood)	1.05
N359DW	Piper PA-30 Twin Comanche	30-770	G-ATET	L W Durrell	Jersey	1.05
			N230ET			
N369AN	Cessna 182S	18280696		Air View Ltd	Jersey	1.05
N370SA	Piper PA-23-250 Aztec F	27-8054005	G-BKVN	Southern Aircraft Consultancy Inc		
			N6959A	*(Op B K Pugh)*	Guernsey/Southend	2.05
N372SA	Cessna 172RG Cutlass II	172RG0550	G-BHVC	Southern Aircraft Consultancy Inc		
			N5515V		Top Farm, Croydon, Royston	12.04
N375SA	Piper PA-34-200T Seneca II	34-7670002	G-BMWP	Southern Aircraft Consultancy Inc	Gamston	10.04
			N3946X			
N380CA	Piper PA-32R-301T Saratoga IITC	3257080		Continental Capital Aviation	Durham Tees Valley	10.04
N382AS	Reims/Cessna F182Q Skylane	F1820049	D-EAAF	Southern Aircraft Consultancy Inc	Barton	11.04
N382RW*	Vickers 361 Spitfire LFXVIe	CBAF.IX4640	G-XVIA	Airframe Assemblies	Sandown	6.04
			8075M, RW382, "RW729", 7245M, RW382 *(Crashed California 3.6.98 - wreck stored)*			
N395TC	Commander Aircraft Commander 114TC	20003		BNZ Aviation Inc	Bembridge	2.05
N400RG	Boeing 727-22	19149	N7085U	Wilmington Trust Co *(Op MBI Aviation)*	Lasham	1.05
N400YY	Extra EA400	019		Tamboti Aviation	Leeds-Bradford	9.04
N401S	Aérospatiale SA.341G Gazelle	1509		MW Helicopters	Stapleford	1.05
N409SA	Reims/Cessna FR182 Skylane RG	FR18200046	G-BJDI	Southern Aircraft Consultancy Inc	Leicester	11.04
			N8062H			
N410RS	Cessna 310R II	310R-1541	G-BTGN	R A Swick & Associates	Kirknewton	11.04
			N410RS, N5331C			
N412MD	Pilatus PC-12/45	412	HB-FSI	Senate Inc	Bournemouth	1.05
N414FZ	Cessna 414RAM	414-0175	G-AZFZ	Lizard Aviation Inc	Jersey	1.05
			N8245Q			
N418WS	Beech 58 Baron	TH-1967	N4467N	Millburn World Travel Services Two Inc	Edinburgh	1.05
				(Op W Scott & Partners Ltd)		
N421CA	Cessna 421C Golden Eagle III	421C0153	XA-RYC	Offshore Marine Inc	Ronaldsway	1.05
			N115JH, N5263J			
N421N	Cessna 421C Golden Eagle III	421C1235		Wix Aviation Services	Humberside	4.04
N423RS	Consolidated-Vultee PBY-5A Catalina	1785	C-FJJG	Southern Aircraft Consultancy Inc	Lee-on-Solent	12.04
			CF-JJG, N4002A, BuAer48423	*(Op Super Catalina Restoration)*		
				(As "JV828" of 210 Sqdn in RAF c/s)		
N425DR	Cessna 425 Conquest I	425-0199	VP-BDR	Intercity Co Inc	Booker	1.05
N425RR	Rockwell Commander 690A	11259	VP-BRR	Rami Aviation Inc *(Op Mann Aviation)*	Fairoaks	1.05
			SE-KYY, OY-BEO, SE-IYX, OY-BEO, N57090			
N425TV	Cessna 425 Corsair	425-0176	ZS-LDR	Conquest Aircraft Leasing	Aberdeen	9.02
			N6873T	*(Op Apex Tubulars Ltd)*		
N438DD	Cessna 310D	39278		Southern Aircraft Consultancy Inc	(Bourn)	1.05
N440GC	Piper PA-44-180T Turbo Seminole		G-GISO	Lucon Chasnais Flying Inc	Coventry	1.05
		44-8107065	D-GISO, N82112, N9602N			
N442BJ	Reims/Cessna F177RG Cardinal RG	F177RG0094	F-BVBC	Southern Aircraft Consultancy Inc	Seething	1.05
N448JC	Cessna 525 CitationJet	525-0448		Wells Fargo Bank Northwest NA	Bournemouth	1.05
N449TA	Piper PA-31 Turbo Navajo	31-480	G-CCRY	William L Shoufler	Elstree	1.05
			F-BTMM, N449TA			
N454CC	Bell UH-1E	6200	Bu155344	S W Firczak	Howth, Dublin	6.04
		(C/n 6199 quoted also)		*(Op Independent Helicopters Ltd)*		

Registration	Type	C/n	Previous identity	Owner/Operator	Base	Date
N456TL	Reims/Cessne FT337GP Super Skymaster	FP3370019	SX-PBA F-ODFY, F-BUDU	CCC Aviation	Coventry	9.04
N458BG	de Havilland DHC.1 Chipmunk 22	C1/0508	WG458	BG Chipmunks Inc *(As "WG458/2")*	Breighton	1.05
N460RB	Piper PA-34-220T Seneca III	34-8133177	G-BYKM HB-LMV	Tyler International School of Aviation *(Oxford Aviation Services c/s)*	Oxford	9.04
N473BS	Piper PA-28RT-201T Turbo Arrow IV	28R-8631003	G-BNYY N25WA, N77860, G-BNYY, N9129X, N9517N	Piper Arrow Inc *(Op I Jacobs t/a Chatham Glyn Fabrics)*	Southend	2.05
N476D	Pilatus PC-12/45	476		HPM Investments	Bournemouth	1.05
N480BB	Enstrom 480B	5056		Eastern Atlantic Helicopters	Shoreham	8.04
N480DS	Enstrom 480	5045		Hughston Aircraft Corp *(Heavy landing Droitwich 12.11.04 - "write off")*	Gloucestershire	6.04
N480E	Enstrom F480	5001	HB-XUX	Eastern Atlantic Helicopters	Shoreham	9.04
N480KP	Enstrom 480B	5053		Chartfleet Helicopters	(Fenland)	12.04
N484CJ	Cessna 525 CitationJet	525-0484	N5211F	Fegotila Inc	Gloucestershire	1.05
N485A	Enstrom F480	5029		Hughston Aircraft Corp	(Gloucestershire)	12.04
N485ED	Piper PA-23-250 Aztec C	27-3864	G-BAED N6567Y	Southern Aircraft Consultancy Inc	Waterford	1.05
N485LT	Hawker 800XP	258485	A7-AAL (HZ-KSRD), N44515	Surewings Inc *(Op Ambrion Aviation)*	Luton	1.05
N492AF	Piper PA-44-180 Seminole	4496170	G-CCDA	Saxon Aviation Inc	Norwich	12.04
N492PA	Beech B90 KingAir	LJ-492		Southern Aircraft Consultancy Inc	Bournemouth	3.04
N494AT	British Aerospace BAe 125 Srs.-800XP	258103 & NA0404		Vodaphone Americas Inc	Farnborough	12.04
N495AF	Piper PA-28-181 Archer III	2843466	G-SUEB N5330M	Saxon Aviation Inc	Norwich	1.05
N496DT	Pilatus PC-12/45	496		Normac Aviation	Lydd	1.05
N498YY	Cessna 525 CitationJet	525-0498	N5201J N498YY, N5223K	John Mills Aviation	Blackpool	1.05
N499MS	Piper PA-28-181 Archer III	2843166	G-EPJM N41268	MS Aviation	Jersey	1.05
N500AV	Piper PA-24-260 Comanche C	24-4805	OO-SAP	Southern Aircraft Consultancy Inc	Blackbushe	1.05
N500CS	Beech B200 Super King Air	BB-773	N83JE N3913U, OY-BEH	Wells Fargo Bank Northwest NA	Weston, Dublin	8.04
N500GV	Gulfstream Gulfstream V	506	N158AF N506GV	Wells Fargo Bank Northwest NA *(Op GAMA Aviation)*	Farnborough	1.05
N500LN	Howard 500 *(Lockheed PV-1 Ventura [5560] conversion)*	500-113	N381RD N206G, N200G, N539N, SAAF 6417, FP579, Bu.34670	Western Aviation Leasing Inc *(Op Baker Petroleum)*	Exeter	11.04
N500TY	MD Helicopters MD.369E	0086E	C-GRVV	Eastern Atlantic Helicopters	Shoreham	10.04
N500XV	Hughes 369D	120-0881D	OO-LVK OE-XBB, N190CA, N5293E, C-GHVK	Southern Aircraft Consultancy Inc	Blackpool	11.04
N502TC	Piper PA-30-160 Twin Comanche	30-881	G-BMSX N502TC, N7802Y	Southern Aircraft Consultancy Inc	Standalone Farm, Meppershall	11.03
N503BA	Beech B58 Baron	TH-862	EI-CPS G-BEUL	Leadbolt Inc	Gloucestershire	1.05
N510W	Bell 222B	47133	N7040Z	Wells Fargo Bank Northwest NA	Naas, County Kildare	7.04
N511TC	Cessna 525 CitationJet	525-0074	N26581	Eagle III LLC	Cambridge	11.04
N519MC	Piper PA-28-140 Cherokee Cruiser	28-7325519	G-BBID	R Lobell	Elstree	8.04
N524SF	Cessna 525 CitationJet	525-0240	N525GM	C P Lockyer Inc	Coventry	1.05
N525AL	Cessna 525 CitationJet	525-0011	(N1327N)	Jetco Holdings Inc	Guernsey	1.05
N525CM	Cessna 525 CitationJet	525-0093	I-IDAG N5151S	Wells Fargo Bank Northwest NA *(Op Ravenair)*	Liverpool	11.04
N525PM	Cessna 525A CitationJet CJ2	525A-0067	N5141F	Wells Fargo Bank Northwest NA	Oxford	12.04
N527EW	Cessna 501 Citation 1	501-0322	(N769EW) (N669DM), N314GS, N374GS, N2663J	Rockville Aero Inc	Jersey	1.05
N535CE	Cessna 560 Citation Ultra	560-0635		Latium Jet Services (IOM)	Gloucestershire	1.05
N554RB	Beech E55 Baron	TE-1141	G-BNRH N7855E	Rodney Badham Inc	Coventry	3.04
N555GS	Agusta A109E Power	11112	N109LF	TJH Helicopters	Blackpool	1.05
N556MA	Beagle B.121 Pup Series 1	B121-013	G-AWEB	George V Crowe	Norwich	12.04
N559C	Piper PA-34-220T Seneca V	3449238		Chiswell Aviation Inc	Fowlmere	9.04
N560S	Cessna 560XL Citation Excel	560-5190		Sisma Aviation	Jersey	1.05
N560TH	Cessna 560XL Citation Excel	560-5215	VP-CPC N5091J, N560TH	TJH Air Inc	Blackpool	12.04
N565F	Aérospatiale SA.341G Gazelle	1182		Transheli Inc	Fairoaks	1.05
N565G	SOCATA TB-20 Trinidad	2140		Ginsberg Aviation	Blackpool	1.05
N573VE	Cirrus Design SR22	1078		Flanes Ltd	Biggin Hill	1.05
N587PB	Beech C90B King Air	LJ-1408	N749RN N749RH	Air Montgomery *(Op Colin Montgomerie)*	Fairoaks	9.04
N590HM	Gulfstream Gulfstream IV	1153	N589HM N110LE, N448GA	Cirrus Gas 4 Llc	Luton	12.04
N598MT	Canadair CL-604 Challenger	5502	HB-JRY D-ACTU, D-ACTO, N502TF	Wells Fargo Bank Northwest NA *(Op Andaman Aviation)*	Northolt	11.04
N600MG	MD Helicopters MD.600N	RN-049	N3266A	Mentor Adi Recruitement Inc	Durham Tees Valley	1.05
N600PV	MD Helicopters MD.600N	RN-048		Southern Aircraft Consultancy Inc	Maryport, Cumbria	1.05
N600RN	MD Helicopters MD.600N	RN-015		Alan Smith Aviation	(Stonehouse, Gloucestershire)	11.04
N600SY	MD Helicopters MD.600N	RN-031	N9211F	Normac Aviation	Gloucestershire	11.04
N601AR	Piper Aerostar 601P	61P-0569-7963247	N3839H F-GKCL, N3839H, G-RACE, N8083J	Southern Aircraft Consultancy Inc *(Op The Fabric Factory)*	Jersey	1.05
N606AT	Cessna 650 Citation VI	650-0225	(N225CV) N1301Z	Longborough Aviation	Gloucestershire	1.05
N620LH	Aérospatiale AS355F Ecureuil 2	5463		MJD Aviation Inc	Leeds-Bradford	1.05
N620PL	Piper PA-32R-301 Saratoga II HP	3213078		Southern Aircraft Consultancy Inc	Lee-on-Solent	9.04
N629RS	Piper PA-44-180T Turbo Seminole	44-8207005		Richard J Schreiber	Ronaldsway	1.05

Registration	Type	C/n	Previous identity	Owner/operator	Base	Date
N637CG	Agusta A109C	7619	D-HARI	Castle Air Services Inc	Biggin Hill	1.05
			TC-HHI, D-HAAX, JA6608			
N638DB	Piper PA-46-350P Malibu Mirage	4636248		WA Aviation	Fowlmere	11.04
N642P	Piper PA-31 Turbo Navajo	31-761	N500UD	Universal Direct Inc	Southend	2.05
			G-EEAC, G-SKKA, G-FOAL, G-RMAE, N7239L *(Op JRB Aviation)*			
N646JR	Piper PA-32RT-300T Turbo Lance II	32R-7987019	PH-LFD	Southern Aircraft Consultancy Inc	Gloucestershire	1.05
			N3032A			
N652NR	Cessna 560 Citation Encore	560-0652		Cross Jet Inc	Kerry, County Kerry	1.05
N652P	Piper PA-18-150 Super Cub	18-7809098	G-BTDX	Southern Aircraft Consultancy Inc	(Ireland)	8.04
			N62595			
N656JM	Reims Cessna FR182 Skylane RGII	FR1820049	G-BHEO	JM Aviation Inc	Old Sarum	10.04
N661KK	Piper PA-28-181 Archer II	2890028	HB-PKN	Southern Aircraft Consultancy Inc	Fairoaks	1.05
			N9104F			
N665CH	Cessna 525 CitationJet	0504		Volante Aviation	Coventry	1.05
N666AW	Piper PA-31 Navajo C	31-7612061		Atlantic International Air Charter Inc	Biggin Hill	12.04
N666BE	Dassault Falcon 2000EX/EASy	032	F-WWGK	Wilmington Trust Co *(Op Bernie Ecclestone)*	Biggin Hill	1.05
N666BM	Aviat Pitts S-1T	1057		Not known, *"Devil Poo"*	Gamston	1.05
N666EX	Piper PA-32R-301T Saratoga II TC	3257241		Wells Fargo Bank Northwest NA *(N9666 reserved)*	Oxford	9.04
N666GA	Gulfstream AA-5B Tiger	AA5B-1136		Southern Aircraft Consultancy Inc	Thruxton	10.04
N669MM	Bellanca 8KCAB-180 Super Decathlon	825-99		Red Kite Inc	White Waltham	1.05
N671B	Raytheon A36 Bonanza	E-3409		N671B Inc	Ronaldsway	12.04
N672LE	Eurocopter EC155B1 Dauphin	6652	F-WQEP	HEC01 LLC *(Op TAG Aviation)*	Blackbushe	11.04
N674BW	Grumman AA-5A Cheetah	AA5A-0674	G-BXOO	Southern Aircraft Consultancy Inc	Denham	12.04
			N26721			
N700EL	SOCATA TBM-700	209	N701AR	Air Twinlite Inc	Weston	1.05
N700S	SOCATA TBM-700	193		Speedbird Aviation	Fairoaks	12.04
N700VA	SOCATA TBM-700	233	F-OIKI	Wells Fargo Bank Northwest NA	Biggin Hill	6.04
			(Overran runway into River Tay, Dundee 24.10.03) (Fuselage stored with Air Touring)			
N700VB	SOCATA TBM-700	237	F-OIKJ	ATM Leasing Inc	Biggin Hill	1.05
N702AR	SOCATA TBM-700	275		Isnet Aviation	Cambridge	11.04
N707LD	Piper PA-E23-250 Aztec C	27-2754	G-JANK	Southern Aircraft Consultancy Inc	Shoreham	11.04
			EI-BOO, G-ATCY, N5640Y	*(Op I A Qureshi in Keenair titles)*		
N707LG	Boeing 707-3M1C	21092	A-7002 (TNAIU)	Omega Air Inc *(Stored)*	Prestwick	1.05
			PK-GAU, A-7002, PK-PJQ, A-7002, PK-PJQ			
N707TJ	Boeing-Stearman A75N1 (N2S-1) Kaydet (Pratt & Whitney R-985 450hp)	75-950	N9PK	M G Plaskett *"Butterly"*	Rendcomb	9.04
			N50057, Bu.3173	*(Op V.S.E.Norman t/a Aerosuperbatics Ltd) (Utterly Butterly titles)*		
N708SP	Learjet Inc. Learjet 45	45-014		Tappetto Magico Inc	Cranfield	1.05
N709AT	Agusta A109E Power	11017	HB-XQM	Associated Technologies	Sywell	12.04
N709EL	Beech 400A Beechjet	RK-52	(N709EW)	GAL Air Inc	Nottingham East Midlands	1.05
			N709JB	*(Op DFS Furniture)*		
N711TL	Piper PA-60 Aerostar 700P	60-8423017	N700SX	Southern Aircraft Consultancy Inc	Biggin Hill	1.05
			N15GK, XB-EXQ, N6906Y			
N715BC	Beech A36 Bonanza	E-782	F-BXOZ	N715BC Inc	Denham	2.05
N719CD	Cirrus Design SR22	0051		Southern Aircraft Consultancy Inc	Exeter	11.04
N719CS	Piper PA-18S-135 Super Cub (L-21C)	18-3569	G-BWUC	Southern Aircraft Consultancy Inc	Cumbernauld	5.02
			SX-ASM, EI-181, I-EIYB, MM54-2369, 54-2369 *(Op Caledonian Seaplanes Ltd) (Under restoration)*			
N720B	Bell 206L-1 LongRanger II	45452	G-DALE	Omega Air Inc	Dublin	10.03
			G-HBUS			
N735BZ	Cessna 182Q Skylane	18265307		Luke Underhill	Sibson	12.04
N735CX	Cessna 182Q Skylane II *(Modified to Advanced Lift 260 STOL)*	18265329		Wilmington Trust Company *(Op B.Holmes)*	Barnard Farm, Thurrock	12.04
N737M	Boeing 737-8EQ	33361	N737SP	Wells Fargo Bank Northwest NA	Luton	1.05
			N737M			
N741CD	Cirrus Design SR22	0137		Southern Aircraft Consultancy Inc	Cambridge	1.05
N745HA	Agusta A109A-II	7413	G-CBDR	Helix Aviation	Prees, Shropshire	11.04
N747MM	Piper PA-28R-200 Arrow II	28R-7335445	PH-MLP	Southern Aircraft Consultancy Inc	Denham	1.05
			N56489			
N747WW	Piper PA-23-250 Aztec D	27-4330	G-AXOG	Southern Aircraft Consultancy Inc	Biggin Hill	12.04
			N6965Y			
N750NS	Cessna 750 Citation X	750-0172	N5066U	Sealpoint Aviation USA	Jersey	1.05
				(Departed 27.1.05 - to be replaced by AMD Falcon 900EX)		
N753RT	Hughes 369D	80-0753D	N111RS	Newtown Aviation	Waterford	1.05
			N1090V			
N754AM	Agusta A109A	7154	I-AVJJ	Capital Helicopters London Inc	Biggin Hill	10.04
			I-CELB, N33SV, N59340			
N758BK	Cessna R172K Hawk XP	R1722963		Eros Inc *(Current status unknown)*	Not known	8.01
N766AM	Aérospatiale AS355N Ecureuil 2	5601		Beacon Aviation *(Op Beacon Energy (Aviation) Ltd)*	Beacon Farm, Leicestershire	11.04
N767CM	Beech A36 Bonanza	E-2723	G-ORSP	Makins Aviation	Leeds-Bradford	1.05
			N56037			
N767CW	SOCATA TBM-700	96		High Sierra Inc *(Current status unknown)*	Biggin Hill	4.01
N770RM	SOCATA TB-9 Tampico	131	PH-CAG	Mary P Sandell	Ronaldsway	8.04
N771SC	Raytheon B200 Super King Air	BB-1693	N773TP	Wal-Mart Leasing	Leeds-Bradford	1.05
N773DC	Beech 58 Baron	TH-755	G-BDWK	DC Aviation Inc	Gamston	8.04
			(G-BEET)	*(Op DC Energy Ltd)*		
N777NG	Cessna 550 Citation Bravo	550-0992		Tazio Aviation Inc	Hawarden	12.04
NX793QG	Bell P-39Q Aircobra	26E-397	N139DP	Steven J Hinton	Duxford	9.04
			42-19993	*(Op The Fighter Collection) "219993" "Brooklyn Bum 2nd"*		
N799JH	Piper PA-28RT-201T Turbo Arrow IV	28R-8231051	HB-PNE	Southern Aircraft Consultancy Inc	North Weald	1.05
			PH-HJM, N8206B	*(Op J Havers)*		
N800C	Cirrus Design SR22	0367		Context GB Inc	Blackpool	9.04

N800HL	Bell 222	47054	N800HH	Yorkshire Helicopters USA Inc	Coney Park, Leeds	9.04
			N8140A, N37VA			
N800MG	Beech 200 Super King Air	BB-1259	D-IDSM	MGI Aviation	Cranfield	1.05
			N734P			
N800UK	Raytheon Hawker 800XP	258577	N51027	Wells Fargo Bank Northwest NA	Leeds-Bradford	1.05
				(Op Liberty Aviation)		
N800VM	Beech 76 Duchess	ME-318	G-BHGM	Southern Aircraft Consultancy Inc	Gloucestershire	12.04
N808NC	Gulfstream 695B Commander 1200	96085		Wilmington Trust Co	Gamston	11.03
				(Op Coopers Aerial Surveys)		
N816RL	Beech E90 King Air	LW-187	N66BP	Springair Inc	Gloucestershire	12.04
			N816EP, N900MH, N2187L	*(Op English Braids Ltd)*		
N818MJ	Piper PA-23-250 Aztec B	27-2486	G-ASNH	Retail Management Associates		
					Charlton Park, Malmesbury	12.04
N818Y	Piper PA-30 Twin Comanche B	30-1458	ZS-CAO	One Eight Yankee Aviation Inc	Guernsey	1.05
			ZS-EYB, A2-ZFE, ZS-EYB, VQ-ZIY, ZS-EYB, N8318Y			
N820CD	Cirrus Design SR22	0180		Southern Aircraft Consultancy Inc	Guernsey	1.05
N829CB	Cessna 550 Citation Bravo	550-0829	N5096S	Wells Fargo Bank Northwest NA	Blackpool	10.04
				(Op JJB Sports)		
N834CD	Cirrus Design SR22	0168		Southern Aircraft Consultancy Inc	Norwich	12.04
N836TP	Beech A36TP Bonanza	E-2124	N6770M	Hastingwood Aviation Inc	Tatenhill	10.04
				(Op Velcourt East plc)		
N840JC	Rockwell RC690C Turbo Commander	11643	(N5895K)	Semitool Inc	Cambridge	7.03
N840LE	Rockwell Commander 690C	11709	N690BA	Wells Fargo Bank Northwest NA	Guernsey	12.04
			ZS-KZM, N5961K	*(Op O.Henriksen)*		
N841WS	Hawker 800XP	258674	N674XP	Millburn World Travel Service Three Inc	Edinburgh	12.04
				(Op Walter Scott & Partners)		
N852CD	Cirrus Design SR22	0219		CS Aviation Europe	Biggin Hill	9.04
N852FT*	Boeing 747-122F	19757	N4712U	PIK Ltd *(Unmarked in fire service use)*	Prestwick	5.04
N853CD	Cirrus Design SR22	0348		Assegai Aviation	Denham	2.05
N866C	Cirrus Design SR22	0397		Aircraft Guaranty Trust	Turweston	1.05
N866LP	Piper PA-46-350P Malibu Mirage	4636130	N666LP	TLP Aviation Inc	Guernsey	1.05
			N92928			
N882	SOCATA TB-20 Trinidad GT	2161	F-OIMA	Harland Aviation	Ronaldsway	10.04
N882JH	Maule M.7-235B	23056C		Everbright Aviation Inc	Exeter	11.04
N883DP	Cessna R182 Skylane RGII	R18201883	G-GOZO	Southern Aircraft Consultancy Inc	Humberside	11.04
			G-BJZO, (G-BJYE), N5521T	*(Op D Pelling)*		
N888KT	Piper PA-60 Aerostar 601P	61P-0783-8063396	N9219Y	Southern Aircraft Consultancy Inc	Bournemouth	1.05
			OO-GMF, OO-HLV, N6072A			
N889VF	Cessna T303 Crusader	T30300102		Simply Living	Barton	1.05
N900CB	Cessna 421C Golden Eagle III	421C0837	VP-CPR	Southern Aircraft Consultancy Inc	Leeds-Bradford	8.04
			VR-CPR, N2659F	*(Op Fifty North)*		
N900RK	Mooney M.20J	24-3402		Romeo Kilo Flying Group	Turweston	12.04
N908W	Sikorsky S-92	920007		Laws Helicopter LLC *(Op Air Harrods)*	Stansted	12.04
N909PS	Cessna 501 Citation I/SP	501-0008	N900PS	Silversteel America Inc	Jersey	1.05
			(N501DB), N6HT, N362CC, N5362J			
N915TC	Aeronca 15AC Sedan	15AC-429	EI-ETC	Southern Aircraft Consultancy Inc	Athboy, County Meath	6.00
			G-CETC, HB-ETC	*(Op H.Moreau: current status unknown)*		
N916CD	Cirrus Design SR22	0318		Farnborough Aircraft Inc	Redhill	12.04
N918Y	Piper PA-30-160 Twin Comanche	30-736	ZS-NLH	Southern Aircraft Consultancy Inc	Shobdon	1.05
			N31RK, ZS-NLH, ZS-FZR, CR-AJR, ZS-EIU, VQ-ZIS, N7658Y			
N937BP	Mooney M.20J	24-3046	G-OOOO	Virginia Aircraft Trust Corp	Little Staughton	3.04
			N205EE			
N950H	Dassault Falcon 50EX	307		Island Aviation Inc	Farnborough	11.04
N951SF	Beech 56TC Baron	TG-83	N23PB	Timcar Inc	Elstree	1.05
N958SD	MD Helicopters MD.600N	RN-018		Eastern Atlantic Helicopters	Shoreham	11.04
N959JB	Piper PA-23-250 Aztec F	27-7754115	G-BSVP	Grand Motte Inc	Shoreham	11.04
			N63787	*(Op J Berry)*		
N966PR	Sikorsky S-76B	76-0354	N421MK	Accorp Helicopters	Dublin	10.04
N971RJ	Piper PA-39 Twin Comanche C/R	39-111	G-AZBC	Aircraft Guaranty Corporation	Biggin Hill	12.04
			N8951Y			
N973BB	Mitsubishi MU-2B-60 Marquise	1509SA		Romeo Aviation Inc	Jersey	1.05
N980HB	Rockwell Commander 695	95006		HBC Aviation Inc	Thurrock	10.04
N994K	Hughes 269A (TH-55A)	0840	67-16733	Southern Aircraft Consultancy Inc	Bournemouth	1.05
N997JB	Partenavia P.68C-TC	288-20-TC	F-GROG	Pangaea Air Service	Little Staughton	2.01
			HB-LSB, F-GEQD, N60CH, YV-2318P	*(Current status unknown)*		
N997JM	SOCATA TBM-700	244	LX-JFG	Blackbrook Aviation	Dunkeswell	1.05
			F-GLBQ			
N999F	Beech F33A Bonanza	CE-1282	OO-OVB	N T N Fox Systems Inc	Newcastle	10.04
N999MH	Cessna 195B	7168	OH-CSE	E Detiger	Compton Abbas	9.04
N1024L	Beech 60 Duke	P-78	C-FOPH	Southern Aircraft Consultancy Inc	North Weald	1.05
			CF-OPH, N1024L, CF-OPH	*(Op R.Ogden)*		
N1027G	Maule M.7-235B	23032C		Southern Aircraft Consultancy Inc *(Op T Clark)*	Elstree	2.05
N1092H	Beech C90A King Air	LJ-1454		Park Close Aviation Inc	Blackbushe	7.04
N1298C	Cirrus Design SR20	1315		Aircraft Guaranty Management and Trust	Denham	1.05
N1325M	Boeing Stearman E75(N2S-5) Kaydet	75-8484	Bu43390	Eastern Stearman Inc	Priory Farm, Tibenham	9.04
				(Op Blackbarn Aviation) (Frame only)		
NC1328	Fairchild F24R-46KS Argus	3310		Eastern Stearman Inc	Priory Farm, Tibenham	9.04
				(Op Blackbarn Aviation) (Frame only)		
N1344	Ryan PT-22-RY Recruit	2086	41-20877	Flying Heritage Inc	RAF Cosford	5.02
				(Op Mrs.H.Mitchell t/a PT Flight)		
N1350J	Rockwell Commander 112B	516		Southern Aircraft Consultancy Inc	Elmsett	6.04
				(Op G.Richards)		
N1363M	Boeing Stearman B75N1	75-7854		Pluto Inc	Priory Farm, Tibenham	9.04
				(Under rebuild - to be G-CCXB (As "996" in error - to be "699")		

N1407J	Rockwell Commander 112A	407		Blue Lake Aviation Inc	Blackbushe	12.04
N1551D	Cessna 190	7773		Southern Aircraft Consultancy Inc *(Under repair)*	East Winch	10.04
N1554E	Cessna 172N	17271044		Robert D Garretson	Studley Barton	8.04
N1711G	Cessna 340	340-0516		Paul Castle Inc	Cranfield	1.05
N1731B	Boeing A75N-1 Stearman	75-5716		Eastern Stearman Inc *(Stored)*	Priory Farm, Tibenham	9.04
N1745M	Cessna 182P Skylane II	18264424		David Thomas	Cardiff	11.04
N1778X	Cessna 210L Centurion	21060798		Central Investment Corporation	Booker	1.05
N1835W	Beech 95-B55 Baron	TC-1513	ZS-ING N1835W	Southern Aircraft Consultancy Inc	Bournemouth	11.04
N1937Z	Cessna 172RG Cutlass RG	172RG0908	EI-BVS	Virginia Aircraft Trust Corp	Ronaldsway	7.04
N1944A	Douglas DC-3C-47A-80-DL	19677	(N5211A) N3239W, RDan AF K-683, RNorAF, 43-15211	Wings Venture Ltd *(As "315211/JB-Z")*	Kemble	9.04
N2061K	Beech 58P Pressurised Baron	TJ-161		R Wagstaff	Oxford	6.04
N2121T	Gulfstream AA-5B Tiger	AA5B-1031		J.Siebols	Southend	2.05
N2195B	Piper PA-34-200T Seneca II	34-7970006		Papa Alpha Inc	Shoreham	9.04
N2216X	Cessna 337 Super Skymaster	3370116		Skymaster Air Services	Lee-on-Solent	8.04
N2273Q	Piper PA-28-181 Cherokee Archer II	28-7790389		Southern Aircraft Consultancy Inc	Panshanger	10.04
N2299L	Beech F33A Bonanza	CE-677		Stafford W Freeborn	Thruxton	10.04
N2326Y	Beech 58P Baron	TJ-83	F-GALL	Southern Aircraft Consultancy Inc	Gamston	2.05
N2341S	Raytheon B300 Super King Air	FL-241		Specsavers Aviation Inc	Guernsey	1.05
N2366D	Cessna 170B	20518		Southern Aircraft Consultancy Inc	Turweston	1.05
N2379C	Cessna R182 Skylane RG	R18200170		West Country Aviation Inc	(Berrow)	1.05
N2401Z	Piper PA-23-250 Aztec F	27-8054034		Pan Maritime Inc	Filton	8.04
N2405Y	Piper PA-28-181 Archer II	28-8590070		A D Bly Aviation	Panshanger	1.05
N2437B	Cessna 172S	172S8704		Southern Aircraft Consultancy Inc	Belle Vue Farm, Yarnscombe	11.04
N2454Y	Cessna 182S	18280918		Seima Aviation	Great Massingham	9.04
N2480X	Piper PA-31T1 Cheyenne I	31T-8104026		Jane Air	Southampton	2.02
N2548T	Navion Model H Rangemaster	NAV-4-2548		Navion Airways Inc	Guernsey	1.05
NC2612	Stinson Junior R	8754		A.L.Young *(Stored)*	Henstridge	4.04
N2652P	Piper PA-22-135 Tri-Pacer	22-2992		Southern Aircraft Consultancy Inc *(Op Anne Lait) "Jeff Jeff"*	Not known	8.01
				(Flipped over after abortive take-off 23.8.02 Hacketstown Airstrip, CountyCarlow: current status unknown)		
N2923N	Piper PA-32-300 Cherokee Six	32-7940207		Southern Aircraft Consultancy Inc	Jersey	1.05
N2929W	Piper PA-28-151 Cherokee Warrior	28-7415457	OO-GPE N9619N	Funair Inc *(Op R.Lobell)*	Elstree	2.05
N2943D	Piper PA-28RT-201 Arrow IV	28R-7918231	G-BSLD N2943D	Southern Aircraft Consultancy Inc *(Op E.Gawronek)*	Barton	10.04
N2967N	Piper PA-32-300 Six	32-7940242		Aerotechnics Aviation Inc	Guernsey	1.05
N2975K*	Luscombe 8E	5702		Not known *(for UK registration)*	Weston Zoyland	10.01
				(Crashed landing 19.2.83 Jackson, MS: cancelled 11.92 by FAA) (Current status unknown)		
N3023W	Beech V35B Bonanza	D-9517		M A Sargent	Cambridge	10.04
N3044B	Piper PA-34-200T Seneca II	34-7970012		Aerotechnics Aviation Inc	Alderney	11.04
N3053R	Piper PA-32R-301 Saratoga II HP	3246217		Meridian Aircraft Ferrying	Bournemouth	1.05
N3109X	Cessna 150F	15064509		Southern Aircraft Consultancy Inc *(Op The Lucy Flying Group)*	Kilrush	8.04
N3456Q	Cessna 340	340-0004		Cari Miller	Filton	12.04
N3586D	Piper PA-31-325 Navajo C/R	31-8012065		L. W.Durrell	Jersey	9.04
N3669D	Beech 60 Duke	P-544	G-CBYK N3669D	Vetch Aviation	Goodwood	9.04
N3864	Ryan Navion B	NAV-4-2285B		Southern Aircraft Consultancy Inc *(Under rebuild by G Spooner)*	(Kelvedon)	11.03
N3922B	Boeing-Stearman E75 (PT-17) Kaydet (Continental W670)	75-5805	42-17642	Eastern Stearman Inc *(Op Peter Hoffmann)*	Priory Farm, Tibenham	8.04
N3995W	Piper PA-32-260 Cherokee Six	32-963		L Major	Bournemouth	5.00
				(Crashed Le Rignolent, France 31.5.98) (Fuselage stored: current status unknown)		
N4085E	Piper PA-18-150 Super Cub	18-7809059		R N Hall	Goodwood	11.03
N4102D	Reims/Cessna FR182 Skylane RG	FR18200029	PH-CTM SE-IBB	Alpine Air Leasing	Top Farm, Croydon, Royston	12.04
N4168D	Piper PA-34-220T Seneca V	3449158		AAL Inc	Shoreham	12.04
N4173T	Cessna 320D Skyknight	320D0073		N4173T Inc *(Op J.Irwin)*	Cranfield	1.05
N4178W	Piper PA-32R-301T Saratoga IITC	3257178		Mistress Two Inc	Blackbushe	1.05
N4232Y	Reima Cessna F150G	F1500098	D-EBYW	F Acevedo	Stapleford	3.03
N4238C	Mudry CAP.10B	155		Southern Aircraft Consultancy Inc *(As "52" in Mexican AF c/s)*	Rotary Farm, Hatch	9.04
N4306Z	Piper PA-28-161 Warrior II	28-8316073		Thomas Stuer Aviation Inc *(Op USAF Flying Club)*	Stapleford	10.02
N4337K	Cessna 150K	15071583	G-BTSA N6083G	T L Crook	Coleman Green	11.04
N4422P	Piper PA-23-160 Geronimo	23-1936		W J Armstrong Inc	Thruxton	10.04
N4500L	Piper PA-31-350 Chieftain	31-8052155		Rosefly Inc	Cambridge	1.05
N4514X	Piper PA-28-181 Cherokee Archer II	28-7690027		Baxter 4514X Holdings Inc	Elstree	1.05
NC4531H	Piper PA-15 Vagabond	15-305		E A Terris	Oxford	9.04
N4575C	Grumman G.21A Goose	B-120		Aerofloat G21A Inc	Weston, Dublin	8.04
N4596N	Boeing-Stearman E75 (PT-13D) Kaydet (Lycoming R680-7)	75-5945	42-17782	Phil Dacy Aviation *(US Mail c/s)*	Shoreham	11.04
N4599W	Rockwell Commander 112TC	13089		Skyfast Inc	Haverfordwest	6.04
N4698W	Rockwell Commander 112TC-A	13274		Syston Aviation Inc I	Denham	1.05
N4712V	Boeing Stearman PT-13D Kaydet	75-5094	42-16931	Southern Aircraft Consultancy Inc	Lower Botrea Farm, Sancreed	8.04
N4770B	Cessna 152	15283626		Walkwitz Aviation	Panshanger	10.04
N4806E	Douglas B-26C Invader	27451	44-34172	A26 Europe Inc *(Stored Hull Aero)*	(Norwich)	8.02

Registration	Type	C/n	Previous identity	Owner/Operator	Location	Date
N5050	Klemm Kl.35D	1979		Lars A G de Jounge	Fairoaks	10.04
N5052P	Piper PA-24-180 Comanche	24-56	G-ATFS	T A G Randell	Farley Farm, Romsey	10.02
			N5052P	*(Under restoration)*		
N5057V	Boeing-Stearman PT-13D Kaydet	75-5598	42-17435	M G Plaskett *"Utterly"*	Rendcomb	9.04
				(Op V.S.E.Norman) (Utterly Butterly titles)		
N5084V	Cirrus Design SR22-G2	0831		BPAD USA Inc	Elstree	12.04
N5120	Bell 430	49095		Wells Fargo Bank Northwest NA	Blackpool	12.04
				(Op JJB Sports)		
N5144Q	McDonnell Douglas MD.369E	0007E		0454E Inc	Shoreham	12.04
N5180Y	Piper PA-23-250 Aztec B	27-2226		Belair Inc *(Current status unknown)*	Glasgow	4.01
N5240H	Piper PA-16 Clipper	16-44		Southern Aircraft Consultancy Inc		
				(Op D.Hillier)	Wellcross Grange, Slinfold	8.04
N5264Q	MD Helicopters MD.369E	0126E		Trafficopters Inc	Donegal	4.04
N5277T	Piper PA-32-260 Cherokee Six	32-7200031		K R Denman	Goodwood	7.04
N5315V	Hiller UH-12C	757		Southern Aircraft Consultancy Inc	Lower Upham	11.04
N5317V	Hiller UH-12C	768		Canaan Helicopters	Sywell	9.04
				(Op Pulse Helicopters)		
N5336Z	Cirrus Design SR20	1413		Southern Aircraft Consultancy Inc	Guernsey	1.05
N5428C	Cessna 170A	19462		Southern Aircraft Consultancy Inc	Audley End	9.04
				(Op P.Norman)		
N5632R	Maule M-5-235C Lunar Rocket	7244C		Southern Aircraft Consultancy Inc		
				(Op RD Group)	Stowes Farm, Tillingham	10.04
N5647S	Maule M-5-235C Rocket	7345C		Virginia Aircraft Trust Corp		
					Yeatsall Farm, Abbotts Bromley	9.03
N5675Z*	Piper PA-22-108 Colt	22-9501		Not known	Kilrush, County.Clare	5.00
				(Cancelled 11.99 by FAA: current status unknown)		
N5730H	Piper PA-16 Clipper	16-342		Southern Aircraft Consultancy Inc		
					Yeatsall Farm, Abbotts Bromley	7.04
N5736	Raytheon Hawker 800XP	258471	N43642	Wells Fargo Bank Northwest NA	Leeds-Bradford	12.04
				(Op A Ogden & Sons Plc)		
N5834N*	Rockwell Commander 114	14383		W F Chmura	(Cardiff)	11.04
				(Crashed 23.10.98 - cancelled 7.99 by FAA - hulk only)		
N5839P	Piper PA-24-180 Comanche	24-920		Aircraft Guaranty Management LLC	(Blackbushe)	12.04
N5880T	Westland WG-30-100	009	G-17-31	Offshore Fire and Survival Training Centre	Norwich	8.04
N5900H	Piper PA-16 Clipper	16-520		Southern Aircraft Consultancy Inc	Shenstone	8.02
				(Damaged landing Saint Girons, France 25.4.03)		
N5915V	Piper PA-28-161 Cherokee Warrior II	28-7716215		Southern Aircraft Consultancy Inc	North Weald	1.05
N5927G*	Cessna 150K	15071427		Not known	Plaistowes Farm, St Albans	8.04
				(Cancelled 4.89 by FAA: stored)		
N6010Y	Commander Aircraft Commander 114B	14589		Camrose Inc	Biggin Hill	1.05
N6039X	Commander Aircraft Commander 114B	14639		Little Beetle Inc	Guernsey	1.05
N6095A	Commander Aircraft Commander 114B	14635		Bonbois Aviation	Guernsey	1.05
N6130X	Maule M6-235C Super Rocket	7497C		Pelmont Aviation	Chilbolton	10.04
N6182G	Cessna 172N Skyhawk II	17273576		Southern Aircraft Consultancy Inc	Duxford	9.04
N6240V	Beech V35A Bonanza	D-8628		Not known	Biggin Hill	1.05
N6302W	Government Aircraft Factory N22B Nomad	F-159	VH-HWB	Chatteris Aviation Inc	Lower Mountpleasant, Chatteris	10.04
				(Op London Parachute Centre)		
N6339U	Piper PA-28-236 Dakota	28-8011089	OO-JFD	Southern Aircraft Consultancy Inc	Wickenby	12.04
			F-GCMU, OO-HLM, N8152S)			
N6438C	Stinson L-5C Sentinel	1428		Eastern Stearman Inc	Priory Farm, Tibenham	9.04
				(Op Paul Bennett and Nik Nice: under rebuild)		
N6498V	Cessna T303 Crusader	T30300313	G-CRUS	Southern Aircraft Consultancy Inc	Guernsey	1.05
			N6498V			
N6593W	Cessna P210N Centurion	P210-00801		Southern Aircraft Consultancy Inc *"Rose Anne"*		
				(Op Pacific Network Air)	Brittas House, Limerick	1.05
N6601Y	Piper PA-23-250 Aztec C	27-3905	XC-DAZ	3 Greens Aviation	Blackpool	12.02
			N6601Y			
N6602Y	Piper PA-28-140 Cherokee	28-21943	G-ATTG	Southern Aircraft Consultancy Inc	Norwich	9.03
			N11C			
N6632L	Beech C23 Musketeer	M-2188		W J Forrest	Enstone	4.03
N6690D	Piper PA-18-135 Super Cub	18-3848	PH-KNK	6688 Delta Inc	Netherley	6.00
				(Op S Gruver: current status unknown)		
N6819F	Cessna 150F	15063419		W J Davis *(On fire dump)*	Shoreham	5.03
N6830B	Piper PA-22-150 Tri-pacer	22-4128		Adam J Wynne	Leicester	11.04
N6907E	Cessna 175A Skylark	56407		Southern Aircraft Consultancy Inc	Popham	8.04
				(Op Colin Webb)		
N6954J	Piper PA-32R-300 Cherokee Lance	32R-7680394		Matrix Aviation Inc	Norwich	12.04
N7027E	Hawker Tempest V	---	EJ693	K Weeks *(Stored)*	Booker	6.03
N7070A	Cessna S550 Citation II	S550-0068	N4049	Omega Air Inc	Dublin	1.05
			N404G, N1272Z			
N7098V	North American TF-51D Mustang	122-40411	IDF/AF	Mustang Air Inc	(Greenham Common)	2.04
			RCAF9245, 44-73871	*(Stored)*		
N7148R	Beech B55 Baron	TC-2028	N2198L	Air Services Holdings Corp	Guernsey	10.04
			C-GWFD, N2198L, D-IGRW, N2198L			
N7205T	Beech A36 Bonanza	E-2182		Minster Enterprises	Tatenhill	9.04
N7214Y	Beech A36 Bonanza	E-2169		Whitsunday Inc	Fairoaks	12.04
N7219L	Beech B55 Baron	TC-717	HB-GBX	Southern Aircraft Consultancy Inc	Kemble	1.05
			OE-FDF, HB-GBX			
N7242N	Agusta A109A-II	7355	ZK-HJC	Cannon Air	Liskeard	12.04
N7263S	Cessna 150H	15067963		Cesna Inc *(Stored)*	Plaistowes Farm, St Albans	11.04
N7348P	Piper PA-24-250 Comanche	24-2526		Southern Aircraft Consultancy Inc	Netherthorpe	8.04
				(Op J.Bown)		
N7374A	Cessna A150M Aerobat 135	A1500726		J A Thomas	Branscombe	8.04
	(Tail-wheel conversion)			*"Turnin' Tricks"*		

N7423V	Mooney M.20E Chapparal	21-1163		Southern Aircraft Consultancy Inc	Hinton-in-the-Hedges	1.05
N7456P	Piper PA-24-250 Comanche	24-2646		Southern Aircraft Consultancy Inc	Gamston	8.04
N7640F	Piper PA-32R-300 Cherokee Lance	32R-7780069	ZS-OGX N7640F	Southern Aircraft Consultancy Inc	Wellesbourne Mountford	12.04
N7801R	Bell 47G-5	7801		Skyman Logistics	Beckley, Oxon	8.04
N7813M	Piper PA-28-180 Cherokee D	28-5227	G-AZYF 5Y-AJK, N7813N	Southern Aircraft Consultancy Inc	Leicester	1.05
N7832P	Piper PA-24-250 Comanche	24-3052		Three Two Papa Inc	White Waltham	6.04
N7976Y	Piper PA-30 Twin Comanche B	30-1075		Southern Aircraft Consultancy Inc	Guernsey	1.05
N8027U	Europa Aviation Europa XL	A048		Nathan E McGuire	Kemble	1.05
N8153E	Piper PA-28RT-201T Turbo Arrow IV	28R-8131185	N9561N N84205	P B Payne	RAF Mona	12.04
N8159Q	Cirrus Design SR20	1388		Lisa Hall	Cambridge	1.05
N8241Z	Piper PA-28-161 Warrior II	28-8316079		Pett Air Inc	Henstridge	1.05
N8258F	Beech B36TC Bonanza	EA-513		Millfore Aviation Inc *(Damaged in aborted takeoff 23.2.03 Tregorras Farm, Llangorren)*	Elstree	2.03
N8403Y	Piper PA-30-160B Twin Comanche	30-1552	4X-AVY N8403Y	D Yoshpe	Elstree	1.05
N8412B	Piper PA-28RT-201T Turbo Arrow IV	28R-8131164		Sweepline Inc	Bourne Park, Hurstbourne Tarrant	10.04
N8471Y	Piper PA-28-236 Dakota	28-8211019		Turbo Arrow Inc	Elstree	1.05
N8754J	Aviat A-1 Husky *(Built Christen Industries)*	1160		Southern Aircraft Consultancy Inc *(Op A.Febrache)*	Guernsey	1.05
N8862V	Bellanca 17-31ATC Turbo Super Viking	31022		Scott B Barber	Popham	10.04
N8911Y	Piper PA-39 Twin Comanche C/R	39-66	G-AYFT N8911Y	Southern Aircraft Consultancy Inc	Farley Farm, Romsey	1.05
N9122N	Piper PA-46-310P Malibu	4608097		Libra Air Inc	Oxford	5.04
N9123X	Piper PA-32R-301T Turbo Saratoga SP	3229003		Vector Sky Service Inc	Shoreham/Hilversum	11.04
N9146N	Cessna 401B *(RAM conversion)*	401B0010		A J Air Ltd Inc *(In derelict condition)*	Weston, Dublin	8.04
N9201U	MD Helicopters MD.900	900-00042		MD Helicopters Inc	Shoreham	1.02
N9208V	MD Helicopters MD.900	900-00010		WA Helicopters	Hexham	12.04
N9325N	Piper PA-28R-200 Cherokee Arrow *(Lopresti version)*	28R-35025		Southern Aircraft Consultancy Inc *(Op Hiam Mercado)*	Panshanger	1.05
N9381P	Piper PA-24-260 Comanche C	24-4882		Southern Aircraft Consultancy Inc	Elstree	1.05
N9533Y	Cessna T.210N Centurion	21064539		Simply Living Ltd	Liverpool	7.04
N9727G	Cessna 180H	18052227	G-FESC N9727G	Wessex Air Services	Thruxton	10.04
N9861M	Maule M.4-210C	1058C		Southern Aircraft Consultancy Inc	Fairoaks	1.05
N9950	Curtiss P-40N Warhawk	33723	44-7983	Ice Strike Corp *(Stored in container)*	(Greenham Common)	2.04
N11824	Cessna 150L	15075652		Southern Aircraft Consultancy Inc *(Stored)*	Manor Farm, Glatton	7.04
N13253	Cessna 172M	17262613		Anglia Aviation Inc *(Stored)*	Plaistowes Farm, St Albans	11.04
N14113	North American T-28B Trojan	174-398	Haiti AF 1236 N14113, FrAF 119, 51-7545	Radial Revelations Ltd *(As "119" inAdlA c/s) "Little Rascal"*	Duxford	9.04
N15750	Beech D.18S	A-850	G-ATUM D-IANA, N20S	S Quinto *(Nose only preserved)*	Strathdon	2002
NC16403*	Cessna C.34 Airmaster	322		Alan House *(Cancelled 5.99 by FAA: stored)*	Lower Wasing Farm, Brimpton	8.04
NC18028	Beech D17S Staggerwing	147	NC18028	P.H.McConnell	Popham	10.04
N19753	Cessna 172L	17260723		Aircraft Guaranty Title and Trust Co	Elstree	1.05
N21381	Piper PA-34-200 Seneca	34-7350274	F-BUTM F-ETAL	Tickton Inc	Dunkeswell	9.04
N23659	Beech B58 Baron	TH-893		S W Freeborn	Guernsey	1.05
N24136	Beech A36 Bonanza	E-1233		Dickens Aviation Inc	North Weald	1.05
N25644	North American B-25D Mitchell	100-20644	G-BYDR N88972, CF-OGQ, KL161, 43-3318	Vulcan Warbirds Inc *(As "KL161/VO-B" in 98 Sqn c/s) "Grumpy"*	North Weald	1.05
N26634	Piper PA-24-250 Comanche	24-3551	G-BFKR PH-BUS, D-ELPY, N8306P	N26634 Inc *(Op Graham Lees)*	Southend	1.05
N29566	Piper PA-28RT-201 Arrow IV	28R-7918146	D-EJLH	Southern Aircraft Consultancy Inc	Denham	2.05
N30593	Cessna 210L Centurion	21059938		Southern Aircraft Consultancy Inc	Cranfield	1.05
N30614	Piper PA-32-301FT 6x	3232021		Senate Inc	Bournemouth	1.05
N31356	Douglas DC-4-1009	42914	C-FTAW EL-ADR, N6404	Aces High US Inc *(As "44-42914")*	North Weald	1.05
N33145	Piper PA-34-200T Seneca II	34-7570095	Z-PTJ G-BPTJ, N33145	Southern Aircraft Consultancy	Goodwood	2.04
N33514	Hiller UH-12B	661	51-16408	Pulse Helicopter Corp	Sywell	9.04
N33870	Fairchild M62A (PT-19-FA) Cornell	T40-237	G-BTNY N33870, US Army	Ice Strike Corp *(Op R.M Lamplough as "02538" in US Army c/s)*	(Greenham Common)	2.04
NC33884	Aeronca 65CA Chief	CA.14101		Southern Aircraft Consultancy Inc	Bodmin	8.04
N36362	Cessna 180 Skywagon	31691	G-BHVZ F-BHMU, N4739B	Southern Aircraft Consultancy Inc *(Op W Burgess)*	Oxford	11.04
N38273	Piper PA-28R-201 Cherokee Arrow III	28R-7737086		S W Freeborn *(Op L.Slater)*	Blackbushe	1.05
N38940	Boeing-Stearman A75N1 (PT-17) Kaydet (Continental R670)	75-1822	(G-BSNK) N38940, N55300, 41-8263	Eastern Stearman Inc *(Op Paul Bennett as "18263/822" in US Army c/s)*	Priory Farm, Tibenham	9.04
N38945	Piper PA-32R-300 Cherokee Lance	32R-7780490		Southern Aircraft Consultancy Inc	Panshanger	11.04
N39132	Piper PA-38-112 Tomahawk II	38-82A0065	(G-NCFD) D-EIIS, N2477V	Not known *(Derelict: current status unknown)*	Swanton Morley	7.01
N39605	Piper PA-34-200T Seneca	34-7870397		Heliquick Aviation *(Op Direct Air)*	Compton Abbas	2.05
N41098	Cessna 421B Golden Eagle	421B0448		M C I Inc	Elstree	1.05
N44914	Douglas C-54D Skymaster	10630	Bu56498 42-72525	Aces High US Inc *(As "56498")*	North Weald	1.05

N47914	Piper PA-32-300 Six	32-7840018		Southern Aircraft Consultancy Inc	Alderney	1.05
N49272	Fairchild M.62/PT-23-HO Cornell	HO-437	42-.....	Flying Heritage Inc	RAF Cosford	5.02
	(Continental W670)			*(Op R.E.Mitchell t/a PT Flight as "23" in USAAC c/s)*		
N50029	Cessna 172	28807	LX-AIB	Southern Aircraft Consultancy Inc	Exeter	11.04
			N6707A	*(Op E.Byrd)*		
N52485	Boeing-Stearman A75N1 (PT-17) Kaydet	75-4494		Roland Stearman Aviation	Rendcomb	9.04
				(Op V S E Norman as "169" in US Navy c/s)		
N53103	Cessna 177RG Cardinal RG	177RG1347		Vectis Aircraft	Sandown	9.04
N53497	Piper PA-34-220T Seneca V	3449279		Jet 1 Services Inc	Bournemouth	1.05
N54211	Piper PA-23-250 Aztec E	27-7554006	G-ITTU	Southern Aircraft Consultancy Inc	Elstree	1.05
			D-IKLW, G-BCSW, N54211			
N54922	Boeing-Stearman A75N1 (N2S-4) Kaydet	75-3491	Bu.30054	M G Plaskett *"Yum"*	Rendcomb	9.04
	(Pratt & Witney 985-14B)			*(Op V.S.E.Norman) (Utterly Butterly titles)*		
N56421	Ryan PT-22-RY Recruit	1539	41-15510	Flying Heritage Inc	RAF Cosford	5.02
				(Op R.E.Mitchell t/a PT Flight as "855" in US Army c/s)		
N56462	Maule M.6-235 Rocket	7409C		Avocet (US) Inc	Old Buckenham	8.04
N56608	Boeing-Stearman A75N1 Kaydet	75-2888		Southern Aircraft Consultancy Inc *"17"*	Leicester	6.03
N56643	Maule M.5-180C	8086C		Southern Aircraft Consultancy Inc	Duxford	9.04
N57783	Stinson L-5 Sentinel	76-511	42-98270 ?	J F Tillman	Priory Farm, Tibenham	8.04
				(Frame - spares for N6438C)		
N58566	Consolidated-Vultee BT-15-VN Valiant	10670	42-41882	Flying Heritage Inc	RAF Cosford	5.02
				(Op R.E.Mitchell t/a PT Flight in US Army c/s)		
N59269	Boeing Stearman A75L3	75-3867		R W Hightower *"817"*	Stock	7.04
N60256	Beech C35 Bonanza	D-3346	OO-DOL	R.M.Hornblower	Southend	1.05
			OO-JAN	*(Damaged at Southend 28.1.05)*		
N60526	Beech E55 Baron	TE-1159		E Walsh	Elstree	12.04
N61787	Piper J-3C-65 Cub	13624	45-4884	Gypsy Fliers	Draycott Farm, Chiseldon	1.05
				(As "54884/57-D")		
N61970	Piper PA-24-250 Comanche	24-3364	OO-GOE	Southern Aircraft Consultancy Inc	Gamston	8.04
			F-OCBM, 5R-MVA, N8198P, N10F	*(Op Nunn & Green)*		
N62171	Hiller Felt UH-12C	GF-2		Southern Aircraft Consultancy Inc	Tuam, Galway	12.04
N62842	Boeing-Stearman PT-17 Kaydet	75-3851		Eastern Stearman Inc	Priory Farm, Tibenham	9.04
				(Frame stored)		
N63560	Piper PA-31 Turbo Navajo	31-188	HB-LFW	Southern Aircraft Consultancy Inc	Norwich	12.04
			N9141Y	*(Op Earl Aviation)*		
N63590	Boeing-Stearman N2S-3 Kaydet	75-7143	Bu.07539	Eastern Stearman Inc	Brock Farm, Brentwood	1.05
				(Op Ian Stockwell as "07539/143" in US Navy c/s)		
N65200	Boeing-Stearman D75N1 Kaydet	75-3817	FJ767	Eastern Stearman Inc	Bagby	9.04
N65565	Boeing Stearman B75N1 Kaydet	75-7463		Ortac Inc	Sywell	1.05
N67548	Cessna 152	15281906		Southern Aircraft Consultancy Inc	Norwich	12.04
N68427	Boeing-Stearman A75N1 (N2S-4) Kaydet	75-5008	Bu 55771	Eastern Stearman Inc	Priory Farm, Tibenham	9.04
				(Op Blackbarn Aviation) (Frame only)		
N70154	Piper J-3C-65 Cub	17139		M G Plaskett *(Current status unknown)*	Rendcomb	7.01
N70526	MD Helicopters MD.369E	0556E		Eastern Atlantic Helicopters	Shoreham	10.04
N71763	Cessna 180K	18053191		China Pilot Inc	(Newmarket)	9.04
N72127	Cessna U206D Skywagon	U2061368	G-AXJY	R D Garretson	Hill Farm, Nayland	9.01
	(Robertson STOL conversion)		N72326	*(Current status unknown)*		
N74189	Boeing Stearman PT-17	75-717		M G Plaskett	Rendcomb	9.04
				(Op Aerosuperbatics Ltd) (Utterly Butterly titles)		
N75048	Piper PA-28-181 Archer II	28-7690286		Southern Aircraft Consultancy Inc	Bodmin	6.04
N75822	Cessna 172N	17267979		Hercmar Inc	Crowfield	1.05
N76402*	Cessna 140	10828	NC76402	(C.Murgatroyd) *(Wreck stored)*	Blackpool	12.02
				(W/O near Meppershall 9.8.98 : cancelled 4.99 by FAA)		
N80533	Cessna 172M Skyhawk	17266640		Southern Aircraft Consultancy Inc	Alderney	10.02
N82507	Piper PA-28RT-201 Arrow IV	28R-8018100		David V Christofferson	Southend	1.05
N84718	Piper PA-28RT-201T Turbo Arrow IV	28R-8231013		Southern Aircraft Consultancy Inc		
					Nottingham East Midlands	12.04
N90704	Grumman AA-5A Cheetah	AA5A-0301	LX-AVC	Aerocolor Inc	Turweston	1.05
N90724	Hiller UH-12C	810	55-4106	Southern Aircraft Consultancy Inc		
					Lower Botrea Farm, Sancreed	11.03
N91384	Rockwell Commander 690A	11118	SE-FLN	Airbourne Data Inc	Gamston	1.05
N92001	MD Helicopters MD.900	900-00040		Latium Helicopter Charters (UK) Inc	(Swettenham)	11.04
N92562	Piper PA-46-350P Malibu Mirage	4636010		Trevair Inc	Ronaldsway	12.04
N93938	Erco 415C Ercoupe	1261		Merkado Holdings	Panshangar	10.04
N96240	Beech D18S (3TM)	CA-159	G-AYAH	M L & C E Edwards	North Weald	1.05
			N6123, RCAF 1559			
N97121*	Embraer EMB-110P1 Bandeirante	110.334	PT-SDK	Guernsey Airport Fire Service *(Hulk only)*	Guernsey	1.05
N97821	Mooney M.20J	24-1080		Southern Aircraft Consultancy Inc	Panshanger	1.05

CZECH REPUBLIC

OK-DUU 15	Urban Air UFM-11 Lambada	3/11		M.Tormey	Abbeyshrule	8.03
OK-EUU 56	Urban Air UFM-11 Lambada	12/11		F Mayhan	Abbeyshrule	9.04
OK-FUA 05	Urban Air UFM-11 Lambada	13/11	OK-EUU-02	T Mackey	Waterford	7.04
OK-FUU 31	Urban Air UFM-10 Samba	03/10		Not known	Abbeyshrule	8.04
OK-GUA 16	Urban Air UFM-10 Samba	NK		Not known	Abbeyshrule	8.03
OK-GUA 19	Urban Air UFM-10 Samba	NK		Not known	Abbeyshrule	9.04
OK-GUA 24	Urban Air UFM-10 Samba	N/K		Not known	Abbeyshrule	8.04
OK-GUA 27	Urban Air UFM-10 Samba	N/K		Not known	Kilrush	6.04
OK-GUA 28	Urban Air UFM-10 Samba	21/10		Not known	Abbeyshrule	8.04

BELGIUM

OO-DHN*	Boeing 727-31	20113	N260NE	European Air Transport/DHL	Lasham	12.04
			N97891	*(Stored)*		

OO-DHR*	Boeing 727-35F	19834	N932FT	European Air Transport/DHL	Lasham	12.04
			(N526FE), N932FT, N1958	*(Stored)*		
OO-MEL*	Focke-Wulf Piaggio FWP.149D	113	90+93	Not known	Draycott Farm, Chiseldon	1.05
			AC+470, JC+394, AS+428	*(Stored dismantled)*		
OO-MHB*	Piper PA-28-236 Dakota	28-8011143	G-BMHB	R W H Watson	Blackpool	12.02
			D6-PAD, N81321, N9593N	*(Damaged Southend 20.10.90: wreck stored)*		
OO-NAT	SOCATA MS.880B Rallye Club	2253	G-BAOK	R W H Watson	Grimmet Farm, Maybole	12.02
				(Fuselage stored)		
OO-RLD	Miles M.65 Gemini 1A	6285	G-AISD	Not known	Spanhoe	8.04
			OO-RLD, (OO-PRD), VP-KDH, G-AISD			
OO-WIO*	Reims/Cessna FRA150L Aerobat	FRA1500183	F-BUMG	Department of Engineering, Salford University	Salford	8.03
				(Instructional Airframe)		

DENMARK

OY-AVT	Piper PA-18 Super Cub 95	18-3202	D-ELFT	Henrik Burkal	Breighton	1.05
			OL-L05, L-128, 53-4802			
OY-BNM*	Embraer EMB.110P2 Bandeirante	110.200	N5071N	Air Salvage International	Alton	7.00
			G-BFZK, PT-GLS	*(Current status unknown)*		
OY-BTZ	Piper PA-31-350 Navajo Chieftain	31-7752031	SE-GPM	Company Flight K/S	Nottingham East Midlands	1.05
OY-DRS	Reims/Cessna F172K	F17200786	LN-LJY	Ulf Odlund *(Stored dismantled)*	Old Buckenham	8.04
OY-EGZ	Cessna F172H	F172-0324	N17013	Peter and Christen Bak	Alderney	1.05
OY-JRR	de Havilland DHC.2 Turbo Beaver III	1632/TB-18	N911CC	Ulrich Steen	Headcorn	11.04
			C-FUKK, CF-UKK	*"Black Beaver"*		
OY-JRW*	GAF N-24A Nomad	117	VH-KNA	Nomad Ap S	RAF Weston on the Green	4.02
			ZK-ECM, VH-PGW, (N415NE), VH-PGW, (VH-AUR)			
				(Damaged on take off 13.4.02: cancelled 1.04)		
OY-LGI	Learjet Inc. Learjet 60	60-243	N50287	Graff Aviation	Luton	1.05
OY-MUB*	Short SD.3-30 Var.200	SH.3069	G-BITX	Air Salvage International	Alton	1.05
			G-14-3069	*(Fuselage only)*		
OY-NMH	GAF N-24A Nomad	74FA	ZK-NMH	Airlog	Peterlee	6.04
			N870US, PH-HAG, (PH-DHL), N5579K, VH-PNF			
OY-PBH	LET L-410UVP-E20	972736	OK-DDC	Alebco Corporation	Aberdeen/Sumburgh	12.04
				(Op Benair A/S)		
OY-PBI	LET L-410UVP-E20	871936	OK-SDM	Alebco Corporation	Aberdeen/Sumburgh	12.04
			Sov.AF 1936	(Op Benair A/S)		

The NETHERLANDS

PH-DUC	Stoddard-Hamilton Glasair IIRG-S	2069	N51DA	R S van Dijk	Little Gransden	1.05
PH-END	Bölkow Bö.208A Junior	515	D-ENDA	M Palfreman	Bagby	9.04
			VH-UES, D-ENDA			
PH-MLB	American General AG-5B Tiger	10141		R K Hyatt	Perranporth	1.05
PH-NLK*	Piper PA-23-160 Apache	23-1694	OY-DCG	Not known	Water Leisure Park, Skegness	4.04
			SE-CKW, N10F	*(Wreck stored for spares)*		
PH-PAB	Neico Lancair 360	766		D C Ratcliffe	Shoreham	7.04
PH-ZZY	SOCATA MS893E Rallye 180GT	12074	OO-AON	P E Mooser	Trevethoe Farm, Lelant	11.03
			F-GACN			
PH-3P3	WDL Fascination D4BK	106		R Simpson	(Longhope, Gloucestershire)	5.04

ARUBA

P4-LJG	Cessna 750 Citation X	750-0228	N5267J	Venair	Dublin	1.05

RUSSIA

FLARF01035	Yakovlev Yak-52	8910106	LY-AIG	Not known	Haverfordwest	6.04
			Ukraine AF 23 yellow			
RA-01274	Yakovlev Yak-55	910103	DOSAAF 03	Not known *"03"*	Wolverhampton	8.04
RA-01370	Yakovlev Yak-18T	Not known		F.M.Govern	Old Sarum	8.01
			(Crashed 24.5.01 one mile south Old Sarum: wreck stored: current status unknown)			
RA-01813	Yakovlev Yak-52	Not known		Not known *"13"*	Sleap	12.04
RA-02933	Yakovlev Yak-18T	22202040425	LY-AOO	W Marriott	Wickenby	7.03
			LY-AOG	*"2"*		
RA-22521	Yakovlev Yak-52	9211612	DOSAAF 04	D.Squires *"04"*	Wellesbourne Mountford	11.04
RA-44476	Yakovlev Yak-52	Not known		Not known	Little Gransden	1.05
	(Also reported as Yak-50)					
RA-44515	Yakovlev Yak-52	9111515	DOSAAF 56	M.Stebbing *"65"*	Poplar Hall Farm, Elmsett	2.03

SWEDEN

SE-BRG*	Fairey Firefly TT.1	F.6071	DT989	ARCO *(Stored dismantled)*	Duxford	9.04
SE-CAU*	Fairey Firefly TT.1	F.6180	PP469	ARCO *(Stored dismantled)*	Duxford	9.04
SE-CEE	Piper PA-18-150 Super Cub	18-5700		T Spurge	Great Oakley	10.04
SE-ETR	Cessna 150F	15062718	N11B	Marham Aero Club	RAF Marham	10.04
			N8618G			
SE-GPU	Piper PA-28-161 Cherokee Warrior II	28-7716200		Not known	East Winch	10.04
SE-GVH	Piper PA-38-112 Tomahawk	38-78A0053		Not known *(Fuselage only)*	Little Staughton	9.03
SE-HXF*	Rotorway Scorpion	SE-1		Not known *(Stored)*	Earls Colne	2.04
SE-IIV	Piper PA-24-260 Comanche C	24-4970	HB-OHZ	Not known	Gamston	8.04
			N9462P			
SE-KBU	Christen A-1 Husky	1038		A Allan	Perth	11.04

POLAND

SP-CHD*	PZL-101A Gawron	74134		T Wood ? *(Stored)*	North Weald	1.05

SUDAN

ST-AHZ	Piper PA-31 Turbo Navajo	31-473	G-AXMR N6558L	Not known *(Fire practice: burnt-out fuselage remains)*	Elstree	10.03

GREECE

SX-BFM*	Piper PA-31-350 Chieftain	31-8052204	N4504J	Not known *(Fuselage stored unmarked and derelict)*	Bournemouth	1.05
SX-BLW	Boeing 757-236	24397	G-OOOS G-BRJD, EC-ESC, EC-349, G-BRJD	Greece Airways *(Op Air Scotland)*	Glasgow	1.05
SX-BNL	Embraer EMB.110P2 Bandeirante	110224	N614KC PT-GMQ	Not known *("Euroair" titles)*	Kemble	1.05
SX-BTA	Cessna 340A	340A0105	C-GTJT N5405J	GT Aviation	Bournemouth	1.05
SX-HCF	Agusta A109A-II	7207	N71PT N4263A	Castle Air Charters Ltd *(Spares use: current status unknown)*	Liskeard	6.01

SLOVENIA

S5-HPC*	Agusta A.109A	7129	SL-HPC	Castle Helicopters *(Slovenian Police titles)*	Liskeard	9.04

TURKEY

TC-ALM*	Boeing 727-230	20431	TC-IKO TC-JUH, TC-ALB, N878UM, D-ABDI	Fire Services *(Trainer)*	Nottingham East Midlands	1.05

ICELAND

TF-ELL	Boeing 737-210C	20138	N41026 F-GGFI, N4906	Ardennes Epsilon Ltd *(Noted in ATA Brasil c/s) (To become PR-CMA)*	Southend	2.05

UKRAINE

UR-67439	LET L-410UVP Turbolet	841204		Not known *(Universal Avia c/s)*	Headcorn	9.04
UR-67477	LET L-410UVP Turbolet	841302	CCCP-67477	Not known *(Universal Avia c/s)*	Sibson	9.04

AUSTRALIA

VH-BGH	de Havilland DH.82A Tiger Moth	915/T230	VH-BGA A17-492	Oscheat P/L *(Dismantled)*	Coventry	9.04
VH-KLN	Gippsland GA-8 Airvan	GA8-04-046		Gippsland Aeronautics Pty Ltd	Oxford	11.04
19-3431	Jodel D18	SAAA 177		L Usherwood *(Dismantled)*	Headcorn	1.05

BERMUDA (current series)

VP-BAA	Boeing 727-51	19123	N727AK TP-05, TP-01, N477US	Marbyia Investments	Lasham	1.05
VP-BAB	Boeing 727-76	19254	N682G N10XY, N8043B, VH-TJD, (N8043B), VH-TJD	Marbyia Investments	Lasham	2.05
VP-BAT	Boeing 747SP-21	21648	VR-BAT N148UA, N539PA	Worldwide Aircraft Holding (Bermuda)	Bournemouth	1.05
VP-BBT	Boeing 737-705	29089	LN-TUB (LN-SUB)	Interlocutory Ltd *(Op Ford Motor Co)*	Stansted	1.05
VP-BBU	Boeing 737-705	29090	LN-TUC (LN-SUC)	Interlocutory Ltd *(Op Ford Motor Co)*	Stansted	1.05
VP-BBW	Boeing 737-7BJ	30076	N737BF P4-CZT, VP-CZT, N737MC, D-AXXL, N374MC, N1784B, N1786B	Altitude 41 Ltd *(Op GAMA Aviation)*	Farnborough	2.05
VP-BCC	Canadair CL600-2B19 *(CRJ200)*	7717	C-GZSQ	CCC (Bermuda) Inc	Farnborough/Athens	1.05
VP-BCE	Eurocopter AS355N Ecureuil 2	5663		Sioux Corporation	Jersey/Athens	1.05
VP-BCI (2)	Canadair CL-600-2B19 *(CRJ-200)*	7351	N351BA ZS-OGH, N351EJ, C-FMLF	CCC (Bermuda) Ltd	Farnborough/Athens	11.04
VP-BCT	Rockwell 695B Turbo Commander	96208	N695BE VH-PJC, VH-LTM, N230GA	Control Techniques (Bermuda) Ltd	Welshpool	12.04
VP-BDL	Dassault Falcon 2000	111	F-WWVF	Sioux Corporation	Luton	1.05
VP-BGE	Cessna 500 Citation I	500-0287	N287AB PT-WHZ, N31LH, OY-CGO, N57MB, N73LL, N287CC, (N5287J)	Ross Aviation	Filton	1.05
VP-BGN	Gulfstream G-550	5011	N991GA (N522QS)	Rockfield Holdings	Luton	1.05
VP-BGO	Canadair CL-604 Challenger	5404	C-GLYO	Sun International Management Ltd	Farnborough	1.05
VP-BHB	British Aerospace BAe 125 Srs 800A	258221 & NA0473	N25WN N25W, N675BA, G-5-731	Freestream Aircraft (Bermuda) Ltd *(Op GAMA Aviation)*	Farnborough	10.04
VP-BHS	Canadair CL-604 Challenger	5505	N505JD C-GLYK	CCC (Bermuda) Ltd	Farnborough/Athens	11.04
VP-BIE	Canadair CL601 Challenger 1A	3016	N601CL N1107Z, N4562Q, C-GLWV	Inflite Aviation	Stansted	10.04
VP-BJT	Cessna 425 Corsair	425-0027	VP-BNM VR-BNM, N181AA, HI-598SP, N97DA, (N711EF), N97DA, N67720	Rig Design Services Group Ltd	Booker	1.05
VP-BKH	Gulfstream Gulfstream IV	1029	VP-BKI VR-BKI, N429GA	Specialised Transportation	Ronaldsway	1.05
VP-BKK	Hawker Siddeley HS.125-Srs.400A/731	25238	VR-BKK N808V, N125GC, G-TOPF, G-AYER, 9K-ACR, G-AYER	Air 125 Ltd/Business Real Estates	Bournemouth	12.04
VP-BKQ	Bell 430	49008	N62833	Arkesden Aviation	Blackbushe	12.04
VP-BKZ	Gulfstream Gulfstream V	602	N602GV N538GA	Dennis Vanguard International (Switchgear)	Birmingham	1.05
VP-BLA (2)	Gulfstream G-550	5024	N924GA	ISPAT Group	Luton	1.05
VP-BLD	Lockheed 329-731 Jetstar	5117/35	(N858SH) VP-BSH, VR-BSH, N310CK, N210EK, N7962S *(Stored)*	Magnair Ltd	Luton	1.05

Reg	Type	C/n	Previous identity	Owner/operator	Base	Date
VP-BLS	Pilatus PC-XII	176	N176BS VP-BLS, HB-FSL	B.L.Schroeder	Fairoaks	1.05
VP-BMD	British Aerospace BAe 125 Srs.700B	257200	VR-BMD G-MSFY, G-5-14	AirVIP	Hawarden	6.04
VP-BMJ	Dassault Falcon 2000EX	1	F-WMEX	Ormond Ltd/Theburton Inc	Stansted	1.05
VP-BMZ	Rockwell Turbo Commander 690D *(Built Gulfstream Aerospace)*	15033	VR-BMZ G-MFAL, N49GA, (N5925N)	Aviatica Trading Co Ltd/Marlborough Fine Art Ltd	Fairoaks	11.04
VP-BNL	Gulfstream Gulfstream V	607	N303K N559GA	Nebula Ltd *(Op GAMA Aviation)*	Farnborough	1.05
VP-BNM	Sikorsky S-76B	760333	N595JS N595ST, N5AY	Nebula II Ltd)	Blackbushe	1.05
VP-BNN	Gulfstream Gulfstream IVSP	1255	N600PM N437GA	Nebula III Ltd *(Op GAMA Aviation)*	Farnborough	1.05
VP-BPS*	Consolidated 28-5ACF (PBY-5A) Catalina	1997	VR-BPS G-BLSC, C-FMIR, N608FF, CF-MIR, N10023, Bu.46633	PS (Bermuda) Ltd *(On rebuild)* *(Op Super Catalina Restoration)*	Lee-on-Solent	10.04
VP-BPW	Dassault Falcon 900B	135	VR-BPW F-WWFJ	Service Aviation	Southampton	1.05
VP-BRD	Eurocopter EC120B	1155	F-WQDK	Specialised Transportation Ltd *(Current status unknown)*	Redhill	7.01
VP-BUS	Gulfstream Gulfstream IV	1127	VR-BUS VR-BLR, N427GA	A E C International Ltd *(Op U Schwarzenbach)*	Farnborough	1.05
VP-BVT	Gulfstream Gulfstream IV	1419	EI-CVT N419GA	A C Executive Aircraft Bermuda Ltd *(Op Jet Club)*	Farnborough	1.05

CAYMAN ISLANDS

Reg	Type	C/n	Previous identity	Owner/operator	Base	Date
VP-CAT	Cessna 501 Citation 1	501-0232	VR-CAT VR-CHF, N35TL, N853KB, N2616C, (N2616G)	Kestrel Aviation/Aviation Jet	Guernsey	1.05
VP-CBM	Cessna 550 Citation II	550-0729	VR-CBM N1210V	Bernard Matthews plc	Norwich	12.04
VP-CBW	Gulfstream Gulfstream IV	1096	VR-CBW (G-...), N17589	TAG Aviation	Farnborough	1.05
VP-CBX	Gulfstream Gulfstream V	511	N511GA	Aravco	Farnborough	1.05
VP-CCO	Cessna 550 Citation II	550-0321	N321GN TC-COY, N321SE, N5430G	Not known	Biggin Hill	12.04
VP-CEB	Bombardier BD.700 Global Express	9083	C-GKLF N700AU, C-GHYT	Silver Arrows SA	Luton	1.05
VP-CED	Cessna 550 Citation Bravo	550-0870	N50612	Iceland Foods	Hawarden	1.05
VP-CEZ	Dassault Falcon 900EX	134	F-WWFL	J. Dyson	Filton	9.04
VP-CFG	Cessna 501 Citation I/SP	501-0176	VR-CFG (VR-CIA), N49LC, N44LC, N6779L	Alpha Golf Aviation Ltd	Biggin Hill	12.04
VP-CFT	Canadair CL-601-3A Challenger	5067	HB-IUF N220TW, 9A-CRT, 9A-CRO, N603CC, C-GLXF	Not known	Farnborough	11.04
VP-CGE	Cessna 650 Citation VII	650-7077	(N582JF) N532JF, N877CM, N5203J	Not known *(Op Grosvenor Estates)*	Hawarden	2.05
VP-CIC	Canadair CL601 Challenger	5011	VR-CIC N602UK, N611MH, JA8283, N603CC, C-GLXD	TGC Aviation Ltd/Fakhar Ltd	Stansted	1.05
VP-CJR	Cessna 550 Citation II	550-0354	VR-CJR N121C, N121CG	Broome and Wellington (Aviation) Ltd	Manchester	11.04
VP-CKA	Boeing 727-82	20489	VR-CKA N727FH, N727KS, N46793, CS-TBP	Samco Aviation *(Op Executive Air Transport)*	Lasham	1.05
VP-CLA	Beech F90	LA-231	N27PA N7220T	Claessons International Ltd	Farnborough	12.04
VP-CLD	Cessna 550 Citation II	550-0323	N323AM TC-YZB, TC-FMB, TC-FAL, OE-GCP, (N5703C)	Pan Maritime Ltd *(Op Dovey Aviation)*	Filton	1.05
VP-CME	Boeing 767-231ER	22567	N604TW	Sheikh Mustafa Edrees *(Stored)*	Filton	1.05
VP-CNF	Cessna 525 CitationJet	525-0153	(N525EF) N551Q, N551G, N5090V	Foster Aviation	Biggin Hill	12.04
VP-CNP	Gulfstream G1159A Gulfstream III	496	N843HS (N99SU), N99SC, N89AB, N89AE, N21NY, N310SL, N327GA	Fitzwilton plc	Dublin	1.05
VP-COM	Cessna 500 Citation I	500-318	VR-COM N944B, N518CC, N5318J	Rapid 3864 Ltd	Biggin Hill	1.05
VP-CPT	British Aerospace BAe 125 Srs.1000B	259004	VR-CPT G-LRBJ, G-5-779	Reno Investments Inc *(Op Avtec)*	Biggin Hill	1.05
VP-CRB	Learjet Inc. Learjet 60	60-125	N60LR	Lisane Ltd	Guernsey	1.05
VP-CSF	Gulfstream IV	1390	N1874M N490GA	MSF Aviation	Luton	1.05
VP-CSN	Cessna 560 Citation Ultra	560-0401	N401CV N5197A	Scottish & Newcastle Breweries Ltd	Edinburgh	11.04
VP-CTJ	Cessna 550 Citation II	F550-0073	F-GBTL N4621G	Flight Consultancy Services	Biggin Hill	10.04
VP-CVL	Learjet Inc. Learjet 45	45-059	N50153	Czar Aviation	Biggin Hill	1.05
VP-CWA	Agusta A109C	7628	JA6610	Williams Grand Prix Engineering Ltd *(Op Alan Mann Helicopters)*	Oxford	7.04

BERMUDA (old series)

Reg	Type	C/n	Previous identity	Owner/operator	Base	Date
VR-BEB*	British Aircraft Corporation One-Eleven 527FK	BAC.226	RP-C1181 PI-C1181	European Aviation Ltd *(Fire Compound - all white and no marks)*	Bournemouth	1.05

INDIA

Reg	Type	C/n	Previous identity	Owner/operator	Base	Date
VT-EKE*	Westland WG.30-160	021	G-BLPR G-17-17	Turbine World *(Current status unknown)*	Honeycrock Farm, Redhill	2000
VT-EKK*	Westland WG.30-160	025	G-17-13	Turbine World *(Current status unknown)*	Honeycrock Farm, Redhill	2000

Reg	Type	C/n	Previous identity	Owner/operator	Location	Date
VT-EKL*	Westland WG.30-160	028	G-17-14	Turbine World	Honeycrock Farm, Redhill	2000
				(Current status unknown)		
VT-EKM*	Westland WG.30-160	027	G-17-15	Turbine World	Honeycrock Farm, Redhill	2000
				(Current status unknown)		
VT-EKT*	Westland WG.30-160	035	G-17-23	Turbine World	Honeycrock Farm, Redhill	2000
				(Current status unknown)		
VT-EKW*	Westland WG.30-160	038	G-17-26	Turbine World	Honeycrock Farm, Redhill	2000
				(Current status unknown)		
VT-EKX*	Westland WG.30-160	039	G-17-27	Turbine World	Honeycrock Farm, Redhill	2000
				(Current status unknown)		

MEXICO

Reg	Type	C/n	Previous identity	Owner/operator	Location	Date
XA-TLJ	Boeing 737-2H6	20926	PK-IJE	European Aviation	Bournemouth	10.04
			9M-MBH, 9M-ASR	*(Scrapping commenced 9.04)*		
XB-RIY	Boeing- Stearman N2S-3 Kaydet	75-7275	N52093	Not known	Rendcomb	4.04
			BuA 07671	*(Stored - composite)*		

LATVIA

Reg	Type	C/n	Previous identity	Owner/operator	Location	Date
YL-CBJ*	Yakovlev Yak-52	790404	DOSAAF	Hawarden Air Services *"20 blue"*	Hawarden	1.05
YL-LEU*	WSK-PZL Antonov An-2R	1G-165-45	CCCP19731	Hawarden Air Services	Hawarden	1.05
			SP-ZFP, CCCP19731	*(As "CCCP-19731": dismantled)*		
YL-LEV*	WSK-PZL Antonov An-2R	1G-148-29	CCCP07268	Hawarden Air Services	Hawarden	7.04
				(As "CCCP07268")		
YL-LEW*	WSK-PZL Antonov An-2R	1G-182-28	CCCP56471	Hawarden Air Services	Hawarden	1.05
				(As "CCCP56471")		
YL-LEX*	WSK-PZL Antonov An-2R	1G-187-58	CCCP54949	Hawarden Air Services	Hawarden	1.05
				(As "CCCP54949")		
YL-LEY*	WSK-PZL Antonov An-2R	1G-173-11	CCCP40784	Hawarden Air Services	Hawarden	1.05
				(As "CCCP40784")		
YL-LEZ*	WSK-PZL Antonov An-2R	1G-165-47	CCCP19733	Hawarden Air Services	Hawarden	1.05
				(As "CCCP19733")		
YL-LFA*	WSK-PZL Antonov An-2R	1G-172-20	CCCP40748	Hawarden Air Services	Hawarden	1.05
				(As "CCCP40748")		
YL-LFB*	WSK-PZL Antonov An-2R	1G-173-12	CCCP40785	Hawarden Air Services	Hawarden	1.05
				(As "CCCP40785")		
YL-LFC*	WSK-PZL Antonov An-2R	1G-206-44	CCCP17939	Hawarden Air Services	Hawarden	1.05
				(As "CCCP17939")		
YL-LFD*	WSK-PZL Antonov An-2R	1G-172-21	CCCP40749	Hawarden Air Services	Hawarden	1.05
				(As "CCCP40749")		
YL-LHN*	Mil Mi-2	524006025	CCCP20320	Hawarden Air Services	Hawarden	1.05
				(As "CCCP20320")		
YL-LHO*	Mil Mi-2	535025126	CCCP20619	Hawarden Air Services	Hawarden	1.05
				(As "CCCP20619")		

NICARAGUA

Reg	Type	C/n	Previous identity	Owner/operator	Location	Date
YN-CCN	Boeing 707-123B	18054	5B-DAO	Omega Air	Shannon	4.03
			G-BGCT, N7526A	*(Stored)*		

REPUBLIC OF SERBIA AND MONTENEGRO

Reg	Type	C/n	Previous identity	Owner/operator	Location	Date
YU-DLG	UTVA 66	0812 ?	JRV51109	Shuttle Air	Biggin Hill	11.04
YU-HEH	Soko/Aérospatiale SA.341G Gazelle	011	JRV12619	Kestrel Shipping	Stapleford	10.04
YU-HEI	Soko/Aérospatiale SA.341G Gazelle	012	JRV12620	Not known	Stapleford	1.05
YU-HES	Aérospatiale SA.342J Gazelle	1057	F-GOSO	MW Helicopters	Stapleford	1.05
			EC-EQU, C-GEJE, (N341NA), N9042U, C-FGCE, CF-GCE			
YU-YAB	SOKO G-2A Galeb	Not known	JRV23170	Shuttle Air	North Weald	1.05

NEW ZEALAND

Reg	Type	C/n	Previous identity	Owner/operator	Location	Date
ZK-AGM*	de Havilland DH.83 Fox Moth	"4085"		Not known	Denford Manor, Hungerford	12.03
				(Unmarked fuselage)		
	(Composite of parts of ZK-ADH c/n 4085 with fuselage produced by de Havilland Technical School apprentices: provenance uncertain as ZK-AGM "destroyed by fire" 27.4.63 after crash near Wanaka)					
ZK-BMI*	Auster B.8 Agricola Srs.1	B.101	(G-ANYG)	D.J.Baker	Carr Farm, Thorney, Newark	1.05
			G-25-3	*(Rear fuselage frame - new fin constructed 2004)*		
ZK-CCU*	Auster B.8 Agricola	B.105	ZK-BMM	D J Baker	Carr Farm, Thorney, Newark	1.05
	(Composite rebuild by Airepair as c/n AIRP/850 after crash)			*(Centre section stored in container)*		
ZK-GIL	Schempp-Hirth Discus 2a	41		B Flewett *"2a"*	Booker	6.04
ZK-KAY	Pacific Aerospace PAC 750XL	107		Pacific Aerospace Corporation	Cark	1.05
				(Op North West Parachute Centre)		

REPUBLIC of SOUTH AFRICA

Reg	Type	C/n	Previous identity	Owner/operator	Location	Date
ZS-MBI	Rockwell Commander 114	14361		F W A Engelbrecht	Oxford	1.05
ZS-RSI	Lockheed L-100-30	4600	F-GIMV	Safair	Shannon	9.03
			ZS-RSI, F-GDAQ, F-WDAQ, ZS-RSI, C-FNWY, ZS-RSI *(Op Air Contractors)*			

EQUATORIAL GUINEA

Reg	Type	C/n	Previous identity	Owner/operator	Location	Date
3C-QSB	Fokker Friendship 200MAR	10612	M-1	Trygon	Southend	2.05
			PH-EXC	*(Stored engineless)*		
3C-QSC	Fokker Friendship 200MAR	10622	M-2	Trygon	Southend	2.05
			PH-FSI, PH-EXD	*(Stored engineless)*		

CYPRUS

5B-CJV	Beech 58 Baron	TH-1341	N6342U	Not known	Southend	1.05

NIGERIA

5N-AAN*	British Aerospace BAe 125 Srs.F3B/RA	25125	F-GFMP G-AVAI, LN-NPA, G-AVAI	Newcastle Aviation Academy *(Instructional use)*	Newcastle	9.04
5N-AJT	Bell 212	30636	G-BCLG EP-HBY, VR-BFK, G-BCLG, 9M-ATV, VR-BFK, G-BCLG, N18091 *(Stored for spares)*	Bristow Helicopters Ltd	Redhill	1.04
5N-AJU	Bell 212	30632	G-BFDJ EP-HCA, VR-BGP, G-BFDJ, 9V-BGE, B-2309, 9V-BGE *(Stored for spares)*	Bristow Helicopters Ltd	Redhill	1.04
5N-AJV	Bell 212	30868	G-BGMK EP-HCC, VR-BGR, N18096	Bristow Helicopters Ltd *(Stored for spares)*	Redhill	1.04
5N-AJW	Bell 212	30601	G-BGML EP-HBL, VR-BEX	Bristow Helicopters Ltd *(Stored as hulk for spares)*	Redhill	1.04
5N-AQV	Bell 212	30782	G-BERF VR-BIJ, G-BERF, 9Y-TFW, G-BERF, N9925K, N18092 *(Stored)*	Bristow Helicopters	Redhill	1.04
5N-AWD	Hawker Siddeley HS.125 Srs.1	25008	G-ASSI	Not known) *(On fire dump: current status unknown)*	Luton	10.00
5N-BHN	Bell 212	32135	G-BJJP N5736D	Bristow Helicopters Ltd *(Stored as hulk for spares)*	Redhill	1.04
5N-HHH	British Aircraft Corporation One-Eleven 401AK	BAC.064	HZ-NB2 N5024	Airport Fire Service *(Rescue training as "G-FIRE")*	Southend	2.05

MALAGASY REPUBLIC

5R-MFT	Boeing 747-2B2M	21614		MK Airlines *(Spares)*	Filton	11.04

MALAWI

7Q-YDF	Piper J3C-65 Cub	18711	5Y-KEV VP-KEV, VP-NAE, ZS-AZT	Not known *(Stored)*	Pent Farm, Postling	4.04

GHANA

9G-BOB	Westland Wessex HC.2	WA/624	G-HANA XV729	Not known *(Africa Gateway titles)*	Honeycrock Farm, Redhill	11.03
9G-LCA	Conroy CL-44-O	16	(P4-GUP) 4K-GUP, EI-BND, N447T	Johnsons Air *(Stored)*	Bournemouth	1.05
9G-MKA	Douglas DC-8F-55	45804	N855BC CX-BLN, C-GMXP, N855BC, HP-927, PH-DCZ, OY-KTC *(Stored)*	MK Airlines	Manston	1.05

ZAMBIA

9J-RBC	Piper PA-28-140 Cherokee	28-20693	N11C	Not known	Biggin Hill	1.05

SIERRA LEONE

9L-LDA	Beech A90 King Air	LJ-281	RP-C3318	Air Leone *(In open store)*	Southend	2.05
9L-LSA	Sud Aviation SA330L Puma	1506	Chilean Army H261	Not known *"301" (Stored)*	Kemble	1.05
9L-LSG	Sud Aviation SA330F Puma	1242	Chilean Army H259	Not known *(Stored)*	Seaforth Docks, Merseyside	11.04

MALAYSIA

9M-BCR	Dassault Falcon 20C	35	N809P (N1777R), N809F, F-WMKG	(FR Aviation) *(Stored)*	Bournemouth	1.05

PART 2 – OVERSEAS CIVIL REGISTERED AIRCRAFT LOCATED IN MUSEUMS AND PRIVATE COLLECTIONS

Registration	Type	Construction No	Previous Identity	Owner(Operator)	Location
CANADA					
CF-BXO	Supermarine 304 Stranraer	CV-209	RCAF 920	RAF Museum *(As "920/QN-" in RCAF c/s)*	Hendon
CF-EQS	Boeing-Stearman A75N1 (PT-17-BW) Kaydet	41-8169	75-1728	American Air Museum *(As "217786/25" in USAAF c/s)*	Duxford
CF-KCG	Grumman TBM-3E Avenger AS.3	2066	RCN326 Bu.69327	American Air Museum *(As "46214/X-3" in USN c/s)*	Duxford
C-GYZI	Cameron O-77 HAB	269		Balloon Preservation Group *"Aeolus"*	(Lancing)
GERMANY					
D-CATA	Hawker Sea Fury T.20S	ES.8503	D-FATA G-9-30, VZ345	Royal Naval Historic Flight *(As "VZ345": crashed 19.4.85 and stored)*	RNAS Yeovilton
D-HGBX	Enstrom F.280	1189	SE-HKX	Aces High Flying Museum *(Gutted pod)*	North Weald
D-HMQV	Bölkow Bö.102 Helitrainer		6216	The Helicopter Museum *(Development aircraft)*	Weston-super-Mare
D-HOAY	Kamov Ka.26	7001309	DDR-SPY DM-SPY	The Helicopter Museum	Weston-super-Mare
D-IFSB	de Havilland DH.104 Dove 6			See G-AMXR in SECTION 3	
D-Opha	Fire Balloons 3000 HAB	057	D-TALCID	Balloon Preservation Group *"Talcid"*	(Lancing)
D-Pamgas	Cameron N-90 HAB	1288		Balloon Preservation Group *"Pamgas"*	(Lancing)
FIJI					
DQ-PBF	Thunder Ax10-180 Srs.2 HAB			See G-WORK in SECTION 3	
LIBERIA (old series)					
EL-WXA	Bristol 175 Britannia 253F			See G-BDUP in SECTION 3	
FRANCE					
F-BDRS	Boeing B-17G-95DL Flying Fortress	Not known	N68269 32376, NL68269, 44-83735	American Air Museum *"Mary Alice"* *(As "231983/IY-G" in 401st BG/615th BS USAAF c/s)*	Duxford
F-BGEQ	de Havilland DH.82A Tiger Moth	86305	Fr.AF NL846	Brooklands Museum *(Under restoration 3.02)*	Denford Manor, Hungerford
F-BGNX	de Havilland DH.106 Comet 1XB			See G-AOJT in SECTION 3	
F-BTGV	Aero Spacelines 377SGT Super Guppy 201	001	N211AS	British Aviation Heritage-Cold War Jets Collection *(As "1")*	Bruntingthorpe
F-BTRP	Sud SA.321F Super Frelon *(Converted from SA.321 c/n 116)*	01	F-WMHC F-BTRP, F-WKQC, F-OCZV, F-RAFR, F-OCMF, F-BMHC, F-WMHC	The Helicopter Museum *(As "F-OCMF" in Olympic Airways c/s)*	Weston-super-Mare
F-WGGM	Thunder & Colt AS-261 HA Airship			See G-BPLD in SECTION 3	
F-WQAP	Aerospatiale SA365N Dauphin 2	6001	F-WZJJ	The Helicopter Museum	Weston-super-Mare
F-HMFI	Farman F.40 *(Modified to F141 Status)*	6799	9204M	RAF Museum (Conservation Centre)	RAF Cosford
SWITZERLAND					
HB-BOU	Brighton MAB-65 HAB			See G-AWJB in SECTION 3	
HB-NAV	Beagle B.121 Pup Series .150			See G-AZCM in SECTION 3	
NORWAY					
LN-BNM*	Noorduyn AT-16-ND Harvard IIB	14-639	31-329 R.Dan AF, FE905, 42-12392	RAF Museum *(As "FE905" in RAF/RCAF c/s)*	Hendon
ARGENTINA					
LQ-BLT*	MBB Bö.105/CBS *(Non-airworthy pod is original airframe which crashed 13.6.96: shipped to UK and rebuilt with airframe c/n S.915)*	S.863		North East Aircraft Museum	Sunderland
UNITED STATES					
N7SY	Hunting Percival P.57 Sea Prince			See G-BRFC in SECTION 3	
N18E	Boeing 247D	1722	NC18E NC18, NC13340	Science Museum Air Transport Collection & Storage Facility	Wroughton
N46EA	Percival P.66 Pembroke C.1 *(Regd with c/n K66-046)*	P66/83	8452M XK885	P.G.Vallance Ltd *(Gatwick Aviation Museum)*	Charlwood, Surrey
N47DD	Republic P47D-30-RA Thunderbolt	399-55731	N47DD Peru AF FAP119, 45-49192	Imperial War Museum Collection/American Air Museum *"Oregon's Britannia" (As "226413/UN-Z")*	Duxford
N51RT	North American F-51D Mustang	122-40949	N555BM YV-508CP, N555BM, N4409, N6319T, RCAF9235, 44-74409	Robert C Tullius *(loaned to RAF Museum)* *(As "413317/VF-B/The Duck" in 336FS/4thFG c/s)*	Hendon
N75	Hanriot HD.1			See G-AFDX in SECTION 3	
N112WG	Westland WG-30-100	012		The Helicopter Museum	Weston-super-Mare
N114WG	Westland WG-30-100			See G-EFIS in SECTION 3	
N118WG	Westland WG-30-100	018		The Helicopter Museum	Weston-super-Mare
N196B	North American F-86A-5-NA Sabre	151-43611	48-0242	American Air Museum *(As "48-0242/242" in USAF c/s)*	Duxford
NC285RS	North American Navion	NAV-4-119		South East Aviation Enthusiasts Group *"My Way"* *(Crashed 11.6.79:cockpit section/tailplane only)*	Dromod, Leitrim, County.Leitrim

Registration	Type	c/n	Previous identity	Owner/Operator	Location
N413JB	Cameron O-84 HAB	723		Balloon Preservation Group *"Autumn Fall"*	(Lancing)
N588NB	Polikarpov PO-2 (CSS-13)			See G-BSSY in SECTION 3	
N2138J	English Electric Canberra TT.18 *(Built A V Roe & Co)*	EEA/R3/EA3/6640	WK126	S D Picatti *(Stored)* *(Loaned to Gloucestershire Aviation Collection as "WK126/843")*	Gloucestershire
N2700	Fairchild C-119G-FA	10689	3C-ABA Belg AF CP-9, 51-2700	Aces High Flying Museum *(Nose only noted 2.04)*	North Weald
N3188H	ERCO 415C Ercoupe	3813	NC3188H	AeroVenture *(Damaged c 7.92: fuselage suspended from roof)*	Doncaster
N4519U	Head AX9-118 HAB	184		Northern Light Balloon Expeditions *(Op Balloon Preservation Group) "Ground Hog"*	(Lancing)
N4565L	Douglas DC-3-201A	2108	(N3TV) LV-GYP, LV-PCV, N129H, N512, N51D, N80C, NC21744	Aero Venture	Doncaster
N4990T	Thunder Ax7-65B HAB	123		British Balloon Museum and Library *"Stormy Weather"*	Newbury
N5023U	Avian Magnum IX HAB	169		Balloon Preservation Group *"Tumbleweed"*	(Lancing)
NC5171N	Lockheed 10A Electra			See G-LIOA in SECTION 3	
N5237V	Boeing B-17G-95-DL Flying Fortress	32509	(N6466D) N5237V, Bu.77233, 44-83868	RAF Museum *(As "44-83868/N" in 94th BG USAAF c/s)*	Hendon
N5419	Bristol Scout D rep *(Built Leo Opdycke 1983)*	01		FAA Museum *(Frame only)*	RNAS Yeovilton
N5820T	Westland WG-30-100			See G-BKFD in SECTION 3	
N5840T	Westland WG-30-100			See G-BKFF in SECTION 3	
N6526D	North American P-51D-25NA Mustang (Composite)	122-39874	RCAF 9289 44-73415	RAF Museum *"Little Friend"* *(As "413573/B6-K" in 361st FS/357th FG USAAF c/s)*	RAF Cosford
N6699D	Piasecki HUP-3 Retriever	51	RCN 622 USN, 51-16622	Not known *(Loaned to The Helicopter Museum) (As "622" in RCN c/s)*	Weston-super-Mare
N7614C	North American B-25J/PBJ-1J Mitchell	108-37246	44-31171	Imperial War Museum/American Air Museum *(As "31171" in US Marines c/s)*	Duxford
N7777G	Lockheed L.749A-79 Constellation			See G-CONI in SECTION 3	
N9050T	Douglas C-47A-10DK Dakota 3			See G-AGYX in SECTION 3	
N9089Z	North American TB-25J-25NC Mitchell			See G-BKXW in SECTION 3	
N9115Z	North American TB-25N-20NC Mitchell	108-32641	44-29366	RAF Museum *"Hanover Street/Catch 22"* *(As "34037" in USAAF c/s: allotted 8838M)*	Hendon
N12006	Raven S.50A HAB	111		R Higbie *"Cheers" (On loan to British Balloon Museum and Library)*	Newbury
N14234	Handley Page HP.137 Jetstream			See G-BBBV in SECTION 3	
N16676	Fairchild F.24CR-C8F Argus	3101	NC16676	A Langendal *(Frame only)* *(On loan to Norfolk and Suffolk Aviation Museum)*	Flixton
N33600	Cessna L-19A-CE Bird Dog	22303	51-11989	Museum of Army Flying *(As "111989" in US Army c/s)*	AAC Middle Wallop
N66630	Schweizer TG-3A	63	42-52983 *(P/i not confirmed)*	Imperial War Museum *(As "252983" in USAAC c/s)*	Duxford
N70457	MD Helicopters MD.600N	RN-057		Aces High *(Cancelled 10.01) (As "511" for film)*	North Weald
N99153	North American T-28C Trojan	252-52 *(FAA quote c/n 226-93)*	Zaire AF FG-289 Congo AF FA-289, Bu.146289	W R Montague *(Crashed Limoges, France 14.12.77)* *(On loan to Norfolk and Suffolk Aviation Museum: fuselage only as "146289/2W")*	Flixton

BELGIUM

Registration	Type	c/n	Previous identity	Owner/Operator	Location
OO-ARK	Cameron N-56 HAB	276		Balloon Preservation Group *"Princess Alex" (Spares use)*	(Lancing)
OO-BDO	Cameron N-90 HAB	1960	(LX-PRO)	Balloon Preservation Group *"Profi 2" (Spares use)*	(Lancing)
OO-BFH	Piccard Gas Balloon	Not known		Science Museum Air Transport Collection and Storage Facility *(Gondola only)*	Wroughton
OO-BRM	Thunder Ax7-77 HAB	1111		Balloon Preservation Group	(Lancing)
OO-JAT	Cameron Zero 25 Airship	1407		Balloon Preservation Group	(Lancing)

DENMARK

Registration	Type	c/n	Previous identity	Owner/Operator	Location
OY-BOB	Omega O-80 HAFB			See G-AWMO in SECTION 3	
OY-BOW	Colting 77A HAB	77A-014	SE-ZVB	British Balloon Museum and Library *"Circus"*	Newbury

RUSSIA

Registration	Type	c/n	Previous identity	Owner/Operator	Location
RA-01277	Sukhoi Su-29	80-02	RA8002	Yorkshire Air Museum	Elvington
RA-01378	Yakovlev Yak-52 *(Composite with c/n 833805/DOSAAF 134 which is now N54GT)*	833004	DOSAAF 14	Wellesbourne Wartime Museum	Wellesbourne Mountford
RA-01641	Antonov An-2R	1G-190-47		Black Country Aircraft Collection *"3":(Crashed Milton 2.11.99) (Forward fuselage stored)*	Wolverhampton

SWEDEN

Registration	Type	c/n	Previous identity	Owner/Operator	Location
SE-AZB*	Cierva C.30A Autogiro (Avro 671)	R3/CA.954	K4232	RAF Museum *(As "K4232")*	Hendon

POLAND

Registration	Type	c/n	Previous identity	Owner/Operator	Location
SP-SAY*	Mil Mi-2	529538125		The Helicopter Museum	Weston-super-Mare

GREECE

Registration	Type	c/n	Previous identity	Owner/Operator	Location
SX-OAD	Boeing 747-212B	21684	9V-SQI	Cold War Jets Collection *"Olympic Flame"*	Bruntingthorpe

ICELAND

TF-ABP	Lockheed L.1011-385-100 Tristar	1045	VR-HOG N323EA	Aces High Flying Museum *(Nose only)*	North Weald
TF-SHC	Miles M.25 Martinet TT.1	Not known	MS902	Museum of Berkshire Aviation *(Crashed 18.7.51: on rebuild with Master components)*	Woodley

AUSTRALIA

VH-ALB	Supermarine 228 Seagull V	Not known	A2-4	RAF Museum *(As "A2-4")*	Hendon
VH-ASM	Avro 652A Anson I	72960	W2068	RAF Museum *(As "W2068/68" in RAF c/s)*	Hendon
VH-AYY	Kavanagh D-77 HAB	KB136		Balloon Preservation Group *"Carlsberg"*	(Lancing)
VH-BRC	Short S.24 Sandringham IV	SH.55C	N158C VP-LVE, N158C, VH-BRC, ZK-AMH, JM715	Science Museum *"Beachcomber"* *(On loan to Hall of Aviation) (Ansett c/s)*	Southampton
VH-SNB	de Havilland DH.84A Dragon	2002	VH-ASK A34-13	National Museums of Scotland/Museum of Flight	East Fortune
VH-UQB	de Havilland DH.80A Puss Moth			See G-ABDW in SECTION 3	
VH-UTH	General Aircraft Monospar ST-12	ST12/36		Newark Air Museum *(On rebuild by Cotswold Aircraft Restoration Group)*	Innsworth
VH-UUP	Short S.16 Scion 1			See G-ACUX in SECTION 3	

BERMUDA (new series)

VP-BDF	Boeing 707-312			See G-AYAG in SECTION 3	

ZIMBABWE formerly SOUTHERN RHODESIA

VP-YKF	de Havilland DH.104 Dove 6			See G-AMDD in SECTION 3	

BERMUDA (old series)

VR-BEP	Westland WS-55 Whirlwind 3			See G-BAMH in SECTION 3	
VR-BEU	Westland WS-55 Whirlwind 3			See G-ATKV in SECTION 3	

KENYA

5Y-SIL	Cameron A-140 HAB			See G-AZUW in SECTION 3	

SENEGAL

6W-SAF	Douglas C-47A-65-DL 19074	42-100611	F-GEFU	Aces High Flying Museum *(As "42-100766") "Lilly Belle" (Nose section only)*	North Weald

MUSCAT AND OMAN

A40-AB	Vickers VC-10 Srs.1103			See G-ASIX in SECTION 3	

PART 3 – ENTRIES REMOVED FROM 2004 EDITION

Regn	Type	C/n	Reason for removal
PORTUGAL			
CS-TPJ	Embraer EMB-145EU	145037	Returned to Portugal 4.04
GERMANY			
D-CAOA	Embraer EMB.120RT Brasilia	120.013	To EC-JBE 8.04
D-CFXD	Short SD.3-60 Variant 300	SH.3749	To N749JT 9.04
D-CPRX	Dornier Do328-110	3101	To G-CCGS 3.04
D-EDUB	Bucker Bu.181 Bestmann	25071	To G-GLSU 7.04
D-EECN	Piper PA-28-235 Cherokee Charger	28-7310156	To G-EWME 9.04
D-EIBV*	Piper PA-28-161 Warrior II	28-8216115	To EI-SKW 2.04
D-EKKO	Reims FR172G Rocket	FR17200185	To G-EKKC 2.04
D-EOLW	Maule MX.7-180 Star Rocket	11050C	To N266MM 5.04 then to G-RAZZ 11.04
D-EPUD	Bölkow Bö.209 Monsun 160FV	196	Returned to Bonn-Hangelar, Germany by 6.03
D-EXLH	Extra EA.400	06	To Switzerland by 3.03
D-EZAP	Cessna 152	15283078	Returned to Fehrbellin, Germany by 5.04
D-FKMA	Antonov An-2T	117411	Returned to Germany by 8.04
D-GDCO	Piper PA-23-160 Apache	23-1800	Returned to Egelsbach, Germany by 5.04
D-HOXQ	Mil Mi-8T	105103	Returned to Berlin by 5.02
SPAIN			
EC-IKV	Eurocopter EC135P2	0191	Delivered to Spanish Police 2003
EC-IKX	Eurocopter EC135P2	0222	Delivered to Spanish Police 2003
FRANCE			
F-BTMM	Piper PA-31 Turbo Navajo	31-480	To G-CCRY 3.04 then to N449TA 6.04 - see SECTION 4 Part 1
F-GCCZ	Aérospatiale SA.342J Gazelle	1393	To YU-HEV 1.05
F-GFRD	Robin ATL	53	To G-GFRD 2.05
F-GFVE	Cessna 305C (L-19E) Bird Dog	24541	To N134TT 3.04 - see SECTION 4 Part 1
F-GGHZ	Robin ATL	123	To G-GGHZ 2.05
F-GKQD	Robin DR400/120 Dauphin 2+2	2044	To G-CCKP 9.04
F-GLIS	Fokker F.28-0070	11540	Returned to service by 3.04
F-GLIT	Fokker F.28-0070	11541	Returned to service by 9.04
F-GLIU	Fokker F.28-0070	11543	Returned to service by 7.04
F-GLIX	Fokker F.28-0070	11558	Returned to service by 3.04
F-GMPA	Aérospatiale AS350B Ecureuil	1749	To G-HRAK by 7.04
F-GNFT	Boeing 737-3YO	23921	To N921NB 2.04
F-GNFU	Boeing 737-3YO	24256	To OM-AAA 3.04
SWITZERLAND			
HB-IBX	Gulfstream Gulfstream IV/SP	1183	Occasional visitor only in 2004
ITALY			
I-ISAK*	SIAI-Marchetti SF.260D	839	To N260AP 2.04 - see SECTION 4 Part 1
I-JULI	Beech 95-B55 Baron	TC-629	Crashed Monte Mommio Fivizzano, Switzerland 26.5.02
I-RALY	Agusta A.119 Koala	14013	To G-CCSZ 4.04
NORWAY			
LN-RCE	Boeing 767-383ER	24846	To N2484B 6.03
LN-RCM	Boeing 767-383ER	26544	To CC-CGN 5.04
LITHUANIA			
LY-ABZ	Yakovlev Yak-52	9611914	To ZK-YNZ 2.04
LY-AMJ	Yakovlev Yak-18T	22202047812	To HA-YAM 8.04 - see SECTION 4 Part 1
LY-AOO	Yakovlev Yak-18T	22202040425	To RA-02933 - see SECTION 4 Part 1
UNITED STATES			
N7LQ	Agusta A 109C	7634	To N109CW 10.04 - then to VH-XNB 12.04
N25PJ	Cessna 340A II	340A0912	To G-CCXJ 7.04
N26PJ	Piper PA-30-160 Twin Comache B	30-1477	To G-ELAM 10.04
N33CJ	Cessna 525 CitationJet	525-0245	To G-SFCJ 7.04
N37WC	Cessna 401	401-0183	Sold - to US 30.7.04
N47FK	Douglas C-47A-35-DL Dakota 3	9700	To Lelystad, The Netherlands 8.04
N63ST	Canadair CL601-3R Challenger	5149	Returned to US 19.3.04
N75TL	Boeing-Stearman A75N1 (N2S-4) Kaydet	75-3616	To G-CCXA 6.04
N83HN	British Aerospace BAe146 Srs.100	E1083	To G-CCXY 7.04
N93GS	Grumman G.21A Goose	B-76	To Ontario, Canada by 10.02
N121ZR	Beech 1900C-1	UC-121	To US 23.3.04 - then to ZS-PKX 12.04
N123SX	Piper PA-46-500TP Malibu Meridian	4697050	To LX-FUN 4.04
N125YY	British Aerospace BAe 125 Srs.700B	257115	To N333MS 3.04
N126SE	Boeing Stearman E75	75-5498	Marked EC-AMD by 11.03 - see SECTION 4 Part 1
N128M	Dassault Falcon 50EX	276	To N789ME 8.04
N139JV	Commander Aircraft Commander 114TC	20034	To US for sale 13.12.04
N142TV	Piper PA-28RT-201T Turbo Arrow IV	28R-8031020	To US late 2003

N154CD	Cirrus Design SR20	1053	Sold in Germany 8.04
N156RH	Cessna 421C Golden Eagle	421C-0008	To HA-FAG 6.04
N210SA	Maule M.7-235B	23062C	To G-TAFC 1.05
N213CT	Beech C90-1 King Air	LJ-1028	Restored as G-BKFY 11.04
N251JS	Gulfstream G1159 Gulfstream II	251	Sold in Illinois, US 2.04
N265CT	Boeing 737-529	26538	To C-GAHB 3.04
N279SA	Enstrom F-28F	734	Restored as G-BXLW 8.04
N285RA	Consolidated PBY-6A Catalina	2087	To France en route Israel 5.04
N295CP	Beech B200 Super King Air	BB-1295	Sold in Virginia, US 8.04
N314BG	North American P-51D-20NA Mustang	Not known--	Crated to US by 8.03
N417RK	Piper PA-46-350P Malibu Mirage	4636249	Sold in Belgium 12.04
N429PK	Cessna 525 CitationJet	525-0429	To EC-IVJ 4.04
N430CE	Bell 430	49064	Sold 10.04 - replaced by N5120
N431WH	Bell 430	49066	Sold in Virginia, US 10.04 - N404DJ reserved.
N432A	Raytheon B36TC Bonanza	EA-666	Sold in Germany 6.04
N435XS	Airbus A310-304	435	Scrapped Kemble by 1.04
N440XS	Airbus A310-304	440	Scrapped Kemble by 4.04
N456JR	Piper PA-46-350P Malibu Mirage	4636311	To N49ET 2.04
N510PS	Cessna 310N	310N0054	To G-YHPV 12.04
N511HE	Boeing 737-4Y0	24511	To PR-BRH 5.04
N511VA	MD Helicopters MD 600N	RN-023	Based in Belgium by 2.04
N585D	Gulfstream Gulfstream IV/SP	1258	Returned to California, US by 1.04
N600HV	MD Helicopters MD.600N	RN-058	To Russia 10.04
N621MM	Agusta A109C	7622	To G-DBOY 10.04
N656AG	Piper PA-34-220T Seneca III	34-8333087	Sold the Netherlands 7.04, restored 9.04 to Iowa, US
N666HM	SOCATA TB-20 Trinidad	1073	To N33NW 2.04 - see SECTION 4 Part 1
N666JH	Cessna 182T	18281025	Destroyed in fatal crash Marlow, Buckinghamshire 1.8.03
N685TT	Rockwell Commander 685	12043	Sold in Texas, US 8.04
N700KH	SOCATA TBM-700	210	Sold in Ohio, US 11.04
N701AR	SOCATA TBM-700	209	To N700EL 3.04 - see SECTION 4 Part 1
N737SP	Boeing 737-8EQ	33361	To N737M 8.04 - see SECTION 4 Part 1
N777AS	Boeing 777-24Q	29271	To Hamburg, Germany by 6.04
N800VP	Cessna 340A	340A1532	To US 22.11.04 - sold in Utah
N812TC	Douglas DC-8F-55	45764	To 9G-MKT 4.04
N841WS	Cessna 550 Citation Bravo	550-0841	To N841W 8.04
N854CC	Piper PA-12 Super Cruiser	12-2907	To G-CDCS 9.04
N909WJ	Grumman FM-2 Wildcat	-	To the US by 8.03
N957JK	Piper PA-24-250 Comanche	24-1931	To N250CC 1.05
N966SW	Cessna 560 Citation V Ultra	560-0284	Sold in North Carolina, US 2.04
N991RV	IAI 1125 Astra	011	To N500FA 6.04
N999BE	Dassault Falcon 2000	147	Sold in Colorado, US 8.04 (N777MN reserved)
N2423C	Piper PA-38-112 Tomahawk	38-79A0177	No reports since 2.00: cancelled 4.04 by FAA as "destroyed"
N2683Y	Cessna 421C	421C1218	Sold in Georgia, US 8.04
N3046P	Piper PA-46-500TP Malibu Meridian	4697170	To EC-IVZ 4.04
N3839H	Piper Aerostar 601P	61P-0569-7963247	To N601AR 1.05
N5064L	Raytheon B90B King Air	LJ-1664	To N789KP 6.04
N5346S	Piper PA-32R-301T Saratoga II TC	3257257	To SX-ACV 5.04
N5361C	Piper PA-46-500TP Malibu Meridian	4697125	To D-EICO 7.04
N5361L	Cessna T206H	T20608401	To G-OLLS 3.04
N6003F	Commander Aircraft Commander 114B	14590	To G-HPSF 11.04
N6107Y	Commander Aircraft Commander 114B	14627	To G-OECM 3.04
N6171N	Raytheon B300 Super King Air	FL-371	Returned to US 19.3.04. Sold in Arizona 6.04
N6315X	Cessna 421C	421C-1003	Sold in Georgia, US 4.04
N6602W	Piper PA-23-250 Aztec C	27-3799	To N250BW 4.04 - see SECTION 4 Part 1
N6834L	Cessna T310R II	310R2137	Crashed near Colne, Lancs 30.3.04
N9308V	Mooney M.20F	69-0086	To US 9.03
N9469P	Piper PA-24-260 Comanche C	24-4979	To G-PETH 10.04
N9999M	Gulfstream Gulfstream IV	1090	Sold in New York, US 6.04
N12739	de Havilland DH.83 Fox Moth	4026	Unmarked frame now reported as ZK-AGM - see SECTION 4 Part 1
N27597	Piper PA-31-350 Navajo Chieftain	31-7852073	To G-OETV 604
N41762	Raytheon Hawker 800XP	258456	To G-JMAX 10.04
N50429	Fairchild M-62A Cornell	T43-5205	Loaned to Red Bull Collection, Salzburg, Austria by 12.03
N53643	Piper PA-34-220T Seneca V	3449273	To G-MDCA 6.04
N53665	Piper PA-28-161 Warrior III	2842178	Sold in Spain 1.04
N58093	Mooney M.20K Srs.231	25-0877	To Germany by 8.03
NC92782	Piper PA-12 Super Cruiser	12-228	Sold in New Zealand 4.04

CZECH REPUBLIC

OK-EUU 24	Urban Air UFM-11 Lambada	Not known	No recent reports - possible misreading ?
OK-EUU 55	Urban Air UFM-11 Lambada	10/11	To EI-DGY 10.04
OK-FUA 09	Urban Air UFM-11UK Lambada	14/11	To EI-DGT 8.04

BELGIUM

OO-JKT	Focke-Wulf FW.44J Stieglitz	183	To G-STIG 2.04
OO-LBC	SEEMS MS.890A Rallye Commodore 150	10283	To G-TLBC 10.04
OO-RTC	Reims FR172H Rocket	FR17200265	To G-CCSB 3.04

DENMARK

OY-ALW	Miles M.28 Mercury 6	6268	Returned to Denmark by 5.03
OY-PBA	Pilatus PC-6/B-4-H2 Turbo-Porter	678	No reports since 7.00 - Offered for sale in Denmark 8.04

The NETHERLANDS

PH-NLH*	Hawker Hunter T.7	41H-695342	To "Netherlands" 2001 but "nose section" at Weelde, Belgium by 2.04
PH-RAR	Beech 1900D	UE-372	Lease terminated - returned to Rossair Europe
PH-RAT	Beech 1900D	UE-350	Lease terminated - returned to Rossair Europe

RUSSIA

RA-01153	Yakovlev Yak-18T	22202047817	To HA-YAV 7.04 - see SECTION 4 Part 1
RA-01294	Yakovlev Yak-50	853104	Now believed to be mispaint of RA-01293 which is now G-YAKK
RA-01333	Yakovlev Yak-55M	920506	To Romilly sur Seine,France by 5.04 when offered for sale
RA-01555	Yakovlev Yak-18T	05-36	To HA-YAG 5.04 - see SECTION 4 Part 1
RA-01607	Sukhoi Su-29	77-02	To Italy by 9.04
RA-01609	Sukhoi Su-29	75-03	To HA-YAR 4.04 - see SECTION 4 Part1
RA-01610	Sukhoi Su-29	78-02	To HA-YAT 6.04
RA-02041	Yakovlev Yak-52	822603	To G-CBRP 9.04
RA-02622	Yakovlev Yak-52	9612001	To G-CCJK 3.04
RA-44465 (2)	Yakovlev Yak-18T	10-34	To HA-YAN 1.04 - see SECTION 4 Part 1
RA-44470	Technoavia Yak-18T	18-33	To HA-YAH 1.04 - see SECTION 4 Part 1
RA-44479	Sukhoi Su-29	N1001.001	To HA-YAO 12.03
RA-44506	Yakovlev Yak-18T	22202034139	Restored as HA-JAC 6.04 - see SECTION 4 Part 1
RA-44508 (2)	Sukhoi Su-31	03-01	To N131BT 4.04
RA-44510	Yakovlev Yak-55M	930707	To RF-3319K 12.04
RA-44527	Yakovlev Yak-18T	15-35	To HA-YAU 6.04 - see SECTION 4 Part 1
RA-44544	Yakovlev Yak-18T	11-33	To Westerstede-Felde, Germany by 7.03
RA-44545	Yakovlev Yak-18T	22202034023	To HA-YAP 4.04 - see SECTION 4 Part 1
RA-44547	Technoavia SP-55M	0101-0007	To US in 8.04

SWEDEN

SE-IYF	Cessna 172M	17262765	Crashed on take off Lundy Island 4.7.04
SE-RAA	Embraer EMB.135ER	145.210	Returned to City Airline

GREECE

SX-BLV	Boeing 757-2G5	26278	Restored as G-JMCG 4.04

TURKEY

TC-FLD	Boeing 757-256	26244	To N262SR 2.04 then HS-KAK 4.04

ICELAND

TF-ATN	Boeing 747-219B	22723	Restored as G-VBEE 5.04
TF-ELC	Boeing 737-3M8	25041	Returned to Islandsflug
TF-ELD	Boeing 737-46B	24124	Returned to Islandsflug
TF-ELQ	Boeing 737-3Q8	23535	Returned to Islandsflug
TF-ELV	Boeing 737-4S3	24796	Returned to Islandsflug

CONGO BRAZZAVILLE

TN-AEE	Boeing 737-2Q5C	21538	Scrapped Shannon 7.7.04

UKRAINE

UR-67199	LET L-410UVP Turbolet	790305	To Piotrkow Trybunalski, Poland by 5.04

BERMUDA (new series)

VP-BCI	Canadair CL601 Challenger	5193	To VP-BIH 1.04
VP-BCO	Canadair CL604 Challenger	5420	To N604TS 4.04
VP-BDB	Cessna 560 Citation V	560-0503	To N204BG 5.04
VP-BLA	Gulfstream Gulfstream V	654	To N404M 8.04
VP-BLK	Rockwell Turbo Commander 690C	11672	T0 N41462 9.04
VP-BME	Gulfstream Gulfstream IV	1058	To G-GMAC 7.04

CAYMAN ISLANDS

VP-CAD	Cessna 525 CitationJet	525-0297	Based at Zurich by 7.04
VP-CAS	British Aerospace BAe 125 Srs.800A	258167	To N825DA 4.04
VP-CCP	Cessna 550 Citation Bravo	550-0857	To N984BK 6.04
VP-CCW	MD Helicopters MD 600N	RN030	To G-GOON 2.04
VP-CGG	Cessna 560XL Citation Excel	560-5361	To G- CFGL 1.05
VP-CGP	Dassault Falcon 900	163	To N25MB 5.04
VP-CIS	Cessna 525 CitationJet	525-0252	To N252RV 4.04
VP-CJP	Canadair CL-601-3A Challenger	5022	To G-SAJP 2.05
VP-CMF	Gulfstream Gulfstream IV	1062	To N104JG 1.04

FALKLANDS ISLANDS AND DEPENDENCIES

VP-FAZ	de Havilland DHC.6-310 Twin Otter	748	Antarctic Winters not spent in UK since 2002. Recently to Cayman Islands/Canada
VP-FBB	de Havilland DHC.6-310 Twin Otter	783	Crashed landing Rothera, British Antarctic Territory 7.11.03
VP-FBC	de Havilland DHC.6-310 Twin Otter	787	See VP-FAZ above
VP-FBL	de Havilland DHC.6-310 Twin Otter	839	See VP-FAZ above
VP-FBQ	de Havilland DHC-7-110	111	See VP-FAZ above

REPUBLIC OF SERBIA AND MONTENEGRO

YU-HCC	Agusta Bell AB.212	5712	Believed returned to Serbia

NEW ZEALAND

ZK-LIX	Lavochkin LA-9	828	Returned to New Zealand 1.04

REPUBLIC of SOUTH AFRICA

ZS-JIY	Lockheed L-100-30	4691	Returned to Safair

CYPRUS

5B-HAC	Rutan Defiant	162	To N765MM 7.04

UGANDA

5X-ONE	Douglas DC-10-30	46952	Scrapped Kemble by 7.04

SECTION 5

PART 1 – BRITISH GLIDING ASSOCIATION

The Register includes the Certificate of Airworthiness (CA) reference number as issued by the British Gliding Association (BGA). This is usually found below the tailplane in small characters. We include, as a primary reference, the corresponding three-letter coding (or Trigraph (T/G) system) frequently marked on the tails. If gliders are known to be wearing their respective tri-graphs then we have indicated this and/or any other identity in the first composite column, for example BGA Competition Numbers are shown if carried with further details contained in Part 3. The official BGA list is extended by including non-current gliders and those with recently lapsed CAs for which no cancellation details are known but which may survive. These are identified by * in the CA expiry column. We include the complete expiry dates for CA and the complete date for the first issue where known. Where a BGA CA number has been reserved for future use, the reservation date is shown prefixed by the letter 'R'. The non-current gliders include examples known to be in storage or under restoration in the ownership of members of the Vintage Glider Club (VGC). Note that we have now deleted a large number of entries with a CA expiry before the year 2000. Similar entries where a reason for the non-renewal of CA is suspected (ie accident details) have been retained but these will be removed in the next edition. Information in the "(Unconfirmed) Base" information is largely related to owners' addresses and feedback we receive from readers as there is no official record made of bases. Many gliders are normally kept at their owners' homes and trailered to a site for use.

Because of the introduction of new, Europe-wide, arrangements for aircraft certification following the establishment of the European Aviation Safety Agency, significant changes can be expected to occur in the regulation of UK glider certification over the next few years. The detailed effects of the new legislation is still under negotiation, but it can be expected that gliders will be divided into three groups:

(i) Gliders exempted from the regulations are expected to be largely vintage designs and will probably remaining under BGA control)
(ii) Other gliders extant in the UK before 28. 9.03 will need to comply with the new regulations by 28. 3.07
(iii) Gliders imported after 28. 9.03 will have to be CAA registered and have an EASA CofA; by definition this means that they have to comply with their Type Certificates in all respects. These rules have not yet been implemented but can be expected to come into force in the near future.

In the past the BGA has given its own technical approvals to various aircraft modifications some of which might fall outside the Type Approval given in the glider's country of origin. For this reason, the BGA is to apply for EASA Design Organisation Approval. Watch this space !

Once again Phil Butler has updated the BGA register, with assistance from Wal Gandy, and we extend our thanks also to Richard Cawsey, Tony Morris, Barry V Taylor and Ged Terry for their additional comments. Special thanks to the BGA's former Secretary Barry Rolfe, and his successor Pete Stratten, also to Colin Childs (the Coaching Secretary) for their assistance. Information is current to mid February 2005.

T/G	BGA	Code	Type	Construction No	Previous Identity	Date	Owner/Operator	(Unconfirmed) Base	CA Expy
	162		Manuel Willow Wren				See SECTION 3 Part 1	Brooklands	
AAA	231		Abbott-Baynes Scud II *(Built Slingsby)*	215B	G-ALOT BGA.231	8.35	L.P.Woodage	Snitterfield	29. 5.04
(AAF)	236		Slingsby T.6 Kite 1	27A	G-ALUD BGA.236, (BGA.222)	14.11.35	Not known *(Stored pending rebuild 12.95)*	Dunstable	*
AAX	255		Slingsby T.6 Kite 1	227A	(ex RAF) BGA.251	30. 3.36	R.Boyd	Rivar Hill	6. 5.04
ABG	260		Schleicher Rhönsperber	32-16		4. 5.36	F.K.Russell tr Rhönsperber Syndicate	Dunstable	13. 8.05
(ABN)	266		Slingsby T.1 Falcon 1 Waterglider				See SECTION 3 Part 1	Windermere	
ABZ	277		Grunau Baby 2 *(Built F.Coleman)*	?	RAFGSA.270 BGA.277, G-ALKU, BGA.277	25. 8.36	J.L.Smoker and Partners	Hinton-in-the-Hedges	22. 7.95*
ACF	283	ACF	Abbott-Baynes Scud III	2	G-ALJR BGA.283	18.12.36	L.P.Woodage	Dunstable	14. 6.03
ACH	285	E	Slingsby T.6 Kite 1	247A	G-ALNH BGA.285	30.12.36	E.B.Scott *(On loan to The Museum of Army Flying as "G285/E" in 1 GTS RAF c/s)*	AAC Middle Wallop	5.99*
ADJ	310		Slingsby T.6 Kite 1	258B	RAFGSA182 VD218, BGA.310	9. 2.37	A.M.Maufe *(Rebuilt 1982 with components ex BGA.327 c/n 285A)*	Tibenham	26. 5.05
	334		Slingsby T.12 Gull 1				See SECTION 3 Part 1	Bull Creek, Western Australia	
AEM	337		Schleicher Rhönbussard	620	G-ALME BGA.337	25. 4.38	C.Wills and S.White	Booker	31. 5.02
(AFW)	370		Grunau Baby 2 *(Built J.Hobson)*	1		11.10.38	Not known *(Under restoration 2000)*	Saltby	5.77
(AGE)	378		Slingsby T.12 Gull I	312A	G-ALPJ BGA.378	14. 9.38	M L Beach	Halton	16. 8.04
AHC	400	F	Slingsby T.6 Kite 1 *(Wings from Special T.6 c/n 355A)*	336A	VD165 BGA.400	6. 5.39	R.Hadlow and Partners *(In 1 GTS RAF c/s)*	Thame	21. 8.01
(AHU)	416	G-ALRD	Scott Viking 1	114	G-ALRD BGA.416	19. 6.39	M.L.Beach	Dunstable	2. 7.03
AJW	442	AJW	Slingsby T.8 Tutor	MHL/RC/8	G-ALMX BGA.442	8.46	M.Hodgson	Dunstable	14. 8.98
(AKC)	448		DFS 108-68 Weihe	000348	G-ALJW BGA.448, LO+WQ	6.47	D.Philips *(Damaged Thun,Switzerland 20.7.79: on rebuild 1994)*	Snitterfield	*
AKD	449		DFS/70 Olympia-Meise	227	LF+VO	7.47	L.S.Phillips *(Stored)*	Perranporth	5.85*
AKW	486		Slingsby T.8 Tutor	MHL/RT/7		11.46	D.Kitchen	Tibenham	6. 7.96
(ALA)	470		Short Nimbus				See SECTION 3 Part 1	Holywood, Belfast	
ALR	485		Slingsby T.8 Tutor	513	G-ALPE BGA.485	11.46	M.H.Birch *(Stored)*	Booker	14. 6.97
ALW	490	G-ALRC	Hutter H-17A *(Built D.Campbell)*		G-ALRK BGA.490	8.48	N.I.Newton	Booker	16. 7.05
(ALX)	491		Hawkridge Dagling	08471		2.47	N.H.Ponsford *(Stored 1.98)*	(Breighton)	*
(ALZ)	493		Hawkridge Nacelle Dagling *(Also allotted BAPC.81)*	10471		7.47	P.and D.Underwood *(On rebuild 2000)*	Eaton Bray	*
AMK	503		EoN AP.5 Olympia 2	EoN/O/003	G-ALJP BGA.503	5.47	D.T.Staff *(Under restoration)*	Booker	28. 5.95*
AMM	505	AMM	EoN AP.5 Olympia 2	EoN/O/006	G-ALJV BGA.505	5.47	I.Hodge	RAF Marham	31. 5.03
AMP	507		EoN AP.5 Olympia 2	EoN/O/008	G-ALJO BGA.507	6.47	M.Briggs	Cranfield	25. 6.95*

AMR	509	AMR	EoN AP.5 Olympia 2	EoN/O/011	G-ALLA BGA.509	5.47	K.J.Nurcombe	Husbands Bosworth	1. 8.05
AMT	511		EoN AP.5 Olympia 2	EoN/O/005	G-ALLM BGA.511	5.47	E.W.Burgess	Lyveden	25. 7.97
AMU	512		EoN AP.5 Olympia 2	EoN/O/012			Not known *(Under rebuild 2000)*	Challock	
AMV	513		EoN AP.5 Olympia 2	EoN/O/014	G-ALNB BGA.513		Not known *(Under refurbishment 2000)*	Camphill	
(AND)	521		Slingsby T.26 Kite II	MHL/RK.5			Not known *(Stored)*	Chalfont St.Giles	
ANW	538	ANW	EoN AP.5 Olympia 2B	EoN/O/040	G-ALNE BGA.538	7.47	E.A.Barnacle	Snitterfield	9. 3.04
(AND)	521		Slingsby T.26 Kite II	MHL/RK.5			Not known *(Stored)*	Chalfont St.Giles	
ANW	538	ANW	EoN AP.5 Olympia 2B	EoN/O/040	G-ALNE BGA.538	7.47	E.A.Barnacle	Snitterfield	9. 3.04
ANZ	541		EoN AP.5 Olympia 2	EoN/O/043		9.47	Not known *(Under restoration)*	RAF Halton	
APC	544	APC	EoN AP.5 Olympia 2	EoN/O/046	G-ALMJ BGA.544	9.47	N.G.Oultram	Seighford	19. 7.03
(APV)	561		EoN AP.5 Olympia 2B	EoN/O/032	G-ALKN BGA.561	6.47	A.Kepley t/a Fenland and West Norfolk Aviation Museum *(Stored 8.97)*	Crowland	8.78*
(APW)	562		EoN AP.5 Olympia 2B	EoN/O/037	G-ALJZ	7.47	*(Wreck at Museum of Berkshire Aviation, Woodley, 8.04)*		
APZ	565		Slingsby T.25 Gull 4	505	G-ALPB (BGA.565)		A.Fidler	Crowland	19. 5.05
AQE	570		Slingsby T.21B	538	G-ALNJ BGA.570		Not known *(Crashed on landing Ridgewell 12.6.80: stored pending rebuild 10.97)*	Camphill	*
AQG	572		Slingsby T.21B Sedbergh TX.1				See SECTION 3 Part 1	RAF Cosford	
AQN	578	AQN	Hawkridge Grunau Baby 2B	G.3348	G-ALSO BGA.578	.48	R G Hood	Lasham	3. 6.05
(AQQ)	580		EoN AP.7 Primary	EoN/P/003	G-ALPS BGA.580		Imperial War Museum *(Stored)*	Duxford	*
(AQY)	588		EoN AP.7 Primary	EoN/P/011			N.H.Ponsford *(Stored 1.98)*	(Breighton)	*
	589		EoN AP.7 Primary	EoN/P/012	G-ALMN BGA.589	19 5.48	Not known *(As "G-ALMN" at Museum of Berkshire Aviation, Woodley)*	(Farnborough)	4.51*
ARK	599	ARK	Slingsby T.30A Prefect	548	PH-1 BGA.599, G-ALLF, BGA.599	.	K.M.Fresson	RNAS Yeovilton	15. 8.04
ARM	601	ARM	Slingsby T.21B	543	G-ALKX BGA.601	8.48	South London Gliding Centre	Kenley	23. 8.05
ASB	614	ASB	Slingsby T.21B	549	RNGSA G-ALLT, BGA.614	9.48	J.L.Rolls and Partners	Talgarth	10. 8.05
	619		Slingsby T.21B				See SECTION 3 Part 1	Alverca, Portugal	
ASR	628		EoN AP.8 Baby	EoN/B/004	G-ALRU BGA.628	3.49	Not known *(Crashed Bardney 28.5.71; stored11.04)*	Aston Down	*
AST	629	G-ALRH	EoN AP.8 Baby	EoN/B/005	G-ALRH BGA.629	3.49	EoN Baby Syndicate *"Liver Bird" (Extant 2000)*	Chipping	9.96
ATH	643		Slingsby T.15 Gull III				See SECTION 3 Part 1	Brooklands	
ATL	646		Slingsby T.21B	536	G-ALKS	6.50	G.Markham *(Stored 2.01)*	Enstone	7.96
ATR	651		Slingsby T.13 Petrel 1	361A	EI-101 IGA.101, IAC.101, BGA.651, G-ALPP	7.50	G.P.Saw	Booker	17. 6.05
ATV	655	OK-8592	Zlin 24 Krajanek	101	G-ALMP OK-8592	4.50	J.Dredge	Booker	21. 5.05
AUD	663	663	Slingsby T.26 Kite 2B	727		1.52	R.S.Hooper	Lasham	15.10.04
AUG	666	AUG	Slingsby T.21B	643		6.51	Cambridge University Gliding Club	Gransden Lodge	19. 5.05
AUP	673	N21	Slingsby T.21B	636		7.51	The Solent T21 Group	Lee-on-Solent	4. 6.05
AUU	678	AUU	EoN AP.5 Olympia	EoN/0/076		4.52	Mrs.B.Lee	Parham Park	15. 7.04
AVA	684		Abbott-Baynes Scud III	3		1.53	E.A.Hull	Dunstable	1. 6.05
AVB	685	AVB	Slingsby T.34 Sky	644	G-644	2.53	R.Moyse	Lasham	25. 7.03
AVC	686	AVC	Slingsby T.34 Sky	670		3.53	P.J.Teagle *"Kinder Scout II"*	Camphill	27. 8.04
AVF	689	AVF	Slingsby T.26 Kite 2A	728	RAFGSA.294 BGA.689	4.53	C.P.Raine	Dunstable	2. 7.05
AVG	690		Slingsby T.31B	717		4.53	A.R.Worters *(Refurbishing: previously reported as becoming G-BMDD)*	North Connel	
AVL	694		Slingsby T.34 Sky	671	G-671	5.53	M.P.Wakem	Long Mynd	12. 7.05
AVQ	698	G46	Slingsby T.34 Sky	645	G-645	8.53	L.M.Middleton *"Gertie"*	Easterton	23. 7.05
AVT	701		Slingsby T.30B Prefect	857	AGA... BGA.701	1.53	Booker Gliding Club	Booker	14. 5.00
AWD	711	AWD	Slingsby T.21B	950		30. 9.54	D.B.Brown tr T.21 Syndicate	Chipping	19. 9.04
AWS	724	AWS	Slingsby T.41 Skylark 2S	997		6.55	D.M.Cornelius	Dunstable	8. 7.05
AWU	726		EoN AP.5 Olympia 2	EoN/O/082		5.55	M J Riley	Sackville Farm, Riseley	14. 4.03
AWX	729	AWX	Slingsby T.41 Skylark 2	946		1.56	A.G.Leach	Cranfield	24. 4.01
AWZ	731	AWZ	Slingsby T.7 Cadet	SSK/FF/169	RA847	8. 1.57	R.Moyse	Lasham	3. 5.05
AXB	733	AXB	Slingsby T.41 Skylark 2	926		2.55	A.L.Shaw	Lyveden	30. 6.02
AXD	735	AXD	Slingsby T.43 Skylark 3	1014		.55	F.G.T.Birlison *(Badly damaged Aston Down 11.7.03: fell off tow bar on tow and whole tail section broke off)*	Aston Down	5. 6.04
AXE	736	AXE	Slingsby T.43 Skylark 3	1029		24. 3.57	C.J.Bushell	Snitterfield	21. 8.01
AXJ	740	AXJ	Slingsby T.42A Eagle 2	994		.55	B.Mossop and Partners	Pocklington	11.10.05
AXL	742	AXL	Slingsby T.43 Skylark 3	1030		6.56	D.Mills and Partners	Kingston Deverill	30. 1.05
AXP	745	AXP	Slingsby T.41 Skylark 2	949		4.55	M.Sanderson	Strathaven	16. 3.99
AXV	751		Slingsby T.26 Kite 2A	.		4.56	Not known *(Refurbishing)*	Booker	
AYD	759	AYD	Slingsby T.41 Skylark 2	1048		.56	B.Milburn	Currock Hill	11. 9.03
AYF	761	AYF	Slingsby T.43 Skylark 3B	1058		9.56	D.Chisholm	Hinton in the Hedges	28. 8.05
AYH	763	AYH	Slingsby T.43 Skylark 3B	1066		.56	A.D.Griffiths	Lyveden	5. 6.05
(AYK)	765		Slingsby T.21B	1080		.	Not known *(Stored 1.04)*	RAF Keevil	
(AYN)	768		Slingsby T.41 Skylark 2	996			Not known *(Stored)*	Not known	

AZA	780	AZA	Slingsby T.42 Eagle 3	1085	RNGSA	2-08	J.M.Crewe	Hinton-in-the-Hedges	23. 7.01
					BGA.780				
AZC	782	782	Slingsby T.21B	1096		5.57	M Schima	(Wiener Neustadt, Austria)	9. 7.05
	787		Slingsby T.42B Eagle				See SECTION 3 Part 1	Hamilton East, New Zealand	
AZQ	794	VM687	Slingsby T.8 Tutor		VM687	.57	D.Gibbs	Lee-on-Solent	26. 2.05
AZR	795	AZR	EoN AP.5 Olympia 2	EoN/O/101		7.58	Not known *(Under restoration)*	Syerston	28. 7.95*
AZT	797	AZT	EoN AP.5 Olympia 2	EoN/O/063	ZS-GCM	3.57	T. Green	Winthorpe	15. 8.05
							(De-Registered 22. 9.04 - sold abroad to New Zealand)		
AZX	801	AZX	Slingsby T.41 Skylark 2	995	BGA.1909	4.57	B.J.Griffin	Saltby	
					AGA.4, BGA.801		*(Under rebuild)*		
BAA	804	BAA	Slingsby T.8 Tutor	931	XE761	5.57	A.Chadwick *(Stored 6.02)*	RAF Keevil	9. 3.97*
					VM589	*(Note Cadet TX.1 as "BGA.804 ex VM589" @ Midland Air Museum, Coventry)*			
BAC	806	BAC	Slingsby T.43 Skylark 3B	1101	RNGSA CU19	6.58	FK Hutchinson	Husbands Bosworth	24. 4.04
					BGA.806				
BAL	813	BAL	Slingsby T.43 Skylark 3B	1111		11.57	D.A.Wilson	Milfield	14. 5.05
BAM	814	BAM	Slingsby T.41 Skylark 2	1108		1.58	R.Boyton	Wormingford	2. 6.02
BAN	815		Slingsby T.30B Prefect	1120		1.58	J.S.Allison	RAF Halton	15. 5.00
BAV	822	BAV	Slingsby T.41 Skylark 2B	1113		2.58	M.H.Simms	Shipdham	16. 4.02
BAW	823	BAW	Slingsby T.43 Skylark 3B	1126		2.58	R.Joy and Partner	Halesland	26. 9.05
BAY	825	BAY	Slingsby T.42B Eagle 3	1116		3.58	M.Lodge	Ringmer	15. 8.00
BAZ	826	BAZ	Slingsby T.41 Skylark 2	1112		3.58	W.Fuller	Currock Hill	17. 5.03
BBB	828	BBB	Slingsby T.42B Eagle 3	1118		4.58	I.K.Mitchell	North Hill	22.10.03
BBD	830		Slingsby T.42B Eagle 3	1119		6.58	D.Williams and M.Lodge *(Stored 12.02)*	Ringmer	
BBG	833	BBG	Slingsby T.8 Tutor	-	VW535	3. 9.57	P.Pearson	Upwood	25. 7.04
BBP	840		Slingsby T.43 Skylark 3	1125		6.58	Not known *(Under repair 2000)*	Not known	
BBQ	841	BBQ	Slingsby T.42B Eagle 3	1115		5.58	Eagle Syndicate	Milfield	3.10.05
BBT	844	BBT	Slingsby T.43 Skylark 3B	1134	RAFGSA.234	.58	J.P.Gilbert	Wormingford	11. 5.02
					BGA.844				
BBU	845	BBU	Slingsby T.41 Skylark 2B	1135		.58	D.A.Bullock	Bicester	23. 7.05
BCB	852	TS291	Slingsby T.8 Tutor				See SECTION 3 Part 1	East Fortune	
BCF	856		Slingsby T.21B	1		10.58	P.Underwood	Eaton Bray	*
			(Built Leighton Park School)				*(Blown over Haddenham 14.6.80: stored 2000)*		
BCH	858	BCH	Slingsby T.8 Tutor	SSK/FF/489	VM547	9.58	K.van Rooy	(Belgium)	24. 1.05
BCP	864	BCP	Slingsby T.43 Skylark 3B	1140		11.58	K.I.Latty	Walney Island	13. 4.02
BCS	867	549	Slingsby T.43 Skylark 3B	1144		12.58	A Towse	Rattlesden	16. 4.05
BCU	869	2	Slingsby T.21B	1148		1.59	Not known *(Stored 2002)*	North Connel	*
BCV	870	BCV/155	Slingsby T.43 Skylark 3B	1195		4.59	J. Turner	Challock	18. 7.05
BCW	871	BCW	Slingsby T.43 Skylark 3B	1147		3.59	B Baker	Crowland	3. 4.05
BCX	872		Slingsby T.41 Skylark 2B	1197		4.59	G W Haworth	Tibenham	20. 5.00
BCY	873	T45	Slingsby T.45 Swallow	1198		4.59	T.Wilkinson	Sackville Farm,Riseley	19. 5.05
BDA	875		Slingsby T.21B	1205	AGA.7	6.59	W Grosskinsky	Dahlemer Binz, Germany	4. 7.02
					BGA.875				
BDF	880	BDF	Slingsby T.42B Eagle 3	1213		9.59	D.C.Phillips	Snitterfield	14. 8.04
BDR	890	BDR	Slingsby T.45 Swallow	1243		6.60	J.Turner	Challock	18. 6.05
							(On rebuild 1.96: probably to be VM637)		
BDX	896	BDX	Slingsby T.41 Skylark 2	CH.095/1		6.59	H D Maddams	Ridgewell	3. 6.05
			(Built C Hurst)						
BEA	899	BEA	Slingsby T.41 Skylark 2	1194		7.59	M.L.Ryan	RAF Keevil	9. 1.04
BED	902		Slingsby T.12 Gull I				See SECTION 3 Part 1	East Fortune	
BEL	909	BEL	EoN AP.5 Olympia 2B	EoN/O/126		12.59	J.G.Gilbert and Partners	Wormingford	11. 5.05
BEM	910	BEM	Slingsby T.45 Swallow	1221		5. 2.60	J.Whithead	Halesland	20.10.05
BER	914	BER	Slingsby T.43 Skylark 3B	1225		3.60	K.V.Payne	Feshiebridge	3. 2.99
							(Take-off accident at Bryn-Gwyb-Bach 23. 6.98: stored 5.03)		
BEX	920	BEX	Slingsby T.43 Skylark 3F	1229		4.60	S Barber	Rivar Hill	27. 9.03
BEY	921	BEY	Slingsby T.45 Swallow	1230		4.60	G.P.Hayes	Kenley	4. 6.05
BEZ	922	BEZ	Slingsby T.43 Skylark 3F	1232		4.60	T.L.Cook	Seighford	7. 3.05
BFC	925	BFC	Slingsby T.43 Skylark 3F	1239		20. 5.60	Strathclyde Gliding Club	Strathaven	.9. 4.03
BFE	927	BFE	Slingsby T.43 Skylark 3F	1244		6.60	Essex Skylark Gliding Syndicate	North Weald	.4. 6.03
							(I.F.Barnes) (Noted 1.05)		
BFG	929	BFG	Slingsby T.43 Skylark 3F	1245		8. 7.60	T.J.Wilkinson	Sackville Lodge, Riseley	4. 4.05
BFN	935		EoN AP.5 Olympia 2B	EoN/O/125		2. 3.60	Not known *(Being refurbished 2000)*	Not known	
BFS	939		SZD-8ter-ZO Jaskolka	235		3.60	Not known	Not known	
							(Stored during 2002; damaged by fire)		
BFY	945	BFY	Slingsby T.21B	1251	RAFGGA.515	9.60	D.M.Hayes and Partners	Sutton Bank	24. 4.05
					RAFGSA.286, BGA.945				
BGB	948		Slingsby T.21B	1274	RAFGSA.282	11.60	J Elliott and H Bosworth	Edgehill	25. 8.05
			(Robin EC-44PM)		BGA.948				
BGD	950	BGD	Slingsby T.43 Skylark 3F	1276		26.11.60	Essex University Gliding Club	Wormingford	21. 5.05
BGL	957	BGL	Slingsby T.43 Skylark 3F	1296		3.61	C.Willey and Partners	Perranporth	11. 7.02
BGP	960	R83	Slingsby T.21B	1297	RAFGSA.283	1.61	Bannerdown Gliding Club	RAF Keevil	16. 4.05
					BGA.960				
BGR	962	BGR	EoN AP.5 Olympia 2B	EoN/O/124		6.60	M.H.Gagg	RAF Cosford	5. 1.05
BGT	964		DFS/30 Kranich II	087	SE-STF	29.10.6	C.Wills	Marpingen, Germany	4. 9.05
			(Built AB Flygplan)		Fv.8226				
BHC	973	BHC	EoN AP.5 Olympia 2B	EoN/O/138		1.61	M.Pedwell	Bidford	15. 5.01
							(Damaged beyond repair in hangar)		
BHQ	985	BHQ	Slingsby T.43 Skylark 3F	1304		4.61	T.Wiseman	Andreas	28.12.05
BHS	987		Slingsby T.43 Skylark 3F	1305	RAFGSA293		Not known	Not known	
					BGA.987		*(Under repair during 2000)*		
BHT	988	BHT	Slingsby T.43 Skylark 3F	1306		4.61	K.Chichester and Partners	Hinton in the Hedges	13. 5.02
BHV	990	BHV	Slingsby T.45 Swallow	1308	NEJSGSA.4	4.61	M.G.Dawson	Spilsby	27. 3.01
					BGA.990				

BJB	996	BJB	Slingsby T.43 Skylark 3F	SSK/JPS/1		4.61	CJ Ferrier	Falgunzeon	1.11.05
			(Built Jones, Pentelow and Saint)						
BJC	997	BJC	EoN AP.5 Olympia 2B	EoN/O/135		4.61	P.J.Devey	Lyveden	29. 9.04
BJD	998		SZD-9 bis Bocian 1D	P-391		5.61	G.Pullen *(Bocian Syndicate)*	Lasham	4. 8.05
BJF	1000	BJF	Slingsby T.21B	1309		6.61	Sedbergh Syndicate	Wormingford	4. 8.05
BJJ	1003		Slingsby T.45 Swallow	1310		6.61	*(Stored at RAF Keevil 10.03)*		
BJK	1004	BJK	Slingsby T.43 Skylark 3F	1311		7.61	R.Skinner	Wormingford	20. 7.05
BJP	1008	BJP	Slingsby T.45 Swallow	1316		9.61	G.Winch	Wormingford	31. 3.05
BJQ	1009	BJQ	Slingsby T.49A Capstan	1314			L.A.Glover and Partners	Carlisle	14. 6.05
BJU	1013	113	Slingsby T.43 Skylark 3G	1320		12.61	*(Stored airport 10.03)*		
BJV	1014		Slingsby T.21B				See SECTION 3 Part 1	East Fortune	
BJY	1017	BJY	Slingsby T.45 Swallow	1324		3.62	D.A.Wiseman	Andreas	28.12.05
BJZ	1018		Slingsby T.45 Swallow	1325		3.62	J.P.Marshall	North Connel	20.11.96*
							(Landing accident at North Connel 8. 5.95)		
BKA	1019	BKA	Slingsby T.50 Skylark 4	1326	EI-117	28. 5.62	Staffordshire Gliding Club	Seighford	28. 9.05
					BGA.1019				
BKC	1021	BKC	DFS 108-68 Weihe	231	SE-SNE	4.61	B.Briggs	RAF Cranwell	10. 6.05
			(Built AB Flygindustri)		Fv.8312				
BKE	1023	BKE	Slingsby T.43 Skylark 3F	1715/CR/1		13. 7.61	T.Linee	Kingston Deverill	15. 5.05
			(Built C Ross)						
BKJ	1027	BKJ/270	Schleicher Ka6CR	565/59	9G-AAR	7.61	Vale of White Horse Gliding Centre		
								Sandhill Farm, Shrivenham	25. 6.05
BKL	1029	BKL	EoN AP.5 Olympia 2B	EoN/O/134		6.61	J.S.Orr	Lasham	6. 6.05
BKN	1031	BKN	Schleicher Ka7 Rhönadler	1091/61		9.61	East Sussex Gliding Club	Ringmer	4. 8.04
BKP	1032	BKP	Slingsby T.45 Swallow	1203		10.61	E.Traynor	Easterton	9. 5.02
BKS	1035	BKS	EoN AP.5 Olympia 2B	EoN/O/144		11.61	N W Woodward	Booker	9. 7.05
BKU	1037	BKU	EoN AP.5 Olympia 2B	EoN/O/153		1.62	D.N.MacKay	Aboyne	25. 3.05
BKW	1039	BKW	Schleicher Ka6 Rhönsegler	295	OH-RSA	10.62	P.Montgomery *(Under restoration)*	(West Sussex))	2. 5.01
BKX	1040	BKX	EoN AP.5 Olympia 2B	EoN/O/148		3.62	D.J.Allibone	Gallows Hill	18. 8.04
BLA	1043	327	Slingsby T.50 Skylark 4	1331		5.62	I.Russell	Portmoak	17.12.05
BLE	1047	BLE	Slingsby T.50 Skylark 4	1335	RNGSA 1-228	6.62	S.Franks	Easterton	11. 4.05
					BGA.1047				
BLH	1050	BLH	Slingsby T.50 Skylark 4	1338	RAFGSA	7.62	K.Moon	Rufforth	19.10.04
					BGA.1050				
BLJ	1051	BLJ	EoN AP.6 Olympia 419X	EoN/4/009		3.62	G.Balshaw and Partners	Lleweni Parc	10.10.96*
							"Big Bird" (Stored and for sale 2003)		
BLK	1052	67	EoN AP.6 Olympia 419X	EoN/4/007	G-APSX	4.62	C.J.Abbott and Partners	Seighford	17. 7.03
BLN	1055	BLN	EoN AP.5 Olympia 2B	EoN/O/152		5.62	M.R.Derwent	RAF Cranwell	27. 3.05
BLP	1056	BLP	EoN AP.5 Olympia 2B	EoN/O/149		3.62	R.E.Wooler and Partners	Chipping	13. 6.05
							(Being refurbished during 2000)		
BLQ	1057	BLQ	EoN AP.5 Olympia 2 Special	EoN/O/042	RAFGSA 145	7.62	K.Wood and Partners.	Winthorpe	2. 6.05
					BGA.540				
BLS	1059		EoN AP.5 Olympia 2B	EoN/O/151		7.62	D.J.Wilson	Seighford	25. 8.04
			(EoN rebuild of BGA.897 [EoN/O/128])						
BLU	1061	BLU	Slingsby T.45 Swallow	1340		7.62	J.M.Brookes	Strubby	22. 8.05
BLW	1063	BLW	Slingsby T.50 Skylark 4	1342		8.62	S.R.A.Trusler	RAF Weston on the Green	21. 6.05
BLZ	1066		Slingsby T.50 Skylark 4	1346		11.62	R.M Lambert	Easterton	12. 8.04
BMM	1078	BMM	SZD-9bis Bocian 1	P-397		8.99	W.Burgess	Lyveden	19. 6.05
BMQ	1081	BMQ	Slingsby T.21B	1351		11.62	W.Masterton	(Blenheim, Jamaica)	25. 1.04
BMU	1085		Slingsby T.21B (T)	1355	9G-ABD	12.62	D.Woolerton and Partners	North Coates	27. 9.97
			(Rotax 503)		BGA.1085		*"Spruce Goose" (Noted dismantled 7.02)*		
BMW	1087	BMW	Slingsby T.50 Skylark 4	1357		12.62	M.Williams	Lasham	9. 2.05
BMX	1088	739	Slingsby T.50 Skylark 4	1358	RAFGSA.308	1.63	S.Stanwix	Gallows Hill	26. 5.05
					BGA.1088		*(739 Syndicate)*		
BNA	1091	-	Shenstone Harbinger Mk.2	1		12.62	S.Edyvean	Winthorpe	7. 5.04
BNC	1093	BNC	DFS 108-68 Weihe	1	SE-SHU	3.63	K.S.Green	Lasham	22. 8.05
			(Built AB Kockums Flygindustri)						
BND	1094	BND	Schleicher Ka6CR	1157		3.63	D.Laidlaw and Partners.	Strubby	3. 4.05
BNE	1095	BNE	Slingsby T.50 Skylark 4	1375		4.63	T.Davies	Talgarth	27.10.05
BNH	1098	BNH	Schleicher Ka6CR	6115		3.63	Bath, Wilts and North Dorset Gliding Club		
								Kingston Deverill	22. 7.05
BNK	1100	BNK	Slingsby T.50 Skylark 4	1362		2.63	E.D.Weekes and Partners	RAF Weston on the Green	21. 4.05
BNM	1102		Slingsby T.50 Skylark 4	1367		3.63	D.Hertzberg *(Reported as BMN)*	North Weald	10. 6.05
BNN	1103	BNN	Slingsby T.50 Skylark 4	1366		3.63	A.Jenkins	Hinton-in-the-Hedges	9. 6.05
BNP	1104	653	Slingsby T.50 Skylark 4	1368		3.63	A.R.Worters	North Connel	22. 7.01
BNR	1106	BNR	Slingsby T.49B Capstan	1370		8.63	R.Hardy and Partners	Gransden Lodge	13. 2.05
BNS	1107	XS652	Slingsby T.45 Swallow	1373	XS652	20. 3.63	N.Ling and Partner.	Rufforth	5. 7.05
					BGA.1107				
BNU	1109		Slingsby T.45 Swallow	1377		5.63	P.Cammish	Sutton Bank	15. 6.02
BPA	1115	BPA	Slingsby T.50 Skylark 4	1383		5.63	E.Gardner and Partner	Halesland	3. 1.06
BPB	1116	255	Slingsby T.50 Skylark 4	1384	RNGSA	6.63	G.Robertson	Lasham	15. 4.05
					BGA.1116		*(Collision with Ventus BGA.3295 near Lasham 26. 4.04)*		
BPC	1117	BPC	Slingsby T.50 Skylark 4	1389		7.63	J.L.Grayer and Partner	Ringmer	29. 5.05
BPD	1118	N55	Slingsby T.49B Capstan	1390		22. 7.63	Seahawk Gliding Club	RNAS Culdrose	19. 3.05
BPE	1119	BPE	Slingsby T.50 Skylark 4	1391		6.63	D.H.Scales	Shipdham	25. 5.03
BPG	1121	741	Slingsby T.50 Skylark 4	1393		7.63	R.M.Neill and Partners	Long Mynd	10.10.05
BPK	1124	BPK	Slingsby T.50 Skylark 4	1381		6.63	D.Crowhurst	Crowland	14. 8.02
BPL	1125	BPL	EoN AP.5 Olympia 2B	EoN/O/136	G-APXC	6.63	D.Harris and Partners *(Stored)*	(Essex)	19. 5.97*
BPN	1127	BPN	Oberlerchner Standard Austria	003	OE-0496	6.63	R.K.Avery and Partners	Eaglescott	2. 7.05
BPS	1131	BPS	Slingsby T.49B Capstan	1399		9.63	Capstan Gliding Group	Aboyne	18. 5.05
BPT	1132		Slingsby T.49B Capstan	1400		10.63	J.E.Neville *(Fuselage stored 2004)*	Bicester	11. 2.95*

BPU	1133		Slingsby T.49B Capstan	1402		11.63	J. Hailey	Dunstable	17. 3.98
							tr Capstan Syndicate *(Under repair 204)*		
BPV	1134	BPV	Slingsby T.49B Capstan	1404		20.12.63	G.L.Barrett	RAF Weston on the Green	2.11.04
BPW	1135	BPW	Slingsby T.49B Capstan	1408		1.64	Ulster Gliding Club	Bellarena	20. 6.04
BPX	1136	859	Slingsby T.45 Swallow	1397	XS859	1. 1.64	F.Pape and Partners	Pocklington	5. 7.98
					BGA.1136		*(Stored 10.02)*		
BPZ	1138	BPZ	Slingsby T.50 Skylark	41406		3.64	A.Pattermore and Partners	Rivar Hill	10. 7.00
BQA	1139		Slingsby T.51 Dart 15	1421		4.64	Not known	RAF Odiham	
							(Stored 2000; being refurbished)		
BQE	1143	RA905	Slingsby T.7 Cadet	.	RAFGSA.273	8.63	M.L.Beach	Halton	14. 3.00
					RA905				
BQF	1144	1	Slingsby T.21B	1168	XN189	10.63	Connel Gliding Club	North Connel	13. 4.02
(BQJ)	1147		DFS/30 Kranich II	821	RAFGSA.215	11.63	M.C.Russell	Bishops Stortford	*
			(Built Schleicher)				*(To Bad Tolz Vintage GC, Germany for restoration 7.02)*		
BQM	1150	BQM	EoN AP.10 460 Srs.1B	EoN/S/002	RAFGSA.276	12.63	A.Duncan	Portmoak	5. 6.01
BQP	1152	BQP	Slingsby T.30B Prefect	646	RAFGSA.159	9.99	A.Downie	Dunstable	9. 9.00
BQQ	1153	-	EoN AP.5 Olympia 2B	EoN/0/121	RAFGSA.244	2.64	P.R.Brinson *"Dopey"*	Nympsfield	18. 5.05
							(Rebuilt 1993 using wings from BGA.678)		
BQT	1156	BQT	EoN AP.10 460 Srs.1				See SECTION 3 Part 1	Manchester	
BQU	1157	BQU	Schleicher Ka7 Rhönadler	7141	RNGSA AR66	4.64	A.J.Pellatt	Llantisilio	31. 3.04
					BGA.1157				
BQZ	1162	BQZ	Slingsby T.50 Skylark 4	1416		4.64	G.Colledge	Edgehill	3. 8.02
BRA	1163	BRA	Slingsby T.49B Capstan	1417		4.64	I.Pattingdale	Odiham	29. 5.04
BRB	1164	T51	Slingsby T.51 Dart 15	1423	RAFGSA.334	4.64	M.Sansom	North Hill	4. 2.05
					BGA.1164				
BRC	1165	BRC	Slingsby T.45 Swallow 1	1407		5.64	J.R.Smalley	Kirton-in-Lindsey	26. 3.05
BRD	1166	BRD	Slingsby T.51 Dart 15	1425	RAFGSA.335	5.64	T.Wilkinson	Sackville Lodge, Riseley	4. 4.05
					BGA.1166				
BRE	1167		Slingsby T.45 Swallow 2	1415		5.64	A.Swannock and Partners	Gamston	19. 5.01
BRG	1169	-	Slingsby T.45 Swallow	1410		5.64	A.W.F.Edwards	Gransden Lodge	16. 5.05
BRH	1170		EoN AP.5 Olympia 2B	EoN/O/154		3.64	A.Shallcrass *(For refurbishment)*	Challock	
BRK	1172	243	EoN AP.10 460 Srs.1A	EoN/S/001	G-APWL	4.99	D.G.Andrew	Eaglescott	26. 4.00
					BGA.1172, G-APWL, RAFGSA.268, G-APWL *(Restored as G-APWL 10.02)*				
BRM	1174	BRM	Schleicher Ka7 Rhönadler	776/60	D-4635	5.64	D.S.Driver	Milfield	12. 8.05
BRQ	1177	BRQ	EoN AP.10 460 Srs.1C	EoN/S/003	G-ARFU	6.64	J.Steel and Partners *(Stored 2002)*	Falgunzeon	4. 8.96*
BRT	1180	BRT	Slingsby T.51 Dart 15	1430		6.64	C.Logue	Cosford	9. 9.05
BRU	1181	BRU	Slingsby T.51 Dart 15	1429		6.64	G.M.Polkinghorne	Currock Hill	22. 8.04
BRW	1183	BRW	Slingsby T.49B Capstan	1413		6.64	A.W.West and Partners	Lasham	27.11.04
BRY	1185	BRY	Slingsby T.51 Dart 15	1434		7.64	D.Knight and Partners	Kirton-in-Lindsey	25. 5.05
BSA	1187	BSA	Slingsby T.51 Dart 15	1405		7.64	N G Oultram	Seighford	25. 3.05
BSC	1189	BSC	Slingsby T.50 Skylark 4	1422		8.64	P.Pain	Bidford	12. 8.05
BSE	1191	BSE	Slingsby T.49B Capstan	1414		9.64	D.A Bullock	Bicester	3. 7.05
BSG	1193	BSG	Slingsby T.50 Skylark 4	1436		9.64	Denbigh Gliding Club	Llantisilio	22. 7.01
BSH	1194	BSH	Slingsby T.50 Skylark 4	1444		11.64	D.Trotter	Long Mynd	22. 6.05
BSK	1196	BSK	Slingsby T.49B Capstan	1418		10.64	S.Whitaker	Parham Park	24. 1.05
BSL	1197	BSL	Slingsby T.51 Dart 17	1445		10.64	C.J.Owles	Tibenham	6. 8.05
BSM	1198	597	Slingsby T.51 Dart 15	1439		10.64	D.Tait	Cross Hayes	10. 4.05
BSQ	1201	463	EoN AP.10 460 Srs.1	EoN/S/014		5.64	K.G.Ashford	Husbands Bosworth	10. 5.05
BSR	1202	BSR	Slingsby T.50 Skylark 4	1443		12.64	D.Johnstone	Rattlesden	13. 6.04
BSS	1203	T49	Slingsby T.49B Capstan	1449		12.64	Black Mountains Gliding Club	Talgarth	23. 8.05
BST	1204	J13	Slingsby T.49 Capstan	1451		1.65	B.Bullimore and J.N.Gale	Gransden Lodge	5.11.05
BSW	1207	BSW	Slingsby T.51 Dart 15	1459		2.65	B.L.Owen	Tibenham	19. 8.01
BSX	1208	BSX	Slingsby T.45 Swallow	1461	OO-ZWC	4.65	P.Brownlow and Partners	Sackville Lodge, Riseley	30. 5.01
					F-OTAN-C5, BGA.1208				
BSZ	1210	BSZ	Slingsby T.50 Skylark 4	1460		4.65	T.Cummins	Llantisilio	12. 4.05
BTA	1211	BTA	Slingsby T.45 Swallow	1473		6.65	L.P.Woodage	RAF Odiham	25. 4.05
BTE	1215	BTE	Slingsby T.21B	557	OH-KSA	1.65	J.Assman	Dulmen, Germany	17. 9.05
					SE-SHL				
BTG	1217	BTG	EoN AP.10 460 Srs.1	EoN/S/024		2.65	J.Labell	Strubby	13.12.04
BTH	1218	BTH	Slingsby T.21B	JHB/2		3.65	Mrs. P.Gilmore	Aston Down	2. 3.05
			(Built J.Hulme) (Restored 1995 with wings from BGA.3238/WB981)						
BTJ	1219	BTJ	Schleicher Ka6CR	6367		3.65	D.Keith	Kingston Deverill	14. 3.05
BTK	1220	BTK	Slingsby T.50 Skylark 4	1364	SE-SZW	3.65	J.M.Hall	Long Mynd	27. 8.03
BTM	1222	211	Schleicher Ka6CR	6174		3.65	A.Childs	Strubby	12. 2.05
BTN	1223	BTN	EoN AP.10 460 Srs.1	EoN/S/022	AGA.15	4.65	S.C.Thompson	Parham Park	17. 8.04
					BGA.1223				
BTQ	1225	BTQ	EoN AP.10 460 Srs.1	EoN/S/029		4.65	D.Street	Burn	9. 5.05
BTV	1230		DFS/68 Weihe	000358	RAFGGA	7. 5.65	B.Briggs *(Being refurbished)*	RAF Cranwell	23. 5.93*
BUC	1237	A23	Slingsby T.49B Capstan	1472		6.65	P.Redshaw	Walney Island	17. 2.05
BUE	1239	BUE	Slingsby T.50 Skylark 4	1468		7.65	I.H.Davies	Seighford	29. 3.05
BUF	1240	366	Slingsby T.51 Dart 17R	1469		7.65	C.H.Brown and Partners	Chipping	19. 3.05
BUL	1245	BUL	Slingsby T.51 Dart 17R	1470		7.65	A.Parrish and Partner	Lyveden	24. 2.04
							(Destroyed by fire at Lyveden 16. 9.03)		
BUP	1247	837	Slingsby T.51 Dart 17R	1478		9.65	D.S.Carter	Enstone	10. 6.99
							(Damaged by aerodynamic flutter, Enstone, 25.7.98)		
BUR	1249	BUR	Slingsby T.49B Capstan	1482		11.65	T.Knight	Llantisilio	29. 7.05
BUT	1251	BUT	Slingsby T.43 Skylark 3F	VRT.1		7.65	I.Bannister	Chipping	30. 4.05
			(Built V.R.Tull and Partners)				tr Sky Syndicate		
BUV	1253		EoN AP.10 460 Srs.1	EoN/S/030		7.65	T. Edwards	Camphill	25. 8.05
BUW	1254	BUW	Slingsby T.21B	?	RAFGSA.242	8.65	J.N.Wardle *"Lucy"*	Lasham	1. 6.05
BUZ	1257	BUZ	Schleicher Ka6CR	6418		8.65	R.Leacroft	Lyveden	7. 4.05

BVB	1259	BVB	Schleicher Ka7 Rhönadler	7230		9.65	Dartmoor GC	Brentor	7. 3.05
			(Modified to ASK 13 standard)						
BVE	1262	61	Slingsby T.51 Dart 17R	1483		11.65	K.Hale and Partner	Sandhill Farm, Shrivenham	19. 6.05
BVF	1263	BVF	Slingsby T.45 Swallow	1481		11.65	Bidford Swallow Syndicate	Bidford	13. 8.03
BVH	1265	BVH	Slingsby T.51 Dart 17R	1485		12.65	D.J.Simpson	Halesland	6.12.05
BVJ	1266	BVJ	Slingsby T.51 Dart 17R	1486		1.66	R.and M.Weaver	Usk	7. 1.05
BVM	1269	150	Slingsby T.51 Dart 17R	1492		1.66	N.H.Ponsford *(Stored 12.99)*	(Breighton)	5.89*
BVN	1270		EoN AP.10 460 Srs.1	EoN/S/023		3.65	F J Clarke and Partners	North Hill	18. 4.04
BVR	1273	BVR	Schleicher Ka6CR	6441		5.10.65	R.C Cannon and Partners	Lasham	9. 3.05
BVS	1274		SZD-9 bis Bocian 1D	F-831		16.10.65	Spilsby Soaring	Spilsby	28. 7.01
BVW	1278		EoN AP.6 Olympia 403	EoN/4/001	RAFGSA.306	8.65	J.B.and K.D.Dumville	Camphill	29. 5.01
					G-APEW				
BVX	1279	BVX	Schleicher Ka6CR	6439		10.65	M.Schlotter	Nympsfield	26. 9.05
BVY	1280	BVY	LET L-13 Blanik	173121		17.10.65	Strathclyde Gliding Club	Strathaven	19. 8.05
BVZ	1281	BVZ	Schleicher Ka6CR	6446		10.65	L.Blair	Bellarena	19. 8.05
BWB	1283	B96	EoN AP.10 460 Srs.1	EoN/S/036		12.65	S Metcalfe	Tibenham	17.12.03
BWC	1284	BWC	Schleicher Ka6CR	6449		12.65	Birmingham University Gliding Club	Snitterfield	26.11.05
BWE	1286	BWE	EoN AP.10 460 Srs.1	EoN/S/035		12.65	C.Hughes	Nympsfield	18. 8.05
BWG	1288	465	EoN AP.10 465 Srs.2	EoN/S/038		7.12.65	K.S.Green and Partner *(Being refurbished)*	Lasham	27. 4.97*
BWJ	1290	377	Slingsby T.51 Dart 17R	1495		10.2.66	D.Godfrey and Partners	Edgehill	8.10.02
BWK	1291		Slingsby T.45 Swallow	1493		2.66	K.Hubbard and Partners	North Hill	5. 6.03
BWM	1293	182	Slingsby T.51 Dart 17R	1500		4.66	P.L. and L.E.Poole	Kenley	9. 6.05
BWP	1295	861	Slingsby T.51 Dart 17R	1501		3.66	R.Johnson	Parham Park	13. 3.05
BWQ	1296	BWQ	Slingsby T.51 Dart 15	1505		3.66	C.Uncles	Halesland	31. 7.05
BWS	1298	517	Slingsby T.51 Dart 17R	1502		4.66	R.D.Broom and E.A.Chalk	Hinton-in-the-Hedges	13. 5.02
BWU	1300	BWU	EoN AP.10 460 Srs.1	EoN/S/034		1.66	P.W.Berringer	Ridgewell	23. 4.05
BWX	1303	BWX	EoN AP.5 Olympia 2B	101		2.66	A.Buxton	Husbands Bosworth	19. 7.05
			(Built from spares)						
BXB	1307		EoN AP.10 460 Srs.1	EoN/S/040		3.66	J.M.Lee	Parham Park	30.10.02
BXE	1310	BXE	Slingsby T.51 Dart 15R	1509		5.66	S.Briggs	Currock Hill	28. 5.05
BXG	1312	BXG	Slingsby T.51 Dart 17R	1512		5.66	J.M.Whelan	Husbands Bosworth	7. 5.05
BXH	1313	BXH	Slingsby T.51 Dart 17R	1516		6.66	C.Rodwell	Husbands bosworth	19. 4.05
BXK	1315		Slingsby T.21B	1510		6.66	Not known	Rufforth	*
							(Damaged Falgunzeon 18.5.80; being refurbished 2000)		
BXL	1316	121	Slingsby T.51 Dart 17R	1517		6.66	W.R.Longstaff and Partner	Feshiebridge	27. 9.04
BXM	1317	9	Slingsby T.51 Dart 17R	1521		7.66	P.R.Davie	Dunstable	21. 8.04
BXP	1319		Slingsby T.45 Swallow 2	1522		7.66	Carlton Moor Gliding Club	Carlton Moor	1. 8.04
BXT	1323	BXT	Schleicher Ka6CR	6492		30.4.66	C Knowles	Dunstable	30.10.05
BXV	1325	G-ATRA	LET L-13 Blanik	173304	G-ATRA	12. 5.66	Blanik Syndicate *(P.Martin)*	Husbands Bosworth	14. 5.04
BXW	1326	BXW	LET L-13 Blanik	173305	G-ATRB	16. 6.66	R Chapman	Bidford	26. 8.05
BXY	1328	BXY	EoN AP.10 460 Srs.1	EoN/S/042		6.66	G.K.Stanford	Brent Tor	26. 7.03
BYA	1330	BYA	Slingsby T.51 Dart 17R	1518		7.66	G.A.Chalmers	Easterton	18. 5.05
							(Overstressed and written-off Easterton 15. 8.04)		
BYC	1332	BYC	Slingsby T.51 Dart 17R	1526		8.66	D.J.Ireland	Snitterfield	11. 4.05
BYE	1334	BYE	EoN AP.10 463 Srs.1	EoN/S/044		9.66	C.J.Bushell	Snitterfield	29. 4.01
BYG	1336	225	Slingsby T.51 Dart 17R	1535	RAFGSA	11.66	W.T.Emery	Rufforth	13. 6.05
					BGA.1336				
BYK	1339	BYK	Slingsby T.45 Swallow	1566		1.67	G.E.Williams	St. Hubert, Belgium	15. 6.04
							(De-Registered 18. 6.04 - presumed sold in Belgium)		
BYL	1340	BYL	Schleicher Ka6CR	6517		7.66	D.Heaton	Seighford	9.12.05
BYM	1341	558	Schleicher Ka6CR	6518	RAFGSA.381	7.66	K.S.Smith	Wormingford	29. 9.05
					BGA.1341				
BYU	1348	350	Schleicher Ka6CR	6525	XW640	9.66	R.N.John	Dunstable	13. 6.05
					BGA.1348				
BYX	1351	BYX	Schleicher Ka6E	4055		12.66	J.Dent and D.B.Andrews	Chipping	6. 4.05
BYY	1352	BYY	Slingsby T.21B	628	RAFGSA.338	11.66	T.Akerman	(Chauvigny,France)	28. 6.02
					WB967				
BZA	1354	BZA	Slingsby T.21B	1162	RAFGSA.318	11.66	A Hill	Bicester	9.11.04
					XN183				
BZB	1355	BZB	EoN AP.10 460 Srs.1	EoN/S/047		10.66	D.C.Ratcliffe Syndicate	Parham Park	21. 5.04
BZC	1356	BZC	Slingsby T.51 Dart 17R	1563		2.67	A.N.Ely	Strubby	28. 3.04
BZF	1359	311	Slingsby T.51 Dart 17R	1570		3.67	P.C.Gill and Partners	Ridgewell	8. 8.05
BZG	1360	N54	Slingsby T.49B Capstan	1581		28. 4.67	Seahawk Gliding Club	RNAS Culdrose	20. 3.05
BZM	1365	BZM	Slingsby T.45 Swallow	1597		7.67	F.Webster	Easterton	4. 6.05
BZP	1367	F4	SZD-24-4A Foka 4	W-301		1.67	J.Mare	Crowland	1. 5.05
BZR	1369	471	EoN AP.10 460 Srs.1	EoN/S/049		2.67	G.Wardell	Camphill	6. 7.02
BZV	1373	BZV	EoN AP.10 460 Srs.1	EoN/S/046		2.67	I.F.Smith	Lasham	21. 2.05
BZX	1375		Schleicher Ka6CR	6571		3.67	Leeds University Gliding Club	Dishforth	6. 6.05
BZY	1376		Slingsby T.31B	SSK/FF1817	BGA.1175	3.67	A.L.Higgins	Thame	3. 5.03
			(Rebuild of BGA.1175)				tr The Blue Brick Syndicate		
BZZ	1377	77	SZD-24-4A Foka 4	W-308		3.67	M.Hudson and Partners	Lasham	25. 5.05
CAC	1380	994	Schleicher Ka6E	4054		3.67	M.Burridge	Crowland	24. 3.05
CAE	1381	575	Schleicher Ka6E	4076		8. 4.67	N.Rolfe	Seighford	4. 4.05
CAF	1382	CAF	EoN AP.5 Olympia 2B	EoN/O/131	RAFGSA.254	5. 4.67	B.Kozuh	Grobnik, Croatia	17. 7.05
CAG	1383	715	Schleicher Ka6E	4080		4.67	S.L.Beaumont	Crowland	6. 6.05
CAK	1386	117	EoN AP.5 Olympia 2B	EoN/O/122	RAFGSA.246	3.67	K.Veyama	Yokohama, Japan	2.10.05
CAN	1389	CAN	EoN AP.10 460 Srs.1	EoN/S/050		3.67	B.Kozuh	Grobnik, Croatia	11. 9.05
CAQ	1391	812	Schempp-Hirth SHK	37		3.67	M.Clarke	Kingston Deverill	17. 7.05
CAR	1392	422	Schempp-Hirth SHK-1	40		4.67	P.Gentil and M.Gresty	Aston Down	12. 4.04
CAS	1393	372	Schleicher Ka6E	4029	RAFGSA.372	5.67	R.F.Tindall	Gransden Lodge	22. 8.02

CAT	1394	CAT	EoN AP.10 460 Srs.1	EoN/S/051		5.67	D.C.Phillips and Partners	Snitterfield	23. 7.05
CAV	1396	CAV	Schleicher ASK13	13015		5.67	M.F.Cuming	Edgehill	24. 6.04
CAX	1398	CAX	Slingsby T.45 Swallow	1598		7.67	G.M.Hicks	Waldershare Park	17. 8.02
CAZ	1400	500	Slingsby T.51 Dart 17WR	1611		7.68	J.S.Halford	Gallows Hill	9. 6.05
CBA	1401	679	Slingsby T.51 Dart 17WR	1612		7.68	D.R.Bennett and P.H.Pickett	Snitterfield	12. 4.05
	1402		Slingsby HP-14C				See SECTION 3 Part 1	Cuatro-Vientos, Madrid, Spain	
CBK	1410		Grunau Baby III	----	RAFGSA.378	5. 9.67	N.H.Ponsford	Breighton	4.83*
			(Built Sfg.Schaffin)		D-4676		*(Preserved at Cavan & Leitrim Railway, Dromod, Ireland 8.04)*		
CBM	1412	343	Schleicher Ka6CR	6607		7.67	F.Ballard	Nympsfield	9. 6.05
CBN	1413	CBN	SZD-30 Pirat	W-320		5.67	B.C.Cooper	Portmoak	12.11.01
							(Crashed at Strahallan 6. 5.01, scrapped at Portmoak)		
CBP	1414	CBP	SZD-24C Foka	W-198	OY-BXR	7.67	G.Sutton	Sutton Bank	18. 6.01
CBR	1416	CBR	Aeromere M.100S	044		7.67	G.Viglione	Rattlesden	7. 3.04
CBW	1421	CBW	Schleicher ASK13	13034		8.67	Stratford-upon-Avon Gliding Club	Snitterfield	22. 1.05
CBY	1423	475	Schleicher Ka6CR	960	RAFGSA.322	10.67	G.Martin	Talgarth	31. 7.05
					D-3222				
(CBZ)	1424		Slingsby T.8 Tutor	SSK/FF27	RAFGSA.214	10.67	*(Preserved at Cavan & Leitrim Railway, Dromod, Ireland 8.04)*		
					RA877				
CCA	1425	CCA	Schleicher Ka6E	4126		10.67	R.K.Forrest	Feshiebridge	6. 12.04
CCB	1426	CCB	Schempp-Hirth SHK-1	52		7.67	R.M.Johnson	Milfield	19. 9.05
CCC	1427	CCC	Schleicher ASK13	13035	RAFGSA.R83	6.99	Shenington Gliding Club	Bicester	20. 3.04
					BGA.1427		*(Dismantled 1.05)*		
CCD	1428	373	Schleicher Ka6E	4127		12.67	M.H.Phelps	Husbands Bosworth	2. 4.05
CCE	1429	CCE	Schleicher ASK13	13047		12.67	Oxford Gliding Club	RAF Weston on the Green	29. 3.05
CCF	1430	CCF	Schleicher ASK13	13042		12.67	Norfolk Gliding Club	Tibenham	10. 4.04
CCG	1431	CCG	Schleicher Ka6E	4125		12.67	R.J.Playle	Edgehill	21. 6.02
CCJ	1433	878	Schleicher Ka6CR	6145	RAFGSA.323	10.67	R.J.Playle	Edgehill	4. 6.05
CCL	1435	47	Schleicher Ka6E	4129		3.68	M.T.Stanley	Sutton Bank	14. 1.05
CCM	1436	CCM	Schleicher ASK13	13053		2.68	Burn Gliding Club	Burn	17. 6.05
CCN	1437	CCN	SZD-9 bis Bocian 1E	P-431		3.68	M and D Morley	Kenley	14. 2.04
							(Scrapped by new owners 11.4.04)		
CCP	1438	L99	Schleicher ASK13	13052		2.68	G.D.Pullen	Lasham	18. 2.05
CCR	1440	CCR	Schleicher Ka6E	4149		2.68	A.Shaw	Edgehill	18. 3.05
CCS	1441	CCS	Slingsby T.41 Skylark 2	1008	PH-230	3.68	S.Longstaff	Aston Down	14. 2.05
CCT	1442	CCT	Schleicher ASK13	13057		3.68	Stratford-upon-Avon Gliding Club	Snitterfield	14. 1.05
CCU	1443	CCU	Schleicher Ka6E	4122		3.68	C.Hocking	RAF Keevil	9. 4.05
CCV	1444	CCV	Schleicher Ka6E	4160		3.68	R.Hood	Lasham	8. 7.04
CCW	1445	CCW	Schleicher ASK13	13051		3.68	Needwood Forest Gliding Club	Cross Hayes	14. 2.05
CCX	1446	CCX	Schleicher ASK13	13054		3.68	Trent Valley Gliding Club	Kirton-in-Lindsey	22. 3.05
CCY	1447	CCY	Schleicher ASK13	13050		3.68	Devon and Somerset Gliding Club	North Hill	3. 2.05
CCZ	1448	CCZ	Schleicher ASK13	13070		3.68	Trent Valley Gliding Club	Kirton-in-Lindsey	6.10.04
CDA	1449	CDA	Schleicher Ka6E	4136		3.68	F.T.Bick and Partners	Aboyne	17. 5.05
CDB	1450	CDB	Schleicher Ka6E	4137		3.68	K.L.Holburn	Currock Hill	7. 5.05
CDC	1451	CDC	Schleicher K8B	8743		3.68	Enstone Eagle Gliding Club *(Stored 7.01)*	Rivar Hill	15. 3.99
CDD	1452	CDD	Schleicher Ka6E	4165		3.68	C.Sherriff	Booker	15. 8.05
CDF	1454	683	Schleicher Ka6E	4162		3.68	K.G.Reid and Partners	Rivar Hill	25. 5.05
CDG	1455	CDG	FFA Diamant 18	35		3.68	J.A.Luck	Hinton-in-the-Hedges	23. 4.05
CDH	1456	619	Schempp-Hirth HS.2 Cirrus	10		4.68	A.K.Moore	Lasham	23. 1.05
CDK	1458	CDK	Schleicher K8B	8747		5.68	Burn Gliding Club	Burn	23. 5.05
(CDN)	1461	CDN	EoN AP.7 Primary				See SECTION 3 Part 1	Flixton, Bungay	
CDQ	1463		Grunau Baby III	R161		6.68	Not known *(Under repair 2000)*	Not known	
CDR	1464	CDR	Scheibe Bergfalke III	5625		8.68	N.M Neil	Eagelscott	2. 4.04
CDU	1467	CDU	Schempp-Hirth SHK-1	V.1	D-8441	5.68	SHK Syndicate	Wormingford	23. 3.05
CDW	1469	CDW	FFA Diamant 18	033		8.68	J.L.McIver	Falgunzeon	2.11.05
CDZ	1472	CDZ	Schleicher Ka6E	4177		5.68	J R Minnis	Wormingford	15. 7.05
CEA	1473	CEA	Schempp-Hirth HS.2 Cirrus	21	XZ405	8.68	M.S.Whitton	Long Mynd	18. 4.05
					BGA.1473, D-8437				
CEB	1474	CEB	SZD-9 bis Bocian 1E	P-433		5.68	Bath, Wilts and North Dorset Gliding Club		
								Kingston Deverill	22. 7.05
CEC	1475	18	Schempp-Hirth HS.2 Cirrus	22		7.68	C.R.Ellis	Long Mynd	9. 2.05
CED	1476	814	Schleicher Ka6E	4196		6.68	H.G.Williams and Partners	Snitterfield	9. 3.05
							(Damaged in field landing, Loxley, Warwicks, 28. 7.04)		
CEG	1479		Schleicher Ka6E	4203		6.68	J.C.Boley	Halesland	21. 3.96*
							(Written off 'in the Alps' 13. 6.95)		
CEH	1480	CEH	Wassmer WA.22 Super Javelot	68	F-OTAN-C6	7.68	N.A.Mills	Lasham	18. 6.05
					F-CCLU				
CEK	1482		Slingsby T.21B (T)	1151	RAFGSA.369	7.68	D.Woolerton	North Coates	7. 5.05
			(Fuji-Robin EC34PM s/n 82-00391)		XN147				
CEL	1483	JD	Schleicher Ka6E	4174		8.68	Essex and Suffolk Gliding Club	Wormingford	26.11.05
CEM	1484	CEM	Schleicher Ka6E	4212		8.68	G.D.Bowes	Kirton-in-Lindsey	1. 6.05
CEN	1485	CEN	SZD-30 Pirat	W-393	SP-2520	7.68	A.Bogan	Kirton-in-Lindsey	28. 9.05
CEQ	1487	458	Schleicher Ka6E	4230		8.68	C.L.Lagden and Partners	Ridgewell	19. 7.05
CEW	1493	CEW	Schleicher Ka6E	4209		8.68	J.W.Richardson	Dunstable	8. 5.05
CEX	1494	CEX	Schleicher ASK13	13108		9.68	Carlton Moor Gliding Club	Carlton Moor	8. 4.05
CEY	1495	CEY	Schleicher Ka6E	4222		8.68	P.A.Pearson	Husbands Bosworth	16. 2.05
CFA	1497	CFA	Schleicher ASK13	13113		10.68	Booker Gliding Club	Booker	11. 8.05
CFC	1499	CFC	Schleicher Ka7 Rhönadler	470	RAFGSA.387	11.68	P.Barnes	Edgehill	13. 4.05
					F-OTAN-C1				
CFF	1502	CFF	Schleicher K8B	8765		10.68	Denbigh Gliding Club	Lleweni Parc	24. 5.05
							(Damaged at Lleweni Parc, 30. 7.04)		
CFG	1503	CFG	Schleicher ASK13	13115		10.68	Staffordshire Gliding Club	Seighford	21. 3.05

CFK	1506	CFK	Schempp-Hirth HS.2 Cirrus	38		11.68	C.V Webb and Partners	Seighford	21.12.05
CFL	1507	CFL	Schleicher Ka6E	4215		10.68	E. Lambert	Aston Down	14. 4.05
CFM	1508	CFM	Schleicher ASK13	13121		7.12.68	Vale of The White Horse Gliding Centre	Sandhill Farm, Shrivenham	24.11.05
CFS	1513	CFS	Glasflugel H.201 Standard Libelle	83		4.70	P.G.F.Steele and Partner.	Shipdham	30. 4.05
CFT	1514	62	Slingsby T.59A Kestrel 17	1729		3.73	J.A.Kane	Sutton Bank	30. 6.05
CFW	1517	-	Glasflugel H.201 Standard Libelle	235		2.72	D.Edwards	Rhigos	26. 5.05
CFX	1518	CFX	Glasflugel H.201 Standard Libelle	274		2.72	K.D.Fishenden	Dunstable	15. 4.05
CFY	1519	862	Glasflugel H.201 Standard Libelle	270		3.72	C.W.Stevens	Rufforth	3. 1.05
CGB	1522	CGB	Schleicher Ka6E	4247		12.68	M.J.Huddart	Winthorpe	16. 5.05
CGD	1524	CGD	Schleicher Ka6E	4202		1.69	I F Smith	Lasham	6. 6.05
CGE	1525	CGE	Schleicher Ka6E	4246		1.69	C.Weir	Bellarena	2. 4.05
CGH	1528	153	Schleicher K8B	8772		23. 2.69	I.G.Brice	Lasham	19. 4.05
CGJ	1529	CGJ	Schleicher K8B	8773		2.69	Nene Valley Gliding Club	Upwood	5. 6.05
CGK	1530	124	Schleicher Ka6E	4261		3.69	J.E.Hampson *(Crashed Wortham near Diss, 28. 7.03)*	Ridgewell	24. 9.03
CGM	1532	CGM	FFA Diamant 18	053		4. 4.69	J.G.Batch	Hinton in the Hedges	13. 2.05
CGN	1533	309	Schleicher Ka6E	4173		3.69	R.Targett	Nympsfield	18. 4.01
CGQ	1535	CGQ	Schleicher ASK13	13153		7. 4.69	Oxford Gliding Club	RAF Weston on the Green	18. 4.05
CGS	1537	CGS	FFA Diamant 18	055		7.69	C.J.Wimbury	Talgarth	18.10.05
CGT	1538	449	Schempp-Hirth SHK-1	38	D-1966	4.69	B.W.Svenson	Pocklington	28.12.05
CGV	1540	CGV	PIK-16C Vasama	48		4.69	D.J.Osborne and Partners	Currock Hill	4. 7.05
CGX	1542	CGX	Bölkow Phoebus C	869		4.69	W.N.Smith and Partners	Sackville Lodge, Riseley	26. 5.03
CGY	1543	CGY	Schempp-Hirth HS.2 Cirrus	51		18. 4.69	R.M.Munday and Partners	Eaglescott	22. 6.05
CGZ	1544	CGZ	Schempp-Hirth SHK	39		5.69	M.C.Ridger	Saltby	3. 3.03
CHB	1546	577	Schleicher Ka6E	4235		5.69	D.L.Jones	RAF Weston on the Green	10. 4.05
CHE	1549	CHE	Slingsby T.41 Skylark 2 *(Built Doncaster Sailplane Services)*	DSS.002		6.69	M.S.Howey	Camphill	30. 5.05
CHG	1551	N52	SZD-30 Pirat	B-294		27. 6.69	SeahawkGliding Club	RNAS Culdrose	19. 3.05
CHJ	1553	CHJ	Bölkow Phoebus C	879		6.69	D.C.Austin	Sutton Bank	30. 4.02
CHK	1554	CHK	EoN AP.5 Olympia 2B	?	RAFGSA	6.69	Oly Gliding Syndicate	Halesland	5. 2.05
CHL	1555	CHL	SZD-30 Pirat	B-295		6.69	T.A.Sage	Dunstable	1. 6.04
CHM	1556		Schleicher Rhonlerche II	860	D-5015	4.69	Not known *(Stored)*	RAF Keevil	
CHT	1562	846	Schleicher ASW15	15013		8.69	D.Dyer and Partners	Hinton-in-the-Hedges	22. 4.05
CHU	1563	CHU	Schleicher K8B	8794		10. 8.69	H B Chalmers *(Highland Gliding Club Syndicate)*	Easterton	14. 5.05
CHW	1565	CHW	Schleicher ASK13	13187		8.69	Dorset Gliding Club *(Damaged Gallows Hill 19.11.04)*	Gallows Hill	2. 4.05
CHY	1567	CHY	Slingsby T.45 Swallow *(Built R.Greenslade from kit)*	RG.103		9.69	J.L.H.Pegman and Partners	Currock Hill	27. 6.02
CHZ	1568	857	Schleicher Ka6E	4153	N6916	9.69	P.Everitt	Parham Park	30. 3.05
CJB	1570	764	Bölkow Phoebus C	919		12.69	D.Clarke	Burn	30. 8.05
CJC	1571		Ginn-Lesniak Kestrel	1		10.69	P.G.Fairness and K Burns	Milfield	25. 3.02
CJD	1572	S14	Schleicher ASK13	13182		19.10.69	Shenington Gliding Club	Edgehill	7. 1.03
CJF	1574	474	Schleicher K8B	8803		29.11.69	Surrey and Hampshire Gliding Club	Lasham	20. 4.05
CJG	1575	CJG	Wassmer WA.21 Javelot II	38	F-OTAN-C4 F-CCEZ	1.70	R S Hanslip	Burn	11. 5.04
CJJ	1577	CJJ	Bölkow Phoebus C	913		15. 1.70	P Maddocks	Portmoak	8. 6.04
CJK	1578	CJK	Schempp-Hirth SHK	35	RAFGSA	25 2.70	R.H.Short	Lyveden	16. 4.05
CJL	1579	222	Schempp-Hirth SHK-1	42	OO-ZLG	2.70	M.F.Brook	Barkston Heath	17. 4.05
CJM	1580	CJM	Schleicher K8B	8814		8. 3.70	Surrey and Hampshire Gliding Club	Lasham	13. 5.05
CJN	1581	CJN	Schempp-Hirth SHK-1	55		3.70	R.Makin	Camphill	25. 6.05
CJP	1582	CJP	Schleicher ASW15	15041		3.70	C.Paine	Ridgewell	10. 6.05
CJR	1584	CJR	Schempp-Hirth HS.2 Cirrus	87		3.70	M and S Tolson and J.H.Stanley	Dishforth	6. 2.05
CJY	1591	CJY	Schleicher Ka6CR	555	(RAFGSA)	4.70	Bristol University Gliding Club	Nympsfield	10. 6.05
CKC	1595	CKC	Bölkow Phoebus C	936	(BGA.1590)	21. 4.70	S.J.Bennett	Bidford	18. 7.02
CKD	1596	CKD	SZD-30 Pirat	B-327		4.70	Borders Gliding Club	Milfield	27. 6.05
CKF	1598	961	Glasflugel H.201 Standard Libelle	101		4.70	S.M.Turner	Crowland	27. 6.04
CKL	1603	CKL	Schleicher Ka6E	4336		4.70	I.Lowes and Partners	Milfield	22. 8.05
CKN	1605	CKN	SZD-9 bis Bocian 1E	P-496		5.70	Lincolnshire Gliding Club *"Enola Gay"*	Strubby	26. 6.04
CKP	1606	CKP	Schleicher ASW15A	15058		7.70	M.G.Shaw and Partners	Portmoak	31. 3.05
CKR	1608		Schleicher ASK13	13247		7.70	Essex Gliding Club *(Noted 1.05)*	North Weald	4.10.04
CKT	1610		Scheibe Bergfalke II	E.03	D-9208	7.70	Not known *(Being restored)*	Thame	
CKU	1611	CKU	Schleicher ASK13	13243		11. 8.70	Essex Gliding Club	Ridgewell	27. 5.05
CKV	1612	T10	Schleicher ASK13	13253		8.70	Black Mountains Gliding Club	Talgarth	19. 7.05
CKY	1615	743	Glasflugel H.201 Standard Libelle	139		22. 8.70	G.Herbert	Bidford	13. 5.03
CKZ	1616	724	Schempp-Hirth HS.4 Standard Cirrus	52	RAFGSA BGA.1616	8.70	R.Johnson	Dunstable	24.11.05
CLA	1617	CLA	Schempp-Hirth HS.4 Standard Cirrus	63		11.70	J.A.Wight and D.Dye	Nympsfield	18.10.05
CLC	1619		Slingsby T.21B	1200	RNGSA 2-07	11.70	Not known *(Stored 2000)*	Not known	
CLF	1622	CLF	Schleicher Ka7 Rhönadler	931	D-5062	5. 1.71	P.M.Morgan and Partners	Tibenham	27. 6.04
CLG	1623	CLG	Schempp-Hirth SHK-1	36	RAFGSA.27	9. 1.71	D. Evans *(Crashed on landing, Bembridge 5.10.03 - wreck exported to Finland)*	Bembridge	17. 5.04
CLH	1624	252	Schempp-Hirth HS.4 Standard Cirrus	77		??	P.C.Bray and Partners *(Crashed Nympsfield 7. 8.04)*	Nympsfield	
(CLJ)	1625		EoN AP.7 Primary	EoN/P/035	WP267	8. 2.71	T Akerman *(On rebuild 2004)*	Poitiers, France	2.72*
CLK	1626	CLK	Schleicher Ka7 Rhönadler *(Partly modified to ASK-13 standard)*	607	D-5714	2.71	Cornish Gliding Club	Perranporth	27. 7.05
CLM	1628	535	Glasflugel H.201 Standard Libelle	178		2.71	J.N.Cochrane	North Hill	15. 4.05
CLN	1629	142	Glasflugel H.201 Standard Libelle	175		4.71	J.N.Wardle *(Crashed near Didcot 11. 8.03 after collision with BGA.2806)*	Lasham	24. 4.04

CLP	1630	948	Glasflugel H.201B Standard Libelle	176		2.71	A.Jelden	Booker	7. 4.05
CLQ	1631	CLQ	Schempp-Hirth HS.2 Cirrus	99		29. 1.71	K.Bastenfield	Brent Tor	11. 3.05
CLR	1632	284	Glasflugel H.201B Standard Libelle	173		4.71	R.A.Christie	Easterton	13. 3.05
CLT	1634		Schleicher Ka7 Rhönadler	251	D-5529	4.71	R.Spencer	Rhigos	7.12.02
			(Partly modified to ASK-13 standard)				tr The Syndicate		
CLV	1636	CLV	Glasflugel H.201 Standard Libelle	180		3.71	J.M.Sherman	Parham Park	17.11.05
CLW	1637	937	Glasflugel H.201 Standard Libelle	174		12. 3.71	N.A.Dean and Partners	Kirton-in-Lindsey	16. 2.05
CLX	1638	CLX	Schleicher K8B	8851		12. 3.71	Midland Gliding Club	Long Mynd	9. 1.03
							(Blown over Long Mynd 8.3.02)		
CLY	1639		Hirth Go.III Minimoa	378	PH-390	20. 3.72	Not known	Dunstable	1.79*
					D-5076		*(On rebuild 2000)*		
CLZ	1640	799	Schleicher Ka6E	4056	AGA.2	2. 4.71	M.Pagram and G. Walker	Pocklington	11. 9.05
CMF	1646	CMF	SZD-32A Foka 5	W-534		7.71	D J Linford	Ringmer	29. 6.05
CMG	1647	CMG	Schleicher Ka7 Rhönadler	462	D-8116	7.71	I.G.Brice *"Fledermaus"*	Lasham	26. 2.04
CMH	1648	165	Glasflugel H.201B Standard Libelle	224		12. 7.71	D.N Greig	North Hill	18. 3.04
CMK	1650	CMK	Schleicher ASK13	13305		8.71	South Wales Gliding Club	Usk	9.12.05
CMN	1653	CMN	Schleicher K8B	8870		23. 8.71	Bristol and Gloucestershire Gliding Club	Nympsfield	20. 5.05
CMQ	1655	CF	Glasflugel H.201 Standard Libelle	233		13. 8.71	P.Britten	Booker	2. 7.05
CMR	1656	CMR	Glasflugel H.201 Standard Libelle	225		15. 8.71	J.A.Dandie and Partners	Portmoak	3. 8.05
CMS	1657	602	Glasflugel H.201 Standard Libelle	234		8.71	D.Manser and Partners	Challock	16. 4.05
CMT	1658		Scheibe Bergfalke II	124	D-6012	22. 8.71	Not known *(Stored 2000)*	Not known	
CMV	1660	184	Glasflugel H.201 Standard Libelle	235		11. 8.71	S.E.Evans and Partners	RAF Weston on the Green	21. 2.05
CMW	1661	CMW	Glasflugel H.201B Standard Libelle	242		9.71	T.E. Rose	Booker	16. 4.05
CMX	1662	226	Glasflugel H.201 Standard Libelle	232		9.71	D.Bell and Partners	Burn	14. 2.05
CMZ	1664	CMZ	Schleicher Ka7 Rhönadler	323	D-5589	11. 6.72	Cornish Gliding Club	Perranporth	27. 7.05
			(Partly modified to ASK-13 standard)						
CNC	1667	273	Schempp-Hirth Standard Cirrus	167		12.71	J.H. Fox and Partners.	Portmoak	5.3.05
CND	1668	CND	SZD-9 bis Bocian 1E	P-428	RAFGSA.392	1.72	Angus Gliding Club	Drumshade	1. 5.04
CNE	1669	525	Glasflugel H.201 Standard Libelle	266		1.72	E.T.Melville	Portmoak	24. 4.05
CNF	1670	709	Glasflugel H.201B Standard Libelle	271		1.72	D.F.Mazingham	Pocklington	23. 3.05
CNG	1671	622	Glasflugel H.201 Standard Libelle	265		2.72	C.F.Smith and Partner	Nympsfield	9.11.05
CNH	1672	442	Glasflugel H.201 Standard Libelle	269		2.72	A.D'Otreppe	Lasham	28. 7.05
CNJ	1673	CNJ	Glasflugel H.201 Standard Libelle	272		9. 2.72	R.Thornley	Crowland	23. 3.05
CNK	1674	CNK	SZD-30 Pirat	B-459		3.72	R.C.T.Birch	Portmoak	25. 4.05
CNM	1676	CNM	SZD-9 bis Bocian 1E	P-551		2.72	G.Morriss	Crowland	30. 5.04
CNN	1677	CNN	Schempp-Hirth HS.4 Standard Cirrus	173		2.72	M.A.Cropper	Brentor	18. 2.05
CNP	1678	CNP	Glasflugel H.201 Standard Libelle	264		3.72	S.and J.McKenzie	Camphill	1. 7.05
CNS	1681		Slingsby T.59A Kestrel 17	1724		4.72	J.R.Greenwell	Carlton Moor	1. 5.02
CNV	1683	229	Slingsby T.59F Kestrel 19	1790		6.72	P.H.Fanshawe and E.A.Smith	Snitterfield	22. 1.05
CNW	1684	625	Slingsby T.59F Kestrel 19	1791		7.72	S.R.Watson	Seighford	16. 3.05
CNX	1685	818	Slingsby T.59F Kestrel 20	1792		27. 7.72	M.Boxall	Rufforth	16. 7.05
CNY	1686	151	Glasflugel H.201 Standard Libelle	322		9.72	B. Cole-Hamilton and Partners	Portmoak	16. 1.05
CPA	1688	466	Glasflugel H.201B Standard Libelle	328		9.72	M.Corrance	Kenley	2. 3.05
CPB	1689	858	Slingsby T.59D Kestrel 19	1796		10.72	A.T.Vidion	Tibenham	27. 4.05
CPD	1691	401	Schleicher ASW17	17026		3.74	D.A.Johnstone	Marham	12. 6.05
CPF	1693	T15	Glasflugel H.201 Standard Libelle	267		3.72	J.M.Norman and P.Elvidge	Pocklington	1. 4.05
CPG	1694	CPG	Schleicher Ka7 Rhönadler	7036	D-4029	15. 4.72	Queens University Gliding Club	Bellarena	10.11.04
CPJ	1696	CPJ	Schleicher Ka6E	4059	OO-ZDA	9. 4.72	A.Forbes and Partners	Chipping	13. 4.05
CPL	1698		Slingsby T.8 Tutor	FF477	RAFGSA183	26. 4.72	Not known *(Being refurbished 2001)*	Lasham	
CPM	1699	CPM	Glasflugel H.201 Standard Libelle	179		4.72	K. Marsden	Upavon	15. 5.05
(CPQ)	1702	695	Torva Sprite Series 2			003	*(Stored 6.03)*	Rufforth	
CPU	1706	761	Schempp-Hirth HS.4 Standard Cirrus	194		4.72	J.P.J.Ketelaar	Feshiebridge	26. 5.05
CPV	1707	P19	SZD-30 Pirat	B-470		3.72	G.Hall	Portmoak	10. 7.05
CPX	1709	CPX	SZD-30 Pirat	B-460		4.72	M.Codd	Talgarth	12. 6.05
	1711		Aachen FVA.10B Rheinland				See SECTION 3 Part 1	Wassererkuppe, Germany	
CQC	1714	CQC	SZD-30 Pirat	B-472		15. 4.72	N.White	Crowland	8. 4.05
CQD	1715	CQD	Schleicher K8B	419/58	D-5625	4.72	E.McCaig	Challock	19. 4.01
CQG	1718	CQG	EoN AP.5 Olympia 2B	EoN/O/044	RAFGSA.206	4.72	L.McKenzie	Sutton Bank	31. 3.05
					BGA.542				
CQJ	1720	K17	Slingsby T.59A Kestrel 17	1727		5.72	A.Shelton	Portmoak	27. 7.02
CQL	1722	339	Schempp-Hirth HS.5 Nimbus 2	11		5.72	M N Erlund	East Kirkby	16. 4.05
CQM	1723	234	Slingsby T.59F Kestrel 19	1765		22. 5.72	J.A.Knowles	Lasham	11. 7.05
CQN	1724	CQN	Schempp-Hirth HS.4 Standard Cirrus	204G		19. 5.72	M.G.Sankey and Partners	Lasham	20. 5.05
CQP	1725	918	Schempp-Hirth HS.5 Nimbus 2	4		4.72	D.Caunt and Partners	Booker	1. 2.05
CQQ	1726	139	Schempp-Hirth HS.5 Nimbus 2	5		4.72	M.D.J.White	Pocklington	25. 3.05
CQR	1727	703	Schempp-Hirth HS.4 Standard Cirrus	220G		28. 5.72	B.E.Richards	Aston Down	1. 5.05
CQW	1732	342	SZD-36A Cobra 15	W-572		6.72	I.Norman	Portmoak	20. 4.05
CQX	1733	789	SZD-30 Pirat	B-483		9. 6.72	J.G.Jones	Lleweni Parc	6. 7.01
CQY	1734	CQY	Schempp-Hirth HS.4 Standard Cirrus	214		16. 6.72	R. Farmer	Husbands Bosworth	6. 6.05
CRA	1736	CRA	Schleicher Ka7 Rhönadler	7009	???	7. 7.72	Welland Gliding Club	Lyveden	19. 5.02
CRB	1737	241	Glasflugel H.201 Standard Libelle	243		6.72	A.I.Mawer	Winthorpe	31. 5.05
CRF	1741	351	Birmingham Guild BG-135	001		2.72	C D Stevens	Lee-on-Solent	15. 4.05
CRH	1743	650	Schempp-Hirth HS.4 Standard Cirrus	233G		8.72	D.Thompson	Portmoak	31. 3.05
CRJ	1744	KT	Slingsby T59A Kestrel 17		1728	7.72	M.Krup	Nordhorn, Germany	23. 2.05
CRK	1745		Slingsby T.8 Tutor	930	XE760	25. 7.72	I.D.Smith	Nympsfield	8.82*
					VM539		*(Stored 7.04)*		
CRL	1746	CRL	Schleicher ASK13	13013	???	4. 8.72	Midland Gliding Club	Long Mynd	20. 2.05
CRN	1748	566	Schempp-Hirth HS.4 Standard Cirrus	234G		4. 8.72	J.Craig and Partner.	Dunstable	20. 2.05
CRQ	1750	CRQ	Glasflugel H.201B Standard Libelle	326		10. 8.72	K.G.Counsell and Partners	Usk	6. 4.05
CRS	1752	707	Glasflugel H.201B Standard Libelle	325		13. 8.72	S.Biggs	Husbands Bosworth	8. 7.05
CRT	1753	CRT	Schleicher ASK13	13396		1. 8.72	Bowland Forest Gliding Club	Chipping	4. 1.04

CRV	1755	CRV	Glasflugel H.201B Standard Libelle	329		22. 8.72	S.Cervantes	Portmoak	5. 6.05
CRW	1756	417	Glasflugel H.201 Standard Libelle	324		10.72	P.Dunster	Nympsfield	21. 5.05
CRX	1757		Slinsgsby T.41 Skylark 2B	1146	RNGSA	10.72	P.N.Fanning (Under restoration 2003)	RAF Wattisham	
CRZ	1759		Slingsby T.8 Tutor				See SECTION 3 Part 1	Wolverhampton	
CSA	1760	182	Slingsby T.59F Kestrel 20	1797		11.72	P.L Poole	Parham Park	9. 5.03
							(Stalled on field landing, Crewkerne, Somerset 24. 7.02)		
CSB	1761	76	Slingsby T.59F Kestrel 19	1798		17.11.72	R and T Linee	Kingston Deverill	31. 8.05
CSD	1763	53	Slingsby T.59D Kestrel 19	1800		1.12.72	J.T.Goodall	Pocklington	25. 2.05
CSF	1765	347	Slingsby T.59F Kestrel 19	1802		1.73	G.R.Glazebrook	Dunstable	18.11.05
CSJ	1768	CSJ	Glasflugel H.201B Standard Libelle	372		26. 1.73	S.N.Kroner	Hinton-in-the-Hedges	24. 2.05
CSK	1769	387	Slingsby T.59D Kestrel 20	1806		3.73	H.A.and J.E.Torode	Lasham	4. 3.05
CSL	1770		Slingsby T.8 Tutor	928	XE758 VF181	15.10.72	W.D.Baars	(The Netherlands)	19. 9.00
CSN	1772	CSN	Pilatus B4 PC-11	021		12.72	D.Brummitt and Partrner.	North Hill	12. 5.05
CSP	1773	CSP	Pilatus B4 PC-11	027		3.73	I.Pendlebury	Chipping	17. 4.05
CSR	1775	808	Glasflugel H.201B Standard Libelle	368		14. 1.73	W.G.Miller and Partners	North Connel	18. 5.02
CSU	1778		Manuel Hawk	1		11.72	Not known *(Stored)*	Sackville Lodge, Riseley	
CSV	1779	CSV	SZD-30 Pirat	B-515		3.12.72	M.Terry and Partners	Gamston	19. 1.05
CSW	1780	CSW	Pilatus B4 PC-11AF	022		12.72	I.H.Keyser	Waldershare Park	31. 3.05
CTA	1784	CTA	EoN AP.5 Olympia 2B	EoN/O/146	RAFGSA.285	12.72	P.N.Tolson	Wormingford	30. 9.00
CTB	1785	579	Schempp-Hirth HS.4 Standard Cirrus	264G		1.73	I.M.Young and Partners	RAF Weston on the Green	4. 2.05
CTD	1787		Yorkshire Saiplanes YS-53	1721		4.74	P.Older	Andreas	
							(Under restoration in 2000 after heavy landing)		
CTE	1788	40	Schleicher ASW17	17012		1.73	D.Edwards and S.Blackmore	Lasham	27. 2.05
CTF	1789		Schleicher Ka4 Rhönlerche	01	D-3574	1.73	M.Goodman	Winthorpe	24. 1.04
			(Owner quotes p/i D-4346, but unconfirmed)						
CTJ	1792	CTJ	Slingsby T.59D Kestrel 19	1810		14. 3.73	W.Waldren and Partners	Dunstable	15.12.05
CTL	1794	CTL	Slingsby T.59D Kestrel 19	1812		28. 3.73	P.G.Codd	Wormingford	15. 4.05
CTM	1795	254	Slingsby T.59D Kestrel 19	1813		3.73	G.P.Emsden	Dunstable	3. 4.04
CTP	1797	49	Slingsby T.59D Kestrel 19	1815		14. 4.73	D.C.Austin	Sutton Bank	24. 5.05
CTQ	1798	CTQ	Slingsby T.59D Kestrel 20	1816		27. 4.73	K.A.Moules	RAF Bicester	1. 5.05
CTR	1799	402	Slingsby T.59D Kestrel 19	1817		5.73	D.J.Marpole	Kingston Deverill	6. 4.05
CTS	1800	CTS	EoN AP.5 Olympia 2B	EoN/O/157	RNGSA	13. 1.73	M.D.Smith	Parham Park	6. 6.04
CTT	1801	873	Schempp-Hirth HS.4 Standard Cirrus	277G		17. 2.73	S.M.L.Young	Nympsfield	20. 4.05
CTU	1802	501	Glasflugel H.201 Standard Libelle	371		2.73	J.R.Humpherson	Camphill	13. 5.05
CTV	1803		SZD-30 Pirat	B-528		2.73	B.Fantham	Rhigos	6. 9.04
CTW	1804	1	SZD-9 bis Bocian 1E	P-598		2.73	Mendip Gliding Club	Halesland	1. 9.01
CTX	1805	CTX	SZD-30 Pirat	B-527		2.73	P.Goulding	Crowland	21. 5.05
CTZ	1807	CTZ	Schleicher K8B	8035/B5	D-KOCU D-5203	3. 4.73	Scottish Gliding Union Ltd	Portmoak	8. 7.05
CUB	1809	CUB	Pilatus B4 PC-11	047		31. 3.73	G.Saunders	Hinton-in-the-Hedges	20. 2.05
CUC	1810	678	Pilatus B4 PC-11	003	HB-1102	5.73	H.M.Pantin and Partners	AAC Dishforth	7. 5.05
CUD	1811	CUD	Yorkshire Sailplanes YS-53 Sovereign	02		7.72	D R Bricknell	Saltby	20. 5.01
			(Built fromSlingsby T.53B XV951 [1574] w/o 11.4.72)						
CUF	1813	331	Yorkshire Sailplanes YS-55 Consort	04		9.11.73	C.G.Taylor and Partners	Sutton Bank	21. 7.00
CUJ	1816	706	Glasflugel H.201B Standard Libelle	370		13. 2.73	T.G.B.Hobbis and Partners	Lasham	14. 4.05
CUK	1817	380	Glasflugel H.201 Standard Libelle	367		3.73	G.R.Brown	Dunstable	9. 8.01
CUM	1819	CUM	SZD-30 Pirat	B-534		24. 2.73	E.Hughes	Pocklington	27. 2.05
CUQ	1821	CUQ	Pilatus B4 PC-11	040		2.73	Cotswold Gliding Club	Aston Down	13. 5.05
CUS	1822	842	Schempp-Hirth HS.2 Cirrus VTC	126Y		3.73	G.F.Wearing	Chipping	19. 6.05
CVA	1830	CVA	LET L-13 Blanik	025418		3.73	Andreas Gliding Club	Andreas	16. 4.05
							"Boggles the Blanik"		
CVB	1831	CVB	LET L-13 Blanik	025419		3.73	W.N.Smith	Sackville Lodge, Riseley	26. 5.03
CVC	1832	CVC	SZD-30 Pirat	B-535		3.73	T.Wilkinson	Sackville Lodge, Riseley	4. 4.05
CVE	1834	BZ	Schempp-Hirth HS.2 Cirrus VTC	127Y		15. 3.73	D.Jamin	Gransden Lodge	20. 3.05
CVG	1836	656	Pilatus B4 PC-11	045		19. 3.73	I H Keyser	Waldershare Park	9. 4.05
CVH	1837	CVH	Schempp-Hirth SHK	34	N6524A	30. 3.73	J.C.Fletcher	Parham Park	21. 8.04
CVJ	1838	CVJ	Breguet Br.905S Fauvette	37	F-CCJH	29. 6.73	I.C.Gutsell	Burn	22. 5.02
CVK	1839	92	Pilatus B4 PC-11AF	048		22. 3.73	R.R.Stoward	Dunstable	5. 8.05
CVL	1840	253	Glasflugel H.201B Standard Libelle	369		2.73	N.A.Dean	Kirton-in-Lindsey	21. 3.05
CVM	1841	CVM	Pilatus B4 PC-11 (Powered)	036		23. 3.73	J.A.Mace	Upavon	22. 4.05
CVP	1843	CVP	SZD-9 bis Bocian 1E	P-597		17. 3.73	M.Boyle	Rufforth	3. 8.04
CVQ	1844	428	Glasflugel H.201 Standard Libelle	374		23. 3.73	C.J.Taunton and Partners	Dunstable	29. 4.04
CVR	1845	CVR	SZD-30 Pirat	B-538		27. 3.73	J.R.Hornby	Gamston	26. 6.04
CVS	1846	CVS	SZD-36A Cobra 15	W-610		3.73	B.Mossop	Pocklington	5. 8.05
CVT	1847	CVT	SZD-36A Cobra 15	W-609		27. 3.73	C. Winch	Ridgewell	18.11.05
CVV	1849	CVV	Pilatus B4 PC-11	028		27. 3.73	F.R.Wolff and Partners	Brent Tor	2. 2.05
CVW	1850	423	Slingsby T.59D Kestrel 19	1818		29. 5.73	P.B.Hogarth	Halesland	29. 6.05
CVY	1852	355	Slingsby T.59D Kestrel 19	1821		4. 7.73	J.Ainsworth	Sleap	25. 9.05
CVZ	1853	269	Slingsby T.59D Kestrel 19	1824	RAFGSA.269 BGA.1853	3. 8.73	T.R.F.Gaunt and Partners	Upavon	20. 3.05
CWA	1854	539	Slingsby T.59D Kestrel 19	1825		9.73	I.B.Kennedy	Usk	17. 1.05
CWB	1855	CWB	Slingsby T.59D Kestrel 19	1833		9. 1.74	K.Fairness	Milfield	7. 6.05
CWD	1857	CWD	Slingsby T.59D Kestrel 19	1835		1.74	J.R.Dransfield	Aboyne	14. 9.05
CWE	1858	468	Glasflugel H.201 Standard Libelle	482		31. 1.74	T.W.S.Stoker	Rufforth	11. 5.05
CWF	1859	CWF	Slingsby T.59D Kestrel 19	1838		2.74	P.F.Nicholson	Thame	1. 4.05
CWG	1860	322	Glasflugel H.201 Standard Libelle	391		7. 4.73	G.Pledger	Currock Hill	11. 8.05
CWH	1861	CWH	Schleicher ASK13	13424		12. 4.73	York Gliding Centre	Rufforth	22. 1.05
CWJ	1862	CWJ	Schleicher Ka7 Rhönadler	630	D-6057 D-5723	27. 4.73	Wolds Gliding Club	Pocklington	8. 9.05
CWN	1866	CWN	Glasflugel H.201B Standard Libelle	386		4.73	Strathclyde Gliding Club	Strathaven	20. 3.05

CWR	1869	917	Schempp-Hirth HS.2 Cirrus VTC	133Y		4.73	D.C.Perkins and Partner	Thame	11. 3.05
CWS	1870	CWS	Schempp-Hirth HS.2 Cirrus VTC	129Y		19. 4.73	R.W.Cassels and Partners	Ridgewell	19. 3.05
CWT	1871	978	Glasflugel H.201B Standard Libelle	384		4.73	C.A.Lawrence	Husbands Bosworth	23. 1.05
CWU	1872		Schleicher Ka4 Rhönlerche II	390	D-5627	22.4.73	Not known *(Under restoration 2001)*	(West Sussex)	
CWV	1873	Z	Schleicher Ka4 Rhönlerche II	123	D-8226	22. 4.73	11th Bristol (Headley Park) Scout Troop	Aston Down	5.94*
							(Stored11.04)		
CWX	1875	832	Glasflugel H.201 Standard Libelle	36	RAFGSA.132	4.73	C.A.Weyman and Partners	Gallows Hill	17. 4.05
CWY	1876	146	Glasflugel H.201 Standard Libelle	387		14. 4.73	J Dixon	Portmoak	17. 2.05
CWZ	1877	CWZ	Glasflugel H.201 Standard Libelle	392		30. 4.73	D.Mackenzie	Camphill	3.12.05
CXH	1885	CXH	SZD-36A Cobra 15	W-619		10. 6.73	C D Street	Parham Park	11. 9.04
CXK	1887	CXK	Glasflugel H.201 Standard Libelle	383		6.73	C.A Turner	Cross Hayes	1. 3.05
CXL	1888	CXL	SZD-30 Pirat B-	548		6.73	N.Clarke	Rattlesden	10. 4.05
CXM	1889	532	Slingsby T.59D Kestrel 19	1820		7.73	R.P.Beck and Partners	AAC Dishforth	15. 4.05
CXN	1890	508	Yorkshire Sailplanes YS-55 Consort	05		21.12.73	A.A.Priestley and Partners	Sutton Bank	13. 3.02
CXP	1891		Yorkshire Sailplanes YS-55 Consort	07	BGA.1892	5.76	A.D.Coles	North Hill	19. 4.04
CXV	1897	CXV	Yorkshire Sailplanes YS-53 Sovereign	03		7.74	P.Myers	Chipping	.10. 7.05
CYA	1902	503	Pilatus B4 PC-11	072		7.73	E.J.Bromwell and Partners	North Hill	18.12.05
CYC	1904	CYC	Pilatus B4 PC-11	029	N47247	7.73	D.F.Barley	Ringmer	30. 7.04
CYD	1905	CYD	SZD-30 Pirat	B-559		25. 7.73	I.Johnstone	Milfield	8. 7.04
CYG	1908	CYG	Glasflugel H.201B Standard Libelle	441		8.73	D.J.Cooke	Husbands Bosworth	12. 2.05
CYJ	1910	CYJ	DFS/49 Grunau Baby 2B	031000	D-6021	11. 8.73	C.Bird	Dunstable	1.90*
			(Built Petera 1943)				*(Under restoration during 2000)*		
CYK	1911	248	Pilatus B4 PC-11	078		8.73	I.M.Trotter tr Pilatus Soaring Syndicate	Portmoak	19. 9.04
CYM	1913	299	Schempp-Hirth HS.4 Standard Cirrus	48	RAFGSA D-0578	9.73	K.Fisher	Husbands Bosworth	10. 2.05
CYN	1914	N4	Slingsby T.59D Kestrel 19	JP.054		10.74	S.J.Cooke and Partners	Gransden Lodge	24. 1.05
			(Built D Jones and T Pentelow)						
CYQ	1916	477	Schempp-Hirth HS.4 Standard Cirrus	364		27.9.73	B.M.Reeves	Nympsfield	21. 2.05
CYT	1919	CYT	Schempp-Hirth HS.4 Standard Cirrus	357G		9.73	G.Royle	Llantisilio	25. 4.05
CYV	1921		Birmingham Guild BG-135	5		9.73	N.Moffatt	Falgunzeon	16. 3.04
							(Ground-looped on landing at Grasmere, 6. 6.03)		
CYW	1922		Birmingham Guild BG-135	6		1.10.73	Not known	Not known	
							(Crashed at Lleweni Parc 16.10.93: stored 2000)		
CYY	1924		Schleicher Ka4 Rhönlerche	Not known	AGA.19	4.11.73	Not known (Stored 2000)	Not known	
CYZ	1925	CYZ	Schleicher K8B	8882	RAFGSA	8. 9.74	Oxford Gliding Club	RAF Weston on the Green	2. 8.04
CZD	1929	CZD	Pilatus B4 PC-11	081		12.73	S.E.Marples	Milfield	16. 9.05
CZE	1930	CZE	SZD-30 Pirat	S-0114		23.12.73	G. Graham	Snitterfield	1. 4.05
CZG	1932	CZG	SZD-30 Pirat	S-0116		31.12.73	J.T.Pajdak	Kenley	10. 6.05
CZJ	1934	CZJ	SZD-30 Pirat	S-0115		29.12.73	C.L Groves and Partners	Husbands Bosworth	15. 4.05
CZL	1936	504	Glasflugel H.201 Standard Libelle	483		31. 1.74	L P Woodage	Dunstable	29.11.02
							(Hit trees on approach to Husbands Bosworth 31. 8.02)		
CZM	1937	CZM	Munchen Mu-13D-III	10/52	D-1488	9.74	H.Chapple	RAF Bicester	18. 6.05
CZN	1938	CZN	Schleicher ASW15B	15329		13. 3.74	P.C.Tuppen and Partners	Bembridge	3. 6.05
CZQ	1940	CZQ	Slingsby T.59D Kestrel 19	1840		5. 4.74	J.P.Walker and Partners	Husbands Bosworth	26. 4.05
CZR	1941	CZR	Slingsby T.59D Kestrel 19	1842		4.74	R.P.Brisbourne	Rufforth	11. 3.05
CZU	1944	826	Slingsby T.59D Kestrel 19	1849		14. 6.74	K.R.Merritt and P.F.Croote	Halesland	14. 6.05
CZV	1945	415	Slingsby T.59D Kestrel 19	1850		16. 7.74	V.F.G.Tull	Dunstable	15. 6.03
CZW	1946	A3	Slingsby T.59D Kestrel 20	1846		17.10.76	C.D Berry	Bidford	21. 4.04
DAA	1950	DAA	SZD-9 bis Bocian 1E	P-639		10. 3.74	Highland Gliding Club	Easterton	24. 9.01
DAC	1952	DAC	SZD-36A Cobra 15	W-656		3.74	R.Colenso	Dunstable	26. 4.05
DAF	1955		LET L-13 Blanik	025825		4.74	Not known *(Stored 2000)*	Not known	
DAJ	1958	14	Schempp-Hirth HS.5 Nimbus 2	50		21. 3.74	J.D.Jones	Nympsfield	5. 5.05
DAL	1960	DAL	EoN AP.6 Olympia 419	EoN/4/010	RAFGSA.301	4.74	D.M.Judd and Partners	Snitterfield	30. 3.05
DAM	1961	675	ICA IS-29D	27		4.74	W.T.Barnard	Strathaven	24. 9.00
DAN	1962		SZD-30 Pirat	S-0145		23. 3.74	W.Pottinger and Partners	Ridgewell	26. 2.05
DAP	1963	DAP	SZD-30 Pirat	S-0147		4.74	T.D.Younger	Currock Hill	11. 5.05
DAR	1965	DAR	Slingsby T.21B	SSK/FF1200	RAFGSA.404	1. 6.74	J. Bennett and Partners.	Upwood	10. 7.05
							(Damaged in heavy landing at Upwood 31. 7.04)		
DAS	1966	DAS	Schempp-Hirth HS.4 Standard Cirrus	378	(BGA.1925)	4.74	G.Goodenough	Burn	9. 4.05
DAU	1968		SZD-30 Pirat	S-0150		4.74	W.Fisher	Winthorpe	3. 4.04
							(Crashed North of Winthorpe, 30. 5.03)		
DAV	1969	240	SZD-38A Jantar-1	B-608		4.74	B.Taylor and Partne.	Brent Tor	8.10.05
DAW	1970	DAW	Schleicher Ka6CR	951	RAFGSA D-2025	8. 6.74	P.S.Holmes	Bellarena	23. 4.05
DBA	1974	207	EoN AP.5 Olympia 2B	EoN/O/156	RNGSA.208	25. 5.74	W.R.Williams	RAF Halton	20. 6.04
DBB	1975	DBB	Slingsby T.51 Dart 17R	DG/51/01		5. 2.76	A. Salisbury	Edgehill	27. 5.05
			(Built Greenfly Aviation)						
DBC	1976	851	Pilatus B4 PC-11	135		6.74	J.H.France and Partners	Shobdon	15. 4.05
DBD	1977	DBD	SZD-30 Pirat	S-0202		6.74	E W Burgess	Lyveden	18. 7.04
DBG	1980	DBG	ICA IS-29D	31		6.74	P.S.Whitehead	Rufforth	26. 6.04
DBJ	1982	691	Slingsby T.59D Kestrel 19	1856	BGA.1892 BGA.1982	19.10.74	T.Barr-Smith	Challock	2. 2.05
DBK	1983	523	Slingsby T.59D Kestrel 19	1861		2.12.74	R.E.Perry and C.Crabb	North Hill	5. 6.05
DBN	1986	617	Slingsby T.59D Kestrel 19	1857		25. 3.75	A.C.Wright	Sutton Bank	31. 1.05
DBP	1987	551	Glasflugel H.205 Club Libelle	51		1. 5.75	N.A.White	Wormingford	14. 4.05
DBQ	1988	DBQ	Slingsby T.59D Kestrel 19	1863		17. 4.75	S.Snaderson	Rufforth	8.10.05
DBR	1989	182	Slingsby T.59D Kestrel 19	1858		22. 4.75	P.L.Poole	Parham Park	26. 4.05
DBS	1990	DBS	Slingsby T.59D Kestrel 19	1864		24. 4.75	J.W.Rice	Gamston	24. 7.05
DBT	1991	DBT	Schempp-Hirth Standard Austria S	35	F-CCPQ	7.74	G.Bambrook	Edgehill	20. 5.05
DBU	1992		Hirth Go IV Goevier 3				See SECTION 3 Part 1	Wolverhampton	
DBV	1993	DBV	SZD-30 Pirat	S-0227		1. 7.74	P.C.Herniman	Usk	2. 5.05

DBX	1995	DBX	SZD-9 bis Bocian 1E	P-642		20. 7.74	Miss A.G.Veitch	Easterton	31. 3.05
							tr Highland Bocian Syndicate		
DCA	1998	DCA	SZD-36A Cobra 15	W-686		18. 9.74	J.Young	Husbands Bosworth	19. 6.05
DCC	2000	324	Glasflugel H.201B Standard Libelle	585		11.74	M.Hutchinson	Shobdon	20. 3.05
DCE	2002	DCE	Slingsby T.41 Skylark 2B	1003	RAFGGA.540	11.74	J Salvin	Winthorpe	14. 5.05
					PH-225				
DCG	2004		Schleicher Ka2B	2	D-7064	12.74	Not known *(Stored)*	Falgunzeon	*
DCM	2009		LET L-13 Blanik	026155		12.74	Not known *(Wreck noted 2001)*	Strathaven	7. 8.92*
	2014	DCS	Slingsby T.45 Swallow	1538	RAFGGA.544	4.74	N. Dickenson	Chipping	23. 8.01
							(Written-off at Denbigh 6. 1.01)		
DCW	2018	DCW	Schleicher Ka6CR	1076	D-5228	2.75	Trent Valley Gliding Club	Kirton-in-Lindsey	11. 7.01
DCY	2020	DCY	Swales SD.3-15V	01		2.75	T.R.Edwards	Chipping	16. 5.97*
			(Rebuild of incomplete Yorkshire Sailplanes YS-55 Consort c/n 09; under restoration 2001)						
DCZ	2021	DCZ	King-Elliott-Street Osprey	1470		10.75	G.R.Burkert	Edgehill	28. 5.00
			(Believed to be converted Slingsby T.51 Dart but c/n conflicts with BGA.1245)						
DDA	2022	DDA	Schempp-Hirth HS.4 Standard Cirrus	532G		2.75	P Berridge	Sandhill Farm, Shrivenham	22. 4.05
DDB	2023	DDB	Schleicher ASK13	13493		2.75	Shenington Gliding Club	Edgehill	24.11.05
DDC	2024	DDC	Slingsby T.21B	1157	RAFGSA.313	3.75	J.S.Shaw and Partners	Perranporth	12. 5.05
					XN153				
DDD	2025	695	Schempp-Hirth HS.5 Nimbus 2	84	G-BKPM	3.75	D.D.Copeland	Lasham	13. 4.05
					BGA.2025				
DDK	2031	DDK	SZD-30 Pirat	S-0408		4.75	P.Jennings	Bembridge	15.10.05
DDL	2032	DDL	Schleicher K8B	218/61	D-5156	4.75	York Gliding Centre	Rufforth	7. 4.05
DDM	2033	959	Schempp-Hirth HS.2 Cirrus VTC	164Y		3.75	L Mundy	Wormingford	19. 3.05
DDN	2034	DDN	SZD-9 bis Bocian 1E	P-429	RAFGSA.393	3.75	Bath, Wilts and North Dorset Gliding Club		
								Kingston Deverill	5. 8.05
DDQ	2036		Slingsby T.21B	630	RAFGSA.247	3.75	Aeroventure	Doncaster	
					WB969		*(On display 7.03)*		
DDR	2037	680	Schempp-Hirth HS.4 Standard Cirrus	531G		3.75	J.Smith	Pocklington	25. 1.05
DDW	2042	DDW	SZD-30 Pirat	S-0433		4.75	C.P.Offen	Parham Park	17. 5.05
DDY	2044	DDY	Schleicher Ka6CR	678	D-8841	4.75	D Bowden	Cross Hayes	14. 5.05
DEB	2047	DEB	Slingsby T.59D Kestrel 19	1866		24.10.75	T.Moss	RAF Weston on the Green	19. 6.05
DEG	2051	DEG	ICA IS-28B2	48			P.S.and H.Whitehead	Sutton Bank	23. 9.05
DEN	2057	588	ICA IS-29D	41		4.75	S.E.Marples	Feshiebridge	13. 3.01
							(Undershot on landing at Milfield, 1. 7.00)		
DEP	2058	DEP	Schleicher Ka6CR	6452	AGA...	4.75	S.J.Aldridge	Saltby	12. 3.05
					BGA.2058, RAFGSA.350				
DEQ	2059	716	Glasflugel H.205 Club Libelle	97		4.75	J.A.Holland and Partners	Kingston Deverill	2. 3.01
DEV	2064	DEV	Schleicher Ka6CR	6453	RAFGSA.354	5.75	P.J.Groves and Partners	Long Mynd	22. 2.05
DEW	2065	DEW	ICA IS-29D	40		12. 6.75	M.D.Smith	Parham Park	26. 3.05
DEX	2066	DEX	LET L-13 Blanik	026348	RAFGSA.R4	1. 7.75	Black Mountains GC	Talgarth	28.10.05
					BGA.2066				
DEY	2067	DEY	LET L-13 Blanik	026352	RAFGSA.R12	10. 7.75	Bath, Wilts and North Dorset Gliding Club		
					BGA.2067			Shipdham	9. 9.03
DFC	2071	128	Schempp-Hirth HS.4 Standard Cirrus	592G		7.75	J.Cornish	Tibenham	10. 2.05
DFE	2073	DFE	Eiriavion PIK-20B	20052		7.75	M.Roff-Jarrett	Lasham	13.12.05
DFK	2078	DFK	Molino PIK-20	20039	OH-500	14. 9.75	M.J Fairclough	North Hill	27. 4.05
DFL	2079	DFL	SZD-38A Jantar-1	B-682		9.75	J.Howlett	Crowland	28. 4.05
DFN	2081		Glaser-Dirks DG-100	30		9.75	D.J.Clarke	Wormingford	2. 8.05
DFP	2082	DFP	Aeromere M.100S	029	I-LSUO	9.75	D.and J.Lee	Pocklington	23. 2.03
DFR	2084	906	Grob G.102 Astir CS	1038		10.75	D.Peggott	Lasham	6.12.05
DFV	2088	164	SZD-38A Jantar-1	B-684		7.12.75	R.R.Rodwell	Bellarena	15.12.01
DFW	2089	DFW	SZD-30 Pirat	S-0545		26.11.75	R.Skerry	Strubby	19. 2.05
DFX	2090	DFX	SZD-41A Jantar Standard	B-691		10.75	J.C.Tait and Partners	Easterton	18. 5.05
DFY	2091		Schempp-Hirth HS.4 Standard Cirrus				See SECTION 3 Part 1	South Kensington, London SW7	
DFZ	2092	774	Eiriavion PIK-20B	20080		27.11.75	M.J.Leach and J.P.Ashcroft	Sandhill Farm, Shrivenham	
							(Crashed on landing at Sandhill Farm, 19. 6.04)		
DGA	2093	DGA	Schleicher K8B	8587	RAFGSA	11.75	Welland Gliding Club	Lyveden	2. 5.05
					BGA.1926/D-...				
DGE	2097	DGE	Schempp-Hirth HS.4 Standard Cirrus 75	606		12.75	T.E.Snoddy	Bellarena	8. 8.05
DGG	2099	DGG	Schleicher Ka6E	4061	RAFGSA.263	15.12.75	N.F.Holmes and Partners	Long Mynd	8. 5.05
DGH	2100	DGH	SZD-30 Pirat	B-533	RNGSA	12.75	T.Bell and V. Turner	RAF Keevill	24. 2.05
					BGA.2100				
DGK	2102	DGK	Schleicher Ka6CR	6287	D-3224	21.12.75	IBM Gliding Club *"Betty Blue"*	Lasham	29. 6.05
DGM	2104		Slingsby T.45 Swallow	1506	RAFGGA	12.75	Not known *(Being refurbished 2000)*	North Hill	8.11.93
DGP	2106	DGP	LET L-13 Blanik	026560		26. 5.76	B.Kozuh	Grobnik, Slovenia	17. 5.05
DGT	2110		Schleicher Ka2B	181	D-5469	5.76	Not known *(Stored 2002)*	Falgunzeon	23.10.93*
DGV	2112		Brequet Br.905S Fauvette	2	HB-632	5.76	R.M.Cust and Partners	Burn	8. 8.05
DGX	2114	610	Schempp-Hirth HS.4 Standard Cirrus 75	619	AGA.3	11. 5.76	G.R.Seaman and Partners	Lasham	28. 5.04
					BGA.2114				
DGY	2115	195	Schempp-Hirth HS.5 Nimbus 2	105		11. 5.76	J.H.Taylor	Nympsfield	14. 1.05
DHA	2117	DHA	Schleicher K8B	1055	D-8848	23. 6.76	Booker Gliding Club	Booker	16.12.04
			(C/n conflicts with D-8616)		D-5148				
DHB	2118	DHB	SZD-30 Pirat	S-0643		9. 6.76	J.T.Winsworth	Tibenham	11. 4.00
DHC	2119	811	SZD-41A Jantar Standard	B-710	(BGA.2109)	9. 6.76	M.C.Burlock	Aston Down	13. 3.05
DHG	2123	517	Schleicher Ka6CR	1131	D-5170	23. 6.76	Angus Gliding Club	Drumshade	3. 8.03
DHH	2124	116	Molino PIK-20B	20124		5.76	D.M.Steed	Dunstable	9. 7.05
DHJ	2125	DHJ	Glaser-Dirks DG-100	48		1.76	H.D.Armitage	Usk	25. 6.05
DHK	2126	A30	Glaser-Dirks DG-100	50		1.76	R.Dell and B.J.Griffin	Saltby	11. 4.05
DHL	2127	DHL	Glaser-Dirks DG-100	52		1.76	K.J.Adam	Aboyne	24. 9.05
DHM	2128	DHM	Schleicher Ka6E	4124	RAFGSA.26	6. 2.76	J.G.Heard	Seighford	23. 3.05

DHN	2129	824	Molino PIK-20B	20082		1.76	G J Bass	Challock	9. 6.03
DHP	2130		Slingsby T.45 Swallow	45176		5. 2.76	Dumfries and Galloway Gliding Club	Falgunzeon	10.89*
			(Components ex BGA.1041 [1329]: c/n = type/year)				*(W/o 3.4.64 with parts from BGA.1032; sold for rebuild 2002)*		
DHR	2132	DHR	Slingsby T.53B	1718		2.76	E.L.Pole	Portmoak	18.11.05
							(Aviation Preservation Society of Scotland)		
DHT	2134	DHT	Schleicher Ka6E	4065	???	2.76	D.Wilkinson	Nympsfield	13. 1.05
DHV	2136	DHV	Molino PIK-20B	20111		3.76	J.D.and G.J.Walker	Booker	11. 7.05
DHW	2137	951	Schempp-Hirth HS.5 Nimbus 2	106		23. 4.76	A.J.Bauld	Portmoak	24. 5.05
DHY	2139	DHY	Schleicher Ka7 Rhönadler	1137	RAFGSA.266 D-5162	14. 4.76	G.Whittaker	Chipping	3. 8.05
DHZ	2140	DHZ	SZD-30 Pirat	S-0641		23. 4.76	Peterborough and Spalding Gliding Club	Crowland	15. 5.05
DJA	2141	DJA	SZD-30 Pirat	S-0642		23. 4.76	J.Young	Milfield	18. 5.03
DJB	2142	N11	Schleicher K8B	8879	(RAFGSA.N11) AGA.17, BGA.2142, RAFGSA.397 (1)	7. 4.76	Portsmouth Naval Gliding Club	Lee-on-Solent	28. 3.05
DJD	2144	DJD	Grob G.102 Astir CS	1226		5. 8.76	P.J.Gascoigne and Partners	Kingston Deverill	19. 3.05
DJE	2145	DJE	Schleicher Ka6CR	6412	D-3682	15. 7.76	R.A. Davenport	Andreas	6. 5.05
DJF	2146	500	Halford JSH Scorpion	001		7.77	R.G.Greenslade	(Doncaster)	
							(Previously at South Yorkshire Air Museum - for restoration)		
DJG	2147	K2	Schleicher Ka2B Rhönschwalbe	231	D-6179	6.76	J.Harmer	Lasham	10. 6.01
							(Withdrawn from use after water-damaged in trailer)		
DJJ	2149	DJJ	Schleicher ASK18	18029		16. 7.76	Mendip Gliding Club	Halesland	31. 7.05
DJK	2150	DJK	Schleicher ASK18	18030		15. 7.76	Booker Gliding Club	Booker	6. 6.05
DJL	2151	DJL	SZD-41A Jantar Standard	B-714		16. 7.76	A.M.Cooper	Llantisilio	22. 2.05
DJM	2152	DJM	SZD-41A Jantar Standard	B-715		17. 8.76	M.Gatehouse	North Hill	5. 7.05
DJN	2153	407	Molino PIK-20B	20140C		16. 7.76	S.L.Cambourne	Kingston Deverill	3. 3.05
DJP	2154	DJP	Schleicher K8B	8588	RAFGSA.335	21. 7.76	Sackville Gliding Club	Sackville Lodge, Riseley	17. 6.03
DJQ	2155	214	Grob G.102 Astir CS	1258		17.10.76	R.D.Slater and N.S. Jones	Usk	12. 2.05
DJR	2156	FGT	Schleicher Ka6CR	680	D-8423	4. 8.76	Cornish Gliding Club	Perranporth	17. 6.05
DJS	2157	DJS	Schempp-Hirth SHK-1	51	SE-TNF OY-MFX, HB-898	16. 7.76	J.Selman	Limerick, County Limerick	3. 8.05
DJT	2158	DJT	Schleicher Ka7 Rhönadler	?	RAFGSA	7.76	Enstone Eagles Gliding Club	Lleweni PArc	1.12.03
							(BGA.4271 is marked "DJT" also)		
DJU	2159		Scheibe Bergfalke II/55	204	D-....	7.76	Not known *(Stored 2000)*	Not known	
DJW	2161		Manuel Condor	1		7.76	C.V. and R.C.Inwood	Booker	18. 5.98
DJX	2162	614	Grob G.102 Astir CS	1259		17.11.76	P. Barnwell and R.Duke	Crowland	16. 4.05
DJZ	2164	989	Eiri PIK-20B	20144		4. 8.76	D.S.Puttock	Talgarth	22. 2.05
DKB	2166	DKB	Schempp-Hirth Standard Austria S	32	F-CCPR	4. 8.76	J R Parr	Burn	25.10.03
DKC	2167	DKC	Schleicher K8B	8261	D-1431	8.76	Yorkshire Gliding Club	Sutton Bank	21. 4.05
DKD	2168	759	Glasflugel H.206 Hornet	67	(BGA.2165)	8.76	I.M.Evans	Long Mynd	19. 8.04
DKE	2169	DKE	Schleicher ASK13	13548		28. 9.76	South Wales Gliding Club	Usk	6. 4.05
DKG	2171	DKG	Schleicher Ka6CR	6233	D-4327	8.76	L. Footring and Partners	Wormingford	26. 3.05
DKH	2172	769	LET L-13 Blanik	026644		8. 9.76	T.Wiltshere	Spilsby	11. 9.04
DKL	2175	444	Schempp-Hirth HS.5 Nimbus 2	086	D-2111	8. 9.76	G.J.Croll	Snitterfield	11. 4.05
DKM	2176	DKM	Glasflugel H.206 Hornet	49	(BGA2213) BGA2176, D-7816	9.76	M.Lee	Rattlesden	10. 4.05
DKN	2177	DKN	Schleicher Ka6CR	6456	D-9358	12 .3.77	C.P.Godfrey and S. Jarvis	Upwood	24. 6.05
DKQ	2179	DKQ	Glaser-Dirks DG-100G	91G11		9.76	G.Peters	North Hill	10. 2.03
							(Crashed Nympsfield 5.12.02; de-registered 1.8.03 - remains sold in Germany)		
DKR	2180	DKR	Grob G.102 Astir CS	1327		9.76	Oxford Gliding Club	RAF Weston on the Green	20. 6.05
DKS	2181	788	Grob G.102 Astir CS	1330		23.12.76	C.K.Lewis	Lasham	9. 5.04
DKT	2182	DKT	Eiri PIK-20B	20155		4.11.76	G.Barnham	Rufforth	15. 4.05
DKU	2183	DKU	Grob G.102 Astir CS	1326		11.76	P. Stapleton	North Hill	4. 3.05
DKW	2185	DKW	Grob G.102 Astir CS	1329		9.76	J.H.C.Friend	Llantisilio	8.10.05
DKX	2186	353	Grob G.102 Astir CS	1331		9.76	L.R.and J.M.Bennett	Usk	4. 3.05
DKY	2187	DKY	Schleicher Ka7 Rhönadler	7187	RAFGSA.342	4.11.76	Defford Aero Club *(Stored 2005)*	Challock	14. 8.02
DLA	2189	DLA	Pilatus B4 PC-11	149	RAFGSA	17.10.76	P.R.Seddon	Chipping	7. 5.05
DLB	2190	DLB	Schleicher ASK18	18040		17.10.76	Vale of The White Horse Gliding Club Centre	Sandhill Farm, Shrivenham	3. 3.05
DLC	2191	C	Schleicher ASK13	13549		4.11.76	Lasham Gliding Society	Lasham	27.10.04
DLD	2192	DLD	Schleicher K8B	8766	RAFGSA.383	17.11.76	Shalbourne Soaring Group	Rivar Hill	22. 2.04
							(Undershot and crashed Rivar Hill 27.8.03)		
DLE	2193	433	Schleicher Ka6E	4074	AGA.8 RAFGSA	10.76	D.C.Unwin	Gamston	28. 4.04
DLG	2195	DLG	Schempp-Hirth HS.4 Standard Cirrus	579	AGA.2	17.10.76	S.Naylor	Burn	30. 7.04
DLH	2196	378	Grob G.102 Astir CS77	1646		2.77	M.Saunders	Lasham	18. 3.05
DLJ	2197	DLJ	Molino PIK-20B	20157		16.12.76	M.S Parkes	Milfield	25. 7.04
DLM	2200	266	Grob G.102 Astir CS	1260	(BGA.2163)	1.12.76	I.A.Davidson	North Hill	6. 6.05
DLP	2202	DLP	Schleicher Ka6CR	6519	RAFGSA.355	11.76	J.R Crosse	Crowland	7. 5.05
							(Damaged Sandhill Farm, Shrivenham 7.8.04)		
DLS	2205	DLS	Schleicher K8B	8650	D-5718	1.12.76	D.B.Rich	Eaglescott	23. 3.05
DLT	2206	DLT	ICA IS-28B2	32		12.76	A.Woodrow	Tibenham	3. 7.03
DLU	2207	R93	ICA IS-28B2	33	RAFGSA.R93 NEJSGSA.3, EI-141, BGA.2207	12.76	Crusaders Gliding Club	Kingsfield, Dhekelia	12.12.04
DLY	2211	DLY	Eiri PIK-20D	20509		5. 1.77	M.Conrad	Dunstable	13. 9.05
DLZ	2212	456	Swales SW.3-15T	03		12.76	R.E.Harris	Rivar Hill	25. 5.05
DMB	2214	831	Schleicher K8B	8209	D-4331	18. 1.77	Crown Services Gliding Club *"Kate"*	Lasham	5.11.05
DMD	2216	251	Glaser-Dirks DG-100	75		12.76	N.Hill	RAF Weston on the Green	25. 1.05
DMF	2218	DMF	Schleicher Ka7 Rhönadler	7073	D-4313	2. 3.77	Staffordshire Gliding Club	Seighford	19. 5.04
DMG	2219	DMG	Schleicher K8B	8763	RAFGSA.382	19. 3.77	Dorset Gliding Club	Gallows Hill	8. 5.05
DMH	2220	DMH	Grob G.102 Astir CS	1511		23. 4.77	Oxford Gliding Club	RAF Weston-on-the-Green	25. 5.05
DMJ	2221	DMJ	Schleicher K8B	8077	PH-290	19. 3.77	Not known *(Stored 10.04)*	Strathaven	23. 9.93*

DMK	2222	593	Schempp-Hirth SHK	25	D-5401	5. 4.77	P.Gray	Aston Down	8. 3.05
DML	2223	DML	Schleicher Ka7 Rhönadler	929	D-6194	4. 2.77	Newark & Notts Gliding Club	Winthorpe	21. 2.05
					D-5005				
DMM	2224	74	Schempp-Hirth HS.5 Nimbus 2	125		4. 2.77	T.E.Linee	Gallows Hill	28. 2.05
DMN	2225	DMN	Glasflugel H.303 Mosquito	20		2.77	R.P.Brecknock	Dunstable	27. 3.03
DMP	2226	233	Grob G.102 Astir CS	1239	ZS-GKF	28. 2.77	D.G.Nisbet	Dunstable	19. 2.05
DMQ	2227	DMQ	Schleicher Ka6E	4062	RAFGSA.264	12. 3.77	A.R.Bushnell and Partner	Lyveden	24. 5.05
DMR	2228	511	Grob G.102 Astir CS	1435		19. 3.77	L.A.Beale and Partners	Parham Park	30. 1.05
DMS	2229	259	Glasflugel H.201B Standard Libelle	385	RNGSA	19. 3.77	D.J.White and Partners	Burn	9. 4.05
DMU	2231	392	Eiri PIK-20D	20524		3.77	P.Warner and Partner	Gransden Lodge	29. 1.05
DMV	2232	DMV	Eiri PIK-20D	20526		12. 4.77	F.S.Parkhill	Bellarena	30. 1.05
DMX	2234	DMX	Schleicher ASK13	13567		5. 4.77	Dartmoor Gliding Society	Brentor	2. 6.05
							(Incorrectly marked as BGA.2294)		
DNB	2238		DFS/49 Grunau Baby 2B	2	RAFGSA.380	2.77	P.Underwood	Eaton Bray	*
			(Built Flg.u.Arbeitsg.Hall)		D-5766		*(On rebuild 2000; to be in Luftwaffe c/s)*		
DNC	2239	588	Grob G.102 Astir CS	1428		13. 4.77	A.J.Carpenter *"Natural High"*	Edgehill	27. 6.99
DND	2240	DND	Pilatus B4 PC-11AF	136		19. 3.77	R J Happs	Lasham	30. 4.05
DNE	2241	DNE	Grob G.102 Astir CS77	1631		3.77	G. Jones	Gransden Lodge	5. 2.05
DNF	2242	DNF	SZD-9 bis Bocian 1D	P-354	HB-657	23. 4.77	M.G.Shaw	Portmoak	31. 5.03
DNG	2243	265	Schempp-Hirth HS.5 Nimbus 2	126		5. 4.77	A.O.Harkins and A Brown	Upavon	27. 3.05
DNJ	2245	DNJ	Schleicher ASK18	18042		13. 4.77	Derbyshire and Lancashire Gliding Club	Camphill	3.12.05
DNK	2246	745	Grob G.102 Astir CS	1434		3. 5.77	745 Syndicate	Rattlesden	22.11.03
DNL	2247	DNL	Glasflugel H.201 Standard Libelle	82	RAFGSA.742	23. 4.77	P.J.Luckhurst	Tibenham	6. 4.02
					RAFGSA 16, G-AXZH		*(Landing accident at Marham 28. 7.01)*		
DNQ	2251	307	Rolladen-Schneider LS-3	3035		1.77	M.Cooper	Challock	20. 2.05
DNT	2254	DNT	SZD-30 Pirat	S-0712		25. 5.77	R.Lashly	Drumshade	20. 2.05
DNU	2255	U2	SZD-42-1 Jantar 2	B-783		4.77	C.D.Rowland and Partners	Booker	23. 8.04
DNV	2256	DNV	Schleicher ASK13	13568		4.77	Buckminster Gliding Club	Saltby	4. 5.05
DNW	2257	DNW	Schleicher Ka6CR	829	???	5.77	F.G.Broom	Rhigos	20. 2.05
DNX	2258		Schleicher Ka6CR	6094Si	D-5107	14. 6.77	C.J.Riley	Burn	15.11.02
DNZ	2260	DNZ	Schleicher K8B	8095	D-1711	5.77	Southampton University Gliding Club	Lee-on-Solent	17. 9.04
DPA	2261	DPA	Schleicher ASK18	18044		25. 5.77	Vectis Gliding Club	Bembridge	8. 5.05
DPD	2264	DPD	LET L-13 Blanik	026860		14. 6.77	E.McCaig	Challock	27. 8.01
			(Incorporates major portions of BGA2121)						
DPG	2267	DPG	Munchen Mu-13D III	005	D-1327	14. 6.77	G.J.Moore	Dunstable	30. 5.05
DPH	2268	287	Schempp-Hirth HS.7 Mini Nimbus	009		5.77	J.W.Murdoch	Portmoak	10. 9.05
DPJ	2269	DPJ	Grob G.102 Astir CS77	1641		8. 7.77	J.Liddiard and Partners	Lasham	26. 4.05
DPK	2270	DPK	Glasflugel H.303 Mosquito	27		6.77	G.Lawley	Cross Hayes	5. 8.05
DPL	2271	437	Eiri PIK-20D	20549		26. 6.77	D.W.Standen	Dunstable	19. 3.05
DPP	2274	DPP	Schleicher Ka2B Rhönschwalbe	105	D-1880	1. 7.77	B.G.Hoekstra	Breda, The Netherlands	8. 7.02
DPQ	2275	DPQ	Grob G.102 Astir CS77	1632		8. 7.77	J.Bone	Wormingford	14. 1.05
DPR	2276	D-1265	Scheibe L-Spatz	05	D-1265	8. 7.77	V.W.Jennings *"Sparrowfahrt"*	Thame	25. 7.01
DPT	2278	DPT	Scheibe L-Spatz 55	01	RAFGSA	25. 8.77	B.V.Smith *(Stored 6.03)*	Rufforth	12. 7.98
DPU	2279	DPU	EoN AP.5 Olympia 2B	EoN/O/142	RAFGSA.274	6. 9.77	J.M.Turner	Challock	8. 4.05
DPX	2282	440	Schleicher ASW19B	19126		7.77	A.J.Wilson	Nympsfield	7. 4.05
							(Damaged landing in field near Stow-on-the-Wold 21.8.04)		
DPY	2283	375	Grob G.192 Astir CS77	1652		8.77	D S Burton	Lasham	12. 6.05
DPZ	2284	DPZ	Slingsby T.34A Sky	822	HB-561	9. 8.77	N McLaughlin	Saltby	3. 7.00
DQA	2285	DQA	Schleicher ASK13	13582		8.77	Essex and Suffolk Gliding Club	Wormingford	28. 3.05
DQB	2286	844	Grob G.102 Astir CS77	1653		8.77	G.R.Davey	Kirton-in-Lindsey	18. 5.05
DQC	2287	DQC	Schleicher Ka6CR	6373Si	D-5725	26. 9.77	P. Morantl	Lasham	25. 5.05
DQD	2288	DQD	Slingsby T.8 Tutor	---		25. 8.77	K.J.Nurcombe	Husbands Bosworth	10.10.03
			(Built F.Breeze from parts)						
DQE	2289	480	Grob G.102 Astir CS77	1636		8.77	Heron Gliding Club	RNAS Yeovilton	7. 4.05
DQF	2290	DQF	Schleicher Ka6CR	6417	D-5827	9.77	P.James	Saltby	26. 3.05
DQG	2291	770	Grob G.102 Astir CS77	1649		6. 9.77	Miss A.G.Veitch	Easterton	10. 4.05
DQH	2292	DQH	Schmetz Condor IV	2	D-8538	7. 7.78	G.Saw and J.Kruse	Uetersen, Germany	6. 8.04
DQJ	2293	DQJ	Schleicher Ka6CR	228	D-5467	7.10.77	K.Whitworth	Kenley	27. 1.02
DQK	2294	542	Schleicher Ka6E	4341	D-0541	15.10.77	S.Y.Duxbury and R.S.Hawley	Long Mynd	28. 6.03
							(See BGA.2234)		
DQL	2295	DQL	Schleicher Ka8	509	D-5675	15.11.77	Lakes Gliding Club	Walney Island	23. 5.03
							(Damaged landing in field Dalton, Cumbria, 1.9.02)		
DQM	2296	DQM	Pilatus B4 PC-11	138	RAFGSA 506	15.10.77	T.Flude and H.Weston	Ringmer	9.10.04
DQP	2298	DQP	Schleicher K8B	1181	???	15.11.77	Soaring Centre	Husbands Bosworth	19. 8.05
DQR	2300	556	Grob G.102 Astir CS77	1667		10.77	N.R.Warren and Partners	Kingston Deverill	26. 1.05
DQS	2301	DQS	Schleicher Ka6CR	1065	D-5144	23.11.77	L.Hill	North Hill	2. 4.05
DQU	2303	DQU	Eiri PIK-20D	20579		10.77	A.B.Duncan	Portmoak	9. 8.05
DQX	2306	DQX	Schleicher Ka7 Rhönadler	743	D-9127	15.11.77	Scottish Gliding Union Ltd	Portmoak	8. 4.04
			(Modified to ASK-13 standard, with fuselage from ASK-13 BGA.1833)						
DQY	2307	DQY	Schleicher K8B	647	D-4375	11.77	Mendip Gliding Club	Halesland	26. 9.05
DRA	2309	904	Schleicher Ka6CR	1118	D-901141	11.77	P Davis	Shobdon	6. 8.04
DRB	2310	86	Glaser-Dirks DG-100	31	PH-532	11.77	J.D.Peck	Bicester	10. 4.05
DRD	2312	DRD	Schleicher Ka6CR	6377Si	D-9080	11.77	Essex and Suffolk Gliding Club	Wormingford	12. 4.05
DRE	2313	DRE	Schleicher Ka6CR	6197	D-8558	25. 1.78	J.H.Jowett	North Hill	3. 3.05
DRG	2315	DRG	Schleicher Ka6CR	6157	D-4090	11.77	WE Smith tr Summer Wine Syndicate	Gallows Hill	14. 8.01
DRJ	2317	D	Schleicher ASK13	13583		11. 1.78	Lasham Gliding Society	Lasham	21.11.05
DRK	2318	NT	Grob G.102 Astir CS77	1686		12.77	N.Toogood	Lasham / Weston-on-the-Green	14. 8.05
DRL	2319	DRL	Scheibe SF-26 Standard	5040	D-7073	15.12.77	T McKinley	Kirton-in-Lindsey	13. 6.04
DRM	2320	DRM	Schleicher Ka7 Rhönadler	7017	D-4666	1. 2.78	L.G.Cross and Syndicate	Dunstable	21. 8.05
							(Damaged landing in field south of Talgarth 13.10.04)		
DRN	2321	821	Glasflugel H.303 Mosquito	082		11. 2.78	A.Roberts	Cross Hayes	16. 4.05

DRP	2322	DRP	Pilatus B4 PC-11	080	RAFGSA	5. 1.78	M.C.Moxon	RAF Weston on the Green	24. 2.05
					BGA.1927				
DRQ	2323	258	Grob G.103 Twin Astir	3027		25. 1.78	V.C.Carr and Partners	Sleap	23. 4.04
DRR	2324	DRR	Schleicher Ka2B Rhönschwalbe	49	D-8108	11. 1.78	Dumfries and District Gliding Club	Falgunzeon	18.11.05
DRS	2325	DRS	SZD-9 bis Bocian 1E	P-783		1. 2.78	Mendip Gliding Club	Halesland	24. 5.04
							(Withdrawn from use and burned Haylesland, 11.03)		
DRT	2326	688	Eiri PIK-20D	20587		5. 1.78	P F Fowler	Long Mynd	18. 4.05
DRU	2327	334	Grob G.102 Astir CS77	1685		5. 1.78	J.R.Goodenough	Rivar Hill	3. 2.05
DRV	2328	DRV	Schleicher K8b	8026	D-6169	1.78	P.G.Clayton	Portmoak	27. 5.05
DRW	2329	798	Grob G.102 Astir CS	1081	D-3311	23. 2.78	P.A.Brooks	Lasham	18. 5.05
DRY	2331		Schleicher Ka6BR	370	D-5533	11. 2.78	A.May	RAF Marham	9. 1.06
DRZ	2332	DRZ	Schleicher K8B	668	D-4622	1. 2.78	East Sussex Gliding Club	Ringmer	1. 3.05
					D-KANB, D-4622				
DSA	2333	DSA	Slingsby T.30 Prefect	575	WE985	11. 2.78	J.Turner	Challock	14. 6.05
DSB	2334	DSB	Schleicher Ka6E	4300	D-0263	15. 3.78	B.Griffith	Ridgewell	25. 1.05
DSF	2338	DSF	Schleicher K8B	8220	D-7114	2.78	Edinburgh University Gliding Club *"Snoopy"*	Portmoak	18.12.04
DSH	2340	648	Grob G.102 Astir CS77	1696		13. 4.78	R B Petrie	Portmoak	24. 4.05
DSJ	2341	DSJ	Grob G.103 Twin Astir	3050		8. 3.78	L J Kaye	Shobdon	29. 7.05
DSL	2343	447	Grob G.103 Twin Astir	3041		2. 3.78	Shenington Gliding Club	Edgehill	20. 2.05
DSN	2345	893	Grob G.102 Astir CS77	1698		23. 3.78	J.J.M.Riach	Feshiebridge	27. 6.04
							(Damaged landing in field near Loch Laggan 22..5.04)		
DSP	2346	270	Schempp-Hirth HS.7 Mini Nimbus	33		9. 3.78	R.I.Hey and Partners	Nympsfield	28. 2.05
DSR	2348	DSR	Schleicher Ka6CR	970	D-5040	31. 5.78	Cornish Gliding Club	Perranporth	21. 4.02
							(Crashed near Perranporth 13. 4.02)		
DST	2350	972	Schleicher ASW20L	20059		24. 8.78	M.Makin Syndicate	Dunstable	4. 2.05
DSU	2351	DSU	Grob G.102 Astir CS77	1663		19. 4.78	Bowland Forest Gliding Club	Chipping	11. 4.05
DSV	2352	DSV	Pilatus B4 PC-11AF	134	RAFGSA.718	23.3.78	R.P.Watson	Dunstable	11. 4.05
					RAFGGA.518				
DSW	2353	533	Schempp-Hirth HS.7 Mini Nimbus	37	RNGSA.N33	3.78	S.C.Fear	Crowland	14. 3.04
							(De-Registered 3.2.04 - sold in Belgium as OO-YGN 3.04)		
DSX	2354	877	Schleicher ASW19	19188		13. 4.78	J.Reynolds and Partners.	Kingston Deverill	26. 3.05
DSY	2355	DSY	Schleicher Ka6CR	561	D-5702	28. 4.78	D.J.L.Smith	Rufforth	30.11.02
DTA	2357	699	Glaser-Dirks DG-200	2-27		9. 5.78	R P Hardcastle and D. Wakefield	Camphill	29. 6.05
DTC	2359	DTC	Schempp-Hirth HS.6 Janus B	63	RAFGSA.R9	4.78	Dukeries Gliding Club	Gamston	16. 5.05
					RAFGSA 16, BGA.2359				
DTD	2360	DTD	Schleicher ASW19	19187		4.78	R.K.Warren	Cross Hayes	20. 3.04
							(Badly damaged in fatal accident Camphil 11.6.03)		
DTE	2361	DTE	Schleicher ASW19B	19185		28. 4.78	G.R.Purcell	Lasham	4. 3.05
DTG	2363	DTG	Schempp-Hirth SHK-1	012	D-2034	9. 5.78	G.Thomas	Wattisham	15. 4.05
DTK	2366	760	Glasflugel H.303 Mosquito B	109		4.78	P.France	Usk	14. 3.05
DTM	2368	DTM	Glaser-Dirks DG-200	2-34		31. 5.78	C.Neil	Wormingford	12. 5.03
DTN	2369	DTN	Schleicher K8B	117/58	???	1. 6.78	F.McKeegan	Wormingford	18. 5.05
DTP	2370	915	Schleicher ASW20	20078		5.78	T.S.Hills and Partners	Lasham	31. 1.05
DTQ	2371	189	Schleicher ASW20L	20054		31. 7.78	J.P.Walsh	(Ireland)	1. 5.04
							(De-Registered 4.5.04 - sold in Ireland as EI-167)		
DTR	2372	DTR	EoN AP.6 Olympia 401	EoN/4/005	NEJSGSA.7	31. 5.78	B.D.Clarke	Ringmer	15. 5.02
					RAFGSA.252, G-APSI				
DTS	2373	DTS	CARMAM M.100S Mésange	031	F-CCST	18. 5.78	R.C.Holmes	Llantisilio	20. 9.01
DTU	2375	DTU	Schempp-Hirth HS.5 Nimbus 2B	167		9. 5.78	R.E.Wooler	Chipping	23. 4.05
DTV	2376	704	Glasflugel H.303 Mosquito B	110		7. 6.78	S.R.Evans	Aston Down	17. 3.05
DTW	2377	DTW	SZD-30 Pirat	S-0711		7. 6.78	C Kaminski	North Hill	30. 7.05
DTX	2378	320	Glasflugel H.303 Mosquito B	111		18. 5.78	T.Pentelow	Nympsfield	23 .1.02
DTY	2379	766	Glasflugel H.303 Mosquito B	112		18. 5.78	B.Bullimore	Gransden Lodge	27. 3.05
DTZ	2380	S30	Slingsby T.30 Prefect	573	WE983	21. 6.78	C.Hughes	Nympsfield	13.12.05
DUB	2382	911	Glasflugel H.303 Mosquito B	113		31. 5.78	F.B.Reilly and Partners	Portmoak	15. 5.05
DUC	2383		CARMAM M.100S Mésange	012	F-CCSA	7. 6.78	Not known *(Stored 1.05)*	Carlton Moor	5.88*
DUD	2384		Grunau Baby III		BGA.2074	7. 6.78	Not known	Flixton	
					RAFGSA.374, D-9142		*(On display Norfolk & Suffolk Aviation Museum 4.04)*		
DUE	2385	A11	Schleicher ASK-13	13591	AGA.15	4. 7.78	Kestrel Gliding Club	RAF Odiham	13. 5.05
					BGA.2385				
DUF	2386		Schleicher K8B	8296A	D-5294	11. 7.78	Essex Gliding Club	Ridgewell	15. 6.04
DUH	2388	DUH	Scheibe L-Spatz	760	Not known	28. 7.78	R.J.Aylesbury	Upwood	12. 6.05
DUK	2390	DUK	Schleicher K8B	752	D-4048	21 6.78	Bristol and Gloucestershire Gliding Club	Nympsfield	7. 7.05
DUL	2391	642	Grob G.102 Astir CS77	1720		21.6.78	J.Warren	Long Mynd	6. 3.05
DUQ	2394	DUQ	Glaser-Dirks DG200	2-43		7.78	D.M.Cottingham	North Hill	5. 5.05
DUR	2395	DUR	Schleicher Ka6CR	6273	OY-DLX	15. 8.78	K.W.Brown	Kirton in Lindsey	3. 8.05
DUS	2396	638	Schleicher Ka6E	4263	OY-XCB	.8.78	R.Bartlett	Tibenham	10.11.05
DUT	2397	T34	Schleicher ASW20	20089		24. 8.78	T.J.Murphy	Portmoak	29. 4.05
DUX	2401	DUX	Grob G.102 Club Astir CS Jeans	2140		7.78	B.T.Spreckley	Le Blanc, France	8. 3.05
DUY	2402	652	Glaser-Dirks DG-100	24	PH-525	13.10.78	A.C.Saxton and Partners	Currock Hill	26. 7.05
DVB	2405	DVB	Schleicher ASK13	13596		8.78	Essex and Suffolk Gliding Club	Wormingford	12. 2.05
			(Components including c/n plate donated to BGA.3493 and possibly discarded parts from crash Dunstable 5.6.82)						
DVC	2406	DVC	Schleicher ASK13	13597		31. 8.78	Southdown Gliding Club	Parham Park	9. 5.05
DVD	2407	DVD	LET L-13 Blanik	027021	RNGSA.N22	9. 9.78	Vectis Gliding Club	Bembridge	4. 3.05
DVE	2408	879	Schleicher Ka6E	4226	RAFGSA.379	15. 8.78	N.Greenwood	Aston Down	6. 3.05
DVG	2410	DVG	Schleicher Ka6CR	003	D-1916	22.11.78	R.F.Warren	Ringmer	1. 4.05
			(Built Holzmann-Drespack)						
DVH	2411	DVH	Schleicher Ka6E	4117	RAFGSA	315. 8.78	R.A.J.Jones	Brentor	14.12.05
DVJ	2412	869	Eiri PIK-20D	20638		9. 9.78	M.C.Hayes	Shobdon	14.12.04
DVK	2413	732	SZD-48 Jantar Standard 2	W-868		22. 9.78	G.P.Nuttall	Booker	5. 7.04
DVL	2414	X96	Schleicher ASW19	19222		10.10.78	P.T Healy and Partners	Lasham	1. 6.05

DVM	2415	DVM	Glasflugel H.205 Club Libelle	52	RAFGGA.581	12. 9.78	M.A. Field	Wormingford	5. 4.05
DVN	2416	DVN	Eiri PIK-20D	20641		22.11.78	J.Sharples and Partner	Burn	25. 4.05
DVP	2417	971	Schleicher ASW19	19220		23. 9.78	E.F.Davies	Booker	17. 2.05
DVS	2420	VS	Schempp-Hirth HS.4 Standard Cirrus	380	RAFGSA.824	26. 9.78	T.and C.Milner	Rufforth	3. 4.05
			(C/n duplicates VH-GGC)						
DVV	2423	810	Schleicher ASW20L	20100		9.78	A.Hooper	Usk	18. 6.05
DVX	2425	S13	Schleicher ASK13	13598		5.10.78	Shenington Gliding Club	Edgehill	16. 6.05
DVY	2426	272	Schempp-Hirth HS.2 Cirrus	52	OO-ZIR	10.78	M.G.Ashton	Perranporth	10. 6.05
DVZ	2427	Z25	Glasflugel H.303 Mosquito B	133		31.10.78	B.H.Shaw	Husbands Bosworth	26. 5.04
DWB	2429	733	Glasflugel H.303 Mosquito B	135		10.11.78	C.G.Salt and Partners *(As "BGA.2924")*	Lasham	6.11.04
DWC	2430	DWC	Schleicher Ka6E	4111	AGA.11	24.10.78	D.Jones	Wormingford	13. 6.05
DWE	2432	DWE	Schleicher Ka7 Rhönadler	7132	D-5427	3.11.78	N.T.Large	Llantisilio	5. 1.04
DWF	2433	DWF	DFS/49 Grunau Baby 2B	----	AGA.16	11.78	L.P.Woodage	RAF Keevil	9. 1.02
			(Built RNAY Fleetlands)		RNGSA 1-13, VW743		*(Stored 2005)*)		
DWG	2434	DWG	Schleicher K8B	165/60	D-5750	11.78	Newark & Notts Gliding Club	Winthorpe	18. 3.02
DWJ	2436	191	Glaser-Dirks DG-200	2-59		11.78	P.R.Desmond	Chipping	11. 4.05
DWL	2438	DWL	Glasflugel H.303 Mosquito B	141		2.12.78	A.Stanford and Partners	Husbands Bosworth	20. 4.05
DWN	2440	DWN	Schleicher Ka7 Rhönadler	7101	D-5360	12.78	L.R.Merritt	Saltby	11. 5.05
DWP	2441	447	Glasflugel H.303 Mosquito B	136		15.12.78	I.Pattingale	Odiham	18. 2.05
DWQ	2442	DWQ	Grob G.102 Astir CS77	1758		9. 1.79	C.Ratcliffe	Sleap	25. 6.05
DWR	2443	P9	Glasflugel H.303 Mosquito B	134	(BGA.2428)	23. 1.79	C.D.Lovell	Lasham	28. 2.05
DWS	2444	728	Eiri PIK-20D	20652		7. 3.79	D.G.Slocombe	Camphill	9. 4.05
DWT	2445	886	Slingsby T.65A Vega	1898		28. 3.79	A.P.Grimley	Husbands Bosworth	28. 9.00
DWU	2446	DWU	Grob G.102 Astir CS	1201	D-7269	30. 1.79	D. Evans	Hinton-in-the-Hedges	21. 6.05
DWW	2448	DWW	Slingsby T.65A Vega	1896		2. 3.79	J.Sorrell	Usk	2. 5.05
DWZ	2451	DWZ	Schleicher ASW19	19243		14. 2.79	J.A.Stirk and Partners	Burn	4. 5.02
							(Damaged landing in field Chelveston 16.8.01)		
DXA	2452	483	Glasflugel H.303 Mosquito B	137		14. 2.79	S H Gibson	Dunstable	31. 3.04
DXB	2453	81	Schleicher ASW20	20142		Not known	J.A.Timpany and Partners	Nympsfield	16. 2.05
DXD	2455	132	Slingsby T.65A Vega 17L	1901		20. 4.79	T.C.Harrington and Partners	Bicester	10. 5.05
DXE	2456	DXE	Slingsby T.65A Vega	1902		16. 5.79	H.K.Rattray	Usk	15.12.05
DXF	2457	DXF	Slingsby T.65A Vega	1903		16. 5.79	P.Goulding	Crowland	20. 5.05
DXG	2458	46	Slingsby T.65A Vega 17L	1906		2. 6.79	M.H.Pope	Bidford	8. 7.05
DXH	2459	DXH	Schleicher Ka6E	4198	RAFGSA.489	28. 3.79	B.Hughes	Bicester	18. 2.05
					D-4093				
DXJ	2460	DXJ	Grob G.102 Astir CS77	1762		2. 3.79	S. Flowitt-Hill	Rattlesden	27. 3.05
DXK	2461	160	Centrair ASW20F	20108		15. 5.79	A.Townsend	Nympsfield	5. 7.05
DXL	2462	DXL	Schempp-Hirth HS.4 Standard Cirrus	203G	AGA.1	6. 3.79	C.J.Button	Aston Down	10. 5.05
DXM	2463	DXM	Schleicher Ka7 Rhönadler	626	RAFGGA.551	20. 3.79	Vale of Neath Gliding Club	Rhigos	26. 4.04
			(Partly modified to ASK-13 standard)		D-5707				
DXN	2464	267	Glaser-Dirks DG-200	2-63		17. 3.79	J.A.Johnston	Gransden Lodge	3. 4.05
DXP	2465	DXP	Schleicher K8B	8646A	D-8537	24. 3.79	Stratford-upon-Avon Gliding Club	Snitterfield	24. 2.02
							(Withdrawn from use)		
DXQ	2466	147	Schempp-Hirth HS.7 Mini Nimbus C	96		13. 3.79	T.Lamb and P.Hawkins	RAF Weston on the Green	26. 4.05
			(Build No.MN97)						
DXT	2469	286	Schempp-Hirth HS.7 Mini Nimbus C	97		14. 3.79	C.Chapman	Booker	19.10.04
DXU	2470	DXU	Slingsby T.59J Kestrel 22	1867	G-BDWZ	11. 4.79	D.R.Wilkinson	Riseley	30. 4.05
DXV	2471	DXV	Schleicher ASK13	13602		16. 3.79	Cambridge University Gliding Club	Gransden Lodge	22. 2.99
							(Damaged on landing Gransden Lodge, 15.8.98)		
DXW	2472	354	Glasflugel H.303 Mosquito B	142		7. 4.79	P.Newmark and Partners	Burn	11. 2.05
DXX	2473	580	Schleicher ASW19B	19245		17. 3.79	P.G.Bateman	Kenley	11. 3.05
DXY	2474	HB-474	Muller Moswey III	---	HB-474	20. 4.79	G.M.Bacon and Partners	Gransden Lodge	9. 8.04
DYB	2477		Schleicher Ka7 Rhönadler	167/59	D-5775	29. 3.79	South London Gliding Centre	Kenley	9. 5.04
							"6" (Reported as "DYN")		
DYC	2478	DYC	Schleicher Ka6CR	6390	D-1545	20. 3.79	F.J.Smith	Burn	15. 6.05
DYE	2479	828	Schleicher ASW20L	20143		27. 3.79	T.A.Sage	Dunstable	14. 2.05
DYF	2480	850	Grob G.102 Astir CS77	1805		7. 4.79	York Gliding Centre	Rufforth	1. 2.05
DYG	2481	592	Slingsby T.59H Kestrel 22	1868	G-BDZG	31. 3.79	R.E.Gretton and R.L.Darby	Crowland	8. 7.05
DYJ	2483	DYJ	Schleicher Ka6CR	6583	D-5838	15. 5.79	R.M.Morris	Dunstable	7. 6.04
DYL	2485	DYL	CARMAM JP-15/36A Aiglon	37		4.79	M.P.Edwards	Crowland	20. 5.05
DYN	2486	DYN	Schleicher Ka6CR	6129Si	D-8458	1. 5.79	C.N.Harder *(See BGA.2477)*	Rivar Hill	24. 4.03
DYQ	2488	DYQ	Schleicher Ka6CR	6178	D-5328	9. 5.79	Dorset Gliding Club	Gallows Hill	20. 8.00
							(Withdrawn from use due to glue failure)		
DYR	2489	DYR	Schleicher Ka7 Rhönadler	766	D-5220	12. 4.79	UWE Gliding Club	Aston Down	17. 3.05
			(Partly modified to ASK-13 standard)						
DYU	2491	742	Schempp-Hirth HS.5 Nimbus 2C	181		18. 4.79	P.Brown	Lasham	31. 3.05
DYX	2494	102	Schleicher ASW20	20135		23. 6.79	J.C.Bailey	Challock	23.11.04
DYZ	2495	943	Schempp-Hirth HS.5 Nimbus 2C	180		24. 4.79	M.Sarel	Rivar Hill	28. 2.05
DZA	2496	DZA	Slingsby T.65A Vega 17L	1907		26. 6.79	K.Kuntz	Dunstable	14. 3.05
DZB	2497	DZB	Slingsby T.65A Vega	1908		10. 7.79	A.Black	Portmoak	11. 9.05
DZC	2498	DZC	Scheibe L-Spatz 55	642	RAFGGA	4. 5.79	G.A.Ford	Nympsfield	23.10.98
					D-5629				
DZD	2499	573	Schleicher ASW19B	19268		30. 5.79	D. HoltJ.Anderson	Llantisilio	10.10.03
							(De-registered 26.10.04 - sold in "Europe")		
DZF	2501	152	Schempp-Hirth HS.4 Standard Cirrus	421G	RAFGSA.27	15. 5.79	L.S.Hood	Bicester	13. 4.05
DZG	2502	909	Schleicher ASW19B	19267		10. 5.79	S.P.Wareham	Kingston Deverill	23. 2.05
DZJ	2504	576	Grob G.102 Club Astir	2230		11. 5.79	D.Bassett and Partners	Halesland	14. 6.05
DZM	2507	DZM	Slingsby T.65A Vega	1909		10. 7.79	D.G.MacArthur	Long Mynd	14. 3.05
DZN	2508	990	Slingsby T.65A Vega 17L	1910		13. 7.79	D.A.White	Aboyne	8. 8.05
DZP	2509	DZP	Slingsby T.65A Vega	1911		17.11.79	M.T.Cruise	Currock Hill / Milfield	3. 5.05
	2510		Kronfeld (BAC) Drone de luxe				See SECTION 3 Part 1	Brooklands	

DZR	2511	DZR	ICA IS-28B2	287		13. 6.79	Lakes Gliding Club	Walney Island	8. 2.04
DZS	2512	DZS	SZD-8bis-0 Jaskolka	183	HB-583	30. 5.79	D.A. Read	Booker	16. 7.05
DZT	2513	106	Eiri PIK-20D	20661		9. 6.79	A.C.Garside	Challock	28. 2.05
DZU	2514	DZU	Grob G.102 Astir CS	1076	D-3308	6. 6.79	P.F.Clarke	Booker	28. 2.05
DZV	2515	839	Scheibe SF-27A Zugvogel V	6065	D-5839	17. 7.79	N. Newham	Usk	8. 5.05
DZW	2516	DZW	Schleicher Ka6CR	6628	D-1045	11. 7.79	A.Wildman	Husband Bosworth	18. 7.03
DZX	2517		Slingsby T.30B Prefect				See SECTION 3 Part 1	Doncaster	
DZY	2518	757	Schleicher ASW19B	19275		13. 6.79	M.C.Fairman and T.Marlow	Dunstable	9. 3.05
EAC	2522	367	Grob G.102 Astir CS77	1803		14. 6.79	I.Pickering	Bidford	1. 8.05
EAD	2523	EAD	Slingsby T.65A Vega	1912		30.11.79	D.Smith	Sutton Bank	21.10.05
EAE	2524	107	Schleicher ASW20L	20224		20. 6.79	A. Smith and Partner	Challock	9. 4.05
EAF	2525	EAF	Grob G.102 Astir CS77	1830		12. 7.79	J.Bell	Milfield	20. 9.05
EAG	2526	EAG	Slingsby T.65A Vega	1913		4. 9.79	J.F.Evans	Usk	9. 4.05
EAH	2527	EAH	Schleicher Ka6E	4085	D-7542 D-7142	12. 7.79	M.Lodge	Lasham	14. 7.05
EAJ	2528	79	Schempp-Hirth HS.5 Nimbus 2	7	D-0699	28.6.79	G.Harvey	Currock Hill	17.11.04
EAK	2529	594	Glasflugel H.303 Mosquito B	155		29.6.79	T.Barnes	Aston Down	25. 3.05
EAL	2530		Schleicher Ka4 Rhönlerche II	3051/BR	PH-331	7.79	Newcastle and Teesside Gliding Club *(Stored 2000)*	Carlton Moor	13. 9.91*
EAM	2531	EAM	Schempp-Hirth HS.5 Nimbus 2B	93	D-2787	10. 7.79	C.H.Brown	Chipping	28. 2.05
EAP	2533	R31	Schleicher ASK-13	13609	RAFGSA.R31 BGA.2533	19. 7.79	RAFGSA Crusaders Gliding Club *(Landing accident Kingsfield 12. 5.04)*	Kingsfield, Dhekelia	22. 2.04
EAR	2535	EAR	Eiri PIK-20D	20550		28. 7.79	D.A.Coker	RAF Syerston	12. 4.05
EAT	2537	786	Eiri PIK-20D	20664		22. 8.79	M.Crawley and Partners	Lasham	17. 5.05
EAU	2538	EAU	Schleicher Ka7 Rhönadler *(Part modified to ASK13 standard)*	7092	PH-304	7.79	Welland Gliding Club	Lyveden	20. 4.05
EAV	2539	360	Schempp-Hirth HS.7 Mini Nimbus C	136		31. 7.79	G.D.Crawford	Weston-on-the-Green	15. 4.04
EAW	2540	EAW	Grob G.102 Astir CS77	1831		21. 7.79	D.M.Smith	Camphill	7. 7.05
EBA	2544	EBA	Slingsby T.65A Vega 17L	1914		1.11.79	F.T.Bick	Aboyne	12. 6.05
EBB	2545	881	Grob G.102 Speed Astir IIB	4040		28. 7.79	M.Malcolm and A.F.Grinter	Pocklington	28. 4.05
EBC	2546	EBC	Slingsby T.30B Prefect *(Built from parts ex BGA.808 and BGA.1618?)*	583	RAFGSA.33 *WE993*	1. 8.79	K.R.Reeves *"Jonathan Livingstone Prefect"*	RAF Syerston	16. 5.04
EBD	2547	EBD	Scheibe Bergfalke IV	5822	D-1005	4. 9.79	D.Clarke	Burn	12. 5.05
EBE	2548	EBE	Issoire E78 Silene	07		20.11.79	B.A.Burgess	Husbands Bosworth	21. 7.05
EBF	2549	EBF	Schempp-Hirth HS.7 Mini Nimbus C	138		17. 8.79	M.Pinell	Talgarth	20.12.05
EBJ	2552	h11	Schleicher ASW19B	19282		21. 8.79	S. Hill	Sutton Bank	10.11.05
EBK	2553	552	Schempp-Hirth HS.7 Mini Nimbus C	139	AGA.2 BGA.2553	17. 8.79	J.B.Burgoyne	Husbands Bosworth	9. 6.05
EBL	2554	EBL	Schleicher ASK13	13610		7. 9.79	Bristol and Gloucestershire Gliding Club	Nympsfield	26.11.05
EBM	2555	807	Grob G.102 Astir CS77	1843		8.79	P.K.Hayward	Ringmer	20. 5.05
EBN	2556	37	Centrair ASW20FL	20118		22. 8.79	S.MacArthur	Camphill	18. 6.05
EBP	2557	EBP	Allgaier Geier I	3/4	D-9025	4. 9.79	D.P.Raffan	RAF Marham	21. 7.01
EBQ	2558		Schleicher Ka6CR	6051Si	D-5237	3.10.79	Not known *(Under repair 2000)*	Tibenham	
EBR	2559	EBR	Glaser-Dirks DG-200/17	2-89/1706	D-6893	6. 9.79	M.D.Parsons	Lee-on-Solent	3. 6.05
EBS	2560	EBS	Scheibe Zugvogel IIIA	1054	LX-CAF D-8363	21.11.79	I.D.McLeod *"Schwarzhornfalke"*	Challock	11. 6.05
EBX	2565	644	Schleicher ASW20	20058	D-7973	21. 9.79	J P Davies	Gransden Lodge	5. 4.05
EBZ	2567	EBZ	Schleicher ASK13	13614		9.79	Booker Gliding Club	Booker	11. 2.05
ECC	2570	ECC	Schleicher Ka6CR	60/01	D-5080	10.10.79	A. Jepson and Partners	Burn	9. 4.05
ECF	2573	ECF	Schleicher Ka6CR	856	D-5808	10.79	S. Buckley and partners	Aston Down	24. 4.05
ECG	2574	ECG	Schempp-Hirth SHK	19	D-5359 D-1329	10.10.79	M.F.Hardy	AAC Upavon	6. 1.05
ECH	2575	ECH	Glasflugel H.303 Mosquito B	173		24. 1.80	A.Walker and Partners	Rattlesden	27. 3.05
ECJ	2576	ECJ	Slingsby T.65A Vega	1916		21.12.79	J.E.B.Hart and Partners	Sutton Bank	21. 4.05
ECK	2577	ECK	Slingsby T.65A Vega	1917		13.12.79	B.Chadwick	Lyveden	4. 4.04
ECL	2578	ECL	Slingsby T.65A Vega 17L	1918		2. 2.80	J.B.Strzebrakowski	Lyveden	20. 6.05
ECM	2579	ECM	Slingsby T.65A Vega	1919		15. 1.80	F.L.Wilson	Aston Down	21. 9.05
ECN	2580	645	Slingsby T.65A Vega 17L	1920		20.11.79	C.Claxton Syndicate *(De-Registered - to ZK-GSV, 11.04)*	Booker	17. 2.04
ECP	2581	ECP	Rolladen-Schneider LS-3-17	3426		26. 3.80	D.Crowhurst	Crowland	3. 5.05
ECQ	2582	ECQ	Grob G.102 Astir CS77	1837		25.10.79	N.Harrison	Tibenham	19. 4.05
ECS	2584	955	Glasflugel H.303 Mosquito B	166		26.10.79	R.C.Adams and P.Robinson	Wormingford	12. 3.05
ECT	2585	604	Glasflugel H.604	2	I-FEVG D-0279	4.10.79	P.T. Nash	Crowland	8. 4.05
ECW	2588	ECW	Schleicher ASK21	21008		2. 3.80	Norfolk Gliding Club	Tibenham	9. 3.05
ECX	2589	600	Schleicher ASW20L	20315		10. 6.80	A.C.Robertson *(De-Registered 16. 2.04 - sold abroad)*	Feshiebridge	17. 5.04
ECY	2590	ECY	Glasflugel H.201B Standard Libelle	530	RAFGGA.557	13.11.79	T.Huttlestone	Bidford	4. 6.05
ECZ	2591	ECZ	Schleicher ASK21	21009		26. 4.80	Booker Gliding Club	Booker	26. 7.05
EDA	2592	647	Slingsby T.65A Vega 17L	1888	G-BFYW	30.11.79	A.R.Worters	North Connel	17. 9.01
EDB	2593	EDB	CARMAM JP-15/36AR Aiglon	40		1. 2.80	P.J.Martin and Partners	Husbands Bosworth	14. 3.05
EDC	2594		Schleicher Ka7 Rhönadler	244	D-8527	13. 2.80	J.C.Shipley	Camphill	8. 7.04
EDD	2595	EDD	Schleicher ASW17	17043	D-6865	12.79	M.Etherington	Husbands Bosworth	14.12.05
EDF	2597	530	Schempp-Hirth HS.7 Mini Nimbus C	149		8. 1.80	C.W.Boutcher	Snitterfield	4.11.05
EDG	2598	EDG	Schleicher Ka6CR	6512	RAFGSA	9. 1.80	S.J.Wood	Saltby	24. 5.05
EDH	2599	EDH	Glasflugel H.303 Mosquito B	184		25. 3.80	D.G.Cooper	Tibenham	24. 7.05
EDJ	2600	EDJ	Glasflugel H.303 Mosquito B	185		4. 4.80	D.H.Martin	Camphill	14. 3.05
EDK	2601	EDK	Schleicher Ka7 *(Partly modified to ASK-13 standard)*	791	D-1633	13. 2.80	North Wales Gliding Club	Llantisilio	29.12.05
EDL	2602		Focke-Wulf Weihe 50	4	D-0893 HB-555	26. 1.80	F.K.Russell *(Being restored during 2000)*	Dunstable	14.10.96*

EDM	2603	EDM	Glaser-Dirks DG-200	2-98		17. 2.80	S. St Pierre	Sutton Bank	12. 3.05
EDN	2604	820	Glaser-Dirks DG-100G Elan	E12G6		14. 2.80	A.P.Scott and Partners	Currock Hill	5. 7.05
EDP	2605	EDP	Glaser-Dirks DG-100G Elan	E19G7		12. 2.80	G.Peters and partner	North Hill	16. 2.05
EDU	2610	EDU	Schleicher ASK13	13613		22. 3.80	Kent Gliding Club	Challock	13.11.05
EDV	2611	541	Slingsby T.65A Vega 17L	1893	G-BGCU	8. 2.80	J.L.Clegg	Aston Down	11. 4.04
EDW	2612	EDW	Schleicher ASK21	21010		5. 4.80	UCLU Gliding Club	RAF Halton	2. 8.04
EDX	2613	EDX	Slingsby T.65D Vega	1928		20. 5.80	G.Kirkham	Camphill	15.12.05
EDY	2614	EDY	Slingsby T.65D Vega	1929		23. 5.80	C.J.Steadman	Husbands Bosworth	29. 3.05
EDZ	2615	EDZ	Slingsby T.65C Sport Vega	1931		18. 6.80	R.C.Copley	Walney Island	6. 3.05
EEA	2616	337	Slingsby T.65C Sport Vega	1932		27. 6.80	M.R.Johnson	Crowland	7. 7.05
EEC	2618	EEC	Schleicher ASW20L	20311	(G-BSTS)	27. 6.80	D.M.Cushway	Challock	10.12.05
					BGA.2618				
EED	2619	R91	Schleicher K8B	590	RAFGSA.R91	1. 3.80	RAFGSA Crusaders Gliding Club	Kingsfield, Dhekelia	9. 2.02
					NEJSGSA, BGA.2619, D-5703 *(Crashed Kingsfield 3.2.02)*				
EEE	2620	EEE	Schleicher ASW20L	20312		4. 4.80	T E MacFadyen	Nympsfield	13. 4.05
EEF	2621	EEF	Rolladen-Schneider LS-3-17	3441		13. 6.80	G.J.Nicholas	Rivar Hill	28.11.05
EEG	2622	EEG	Slingsby T.65C Sport Vega	1922	EI-129	3.80	G.Harris	Rufforth	13. 4.05
					BGA.2622				
EEH	2623	166	Schleicher ASW19	19042	RAFGGA.166	27. 2.80	K Kiely	AAC Dishforth	14. 6.05
EEJ	2624	EEJ	Schleicher ASW20L	20314		20. 9.80	R.R.Stoward	Dunstable	25. 7.04
EEK	2625	996	Schempp-Hirth HS.5 Nimbus 2C	201		23. 2.80	M. Erland	Kirton-in-Lindsey	11.12.04
EEM	2627	EEM	Schleicher K8B	8688AB	D-0254	28. 2.80	South Wales Gliding Club	Usk	20. 6.05
EEN	2628	EEN	Schempp-Hirth HS.4 Standard Cirrus	75 621	RAFGSA 87	1. 3.80	J.Hanlon	RAF Weston on the Green	5. 3.05
					(BGA.2609)				
EEP	2629	EEP	Wassmer WA.26P Squale	36	F-CDSX	3.80	B.Key and Partners	Aston Down	21. 3.05
EEQ	2630	EEQ	Grob G.102 Standard Astir II	5015S	RNGSA.N12	3.80	M.Abbott	Lleweni Parc	11.10.05
EER	2631	EER	Schempp-Hirth HS.7 Mini Nimbus C	150		14. 3.80	D.J.Uren	Perranporth	5. 3.05
EES	2632	50	Rolladen-Schneider LS-3-17	3248		26. 3.80	J.Illidge and Partners	Camphill	28.10.05
EEV	2635		Centrair ASW20FL	20145		15. 5.80	R.Baez	(Wittstock-Berlinchen, Germany)	5. 8.05
EEW	2636	EEW	Schleicher Ka6CR	6188	RAFGGA	26. 3.80	Buckminster Gliding Club	Saltby	7. 6.05
					D-6151				
EEX	2637	EEX	Rolladen-Schneider LS-3-17	3442		5. 7.80	W.A.Dallimer and Partner	Aston Down	1. 7.05
EEZ	2639	157	Rolladen-Schneider LS-3A	3458		10. 6.80	J.P.Gilbert	Wormingford	22. 1.05
EFA	2640	470	Schleicher ASW20L	20326		2. 7.80	R.Cooper	Dunstable	25. 7.05
EFB	2641	EFB	Schempp-Hirth HS.5 Nimbus 2C	216		3. 4.80	N.Revell and Partners	Gamston	23. 6.05
EFC	2642	EFC	Siebert Sie-3	3018	D-0811	3.4.80	M.S.A.Skinner	Sandhill Farm, Shrivenham	28.12.05
EFD	2643	EFD	Schleicher Ka7 Rhönadler	7007	PH-277	15.7.80	South London Gliding Centre	Kenley	19. 1.99
							(Blown over Kenley 21.4.99)		
EFE	2644	586	Centrair ASW20F	20139		1. 5.80	D. Salmon and Partners	Camphill	22. 4.05
							(Damaged Camphill 31.7.04)		
EFF	2645	737	Schempp-Hirth HS.5 Nimbus 2C	208		10. 4.80	D.L.Jobbins	Usk	14. 4.05
EFG	2646	EFG	Schleicher K8B	?	RAFGGA	10. 4.80	Essex Gliding Club	Ridgewell	7. 5.05
EFJ	2648	EFJ	Centrair ASW20F	20127		12. 4.80	D.E.Ball	Booker	17. 5.05
EFK	2649	643	Centrair ASW20FL	20140		15. 5.80	G.B.Monslow	Snitterfield	14. 3.05
EFL	2650	EFL	Centrair ASW20FL	20133		15. 5.80	T.Dews	Kingston Deverill	1.11.05
EFM	2651	GAZ	Schleicher Ka6E	4103	RAFGSA	4. 6.80	G.S.Foster	Parham Park	13. 9.05
EFN	2652	EFN	Scheibe L-Spatz 55	635	D-1617	17. 5.80	J.A.Halliday	Aston Down	15. 3.05
EFT	2657	J45	Schempp-Hirth HS.5 Nimbus 2B	26	HB-1160	23. 4.80	S.A.Adlard	Long Mynd	16. 5.05
EFV	2659	EFV	Schleicher ASW20	20041	OE-5162	12. 6.80	A.R McKillen	Bellarena	11.12.04
EFW	2660	EFW	Slingsby T.65C Sport Vega	1938		18. 7.80	Dukeries Gliding Club	Gamston	6. 8.05
EFZ	2663	EFZ	Rolladen-Schneider LS-3A	3273		21.7.80	D.H.Gardner and J.Higgins	Aston Down	13. 4.05
EGD	2667	D3	Schleicher ASW17	17028	D-2343	25. 6.80	W.J.Dean	Long Mynd	1. 6.05
EGE	2668	EGE	Rolladen-Schneider LS-3A	3465		31. 7.80	D.Barker	Nympsfield	9. 1.06
EGF	2669	EGF	Slingsby T.65C Sport Vega	1936		28. 6.80	S.Margant and Partner	Ringmer	2. 2.05
EGG	2670	JH	Slingsby T.65C Sport Vega	1939		23. 9.80	R.Robbins	Usk	22.12.05
EGH	2671	EGH	Slingsby T.65C Sport Vega	1943		28.11.80	M.J Davies and Partners	Winthorpe	22. 5.05
EGJ	2672	672	Slingsby T.65C Sport Vega	1944		12.12.80	J. Bradbury and M. Hannigan	Lee-on-Solent	27. 3.05
EGK	2673	569	Schempp-Hirth HS.4 Standard Cirrus	542G	RAFGSA.569	1. 7.80	I.M.Deans and Partners	Lasham	19.12.05
					RAFGSA.R2 (2)				
EGN	2676	EGN	Grob G.103 Twin II	3542		19. 8.80	J.F. Rogers	Edgehill	13. 5.04
							(Crashed south of Malton, Yorks, 27.8.04 - to Rufforth for repair)		
EGP	2677	BS1	Schleicher ASW20L	20336		17. 9.80	B.K.Scaysbrook	Husbands Bosworth	29. 3.05
EGR	2679	EGR	Breguet Br.905SA Fauvette	18	F-CCGT	22. 8.80	P.Parker	Dunstable	15. 2.03
EGS	2680	2CS	Schempp-Hirth HS.5 Nimbus 2CS	192	D-2111	21. 7.80	P.G.Myers	Long Mynd	26. 3.05
EGT	2681	EGT	Slingsby T.65D Vega	1933		28. 7.80	D.M.Badley and Partners	Sleap	4. 3.05
EGU	2682	EGU	Slingsby T.65A Vega	1921		28. 7.80	M.N.Bishop	Challock	22. 8.05
EGW	2684	EGW	Schempp-Hirth HS.7 Mini Nimbus B	78	HB-1447	1. 8.80	I F Barnes	Ridgewell	28. 8.04
			(Modified to Mini Nimbus C?)						
EGX	2685	EGX	Slingsby T.65C Sport Vega	1937	RAFGSA.R23	15.10.80	O. Raine	Thame	9. 2.05
					BGA.2685				
EGZ	2687	EGZ	Schleicher ASK21	21030		28.10.80	Needwood Forest Gliding Club	Cross Hayes	5. 3.05
EHA	2688	D5084	Schleicher K8B	136/59	D-5084	6. 9.80	Not known (Stored 6.04)	Fairwood Common	*
EHB	2689	K3	Schleicher Ka3	3	????	23.10.80	L.S.Hood	Cranwell	28. 5.05
EHC	2690	EHC	Eichelsdorfer SB-5B	5017	D-9310	14. 8.80	R.I.Davidson	Husbands Bosworth	12. 5.05
EHD	2691	891	Schleicher ASW20L	20386		10.80	B.Lumb	Burn	9. 4.05
EHE	2692	WE992	Slingsby T.30B Prefect	582	WE992	29. 9.80	A.P.Stacey	RAF Keevil	5. 3.98
							tr A.T.C.Syndicate *(Noted 1.04)*		
EHG	2694	453	Slingsby T.65C Sport Vega	1940		21.10.80	A. Smith Partners	Lasham	19. 1.05
EHH	2695	V7	Schempp-Hirth Ventus A	07		5.11.80	P.G.Sheard and A.Stone	Lasham	11. 3.04
EHK	2697	490	Rolladen-Schneider LS-4	4068		15. 3.81	R.Starling	Nympsfield	4. 2.05
EHL	2698	138	Rolladen-Schneider LS-4	4024		24. 4.81	C.J.Evans	Booker	14. 3.05

EHM	2699	EHM	Schleicher Ka6E	4118	RAFGSA.318	28. 3.81	J. Symonds	Kingston Deverill	12. 8.05
EHN	2700	EHN	Slingsby T.65C Sport Vega	1942	G-BILH	17.12.80	P.Cheraten and Partners	Challock	1. 9.05
					BGA.2700				
EHP	2701	EHP	Schempp-Hirth HS.5 Nimbus 2C	234		11.80	R.Hudson	Sutton Bank	19. 2.05
EHQ	2702	431	Schleicher ASK21	21035		21.11.80	University of Surrey Gliding Club	Lasham	14.12.05
EHS	2704	EHS	ICA IS-28B	289		3.12.80	M.Terry	Gamston	1. 9.03
EHT	2705	EHT	Schempp-Hirth HS.5 Nimbus 2C	235		12.80	S.Codd	Edgehill	29. 4.05
EHU	2706	849	Glasflugel H.304	209		5.11.80	F.and J.M.Townsend	Bidford	3.12.04
EHV	2707	481	Schleicher ASW20L	20385		6. 1.81	M.A.Roberts	Rattlesden	2. 1.06
EHW	2708	EHW	ICA IS-28B2	86		9. 1.81	J.P.Bakker	Talgarth	13.12.04
EHY	2710	EHY	Slingsby T.65D Vega	1941		8. 1.81	B.G.Skilton	Ringmer	30. 3.05
EHZ	2711	413	Schleicher ASW20L	20388		29. 1.81	J.B.Hoolahan	Challock	29. 6.05
EJA	2712	EJA	ICA IS-28B2	88		16. 1.81	Not known *(Stored 11.04)*	Aston Down	15.12.00
EJB	2713	EJB	Slingsby T.65C Sport Vega	1945		23. 1.81	I.G.Walker and Partners	Cross Hayes	23. 4.05
EJC	2714	EJC	Slingsby T.65C Sport Vega	1946		9. 2.81	D.Redfearn	Rattlesden	17. 4.05
EJD	2715	261	Slingsby T.65D Vega 17L	1930	RAFGGA.510	29. 6.81	A.J.French	Rufforth	28. 5.97*
							(Damaged Dunstable 28.3.97: wreck noted.03)		
EJE	2716	EJE	Slingsby T.65C Sport Vega	1947		16. 2.81	Crown Service Gliding Club	Lasham	7. 2.05
EJF	2717	EJF	Schleicher K8B	8966	D-2328	15. 1.81	Cotswold Gliding Club	Aston Down	24. 1.05
EJH	2719	EJH	Eichelsdorfer SB-5E	5041A	D-5430	14. 1.81	H.J.McEvaddy	Husbands Bosworth	11. 6.01
					D-0087				
EJJ	2720	EJJ	Slingsby T.21B	618	RAFGSA.120	1. 3.81	N.P.Marriott	Parham Park	4. 5.05
					BGA.662, WB957				
EJL	2722	904	Centrair ASW20FL	20183		27. 4.81	S.R.Jarvis	Bidford	22. 6.04
EJQ	2726	EJQ	Centrair ASW20FL	20184		20. 4.81	S. Foster and Partners	Gransden Lodge	18. 1.05
EJR	2727	193	Schleicher ASW19B	19334		12. 4.81	M.Thompson	Nympsfield	25.12.05
EJS	2728	319	Slingsby T.65C Sport Vega	1948		20. 3.81	A.D.McLeman	Portmoak	4. 9.05
EJT	2729	890	Slingsby T.65A Vega	1889	G-VEGA	5. 3.81	W.A.Sanderson	Wormingford	3.12.03
					(G-BFZN)				
EJY	2734	EJY	SZD-9 bis Bocian 1D	P-351	D-1587	13. 4.81	J.A.Stephen	Aboyne	15. 6.05
EKA	2736	EKA	Glaser-Dirks DG-200/17	2-128/1730		3. 8.81	M.J.Lindsey	Tibenham	25. 1.05
EKC	2738	EKC	Schleicher Ka6E	4079	OO-ZDV	18. 6.81	S.L.Benn	RAF Cranwell	21. 3.05
					OE-0813				
EKD	2739	EKD	Schleicher ASK13	13539	OH-494	21. 4.81	Bristol and Gloucestershire Gliding Club	Nympsfield	30. 4.05
EKE	2740	20L	Schleicher ASW20L	20387		15. 4.81	A.G.Mackenzie	Burn	25. 4.05
EKF	2741	EKF	Grob G.102 Club Astir III	5519C		14. 6.81	Bristol and Gloucestershire Gliding Club	Nympsfield	10. 4.05
EKG	2742	EKG	Schleicher ASK21	21067	AGA.8	5.81	Wyvern Gliding Club	AAC Upavon	4.12.05
					BGA.2742				
EKH	2743	714	Schempp-Hirth Ventus B	32		1. 5.81	R Johnson	Parham Park	10. 2.05
EKJ	2744	EKJ	Schempp-Hirth Ventus B	36		7. 5.81	I.J.Metcalfe	Nympsfield	20. 9.05
EKK	2745	EKK	SZD-48 Jantar Standard 2	W-853		9. 5.81	K.A. Harrison	Hinton-in-the-Hedges	19. 5.05
EKM	2747	EKM	Schleicher K8B	647	PH-258	28. 5.81	Not known *(Wreck stored 11.04)*	Aston Down	
EKP	2749	EKP	Glaser-Dirks DG-100G Elan	E71G46		3.10.81	P.J.Masson	Lasham	12. 3.05
EKR	2751	117	Schempp-Hirth HS.5 Nimbus 2C	195	D-4904	25. 6.81	K.Wells and Partners	Lyveden	8. 2.05
EKS	2752	EKS	Scheibe SF-27A Zugvogel V	6096	D-8166	28. 4.81	J.C.Johnson	Parham Park	1. 3.05
EKT	2753		Wassmer WA30 Bijave	241	F-CDML	11. 5.81	D.C.Reynolds	Carlton Moor	23. 6.05
EKU	2754	408	Schleicher ASW20L	20384		19. 5.81	A.Gilson	Sleap	29. 5.05
EKV	2755	EKV	Rolladen-Schneider LS-4	4102		12. 7.81	M.Ray	Lasham	25. 5.05
EKW	2756	430	Schempp-Hirth HS.5 Nimbus 2B	111	D-7245	7. 6.81	R.S.Jobar	Lasham	24. 6.05
EKX	2757	EKX	Schleicher Ka6E	4027	D-1221	20. 6.81	A.Coatsworth	Gallows Hill	15. 8.04
EKY	2758	EKY	Slingsby T.65C Sport Vega	1949		17. 6.81	Essex and Suffolk Gliding Club	Wormingford	26. 3.03
							(Crashed Wormingford 23.6.02)		
ELA	2760	ELA	Schleicher ASW19B	19346		28. 7.81	A.G.Stark	Aboyne	15. 2.05
ELC	2762	ELC	Slingsby T.45 Swallow	1474	AGA	25. 5.81	J.Povall	AAC Dishforth	2. 5.05
					RAFGSA.346				
ELD	2763	ELD	Slingsby T.65C Sport Vega	1950		17. 8.81	D.J Clark and Partners	Challock	13. 2.05
ELE	2764	ELE	Schleicher ASK21	21065		1. 7.81	Midland Gliding Club	Long Mynd	28. 3.05
ELG	2766	ELG/931	Schempp-Hirth Ventus B	46		19. 8.81	H.Forshaw	Burn	1. 5.05
ELH	2767	ELH	Slingsby T.21B	---	RAFGSA.314	16. 9.81	Not known	Enstone	7.91*
			(Possibly ex WB966 [627])		(RAF)		*(Stored 2004)*		
ELJ	2768	ELJ	Breguet Br.905SA Fauvette	21	F-CCGU	20. 8.81	E.A.Hull	Dunstable	10. 2.05
ELL	2770	L01	Vogt Lo-100 Zwergreiher	25	HB-591	27. 7.81	I.E.Tunstall	RAF Syerston	14. 5.01
ELN	2772	ELN	Grob G.102 Astir CS Jeans	2024	???	12. 8.81	I.Hammond Syndicater	Hinton-in-the-Hedges	3. 3.05
ELQ	2774	ELQ	Slingsby T.65D Vega	1934		3. 9.81	I.Sim and Partners	Milfield	29.10.05
ELR	2775	188	Schempp-Hirth Ventus B	45		19. 8.81	I.D.Smith	Nympsfield	6. 3.05
ELS	2776	ELS	EoN AP.10 460 Srs.1	EoN/S/020	RAFGGA.530	8.81	D.G.Shepherd	Easterton	8. 4.05
ELT	2777	ELT	Rolladen-Schneider LS-4	4186		9.81	P.D.MaCarthy	Lasham	29. 2.04
ELU	2778	696	Schleicher ASW20L	20462		9.81	D.Briggs	Aston Down	8. 2.05
ELV	2779	ELV	Scheibe Zugvogel IIIB	1088	F-CCPX	5. 9.81	C.R.W.Hill	Upwood	16. 4.05
ELX	2781	ELX	Schleicher Ka7 Rhönadler	928	D-4023	18. 9.81	Nene Valley Gliding Club	Upwood	12. 3.05
ELY	2782	ELY	Schleicher Ka6CR	6485Si	D-5172	28. 9.81	P.F.Richardson and Partners	Bellarena	7. 3.04
ELZ	2783	719	Schleicher ASW20L	20310	RAFGGA,569	4.10.81	D.A.Fogden	Booker	12. 3.05
EMB	2785	RH	Rolladen-Schneider LS-4	4185		10.81	R.Hine	Winthorpe	13. 3.05
EME	2788	515	Glaser-Dirks DG-202/17C	2-176CL18		11.81	E.D.Casagrande	Usk	15. 3.05
EMF	2789	452	Rolladen-Schneider LS-4	4187		11.81	G.R.Nunan	Nympsfield	9. 4.05
EMG	2790	EMG	Rolladen-Schneider LS-4	4242		8. 5.82	R C Bowsfield	Aston Down	16. 3.05
EMJ	2792	EMJ	Slingsby T.65C Sport Vega	1951		1. 2.82	Staffordshire Gliding Club	Seighford	25. 5.05
EMK	2793	EMK	Slingsby T.45 Swallow	1514	RAFGGA.545	12.81	J.Wilton	RAF Syerston	18. 4.05
EML	2794	EML	Slingsby T.65A Vega	1892	G-BGCB	8.12.81	P.W.Williams	Brentor	19. 8.05
EMN	2796	EMN	Slingsby T.65D Vega	1935		25. 1.82	C.D Sword	Milfield	25. 7.05
EMP	2797	EMP	Slingsby T.65C Sport Vega	1952		2. 2.82	D.R Freehold	Kenley	17. 8.05

EMR	2799	EMR	Slingsby T.65C Sport Vega	1954		10. 2.82	P.Greenway and Partners	Shobdon	12. 8.05
EMS	2800	T65	Slingsby T.65A Vega 17L	1890	G-BGBV	26. 1.82	W.P.Day	Kenley	16. 4.05
EMT	2801	MF	Rolladen-Schneider LS-4	4243		12.81	M.Fox	Rufforth	31. 3.05
EMU	2802	606	Glaser-Dirks DG-202/17	2-162/1753		1.82	P.B.Gray and Partners	Camphill	27. 5.05
EMV	2803	EMV	Schleicher Ka7 Rhönadler	---	AGA.13	1. 1.82	Shalbourne Soaring Group	Rivar Hill	11. 7.04
EMY	2806	264	Rolladen-Schneider LS-4	4189		15. 4.82	N.V.Parry	Nympsfield	20. 3.05
EMZ	2807	EMZ	Slingsby T.65A Vega	1891	G-BGCA	5. 2.82	F.S.Smith	Portmoak	11. 2.05
ENA	2808	288	Rolladen-Schneider LS-4	4191		31. 5.82	A.J.Danbury	Long Mynd	4. 4.05
ENC	2810	ENC	Schleicher Ka7 Rhönadler	384	D-8111	2. 3.82	I.H.Keyser	Waldershare Park	23. 4.02
			(Modified to ASK13 status)						
ENE	2812	281	Rolladen-Schneider LS-4	4271		1. 6.82	A. Samson	Dunstable	9.12.04
ENG	2814	ENG	Focke-Wulf Kranich III	79	D-5420	8. 3.82	P.R.Davie and Partners	Dunstable	17. 8.04
ENJ	2816	771	Schempp-Hirth Ventus B	62		25. 3.82	S.J.Boyden	Lasham	20. 4.05
ENK	2817	ENK	Schleicher ASK21	21106		12. 4.82	H Jakeman	Aston Down	13. 4.05
ENN	2820	345	Schempp-Hirth Nimbus 3	9		5. 4.83	R.Kalin and Partners	Rufforth	17. 1.05
ENP	2821	295	Schempp-Hirth Nimbus 3	10		12.11.82	R.M.Hitchin	Kingston Deverill	23. 1.05
ENT	2825	902	Glasflugel H.304	210		13. 5.82	M.J.Hastings and Partners	RAF Weston-on-the-Green	19. 5.05
ENU	2826	435	Glaser-Dirks DG-100G Elan	E108G78		8. 8.82	R.D.Platt	Long Mynd	26. 4.05
ENV	2827	181	Schleicher ASW20L	20554		27. 5.82	R.D.Hone	Booker	27. 7.04
ENW	2828	ENW	Schleicher ASW20L	20567		28. 5.82	A.Hunter	Pocklington	17. 3.05
ENX	2829	276	SZD-48 Standard Jantar 2	W857	(BGA.2746)	6.82	J.M.Hire	Currock Hill	12. 4.05
ENY	2830	ENY	Schleicher ASK13	13606	RAFGSA.R17	22. 7.82	Aquila Gliding Club	Hinton in the Hedges	2. 6.04
ENZ	2831	RNT	Schleicher ASW19B	19366		29. 6.82	O.Pugh	Booker	13. 5.05
EPD	2835	EPD	Schleicher ASK21	21119		29. 8.82	J.E.Ashcroft	Chipping	22. 4.05
EPE	2836	EPE	Schleicher ASW19B	19335	RAFGSA.R18	29. 6.82	R.Witty	Pocklington	17. 3.05
					BGA.2836, RAFGSA.R18				
EPF	2837	323	Centrair ASW20FLP	20515		1. 7.82	D.J.Howse	Gransden Lodge	6. 3.05
EPJ	2840		Nord 2000 (Olympia)	10399/69	F-CACX	8.82	B.V.Smith	Sutton Bank	18. 3.04
EPK	2841	742	Centrair 101A Pégase	101-012		16. 1.83	742 Syndicate	Bicester	14. 3.02
EPM	2843	EPM	Scheibe SFH-34 Delphin	5115		22.10.82	Angus Gliding Club	Drumshade	29. 8.05
EPP	2845	EPP	Schleicher ASK13	1609		28.12.82	Mendip Gliding Club	Halesland	22. 4.05
			(Rebuild of PH-368 c/n 13064: c/n is spare fuselage no.)						
EPR	2847	EPR	Hutter H-17	---	(Kenya)	30. 9.82	B.Molineux	Halesland	19. 3.05
					PH-269				
EPS	2848	765	Schleicher ASW20L	20245	RAFGSA.87	5. 7.85	D.Richardson	Booker	26. 7.05
EPT	2849	EPT	Schleicher K8B	---	RAFGGA.504	14. 9.82	Trent Valley Gliding Club	Kirton-in-Lindsey	28. 2.05
EPU	2850	EPU	Glaser-Dirks DG-100G Elan	E116G85	(BGA.2833)	31.10.82	J.F.Rogers	Booker	18. 4.05
EPV	2851	EPV	Schleicher Ka7 Rhönadler	7148	D-5468	8.10.82	Surrey Hills Gliding Club	Kenley	16. 8.05
							(Landing accident Kenley 5.10.04)		
EPW	2852	EPW	Schleicher Ka6CR	6537	(Kenya)	21. 3.83	J.Kitchen	Strubby	7. 2.05
EPX	2853	EPX	Schempp-Hirth Ventus B/16.6	107		20.10.82	D.L. Slobom	Dunstable	29.12.04
EPZ	2855		Scheibe Bergfalke II/55	370	D-4012	15. 1.83	G.W.Sturgess *(Being refurbished 2000)*	Upavon	11. 8.96*
EQA	2856	279	Rolladen-Schneider LS-4	4259		24. 6.83	R.L.Smith	Booker	13. 4.05
EQB	2857	EQB	SZD-30 Pirat	S-0648	D-2702	31.10.82	R.Firman	Booker	23. 8.04
EQD	2859	EQD	Grob G.102 Astir CS77	1614	PH-570	14. 1.83	D.S.Fenton and Partners	Usk	7. 5.05
EQE	2860	EQE	Schleicher ASK13	13627		14. 4.83	Essex Gliding Club	Ridgewell	14. 5.05
EQF	2861	EQF	Schleicher ASK13	13626		8. 3.85	Essex Gliding Club	North Weald	3. 7.05
EQG	2862	239	Schleicher ASW19B	19265	PH-665	16.12.82	M. Bowyer and Partners	Challock	13. 4.05
EQJ	2864	968	Centrair ASW20FL	20512		21. 1.83	R.Grey	Nympsfield	28. 3.05
EQK	2865	EQK	Centrair 101A Pégase	101-054		30. 5.83	F.G.Irving and Partners	Lasham	4. 5.04
EQL	2866		Avialsa (Rocheteau) CRA-60 Fauconnet						
				03K	F-CDNR	18. 1.86	J.James	Saltby	30. 6.01
EQM	2867		CARMAM M.100S Mésange	81	F-CDKQ	20. 3.83	R.Boyd	Rivar Hill	27. 8.04
EQN	2868	340	Schempp-Hirth Nimbus 3	31		21. 2.83	D. Smith	Snitterfield	1.12.05
EQP	2869	158	Glaser-Dirks DG-202/17	2-187/1761		3.83	Not known *(Wreck noted 5.04)*	Aston Down	
EQQ	2870	451	Schleicher Ka6CR	6541	AGA.24	6. 3.83	C.R. Reese and Partners	Challock	23. 2.05
					BGA.1353				
EQR	2871	EQR	Schleicher ASK21	21157		25. 4.83	London Gliding Club	Dunstable	16.12.04
EQT	2873	EQT	Grob G.103A Twin II Acro	3787-K-65	RAFGSA.R58	15. 4.83	R.Tyrell	Edgehill	15.10.05
					BGA.2873				
EQU	2874	EQU	Pilatus B4 PC-11	201	PH-535	2. 4.83	G.A.Settle	Chipping	17. 4.05
EQV	2875	EQV	Schempp-Hirth Janus C	169	ZD974	8. 3.83	Burn Gliding Club	Burn	10.12.05
					BGA.2875				
EQW	2876	383	Schempp-Hirth Janus C	171	ZD975	24. 4.83	G.R.Seaman	Lasham	4. 4.05
					BGA.2876				
EQX	2877	EQX	CARMAM M.200 Foehn	54	F-CDKR	11.4.83	R.Pettifer and Partners	Chipping	30. 4.05
EQY	2878		BAC.VII rep	01		8. 9.91	D.Rogers	(Middlesex)	21. 5.96P*
			(Rebuild of BAC Drone using wings of G-AEJR and new fuselage)				*(Being refurbished as powered aircraft)*		
EQZ	2879	EQZ	Schleicher K8B	8113A	D-8763	13. 4.83	Cotswold Gliding Club	Aston Down	1. 5.05
ERA	2880	283	Centrair ASW20FL	20526		4.83	R.Lockett	Seighford	17. 3.05
ERB	2881	ERB	Slingsby T.50 Skylark 4 Special	001		25. 4.83	B.V.Smith	Sutton Bank	22. 5.05
			(Built C.Almack)						
ERH	2887	ERH	Schleicher ASK21	21147	ZD647	28. 4.83	Burn Gliding Club	Burn	3. 4.05
					BGA.2887				
ERJ	2888	R35	Schleicher ASK21	21148	RAFGSA.R35	28.4.83	Cranwell Gliding Club	RAF Cranwell	19.12.04
					ZD648, BGA.2888				
ERP	2893	SH5	Schleicher ASW-19B	19348	ZD657	28. 4.83	Surrey and Hampshire Gliding Club	Lasham	18. 1.05
					BGA.2893, BGA.2773				
ERQ	2894	298	Schleicher ASW-19B	19381	(BGA.4849)	28. 4.83	G.Lane	Gransden Lodge	5. 5.05
					ZD658, BGA.2894				

ERR	2895	ERR	Schleicher ASW-19B	19382	ZD659 BGA.2895	28.4.83	D.M. Hook	Lasham	29. 7.05
ERS	2896	ERS	Schleicher ASW-19B	19383	ZD660 BGA.2896	28. 4.83	J.C.Marshall	Kingston Deverill	18. 3.05
ERU	2898	3	Schempp-Hirth Nimbus 3	13	RAFGSA.R26 D-6330	4. 5.83	L.Urbani	Rieti, Italy	28. 2.05
ERV	2899	854	Rolladen-Schneider LS-4	4257		19. 5.83	R E Francis	Nympsfield	10.10.05
ERW	2900	ERW	Slingsby T.21B	1130	RAFGSA.237 BGA.842	24. 5.83	High Moor Gliding Club *"The Spruce Goose"*	Hafotty Bennett	11. 6.05
ERX	2901	180	Centrair 101A Pégase	101-058		3. 8.83	C.N.Harder	Rivar Hill	21. 5.05
ERY	2902	983	Slingsby T.59D Kestrel 19	1839	EI-125 D-9253	14. 6.83	S.W. Bradford	Tibenham	3.12.05
ERZ	2903	-	Oberlerchner Mg19a Steinadler	015	OE-0324	1. 6.83	C.Wills	Booker	14. 9.04
ESA	2904	ESA	SZD-9 bis Bocian 1E	P-750		21. 6.83	F.Wevers	Amersfoort, The Netherlands	17.10.03
ESB	2905	ESB	Schleicher ASK21	21176		2. 9.83	A.L.Garfield	Dunstable	28.10.05
ESC	2906	379	Rolladen-Schneider LS-4	4261		25. 6.83	J.M.Staley	Edgehill	18. 7.05
ESE	2908	LS4	Rolladen-Schneider LS-4	4260		26. 6.83	M.Platt	Aston Down	1. 1.06
ESH	2911	118	Centrair 101A Pégase	101-069		2. 7.83	J.E.Moore	Booker	23. 1.05
ESJ	2912	ESJ	Schleicher K8B	8730	D-5010	13. 7.83	Bowland Forest Gliding Club	Chipping	9.12.05
ESK	2913		Schleicher Ka2B	697	RAFGGA.594 D-5947	23. 7.83	W.R.Williams	RAF Halton	20. 5.00
ESM	2915	ESM	Breguet Br.905SA Fauvette	30	F-CCJA	14.10.83	J. Doppelbauer	Heidenheim, Germany	16.10.05
ESP	2917	ESP	SZD-48-3 Jantar Standard 3		B-1294	19. 4.84	R.Kirbitson	Pocklington	14. 3.05
ESQ	2918	231	Glaser-Dirks DG-300 Elan	3E10		4.84	G.R.Brown	Sandhill Farm, Shrivenham	23. 6.05
ESU	2922	ESU	Schleicher ASK21 *(Composite with RAFGSA.R28 c/n 21154)*	21180	RAFGSA.R40 BGA.2922	12.11.83	Aquila Gliding Club *(Ground-looped landing Hinton 13.6.04)*	Hinton in the Hedges	2. 4.05
ESW	2924	590	Centrair 101A Pégase	101-068		20. 3.84	D.A.Brown	Usk	13. 4.05
ESX	2925	ESX	Schleicher K8B	8805	RAFGGA.553	10. 9.83	Wolds Gliding Club *(W/o Pocklington 2. 2.96 ; stored 6.04)*	Aston Down	31.12.96*
ESY	2926	ESY	Rolladen-Schneider LS-4	4334		4. 2.84	K.Jenkins and Partners	North Hill	27. 2.05
ETA	2928	ETA	Schleicher ASK21	21181		19.11.83	P.Hawkins	Hinton-in-the-Hedges	9. 8.05
ETB	2929	ETB	Schleicher Ka6E	4365	HB-1021	20.11.83	A.J.Padgett	Marham	22. 5.05
ETD	2931	R44	Schleicher K8B	8918	RAFGSA.R44 BGA2931, RAFGSA 244	30.12.83	RAFGSA Wrekin Gliding Club	RAF Cosford	19.11.04
ETG	2934	ETG	Rolladen-Schneider LS-4	4349		7. 2.84	N. Woods	Crowland	15.11.05
ETH	2935	ETH	Schleicher K8B	120	D-5755	23. 3.84	North Wales Gliding Club	Llantisilio	3. 9.03
ETJ	2936	223	Centrair 101A Pégase	101A-0110		8. 8.84	K.J.Bye	Wormingford	17. 1.05
ETK	2937	215	SZD-48 Jantar Standard 2	W-876	OY-XJO	14. 2.84	J.A.Cowie	Portmoak	15. 7.05
ETM	2939	M7	Centrair 101A Pégase	101A-0111		11. 4.84	M.D.Evans	Booker	24. 6.05
ETP	2941	WB943	Slingsby T.21B	610	WB943	7.84	P.Hepworth tr Ouse T.21 Syndicate	Rufforth	30.10.05
ETR	2943	S7	Schleicher Ka7 Rhönadler	3	D-8339	21. 2.84	Shenington Gliding Club	Edgehill	26. 7.04
ETS	2944	ETS	Schleicher ASK13	13635AB		3. 3.84	Upward Bound Trust	Thame	7. 2.05
ETU	2946	ETU	Schleicher Ka7 Rhönadler	---	RAFGSA R.8	24. 3.84	M.J.Libelle	Strubby	25. 5.04
ETV	2947	ETV	Rolladen-Schneider LS-4	4314	(BGA.2919)	23. 3.84	T.A.Meaker	Kirton-in-Lindsey	27. 9.04
ETY	2950	249	Rolladen-Schneider LS-4	4368		16. 4.84	R.Harris *(Carries "BGA.2350" on fin)*	Booker	15. 1.05
ETZ	2951	20	Schleicher ASW20CL	20730		20. 3.84	N.L.Clowes	Tibenham	19. 3.05
EUC	2954	EUC	Schleicher ASK13	13104	AGA.12	15. 4.84	Bristol and Gloucestershire Gliding Club	Nympsfield	28. 3.05
EUD	2955	X56	Schleicher ASW20C	20734		3. 6.84	R.Fletcher and Partners	Lasham	9. 3.05
EUE	2956	EUE	Scheibe SF-27A Zugvogel V	6106	D-5342	6. 4.84	Newark & Notts Gliding Club	Winthorpe	6. 2.05
EUF	2957		SZD-50-3 Puchacz	B-1090		30. 5.84	A.J.Pettitt	Lasham	12. 6.05
EUH	2959	446	Rolladen-Schneider LS-4	4382		14. 4.84	A.R.Turner	Nympsfield	7. 4.05
EUJ	2960	217	Schempp-Hirth Ventus B/16.6	162		9. 4.84	M. Millar	Parham Park	15. 2.05
EUK	2961	992	Centrair ASW20FL	20530		20. 5.84	J.L.Caton	Lasham	20.10.05
EUM	2963		Scheibe SF-26A Standard	5039	RAFGSA	20. 4.84	Vale of Neath Gliding Club *(Beyond repair due to water damage - fuselage to local ATC unit)*	(Blackpool)	14. 7.97*
EUN	2964		Slingsby T.21B	588	RAFGSA.R92 RAFGSA.212, WB925	20. 4.84	Booker Gliding Club	Booker	27. 6.04
EUQ	2966	EUQ	Schleicher Ka7 Rhönadler	863	D-4639	8. 5.84	J.Powell	Shobdon	21. 2.05
EUS	2968	443	Schempp-Hirth Ventus B/16.6	192		19. 5.84	R.Whitaker	Lasham	14.12.05
EUX	2973	EUX	Schleicher ASK18	18005	D-3988	28. 5.84	Southdown Gliding Club	Parham Park	2.12.04
EUY	2974	88	Schleicher ASW20BL	20645		8. 6.84	D.G.Roberts and Partners	Aston Down	11.12.05
EUZ	2975		Slingsby T.21B	620	WB959	24. 6.84	Dartmoor Gliding Association *(Damaged in heavy landing Long Mynd 11.5.03)*	Brent Tor	31. 8.04
EVB	2977	EVB	Schleicher Ka7 Rhönadler	7004	D-5109	12. 6.84	R.Armitage t/a Channel Gliding Club	Waldershare Park	11. 5.05
EVC	2978	EVC	CARMAM M.200 Foehn	55	F-CDKT	20. 6.84	W.Young and Partners	Pocklington	13. 8.04
EVD	2979	382	Rolladen-Schneider LS-3	3024	N63LS D-7914	11. 8.84	C.H.Appleyard	Lasham	26. 2.05
EVE	2980	491	Centrair 101A Pégase	101A-0141		19. 6.84	J.Veraing and J.North	Lasham	12. 2.05
EVF	2981	90	Schempp-Hirth Nimbus 3T	15/76	D-KHIJ	19. 3.85	R.A.Foot and Partners	Lasham	16. 2.05
EVG	2982	EVG	Schleicher Ka7 Rhönadler	396	D-0018	17. 7.84	Derbyshire and Lancashire Gliding Club	Camphill	6. 3.05
EVH	2983	EVH	Schleicher Ka10	10008	HB-791	26. 5.86	J.W Bolt	Brent Tor	21. 8.03
EVJ	2984	H	Schleicher ASK13	13637AB		14. 7.84	Lasham Gliding Society	Lasham	18. 6.05
EVK	2985	EVK	Grob G.102 Astir CS	1397	PH-546	20.12.86	Peterborough and Spalding Gliding Club	Crowland	1. 4.05
EVL	2986	SH1	Grob G.102 Astir CS77	1638	PH-575	26. 7.84	J.R.Bates	Lasham	3. 6.05
EVM	2987	N51	Centrair 101A Pégase	101A-0157		17. 8.84	Seahawk Gliding Club	RNAS Culdrose	20. 3.05
EVN	2988		Monnett Monerai		312	R	*(Noted 11.03)*	Bellarena	
EVP	2989	K	Schleicher ASK13	13638AB		23. 8.84	Lasham Gliding Society	Lasham	22. 4.05
EVQ	2990	682	Centrair 101A Pégase	101A-0149		19. 8.84	Essex & Suffolk Gliding Club	Wormingford	13. 3.05
EVR	2991	EVR	LET L-13 Blanik	172604	G-ASVS OK-3840	30. 8.84	D.Latimer	Hinton-in-the-Hedges	14.11.04

EVS	2992	EVS	SZD-50-3 Puchacz	B-1091		6. 9.84	Deeside Gliding Club	Aboyne	25. 3.05
EVT	2993	EVT	Scheibe Bergfalke IV	5807	D-0730	24. 9.84	J.Selman	(Limerick)	18.11.01
EVV	2995	EVV	Schleicher ASK23	23004		21.11.84	Midland Gliding Club	Long Mynd	13. 1.05
							(Crashed Wentnor near Long Mynd 2..7.04)		
EVW	2996	EVW	Schleicher ASK23	23006		4. 1.85	London Gliding Club	Dunstable	21.12.04
EVX	2997	EVX	Schleicher ASK23	23007		7. 1.85	London Gliding Club	Dunstable	10.12.04
EVY	2998	EVY	Schleicher ASK23	23008		31. 1.85	London Gliding Club	Dunstable	25. 5.05
EWG	3006	VE	Grob G.103A Twin II Acro	33885-K-124	ZE501 BGA.3006	10.84	N.Bigrigg	Syerston	16. 6.05
EWP	3013	EWP	Grob G.103A Twin II Acro	33892-K-130	ZE523 BGA.3013	3.11.84	Cambridge University Gliding Club	Gransden Lodge	20. 2.05
EWR	3015	R70	Grob G.103A Twin II Acro	33894-K-132	RAFGSA.R70 ZE525, BGA.3015	9.11.84	Anglia Gliding Club	RAF Wattisham	1. 4.05
EXA	3024	VW	Grob G.103A Twin II Acro	33908-K-143	ZE534 BGA.3024	12.84	Trent Valley Gliding Club	Kirton-in-Lindsey	20. 5.05
EYS	3064	R71	Grob G.103A Twin II Acro	33961-K-194	RAFGSA.R71 ZE612, BGA.3064	3.99	Fenland Gliding Club	RAF Marham	9. 4.05
EZE	3076		Grob G.103A Twin II Acro	33981-K-214	ZE634 BGA.3076	3. 5.85	Oxford Gliding Club	RAF Weston on the Green	25. 1.00
							(Damaged in field landing Turkdean, Glos. 11.7.99)		
EZF	3077	XY	Grob G.103A Twin II Acro	33982-K-215	ZE635 BGA.3077	5.85	N.Hatton	Winthorpe	18. 4.05
FAF	3101	271	Schleicher ASW20	20214	RAFGSA.271 RAFGSA.R27	5.10.84	M.S.Armstrong	Gallows Hill	30. 6.04
FAJ	3103	FAJ	Glaser-Dirks DG-300 Elan	3E50		3.10.84	B.A.Brown	Milfield	26. 2.05
FAK	3104	FAK	Avialsa A.60 Fauconnet	104K	F-CDFG	12. 5.85	I.D.Gumbrell	Kingston Deverill	9. 5.01
FAM	3106	J15	Schempp-Hirth Nimbus 3/24.5	79		16. 3.85	K.Hartley	Bicester	31. 5.05
FAN	3107	202	Centrair 101A Pégase	101A-0161		25.10.84	J.Rayner	Parham Park	6.12.05
FAQ	3109	631	Rolladen-Schneider LS-4	4465		9. 3.85	C.H.Meir	Camphill	24. 2.05
FAR	3110	FAR	Glasflugel H.205 Club Libelle	58	HB-1262	5. 6.85	G.A.Gair	Ringmer	5. 4.05
FAT	3112	FAT	Schleicher ASK13	13528	PH-456	22. 1.85	Dorset Gliding Club	Gallows Hill	23.11.04
FAV	3114	FAV	ICA IS-32A	05		27.12.84	Black Mountains Gliding Club	Talgarth	9. 6.05
FAW	3115	333	Schempp-Hirth Ventus B/16.6	26	D-6768	23. 3.85	P.R.Stafford-Allen	Marham	22. 3.05
FBA	3119	178	Schleicher ASW20BL	20665		23. 1.85	R.H.Prestwich	Long Mynd	28. 1.05
FBB	3120	822	Schempp-Hirth HS.4 Standard Cirrus	327G	RAFGGA.312	1.85	R.Andrew	Nympsfield	17. 1.05
FBC	3121	FBC	Schleicher ASW15B	15356	OH-439	19. 5.85	C.Knock	Sandhill Farm, Shrivenham	14. 3.05
FBD	3122	FBD	Schleicher ASW15B	15407	OH-445	19. 5.85	R.Coleman	Strubby	10. 8.05
FBE	3123	1	Rolladen-Schneider LS-6	6028	D-9384	2. 6.85	T.J.Wills	New Zealand / Booker	27. 8.05
FBF	3124	175	Glaser-Dirks DG-300 Elan	3E9	BGA.2952	16. 1.85	P.J.Machacek	Husbands Bosworth	21. 5.05
FBH	3126	177	Glaser-Dirks DG-100G Elan	E156G123		1. 4.85	IBM (South.Hants) Gliding Club	Lasham	29. 6.05
FBJ	3127	FBJ	Schleicher K8B	8221	D-6340	18. 2.85	Bidford Gliding Club	Bidford	14. 7.05
FBM	3130	727	Schempp-Hirth Nimbus 3/24.5	73		28. 2.85	D.K.Gardiner	Aboyne / Portmoak	27. 7.05
FBN	3131	FBN	Glasflugel H.303 Mosquito B	167	D-6364	1. 5.85	S.R.Nash	Sandhill Farm, Shrivenham	12. 2.05
FBQ	3133	464	Schleicher ASW20BL	20669		16. 4.85	R.A. Robertson	Talgarth	30. 3.05
FBR	3134	773	Grob G.102 Astir CS77	1701	SE-TSV	5. 3.89	R. Robinson	Wormingford	. 6. 5.05
FBT	3136	488	Schempp-Hirth Ventus BT	218/35		12. 3.85	C.J.Pollard	Tibenham	28.11.05
FBV	3138	FBV	Schleicher ASK21	21223		2. 5.85	London Gliding Club	Dunstable	22. 1.05
FBW	3139	395	Glaser-Dirks DG-101G Elan	E174G140		11. 4.85	C.Tye	Lasham	29. 3.05
FBY	3141	780	Schempp-Hirth Discus B	20		12. 4.85	D Latimer	Hinton-in-the-Hedges	20. 8.05
FBZ	3142	D4667	Schleicher Ka6CR	6016	D-4667 D-KIMN, D-4667	12. 5.85	R.H.Martin and Partners	Booker	8. 2.05
FCB	3144	FCB	Centrair 101 Pégase	101-0178	F-CGEA	15. 4.85	N.Stratton	Portmoak	16.12.05
FCC	3145	XN243	Slingsby T.31B	1182	XN243	6. 5.85	R. Linde	Emmerich, Germany	21. 9.05
FCD	3146	841	Centrair 101A Pégase B	101A-0207		1. 5.85	G.J.Bass	Challock	23. 6.05
FCF	3148	993	Slingsby T.21B	MHL.017	WB990	12. 5.85	G.Pullen	Lasham	16. 4.05
FCG	3149	WT871	Slingsby T.31B	681	WT871	5.85	C.Wevers	Amersfoort, The Netherlands	21. 2.04
FCH	3150	FCH	CARMAM M.100S Mésange	72	F-CDKD	29. 5.85	D.Laidlaw	Strubby	26. 1.03
							(Damaged in take-off accident Strubby 27.3.02)		
FCJ	3151	571	Grob G.102 Astir CS	1231	D-4205	19. 5.85	M.Levitt and G.Fellows	Aston Down	13. 3.05
FCK	3152	671	Schempp-Hirth Ventus B/16.6	241		6. 5.85	L.J.Scott	Sleap	21. 6.05
FCM	3154	411	Glaser-Dirks DG-300 Elan	3E94		17. 5.85	R.B.Coote	Parham Park	21.12.05
FCN	3155	920	Schempp-Hirth HS.4 Standard Cirrus	131	D-0191	18. 6.85	S.M.Robinson	Nympsfield	.4 2.05
FCP	3156	721	Rolladen-Schneider LS-6a	6030		2. 7.85	R.E.Robertson	Dunstable	13.12.04
FCR	3158	113	Schleicher Ka6E	4223	OH-375 OH-REC	10. 6.85	R.F.Whitaker and Partners	Lasham	16. 2.05
FCS	3159	2R	Schempp-Hirth HS.5 Nimbus 2C	233	D-5993	7. 7.85	R.W.Hawkins	Parham Park	20. 4.05
FCT	3160	WB944	Slingsby T.21B	611	WB944	16.12.86	A.Dyer	Bicester	12.11.05
FCV	3162	FCV	Schleicher ASW20	20076	RAFGSA.R24	7. 6.85	M.J.Davis and Partners	RAF Cosford	24.10.05
FCW	3163	L	Schleicher ASK13	13642AB		27. 6.85	Lasham Gliding Society	Lasham	4. 2.05
FCY	3165	FCY	Schleicher ASW15	15122	D-0748	29. 6.85	M. Shaw	Ridgewell	16. 8.05
FCZ	3166		Slingsby T.1 Falcon 1 rep *(Built Southdown Aero Services)*	---		7.85	D.D.Knight and J.Harber *(Stored 8.04)*	RAF Halton	
FDA	3167	7D	Schleicher ASW15	15050	D-0511	2. 7.85	N W Woodward	Booker	14.12.05
FDB	3168		ICA IS-30	07		23. 9.85	Black Mountains Gliding Club	Talgarth	19. 6.00
FDC	3169	FDC	CARMAM JP-15/34 Kit Club	TAH.50/60		10. 3.87	T.A.Hollings	Rufforth	20. 7.03
FDD	3170	FDD	Schleicher K8B	8972	AGA.5	11. 7.85	Shalbourne Soaring Society	Rivar Hill	21. 4.05
FDE	3171	510	Schempp-Hirth Ventus BT	256/53		8.85	P.Clay	Camphill	29. 3.05
FDF	3172	FDF	Grob G.102 Astir CS	1321	D-7338	2. 9.85	R.H.Davies	Nympsfield	26. 7.05
FDG	3173	FDG	ICA IS-29D2 Club	02		23. 8.85	D.Mole	Sackville Lodge, Riseley	1. 6.02
FDK	3176	FDK	Slingsby T.59 Kestrel 19	1832	(BGA.4927) BGA.3176, G-BBVC	8.85	T.Gauder	Speyer, Germany	20. 4.05
FDP	3180	FDP	ICA IS-30	08		24. 5.86	M.H.Simms	Llantisilio	18. 9.04

FDQ	3181	FDQ	Slingsby T.31B	710	WT915	19. 9.85	J.F.J.M.Forster *"Chris Wills"*	Maastricht	21. 5.05
FDR	3182	FDR	Schleicher Ka6CR	6119	D-8456	18.11.85	Dartmoor Gliding Club	Brentor	22. 3.05
FDU	3185	H20	Schempp-Hirth Discus B	87		23. 5.86	J.L.Whiting	Edgehill	9. 3.05
FDW	3187	FDW	Glaser-Dirks DG-300 Elan	3E143		26.1.86	C.M.and A.J. Hadley	North Hill	14. 3.05
FDX	3188	FDX	SZD-48-1 Jantar Standard 2	B-1251	(BGA.2916)	22.12.85	F.Bishop and Partners	Ringmer	5.10.05
FDY	3189	FDY	Slingsby T21B	MHL.005	WB978	26. 1.86	R. Lloyd	Challock	26.12.05
FEA	3191	FEA	Grob G.103 Twin Astir	3151	RAFGSA.R83 RAFGSA 833	6.12.85	M.Boyle	Rufforth	11. 5.05
FEB	3192	SH8	Grob G.102 Club Astir III	5643C		15.11.85	Surrey and Hampshire Gliding Club	Lasham	2. 2.05
FEE	3195	FEE	Slingsby T.21B	MHL.016	WB989	20. 1.86	K.Schickling	Aschaffenburg, Germany	9. 7.05
FEF	3196	FEF	Grob G.102 Astir CS	1164	OY-XGC	9. 2.86	Oxford University Gliding Club	Bicester	28. 2.05
FEG	3197	120	Schempp-Hirth Ventus B/16.6	279		12. 2.86	K.Moorhouse and Partner	Rivar Hill	31. 3.05
FEH	3198	318	Centrair 101A Pégase Club	101A-0268		1. 5.86	Booker Gliding Club	Booker	3. 3.05
			(Rebuilt with new fuselage c/n 01304: original fuselage rebuilt as BGA.3560)						
FEJ	3199	538	Schempp-Hirth Discus B	76		22. 2.86	J.W.White	Booker	7. 1.05
FEL	3201	FEL	Schleicher Ka7 Rhönadler (See BGA.3231)	7231	RAFGGA.. D-???	22. 9.86	Dukeries Gliding Club	Gamston	19.11.05
FEN	3203	FEN	SZD-50-3 Puchacz	B-1326		24. 3.86	Northumbria Gliding Club	Currock Hill	17. 5.05
FEQ	3205	M	Schleicher ASK13	13650AB		7. 4.86	Lasham Gliding Society	Lasham	14. 5.05
FER	3206	FER/370	Schempp-Hirth Discus B	75		27. 3.86	R.Acreman	Halesland	25. 3.05
FES	3207	564	Schempp-Hirth Discus B	88		4. 4.86	N.G.Storer	Booker	20. 2.05
FEX	3210	ZS-GFZ	Glasflugel H.301B Libelle	100	ZS-GFZ	5.86	T.J.Wills	(New Zealand)	25.10.04
							(De-Registered 17.11.04 - sold in New Zealand as ZK-GFZ)		
FEX	3212	FEX	Grob G.102 Astir CS77	1660	D-7492	6. 4.86	J.Taylor	Upwood	7. 5.04
FEZ	3214	FEZ	EoN AP.7 Primary	EoN/P/037	RAFGSA.R13 RAFGSA 113, WP269	19. 9.86	G.J.Moore *(For restoration)*	RAF Keevil	3. 6.01
FFA	3215	FFA	Schleicher ASK13	13651AB		15. 5.86	Staffordshire Gliding Club	Seighford	9.12.05
FFB	3216	R9	Grob G.102 Astir CS	1123	RAFGSA.R9 RAFGSA.R97, BGA.3216, D-6977	10. 6.86	Chilterns Gliding Club	RAF Halton	5.11.05
FFC	3217	FFC	Centrair 101A Pégase	101A-0255		30. 5.86	J.Guy	Milfield	27. 3.05
FFG	3221	WB920	Slingsby T.21B	559	WB920	2. 6.86	J.H.Wisselink	Roosendaal, The Netherlands	18. 6.04
FFH	3222	FFH	Schleicher ASW20	20037	D-7947	4. 4.87	J.Hayes	AAC Dishforth	13. 7.05
FFK	3224	FFK	Schempp-Hirth Nimbus 3	87		11. 4.87	Dr. D.Brennig-James	Booker	4. 5.04
FFL	3225	FFL	Slingsby T.21B	MHL.020	WB993	28. 6.87	B. van Aalst	Zwolle, The Netherlands	30. 4.05
FFN	3227	987	ICA IS-29D	21	D-9223	13. 8.86	K.J.Sleigh *(Blown over Rattlesden 22.9.03)*	Rattlesden	11. 6.00
FFP	3228	93	Schleicher ASW19B	19317	RAFGSA.R19	12. 6.86	J.M.Hutchinson	Booker	28. 2.05
FFQ	3229	FFQ	Slingsby T.31B	913	XE800	18. 8.86	K.J.Grosse	Odiham	26. 3. 03
FFS	3231	FFS	Centrair 101A Pégase	101A-0265		27. 6.86	W Murray *(Carries "BGA.3201")*	Gransden Lodge	21. 4.05
FFT	3232	FFT	Schempp-Hirth Discus B	110		30. 6.86	R.Maskell and Partners	Gransden Lodge	4. 3.05
FFU	3233	FFU	Glaser-Dirks DG-100G Elan	E200G166		11. 1.87	T. Tuthill and partners *(Damaged in ground accident Dishforth 10.1.04)*	Dishforth	7. 4.05
FFV	3234	FFV	SZD-51-1 Junior	B-1616	F-WGJA	26. 8.86	Herefordshire Gliding Club	Shobdon	13. 1.04
FFW	3235		Slingsby T21B	1155	XN151	2.10.87	R.Schmid	Southeim, Germany	9. 9.05
FFX	3236	627	Schempp-Hirth Discus B	109		12. 7.86	P.J.Tiller	Husbands Bosworth	29. 3.05
FFY	3237	FFY	SZD-51-1 Junior	W-938		24.11.86	Cornish Gliding Club	Perranporth	11. 6.05
FFZ	3238	WB981	Slingsby T.21B	MHL.008	WB981	21. 8.86	A P Stacey	RAF Keevil	16. 4.05
FGA	3239	WT913	Slingsby T.31B				See SECTION 3 Part 1	Doncaster	
FGB	3240	FGB/ WJ306	Slingsby T.21B	654	WJ306	23. 8.86	Oxford Gliding Club	RAF Weston on the Green	4. 8.05
FGC	3241	WT918	Slingsby T.31B	713	WT918	24. 8.86	E.Woefeel	Jena, Germany	18. 6.05
	3242		Slingsby T.21B Sedbergh TX.1				See SECTION 3 Part 1	Ekeby, Eskiltuna, Sweden	
FGF	3244	141	Schempp-Hirth Nimbus 3T	25/91		16. 8.86	R.E.Cross	Lasham	28. 2.05
FGG	3245	WG498	Slingsby T.21B	665	WG498	29. 9.86	G.A.Ford and Partners	Aston Down	28. 9.03
FGJ	3247	FGJ	Schleicher Ka6CR	6634	D-1041	22. 9.86	JCB Syndicate	Lleweni Parc	1.11.05
FGK	3248	FGK	Grob G.102 Astir CS	1323	RAFGSA.R61 RAFGSA.316	9.86	P.Allingham	Rivar Hill	2. 4.05
FGM	3250	FGM	Slingsby T.21B *(No longer motorised)*	1160	XN156	19. 7.87	R.B.Petrie	Strathaven	14. 8.05
FGP	3252	FGP	Schleicher ASW19	19121	C-GJXG	1.11.86	N.R.Foreman	Gransden Lodge	8. 1.05
FGR	3254	N29	Schleicher ASK13 *(Built Jubi)*	13655AB		24.10.86	Portsmouth Naval Gliding Club	Lee-on-Solent	14.11.04
FGT	3256	FGT	Glaser-Dirks DG-300 Elan	3E217		6. 3.87	S.C.Williams	Booker	5. 1.06
FGU	3257	806	Schempp-Hirth HS.4 Standard Cirrus	147	D-0193	27. 4.87	L.E.Ingram	Snitterfield	31. 3.05
FGV	3258	FGV	Schleicher Ka7 Rhönadler	Not known	OO-Z..	15.12.86	Nene Valley Gliding Club	Upwood	6. 4.03
			(Hybrid using ex Belgian Ka7 fuselage and wings from Ka2 BGA.2662)						
FGW	3259	701	Centrair 101A Pégase	101A-0275		23. 6.87	L.P.Smith	Kingston Deverill	29. 3.05
FGY	3261	527	Schleicher ASW22	22027	D-3527	3.12.86	P. Shrosbree	Dunstable	5. 3.05
FHB	3264		Slingsby T.21B(T) (Fuji-Robin EC-34PM)	MHL.018	WB991	17. 2.87	G.Traves	East Kirkby	12. 9.04
FHD	3266	196	Schleicher ASW20BL	20694	RAFGGA..	15. 2.87	K.J.Hartley	RAF Bicester	26. 6.05
FHF	3268	FHF	SZD-51-1 Junior	W-952		20. 3.87	Black Mountains Gliding Club	Talgarth	7. 5.05
FHG	3269	187	Schempp-Hirth HS.7 Mini Nimbus C	140	(BGA.3213) ZS-GNI	22. 3.87	R.W.Weaver	Usk	29. 1.05
FHJ	3271	987	Centrair 101A Pégase	101A-0278		13. 5.87	Booker Gliding Club	Booker	4. 3.05
FHK	3272	FHK	Slingsby T.31B	695	WT900	22. 4.87	N.A.Scully and Partners *"Tweety"*	Saltby	9. 6.01
							(Crashed at Tibenham 31.7.00 - under restoration Lee-on-Solent 2004)		
FHL	3273	136	Rolladen-Schneider LS-4	4633		17. 4.87	I.P.Hicks	Dunstable	25. 3.05
FHM	3274	D	Schleicher ASK13	13662AB		7. 6.87	Lasham Gliding Society	Lasham	19. 8.05
FHN	3275	FHN	Schleicher K8B	Not known	RAFGSA.R85 RAFGSA.385, RAFGSA.360	5. 6.87	B.F.Cracknell	Upwood	25. 3.05

FHQ	3277		Hols-der-Teufel rep *(Built M.L.Beach)*	---		6.87	M.L.Beach *(Sold to Germany in 1998- displayed Deutsches Segelflugmuseum, Wasserkuppe 5.02)*	Brooklands	N/E
FHR	3278	Q5	Schempp-Hirth Discus B	152		8. 6.87	J.Vella Gretch	Long Mynd	20. 1.05
FHS	3279	154	Schempp-Hirth Ventus CT	326/82		11. 6.87	R.Andrews	Long Mynd	28. 7.05
FHT	3280	FHT	Grob G.102 Astir CS	1234	D-4208	12. 6.87	A.C.Howells	Rattlesden	29. 1.05
FHU	3281	FHU	Schleicher Ka7 Rhönadler *(Modified to ASK13 standard)*	629	RAFGSA.R15 RAFGGA, D-5722	17. 6.87	Dartmoor Gliding Club	Brentor	4. 3.05
FHV	3282	FHV	SZD-48-1 Jantar Standard 2	B-1036	D-4516	25. 6.87	R.A.Williams and Partners	Long Mynd	5. 5.05
FHW	3283	698	Grob G.102 Astir CS	1087	D-6987	30. 6.87	P.R.J.Halliday	Lasham	10. 2.05
FHY	3285	H5	Scheibe SF-27A Zugvogel V	6045	D-1868	28. 6.87	J.M.Pursey	North Hill	26.11.04
FHZ	3286	FHZ	Schleicher Ka6CR	949	D-4661	20. 8.87	J.Hiley	Husbands Bosworth	6. 7.05
FJA	3287	FJA	Slingsby T.21B	1152	XN148	8. 7.87	M.Steiner *(Damaged in take-off accident Greiling, Germany 14.7.04)*	Lachen-Speyerdorf, Germany	7. 5.05
FJB	3288	FJB	Slingsby T.21B	MHL.002	WB975	8. 7.87	Angus Gliding Club *(As "WB975")*	Drumshade	17. 6.05
FJD	3290	T21	Slingsby T.21B	MHL.007	WB980	29. 8.87	R. Payne and Partners	Husbands Bosworth	12. 4.05
FJE	3291	744	Schleicher ASW20BL	20953		1. 8.87	B.Pridal	Booker	3.11.05
FJF	3292	FJF	Slingsby T.21B *(Frame No.SSK/FF 1085)*	586	WB923	7. 9.87	Sedbergh Syndicate	Snitterfield	5. 6.05
FJH	3294	FJH	Grob G.102 Astir CS77	1763	AGA.7	11. 7.87	Shalborne Soaring Society	Rivar Hill	13. 3.05
FJJ	3295	AL	Schempp-Hirth Ventus BT	344/93		20. 8.87	A.D.Purnell *(Collision with Skylark 4 BGA.1116 Lasham 26.4.04)*	Lasham	6. 2.05
FJK	3296	FJK	Centrair 101A Pégase	101-070	N4429W	30. 4.88	D.J.Ingledew	Lee on Solent	11. 9.04
FJM	3298	143	Rolladen-Schneider LS-4A	4665	D-1431	4.12.87	G.C.Beardsley and Partner	Dunstable	28. 1.04
FJN	3299	903	Slingsby T.31B	698	WT903	17. 2.88	B.Kozuh	Grobnik, Croatia	17. 7.05
FJQ	3301	FJQ	Schempp-Hirth Ventus CT	104/365		24. 3.88	B.Rood	Hinton-in-the-Hedges	5. 8.05
FJR	3302	950	Glaser-Dirks DG-300 Club Elan	3E270C2		12. 2.88	G Smith	(France)	8. 3.05
FJS	3303	257	Glaser-Dirks DG-300 Club Elan	3E271C3		28. 5.88	Yorkshire Gliding Club	Sutton Bank	22.12.05
FJT	3304	997	Centrair 101A Pégase	101A-0284		20. 2.88	D.M.Smith and A.Marlow	Booker	13. 5.05
FJV	3306	FJV	Schleicher ASW15	15109	D-0710	3.11.87	M.H.Simms	Shipdham	12. 9.05
FJW	3307	FJW	Schleicher Ka7	980	OH-241 OH-KKF	15.11.88	A.J.Pettitt and Syndicate	Rivar Hill	10.10.04
FJX	3308	FJX	Glaser-Dirks DG-300 Elan	3E261		6. 2.88	Crown Service Gliding Club	Lasham	12.12.05
	3309		Slingsby T.21B Sedbergh TX.1				See BGA.3242		
FJZ	3310	FJZ	Schempp-Hirth SHK	14	D-9330	2. 4.88	R H.Hanna and A.and R.Willis	Bellarena	30. 1.05
FKA	3311	FKA	Schleicher Ka6CR	6239	D-7037 D-5435	21. 2.88	E.Drake	Kingston Deverill	3. 6.05
FKB	3312	FKB	Glaser-Dirks DG-600	6-08		10.88	J.A.Watt	Dunstable	6. 12.05
FKE	3315	G2	Schleicher ASW15	15146	D-0794	6. 3.88	J.Watson	Bidford	13. 3.05
FKG	3317	125	Rolladen-Schneider LS-4A	4673		18. 5.88	B.A.Pocock	Kingston Deverill	12. 3.05
FKH	3318	FKH	Schleicher Ka6CR	6343	EI-109 IGA.106	3.88	Ulster Gliding Club	Bellarena	8. 8.05
FKK	3320	406	Schempp-Hirth Discus B	219		12. 3.88	D.J.Eade *(To EI-149 2003)*	Lasham	15. 1.04
FKL	3321	159	Schleicher ASW20BL	20954		21. 3.88	J.M.Ley and J.Rollason	Ridgewell	2. 3.05
FKM	3322	SH3	Schempp-Hirth Discus B	212		19. 3.88	Surrey and Hampshire Gliding Club	Lasham	13. 5.05
FKN	3323	13	Schleicher ASH25	25042	(BGA.3491) BGA.3323	19. 7.88	M Bird *(De-Registered 20.12.04 - sold abroad)*	Dunstable	25. 3.05
FKP	3324	WB971	Slingsby T.21B	632	WB971	28. 2.88	M.Powell *(Ground-looped at Camphill, 1. 7.04)*	Tibenham	13. 6.05
FKT	3328	FKT	Schleicher K8B	8382	D-5366	8. 5.88	L.Shepherd and partners	Lyveden	24. 5.05
FKU	3329	FKY	Schleicher Ka6CR	822	D-0025	6. 4.88	J.A.Timmis	Camphill	24. 5.05
FKW	3331	FKW	Schleicher Ka7 Rhönadler	7145	OH-302 OH-KKJ	2. 4.88	Nene Valley Gliding Club	Upwood	20.10.05
FKX	3332	FKX	Schleicher Ka6CR	6433	D-4316	30. 4.88	J.Bates and Partners	Lasham	1. 7.05
FLB	3336		Slingsby T.31B	837	XA295	23 .8.88	R.Birch *(Under restoration 6.04)*	Aston Down	
FLC	3337	368	Glaser-Dirks DG-300 Elan	3E310		26. 9.88	J.L.Hey	Camphill	9. 3.05
FLE	3339	314	Schempp-Hirth Discus B	207		7. 5.88	Booker Gliding Club	Booker	5.12.05
FLF	3340	Z4	Rolladen-Schneider LS-4A	4694		20. 3.88	D.E.Lamb	Booker	3. 3.05
FLG	3341	A25	Schleicher ASH25E (Turbo)	25044		6.88	R.Matthews	North Hill	4. 3.05
FLH	3342	FLH	Schleicher K8B (Built KK Lehtovaara O/Y)	22	OH-361 OH-RTW	12. 5.88	South Wales Gliding Club	Usk	9. 7.03
FLK	3344	FLK	Schleicher Ka7 Rhönadler	985	D-5047	12. 1.89	Dukeries Gliding Club	Gamston	2. 5.04
FLL	3345	FLL	SZD-9 bis Bocian 1D	F-877	OH-336 OH-KBP	30. 7.88	Bath, Wilts and North Dorset Gliding Club	Kingston Deverill	25. 5.04
FLP	3348	FLN	Schleicher K8B (Built KK Lehtovaara O/Y)	07	OH-316 OH-RTP	7. 5.88	Bath, Wilts and North Dorset Gliding Club	Shipdham	4. 5.02
FLQ	3349	FLQ	Schleicher K8B	8195A	D-8887	15. 8.88	F.J.Glanville	Long Mynd	15. 4.04
FLS	3351	FLS	Schleicher Ka6CR	6180	D-4001	16.11.88	P.B.Arms	RAF Halton	3.12.04
FLT	3352		Glasflugel H.201B Standard Libelle	41	D-0211	11.12.88	C.Glover	Husbands Bosworth	20. 8.00
FLU	3353	DJ2	Glasflugel H.201B Standard Libelle	52	D-0298	22. 6.88	C.D.Duthy-James	Talgarth	3. 6.03
FLW	3355	127	Schempp-Hirth HS.4 Standard Cirrus	75 656	F-CEMT	3. 7.88	D.Wilson	Long Mynd	15.12.04
FLX	3356	FLX	Glaser-Dirks DG-300 Club Elan	3E304C19		19.10.88	R.Emms	Upwood	24. 3.05
FLZ	3358	FLZ	Scheibe SF-27A Zugvogel V	6061	D-5378	14. 7.88	R Russon	Cosford	19. 4.03
FMC	3361	68	Rolladen-Schneider LS-6B	6184		27. 7.88	B.L.Cooper	Booker	24. 3.05
FMD	3362	FMD	Schleicher Ka7 Rhönadler	343	D-2877 HB-603	23. 7.88	Not known *(Damaged Ringmer 6.5.92: stored 7.04)*	Nympsfield	1.12.92*
FME	3363	927	Schleicher ASW15A	15164	D-0825	1. 8.88	T.J.Stanley	Sutton Bank	29. 1.05
FMG	3365	969	Schempp-Hirth Discus B	242		3. 8.88	J.Melvin	Dunstable	4. 3.05
FMH	3366	B	Schleicher ASK13	13673AB		22. 8.88	Lasham Gliding Society	Lasham	8. 7.05
FMK	3368	FMK	Centrair 101 Pégase *(Model 101B?)*	101-0293		20. 1.90	A.Bailey	Bidford	23. 4.05
FML	3369	FML	Schleicher ASW15B	15294	F-CEGR	21.11.89	G.Macmillan	Snitterfield	28. 2.05

FMM	3370	FMM	Schleicher Ka6CR	6328	D-1260	20.10.88	D.Hall	Burn	21. 5.05
FMN	3371	FMN	Schempp-Hirth Ventus CT	123/397		12. 9.88	J.Eccles	Lasham	18.12.04
FMP	3372	328	Schleicher ASW24	24023		21. 1.89	A.Harrison and G.Piersey	Ridgewell	28. 1.05
FMQ	3373	158	Schempp-Hirth Discus B	243		1.10.88	A.L Harris and A.Price	Nympsfield	4. 4.05
FMR	3374	FMR	Neukom Standard Elfe S-2	05	HB-801	8.11.88	M.Powell and Partners	Camphill	7. 6.05
FMS	3375	519	Schleicher ASW15	15061	N111SP	11.88	A.Brind	Rivar Hill	6. 4.05
FMT	3376	FMT	Schempp-Hirth HS.4 Standard Cirrus	249	N2HM	10.89	S.R.Westlake	North Hill	4. 6.05
FMU	3377	FMU	Schempp-Hirth HS.4 Standard Cirrus	236	N3LB	14. 7.90	A.Harrison and Partner	Aston Down	8. 2.05
FMX	3380	FMX	Schleicher ASW24	24014		5. 3.90	D.T.Reilly	North Hill	29. 3.05
							(De-Registered 22. 6.04, sold abroad as OY-RXR)		
FMY	3381	371	Rolladen-Schneider LS-7	7004	D-1256	18.12.88	M.Newman	Camphill	20. 4.05
FMZ	3382	FMZ	Schleicher Ka7 Rhönadler	7018	D-6035	22.11.88	Nene Valley Gliding Club	Upwood	17. 7.04
FNA	3383	FNA	Schleicher K8B	8499	D-5670	13.11.88	Bowland Forest Gliding Club	Chipping	17. 4.05
FNC	3385		Slingsby T.21B	601	WB934	5.11.88	P.Hoffmann	Oberschleissheim, Germany	8. 6.02
FND	3386	FND	Schleicher Ka6E	4069	PH-366	14.11.88	I.Kerby	Rhigos	30.12.05
FNE	3387	FNE	SZD-38A Jantar-1	B-612	HB-1215	20.12.88	D.A Salmon and Partners	Camphill	14. 6.04
FNF	3388	461	Schleicher ASW22B	22053		20.12.88	T.J.Parker	Dunstable	9.12.04
FNG	3389	163	Schleicher ASW24	24015		10. 5.89	P.H.Pickett	Snitterfield	9. 1.06
FNH	3390	A19	Schleicher ASW19	19174	D-7969	9. 2.89	S.N.Longland	Gransden Lodge	15. 4.05
FNK	3392	FNK	Slingsby T.65A Vega	1897	N9023H	10.12.88	P.Goldstraw and Partner	Dunstable	5. 3.05
FNL	3393	705	Schempp-Hirth Discus B	253		28.11.88	P.Musto	Husbands Bosworth	22. 2.05
FNM	3394	FNM	Centrair 101B Pégase	101B-0289	F-CGSE	30. 3.89	D.Hartley	Husbands Bosworth	16.11.04
FNN	3395		Schempp-Hirth Ventus CT	130/407		8.12.88	C.A.Marren	Aston Down	20. 1.03
FNP	3396	FNP	Schleicher Ka6CR	567	D-4657	16. 1.89	Trent Valley Gliding Club	Kirton-in-Lindsey	13. 7.05
FNQ	3397	282	Schempp-Hirth Discus B	259		18.12.88	C.Huck	Aston Down	29. 1.05
FNR	3398	330	Schempp-Hirth Discus B	255		20. 3.89	P.Gelsthyorpe	Lasham	11. 3.05
FNS	3399	FNS	Glaser-Dirks DG-300 Club Elan	3E314C23		2. 4.89	P.E.Williams	Portmoak	2. 7.05
FNT	3400	674	Glaser-Dirks DG-600	6-12		12.88	D.M.Hayes	Rufforth	12. 3.05
FNU	3401	190	Rolladen-Schneider LS-4A	4732	D-1376	9. 4.89	R J Simpson	Nympsfield	10. 4.05
FNW	3403	FNW	Schleicher Ka6CR	598	HB-634	20. 3.89	Cotswold Gliding Club	Aston Down	17. 1.99
							(Crashed near Aston Down 3.6.98)		
FNX	3404	FNX	Wassmer WA.30 Bijave	84	F-CCTJ	2. 1.89	A.Walker	Milfield	17. 1.05
FPB	3408	FPB	Schleicher ASW15B	15243	D-2068	18.12.88	R C Tatlow	Winthorpe	7. 7.05
FPD	3410	973	Rolladen-Schneider LS-7	7033	D-5178	14. 1.89	P.H.Rackham	Dunstable	22.12.05
FPE	3411	238	Schempp-Hirth Ventus CT	131/408		14. 1.89	P.Whitt and N.Francis	Shobdon	21. 2.05
FPF	3412	FPF	Scheibe L-Spatz 55	2720	RAFGGA...	11. 2.89	P.Saunders	Usk	18. 5.01
FPH	3414	D5	Centrair ASW 20F	20132	F-CFFX	29. 1.89	R.Gibson and Partners	Bidford	3. 6.04
FPJ	3415	459	Schleicher ASW19	19001	D-1909	29. 1.89	F.W.Pinkerton	Lyveden	30. 4.01
FPK	3416	Y1	Glaser-Dirks DG-300 Elan	3E6	D-1233	21. 1.89	G.C.Keall and Partners	Husbands Bosworth	7. 2.04
							(Crashed in field landing, Clapton on the Hill, Glos., 21.6.03)		
FPL	3417	242	Schempp-Hirth Ventus C	409		27. 1.89	R.V Barrett	Nympsfield	7. 2.05
FPM	3418	FPM	SZD-51-1 Junior	B-1788		6. 3.89	Kent Gliding Club	Challock	13. 1.06
FPN	3419	69	Schleicher ASW20	20376	RAFGGA.545 D-8780	5. 3.89	M.Rayner	Ringmer	29. 6.05
FPP	3420	N2	Schempp-Hirth HS.5 Nimbus 2B	142	D-6779 D-2111	26. 3.89	R.Jones	Walney Island	5. 6.05
FPQ	3421	FPQ	Schleicher Ka7 Rhönadler	EB180/61	D-5184	15. 2.89	East Sussex Gliding Club	Ringmer	24. 2.05
FPS	3423		Slingsby T.21B	MHL.001.	OH-914X LN-GAO, BGA.3423, WB974	2.89	M.Schopka	Brandenburg-Muhlenfeld, Germany	6.05
FPT	3424	574	Schleicher ASW20	20007	D-7574	18. 2.89	L.Hornsey and Partners	RAF Halton	19. 5.05
FPU	3425	FPU	Schleicher Ka2B Rhönschwalbe *(Built Segelfluggruppe Zwingen)*	NK	HB-698	17. 2.89	T.J.Wilkinson	Sackville Lodge, Riseley	4. 4.05
FPV	3426	FPV	Schleicher Ka6E	4123	N29JG G-AWTP, RAFGSA '29'	10. 3.89	J.E.Stewart	Bembridge	16. 4.05
FPW	3427	39	Glaser-Dirks DG-600	6-17		13. 4.89	W.S.Stephen	Aboyne	9. 1.05
FPX	3428	FPX	Schleicher ASK13	13325	F-CDYR	21. 6.89	Channel Gliding Club	Waldershare Park	9. 6.05
FQB	3432	FQB	Schleicher ASW15B	15340	D-2345	10. 8.89	J.Knowles	Dunstable	8. 3.05
FQC	3433	201	Glaser-Dirks DG-202/17c	2-178CL19	HB-1645	8. 3.89	A.T.MacDonald	Wormingford	16. 3.05
FQD	3434	FQD	Schleicher K8B	8289	D-1908	6. 3.89	Kent Gliding Club	Challock	7. 1.05
FQE	3435	FQE	Schleicher K8	3	D-6329	3. 4.89	Cotswold Gliding Club *(Stored 11.04)*	Aston Down	9. 4.01
FQF	3436	FQF	Scheibe SF-27A Zugvogel V	6025	D-0009	19. 2.89	S.Maddox	Winthorpe	18. 4.05
FQG	3437	952	Rolladen-Schneider LS-7	7050	D-1712	4. 6.89	D.W. Smith	Sutton Bank	24. 4.05
FQH	3438	A98	Rolladen-Schneider LS-7	7029	D-1316	15. 4.89	P.J.Lazenby	Sutton Bank	9.12.05
FQK	3440	601	Grob G.103C Twin III Acro	34123		15. 8.89	P.O'Donald	Gransden Lodge	4. 1.05
							(Crashed Husbands Bosworth 5. 8.04)		
FQL	3441	772	Schleicher Ka6CR	6235	HB-772	18. 3.89	P.R. Alderson	Lasham	4. 3.05
FQM	3442	FQM	Scheibe SF-27A Zugvogel V	6098	D-9421	19. 2.89	R.D.Noon	Winthorpe	5. 6.05
FQN	3443	479	Schempp-Hirth Ventus B/16.6	141	D-8772 D-KHIB	26. 3.89	R P S Montague-Scott	Lasham	28. 1.05
			(Composite rebuild of D-8772 - ex Ventus BT D-KHIB (10/141): w/o 27.5.85 and possibly HB-1626 (91) as BGA.3443 has build plate V91)						
FQQ	3445	656	Glaser-Dirks DG-600	6-11		3.89	P.Manley	Ridgewell	11. 6.05
FQR	3446	FQR	Schleicher K8B	8537	PH-349	9. 3.89	Dorset Gliding Club	Gallows Hill	29. 3.05
FQT	3448	484	SZD-48-3 Jantar Standard 3	B-1891	(BGA.3409)	28. 3.90	T H Greenwood	Sandhill Farm, Shrivenham	18. 4.05
FQU	3449	FQU	Schleicher Ka7 Rhönadler	1139	D-8614 HB-709	13. 3.89	S.Waters	Challock	21. 1.05
FQY	3453	785	Schempp-Hirth Discus B	274		16. 4.89	P.Studer	Sleap	30.11.05
FQZ	3454	L51	Rolladen-Schneider LS-1F	391	F-CEKH	12. 6.89	G.M.Ariss	Husbands Bosworth	19. 3.05
FRA	3455	79	Rolladen-Schneider LS-6B	6151	D-8081	20. 4.89	M.Randle	Aston Down	29. 3.05
FRB	3456	FRB	Schempp-Hirth Ventus C	404		4 .3.89	C.J.Ratcliffe	Sleap	7. 3.05
FRC	3457	988	Schempp-Hirth HS.5 Nimbus 2B	151	D-4980	15. 5.89	C.F.Whitbread	Challock	29. 3.05
FRD	3458	JPB	Centrair 101A Pégase	101A-0311		15. 4.89	A.Kangars	Husbands Bosworth	16. 6.04
							(De-Registered 26. 5.04, sold to Ireland as EI-162)		

FRE	3459	FRE	Schleicher Ka6E	4349	F-CDTL	13. 4.89	D.J.Stewart	Parham Park	30. 5.05
FRF	3460	FRF	Schleicher Ka7 Rhönadler	450/58	D-5653	7. 4.89	P.Roberts and Co	Dunstable	6. 9.01
FRG	3461	FRG	Siebert Sie-3	3009	D-0739	7. 4.89	A. Cridge	Talgarth	14. 9.05
FRH	3462	634	Schleicher ASW20CL	20740	D-9229	2. 4.89	J.N.Wilton and Partner	Husbands Bosworth	19. 3.05
FRJ	3463	RJ	Schempp-Hirth HS.4 Standard Cirrus	103	HB-1041	23. 4.89	P.D.Oswald and Partners	Portmoak	26. 1.05
FRK	3464	FRK	Schleicher ASW15B	15214	D-0941	21. 3.89	M.Hill	Snitterfield	24.10.05
FRL	3465	609	Grob G.102 Astir CS	1373	D-7402	22. 4.89	South Wales Gliding Club	Usk	11. 9.05
FRQ	3469	XT653	Slingsby T.45 Swallow	1420	XT653	27. 4.89	D.Shrimpton	RAF Keevil	3. 7.03
FRR	3470	495	Centrair 101A Pégase	101A-0034	(BGA.3451) F-CFQA	16. 4.89	P.A.Lewis *"Scoundrel"*	Walney Island	7. 5.05
FRS	3471	FRS	Scheibe Zugvogel IIIB	1097	D-2171 HB-749	27. 4.89	S.W.Vallei	Rivar Hill	1. 7.05
FRT	3472	S9	Schempp-Hirth Ventus CT	137/421		26. 4.89	S.Edwards	Dunstable	12. 3.04
							(De-Registered 17.11.03 - sold to the Netherlands as PH-1299)		
FRV	3474	FRV	Centrair 101A Pégase	101A-0325		28.10.89	D.G.Every	Gallows Hill	20. 1.05
FRW	3475	268	Schleicher ASW20L	20202	D-5981	5. 5.89	D.Cooper	Booker	24. 1.05
FRX	3476	FRX	Centrair 101A Pégase	101A-0315		30. 5.89	L.Sparrow	Rivar Hill	10. 2.05
FRZ	3478	H6	Schempp-Hirth HS.4 Standard Cirrus	348G	HB-1194 D-2172	15. 5.89	N.A.Maclean	Lasham	16. 3.04
FSA	3479	498	Grob G.102 Astir CS	1277	D-7371	9. 4.89	J.R.Carpenter	Lasham	1 .4.05
FSC	3481		Slingsby T.38 Grasshopper	751	WZ755	27. 4.90		Wolverhampton	30. 4.93
							(To Boulton Paul Heritage Project 2004)		
FSD	3482	N28	Schleicher ASK13	13367	D-0863	24. 5.89	Portsmouth Naval Gliding Club	Lee-on-Solent	8. 4.05
FSE	3483	FSE	Schleicher Ka6CR	6021	D-1946	19. 8.89	G.W.Lobb	North Hill	28. 7.04
FSH	3486	FSH	Grob G.102 Astir CS Jeans	2090	D-7532	9. 5.89	Buckminster Gliding Club	Saltby	14. 5.05
FSR	3494	FSR	Glaser-Dirks DG-300 Elan	3E343		24. 8.89	E.J.Dent	Nympsfield	24.11.05
FSS	3495	FSS	Schleicher Ka6E	4019	D-5260	19. 8.89	P.Roby and P. Lloyd	Lasham	9. 6.05
FST	3496	FST	Schleicher ASH25E	25073	(BGA.3530) (BGA.3496)	12.10.89	K.H.Lloyd and Partners	Aston Down	13. 3.05
FSU	3497	FSU/55	Scheibe Zugvogel IIIA	1060	D-9055	30. 6.89	P.W.Williams	Brent Tor	18. 6.04
FSV	3498	WZ819	Slingsby T.38 Grasshopper	800	WZ819	26. 6.89	P.D.Mann	RAF Halton	6. 6.04
FSX	3500	405	Glaser-Dirks DG-300 Elan	3E344		11. 7.89	C.Hyett	Lasham	20. 1.05
FSY	3501	162	Schleicher ASH25	25064	D-1578	14. 7.89	B.T.Spreckley	Le Blanc, France	14.10.03
							(De-Registered 23. 6.04, sold to South Africa as ZS-GXX 8.04)		
FSZ	3502	FSZ	Grob G.102 Astir CS77	1841	D-2908	29. 7.89	R.Birch	Aston Down	14. 3.05
FTB	3504	FTB	Schleicher Ka6CR *(Built Bitz)*	019	D-8900	22. 7.89	P.J.Blair	Bidford	15. 4.04
FTC	3505	N56	SZD-51-1 Junior	B-1860		23. 7.89	Seahawk Gliding Club	RNAS Culdrose	20. 3.05
FTD	3506	FTD	Schleicher ASW15B	15191	D-0872	23. 8.89	A.E.Stephenson	Walney Island	22. 4.05
FTF	3508	FTF	Schleicher Ka6CR	6294	D-6081	5. 9.89	T.Delap	Parham Park	13. 6.05
FTG	3509	FTG	Schleicher Ka7 Rhönadler	535	D-8321	18.10.89	Angus Gliding Club *(Stored)*	Drumshade	26. 3.00
FTH	3510	FTH	SZD-50-3 Puchacz	B-1881		24. 8.89	Buckminster Gliding Club	Saltby	21.12.05
FTJ	3511	FTI	SZD-48 Jantar Standard 2	W-889	HB-1472	25. 8.89	D.P.Bieniasz	Kirton-in-Lindsey	21. 3.05
FTK	3512	518	Grob G.102 Astir CS Jeans	2059	OE-5152	10.89	Ulster Gliding Club	Bellarena	16. 7.05
FTL	3513	FTL	Schleicher ASW20CL	20751	D-3564	2. 9.89	J.S.Shaw	Perranporth	17. 5.05
FTM	3514	-	Schleicher K8B	513	D-5708	30. 8.89	West Wales Gliding Club	Usk	28. 4.04
FTN	3515	853	Schleicher K8B	996	D-8539 D-KAEL, D-8539	11. 3.89	Mendip Gliding Club	Halesland	12. 5.05
FTP	3516	332	Schleicher ASW20CL	20733	D-3640	4. 1.90	Crabb Computing	Dunstable	24. 5.05
FTR	3518	FTR	Grob G.102 Astir CS77	1606	D-4807	6.10.89	Lakes Gliding Club	Walney Island	19. 1.05
FTS	3519	FTS	Glaser-Dirks DG-300 Club Elan	3E349C38		12.10.89	A.J.Taylor	Parham Park	25. 5.05
FTU	3521	FTU	Schleicher Ka7 Rhönadler	302	HB-599	25. 9.89	Dartmoor Glidng Society *"Fondue"*	Brentor	25. 3.03
FTV	3522	944	Rolladen-Schneider LS-7	7073		9.10.89	D.Hilton and S.White	Booker	5. 3.05
FTW	3523	230	Schempp-Hirth Discus B	292		4.10.89	P.Startup	North Hill	17. 2.06
			(Rebuilt with new fuselage after accident 21.6.91; original fuselage rebuilt as BGA.3879)						
FTY	3525	753	Rolladen-Schneider LS-7	7075		8.10.89	A.M.Burgess	Easterton	5. 3.05
FUB	3528	-	Schleicher Ka6CR	6007	D-8573	6.11.89	D.E.Hooper	Brent Tor	26. 5.05
FUD	3529	2	SZD-9 bis Bocian 1E	P-689	SP-2807	2.11.89	Mendip Gliding Club	Halesland	16. 4.04
FUF	3531	FUF	Scheibe SF-27A Zugvogel V	6089	D-6068	30. 9.89	East Sussex Gliding Club	Ringmer	22. 4.04
FUG	3532	BB	Schleicher ASH25	25074	(BGA.3526)	20.10.89	J.P.Gorringe and D.S.Hill	Lasham	9.12.04
FUH	3533	192	Schempp-Hirth Ventus C	438		12.10.89	M.A.Gale and Partners	Gallows Hill	14. 3.05
FUJ	3534	FUJ	Glaser-Dirks DG-300 Elan	3E353		5.12.89	J.D.Cook and Partners	Portmoak	28. 2.05
FUL	3535	803	Schempp-Hirth Discus B	293		14. 3.90	P.Warner	Dunstable	8.12.05
FUM	3536	FUM	Schleicher Ka6CR	808	D-6289	25. 3.90	S.Stanley	Dunstable	25. 5.03
							(De-Registered 5.4.04, sold to Ireland as EI-161)		
FUN	3537	FUN	Schleicher ASW20CL	20813	D-3432	5. 4.91	W.H.Parker	Booker	2. 3.05
FUP	3538	397	Schempp-Hirth Discus B	291		18.10.89	Surrey and Hampshire Gliding Club	Lasham	4. 2.05
FUQ	3539	FUQ	Scheibe SF-27A Zugvogel V	6090	D-5196	1. 4.90	G.Elliott and Partners	Ringmer	10. 5.03
FUR	3540	256	Schempp-Hirth Ventus CT	145/446		15. 3.90	K.Martin	Shobdon	15. 4.05
FUS	3541	FUS	SZD-51-1 Junior	B-1912		20.11.89	Scottish Gliding Union Ltd	Portmoak	4. 2.05
FUT	3542	612	Glaser-Dirks DG-300 Club Elan	3E350C39		11. 3.90	A.Eltis	Gransden Lodge	20. 6.05
FUU	3543	FUU	Glaser-Dirks DG-300 Club Elan	3E360C45		4. 3.90	S. Moss	Nympsfield	17. 6.05
FUV	3544	194	Rolladen-Schneider LS-7	7068		4.11.89	E.Alston	North Hill	16.12.05
FUW	3545	XE807	Slingsby T.31B Cadet TX.3	920	XE807	20.11.89	D.Shrimpton	Halesland	10. 1.05
FUY	3546	FUY	SZD-50-3 Puchacz	B-1983		30.11.89	Bath, Wilts and North Dorset Gliding Club	Kingston Deverill	10. 5.05
FVA	3548	N15	Schleicher K8B	1051	D-5117	13. 4.90	Portsmouth Naval Gliding Club	Lee-on-Solent	22. 7.05
FVB	3549	228	Schempp-Hirth Ventus CT	144/445		13. 1.90	R.Harraway	Camphill	29.11.05
FVC	3550	FVC	Schleicher ASK13 *(Built Jubi)*	13682AB		11.12.89	Mendip Gliding Club	Halesland	26. 1.05
FVD	3551	FVD	Scheibe Bergfalke IV	5806	D-0729	9.12.89	North Wales Gliding Club	Llantisilio	26.11.03
							(Damaged in field landing near Llandegla 30. 8.03)		

FVE	3552	7D	Rolladen-Schneider LS-4	4190	RAFGSA232	12. 1.90	R.May and Partners	Thame	19. 2.05
					RAFGSA R30, D-4542				
FVF	3553	FVF	Schempp-Hirth HS.5 Nimbus 2C	202	D-2880	3. 3.90	L.C.Mitchell and J.Wood	Chipping	6. 4.05
FVG	3554	660	Glaser-Dirks DG-600	6-41		17.12.89	R.G.Tomlinson	Winthorpe	23. 4.03
							(De-Registered 27. 5.03, sold to Germany as D-2224)		
FVH	3555	246	Rolladen-Schneider LS-7	7067	(BGA.3527)	18.12.89	B.R.Forrest and A.Hallum	Booker	28. 3.05
FVL	3558	FVL	Scheibe Zugvogel IIIB	1082	D-5224	29.12.89	T.G.Homan	Kirton-in-Lindsey	31. 8.05
FVM	3559	369	Centrair 101A Pégase	101A-0345		15. 3.90	S.H.North	RNAS Yeovilton	10. 4.05
FVN	3560	FVN	Centrair 101A Pégase	101A-0268/2		16. 1.90	G.G.Butler	Snitterfield	1. 5.05
			(Rebuild of BGA.3198 and carries c/n 10100268)				*(Damaged Snitterfield 22.5.04)*		
FVP	3561	FVP	Centrair 101A Pégase	101A-0350		22. 4.90	J.R.Parry and Partner	Long Mynd	10. 2.05
FVQ	3562	FVQ	Rolladen-Schneider LS-7	7079		18. 1.90	P.Harvey	Gransden Lodge	15. 4.05
FVS	3564	FVS	Schempp-Hirth HS.4 Standard Cirrus	359G	D-2168	25. 3.90	P.A Clark	Lasham	10. 2.05
FVU	3566	FVU	Schleicher ASK13	13062	D-1348	17. 4.90	Edinburgh University Gliding Club	Portmoak	30. 6.05
FVV	3567	FVV	Centrair 101A Pégase	101A-0353		27. 4.90	Cambridge University Gliding Club	Gransden Lodge	26. 4.05
FVW	3568	FVW	Schempp-Hirth Ventus BT	252/51	D-KORN	26. 1.90	I.Champness	Lasham	11. 3.05
FVZ	3571	PS	Schleicher Ka6E	4007	D-4104	27. 2.90	R.C.Fisher	Booker	10. 6.05
FWA	3572	FWA	Schleicher Ka6CR	6227	D-1062	26. 3.90	D.Cousins	Usk	29. 9.05
FWB	3573	FWB	Schleicher ASK13	13224	HB-989	2. 4.90	Cotswold Gliding Club	Aston Down	2.12.05
FWC	3574	45	Grob G.103C Twin III Acro	34154		5. 4.90	Lasham Gliding Society Ltd	Lasham	6. 1.06
FWD	3575	FWD/888	Schempp-Hirth Ventus CT	148/468		10. 5.90	R.S.Maxwell-Fendt	Lasham	20. 4.05
FWE	3576	FWE	SZD-50-3 Puchacz	B-1984	(BGA.3547)	5. 2.90	Deeside Gliding Club	Aboyne	4. 5.05
FWF	3577	L57	Rolladen-Schneider LS-7	7097		28. 2.90	G.P.Hibberd	Husbands Bosworth	1. 4.05
FWG	3578	FWG	Centrair 101A Pégase	101A-0252	PH-793	22. 2.90	Devon and Somerset Gliding Club	North Hill	27. 2.04
							(Dived into the ground near North Hill 17. 4.03, fatal accident)		
FWH	3579	FWH	Scheibe SF-27A Zugvogel V	6024	D-4733	10. 2.90	R.Sampson	Husbands Bosworth	28. 6.05
FWJ	3580	S3	Rolladen-Schneider LS-7WL	7078		22. 3.90	J.P.Popika	Gransden Lodge	13. 6.05
FWK	3581	29	Schempp-Hirth Nimbus 3DT	32		22. 3.90	J.D.Glossop	Gransden Lodge	11.12.04
FWL	3582	FWL	Schleicher K8B	106/58	D-7151	24. 2.90	Dukeries Gliding Club	Gamston	27. 5.05
FWM	3583	FWM	Glaser-Dirks DG-300 Club Elan	3E373C50		15. 6.90	P.J.Groves	Long Mynd	3. 3.05
FWN	3584	FWN	Schleicher ASK13	13285	HB-1023	20. 4.90	Booker Gliding Club	Booker	24. 2.05
							(Crashed Booker, 6. 8.04)		
FWP	3585	980	Schleicher ASW19B	19262	D-5980	4. 4.90	A.Kefford and Partners	Dunstable	11. 4.05
							(Damaged in take-off accident Dunstable 24. 9.04)		
FWQ	3586	FWQ	Schleicher ASK21	21460		15. 5.90	Midland Gliding Club	Long Mynd	18. 4.05
FWR	3587	277	Glasflugel H.303 Mosquito	34	N77RL	26. 3.90	S.J.Ferguson	Aston Down	26. 9.04
FWS	3588	662	Schleicher ASW20C	20765	D-6623	18. 2.90	P.C.Gill	North Weald	3. 2.05
FWT	3589	FWT	SZD-50-3 Puchacz	B-1988		24. 3.90	The Soaring Centre	Husbands Bosworth	27. 1.05
FWU	3590	768	Rolladen-Schneider LS-7	7080		19. 3.90	C.Brown	Husbands Bosworth	8. 3.05
FWW	3592	FWW	Schleicher ASH25E	25093		19. 6.90	A.T.Farmer	RAF Weston on the Green	28. 2.05
FWX	3593	FWX	Centrair 101A Pégase	101A-033	F-CFRZ	27. 3.90	I.R.Stanley	Booker	23. 3.05
FWY	3594	FWY	Centrair 101A Pégase	101A-071	F-CFXE	23. 3.90	I.P.Johnson	Lasham	28. 4.05
FWZ	3595	FWZ	Schleicher ASW19B	19342	D-2603	14. 4.90	C.Fowler	Camphill	17.12.05
FXA	3596	567	Grob G.102 Speed Astir IIB	4083	D-2671	16. 4.90	A.D.Duke	Nympsfield	17.11.05
FXB	3597	FXB	Schleicher K8B	8193/A	D-5597	29. 3.90	R.J.Morris	Brent Tor	6. 7.04
FXC	3598	FXC	Schleicher Ka6E	4268	D-0150	9. 8.90	J.Wilson	Kingston Deverill	19. 5.05
FXD	3599	285	Centrair 101A Pégase	101A-0346	(BGA.3563)	31. 3.90	The Soaring Centre	Husbands Bosworth	21.12.05
FXE	3600	35	Rolladen-Schneider LS-7	7090		23. 3.90	J.C.Kingerlee	RAF Weston on the Green	28. 9.05
FXF	3601	FXF	Slingsby T.50 Skylark 4	1455	HB-812	7. 5.90	S.White	Booker	13. 5.02
FXG	3602	FXG	Schempp-Hirth HS.2 Cirrus	23	N1216	23. 8.90	G.F.King	Rufforth	15. 2.00
							(Crashed North Hill 15.9.99; noted 5.01 for potential rebuild)		
FXH	3603	-	Schleicher Ka7 Rhönadler	353	D-4040	10. 4.90	Vale of Neath Gliding Club	Rhigos	27. 5.05
FXJ	3604	247	Schleicher ASW24	24086		4. 5.90	A.K.Laylee	Lasham	23. 6.05
FXL	3606	108	Schleicher ASH25	25088		11. 4.90	C.Simpson and Partners	Husbands Bosworth	23.12.05
FXM	3607	173	Schempp-Hirth Discus BT	16/301	D-KHIA	12. 4.90	R.J.H.Fack	Long Mynd	15. 7.05
FXN	3608	FXN	CARMAM M.200 Foehn	4	OO-ZNI	14. 4.90	I.C.Gutsell and Partners	Burn	9. 9.01
					(OO-ZXS), F-CCXS				
FXP	3609	FXP	LET L-23 Super Blanik	907609		17. 7.90	Dill Faulke Education Trust	Dunstable	26. 6.05
FXQ	3610	954	Schempp-Hirth Nimbus 3DT	31		21. 4.90	D.G.Tanner	Lasham	20. 4.05
FXR	3611	L12	LAK-12 Lietuva	6162		9.90	S R Blackmore	Hinton-in-the-Hedges	23. 4.05
FXS	3612	FXS	Schleicher Ka6E	4228	D-0073	7. 5.90	R.Woodhouse and B.C.Wade	Tibenham	28. 5.05
FXT	3613		Centrair 101A Pégase	101A-0056	F-CFQV	4.90	K.Ludlow	Viterbo, Italy	4. 8.02
FXU	3614	FXU	Schleicher Ka6E	4071	OH-343	8. 6.90	M E Mann Syndicate	Lasham	14. 3.05
					OH-RSY				
FXW	3616	FXW	Schleicher K8B	8651	D-7203	5. 4.90	South Wales Gliding Club	Usk	26.11.05
					D-KOLA, D-7203				
FXX	3617		Scheibe L-Spatz 55	756	D-3598	1. 9.91	P.Brown	Ridgewell	19. 6.00
FXY	3618	723	Schleicher ASW15B	15348	F-CEJL	21. 5.90	C.I.Willey	Brentor	22. 4.05
FYA	3620	FYA	SZD-50-3 Puchacz	B-2022		9. 5.90	Cairngorm Gliding Club	Feshiebridge	5. 3.05
FYB	3621	779	Rolladen-Schneider LS-7	7102		2. 5.90	V.Haley and Partner	Wormingford	20. 2.05
FYC	3622	A10	Schempp-Hirth Ventus B	83	F-CEDR, F-WEDR	2. 5.90	D.B.Meeks	Sutton Bank	25. 9.04
FYD	3623	942	Schleicher ASH25	25095		19. 5.90	C.C.Lyttelton	Lasham	9. 2.05
FYE	3624	FYE	Scheibe Zugvogel IIIB	1067		20. 5.90	G.Pearce and Partners	Brent Tor	15. 6.05
			(Not the same aircraft as OY-MHX although given as c/n 1067)						
FYF	3625	FYF	Schleicher ASK21	21470		4. 8.90	London Gliding Club	Dunstable	3. 5.05
FYG	3626	FYG	Glasflugel H.205 Club Libelle	22	OH-545	13. 5.90	I.H Shattock	Usk	27. 6.05
FYH	3627	224	Rolladen-Schneider LS-4A	4804		4. 7.90	G.W.Craig	RAF Weston on the Green	11. 8.05
FYJ	3628	FYJ	Schempp-Hirth HS.4 Standard Cirrus	581G	D-8931	12. 7.90	S.Gibson	Pocklington	30.11.05
FYK	3629	34	Rolladen-Schneider LS-7	7108		1. 6.90	R.Ward	Gransden Lodge	27. 3.05
FYL	3630	FYL	SZD-50-3 Puchacz	B-1990		6.90	Deeside Gliding Club	Aboyne	21.10.05
FYM	3631	326	Schempp-Hirth Discus BT	31/328		1. 6.90	J.A.Denne	RAF Weston-on-the-Green	27. 5.05

Reg	BGA No	Comp No	Type	C/n	Previous identity	Date	Owner	Base	CofA expiry
FYN	3632	J3	Schempp-Hirth Discus B	179	N75J	14. 7.90	P.Foulger	Wormingford	7. 1.05
FYP	3633	FYP	LET L-23 Super Blanik	907620		4. 8.90	Needwood Forest Gliding Club *(Crashed on landing, Cross Hayes 27. 8.03)*	Cross Hayes	29. 2.04
FYR	3635	FYR	LET L-23 Super Blanik	917816		2. 7.92	North Wales Gliding Club	Llantisilio	31.12.04
FYU	3638	DG	Glaser-Dirks DG-100 Elan	E111	OY-XMR SE-TYO	28. 6.90	I.Shepherd	RAF Weston-on-the-Green	20. 5.05
FYV	3639	FYV	Schleicher ASK21	21468		25. 7.90	Booker Gliding Club	Booker	25. 1.05
FYW	3640	C1	Rolladen-Schneider LS-7	7111		6. 6.90	D.Tagg	Rufforth	6. 1.06
FYX	3641	208	Schempp-Hirth Discus bT	32/333		3. 7.90	M.P.Brockington	Talgarth	18. 4.05
FYY	3642	S	Schleicher ASK13	13685AB		9. 7.90	Lasham Gliding Society	Lasham	20.10.05
FYZ	3643	171	Schleicher ASH25	25097		18. 7.90	M.G.Thick	Sutton Bank	10. 6.04
FZA	3644	FZA	SZD-51-1 Junior	B-1913		23. 7.90	Booker Gliding Club	Booker	19. 1.06
FZB	3645	669	Glasflugel H.201B Standard Libelle	112	OH-388 OH-GLA	31. 7.90	C.Thomas and J.E.Herring	Lasham	22. 2.04
FZC	3646	FZC	Schempp-Hirth SHK-1	58	OH-357 OH-SHA	30. 8.91	J.F Mills	RAF Cranwell	28. 5.05
FZF	3649	FZF	SZD-51-1 Junior	B-1861		21. 7.90	Devon and Somerset Gliding Club	North Hill	26. 2.05
FZG	3650	FZG	SZD-9 bis Bocian 1D	F-859	SP-2450	24. 9.90	The Borders Gliding Club	Milfield	16. 6.02
FZH	3651	FZH	Schempp-Hirth Ventus C	455		26. 7.90	G.D.Clack	Lasham	16. 2.05
FZK	3653	FZK	Schempp-Hirth HS.4 Standard Cirrus	81	HB-967	2. 9.90	R.S.Burgoyne and S. Lucas	Aston Down	29. 9.05
FZL	3654	Z6	Schleicher ASW20CL	20764	D-5937	12. 7.90	R.M.Housden	Aboyne	17. 4.05
FZM	3655	FZM	Scheibe SF-27A Zugvogel V	6103	D-1772	7. 8.90	N.Dickenson	Chippingl	31. 7.04
FZN	3656	K13	Schleicher ASK13	13045	D-5759	9. 8.90	Black Mountains Gliding Club	Talgarth	16. 6.05
FZP	3657	N16	SZD-51-1 Junior	B-1926		9. 8.90	Portsmouth Naval Gliding Club	Lee-on-Solent	27. 3.05
FZQ	3658	FXQ	SZD-50-3 Puchacz	B-2024	(BGA.3637)	9. 8.90	The Soaring Centre	Husbands Bosworth	19. 1.05
FZR	3659	FZR	Schleicher Ka6CR	6136	D-8459	17.12.90	D.Albasing	North Hill	13.12.05
FZV	3663	480	Rolladen-Schneider LS-7	7116		16.12.90	R.N.Boddy	Booker	10.11.05
FZW	3664	FZW	Glaser-Dirks DG-300 Club Elan	3E378C53		23. 9.90	Mr and Mrs S.L.Barter	Ringmer	9. 2.05
FZX	3665	FZX	SZD-51-1 Junior	B-1925		18. 9.90	Nene Valley Gliding Club *(Ground accident Upwood 5. 2.04; de-registered 19.5.04 - remains sold in Poland, to SP-3696)*	Upwood	12. 3.04
FZZ	3667	FZZ	LET L-33 Solo	940220		28. 4.95	D.A Wiseman	Andreas	16. 4.05
GAB	3669	GAB	LAK-12 Lietuva	6170		12. 1.91	M.J. Wilshere	RAF Halton	5. 4.04
GAD	3671	L5	Rolladen-Schneider LS-3	3032	HB-1363	19.11.90	P.Turner	Challock	1. 8.05
GAF	3673	778	Schleicher ASK21	21152	ZD652 BGA.2892	8.11.90	Lasham Gliding Society	Lasham	25. 3.05
GAG	3674	GAG	Schleicher ASK21	21143	ZD645 BGA.2885	24. 1.91	Stratford-upon-Avon Gliding Club	Snitterfield	29.11.04
GAH	3675	GAH	Schempp-Hirth HS.4 Standard Cirrus	572	HB-1240	3.12.90	E.Wright	Nympsfield	16.11.05
GAJ	3676	GAJ	Glaser-Dirks DG-300 Club Elan	3E385C56		10.12.90	S.M. Lewis	Long Mynd	12. 5.04
GAK	3677	GAK	LET L-13 Blanik	174522	2-84 (Lithuania)	7.97	North Wales Gliding Club	Llantisilio	5.10.05
GAM	3679	GAM	Schleicher ASK21	21144	ZD646 BGA.2886	21.11.90	Oxford University Gliding Club	Bicester	21.12.04
GAN	3680	83	Glasflugel H.301 Libelle	8	D-4111	12.12.90	W.J.Dean	Long Mynd	24. 8.04
GAP	3681	GAP	Schempp-Hirth Ventus bT	14/150	OH-774 N416DP	28. 4.91	J.R.Greenwell	Milfield	21. 5.05
GAQ	3682	GAQ/K7	Schleicher Ka7 Rhönadler	3	PH-788 D-5550	19. 4.91	York Gliding Centre	Rufforth	24. 2.05
GAR	3683	AW	Rolladen-Schneider LS-6C	6205		2.11.90	A. Warbrick	Aboyne	21. 7.05
GAS	3684	GAS	Schempp-Hirth Ventus CT	157/509		30. 5.91	M W Edwards	Kingston Deverill	30. 1.05
GAT	3685	GAT	Grob G.102 Astir CS	1130	D-4176	23.11.90	R.S.Scott	Lasham	26. 7.03
GAU	3686	725	Glasflugel H.201B Standard Libelle	498	F-CELA	11. 6.93	D.R.Pickett	Crowland	22. 5.05
GAV	3687	GAV	Scheibe SF-27A Zugvogel V	6073	D-5287	18.11.90	W.Waite	Lleweni Parc	17. 6.04
GAW	3688	GAW	Schleicher Ka6CR	61/08	D-6320	30.12.90	H.C.Yorke	Snitterfield	22. 3.05
GAX	3689	302	SZD-55-1	551190008		30. 4.91	P.Gold	Rattlesden	21. 4.04
GBA	3692	GBA	Schleicher ASK13	13417	D-2114	4.12.90	Burn Gliding Club	Burn	5. 9.05
GBB	3693	GBB	Schleicher ASK21	21073	D-3239	11.12.90	B.T.Spreckley	Le Blanc, France	6. 3.05
GBD	3695	GBD	SZD-50-3 Puchacz	B-2028		27. 4.91	Northumbria Gliding Club	Currock Hill	11.12.04
GBE	3696	-	Schleicher Ka6CR (Pe)	6133A	D-4085	23.12.90	T.H.B.Bowles	Gamston	11. 6.05
GBF	3697	GBF	Schleicher ASK21	21142	ZD644 BGA.2883	3. 2.91	BBC Gliding Group	Booker	13. 1.06
GBG	3698	S21	Rolladen-Schneider LS-6c	6214		12.12.90	C.M.Greaves	Rufforth	20. 1.05
GBJ	3700	GBJ	Grob G.102 Astir CS	1107	D-4167	5. 1.91	Aquila Gliding Club	Hinton-in-the-Hedges	6. 6.04
GBK	3701	GBK	Grob G.102 Astir CS	1461	D-7451	5. 1.91	B.J. and A.R.Griffiths *"Mountain Man"*	Cross Hayes	16. 2.05
GBL	3702	720	Rolladen-Schneider LS-7WL	7119		12.11.90	C.R.Sutton	Saltby	17. 3.05
GBM	3703	GBM	Scheibe SF-27A Zugvogel V	6060	RAFGGA D-5409	2. 1.91	G.Cook tr BFMT Syndicate	North Hill	19.11.05
GBN	3704	843	Schleicher ASK21	21141	ZD643 BGA.2884	14.3.91	Essex and Suffolk Gliding Club	Wormingford	5.12.05
GBQ	3706	630	Rolladen-Schneider LS-6	6082	D-3725	6. 2.91	A.and P.R.Pentecost	Lasham	1.12.05
GBR	3707	218	Rolladen-Schneider LS-6C	6196	D-3482	25.11.90	S.Hurd	Dunstable	21. 2.05
GBS	3708	206	Glaser-Dirks DG-300 Club Elan	3E389C58		15. 3.91	Yorkshire Gliding Club	Sutton Bank	26. 1.05
GBT	3709	IV	Rolladen-Schneider LS-4A	4355	N220BB	8. 4.91	S.A.Adlard	Long Mynd	4. 3.05
GBU	3710	922	Centrair 101A Pégase	101A-0394		3. 4.91	S.I.Ross	Parham Park	3. 3.05
GBV	3711	GBV	Schleicher ASK21	21149	ZD649 BGA.2889	23. 4.91	Wolds Gliding Club	Pocklington	5. 5.05
GBX	3713	52	Schleicher ASW22	22029	D-4325	27. 2.91	I.Ashdown	Parham Park	7. 2.05
GBY	3714	425	Rolladen-Schneider LS-7	7121		21. 1.91	W.J.Morecraft and Partners *(Crashed at Javierregay, Spain 20. 7.04)*	Saltby	27.12.04
GBZ	3715	GBZ	Glaser-Dirks DG-500 Elan Trainer	5E34T10		10. 8.91	Needwood Forest Gliding Club	Cross Hayes	20. 3.05
GCA	3716	GCA	Schleicher ASW19B	19281	D-3179	2. 3.91	Deeside Gliding Club	Aboyne	12. 2.05

GCB	3717	236	LAK-12 Lietuva	647		29. 3.91	B.Middleton	Easterton	11. 2.05
GCC	3718	GCC	SZD-51-1 Junior	B-1928		10. 3.91	The Soaring Centre	Husbands Bosworth	12.5.05
GCD	3719	507	Schempp-Hirth HS.4 Standard Cirrus	476	PH-507	21. 2.91	B.Van Woerden	Feshiebridge	17. 7.05
GCE	3720	8	Schleicher ASH25	25105		19. 2.91	C.L.Withall	Dunstable	24. 2.05
GCF	3721	GCF	Schleicher ASK23	23010	AGA.9	8. 2.91	Needwood Forest Gliding Club	Cross Hayes	16. 1.06
GCG	3722	S81	Schleicher K8B	8186	D-5227	5. 2.91	Shenington Gliding Club	Edgehill	24.10.04
GCH	3723	438	Schleicher ASW15B	15212	PH-438 D-0950	17. 4.91	M.D.Woodman-Smith and Partner *(Under restoration)*	Dunstable	24. 4.99
GCJ	3724	GCJ	LAK-12 Lietuva *(New wings with reconditioned 1982-built fuselage)*	626		30. 3.91	P.Crowhurst	Crowland	27. 5.05
GCK	3725	GCK	SZD-50-3 Puchacz	B-2025	(G-BTJV) BGA.3725	8. 3.91	Kent Gliding Club	Challock	11. 5.05
GCL	3726	GCL	Grob G.102 Astir CS	1194	D-7311	10. 3.91	C.P.Offen	Burn	8. 1.06
GCM	3727	Z29	Rolladen-Schneider LS-6C	6216		12. 3.91	M.H.Hardwick	Booker	13.12.05
GCP	3729	GCP	Schleicher Ka6CR	6416	D-6369	3. 5.91	D.Clarke	Burn	18. 4.05
GCQ	3730	GCQ/845	Schempp-Hirth HS.2 Cirrus VTC	135Y	D-2945	2. 4.91	Dumfries and District Gliding Club	Falgunzeon	25.11.05
GCR	3731	748	Schleicher ASW15B	15447	D-6887	23. 3.91	H.McLean	Aboyne	9. 4.05
GCS	3732	H12	Glasflugel H.205 Club Libelle	159	F-CEQL	7. 7.91	D.Coates	Gamston	23. 5.05
GCT	3733	GCT	Schempp-Hirth Discus B	360		22. 3.91	J.C.Leonard	Bembridge	9. 5.05
GCU	3734	GCU	SZD-50-3 Puchacz	B-2023	(BGA.3619)	19. 3.91	Buckminster Gliding Club	Saltby	14. 3.05
GCX	3736	N6	Schleicher ASW15	15034	D-0420	21. 5.91	P.Collier and partner	Husbands Bosworth	15. 3.05
GCY	3737	GCY	Centrair 101A Pégase	101A-0392		22. 4.91	G.R.Hudson	Lasham	6. 1.05
GDA	3739	546	Rolladen-Schneider LS-3-17M	3448	RAFGGA.546	20. 5.91	A.Cluskey	Saltby	11. 2.05
GDB	3740	GDB	Schleicher K8B	8152	HB-738	23. 3.91	Welland Gliding Club	Lyveden	1. 5.05
GDD	3742	GDD	Bölkow Phoebus 17C	836	D-0060	18. 4.91	I.D.McLeod	Challock	2. 6.02
GDE	3743	GDE	Schleicher Ka6CR	6570Si	D-5306	26. 4.91	R.Hills	North Hill	17. 6. 05
GDF	3744	GDF	Schleicher Ka6BR	389	D-8544	17. 4.91	P.Kent	Cross Heyes	12. 6.05
GDJ	3747	450	Rolladen-Schneider LS-4A	4832		27. 4.91	A Clark	Lee-on-Solent	10. 4.05
GDK	3748	GDK	Schleicher K8B	8240	D-5381 D-KANU, D-5381	15. 4.91	Vale of Neath Gliding Club	Rhigos	14. 3.05
GDM	3750	668	Glasflugel H.201B Standard Libelle	597	D-6666	29. 4.91	K.Fear and Syndicate	Crowland	27. 3.05
GDN	3751	294	Rolladen-Schneider LS-3-17M	3291	D-6932	28. 4.91	S.J Pepler	Sandhill Farm, Shrivenham	29. 9.05
GDP	3752	GDP	Schleicher ASW19B	19285	D-3160	2. 5.91	P.Heywood	Lyveden	23. 4.05
GDQ	3753	GDQ	Grob G.102 Astir CS	1145	D-7229	11. 5.91	P. Lowe and R. Bostock	Seighford	27. 8.05
GDR	3754	GDR	Schempp-Hirth Discus CS *(Built Orlican)*	016CS		5. 5.91	T.Limb	Husbands Bosworth	21. 7.05
GDS	3755	GDS	Schleicher ASW15B	15205	D-0902	20. 6.91	J.Edwards	Dunstable	10. 9.04
GDT	3756	T54	Schleicher ASW24	24120		10. 5.91	A.Ditchfield	Camphill	1. 4.05
GDU	3757	801	Schleicher ASW24	24118		8. 6.91	G.J.Moore	Dunstable	14.12.05
GDV	3758	GDV	Schleicher Ka6E	4099	OO-ZWQ I-NEST, OE-0807	20. 6.91	A.Ruddle and D. Parkes	North Hill	22. 4.05
GDW	3759	GDW	Scheibe SF-27A Zugvogel V	6116	D-1997	16. 5.91	M.W.Hands	Saltby	21. 5.05
GDX	3760	HB2	Schempp-Hirth Discus CS	023CS		2. 7.91	The Soaring Centre	Husbands Bosworth	19.12.05
GDY	3761	914	Schleicher ASW15B	15220	D-0947	6. 5.91	J Archer	Bidford	11. 2.05
GDZ	3762	524	Schleicher ASW24	24116		17. 5.91	M.Haddon	Pocklington	14. 2.05
GEA	3763	GEA	Schleicher Ka6CR	849	(BGA.3605) D-5801	7. 6.91	C.Deane	Rufforth	9. 1.06
GEB	3764	GEB	Grob G.102 Astir CS77	1628	PH-576	7. 6.91	J.O.Lavery	Bellarena	9. 7.05
GEE	3767	928	Glasflugel H.201B Standard Libelle	94	D-0928	6. 6.91	C.Metcalfe	Gamston	20. 6.05
GEG	3769	GEG	Schleicher K8B	689	HB-639	27. 4.91	Newark & Notts Gliding Club	Winthorpe	6. 2.05
GEH	3770	219	Schleicher ASW15B	15276	D-2124	9. 7.91	K.G.Vincent and Partners	Challock	28. 5.05
GEL	3772	GEL	SZD-50-3 Puchacz	B-2030		29. 5.91	Northumbria Gliding Club	Currock Hill	23. 4.05
GEM	3773	GEM	Schleicher Ka6CR	6249	D-8486	4. 6.91	D.Perkins and partners	Thame	16. 6.05
GEN	3774	GEN	Slingsby T.21B	1154	RAFGGA.550 XN150	16. 5.92	A.Harris	RAF Bruggen	23. 7.05
GEP	3775	GEP	Schempp-Hirth HS.4 Standard Cirrus	205G	D-0917	10. 6.91	G.S.Wadforth *(Crashed in field landing between Towcester and Weedon 23. 7.04)*	Pocklington	28. 4.05
GEQ	3776	2001	SZD-12A Mucha 100A	462	SP-2001	6. 6.91	R.A. Earnshaw-Fretwell	RAF Keevil	10. 7.05
HAA	3777	263	Glasflugel H.201B Standard Libelle	356	HB-1090	20. 6.91	D. Jones and Partners	Gransden Lodge	3. 3.05
HAB	3778	HAB	Schleicher Ka6CR	6596	D-1596	1.92	M Greenwood	Long Mynd	19. 4.05
HAC	3779	HAC	SZD-50-3 Puchacz	B-2035		29. 6.91	Peterborough and Spalding Gliding Club	Crowland	8. 4.05
HAD	3780	429	Glasflugel H.201 Standard Libelle	3	D-8914	5. 7.91	A.Barker-Mill	Lasham	10. 3.05
HAE	3781	HAE	Glasflugel H.205 Club Libelle	75	D-8687	5. 7.91	J.P.Kirby	Bembridge	3. 5.05
HAF	3782	HAF	SZD-50-3 Puchacz	B-2031		4. 7.91	Seahawk Gliding Club	RNAS Culdrose	5. 3.05
HAG	3783	HAG	Schleicher Ka7 Rhönadler	834	D-5795	21. 5.92	Denbigh Gliding Club	Lleweni Parc	3. 8.05
HAJ	3785	391	Schempp-Hirth Ventus C	517		19. 7.91	Surrey and Hampshire Gliding Club	Lasham	13. 1.06
HAL	3787	HAL	Schleicher ASK13 *(Built Jubi)*	13690AB		7. 9.91	Cotswold Gliding Club	Aston Down	10.11.05
HAN	3789	278	Schempp-Hirth HS.4 Standard Cirrus	130	D-0326	14. 8.91	I.Ashdown	Parham Park	8. 4.05
HAP	3790	HAP	Schleicher Ka6E	4335	HB-985	29. 8.91	T.Turner	Dunstable	6. 9.05
HAQ	3791	114	Rolladen-Schneider LS-6B	6150	D-8079	2. 9.91	A R Hughes	Booker	7. 4.05
HAR	3792	HAR	Schleicher K8B	8151	D-8453	4. 9.91	T.W.Roberts *(For restoration)*	Brentor	4. 7.97*
HAS	3793	HAS	SZD-50-3 Puchacz	B-2043		17. 8.91	The Soaring Centre	Husbands Bosworth	16.12.05
HAT	3794	HAT	Glaser-Dirks DG-200/17	2-93/1709	D-6843	26. 8.91	D.Simon	Aboyne	5. 1.05
HAU	3795	HAU	Grob G.102 Astir CS Jeans	2043	D-3887	11. 9.91	Yorkshire Gliding Club	Sutton Bank	1. 3.04
HAV	3796	HAV	Glasflugel H.201B Standard Libelle	40	HB-950	25. 8.91	P.W.Andrews	Husbands Bosworth	5. 7.05
HAX	3798	HAX	Schempp-Hirth HS.4 Standard Cirrus	02	ZS-GHZ ZS-TIM, ZS-GGR, D-0302	10.10.91	R.Jarvis	Rivar Hill	10. 4.05
HAY	3799	778	Rolladen-Schneider LS-7	7154		15.10.91	N.Leaton and Partners	Gransden Lodge	6. 5.04
HBA	3801	729	Rolladen-Schneider LS-7	7156	D-6041	12.10.91	P.O'Donald	Gransden Lodge	15.11.05
HBB	3802	HBB	Schleicher ASW24	24132		19. 9.91	London Gliding Club	Dunstable	25.11.05

HBC	3803	HBC	Rolladen-Schneider LS-6C	6209	D-....	20. 9.91	A.Crowden	Usk	1. 4.05
HBD	3804	HBD	Glaser-Dirks DG-2002-	12	HB-1384	5.10.91	D.A.Clempson	Portmoak	16. 9.05
HBE	3805	356	Glaser-Dirks DG-300 Elan	3E237	SE-UFB	21. 5.92	A.W.Cox and Partners	Aston Down	9. 1.05
HBF	3806	HBF	Schempp-Hirth HS.5 Nimbus 2C	191	D-3369	5.10.91	T.Cauldwell	Sackville Lodge, Riseley	21. 7.00
HBG	3807	96	Schleicher ASW24	24133		10.12.91	Imperial College Gliding Club	Lasham	13. 4.05
HBH	3808	496	Grob G.103C Twin III	36006		14.10.91	Imperial College Gliding Club	Lasham	9. 1.06
HBJ	3809	949	Rolladen-Schneider LS-6C-18	6230		26. 9.91	G.Lyons	Booker	13. 1.06
HBK	3810	HBK	Grob G.103 Twin Astir	3254-T-31	RAFGGA.550 D-2389	29. 9.91	A.G. Machin	Burn	2. 5.05
HBL	3811	HBL	Grob G.102 Astir CS77	1626	RAFGSA R78 RAFGSA.778	17.10.91	J.McCormick	Bidford	4. 6.05
HBM	3812	755	Grob G.102 Astir CS77	1633	RAFGSA R65 RAFGSA.R66, RAFGSA.546	3.12.91	A.Fox	RAF Syerston	17. 3.05
HBP	3814	522	Glaser-Dirks DG-500/22 Elan	5E36S8		10.91	A.Taverna	Borgo San Lorenzo, Italy	13. 6.04
HBQ	3815	HBQ	Schleicher Ka6CR	6611	D-5616	22.11.91	P.N.Jones	Dunstable	14. 9.04
HBR	3816	PH	Schempp-Hirth Nimbus 4T	3/6	(BGA.3784)	1. 8.92	P S Hawkins	Snitterfield	12. 5.05
HBS	3817	HBS	SZD-41A Jantar Standard	B-852	D-4160	2.12.91	G.King	Dunstable	11. 7.05
HBT	3819	HBT	Grob G.102 Club Astir	2235	PH-675	9. 2.92	M.D.Evans	Winthorpe	19. 3.05
HBU	3820	605	Centrair ASW20F	20527	F-CFSI	22.11.91	N.H.Cotterell	Ringmer	4. 2.05
HBV	3821	667	Schempp-Hirth HS.5 Nimbus 2B	143	D-7850	11.91	G.Evison and partners	Sutton Bank	15. 2.05
HBW	3822	829	Glaser-Dirks DG-300 Club Elan	3E405C64		15.12.91	D.R.Smith *(De-Registered 26. 1.03 - sold in Japan as JA-2611)*	Halesland	4. 5.04
HBX	3823	HBX	Slingsby T.45 Swallow	1386	8801M XS650	16. 5.93	C.D.Street and Partners	Lasham	20. 5.05
HBY	3824	664	Rolladen-Schneider LS-7	7148		7.11.91	P O'Donovan *(Crashed on launch Dunstable 7. 8.04)*	Dunstable	12. 5.05
HBZ	3825	HBZ	Slingsby T.15 Gull III rep	---		28. 6.92	P.R.Philpot	Chipping	12. 7.05
HCA	3826	HCA	Grob G.103 Twin Astir	3289	D-0094 OO-ZOH/D-3063	24.12.91	D.Munroe and Partners	Tibenham	26. 4.05
HCB	3827	754	Schempp-Hirth Nimbus 3DT	47		24.12.91	P.A.Green	Lasham	2. 4.05
HCC	3829	HCC	SZD-50-3 Puchacz	B-2048		4. 1.92	Heron Gliding Club	RNAS Yeovilton	1. 4.05
HCD	3830	HCD	SZD-50-3 Puchacz	B-2049		7. 1.92	The Soaring Centre *(Crashed Husbands Bosworth 18.1.04 and destroyed, wreck to AAIB)*	Farnborough	15. 6.04
HCE	3831	346	Schleicher ASW19B	19305	D-6527	6. 1.92	R.Halliburton	Pocklington	25.11.05
HCF	3832	HCF	SZD-50-3 Puchacz	B-2047		20.12.91	Shalbourne Soaring Society	Rivar Hill	12. 3.05
HCG	3833	HCG	Maupin Woodstock One *(Built R.Harvey)*	---		10.92	R.Harvey	Tibenham	30.12.05
HCH	3834	355	Centrair ASW20FP	20178	F-CEUL	20. 3.92	Not known	Booker	18. 9.05
HCJ	3835	HCJ	Grob G.103 Twin II	3709	D-2611	24. 1.92	Peterborough and Spalding Gliding Club	Crowland	28. 5.05
HCK	3836	WB962	Slingsby T.21B	623	RAFGGA5.. WB962	2. 1.92	V.Mallon	Kleve/Wissler Dunen, Germany	21. 9.05
HCL	3837	HCL	Schempp-Hirth Discus B	136	D-4682	13. 3.92	C.E.Broom	Usk	13. 3.05
HCM	3838	HCM	Schleicher Ka7 Rhönadler	498	D-5669	4. 3.92	M.Barnard	Edgehill	29. 8.02
HCN	3839	HCN	CARMAM M.200 Foehn	24	F-CDDR	21.12.92	J S Shaw	Perranporth	28. 6.05
HCP	3840	HCP	Avialsa A.60 Fauconnet	123K	F-CDLA	3.93	C.Kaminski *(Being refurbished)*	Eaglescott	31. 7.95*
HCQ	3841	HCQ	Glasflugel H.201B Standard Libelle	197	HB-999	28. 1.92	E.K.Harris	Dunstable	14. 3.04
HCR	3842	394	SZD-51-1 Junior	B-2003		28. 4.92	East Sussex Gliding Club *(Landing accident Ringmer 10.11.04)*	Ringmer	12.12.04
HCU	3845	78	Glaser-Dirks DG-300 Club Elan	3E407C66		7. 2.92	M.S Smith and Partners	Kingston Deverill	21. 3.05
HCV	3846	X19	Schleicher ASW19B	19084	D-4486	14. 5.93	A.B.Laws	Syerston	29. 4.05
HCW	3847	HCW	SZD-51-1 Junior	B-2002	(BGA.3844)	1. 2.92	Deeside Gliding Club	Aboyne	4.12,04
HCX	3848	HCX	Schleicher ASK21	21541		16. 5.92	Devon and Somerset Gliding Club	North Hill	9. 5.05
HCY	3849	HCY	Glaser-Dirks DG-300 Club Elan	3E413C67		10. 5.94	S.T.Dry	RAF Keevil	28. 8.05
HCZ	3850	HCZ	Schleicher K8B	8114A	D-4675	21. 2.92	South London Gliding Centre	Kenley	19. 8.05
HDA	3851	HDA	Pilatus B4 PC-11AF	017	D-0964	18. 3.92	P.Bois	(Jersey)	7.10.05
HDB	3852	HDB	SZD-51-1 Junior	B-1997		4. 3.92	Stratford-upon-Avon Gliding Club	Snitterfield	9. 1.05
HDC	3853	HDC	Schleicher ASK13	13308	D-0750	19. 3.93	Bowland Forest Gliding Club	Chipping	29.11.05
HDD	3854	591	Centrair 101B Pégase	101B-0425		5. 4.92	Scottish Gliding Union	Portmoak	28.12.05
HDE	3855	HDE	Pilatus B4 PC-11AF	223	VH-XOZ VH-WQP	12. 4.92	A.Jenkins	Lasham	10. 6.05
HDF	3856	910	Schempp-Hirth Discus B	404		21. 2.92	B.Laverick-Smith	Challock	26.12.05
HDH	3858	991	Glaser-Dirks DG-202-15	2-197	???	24. 5.92	A.J.Millson	Rattlesden	23. 2.05
HDJ	3859	HDJ	Schleicher ASW20CL	20828	D-8442	4. 3.92	G.E.Lambert	Weelde, The Netherlands	18. 4.05
HDL	3861	137	Schleicher ASW20	20082	D-1617 OH-495	13. 4.92	S.Thackray	Rivar Hill	20. 5.05
HDM	3862	HDH	SZD-12A Mucha 100A	448	SP-1987	15. 4.92	T.J.Wilkinson	Sackville Lodge, Riseley	19. 5.05
HDN	3863	HDN	Schleicher K8B	2	D-8017	17. 3.92	Upward Bound Trust	Thame	9. 5.05
HDP	3864	N36	SZD-50-3 Puchacz	B-2050		3.92	Heron Gliding Club	RNAS Yeovilton	23. 1.05
HDR	3866	467	Glaser-Dirks DG-300 Elan	3E95	RAFGSA R30	14. 3.92	R.Starmer	Bidford	6. 4.05
HDT	3868	291	Schempp-Hirth Discus BT	76/405		18. 3.92	J.D.J.Glossop and Partners	Gransden Lodge	7. 3.05
HDU	3869	HDU	SZD-51-1 Junior	B-1996		25. 3.92	Cambridge University Gliding Club	Gransden Lodge	26.11.05
HDV	3870	882	Schleicher ASW19B	19345	D-2876	20. 4.92	R.J.Hinley	Long Mynd	19. 4.05
HDW	3871	HDW	Centrair 101A Pégase	101A-0179	F-CGEE	21. 3.92	T.Head	Husbands Bosworth	21. 5.05
HDX	3872	A2	Rolladen-Schneider LS-7	7161		27. 3.92	G.Davison	Saltby	28. 5.05
HDY	3873	HDY	Schleicher K8B	8277	D-4094	24. 3.92	M.A Everett	Crowland	21. 9.05
HDZ	3874	W4	Schempp-Hirth Discus CS	078CS		7. 7.92	D.Keith and partners	Dunstable	22. 2.05
HEA	3875	HEA	Slingsby T.38 Grasshopper	'SSK/FF529'	(ex RAF)	4. 7.92	R.L.McLean	Rufforth	19.11.01
HEB	3876	HEB	Schleicher Ka6CR	6289	HB-773	6. 5.92	Devon and Somerset Gliding Club	North Hill	26. 2.05
HEC	3877	308	SZD-55-1	551191019		10. 5.92	G.P Davis	Nympsfield	6. 4.05
HED	3878	840	Schempp-Hirth Ventus A	17	D-2524	18. 4.92	S. Foster	RAF Keevil	9. 5.05

HEE	3879	316	Schempp-Hirth Discus B	292		19. 4.92	Booker Gliding Club	Booker	6. 1.06
			(Rebuild of BGA.3523 after accident 21.6.91 but see BGA.4047)						
HEF	3880	HEF	Glaser-Dirks DG-500 Elan Trainer	5E53T20		24. 5.92	Yorkshire Gliding Club	Sutton Bank	16. 4.05
HEG	3881	HEG	LAK-12 Lietuva	6206		27. 6.92	R.Kmita and Partners	Kirton-in-Lindsey	17. 3.04
HEH	3882	795	Rolladen-Schneider LS-7WL	7163	D-6078	24. 6.92	P.D.Chandler	Dunstable	5. 1.06
HEJ	3883	687	Schleicher ASW15B	15441	D-6871	5.92	A.Webb	Booker	12. 1.06
HEK	3884	HEK	SZD-51-1 Junior	B-2009	BGA.3893 (BGA.3884)	29. 5.94	Cambridge University Gliding Club	Gransden Lodge	3.11.05
HEL	3885	DZ	Rolladen-Schneider LS-4	4027	(BGA.3896) BGA.3885, D-6431	26. 5.92	J. Ballard	Long Mynd	30. 3.05
HEM	3886	473	Schempp-Hirth Discus CS	073CS		22. 5.92	G.G. Lee	Lasham	3. 3.05
HEN	3887	735	Schempp-Hirth Discus B	422		5. 6.92	A.R.Verity and Partners	Challock	6. 1.05
HEP	3888	HEP	SZD-50-3 Puchacz	B-2057		30. 5.92	Peterborough and Spalding Gliding Club	Crowland	26. 3.05
HEQ	3889	611	Schleicher ASW20L	20410	D-6747	6. 6.92	M.Chant	North Hill	1. 4.05
HER	3890	HER	Schleicher ASW19B	19240	F-CERR	1. 4.93	B.T.Spreckley	Le Blanc, France	12. 3.04
HES	3891	HES	Centrair 101A Pégase	101A-039	F-CFQF	13. 2.93	B.T.Spreckley	Le Blanc, France	12. 3.05
HET	3892	335	Rolladen-Schneider LS-6C	6263		26. 5.92	M.P.Brooks	Lasham	22. 2.05
HEV	3894	HEV	Schempp-Hirth HS.2 Cirrus	41	OO-ZXY (OO-ZOZ), D-0104	26. 5.92	R.R.Welch	Portmoak	17. 4.05
HEW	3895	486	Rolladen-Schneider LS-6-18W	6250		29. 4.92	R.M.Underhill	RAF Weston-on-the-Green	8. 4.05
			(Same c/n quoted for BGA.4952; above possibly rebuild retaining original c/n ?)						
HEY	3897		Hutter H-17A	02		6.92	J.M.Lee	Parham Park	17. 5.00
			(Built J.M.Lee) (Is NOT ex BGA.3661)				*(De-Registered 21. 8.03 - sold in the USA as N17HU 11.03)*		
HEZ	3898	607	Rolladen-Schneider LS-6C	6264		16. 7.92	J.E.Cruttenden	Lasham	4. 6.05
HFB	3900	HFB	Schleicher Ka6CR	6344Si	D-5825	13. 7.92	P.J.Galloway	Talgarth	26. 8.05
HFC	3901	WB924	Slingsby T.21B	587	WB924	25.7.92	T.Rose and partners	Dunstable	10. 7.05
HFE	3903	XN187	Slingsby T.21B	1166	XN187	23. 6.92	N.G.Oultram	Seighford	18. 4.05
HFF	3904	870	Schempp-Hirth Standard Cirrus	539	D-8916	14. 2.93	R.S.Morrisroe	Upwood	24. 3.05
HFG	3905	HFG	Slingsby T.21B	1165	XN186	28. 6.92	M.H.Simms	Shipdham	27. 8.05
HFH	3906	HFH	SZD-50-3 Puchacz	B-2059		4. 8.92	Trent Valley Gliding Club	Kirton-in-Lindsey	5. 5.05
HFL	3909	925/SSC	Schleicher ASH25	25147		18. 7.92	Scottish Gliding Union	Portmoak	29. 1.05
HFM	3910	747	Rolladen-Schneider LS-6C	6266		8. 7.92	F.J.Sheppard	Booker	1. 6.05
HFQ	3913	126	Rolladen-Schneider LS-6C	6260	(BGA.3908)	23. 6.92	B.McKnight	Gamston	30. 1.05
HFV	3918	F2	Schempp-Hirth Ventus B 16.6	204	D-5235	1.10.92	A.Cliffe	Camphill	26. 4.05
HFW	3919	HFW	Schleicher K8B	8108	HB-705	24. 9.92	Oxford Gliding Club	RAF Weston-on-the-Green	24. 2.05
HFX	3920	82	Schempp-Hirth Nimbus 4T	12		3. 7.92	R.Jones	Lasham	7. 4.05
HFY	3921	940	Schempp-Hirth Ventus CT	168/554	(BGA.3916) (BGA.3867)	21. 7.92	M.T.Day and D.J.Ellis	Lasham	16. 2.05
HFZ	3922	HFZ	Abbott-Baynes Scud I rep.				See SECTION 3, Part 1	Brooklands	
HGA	3923	HGA	Wassmer WA-26P Squale	43	F-CDUH	30. 3.93	E.C.Murgatroyd	Sackville Lodge, Riseley	15. 5.04
HGB	3924	509	Grob G.102 Astir CS	1356	D-7386	16.11.92	P.J.Hollamby and Partners	Lee-on-Solent	20. 9.04
			(Rebuilt with wings and components from RAFGGA.587)						
HGC	3925	HGC	Schleicher Ka7 Rhönadler	540	D-5689	6. 3.94	T.A.Joint	Lasham	27. 8.00
HGF	3928	HGF	Schleicher ASW15B	15264	D-2128	25. 8.92	D.Burton and Partners	Camphill	29. 3.05
HGG	3929	HGG	Schempp-Hirth HS.4 Standard Cirrus	362	HB-1172	31.12.92	P.W.Cooper	Seighford	7. 3.05
HGH	3930	HGH	Schleicher ASW19B	19351	D-1199	26. 8.92	A.Wood	Brent Tor	19. 1.04
HGK	3932	HGK	Schempp-Hirth Discus BT	96/435		30.10.92	C.T.Skeate	Lasham	17.11.05
HGL	3933	583	Schempp-Hirth Discus B	431		30. 7.92	D.Nicholls	Aston Down	19. 3.05
HGM	3934	HGM	Scheibe SF-27A Zugvogel V	6017	D-9351	26. 9.92	S.R.Algeo	Lyveden	11. 6.05
HGN	3935	D9	Schempp-Hirth Ventus CT	172/562		18. 9.92	D.J.Scholey	Lasham	11. 3.05
HGP	3936	HGP	Rolladen-Schneider LS-6C	6270		3.11.92	D.Elrington	Camphill	18. 2.05
HGQ	3937	637	LAK-12 Lietuva	6208		1.12.92	R.Perry	Dunstable	13. 6.05
HGR	3938	HGR	LAK-12 Lietuva	6186		22. 3.93	R.G.Stevens	Husbands Bosworth	31.10.05
HGS	3939	730	Schempp-Hirth Discus B	439		6.11.92	P.J.Bramley	Challock	25. 1.05
HGT	3940		FFA Diamant 16.5	40	HB-929	12. 4.94	R.W.Collins	Burn	29. 3.05
HGV	3942	HGV	Glaser-Dirks DG-500/22 Elan	5E70S11		26. 2.93	B.H.Bryce-Smith	Gransden Lodge	26. 1.05
HGW	3943	HGW	Centrair ASW20F	20102	F-CFFB	1. 1.93	C.Smith	Husbands Bosworth	5.12.05
HGX	3944	783	LAK-12 Lietuva	6201		3. 5.93	M.Jenks	Kingston Deverill	9. 1.06
HGY	3945		SZD-24C Foka	W-180	SP-2385	16.12.92	Peterborough and Spalding Gliding Club	Crowland	19. 6.05
HGZ	3946	502	Schempp-Hirth Discus BT	95/434		18.12.92	G.D.Coppin	Lasham	11. 3.05
HHA	3947	HHA	SZD-50-3 Puchacz	B-2058		18. 2.93	Derbyshire and Lancashire Gliding Club	Camphill	15. 2.04
HHC	3949	HHC	SZD-50-3 Puchacz	B-2080		16. 4.93	Derbyshire and Lancashire Gliding Club	Camphill	12. 5.05
							(Crashed on ridge Camphill, 26. 8.04)		
HHD	3950	HHD	SZD-51-1 Junior	B-2010		19. 3.93	Derbyshire and Lancashire Gliding Club	Camphill	6. 1.06
HHE	3951	HHE	SZD-51-1 Junior	B-2008		30. 6.93	Bowland Forest Gliding Club	Chipping	6. 4.05
HHG	3953	WT910	Slingsby T.31B	705	WT910	9. 1.93	P.Wickwar and Partner	Bruntingthorpe	30. 5.97*
							(Under restoration 2004)		
HHH	3954	963	Rolladen-Schneider LS-6C	6289		11.12.92	B.R.Wise	Booker	31.12.05
HHJ	3955	X97	Glaser-Dirks DG-500/22 Elan	5E71S12		9. 2.93	N.A.Kelly	Bicester	6. 4.05
HHK	3956	838	Schleicher ASW19B	19384	ZD661 BGA.2897	14. 3.93	A.J.Peters Syndicate	Lasham	8. 5.05
HHL	3957	HHL	Schleicher Ka7 Rhönadler	446	OY-XCK D-5619	30. 7.93	Lincolnshire Gliding Club	Strubby	2. 9.04
							"Buttercup" (Crashed Strubby 26. 5.04)		
HHM	3958	HHM	LAK-12 Lietuva	6195		9. 8.93	D. Martin	Bordeaux, France	8. 3.05
HHN	3959	979	Schempp-Hirth Ventus B/16.6	205	RAFGSA.R27	6. 2.93	N.A.C.Norman	Feshiebridge	10. 5.05
			(Build No. V-204)						
HHP	3960	KL	Schempp-Hirth Discus B	399	SE-UKL	12. 2.93	D.J.Knowles	Easterton	20.10.05
HHQ	3961	97Z	Schempp-Hirth Discus BT	106/453		12. 2.93	P.Tratt and Partners	Ringmer	3. 4.05
HHR	3962	100	SZD-55-1	551191020		18. 4.93	R.T.Starling	Nympsfield	31. 1.05
HHS	3963	HHS	Schleicher ASW20	20008	SE-TTU	3. 3.93	P.J.Rocks	Kirton-in-Lindsey	12. 9.04
HHT	3964	855	Rolladen-Schneider LS-6C	6292		18. 7.93	G.O.Humphries	Kingston Deverill	9. 3.05

HHU	3965	445	Rolladen-Schneider LS-6C	6296		15. 2.93	J.S.Weston	Bellarena	2. 1.06
HHW	3967	237	LAK-12 Lietuva	6212		19. 3.93	A.J.Dibdin	Dunstable	17. 2.05
HHX	3968	HHX	Wassmer WA-26P Squale	14	F-CDQJ	27. 2.93	M.H.Gagg	Chauvigny, France	25. 2.05
HHY	3969	RT	Glasflugel H.201B Standard Libelle	119	SE-TIU	1. 5.93	R.Tietma	Husbands Bosworth	14. 3.05
HJA	3971	HJA	VFW-Fokker FK-3	0008	D-0409	9. 8.93	M.A Johnson and Partners	Sackville Lodge, Riseley	26. 3.05
HJC	3973	25	Rolladen-Schneider LS-6C	6290		17. 3.93	F.J Davies and I.C.Woodhouse	Husbands Bosworth	18.12.05
HJD	3974	HJD	Schleicher Ka6E	4141	D-....	4. 2.94	J.P.Stafford	Snitterfield	16. 7.04
					OH-505/SE-TFM				
HJE	3975	505	Schleicher K8B	8259	(BGA.3926)	16. 4.93	Denbigh Gliding Club	Lleweni Parc	28.11.04
					RAFGGA.505, RAFGGA.971				
HJF	3976	245	Rolladen-Schneider LS-6C	6291		29. 4.93	J.L.Bridge	Gransden Lodge	25. 2.05
HJH	3978	HJH	Schempp-Hirth Discus BT	65/391	N224WT	22. 4.93	P.J.Goulthorpe	Crowland	5.12.03
							(Crashed at Wyton 31. 8.03)		
HJJ	3979		Slingsby T.38 Grasshopper	797	WZ816	R	J.Wilkins *(On rebuild 2000)*	Redhill	
HJK	3980	HJK	Schleicher Ka7 Rhönadler	795	RAFGSA.R5	14. 6.93	Leeds University Gliding Club	Dishforth	1. 5.03
					D-5791				
HJL	3981	306	Schempp-Hirth Discus BT	105/451		3. 5.93	A.R.MacGregor	Kingston Deverill	7. 1.05
HJM	3982	HJM	Hutter H.28-III rep	ED.02		25. 5.93	E.R.Duffin	Nympsfield	12. 6.99
			(Built E.R.Duffin)				*(Being refurbished)*		
HJN	3983	HJN	Grob Standard Cirrus	440G	HB-1206	2. 6.93	D.F Marlow	Aston Down	24. 8.05
HJR	3986	B9	Glasflugel H.201B Standard Libelle	102	SE-TIO	26. 5.95	B.Magnani	Wormingford	8. 4.05
HJT	3988	292	Centrair ASW20F	20115	F-CFFL	3. 6.93	C.I Roberts and Partners	Booker	7. 6.05
HJV	3990	HJV	Grob G.102 Astir CS	1007	D-7000	7. 6.93	Cotswold Gliding Club	Aston Down	22.12.05
HJX	3991	203	Rolladen-Schneider LS-6C	6271		28. 5.93	R.S.Hatwell and M.Haynes	Tibenham	3. 3.05
HJY	3992	HJY	Schempp-Hirth Standard Cirrus	459	HB-1207	23. 6.93	A.Hogbin	Currock Hill	21. 7.05
HJZ	3993	865	Schleicher ASW15B	15190	OH-408	15. 5.94	R.R.Beezer	Camphill	29. 9.05
HKA	3994	860	Schempp-Hirth Discus CS	120CS		20. 5.93	M. and R.Weaver	Lasham	16.12.05
HKB	3995	HKB	Grob G.102 Astir CS77	1658	D-7491	26.10.93	R. Peach	RAF Keevil	20. 2.05
HKC	3996	HKC	Grob Standard Cirrus	520G	D-3268	28. 8.93	Welland Gliding Club	Lyveden	22. 4.05
HKD	3997	C34	Grob Standard Cirrus	576G	F-CEMF	7. 7.93	J.A.Clark	Talgarth	24. 5.05
HKJ	4002		Penrose Pegasus 2	001		7.93	J.M.Lee	Flixton	15. 9.98
			(Built J.M.Lee)				*(On display at Norfolk & Suffolk Aviation Museum4.04).*		
HKK	4003	HKK	Schleicher K8B	8886	D-0866	16. 1.94	R.K.Lashly	Drumshade	18. 4.05
HKL	4004	919	Schempp-Hirth Discus bT	120/476		2.94	M.A.Thorne	Kingston Deverill	17. 3.05
HKM	4005	HKM	Grob G.102 Astir CS Jeans	2108	D-7636	3.94	G.Wright	Wormingford	31. 3.05
HKP	4007	HKP	Schleicher ASK23B	23100	D-2935	9. 8.93	Midland Gliding Club	Long Mynd	10. 1.06
					HB-1935				
HKQ	4008	970	Schempp-Hirth Nimbus 3DT	63		7. 8.93	R.I.Hey and Syndicate	Nympsfield	23.11.05
HKR	4009	985	Jastreb Standard Cirrus G/81	276	OH-663	15.10.93	S.Crozier	Strubby	3. 4.05
HKS	4010	HKS	Jastreb Standard Cirrus G/81	361	SE-TZS	11.11.93	P.Uden	Gamston	17. 4.05
HKT	4011	HKT	Schleicher ASW19	19168	D-7958	4.10.93	P.Clayton	Burn	7. 9.05
			(C/n conflicts with OE-5174 but believed correct)						
HKU	4012	C29	Grob Standard Cirrus	513G	F-CEMA	5.12.93	T.J Wheeler and Partner	Lyveden	24. 5.05
HKV	4013	HKV	Scheibe Zugvogel IIIA	1034	D-8294	6.10.93	Dartmoor Gliding Club	Brentor	31. 5.05
HKW	4014	HKW	Marco J-5	009	G-BSBO	2. 6.94	G.K Owen	Bidford	19. 3.04
			(Built D.Austin - regd with c/n 001)				*"Flying Penguin II"*		
HKX	4015	HKX	Rolladen-Schneider LS-4B	4933		18.12.93	D.J Hughes	Long Mynd	19. 2.05
HKY	4016	JA	Schempp-Hirth Discus B	461		14.10.93	J.G Arnold	RAF Keevil	3. 2.05
HKZ	4017	P31	CARMAM JP-15/36AR Aiglon	31	F-CFGA	27. 9.93	R.Borthwick	Milfield	9.11.04
HLB	4019	365	Rolladen-Schneider LS-4B	4935		27. 4.94	E.G.Leach	Wormingford	1. 9.05
HLC	4020	HLC	Pilatus B4 PC-11AF	177	SE-UFX	10. 3.94	E.A.Lockhart	Lasham	11. 5.05
					OH-455				
HLG	4024	HLG	Schleicher ASK21	21596		1. 4.95	London Gliding Club	Dunstable	18. 2.04
HLH	4025	HLH	Schleicher K8B	8637	RAFGGA.569	26. 2.94	R.Das	Usk	18.12.05
					D-5691				
HLK	4027	HLK	Glasflugel H.301 Libelle	85	SE-TFS	11. 4.95	E.Sweetland	Dunstable	21. 4.05
HLM	4029	819	Schleicher ASW19B	19269	OH-538	10. 2.94	R.A.Colbeck	Booker	24. 5.05
HLN	4030	805	Schempp-Hirth Discus CS	143CS		18. 1.94	Portsmouth Naval Gliding Club	Lee-on-Solent	23. 4.05
HLP	4031	HLP	Schleicher ASK21	21597		24. 3.94	Yorkshire Gliding Club	Sutton Bank	25. 3.05
HLQ	4032	381	Schempp-Hirth Discus bT	128/490		22.12.93	J.F.Goudie	Portmoak	13.12.05
HLR	4033	XE786	Slingsby T31B	899	XE786	18.12.93	D.Thomson	Arbroath	15. 4.04
HLS	4034	V5	Schempp-Hirth Discus B	114	RAFGSA.R11	28. 1.94	R.A.Lennard	Dunstable	13. 1.06
HLU	4036	HLU	Scheibe SF-27A Zugvogel V	6101	SE-TGP	22. 2.94	G.Dennis	Edgehill	2. 7.05
HLW	4038	"HLV"	Schleicher ASW19B	19325	D-8799	3. 4.94	P.Belcher and partner	Gransden Lodge	29. 3.05
HLX	4039	260	Schleicher ASH25	25124	D-3988	27. 2.94	P.Pozerskis	Husbands Bosworth	28.11.05
HLY	4040	Z45	Schempp-Hirth Discus CS	161CS		15. 6.94	S.Clark	North Hill	24. 3.05
HLZ	4041	359	Schleicher ASW20BL	20951	D-8188	19. 3.94	T Vines	Dunstable	5. 5.05
HMA	4042	HMA	SZD-51-1 Junior	B-2132		30. 3.94	The Soaring Centre	Husbands Bosworth	23. 2.05
HMB	4043	HMB	Glaser-Dirks DG-300 Elan	3E105	D-4676	31. 3.94	A.D.Langlands	Edgehill	26. 1.05
HMG	4044	HMG	ICA IS-28B2	353	HA-....	20. 4.94	A.Sutton and Partners	Husbands Bosworth	12. 4.05
HMH	4045	S82	Schleicher K8B	5	D-5735	15. 4.94	Shenington Gliding Club	Edgehill	22. 7.03
			(Officially regd as c/n 2330)						
HMK	4046	122	Rolladen-Schneider LS-6-18W	6324	D-1245	18. 3.94	A.S.Decloux	Gransden Lodge	18. 1.06
HML	4047	38	Schempp-Hirth Discus CS	114CS	OO-ZTU	16. 3.94	M.E.Hahnefeld	Parham Park	5.11.05
			(Composite with wings from BGA.3879)						
HMM	4048	514	Glasflugel H.304B	322	SE-UGZ	31. 3.94	S.Hogg and Partners	Husbands Bosworth	3. 5.05
					D-1005				
HMP	4050	297	Schempp-Hirth Discus B	497		13. 3.94	P.Charatan	Challock	14. 3.05
HMQ	4051	364	Schempp-Hirth Discus CS	099CS	D-7160	8. 3.94	S.A.Crabb	Edgehill	30. 4.04
HMS	4053	HMS	Glaser-Dirks DG-100	40	D-2579	8. 4.94	B.Walton-Knight	Shobdon	18.11.03
HMT	4054	380	Glasflugel H.303 Mosquito B	153	F-CEDY	20. 3.94	R.J.Pirie and Partners	North Hill	7. 4.05

HMU	4055	HMU	CARMAM JP-15/36AR Aiglon	22	F-CETT	7. 5.94	J.R.Holmes	Kingston Deverill	30. 1.05
HMV	4056	N26	Schleicher ASK13	13177	D-0268	9. 5.94	Portsmouth Naval Gliding Club	Lee-on-Solent	9. 4.05
HMX	4058	PZ	Rolladen-Schneider LS.4B	4230	OO-ZNN	14. 5.94	J.Hall	Nympsfield	9. 4.05
					F-CEIO				
HMY	4059	HMY	Schempp-Hirth HS.4 Standard Cirrus	121	HB-1034	29. 4.94	C.P.Woodcock and B.J.Thomas		
								RAF Weston on the Green	26. 1.05
HMZ	4060	469	Federov Me-7 Mechta	M.004		4.94	R.Ellis	Rufforth	12. 2.96*
							(Major components stored in Wales 2003)		
HNA	4061	HNA	Glaser-Dirks DG-500/20 Elan	5E128W3		14. 7.94	J.P.Boneham	Camphill	20. 8.05
HNB	4062	563	Schempp-Hirth HS.6 Janus C	215	D-4149	16. 4.94	C.M.Fox	Sleap	16. 2.05
HNC	4063	HNC	Schleicher ASW19B	19297	OH-515	18. 4.94	J.Lawn	Tibenham	16. 3.05
HND	4064	HND	Scheibe Zugvogel IIIA	1044	HB-735	23. 5.94	D.Spillane	Husbands Bosworth	23. 5.04
					D-9119				
HNE	4065	708	Schempp-Hirth HS.5 Nimbus 2B	91	D-2786	10. 5.94	S.Noad and Partners	Challock	23. 6.05
HNF	4066	315	Schempp-Hirth Duo Discus	11		11. 5.94	Booker Gliding Club	Booker	5.12.05
HNG	4067	HNG	Schleicher K8B	132/59	D-8378	5. 5.94	Bidford Gliding Club *(Also as "8378")*	Bidford	5. 8.05
HNH	4068	Z99	Schempp-Hirth HS.5 Nimbus 2C	187	D-2830	31. 3.94	R.J.Hart	Tibenham	2. 4.05
HNJ	4069	HNJ	Schleicher Ka7 Rhönadler	7031	D-1667	6. 5.94	N.J.Orchard-Armitage	Waldershare Park	5. 4.05
					RAFGGA??, D-6233				
HNK	4070	HNK	SZD-51-1 Junior	B-1496	SP-3299	20. 5.94	Booker Gliding Club	Booker	11. 1.06
					(SP-3290)				
HNM	4072	HNM	Jastreb Standard Cirrus G/81	360	SE-TZT	2. 7.94	J.Phillips	Usk	13. 4.05
HNN	4073	HNN	Schempp-Hirth Duo Discus	21		15. 9.94	B.T.Spreckley	Le Blanc	12. 3.05
HNS	4077	XN185	Slingsby T.21B	1164	8942M	21. 6.94	B.Walker	RAF Syerston	12. 4.04
					XN185		*(Donated to RAF Museum)*		
HNT	4078	105	Schleicher ASW15	15167	F-CEAQ	27. 4.94	M.C.Moulang	Challock	5. 5.05
HNU	4079	48	Schempp-Hirth Nimbus 4DT	3/5	D-KHIA	25. 5.94	D.E.Findon	Bidford	22. 3.05
HNV	4080	692	Rolladen-Schneider LS-4B	4960		11.12.94	P.W.Armstrong	Kirton-in-Lindsey	8. 5.05
HNW	4081	220	Schempp-Hirth Duo Discus	25		21.11.94	J.L.Birch	Dunstable	28. 2.05
HNX	4082	585	Rolladen-Schneider LS-4B	4937		6. 7.94	C.S.Crocker	Long Mynd	30. 3.05
HNY	4083	HNY	Centrair 101A Pégase	101A-020	F-CFRP	12.10.95	M.O.Breen	Booker	10. 8.05
HNZ	4084	RY	Centrair 101A Pégase	101A-032	F-CFRY	14. 7.94	R.H.Partington	Milfield	11. 1.05
HPA	4085		Issoire E78 Silene	4	F-CFEA	R	T.M.Perkins	Dunstable	
HPC	4087	HPC	Schleicher ASW20CL	20787	D-3424	20 .7.94	D.R.Sutton	Pocklington	13. 5.05
HPD	4088	62	Rolladen-Schneider LS-6C-18	6331	D-1054	24.10.94	S.A.Hughes and partner	Sutton Bank	27.11.04
HPE	4089	HPE	Schleicher ASK13	13510	D-3992	2.10.94	Nottingham University Gliding Club	RAF Syerston	11. 1.05
HPF	4090	HPH	SZD-9 bis Bocian 1E	P-740	OH-508	3. 8.94	Bath, Wilts and North Dorset Gliding Club		
							(Overshot landing 24.4.00 - stored 6.04)	Kingston Deverill	23. 2.01
HPG	4091	HPG	Maupin Woodstock	551	VR-HKI	8.94	J.M.Stockwell	Perranporth	27. 5.05
			(Built J.M.Stockwell)						
HPH	4092	73	Schempp-Hirth Discus CS	174CS		21. 9.94	M.T.Burton	Ridgewell	19. 9.03
							(Crashed after collision with ASW-27 BGA.4338 near Lasham 4. 9.03)		
HPJ	4093	HPJ	Edgley EA9 Optimist	EA9/001		5.94	Edgley Aeronautics Ltd	Lasham	11. 8.01
			(C/n reported by John Edgley as "004")						
HPL	4095	655	Rolladen-Schneider LS-4B	4959	(BGA.4071)	28. 7.94	P.G.Mellor	Booker	31. 3.05
HPM	4096	HPM	Grob G.102 Astir CS	1072	D-3304	21.11.94	P.Turner and Partners	North Hill	13. 3.05
HPP	4098		Slingsby T.38 Grasshopper	863	XA230	5. 2.95	S.Butler	Gransden Lodge	6. 6.05
HPQ	4099	HPQ	Schleicher Ka6CR	6200	D-1933	5.10.94	M.Ewer	Crowland	24. 3.05
HPR	4100	150	Schempp-Hirth Discus B	532		20. 2.95	J.S.McCullagh	Lasham	18.10.05
HPT	4102	HPT	Federov Me-7 Mechta	M.006		29. 3.96	A E.Griffiths	Long Mynd	4. 3.05
HPV	4104	HPV	Schleicher ASK21	21608		13.10.94	Scottish Gliding Union Ltd	Portmoak	19. 1.05
HPW	4105	HPW	Schleicher ASK21	21609		25.11.94	Scottish Gliding Union Ltd	Portmoak	9. 4.05
HPX	4106	693	Schempp-Hirth Discus CS	177CS		12. 4.95	P.C.Whitmore and M.A.Whitehead	Gransden Lodge	11. 2.05
HQB	4110	HQB	Slingsby T.21B	602	WB935	1.10.94	C.E.Anson	Hahnweide, Germany	13. 2.05
			(Officially regd with c/n 1099 which is a corruption of fuselage no. SSK/FF/1099)						
HQD	4112	A20	Schleicher ASW20	20288	SE-ULA	1.11.94	R.Hodge	Dunstable	12.12.04
					OH-548				
HQE	4113	3D	Schempp-Hirth Duo Discus	29		19. 2.95	3D Syndicate	Aboyne	21. 3.05
HQF	4114	HQF	CARMAM M.100S Mésange	26	F-CCSO	3.11.94	R.E.Stokes	Long Mynd	12.11.04
HQG	4115	HQG	LAK-12 Lietuva	6222		30. 4.95	J.M. Pursey	Brentor	1. 6.05
HQH	4116	HQH	Schleicher Ka4 Rhönlerche II	3072/Br	(BGA.4097)	5.95	D.Fulchiron	Bellechasse, France	24. 7.03
					HB-877		*(D-4116 reserved 12.02, not yet de-registered in UK)*		
HQJ	4117	762	Schempp-Hirth Discus B	336	D-1762	10. 2.95	D.G.Lingafelter	Lasham	5. 2.05
HQK	4118	S2	Schleicher ASW20CL	20854	D-3366	18. 1.95	S.D.Minson	North Hill	9. 4.05
HQL	4119	LS6	Rolladen-Schneider LS-6C-18W	6352	D-0794	3. 3.95	D.P.Masson	Lasham	5. 5.05
HQM	4120	HQM	Schempp-Hirth Discus B	44	RAFGSA.R10	23. 1.95	Cambridge University Gliding Club	Gransden Lodge	9.12.05
HQN	4121	D64	Schempp-Hirth HS.5 Nimbus 2B	139	D-6494	29. 1.95	A.J.Nurse	Nympsfield	10. 3.05
HQR	4123	19	Schempp-Hirth Discus B	531		26. 4.95	British Gliding Association	RAF Bicester	23. 2.05
HQS	4124	HQS	Grob G.103 Twin Astir	3155	OO-ZEG	26. 2.95	Essex and Suffolk Gliding Club	Wormingford	27. 5.05
HQT	4125	A77	Grob G.102 Astir CS77	1678	RAFGGA.561	12. 2.95	Crown Services Gliding Club	Lasham	2. 7.04
HQV	4127	HQV	SZD-51-1 Junior	B-2139		20. 3.95	The Soaring Centre	Husbands Bosworth	25. 6.05
HQW	4128	W3	Schempp-Hirth Discus B	538		9. 3.95	P. Turner	Halesland	27.12.05
HQX	4129	HQX	Schleicher ASW15B	15326	D-2315	13. 3.95	P. Ridgill	Upwood	20.11.04
HQY	4130	HQY	Schempp-Hirth HS.7 Mini Nimbus C	113	D-3364	14. 3.95	A.H Sparrow	Lasham	11. 2.05
HQZ	4131	U2	Rolladen-Schneider LS-6C-18W	6353	D-1486	6. 3.95	D. Champion	Lasham	26. 5.05
HRA	4132	N19	Grob G.102 Astir CS	1109	D-4169	21. 4.95	Portsmouth Naval Gliding Club	Lee-on-Solent	17. 4.04
HRB	4133	HRB	LAK-12 Lietuva	6223		3.95	J.E.Neville	Aboyne	18. 1.05
HRC	4134	390	Glaser-Dirks DG-500-20 Elan Trainer			15. 5.95	D.Rhys-Jones	Parham Park	19. 2.04
				5E136W5			*(Crashed Parham Park 18.10.03)*		
HRD	4135		Slingsby T.21B	634	WB973	18. 3.95	C.Langenau	Aukrug, Germany	18. 6.05
HRE	4136	HRE	Schleicher Ka6CR	572	D-9326	26. 4.95	R.J.Playle	Edgehill	19. 6.04

HRF	4137	JM	Schleicher Ka6E	4272	OO-ZJR	12. 5.95	J.F.Morris	Gransden Lodge	7. 3.05
					D-0165				
HRG	4138	HRG	SZD-51-1 Junior	B-2013		25. 4.95	Scottish Gliding Union Ltd	Portmoak	18. 1.05
							(Crashed Portmoak 20.4.04)		
HRJ	4139	HRJ	Schleicher K8B	8093Ei	D-5048	26. 4.95	Shenington Gliding Club	Edgehill	30. 6.05
HRK	4140	HRK	Centrair 101A Pégase	101A-048	F-CFQJ	16. 5.95	P. Bushill	Dunstable	1. 6.05
HRL	4141	HRL	Schempp-Hirth HS.4 Standard Cirrus	525	D-3099	26. 4.95	M.Harbour	Camphill	3. 4.05
HRN	4143	HRN	Schleicher ASK18	18026	HB-1308	7. 4.95	Stratford-upon-Avon Gliding Club	Snitterfield	19. 2.05
HRP	4144	HRP	SZD-51-1 Junior	B-1807	SP-3442	19. 5.95	Wolds Gliding Club	Pocklington	11. 3.04
							(Spun in Pocklington 1.8.03)		
HRQ	4145	169	Schempp-Hirth HS.7 Mini Nimbus	C123	(BGA.4122)	17.4.95	C.Buzzard	Husbands Bosworth	1. 4.05
					SE-TVB				
HRR	4146	D70	Schleicher ASK21	21033	D-7083	2. 2.95	Lakes Gliding Club	Walney Island	19. 1.05
HRS	4147	B33	Schempp-Hirth Discus CS	100CS	D-5100	1. 5.95	M.E.Hughes	Husbands Bosworth	24. 4.05
HRT	4148	HRT	Schleicher K8B	8390A	D-5599	9. 3.96	Heron Gliding Club	RNAS Yeovilton	20. 6.04
HRV	4150		SZD-55-1	551195076		2.11.95	R.W.Southworth	Warsaw, Poland	24.11.01
							(De-Registered 7. 8.03, sold in Germany as D-2598)		
HRW	4151	802	Schempp-Hirth Duo Discus	43	(BGA.4160)	22. 6.95	A.J.Davis	Nympsfield	22.11.05
					BGA.4151				
HRX	4152	P5	Schempp-Hirth Discus A	545		24. 5.95	N.Worrell	Lasham	19. 2.05
HRY	4153	L8	Rolladen-Schneider LS-6C-18W	6362		2. 6.95	F.K.Russell	Dunstable	29. 1.05
HSA	4155	A1	Rolladen-Schneider LS-6C-18W	6361		8. 8.95	D.A.Benton	Snitterfield	26. 3.05
HSB	4156	HSB	Glaser-Dirks DG-300 Elan	3E461		20. 7.95	J.S.Foster	Parham Park	11. 3.05
HSC	4157	99	SZD-50-3 Puchacz	B-2079		15. 8.95	British Gliding Association	RAF Bicester	1. 3.05
HSD	4158	D15	Schempp-Hirth Discus B	258/1	(BGA.4142)	19. 6.95	J.R.Reed	Dunstable	5. 4.05
			(Rebuild of BGA.3406 c/n 258 w/o 26.8.94)						
HSE	4159	HSE	Grob G.102 Astir CS77	1635	RAFGSA.R68	2. 9.95	R.G.Tait	Easterton	28. 5.05
					RAFGSA.548				
HSG	4161	HSG	Scheibe SF-27A Zugvogel V	1705/E	D-7827	11. 7.95	Welland Gliding Club	Lyveden	30. 7.05
					OE-0827				
HSH	4162	HSH	Scheibe Zugvogel IIIB	7/1041	D-6558	12. 7.95	Zugvogel Syndicate *"Brigitta"*	Wormingford	18.12.05
HSJ	4163	D54	Schempp-Hirth Discus B	546		3. 7.95	A.A.Jenkins	Dishforth	1. 3.05
HSK	4164	751	Schleicher ASW20CL	20827	D-3499	18. 7.95	C. Ramshorn	Gransden Lodge	5. 4.05
HSL	4165	213	Schempp-Hirth Ventus 2C	1/2	(BGA.4154)	4. 7.95	H.G.Woodsend	Aston Down	1. 3.05
			(Incomplete airframe assembled by Southern Sailplanes)				*(Crashed at Nympsfield 7. 8.04)*		
HSM	4166	HSM	Schleicher ASK13	13145	D-0168	18. 7.95	Stratford-upon-Avon Gliding Club	Snitterfield	31. 5.05
HSN	4167	HSN	Schleicher Ka6CR	6218	OO-ZZF	1. 8.95	Needwood Forest Gliding Club	Cross Hayes	20. 2.05
					D-8546				
HSP	4168	385	Schempp-Hirth HS.6 Janus C	112	RAFGSA.R1	9. 5.04	H.A. Torode	Lasham	8. 5.05
					BGA.2723, /D-7013				
HSQ	4169	493	Schempp-Hirth Discus B	99	D-2943	8. 8.95	Midland Gliding Club	Long Mynd	9. 1.06
HSR	4170	313	LAK-12 Lietuva	6178		3. 8.95	G.Forster	Currock Hill	19. 3.05
HSU	4173	K18	Schleicher ASK18	18025	AGA.16	R	R.C.Martin *(Being refurbished)*	Not known	
HSV	4174	HSV	Schempp-Hirth HS.4 Standard Cirrus	195	D-0785	14. 3.96	C.D.Morrow	Rivar Hill	22. 2.05
HSW	4175	895	Schempp-Hirth Duo Discus	48		25. 8.95	C.R.Simpson	Husbands Bosworth	28. 3.05
HSX	4176	HSX	Scheibe SF-27A Zugvogel V	6031	SE-TDT	26. 9.95	C.Smith	Wormingford	9. 5.04
HSY	4177	A15	Pilatus B4 PC-11	050	SE-UFF	22. 9.95	A.L.Dennis	Walney Island	19. 9.00
					OH-431				
HSZ	4178	497	Rolladen-Schneider LS-8a	8030		9.95	A.Green	Dunstable	9.10.05
HTA	4179		Centrair C-201B1 Marianne	201-014	F-CGMM	21.10.95	E.Crooks	Kirton-in-Lindsey	31.10.00
							(Undershot in field landing near Denbigh 27. 9.00)		
HTB	4180	HTB	Schempp-Hirth HS.6 Janus A	007	D-3114	27. 7.95	J.B.Madison	Kirton in Lindsey	29. 3.05
HTC	4181	HTC	Schleicher ASW15B	15188	OE-0930	26.10.95	J. Klunder	Camphill	17. 6.05
HTD	4182	VMC	Grob G.102 Astir CS	1012	D-6508	20. 5.96	G.V.McKirdy	Edgehill	15. 4.00
HTE	4183	HTE	Grob G.102 Astir CS77	1716	RAFGSA.R82	23.11.95	J.R.Whittington	Challock	9. 4.05
					RAFGSA.882				
HTF	4184	HTF	LAK-12 Lietuva	6180		1. 6.96	D.Stidwell	Cross Hayes	7. 4.05
HTG	4185	HTG	Grob G.102 Astir CS	1510	RAFGSA.R59 (2)	30.10.95	Trent Valley Gliding Club	Kirton-in-Lindsey	16. 8.05
					RAFGSA.R69 (2), RAFGSA.519				
HTH	4186	C4	Schempp-Hirth Janus CT	185/2	N137DB	13.11.95	D. Bramwell	Thame	5. 7.05
					D-KHIE				
HTJ	4187	HJT	Schleicher ASK13	13125	D-6048	2.12.95	Ulster Gliding Club	Bellarena	27. 3.05
HTL	4189	LS	Rolladen-Schneider LS-8-18	8038	D-3156	10.95	A.and L.Wells	Hinton-in-the-Hedges	23.12.05
HTM	4190	Z8	Rolladen-Schneider LS-8-18	8036		10.95	W.Payton	Sutton Bank	6. 4.05
HTN	4191	S22	Schleicher ASW22	22013	ZS-GLN	16. 4.96	J.B.Giddins	Hinton-in-the-Hedges	25.10.05
HTP	4192	L58	Rolladen-Schneider LS-8-18	8039	D-3175	11.95	R.A.Browne	Crowland	25. 2.05
HTR	4194	HTR	Grob G.102 Astir CS	1190	D-7307	20.11.95	I.Wright	Kingston Deverill	11. 3.05
HTS	4195	H8	Rolladen-Schneider LS-8-18	8040		22.11.95	A.Head and Partner	Gransden Lodge	3. 4.05
HTT	4196	HTT	Schleicher ASW20CL	20627	D-2410	24.11.95	J.Potter	Camphill	25. 3.05
HTU	4197	HTU	Schempp-Hirth HS.2 Cirrus	88	D-0478	29.12.95	R.Salmon	Burn	7. 4.05
HTV	4198	HTV	Schleicher ASK21	21624	D-8355	3. 3.96	Cambridge University Gliding Club	Gransden Lodge	24. 3.05
HTX	4200	900	Schleicher ASW20	20239	D-3180	3. 3.96	C.Ramshorn	Gransden Lodge	18. 2.02
HTY	4201	HTY	LET L-13 Blanik	026318	LY-GDT	1. 4.96	North Devon Gliding Club	Eaglescott	20. 5.05
					DOSAAF				
HUA	4203	HUA	Schleicher ASW19	19091	D-3840	29.12.95	S.Glazzard	Long Mynd	25. 5.05
HUB	4204	HUB	SZD-48-3 Jantar Standard	B-1527	DOSAAF	17. 3.96	C.F.Sermanni	Portmoak	18. 1.05
HUC	4205	HUC	Schempp-Hirth Janus C	170	(BGA.4188)	24. 2.96	C.W.Price	Wormingford	12.12.04
					D-3189				
HUD	4206	HUD	Schleicher ASK13	13018	D-9203	21. 2.96	London Gliding Club	Dunstable	19. 5.05
HUE	4207	N5	Schleicher ASW27	27022		10.96	E.H.Downham	Dunstable	10. 3.05
HUF	4208	HUF	Schleicher ASK13	13109	OO-ZWE	10. 3.96	Welland Gliding Club	Lyveden	5. 2.05HUH

	4210	D31	Schempp-Hirth HS.6 Janus	15	D-3116	9. 5.96	B.A.Fairston	Husbands Bosworth	16. 3.05
HUJ	4211	X70	Centrair ASW20F	20170	F-CFLY	5. 3.96	M.Manning	Rattlesden	11. 5.05
HUK	4212	HUK	Schleicher Ka6CR	6385	SE-TCN	15. 4.96	T.J Donovan and Partner	Lyveden	9. 8.04
HUL	4213	624	Schempp-Hirth HS.2 Cirrus	V3	HB-900	26. 2.96	I.Ashton and Partners	Chipping	1. 5.05
HUM	4214	241	Rolladen-Schneider LS-6C	6267	OO-ZXS D-4350	25. 3.96	A.Hall	Lasham	15. 4.05
HUN	4215	HUN	Grob G.102 Astir CS Jeans	2089	D-7531	27. 2.96	D P.Manchett	Lleweni Parc	18.12.05
HUP	4216	170	Schempp-Hirth Ventus 2CT	4/11		16. 2.96	C G Corbett	Dunstable	26. 2.05
HUQ	4217	HUQ	Federov Me-7 Mechta	007		21. 6.96	J S Fielden	Brent Tor	20. 2.04
HUR	4218	HUR	Schempp-Hirth HS.2 Cirrus	12	HB-927	20. 4.96	M.Rossiter	Talgarth	23. 4.05
HUS	4219	HUS	Scheibe SF-27A Zugvogel V	6010	D-1035	6. 8.97	G.Jackson	Edgehill	22. 3.05
HUT	4220	HUT	Centrair ASW20F	20187	F-CEUQ	14. 4.96	G.A MacFadyen	Nympsfield	13. 4.05
HUU	4221	HUU	Schleicher ASK13	13527AB	D-7506 D-8945	15. 4.96	Upward Bound Trust	Thame	16. 4.05
HUV	4222	64	Rolladen-Schneider LS-8-18	8056	D-3823	4.96	C.P.Jeffery	Gransden Lodge	23. 4.05
HUW	4223	S8	Rolladen-Schneider LS-8A	8058		3.96	S.E.Bort	Challock	28. 3.05
HUY	4225	HUY	Schempp-Hirth Ventus CT	84/329	D-KILZ	3. 4.96	M A Challans	Lasham	16. 1.06
HUZ	4226	200	Schempp-Hirth Discus BT	158/559		2. 3.96	Smith & Farr Syndicate	Keevil	26. 7.05
HVA	4227	HVA	SZD-59 Acro	B-2169		2. 5.96	T.Williams	Lasham	17. 6.03
							(Sold as N459TW 14.11.02, not yet de-registered in UK)		
HVB	4228	HVB	Slingsby T.31B	850	(BGA.3249) XA308	27. 4.96	M.Hoogenbosch *"Top Less"*	Hilversum, The Netherlands	21. 5.05
HVC	4229	HVC	Slingsby T.38 Grasshopper	766	WZ770	4. 5.01	J.F.Forster	Hilversum, The Netherlands	21. 9.05
HVD	4230	304	SZD-55-1	551193052		23. 4.96	M.Drecka	Booker	15. 3.05
HVE	4231	W54	Schempp-Hirth Ventus 2CT	8/19	D-KHIA	28. 3.96	Glyndwr Soaring Club	Lleweni Parc	9. 5.05
HVF	4232	B8	Rolladen-Schneider LS-8-18	8059	D-1683	4.96	B.Scougall	Aboyne	11.11.05
HVG	4233	RP1	Schleicher ASK21	21062	D-2606	17. 4.96	Rattlesden Gliding Club	Rattlesden	14. 1.05
							(Crashed on take-off Rattlesden 18. 2.04)		
HVH	4234	HVH	Pilatus B4 PC-11AF	067	D-2156	21. 5.96	London Gliding Club	Dunstable	27. 5.05
HVJ	4235	962	Scheibe SF-27A Zugvogel V	6012	OE-0762	1. 6.96	C.P.Bleaden	Kirton-in-Lindsey	27. 7.04
HVK	4236	HVK	Grob G.102 Astir CS	1161	D-4182	26. 4.96	P.A.Davey	Crowland	22. 4.05
HVL	4237	LS8	Rolladen-Schneider LS-8-18	8060		16.4.96	D.W.Allison and S.Thompson	Ringmer	27. 1.05
HVM	4238	393	Glaser-Dirks DG-300 Elan	3E177	D-4314	17. 5.96	Surrey and Hampshire Gliding Club	Lasham	15. 1.05
HVP	4240	930	Schleicher ASW20	20374	D-1961 BGA.4076, EC-DLN	3. 5.96	E.J.Smallbone	Lasham	21. 3.05
HVQ	4241	HVQ	Schleicher ASK13	13251	D-0605	27. 4.96	R.B.Brown	Edgehill	17. 4.05
HVR	4242	HVR	Schempp-Hirth Discus B	560		3. 5.96	Yorkshire Gliding Club (Pty) Ltd	Sutton Bank	10. 3.05
							(Crashed Sutton Bank 4. 6.04)		
HVT	4244	A9	Schempp-Hirth Ventus 2B	37		10. 5.96	G.Allison	Booker	14.12.05
HVU	4245	59	Rolladen-Schneider LS-8-18	8066		4.96	European Soaring	Le Blanc, France	19. 4.05
HVV	4246	HVV	Rolladen-Schneider LS-4B	41009		5. 1.97	A.J.Bardgett	Milfield	3.12.05
HVW	4247	HVW	Schleicher ASK13	13431	D-2140	14. 4.96	Rattlesden Gliding Club	Rattlesden	30. 7.05
HVX	4248	F20	Centrair ASW20F	20528	F-CFSJ	13. 6.96	L.Clarke	Rivar Hill	14. 5.05
HVY	4249	584	Schempp-Hirth Ventus 2C	9/21		17. 5.96	R.Ashurst	Lasham	16. 3.05
HVZ	4250	HVZ	Schempp-Hirth HS.4 Standard Cirrus	567G	HB-1269	5. 6.96	J.Bennett	Upwood	24. 4.05
HWA	4251	31	Schempp-Hirth Ventus 2C	8/20		7. 6.96	C.Garton	Dunstable	9. 9.05
HWB	4252	775	Schempp-Hirth Duo Discus	84		25. 5.96	Lasham Gliding Society	Lasham	28. 2.05
HWC	4253	L18	Glasflugel H.201B Standard Libelle	310	HB-1076	7. 7.96	J.C.Rogers	Winthorpe	23. 4.05
HWD	4254	HWD	Schempp-Hirth HS.4 Standard Cirrus	97	HB-987	3. 6.96	M.R.Hoskins	Rivar Hill	26. 4.05
HWE	4255	HWE	Schleicher K8B	1151	HB-700	31. 5.96	J.P.Brady	Brent Tor	9. 2.04
HWF	4256	ZC	Jastreb Standard Cirrus G/81	281	SE-TZC	12. 6.96	M.D.Langford and Partner	Talgarth	30. 5.05
HWG	4257	HWG	Glasflugel H.201B Standard Libelle	259	HB-1051	19. 6.96	T.Kendall	Pocklington	19.11.05
HWH	4258	712	Schempp-Hirth Ventus CT	182/599	RAFGGA.506	7. 7.96	H.R.Browning	Lasham	19.10.05
HWK	4260		Grob G.104 Speed Astir IIB	4070	OO-ZVQ LX-CRT	14.7.96	P.Gilbert *(De-Registered 15. 9.04)*	Tours, France	13. 7.97*
HWL	4261	84	Rolladen-Schneider LS-8A	8076		15. 6.96	M.Coffee	Snitterfield	19. 3.05
HWM	4262	D7	Rolladen-Schneider LS-8A	8079		7.96	A.R.Paul	Dunstable	16. 3.05
HWN	4263	598	Schempp-Hirth Nimbus 3T	8/60	D-KHIF	5. 7.96	H.Hampel	Michelbach,Germany	24. 6.05
HWP	4264	HWP	Glaser-Dirks DG-100G Elan	E24G13	D-3772	12. 7.96	R.L.Fox	SuttonBank	3. 6.05
HWQ	4265	HWQ	Scheibe L-Spatz 55	607	D-6195	7.96	A.Gruber *"Heisse Kartoffel"*	Usk	9. 4.05
HWS	4267	75	Rolladen-Schneider LS-8-18	8080		14. 2.97	E.A.Coles	Dunstable	13. 1.05
HWT	4268	S83	Schleicher K8B	8780	HB-958	13. 7.96	Shenington Gliding Club	Edgehill	25.10.05
HWV	4270	526	Schempp-Hirth Discus B	561		12. 8.96	J.R.Martindale	Walney Island	30. 4.04
			(Rebuild of fuselage ex AGA.4 [206] with new wings [561])				*(Badly damaged near Masham29.6.03; de-registered 10.6.04 - wreckage to New Zealand)*		
HWW	4271	DJT	Grob G.103 Twin II Acro	3658-K-27	OE-5285	23. 8.96	T.Gage	Lasham	11. 3.05
HWX	4272	HWX	SZD-59 Acro	B-2170		3. 9.96	D.W.Gosden	Talgarth	1. 8.05
HWY	4273	168	Jastreb Standard Cirrus VTC G/81	359	LN-GAL	6. 9.96	D.D.Copeland	Lasham	7.11.04
HWZ	4274	HWZ	Schleicher ASW19B	19316	HB-1524	6. 9.96	K.Commins and L. Keegan	Dublin	29. 5.02
							(To owners in Ireland as EI-153, and de-registered on 22. 1.03)		
HXA	4275	HXA	Scheibe Zugvogel IIIB	1107	D-2005	17. 3.97	B W Millar	North Connel	25. 5.02
HXB	4276	HXB	Grob G.102 Astir CS77	1819	D-6755	29. 9.96	M.Rushton	Lyveden	29. 8.04
HXC	4278	M8	Rolladen-Schneider LS-8-18	8094		24. 2.97	S.M.Smith	Gransden Lodge	23.11.05
HXD	4279	HXD	Schleicher ASW27	27030		8. 3.97	M.Jerman	Wormingford	5. 2.05
HXE	4280	Y4	Schleicher ASW19B	19053	D-6699	25. 9.96	M.Lloyd-Owen	Lasham	21. 4.05
HXH	4283	HXH	Schempp-Hirth Discus B	573	BGA.4375 (BGA.4283)	28.6.98	Deeside Gliding Club	Aboyne	7.12.05
							(NTU then re-allotted as BGA.4375 and finally reverted to BGA.4283)		
HXJ	4284	HXJ	Schleicher ASK13	13216	D-0417	29.11.96	Cotswold Gliding Club	Aston Down	23. 1.05
HXL	4286	OK-0927	Letov LF107 Lunak	39	OK-0927 OK-0827	1.11.96	G.P.Saw	Booker	25. 2.04
HXM	4287	HXM	Grob G.102 Astir CS	1272	D-7367	15.11.96	C.Fretwell	Challock	3. 6.05
HXN	4288	JS8	Rolladen-Schneider LS-8-18	8095		28. 1.97	D.Pitman	Gransden Lodge	11.10.05

HXP	4289	856	Schleicher ASK13	13023	D-3656	19.11.96	Vale of White Horse Gliding Club		
								Sandhill Farm, Shrivenham	1. 4.05
HXQ	4290	711	Schleicher ASH25B	25187	OH-874	23.11.96	M.Chant	North Hill	10. 2.05
HXR	4291	560	Schempp-Hirth Ventus CT	88/333	D-KESH	20. 2.97	J.W.A'Court	Lasham	24.11.05
HXS	4292	V11	Schempp-Hirth Ventus 2CT	10/41		27.11.96	I.R.Cook	Lasham	7. 1.05
HXT	4293	S4	Rolladen-Schneider LS-4A	4325	ZS-GNV	21. 3.97	B.T.Spreckley	Le Blanc, France	6. 3.05
HXU	4294	HXU	Schleicher ASW19B	19359	SE-TXN	14. 4.97	R.Whittle	Camphill	6. 8.05
				(Crashed on failed launch, 5. 7.03. De-Registered 16.2.04 pending foreign sale, then re-registered 7. 8.04)					
HXV	4295	HXV	Schleicher ASK13	13080	D-5462	5.12.96	Aquila Gliding Club	Hinton-in-the-Hedges	1. 1.06
HXW	4296	325	Rolladen-Schneider LS-8-18	8097		7. 3.97	W.Aspland	Dunstable	14. 3.05
HXX	4297	HXX	Schempp-Hirth HS.4 Standard Cirrus	154G	D-0363	29. 1.97	J.F.Anscombe	Lasham	8. 2.05
			(Built Grob)						
HXY	4298	HXY	Grob G.102 Astir Jeans	1781	D-7689	8.12.96	G. Powelll	Edgehill	8. 3.05
HXZ	4299	S5	Rolladen-Schneider LS-4	4249	SE-TXF	30. 3.97	E.J.Foggin	Sandhill Farm, Shrivenham	9. 2.05
HYA	4300	DD	Rolladen-Schneider LS-6-18	634B		28. 2.97	R.H.Dixon	Parham Park	8. 3.05
			(Probably c/n 6349 ex D-2162)						
HYB	4301	T5	Schempp-Hirth Discus B	140	D-4684	19. 4.97	K.A.Boost	Wormingford	8. 3.05
HYD	4303	HYD	Schleicher ASW24	24039	OE-5460	13. 2.97	S. Adlard	Long Mynd	23. 2.05
HYE	4304	913	Glaser-Dirks DG-505 Elan Orion	5E167X22		22.12.96	Bristol and Gloucestershire Gliding Club	Nympsfield	14.12.05
HYF	4305	AJ	Rolladen-Schneider LS-8-18	8106		15. 3.97	A.T.Johnstone	Dunstable	12. 1.06
HYH	4307	HYH	Rolladen-Schneider LS-3-17	3186	D-6650	18. 1.97	J.Lamb	Bellarena	9. 4.05
HYJ	4308	HYJ	Schleicher ASK21	21066	D-2724	17. 1.97	Highland Gliding Club	Easterton	10. 4.05
HYK	4309	HYK	Centrair ASW20FLP	20176	F-CEUN	25. 9.97	J.C.Riddell	Rufforth	27. 9.05
HYL	4310	K4	Schempp-Hirth Ventus 2A	44		31. 5.97	A J Stone	Booker	31. 1.05
HYM	4311	PW5	DWLKK PW-5 Smyk	17.06.020		28. 2.97	T.A.Joint	Lasham	11. 5.02
			(C/n is officially recorded as "100")				*"Iceman"*		
HYN	4312	HYN/1018	Schleicher K8B	8310A	D-1018	11.97	G.Brook	Upwood	9. 4.05
HYP	4313	HYP	SZD-50-3 Puchacz	B-2082		22. 2.97	Rattlesden Gliding Club	Rattlesden	18. 5.05
HYR	4315	432	Schleicher ASW27	27013	D-8733	11. 3.97	A.R.Hutchings	Dunstable	26. 2.05
HYS	4316	A14	Schleicher ASK21	21519	RAFGGA.514	14. 4.98	AGA Kestrel Gliding Club	RAF Odiham	8. 4.05
HYT	4317	HYT	Schleicher ASK21	21568	AGA.20	23.1.99	AGA Wyvern Gliding Club	AAC Upavon	6. 1.05
					RAFGGA.515				
HYU	4318	A61	Schempp-Hirth Discus CS	192CS	RAFGSA.R61	29.6.99	Anglia Gliding Club	RAF Wattisham	8.12.05
					RAFGGA.561				
HYW	4320	HYW	Schleicher K8B	8163A	D-5316	13. 3.97	Lincolnshire Gliding Club	Strubby	13. 2.05
					D-3202				
HYX	4321	HYX	Schleicher K8B	686	D-5742	???	Oxford University Gliding Club	Bicester	5. 6.05
HYY	4322	A26	Schempp-Hirth Nimbus 3DT	21	RAFGSA.R26	1. 3.97	A.Broadbridge	Bidford	21. 4.05
					D-KAFA				
HYZ	4323	L88	Rolladen-Schneider LS-8-18	8104		12. 4.97	P J Coward	Crowland	19. 8.03
							(Sold in Australia as VH-PNL, 10.02)		
HZA	4324	374	Schempp-Hirth Nimbus 3/24.5	94	SE-UFO	27. 4.97	P.J.Kite	Lasham	24. 2.05
HZB	4325	HZB	PZL Swidnik PW-5 Smyk	17.06.021		28. 2.97	Burn Gliding Club	Burn	3. 9.05
HZC	4326	216	Grob G.102 Astir CS	1092	D-6991	24. 3.97	K.G.Laws	Lasham	15.12.05
HZD	4327	HZD	Schleicher ASW15B	15327	D-2191	20. 3.97	C.P.Ellison	Booker	25. 4.05
HZE	4328	393	Schempp-Hirth Discus CS	121CS	D-6946	24. 3.97	Surrey and Hants Gliding Club	Lasham	13.12.05
HZF	4329	G7	Centrair 101A Pégase	101A-0262	PH-796	6. 5.97	J.Dadson	Gransden Lodge	7. 3.05
HZG	4330	X7	Rolladen-Schneider LS-8-18	8118		10. 3.97	N.G.Hackett	Husbands Bosworth	19. 2.05
HZH	4331	HZH	Schleicher Ka6CR	6461	HB-836	1. 6.97	M.E.de Torre	Gamston	30. 5.04
HZJ	4332	HZJ	Schempp-Hirth HS.4 Standard Cirrus	23	HB-981	18. 3.97	A.B.Stokes	Edgehill	19. 1.05
HZL	4334	HZL	Schempp-Hirth HS.4 Standard Cirrus	304	D-2060	5. 4.97	P.Conran	Lasham	14. 4.01
							(De-registered 22 .6.04, sold in Ireland as EI-148)		
HZM	4335	U1	Rolladen-Schneider LS-4A	4762	D-1394	14. 4.97	J.M.Bevan	Husbands Bosworth	13. 3.05
HZN	4336	-	Schleicher Ka2B Rhönschwalbe	195	D-6173	28. 3.97	R.A.Willgoss	Booker	12. 7.04
HZP	4337	56	Rolladen-Schneider LS-8-18	8117		12. 3.97	S.J.Redman	Gransden Lodge	25. 4.05
HZQ	4338	LZ	Schleicher ASW27	27018	D-4499	10. 3.97	S.Turner	Cosford	1. 9.05
							(Collision with Discus CS BGA.4092 near Lasham 4. 9.03)		
HZR	4339	HZR	Schleicher ASK21	21079	D-4491	1. 5.97	A.Roseberry	Aston Down	15. 2.05
HZS	4340	K1	Schempp-Hirth Ventus 2A	43		27.3.97	A.E.Kay	RAF Weston on the Green	20. 3.05
HZT	4341	X50	Centrair ASW20F	20150	F-CFLL	19. 8.97	T.J.Banks	Ringmer	2. 3.05
HZU	4342	B11	Schempp-Hirth HS.4 Standard Cirrus	366	HB-1258	30. 4.97	D.R.Piercy	Gallows Hill	19. 2.05
					N71KW				
HZV	4343	P61	Schempp-Hirth HS.4 Standard Cirrus	305	VH-GFZ	30. 4.97	P.R.Cox	Tibenham	21. 2.05
					BGA.4343, HB-1457, D-2061		*(Restored 5.8.02 ex VH-GFZ)*		
HZW	4344	112	Schempp-Hirth Nimbus 3T	22/88	D-KILO	1. 6.97	J.Ellis	Sutton Bank	18. 5.05
HZX	4345	476	Schleicher K8B	8257	D-8476	12.4.97	G.E.W.Woodward	Upwood	4. 6.05
HZY	4346	EN	Rolladen-Schneider LS-4A	4479	D-3458	19.4.97	N.P.Wedi	Booker	3. 2.05
HZZ	4347	LD	Schleicher ASW20L	20273	N727AM	27.4.97	P.E.Rice	Wormingford	18. 6.05
JAA	4348	JAA	Schempp-Hirth HS.6 Janus B	163	D-3147	18. 4.97	A.A.Baker	Parham Park	5. 2.05
							(Crashed Parham Park 8. 8.04)		
JAB	4349	JAB	Glaser-Dirks DG-300 Elan	3E320	OY-XTC	24. 4.97	I.G. Johnston	Sutton Bank	3. 3.05
JAC	4350	98	Schempp-Hirth Duo Discus	128		24. 5.97	British Gliding Association	RAF Bicester	14. 1.06
JAD	4351	JAD	Schleicher ASK21	21659		10.97	Borders Gliding Club	Milfield	3.10.05
JAE	4352	N8	Glaser-Dirks DG-200/17C	2-62	HB-1443	21. 4.97	S.A.White	Hinton-in-the-Hedges	18. 5.05
JAF	4353	620	Schempp-Hirth Ventus 2B	33	(BGA.4306)	24. 4.97	D.K.McCarthy	Gransden Lodge	11. 2.04
							(De-Registered 17.12.03, sold in Sweden as SE-UTN)		
JAG	4354	JAG	Schleicher ASW20L	20136	HB-1474	15. 7.97	K.J.McPhee	Keevil	18. 5.05
JAH	4355	921	Schempp-Hirth Discus B	572		30. 5.97	K.Neave	Nympsfield	26. 6.05
JAJ	4356	Y1	Glaser-Dirks DG-202/17	2-150/1744	D-4154	8. 5.97	M.Edmonds	Talgarth	28. 1.05
JAK	4357		Schleicher Ka6E	4301	F-CDRJ	R			
JAL	4358	JAL	Schleicher Ka6E	4360	F-CDTX	19. 6.97	N.Gilkes	Lee-on-Solent	13. 5.05

JAM	4359	777	Schleicher ASW15B	15353	D-2360	5.97	T.J.Beckwith	Sackville Lodge, Riseley	20. 1.00
JAN	4360	JAN	Schempp-Hirth Discus B	575		10.97	Wolds Gliding Club	Pocklington	18.12.04
JAP	4361		Slingsby T.38 Grasshopper	779	WZ783	R	R.H.Targett	Nympsfield	
			(Thought to be ex WZ818 [799])						
JAQ	4362	823	Schempp-Hirth Discus B	190	D-0960	2. 6.97	S.Smithers	Lasham	21. 5.05
JAR	4363	P3	Schempp-Hirth Discus BT	83/417	D-KHEI	16. 5.97	C.J.Partridge	Lasham	14. 3.05
JAS	4364	7Q	Glasflugel H.201 Standard Libelle	109	SE-TIS	29. 5.97	M. Collett	Booker	25. 7.05
JAT	4365	JAT	Schleicher K8B	8150	D-4390	6. 6.97	Wolds Gliding Club	Pocklington	9. 5.05
JAU	4366	WB922	Slingsby T.21B	585	WB922	27. 5.97	A.Clarke	RAF Hullavington	7. 7.05
JAV	4367	JAV	Schleicher ASK21	21662		1.97	Wolds Gliding Club	Pocklington	28.12.05
JAW	4368	M4	Glaser-Dirks DG-200/17	2-180/1759	D-5618	12. 6.97	P.W.Roberts	Lasham	19. 4.05
JAX	4369	JAX	Schleicher ASK21	21665		1.98	Wolds Gliding Club	Pocklington	6. 1.06
JAY	4370	123	Schleicher ASW20	20034	D-7941	19. 6.97	M.Gregorie	Gransden Lodge	19. 2.05
							(Crashed in field near Winslow, Bucks, 13. 9.04)		
JAZ	4371	JAZ	Grob G.102 Astir CS Jeans	2073	D-7586	28. 8.97	Bath, Wilts and North Dorset Gliding Club		
								Kingston Deverill	3.11.05
JBA	4372	JBA	Slingsby T.38 Grasshopper	1262	XP463	6.98	M.Morley	Challock	3.10.04
			(Assembled from components; p/i is starboard wing only)						
JBB	4373	P8	Rolladen-Schneider LS-8	8003	D-8023	22. 8.97	P. Whitehouse	Challock	13. 4.05
JBC	4374	2B	Schempp-Hirth Ventus 2CT	15/61		20. 6.97	B.A.Bateson	Ringmer	7. 5.01
							(Spun in Benalla, Australia, 3. 2.01)		
JBE	4376	LX	ISF Mistral C	MC.020/79	OY-XLX	4. 7.97	H.H.Crowther	RNAS Yeovilton	8. 4.05
					PH-667				
JBF	4377	JBF	Glasfugel H.201 Standard Libelle	246	F-CDPV	22. 7.97	K.Johnson	Booker	9. 5.05
JBG	4378	U9	Schempp-Hirth Ventus B	135	OE-5315	20. 7.97	W.R.Longstaff	Feshiebridge	20.12.05
JBH	4379	537	Eiri PIK-20D	20621	OH-529	6. 7.97	537 Syndicate (P.J.Holloway)	Parham Park	8. 1.05
JBJ	4380	G81	Jastreb Standard Cirrus G/81	280	SE-TZD	8. 7.97	T.Rendell	Lasham	30. 7.03
JBK	4381	L3	Schleicher ASW19B	19204	D-4099	11.97	P.M.Sharpe	Dunstable	27. 1.05
					PH-602				
JBM	4383	S21	Schleicher ASK21	21089	D-6391	26. 7.97	Staffordshire Gliding Club	Seighford	19. 6.05
JBN	4384		Schempp-Hirth Discus B	94	D-7175	R	*(or Discus bT c/n 94/433 ex D-KCFL ?)*		
JBP	4385	B2	Rolladen-Schneider LS-6-18W	6378		25. 7.97	I.C.Baker	Nympsfield	17.12.04
JBQ	4386	LH7	Rolladen-Schneider LS-8-18	8148		24. 9.97	L.Hill	North Hill	26. 9.05
JBR	4387	AC	Schempp-Hirth Discus B	90	F-CGGD	8. 8.97	A.A.Baker	RAF Odiham	6.12.04
					F-WGGD				
JBS	4388	JBS	LAK-12 Lietuva	6115	???	10. 8.97	I.G.Smith and Partners	Ringmer	8. 4.04
JBT	4389	JBT	Schleicher ASW19B	19075	D-4477	2. 8.97	Aquila Gliding Club	Hinton-in-the-Hedges	10. 4.05
JBU	4390	HL	Rolladen-Schneider LS-6-18W	6350	D-0462	23. 9.97	J.Gorringe	Lasham	12. 9.05
JBW	4392	710	Schempp-Hirth Discus BT	34/337	D-KBJR	8.97	N.C.Pringle	Lasham	29. 1.05
JBX	4393	JBX	Rolladen-Schneider LS-4A	4293	D-9111	4.98	P.W.Lee and Partner	Nympsfield	28. 3.05
JBY	4394	960	LAK-12 Lietuva	6185		12. 8.97	C.Davison	Gamston	17. 4.05
JBZ	4395	JBZ	Grob G.102 Astir CS	1492	D-4794	30. 8.97	M.Ritson	RAF Keevil	24. 4.05
JCA	4396	JCA	Schleicher ASW15B	15202	(BGA.4049)	11. 9.97	S.Briggs and Partners	Currock Hill	19.11.05
					OH-410				
JCB	4397	JCB	Rolladen-Schneider LS-6C-18WL	6234	D-6116	4.98	R. Bradley	Pretoria	16.12.03
					(De-Registered 23. 6.04, sold in South Africa as (ZS-GXX), became ZS-GXY 8.04)				
JCD	4399	H5	Schleicher ASW24	24101	D-6091	10.97	M.D.Evershed	Gransden Lodge	25. 2.05
JCE	4400	911	Schempp-Hirth Ventus BT	46/240	D-KFMS	15. 9.97	A.G.Reid	Bidford	24. 3.05
JCF	4401	JCF	Grob G.102 Astir CS77	1705	PH-1012	23. 9.97	Northumbria Gliding Club	Currock Hill	6. 1.05
					D-7634				
JCG	4402	JCG	DWLKK PW-5 Smyk	17.09.003		9.97	H.V.Jones	Crowland	28. 5.03
JCJ	4404	C7	Grob Standard Cirrus	434G	SE-TNC	5.98	R.P.Carter	Husbands Bosworth	31. 5.05
JCK	4405	DC	Schempp-Hirth Discus BT	92/430	D-KIDE	10.97	D.Coppin	Lasham	27. 9.05
JCL	4406	T2	Rolladen-Schneider LS-8-18	8147		10.97	T.W.Slater	Aboyne	7. 4.05
JCM	4407	W27	Schleicher ASW27	27064		12.97	M.J.Clayton	Bidford	9. 7.05
							(Damaged in take-off accident at Camphill, 8.10.04)		
JCN	4408	JCN	Schempp-Hirth HS.4 Standard Cirrus	646	D-7247	10.97	P.W.Reavill	Camphill	30.11.05
JCP	4409	F2	Rolladen-Schneider LS-8-18	8146		10.97	D.Francis	Weston-on-the-Green	27.12.05
JCQ	4410	W19	Schleicher ASW19B	19086	PH-562	12.97	A.J.Preston	Hinton-in-the-Hedges	23. 4.05
JCR	4411	JCR	Grob G.102 Astir CS	1181	OE-5188	10.97	B Harrison	Kingston Deverill	11.11.05
JCS	4412	WT898	Slingsby T.31B	693	BGA.3284	10.97	G.Schwab	Graz, Austria	28. 6.04
					WT898				
JCT	4413	176	Schempp-Hirth Nimbus 4T	21	D-KKKL	10.97	D S Innes	Lasham	29.10.04
JCU	4414	616	Schempp-Hirth HS.4 Std Cirrus 75	688	D-6604	10.97	N.Swinton	Weston-on-the-Green	15. 1.06
JCV	4415	IM	Schleicher ASH25E	25069	D-KAIM	11.97	R.Verdier	Gransden Lodge	31. 3.05
JCW	4416	JCW	Grob G.102 Astir CS77	1612	PH-573	10.97	C.R.Phipps	Llantisilio	22. 4.05
JCX	4417	JCX	Schempp-Hirth Discus BT	93/432	D-KJOB	10.97	R.Starmer	Bidford	9.11.05
JCY	4418	F3	Rolladen-Schneider LS-8-18	8171		12.97	R.Starey	Lasham	18.12.03
JCZ	4419	JCZ	Schleicher Ka6CR	6108	D-7152	11.97	R.Monk	Kingston Deverill	4. 5.05
JDA	4420	GR	Schempp-Hirth Nimbus 3/24.5	8	D-1788	11.97	G.R.Ross	Lasham	19. 3.04
JDB	4421	WZ828	Slingsby T.38 Grasshopper	809	WZ828	11.97	H.Chapple	Bicester	23. 4.03
JDC	4422	A27	Schleicher ASW27	27010	D-6209	12.97	J.Marshall	Dunstable	14.11.05
JDD	4423	JDD	Glaser-Dirks DG-200/17C	2-171/CL17	PH-717	12.97	D.Littler	Chipping	25. 2.05
JDE	4424	543	Rolladen-Schneider LS-8-18	8151		5.98	M.N.Davies	Snitterfield	16. 4.05
JDF	4425	907	Schleicher ASH25E	25150	D-KPAS	11.97	W.Young	Pocklington	1.10.04
JDG	4426	KW	Rolladen-Schneider LS-6B	6145	D-5675	2.98	A.Moss	Nympsfield	20. 2.05
JDH	4427	2F	Rolladen-Schneider LS-6C-18W	6287	D-9128	2.98	F.Schlafke	Wilsche, Germany	14.10.03
							(De-Registered 28.10.04, sold abroad to Germany)		
JDJ	4428	434	Rolladen-Schneider LS-3	3010	D-7729	11.97	J.C.Burdett	Walney Island	3. 2.05
JDK	4429	841	Rolladen-Schneider LS-8-18	8153		3.98	G.K Drury	Challock	15. 3.05
JDL	4430	SR	Schempp-Hirth Discus BT	165/578		4.98	S Robinson	Chipping	14. 5.05

JDM	4431	JDM	Schleicher ASW15B	15280	F-CEGL	1.98	D C Blyth	Tibenham	8. 4.05
JDN	4432	JDN	Glaser-Dirks DG-505 Elan Orion	5E180X31		3.98	Devon and Somerset Gliding Club	North Hill	15. 2.05
JDP	4433	JDP	Glaser-Dirks DG-200/17	2-136/1734	D-0152	12.97	R.Fielding	Camphill	7. 1.06
JDR	4435	JDR	Schleicher ASW15A	15053	D-6910	2.98	R.Briggs	Wormingford	13. 1.06
JDS	4436	JDS	Schempp-Hirth HS.4 Standard Cirrus 75	638	D-4057	3.98	I.P. Santos	Usk	18. 2.05
					OY-XCZ				
JDT	4437	232	Rolladen-Schneider LS-8a	8172		1.98	R.J.Rebbeck	Dunstable	15. 4.05
JDU	4438	JDU	LET L-13 Blanik	026303	D-8919	12.97	Herefordshire Gliding Club	Shobdon	27. 3.05
JDV	4439	JDV	DG Flugzeubau DG-303 Elan Acro			3.98	I.N.Busby	Booker	25. 3.05
				3E481A24					
JDX	4441	JDX	Wassmer WA-28F Espadon	101	F-CDZU	12.97	P.J.Martin	Husbands Bosworth	26. 5.05
JDY	4442	P2	Rolladen-Schneider LS-8-18	8173		2.98	P.O.Paterson	Lasham	14. 3.05
JDZ	4443	N1	Schempp-Hirth Nimbus 4T	18	D-KOLF	1.98	A.J.French	Lasham	1. 4.05
JEA	4444	D4	Rolladen-Schneider LS-8-18	8159	D-2411	3.98	R.I.Davidson	Husbands Bosworth	19. 4.05
JEB	4445	JEB	Schleicher ASW24	24172	D-9344	2.98	M.A.and J.Taylor	Rattlesden	11.11.05
JEC	4446	JEC	SZD-50-3 Puchacz	B-2197		4.98	Cambridge University Gliding Club	Gransden Lodge	31.12.05
JED	4447		Schleicher ASW15B	15427	D-3976	1.98	J.T.Leppanen	(Portugal)	26. 2.05
JEE	4448	M5	Schleicher ASW20L	20073	(BGA.4456)	2.98	J.L.Bugbee	North Hill	20. 5.05
					D-7666		*(Undershot T North Hill 24. 7.04)*		
JEF	4449	M2	Schempp-Hirth Ventus CT	126/400	D-KFWH	1.98	M. Powell-Brett	Snitterfield	26. 1.05
JEG	4450	685	Rolladen-Schneider LS-8-18	8150		1.98	J.R.Luxton	Booker	28. 3.05
JEH	4451	KE	Glasflugel H.303 Mosquito B	172	OY-XKE	2.98	I.T.Vickery	Lasham	27. 1.05
JEJ	4452	F1	Rolladen-Schneider LS-7WL	7074	OE-5477	3.98	I.Mountain	Cranwell	31. 5.05
JEL	4454	JEL	Schleicher ASW24	24044	PH-866	2.98	D.Robson	Milfield	15. 4.05
JEM	4455	572	Schempp-Hirth Duo Discus	146		3.98	D.W.Briggs	Aston Down	15. 1.05
JEP	4457	JEP	Rolladen-Schneider LS-4B	41021		3.99	C.F.Carter	Long Mynd	13. 4.05
JEQ	4458	OZ	Schempp-Hirth Nimbus 3DT	6	OO-ZOZ	2.98	R.Lynch and Partners	Kingston Deverill	29.10.05
					HB-1921, D-7695				
JER	4459	JER	Schempp-Hirth Standard Cirrus 75	654	D-6475	4.98	S. Brown	Snitterfield	10. 3.05
					OO-ZBM				
JES	4460	V4	Schleicher ASW19B	19119	SE-TTV	3.98	J.A.Crowhurst	Dubbo, Australia	8. 1.06
JET	4461	JET	Schempp-Hirth Ventus cT	161/521	RAFGSA.R38	2.98	R.Jarvis	Bidford	27. 4.05
JEU	4462	JEU	Glasflugel H201 Standard Libelle	55	SE-TIC	4.98	D.Johns	Bidford	14. 3.05
JEV	4463	JEV	Schempp-Hirth Standard Cirrus B	650	OE-5072	2.98	G.F.King	North Hill	11.12.04
JEW	4464	D41	Schleicher Ka6CR	6493	D-4116	2.98	A.Turk	Seighford	25. 4.05
JEX	4465	DW	Schempp-Hirth Ventus 2A	64		3.98	D.S.Watt	Booker	8. 3.05
JEY	4466	T7	Schempp-Hirth Standard Cirrus	456G	D-3255	3.98	R.J.Lockett	Ridgewell	21. 2.05
JEZ	4467	274	Glaser-Dirks DG-100	3	PH-792	3.98	M. Lavender	Booker	9. 2.05
					D-3721				
JFA	4468	JFA	Schempp-Hirth Standard Cirrus	225	D-0974	4.98	P.M.Sheahan	Lasham	19. 3.05
JFB	4469	S6	Rolladen-Schneider LS-8-18	8152		3.98	S.Turner and Partners	Husbands Bosworth	4. 3.05
							(Crashed in field, Little Linford, Buckinghamhire 13 .8.04)		
JFC	4470	R55	Schempp-Hirth Discus CS	054CS	RAFGSA.R55	3.98	RAFGSA Fenland Gliding Club	RAF Marham	19. 4.04
JFD	4471	R8	Grob G.102 Astir CS	1379	RAFGSA.R8	3.98	RAFGSA Centre	RAF Bicester	30.10.03
					OY-XGE				
JFE	4472	16	Schempp-Hirth Janus Ce	21/299	RAFGSA.R16	3.98	Bannerdown Gliding Club	RAF Keevil	24. 1.05
							(Crashed Keevil 7. 4.04)		
JFF	4473	26	Schempp-Hirth Duo Discus	131	RAFGSA.R26	3.98	RAFGSA Centre	RAF Halton	24.11.05
JFH	4475	R1	Schempp-Hirth Duo Discus	118	RAFGSA R1	3.98	Four Counties Gliding Club	Barkston Heath	8. 4.05
JFJ	4476	JFJ	Schleicher ASW20CL	20830	D-8307	4.98	J.A.Stirk	Burn	10. 3.05
					F-CGCS				
JFK	4477	JFK	Schleicher ASW20L	20201	D-5979	4.98	M.K.Field	Sleap	9. 6.05
JFL	4478	42	Rolladen-Schneider LS-8-18	8178		3.98	G.N.Smith	Dunstable	15. 1.06
JFM	4479	JFM	Schleicher ASK13	13222	D-0396	3.98	Newark & Notts Gliding Club	Winthorpe	30. 1.05
JFN	4480	M25	Schleicher ASH25E	25060	D-KCOH	6.98	R.J.Baker	Gransden Lodge	7. 3. 05
JFP	4481	HB	Schempp-Hirth Ventus A	19	PH-707	3.98	J. Dawson	Keevil	28. 2.05
JFQ	4482	651	Schempp-Hirth Nimbus 4DT	9/40		5.98	P.Whitt	Shobdon	1. 4.05
JFR	4483	221	Schempp-Hirth Ventus cT	170/560	RAFGSA.R24	12.97	J.G.Allen	RAF Bicester	18. 1.06
JFS	4484	528	Schempp-Hirth Ventus cT	147/456	RAFGSA.R28	7.98	M.J.Towler	Bidford	28. 7.00
JFT	4485	JFT	Schleicher K8B	8451	D-1883	3.98	South London Gliding Centre	Kenley	16. 6.05
JFU	4486	JFU	Schleicher ASW19	19038	D-4531	3.98	East Sussex Gliding Club	Ringmer	2. 2.00
							(Crashed on landing Ringmer 10.11.99)		
JFV	4487	WA1	Schleicher ASK21	21675		6.98	Scottish Gliding Union	Portmoak	25. 6.05
JFW	4488	JFW	LAK-12 Lietuva	6192	LX-CDM	4.98	K.Burgess	Lyveden	9. 7.05
JFX	4489	144	Rolladen-Schneider LS-8A	8174		3.98	P.E.Baker	Gransden Lodge	4. 3.05
JFY	4490	JFY	Federov Me-7b	8		10.98	D.S.Adams	Booker	22. 1.04
JFZ	4491		Federov Me-7b	9		10.98	M.Powell-Brett	Long Mynd	16.11.00
JGA	4492	JGA	Federov Me-7b	10		10.98	N.Wilkinson	Challock	19.12.03
JGB	4493	CU	Schleicher K8B	AB.02	D-8868	5.98	Cambridge University Gliding Club	Gransden Lodge	7. 3.05
JGC	4494	CC	Rolladen-Schneider LS-6A	6031	D-6699	4.98	Gliding Expeditions Ltd	Les Ages, France	11. 5.03
					PH-763		*(To ZS-GXV 8.03)*		
JGD	4495	JGD	Schleicher K8B	8214A-SH	D-??..	4.98	R.E.Pettifer	Chipping	2. 5.05
JGE	4496	K21	Schleicher ASK21	21068	RAFGSA.R21	2.98	N.Wall	Bidford	26. 5.05
JGF	4497	JGF	Neukom Elfe S4D	416	BGA.3316	5.98	C.V.Inwood	Lasham	14. 5.02
					D-4820				
JGG	4498	JGG	Schleicher ASW15B	15332	D-2325	4.98	L.R.Groves	Ringmer	1. 3.05
JGH	4499	JGH	Schempp-Hirth Nimbus 2c	188	OO-ZZM	4.98	J.Swannack	Gamston	13. 9.04
					D-2834				
JGJ	4500	JGJ	Schleicher ASK21	21039	RAFGSA.R22	4.98	Midland Gliding Club	Long Mynd	15. 3.05
JGK	4501	JGK	Molino Pik-20D	20571	OO-ZDL	4.98	R.Cassidy	Milfield	19. 4.05
					D-6707				

JGL	4502	27	Schempp-Hirth Discus CS	148CS	RAFGSA.R27	4.98	Chilterns Glding Club	RAF Halton	11. 2.05
JGM	4503	R53	Schempp-Hirth Discus CS	036CS	RAFGSA.R53	4.98	Chilterns Gliding Club	RAF Halton	8. 4.05
JGN	4504	867	Schempp-Hirth Standard Cirrus	554	D-8674	4.98	C.J.Sturdy	Rufforth	11. 5.05
JGP	4505	E8	Schempp-Hirth Ventus 2cT	3/10	N200EE	4.98	C.Morris	Bidford	28. 2.05
					D-KHIA				
JGQ	4506	JGQ	LET L-13 Blanik	026224	HB-1282	6.98	Joint Aviation Services	Shipdham	19. 3.02
JGR	4507	BT	Schempp-Hirth Discus bT	10/275	D-KGPS	5.98	J.A.Horne	Wormingford	11. 3.05
					D-5461				
JGS	4508	L2	Rolladen-Schneider LS-8-18	8180		5.98	M.D.Allan	Shobdon	30. 1.05
JGU	4510	JGU	Schempp-Hirth Mini Nimbus B	69	HB-1427	5.98	M.Burrows	Husbands Bosworth	5. 4.05
JGV	4511	LGC	Schempp-Hirth Duo Discus	173	(D-4020)	4.00	London Gliding Club	Dunstable	29. 1.05
JGW	4512	JGW	Schleicher ASK13	13146	D-0169	5.98	Newark & Notts Gliding Club	Winthorpe	31. 1.05
JGX	4513	JGX	Schleicher K8B	753	D-1878	5.98	J.Fisher	Andreas	16. 4.05
JGY	4514	C3	Schempp-Hirth Standard Cirrus	333	SE-TMU	5.98	P.Short	Husbands Bosworth	19. 4.05
JGZ	4515	JGZ	Glasflugel H-201 Standard Libelle	193	D-0697	5.98	J.H.Edwards	Pocklington	16. 3.054
JHA	4516	EU	Schempp-Hirth Standard Cirrus 75	645	D-4240	5.98	G.Carruthers	Sutton Bank	15. 2.05
JHB	4517	JHB	Scheibe L-Spatz 55	552	D-1618	8.98	A.Gruber	Usk	9. 4.05
JHD	4519	JHD	Schleicher Ka6E	4307	OY-XGS	5.98	M.A.King	Lyveden	5. 4.05
					D-0272				
JHE	4520	JHE	Grob G.102 Astir CS Jeans	2189	CS-PBI	5.98	AC de Portugal	Lisbon	5. 4.05
					BGA3977/D-7764				
JHF	4521	VW	Schempp-Hirth Nimbus 4T	30		5.98	P.S.Kurstjens-Hawkins	Bidford	23. 2.05
JHG	4522	51	Grob G.102 Astir CS	1084	D-6984	5.98	J.K.G.Pack	Rivar Hill	26. 3.05
JHH	4523	986	Schempp-Hirth Standard Cirrus	349G	D-3006	5.98	R.J.Lodge	Dunstable	23. 3.05
JHJ	4524	JHJ	Glasflugel H-201 Standard Libelle	495	HB-1187	30. 6.01	A.C.Jarvis	Parham	6. 2.05
JHK	4525	JHK	Schleicher K8B	558	(BGA4319)	8.98	Stratford Gliding Club	Snitterfield	4. 4.05
					AGA.21, RAFGGA.558				
JHL	4526	JHL	Schleicher Ka6E	4073	SE-TFB	6.98	M.R.Doran	Wormingford	20. 5.05
JHM	4527	JHM	Schempp-Hirth Discus b	373	???	6.98	J.H.May	Sutton Bank	8. 6.05
JHN	4528	JHN	Grob G.102 Astir Jeans CS	2110	D-7638	5.98	A. Percival	Kingston Deverill	27. 1.05
JHP	4529	JHP	Valentin Mistral C	MC048-82	D-4948	6.98	R.H.Targett	Nympsfield	13. 2.01
JHQ	4530	R43	Schleicher ASK18	18021	RAFGSA.R43	9.98	Wrekin Gliding Club	RAF Cosford	13.12.04
					RAFGSA.713, RAFGSA.113				
JHR	4531	JHR/A34	Centrair Alliance SNC-34c	34026		6.98	Borders Gliding Club	Milfield	20. 6.05
JHS	4532	JHS	Schleicher ASW19B	19047	D-6716	3.99	B.Crow	Usk	17.12.04
JHT	4533	D2	Schempp-Hirth Discus 2A	2		5.98	H.Jones	Lasham	7. 4.05
JHU	4534	868	Rolladen-Schneider LS-8-18	8197		7.98	S.Ball and partners	Pocklington	19. 5.05
JHW	4536	JHW	Glaser-Dirks DG-200	2-19	HB-1400	6.98	A.Thornhill	Burn	6. 2.05
JHX	4537	JHX	Bölkow Phoebus C	930	OO-ZYN	6.98	M.Dunlop	Usk	19. 2.03
					F-CDON				
JHY	4538	LT	Rolladen-Schneider LS-8a-18	8181	D-9988	7.98	L.E.N.Tanner	Nympsfield	28. 2.05
JHZ	4539	G41	Schleicher ASW20	20313	D-6532	6.98	T.J.Stanley	Sutton Bank	18.10.05
JJB	4541	615	Rolladen-Schneider LS-4	4542	D-2397	6.98	M.Tomlinson	Talgarth	12. 4.05
JJD	4543	K11	Schempp-Hirth Discus bT	5/262	D-KIHS	7.98	D.Wilson	Burn	2. 9.05
JJE	4544	JO1	Schempp-Hirth Discus a	379	OE-5530	7.98	N.Braithwaite	Walney Island	12. 7.05
					VH-XQT				
JJF	4545	G1	Schleicher ASW27	27086		6.98	G.F.Read	Booker	24.11.05
JJG	4546	V1	Schempp-Hirth Nimbus 4T	3	D-KIXL	7.98	P.G.Sheard	Dunstable	13. 5.05
JJH	4547	899	Glaser-Dirks DG-800S	8-137S30		4.99	W.R.Brown	Husbands Bosworth	9. 4.05
JJJ	4548	JJJ/2946	Schempp-Hirth Standard Cirrus	284	D-2946	7.98	F.R.Stevens	Husbands Bosworth	22.11.05
JJK	4549	K8	Rolladen-Schneider LS-8-18	8199		8.98	J.E.C.White	Dunstable	18. 3.05
JJL	4550	JJL	Schleicher ASW19B	19302	D-4227	4.99	Newark & Notts Gliding Club	Winthorpe	3. 2.05
							(Damaged in take-off accident Winthorpe 5.6.04)		
JJM	4551	JJM	Schempp-Hirth Standard Cirrus	403G	D-2933	7.98	S.A.Young	Husbands Bosworth	12. 7.05
JJN	4552	JJN	Slingsby T.38 Grasshopper	1267	XP490	7.98	611 VGS Collection	RAF Watton	21. 7.99
			(Regd as '2067', from Frame no.SSK/FF/2067)						
JJP	4553	494	Schempp-Hirth Duo Discus	180		7.98	R.J.Fack	Long Mynd	9. 3.05
JJQ	4554	JJQ	SZD-51-1 Junior	B-2191		9.98	Norfolk Gliding Club	Tibenham	22. 9.01
							(Spun in Tharston near Long Stratton, Norfolk, 1. 4.01)		
JJR	4555	R73	Schleicher ASK21	21054	RAFGGA513	8.98	RAFGSA Centre	RAF Halton	29. 1.05
JJS	4556	XA240	Slingsby T.38 Grasshopper	873	XA240	R	A.Stacey *(Stored 6.02)*	RAF Keevil	
JJT	4557	933	Schleicher ASW27	27070	D-6209	8.98	T.M.World	Lee-on-Solent	31. 3.05
JJU	4558	H2	Rolladen-Schneider LS-8a	8200		8.98	P.J.Harvey	Dunstable	6. 4.05
JJV	4559	JJV	Schleicher Ka6CR	1001	D-1719	8.98	C.J.Huck	Aston Down	7. 4.03
							(Crashed on ridge, Bishop Hill near Portmoak, 30.10.02)		
JJX	4561	T9	Schleicher ASW15B	15323	D-2312	9.98	T.J.Davies	Portmoak	6. 1.06
JJY	4562	SW	Schempp-Hirth Ventus bT	273/61	PH-981	9.98	S.Way	Parham Park	15.12.05
					D-KMIH				
JJZ	4563	15	Schempp-Hirth Discus bT	156/556	OO-ZQX	9.98	S.J.C.Parker	Nympsfield	10. 1.06
JKA	4564	JKA	Schleicher ASK21	21059	D-8835	10.98	East Sussex Gliding Club	Ringmer	18. 1.05
JKB	4565	JKB	DWLKK PW-5	17.10.08		9.98	J.C.Gibson	Chipping	4. 9.03
JKC	4566	-	MDM-1 Fox	224	SP-P632	9.98	G.C.Westgate and Partner	Parham Park	10.11.05
JKD	4567	P1	Rolladen-Schneider LS-8-18	8215		10.98	R.J.Large	Husbands Bosworth	12. 3.05
JKE	4568	JKE	DWLKK PW-5	17.11.025		10.98	Burn Gliding Club	Burn	31. 5.05
JKF	4569	JKF	Glaser-Dirks DG-200	2-35	D-6069	10.98	M.Barrett	North Hill	9.11.05
JKG	4570	R48	Schleicher ASK18	18036	RAFGSA.R48	3.99	RAFGSA Centre	RAF Halton	9.12.05
					RAFGSA.448				
JKH	4571	P30	Schempp-Hirth Ventus cT	174/566	RAFGSA.R30	10.98	E.Fitzgerald	Usk	5.12.05
JKJ	4572	R21	Schleicher ASK21	21679		10.98	RAFGSA Centre	RAF Halton	25.11.05
JKK	4573	A7	Schleicher ASK21	21182	AGA.11	10.98	AGA Kondor Gliding Club	Bruggen	6. 8.05
JKL	4574	W8	Rolladen-Schneider LS-8-18	8218		4.99	R.J.Welford	Gransden Lodge	19. 3.05

JKM	4575	Z10	Glaser-Dirks DG-202-17M	2-148/1746	D-4155	10.98	A.H.Brown	Portmoak	11. 2.05
JKN	4576	790	Rolladen-Schneider LS-8-18	8214		10.98	D.A.Booth	Husbands Bosworth	30.12.04
JKP	4577	PH1	Rolladen-Schneider LS-4B	41000	PH-1089	12.98	D.M.Hope	Booker	18. 2.05
JKQ	4578	R20	Schleicher ASK21	21098	RAFGSA.R20	1.99	RAFGSA Bannerdown Gliding Club	RAF Keevil	23. 4.05
JKR	4579	P12	Schempp-Hirth Discus b	151	RAFGSA.R12	11.98	T.Wright	Husbands Bosworth	5. 4.05
JKS	4580	S19	Schleicher ASW19B	19362	D-1273	11.98	D.M.Brown	Dunstable	13.10.05
JKT	4581	R7	Schleicher ASK13	13615	RAFGSA.R7	8.99	Clevelands Gliding Club	AAC Dishforth	16. 5.05
JKU	4582	R33	Schleicher ASK18	18022	RAFGSA.R33	6.99	Clevelands Gliding Club	AAC Dishforth	30. 7.05
					RAFGSA.223				
JKV	4583	R52	Grob G103A Twin II Acro	34042-K-273	RAFGSA.R52	5.99	Clevelands Gliding Club	AAC Dishforth	23. 5.04
JKW	4584	JKW	Grob G.102 Astir CS 77	1666	RAFGSA.R60	1.99	Bath, Wilts and North Dorset Gliding Club		
					RAFGSA.560			Kingston Deverill	19. 5.05
JKX	4585	R17	Schempp-Hirth Discus B	247	RAFGSA.R17	8.99	Clevelands Gliding Club	AAC Dishforth	10. 7.05
JKY	4586	243	Schempp-Hirth Ventus cT	181/597	RAFGSA.R24	11.98	M.P.Osborne	Long Mynd	6.12.05
					RAFGGA.557				
JKZ	4587	R25	Schleicher ASK21	21123	RAFGSA.R25	12.98	RAFGSA Centre	RAF Halton	14. 1.05
JLA	4588	JLA	Schempp-Hirth Ventus 2cT	26/94	PH-1129	11.98	E.C.Neighbour	Camphill	14. 9.05
JLC	4590	R10	Schempp-Hirth Discus CS	193CS	RAFGSA.R10	11.98	RAFGSA Four Counties Gliding Club		
								RAF Barkston Heath	29.11.05
JLE	4592	JLE	Schleicher ASK13	13245	RAFGSA.R90	10.98	AGA Kondor Gliding Club	Bruggen	22. 2.05
					NEJSGSA.1				
JLF	4593	JLF	Schleicher ASK13	13150	AGA.14	12.98	Wyvern Gliding Club	AAC Upavon	28. 9.05
JLG	4594	JLG	SZD-51-1 Junior	B.1933	AGA.5	2.99	Wyvern Gliding Club	AAC Upavon	20. 2.05
					BGA.3699				
JLH	4595	JLH	Rolladen-Schneider LS-4	4256	AGA.1	3.99	Wyvern Gliding Club	AAC Upavon	30. 1.05
JLJ	4596	A8	Rolladen-Schneider LS-4B	4997	AGA.2	12.98	Wyvern Gliding Club	AAC Upavon	1. 5.05
JLK	4597	LS7	Rolladen-Schneider LS-7	7112	AGA.3	4.99	A.R.Mountain	Llantisilio	2. 4.05
JLL	4598	N25	Schleicher ASK13	13144	HB-952	11.98	Portsmouth Naval Gliding Club	Lee-on-Solent	5. 11.05
JLM	4599	R2	Schempp-Hirth HS.6 Janus C	210	RAFGSA.R2	5.99	RAFGSA Cranwell Gliding Club	RAF Cranwell	23.11.03
							(De-Registered 25. 7.03)		
JLN	4600	R4	Rolladen-Schneider LS-8-18	8169	RAFGSA.R4	3.99	RAFGSA Cranwell Gliding Club	Cranwell	6. 3.05
JLP	4601	R39	Schempp-Hirth Discus CS	034CS	RAFGSA.R39	4.99	RAFGSA Cranwell Gliding Club	RAF Cranwell	13. 2.05
JLQ	4602	JLQ	Schleicher ASK13	13608	RAFGSA.R40	4.99	Bowland Forest Gliding Club	Chipping	16. 4.05
					RAFGSA.R4				
JLR	4603	R57	Grob G.102 Astir CS	1509	RAFGSA.R57	2.99	RAFGSA Cranwell Gliding Club	RAF Cranwell	26.11.05
					RAFGSA.507				
JLS	4604	R75	Schleicher K8B	8950	RAFGSA.R75(2)	3.99	RAFGSA Crusaders Gliding Club	Kingsfield, Dhekelia	16. 2.05
					RAFGSA.285(2)				
JLT	4605	JLT	Schleicher Ka6E	4115	D-6082	11.98	M.Thompson	Husbands Bosworth	4.12.05
JLU	4606	11	Schempp-Hirth Ventus 2cT	37/126		3.99	J.C.Mitchell	Chipping	29. 3.05
JLV	4607	JLV	Schleicher Ka6E	4192	OY-XEU	1.99	M.S.Neal	Lyveden	30. 1.05
					D-4424				
JLW	4608	87	Schempp-Hirth Discus CS	033CS	RAFGSA.R87	12.98	Wrekin Gliding Club	RAFCosford	15. 3.04
JLX	4609	JLX	Schempp-Hirth Standard Cirrus	279G	OO-ZGL	11.98	M.W.Fisher	Talgarth	31. 3.05
					D-1985				
JLY	4610	JLY	Schleicher ASW27	27111		6.99	P.C.Piggott	Husbands Bosworth	1. 6.05
JLZ	4611	21	Grob G.103A Twin II Acro	3633-K-15	D-7912	12.98	B.L. Eppy	Lasham	27. 1.05
JMA	4612	R36	Schleicher ASK18	18038	RAFGSA.R36	6.99	Four Counties Gliding Club	RAF Barkston Heath	18. 3.05
					RAFGSA.236				
JMB	4613	R5	Rolladen-Schneider LS-8-18	8130	RAFGSA.R5	2.99	RAFGSA Centre	RAFHalton	10. 2.02
							(Crashed on winch launch, Syerston, 26.8.01, w/o)		
JMC	4614	R22	Schleicher ASK21	21681	(RAFGSA.R22)	12.98	RAFGSA Wrekin Gliding Club	RAF Cosford	28. 2.05
JMD	4615	P23	Schempp-Hirth Discus b	241	RAFGSA.R23	4.99	R.C.Oliver	Kenley	16. 3.05
JME	4616	JME	Schleicher Ka7	5	D-8867	12.98	W.E.Masterson	Blenheim, Jamaica	25. 1.04
JMG	4618	JMG	SZD-51-1 Junior	B.2192		4.99	Kent Gliding Club	Challock	15. 1.05
							(Crashed at Challock 25.7.04)		
JMH	4619	JMH	Schempp-Hirth Standard Cirrus	571	HB-1263	12.98	G.J.McCann	Pocklington	8. 5.05
JMJ	4620	R46	Schleicher ASK13	13616	RAFGSA.R46(2)	9.99	RAFGSA Fenland Gliding Club	RAF Marham	19.11.05
					RAFGSA.R16				
JMK	4621	R49	Schleicher ASK18	18023	RAFGSA.R49	1.99	RAFGSA Fenland Gliding Club	RAF Marham	11. 3.05
					RAFGSA.318(2)				
JML	4622	R63	Grob G.102 Astir CS 77	1718	RAFGSA.R63	1.99	RAFGSA Centre	RAF Halton	20. 4.05
					RAFGSA.883				
JMM	4623	JMM	Schempp-Hirth Discus b	254	BGA.4474	12.98	R.Bottomley	Pocklington	7. 4.05
					RAFGSA.R15				
JMN	4624	636	Schempp-Hirth Nimbus 2	38	D-1129	12.98	R.A.Holroyd	Pocklington	14. 3.05
					HB-1159				
JMO	4625	781	Rolladen-Schneider LS-8-18	8225		12.98	D.J.Langrick	Husbands Bosworth	19. 4.05
JMP	4626	JMP	Schleicher ASK13	13436	D-2984	1.99	East Sussex Gliding Club	Ringmer	15. 2.05
JMQ	4627	KR	Schleicher ASW20L	20499	D-2026	1.99	S.G.Back	Feshiebridge	2. 1.05
					F-CADB				
JMR	4628	628	Rolladen-Schneider LS-8-18	8198	D-0280	12.98	R.J.Rebbeck	Dunstable	6. 5.05
JMS	4629	R23	Schleicher ASK21	21212	RAFGGA.521	11.98	Anglia Gliding Club	Wattisham	5. 3.05
JMT	4630	301	Rolladen-Schneider LS-8-18	8223		2.99	D and K Draper	Kingston Deverill	15. 1.05
JMU	4631	Z7	Rolladen-Schneider LS-8-18	8246		3.99	J.Williams	Aboyne	17. 2.05
JMV	4632	EW	Schempp-Hirth HS.5 Nimbus 2C	179	D-6738	1.99	K.R.Walton	Lee-on-Solent	16. 2.05
JMW	4633	R61	Schleicher ASK13	13688AB	RAFGGA.567	1.99	RAFGSA Bannerdown Gliding Club	RAF Keevil	19. 3.05
JMX	4634	JMX	Schleicher ASK13	13107	RAFGSA.R86	2.99	Shalbourne Soaring Group	Rivar Hill	26. 9.05
					RAFGSA.386				
JMY	4635	JMY	SZD-51-1 Junior	W-959	OO-ZRH	3.99	Highland Gliding Club	Easterton	5. 4.05

JMZ	4636	R37	Schleicher ASK13	13099	RAFGSA.R37(2)	1.99	RAFGSA Wrekin Gliding Club	RAF Cosford	17. 4.05
					RAFGSA.378(2)				
JNA	4637	JNA	Grob G.102 Astir CS Jeans	2160	D-4556	2.99	Shenington Gliding Club	Edgehill	17. 4.05
JNB	4638	D1	Rolladen-Schneider LS-8-18	8227		2.99	P.M. Shelton	Husbands Bosworth	23. 1.05
JNC	4639	Z22	Schempp-Hirth Standard Cirrus	322	HB-1157	2.99	S.M.Veness	Bathurst, NSW, Australia	23. 5.04
							(De-Registered 5. 8.04, sold Australia, to VH-VKC 26.10.04)		
JNE	4641	JNE	Schempp-Hirth Discus 2A	18	D-4499	2.99	J.R.W.Kronfeld	Lasham	18.11.05
JNF	4642	80	Schempp-Hirth Discus 2A	12		2.99	A.J.Davis	Nympsfield	6. 4.05
JNG	4643	JNG	Glasflugel H.201B Standard Libelle	6	SE-TFU	3.99	P.K.Spencer	Edgehill	9. 8.05
JNJ	4645	601	Rolladen-Schneider LS-8-18	8226		2.99	J.D.Spencer	Hinton-in-the-Hedges	24. 5.05
							(Damaged Husbands Bosworth 5. 8.04, undercarriage collapsed on landing)		
JNK	4646	676	Rolladen-Schneider LS-8-18	8244		2.99	Wyvern Gliding Club	Upavon	2. 4.05
JNL	4647	DR7	Schempp-Hirth HS.6 Janus	25	HB-1313	2.99	D.M.Ruttle	Strubby	29. 4.04
JNM	4648	P4	Rolladen-Schneider LS-8-18	8232	(BGA.4617)	3.99	P.Onn	Dunstable	12. 1.05
					D-8217				
JNN	4649	JNN	Schleicher K-8B	8744	D-8583	3.99	Buckminster Gliding Club	Saltby	18. 4.05
JNP	4650	EF	Rolladen-Schneider LS-6B	6109	D-5853	20. 7.03	P.S.Fink	Lasham	26. 6.05
JNQ	4651	441	Glaser-Dirks DG-300 Elan	3E-341	SE-UHO	4.99	F.C.Roles	Camphill	11. 3.05
JNR	4652	JNR	Glasflugel H.303 Mosquito B	159	D-5908	3.99	I.Agutter	Wormingford	27. 1.05
JNS	4653	D48	Schleicher ASW27	27103		3.99	M.Morley	Lasham	8. 4.05
JNT	4654	SC	Schleicher ASW19B	19371	D-2233	4.99	S.Cheshire	Lasham	16.11.04
JNU	4655	R69	Rolladen-Schneider LS-6-18W	6345	RAFGSA.R69(3)	3.99	RAFGSA Chilterns Gliding Club	RAF Halton	24. 4.05
					RAFGGA.553, D-8037				
JNV	4656	E2	Schleicher ASW22BL	22079		4.99	R.A.Cheetham	Husbands Bosworth	8. 4.05
							(Re-regd G-CDII 17.1.05)		
JNW	4657	L4	Rolladen-Schneider LS-8-18	8217		3.99	P.C.Fritche	Parham Park	17.12.05
JNX	4658	JNX	LET L-13A Blanik	827408	OK-2712	5.99	Vectis Gliding Club	Bembridge	4. 3.05
JNY	4659	CZ	Schempp-Hirth Discus 2B	17	D-4498	3.99	D.H.Conway	RAF Keevil	4. 3.05
							(Marked on fin as 'BGA.4569')		
JNZ	4660	813	Glaser-Dirks DG-100	70	(D-7324)	3.99	R.Jones	Rufforth	2. 8.05
					HB-1324				
JPA	4661	HB1	Schempp-Hirth Duo Discus	201		3.99	The Soaring Centre	Husbands Bosworth	3. 2.05
JPC	4663	JPC	Schleicher ASK13	13256	RAFGSA.R51	3.99	Shalbourne Soaring Group	Rivar Hill	15. 7.05
JPD	4664	V2T	Schempp-Hirth Ventus 2cT	39/129		3.99	P.A.Hearne	Challock	28.10.05
JPE	4665	LS1	Rolladen-Schneider LS-1f	488	F-CESC	5.99	I.Craigie	Lasham	12. 9.05
JPF	4666	JPF	Glaser-Dirks DG-100	22	D-3735	5.99	H.G.Burkert	Bidford	10. 6.05
JPH	4668	CB	Rolladen-Schneider LS-8-18	8259		7.99	J.P.Ben-David	Lasham	15. 2.05
JPJ	4669	JPJ	Grob G.104 Speed Astir IIB	4089	OE-5352	4.99	R.J.Maison-Pierre	Rattlesden	26. 3.05
JPK	4670		Slingsby T34 Sky	672	RAFGSA.876	5.99	J.Tournier	Booker	21. 5.05
					XA876/G-672				
JPL	4671	RW	Rolladen-Schneider LS-8-18	8249	D-2562	4.99	I.Reekie	Dunstable	9. 3.05
JPM	4672	JPM	Grob G.102 Astir CS Jeans	2209	D-3825	5.99	J.Thorpe	Cross Hayes	14. 4.05
JPP	4674	388	Schempp-Hirth Discus B	206	AGA.4	4.99	Kestrel Gliding Club	RAF Odiham	20. 4.05
JPQ	4675	R32	Schleicher ASK18	18002	RAFGSA.R32	5.99	RAFGSA Fulmar Gliding Club	Easterton	12.12.04
					RAFGSA.213, D-3978				
JPR	4676	161	Rolladen-Schneider LS-8-18	8245		4.99	D.M.Byass	Booker	6. 4.05
JPS	4677	CL	Schleicher ASW27	27108		5.99	C.C.Lyttelton	Lasham	21. 2.05
JPT	4678	JPT	Schleicher ASW27	27113		10.99	R and W Willis-Fleming	North Hill	27. 8.05
JPU	4679	TL2	Schempp-Hirth Discus CS	257CS		4.99	K.Armitage	Camphill	19. 2.05
JPV	4680	R88	Schleicher ASK13	13312	RAFGSA.R88	5.99	RAFGSA Crusaders Gliding Club	Kingsfield, Dhekelia	29. 5.05
					RAFGSA.186				
JPW	4681	JPW	Glaser-Dirks DG-200	2-48	D-2201	5.99	J.P.Goodison	Burn	20. 7.05
JPX	4682	JPX	Schleicher ASW15	15160	D-0823	5.99	P.Seymour	Upwood	6. 3.05
JPY	4683	R59	Schleicher ASK13	13653AB	RAFGGA.509	5.99	RAFGSA Cranwell Gliding Club	Cranwell	1. 1.06
JPZ	4684	R56	Schleicher ASK18	18027	RAFGGA.563	4.99	RAFGSA CranwellGliding Club	Cranwell	20. 3.05
JQA	4685	547	Schempp-Hirth Discus b	265	BGA.4535	4.99	Bannerdown Gliding Club	RAF Keevil	16. 1.05
					RAFGGA.547/RAFGGA.500		*(Take-off accident at Keevil 18. 2.04)*		
JQB	4686	JQB	Schleicher K8B	8880	RAFGSA.R98	5.99	A.J.Taylor and Partners	Lee on Solent	19. 5.05
					RAFGSA.398				
JQC	4687	208	Schempp-Hirth Discus bT	127/488	D-KITT(3)	5.99	R.Neil	Booker	14. 1.06
JQD	4688	R3	Rolladen-Schneider LS-8-18	8224		5.99	RAFGSA Bannerdown Gliding Club	RAF Keevil	18.11.05
JQE	4689	JQE	Schempp-Hirth HS4 Standard Cirrus 75	25	OO-ZRS	6.99	C.Nunn	Wormingford	7. 5.05
					D-0483				
JQF	4690	5GC	Glaser-Dirks DG-505 Elan Orion	5E194X38	S5-7516	6.99	Scottish Gliding Union	Portmoak	19. 4.05
JQG	4691	R50	Grob G.103A Twin II Acro	33964-K-197	RAFGSA.R50	5.99	RAFGSA Fulmar Gliding Club	Easterton	3. 4.05
JQH	4692	JQH	LAK-12 Lietuva	6188		5.99	D.J.Lee	Pocklington	5. 8.05
JQJ	4693	JQJ	Schleicher K8B	8795	RAFGSA.R47	5.99	Staffordshire Gliding Club	Seighford	9.12.04
					BGA.1564				
JQK	4694	506	Schempp-Hirth Discus CS	075CS	RAFGGA.501	5.99	RAFGSA Chilterns Gliding Club	Halton	27. 1.05
JQL	4695	W2	Schempp-Hirth Ventus 2A	79		5.99	M.L.Dawson	RAF Keevil	6. 4.05
JQM	4696	Z9	Schleicher ASW-27	27114		6.99	A.Heynor	Nympsfield	20.12.05
JQN	4697	R67	Grob G.102 Astir CS 77	1634	RAFGSA.R67	6.99	RAFGSA Chilterns Gliding Club	Halton	28. 6.05
					RAFGSA.547				
JQP	4698	eb	Centrair 101A Pégase	101-066	F-CFQY	6.99	J.Rees	Nympsfield	8. 3.05
JQQ	4699	185	Schempp-Hirth Duo Discus	227		3.00	K.G.Reid and Partners	RAF Keevil	13. 3.05
JQR	4700	JQR	Schempp-Hirth Ventus 2cT	49/152		10.99	M.C.Costin	Husbands Bosworth	4. 1.06
JQS	4701	JQS	Schempp-Hirth Standard Cirrus	251G	D-1147	6.99	R.K.Arkley and Partners	Milfield	5. 7.05
			(Built Grob)						
JQT	4702	JQT	Grob G.102 Astir CS Jeans	2076	D-7589	6.99	Southdown Gliding Club	Parham Park	11. 3.05
JQU	4703	L17	LAK-17A	102		6.99	A Pozerskis	Husbands Bosworth	8. 6.05

JQV	4704	Z12	Schleicher ASW27	27112		7.99	J.White	Booker	16.11.05
JQW	4705	GS	Schempp-Hirth HS.2 Cirrus 18	47	D-0186	7.99	G.E.Smith	Parham Park	21. 2.05
JQX	4706	JQX	Schleicher ASK21	21702		9.99	Southdown Gliding Club	Parham Park	21.11.05
JQY	4707	R92	Slingsby T21B Sedbergh	666	RAFGSA.R92	6.99	RAFGSA Crusaders Gliding Club	Kingsfield, Dhekelia	18. 6.05
					NEJSGSA.4, WG499				
JQZ	4708	JQZ	Schleicher K8B	8854	RAFGSA.R42	7.99	Derbyshire and Lancashire Gliding Club	Camphill	8. 6.05
					RAFGSA.323				
JRA	4709	E11	Rolladen-Schneider LS-8-18	8263		7.99	S.R.Ell	Sutton Bank	13. 1.05
JRB	4710	S33	Schleicher ASW19B	19227	D-2713	7.99	J. Barter	Lasham	4. 1.05
JRC	4711	JRC	Glaser-Dirks DG300 Elan	3E-20	HB-1718	8.99	D.P.Sillett	Rattlesden	15. 8.05
JRD	4712	JRD	Grob G102 Astir CS	1487	RAFGSA.R18	6.99	RAFGSA Centre	RAF Halton	22. 1.05
					RAFGGA.540, D-4791				
JRE	4713	JP	Schleicher ASW-15	15048	LN-GGL	8.99	J.A. Woodforth	Bidford	5. 9.05
					OH-391, OH-RWA, D-4391				
JRF	4714	JRF	SZD-50-3 Puchacz	B-1395	OO-ZTX	8.99	Wolds Gliding Club	Pocklington	22. 1.05
					D-8213, SP-3285				
JRG	4715	JRG	Schempp-Hirth Standard Cirrus	146	D-0297	8.99	D.Draper	Rivar Hill	21. 2.05
JRH	4716	T27	Schleicher ASW27	27118		8.99	P.C.Jarvis	Lasham	3.11.05
JRJ	4717	JRJ	SZD-50-3 Puchacz	503199327		8.99	Bidford Gliding Centre	Bidford	21. 7.05
JRK	4718	618	Rolladen-Schneider LS-8-18	8267		9.99	D.King	Snitterfield	29. 9.05
JRL	4719	JRL	Glaser-Dirks DG-100G Elan	E185G151	D-1246	2.00	P.Lazenby	Aston Down	1. 4.05
JRM	4720	212	Grob G102 Astir CS	1332	BGA.4314	9.99	Kestrel Gliding Club	RAF Odiham	12. 6.05
					AGA.6				
JRN	4721	PT	Glaser-Dirks DG-202/17C	2-118CL01	D-7267	9.99	T.G.Roberts	Lasham	16. 5.05
			(C/n 2-118CL04?)						
JRP	4722	JRP	Grob G102 Astir CS Jeans	2244	D-5951	9.99	Borders Gliding Club	Milfield	24.10.05
JRQ	4723	PM3	Pfenninger-Markwalder Elfe PM3	001	N6351U	11.99	G.Mclean	Lleweni Parc	23. 5.05
					N63514, HB-526				
JRR	4724	LA	Schempp-Hirth Discus bT	50/367	PH-1087	3.00	C.R.Lear	RAF Keevil	3. 3.05
					D-KBHM				
JRS	4725	AT	Valentin Mistral C	MC021/79	D-4921	3.00	A.Towse	RAF Wattisham	10. 2.05
JRT	4726	JRT	Schempp-Hirth Standard Cirrus	99	D-0734	10.99	S.Hutchinson	Husbands Bosworth	18.10.05
JRU	4727	SK	Schleicher ASW-24	24168	D-7085	1.00	S.Kerby	Snitterfield	17. 3.05
JRV	4728	B19	Schleicher ASW19B	19233	D-2644	10.99	M.Roome	Lasham	27. 1.05
JRW	4729	R15	Grob G103A Twin II Acro	34040-K-271	RAFGGA.556	10.99	Four Counties Gliding Club	Barkston Heath	27.11.05
JRX	4730	R41	Schleicher ASK13	13375	RAFGSA.R41	10.99	Chilterns Gliding Club	RAF Halton	18. 3.05
					RAFGSA.241 (2)				
JRZ	4732		Colditz Cock rep	Not known			R M.Francis	Camphill	
			(Built M.Francis)						
JSB	4734	-	Rolladen-Schneider LS-4	4424	D-4541	2.00	H.Vare	Speyer, Germany	11. 3.05
JSC	4735	827	Schempp-Hirth Nimbus 3dT	10	F-CFUE	11.99	M.J.Aldridge	Tibenham	26. 3.05
					F-WFUE/D-KFUE				
JSD	4736	R77	Grob G102 Astir CS	1133	RAFGSA.R77	2.00	RAFGSA Wrekin Gliding Club	RAF Cosford	8. 4.04
					D-4177				
JSE	4737	296	Schempp-Hirth Discus b	365	PH-918	3.00	Imperial College Gliding Club	Lasham	29. 3.05
JSF	4738	414	Rolladen-Schneider LS-1f	383	LN-GGE	3.00	R.C.Godden	Wormingford	13. 5.05
					SE-TOU				
JSG	4739	36	Schleicher K-6E	4248	D-0090	2.00	A.J.Emck	Lasham	15.12.04
JSH	4740	SH9	Grob G102 Club Astir IIIB	5504CB	D-6470	11.99	Surrey and Hampshire Gliding Club	Lasham	11. 2.05
JSJ	4741	7X	Rolladen-Schneider LS-7WL	7058	D-5774	10.99	S.P.Woolcock	Gransden Lodge	12. 1.05
JSK	4742	JSK	Grob G102 Astir CS	1521	D-7455	11.99	J.W.Bolt	Brentor	29. 4.05
JSL	4743	JSL	Schempp-Hirth Ventus cT	121/395	D-KIFL	11.99	W.D.Ingles	Bidford	27.12.04
JSN	4745	R45	Schleicher K8B	8916	RAFGSA.R45	11.99	Chilterns Gliding Club	RAF Halton	9.11.04
					RAFGSA.245				
JSP	4746	JSP	Slingsby T31B	1189	BGA.2724	4. 7.04	R. Wulfers	Deelen, The Netherlands	3. 7.05
			(Restoration using wings of XE798).		XN250		*(Marked as "BGA. 4756")*		
JSQ	4747	T8	Rolladen-Schneider LS-8T	8301	D-KKAF	5.00	Strong Words Ltd.	Gransden Lodge	24. 4.04
JSR	4748	JSR	SZD-50-3 Puchacz	B-1386	OY-XRV	3.00	Bidford Gliding Club	Bidford	15. 4.05
					SP-3283				
JSS	4749	JB	Schleicher ASW27B	27121		12.99	J.H.Belk	Dunstable	8. 3.05
JST	4750	X5	Rolladen-Schneider LS-1c	86	OO-ZPA	1.00	L.Gerrard	Saltby	16. 6.05
					D-0766				
JSU	4751	95	Rolladen-Schneider LS-8-18	8297	D-0543	3.00	J.G.Bell	Parham Park	13. 1.05
JSV	4752	R80	Schleicher ASK13	13127	RAFGSA.R80	1.00	RAFGSA Centre	Bicester	1.12.04
					BGA.1509		*(Dismantled 1.05)*		
JSW	4753	H4	Rolladen-Schneider LS-4	4262	ZS-GOP	10.00	European Soaring	Le Blanc, France	8. 3.05
JSX	4754	JSX	Glaser-Dirks DG-505 Elan Orion	5E200X44		5.00	Oxford Gliding Club	RAF Weston on the Green	19. 2.05
JSY	4755		Pilatus B4 PC-11AF	127	D-3055	1.00	A.de Tourboulon and Partner	Wormingford	16. 2.04
					D-5787, PH-578		*(De-Registered 11.12.03, sold to Germany as D-1315)*		
JSZ	4756	JSZ	Schleicher ASK-18	18012	D-6878	4.00	C.J.N.Weston	Challock	11. 7.04
JTA	4757	-	Colditz Cock rep				See SECTON 3 Part 1	Flixton, Bungay	
JTB	4758	V17	Schleicher ASW-24	24017	D-3465	2.00	M.E. Knell	Booker	10. 2.05
JTC	4759	JTC	Glaser-Dirks DG-100G	84G5	(BGA.4744)	5.00	A.Burger	Nunstadt, Germany	5. 7.05
					HB-1335				
JTD	4760	103	Rolladen-Schneider LS-6-18W	6371	D-2571	2.00	C.J.Mayhew	Lasham	23. 1.05
							(De-Registered 5. 4.04, sold in the Netherlands as PH-1303)		
JTE	4761	JTE	Schempp-Hirth Standard Cirrus	470G	D-3718	2.00	T.and R.Scott	Ringmer	7. 4.05
JTF	4762	Z3	Schleicher ASW-27B	27125		3.00	P.M.Wells	Booker	21. 3.05
JTG	4763	D55	SZD-55	551197100	N4364R	3.00	N.A.McLean	Lasham	14. 3.05
JTH	4764	H3	Schleicher ASW-24	24218	D-7681	3.00	R.H.Yarney	Lasham	29. 3.05

JTJ	4765	JTJ	Schempp-Hirth Mini Nimbus b	73	D-7620	3.00	A.Richards	North Hill	6. 3.05
JTK	4766	JTK	DG Flugzeubau DG-303 Elan Acro	3E487A28		5.00	A. Jorgensen	Booker	28. 2.05
JTL	4767	352	Rolladen-Schneider LS-8-18	8317		3.00	L.S.Hood	RAF Barkston Heath	3. 4.05
JTM	4768	205	Rolladen-Schneider LS-8-18	8268		2.00	J.T. Birch	Gransden Lodge	11. 2.05
JTN	4769	E5	Glaser-Dirks DG-300 Elan	3E19	HB-1717	3.00	M.J.Chapman	Seighford	24. 3.05
JTP	4770	JTP	Schleicher ASW-20L	20569	D-4688	3.00	C.A.Sheldon	Pocklington	6. 3.05
JTQ	4771	JTQ	Glasflugel H303 Mosquito	88	OO-ZYL	4.00	I.Hamilton	Chipping	15. 7.05
JTR	4772	D-53	Rolladen-Schneider LS-7	7104	D-5309	3.00	B.L.Anson	Halton	23. 2.05
JTS	4773	JTS	Schempp-Hirth Cirrus VTC	108	S5-3059 SL-3059/, U-4200	3.00	P.V. Skinner	Not known	7. 3.05
JTU	4775	377	Schempp-Hirth Duo Discus	4/234		3.00	R.Starmer Syndicate	Bidford	6. 4.05
JTV	4776	666	Schempp-Hirth Ventus 2cT	52/173		3.00	A.P.Moulang	Challock	24. 2.05
JTW	4777	AV8	Glasflugel H303 Mosquito	199	F-CELX	4.00	M.Wright	Rattlesden	8. 2.05
JTX	4778	JTX	Start+Flug H101 Salto	47	D-9260	3.00	C.Schneeberger	Maxdorf, Germany	1. 3.05
JTY	4779	JTY	Rolladen-Schneider LS-8A	8102	SE-USA	4.00	BBC Club	Booker	12. 1.05
JTZ	4780	GC	Schempp-Hirth Ventus 2B	23	OO-ZQS	2.00	G.Costacurta	Asiago, Italy	12. 6.04
JUA	4781	Y2K	Schempp-Hirth Discus 2B	48		2.00	C.Costa *(De-Registered 5.12.02, sold in Germany as D-3125)*	(Italy)	30. 6.03
JUB	4782	894	Schempp-Hirth Discus CS	268CS		4.00	D.F.Wass	Gamston	10. 2.05
JUC	4783	X1	Rolladen-Schneider LS-8-18	8305		4.00	G.P.Stingemore	Winthorpe	6. 4.05
JUD	4784	Z19	Rolladen-Schneider LS-8-18	8309		3.00	D.S.Haughton	Long Mynd	12. 2.05
JUE	4785	X15	Rolladen-Schneider LS-8-18	8295		3.00	G.A.Chalmers	Easterton	2. 9.05
JUF	4786	46	Schempp-Hirth Ventus 2cT	53/174		4.00	M.H.Pope	Bidford	20. 3.05
JUG	4787	JUG	Issoire E78 Silene	9	F-CFED	7.00	J.Sanders	Ingolstadt	11.11.05
JUH	4788	R6	Schleicher ASW-27B	27129		4.00	RAFGSA Centre	RAF Halton	6. 1.06
JUJ	4789	Z1	Schleicher ASW-27B	27127		4.00	D.R.Campbell	Booker	21. 3.05
JUK	4790	554	Grob G102 Astir CS	1430	PH-552	6.00	P.G.Freer	Booker	3. 5.05
JUL	4791	621	Schleicher ASW-27B	27132		4.00	T.Stuart	Nympsfield	12. 2.05
JUM	4792	2UP	Schempp-Hirth Duo Discus T	5/243		4.00	B.A.Bateson	Parham Park	12. 4.05
JUN	4793	M19	Schleicher ASW-19B	19096	D-3844	5.00	M.P.Roberts	Gransden Lodge	5. 5.05
JUP	4794	183	Schempp-Hirth Discus 2b	60		5.00	P.J.Ward	Aston Down	7. 3.05
JUQ	4795	W1	Schempp-Hirth Ventus 2cT	55/179		5.00	J.Wright	Rivar Hill	6. 4.05
JUR	4796	JUR	Valentin Mistral C	MC042/81	HB-1596	5.00	Essex and Suffolk Gliding Club	Wormingford	11. 2.05
JUS	4797	JUS	Grob G102 Astir CS	1403	(BGA.4774) D-4269	5.00	Rattlesden Gliding Club	Rattlesden	23. 2.05
JUU	4799	JUU	Schempp-Hirth Standard Cirrus	450	PH-500	5.00	S.K. Ruffell	Rufforth	17. 6.05
JUV	4800	SH2	Schempp-Hirth Discus b	551	D-8257	6.00	Surrey and Hampshire Gliding Club	Lasham	24. 3.05
JUW	4801	4T	Schleicher ASW-19B	19074	D-4476	5.00	E.Cole	RAF Keevil	29.12.04
JUX	4802	CD1	Avia Stroitel AC-4c *(Formerly known as Federov Me-7)*	051		5.00	R.J. Walton	Winthorpe	18. 2.05
JUY	4803	JUY	Valentin Mistral C	MC041/81	D-4941	5.00	C.A.Pennifold	Lee-on-Solent	5. 6.04
JUZ	4804	SH6	Schleicher ASW-19B	19146	D-7932	6.00	J.P.Eldem	Walney Island	7. 1.05
JVA	4805	JVA	Schempp-Hirth Ventus 2cT	66/203		8. 1.02	N.A.Leigh	Camphill	11. 2.05
JVB	4806	DF	Schempp-Hirth Discus bT	111/462	D-KUNK	5.00	C.J.Edwards	Nympsfield	11.12.05
JVC	4807	JVC	SZD-51-1 Junior	B-1799	SP-3434	6.00	Yorkshire Gliding Centre	Rufforth	19. 2.05
JVE	4809	JVE	Eiri PiK-20D	20631	OY-XJC	6.00	S.R.Wilkinson	Kirton-in-Lindsey	28. 4.05
JVF	4810	JH1	Schempp-Hirth Discus CS	271CS		6.00	J.Hodgson	Wormingford	26. 5.05
JVG	4811	420	Schempp-Hirth Discus bT	121/477	D-KSOP	6.00	P.J.Charnell	Lasham	15. 6.05
JVH	4812	AC4	Avia Stroitel AC-4c	052		6.00	G.Bowser	Shobdon	19. 2.05
JVJ	4813	JS	LAK-17A	108		6.00	J.A.Sutton	Milfield	24. 2.05
JVK	4814	CP	Rolladen-Schneider LS-6c-18	6236	D-6417	7.00	M.F.Collins	Lasham	27. 7.05
JVL	4815	JVL	Glaser-Dirks DG-300 Elan	3E158	HB-1833	7.00	A.M.Blackburn	Camphill	13. 3.05
JVM	4816	GP	Schleicher ASW-27B	27138		6.00	G.K.Payne	Dunstable	1.12.05
JVN	4817	T4	Schleicher ASW-27B	27134		6.00	N and R Tillett *(Hit hedge on approach to Booker, 25 .8.03)*	Dunstable	13. 1.04
JVP	4818	D8	Glaser-Dirks DG-200	2-1	D-8200	7.00	M.P.Ellis and Partners	Burn	31. 5.05
JVQ	4819	275	Schleicher ASW-27B	27136		7.00	M.R.Fountain	Booker	11. 4.05
JVR	4820	540	Schempp-Hirth Discus 2b	72		7.00	P.Davies	Lasham	2. 7.05
JVS	4821	S1	Schleicher ASW-28	28003	D-4008	9.00	S.J.Kelman	Gransden Lodge	10. 1.06
JVT	4822	JED	Schempp-Hirth Nimbus 3-25	5 37	D-3176	7.00	J.R.Edyvean	Winthorpe	17. 5.05
JVU	4823	C74	Lanverre Cirrus CS11-75L	28	F-CEVT	7.12.00	C.R.Coates	Snitterfield	23. 1.05
JVV	4824	J50	Sch-Hirth Janus Ce	176	D-4150	8.00	G.R.Jenkins	Lasham	3. 2.05
JVW	4825	15A	Schleicher ASW-15A	15042	HB-992	20. 2.01	G.J.Armes	Rattlesden	15. 4.05
JVX	4826	163	Schempp-Hirth Discus CS	087CS	D-0263	8.00	M.A.Gatehouse	North Hill	
JVY	4827	F6	Schempp-Hirth Discus b	175	RAFGSA.R6	8.00	M.R.Garwood	Husbands Bosworth	7.12.05
JVZ	4828	JVZ	Schleicher ASK-21	21721		12.00	Yorkshire Gliding Club *"Sharpe's Classique"*	Sutton Bank	14.12.05
JWA	4829	E1	Schleicher ASW-28	28005		10.00	R.A.Cheetham	Winthorpe	24.10.05
JWB	4830	JWB	Schleicher ASK-13	13671AB	D-1066	12.00	East Sussex Gliding Club	Ringmer	24. 2.05
JWC	4831	900	Schleicher ASW-27B	27142		10.00	C.G.Starkey	Lasham	2.12.05
JWD	4832	JWD	Schleicher ASK-21	21724		16. 2.01	London Gliding Club	Dunstable	2. 3.05
JWE	4833	JWE	Slingsby T.21B	1159	XN155	1. 7.00	M.Selss	Bad Tolz	13. 7.05
JWF	4834	S27	Schleicher ASW-27B	27144		11.00	B.A.Fairston	Husbands Bosworth	8. 9.05
JWG	4835	880	Schempp-Hirth Nimbus 3DT	11	D-KMGD	11. 2.01	T.P.Browning	Lasham	8.12.05
JWH	4836	C2	Schempp-Hirth Standard Cirrus	256	SE-TMZ	14. 6.01	R.J.Griffin	Edgehill	20. 3.05
JWJ	4837	R38	Schleicher ASK-13	13599	RAFGSAR38 RAFGSA R3	11.00	Wrekin Gliding Club	RAF Cosford	7.12.05
JWK	4838	722	Schempp-Hirth Discus bT *(Not ex HB-1680; rebuilt from parts)*	106		11.00	D.E.Barker	Nympsfield	13.12.05
JWL	4839		Schempp-Hirth Ventus B	125	D-6667	11.00	S Weber	Dettingen,Germany	26.11.05
JWM	4840	T12	Grob G.103 Twin II	3536	D-8730	11.00	Norfolk Gliding Club	Tibenham	25. 1.05

JWN	4841	UY	Rolladen-Schneider LS.4	4728	D-7008	24. 2.01	D.Bartek	Speyer, Germany	11. 3.05
JWQ	4843	310	Schempp-Hirth Discus 2A	82		11.00	P.G.Sheard	Dunstable	11.11.05
JWR	4844	NJ1	Grob G.102 Astir CS	1271	D-7366	1.01	N.J.Irving	Portmoak	3. 1.06
JWS	4845	JWS	Schleicher ASW-15B	15098	D-4656	14. 4.01	B. Pridgeon	Burn	26. 6.05
JWT	4846	JWT	Glaser-Dirks DG-200	2-42	D-6560	11.00	K.R.Nash	Aston Down	30. 5.05
					D-6660				
JWU	4847	GA	Schempp-Hirth Ventus bT	19/159	ZS-GOW	10.00	G Tabbner	Gransden Lodge	24. 2.05
JWV	4848	S60	Glasflugel 201B Standard Libelle	411	OY-XBG	14. 3.01	G.K.Drury	Challock	8. 1.05
JWX	4850	63	Schempp-Hirth Ventus 2cT	63/198		20.12.00	S.G.Olender	Santo Tome del Puerto, Spain	13. 4.05
JWZ	4852	W22	Schleicher ASW-22	22037	D-3422	22. 1.01	D.Prosolek	Saltby	16.12.05
JXA	4853	Y44	Schempp-Hirth Nimbus 3dT	9	D-KKYY	3. 2.01	B.C.Morris	Booker	6. 2.05
					D-4444				
JXB	4854	JXB	Centrair 201 Marianne	201-015	F-CGMN	27. 1.01	C.A.Sheldon	Pocklington	21. 2.05
JXC	4855	JXC	Wassmer WA.28F	102	F-CDZV	27. 3.01	A.P.Montague	Nympsfield	10. 6.05
JXD	4856	JXD	Slingsby T.21B	622	WB961	20. 1.01	F.Brune	Eudenbach, Germany	11. 5.05
JXE	4857	JXE	SZD-22C Mucha Standard	F-717	SP-2330	12. 8.01	C.E.Harwood	Challock	11. 6.05
JXF	4858		Bölkow Phoebus B1	875	(BGA.4842)	R18. 1.01	P.A.Hearne	Not known	
					D-0128				
JXG	4859	W5	Eiri PIK-20D	20660	PH-670	23. 2.01	T.J.Clubb	Lee-on-Solent	19. 3.05
JXH	4860	DH2	Schleicher ASW-20L	20067	D-7657	17. 2.01	G.M.Brightman	Edgehill	2. 5.05
JXJ	4861	W7	Schleicher ASW-28	28012		29. 1.01	E.W.Johnston	Dunstable	4. 2.05
JXK	4862	W3	Schempp-Hirth Ventus bT	49/247	D-KLOE	25. 1.01	P.Turner	North Hill	27.11.05
JXL	4863	SG1	Schempp-Hirth Discus CS	278CS		5. 4.01	Southdown Gliding Club	Parham Park	15. 1.06
JXM	4864	R34	Schleicher ASK-13	13542	RAFGSA.R34	25. 2.01	RAFGSA Chilterns Gliding Club	Bicester	16. 5.04
					F-CERF		*(Dismantled 1.05)*		
JXN	4865	Z35	Centrair 201B Marianne	201B-035	F-CBLI	21. 2.01	E.Crooks	Kirton-in-Lindsey	14. 3.05
JXP	4866	JXP	Glaser-Dirks DG-100	18	PH-520	16. 5.01	P.W. Butcher	Husbands Bosworth	1. 6.05
JXQ	4867	JXQ	Wassmer WA.26P	19	F-CDQQ	26. 9.01	J.A.French	Nympsfield	25. 9.02
JXR	4868	DM	Schempp-Hirth Discus B	540	D-9152	8. 4.01	Cambridge Gliding Club	Gransden Lodge	11. 2.05
JXS	4869	JXS	Schleicher K8B	8778	RAFGSA.R95	7. 3.04	Stratford Gliding Club	Snitterfield	6. 3.05
					RAFGSA.395				
JXT	4870	CT	Schleicher ASW-24B	24233	D-6706	9. 3.01	C.Thwaites	Rufforth	12.11.05
JXU	4871	646	Rolladen-Schneider LS-8-18	8354		10. 3.01	C.J.Alldis	Long Mynd	10. 4.04
JXV	4872	JXV	Glaser-Dirks DG-100	63	PH-543	24. 3.01	R.Vos	Challock	4. 4.04
							(De-Registered 20. 5.04, sold in Germany as D-3812)		
JXW	4873	871	Schempp-Hirth Duo Discus T	7/250	D-KOZX	11. 5.01	C.Bainbridge	Wormingford	5. 5.05
JXX	4874	JXX	Pilatus B4 PC-11AF	13	HB-1112	28. 5.01	N.Buckenham	Rattlesden	20. 5.04
JXY	4875	JXY	Neukom Elfe S4A	68	HB-1267	29. 5.01	D.V. Wilson	Parham ParkRattlesden	27. 4.05
JXZ	4876	27B	Schleicher ASW-27B	27152		16. 3.01	M.Fryer	Rufforth	27.11.05
JYA	4877	WB988	Slingsby T.21B	MHL015	WB988	31. 3.01	C.Bravo	Santo Tome del Puerto, Spain	16. 4.05
JYB	4878	IZ	Glaser-Dirks DG-202/17	2-143/1738	D-1086	1. 5.01	N.Wood	Rufforth	27. 5.05
JYC	4880	JYC	Grob G102 Astir CS	1429	RAFGSA.R19	9. 4.01	R.A.Scothern	Easterton	31.10.05
					RAFGGA 742/D-7425				
JYD	4881	T6	Schleicher ASW-27B	27155		25. 3.01	J.E.Gatfield	Booker	21.10.05
JYE	4882	JYE	Schleicher ASK-13	13191	D-0347	11. 4.01	Ulster Gliding Club	Bellarena	8. 4.05
JYF	4883	N55	Schempp-Hirth Discus CS	281CS		11. 4.01	D.Bradley	Rufforth	6. 4.05
JYG	4884	OK-0833	Letov LF107 Lunak	49	OK-0833	21. 9.02	M.Launer	Rossfeld. Germany	24. 9.05
JYJ	4886	2	Schempp-Hirth Ventus 2cT	64/199		1. 5.01	P.G.Myers	Chipping	24. 4.05
JYL	4888		LAK-12 Lietuva	6197	(Slovenia)	9. 5.01	C.Arrigo	Udine, Italy	11. 6.04
JYM	4889	LR	Schempp-Hirth Discus 2A	86		9. 5.01	S.Meriziola	Rieti, Italy	26. 2.05
JYN	4890		Schempp-Hirth Discus 2B	94		9. 5.01	A.Carugati	Saronno,, Italy	12. 6.04
JYP	4891	B12	Grob G102 Club Astir II	5018C	D-8743	19. 5.01	Norfolk Gliding Club	Tibenham	11. 9.04
JYQ	4892	485	Glaser-Dirks DG-101G Elan	E181G147	D-1485	15. 6.02	A.M.Booth	Wormingford	13. 7.05
JYR	4893	B20	Schempp-Hirth Duo Discus T	16/267	D-KOZX	3. 7.01	B.F.Walker	Nympsfield	6. 7.05
JYS	4894	878	Schempp-Hirth Mini Nimbus C	106	HB-1437	6. 6.01	A.Jenkins	Shobdon	10. 6.05
JYT	4895	E4	Schempp-Hirth Ventus 2B	114	(BGA.4885)	15. 6.01	J.C.Bastin	Booker	22. 3.05
JYU	4896	R11	Schempp-Hirth Ventus 2cT	70/216		15. 6.01	RAFGSA Centre	RAF Halton	21. 1/05
JYV	4897	JYV	Schleicher K8B	133	D-8395	18. 6.01	European Soaring Club	Le Blanc, France	4. 7.04
JYW	4898	JYW	Schleicher K8B	8432A	D-5682	18. 6.01	European Soaring Club	Le Blanc, France	4. 7.04
JYX	4899	JYX	Rolladen-Schneider LS-3-17	3289	D-3517	18. 6.01	D.M.Beck	Ad Ontur, Spain	24. 6.05
JZA	4902	JZA	Start+Flug H101 Salto	27	D-2997	10. 7.01	B. Merieau	(France)	17. 7.04
JZB	4903	JZB	DG Flugzeugbau DG-505 Elan Orion	5E223X61		28. 7.01	Faulkes Flying Foundation	Long Mynd	13. 5.05
							"Jezabel"		
JZC	4904	M9	Rolladen-Schneider LS-8-18	8395		11.3.02	M.H.Patel	Booker	2. 3.05
JZD	4905	JZD	Dittmar Condor IV	018	(D-0125)	7.3.02	P.J.Underwood	Dunstable	3. 7.05
					LV-DHV				
JZE	4906	SG2	Schleicher ASK-13	13423	OY-XPJ	11.11.01	Southdown Gliding Club	Parham Park	1. 3.05
					D-2125				
JZG	4908	SM	Schempp-Hirth Discus bT	9/272	D-KISM	9. 7.01	R.J.Middleditch	Nympsfield	7. 8.04
JZH	4909	JZH	Schleicher ASW-20CL	20754	D-5932	10. 7.01	C.P.Gibson	Lasham	15.12.05
JZJ	4910		Schleicher ASW-15	15147	D-0791	R12. 7.01	K.Sleigh	Rattlesden	
							(Not taken up ; to OM-1515 3.02)		
JZK	4911	JZK	DG Flugzeugbau DG-505 Elan Orion	5E225X63		1. 9.01	Faulkes Flying Foundation	Gransden Lodge	12. 3.05
JZL	4912	BS	Schempp-Hirth Mini Nimbus B	92	HB-1453	25. 8.01	S.J.Ware and Partner	Kirton-in-Lindsey	17. 3.05
JZM	4913	110	Schempp-Hirth Ventus 2Ax	117		24. 8.01	R.Jones	Lasham	7. 4.05
JZN	4914	JZN	Schleicher ASW-28	28038		19.12.01	I.C.Lees	Rufforth	9. 1.05
JZP	4915	JZP	Marganski Swift S1	119	F-CIAB	26. 9.01	I.E.Tunstall	Winthorpe	19. 5.05
JZQ	4916		Edgley EA9 Optimist	006		R3. 9.01	T.Henderson	Not known	
JZR	4917		Edgley EA9 Optimist	008	(BGA.4731)	R3. 9.01	University of London	Not known	
JZV	4921	V2C	Schempp-Hirth Ventus 2cT	72/225		14. 9.01	P.McLean	Tibenham	16. 4.05

JZW	4922	JZW	Grob G102 Astir CS	1208	D-7281	27. 9.01	R.Theil	RAF Cranwell	26. 9.03
							(De-Registered 2.12.02, sold in Japan as JA-00YH)		
JZX	4923	P10	Schleicher ASW-27	27166		22. 9.01	P.R.Barley	Halton	1.11.05
JZY	4924	SH7	Grob G102 Standard Astir III	5600S	D-6951	2.10.01	Surrey and Hampshire Gliding Club	Lasham	9. 1.05
JZZ	4925	JZZ	Rolladen-Schneider LS-7WL	7128	SE-UIU	28.3.02	J.H.Tucker	Crowland	3. 3.05
KAA	4926	XE790/KAA	Slingsby T31B	903	XE790	14. 9.02	N.Stalpers	Castricum, The Netherlands	25. 9.05
KAC	4928	KAC	Glaser-Dirks DG-200	2-159	HB-1611	3.10.01	C.Morton-Fincham	Kirton-in-Lindsey	19. 7.05
KAD	4929		Glaser-Dirks DG-300 Elan	3E259	D-8411 OE-5420	5.10.01	N.G.Maxey *(De-Registered 11. 8.03, sold in New Zealand as ZK-GBA)*	Challock	16.12.03
KAE	4930	KAE	Centrair 101A Pégase	101A-0152	F-CGBN	4.6.02	Rattlesden Gliding Club	Rattlesden	12. 1.05
KAF	4931	DD2	Schempp-Hirth Duo Discus T	49/330		24. 7.02	M.R.Smith	Aboyne	20. 8.05
KAG	4932	EE	Schempp-Hirth Nimbus 3T	23/89	D-KMHF	15.11.01	K.Engelhardt	Lusse, Germany	15. 9.05
KAH	4933	KAH	Schempp-Hirth Discus Bt	464/112	D-KNZZ	13.11.01	R.M.Brown	Bicester	7. 1.05
KAJ	4934	KAJ	Schempp-Hirth Ventus 2cT	86		18.5.02	A.N.Reddington	Perranporth	3. 5.05
KAK	4935	J1	Schleicher ASW-28	28032		1.11.01	R.A.Johnson	Husbands Bosworth	7. 4.05
KAL	4936	A28	Schleicher ASW-28	28031		13.11.01	P.A.Ivens	Aston Down	1. 4.05
KAM	4937	KAM	Glasflugel H205 Club Libelle	83	D-8928	10.12.02	M.W.Black	Easterton	16. 3.05
KAN	4938	KAN	SZD-50-3 Puchacz	B2106	PH-1104	10.12.01	Bath, Wilts and North Dorset Gliding Club	Kingston Deverill	4. 4.05
KAP	4939	KAP	Schempp-Hirth Discus CS	290CS		16.2.02	A.A.Stewart	Portmoak	8. 1.06
KAQ	4940	BG	LAK-17A	125		8.12.01	LAK Deutschland	Musbach, Germany	7.12.02
							(De-Registered 4. 3.03, became D-7417 with same owner)		
KAR	4941	977	Schempp-Hirth Duo Discus T	35/309	D-KHAF	3.12.01	J.P.Galloway	Portmoak	13. 2.05
KAS	4942	KAS	Schempp-Hirth Ventus 2cT	93	(BGA.4995)		J.E.Bowman *(See BGA.4995)*	Bidford	14. 4.05
KAT	4943	520	Schempp-Hirth Ventus 2cT	82/249		22.3.02	P.C.Naegeli	Lasham	4. 4.05
KAU	4944	KAU	DG Flugzeugbau DG-303 Elan Acro	3E500A35		20.4.02	G.Earle	Lasham	15. 4.05
KAV	4945	GG	Rolladen-Schneider LS-4A	4696	D-1055	11. 2.02	G.S.Goudie	Gransden Lodge	6. 1.06
KAW	4946	KAW	DG FlugzeugbauDG-505 Elan Orion	5E228X66		10.3.02	Faulkes Flying Foundation	Not known	11. 2.05
KAX	4947	KAX	DG FlugzeugbauDG-505 Elan Orion	5E229X67		12.4.02	Faulkes Flying Foundation	Rufforth	30. 4.05
KAY	4948	KAY	Grob G102 Astir CS	1452	D-7433	7.2.02	D.A.Woodforth	Saltby	26. 1.04
							(As "D-7433" under wing and "BGA.4847" on fin)		
KAZ	4949	32	Glaser-Dirks DG-600	6-5	VH-GHS D-1666	19. 3.02	M.Geisen	Bremerhaven	14. 1.05
KBA	4950	KBA	Centrair 101 Pégase	101A-0435	HB-3096	28.3.02	K.J.Sleigh	Rattlesden	3. 3.05
KBB	4951	KBB	Schempp-Hirth Mini Nimbus C	147	HB-1508	12.2.02	K.E.Ballington	Cross Hayes	13. 7.04
KBC	4952	C66	Rolladen-Schneider LS-6c *(Rebuild of c/n 6250 in Poland)*	6250-1			A.P.Hatton *(See BGA.3895, still current as c/n 6250)*	Husbands Bosworth	1. 3.05
KBD	4953	NU	Grob G102 Astir CS	1217	D-7290	4. 4.02	Notts University Gliding Club	Syerston	12. 2.05
KBE	4954	SP	LAK-17A	130		14. 3.02	C.Triebel	(Germany)	13. 3.03
							(De-Registered 14. 3.03, became D-3117 with same owner)		
KBF	4955	AG	DG Flugzeugbau DG-303 Elan	3E498		10. 4.02	A.L.Garfield	Dunstable	15. 4.05
KBG	4956	71	Schempp-Hirth Ventus 2cT	83/249		15. 3.02	J.F.D'Arcy	Strathaven	7. 4.05
KBH	4957	H1	Rolladen-Schneider LS-6	6072	D-7798	4. 4.02	J.A.French	Nympsfield	6. 4.05
KBJ	4958	R	Rolladen-Schneider LS-8-18	8422		8. 4.02	W.Schneider	Speyer, Germany	21. 4.05
KBK	4959	409	Schempp-Hirth Ventus 2cT	79/244		15. 4.02	R.J.Nicholls	Husbands Bosworth	29. 4.05
KBL	4960	N12	Grob G102 Astir CS	1464	D-7436	13. 4.02	Norfolk Gliding Club	Tibenham	30.12.05
KBM	4961	73	Schleicher ASW-28	28046	D-0001	29. 3.02	M.T.Burton	Dunstable	4.11.05
KBN	4962	NH	SZD-55-1	551190004	SE-ULV	26. 3.02	N.D.Pearson	Parham Park	29. 1.05
KBP	4963	KBP/ XA310	Slingsby T.31B	852	XA310	20. 4.02	A.P.Stacey	RAF Keevil	19. 4.03
KBQ	4964	Y2	Schempp-Hirth Discus A	279	OH-807	5. 4.02	R.Priest	Booker	30. 3.05
KBR	4965		Glasflugel Libelle	57		R15.4.02	M.Geisen	Not known	
KBS	4966	KBS	Glaser-Dirks DG-600	6-42	OO-YPH D-4882	20. 4.02	H.Altmann	Booker	5. 4.05
KBT	4967	633	Schempp-Hirth Standard Cirrus	561G	D-4755	3. 6.02	P R Johnson	Rattlesden	2. 6.04
KBU	4968	104	Schleicher ASW-28	28040		18. 4.02	G C Metcalfe	Lasham	12.10.05
KBV	4969	MM	Schleicher ASW-28	28045		24. 4.02	M P Mee	Booker	12. 3.05
KBW	4970	OM-0973	Letov LF107 Lunak	22	OM-0973 OK-0973	9. 6.02	A.P.Middleton *(As "OM-0973")*	Saltby	4. 9.05
KBX	4971	421	Schleicher ASW-27	27092	PH-1146	11. 5.02	C Colton	Gransden Lodge	7. 4.05
KBZ	4973	C64	Rolladen-Schneider LS-8B	8425	D-3229	12. 5.02	P G Crabb	Husbands Bosworth	10. 5.05
KCA	4974	C65	Rolladen-Schneider LS-8B	8424	D-9503	3. 5.02	S J Crabb	Wroclaw, Poland	22. 4.05
KCB	4975	MY	Rolladen-Schneider LS-4A	4776	PH-887 D-1597	7. 6.02	Bristol and Gloucestershire Gliding Club	Nympsfield	24. 4.05
KCC	4976	V26	Schempp-Hirth Ventus 2cT	90/266		13. 5.02	D.J. Fawcett	Lee-on-Solent	10. 3.05
KCD	4977	B6	Schempp-Hirth Ventus 2cT	91/267		20. 5.02	J.Delafield	RAF Bicester	9. 5.05
KCE	4978	24	Schempp-Hirth Ventus 2cT	85/256		28. 5.02	RAFGSA Centre	Halton	1. 2.05
KCF	4979	72	Schempp-Hirth Duo Discus T	48/328		13. 5.02	J.L.Birch	Gransden Lodge	10. 3.05
KCG	4980	T99	Schleicher ASW-27B	27182		31. 5.02	T.N.McGee	Dunstable	22.12.05
KCH	4981	EA	Schempp-Hirth Ventus 2cT	56/186	PH-1191	12. 7.02	L.Marks and W.Pridell	Lasham	2. 6.05
KCJ	4982	V8	Schleicher ASW-28	28051		16. 7.02	S.L.Withall	Dunstable	15. 4.05
KCK	4983	RA	Schempp-Hirth Ventus 2cT	.94/272		5. 7.02	R.F.Aldous	Booker	18. 7.05
KCL	4984	KCL	Rolladen-Schneider LS-4	4123	D-4239	12. 7.02	R.L.Fox	Rufforth	24. 4.05
KCM	4985	J8	Glasflugel H201 Standard Libelle	104	HB-968	17. 7.02	G.A.Cox	Kingston Deverill	10. 7.05
KCN	4986	700	Schleicher ASW-27B	27188	D-0001	11. 7.02	W.J.Head and Partner	Gransden Lodge	22.12.05
KCP	4987	NG1	Grob G102 Astir CS	1094	OY-XDB	7. 8.02	Norfolk Gliding Club	Tibenham	9.12.05
KCQ	4988	KCQ	DWLKK PW-5	17-04-010	OY-XYE	R15. 7.02	P.H. Young	Chipping	11. 2.05
KCR	4989	RB	LAK-17A	132		3. 6.02	R.A.Bickers	Alzate Brianza, Italy	14. 6.04

KCS	4990	PK	Grob G103A Acro	3691-K-42	D-6940 OH-645	7. 7.02	Cairngorm Gliding Club	Feshiebridge	12.12.05
KCT	4991	KCT	Schleicher ASK-21	21751		R18. 7.02	Kent Gliding Club	Challock	25.12.05
KCU	4992	KCU	Schempp-Hirth Standard Cirrus	100	D-0741	7. 7.02	G.Carruthers *(Crashed on launch Pocklington, 6.3.03)*	Sutton Bank	6. 8.03
KCV	4993	WE4	Schempp-Hirth Duo Discus T	54/339	D-KOZZ	7. 8.02	T.W.Slater	Wormingford	10. 8.05
KCW	4994	KCW	Glaser-Dirks DG-200-17	2-137/1735	D-0153	6. 9.02	R.W.Adamson	Portmoak	19.11.05
KCY	4996	AV	Schleicher ASW-20	20068	PH-597 OY-XTM/PH-597	14.11.02	J.R.Epple	Camphill	17. 3.05
KCZ	4997	KCZ	Schleicher ASK-21	21749		9. 8.02	Booker Gliding Club	Booker	5. 3.05
KDA	4998	406	Schempp-Hirth Ventus 2BX	138	D-4999	21. 2.03	D.J.Eade *(See BGA.5008)*	Lasham	1. 4.05
KDB	4999	KDB	Schleicher Ka-6CR	6431	HB-805	20. 9.02	Aquila Gliding Club	Hinton-in-the-Hedges	25.10.05
KDC	5000	SK	Schleicher ASW-20	20254	F-CFSF	R13. 9.02	A.R.Blanchard	RAF Wattisham	15. 3.04
KDD	5001		Avia Stroitel AC-5T	027		26. 9.02	Russia Sailplanes	Not known	25. 9.03
					(Crashed Nympsfield during flight testing 19.12.02 and written off -suspected structural failure)				
KDE	5002	SI	Schempp-Hirth Duo Discus	193	PH-1141	29.10.02	M.G.Stringer	Dunstable	10.11.05
KDF	5003	E7	Schleicher ASK-21	21006	D-6539	29. 9.02	Portsmouth Naval Gliding Club	Lee-on-Solent	17.10.04
KDG	5004	6S	Glasflugel H205 Club Libelle	119	OO-YHL D-2473	22.10.02	R Southern and N Hyde *(Crashed Advie Bridge 27.7.03; wreck noted 10.04)*	Perth	21.10.03
KDH	5005	KDH	Schleicher K-8B	E4	D-8428	3. 1. 03	Midland Gliding Club	Long Mynd	16.12.04
KDJ	5006	6	Schempp-Hirth Ventus 2cT	102/293		21. 3.03	M.Nash-Wortham	Lasham	17. 3.05
KDK	5007	KDK	Rolladen-Schneider LS-4A	4352	D-4106	9.11.02	M.C.Ridger	Long Mynd	9.11.04
KDL	5008		Schempp-Hirth Ventus 2BX	138	(BGA4998)	R11.11.02	D.Eade *(Not taken up, remained as BGA.4998)*		
KDM	5009	RC	Schleicher ASH-25WL	25139	F-CHAY	15. 2.03	R.Browne	Cranwell	15.3.05
KDN	5010	140	Schleicher ASW-27B	27208		3. 1.03	R.D.Payne	Nympsfield	29.12.04
KDP	5011	KDP	Schleicher ASK-21	21760		23. 5.03	Kent Gliding Club	Challock	22.5.05
KDQ	5012	808	Schempp-Hirth Ventus 2cT	97		29.11.02	A.Milne	Kingston Deverill	15.11.05
KDR	5013	KDR	SZD-48-3 Standard Jantar 3	B1642	DOSAAF	22.10.03	S.J.Kochanowski	Gamston	21.10.04
KDS	5014	172	Schleicher ASW-27B	27202		20. 2.03	G.D.Morris	Nympsfield	17. 2.05
KDT	5015		Letov LF107 Lunak	12	OK-0975	.... 12.02	I.H.Keyser *(As "OK-0975")*	(Switzerland)	9. 9.04
KDU	5016	KDU	Glaser-Dirks DG-202-17	2-161/1752	D-6000	10. 3.03	P.G.Noonan	Edgehill	7. 3.05
KDV	5017	KDV	Schempp-Hirth Ventus B	224	HB-1770	1. 2.03	R.Theil	Crowland	27.11.04
KDW	5018	292	Schleicher ASW-27B	27196		28. 3.03	A.J.Kellerman	Gransden Lodge	7. 3.05
KDX	5019	KDX	Glaser-Dirks DG-200	2-11	D-7218	5. 2.03	B.Searle	Long Mynd	26. 1.05
KDY	5020	503	Glaser-Dirks DG-100	78	D-2591	21. 3.03	N.M.L.Claiden	Dunstable	6. 3.05
KDZ	5021	C75	Schempp-Hirth Standard Cirrus 75	696	OO-ZKE	2. 2.03	H.E.L.Williams	Syerston	18. 2.05
KEA	5022	KEA	Schempp-Hirth Cirrus 18	11	D-8807, HB-911	11. 3.03	C.M.Reed	Rattlesden	3. 4.05
KEB	5023	N	Schempp-Hirth Standard Cirrus	436G	F-CEFN	13. 3.03	P.Smith	Rivar Hill	19. 3.05
KEC	5024		Rolladen-Schneider LS-4	4469	D-5170	13. 3.03	European Soaring Centre	Le Blanc	23. 3.05
KED	5025	427	Schleicher ASW-27B	27203	D-0001	14. 2.03	P F.Brice	Booker	24. 1.05
KEE	5026	KEE	Grob G102 Astir CS	1135	D-4179	30. 3.03	T. Barton	Talgarth	27 3.05
KEF	5027		Schempp-Hirth Ventus 2BX	137			*(Reservation cancelled - to N57LK 11.02)*		
KEG	5028		Slingsby T45 Swallow	1541	EC-BHG	R24. 2.03	C Bravo	Madrid, Spain	
KEH	5029		Schempp-Hirth Duo Discus			R27. 2.03	*(Probably became BGA.5033)*		
KEJ	5030	KEJ	Schleicher ASK-21	21765		23. 7.03	London Gliding Club	Dunstable	2. 8.05
KEK	5031	KEK	Schleicher ASK-21	21767		12. 9.03	Devon and Somerset Gliding Club	North Hill	27. 3.05
KEL	5032	SK1	Rolladen-Schneider LS-8-18	8454		16. 3.03	P.D.Kaye	Gransden Lodge	15. 3.04
KEM	5033	570	Schempp-Hirth Duo Discus T	69/364		13. 3.03	H Kindell	Lasham	4. 2.05
KEN	5034	CH	Rolladen-Schneider LS-4A	4172	OO-ZSM (OO-ZDG)	20. 3.03	Miss H.R.Hay	Kingston Deverill	22.12.05
KEP	5035	KEP	Rolladen-Schneider LS-6B	6188	F-CGUI F-WGUI, D-5017	31. 3.03	P.J.Coward	Husbands Bosworth	19.9.05
KEQ	5036	866	Schleicher ASW-28	28058		6. 5.03	H Franks *(To VH-GHS 6.7.04)*	Thame	5. 5.04
KER	5037	KER	Schleicher ASW-19B	19224	OY-XJI	6. 5.03	C Osborne	Feshie Bridge	21. 3.05
KES	5038	KES	Schempp-Hirth Cirrus 18	46	D-6955 HB-955	6. 5.03	N H Hawley	Sleap	28. 4.05
KET	5039	456	Rolladen-Schneider LS-8B	8459		29. 4.03	M.B.Jefferyes	Dunstable	21. 5.05
KEU	5040	ED	Grob G102 Astir CS	1112	OY-XDE	30. 4.03	E Weaver	Wattisham	7. 5.05
KEV	5041	R2	Schempp-Hirth Duo Discus	368		15. 4.03	RAFGSA	Cranwell	27. 2.05
KEW	5042	FB	Rolladen-Schneider LS-7WL	7009	F-CGYA F-WGYA, D-1272	22. 5.03	J J M Riach	Feshiebridge	29. 5.05
KEX	5043	KEX	Schleicher ASW-19B	19115	I-IUUH	18. 5.03	P R Porter	Lyveden	19. 5.05
KEY	5044	KEY	Schleicher ASW-20CL	20840	D-3171	23. 5.03	K W Payne	Husbands Bosworth	10. 5.05
KEZ	5045	EZ	Rolladen-Schneider LS-8T	8464	D-KSAB	20. 5.03	I.M.Evans	Long Mynd	20. 5.05
KFA	5046	C75	Schempp-Hirth Standard Cirrus 75	644	D-2124 F-CEMQ	9. 5.03	S Withey	Talgarth	8. 5.04
KFB	5047	BK	Schempp-Hirth Discus 2T	30	D-KOZZ	7. 5.03	M Holmes	Booker	5.11.05
KFC	5048	M2	Schempp-Hirth Ventus 2cT	105	D-KOZZ	17. 5.03	M Emmett	Booker	28. 3.05
KFD	5049	906	Schleicher ASW-27B	27211		12. 5.03	W T Craig	Dunstable	28. 2.05
KFE	5050	61	Eiri PiK-20D	20520	D-8103 OE-5103	19. 7.03	M. McSorley	Bellarena	13. 7.05
KFF	5051	KFF	SZD-50-3 Puchacz	B2111	(N806JH) N503SZ	11. 6.03	Derbyshire and Lancashire Gliding Club	Camphill	18. 4.05
KFG	5052	KFG	Grob G103A Twin II Acro	3771-K-57	D-1339	25. 6.03	Surrey Hills Gliding Club	Kenley	24. 6.05
KFH	5053	PC	Schempp-Hirth Mini Nimbus	15	D-4819	12. 7.03	P Purdie	Lasham	17. 7.05
KFJ	5054	KFJ	Schleicher ASK-13	13136	PH-383	15. 7.03	York Gliding Centre	Rufforth	8. 6.05
KFK	5055	2W	VTC Standard Cirrus	203	S5-3058 SL-3058, YU-4295	8. 7.03	J Wingett	North Hill	7. 7.05
KFL	5056	Y11	Rolladen-Schneider LS-4	4080	HB-1619	1. 8.03	D A O'Brien	Tibenham	25. 2.05
KFM	5057	149	Rolladen-Schneider LS-8-18	8475	D-9502	28. 6.03	A J H Smith	Tibenham	10. 5.05

KFN	5058	DS2	DG Fzb DG-1000S	10-29S28		5. 7.03	Yorkshire Gliding Club	Sutton Bank	30. 4.05
KFP	5059	210	Schempp-Hirth Ventus 2cxT	110	D-KKAH	18. 6.03	Southern Sailplanes	Membury	5. 4.05
KFQ	5060	250	Schempp-Hirth Ventus 2cxT	108	D-KXIK	3. 7.03	Southern Sailplanes	Membury	13.10.05
KFR	5061	KFR	Schleicher ASK-13	13433	D-3536	15. 9.03	European Soaring Centre	Le Blanc	14 .9.04
					RAFGGA.535, D-3535				
KFS	5062	M3	Rolladen-Schneider LS-8A	8482		2. 7.03	I M Evans	Long Mynd	1. 7.04
							(De-Registered 31.10.03 - sold in Japan as JA-221A)		
KFT	5063	UP2	Schempp-Hirth Duo Discus T	77	D-KIIH	31. 7.03	L Merritt	Saltby	30. 7.05
KFU	5064	X11	Schempp-Hirth Ventus 2cT	114/311	D-KOAX	4. 8.03	FB Jeynes	Bidford	4. 8.05
KFV	5065	111	Rolladen-Schneider LS-8T	8476	D-KOBP	19. 8.03	M Lassan	Snitterfield	9. 8.05
KFW	5066		Grunau Baby IIB	5		R16. 8.03	R Slade	Not known	
			(Possibly ex D-5149, to be confirmed)						
KFX	5067	T11	Centrair 101AP Pégase	101-029	HB-1664	18.11.03	K.J.Sleigh	Sandhill Farm, Shrivenham	17.11.04
KFY	5068		Schleicher ASK-21	21776		20. 3.04	Cambridge Gliding Club	Gransden Lodge	19. 3.05
KFZ	5069		Grob G103A Twin II Acro	34012-K-245	ZE659	R8. 9.03	T Dews	Kingston Deverill	
					BGA.3089				
KGA	5070	370	Schempp-Hirth Ventus 2cxT	115/312	D-KIBL	13. 4.04	D Campbell	Booker	12.4.05
KGB	5071	T3	Schempp-Hirth Ventus 2cxT	116	D-KDAH	4. 1.04	D.Irving	Portmoak	3.1.05
KGC	5072	J6	Schempp-Hirth Ventus 2cxT	118	D-KMAF	R8. 9.03	C Marren	Not known	18.12.04
KGD	5073	V9	Schempp-Hirth Ventus 2cxT	119/318	D-KEAD	27. 2.04	C.Morris and H Bosworth	Not known	26.2.05
KGE	5074		Slingsby T38 Grasshopper	'785'		R11. 9.03	W den Baars	Not known	
			[Possibly c/n 778 ex WZ782 reported with W den Baars, Haamstede]						
KGF	5075	233	Schempp-Hirth Duo Discus T	84		11. 2.04	J.McNamee	Bicester	10. 2.05
KGG	5076		Schleicher Ka-6CR	6646	HB-924	R22.9.03	K Ballington	(Cross Hayes	
KGH	5077	KGH	Grob G102 Club Astir II	5057C	OO-ZVS	8.10.03	Wolds Gliding Club	Pocklington	3.11.05
KGJ	5078	N3	Schleicher ASK-21	21770		2.11.03	Portsmouth Naval Gliding Club	Lee-on-Solent	22.10.05
KGK	5079	R28	Schleicher ASK-21	21766		24.10.03	RAFGSA	Bicester	6.11.05
KGL	5080	X4	Schempp-Hirth Ventus 2cT	88/263	EI-152	6. 3.04	W.D.Ingles	Bidford	5.3.05
KGM	5081		Centrair Pegase	'10155'		R11.11.03	European Soaring Centre	Le Blanc, France	
KGN	5082		Schleicher ASW-28	28505	D-3063	R21.11.03	M.German	Not known	
KGP	5083	GB	Schempp-Hirth Nimbus 3T	16/77	OO-YEN	5. 6.04	I Molesworth	Challock	4. 6.05
KGQ	5084	565	Schempp-Hirth Ventus 2c	28/75	D-0602	19. 1.04	F.Birlison	Aston Down	13.12.05
KGR	5085	608	Rolladen-Schneider LS.4a	4301	D-6086	26. 1.04	R.de Abaffy	Seighford	25.1.05
					D-KOTT				
KGS	5086	C9	Eiri PiK-20	20058	OY-XCP	16. 4.04	K Sleigh	Upwood	15. 4.05
KGT	5087		VTC Standard Cirrus	294	HA-4283	R1.04	R.A.Sheard	Not known	
KGU	5088		Schleicher ASW-19B	19208	PL-68	R1.04	K.Sleigh	Rattlesden	
KGV	5089	KGV	Schleicher ASW-28-18	28512	D-7062	4.11.04	A.Reynolds	Long Mynd	13. 7.05
KGW	5090	SF	Rolladen-Schneider LS-3a	3166	I-MMST	28. 2.04	A.Birardi	Winthorpe	27.2.05
					HB-1362				
KGX	5091	KGX	Schleicher ASK-21	21782		2. 7.04	The Soaring Centre	Husbands Bosworth	1. 7.05
KGY	5092	KGY	Scheibe Bergfalke IV	5839	SE-TLL	9. 4.04	North Devon Gliding Club	Eaglescott	8. 4.05
KGZ	5093	57	Schempp-Hirth Discus 2A	196		18. 3.04	M.Young	(France)	17.3.05
KHA	5094	KHA	SZD-51-1 Junior	B-1918	(SP-3691)	16. 6.04	Devon & Somerset Gliding Club	North Hill	15. 6.05
					D-2843, DDR-2803				
KHB	5095	635	Rolladen-Schneider LS-3	3316	D-2635	7. 5.04	P.Duntthorne	Aston Down	6.5.05
KHC	5096	KHC	DG Fzb DG-505/20 WL	5E178W11	D-6401	22. 5.04	P.Holland	Kirton-in Lindsey	21. 5.05
KHD	5097	T4	Schleicher ASW-27B	27222		7. 4.04	N.Tillett	Dunstable	13.1.06
KHE	5098	0118	LAK-17A	122	OM-0118	9. 5.04	N.Gough	Barkston Heath	8.5.05
KHF	5099	XDZ	Schleicher ASW-20	20229	D-3162	16. 5.04	C.Brown	Chipping	15. 5.05
KHG	5100	K5	Schleicher ASW-27B	27223		7. 5.04	B.King	Booker	6. 5.05
KHH	5101	KHH	Schleicher ASK-13	13171	LN-GAX	4. 4.04	P.Carrington	Strubby	3. 4.05
KHJ	5102	KM	Schempp-Hirth Std. Cirrus	577G	F-CEMG	16. 5.04	K.Barker	(France)	15. 5.05
							(Registered as G-CDDB 10.04)		
KHK	5103	KHK	Schempp-Hirth Duo Discus T	426/100		28. 8.04	B.Parslow and Partners	Chipping	27.8.05
KHL	5104		Rolladen-Schneider LS-1f	452	F-CEKZ	R17. 5.04	K.Sleigh	Rattlesden	
KHM	5105		Centrair Pegase	101A-0359	F-CHDE	R17. 5.04	K.Sleigh	Rattlesden	
KHN	5106	KHN	SZD-51-1 Junior	B-2142	OE-5614	10. 6.04	Nene Valley Gliding Club	Upwood	9. 6.05
					SP-3612				
KHP	5107	70	Rolladen-Schneider LS-8-18	8337	N818FD	2. 6.04	S.May	Dunstable	1. 6.05
KHQ	5108		Schempp-Hirth Ventus 2cXT	134		16.12.04	R.Wilson	Aboyne	15.12.05
KHR	5109	KHR	SZD-51-1 Junior	B-1775	HB-1928	27. 6.04	Wolds Gliding Club	Pocklington	26. 6.05
KHS	5110	KO	Rolladen-Schneider LS-7WL	7043	PH-862	26.10.04	G.Coles	Wormingford	25.10.05
					D-5157				
KHT	5111	424	Schempp-Hirth Std. Cirrus	259G	D-1139	1. 9.04	M.Holder	Lasham	31. 8.05
KHU			Not allotted						
KHV	5112	AD	Glaser-Dirks DG-100	77	HB-1331	26. 7.04	A.Roberts	Not known	25. 7.05
KHW	5113	KHW	SZD-50-3 Puchacz	503A-004-001		17. 8.04	Derbyshire and Lancashire Gliding Club	Camphill	16. 8.05
KHX	5114		Schleicher ASW-28-18E	28720	D-KOZZ	R12 .7.04	Forman	Not known	
KHY	5115	620	Schempp-Hirth Duo Discus T	95	D-KOZZ	9. 7.04	D.McCarthy	Lasham	8. 7.05
KHZ	5116	PE	Schempp-Hirth Nimbus 3DT	25	D-KEPE	11. 8.04	T.Salter	Lasham	10. 8.05
KJA	5117		Schleicher ASW-28-18	28712	D-6051	R28. 7.04	M.Fryer	Not known	
KJB	5118	LW	Schempp-Hirth Ventus bT	191/26	D-KBST	15. 9.04	G.Barley	Usk	14. 9.05
KJC	5119	617	Schempp-Hirth Nimbus 3T	57/6	D-KUPA	14. 9.04	A.Wright	Camphill	13. 9.05
					OY-KHX, D-KHXB				
KJD	5120		Schempp-Hirth Cirrus	241	F-CDOZ	R7. 9.04	Rebbeck	Not known	
KJE	5121	321	Rolladen-Schneider LS-8-18	8498		24.11.04	M Wells	Not known	23.11.05
KJF	5122		Schempp-Hirth Std. Cirrus	414G	OY-XGW	R22. 9.04	O Walters	Not known	
					D-9247				
KJG	5123		Schempp-Hirth Cirrus	153	EC-CKO	R24. 9.04	S.Wright		
KJH	5124	KJH	Glaser-Dirks DG-303	3E-506		8.12.04	Yorkshire Gliding Club	Sutton Bank	7.12.05

KJJ	5125	KJJ	DG Fzb DG-505 Orion	5E249X79		12.12.04	Ulster Gliding Club	Bellarena	11.12.05
KJK	5126	C66	Schempp-Hirth Janus Ce	283	D-3666	7.1.05	A.McWhirter	Not known	6.1.06
KJL	5127		Schleicher ASK-13	13468	D-2338	8. 9.04	Lincolnshire Gliding Club	Strubby	7. 9.05
KJM	5128	7A	Schempp-Hirth Ventus cT	394/120	D-KAHE OH-781	5.11.04	J.Ferguson	Portmoak	4.11.05
KJN	5129	9E	Schleicher ASW-20	20052	D-7964	15.12.04	R Logan	Bellarena	14.12.05
KJP	5130	R12	Schleicher ASK-21	21783	D-0001	9.11.04	Bannerdown Gliding Club	RAF Keevil	8.11.05
KJQ			Omitted in error						
KJR	5131		Schleicher ASW-28T	28723		R8.11.04	B.Bateson	Not known	
KJS	5132		Schleicher ASW-28-18E	28713	D-KHJW	6.11.04V	J.Warren	Not known	11.12.04
KJT	5133	AA	Schempp-Hirth Janus A	Not known	N2AA	11.12.04	D.Cott	Bicester	10.12.05
	5134 to 5143		Omitted in error						
KJU	5144		Rolladen-Schneider LS-3-17	'318'	D-2637?	R24.11.04	M Bewley	Not known	
KJV	5145		Schleicher ASW-28-18E	28725		R24.11.04	A.Price	Not known	
KJW			Omitted in error						
KJX	5146		Schempp-Hirth Ventus cT	332/87	D-KCHG	R3.12.04	D.Jones	Not known	
KJX	5147		Schempp-Hirth Ventus 2CxT	'137-04'		R26.1.05	D.Zarb	Not known	

PART 2 (i) – BGA COMPETITION NUMBERS

BGA Competition Numbers are issued to members and/or pilots and not to individual gliders. They change frequently. There is no formal list of Competition Numbers or control over other gliders wearing similar or previous numbers or other (tail) codes- see Part 2 (ii) below. Therefore, this listing is a composite one based on BGA information and reported sightings. Competition Numbers marked with an asterisk indicate where gliders have been observed with the numbers shown although not listed in the current BGA record. In some cases because joint (syndicate) ownership is common this means that the Competition Number belongs to a member of a syndicate other than the one whose name appears as owner in the BGA's records. The member's name is shown where the glider concerned is not identified. The missing numbers are not allocated.

No.	BGA No.
1	1144*, 1804*, 3123
2	869*, 3529*, 4886
3	2898*
4	(W A Kahn)
5	(G.D.Green)
6	5006
8	3720
9	1317*, (R.C.Ellis)
11	4606*
13	3323*
14	1958
15	4563*
16	4472
18	1475
19	4123
20	2951
21	(M I Gee), 4611*
22	T S Zealley
23	3965*
24	4978
25	3973
26	4473
27	4502
29	3581
31	4251
32	4949
34	3629*
35	3600
36	4739
37	2556*
38	4047*
39	3427
40	1788*
41	R F Aldous
42	4478
44	(T B Sarjeant)
45	3574
46	2458, 4786
47	1435*
48	4079
49	1797
50	2632
51	4522
52	3713
53	1763
54	(R Jones)
55	3497*
56	4337
57	5093
58	(J.F.D'Arcy)
59	4245
60	(S H Marriott)
61	1262, 5050
62	1514, 4088*
63	4850
64	4222
65	(A K Lincoln)
66	4977
67	1052
68	3361
69	3419
70	5109
71	4956
72	4979
73	4092*, 4961*
74	2224
75	4267*
76	1761
77	1377
78	3845
79	2528*, 3455
80	4642
81	2453*
82	3920
83	3680
84	4261
85	(D J Robertson)
86	2310
87	4608
88	2974
89	(J A Millar)
90	2981
91	(G P Stingemore)
92	1839
93	3228
94	(R C Bromwich)
95	1989, 4751
96	3807
97	(BGA)
98	4350
99	4157
100	3962
102	2494
103	4760
104	4968
105	4078
106	2513
107	(G O Avis), 2524*
108	3606
109	(M J Botwinski)
110	5060
111	5065
112	4344
113	1013*, 3158
114	3791
116	2124
117	1386*, 2751*
118	2911
120	3197
121	1316
122	4046
123	4370
124	1530*
125	3317
126	3913
127	3355*
128	2071
130	(R Lewin)
131	(W J Dean)
132	2455
134	(P.G. Purdie)
136	3273
137	(S J Parsonage), 3861*
138	2698
139	1726
140	5010
141	3244
142	1629
143	3298*
144	4489
145	(N L Jennings)
146	1876
147	2466
148	(S.Burton)
149	5057
150	1269*, 4100
151	1686
152	2501
153	1528
154	3279
155	870*
157	2639*
158	2869*, 3373
159	3321
160	2461
161	(J A McCoshim), 4676*
162	3501*, 4653
163	3389, 4826
164	2088*
165	1648*
166	2623
167	(V.L.Brown)
168	4273
169	4145
170	4216
171	3643
172	5014
173	3607
175	3124
176	4413
177	3126
178	3119
180	2901
181	2827
182	1293, 1760, 1989
183	4794
184	1660
185	4699
187	3269
188	2775
189	2371*
190	3401
191	2436
192	3533
193	2727
194	3544
195	2115
196	3266
199	(J Fuchs)
200	4226
201	3433*
202	3107
203	3991
205	4768
206	3708
207	1974*
208	3641, 4687
210	5059
211	1222*
212	4720
213	4165
214	2155*
215	2937*, (M H Jones)
216	4326
217	2960
218	3707
219	(C J Ireland), 3770*
221	4483
222	1579
223	2936
224	3627
225	1336
226	(D M Bellamy), 1662*
227	(A T Farmer)
228	3549
229	1683
230	3523
231	2918*
232	4437
233	2226*, 5075
234	1723
236	3717
237	3967
238	3411
239	2862
240	1969*
241	1737*, 4214
242	3417
243	1172*, 4586
244	(S G Olender)
245	3976
246	3555
247	3604*, (G G Dale)
248	1911
249	2950*, (J A Johnston)
250	4795, 5060
251	2216
252	1624
253	1840
254	1795
255	1116*
256	3540
257	3303
258	2323
259	2229*, (B I Stoddart)
260	4039
261	2715*
262	(D W Allison)
263	(R N Turner), 3777*
264	2806
265	2243
266	2200
267	2464
268	3475
269	1853
270	1027, 2346
271	3101
272	2426*
273	1667
274	4467
275	4819
276	2829
277	3587*
278	3789
279	2856
280	
281	2812*
282	3397
283	2880*
284	1632
285	3599
286	(J M Beattie), 2469*
287	2268
288	2808
289	
290	
291	3868
292	3988*, 5018
294	(S C Foggin), 3751
295	2821
296	4737
297	4050
298	2894
299	1913
301	4630*
302	3689
303	(S G Olender)
304	4230*, Z. Marcynski
306	3981
307	2251
308	3877
309	1533
310	4843
311	1359
312	(Booker GC)
313	4170
314	3339
315	4066
316	3879
317	(P Bell)
318	3198
319	2728

320 2378
321 5121
322 1860*
323 2837
324 2000
325 4296
326 3631
327 1043*
328 3372
330 3398*
331 1813*
332 3516
333 3115
334 2327
335 3892
337 2616
339 1722*, (L I Rigby)
340 2868
342 1732*
343 (I R Starfield), 1412*
345 2820
346 (G G Pursey), 3831*
347 1765
350 1348
351 1741
352 4767
353 2186
354 2472
355 1852*, 3834
356 3805
357 (M P Brooks)
359 (C C Watt), 4041*
360 2539
363 (D K Gardiner)
364 4051
365 4019
366 1240
367 2522
368 3337
369 3559
370 3206*, 5070
371 3381
372 1393
373 1428
374 4324
375 2283
376 (C J Short)
377 1290*, 4775
378 2196
379 2906
380 1817*, 4054*
381 4032
382 2979*
383 2876*
385 4168*
387 1769
388 4674
390 4134*
391 3785
392 2231
393 4238, 4328
394 3842
395 3139*
397 3538
401 1691
402 1799
405 3500
406 3320*, 4998
407 2153
408 2754*
409 4959
410 (R Jones)
411 3154
413 2711
414 4738
415 1945
417 1756*
419 (M L Boxall)
420 4811
421 4971
422 1392
423 1850
424 5111
425 3714
427 5027
428 1844
429 (S M Sagun), 3780
430 2756
431 2702*
432 4315
433 2193
434 4428
435 2826
437 2271*, (H A Schuricht)
438 3723*
440 2282*
441 4651
442 1672
443 2968*
444 2175
445 3965
446 2959
447 2343*, 2441
449 1538
450 3747
451 2870*
452 2789
453 2694
456 2212*, 5039
458 1487
459 3415
461 3388*
463 1201
464 3133
465 1288
466 1688*
467 3866*
468 1858
469 4060*
470 2640*
471 1369
473 3886
474 1574
475 1423
476 4345*
477 1916
479 3443
480 2289*, 3663
481 2707
483 (K S Whiteley), 2452*
484 3448
485 4892
486 3895
488 3136
490 2697*
491 2980*
493 4169
494 4553
495 3470
496 3808
497 4178
498 3479
500 1400, 2146*
501 (A B Dickinson), 1802*
502 3946
503 1902*, 5020
504 (N M Claiden), 1936*
505 3975*
506 4694
507 3719
508 1890*
509 3924
510 3171
511 (M D Lawrence), 2228*
514 4738*
515 2788
517 1298, 2123*
518 3512*, (T. Busby)
519 3375
520 4943
521 (C M Greaves)
522 3814
523 1983*
524 3762*
525 1669
526 4270
527 3261*
528 4484
530 (A A Maitland), 2597*
532 1889
533 2353
535 1628*
537 4379
538 3199*
539 (I.B.Kennedy)
540 (M F Evans), 4820*
541 2611
542 2294
543 4424
546 3739
547 4685
549 867
550 (L.G.Watts)
551 1854, 1987
552 2553
554 4790
555 (R M Fendt)
556 (R G Wardell), 2300*
558 1341
560 4291
561 4641*, (J A Hallam)
563 4062
564 3207
565 5084
566 1748
567 3596
569 2673
570 5033
571 3151
572 4455
573 2499*, (J Farley)
574 3424
575 1381*
576 2504*
577 1546
579 1785
580 2473*
581 (P T Worth)
583 3933, (D Nichols)
584 4249
585 4082
586 2644
588 2057*
590 2924
591 3854
592 2481
593 2222
594 2529
597 1198
598 4263*
600 916*, 2589
601 3440*, 4645
602 1657
604 2585
605 3820
606 2802
607 3898
608 5085
609 3465
610 2114
611 3889
612 3542*
614 2162*
615 4541
616 4414
617 1986, 5119
618 4718
619 1456
620 4353, 5115
621 4791
622 1671
624 4213
625 1684
627 3236
628 4628
630 3706
631 3109
633 4967
634 3462
635 5095
636 4624
637 3937*
638 2396*, (M Toon)
642 2391*, (W J Tolley)
643 2649
644 2565
645 2580*, (B A Walker)
646 4871
647 2592*
648 2340
650 1743
651 4482
652 2402*, (M A Fellis)
653 1104*
655 4095
656 1836*, 3445
660 3554*
661 (J R W Kronfeld)
662 3588
663 663*
664 3824
665 (J B Dalton)
666 4776
667 (R J Strarup), 3821
668 3750
669 3645
671 3152
672 2672
674 3400
675 1961
676 4646
678 1810
679 1401
680 2037
682 2990
683 1454
685 4450
687 3883
688 (L Dent), 2326
690 (D M Ruttle)
691 1982
692 4080
693 4106*
695 1702*, 2025
696 2778*
698 3283
699 2357
700 4986
701 3259
703 1727
704 2376
705 3393
706 1816
707 1752
708 4065*
709 1670
710 (M Dowding), 4392
711 4290
712 4258
714 2743*
715 1383
716 2059*
719 2783
720 3702
721 3156
722 4838
723 (N Climpson), 3618
724 1616*
725 3686
727 3130*, (T P Docherty)
728 2444
729 3801
730 3939
732 2413*
733 2429
734 (F G Bradney)
735 3887
737 2645
739 1088*
741 1121
742 2491, 2841
743 1615
744 3291
745 2246
746 (D P Holdcroft)
747 3910

748	3731*
751	4164
753	3525
754	3827
755	3812
757	2518
758	(J Nash)
759	2168*
760	985*, 2366
761	1706
762	4117
764	1570*
765	2848*
766	2379
768	3590*
769	2172*
770	2291
771	(C E Wick), 2816*
772	3441
773	3134
774	2092
775	4252
777	4359*, (M.P.Roberts)
778	3673, 3799
779	3621
780	3141
781	4625
783	3944
785	3453
786	2537
788	2181
789	1733
790	4576
795	3882
797	(J P Galloway)
798	2329
799	1640
800	(J Bradley)
801	3757
802	4151
803	3535
805	4030
806	3257
807	2555
808	1775*, 5012
810	2423*
811	2119
812	1391*
813	4660
814	1476
818	1685
819	4029
820	2604*
821	2321
822	3120
823	4362
824	2129
826	1944
827	4735
828	2479
829	3822
831	2214
832	1875
837	1247
838	3956
839	2515
840	3878*
841	3146*, 4429
842	1822
843	3704
844	2286
845	3730*
846	1562
849	2706
850	2480
851	1976
853	3515*
854	2899
855	3964*, (G O Humphries)
856	4289
857	1568*
858	1689
859	1136*
860	3994
861	1295
862	1519
865	3993*
866	5036
867	4504
868	4534
869	2412
870	3904
871	4873
873	(S.K.Anksell), 1801*
877	2354
878	1433*, 4894
879	2408
880	4835
881	2545
882	3870
886	2445
888	3575
890	2729
891	2691
893	2345
894	4782
895	4175
899	4547
900	4200*, 4831
902	2825
903	3299
904	2722
906	2084*, 5049
907	4425*
909	2502
910	3856
911	2382, 4400
912	(F B Reilly)
913	1536, 4304
914	3761)
915	2370
916	(P I Fenner)
917	(PJ Concannon), 1869*
918	1725
919	4004
920	3155*, (K Neave)
921	4355
922	3710
925	3909
927	3363
928	3767
930	4240
931	2766*
933	4164, 4557
937	(J Williams), 1637*
940	3921
941	(R J Smith)
942	3623
943	2495*, (M J Slade)
944	3522
948	1630*
949	3809*
950	3302*, (W J Palmer)
951	2137*, (M J Carruthers)
952	3437*
954	3610*, (M C Foreman)
955	2584
959	2033
960	4394
961	1598
962	4235*
963	3954
968	2864
969	3365
970	(D A Vennard), 4008
971	2417
972	2350*
973	3410
977	4941
978	1871
979	3959
980	3585*
983	2902*
985	4009*
986	4523
987	3227*, 3271
988	3457
989	2164
990	2508
991	3858
992	2961
993	3148
994	1380
996	2625
997	3304
2001	3299

PART 2 (ii) – ALPHA / NUMERIC TAIL CODES

0D (R B Witter)
1Z 4878
2A (B Flewett)
2B 4374*
2C (D Heslop)
2CS 2680
2F 4427*
2R 3159
2UP 4792
2W 5055
3D 4113
4T 4801
5GC 4690
5UE (S K Armstrong)
6S 5004
7A (J C Ferguson)
7D 3167*, 3552*
7Q 4364*
7X 4741
9E 5129
15A 4825
17K 4720
20L 2740
21C 4244
27B 4876
28E (J M Fryer)
97Z 3961
A1 4155
A2 3872
A3 1946
A7 4573
A8 4596
A9 4244
A10 3622
A11 2385
A14 4316
A15 4177
A19 3390
A20 4112
A23 1237
A25 3341
A26 4322
A27 4422
A28 4936
A30 2126
A34 4531
A61 4318*
A77 4125*
A98 3438
AA 5133
AC 4387
AD 5112
AC4 4812
AC5 (S Kotomin)
AG 4955*
AJ 4305
AL 3295
AT 4725
AV 4996
AV8 4777
AW 3683
B 3366
B2 4385
B3 4373
B6 (J H Russell)
B9 (B O Marcham)
B11 4342
B12 4891*
B19 4728
B20 4893
B33 4147
B96 1283
BB (M J Wells), 3532
BG 4940
BJ (J A Hallam)
BK 5047
BR 2487*
BS 4912
BS1 2677
BT 4507
BZ 1834*
C 2191
C1 3640
C2 4836
C3 4514
C4 4186
C7 4404
C8 (C Bradley)
C9 (5086)
C29 4012
C34 3997
C64 4973
C65 4974
C66 4952, 5126
C74 4823
C75 5021, 5046
CB 4668
CC 4494
CD1 4802
CF 1655
CH 5034
CJ (M Wright)
CL 4677
CP (L G Blows), 4814*
CT 4870
CU 4493
CZ 4659
D 2317
D1 4638
D2 4533
D3 2667
D4 4444
D5 (D H Smith), 3414*
D6 (R J Large)
D7 4262
D8 4818
D9 (D J Scholey), 3935
D15 4158
D31 (S A Young), 4210*
D41 4464
D48 4653
D49 1734
D53 4772
D54 4163
D55 4763
D64 4121
D70 4146
DC 4405
DD 4300
DD2 4931
DF 4806
DG 3638
DH2 4860
DJ2 3352*
DM 4868
DR7 4647
DS2 5058
DT (R B Witter)
DV8 (P Thelwall)
DW 4465
DZ 3885
E 285*
E1 4829
E2 4656
E4 4895
E5 4769
E7 5003
E8 4505
E11 4709
EA 4981
EB 4698
ED 5040
EE 4932
EF 4650
EN 4346
EU 4516
EW 4632
EZ 2168, 5045
F 400*
F1 (A J Clarke), 4452
F2 3918*, 4409
F3 4418, (R Zaccour)
F4 1367
F6 4827
F20 4248
FB 5042
G1 4545
G2 3315
G7 4329
G41 4539
G46 698
G81 4380
G91 (S Barber)
GA 4847
GB 5083
GC 4780*
GG 4945
GP 4816
GR 4420
GS 4705
H 2984
H1 4957
H2 4558
H3 4764
H4 4753*
H5 3285*
H6 3478
H8 4195
H11 2552
H12 3732
H20 3185
H23 1189*
HB 4481
HB1 4661
HB2 3760
HL 4390
HS 4399
IM 4415
IV 3709
IZ (see 1Z)*
J1 4935
J2 (M Geisen)
J3 3632
J4 5087
J6 5072
J8 4985
J13 1204
J15 3106
J45 2657
J50 4824
JA 4016
JB 4749
JD 1483
JG (J Gorringe)
JH 2670
JH1 4810
JJ 4562
JM 4137
JO1 4544
JP 4713
JS 4813
JT 4687
JW (J R Warren)
K 2989
K1 4340
K2 2147
K3 2689*
K4 4310
K5 5100
K7 3682*
K8 4549
K11 4543
K13 3656
K17 1720*
K18 4173
K21 4496
KE 4451
KL 3960
KM 5102
KO 5110
KR 4627
KT 1744
KW 4426
L 3163
L2 4508
L3 4381
L4 4657
L5 3671
L8 4153
L10 (P H Young)
L12 3611
L17 4703
L18 4253
L51 3454
L57 3577
L58 4192
L88 4323
L99 1438
L01 2770
LA 4724
LD 4347
LGC 4511
LH7 4386
LR 4889*
LS 4189
LS1 4665
LS4 2908
LS6 4119
LS7 4597
LS8 4237
LT 4538
LW 5118
LX 4376*
LZ 4338*
M 3205
M2 4449, 5048
M3 5062
M4 4368
M5 4448
M6 (P Richer)
M7 2939
M8 4278
M9 4904
M19 4793
M25 4480
MF 2801
ML 2893
MM 4969
MY 4975
N 5023
N1 4443
N2 (R Murfitt), 3420*
N3 5078
N4 1914
N5 4207
N6 3736*
N8 4352
N11 2142
N12 4960
N15 3548
N16 3657
N19 4132
N21 673
N23 3772
N25 4598
N26 4056
N28 3482
N29 3254
N36 3864*
N51 2987
N52 1551*
N54 1360*
N55 1118*, 4883
N56 3505
NG1 4987
NH 4962
NJ1 4844
NT 2318
NU 4953
OD see 0D

OE	(M A Gatehouse)
OZ	(R Lynch), 4458
P	3274
P1	4567
P2	4442
P3	4363
P4	4648
P5	4152
P8	4373
P9	2443
P10	4923
P12	4579*
P19	1707
P23	4615
P30	4571
P31	4017
P61	4343
PB	4900
PC	5053
PE	5116
PH	3816
PH1	4577
PK	4990
PM3	4723
PS	3571
PT	4721
PZ	4058
Q5	3278
R	4958*
R1	4475
R2	5041
R3	4688
R4	4600
R5	4613
R6	4788
R7	4581
R8	4471
R9	3216
R10	4590
R11	4896
R12	5130
R15	4729
R17	4585
R20	4578*
R21	4572
R22	4614
R23	4629
R25	4587
R28	5079
R31	2533*
R32	4675*
R33	4582*
R34	4864
R35	2888*
R36	4612*
R37	4636
R38	4837
R39	4601
R41	4730*
R43	4530*
R44	2931*
R45	4745*
R46	4620*
R48	4570
R49	4621*
R50	4691*
R52	4583*
R53	4503
R55	4470
R56	4684*
R57	4603
R59	4683
R61	4633
R63	4622
R67	4697
R69	4655
R70	3015*
R71	3064*
R73	4555
R75	4604*
R77	4736
R80	4752*
R83	960
R88	4680*

R91	2619*
R92	4707*
R93	2207*
RA	4983
RB	4989
RC	5009
RH	2785
RJ	(M J Cook)
RNT	2831
RP1	4233
RT	3969
RW	4671
RY	4084
S	3642
S1	4821
S2	4118
S3	3580
S4	4293
S5	4299
S6	4469
S7	2943*
S8	4223
S9	3472
S13	2425
S14	1572*
S19	4580
S21	3698*, 4383*
S22	4191
S27	4834
S30	2380
S33	4710
S60	4848
S81	3722
S82	4045
S83	4268
SC	4654
SF	5090
SG1	4863
SG2	4906
SH1	2986
SH2	4800
SH3	3322
SH4	Surrey and Hampshire GC
SH5	Surrey and Hampshire GC
SH6	4804
SH7	4924
SH8	3192
SH9	4740
SH10	Surrey and Hampshire GC
SI	5002
SK	4727, 5000
SK1	5032
SM	4908
SP	4954
SR	4430
SSC	3909
T1	(P H Turner)
T2	4406
T3	5071
T4	4817, 5097
T5	4301
T6	4881
T7	4466
T8	4747
T9	4561
T10	1612
T11	5067
T12	4840
T15	1693
T21	3290
T27	4716
T34	2397
T45	873
T49	1203
T51	1164
T54	3756
T65	2800
T99	4980
TL2	4679
U1	4335
U2	4131, 2255*
U9	4378
UP2	5063
UY	4841

V1	4546
V2	(R Jones)
V2C	4921
V2T	4664
V4	4460
V5	4034
V6	(M Nash-Wortham)
V7	2695
V8	4982
V9	5073
V11	4292
V17	4758
V26	4976
VE	3006*
VMC	4182*
VS	2420*
VW	4521*, 3024*
W1	4795
W2	4695
W3	4128, 4862
W4	3874
W5	4859
W7	4861
W8	4574
W9	(K Sleigh)
W19	4410
W22	4852
W27	4407
W54	4231
WA1	4487
WE4	4993
X1	4783
X4	5080
X5	4750
X7	4330
X11	5064
X15	4785
X19	3846
X50	4341
X56	2955
X70	4212
X96	2414
X97	3955
XL5	(S R Ell)
XY	3077*
Y1	3416, 4356
Y2	4964
Y2K	4781
Y4	4280
Y11	5056
Y44	4853
Z	1873
Z1	4789
Z2	(P Wells)
Z3	4762
Z4	3340
Z6	3654*
Z7	4631
Z8	4190
Z9	4696
Z10	4575
Z11	1374
Z12	4704
Z19	4784
Z22	4639
Z25	2427
Z29	3727
Z35	4865
Z45	4040
Z99	4068
ZC	4256*

Notes:

(i) Codes "SY" to "ZZ" are allocated to Air Cadets Central Gliding School

(ii) Some gliders, mostly imported vintage specimens, still carry their previous civil registrations for authenticity. Those examples known will be found at the end of the Military Markings decode, Section 6, Part 2.

(ii) Ex British military gliders which still carry their former serials will be found listed with other military serials in Section 6, Part 2.

PART 3 – IRISH GLIDING AND SOARING ASSOCIATION REGISTER

The system is similar to the British Glider Association and the register is maintained by the Irish Gliding & Soaring Association. Originally this was kept by the Irish Aviation Club using the prefix "IAC". From 1960 until 1967 gliders were allocated with a number prefixed "IGA". The IGSA listing is updated from Air-Britain sources. We have decided to expand the data this year and include some historical entries. Thanks to Richard Cawsey for this and Further new information.

No	Code	Type	Construction No	Previous Identity	Date	Owner*(Operator)*	(Unconfirmed) Base	CA Expy
IGA.6		Slingsby T.8 Tutor	-	IAC.6	.56	Meath Aero Museum	Ashbourne, County.Meath	
				VM657		*(Noted 3.01)*		
EI-100		SZD-12A Mucha 100A	494	OY-XAN	.95	J.Finnan and M.O'Reilly	Gowran Grange	1. 7.97
				(Withdrawn from use due to glue failure - preserved at Cavan & Leitrim Railway, Dromod, 8.04)				
EI-101		Slingsby T.13 Petrel	361A	IGA.101	54	*To BGA 651 1973*		
				IAC.101, BGA 651, G-ALPP				
EI-102		Slingsby T.26 Kite 2	?	IGA.102	.54	Dublin Gliding Club	Gowran Grange	
				IAC.102, BGA...		*(Stored 5.99)*		
EI-103		Chilton Olympia	CAH/OL/7	IGA 103	.61	*EI- marks not taken up - damaged 1966 & scrapped*		
				BGA434, G-ALJN, BGA 434				
EI-104		Schleicher Ka4 Rhönlerche II	?	IGA 104?	.61	*Fate unknown*		
EI-105		Schleicher Ka7 Rhönadler	775	IGA.7	.60	Dublin Gliding Club *(Noted 5.99)*	Gowran Grange	14. 6.97
EI-106		Schleicher K 8B	1099	IGA 8	.61	*To BGA 1695 4.72*		
EI-107		Scheibe Bergfalke II	190	IGA	.61	*To BGA 2571 10.79*		
				OY-AXP				
EI-108	08	Schleicher K8B	8486		.65	Dublin Gliding Club	Gowran Grange	13. 4.97
			(Logbook shows c/n 8468)			*(Active 10.01)*		
EI-109		Schleicher Ka 6CR	6343	IGA 106	.64	*To Ulster Gliding Club as BGA 3318 3.88*		
EI-110		LET L-13 Blanik	173328	IGA 110	.66	*To BGA 2923 4.84*		
				G-ATWW				
EI-111	11	Schleicher Ka6CR	6565	IGA.9	.67	Not known *(Noted 5.99)*	Gowran Grange	30. 3.97
EI-112		Schleicher ASK13	13131		.69	Dublin Gliding Club *(Noted 5.99)*	Gowran Grange	9. 3.97
EI-113		Schleicher ASK13	13189		.69	Clonmel Gliding Club	Kilkenny	17. 7.97
EI-114		Schleicher ASK14	14008	EI-APS	.69	SLG Group	Gowran Grange	
				G-AWVV, D-KOBB		*(Not used became EI-APS)*		
EI-115		EoN AP.5 Olympia 2B	EoN/O/155	BGA.1097	.70	S.Cashin *(Active 5.99)*	Kilkenny	
EI-116		Scheibe Mü 13E	305	D-1127	.71	Kerry Gliding Club; *fate unknown*		
EI-117		Slingsby T.50 Skylark 4	1326	BGA 1019	.73	*To BGA 1019 5.84*		
EI-118		EoN AP.8 Baby	EoN/B/001	BGA.608	.73	B.Douglas	Gowran Grange	
				RAFGSA.217, BGA.608, G-ALLU, BGA.608 (Stored 6.95)				
EI-119		Schleicher ASK16	16022	EI-AYR	??	*(Not used - remained as Ei-AYR)*		
EI-120		LET L-13 Blanik	175205	RAFGSA	.75	Private Syndiate	Gowran Grange	
				BGA.1730		*(Active 10.01)*		
EI-121		Pilatus B4 PC11AF	199		.77	Clonmel Gliding Club *(Stored 6.95)*	Kilkenny	
EI-122		Scheibe SF 28A Tandem Falke	5739	G-BBGA	.78	*Not used - to EI-BFD 2.80*		
				D-KOEI				
EI-123		Bölkow Phoebus C	840	D-0057	??	*Crashed*		
EI-124		Grob G.102 Astir Standard CS 77	1761	D-...	.80	Nutgrove Shopping Centre	Churchtown, Dublin	
						(Noted 5.99)		
EI-125		Slingsby T.59D Kestrel 19	1839	D-9253	.80	*To BGA 2902 6.83*		
EI-126		Centrair ASW 20FL	20107	F-CFFG	.81	*Damaged 27 May 82 - to F-CFTA*		
EI-127		Schleicher Ka6CR	662	PH-259	??	N van Kuyk	Not known	
EI-128		Schleicher Ka6CR	6649	D-1393	??	Dublin Gliding Club *(On rebuild 4.96)*	Gowran Grange	
EI-129		Slingsby T.65C Sport Vega	1922	BGA 2622	.87	*To BGA 2622 9.94*		
EI-130		Scheibe L-Spatz	200	BGA.2199	??	J.J.Sullivan	Gowran Grange	
				D-4707		*"White Cloud"*		
EI-131		Schleicher Ka 2	133	BGA 1418	.90	*Crashed near Bellarena 3.6.95*		
				RAFGSA, D-4327				
EI-132	TK	Schleicher ASW17	17031	D-2365		Not known	Gowran Grange	17. 6.97
EI-133	33	Schleicher K8B	8557	D-8517	.91	Dublin Gliding Club	Gowran Grange	14. 6.97
				D-9367		*(Active 10.01)*		
EI-134	34	Schleicher ASW15B	15249	D-1087	.91	Not known	Gowran Grange	7. 5.97
EI-135		Slingsby T.38 Grasshopper	758	WZ762	.91	(Syndicate)	Gowran Grange	
		(Wings from WZ756 or WZ768)				*(Stored as "WZ762" 5.99)*		
EI-136		Schleicher ASK18	18007	BGA.2945	.91	Dublin Gliding Club	Gowran Grange	30. 7.97
				D-6868		*(Noted 5.99)*		
EI-137		Rolladen-Schneider LS3-17	3308	D-3521	.92	Not known	Gowran Grange	1. 3.97
						(Sold to US 1.02)		
EI-138	BR	Schempp-Hirth Discus CS	089CS		.92	B.Ramseyer	Gowran Grange	
						(Sold as N189HH 12.00)		
EI-139		Slingsby T.31B	902	BGA.3485	.93	P.Bedford Syndicate	Dromod	2. 8.97
				G-BOKG, XE789		*(Preserved at Cavan & Leitrim Railway 8.04)*		
EI-140		SZD-12A Mucha 100A	491	HB-647	.93	D.Mongey	Gowran Grange	
EI-141		ICA IS-28B	233	BGA 2207	.93	*To Crusaders Gliding Club, Cyprus 6.95*		
EI-142		Scheibe SF-27A Zugvogel V	6049	D-1444	.94	Not known	Gowran Grange	19. 7.97
						(Active 5.99, to EI-144 qv)		
EI-143		Schleicher ASK13	13112	BGA.1501	.94	Dublin Gliding Club	Gowran Grange	
EI-144		Scheibe SF-27A Zugvogel V	6049	(EI-142)	.94	Not known	Gowran Grange	
				D-1444				
EI-145		Glaser-Dirks DG-200	2-88	PH-930	.95	Not known	Gowran Grange	
				D-7610		*(Noted 5.99)*		
EI-146	TK	Scheibe Zugvogel IIIB	1085	D-4096	.96	N.Short and T.Daly *(Active 10.01)*	Gowran Grange	
EI-147	DS	Glaser-Dirks DG-200	2-22	D-6760	.97	Not known *(Active 6.00)*	Gowran Grange	

EI-148	Schempp-Hirth Std. Cirrus	304	BGA.4334 D-2060		Not known	
EI-149 49	Schempp-Hirth Discus b	219	BGA.3320	.03	S.Kinnear	Gowran Grange
EI-150	Schleicher ASK-21	21002	D-6957	12.02	Dublin Gliding Club	Gowran Grange
EI-151 51	Schleicher ASW-27B	27174		.02	K.Houlihan	Gowran Grange
EI-152 52	Schempp-Hirth Ventus 2cT	88/263		.02	B.Ramseyer *(To BGA.5080 3.04)*	Gowran Grange
EI-153	Schleicher ASW-19B	19316	BGA.4274/HWZ HB-1524	.02	K.Commins	Gowran Grange
EI-154	LET L-13 Blanik	173214	BGA.1500/CFD G-ATCG	6.02	J.Selman	Ardagh
EI-155	Not yet allocated					
EI-156	Not yet allocated					
EI-157	Slingsby T.21B	1158	BGA.1465 RAFGSA.333, XN154	.02	Dublin Gliding Club *(Active 10.01)*	Gowran Grange
EI-158	Bölkow Phoebus C	908	BGA.4202 OO-ZDJ, BGA.1573	.02	Not known	Gowran Grange
EI-159 BR	Schempp-Hirth Duo Discus T	106/435		10.04	B.Ramseyer	Gowran Grange
EI-160	Grob G.103 Twin Astir	Not known			Not known	Kilkenny
EI-161	Schleicher Ka-6CR	808	BGA.3536 D-6289		Not known	Not known
EI-162	Centrair 101A Pegase	101A-0311	BGA.3458		Not known	Not known
EI-167 189	Schleicher ASW-20L	20054	BGA.2371		J.P.Walsh	Gowran Grange

Note: Ulster Gliding Club aircraft are registered with the British Gliding Association.

SECTION 6

PART 1 – OVERSEAS REGISTRATIONS DECODE

To help the discerning reader decipher the many and varied origins of UK and Irish registered aircraft listed above we include both an historical and a current list of overseas aircraft registrations. Both lists were originally based on member Roger House's web-site and expanded with reference to Tony Pither's "Airline Fleets" and Ian Burnett's Overseas Registers section in "Air-Britain News" with further data added by John Davis and Dave Partington. Any further amendments are always welcomed.

HISTORICAL Prefix	Country	Period	Remarks
A-	Austria	1929-1939	Changed to OE-
AN-	Nicaragua	1936-	Changed to YN-
BR-	Burundi	1962-1965	Changed to 9U-
C-	Colombia	1929-1946	Changed to HK-
CB-	Bolivia	1929-1954	Changed to CP-
CCCP-	Soviet Union	1929-	Changed to RA-
CF-	Canada	1929-1974	Changed to C-
CH-	Switzerland	1929-1936	Changed to HB-
CR-A	Mozambique	1929-1975	Changed to C9-
CR-B	Mozambique	1971-1975	Changed to C9-
CR-C	Cape Verde Islands	1929-	Changed to D4-
CR-G	Portuguese Guinea/Guinea Bissau	1929-1975	Changed to J5-
CR-I	Portuguese India	1929-1961	Changed to VT-
CR-L	Angola	1929-1975	Changed to D2-
CR-S	Sao Tome and Principe	1929-	Changed to S9-
CR-T	Timor	1929-1975	Changed to PK-
CV-	Romania	1929-1936	Changed to YR-
CY-	Ceylon	1948-1954	Changed to 4R-
CZ-	Canal Zone	1930s	
CZ-	Monaco	1929-1949	Changed to MC-
DDR-	East Germany	1981-1990	Changed to D-
DM-	East Germany	1955-1981	Changed to DDR-
ES-	Estonia	1929-1939	Merged into Soviet Union CCCP-
EZ-	Saar Territory	1929-1933	Changed to SL-
FC-	Free French	1940-1944	Changed to F-
F-D	French Morocco	1929-1952	Changed to CN-
F-KH	Cambodia	1945-1954	Changed to XU-
F-L	Laos	1945-1954	Changed to XW-
F-O	Dahomey/Benin	1929-1960	Changed to TY-
F-O	Algeria	1929-1962	Changed to 7T-
F-O	Upper Volta	1929-1960	Changed to XT-
F-O	Ubangi-Shari	1929-1960	Changed to TL-
F-O	Tunisia	1929-1956	Changed to TS-
F-O	Cameroon	1929-1960	Changed to TJ-
F-O	Chad	1929-1960	Changed to TT-
F-O	Senegal	1929-1960	Changed to 6V-
F-O	Congo	1929-1960	Changed to TN-
F-O	Mauritania	1929-1960	Changed to 5T-
F-O	Gabon	1929-1960	Changed to TR-
F-O	Guinea	1929-1958	Changed to 3X-
F-O	Cote D'lvoire	1929-1960	Changed to TU-
F-O	Madagascar	1929-1960	Changed to 5R-
F-O	Niger	1929-1960	Changed to 5U-
F-O	Mali	1929-1960	Changed to TZ-
F-VN	French Indo-china (Vietnam)	1929-1954	Changed to XV-
G-AU	Australia	1921-1928	Changed to VH-
G-CA	Canada	1920-1928	Changed to CF-
G-CY	Canada	1920-1931	Military
G-IA	India	1919-1928	Changed to VT-
G-K	Kenya	1928-1928	Changed to VP-K
G-NZ	New Zealand	1921-1928	Changed to ZK-
G-UA	Union of South Africa	1927-1928	Changed to ZS-
J-	Japan	1929-1945	Changed to JA
JZ-	Dutch East Indies	1945-	
K-	Kuwait	1967-1968	Interim - Changed to 9K-
KA-	Katanga	1961-1963	Unoffical - Changed to 9O-
KW-	Cambodia	1954-	Changed to XU-
LG-	Guatemala	1936-1948	Changed to TG
LI-	Liberia	1929-1952	Changed to EL-, now A8-
LR-	Lebanon	1944-1954	Changed to OD-
LY-	Lithuania	1929-1939	Merged into Soviet Union CCCP-
M-	Spain	1929-1933	Changed to EC-
MC-	Monaco	1949-1959	Changed to 3A-
OA	Peru	1929-1939 or 1940	Changed to OB-
OO-C	Belgium Congo	1929-1960	Changed to 9O-
P-B	Brasil	1927-1932	Changed to PP-
P-P	Poland	1929-1929	Changed to SP-
PI-	Philippines	1941-1975	Changed to RP-
R-	Argentina *(three digits)*	1928-1937	Changed to LV-

R-	Argentina *(four letters)*	1927-1931	
R-	Panama *(two digits)*	1929-1943	Changed to RX-
RR	Russia	1926-1929	Changed to CCCP-
RV-	Persia (Iran)	1929-1944	Changed to EP-
RX-	Panama	1943-1952	Changed to HP-
RY-	Lithuania	1929-1939	Merged into Soviet Union CCCP-
SL-	Saar Territory	1947-1959	Changed to D-
SN-	Sudan	1929-1959	Changed to ST-
TJ-	Transjordan	1946-1954	Changed to JY-
TS-	Saar Territory	1930-1931	Unoffical - Changed to EZ-
UH-	Saudi Arabia	1945	Interim - Changed to HZ-
UL-	Luxembourg	1929-1939	Changed to LX-
UN-	Yugoslavia	1929-1935	Changed to YU-
VO-	Newfoundland	1929-1939	Merged into Canada CF-
VP-A	Gold Coast (Ghana)	1929-1957	Changed to 9G-
VP-B	Bahamas	1929-1975	Changed to C6-
VP-C	Ceylon	1929-1948	Changed to CY-
VP-G	British Guiana	1929-1967	Changed to 8R-
VP-H	British Honduras (Belize)	1947-1983	Changed to V3-
VP-J	Jamaica	1930-1964	Changed to 6Y-
VP-K	Kenya	1929-1963	Previously G-K.. 9.28. - changed to 5Y-K..
VP-L	Leeward and Windward Islands	1929-	Now Antigua V2-, Anguila VP-A and St. Kitts and Nevis VP-LK+
VP-LKA-LLZ	St. Kitts and Nevis		Changed to V4-
VP-M	Malta	1929-1968	Changed to 9H-
VP-N	Nyasaland	1929-1953	Changed to VP-Y
VP-P	Western Pacific Islands	1929-	Now Solomon Islands H4-
VP-R	Northern Rhodesia	1929-1953	Changed to VP-Y
VP-S	Somliland	1929-1960	Changes to 6OS-
VP-T	Trinidad and Tobago	1931-1965	Changed to 9Y-
VP-U	Uganda	1929-1962	Changed to 5X-
VP-V	St. Vincent and Grenadines	1959-	Changed to J8-
VP-W	Rhodesia	1971-	Changed to Z-
VP-W	China (Wei-Hai-Wei)	1929-1939	
VP-X	Gambia	1929-1945	
VP-Y	Southern Rhodesia, Rhodesia and Nyasaland	1929-1964	Changed to 7Q-, 9J-
VP-Z	Zanzibar	1929-1964	Changed to 5H-
VQ-B	Barbados	1952-1968	Changed to 8P-
VQ-C	Cyprus	1952-1960	Changed to 5B-
VQ-F	Fiji/Tonga/Friendly Isles	1929-1971	Changed to DQ-
VQ-G	Grenada	1962-	Changed to J3-
VQ-L	St. Lucia	1965-1981	Changed to J6-
VQ-M	Mauritius	1929-1968	Changed to 3B-
VQ-P	Palestine	1930-1948	Changed to either TJ- or 4X-
VQ-S	Seychelles	1929-1977	Changed to S7-
VQ-ZA, -Z	Basutoland	1929-1967	Changed to 7P-
VQ-ZE, -ZH	Bechuanaland	1929-1968	Changed to A2-
VQ-ZI	Swaziland	1929-1975	Changed to 3D-
VR-A	Aden	1939	
VR-B	Bermuda	1931-	Changed to VP-B
VR-C	Cayman Islands	1968-	Changed to VP-C
VR-G	Gibraltar	1929-1939	Changed to G-
VR-H	Hong Kong	1929-	Changed to B-H
VR-J	Johore	1929-1963	Changed to 9M-
VR-L	Sierra Leone	1929-1961	Changed to 9L-
VR-N	British Cameroons	1929-1958	Changed to either TJ- or 5N-
VR-O	Sabah (North Borneo)	1929-1963	Changed to 9M-
VR-R	Malaya	1929-1963	Changed to 9M-
VR-S	Singapore	1929-1965	Changed to 9V-
VR-U	Brunei	1929-	Changed to V8-
VR-W	Sarawak	1929-	
XH-	Honduras	1929-1960	Changed to HR-
XT-	China	1929-1949	
XV-	South Vietnam	1959-1975	
XW-	Laos	1954-	Changed to RDPL-
YE-	Yemen	1955-1969	Changed to 4W-
YL-	Latvia	1929-1939	Merged in Soviet Union CCCP-
YM-	Danzig Free State	1929-1939	Changed to D-
YN-	Nicaragua	1929-1936	Changed to AN-
ZM-	New Zealand	1929-1939	
3W-	Vietnam	1954-1959	Interim - Changed to XV-
4W-	Yemen	1969-	Merged into 7O-
6OS-	Somalia	1960-1969	Changed to 6O-
9O-	Zaire	1960-1966	Changed to 9Q-

CURRENT Prefix	Previous Prefix	Country	Period
AP-		Pakistan	1947-
A2-	VQ-ZE, -ZH	Botswana	1972-
A3-		Tonga	
A40-		Oman	1974-
A5-		Bhutan	
A6-		United Arab Emirates	1977-
A7-		Qatar	1975-
A8--	LI-, EL-	Liberia	2003-
A9C-		Bahrain	1977-
B-	XT-	Peoples' Republic of China	1975-
B-	XT-	Taiwan	1949-
B-H	VR-H	Hong Kong, China	1997-
C-	G-C, CF-	Canada	1974-
CC-		Chile	1929-
CN-	F-D	Morocco	1952-
CP-	CB-	Bolivia	1954-
CS-	CR-	Portugal	1929-
CU-		Cuba	1947-
CX-		Uruguay	1929-
C2-	VH-	Nauru	1971-
C3-		Andorra	
C5-	VP-X	The Gambia	1978-
C6-	VP-B	Bahamas	1975-
C9-	CR-A, CR-B	Mozambique	1975-
D-		(West) Germany	1929-
D-	DM-, DDR-	(East) Germany	Used by GDR 1955-89
D2-	CR-L	Angola	1975-
D4-	CR-C	Cape Verde Islands	1975-
D6-		Comoro Islands	1977-
DQ-	VQ-F	Fiji	1971-
EC-	M	Spain	1929-
EI-		Ireland	1928-
EK-	CCCP-	Armenia	1991-
EP-	RV-	Iran	1944-
ER-	CCCP-	Republic of Moldova	1991-
ES-	ES-, CCCP-	Estonia	1929-
ET-		Ethiopia	1945-
EW-	CCCP-	Belarus	1991-
EX-	CCCP-	Kyrgyzstan	1991-
EY-	CCCP-	Tajikistan	1991-
EZ-	CCCP-	Turkmenistan	1993-
E3-	ET-	Eritrea	1994
F-		France	1919-
F-O		French Overseas Territories	1929-
G-		United Kingdom	1919-
HA-	H-M	Hungary	1935-
HB-	CH-	Switzerland	1935-
HC-		Ecuador	1929-
HH-		Haiti	1929-
HI-		Dominican Republic	1929-
HK-	C-	Republic of Colombia	1946-
HL		Republic of Korea	1948-
HP-	RX-	Panama	1952-
HR-	XH-	Honduras	1961-
HS-	H-S	Thailand	1929-
HV-		Vatican City	
HZ-	UH-	Saudi Arabia	1945-
H4-	VP-P	Solomon Islands	1978-
I-		Italy	1929-
JA	J-	Japan	1948-
JU-	MT-, HMAY-	Mongolia	
JY-	TJ-	Jordan	1954-
J2-	F-O	Djibouti	1977-
J3-	VQ-G	Grenada	
J5-	CR-G	Guinea Bissau	1979-
J6-	VQ-L	St. Lucia	1981-
J7-		Dominica	
J8-	VP-V	St. Vincent and Grenadines	
LN-	N-	Norway	1931-
LQ-		Argentina (Government)	1932-
LV-	R-	Argentina	1932-
LX-	UL-	Luxembourg	1935-
LY-	LY-, CCCP-	Lithuania	1936-
LZ-		Bulgaria	1929-
N		United States (K-, W- 1927, unused)	1927-
N	NC, NL, NR, NS, NX	Used pre-1949 to denote category of use, can now be applied to vintage aircraft for historical accuracy	
OB-	OA-	Peru	1940-
OD-	F-, LR-	Lebanon	1951-
OE-	A-	Austria	1936-
OH	K-S	Finland	1931-

OK-	L-B	Czech Republic	1929-	
OM-	OK-	Slovakia		
OO-	O-B	Belgium	1929-	
OY-	T-D	Denmark	1929-	
P-		Korea, Democratic Peoples Republic (North Korea)		
PH-		The Netherlands	1929-	
PJ-		The Netherlands Antilles	1929-	
PK-		Indonesia	1929-	
PP-	P-B	Brasil	1932-	
PR-		Brasil	1950-	
PT-		Brasil	1946-	
PZ-		Suriname	1929-	
P2-	VH-	Papua New Guinea	1974-	
P4-	PJ-	Aruba		
RA-	CCCP-	Russian Federation	1991-	
RDPL-	F-L, XW-	Laos	1975-	
RP-	PI-	Philippines	1975-	
SE-	S-A	Sweden	1929-	
SP-	P-P	Poland	1929-	
ST-	SN-	Sudan	1959-	
SU-		Egypt	1931-	
SX-		Greece	1929-	
S2-	AP-	Bangladesh	1972-	(former East Pakistan)
S3-		Bangladesh	1976-	
S5-	YU-, SL-	Slovenia	1993-	
S7-	VQ-S	Seychelles	1976-	
S9-	CR-S	Sao Tome Island	1977-	
TC-		Turkey	1929-	
TF-		Iceland	1937-	
TG-	LG-	Guatemala	1948-	
TI-		Costa Rica	1927-	
TJ-	F-O, VR-N	Cameroon	1960-	
TL-	F-O	Central African Republic	1961-	
TN-	F-O	Congo Brazzaville	1960-	
TR-	F-O	Gabon	1960-	
TS-	F-O	Tunisia	1956-	
TT-	F-O	Chad	1960-	
TU-	F-O	Ivory Coast	1960-	
TY-	F-O	Benin (Dahomey)	1960-	
TZ-	F-O	Mali	1960-	
T3-	VP-P	Kiribati	1983-	
T7-	(9A-)	San Marino		
T8A-		Republic of Palau		
T9-	YU-	Bosnia Herzegovina	1992-	
UK-	CCCP-	Uzbekistan	1991-	
UN-	CCCP-	Kazakstan	1991-	
UR-	CCCP-	Ukraine	1991-	
VH-	G-AU	Australia	1929-	
VN-	XV-	Vietnam		
VP-A	VP-L	Anguilla	1997-	
VP-B	VR-B	Bermuda	1997-	
VP-C	VR-C	Cayman Islands	1997-	
VP-F		Falkland Islands & Dependencies	1929-	
VP-LVA-LZZ		British Virgin Islands	1971-	
VQ-H		St. Helena	1929-	
VQ-T	VP-J	Turks and Caicos Islands		
VT-	G-I	India	1930-	
V2-	VP-L	Antigua		
V3-	VP-H	Belize	1983-	
V4-	VP-LKA-LLZ	St.Kitts and Nevis		
V5-	ZS-	Namibia	1990-	
V6-		Micronesia		
V7-		Marshall Islands		
V8-	VR-U	Brunei	1984-	
XA-	X-	Mexico (Commercial)	1929-	
XB-	X-	Mexico (Private)	1929-	
XC-	X-	Mexico (Government)	1929-	
XT-	F-O	Burkina Faso (Upper Volta)	1984-	
XU-	F-KH, KW-	Kampuchea (Cambodia)	1954-	
XY-	VT-	Myanmar (Burma)	1938-	
YA-		Afghanistan	1929-	
YI-		Iraq	1931-	
YJ-		Vanuatu (New Hebrides)	1933-	Used F-O or VP-P marks until c.1980
YK-	F-, SR-	Syria	1949-	
YL-	YL-, CCCP-	Latvia	1929-	
YN-	AN-	Nicaragua	1981-	
YR-	CV-	Romania	1936-	
YS-		Republic of El Salvador	1939-	
YU-	X-S, UN-	Yugoslavia	1934-	Now Serbia and Montenegro
YV-		Venezuela	1929-	
Z-	VP-W, VP-Y	Zimbabwe	1980-	
Z3-	YU-	Macedonia	1994-	
ZA-		Albania	1946-	

ZK-	G-NZ	New Zealand	1929-	
ZP-		Paraguay	1929-	
ZS-	G-U	Republic of South Africa	1929-	
ZT-		Republic of South Africa		
ZU-		Republic of South Africa		
3A-	CZ, MC	Monaco	1959-	
3B-	VQ-M	Mauritius	1968-	
3C-	EC-	Equatorial Guinea	1975-	
3D-	VQ-ZIA-ZLZ	Swaziland	1971-	
3X-	F-O	Guinea	1958-	
4K-	CCCP-	Azerbaijan	1991-	
4L-	CCCP-	Georgia	1991-	
4R-	VP-C, CY-	Sri Lanka	1954-	
4X-	VQ-P	Israel	1948	
5A-	I-	Libya	1959-	
5B-	VQ-C	Cyprus	1960-	
5H-	VP-Z	Tanzania	1964-	
5N-	VR-N	Nigeria	1960-	
5R-	F-O	Malagasy Republic	1960-	
5T-	F-O	Mauritania	1960-	
5U-	F-O	Niger	1960-	
5V-	F-O	Togo	1976-	
5W-	ZK-	Western Samoa	1962-	
5X-	VP-U	Uganda	1962-	
5Y-	VP-K	Kenya	1963-	
6O-	I-, 6OS-	Somalia	1969-	
6V-	F-O	Senegal	1960-	
6Y-	VP-J	Jamaica	1964-	
7O-	YE-, 4W-	Yemen	1974-	
7P-	VQ-ZA,-ZD	Lesotho	1967-	
7Q-	VP-Y	Malawi	1964-	
7T-	F-O	Algeria	1962-	
8P-	VQ-B	Barbados	1968-	
8Q-	VP-C	Maldives	1976-	
8R-	VP-G	Guyana	1967-	
9A		San Marino	1959	Not used - see T7-
9A-	YU-, RC-	Croatia	1992-	
9G-	VP-A	Ghana	1957-	
9H-	VP-M	Malta	1968-	
9J-	VP-Y	Zambia	1964-	
9K-	K-	Kuwait	1960-	
9L-	VR-L	Sierra Leone	1961	
9M-	VR-J, -O, -R	Malaysia	1963-	
9N-		Nepal	1960-	
9Q-	OO-C, 9O-	Democratic Republic of Congo	1962-	Formerly known as Zaire
9U-	OO-C	Burundi	1966-	
9V-	VR-S, 9M-	Singapore	1966-	
9XR-	OO-C	Rwanda	1966-	
9Y-	VP-T	Trinidad and Tobago	1965-	

PART 2 – MILITARY SERIALS DECODE

In certain circumstances the Civil Aviation Authority may permit the operation of aircraft without the need to carry regulation size national registration letters. These conditions are referred to as "exemptions". The CAA issues to operators Exemption Certificates which are usually valid for two years. The basic requirements are that the owner undertakes to notify the CAA of the markings carried and may not fly overseas, without specific permission of the overseas country. In the case of aircraft wearing military marks the authority of the relevant department at the Ministry of Defence is required for UK markings whilst an equivalent establishment must sanction any overseas markings to be carried.

Below are current details of those aircraft and gliders which are known to be wearing military or, in a very few cases, "B" Conditions markings, see Part 6. The information is compiled from member's observations and includes any BAPC and "B" Conditions identitites and overseas registered aircraft known to be based in the UK and Ireland. Full details of BAPC markings are carried in SECTION 3 and c/ns for the others can be found in their respective Sections. We should point out that most of the serials used are spurious.

Country	Serial	Code	Regn	Type
UNITED KINGDOM (RAF unless otherwise shown)				
	4		BAPC.11	English Electric Wren composite
	10		G-BPVE	Bleriot Type XI 1909 rep
	12A		BAPC.2	Bristol Boxkite reconstruction
	14		BAPC.6	Roe Triplane Type I
	168		G-BFDE	Sopwith Tabloid Scout rep *(Royal Navy Air Service)*
	304		BAPC.62	Cody Type V Biplane *(Royal Flying Corps)*
	687		BAPC.181	Royal Aircraft Factory BE.2b rec *(Royal Flying Corps)*
	1701		BAPC.117	Royal Aircraft Factory BE.2c rep *(Royal Flying Corps)*
	1881		BAPC.122	Avro 504 fsm
	2345		G-ATVP	Vickers FB.5 Gunbus rep *(Royal Flying Corps)*
	2882		BAPC.234	Vickers FB.5 Gunbus rep *(Royal Flying Corps)*
	3066		G-AETA	Caudron G.III *(RNAS)*
	5964		BAPC.112	AirCo DH.2 fsm *(Royal Flying Corps)*
	5964		G-BFVH	AirCo DH.2
	6232		BAPC.41	Royal Aircraft Factory BE.2c rep *(Royal Flying Corps)*
	A1742		BAPC.38	Bristol Scout D fsm *(Royal Flying Corps)*
	A4850		BAPC.176	Royal Aircraft Factory SE.5a rep *(Royal Flying Corps)*
	A7317		BAPC.179	Sopwith Pup fsm *(Royal Flying Corps)*
	A8226		G-BIDW	Sopwith "1½" Strutter rep *(Royal Flying Corps)*
	B-415		BAPC.163	AFEE 10/42 Rotachute Rotabuggy reconstruction
	B595	W	G-BUOD	SE.5A rep *(Royal Flying Corps)*
	B1807	A7	G-EAVX	Sopwith Pup *(Royal Flying Corps)- intended marks*
	B2458	R	G-BPOB	Sopwith F1 Camel rep *(Royal Flying Corps)*
	B3459	2	G-BWMJ	Nieuport Scout 17/23 rep *(Royal Flying Corps)*
	B4863		BAPC.113	Royal Aircraft Factory SE.5a fsm
	B5577		BAPC.59	Sopwith Camel fsm *(Royal Flying Corps)*
	B6401		G-AWYY	Sopwith F1 Camel rep *(Royal Flying Corps)*
	B7270		G-BFCZ	Sopwith F1 Camel rep *(Royal Flying Corps)*
	C1904	Z	G-PFAP	Royal Aircraft Factory SE.5a (Currie Wot) *(RFC)*
	C3011	S	G-SWOT	Royal Aircraft Factory SE.5a (Currie Wot) *(RFC)*
	C4451		BAPC.210	Avro 504J *(Royal Flying Corps)*
	C4912		BAPC.135	Bristol 20 M.1C Monoplane fsm
	C4918		G-BWJM	Bristol 20 M.1C Monoplane rep
	C4994		G-BLWM	Bristol 20 M.1C Monoplane rep *(Royal Flying Corps)*
	C9533	M	G-BUWE	Royal Aircraft Factory SE.5a rep *(Royal Flying Corps)*
	D276	A	BAPC.208	Royal Aircraft Factory SE.5a rep *(Royal Flying Corps)*
	D7889		G-AANM	Bristol F.2b Fighter
	D8084	S	G-ACAA	Bristol F.2b Fighter
	D8096		G-AEPH	Bristol F.2b Fighter
	E449		G-EBJE	Avro 504K
	E2466		BAPC.165	Bristol F.2b Fighter
	F141	G	G-SEVA	Royal Aircraft Factory SE.5a rep *(Royal Flying Corps)*
	F235	B	G-BMDB	Royal Aircraft Factory SE.5a rep *(Royal Flying Corps)*
	F904	H	G-EBIA	Royal Aircraft Factory SE.5a rep *(Royal Flying Corps)*
	F938		G-EBIC	Royal Aircraft Factory SE.5a rep *(Royal Flying Corps)*
	F943		G-BIHF	Royal Aircraft Factory SE.5a rep *(Royal Flying Corps)*
	F943		G-BKDT	Royal Aircraft Factory SE.5a rep *(Royal Flying Corps)*
	F5447	N	G-BKER	Royal Aircraft Factory SE.5a rep *(Royal Flying Corps)*
	F5459	Y	G-INNY	Royal Aircraft Factory SE.5a rep *(Royal Flying Corps)*
	F5459	Y	BAPC.142	Royal Aircraft Factory SE.5a rep *(Royal Flying Corps)*
	F5475	A	BAPC.250	Royal Aircraft Factory SE.5a rep *(Royal Flying Corps)*
	F8010	Z	G-BDWJ	Royal Aircraft Factory SE.5a rep *(Royal Flying Corps)*
	F8614		G-AWAU	Vickers FB.27A Vimy rep
	H1968		BAPC.42	Avro 504K rep
	H3426		BAPC.68	Hawker Hurricane rep
	H5199		G-ADEV	Avro 504K
	J7326		G-EBQP	de Havilland DH.53 Humming Bird - *intended marks*
	J9941		G-ABMR	Hawker Hart II
	K1786		G-AFTA	Hawker Tomtit
	K1930		G-BKBB	Hawker Fury II
	K2048		G-BZNW	Hawker (Isaacs) Fury
	K2050		G-ASCM	Hawker (Isaacs) Fury
	K2059		G-PFAR	Hawker (Isaacs) Fury
	K2075		G-BEER	Hawker (Isaacs) Fury
	K2227		G-ABBB	Bristol Bulldog IIA
	K2567		G-MOTH	de Havilland DH.82 Tiger Moth

K2572		G-AOZH	de Havilland DH.82A Tiger Moth
K2587		G-BJAP	de Havilland DH.82A Tiger Moth
K3241		G-AHSA	Avro Tutor
K3731		G-RODI	Hawker (Isaacs) Fury
K4232		SE-AZB	Cierva C.30A
K-4259	71	G-ANMO	de Havilland DH.82A Tiger Moth
K5054		G-BRDV	Supermarine Spitfire Prototype rep
K5054		BAPC.214	Supermarine Spitfire Prototype fsm
K5414	XV	G-AENP	Hawker Hind
K5600		G-BVVI	Hawker Audax
K5673		BAPC.249	Hawker Fury I fsm
K5673		G-BZAS	Hawker Fury I rep
K7271		BAPC.148	Hawker Fury II fsm
K8203		G-BTVE	Hawker Demon I
K8303	D	G-BWWN	Hawker (Isaacs) Fury
K9926	JH-C	BAPC.217	Supermarine Spitfire fsm
L1070	XT-A	BAPC.227	Supermarine Spitfire fsm
L1679	JX-G	BAPC.241	Hawker Hurricane 1 fsm
L1710	AL-D	BAPC.219	Hawker Hurricane fsm
L2301		G-AIZG	Supermarine Walrus 1 *(RN)*
L6906		G-AKKY	Miles Magister
L7005	PS-B	BAPC.281	Boulton Paul Defiant fsm
N500		G-BWRA	Sopwith Triplane rep *(Royal Navy Air Service)*
N185		G-AIBE	Fairey Fulmar 2 *(RN)*
N248		BAPC.164	Wight Quadraplane Type 1 fsm
N2276		G-GLAD	Gloster Gladiator II
N3194	GR-Z	BAPC.220	Supermarine Spitfire fsm
N3289	DW-K	BAPC.65	Supermarine Spitfire fsm
N3313	KL-B	BAPC.69	Supermarine Spitfire fsm
N3317	AI-A	BAPC.268	Supermarine Spitfire fsm
N3788		G-AKPF	Miles M.14A Hawk Trainer
N4877	MK-V	G-AMDA	Avro 652A Anson 1
N5182		G-APUP	Sopwith Triplane fsm*(Royal Navy Air Service)*
N5195		G-ABOX	Sopwith Pup *(Royal Navy Air Service)*
N5199		G-BZND	Sopwith Pup rep
N5492	B	BAPC.111	Sopwith Triplane fsm *(Royal Navy Air Service)*
N6181		G-EBKY	Sopwith Pup *(Royal Navy Air Service)*
N6290		G-BOCK	Sopwith Triplane rep *(Royal Navy Air Service)*
N6452		G-BIAU	Sopwith Pup rep *(Royal Navy Air Service)*
N6-466		G-ANKZ	de Havilland DH.82A Tiger Moth
N6720	VX	G-BYTN	de Havilland DH.82A Tiger Moth
N6740		G-AISY	de Havilland DH.82A Tiger Moth
N-6797		G-ANEH	de Havilland DH.82A Tiger Moth
N6847		G-APAL	de Havilland DH.82A Tiger Moth
N6965	FL-J	G-AJTW	de Havilland DH.82A Tiger Moth
N6985		G-AHMN	de Havilland DH.82A Tiger Moth
N9191		G-ALND	de Havilland DH.82A Tiger Moth *(RN)*
N9192	RCO-N	G-DHZF	de Havilland DH.82A Tiger Moth
N9389		G-ANJA	de Havilland DH.82A Tiger Moth
P641		BAPC.123	Vickers FB.5 Gunbus fsm
P2793	SD-M	BAPC.236	Hawker Hurricane fsm
P2902	DX-X	G-ROBT	Hawker Hurricane I
P2970	US-X	BAPC.291	Hawker Hurricane I fsm
P3059	SD-N	BAPC.64	Hawker Hurricane fsm
P3208	SD-T	BAPC.63	Hawker Hurricane fsm
P3873	YO-H	BAPC.265	Hawker Hurricane fsm
P6382	C	G-AJRS	Miles Magister
P7350	XT-D	G-AWIJ	Supermarine Spitfire F.IIA
P8140	ZP-K	BAPC.71	Supermarine Spitfire fsm
P8448	UM-D	BAPC.225	Supermarine Spitfire rep
P9338		BAPC.299	Supermarine Spitfire I fsm
R1914		G-AHUJ	Miles Magister
R3821	UX-N	G-BPIV	Bristol Blenheim IV
R4115	LE-X	BAPC.267	Hawker Hurricane fsm
R4118	UP-W	G-HUPW	Hawker Hurricane I
R4897		G-ERTY	de Havilland DH.82A Tiger Moth
R4959	59	G-ARAZ	de Havilland DH.82A Tiger Moth
R5136		G-APAP	de Havilland DH.82A Tiger Moth
R5172	FIJ-E	G-AOIS	de Havilland DH.82A Tiger Moth
R6690	PR-A	BAPC.254	Supermarine Spitfire 1 fsm
S1287	5	G-BEYB	Fairey Flycatcher rep *(FAA)*
S1595		BAPC.156	Supermarine S.6B fsm
S1579	571	G-BBVO	Hawker Nimrod (Isaacs Fury) *(RN)*
S1581	573	G-BWWK	Hawker Nimrod 1 *(FAA)*
T5672		G-ALRI	de Havilland DH.82A Tiger Moth
T5854		G-ANKK	de Havilland DH.82A Tiger Moth
T-5879	RUC-W	G-AXBW	de Havilland DH.82A Tiger Moth
T6313		G-AHVU	de Havilland DH.82A Tiger Moth
T6818		G-ANKT	de Havilland DH.82A Tiger Moth
T-6953		G-ANNI	de Havilland DH.82A Tiger Moth
T-7230		G-AFVE	de Havilland DH.82A Tiger Moth
T7281		G-ARTL	de Havilland DH.82A Tiger Moth
T7328		G-APPN	de Havilland DH.82A Tiger Moth
T7404	04	G-ANMV	de Havilland DH.82A Tiger Moth

T7909		G-ANON	de Havilland DH.82A Tiger Moth
T8191		G-BWMK	de Havilland DH.82A Tiger Moth
T9707		G-AKKR	Miles M.14A Hawk Trainer
T9738		G-AKAT	Miles M.14A Hawk Trainer
V3388		G-AHTW	Airspeed Oxford 1
V6028	GB-D	G-MKIV	Bristol Blenheim IV
V6799	SD-X	BAPC.72	Hawker Hurricane fsm
V7467	LE-D	BAPC.223	Hawker Hurricane fsm {1}
V7467	LE-D	BAPC.288	Hawker Hurricane fsm {2}
V9367	MA-B	G-AZWT	Westland Lysander IIIA
V9545	BA-C	G-BCWL	Westland Lysander IIIA
V9673	MA-J	G-LIZY	Westland Lysander III
W2718	AA5Y	G-RNLI	Supermarine Walrus *(RN)*
W5856	A2A	G-BMGC	Fairey Swordfish II
W9385	YG-L	G-ADND	de Havilland DH.87B Hornet Moth
Z2033	N/275	G-ASTL	Fairey Firefly TT.1
Z5053		G-BWHA	Hawker Hurricane IIB
Z5140	HA-C	G-HURI	Hawker Hurricane IIB
Z5252	GO-B	G-BWHA	Hawker Hurricane IIB
Z7015	7-L	G-BKTH	Hawker Sea Hurricane IB *(RN)*
Z7197		G-AKZN	Percival Proctor III
AB550	GE-P	BAPC.230	Supermarine Spitfire fsm
AB910	ZD-C	G-AISU	Supermarine Spitfire LF.Vb
AP507	KX-P	G-ACWP	Cierva C.30A
AR213	PR-D	G-AIST	Supermarine Spitfire IA
AR501	NN-A	G-AWII	Supermarine Spitfire Vc
AR614	DU-Z	G-BUWA	Supermarine Spitfire Vc
BB807		G-ADWO	de Havilland DH.82A Tiger Moth
BE421	XP-G	BAPC.205	Hawker Hurricane IIc fsm
BL924	AZ-G	BAPC.242	Supermarine Spitfire Vb fsm
BM597	JH-C	G-MKVB	Supermarine Spitfire Vb
BM631	XR-C	BAPC.269	Supermarine Spitfire V fsm
BN230	FT-A	BAPC.218	Hawker Hurricane fsm
BR600	SH-V	BAPC.222	Supermarine Spitfire fsm
BR600		BAPC.224	Supermarine Spitfire fsm
BW881		G-KAMM	Hawker Hurricane XIIA
CB733		G-BCUV	Scottish Aviation Bulldog
DE208		G-AGYU	de Havilland DH.82A Tiger Moth
DE470	16	G-ANMY	de Havilland DH.82A Tiger Moth
DE623		G-ANFI	de Havilland DH.82A Tiger Moth
DE673		G-ADNZ	de Havilland DH.82A Tiger Moth
DE730		G-ANFW	de Havilland DH.82A Tiger Moth
DE970		G-AOBJ	de Havilland DH.82A Tiger Moth
DE992		G-AXXV	de Havilland DH.82A Tiger Moth
DF112		G-ANRM	de Havilland DH.82A Tiger Moth
DF128	RCO-U	G-AOJJ	de Havilland DH.82A Tiger Moth
DF155		G-ANFV	de Havilland DH.82A Tiger Moth
DG590		G-ADMW	Miles Hawk Major
DR628	PB-1	N18V	Beech Traveler
EM720		G-AXAN	de Havilland DH.82A Tiger Moth
EN224		G-FXII	Supermarine Spitfire XII – *intended marks*
EN343		BAPC.226	Supermarine Spitfire fsm
EN398		BAPC.184	Supermarine Spitfire IX fsm
EP120	AE-A	G-LFVB	Supermarine Spitfire Vb
FB226	MT-A	G-BDWM	North American Mustang *(Bonsall Mustang)*
FE695	94	G-BTXI	North American Harvard IIB
FE788		G-CTKL	Noorduyn AT-16 Harvard IIB
FE905		LN-BNM	North American Harvard IIB
FJ777		G-BIXN	Boeing Stearman Kaydet
FR886		G-BDMS	Piper Cub
FS628		G-AIZE	Fairchild Argus
FT323	GN	FAP 1513	North American Harvard III
FT375		G-BWUL	North American Harvard IIB
FT391		G-AZBN	North American Harvard IIB
FX301	FD-NQ	G-JUDI	North American Harvard III
FZ625		G-AMPO	Douglas Dakota 3
HB275		G-BKGM	Beech Expeditor
HB751		G-BCBL	Fairchild Argus III
HG691		G-AIYR	de Havilland DH.89A Dragon Rapide
HH268		BAPC.261	General Aircraft Hotspur replica
HM580	KX-K	G-ACUU	Cierva C.30A
HS503		BAPC.108	Fairey Swordfish IV
JV828		N423RS	Consolidated-Vultee PBY-5A Catalina
KB889	NA-I	G-LANC	Avro Lancaster X
KD345	130	G-FGID	Vought FG-1D Corsair *(RN)*
KF584	RAI-X	G-RAIX	North American Harvard IV
KJ351		BAPC.80	Airspeed AS.58 Horsa II
KL161	VO-B	G-BYDR	North American B-25D Mitchell II
KZ321	JV-N	G-HURY	Hawker Hurricane IV
LB312		G-AHXE	Taylorcraft Plus D *(Auster I)*
LB367		G-AHGZ	Taylorcraft Plus D *(Auster I)*
LB375		G-AHGW	Taylorcraft Plus D *(Auster I)*
LF789	R2-K	BAPC.186	de Havilland DH.82B Queen Bee composite
LF858		G-BLUZ	de Havilland DH.82B Queen Bee

VP981		G-DHDV	de Havilland DH.104 Devon C.2/2
VR192		G-APIT	Percival Prentice T.1
VR249	FA-EL	G-APIY	Percival Prentice T.1
VR259	M	G-APJB	Percival Prentice T.1
VS356		G-AOLU	Percival Prentice T.1
VS610	K-L	G-AOKL	Percival Prentice T.1
VS623		G-AOKZ	Percival Prentice T.1
VT871		G-DHXX	de Havilland DH.100 Vampire FB.6
VV612		G-VENI	de Havilland DH.112 Venom FB.1
VX118		G-ASNB	Auster AOP.6
VX147		G-AVIL	Ercoupe 415
VX653		G-BUCM	Hawker Sea Fury FB.11
VX927		G-ASYG	Beagle A 61 Terrier 2
VZ345		D-CATA	Hawker Sea Fury T.20 *(RN)*
VZ467	A	G-METE	Gloster Meteor F.8
VZ638		G-JETM	Gloster Meteor T.7 *(RN/FRU)*
VZ728		G-AGOS	Reid and Sigrist Bobsleigh
WA591		G-BWMF	Gloster Meteor T.7 - *intended marks*
WB188		G-HUNT	Hawker Hunter F.51
WB188		G-BZPB	Hawker Hunter GA.Mk.11
WB188		G-BZPC	Hawker Hunter GA.Mk.11
WB531		G-BLRN	de Havilland DH.104 Devon C.2/2
WB533		G-DEVN	de Havilland DH.104 Devon C.2/2
WB565	X	G-PVET	de Havilland DHC-1 Chipmunk T.10 *(Army)*
WB569	R	G-BYSJ	de Havilland DHC-1 Chipmunk T.10
WB571	34	G-AOSF	de Havilland DHC-1 Chipmunk T.10
WB585	M	G-AOSY	de Havilland DHC-1 Chipmunk T.10
WB588	D	G-AOTD	de Havilland DHC-1 Chipmunk T.10
WB615	E	G-BXIA	de Havilland DHC-1 Chipmunk T.10
WB652		G-CHPY	de Havilland DHC-1 Chipmunk T.10
WB654	U	G-BXGO	de Havilland DHC-1 Chipmunk T.10
WB671	910	G-BWTG	de Havilland DHC-1 Chipmunk T.10 *(RN)*
WB697	95	G-BXCT	de Havilland DHC-1 Chipmunk T.10
WB702		G-AOFE	de Havilland DHC-1 Chipmunk T.10
WB703		G-ARMC	de Havilland DHC-1 Chipmunk T.10
WB711		G-APPM	de Havilland DHC-1 Chipmunk T.10
WB726	E	G-AOSK	de Havilland DHC-1 Chipmunk T.10
WB763		G-BBMR	de Havilland DHC-1 Chipmunk T.10
WB920		BGA.3221	Slingsby T.21B
WB922		BGA.4366	Slingsby T.21B
WB924		BGA.3901	Slingsby T.21B
WB943		BGA.2941	Slingsby T.21B
WB962		BGA.3836	Slingsby T.21B
WB971		BGA.3324	Slingsby T.21B
WB975		BGA.3290	Slingsby T.21B
WB981		BGA.1218	Slingsby T.21B
WB985	FJB	BGA.3288	Slingsby T.21B
WD286	J	G-BBND	de Havilland DHC-1 Chipmunk T.10
WD292		G-BCRX	de Havilland DHC-1 Chipmunk T.10
WD310	B	G-BWUN	de Havilland DHC-1 Chipmunk T.10
WD331		G-BXDH	de Havilland DHC-1 Chipmunk T.10
WD347		G-BBRV	de Havilland DHC-1 Chipmunk T.10
WD363		G-BCIH	de Havilland DHC-1 Chipmunk T.10
WD373	12	G-BXDI	de Havilland DHC-1 Chipmunk T.10
WD379	K	G-APLO	de Havilland DHC-1 Chipmunk T.10
WD390		G-BWNK	de Havilland DHC-1 Chipmunk T.10
WD413		G-VROE	Avro 652A Anson T.21
WE569		G-ASAJ	Beagle Terrier (Auster T.7)
WE591	Y	G-ASAK	Beagle Terrier (Auster T.7)
WE992		BGA.2692	Slingsby T.30B Prefect
WF118		G-DACA	Percival P.57 Sea Prince T.1
WG307		G-BCYJ	de Havilland DHC-1 Chipmunk T.10
WG316		G-BCAH	de Havilland DHC-1 Chipmunk T.10
WG321		G-DHCC	de Havilland DHC-1 Chipmunk T.10
WG348		G-BBMV	de Havilland DHC-1 Chipmunk T.10
WG350		G-BPAL	de Havilland DHC-1 Chipmunk T.10
WG407	67	G-BWMX	de Havilland DHC-1 Chipmunk T.10
WG422	16	G-BFAX	de Havilland DHC-1 Chipmunk T.10
WG465		G-BCEY	de Havilland DHC-1 Chipmunk T.10
WG469		G-BWJY	de Havilland DHC-1 Chipmunk T.10
WG472		G-AOTY	de Havilland DHC-1 Chipmunk T.10
WG498		BGA.3245	Slingsby T.21B
WG719		G-BRMA	Westland Dragonfly HR.5
WJ306		BGA.3240	Slingsby T.21B
WJ358		G-ARYD	Auster AOP.6
WJ680	CT	G-BURM	English Electric Canberra TT.18
WJ945	21	G-BEDV	Vickers Varsity T.1
WK126	843	N2138J	English Electric Canberra TT.18
WK163		G-BVWC	English Electric Canberra B.2
WK436		G-VENM	de Havilland DH.112 Venom FB.50 (FB.1)
WK512	A	G-BXIM	de Havilland DHC-1 Chipmunk T.10 *(Army)*
WK514		G-BBMO	de Havilland DHC-1 Chipmunk T.10
WK517		G-ULAS	de Havilland DHC-1 Chipmunk T.10

WK522		G-BCOU	de Havilland DHC-1 Chipmunk T.10
WK549		G-BTWF	de Havilland DHC-1 Chipmunk T.10
WK577		G-BCYM	de Havilland DHC-1 Chipmunk T.10
WK585		G-BZGA	de Havilland DHC-1 Chipmunk T.10
WK586	V	G-BXGX	de Havilland DHC-1 Chipmunk T.10 *(Army)*
WK590	69	G-BWVZ	de Havilland DHC-1 Chipmunk T.10
WK609	93	G-BXDN	de Havilland DHC-1 Chipmunk T.10
WK611		G-ARWB	de Havilland DHC-1 Chipmunk T.10
WK622		G-BCZH	de Havilland DHC-1 Chipmunk T.10
WK624	M	G-BWHI	de Havilland DHC-1 Chipmunk T.10
WK628		G-BBMW	de Havilland DHC-1 Chipmunk T.10
WK630		G-BXDG	de Havilland DHC-1 Chipmunk T.10 *(RAF)*
WK633	B	G-BXEC	de Havilland DHC-1 Chipmunk T.10
WK640	C	G-BWUV	de Havilland DHC-1 Chipmunk T.10
WK642		G-BXDP	de Havilland DHC-1 Chipmunk T.10
WL505		G-FBIX	de Havilland DH.100 Vampire FB.9
WL505		G-MKVI	de Havilland DH.100 Vampire FB.6
WL626	P	G-BHDD	Vickers Varsity T.1
WM167		G-LOSM	Armstrong-Whitworth Meteor NF.11
WP308	572	G-GACA	Hunting Percival P.57 Sea Prince T.1
WP788		G-BCHL	de Havilland DHC-1 Chipmunk T.10
WP790	T	G-BBNC	de Havilland DHC-1 Chipmunk T.10
WP795	901	G-BVZZ	de Havilland DHC-1 Chipmunk T.10 *(RN)*
WP800	2	G-BCXN	de Havilland DHC-1 Chipmunk T.10
WP803		G-HAPY	de Havilland DHC-1 Chipmunk T.10
WP805		G-MAJR	de Havilland DHC-1 Chipmunk T.10
WP808		G-BDEU	de Havilland DHC-1 Chipmunk T.10
WP809	78	G-BVTX	de Havilland DHC-1 Chipmunk T.10 *(RN)*
WP833		G-BZDU	de Havilland DHC-1 Chipmunk T.10 *(RAF)*
WP840	9	G-BXDM	de Havilland DHC-1 Chipmunk T.10
WP844		G-BWOX	de Havilland DHC-1 Chipmunk T.10
WP856	904	G-BVWP	de Havilland DHC-1 Chipmunk T.10 *(RN)*
WP857	24	G-BDRJ	de Havilland DHC-1 Chipmunk T.10
WP859		G-BXCP	de Havilland DHC-1 Chipmunk T.10
WP860	6	G-BXDA	de Havilland DHC-1 Chipmunk T.10
WP870	12	G-BCOI	de Havilland DHC-1 Chipmunk T.10 *(RAF)*
WP896		G-BWVY	de Havilland DHC-1 Chipmunk T.10
WP901		G-BWNT	de Havilland DHC-1 Chipmunk T.10
WP903		G-BCGC	de Havilland DHC-1 Chipmunk T.10 *(Queens Flight)*
WP920		G-BXCR	de Havilland DHC-1 Chipmunk T.10
WP925	C	G-BXHA	de Havilland DHC-1 Chipmunk T.10 *(Army)*
WP928	D	G-BXGM	de Havilland DHC-1 Chipmunk T.10 *(Army)*
WP929		G-BXCV	de Havilland DHC-1 Chipmunk T.10
WP930	J	G-BXHF	de Havilland DHC-1 Chipmunk T.10 *(Army)*
WP971		G-ATHD	de Havilland DHC-1 Chipmunk T.10
WP977		G-BHRD	de Havilland DHC-1 Chipmunk T.10
WP983	B	G-BXNN	de Havilland DHC-1 Chipmunk T.10
WP984	H	G-BWTO	de Havilland DHC-1 Chipmunk T.10
WR360		G-DHSS	de Havilland DH.112 Venom FB.1
WR410		G-DHUU	de Havilland DH.112 Venom FB.1
WR421		G-DHTT	de Havilland DH.112 Venom FB.1
WR470		G-DHVM	de Havilland DH.112 Venom FB.1
WT327		G-BXMO	English Electric Canberra B.6
WT333		G-BVXC	English Electric Canberra B(I).8
WT722	878/VL	G-BWGN	Hawker Hunter T.8C *(RN)*
WT723		G-PRII	Hawker Hunter PR.11 *(RN)*
WT871		BGA.3149	Slingsby T.31B
WT898		BGA.4412	Slingsby T.31B
WT903		BGA.3299	Slingsby T.31B
WT908		BGA.3487	Slingsby T.31B
WT910		BGA.3953	Slingsby T.31B
WT913		BGA.3239	Slingsby T.31B
WT918		BGA.3241	Slingsby T.31B
WV198	K	G-BJWY	Sikorsky Whirlwind HAR.21
WV318	D	G-FFOX	Hawker Hunter T.7B
WV322		G-BZSE	Hawker Hunter T.11 *(RAF)*
WV372	R	G-BXFI	Hawker Hunter T.7
WV493	29/A-P	G-BDYG	Percival Provost T.1
WV666	O-D	G-BTDH	Percival Provost T.1
WV740		G-BNPH	Hunting Percival Pembroke C.1
WW421	P	G-BZRE	Percival Provost T.1
WW453	W-S	G-TMKI	Percival Provost T.1
WW499	P-G	G-BZRF	Percival Provost T.1
WZ507		G-VTII	de Havilland DH.115 Vampire T.11
WZ553	40	G-DHYY	de Havilland DH.115 Vampire T.11
WZ589		G-DHZZ	de Havilland DH.115 Vampire T.55
WZ662		G-BKVK	Auster AOP.9 (Army)
WZ706		G-BURR	Auster AOP.9 (Army)
WZ765		BGA.3662	Slingsby T.38 Grasshopper
WZ819		BGA.3498	Slingsby T.38 Grasshopper
WZ828		BGA.4421	Slingsby T.38 Grasshopper
WZ847		G-CPMK	de Havilland DHC-1 Chipmunk T.10
WZ868	H	G-ARMF	de Havilland DHC-1 Chipmunk T.10
WZ879	X	G-BWUT	de Havilland DHC-1 Chipmunk T.10

WZ882	K	G-BXGP	de Havilland DHC-1 Chipmunk T.10 *(Army)*
XA229		BGA.3379	Slingsby T.38 Grasshopper
XA240		BGA.4556	Slingsby T.38 Grasshopper
XA302		BGA.3786	Slingsby T.31B
XA880		G-BVXR	de Havilland DH.104 Devon C.2 *(RAE)*
XD693	Z-Q	G-AOBU	Percival Jet Provost T.1
XE489		G-JETH	Armstrong-Whitworth Sea Hawk FGA.6
XE665	876/VL	G-BWGM	Hawker Hunter T.8C *(RN)*
XE689	864/VL	G-BWGK	Hawker Hunter GA.11 *(RN)*
XE786		BGA.4033	Slingsby T31B
XE807		BGA.3545	Slingsby T.31B Cadet TX.3
XE897		G-DHVV	de Havilland DH.115 Vampire T.55
XE920	A	G-VMPR	de Havilland DH.115 Vampire T.11
XE956		G-OBLN	de Havilland DH.115 Vampire T.11
XF515	R	G-KAXF	Hawker Hunter F.6A
XF516	19	G-BVVC	Hawker Hunter F.6A
XF597	AH	G-BKFW	Percival Provost T.1
XF603	H	G-KAPW	Percival Provost T.1
XF690		G-MOOS	Percival Provost T.1
XF836		G-AWRY	Percival Provost T.1
XF877	JX	G-AWVF	Percival Provost T.1
XF995		G-BZSF	Hawker Hunter T.8B *(RAF)*
XG160	U	G-BWAF	Hawker Hunter F.6A
XG232		G-HHAC	Hawker Hunter F.6
XG452		G-BRMB	Bristol Belvedere HC.1
XG547	T-S/S-T	G-HAPR	Bristol Sycamore HR.14
XG775		G-DHWW	de Havilland DH.115 Vampire T.11 *(RN)*
XH558		G-VLCN	Avro Vulcan B.2
XH568		G-BVIC	English Electric Canberra B.2/B.6
XJ389		G-AJJP	Fairey Jet Gyrodyne
XJ615		G-BWGL	Hawker Hunter T.8C *(As T.7 prototype)*
XJ729		G-BVGE	Westland Whirlwind HAR.10
XJ763	P	G-BKHA	Westland Whirlwind HAR.10
XJ771		G-HELV	de Havilland DH.115 Vampire T.55
XK417		G-AVXY	Auster AOP.9
XK895	CU-19	G-SDEV	de Havilland DH.104 Sea Devon C.20 *(RN)*
XK940		G-AYXT	Westland Whirlwind HAS.7
XL426		G-VJET	Avro Vulcan B.2
XL502		G-BMYP	Fairey Gannet AEW.3 *(RN)*
XL571	V	G-HNTR	Hawker Hunter T.7 *(Blue Diamonds c/s)*
XL573		G-BVGH	Hawker Hunter T.7
XL577	V	G-BXKF	Hawker Hunter T.7 *(Blue Diamonds c/s)*
XL602		G-BWFT	Hawker Hunter T.8M
XL613		G-BVMB	Hawker Hunter T.7A
XL616	D	G-BWIE	Hawker Hunter T.7A
XL621		G-BNCX	Hawker Hunter T.7
XL714		G-AOGR	de Havilland DH.82A Tiger Moth
XL-716		G-AOIL	de Havilland DH.82A Tiger Moth
XL809		G-BLIX	Saro Skeeter AOP.12 *(Army)*
XL812		G-SARO	Saro Skeeter AOP.12
XL954		G-BXES	Percival Pembroke C.1
XM223		G-BWWC	de Havilland DH.104 Devon C.2
XM365		G-BXBH	Hunting Jet Provost T.3A
XM376	27	G-BWDR	Hunting Jet Provost T.3A
XM424		G-BWDS	Hunting Jet Provost T.3A
XM478		G-BXDL	Hunting Jet Provost T.3A
XM479	54	G-BVEZ	Hunting Jet Provost T.3A
XM553		G-AWSV	Saro Skeeter AOP.12
XM575		G-BLMC	Avro Vulcan B.2A
XM655		G-VULC	Avro Vulcan B.2A
XM685	PO/513	G-AYZJ	Westland Whirlwind HAS.7
XM819		G-APXW	Lancashire Aircraft EP.9 *(Army)*
XN157		BGA.3255	Slingsby T.21B
XN185		BGA.4077	Slingsby T.21B
XN187		BGA.3903	Slingsby T.21B
XN243		BGA.3145	Slingsby T.31B
XN351		G-BKSC	Saro Skeeter AOP.12 *(Army)*
XN441		G-BGKT	Auster AOP.9 *(Army)*
XN459		G-BWOT	Hunting Jet Provost T.3A
XN470		G-BXBJ	Hunting Jet Provost T.3A
XN498	16	G-BWSH	Hunting Jet Provost T.3A
XN510		G-BXBI	Hunting Jet Provost T.3A
XN629	49	G-KNOT	Hunting Jet Provost T.3A
XN637	03	G-BKOU	Hunting Jet Provost T.3
XP242		G-BUCI	Auster AOP.9 *(Army)*
XP254		G-ASCC	Auster AOP.11 *(Army)*
XP279		G-BWKK	Auster AOP.9 *(Army)*
XP355	A	G-BEBC	Westland Whirlwind HAR.10
XP672	03	G-RAFI	Hunting Jet Provost T.4
XP772		G-BUCJ	de Havilland DHC.2 Beaver AL.1 *(Army)*
XP907		G-SROE	Westland Scout AH.1
XR240		G-BDFH	Auster AOP.9 *(Army)*
XR241		G-AXRR	Auster AOP.9 *(Army)*
XR246		G-AZBU	Auster AOP.9

XR486		G-RWWW	Westland Whirlwind HCC.12 *(Queens Flight)*
XR537	T	G-NATY	Folland Gnat T.1
XR538	01	G-RORI	Folland Gnat T.1 *(RAF Training c/s)*
XR595	M	G-BWHU	Westland Scout AH.1 *(Army)*
XR673		G-BXLO	Hunting Jet Provost T.4
XR944		G-ATTB	Wallis WA.116
XR991		G-MOUR	Folland Gnat T.1 *(Yellowjacks c/s)*
XR993		G-BVPP	Folland Gnat T.1 *(Red Arrows c/s)*
XS101	1	G-GNAT	Folland Gnat T.1 *(Red Arrows c/s)*
XS111		G-TIMM	Folland Gnat T.1
XS165	37	G-ASAZ	Hiller UH-12E-4
XS587		G-VIXN	de Havilland DH.110 Sea Vixen FAW.2 *(RN*
XS652		BGA.1107	Slingsby T.45 Swallow
XS765		G-BSET	Beagle Basset CC.1
XS770		G-HRHI	Beagle Basset CC.1 *(Queens Flight)*
XT223		G-XTUN	Westland Sioux AH.1 *(Army)*
XT435	430	G-RIMM	Westland Wasp HAS.1 *(RN)*
XT634		G-BYRX	Westland Scout AH.1 *(Army)*
XT653		BGA.3469	Slingsby T.45 Swallow
XT781	426	G-KAWW	Westland Wasp HAS.1 *(RN)*
XT787		G-KAXT	Westland Wasp HAS.1 *(RN)*
XT78?	316	G-BMIR	Westland Wasp HAS.1 *(RN)*
XV121		G-BYKJ	Westland Scout AH.1 *(Army)*
XV130	R	G-BWJW	Westland Scout AH.1 *(Army)*
XV134		G-BWLX	Westland Scout AH.1 *(Army)*
XV137		G-CRUM	Westland Scout AH.1
XV140	K	G-KAXL	Westland Scout AH.1 *(Army)*
XV238	41	G-ALYW(2)	de Havilland DH.106 Comet 1
XV268		G-BVER	de Havilland DHC.2 Beaver *(Army)*
XW281	U	G-BYNZ	Westland Scout AH.1 *(Royal Marines)*
XW289	73	G-JPVA	BAC Jet Provost T.5A
XW293	Z	G-BWCS	BAC Jet Provost T.5
XW324		G-BWSG	BAC Jet Provost T.5
XW325	E	G-BWGF	BAC Jet Provost T.5A
XW333		G-BVTC	BAC Jet Provost T.5A
XW422		G-BWEB	BAC Jet Provost T.5A
XW423	14	G-BWUW	BAC Jet Provost T.5A
XW431	A	G-BWBS	BAC Jet Provost T.5A
XW433		G-JPRO	BAC Jet Provost T.5A *(CFS)*
XW635		G-AWSW	Beagle Husky
XW784	VL	G-BBRN	Mitchell-Procter Kittiwake *(RN)*
XW858		G-DMSS	Westland Gazelle HT.3
XW861	CU-52	G-BZFJ	Westland Gazelle HT.2
XW895	51	G-BXZD	Westland Gazelle HT.2 *(RN)*
XW898	G	G-CBXT	Westland Gazelle HT.3 *(RN)*
XX110		BAPC.169	Sepecat Jaguar GR.1 fsm
XX226	74	BAPC.152	BAe Hawk T.1A fsm
XX253		BAPC.171	BAe Hawk T.1 *(Red Arrows c/s)*
XX467	86	G-TVII	Hawker Hunter T.7 *(TWU)*
XX513	10	G-CCMI	Scottish Aviation Bulldog
XX514		G-BWIB	Scottish Aviation Bulldog
XX515	4	G-CBBC	Scottish Aviation Bulldog
XX518	S	G-UDOG	Scottish Aviation Bulldog
XX521	H	G-CBEH	Scottish Aviation Bulldog
XX522	06	G-DAWG	Scottish Aviation Bulldog
XX524	04	G-DDOG	Scottish Aviation Bulldog
XX525	6	G-CBJJ	Scottish Aviation Bulldog
XX528	X	G-BZON	Scottish Aviation Bulldog
XX534	B	G-EDAV	Scottish Aviation Bulldog
XX537	C	G-CBCB	Scottish Aviation Bulldog
XX538	O	G-TDOG	Scottish Aviation Bulldog
XX543	F	G-CBAB	Scottish Aviation Bulldog
XX546	03	G-CBCO	Scottish Aviation Bulldog
XX549	6	G-CBID	Scottish Aviation Bulldog
XX550	Z	G-CBBL	Scottish Aviation Bulldog
XX551	E	G-BZDP	Scottish Aviation Bulldog
XX554	09	G-BZMD	Scottish Aviation Bulldog
XX561	7	G-BZEP	Scottish Aviation Bulldog
XX611		G-CBDK	Scottish Aviation Bulldog
XX612	A03	G-BZXC	Scottish Aviation Bulldog
XX614	V	G-GGRR	Scottish Aviation Bulldog
XX619	T	G-CBBW	Scottish Aviation Bulldog
XX621	H	G-CBEF	Scottish Aviation Bulldog
XX622	B	G-CBGX	Scottish Aviation Bulldog
XX624	E	G-KDOG	Scottish Aviation Bulldog
XX625	01	G-CBBR	Scottish Aviation Bulldog
XX628	9	G-CBFU	Scottish Aviation Bulldog
XX629	V	G-BZXZ	Scottish Aviation Bulldog
XX630	5	G-SIJW	Scottish Aviation Bulldog
XX631	W	G-BZXS	Scottish Aviation Bulldog
XX636	Y	G-CBFP	Scottish Aviation Bulldog
XX638		G-DOGG	Scottish Aviation Bulldog
XX658	03	G-BZPS	Scottish Aviation Bulldog
XX667	16	G-BZFN	Scottish Aviation Bulldog

	XX668	I	G-CBAN	Scottish Aviation Bulldog
	XX692	A	G-BZMH	Scottish Aviation Bulldog
	XX693	07	G-BZML	Scottish Aviation Bulldog
	XX694	E	G-CBBS	Scottish Aviation Bulldog
	XX695	3	G-CBBT	Scottish Aviation Bulldog
	XX698	9	G-BZME	Scottish Aviation Bulldog
	XX699	F	G-CBCV	Scottish Aviation Bulldog
	XX700	17	G-CBEK	Scottish Aviation Bulldog
	XX702		G-CBCR	Scottish Aviation Bulldog
	XX707	4	G-CBDS	Scottish Aviation Bulldog
	XX711	X	G-CBBU	Scottish Aviation Bulldog
	XX713	2	G-CBJK	Scottish Aviation Bulldog
	XX725	GU	BAPC.150	Sepecat Jaguar GR.1
	XZ329		G-BZYD	Westland Gazelle AH.1
	XZ363	A	BAPC.151	Sepecat Jaguar GR.1A
	ZA556		BAPC.155	Panavia Tornado GR.1 fsm
	ZA634	C	G-BUHA	Slingsby T-61F Venture T.2
	ZB500		G-LYNX	Westland Lynx 800 *(Army)*
	ZB627	A	G-CBSK	Westland SA.314G Gazelle HT.3
	ZB629	CU	G-CBZL	Westland SA.314G Gazelle HT.3
	ZH139	01	BAPC.191	BAe Harrier GR.5
		AL-K	G-HURR	Hawker Hurricane XII
		F	G-RUMW	Grumman FM-2 Wildcat *(RN/FAA)*
	"B" Conditions markings			
	G-17-3		G-AVNE	Westland Wessex 60
	G-29-1		G-APRJ	Avro Lincoln
	G-48/1		G-ALSX	Bristol Sycamore
	U-0247		G-AGOY	Miles Messenger - *intended marks*
	W-2		BAPC.85	Weir W-2
	X-25		BAPC.274	Boulton and Paul P.6 fsm
	Other markings			
	SR-XP020		G-BZUG	Tiger Cub RL7A XP Sherwood Ranger
OTHER ARMED FORCES				
AUSTRALIA				
	A2-4		VH-ALB	Supermarine Seagull
	A16-199	SF-R	G-BEOX	Lockheed Hudson IIIA
	A17-48		G-BPHR	DH.82A Tiger Moth
	N6-766		G-SPDR	DH.115 Sea Vampire T.55 *(Navy)*
BELGIUM				
	HD-75		G-AFDX	Hanriot HD.1
BOLIVIA				
	FAB-184		G-SIAI	SIAI-Marchetti SF.260W
BOTSWANA				
	OJ-1		G-BXFU	BAC.167 Strikemaster 83
	OJ-7	Z-28	G-BXFX	BAC.167 Strikemaster 83
	OJ-8		G-BXFV	BAC.167 Strikemaster 83
BURKINA FASO				
	BF8431	31	G-NRRA	SIAI-Marchetti SF.260W
CANADA				
	622		N6699D	Piasecki HUP-3 Retreiver *(RCN)*
	920	QN-	CF-BXO	Supermarine Stranraer
	16693	693	G-BLPG	Auster J/1N *(AOP.6 c/s)*
	18013	013	G-TRIC	de Havilland DHC-1 Chipmunk
	18393		G-BCYK	Avro Canada CF.100 Canuck IV
	18671		G-BNZC	de Havilland DHC-1 Chipmunk
	20310	310	G-BSBG	North American Harvard IV
	21261		G-TBRD	Lockheed T-33A
PEOPLES' REPUBLIC OF CHINA (including HONG KONG)				
	663/P11151	88	ZK-RMH	Curtiss P-40E Kittyhawk *(Chinese AF)*
		68	G-BVVG	Nanchang CJ-6A
	HKG-6		G-BPCL	Scottish Aviation Bulldog
	HKG-11		G-BYRY	Slingsby T.67M-200 Firefly
	HKG-13		G-BXKW	Slingsby T.67M-200 Firefly
ECUADOR				
	259	T59	G-UPPI	BAC 167 Strikemaster Mk.80A
FRANCE				
	124		G-BOSJ	Nord 3400
	143		G-MSAL	Morane-Saulnier MS.733 *(Aeronavale)*
	185	44-CA	G-BWLR	Max Holste Broussard
	316	315-SN	F-GGKR	Max Holste Broussard
	394		G-BIMO	Stampe SV-4C
	MS.824		G-AWBU	Morane-Saulnier N rep *(French AF)*

	1 4513	S	G-BFYO	SPAD XII rep
		315-SQ/20	G-BWGG	Max Holste Broussard
	F-OTAN-6		G-ATDB	SNCAN 1101 Noralpha
	F-OTAN-6		G-BAYV	SNCAN 1101 Noralpha
GERMANY				
	1		G-BWUE	Hispano HA.1112-MIL Buchon
	1+4		G-BSLX	WAR FW190 scale rep
	2+1	7334	G-SYFW	WAR FW190 scale rep
	3		G-BAYV	Nord 1101 *(Messerschmitt guise)*
	8+--		G-WULF	WAR FW190 scale rep
	- + 9		G-CCFW	WAR FW190 scale rep
	10	KG+EM	G-ETME	Nord 1002 Pingouin
	14		BAPC.67	Messerschmitt Bf.109 fsm
	+14		G-BSMD	Nord 1101 *(Messerschmitt guise)*
	17+TF		G-BZTJ	Bücker Bü.133C Jungmeister
	28+10		G-BWTT	Aero L-39ZO Albatros
	50	CW+BG	G-BXBD	CASA 1131 Jungmann
	97+04		G-APVF	Putzer Elster B
	99+32		G-BZGK	North American OV-10B Bronco
	99+26		G-BZGL	North American OV-10B Bronco
	102/17		BAPC.88	Fokker Dr.1 5/8th scale model
	152/17		G-ATJM	Fokker Dr.1 rep
	210/16		BAPC.56	Fokker E.III
	422/15		G-AVJO	Fokker E-III rep
	425/17		BAPC.133	Fokker Dr.1 rep
	450/17		G-BVGZ	Fokker Dr.1 rep *(German Army Air Service)*
	626/8		N6268	Fokker D.VII (Travel Air 2000)
	694		BAPC.239	Fokker D.VIII 5/8th scale model
	1227	DG+HO	G-FOKW	Focke-Wulf FW190A-5
	1480	6	BAPC.66	Messerschmitt Bf.109 rep
	CL.1 1803/18		G-BUYU	Bowers Fly Baby *(German Army Air Service)*
	2292		BAPC.138	Hansa Brandenberg W.29 fsm
	4477	GD+EG	G-RETA	CASA 1131E Jungmann Srs.2000
	6357	6	BAPC.74	Messerschmitt Bf.109 rep
	7198/18		G-AANJ	LVG C.VI
	10639	6 (Black)	G-USTV	Messerschmitt Bf.109G-2
	191454		BAPC.271	Messerschmitt Me 163B Komet fsm
	C19/15		BAPC.118	Albatros D.Va fsm
	D604		G-FLIZ	Staaken Flitzer
	D692		G-BVAW	Staaken Flitzer
	D5397/17		G-BFXL	Albatros D.VA rep
	DR1/17		BAPC.139	Fokker Dr.1 Triplane fsm
		6G+ED	G-BZOB	Slepcev Storch
		BU+CC	G-BUCC	CASA 1131E Jungmann
		BU+CK	G-BUCK	CASA 1131E Jungmann
		CC+43	G-CJCI	Pilatus P.2 *(Arado Ar.96B guise)*
		CF+HF	EI-AUY	Morane-Saulnier MS.502 Criquet
		F+IS	G-BIRW	Morane-Saulnier MS.505 Criquet
		JU.FB. D.I	G-BNPV	Bowers Fly Baby *(German Army Air Service)*
		LG+0I	G-AYSJ	Bücker Bü.133 Jungmeister
		LG+03	G-AEZX	Bücker Bü.133 Jungmeister
		NJ+C11	G-ATBG	Messerschmitt Bf.108 *(Nord 1002)*
		S4+A07	G-BWHP	CASA 1131E Jungmann
		S5+B06	G-BSFB	CASA 1131E Jungmann
		TA+RC	G-BPHZ	Morane-Saulnier MS.505 Criquet
		VZ-NK	G-BFHG	CASA 352L
HUNGARY				
	503		G-BRAM	MiG 21PF *(Russian c/s)*
IRELAND				
	161		G-CCCA	Vickers Supermarine 509 Spitfire T.IX
	177		G-BLIW	Percival Provost T.51
ITALY				
		W7	G-AGFT	Avia FL.3
	MM12822	20	G-FIST	Fiesler Fi.156C-3 Storch
	MM53211	ZI-4	BAPC.79	Fiat G.46-4b
JAPAN				
	15-1585		BAPC.58	Yokosuka MXY-7 Ohka II
	24		BAPC.83	Kawasaki Ki 100-1b
	997		BAPC.98	Yokosuka MXY-7 Ohka II
THE NETHERLANDS				
	BI-005		G-BUVN	CASA 1131E Jungmann
	E-15		G-BIYU	Fokker S.11 Instructor
	K-174		G-BEPV	Fokker S.11 Instructor *(Navy)*
	R-55	*(Also carries 54-2466)*	G-BLMI	Piper L-18C Super Cub *(Air Force)*
	R-151		G-BIYR	Piper L-21B Super Cub
	R-156	*(Also carries 54-2446)*	G-ROVE	Piper L-21B Super Cub *(Air Force)*
	R-163	*(Also carries 54-2453)*	G-BIRH	Piper L-21B Super Cub *(Air Force)*
	R-167		G-LION	Piper L-21B Super Cub *(Air Force)*

NORTH KOREA				
	01420		G-BMZF	MiG-15
NORTH VIETNAM				
	1211		G-MIGG	WSK PZL-Mielec Lim-5 *(MiG-17F)*
NORWAY				
	321		G-BKPY	Saab Safir
	423 and 427		G-AMRK	Gloster Gladiator
PORTUGAL				
	85		G-BTPZ	Hawker (Isaacs) Fury
	1365		G-DHPM	de Havilland DHC-1 Chipmunk
	1377		G-BARS	de Havilland DHC-1 Chipmunk
	1747		G-BGPB	CCF Harvard 4
RUSSIA				
	01		G-YKSZ	Yakovlev Yak-52
	07 *(Yellow)*		G-BMJY	Yakovlev Yak-18
	09		G-BVMU	Yakovlev Yak-52 *(DOSAAF)*
	10?		G-BTZB	Yakovlev Yak-50 *(DOSAAF) (but see below)*
	1 *(White)*		G-BZMY	Yakovlev Yak C-11
	11 *(Yellow)*		G-YCII	LET Yakovlev C.11
	12 *(Red)*		G-DELF	Aero L-29A Delfin
	26		G-BVXK	Yakovlev Yak-52 *(DOSAAF)*
	27		G-OYAK	Yakovlev Yak-11
	27 *(Red)*		G-YAKX	Yakovlev Yak-52 *(DOSAAF)*
	36 *(White)*		G-KYAK	Yakovlev Yak C-11 *(DOSAAF)*
	39		G-XXVI	Sukhoi Su-26M
	42 *(White)*		G-CBRU	Yakovlev Yak-52 *(DOSAAF)*
	49		G-YAKU	Yakovlev Yak-50 *(DOSAAF)*
	50		G-CBRW	Yakovlev Yak-52 *(DOSAAF)*
	55		G-BVOK	Yakovlev Yak-52 *(DOSAAF)*
	61		G-YAKM	Yakovlev Yak-50 *(DOSAAF)*
	66		G-YAKN	Yakovlev Yak-52 *(DOSAAF)*
	69		G-BTZB	Yakovlev Yak-50 *(DOSAAF)*
	72		G-BXAV	Yakovlev Yak-52 *(DOSAAF)*
	139		G-BWOD	Yakovlev Yak-52 *(DOSAAF)*
Republic of **SOUTH AFRICA**				
	92		G-BYCX	Westland Wasp HAS.Mk.1 *(Navy)*
SAUDI ARABIA				
	1108		G-CBPB	BAC 167 Strikemaster Mk.80A
	1133		G-BESY	BAC 167 Strikemaster Mk.80A
SPAIN				
	E3B-153	781-75	G-BPTS	CASA 1131 Jungmann
	E3B-350	05-97	G-BHPL	CASA 1131 Jungmann
	E3B-369	781-32	G-BPDM	CASA 1131 Jungmann
		781-25	G-BRSH	CASA 1131 Jungmann
SWITZERLAND				
	A 10		G-BECW	CASA 1131E Jungmann
	A 12		G-CCHY	Bücker Bü.131 Jungmann
	A 50		G-CBCE	CASA 1131E Jungmann rep
	A 57		G-BECT	CASA 1131E Jungmann
	A 125		G-BLKZ	Pilatus P.2-05
	A 806		G-BTLL	Pilatus P.3
	C-552		G-DORN	EKW C-3605
	J-1149		G-SWIS	de Havilland DH.100 Vampire FB.6
	J-1573		G-VICI	de Havilland DH.112 Venom FB.50
	J-1605		G-BLID	de Havilland DH.112 Venom FB.50
	J-1611		G-DHTT	de Havilland DH.112 Venom FB.50
	J-1758		G-BLSD	de Havilland DH.112 Venom FB.50
	J-4021		G-HHAC	Hawker Hunter F.58
	J-4031		G-BWFR	Hawker Hunter F.58
	J-4058		G-BWFS	Hawker Hunter F.58
	J-4066		G-BXNZ	Hawker Hunter F.58
	J-4090		G-SIAL	Hawker Hunter F.58
	U-80		G-BUKK	Bücker Bü.133 Jungmeister
	U-95		G-BVGP	Bücker Bü.133 Jungmeister
	U-99		G-AXMT	Bücker Bü.133 Jungmeister
	U-110		G-PTWO	Pilatus P.2
	U-1234		G-DHAV	de Havilland DH.115 Vampire T.11
	V-54		G-BVSD	SE.3130 Alouette II
UNITED NATIONS				
	001		G-BFRI	Sikorsky S-61N Mk.II
UNITED STATES				
	2		G-AZLE	Boeing Stearman Kaydet *(Army)*
	5		G-BEEW	Taylor Monoplane (Boeing P-26A) *(Army)*
	14		G-ISDN	Boeing Stearman Kaydet *(Army)*

LS326	L/2	G-AJVH	Fairey Swordfish II
LZ766		G-ALCK	Percival Proctor III
MAV467	R-O	BAPC.202	Supermarine Spitfire V fsm
MH415	FU-N	BAPC.209	Supermarine Spitfire LF.IXC fsm
MH434	ZD-B	G-ASJV	Supermarine Spitfire IXB
MH486	FF-A	BAPC.206	Supermarine Spitfire IX fsm
MH777	RF-N	BAPC.221	Supermarine Spitfire rep
MJ627	9G-P	G-BMSB	Supermarine Spitfire IX
MJ730	GZ-?	G-HFIX	Supermarine Spitfire IXe
MJ832	DN-Y	BAPC.229	Supermarine Spitfire fsm
MK732	3W-17	G-HVDM	Supermarine Spitfire IXc
MK805	SH-B	----	Supermarine Spitfire IX rep
(Built TDL Reproduction Aircraft)			*"Peter John III" in 64 Sqn c/s*
ML407	OU-V/NL-D	G-LFIX	Supermarine Spitfire IX
ML417	2I-T	G-BJSG	Supermarine Spitfire IXe
MP425		G-AITB	Airspeed Oxford I
MT438		G-AREI	Auster III
MT928	ZX-M	G-BKMI	Supermarine Spitfire VIIIc
MV268	JE-J	G-SPIT	Supermarine Spitfire XIVe
MV370		G-FXIV	Supermarine Spitfire XIVc
MW763	HF-A	G-TEMT	Hawker Tempest II
MW800	HF-V	G-BSHW	Hawker Tempest II
NH238	D-A	G-MKIX	Supermarine Spitfire IX
NJ633		G-AKXP	Auster 5
NJ673		G-AOCR	Auster 5
NJ695		G-AJXV	Auster 4
NJ703		G-AKPI	Auster 5
NJ719		G-ANFU	Auster 5 - *intended marks*
NL750		G-AOBH	de Havilland DH.82A Tiger Moth
NL985		G-BWIK	de Havilland DH.82A Tiger Moth
NM181		G-AZGZ	de Havilland DH.82A Tiger Moth
NX534		G-BUDL	Auster III
NX611	LE-C/DX-C	G-ASXX	Avro Lancaster B.VII
PL344	Y2-B	G-IXCC	Supermarine Spitfire IXe
PL965	R	G-MKXI	Supermarine Spitfire PR.XI
PL983	JV-F	G-PRXI	Supermarine Spitfire XI
PR772		G-BTTA	Hawker Iraqi Fury FB.11
PS853	C	G-RRGN	Supermarine Spitfire PR.XIX
PT462	SW-A	G-CTIX	Supermarine Spitfire IX
PV202	5R-Q	G-TRIX	Supermarine Spitfire IX
PZ865	Q	G-AMAU	Hawker Hurricane IIc
RA905		BGA.1143	Slingsby T.7 Cadet
RG333		G-AIEK	Miles Messenger
RM221		G-ANXR	Percival Proctor IV
RN201		G-BSKP	Supermarine Spitfire XIV
RN218	N	G-BBJI	Isaacs Spitfire
RR232		G-BRSF	Supermarine Spitfire IXc
RT486	PF-A	G-AJGJ	Auster 5
RT610		G-AKWS	Auster 5A
RX168		G-BWEM	Supermarine Seafire L.III - *intended marks*
SM520		G-BXHZ	Supermarine Spitfire HF.IX
SM845	GZ-J	G-BUOS	Supermarine Spitfire XVIIIe
SM969	D-A	G-BRAF	Supermarine Spitfire XVIII
SX336		G-BRMG	Supermarine Seafire XVII
TA634	8K-K	G-AWJV	de Havilland DH.98 Mosquito TT.35
TA719	6 T	G-ASKC	de Havilland DH.98 Mosquito TT.35
TA805		G-PMNF	Supermarine Spitfire IX
TB252	GW-H	G-XVIE	Supermarine Spitfire XVIe
TD248	D	G-OXVI	Supermarine Spitfire XVIe
TE184	D	G-MXVI	Supermarine Spitfire XVIe
TJ398		BAPC.70	Auster AOP.5
TJ534		G-AKSY	Auster 5
TJ565		G-AMVD	Auster 5
TJ569		G-AKOW	Auster 5
TJ672		G-ANIJ	Auster 5
TS291		BGA.852	Slingsby T.8 Tutor
TS423	YS-L	G-DAKS	Douglas Dakota 3
TS798		G-AGNV	Avro 685 York C.1
TW439		G-ANRP	Auster 5
TW467	ROD-F	G-ANIE	Auster 5
TW511		G-APAF	Auster 5 *(Army)*
TW536	T-SV	G-BNGE	Auster AOP.6
TW591		G-ARIH	Auster AOP.6 *(Army)*
TW641		G-ATDN	Auster AOP.6
TX183		G-BSMF	Avro Anson C.19
VF512	PF-M	G-ARRX	Auster AOP.6
VF516		G-ASMZ	Auster AOP.6
VF526	T	G-ARXU	Auster AOP.6 *(Army)*
VF581		G-ARSL	Auster AOP.6
VL348		G-AVVO	Avro Anson C.19/2
VL349		G-AWSA	Avro Anson C.19/2
VM360		G-APHV	Avro Anson C.19/2
VM687		BGA.794	Slingsby T.8 Tutor
VP955		G-DVON	de Havilland DH.104 Devon C.2/2

23		N49272	Fairchild PT-23 Cornell *(Army Air Corps)*
26		G-BAVO	Boeing Stearman Kaydet *(Army)*
27		G-AGYY	Ryan PT-21 *(Army Air Corps)*
28		N8162G	Boeing Stearman Kaydet *(Army)*
33		G-THEA	Boeing Stearman Kaydet *(Navy)*
43	SC	G-AZSC	North American AT-16 Texan *(Army Air Force)*
44		G-RJAH	Boeing Stearman Kaydet *(Army Air Corps))*
49		G-KITT	Curtiss TP-40M Kittyhawk *(Army Air Corps))*
54		G-BCNX	Piper L-4H *(Air Force)*
85		G-BTBI	Republic P-47 Thunderbolt scale rep *(Air Force)*
112		G-BSWC	Boeing Stearman Kaydet *(Army Air Corps))*
118		G-BSDS	Boeing Stearman Kaydet *(Army Air Corps)*
379		G-ILLE	Boeing Stearman Kaydet *(Army Air Corps))*
441		G-BTFG	Boeing Stearman Kaydet *(Navy)*
526		G-BRWB	North American T-6G Texan *(Air Force)*
624	D-39	G-BVMH	Piper L-4 *(Wag-Aero Cuby) (Army Air Corps)*
669		G-CCXA	Boeing Stearman A75N1 *(Army Air Corps)*
854		G-BTBH	Ryan PT-22 *(Army Air Corps)*
855		N56421	Ryan PT-22 *(Army Air Corps)*
897E		G-BJEV	Aeronca Chief *(Navy)*
1164		G-BKGL	Beech C-45 *(Army)*
2807	103	G-BHTH	North American T-6G Texan *(Navy)*
7797		G-BFAF	Aeronca L-16A *(Army)*
8178	FU-178	G-SABR	North American F-86A Sabre *(Air Force)*
8242	FU-242	N196B	North American F-86A Sabre *(Air Force)*
02538		N33870	Fairchild PT-19 Cornell *(Army Air Corps)*
07539	143	N63590	Boeing Stearman Kaydet *(Navy)*
14863		G-BGOR	North American AT-6D Texan *(Army Air Force)*
15445		G-BLGT	Piper L-18C-PI
16136	205	G-BRUJ	Boeing Stearman Kaydet *(Navy)*
18263	822	N38940	Boeing Stearman Kaydet *(Army Air Corps)*
21714	201B	G-RUMM	Grumman F8F-2P Bearcat *(Navy)*
28521	TA-521	G-TVIJ	North American T-6J Harvard *(Air Force)*
29261		G-CDET	Culver Cadet *(Army Air Force)*
30274		N203SA	Piper AE-1 Cub Cruiser
31145	G-26	G-BBLH	Piper L-4B *(Army)*
31171		N7614C	North American B-25J Mitchell *(Marines)*
31952		G-BRPR	Aeronca L-3C Grasshopper *(Army)*
38674		G-MTKM	Thomas-Morse S4 Scout scale rep *(Army Air Corps)*
40467	19	G-BTCC	Grumman F6F Hellcat *(Navy)*
41386		G-MJTD	Thomas-Morse S4 Scout scale rep *(Army Air Corps)*
46214	X-3	CF-KCG	Grumman TBM-3E Avenger *(Navy)*
53319	RB 319	G-BTDP	Grumman TBM-3R Avenger *(Navy)*
54884	57-H	G-AKAZ	Piper L-4A *(Army Air Corps)*
56498		N44914	Douglas C-54D Skymaster
80105	19	G-CCBN	Replica SE5A rep (*US Air Service)*
80425	4-WT	G-RUMT	Grumman F7F-3P Tigercat *(Navy)*
90678	27	G-BRVG	North American SNJ-7 Texan *(Navy)*
91007	TR-007	G-NASA	Lockheed T-33A *(Air Force)*
92844	8	G-BXUL	Vought FG-1D Corsair *(Navy)*
93542	LTA-542	G-BRLV	North American T-6 Texan *(Air Force)*
111836	JZ-6	G-TSIX	North American AT-6C Texan *(Navy)*
111989		N33600	Cessna L-19A Bird Dog *(Army)*
115042	TA-042	G-BGHU	North American T-6G Texan *(Air Force)*
115302	TP	G-BJTP	Piper L-18C Super Cub *(Marines)*
115684	VM	G-BKVM	Piper L-21A Super Cub *(Army)*
124485	DF-A	G-BEDF	Boeing B-17G Flying Fortress *(Army Air Corps)*
122351		G-BKRG	Beech C-45G
126603		G-BHWH	Weedhopper JC-24C *(Navy)*
126922	AK 402	G-RADR	Douglas AD-4NA Skyraider *(Navy)*
151632		G-BWGR	North American TB-25N Mitchell *(Air Force)*
217786	25	CF-EQS	Boeing Stearman Kaydet *(Army Air Force)*
224319	L4-D	N147DC	Douglas C-47A-75-DL Dakota *(Army Air Force)*
226413	ZU-N	N47DD	Republic P-47D Thunderbolt *(Army Air Force)*
226671	MX-X/LH-X	G-THUN	Republic P-47D Thunderbolt *(Army Air Force)*
231983	IY-G	F-BDRS	Boeing B-17G Flying Fortress *(Army Air Force)*
236657	72-D	G-BGSJ	Piper L-4A-PI *(Army Air Corps)*
237123		BAPC.157	Waco CG-4A Hadrian
238410	44-A	G-BHPK	Piper L-4A *(Army Air Corps)*
243809		BAPC.185	WACO CG-4A Hadrian Glider
252983		N66630	Schweizer TG-3A
292912	LN-F	N47FK	Douglas C-47A Dakota 3 *(Army Air Force)*
314887		G-AJPI	Fairchild UC-61 Forwarder *(Army Air Force)*
315211	JB-Z	N1944A	Douglas C-47A
315509	W7-S	G-BHUB	Douglas C-47A Dakota *(Army Air Force)*
329405	23-A	G-BCOB	Piper L-4H *(Army Air Corps)*
329417		G-BDHK	Piper L-4A *(Army Air Corps)*
329471	44-F	G-BGXA	Piper L-4H *(Army Air Corps)*
329601	44-D	G-AXHR	Piper L-4H *(Army Air Corps)*
329854	44-R	G-BMKC	Piper L-4H *(Army Air Corps)*
329934	72-B	G-BCPH	Piper L-4H (*(Army Air Corps)/French)*
330238	24-2	G-LIVH	Piper L-4H *(Army Air Corps)*
330485	44-C	G-AJES	Piper L-4H *(Army Air Corps)*
343251	27	G-NZSS	Boeing Stearman Kaydet *(Army Air Corps)*

	413573	B6-V	N6526D	North American P-51D Mustang *(Army Air Corps)*
	413704	B7-H	G-BTCD	North American P-51D Mustang *(Army Air Corps)*
	414419	LH-F	G-MSTG	North American P-51D Mustang *(Air Force)*
	436021		G-BWEZ	Piper L-4 *(Army Air Corps))*
	454467	44-J	G-BILI	Piper L-4J *(Army Air Corps)*
	454537	04-J	G-BFDL	Piper L-4J *(Army Air Corps)*
	461748	Y	G-BHDK	Boeing B-29A Superfortress *(Air Force)*
	463209	WZ-S	BAPC.255	North American P-51D Mustang *(Army Air Corps)*
	463864	HL-W	G-CBNM	North American P-51D Mustang *(Army Air Corps)*
	472035		G-SIJJ	North American P-51D Mustang
	472216	HO-M	G-BIXL	North American P-51D Mustang *(Army Air Corps)*
	472218	WZ-I	G-HAEC	North American P-51D Mustang *(Army Air Corps)*
	473877		N167F	North American P-51D Mustang *(Army Air Corps)*
	479609	PR-L4	G-BHXY	Piper L-4H *(Army Air Corps)*
	479744	49-M	G-BGPD	Piper L-4H *(Army Air Corps)*
	479766	63-D	G-BKHG	Piper L-4H *(Army Air Corps)*
	479781		G-AISS (2)	Piper L-4H *(Army Air Corps)*
	479897	JD	G-BOXJ	Piper L-4H *(Army Air Corps)*
	480015	44-M	G-AKIB	Piper L-4H *(Army Air Corps)*
	480133	44-B	G-BDCD	Piper L-4J *(Army Air Corps)*
	480321	44-H	G-FRAN	Piper L-4J *(Army Air Corps)*
	480480	44-E	G-BECN	Piper L-4J *(Army Air Corps)*
	480636	58-A	G-AXHP	Piper L-4J *(Army Air Corps)*
	480752	39-E	G-BCXJ	Piper L-4J *(Army Air Corps)*
	483868	N	N5237V	Boeing B-17G Flying Fortress *(Air Force)*
	493209		G-DDMV	North American T-6G Texan *(California Air Nat.Guard)*
	517962		G-TROY	North American T-28B Trojan
	607327	09-L	G-ARAO	Piper L-21B Super Cub *(Army Air Corps)*
	00195700		G-OIDW	Cessna F150G *(Pseudo Air Force)*
	3-1923		G-BRHP	Aeronca O-58B Grasshopper *(Army)*
	18-2001		G-BIZV	Piper L-18C Super Cub *(Army)*
	18-5395	CDG	G-CUBJ	Piper L-18C Super Cub (ALAT)
	41-33275	CE	G-BICE	North American AT-6C Texan *(Army Air Corps)*
	42-35870	129	G-BWLJ	Taylorcraft DCO-65 O *(Army Air Corps)*
	42-42914		N31356	Douglas DC-4-1009
	42-58678	IY	G-BRIY	Taylorcraft L-2A *(Army Air Corps)*
	42-78044		G-BRXL	Aeronca L-3F *(Army)*
	42-84555	EP-H	G-ELMH	North American AT-6D Harvard *(Army Air Corps)*
	44-30861		N9089Z	North American B-25J Mitchell *(Army Air Corps)*
	44-63507		NL51EA	North American P-51D Mustang
	44-80594		G-BEDJ	Piper L-4J *(Army Air Corps)*
	44-80723	E5-J	G-BFZB	Piper L-4J *(Army Air Corps)*
	44-83184	7	G-RGUS	Fairchild UC-61K Forwarder *(Army Air Corps)*
	51-7545		N14113	North American T-28B Trojan
	51-11701	AF258/AF313	G-BSZC	Beech C-45H *(Air Force)*
	51-15227	10	G-BKRA	North American T-6G Texan *(Navy)*
	51-15319	319-A	G-FUZZ	Piper L-18C Super Cub *(Army)*
	54-2447		G-SCUB	Piper L-21B Super Cub *(Army)*
	146-11042	7	G-BMZX	SPAD rep *(Wolf W.II)* *(Army/Allied Expeditionary.Force)*
	146-11083	5	G-BNAI	SPAD rep *(Wolf W.II)* *(Army/Allied Expeditionary.Force)*
YUGOSLAVIA				
	30140		G-RADA	Soko Kraguj
	30146		G-BSXD	Soko Kraguj
	30149		G-SOKO	Soko Kraguj
UNATTRIBUTED				
	001		G-BYPY	Ryan ST3-KR
	52024	LK-A	G-HURR	Hawker Hurricane XII (IIB)
	442795		BAPC.199	Fieseler Fi 103 (V-1)
CIVIL GLIDERS				
	G-ALRD		BGA.416	Slingsby T.6 Kite 1
	G-ALRH		BGA.629	EoN AP.8 Baby
	G-ALRK		BGA.490	Hutter H-17A
	G-ATRA		BGA.1325	LET L-13 Blanik
	D-1221		BGA.2757	Schleicher Ka6E
	D-1265		BGA.2276	Scheibe L-Spatz
	D-4667		BGA.3142	Schleicher Ka6CR
	D-5084		BGA.2688	Schleicher K8B
	D-6173		BGA.4336	Schleicher Ka2B Rhönschwalbe
	HB-474		BGA.2474	Muller Moswey III
	OK-0833		BGA.4884	Letov LF107 Lunak
	OK-0927		BGA.4286	Letov LF107 Lunak
	OK-8592		BGA.655	Zlin 24 Krajanek
	ZS-GFZ		BGA.3210	Glasflugel H.301B Libelle

PART 3 – "B CONDITIONS" MARKINGS

Air Navigation Order (ANO2000) promulgateS the specific circumstances under which aerospace maufacturers can pursue the conduct of aircraft trials without the need for valid Certificates of Airworthiness. ANO2000 establishes both "A" and "B" conditions but we are only concerned here with the latter requirements which stipulate the use of identity marks as approved by the CAA for the purposes of "B Conditions" flight.

In brief, under "B Conditions" an aircraft must fly only for the purpose of:

(a) experimenting with or testing the aircraft (including any engines installed thereon) or any equipment installed or carried in the aircraft;
(b) enabling it to qualify for the issue of a certificate of airworthiness or the validation thereof or the approval of a modification of the aircraft or the issue of a permit to fly;
(c) demonstrating and displaying the aircraft, any engines installed thereon or any equipment installed or carried in the aircraft with a view to the sale thereof or of other similar aircraft, engines or equipment;
(d) demonstrating and displaying the aircraft to employees of the operator;
(e) the giving of flying training to or the testing of flight crew employed by the operator or the training or testing of other persons employed by the operator; or
(f) proceeding to or from a place at which any experiment, inspection, repair, modification, maintenance, approval, test or weighing of the aircraft, the installation of equipment in the aircraft, demonstration, display or training is to take place or at which installation of furnishings in, or the painting of, the aircraft is to be undertaken.

The flight must be operated by a person approved by the CAA for the purposes of these Conditions and subject to any additional conditions which may be specified in such an approval. If not registered in the United Kingdom the aircraft must be marked in a manner approved by the CAA for the purposes of these Conditions. The aircraft must carry such flight crew as may be necessary to ensure the safety of the aircraft. No person can act as pilot in command of the aircraft except a person approved for the purpose by the CAA.

Prompted by the SBAC a radically new system was introduced in 1948. In essence, this remains in existence today. Whilst deemed "current" many Companies included have long sinced merged or ceased to trade and so, by coincidence, the table below enscapsulates in miniature the absorbing changes within the UK aircraft industry which have occured during the period 1948 to date. The same can also be said for the original series covering the period from 1929 to 1948.

ORIGINAL SERIES

Prefix	Company	Period	Issued	Remarks
A	The Sir W G Armstrong Whitworth Aircraft Ltd	1929-1948	23.12.29	
B	Blackburn Aeroplane and Motor Co.Ltd	1929-1948	23.12.29	
C	Boulton and Paul Ltd	1929-1948	23.12.29	
D	Bristol Aeroplane Co.		Not taken up	
D	Cunliffe Owen Aircraft		Not taken up	
D	Portmouth Aviation Ltd	1947-1948	1947	
E	de Havilland Aircraft Co Ltd	1929-1948	23.12.29	
F	The Fairey Aviation Co Ltd	1929-1948	23.12.29	
G	Gloster Aircraft Ltd	1929-1948	23.12.29	
H	Handley Page Ltd	1929-1948	23.12.29	
I	H G Hawker Engineering Co Ltd	1929-1948	23.12.29	
J	George Parnall and Co *(became Parnall Aircraft Ltd)*	1929-1946	23.12.29	Cancelled 1946
J	Reid and Sigrist Ltd	1947-1948	1947	
K	A V Roe and Co Ltd	1929-1948	23.12.29	
L	Saunders-Roe Ltd	1929-1948	23.12.29	
M	Short Bros (Rochester & Bedford) Ltd	1929-1948	23.12.29	
N	Supermarine Aviation Works (Vickers) Ltd	1929-1948	23.12.29	
O	Vickers (Aviation) Ltd	1929-1948	23.12.29	
P	Westland Aircraft Works	1929-1948	23.12.29	
R	The Bristol Aeroplane Co.Ltd	1929-1948	23.12.29	
S	Spartan Aircraft Ltd	1930-1936	30. 8.30	Cancelled 29. 2.36
S	Heston Aircraft Ltd	1936-1948	15. 5.36	
S	Comper Aircraft		Not taken up	
T	General Aircraft Ltd	1933-1948	8. 5.33	
U	Phillips & Powis Aircraft Ltd	1934-1948	5. 2.34	
V	Airspeed (1934) Ltd	1934-1948	27. 6.34	
W	G & J Weir Ltd	1933-1948	8. 5.33	Cancelled 1946
X	Percival Aircraft Ltd	1936-1948	21.1.36	
Y	The British Aircraft Manufacturing Co.Ltd	1936-1938	1936	Cancelled 1938
Y	Cunliffe-Owen Aircraft Ltd	1940-1948	28.10.40	
Z	Taylorcraft Aeroplanes (England) Ltd	1946-1948	12.1.46	
AA	Believed not allocated			
AB	Slingsby Sailplanes Ltd	1947-1948	1947	

CURRENT SERIES

G-1-	Sir W G Armstrong-Whitworth Aircraft Ltd	1948-1967	1.1.48	Cancelled
G-1-	Rolls-Royce Ltd (Bristol Engines Division)	1949-19xx	9.4.69	Cancelled
G-2-	Blackburn Aircraft Ltd	1949-1967	1.1.48	Cancelled
G-3-	Boulton Paul Aircraft Ltd	1948-1973	1.1.48	Cancelled 31.12.73
G-03	BAE Systems (Operating) Ltd	19xx- Current		
G-4-	Portsmouth Aviation Ltd	1948-1949	1.1.48	Cancelled 23.5.49
G-4-	Miles Aviation and Transport (R & D) Ltd	1969-19xx	1.5.69	Cancelled
G-04	BAe Systems (Operating) Ltd	19xx- Current		
G-5-	The de Havilland Aircraft Co Ltd *(became Raytheon Services Ltd)*	1948- Current	1.1.48	
G-6-	Fairey Aviation Ltd	1948-1969	1.1.48	Cancelled 17.1.69
G-7-	Gloster Aircraft Ltd	1948-1961	1.1.48	Cancelled
G-7-	Slingsby Sailplanes *(became Slingsby Aviation Ltd)*	1971- Current	21.10.71	
G-8-	Handley Page Ltd	1948-1970	1.1.48	Cancelled 28.2.70
G-08	BAE Systems (Operating) Ltd	19xx- Current		
G-9-	Hawker Aircraft Ltd *(became British Aerospace Defence Ltd)*	1948-1996	1.1.48	Cancelled
G-10-	Reid and Sigrist Ltd	1948-1953	1.1.48	Cancelled 1.4.53
G-11-	A.V.Roe and Co. Ltd *(became BAe Systems (Operations) Ltd)*	1948- Current	1.1.48	
G-12-	Saunders-Roe Ltd	1948-1967	1.1.48	Cancelled 9.6.67
G-13-	Not allocated			
G-14-	Short Brothers and Harland Ltd *(became Short Brothers plc)*	1948- Current	1.1.48	
G-15-	Vickers Armstrong Ltd, Supermarine Division	1948-1968	1.1.48	Cancelled 17.10.68
G-16-	Vickers Armstrong Ltd, Weybridge Division *(became British Aerospace Airbus Ltd)*	1948-1999	1.1.48	Cancelled
G-17-	Westland Aircraft Ltd *(became GKN Westland Helicopters Ltd)*	1948- Current	1.1.48	
G-18-	The Bristol Aeroplane Co.Ltd	1948-1975	1.1.48	Cancelled 31.7.75
G-19-	Heston Aircraft Ltd	1948-1960	1.1.48	Cancelled 4.2.60
G-20-	General Aircraft Ltd	1948-1949	1.1.48	Cancelled 23.5.49
G-21-	Miles Aircraft Ltd (H P Reading Ltd)	1948-1963	1.1.48	Cancelled 11.2.63
G-22-	de Havilland Aircraft Co Ltd, Airspeed Division	1948-1952	1.1.48	Cancelled 23.5.52
G-23-	Percival Aircraft Ltd	1948-1966	1.1.48	Cancelled 31.5.66
G-24-	Cunliffe-Owen Aircraft Ltd	1948-1949	1.1.48	Cancelled 23.5.49
G-25-	Auster Aircraft Ltd	1948-1962	1.1.48	Cancelled
G-26-	Slingsby Sailplanes Ltd	1948-1949	1.1.48	Cancelled 19.12.49
G-27-	English Electric Co Ltd Aircraft Division *(became British Aerospace Military Aircraft Division Ltd)*	1948-1991	1.1.48	Cancelled
G-28-	British European Airways Corporation (Helicopters) *(became Brintel Helicopters Ltd)*	1948- Current	1.1.48	
G-29-	D Napier and Son Ltd	1948-1962	1.1.48	Cancelled 9.11.62
G-30-	Pest Control Ltd	1952-1957	1.1.48	Cancelled 4.3.57
G-31-	Scottish Aviation Ltd *(became BAe Systems (Operations) Ltd)*	1948- Current	1.1.48	
G-32-	Cierva Autogiro Co. Ltd	1948-1951	1.1.48	Cancelled 9.3.51
G-33-	Flight Refuelling Ltd	Not known		
G-34-	Chrislea Aircraft Ltd	1948-1952	1.1.48	Cancelled
G-35-	F.G.Miles Ltd *(became Beagle Aircraft Ltd)*	1951-1970	9.8.54	Cancelled 29.6.70
G-36-	College of Aeronautics *(became Cranfield Aerospace Ltd)*	1954- Current	9.8.54	
G-37-	Rolls-Royce Ltd	1954-1971	9.8.54	Cancelled 17.9.71
G-38-	de Havilland Propellers Ltd *(became Hawker Siddely Dynamics)*	1954-1975	9.8.54	Cancelled 22.10.75
G-39-	Folland Aircraft Ltd	1954-1965	9.8.54	Cancelled 2.4.65
G-40-	Wiltshire School of Flying Ltd	Not known		Not taken up
G-41-	Aviation Traders (Engineering) Ltd	1956-1976	3.10.56	Cancelled 27.2.76
G-42-	Armstrong Siddeley Motors Ltd	1956-1959	13.11.56	Cancelled 28.8.59
G-43-	Edgar Percival Aircraft Ltd	1956-1959	21.11.56	Cancelled 26.6.59
G-44-	Agricultural Aviation Ltd	1959-1959	17.4.59	Cancelled 14.12.59
G-45-	Bristol Siddeley Engines Ltd	1959-1969	27.5.59	Cancelled 8.4.69
G-46-	Saunders-Roe Ltd, Helicopter Division	1959-1962	15.6.59	Cancelled 3.5.62
G-47-	Lancashire Aircraft Co. Ltd	1960-19xx	8.2.60	Cancelled
G-48-	Westland Aircraft Ltd, Bristol Division	1960-1969	7.7.60	Cancelled 28.2.69
G-49-	F.G.Miles Engineering Ltd	1965-1969	23.7.65	Cancelled c.1969
G-50-	Alvis Ltd	1967-1975	16.2.67	Cancelled
G-51-	Britten-Norman Ltd *(became BN Group Ltd)*	1967- Current	20.9.67	
G-52-	Marshalls of Cambridge (Engineering) Ltd *(became Marshalls of Cambridge Aerospace Ltd)*	1968- Current	18.1.68	
G-53-	Norman Aeroplane Co. Ltd	1977-19xx	23.5.77	Cancelled
G-54-	Cameron Balloons Ltd	197x-	Not known	
G-55-	W.Vinten Ltd	19xx-19xx	Not known	Cancelled
G-56-	Edgley Aircraft Ltd	19xx-19xx	Not known	Cancelled
G-57-	Airship Industries (UK) Ltd	19xx-19xx	Not known	Cancelled
G-58-	ARV Aviation *(became Island Aircraft)*	19xx-19xx	Not known	Cancelled
G-59-	Mainair Sports	19xx-19xx	Not known	
G-60-	FR Aviation Ltd	19xx-	Not known	*(Current together with G-71)*
G-61-	Aviation Enterprises Ltd	19xx-	Not known	
G-62-	Curtiss and Green Ltd	19xx-19xx	Not known	Cancelled
G-63-	Thunder and Colt Balloons	1994-19xx	c.1994	Cancelled
G-64-	Brooklands Aerospace Group plc	19xx-19xx	Not known	Cancelled
G-65-	Solar Wings Aviation Ltd	19xx-1995	Not known	Cancelled

G-66-	Aerial Arts Ltd	19xx-19xx	Not known	Cancelled
G-67-	Atlantic Aerengineering Ltd	19xx	Not known	
G-68-	Medway Microlights	19xx	Not known	
G-69-	Cyclone Airsports Ltd	19xx	Not known	
G-70-	FLS Aerospace (Lovaux) Ltd	19xx	1997	Cancelled
G-71-	FR Aviation Ltd	19xx	Not known	
G-72-	Lindstrand Balloons Ltd	19xx	Not known	
G-73-	Aviation (Scotland) Ltd	19xx	1995	Cancelled
G-74-	Fleaplanes UK Ltd	19xx	2001	Cancelled
G-75-	Chichester Miles Consultants	19xx	Not known	
G-76-	Police Aviation Services Ltd	19xx	Not known	
G-77-	Thruster Air Services Ltd	19xx	Not known	
G-78-	Bristow Helicopters	19xx	Not known	
G-79-	McAlpine Helicopters Ltd	19xx	Not known	
G-80-	British Microlight Aircraft Association	19xx	Not known	
G-81-	Cooper Aerial Services	19xx	Not known	Cancelled
G-82-	European Helicopters Ltd	19xx	1999	Cancelled
G-83-	Mann Aviation Group (Engineering) Ltd	19xx	Not known	
G-84-	Intora-Firebird plc	19xx	Not known	Cancelled
G-85-	CFM Aircraft Ltd	19xx	2004	
G-86-	Advanced Technologies Group Ltd	19xx	Not known	
G-87-	CHC Scotia Ltd	19xx	Not known	
G-88-	Air Hanson Engineering	19xx	Not known	
G-89	Not known			
G-90	Not known			
G-91	Bella Aviation	2004		

PART 4 – FICTITIOUS MARKINGS

Registration	Type	Comments
"K.158"	Austin Whippet rep	See BAPC.207 in SECTION 3 Part 2
"EI-ABH"	HM.14 Pou-du-Ciel rep (1)	Under construction @ Meath Aero Museum 2001
"F-OCMF"		See F-BTRP in SECTION 4 Part 2
"G-EASQ"	Bristol 30/46 Babe III rep	See BAPC.87 in SECTION 3 Part 2
"G-EBED"	Vickers 60 Viking IV rep	See BAPC.114 in SECTION 3 Part 2
"G-EBVO"	Blackburn Lincock fsm	See BAPC.287 in SECTION 3 Part 2
"G-AAAH"	de Havilland DH.60 Moth rep	Located at Yorkshire Aircraft Museum, Elvington
"G-AAAH"	de Havilland DH.60G Moth rep	See BAPC.168 in SECTION 3 Part 2
"G-AACA"	Avro 504K rep	See BAPC.177 in SECTION 3 Part 2
"G-ABUL"	de Havilland DH.82A Tiger Moth	See G-AOXG in SECTION 1 Part 2
"G-ACDR"	de Havilland DH.82A Tiger Moth	US registered as N9295 [c/n 86536]
"G-ACSS"	de Havilland DH.88 Comet model	See BAPC.216 in SECTION 3 Part 2
"G-ACSS"	de Havilland DH.88 model	See BAPC.257 in SECTION 3 Part 2
"G-ADRG"	Mignet HM.14 Pou-Du-Ciel	See BAPC.77 in SECTION 3 Parts 1 and 2
"G-ADRX"	Mignet HM.14 Pou-Du-Ciel	See BAPC.231 in SECTION 3 Part 2
"G-ADRY"	Mignet HM.14 Pou-Du-Ciel	See BAPC.29 in SECTION 3 Part 2
"G-ADVU"	Mignet HM.14 Pou-Du-Ciel	See BAPC.211 in SECTION 3 Part 2
"G-ADYV"	Mignet HM.14 Pou-Du-Ciel	See BAPC.243 in SECTION 3 Part 2
"G-ADZW"	Mignet HM.14 Pou-Du-Ciel	See BAPC.253 in SECTION 3 Part 2
"G-AEAJ"	de Havilland DH.89 Dragon Rapide rep	See BAPC.280 in SECTION 3 Part 2
"G-AEOF"	Mignet HM.14 Pou-Du-Ciel	See BAPC.22 in SECTION 3 Part 2
"G-AFAP"	CASA 352L *(Junkers Ju 52/3m)*	Ex Spanish AF T2B-272 [c/n 163] @ RAF Museum, Cosford *(Original British Airways titles)*
"G-AFFI"	Mignet HM.14 Pou-Du-Ciel	See BAPC.76 in SECTION 3 Part 2
"G-AFUG"	Luton LA.4 Minor	See BAPC 97 in SECTION 3 Part 2
"G-AJOV"	Westland WS-51 Dragonfly HR.3	Ex WP495 [c/n WA/H/80] @ RAF Museum, Cosford *(BEA titles)*
"G-AJOZ"	Fairchild F.24W-41A Argus 1	The Thorpe Camp Preservation Group Woodhall Spa *(Full-scale replica)*
"G-ALVD"	de Havilland DH.104 Dove 2B	See G-ALCU in SECTION 3 Part 1
"G-AMAF"	Cessna 150J	See G-BOWC in SECTION 1 Part 2
"G-AMSU"	Douglas C-47A Dakota 3	See G-AMPP in SECTION 1 Part 2
"G-AMZZ" (2)	Douglas C-47A-DK Dakota [12254,] (ex 42-92452 - to RCAF FZ669, CAF12943, C-GCXE, HI-502, N688EA)	Al Mahata Museum/The Sharjah Avin Museum, Sharjah,UAE
"G-AOXL"	de Havilland DH.114 Heron 2	See G-ANUO in SECTION 1 Part 2
"G-ASOK"	Cessna F172E Rocket	See D-ECDU in SECTION 4 Part 1
"G-CARS"	Pitts S-2A Special	See BAPC.134 in SECTION 3 Part 2
"G-CDBS"	MBB Bo.105D	See G-BCXO in SECTION 3 Part 1
"G-DRNT"	Sikorsky S-76A	Petak Offshore Industry Training Centre, Norwich
"G-ESKY"	Piper PA-23-250 Aztec	Is G-BADI (qv)- used 1999 for TV work as "G-BADF"
"G-FIRE"	British Aircraft Corporation One-Eleven 401AK	See 5N-HHH in SECTION 4 Part 1 *(Note allocated G-FIRE in SECTION 3 Part 1)*
"G-MAZY"	de Havilland DH.82A Tiger Moth *(On loan from Cotswold Aircraft Restoration Group) (Rebuilt for static display and loaned Newark Air Museum)* *Composite ex Newark components and G-AMBB/T6801; also reported as ex DE561 [lost at sea 1942]*	H.Hodgson, Winthorpe *"Maisie"*
"G-RAFM"	Robinson R22 Beta	See G-OTHL in SECTION 3 Part 1
"G-SHOG"	Colomban MC-15 Cri-Cri	V.S.E.Norman, Rendcomb *(Static model 1999)*

PART 5 – NO EXTERNAL MARKINGS

These are listed by Type to ease identification! Amendments and alterations are always welcome.

Registration	Type	Comments
G-AANI	Blackburn Monoplane	See SECTION 1, Part 2
G-AANG	Bleriot XI	See SECTION 1, Part 2
G-BWJM	Bristol M.1C rep	See SECTION 1, Part 2
G-MYBL	CFM Shadow CD	See SECTION 1, Part 2
G-AANH	Deperdussin Monoplane	See SECTION 1, Part 2
G-EBNV	English Electric Wren	See SECTION 1, Part 2
G-BAAF	Manning-Flanders MF.1	See SECTION 1, Part 2
G-ARSG	Roe Triplane IV rep	See SECTION 1, Part 2
G-BFIP	Wallbro Monoplane	See SECTION 1, Part 2

PART 6 – GLIDER TYPE INDEX (UK AND IRELAND)

Below is a summary of BGA Number/Tri-graph tie-ups. Th letters "I" and "O" are not used except in the case of JMO. Many of these have now been replaced by other codes. Tri-graphs not used are shown also.

BGA Number.	Tri-graph	BGA No.	Tri-graph
101 - 230	*None*	3535 - 3545	FUL - FUW *(FUX not used)*
231 - 246	AAA - AAR *(AAS not used)*	3546 - 3735	FUY - GCV *(GCW not used)*
247 - 605	AAT - ARR *(ARS not used)*	3736 - 3770	GCX - GEH *(GEJ not used)*
606 - 628	ART - ASR *(ASS not used)*	3771 - 3776	GEK - GEQ
629 - 806	AST - BAC *(BAD not used)*	3777 - 3817	HAA - HBS *(3818 not used)*
807 - 1245	BAE - BUL *(BUM not used)*	3819 - 3827	HBT - HCB *(3828 not used)*
1246 - 1380	BUN - CAC *(CAD/CAE/CAF not used)*	3829 - 3990	HCC - HJV *(HJW not used)*
1381 - 1681	CAG - CNS *(CNT not used)*	3991 - 4043	HJX - HMB *(HMC to HMF not used)*
1682 - 1819	CNU - CUM *(CUN not used)*	4044 - 4045	HMG - HMH *(HMJ not used)*
1820 - 1821	CUP - CUQ *(CUR not used)*	4046 - 4122	HMK - HQP *(HQQ not used)*
1822 - 2048	CUS - DEC *(DED not used)*	4123 - 4138	HQR - HRG *(HRH not used)*
2049 - 2391	DEE - DUL *(DUM not used)*	4139 - 4276	HRJ - HXB *(4277 not used)*
2392 - 2478	DUN - DYC *(DYD not used)*	4278 - 4878	HXC - JYB *(4879 not used)*
2479 - 2485	DYE - DYL *(DYM not used)*	4880 - 5111	JYC - KHT *(KHU not used)*
2486 - 2489	DYN - DYR *(DYS not used)*	5112 - 5130	KHV - KJP *(KJQ not used)*
2490 - 2494	DYT - DYX *(DYY not used)*	5131 - 5133	KJR - KJT
2495 - 3101	DYZ - FAF *(FAG not used)*	5134 - 5143	*Not used*
3102 - 3528	FAH - FUB *(FUC not used)*	5144 - 5145	KJU - KJV *(KJW not used)*
3529 - 3534	FUD - FUJ *(FUK not used)*	5146 -	KJX onwards

ABBOTT-BAYNES SAILPLANES LTD
SCUD I
HFZ
SCUD II
AAA
SCUD III
ACF AVA

AEROMERE see CARMAM

ALLGAIER
GEIER
EBP

ASC
FALCON
HPZ
SPIRIT
HPY

AVIA
40P
AUW

AVIASTROITEL see FEDEROV
AC-4
JUX JVH
AC-5
KDD

AVIALSA see SCHEIBE

AVIONAUTICA RIO see CARMAM

BAC
VII rep
EQY

BIRMINGHAM GUILD LTD see SWALES and YORKSHIRE SAILPLANES
BG.135
CRF CUF CYV CYW CXN CXP DCY DLZ

BÖLKOW
PHOEBUS C
CGX CHJ CJB CJJ CKC GDD JHX JVH JWP EI-158

BREGUET
905 FAUVETTE
CVJ DGV EGR ELJ EPN ESM

SOCIETE CARMAM see AEROMERE/AVIONAUTICA RIO
M.100S MESANGE
CBR CLU DFP DTS DUC EQM FCH HQF
M.200 FOEHN
EQX EVC FXN HCN
JP-15/34 KIT-CLUB & JP-15/36A AIGLON
DYL EDB FDC HKZ HMU

CAUDRON
C.801
EHF

SA **CENTRAIR** see SCHLEICHER
101 PÉGASE
EPK EQK ERX ESH ESW ETJ ETM EVE EVM EVQ FAN FCB FCD FEH FFC FFS FGW FHJ FJK FJT FMK FNM FRD FRR FRV FRX FVM FVN FVP FVV FWG FWX FWY FXD FXT GBU GCY HDD HDW HES HNY HNZ HRK HZF JQP KAE KBA KFX KGM KJX EI-162
201 MARIANNE
HTA JXB JXN
ALLIANCE SNC-34
JHR

CHARD see KING-ELLIOTT-STREET

COLDITZ - *two unrelated designs*
COCK rep
JRZ JTA

DFS see GRUNAU, EoN, NORD, FOCKE-WULF, WEIHE and SCHLEICHER production
KRANICH
BGT BQJ
OLYMPIA-MEIS
AKD
108-68 WEIHE
AKC BKC BNC BTV BWR EDL

DG FLUGZEUGBAU - see GLASER-DIRKS

DITTMAR
CONDOR IV
JZD

DWLKK
PW-5 SMYK
HYM HZB JCG JDW JKB JKE KCQ

EDGLEY
EA.9
HPJ JZQ JZR

EICHELSDORFER
SB.5
EHC EJH

EIRI
PIK-20
DFE DFK DFZ DHH DHN DHV DJN DJZ DKT DLJ DLY DMU DMV DPL DQU DRT DVJ DVN DWS DZT EAR EAT JBH JGK JVE JXG KFE KGS

EoN see DFS/NORD
AP.5 OLYMPIA
AMK AMM AMP AMR AMT AMU AMV AMW ANW ANZ APC APV APW AUU AWU AZR AZT BEL BFN BGR BHC BJC BKL BKS BKU BKX BLN BLP BLQ BLS BNG BPL BQQ BRH BWX CAF CAK CHK CQG CTA CTS DBA DPU EI-115
AP.6 OLYMPIA 401/403/419
BLJ BLK BVW CDW DAL DTR
AP.7 PRIMARY/ETON TX.1
AQQ AQY AQZ CLJ FEZ
AP.8 BABY
ASS AST EI-118
AP.10 460/463/465
BQM BQT BRK BRQ BSQ BTG BTN BTQ BUV BVN BWB BWE BWG BWU BXB BXY BYE BZB BZR BZV CAN CAT ELS

FEDEROV see AVIA STROITEL
Me-7 MECHTA
HMZ HPT HUQ JFY JFZ JGA

FFA FLUGZEUGWERKE AG
DIAMANT
CDG CDW CGM CGS HGT

FOCKE-WULF see DFS WEIHE
KRANICH III
ENG

GINN-LESNIAK
KESTREL
CJC

GLASER-DIRKS INCLUDING DG FLUGZEUGBAU
DG-100/DG-101
DFN DHJ DHK DHL DKQ DMD DRB DUY EDN EDP EKP ENU EPU FBH FBW FFU FYU HMS HWP JEZ JNZ JPF JRL JSM JTC JXP JXV JYQ KDY KHV
DG-200/DG-202
DTA DTM DUQ DWJ DXN EBR EDM EKA EME EMU EQP FQC HAT HBD HDH JAE JAJ JAW JDD JDP JHW JKF JKM JPW JRN JVP JXG JYB KAC KCW KDU KDY EI-145 EI-147
DG-300/DG-303 ELAN
ESQ FAJ FBF FCM FDW FGT FJR FJS FJX FLC FLX FNS FPK FSR FSX FTS FUJ FUT FUU FWM FZW GAJ GBS HBE HBW HCU HCY HDR HMB HSB HVM JAB JDV JNQ JRC JTK JTN JVL KAD KAU KBF KJH
DG-500/DG-505 ELAN
GBZ HBP HEF HGV HHJ HNA HRC HYE JDN JQF JSX JZB JZK KAW KAX KJJ
DG-600
FKB FNT FPW FQQ FVG KAZ KBS
DG-800
JBL JJH
DG-1000
KFN

GLASFLUGEL
H.201 STANDARD LIBELLE
CFS CFW CFX CFY CKF CKY CLM CLN CLP CLR CLV CLW CMH CMQ CMR CMS CMV CMW CMX CNE CNF CNG CNH CNJ CNP CNY CPA CPF CPM CRB CRQ CRS CRV CRW CSJ CSR CTU CUJ CUK CVL CVQ CWE CWG CWN CWT CWX CWY CWZ CXK CYG CZL DCC DMS DNL ECY FLT FLU GAU GDM GEE HAA HAD HAV HCQ HHY HJR HWC HWG JAS JBF JEU JGZ JHJ JNG JVW KBR KCM
H.205 CLUB LIBELLE
DBP DEQ DVM FAR FYG GCS HAE KAM KDG
H.206 HORNET
DKD DKM
H.301 LIBELLE
FEV GAN HLK
H.303 MOSQUITO
DMN DPK DRN DTK DTV DTX DTY DUB DVZ DWB DWL DWO DWR DXA DXW EAK ECH ECS EDH EDJ FBN FWR HMT JEH JNR JTQ JTW
H.304
EHU ENT HMM
H.604
ECT

GROB (BURKHART GROB LUFT und RAUMFAHRT GmbH) see SCHEMPP-HIRTH
G.102 ASTIR & G.104 SPEED ASTIR
DFR DJD DJQ DJX DKR DKS DKU DKW DKX DLH DLM DMH DMP DMR DNE DNK DPJ DPQ DPY DQB DQE DQG DQR DRK DRU DRW DSH DSN DSU DUL DUX DWQ DWU DXJ DYF DZJ DZU EAC EAF EAW EBB EBM ECQ EEQ EKF ELN EQD EVK EVL FBR FCJ FDF FEB FEF FEX FFB FGK FHT FHW FJH FRL FSA FSH FSZ FTK FTR FXA GAT GBJ GBK GCL GDQ GEB HAU HBL HBM HBT HGB HJV HKB HKM HPM HQT HRA HSE HTD HTE HTG HTR HUN HVK HWK HXB HXM HXY HYQ HZC JAZ JBZ JCF JCR JCW JFD JHE JHG JHN JKW JLR JML JNA JPJ JPM JQN JQT JRD JRM JRP JSD JSH JSK JTT JUK JUS JWR JYC JYP JZW JZY KAY KBD KBL KCP KEE KEU KGH EI-124
G.103 TWIN ASTIR/ACRO
DRQ DSJ DSL EGN EQT EWG EWP EWR EXA EYS EZE EZF FEA FQK FWC HBH HBK HCA HCJ HQS HWW JKV JLZ JQG JRW KCS KFG KFZ EI-160

GRUNAU including DFS,FOKKER and HAWKRIDGE production
BABY
ABZ AFY AQN CBK CDQ CYJ DNA DNB DUD DWF HJB KFW

HALFORD
JSH SCORPION
DJF

HAWKRIDGE
DAGLING
ALX ALZ

HIRTH
Go.III MINIMOA
CLY
GOEVIER
DBU

HOLS-DER-TEUFEL
REPLICA
FHQ

HUTTER
H.17
ALW EPR HEY
H.28
HJM

ICA (INTREPRINDERA DE CONSTRUCTII AERONAUTICE OF CIAR 1968)
IS-28B2
DEG DLT DLU DZR EHS EHW EJA HMG
IS-29D
DAM DBG DEN DEW FDG FFN
IS-30
FDB FDP
IS-32A
FAV

ISF
MISTRAL C
JBE JRS

SOCIÉTÉ ISSOIRE AVIATION
D77 IRIS
EET
E78 SILENE
EBE HPA JUG

JASTREB see SCHEMPP-HIRTH

KING-ELLIOTT-STREET
OSPREY
DCZ

LAK
LAK-12 LIETUVA
FXR GAB GCB GCJ HEG HGQ HGR HGX HHM HHW HQG HRB HSR HTF JBS JBY JFW JQH JYL

<u>*LAK-17*</u>
JQU JVJ KAQ KBE KCR

LANAVERRE see SCHEMPP-HIRTH

LET
<u>*L-13 BLANIK*</u>
BVY BXV BXW CVA CVB DAF DCL DEX DEY DGP DKH DPD DVD EVR GAK HTY JDU JGQ JNX EI-120 EI-154
<u>*L-23 SUPER BLANIK*</u>
FXP FYP FYR
<u>*L-33 SOLO*</u>
FZZ

LETOV (VOJENSKÁ továrna na letadla LETOV)
<u>*LF-107 LUNAK*</u>
HXL JYF KBW KDT

MANUEL
<u>*CONDOR*</u>
DJW
<u>*HAWK*</u>
CSU
<u>*WILLOW WREN*</u>
BGA.162

MARCO
<u>*J-5*</u>
HKW

MARGANSKI
<u>*SWIFT S1*</u>
JZP

MAUPIN
<u>*WOODSTOCK*</u>
HCG HPG

MDM
<u>*MDM-1 FOX*</u>
JKC

MOLINO see EIRI

MONNETT
<u>*MONERAI*</u>
EVN

MÜLLER
<u>*MOSWEY III*</u>
DXY

MÜNCHEN
<u>*MÜ-13D*</u>
CZM DPG

NEUKOM (includes Pfenninger-Markwalder)
<u>*STANDARD ELFE S-2*</u>
FMR JGF JXY
<u>*ELFE PM3*</u>
JRQ

NORD see EoN
<u>*2000*</u>
EPJ

OBERLERCHNER see SCHEMPP-HIRTH
<u>*Mg19a STEINADLER*</u>
ERZ

PENROSE
<u>*PEGASUS*</u>
HKJ

PIK see EIRI
<u>*PIK-16C VASAMAss*</u>
CGV

PILATUS FLUGZEUGWERKE
<u>*B4 PC-11*</u>
CSN CSP CSW CUB CUC CUQ CVG CVK CVM CVV CYA CYC CYK CZD DBC DLA DND DQM DRP DSV EQU HDA HDE HLC HSY HVH JSY JXX EI-121

POTTIER see CARMAM

ROLLADEN-SCHNEIDER
<u>*LS1F*</u>
FQZ JPE JSF JST KHL
<u>*LS3*</u>
DNQ ECP EEF EES EEX EEZ EFZ EGE EVD GAD GDA GDN HYH JDJ JYX KGW KHB KJU EI-137
<u>*LS4*</u>
EHK EHL EKV ELT EMB EMF EMG EMT EMY ENA ENE EQA ERV ESC ESE ESY ETG ETV ETY EUH FAQ FHL FJM FKG FLF FNU FVE FYH GBT GDJ HEL HKX HLB HMX HNV HNX HPL HVV HXF HXT HXZ HZM HZY JBX JEP JJB JKP JLH JLJ JSB JSW JWN KAV KCB KCL KDK KEC KEN KFL KFR
<u>*LS6*</u>
FBE FCP FMC FRA GAR GBG GBQ GBR GCM HAQ HBC HBJ HET HEW HEZ HFM HFQ HGP HHH HHT HHU HJC HJF HJX HMK HPD HQL HQZ HRY HSA HUM HYA JBP JBQ JBU JCB JDG JDH JGC JNP JNU JTD JVK KBC KBH KEP
<u>*LS7*</u>
FMY FPD FQG FQH FTV FTY FUV FVH FVQ FWF FWJ FWU FXE FYB FYK FYW FZV GBL GBY HAY HBA HBY HDX HEH JEJ JLK JSJ JTR JZZ KEW KHS
<u>*LS8*</u>
HSZ HTL HTM HTP HTS HUG HUV HUW HVF HVL HVU HWL HWM HWS HXC HXN HXW HYF HYZ HZG HZP JBB JCL JCP JCY JDE JDK JDT JDY JEA JEG JFB JFL JFX JGS JHU JHY JJK JJU JKD JKL JKN JLN JMB JMO JMR JMT JMU JMW JNB JNJ JNK JNM JNW JPH JPL JPR JQD JRA JRK JSQ JSU JTL JTM JTY JUC JUD JUE JXU JZC KBJ KBZ KCA KEL KET KEZ KFM KFS KFV KHP KJE

SCHEIBE-FLUGZEUBAU including AVIALSA/ROCHETAU production
<u>*BERGFALKE*</u>
CDR CKT CMT DJU EBD EPZ EVT FVD KGY
<u>*L-SPATZ*</u>
DPR DPT DUH EFN EQL FAK FPF FXX HCP HWQ JHB EI-130
<u>*ZUGVOGEL III*</u>
EBS ELV FRS FSU FVL FYE HKV HND HSH HXA EI-146
<u>*SF-26 STANDARD*</u>
DRL EUM
<u>*SF-27A ZUGVOGEL V*</u>
DZV EKS EUE FHY FLZ FQF FQM FUF FUQ FWH FZM GAV GBM GDW HGM HLU HSG HSX HUS HVJ EI-142
<u>*SFH-34 DELPHIN*</u>
EPM

SCHEMPP-HIRTH OHG including GROB, JASTREB and OBERLERCHNER production
<u>*STANDARD AUSTRIA*</u>
BPN DBT DKB
<u>*SHK/SHK-1*</u>
CAQ CAR CCB CDU CGT CGZ CJK CJL CJN CLG CVH DJS DMK DTG ECG FJZ FZC
<u>*HS.2 CIRRUS/CIRRUS VTC*</u>
CDH CEA CEC CFK CGY CJR CLQ CUS CVE CWR CWS DDM DVY FXG GCQ HEV HTU HUL HUR JQW JTS KEA KES KJG
<u>*HS.4 STANDARD CIRRUS*</u>
CKZ CLA CLH CNC CNN CPU CQN CQR CQY CRH CRN CTB CTT CYM CYQ CYT DAS DDA DDR DFC DFY DGE DGX DLG DVS DXL DZF EEN EGK FBB FCN FGU FLW FMT FMU FRJ FRZ FVS FYJ FZK GAH GCD GEP HAN HAX HFF HGG HJN HJY HKC HKD HKR HKS HKU HMY HNM HRL HSV HVZ HWD HWF HWY HXX HZJ HZL HZU HZV JBJ JCJ JCN JCU JDS JER JEV JEY JFA JGN JGY JHA JHH JJJ JJM JLX JMH JNC JQS JRG JRT JTE JUU JVU JWH KBT KCU KDZ KEB KFA KFK KGT KHJ KHT KJD KJF EI-148
<u>*HS.5 NIMBUS 2*</u>
CQL CQP CQQ DAJ DDD DGY DHW DKL DMM DNG DTU DYU DYZ EAJ EAM EEK EFB EFF EFT EGS EHP EHT EKR EKW FCS FPP FRC FVF HBF HBV HNE HNH HQN JGH JMN JMV JQE
<u>*HS.6 JANUS*</u>
DTC EQV EQW HNB HSP HTB HTH HUC HUH JAA JFE JLM JNL JVV KJK KJT
<u>*HS.7 MINI NIMBUS*</u>
DPH DSP DSW DXQ DXT EAV EBF EBK EDF EER EGW FHG HQY HRQ JGU JTJ JYS JZL KBB KFH

<u>NIMBUS 3/4</u>
ENN ENP EQN ERU EVF FAM FBM FFK FGF FRP FWK FXQ HBR HCB HFX HKQ HNU HWN HYY HZA HZW JCT JDA JDZ JEQ JFQ JHF JJG JSC JUT JVT JWG JXA KAG KGP KHZ KJC
<u>DISCUS A/B/CS</u>
FBY FDU FEJ FER FES FFT FFX FHR FKK FKM FLE FMG FMQ FNL FNQ FNR FQY FTW FUL FUP FXM FYM FYN FYX GCT GDR GDX HCL HDF HDT HDZ HEE HEM HEN HGK HGL HGS HGZ HHP HHQ HJH HJL HKA HKL HKY HLN HLQ HLS HLY HML HMP HMQ HPH HPX HQJ HQM HQR HQW HRS HRX HSD HSJ HSQ HUZ HVR HWV HXH HYB HYU HZE JAH JAN JAQ JAR JBD JBN JBR JBW JCK JCX JDL JFC JFG JGL JGM JGR JHM JHT JHV JJD JJE JJZ JKR JKX JLC JLP JLW JMD JMM JPP JPU JQA JQC JQK JRR JSE JUB JUV JVB JVF JVG JVX JVY JWK JXL JXR JYE JZG KAH KAP KBQ EI-138 EI-149
<u>DISCUS 2</u>
JNE JNF JNY JUA JUP JVR JWQ JYM JYN KFB KGZ
<u>DUO DISCUS</u>
HNF HNN HNW HQE HRW HSW HWB JAC JEM JFF JFH JGV JJP JPA JQQ JTU JUM JXW JYR KAF KAR KCF KCV KDE KEH KEM KEV KFT KGF KHK KHY EI-159
<u>VENTUS A/B/C</u>
EHH EKH EKJ ELG ELR ENJ EPX EUJ EUS FAW FBT FCK FDE FEG FHS FJJ FJQ FMN FNN FPE FPL FQN FRB FRT FUH FUR FVB FVW FWD FYC FZH GAP GAS HAJ HED HFV HFY HGN HHN HUY HVE HWH HXR HYG JBG JCE JEF JET JFP JFR JFS JJY JKH JKY JLA JLU JSL JWL JWU JXK KAS KDV KJB KJM KJX
<u>VENTUS 2</u>
HSL HUP HVE HVT HVY HWA HXS HYL HZS JAF JBC JEX JGP JLU JPD JQL JQR JTV JTZ JUF JUQ JVA JWX JYH JYJ JYT JYU JZM JZV KAJ KAT KBG KBK KCC KCD KCE KCH KCK KDA KDJ KDL KDQ KFC KFP KFQ KFU KGA KGB KGC KGD KGL KGQ KHQ KJY EI-152

SCHLEICHER including CENTRAIR production
<u>RHÖNBUSSARD</u>
AEM
<u>RHÖNSPERBER</u>
ABG
<u>Ka2B RHÖNSCHWALBE</u>
DCG DGT DJG DPP DRR ESK FPU HZN
<u>Ka3</u>
EHB
<u>Ka4 RHÖNLERCHE II</u>
CHM CTF CVY CWU CWV EAL HQH EI-104
<u>Ka6/BR/CR RHÖNSEGLER</u>
BKJ BKW BND BNH BTJ BTM BUZ BVR BVX BVZ BWC BXT BYL BYM BYU BZX CBM CBY CCJ CJY DAW DCW DDY DEP DEV DGK DHG DJE DJR DKG DKN DLP DNW DNX DQC DQF DQJ DQS DRA DRD DRE DRG DRY DSR DSY DUR DVG DYC DYJ DYN DYQ DZW EBQ ECC ECF EDG EEW ELY EPW EQQ FBZ FDR FGJ FHZ FKA FKH FKU FKX FLS FMM FNP FNW FQL FSE FTB FTF FUB FUM FWA FZR GAW GBE GCP GDE GDF GEA GEM HAB HBQ HEB HFB HPQ HRE HSN HUK HZH JCZ JEW JJV KDB KGG EI-111 EI-127 EI-128 EI-161
<u>Ka6E</u>
BYX CAC CAE CAG CAS CCA CCD CCG CCL CCR CCU CCV CDA CDB CDD CDF CDZ CED CEG CEL CEM CEQ CEW CEY CFL CGB CGD CGE CGK CGN CHB CHZ CKL CLZ CPJ DGG DHM DHT DLE DMQ DQK DSB DUS DVE DVH DWC DXH EAH EFM EHM EKC EKX ETB FCR FND FPV FRE FSS FVZ FXC FXS FXU GDV HAP HJD HRF JAK JAL JHD JHL JLT JLV JSG
<u>Ka7 RHÖNADLER</u>
BKN BQU BRM BVB CFC CLF CLK CLT CMG CMZ CPG CRA CWJ DHY DJT DKY DMF DML DQX DRM DWE DWN DXM DYB DYR EAU EDC EDK EFD ELX EMV ENC EPV ETR ETU EUQ EVB EVG FEL FGV FHU FJW FKW FLK FMD FMZ FPQ FQU FRF FTG FTU FXH GAQ HAG HCM HGC HHL HJK HNJ JME EI-105
<u>K8B</u>
CDC CDK CFF CGH CGJ CHU CJF CJM CLX CMN CQD CTZ CYZ DDL DFQ DGA DHA DJB DJP DKC DLD DLS DMB DMG DMJ DNZ DQL DQP DQY DRV DRZ DSF DTN DUF DUK DWG DXP EAZ EED EEM EFG EHA EJF EKM EPT EQZ ESJ ESX ETD FBJ FDD FHN FKT FLH FLP FLQ FNA FQD FQE FQR FTM FTN FVA FWL FXB FXW GCG GDB GDK GEG HAR HCZ HDN HDY HFW HJE HKK HLH HMH HNG HRJ HRT HWE HWT HYN HYV HYW HYX HZX JAT JFT JGB JGD JGX JHK JLS JNN JQB JQJ JQZ JSN JXS JYV JYW KDH EI-108 EI-133
<u>Ka10</u>
EVH
<u>ASK 13</u>
CAV CBW CCC CCE CCF CCM CCP CCT CCW CCX CCY CCZ CEX CFA CFG CFM CGQ CHW CJD CKR CKU CKV CMK CRL CRT CWH DDB DKE DLC DMX DNV DQA DRJ DUE DVB DVC DVX DXV EAP EBL EBZ EDU EKD ENY EPP EQE EQF ETS EUC EVJ EVP FAT FCW FEQ FFA FGR FHM FHU FMH FPX FSD FVC FVU FWB FWN FYY FZN GBA HAL HDC HMV HPE HSM HTJ HUD HUF HUU HVQ HVW HXJ HXP HXV JFM JGW JKT JLE JLF JLL JLQ JMJ JMP JMW JMX JMZ JPC JPV JPY JRX JSV JWB JWJ JXM JYD JZE KFJ KFR KJL EI-112 EI-113 EI-143
<u>ASW 15</u>
CHT CJP CKP CZN FBC FBD FCY FDA FJV FKE FME FML FMS FPB FQB FRK FTD FXY GCH GCR GCX GDS GDY GEH HEJ HGF HJZ HNT HQX HTC HZD JAM JCA JDM JDR JED JGG JJX JPX JRE JVW JWS JZJ EI-134
<u>ASW 17</u>
CPD CTE EDD EGD EI-132
<u>ASK 18</u>
DJJ DJK DLB DNJ DPA EUX HRN HSU JHQ JKG JKU JMA JMK JPQ JPZ JSZ EI-136
<u>ASW 19</u>
DPX DSX DTD DTE DVL DVP DWZ DXX DZD DZG DZY EBJ EEH EJR ELA ENZ EPE EQG ERP ERQ ERS FFP FGP FNH FPJ FWP FWZ GCA GDP HCE HCV HDV HER HGH HHK HKT HLM HLW HNC HUA HWZ HXE HXU JBK JBT JCQ JES JFU JHS JJL JKS JNT JRB JRV KER KEX KGU EI-153
<u>ASW 20</u>
DST DTP DTQ DUT DVV DXB DXK DYE DYX EAE EBN EBX ECX EEC EEE EEJ EEV EFA EFE EFJ EFK EFL EFV EGP EHD EHV EHZ EJL EJQ EKE EKU ELU ELZ ENV ENW EPF EPS EQJ ERA ETZ EUD EUK EUY FAF FBA FBQ FCV FFH FHD FJE FKL FPH FPN FPT FRH FRW FTL FTP FUN FWS FZL HBU HCH HDJ HDL HEQ HGW HHS HJT HLZ HPC HQD HQK HSK HTT HTX HUJ HUT HVP HVX HYK HZT HZZ JAG JAY JEE JEN JFJ JFK JHZ JMQ JTP JXH JZH KCY KDC KEY KJN EI-167
<u>ASK 21</u>
ECW ECZ EDW EGZ EHQ EKG ELE ENK EPD EQR ERH ERJ ESB ESU ETA FBV FWQ FYF FYV GAF GAG GAM GBB GBF GBN GBV HCX HLG HLP HPV HPW HRR HTV HVG HYJ HYS HYT HZR JAD JAV JAX JBM JFV JGE JGJ JJR JKA JKJ JKQ JKZ JMC JMS JQX JVZ JWD KCT KCZ KDF KDP KEJ KEK KFY KGJ KGK KGX KJP EI-150
<u>ASW 22</u>
FEU FGY FNF GBX HTN JNV JWZ
<u>ASK 23</u>
EVV EVW EVX EVY GCF HKP
<u>ASW 24/E</u>
FMP FMX FNG FXJ GDT GDU GDZ HBB HBG HYD JCD JEB JEL JRU JTC JXT
<u>ASH 25</u>
FKN FLG FST FSY FUG FWW FXL FYD FYZ GCE HFL HLX HXQ JCV JDF JFN KDM
<u>ASW 27</u>
HUE HXD HYR HZQ JCM JDC JJF JJT JLY JNS JPS JPT JQM JQV JRH JSS JTF JUH JUJ JUL JVM JVN JVQ JWC JWF JXZ JYD JZX KBX KCG KCN KDN KDS KDW KED KFD EI-151
<u>ASW 28</u>
JVS JWA JXJ JZN KAK KAL KBM KBU KBV KCJ KEQ KGN KGV KHX KJA KJR KJS KJV

SCHMETZ
<u>CONDOR</u>
DQH

SCOTT
<u>VIKING</u>
AHU

SHENSTONE
<u>HARBINGER</u>
BNA

SHORT
<u>NIMBUS</u>
ALA

SIEBERT
<u>SIE 3</u>
EFC FRG

SLINGSBY SAILPLANES LTD including YORKSHIRE SAILPLANES production
<u>T.1 FALCON 1</u>
ABN FCZ
<u>T.6 KITE 1</u>
AAF AAX ACH ADJ AHC
<u>T.7 CADET</u>
AWZ BQE
<u>T.8 TUTOR</u>
AJW AKW ALR AZQ BAA BBG BCB BCH CBZ CPL CRK CRZ CSL DQD IGA.6
<u>T.12 GULL I</u>
AGE BED
<u>T.13 PETREL</u>
ATR
<u>T.15 GULL III</u>
ATH HBZ

PART 7 - AIRCRAFT TYPE INDEX (OVERSEAS CIVIL) (COVERS SECTION 4, PART 1)

ADAM
RA.14 LOISIR
F-PFUG

AERO COMMANDER INC including ROCKWELL INTERNATIONAL CORPORATION production
680/685/690
N60B N67TC N71VE N425RR N808NC N840JC N840LE N980HB N91384 VP-BCT VP-BMZ

AERO VODOCHODY NÁRODNÍ PODNIK see LET
L-29 DELFIN
ES-YLK

AERONAUTICAL CORP OF AMERICA (AERONCA) see CHAMPION
15AC SEDAN
N915TC
A65TAC/65C SUPER CHIEF/O-58B(L-3) GRASSHOPPER
NC33884

AÉROSPACELINES
377SGT SUPER GUPPY
F-BTGV

AÉROSPATIALE including EUROCOPTER production and see SUD-AVIATION
AS350B ECUREUIL
N350UK
AS355 ECUREUIL 2
N620LH N766AM VP-BCE
SA365 DAUPHIN 2
F-WQAP

COSTRUZIONI AERONAUTICHE GIOVANNI AGUSTA S.p.A
A109
D-HCKV N2NR N7AG N9VL N74PM N109AB N109AN N109AR N109GR N109KH N109TF N109TK N109UK N109WF N188S N555GS N637CG N709AT N745HA N754AM N7242N SX-HCF S5-HPC VP-CWA
A119
N119BM

AIRBUS INDUSTRIE/SAS
A319
A6-ESH

AMERICAN BLIMP CORPORATION
A-1-50/A-60+ AIRSHIP
N11ZP N12ZP N156LG

ANTONOV
AN-2
HA-MKE HA-MKF LY-AFO LY-BIG LY-MHC RA-01641 YL-LEU YL-LEV YL-LEW YL-LEX YL-LEY YL-LEZ YL-LFA YL-LFB YL-LFC YL-LFD

AUSTER AIRCRAFT LTD including TAYLORCRAFT production
Model 5
F-BBSO
J/1 AUTOCRAT
F-PAGD
B.8 AGRICOLA
ZK-BMI ZK-CCU

AVIAN
MAGNUM HAB
N5023U

AVIATION TRADERS ENGINEERING LIMITED
ATL98 CARVAIR
CF-EPV

A V ROE and CO LTD
652A ANSON
VH-ASM
671 - see CIERVA
RJ SERIES - see BRITISH AEROSPACE 146

BAE SYSTEMS (OPERATIONS) LTD including BRITISH AEROSPACE plc, BRITISH AEROSPACE (REGIONAL AIRCRAFT) Ltd, HANDLEY-PAGE and SCOTTISH AVIATION production
JETSTREAM variants to Series 32
N14234
146 including Avro variants
A6-SHK N76HN

BEAGLE AIRCRAFT LTD
B.121 PUP
HB-NAV N556MA
B.206
N181WW

BEECH AIRCRAFT CORPORATION
17 TRAVELER
N18V NC18028
18/3NM, 3TM and C-45
N15750 N96240
23/24 MUSKETEER/SUNDOWNER/SIERRA
D-EDEQ N39TA N6632L
33 DEBONAIR/33 BONANZA
N42FW N232N N999F N2299L
35 BONANZA/("V" tail)
D-ENCA N35ZS N59VT N150JC N3023W N60256
36 BONANZA
D-EKDN N36NB N250TP N345SF N671B N715BC N767CM N836TP N7205T N7214Y N8258F N24136
55/56/58 BARON
D-IBPN N27MW N28TE N55AE N55BN N55EN N58GT N58YD N64VB N65MJ N79AP N80HC N142TW N418WS N503PA N554RB N773DC N951SF N1835W N2061K N2326Y N7148R N7219L N23659 N60526 5B-CJV
60 DUKE
N80RF N322RJ N1024L N3669D
65/70/80 QUEEN AIR
N70AA
76 DUCHESS
N800VM
90 KING AIR
F-GFLD N41AK N146FL N492PA N587PB N816RL N1092H VP-CLA 9L-LDA
200/300/350 SUPER KING AIR
N64GG N73MW N250TM N500CS N771SC N800MG N2341S
400 BEECHJET
N79EL N709EL

BELL
P-39 AIRCOBRA
NX793QG

BELL HELICOPTER TEXTRON INC including AGUSTA, BELL HELICOPTER CO, BELL HELICOPTER TEXTRON CANADA, IPTN (412) AND WESTLAND production
47G
N7801R
206A/B JETRANGER I/II/III
I-VFAN N64JG N206GF N206HE
206L LONG RANGER
N720B
212
5N-AJT 5N-AJU 5N-AJV 5N-AJW 5N-AQV 5N-BHN
222
N153H N222WX N510W N800HL
430
N5120 VP-BKQ
UH-1 IROQUOIS
N116HS N454CC

BELLANCA-AIRCRAFT CORPORATION
DECATHLON
N669MM
SUPER VIKING
N8862V

BOEING AIRCRAFT CO including BOEING COMPANY
B-17G FORTRESS
F-BDRS N5237V

247
N18E
707
EL-AKJ EL-AKL HZ-123 N707LG VP-BDF YN-CCN
727-100 series
N400RG OO-DHN OO-DHR VP-BAA VP-BAB VP-CKA
727-200 series
EC-CFA EC-CFI EC-DDX HZ-AB3 TC-ALM
737-200 series
C-GWJO TF-ELL XA-TLJ
737-700 series
A6-HRS VP-BBT VP-BBU VP-BBW
737-800 series
A6-MRM N737M
747-100/200 series
N218BA N852FT SX-OAD 5R-MFT
747SP
VP-BAT
757
SX-BLW
767
VP-CME

BOEING AIRPLANE CO
STEARMAN 75 KAYDET/N2S/PT-13/PT-17:
CF-EQS EC-AMD N43SV N126SE N707TJ N1325M N1363M N1731B N3922B N4596N N4712V N5057V N38940 N52485 N54922 N56608 N59269 N62842 N63590 N65200 N65565 N68427 N74189 XB-RIY

BÖLKOW APPARATEBAU GmbH including MALMO and MBB production
Bö.102 HELITRAINER
D-HMQV
Bö.105
LQ-BLT
Bö.207
D-EBLI D-EBLO D-EFQE D-EFTI D-EHKY D-EHLA D-EHUQ D-EHYX D-EJBI
Bö.208 JUNIOR
D-ECGI D-EEAH D-EFNO D-EMFU D-EMUH F-BRHN PH-END
Bö.209 MONSUN
D-EFJD D-EFJG

BOMBARDIER INC including CANADAIR production
CL600/604 CHALLENGER
HB-IVR HZ-SJP3 N328BX N598MT VP-BGO VP-BHS VP-BIE VP-CFT VP-CIC
CRJ 200 REGIONAL JET
VP-BCC VP-BCI
BD-700 GLOBAL EXPRESS
VP-CEB

BRANTLY HELICOPTER CORPORATION
B.2
N19GL N276SA

BRIGHTON
MAB-65 HAB
HB-BOU

BRISTOL AEROPLANE CO LTD
SCOUT D REP
N5419
175 BRITANNIA
EL-WXA

BRITISH AIRCRAFT CORPORATION (BAC)
ONE-ELEVEN
VR-BEB 5N-HHH

BRITTEN-NORMAN LTD
BN.2A/B/T ISLANDER/DEFENDER
A40-CT N121MT

BROCHET
MB.100
F-BBGH

CAMERON BALLOONS LTD see CAMERON-COLT and CAMERON-THUNDER
Hot Air Airship:
Zero 25
OO-JAT
Hot Air Balloon:
56 variants
OO-ARK
77 variants
C-GYZI
84 series
N413JB
90 variants
OO-BDO
140 variants
5Y-SIL

CANADA BENJAMIN CO
TWINSTARR GYROCOPTER
31-WI

CESSNA AIRCRAFT COMPANY including REIMS AVIATION SA production (F.prefix)
C.34 AIRMASTER
NC16403
120/140
N76402
150
D-EALX D-EAWD N3109X N4232Y N4337K N5927G N6819F N7263S N11824 SE-ETR
A150 AEROBAT
N7374A OO-WIO
152
N4770B N67548
170
D-EEVY N170AZ N2366D N5428C
172/SKYHAWK
D-EAGC D-ECDU D-EFZO N90SA N172AM N259SA N1554E N2437B N6182G N13253 N19753 N50029 N75822 N80533 OY-DRS OY-EGZ
172RG CUTLASS
N372SA N1937Z
R172 HAWK/REIMS FR172 ROCKET
D-EADD D-EBXR D-EKJD D-EMZC N758BK
175/SKYLARK
N6907E
177(RG) CARDINAL
N56PZ N177SA N278SA N442BJ N53103
180/SKYWAGON
N180BB N180FN N9727G N36362 N71763
182/SKYLANE variants
N14MT N25XZ N182PN N369AN N382AS N409SA N656JM N735BZ N735CX N883DP N1745M N2379C N2454Y N4102D
185 SKYWAGON/AG CARRYALL
N185UK
190/195
N999MH N1551D
206 SUPER SKYLANE/SUPER SKYWAGON/STATIONAIR
F-BRII N191ME N72127
208/B (GRAND) CARAVAN
D-FBPS D-FLOH N208KP N208NJ N208ST
210 CENTURION
D-EEHW D-EMLS HB-CII N210AD N210CP N210NM N249SP N277CD N1778X N6593W N9533Y N30593
T303 CRUSADER
N11FV N57MT N154DJ N889VF N6498V
305 BIRD DOG (L-19)
N134TT N33600
310
N218Y N234SA N310QQ N310WT N410RS N438DD
320
N4173T
337 SUPER SKYMASTER
N19F N337UK N456TL N2216X
340
N3HK N27BG N66SW N340DW N340GJ N340SC N340YP N1711G N3456Q SX-BTA
401/402
N9146N
414/CHANCELLOR
N41TT N414FZ
421/GOLDEN EAGLE
N1FY N9AY N26HE N37VB N60GM N132CK N202AA N421CA N421N N900CB N41098
425/441 CONQUEST I/II
N22CG N425DR N425TV VP-BJT
500/501 CITATION 1
N324JC N527EW N909PS VP-BGE VP-CAT VP-CFG VP-COM

FARMAN
F.40
F-HMFI

FISHER
FP202 SUPER KOALA
50-BH

FOKKER AIRCRAFT BV
F.27 FRIENDSHIP
3C-QSB 3C-QSC
F.27-050 (FOKKER 50)
LN-KKA

FOUGA
CM170 MAGISTER
F-GJJI

AVIONS FOURNIER
RF-3
D-KMDP
RF-9
F-CARF

GENERAL AIRCRAFT LTD
MONOSPAR ST.12
VH-UTH

GIPPSLAND AERONAUTICS
GA-8 AIRVAN
VH-KLN

GOVERNMENT AIRCRAFT FACTORY
N22 NOMAD
N6302W OY-JRW OY-NMH

GRUMMAN AIRCRAFT ENGINEERING
G.21 GOOSE
N4575C
TBM-3 AVENGER
CF-KCG

GRUMMAN AIRCRAFT CORPORATION including AMERICAN AVIATION, AMERICAN-GENERAL and GULFSTREAM-AMERICAN CORPORATION production
AA-5/AG-5 TRAVELER/CHEETAH/TIGER
N31RB N136SA N666GA N674BW N2121T N90704 PH-MLB

GULFSTREAM AEROSPACE CORPORATION
G1159 GULFSTREAM III
VP-CNP
GULFSTREAM IV/V/G550
A6-HHH HB-IIY HZ-KAA N500GV N590HM VP-BGN VP-BKH VP-BKZ VP-BLA VP-BNL VP-BNN VP-BUS VP-BVT VP-CBW VP-CBX VP-CSF

HANRIOT
HD.1
N75

HAWKER AIRCRAFT LTD
TEMPEST
N7027E
SEA FURY
D-CATA

HAWKER SIDDELEY AVIATION including BAe, CORPORATE JETS LTD, DE HAVILLAND and RAYTHEON-HAWKER production
HS.125
N82GK N125XX N125ZZ N228TM N485LT N494AT N800UK N841WS N5736 VP-BHB VP-BKK VP-BMD VP-CPT 5N-AAN 5N-AWD

HEAD
Ax9-118 HAB
N4519U

HILLER
UH-12 (360)
N21UH N57CR N5315V N5317V N33514 N62171 N90724

HOWARD
500
N500LN

HUGHES HELICOPTER INC including SCHWEIZER AIRCRAFT CORPORATION (269 wef 1986) and McDONNELL DOUGLAS (369 wef 1983) production
269
N333UC N994K
369
CS-HBK CS-HBL CS-HCE N322MC N500TY N500XV N753RT N5144Q N5264Q N70526

INTORA FIREBIRD
HELI ATLAS
F-WGTX F-WGTY

JABIRU AIRCRAFT (PTY) LTD
JABIRU UL
I-6052

JODEL including CEA
D.1190S
EC-AOY
D.18
19-3431
DR.250
D-EIAR

KAMOV
Ka.26
D-HOAY

KAVANAGH
D-77 HAB
VH-AYY

KLEMM
KL.35
N5050

LAKE AIRCRAFT CORPORATION INC
LA-4 BUCCANEER
C-FQIP

LATULIP
LM-3X
N6FL

LEARJET CORPORATION INC
LEARJET 25
N121EL N309LJ
LEARJET 31
N2FU N125GP
LEARJET 45
N66SG N125GW N708SP VP-CVL
LEARJET 60
N69LJ OY-LGI VP-CRB

LET NARODNI PODNIK KUNOVICE
L-410 TURBOLET
HA-LAQ HA-YFC OY-PBH OY-PBI UR-67439 UR-67477

LOCKHEED AIRCRAFT CORPORATION including LOCKHEED-CALIFORNIA CO and CANADAIR production
10 ELECTRA
NC5171N
L.100 HERCULES
ZS-RSI
L.188 ELECTRA
LN-FOI N2RK N4HG
L.749 CONSTELLATION
N7777G
L.1011 TRISTAR
TF-ABP
JETSTAR
N6NE VP-BLD

LUSCOMBE AIRPLANE CORPORATION
8 SILVAIRE
N2975K

McDONNELL DOUGLAS CORPORATION
MD-80
EC-GQZ LV-WTY

MAULE AIRCRAFT CORPORATION
M-4
N9861M
M-5 LUNAR ROCKET
N346X N5632R N5647S N56643
M-6 SUPER ROCKET
N6130X N56462
M(XT)-7 SUPER/STAR ROCKET/STARCRAFT
N14HF N15CK N280SA N882JH N1027G

MD HELICOPTERS INC
MD.600N
N600MG N600PV N600RN N600SY N958SD N70457
MD.900
N9201U N9208V N92001

MIL
Mil-2
SP-SAY YL-LHN YL-LHO

MILES AIRCRAFT LTD
M.25 MARTINET
TF-SHC
M.65 GEMINI
OO-RLD

MITSUBISHI
MU-2
N33EW N60NB N80JN N202MC N973BB

MOONEY AIRCRAFT CORPORATION
M.20/M.252
F-BOJP HB-DFT N20UK N61MF N90BE N90TK N101UK N123UK N201YK N305RD N305SE N321KL N900RK N937BP N7423V N97821

MORANE-SAULNIER including GEMS, MORANE, SEEMS and SOCATA production
MS.880/885/887/892/894 RALLYE/GALERIEN/GALOPIN/MINERVA
D-EAXX F-GFGH OO-NAT PH-ZZY

AVIONS MUDRY AND CIE including AKROTECH EUROPE and CONSTRUCTIONS AÉRONAUTIQUES DE BOURGOGNE
CAP.10
N4238C
CAP.222/231/232
F-GKKI F-GKMZ F-GOTC F-GUJM F-WWMX F-GXDB F-GYRO

NARDI
FN.305
I-TOMI

NEICO
LANCAIR 200/235/320/IV
N250JF PH-PAB

NOORDUYN AVIATION LTD see NORTH AMERICAN

NORD
1002 PINGOUIN
LV-RIE

NORTH AMERICAN AVIATION INC and including CAC, CCF, NOORDUYN AVIATION LTD and NORTH AMERICAN ROCKWELL production
B-25 MITCHELL
N7614C N9089Z N9115Z N25644
F-86 SABRE
N196B
P-51 MUSTANG
N51RT N6526D N7098V
T-6/AT-16 HARVARD
LN-AMY LN-BNM
T-28 TROJAN
N14113 N99153

OMEGA BALLOONS
80
OY-BOB

PACIFIC AEROSPACE
PAC750
ZK-KAY

PARTENAVIA COSTRUZIONI AERONAUTICHE S.p.A
P.68
N997JB

PERCIVAL AIRCRAFT CO LTD
P.57 SEA PRINCE
N7SY
P.66 PEMBROKE
N46EA

PIAGGIO including FOCKE WULF production
P.149
D-EHJL OO-MEL
P.180
I-FXRF

PIASECKI
HUP-3 RETRIEVER
N6699D

PICCARD
GAS BALLOON
OO-BFH

PILATUS AIRCRAFT LTD
PC.6 PORTER
F-GODZ
PC.12
N234RG N412MD N476D N496DT VP-BLS

PIPER AIRCRAFT CORPORATION including TAYLOR AIRCRAFT CO LTD and THE NEW PIPER AIRCRAFT INC production and see TED SMITH
J-3C CUB (L-4/O-59)
F-BMHM N2MD N61787 N70154 7Q-YDF
J-5A CUB CRUISER
N203SA
PA-15 VAGABOND
NC4531H
PA-16 CLIPPER
N5240H N5730H N5900H
PA-18/PA-18A SUPER CUB (L-18/L-21)
I-EIXM N123SA N652P N719CS N4085E N6690D OY-AVT SE-CEE
PA-22 TRI-PACER/CARIBBEAN/COLT
D-EGEU N209SA N2652P N5675Z N6830B
PA-23/PA-27 APACHE/AZTEC
N139DB N250BW N250TB N370SA N485ED N707LD N747WW N818MJ N959JB N2401Z N4422P N5180Y N6601Y N54211 PH-NLK
PA-24/PA-26 COMANCHE
N25KB N218SA N250CC N500AV N5052P N5839P N7348P N7456P N7832P N9381P N26634 N61970 SE-IIV
PA-28-140/160 CHEROKEE/CHALLENGER/CRUISER/FLITE-LINER
N277SA N519MC N6602Y 9J-RBC
PA-28-151/161 CHEROKEE WARRIOR/CADET
D-EIAL N45CD N123DU N2929W N4306Z N5915V N8241Z SE-GPU
PA-28-180/181 CHEROKEE/CHALLENGER/ARCHER
N65JF N115RT N180LK N495AF N499MS N661KK N2273Q N2405Y N4514X N7813M N75048
PA-28-235/236 CHEROKEE/DAKOTA
D-EBWE N235PF N6339U N8471Y N81188 OO-MHB
PA-28R/PA-28RT CHEROKEE ARROW
D-EAWW N2CL N45AW N171JB N187SA N216GC N473BS N747MM N799JH N2943D N8153E N8412B N9325N N29566 N38273 N82507 N84718
PA-30/PA-39 TWIN COMANCHE
D-GPEZ N8MZ N25PR N30NW N35AD N41FT N230MJ N320MR N359DW N502TC N818Y N918Y N971RJ N7976Y N8403Y N8911Y
PA-31/PA-31T NAVAJO/CHIEFTAIN/CHEYENNE
N5LL N31NB N76JN N77XB N95TA N103ZZ N160TR N189SA N250AC N250MD N350PB N449TA N642P N666AW N2480X N3586D N4500L N63560 OY-BTZ ST-AHZ SX-BFM
PA-32/PA-32R/PA-32RT CHEROKEE SIX/LANCE/SARATOGA
D-EDEL D-EDYQ F-BCJJ N8YG N32LE N48HB N51AH N61DE N77YY N101DW N123AX N129SC N132LE N145AV N257SA N380CA N620PL N646JR N666EX N2923N N2967N N3053R N3995W N4178W N5277T N6954J N7640F N9123X N30614 N38945 N47914
PA-34 SENECA
N35AL N37US N43GG N61HB N95D N145DR N245CB N375SA N460RB N559C N2195B N3044B N4168D N21381 N33145 N39605 N53497

PART 9 – AIRCRAFT TYPE INDEX (IRISH REPUBLIC) (COVERS SECTION 2 Part 1)

AERONCA
7AC CHAMPION
EI-ATL AVB BJB BJC DDD
11AC CHIEF
EI-CCF CRR
15AC SEDAN
EI-BJJ BKC

AÉROSPATIALE
AS.350B ECUREUIL
EI-CGQ IRV PJD
SA.365 DAUPHIN 2
EI-MIP

AÉROSPATIALE-ALENIA
ATR-42
EI-BYO CBK CPT CVR CVS SLB SLC SLD SLE
ATR-72
EI-CLB CLC CLD CMJ REA REB RED REE REF SLF

AGUSTA S.p.A
A109
EI-CHV DLP DUN ECA JBC MEL MIT NPG SBM SQG

AIR and SPACE
18A
EI-CNG

AIRBUS INDUSTRIE/SAS
A300
EI-OZA,OZB OZC
A319
EI-DEY DEZ DFA DFP
A320
EI-CTD CUQ CVA CVB CVC CVD DEA DEB DEC DEE DEF DEG DEH DEI DEJ DEK DEL DEM DEN DEO DEP DER DES DFN DFO DIG DIH DIJ DIU DIV DIW DIX DJH DJI LTE TAB TAC TAD TAE
A321
EI-CPC CPD CPE CPF CPG CPH
A330
EI-CRK DAA DIP DIR DUB EWR JFK LAX ORD

AIR CRÉATION
KISS
EI-DBO DFX DIB

AMF MICROFLIGHT LTD
CHEVVRON
EI-BVJ

ATEC
2000 ZEPHYR
EI-CZA DGS DGV

AUSTER AIRCRAFT LTD
J/1 AUTOCRAT & J/1N ALPHA
EI-AGJ AMK AUM AMY
J/4
EI-CPN
J/5F AIGLET TRAINER
EI-AUS

AVIAMILANO SRL
F.8L FALCO
EI-BCJ BMF

AVID AIRCRAFT INC
AVID SPEEDWING
EI-CIM

A V ROE and CO LTD
643 CADET
EI-ALP
748
EI-BSF

BA
SWALLOW 2
EI-AFF

BAE SYSTEMS (OPERATIONS) LTD
BAe 146 including Avro variants
EI-CLG CLH CLI CMS CMY CNB CNI CNJ CNQ COF COQ CPJ CPK CPL CPY CSK CSL CTM CTN CTO CWA CWB CWC CWD CZO DBX DDE DDF DEV DEW DEX PAT

BEAGLE AIRCRAFT LTD
B.121 PUP
EI-ATJ

BEAGLE-AUSTER AIRCRAFT LTD
A.61 TERRIER
EI-ASU

BEECH AIRCRAFT CORPORATION
23 MUSKETEER
EI-BFF
76 DUCHESS
EI-BUN
77 SKIPPER
EI-BHT

BELL HELICOPTER TEXTRON INC
206A/B JETRANGER I/II
EI-BHI BIJ BKT BYJ CAW CLT CUG CUS HER HXM LAF MER PCI PKS PMI PRI RMC WSH WSN
206L LONG RANGER
EI-BYR CHL CIO LHD
407
EI-STR DBN GAN SNJ
430
EI-TIP WAV

BENSEN AIRCRAFT CORPORATION including MONTGOMERIE variants
B.8 GYROCOPTER
EI-BCF BSG
MERLIN
EI-DDO

BOEING AIRCRAFT CO including THE BOEING COMPANY
737-200 series
EI-CJC CJD CJE CJF CJG CJH CJI CKP CKS CNT CNV CNW CNX CNZ COA COB CON COX
737-300 series
EI-BZE BZF BZJ BZL BZN CHH CLW CLZ CRZ CXN CXR DJK
737-400 series
EI-BXI BXK COH COI COJ COK CUA CUD CUN CVN CVO CVP CWE CWF CWW CWX ,CXI CXJ CXK CXL CXM CZG CZK DDK DDY DGD DGL DGM DGN
737-500 series
EI-CDC CDD CDE CDF CDH DDT
737-700 series
EI-CXD,CXE
737-800 series
EI-CSA CSB CSC CSD CSE CSF CSG CSH CSI-CSJ CSM CSN CSO CSP CSR CSS CST CSU CSV CSW CSX CSY CSZ CTA CTB CXP CXT CXV DAC DAD DAE DAF DAG DAH DAJ DAK DAL DAM DAN DAO DAP DAR DAS DAT DAV DAW DAX DAY DAZ DCB DCC DCD DCE DCF DCG DCH DCI DCJ DCK DCL DCM DCN DCO DCP DCR DCS DCT DCV DCW DCX DCY DCZ DFD DFE DFF DGZ DHA DHB DHC DHD DHE DHF DHG DHI DHJ DHK DHM
757-200 series
EI-CEY CEZ
767-200 series
EI-CXZ CZD DBW GAA GBA
767-300 series
EI-CRD CRF CRL CRO CRM CXO CZH DBF DBG DBU DDW DFS
777-200 series
EI-DBK DBL DBM DDH

BOX
DUET
EI-BOX

FOUGA
CM-170 MAGISTER
EI-BXO

GARDAN
GY-80 HORIZON
EI-AYB

GROB-WERKE GmbH and CO KG
G-109
EI-HCS
G-115
EI-CAC CAD CAE CCD

GRUMMAN-AMERICAN AVIATION CORPORATION
AA-5 TIGER/TRAVELER
EI-AYD BJS BMV

GULFSTREAM AEROSPACE CORPORATION
GULFSTREAM
EI-WGV

GYROSCOPIC ROTORCRAFT
GYROPLANE
EI-COG

HAWKER-SIDDELEY AVIATION
HS.125
EI-WJN

HIWAY HANG GLIDERS LTD
DEMON (Wing) SKYTRIKE
EI-BPU BRV

HOAC FLUGZEUGWERKE
DV-20 KATANA
EI-CLA

HOFFMANN FLUGZEUGBAU FRIESACH
H-36 DIMONA
EI-CRV

HOVEY
DELTA BIRD
EI-BRW

HUGHES TOOL CO/HELICOPTER INC including SCHWEIZER AIRCRAFT CORPORATION
269 (Srs 300)
EI-CVM CZL CZP DDI

HUNT
AVON
EI-CKG

III
SKY ARROW
EI-CPX

JODEL including CEA, SAN and WASSMER production
D.9 BÉBÉ
EI-BUC
D.112
EI-BSB
D.120 PARIS-NICE
EI-CJS
D.18
EI-CKZ
DR.1050
EI-ARW

JORA AIRCRAFT
JORA
EI-NVL

LAKE AIRCRAFT CORPORATION
LA-4 BUCCANEER
EI-BUH CEX

LEARJET CORPORATION INC
LEARJET 31
EI-MAX
LEARJET 60
EI-IAW

LETOV AIR
LK-2M SLUKA
EI-CNA

LINDSTRAND BALLOONS LTD
LBL-90A
EI-CRB

LOCKHEED-CALIFORNIA CO
L.382 HERCULES
EI-JIV
L.1011 TRISTAR
EI-CNN

McCANDLESS
M.4 GYROPLANE
EI-ASR

McDONNELL DOUGLAS CORPORATION
DC-9-82/83
EI-BTX BTY BWD CBR CBS CBY CBZ CCC CCE CDY CEK CEP CEQ CFZ CIW CKM CMZ CPB CRE CRH CRJ CRW CTJ

MAGNI GYRO of ITALY
SPARTAN
EI-DBX

MAINAIR SPORTS LTD
GEMINI/FLASH
EI-CKT
MERCURY
EI-CMU

MALMO
MFI-9 JUNIOR
EI-AWR

MAULE AIRCRAFT CORPORATION
MX-7
EI-CUT GER

MBB
Bö.105
EI-BLD LIT

MEDWAY MICROLIGHTS LTD
ECLIPSE R
EI-CRY CTC ELL

MONNETT
MONI
EI-BMU

MOONEY AIRCRAFT CORPORATION
M.20
EI-CAY

MORANE-SAULNIER
MS.500
EI-AUY
MS.880/885/887/892/894 RALLYE variants
EI-ATS AUE AUG AUJ AWU AYA AYI AYT BBG BBI BBJ BBO BCS BCU BCW BDH BDK BEA BEP BFI BFP BFR BFV BGA BGB BGC BGG BGS BGU BHF BHK BHN BHP BHY BIM BIW BIT BIW BJK BKE BKN BKU BMA BMB BMJ BNG BNU BUJ BUT CEG CIA

MOSKITO
MOSKITO 2
EI-CJV

PARTENAVIA COSTRUZIONI AERONAUTICHE S.p.A
P.68
EI-BWH

PEGASUS AVIATION
QUANTUM
EI-CNU

PHOENIX
LUTON LA-5A MAJOR
EI-CGF

PIEL
CP.301 EMERAUDE
EI-CFG

PIPER AIRCRAFT CORPORATION
J-3C CUB
EI-AFE AKM BBV BCM BCN BCO BEN BFO BIO BSX BYY CFO CHK COY CPP
J-5A CUB CRUISER
EI-CGV
PA-12 SUPER CRUISER
EI-ADV CFF CFH
PA-18 SUPER CUB
EI-ANY BID BIK CIG CKH
PA-22 TRI-PACER/COLT
EI-BAV UFO
PA-23 AZTEC
EI-BDM EEC WAC WMN
PA-28-140/160 CHEROKEE/CRUISER
EI-AOB ATK BSO CGP CIV CMB COZ
PA-28-151/161 CHEROKEE/WARRIOR
EI-CTT DJM SKW WRN
PA-28-180/181 CHEROKEE
EI-BBC BDR BGF CIF DDZ
PA-28R CHEROKEE ARROW
EI-EDR SKU
PA-31 NAVAJO CHIEFTAIN
EI-CNM
PA-34 SENECA
EI-BSL CMT
PA-38 TOMAHAWK
EI-BJT BVK
PA-44 SEMINOLE
EI-SKT

PTERODACTYL LTD
MICROLIGHT
EI-BOA

RAJ HAMSA
X'AIR
EI-CXC DBI DBV DCA DDJ DFY DGG DGH DGJ DGK KEV

RAND-ROBINSON
KR-2
EI-BNL BOV

AVIONS PIERRE **ROBIN**
DR400/180R REMORQUER
EI-CRG
R.1180T AIGLON
EI-BIS
R.2160D
EI-SKS SKV

ROBINSON HELICOPTER CO INC
R22 BETA
EI-CFE CFX CPO CWL CWP CWR DIZ EHB EHC EHD EHE EHG EWM EXG GAV GKL GPT JWM MAG MUR OBJ TKI
R44 ASTRO
EI-CZM DDA DFW DOC EBJ EGG EXC GPZ JAL JFD MCC MJR MMO MOR MUL PEC SGF TMH TOY

ROTORWAY
EXECUTIVE
EI-BNP CMW

RUTAN
LONG-EZE
EI-CMR CPI

SCHLEICHER
ASK14
EI-APS
ASK16
EI-AYR

SHORT BROTHERS LTD
SD.3-60
EI-SMA SMB

SIKORSKY AIRCRAFT
S-61
EI-BLY CNL CXS CZN MES RCG SAR

SKYHOOK
SABRE
EI-BPT

SKYTRIKE/HIWAY
Vulcan
EI-BMW BNH

SLINGSBY SAILPLANES LTD
T.21 CADET
EI-CJT
T.56 SE.5 rep
EI-ARH ARM

SNCAN STAMPE including AIA production
STAMPE SV.4A/C
EI-BAJ BLB CJR

SOCATA
ST.10 DIPLOMATE
EI-BUG
TB-9 TAMPICO
EI-BMI BSK BYG CRX GFC
TB-10 TOBAGO
EI-BOE
TB-20 TRINIDAD
EI-BSV

SOLAR WINGS LTD
XL-R
EI-BSW CGJ CGM CGN CHT CKU

SOUTHDOWN SAILWINGS LTD
PUMA
EI-BPO

STEEN AERO LAB.INC
SKYBOLT
EI-CIZ SAT

STINSON AIRCRAFT CORPORATION
108 STATION WAGON
EI-CGC

STODDARD-HAMILTON
GLASAIR
EI-CTG

STOLP
SA.300 STARDUSTER TOO
EI-CDQ

TAYLOR
JT.1 MONOPLANE
EI-BKK

TAYLORCRAFT AIRCRAFT CORPN
PLUS D
EI-AGD ALH AMF
BC-65
EI-CES

TEAM
MINI-MAX
EI-CNC CGB

PART 9 – AIRCRAFT TYPE INDEX (UK) (COVERS SECTION 1 Parts 1 and 2, SECTION 3 and includes selected default ENGINE details)

ACRO
ADVANCED
G-BPAA

ADAM
RA.14 LOISIR
G-BHIK

ADVANCED AIRSHIP CORPORATION
ANR-1
G-MAAC

ADVANCED TECHNOLOGIES GROUP LTD
AIRSHIP (GAS) AT-10
G-OATG

ADVANCED TECHNOLOGIES INC
FIREBIRD CH1 ATI
G-BXZN

AERIAL ARTS (LTD)
CHASER (Rotax 377) with Aerial Arts 110SX and 130SX wings
G-MNTD MNYD MNYE MNYF MTCP MTDD MTDE MVOD MWGO
CHASER S (Rotax 377)
G-MVDK MVDL MVDP MVDR MVGA MVGF MVGG MVGH MVGI MVHA MVID MVIE MVJF MVJG MVJH MVJI MVJJ MVJK MVKY MVKZ MVLA MVLB MMLB MVLC MVLD MVLE MVLF MVLG MVLH MVLS MVLT MVLW MVML MVMM MVOP MVRG MVRL MVSG MVTF MVTL MVTM MVUS MVUT MVVU MVXG MVYY MVZM MWWZ MWXW MWXX MWXY MWXZ MWYM MYBU MYCB MYEI MYEJ MYFO MYGK MYIL MYIT MYJO MYJW MYKD MYLJ MYMY MYRE MYSA MYSV MYWN MYWS MYYD MYZX MZCB MZTS REGI

AERO SP ZOO
AT-3
G-SPAT UKAT

AERO COMMANDER INC including ROCKWELL INTERNATIONAL CORPORATION production
680/685 COMMANDER
G-AWOE OMAP

AERO DESIGNS
PULSAR X (Rotax 582)
G-BSFA BTDR BTRF BTWY BUDI BUJL BULM BUSR BUYB BVJH BVSF BVTW BXDU CCBZ CCIG IIAN LUED LWNG MCMS RMAN
PULSAR XP (Rotax 912/912-UL)
G-BUOW BUZB CBLA EPOX LEEN OOXP PLSA WYNS XPXP
PULSAR 3 (Rotax 912-ULS)
G-BYJL

AERO DIFUSIÓN see JODEL

AERO DYNAMICS LTD
SPARROWHAWK
G-BOZU

AERO VODOCHODY NÁRODNÍ PODNIK see CZL/LET
C-104
G-CCOB
L-29 DELFIN
G-BYCT BZNT DELF DLFN ODAT
L-39 ALBATROS
G-BZDI CCWD OALB
SB LIM-2
G-OMIG

AEROCAR
TAYLOR MINI IMP
G-BLWW
TAYLOR SOOPER COOT
G-COOT

AERODYNE SYSTEMS INC
VECTOR
G-MBTW MJAZ

AEROFAB see LAKE

AEROMERE see AVIAMILANO

AEROMOT INDUSTRIA MECANICO
AMT-200 SUPER XIMANGO
G-BWNY JTPC KHOM LLEW MOAN RFIO XMGO

AERONAUTICAL CORP OF AMERICA (AERONCA) see CHAMPION
C-3/100
G-ADRR ADYS AEFT AESB AEVS AEWU AEXD
K
G-ONKA
11AC CHIEF/11CC SUPER CHIEF
G-AKTK AKUO AKVN BJEV BJNY BPRA BPRX BPXY BRCW BRFJ BRWR BRXF BRXL BSTC BTFL BTRI BTSR BUAB BUTF IIAC IVOR
15AC SEDAN
G-AREX
A65TAC/65C SUPER CHIEF/O-58B(L-3) GRASSHOPPER
G-BRHP BRPR BTRG BTUV

AÉRONAUTIQUE HAVRAISE see GARDAN

AEROPHILE
Gas Balloon:
AEROPHILE 5500
G-CCYF

AEROPRAKT
A22 FOXBAT (Rotax 912-ULS)
G-CBGJ CBJH CBYH CCCE CCJV CDDW CHAD COXS CWTD FBAT FJTH FOXB FXBT MRAF SDOI TADC VROD

AÉROSPATIALE including EUROCOPTER production also see AÉROSPATIALE, ALENIA, ATR, SOCATA and SUD-AVIATION
AS332 SUPER PUMA
G-BKZE BKZG BLPM BLRY BLXR BMCW BMCX BUZD BWWI BWZX CHCF CHCG CHCH JSAR PUMA PUMB PUMD PUME PUMH PUML PUMN PUMO PUMS REDJ REDK REDL REDN TIGB TIGC TIGE TIGF TIGG TIGH TIGI TIGJ TIGO TIGP TIGR TIGS TIGT TIGV TIGZ
AS350B ECUREUIL
G-BMAV BRVO BVJE BVXM BXGA BYZE BZVG CBHL DEMM DOIT EFTF EJOC FIBS FROH HRAK IANW IIPM JBBZ JESI MURP ODMC OGOA OMCC OOIO NUTY PLMB PLMH PROB REAL RICC SCHI SMDJ TATS WHAM WHST WKRD XMEN
AS355 ECUREUIL 2
G-BOOV BPRI BPRJ BPRL BSTE BSYI BUFW BVLG BYPA BYZA BZGC BZVZ CAMB CCAO CCWK CDFV CPOL DANZ DOOZ EMAN EMHH EPOL FFRI FTWO GMPA GONN GRID HARO HOOT ICSG IFBP JARV JETU JPAL LECA LENI LHEL LINE LNTY LUVY NAAS OASP OHCP OHMS OITN OLCP OROM ORMA OTSP PASF PASH RANI REEM SASU SEPA SEPB SEPC SEWP SKYN SKYW STRL SYPA TOPC TOPS ULES VGMC VONE VONF WENA WIRE WMPA XCEL XOIL ZITZ
SA365 DAUPHIN 2
G-MLTY PLMI

AÉROSPATIALE-ALENIA
ATR-42
G-BVJP BXEH CDFF DRFC IONA KNNY SSEA TAWE WLSH
ATR-72
G-BWDA BWDB BXTN

AEROSPORT
SCAMP
G-BKFL BKPB BOOW
WOODY PUSHER
G-AWWP AYVP BSFV SHUV

AEROSTAR SA see TED SMITH and YAKOVELEV

AEROTEC/AEROTEK INC see PITTS

AEROTECH see WHITTAKER

AESL see VICTA

COSTRUZIONI AERONAUTICHE GIOVANNI AGUSTA S.p.A see BELL HELICOPTER
A109
G-BWNZ BWZI CHSL DATE DBOY DPPF ECMM ESLH HDTV HIMJ JERL JMXA JODI MDPI MOMO OCMM PBEK RCMS RFDS SCOI TBGL TELY TGRA TYCN USTB USTC USTS VIPH WCIL WEST WNAA

<u>A119 KOALA</u>
G-CCSZ

AHERNE
<u>BARRACUDA</u>
G-BZSV

AIR & SPACE
<u>18A GYROPLANE</u>
G-BVWK BVWL

AIR COMMAND MANUFACTURING INC
<u>503 COMMANDER</u>
G-BOAS BOIK KENB
<u>532 ELITE</u>
G-BOGV BOKF BOOJ BPAO BPPU BPUE BPUG BPUI BRGO BRLB BRSP BSCB BSND BSRZ OGTS TFRB YROI
<u>582 SPORT</u>
G-BTCB URRR

AIR CRÉATION
<u>503/FUN 18 GT</u>
G-MYMM MYOL MYTZ MYUA MYVI MYXF
<u>582/KISS</u>
G-BZXP CBEB CBJA CBJL CBKE CBKS CBLX CBMX CBNY CBRZ CBSX CCEK CCFA CCGM CCHJ CCHM COXY CCPA CHKN KIZZ PGHM SNOG TEDW TRYK
<u>IXESS 912</u>
G-DJST IXES

AIR NAVIGATION AND ENGINEERING CO
<u>ANEC II</u>
G-EBJO
<u>ANEC IV MISSEL THRUSH</u>
G-FBPI

AIRBORNE WINDSPORTS PTY LTD
<u>XT912/STREAK III</u>
G-CDGE LVRL XTEE

AIRBUS INDUSTRIE/SAS
<u>A300</u>
G-CEXH CEXI CEXJ CEXK MAJS MONR MONS OJMR
<u>A319</u>
G-CBTU DBCA DBCB DBCC DBCD DBCE DBCF EJAR EJJB EUOA EUOB EUOC EUOD EUOE EUOF EUOG EUOH EUOI EUOJ EUOK EUOL EUPA EUPB EUPC EUPB EUPC EUPD EUPE EUPF EUPG EUPH EUPJ EUPK EUPL EUPM EUPN EUPO EUPP EUPR EUPS EUPT EUPU EUPV EUPW EUPX EUPY EUPZ EZAM EZBS EZDC EZEA EZEB EZEC EZED EZEF EZEG EZEJ EZEK EZEO EZEP EZET EZEU EZEV EZEW EZEX EZEY EZEZ EZIA EZIB EZIC EZID EZIE EZIF EZIG EZIH EZII EZIJ EZIK EZIL EZIM EZIN EZMH EZMK EZMS EZNC EZNM EZPG EZSM HMCC OMAK
<u>A320</u>
G-BXKA BXKB BUSB BUSC BUSD BUSE BUSF BUSG BUSH BUSI BUSJ BUSK BVYA BVYC BXKC BXKD CRPH DJAR EUUA EUUB EUUC EUUD EUUE EUUF EUUG EUUH EUUI EUUJ EUUK EUUL EUUM EUUN EUUO EUUP EUUR EUUS EUUT EUUU FHAJ MEDA MEDE MEDH MEDJ MEDK MIDK MIDO MIDP MIDR MIDS MIDT MIDU MIDV MIDW MIDX MIDY MIDZ MONX MPCD OOAP OOAR OOAU OOAW OOAX OUZO OZBB SSAS SUEW TCKE TTOA TTOB TTOC TTOD TTOE TTOF TTOG TTOH TTOI TTOJ UNIH VCED
<u>A321</u>
G-CTLA DHJH EFPA EUXC EUXD EUXE EUXF EUXG EUXH EUXK MEDF MEDG MEDJ MIDA MIDC MIDE MIDF MIDH MIDI MIDJ MIDK MIDL MIDM NIKO OOAE OOAF OOAH OOAV OOAX OZBE OZBF OZBG OZBH OZBI SMTJ TTIA TTIB TTIC TTID TTIE TTIF
<u>A330</u>
G-EOMA MDBD MLJL OJMB OJMC OMYT SMAN WWBB WWBC WWBD WWBM
<u>A340</u>
G-VAEL VAIR VATL VBUS VEIL VELD VFAR VFLY VFOX VGAS VGOA VHOL VMEG VNAP VOGE VSEA VSHY VSSH VSUN VWKD

AIRCRAFT MANUFACTURING CO (AIRCO)
<u>DH.2 rep</u>
G-BFVH
<u>DH.6</u>
G-EAML
<u>DH.9</u>
G-EAQM

AIRSHIP INDUSTRIES including AEROSPACE DEVELOPMENTS
<u>Skyship 500</u>
G-BIHN

AIRSPEED LTD
<u>AS.40 OXFORD</u>
G-AHTW AITB AITF
<u>AS.57 AMBASSADOR</u>
G-ALZO

AIRTOUR BALLOONS
<u>Hot Air Balloon:</u>
<u>AH-31</u>
G-BKVY
<u>AH-56</u>
G-BKVW BKVX BLVA BLVB BSGH BWPL OAFC
<u>AH-77</u>
G-BLYT BOBH IVAC OAAC

AIRWAVE see CHARGUS

ALLPORT
<u>MLB variants</u>
G-BJIA BJSS

ALON INC see ERCOUPE

SOCIÉTÉ **ALPAVIA** see FOURNIER

ALPHA
<u>Trike (Rotax 277) with Aerial Arts 130SX wing</u>
G-MNIT MNZS

ALPI
<u>PIONEER 300</u>
G-DEBT EWES IPKA OPFA PCCC TREX XCIT YVES

AMD-BA see DASSAULT

AMERICAN AEROLIGHTS INC
<u>EAGLE</u>
G-MBCU MBEP MBFS MBHE MBJD MBJK MBKY MBNK MBRD MBRS MBWE MBZV MJAE MJBL MJBV MJEO MJNM MJNO MMTV

AMERICAN AVIATION CORPORATION see GRUMMAN-AMERICAN

AMERICAN BLIMP CORPORATION
<u>A-60+ AIRSHIP</u>
G-OLEL TLEL

AMERICAN CHAMPION AIRCRAFT CORPORATION
<u>7ECA CITABRIA AURORA</u>
G-CDGJ
<u>7GCAA CITABRIA ADVENTURE</u>
G-IGLS
<u>7GCBC CITABRIA EXPLORER</u>
G-EXPL
<u>8KCAB</u>
G-IGLZ IZZZ YZMO

AMF MICROFLIGHT LTD including AMF AVIATION ENTERPRISES LTD
<u>CHEVVRON (Konig SD570)</u>
G-MNFL MTFG MVGC MVGD MVGE MVIP MVOO MVUO MVVV MVXX MVZZ MWNO MWNP MWPW MWRZ MWUI MWZB MYGN MYYP MZCK MZDP MZFH MZMK

ANDERSON
<u>EA-1 KINGFISHER AMPHIBIAN</u>
G-BUTE BXBC

ANDREASSON including CROSBY
<u>BA.4B</u>
G-AWPZ AYFV BEBS BFXF JEDS YPSY

ANONIMA POIAZIONARA VERCELLESE IND. AERONAUTICHE
<u>AVIA FL.3</u>
G-AGFT

ARBITER SERVICES
<u>Trike with Aerial Arts 130SX wing</u>
G-MNWL

BENSEN AIRCRAFT CORPORATION including CAMPBELL-BENSEN and MONTGOMERIE-BENSEN production

B.7/B.8 GYROCOPTER

G-APSY APUD ARBF ARTJ ASCT ASME ASWN ASYP ATLP ATOZ AWDW AWPY AXBG AZAZ BCGB BGIO BHEM BIFN BIGP BIGU BIHX BIPY BIVK BJAO BKBS BLLA BLLB BMBW BMOT BMYF BMZW BNBU BNJL BOUV BOWZ BOZW BPCV BPIF BPNN BPOO BPSK BPTV BRBS BRCF BREA BREU BRFW BRHL BRXN BSBX BSJB BSMG BSMX BSNL BSNY BSZM BTAH BTBL BTFW BTIG BTJN BTJS BTST BTTD BUJK BUPF BVAZ BVIF BVJF BVKJ BVMG BVPX BWAH BWEY BWJN BWSZ BXCL BYTS BZID BZJR CBFW CBNX CBSV CCXS CDBE HAGS JOEL OOJC OTIM SCUD YJET YROS YROY

BEST OFF

SKYRANGER

G-CBIV CBVR CBVS CBWW CBXS CCAF CCBA CCBG CCBJ CCCK CCCM CCCR CCCY CCDG CCDH CCDW CCDY CCEH CCIK CCIO CCIY CCJA CCJT CCJW CCKF CCKG CCLF CCLU CCMX CCMZ CCNJ CCNR CCNS CCNU CCPF CCPL CCRR CCRV CCSX CCTR CCUC CCUD CCUF CCVR CCWC CCWU CCXH CCXL CCXM CCXN CCYM CCZM CDAY CDBA CDBO CDBV CDCH CDDR CDDU CDFJ CDFP CDHA CDHE CDIP CDJP CFWR CRAB CUBE CZMI DOIN FRNK FLDG FONZ GHLI INCE ISEL JAYS JEZZ KULA LASN LDAH MARO OBAZ OKIM OMSS OSKR OTCV POLL PSKY RAFR REVO RSKY SKPG SKRG SOPH TEDI TMCB TYGR UACA VVVV WAZP XYJY

BETTS

TB.1

G-BVUG

BINDER AVIATIK GmbH see PIEL

BIRDMAN ENTERPRISES LTD

WT-11 CHINOOK

G-MMKE

BLACKBURN AEROPLANE and MOTOR CO LTD

MONOPLANE

G-AANI

B.2

G-AEBJ

BLAKE

BLUETIT

G-BXIY

BLERIOT

XI

G-AANG AVXV BLXI BPVE BWRH LOTI

BOEING AIRCRAFT CO including BOEING COMPANY

B-17G FORTRESS

G-BEDF FORT

B-29 SUPERFORTRESS

G-BHDK

707-300 series

G-AYAG

707-400 series

G-APFG APFJ

737-200 series

G-BYYK CEAC CEAD CEAE CEAF CEAG CEAH CEAI CEAJ GPFI

737-300 series

G-BYZJ CELA CELB CELC CELD CELE CELF CELG CELH CELI CELJ CELK CELP CELR CELS CELU CELV CELW CELX CELY CELZ ECAS EZYD EZYH EZYI EZYJ EZYK EZYL EZYM EZYN EZYO EZYP EZYR EZYS EZYT GSPN IGOB IGOC IGOG IGOH IGOI IGOJ IGOK IGOL IGOM IGOO IGOP IGOR IGOS IGOT IGOU IGOV IGOW IGOY IGOZ LGTE LGTF LGTG LGTH LGTI OBWZ ODSK ODUS OFRA OGBD OGBE OHAJ OJTW OMUC OTDA STRA STRB STRE STRF STRH THOE THOF THOG THOH THOI THOJ TOYA TOYB TOYC TOYD ZAPM ZAPV ZAPW

737-400 series

G-DOCA DOCB DOCE DOCF DOCG DOCH DOCL DOCN DOCO DOCP DOCR DOCS DOCT DOCU DOCV DOCW DOCX DOCY DOCZ GBTA GBTB

737-500 series

G-BVKA BVKB BVKD BVZE BVZG BVZH BVZI GFFA GFFB GFFC GFFD GFFE GFFF GFFG GFFH GFFI GFFJ THOA THOB THOC THOD

737-700 series

G-EZJA EZJB EZJC EZJD EZJE EZJF EZJG EZJH EZJI EZJJ EZJK EZJL EZJM EZJN EZJO EZJP EZJR EZJS EZJT EZJU EZJV EZJW EZJX EZJY EZJZ EZKA EZKB EZKC EZKD EZKE EZKF EZKG

737-800 series

G-OBBJ OBMP THOK XLAA XLAB XLAE XLAF XLAG XLAH

747-200 series

G-BDXE BDXF BDXG BDXH BDXI BDXJ BDXO CCMA INTL VIBE VPUF VSSS VZZZ

747-400 series

G-BNLA BNLB BNLC BNLD BNLE BNLF BNLG BNLH BNLI BNLJ BNLK BNLL BNLM BNLN BNLO BNLP BNLR BNLS BNLT BNLU BNLV BNLW BNLX BNLY BNLZ BYGA BYGB BYGC BYGD BYGE BYGF BYGG CIVA CIVB CIVC CIVD CIVE CIVF CIVG CIVH CIVI CIVJ CIVK CIVL CIVM CIVN CIVO CIVP CIVR CIVS CIVT CIVU CIVV CIVW CIVX CIVY CIVZ GSSA GSSB GSSC VAST VBEE VBIG VFAB VGAL VHOT VLIP VROC VROM VROS VROY VTOP VWOW VXLG

757-200 series

G-BIKC BIKF BIKG BIKI BIKJ BIKK BIKM BIKN BIKO BIKS BIKU BIKV BIKZ BMRA BMRB BMRC BMRD BMRE BMRF BMRH BMRJ BPEC BPED BPEE BPEI BPEJ BYAD BYAE BYAF BYAH BYAI BYAJ BYAK BYAL BYAN BYAO BYAP BYAR BYAS BYAT BYAU BYAW BYAX BYAY CDUO CDUP CPEL CPEL CPEM CPEN CPEO CPEP CPER CPES CPET CPEU CPEV DAJB FCLA FCLB FCLC FCLD FCLE FCLF FCLG FCLH FCLI FCLJ FCLK FJEA FJEB JALC JMAA JMAB JMCD JMCE JMCF JMCG MCEA MONB MONC MOND MONE MONJ MONK OAVB OOBA OOBC OOBD OOBE OOBF OOBI OOBJ OOOB OOOG OOOX OOOY PIDS RJGR ZAPU

767-200 series

G-BNYS BOPB BRIF BRIG BYAA BYAB

767-300 series

G-BNWA BNWB BNWC BNWD BNWH BNWI BNWM BNWN BNWO BNWR BNWS BNWT BNWU BNWV BNWW BNWX BNWY BNWZ BRIF BRIG BZHA BZHB BZHC DAJC DIMB OBYB OBYC OBYE OBYG OBYH OBYI OBYJ OOAL OOAN OOBK SJMC VKNG

777-200 series

G-RAES VIIA VIIB VIIC VIID VIIE VIIF VIIG VIIH VIIJ VIIK VIIL VIIM VIIN VIIO VIIP VIIR VIIS VIIT VIIU VIIV VIIX VIIY VIIZ YMMA YMMB YMMC YMMD YMME YMMF YMMG YMMH YMMI YMMJ YMMK YMML YMMM YMMN YMMO YMMP ZZZA ZZZB ZZZC

BOEING AIRPLANE CO

STEARMAN 75 KAYDET/N2S/PT-13/PT-17:

G-AROY AWLO AZLE BAVN BAVO BHUW BIXN BNIW BPEX BRHB BRSK BRTK BRUJ BSDS BSGR BSWC BTFG CCXA CCXB IIIG ILLE ISDN NZSS RJAH

BOLAND

52 HAB

G-BYMW

BÖLKOW APPARATEBAU GmbH including MALMO and MBB production

Bö.207

G-EFTE

Bö.208 JUNIOR

G-ASFR ASZD ATDO ATRI ATSI ATSX ATTR ATUI ATVX ATXZ AVKR AVLO AVZI BIJD BSME CLEM ECGO

Bö.209 MONSUN

G-AYPE AZBB AZDD AZOA AZOB AZRA AZTA AZVA AZVB BLRD

BOMBARDIER INC

CANADAIR CL600/604 CHALLENGER

G-ELNX FBPI FTSL LVLV REYS

CANADAIR CRJ 700 REGIONAL JET

G-DUOA DUOC DUOD DUOE

BD-700 GLOBAL EXPRESS

G-LOBL XPRS

BONSALL

DB-1 MUSTANG

G-BDWM

BOULTON PAUL AIRCRAFT LTD

P.108 BALLIOL

G-APCN

BOWERS

FLY BABY

G-BFRD BNPV BUYU

BRADSHAW

HAB-76

G-AXXP

BRANDLI

BX-2 CHERRY

G-BXUX

BRANTLY HELICOPTER CORPORATION
B.2
G-ASHD ASXD ATFG AVIP AWDU BPIJ OAPR OMAX ROTR
305
G-ASXF

BREMNER see MITCHELL wing

BRIGHTON
Ax7-65 HAB
G-AVTL AWJB

BRISTOL AEROPLANE CO LTD
BOXKITE
G-ASPP
BABE
G-EASQ
F.2B FIGHTER
G-AANM ACAA AEPH
20 M.1C REP
G-BLWM BWJM BPLT
105 BULLDOG
G-ABBB
149 BOLINGBROKE (BLENHEIM)
G-BPIV MKIV
156 BEAUFIGHTER
G-AJMC DINT
170 FREIGHTER
G-AICH AICT AIMI
171 SYCAMORE
G-ALSX AMWI HAPR
173
G-ALBN
175 BRITANNIA
G-ANCF AOVF AOVT
192 BELVEDERE
G-BRMB

BRITISH AIRCRAFT CORPORATION/AÈROSPATIALE
CONCORDE
G-AXDN BBDG BOAA BOAB BOAC BOAD BOAE BOAF BOAG SSST

BRITISH AIRCRAFT CORPORATION (BAC) see HUNTING
ONE-ELEVEN
G-ASYD AVMJ AVMN AVMO AVMT AVMU AVMZ AWYV AZMF MAAH OBWD

BRITTEN
SHERIFF
G-FRJB

BRITTEN-NORMAN LTD including FAIREY BRITTEN-NORMAN LTD, IRMA and PILATUS (BN-2) production
BN.1F
G-ALZE
BN.2A/B/T ISLANDER/DEFENDER
G-AWNT AWID AXUB AXZK AYAZ AYRU AZBV BCEN BCWO BCWR BCZS BDPP BDPU BEEG BELF BFNU BIIP BJEC BJEE BJEF BJEJ BJOP BJWO BLDV BLNJ BOMG BPCA BPLR BSWR BUBN BVFK BVHY BWPV BWPX CCUX CDCJ CHES CHEZ CIAS CZNE GMPB ISLA JSAT JSPC LEAP MAFF NESU OBNC OBNG ORED OSEA PASV RAPA RLON SBUS SJCH SSKY WOTG XAXA
BN.2A/III TRISLANDER
G-BBYO BDOT BDTN BDTO BEDP BEVR BEVT BEVV FTSE JOEY LCOC OJAV PCAM RBCI RHOP XTOR

BROCHET
MB.50 PIPISTRELLE
G-AVKB BADV

BROOKLAND
MOSQUITO
G-AWIF

BROOKLANDS AIRCRAFT CO see OPTICA

BROOKMOOR BEDE AIRCRAFT
BD-4
G-BEKL BKZV BOPD BYLS
BD-5
G-BCOX BGLB BJPI

BRÜGGER
MB.2/MB.3 COLIBRI
G-BKCI BKRH BNDP BNDT BPBP BRWV BSUJ BUDW BUTY BVIS BVVN BXVS HRLM KARA PRAG

BÜCKER including CASA and DORNIER-WERKE AG production
Bü.131 JUNGMANN (CASA 1-131)
G-BECT BECW BHPL BIRI BJAL BPDM BPTS BPVW BRSH BSAJ BSFB BSLH BTDT BTDZ BUCC BUCK BUOR BUTA BUVN BVPD BWHP BXBD BYIJ BZJV BZVS CBCE CCHY CDRU DUDS EHBJ EMJA JGMN JUNG JWJW RETA TAFF WIBS
Bü.133 JUNGMEISTER
G-AEZX AXMT AYFO AYSJ BSZN BUKK BUTX BVXJ BZTJ TAFI
Bü 181 BESTMANN
G-CBKB GLSU

BUSHBY-LONG see LOEHLE
MIDGET MUSTANG
G-AWIR BXHT IIMT MIDG

BYRNE
ODYSSEY 4000 MLB
G-BCFE

CAB see GARDAN

CALL AIRCRAFT CO see IMCO

CAMBRIDGE HOT-AIR BALLOONING ASSOCIATION
HAB
G-BBGZ

CAMERON BALLOONS LTD see CAMERON-COLT and CAMERON-THUNDER
Gas Airship:
DG-19
G-BKIK
Gas Balloon:
R-15
G-CICI
R-36
G-ROZY
R-42
G-BLIO
R-77
G-BUFA BUFC BUFE
R-150
G-BVUO
Hot Air Airship:
Zero 25
G-BMUR
DP-50
G-BMEZ
DP-70
G-BNXG-BPFF BRDT
DP-80
G-UPPY VIBA
DP-90
G-BXKY
D-96
G-BAMK BEPZ
Hot Air Balloon:
20 variants
G-BIBS BJUV BOYO BPRU
24 series
G-BSCK BVCY
31 variants
G-BEJK BEUY BGHS BMST BPUB BRMT BVFB BZYR CBIH CBLN CCHP CDFI LEAU NOMO PRTT RBMV TOHS WETI
34 variants
G-BRKL BUCB BVZX BYNW BZBT EROS EZER FZZI IAMP OBLU RAPP
42 variants
G-BZER BCDL BISH BKNB BMWU BPHD BUPP BWEE BWGX BXJH BXTG BYRK CCAY CCJY CCSI HOPI SKOT
56 variants
G-AZKK BBYU BCOJ BCXZ BDPK BDSF BDUI BDUZ BDYH BECK BEEH BEND BENN BERT BEXX BEXZ BFAB BFFT BFKL BGLX BGOI BHGF BHSN BICU BKRS BKZF BNIF BNZN BOWM BRIR BRSA BTHZ BUVG BZKK HOFM HOOV LENN ODAY OVET SWPR WAAC WYNT
60 series
G-BTZU BVDM BVDY BWRT BXJZ CBJS CBVD ROGY

65 variants
G-AZIP AZUP AZUV AZXB BAOW BBGR BBYR BCAP BCFN BCRI BDRK BDSK BEIF BGJU BHKH BHNC BHND BHOT BHOU BIBO BIGL BIWK BIWU BIYI BJAW BJWJ BJZA BKGR BKWR BLEP BLJF BLZB BMCD BMJN BMKY BMPD BMVW BMYJ BNAN BNAU BNAW BOAL BOOB BOWV BPGD BPPA BPXF BREH BRMI BROE BROG BSAS BSGP BTUH BWBA BWHB BWHG BXGY BXUU BYZL GLUE HENS KAFE MUIR NATX OERX PMAM PYRO RUDD SMIG WELS

70 variants
G-BXOT BYJX BZEK CCCX CCPX CCTY

77 variants
G-BAXF BBCK BBOC BCNP BCRE BCZO BDBI BDSE BEPO BFUZ BFYK BGAZ BGHV BHDV BHHB BHHK BHHN BHII BHYO BIEF BIET BIRY BJGK BKNP BKPN BKTR BKWW BKZB BLFY BLIP BLJH BLLD BLPP BLSH BLZS BMAD BMKJ BMKP BMKW BMLJ BMLW BMOH BMPP BMTN BMTX BMZB BNCB BNCJ BNDN BNDV BNEO BNES BNFG BNFO BNGJ BNGN BNHI BNIN BNIU BNJG BNKT BNMA BNMG BNNE BNPE BNTW BNTZ BNUC BOAU BOBR BOEK BOFF BOGP BOJB BOJD BOJU BOOZ BORB BORN BOTW BOVV BOWB BOWL BOXG BOYS BOZN BPBV BPBY BPDF BPDG BPHH BPHJ BPLV BPPP BPSH BPSR BPTD BPVC BPVM BPWC BPYI BPYS BPYT BPYV BRBO BRFE BRFO BRIE BRKW BRLX BRMU BRMV BRNW BROB BRRF BRRO BRRR BRRW BRSD BRTV BRUV BRZA BRZT BSBI BSBM BSBR BSDX BSEV BSGY BSHO BSHT BSIC BSIJ BSLI BSKD BSMS BSUV BSWV BSWY BSXM BTAG BTIX BTJH BTKZ BTOI BTOP BTPT BTRX BTWJ BTWM BTXW BTZV BUAF BUAM BUDU BUEV BUGD BUGP BUGS BUHM BUNG BUOX BUPI BUTJ BUWU BUZK BVBS BVBU BVDR BVFF BVHK BVLI BVMF BVUK BVXB BWAJ BWAN BWHC BWKV BWPB BWPC BWTJ BWYN BXAX BXSX BXTJ BXVT BYBN BYHY BYLY BYNJ BZPU BZPW CBHX CBKV CCAR CCHW CCPO CCSP CEJA CGOD CHOK CHUK CTGR CXHK DASU DRYI ENNY EPDI ERIK FABB FELT FUZY GEES GEEZ GEUP GUNS HARE HENY HORN JLMW KEYY KODA KTEE LAZR LEGO LEND LEXI LIDD LIOT LOAG LOAN LOLL LUBE MAMO MILE MOFF MOKE MRTY NWPR OATH OCND OEDP OHSA OJEN OKYA OMRB ONZO ORPR PADI POLY PUSS RAPH RCMF RONI SAFE SAIX SCAH SCFO SNOW SUCH SUSI TECK TUDR UPUP ULIA VODA VOLT WAIT WBEV WELI XSKY ZINT ZUMP

80 series
G-BUYC BVEK BVEN BVGJ BVUU BVZN BWAO BWGP BXJP BXLG BXSC BXSJ BYER BYJJ BYTJ BZMV BZPK CBEY CCYN EVET LIMP MCAP NMOS OARG OBTS OGJM ONIX RMAX SLCE UPHL

84 series
G-AYAJ AYJZ AYVA AZBH AZNT AZRN AZSP BZVT BAGY BAKO BALD BAND BAST BBLL BCEZ BNET BNFP BNXR BOWU BOYM BRGD BSKE BSKU BSMK BUYN BVXD BWLN KEYB MOSY

90 variants
G-BMFU BMJZ BNII BOOP BPSO BPUJ BROH BROY BRPJ BRZC BSCA BSNJ BSSO BSWX BTBP BTCM BTFU BTHF BTJU BTTB BTTL BTWV BTXF BUAJ BUFJ BUFX BUGY BUIE BUIU BUIZ BUOE BUUO BUVW BVBX BVDX BVFP BVHO BVHR BVMR BVOC BVOP BVPK BVTN BWAU BWBC BWDU BWIP BWJI BWNO BWNS BWPT BWUU BWVU BXAM BXCS BXJO BXVV BYDT BYHC BYIU BYJC BYKX BYMY BYNN BYOK BYOX BYTW BYZX BZFD BZIX BZJH BZKX BZMX BZOX BZRU BZTK BZUU BZXR BZYW BZYY CBAT CBED CBJI CBJK CBKK CBLU CBRV CBUO CCBB CCBT CCEX CCJF CCMN CCNN CCPT CCUJ CCXF COMP CONC CPSF CTEL CXCX DEKA DHLB DIAL DRYS ELLE FBNW FOGG FVEL GLAW GOCX GOGW HBUG IBEV IBLU IGEL IGLE INSR ITOI IWON JMJR JULU LAGR LTSB MANI MFLI MOFZ OJBM OJBW OMEN OXBY PATG PERC PINC PKCC PRIT PRNT RECE RISE RIZE RIZI SBIZ SIAM SLII SORT SRVO STRM SWEB TANK TEDF TEEL TETI TINS TMCC TRIG VRVI WIFI YUMM YUPI YVET

100 series
G-NPWR

105 variants
G-BAVU BMEE BMOV BNFN BOTD BOTK BPBW BPJE BRFR BRLL BRZB BSRD BSNZ BTEA BTFM BTIZ BTKW BTPB BTRL BTOU BUAV BUHU BULD BUPT BUWF BVCA BVEU BVHV BVNR BVUA BVXA BWDH BWEW BWKF BWOW BWPZ BWRY BWSU BXBM BXBR BXBY BXEN BXGC BXXG BXXL BYFJ BYHU BYIL BYKI BYMX BYNX BYPD BZDJ BZKU BZVU BZXO CAMP CBEC CBHW CBMC CBNW CCFN CCGY CCIU CCOT CCVD CCYI CITR CLIC DRGN ELEE ENRY FOWS FUSE GFAB HONK JOSH JSON KSKS LOSI MHBD NPNP NYLB NZGL OAML OJBS OSET OUCH OUVI SAXO SDLW SEDO SEPT SERV SFSL SSTI SUCK TUTU ULTR VSGE WNGS YLYB

120 variants
G-BNEX BOBB BOHL BOZY BPSS BPTX BPZK BRXA BSRD BSYB BTEE BTKN BTUU BTXS BUDV BUFT BURN BVSO BVXF BWAG BWKD BWLD BWYS BXVJ BXWI BYSV CBFF CBMK CBOW CBVV CCEN CCTS CCVZ CCXE FLOA GHIA HOTT LOBO MEUP MOFB OMFG TING VIKY VALZ

133 series
G-BWAA BZVE CBUW

140 variants
G-AZUW BVYU BWTE CCCS DSPK FLTG OXBC

145 series
G-DENT HIBM

160 variants
G- BPCN BPLE CBHD TGAS

165 series
G-BIAZ

180 variants
G-BRTH BRVC BRZI BSWD BSWZ BSYD BTYE BUJR BUKC BVKL BWBR BWHW BXMM OBRY RWHC SKYR

200 series
G-BXOS BZJU CCJG

210 series
G-BTXV BUHY BVBN BWZK BXBA BXJC BXRM BXZG BXZH BYDI BYJV BYMG BYSM BZBE BZXF CBFY CCNV CVBF FLYE JHNY JOJO ORGY RPBM SIAX SKYU SKYX TRAV YTUK

225 series
G-CCBV CCPZ

250 series
G-BUBR BUXE BUXR BUZY BVIG BVYR BWKU BWKX BWZJ BXPK BYHX BYYD BZIK CCGZ HIUP MOLI OBUN OVBF OVIC SCRU SKYY STPI

275 series
G-BWML BXIC BXKJ BXTE BYSK BYZG BZTE BZTT CBYC CBZZ CCNC CCSG CCSJ OOTW SKYK TCAS

300 series
G-BZSU CBAW OXXL SIMI

315 series
G-BZNU CCRH

340 series
G-BZUO KVBF KYBF RANG RVBF

350 series
G-CCOS CCSA

400 series
G-ZVBF

425 series
G-CCGT

SPECIAL SHAPES:

SHAPE	REGISTRATION(S)
ACTION MAN PARACHUTIST	G-RIPS
APPLE	G-BWSO
BALL	G-RNIE
BEER BARREL	G-PINT
BEER CAN	G-IBET
BEER CRATE	G-FGSK
BEETHOVEN BUST	G-BNJU
BELLS WHISKY BOTTLE	G-BUUU
BENIHANA	G-BMVS
BERENTZEN BOTTLE	G-KORN
BERTIE BASSETT	G-BXAL BZTS
BIBENDUM	G-GRIP PNEU TRED
BIERKRUG	G-BXFY
BOWLER	G-OPKF
BRADFORD AND BINGLEY	G-BWMY
BUDWEISER CAN	G-BPFJ
BULB	G-BVWI
BULL	G-CBON
BUS	G-BUSS
CAN	G-BXPR OFIZ
CARROTS	G-BWSP HUCH
CHAMPION SPARK PLUG	G-BETF
CHATEAU DE BALLEROY	G-BKBR BTCZ
CHESTIE	G-USMC
CHICK	G-BYEI
CLUB	G-BWNP
COCA COLA BOTTLE	G-BXSB BYIV BYIW BYIX
COOLING TOWER	G-WATT
COTTAGE	G-COTT
DOUGLAS LURPAK BUTTERMAN	G-BXCK
DUDE	G-OIFM
EAGLE	G-BVMJ
ELEPHANT	G-BLRW BMKX BPRC
FABERGE EGG	G-BNFK
FIRE EXTINGUISHER	G-BZJA
FLAME	G-CBIU
FORBES' MAGAZINE	G-BPOV
FURNESS BUILDING	G-BSIO
GOLFBALL	G-PUTT
GOLLY	G-OLLI
GRAND ILLUSION	G-MAGC
HELIX OILCAN	G-HLIX
HOFMEISTER LAGER BEAR	G-HEYY
HOME SPECIAL	G-BWZP
HORSE	G-HORS
ICE CREAM CONE	G-BZTL
IKEA	G-IKEA
JAGUAR XK8 SPORTS CAR	G-OXKB

KATALOG G-OTTO
KINDERNET DOG G-LKET
KP CHOC DIPS TUB G-DIPI
LIGHTBULB G-BVWH LAMP
LIPS G-LIPS
LOCOMOTION G-LOKO
MACAW G-BRWZ
MINAR E PAKISTAN G-BKNN
MONSTER TRUCK G-BWMU
MUG G-RMUG
N ELE G-WBMG
OIL CAN G-UNIP
OTTI G-OTTI
PERRIER BOTTLE G-PERR
PIG G-HOGS
POT G-CHAM
PRINTER G-BYFK
ROBINSON'S BARLEY WATER G-BKES
RUPERT BEAR G-BTML
RUSSIAN DOLL G-USSR
SAMSUNG COMPUTER G-SEUK
SANTA MARIA SHIP G-BPSP
SATURN G-DREX
SAUCER G-GUFO
SCOTTISH PIPER G-PIPY
SIGN G-UCCC
SONIC THE HEDGEHOG G-SEGA
SPARKASSE BOX G-BXKH
SPHERE G-BVFU BYJW IBBC SATL
SPHINX G-BLFE
STRAWBERRY G-BXTF SAMI
SUGAR BOX G-BZDX BZDY BZKR
TEMPLE G-BMWN
TENNENT'S LAGER GLASS G-BTSL
TRAIN G-BTMY
TRAINER'S SHOE G-BUDN
TRUCK G-BLDL DERV
TV G-TVTV
UNCLE SAM G-USAM
VAN G-ORAC
WITCH G-WYCH
WINE BOX G-STOW

CAMPBELL AIRCRAFT including BENSEN and EVERETT GYROPLANES LTD production
COUGAR
G-BAPS
CRICKET
G-AXPZ AXRC AXVK AXVM AYBX AYCC AYHI AYPZ BHBA BKVS BORG BRLF BSPJ BSRL BTEI BTMP BUIG BVDJ BVIT BVLD BVOH BWSD BWUA BWUZ BXCJ BXHU BXUA BYMO BYMP BZKN CBWN CCPD GYRO KGED RUGS

CANADIAN HOME ROTORS
SAFARI
G-CCKJ

CARLSON
SPARROW
G-BSUX BVVB

CASA see BUCKER (1131), HEINKEL (2111) and JUNKERS (352L)

CASSUTT including MUSSO SPECIAL
RACER
G-BEUN BNJZ BOMB BOXW BPVO BXMF CXDZ FRAY NARO RUNT

CAUDRON
G.III
G-AETA

CCF see HAWKER and NORTH AMERICAN

CEA see JODEL and ROBIN

CENTRAIR
MOTO-DELTA
G-MBPJ

CESSNA AIRCRAFT COMPANY including REIMS AVIATION SA production (F.prefix)
C.165 AIRMASTER
G-BTDE

120/140
G-AHRO AJJS AJJT AKTS AKUR AKVM ALOD ALTO ANGK BHLW BJML BOCI BPHW BPHX BPKO BPUU BPWD BPZB BRJC BRPE BRPF BRPG BRPH BRUN BRXH BSUH BTBW BTEW BTOS BTYW BTYX BUHO BUHZ BUJM BUKO BVUZ BYCD GAWA HALJ JOLY OVFM
150
G-APXY APZR ARFI ARFO ASMS ASMW ASST ASUE ASYP ASZB ASZU ATEF ATHV ATHZ ATKF ATMC ATML ATMM ATMY ATNE ATNL ATRK ATRL ATRM ATRM ATUF ATYM ATZY AVAA AVAR AVCU AVEM AVEN AVER AVGU AVHM AVIA AVIB AVIT AVJE AVMD AVMF AVNC AVUG AVUH AVVL AVVX AVZU AWAW AWAX AWBX AWCP AWES AWFF AWGK AWLA AWOT AWPJ AWPU AWRK AWTJ AWTX AWUG AWUH AWUJ AWUK AWUL AWUN AWUO AWUT AWUU AXGG AXPF AYBD AYGC AYKL AYRF AZLH AZLY AZXC BABB BABC BABH BAEU BAHI BAIK BAIP BAMC BAXU BAXV BAYO BAYP BAZS BBBC BBCI BBDT BBJX BBKA BBKB BBKE BBKY BBNJ BBTZ BCBX BCCC BCRT BCUH BCUJ BCZN BDBU BDFJ BDFZ BDOD BDSL BDTX BDUM BDUO BDZC BEIG BELT BEOK BEWP BFFY BFGW BFIY BFOG BFSR BFVU BFWL BGBI BGOJ BHIY BIFY BIOC BJOV BLVS BMBB BMLX BMXJ BNFI BOBV BOIV BOMN BORY BOTP BOUJ BOUZ BOVS BOVT BOWC BPAB BPAW BPAX BPCJ BPEM BPGZ BPNA BPOS BPWG BPWM BPWN BRBH BRJT BRLR BRNC BRTJ BSBZ BSEJ BSJU BSJZ BSKA BSSB BSYV BSYW BSZU BSZV BTES BTGP BTHE BTIN BTSN BTTE BTYC BUCS BUCT BUGG BURH BWII BWVL BZTE CSBM CSFC DENA DENB DENC DEND ECBH EJMG FAYE FFEN FINA GBLR GCNZ GFLY GLED HFCB HFCI HIVE HULL IANJ IAWE JHAC JWDS LUCK MABE NSTG OIDW OJVH OKED OSTY PHAA PLAN SADE SALL SAMZ SCAT TAIL TEDB UFLY YIII
A150 AEROBAT
G-AXRT AXSW AXUF AYCF AYOZ AYRO AZID AZJY AZOZ AZUZ AZZX BACC BACN BABD BACO BACP BAEP BAEV BAEZ BAII BAIN BAOP BAPI BAPJ BBKF BBKU BBNY BBTB BBXB BCDY BCFR BCKU BCKV BCTU BCUY BCVG BCVH BDAI BDEX BDOW BDRD BEIA BEKN BEMY BEOE BEOY BFGG BFGX BFGZ BFIE BFRR BHRH BIBN BJTB BMEX BOFW BOFX BOYU BPJW BTFS BUCA CLUB FMSG HFCA JAGS OISO OPIC OSND PNIX
152
G-BFEK BFFC BFFE BFFW BFHT BFHU BFHV BFKH BFLU BFOE BFOF BGAA BGAB BGAE BGFX BGGO BGGP BGHI BGIB BGLG BGNT BHAA BHAI BHAV BHCP BHDM BHDR BHDS BHDU BHDW BHEC BHFC BHFI BHHG BHIN BHNA BHPY BHRB BHRM BHRN BHUI BHWA BHWB BHYX BHZH BICG BIDH BIJV BIJW BILR BILS BIOK BITF BIUM BIXH BIZG BJKY BJNF BJVJ BJVT BJWH BJYD BKAZ BKFC BKGW BKTV BKWY BLJO BLWV BLZE BLZH BLZP BMCN BMCV BMFZ BMGG BMJB BMJC BMJD BMMM BMSU BMTA BMTB BMTJ BMVB BMXA BMXB BMXC BMXX BNAJ BNDO BNFR BNFS BNHJ BNHK BNID BNIV BNJB BNJC BNJH BNKC BNKI BNKP BNKR BNKS BNKV BNMC BNMD BNME BNMF BNNR BNOZ BNPY BNPZ BNRK BNRL BNSI BNSM BNSN BNSU BNSV BNUL BNUS BNUT BNXC BNYL BNYN BOAI BODO BOFL BOFM BOGC BOGG BOHI BOHJ BOIO BOIP BOIR BOKY BOLV BOLW BONW BOOI BORI BORJ BORO BOTG BOYL BOZR BPBG BPBJ BPBK BPEO BPFZ BPHT BPIO BPME BPTU BRBP BRND BRNE BRNK BRNN BRPV BRTD BRTP BRUA BSCP BSCZ BSDO BSDP BSFP BSFR BSRC BSTO BSTP BSWH BSZI BSZO BSZW BTAL BTCE BTDW BTFC BTGH BTGR BTGW BTGX BTIK BTVW BTVX BTYT BUEF BUEG BVTM BWEU BWEV BWNB BWNC BWND BXJM BXTB BXVB BXVY BXWC BXYU BYFA BYMH BYMJ BZAD BZAE BZEB BZEC BZWH CCHT CCTW CHIK CPFC CWFY DACF DRAG ENTT ENTW FIGA FIGB HART HFCL HFCT IAFT IBRO IRAN KATT LAMS LSMI MASS OAFT OBEN ODAC OFRY OIMC OLEE OPAM OPJC OSFC OVMC OWAC OWAK OWFS OWOW PFSL POCO RICH SACB SACF SHAH TAYS TFSA WACB WACE WACF WACG WACT
A152 AEROBAT
G-BFGL BFKF BFMK BFRV BFZN BFZT BFZU BGAF BGLN BHAC BHAD BHED BHEN BHMG BILJ BIMT BLAC BLAX BMUO BMYG BOPX BOSO BOYB BRCD BRUM BZEA FIFE FLAP FLIP JEET JONI LEIC MPBH OCPC RLFI TFCI WACH WACU ZOOL
170
G-AORB APVS AWOU BCLS MDAY
172/SKYHAWK
G-ARID ARMO ARMR AROA ARWH ARWO ARWR ARYI ARYK ARYS ASFA ASIB ASMJ ASNW ASOK ASSS ASUP ASVM ASWL ATAF ATFY ATGO ATKT ATKU ATLM ATSL ATWJ AVEC AVHH AVIC AVIS AVJF AVKG AVPI AVTP AVVC AVZV AWGD AWLF AWMP AWMZ AWUX AWUZ AWVA AXBH AXBJ AXDI AXSI AXVB AYCT AYRG AYRT AYUV AZDZ AZJV AZKW AZKZ AZLM AZLV AZTK AZTS AZUM AZXD AZZV BAEO BAEY BAIW BAIX BANX BAOB BAOS BAVB BAXY BAZT BBDH BBJY BBJZ BBKI BBKZ BBNZ BBOA BBTG BBTH BCCD BCHK BCOL BCPK BCRB BCUF BCVJ BCYR BCZM BDNU BDZD BEHV BEMB BENK BEUX BEWR BEZK BEZO BEZR BEZV BFGD BFKB BFMX BFOV BFPH BFPM BFRS BFTH BFZV BGAG BGBR BGHJ BGIU BGIY BGLO BGMP BGND BGRO BGSV BHAW BHCC BHCM BHDX BHDZ BHMI BHPZ BHSB BHUG BHUJ BHVR BHYP BHYR BIBW BIDF BIGJ BIHI BIIB BIIE BING BIOB BITM BIZF BJDE BJDW BJGY BJVM BJWI BJWW BJXZ BKCE BKEP BKEV BKII BKIJ BKLO BLHJ BLVW BMCI BMHS BMIG BMTS BNKD BNKE BNRR BNST BNTP BNXD BNYM BOEN BOHH BOIL BOIX BOIY BOJR BOJS

BOLI BOLX BOLY BOMS BONO BONR BONS BOOL BORW BOUE BOUF BOVG BOYP BPML BPRM BPTL BPVA BPVY BPWS BRAK BRBI BRBJ BRCM BRWX BRZS BSEP BSNG BSOG BSOO BSPE BSTM BTMA BTMR BUAN BUJN BULH BUOJ BURD BUZN BWJP BXGV BXHG BXOI BXSD BXSE BXSM BXSR BXXK BYBD BYEA BYEB BYES BYET BYNA BZBF BZZD BZGH BZPM CBFO CBME CBOR CBXJ CCCC CCTT CDDK CLUX COCO CSCS CURR DCKK DEMH DENR DODD DRAM DRBG DUVL ECGC EGEG ECON ENII ENNK ENOA EOFM ETDC FACE FLOW FNLD FNLY GBFF GBLP GEHL GFMT GRAY GWYN GYAV GZDO HERC HILS ICOM IZSS IZZY JFWI-JMKE JONZ LANE LAVE LENX LICK LSCM MALK MELT MFAC MICK MILA OAKR OBMS OFCM OOLE OPFT OPYE ORMG OSII OSKY OTAM OVFR OZOO PDSI PFCL PLBI RARB ROLY ROOK ROUP RSWO RUIA SACD SBAE SEVE SEXI SHSP SHWK SKAN SKYH TAAL TAIT TASH TOBI TOBY TRIO TYRE UFCB UFCC UFCD UFCE UFCF UFCG UFCH WABH WACL WACM WACW WACY WACZ YFZT YORK YSPY ZACE

172RG CUTLASS
G-BHYC BILU BIXI GKRG PARI

R172 HAWK XP/REIMS FR172 ROCKET
G-AWCN AWDR AWWU AWYB AYGX AYJW BARC BBKG BBXH BCTK BDOE BEZS BFIG BFIU BFSS BHYD BLPF BPCI BPWR BTMK BZVB CCSB DAVD DIVA DRID EDTO EFBP EKKC FANL JANS LOYA MFEF OMAC PJTM RABA ROKT STAY THIN YBAA YIPI

175/SKYLARK
G-ARCV ARFG ARML ARMN AROC ARRI ARUZ ARWS OTOW

177(RG) CARDINAL
G-AYPG AYPH AYSX AYSY AZTF AZTW AZVP BAGN BAIS BAJA BAJB BAJE BBHI BBJV BEBN BFIV BFMH BPSL BRDO BRPS BTSZ BUJE FIJJ FNEY GBFR LNYS OAMP OSFS TOTO

180/SKYWAGON
G-ARAT ASIT AXZO BETG BNCS BOIA BTSM BUPG CCYK CIBO DAPH

182/SKYLANE variants
G-ARAW ASLH ASSF ASXZ ATCX ATLA ATPT ATTD AVCV AVDA AVGY AVID AWJA AXNX AXZU AYOW AYWD AZNO BAAT BAFL BAHD BAHX BAMJ BBGX BBYH BBYS BCWB BDBJ BDIG BEKO BFOD BFSA BFZD BGAJ BGFH BGPA BHDP BHIB BHIC BHVP BHYA BJVH BKHJ BKKN BKKO BMMK BMUD BNMO BNRY BOPH BOTH BOWO BPUM BRKR BRRK BSDW BSRR BTHA BUVO BWMC BWRR BXEZ BXZM BYEM BZVF BZVO CCAN CBIL CBMP CBVX CCYS DATG DOVE DRGS EEZS EIRE EIWT ENRM EOHL ESME FAUX GBUN GCYC GHOW GUMS HRNT HUFF IATU IBZS IFAB IOPT IRPC ISEH IZZI JBRN JENI JOON KEMY KWAX LEGG LVES MICI MILN MISH MOUT NLEE NOCK NYZS OBBO OHAC OJRM OLDG OPCG OPST ORAY OTRG OWCS OWRT OZOI PDHJ PLEE POWL PUGS RACY ROWE SAAM SKYL THRE TPSL VIPA WARP WHDP WIFE WMLT XLTG ZBLT

185 SKYWAGON/AG CARRYALL
G-AYNN BBEX BDKC BKPC BLOS BWWF BXRH BYBP RNRM

(T)188 AG WAGON/AG TRUCK/AG HUSKY
G-AZZG

190/195
G-BSPK BTBJ

205
G-ASNK ASOX

206 SUPER SKYLANE/SUPER SKYWAGON/STATIONAIR
G-ASVN ATCE ATLT AYCJ AZRZ BAGV BFCT BGWR BMHC BMOF BOFD BPGE BSMB BSUE BXDB BXRO BYIC CCSN CTFF DROP EESE GNMG LIPA OLLS OSSA PEPA SEAI SKYE STAT

207 SKYWAGON/STATIONAIR 8
G-NJAG PARA

208/B (GRAND) CARAVAN
G-BZAH EELS EORD ETHY OCIT WIKY

210 CENTURION
G-ASXR BEYV BNZM BSGT DECK MANT MPRL OFLY PIIX SEEK TOTN VMDE

T303 CRUSADER
G-BSPF CRUZ CYLS DOLY GAME IKAP INDC JUIN OAPE PTWB PUSI ROCH UILT

305 BIRD DOG (L-19)
G-PDOG

310
G-APNJ AXLG AYGB AYND AZRR AZUY BALN BARG BARV BBBX BBXL BCTJ BGTT BIFA BJMR BKSB BMMC BODY BPIL BRIA BTFF BTYK BWYE BWYG BWYH BXUY BXYG EGEE EGLT FFWD FISH IMLI MIWS MPBI ODLY REDB REDD RIST RODD SOUL TKPZ TROP VDIR

335
G-FITZ

336 SKYMASTER
G-PIXS

337 SUPER SKYMASTER
G-ATCU ATSM AXHA AZKO BCBZ BFGH BFJR BMJR BOYR BTVV HIVA RGEN

340
G-BVES CCXJ FEBE HAFG LIZA LUND OPLB PUFN REEN SAMM

401/402
G-AVKN AWWW AZFR AZRD BXJA DACC EYES MAPP NOSE OVNE ROAR

404 TITAN/AMBASSADOR/COURIER
G-BWLF BYLR EXEX MIND OOSI TASK TVIP

406 CARAVAN II
G-BVJT LEAF MAFA MAFB SFPA SFPB TURF TWIG

414/CHANCELLOR
G-DYNE SMJJ

421/GOLDEN EAGLE
G-BAGO BBUJ BDZU BFTT BHKJ BKNA BLST CSNA CJEA FTAX GILT HASI HIJK JACK KWLI MUVG OSCH PVIP TAMY TREC UVIP VVIP

425/441 CONQUEST I/II
G-BNDY FPLC FRAX

500/501 CITATION 1
G-DJAE JTNC LOFT

525 CITATIONJET
G-BVCM CJAD HMMV IUAN SFCJ ZIZI

550/CITATION BRAVO/551 CITATION II
G-BJIR CBTU EJEL ESTA FIRM FCDB FJET FLVU GHPG IDAB IKOS IPAL JETC JETJ JMDW ORDB RDBS SPUR VUEA

560/XL CITATION EXCEL/ULTRA
G-CBRG CFGL CJAE IAMS IPAX KDMA NETA OGRG REDS SIRS TTFN WINA XLMB

650 CITATION III
G-HNRY

750 CITATION X
G-CDCX

CFM METAL-FAX including CFM AIRCRAFT LTD

IMAGE
G-BTUD

SHADOW SERIES B/BD (Rotax 447)
G-MNZP MNZR MTGN MTHV MTTH MTWL MVBB MVCW MVIG MVRO MVRS MVRT MVYZ MWJF MYKE

SHADOW SERIES C/CD (Rotax 503)
G-BZLF CAIN MGUY MJVF MNCM MNER MNIS MNTK MNTP MNVJ MNVK MNWK MNWY MNXX MNZJ MNZZ MTBE MTCA MTCT MTDU MTDX MTFZ MTGV MTGW MTHT MTKR MTMX MTMY MTSG MTWH MTWK MTXR MVAC MVAM MVAN MVCC MVEI MVEN MVFH MVHD MVLJ MVLP MVOH MVPK MVRE MVRP MVRR MVVT MWAE MWDB MWDN MWEN MWEZ MWFB MWIZ MWJZ MWLD MWMU MWON MWPN MWRL MWRY MWSZ MWTJ MWTN MWTP MWUA MWVG MWVH MWYD MYBC MYBL MYCM MYDD MYDE MYEP MYGO MYIF MYIP MYLV MYNA MYOH MYON MYOS MYPL MYPT MYSM MYTH MYUS MYWF MYWM MYXY MZBN MZBS MZCT MZRS MZLP PSUE THAI

SHADOW SERIES DD (Rotax 582)
G-BXZY BYCJ CCMW DARK DMWW LYNK MGTW MYZP MZGS MZKH MZNH MZOM ODVB PBEL

STREAK SHADOW
G-BONP BROI BRSO BRWP BRZZ BSMN BSOR BSPL BSRX BSSV BTDD BTEL BTGT BTKP BTZZ BUGM BUIL BULJ BUOB BUTB BUVX BUWR BUXC BVDT BVFR BVLF BVOR BVPY BVTD BWAI BWCA BWOZ BWPS BXFK BXVD BXWR BYAZ BYFI BYOO BZEZ BZMZ BZWJ BZWY CBCZ CBGI CBNO CZBE DOTT ENEE FAME GORE HLCF MEOW MGGT MGPH MWPP MYNX MYTY OLGA OPIT ORAF OTCH RINT ROTS SHIM SNEV STRK TEHL TTOY WESX WHOG WYAT

CGS including ARROWFLIGHT AVIATION LTD

HAWK
G-MWYS MYTP MZGU

CHAMPION AIRCRAFT CORPORATION- see BELLANCA

7AC/7DC CHAMPION
G-AJON AKTO AKTR AOEH ATHK AVDT AWVN BGWV BPFM BPGK BRAR BRCV BRER BRFI BRWA BRXG BTGM BTNO BVCS HAMP JTYE LEVI OTOE TECC

7BCM (L-16)
G-BFAF TIMP

7FC TRI-TRAVELER
G-APYT ARAS

CITABRIA
G-BFHP BIZW BPMM BSLW BUGE

CHANCE-VOUGHT see VOUGHT

CHARGUS GLIDING CO LTD see HIWAY

T.250
G-MBJG

TITAN 38
G-MYZH

CHICHESTER-MILES
LEOPARD
G-BRNM

CHILTON AIRCRAFT
DW.1/1A/1B/2
G-AESZ AFGH AFGI AFSV AFSW BWGJ DWIA DWIB

CHOWN
OSPREY MLB variants
G-BJID BJND BJNH BJPL BJRA BJRG

CHRISLEA AIRCRAFT CO LTD
LC.1 AIRGUARD
G-AFIN
CH.3 SUPER ACE
G-AKUW AKVD AKVF
CH.3 series4 SKYJEEP
G-AKVR

CHRISTEN INDUSTRIES INC including AVIAT AIRCRAFT INC and see PITTS
EAGLE
G-CCYO EEGL EGAL EGEL EGLE EGUL ELKA IXII OEGL
A-1/1B HUSKY
G-BUVR HAIB HUSK OGGY USKY WATR

CHRIS TENA
MINI COUPE
G-BPDJ

CIERVA
C.19 (AVRO 620)
G-ABXH
C.24
G-ABLM
C.30A (AVRO 671)
G-ACUU ACWM ACWP ACXA AHMJ

CIRRUS DESIGN CORPORATION
SR20
G-OPSS

CIVILIAN AIRCRAFT CO
CAC.1 COUPE
G-ABNT

CLUTTON
FRED (Volkswagen 1834)
G-BBBW BDBF BGFF BGHZ BISG BITK BKAF BKDP BKEY BKVF BKZT BLNO BMAX BMMF BMSL BNZR BOLS BVCO BWAP BYLA MANX OLVR ORAS PFAF PFAL RONW USTY

COLT BALLOONS LTD including CAMERON BALLOONS, THUNDER and COLT BALLOONS production and see COLTING
Gas Airship:
GA-42
G-MATS ZEPI
Hot Air Airship:
AS-42 (Cuyana 430R)
G-WZZZ
AS-56
G-BNKF BTXH NOVO
AS-80 (Rotax 462)
G-BPCF BPGT BPKN
AS-105 (Rotax 462)
G-BTSW BUKV BWKE BWMV BXEY BXNV BXYF RBOS OVAX
AS-120 (Rotax 582)
G-BXKU BZWF
AS-261
G-BPLD
Hot Air Balloon:
12A Cloudhopper
G-BHOJ
14A Cloudhopper
G-BHKN BHKR BHPN BVKX
17A Cloudhopper
G-BIDV BIYT BJWV BKBO BKXM BLHI BONV BPXH HELP HEXE
21A Cloudhopper
G-BLVY BLXG BMKI BNFM BOLN BOLP BOLR BPFX BSAK BSIG BTNN BUEU BWBJ LLAI MLGL SOOS
H-24
G-BZUV
25A Sky Chariot
G-BSOF BUPH BVAO OKBT
31A Air Chair
G-BHIG BLOB BROJ BSDV BSMM BVTL BXXU DNGR DHLZ DOWN HOUS IMAN MUTE PIXE PONY
42 variants
G-BJZR BVHP
56 variants
G-BGIP BHEX BHRY BICM BISX BIXW BJXP BJYF BKSD BLCH BLLW BLOT BTZY BUGO BVOZ BVUC BVYL CCYP CFBI EZXO ILEE FZZZ MERC POSH TTWO WIMP
69 series
G-BLEB BLUE BOVW BPAH BSHC BSHD BTMO BVDD COLR FZZY JBJB OABC OBUD ODIY TCAN TWEY
77 variants
G-BGOD BIGT BKOW BLTA BOCF BOGT BOHD BORA BORE BORT BPEZ BPFB BPJK BRLT BRVF BRVU BSCI BSUB BSUK BTDS BTTS BTXB BTZR BTZS BUJH BUKS BULF BURG BUVE BUVS BUVT BUYO BUZF BVAX BXFN BXIE BYFX CHEL CURE DING DRAW DURX EZVS FLAG GGOW GOBT HOME HOTI HOTZ HRZN IMAG JONO LSHI MAUK MKAK OAWS OCAR ODAD OLPG ONCL ORON OSST READ SGAS SIXX STOK TRUX UPPP UZLE WHAT WHIM WOOL WRIT
90 series
G-BPUW BRHG BRRU BSIU BTCS BTMH BTPV BVEI BXUW FOWL IRLY JNNB OBBC OMMM PEGG PHSI SAUF SEND TOFT
105 series
G-BGAS BLHK BLMZ BMBS BNAG BPZS BRUH BSCC BSHS BSNU BTHX BURL BUSV BWMA BWRM BXOW BYIO DYNG HSHS PLUG RAIL TIKI
120 series
G-BXAI BXCO BYDJ BYPV BZIL BZNF CBEJ OBIB VYGR
180 series
G-BMXM BOGR BONK CUCU
210 series
G-BTYZ BULN BUXA BVFY BZYO
240 series
G-BNAP IGLA LCIO
260 series
G-HUGO
300 series
G-RAPE
315 series
G-CCIE VVBF
2500 series
G-USUK

SPECIAL SHAPES:

SHAPE	REGISTRATION(S)
AGFA FILM CASSETTE	G-OHDC
APPLE	G-BRZV
ARIEL BOTTLE	G-BNHN
BEER GLASS	G-BNHL PUBS
BLACK KNIGHT	G-BNMI
BOTTLE	G-BOTL
BUDWEISER CAN	G-BUET BVIO
CLOWN	G-GWIZ
DRACHENFISCH	G-BMUJ
EGG	G-BWWL
FINANCIAL TIMES	G-ETFT FTFT
FIRE EXTINGUISHER	G-CHUB
FLAME	G-BLKU
FLYING MITT	G-WCAT
FLYING YACHT	G-BXXJ
GAS FLAME	G-BGOO
HOP	G-MALT
HUT	G-SMTC
ICE CREAM CONE	G-BWBE BWBF OJHB
J and B WHISKY BOTTLE	G-JANB
JUMBO	G-OVAA
KINDERMOND	G-BMUL
MAXWELL HOUSE COFFEE JARS	G-BVBJ BVBK
MICKEY MOUSE	G-BTRB
OLD PARR WHISKY BOTTLE	G-PARR
PANASONIC BATTERY	G-PSON
PIG	G-BUZS
PIGGY BANK	G-BXVW
SANTA CLAUS	G-HOHO
SATZENBRAU BOTTLE	G-BIRE
SHUTTLECOCK	G-OOUT
STORK	G-BRGP
UFO	G-BMUK
WORLD	G-DHLI

COLOMBAN
MC-12/15 CRI-CRI
G-BOUT BWFO CRIC CRIK MCXV OCRI SHOG

COMCO IKARUS GmbH including FLYBUY ULTRALIGHTS Ltd production
IKARUS C42 variants
G-CBFV CBGP CBIJ CBJW CBKU CBOD CBPD CBRF CBTG CBVY CBXC CCCT CCFZ CCLS CCNT CCPS CCYR CCZL CDBU CDCG CDCM CDCO DTOY DUGE EGGI FBII FIFT FLYB FROM GNJW GRPA GSCV HIJN IBAZ ICRS IKRS IKUS ILRS INJA MGPA MROY MSKY NPPL NSBB OAJL OFBU OJDS OOMW OVAL OVLA RBSN ROZZ SGEN SIMM

COMMANDER AIRCRAFT COMPANY see ROCKWELL

COMMONWEALTH see NORTH AMERICAN

COMPER AIRCRAFT CO
CLA.7 SWIFT
G-ABTC ABUS ABUU ACTF LCGL

CONSOLIDATED VULTEE AIRCRAFT see STINSON AIRCRAFT CORPORATION
CV.32-3 LIBERATOR
G-AGZI
PBY-5A CATALINA
G-PBYA

CONVAIR see GENERAL DYNAMICS CORPORATION

COOK
ARIES P
G-MYXI

COPE
BUG
G-BXTV

CORBEN
BABY ACE
G-BTSB BUAA
JUNIOR ACE
G-BSDI

CORBETT FARMS LTD see MICROFLIGHT AIRCRAFT LTD

CORBY
CJ-1 STARLET
G-BVVZ CBHP CCHN CCXO ILSE

COSMOS
TRIKE
G-MVCK

ETS **COUESNON** - see JODEL

COUGAR
(Wing)
G-MMUJ

CRANFIELD
A.1-400 EAGLE
G-COAI

CREMER
MLB variants
G-BJLX BJLY BJRP BJRR BJRV

CROSBY see ANDREASSON

CSS see POLIKARPOV

CUB PROSPECTOR see PIPER

CULVER
LCA CADET
G-CDET

CURRIE including TURNER
WOT
G-APNT ARZW ASBA AVEY AYMP AYNA BANV BDFB BFAH BFWD BKCN BLPB BXMX CWBM CWOT PFAP SWOT

CURTISS-WRIGHT CORPORATION
TRAVEL AIR 12Q
G-AAOK
P-40 KITTYHAWK/TOMAHAWK
G-KITT TOMA

CURTISS ROBERTSON
ROBIN C.2
G-BTYY HFBM

CUSTOMCRAFT BALLOON SERVICES
Hot Air Balloon:
A25
G-CCKZ DUMP

CVJETKOVIC
CA-65 SKYFLY HAB
G-BWBG

CYCLONE AIRSPORTS LTD see PEGASUS AVIATION and SOLAR WINGS (AVIATION)
70 trike with Aerial Arts 110SX/130SX wings (Rotax 377/277)
G-MMYL-MNMY
AX3
G-BVJG MYFI MYFV MYFW MYFY MYFZ MYGD MYHG MYHH MYHJ MYHM MYHR MYIJ MYIU MYKA MYKF MYKT MYME MYMF MYMW MYMZ MYOY MYPM MYPR MYRO MYRU MYRV MYSO MYTM MYUI MYVN MYXH MYYL MYZC MYZF MYZG MZDO MZDS ZELE
AX2000
G-BYJM CBGS CBHC CBIT CBMB CBUX JONY MGUN MYER MZER MZFA MZFX MZGA MZGB MZGC MZGM MZGP MZHR MZIV MZJF MZJL MZJR MZKC MZLS MZLU MZMX MZOE OAJB ROMW STRG WAKY YROO

CZAL
AERO 45/145
G-APRR ATBH AYLZ

DAN RIHN
DR.107 ONE DESIGN
G-IDDI IIID LOAD RIHN

DART AIRCRAFT LTD
KITTEN
G-AEXT

DASSAULT
FALCON 20/200
G-FFRA FRAF FRAH FRAI FRAJ FRAK FRAL FRAM FRAO FRAP FRAR FRAS FRAT FRAU FRAW FRBA
FALCON 900
G-CBHT DAEX HMEI JCBG JJMX JPSX LYCA OPWH RBSG
FALCON 2000
G-GEDY

DAVIS
DA-2
G-BPFL

DE HAVILLAND AIRCRAFT LTD including AMERICAN MOTH CORPORATION, F + W, MORANE-SAULNIER, MORRIS MOTORS and OGMA production - also see AIRCO and HAWKER SIDDELEY AVIATION
DH.51
G-EBIR
DH.53 HUMMING BIRD
G-EBHX EBQP
DH.60/60G/60M/60MW/60X MOTH
G-EBLV EBNO EBWD EBZN
G-AAAH AADR AAEG AAHI AAHY AAMX AALY AAMY AANF AANL AANO AANV AAOR AAWO AAXG AAZG ABAG ABDX ABEV ABSD ABYA ATBL
DH.60GIII MOTH MAJOR
G-ABZB ACGZ ACNS ACXB ADHD BVNG
DH.71 TIGER MOTH
G-ECDX
DH.80A PUSS MOTH
G-AATC AAZP ABHC ABIH ABLS AEOA AHLO FAVC
DH.82A TIGER MOTH
G-ABUL ACDA ACDC ACDI ACDJ ACMD ADCG ADGT ADGV ADGZ ADIA ADJJ ADLV ADNZ ADPC ADWJ ADWO ADXT AFGZ AFSH AFVE AFWI AGEG AGHY AGNJ AGPK AGYU AGZZ AHAN AHIZ AHLT AHOO AHPZ AHUF AHUV AHVU AHVV AIDS AIRI AIRK AIXJ AJHR AJHS AJHU AJOA AJTW AJVE AKEE AKUE AKXS ALAD ALBD ALIW ALJL ALNA ALND ALRI ALUC ALVP ALWS ALWW ALZA AMBB AMCK AMCM AMEX AMHF AMIU

AMJD AMMK AMNN AMOU AMPM AMPN AMRM AMTK AMTL AMTP AMTV AMVS ANCS ANCX ANCY ANDM ANDP ANEF ANEH ANEJ ANEL ANEM ANEN ANEW ANEZ ANFC ANFI ANFL ANFM ANFP ANFV ANFW ANHK ANJA ANJD ANJG ANKK ANKN ANKT ANKV ANKZ ANLD ANLS ANMO ANMV ANMY ANNB ANNE ANNG ANNI ANNK ANOH ANOM ANON ANOO ANOS ANPE ANPK ANRF ANRM ANRN ANRX ANSE ANSJ ANSM ANSU ANTE ANTV ANZU ANZZ AOAA AOAF AOBH AOBO AOBX AODR AODT AOEI AOEL AOES AOET AOFR AOGI AOGR AOHY AOIL AOIM AOIS AOJJ AOJK AOJX AOUR AOYU AOZH APAL APAM APAO APAP APBI APCC APCU APFU APGL APGM APIH APJO APJP APLI APLR APLU APMX APOU APPN ARAZ AREH ARTL ASKP ASPV AVPJ AXAN AXBW AXBZ AXTY AXXV AYDI AYIT AYUX AZDY AZGZ AZZZ BABA BACK BAFG BBRB BEWN BFHH BHLT BHUM BJAP BJZF BMPY BNDW BPAJ BPHR BRHW BTOG BWIK BWMK BWMS BWVT BYTN DHTM DHZF EMSY ERDS ISIS MOTH OOSY PWBE TIGA YVFS

DH.82B QUEEN BEE
G-BLUZ

DH.83/83C FOX MOTH
G-ACCB ACEJ ADHA AOJH

DH.84 DRAGON
G-ACET ACIT AJKF ECAN

DH.85 LEOPARD MOTH
G-ACLL ACMA ACMN ACOJ ACTH ACUS AIYS ATFU

DH.87B HORNET MOTH
G-ADKC ADKK ADKL ADKM ADLY ADMT ADND ADNE ADOT ADRH ADSK ADUR AELO AESE AHBL AHBM AMZO

DH.88 COMET
G-ACSP ACSS

DH.89A DRAGON RAPIDE
G-ACPP ACYR ACZE ADAH AEML AERN AGJG AGSH AGTM AHAG AHED AHGD AHXW AIDL AIWY AIYR AJBJ AJCL AKDW AKGV AKIF AKNV AKOE AKRP ALGB ALXT ALZF AOAO

DH.90 DRAGONFLY
G-AEDT AEDU

DH.94 MOTH MINOR
G-AFNG AFNI AFNJ AFOB AFOJ AFON AFOR AFOW AFPN AFUU

DH.98 MOSQUITO
G-ASKB ASKC AWJV MOSI

DH.100 VAMPIRE FIGHTER BOMBER
G-BVPD DHXX FBIX MKVI

DH.104 DOVE/(SEA) DEVON
G-AHRI ALCU ALFT ALFU AMDD AMXS AMXT AMXX ANOV ANUW ANVU AOUF APSO ARBE ARDE AREA ARHW ARHX ARJB AROI ARUE ATGJ ATGK AWFM BAHB BLPZ BLRN BVXR DHDV DVEN DVON HBBC KOOL OEWA OPLC RNAS SDEV

DH.106 COMET
G-AOJT AOVU APAS APDB APDF APYD BDIW BDIX BEEX CPDA

DH.110 SEA VIXEN
G-CVIX VIXN

DH.112 VENOM
G-BLID BLKA BLSD DHSS DHTT DHUU DHVM VENI VENM VICI VNOM

DH.114 HERON
G-ALZL ANPV ANUO ANXB AOGW AORG AOTI AOXL APRK APXG ASCX ASFI HRON

DH.115 VAMPIRE TRAINER
G-BZRC BZRD BVLM DHVV DHWW DHZZ DUSK HELV OBLN SPDR VMPR VTII

DH.121 TRIDENT – see HAWKER SIDDELEY

DH.125 – see HAWKER SIDDELEY

DE HAVILLAND (AUSTRALIA)

DHA.3 DROVER
G-APXX

DE HAVILLAND (CANADA) including BOMBARDIER INC and OGMA production

DHC-1 CHIPMUNK
G-AKDN ALWB AMMA AMUF AMUH ANWB AOFE AOJR AORW AOSF AOSK AOSL AOSP AOSU AOSW AOSY AOTD AOTF AOTR AOTY AOUO AOUP AOZP APLO APPA APPM APTS APYG ARGG ARMC ARMD ARMF ARMG ARWB ATHD ATVF BAPB BARS BAVH BBMN BBMO BBMR BBMT BBMV BBMW BBMX BBMZ BBNA BBNC BBND BBRV BBSS BCAH BCCX BCEY BCGC BCHL BCHV BCIH BCKN BCOI BCOO BCOU BCOY BCPU BCRX BCSA BCSL BCXN BCYM BCZH BDCC BDDD BDEU BDRJ BFAW BFAX BFDC BHRD BNZC BPAL BTWF BVTX BVWP BVZZ BWHI BWJY BWMX BWNK BWNT BWOX BWTG BWTO BWUN BWUT BWUV BWVY BWVZ BXCP BXCT BXCV BXDA BXDG BXDH BXDI BXDM BXDP BXEC BXGL BXGM BXGO BXGP BXGX BXHA BXHF BXIA BXIM BXNN BYHL BYSJ BYYW BZDU BZGA BZGB BZXE CBJG CHPY CPMK DHCC DHPM HAPY HDAE ITWB MAJR OACP PVET TRIC ULAS

DHC-2 BEAVER
G-ARTR BUCJ BVER

DHC-6 TWIN OTTER
G-BIHO BVVK BZFP CBML

DHC-8 DASH EIGHT variants
G-BRYI BRYJ BRYU BRYV BRYW BRYY BRYX BRYZ JECE JECF JECG JEDE JEDF JEDI JEDJ JEDK JEDL JEDM JEDN JEDO JEDP JEDR JEDT JEDU JEDV JEDW NVSA NVSB WOWA WOWB WOWC

DEMON see HIWAY

DENNEY AEROCRAFT COMPANY including SKYSTAR

KITFOX (Rotax 582)
G-BNYX BONY BPII BPKK BRCT BSAZ BSCG BSCH BSCM BSFX BSFY BSGG BSHK BSIF BSIK BSMO BSRT BSSF BSUZ BSVK BTAT BTBG BTBN BTDC BTDN BTIF BTIR BTKD BTMT BTMX BTNR BTOL BTSV BTTY BTVC BTWB BUDR BUIC BUIP BUKF BUKP BULZ BUNM BUOL BUPW BUWS BUYK BUZA BVAH BVCT BVEY BVGO BWAR BWHV BWSJ BWSN BWWZ BWYI BXCW BXWH BZAR BZLO CBDI CBTX CJUD CRES CTOY DJNH ELIZ EYAS FOXC FOXD FOXF FOXG FOXI FOXS FOXX FOXZ FSHA HOBO HUTT KAWA KFOX KITF KITY LACR LEED LESJ LESZ LOST OFOX OPDS PHYL PPPP RAYA ROXY RWSS RSSF RWSS TFOX TOMZ TWTW

DEPERDUSSIN CIE

MONOPLANE
G-AANH

DESIGNABILITY see JORDAN

DESOUTTER AIRCRAFT COMPANY

DESOUTTER 1
G-AAPZ

CONSTRUCTIONS AÉRONAUTIQUES EMILE DEWOITINE including EKW

D.26
G-BBMI

DIAMOND AIRCRAFT INDUSTRIES

DA.20 KATANA
G-BWTA BXGH BXJV BXJW BXMZ BXOF BXPC BXPD BXPE BXTP BXTS BYFL BYMB IKAT OBDA RIBS

DA.40(D) STAR
G-CBFA CBFB CCFP CCFR CCFS CCFU CCHA CCHB CCHC CCHD CCHE CCHF CCHG CCHK CCKH CCKI CCLB CCLC CCLV CCLW CCLZ CCMF CCPX CCUS CCXU CCZU CDEJ CDEK CDEL EMDM HASO JKMF MOPB OPHR OTDI SFLY SOHO WBVS ZANY

DORNIER including AG fur DORNIER-FLUGZEUGE, DORNIER GmbH, DORNIER-WERKE AG, DORNIER LUFTFAHRT GmbH and see BUCKER

DO.27
G-BMFG BNMK

DO.28 SKYSERVANT
G-ASUR BWCO

228
G-MAFE MAFI OMAF

328
G-BWIR BWWT BYHG BYMK BYML BYTY BZOG CCGS

DOUGLAS AIRCRAFT COMPANY INC including DOUGLAS AIRCRAFT CORPORATION and see McDONNELL DOUGLAS

AD-4 SKYRAIDER
G-BMFB RADR

DC-3/C-47 DAKOTA/SKYTRAIN
G-AGIP AHLX AIBA AIYT AJAV AJAZ AKLL ALWC AMCA AMHJ AMJX AMNL AMPO AMPP AMPY AMRA AMSM AMSN AMSV AMYJ AMZH AMZW ANAF APML BGCF BHUB BLFL BLYA BPMP BVOL DAKK DAKS

DC-6
G-APSA SIXC

DC-7C
G-AOIE

DRAGON BALLOONS

G77
G-BKRZ

DRAGON LIGHT AIRCRAFT CO LTD

DRAGON series150/200/250
G-MJSL MJVY MMAC MMAE MMAI MMML MMNH MMPR MNJF

DRAYTON BALLOONS

DRAYTON B-56 HAB
G-BITS

DRUINE including ROLLASON AIRCRAFT and ENGINES LTD (D.31/D.62) production

D.5 TURBI
G-AOTK APBO APFA

D.31 TURBULENT
G-AJCP APIZ APNZ APOL APTZ APUY APVN APVZ APWP ARBZ ARGZ ARIM ARLZ ARMZ ARNZ ARRU ARRZ ASFX ASHT ASMM ASPU ASSY ASTA ATBS AWBM AWDO AWMR AWWT BFXG BGBF EGMA BKXR BLTC BUKH BVLU BWID OJJF
D.62 CONDOR (Continental O-200-A)
G-ARHZ ARVZ ASEU ASRB ASRC ATAU ATAV ATOH ATUG ATVW AVAW AVEX AVJH AVMB AVOH AVXW AWAT AWEI AWFN AWFO AWFP AWSN AWSP AWSS AWST AXGS AXGV AXGZ AYFC AYFD AYFE AYFF AYFG AYZS BADM BUOF OPJH YNOT

DYKE
DELTA
G-DYKE

SOCIÉTÉ **DYN'AÉRO**
CR100
G-BZGY
MCR-01 CLUB
G-BYEZ BYTM BZXG CBNL CBZX CCMM CCFG CCPN CCTE CCWH CDBY CDGG CUTE DECO DGHI KARK LARS LMLV PGAC POOP RESG TBEE TDVB TOOT

EAA
ACROSPORT
G-BJHK BKCV BLCI BPGH BSHY BTAK BTWI BVVL CCFX NEGG OJDA TANY TSOL VCIO
BIPLANE
G-ATEP AVZW BBMH BPUA BRUU PFAA PFAY

EAGLE see AMERICAN AEROLIGHTS

EAVES
DODO MLB
G-BJIC
EUROPEAN MLB variants
G-FYDN FYFI

ECLIPSE
SUPER EAGLE
G-BGWZ

EDGAR PERCIVAL including LANCASHIRE AIRCRAFT COMPANY production
EP.9 PROSPECTOR
G-APAD APXW ARDG

EDWARDS
GYROCOPTER
G-ASDF

EH INDUSTRIES
EH-101
G-EHIL

EIPPER AIRCRAFT INC
QUICKSILVER
G-MBBM MBCK MBFO MBYM MJAM MJBT MJDW MJHX MJIR MJJK MJPV MJUO MJVP MJVU MJZL MMAH MMBU MMMG MMNA MMNB MMNC MMWC MMYR MNCO MTDO MWDZ

EIRI.EINO RHIELA KY
PIK 20E
G-BHFR BHNP OPIK POPE

EKW see DEWOITINE and DORNIER
C-3605
G-CCYZ DORN

ELECTRA FLYING CORPORATION
EAGLE
MBRB

ELLIOTTS OF NEWBURY
EoN AP.10 460
G-APWL

ELMWOOD
CA-05 CHRISTAVIA
G-MRED

EMBRAER
EMB-110 BANDEIRANTE
G-BGYT FLTY TABS
EMB-135 variants
G-CDFS RJXL
EMB-145 variants
G-CCYH EMBC EMBD EMBE EMBF EMBG EMBH EMBI EMBJ EMBK EMBL EMBM EMBN EMBO EMBP EMBS EMBT EMBU EMBV EMBW EMBX EMBY ERJA ERJB ERJC ERJD ERJE ERJF ERJG RJXA RJXB RJXC RJXD RJXE RJXF RJXG RJXH RJXI RJXJ RJXK RJXL RJXM
EMB-312 TUCANO
G-BTUC

ENGLISH ELECTRIC CO LTD including AVRO production
WREN
G-EBNV
CANBERRA
G-BVWC BVXC
LIGHTNING
G-AXEE FSIX LNTG OPIB

ENSTROM HELICOPTER CORPORATION
F-28
G-BBHE BBIH BBPN BBPO BDKD BHAX BONG BRZG BSHZ BURI BWOV BXLW BXXB BXXW BYKF BZHI MHCE MHCJ OABO SERA WSEC
280 SHARK
G-BEYA BGWS BIBJ BPXE BSDZ BSLV BXEE BXFD BXRD BYSW CBYL CKCK COLL GKAT HDIX HYST IDUP MHCB MHCD MHCF MHCG MHCI MHCK MHCL OGES OITV OJBB OJMF ONUP OTHE PALS PBYY SHAA SHRK SHUU SOPP VETS VRTX VVWW WIZI WRFM WSKY ZZWW
480
G-GUAY HADA IGHH IJBB JPTT LADD OSKP OZAR PBTT RIBZ TRUD UZZY WOOF

ERCO including ALON and FORNEY production
ERCOUPE 415
G-ARHB ARHC AROO AVIL AVTT BZKS BZNO COUP EGHB ERCO HARY ONHH

EUROCOPTER see AÉROSPATIALE, SUD-AVIATION and MBB
EC120 COLIBRI
G-BXYD CBNB DEVL EIZO FEDA IBRI IGPW ISSY MODE PDGE RCNB SCUR TBLY WUSH ZZOE

EUROCOPTER DEUTSCHLAND GmbH
EC135
G-BZRS CCAU CHSU CPSH DAAT EMAS ESEX ETHU EWRT FEES IWRC KRNW LASU NEAU NESV NMID NWPS SASA SASB SPHU SSXX SUFF SURY TAGG WCAO WMAS XMII

EUROPA AVIATION
EUROPA (Rotax 912)
G-BVGF BVIZ BVJN BVKF BVLV BVOS BVOW BVRA BVUV BVVH BVVP BVWM BWCV BWDP BWDX BWEG BWFH BWFX BWGH BWIJ BWIV BWJH BWKG BWNL BWON BWRO BWUP BWVS BWWB BWYD BWRA BWZT BXCH BXDY BXEF BXFG BXGG BXHY BXII BXIJ BXLK BXNC BXOB BXTD BXUM BYFG BYIK BYPM BYJI BYSA BZAM BZNY BZTH BZTI BZTN CBES CBHI CBOF CBWF CBWP CBXW CBYN CCEF CCFK CCGV CCJX CCOV CCUY CDBX CDEX CDGH CERI CHAH CHAV CHEB CHET CHOX CHUG CLAV COPY CROB CROY CUTY DAMY DAYI DAYS DDBD DEBR DLCB DONZ DRMM EENI EESA EIKY EMIN EMSI EOFS EORJ EURX EXES EZZA FELL FIZY FLOR FLOX FLRT FLYT FOGI GCAC GIDY GIWT HOFC HUEW IANI IBBS IGII IKRK ILUM INAV IOWE IRON IVER IVET JAMY JAPS JERO JHKP JHYS JIMM JOST JULZ JXWS KIMM KITS LABS LACE LAMM LEBE LILP LINN MAAN MFHI MIME MEGG NDOL NEAT NESA NHRJ NIGL OBDM OBEV OBJT ODJG ODTW OEZY OGAN OIZI OJHL OKEV OMIK OPJK OPRC ORPC OSJN OSLD OUHI OURO OWWW OZEF PATF PATS PATZ PEGY PHXS PLPM PTAG PTYE PUDS PUPY RATZ RBBB RBJW RDHS RICS RIKS RIXS RJWX RMAC RMMT ROBD RONA ROOV ROWI RPAF RWLY SAMY SELF SHSH SMDH SNOZ SRYY SUES SYCO TAGR TERN TKAY TOPK VKIT VPSJ WEEM WUFF WWWG XSDJ YURO ZORO ZTED

EUROWING LTD
GOLDWING
G-MBDG MBFZ MBPM MBPX MBZH MJAJ MJAY MJDP MJEG MJOE MJRL MJRO MJRS MJSY MJUU MJWB MJWS MMBN MMLE MMTZ MMWL MNNS MNZU

EVANS
VP-1
G-AYUJ AYXW BAAD BAJC BBXZ BCTT BDAR BDTB BDTL BDUL BEIS

<u>*AA-5/5B/AG-5 TRAVELER/CHEETAH/TIGER*</u>
G-AZMJ AZVG BAFA BAJN BAJO BAOU BASH BAVR BBBI BBCZ BBDL BBDM BBLS BBRZ BBSA BBUE BBUF BCCJ BCCK BCEE BCEF BCEO BCEP BCIJ BCIK BCLI BCLJ BCPN BCRR BDCL BDFY BDLO BEBE BEZC BEZF BEZG BEZH BEZI BFIJ BFIN BFLW BFLX BFPB BFTF BFTG BFVS BFXW BFXX BFZO BFZR BGCM BGFG BGFI BGPH BGVV BGVY BHLX BHZK BHZO BIAY BIBT BIPA BIPV BIVV BIWW BJAJ BJDO BKPS BLFW BLSF BMYI BNVB BOXU BOZO BOZZ BPIZ BSTR BTII BTUZ BXHH BXOX BXTT BYDX CCAT CCXX CHTA DAVO DINA DOEA DONI ERRY GAJB GIRY IFLI IRIS JAJB JAZZ JENN JNAS JUDY KINE LSFI MALC MILY MOGI MSTC NODE NODY NONI OABR OBMW OCAM OECH OMOG OPPL OPWK OSSF OSTC OSTU OTIG PADD PAWS PING PORK PROP PURR RATE REEK RICO ROWL RUBB TGER TYGA WINK WMTM ZARI ZERO

<u>*GA-7 COUGAR*</u>
G-BGNV BGON BGSY BLHR BOGS BOOE BOXR CYMA EENY FLII GABD GENN GOTC HIRE OOGA OOGI OOGO REAT TANI

GRYPHON SAILWINGS see WASP and WILLGRESS
<u>*GRYPHON*</u>
G-MMYC

GULFSTREAM AEROSPACE CORPORATION
<u>*G159 GULFSTREAM I*</u>
G-BNCE
<u>*GULFSTREAM IV/V*</u>
G-DNVT EVLN GMAC HARF HRDS JCBV

GULFSTREAM-AMERICAN CORPORATION see GRUMMAN-AMERICAN

GYROFLIGHT
<u>*BROOKLAND HORNET*</u>
G-BRPP MIKE PHIL

HADLAND
<u>*WILLOW*</u>
G-MMMH

HALLAM
<u>*FLECHE*</u>
G-FLCT

HANDLEY PAGE LTD
<u>*O/400 Rep*</u>
G-BKMG
<u>*HP.39 GUGNUNC*</u>
G-AACN
<u>*HP.81 HERMES*</u>
G-ALDG

HANDLEY PAGE (READING) LTD
<u>*HPR.7 DART HERALD*</u>
G-APWA APWJ ASKK ASVO ATIG AVEZ AVPN BAZJ BEYF BEYK CEXP

HANRIOT
<u>*HD.1*</u>
G-AFDX

HAPI
<u>*CYGNET SF-2A*</u>
G-BRZD BWFN BXCA BXHJ BYYC

HARKER
<u>*DH/WASP*</u>
G-MJSZ

HARMON
<u>*ROCKET*</u>
G-RCKT

HATZ
<u>*CB-1*</u>
G-BRSY BXXH CBYW HATZ TIKO

HAWKER AIRCRAFT LTD including AVRO and CCF production and see W.A.R
<u>*CYGNET*</u>
G-CAMM EBJI EBMB
<u>*AUDAX*</u>
G-BVVI
<u>*DEMON*</u>
G-BTVE
<u>*FURY (Biplane)*</u>
G-BKBB
<u>*TOMTIT*</u>
G-AFTA
<u>*HART*</u>
G-ABMR
<u>*HIND*</u>
G-AENP CBLK
<u>*NIMROD*</u>
G-BURZ BWWK
<u>*HURRICANE*</u>
G-AMAU BKTH BYDL CBOE HRLI HUPW HURI HURR HURY KAMM ORGI ROBT
<u>*TEMPEST*</u>
G-PEST TEMT
<u>*FURY/SEA FURY*</u>
G-AGHB BUCM BWOL CBEL CBZP
<u>*HUNTER*</u>
G-BNCX BOOM BUEZ BVGH BVWG BVWV BWAF BWFR BWFT BWGK BWGL BWGM BWGN BWKA BWKC BWOU BXFI BXKF BZPB BZPC BZSE BZSF BZSR EGHH ETPS FFOX HHAB HHAC HHAD HHAE HHAF HNTR HPUX HUNT HVIP KAXF PRII PSST SIAL TVII VETA

HAWKER SIDDELEY AVIATION including BAe, CORPORATE JETS LTD, DE HAVILLAND and RAYTHEON-HAWKER production
<u>*DH/HS.121 TRIDENT*</u>
G-ARPH ARPK ARPO AVFB AVFE AVFG AVFH AVFJ AWZI AWZJ AWZM AWZO AWZP AWZR AWZS AWZU AZFW BAJJ BBVU BBVV BBVW BBVZ BBWG
<u>*DH/HS.125 and subsequent (Hawker) variants*</u>
G-ARYB ARYC ASSM ATPD AWYE BOCB BTAB BYHM BZNR CJAA DEZC ETOM FANN GDEZ GIRA GMAB IFTE JETI JJSI JMAX LAOR MKSS OCBA OHEA OLDD OSPG OURB OWDB RCEJ SUFC SVLB TCAP VIPI
<u>*390 PREMIER*</u>
G-FRYL OMJC
<u>*BUCCANEER*</u>
G-HHAA
<u>*HARRIER*</u>
G-CBCU CBGK VTOL

HEAD
<u>*Ax8-105 HAB*</u>
G-UKUK

HEINKEL
<u>*CASA 2111 (He.111H-16)*</u>
G-AWHA AWHB BDYA

HEINTZ see ZENAIR

HELIO
<u>*SUPER COURIER*</u>
G-BAGT BGIX

HELTON
<u>*LARK 95*</u>
G-LARK

HILL see MAXAIR

HILLER
<u>*UH-12 (360)*</u>
G-ASAZ ASTP

HINDUSTAN
<u>*PUSHPAK*</u>
G-AVPO BXTO

HISPANO see MESSERSCHMITT

HIWAY HANG GLIDERS LTD
<u>*SKYTRIKE with HIWAY and SOLAR WINGS sail-wings*</u>
G-MBAA MBCL MBDD MBEU MBFK MBIA MBIT MBJF MBKZ MBLM MBPU MBTE MBUA MBVV MBXF MBXJ MJAN MJAV MJDJ MJDR MJFZ MJHV MJMA MJMD MJMS MJMT MJMU MJNK MJNT MJOU MJPE MJPP MJSO MJUM MJXY MMBS MMCN MMCV MMEF MMEI MMHG MMHK MMHL MMHP MMLH MMOO MMRH MMUR MNME MVDS MYBN
<u>*DEMON (Wing only)*</u>
G-MJKF

HOAC FLUGZEUGWERKE (HOFFMANN FLUGZEUGBAU FRIESACH) including DIAMOND AIRCRAFT INDUSTRIESGmbH production
<u>*DV.20 KATANA*</u>
G-BWEH BWFI BWFV BWGY BWGZ BWLP BWLS BWLV BWPY BWYM TENS

LBL-35 CLOUDHOPPER
G-CCOE HOPA
LBL-42
G-BWCG CBLO
LBL-56
G-COSY DBAT
LBL-60
G-CCBP OERR
LBL-69
G-BVDS BVGG BVIR BWLA BYKA BZJY CBBX LBLI REAR
LBL-77
G-BUBS BUWI BUZR BVPV BVRR BWAW BWBO BWFK BWKZ BWMH BXDR BYJG BYKW BYLW BYRZ BYYE BYYR BZBJ BZKE CCFV ERRI HERD HUNK ICEY ICKY LBUK MERE MUCK PATP PSAX TAJF
LBL-90/A
G-BVAG BVWW BVZT BWBT BWRV BWTN BWWE BWZU BXLF BXXO BXZF BXZI BYEP BZLU CBIM CBNI CCJH CCOI CCSS CDDM CDDN CHLL DUGI FLEW FWAY GOGB JEMY JIGS MINN MRKT OBJB OSUP PATX PROF RAYO SJKR ULLS UNER UNGE
LBL-105/A
G-BUYJ BUZJ BVDO BVON BVOO BVRU BWGA BWOK BWRZ BWSB BWTB BWWY BXHE BXJG BXUO BYFU BYIY BYJN BYJZ BYLX BZAG BZPV BZUD CBPH CBPW CCIA CCSM CCVF CCXD CDHP ENRI FLGT GOAL GULF HAPI ICOI ITVM JENO LPAD OAER ODDY OPMT OUMC PIZZ ROMS RIMB RXUK SNAK VITL ZETA
LBL-120/A
G-BVLZ BWEA BZBL CBTR CBVH OGSS UPUZ
LBL-150
G-BVEW BXCM CCFF CCGV IRTH
LBL-180
G-BVBM BVIX EVNT CBZU KNOB OTUP WIZD
LBL-203
G-BXGK
LBL-210
G-BVLL BVML BXNX BZDE CCKX CCRS DVBF FVBF HVBF JVBF NVBF OCBS SSLF WVBF
LBL-240
G-BXBL CCKY
LBL-260/A
G-CDDZ PVBF
LBL-310
G-BZPE CBIW TVBF
LBL-317
G-YVBF
LBL-330
G-BXVE CCWE LRGE LVBF XVBF

SPECIAL SHAPES:

SHAPE	REGISTRATION(S)
ARMCHAIR	G-LAZY
BABY BEL	G-BXUG
BEAR	G-BWTF
BIRTHDAY CAKE	G-WISH
BUDWEISER CAN	G-BXHN
BUNNY	G-FLUF
CAKE	G-BZNZ
DOG	G-CDOG
FLOWERS	G-ODBN
HUMPTY DUMPTY	G-EGGG
J and B BOTTLE	G-OJBW
NEWSPAPER	G-FFTT
PIG	G-PIGG
PINK PANTHER	G-PINX
RACING CAR	G-TKGR
SUN	G-BZIC
SYRUP BOTTLE	G-BXUB
TELEWEST SPHERE	G-BXHO
TULIPS	G-TULP

LOCKHEED AIRCRAFT CORPORATION including LOCKHEED-CALIFORNIA CO and CANADAIR production
10 ELECTRA
G-LIOA
18 LODESTAR
G-AGCN AGIJ BMEW
414 HUDSON
G-BEOX
L.188 ELECTRA
G-FIJR FIJV FIZU LOFB LOFC LOFD LOFE LOFF LOFG OFRT
L.749 CONSTELLATION
G-CONI
L.1011 TRISTAR
G-IOIT
T-33A
G-BYOY TBRD WGHB

LORIMER
IOLAIRE
G-MZFI

LOVEGROVE see BENSEN
AV-8 GYROPLANE
G-BXXR

LUSCOMBE AIRPLANE CORPORATION
8 SILVAIRE/MASTER/RATTLER
G-AFUP AFYD AFZK AFZN AGMI AHEC AICX AJAP AJJU AJKB AKPG AKTI AKTN AKTT AKUF AKUG AKUH AKUI AKUJ AKUK AKUL AKUM AKUP AKVP BNIO BNIP BPOU BPPO BPVZ BPZA BPZC BPZE BRDJ BRGF BRGG BRHX BRHY BRJA BRJK BROO BRPZ BRRB BRSW BRUG BSHH BSHI BSNE BSNT BSOE BSOX BSSA BSTX BSUD BSYF BSYH BTCH BTCJ BTDF BTIJ BTJA BTJB BTJC BUAO BUKT BUKU BULO BVEP BVGW BVGY BVMD BWOB CCRK DAIR EITE KENM LUSC LUSI LUST NIGE ROTI SAGE YRIL

LVG
C.VI
G-AANJ

LYNDEN
AURORA
G-CBZS

McCANDLESS
M.4 GYROPLANE
G-ARTZ ATXX AXVN BVLE

McCULLOGH
J.2
G-ORVB

McDONNELL DOUGLAS CORPORATION
DC-10
G-BYDA

McDONNELL DOUGLAS HELICOPTER CO see HUGHES and MD HELICOPTERS

MACAIR
MERLIN
G-BWEN

MAINAIR SPORTS LTD see PEGASUS/FLASH and SOUTHDOWN INTERNATIONAL
BLADE (Rotax 912)
G-BYCW BYHN BYHO BYHS BYJB BYKC BYKD BYNM BYON BYOS BYOW BYRO BYRP BYRR BYTL BYTU BZAA BZAL BZDC BZDD BZEG BZEL BZHY BZFO BZFS BZGW BZIR BZJL BZJN BZLM BZMS BZNS BZPA BZPN BZPZ BZRB BZRW BZTM BZTR BZTV BZTX BZUB BZUM BZUN BZWB BZXM BZXT CBAD CBBG CBDD CBDL CBDN CBDP CBEM CBET CBGM CBGT CBGY CBHG CBHJ CBHM CBJT CBKM CBKN CBKO CBLD CBLM CBLT CBMM CBNC CBOG CBOL CBOM CBOO CBOV CBRE CBRJ CBRM CBSM CBSZ CBTE CBTM CBTW CBVG CBWM CBXM CBXV CBYF CBYM CBZA CBZB CBZD CCAB CCAG CCAM CCAW CCDM CCFM CCGK CCIF CCPM CCTM CCZW CDCU CLFC CCPB CCWL CCXR CDAG EEYE ENVY FERN JAIR JBEN JMAN JOOL LENF MAIN MYLT MYRC MYRD MYTD MYTG MYTL MYTU MYUC MYUN MYVB MYVE MYVH MYVO MYVY MYVZ MYXJ MYXM MYXN MYYA MYYG MYYH MYYW MYYY MZAA MZAB MZAE MZAF MZAG MZAJ MZAM MZAP MZAR MZAS MZAT MZAY MZAZ MZBA MZBL MZCC MZCD MZCE MZCG MZCN MZCU MZDF MZDK MZDT MZEB MZED MZEG MZEJ MZEW MZFB MZFS MZFZ MZGI MZGW MZIH MZIR MZIS MZIT MZIW MZJA MZJD MZJK MZJV MZJX MZJZ MZKG MZKJ MZKK MZKM MZKO MZKV MZKZ MZLC MZLZ MZMB MZMD MZMJ MZML MZMM MZMP MZMV MZMY MZNC MZNI MZNJ MZNK MZNL MZNO MZOC MZOF MZOP MZOR MZPH MZSD MZSM MZZY NNON NOOK OBMI OHVA ORBS OSEP OYES REED REEF RIKI RINN RUFF RYPH SHUF WAKE WLMS YZYZ
GEMINI (Trike)
G-JESA MBST MBTF MBTG MJYP MMAJ MMAR MMDP MMIR MMIV MMJT MMKM MMMD MMOB MMPJ MMRP MMRW MMSC MMSO MMTG MMTL MMTX MMUX MMXW MNGB MNJG MNMC MTBY RAVE
GEMINI/FLASH variants (Rotax 503)
G-CCER MJZD MMDP MMKL MMPO MMSP MMTG MMUO MMUT MMUW MMVP MMWA MMXD MMXG MMXJ MMXL MMXU MMXV MMZA MMZB MMZC MMZF MMZJ MMZK MMZM MMZV MNAC MNAE MNBD MNBF MNBG MNBN MNBP MNBR MNBS MNBT MNBV MNBW MNCF MNCG

MNCJ MNDC MNDF MNEF MNEG MNEH MNET MNEV MNEY MNFE MNFF MNFH MNFM MNFN MNFP MNGK MNGM MNGT MNGU MNGW MNHZ MNIA MNID MNIE MNIF MNIG MNIH MNII MNIX MNIZ MNJU MNLI MNLX MNLY MNMG MNMI MNMV MNNF MNNI MNNJ MNNL MNNR MNNV MNPC MNPG MNRW MNRX MNSA MNSI MNSJ MNSR MNTI MNTS MNTU MNTV MNTX MNTZ MNUA MNUF MNUG MNUO MNUR MNUY MNVT MNVU MNVV MNVW MNWD MNWI MNWZ MNXS MNXU MNYJ MNZB MNZC MNZD MNZF MTAB MTAE MTAF MTAG MTAH MTAR MTBD MTBH MTBJ MTBX MTBY MTCC MTCE MTCU MTCW MTDF MTDR MTDW MTDY MTEJ MTEK MTEN MTEY MTFF MTFI MTFJ MTGA MTGH MTGO MTHW MTHZ MTIA MTIB MTIL MTIM MTIN MTJA MTJB MTJC MTJD MTJE MTJL MTJM MTJT MTJV MTJW MTJZ MTKN MTKV MTKW MTKX MTKZ MTLB MTLC MTLD MTLL MTMA MTMC MTML MTMT MTMV MTMW MTNC MTNG MTNH MTNI MTNJ MTNL MTNM MTNX MTNY MTPA MTRA MTRZ MTSC MTTI MTTM MTTP MTTR MTTW MTUU MTUV MTVG MTVH MTVI MTVJ MTWF MTWG MTWR MTWS MTWX MTXM MTXP MTXS MTXZ MTZG MTZH MTZL MTZM MTZO MTZV MTZW MTZX MTZY MTZZ MVAA MVAB MVAD MVAO MVAP MVBD MVBF MVBG MVBI MVBK MVBL MVBM MVBN MVBO MVCE MVCF MVCY MVCZ MVDA MVDT MVEH MVEJ MVEK MVEL MVEO MVEP MVER MVES MVET MVEV MVEW MVGM MVHE MVHF MVHG MVHH MVIB MVIX MVIY MVIZ MVJA MVJC MVJE MVJL MVKC MVLL MVLR MVMO MVMR MVMT MVMU MVMV MVMX MVMY MVMZ MVNM MVNW MVNX MVNY MVNZ MVOB MVOF MVON MVOR MVPA MVPB MVPC MVPD MVPE MVPI MVRA MVRB MVRC MVRD MVRM MVSN MVSO MVSP MVST MVSV MVTC MVUA MVXB MVXC MVXR MVXS MVYS MVZS MWAB MWAU MWCE MWCW MWDJ MWEL MWGG MWHI MWHO MWHR MWIA MWIG MWIH MWIV MWJY MWLP MWLT MWLX MWMM MWMS MWMT MWMX MWMY MWNE MWNS MWNT MWNU MWOJ MWOK MWPB MWPC MWPD MWPF MWPO MWRB MWRC MWRD MWRE MWRF MWRG MWRH MWRJ MWRR MWSL MWSM MWTG MWTH MWTO MWTR MWTY MWTZ MWVN MWVO MWVR MWVS MWVT MWVW MWVY MWWB MWWC MWWI MWWJ MWWK MWWN MWXB MWXC MWXL MWXN MWXN MWXO MWXU MWXV MWYA MWYG MWYH MWYL MWYT MWYV MWZC MWZG MWZL MWZN MYAO MYAS MYAU MYBJ MYCK MYCR MYCS MYDV MYEU MYFP MYFR MYFU MYGZ MYHF MYHL MYHN MYHX MYIH MYIV MYIY MYJB MYJC MYJM MYKC MYKG MYKH MYKV MYLG MYLR MYMK MYMO MYMV MYND MYOM MYOW MYPE MYPW MYSJ MZCF OLJT

MERCURY (Rotax 503)
G-MWVK MWXF MWXJ MWXK MWZA MYAI MYAV MYCJ MYCL MYCN MYCV MYDC MYGJ MYJR MYKI MYKW MYKX MYKY MYLS MYML MYMT MYNC MYNF MYNJ MYOB MYOF MYOV MYOX MYPD MYPV MYRW MYSG MYSZ MYTB MYTK MYTX MYUB MYUD MYUE MYUK MYUW MYVL MYVS MYWA MYYU MZAK MZCO

RAPIER (Rotax 503)
G-BYBV BYOZ BZAB BZUF BZWR CCHV MFLY MJYV MZEP MZEV MZFD MZGL MZHJ MZHL MZIL MZIM MZJE MZKN MZND MZNU MZON YARR

SCORCHER SOLO (Rotax 447)
G-MNDD MNNM MNPV MNPY MNPZ MNRE MNRZ MVBE MYFT MZKI MZKN

TRI-FLYER Trike with various wings
G-CCVX MBCJ MBGA MBHK MBIZ MBMT MBPG MBPZ MBUK MBZA MBZO MJEE MJEY MJFK MJHR MJIF MJJO MJMN MJMR MJPE MJTP MJXE MJYV MJYX MJZU MMAL MMCZ MMDE MMDK MMDN MMDT MMEJ MMFD MMFE MMFK MMFV MMJG MMKR MMLI MMMB MMNW MMTD MMUH MMWG MMWN MMYV MNJD MNJG MNUI MNXB MVBC

MALMO see BÖLKOW

MANNING-FLANDERS

MF.1
G-BAAF

MANTA

PFLEDGE (Wing)
G-MBNY

MANUEL

LADYBIRD
G-MJPB

MARQUART

MA.5 CHARGER
G-BHBT BVJX

MASQUITO

M.58
G-MASX MASY MASZ

MAULE AIRCRAFT CORPORATION

M-5 LUNAR ROCKET
G-BHJK BICX BIES BPMB BVFT BVFZ CCBF FMGG KRIS OJGT RAIN RJWW

M-6 SUPER ROCKET
G-BKGC MOUL

M(XT)-7 SUPER/STAR ROCKET/STARCRAFT
G-BSKG BSKO BTXT BUEP BUXD BVIK BVIL BZDT CROL GROL HIND ITON JREE LOFM MLHI OMOL RAZZ TAFC URUS WALY

MAX HOLSTE

MH.1521M BROUSSARD
G-BWGG BWLR CBGL YYYY

MAXAIR

DRIFTER
G-MYBB

HUMMER
G-MBYH MJCF MMZZ MNIM

MBB including EUROCOPTER production

Bö.105
G-BAMF BATC BCXO BFYA BTHV BTKL BUXS CDBS DNLB EYNL NAAA NAAB PASB PASG PASX TVAM WAAN WAAS WMAA WYPA

BK.117
G-CDGM DCPA

MD HELICOPTERS INC

MD.500N
G-NOTR SIVN

MD.600N
G-BZTZ GOON ODOD PEPL

MD.900
G-BXZK EHMS GMPS GNAA HPOL KAAT LNAA SIVR SUSX SYPS WMID WPAS YPOL

MEA

MISTRAL TRAINER
G-MBET MBOH MBUS

MEDWAY MICROLIGHTS LTD see RAVEN/SOUTHDOWN

AV8R
G-CCGO

ECLIPSER (Rotax 912-UL)
G-BYBO BYXV BYXW BZGE BZWI CBMR CBMS CCCI CCGA OBRI

HALF PINT trike (JPX PUL 425) with Aerial Arts 130SX wing
G-MMZI MNDE MNEK MNTT MNVL

HYBRED 44XL (Fuji-Robin EC-44)
G-MJVE MMEK MMKG MMKH MNCU MNCV MNEI MNFW MNJK

HYBRED 44XLR (Rotax 447)
G-BYBJ BYRH MNMN MNXO MTFC MTJG MTJP MTLX MTNE MTNF MTUX MVCD MVDJ MVEE MVGB MVGL MVGY MVKB MVMK MVPF MVPG MVPL MVRY MVRZ MVSI MVUD MVVH MVVI MVVR MVWV MVXD MVXE MVXI MVXJ MVXM MVYP MVYR MVZO MWBJ MWCX MWCY MWCZ MWGC MWIL MWJP-MWJR MWJX MWLB MWLS MWRM MWSS MWST MWSU MWVU MYRI MYVV MYVX MZME

REBEL
G-BYPP BYSS CCPK

SLA 100 EXECUTIVE
G-CCJJ

SOCIÉTÉ MENAVIA see PIEL

MERIDIAN ULTRALIGHTS LTD

RENEGADE SPIRIT (Rotax 582)
G-BTHN BTKB BWPE BYBU FIRZ MGOO MVZP MVZX MWAJ MWDM MWKA MWMW MWNF MWNR MWOO MWPS MWPZ MWUH MWVP MWWD MYAM MYAZ MYCO MYFM MYRK MYUF MYXR MZIZ NINE RCMC RENE TBAG TBMW

MESSERSCHMITT AG including HISPANO production and see NORD

Bf.108D-1 TAIFUN
G-AKZY

Bf.109/HA.1112
G-AWHE AWHG AWHJ AWHL AWHN AWHO AWHS BWUE EMIL HUNN SMIT

MICKLEBURGH

L107
G-BZVC

MICRO AVIATION

B-22 BANTAM (Rotax 582)
G-BXZU BZYS MZEY MZLX

MICRO BIPLANE AVIATION
TIGER CUB 440 (Fuji-Robin EC-44)
G-MJRU MJSP MJSU MJSV MJUC MJUF MJUW MJWF MJWJ MJWW MJXD MJXF MJYD MJZE MMAG MMBE MMBH MMBT MMCX MMFS MMGF MMGL MMHN MMIE MMIH MMIX MMJV MMKP MMLB MMUM MNJC MNKM MWFT

MICROFLIGHT AIRCRAFT LTD including CORBETT FARMS production
SPECTRUM (Rotax 503)
G-MVJM MWCG MWKW MWKX MWOF MWPG MWPH MWTD MWTE MWWR MYAY

MIDLAND ULTRALIGHTS LTD
SIROCCO
G-MNDU MNRT MTJN MTRC MVSM

MIGNET
HM.14/HM.19 POU-DU-CIEL
G-ADRG ADRX ADRY ADVU ADYV ADZW AEBB AEEH AEFG AEGV AEHM AEJZ AEKR AEOH AFFI MYSI
HM.293
G-AXPG

SOCIÉTÉ D'EXPLOITATION DES AERONEFS HENRI **MIGNET**
HM-1000 BALERIT (Rotax 582)
G-MRAM MYDZ MYXL MZIX MZLI MZMW MZPB MZTA

MIKOYAN AVIATION including WSK-PZL
MiG-15 (Lim-2)
G-BMZF
MiG-17 (Lim 5)
G-MIGG
MiG-21
G-BRAM

MILES AIRCRAFT LTD including Phillips and Powis Aircraft Ltd
M.2H HAWK MAJOR
G-ACYZ ADMW CCMH
M.2L HAWK SPEED SIX
G-ADGP
M.3 FALCON
G-ADLS AEEG
M.5 SPARROWHAWK
G-ADNL
M.11A WHITNEY STRAIGHT
G-AERV AEUJ AFGK
M.12 MOHAWK
G-AEKW
M.14A HAWK TRAINER 3
G-AETL AFBS AHUJ AIUA AJRS AKAT AKKR AKKY AKPF AMMC ANWO
M.17 MONARCH
G-AFJU AFLW AFRZ
M.18
G-AHKY
M.28 MERCURY
G-AHAA
M.38/48 MESSENGER
G-AGOY AIEK AJOC AJOE AJVL AJWB AKAO AKBO AKEZ AKIN AKIS AKVZ ALBP
M.65 GEMINI
G-AISD AJTI AKDK AKEL AKER AKGE AKHP AKHW AKKA AKKB AKKH ALCS
M.75 ARIES
G-AOGA
M.100 STUDENT
G-APLK MIOO

MILLS
MH-1
G-OMHI

MIRAGE see ULTRAFLIGHT

MITCHELL
WING B-10
G-MMJA
WING U-2
G-MMNS

MITCHELL-PROCTER see PROCTER
KITTIWAKE
G-ATXN BBRN BBUL

MONNETT
MONI
G-BMVU CBTL NOW MONI TRIM
SONERAI
G-BGEH BGLK BICJ BJBM BJLC BKDC BKFA BKNO BLAI BMIS BOBY BSGJ BVCC CCOZ PFAT

MONOCOUPE CORPORATION
90A
G-AFEL

MONTGOMERIE-BENSEN see BENSEN and PARSONS

MOONEY AIRCRAFT CORPORATION
M.20/M.252
G-APVV ASUB ATOU AWLP BCJH BDTV BHBI BHJI BIBB BIWR BJHB BKMA BKMB BPCR BPFC BSXI BVZY BWJG BWTW BXML BYEE CERT DBYE DESS DEST EKMW FLYA GCKI GJKK JAKI JAST JDIX JENA MALS MUNI NYRL OBAL ODJH OEAC OJAC OJJB OONE OPWS OSUS RAFW

MORANE-SAULNIER including GEMS, MORANE, SEEMS and SOCATA production and see DE HAVILLAND and FIESELER
TYPE N
G-AWBU
MS.230 PARASOL
G-BJCL
MS.315
G-BZNK
MS.502/505
G-BIRW BPHZ
MS.733 ALCYON
G-MSAL
MS.880/885/887/892/894 RALLYE/GALERIEN/GALOPIN/MINERVA
G-ARXW ASAT ASAU AVIN AVTV AVVJ AVZX AWKT AWOA AWYX AXCM AXGE AXHS AXHT AXOH AXOS AXOT AYRH AYTA AYYX AZEE AZGI AZGL AZKC AZKE AZMZ AZUT AZVF AZVH AZVI AZYD BAAI BAOG BAOH BAOJ BAOM BBED BBGC BBLM BCLT BCOR BCST BCUL BCVC BCXB BDEC BDWH BECA BECB BECC BEIL BERA BERC BETO BEVB BEVC BEVW BFAK BFDF BFGS BFTZ BGKC BGMT BGSA BHWK BIAC BIIK BIRB BJDF BKBF BKGA BKGT BKJF BKOA BKVA BKVB BLGS BLIY BOJL BPJD BRDN BTIU BTOW BTUG BUKR BVAN BWWG BXZT BYPN BZNX EISO EXIT FARM FOSY GIGI HENT HHAV KHRE MELV OACI OMIA PIGS TLBC WCEI

MORAVA see LET NARODNI PODNIK KUNOVICE

MORAVAN NARODNI PODNIK
ZLIN Z-226T
G-EJGO
ZLIN Z-326 TRENER MASTER
G-BEWO BKOB
ZLIN Z-242L
G-BWTC BWTD EKMN
ZLIN Z-50LX
G-MATE
ZLIN Z-526 TRENER MASTER (Walter Minor 6-3)
G-AWJX AWJY AWSH BLMA BPNO PCDP TINY ZLIN ZLYN

MORETON
ORIENTAL MLB
G-BINY

MORRIS
SCRUGGS MLB
G-BILE BILG BINL BINM BINX BIPH BISL BISM BISS BIST BIWB BIWC

MORRIS MOTORS LTD see DE HAVILLAND

MOSS BROS AIRCRAFT LTD
MOSS MA.1
G-AFHA
MOSS MA.2
G-AFJV

MOTH CORPORATION see DE HAVILLAND

MOTO-DELTA see CENTRAIR

MOULT
(Trike)|
G-MTFK

PAZMANY
PL-1
G-BDHJ
PL-2
G-OPAZ
PL-4/A
G-BMMI BRFX FISK PAZY PLIV

PEARSON
MLB
G-BIXX

PEGASUS AVIATION including MAINAIR SPORTS LTD production and see CYCLONE AIRSPORTS LTD
QUANTUM (Rotax 912)
G-BYDM BYDZ BYEW BYFF BYFG BYIS BYIZ BYJK BYKT BYLC BYMF BYMI BYMT BYND BYNO BYOG BYOV BYPB BYPJ BYPL BYRJ BYRU BYSR BYSX BYTC BYYN BYYP BYYY BZAI BZBR BZED BZGZ BZHN BZHO BZIM BZIW BZJF BZJO BZJZ BZKT BZLL BZLX BZLZ BZMI BZMW BZNB BZNC BZNM BZOD BZOE BZOO BZOU BZOV BZRJ BZRP BZSA BZSG BZSI BZSM BZSS BZSX BZUC BZUE BZUI BZUX BZVJ BZVV BZWS BZWU BZXV BZXX BZYN CBAY CBBB CBBN CBBP CBBZ CBCD CBCF CBCX CBDX CBDZ CBEN CBEU CBEV CBGG CBHK CBHN CBHY CBIZ CBJO CBKW CBLL CBMV CBNT CBOY CBSP CBTD CBTZ CBUD CBUS CBUU CBUZ CBYI CBYV CCCD CCDK CCDZ CCFT CCIH CCJD CCNE CCNW CCOC CCRF CCRT CCUR CCWN CCWO CCWW CCYL CCZB CDAA CDAO CDCY CDCZ CDDF CDEN CDGX CDHM CDIL DINO DSLL EDMC EMLY EOFW FFUN GBJP JAWC JGSI KICK MCEL MCJL MDBC MGDL MGEF MGFK MGFO MGGG MGGV MGMC MGTG MROC MSPY MYLC MYLE MYLH MYLI MYLK MYLL MYLM MYLZ MYMB MYMC MYMD MYMX MYNB MYNK MYNL MYNN MYNO MYNP MYNR MYNS MYNT MYNV MYNZ MYOU MYPH MYPI MYPN MYPX MYPY MYRF MYRM MYRN MYRS MYRT MYRY MYRZ MYSB MYSC MYSR MYSW MYSX MYSY MYTI MYTJ MYTN MYUO MYUU MYUV MYVC MYVJ MYVK MYVM MYVR MYWG MYWI MYWJ MYWK MYWL MYWO MYWR MYWT MYWU MYWW MYWX MYWY MYXE MYXT MYXW MYXX MYXZ MYYB MYYC MYYI MYYK MYYN MYYX MYZB MYZJ MYZK MYZL MYZM MYZY MZAN MZAW MZAX MZBB MZBC MZBI MZBM MZBO MZBT MZBY MZCI MZCJ MZCM MZCR MZCV MZCY MZDB MZDC MZDD MZDE MZDH MZDN MZDU MZDV MZDY MZEC MZEE MZEH MZEM MZEX MZEZ MZFG MZFM MZFV MZGG MZGK MZGN MZGO MZGV MZHI MZHK MZHN MZHP MZIB MZIC MZIE MZIF MZIJ MZIK MZIU MZJG MZJH MZJN MZJO MZJT MZJW MZJY MZKA MZKD MZKF MZKL MZKX MZKY MZLA MZLD MZLF MZLH MZLJ MZLN MZLT MZLV MZLW MZMC MZMF MZMG MZMH MZNP MZMN MZMT MZNB MZNG MZNR MZNS MZNT MZOD MZOG MZOJ MZOS MZOV MZOW MZPD MZRC MZRH MZRM MZSC NAPO OAKS OAMF OATE OBJP OELD OLDM OLFB OTJH PEGA PIXI PRSI REDC REPH RUSA SILY SITA SMBM TBBC TFIX TRAM TUSA WHEE
QUASAR (Rotax 503)
G-MWHT MWHU MWIM MWIU MWIW MWIX MWIY MWJD MWJH MWJI MWJJ MWJK MWJS MWJT MWJV MWLH MWLJ MWLK MWMI MWMJ MWMK MWML MWNK MWNL MWOM MWOP MWPU MWSH MWSI MWTK MWTL MWVM MWXG MWXH MWYI MWYJ MWZD MWZE MWZF MWZO MWZP MWZR MWZS MYAK MYBD MYBE MYBT MYCE MYEK MYEM MYEN MYEO MYFK MYFL MYIM MYIN MYIO MYJJ MYJK MYJS MYJT MYJU MYKP MYKR MYKS MYTR MYXD MZMA MZPW REKO
QUIK (Rotax 912-UL(S))
G-CBRY CBVN CBYE CBYO CBZH CBZT CCAD CCAS CCAW CCAZ CCCG CCDB CCDD CCDF CCDO CCEA CCEW CCFB CCFL CCGC CCGI CCHH CCHI CCHO CCIV CCJM CCKM CCKO CCLM CCLX CCMD CCME CCML CCMS CCNM CCOG CCOK CCOU CCOW CCPC CCPG CCRW CCSD CCSF CCSH CCSL CCSY CCTC CCTD CCTU CCTZ CCUA CCVB CCWR CCWV CCXT CCXZ CCYE CCYJ CCZO CDAR CDAX CDBB CDCF CDCI CDCK CDEC CDEW CDFG CDFO CDGC CDGD CDGO CDKM CDPD CJAY CWIC CWIK CWVY DCMI EZAR FLEX GBEE HALT KUIK KWIC LUNE MRJJ OKEM OLDP OMPW OUIK RIKY SHEZ SOCK SYTX TBJP TCNY TERR TJAV TONN WBLY WFLY WIZS YSMO

PENN-SMITH
GYROPLANE
G-AXOM

PERCIVAL AIRCRAFT CO/LTD see HUNTING
P.1 GULL FOUR
G-ACGR
P.3 GULL SIX
G-ACUP ADPR AERD
P.6 MEW GULL
G-AEXF
P.10 VEGA GULL
G-AEZJ
P.16 Q SIX
G-AFFD
P.28 PROCTOR I
G-AHDI AHTN
P.30 PROCTOR II
G-AHVG
P.31 PROCTOR IV
G-ANXR
P.34 PROCTOR III
G-AKZN ALCK ALIS ALJF AOGE
P.44 PROCTOR V
G-AGTB AHTE AHZY AKIU
P.40 PRENTICE
G-AOKL AOKO AOKZ AOLK AOLP AOLU AOMF AOPL AOPO APIT APIY APJB APPL
P.50 PRINCE
G-AMLZ
P.56 PROVOST
G-AWPH AWRY AWTD AWVF BDYG BGSB BKFW BLIW KAPW MOOS TMKI
P.57 SEA PRINCE
G-BRFC DACA GACA RACA
P.66 PEMBROKE/PRESIDENT
G-AOJG BNPG BNPH BXES

PEREIRA
OSPREY
G-BEPB BVGI CCCW GEOF PREY

PHANTOM see SKYRIDER

PHILLIPS - see SPEEDTWIN DEVELOPMENTS LTD
ST.1 SPEEDTWIN
G-DPST GPST

PHILLIPS AND POWIS AIRCRAFT LTD - see MILES AIRCRAFT LTD

PHOENIX see CURRIE WOT and ROLLASON
LUTON LA-4/A MINOR/PARKER CA-4/PHOENIX DUET
G-AFIR AMAW ARIF ARXP ASAA ASEA ASEB ASML ASXJ ATCJ ATFW ATKH ATWS AVDY AVUO AWIP AWMN AXGR AXKH AYDY AYSK AYTT AZHU AZPV BANF BBCY BBEA BCFY BDJG BIJS BKHR BRWU
LA-5A MAJOR
G-ARAD

PIAGGIO including FOCKE WULF production
P.149
G-RORY
P.166
G-APWY

PICCARD
Hot Air Balloon
G-ATTN
AX6
G-AWCR AZHR

PIEL including COOPAVIA, MENAVIA, ATELIERS AERONAUTIQUE ROUSSEAU, SCINTEX production and BINDER/FAIRTRAVEL variants
CP.301/328 EMERAUDE
G-APNS ARDD ARRS ARUV ASCZ ASLX ASMT ASVG ASZR AXXC AYCE AYEC AYTR AZGY AZYS BBKL BCCR BDCI BDDZ BDKH BHRR BIDO BIJU BIVF BKFR BKNZ BKUR BLRL BPRT BSVE BXAH BXYE DENS PIEL SAZZ
CP.1310/1315/1320 SUPER EMERAUDE
G-ASNI BANW BCHP BGVE BHEK BJCF BJVS BLXI BXRF SAFI

PIETENPOL
AIRCAMPER
G-ADRA BBSW BKVO BMDE BMLT BNMH BPOL BRXY BSVZ BUCO BUXK BUZO BVYY BWAT BWVB BXZO BYFT BYKG BYLD BYZY CCKR DAYZ ECOX EDFS IMBY OFFA OHAL OPJS PCAF PIET RAGS SILS SLOW TARN UNGO VALS

PIK see EIRI and SIREN

PILATUS AIRCRAFT LTD
P.2
G-BLKZ CJCI PTWO
P.3
G-BTLL
PC.6 PORTER
G-BYNE WGSC
PC.12/45
G-CCPU CCWY

PIPER AIRCRAFT CORPORATION including TAYLOR AIRCRAFT CO LTD and THE NEW PIPER AIRCRAFT INC production and see TED SMITH

J-2 CUB
G-AEXZ AFFH JTWO

J-3C CUB (L-4/O-59)
G-AFDO AGAT AGIV AGVV AHIP AIIH AISS AISX AJAD AJES AKAA AKAZ AKBT AKIB AKRA AKTH AKUN ASPS ATKI ATZM AXGP AXHP AXHR AYCN AYEN BAET BBHJ BBLH BBUU BBXS BCNX BCOB BCOM BCPH BCPJ BCUB BCXJ BDCD BDEY BDEZ BDHK BDJP BDMS BDOL BECN BEDJ BEUI BFBY BFDL BFHI BFZB BGPD BGSJ BGTI BGXA BHPK BHVV BHXY BHZU BIJE BILI BJAF BJAY BJSZ BJTO BKHG BLPA BMKC BOTU BOXJ BPCF BPUR BPVH BPYN BREB BROR BSBT BSFD BSNF BSTI BSVH BSYO BTBX BTET BTSP BTUM BTZX BVAF BVPN BWEZ CCOV CCUB COPS CUBS CUBY FRAN HEWI KIRK LIVH LOCH NCUB OCUB OINK OLEZ POOH RAMP SEED TCUB

J-4A CUB COUPÉ
G-AFGM AFWH AFZA BRBV BSDJ BUWL

J-5A CUB CRUISER
G-BRIL BRLI BSDK BSXT BTKA

PA-12 SUPER CRUISER
G-AMPG ARTH AWPW AXUC BCAZ BOWN BSYG CDCS PAIZ

PA-15/PA-17 VAGABOND
G-AKTP ALEH ALGA ALIJ AMYL ASHU AWKD AWOF AWOH BCVB BDVA BDVB BDVC BIHT BLMP BOVB BRJL BRPY BRSX BSFW BSMV BSWG BTBY BTCI BTFJ BTOT BUKN BUXX CCEE FKNH

PA-16 CLIPPER
G-BAMR BBUG BIAP BSVI BSWF

PA-18/PA-18A SUPER CUB (L-18/L-21)
G-AMEN APZJ ARAM ARAN ARAO ARCT AREO ARGV ARVO ASCU ATRG AVOO AWMF AXLZ AYPM AYPO AYPR AYPS AYPT AZRL BAFT BAFV BAKV BBOL BBYB BCFO BCMD BEOI BEUA BEUU BFFP BGPN BGWH BGYN BHGC BHOM BHPM BIDJ BIDK BIID BIJB BIMM BIRH BITA BIYJ BIYR BIYY BIZV BJBK BJCI BJEI BJFE BJIV BJTP BJWX BJWZ BKET BKJB BKRF BKTA BKVM BLGT BLHM BLIH BLLN BLLO BLMI BLMR BLMT BLPE BLRC BMAY BMEA BMKB BNXM BOOC BPJG BPJH BPUL BROZ BRRL BSHV BTBU BTUR BVIE BVIW BVMI BVRZ BWOR BWUB BZHT CCKW CUBB CUBI CUBJ CUBP DADG ECUB FUZZ HACK HELN JCUB KAMP LION MGMM NESY NETY NNAC OFER OROD OSPS OTAN OTUG PIPR PUDL ROVE SCUB SUPA TUGG WCUB WGCS WLAC XCUB YCUB ZAZA

PA-20 PACER/PA-22 conversions
G-APYI ARBS ARGY ARNK ATBX AVDV ATXA BFMR BIYP BSED BTLM BUDE BUOI BUXV BWWU BXBB GGLE PAXX

PA-22 TRI-PACER/CARIBBEAN/COLT
G-APUR APXR APXT APXU APYN APZL APZX ARAI ARAX ARBV ARCC ARCF ARDS ARDT ARDV AREL ARET AREV ARFB ARFD ARGO ARHN ARHP ARHR ARIK ARIL ARJE ARJF ARJH ARKK ARKM ARKN ARKP ARKS ARND ARNE ARNG ARNJ ARNL ARON ARSU ARYH AWLI AZRS BMCS BNED BRNX BTKV BTWU BUVA CBEI EMSB HALL TJAY TLDK

PA-23/PA-27 APACHE/AZTEC
G-APMY ARCW ARJS ARJT ARJU ARJV ARYF ASEP ASHH ASMY ATFF ATOA AXDC AXZP AYBO AYMO AZRG AZSZ AZXG AZYU BADI BADJ BAPL BATN BAUJ BAUW BAVL BBCC BBDO BBEY BBHF BBIF BBMJ BBRA BBTJ BCBG BCBM BCCE BCEX BCRP BEXO BFBB BFWE BGTG BGWW BICY BJNZ BJXX BKJW BKVT BMFD BNUV BRAV BXPS BYRW CALL CSFT EEVA ESKY FOTO HFTG JTCA KEYS LIZZ MLFF MOLY NRSC OART OPME OSNI OXTC RVRC RVRD RVRJ RVRW SFHR TAPE TAXI UNDO XSFT

PA-24/PA-26 COMANCHE
G-APUZ APXJ ARBO ARDB ARFH ARHI ARLB ARLK ARUO ARXG ARYV ASCJ ASEO ASYK ATFK ATIA ATJL ATNV ATOY AVCM AVGA AXMA AXTO AZKR AZWY BAHJ BRDW BRXW BWNI BYTI DISK KSVB MOTO PETH UVNA

PA-25 PAWNEE
G-ASIY ASVP ATFR AVPY AVXA AXED AZPA BAUC BCBJ BDDS BDPJ BEII BETL BETM BFBP BFEV BFEW BFEY BFPR BFPS BFRY BFSC BFSD BHUU BILL BLDG BNZV BPWL BSTH BUXY BVYP BXST CCUV CCUW CMGC CTUG DSGC LYND NYMF PAWN PSGC TOWS

PA-28-140/160 CHEROKEE/CHALLENGER/CRUISER/FLITE-LINER
G-ARVT ARVU ARVV ASLV ASPK ASSW ASVZ ATDA ATEZ ATIS ATJG ATMW ATOI ATOJ ATOK ATOL ATOM ATON ATOO ATOP ATOR ATOS ATPN ATRO ATRP ATRR ATTI ATTK ATTV ATUB ATUD ATVK ATVO AVFP AVFR AVFX AVFZ AVGC AVGD AVGE AVGI AVLB AVLC AVLE AVLF AVLG AVLH AVLI AVLJ AVLT AVSI AVUS AVUT AVUU AVUT AVUU AVWA AVWD AVWE AVWI AVWJ AVWL AVWM AVYP AVYR AWBE AWBG AWBH AWBS AWEV AWEX AWPS AWSM AWTM AXAB AXIO AXJV AXJX AXRL AXSZ AXTA AXTC AXTJ AXTL AYAT AYIG AYJP AYJR AYKW AYKX AYMK AYNF AYNJ AYPV AYRM AYWE AZEG AZFC AZRH AZWB AZWD AZWE AZZO BAFU BAFW BAGX BAHE BAHF BAKH BAMM BASL BATW BAWK BAXZ BBBK BBBY BBDC BBEF BBEV BBHY BBIL BBIX BBYP BBZF BCDJ BCGI BCGJ BCGN BCGT BCJM BCJN BCJP BDSH BDWY BEAC BEEU BEFF BEYO BEYT BFXK BGAX BGPU BGRC BHXK BIFB BIYX BOFY BOSR BOSU BRBW BRPK BRPL BRWO BSER BSLM BSLU BSSE BSTZ BTEX BTGO BTON BTVR BULR BWYB BXPL BXVU BXYM BYCA BZWG CCLJ CGHM COLH DAKS DENE DIAT FIAT GCAT JAKS JDJM KATS LFSC LFSI LIZI LTFC MATZ MIDD MKAS NHRH OFTI OKYM OMAT PAWL PETR PIKK RECK SCPL SMTH TEFC TEWS UANT WOLF ZANG ZEBY

PA-28-151/161 CHEROKEE WARRIOR/CADET
G-BBXW BCIE BCIR BCRL BCTA BCTF BDGM BDPA BEBZ BEFA BELP BFBR BFDK BFMG BFNI BFNJ BFNK BFWB BFWK BGKS BGOG BGPJ BGPL BGVK BHFK BHIL BHJO BHOR BHRC BHVB BICW BIEY BIIT BIUW BJBW BJBX BJCA BJSV BJYG BLEJ BLVL BMFP BMKR BMTR BMUZ BNCR BNEL BNJM BNJT BNMB BNNO BNNS BNNT BNNY BNNZ BNOE BNOF BNOG BNOH BNOJ BNOK BNOM BNON BNOP BNOT BNRG BNSY BNSZ BNTD BNXE BNXT BNXU BNZB BNZZ BOAH BODB BODC BODD BODE BODR BOER BOFZ BOHA BOHO BOHR BOIG BOJW BOJZ BOKB BOKX BOMY BOPC BORK BORL BOSP BOTF BOTI BOTN BOUP BOUR BOVK BOXA BOXB BOXC BOYH BOYI BOZI BPAF BPBM BPCK BPDT BPEL BPFH BPHB BPHL BPID BPIU BPJO BPJP BPJR BPJU BPKM BPKR BPMF BPMR BPOM BPPK BPRN BPRY BPWE BRBA BRBB BRBD BRBE BRDF BRDG BRDM BRFM BRJV BRRN BRSE BRTM BRTX BRUB BRXC BSAW BSBA BSCV BSCY BSFK BSGL BSHP BSJX BSLE BSLK BSLT BSMZ BSOK BSOZ BSPI BSPM BSSC BSSW BSSX BSVG BSVM BSXA BSXB BSXC BSYZ BSZT BTAW BTBC BTDV BTFO BTGY BTID BTIM BTIV BTKT BTNE BTNH BTNT BTNV BTRK BTRS BTRY BTSJ BTUW BUFH BUFY BUIF BUIJ BUIK BUJO BUJP BUKX BURT BVJZ BVTO BWOI BWOJ BXAB BXJJ BXJX BXLY BXNH BXTX BXTY BXTZ BYHH BYHI BYKN BYKO BYKR BYXU BYZM BZBS BZDA BZIO BZLH BZMT CBAL CBKR CBWD CBYU CCYY CCZV CDCL CDDG CDEF CDGW CDON CLAC CLEA CPTM CWFZ DOME EDGI EEGU EGTB EGTR EJRS EKIR EKKL ELZY EMSL EOLD ERFS ESFT ESSX ETDA EVIE EXON EXXO FIZZ FLAV FLEN FMAM FNPT FOXA FPIG FPSA GALB GBRB GFCA GFCB GFTA GFTB GHKX GOTH GRRC GURU GUSS GYTO HMED IKBP ISDB ISHA JACA JAMP JASE JAVO KART KBPI KDET KNAP LACA LACB LAZL LBMM LFSJ LFSK LORC LSFT LUSH MAYO MSFT NINA NSFT OAAA OANI OBDN OBFC ODEN OJWS OMST OOFT OONY OTYJ OWAR OXOM PJCC PSRT RIZZ RSKR ROWS SACI SACO SACR SACS SACT SACZ SARH SEJW SEXX SLYN SNUZ SRWN SUZN TAGS TLET VICC WARB WARC WARE WARH WARR WARS WARV WARW WARX WARY WARZ WFFW XENA XINE ZULU

PA-28-180/181 CHEROKEE/CHALLENGER/ARCHER
G-ARYR ASFL ASHX ASII ASIJ ASIL ASKT ASRW ASUD ASWX ATAS ATEM ATHR ATNB ATOT ATTX ATUL ATVS ATXM ATYS ATZK AVAX AVBG AVBH AVBS AVGK AVNN AVNO AVNP AVNS AVNU AVNW AVOZ AVPV AVRK AVRU AVRY AVRZ AVSA AVSB AVSC AVSD AVSE AVSF AVSP AVYL AVYM AVZR AWDP AWET AWIT AWSL AWTL AWXR AWXS AXSG AXTP AXZD AXZF AYAB AYAR AYAS AYAW AYEE AYEF AYPJ AYUH AYUI AZDX AZLN BABG BAJR BASJ BATV BBBN BBEC BBKX BBPY BCCF BCLL BDSB BEIP BEMW BEXW BEYL BFDI BFSY BFVG BGBG BGTJ BGVZ BGWM BHNO BHWZ BHZE BIIV BIUY BJAG BJOA BKCC BLFI BMIW BMPC BMSD BNGT BNPO BNRP BNVE BNYP BOBZ BODM BOEE BOHM BOJM BOMP BOMU BOOF BOPA BORS BOSE BPAY BPFI BPGU BPOT BPTE BPXA BPYO BRBG BRBX BRGI BRME BRNV BRUD BRXD BSCS BSEF BSEU BSGD BSIM BSIZ BSKW BSNX BSVB BSXS BSZJ BTAM BTGZ BTKX BTYI BUMP BUTZ BUUX BUYY BVNS BVOA BWPH BWUH BXEX BXIF BXJD BXOZ BXRG BXRJ BXTW BXWO BYHK BYKL BYSP BZHK BZHV CBMO CBSO CBTT CCAV CCDN CCHL CCWA CHAS CHIP CIFR CJBC DEVS DIXY DJJA DLTR EFIR EGLS EHGF EHLX EMAZ ERNI FBRN FEAB FLUX GALA GASP GBRB HARN HOCK IDPH ILLY ISAX ITUG JACB JACC JACS JADJ JANA JANT JCAS JJAN JJEN JOYT JOYZ KAIR KEES KEMI KEVB KITE LACD LFSG LKTB LORR MALA MASF MDAC MERI NIKE NINB NINC NITA NOTE NUKA OBFS ODUD OGEM OIBO OJEH ONET OODW OPET ORAR OTYP PACT PEJM PIPA RADI RAZY SARA SGSE SHED SOBI SOOT SVEA SYDD TEMP TERY TIMK TSGJ TWEL USSY VOAR VSPN WACP WACR WWAY YANK YULL ZMAM

PA-28-235/236 CHEROKEE/DAKOTA
G-BGXS BHTA BNYB BOKA BPCX BRKH BXCC BZEH CCBH DAKO EWME FRGN FWPW KOTA LEAM ODAK TART

PA-28R/PA-28RT (CHEROKEE) ARROW
G-AVWN AVWO AVWR AVWT AVWU AVWV AVXF AVYS AVYT AWAZ AWBA AWBB AWBC AWEZ AWFB AWFC AWFD AWFJ AXCA AXWZ AYAC AYII AYPU AYRI AZAJ AZDE AZFI AZFM AZNL AZOG AZSF AZRV AZWS BAHS BAIH BAMY BAPW BAWG BBDE BBEB BBFD BBIA BBZH BBZV BCGS BCJO BCPG BDKV BEOH BEWX BFDO BFLI BFTC BFZH BGKU BGKV BGOL BGVN BHAY BHEV BHFJ BHGY BHIR BHWY BIDI BIKE BIZO BKFZ BKXF BLXP BMGB BMIV BMKK BMLS BMNL BMOE BMPR BNEE BNJR BNNX BNSG BNTC BNTS BNVT BOBA BOET BOGM BOIC BOJI BONC BOOG BOWY BOYV BPBO BPXJ BPZM BRLG BRMS BRRJ BSNP BSPN BTLG BTRT BUND BUNH-BUUM BVDH BWMI BWNM BXYO BXYP BXYR BXYT BYHJ BYKP BZDH BZKL CBEE CBPI CBVU CBZR CCIJ CSWH DAAH DAAZ DDAY DIZY DMCS DNCS DONS DORA DSFT ECJM EDVL ELUT EPTR FBWH FULL GDOG GEHP GHRW GPMW GYMM HALC HERB IBFW IJOE IRKB ISCA JANG JANO JESS JMTT LAOL LBRC LZZY MACK MEAH MEGA MEME MERL MRST NELI NIJM OARC OARI OARO OARU OBAK ODOG OJIM OKAG OKEN OMHC OMNI ONSF OOTC OPEP

OPJD OTGA RACO RJMS RONG RUBY SABA SBMM SHAY SHUG SKYV STEA TAPS TCTC TEBZ THSL TOLL TORC UTSY VOID WAMS WEND WILS WWAL YAWW

PA-30/PA-39 TWIN COMANCHE

G-ASMA ASRO ASSB ASSP ASWW ATEW ATMT ATSZ ATWR ATXD AVAU AVCY AVJJ AVKL AVPS AVUD AVVI AWBN AYSB AYZE AZAB BAKJ BAWN BKCL BLOR BZRO COMB ELAM LADI LARE OAJS OGET RROD SIGN SURG TCOM UAVA

PA-31/PA-31T NAVAJO/CHIEFTAIN/CHEYENNE

G-BBDS BBZI BEZL BFAM BFIB BFOM BLFZ BPYR BVYF BWDE BWHF BXKS CBGF CBTN CITY EEJE EHJM EMAX EPED FILL GLTT GLUG GURN HVRD IFIT ILEA ISFC JAJK LIDE LYDD MOHS MRMR NERC NEWR OBNW OETV OJIL ONAV ONPA OSGB OWLC PLAC PMAX PZAZ PZIZ RHYM SASK UMMI VICT VIPP YEOM

PA-32/PA-32R/PA-32RT CHEROKEE SIX/LANCE/SARATOGA

G-ATES ATJV ATRW ATRX AVFU AVUZ AZDJ AZTD BAGG BAXJ BBFV BBSM BDWP BEHH BEZP BFUB BFYC BGUB BHBG BHGO BIWL BJCW BKEK BKMT BMDC BMJA BOGO BOON BOTV BPVI BPVN BRGT BRHA BRNZ BSTV BSUF BSYC BTCA BVWZ BXWP BYFR BYPU CBCA CCFI CCST CDUX CEYE CSIX DCAV DENI DIGI DIWY EENA ELDR ELLA ETBY FRAG GOTO HDEW HERO HYLT IFFR ILTS JPOT JUPP KFRA KNOW LUNA MAIE MOLL MOVI NEAL NIOS NROY OARW OCPF OCTI OEVA OJCW OSCC OSIX OTBY PAPS PECK PUSK RAMS RAYE RHHT RIGH ROLF SALA SAWI SIMY SIXD TFIN VERN VONS WAIR WINS

PA-34 SENECA

G-AZIK AZOL AZOT AZVJ BABK BACB BADL BAIG BAKD BASM BATR BBLU BBNH BBNI BBPX BBXK BBZJ BCID BCVY BDUN BEAG BEHU BEJV BETT BEVG BFKY BFLH BGFT BGLW BHFH BHYE BHYF BHYG BLWD BLYK BMDK BMJO BMUT BNEN BNRX BOCG BOCS BOFE BOIZ BOJK BORH BOSD BOUK BOUL BOUM BOWE BPON BPXX BRHO BRXO BSDN BSGK BSHA BSII BSPG BSUW BTGU BTGV BUBU BVDN BVEV BWDT BXPV BXPW BYBH BZTG CAHA CBWB CEGA CHEM CLOS CLUE CTWW CVLH DARA DAZY DCEA DSID ELIS EMER EXEC EZYU FILE FLYI GAFA GFEY GFCD GOAC GOGS GUYS HCSL HMJB HTRL IEIO IFLP JANN JDBC JLCA LENY LORD MAIK MAIR MAXI MDCA MPWT NESW NSUK OACG OBNA OCFM OOON OPAG OWAL PEGI POPS ROUS RVRB SENE SENX TAIR TEST VASA VVBK WIZO WWAS XKEN

PA-38 TOMAHAWK

G-BGBK BGBN BGBW BGBY BGEK BGGE BGGG BGGI BGGL BGGM BGGN BGIG BGKY BGLA BGRM BGRN BGRR BGRX BGSH BGSI BGWN BGWU BGXB BGXO BGZF BGZW BHCZ BJNN BJUR BJUS BJYN BKAS BLWP BMKG BMML BMSF BMTO BMVL BMVM BMXL BNCO BNDE BNFR BNEK BNGR BNGS BNHG BNIM BNKH BNNU BNPL BNPM BNSL BNUY BNVD BNXV BNYK BNYV BOBL BOCC BODP BODS BOHS BOHT BOHU BOLD BOLE BOLF BOMO BOMZ BOUD BPER BPES BPHI BPIK BPJF BPPD BPPE BPPF BRFL BRFN BRHR BRHT BRJR BRLO BRLP BRML BRNJ BRSJ BSFE BSKK BSKL BSOT BSOU BSVW BSYL BSYM BTAP BTAS BTFP BTIL BTJK BTJL BTND BVHM BVLP BWNU BWSC BXET BXZA BYLE BYMC BYMD CHER CWFA CWFB CWFD CWFE DFLY-DTOO EDNA EMMS EORG EGNR GALL GTHM LFSA LFSB LFSD LFSH LFSM MSFC NCFC NCFE OATS OEDB OLFC OPSF OTFT PRIM RECS PVRF PVRG SUKI TOMS VMCG

PA-44 SEMINOLE

G-BGCO BGJB BGTF BHFE BHRP BOHX BRUI BRUX DENZ GAFT GHSI OACA OACC PDOC SEMI TWIN

PA-46 MALIBU/MERIDIAN

G-BZSD DERI DERK DIPM DNOP DODI GOJP GREY HITS JCAR RKJT VRST WADI

PIPER

CP.1 METISSE

G-BVCP

PITTS including AEROTECK, AVIAT and CHRISTEN INDUSTRIES INC

S-1/S-2 (Lycoming O-360 variants)

G-AXNZ AZCE AZPH BADW BADZ BBOH BETI BHSS BIRD BKDR BKPZ BKVP BLAG BMTU BOEM BOXH BOXV BOZS BPDV BPLY BPRD BPZY BRAA BRBN BRCE BRCI BRJN BRRS BRVL BRVT BRZL BRZX BSDB BSRH BTEF BTOO BTTR BTUK BTUL BUWJ BVSZ BXAF BXAU BXFB BXTI BYIP BYIR BYJP BZSB CCFO CCTF CCXK CDBH EWIZ FARL FCUK FDPS FLIK FOLY FORZ HISS ICAS IICI IIDY IIIE IIII IIIL IIIR IIIT IIIX ITII JAWZ KITI LITZ LOOP MAGG MAXG MINT OKAY OODI OSIC OSIS OSIT OSZB OWAZ PARG PITS PITZ PTTS REAP ROLL SIIA SIIB SIIE SIIS SKNT SOLO SPIN STUA STUB STYL SWUN TIII VOOM WAZZ WILD WREN XATS YOYO

PLUMB

BGP.1 BIPLANE

G-BGPI FUNN

PMPS

DRAGONFLY MPA Mk 1

G-BDFU

POBER

P-9 PIXIE

G-BUXO

POLIKARPOV

Po-2 (CSS-13)

G-BXYA

PORTERFIELD

CP-50

G-AFZL

CP-65

G-BVWY

PORTSLADE SCHOOL

HAB

G-AZYL

PORTSWOOD

HAB

G-FYBS FYBX

AVIONS **POTTIER**

P.80S

G-BTYH

POWERCHUTE SYSTEMS INTERNATIONAL LTD

KESTREL (Rotax 503

G-CCLT MVRV MWCI MWCJ MWCK MWCM MWCN MWCO MWCP MWCS MWFG MWFI MWFL MWGU MWGW MWGZ MWMB MWMC MWMD MWMG MWMH MWNV MWNX MWOC MWOD MWOE MYCW MYCX MYCY MYDA MYEW MYEX MYFA MYHS

RAIDER (Rotax 447)

G-MVHB MVHC MVMD MVNA MVNB MVNC MVNI MVNK MVNL MVNM MVVZ MVWJ

PRACTAVIA

PILOT SPRITE

G-AZZH BALY BCVF BCWH ROSS

PRICE

AX7-77 HAB

G-BMDJ

PRIVATEER see SLINGSBY

PROCTER see MITCHELL

PETREL

G-AXSF

PROTECH

PT-2C SASSY

G-EWAN

PTERODACTYL LTD see SOLEAIR

PFLEDGLING/PTRAVELER

G-MBAW MBHZ MBPB MJST MMPI

ALFONS **PUTZER KG**

ELSTER B

G-APVF BMWV LUFT

PZL-BIELSKO

SZD-45A OGAR

G-BEBG BKTM BMFI OGAR

PZL WARSZAWA-OKECIE SA

PZL-104 WILGA variants

G-BUNC BWDF BXBZ RIIN WILG WLGA

PZL-110 KOLIBER 150A/160A

G-BUDO BVAI BXLR BXLS BYSI BZAJ BZLC CBGA CDDE KOLI LOKM

QAC

QUICKIE/TRI-Q

G-BKFM BKSE BMFN BMVG BNCG BNJO- BPMW BSPA BSSK BXOY CUIK FARY IMBI KUTU KWKI OSAW WAHL

QUAD CITY including BFC kits

CHALLENGER (Rotax 503)

G-BYKU CAMR CBDU CCFD IBFC MGAA MGRH MVZK MWFU MWFV MWFX MWFY MWFZ MYAG MYDN MYDS MYFH MYGM MYIA MYIX

MYOZ MYPZ MYRH MYRJ MYSD MYTO MYTT MYUL MYXC MYXK MYXV MYYF MZAC MZBW MZBZ MZEA MZHO MZKW MZNA

ROYAL AIRCRAFT FACTORY-see REPLICA PLANS and SLINGSBY
_SE-5_
G-EBIA EBIB EBIC
G-BKDT

RAJ HAMSA
X'AIR variants
G-BYCL BYHV BYJU BYLT BYMM BYMR BYNT BYOH BYOJ BYOR BYPO BYPW BYRV BYSY BYTW BYTT BYTZ BYYM BYYR BYZF BYZW BZAF BZAK BZBP BZDK BZEJ BZER BZEX BZFF BZGX BZIA BZIS BZIY BZKC BZLT BZMR BZNG BZUP BZVH BZVK BZVR BZWC BZYM BZYX CBAH CBAV CBBH CBCI CBCM CBDO CBDV CBDY CBFE CBFT CBHB CBHV CBIC CBII CBIS CBJX CBKL CBLF CBLH CBLP CBLW CBMA CBNJ CBOC CBPU CBTK CBTY CBUC CBUJ CBVC CBVE CBVO CBWY CBXA CCAX CCBI CCBU CCBX CCCV CCCZ CCDJ CCDL CCDP CCDR CCEP CCES CCEY CCHS CCIW CCKJ CCMK CCNF CCNL CCNZ CCOH CCOO CCSO CCVJ CCWF CCWZ CCZJ CCZS CDDH CDDO CDEM CDFB CDFM CWAL HARI HITM MITE NEMO ODJD OHWV RAJA TANJ THAT TKSD UFAW XAYR XIOO XRXR

RAND-ROBINSON
_KR-2_
G-BETW BLDN BNAD BNML BOLZ BOUN BPRR BRJX BRJY BRSN BTGD BUDF BUDS BURF BUWT BVIA BVZJ BXXE BYLP BSTL CBAU CBPK DGWW JCMW KISS KRII OFMB UTSI WYNN XRAY

RANGO BALLON and KITE COMPANY including RANGO-SAFFERY
MLB variants
G-BJAS BJRH FYEU FYFW FYFY FYGI FYGK

RANS
S-4 COYOTE (Rotax 447)
G-MVXW MWBO MWEP MWES MWFW MWGN MWIO MWLA MWLZ MWWP MYWV
S-5 COYOTE (Rotax 447)
G-MVPJ MWFF MWGA MYDO MYFN MZGD
S-6 COYOTE variants (Rotax 503)
G-BSMU BSSI BSTT BSUA BSUT BTNW BTXD BUEW BUOK BUTM BUWK BVCL BVFM BVIN BVOI BVPW BVRK BVUM BVZO BVZV BWHK BWWP BWYR BXCU BXRZ BXWK BYBR BYCM BYCN BYCO BYIB BYID BYJO BYKE BYMN BYMU BYMV BYNP BYOT BYOU BYPZ BYRG BYRS BYSN BYZO BZBC BZBX BZEW BZKD BZKO BZLE BZMJ BZNH BZNJ BZRA BZRY BZUH BZVM BZYA BZYL CBAS CBAZ CBFX CBNV CBOK CBOS CBTO CBUY CBXZ CBYD CBZG CBZN CCDC CCEG CCJN CCLH CCNB CCNH CCOF CCTV CCTX CCZN CDFU CDGB CDGE CLEE IZIT KEPP MGEC MGND MWCH MWHP MWIF MWSC MWTT MWUK MWUL MWUN MWVL MWWL MWYE MWYN MYAJ MYBA MYBI MYDK MYDX MYES MYGH MYGP MYGR MYHI MYHK MYHP MYIR MYIS MYJD MYJY MYKN MYLD MYLF MYLO MYLW MYMH MYMP MYMR MYMS MYNE MYNH MYOA MYOI MYOT MYPA MYPJ MYSP MYSU MYTE MYUZ MYVP MYXB MYXG MYXP MYYV MYZR MZAH MZBD MZBH MZBU MZBV MZCA MZDA MZDG MZDM MZDR MZEN MZEO MZEU MZFL MZFN MZFY MZIY MZJI MZJM MZKE MZLG MZLL MZMP MZMS MZMU MZNV MZOZ MZUB RDNS RINS RTMS SAUK SOOZ SSIH TIVS TSOB WYLE
S-7 COURIER
G-BVNY BWKJ BWMN CBNF KATI OJKM
S-9 CHAOS
G-BSEE
S-10 SAKOTA
G-BRPT BRSC BRZW BSBV BSGS BSMT BSNN BSWB BTCR BTGG BTJX BTWZ BUAX BUKB BVFA BVHI BWIA BWIL BYREJSCL OEYE RANS RANZ
S-12 AIRAILE
G-BZAO

RAYTHEON HAWKER see HAWKER SIDDELEY AVIATION

REALITY AIRCRAFT LTD - see JUST AIRCRAFT
EASY RAIDER
G-CBKF CBXE CBXF CCEZ CCHR CCJS CCMJ OESY OEZI SLIP SRII

REARWIN AIRCRAFT and ENGINES INC
175 SKYRANGER
G-BTGI RWIN
8125 CLOUDSTER
G-EVLE
8500 SPORTSTER
G-AEOF
9000L SPORTSTER
G-BGAU

REECE
SKY RANGER
G-MJRR

REID and SIGRIST
RS.4 DESFORD
G-AGOS

REIMS AVIATION SA see CESSNA

RENEGADE see MURPHY

REPLICA PLANS
_SE-5A_
G-BDWJ BIHF BKER BMDB BUOD BUWE CCBN CCXG INNY SEVA

REPUBLIC AVIATION CORPORATION
P-47 THUNDERBOLT
G-BLZW THUN

REVOLUTION HELICOPTERS
MINI-500
G-BWCZ OREV PDWI

RIDOUT
MLB variants
G-BIWA BJDK BJFC G-BJMX BJMZ BJNA FYDN FYFI

RIGG
MLB variants
G-BIAR

AVIONS PIERRE **ROBIN** including CONSTRUCTIONS AÉRONAUTIQUES DE BOURGOGNE
DR400/500 variants
G-BAEB BAEM BAEN BAFP BAFX BAGC BAGR BAGS BAHL BAJY BAJZ BAKM BALF BALG BALH BALI BALJ BAMS BAMT BAMU BAMV BANB BAPV BAPX BAZC BBAX BBAY BBCH BBCS BBDP BBJU BBMB BCXE BDUY BEUP BFJZ BGRH BGWC BHAJ BHFS-BHJU BHLE BHLH BHOA BIHD BIZI BJUD BKDH BKDI BKDJ BKVL BNFV BNXI BOGI BPHG BPZP BRBK BRBL BRBM BRNT BRNU BSDH BSFF BSLA BSSP BSVS BSYU BSZD BTRU BUGJ BUYS BXRT BYHT BYIT BZIJ BZMM CBBA CBEZ CBMT CBVB CBZK CCKP CCWM CCZX CDAI CDBM CHIX CONB DUDZ EGGS EHMM ELEN ELUN ETIV EUSO EYCO FCSP FTIL FTIM FTIN FUEL GBUE GOSL HAIR HANS-HXTD IEYE IOOI IYCO JBDH JEDH JMTS JUDE KENW KIMY LEKT LARA LEOS LGCA LISE MIFF MOTI NBDD NFNF OACF ONGC PAYD PREZ RNDD RONS SELL TUGY UAPA XLXL YOGI ZACH ZIGI ZIPI
HR100 ROYALE/SAFARI/TIARA
G-AZHB AZHK BAEC BAPY BAWR BAYR BBAW BBCN BBIO BEUD BLHN BLWF BXWB CBFN HRIO MPWI RUES
HR200/CLUB
G-BBOE BCCY BETD BFBE BGXR BLTM BNIK BUWZ BVMM BWFG BWPG BXDT BXGW BXOR BXVK BYLG BYLH BYNK BYSG BZLG BZXK GORF JPAT NSOF WAVA WAVI WAVN
R1180T AIGLON
G-BGHM BIRT BJVV GBAO GDER GEEP PACE ROBN VECD VITE
R2100/2112/2120/2160 variants
G-BGBA BICS BIVA BKXA BLWY BVYO BWZG BYBF BYOF BZFB CBLE CBNG EWHT MATT OCFC PGSI PLAY RAFC SACK SBMO TOUR VECE VECG WAVT
_R3000_
G-BLYP BOLU BZOL CCCN ENNI PAVL
ATL
G-GFRD GGHZ

ROBINSON AIRCRAFT CO
REDWING
G-ABNX

ROBINSON HELICOPTER CO INC
R22 ALPHA/BETA/MARINER
G-BJUC BLDK BLME BLTF BOAM BOCN BODZ BOEW BOEZ BOVR BOYC BOYX BPGV BPIT BPNI BPTZ BRBY BROX BRRY BRVI BRWD BRXV BSCE BSEK BSGF BTBA BTHI BTNA BTOC BTVU BUBW BVGS BVPR BWHY BWTH BXOA BXSG BXSY BXUC BXXN BXYK BYCF BYCK BYCU BYHE BYTD BYTE BYZP BYZZ BZBU BZJJ BZMO BZYE CBOT CBPT CBVL CBWZ CBXK CCAP CBXN CBZF CCDE CCGC CCGE CCGF CCHZ CCMR CCVU CCVY CCYW CDAW CDED CDBF CDBG CDDD CHIS CHYL CHZN CMSN CRAY CTRL DAAM DABS DEER DEFY DELT DERB DLDL DMCD DODB

DODR EFGH EFOF EGGY EIBM EPAR ERBL EROM ETIN FEBY FIRS FLYH FOGY FOLI GEGE GJCD GOUP HARR HBMW HIEL HIPO HIZZ HONI HRBS HRHE HSLA HUGS HURN IAGD IBED ICCL IHSB IIFR IIPT INIS IORG ISMO JARA JATD JBII JCAP JERS JHEW JONH JSAK JWFT KNIB KNOX KUKI LAIN LHCA LHCB LIPE LNIC LYNC MATY MAVI MDGE MDKD MFHT MICH MIGK MOGY MRSN MUFY NJSH NORT OASH OAVA OBIL OBIO OCOV ODCS ODJB ODOT OEAT OFAS OGOH OHFT OHSL OIIO OJAN OKEY OLAU OLIZ OLRT OMMG ONMT OPAL ORMB OROB OSEE OSHL OSIP OSMS OTHL OTOY OVNR OZZY PACL PERE RALD REDY RENT RIAT RICE RIDD RIDL RNGO ROUT ROVY RSVP RSWW SBUT SIMN SIMS SPEE SUMT SUMX TGRD TGRE TGRR TGRS THLA TINK TILE TOLY TOMM TORS TOSH TTHC TUNE UDAY UESY UNYT VEEE VEYE VMSL WADS WAGG WFOX WIRL WIZA WIZR WIZY WRWR YACB YMBO ZAPY

R44 ASTRO/CLIPPER/RAVEN

G-BVMC BWVH BXPY BXUK BYCE BYKK BZIN BZLP BZMG BZOP BZPL BZRN BZTA BZXY CBAK CBEG CBFJ CBRO CBVI CBYY CBZC CCFC CCNK CCNO CCNY CCRD CCTL CCWD CCWI CCWJ CCYC CCYG CCYT CCZG CCZH CDCA CDCB CDCV CDEY CDXX CEEE CHAP CHUM CLKE CRIB DAVG DBUG DCSE DGHD DMRS DRIV DSPI DSPZ DYCE EDES EEZA EJTC EKKO EKYD ELMO EMMI ENYA EPMI EPPO ESSY EUGN EYET FABI FARE FLBI FOFO GACB GATE GATT GILI GLIB GRWW HALE HEPY HHOG HMPF HRHS HRPN HTEL ICAB IFDM IFTS ILTY INDX IVEN IVIV JBKA JEFA JILY JWEB KAZZ KPAO KTOL KYDD KYNT LARY LFBW LOTA LRSN LUKI LUKY LWAY LWUK MAMK MAPL MAYB MCAI MGAN MGWI MIKS MRKS MURY MUSH NIOG NIOL NOSY NSEW OAJC OBBY OCHM ODHG ODOC OFIL OJRH OKES OLFO OLOW OMCD OMEL OMGH OMMT ONGA OPHA OPTF OSSI OTJB OWND PBEE PFML PIDG PIXX POTT PPTS PRET RAVN RAYC RDEL REDI RFUN RGNT RONN ROZI RROB RTWO RTWW RUZZ SBRA SEFI SHAN SHRT SJDI SPAL SSJP STOT STUY SUMZ SUNN SWAT SYTN TAND TATY TEXT THEL TRAC TRCY TRNT TRYG UTSS UTZI VALV VEIT WAFU WEGO WMBT WYSP XLIV XTEK YEAH YIIK

ROCKWELL INTERNATIONAL CORPORATION including COMMANDER AIRCRAFT COMPANY **production and** see AERO COMMANDER

COMMANDER 112/114

G-BDAK BDFW BDIE BDKW BDLT BDYD BEBU BEDG BENJ BEPY BERI BERW BFAI BFPO BFXS BFZM BGBZ BHRO BHSE BIOJ BKAY BLTK BMJL BMWR BOLT BPTG BUSW BVNL BYKB CCDT CRIL DANT DASH DIME EHXP EMCA ERIC FATB FLPI HILO HMBJ HPSB HPSE HPSF HPSL HROI IMPX JILL JURG KEEF LADS LITE NATT NOOR OECM OIBM OLFT OMUM OOJP OVIN PJNZ PLAZ RCED SAAB TECH TWIZ VICS ZIPA

ROE

TRIPLANE

G-ARSG

ROGER HARDY

RH7B TIGER LIGHT

G-MZGT

ROGERSON

HORIZON 1

G-DOGZ

ROLLASON AIRCRAFT and ENGINES LTD see DRUINE

BETA

G-AWHX BADC BETE BUPC

ROMAIN

COBRA BIPLANE

G-MNLH

ROOSTER see LIGHTWING

ROTARY AIR FORCE INC

RAF 2000 variants (Subara EJ22)

G-BUYL BVSM BWAD BWHS BWTK BWWS BXAC BXDD BXEA BXEB BXGS BXKM BXMG BYDW BYIN BYJA CBCJ CBHZ CBIT CBJE CBJN CBMJ CCEU HOWL IRAF JEJE ONON PHLB RAFZ REBA SAYS YRAF YROJ

ROTEC ENGINEERING INC

RALLY 2B

G-MBAZ MBGS MBMG MJPA MVRF

ROTORWAY

SCORPION/EXECUTIVE

G-BNZL BNZO BPCM BRGX BSRP BUJZ BURP BUSN BVOY BVTV BWLY BWUJ BYNI BYNJ BZBW BZES BZOM BZXD CBWO CBWU CBYB CBZI CCFY CCMU CDBK CHTG ESUS FLIT JONG KENI MAMC NEEL OHOV PILE PURS RATH RAWS RHYS RWAY SFOX VART WHOO YEWS ZHWH

ATELIERS AERONAUTIQUE **ROUSSEAU** see PIEL

ROYAL AIRCRAFT FACTORY

BE.2C

G-AWYI

SE.5A

G-BLXT

RUSCHMEYER LUFTFAHRTTECHNIK GmbH

RUSHMEYER R90

G-EERH TODE UAPO

RUTAN

COZY

G-BXDO BXVX COZI OGJS SCUL SPFX

DEFIANT

G-OTWO

LONG-EZ

G-BKXO BLLZ BLMN BLTS BMHA BMIM BMUG BNCZ BOOX BPWP BRFB BSIH BUPA BZMF CBLZ HAIG ICON LEZE LUKE MUSO OMJT PUSH RAEM RAFT RPEZ SENA WILY

VARIEZE

G-BEZE BEZY BIMX BKST BVAY BVKM EMMY EZOS IPSY KENZ LASS OOSE SKCI TIMB VEZE

RYAN AERONAUTICAL CORPORATION

ST3KR/PT-22

G-AGYY BTBH BYPY

SAAB-SCANIA AB including SVENSKA AEROPLAN AB (SAAB)

32 LANSEN

G-BMSG

91 SAFIR

G-ANOK BCFW BKPY HRLK SAFR

SF.340

G-GNTB GNTC LGNA LGNB LGNC LGND LGNE LGNF LGNG LGNH RUNG

2000

G-CDEA CDEB

SAFFERY MODEL BALLOONS including CUPRO SAPHIRE LTD

S.200 MLB

G-BERN BFBM

S.330 MLB

G-BIHU

Princess MLB

G-FYGM

Rigg Skyliner MLB

G-BHLJ

SAN see JODEL

SAUNDERS-ROE (SARO)

A.19 CLOUD

G-ACGO

P.531-2

G-APVL

SKEETER

G-APOI AWSV BLIX HELI SARO

SCALLAN

EAGLE MLB

G-FYEO

FIREFLY MLB

G-FYEZ

SCHEIBE-FLUGZEUGBAU GmbH

SF23 SPERLING

G-BCHX

SF24 MOTORSPATZ

G-BBKR BZPF

SF25 FALKE including SLINGSBY T.61 variants

G-AVIZ AXEO AXIW AXJR AYBG AYSD AYUM AYUN AYUP AYUR AYZU AYZW AZHD AZIL AZMC AZMD AZPC AZYY BADH BAIZ BAKY BAMB BDZA BECF BEGG BFPA BFUD BGMV BHSD BIGZ BKVG BLCU BLTR BLZA BMBZ BMVA BODU BPIR BPZU BRRD BRWT BSEL BSUO BSWL BSWM BTDA BTRW BTTZ BTUA BTWC BTWD BTWE BUDA BUDB BUDC BUDT BUED BUEK BUFG BUFN BUFR BUGL BUGT BUGV BUGW BUGZ BUHA BUHR BUIH BUJA BUJB BUJI BUJX BUNB BUXJ BVKK BVKU BVLX BWTR BXAN BXMV BYOA CCHX CDFD FEFE FHAS FLKE FLKS GBGA HBOS KAOM KDEY KFAN KGAO KWAK MFMM OWGC SEXE

STRIKER see FLEXIFORM

STRIPLIN
LONE RANGER
G-MBDL MBJM
SKY RANGER
G-MJKB MMFZ

STROJIRNY PRVNI PETILESKY see MORAVA

STROJNIK
_S-2A_
G-BMPS

SUD-AVIATION see GARDAN and SOCATA and including AéROSPATIALE, SOKO and WESTLAND HELICOPTERS LTD production
SE.313x ALOUETTE II
G-BVSD BZGG UGLY
SA.315 LAMA
G-LAMA
SA.341 GAZELLE
G-BAGL BCHM BXTH BXZD BZDV BZLA BZYB BZYD CBBV CBGZ CBJZ CBKA CBKC CBKD CBSB CBSC CBSD CBSE CBSF CBSH CBSI CBSJ CBSK CBXT CBZL DFKI DMSS EHUP EROL EZEL FUKM GAZA GAZI GAZL GAZZ GZLE LOYD MANN OGAZ OGEO OLDH PAGS SFTA SIVJ SKUL SWWM TURP UZEL WCRD WDEV WMAN ZELE ZLLE ZZEL
SA.365 DAUPHIN 2
G-BKXD BLEZ BLUM BLUN BTEU BTNC BXLL BXPA MLTY PDGN PLMI

SUKHOI
_Su-26M_
G-IIIS IIIZ SIID XXVI

SUPER SCORPION see HIWAY

SUPER MARINE AIRCRAFT (PTY)
SPITFIRE Mk.26
G-CCGH CCJL CCZP

SUSSEX
GAS BALLOON
G-AWOK

SWALLOW AEROPLANE CO
SWALLOW B
G-MJBK

SWEARINGEN
SA-227 METRO III
G-BUKA

SZD see PZL

TARJANI
(Trike)
G-MJCU

TAYLOR
JT.1 MONOPLANE (Volkswagen 1600)
G-APRT AWGZ AXYK AYSH AYUS BBBB BDAD BDAG BDNC BDNG BDNO BEUM BEVS BEYW BFBC BFDZ BFOU BGCY BGHY BILZ BJMO BKEU BKHY BLDB BMAO BMET BRUO BUXL BXTC BYAV CDGA CRIS DIPS DRAY SUZY WARD
JT.2 TITCH
G-BABE BARN BCSY BDRG BFID BGMS BIAX BKWD BVNI BZJS MISS MOLE OJON RKET TICH VIVI

TAYLOR see AEROCAR

TAYLOR-WATKINSON
DINGBAT
G-AFJA

TAYLOR AIRCRAFT CO INC see PIPER

TAYLORCRAFT see AUSTER

TAYLORCRAFT AIRCRAFT CORPORATION
BC-12D/BL-65/DF-65/DCO-65
G-AHNR AKVO BIGK BOLB BPHO BPHP BPPZ BREY BRIH BRIY BRPX BRXE BSCW BSDA BTFK BVDZ BVRH-BVXS BWLJ
_F-19/F-21/F-22_
G-BPJV BRIJ BVOX BWBI

TEAM
HI-MAX (Rotax 447)
G-CBNZ CCAJ MZHM MZIA
MINI-MAX (Rotax 447)
G-BVSB BVSX BVYK BXCD BXSU BYBW BYFV BYII BYJE BYYX BZDR BZTC CBIN CBPL CBXU CCGB MWFC MWFD MWHH MWLW MWSA MWWE MWZM MYAT MYBM MYCT MYDF MYGF MYGL MYII MYIZ MYKJ MYKZ MYLB MYNI MYRG MYRL MYSK MYXA MYYR MYYS MYZE MZCS MZII MZMO MZNM MZNN MZOY MZPJ NADS OJLH OSCO THEO

TECNAM (CONSTRUZIONI AERONAUTICHE TECNAM SRL)
P92 ECHO variants (Jabiru 2200)
G-BZHG BZWT CBAX CBDM CBGE CBLB CBUG CBYZ CCAL CCDU DWPF OALH PGFG PLOD SDOZ TECM TECS TCNM WHEN

TED SMITH including PIPER production
AEROSTAR 601
G-MOVE RIGS TIME

TEMAN
MONO-FLY
G-MMJX MMPZ

TEVERSON
BISPORT
G-CBGH

THORN
COAL GAS BALLOON
G-ATGN

THORP including VENTURE
_T-18_
G-BLIT BSVN BYBY HATF
_T-211_
G-BTHP BXPF BXPO BYJF CCXI TZII

THRUSTER AIR SERVICES LTD including THRUSTER AIRCRAFT (UK) LTD
TST Mk.1 (Rotax 503)
G-MCCF MGTV MTGB MTGC MTGD MTGE MTGF MTGP MTGR MTGS MTGT MTGU MTKA MTKB MTKD MTKE MTLM MTLN MTNR MTNS MTNT MTNU MTNV MTPT MTPU MTPW MTPX MTPY MTSH MTSJ MTSK MTSM MTUB MTUC MTUD MTUE MTUF MTVP MTVR MTVS MTVT MTVV MTWY MTWZ MTXA MTXB MTXC MTXD MTZA MTZB MTZC MTZD MTZF MVAG MVAH MVAI MVAJ MVAK MVAL MVBP MVBT MVDD MVDE MVDF MVDG MVDH MVFJ MVFK MVFL MVFM MVFN MVFO MVHI MVHJ MVHK MVHL MVIR MVIT MVIU MVIV MVIW MVME MVMG MVMH MVMI MVOT MVOU MVOV MVOW MVOX MVOY MVXL-MVYE MWDP MYWZ NDOT OASJ OBAX OJSH OMAL PYNE RAFS
_T.300_
G-MGWH MVUB MVWN MVWR MVWS MVZA MVZC MVZD MVZG MVZI MWAN MWAP MWAR MWDS MWWS MYAP MYAR MYDR MYDU MYJF MYJG MYXU
T.600 (Jabiru 2200A)
G-BYPF BYPG BYPH BYPI BZDB BZJC BZJD BZNP BZTD CBGU CBGV CBGW CBIO CBIP CBIR CBKG CBPN CBVA CBWI CBWJ CBXG CBXH CBYT CCBC CCCB CCCF CCCH CCCU CCDV CCEB CCIC CCMT CCRN CCUZ CCXV CCXW CDBN CDBZ CDDI CDDX CDGI CSAV CXIP EVEY FJCE INGE IRAL KIPP MARZ MOMA MYWD MYWE MZFO MZFP MZFR MZFU MZGX MZGY MZGZ MZHA MZHC MZHD MZHE MZHF MZHS MZHU MZHV MZHW MZHY MZHZ MZKP MZKR MZKS MZKT MZKU MZNX MZNY OHYE OOFE ORDS ORUG OWMC PGSA PSUK PVST RAFH REDZ RIVR UDGE ULLY WORM

THRUXTON see JACKAROO AIRCRAFT
JACKAROO
G-ANZT AOEX AOIR APAJ APHZ

THUNDER BALLOONS LTD including THUNDER and COLT LTD
Airship:
THUNDER AS-33
G-ERMS
Gas Balloon:
_AA-1050_
G-BSRJ
Hot Air Balloon:
Ax3 SKY CHARIOT
G-BHUR BJGE BKBD BKFG BKIY BKMR NEIL

<u>*Ax4 series*</u>
G-LORY
<u>*Ax5 series*</u>
G-BDAY BEEP BEMU BLOV
<u>*Ax6 series*</u>
G-BBDJ BBOO BCCH BDVG BEEE BEJB BERD BECS BETH BFIT BFOS BFOZ BGPF BGZZ BHTG BIIG BIIL BJVU BLWB BPSJ BPUF BUSY BVRI BVUH DICK LDYS LIFE RTBI THOM TNTN
<u>*Ax7 series*</u>
G-BAXK BBDJ BBOX BCAR BCAS BCCG BCIN BCSX BCZI BDGH BDGO BDMO BDON BEVI BFIX BGRS BGST BHAT BHEU BHHH BHIS BHSP BHZX BJSW BJZC BKDK BKUU BLAH BLCC BLCY BLET BLGX BLKJ BLTN BLUI BLZF BMCC BMHJ BMMW BMMY BMOG BMUU BMVT BMYS BNBL BNBV BNBW BNCC BNCU BNGO BNMX BNXZ BNZK BOAO BOIJ BORD BOSB BPGF BPHU BPNU BPVU BPYK BPYZ BRDE BRLS BROA BRVN BRXB BRZE BSAV BSBN BSCF BSCO BSOJ BSZH BTAN BTAU BTHK BTRR BTSX BTTW BTVA BTXK BUDK BUIN BUKI BULB BUNV BUPU BVDB BYNU BZBH CBWR GASS GGGG GHIN HOWE LENS LYTE MLWI NEGS NIGS NWPB OFBJ OJDC OONI ORDY PIAF PUFF RAFE RBOW RIGB RINO ROCK ROSI RUBI SFRY SOFT THOS USIL VIVA WDEB WINE
<u>*Ax8 series*</u>
G-BJMW BOHF BORR BOTE BPZZ BRTT BRVY BSCX BSKI BSPB BSTK BSTY BTBB BTHM BTJD BTPX BTRO BUBL BUBY BUEI BUJW BUXW BUYD BVDW BVGB BVKH BVPA BVWB BWKW BYLV CBFH GEMS FABS HAZE INGA ISTT KBKB OMDD OTEL PUNK SUED THOR TOOL ZEBO
<u>*Ax9 series*</u>
G-BGHW BTJO BTMN BTOZ BTRN BTUJ BUAT BULK BVKZ BVSY BZBL IOAZ
<u>*Ax10 series*</u>
G-BTJF BTYF BUNZ BUOZ BUVZ BWUR CCEO OLEO WORK
<u>*Ax11 series*</u>
G-BXAD BXVF BZHX BZRZ
<u>***MLB***</u>
<u>*O.5*</u>
G-BBOD

<u>***SPECIAL SHAPES:***</u>

SHAPE	REGISTRATION(S)
FILM CASSETTE	G-PHOT
ICE CREAM	G-ICES
JUMBO JET	G-VJIM
WHISKY BOTTLE	G-RARE

THURSTON
<u>*TEAL*</u>
G-OWET

TIGER CUB see MICRO BIPLANE AVIATION
<u>*RL5A SHERWOOD RANGER*</u>
G-BZUG CBHU CCBW GKFC HVAN MWND WZOL PUSY

TIPSY including AVIONS FAIREY, COBELAVIA, SLINGSBY and homebuild production
<u>*TRAINER/B/BELFAIR*</u>
G-AFJR AFRV AFSC AFVH AFVN AFWT AISA AISB AISC APIE APOD
<u>*JUNIOR*</u>
G-AMVP
<u>*T.66 NIPPER*</u>
G-APYB ARBG ARBP ARDY ARFV ASXI ASZV ATBW ATUH AVKI AVKK AVKT AVTC AVXC AVXD AWDA BWHR AWJE AWJF AWLR AWLS AXLI AXZM AZBA BLMW BRIK BRPM BWCT BYLO CBCK CCFE CORD ENIE NIPA NIPP OVAG TIPS

TRAGO MILLS see FLS

TRI-R TECHNOLOGIES
<u>*KIS*</u>
G-BVTA BVZD BXJI BZDR CBTV MANW OKIS OKMA OKPW TKIS
<u>*KIS CRUISER*</u>
G-BYZD

TRIDENT see SHEFFIELD

TROTTER
<u>*AX3-90 HAB*</u>
G-BRBT

TURNER
<u>*SUPER T-40A*</u>
G-BRIO

TURNER see CURRIE

TWAMLEY
<u>*TRIKE*</u>
G-MBGF MJWI

ULTRAFLIGHT LTD including AMF Microflight Ltd production
<u>*LAZAIR*</u>
G-MBYI MNRD MTDN MTFL MVGZ

ULTRAFLIGHT FLIGHT INC
<u>*MIRAGE (Rotax 447)*</u>
G-MBRH MBSX MBXX

ULTRALIGHT AVIATION SYSTEMS
<u>*STORM BUGGY (Trike) with LIGHTNING and SOLAR wings*</u>f
G-MBGX MJBS

ULTRAMAGIC SA
<u>***Hot Air Balloon:***</u>
<u>*H-31*</u>
G-BZIZ BZPY
<u>*77 variants*</u>
G-BXPT BZKW BZSH BZSO CBRK CBWK CCLO CCRG DAIV DWPH KNEE RWRW VOTE
<u>*90 variants*</u>
G-CCYU CSFD KEWT
<u>*105 variants*</u>
G-BZPX BZRX CBRB CCOP GBGB
<u>*S-130*</u>
G-CBKK CBZV
<u>*M-145*</u>
G-BZGI
<u>*T-180*</u>
G-CCTN CCUE
<u>*N-210*</u>
G-BZPR BZPT
<u>*N-250*</u>
G-BZJX CBUE
<u>*N-300*</u>
G-CBPZ

ULTRASPORTS
<u>*PANTHER/FLEXIFORM DUAL STRIKER*</u>
G-MJVR MMDY
<u>*PUMA SPRINT X*</u>
G-MJCE MMCI
<u>*TRI-PACER with Flexiform, Hiway, Moyes, Solar Wings, Southdown and Wasp sail-wings*</u>
G-MBIY MBLK MBLU MBPY MBTJ MJER MJFB MJHC MJIA MJIC MJIY MJOI MJTC MJVN MMBL MMEO MMMR MMRK MMRL MMTT MMWS MNGD MNSD MTTL MZLK

ULTRAVIA
<u>*SUPER PELICAN*</u>
G-MPAC MWRS
<u>*PELICAN CLUB*</u>
G-BWWA

UNICORN GROUP
<u>*MLB variants*</u>
G-BINR BINS BINT BIWJ BJGM BJLF BJLG FYEK

VAHDAT
<u>*SEMICOPTER GYROPLANE*</u>
G-BZEV

VALENTIN
<u>*TAIFUN 17E*</u>
G-BMSE <u>OACE</u> TFUN

VALTION LENTOKONETEHDAS
<u>*VIIMA II*</u>
G-BAAY

VAN'S
<u>*RV-3*</u>
G-BVDC CCTG RVRH
<u>*RV-4/4A*</u>
G-BOHW BROP BULG BVDI BVLR BVRV BVUN BVVS BXPI BXRV BZPH CBGN CDJB FTUO IIGI MARX MAXV MUMY ORCA PIPS RMIT RVMJ RVDP RVDS RVRV SARV VANS

AIR-BRITAIN MEMBERSHIP

Join on-line at www.air-britain.co.uk

If you are not currently a member of Air-Britain, the publishers of this book, you may be interested in what we have on offer to provide for your interest in aviation.

About Air-Britain

Formed over 50 years ago, we are the world's most progressive aviation society, and exist to bring together aviation enthusiasts with every type of interest. Our members include aircraft historians, aviation writers, spotters and pilots – and those who just have a fascination with aircraft and aviation. Air-Britain is a non-profit organisation, which is independently audited, and any financial surpluses are used to provide services to the ever-growing membership. Our current membership now stands at over 4,000 some 700 of whom live overseas.

Membership of Air-Britain

Membership is open to all. A basic membership fee is charged and every member receives a copy of the quarterly house magazine, Air-Britain Aviation World, and is entitled to use all the Air-Britain specialist services and to buy **Air-Britain publications at discounted prices**. A membership subscription includes the choice to add any or all of our other 3 magazines, News and/or Archive and/or Aeromilitaria. Air-Britain publishes 10-20 books per annum (around 70 titles in stock at any one time). Membership runs January - December each year, but new members have a choice of options periods to get their initial subscription started.

Air-Britain Aviation World is the quarterly 48-page house magazine containing not only news of Air-Britain activities, but also a wealth of features, often illustrated in colour, on many different aviation subjects, contemporary and historical, contributed by our 4,000 members.

Air-Britain News is the world aviation news monthly, containing data on Aircraft Registrations worldwide, and news of Airlines and Airliners, Business Jets, Local Airfield News, Civil and Military Air Show Reports and International Military Aviation. An average 160 pages of lavishly–illustrated information for the dedicated enthusiast.

Air-Britain Archive is the quarterly 48 page specialist journal of civil aviation history. Packed with the results of historical research by Air-Britain specialists into aircraft types, overseas registers and previously unpublished photographs and facts about the rich heritage of civil aviation. Around 100 photographs per issue, some in colour.

Air-Britain Aeromilitaria is the quarterly 48-page unique source for meticulously researched details of military aviation history edited by the acclaimed authors of Air-Britain's military monographs, featuring British, Commonwealth, European and U.S. Military aviation articles. Illustrated in colour and black and white.

Other Benefits

Additional to the above, members have exclusive access to the Air-Britain e-mail Information Exchange Service (ab-ix) where they can exchange information and solve each other's queries, and to an on-line UK airfield residents database. Other benefits include numerous Branches, use of the Specialists' Information Service; Air-Britain trips and access to black and white and colour photograph libraries. During the summer we also host our own popular FLY-IN. Each autumn, we host an Aircraft Recognition Contest.

Membership Subscription Rates – from £18 per annum.

Membership subscription rates start from as little as £18 per annum, and this amount provides a copy of 'Air-Britain Aviation World' quarterly as well as all the other benefits covered above. Subscriptions to include any or all of our other three magazines vary between £25 and £62 per annum (slightly higher to overseas).

Join on-line at www.air-britain.co.uk or, write to 'Air-Britain' at 1 Rose Cottages, 179 Penn Road, Hazlemere, High Wycombe, Bucks HP15 7NE, UK, or telephone/fax on 01394 450767 (+44 1394 450767) and ask for a membership pack containing the full details of subscription rates, samples of our magazines and a book list.